neuro- *nerve* neurophysiology, the physiology of the nervous system
ob- *before, against* obstruction, impeding or blocking up
oculo- *eye* monocular, pertaining to one eye
odonto- *teeth* orthodontist, one who specializes in proper positioning of the teeth in relation to each other
ophthalmo- *eye* ophthalmology, the study of the eyes and related disease
ortho- *straight, direct* orthopedic, correction of deformities of the musculoskeletal system
osteo- *bone* osteodermia, bony formations in the skin
oto- *ear* otoscope, a device for examining the ear
oxy- *oxygen* oxygenation, the saturation of a substance with oxygen
pan- *all, universal* panacea, a cure-all
para- *beside, near* paraphrenitis, inflammation of tissues adjacent to the diaphragm
peri- *around* perianal, situated around the anus
phago- *eat* phagocyte, a cell that engulfs and digests particles or cells
phleb- *vein* phlebitis, inflammation of the veins
pod- *foot* podiatry, the treatment of foot disorders
poly- *multiple* polymorphism, multiple forms
post- *after, behind* posterior, places behind (a specific) part
pre-, pro- *before, ahead of* prenatal, before birth
procto- *rectum, anus* proctoscope, an instrument for examining the rectum
pseudo- *false* pseudotumor, a false tumor
psycho- *mind, psyche* psychogram, a chart of personality traits
pyo- *pus* pyocyst, a cyst that contains pus
ren- *kidney* renal capsule, capsule around the kidney
retro- *backward, behind* retrogression, to move backward in development
sclero- *hard* sclerodermatitis, inflammatory thickening and hardening of the skin
semi- *half* semicircular, having the form of half a circle
steno- *narrow* stenocoriasis, narrowing of the pupil
sub- *beneath, under* sublingual, beneath the tongue
super- *above, upon* superior, quality or state of being above others or a part
supra- *above, upon* supracondylar, above a condyle
sym-, syn- *together, with* synapse, the region of communication between two neurons
tachy- *rapid* tachycardia, abnormally rapid heartbeat
therm- *heat* thermometer, an instrument used to measure heat
tox- *poison* antitoxic, effective against poison
trans- *across, through* transpleural, through the pleura
tri- *three* trifurcation, division into three branches
viscero- *organ, viscera* visceroinhibitory, inhibiting the movements of the viscera

Suffixes

-able *able to, capable of* viable, ability to live or exist
-ac *referring to* cardiac, referring to the heart
-algia *pain in a certain part* neuralgia, pain along the course of a nerve
-ary *associated with, relating to* coronary, associated with the heart
-cide *destroy, kill* germicide, an agent that kills germs
-ectomy *cutting out, surgical removal* appendectomy, cutting out of the appendix
-emia *condition of the blood* anemia, deficiency of red blood cells
-ferent *carry* efferent nerves, nerves carrying impulses away from the CNS
-fuge *driving out* vermifuge, a substance that expels worms of the intestine
-gen *an agent that initiates* pathogen, any agent that produces disease
-gram *data that are systematically recorded, a record* electrocardiogram, a recording showing action of the heart
-graph *an instrument recording data or writing* electrocardiograph, an instrument used to make an electrocardiogram
-ia *condition* insomnia, condition of not being able to sleep
-iatrics *medical specialty* geriatrics, the branch of medicine dealing with disease associated with old age
-itis *inflammation* gastritis, inflammation of the stomach
-logy *the study of* pathology, the study of changes in structure and function brought on by disease
-lysis *loosening or breaking down* hydrolysis, chemical decomposition of a compound into other compounds as a result of taking up water
-malacia *soft* osteomalacia, a process leading to bone softening
-mania *obsession, compulsion* erotomania, exaggeration of the sexual passions
-odyn *pain* coccygodynia, pain in the region of the coccyx
-oid *like, resembling* cuboid, shaped as a cube
-oma *tumor* lymphoma, a tumor of the lymphatic tissues
-opia *defect of the eye* myopia, nearsightedness
-ory *referring to, of* auditory, referring to hearing
-pathy *disease* osteopathy, any disease of the bone
-phobia *fear* acrophobia, fear of heights
-plasty *reconstruction of a part, plastic surgery* rhinoplasty, reconstruction of the nose through surgery
-plegia *paralysis* paraplegia, paralysis of the lower half of the body or limbs
-rrhagia *abnormal or excessive discharge* metrorrhagia, uterine hemorrhage
-rrhea *flow, discharge* diarrhea, abnormal emptying of the bowels
-scope *instrument used for examination* stethoscope, instrument used to listen to sounds of various parts of the body
-stasis *arrest, fixation* hemostasis, arrest of bleeding
-stomy *establishment of an artificial opening* enterostomy, the formation of an artificial opening into the intestine through the abdominal wall
-tomy *to cut* appendectomy, surgical removal of the appendix
-ty *condition of, state* immunity, condition of being resistant to infection or disease
-uria *urine* polyuria, passage of an excessive amount of urine

Additional resources included with your textbook purchase.

As the owner of this book, you can get the most from your financial investment and study time by using a variety of study tools that accompany the book at no additional charge.

Each new copy of the text includes the following items:

- A 12-month, prepaid subscription to **The A&P Place** including **Adam Online Anatomy (AOA)**
- **Study Partner CD-ROM***
- ***Atlas of the Human Skeleton***

How to activate your prepaid subscription to The A&P Place:

1. Point your web browser to **www.anatomyandphysiology.com**
2. Click on the ***Human Anatomy,* Third Edition** book cover
3. Select the "Register Here" link
4. Enter your pre-assigned activation ID and password exactly as they appear below in the activation ID and password fields:

Activation ID

Password

5. Select "Enter"
6. Complete the online registration form to choose your own personal user ID and password.
7. Once your personal ID and password are confirmed, go back to **www.anatomyandphysiology.com** to enter the site with your new ID and password.

Your activation ID and password can be used only once to establish your subscription, which is not transferable.

If you have purchased a used copy of this text, this activation ID and password may not be valid. However, if your instructor is recommending or requiring use of **The A&P Place**, or if you would like to have access to the additional study tools found on this website, you can find more information about purchasing a subscription directly on-line at **www.anatomyandphysiology.com**. You also have the option of purchasing a subscription at your local college bookstore by placing an order for ISBN: 0-8053-4935-9.

The A&P Place contains several tools to help you prepare for exams, including 50 review questions per chapter, interactive clinical case studies, bone review anatomy quizzing, web research activities, on-line histology, a glossary with audio pronunciation guide, and A&P-related news articles. Also features access to **Adam Online Anatomy (AOA)**, which includes over 20,000 dissectible, atlas, and 3D anatomy images.

Turn the page to learn more...

* These items are available only with **NEW** copies of ***Human Anatomy,* Third Edition**, and are not sold separately. Please ask your textbook retailer for more information.

How to use the Study Partner CD-ROM:

The **Study Partner CD-ROM** includes multiple-choice review questions and practice testing for each chapter, interactive case studies and art exercises, and a glossary with pronunciation guide. To help you locate relevant Study Partner review materials, each chapter summary in the textbook includes Study Partner references which are signaled by an **SP** icon.

How to contact Customer Service and Technical Support:

To purchase additional study tools, including Interactive Physiology CD-ROMs, Student Video Series, and coloring books, please contact your local textbook retailer, visit our website www.awl.com/bc, or call Addison Wesley Longman Customer Service at 800-282-0693.

For technical support for **The A&P Place**, **Study Partner CD-ROMs**, or **Adam Online Anatomy**, please visit the technical support website at **www.awlonline.com/techsupport**, or send an e-mail to **online.support@pearsoned.com** with a detailed description of your computer system and the technical problem. For assistance with the **Study Partner CD-ROM**, please send an e-mail to **media.support@pearsoned.com**.

How can we be of further assistance?

We invite your comments and suggestions via email at **question@awl.com**, or, as you complete your Anatomy course, please take a few minutes to fill out the postage-paid questionnaire at the end of the text.

Allied Health Student Scholarship

Each year, Benjamin Cummings offers a scholarship award to five students who are enrolled in allied health programs in the United States. For more information and application instructions, please visit our website at **www.awl.com/bc** and select "Student Resources."

In the meantime, we wish you great success in your anatomy course!

Human Anatomy

Third Edition

Elaine N. Marieb, R.N., Ph.D.
Holyoke Community College

Jon Mallatt, Ph.D.
Washington State University

An imprint of Addison Wesley Longman, Inc.
San Francisco Boston New York
Capetown Hong Kong London Madrid Mexico City
Montreal Munich Paris Singapore Sydney Tokyo Toronto

Publisher: Daryl Fox
Senior Project Editor: Kay Ueno
Senior Developmental Editor: Mark Wales
Managing Editor: Wendy Earl
Associate Project Editor: Tiffany Barnes
Associate Editor: Susan Teahan
Publishing Assistant: Richard Gallagher
Production Supervisor: Karen Gulliver
Art and Design Director: Bradley Burch
Art Manager: Michele Mangelli
Developmental and Copy Editor: Alan Titche
Art and Photo Coordinator: David Novak
Assistant Art Coordinator: Anne Hikido
Photo Research: Stuart Kenter Associates and Bradley Burch
Text Designer and Cover Design: F. tani Hasegawa
Manufacturing Coordinator: Stacey Weinberger
Senior Marketing Manager: Lauren Harp

Photography and illustration credits appear following the Appendices.

Library of Congress Cataloging-in-Publication Data

Marieb, Elaine Nicpon
Human anatomy / Elaine N. Marieb, Jon Mallatt.—3rd ed.
p. cm.
Includes index.
ISBN 0-8053-4920-0 — ISBN 0-8053-4925-1
1. Human anatomy. I. Mallatt, Jon.

QM23.2 .M348 2001
611—dc21 00-034063

ISBN 0-8053-4920-0 (For sale copy with CD-ROM)
ISBN 0-8053-4925-1 (Professional copy with CD-ROM)

10 9 8 7 6 5 4 3 2 1–QWV–04 03 02 01 00

ABOUT THE AUTHORS

Elaine N. Marieb and Jon Mallatt are keenly aware of the challenges faced by all anatomy instructors. It is important to communicate a vast amount of information to students in a way that stimulates their interest. Forty-nine combined years of teaching experience in both the laboratory and the classroom have enhanced the authors' sensibilities about pedagogy and presentation. The insights gained from this experience are distilled in this thoroughly revised Third Edition of *Human Anatomy*—a book that aims to address the problems students encounter in this course.

Elaine N. Marieb

By any measure, Elaine N. Marieb is an accomplished educator and author. Her academic career began at Springfield College, where she taught anatomy and physiology to physical education majors. She joined the faculty of the Biological Science Division of Holyoke Community College in 1969, with a Ph.D. in Zoology from the University of Massachusetts at Amherst. Like other fine, inquisitive teachers, however, Dr. Marieb found there was more to learn—and more to understand—in order to meet the needs of her students, many of whom were pursuing degrees in nursing. To that end, Dr. Marieb, while teaching full time, continued her education, which culminated in both a bachelor and a master of science degree with a clinical specialization in gerontology from the University of Massachusetts. As a result, she gained an understanding of the relationship between the scientific study of the human body and the clinical aspects of nursing practice. From this experience and stories from former students in health careers, Dr. Marieb developed the unique perspective for which her widely used texts and laboratory manuals are known.

While actively engaged as an author, Dr. Marieb also serves as series consultant for the Benjamin Cummings–A.D.A.M.® CD-ROM series Interactive Physiology, and she is an active member of the Human Anatomy and Physiology Society (HAPS). In 1994 Dr. Marieb received the Benefactor Award from the National Council for Resource Development, American Association of Community Colleges in recognition of her ongoing sponsorship of student scholarships, faculty teaching awards, and other academic contributions to her community college. In May, 2000, Holyoke Community College renamed the Science Building in Dr. Marieb's honor.

Jon Mallatt

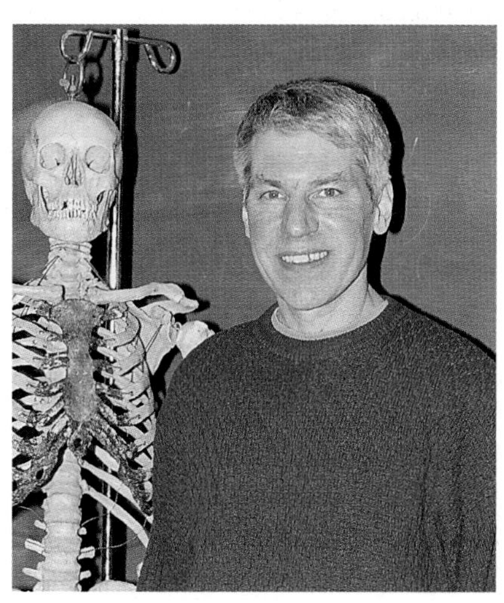

With a Ph.D. in Anatomy from the University of Chicago, Dr. Mallatt is currently an Associate Professor of Biological Sciences at Washington State University, where he has been teaching human anatomy to undergraduates of all backgrounds for 20 years. He is also a member of the department of Basic Medical Sciences, where he teaches courses in Histology and Anatomy of the Trunk in the WWAMI Medical Program. WWAMI honored him with their "Excellence in Teaching Award" in 1992, 1993, and 1995. Additionally Dr. Mallatt holds a position as adjunct Associate Professor in the department of Biological Structure at the University of Washington. His particular areas of expertise in the study of anatomy are histology, comparative anatomy, and anatomical drawing, although his research now focuses on the origin of vertebrate animals and molecular phylogeny. Dr. Mallatt is an accomplished researcher with 30 publications in the fields of anatomy and molecular phylogeny to his credit.

PREFACE

The general philosophy behind this revised third edition of *Human Anatomy* remains the same as in the previous editions. As an instructor, you know that teaching anatomy is not just the presentation of facts. You must provide information in a framework that encourages genuine understanding, devise new presentations to help students remember large amounts of material, and help students apply what they have learned to new situations. All the while you hope that you inspire in the students a love of the subject.

After many years of teaching human anatomy, we became convinced that new approaches to the subject could excite and challenge the students' natural curiosity. That is why we decided to write this book. We are fortunate to have collaborated with Benjamin Cummings, a publisher that shares our goal: to set a new standard for pedagogical and visual effectiveness in an anatomy text.

This book is designed for one-semester or one-quarter introductory anatomy courses that serve students in prenursing, premedical, prephysical therapy, radiological technology, physician assistant training, predentistry, pharmacy, and other allied-health fields, as well as physical education, athletic training, and nutrition.

This third edition has been extensively revised, its subject matter updated with the newest research findings, its art greatly improved, and the amount of clinical information increased. Additionally, the textbook is now linked to a suite of powerful new electronic-media resources.

Unique Approach to Anatomy

Since its inception, we have worked diligently to distinguish *Human Anatomy* from the many other anatomy books currently available. This book explains anatomy thoroughly, and its discussions are not merely brief summaries of the art. We have striven to present the basic concepts of anatomy—gross, microscopic, developmental, and clinical—in a manner that is clearly written, effectively organized, up to date, and well illustrated. We realize that learning anatomy involves assimilating gargantuan amounts of material, and we have tried to make our presentation as logical and accessible as possible. To this end, we present anatomy as a "story" that can be explained and understood—convincing the students that the structure of the body makes sense.

It is sometimes claimed that anatomy is a relatively static science and that anatomical knowledge does not grow much from year to year. Although this may be true for descriptive gross anatomy, it is not true for functional anatomy, neuroanatomy, developmental anatomy, nor for the functional aspects of tissue and cellular anatomy, subfields in which knowledge is growing quickly. This edition not only strives to keep up with the knowledge explosion in each of these subfields, but presents anatomy in a way that allows modern biology students, whose training is becoming ever more molecular and cellular, to anchor their biochemical and medical training in the physical context of the human body. Here, then, are the distinguishing features of our book.

Functional Approach

First, we emphasize strongly the functional anatomy theme, giving careful consideration to the adaptive characteristics of the anatomical structures of the body. Wherever possible, we explain how the shape and composition of the anatomical structures allow them to perform their functions. Such functional anatomy is not physiology (which focuses on biological mechanisms), but is more akin to "design analysis." This approach is not used by other texts at this level.

Emphasis on Current Research

This edition has been thoroughly updated. Both authors read extensively in current clinical and research journals to provide a wealth of cutting-edge information for this book. The best examples of this updating are found in the expanded coverage of the brain in Chapter 13 and of the immune system in Chapter 20, but all chapters have been updated to include information that is not present in other textbooks at this level. Other, more specific, examples of new information include:

- **Chapter 2:** Revised consideration of the cytoskeleton
- **Chapter 10:** Coverage of the functional ultrastructure of smooth muscle cells
- **Chapter 11:** The role of somitomeres in muscle development
- **Chapter 23:** Nervous control of urination
- **Chapter 24:** The functional anatomy of the placenta and labor
- **Chapter 25:** Updated effects of aging on the endocrine glands

Moreover, to emphasize some of the most exciting new information, certain findings are highlighted and labeled "Advances in Understanding," a new feature that is spread throughout the chapters of this edition.

Microscopic Anatomy

We have worked to provide an especially effective treatment of microscopic anatomy. Many undergraduate texts treat histology as a specialized and minor subfield that takes a back seat to gross anatomy. This is unfortunate, because most physiological and disease processes take place at the cell and tissue level, and most allied-health students require a solid background in histology and subcellular structure to prepare them for their physiology courses. Microscopic anatomy is one of Dr. Mallatt's research areas, and he has taught histology to University of Washington medical students for 20 years.

Embryology

Our text is designed to present embryology in the most effective and logical way. Currently, all competing texts describe the embryology of each organ system in the relevant chapters, yet do not provide the fundamentals of embryology until the very end of the book, in the section on the female reproductive organs. This approach seems backward to us. We are convinced that the fundamentals should be presented early in the text, before the more advanced discussions of the developing organ systems in the relevant chapters. Therefore, we wrote Chapter 3 as a basic introduction to embryology. Because a comprehensive presentation of embryology early in the book could be intimidating to some students, we have used a "velvet glove approach," providing only the most important concepts in a concise, understandable way, and visually reinforced with exceptionally clear new art.

Life Span Approach

Most chapters in this book close with a "Throughout Life" section that first summarizes the embryonic development of organs of the system and then examines how these organs change across one's life span. Diseases particularly common during certain periods of life are pointed out, and effects of aging are considered. The implications of aging are particularly important to students in the health-related curricula because many of their patients will be in the senior age group. Because aging is a fast-moving field of research, the "Throughout Life" sections are some of the most extensively updated parts of this edition.

Interactive Learning

This book is an interactive learning tool that encourages understanding; it is not an encyclopedia that encourages rote memorization. Whereas many scientific textbooks read like long lists of terms, this book reads like an instructor talking to, explaining to, and challenging the student. Many of the chapters include special topic boxes entitled "A Closer Look," which present clinical and topical information in a lighter, more inviting format. For those organs and systems with especially many components (e.g., skeletal muscles, cranial nerves, sensory receptors), we present these components in illustrated tables that organize the information in a logical way. In-text Critical Reasoning questions and Clinical Applications boxes are provided to challenge the student and to encourage synthesis of information.

Organization

This text conforms to the organization of most anatomy courses, starting with several introductory chapters followed by skeletal, muscular, nervous, and circulatory chapters, and then chapters on the various systems of visceral organs (see the Table of Contents). We also retain the following unique organizational features of previous editions of this book.

- As mentioned, we introduce basic embryology early (Chapter 3), a position that provides students with the background they need to understand the development of specific organ systems in later chapters.
- We present the endocrine system late (Chapter 25), after students have learned the visceral systems that the endocrine system controls. Nonetheless, we realize that many courses teach endocrine system anatomy with the nervous system, so we designed the endocrine system chapter so that it can also be interjected and understood at an earlier point in the course.
- We cover surface anatomy last (Chapter 26) so that we can discuss more of the clinical information related to body surface features. Books that place this chapter earlier, after bone and muscles, cannot logically discuss the surface anatomy of blood vessels (pulse points), lymph nodes, nerves, and the visceral organs. Our coverage of surface anatomy is the book's capstone, offering students a uniquely practical summary of the entire body.

Outstanding Illustration Program

Art plays a critical role in helping students visualize anatomical structures and concepts, and over the years we have diligently striven to maintain the hallmark accuracy of *Human Anatomy*'s art. Hence, when preparing the Third Edition, we asked reviewers from colleges and universities across the country to suggest how we might further refine our illustration program. Their mandate was clear: Provide anatomy art that is larger, sharper, and richer looking. We responded by undertaking the most thorough revision and updating of the illustration program since the book's inception. Here are some of the most noteworthy art changes:

- **Enhanced clarity and color.** We incorporated into the art the latest generation of digital drawing techniques and commissioned classic painted images that have been digitally enhanced for greater clarity. What is more, the colors in every figure are deeper and more vivid, creating a more pleasing three-dimensional appearance.
- **Larger illustrations of bones, muscles, and joints.** Virtually every illustration in the musculoskeletal system chapters of the Third Edition of *Human Anatomy* has been enlarged a minimum of 20 percent—and sometimes considerably more! The generous new illustration size, in conjunction with the aforementioned digital enhancement techniques, allows even the smallest details of this complex anatomy to be easily seen.
- **New Art.** Some chapters feature entirely new illustration programs. See, for example, the beautiful new art in Chapter 2, on the cell, drawn by the noted science illustrator Tomo Narashima. Also, all of the embryology art in Chapter 3 is new, featuring clearer layouts and greatly enhanced three-dimensional details. In some chapters, such as Chapter 13: The Central Nervous System and Chapter 20: The Lymphatic and Immune Systems, we substantially augmented the illustration program to match our updated and expanded coverage. What is more, in almost every chapter we either added an assortment of new illustrations or we reconceived and redrew existing illustrations using a contemporary, three-dimensional style. For example, compare Figure 16.8, Structures in the anterior region of the eye, and Figure 22.6, Histological layers of the alimentary canal, to their counterparts in the Second Edition. We trust that you will find the new illustrations to be more attractive and pedagogically effective.

We have continued to refine other aspects of our illustration program, too. As in previous editions, we use color not only for aesthetic purposes but also to denote the functional characteristics of certain anatomical structures. More illustrations than ever feature orientation diagrams, simple "thumbnail" sketches that clarify an illustration's anatomical perspective or pinpoint the location of an organ within the body. Not only are light and scanning electron micrographs used abundantly, but images produced by modern medical scanning techniques (CT, PET, MRI) are included where appropriate to enhance the student's grasp of anatomical structures. We have continued to augment key illustrations with the highly acclaimed Bassett photographs of dissected cadavers. Moreover, the skeletal system chapters now include an assortment of newly prepared bone photographs provided by the distinguished anatomy photographer Ralph T. Hutchings.

Improved Coverage for the Third Edition

Besides the extensive updating of the text and art described above, the Third Edition of *Human Anatomy* has some additional improved features.

More Clinical Information For this edition, the amount of clinical information has been expanded by about 50%. Students entering the allied-health fields always enjoy such information because it teaches them anatomy in the context in which they are most interested. Not only has the clinical information been increased, but it has been more clearly marked for easier identification. Disorders discussions now appear near the end of most chapters, and highlighted Clinical Applications paragraphs are woven into the text at appropriate places. Noteworthy among all the new clinical information are an expanded coverage of traumatic and sports-related injuries in the bone and joint chapters, and of cancers in the chapters on visceral organs.

Improved Presentation of the Skeleton The presentation of the bones of the skeleton, a single chapter in previous editions, has been split into two chapters: The Axial Skeleton (Chapter 7) and The Appendicular Skeleton (Chapter 8). This dual-chapter approach better fits the way in which bones are taught in most anatomy courses, and it provides students with twice the amount of end-of-chapter review material and questions on the skeleton. Equally important, we have added a number of new full-page bone illustrations: the bones of the elbow and posterior views of the hand. Given the frequency of injuries to the ankle, we also added a new fully illustrated consideration of the ankle joint to Chapter 9.

New and Revised Hallmark Features

Spotlighting Special Information We took care to make important information and subsidiary themes stand out in each chapter's coverage. As mentioned previously, recent research findings now are highlighted via the new Advances in Understanding boxes found throughout most chapters. To provide even more topical and clinical information, we added five new A CLOSER LOOK boxes to the Third Edition, bringing the total number of these popular boxes to 21. The new boxes cover the use of stem cells in tissue engineering (Chapter 4); tattoos (Chapter 5); harvesting blood stem cells from umbilical cords (Chapter 17), polycystic kidney disease (Chapter 23), and sexually transmitted diseases (Chapter 24). Along with these additions, we revised and updated all of the previous A CLOSER LOOK boxes.

Clinical Applications and Critical Reasoning In keeping with the greatly augmented clinical emphasis of the Third Edition, we edited and redesigned our popular Clinical Applications boxes so that they are more timely than ever, succinct, and easy to spot. Also, as part of the book's interactive pedagogy, a number of Critical Reasoning questions can be found in the text columns of most chapters, with an answer following each question. This feature asks students to synthesize information from previous reading, or to draw on their own experience. Although some of the Critical Reasoning questions are anatomically oriented, most are clinical in nature. Additional Critical Reasoning questions are included in the Review Questions found at the end of each chapter.

Helpful Presentation of Terminology The complex terminology of anatomy is one of the most difficult aspects of the subject to make interesting and accessible. To this end, we highlight important terms in **boldfaced type,** and we provide the pronunciations of more terms than do many competing texts. Also, we include the Latin or Greek translations of almost every term at the point where the term is introduced in the text. This promotes learning by showing students that difficult terms actually have simple, logical derivations. Furthermore, we call attention to clinical terminology in Related Clinical Terms, a section found at the conclusion of most chapters. A helpful glossary and pronunciation guide is located at the end of the text.

Practical Study Tools On each chapter's vividly designed opening pages, readers will find student objectives amalgamated with a basic chapter outline. The popular chapter summaries now feature special icons that direct students to related interactive art exercises on the Study Partner CD SP or to supplementary illustrations found in the A.D.A.M.® Interactive Anatomy Student Package CD AIA. Students have more testing opportunities than ever before, too. Chapters conclude with our three-part review question section offering questions in multiple choice and matching, short answer and essay, and critical reasoning and clinical applications formats. Many chapters also feature questions linked to the A.D.A.M.® Interactive Anatomy Student Package CD. What is more, a 35-question quiz for each chapter can be found on the Study Partner CD, and students can access a 50-question exam for each chapter on the Anatomy and Physiology Place website. (Please see the discussion of electronic media supplements for more information.)

A Comprehensive Teaching and Learning Package

The supplementary material developed by Benjamin Cummings for the Third Edition of *Human Anatomy* is more innovative and thorough than ever before—in print, on CD-ROM, and on the Web—for instructor and student alike.

New Electronic Media Supplements

The Anatomy & Physiology Place From the home page of the Anatomy & Physiology Place (www.anatomyandphysiology.com), you and your students can access a Special Edition for *Human Anatomy,* Third Edition, which features the following resources and activities:

- 50-question practice quizzes in every chapter
- Interactive clinical case studies with labeling and matching exercises
- Illustrated bone anatomy quiz
- Web research activities
- On-line histology tutor
- Pronunciation glossary with an audio feature
- Anatomy-related news articles

Instructors will find a useful Syllabus Manager to help organize the course, plus access to the Benjamin Cummings Digital Library (discussed next). A free 12-month subscription to this powerful website is included with each new student copy of the text. Instructors receive a 6-month subscription, which can be renewed free of charge by adopters. A 7-day trial option is also available at the website. Access directions, along with password and PIN code numbers, are attached inside the front cover.

Benjamin Cummings Digital Library for Human Anatomy, Third Edition The Digital Library is a CD-ROM featuring all of the illustrations (excluding photomicrographs) from the book—approximately 650 images in all. This important new teaching tool includes a presentation program that can be used for designing a customized slide show of images. You may edit labels, import illustrations and photos from other sources, and export figures to other software programs such as PowerPoint. For increased flexibility, the Digital Library presents art in three formats: art with leaders and labels, art with leaders only, and art without leaders and labels. A helpful topic and key word search function makes it easy to locate the illustrations you need without consulting figure numbers in the text.

Study Partner CD-ROM This helpful new interactive learning tool is designed to engage your students as they study and review the main concepts of anatomy. The Study Partner contains 53 interactive illustrated exercises along with quizzes for each chapter. A random testing feature allows students to construct tests that cover multiple chapters, with the total score revealed at the end of the test. Students can also access a comprehensive Pronunciation Glossary with audio feature. Special SP icons appearing within the book's end-of-chapter summaries alert students to specific, related exercises on the Study Partner. A copy of the Study Partner is attached to the inside back cover of the student and instructor versions of each textbook.

CD-ROM Test Bank Revised by Robin McFarland, Cabrillo College; Temma al Muhktar, San Diego Mesa College; James T. Wetzel, Presbyterian College; and Jon Mallatt, Washington State University. The revised test bank accompanying the Third Edition offers two 25-question multiple-choice exams per chapter as well as two final exams, each consisting of over 200 multiple-choice questions. Approximately one third of the questions are new to this edition. The electronic test bank is formatted on a cross-platform CD-ROM in TestGen-EQ. This versatile new software allows instructors to generate tests via a user-friendly interface in which they can easily view, edit, sort, and add questions. QuizMaster-EQ makes testing even easier for instructors. This free and fully networkable program enables the instructor to create and save tests and quizzes using TestGen-EQ so that students can take tests or quizzes on a computer network. The program automatically grades exams and allows the instructor to view or print reports for individual students, classes, or courses. The test bank also is available in print.

A.D.A.M.® Interactive Anatomy Student Package A.D.A.M.® Interactive Anatomy (AIA), version 3.0, is now available packaged with Lafferty and Panella's A.D.A.M.® Interactive Anatomy Student Lab Guide for a very special price. With over 20,000 dissectible, atlas, and 3D anatomy images, AIA also features female and male anatomical structures, cadaver dissections, and 3D models of the heart, skull, and lungs. Special AIA icons appearing within most of the end-of-chapter summaries provide specific directions for locating related images on the A.D.A.M.® Interactive CD-ROM. What is more, AIA-related anatomy review questions are located at the conclusion of most chapters of the text.

A.D.A.M.® Anatomy Practice This CD-ROM is designed to augment your students' laboratory experience. Anatomy Practice features 500 pinned anatomical images, and it includes more than 15,000 practice test questions. Students will find a wide variety of options to review anatomical images and take randomly generated tests with time limits, just as in the lab.

Course Management Systems and Special Student Services

WebCT and Blackboard Both course management systems are useful for developing and managing on-line distance learning courses. They are available in stand-alone form or packaged with the text. The content provided with either system includes selections from the Anatomy & Physiology Place as well as the entire computerized test bank.

AWL Tutoring Center

Free tutoring is available to students who purchase a copy of *Human Anatomy,* Third Edition. The AWL Tutoring Center is staffed by qualified anatomy instructors who can tutor students on all the material covered in the text, including the art and multiple choice/matching questions at the end of each chapter. Tutoring content is restricted to text material, and tutors will only discuss answers that are provided in the textbook. Students can contact the Tutoring Center by phone, fax, e-mail, or the Web. The Tutoring Center is open Sunday through Thursday, 5:00 PM to 12:00 AM EST. Instructors should contact their local Benjamin Cummings sales representative for instructions about how to authorize student access to the Tutoring Center.

New and Revised Printed Supplements

Full-Color Transparency Acetates We are greatly pleased to have more than doubled the number of transparency acetates available with the Third Edition—approximately 450 in all—which has enabled us to include every key illustration in the book (except for photomicrographs). The illustrations in every transparency have been color enhanced and then enlarged for easy viewing in the classroom or lecture hall. The acetate package is available free to adopters of the text.

Atlas of the Human Skeleton Prepared by Matt Hutchinson, Jon Mallatt, and Elaine N. Marieb, this new photographic atlas of the human skeleton is an indispensable reference for the laboratory and the classroom. You will find approximately 102 four-color images, with labels and leaders, all beautifully photographed by Ralph T. Hutchings. The Atlas is packaged free with each new textbook.

Instructor's Guide Revised by Gail C. Turner, Virginia Commonwealth University, the Instructor's Guide to the Third Edition of *Human Anatomy* features a number of noteworthy changes. An innovative Teaching with Art feature explains how to interpret a key illustration in each chapter during lecture. This feature also provides an art-related exercise and a visual critical reasoning challenge for students. As before, the Instructor's Guide offers a host of useful features such as student objectives and suggested lecture outlines, which now are augmented with lecture hints. Each chapter lists the available transparencies and provides lists of media resources, suggested readings, and classroom discussion topics and activities. You will also find answers to the text's short answer and essay questions and critical reasoning questions. Each chapter concludes with a Supplementary Student Materials section that contains study-tips pages that you may elect to photocopy and distribute to your students.

Human Anatomy Laboratory Manual Elaine N. Marieb's widely used Human Anatomy Laboratory Manual with Cat Dissection, Third Edition, accompanies this text (available January 2001 for Spring classes). The manual contains 29 gross anatomy and histology exercises for all major body systems. Illustrated in full color, with a convenient spiral binding, the lab manual has an accompanying *Instructor's Guide* by Linda Kollett of Massasoit Community College.

Additional Supplements Available from Benjamin Cummings

Human Cadaver Dissection Videos
By Rose Leigh Vines, et al.

Anatomy Flashcards
By Glenn Bastian

The Anatomy Coloring Book, Second Edition
By Kapit and Elson

A Color Atlas of Histology
By Dennis Strete

Histology for the Life Sciences
By Allen Bell and Victor Eroshenko

Contact your Benjamin Cummings sales representative or your campus bookstore for more information.

ACKNOWLEDGMENTS

An excellent group of people has helped us with this major revision of *Human Anatomy.* First, we thank Daryl Fox, Vice President for Anatomy and Physiology, for launching the project. Kay Ueno, as Marieb project manager, oversaw the revision process; her calm and fair manner, confidence, and encouragement guided us through some hectic times and are greatly appreciated. Amy Folsom, the sponsoring editor, and Claire Brassert, the developmental editor for the first seven months of the project, were extremely kind, pleasant, and fair, and we appreciate the free hand that Claire gave Dr. Mallatt in updating the chapters. Mark Wales, a remarkably talented individual, served as developmental editor for all the art and for the text of Chapters 1–4, for the final eleven months of the project. We appreciate his vast experience, his dedication to detail, hard work, and his many creative ideas. Mark was involved with all aspects of the project and tied it all together, while bearing the brunt of the heaviest deadline pressures, as he also assembled the extensive package of supplements. He taught us much about art presentation, and Mark and Dr. Mallatt developed a special synergy that greatly improved the art and the book as a whole. Alan Titche, who was the developmental editor for the text of Chapters 5–26, brought an extremely logical mind to the project. He increased the clarity of many written passages, improved the organization and explanatory effectiveness of many sections, helped with the reorganization of the clinical material, and revealed a talent for excellent scientific writing in his own right. Tiffany Barnes and Richard Gallagher were editorial assistants who kept the project moving through endless copying and mailing, all with flawless efficiency.

Wendy Earl and her team of experts at Wendy Earl Productions produced this edition of *Human Anatomy.* This group also produced previous editions of this book, and when we heard Wendy Earl Productions would be back for this edition, we knew the book would look good. Bradley Burch, who served as art and design director (and incidentally, chose the fine new color palette for the art), and David Novak, who was the photo coordinator and coordinator of supplements, are both extremely friendly and were a joy to work with. Special thanks, too, to Michele Mangelli, our art manager, whose ability to work hard while remaining helpful and pleasant are, frankly, astounding; she laughed off the pressure and pushed onward. Anne Hikido did a fine job as assistant art coordinator, mailing all the art back and forth across country to the bicoastal coauthors again and again. Art coordination interns from "across the pond" Charlotte Ross and Olivia Ross kept everyone's spirits up.

We also thank Stuart Kenter, Rachael Epstein, and Bradley Burch, who as photo researchers found many fine new photos, including some especially good clinical shots and photomicrographs.

The talents of our artists are also appreciated, as we worked them quite hard! Tomo Narashima painted several showcase pieces of reflective cell biology and anatomical art, and the team of illustrators at Imagineering produced the large amount of computer-rendered art that is new to this edition. Anatomical art must be some of the most difficult art to render digitally, given its complexity, detail, and three-dimensional nature, yet these computer artists showed remarkable ability. We especially appreciate how well they were able to follow our suggestions for polishing the rough art into accurate, finished pieces. Our thanks as well to Wendy Hiller Gee and to Kristin Mount, who breathed new life into the bone and muscle art in this edition, Karl Miyajima, who acted as our production artist, and Jack Haley, project manager at Imagineering. Finally we want to thank F. tani Hasegawa, who designed the impressive cover and vivid chapter openers, and who is responsible for the Third Edition's spacious new look.

Karen Gulliver served as the production supervisor, directing the transformation of our messy manuscript into final pages. As a calm, competent, and supportive person, she was a joy to talk to. The professionals at GTS Publishing Services assembled the dummy and final pages under a tight schedule, and we were most impressed at how they set so much text with so few errors from such a heavily revised manuscript. This book also benefited from the fine work of Kristin Barendesen as the copy editor, and we are grateful for the work of our eagle-eyed proofreaders, Dave Rich and Anita Wagner. Finally our thanks to Shane-Armstrong Information Systems for a splendid new index.

We would also like to thank our reviewers, colleagues from across the United States and Canada who devoted much effort to this book, pointed out mistakes, made helpful suggestions, and believed in us. A complete list of these reviewers appears on the next page.

Lastly, we would like to thank our spouses, Marisa de los Santos and Harvey Howell, who supported us through the entire project. We also treasure the memory of "Zeb" Marieb, beloved husband, who was an inspiration for the early versions of this book.

Jon Mallatt
Elaine Marieb

REVIEWERS

Text and Art Reviewers

Kathleen Andersen
University of Iowa College of Medicine

Steven Basset
Southeast Community College

Robert Bauman
Amarillo College

Leann Blem
Virginia Commonwealth University

Kishori Chaubal
University of the Pacific

Joe Coelho
Culver-Stockton College

Danielle Desroches
William Patterson University

David Deutsch
Kent State University

Roisheen Doherty
University College of Fraser Valley

George S. Dougherty
Indiana University

Patricia B. Fulks
Bakersfield College

Larry Ganion
Ball State University

Glenn Gorelick
Citrus College

Frank Hagy
Wright State University

Christine Iltis
Salt Lake Community College

Anne Johnson
Southern Illinois University

Jamie King
Craven Community College

Karen LaFleur
Greenville Technical College

Steven Lewis
Penn Valley Community College

Jerri K. Lindsey
Tarrant County Junior College

Theresa Martin
College of San Mateo

Robin McFarland
Cabrillo College

Patricia Munn
Longview Community College

Gail Patt
Boston University

John O. Reiss
Humboldt State University

Bill Sattarelli
Central Michigan University

Robert Seegmiller
Brigham Young University

Mary Lynne Stephanou
Santa Monica College

Deborah Taylor
Kansas City Community College

Gail C. Turner
Virginia Commonwealth University

Christine K. Wade
University of Wyoming

Jeff Wanderer
West Hills Community College

Ophelia I. Weeks
Florida International University

Toni Yates
Seminole State College

Joan E. Zuckerman
Long Beach City College

Instructor Art Focus Group

Billie S. Lane
Chattanooga State Technical College

Gail O. McKenzie
Samford University

Robert J. McDonough
Georgia Perimeter College

Andrew J. Penniman
Georgia Perimeter College

Benjamin R. Wall, Jr.
Bevill State College, Walker College Campus

Anatomy Art Advisory Board

Linda Adamchak
Hudson Valley Community College

Patricia Ahanotu
Georgia Perimeter College, Central

Roger Biduaka
St. Phillips College

Ed Bradel
Raymond Walters College

Kenneth Bynum
University of North Carolina, Chapel Hill

Russell Garcia
San Antonio College

Delaine Gilcrease
Mesa Community College

Edwin Gines-Candelaria
Miami Dade Community College, Wolfson

James Horwitz
Palm Beach Community College

Christine Ilitis
Salt Lake Community College

Bonnie Kallison
Mesa Community College

Alice Mills
Middle Tennessee State University

Betsy Ott
Tyler Junior College, Texas

Linda Powell
Community College of Philadelphia

Robert Seegmiller
Brigham Young University

Sharon Simpson
Broward Community College

R. W. Stevens
Old Dominion University

Pamela Vaughn
Laredo Community College

Michael Vitale
Daytona Beach Community College

Deloris Wenzel
University of Georgia

Art Zeitlin
Kingsborough Community College

HAPS 1999 Focus Group

Steven Lewis
Penn Valley Community College

Gail Turner
Virginia Commonwealth University

Bobby Baldridge
Ashbury College

BRIEF CONTENTS

1 The Human Body: An Orientation 1

2 Cells: The Living Units 27

3 Basic Embryology 57

4 Tissues 75

5 The Integumentary System 111

6 Bones and Skeletal Tissues 129

7 Bones: The Axial Skeleton 151

8 Bones: The Appendicular Skeleton 185

9 Joints 211

10 Muscle Tissue 243

11 Muscles of the Body 265

12 Fundamentals of the Nervous System and Nervous Tissue 335

13 The Central Nervous System 359

14 The Peripheral Nervous System 409

15 The Autonomic Nervous System and Visceral Sensory Neurons 443

16 The Special Senses 465

17 Blood 503

18 The Heart 521

19 Blood Vessels 543

20 The Lymphatic and Immune Systems 583

21 The Respiratory System 603

22 The Digestive System 633

23 The Urinary System 675

24 The Reproductive System 701

25 The Endocrine System 743

26 Surface Anatomy 769

Appendix A The Metric System 799

Appendix B Answers to Multiple Choice and Matching Questions 801

Glossary 803

Photography and Illustration Credits 815

Index 819

DETAILED CONTENTS

1 The Human Body: An Orientation 1

AN OVERVIEW OF ANATOMY 2

Anatomical Terminology 2

Branches of Anatomy 2

Gross Anatomy 2

Microscopic Anatomy 2

Other Branches of Anatomy 2

The Hierarchy of Structural Organization 3

Scale: Length, Volume, and Weight 4

Anatomical Variability 4

GROSS ANATOMY: AN INTRODUCTION 4

The Anatomical Position 4

Directional and Regional Terms 8

Body Planes and Sections 10

The Human Body Plan 12

Body Cavities and Membranes 13

Dorsal Body Cavity 13

Ventral Body Cavity 13

Serous Cavities 14

Other Cavities 14

Abdominal Regions and Quadrants 15

MICROSCOPIC ANATOMY: AN INTRODUCTION 16

Light and Electron Microscopy 16

Preparing Human Tissue for Microscopy 17

Scanning Electron Microscopy 18

Artifacts 18

CLINICAL ANATOMY: AN INTRODUCTION TO MEDICAL IMAGING TECHNIQUES 19

X Rays 19

Advanced X-Ray Techniques 19

Computed Tomography 19

Digital Subtraction Angiography Imaging 20

Positron Emission Tomography 21

Sonography 21

Magnetic Resonance Imaging 22

CHAPTER SUMMARY 23

REVIEW QUESTIONS 24

CRITICAL REASONING AND CLINICAL APPLICATION QUESTIONS 25

A.D.A.M.® INTERACTIVE ANATOMY QUESTIONS 26

2 Cells: The Living Units 27

INTRODUCTION TO CELLS 28

THE PLASMA MEMBRANE 29

Structure 29

Functions 30

THE CYTOPLASM 33

Cytosol 33

Cytoplasmic Organelles 33

Mitochondria 33

Ribosomes 34

Endoplasmic Reticulum 35

Rough Endoplasmic Reticulum 35 Smooth Endoplasmic Reticulum 36

Golgi Apparatus 36

Lysosomes 38

Peroxisomes 39

Cytoskeleton 40

Centrosome and Centrioles 40

Vaults 40

Cytoplasmic Inclusions 40

THE NUCLEUS 41

Nuclear Envelope 42

Chromatin and Chromosomes 42

Nucleoli 45

THE CELL LIFE CYCLE 45

Interphase 45

Cell Division 45

Mitosis 48

Cytokinesis 48

CELLULAR DIVERSITY 48

DEVELOPMENTAL ASPECTS OF CELLS 52

Youth 52

Aging 52

A CLOSER LOOK

Cancer—The Intimate Enemy 50

CLINICAL APPLICATIONS
Hypercholesterolemia 32 Tay-Sachs Disease 38

RELATED CLINICAL TERMS 53

CHAPTER SUMMARY 53

REVIEW QUESTIONS 55

CRITICAL REASONING AND CLINICAL APPLICATION QUESTIONS 56

3 Basic Embryology 57

THE BASIC BODY PLAN 59

THE EMBRYONIC PERIOD 59

Week 1: From Zygote to Blastocyst 59

Week 2: The Two-Layered Embryo 61

Week 3: The Three-Layered Embryo 62

The Primitive Streak and the Three Germ Layers 62

The Notochord 62

Neurulation 62

The Mesoderm Begins to Differentiate 62

Week 4: The Body Takes Shape 66

Folding 66

Derivatives of the Germ Layers 66

Derivatives of Ectoderm 66 Derivatives of Endoderm 66 Derivatives of Mesoderm and Notochord 67

Weeks 5–8: The Second Month of Embryonic Development 68

THE FETAL PERIOD 68

A CLOSER LOOK
Focus on Birth Defects 70

RELATED CLINICAL TERMS 72

CHAPTER SUMMARY 72

REVIEW QUESTIONS 73

CRITICAL REASONING AND CLINICAL APPLICATIONS QUESTIONS 74

4 Tissues 75

EPITHELIA AND GLANDS 76

Special Characteristics of Epithelia 77

Classification of Epithelia 77

Simple Epithelia 78

Simple Squamous Epithelium 78 Simple Cuboidal Epithelium 78 Simple Columnar Epithelium 78 Pseudostratified Columnar Epithelium 78

Stratified Epithelia 78

Stratified Squamous Epithelium 78 Stratified Cuboidal and Columnar Epithelium 83 Transitional Epithelium 83

Epithelial Surface Features 83

Apical Surface Features: Microvilli and Cilia 83

Lateral Surface Features: Cell Junctions 84

Desmosomes 84 Tight Junctions 85 Gap Junctions 86

Basal Feature: The Basal Lamina 86

Glands 86

Exocrine Glands 87

Unicellular Exocrine Glands 87 Multicellular Exocrine Glands 87

Endocrine Glands 87

CONNECTIVE TISSUE 88

Connective Tissue Proper 88

Areolar Tissue: A Model Connective Tissue 89

Fibers Provide Support 90 Ground Substance Holds Fluid 91 Defense Cells Fight Infection 91 Fat Cells Store Nutrients 97

Other Loose Connective Tissues 97

Adipose Tissue 97 Reticular Connective Tissue 98

Dense Connective Tissue 98

Dense Irregular Connective Tissue 98 Dense Regular Connective Tissue 98

Other Connective Tissues: Cartilage, Bone, and Blood 98

Cartilage 99

Bone Tissue 99

Blood 99

COVERING AND LINING MEMBRANES 99

MUSCLE TISSUE 99

NERVOUS TISSUE 102

TISSUE RESPONSE TO INJURY 102

Inflammation 102

Repair 103

THE TISSUES THROUGHOUT LIFE 106

A Closer Look
Stem Cells: Medical Miracle of the Twenty-First Century? 105

Clinical Applications
Kartagener's Syndrome 83 Basement Membranes and Diabetes 86 Scurvy 91

RELATED CLINICAL TERMS 106

CHAPTER SUMMARY 107

REVIEW QUESTIONS 109

CRITICAL REASONING AND CLINICAL APPLICATIONS QUESTIONS 110

5 The Integumentary System 111

THE SKIN AND THE HYPODERMIS 112

Epidermis 113

Layers of the Epidermis 114

Stratum Basale 114 Stratum Spinosum 114 Stratum Granulosum 114 Straum Lucidum 115 Stratum Corneum 115

Dermis 115

Hypodermis 115

Skin Color 116

APPENDAGES OF THE SKIN 117

Hair and Hair Follicles 117

Hair 117

Hair Follicles 117

Types and Growth of Hair 119

Hair Thinning and Baldness 120

Sebaceous Glands 120

Sweat Glands 121

Nails 121

DISORDERS OF THE INTEGUMENTARY SYSTEM 122

Burns 122

Skin Cancer 123

Basal Cell Carcinoma 123

Squamous Cell Carcinoma 124

Melanoma 124

THE SKIN THROUGHOUT LIFE 124

A Closer Look
Tattoos 116

Clinical Applications
Skin's Response to Friction 113 Cyanosis 117 Chemotherapy and Hair Loss 120 Acne 121

RELATED CLINICAL TERMS 125

CHAPTER SUMMARY 125

REVIEW QUESTIONS 127

CRITICAL REASONING AND CLINICAL APPLICATIONS QUESTIONS 128

A.D.A.M.® INTERACTIVE ANATOMY QUESTIONS 128

6 Bones and Skeletal Tissues 129

CARTILAGES 130

Location and Basic Structure 130

Types of Cartilage 130

Hyaline Cartilage 130

Elastic Cartilage 130

Fibrocartilage 130

Growth of Cartilage 130

BONES 132

Functions of Bones 132

Classification of Bones 133

Gross Anatomy of Bones 134

Compact and Spongy Bone 134

Structure of a Typical Long Bone 134

Diaphysis and Epiphysis 134 Blood Vessels 134 The Medullary Cavity 134 Membranes 134

Structure of Short, Irregular, and Flat Bones 134

Bone Design and Stress 135

Microscopic Structure of Bone 136

Compact Bone 137

Spongy Bone 137

Chemical Composition of Bone Tissue 137

Bone Development 139

Intramembranous Ossification 139

Endochondral Ossification 139

Anatomy of Epiphyseal Growth Areas 140

Postnatal Growth of Endochondral Bones 141

Bone Remodeling 142

Repair of Bone Fractures 143

DISORDERS OF BONES 143

Osteoporosis 143

Osteomalacia and Rickets 145
Paget's Disease 146
Osteosarcoma 146
THE SKELETON THROUGHOUT LIFE 147
A CLOSER LOOK
The Marvelous Properties of Cartilage 132
RELATED CLINICAL TERMS 147
CHAPTER SUMMARY 148
REVIEW QUESTIONS 149
CRITICAL REASONING AND CLINICAL APPLICATIONS QUESTIONS 150
A.D.A.M.® INTERACTIVE ANATOMY QUESTIONS 150

7 Bones, Part I: The Axial Skeleton 151

THE SKULL 154
Overview of Skull Geography 154
Cranial Bones 154
Frontal Bone 154
Parietal Bones and the Major Sutures 154
Sutural Bones 157
Occipital Bone 157
Temporal Bones 157
Sphenoid Bone 160
Ethmoid Bone 162
Facial Bones 162
Mandible 162
Maxillary Bones 163
Zygomatic Bones 163
Nasal Bones 163
Lacrimal Bones 163
Palatine Bones 163
Vomer 163
Inferior Nasal Conchae 163
Special Parts of the Skull 165
Orbits 165
Nasal Cavity 165
Paranasal Sinuses 167
The Hyoid Bone 167
THE VERTEBRAL COLUMN 170
Intervertebral Discs 170
Regions and Normal Curvatures 172
General Structure of Vertebrae 172
Regional Vertebral Characteristics 173
Cervical Vertebrae 173
Thoracic Vertebrae 173
Lumbar Vertebrae 175
Sacrum 175
Coccyx 177
THE BONY THORAX 177
Sternum 177
Ribs 178
DISORDERS OF THE AXIAL SKELETON 179
Abnormal Spinal Curvatures 179
Stenosis of the Lumbar Spine 179
THE AXIAL SKELETON THROUGHOUT LIFE 179
CLINICAL APPLICATIONS
Fractured Mandible 163 Herniated Disc 171
The Dens and Fatal Trauma 173
RELATED CLINICAL TERMS 182
CHAPTER SUMMARY 182
REVIEW QUESTIONS 183
CRITICAL REASONING AND CLINICAL APPLICATIONS QUESTIONS 184
A.D.A.M.® INTERACTIVE ANATOMY QUESTIONS 184

8 Bones, Part II: The Appendicular Skeleton 185

THE PECTORAL GIRDLE 186
Clavicles 186
Scapulae 186
THE UPPER LIMB 189
Arm 189
Forearm 190
Ulna 190
Radius 190
Hand 190
Carpus 190
Metacarpus 193
Phalanges of the Fingers 196
THE PELVIC GIRDLE 196
Ilium 196
Ischium 196

Pubis 198

True and False Pelves 198

Pelvic Structure and Childbearing 199

THE LOWER LIMB 199

Thigh 199

Leg 202

Tibia 202

Fibula 202

Foot 202

Tarsus 202

Metatarsus 202

Phalanges of the Toes 205

Arches of the Foot 205

DISORDERS OF THE APPENDICULAR SKELETON 205

THE APPENDICULAR SKELETON THROUGHOUT LIFE 205

A CLOSER LOOK

Clinical Advances in Bone Repair 206

CLINICAL APPLICATIONS

Fractures of the Clavicle 186 Carpal Tunnel Syndrome 191 Hip Pointer 191 Metatarsal Stress Fracture 202

RELATED CLINICAL TERMS 207

CHAPTER SUMMARY 208

REVIEW QUESTIONS 209

CRITICAL REASONING AND CLINICAL APPLICATIONS QUESTIONS 210

A.D.A.M.® INTERACTIVE ANATOMY QUESTIONS 210

9 Joints 211

CLASSIFICATION OF JOINTS 212

FIBROUS JOINTS 212

Sutures 212

Syndesmoses 212

Gomphoses 213

CARTILAGINOUS JOINTS 213

Synchondroses 213

Symphyses 213

SYNOVIAL JOINTS 214

General Structure of Synovial Joints 214

How Synovial Joints Work 217

Bursae and Tendon Sheaths 219

Factors Influencing the Stability of Synovial Joints 219

Articular Surfaces 219

Ligaments 219

Muscle Tone 220

Movements Allowed by Synovial Joints 220

Gliding 220

Angular Movements 220

Flexion 221 Extension 221 Abduction 221 Adduction 221 Circumduction 221

Rotation 221

Special Movements 222

Supination and Pronation 222 Dorsiflexion and Plantar Flexion 222 Inversion and Eversion 222 Protraction and Retraction 222 Elevation and Depression 222 Opposition 222

Synovial Joints Classified by Shape 224

Plane Joints 224

Hinge Joints 224

Pivot Joints 225

Condyloid Joints 225

Saddle Joints 226

Ball-and-Socket Joints 226

Selected Synovial Joints 226

Temporomandibular Joint 226

Shoulder (Glenohumoral) Joint 227

Elbow Joint 227

Hip Joint 229

Knee Joint 231

Ankle Joint 233

DISORDERS OF JOINTS 234

Joint Injuries 234

Sprains 234

Dislocations 234

Torn Cartilage 236

Inflammatory and Degenerative Conditions 236

Bursitis and Tendonitis 236

Arthritis 236

Osteoarthritis (Degenerative Joint Disease) 236 Rheumatoid Arthritis 237 Gouty Arthritis (Gout) 237

Lyme Disease 237

THE JOINTS THROUGHOUT LIFE 238

A CLOSER LOOK
The Development of Artificial Joints 238

CLINICAL APPLICATIONS
Temporomandibular Disorder 227
Shoulder Dislocations 227 Elbow Trauma 227
Hip Dislocation 231 Knee Injuries 232
Ankle Sprains 234

RELATED CLINICAL TERMS 239

CHAPTER SUMMARY 239

REVIEW QUESTIONS 241

CRITICAL REASONING AND CLINICAL APPLICATION QUESTIONS 242

A.D.A.M.® INTERACTIVE ANATOMY QUESTIONS 242

10 Muscle Tissue 243

OVERVIEW OF MUSCLE TISSUE 244

Functions 244

Types of Muscle Tissue 244

SKELETAL MUSCLE 244

Basic Features of a Skeletal Muscle 244

Connective Tissue and Fascicles 244

Nerves and Blood Vessels 247

Muscle Attachments 247

Microscopic and Functional Anatomy of Skeletal Muscle Tissue 248

The Skeletal Muscle Fiber 248

Myofibrils and Sarcomeres 248

Mechanism of Contraction 248

Muscle Extension 248

Muscle Fiber Length and the Force of Contraction 250

The Role of Titin 250

Sarcoplasmic Reticulum and T Tubules 251

Types of Skeletal Muscle Fibers 252

Red Slow-Twitch Fibers 252 White Fast-Twitch Fibers 252
Intermediate Fast-Twitch Fibers 252

CARDIAC MUSCLE 254

SMOOTH MUSCLE 254

DISORDERS OF MUSCLE TISSUE 257

Muscular Dystrophy 258

Myofascial Pain Syndrome 258

Fibromyalgia 259

MUSCLE TISSUE THROUGHOUT LIFE 259

A CLOSER LOOK
Anabolic Steroid Abuse 260

CLINICAL APPLICATIONS
Rhabdomyolysis 252

RELATED CLINICAL TERMS 261

CHAPTER SUMMARY 262

REVIEW QUESTIONS 263

CRITICAL REASONING AND CLINICAL APPLICATION QUESTIONS 264

A.D.A.M.® INTERACTIVE ANATOMY QUESTIONS 264

11 Muscles of the Body 265

LEVER SYSTEMS: BONE-MUSCLE RELATIONSHIPS 266

ARRANGEMENT OF FASCICLES IN MUSCLES 268

INTERACTIONS OF SKELETAL MUSCLES IN THE BODY 268

NAMING THE SKELETAL MUSCLES 270

A DEVELOPMENT-BASED ORGANIZATION OF THE MUSCLES 270

MAJOR SKELETAL MUSCLES OF THE BODY 272

Muscle Galleries

Muscle Gallery: Table 11.1
Muscles of the Head, Part I: Facial Expression 276

Muscle Gallery: Table 11.2
Muscles of the Head, Part II: Mastication and Tongue Movement 278

Muscle Gallery: Table 11.3
Muscles of the Anterior Neck and Throat: Swallowing 280

Muscle Gallery: Table 11.4
Muscles of the Neck and Vertebral Column: Head Movements and Trunk Extension 283

Muscle Gallery: Table 11.5
Deep Muscles of the Thorax: Breathing 287

Muscle Gallery: Table 11.6
Muscles of the Abdominal Wall: Trunk Movements and Compression of Abdominal Viscera 289

Muscle Gallery: Table 11.7
Muscles of the Pelvic Floor and Perineum: Support of Abdominopelvic Organs 292

Muscle Gallery: Table 11.8
Superficial Muscles of the Anterior and Posterior Thorax: Movements of the Scapula 294

Muscle Gallery: Table 11.9
Muscles Crossing the Shoulder Joint: Movements of the Arm (Humerus) 297

Muscle Gallery: Table 11.10
Muscles Crossing the Elbow Joint: Flexion and Extension of the Forearm 300

Muscle Gallery: Table 11.11
Muscles of the Forearm: Movements of the Wrist, Hand, and Fingers 301

Muscle Gallery: Table 11.12
Summary of Actions of Muscles Acting on the Arm, Forearm, and Hand 307

Muscle Gallery: Table 11.13
Intrinsic Muscles of the Hand: Fine Movements of the Fingers 309

Muscle Gallery: Table 11.14
Muscles Crossing the Hip and Knee Joints: Movements of the Thigh and Leg 312

Muscle Gallery: Table 11.15
Muscles of the Leg: Movements of the Ankle and Toes 319

Muscle Gallery: Table 11.16
Summary of Actions of Muscles Acting on the Thigh, Leg, and Foot 326

Muscle Gallery: Table 11.17
Intrinsic Muscles of the Foot: Toe Movement and Foot Support 329

CLINICAL APPLICATIONS
Torticollis 283 Hernia 289 Tennis Elbow 301

RELATED CLINICAL TERMS 332

CHAPTER SUMMARY 332

REVIEW QUESTIONS 333

CRITICAL REASONING AND CLINICAL APPLICATION QUESTIONS 334

A.D.A.M.® INTERACTIVE ANATOMY QUESTIONS 334

12 Fundamentals of the Nervous System and Nervous Tissue 335

BASIC DIVISIONS OF THE NERVOUS SYSTEM 336

NERVOUS TISSUE 338

The Neuron 338

The Cell Body 338

Neuron Processes 338

Dendrites 340 Axons 340

Synapses 340

Signals Carried by Neurons 342

Action Potentials on Axons 342 Graded Potentials on Axons 342 Synaptic Potentials 342

Classification of Neurons 343

Structural Classification of Neurons 343 Functional Classification of Neurons 343

Supporting Cells 346

Supporting Cells in the CNS 346

Supporting Cells in the PNS 347

Myelin Sheaths 347

Myelin Sheaths in the CNS 347 Myelin Sheaths in the PNS 348

Nerves 348

BASIC NEURONAL ORGANIZATION OF THE NERVOUS SYSTEM 350

Reflex Arcs 350

Simplified Design of the Nervous System 351

DISORDERS OF NERVOUS TISSUE 353

NERVOUS TISSUE THROUGHOUT LIFE 354

CLINICAL APPLICATIONS
Tic Douloureux 346 Regeneration and Spinal Cord Injuries 356

RELATED CLINICAL TERMS 356

CHAPTER SUMMARY 356

REVIEW QUESTIONS 358

CRITICAL REASONING AND CLINICAL APPLICATION QUESTIONS 358

13 The Central Nervous System 359

THE BRAIN 360

Embryonic Development of the Brain 360

Basic Parts and Organization of the Brain 362

Ventricles of the Brain 363

The Cerebral Hemispheres 363

Cerebral Cortex 365

Motor Areas 366 Sensory Areas 369 Association Areas 370 Lateralization of Cortical Functioning 371

Cerebral White Matter 372

Basal Nuclei (Basal Ganglia) 372

The Diencephalon 374

The Thalamus 375
The Hypothalamus 377
The Epithalamus 379
The Brain Stem 379
The Midbrain 379
The Pons 381
The Medulla Oblongata 381
The Cerebellum 382
Cerebellar Peduncles 385
Functional Brain Systems 385
The Limbic System 385
The Reticular Formation 387
Protection of the Brain 388
Meninges 388
Dura Mater 388 Arachnoid Mater 389
Pia Mater 389
Cerebrospinal Fluid 391
Blood-Brain Barrier 391

THE SPINAL CORD 391
Gray Matter of the Spinal Cord and Spinal Roots 395
White Matter of the Spinal Cord 395
Sensory and Motor Pathways 397
Ascending (Sensory) Pathways 397
Descending (Motor) Pathways 397

DISORDERS OF THE CENTRAL NERVOUS SYSTEM 401
Brain Dysfunction 401
Traumatic Brain Injuries 402
Degenerative Brain Diseases 402
Cerebrovascular Accidents 402 Alzheimer's Disease 402
Spinal Cord Damage 403

THE CENTRAL NERVOUS SYSTEM THROUGHOUT LIFE 403

A CLOSER LOOK
Sex-Related Differences in Brain Structure 373

CLINICAL APPLICATIONS
Dyskinesia 372 Hydrocephalus 391

RELATED CLINICAL TERMS 404
CHAPTER SUMMARY 405
REVIEW QUESTIONS 407
CRITICAL REASONING AND CLINICAL APPLICATION QUESTIONS 408
A.D.A.M.® INTERACTIVE ANATOMY QUESTIONS 408

14 The Peripheral Nervous System 409

PERIPHERAL SENSORY RECEPTORS 411
Classification by Location 411
Classification by Stimulus Detected 411
Classification by Structure 413
Free Dendritic Endings 413
Encapsulated Dendritic Endings 413
Meissner's Corpuscles 414 Krause's End Bulbs 414
Pacinian Corpuscles 414 Ruffini's Corpuscles 414
Proprioceptors 414

PERIPHERAL MOTOR ENDINGS 416
Innervation of Skeletal Muscle 416
Innervation of Visceral Muscle and Glands 417

CRANIAL NERVES 418
I The Olfactory Nerves 420
II The Optic Nerves 420
III The Oculomotor Nerves 420
IV The Trochlear Nerves 421
V The Trigeminal Nerves 421
VI The Abducens Nerves 422
VII The Facial Nerves 423
VIII The Vestibulocochlear Nerves 424
IX The Glossopharyngeal Nerves 424
X The Vagus Nerves 425
XI The Accessory Nerves 425
XII The Hypoglossal Nerves 426

SPINAL NERVES 426
Innervation of the Back 428
Innervation of the Anterior Thoracic and Abdominal Wall 428
Introduction to Nerve Plexuses 428
Cervical Plexus and the Neck 428
Brachial Plexus and Innervation of the Upper Limb 429
Lumbar Plexus and Innervation of the Lower Limb 432
Sacral Plexus and Innervation of the Lower Limb 432
Innervation of Joints of the Body 435
Innervation of the Skin: Dermatomes 435

DISORDERS OF THE PERIPHERAL NERVOUS SYSTEM 437

THE PERIPHERAL NERVOUS SYSTEM THROUGHOUT LIFE 437

A CLOSER LOOK
Post-Polio Syndrome—The Plight of Some "Recovered" Polio Victims 438

CLINICAL APPLICATIONS
Anosmia 420 Optic Nerve Damage 420 Oculomotor Nerve Paralysis 420 Tic Douloureux 421 Trochlear Nerve Damage 421 Abducens Nerve Paralysis 422 Bell's Palsy 423 Vestibulocochlear Nerve Damage 424 Glossopharyngeal Nerve Damage 424 Vagus Nerve Damage 425 Damage to Accessory Nerves 425 Hypoglossal Nerve Damage 426 Brachial Plexus Injuries 431 Sciatic Nerve Injuries 434 Clinical Importance of Dermatomes 436

RELATED CLINICAL TERMS 439

CHAPTER SUMMARY 439

REVIEW QUESTIONS 441

CRITICAL REASONING AND CLINICAL APPLICATION QUESTIONS 442

A.D.A.M.® INTERACTIVE ANATOMY QUESTIONS 442

15 The Autonomic Nervous System and Visceral Sensory Neurons 443

INTRODUCTION TO THE AUTONOMIC NERVOUS SYSTEM 444

Comparison of the Autonomic and Somatic Motor Systems 444

Divisions of the Autonomic Nervous System 445

THE PARASYMPATHETIC DIVISION 447

Cranial Outflow 448

Outflow via the Oculomotor Nerve (III) 448

Outflow via the Facial Nerve (VII) 448

Outflow via the Glossopharyngeal Nerve (IX) 448

Outflow via the Vagus Nerve (X) 449

Sacral Outflow 449

THE SYMPATHETIC DIVISION 452

Basic Organization 452

Sympathetic Trunk Ganglia 452

Prevertebral Ganglia 452

Sympathetic Pathways 452

Sympathetic Pathways to the Body Periphery 454

Sympathetic Pathways to the Head 454

Sympathetic Pathways to the Thoracic Organs 455

Sympathetic Pathways to the Abdominal Organs 456

Sympathetic Pathways to the Pelvic Organs 456

The Role of the Adrenal Medulla in the Sympathetic Division 456

CENTRAL CONTROL OF THE AUTONOMIC NERVOUS SYSTEM 457

Control by the Brain Stem and Spinal Cord 457

Control by the Hypothalamus and Amygdala 458

Control by the Cerebral Cortex 458

VISCERAL SENSORY NEURONS 458

VISCERAL REFLEXES 459

DISORDERS OF THE AUTONOMIC NERVOUS SYSTEM 459

THE AUTONOMIC NERVOUS SYSTEM THROUGHOUT LIFE 460

RELATED CLINICAL TERMS 461

CHAPTER SUMMARY 461

REVIEW QUESTIONS 463

CRITICAL REASONING AND CLINICAL APPLICATION QUESTIONS 464

A.D.A.M.® INTERACTIVE ANATOMY QUESTIONS 464

16 The Special Senses 465

THE CHEMICAL SENSES: TASTE AND SMELL 466

Taste (Gustation) 466

Taste Buds 466

Taste Sensations and the Gustatory Pathway 466

Smell (Olfaction) 468

Disorders of the Chemical Senses 470

Embryonic Development of the Chemical Senses 470

THE EYE AND VISION 470

Accessory Structures of the Eye 470

Eyebrows 470

Eyelids 470

Conjunctiva 471

Lacrimal Apparatus 471

Extrinsic Eye Muscles 472

Anatomy of the Eyeball 474

The Fibrous Tunic 474

The Vascular Tunic 476

The Sensory Tunic (Retina) 477

Photoreceptors 478 Regional Specializations of the Retina 479 Blood Supply of the Retina 479

Internal Chambers and Fluids 479

Lens 480

The Eye as an Optical Device 480

Visual Pathways 481

Visual Pathway to the Cerebral Cortex 481

Visual Pathways to Other Parts of the Brain 484

Disorders of the Eye and Vision 484

Embryonic Development of the Eye 485

THE EAR: HEARING AND EQUILIBRIUM 486

The Outer (External) Ear 486

The Middle Ear 486

The Inner (Internal) Ear 489

The Vestibule 489

The Semicircular Canals 491

The Cochlea 491

Equilibrium and Auditory Pathways 494

Disorders of Equilibrium and Hearing 495

Motion Sickness 495

Meniere's Syndrome 495

Deafness 495

Embryonic Development of the Ear 496

THE SPECIAL SENSES THROUGHOUT LIFE 496

A CLOSER LOOK

Eye-Focusing Disorders 482

CLINICAL APPLICATIONS

Strabismus 473 Corneal Transplants 476 Detached Retina 479 Glaucoma 480 Cataract 480 Perforated Eardrum 486 Middle Ear Infections 488

RELATED CLINICAL TERMS 497

CHAPTER SUMMARY 498

REVIEW QUESTIONS 500

CRITICAL REASONING AND CLINICAL APPLICATION QUESTIONS 501

A.D.A.M.® INTERACTIVE ANATOMY QUESTIONS 501

17 Blood 503

OVERVIEW: COMPOSITION OF BLOOD 504

BLOOD PLASMA 504

FORMED ELEMENTS 505

Erythrocytes 505

Leukocytes 506

Granulocytes 507

Neutrophils 507 Eosinophils 508 Basophils 508

Agranulocytes 508

Lymphocytes 508 Monocytes 509

Platelets 510

BLOOD CELL FORMATION 512

Bone Marrow as the Site of Hematopoiesis 512

Cell Lines in Blood Cell Formation 513

Genesis of Erythrocytes 513

Formation of Leukocytes and Platelets 513

BLOOD DISORDERS 515

Disorders of Erythrocytes 515

Disorders of Leukocytes 515

Platelet Disorders 515

THE BLOOD THROUGHOUT LIFE 516

A CLOSER LOOK

Bone Marrow and Cord-Blood Transplants 516

CLINICAL APPLICATIONS

Complete Blood Count 510

RELATED CLINICAL TERMS 517

CHAPTER SUMMARY 517

REVIEW QUESTIONS 519

CRITICAL REASONING AND CLINICAL APPLICATION QUESTIONS 520

18 The Heart 521

LOCATION AND ORIENTATION WITHIN THE THORAX 522

STRUCTURE OF THE HEART 524

Coverings 524

Layers of the Heart Wall 524

Heart Chambers 525

Right Atrium 525

Right Ventricle 525

Left Atrium 525

Left Ventricle 528

PATHWAY OF BLOOD THROUGH THE HEART 528

HEART VALVES 529
Valve Structure 529
Valve Function 529
Heart Sounds 530
FIBROUS SKELETON 532
CONDUCTING SYSTEM AND INNERVATION 532
Conducting System 532
Innervation 533
BLOOD SUPPLY TO THE HEART 534
DISORDERS OF THE HEART 535
Coronary Artery Disease 535
Heart Failure 535
Disorders of the Conduction System 537
THE HEART THROUGHOUT LIFE 537
Development of the Heart 537
The Heart in Adulthood and Old Age 539
A CLOSER LOOK
Treating Heart Disease 536
CLINICAL APPLICATIONS
Pericarditis and Cardiac Tamponade 524 Valve Disorders 532 Damage to the Conducting System 533
RELATED CLINICAL TERMS 540
CHAPTER SUMMARY 540
REVIEW QUESTIONS 541
CRITICAL REASONING AND CLINICAL APPLICATION QUESTIONS 542
A.D.A.M.® INTERACTIVE ANATOMY QUESTIONS 542

19 Blood Vessels 543

Part I: General Characteristics of Blood Vessels
STRUCTURE OF BLOOD VESSEL WALLS 544
TYPES OF BLOOD VESSELS 544
Arteries 545
Elastic Arteries 546
Muscular Arteries 546
Arterioles 546
Capillaries 547
Capillary Beds 547
Capillary Permeability 547
Routes of Capillary Permeability 549 Low-Permeability Capillaries: The Blood-Brain Barrier 549
Sinusoids 549
Veins 550
VASCULAR ANASTOMOSES 550
VASA VASORUM 551

Part II: Blood Vessels of the Body
THE PULMONARY CIRCULATION 552
THE SYSTEMIC CIRCULATION 553
Systemic Arteries 553
Aorta 553
Ascending Aorta 553 Aortic Arch 553 Descending Aorta 553
Arteries of the Head and Neck 553
Common Carotid Arteries 553 Vertebral Arteries 555 Thyrocervical and Costocervical Trunks 557
Arteries of the Upper Limb 557
Axillary Artery 557 Brachial Artery 557 Radial Artery 557 Ulnar Artery 557 Palmar Arches 557
Arteries of the Thorax 557
Arteries of the Thoracic Wall 557 Arteries of the Thoracic Visceral Organs 559
Arteries of the Abdomen 559
Inferior Phrenic Arteries 559 Celiac Trunk 559 Superior Mesenteric Artery 560 Suprarenal Arteries 560 Renal Arteries 560 Gonadal Arteries 560 Inferior Mesenteric Arteries 561 Lumbar Arteries 561 Median Sacral Artery 561 Common Iliac Arteries 561
Arteries of the Pelvis and Lower Limbs 561
Internal Iliac Arteries 561 External Iliac Artery 562 Femoral Artery 562 Popliteal Artery 563 Anterior Tibial Artery 563 Posterior Tibial Artery 563
Systemic Veins 563
Venae Cavae and Their Major Tributaries 564
Superior Vena Cava 564 Inferior Vena Cava 564
Veins of the Head and Neck 566
Dural Sinuses 566 Internal Jugular Veins 566 External Jugular Veins 567 Vertebral Veins 567
Veins of the Upper Limbs 567
Deep Veins 567 Superficial Veins 567
Veins of the Thorax 568
Azygos Vein 569 Hemiazygos Vein 569 Accessory Hemiazygos Vein 569
Veins of the Abdomen 569
Lumbar Veins 569 Gonadal Veins 569 Renal Veins 569 Suprarenal Veins 569 Hepatic Veins 570 Hepatic Portal System 570

Veins of the Pelvis and Lower Limbs 571
Deep Veins 571 Superficial Veins 571
Portal-Systemic Anastomoses 572

DISORDERS OF BLOOD VESSELS 573

THE BLOOD VESSELS THROUGHOUT LIFE 576

Fetal Circulation 576
Vessels to and from the Placenta 576
Shunts Away from the Pulmonary Circuit 576

Blood Vessels in Adulthood 578

A CLOSER LOOK
Atherosclerosis—Controlling the Silent Killer 574

CLINICAL APPLICATIONS
Varicose Veins 550 Medical Importance of the Cavernous Sinus 566 Medical Importance of the Saphenous Vein 572

RELATED CLINICAL TERMS 578

CHAPTER SUMMARY 579

REVIEW QUESTIONS 581

CRITICAL REASONING AND CLINICAL APPLICATION QUESTIONS 582

A.D.A.M.® INTERACTIVE ANATOMY QUESTIONS 582

20 The Lymphatic and Immune Systems 583

THE LYMPHATIC SYSTEM 584

Lymph Capillaries 585

Lymphatic Collecting Vessels 585

Lymph Nodes 586

Lymph Trunks 587

Lymph Ducts 589

THE IMMUNE SYSTEM 589

Lymphocytes 589

Lymphocyte Activation 589

Lymphoid Tissue 593

Lymphoid Organs 595
Lymph Nodes 595
Spleen 595
Thymus 597
Tonsils 597
Aggregated Lymphoid Nodules and the Appendix 598

DISORDERS OF THE LYMPHATIC AND IMMUNE SYSTEMS 598

THE LYMPHATIC AND IMMUNE SYSTEMS THROUGHOUT LIFE 599

A CLOSER LOOK
AIDS: The Modern Day Plague? 592

CLINICAL APPLICATIONS
The Lymph Vessels and Edema 586 Swollen Lymph Nodes 587 Splenectomy 596

RELATED CLINICAL TERMS 600

CHAPTER SUMMARY 600

REVIEW QUESTIONS 601

CRITICAL REASONING AND CLINICAL APPLICATION QUESTIONS 602

A.D.A.M.® INTERACTIVE ANATOMY QUESTIONS 602

21 The Respiratory System 603

FUNCTIONAL ANATOMY OF THE RESPIRATORY SYSTEM 604

The Nose and the Paranasal Sinuses 604
The Nose 604
The Nasal Cavity 604 The Respiratory Mucosa 607
The Nasal Conchae 607
The Paranasal Sinuses 608

The Pharynx 608
The Nasopharynx 608
The Oropharynx 608
The Laryngopharynx 608

The Larynx 609
Voice Production 610
Sphincter Functions of the Larynx 611
Innervation of the Larynx 611

The Trachea 612

The Bronchi and Subdivisions: The Bronchial Tree 613
Bronchi in the Conducting Zone 613
The Respiratory Zone 614

The Lungs and Pleurae 615
The Pleurae 615
Gross Anatomy of the Lungs 617

Blood Supply and Innervation of the Lungs 619

VENTILATION 621

The Mechanism of Ventilation 621

Inspiration 621

Expiration 622

Neural Control of Ventilation 623

DISORDERS OF THE RESPIRATORY SYSTEM 623

Disorders of Lower Respiratory Structures 624

Bronchial Asthma 624

Chronic Obstructive Pulmonary Disease 624

Pneumonia 625

Tuberculosis 625

Cystic Fibrosis 626

Disorders of the Upper Respiratory Structures 627

Epistaxis 627

Epiglottitis 628

THE RESPIRATORY SYSTEM THROUGHOUT LIFE 628

A CLOSER LOOK
Lung Cancer—The Facts Behind the Smoke Screen 626

CLINICAL APPLICATIONS
Rhinitis 607 Sinusitis 608 Pleurisy and Pleural Effusion 617 Collapsed Lung 623 Respiratory Distress Syndrome 623

RELATED CLINICAL TERMS 629

CHAPTER SUMMARY 629

REVIEW QUESTIONS 631

CRITICAL REASONING AND CLINICAL APPLICATION QUESTIONS 632

A.D.A.M.® INTERACTIVE ANATOMY QUESTIONS 632

22 The Digestive System 633

OVERVIEW OF THE DIGESTIVE SYSTEM 634

Digestive Processes 635

Abdominal Regions and Quadrants 637

ANATOMY OF THE DIGESTIVE SYSTEM 637

The Peritoneal Cavity and Peritoneum 637

Histology of the Alimentary Canal Wall 638

The Mucosa 638

The Submucosa 638

The Muscularis Externa 638

The Serosa 638

Nerve Plexuses 639

The Mouth and Associated Organs 640

The Mouth 640

Histology of the Mouth 640 Lips and Cheeks 640 The Palate 641

The Tongue 641

The Salivary Glands 642

The Teeth 643

Dentition and the Dental Formula 644 Tooth Structure 644

The Pharynx 645

The Esophagus 646

Gross Anatomy 646

Microscopic Anatomy 646

The Stomach 646

Gross Anatomy 646

Microscopic Anatomy 648

The Small Intestine 652

Gross Anatomy 652

Microscopic Anatomy 653

Modifications for Absorption 653 Histology of the Wall 654

The Large Intestine 654

Gross Anatomy 656

Cecum and Vermiform Appendix 656 The Colon 656 The Rectum 656 The Anal Canal 657 Vessels and Nerves 657 Defecation 657

Microscopic Anatomy 657

The Liver 658

Gross Anatomy 658

Microscopic Anatomy 660

The Gallbladder 662

The Pancreas 663

Mesenteries 663

DISORDERS OF THE DIGESTIVE SYSTEM 666

Intestinal Obstruction 666

Inflammatory Bowel Disease 666

Viral Hepatitis 667

Cystic Fibrosis and the Pancreas 667

THE DIGESTIVE SYSTEM THROUGHOUT LIFE 667

Embryonic Development 667

The Digestive System in Later Life 668

A CLOSER LOOK

Peptic Ulcers 650

CLINICAL APPLICATIONS

Peritonitis 638 Gum Disease 645 Hiatal Hernia and Gastroesophageal Reflux Disease 646 Appendicitis 656 Diverticulosis and Diverticulitis 656 Hemorrhoids 657 Cirrhosis 662 Gallstones 662 Pancreatitis 663

RELATED CLINICAL TERMS 668

CHAPTER SUMMARY 669

REVIEW QUESTIONS 672

CRITICAL REASONING AND CLINICAL APPLICATION QUESTIONS 673

A.D.A.M.® INTERACTIVE ANATOMY QUESTIONS 673

23 The Urinary System 675

KIDNEYS 676

Gross Anatomy of the Kidneys 676

Location and External Anatomy 676

Internal Gross Anatomy 678

Gross Vasculature and Nerve Supply 678

Microscopic Anatomy of the Kidneys 681

Mechanisms of Urine Production 681

The Nephron 682

Renal Corpuscle 682 Tubular Section of the Nephron 682 Classes of Nephrons 684

Collecting Tubules 684

Microscopic Blood Vessels Associated with Uriniferous Tubules 685

Juxtaglomerular Apparatus 686

Interstitial Connective Tissue 686

URETERS 686

Gross Anatomy 686

Microscopic Anatomy 687

URINARY BLADDER 687

URETHRA 689

MICTURITION 691

DISORDERS OF THE URINARY SYSTEM 692

Urinary Tract Infections 692

Renal Calculi 693

Cancer of the Urinary Organs 693

THE URINARY SYSTEM THROUGHOUT LIFE 693

A CLOSER LOOK

ADPKD—The Most Interesting Disease You've Never Heard Of? 696

CLINICAL APPLICATIONS

Surgical Approach to the Kidney 678 Pyelonephritis 678 Pyelography 678 Micturition Dysfunctions 691 Kidney Transplant 693

RELATED CLINICAL TERMS 695

CHAPTER SUMMARY 696

REVIEW QUESTIONS 699

CRITICAL REASONING AND CLINICAL APPLICATION QUESTIONS 699

A.D.A.M.® INTERACTIVE ANATOMY QUESTIONS 700

24 The Reproductive System 701

THE MALE REPRODUCTIVE SYSTEM 702

The Scrotum 703

The Testes 703

Gross Anatomy 703

The Seminiferous Tubules and Spermatogenesis 705

Spermatogenic Cells 705 Sustentacular Cells 705 Myoid Cells 707 Interstitial Cells 707

The Reproductive Duct System in Males 708

The Epididymis 708

The Ductus Deferens 709

The Spermatic Cord 709

The Urethra 710

Accessory Glands 710

The Seminal Vesicles 710

The Prostate Gland 711

The Bulbourethral Glands 712

The Penis 712

The Male Perineum 715

THE FEMALE REPRODUCTIVE SYSTEM 715

The Ovaries 715

The Ovarian Cycle 716

Follicular Phase 717 Ovulation 718 Luteal Phase 718

Oogenesis 718

The Uterine Tubes 718

The Uterus 720

Supports of the Uterus 720

The Uterine Wall 721
The Uterine Cycle 722
The Vagina 724
The External Genitalia and Female Perineum 724
The Mammary Glands 725
PREGNANCY AND CHILDBIRTH 726
Pregnancy 727
Events Leading to Fertilization 727
Implantation 728
Formation of the Placenta 728
Anatomy of the Placenta 729
Childbirth 729
DISORDERS OF THE REPRODUCTIVE SYSTEM 732
Reproductive System Cancers in Males 733
Testicular Cancer 733
Prostate Cancer 733
Reproductive System Cancers in Females 733
Ovarian Cancer 733
Endometrial Cancer 733
Cervical Cancer 734
Breast Cancer 734
THE REPRODUCTIVE SYSTEM THROUGHOUT LIFE 734
Embryonic Development of the Sex Organs 734
Descent of the Gonads 737
Puberty 738
Menopause 738
A CLOSER LOOK
Sexually Transmitted Diseases 712
CLINICAL APPLICATIONS
Vasectomy 709 Inguinal Hernia 710 Benign Prostatic Hyperplasia 711 Circumcision 712 Pelvic Inflammatory Disease 720 Endometriosis 724 Episiotomy 725 Displaced Placentas 729 Hypospadias 737 Cryptorchidism 737
RELATED CLINICAL TERMS 738
CHAPTER SUMMARY 739
REVIEW QUESTIONS 741
CRITICAL REASONING AND CLINICAL APPLICATION QUESTIONS 742
A.D.A.M.® INTERACTIVE ANATOMY QUESTIONS 742

25 The Endocrine System 743
THE ENDOCRINE ORGAN SYSTEM AND HORMONES: AN OVERVIEW 744
Endocrine Organs 744
Hormones 745
Classes of Hormones 745
Basic Hormone Action 745
Control of Hormone Secretion 745
THE MAJOR ENDOCRINE ORGANS 745
The Pituitary Gland 745
The Adenohypophysis 747
Hypothalamic Control of Hormone Secretion from the Adenohypophysis 750
The Neurohypophysis 751
The Thyroid Gland 753
The Parathyroid Glands 754
The Adrenal (Suprarenal) Glands 754
The Adrenal Medulla 755
The Adrenal Cortex 755
Structure of Steroid-Secreting Cells 757
The Pineal Gland 758
The Pancreas 758
The Thymus 759
The Gonads 759
OTHER ENDOCRINE STRUCTURES 759
DISORDERS OF THE ENDOCRINE SYSTEM 760
Pituitary Disorders 760
Disorders of the Thyroid Gland 760
Disorders of the Adrenal Cortex 760
Disorders of the Pancreas: Diabetes Mellitus 761
THE ENDOCRINE SYSTEM THROUGHOUT LIFE 762
A CLOSER LOOK
Potential Uses for Growth Hormone 752
CLINICAL APPLICATIONS
Therapeutic Uses of Oxytocin 753
RELATED CLINICAL TERMS 763
CHAPTER SUMMARY 763
REVIEW QUESTIONS 765
CRITICAL REASONING AND CLINICAL APPLICATION QUESTIONS 767
A.D.A.M.® INTERACTIVE ANATOMY QUESTIONS 767

26 Surface Anatomy 769

THE HEAD 770
Cranium 770
Face 770
THE NECK 774
Skeletal Landmarks 774
Muscles of the Neck 774
Triangles of the Neck 777
THE TRUNK 777
The Back 777
Bones of the Back 777
Muscles of the Back 779
The Thorax 779
Bones of the Thorax 779
Muscles of the Thorax 780
The Abdomen 781
Bony Landmarks 781
Muscles and Other Abdominal Surface Features 781
The Pelvis and Perineum 783
UPPER LIMB AND SHOULDER 783
Axilla 783
Shoulder 783
Arm 783
Elbow Region 783
Forearm and Hand 787
LOWER LIMB AND GLUTEAL REGION 789
Gluteal Region 789
Thigh 791
Leg and Foot 791

CLINICAL APPLICATIONS
Triangle of Auscultation 779 Diagnosing Inguinal Hernias 781 Umbilical Hernias and McBurney's Point 781 Bowel Sounds 783 Palpation of a Colles' Fracture 787 Gluteal Intramuscular Injections 790

CHAPTER SUMMARY 794
REVIEW QUESTIONS 796
CRITICAL REASONING AND CLINICAL APPLICATION QUESTIONS 796
A.D.A.M.® INTERACTIVE ANATOMY QUESTIONS 797

Appendices

Appendix A The Metric System 799
Appendix B Answers to Multiple Choice and Matching Questions 801
Glossary 803
Photography and Illustration Credits 815
Index 819

CHAPTER 1

The Human Body: An Orientation

Illumination of the body

CHAPTER OUTLINE + STUDENT OBJECTIVES

An Overview of Anatomy 2–4

1. Define anatomy and physiology, and describe the subdivisions of anatomy.
2. Name the levels of structural organization in the human body, and explain their relationships.
3. List the organ systems of the body, and briefly state their functions.
4. Classify the levels of structural organization in the body according to relative and actual size.
5. Give an example of anatomical variation.

Gross Anatomy: An Introduction 4–16

6. Define the anatomical position.
7. Use anatomical terminology to describe body directions, regions, and planes.
8. Describe the basic structures that humans share with other vertebrates.
9. Locate the major body cavities and their subdivisions.
10. Name the nine regions and four quadrants of the abdomen, and name the visceral organs associated with these regions.

Microscopic Anatomy: An Introduction 16–18

11. Explain how human tissue is prepared and examined for its microscopic structure.
12. Distinguish tissue viewed by light microscopy from that viewed by electron microscopy.

Clinical Anatomy: An Introduction to Medical Imaging Techniques 19–23

13. Describe the medical imaging techniques that are used to visualize structures inside the body.

AS YOU READ THIS BOOK, YOU WILL BE LEARNING about one of the most fascinating subjects there is—your own body. The study of human anatomy is not only an interesting and highly personal experience, but it is timely as well. Almost every week, the news media report advances in medical science. Understanding how your body is built and how it works enables you to appreciate newly developed techniques for detecting and treating disease and to apply guidelines for staying healthy. For those of you preparing for a career in the health sciences, your knowledge of human anatomy is the foundation of your clinical practice.

AN OVERVIEW OF ANATOMY

Anatomy is the study of the structure of the human body. It is also called **morphology** (mor″fol′o-je), the science of form. An old and proud science, anatomy has been a field of serious intellectual investigation for at least 2300 years. It was the most prestigious biological discipline of the 1800s and is still dynamic and growing.

Anatomy is closely related to **physiology**, the study of body function. Although you may be studying anatomy and physiology in separate courses, the two are truly inseparable, because structure tends to reflect function. For example, the lens of the eye is transparent and curved; it could not perform its function of focusing light if it were opaque and uncurved. Likewise, the thick, long bones in our legs could not support our weight if they were soft and thin. This textbook stresses the closeness of the relationship between structure and function. In almost all cases, a description of the anatomy of a body part is accompanied by an explanation of its function, emphasizing the structural characteristics that contribute to that function. This approach is called *functional anatomy.*

Anatomical Terminology

Most anatomical terms are based on ancient Greek or Latin words. For example, the arm is the brachium (bra′ke-um; Greek for "arm"), and the thigh bone is the femur (fe′mer; Latin for "thigh"). This terminology, which came into use when Latin was the official language of science, provides a standard nomenclature that scientists can use worldwide, no matter what language they speak. Learning anatomical terminology can be difficult, but this text will help you by explaining the origins of selected terms as you encounter them. For further help, see the Glossary in the back of the book, and the list of word roots located inside the covers of the book.*

Branches of Anatomy

Anatomy is a broad field of science consisting of several subdisciplines or branches. Each branch of anatomy studies the body's structures in a specialized way.

Gross Anatomy

Gross anatomy (*gross* = large) is the study of body structures that can be examined by the naked eye—the bones, lungs, and muscles, for example. An important technique for studying gross anatomy is **dissection** (dĭ-sek′shun; "cut apart"), in which connective tissue is removed from between the body organs so that the organs can be seen more clearly. Then the organs are cut open for viewing. The term *anatomy* is derived from Greek words meaning "I dissect."

Studies of gross anatomy can be approached in several different ways. In **regional anatomy,** all structures in a single body region, such as the abdomen or head, are examined as a group. In **systemic** (sis-tem′ik) **anatomy,** by contrast, all the organs with related functions are studied together. For example, when studying the muscular system, you consider the muscles of the entire body. The systemic approach to anatomy is best for relating structure to function. Therefore, it is the approach taken in most college anatomy courses and in this book. Medical schools, however, favor regional anatomy because many injuries and diseases involve specific body regions (sprained ankle, sore throat, heart disease); furthermore, surgeons must know the structure of each body region in depth.

Another subdivision of gross anatomy is **surface anatomy,** the study of shapes and markings (called "landmarks") on the surface of the body that reveal the underlying organs. This knowledge is used to identify the muscles that bulge beneath the skin in weight lifters, and clinicians use it to locate blood vessels for placing catheters, feeling pulses, and drawing blood. (Chapter 26 will introduce you to clinically useful surface anatomy landmarks.)

Microscopic Anatomy

Microscopic anatomy, or **histology** (his-tol′o-je; "tissue study"), is the study of structures that are so small they can only be seen with a microscope. These structures include cells and cell parts; groups of cells, called tissues; and the microscopic details of the organs of the body (stomach, spleen, and so on). A knowledge of microscopic anatomy is important because physiological and disease processes occur at the cellular level.

Other Branches of Anatomy

Two branches of anatomy explore how body structures form, grow, and mature. **Developmental anatomy** traces the structural changes that occur in the body throughout the life span and the effects of aging. **Embryology** is the study of how body structures form and develop before birth. A knowledge of embryology will help you understand the complex design of the adult

*For a guide to pronunciation, see the Glossary.

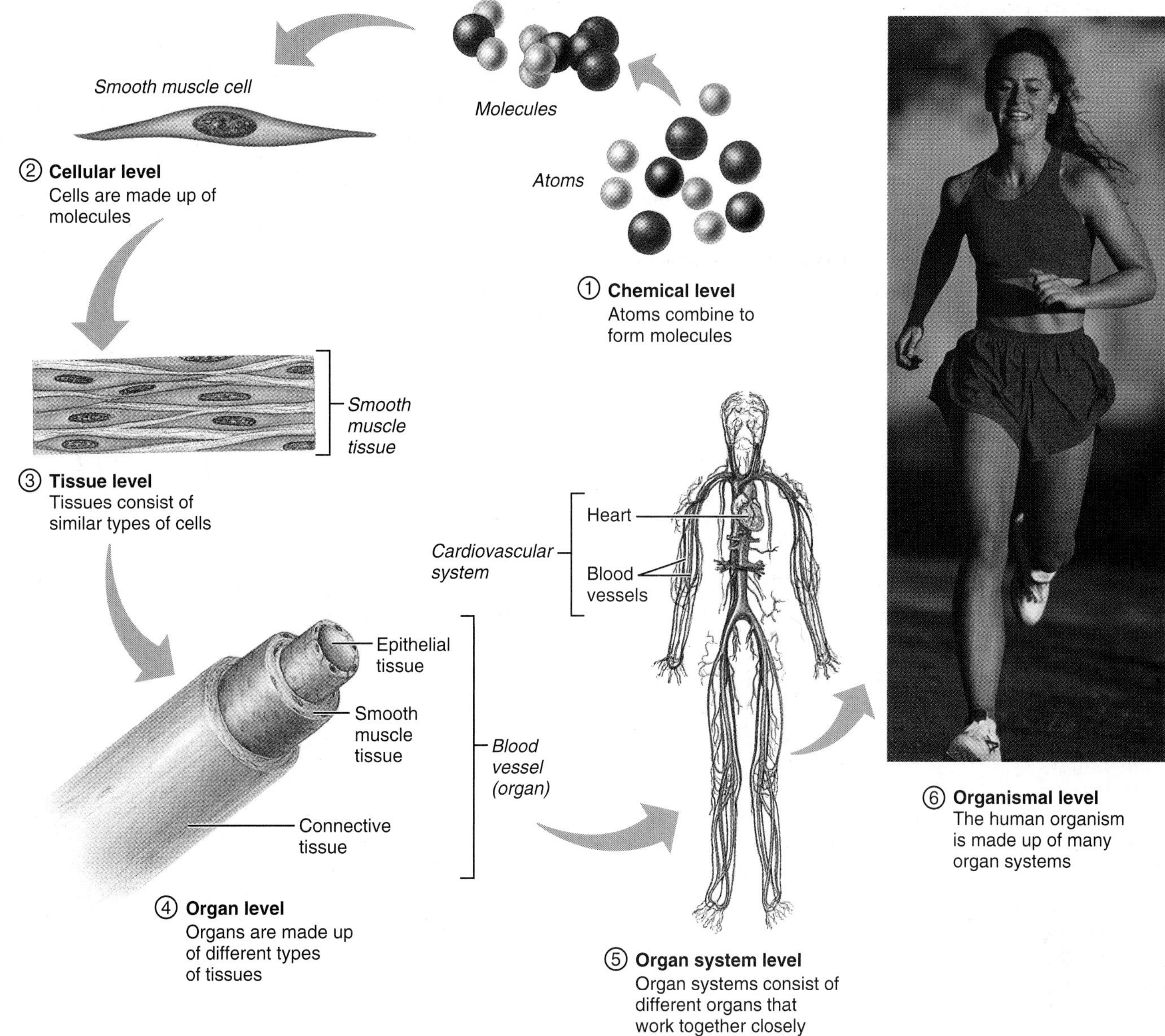

FIGURE 1.1 Levels of structural complexity. In this diagram, components of the cardiovascular system illustrate the various levels of structural complexity in the human body, from the chemical (1) to the organismal (6).

human body and helps to explain birth defects, which are anatomical abnormalities that form during development and are evident after birth.

Some specialized branches of anatomy are used primarily for medical diagnosis and scientific research. **Pathological** (pah-tho-loj′ĭ-kal) **anatomy** deals with the structural changes in cells, tissues, and organs caused by disease. (**Pathology** is the study of disease.) **Radiographic** (ra″de-o′graf′ic) **anatomy** is the study of internal body structures by means of X rays and other forms of radiation (see pp. 19–23). **Functional morphology** explores the functional properties of body structures and assesses the efficiency of their design.

The Hierarchy of Structural Organization

The human body has many levels of structural complexity (Figure 1.1). At the **chemical level**, *atoms*, the tiny building blocks of matter, combine to form small *molecules*, such as carbon dioxide (CO_2) and water (H_2O), and larger *macromolecules* (*macro* = big). There are four classes of macromolecules that are found in the body: carbohydrates (sugars), lipids (fats), proteins, and nucleic acids (DNA, RNA). These are the building blocks of the structures at the **cellular level**: the *cells* and their functional subunits, called *cellular organelles.* Cells are the smallest living things in our body, and we have trillions of them.

The next level is the **tissue level**. A tissue is a group of cells that together perform a common function. Only four tissue types together make up all organs of the human body: epithelial tissue (epithelium), connective tissue, muscle tissue, and nervous tissue. Each tissue plays a characteristic role in the body (see Chapter 4). Briefly, epithelium (ep″ĭ-the′le-um) covers the body surface and lines its cavities; connective tissue supports the body and protects its organs; muscle tissue provides movement; and nervous tissue provides fast internal communication by transmitting electrical impulses.

Extremely complex physiological processes occur at the **organ level**. An organ is a discrete structure made up of more than one tissue. In fact, most organs contain all four tissues. The liver, brain, femur, and heart are good examples. You can think of each organ in the body as a functional center responsible for an activity that no other organ can perform.

Organs that work closely together to accomplish a common purpose make up an **organ system**, the next level. For example, organs of the cardiovascular system—the heart and blood vessels (Figure 1.1)—transport blood to all body tissues. Organs of the digestive system—the mouth, esophagus, stomach, intestine, and so forth—break down the food we eat so that the nutrients can be absorbed into the blood. The body's organ systems are the *integumentary* (skin), *skeletal, muscular, nervous, endocrine, cardiovascular, lymphatic, immune, respiratory, digestive, urinary* and *reproductive* systems.* Figure 1.2 provides a brief overview of the organ systems and their basic functions.

The highest level of organization is the **organismal level;** for example, the human organism is a whole living person (see Figure 1.1). The organismal level is the result of all of the simpler levels working in unison to promote life.

Scale: Length, Volume, and Weight

To describe the dimensions of cells, tissues, and organs, anatomists need a precise system of measurement. The **metric system** provides such precision. Familiarity with this system enables you to understand the sizes, volumes, and weights of body structures. Metric units are described in detail in Appendix A, and you will find yourself referring to that appendix often as you read through this book.

An important unit of *length* is the **meter** (m), which is a little longer than a yardstick. If you are 6 feet tall, your metric height is 1.83 meters. Most adults are between 1.5 and 2 meters tall. A **centimeter** (cm) is a hundredth of a meter (*cent* = hundred). You can visualize this length by remembering that a nickel is about 2 cm in diameter. Many of our organs are several centimeters in height, length, and width. A **micrometer** (μm) is a millionth of a meter (*micro* = millionth). Cells, organelles (structures found inside cells), and tissues are measured in micrometers. Human cells average about 10 μm in diameter, although they range from 5 μm to 100 μm. The human cell with the largest diameter, the egg cell (ovum), is about the size of the tiniest dot you could make on this page with a pencil.

The metric system also measures *volume* and *weight* (mass). A **liter** (L) is a volume slightly larger than a quart; think of a quart container of milk. A **milliliter** (mL) is one thousandth of a liter (*milli* = thousandth). A **kilogram** is a mass equal to about 2.2 pounds, and a **gram** (g) is a thousandth of a kilogram (*kilo* = thousand).

Anatomical Variability

We all know from looking at the faces and body shapes of the people around us that humans differ in their external anatomy. The same kind of variability holds for our internal organs as well. Thus, not every structural detail described in an anatomy book is true of all people or of all the cadavers (dead bodies) you observe in the anatomy lab. In some bodies, for example, a certain blood vessel may branch off higher than usual, a nerve or vessel may be somewhat "out of place," or a small muscle may be missing. Such minor variations are unlikely to confuse you, however, because well over 90% of all structures present in any human body match the textbook descriptions. Extreme anatomical variations are seldom seen, because they are incompatible with life. For example, no living person could be missing the blood vessels to the brain.

Critical Reasoning

What are some easily observed or commonplace physical traits that vary noticeably among different people?

Your examples may include eye color, curly or straight hair, presence or absence of ear lobes, presence or absence of wisdom teeth, and the individual nature of fingerprints.

GROSS ANATOMY: AN INTRODUCTION

The Anatomical Position

To accurately describe the various body parts and their locations, we need to use a common visual reference point. This reference point is the **anatomical position**

* The cardiovascular and lymphatic systems are collectively known as the *circulatory system* due to their interrelated roles in circulating fluids (blood and lymph) through the body.

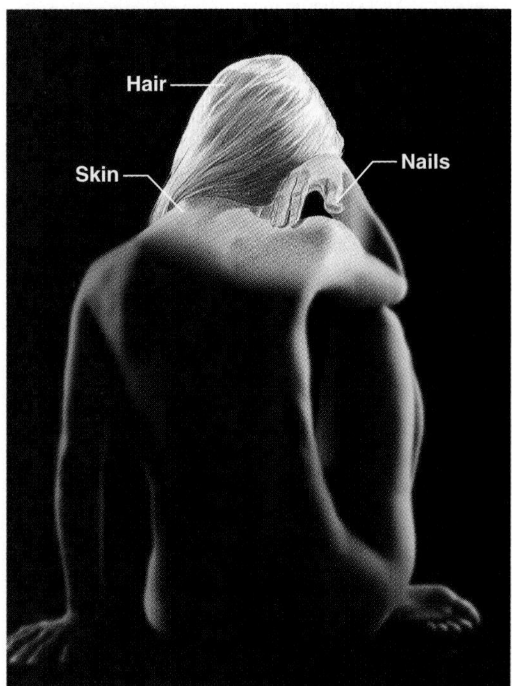

(a) Integumentary System
Forms the external body covering; protects deeper tissues from injury; synthesizes vitamin D; site of cutaneous receptors (pain, pressure, etc.), and sweat and oil glands.

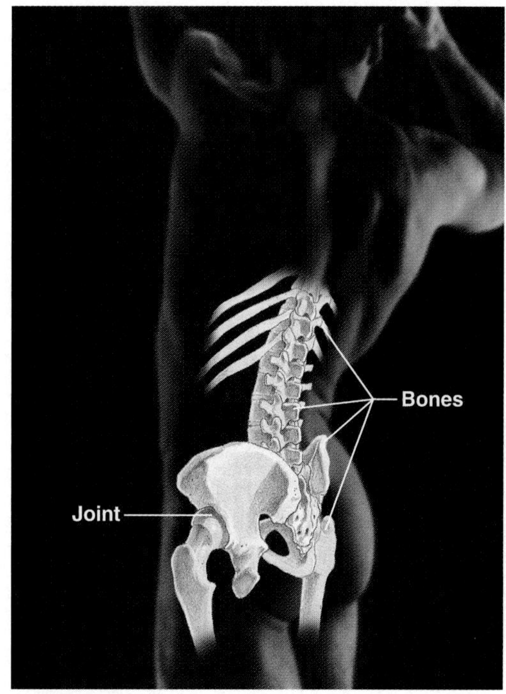

(b) Skeletal System
Protects and supports body organs; provides a framework the muscles use to cause movement; blood cells are formed within bones; stores minerals.

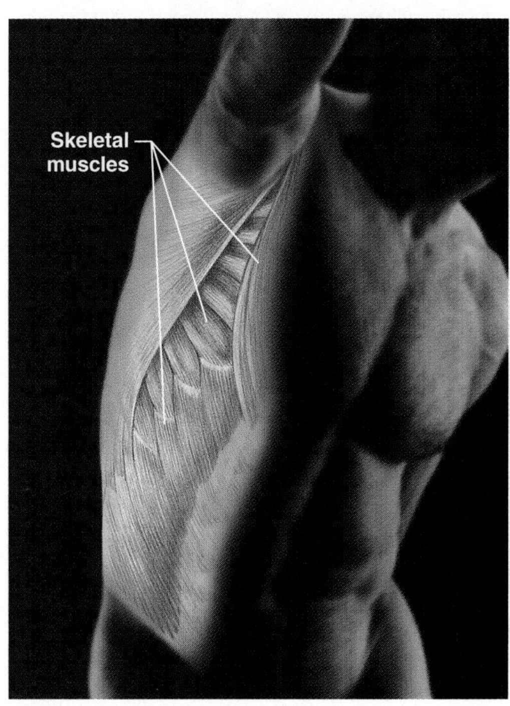

(c) Muscular System
Allows manipulation of the environment, locomotion, and facial expression; maintains posture; produces heat.

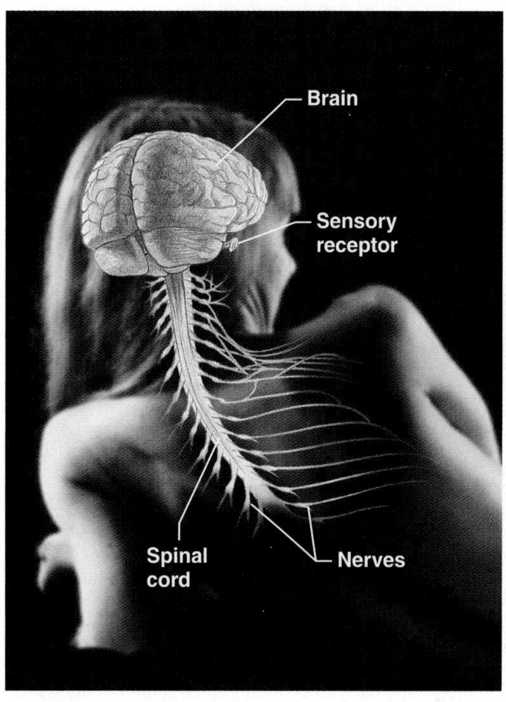

(d) Nervous System
Fast-acting control system of the body; responds to internal and external changes by activating appropriate muscles and glands.

FIGURE 1.2 Summary of the body's organ systems. The structural components of each organ system are illustrated in the diagrammatic view. The major functions of the organ system are listed beneath each illustration.

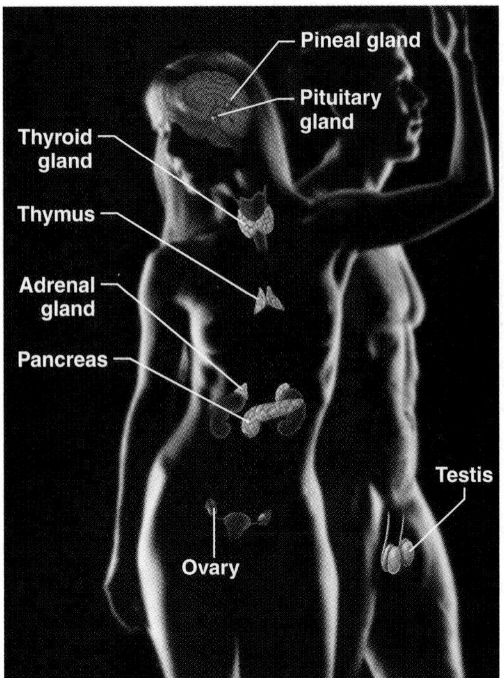

(e) Endocrine System
Glands secrete hormones that regulate processes such as growth, reproduction and nutrient use (metabolism) by body cells.

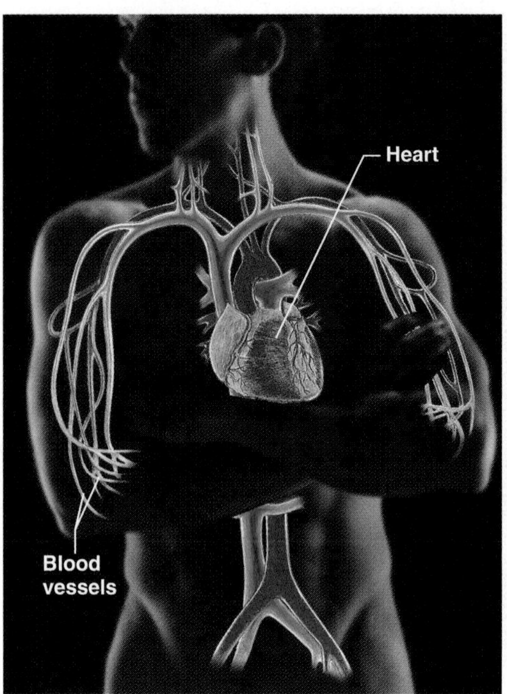

(f) Cardiovascular System
Blood vessels transport blood, which carries oxygen, carbon dioxide, nutrients, wastes, etc.; the heart pumps blood.

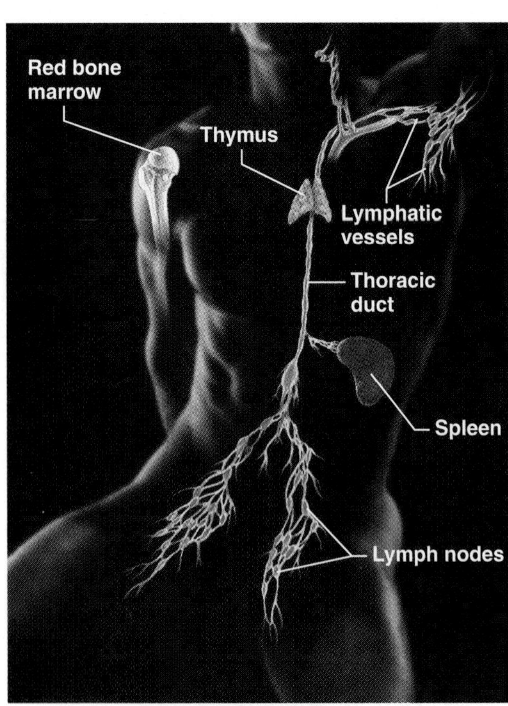

(g) Lymphatic System/Immunity
Picks up fluid leaked from blood vessels and returns it to blood; disposes of debris in the lymphatic stream; houses white blood cells (lymphocytes) involved in immunity. The immune response mounts the attack against foreign substances within the body.

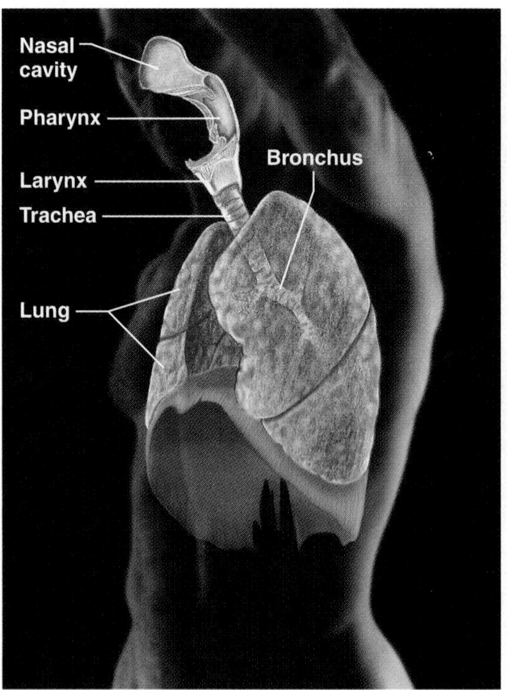

(h) Respiratory System
Keeps blood constantly supplied with oxygen and removes carbon dioxide; the gaseous exchanges occur through the walls of the air sacs of the lungs.

FIGURE 1.2 Summary of the body's organ systems, *continued.*

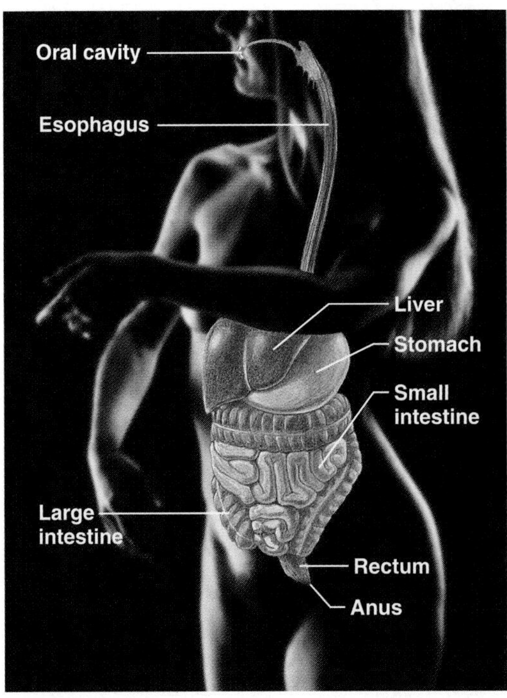

(i) Digestive System
Breaks down food into absorbable units that enter the blood for distribution to body cells; indigestible foodstuffs are eliminated as feces.

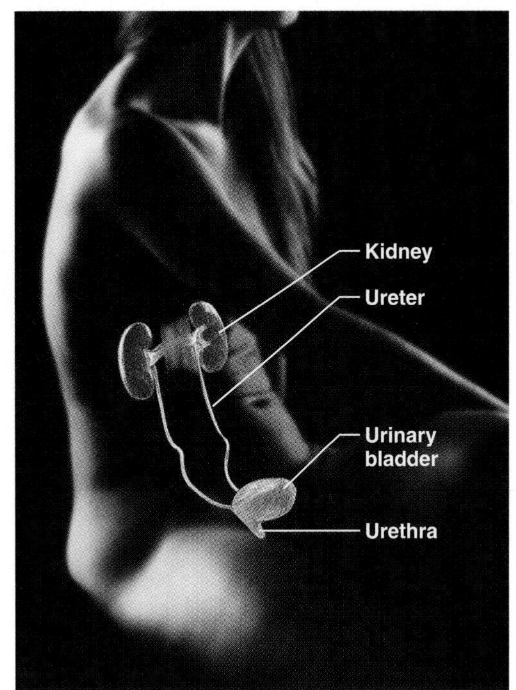

(j) Urinary System
Eliminates nitrogenous wastes from the body; regulates water, electrolyte and acid-base balance of the blood.

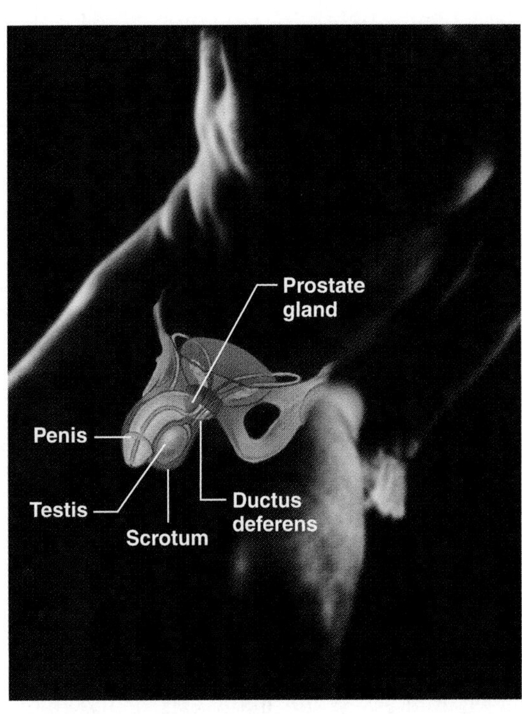

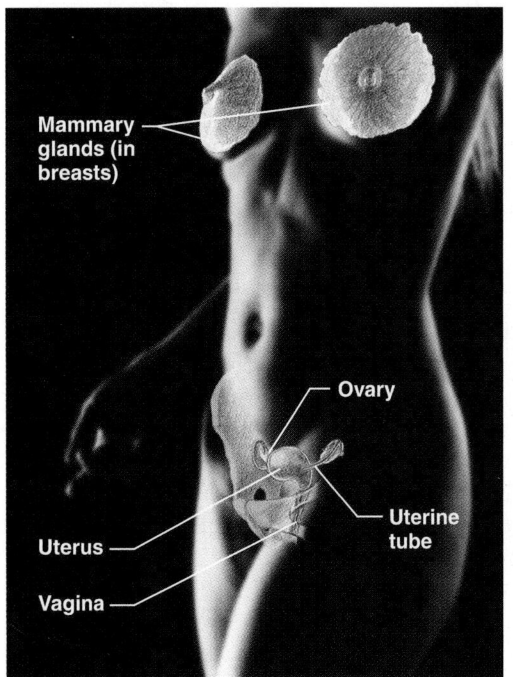

(k) Male Reproductive System **(l) Female Reproductive System**
Overall function is production of offspring. Testes produce sperm and male sex hormone; ducts and glands aid in delivery of sperm to the female reproductive tract. Ovaries produce eggs and female sex hormones; remaining structures serve as sites for fertilization and development of the fetus. Mammary glands of female breasts produce milk to nourish the newborn.

FIGURE 1.2 *continued*

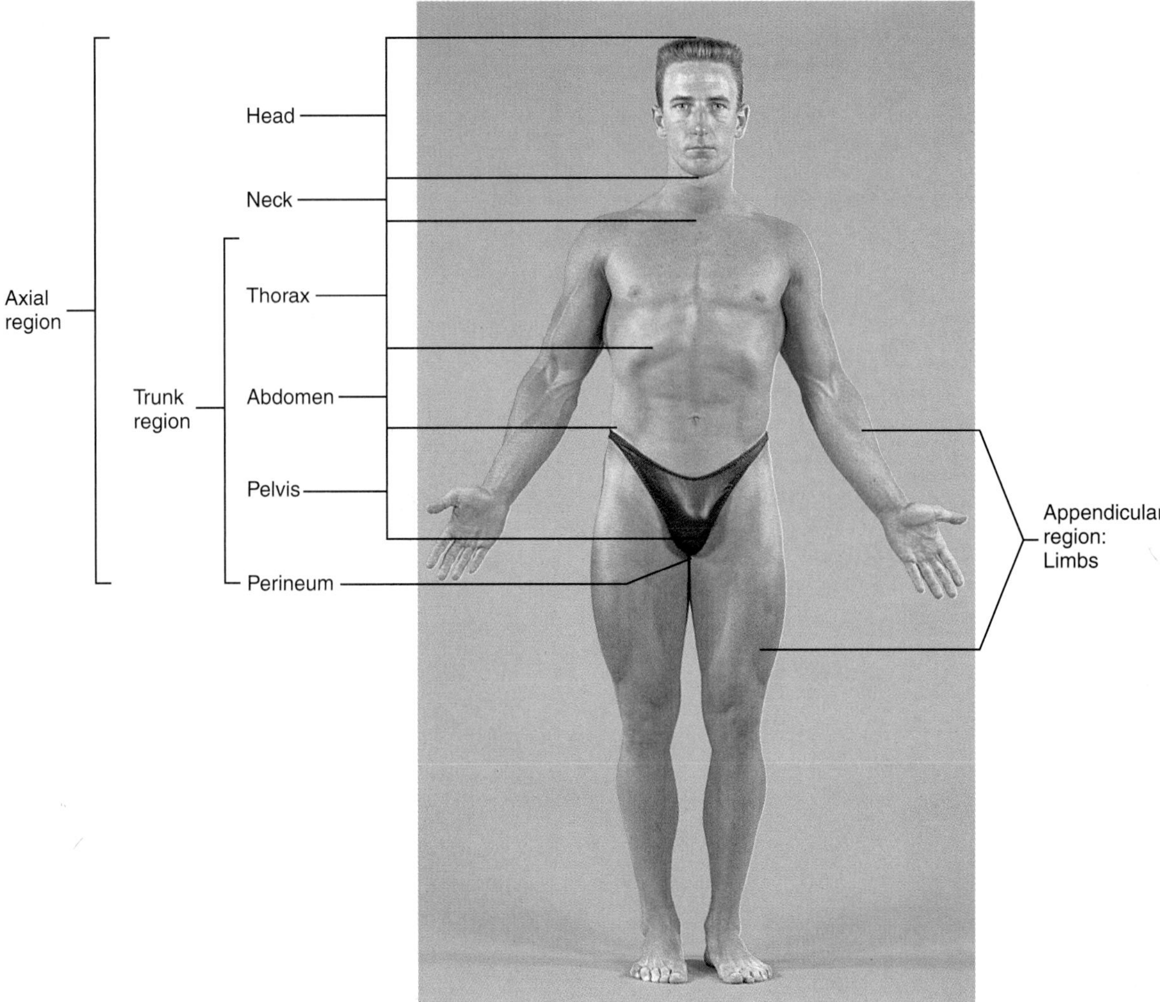

FIGURE 1.3 The anatomical position. The largest divisions of the body are indicated.

(Figure 1.3). In this position, a person stands erect with feet together and eyes forward. The palms face anteriorly with the thumbs pointed away from the body. It is essential to learn the anatomical position because most of the directional terminology used in anatomy refers to the body in this position. Additionally, the terms *right* and *left* always refer to those sides belonging to the person or cadaver being viewed—not to the right and left sides of the viewer.

Directional and Regional Terms

Medical personnel and anatomists use certain directional terms to explain precisely where one body structure lies in relation to another. For example, we could describe the relationship between the eyebrows and the nose informally by stating, "The eyebrows are located on each side of the face to the right and left of the nose and higher than the nose." In anatomical terminology, this condenses to, "The eyebrows are lateral and superior to the nose." Clearly, the anatomical terminology is less wordy and confusing. The standardized terms of direction are defined and illustrated in Table 1.1. Most often used are the paired terms **superior/inferior, anterior (ventral)/posterior (dorsal), medial/lateral,** and **superficial/deep.**

Regional terms are the names of specific body areas. These areas are shown in Figure 1.3. The fundamental divisions of the body are the *axial* and *appendicular* (ap″en-dik′u-lar) *regions.* The **axial region,** so named because it makes up the main axis of the body, consists of the *head, neck,* and *trunk.* The trunk, in turn, is divided into the *thorax* (chest), *abdomen,* and *pelvis;* the trunk also includes the region around the anus and external genitals, called the *perineum* (per″ĭ-ne′um; "around the anus"). The **appendicular region** of the body consists of the limbs, which are also called *appendages* or *extremities.*

Table 1.1 Orientation and Directional Terms

Term	Definition	Example
Superior (cranial)	Toward the head end or upper part of a structure or the body; above	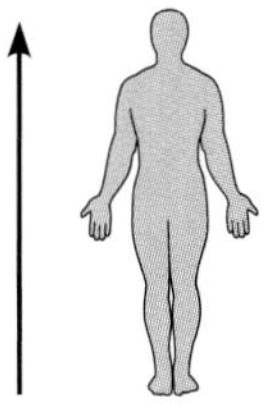 The head is superior to the abdomen
Inferior (caudal)	Away from the head end or toward the lower part of a structure or the body; below	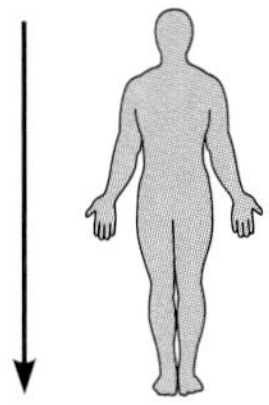 The navel is inferior to the chin
Anterior (ventral)*	Toward or at the front of the body; in front of	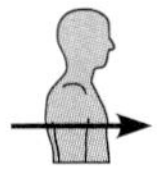 The breastbone is anterior to the spine
Posterior (dorsal)*	Toward or at the back of the body; behind	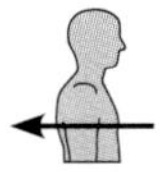 The heart is posterior to the breastbone
Medial	Toward or at the midline of the body; on the inner side of	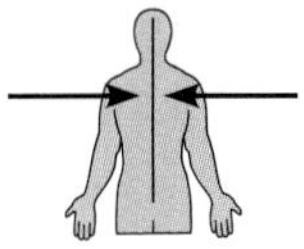 The heart is medial to the arm
Lateral	Away from the midline of the body; on the outer side of	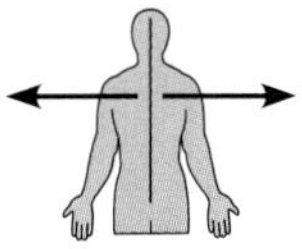 The arms are lateral to the chest
Intermediate	Between a more medial and a more lateral structure	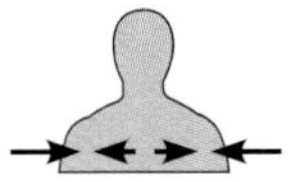 The collarbone is intermediate between the breastbone and shoulder
Proximal	Closer to the origin of the body part or the point of attachment of a limb to the body trunk	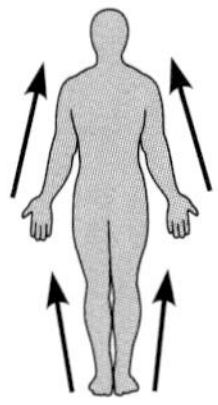 The elbow is proximal to the wrist
Distal	Farther from the origin of a body part or the point of attachment of a limb to the body trunk	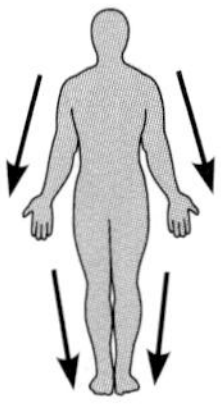 The knee is distal to the thigh
Superficial (external)	Toward or at the body surface	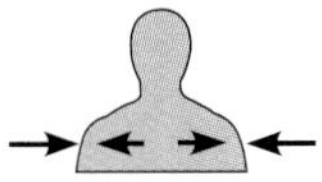 The skin is superficial to the skeletal muscles
Deep (internal)	Away from the body surface; more internal	 The lungs are deep to the skin

*Whereas the terms *ventral* and *anterior* are synonymous in humans, this is not the case in four-legged animals. *Ventral* specifically refers to the "belly" of a vertebrate animal and thus is the inferior surface of four-legged animals. Likewise, although the dorsal and posterior surfaces are the same in humans, the term *dorsal* specifically refers to an animal's back. Thus, the dorsal surface of four-legged animals is their superior surface.

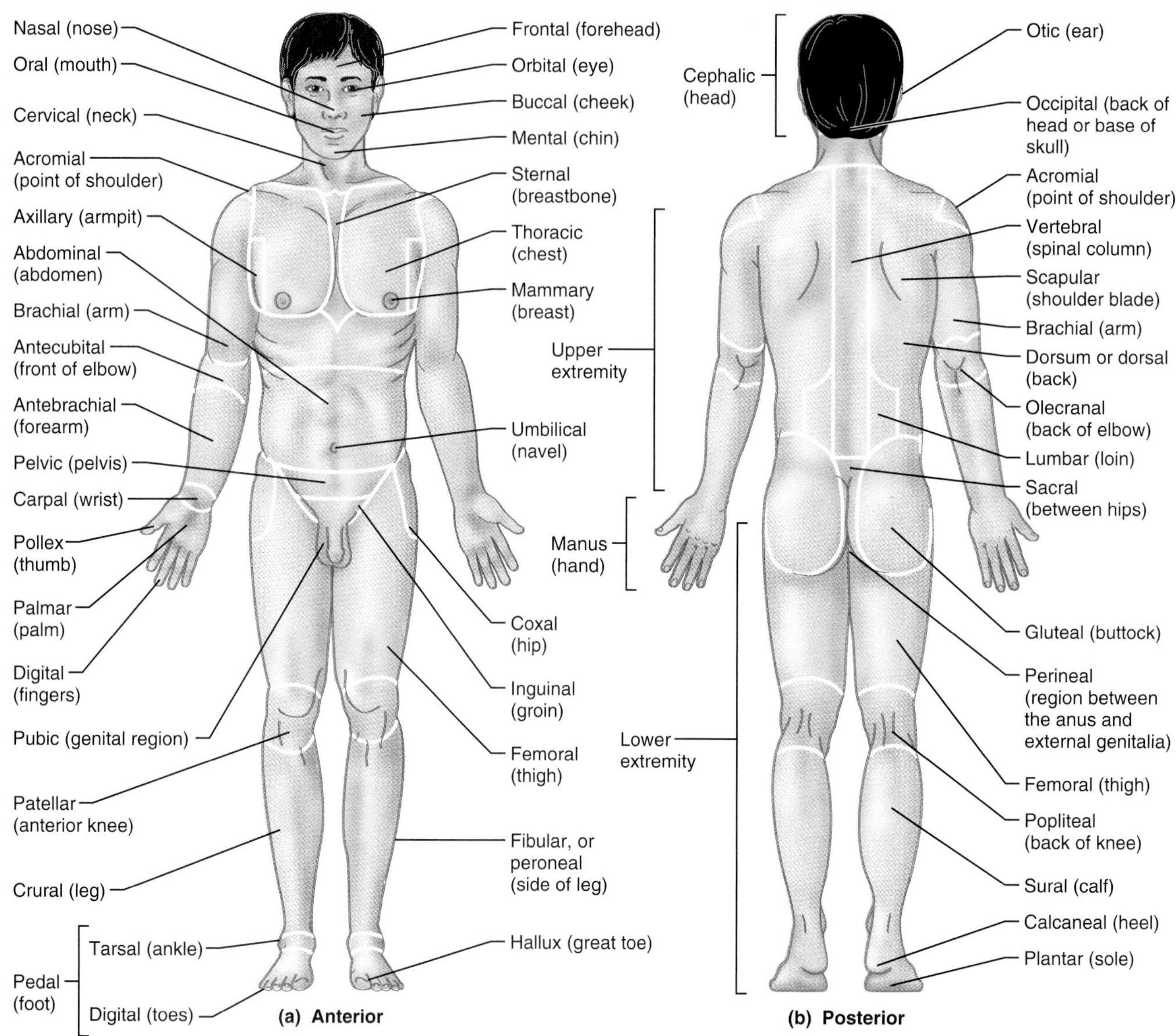

FIGURE 1.4 Regional terms: names of specific body areas.

The fundamental divisions of the body are further divided into smaller regions, as shown in Figure 1.4. For each of these regions, the figure gives the anatomical name, followed by the common name; for example, *oral (mouth)* and *gluteal (buttock)*.

Critical Reasoning

Referring to Figure 1.4, where you would be injured if you pulled a muscle in your axillary *region, cracked a bone in your* occipital *region, or received a cut on a* digit*?*

These injuries would be to the armpit, base of the skull (posteriorly), and a finger or toe, respectively.

Body Planes and Sections

In the study of anatomy, the body is often *sectioned* (cut) along a flat surface called a plane. The most frequently used body planes are sagittal, frontal, and transverse planes, which lie at right angles to one another (Figure 1.5). A section bears the name of the plane along which it is cut. Thus, a cut along a sagittal plane produces a sagittal section.

A **frontal (coronal) plane** lies vertically and divides the body into anterior and posterior parts (Figure 1.5a). A **transverse (horizontal) plane** runs horizontally from right to left, dividing the body into superior and inferior parts (Figure 1.5b). Transverse sections are also called

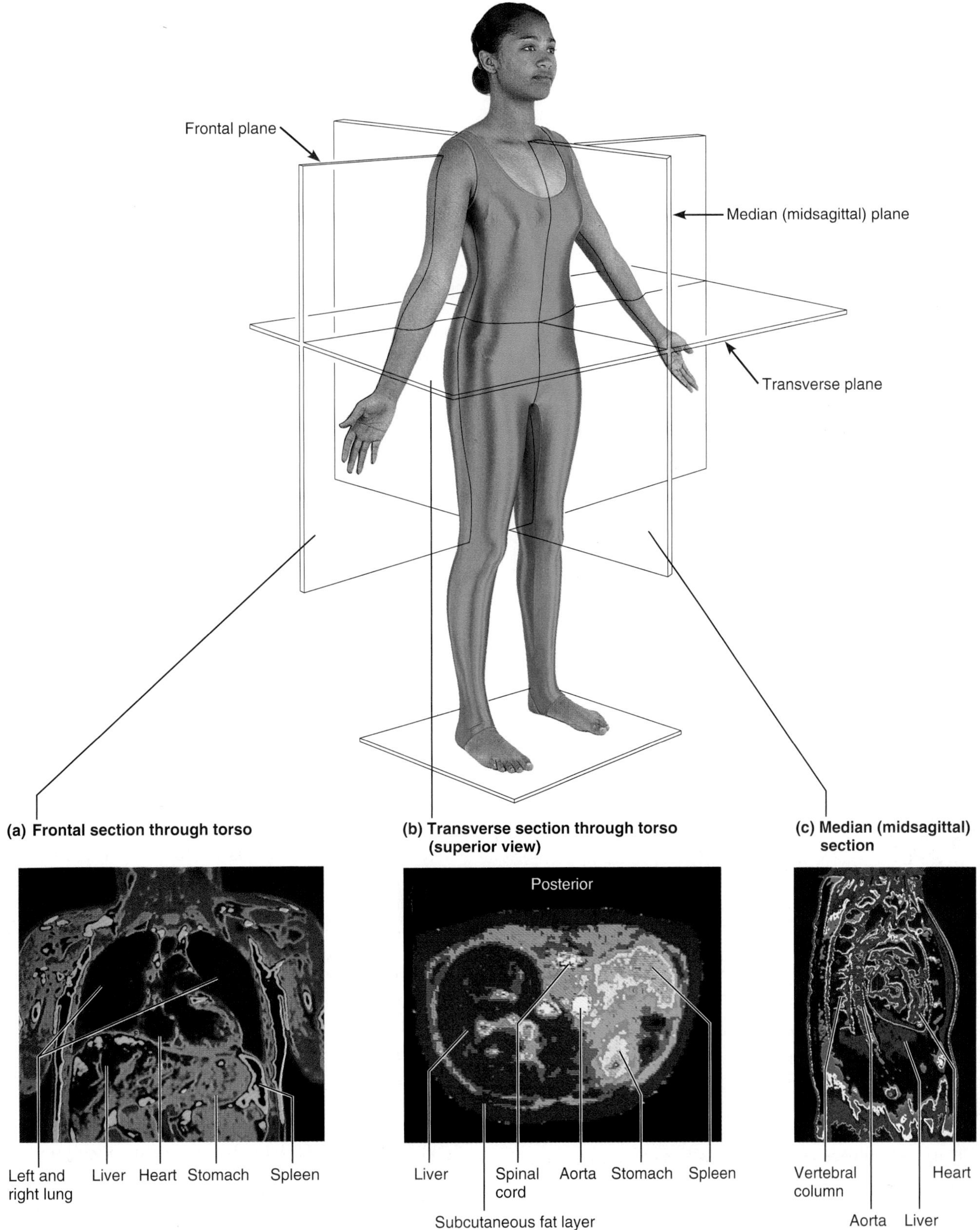

FIGURE 1.5 Planes of the body. Three major types of planes (frontal, transverse, and median [midsagittal]) are superimposed on the photograph of the young woman. Sections taken in each plane are shown below, as visualized by the medical imaging technique called magnetic resonance imaging (MRI).

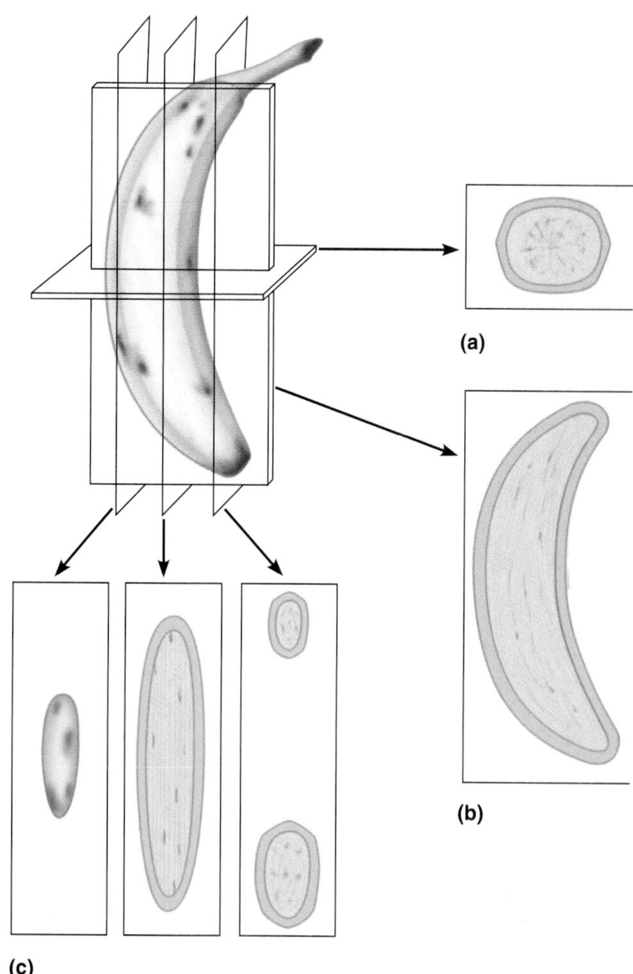

FIGURE 1.6 Objects can look odd when viewed in section. This banana has been sectioned in three different planes **(a–c)**, and only in one of these planes is it easily recognized as a banana (plane b). In order to recognize human organs in section, one must anticipate how the organs will look when cut that way. If one cannot recognize a sectioned organ, it is possible to reconstruct its shape from a series of successive cuts, as from three serial sections in **(c)**.

cross sections. **Sagittal planes** (sag′ĭ-tal; "arrow") are vertical, like frontal planes, but divide the body into right and left parts (Figure 1.5c). The specific sagittal plane that lies exactly in the midline is the **median plane,** or **midsagittal plane.** All other sagittal planes, offset from the midline, are **parasagittal** (*para* = near).

Cuts made along any plane that lies diagonally between the horizontal and the vertical are called **oblique sections**. Neither frontal, transverse, or sagittal, such oblique sections are difficult to interpret because the orientation of the view isn't obvious. For this reason, oblique sections are not used very much.

The ability to interpret sections through the body, especially transverse sections, is increasingly important in the clinical sciences. New medical imaging devices described on pp. 19–23 produce sectional images rather than three-dimensional images. It can be difficult, however, to decipher an object's overall shape from a sectional view alone. A cross section of a banana, for example, looks like a circle and gives no indication of the whole banana's crescent shape (Figure 1.6). Sometimes, one must mentally assemble a whole series of sections to understand the true shape of an object. It will take practice, but you will gradually learn to relate two-dimensional sections to three-dimensional shapes.

The Human Body Plan

Humans belong to the group of animals called *vertebrates.* This group also includes cats, rats, birds, lizards, frogs, and fish. All vertebrates share the following basic features (Figure 1.7):

1. **Tube-within-a-tube body plan.** Our digestive organs form a tube that runs through the axial region of the body, which itself can be viewed as forming a larger tube.
2. **Bilateral symmetry.** The left half of our body is essentially a mirror image of the right half. Most body structures occur in pairs, such as the right and left hands, eyes, and ovaries. Structures in the median plane are unpaired, but they tend to have identical right and left sides (the nose is an example).
3. **Dorsal hollow nerve cord.** All vertebrate embryos have a hollow nerve cord running along their back in the median plane. This cord develops into the brain and spinal cord of adults.
4. **Notochord and vertebrae.** The **notochord** (no′to-kord; "back string") is a stiffening rod in the back just deep to the spinal cord. In humans, a complete notochord forms in the embryo, though most of it is quickly replaced by the development of vertebrae (ver′tĕ-bre), the bony pieces of the vertebral column, or backbone. Still, some of the notochord persists throughout life as the cores of the discs between the vertebrae (see "nucleus pulposus," p. 168).
5. **Segmentation.** The "outer tube" of our body shows evidence of segmentation. Segments are repeating units of similar structure that run from the head along the full length of the trunk. In humans, the ribs and the muscles between the ribs are evidence of segmentation, as are the many nerves branching off our spinal cord. The bony vertebral column, with its repeating vertebrae, is also segmented.
6. **Pharyngeal pouches.** Humans have a **pharynx** (far′ingks), which is the throat region of the digestive tube. In the embryonic stage, our pharynx has a set of outpocketings called pharyngeal (far-rin′je-al) pouches that correspond to the clefts between the gills of fish. Such pouches give rise to some struc-

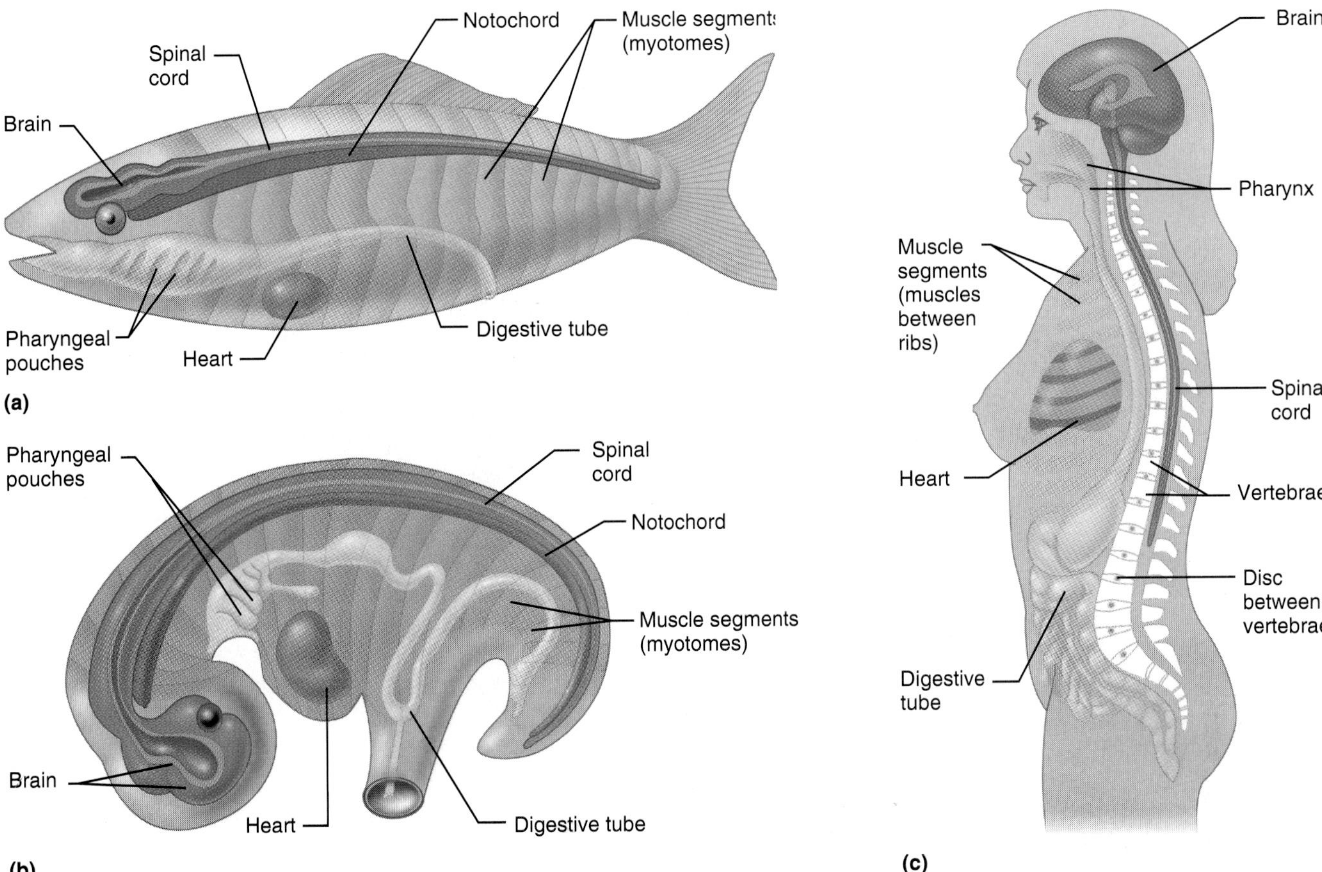

FIGURE 1.7 Basic human body plan, indicated by structures shared with all vertebrates. The bodies are shown as semi-transparent, in order to reveal the internal organs. **(a)** Generalized vertebrate, as represented by a primitive fish (highly simplified). **(b)** Human embryo 5 weeks after conception; note the features shared with the fish. **(c)** Adult human.

tures in our head and neck. An example is the middle ear cavity, which runs from the eardrum to the pharynx.

Body Cavities and Membranes

Within the body are two large cavities called the dorsal and ventral cavities (Figure 1.8). These are closed to the outside, and each contains internal organs. Think of them as filled cavities, like toy boxes containing toys.

Dorsal Body Cavity

The **dorsal body cavity** is subdivided into a **cranial cavity,** which lies in the skull and encases the brain, and a **vertebral cavity,** which runs through the vertebral column to enclose the spinal cord. The hard, bony walls of this cavity protect the contained organs.

Ventral Body Cavity

The more anterior and larger of the closed body cavities is the **ventral body cavity** (Figure 1.8). The organs it contains, such as the lungs, heart, intestines, and kidneys, are called **visceral organs** or **viscera** (vis′er-ah). The ventral body cavity has two main divisions: (1) a superior **thoracic cavity,** surrounded by the ribs and the muscles of the chest wall; and (2) an inferior **abdominopelvic** (ab-dom″ĭ-no-pel′vic) **cavity** surrounded by the abdominal walls and pelvic girdle. The thoracic and abdominal cavities are separated from each other by the diaphragm, a dome-shaped muscle used in breathing.

The *thoracic cavity* is subdivided into three parts: (a) two lateral parts, each of which contains a lung surrounded by a **pleural cavity** (ploo′ral; "the side, a rib"), and (b) a central band of organs called the **mediastinum** (me″de-ah-sti′num; "in the middle"). The mediastinum contains the heart surrounded by a **pericardial cavity** (per″ĭ-kar′de-al; "around the heart"). It also houses other major thoracic organs, such as the esophagus (gullet) and trachea (windpipe).

The *abdominopelvic cavity* is divided into two parts. The superior part, called the **abdominal cavity,** contains

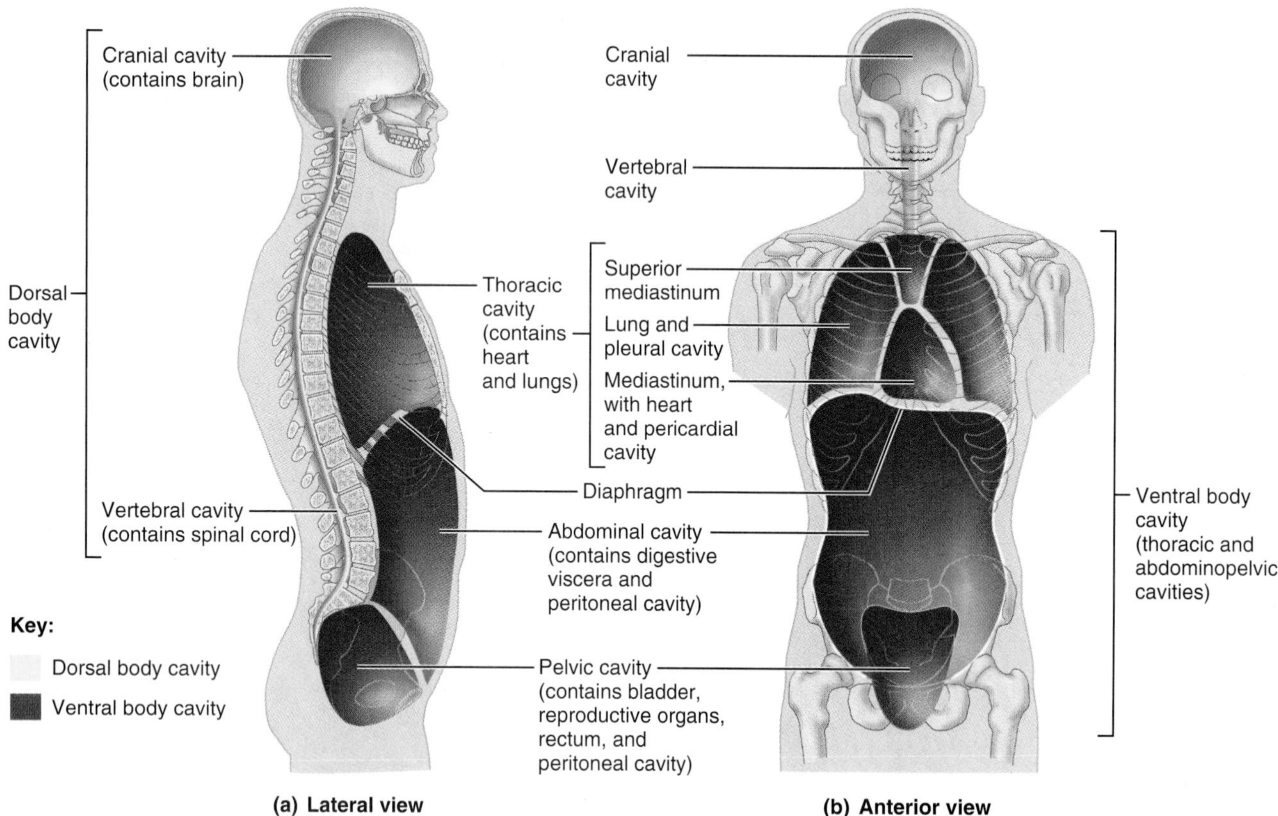

FIGURE 1.8 Dorsal and ventral body cavities and their subdivisions. The dorsal body cavity has cranial and vertebral subdivisions. The ventral body cavity has thoracic, abdominal, and pelvic subdivisions.

the liver, stomach, kidneys, and other organs. The inferior part, or **pelvic cavity,** contains the bladder, some reproductive organs, and the rectum. These two parts are continous with each other, not separated by any muscular or membranous partition. Many organs in the abdominopelvic cavity are surrounded by a **peritoneal** (per″ĭ-to-ne′al) **cavity** (Figure 1.9d).

Serous Cavities

The previous section mentioned the *pericardial cavity* around the heart, *pleural cavity* around the lung, and the *peritoneal cavity* around the viscera in the abdominopelvic cavity. As shown in Figure 1.9, each of these serous cavities is a slit-like space lined by a **serous** (se′rus) **membrane,** or **serosa** (se-ro′sah). Indicated by the blue lines in Figure 1.9, these serous membranes are named **pleura,** serous **pericardium,** and **peritoneum,** respectively. The part of a serosa that forms the outer wall of the cavity is called **parietal serosa** (pah-ri′ĕ-tal; "wall"). The parietal serosa is continuous with the inner, **visceral serosa,** which covers the visceral organs. You can visualize this relationship by pushing your fist into a limp balloon (Figure 1.9a). The part of the balloon that clings to your fist represents the visceral serosa on the organ's (your fist's) outer surface, the outer wall of the balloon represents the parietal serosa, and the balloon's thin airspace represents the serous cavity itself. Serous cavities do not contain air, however, but a thin layer of **serous fluid** (serous = watery). This fluid is secreted by both serous membranes.

The slippery serous fluid allows the visceral organs to slide without friction across the cavity walls as they carry out their routine functions. This freedom of movement is extremely important for organs that move or change shape, such as the pumping heart and the churning stomach.

Other Cavities

In addition to the large, closed body cavities, there are several types of smaller body cavities (Figure 1.10). Many of these are in the head, and most open to the body exterior:

1. **Oral cavity.** The oral cavity, commonly called the mouth, contains the tongue and teeth. It is continuous with the rest of the digestive tube, which opens to the exterior at the anus.

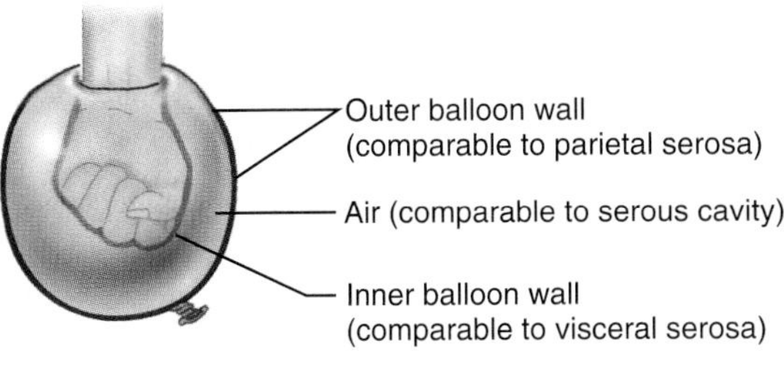

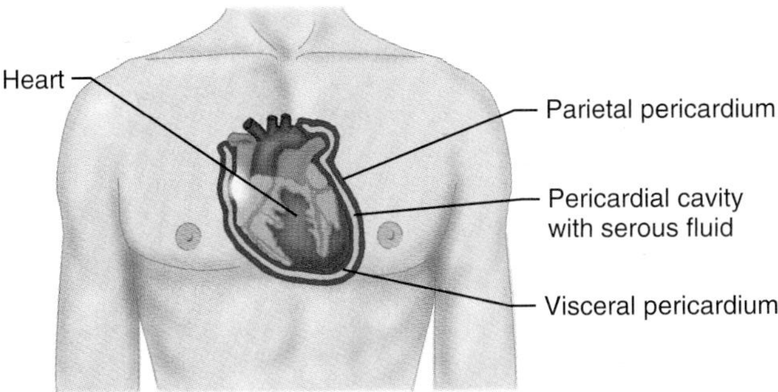

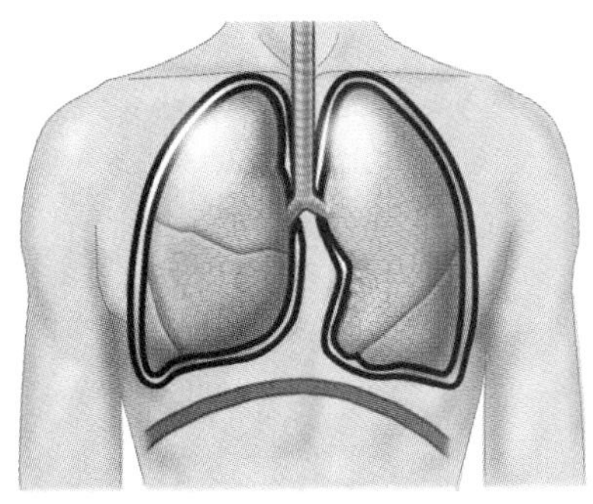

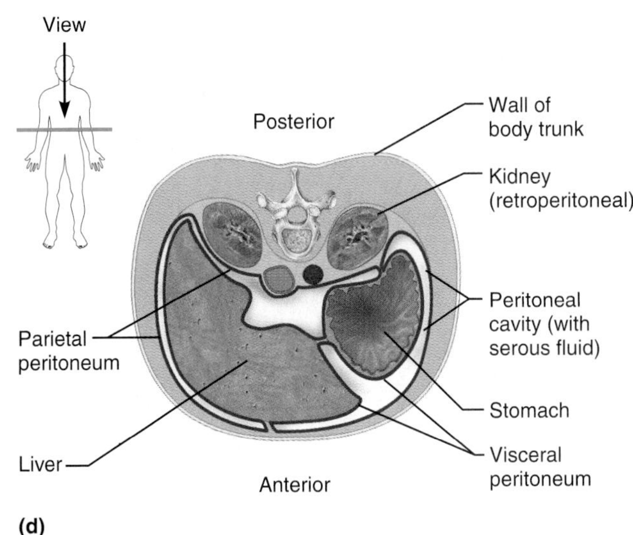

FIGURE 1.9 The serous cavities: pericardial, pleural, and peritoneal. (a) A serous cavity, lined by serous membrane (blue), surrounds visceral organs in the same way that a balloon surrounds a fist that has been thrust into it. The outer part of the membrane is parietal serosa, the inner part is visceral serosa, and the intervening space is the cavity itself. The serous cavities are actually narrow and slit-like, but have been widened in this figure to make them easier to see. **(b)** The pericardial cavity, associated with the heart: The parietal pericardium is the outer lining; the visceral pericardium clings to the heart. **(c)** The pleural cavities, around the lungs: The parietal pleura covers the walls of the thoracic cavity; the visceral pleura covers each lung. **(d)** The peritoneal cavity, around some abdominopelvic organs. This is a transverse section through the abdomen at the level of the liver and kidneys. The parietal peritoneum covers the wall of the abdomen; the visceral peritoneum covers the organs of the peritoneal cavity. However, some abdominopelvic viscera, such as the kidneys, are retroperitoneal (fully behind the peritoneum).

2. **Nasal cavity.** Located within and posterior to the nose, the nasal cavity is part of the passages of the respiratory system.
3. **Orbital cavities.** The orbital cavities (orbits) in the skull house the eyes and present them in an anterior position.
4. **Middle ear cavities.** Each middle ear cavity lies just medial to an ear drum and is carved into the bone of the skull. These cavities contain tiny bones that transmit sound vibrations to the organ of hearing in the inner ears.
5. **Synovial** (sĭ-no′ve-al) **cavities.** Synovial cavities are joint cavities: They are enclosed within fibrous capsules that surround the freely movable joints of the body such as the knee joint and hip joint. Like the serous membranes of the ventral body cavity, membranes lining the synovial cavities secrete a lubricating fluid that reduces friction as the bones move across one another.

Abdominal Regions and Quadrants

Because the abdominopelvic cavity is large and contains many organs, it is helpful to divide it into smaller areas for study. In one method of division, two transverse planes and two parasagittal planes divide the abdomen

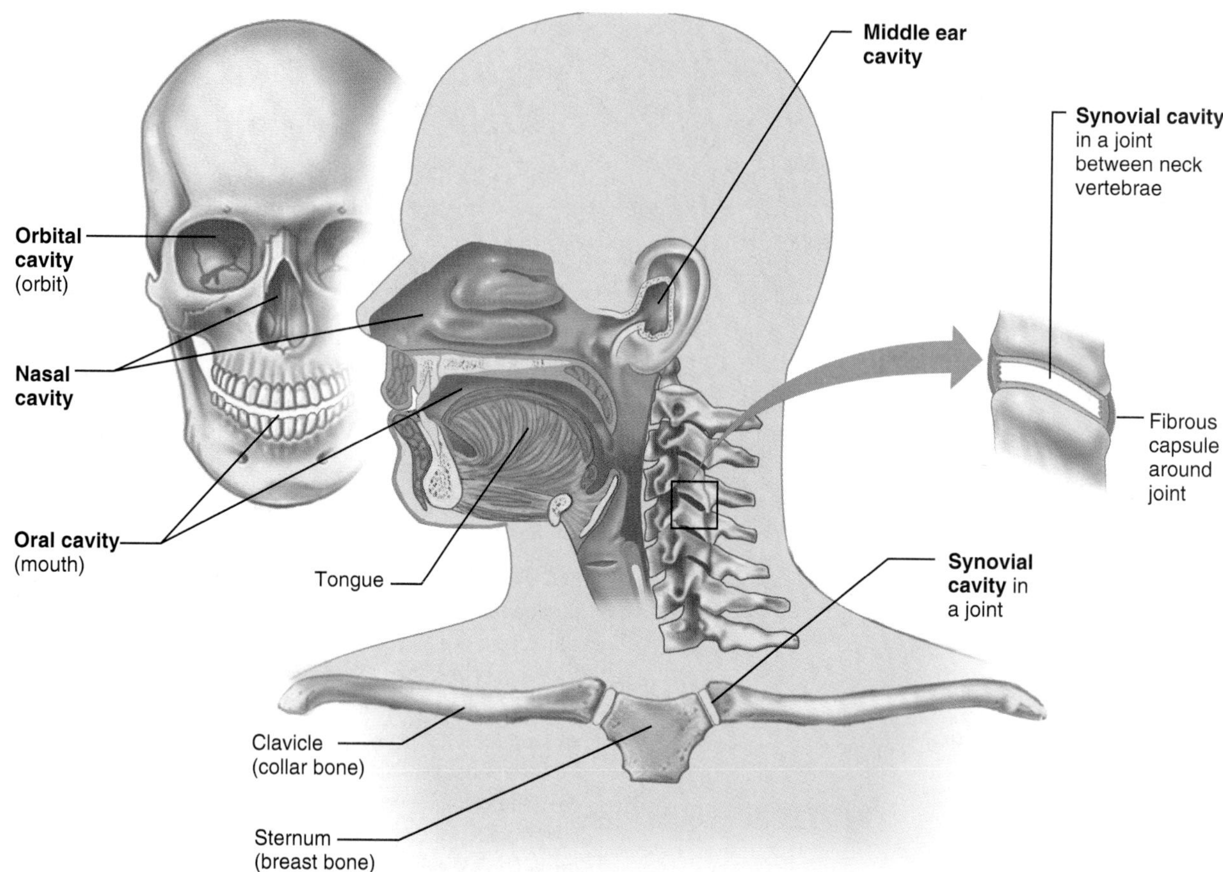

FIGURE 1.10 **Other body cavities.** The oral, nasal, orbital, and middle ear cavities are located in the head and open to the body exterior. Synovial cavities are found in the joints between many bones such as the vertebrae of the spine.

into nine **regions** (Figure 1.11): the right and left hypochondriac regions, the epigastric region, the right and left lumbar regions, the umbilical region, the right and left iliac regions, and the hypogastric region. These nine regions are discussed in more detail on p. 637. For localizing organs in a more general way, an alternative scheme divides the abdomen into just four **quadrants** ("quarters") by drawing one vertical and one horizontal plane through the navel (Figure 1.12). You can use Figures 1.11b and 1.12 to become familiar with the locations of some of the major viscera, such as the stomach, liver, and intestines.

Critical Reasoning

As shown in Figure 1.11b, the most superior part of the abdominal cavity is protected by the rib cage. This protection is especially important for a large visceral organ on the right side of the body, which could bleed profusely if ruptured by a blow. Can you deduce which organ we are referring to?

It is the liver. Similarly, the spleen, another abdominal organ subject to rupture, is protected by the rib cage on the left (see Figure 1.5a).

MICROSCOPIC ANATOMY: AN INTRODUCTION

Light and Electron Microscopy

Microscopy is the examination of small structures with a microscope. When microscopes were introduced in the early 1600s, they opened up a tiny new universe whose existence was unsuspected before that time. Two main types of microscopes are now used to investigate the fine structure of organs, tissues, and cells: the **light microscope (LM)** and the **transmission electron microscope (TEM or just EM).** Light microscopy illuminates body tissue with a beam of light, whereas electron microscopy uses a beam of electrons. LM is used for lower-magnification viewing, EM for higher magnification (Figure 1.13a and b). Light microscopy can produce sharp, detailed images of tissues and cells, but not of the small structures within cells (organelles). Most intracellular structures are too small to alter the path of light waves enough to appear on the image. A light microscope's *low resolution*—its inability to reveal small structures

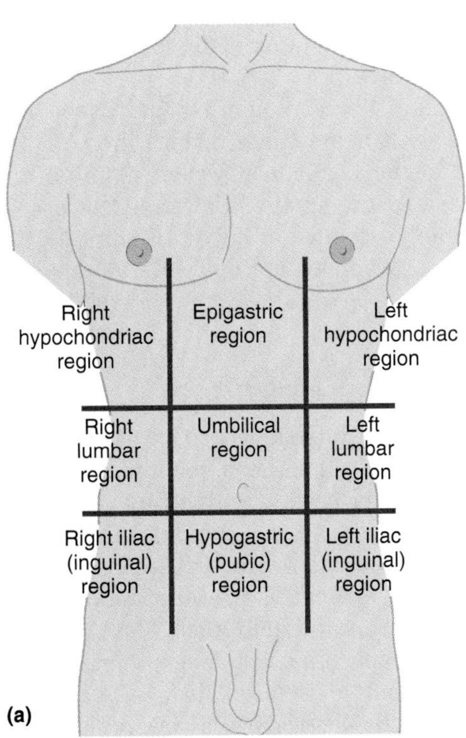

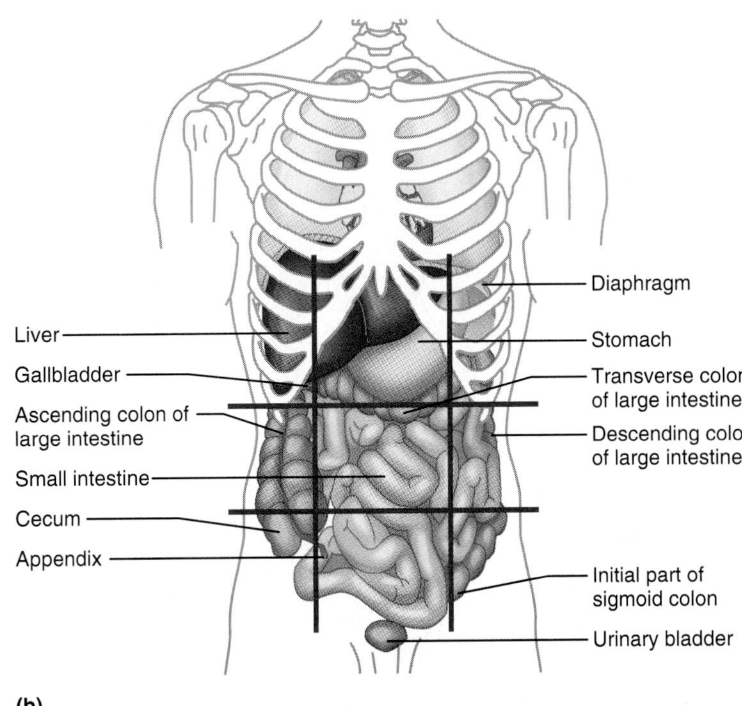

FIGURE 1.11 The nine abdominal regions. (a) Division of the abdomen into nine regions delineated by four planes. The superior horizontal plane is just inferior to the rib cage; the inferior horizontal plane is near the top of the hip bones (see p. 637). The two parasagittal planes lie just medial to each nipple. (b) Anterior view of the abdominopelvic cavity, showing many of its visceral organs.

clearly—remains its basic limitation, despite technical advances that have greatly improved the quality of LM images. EM, on the other hand, uses electron beams of much smaller wavelength, so it can produce sharp images at much greater magnification. The fine details of cells and tissues, as viewed by electron microscopy, are said to comprise their **ultrastructure.**

Preparing Human Tissue for Microscopy

Elaborate steps are taken to prepare human or animal tissue for microscopic viewing. The specimen must be **fixed** (preserved) and then cut into **sections** (slices) thin enough to transmit light or electrons. Finally, the specimen must be **stained** to enhance contrast. The stains used in light microscopy are beautifully colored organic dyes, most of which were originally invented by the clothing industry in the mid 1800s. These dyes helped to usher in the golden age of microscopic anatomy from 1860 to 1900. The stains come in almost all colors of the rainbow. Many consist of charged dye molecules (negative or positive molecules) that bind within the tissue to macromolecules of the opposite charge. This electrical attraction is the basis of the staining ability. Stains with negatively charged dye molecules are called **acidic stains.** Positively charged dyes, by contrast, are called **basic stains.** Because different parts of cells and tissues

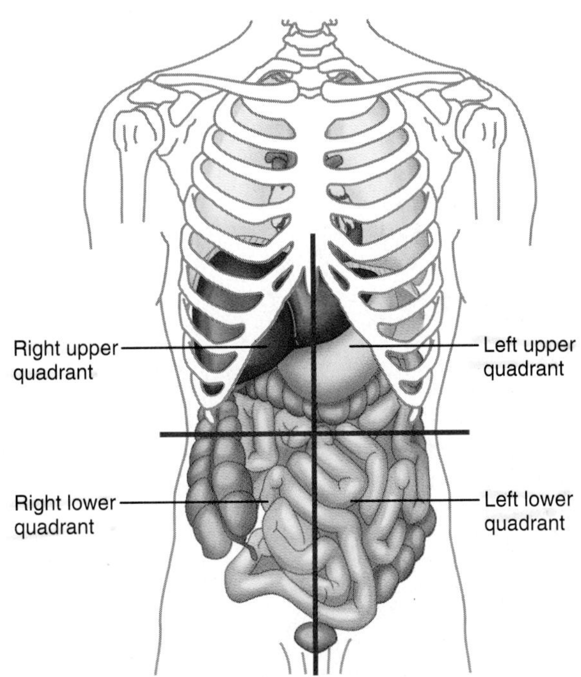

FIGURE 1.12 The four abdominal quadrants. In this scheme, two planes divide the abdominopelvic cavity into four quarters. The planes intersect at the navel.

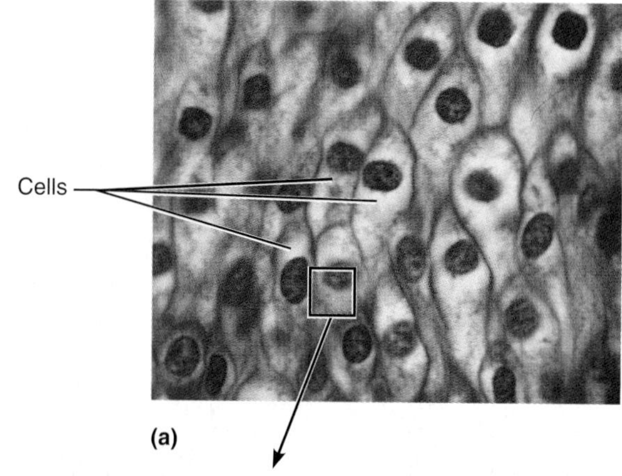

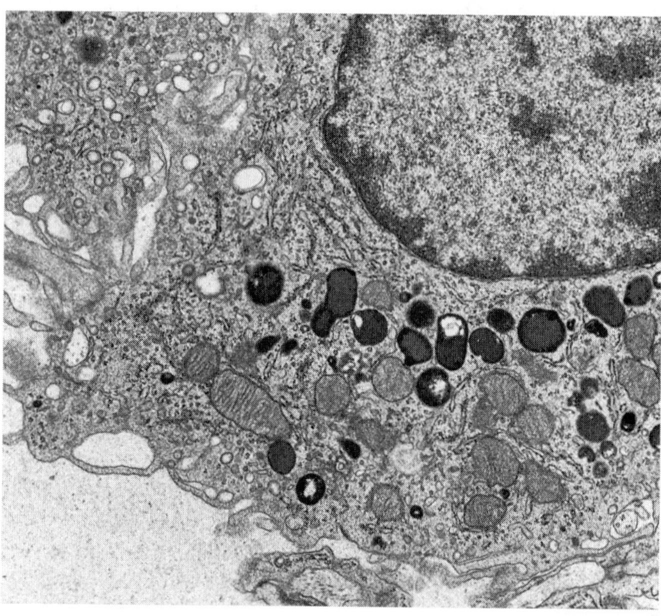

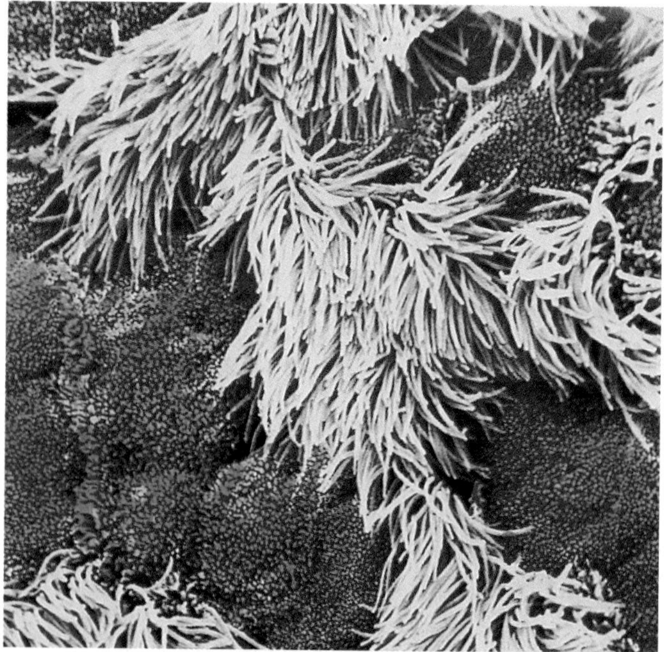

take up different dyes, the stains distinguish the different anatomical structures.

For transmission electron microscopy, tissue sections are stained with heavy-metal salts. These metals deflect electrons in the beam to different extents, thus providing contrast in the image. Electron-microscope images contain only shades of gray (Figure 1.13b), because color is a property of light, not of electron waves.

Scanning Electron Microscopy

The types of microscopy introduced so far are used to view cells and tissue that have been sectioned. Another kind of electron microscopy, **scanning electron microscopy (SEM),** provides three-dimensional pictures of whole, unsectioned surfaces with striking clarity (Figure 1.13c). First, the specimen is preserved and coated with fine layers of carbon and gold dust. Then, an electron beam scans the specimen, causing other, secondary, electrons to be emitted from its surface. A detector captures these emitted electrons and assembles them into a three-dimensional image on a video screen (cathode ray tube), based on the principle that more electrons are produced by the high points on the specimen surface than by the lower points. Although artificially constructed, the SEM image is accurate and looks very real. Like all electron-microscopy images, the original is in black and white, although the photo can be colored artificially to make it look more pleasing to the eye (Figure 1.13c).

Artifacts

The preserved tissue we see under the microscope has been exposed to many procedures that alter its original condition. Because each preparatory step introduces minor distortions, called **artifacts**, most microscopic structures viewed by anatomists are not exactly like those in living tissue. Furthermore, the human and animal corpses studied in the anatomy laboratory have also been preserved, so their organs have a drabber color and a different texture than living organs. Keep these principles in mind as you look at the micrographs (microscope pictures) and the photos of human cadavers in this book.

FIGURE 1.13 Cells viewed by three types of microscopy. **(a)** Light micrograph (cells from the epithelium lining the bladder). **(b)** Transmission electron micrograph (enlarged area of the cell region that is indicated in a box in part a). **(c)** Scanning electron micrograph: surface view of cells lining the trachea, or windpipe. The long, grass-like structures on the surfaces of these cells are cilia, and the tiny knob-like structures are microvilli (see Chapter 4, p. 83). This scanning electron micrograph is artificially colored.

CLINICAL ANATOMY: AN INTRODUCTION TO MEDICAL IMAGING TECHNIQUES

Physicians have long sought ways to examine the body's internal organs for evidence of disease without subjecting the patient to the shock and pain of exploratory surgery. Physicians can identify some diseases and injuries by feeling the patient's deep organs through the skin, or by using traditional X rays—but these are relatively crude methods. Modern medical science is developing powerful new techniques for viewing the internal anatomy of living people. These imaging techniques not only reveal the structure of our functioning internal organs, but also can extract information about the workings of our molecules. The new techniques all rely on powerful computers to construct images from raw data transmitted by electrical signals.

X Rays

Before considering the new techniques, we must discuss traditional X rays, because these still play the major role in medical diagnosis (Figure 1.14). X rays are best for visualizing bones and locating abnormal dense structures, such as some tumors and tuberculosis nodules in the lungs. Discovered in 1895 and used in medicine ever since, X rays are electromagnetic waves of very short wavelength. When X rays are directed at the body, some are absorbed. The amount of absorption depends on the density of the matter encountered. X rays that pass through the body expose a piece of film behind the patient. The resulting radiograph is a negative: The darker, exposed areas on the film represent soft organs that are easily penetrated, whereas light, unexposed areas correspond to denser structures, like bones, that absorb most X rays.

Exposure to high or prolonged levels of X radiation can damage tissue, cause hereditary disorders, or predispose one to cancer. Therefore, modern X-ray techniques try to minimize the exposure of the patient to radiation. For example, new digital detectors are more sensitive than traditional X-ray film, so images can be obtained using just half the radiation.

In a variation of radiography, X-ray images are viewed on a fluorescent screen, or **fluoroscope,** as they are being generated. Although not as sharp as the film image, the fluoroscope image enables us to view internal organs as they move. Alternatively, movements can be recorded with X-ray cinema film, a technique known as **cineradiography.**

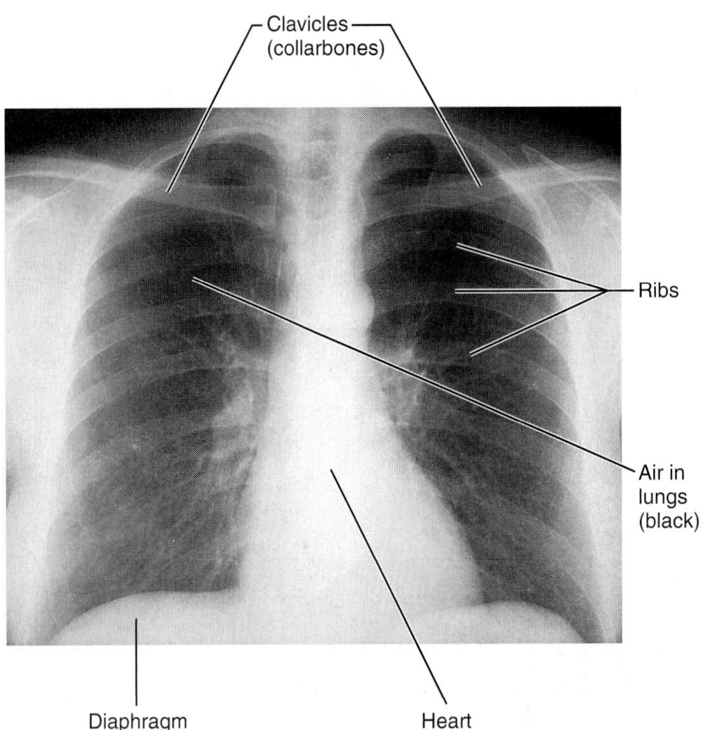

FIGURE 1.14 An X-ray image (radiograph) of the chest.

Critical Reasoning

Conventional X rays have several limitations that have prompted clinicians to seek more advanced imaging techniques. Looking at Figure 1.14, can you deduce some of the problems clinicians face when they try to decipher standard X-ray images?

First, X-ray images can be blurry, especially those of soft tissues. Second, conventional X rays flatten three-dimensional body structures into two dimensions. Consequently, organs appear stacked one on top of another. Worse yet, denser organs block the less dense organs that lie in the same path.

A variety of strategies and technologies have been developed to overcome the limitations of conventional X rays. Soft but hollow organs can be filled or coated with a liquid called **contrast medium,** which contains atoms of heavy elements such as barium that absorb more of the passing X rays. Sharper images are also produced by a new technique called *phase contrast X-ray imaging,* which distinguishes body structures by how much they slow down X rays, as well as by how much they absorb them. However, to overcome the confusion caused by superimposed images, clinicians use newer computer-assisted imaging technologies that produce images of the body's interior in sections.

Advanced X-Ray Techniques

Computed Tomography

The best known of the modern imaging techniques is a refined X-ray technology called **computed tomography (CT),** or **computed axial tomography (CAT)** (Figure 1.15). A CT scanner is shaped like a square metal nut

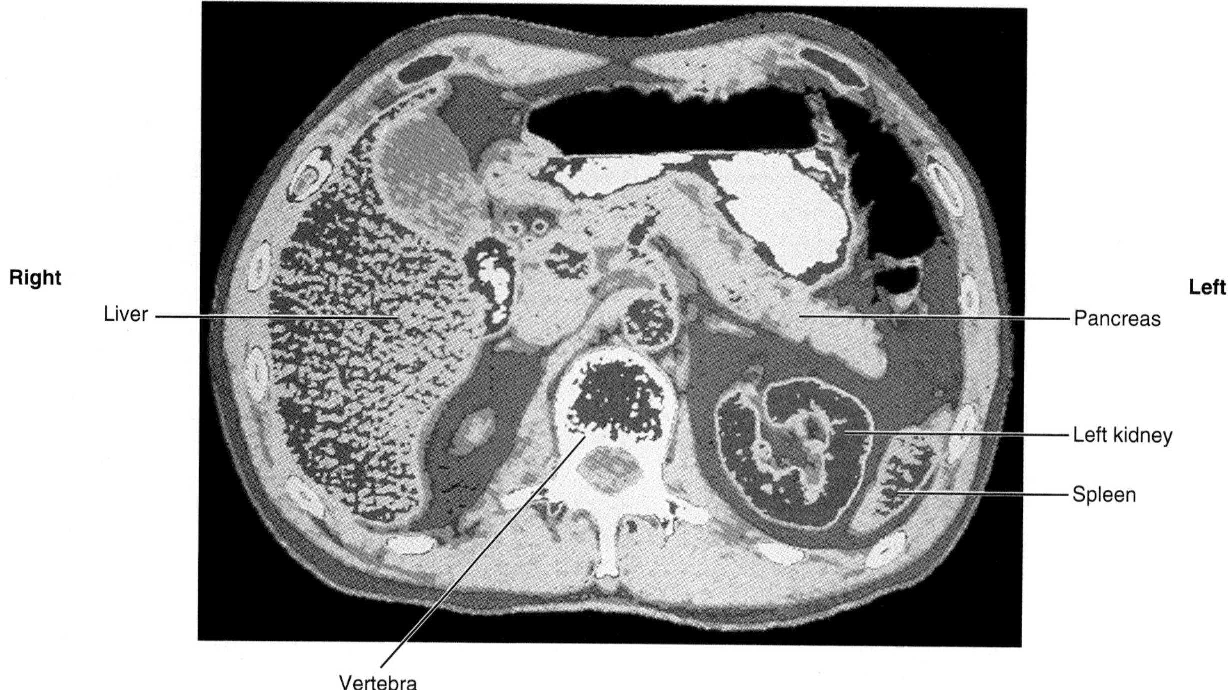

FIGURE 1.15 Computed tomography (CT). CT scan through the superior abdomen. CT sections are conventionally oriented as if viewed from an inferior direction, with the posterior surface of the body directed toward the inferior part of the picture; therefore, the patient's right side is at the left side of the picture. Like all x-ray images, this scan was originally produced in black and white; it has been artificially colored to enhance its visual appeal.

(as in "nuts and bolts") standing on its side. The patient lies in the central hole, situated between an X-ray tube and a recorder, both of which are in the scanner. The tube and recorder rotate to take about 12 successive X rays around the person's full circumference. The fan-shaped X-ray beam is thin, so all pictures are confined to a single transverse body plane about 0.3 cm thick. This explains the term *axial tomography*, which literally means "pictures of transverse sections taken along the body axis." Information is obtained from all around the circumference so that every organ is recorded from its best angle, with the fewest structures blocking it. The computer translates all the recorded information into a detailed picture of the body section, which it displays on a viewing screen. Soft structures are represented more clearly in CT scans than in conventional radiographs, because computerized image-enhancement techniques are used.

Special ultrafast CT scanners are used in a technique called **dynamic spatial reconstruction (DSR).** This process assembles a series of CT sections into three-dimensional images of body organs, which can be rotated and seen from every angle; and movements of these organs can be scrutinized at normal speed, in slow motion, and at a specific moment in time. Its greatest value has been to view the heart beating and blood flowing through blood vessels, allowing heart defects, constricted or blocked vessels, and the status of coronary bypass grafts to be evaluated.

Xenon CT is used to diagnose a stroke (that is, a blockage or cut-off of the flow of blood to the brain). Xenon is an inert gas that the patient breathes, so it enters the bloodstream and distributes to the different body tissues in proportion to their blood flow. It shows up on CT scans, and its absence from a part of the brain indicates that a stroke is occurring there.

Digital Subtraction Angiography Imaging

Another computer-assisted X-ray technique is **digital subtraction angiography (DSA)** (*angiography* = "vessel pictures"). This technique provides an unobstructed view of small arteries (Figure 1.16). Its principle is simple: Conventional radiographs are taken before and after a contrast medium is injected into an artery. Then the computer subtracts the "before" image from the "after" image, eliminating all traces of body structures that obscure the vessel. DSA is often used to identify blockage of the arteries that supply the heart wall and the brain.

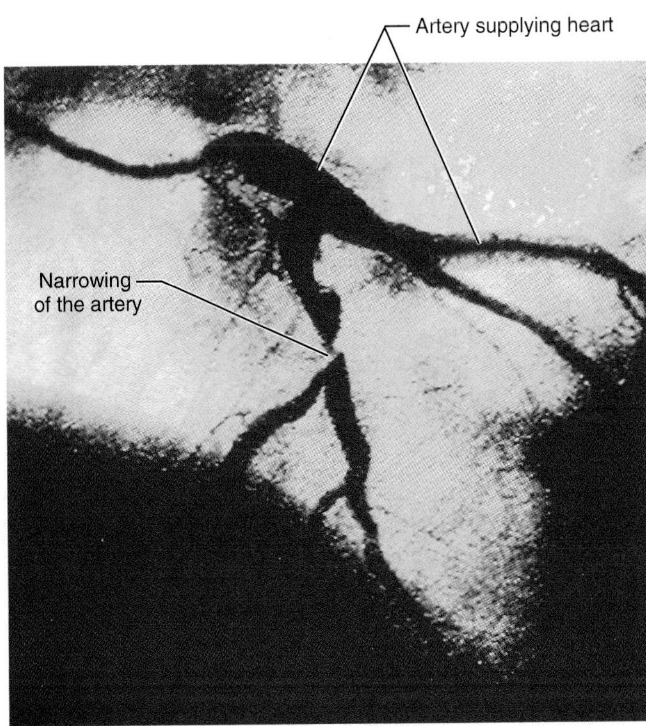

FIGURE 1.16 Digital subtraction angiography (DSA). A DSA image of the arteries that supply the heart.

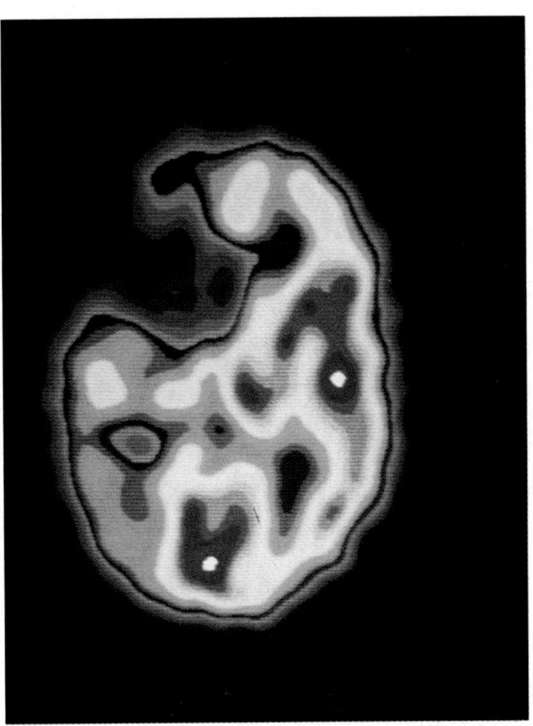

FIGURE 1.17 Positron emission tomography (PET). A PET scan of the brain, in transverse section, shows an area with no neural activity (the frontal region of the brain at upper left). This area was destroyed by a stroke.

Positron Emission Tomography

Positron emission tomography (PET) (Figure 1.17) is an advanced procedure that forms images by detecting radioactive isotopes that are injected into the body. The special advantage of PET is that its images contain messages about the chemical processes of life. For example, radioactively tagged sugar or water molecules are injected into the bloodstream and followed to the body areas that take them up most avidly. This procedure locates our body's most active cells and pinpoints the body regions that receive the greatest supply of blood. As the isotopes decay, they emit particles called positrons, which indirectly lead to the production of gamma rays. Sensors within the doughnut-shaped scanner pick up the emitted gamma rays, which are translated into electrical impulses and sent to the computer. A picture of the isotope's location is then constructed on the viewing screen in vivid colors (Figure 1.17).

PET is used to assess the functional flow of blood to the heart and brain. Furthermore, by mapping increases in blood flow, it can determine which areas of the normal brain are most active during certain tasks (speech, seeing, comprehension), thereby providing direct evidence for the functions of specific brain regions. The resolution of a PET image is low, however, and the image takes a relatively long time to form, so PET cannot record fast changes in brain activity. Additionally, it requires an expensive cyclotron machine on site to make the isotopes. For these reasons, PET is gradually being eclipsed by other techniques, such as functional MRI (see p. 23).

Sonography

In **sonography,** or **ultrasound imaging** (Figure 1.18), the body is probed with pulses of high-frequency (ultrasonic) sound waves that reflect (echo) off the body's tissues. A computer analyzes the echoes to construct sectional images of the outlines of organs. A hand-held device that looks something like a microphone emits the sound and picks up the echoes. The device is moved across the surface of the body, so organs can be examined from the perspective of many different body planes.

Sonography has some distinct advantages over other imaging techniques. First, the equipment is relatively inexpensive. Second, ultrasound seems to be safer than ionizing forms of radiation, with fewer harmful effects on living tissues.

Because of its apparent safety, ultrasound is the imaging technique of choice for determining the age and health of a developing fetus. It is also used to image the gallbladder and other viscera, and increasingly, the arteries in search of atherosclerosis (thickening and hardening of the arterial walls).

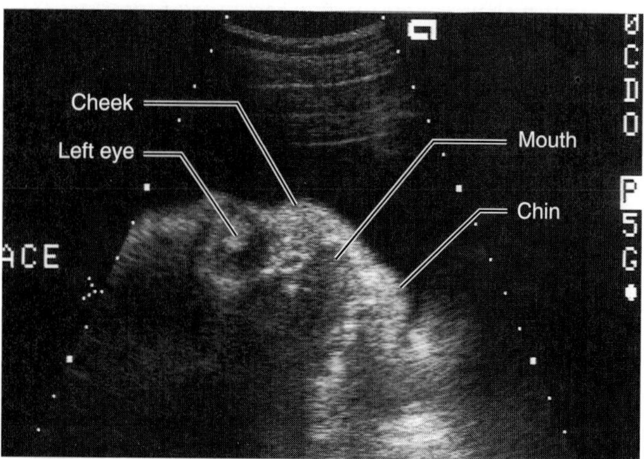

FIGURE 1.18 Ultrasound image of the face of a fetus in the uterus.

Ultrasound images are somewhat blurry, although their sharpness is being improved by using higher-frequency sound waves. Liquid contrast media containing sound-reflecting bubbles can be injected into the bloodstream to better reveal the vessels and heart. Sonography is of little value for viewing air-filled structures (lungs) or structures surrounded by bone (brain and spinal cord) because sound waves do not penetrate hard objects well and rapidly dissipate in air. However, special gases that do reflect sound can now be breathed into the lungs for improved images.

Magnetic Resonance Imaging

Magnetic resonance imaging (MRI) is a technique with tremendous appeal because it produces high-contrast images of our *soft* tissues (Figure 1.19), an area in which X-ray imaging is weak. In its present form, MRI primarily detects the levels of the element hydrogen in our body, most of which is in water. Thus, MRI tends to distinguish body tissues from one another on the basis of differences in water content. For example, MRI can distinguish the fatty white matter from the more watery gray matter of the brain. Bones, which contain less water than other tissues, do not show up at all, so MRI peers easily through the skull to reveal the brain. The joints also image well. Many tumors show up distinctly, and MRI has even revealed brain tumors that had been missed by direct observation during exploratory surgery!

The technique subjects the patient to magnetic fields up to 60,000 times stronger than that of the earth. The patient lies in a chamber, with his or her body surrounded by a huge magnet. When the magnet is turned on, the nuclei of the body's hydrogen atoms—single protons that spin like tops—line up parallel to the strong magnetic field. The patient is then exposed to a brief pulse of radio waves just below the frequency of FM radio, which knock the spinning protons out of alignment. When the radio waves are turned off, the protons return to their alignment in the magnetic field, emitting their own faint radio waves in the process. Sensors detect these waves, and the computer translates them into images.

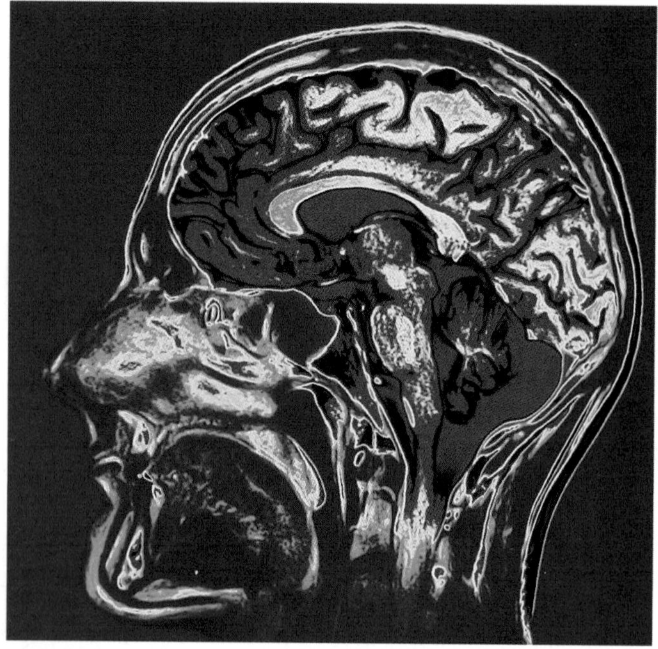

(a)

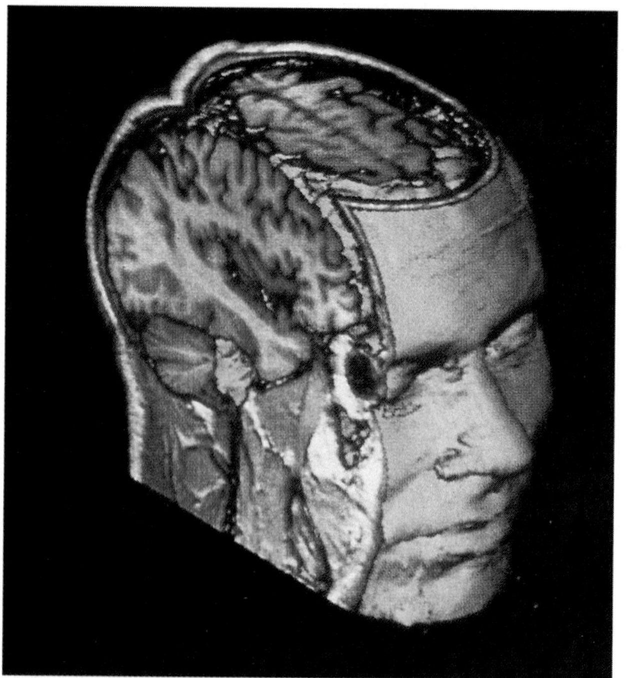

(b)

FIGURE 1.19 Magnetic resonance image (MRI) of a sagittal section of the head. (a) The MRI technique produces great clarity and contrast, in (artificial) colors. Unlike CT, its primary images are not limited to the transverse plane. **(b)** Volume rendering of an MRI of the head. The flat surfaces show the original MRI data.

Advances in Understanding In the early 1990s, MRI technology leaped forward with the development of **functional MRI (fMRI).** This technique measures blood oxygen, so it reveals the amount of oxygenated blood flowing to specific body regions. It can therefore determine which parts of the brain are active during various mental tasks. Functional MRI can pinpoint much smaller brain areas than can PET, it works much faster, and it uses no messy radioactive tracers. For these reasons, it is replacing PET in the study of brain function. ✷

Another type of MRI, called **magnetic resonance spectroscopy (MRS),** maps the body distribution of elements other than hydrogen to reveal more about how diseases change body chemistry. MRS could one day allow scientists to track the signals passing along neural circuits during thought, by measuring changes in *sodium* concentration that occur wherever nerve fibers carry impulses.

Computer scientists are now using advanced volume rendering techniques to assemble multiple MRI scans into three-dimensional reconstructions (Figure 1.19b). With the availability of ever more powerful computers, the speed of these reconstructions is increasing rapidly.

Despite the advantages of MRI, its magnets present problems. They are expensive, costing several million dollars per machine, and they bang with a hammering noise when operating. The magnetic chamber is tight-fitting and may induce claustrophobia, and the patient must stay completely still while being scanned. The magnets interfere with any implanted electronic devices, so people with pacemakers or implanted hearing aids cannot use MRI. Moreover, there is lingering concern that strong magnetic fields can cause harm, although the MRI procedure is currently considered to be safe.

Advances in Understanding Many of the problems with MRIs will be solved by the use of new, more-sensitive detectors of the MRI signals. Called SQUIDS (superconducting quantum interference devices), these detectors not only allow the use of cheaper, less-powerful magnets, but also improve the speed and clarity of MRI imaging. ✷

In conclusion, the images formed by computerized imaging techniques can be quite stunning. Keep in mind, however, that the images are abstractions assembled within the "brain" of a computer. They are artificially enhanced for sharpness and artificially colored to increase contrast or to highlight areas of interest. Although computer-based images are not inaccurate, they are several steps removed from direct visual observation.

CHAPTER SUMMARY

Media study tools that could provide you additional help in reviewing specified key topics of Chapter 1 are referenced below:

SP = Study Partner; AIA = A.D.A.M.® Interactive Anatomy

AN OVERVIEW OF ANATOMY (pp. 2–4)

1. Anatomy is the study of body structure. In this book, structures are considered in terms of their function.

Anatomical Terminology (p. 2)

2. Because most structures in the body have formal Greek and Latin names, learning anatomy is similar to learning a new language.

Branches of Anatomy (pp. 2–3)

3. Subdivisions of anatomy include gross anatomy, microscopic anatomy (histology), and developmental anatomy.

The Hierarchy of Structural Organization (pp. 3–4)

4. The levels of structural organization of the body, from simplest to most complex, are chemical, cellular, tissue, organ, organ system, and the human organism itself.

5. The organ systems in the body are the integumentary (skin), skeletal, muscular, nervous, endocrine, circulatory (cardiovascular and immune), respiratory, digestive, urinary, and reproductive systems.

AIA Link: Atlas; System: Integumentary; Surface anatomy of head (anterior). (To continue, choose another system, then choose an image within that system from the list provided on the screen.)

Scale: Length, Volume, and Weight (p. 4)

6. Important units of length measurement are meters (m) for the organism, centimeters (cm) for the organs, and micrometers (μm) for cells. For other units of size, see Appendix A The Metric System.

Anatomical Variability (p. 4)

7. Bodies vary, so anatomical structures do not always match textbook descriptions.

GROSS ANATOMY: AN INTRODUCTION (pp. 4–16)

The Anatomical Position (p. 4)

1. In the adult anatomical position, the body stands erect, facing forward with legs together. The arms are at the sides, with the palms forward.

Directional and Regional Terms (pp. 8–10)

2. Directional terms allow anatomists to describe the location of body structures with precision. Important terms include superior/inferior; anterior/posterior; ventral/dorsal; medial/lateral; proximal/distal; and superficial/deep (see Table 1.1).

3. Regional terms are used to designate specific areas of the body (see Figures 1.3 and 1.4).

AIA Link: Atlas; Region: Upper limb; Deep veins of upper limb (anterior). (To continue, choose another region, then choose the image within that region from the list provided on the screen.)

Body Planes and Sections (pp. 10–12)

4. The body or its organs may be cut along planes to produce different types of sections. Frequently used are sagittal, frontal, and transverse planes.

SP Exercise: Chapter 1, Body Planes.

AIA Link: Atlas; View: Anterior; Planes of trunk (anterior).

The Human Body Plan (pp. 12–13)

5. The basic structures we share with all other vertebrate animals are the tube-within-a-tube body plan, bilateral symmetry, a dorsal hollow nerve cord, notochord and vertebrae, segmentation, and pharyngeal pouches.

AIA Link: Dissectible anatomy; Male; Medial.

Body Cavities and Membranes (pp. 13–15)

6. The body contains two major closed cavities: the dorsal cavity, subdivided into the cranial and vertebral cavities; and the ventral body cavity, subdivided into the thoracic and abdominopelvic cavites.

7. Within the ventral cavity are the visceral organs (heart, lungs, intestines, kidneys, etc.) and three serous cavities: pleural, pericardial, and peritoneal cavities. These slit-like cavities are lined by thin membranes, the parietal and visceral serosae (see Figure 1.9). The serosae produce a thin layer of lubricating fluid that decreases friction between moving organs.

SP Exercise: Chapter 1, Dorsal and Ventral Body Cavities.

8. There are several smaller body cavities: oral, nasal, orbital, middle ear, and synovial.

AIA Link: Atlas; View: Lateral; Lateral wall of the nasal cavity.

Abdominal Regions and Quadrants (pp. 15–16)

9. To map the visceral organs in the abdominopelvic cavity, physicians divide the abdomen into nine regions or four quadrants.

AIA Link: Atlas; Region: Abdomen; Abdominoplevic regions.

MICROSCOPIC ANATOMY: AN INTRODUCTION (pp. 16–18)

Light and Electron Microscopy (pp. 16–17)

1. To illuminate cells and tissues, the light microscope (LM) uses light beams and the transmission electron microscope (TEM or EM) uses electron beams. EM produces sharper images than LM at higher magnification.

Preparing Human Tissues for Microscopy (pp. 17–18)

2. The preparation of tissues for microscopy involves preservation (fixation), sectioning, and staining. Stains for LM are colored dyes, while stains for TEM are heavy-metal salts.

Scanning Electron Microscopy (p. 18)

3. Scanning electron microscopy (SEM) provides sharp, three-dimensional images at high magnification.

Artifacts (p. 18)

4. All steps in the preparation of tissue for microscopy produce minor distortions called artifacts.

CLINICAL ANATOMY: AN INTRODUCTION TO MEDICAL IMAGING TECHNIQUES (pp. 19–23)

X Rays (p. 19)

1. In conventional radiographs, X rays are used to produce negative images of internal body structures. Denser structures in the body appear lighter (whiter) on the X-ray film.

AIA Link: Type; Radiograph; Flexed elbow (lateral).

Advanced X-Ray Techniques (p. 19)

2. CT produces improved X-ray images that are computer enhanced for clarity, and are taken in cross section to avoid overlapping images of adjacent organs. DSA produces sharp X-ray images of blood vessels injected with a contrast medium.

AIA Link: Type; Radiograph; Carotid angiogram (anterior).

3. PET tracks radioisotopes in the body, locating areas of high energy consumption and high blood flow.

4. Ultrasonography provides sonar images of developing fetuses and internal body structures.

5. MRI subjects the body to strong magnetic fields and radio waves, producing high-contrast images of soft body structures.

SP Exercise: Chapter 1, Imaging Techniques.

REVIEW QUESTIONS

Multiple Choice/Matching Questions

1. The correct sequence of levels forming the body's structural hierarchy is (a) organ, organ system, cellular, chemical, tissue; (b) chemical, tissue, cellular, organismal, organ, organ system; (c) chemical, cellular, tissue, organ, organ system, organismal; (d) organismal, organ system, organ, chemical, cellular, tissue.

2. Using what you learned about scale in this chapter, match each structure at left with the letter indicating its correct diameter or height, at right.

____ **(1)** white blood cell	**(a)** 2m
____ **(2)** stomach	**(b)** 25 cm
____ **(3)** professional basketball player	**(c)** 1 μm
	(d) 10 μm

3. For each part (a–e), circle the structure or organ that matches the given directional term.
   (a) distal: the elbow/the wrist
   (b) lateral: the hip bone/the navel
   (c) superior: the nose/the chin
   (d) anterior: the toes/the heel
   (e) superficial: the scalp/the skull

4. Which of these organs would not be cut by a section through the midsagittal plane of the body? (Hint: See Figure 1.11.) (a) urinary bladder, (b) gallbladder, (c) small intestine, (d) heart.

5. Relate each of the following conditions or statements to either the dorsal body cavity (D) or the ventral body cavity (V).
   (a) surrounded by the skull and the vertebral column
   (b) includes the thoracic and abdominopelvic cavities
   (c) contains the brain and spinal cord
   (d) located more anteriorly
   (e) contains the heart, lungs, and many digestive organs

6. Radiographs. List the following structures, from darkest (black) to lightest (white), as they would appear on an X-ray film. Number the darkest one 1, the next darkest 2, and so on.

___ **(a)** soft tissue

___ **(b)** femur (bone of the thigh)

___ **(c)** air in lungs

___ **(d)** gold (metal) filling in a tooth

7. The ventral surface of the body is the same as its ___ surface. (a) medial, (b) superior, (c) superficial, (d) anterior, (e) distal.

8. Twenty-year-old Joey has a small streak of white hair on his head, although the rest of his hair is black. Joey exhibits (a) a rare disease, (b) a serious birth defect, (c) an example of anatomical variation, (d) an artifact.

9. Which of the following relationships is incorrect? (a) visceral peritoneum/surface of small intestine, (b) parietal part of pericardium/surface of heart, (c) parietal pleura/wall of thoracic cavity.

10. Which microscopic technique provides sharp pictures of three-dimensional structures at high magnification? (a) light microscopy, (b) X-ray microscopy, (c) scanning electron microscopy, (d) transmission electron microscopy.

11. Histology is the same as (a) pathological anatomy, (b) ultrastructure, (c) functional morphology, (d) surface anatomy, (e) microscopic anatomy.

Short Answer Essay Questions

12. Describe the anatomical position, and then assume this position.

13. Describe two ways you could distinguish a transmission electron micrograph from a light micrograph.

14. (a) Define bilateral symmetry. (b) Give an example of segmentation in the human body. (c) Define dissection.

15. The following advanced imaging techniques are discussed in the text: CT, DSA, PET, sonography, and MRI. (a) Which of these techniques uses X rays? (b) Which uses radio waves and magnetic fields? (c) Which uses radioisotopes? (d) Which displays body regions in sections? You may have more than one answer to each question.

16. Give the formal regional term for each of these body areas: (a) arm, (b) chest, (c) groin, (d) armpit, (e) limbs.

17. Look ahead in this book to Figure 4.3 (pp. 79–82). Do you think that the micrographs in the figure are viewed by light microscopy, transmission electron microscopy, or scanning electron microscopy? (Hint: Color is a property of light, not of electron beams.) Please explain your answer as fully as possible.

18. Although many body structures are bilaterally symmetrical, much of the *abdomen* is not. Find a picture that demonstrates this lack of symmetry, and name some abdominal organs that are not symmetrical.

19. Construct sentences that use the following directional terms: superior/inferior; dorsal/ventral; medial/lateral; and superficial/deep. An example would be "The forehead is superior to the nose" provided in Table 1.1, but please use different examples than those in the table. For ideas, it will help to view whole-body diagrams that show many structures, such as Figures 1.4 and 1.12.

20. The main cavities of the body include the abdominal and pelvic cavities. List three organs in each of these cavities.

CRITICAL REASONING + CLINICAL APPLICATION QUESTIONS

1. Dominic's doctors strongly suspect he has a tumor in his brain. Which of the following medical imaging devices—conventional X ray, DSA, PET, sonography, or MRI—would probably be best for precisely locating the tumor?

2. The Nguyen family was traveling in their van and had a minor accident. The children in the back seat were wearing lap belts, but they still sustained bruises around the abdomen and had some internal injuries. Why is the abdominal area more vulnerable to damage than others?

3. A patient had a hernia in his inguinal region, pain from an infected kidney in his lumbar region, and hemorrhoids in his perineal region. Explain where each of these regions is located.

4. A woman fell off a motorcycle and tore a nerve in her axillary region, and broke bones in her coxal, sacral, acromial, and peroneal regions. Explain where each of these regions is located.

5. New anatomy students often mix up the terms *spinal cord, spinal column,* and *spine.* Look up these words elsewhere in this book or in a medical dictionary, define each one, and tell whether they are the same or different.

A.D.A.M.® INTERACTIVE ANATOMY QUESTIONS

1. Link: Atlas; Type: Illustration; Abdominopelvic regions. (a) Name the organs observed in the umbilical area. (b) The right lobe of the liver extends along the transpyloric plane through which two areas?

2. Link: Atlas; Region; Upper arm: Bones of the arm and shoulder (anterior). (a) On which end of the humerus can one observe the capitulum? (b) Name the bone of the forearm that articulates with the capitulum.

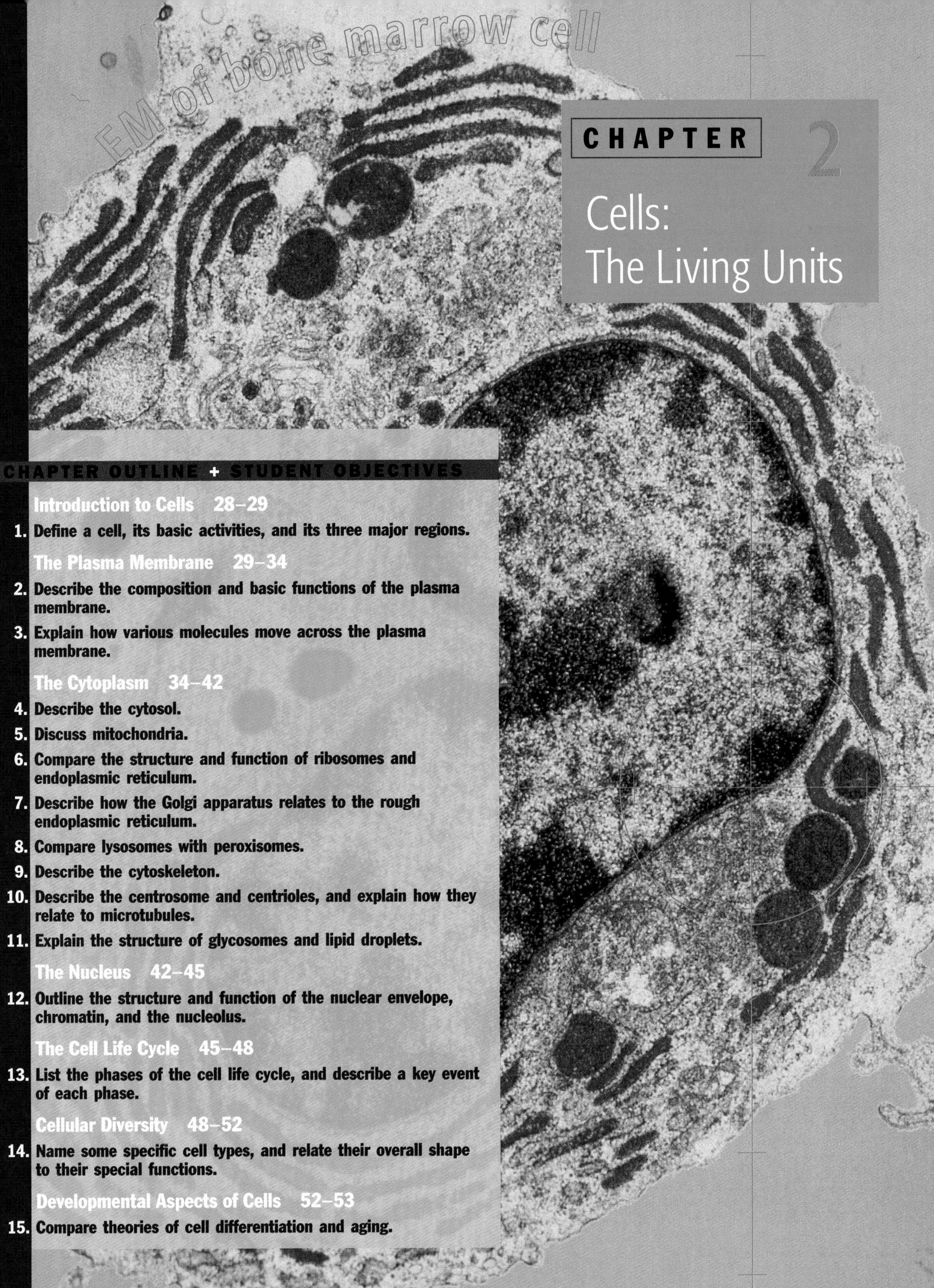

CHAPTER 2

Cells: The Living Units

CHAPTER OUTLINE + STUDENT OBJECTIVES

Introduction to Cells 28–29

1. Define a cell, its basic activities, and its three major regions.

The Plasma Membrane 29–34

2. Describe the composition and basic functions of the plasma membrane.
3. Explain how various molecules move across the plasma membrane.

The Cytoplasm 34–42

4. Describe the cytosol.
5. Discuss mitochondria.
6. Compare the structure and function of ribosomes and endoplasmic reticulum.
7. Describe how the Golgi apparatus relates to the rough endoplasmic reticulum.
8. Compare lysosomes with peroxisomes.
9. Describe the cytoskeleton.
10. Describe the centrosome and centrioles, and explain how they relate to microtubules.
11. Explain the structure of glycosomes and lipid droplets.

The Nucleus 42–45

12. Outline the structure and function of the nuclear envelope, chromatin, and the nucleolus.

The Cell Life Cycle 45–48

13. List the phases of the cell life cycle, and describe a key event of each phase.

Cellular Diversity 48–52

14. Name some specific cell types, and relate their overall shape to their special functions.

Developmental Aspects of Cells 52–53

15. Compare theories of cell differentiation and aging.

ALL LIVING ORGANISMS ARE CELLULAR IN nature, from one-celled "generalists" like amoebas to complex multicellular organisms such as trees, dogs, and humans. Just as bricks and timbers are the structural units of a house, cells are structural units of all living things. The human body has about *50 to 100 trillion* cells. This chapter examines the structures and functions that are shared by the different cells of the body. Specialized cells and their unique functions are addressed in detail in later chapters.

INTRODUCTION TO CELLS

The scientist Robert Hooke first observed plant cells with a crude microscope in the late 1600s. However, it was not until the 1830s that two scientists, Matthias Schleiden and Theodor Schwann, were bold enough to assert that all living things are composed of cells. Shortly thereafter, the pathologist Rudolf Virchow extended this idea by contending that cells arise only from other cells. Virchow's thesis was revolutionary because it challenged the prevailing theory of spontaneous generation, which held that organisms can arise from nonliving matter. Since the late 1800s, cell research has been exceptionally fruitful. Currently, scientific knowledge about the cell doubles every 7 years or so!

Cells are the smallest living units in our body. Each cell performs all the functions necessary to sustain life: It can obtain nutrients and other essential substances from the surrounding body fluids and use these nutrients to make the molecules it needs to survive. Each cell also disposes of its wastes and maintains its shape and integrity. Finally, cells can replicate themselves. These functions are carried out by the cell's many subunits, most of which are called **organelles** ("little organs"). Most chemical reactions in cells are directed and con-

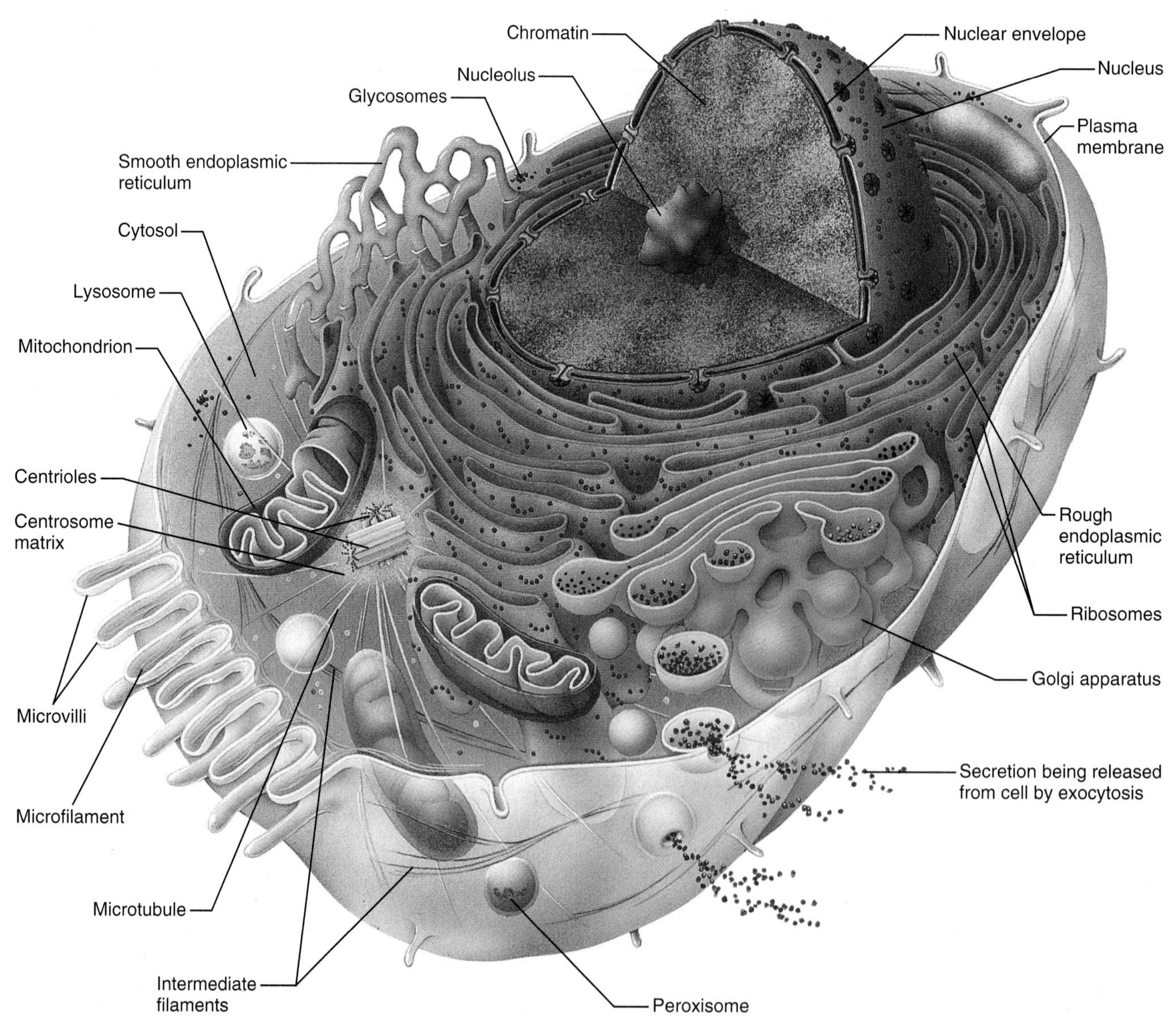

FIGURE 2.1 Structure of a generalized cell. No cell type is exactly like this one, but virtually all the features shown occur in every animal cell. Exceptions are the microvilli, which occur only in certain cells (discussed in Chapter 4).

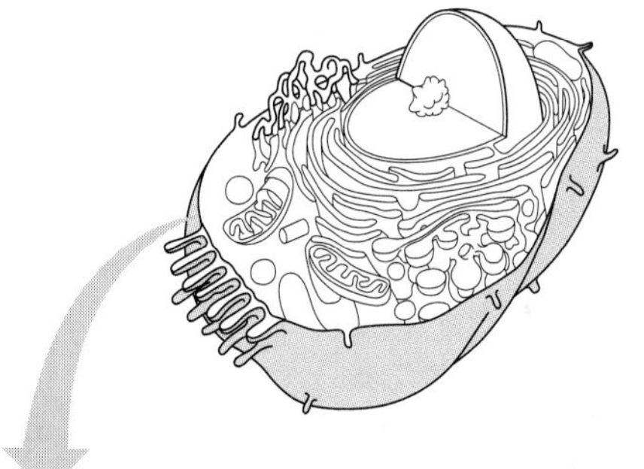

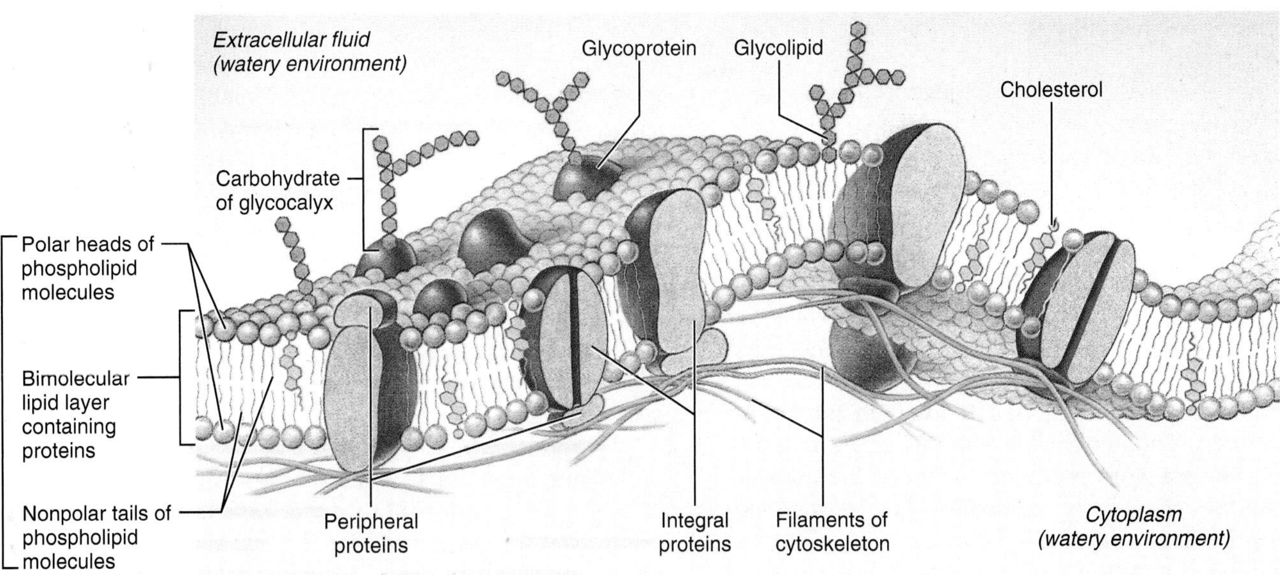

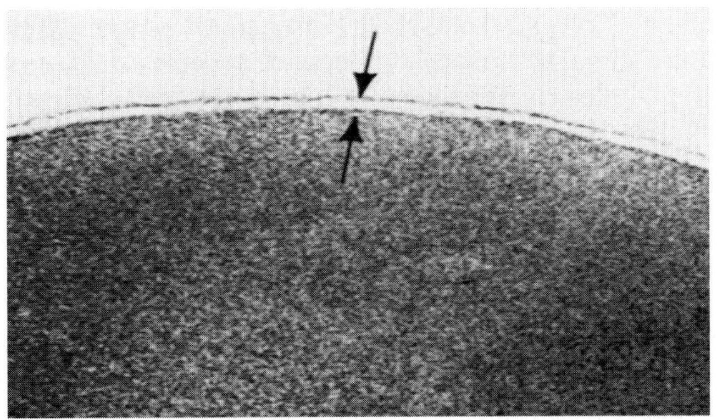

FIGURE 2.2 The plasma membrane. (a) Structure of the plasma membrane according to the fluid mosaic model. **(b)** Electron micrograph of plasma membrane of a red blood cell (220,000 ×). Here, the membrane has a three-layered appearance of two dark bands (arrows) separated by a light band.

trolled by **enzymes** (en′zīmz), catalyst molecules made of protein. The sum of all the chemical reactions in the cell, and thus in the whole body, is called **metabolism** (mĕ-tab′o-lizm; "a state of change").

Although different cell types perform different functions, virtually all human cells contain the same basic parts and can be described in terms of a generalized cell (Figure 2.1). Human cells have three main parts: the plasma membrane, the cytoplasm, and the nucleus. The *plasma membrane* is the outer boundary. Internal to this membrane is the *cytoplasm* (si′to-plazm), which makes up the bulk of the cell and surrounds the nucleus. The *nucleus* (nu′kle-us) controls cellular activities and lies near the cell's center. These cell structures are discussed next, and are summarized in Table 2.1 on page 44.

THE PLASMA MEMBRANE

The outer cell membrane (Figure 2.2) is called the **plasma membrane** or **plasmalemma** (plaz″mah-lem′ah; *lemma* = sheath, husk). This thin, flexible layer defines the extent of the cell, thereby separating two of the body's major fluid compartments: the *intracellular fluid* within the cells and the *extracellular fluid* that lies outside and between cells.

Structure

The *fluid mosaic model* of membrane structure depicts the plasma membrane as a double layer, or bilayer, of lipid molecules with protein molecules dispersed within it (Figure 2.2a). The most abundant lipids in the plasma membrane are phospholipids. Like a lollipop on two sticks, each phospholipid molecule has a polar "head" that is charged, and an uncharged, nonpolar "tail" made of two chains of fatty acids. The polar heads are attracted to water—the main constituent of both the cytoplasm and the fluid external to the cell—so they lie on both the inner and outer faces of the membrane. The nonpolar tails, by contrast, avoid water and line up in the center of the membrane. The result is two parallel sheets of phospholipid molecules lying tail-to-tail, forming the membrane's basic bilayered structure.

The inner and outer layers of the membrane differ somewhat in the specific kinds of lipids they contain. Sugar groups are attached to about 10% of the outer lipid molecules, making them "sugar-fats," or glycolipids (gli″ko-lip′ids). The plasma membrane also contains substantial amounts of cholesterol, another lipid. Cholesterol improves the rigidity of the membrane and increases its impermeability to water and water-soluble molecules.

Even though we know that the plasma membrane is bilayered, in electron micrographs it appears to have *three* layers—two dark layers sandwiching a light one (Figure 2.2b). How can this discrepancy between molecular structure and visualized structure be explained? Evidently, the metal ions of the stain used for electron microscopy (osmium tetroxide) attach to the polar heads of the phospholipid molecules, staining both surfaces of the membrane but leaving the central part (the tails) unstained.

Proteins make up about half of the plasma membrane by weight. The membrane proteins are of two distinct types: integral and peripheral (Figure 2.2a). **Integral proteins** are firmly embedded in or strongly attached to the lipid bilayer. Some integral proteins protrude from one side of the membrane only, but most are *transmembrane proteins* that span the whole width of the membrane and protrude from both sides (*trans* = across). **Peripheral proteins,** by contrast, are not embedded in the lipid bilayer at all. Instead, they attach rather loosely to the membrane surface, usually the inner surface. The peripheral proteins include a network of filaments that helps support the membrane from its cytoplasmic side. Without this strong, supportive base, the plasma membrane would tear apart easily.

Short chains of carbohydrate molecules attach to the integral proteins. These sugars project from the external cell surface, forming the *cell coat* or *glycocalyx* (gli″ko-kal′iks; "sugar covering"). Also contributing to the glycocalyx are the sugars of the membrane's glycolipids. You can therefore think of your cells as "sugar-coated." The glycocalyx is sticky and may help cells to bind when they come together. Because every cell type has a different pattern of sugars that make up its glycocalyx, the glycocalyx is also a distinctive biological marker by which approaching cells recognize each other. For example, a sperm recognizes the ovum (egg cell) by the distinctive composition of the ovum's cell coat.

Functions

The functions of the plasma membrane relate to its location at the interface between the cell's exterior and interior. First, it provides a protective barrier against substances and forces outside the cell. Second, some membrane proteins act as **receptors:** that is, they have the ability to bind to specific molecules arriving from outside the cell. Once bound to the receptor, the molecule can induce a change in the cell. Membrane receptors, therefore, are part of the body's cellular communication system.

Third, and perhaps most importantly, the plasma membrane determines which substances can enter and leave the cell. The membrane is a selectively permeable barrier, meaning that it allows some substances to pass between the intracellular and extracellular fluids while preventing others from doing so. Only small, uncharged molecules like oxygen, carbon dioxide, and fat-soluble drugs can pass freely through the lipid bilayer. This free passage of molecules is accomplished through the process called diffusion. *Diffusion* is the tendency of molecules in a solution to move down their concentration gradient; that is, the molecules move from a region where they are more concentrated to a region where they are less concentrated. Water, like other molecules, diffuses down its concentration gradient. The diffusion of water molecules across a membrane is called *osmosis* (oz-mo′sis).

Water-soluble or charged molecules, such as sugar molecules and ions, cannot pass through the lipid bilayer. Such substances can cross the plasma membrane only by means of specific *transport mechanisms.* These mechanisms carry or pump molecules across the membrane or form channels through which specific molecules pass. All these transport mechanisms are formed by the intrinsic membrane proteins.

The largest molecules (macromolecules) and large solid particles are transported through the plasma membrane by another set of processes, called *vesicular,* or *bulk transport.* Knowledge of the two general types of bulk transport, *exocytosis* and *endocytosis,* is essential to the understanding of basic functional anatomy.

Exocytosis (ek″so-si-to′sis; "out of the cell") is a mechanism that moves substances from the cytoplasm to the outside of the cell (Figure 2.3). Exocytosis accounts for most secretion processes, such as the release of mucus or protein hormones from the gland cells of the body. In exocytosis the substance or cell product to be released from the cell is first enclosed in a membrane-walled sac, or *vesicle,* in the cytoplasm. The vesicle migrates to the plasma membrane and fuses with it, releasing the contents of the sac into the space outside the cell.

Endocytosis (en″do-si-to′sis; "into the cell") is the mechanism by which large particles and macromolecules *enter* cells (Figure 2.4 and Figure 2.5). The substance to be taken into the cell is enclosed by an infolding part of the plasma membrane. This forms a membranous vesicle, which pinches off from the plasma membrane and moves into the cytoplasm, where its contents are digested. Three types of endocytosis are recognized: phagocytosis, pinocytosis, and receptor-mediated endocytosis.

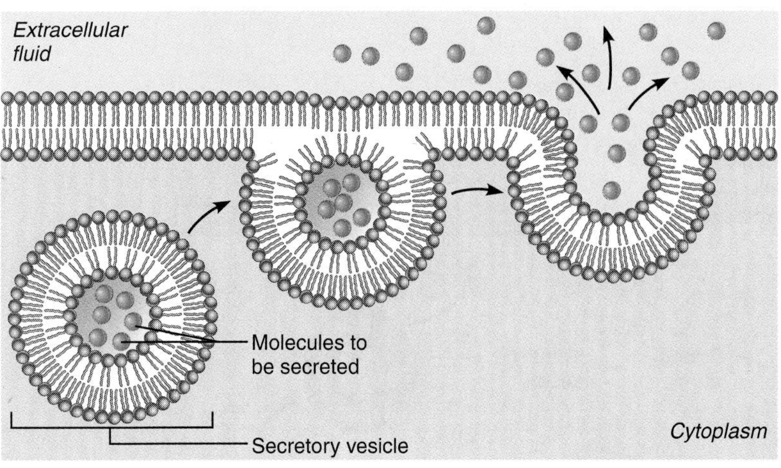

(a)

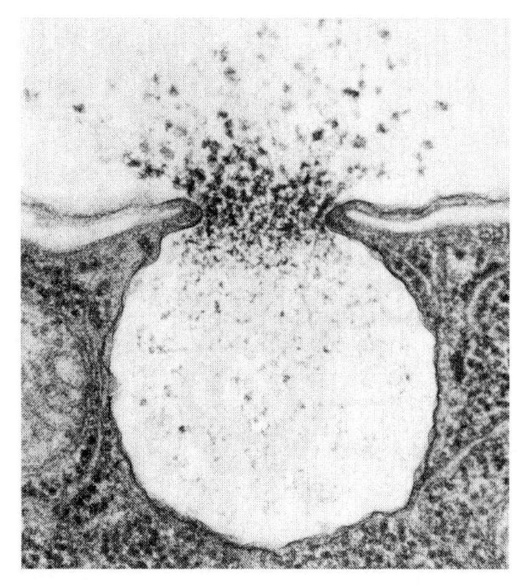

(b)

FIGURE 2.3 Exocytosis. **(a)** A membrane-walled secretory vesicle containing the substance to be secreted migrates to the plasma membrane, and the two membranes fuse. The fused site opens and releases the contents of the vesicle into the extracellular fluid. **(b)** Electron micrograph of a vesicle in exocytosis (100,000 ×).

Phagocytosis (fag″o-si-to′sis) is literally "cell eating." In this process, parts of the plasma membrane and cytoplasm protrude and flow around some relatively large material, such as a clump of bacteria or cellular debris, and engulf it (Figure 2.4a). The membranous vesicle thus formed is called a **phagosome** (fag′o-sōm; "eaten body"). In most cases, the phagosome then fuses with *lysosomes* (li′so-sōmz), organelles containing digestive enzymes that break down the contents of the phagosome. (For more on lysosomes, see p. 38.) Some cells—most white blood cells, for example—are experts at phagocytosis. Such cells help to police and protect the body by ingesting bacteria, viruses, and other foreign substances. They also "eat" the body's dead and diseased cells.

Critical Reasoning

Phagocytic cells gather in the air spaces of the lungs—especially in the lungs of smokers. Can you explain this connection?

Phagocytic cells in the lungs serve to remove bacteria and dust particles from the inhaled air. In smokers, these cells are packed with indigestible particles of carbon.

Just as we can say that cells eat, we can also say that they drink. **Pinocytosis** (pin″o-si-to′sis) is "cell drinking" (Figure 2.4b). In pinocytosis, a bit of infolding plasma membrane surrounds a tiny quantity of extracellular fluid containing dissolved molecules called

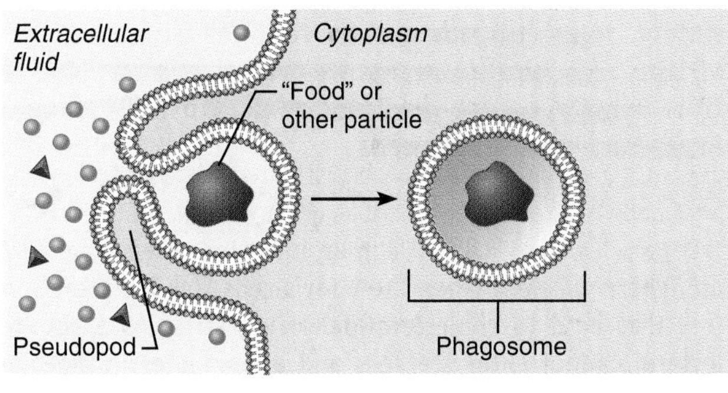

(a) Phagocytosis

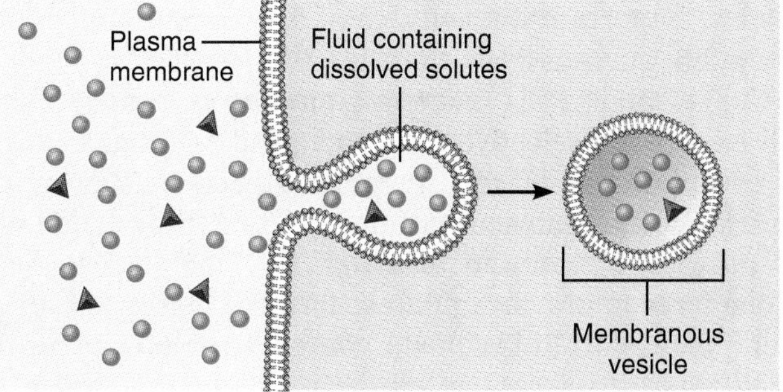

(b) Pinocytosis

FIGURE 2.4 Phagocytosis and pinocytosis—two types of endocytosis. **(a)** Phagocytosis draws particles that are larger than molecules into the cell. **(b)** Pinocytosis draws fluids and dissolved solutes into the cell.

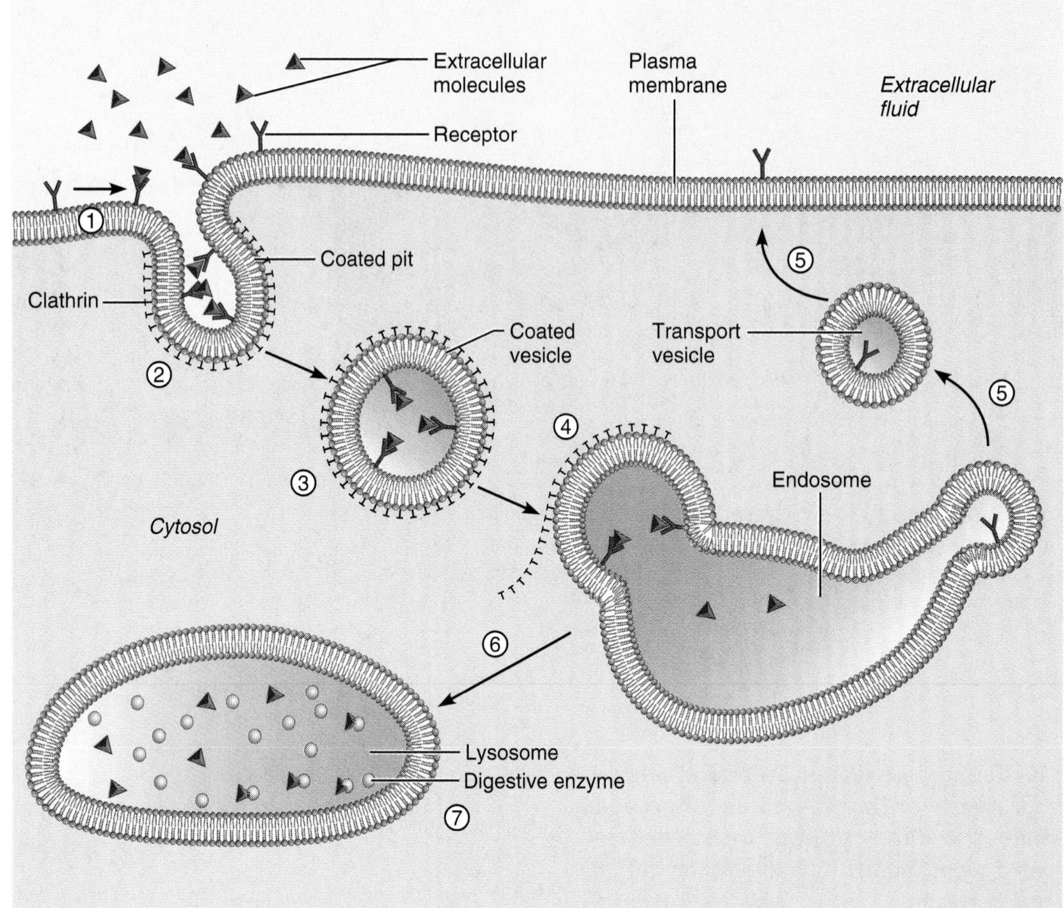

FIGURE 2.5 Receptor-mediated endocytosis. Stages in the processing of clathrin-coated vesicles and endosomes: 1) Extracellular molecule binds to a receptor; 2) Receptors cluster in a clathrin-coated pit; 3) Coated pit invaginates to become a coated vesicle; 4) Vesicle uncoats and fuses to an endosome; 5) Receptors, in transport vesicles, recycle back to the plasma membrane; 6) Endosome fuses with a lysosome; 7) Lysosome digests the molecules.

solutes. This fluid enters the cell in a tiny membranous **pinocytotic vesicle.** Unlike phagocytosis, pinocytosis is a routine activity of most cells. It is an unselective way of sampling the extracellular fluid. This process is particularly important in cells that function in nutrient absorption, such as cells that line the intestines.

Unlike pinocytosis, **receptor-mediated endocytosis** is an exquisitely selective transport process (Figure 2.5). The receptors are plasma-membrane proteins that bind only with certain molecules. Upon binding, the portion of the plasma membrane bearing the molecules and attached receptors invaginates, forming a **coated pit,** which pinches off to become a **coated vesicle.** The term *coated* refers to a coat of **clathrin** (klath′rin; "lattice") protein around the vesicle. Once the coated vesicle is inside the cell, it loses its coat and fuses to a bigger vesicle in the cytoplasm called an **endosome.** This in turn carries the receptor-bound molecules to a lysosome for digestion. Along the way, the attached receptors are pinched off and returned to the plasma membrane.

Molecules taken up by receptor-mediated endocytosis include hormones such as insulin, and *low-density lipoproteins (LDLs),* the molecules that carry cholesterol through the bloodstream to the body's cells. Unfortunately, harmful substances such as some toxins and viruses also bind to receptors on the plasma membrane, thus using receptor-mediated endocytosis as a means to enter and attack our cells.

Clinical Applications

HYPERCHOLESTEROLEMIA In an inherited disease called *familial hypercholesterolemia,* the body's cells lack the protein receptors that bind to cholesterol-delivering LDLs. As a result, cholesterol cannot enter the cells and accumulates in the blood. If untreated, hypercholesterolemia causes atherosclerosis, also known as "hardening of the arteries," a condition that places the individual at high risk of stroke (blockage of a blood vessel in the brain) or of coronary artery disease and heart attack. ✚

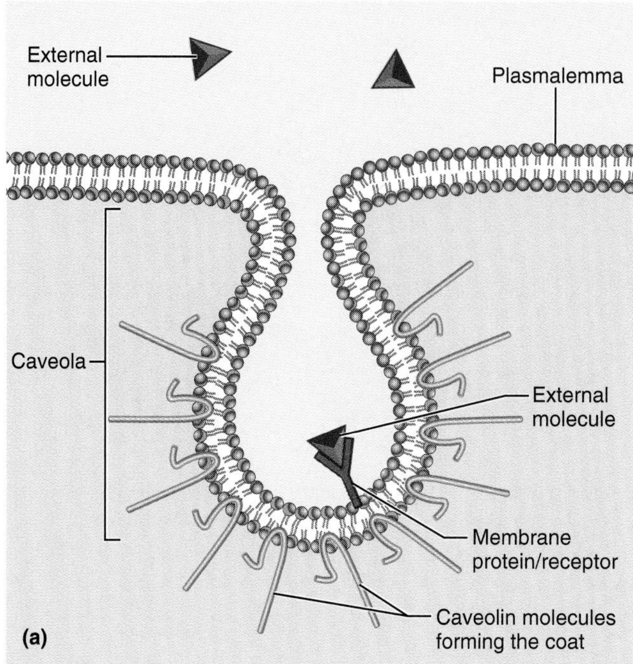

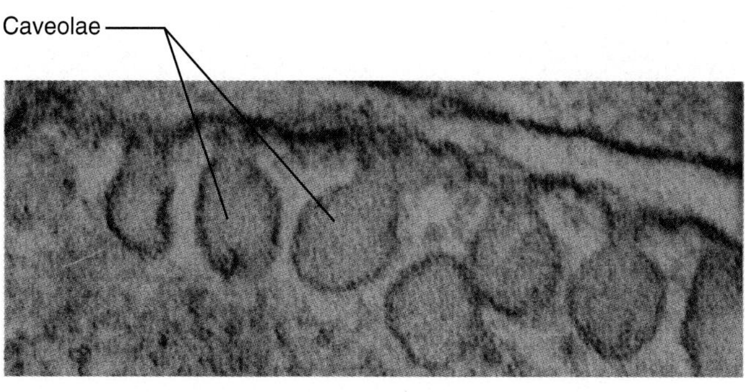

FIGURE 2.6 **Caveolae.** These are flask-shaped pits that perform a type of receptor-mediated endocytosis. **(a)** Diagram of a caveola, showing the caveolin coat. **(b)** Electron micrograph of several caveolae (80,000 ×).

Advances in Understanding It was recently discovered that another kind of receptor-mediated endocytosis occurs, involving **caveolae** (ka″ve-o′le; "little caves"). As shown in Figure 2.6, caveolae are flask-shaped inpocketings of the plasma membrane present in many cell types. Like clathrin-coated pits, they capture specific types of proteins and other molecules from the extracellular fluid, then form vesicles to carry these molecules through the cytoplasm. However, caveolae are smaller than clathrin-coated pits, their "coat" is thinner and composed of different proteins (called caveolins), and they transport different classes of proteins than do clathrin-coated pits. Although their precise role is not well understood in most cell types, caveolae are known to move proteins across the cells that form the walls of capillaries (p. 549) ✱

THE CYTOPLASM

Cytoplasm, literally "cell-forming material," is the part of the cell that lies internal to the plasma membrane and external to the nucleus. Most cellular activities are carried out in the cytoplasm, which consists of three major elements: *cytosol, organelles,* and *inclusions* (see Table 2.1 on p. 44).

Cytosol

The **cytosol** (si′to-sol), or **cytoplasmic matrix,** is the jelly-like, fluid-containing substance within which the other cytoplasmic elements are suspended (see Figure 2.1). It consists of water, ions, and many enzymes. Some of these enzymes start the breakdown of nutrients (sugars, amino acids, lipids) that provide the raw material and source of energy for cell activities. In many cell types, the cytosol makes up about half the volume of the cytoplasm.

Cytoplasmic Organelles

Typically, the cytoplasm contains about nine types of organelles, each of which has a different function and is essential for the survival of the cell. As discrete units, the organelles compartmentalize the cell's many biochemical reactions, thus preventing reactions from interfering with one another and promoting maximal functional efficiency. The organelles include mitochondria, ribosomes, rough and smooth endoplasmic reticulum, Golgi apparatus, lysosomes, peroxisomes, the cytoskeleton, and centrioles (see Figure 2.1). As you will learn, most organelles are bounded by a membrane that is similar in composition to the plasma membrane but lacks a glycocalyx.

With very few exceptions, all cells of the human body share the same kinds of organelles. However, when a cell type performs a special body function, the particular organelles that contribute to that function are especially abundant in that cell. Thus, certain organelles are better developed in some cells than in others. Examples of this principle will be provided as we explore the organelles and their roles.

Mitochondria

Mitochondria (mi′to-kon″dre-ah) are the main "power plants" of the cells. These organelles usually are depicted as being bean-shaped, reflecting their appearance in sections under the microscope (Figure 2.7), but mitochondria are actually long and thread-like (*mitos* = thread). In living cells, they squirm about and change shape as they move through the cytoplasm. Most organelles are surrounded by a membrane, but mitochondria are actually enclosed by *two* membranes: The outer membrane is smooth and featureless, but the inner membrane folds inward to produce shelf-like **cristae** (krĭ′ste; "crests"). These protrude into the

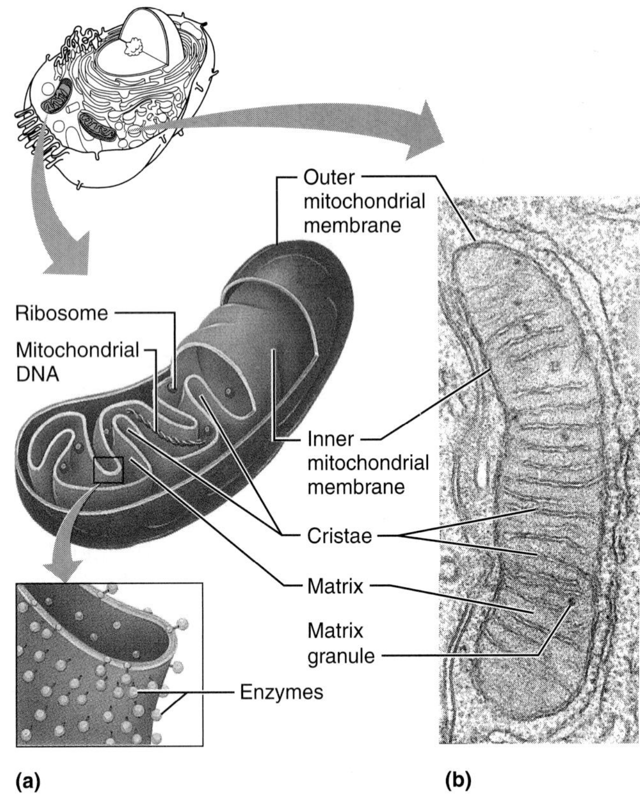

FIGURE 2.7 Mitochondria. **(a)** Diagram of a longitudinally opened mitochondrion. In the inset at lower left, the stalked spheres on the inner mitochondrial membrane (including on the cristae) are the enzyme complexes that make ATP molecules. **(b)** Electron micrograph of a mitochondrion (approx. 36,000 ×).

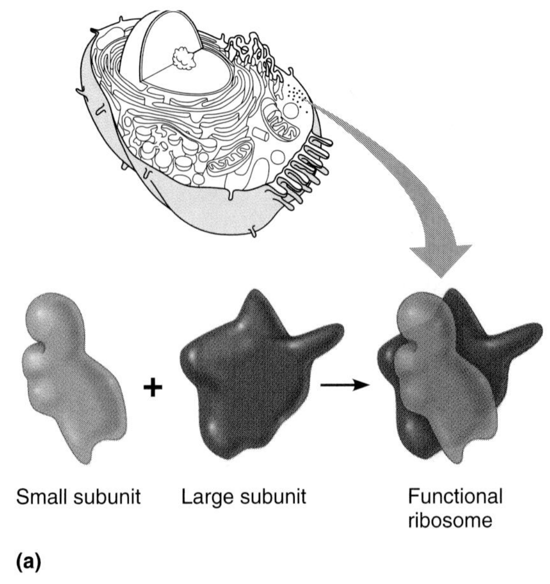

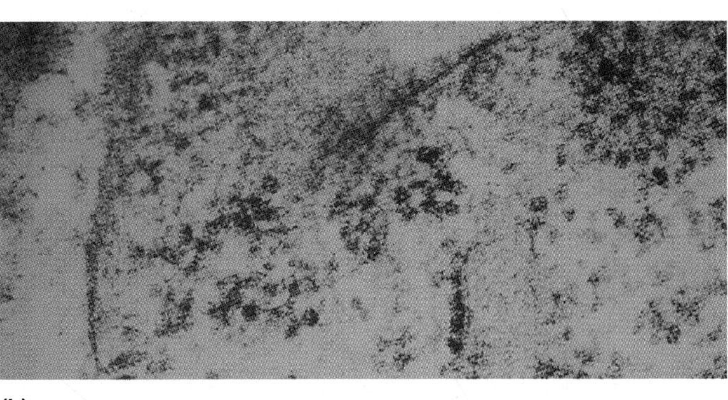

(b)

FIGURE 2.8 Ribosomes. **(a)** Diagram of a ribosome, made from small and large subunits. **(b)** Electron micrograph of free ribosomes (the dark dots) in the cytosol (100,000 ×).

matrix, the jelly-like substance within the mitochondrion. The matrix also contains scattered, dark spheres, the **matrix granules** (Figure 2.7b).

Mitochondria generate most of the energy the cell uses to carry out work. They do this by systematically releasing the energy stored in the chemical bonds of nutrient molecules and then transferring this energy to **ATP (adenosine triphosphate),** the high-energy molecules that cells use to power chemical reactions. Within the mitochondrion, the ATP-generating process starts in the matrix (by a process called the citric acid cycle) and is completed on the inner membrane of the cristae (by the processes called oxidative phosphorylation and electron transport). Those cell types with the highest energy requirements have the largest number of cristae in their mitochondria.

The function of the matrix granules is uncertain. Some consist of lipids (fats) that may provide raw materials for building the mitochondrial membranes. Other matrix granules may bind and store calcium ions that are taken up by mitochondria.

Mitochondria are far more complex than any other organelle. They even contain some genetic material (DNA) and divide to form new mitochondria, as if they were miniature cells themselves. Intriguingly, mitochondria are very similar to a group of bacteria, the purple bacteria phylum. It is now widely believed that mitochondria arose from bacteria that invaded the ancient ancestors of animal and plant cells.

Critical Reasoning

If a particular cell type has a large number of mitochondria in its cytoplasm, what can you say about the activity level of that cell?

The cell performs some process that requires a great deal of energy. The muscle cell, which requires large amounts of energy for contraction, is an example of a mitochondria-rich cell.

Ribosomes

Ribosomes (ri′bo-sōmz) are small, dark-staining granules (Figure 2.8). Unlike most organelles, they are not surrounded by a membrane, but are constructed of proteins plus **ribosomal RNA** (*RNA* = ribonucleic acid). Each ribosome consists of two subunits that fit together like the body and cap of an acorn (Figure 2.8a).

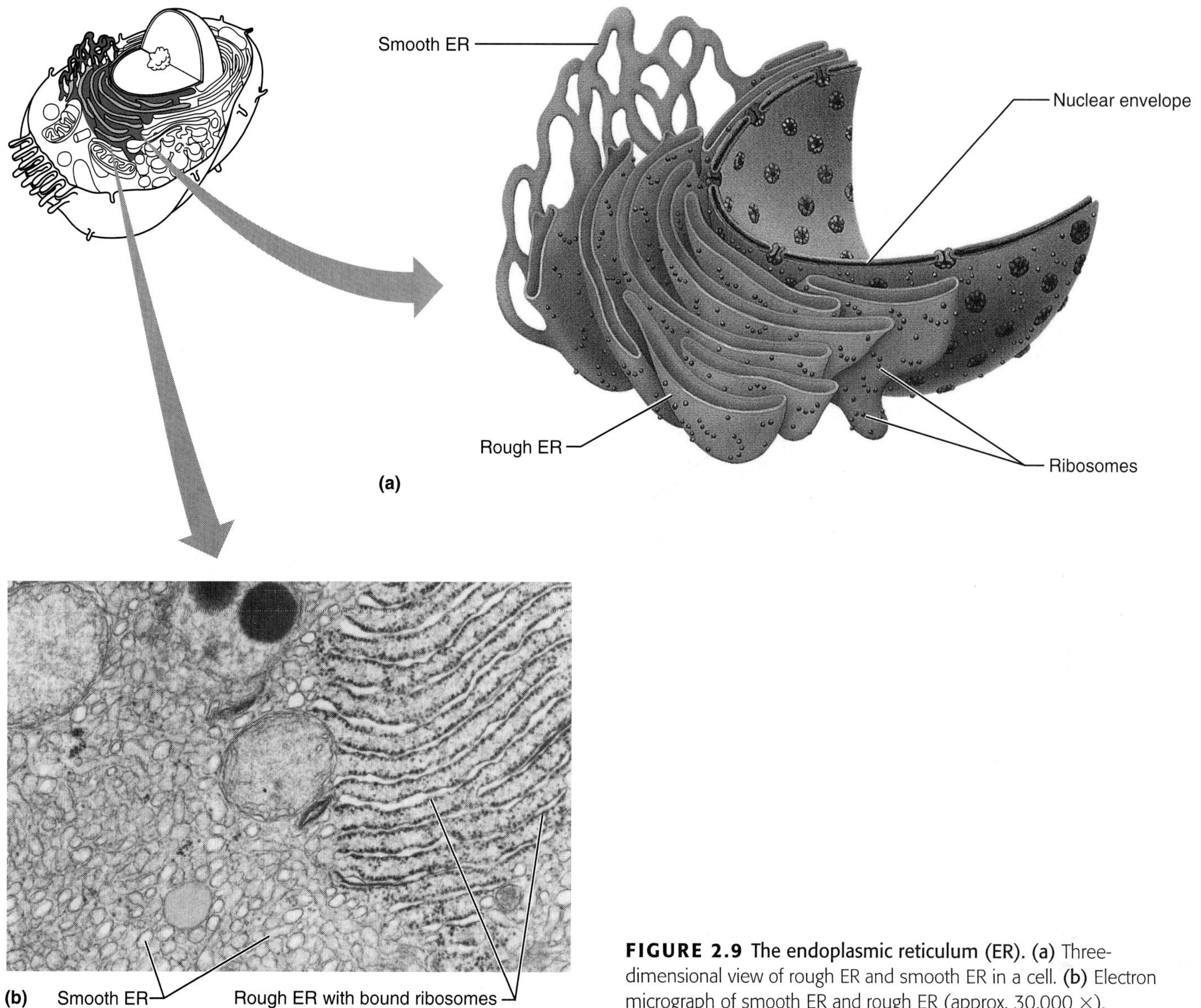

FIGURE 2.9 The endoplasmic reticulum (ER). **(a)** Three-dimensional view of rough ER and smooth ER in a cell. **(b)** Electron micrograph of smooth ER and rough ER (approx. 30,000 ×).

Almost all cells make large amounts of protein, and ribosomes are the site of protein synthesis. On the ribosomes, building blocks called amino acids are linked together to form protein molecules. This assembly process is called *translation*. It is dictated by the genetic material in the cell nucleus (DNA, p. 41), whose instructions are carried to the ribosomes by messenger molecules called **messenger RNA (mRNA).**

Many ribosomes float freely within the cytosol. Such **free ribosomes** (Figure 2.8b) make the soluble proteins that function within the cytosol itself. By contrast, ribosomes attached to the membranes of the rough endoplasmic reticulum (discussed next) synthesize proteins with somewhat different destinies. In either case, some ribosomes exist individually while others line up in chains called polyribosomes. These polyribosomes make multiple copies of the same protein in assembly-line fashion.

Endoplasmic Reticulum

The **endoplasmic reticulum** (en″do-plaz′mik ret-tik′u-lum), or **ER,** is literally the "network within the cytoplasm." As shown in Figure 2.9, the ER is an extensive system of membrane-walled envelopes and tubes that twists through the cytoplasm. It comprises more than half of the membranous surfaces inside an average human cell. There are two distinct varieties of ER: *rough ER* and *smooth ER*. Either variety may predominate in a given cell type, depending on the specific functions of the cell.

Rough Endoplasmic Reticulum The **rough endoplasmic reticulum (rough ER)** consists mainly of stacked envelopes called **cisternae** (sis-ter′ne; "fluid-filled cavities"). Ribosomes stud the external faces of the membranes of the rough ER, assembling proteins. The ribosomes attach to the membrane when the protein is being made, then detach when the protein is completed.

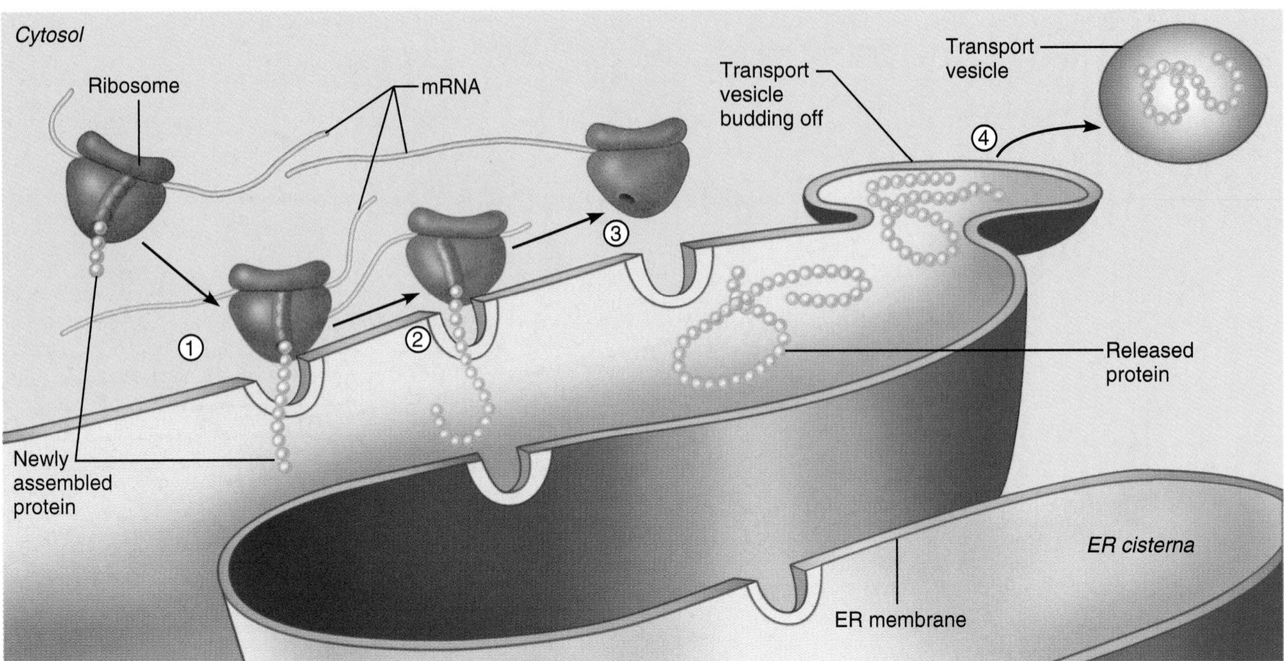

FIGURE 2.10 Assembly of proteins at the rough endoplasmic reticulum. 1) Ribosome binds, and its protein enters the cisterna of the rough ER; 2) Protein elongates; 3) Ribosome detaches; 4) Transport vesicle buds off, taking the protein to the Golgi apparatus.

Figure 2.10 shows how such protein assembly works. 1) A ribosome, translating a strand of mRNA, attaches to the rough ER and its protein enters the ER cisterna. 2) The protein elongates within the cisterna. 3) The ribosome detaches, leaving the completed protein behind. 4) The protein becomes encapsulated by ER membrane, buds off in transport vesicles, and migrates through the cytoplasm to the Golgi apparatus, which sorts and packages the product (see p. 38).

The rough ER has several functions. Its ribosomes make all proteins that are secreted from cells; thus, rough ER is especially well-developed in gland cells that secrete a large amount of protein (mucous cells, for example). It also makes the digestive enzymes that will be contained in lysosomes. More significantly, the rough ER makes both the integral proteins and the phospholipid molecules of the cell's membranes. In other words, all cell membranes start out as rough ER membrane. The rough ER can therefore be considered the cell's "membrane factory."

Smooth Endoplasmic Reticulum The **smooth endoplasmic reticulum (smooth ER)** is continuous with the rough ER (Figure 2.9a and b). It consists of tubules arranged in a branching network. Because no ribosomes are attached to its membranes, the smooth ER is not a site of protein synthesis. It performs different functions in different cell types, but most of these relate to lipid metabolism, the making or breaking down of fats. For example, smooth ER is abundant in cells that make lipid steroid hormones from cholesterol and in liver cells that detoxify lipid-soluble drugs. Most cell types, however, have little smooth ER.

Critical Reasoning

The sedative phenobarbitol is a lipid-soluble drug. What may happen to the smooth ER in liver cells of people who use this drug for a period of time?

The amount of smooth ER in liver cells increases markedly, and the liver becomes more efficient at detoxifying the drug. For this reason, people who take phenobarbitol tend to acquire a tolerance to it with prolonged usage. When this happens, ever higher doses must be taken to achieve the same benefit.

Both smooth and rough ER perform another important function, storing calcium ions. Ionic calcium is a signal for the beginning of many cellular events, including muscle contraction and glandular secretion. The calcium concentration in the cytosol is kept low when such events are not occurring, because most calcium ions are pumped into the ER and held there until the cell needs them.

Golgi Apparatus

The **Golgi** (gōl'je) **apparatus** is a stack of three to ten disc-shaped envelopes (cisternae), each bound by a

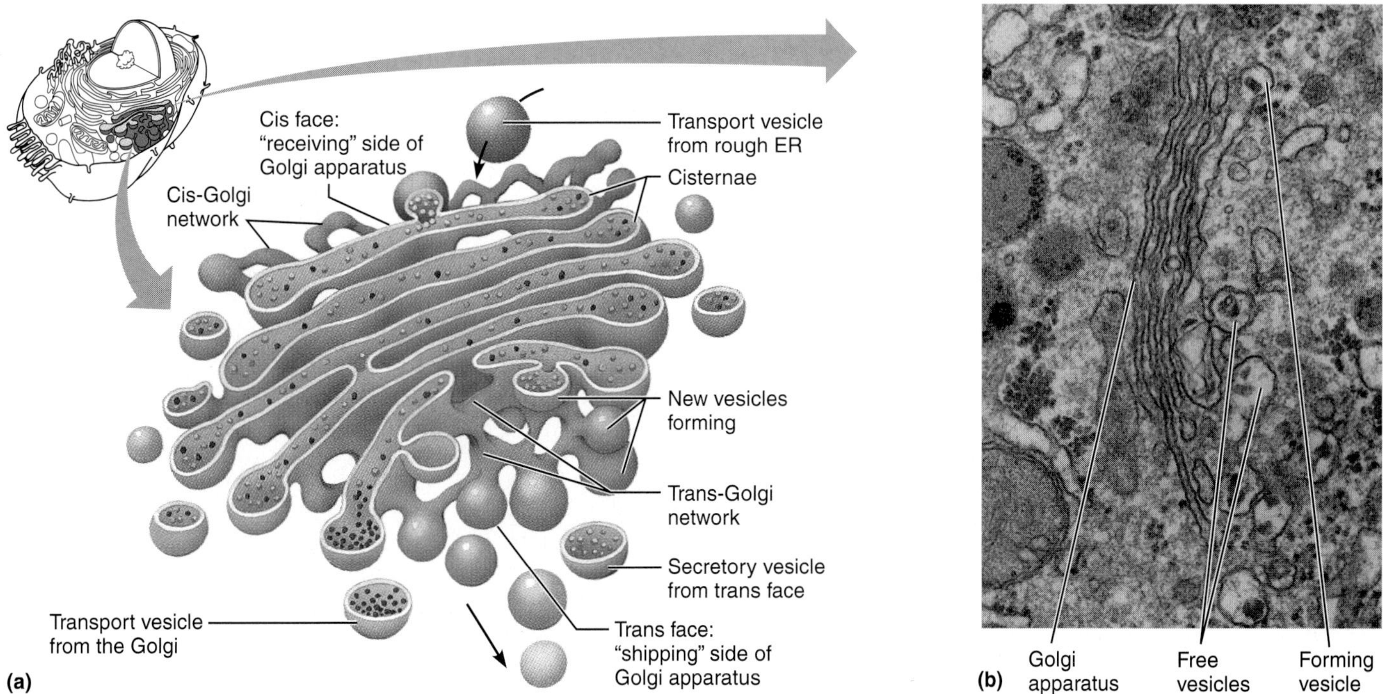

FIGURE 2.11 Golgi apparatus. **(a)** Three-dimensional view of a Golgi apparatus that has been cut in half. The cis face (above) receives small transport vesicles. Larger vesicles leave the trans face. **(b)** Electron micrograph of a Golgi apparatus tilted vertically compared to (a) (approx. 41,000 ×).

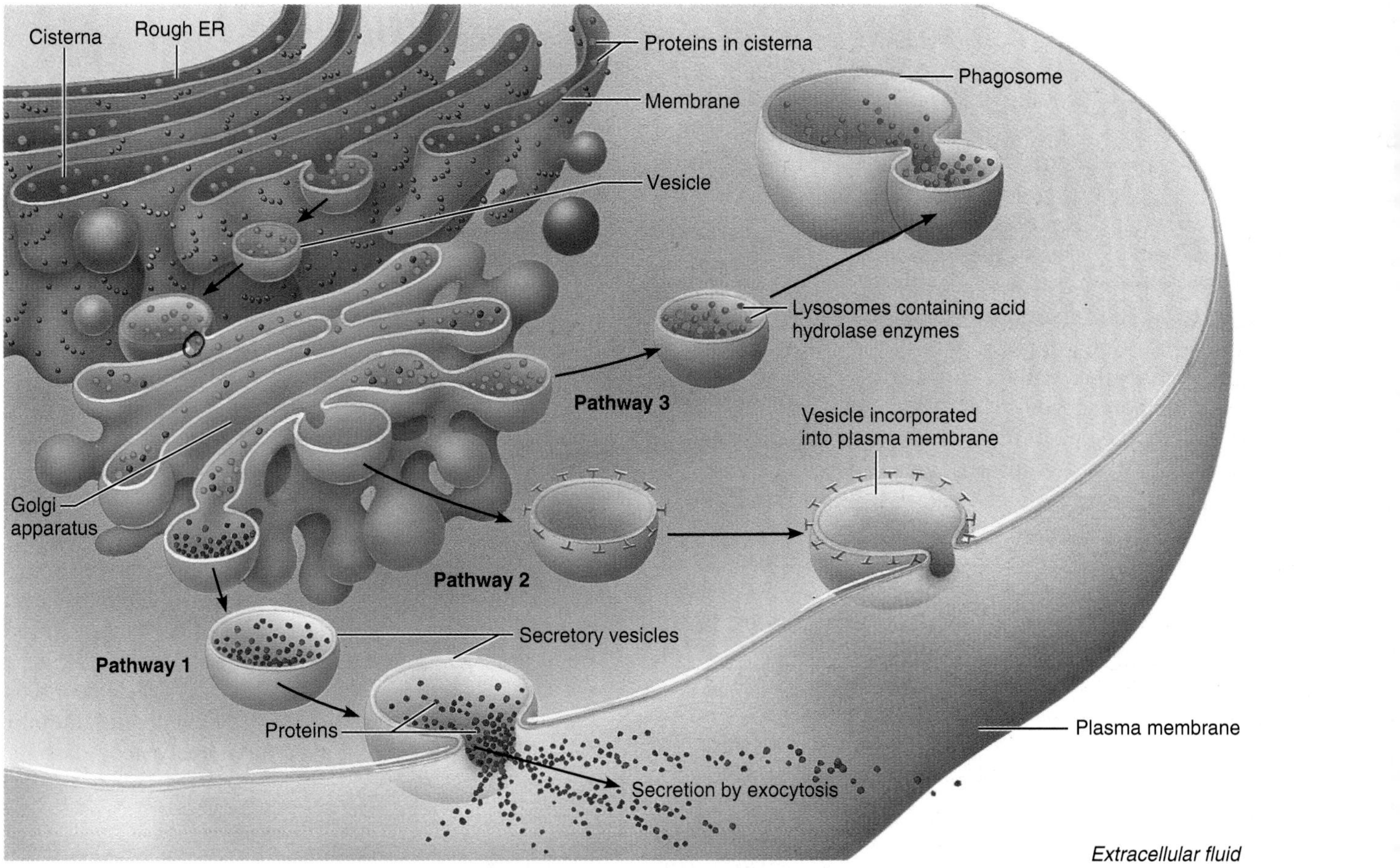

Cisterna
Rough ER
Proteins in cisterna
Membrane
Phagosome
Vesicle
Lysosomes containing acid
hydrolase enzymes
Pathway 3
Vesicle incorporated
into plasma membrane
Golgi
apparatus
Pathway 2
Secretory vesicles
Pathway 1
Proteins
Plasma membrane
Secretion by exocytosis
Extracellular fluid

FIGURE 2.12 Role of the Golgi apparatus in packaging the products of the rough ER for use in the cell and for secretion. Starting at upper left, protein-containing transport vesicles pinch off the rough ER and migrate to fuse with the cis face of the Golgi apparatus. As it passes through the Golgi apparatus, the product is sorted and its chemistry is slightly modified. The product then leaves the trans face of the Golgi apparatus as the contents and membranes of vesicles, which travel by different pathways to different destinations. These pathways 1–3 are explained in the text.

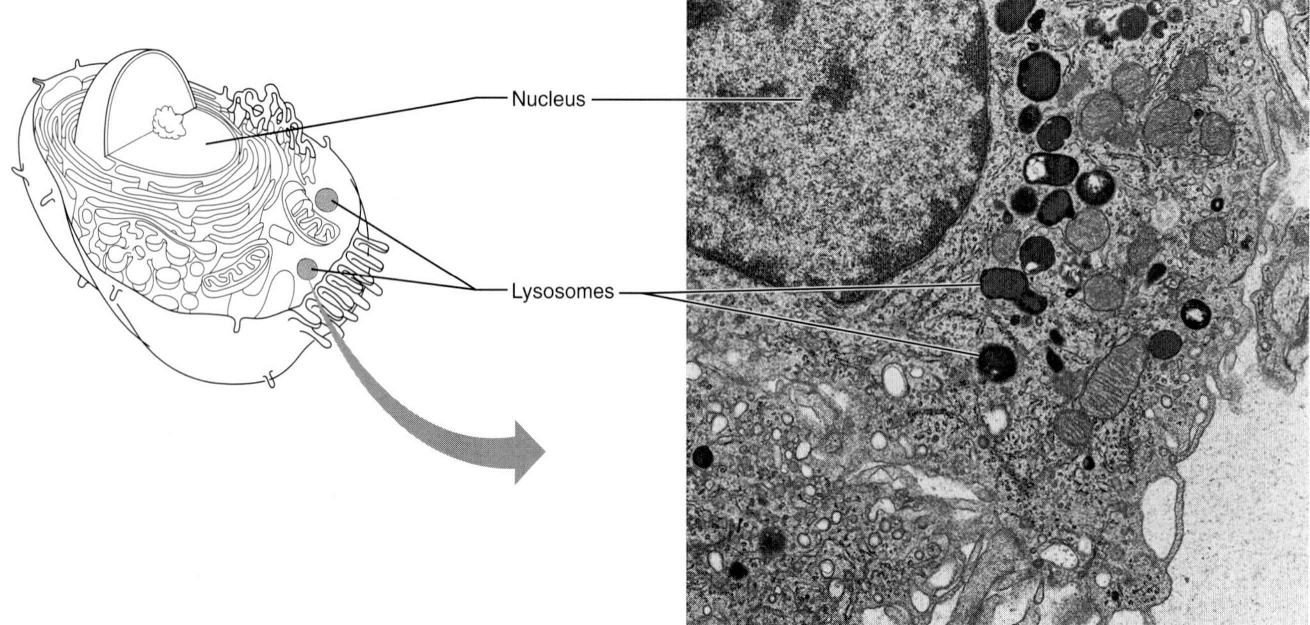

FIGURE 2.13 Lysosomes. Electron micrograph of a cell containing lysosomes. Lysosomes often stain darkly, as shown (10,000 ×).

membrane (Figure 2.11). It resembles a stack of hollow saucers, one cupped inside the next. The products of the rough ER move through the Golgi stack from the convex *(cis)* to the concave *(trans)* side. More specifically, a special compartment called the **cis-Golgi network** receives spherical, membranous **transport vesicles** from the rough ER; and other vesicles bud off a **trans-Golgi network** to leave the apparatus.

The Golgi apparatus sorts, processes, and packages the proteins and membranes made by the rough ER. For example, the Golgi apparatus distinguishes which newly made membranes will become part of the lysosomes (discussed shortly), and which ones are destined for the plasma membrane. It then sends these membranes to their correct destinations via the vesicles that leave the trans face. Thus, the Golgi apparatus is like a post office: It receives "letters" (products of the rough ER) that are "addressed" to different locations within the cell and ensures that they get to their destinations.

Figure 2.12 shows the specific pathways by which products leave the Golgi apparatus. In Pathway 1, which occurs in gland cells, the protein product is contained in **secretory vesicles (secretory granules);** these vesicles ultimately release their contents to the cell's exterior by exocytosis. In Pathway 2, prevalent in all cells, the membrane of the vesicle fuses to and contributes to the plasma membrane, whose components are constantly being renewed and recycled. In Pathway 3, also prevalent in all cells, the vesicle leaving the Golgi apparatus is a lysosome, a sac filled with digestive enzymes that remains inside the cell.

Lysosomes

Lysosomes, literally "disintegrator bodies," are spherical, membrane-walled sacs containing many kinds of digestive enzymes (Figure 2.13). These enzymes, called acid hydrolases, can digest almost all types of large biological molecules. Lysosomes can be considered the cell's "demolition crew," because they digest unwanted substances. For example, they fuse with phagosomes and endosomes, emptying their enzymes into these vesicles and breaking down their contents (see Figure 2.5).

Critical Reasoning

Can you think of a cell type that may have an exceptional number of lysosomes?

Any type of phagocytic cell, such as some white blood cells, use their many lysosomes to degrade ingested bacteria and viruses.

When a cell's own internal membranes, proteins, or organelles are damaged or wear out, these become encircled by a new membrane from the rough ER, forming a vesicle. Then, nearby lysosomes fuse with this vesicle to digest its contents. Within such vesicles, digestion can proceed safely, because the enclosing membrane keeps the destructive enzymes away from other cell components.

CLINICAL APPLICATIONS

TAY-SACHS DISEASE In an inherited condition called Tay-Sachs disease, an infant's lysosomes lack a specific enzyme that breaks down certain glycolipids in the normal recycling of worn-out cellular membranes. Such glycolipids are especially abundant in the membranes of nerve cells. The undigested glycolipids build

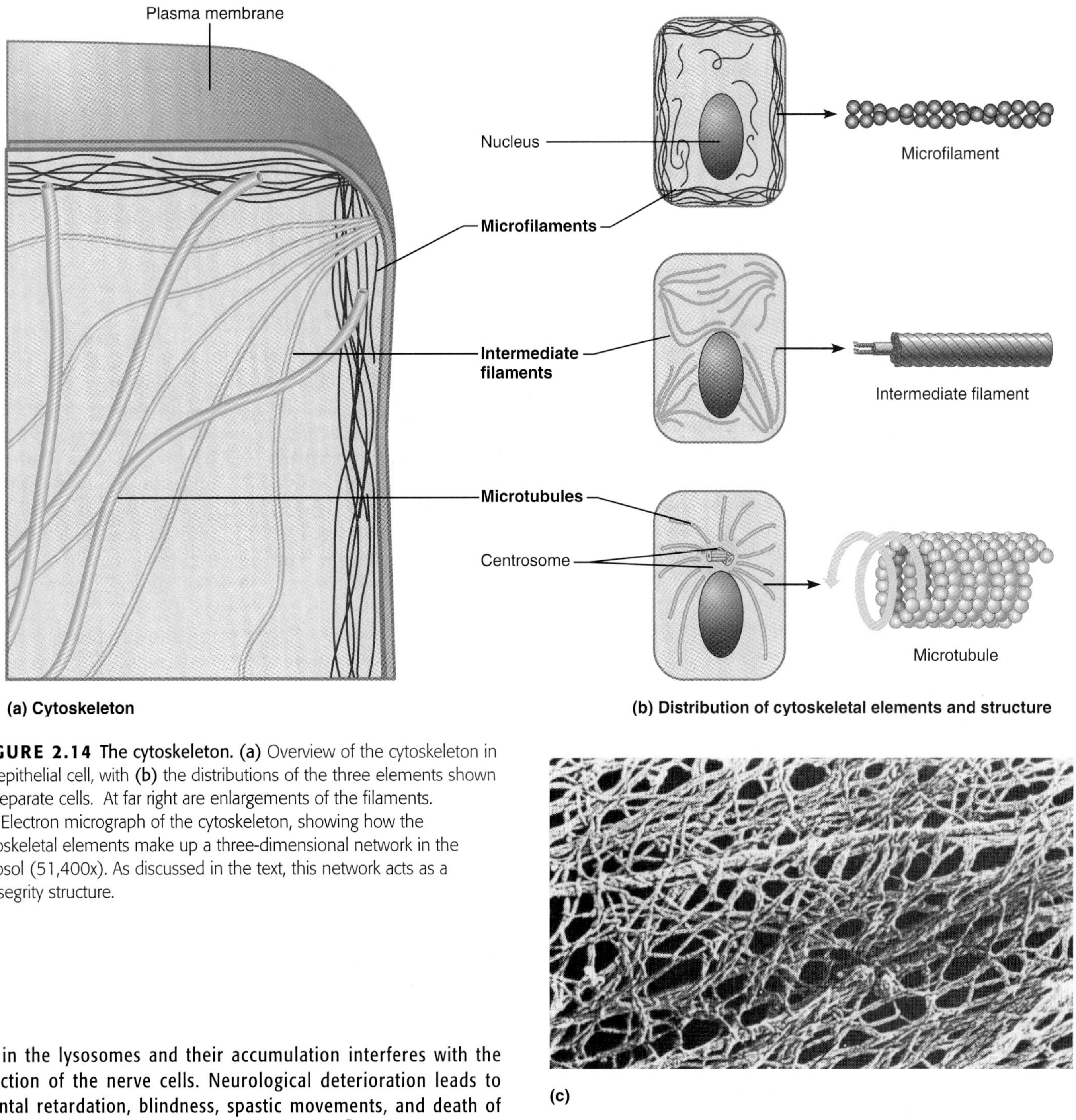

Plasma membrane
Nucleus
Microfilaments
Intermediate filaments
Microtubules
Centrosome
Microfilament
Intermediate filament
Microtubule

(a) Cytoskeleton

(b) Distribution of cytoskeletal elements and structure

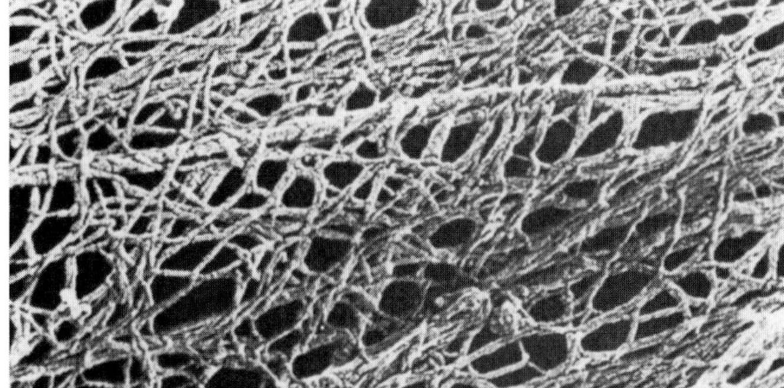

(c)

FIGURE 2.14 The cytoskeleton. **(a)** Overview of the cytoskeleton in an epithelial cell, with **(b)** the distributions of the three elements shown in separate cells. At far right are enlargements of the filaments. **(c)** Electron micrograph of the cytoskeleton, showing how the cytoskeletal elements make up a three-dimensional network in the cytosol (51,400x). As discussed in the text, this network acts as a tensegrity structure.

up in the lysosomes and their accumulation interferes with the function of the nerve cells. Neurological deterioration leads to mental retardation, blindness, spastic movements, and death of the child within a year and a half of birth. ✚

Peroxisomes

Peroxisomes (pĕ-roks′ĭ-sōmz; "peroxide bodies") are membrane-walled sacs that resemble small lysosomes in typical microscopic sections (see Figure 2.1). They contain a variety of enzymes, the most important of which are oxidases and catalases. Oxidases use oxygen to neutralize aggressively reactive molecules called **free radicals,** by converting these to hydrogen peroxide. (Free radicals are normal byproducts of cellular metabolism, but if allowed to accumulate can destroy the cell's proteins, membranes, and DNA.) The hydrogen peroxide is also reactive and dangerous, but it is converted by catalase into harmless water and oxygen. This catalase-driven reaction breaks down poisons that have entered the cell, such as alcohol, formaldehyde, and phenol. Peroxisomes are numerous in liver and kidney cells, which play a major role in removing toxic substances from the body. Peroxisomes also perform many other metabolic reactions, such as breaking down long chains of fatty acids in lipid metabolism.

Cytoskeleton

The **cytoskeleton,** literally "cell skeleton," is an elaborate network of rods running throughout the cytosol (Figure 2.14). This network acts as a cell's "bones," "muscles," and "ligaments" by supporting cellular structures and generating various cell movements. The three types of rods in the cytoskeleton are *microtubules, microfilaments,* and *intermediate filaments,* none of which is covered by a membrane.

Microtubules, the elements with the largest diameter, are hollow tubes made of spherical protein subunits called tubulins. They are stiff but bendable. All microtubules radiate from a small region of cytoplasm near the nucleus called the *centrosome* or *cell center* (discussed shortly). This radiating pattern of stiff microtubules determines the overall shape of the cell, as well as the distribution of cellular organelles. Mitochondria, lysosomes, and secretory granules attach to the microtubules like ornaments hanging from the limbs of a Christmas tree. Organelles move within the cytoplasm, pulled along the microtubules by small *motor proteins* that act like train engines on the microtubular railroad tracks. The specific motor proteins associated with microtubules are *kinesins* (ki-ne'sinz) and *dyneins* (di'ne-inz). Microtubules are remarkably dynamic organelles, constantly growing out from the cell center, disassembling, then reassembling again.

Microfilaments, the thinnest elements of the cytoskeleton, are strands of the protein **actin** (ak'tin). Also called **actin filaments,** they concentrate most heavily in a layer just deep to the plasma membrane. Actin filaments interact with another protein called **unconventional myosin** (mi'o-sin) to generate contractile forces within the cell. Actin and myosin interact to squeeze one cell into two during cell division (cytokinesis: page 48), and to cause the membrane changes that accompany endocytosis and exocytosis. Actinomyosin also enables some cells to send out then retract extensions called *pseudopods* (soo'do-pods; "false feet"), in a crawling action called amoeboid motion (ah-me'boid; "changing shape"). Additionally, myosin acts as a motor protein to move some organelles along the actin filaments. Except in muscle cells where they are stable and permanent, the actin microfilaments are labile; that is, they are constantly breaking down and re-forming from smaller subunits.

Intermediate filaments are tough, insoluble protein fibers, with a diameter between those of microfilaments and microtubules. Intermediate filaments are the most stable and permanent of the cytoskeletal elements. Their most important property is high tensile strength; that is, they act like strong guy wires to resist *pulling* forces that are placed on the cell.

Advances in Understanding In the cytosol, the microtubules, microfilaments, and intermediate filaments are linked together into a three-dimensional network (Figure 2.14c) called a *tensegrity structure.* Tensegrity structures are pre-stressed in both tension and compression, which gives them their strength and shape. Besides stabilizing cells against external and internal forces, the cytoskeleton acts as a force-carrying network, allowing mechanical forces outside the cell to be transmitted from the plasma membrane through the cytoplasm to the nucleus, which is the cell's control center. In this way, the cytoskeleton forms a mechanical communication system between the extracellular environment and the inner workings of the cell. ✷

Centrosome and Centrioles

The **centrosome** (sen'tro-sōm) is a spherical structure in the cytoplasm near the nucleus (Figure 2.15). It contains no membranes. Instead, it consists of an outer cloud of protein called the **centrosome matrix** and an inner pair of **centrioles** (sen'tre-ōlz). The matrix protein seeds the growth and elongation of microtubules, which explains why the long microtubules of the cytoskeleton radiate from the centrosome in nondividing cells (Figure 2.14b), and why a mitotic spindle of microtubules radiates from it in dividing cells (see Figure 2.21, p. 46).

In the core of the centrosome, the two barrel-shaped centrioles lie perpendicular to one another. The wall of each centriole consists of 27 short microtubules, arranged in nine groups of three. Unlike most other microtubules, those in centrioles are stable and do not disassemble. Functionally, centrioles act in forming cilia (p. 83 in Chapter 4).

Vaults

The high magnification of electron microscopy was first used to study cell structure in the 1950s, and by the late 1960s most scientists assumed that all organelles had been discovered. They were proved wrong when a new type of organelle, **vaults** (Figure 2.16) was detected in the late 1980s. The reason vaults had been missed is that these particles do not take up the traditional stains, so they can only be seen as whitish "ghosts" against an overstained cytosol background. Another reason is their relatively small size, little larger than a ribosome. A vault consists of protein and is shaped like two cups joined brim-to-brim to form a barrel, except each cup is a cupped flower with eight petals. Their function is unknown, but they sometimes concentrate in the cytosol just outside the nucleus. Perhaps they act as containers to shuttle large molecules from the nucleus to the cytoplasm.

Cytoplasmic Inclusions

Inclusions are temporary structures in the cytoplasm that are not present in all cell types. Inclusions may consist of pigments, crystals of protein, and food stores. The

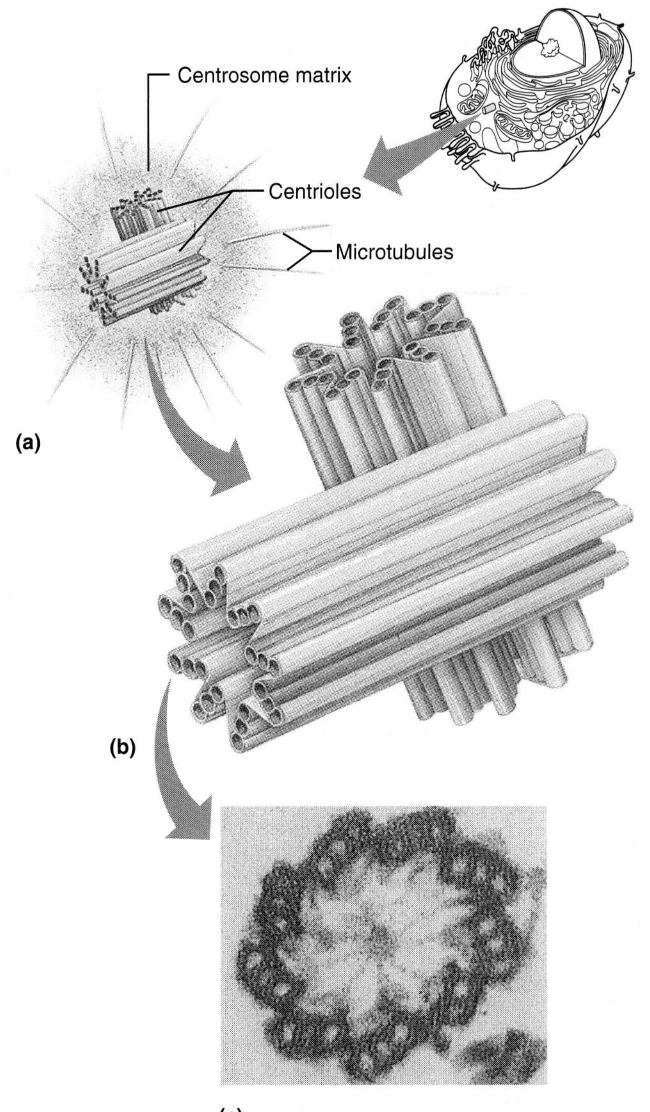

FIGURE 2.15 **Centrosome and centrioles.** **(a)** A centrosome. **(b)** A pair of centrioles oriented at right angles to each other. **(c)** Electron micrograph of a centriole in cross section. Its wall consists of nine groups of three microtubules (165,000 ×).

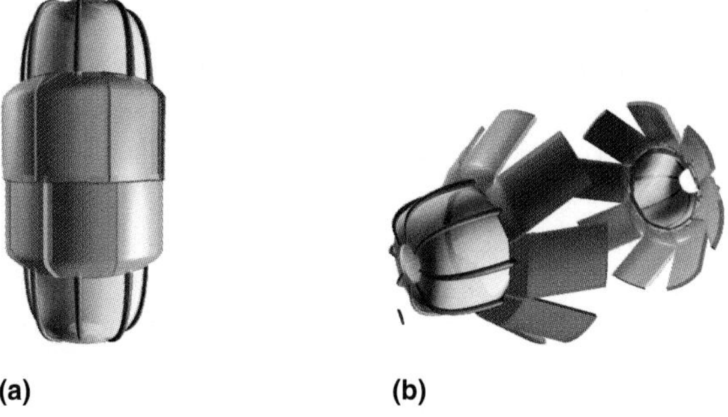

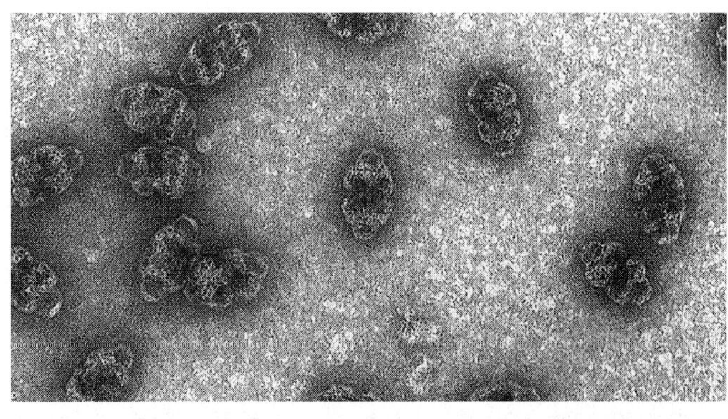

FIGURE 2.16 **Vaults.** **(a)** A complete, barrel-shaped vault. **(b)** Vault opened into its two petal-shaped parts. **(c)** Electron micrograph stained specially to reveal the vaults (250,000 ×).

food stores, by far the most important kind, are lipid droplets and glycosomes. **Lipid droplets** are spherical drops of stored fat. They can have the same size and appearance as lysosomes but can be distinguished by their lack of a surrounding membrane. Only a few cell types contain lipid droplets: Small lipid droplets are found in liver cells, large ones in fat cells. **Glycosomes** ("sugar-containing bodies") store sugar in the form of glycogen (gli′ko-jen), which is a long branching chain of glucose molecules, the cell's main energy source. They also contain enzymes that make and degrade the glycogen into its glucose subunits. Structurally, glycosomes are dense, spherical granules, which resemble ribosomes but their diameter is twice as large.

Advances in Understanding For a long time, the glycosomes dectected by electron microscopy were called *glycogen granules*. Eventually it was realized that their glycogen does not stain in typical electron micrographs and only their enzyme component is visible. Because no glycogen is actually visible, a new name was selected for these inclusions that did not explicitly contain the word "glycogen." ✱

THE NUCLEUS

The **nucleus,** literally a "central core" or "kernel," is the control center of the cell. Its genetic material, **deoxyribonucleic acid (DNA),** directs the cell's activities by

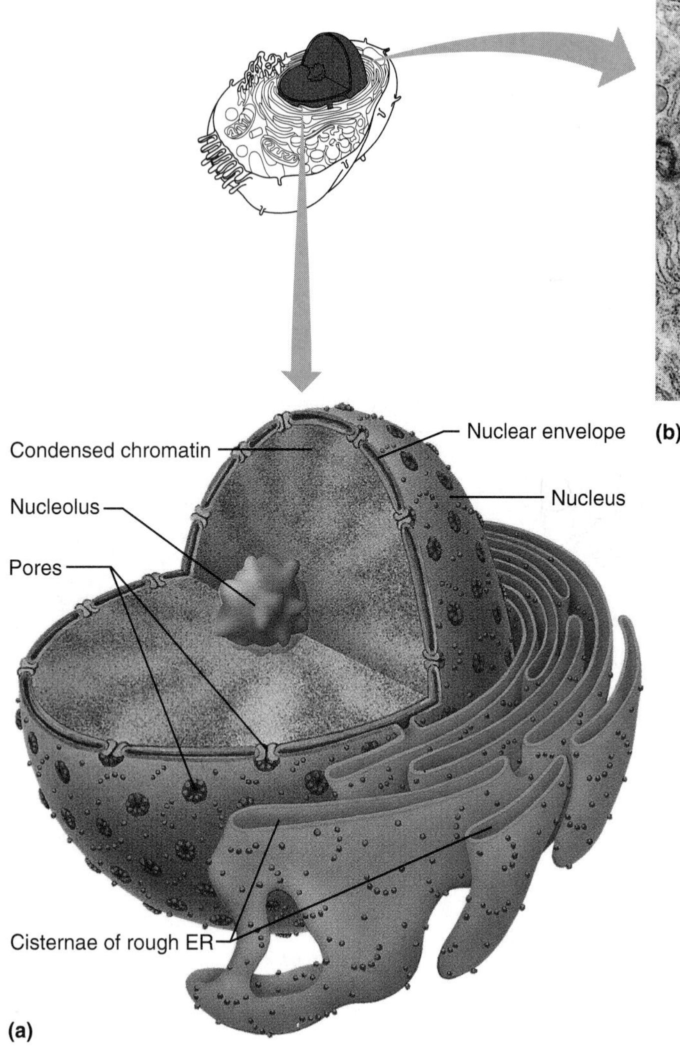

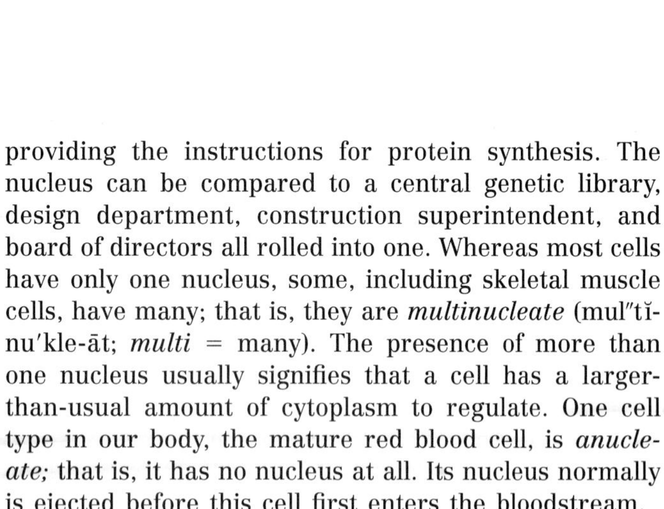

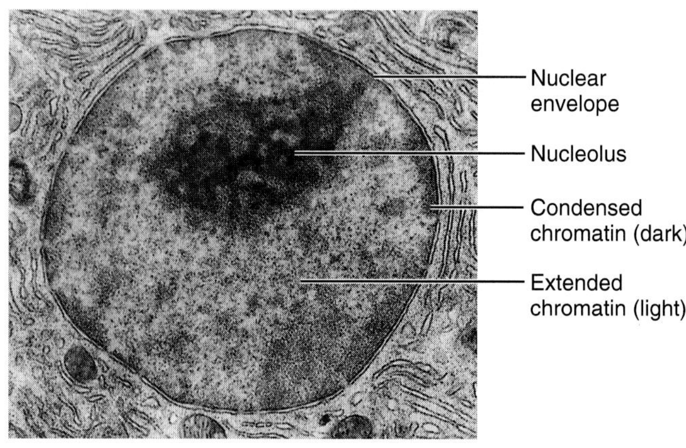

FIGURE 2.17 The nucleus. **(a)** Three-dimensional view of the nucleus, showing the continuity of the nuclear envelope (purple) with the rough ER (blue). **(b)** Electron micrograph of the nucleus, showing the nuclear envelope, chromatin, and the nucleolus (8750 ×).

providing the instructions for protein synthesis. The nucleus can be compared to a central genetic library, design department, construction superintendent, and board of directors all rolled into one. Whereas most cells have only one nucleus, some, including skeletal muscle cells, have many; that is, they are *multinucleate* (mul″tĭ-nu′kle-āt; *multi* = many). The presence of more than one nucleus usually signifies that a cell has a larger-than-usual amount of cytoplasm to regulate. One cell type in our body, the mature red blood cell, is *anucleate;* that is, it has no nucleus at all. Its nucleus normally is ejected before this cell first enters the bloodstream.

The nucleus, which averages 5 μm in diameter, is larger than any of the cytoplasmic organelles (see Figure 2.1). Although it is usually spherical or oval, it generally conforms to the overall shape of the cell. If a cell is elongated, for example, the nucleus may also be elongated. The main parts of the nucleus are the *nuclear envelope, chromatin and chromosomes,* and *nucleoli.*

Nuclear Envelope

The nucleus is surrounded by a **nuclear envelope** that consists of two parallel membranes separated by a fluid-filled space (Figure 2.17a). The envelope is continuous with the rough ER and has ribosomes on its outer face. It forms anew from rough ER after every cell division, so it is evidently a specialized part of the rough ER.

At various points, the two layers of the nuclear envelope fuse, and **nuclear pores** penetrate the fused regions. Each pore is formed by a bracelet-shaped complex of over one hundred proteins, and there are several thousand pores per nucleus. Like other cellular membranes, the membranes of the nuclear envelope are only selectively permeable, but the pores allow large molecules to pass in and out of the nucleus as necessary. For example, protein molecules imported from the cytoplasm and RNA molecules exported from the nucleus routinely travel through the pores.

The nuclear envelope encloses a jelly-like fluid called *nucleoplasm* (nu′kle-o-plazm″), within which the chromatin and nucleoli are suspended. Like the cytosol, the nucleoplasm contains salts, nutrients, and other essential chemicals. It may even contain a supportive nucleoskeleton comparable to the cytoskeleton.

Chromatin and Chromosomes

Most cells undergo brief stages of cell division, followed by longer stages during which they carry out normal cellular activities. When cells in the nondividing stage are examined by electron microscopy, some regions of the nucleus stain darkly (Figure 2.17b). These dark regions, called **condensed chromatin** (kro′mah-tin; "colored

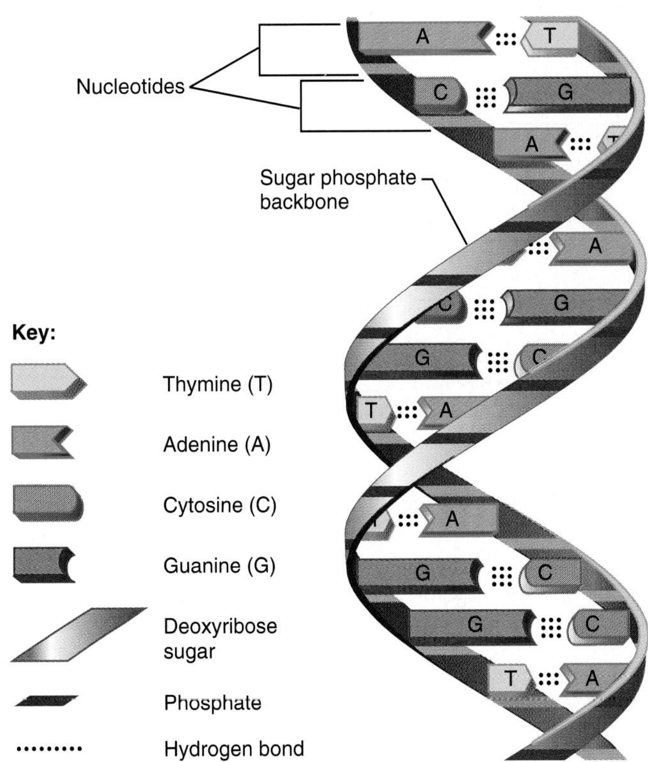

FIGURE 2.18 **Molecular structure of DNA.** DNA is a double helix constructed of chains of nucleotide molecules. Each nucleotide consists of a sugar, phosphate, and one of four bases: thymine (T), adenine (A), cytosine (C), or guanine (G).

substance"), contain tightly coiled strands of DNA. Other regions of the nucleus stain poorly. These light regions, called **extended chromatin,** contain fine uncoiled strands of DNA. While the DNA of condensed chromatin is inactive, that of extended chromatin directs the synthetic activities of the cell during a cell's nondividing stage. More specifically, extended chromatin is where DNA's genetic code is being copied onto messenger RNA molecules in a process called **transcription.** Understandably, the most active cells of the body tend to have a large amount of extended chromatin and little condensed chromatin. In general, most of the condensed chromatin is near the periphery of the nucleus, just internal to the nuclear envelope.

The DNA in chromatin is a long double helix that resembles a spiral staircase (Figure 2.18). This double helix is in turn composed of four kinds of subunits called *nucleotides,* each of which contains a distinct base. These bases—thymine (T), adenine (A), cytosine (C), and guanine (G)—bind to form the "stairs" of the "staircase" and to hold the DNA helix together.

The long DNA helix wraps around clusters of eight spherical protein molecules called **histones** (his′tōnz) (Figure 2.19). An effective packing material, the histones may also help to regulate transcription and gene expression by opening up and condensing the coils of chromatin. Each cluster of DNA and histones is called a **nucleosome,** and nucleosomes appear under the electron microscope as beads on a string (Figure 2.19a). Chromatin is thus made up of strands of DNA plus their associated nucleosomes.*

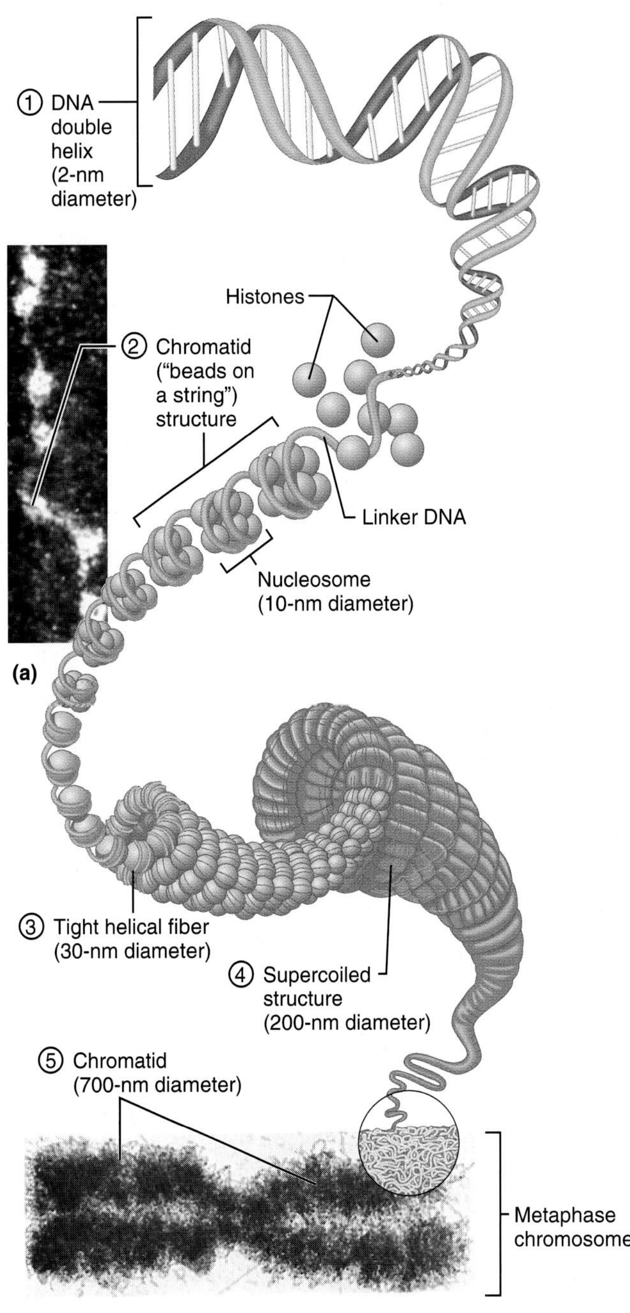

FIGURE 2.19 **Chromatin.** This figure shows the arrangement of DNA and histones in chromatin, from its most extended state to its most condensed state, in a chromosome. Part **(a)** is a high-power electron micrograph of extended chromatin (500,000 ×), whereas **(b)** is a diagram.

Text continues on page 45

*The term *chromatin,* which technically means "DNA and nucleosomes (or histones)," can also be used as an abbreviation for condensed chromatin (see Figure 2.1).

Table 2.1 Parts of the Cell: Structure and Function

Cell Part	Structure	Functions
PLASMA MEMBRANE (Figure 2.2)	Membrane made of a double layer of lipids (phospholipids, cholesterol, etc.) within which proteins are embedded; proteins may extend entirely through the lipid bilayer or protrude on only one face; externally facing proteins and some lipids have attached sugar groups	Serves as an external cell barrier; acts in transport of substances into or out of the cell; externally facing proteins act as receptors (for hormones, neurotransmitters, etc.) and in cell-to-cell recognition
CYTOPLASM	Cellular region between the nuclear and plasma membranes; consists of fluid **cytosol** containing dissolved solutes, **inclusions** (stored nutrients, pigment granules), and **organelles,** the metabolic machinery of the cytoplasm	
Cytoplasmic organelles		
• Mitochondria (Figure 2.7)	Rod-like, double-membrane structures; inner membrane folded into projections called cristae	Site of ATP synthesis; powerhouse of the cell
• Ribosomes (Figures 2.8, 2.9)	Dense particles consisting of two subunits, each composed of ribosomal RNA and protein; free or attached to rough ER	The sites of protein synthesis
• Rough endoplasmic reticulum (Figures 2.9, 2.10)	Membrane system enclosing a cavity, the cisterna, and coiling through the cytoplasm; externally studded with ribosomes	Makes proteins that are secreted from the cell; makes the cell's membranes
• Smooth endoplasmic reticulum (Figure 2.9)	Membranous system of sacs and tubules; free of ribosomes	Site of lipid and steroid synthesis, lipid metabolism, and drug detoxification
• Golgi apparatus (Figures 2.11, 2.12)	A stack of smooth membrane sacs close to the nucleus	Packages, modifies, and segregates proteins for secretion from the cell, inclusion in lysosomes, and incorporation into the plasma membrane
• Lysosomes (Figure 2.13)	Membranous sacs containing acid hydrolases	Sites of intracellular digestion
• Peroxisomes (Figure 2.1)	Membranous sacs of oxidase enzymes	The enzymes detoxify a number of toxic substances; the most important enzyme, catalase, breaks down hydrogen peroxide
• Microtubules (Figure 2.14)	Cylindrical structures made of tubulin proteins	Support the cell and give it shape; involved in intracellular and cellular movements; form centrioles
• Microfilaments (Figure 2.14)	Fine filaments of the contractile protein actin	Involved in muscle contraction and other types of intracellular movement; help form the cell's cytoskeleton
• Intermediate filaments (Figure 2.14)	Protein fibers; composition varies	The stable cytoskeletal elements; resist tension forces acting on the cell
• Centrioles (Figure 2.15)	Paired cylindrical bodies, each composed of nine triplets of microtubules	Organize a microtubule network during mitosis to form the spindle and asters; form the bases of cilia and flagella
• Vaults (Figure 2.16)	Barrel-shaped protein structures; in cytosol adjacent to nuclear pores	Unknown; may shuttle large molecules from nucleus to cytoplasm
NUCLEUS (Figure 2.17)	Surrounded by the nuclear envelope; contains fluid nucleoplasm, nucleoli, and chromatin	Control center of the cell; responsible for transmitting genetic information and providing the instructions for protein synthesis
• Nuclear envelope (Figure 2.17)	Double-membrane structure; pierced by the pores; continuous with the cytoplasmic ER	Separates the nucleoplasm from the cytoplasm and regulates passage of substances to and from the nucleus
• Nucleoli (Figure 2.17)	Dense spherical (non-membrane-bounded) bodies	Site of ribosome subunit manufacture
• Chromatin (Figures 2.17, 2.19)	Granular, thread-like material composed of DNA and histone proteins	DNA constitutes the genes

The highest level of organization of the chromatin is into **chromosomes** (kro′mo-sōmz; "colored bodies"). Each chromosome contains a single, very long molecule of DNA, and there are 46 chromosomes in a typical human cell. When a cell is dividing, its chromosomes are maximally coiled, so they appear as thick rods (Figure 2.19b). Chromosomes are moved extensively during cell division (p. 46), and their compact nature helps to keep the delicate chromatin strands from tangling and breaking as the chromosomes move. When cell division stops, many parts of the chromosome uncoil to form the extended chromatin, thereby allowing transcription to occur.

Advances in Understanding Due to the diffuse nature and great length of the extended chromatin, it had been very difficult for researchers to pinpoint the locations of individual chromosomes in the nucleus of nondividing cells. It is now possible, however, to label whole chromosomes with fluorescent dyes, revealing that each chromosome occupies a discrete territory in the nucleus. The different territories are divided by channels called interchromosomal domains. An even higher degree of compartmentalization is likely to be revealed in the future. ✱

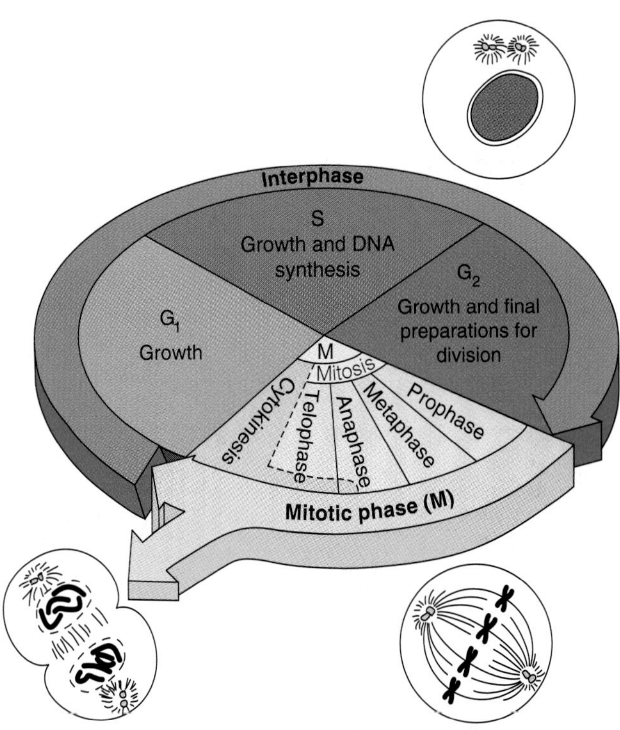

FIGURE 2.20 The cell cycle. The two basic phases in the life and reproduction of each cell are interphase and the mitotic (M) phase. The length of the cell cycle varies in different cell types, but the G_1 stage of interphase tends to be the longest and the most variable in duration. See the text for details.

Nucleoli

As shown in Figure 2.17, the cell nucleus contains one or several dark-staining bodies called **nucleoli** (nu-kle′o-li; singular, **nucleolus:** "little nucleus"). A nucleolus contains parts of several different chromosomes and serves as the cell's "ribosome-producing machine." Specifically, it has hundreds of copies of the genes that code for ribosomal RNA and serves as the site where the large and small subunits of ribosomes are assembled. These subunits leave the nucleus through the nuclear pores and join within the cytoplasm to form complete ribosomes.

THE CELL LIFE CYCLE

The cell life cycle is the series of changes a cell undergoes from the time it forms until it reproduces itself. This cycle can be divided into two major periods: *interphase,* in which the cell grows and carries on its usual activities; and *cell division,* or the *mitotic phase,* during which it divides into two cells (Figures 2.20 and 2.21).

Interphase

In addition to carrying on its life-sustaining activities, an interphase cell prepares for the next cell division. Interphase is divided into G_1, S, and G_2 subphases. During **G_1 (growth 1),** the first part of interphase, cells are metabolically active, make proteins rapidly, and grow vigorously. This is the most variable phase in terms of duration. In cells with fast division rates, G_1 lasts several hours; in cells that divide slowly, it can last days or even years.

For most of G_1, no activities directly related to cell division occur. However, near the end of G_1, the centrioles start to replicate in preparation for cell division. During the next stage, the **S (synthetic)** phase, DNA replicates itself, ensuring that the two daughter cells will receive identical copies of the genetic material.

The final part of interphase, called **G_2 (growth 2),** is brief. In this period, the enzymes needed for the cell-division process are synthesized. Centrioles finish copying themselves at the end of G_2. Throughout the S and G_2 phases, the cell continues to grow and to carry out its normal metabolic activities.

Cell Division

Cell division is essential for body growth and tissue repair. Short-lived cells that continuously wear away, such as cells of the skin and the intestinal lining, reproduce themselves almost continuously. Others, such as liver cells, reproduce slowly (replacing those cells that gradually wear out), but can divide quickly if the organ is damaged. Cells of nervous tissue and skeletal muscle are unable to divide once they are fully mature; repair

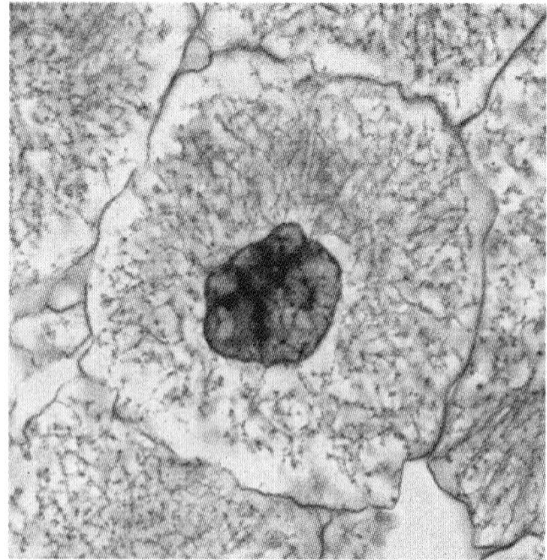

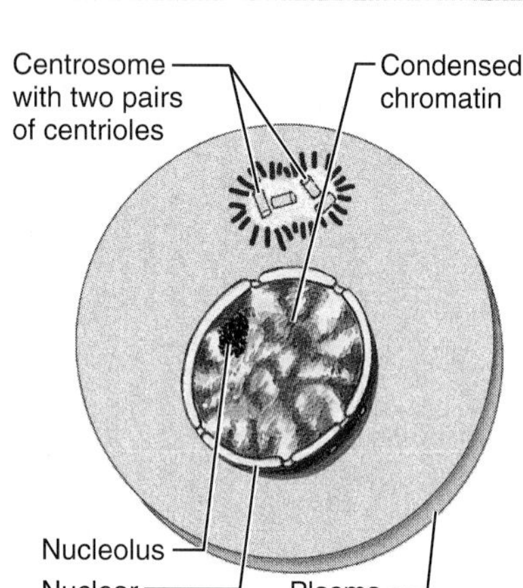

Interphase

Interphase is the period of a cell's life when it is carrying out its normal metabolic activities and growing. During interphase, the chromosomal material is seen in the form of extended and condensed chromatin, and the nuclear membrane and nucleolus are intact and visible. During various periods of this phase, the centrioles begin replicating (G_1 through G_2), DNA is replicated (S), and the final preparations for mitosis are completed (G_2). The centriole pair finishes replicating into two pairs during G_2.

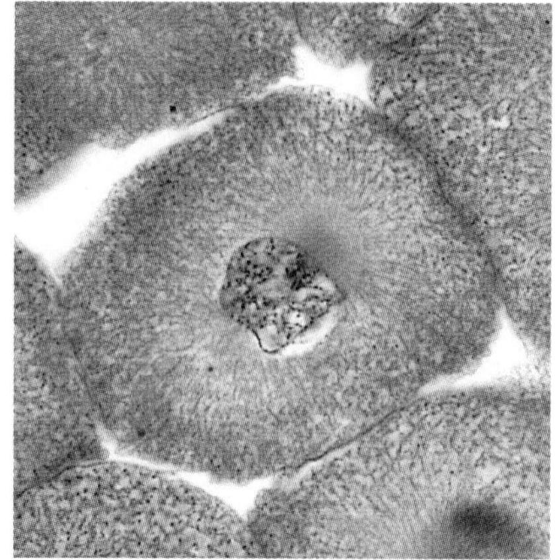

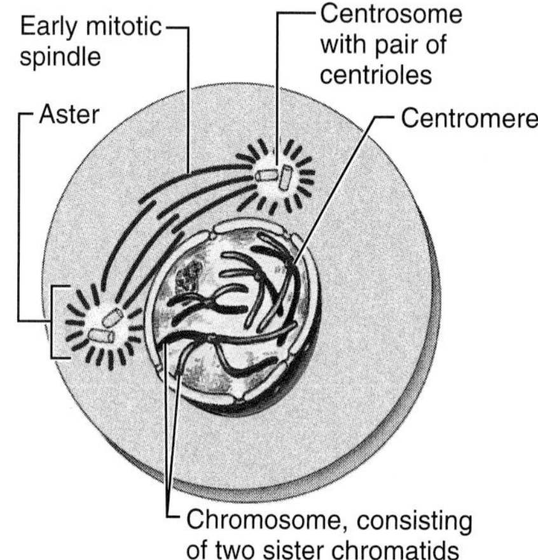

Early prophase

As mitosis begins, microtubule arrays called *asters* ("stars") are seen extending from the centrioles. Early in *prophase,* the first and longest phase of mitosis, the chromatin threads coil and condense,forming barlike *chromosomes* that are visible with a light microscope. Since DNA replication has occurred during interphase, each chromosome is actually made up of two identical chromatin threads, now called *chromatids.* The chromatids of each chromosome are held together by a small, buttonlike body called a *centromere.* Later, after the chromatids separate, each will be considered a new chromosome.

As the chromosomes appear, the nucleoli disappear, and the cytoskeletal microtubules disassemble. The centriole pairs separate from one another. The centrosomes around the centrioles act as focal points for growth of a new assembly of microtubules called the **mitotic spindle.** As these microtubules lengthen, they push the centrioles farther and farther apart, propelling them toward opposite ends (poles) of the cell.

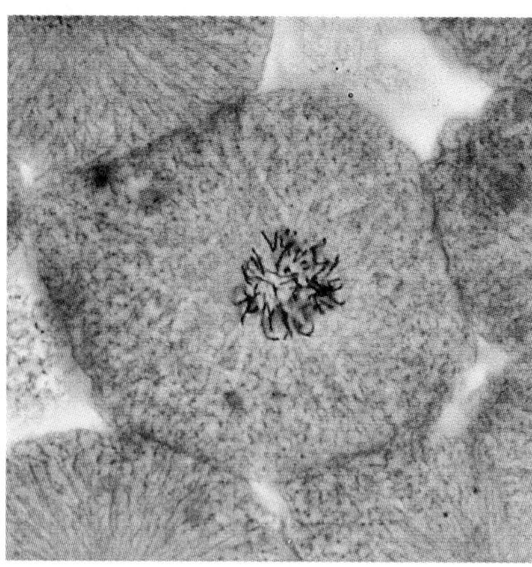

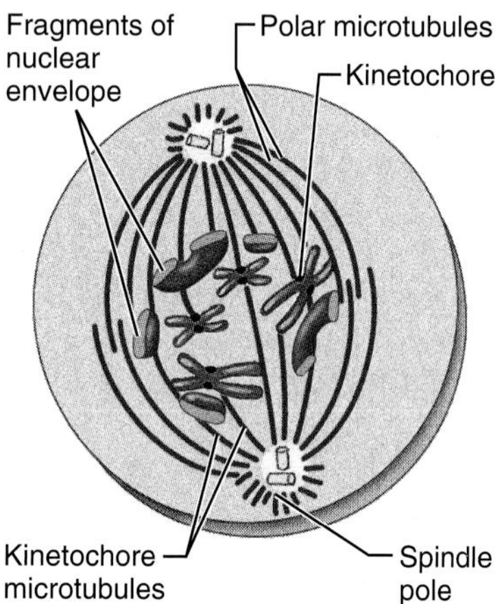

Late prophase

While the centrioles are still moving away from each other, the nuclear membrane fragments, allowing the spindle to occupy the center of the cell and to interact with the chromosomes. Meanwhile, some of the growing spindle microtubules attach to special protein–DNA complexes, called *kinetochores* (ki-ne′to-korz), on each chromosome's centromere. Such microtubules are called *kinetochore microtubules.* The remaining spindle microtubules, which do not attach to any chromosomes, are called *polar microtubules.* The tips of the polar microtubules are linked near the center; these push against each other forcing the poles apart. The kinetochore microtubules, on the other hand, pull on each chromosome from both poles, resulting in a tug-of-war that ultimately draws the chromosomes to the middle of the cell.

FIGURE 2.21 The stages of mitosis. The cells shown are from an early embryo of a whitefish. (Micrographs approx. 600 ×.)

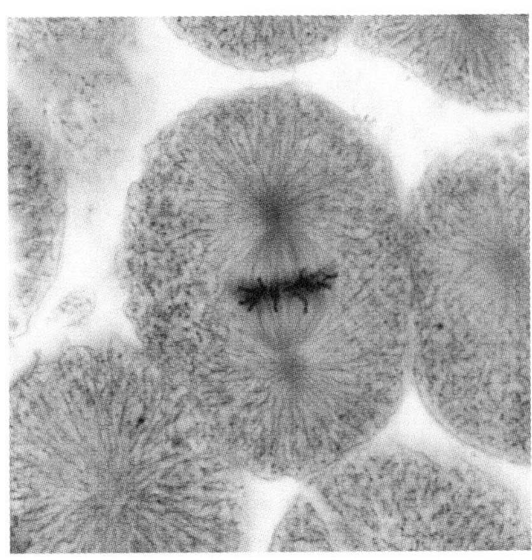

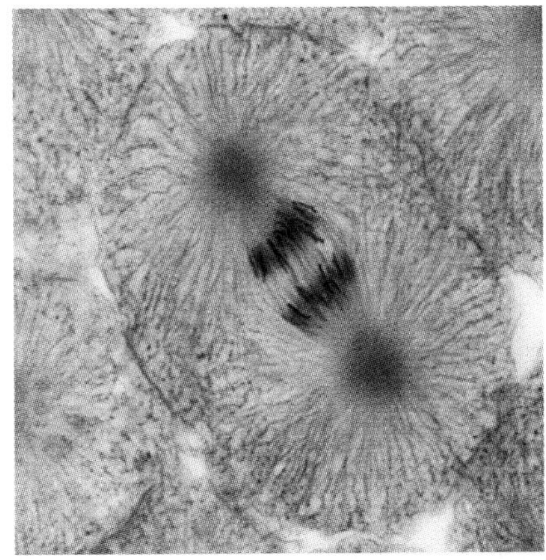

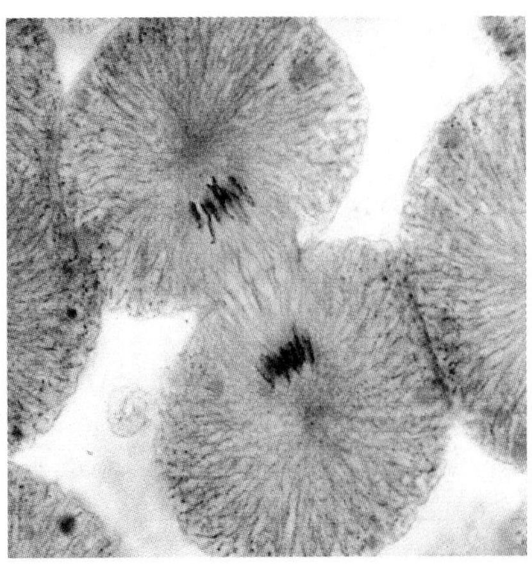

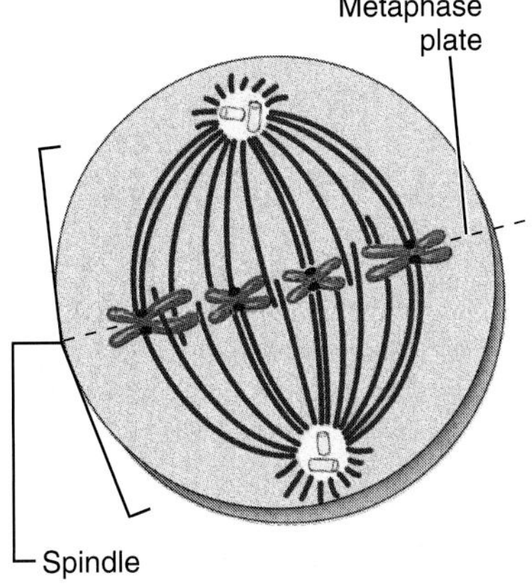

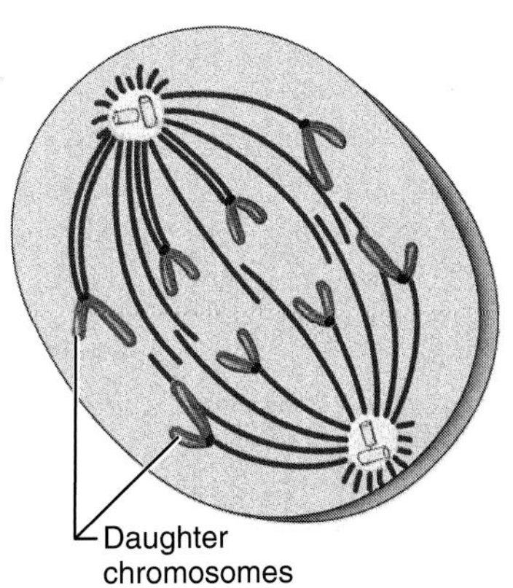

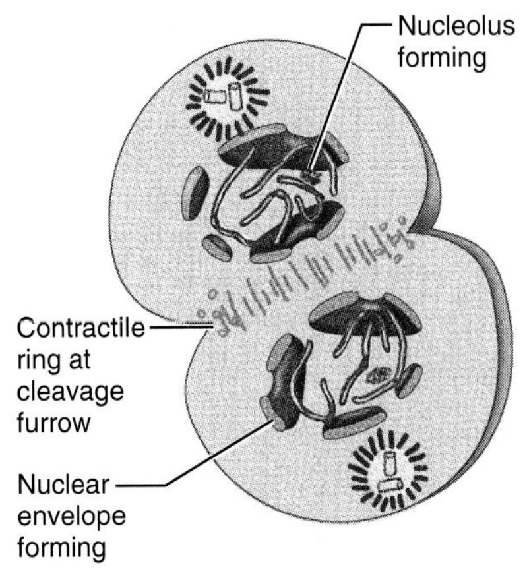

Metaphase

Metaphase is the second phase of mitosis. The chromosomes cluster at the middle of the cell, with their centromeres precisely aligned at the exact center, or *equator,* of the spindle. This arrangement of the chromosomes along a plane midway between the poles is called the *metaphase* plate.

Anaphase

Anaphase, the third phase of mitosis, begins abruptly as the centromeres of the chromosomes split, and each chromatid now becomes a chromosome in its own right. The kinetochore fibers rapidly disassemble at their kinetochore ends by removing tubulin subunits, and gradually pull each chromosome toward the pole it faces. By contrast the polar microtubules slide past each other and lengthen (a process presumed to be driven by kinesin motor molecules), and push the two poles of the cell apart, causing the cell to elongate. Anaphase is easy to recognize because the moving chromosomes look V-shaped. The centromeres, which are attached to the kinetochore microtubules, lead the way, and the chromosomal "arms" dangle behind them. Anaphase is the shortest stage of mitosis; it typically lasts only a few minutes.

This process of moving and separating the chromosomes is helped by the fact that the chromosomes are short, compact bodies. Diffuse threads of extended chromatin would tangle, trail, and break, which would damage the genetic material and result in its imprecise "parceling out" to the daughter cells.

Telophase and cytokinesis

Telophase begins as soon as chromosomal movement stops. This final phase is like prophase in reverse. The identical sets of chromosomes at the opposite poles of the cell uncoil and resume their threadlike extended–chromatin form. A new nuclear membrane, derived from the rough ER, reforms around each chromatin mass. Nucleoli reappear within the nuclei, and the spindle breaks down and disappears. Mitosis is now ended. The cell, for just a brief period, is binucleate (has two nuclei) and each new nucleus is identical to the original mother nucleus.

As a rule, as mitosis draws to a close, *cytokinesis* completes the division of the cell into two daughter cells. Cytokinesis occurs as a contractile ring of peripheral microfilaments forms at the *cleavage furrow* and squeezes the cells apart. Cytokinesis actually begins during late anaphase and continues through and beyond telophase.

is carried out by scar tissue (a fibrous type of connective tissue).

Cells divide in the **M (mitotic) phase** of their life cycle, which follows interphase (Figure 2.20). In most cell types, division involves two distinct events: *mitosis* (mi-to′sis), or division of the nucleus, and *cytokinesis* (si″to-ki-ne′sis), or division of the entire cell into two cells.

Mitosis

Mitosis is the series of events that parcel out the replicated DNA of the original cell into two new cells, culminating in the division of the nucleus. Throughout these events, the chromosomes are evident as thick rods or threads . Indeed, *mitosis* literally means "the stage of threads." Mitosis is said to have four consecutive phases: prophase, metaphase, anaphase, and telophase. However, it is actually a continuous process, with each phase merging smoothly into the next. Its duration varies according to cell type, but it typically lasts about 2 hours. Mitosis is described in detail in Figure 2.21.

Cytokinesis

The separation of one cell into two at the end of the cell cycle is called **cytokinesis,** literally "cells moving (apart)." It begins during anaphase and is completed after mitosis ends (Figure 2.21). Essentially, a ring of contractile actin and myosin filaments in the center of the original cell constricts to pinch that cell in two. The two new cells, called daughter cells, then enter the interphase part of their life cycle.

Critical Reasoning

The drug vinblastine *is used in cancer therapy to stop the runaway division of cancer cells. Vinblastine inhibits the assembly and growth of microtubules. Referring to Figure 2.21, explain how the action of this drug prevents mitosis.*

Without microtubules, the mitotic spindle cannot form, and without the mitotic spindle, mitosis and cell division are impossible.

CELLULAR DIVERSITY

So far in this chapter, we have focused on an "average" human cell. The trillions of cells in the human body are made up of 200 different cell types that vary greatly in size, shape, and function. They include sphere-shaped fat cells, disc-shaped red blood cells, branching nerve cells, and cube-shaped cells of kidney tubules. Figure 2.22 illustrates how the shapes of cells and their arrangement of organelles relate to the specialized function of these cells. The functional groups are:

1. *Cells that connect body parts or cover and line organs*

   Fibroblast. The elongated shape of this cell extends along the cable-like fibers that it secretes. It also has an abundant rough ER and a large Golgi apparatus, to make and secrete the protein components of these fibers.

   Erythrocyte (red blood cell). This cell carries the respiratory gases, oxygen and carbon dioxide. Its concave disc shape provides extra surface area for the uptake of respiratory gases and also provides streamlining so the cell flows easily through the bloodstream. So much oxygen-carrying pigment is packed in erythrocytes that all other organelles have been shed to make room.

   Epithelial cell. The hexagonal shape of this cell is exactly like a "cell" in a honeycomb of a beehive. This shape allows the maximum number of epithelial cells to pack together in a sheet called epithelium. An epithelial cell has abundant intermediate filaments that resist tearing when the epithelium is rubbed or pulled. Some epithelial cells are gland cells, with an abundant rough ER, Golgi apparatus, and secretory granules.

2. *Cells that move organs and body parts*

   Skeletal muscle and *smooth muscle cells.* These cells are elongated and filled with abundant actin and myosin filaments, so they can shorten forcefully.

3. *Cell that stores nutrients*

   Fat cell. The huge spherical shape of a fat cell is produced by a large lipid droplet in its cytoplasm.

4. *Cell that fights disease*

   Macrophage (a phagocytic cell). This cell extends long pseudopods to crawl through tissue to reach infection sites. The many lysosomes within the cell digest the infectious microorganisms it takes up.

5. *Cell that gathers information and controls body functions*

   Nerve cell (neuron). This cell has long processes for receiving messages and transmitting them to other structures in the body. The processes are covered with an extensive plasma membrane, whose components are continually recycled; a large rough ER is present to synthesize membrane components.

6. *Cells of reproduction*

   Oocyte (female). The largest cell in the body, this egg cell contains yolk and many copies of all organelles, for distribution to the daughter cells that arise when the egg divides to become an embryo.

Text continues on p. 52

FIGURE 2.22 Cellular diversity. The shape of human cells and the relative abundances of their organelles relate to their function in the body.

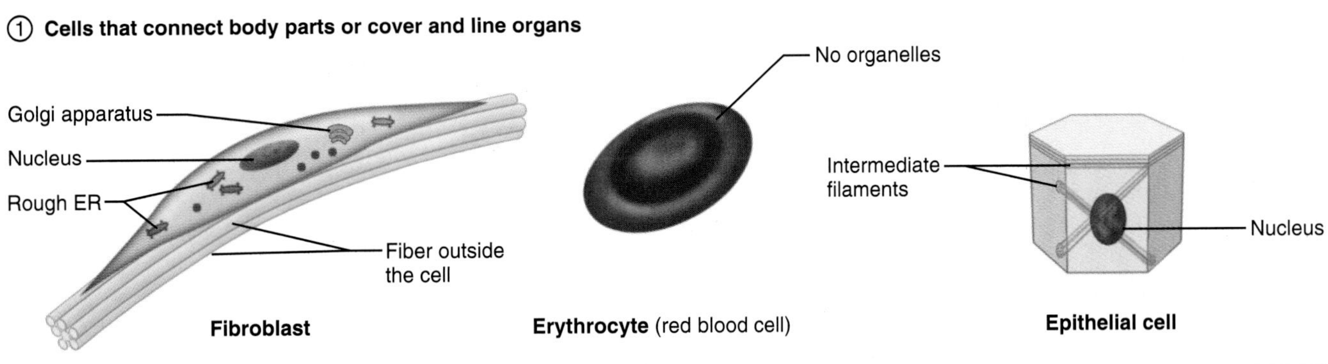
① Cells that connect body parts or cover and line organs
Golgi apparatus
Nucleus
Rough ER
Fiber outside the cell
Fibroblast
No organelles
Erythrocyte (red blood cell)
Intermediate filaments
Nucleus
Epithelial cell

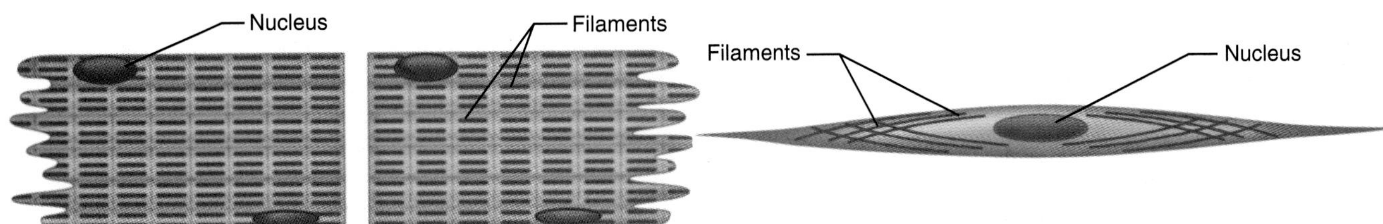
② Cells that move organs and body parts
Nucleus
Filaments
Skeletal muscle cell
Filaments
Nucleus
Smooth muscle cell

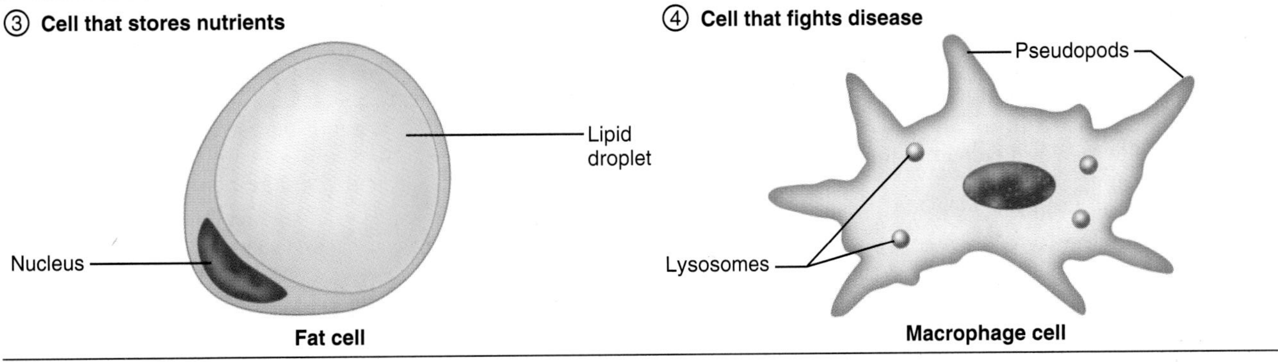
③ Cell that stores nutrients
Lipid droplet
Nucleus
Fat cell
④ Cell that fights disease
Pseudopods
Lysosomes
Macrophage cell

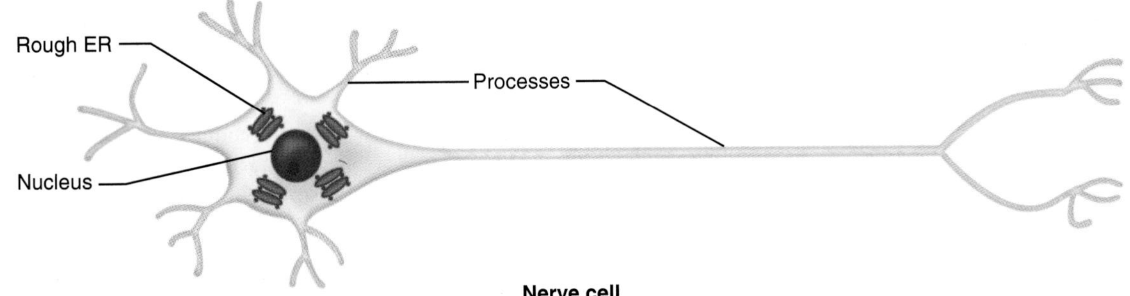
⑤ Cell that gathers information and controls body functions
Rough ER
Processes
Nucleus
Nerve cell

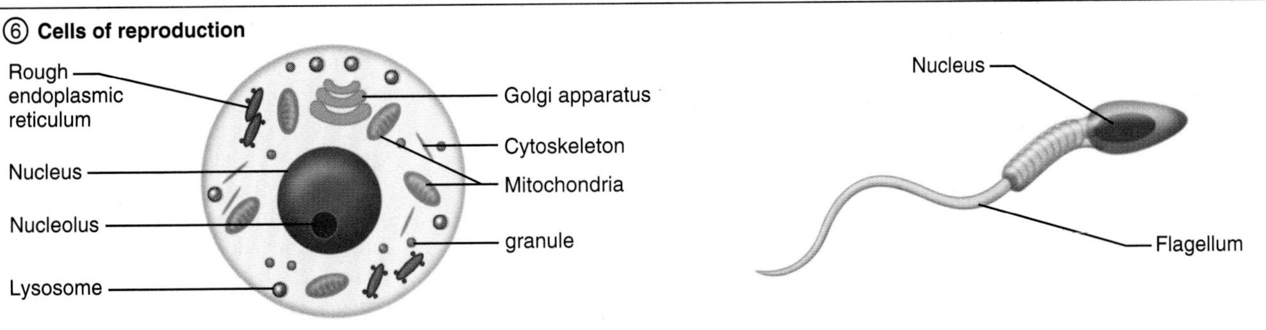
⑥ Cells of reproduction
Rough endoplasmic reticulum
Nucleus
Nucleolus
Lysosome
Golgi apparatus
Cytoskeleton
Mitochondria
granule
Ovum (egg)
Nucleus
Flagellum
Sperm

A CLOSER LOOK

Cancer—The Intimate Enemy

The clinical definition of cancer—the uncontrolled division and spread of cells within the body—gives no hint of its status as one of the most feared diseases. Almost *half* of Americans develop cancer, and a fifth die from it. Cancer can arise from almost any cell type, but the most common cancers arise in the skin, lung, large intestine, breasts, and the prostate gland below the male bladder (in that order).

Cancer is a type of **neoplasm** (ne′o-plazm; "new growth"), an abnormal mass of proliferating cells. Neoplasms, also called **tumors,** are classified as benign (literally, "kindly") or malignant ("bad"). A **benign tumor** is a localized mass of compacted cells, often enclosed in a fibrous capsule. Benign tumors tend to grow slowly, and they seldom kill if they are removed before they compress vital organs.

By contrast, cancers are **malignant neoplasms,** nonencapsulated "killers" that grow relentlessly. Their invasive nature is reflected in the word **cancer,** from the Latin meaning a creeping "crab." Worse, the cancer attracts a rich supply of vessels, and cancer cells enter these vessels and then travel to other body organs, where they form secondary growths. This travel is called **metastasis** (mĕ-tas′tah-sis; "beyond standing"). As you can see, the main properties that distinguish malignant cancer from benign neoplasms are the invasive and metastatic properties of cancer cells.

Malignancies can kill their victims by pressing on or disrupting the funcions of vital organs. Moreover, cancer cells consume many nutrients and thus cause tremendous weight loss. Such wasting weakens the patient and is the main contributor to death in many cases.

Carcinogenesis

Cancer begins with just one renegade cell that has accumulated a series of damaging changes in certain genes through exposure to carcinogenic (cancer-causing) chemicals or other factors. Some of the affected genes are called **oncogenes** (*onco* = tumor, cancer), and these dictate the uncontrolled division of cancer cells. Cancer can also result from damage to **tumor-suppressor genes.**

Increasingly, cancer is viewed as a disease of the cell cycle (p. 45). Many oncogenes are mutated, overactive versions of genes that normally direct cells through the cell cycle or promote cell proliferation. Many tumor-supressor genes act as gatekeepers along "checkpoints" that exist at each important stage of the cell cycle; that is, these genes normally delay or halt mitosis in cells that have defective chromosomes and damaged DNA. The delay allows time for DNA repair, and if the genetic damage is too great, halting mitosis lets the cell die (see *apoptosis,* p. 53). Removing these normal checks and balances leads to the wild proliferation and "immortality" of cancer cells. In fact, loss or malfunction of just two tumor-suppressor genes, called p 53 and p16, has been found in over half of all cancers.

A model of some of the mutations known to occur in colorectal cancer, one of the best-understood human cancers, appears in the accompanying illustration. As with most cancers, the development of metastasizing colon cancer is a gradual process. One of the first signs is the appearance of a polyp, a small benign growth consisting of apparently normal cells that line the intestine. As cell division continues, the growth enlarges, becoming an adenoma. As various tumor-suppressor genes are inactivated and an oncogene called k-*ras* is mobilized, the mutations accumulate and the cells in the adenoma become increasingly abnormal. The final consequence is colon carcinoma, a form of cancer that metabolizes quickly.

Prevalence and Causes of Cancer

Although the incidence of some cancers is down (stomach, colon, and cervical cancers), the incidence of others is up (skin, lymphoid, prostate, and brain cancers), and the overall rate of cancer deaths has actually increased over the past few decades despite a great rise in expenditures for research and treatments. Some good news has appeared, however. When records for the first half of the 1990s became available, they showed that cancer deaths in the United States declined for the first time ever. However, the decline was small (4%) and was influenced heavily by lower rates of tobacco use and lung cancer since the 1960s.

The American Medical Association makes the following rough estimates of the most common cancer-causing agents: Natural chemicals in food seem to cause 35% of all cancers; carcinogens in tobacco (smoking) cause 30%; infections may account for about 10% (ulcer bacteria may cause stomach cancer); sexual behavior apparently contributes 7% (for example, papilloma virus transmitted through sex is associated with cancer of the uterus); occupational hazards may cause 4%; and alcohol, 3%. Environmental pollutants perhaps account for 2%. Other, less common factors would account for the remaining 9%.

Cancer Management

Cancer management techniques vary, but the usual sequence is:

1. *Diagnosis.* Screening procedures, such as having a mammogram or examining one's breasts or testicles for lumps, can detect some cancers very early and have saved many lives. Even so, most tumors are recognized only after they cause symptoms. A **biopsy** (bi′op-se; "life viewing") is another diagnostic procedure: A tissue sample is surgically removed and examined under the microscope for malignant cells. Increasingly, diagnosis is made through a chemical or genetic analysis of the sample. Large tumors may be recognized by surgical exposure or by medical imaging techniques (X rays, MRI, CT).
2. *Staging.* This involves several methods (physical exam, histological examination, lab tests, and imaging techniques) to determine the extent of the disease (such as size of the neoplasm and degree of metastasis). The patient

is placed in a group according to the probability of cure: Stage 1 has the best chance, stage 4 the least.

3. *Treatment.* Most cancers are removed by surgery. To destroy the cells that have metastasized, surgery is often followed by radiation therapy (X-irradiation of the primary tumor or treatment with radioactive isotopes) and by chemotherapy (treatment with anticancer drugs). Chemotherapy is beset with the problem of resistance. That is, a few cancer cells can eject the toxins, and these cells proliferate and metastasize until they form new tumors, which are invulnerable to further chemotherapy. Furthermore, anticancer drugs have unpleasant side effects because they kill all types of rapidly dividing cells, including normal ones. Their side effects include nausea, diarrhea, fatigue, hair loss, and weakened resistance to disease. X rays also have side effects because, in passing through the body, they kill healthy cells that lie in the path of the cancer cells.Treatment is able to arrest and cure about half of all cancer cases. Certain cancers, however, have low survival rates, including cancers of the lungs, liver, and ovaries.

Current cancer treatments—"cut, burn, and poison"—are widely recognized to be too crude and painful. Promising new methods focus on delivering anticancer drugs more exclusively and precisely to the cancer, and on increasing the immune system's effectiveness against cancer cells. The most recent research seeks to fix defective tumor-suppressor genes and oncogenes in cancer cells, to choke off tumors by inhibiting their ability to attract a blood supply, to better understand metastasis, to use viruses to target and destroy cancer cells, to overcome the problem of resistance to chemotherapy, and to signal cancer cells to commit suicide by apoptosis. Much excitement was generated in mid-1999, when scientists finally were able to cause normal human cells in laboratory cultures to become cancerous by manipulating a series of specific genes; this indicated that we finally understand the genesis of cancer, and it could soon lead to better treatments.

We end on a positive note. Although average survival rates have not increased, the quality of life of cancer patients has improved in the last decade. The treatment of cancer-associated pain is better, and discomfort from the side effects of chemotherapy has been reduced by antinausea drugs and other helpful medicines.

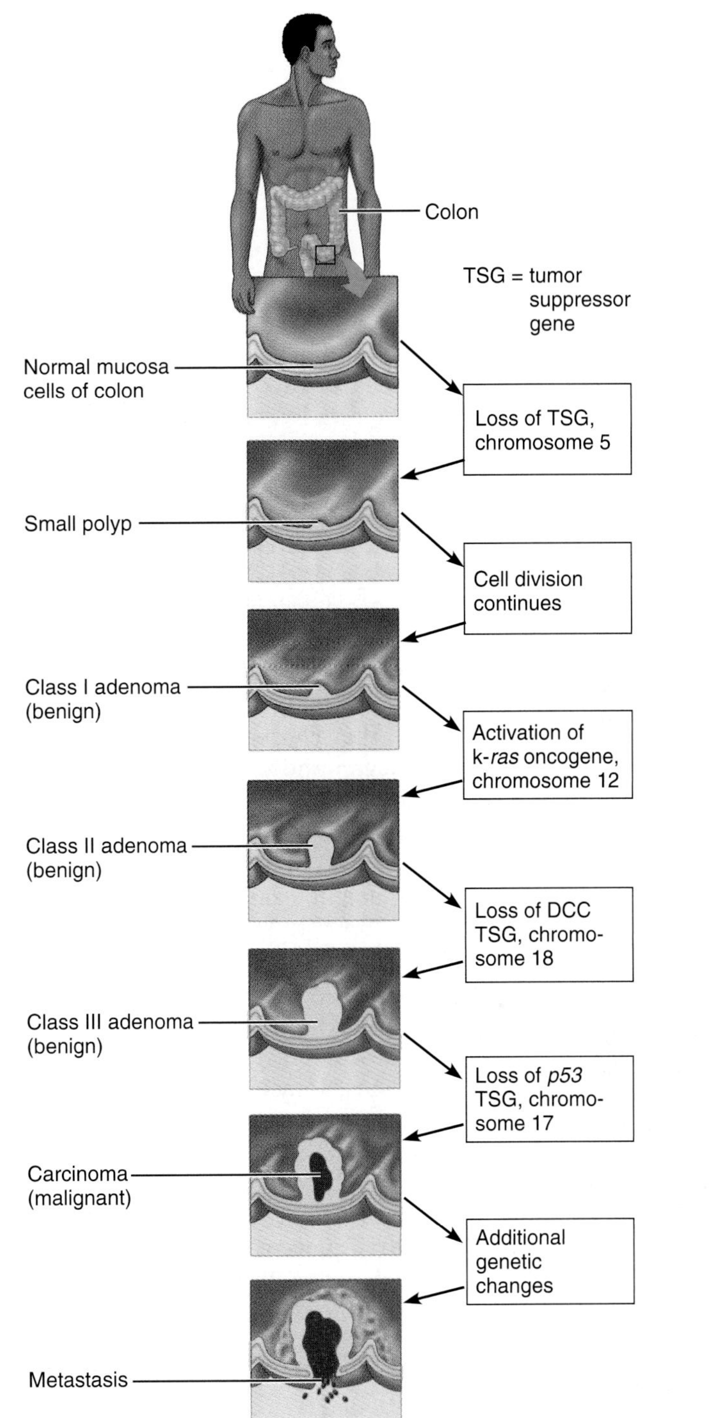

Development of colon cancer.

Sperm (male). This cell is long and streamlined, for swimming to the egg for fertilization. The swimming tail is a motile whip called a flagellum (p. 83 in Chapter 4).

DEVELOPMENTAL ASPECTS OF CELLS

Youth

We all begin life as a single cell, the fertilized egg, from which arise all the cells of our bodies. Early in embryonic development, the cells begin to specialize: Some become liver cells; some become nerve cells; others become the transparent lens of the eye. Every cell in the body carries the same genes. (A *gene,* simply speaking, is a segment of DNA that dictates a specific cell function, usually by coding for a specific protein.) If all our cells have identical genes, how do cells differentiate and take on specialized structures and functions?

This question is currently the subject of extensive research. Apparently, cells in various regions of the developing embryo are exposed to different chemical signals that channel the cells into specific pathways of development. In a young embryo that consists of just a few cells, slight differences in oxygen and carbon dioxide concentrations between the more superficial and the deeper cells may be the major signals. But as development continues, cells begin to release chemicals that influence the development of neighboring cells by switching some of their genes "on" or "off." Some genes are active in all our cells; for example, all cells must carry out protein synthesis and make ATP. However, the genes for enzymes that catalyze the synthesis of specialized proteins such as hormones or mucus are activated only in certain cell populations. Hence, the secret of cell specialization lies in the kinds of proteins made and reflects differential gene activation in the different cell types.

Cell specialization leads to *structural* variation among the cell types in the body. Different organelles come to predominate in different cells. For example, muscle cells make tremendous quantities of actin and myosin proteins, and their cytoplasm fills with actin microfilaments. Lipid accumulates in fat cells. High-energy cells, like those in the liver, make more enzymes for the manufacture of ATP and have more mitochondria than other cells. Phagocytic cells produce more lysosomal enzymes, and contain many lysosomes. Through such events arise the 200 or so different cell types of our body. The development of specific and distinctive features within cells is called **cell differentiation.**

Most organs are well formed and functional long before birth, but the body continues to grow by forming more cells throughout childhood and adolescence. Once adult size is reached, cell division slows considerably and occurs primarily to replace short-lived cell types and to repair wounds.

As you can see, cell proliferation and differentiation are precisely orchestrated events. This exact coordination is disrupted in cancer, a disease in which cells multiply uncontrollably and lose some of their differentiated features. Cancer is discussed in the box on pages 50–51.

Aging

There is no doubt that cellular aging occurs and that it accounts for most problems associated with old age. While aging is very complex and certainly is caused by many mechanisms, the best-documented theory says that it results from *free radicals.* These destructive molecules mostly form as byproducts of normal cellular metabolism, although they also form when external insults such as radiation and chemical pollutants enter our body. The theory proposes that the radicals build up and progressively damage the essential molecules of our cells, gradually weakening us as we age. Most evidence for this comes from experiments on less complex animals such as worms and fruit flies, where neutralizing the radicals and repairing their damage have been shown to increase life span. Vitamins E and C appear to act as antioxidants in our own bodies, and may help to prevent excessive free-radical formation.

Most radicals are produced in our mitochondria, because this is the organelle with the highest rate of metabolism. This has led to the proposal that a decrease in the production of energy by radical-damaged mitochondria weakens and ages our cells. This is called the *mitochondrial theory of aging.*

It has long been known that when laboratory rats and mice are slightly undernourished, their life span increases by up to thirty percent. The same thing has recently been demonstrated in the more human-like monkeys. This phenomenon is explained by the free-radical theory: Because caloric restriction lowers the metabolic rate and makes metabolism more efficient, fewer of the destructive radicals are produced, and aging slows.

Genetic theories of aging propose that aging is programmed into our genes. These theories originated from the observation that the body structure ages in a predictable, patterned way, as if aging were a normal part of human development (and development is known to be controlled by genes). Rats and fruit flies can be bred to live longer than usual, and genes that increase and decrease longevity have been identified in animals. While some of these genes have been shown to work by influencing free radicals, others act in different but less-understood ways.

The closest evidence we have for any kind of planned senescence involves **telomeres,** structures that limit the maximum number of times cells can divide. Telomeres are repeating, seemingly nonsensical stretches of DNA that cap the ends of chromosomes (*telo* = end; *mere* = piece). Though they carry no genes, they appear to be vital for chromosomal survival: Each time DNA is replicated, 50 to 100 of the end nucleotides are lost and the telomeres get a bit shorter. When the telomeres reach a certain minimal length, the "stop-division" signal is given. The idea that cell longevity depends on telomere length was supported by the 1994 discovery of *telomerase,* an enzyme that prevents telomeres from degrading by adding more repeating DNA to the ends. Pegged as the "immortality enzyme," telomerase occurs in our endlessly replicating germ-line cells and in cancer cells, but not in other cell types. Great excitement was generated in 1998 when it was reported that adding telomerase to laboratory-grown human cells greatly increased the number of times these cells could divide, without showing any signs of senescence.

RELATED CLINICAL TERMS

Anaplasia (an″ah-pla′ze-ah) (*an* = without, not; *plas* = form) An abnormality in cell structure. For example, some cancer cells resemble not their parent cells but undifferentiated or embryonic cells.

Apoptosis (ap″o-to′sis; "falling away") Programmed cell death. This process of controlled cellular suicide eliminates cells that are stressed, unneeded, excessive, or aged. In response to damaged macromolecules within the cell or to some extracellular signal, a series of intracellular enzymes is activated. These enzymes, called caspases, destroy the cell's DNA, cytoskeleton, and other structures, producing a quick, neat death. The apoptotic cell shrinks without leaking its contents into the surrounding tissue, its surface forms blebs, and its chromatin condenses as the nucleus shrinks. It detaches from other cells, rounds up, and is immediately phagocytized by nearby cells. This tidy sequence avoids inflammation (p. 102), and therefore minimizes tissue injury. Cancer cells fail to undergo apoptosis, but oxygen-starved cells do so excessively (heart-muscle and brain cells during heart attacks and strokes, for example).

Dysplasia (dis-pla′ze-ah) (*dys* = abnormal) A change in cell size, shape, or arrangement due to long-term irritation or inflammation (from infections, for example).

Hyperplasia (hi″per-pla′ze-ah; "excess shape") Excessive cell proliferation. Unlike cancer cells, hyperplastic cells retain their normal form and arrangement within tissues.

Hypertrophy (hi-per′tro-fe; "excess growth") Growth of an organ or tissue due to an increase in the size of its cells. Hypertrophy is a normal response of skeletal muscle cells to exercise. Hypertrophy differs from hyperplasia, the condition in which cells increase in number but not in size.

Necrosis (ne-kro′sis) (*necros* = death; *osis* = process; condition) Death of a cell or group of cells due to injury or disease. Acute injury causes the cells to swell and burst, and they induce an inflammatory response. This is accidental, *uncontrolled* cell death, in contrast to apoptosis.

CHAPTER SUMMARY

Media study tools that could provide you additional help in reviewing specific key topics of Chapter 2 are referenced below.

SP = Study Partner; AIA = A.D.A.M.® Interactive Anatomy

INTRODUCTION TO CELLS (pp. 28–29)

1. Cells are the basic structural and functional units of life.
2. There are about 200 cell types in the human body, which vary widely in size, shape, and function. In this chapter, the features shared by all cells are emphasized.
3. Each cell has three main regions: plasma membrane, cytoplasm, and nucleus.

THE PLASMA MEMBRANE (pp. 29–33)

Structure (pp. 29–30)

1. The plasma membrane defines the cell's outer boundary. The fluid mosaic model interprets this membrane as a flexible bilayer of lipid molecules (phospholipids, cholesterol, and glycolipids), with proteins in this layer. When viewed by electron microscopy, the membrane has two dark layers (phospholipid heads) that sandwich an inner, light layer (phospholipid tails).
2. Most proteins in the membrane are integral proteins and extend entirely through the membrane. Peripheral proteins, by contrast, are on its cytoplasmic side, helping to support the membrane along with other functions.
3. Sugar groups of membrane glycoproteins and glycolipids project from the cell surface and contribute to the cell coat (glycocalyx), which functions in cell-to-cell binding and recognition.

SP Exercise: Chapter 2, Structure of the Plasma Membrane.

Functions (pp. 30–33)

4. Besides acting as a fragile barrier to protect the cell contents, the plasma membrane determines what enters and leaves the cell. Only small, uncharged molecules may diffuse freely through the membrane; larger or charged molecules must pass by transport mechanisms that involve the integral proteins. Some membrane proteins also serve as receptors for extracellular signal molecules that bind to the plasma membrane.
5. Large particles and macromolecules pass through the membrane by vesicular, or bulk, transport (exocytosis and endocytosis). In exocytosis, membrane-lined cytoplasmic vesicles fuse with the plasma membrane and release their contents to the outside of the cell.
6. Endocytosis brings large substances into the cell, as packets of plasma membrane fold in to form cytoplasmic vesicles. If the substance is a particle, the process is called phagocytosis; if the substance is dissolved molecules in the extracellular fluid, the process is known as pinocytosis. Receptor-mediated endocytosis is selective: Specific molecules attach to receptors on the membrane before being taken into the cell in clathrin-coated vesicles or caveolae.

THE CYTOPLASM (pp. 33–42)

Cytosol (p. 33)

1. The cytosol, or cytoplasmic matrix, is the viscous, fluid-containing substance in which cytoplasmic organelles and inclusions are embedded.

Cytoplasmic Organelles (pp. 33–40)

2. Each organelle performs specific functions. The various cell types in the body have different numbers of each organelle type.

3. Mitochondria are thread-like organelles covered by two membranes, the inner of which forms shelf-like cristae. Mitochondria are the main sites of ATP synthesis, the cell's main energy generators.

4. Ribosomes are dark-staining granules that consist of two subunits, each made of protein and ribosomal RNA. Ribosomes are the sites of protein synthesis (translation). Free ribosomes make proteins used in the cytosol.

5. The rough endoplasmic reticulum is a ribosome-studded system of membrane-walled envelopes (cisternae). Its ribosomes make proteins, which enter the cisternae, and which may ultimately be secreted by the cell. The rough ER also makes all the cell's membranes.

6. The smooth endoplasmic reticulum, a network of membrane-walled tubes containing no ribosomes, is involved in the metabolism of lipids. All ER stores calcium ions.

7. The Golgi apparatus is a stack of disc-shaped envelopes that has a cis (convex) and a trans (concave) face. It sorts the products of the rough endoplasmic reticulum and then sends these products, in membrane-bound vesicles, to their proper destination. Lysosomes and secretory granules arise from the Golgi apparatus.

8. Lysosomes are spherical, membrane-walled sacs of digestive enzymes. They digest deteriorated organelles and endocytosed substances.

9. Peroxisomes are membrane-walled, enzyme-containing sacs that perform several metabolic processes. They protect the cell from free radicals and hydrogen peroxide. They also use hydrogen peroxide to break down some organic poisons and carcinogens.

10. The cytoskeleton includes protein rods of three distinct types—microtubules, actin microfilaments, and intermediate filaments—all in the cytosol. Microtubules, which radiate out from the centrosome region, give the cell its shape; they also organize the distribution and the transport of various organelles within the cytoplasm. Actin microfilaments interact with myosin to produce contractile forces. Both microtubules and microfilaments tend to be labile, but intermediate filaments, which act to resist tension placed on the cell, are stable.

11. The centrosome is a spherical region of cytoplasm near the nucleus. It consists of a cloud-like matrix surrounding a pair of centrioles. Proteins in the matrix anchor long microtubules of the cytoskeleton and mitotic spindle. The centrioles are barrel-shaped structures with walls of short microtubules.

12. Vaults are protein "barrels" of uncertain function.

Cytoplasmic Inclusions (pp. 40–41)

13. Inclusions are impermanent structures in the cytoplasm. Examples include food stores, such as lipid droplets and glycogen-containing glycosomes.

THE NUCLEUS (pp. 41–45)

1. The nucleus contains genetic material (DNA) and is the control center of the cell. Most cells have one centrally located nucleus shaped like a sphere or an egg.

Nuclear Envelope (p. 42)

2. The nucleus is surrounded by a nuclear envelope, which is penetrated by nuclear pores. The nuclear envelope is continuous with the rough endoplasmic reticulum.

Chromatin and Chromosomes (pp. 42–45)

3. Chromatin is strand-like material (DNA and histones) in the nucleus. This chromatin is distributed in chromosomes. During cell division, all chromatin is highly coiled, making the chromosomes appear as thick rods. In nondividing cells, the chromatin is a mixture of dark-staining, coiled regions (condensed chromatin) and light-staining, uncoiled regions (extended chromatin).

4. The DNA molecule is a double helix consisting of four types of nucleotides, with bases of thymine, adenine, cytosine, and guanine.

5. A structural subunit of chromatin is the nucleosome, a cluster of eight histone proteins around which DNA coils.

Nucleoli (p. 45)

6. A nucleolus is a dark-staining body within the nucleus, associated with parts of several chromosomes. Nucleoli make the subunits of ribosomes.

THE CELL LIFE CYCLE (pp. 45–48)

1. The cell life cycle is the series of changes a cell experiences from the time it forms until it divides.

Interphase (p. 45)

2. Interphase is the nondividing phase of the cell life cycle. It consists of the subphases G_1, S, and G_2. During G_1, the cell grows; during S, DNA replicates; and during G_2, the final preparations for division are made.

Cell Division (pp. 45–48)

3. Cell division, essential for growth and repair of the body, occurs during the M (mitotic) phase. Cell division has two distinct aspects: mitosis and cytokinesis.

4. Mitosis, the division of the nucleus, consists of (1) *prophase*, when chromatids appear, the nuclear membrane is lost, and the mitotic spindle forms; (2) *metaphase*, when the chromatids line up at the cell's equator; (3) *anaphase*, when the V-shaped chromatids are pulled apart; and (4) *telophase*, when the chromatin extends and the nucleus reassembles. Mitosis parcels out the replicated chromosomes to two daughter nuclei. Cytokinesis, completed after mitosis, is the division of the cell into two cells.

CELLULAR DIVERSITY (pp. 48–52)

1. The cells of the body have a variety of shapes, which reflect their functions; different organelles dominate in different cell types (see Figure 2.22).

DEVELOPMENTAL ASPECTS OF CELLS (pp. 52–53)

Youth (p. 52)

1. The first cell of a human is the fertilized egg. Cell differentiation begins early in development and is thought to reflect differential gene activation. Different organelles come to predominate in different cell types.
2. During adulthood, cell numbers remain fairly constant, and cell division occurs primarily to replace lost cells.

Aging (pp. 52–53)

3. Aging of cells (and of the whole body) may reflect accumulated damage from free radicals, or it may be a genetically influenced process, or both. It may also reflect a loss of the capacity for cell division over time.

REVIEW QUESTIONS

Multiple Choice/Matching Questions

1. The endocytotic process in which particulate matter is brought into the cell is called (a) phagocytosis, (b) pinocytosis, (c) exocytosis.
2. The nuclear substance composed of histone proteins and DNA is (a) chromatin, (b) the nuclear envelope, (c) nucleoplasm, (d) nuclear pores.
3. Final preparations for cell division are made during this stage of the cell life cycle: (a) G_1, (b) G_2, (c) M, (d) S.
4. The fundamental bilayered structure of the plasma membrane is determined almost exclusively by (a) phospholipid molecules, (b) peripheral proteins, (c) cholesterol molecules, (d) integral proteins.
5. Identify the cell structure or organelle described by each of the following statements.
   (a) A stack of 3–10 membranous discs, with vesicles.
   (b) A continuation of the nuclear envelope forms this ribosome-covered cytoplasmic organelle.
   (c) In nondividing cells, the highly coiled parts of chromosomes form this type of chromatin.
   (d) This type of endoplasmic reticulum is a network of hollow tubes, not envelopes.
   (e) This is a spherical cluster of eight histone molecules wrapped by a DNA strand.
   (f) These are the cytoskeletal rods with the thickest diameter (choose from microtubules, microfilaments, intermediate filaments).
   (g) The only organelle with DNA, cristae, and matrix granules.
   (h) This energy-producing organelle is probably descended from bacteria.
6. Circle the false statement about centrioles. (a) They start to duplicate in G_1; (b) they lie in the centrosome; (c) they are made of microtubules; (d) they are membrane-walled barrels lying parallel to each other.
7. The trans face of the Golgi apparatus (a) is its convex face, (b) is where products leave the Golgi apparatus in vesicles, (c) receives transport vesicles from the rough ER, (d) is in the very center of the Golgi stack, (e) is the same as the cis-Golgi network.
8. Circle the false statement about lysosomes. (a) They have the same structure and function as peroxisomes; (b) they form by budding off the Golgi apparatus; (c) lysosomal enzymes do not occur freely in the cytosol in healthy cells; (d) they are abundant in phagocytic cells.
9. Name the appropriate stage of mitosis (telophase, anaphase, metaphase, or prophase) for each of the following.
   (a) the chromosomes are lined up in the middle of the cell
   (b) the nuclear membrane fragments
   (c) the nuclear membrane re-forms
   (d) the mitotic spindles form
   (e) the chromosomes (chromatids) are V-shaped
10. Name the cytoskeletal element (microtubules, actin microfilaments, or intermediate filaments) for each of the following.
    (a) give the cell its shape
    (b) resist tension placed on a cell
    (c) radiate from the centrosome
    (d) interact with myosin to produce contraction force
    (e) are the most stable
    (f) associated with kinesins and dyneins
    (g) associated with the motor protein myosin
11. Different organelles are abundant in different cell types. Match the cell types with their abundant organelles by placing the correct letter from the column at right into each blank at left. Follow the hints provided in parentheses.

Cell Type	**Abundant Organelle**
____ **(1)** cell lining the small intestine (assembles fats)	(a) mitochondria
____ **(2)** white blood cell (phagocytic)	(b) smooth ER
____ **(3)** liver cell (detoxifies carcinogens)	(c) peroxisomes
____ **(4)** muscle cell (highly contractile)	(d) microfilaments
____ **(5)** mucous cell (secretes protein product)	(e) rough ER
____ **(6)** epithelial cell in the outer layer of skin (withstands tension)	(f) intermediate filaments
____ **(7)** kidney tubule cell (makes and uses large amounts of ATP)	(g) lysosomes

12. Cancer is the same as (a) all tumors, (b) all neoplasms, (c) all malignant neoplasms, (d) benign tumors, (e) AIDS.

13. Which of these processes involves endosomes and clathrin-coated pits? (a) phagocytosis, (b) receptor-mediated endocytosis, (c) exocytosis, (d) pinocytosis.

Short Answer Essay Questions

14. Some of the intrinsic proteins and lipids in the plasma membrane have sugar chains attached to their external side. What role does this sugar coat play in the life of the cell?

15. List all the cytoplasmic organelles that are composed (at least in part) of lipid-bilayer membranes. Then list all the cytoplasmic organelles that are not membranous.

16. Martin missed a point on his anatomy test because he thought *nucleus* and *nucleolus* were the same word and the same structure. Distinguish the nucleus from the nucleolus.

17. (a) What is chemotherapy for cancer (as currently employed)? (b) How does it work? (c) What are some of its side effects?

18. The students in Anatomy 315 class were confused by the similarity between lysosomes and peroxisomes. Then Sue helped out by saying, "Lysosomes break down big things, but peroxisomes break down little things like chemicals." Can you explain what Sue meant?

19. In this chapter, we claimed that mitochondria are the most complex organelles in the cytoplasm. What evidence supports this claim?

20. Define a chromosome. Then compare the arrangement of the chromatin in a dividing versus a nondividing cell.

CRITICAL REASONING + CLINICAL APPLICATION QUESTIONS

1. Chrissy, who is 30 years old, stays 5 kg (11 pounds) underweight because she thinks being thin will slow her aging process. After warning her not to lose any more weight or to become anorexic, her doctor admitted that there is at least some scientific evidence to support her view. What is that evidence?

2. After studying the information box on cancer in this chapter, deduce which of the various types of cancer causes the most deaths in the United States. (Hint: Most skin cancers are easily cured.)

3. Kareem had a nervous habit of chewing on the inner lining of his lip with his front teeth. The lip grew thicker and thicker from years of continual irritation. Kareem's dentist noticed his greatly thickened lip, then told him to have it checked to see if the thickening was a tumor. A biopsy revealed hyperplasia and scattered areas of dysplasia, but no evidence of neoplasia. What do these terms mean? Did Kareem have cancer of the mouth?

4. The normal function of one tumor-suppressor gene is to prevent cells with damaged chromosomes and DNA from "progressing from G_1 to S," whereas another tumor-suppressor gene prevents "passage from G_2 to M." When these tumor-suppressor genes fail to work, cancer can result. Explain what the phrases in quotations mean.

5. In their anatomy labs, many students are exposed to the chemical preservatives phenol, formaldehyde, and alcohol. Our cells break down these toxins very effectively. What cellular organelle is responsible for this?

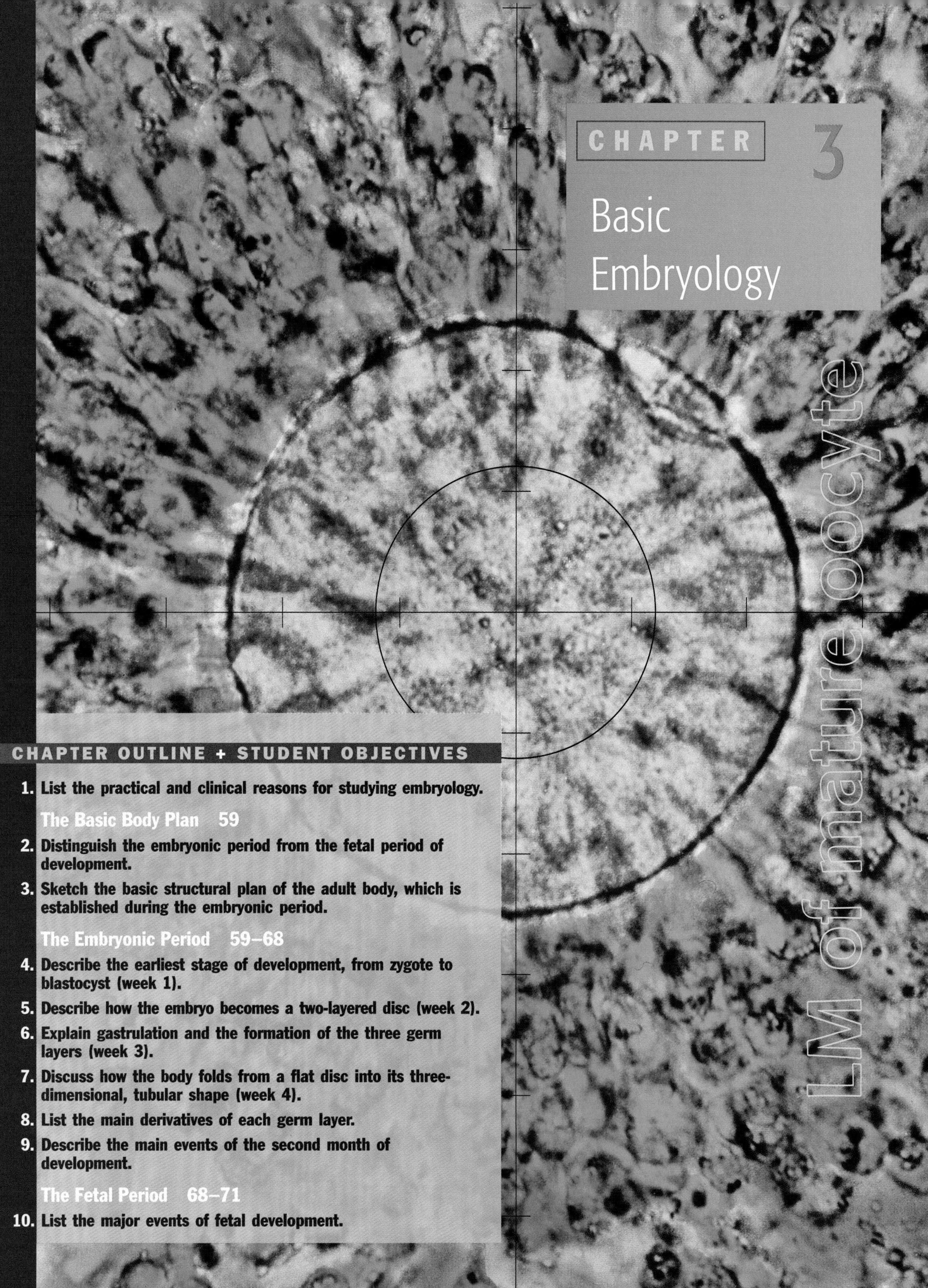

CHAPTER 3
Basic Embryology

CHAPTER OUTLINE + STUDENT OBJECTIVES

1. List the practical and clinical reasons for studying embryology.

The Basic Body Plan 59

2. Distinguish the embryonic period from the fetal period of development.
3. Sketch the basic structural plan of the adult body, which is established during the embryonic period.

The Embryonic Period 59–68

4. Describe the earliest stage of development, from zygote to blastocyst (week 1).
5. Describe how the embryo becomes a two-layered disc (week 2).
6. Explain gastrulation and the formation of the three germ layers (week 3).
7. Discuss how the body folds from a flat disc into its three-dimensional, tubular shape (week 4).
8. List the main derivatives of each germ layer.
9. Describe the main events of the second month of development.

The Fetal Period 68–71

10. List the major events of fetal development.

IN JUST 38 WEEKS, FROM CONCEPTION TO birth, a single fertilized egg cell develops into a fully formed human being. The body will not change this much again during its remaining life span of 70 to 90 years. This chapter will introduce you to human **embryology,** the study of the origin and development of an individual person. A knowledge of basic embryological events and structures is especially valuable as you begin your study of human anatomy. By knowing how the body methodically assembles itself, you will better understand the anatomy of an adult. Moreover, embryology helps to explain the origin of many birth defects, anatomical abnormalities that are evident in about 3% of live births.

The time between conception (the union of an egg and a sperm cell) and birth is the **prenatal period** (pre-na′tal; "before birth"), which is divided into two stages (Figure 3.1). The **embryonic period** spans the first 8 weeks, and the **fetal** (fe′tal) **period** encompasses the remaining 30 weeks. The embryonic period is an exceptionally "busy" one: By its end, all major organs are in place, and the **embryo,*** the early form of the body, looks distinctly human. In the longer fetal period that follows,

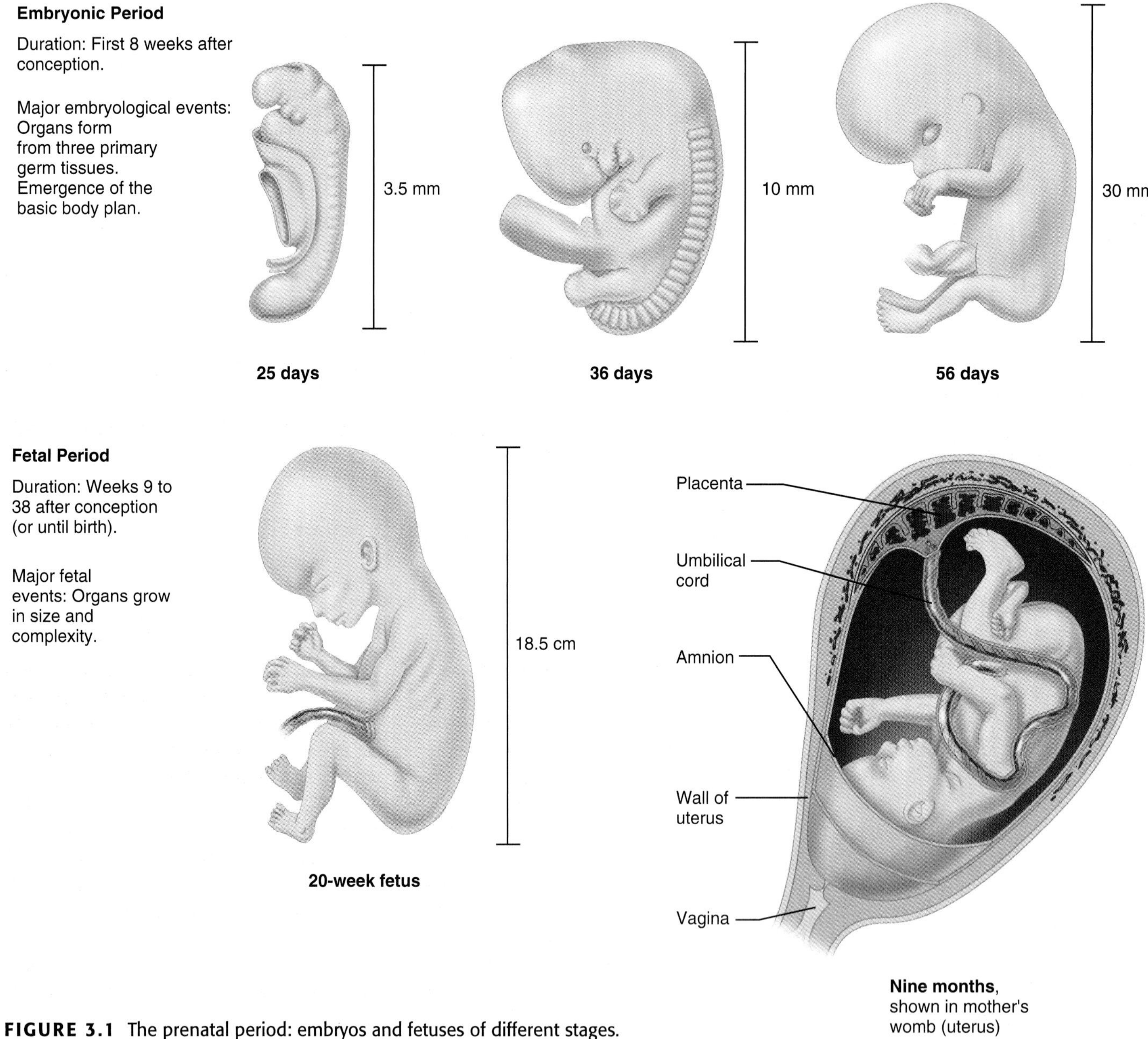

FIGURE 3.1 The prenatal period: embryos and fetuses of different stages.

*Technically, the embryo is the stage in prenatal development between the third or fourth and eighth weeks, inclusive. However, the term *embryo* can be used informally to encompass all stages in the embryonic period.

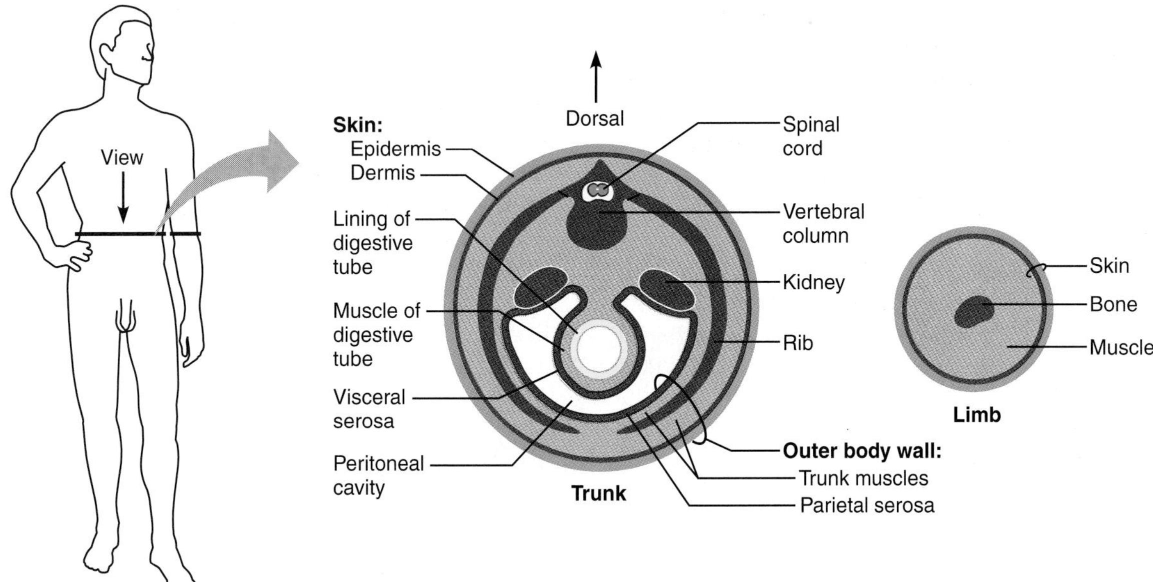

FIGURE 3.2 The adult human body plan (simplified cross section). The development of this body plan is traced in this chapter. The blue, red, and yellow colors denote derivation from the three basic embryonic germ layers. See the discussion on p. 66 and Figure 3.9 for more information.

the organs of the **fetus** (fe′tus; "the young in the womb") grow larger and more complex.

THE BASIC BODY PLAN

To simplify the treatment of embryology in this chapter, we will limit our discussion to the derivation of the basic adult body plan. This plan, which is evident by month 2 of development, is shown in Figure 3.2. In the figure, note the following adult structures:

1. **Skin.** The skin has two layers: an outer layer called the *epidermis* and an inner, leathery layer called the *dermis.*
2. **Outer body wall**. The outer body wall consists mostly of trunk muscles. Dorsally, it also contains the vertebral column, through which runs the spinal cord. Ribs attach to each bony vertebra in the thorax region of the trunk wall.
3. **Body cavity and digestive tube.** The peritoneal cavity, lined by visceral and parietal serosae, surrounds the digestive tube (stomach, intestines, and so on) in the trunk region. The digestive tube has a muscular wall and is lined internally by a sheet of cells. This lining is identified in the figure.
4. **Kidneys and gonads.** The kidneys lie directly deep to the dorsal body wall, in the lumbar region of the back. The gonads (male testes and female ovaries) originate in a similar position but migrate to other parts of the body during the fetal period.
5. **Limbs.** The limbs consist mostly of bone, muscle, and skin.

We can see how this adult body plan takes shape by following the events of month 1 of human development.

THE EMBRYONIC PERIOD

Week 1: From Zygote to Blastocyst

Each month, one of a fertile woman's two ovaries releases an immature egg, called an *oocyte* (Figure 3.3). This cell is normally drawn into a *uterine tube,* which provides a direct route to the uterus (womb). **Conception,** the fertilization of an oocyte by a sperm, generally occurs in the lateral third of the uterine tube. The fertilized oocyte, now called a **zygote** (zi′gōt; "a union"), moves toward the uterus. Along the way, it divides repeatedly to produce two cells, then four cells, then eight cells, and so on. These daughter cells are called **blastomeres** (blas′to-mērz; "generating pieces"). Because there is not much time for cell growth between divisions, the resulting blastomere cells become smaller and smaller. This early division sequence, called **cleavage,** provides the large number of cells needed as building blocks for the embryo.

About 72 hours* after fertilization, the cleavage divisions have generated a solid cluster of 12–16 blastomere cells called a **morula** (mor′u-lah; "mulberry"). During

*All dates given for the developmental events in this chapter are *average* times. Actual dates vary by 1–2 days or more among different pregnancies.

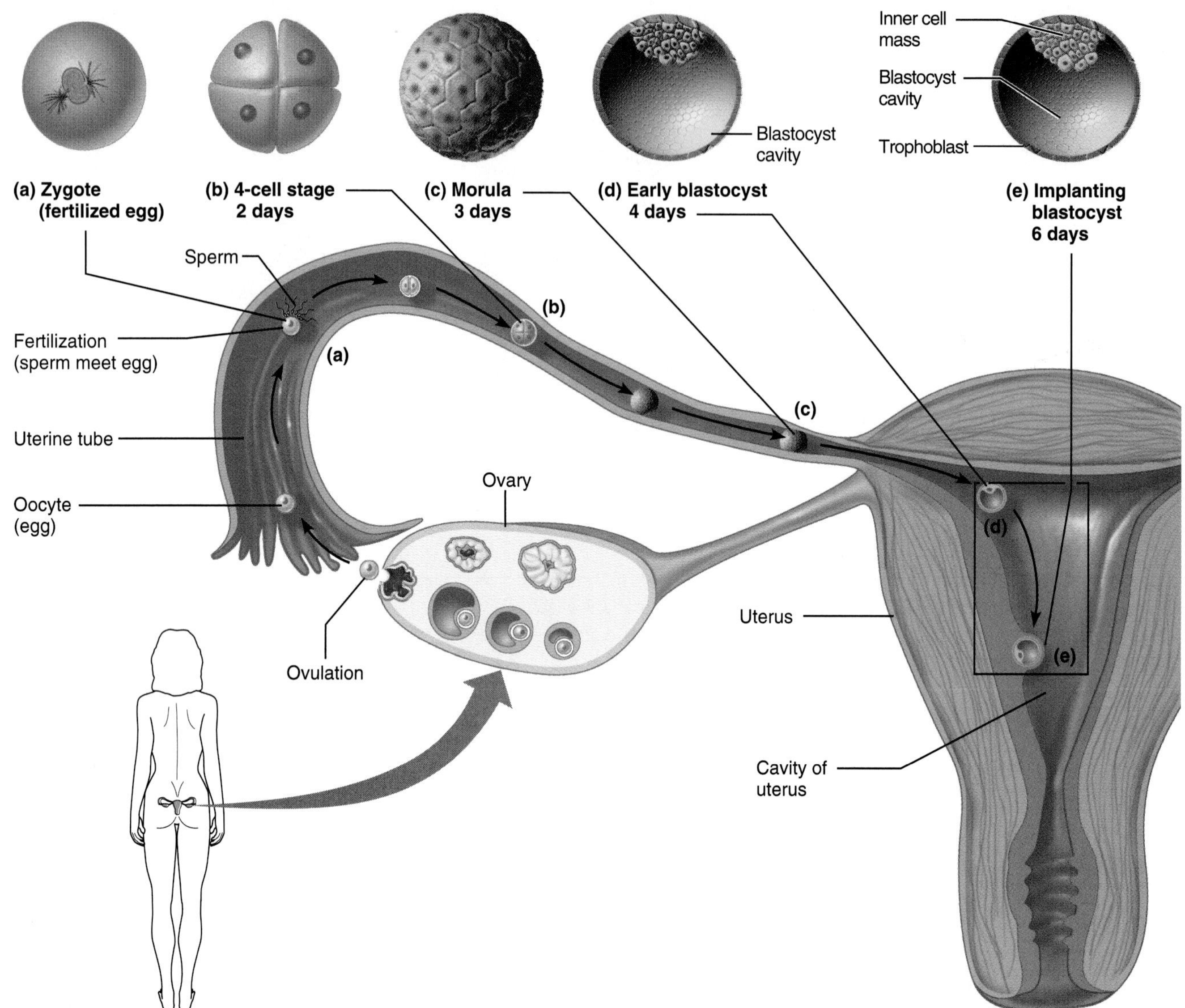

FIGURE 3.3 Fertilization and the events of the first 6 days of development. The ovary, uterus, and uterine tubes, which lie in the mother's pelvis, are shown in posterior view. An egg (oocyte) is released into the peritoneal cavity (ovulation) and taken up by the uterine tube, where it undergoes fertilization to become a zygote **(a)**. Then, as it moves through the uterine tube and into the uterus, it passes through the 4-cell stage **(b)**, morula **(c)**, and blastocyst **(d)** stages. Finally, the blastocyst implants into the wall of the uterus **(e)**, as shown in detail in the next figure. In (d)–(e), the blastocyst has been cut in half so one can see inside.

day 4, the late morula—now consisting of about 60 cells—enters the uterus. It takes up fluid, which gathers into a central cavity. This new fluid-filled structure is called a **blastocyst** (blas′to-sist; *blasto-* = bud or sprout; *-cyst* = bag).

Two distinct types of cells are evident in the morula, and are especially obvious in the blastocyst stage (Figure 3.3e and Figure 3.4b). A cluster of cells on one side of the blastocyst cavity is called the **inner cell mass,** while the layer of cells around the outside of the cavity is called the **trophoblast** (trōf′o-blast; *tropho-* = nourishment). The inner cell mass will form the embryo, while the trophoblast will help form the placenta, the structure by which the fetus obtains nutrients from the mother. In this chapter, we will focus on the inner cell mass and the embryo, leaving the trophoblast and placenta for the discussion of the female reproductive system (Chapter 24).

The blastocyst stage lasts about 3 days, from day 4 to day 7. For most of this time, the blastocyst floats freely in the cavity of the uterus, but on day 6, it starts to burrow into the wall of the uterus (Figure 3.4b). This process, called **implantation,** takes about a week to complete. In implantation, the trophoblast layer erodes inward until the entire embryo is imbedded in the uterine wall.

Critical Reasoning

In some pregnancies, the inner cell mass of a single blastocyst splits into two at the end of week 1. Can you deduce the consequence of this occurrence?

This produces identical twins, also called monozygotic twins (monozygotic = from one zygote). Less often, monozygotic twins result from a splitting that occurs earlier in week 1 during the early stages of cleavage, or else later, in week 2. Overall, twins account for 1 in 90 births in the United States.

Week 2: The Two-Layered Embryo

Figure 3.4d shows the embryo about 9 days after fertilization. The inner cell mass has now divided into two sheets of cells, the **epiblast** (ep′ĭ-blast; *epi-* = over, upon) and the **hypoblast** (hi′po-blast; *hypo-* = under, beneath). Extensions of these cell sheets form two fluid-filled sacs (Figure 3.4d–e). This arrangement resembles two balloons touching one another, with the epiblast and hypoblast at the area of contact. Together, the epiblast and hypoblast make up the **bilaminar** ("two-layered") **embryonic disc,** which will give rise to the whole body.

The sac formed by an extension of the epiblast is the *amniotic* (am″ne-ot′ik) *sac,* and the sac formed by an extension of the hypoblast is the *yolk sac.* The outer membrane of the amniotic sac is called the **amnion,** and the internal **amniotic sac cavity** is filled with *amniotic fluid.* This fluid buffers the developing embryo and fetus from physical shock until the time of birth. The **yolk sac** holds a very small amount of yolk, which is insignificant as a food source. The human yolk sac, however, is important because the digestive tube forms from part of it (see p. 66). Furthermore, tissue around the yolk sac gives rise to the earliest blood cells and blood vessels (see Chapter 17, p. 516).

Critical Reasoning

You may have heard the expression, "The mother's water broke just before she gave birth." Can you determine what that means?

The amniotic sac ruptured and released its amniotic fluid. "Breaking water" occurs near the start of labor, the process that expels the mature fetus from the uterus.

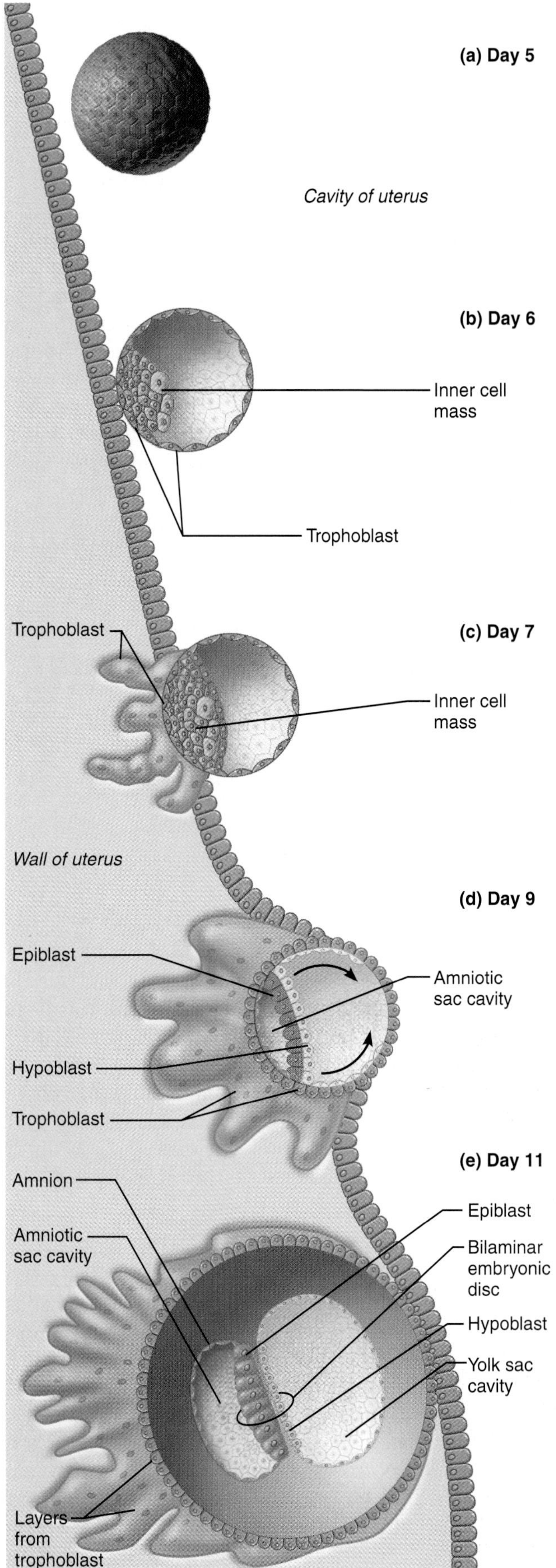

FIGURE 3.4 Implantation of the blastocyst during week 2 of development. For orientation, this is the area in the rectangle in the uterus in Figure 3.3. The embryos are shown in section (cut in half). **(a)** Blastocyst floating in the cavity of the uterus (day 5). **(b)** Blastocyst that has just adhered to the wall of the uterus (day 6). **(c)** Implanting embryo (day 7). **(d)** Slightly later stage in which the epiblast and hypoblast layers appear in the embryo (day 9). **(e)** Embryo is almost completely implanted in the wall of the uterus (day 11). It is now a bilaminar embryonic disc of epiblast and hypoblast situated between an amniotic sac and a yolk sac.

Week 3: The Three-Layered Embryo

The Primitive Streak and the Three Germ Layers

During week 3, the embryo grows from a two-layered disc to a three-layered (trilaminar) disc. In so doing, it establishes the three *primary germ* layers—ectoderm, endoderm, and mesoderm—the layers from which all body tissues develop. Germ-layer formation starts on days 14–15, when a raised groove called the **primitive streak** appears on the dorsal surface of the epiblast (Figure 3.5). Epiblast cells move inward (invaginate) at this streak in a process called **gastrulation** (gas″troo-la′shun). On days 14–15 the first cells that migrate through the primitive streak replace the underlying hypoblast to become the **endoderm** (Figure 3.5g). Then, starting on day 16, the ingressing epiblast cells form a new layer between epiblast and endoderm, the **mesoderm** (Figure 3.5h). The epiblast cells that stay on the embryo's dorsal surface make up the **ectoderm.** In this way, the three primary germ layers of the body are established, all derived from epiblast cells.

The three germ layers differ in their tissue structure. (Recall from Chapter 1 that a tissue is a collection of similar cells occuring in a unique arrangement.) Ectoderm and endoderm are *epithelial tissues,* or *epithelia*—sheets of tightly joined cells (p. 4). Mesoderm, by contrast, is a *mesenchyme tissue* (mes′eng-kīm; *mesen-* = middle; *-chyme* = fluid). A mesenchyme is any embryonic tissue with star-shaped cells that do not attach to one another. Thus, mesenchyme cells are free to migrate widely within the embryo.

The Notochord

At one end of the primitive streak is a swelling called the **primitive node** (Figure 3.5f). The epiblast cells that invaginate at this node migrate straight anteriorly, and together with a few cells from the underlying endoderm, form a rod called the **notochord** (Figure 3.6a). The notochord defines the body *axis* (the midline that divides the left and right sides of the body) and is the site of the future vertebral column. The notochord appears on day 16, and by day 18 it reaches the future head region.

Neurulation

As the notochord develops, it signals the overlying ectoderm to start forming the spinal cord and brain (central column of Figure 3.7). This event is called **neurulation** (nu″roo-la′shun). Specifically, the ectoderm in the dorsal midline thickens into a **neural plate,** and then starts to fold inward as a **neural groove.** This groove deepens until a hollow **neural tube** pinches off into the body. Closure of the neural tube begins at the end of week 3 in the region that will become the neck, and then proceeds both cranially and caudally. Complete closure occurs by the end of week 4. The cranial part of this neural tube becomes the brain, and the rest becomes the spinal cord.

As shown in Figure 3.7, **neural crest** cells are pulled into the body along with the invaginating neural tube. The neural crest cells originate from ectodermal cells on the lateral ridges (neural folds) of the neural plate, and they come to lie just external to the closed neural tube. The neural crest forms the sensory nerve cells and some other important structures, as described on page 66.

The Mesoderm Begins to Differentiate

In the middle of week 3, the mesoderm lies lateral to the notochord on both sides of the body (see the right side of Figure 3.7) and extends from cranially to caudally (from head to tail, or crown to rump). By the end of this week, the mesoderm has divided into three regions.

1. **Somites** (so′mītz; "bodies"). The mesoderm closest to the notochord begins as **paraxial mesoderm** (par″ak′se-al; "near the body axis") (Figure 3.7b). Starting cranially and proceeding caudally, the paraxial mesoderm divides into a series of blocks called *somites* (Figure 3.7c). These raise the overlying ectoderm so they are visible in surface view as a row of bulges on each side of the back (see the left, dorsal views in Figure 3.7c and d). The somites are our first body segments, and about 40 pairs develop by the end of week 4.
2. **Intermediate mesoderm.** This begins as a continuous strip of tissue just lateral to the paraxial mesoderm, then divides into spherical segments in a cranial-to-caudal sequence (Figure 3.7c). Each segment of intermediate mesoderm attaches to a somite.
3. **Lateral plate.** This is the most lateral part of the mesoderm, and it never divides into segments (Figure 3.7c and d). The lateral plate begins as one layer, but soon divides to form a wedge-like space called the **coelom** (se′lum; "cavity"). The two resulting divisions of the lateral plate are the **somatic mesoderm** (so-mat′ik; "body"), apposed to the ectoderm, and the **splanchnic mesoderm** (splangk′nik; "viscera"), apposed to the endoderm. The coelom that intervenes between the splanchnic and somatic mesoderm will become the serous cavities of the ventral body cavity, namely the peritoneal, pericardial, and pleural cavities (Figure 1.9).

FIGURE 3.5 The primitive streak stage. The diagrams in parts **(a)** and **(b)** reorient the embryo from the previous figure, **(c)** shows this in three dimensions as resembling two balloons, and **(d)** removes the top of the amniotic sac so that one looks down on the epiblast of the embryonic disc **(e)**. The primitive streak appears on the epiblast on about day 15. Part **(f)** is the same view as (e), but rotated by 90 degrees. Parts (g) and (h) are sections through the embryonic disc of (f). As shown in **(g)**, the first epiblast cells that migrate medially into the primitive streak become endoderm. In **(h)**, a later stage, the invaginating epiblast cells form mesoderm. The epiblast on the surface is now called ectoderm.

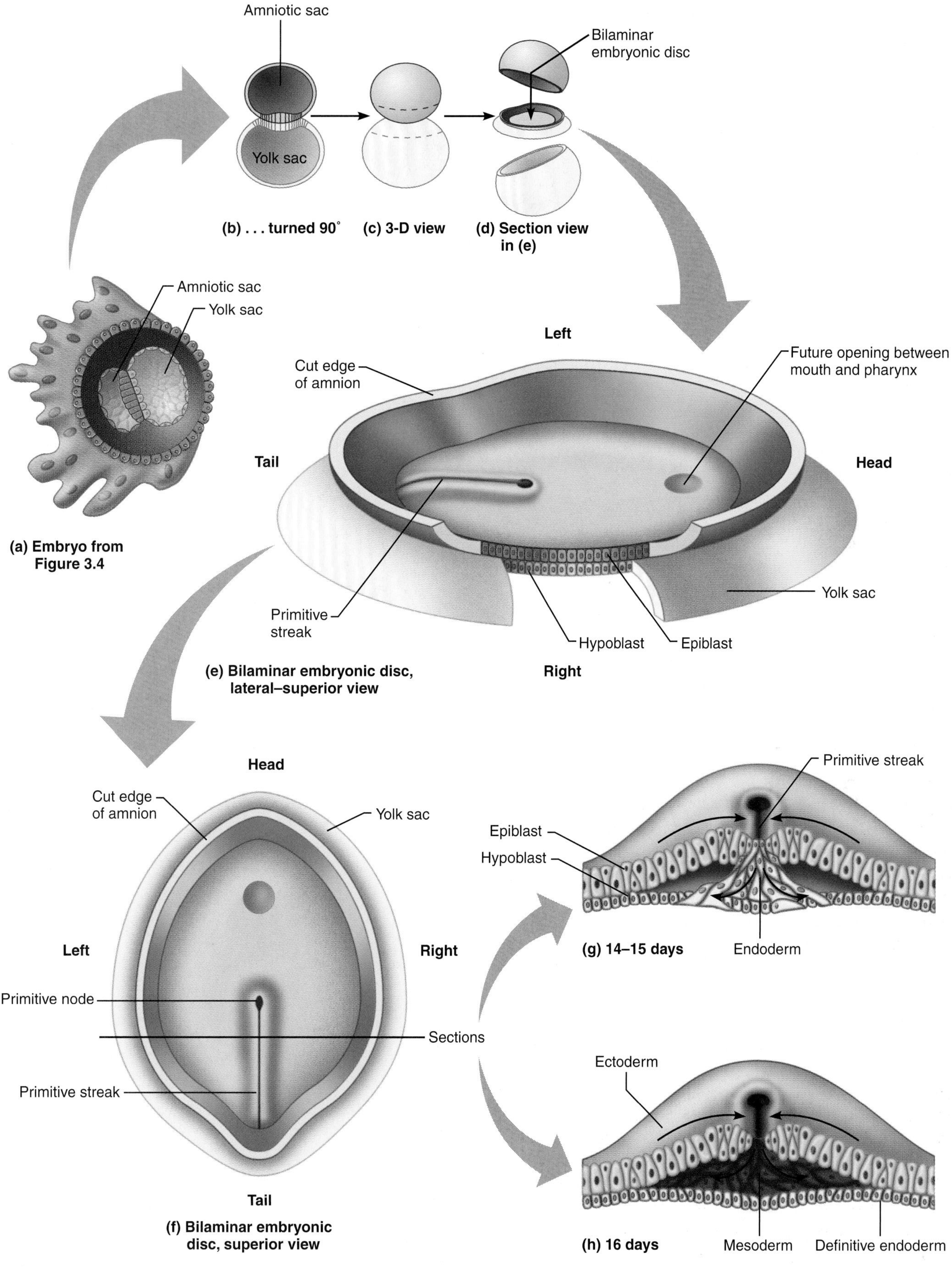
Amniotic sac
Yolk sac
Bilaminar embryonic disc
(b) . . . turned 90°
(c) 3-D view
(d) Section view in (e)
Amniotic sac
Yolk sac
(a) Embryo from Figure 3.4
Left
Cut edge of amnion
Future opening between mouth and pharynx
Tail
Head
Primitive streak
Yolk sac
Hypoblast
Epiblast
Right
(e) Bilaminar embryonic disc, lateral–superior view
Head
Cut edge of amnion
Yolk sac
Left
Right
Primitive node
Sections
Primitive streak
Tail
(f) Bilaminar embryonic disc, superior view
Primitive streak
Epiblast
Hypoblast
(g) 14–15 days
Endoderm
Ectoderm
(h) 16 days
Mesoderm
Definitive endoderm

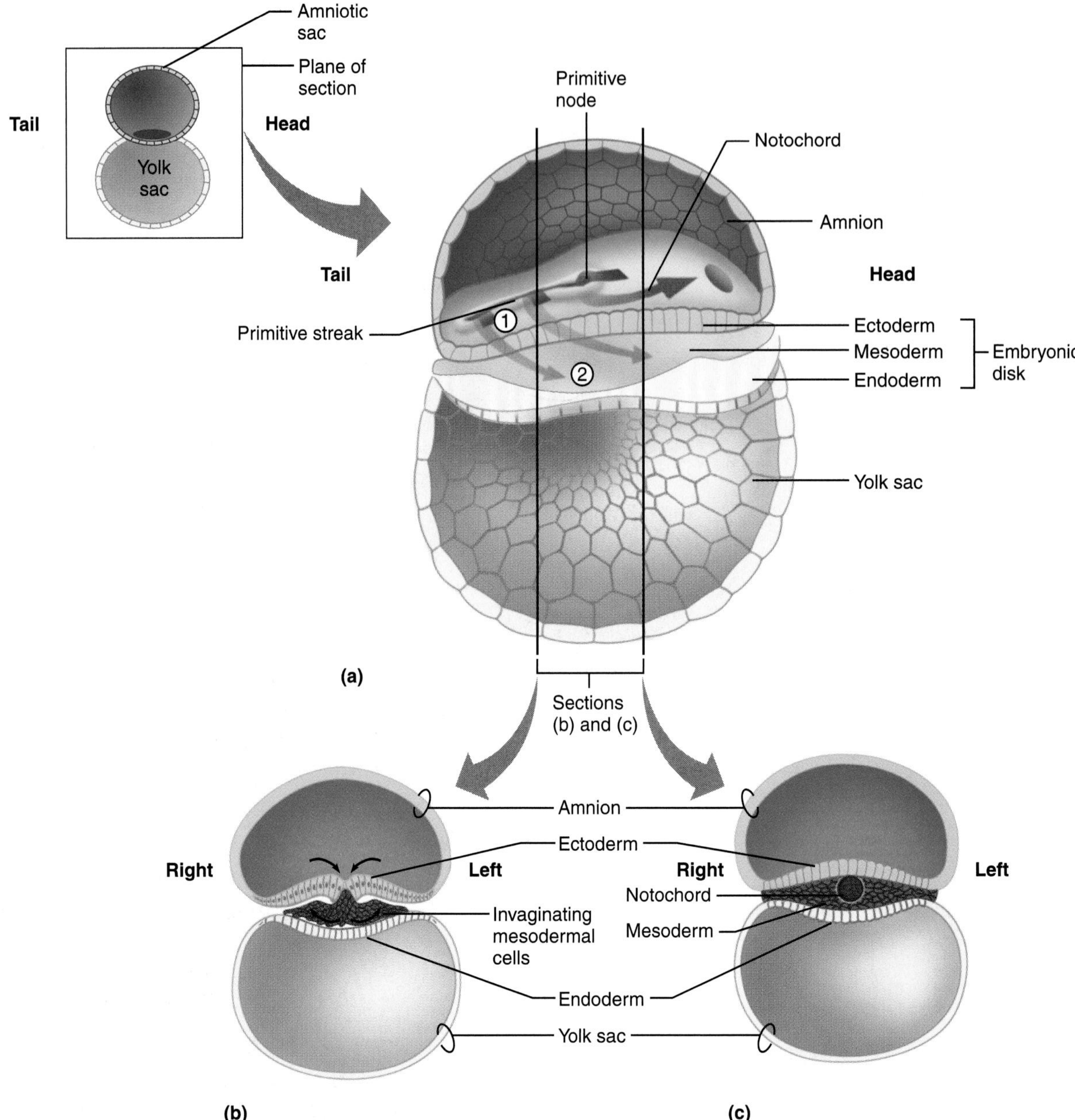

FIGURE 3.6 Formation of the mesoderm and the notochord in a 16-day embryo. **(a)** The three-layered embryonic disc sectioned in side view (see the inset above for orientation). Arrows show how ectoderm cells that migrate inward at the primitive streak (1) form the mesoderm layer (2), and how ectoderm cells that migrate inward at the primitive node form the notochord. **(b)** Cross-sectional view of the embryonic disc, through the primitive streak. **(c)** Cross section taken anterior to the primitive streak, in the future thorax. This is the same cross-sectional view as that of the adult body plan in Figure 3.2.

When you compare the cross sections in the right side of Figure 3.7 with the adult body plan in Figure 3.2, you might begin to see a few similarities, especially if you match structures according to their colors. For example, the blue ectoderm becomes the epidermis of the skin, the neural tube becomes the spinal cord, and the yellow endoderm becomes the lining of the digestive tube. The main difference between the 3-week embryo and the adult body is that the embryo is still a flat disc. The three-dimensional, cylindrical body shape forms during week 4.

FIGURE 3.7 Changes in the embryo near the end of week 3: neurulation and subdivision of the mesoderm. At left, four successive embryonic stages a–d are shown in dorsal view, oriented as in Figure 3.5f. The inset of the backside of the child in part (d) provides further orientation to these embryos. At center is a series of transverse sections showing the formation of the neural tube from ectoderm (neurulation). At right are transverse sections through the entire embryo, showing how the mesoderm forms somites, intermediate mesoderm, and somatic and splanchnic mesoderm.

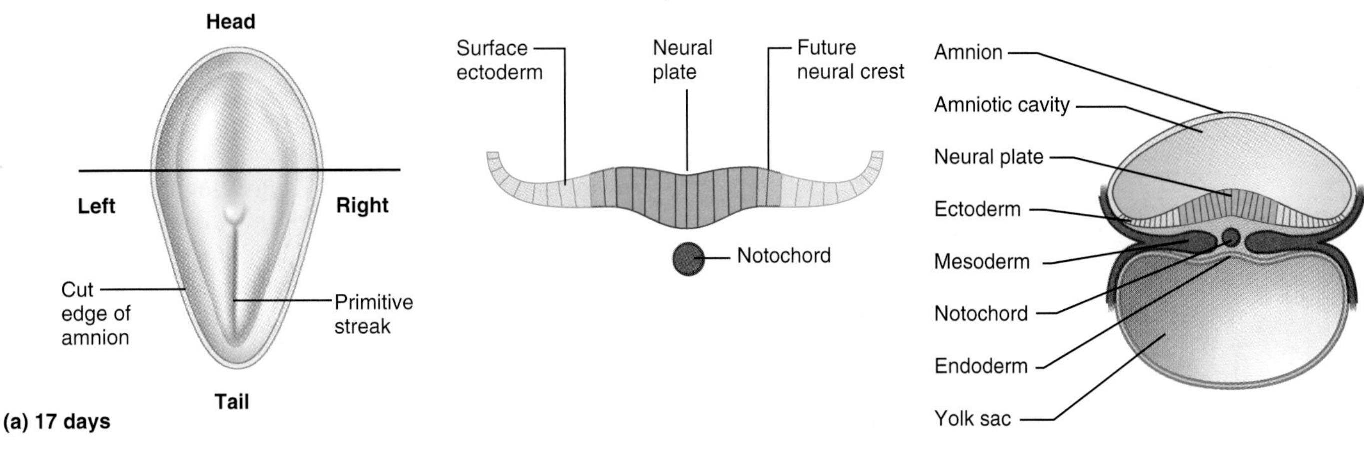
Head
Left
Right
Cut edge of amnion
Primitive streak
Tail
(a) 17 days
Surface ectoderm
Neural plate
Future neural crest
Notochord
Amnion
Amniotic cavity
Neural plate
Ectoderm
Mesoderm
Notochord
Endoderm
Yolk sac

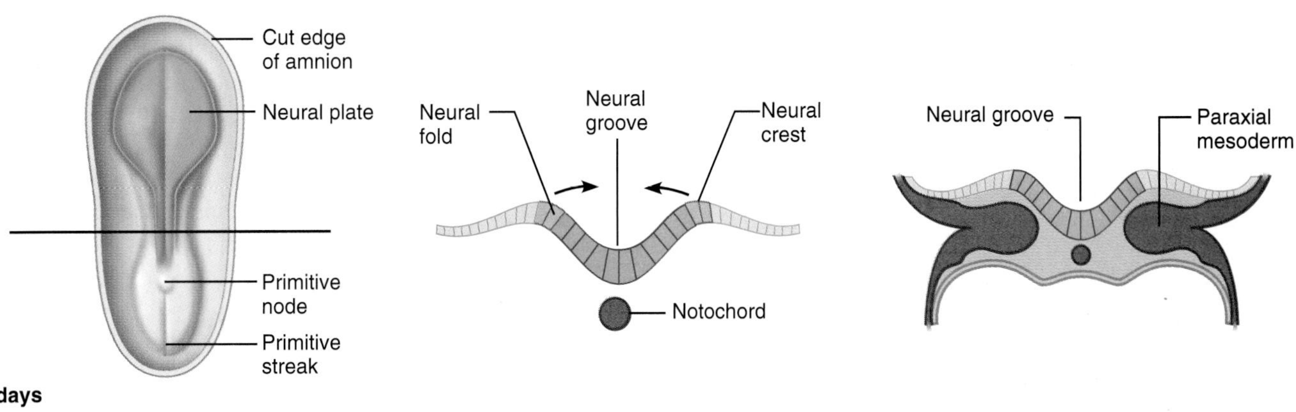
Cut edge of amnion
Neural plate
Primitive node
Primitive streak
(b) 19 days
Neural fold
Neural groove
Neural crest
Notochord
Neural groove
Paraxial mesoderm

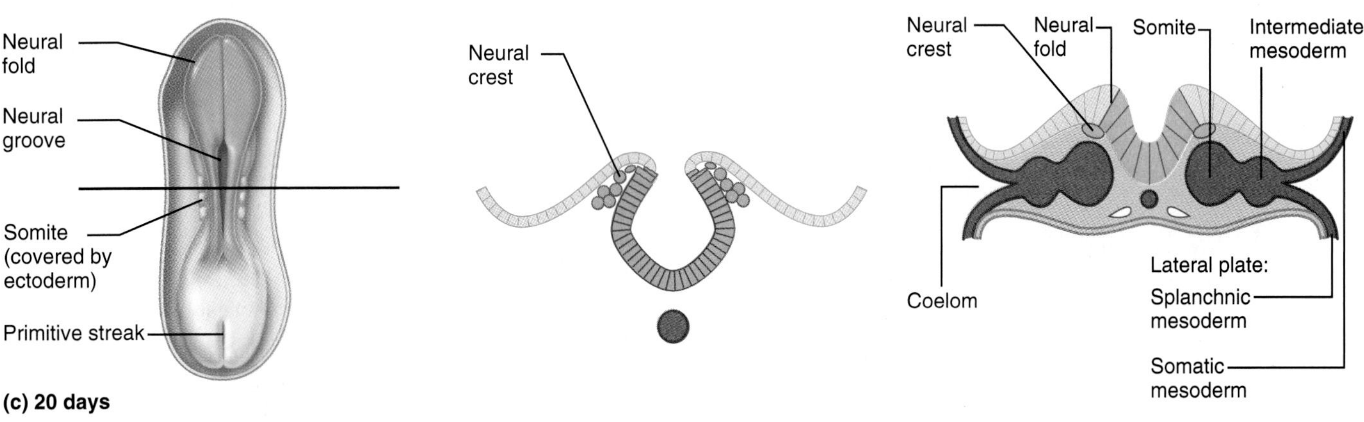
Neural fold
Neural groove
Somite (covered by ectoderm)
Primitive streak
(c) 20 days
Neural crest
Neural crest
Neural fold
Somite
Intermediate mesoderm
Coelom
Lateral plate:
Splanchnic mesoderm
Somatic mesoderm

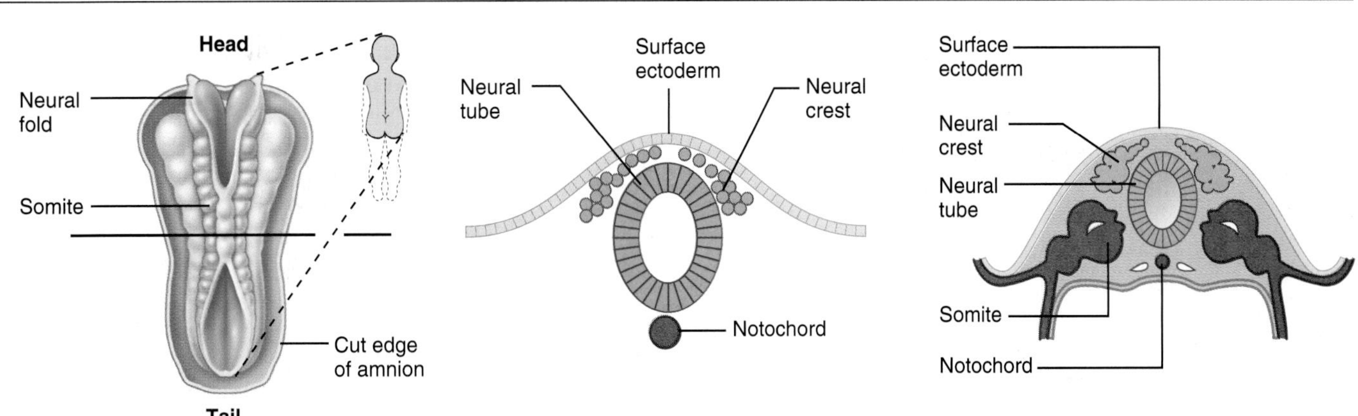
Head
Neural fold
Somite
Cut edge of amnion
Tail
(d) 22 days
Neural tube
Surface ectoderm
Neural crest
Notochord
Surface ectoderm
Neural crest
Neural tube
Somite
Notochord

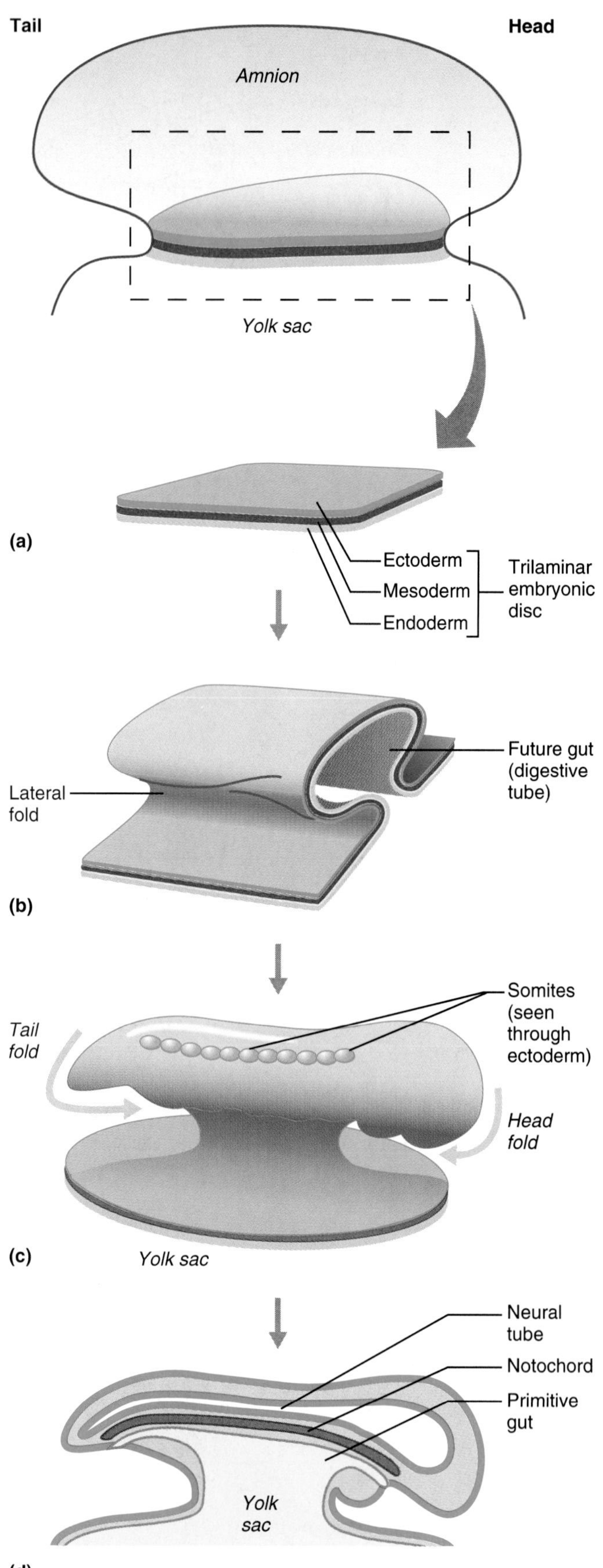

FIGURE 3.8 Simplified diagrams showing the flat embryo folding into a tadpole shape in week 4. These are lateral views, with the head to the right. **(a)** Model of the flat, three-layered embryo as three sheets of paper. **(b)** The sheets fold laterally, and **(c)** at the head and tail. **(d)** A 24-day embryo in sagittal section. Note the primitive gut, which derives from the yolk sac. Dorsal to the gut are the notochord and neural tube.

Week 4: The Body Takes Shape

Folding

The embryo achieves a cylindrical body shape when its sides fold medially and it lifts off the yolk sac and protrudes into the amniotic cavity (Figure 3.8). In the simplest sense, this resembles three stacked sheets of paper folding into a tube. At the same time, the head and tail regions fold under, as shown in Figure 3.8c. The reason the embryonic disc bulges upward is that it is growing so much faster than the yolk sac below it. Its lateral folding is caused by the fast growth of the somites, whereas the folding at the head and tail is caused by expansion of the brain and lengthening of the spinal cord.

As a result of this folding, the embryo acquires a tadpole shape by day 24. As the embryo becomes cylindrical, it encloses a tubular part of the yolk sac, called the **primitive gut** (future digestive tube: Figure 3.8d). This tube is lined by endoderm. The embryo remains attached to the yolk sac below by a duct located at the future navel. This duct becomes incorporated into the umbilical cord.

Derivatives of the Germ Layers

By day 28, the basic human body plan has been attained (Figure 3.9c). Figure 3.9c is a cross section through the trunk of a 1-month-old embryo that can be compared directly to the adult body section in Figure 3.2. We can use such a comparison to explain the adult derivatives of the germ layers. These derivatives are summarized in Table 3.1.

Derivatives of Ectoderm The ectoderm becomes the brain, spinal cord, and epidermis of the skin. The early epidermis, in turn, produces the hair, fingernails, toenails, sweat glands, and the skin's oil glands. Neural crest cells, from ectoderm, give rise to the sensory nerve cells. Furthermore, much of the neural crest breaks up into a mesenchyme tissue, which wanders widely through the embryonic body. These wandering neural-crest derivatives produce such varied structures as the pigment cells and the bones of the face.

Derivatives of Endoderm The endoderm becomes the inner epithelial lining of the digestive tube and its derivatives, such as the respiratory tubes and bladder. It also gives rise to the secretory cells of glands that develop from gut-lining epithelium, such as the liver and pancreas.

Derivatives of the Mesoderm and Notochord The mesoderm has a complex fate, so you may wish to start by reviewing its basic parts in Figure 3.9: the somites, intermediate mesoderm, and somatic and splanchnic mesoderm. Also note the location of the notochord.

The *notochord* gives rise to an important part of our spinal column, the springy cores of the discs between our vertebrae. These spherical centers, each called a *nucleus pulposus* (pul-po′sus), give our vertebral column some bounce as we walk.

Each of the *somites* divides into three parts. One part is the **sclerotome** (skle′ro-tōm; "hard piece"). Its cells migrate medially, gather around the notochord and the neural tube, and produce the vertebra and rib at the associated level. The most lateral part of each somite is a **dermatome** ("skin piece"). Its cells migrate externally until they lie directly deep to the ectoderm, where they form the dermis of the skin in the dorsal part of the body. The third part of each somite is the **myotome** (mi′o-tōm; "muscle piece"), which stays behind after the sclerotome and dermatome migrate away. Each myotome grows ventrally until it extends the entire dorsal-to-ventral height of the trunk. Myotomes become the segmented trunk musculature of the body wall (Figures 1.7b and 3.10). Additionally, the ventral parts of myotomes grow into the limb buds (see p. 68) and form the muscles of the limbs.

The *intermediate mesoderm,* lateral to each somite, forms the kidneys and the gonads. By comparing Figures 3.9 and 3.2, you can see that the intermediate mesoderm lies in the same relative location as the adult kidneys.

We now turn to the splanchnic and somatic mesoderm, which are separated by the coelom body cavity. By now, the *splanchnic mesoderm* surrounds the gut (Figure 3.9c). It gives rise to the entire wall of the digestive tube, except the inner epithelial lining; that is, it forms the musculature and the slippery visceral serosa of the digestive tube. Splanchnic mesoderm also gives rise to the heart and most blood vessels.

Somatic mesoderm, just external to the coelom, produces the parietal serosa and the dermis of the skin in the ventral body region. Its cells migrate into the forming limbs and produce the bone, ligaments, and dermis of each limb.

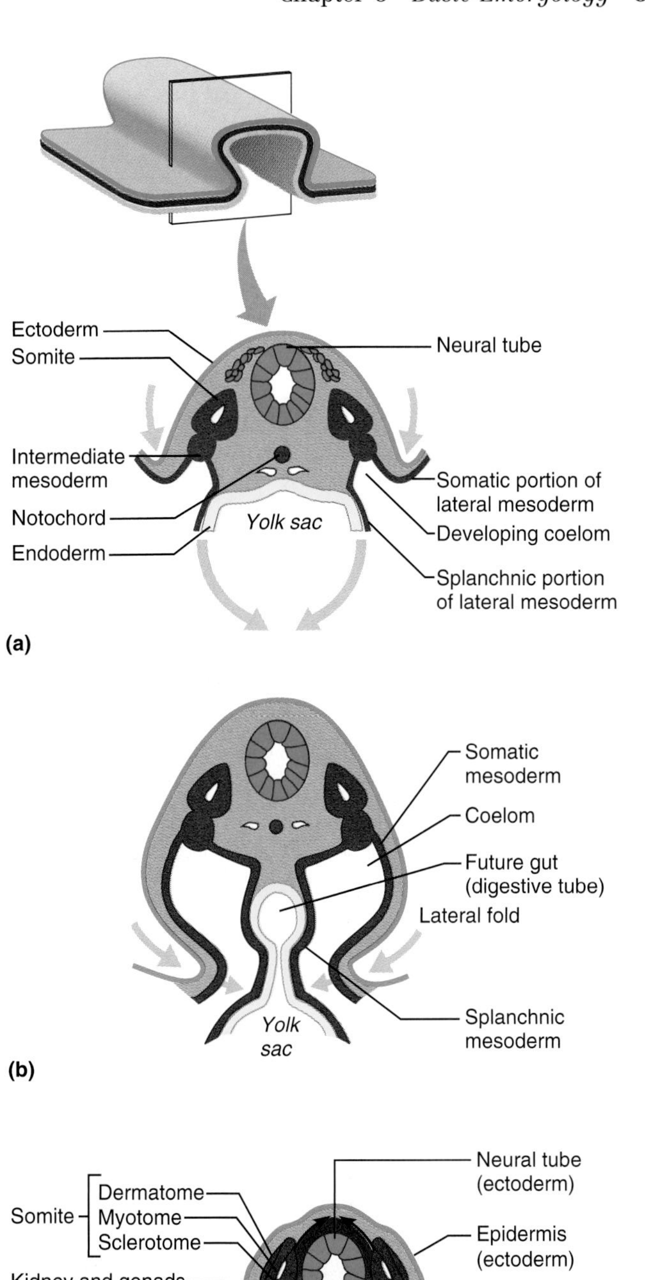

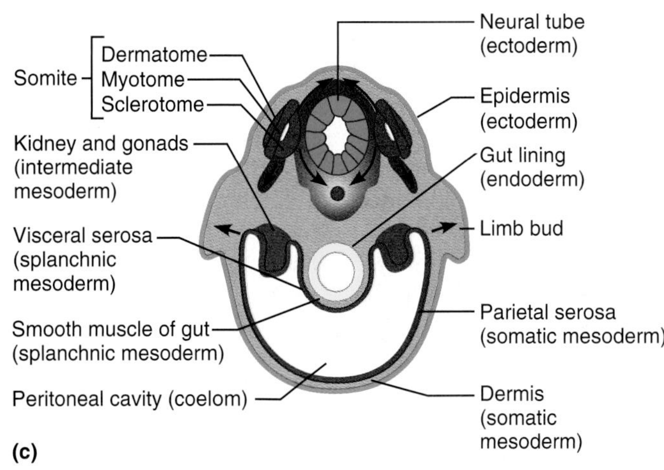

FIGURE 3.9 The germ layers in week 4 and their adult derivatives. These cross-sectional views should be compared to Figure 3.2. **(a)** The embryo on day 23, just as it begins to fold into a tube. The arrows show the direction that folding will take. **(b)** Folding continues, as the embryo forms into a cylinder and lifts up off the yolk sac. The right and left lateral folds will join ventrally. **(c)** The cylindrical human body plan on day 28. This cross section is taken from the trunk region, where the posterior limb buds (future legs) attach.

Table 3.1 Major Derivatives of the Embryonic Germ Layers

ECTODERM

- Epidermis, hair, nails, glands of skin
- Brain and spinal cord
- Neural crest: sensory nerve cells and some nervous structures; pigment cells; portions of skeleton; blood vessels in head and neck

ENDODERM

- Inner lining and glands of digestive tube and its derivatives

MESODERM AND NOTOCHORD

- Mesoderm

  Somite

    Sclerotome: vertebrae and ribs

    Dermatome: dermis of dorsal body region

    Myotome: trunk and limb musculature

  Intermediate mesoderm: kidneys and gonads

  Lateral plate

    Splanchnic mesoderm: wall of digestive tube (except epithelial lining); visceral serosa; heart; blood vessels

    Somatic mesoderm: parietal serosa; dermis of ventral body region; connective tissues of limbs (limb bones, ligaments, joints)

- Notochord: nucleus pulposus of intervertebral discs

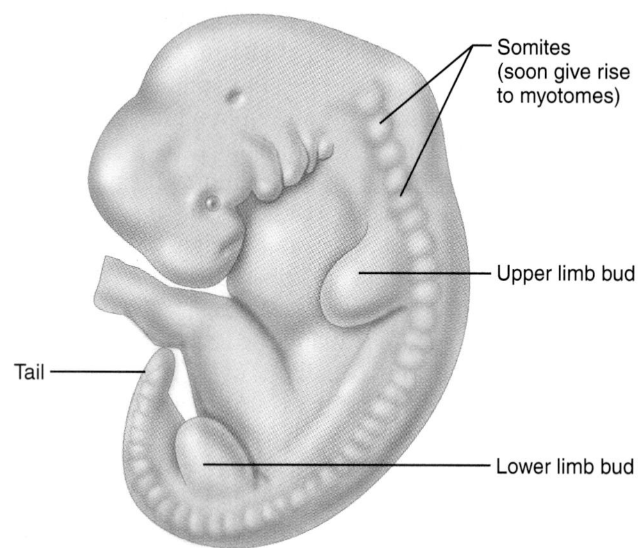

FIGURE 3.10 A 29-day embryo. Note the limb buds. Also note the segmented somites, which soon give rise to segmented myotomes.

Weeks 5–8: The Second Month of Embryonic Development

Figure 3.10 shows an embryo at the start of the second month. At this time, the embryo is only about a half-centimeter long. Around day 28, the first rudiments of the limbs appear as **limb buds.** The upper limb buds appear slightly earlier than the lower limb buds.

One can think of month 2 as the time when the body changes from tadpole-like to recognizably human in form (see Figure 3.1). The limbs grow from rudimentary buds to fully formed extremities with fingers and toes. The head enlarges quickly and occupies almost half the volume of the entire body. The eyes, ears, and nose appear, and the face gains a human appearance as the embryonic period draws to a close. The protruding tail of the 1-month-old embryo disappears at the end of week 8. All major organs are in place by the end of month 2, at least in rudimentary form. In successive chapters we discuss the development of each organ system, so we often will return to the events of month 2.

THE FETAL PERIOD

The main events of the fetal period—weeks 9 through 38—are listed chronologically in Table 3.2. As we stated earlier, the fetal period is a time of maturation and rapid growth of the body structures that were established in the embryo. The fetal period is more than just a time of growth, however. During the first half of this period, cells are still differentiating into specific cell types to form the body's distinctive tissues and are completing the fine details of body structure. The fetus at 3 months and 5 months is presented in Figure 3.11 on page 71.

Normal births typically occur 38 weeks after conception. A **premature birth** is one that occurs before that time. Infants born as early as week 30 (7 months) usually survive without requiring life-saving measures. Medical technology can save many fetuses born prematurely, sometimes as early as 22 weeks (in month 6). However, the lungs of premature infants are not fully functional, and death may result from respiratory distress (see p. 623 for more details). Among those that survive, rates of visual problems and mental retardation are elevated.

Table 3.2 Developmental Events of the Fetal Period

Time after fertilization	Events
8 weeks (end of embryonic period) 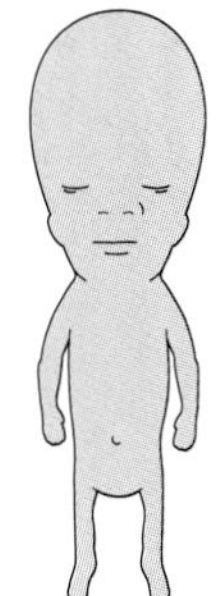 8 weeks	Eyes and ears first look human; neck region first becomes evident Head is nearly as large as rest of body; all major divisions of the brain are present; first brain waves occur in brain stem Liver is disproportionately large Bone formation has just begun; weak muscle contractions occur Limbs are complete; digits are initially webbed, but fingers and toes are free by end of week 8 The cardiovascular system is fully functional; the heart has been pumping blood since week 4 Approximate crown-to-rump length: 3 cm (1.2 inches); weight: 2 g (0.06 ounce)
9–12 weeks (month 3) 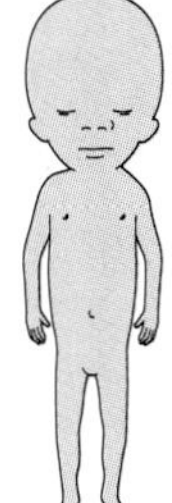 12 weeks	Head is still dominant, but body is elongating; brain continues to enlarge; retina of eye is present Skin epidermis and dermis are obvious Liver is still prominent; right and left halves of palate (roof of mouth) are fusing; walls of hollow visceral organs are gaining smooth muscle Blood cell formation begins in bone marrow (also occurs in liver and spleen) Notochord is degenerating and bone formation is accelerating Sex can be determined from the genitals Approximate crown-to-rump length at end of 12 weeks: 9 cm
13–16 weeks (month 4) 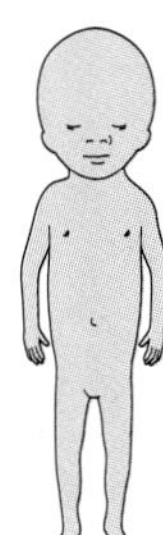 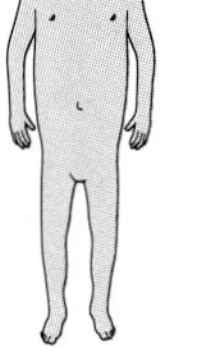 16 weeks	Sucking motions of the lips occur, and eyes can flinch if stimulated (but eyes are closed, not open) Body begins to outgrow head; limbs are no longer so disproportionately small Hard palate is fused Kidneys attain their typical structure Most bones are now distinct, and joint cavities are apparent Approximate crown-to-rump length at end of 16 weeks: 14 cm
17–20 weeks (month 5)	Eyelashes and eyebrows are present; a fatty skin secretion covers the body; silk-like hair, called lanugo, covers the skin Quickening occurs (mother feels fetus moving) Body first bends forward into the fetal position because of space restrictions in the uterus Limbs achieve near-final proportions Approximate crown-to-rump length at end of 20 weeks: 19 cm 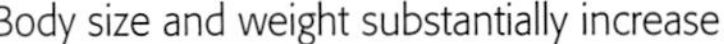
21–30 weeks (months 6 and 7)	Body size and weight substantially increase Eyes open Fingernails and toenails are complete Skin is wrinkled and red; the fatty layer under the skin (hypodermis) is just starting to gain fat, so body is lean Bone marrow becomes the only site of blood cell formation Testes reach the scrotum in month 7 (males) Approximate crown-to-rump length at end of 30 weeks: 28 cm
30–38 weeks (months 8 and 9) At birth 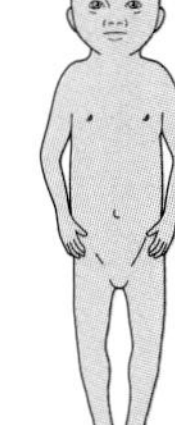	Fat accumulates in hypodermis below skin Approximate crown-to-rump length at end of 38 weeks: 36 cm; weight 2.7–4.1 kg (6–10 pounds)

A Closer Look

Focus on Birth Defects

About 3% of all newborns exhibit some significant birth defect or **congenital abnormality** (*congenital* = present at birth). This figure increases to 6% by 1 year of age, when initially unrecognized defects become evident. The 12 most common and serious defects (see the related list) alone affect almost 3 in 100 children. We will consider some of these birth defects in later chapters. Here, we provide general information on congenital abnormalities and explore their causes.

Birth defects have many causes, including defective chromosomes or genes, exposure to chemicals and harmful radiation, or viruses passed from the mother to the embryo. Chemical, physical, and biological factors that disrupt development and cause birth defects are called **teratogens** (ter'ah-to-jenz; "monster formers"). Teratogens include many drugs, some organic solvents and pollutants, toxic metals, various forms of radiation such as X rays, and infectious agents, particularly German measles. Although many teratogens have been identified, it must be emphasized that most incidences of birth defects cannot be traced to any known teratogen or genetic disorder. That is, for any infant with a birth defect, it is usually impossible to determine what caused the anomaly. Most birth defects are caused by many factors.

When a pregnant woman drinks alcohol, she increases the likelihood of birth defects. Excessive alcohol consumption causes **fetal alcohol syndrome (FAS),** typified by microcephaly (small head and jaw), growth retardation, and distinctive facial features (see the photo on p. 71). FAS is the most common preventable cause of mental retardation in the United States. Among the children of women who drink heavily, the incidence of severe FAS is 6% and a milder form called *fetal alcohol effect* is even more common. There is now evidence that even moderate drinking, one to three drinks per day, adversely affects the development of the brain, lessening the child's ability to learn and remember. Therefore, physicians advise pregnant women not to drink any alcohol at all.

Some prescription drugs can cause catastrophic birth defects. Thalidomide, for example, was used as a sedative and to relieve morning sickness by many pregnant women in Europe from about 1958 to 1962. When taken during the period of limb differentiation (days 24–36), it sometimes results in tragically deformed infants with short, flipper-like limbs. Similarly, a drug called diethylstilbestrol (DES) was commonly given to pregnant women from the 1940s to 1970, because it was wrongly thought to prevent miscarriages. Instead, it caused cancer of the vagina and uterus in daughters exposed to DES in the womb.

An expectant mother must be particularly careful of her health and chemical intake during the third to eighth weeks after conception. It is easy to understand why weeks 3–8 are the "danger period": Insults delivered *before* week 3 tend to destroy critical early cell lines so that development cannot proceed, and a natural abortion occurs. After week 8, almost all organs are already in place, so teratogens may produce only minor malformations. However, this does not imply that the subsequent fetal period is risk-free; exposure of the fetus to teratogens can still affect body growth and organ

Most Common Birth Defects*

Birth defect	Number per 1000 births
1. Heart defects	7
2. Mental retardation	3
3. Pyloric stenosis (a valve in the stomach is too narrow)	3
4. Anencephaly (most of the brain is absent)	2
5. Spina bifida (vertebrae are incomplete, leading to a damaged spinal cord)	1.5
6. Down syndrome† (child has a distinct facial appearance and mild mental retardation)	1.5
7. Cleft palate and cleft lip	1.5
8. Hypospadias (urine leaks from a slit in underside of penis)	>1.5
9. Clubfoot (feet turn inward)	1
10. Congenital dislocation of hip	1
11. Congenital deafness	0.7
12. Cystic fibrosis† (thick mucus and other factors cause severe health problems)	0.5

*SOURCE: Data from Clayman, C.B. (editor): *American Medical Association Encyclopedia of Medicine.* New York: Random House, 1989.

†These two abnormalities result from well-understood chromosomal and genetic defects. The other birth defects on the list have many different causes, which may include exposure to teratogens.

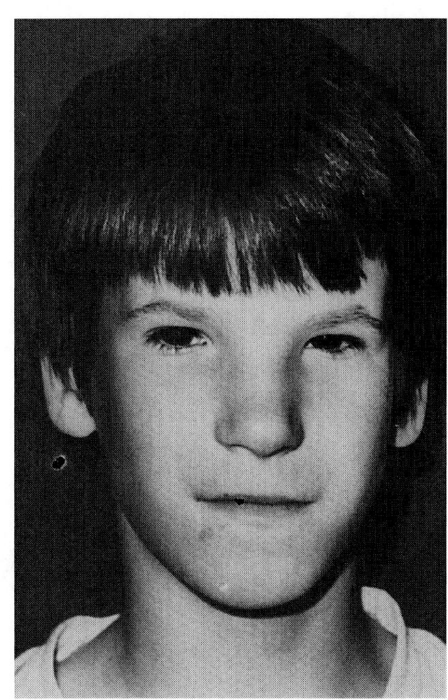

A child with fetal alcohol syndrome. Typical features are unusually shaped eyes and lips (including a thin upper lip), short nose, receding chin, defects in vision and hearing, and possible heart defects. Behavioral traits include mental retardation, difficulty in concentrating, impulsiveness, and speech problems.

functions, produce abnormalities of the eyes, and lead to mental retardation.

Surprisingly, most human embryos die in the womb. About 15 to 30% of all fertilized zygotes never implant in the uterus, either because they have lethal chromosomal defects or the uterine wall is unprepared to receive them. Of the embryos that do implant, about one-third are so malformed that they die, most often during week 2 or 3. Smaller numbers are miscarried later in development. Overall, at least half of all fertilized ova fail to produce living infants—some studies say 75%. We fail to realize how many embryos die because those lost in the first 2 weeks are shed without notice in the next menstrual flow. Perhaps this great loss of embryos is a biological screening process that promotes the birth of healthy infants.

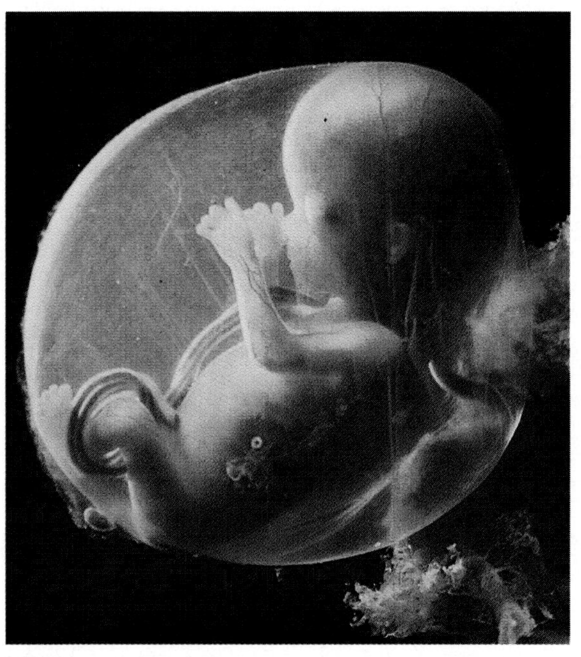

(a)

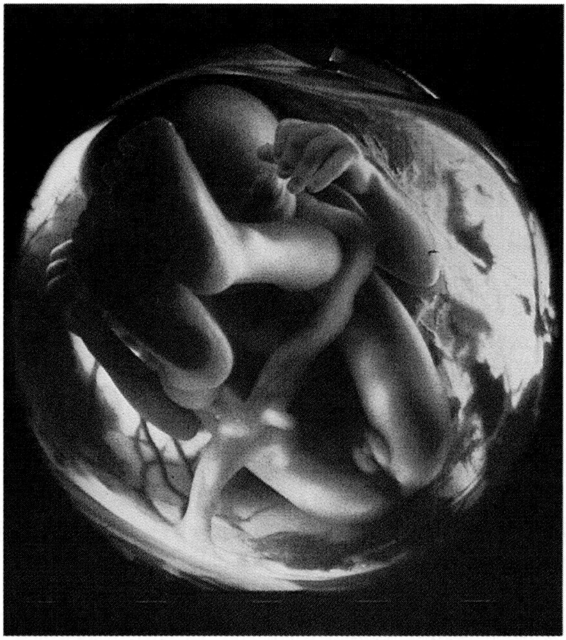

(b)

FIGURE 3.11 Photographs of a developing fetus. The major events of fetal development are growth and tissue specialization. **(a)** Fetus in month 3, about 6 cm long. **(b)** Fetus late in month 5, about 19 cm long. By birth, the fetus is typically 35 cm long from crown to rump.

RELATED CLINICAL TERMS

Abortion (*abort* = born prematurely) Premature removal of the embryo or fetus from the uterus. Abortions can be spontaneous (resulting from fetal death) or induced.

Amniocentesis (am″ne-o-sen-te′sis; "puncture of the amnion") In this procedure, a syringe needle is inserted through the abdominal and uterine walls of a pregnant woman, and a small amount of amniotic fluid is obtained. Shed cells from the fetus can be studied for chromosomal and genetic abnormalities, and chemicals in the fluid can reveal other disorders. Conditions such as Down syndrome and spina bifida can be diagnosed. The procedure can be performed as early as the eleventh week of pregnancy.

Conjoined (Siamese) twins Identical twins that are born joined together. This phenomenon is caused by incomplete division of the inner cell mass or embryonic disc during the twinning process. The twins may be joined at any body region and often share organs. Some can be separated successfully by surgery.

Ectopic pregnancy (ek-top′ik) (*ecto* = outside) A pregnancy in which the embryo implants in any site other than in the uterus. Ectopic pregnancy can occur in the body wall or on the outside of the uterus, but it usually occurs in a uterine tube, a condition known as a tubal pregnancy. Because the uterine tube is unable to establish a placenta or accommodate growth, it ruptures unless the condition is diagnosed early. Once rupture occurs, internal bleeding may threaten the woman's life. For obvious reasons, ectopic pregnancies almost never result in live births.

Teratology (ter″ah-tol′o-je) (*terato* = a monster; *logos* = study of) The scientific study of birth defects and of fetuses with congenital deformities.

CHAPTER SUMMARY

Media study tools that could provide you additional help in reviewing specific key topics of Chapter 3 are referenced below.

SP = Study Partner; AIA = A.D.A.M.® Interactive Anatomy

1. The study of the 38-week prenatal period is called embryology. Embryology helps one understand the basic organization of the human body, as well as the origin of birth defects (such congenital abnormalities are considered in the box on pp. 70–71).

THE BASIC BODY PLAN (p. 59)

1. The embryonic period, during which the basic body plan is established, is the first 8 weeks of prenatal development. The fetal period, primarily a time of growth and maturation, is the remaining 30 weeks.

2. Basic features of the adult body that are established in the embryonic period include the skin, trunk muscles, vertebral column, spinal cord and brain, digestive tube, peritoneal cavity, heart, kidneys, gonads, and limbs (see Figure 3.2).

THE EMBRYONIC PERIOD (pp. 59–68)

Week 1: From Zygote to Blastocyst (pp. 59–61)

1. After fertilization occurs in the uterine tube, the zygote undergoes cleavage (early cell division without growth) to produce a solid ball of cells called the morula. Around day 4, the morula enters the uterus.

2. The morula gains an internal blastocyst cavity and becomes a blastocyst, which consists of an inner cell mass and a trophoblast. The inner cell mass becomes the embryo, and the trophoblast forms part of the placenta. The blastocyst implants in the uterine wall from about day 6 to day 13.

SP Exercise: Chapter 3, Structure of the Embryo.

Week 2: The Two-Layered Embryo (p. 61)

3. By day 7, the inner cell mass has formed two touching sheets of cells, the epiblast and hypoblast. These layers form the bilaminar embryonic disc.

4. The amniotic sac, whose wall is an extension of the epiblast, contains amniotic fluid that buffers the growing embryo from physical trauma. The yolk sac, whose wall is an extension of the hypoblast, later contributes to the digestive tube.

Week 3: The Three-Layered Embryo (pp. 62–65)

5. At the start of week 3, gastrulation occurs. The primitive streak appears on the posterior part of the epiblast layer. Epiblast cells invaginate here to form the mesoderm, which fills the space between the other two layers. The embryonic disc now has three layers: ectoderm, mesoderm, and endoderm, all of which are derived from epiblast.

6. Ectoderm and endoderm are epithelial tissues (sheets of closely attached cells). Mesoderm, by contrast, is a mesenchyme tissue (embryonic tissue with star-shaped, migrating cells that remain unattached to one another).

7. The primitive node, at the anterior end of the primitive streak, is the site where the notochord originates. The notochord defines the body axis and later forms each nucleus pulposus of the vertebral column.

8. The notochord signals the overlying ectoderm to fold into a neural tube, the future spinal cord and brain. This infolding carries other ectoderm cells, neural crest cells, into the body.

SP Exercise: Chapter 3, Derivation of Primary Layers.

9. By the end of week 3, the mesoderm on both sides of the notochord has condensed into three parts. These are, from medial to lateral, somites (segmented blocks that derive from nonsegmented paraxial mesoderm), intermediate mesoderm (segmented balls), and the lateral plate (an unsegmented layer). The coelom divides the lateral plate into somatic and splanchnic mesoderm layers.

Week 4: The Body Takes Shape (pp. 66–68)

10. The flat embryo folds inward at its sides and at its head and tail, taking on a tadpole shape. As this occurs, a tubular part of the yolk sac becomes enclosed in the body as the primitive gut.

11. The basic body plan is attained as the body achieves this tadpole shape. This development can be demonstrated by

comparing Figures 3.2 and 3.9. The major adult structures derived from ectoderm, endoderm, and mesoderm are summarized in Table 3.1.

12. The limb buds appear at the end of week 4.

Weeks 5–8: The Second Month of Embryonic Development (p. 68)

13. All major organs are present by the end of month 2. During this month, the embryo changes from tadpole-like to human-like as the eyes, ears, nose, and limbs form and take on their human shape.

THE FETAL PERIOD (pp. 68–71)

1. Most cell and tissue differentiation is completed in the first months of the fetal period. The fetal period is a time of growth and maturation of the body.

2. The main events of the fetal period are summarized in Table 3.2.

REVIEW QUESTIONS

Multiple Choice/Matching Questions

1. Indicate whether each of the following describes (a) cleavage or (b) events at the primitive streak.

____ **(1)** period when a morula forms

____ **(2)** period when the notochord forms

____ **(3)** period when the three embryonic germ layers appear

____ **(4)** period when individual cells become markedly smaller

2. The outer layer of the blastocyst, which attaches to the uterus wall, is the (a) yolk sac, (b) trophoblast, (c) amnion, (d) inner cell mass.

3. Most birth defects can be traced to disruption of the developmental events during this part of the prenatal period: (a) first 2 weeks, (b) second half of month 1 and all of month 2, (c) month 3, (d) end of month 4, (e) months 8 and 9.

4. The primary germ layer that forms the trunk muscles, heart, and skeleton is (a) ectoderm, (b) endoderm, (c) mesoderm.

5. Match each embryonic structure in column A with its adult derivative in column B.

Column A	Column B
____ **(1)** notochord	**(a)** kidney
____ **(2)** ectoderm (not neural tube)	**(b)** peritoneal cavity
____ **(3)** intermediate mesoderm	**(c)** pancreas secretory cells
____ **(4)** splanchnic mesoderm	**(d)** parietal serosa
____ **(5)** sclerotome	**(e)** nucleus pulposus
____ **(6)** coelom	**(f)** visceral serosa
____ **(7)** neural tube	**(g)** hair and epidermis
____ **(8)** somatic mesoderm	**(h)** brain
____ **(9)** endoderm	**(i)** ribs

6. Match each date in column A (approximate time after conception) with a developmental event or stage in column B.

Column A	Column B
____ **(1)** 38 weeks	**(a)** blastocyst
____ **(2)** 15 days	**(b)** implantation
____ **(3)** 4–7 days	**(c)** first few somites
____ **(4)** about 21 days	**(d)** flat embryo becomes cylindrical
____ **(5)** about 21–24 days	**(e)** embryo/fetus boundary
____ **(6)** about 60 days	**(f)** birth
____ **(7)** days 6–13	**(g)** primitive streak appears

7. The youngest premature babies that can currently be saved by modern technology are born in week ____ of development. (a) 8, (b) 16, (c) 22, (d) 38, (e) 2.

8. Somites are evidence of (a) a structure from ectoderm, (b) a structure from endoderm, (c) a structure from lateral plate mesoderm, (d) segmentation in the human body.

9. Which of the following embryonic structures are segmented? (More than one is correct.) (a) intermediate mesoderm, (b) lateral plate, (c) endoderm, (d) myotomes.

10. Deduce what the primitive streak becomes in the adult body (hint: see Figure 3.7c and d): (a) the navel, (b) the mouth opening, (c) skin epidermis over the tailbone, (d) the cleft between the index finger and thumb.

11. Endoderm gives rise to (a) the entire digestive tube, (b) the skin, (c) the neural tube, (d) the kidney, (e) none of the above.

12. When it is 1.5 months old, an average embryo is the size of (a) Lincoln on a penny, (b) an adult fist, (c) its mother's nose, (d) the head of a common pin.

13. Gastrulation is the (a) formation of three germ layers by epiblast at the primitive streak, (b) formation of the placenta, (c) same as cleavage, (d) folding of the gut into a tube during week 4.

Short Answer Essay Questions

14. What important event occurs at the primitive streak? At the primitive node?

15. What is the function of the amniotic sac and the fluid it contains?

16. (a) What is mesenchyme? (b) How does it differ from an epithelium?

17. Explain how the flat embryonic disc takes on the cylindrical shape of a tadpole.

18. In anatomy lab, Thaya pointed to the vertebrae of a cadaver and said "sclerotome." He pointed to a kidney and said, "intermediate mesoderm," to the biceps muscle in the arm and said, "splanchnic mesoderm," to the inner lining of the stomach and said, "endoderm," and to the brain and said, "ectoderm." Point out Thaya's *one* mistake.

19. Recently, the neural crest has aroused much scientific interest because it is the one embryonic tissue that is strictly unique to vertebrate animals. List some adult structures that develop from the neural crest.

20. Many embryonic events first occur cranially then proceed caudally. Give three examples of structures that develop cranio-caudally like this.

CRITICAL REASONING + CLINICAL APPLICATION QUESTIONS

1. A friend in your dormitory, a freshman, tells you she just discovered she is 3 months pregnant. She recently bragged that since she came to college she has been drinking alcohol heavily and experimenting with every kind of recreational drug she can find. Circle the best advice you could give her, and explain your choice. (a) She must stop taking the drugs, but they could not have affected her fetus during these first few months of her pregnancy. (b) Harmful substances usually cannot pass from mother to embryo, so she can keep using drugs. (c) There could be defects in the fetus, so she should stop using drugs and visit a doctor as soon as possible. (d) If she has not taken any drugs in the last week, she is okay.

2. The citizens of Nukeville brought a class-action lawsuit against the local power company, which ran a nuclear power plant on the outskirts of town. The people claimed that radiation leaks were causing birth defects. As evidence, they showed that of the 998 infants born in town since the plant opened, 20 had congenital abnormalities. Of these 20 infants, 6 had heart defects, 3 were mentally retarded, 2 had pyloric stenosis, 2 were born without a brain, 2 had spina bifida, 2 had Down syndrome, and 3 had cleft palates. Assuming the jury was rational, did the citizens win or lose their case? Explain.

3. Before Delta studied embryology in her anatomy course, she imagined that a developing human is a shapeless mass of indistinct tissues until about halfway through pregnancy. Was Delta correct? At what stage does the embryo or fetus really start to look like a developing human?

4. Germaine took her grandfather to a bar-and-grill restaurant, where they saw a new sign that strongly warned pregnant women against drinking any alcohol. The grandfather was surprised at this and said, “In my day, pregnant women drank freely, because we were told that nothing crosses the placenta to reach the developing baby.” Do you think the warning sign was necessary? Explain.

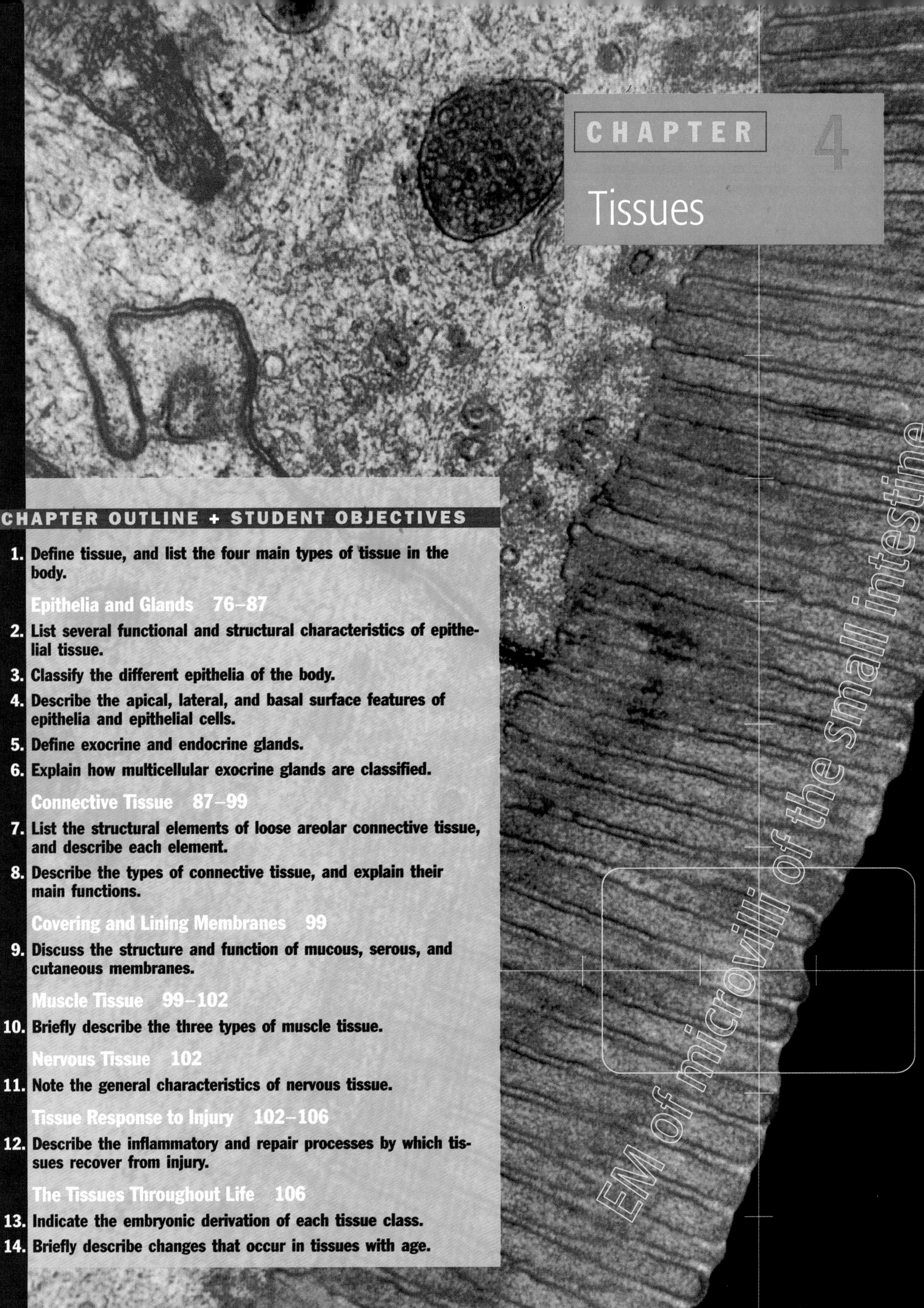

CHAPTER 4

Tissues

CHAPTER OUTLINE + STUDENT OBJECTIVES

1. Define tissue, and list the four main types of tissue in the body.

Epithelia and Glands 76–87

2. List several functional and structural characteristics of epithelial tissue.
3. Classify the different epithelia of the body.
4. Describe the apical, lateral, and basal surface features of epithelia and epithelial cells.
5. Define exocrine and endocrine glands.
6. Explain how multicellular exocrine glands are classified.

Connective Tissue 87–99

7. List the structural elements of loose areolar connective tissue, and describe each element.
8. Describe the types of connective tissue, and explain their main functions.

Covering and Lining Membranes 99

9. Discuss the structure and function of mucous, serous, and cutaneous membranes.

Muscle Tissue 99–102

10. Briefly describe the three types of muscle tissue.

Nervous Tissue 102

11. Note the general characteristics of nervous tissue.

Tissue Response to Injury 102–106

12. Describe the inflammatory and repair processes by which tissues recover from injury.

The Tissues Throughout Life 106

13. Indicate the embryonic derivation of each tissue class.
14. Briefly describe changes that occur in tissues with age.

THE CELLS OF THE HUMAN BODY DO NOT OPERATE independently of one another. Instead, related cells live and work together in cell communities called **tissues.** A tissue is more precisely defined as a group of closely associated cells that perform related functions and are similar in structure. However, tissues do not consist entirely of cells: Between the cells is nonliving *extracellular material.*

The word *tissue* derives from the Old French word meaning "to weave," reflecting the fact that the different tissues weave together to form the "fabric" of the human body. The four basic types of tissue are epithelium, connective tissue, muscle tissue, and nervous tissue. If we were to assign a single, functional term to each basic tissue, the terms would be *covering* (epithelium), *support* (connective), *movement* (muscle), and *control* (nervous). However, these terms reflect only a fraction of the functions that each tissue performs.

As we explained in Chapter 1, tissues are the building blocks of the body's organs. Because most organs contain all four tissue types, learning about tissues will make it easier for you to understand the structure of the organs discussed in the remaining chapters of this book.

EPITHELIA AND GLANDS

An **epithelium** (ep″ĭ-the′le-um; "covering") is a sheet of cells that covers a body surface or lines a body cavity. With minor exceptions, all of the outer and inner surfaces of the body are covered by epithelia. Examples include the outer layer of the skin, the inner lining of all hollow viscera such as the stomach and respiratory tubes, the lining of the peritoneal cavity, and the inner lining of all blood vessels. Epithelia also form most of the body's glands.

Epithelia occur at the interfaces between two different environments. The epidermis of the skin, for example, lies between the inside and outside of the body. Therefore, all functions of epithelia reflect their role as interface tissues and boundary layers: New stimuli, including harmful ones, are experienced at body interfaces, so epithelia both *protect* the underlying tissues and contain nerve endings for *sensory reception*. Nearly all substances that are received or given off by the body must pass across an epithelium, so epithelia function in *secretion* (the release of molecules from cells), *absorption* (bringing small molecules into the body), and *ion transport* (moving ions across the interface). Further-

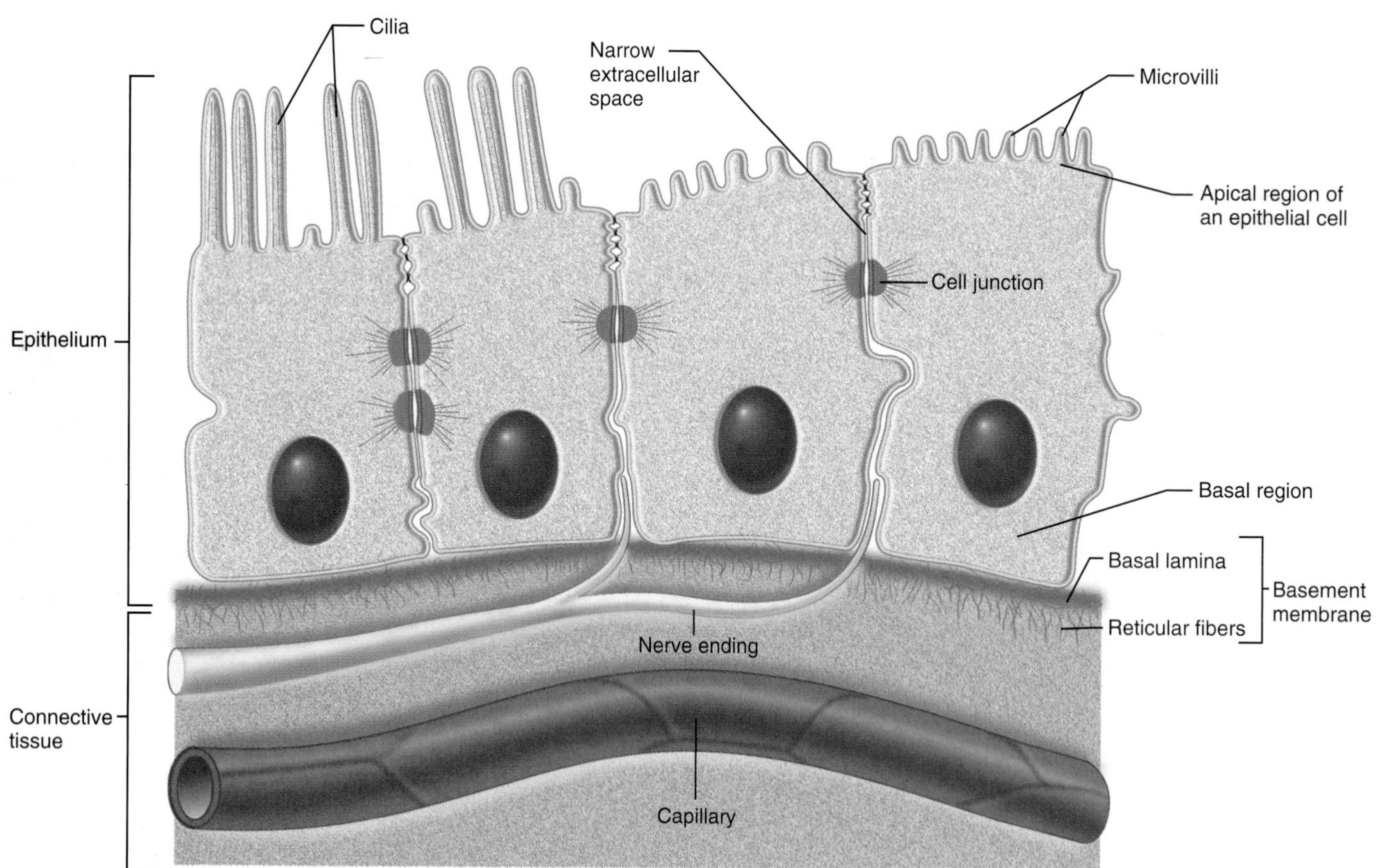

FIGURE 4.1 Special characteristics of epithelium. A sheet of closely joined epithelial cells rests on connective tissue proper. Epithelia contain nerve endings but no blood vessels. Note the special features on the epithelial cell surfaces: cilia, microvilli, cell junctions, and basal lamina.

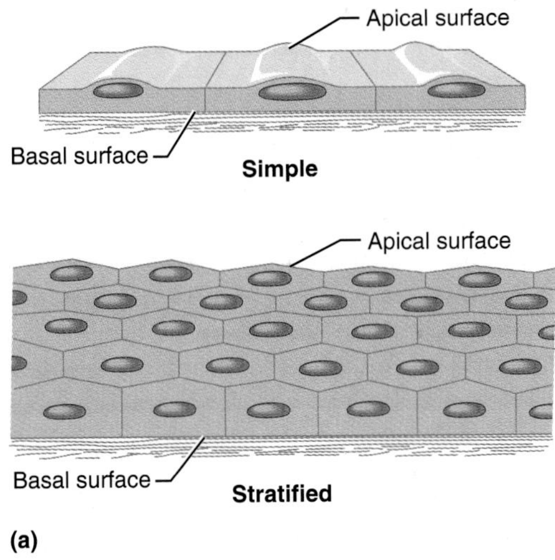

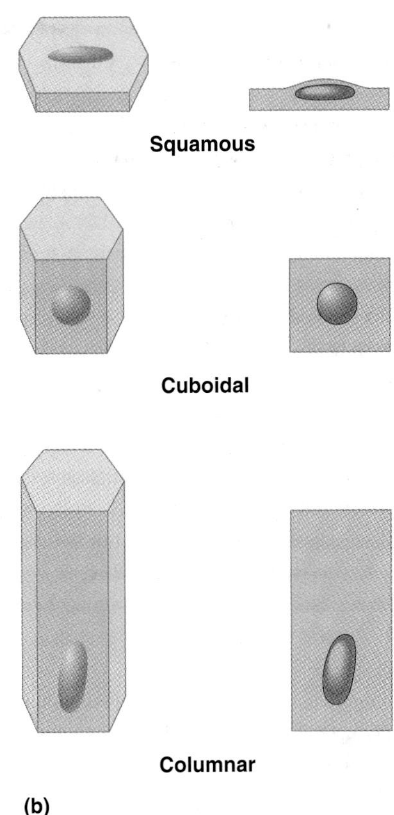

FIGURE 4.2 Classification of epithelia. **(a)** Classification based on number of cell layers. **(b)** Classification based on cell shape. For each category in part (b), a whole cell is shown on the left and a sectioned cell, viewed from the side, is shown on the right.

more, body fluids can be *filtered* across thin epithelia, and some epithelia form *slippery surfaces* along which substances move (food glides along the intestinal lining, for example).

Special Characteristics of Epithelia

Epithelial tissues have many characteristics that distinguish them from other tissue types (Figure 4.1):

1. **Cellularity.** Epithelia are composed almost entirely of cells. These cells are separated by a minimal amount of extracellular material, mainly projections of their integral membrane proteins into the narrow spaces between the cells.
2. **Specialized contacts.** Adjacent epithelial cells are directly joined at many points by special cell junctions.
3. **Polarity.** All epithelia have a free upper (apical) surface and a lower (basal) surface. They exhibit polarity, a term meaning that the cell regions near the apical surface differ from those near the basal surface. As shown in Figure 4.1, the basal surface of an epithelium lies on a thin sheet called a *basal lamina,* which is part of a *basement membrane.* This is explained further on p. 86.
4. **Support by connective tissue.** All epithelial sheets in the body are supported by an underlying layer of connective tissue.
5. **Avascular but innervated.** Whereas most tissues in the body are *vascularized* (contain blood vessels), epithelium is *avascular* (a-vas′ku-lar), meaning it lacks blood vessels. Epithelial cells receive their nutrients from capillaries in the underlying connective tissue. Although blood vessels do not penetrate epithelial sheets, nerve endings do; that is, epithelium is *innervated.*
6. **Regeneration.** Epithelial tissue has a high regenerative capacity. Some epithelia are exposed to friction, and their surface cells rub off. Others are destroyed by hostile substances in the external environment such as bacteria, acids, and smoke. As long as epithelial cells receive adequate nutrition, they can replace lost cells quickly by cell division.

Classification of Epithelia

Many kinds of epithelia exist in the body. In classifying them, each epithelium is given two names. The first name indicates the number of cell layers in the epithelium, and the last name describes the shape of the cells (Figure 4.2). In classifying by cell layers, an epithelium is called **simple** if it has just one cell layer or **stratified** if it has more than one layer. In classifying epithelia by cell shape, the cells are called squamous, cuboidal, or columnar. **Squamous** (sqwa′mus; "plate-like") cells are wider than they are tall; **cuboidal** cells are about as tall as they are wide, like cubes; and **columnar** cells are taller than they are wide, like columns. In each case, the shape of the nucleus conforms to that of the cell: The nucleus of a squamous cell is disc-shaped, that of a cuboidal cell is spherical, and that of a columnar cell is an oval elongated from top to bottom. The shape of the nucleus is an important feature to keep in mind when distinguishing epithelial types.

Simple epithelia are easy to classify by cell shape because all cells in the layer usually have the same shape. In stratified epithelia, however, the cell shapes usually differ among the different cell layers. To avoid ambiguity, stratified epithelia are named according to the shape of the cells in the *apical* layer. This naming system will become clearer as we explore the specific epithelial types.

As you read about the epithelial classes, follow Figure 4.3. Using the photomicrographs, try to pick out the individual cells within each epithelium. This is not always easy, because the boundaries between epithelial cells often are indistinct. Furthermore, the nucleus of a particular cell may or may not be visible, depending on the plane of the cut made to prepare the tissue slides.

Simple Epithelia

Simple Squamous Epithelium (Figure 4.3a). A simple squamous epithelium is a single layer of flat cells. When viewed from above, the closely fitting cells resemble a tiled floor. When viewed in lateral section, they resemble fried eggs seen from the side. Thin and often permeable, this type of epithelium occurs wherever small molecules pass through a membrane quickly, by a process called passive diffusion. The walls of capillaries consist exclusively of this epithelium, whose exceptional thinness encourages efficient exchange of nutrients and wastes between the bloodstream and surrounding tissue cells. In the lungs, this epithelium forms the thin walls of the air sacs, where gas exchange occurs.

Some simple squamous epithelia are slippery, and two examples have special names. **Endothelium** (en″do-the′le-um; "inner covering") provides a slick, friction-reducing lining of the hollow organs of the circulatory system—blood vessels, lymphatic vessels, and the heart (see Figure 4.4 on p. 83). **Mesothelium** (mez″o-the′le-um; "middle covering") is the epithelium that lines the peritoneal, pleural, and pericardial cavities and covers the visceral organs in these cavities (serosae; see pp. 14 and 99).

Simple Cuboidal Epithelium (Figure 4.3b). Simple cuboidal epithelium consists of a single layer of cells as tall as they are wide. This epithelium forms the walls of the smallest ducts of glands, and of many tubules in the kidney. Its functions are the same as those of simple columnar epithelium.

Simple Columnar Epithelium (Figure 4.3c). Simple columnar epithelium is a single layer of tall cells aligned like soldiers in a row. It lines the digestive tube from the stomach to the anal canal. It functions in the active movement of molecules, namely in absorption, secretion, and ion transport. The structure of simple columnar epithelium is ideal for these functions: It is thin enough to allow large numbers of molecules to pass through it quickly, yet thick enough to house the cellular machinery needed to perform the complex processes of molecular transport.

Some simple columnar epithelia bear *cilia* (sil′e-ah; "eyelashes"), whip-like bristles on the apex of epithelial cells that beat rhythmically to move substances across certain body surfaces (see Figure 4.1). A simple ciliated columnar epithelium lines the inside of the uterine tube. Its cilia help move the ovum to the uterus, a journey we traced in Chapter 3. Cilia are considered in detail later in this chapter.

Pseudostratified Columnar Epithelium (Figure 4.3d). The cells of pseudostratified (soo-do-strat′ĭ-fīd) columnar epithelium are varied in height. While all of its cells rest on the basement membrane, only the tall cells reach the apical surface of the epithelium. The short cells are undifferentiated and continuously give rise to the tall cells. The cell nuclei lie at several different levels, giving the false impression that this epithelium is stratified (*pseudostratified* = false stratified).

Pseudostratified columnar epithelium, like simple columnar epithelium, functions in secretion or absorption. A ciliated type lines the interior of the respiratory tubes. Here, the cilia propel sheets of dust-trapping mucus out of the lungs.

Stratified Epithelia

Stratified epithelia contain two or more layers of cells. They regenerate from below; that is, the basal cells divide and push apically to replace the older surface cells. Stratified epithelia are more durable than simple epithelia, and their major (but not their only) role is protection.

Stratified Squamous Epithelium (Figure 4.3e). As you might expect, stratified squamous epithelium consists of many cell layers whose surface cells are squamous in shape. In the deeper layers, the cells are cuboidal or columnar. Of all the epithelial types, this is thickest and best adapted for protection. It covers the often-abraded surfaces of our body, forming the epidermis of the skin and

Text continues on page 83

(a) Simple squamous epithelium

Description: Single layer of flattened cells with disc-shaped central nuclei and sparse cytoplasm; the simplest of the epithelia.

Function: Allows passage of materials by diffusion and filtration in sites where protection is not important; secretes lubricating substances in serosae.

Location: Kidney corpuscles; air sacs of lungs; lining of heart, blood vessels, and lymphatic vessels; lining of ventral body cavity (serosae).

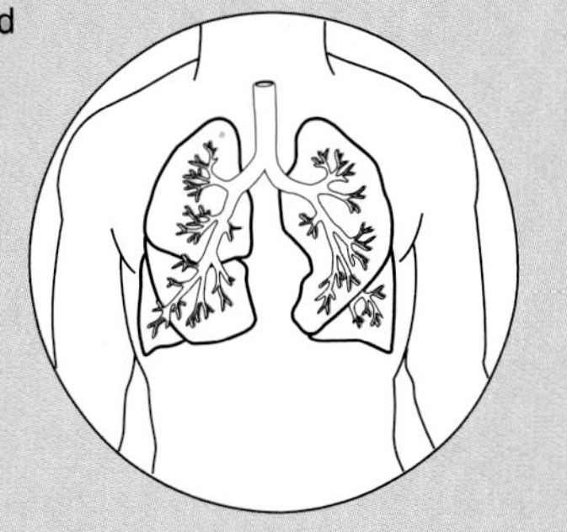

Photomicrograph: Simple squamous epithelium forming part of the alveolar wall (240×).

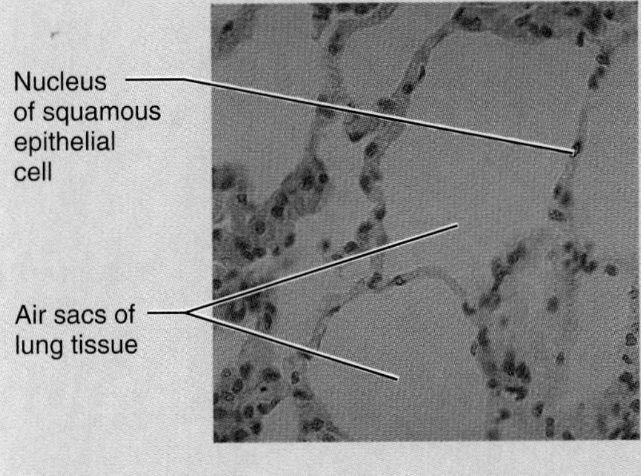

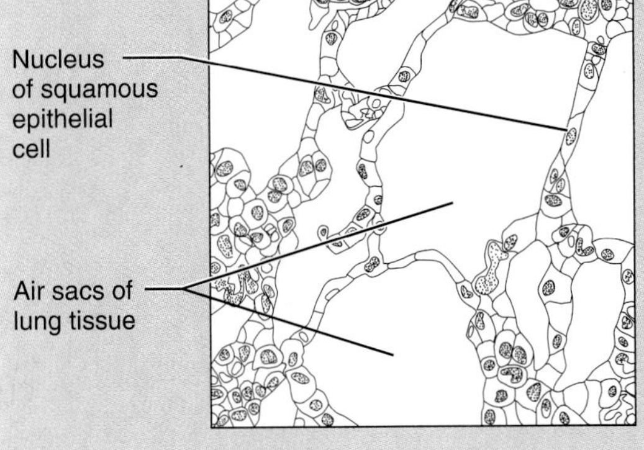

(b) Simple cuboidal epithelium

Description: Single layer of cubelike cells with large, spherical central nuclei.

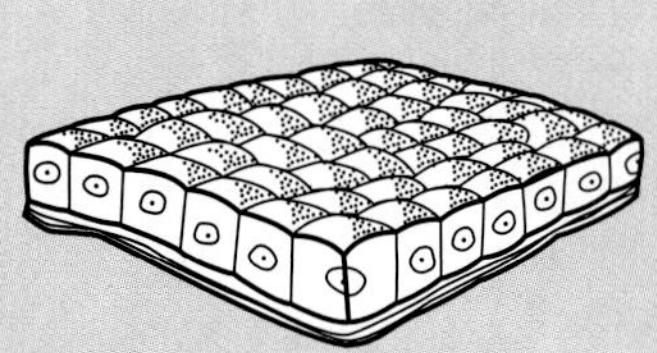

Function: Secretion and absorption.

Location: Kidney tubules; ducts and secretory portions of small glands; ovary surface.

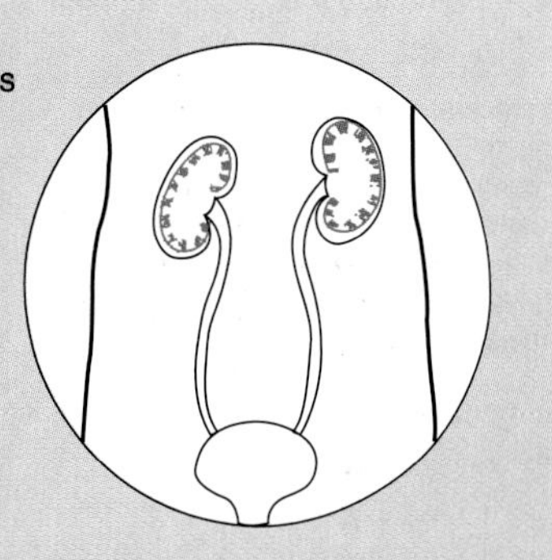

Photomicrograph: Simple cuboidal epithelium in kidney tubules (270×).

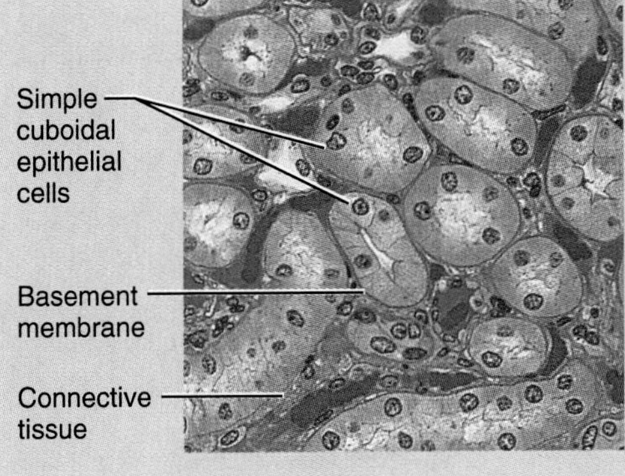

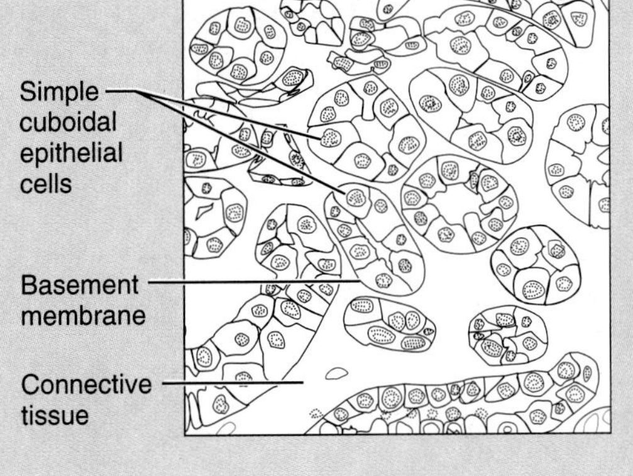

FIGURE 4.3 Epithelial tissues.

(c) Simple columnar epithelium

Description: Single layer of tall cells with *oval* nuclei; some cells bear cilia; layer may contain mucus-secreting unicellular glands (goblet cells).

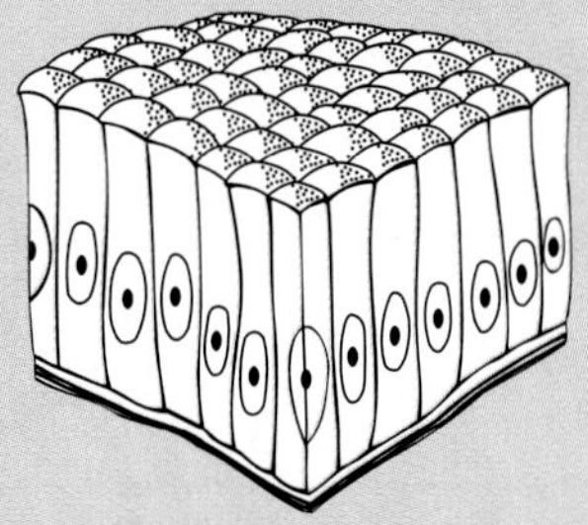

Function: Absorption; secretion of mucus, enzymes, and other substances; ciliated type propels mucus (or reproductive cells) by ciliary action.

Location: Nonciliated type lines most of the digestive tract (stomach to anal canal), gallbladder and excretory ducts of some glands; ciliated variety lines small bronchi, uterine tubes, and the uterus.

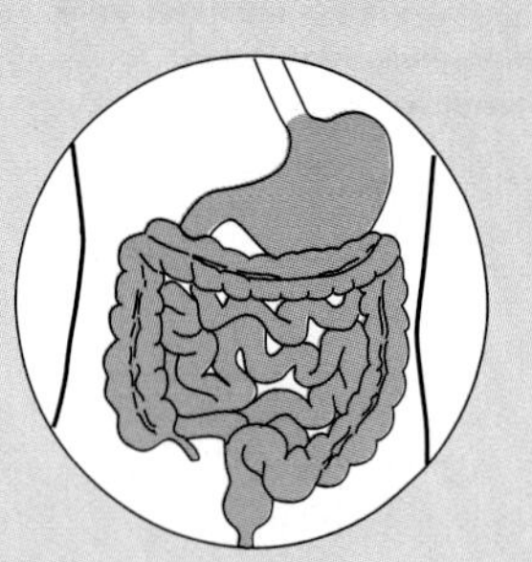

Photomicrograph: Simple columnar epithelium of the stomach mucosa (500×).

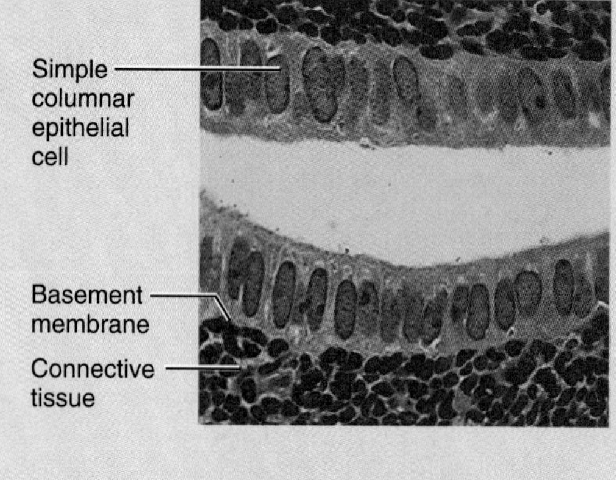

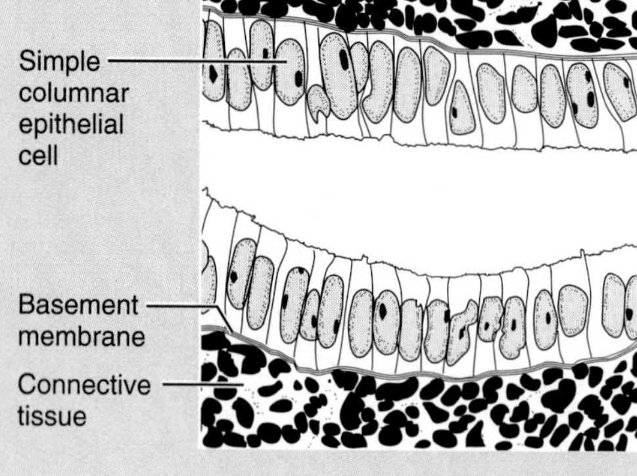

(d) Pseudostratified columnar epithelium

Description: Single layer of cells of differing heights, some not reaching the free surface; nuclei seen at different levels; may contain goblet cells and bear cilia.

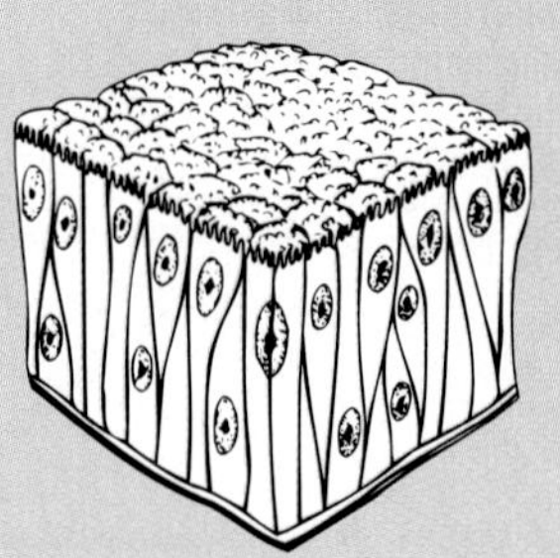

Function: Secretion, particularly of mucus; propulsion of mucus by ciliary action.

Location: Nonciliated type in male's sperm–carrying ducts and ducts of large glands; ciliated variety lines the trachea, most of the upper respiratory tract.

Trachea

Photomicrograph: Pseudostratified ciliated columnar epithelium lining the human trachea (800×).

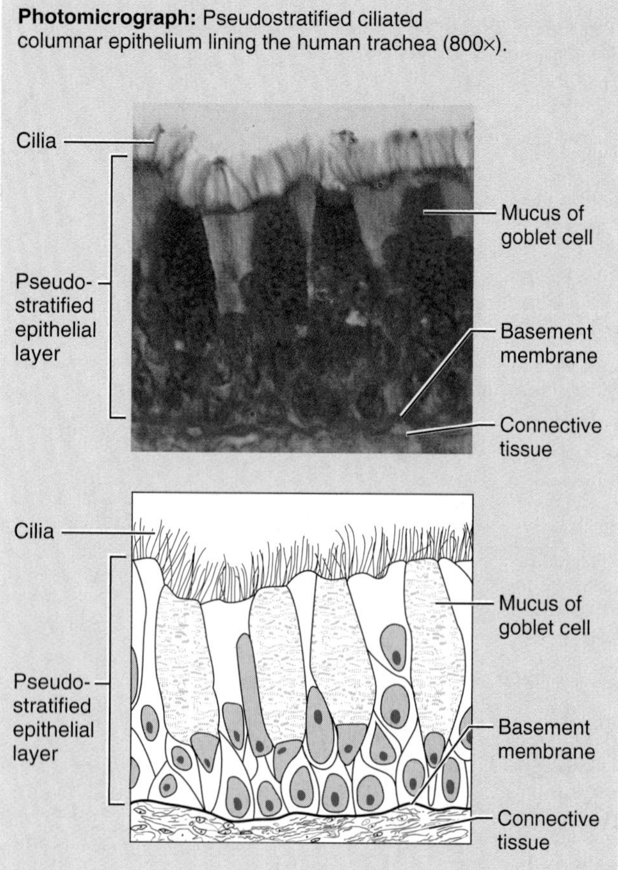

FIGURE 4.3 Epithelial tissues, continued.

(e) Stratified squamous epithelium

Description: Thick membrane composed of several cell layers; basal cells are cuboidal or columnar and metabolically active; surface cells are flattened (squamous); in the keratinized type, the surface cells are full of keratin and dead; basal cells are active in mitosis and produce the cells of the more superficial layers.

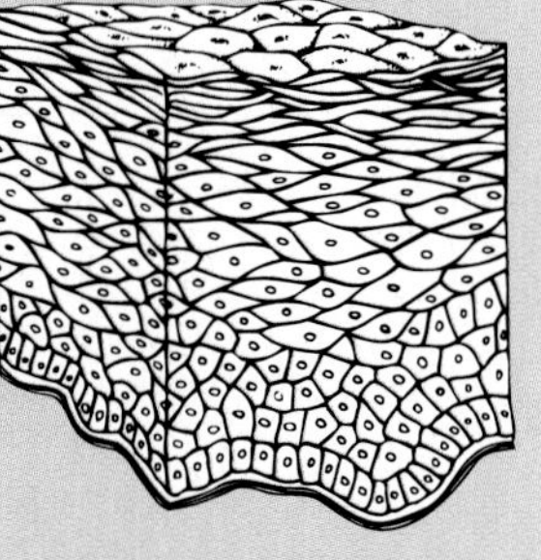

Function: Protects underlying tissues in areas subjected to abrasion.

Location: Nonkeratinized type forms the moist linings of the esophagus, mouth, and vagina; keratinized variety forms the epidermis of the skin, a dry membrane.

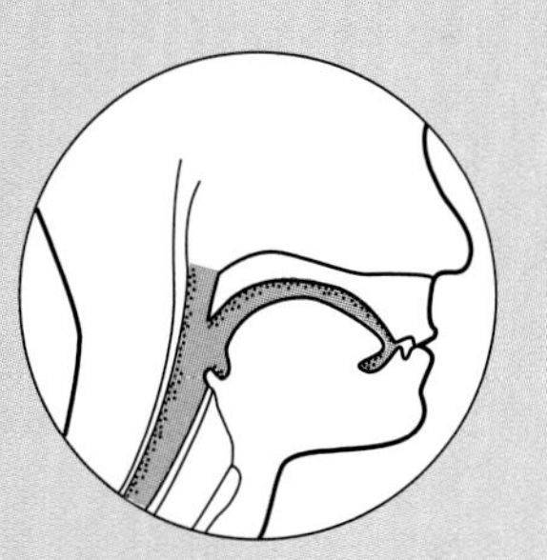

Photomicrograph: Stratified squamous epithelium lining of the esophagus (158×).

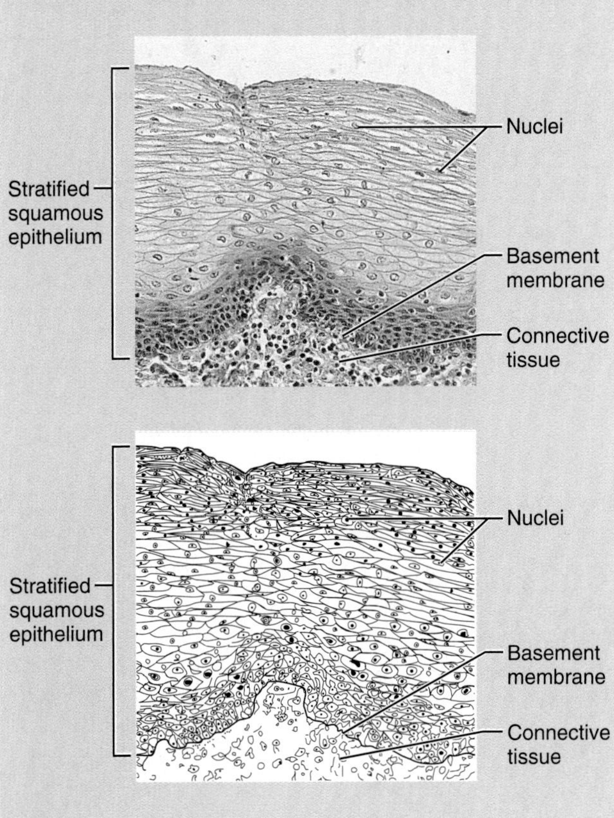

(f) Stratified cuboidal epithelium

Description: Generally two layers of cube-like cells.

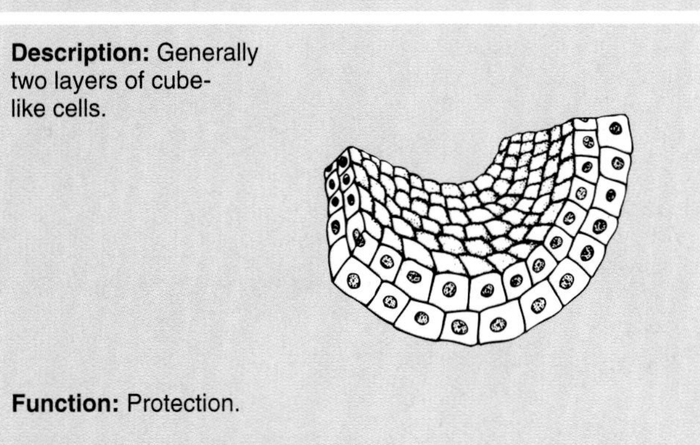

Function: Protection.

Location: Largest ducts of sweat glands, mammary glands, and salivary glands.

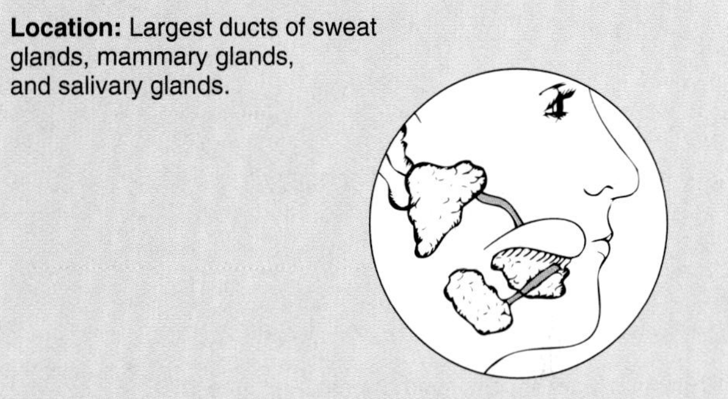

Photomicrograph: Stratified cuboidal epithelium forming a salivary gland duct (150×).

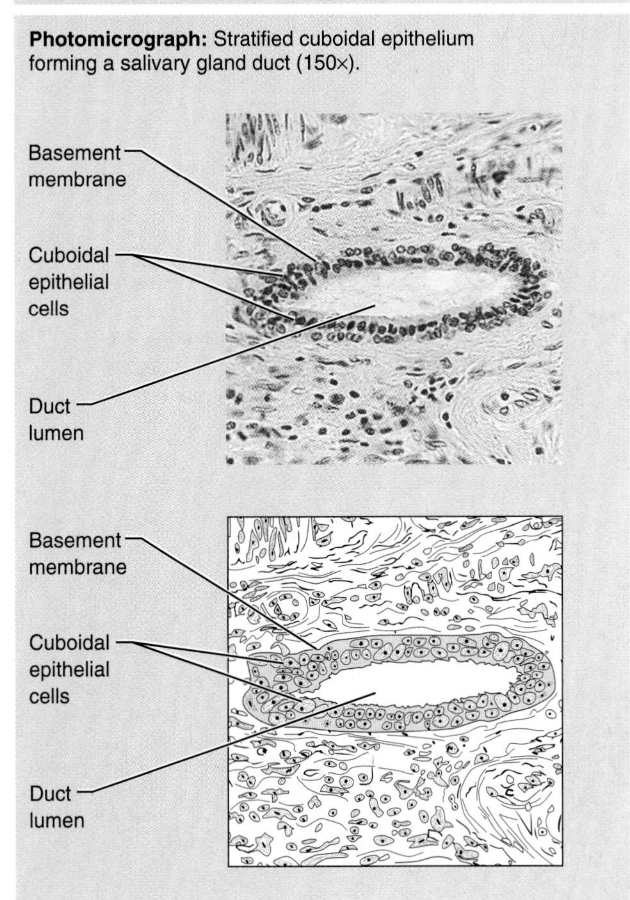

FIGURE 4.3 continued.

(g) Stratified columnar epithelium

Description: Several cell layers; basal cells usually cuboidal; superficial cells elongated and columnar.

Function: Protection; secretion.

Location: Rare in the body; small amounts in male urethra and in large ducts of some glands.

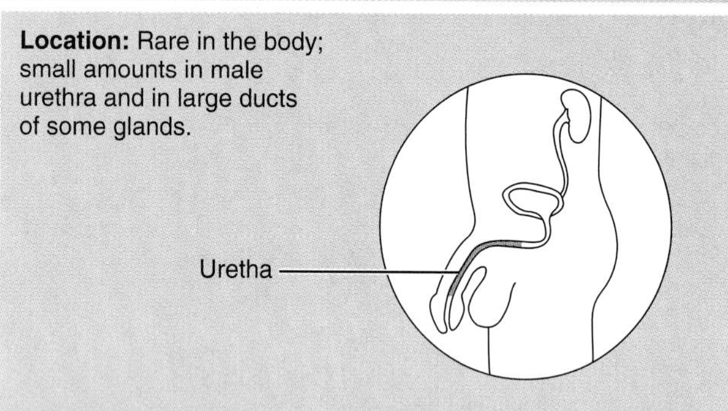

Photomicrograph: Stratified columnar epithelium lining of the male urethra (700×).

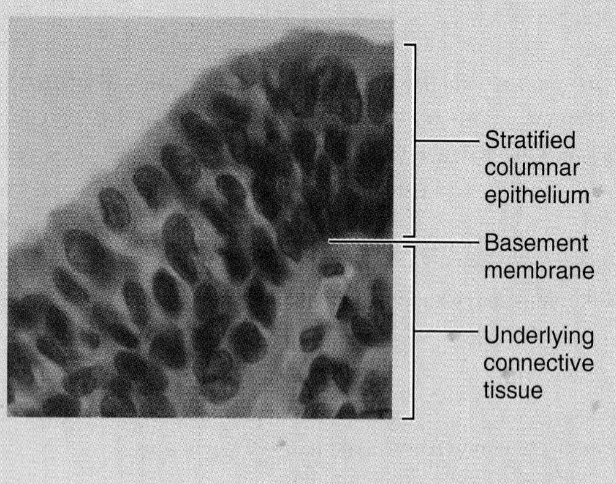

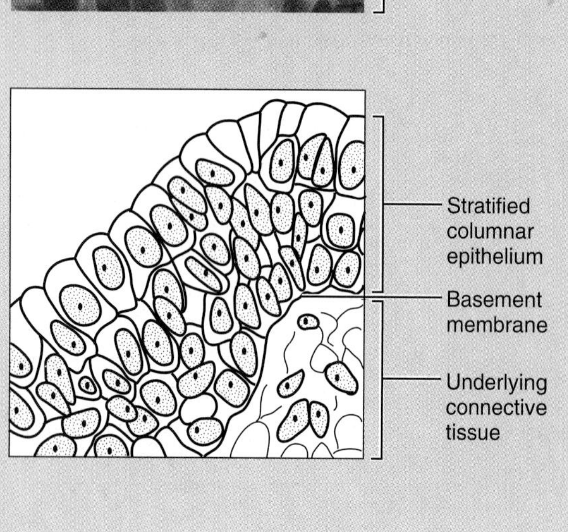

(h) Transitional epithelium

Description: Resembles both stratified squamous and stratified cuboidal; basal cells cuboidal or columnar; surface cells dome-shaped or squamous-like, depending on degree of organ stretch.

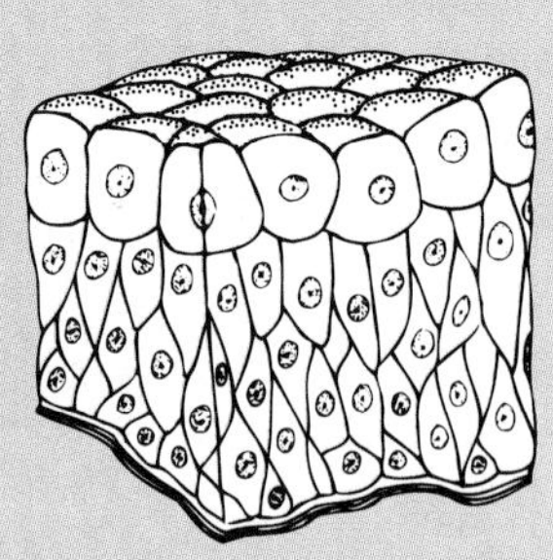

Function: Stretches readily and permits distension of urinary organ by contained urine.

Location: Lines the ureters, bladder, and part of the urethra.

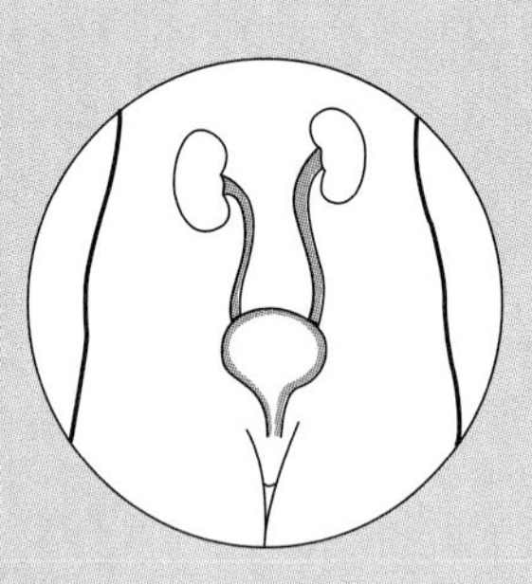

Photomicrograph: Transitional epithelium lining of the bladder, relaxed state (277×); note the bulbous, or rounded, appearance of the cells at the surface; these cells flatten and become elongated when the bladder is filled with urine.

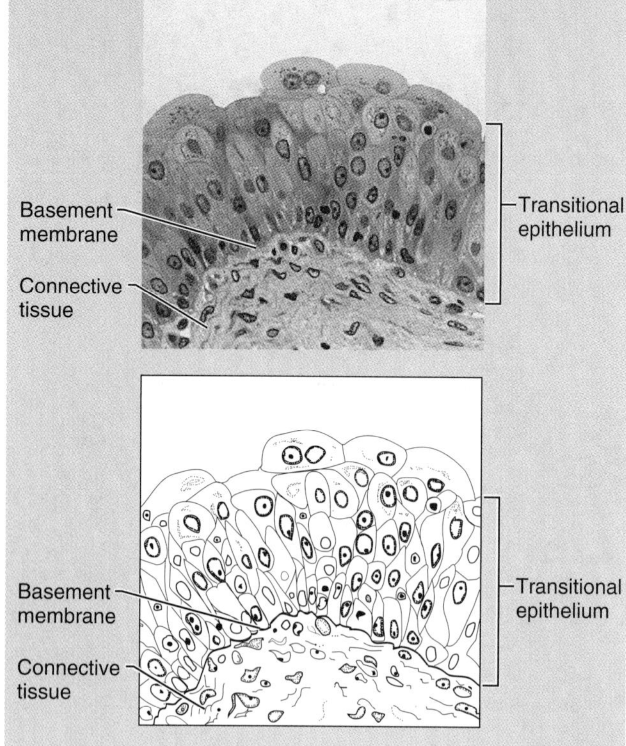

FIGURE 4.3 Epithelial tissues, continued.

the inner lining of the mouth, esophagus, and vagina. To avoid memorizing all its locations, simply remember that this epithelium covers the skin and extends a certain distance into every body opening that is directly continuous with the skin.

The epidermis of the skin is *keratinized*, meaning that its surface cells contain an especially tough protective protein called *keratin*. The other stratified squamous epithelia of the body lack keratin and are *nonkeratinized*. Keratin is explained in detail in Chapter 5 (p. 113).

Stratified Cuboidal and Columnar Epithelia (Figure 4.3f–g). Stratified cuboidal and stratified columnar epithelia are rare types of tissue, forming little else than the large ducts of some glands.

Transitional Epithelium (Figure 4.3h). Transitional epithelium lines the inside of the hollow urinary organs. Such organs (the urinary bladder, for example), stretch as they fill with urine. As the transitional epithelium stretches, it thins from about six cell layers to three, and its apical cells flatten from a rounded shape to a squamous shape. Thus, this epithelium undergoes "transitions" in shape. It also forms an impermeable barrier that keeps urine from passing through the wall of the bladder.

Epithelial Surface Features

The apical, lateral, and basal cell surfaces of epithelia have special features.

Apical Surface Features: Microvilli and Cilia

Microvilli (mi″cro-vĭ′li; "little shaggy hairs") are finger-like extensions of the plasma membrane of apical epithelial cells (see Figure 4.1 and Figure 1.13c on p. 18). They occur on almost every moist epithelium in the body but are longest and most abundant on epithelia that absorb nutrients (in the small intestine) or transport ions (in the kidney). In such epithelia, microvilli maximize the surface area across which small molecules enter or leave cells. Microvilli are also abundant on epithelia that secrete mucus, where they help anchor the mucous sheets to the epithelial surface. Finally, microvilli may act as "stiff knobs" to resist abrasion.

Cilia, you will recall, are whip-like, highly motile extensions of the apical surface membranes of certain epithelial cells (see Figure 4.1). Each cilium contains a core of microtubules, nine pairs of which encircle one middle pair (Figure 4.5). This entire set of microtubules is an **axoneme** (ak′so-nēm; "axis thread"), and each pair is a *doublet*. Ciliary movement is generated when adjacent doublets grip one another with side arms made of the motor protein *dynein* (p. 40) and these arms start to oscillate. This causes the doublets to slide along the length of each other, like centipedes trying to run over each other's backs. As a result, the cilum bends.

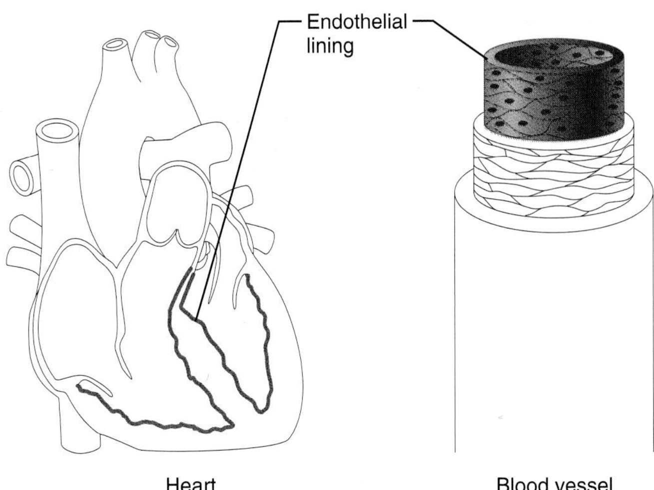

FIGURE 4.4 Endothelium. Endothelium is a simple squamous epithelium that lines the interior of the circulatory vessels and the heart.

The microtubules in cilia are arranged in much the same way as in the cytoplasmic organelles called centrioles (p. 40). The strong resemblence can be seen in Figure 4.5b and c, and it is no coincidence. Indeed, cilia originate as their microtubules assemble around centrioles that have migrated from the centrosome to the apical plasma membrane. The centriole at the base of each cilium is called a **basal body** (Figure 4.5a).The cilia on an epithelium bend and move in coordinated waves, like the waves passing across a field of grass on a windy day. These waves push mucus and other substances over the epithelial surface. Each cilium executes a propulsive *power stroke,* followed by a nonpropulsive *recovery stroke* (see Figure 4.6). This sequence ensures that fluid is moved in one direction only.

CLINICAL APPLICATIONS

KARTAGENER'S SYNDROME A type of *immotile cilia syndrome,* Kartagener's syndrome is an inherited disease in which the dynein arms within the cilia fail to form. This condition leads to frequent respiratory infections because the nonfunctional cilia cannot sweep inhaled bacteria out of the respiratory tubes. People with this syndrome often have a mirror-image reversal of their right and left viscera, meaning their liver is on the left and their heart points to the right, and so on. This suggests that in the normal embryo, the beating action of cilia is necessary for the development of the asymmetry of our viscera. Perhaps the anti-clockwise movement of cilia concentrates left- or right-determining chemicals on one side of the embryo. ✚

An extremely long, isolated cilium is called a *flagellum* (flah-jel′um; "whip"). The only flagellated cells in the human body are sperm, which use their flagella to swim through the female reproductive tract.

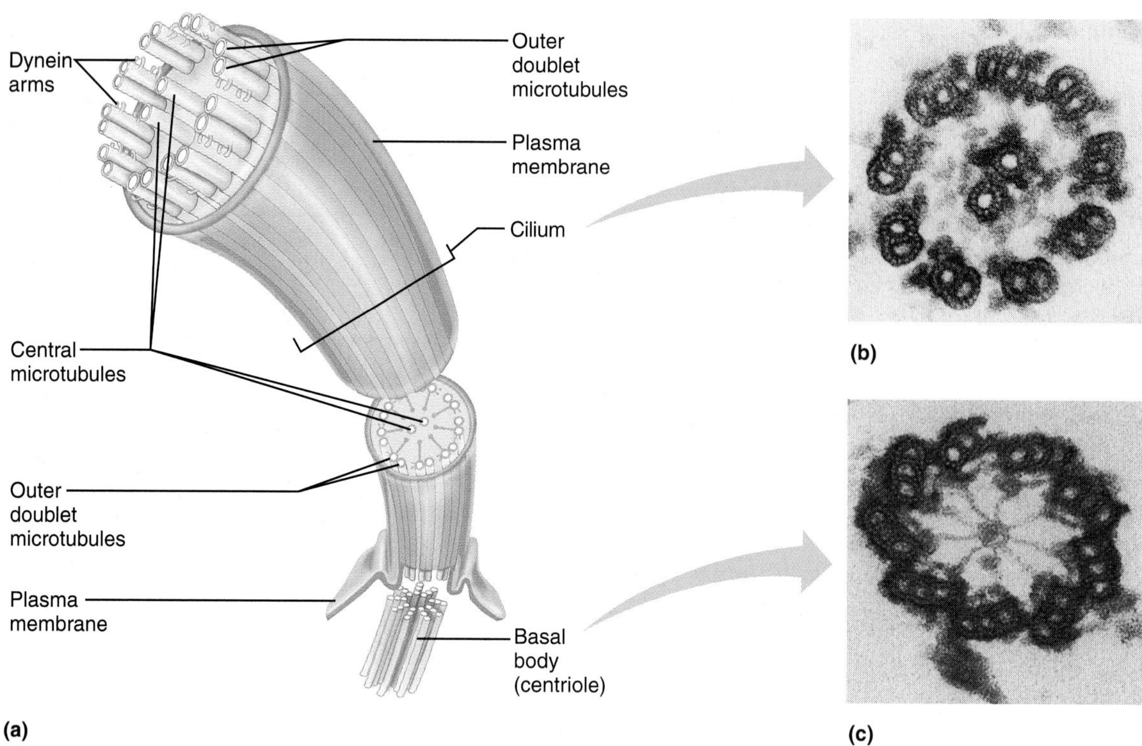

FIGURE 4.5 A cilium. **(a)** Three-dimensional view of a cilium, sectioned to show the arrangement of microtubules in its core. Note the basal body below, which lies in the apical cytoplasm of the cell. **(b)** Electron micrograph of a cilium in cross section (approx. 150,000 ×). **(c)** Electron micrograph of the basal body (centriole) in cross section (approx. 150,000 ×).

Critical Reasoning

All males with immotile cilia syndrome are sterile (cannot father children). Why?

The flagella in their sperm do not function, so the sperm cannot swim to meet and fertilize an ovum.

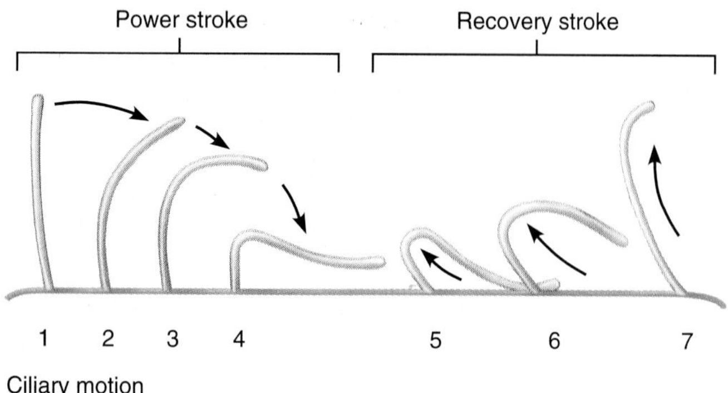

FIGURE 4.6 Ciliary movement. In the initial power stroke, a cilium moves in an arc. This movement is followed by a recovery stroke, in which the cilium bends and pulls back, keeping its tip low. This sequence acts to move mucus in a single direction.

Lateral Surface Features: Cell Junctions

Three factors act to hold epithelial cells to one another: (1) ***immunoglobulin-like adhesion proteins*** in the plasma membranes of the adjacent cells are linked together in the narrow extracellular space; (2) the wavy contours of the membranes of adjacent cells join in a tongue and groove fashion; and (3) there are special cell junctions (Figures 4.7–4.9). **Cell junctions,** the most important of the factors, are most characteristic of epithelial tissue but are occasionally found in other tissue types as well.

Desmosomes The main junctions for binding cells together are called **desmosomes** (dez′mo-sōmz; "binding bodies"). These adhesive spots are scattered like rivets along the abutting sides of adjacent cells (Figure 4.7). Desmosomes have a complex structure: On the cytoplasmic face of each plasma membrane is a circular plaque. The plaques of neighboring cells are joined by linker proteins called **cadherins.** These project from both the cell membranes and interdigitate like the teeth of a zipper in the extracellular space. In addition, intermediate filaments (the cytoskeletal elements that resist tension) insert into each plaque from its inner, cytoplasmic side. Bundles of these filaments extend across

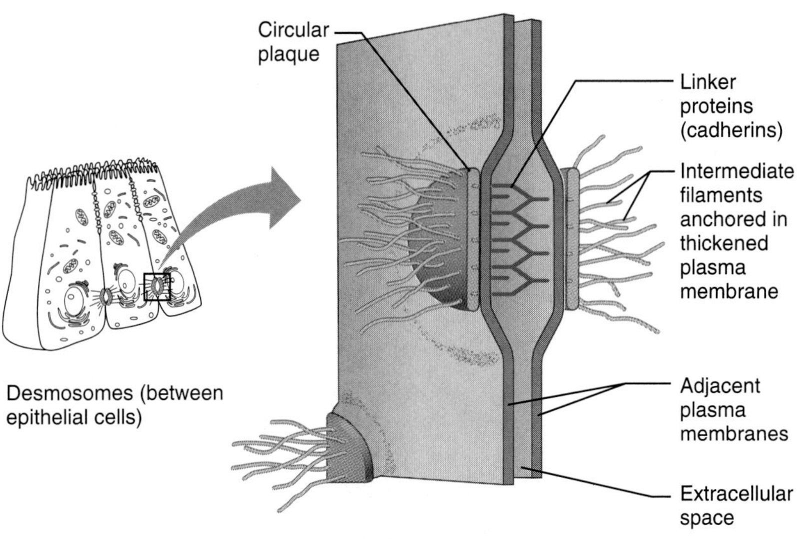

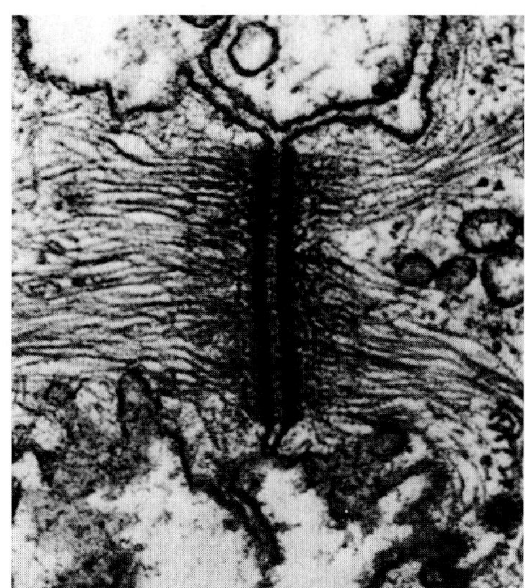

FIGURE 4.7 Desmosome. The diagrams at left and center show the location and structure of a desmosome, a spot junction that connects adjacent epithelial cells. At right is a desmosome viewed by electron microscopy (58,000 ×).

the cytoplasm and anchor at other desmosomes on the opposite side of the same cell. Overall, this arrangement not only holds adjacent cells together but also interconnects all intermediate filaments of the entire epithelium into one continuous network of strong "guy wires." The epithelium is thus less likely to tear when pulled upon, because the pulling forces are distributed evenly throughout the sheet.

Tight Junctions In the apical region of most epithelial types, a belt-like junction extends around the periphery of each cell (Figure 4.8). This is a **tight junction** or a *zonula occludens* (zōn′u-lah o-klood′enz; "belt that shuts off"). At tight junctions, the adjacent cells are so close that some proteins in their plasma membranes are fused. This fusion forms a seal that closes off the extracellular space. Thus, tight junctions prevent molecules from passing between the cells of epithelial tissue. For example, the tight junctions in the epithelium lining the digestive tract keep digestive enzymes, ions, and microorganisms in the intestine from seeping into the bloodstream. Tight junctions need not be entirely impermeable; some are more leaky than others and may let certain types of ions through.

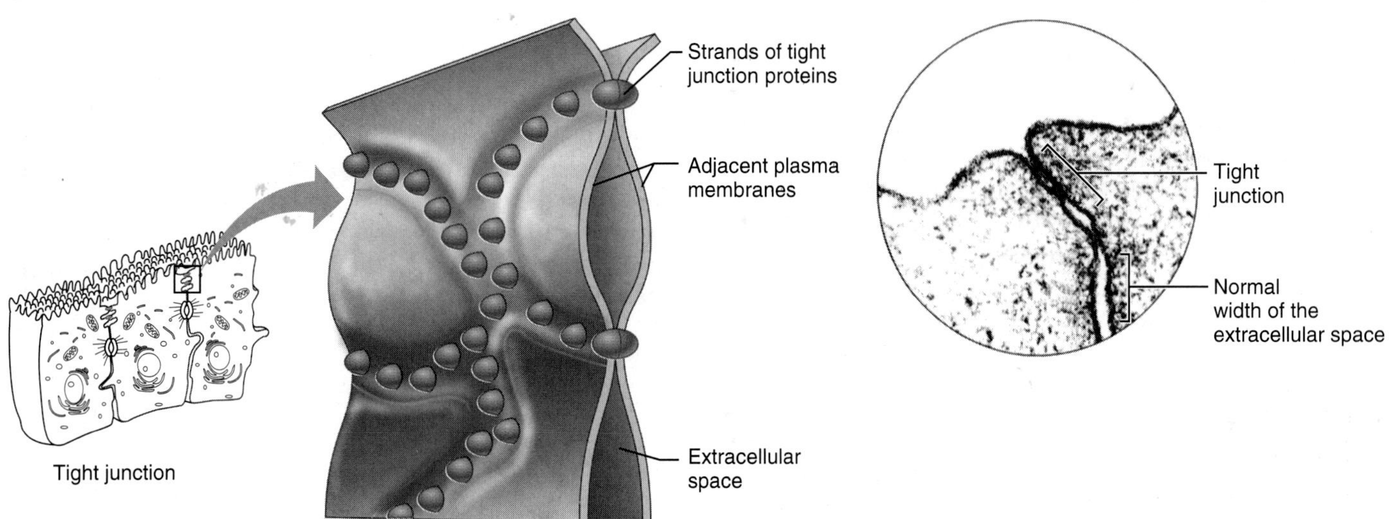

FIGURE 4.8 Tight junction between epithelial cells. Drawing at left and center, and electron micrograph at right (approx. 40,000 ×).

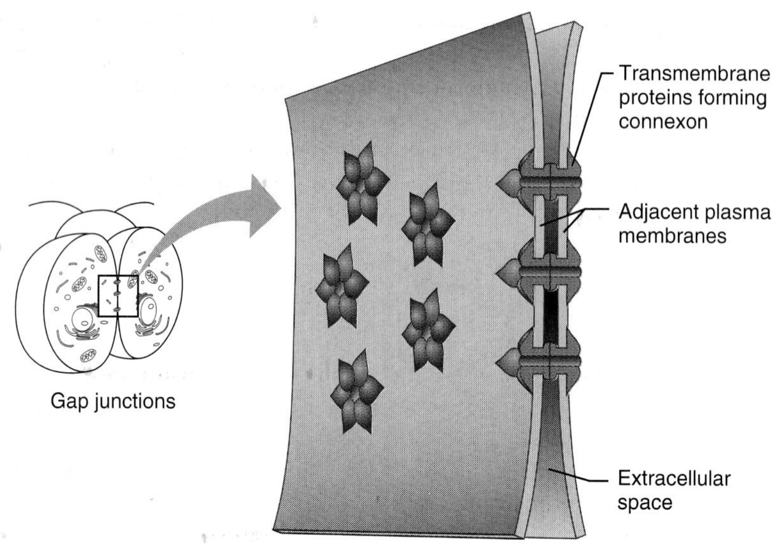

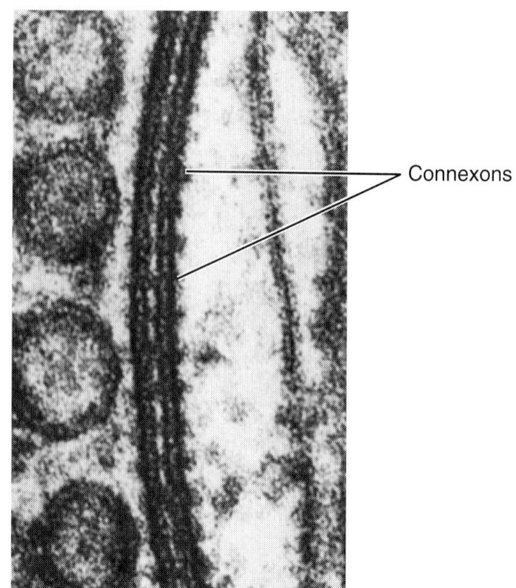

FIGURE 4.9 Gap junction between epithelial cells. Drawing (left) and electron micrograph (right; approx. 200,000 ×).

Gap Junctions A **gap junction,** or *nexus* (nek′sus; "bond"), is a spot-like junction that can occur anywhere along the lateral membranes of adjacent cells (Figure 4.9). Gap junctions let small molecules move directly between neighboring cells. At such junctions, the adjacent plasma membranes are very close, and the cells are connected by hollow cylinders of protein (connexons). Ions, simple sugars, and other small molecules pass through these cylinders from one cell to the next.

Basal Feature: The Basal Lamina

At the border between the epithelium and the connective tissue deep to it is a supporting sheet called the **basal lamina** (lam′ĭ-nah; "sheet"), shown in Figure 4.1. This thin, noncellular sheet consists of proteins secreted by the epithelial cells. Functionally, the basal lamina acts as a selective filter; that is, it determines which molecules from capillaries in the underlying connective tissue will be allowed to enter the epithelium. The basal lamina also acts as scaffolding along which regenerating epithelial cells can migrate. Luckily, infections and toxins that destroy epithelial cells usually leave the basal lamina in place, for without this lamina, epithelial regeneration is more difficult.

Directly deep to the basal lamina is a layer of reticular fibers (defined shortly) belonging to the underlying connective tissue. Together, these reticular fibers plus the basal lamina form the **basement membrane** (see Figure 4.1). The thin basal lamina can be seen only by electron microscopy, but the thicker basement membrane is visible by light microscopy (see Figure 4.3).

While we have been careful to distinguish between *basal lamina* and *basement membrane,* we should point out that many scientists use these two terms interchangeably.

CLINICAL APPLICATIONS

BASEMENT MEMBRANES AND DIABETES In untreated cases of diabetes mellitus, the basement membranes around the endothelium of capillaries thicken over time, perhaps because they take on sugar (glucose), which is present in high concentrations throughout the bodies of diabetics. The capillaries in the kidneys and retina of the eye thicken the most and can become nonfunctional. The resulting kidney failure and blindness are major symptoms of advanced diabetes. +

Glands

Many epithelial cells make and secrete a product. Such cells constitute **glands.** The products of glands are aqueous (water-based) fluids that usually contain proteins. *Secretion* is the process whereby gland cells obtain needed substances from the blood and transform them chemically into a product that is then discharged from the cell. More specifically, the protein product is made in the rough ER, then packaged into secretory granules by the Golgi apparatus and ultimately released from the cell by exocytosis (see p. 30 and Figure 2.12). These organelles are well developed in most gland cells that secrete proteins.

Glands are classified as **exocrine** (ek′so-krin; "external secretion") or **endocrine** (en′do-krin; "internal secretion"), depending on where they release their

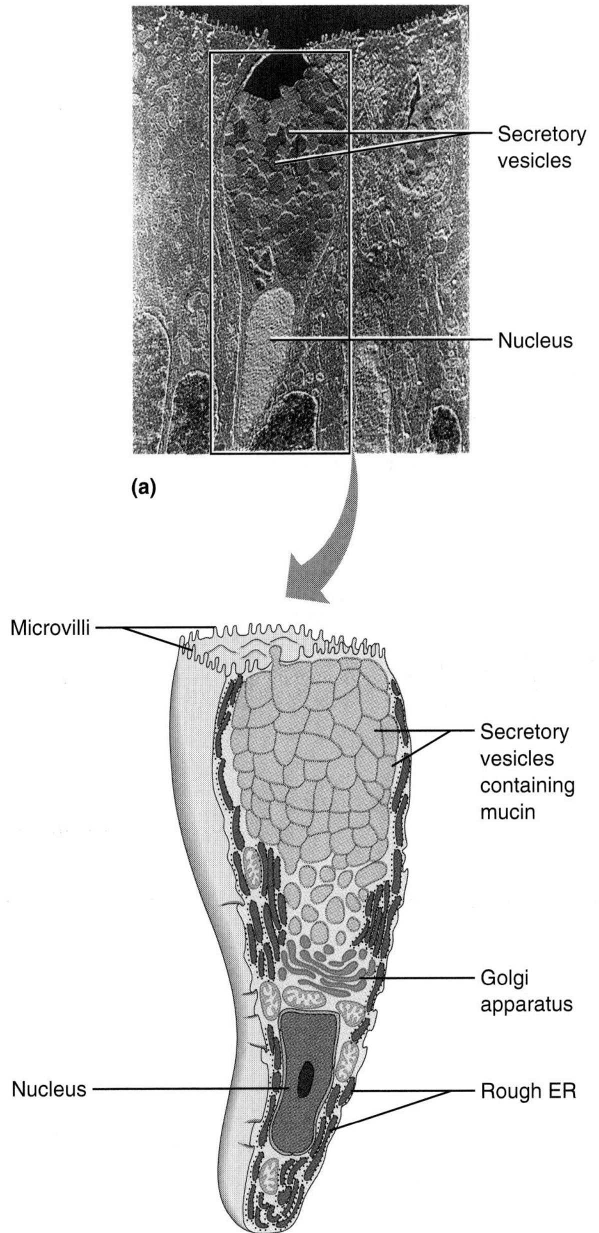

FIGURE 4.10 Goblet cells (unicellular exocrine glands).
(a) Photomicrograph of a mucus-secreting goblet cell in the simple columnar epithelium that lines the small intestine (approx. 1300 ×). (b) Diagram of a goblet cell. Note the secretory vesicles (secretory granules) and the well-developed rough ER and Golgi apparatus.

product, and as **unicellular** ("one-celled") or **multicellular** ("many-celled") on the basis of cell number. Unicellular glands are scattered within epithelial sheets, whereas most multicellular glands develop by invagination of an epithelial sheet into the underlying connective tissue.

Exocrine Glands

Exocrine glands are numerous, and many of their products are familiar ones. All exocrine glands secrete their products onto body surfaces (skin) or into body cavities (like the digestive tube), and multicellular exocrine glands have **ducts** that carry their product to the epithelial surfaces. Exocrine glands are a diverse group: They include many types of mucus-secreting glands, the sweat glands and oil glands of the skin, salivary glands of the mouth, the liver (which secretes bile), the pancreas (which secretes digestive enzymes), mammary glands (which secrete milk), and many others.

Unicellular Exocrine Glands The only important example of a one-celled exocrine gland is the **goblet cell** (Figure 4.10). True to its name, a goblet cell is indeed shaped like a goblet, a drinking glass with a stem. Goblet cells are scattered within the epithelial lining of the intestines and respiratory tubes, between columnar cells with other functions. They produce **mucin** (mu′sin), a glycoprotein (sugar-protein) that dissolves in water when secreted. The resulting complex of mucin and water is viscous, slimy **mucus.** Mucus covers, protects, and lubricates many internal body surfaces.

Multicellular Exocrine Glands Each multicellular exocrine gland has two basic parts: an epithelium-walled *duct* and a *secretory unit* consisting of the secretory epithelium (Figure 4.11). Also, in all but the simplest glands, a supportive connective tissue surrounds the secretory unit and supplies it with blood vessels and nerve fibers. Often, the connective tissue forms a *fibrous capsule* that extends into the gland proper and partitions the gland into subdivisions called *lobes* (not illustrated).

Multicellular glands are classified by the structure of their ducts (Figure 4.11). **Simple** glands have an unbranched duct, whereas **compound** glands have a branched duct. The glands are further categorized by their secretory units: They are **tubular** if their secretory cells form tubes and **alveolar** (al-ve′o-lar) if the secretory cells form spherical sacs (an *alveolus* is literally a small hollow cavity). Furthermore, some glands are **tubuloalveolar;** that is, they contain both tubular and alveolar units. It should also be noted that another word for *alveolar* is *acinar* (as′ĭ-nar) (an *acinus* is literally a grape or berry).

Endocrine Glands

Endocrine glands have no ducts, so they are often referred to as *ductless glands*. They secrete directly into the bloodstream rather than onto an epithelial surface. More specifically, endocrine glands produce messenger molecules called **hormones** (hor′mōnz; "exciters"), which they release into the extracellular space. These hormones then enter nearby capillaries and travel through the bloodstream to specific *target organs*. Each hormone signals its target organs to respond in some characteristic way. For example, endocrine cells in the intestine secrete a hormone that signals the pancreas to release the enzymes that help digest a meal.

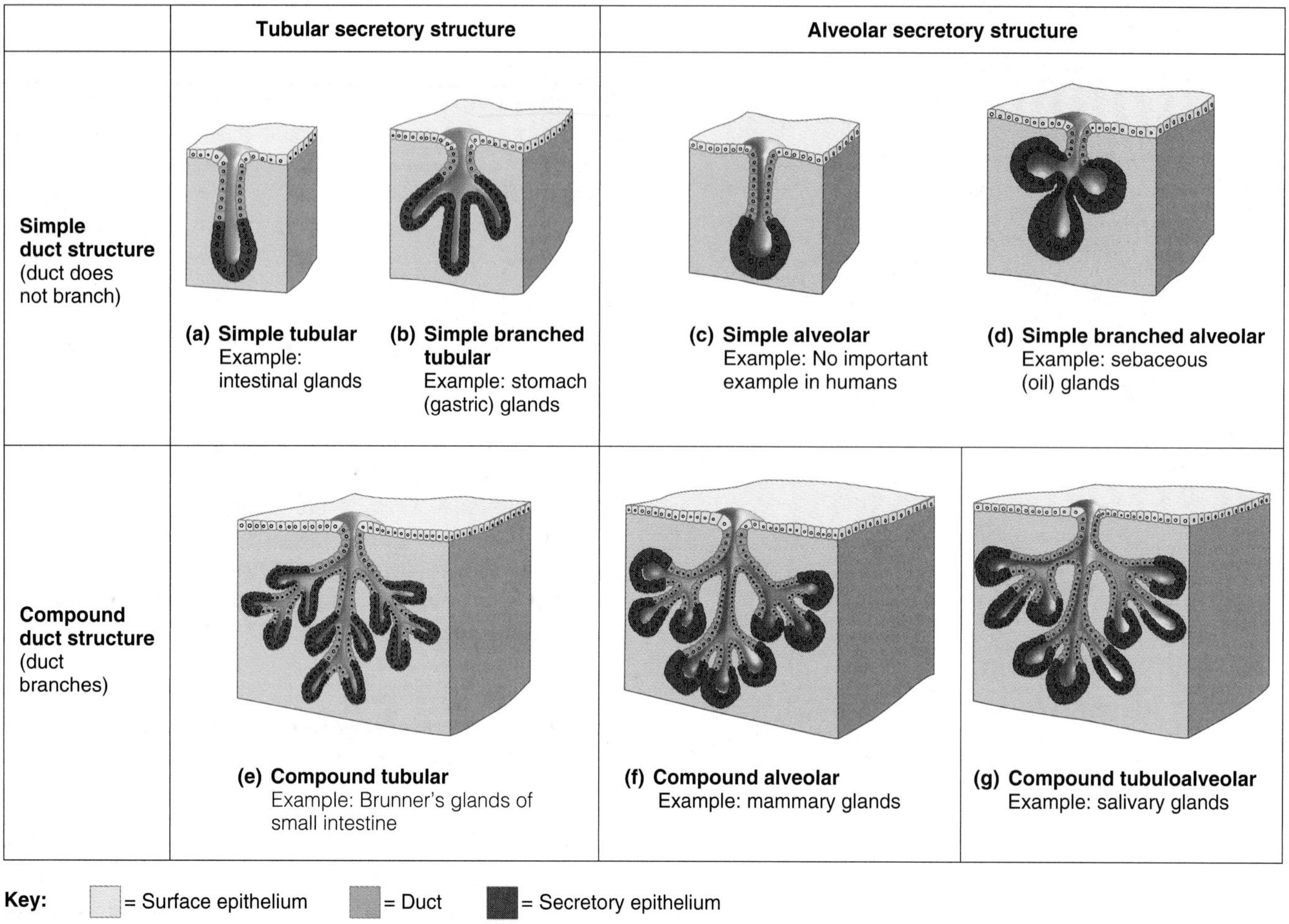

FIGURE 4.11 **Types of multicellular exocrine glands.** These glands are classified according to their ducts (simple or compound) and the structure of their secretory units (tubular, alveolar, tubuloalveolar). The word *acinar* can be substituted for *alveolar*. The largest compound glands, such as the mammary glands and saliva-producing glands, contain many copies of the gland structures illustrated here.

Although most endocrine glands derive from epithelia, some derive from other tissues. The endocrine system is discussed in detail in Chapter 25.

CONNECTIVE TISSUE

Connective tissue is the most diverse and abundant type of tissue. There are four main classes of connective tissue and many subclasses (Figure 4.12). The main classes are: (1) *connective tissue proper,* familiar examples of which are fat tissue and the fibrous tissue of ligaments; (2) *cartilage;* (3) *bone tissue;* and (4) *blood.* Connective tissues do far more than just connect the cells and organs of the body together. They also form the basis of the skeleton (bone and cartilage), store and carry nutrients (fat tissue and blood), surround all the blood vessels and nerves of the body (connective tissue proper), and lead the body's fight against infection. The great variety of connective tissues may seem bewildering at first, but all connective tissues share the same simple structural plan (Figure 4.13): Their cells are always separated from one another by a large amount of nonliving extracellular material called the **extracellular matrix** (ma′triks; "womb"). This differs markedly from epithelial tissue, whose cells crowd closely together. Another feature that unifies the many connective tissues is that they all originate from the embryonic tissue called **mesenchyme** (p. 62). Mesenchyme is shown in Figure 4.15a on p. 92.

Connective Tissue Proper

Connective tissue proper has two subclasses: *loose connective tissue* and *dense connective tissue*. Along with the locations listed below, connective tissue proper also forms the supporting framework, or *stroma* (stro′mah; "mattress"), of many large organs in the body.

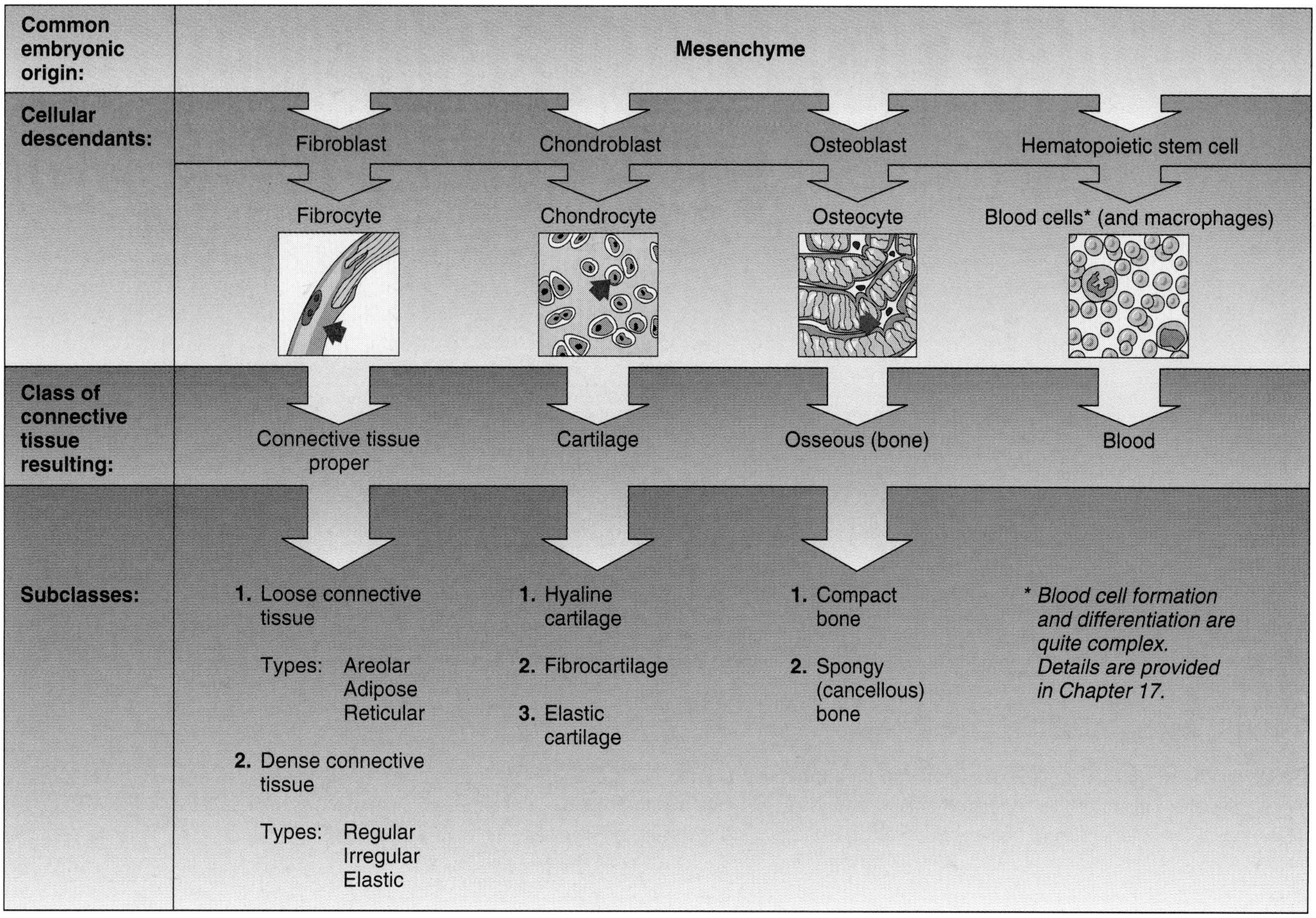

Common embryonic origin:
Mesenchyme
Cellular descendants:
Fibroblast
Chondroblast
Osteoblast
Hematopoietic stem cell
Fibrocyte
Chondrocyte
Osteocyte
Blood cells* (and macrophages)
Class of connective tissue resulting:
Connective tissue proper
Cartilage
Osseous (bone)
Blood
Subclasses:
1. Loose connective tissue
Types: Areolar
Adipose
Reticular
2. Dense connective tissue
Types: Regular
Irregular
Elastic
1. Hyaline cartilage
2. Fibrocartilage
3. Elastic cartilage
1. Compact bone
2. Spongy (cancellous) bone
* Blood cell formation and differentiation are quite complex. Details are provided in Chapter 17.

FIGURE 4.12 Classes of connective tissue. All connective tissues arise from the same embryonic tissue, mesenchyme. Additionally, a stem cell for all mesenchyme derivatives has recently been found in the bone marrow of adults. Such a stem cell forms fibroblasts, chondroblasts, osteoblasts, fat cells, blood cells, and muscle cells.

Areolar Tissue: A Model Connective Tissue

The connective tissues are so diverse and abundant that we need a "model" type, or "prototype," against which the other connective tissues can be compared. For our prototype, we will use the most widespread type of connective tissue proper, called loose **areolar** (ah-re′o-lar) **connective tissue** (Figures 4.14 and 4.15b). This is the connective tissue that underlies almost all the epithelia of the body and surrounds almost all the small nerves and blood vessels, including the capillaries. Because no cell can lie very far from a nutrient-carrying capillary, no cell can lie far from areolar connective tissue. In fact, one might say that all other tissues and cells in the body either border areolar connective tissue or are embedded in it.

We will describe the structure of areolar connective tissue in terms of its basic functions. These functions, which are shared by many other types of connective tissue, are as follows:

1. Support and binding of other tissues
2. Holding body fluids

Fibers
Ground substance
Extracellular matrix
Cells

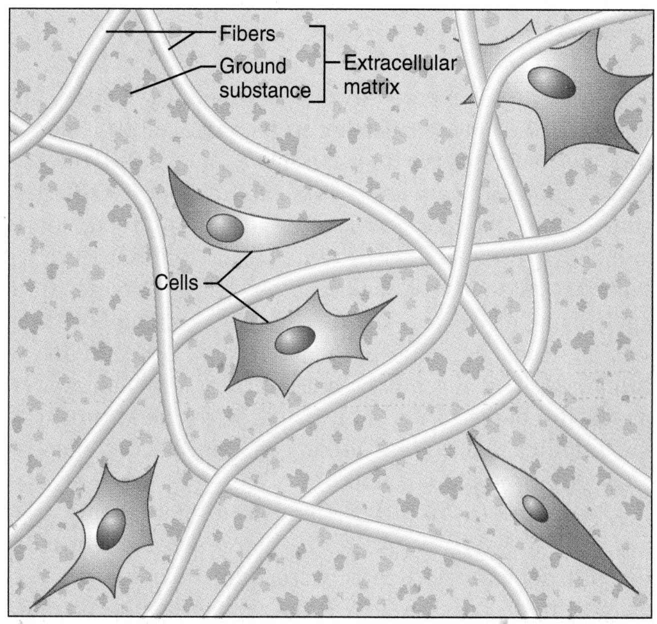

FIGURE 4.13 The basic organization of connective tissues (simplified). Cells are separated by an abundant extracellular matrix. The matrix consists of fibers and a jelly-like ground substance (except in blood tissue).

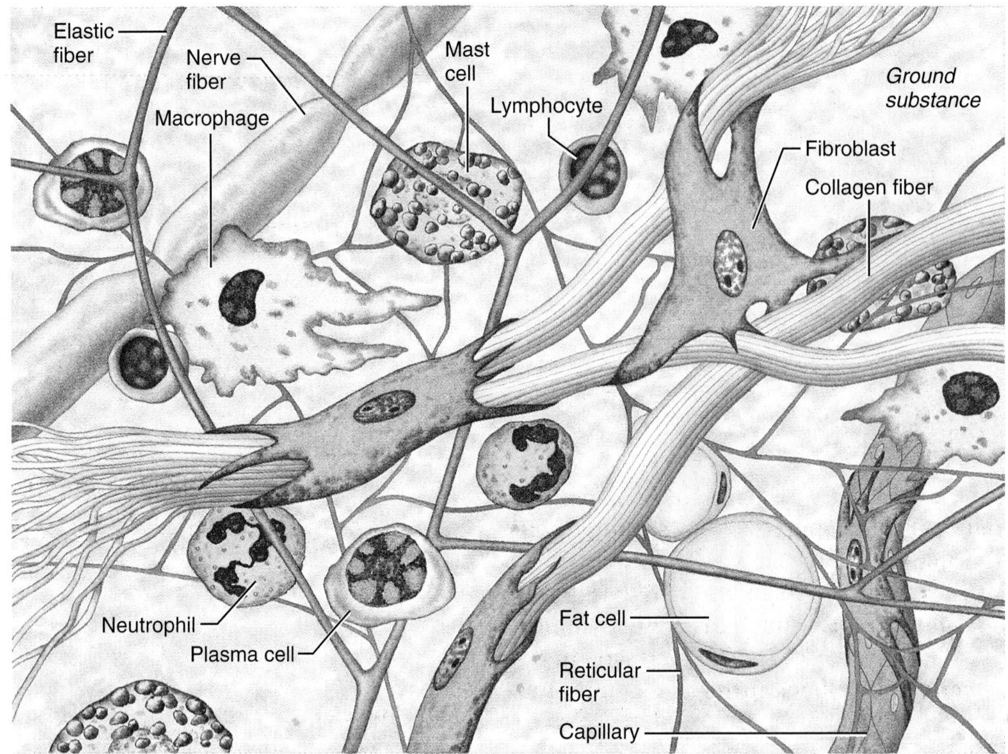

FIGURE 4.14 Areolar connective tissue: A model connective tissue. This tissue underlies epithelia and surrounds capillaries. Note the various cell types and the three classes of fibers (collagen, reticular, elastic) embedded in the ground substance.

3. Defending the body against infection
4. Storing nutrients as fat

Each of these functions is performed by a different structural part of the areolar tissue.

Fibers Provide Support Areolar connective tissue has three types of protein **fibers** in its extracellular matrix that give this tissue its supportive properties (Figure 4.14): *collagen, reticular,* and *elastic fibers.* **Collagen fibers,** the strongest and most abundant type, allow connective tissue to withstand tension (pulling forces). Pulling tests show that collagen fibers are stronger than steel fibers of the same size! The thick collagen fibers that one sees with the light microscope are bundles of thinner, striped threads called *unit fibrils,* which consist of still thinner strands that are strongly cross-linked to one another. This cross-linking is the source of collagen's great tensile strength.

Reticular (re-tik′u-lar) **fibers** are bundles of a special type of collagen unit fibril. These short fibers cluster into delicate networks (*reticulum* = network) that cover and support all structures bordering the connective tissue. For example, capillaries are coated with fuzzy nets of reticular fibers, and these fibers also border the nearby epithelia. Individual reticular fibers glide freely across one another whenever the network is pulled, so they allow more "give" than do collagen fibers. Thus, capillaries can expand a bit without being choked by their collar of reticular fibers.

Areolar connective tissue also contains **elastic fibers.** Long and thin, these fibers branch to form wide networks within the extracellular matrix. Elastic fibers contain a rubber-like protein called **elastin** (e-las′tin), which allows them to function like rubber bands. Connective tissue can stretch only so much before its thick, rope-like collagen fibers become taut. The elastic fibers then recoil so that the stretched tissue springs back to shape.

A single kind of cell produces all the fibers of areolar connective tissue: the **fibroblast** (fi′bro-blast; "fiber-generator"). Fibroblasts are the most abundant cell type in connective tissue proper. They are shaped either like spindles (elongated to a point at both ends) or like stars. To be precise, fibroblasts make and secrete the protein subunits of fibers, which assemble into the fibers upon entering the matrix. Resting fibroblasts that are not actively secreting their product are sometimes called *fibrocytes* (fi′bro-sīts; "fiber cells"). Furthermore, undifferentiated, mesenchyme-like cells that become fibroblasts lie near the small blood vessels, and are thus named *perivascular cells.*

CLINICAL APPLICATIONS

SCURVY Vitamin C, which is abundant in citrus fruits such as oranges and lemons, is necessary for the proper cross-linking of the molecules that make up collagen fibers. A deficiency of this vitamin in the diet can lead to **scurvy,** a weakening of collagen and connective tissue throughout the body. As we will explain later in this text, strong collagen is necessary for holding teeth in their sockets, for reinforcing the walls of blood vessels, and for healing wounds and forming scar tissue. ✚

Critical Reasoning

Can you deduce some of the symptoms of scurvy?

The teeth fall out, blood vessels rupture, and wounds fail to heal.

Ground Substance Holds Fluid Recall that areolar connective tissue lies between the capillaries and all other cells and tissues in the body, such as epithelium and muscle. Nutrients and oxygen diffuse out of the capillaries and travel through a watery fluid in the extracellular matrix to reach the surrounding cells. Likewise, waste molecules from these cells diffuse back through this fluid into the capillaries, to be taken away by the bloodstream. The fluid that occupies areolar connective tissue is called **tissue fluid** or *interstitial fluid,* and it derives from the blood itself. That is, all the small molecules of blood (ions, water, and so on) are slowly pushed out through the capillary walls to form tissue fluid, which gradually returns to the bloodstream at about the same rate it forms. In your body, there is as much tissue fluid in the matrix of areolar connective tissue as there is blood in all your blood vessels. The part of the extracellular matrix that holds this tissue fluid is called **ground substance** (Figure 4.14). This jelly-like material consists of large sugar and sugar-protein molecules that soak up fluid like a sponge. These molecules are called glycosaminoglycans and proteoglycans. The molecules of ground substance are made and secreted by the nearby fibroblasts.

So far, we have established that the extracellular matrix of connective tissue is a combination of (1) fibers and (2) fluid-holding ground substance. We should now add that fibroblasts attach to the matrix components through integral proteins in their plasma membranes, called *integrins.*

Defense Cells Fight Infection Areolar connective tissue is the main battlefield in the body's war against infectious microorganisms, such as bacteria, viruses, fungi, and parasites. It is easy to see why this is so, for microorganisms that invade the body always enter areolar tissue after penetrating the body's epithelial surfaces. Every effort is made to destroy the microorganisms at their entry site in the areolar tissue—otherwise they enter the capillaries and use the circulatory vessels to spread throughout the body. Areolar connective tissue contains a variety of defense cells, all of which originate as blood cells and migrate to the connective tissue by leaving the capillaries. These defenders gather at infection sites in large numbers, where each uses its own strategy to destroy the invaders. The defense cells include the following:

1. **Macrophages** (mak′ro-fāj-es). These "big eaters" are oval cells whose surface is ruffled by pseudopods (Figure 4.14). Macrophages are the nonspecific phagocytic cells of our body; that is, they engulf and devour a wide variety of foreign materials, ranging from whole bacteria to foreign molecules to dirt particles. Macrophages also dispose of dead tissue cells.
2. **Plasma cells.** These egg-shaped cells secrete protein molecules called *antibodies.* Antibodies bind to foreign molecules and microorganisms, marking them for destruction. For example, bacteria covered with antibodies are much more likely to be eaten by macrophages than those that are not. Antibodies and plasma cells are discussed in more detail in Chapter 17 (p. 509).
3. **Mast cells.** These oval cells lie near small blood vessels and possess many large secretory granules. Indeed, *mast* means "stuffed full (of granules)." The granules contain many chemicals that mediate inflammation, especially in severe allergies. Such chemical mediators include histamine, heparin, and proteases (protein-degrading enzymes), and they are secreted in response to infections and to IgE, the type of antibody we produce in the presence of allergy-inducing substances. Histamine, the most important mediator, increases the permeability of the nearby capillaries, causing more tissue fluid to leave the bloodstream. The consequent swelling of the areolar tissue with fluid is a major characteristic of inflammation (p. 102). Heparin in mast cells was recently found to bind and store the other mast-cell molecules, and to regulate their action.

   Besides mediating inflammation, mast cells also seem to play a role in our defenses against parasitic worms, our natural immunity against bacteria, and the normal repair of fibers, ground substance, and blood vessels in connective tissues.
4. **Neutrophils, lymphocytes, and eosinophils.** These are white blood cells that leave the bloodstream to fight infection (Chapter 17, p. 507). *Neutrophils* (nu′tro-filz) enter infected areas quickly and are experts at phagocytizing bacteria.

Cellular defenses are not the only means by which areolar connective tissue fights infection. Both the viscous ground substance and the dense networks of collagen fibers in the extracellular matrix slow the progress of invading microorganisms. Some bacteria, however, secrete enzymes that rapidly break down ground substance or collagen.

Text continues on page 97.

Embryonic connective tissue

(a) Mesenchyme

Description: Embryonic connective tissue; gel-like ground substance containing fibers; star-shaped mesenchymal cells.

Function: Gives rise to all other connective tissue types.

Location: Primarily in embryo.

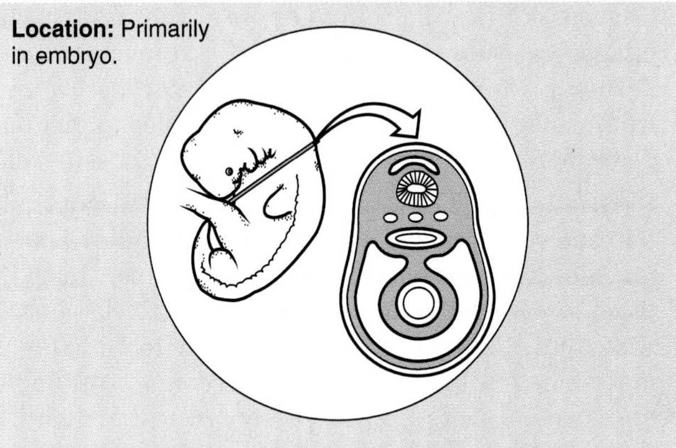

Photomicrograph: Mesenchymal tissue, an embryonic connective tissue (523×); the clear-appearing background is the fluid ground substance of the matrix; notice the fine, sparse fibers.

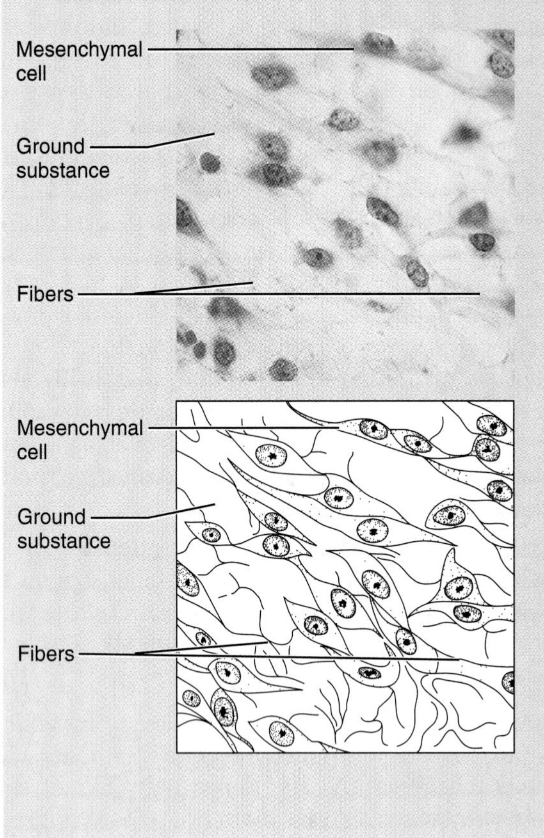

Connective tissue proper: Loose connective tissue (b to d)

(b) Areolar connective tissue

Description: Gel-like matrix with all three fiber types; cells: fibroblasts, macrophages, mast cells, and some white blood cells.

Function: Wraps and cushions organs; its macrophages phagocytize bacteria; plays important role in inflammation; holds and conveys tissue fluid.

Location: Widely distributed under epithelia of body, e.g, forms lamina propria of mucous membranes; packages organs; surrounds capillaries.

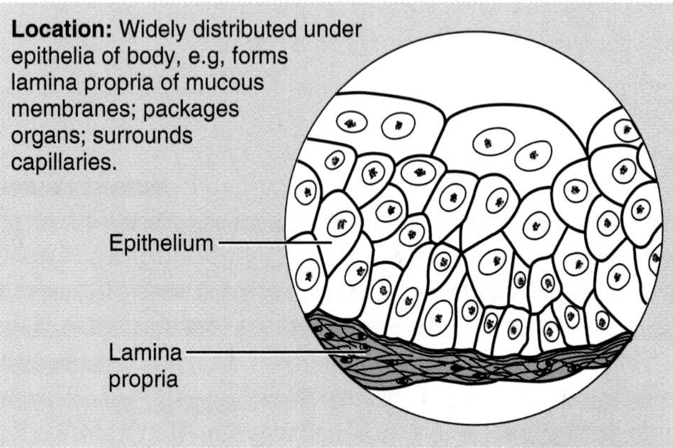

Photomicrograph: Areolar connective tissue, a soft packaging tissue of the body (200×).

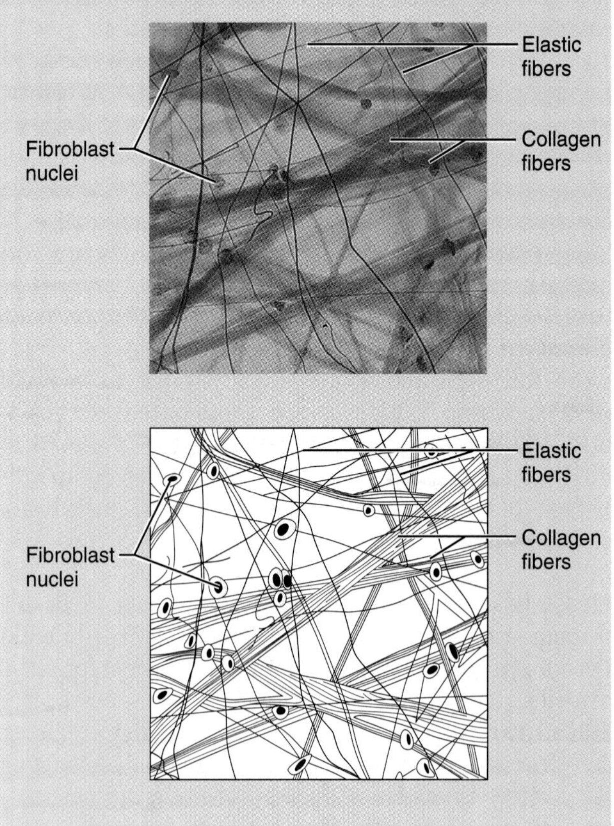

FIGURE 4.15 Connective tissues.

Loose connective tissue (*continued*)

(c) Adipose tissue

Description: Matrix as in areolar, but very sparse; closely packed adipocytes, or fat cells, have nucleus pushed to the side by large fat droplet.

Function: Provides reserve food fuel; insulates against heat loss; supports and protects organs.

Location: Under skin; around kidneys and eyeballs; within abdomen; in breasts.

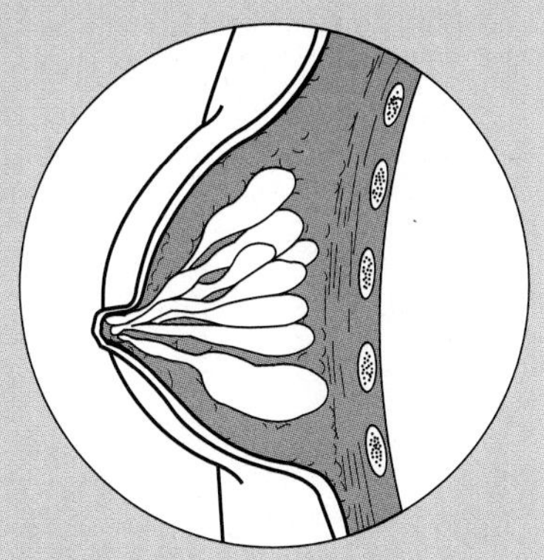

Photomicrograph: Adipose tissue from the subcutaneous layer under the skin (170×).

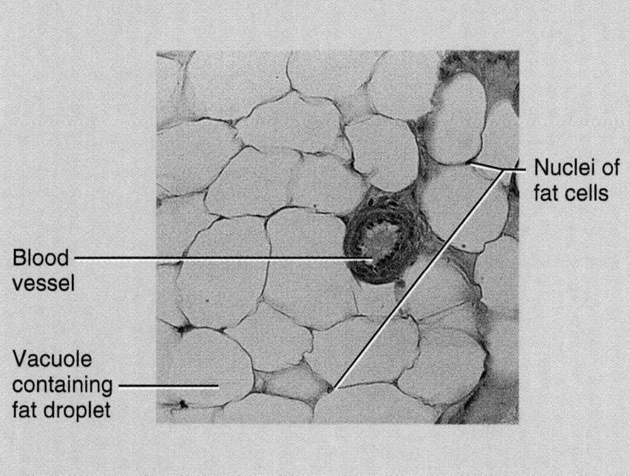

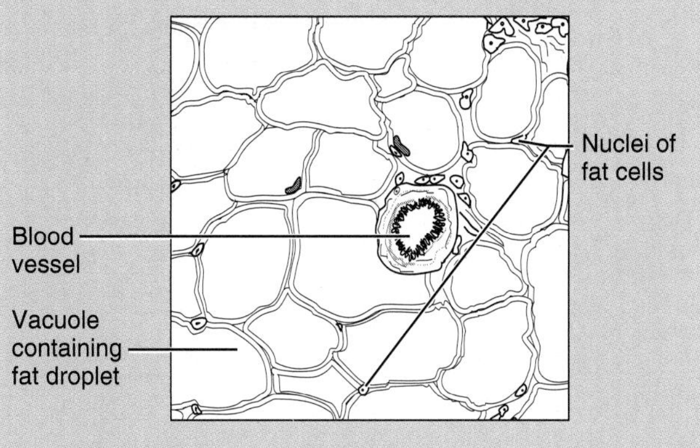

Loose connective tissue (*continued*)

(d) Reticular connective tissue

Description: Network of reticular fibers in a typical loose ground substance; reticular cells lie on the network.

Function: Fibers form a soft internal skeleton (stroma) that supports other cell types.

Location: Lymphoid organs (lymph nodes, bone marrow, and spleen).

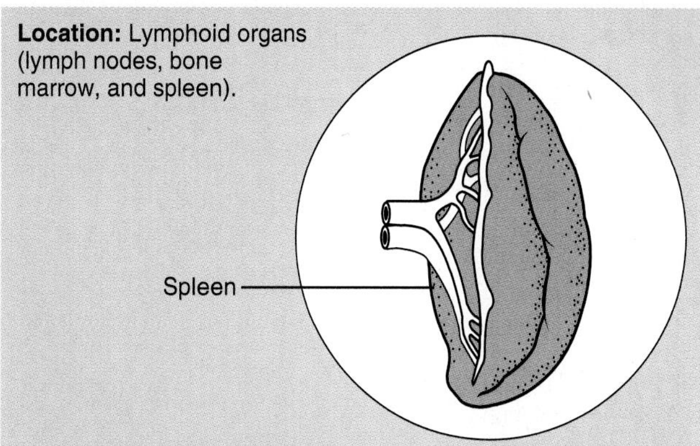

Photomicrograph: Dark-staining network of reticular connective tissue fibers forming the internal skeleton of the spleen (650×).

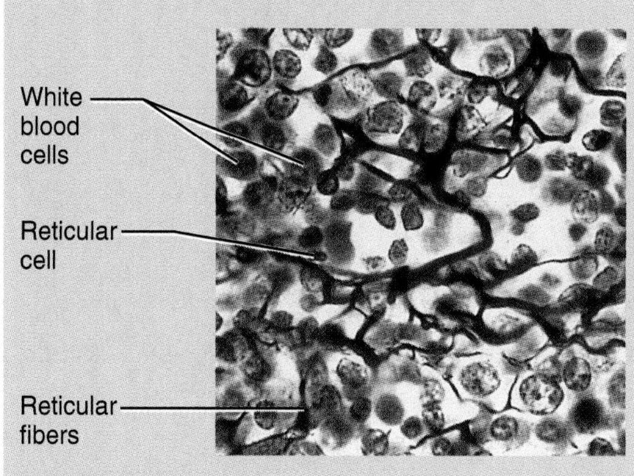

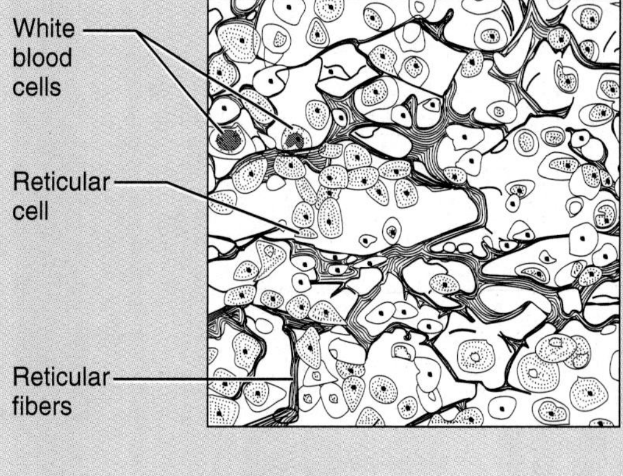

FIGURE 4.15 continued.

Connective tissue proper: Dense connective tissue (e and f)

(e) Dense irregular connective tissue

Description: Primarily irregularly arranged collagen fibers; some elastic fibers; major cell type is the fibroblast.

Function: Able to withstand tension exerted in many directions; provides structural strength.

Location: Dermis of the skin; submucosa of digestive tract; fibrous capsules of organs and of joints.

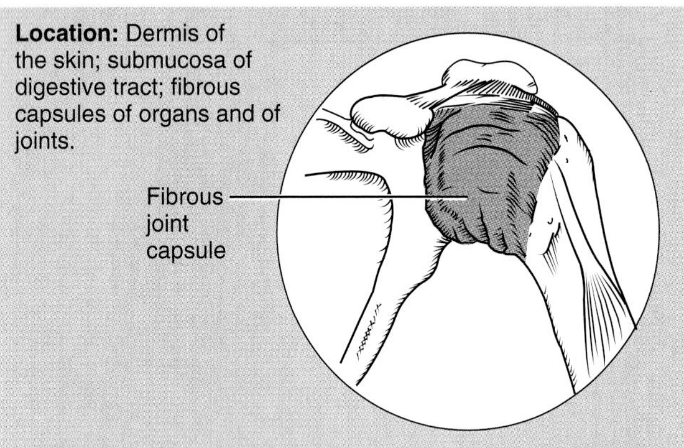

Photomicrograph: Dense irregular connective tissue from the dermis of the skin (550×).

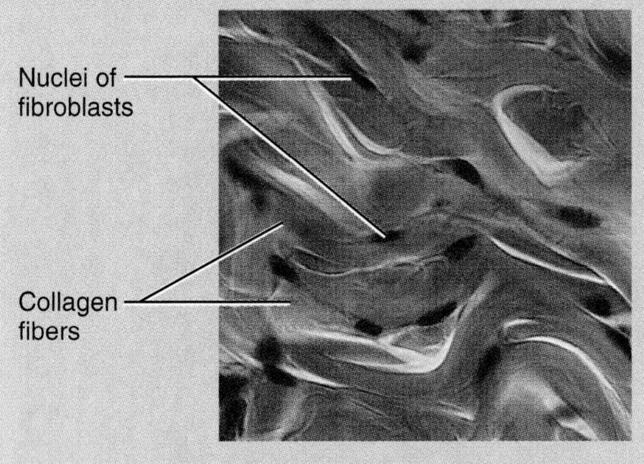

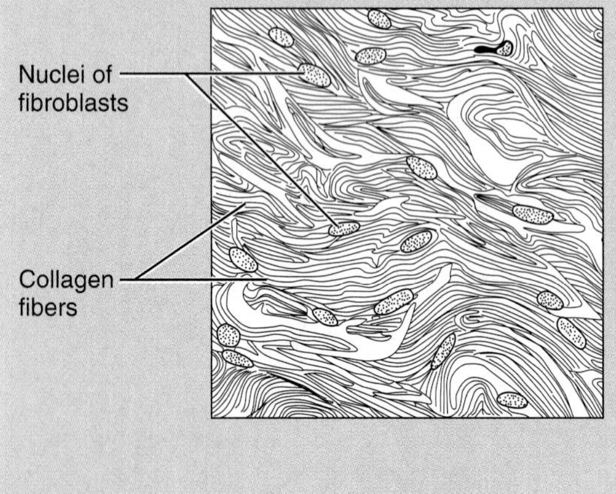

Dense connective tissue *(continued)*

(f) Dense regular connective tissue

Description: Primarily parallel collagen fibers; a few elastin fibers; major cell type is the fibroblast.

Function: Attaches muscles to bones or to muscles; attaches bones to bones; withstands great tensile stress when pulling force is applied in one direction.

Location: Tendons, most ligaments, aponeuroses.

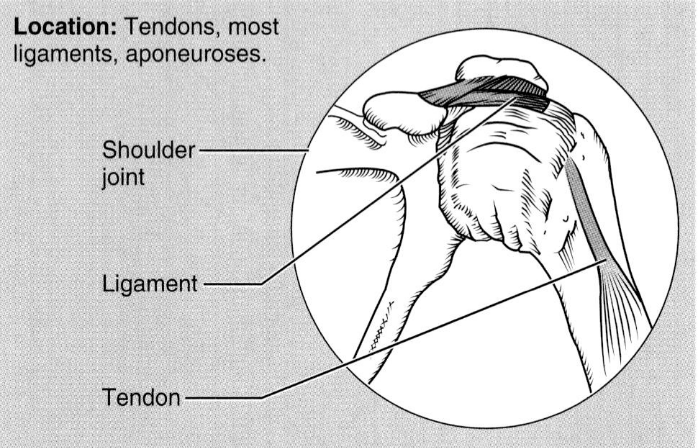

Photomicrograph: Dense regular connective tissue from a tendon (200×).

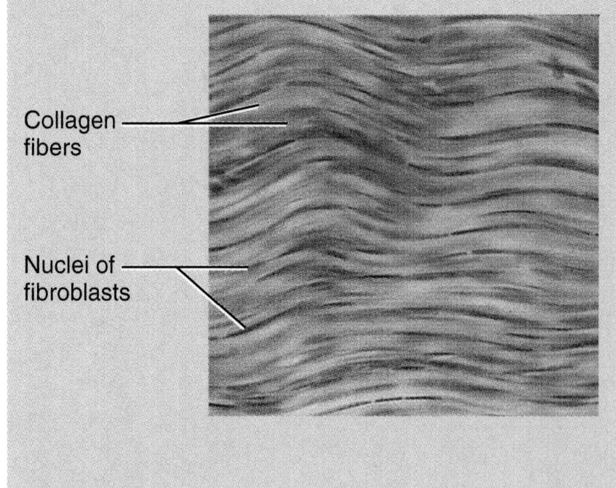

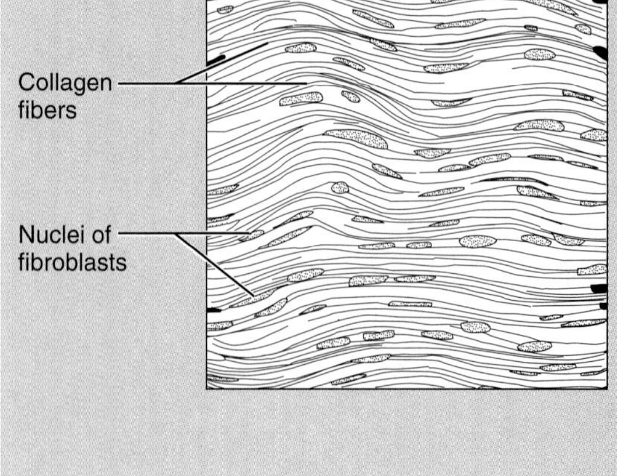

FIGURE 4.15 Connective tissues, continued.

Cartilage: (g to i)

(g) Hyaline cartilage

Description: Amorphous but firm matrix; collagen fibers form an imperceptible network; chondroblasts produce the matrix and when mature (chondrocytes) lie in lacunae.

Function: Supports and reinforces; has resilient cushioning properties; resists compressive stress.

Location: Forms most of the embryonic skeleton; covers the ends of long bones in joint cavities; forms costal cartilages of the ribs; cartilages of the nose, trachea, and larynx.

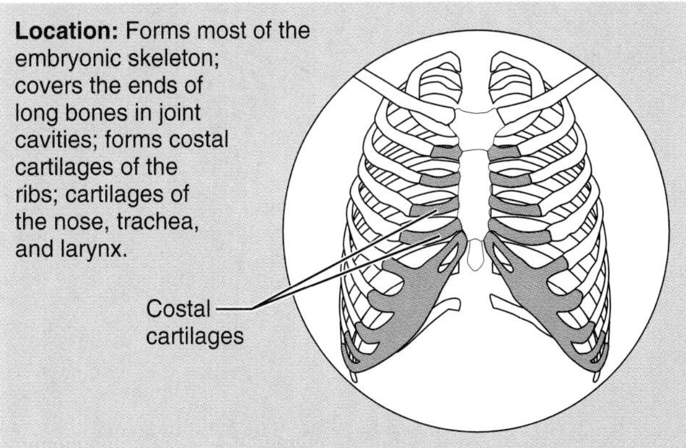

Photomicrograph: Hyaline cartilage from the trachea (375×).

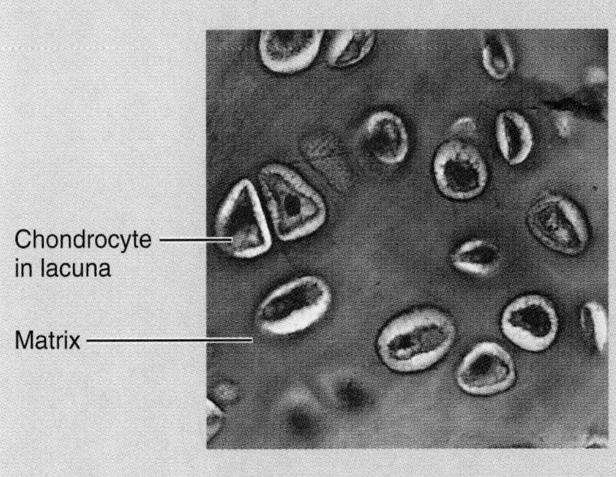

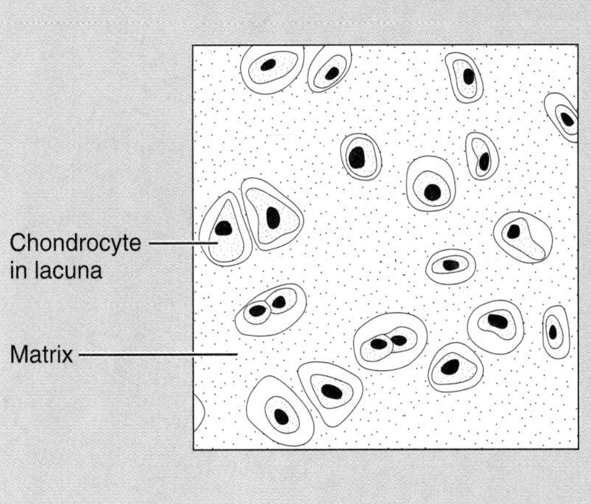

Cartilage *(continued)*

(h) Elastic cartilage

Description: Similar to hyaline cartilage, but more elastic fibers in matrix.

Function: Maintains the shape of a structure while allowing great flexibility.

Location: Supports the external ear (pinna); epiglottis.

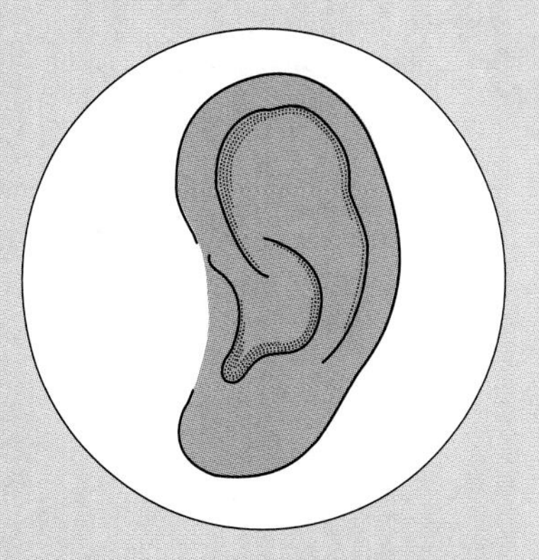

Photomicrograph: Elastic cartilage from the human ear pinna; forms the flexible skeleton of the ear (158×).

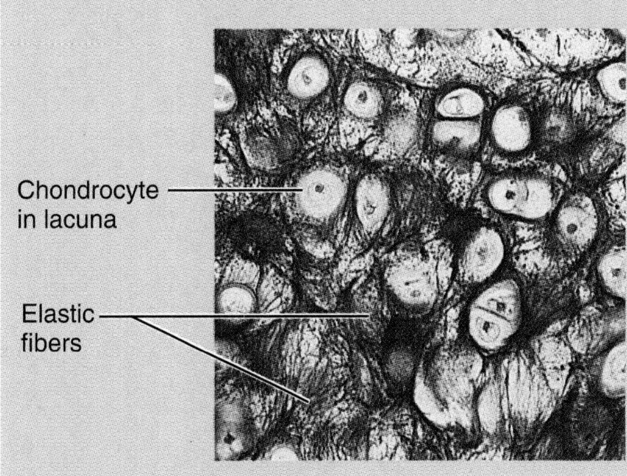

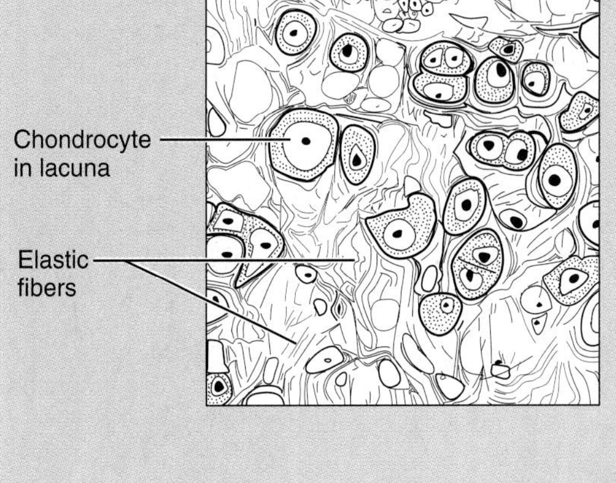

FIGURE 4.15 continued.

Cartilage (*continued*)

(i) Fibrocartilage

Description: Matrix similar but less firm than in hyaline cartilage; thick collagen fibers predominate.

Function: Tensile strength with the ability to absorb compressive shock.

Location: Intervertebral discs; pubic symphysis; discs of knee joint.

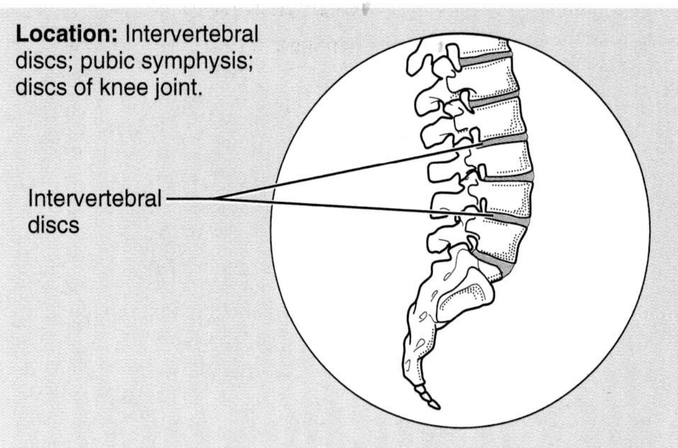

Photomicrograph: Fibrocartilage of an intervertebral disc (200×).

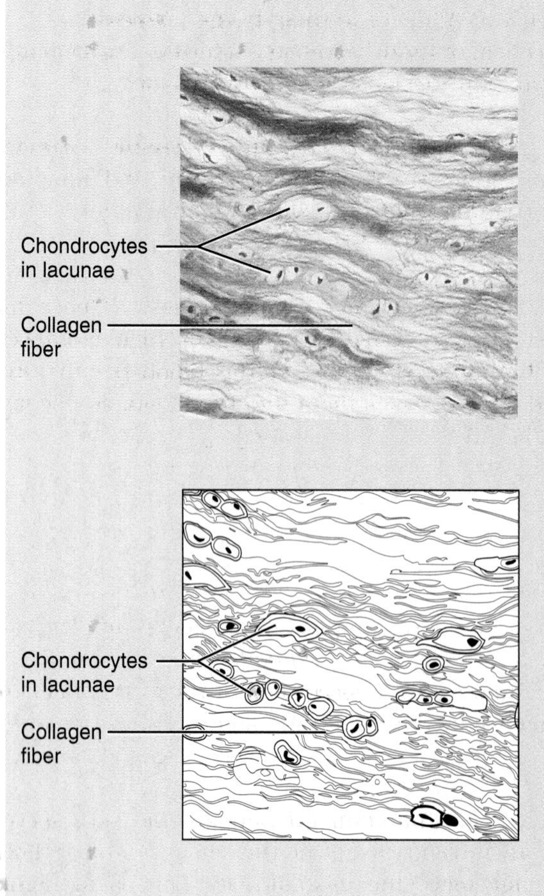

Others: (j and k)

(j) Bone (osseous tissue)

Description: Hard, calcified matrix containing many collagen fibers; osteocytes lie in lacunae. Very well vascularized.

Function: Bone supports and protects (by enclosing); provides levers for the muscles to act on; stores calcium and other minerals and fat; marrow inside bones is the site for blood cell formation (hematopoiesis).

Location: Bones

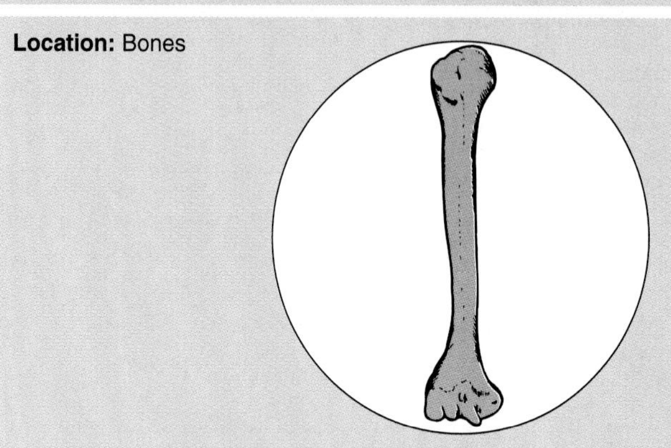

Photomicrograph: Cross-sectional view of bone (110×).

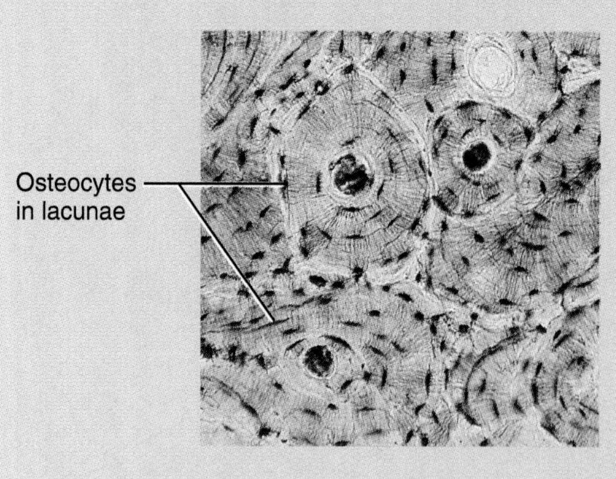

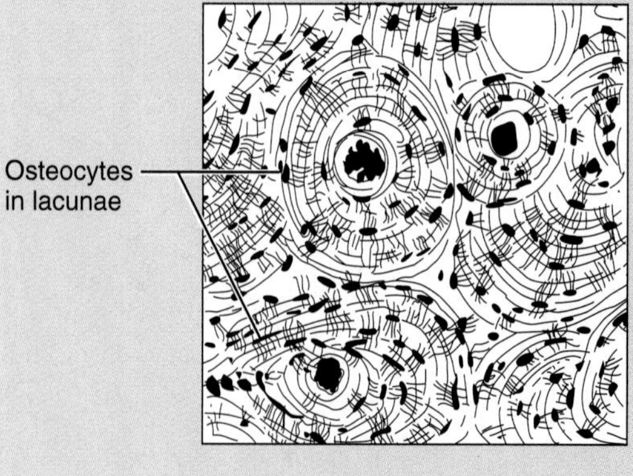

FIGURE 4.15 Connective tissues, continued.

Others *(continued)*

(k) Blood

Description: Red and white blood cells in a fluid matrix (plasma).

Function: Transport of respiratory gases, nutrients, wastes and other substances.

Location: Contained within blood vessels.

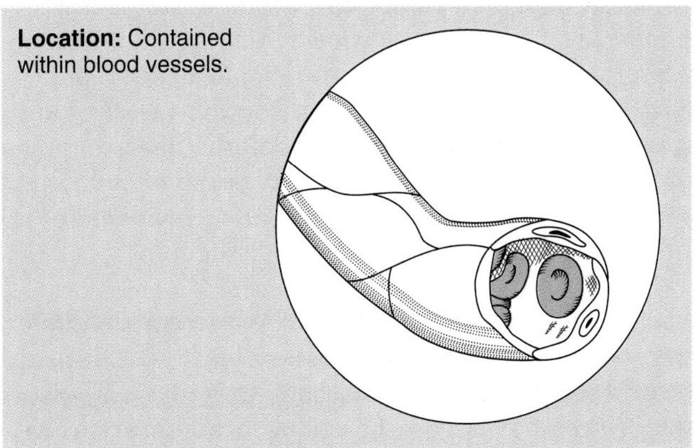

Photomicrograph: Smear of human blood (1100×); two white blood cells (neutrophil in upper left and lymphocyte in lower right) are seen surrounded by red blood cells.

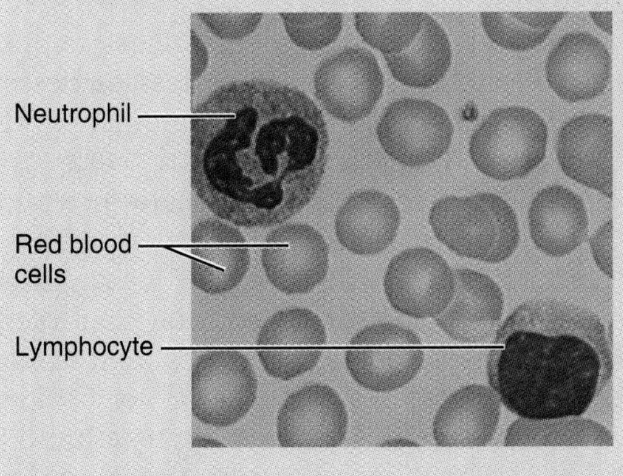

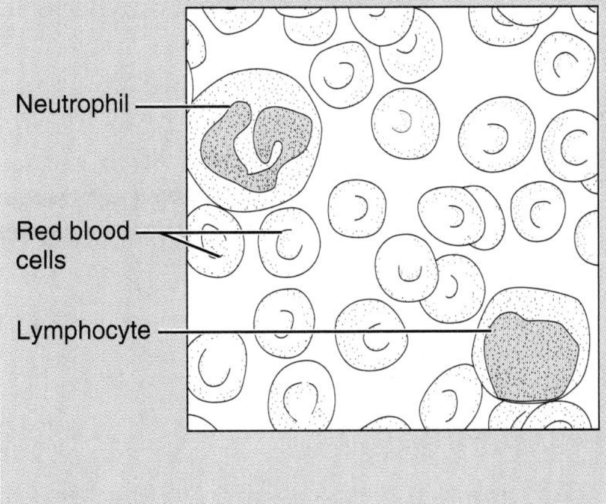

FIGURE 4.15 continued.

Critical Reasoning

What can you deduce about the virulence of matrix-degrading bacteria?

Matrix-degrading bacteria are highly invasive; that is, they spread rapidly through the connective tissues and are especially difficult for the body's defenses to control. An example of a bacterium that degrades ground substance is the Streptococcus *responsible for strep throat.*

Fat Cells Store Nutrients A minor function of areolar connective tissue is to store energy reserves as fat. The large, fat-storing cells are **fat cells,** also called *lipid cells, adipose* (ad'ĭ-pōs) *cells,* and *adipocytes* (Figure 4.14). Fat cells are egg-shaped, and their cytoplasm is dominated by a single, giant lipid droplet that flattens the nucleus and cytoplasm at one end of the cell. Mature fat cells are among the largest cells in the body and cannot divide. In areolar connective tissue, fat cells occur singly or in small groups.

Other Loose Connective Tissues

Now that we have established the structure of a "model" (areolar) connective tissue, we can consider the other connective tissues. We begin with those that most closely resemble areolar connective tissue, then proceed to those that are less similar. Two connective tissues are specialized variations of areolar tissue: adipose (fat) tissue and reticular tissue. Areolar, adipose, and reticular tissues make up the loose connective tissues.

Adipose Tissue (Figure 4.15c). **Adipose tissue** is similar to areolar connective tissue in structure and function, but its nutrient-storing function is much greater. Correspondingly, adipose tissue is crowded with fat cells, which account for 90% of its mass. These fat cells are grouped into large clusters called lobules. Adipose tissue is richly vascularized, reflecting its high metabolic activity. It removes lipids from the bloodstream after meals and later releases them into the blood, as needed. Without the fat stores in our adipose tissue, we could not live for more than a few days without eating. Adipose tissue is certainly abundant: It constitutes 18% of an average person's body weight (15% in men and 22% in women). Indeed, a chubby person's body may be 50% fat without being considered morbidly obese.

Much of the body's adipose tissue occurs in the layer beneath the skin called the ***hypodermis***. Adipose tissue also is abundant in the mesenteries, which are sheets of serous membranes that hold the stomach and intestines in place. Additionally, fat forms cushioning pads around the kidneys and behind the eyeballs in the orbits.

Whereas the abundant fat beneath the skin serves the general nutrient needs of the entire body, smaller depots of fat serve the local nutrient needs of highly active organs. Such depots occur around the hard-working heart and around lymph nodes (where cells of the immune system are furiously fighting infection), within some muscles, and as individual fat cells in the bone marrow, where new blood cells are produced at a frantic rate. Many of these local depots offer special lipids that are exceptionally highly enriched.

The typical, nutrient-storing fat we have considered so far is *white adipose tissue* or *white fat*. Another type, called **brown adipose tissue,** produces heat and is a nutrient consumer. Such brown fat occurs only in babies, who cannot yet warm themselves by shivering. It is located in the hypodermis between the two scapulae (shoulder blades) in the center of the back, on the side of the anterior neck, and on the anterior abdominal wall. It is even more richly vascularized than white fat. Each brown-fat cell contains many lipid droplets and numerous mitochondria, which use the lipid fuel to heat the bloodstream rather than to produce ATP molecules.

Advances in Understanding Adipose tissue is the subject of much research because increasing levels of obesity are causing health problems in the industrialized countries. Here are some recent findings. (1) White fat cells secrete a hormone called *leptin* ("slenderer"), which signals the brain to maintain a constant amount of adipose tissue in our body; that is, an increase in body fat leads to increased leptin secretion, which tells the brain to suppress our appetite and increase our metabolic rate until the excess fat is lost. Disruption of this mechanism may be a cause of obesity. (2) It was shown that weight gain in adults involves the development of new adipose cells from undifferentiated precursors called preadipocytes, as well as enlargement of existing fat cells; an understanding of how fat cells develop could lead to strategies for weight loss. (3) Unexpectedly, brown and white fat cells seem to differ in the activity of just one gene. Through genetic therapy, we may one day "burn off" our excess weight by converting fat-storing white fat to fat-consuming brown fat. ✷

Reticular Connective Tissue
(Figure 4.15d) **Reticular connective tissue** resembles areolar tissue, but the only fibers in its matrix are reticular fibers. These fine fibers form a broad, three-dimensional network, a labyrinth of caverns that hold many free cells. The bone marrow, spleen, and lymph nodes, which house many free blood cells outside of their capillaries, consist largely of reticular connective tissue. Fibroblasts called **reticular cells** lie along the reticular network of this tissue. Reticular tissue is discussed further in Chapters 17 (p. 512) and 20 (p. 593).

Dense Connective Tissue

Dense connective tissue, or *fibrous connective tissue,* contains more collagen than does areolar connective tissue. With its thick collagen fibers, it can resist extremely strong pulling (tensile) forces. The two main kinds of dense connective tissue are *irregular* and *regular*.

Dense Irregular Connective Tissue
Dense irregular connective tissue (Figure 4.15e) is similar to areolar tissue, but its collagen fibers are much thicker. These fibers run in different planes, allowing this tissue to resist strong tensions from different directions. This tissue dominates the leathery dermis of the skin, which is stretched, pulled, and hit from various angles. This tissue also makes up the *fibrous capsules* that surround certain organs in the body, such as kidneys, lymph nodes, and bones. Its cellular and matrix elements are the same as in areolar connective tissue.

Dense irregular connective tissue is confusing, even for experts. The dermis obviously fits the name because its fibers run randomly, as one would expect for a tissue called "irregular." However, the fibrous capsules of organs consist of two layers, with all the fibers in a layer running parallel to each other but perpendicular to the fibers in the other layer. That is, two perpendicular "regular" layers can make the tissue "irregular."

Dense Regular Connective Tissue
All collagen fibers in **dense regular connective tissue** (Figure 4.15f) usually run in the same direction, parallel to the direction of pull. Crowded between the collagen fibers are rows of fibroblasts, which continuously manufacture the fibers and a scant ground substance. When this tissue is not under tension, its collagen fibers are slightly wavy. Unlike areolar connective tissue, dense regular connective tissue is poorly vascularized and contains no fat cells or defense cells.

With its enormous tensile strength, dense regular connective tissue is the main component of *ligaments,* bands or sheets that bind bones to one another. It also is the main tissue in *tendons,* which are cords that attach muscles to bones, and *aponeuroses* (ap″o-nu-ro′sēs), which resemble sheet-like tendons.

Dense regular connective tissue also forms **fascia** (fash′e-ah; "a band"), a fibrous membrane that wraps around muscles, around groups of muscles, and around large vessels and nerves. Many sheets of fascia occur throughout the body, binding structures together like plastic sandwich wrap. When the word *fascia* is used alone, it is understood to mean *deep fascia. Superficial fascia,* something else entirely, is the fatty hypodermis below the skin.

In a few ligaments, bundles of elastic fibers outnumber the collagen bundles. Examples are in the *ligamentum nuchae* and *ligamentum flava*, which connect successive vertebrae (p. 168). The content of elastic fibers is so high that this is called **elastic connective tissue.**

Other Connective Tissues: Cartilage, Bone, and Blood

As we have seen, connective tissue proper has the ability to resist tension (pulling). Cartilage and bone are the firm connective tissues that resist *compression* (pressing) as well as tension. Like areolar tissue, they consist of cells separated by a matrix containing fibers, ground substance, and tissue fluid. However, these skeletal tissues exaggerate the supportive functions of connective tissue and play no role in fat storage or defense against disease. We will discuss cartilage and bone, then introduce blood, the most unusual connective tissue.

Cartilage

Cartilage, a firm but flexible tissue, makes up several parts of the skeleton. For example, it forms the supporting rings of the trachea (windpipe) and gives shape to the nose and ears. Like all connective tissues, cartilage consists of cells separated by an abundant extracellular matrix (Figure 4.15g). This matrix contains thin collagen fibrils, a ground substance, and an exceptional quantity of tissue fluid; in fact, cartilage consists of up to 80% water! The arrangement of water in its matrix enables cartilage to spring back from compression, as we will explain in Chapter 6.

Cartilage is simpler than other connective tissues: It contains no blood vessels or nerves and just one kind of cell, the **chondrocyte** (kon′dro-sīt; "cartilage cell"). Each chondrocyte resides in a cavity called a lacuna, surrounded by the firm matrix. Immature chondrocytes are *chondroblasts,* cells that actively secrete the matrix during cartilage growth. Cartilage is found in three varieties, each dominated by a particular fiber type: hyaline cartilage, elastic cartilage, and fibrocartilage. These are introduced in Figure 4.15 g–i and discussed in depth in Chapter 6.

Bone Tissue

Because of its rock-like hardness, **bone tissue** has a tremendous ability to support and protect body structures (Figure 4.15j). Bone matrix contains inorganic calcium salts (bone salts), which enable bone to resist compression, and an abundance of collagen fibers, which allow bone to withstand strong tension.

Immature bone cells, called *osteoblasts* (os″te-o-blasts; "bone formers"), secrete the collagen fibers and ground substance of the matrix. Then, bone salts precipitate on and between the collagen fibers, hardening the matrix. The mature bone cells, called **osteocytes,** inhabit cavities called lacunae in this hardened matrix. Bone is a living and dynamic tissue, well-supplied with blood vessels. It is discussed further in Chapter 6.

Blood

Blood, the fluid in the blood vessels, is the most atypical connective tissue. It does not connect things or give mechanical support. It is classified as a connective tissue because it develops from mesenchyme and consists of blood *cells* surrounded by a nonliving matrix, the liquid blood *plasma* (Figure 4.15k). Its cells and matrix are very different from those in other connective tissues. Blood functions as the transport vehicle for the cardiovascular system, carrying nutrients, wastes, respiratory gases, and many other substances throughout the body. Blood is discussed in detail in Chapter 17.

COVERING AND LINING MEMBRANES

Now that we have discussed connective and epithelial tissues, we can consider the **covering** and **lining membranes** that combine these two tissue types (Figure 4.16, p. 100). These membranes, which cover broad areas within the body, consist of an epithelial sheet plus the underlying layer of connective tissue proper. These membranes are of three types: cutaneous, mucous, and serous.

The **cutaneous membrane** (ku-ta′ne-us; "skin") is the skin, covering the outer surface of the body (Figure 4.16a). Its outer epithelium is the thick epidermis, and its inner connective tissue is the dense dermis. It is a dry membrane. The skin is discussed further in Chapter 5.

A **mucous membrane,** or **mucosa** (mu-ko′sah), lines the inside of every hollow internal organ that opens to the outside of the body. More specifically, mucous membranes line the tubes of the respiratory, digestive, reproductive, and urinary systems (Figure 4.16b). Although different mucous membranes vary widely in the types of epithelia they contain, all are wet or moist. As their name implies, many mucous membranes secrete mucus. Not all of them do so, however.

All mucous membranes consist of an epithelial sheet directly underlain by a layer of loose connective tissue called the *lamina propria* (lam′ĭ-nah pro′pre-ah; "one's own layer"). In the mucous membranes of the digestive system, the lamina propria rests on a third layer of smooth muscle cells. We will discuss the specific mucous membranes of the body in later chapters.

Serous membranes, or **serosae,** introduced in Chapter 1, are the slippery membranes that line the closed pleural, pericardial, and peritoneal cavities (Figures 4.16c and 1.9). A serous membrane consists of a simple squamous epithelium (mesothelium) lying on a thin layer of areolar connective tissue. A slippery *serous fluid* is produced by this membrane, beginning as a filtrate of the blood in capillaries in the connective tissue, with the addition of lubricating molecules by the mesothelium.

MUSCLE TISSUE

The two remaining tissue types are muscle and nervous tissues. These are sometimes called *composite tissues* because, along with their own muscle or nerve cells, they contain small amounts of areolar connective tissue. (Areolar connective tissue surrounds all blood vessels, and both muscle and nervous tissue are richly vascularized.)

Muscle tissues (Figure 4.17) bring about most kinds of body movements. Most muscle cells are called **muscle fibers.** They have an elongated shape and contract

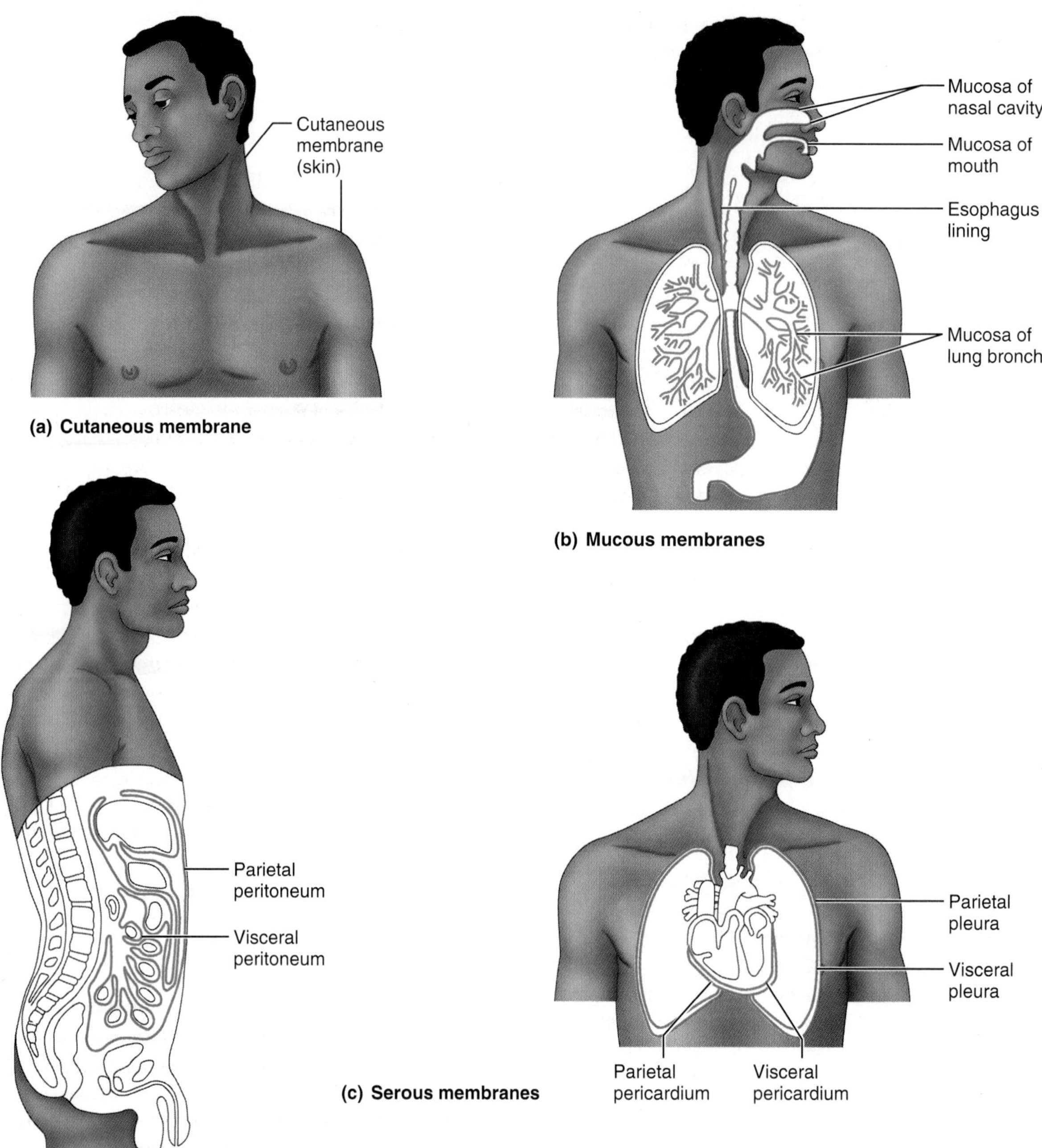

FIGURE 4.16 Covering and lining membranes. Each of these membranes contains an epithelium and an underlying connective tissue. **(a)** The cutaneous membrane is the skin. **(b)** Mucous membranes line body cavities that are open to the exterior. **(c)** Serous membranes line the peritoneal, pericardial, and pleural cavities, which are closed to the exterior.

forcefully as they shorten. These cells contain many myofilaments (mi″o-fil′ah-ments; "muscle filaments"), elaborate versions of the actin and myosin filaments that bring about contraction in all cell types (p. 40). There are three kinds of muscle tissue: skeletal, cardiac, and smooth.

Skeletal muscle tissue (Figure 4.17a) is the major component of organs called skeletal muscles, which pull on bones to cause body movements. Skeletal muscle cells are long, large cylinders that contain many nuclei. Their obvious banded, or striated, appearance reflects a highly organized arrangement of their myofilaments.

Cardiac muscle tissue (Figure 4.17b) occurs in the wall of the heart. It contracts to propel blood through the blood vessels. Like skeletal muscle cells, cardiac muscle cells are striated. However, they differ in two ways: (1) each cardiac cell has just one nucleus, and (2) cardiac cells branch and join at special cellular junctions called *intercalated* (in-ter′kah-la″ted) *discs*.

(a) Skeletal muscle

Description: Long, cylindrical, multinucleate cells; obvious striations.

Function: Voluntary movement; locomotion; manipulation of the environment; facial expression; voluntary control.

Location: In skeletal muscles attached to bones or occasionally to skin.

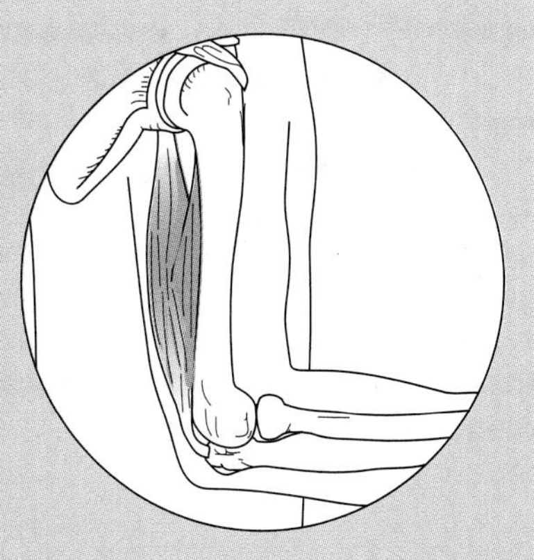

Photomicrograph: Skeletal muscle (approx. 130×). Notice the obvious banding pattern and the fact that these large cells are multinucleate.

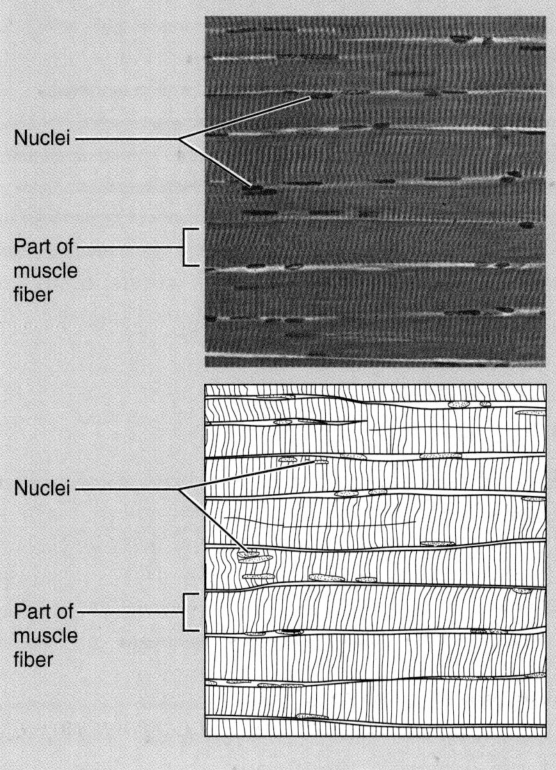

(b) Cardiac muscle

Description: Branching, striated, generally uninucleate cells that interdigitate at specialized junctions (intercalated discs).

Function: As it contracts, it propels blood into the circulation; involuntary control.

Location: The walls of the heart.

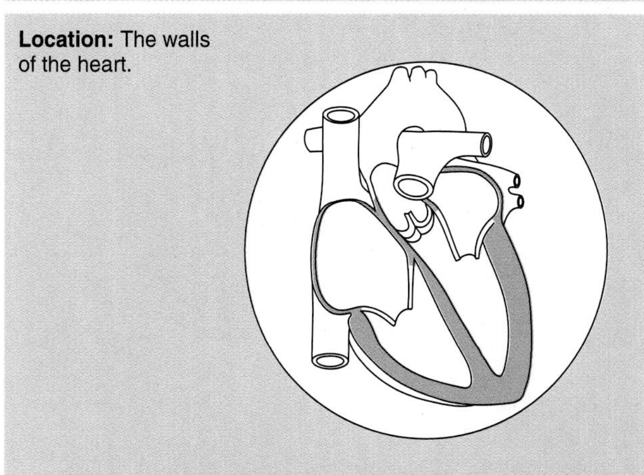

Photomicrograph: Cardiac muscle (275×); notice the striations, branching of cells, and the intercalated discs.

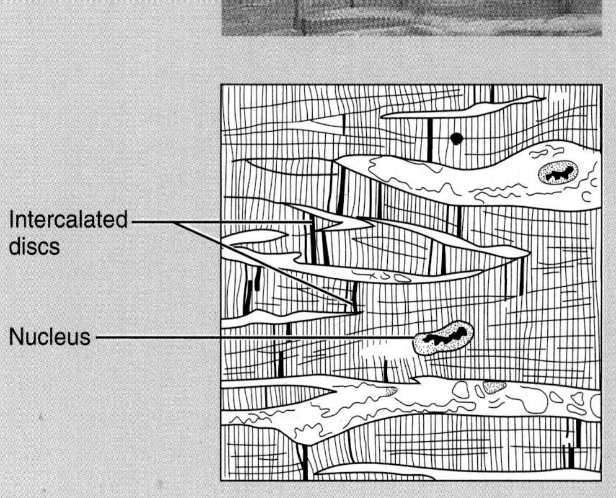

FIGURE 4.17 Muscle tissues (Figure continues on next page).

(c) Smooth muscle

Description: Spindle-shaped cells with central nuclei; cells arranged closely to form sheets; no striations.

Function: Propels substances or objects (foodstuffs, urine, a baby) along internal passageways; involuntary control.

Location: Mostly in the walls of hollow organs.

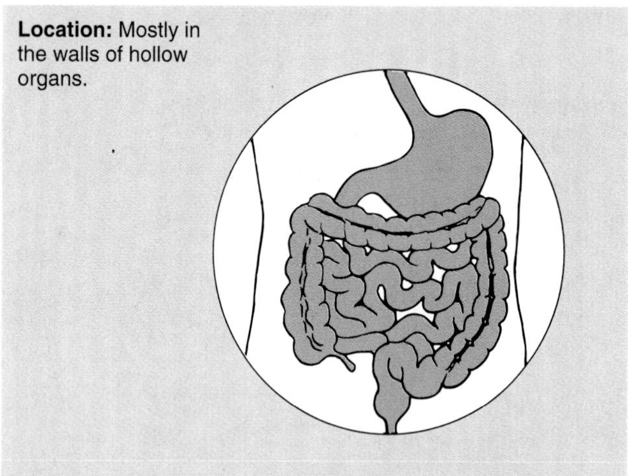

Photomicrograph: Sheet of smooth muscle (approx. 300×).

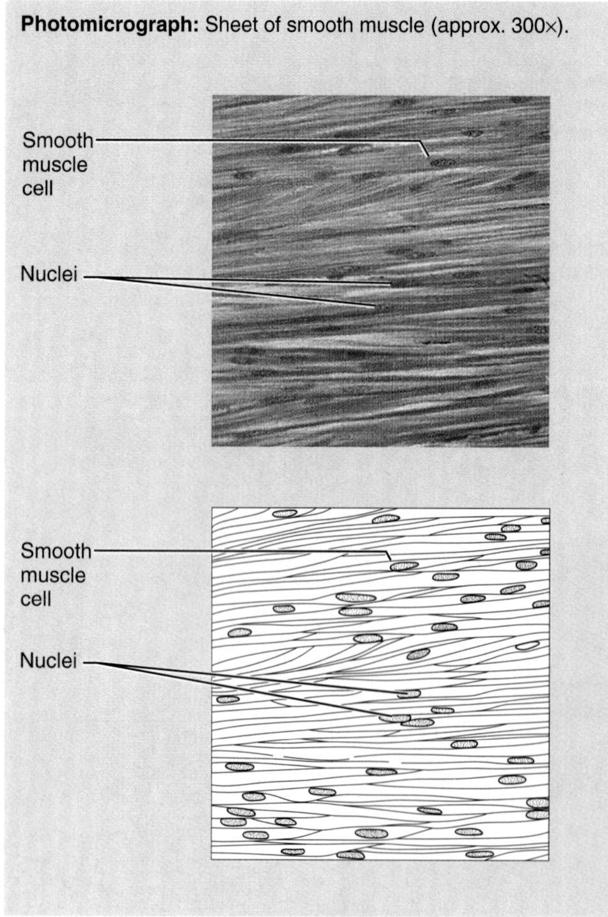

FIGURE 4.17 Muscle tissues, continued.

Smooth muscle tissue (Figure 4.17c) is so named because there are no visible striations in its cells. These cells are spindle-shaped and contain one centrally located nucleus. Smooth muscle primarily occurs in the walls of hollow viscera such as the digestive and urinary organs, uterus, and blood vessels. It generally acts to squeeze substances through these organs by alternately contracting and relaxing. Muscle tissue is described in detail in Chapter 10.

NERVOUS TISSUE

Nervous tissue is the main component of the nervous organs—the brain, spinal cord, and nerves—which regulate and control body functions. It contains two types of cells, neurons and supporting cells. **Neurons** are the highly specialized nerve cells (Figure 4.18) that generate and conduct electrical impulses. They have extensions, or processes, that allow them to transmit impulses over substantial distances within the body. **Supporting cells** are nonconducting cells that nourish, insulate, and protect the delicate neurons. A more complete discussion of nervous tissue appears in Chapter 12.

TISSUE RESPONSE TO INJURY

The body has many mechanisms for protecting itself from injury and invading microorganisms. Intact epithelia act as a physical barrier, but once that barrier has been penetrated, protective responses are activated in the underlying connective tissue proper. These are the *inflammatory* and *immune responses*. Inflammation is a nonspecific, local response that develops quickly and limits the damage to the injury site. The immune response, by contrast, takes longer to develop and is highly specific. It destroys particular infectious microorganisms and foreign molecules at the site of infection and throughout the body. We will concentrate on inflammation here, saving the immune response for Chapter 20.

Inflammation

Almost any injury or infection will lead to an inflammatory response. For example, let us assume the skin is cut by a dirty piece of glass, crushed by a blow in football, or has an infected pimple. As short-term or acute inflammation develops in the connective tissue, it produces four symptoms: *heat, redness, swelling,* and *pain.* Let us trace the source of each symptom.

The initial insult induces the release of *inflammatory chemicals* into the nearby tissue fluid. Injured tissue cells, macrophages, mast cells, and proteins from blood all serve as sources of these inflammatory mediators. These chemicals signal nearby blood vessels to dilate (widen), thus increasing the flow of blood to the injury site. The increase in blood flow is the source of the *heat* and *redness* of inflammation. Certain inflammatory chemicals, such as histamine (p. 91), increase the permeability of the capillaries, causing large amounts of tissue fluid to leave the bloodstream. The resulting accumulation of fluid in the connective tissue, called **edema** (ĕ-de′mah; "swelling"), causes the swelling of inflammation. The excess fluid presses on nerve endings, contributing to the sensation of *pain*. Some of the inflammatory chemicals also cause pain by affecting the nerve endings directly.

Although at first glance inflammatory edema seems detrimental, it is actually beneficial. The entry of blood-derived fluid into the injured connective tissue (1) helps to dilute toxins secreted by bacteria, (2) brings in oxygen and nutrients from the blood, necessary for tissue repair, and (3) brings in antibodies from the blood to fight infection. In very severe infections and in all wounds that sever blood vessels, the fluid leaking from the capillaries contains *clotting* proteins. In these cases, clotting occurs in the connective tissue matrix. The fibrous clot isolates the injured area and "walls in" the infectious microorganisms, preventing their spread.

The next stage in inflammation is *stasis* ("standing"). This is the slowdown in local blood flow that necessarily follows a massive exit of fluid from the capillaries. At this stage, white blood cells begin to leave the small vessels. First to appear at the infection site are neutrophils, then macrophages. These cells devour the infectious microorganisms and the damaged tissue cells as well.

Repair

Even as inflammation proceeds, repair begins. Tissue repair can occur in two major ways: by regeneration and by fibrosis. **Regeneration** is the replacement of a destroyed tissue by new tissue of the same kind, whereas **fibrosis** involves the proliferation of a fibrous connective tissue called **scar tissue.** Tissue repair in a skin wound involves both regeneration and fibrosis. Once the blood within the cut has clotted, the surface part of the clot dries to form a scab (Figure 4.19a and b). At this point, the repair begins with a step called organization.

Organization is the process by which the clot is replaced by granulation tissue (Figure 4.19b). **Granulation tissue** is a delicate pink tissue made of several elements. It contains capillaries that grow in from nearby areas, as well as proliferating fibroblasts that produce new collagen fibers to bridge the gash. Some of its fibroblasts have contractile properties that pull the margins of the wound together. As organization proceeds, macrophages digest the original clot, and the deposit of collagen continues. As more collagen is made, the granulation tissue gradually transforms into fibrous scar tissue. (Figure 4.19c).

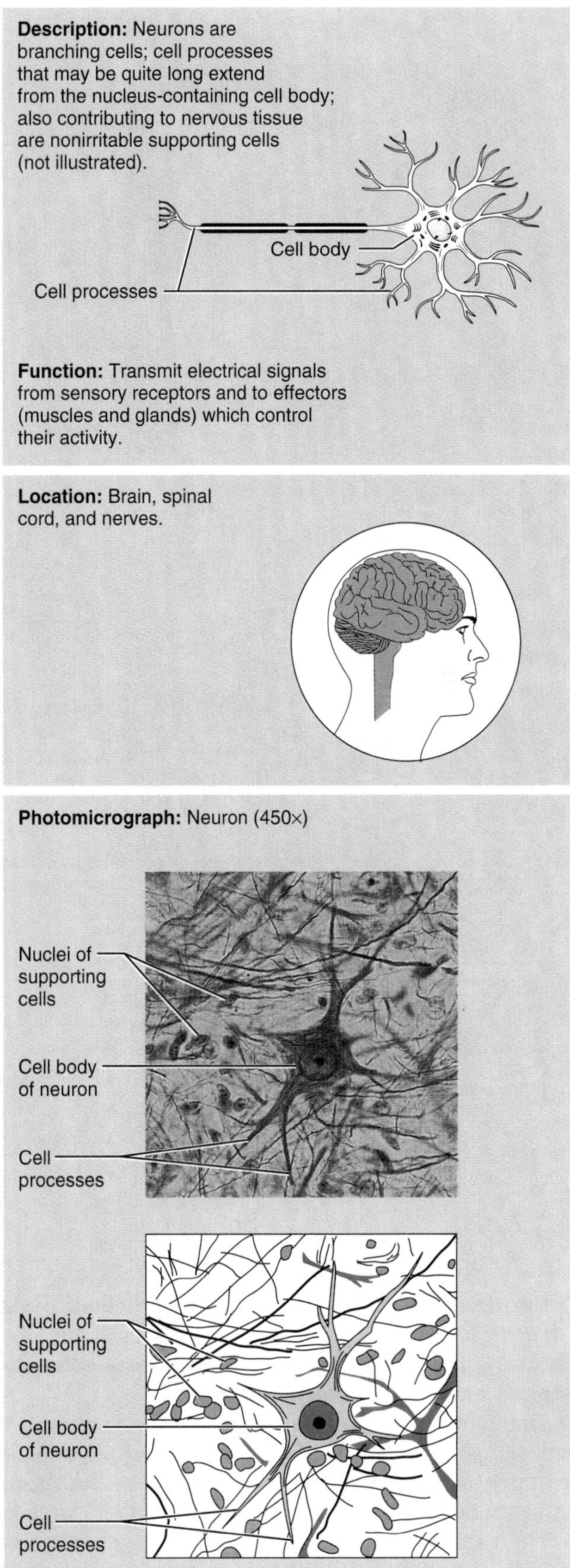

FIGURE 4.18 Nervous tissue.

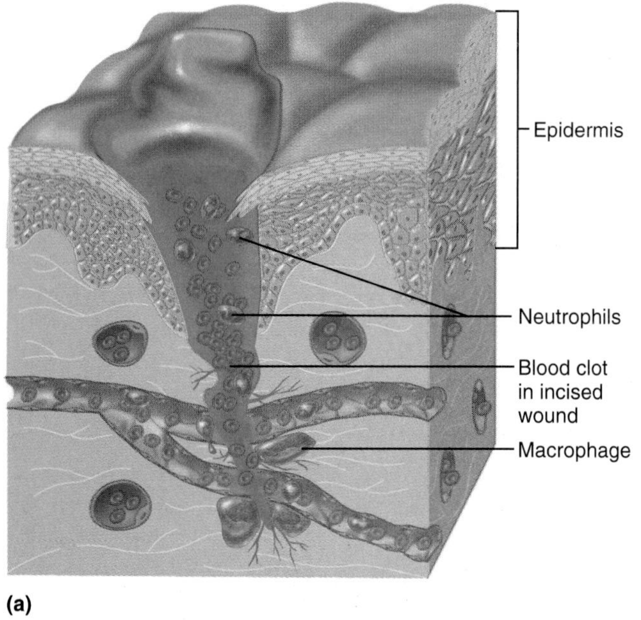

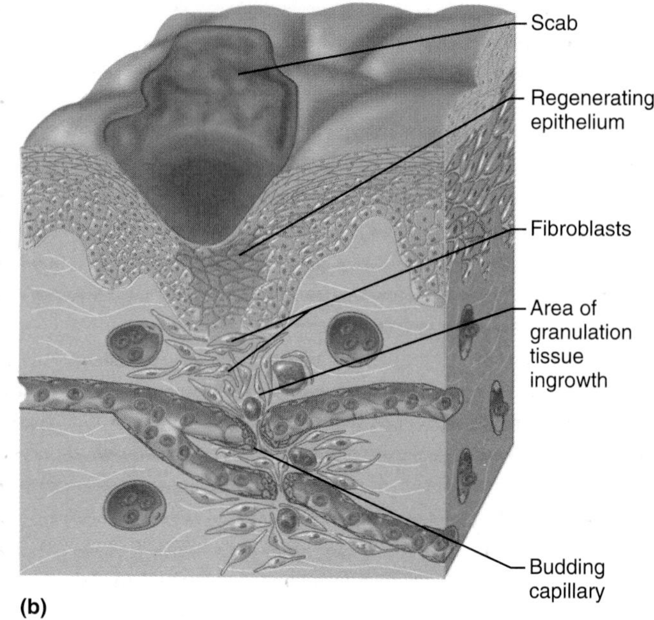

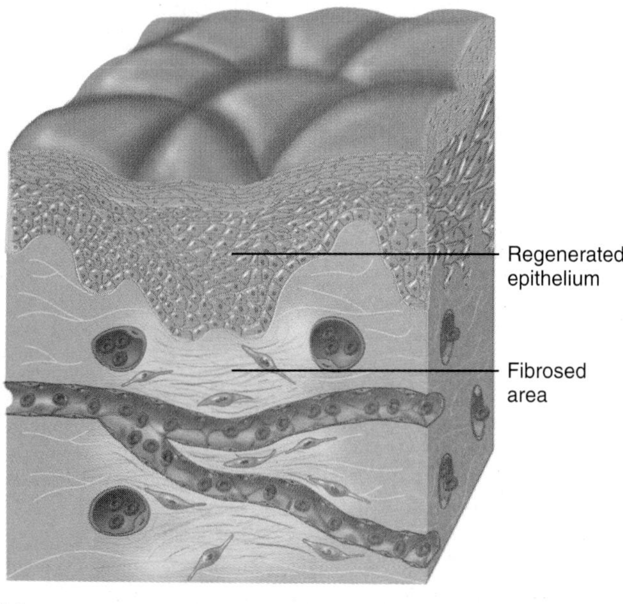

FIGURE 4.19 Tissue repair of a skin wound. (a) At the time of injury, inflammatory chemicals are released, causing local blood vessels to dilate and become more permeable. Fluid, white blood cells, and blood proteins can then enter the injured site. Cut vessels bleed, blood clots within the wound, and the surface of the clot dries to become a scab. **(b)** Organization, or the formation of granulation tissue. Capillary buds invade the clot, restoring a vascular supply. Fibroblasts enter the region and secrete collagen in abundance to bridge the wound. Macrophages dispose of dead cells and other debris. The wound contracts. Surface epithelial cells proliferate and migrate under the scab to cover the granulation tissue. **(c)** About a week after injury, the fibrosed area (scar) is still contracting, and regeneration of the epithelium is almost complete. Depending on the severity of the original wound, the scar may or may not be visible beneath the epidermis.

During organization, the surface epithelium begins to *regenerate,* growing under the scab until the scab falls away. The end result is a fully regenerated epithelium and an underlying area of scar.

The above describes the healing of a wound such as a cut, scrape, or puncture. In pure *infections* (a pimple or sore throat), by contrast, there is usually no clot formation or scarring. Only severe infections lead to scarring.

The capacity for regeneration varies widely among the different tissues. Epithelia regenerate extremely well, as do bone, areolar connective tissue, dense irregular connective tissue, and blood-forming tissue. Smooth muscle and dense regular connective tissue have a moderate capacity for regeneration, but skeletal muscle and cartilage have only a weak capacity. Cardiac muscle and the nervous tissue in the brain and spinal cord have no functional regenerative capacity. However, recent studies have shown that some unexpected cellular division occurs in both these tissues after damage, and efforts are under way to coax them to regenerate better.

In nonregenerating tissues and in exceptionally severe wounds, fibrosis totally replaces the lost tissue. The resulting scar appears as a pale, often shiny area and shrinks during the months after it first forms. A scar consists mostly of collagen fibers and contains few cells or capillaries. Although it is very strong, it lacks the flexibility and elasticity of most normal tissues. Nor can it perform the normal functions of the tissue it has replaced.

A Closer Look

Stem Cells: Medical Miracle of the Twenty-First Century?

Human tissues and organs often fail, become damaged, or wear out and are unable to repair themselves. Common examples are the loss of heart muscle following heart attacks, brain degeneration in Parkinson's disease, the loss of pancreatic insulin-producing cells in diabetes, serious burns, spinal cord injuries, and damage to the joints in arthritis.

Some of this damage can be compensated for by surgical repair, organ transplants, and bionic devices like artificial hips, but these repairs and replacements do not work as well as original healthy tissues and organs. Regeneration of the damaged tissues would be vastly preferable. Ways to encourage such regeneration within the body itself are currently being investigated, but most progress in "tissue engineering" has been made in culturing human cells and organs in the laboratory for return to the body. For example, cultured skin is now available and a laboratory-grown urinary bladder has been tested successfully in dogs. Tests of lab-grown arteries are also under way.

A cluster of living embryonic stem cells

A problem with culturing adult tissues is that most types of cells from fully differentiated organs are difficult to grow in large quantities in the laboratory. However, immature, **tissue-specific stem cells** can be grown in cell culture for months, and animal studies have shown that when returned to their source they can regenerate all cell types of their parent tissue. Once thought to occur in only a few tissue types, such stem cells are now known to be more widespread—perhaps in ***every*** kind of tissue. Stem cells have been isolated and cultured from nervous tissue of the brain, bone, cartilage, fat, muscle, and blood-forming tissue in bone marrow; and they probably also exist in the liver and pancreas.

Despite their advantages, these tissue-specific stem cells are partially differentiated, meaning they can undergo only a limited number of divisions (p. 53) in culture and when reintroduced into the body. This could limit their capacity for regenerating tissues. Therefore, researchers have sought to use immortal cells that divide indefinitely and have unlimited potential, such as the earliest versions of the sex cells (primordial germ cells) and cells from early embryos. Much excitement was generated among scientists in 1998 when such **embryonic stem cells** were isolated from the inner cell mass of human blastocysts (p. 60) and shown to be capable of long-term culture and able to produce a wide range of tissue types when grafted into a developing mouse embryo. (Research on human embryos is controversial and some say unethical; however, it should be noted that the blastocysts were left over after fertility treatments and would have been destroyed if not used. Also, individual embryonic stem cells of blastocysts cannot grow into entire embryos, so their use does not violate current laws that limit research on human embryos.)

The isolation of human embryonic stem cells is viewed as a major advance for tissue transplant, perhaps the breakthrough that will one day allow the replacement of any organ. More immediate goals are to use such cells to produce new neurons for victims of Parkinson's disease or spinal cord injury, new insulin-secreting cells for diabetics, and blood-forming cells for victims of bone marrow cancer. Before the long-term goals can become a reality, however, more research must be done. For example, much more must be learned about how to direct cultured embryonic stem cells to differentiate into specific, desired cell types. So far, only neurons, cardiac muscle cells, and blood-forming cells have been obtained directly from cultured stem cells—and only from ***mouse*** stem cells. Medicine is a long way from growing a many-tissue organ like the brain, the lung, or a bone of the skeleton. Furthermore, before they can be used efficiently, the stem-cell derivatives will probably have to be genetically engineered so the recipient's immune system does not reject them. To avoid this rejection problem, some researchers are seeking ways to reprogram a patient's ***own*** adult cells back to an undifferentiated state, and use these cells to grow personal organs that will not be rejected.

Critical Reasoning

A patch of scar tissue that forms in the wall of the urinary bladder, the heart, or another hollow organ may severely hamper the function of that organ. Why do you think this is so?

The normal shrinking of the scar reduces the internal volume of the organ, possibly hindering or blocking the movement of substances through that organ.

Irritation of visceral organs can cause them to adhere to one another or to the body wall as they scar. Such **adhesions** can prevent the normal churning actions of loops of the intestine, dangerously halting the movement of food through the digestive tube. They can also restrict the movement of the heart and lungs and immobilize the joints. After almost all abdominal surgeries, adhesions form between the body wall and the abdominal viscera, making subsequent surgery in that region more difficult.

THE TISSUES THROUGHOUT LIFE

The embryonic derivations of the four basic tissues are as follows: Connective and muscle tissues derive from mesenchyme, mostly of the mesoderm germ layer. Most epithelial tissues develop from embryonic epithelium, of the ectoderm and endoderm layers. A few epithelia, however, derive from mesodermal mesenchyme, namely the endothelium lining the vessels and the mesothelium lining the ventral body cavity. The nervous tissue in the brain and spinal cord derives from ectodermal epithelium (recall Chapter 3, p. 62).

By the end of the second month of development, the primary tissues have appeared, and all major organs are in place. In virtually all tissues, the cells continue to divide throughout the prenatal period, providing the rapid body growth that occurs before birth. The division of *nerve* cells, however, stops or nearly stops during the fetal period. After birth, the cells of most other tissues continue to divide until adult body size is reached. Cellular division then slows greatly, although many tissues retain a regenerative capacity. In adulthood, only the epithelial tissues and blood cell–forming tissues are highly mitotic.

Some tissues that regenerate throughout life do so through the division of their mature, differentiated cells (the gland cells in the liver and endothelial cells are examples). However, many tissues contain populations of **stem cells,** relatively undifferentiated cells that renew themselves continually and divide to produce new tissue cells as needed.

Advances in Understanding Stem cells have long been known to exist in the rapidly replacing tissues, such as the epidermis, the lining of the digestive tube, some connective tissues, and blood-forming tissue. Now, however, they have been found elsewhere: in the brain, adipose tissue, and probably, liver and pancreas. Stem cells are discussed further in the A Closer Look box on page 105. ✷

Given good nutrition, good circulation, and relatively infrequent wounds and infections, our tissues normally function well through youth and middle age. The importance of nutrition for tissue health cannot be overemphasized. For example, vitamin A is needed for the normal regeneration of epithelium (liver and carrots are rich in this vitamin). Because proteins are the structural material of the body, an adequate intake of protein is essential for the tissues to retain their structural integrity.

With increasing age, the epithelia thin and are more easily breached. The amount of collagen in the body declines, making tissue repair less efficient. Wounds do not heal as fast as in youth. Bone, muscle, and nervous tissues begin to atrophy. These events are due in part to a decrease in circulatory efficiency, which reduces the delivery of nutrients to the tissues, but in some cases diet is a contributing factor. As their income declines or as chewing becomes more difficult, older people tend to eat soft foods, which may be low in protein and vitamins. As a result, the health of the tissues suffers.

RELATED CLINICAL TERMS

Adenoma (ad″ĕ-no′mah) (*aden* = gland; *oma* = tumor) Any neoplasm of glandular epithelium, benign or malignant. The malignant type is more specifically called *adenocarcinoma*.

Carcinoma (kar″sĭ-no′mah) (*karkinos* = crab, cancer) Cancer arising in an epithelium. Ninety percent of all human cancers are of this type: lung, breast, prostate, colon, and others.

Epithelial-connective tissue interactions Epithelia and the underlying connective tissues exchange many chemical signals, especially to control the events of embryonic development and in the generation of cancer as epithelial tumor cells gain the ability to metastasize. These exchanges are also called epithelial-stromal interactions or, in embryos, epithelial-mesenchymal interactions. This is currently a very active area of research.

Healing by first and second intention These are two types of wound healing. Healing by first intention is the simplest type, occurring after the edges of the wound are brought together by sutures, staples, or other means to close surgical incisions. Only small amounts of granulation tissue usually form. In healing by second intention, the edges of the wound stay separated, and the gap is bridged by large amounts of granulation tissue. This is the way in which serious, unattended wounds heal. It is slow and leads to larger scars.

Lesion (le′zhun; "wound") Any injury, wound, or infection that affects tissue over an area of a definite size, as opposed to being widely spread throughout the body.

Sarcoma (sar-ko′mah) (*sarkos* = flesh; *oma* = tumor) Cancer arising in the mesenchyme-derived tissues; that is, in connective tissues and muscle.

CHAPTER SUMMARY

Media study tools that could provide you additional help in reviewing specific key topics of Chapter 4 are referenced below.

SP = Study Partner; AIA = A.D.A.M.® Interactive Anatomy

1. Tissues are collections of structurally similar cells with related functions. They also contain nonliving material between their cells. The four primary tissues are epithelium, connective tissue, muscle, and nervous tissue.

EPITHELIA AND GLANDS (pp. 76–88)

1. Epithelia are sheets of cells that cover body surfaces and line body cavities. Their functions include protection, sensory reception, absorption, ion transport, and filtration. Glandular epithelia function in secretion.

Special Characteristics of Epithelia (p. 77)

2. Epithelia exhibit a high degree of cellularity, little extracellular material, specialized cell junctions, polarity, avascularity, and the ability to regenerate. They are underlain by loose connective tissue.

Classification of Epithelia (pp. 77–83)

3. Epithelia are classified by cell shape as squamous, cuboidal, or columnar, and by the number of cell layers as simple (one layer) or stratified (more than one layer). Stratified epithelia are named according to the shape of their apical cells.

4. Simple squamous epithelium is a single layer of flat cells. Its thinness allows molecules to pass through it rapidly by passive diffusion. It lines the air sacs of the lungs, the interior of blood vessels (endothelium), and the ventral body cavity (mesothelium).

5. Simple cuboidal epithelium occurs in kidney tubules and the small ducts of glands.

6. Simple columnar epithelium lines the stomach and intestines, and a ciliated version lines the uterine tubes. Like simple cuboidal epithelium, it is active in secretion, absorption, and ion transport.

7. Pseudostratified columnar epithelium is a simple epithelium that contains both short and tall cells. A ciliated version lines most of the respiratory passages.

8. Stratified squamous epithelium is multilayered and thick. Its apical cells are flat, and it resists abrasion. Examples are the epidermis and the lining of the mouth, esophagus, and vagina.

SP Exercise: Chapter 4, Classes of Epithelial Tissue.

9. Transitional epithelium is a stratified epithelium that thins when it stretches. It lines the hollow urinary organs.

Epithelial Surface Features (pp. 83–86)

10. Features on apical epithelial surfaces: Microvilli occur on most moist epithelia. They increase the epithelial surface area and may anchor sheets of mucus. Cilia are whip-like projections that beat to move fluid (usually mucus). Microtubules in the cores of cilia generate ciliary movement.

11. Features on lateral cell surfaces: The main types of cell junctions are desmosomes that bind cells together, tight junctions that close off the extracellular spaces, and gap junctions through which small molecules pass from cell to cell.

12. Feature on the basal epithelial surface: Epithelial cells lie on a protein sheet called the basal lamina. This acts as a filter and a scaffolding on which regenerating epithelial cells can grow. The basal lamina, plus some underlying reticular fibers, form the thicker basement membrane.

Glands (pp. 86–88)

13. A gland is one or more cells specialized to secrete a product. Most glandular products are proteins released by exocytosis.

14. Exocrine glands secrete onto body surfaces or into body cavities. Mucus-secreting goblet cells are unicellular exocrine glands. Multicellular exocrine glands are classified by the structure of their ducts as simple or compound and by the structure of their secretory units as tubular, alveolar (acinar), or tubuloalveolar.

15. Endocrine (ductless) glands secrete hormones, which enter the circulatory vessels and travel to target organs, from which they signal a response.

CONNECTIVE TISSUE (pp. 88–99)

1. Connective tissue is the most diverse and abundant class of tissues. Its four basic subtypes are connective tissue proper (loose and dense), cartilage, bone, and blood. Despite their diversity, all derive from mesenchyme and consist of cells separated by abundant extracellular matrix. In all connective tissues except blood, the matrix consists of fibers, ground substance, and tissue fluid.

Connective Tissue Proper (pp. 88–98)

2. Loose areolar connective tissue can be considered a prototype connective tissue against which the other types are compared. It surrounds capillaries and underlies most epithelia. Its main functions are to (1) support and bind other tissues with its fibers (collagen, reticular, elastic), (2) hold tissue fluid in its jelly-like ground substance, (3) fight infection with its many blood-derived defense cells (macrophages, plasma cells, neutrophils, etc.), and (4) store nutrients in fat cells.

3. The fibroblast, the most abundant cell type in connective tissue proper, produces both the fibers and the ground substance of the extracellular matrix. The function of collagen fibers is to resist tension placed on the tissue.
4. Adipose connective tissue is similar to areolar connective tissue but is dominated by nutrient-storing fat cells. It is plentiful in the hypodermis below the skin. This is white fat, but babies also have brown fat that heats their blood and body.
5. Reticular connective tissue resembles areolar tissue, except that its only fibers are reticular fibers. These form networks of caves that hold free blood cells. Reticular tissue occurs in bone marrow, lymph nodes, and the spleen.
6. Dense connective tissue contains exceptionally thick collagen fibers and resists tremendous pulling forces. In dense irregular connective tissue, the collagen fibers run in various directions. This tissue occurs in the dermis and in organ capsules. Dense regular connective tissue contains bundles of collagen fibers that all run in the same direction and are separated by rows of fibroblasts. This tissue, which is subject to high tension from a single direction, is the main tissue in tendons, ligaments, and fascia.

Other Connective Tissues: Cartilage, Bone, and Blood (pp. 98–99)

7. Cartilage and bone have the basic structure of connective tissue (cells and matrix), but their stiff matrix allows them to resist compression. Cartilage is springy and avascular. Its matrix contains mostly water.
8. Bone tissue has a hard, collagen-rich matrix embedded with calcium salts. This mineral gives bone compressive strength.
9. Blood consists of red and white blood cells in a fluid matrix called plasma. It is the most atypical connective tissue.

SP Exercise: Chapter 4, Tissue Types, Blood, Connective Tissue.

COVERING AND LINING MEMBRANES (p. 99)

1. Membranes, each consisting of an epithelium plus an underlying layer of connective tissue, cover broad surfaces in the body. The cutaneous membrane (skin), which is dry, covers the body surface. Mucous membranes, which are moist, line the hollow internal organs that open to the body exterior. Serous membranes, which are slippery, line the pleural, pericardial, and peritoneal cavities.

MUSCLE TISSUE (pp. 99–102)

1. Muscle tissue consists of long muscle cells that are specialized to contract and generate movement. A scant extracellular matrix separates the muscle cells.
2. There are three types of muscle tissue:

   Skeletal muscle is in the muscles that move the skeleton. Its cells are cylindrical and striated.

   Cardiac muscle is in the wall of the heart, and it pumps blood. Its cells branch and have striations.

   Smooth muscle is in the walls of hollow organs, and it usually propels substances through these organs. Its cells are spindle-shaped and lack striations.

NERVOUS TISSUE (p. 102)

1. Nervous tissue, the main tissue of the nervous system, is composed of neurons and supporting cells.
2. Neurons are branching cells that receive and transmit electrical impulses. Nervous tissue regulates body functions.

TISSUE RESPONSE TO INJURY (pp. 102–106)

Inflammation (pp. 102–103)

1. Inflammation is a response to tissue injury and infection. It is localized to the connective tissue and the vessels of the injury site. The inflammatory response begins with dilation of blood vessels (causing redness and heat), followed by edema (causing swelling and pain). Stasis results, and white blood cells migrate into the injured tissue.

Repair (pp. 103–106)

2. Tissue repair begins during inflammation and may involve tissue regeneration, fibrosis (scarring), or both. Repair of a cut begins with organization, during which the clot is replaced with granulation tissue. Collagen deposition replaces granulation tissue with scar tissue.
3. Certain tissues, like cardiac muscle and most nervous tissue, do not regenerate naturally and are replaced by scar tissue.

THE TISSUES THROUGHOUT LIFE (p. 106)

1. Most epithelial and nervous tissues develop from embryonic epithelia, the ectoderm and endoderm. Exceptions are endothelium and mesothelium, which arise from mesodermal mesenchyme. Connective and muscle tissues develop from mesenchyme.
2. Functions of tissues decline with age. The decrease in mass and viability seen in most tissues during old age partially reflects circulatory deficits or poor nutrition.

REVIEW QUESTIONS

Multiple Choice/Matching Questions

1. Use the key to match each basic tissue type with a description below.

Key: **(a)** connective tissue **(b)** epithelium **(c)** muscle tissue **(d)** nervous tissue

____ **(1)** composed largely of nonliving extracellular matrix; important in protection, support, defense, and holding tissue fluid

____ **(2)** the tissue immediately responsible for body movement

____ **(3)** the tissue that provides an awareness of the external environment and enables us to react to it

____ **(4)** the tissue that lines body cavities and covers surfaces

____ **(5)** the tissue that includes most glands

2. An epithelium that has several cell layers, with flat cells in the apical layer, is called (choose all that apply): (a) ciliated, (b) columnar, (c) stratified, (d) simple, (e) squamous.

3. Match the epithelial type named in column B with the appropriate location in column A.

Column A	Column B
____ **(1)** lines inside of stomach and most of intestines	**(a)** pseudostratified ciliated columnar
____ **(2)** lines inside of mouth	**(b)** simple columnar
____ **(3)** lines much of respiratory tract (including the trachea)	**(c)** simple cuboidal
____ **(4)** endothelium and mesothelium	**(d)** simple squamous
____ **(5)** lines inside of urinary bladder	**(e)** stratified columnar
	(f) stratified squamous
	(g) transitional

4. The type of gland that secretes products such as milk, saliva, bile, or sweat through a duct is (a) an endocrine gland, (b) an exocrine gland, (c) a goblet cell.

5. In connective tissue proper, the cell type that secretes the fibers and ground substance is the (a) fibroblast, (b) plasma cell, (c) mast cell, (d) macrophage, (e) chondrocyte.

6. Identify the cell surface features described below.

(1) whip-like extensions that move fluids across epithelial surfaces

(2) little "fingers" on apical epithelial surfaces that increase cell surface area and anchor mucus

(3) the spot-like cell junction that holds cells together so that they are not pulled apart

(4) a cell junction that closes off the extracellular space

(5) of the basement membrane and the basal lamina, the one that can be seen by light microscopy

Short Answer Essay Questions

7. Define tissue.

8. Explain the classification of multicellular exocrine glands, and supply an example of each class.

9. Name the specific type of connective tissue being described: (1) around the capillaries; (2) in the ligamentum flavum between vertebrae; (3) the original, embryonic connective tissue; (4) hard tissue of the skull; (5) main tissue in ligaments; (6) dominates the dermis; (7) dominates the hypodermis.

10. Name four functions of areolar connective tissue, and relate each function to a specific structural part of this tissue.

11. (a) Where does tissue fluid come from? (b) What is the function of tissue fluid?

12. What is the function of a macrophage?

13. Name the four classic symptoms of inflammation, and explain what each symptom represents in terms of changes in the injured tissue.

14. (a) Define endocrine gland. (b) What is a hormone?

15. Name two specific tissues that regenerate well and two that regenerate poorly.

16. Of the four basic tissue types, which two develop from mesenchyme?

17. What is fascia?

18. A fourth-grade teacher told his science class that "the body consists entirely of cells." Hearing this, a pupil raised her hand and said her mother (an anatomist) told her the body does *not* entirely consist of cells, and, in fact, a noncellular material probably makes up more of the body than the cells do. What is this noncellular material?

19. What are the differences between a mucous membrane and a serous membrane?

20. What is the main type of tissue in the following structures? (a) a ligament or tendon, (b) a bone of the leg, (c) a muscle like the biceps of the arm, (d) the brain, (e) the flexible skeleton in the outer ear, (f) the contractile wall of the heart, (g) kidney tubules.

CRITICAL REASONING + CLINICAL APPLICATION QUESTIONS

1. Systemic lupus erythematosus, or lupus, is a condition that sometimes affects young women. It is a chronic (persistent) inflammation that affects all or most of the connective tissue proper in the body. Suzy is told by her doctor that she has lupus, and she asks if it will have widespread or merely localized effects within the body. What would the physician answer?

2. Sailors who made long ocean journeys in the time of Christopher Columbus ate only bread, water, and salted meat on their journeys. They often suffered from scurvy. Eventually the problem was solved, and ocean sailors no longer developed scurvy. Try to deduce how the problem was solved. (Hint: It was not by inventing faster boats!)

3. Three patients in an intensive care unit have sustained damage and widespread tissue death in three different organs. One patient has brain damage from a stroke, another had a heart attack that destroyed cardiac muscle, and the third injured much of her liver (a gland) in a crushing car accident. All three patients have stabilized and will survive, but only one will gain full functional recovery through tissue regeneration. Which one, and why?

4. In adults, over 90% of all cancers are either adenomas (adenocarcinomas) or carcinomas. In fact, cancers of the skin, lung, colon, breast, and prostate are all in these categories. Which one of the four basic tissue types gives rise to most cancers?

CHAPTER 5

The Integumentary System

PM of epithelial cells

CHAPTER OUTLINE + STUDENT OBJECTIVES

The Skin and the Hypodermis 112–117

1. Name the tissue types that compose the epidermis and dermis.
2. Name and describe the functions of the major layers of the epidermis and dermis.
3. Describe the structure and function of the hypodermis.
4. Describe the factors that contribute to skin color.

Appendages of the Skin 117–122

5. List the parts of a hair and a hair follicle, and explain the function of each part.
6. Explain the basis of hair color.
7. Describe the distribution, growth, and replacement of hairs and how hair changes throughout life.
8. Compare the structure and location of oil and sweat glands.
9. Compare eccrine and apocrine sweat glands.
10. Describe the structure of nails.

Disorders of the Integumentary System 122–124

11. Explain why serious burns are life-threatening and how burns are treated.
12. Differentiate first-, second-, and third-degree burns.
13. Summarize the characteristics and warning signs of skin cancers, especially melanoma.

The Skin Throughout Life 124–125

14. Briefly explain the changes that occur in the skin from birth to old age.

WOULD YOU BE ENTICED BY AN ADVERTISEMENT for a coat that is waterproof, stretchable, washable, and permanent-press, that automatically repairs small rips and burns, and is guaranteed to last a lifetime with reasonable care? Sounds too good to be true, but you already have such a coat—your skin. The skin and its appendages (sweat glands, oil glands, hair, and nails) serve a number of functions. Together, these organs make up the **integumentary system** (in-teg″u-men′tar-e; "covering").

THE SKIN AND THE HYPODERMIS

The **skin (integument)** and its appendages are the first *organs* discussed in this book. Recall from Chapter 1 that an organ consists of tissues working together to perform certain functions. Although the skin is less complex than most other organs, it is still an architectural marvel. What is more, it is the largest of all our organs, accounting for about 7% of our total body weight.

The skin, which varies in thickness from 1.5 to 4 mm or more in different regions of the body, has two distinct layers (Figure 5.1). The superficial layer is the *epidermis*, a thick epithelium. Deep to it is the *dermis*, a fibrous connective tissue. Just deep to the skin lies a fatty layer called the *hypodermis*. Even though the hypodermis is not part of the integumentary system, it shares some of the skin's functions and is thus described in this chapter.

The skin performs many functions, most but not all of which are protective. It cushions and insulates the deeper body organs and protects the entire body from bumps, scrapes, and cuts. It also offers protection from harmful chemicals, heat and cold, and invading bacteria. The epidermis is waterproof, preventing unnecessary loss of water across the body surface. The skin's

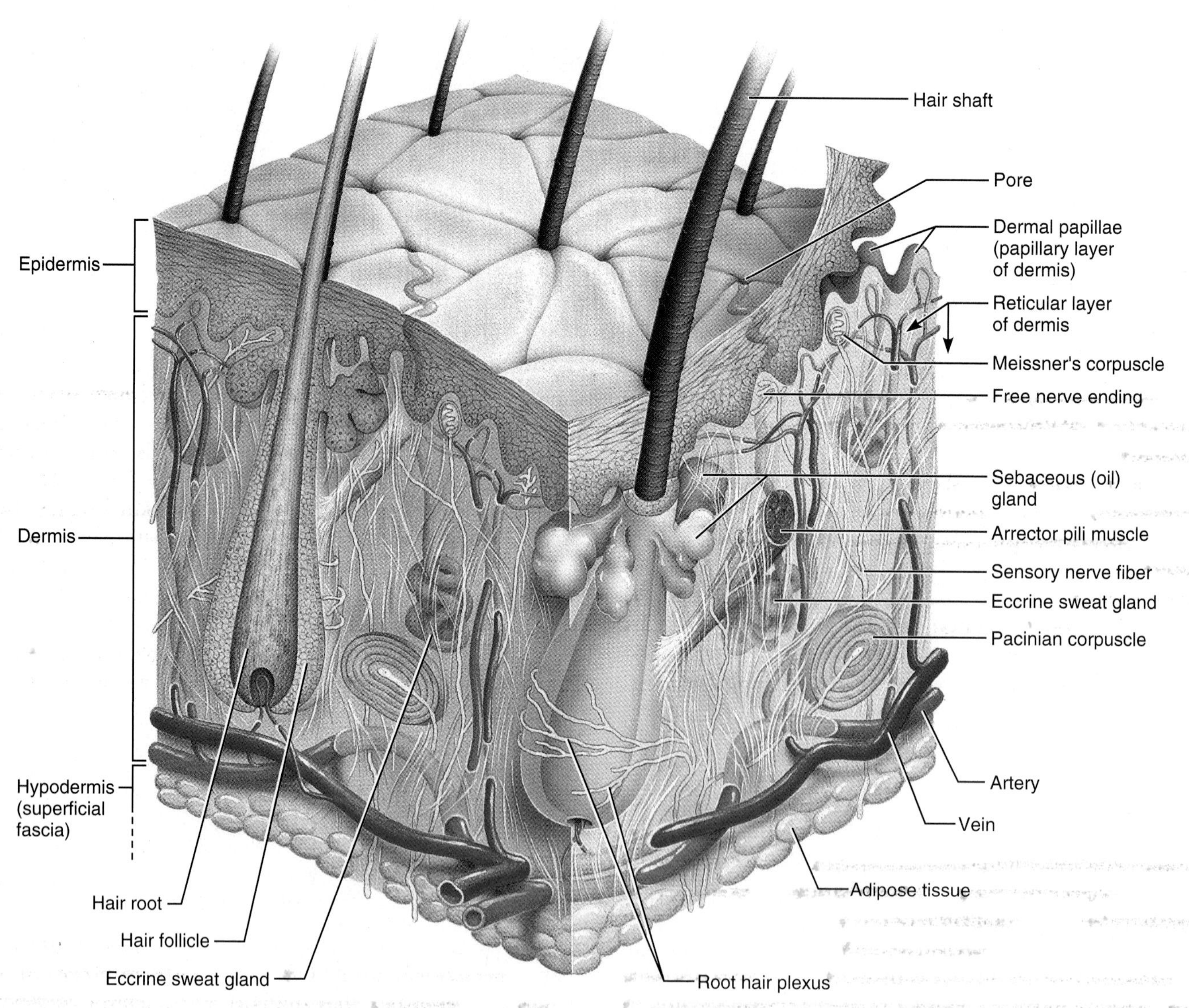

FIGURE 5.1 Skin structure. Three-dimensional view of the skin and the underlying hypodermis. The various nerve endings are sensory receptors, discussed in Chapter 14.

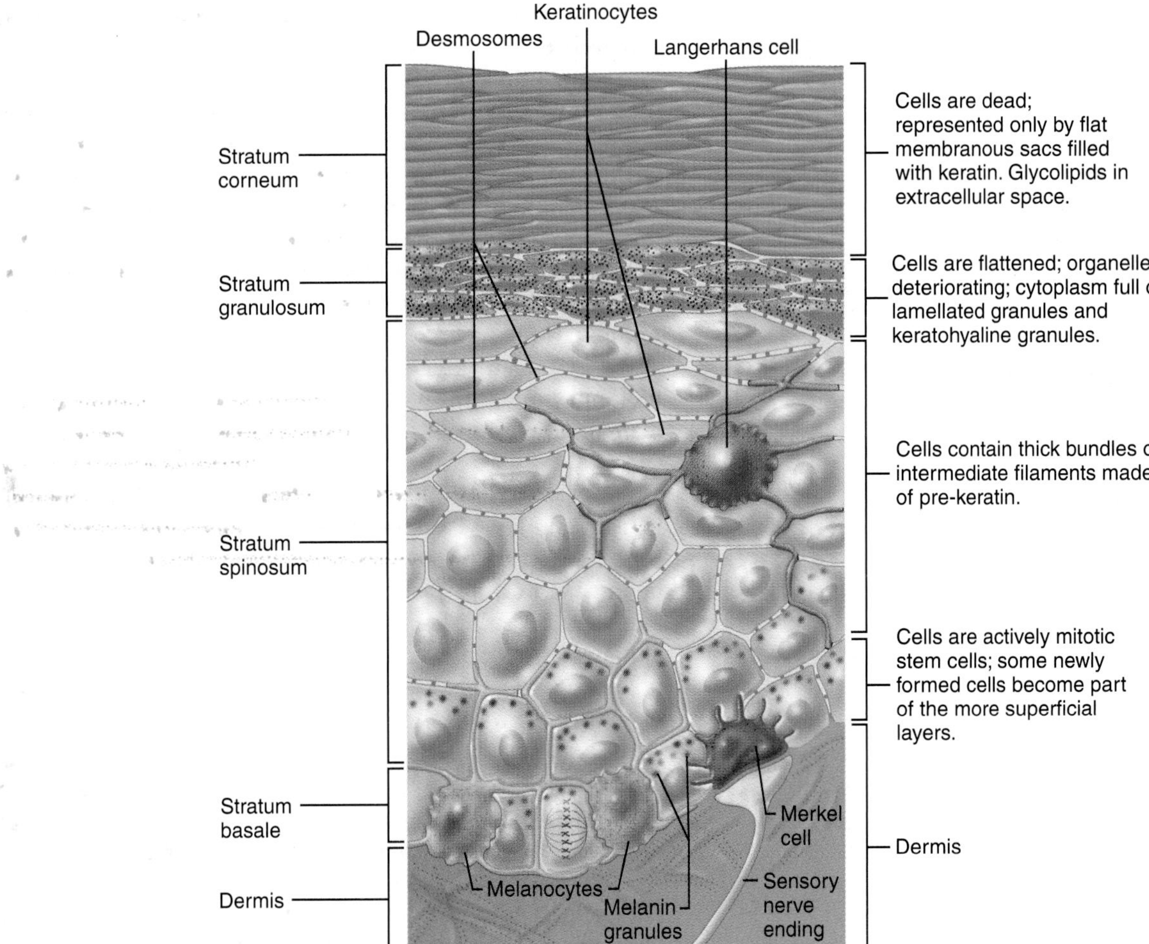

FIGURE 5.2 Epidermal cells and the layers of the epidermis in thin skin. The four cell types are keratinocytes (orange), melanocytes (gray), Langerhans cell (blue), and Merkel cell (purple).

rich capillary networks and sweat glands regulate the loss of heat from the body, helping to control body temperature. Skin acts as a miniature excretory system when urea, salts, and water are lost as we sweat. It screens out harmful ultraviolet (UV) rays from the sun, and its epidermal cells use these rays to synthesize vitamin D. Finally, the skin contains sense organs called *sensory receptors* that are associated with nerve endings. By sensing touch, pressure, temperature, and pain, these receptors keep us aware of conditions at the body surface. We will explore the skin's functions in greater detail as we explore its anatomy.

Epidermis

The **epidermis** (ep″ĭ-der′mis; "on the skin") is a keratinized stratified squamous epithelium that contains four distinct types of cells: *keratinocytes, melanocytes, Merkel cells,* and *Langerhans cells* (Figure 5.2).

The chief role of **keratinocytes** (ke-rat′ĭ-no-sīts), by far the most abundant epidermal cell, is to produce **keratin,** a tough fibrous protein that gives the epidermis its protective properties. Tightly connected to one another by a large number of desmosomes, keratinocytes arise in the deepest part of the epidermis from cells that undergo almost continuous mitosis. As these cells are pushed up by the production of new cells beneath them, they make the keratin that eventually fills their cytoplasm. By the time they approach the skin surface, they are dead, flat sacs completely filled with keratin. Millions of these dead cells rub off every day, giving us an entirely new epidermis every 35 to 45 days—the average time from the birth of a keratinocyte to its final wearing away. Where the skin experiences friction, both cell production and keratin formation are accelerated.

Clinical Applications

SKIN'S RESPONSE TO FRICTION Persistent friction (from a poorly fitting shoe, for example) causes a thickening of the epidermis called a **callus.** Short-term but severe friction (from using a hoe, for example) can cause a **blister,** the separation of the epidermis from the dermis by a fluid-filled pocket. A large blister is called a **bulla** (bul′ah; "bubble"). ✚

Keratinocytes do more than just provide physical and mechanical protection. They also produce antibiotics and enzymes that detoxify the harmful chemicals to which our skin is exposed.

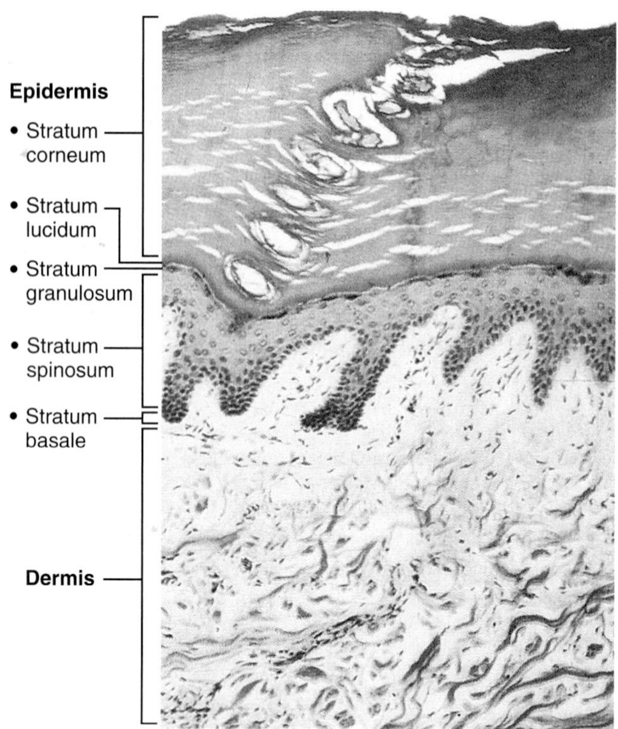

FIGURE 5.3 The epidermis of thick skin. This photomicrograph is magnified 150×.

We will discuss the other epidermal cell types as we examine the various layers of the epidermis.

Layers of the Epidermis

Variation in the thickness of the epidermis determines whether our skin is thick or thin. In **thick skin,** which covers the palms and soles, the thickened epidermis consists of five layers, or *strata* (stra′tah; "bed sheets") (Figure 5.3). In **thin skin,** which covers the rest of the body, only four strata are present (see Figure 5.2).

Stratum Basale (Basal Layer) The **stratum basale** (stra′-tum ba-sal′e), the deepest epidermal layer, is firmly attached to the underlying dermis along a wavy borderline. Also called the *stratum germinativum* (jer-mĭ-na″te′vum; "germinating layer"), it consists of a single row of cells, mostly stem cells representing the youngest keratinocytes (see Figure 5.2). These cells divide rapidly, and many mitotic nuclei are visible. Occasional **Merkel cells** are seen amid the keratinocytes. Each hemisphere-shaped Merkel cell is intimately associated with a disc-like sensory nerve ending and may serve as a receptor for touch.

Between 10% and 25% of the cells in the stratum basale are spider-shaped **melanocytes** (mel′ah-no-sīts; "melanin cells"), which make the dark skin pigment **melanin** (mel′ah-nin; "black"). Melanin is made in membrane-walled granules and then transferred through the cell processes (the "spider-legs") to nearby keratinocytes. Consequently, the basal keratinocytes contain more melanin than do the melanocytes themselves. This melanin shields the cell nuclei from UV rays, which can damage DNA and cause cancer (see p. 123). In Caucasians, the melanin is digested away by lysosomes a short distance above the basal layer. In black people, no such digestion occurs, so melanin occupies keratinocytes throughout the epidermis. Although black people have a darker melanin, more granules, and more pigment in each melanocyte, they do *not* have more melanocytes in their skin. In all but the darkest people, melanin builds up in response to the sun's UV rays, the protective response we know as suntanning.

Stratum Spinosum (Spiny Layer) The **stratum spinosum** (spi-no′sum) is several cell layers thick (see Figure 5.2). Mitosis occurs here, but less often than in the basal layer. This layer is so-named because in typical histological slides, the keratinocytes in this layer have many spine-like extensions. However, these spines do not exist in living cells: They are artifacts created during tissue preparation, when the cells shrink while holding tight at their many desmosomes. Cells of the stratum spinosum contain thick bundles of intermediate filaments, called *tonofilaments* ("tension filaments"), which consist of a tension-resisting protein called prekeratin.

Scattered among the keratinocytes of the stratum spinosum are **Langerhans cells.** These star-shaped cells belong to a class of macrophage-like *dendritic cells* (see Chapter 20). As part of our immune system, Langerhans cells police our outer body surface, using receptor-mediated endocytosis to take up foreign proteins (antigens) that have invaded the epidermis. Then they leave the skin and travel to a nearby lymph node, where they present the antigens to a type of immune cell called a *killer T lymphocyte,* which proceeds to attack all foreign cells that carry the antigen (see p. 592).

Stratum Granulosum (Granular Layer) The thin **stratum granulosum** (gran″u-lo′sum) consists of three to five layers of flattened keratinocytes. Along with abundant tonofilaments, these cells contain *keratohyaline granules* and *lamellated granules*. The keratohyaline (ker″ah-to-hi′ah-lin) granules help form keratin in the higher strata, as described shortly. The lamellated granules (lam′ĭ-la-ted; "plated") contain a waterproofing glycolipid that is secreted into the extracellular space and is the major factor for slowing water loss across the epidermis. Furthermore, the plasma membranes of the cells thicken so that they become more resistant to destruction. It is as if the keratinocytes are "toughening up" to make the outer strata the strongest skin region.

Like all epithelia, the epidermis relies on capillaries in the underlying connective tissue (the dermis) for its nutrients. Above the stratum granulosum, the epidermal cells are too far from the dermal capillaries, so they die, a completely normal occurrence.

Stratum Lucidum (Clear Layer) The **stratum lucidum** (lu′sĭ-dum) occurs in thick skin (see Figure 5.3) but not in thin skin. It is also named the *transition zone.* Appearing through the light microscope as a thin translucent band, it consists of a few rows of flat dead keratinocytes. Electron microscopy reveals that its cells are identical to those at the bottom of the next layer, the stratum corneum.

Stratum Corneum (Horny Layer) The most external part of the epidermis, the **stratum corneum** (kor′ne-um), is many cells thick. It is much thicker in thick skin than in thin skin. Its dead keratinocytes are flat sacs completely filled with keratin because their nuclei and organelles disintegrated upon cell death. Keratin consists of tonofilaments embedded in a "glue" from the keratohyaline granules. Both the keratin and the thickened plasma membranes of cells in the stratum corneum protect the skin against abrasion and penetration. Additionally, the glycolipid between its cells keeps this layer waterproof. It is amazing that a layer of dead cells can still perform so many functions!

The cells of the stratum corneum are referred to as *cornified,* or *horny* cells (*cornu* = horn). These cells are the dandruff shed from the scalp and the flakes that come off dry skin. The average person sheds 18 kg (40 pounds) of these skin flakes in a lifetime. The common saying "Beauty is only skin deep" is ironic, given that nearly everything we see when we look at someone is dead!

Dermis

The dermis, the second major layer of the skin, is a strong, flexible connective tissue. Its cells are those of any connective tissue proper: fibroblasts, macrophages, mast cells, and scattered white blood cells (see pp. 90–91). The fiber types—collagen, elastic, and reticular—also are typical. The dermis binds the entire body together like a body stocking. It is your "hide" and corresponds to animal hides used to make leather products.

The dermis is richly supplied with nerve fibers and blood vessels. The nerve supply of the dermis (and epidermis) is discussed in Chapter 14, p. 411. Dermal blood vessels do more than just nourish the dermis and overlying epidermis: They perform a role in temperature regulation. These vessels are so extensive that they can hold 5% of all blood in the body. When internal organs need more blood or more heat, nerves signal the dermal vessels to constrict, which shunts more blood into the general circulation, making it available to the internal organs. Conversely, on hot days the dermal vessels engorge with warm blood, allowing heat to radiate from the body—a cooling effect. As we will see, humans have a fantastic ability to dissipate excess body heat, which is probably why our dermal blood vessels are more extensive than in any other animal.

The dermis has two layers: papillary and reticular (see Figure 5.1). The **papillary** (pap′ĭ-lar-e) **layer,** the superficial 20% of the dermis, is areolar connective tissue. It includes the **dermal papillae** (pah-pil′le; "nipples"), finger-like pegs that indent the overlying epidermis. On the palms of the hands and the soles of the feet, the dermal papillae lie atop larger mounds called *dermal ridges.* These elevate the overlying epidermis into *epidermal ridges (friction ridges),* which are responsible for fingerprints, palmprints, and footprints. Epidermal ridges increase friction and enhance the gripping ability of the hands and feet. Patterns of these ridges are genetically determined and unique to each person. Because *sweat pores* open along the crests of the epidermal ridges, they leave distinct *fingerprints* on almost anything they touch. Thus, fingerprints are "sweat films."

The deeper **reticular layer,** which accounts for about 80% of the thickness of the dermis, is dense irregular connective tissue. Its extracellular matrix contains thick bundles of interlacing collagen fibers that run in many different planes. However, most run parallel to the skin surface. The reticular layer is so called because of its networks of collagen fibers (*reticulum* = network), not because it has any special abundance of reticular fibers. Separations or less-dense regions between the collagen bundles form the *lines of cleavage* or *tension lines of the skin.* These invisible lines occur over the entire body: They run longitudinally in the skin of the limbs and head and in circular patterns around the neck and trunk. A knowledge of cleavage lines is important to surgeons: When an incision is made *parallel* to these lines, the skin tends to gape less and heals more readily than when the incision is made *across* cleavage lines.

The collagen fibers of the dermis give skin its strength and resilience. Thus, many jabs and scrapes do not penetrate this tough layer. Furthermore, elastic fibers in the dermis provide the skin with stretch-recoil properties. The dermis is also the site of tatoos (see A Closer Look, page 116).

The deep part of the dermis is responsible for skin surface markings called *flexure lines.* Observe, for example, the deep skin creases on your palm. These result from a continual folding of the skin, often over joints, where the dermis attaches tightly to underlying structures. Flexure lines are also visible on the wrists, soles, fingers, and toes.

Critical Reasoning

How might stretch marks *on the skin, commonly seen when one gains a lot of weight, relate to the dermis?*

Extreme stretching of the skin, as occurs in obesity and pregnancy, can tear the collagen in the dermis. Such dermal tearing results in silvery white scars termed striae *(stri′e; "streaks"), which we recognize as "stretch marks."*

Hypodermis

Just deep to the skin is the fatty **hypodermis** (hi″po-der′mis), from Greek words meaning "below the skin"

A Closer Look

Tatoos

Tatoos are made by using a needle to deposit pigment within the dermis. Tatooing is an ancient practice, believed to have originated around 10,000 years ago. These days, tatoos are symbols of group membership for some males (street gangs, the military, fraternities), while other people simply view them as symbols of individuality. In recent years, females have increasingly acquired tatoos as a means of expression and for cosmetic purposes; permanent eyeliner and tatooed liplines now account for over 125,000 tatoos a year.

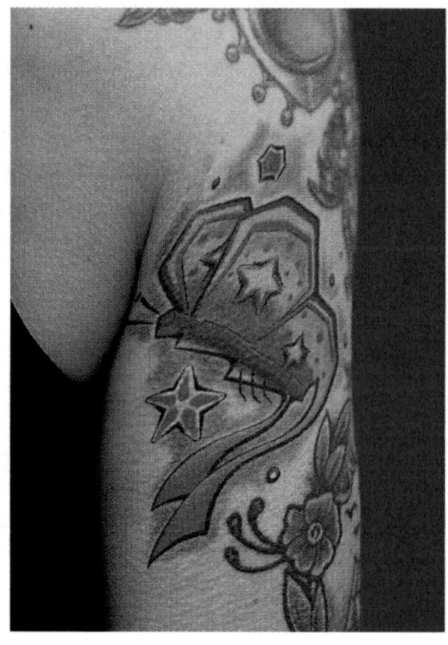

But what if a tatoo becomes unfashionable or the pigment migrates? Until recently, once you had one, you were essentially stuck with it, because attempts at removal—dermabrasion, cryosurgery (freezing), or applying caustic chemicals—left nasty scars. Using new laser-based technologies, dermatologists have no problem destroying the black or blue pigments in tatoos applied a generation ago, but newer, multicolored tatoos post a larger problem. The multitude of pigments in tatoos today require that several different lasers be used over seven to nine treatments spaced about a month apart, at a cost in pain roughly equal to that experienced getting tatooed in the first place.

Tatoos present some other risks. Even though the FDA has some regulations concerning the composition of tatoo pigments, their safety is not well established. Statutory regulations vary widely (from none to complete prohibition) from state to state. Still, in each case, needles are used and bleeding occurs, and practitioners' competence varies significantly. If strict sterile procedures are not adhered to, tatooing can spread infections such as hepatitis.

So if you're thinking about getting a tatoo, look into it carefully. Even with the availability of laser removal techniques, it makes sense to consider the risks.

(see Figure 5.1). This layer is also called the **superficial fascia** and the *subcutaneous layer* (sub″ku-ta′ne-us), from Latin words meaning "below the skin." The hypodermis consists of both areolar and adipose connective tissue, although adipose tissue normally dominates. Besides storing fat, the hypodermis anchors the skin to the underlying structures (mostly to muscles), but loosely enough that skin can slide relatively freely over those structures. Such sliding ensures that many blows just glance off our bodies. The hypodermis is also an insulator: Because fat is a poor conductor of heat, it helps prevent heat loss from the body. The hypodermis thickens markedly with weight gain, but this thickening occurs in different body areas in the two sexes. In females, subcutaneous fat accumulates first in the thighs and breasts, whereas in males it first accumulates in the anterior abdomen as a "beer belly."

Skin Color

Three pigments contribute to skin color: melanin, carotene, and hemoglobin. **Melanin,** the most important, is made from an amino acid called tyrosine. Present in several varieties, melanin ranges from yellow to reddish to brown to black. Its manufacture depends on an enzyme in melanocytes called *tyrosinase* (ti-ro′sĭ-nās). As we noted earlier, melanin passes from melanocytes to keratinocytes in the stratum basale of the epidermis. Freckles and pigmented moles are localized accumulations of melanin.

Carotene (kar′o-tēn) is a yellow to orange pigment that our body obtains from such plant foods as carrots and tomatoes. It tends to accumulate in the stratum corneum of the epidermis and in fat of the hypodermis. It should be noted that the yellowish tinge of the skin of Asian peoples is due to variations in melanin, not to carotene.

The pink hue of Caucasian skin reflects the crimson color of oxygenated **hemoglobin** (he′mo-glo″bin) in the capillaries of the dermis. Because Caucasian skin contains little melanin, the epidermis is nearly transparent and allows the color of blood to show through. Black-and-blue marks represent discolored blood that is visible through the skin. Such bruises, usually caused by blows, reveal sites where blood has escaped from the circulation and clotted below the skin. The general term for a clotted mass of escaped blood anywhere in the body is a **hematoma** (he″mah-to′mah; "blood swelling").

CLINICAL APPLICATIONS

CYANOSIS When hemoglobin is poorly oxygenated, both the blood and the skin of Caucasians appear blue, a condition called **cyanosis** (si"ah-no'sis; "dark blue condition"). Skin often becomes cyanotic during heart failure or in severe respiratory disorders. In black people, the skin may be too dark to reveal the color of the underlying vessels, but cyanosis can be detected in the mucous membranes and nail beds. ✚

The reasons for variations in the skin colors of different human populations are not fully understood. According to one theory, this variation reflects the fact that the sun's UV rays are both dangerous and essential. On the one hand, the intense rays of the tropical sun pose the greatest threat of skin cancer, and the inhabitants of equatorial regions evolved a thick screen of dark melanin that eliminates this danger. (For example, skin cancer is relatively rare among blacks.) On the other hand, UV rays stimulate the deep epidermis to produce vitamin D, a vital hormone that is required for the uptake of calcium from the diet and is essential for healthy bones. Producing vitamin D poses no problem in the sunny tropics, but Caucasians in northern Europe receive little sunlight during the long, dark winter. Therefore, their colorless epidermis ensures enough UV penetration for vitamin D production. Most human skin color types evolved at intermediate latitudes (China, the Middle East, and so on) and are characterized by moderately brown skin tones that are neither "white" nor "black." Such skin is dark enough to provide some protection in the summer, especially when it tans, yet light enough to allow vitamin D production in the moderate winters of intermediate latitudes.

APPENDAGES OF THE SKIN

Along with the skin itself, the integumentary system includes several derivatives of the epidermis. These **skin appendages** include the hair and hair follicles, sebaceous (oil) glands, sweat glands, and nails. Although they derive from the epidermis, they all extend into the dermis.

Hair and Hair Follicles

Together, hairs and their follicles form complex structural units (Figure 5.4). In these units, the **hairs** are the long filaments, and the **hair follicles** are tubular invaginations of the epidermis from which the hairs grow.

Hair

Although hair serves other mammals by keeping them warm, human body hair is far less luxuriant and less useful. Even so, hair is distributed over all our skin surface, except on the palms, soles, nipples, and parts of the external genitalia (the head of the penis, for example). The main function of our sparse body hair is to sense things that lightly touch our skin. The hair on the scalp protects the head against direct sunlight in summer and against heat loss on cold days. Eyelashes shield the eyes, and nose hairs filter large particles such as insects and lint from the air we inhale.

A hair is a flexible strand made of dead, keratinized cells. The **hard keratin** that predominates in hairs confers two advantages over the **soft keratin** that is found in typical epidermal cells: (1) It is tougher and more durable, and (2) the cells of hard keratin do not flake off. The chief parts of a hair are the **root,** the part embedded in the skin, and the **shaft,** the part that projects above the skin surface (Figure 5.4). If its shaft is flat and ribbon-like in cross-section, the hair is kinky; if oval in section, the hair is wavy; if perfectly round, the hair is straight.

A hair consists of three concentric layers of keratinized cells (Figure 5.4c). Its central core, the **medulla** (mĕ-dul'ah; "middle"), consists of large cells and air spaces. The medulla is absent in fine hairs. The **cortex,** surrounding the medulla, consists of several layers of flattened cells. The outermost **cuticle** is a single layer of cells that overlap one another from below like shingles on a roof (Figure 5.5). This shingle pattern helps to keep neighboring hairs apart so that hair does not mat. Hair conditioners smooth out the rough surface of the cuticle to make our hair look shinier. The cuticle is the most heavily keratinized part of the hair, providing strength and keeping the inner layers tightly compacted.

Critical Reasoning

Because it is subjected to the most abrasion, the oldest part of the hair cuticle tends to wear away at the tip of the hair shaft. What is the common name for this phenomenon?

The abrasion causes the keratin fibrils in the cortex and medulla to frizz out, a phenomenon called "split ends."

Hair pigment is made by melanocytes at the base of the hair follicle and is transferred to the cells of the hair root. Different proportions of two types of melanin (black-brown colored and yellow-rust colored) combine to produce all the common hair colors—black, brown, red, and blond. Graying or whitening of hair results from a decrease in the production of melanin and from the replacement of melanin by colorless air bubbles in the hair shaft.

Hair Follicles

Hair follicles extend from the epidermal surface into the dermis. The deep end of the follicle is expanded, forming a **hair bulb** (Figure 5.4c). A knot of sensory nerve endings wraps around each hair bulb to form a **root hair plexus** (see Figure 5.1). Bending the hair shaft stimulates these nerve endings. Hairs are thus exquisite touch receptors. (Feel the tickle as you lightly run your hand over the hairs on your forearm.)

A nipple-like bit of the dermis, the **connective tissue papilla (hair papilla),** protrudes into each hair bulb. This papilla contains a knot of capillaries that

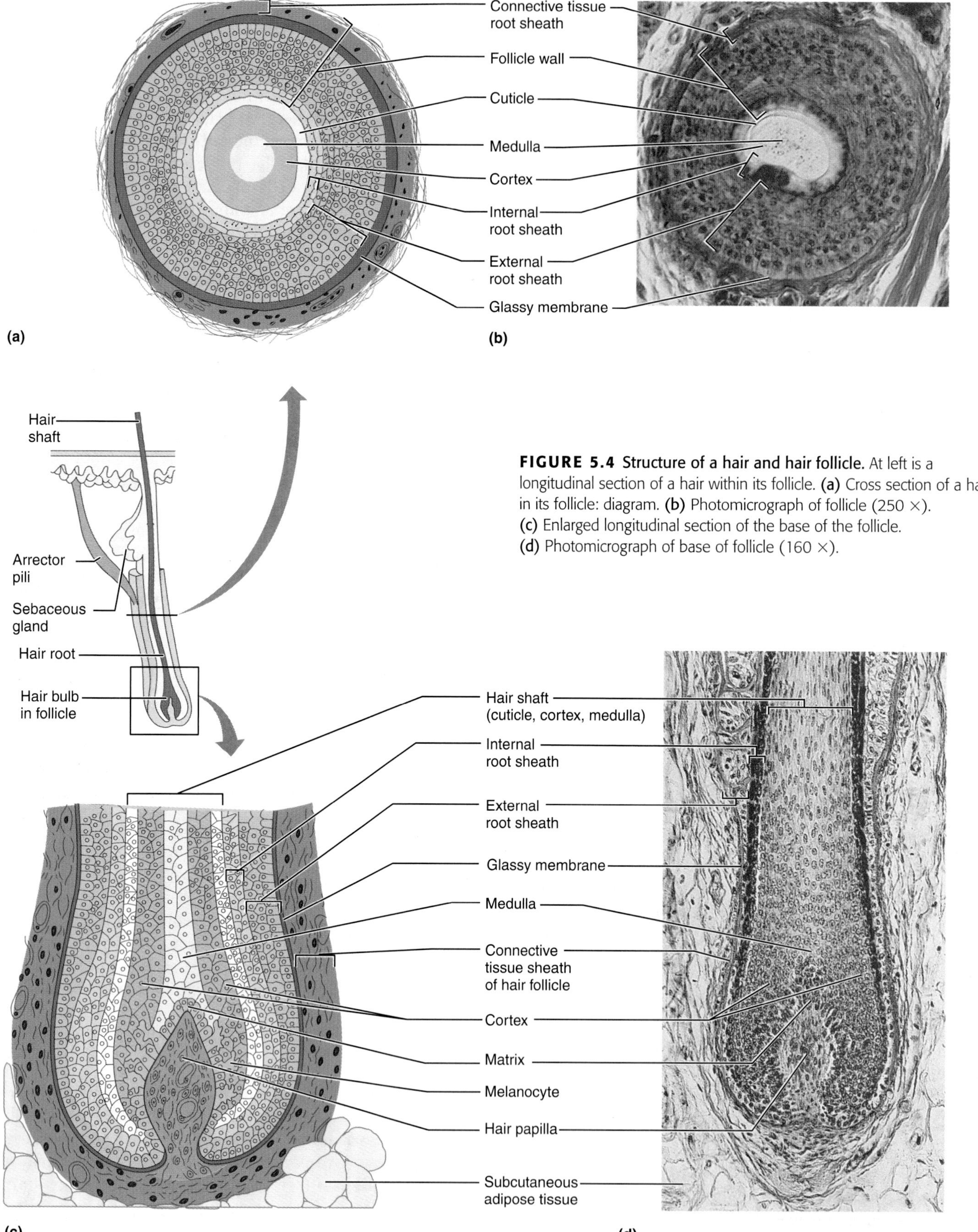

FIGURE 5.4 Structure of a hair and hair follicle. At left is a longitudinal section of a hair within its follicle. **(a)** Cross section of a hair in its follicle: diagram. **(b)** Photomicrograph of follicle (250 ×). **(c)** Enlarged longitudinal section of the base of the follicle. **(d)** Photomicrograph of base of follicle (160 ×).

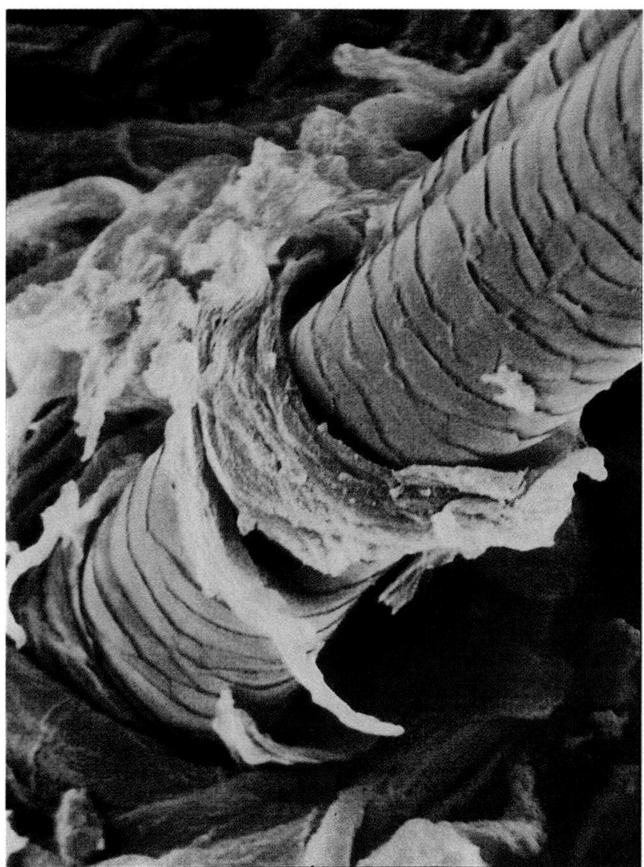

FIGURE 5.5 Scanning electron micrograph of a hair shaft emerging from a follicle at the epidermal surface. Note how the scale-like cells of the cuticle overlap one another (1500 ×; artificially colored).

deliver substances that signal hair growth and supply nutrients to the growing hair. If the hair papilla is destroyed by trauma, the follicle permanently stops producing hair.

Advances in Understanding The epithelial cells in the hair bulb just above the papilla make up the **hair matrix.** The traditional view has been that the matrix cells are the stem cells that produce the growing hair. While these matrix cells certainly do proliferate to form the hairshaft, recent studies indicate they are not the ultimate stem cells after all. Instead, the actual stem cells (from which matrix cells arise) lie more superficially, probably in a bulge in the *external root sheath* near the insertion of the *arrector pili* muscle (these latter structures are defined in the following paragraphs). ✱

The wall of a hair follicle is actually a compound structure composed of an outer **connective tissue root sheath,** derived from the dermis, and an inner **epithelial root sheath,** derived from the epidermis (Figure 5.4a–d). The epithelial root sheath has two parts: Of these, the **internal root sheath** derives from the matrix cells, which it surrounds. The **external root sheath,** by contrast, is a direct continuation of the epidermis.

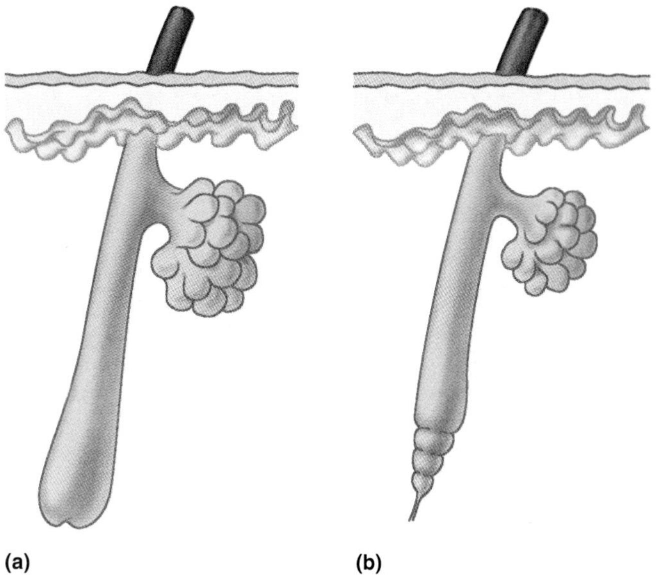

FIGURE 5.6 Shapes of hair follicles during a growth cycle. **(a)** An active follicle. **(b)** A resting follicle. The hair is shed just after each resting phase.

Associated with each hair follicle is a bundle of smooth muscle cells called an **arrector pili** muscle (ah-rek′tor pi′li; "raiser of the hair") (see Figures 5.1 and 5.4). Each arrector pili runs from the most superficial part of the dermis to a deep-lying hair follicle. When their arrector pili is relaxed, most hairs lie flat because most follicles lie at an oblique angle to the skin surface. Then, contraction of this muscle in response to cold or fear causes the hair to stand erect and dimples the skin surface to produce goose bumps. Even though such a "hair-raising" response is not very useful to humans, with our relatively sparse hair, it does keep fur-bearing animals warmer by trapping a layer of insulating air in their fur. Moreover, a frightened animal with its hair on end looks more formidable to an enemy.

Types and Growth of Hair

Hairs come in various sizes and shapes, but as a rule they can be classified as either **vellus** (vel′us; *vell* = wool, fleece) or **terminal.** The body hair of children and women is of the fine, short, vellus variety. The longer, coarser hair of everyone's scalp is terminal hair. Terminal hairs also appear at puberty in the axillary (armpit) and pubic regions in both sexes and on the face, chest, arms, and legs of men. These terminal hairs grow under the influence of male sex hormones called *androgens,* of which *testosterone* is the most important type.

Hairs grow an average of 2 mm per week, although this rate varies widely among body regions and with sex and age. Each follicle goes through growth cycles (Figure 5.6). In each cycle, an active growth phase is followed by a resting phase, in which the hair matrix is inactive and the follicle atrophies somewhat. At the start

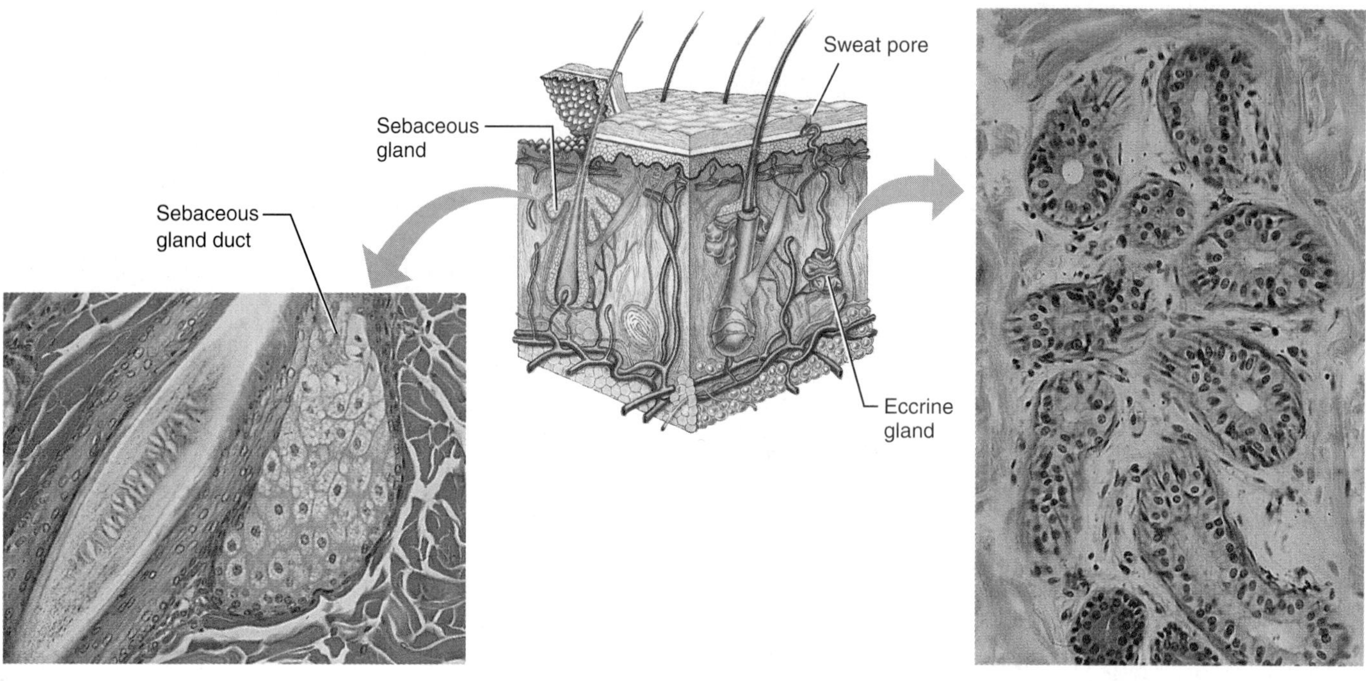

FIGURE 5.7 Sebaceous and sweat glands. **(a)** A sebaceous gland, to the right of a hair follicle, into which the sebum is secreted (micrograph 170 ×). **(b)** Section through the coiled basal part of an eccrine sweat gland (micrograph 200 ×).

of each active phase, the newly growing hair pushes out the old hair, causing it to be shed. The life span of hairs varies: On the scalp, the follicles stay active for an average of 4 years, so individual hairs grow quite long before shedding. On the eyebrows, by contrast, the follicles are active for only a few months, so the eyebrows never grow very long. Fortunately, all the hairs on the head do not fall out simultaneously every few years because the cycles of adjacent hair follicles are not in synchrony. Thus, only a small percentage of the hairs on our head sheds at any one time.

Hair Thinning and Baldness

Given ideal conditions, hair grows fastest from the teen years to the 40s. When hairs are no longer replaced as quickly as they are shed, the hair thins. By age 60 to 65, both sexes usually experience some degree of balding. Coarse terminal hairs are replaced by vellus hairs, and the hair becomes increasingly wispy.

True baldness is different. The most common type, **male pattern baldness,** is a genetically determined, gender-influenced condition. It is thought to be caused by a gene that is not expressed until adulthood, at which time it changes the response of hair follicles to androgens. Now the hairs respond to androgens by increasingly shortening their growth cycles. The cycles become so short that many hairs never even emerge from their follicles before shedding, and those that do emerge are fine vellus hairs that look like peach fuzz in the "bald" area. The drugs used to treat male pattern baldness either inhibit the production of androgens or increase blood flow to the skin and hair follicles. These treatments are only partly successful.

CLINICAL APPLICATIONS

CHEMOTHERAPY AND HAIR LOSS Chemotherapy drugs used to treat cancer target the most rapidly dividing cells in the body, thereby destroying many hair stem cells and causing hair loss. After chemotherapy stops, the hair can recover and regrow. However, hair loss due to severe burns, excessive radiation, or other factors that eliminate the follicles is permanent. ✚

Sebaceous Glands

The **sebaceous glands** (se-ba′shus; "greasy") are the skin's oil glands (Figures 5.1 and 5.7a). They occur over the entire body, except on the palms and soles. They are simple alveolar glands with several alveoli opening into a single duct (see Figure 4.11d on page 88 for their basic structure), but the alveoli are actually filled with cells, so there is no lumen. Their oily product, called **sebum** (se′bum; "animal fat"), is secreted in a most unusual way: The central cells in the alveoli accumulate oily lipids until they become engorged and burst apart. This process is called **holocrine secretion** (hol′o-krin; *holos* = whole), because *whole* cells break up to form the product. Most sebaceous glands are associated with hair follicles, emptying their sebum into the upper third of the follicle. From there, the sebum flows superficially to cover the skin. In addition to making our skin and hair greasy, sebum collects dirt, softens and lubricates the hair and skin, prevents hair from becoming brittle, and keeps the epidermis from cracking. It also helps to slow water loss across the skin and to kill bacteria.

The secretion of sebum is stimulated by hormones, especially androgens. The sebaceous glands are relatively inactive during childhood but are activated in both sexes during puberty, when the production of androgens begins to rise.

CLINICAL APPLICATIONS

ACNE In some teenagers, so much sebum is produced that it cannot be ducted from the glands quickly enough. When a sebaceous gland becomes blocked by sebum, a whitehead appears on the skin surface. If the material oxidizes and dries, it darkens to form a blackhead (note that the dark color of a blackhead is *not* due to dirt). Blocked sebaceous glands are likely to be infected by bacteria, producing a pimple. The bacteria break down the sebum into irritating fatty acids. These acids, along with the bacterial products themselves, induce inflammation, especially when the entire infected mass bursts out of the follicle into the surrounding dermis. The *acne* that results can be mild or extremely severe, leading to permanent scarring. ✚

Sweat Glands

Humans have more than 2.5 million **sweat glands (sudoriferous glands).** Sweat glands are distributed over the entire skin surface, except on the nipples and parts of the external genitalia. We normally produce about 500 ml of sweat per day, but this amount can increase to 12 liters (over 3 gallons!) on hot days during vigorous exercise. Sweating prevents overheating of the body, because its evaporation from the skin is a cooling process. Only mammals have sweat glands, and humans have the most of all, so we can run and work in the hot sun better than any other animal. Hair would interfere with the evaporation of sweat and the ability to cool the body, so our need for increased sweating ability led to a reduction of hairiness.

Sweat is an unusual secretory product in that it is primarily a filtrate of blood that passes through the secretory cells of the sweat glands and is released by exocytosis. Sweat is 99% water, with some salts (mostly sodium chloride) and traces of metabolic wastes (urea, ammonia, uric acid). It is acidic, so it retards the growth of bacteria on the skin.

There are two types of sweat glands, both of which increase their secretion in response to stress as well as to heat: eccrine and apocrine glands. **Eccrine glands** (ek′rin; "secreting") are by far the more numerous type and produce true sweat (see Figure 5.7b). They are most abundant on the palms, soles, and forehead. Each is a coiled version of a simple tubular gland (see Figure 4.11a on p. 88). The coiled, secretory base lies in the deep dermis and hypodermis, and the duct runs superficially to open at the skin surface through a funnel-shaped **pore.** (Although most pores on the skin surface are sweat pores, the "pores" seen on the face are openings of hair follicles.)

Apocrine (ap′o-krin) **glands** are mostly confined to the axillary, anal, and genital areas. They are larger than eccrine glands, and their ducts open into hair follicles. Apocrine glands produce a special kind of sweat consisting of fatty substances and proteins, in addition to the components of true sweat. For this reason, apocrine sweat is viscous and sometimes has a milky or yellow color. This product starts out odorless, but when its organic molecules are decomposed by bacteria on the skin, it takes on a musky smell. This is the basis of body odor.

Apocrine glands start to function at puberty under the influence of androgens. Because their activity is increased by sexual foreplay and they enlarge and recede with the phases of a woman's menstrual cycle, apocrine glands may be analogous to the sexual scent glands of other animals.

The skin forms several types of modified sweat glands. *Ceruminous glands* (sĕ-roo′mĭ-nus; "waxy") are modified apocrine glands in the lining of the external ear canal. Their product is a component of earwax (see p. 486). The *mammary glands* are specialized sweat glands highly modified to secrete milk. Although they are part of the integumentary system, we will discuss the mammary glands with the female reproductive system in Chapter 24.

Nails

A **nail** (Figure 5.8) is a scale-like modification of the epidermis that corresponds to the hoof or claw of other mammals. Fingernails are built-in tools that enable us to pick up small objects and scratch our skin when it itches. Like hairs, nails are made of hard keratin. Each nail has a distal **free edge,** a **body** (the visible attached part), and a **root** (the proximal part embedded in the skin). The nail rests on a bed of epidermis called the **nail bed.** This bed contains only the deeper layers of the epidermis, because the nail itself corresponds to the superficial keratinized layers.

Nails look pink because of the rich network of capillaries in the underlying dermis. At the root and the proximal end of the nail body, the bed thickens to form the **nail matrix,** the actively growing part of the nail. The matrix is so thick that the pink dermis cannot show through it. Instead, we see a white crescent, the **lunula** (lu′nu-la; "little moon"), under the nail's proximal region. The lateral and proximal borders of the nail are overlapped by skin folds called **nail folds.** The proximal nail fold projects onto the nail body as the cuticle or **eponychium** (ep″o-nik′e-um; "on the nail").

An *ingrown toenail* is just what its name implies, a nail whose growth pushes it painfully into the lateral nail fold. This happens when the nail grows crookedly,

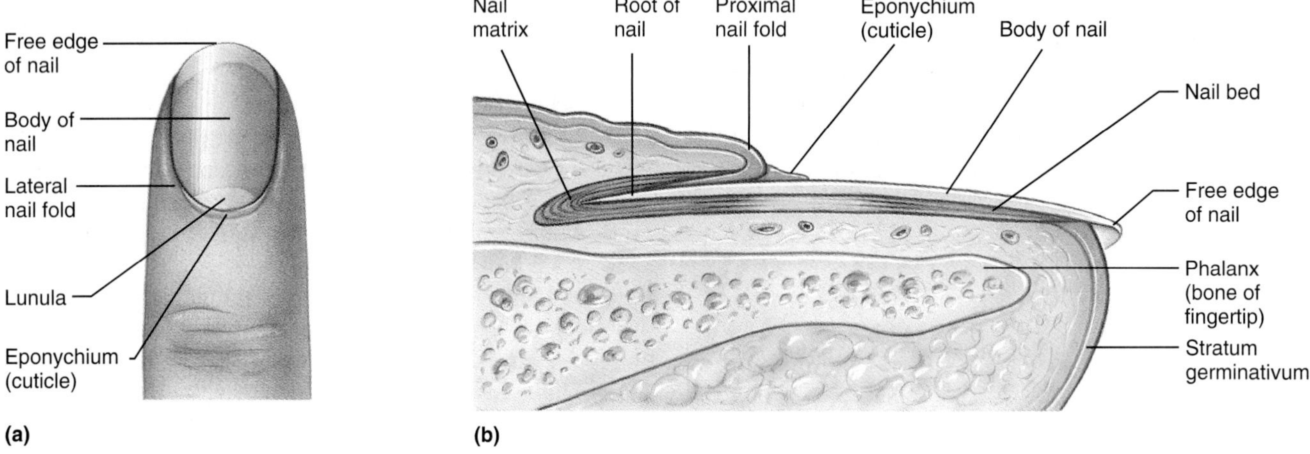

FIGURE 5.8 Structure of a nail. (a) Surface view of the distal part of a finger. (b) Longitudinal section through a fingertip.

most commonly from the pressure of a badly fitting shoe.

DISORDERS OF THE INTEGUMENTARY SYSTEM

Exposed directly to the dangers and germs of the outside world, skin can develop more than a thousand different conditions and ailments. The most common skin disorders are bacterial, viral, or yeast infections, some of which are described in the Clinical Terms at the end of this chapter. The most severe threats to the skin are burns and skin cancer, which we consider next.

Burns

Burns are a devastating threat to the body, primarily because of their effects on the skin. A burn is tissue damage inflicted by heat, electricity, radiation, extreme friction, or certain harmful chemicals.

The immediate threat to life from serious burns is a catastrophic loss of body fluids. Inflammatory edema (p. 103) is severe. As fluid seeps from the burned surfaces, the body quickly loses water and essential salts. This dehydration in turn leads to fatal circulatory shock; that is, inadequate blood circulation caused by the reduction in blood volume. To save the patient, the lost fluids must be replaced immediately. After the initial crisis has passed, infection becomes the main threat. Pathogens can easily invade areas where the skin barrier is destroyed.

Burns are classified by their severity (depth) as first-, second-, or third-degree burns (Figure 5.9). Third-degree burns are the *most* severe. In **first-degree burns,** only the epidermis is damaged. Symptoms include redness, swelling, and pain. Generally, first-degree burns heal in a few days without special attention. Sunburn is usually a first-degree burn. **Second-degree burns** involve injury to the epidermis and the upper part of the dermis. Symptoms resemble those of first-degree burns, but blisters also appear. The skin regenerates with little or no scarring in 3 to 4 weeks if care is taken to prevent infection. First- and second-degree burns are considered **partial-thickness burns.**

Third-degree burns consume the entire thickness of the skin, and thus are **full-thickness burns.** The burned area appears white, red, or blackened. Although the skin might eventually regenerate, it is usually impossible to wait for this because of fluid loss and infection. Therefore, skin from other parts of the patient's body must be grafted onto the burned area. Such a graft, in which an individual is both donor and recipient, is called an *autograft*.

In general, burns are considered critical if any of the following conditions exists: (1) Over 10% of the body has third-degree burns; (2) 25% of the body has second-degree burns; (3) there are third-degree burns on the face, hands, or feet. A quick way to estimate how much surface is affected by a burn is to use the **rule of nines** (see Figure 5.9). This method divides the body surface into 11 regions, each accounting for 9% (or a multiple of 9%) of total body area. This method is only roughly accurate, so special tables are used when greater accuracy is needed.

For patients whose burns are too large or who otherwise are poor candidates for autografts, good artificial coverings are now available. One, called INTEGRA Artificial Skin, consists of a "dermis" made from bovine collagen and an "epidermis" made of silicone. The patient's own dermis gradually replaces and reabsorbs the artificial one, and the silicone sheet is later peeled off and replaced with a network of epidermal cells cultured from the patient's own skin. The artificial skin is not rejected by the body, it saves lives, and it results in min-

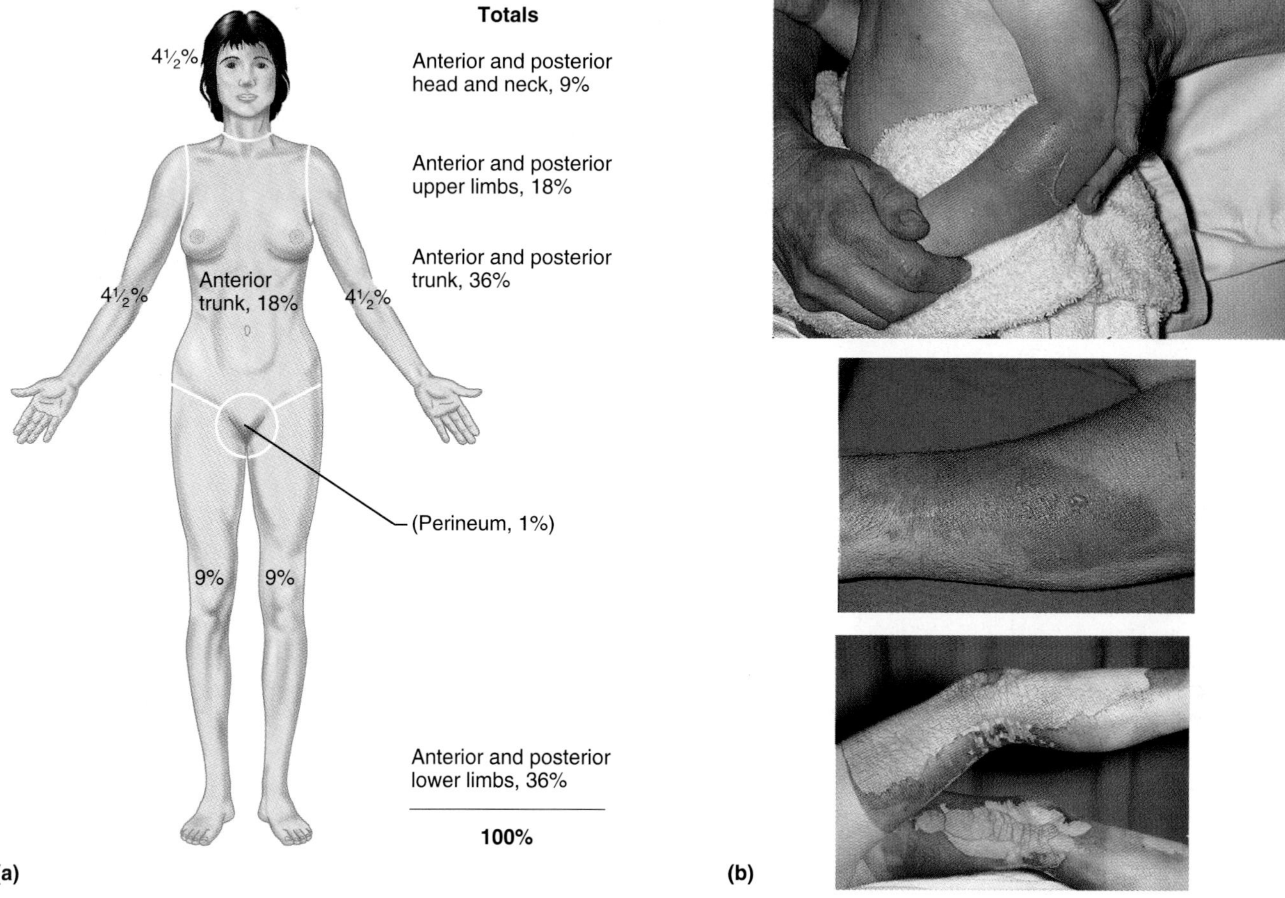

FIGURE 5.9 Burns. **(a)** Estimating the extent of burns using the rule of nines. The surface areas for the anterior body surface are indicated on the human figure. Total surface area (anterior and posterior body surfaces) is tabulated to the right of the figure. **(b)** Burns of increasing severity, from top to bottom: first-degree, second-degree, third-degree.

imal scarring. However, it is more likely to become infected than is an autograft.

Another lab-grown skin is cultured from discarded foreskins of circumcised baby boys. The paper-thin discs of cultured skin are used to treat both burns and leg ulcers.

Skin Cancer

Many types of tumors arise in the skin. Most are benign (warts, for example), but some are malignant. Skin cancer is the most common type of cancer, with about a million new cases appearing each year in the United States. As mentioned earlier, the most important risk factor for skin cancer is overexposure to the UV rays in sunlight. Here we discuss three important types of skin cancer.

Basal Cell Carcinoma

Basal cell carcinoma is the least malignant and most common of the skin cancers (Figure 5.10a). Over 30% of all Caucasians get it in their lifetimes! Cells of the stratum basale proliferate, invading the dermis and hypodermis and causing tissue erosions there. The most common lesions of this cancer are dome-shaped, shiny nodules on sun-exposed areas of the face. These nodules later develop a central ulcer and a "pearly," beaded edge. They often bleed. Basal cell carcinoma grows

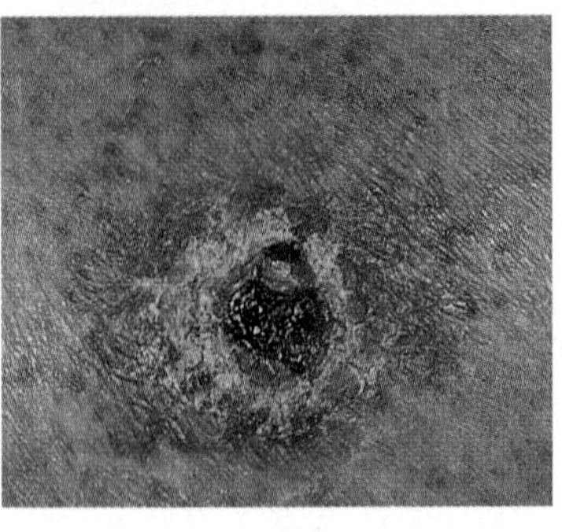

(a)

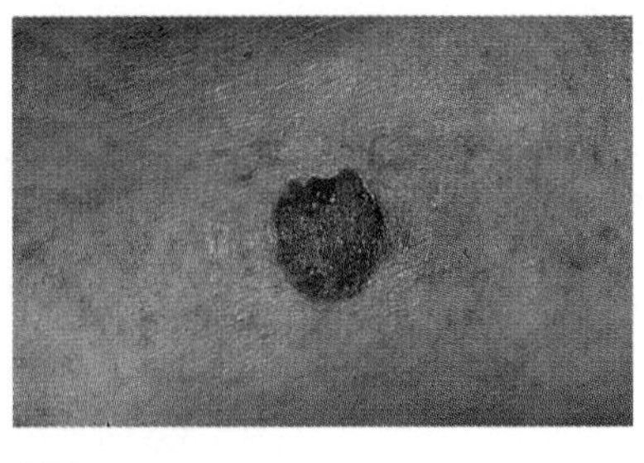

(b)

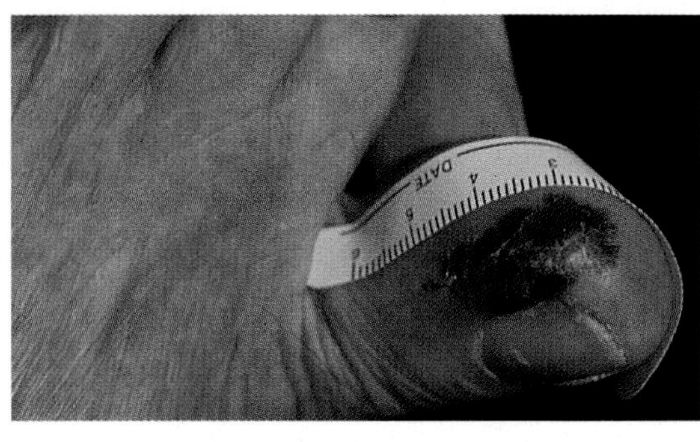

(c)

FIGURE 5.10 Photographs of skin cancers. **(a)** Basal cell carcinoma. **(b)** Squamous cell carcinoma. **(c)** Malignant melanoma. This begins as a multicolored dark lesion with an irregular shape.

relatively slowly, and metastasis seldom occurs. Full cure by surgical excision or other methods of removal is the rule in 99% of cases.

Squamous Cell Carcinoma

Squamous cell carcinoma (Figure 5.10b) arises from the keratinocytes of the stratum spinosum. The lesion appears as a scaly, irregular, reddened papule (small, rounded elevation) that tends to grow rapidly and metastasize if not removed. If treated early, however, the chance of a complete cure is good, and the overall cure rate is 99%. The carcinoma can be removed by radiation therapy, by surgery, or by skin creams containing anticancer drugs.

Melanoma

Melanoma (mel″ah-no′mah), a cancer of melanocytes, is the most dangerous kind of skin cancer (Figure 5.10c). It accounts for only about 1 of every 20 skin cancers, but it is increasing rapidly in countries with light-skinned populations—by 3% to 8% *per year* in the United States! This increase is probably due to the recent increase in recreational suntanning for cosmetic reasons. Melanoma can originate wherever there is pigment, but often arises from existing moles, usually appearing as an expanding dark patch. Because melanoma cells metastasize rapidly into surrounding circulatory vessels, the key to surviving melanoma is early detection. Most people do not survive this cancer if the lesion has grown over 4 mm thick. Melanoma is resistant to chemotherapy and current immunotherapy treatment, although vaccines are being tested.

The American Cancer Society suggests that sun worshippers regularly examine their skin for moles and new pigment spots, applying the **ABCD rule** for recognizing melanoma: **A.** *Asymmetry:* The two halves of the spot or mole do not match; **B.** *Border irregularity:* The borders have indentations and notches; **C.** *Color:* The pigment spot contains several colors, including blacks, browns, tans, and sometimes blues and reds; **D.** *Diameter:* larger than 6 mm (larger than a pencil eraser). Some experts have found that adding an **E,** for *Elevation* above the skin surface, improves diagnosis, so they use the **ABCD(E) rule.**

THE SKIN THROUGHOUT LIFE

The epidermis develops from embryonic ectoderm, and the dermis and hypodermis develop from mesoderm (see Chapter 3, pp. 66–67). Melanocytes, however, develop from neural crest cells that migrate into the ectoderm during the first 3 months of prenatal development. By the end of the fourth month, the skin is fairly well formed: The epidermis has all its strata, dermal papillae are evident, and the epidermal appendages are present in rudimentary form. During the fifth and sixth months, the fetus is covered with a downy coat of delicate hairs called the *lanugo* (lah-nu′go; "wool or down"). This cloak is shed by the seventh month, and vellus hairs make their appearance.

When a baby is born, its skin is covered with *vernix caseosa* (ver′niks ka′se-o-sah; "varnish of cheese"), a cheesy-looking substance produced by sebaceous glands. It protects the skin of the fetus from constant contact with amniotic fluid. A newborn's skin and hypodermis are very thin, but they thicken during infancy and childhood.

With the onset of adolescence, acne may appear. Acne generally subsides in early adulthood, and the skin reaches its optimal appearance in one's 20s and 30s. Thereafter, the skin starts to show the harmful effects of continued environmental assaults, such as abrasion, wind, sun, and chemicals. Scaling and various kinds of skin inflammation, called **dermatitis** (der″mah-ti′-tis), become more common.

In middle age, the lubricating and softening substances produced by the glands start to become deficient. As a result, the skin becomes dry and itchy. People with naturally oily skin may avoid this dryness, so their skin often ages well. Therefore, young people who suffer the humiliation of oily skin and acne may be rewarded later in life by looking younger longer.

Advances in Understanding Recently, it has been discovered that most aspects of skin aging are not intrinsic but are caused by sunlight in a process called "photoaging." Aged skin that has been protected from the sun has lost some elasticity and is thinner (as is the hypodermis under it), but it remains unwrinkled and unmarked. Sun-exposed skin, by contrast, is wrinkled, loose, inelastic, leathery, and has pigmented spots called "liver spots." In the dermis of such sun-aged skin, the amount of collagen has declined, and an abnormal, elastin-containing material has accumulated. Much of this change is due to UV-induced activation of enzymes called matrix metalloproteinases, which degrade collagen and other components of the dermis. ✷

By blocking UV rays, large amounts of melanin protect the skin from photoaging. This is why fair-skinned individuals, who have little melanin to begin with, show age-related changes at a younger age than do people with darker skin and hair. For the same reason, black people look youthful much longer than Caucasians.

RELATED CLINICAL TERMS

Alopecia (al″o-pe′she-ah) Any condition involving absence or loss of hair. Male pattern baldness, the most common type, is technically named *androgenic alopecia.*

Athlete's foot Itchy, red, peeling condition of the skin between the toes resulting from fungal infection.

Boils and carbuncles (kar′bung″k′ls; "little glowing embers") Infection and inflammation of hair follicles and sebaceous glands that has spread into the underlying hypodermis. A boil can be likened to a giant pimple; carbuncles are composite boils. The common cause is bacterial infection.

Cold sores (fever blisters) Small, painful, fluid-filled blisters that usually occur around the lips and in the mucosa in the mouth. They are caused by the herpes simplex virus, which localizes in nerve cells that supply the skin, where it remains dormant until activated by emotional upset, fever, or UV radiation.

Decubitus ulcer (de-ku′bĭ-tus ul′ser; "lying-down opensore") A bedsore, which is a localized breakdown of skin due to a reduction in blood supply resulting from prolonged confinement in bed. A decubitus ulcer usually occurs over a bony prominence, such as the hip or the heel.

Impetigo (im″pĕ-ti′go; "an attack") Pink, fluid-filled, raised lesions that are common around the mouth and nose. They develop a yellow crust and eventually rupture. This contagious condition, caused by a staphylococcus infection, is common in school-age children.

Psoriasis (so-ri′ah-sis; "an itching") A recurrent, chronic inflammatory condition characterized by reddened epidermal papules covered with dry, silvery scales. The scales result from an overproliferation of the epidermis, and the pink color is due to widened capillaries in the dermis. Relatively common, it affects 2% of Americans. May be an autoimmune condition triggered by bacterial products. One effective treatment for psoriasis employs a drug that slows epidermal growth and which is activated by exposure to UV light.

Vitiligo (vit″il-i′go; "blemish") An abnormality of skin pigmentation characterized by light spots surrounded by areas of normally pigmented skin. It can be cosmetically disfiguring, especially in dark-skinned people.

CHAPTER SUMMARY

Media study tools that could provide you additional help in reviewing specific key topics of Chapter 5 are referenced below.

SP = Study Partner; AIA = A.D.A.M.® Interactive Anatomy

1. The integumentary system consists of the skin and its appendages (hair, sebaceous glands, sweat glands, and nails). The hypodermis, although not part of the integumentary system, is also considered in this chapter.

THE SKIN AND THE HYPODERMIS (pp. 112–117)

Epidermis (pp. 113–115)

1. The epidermis is a keratinized stratified squamous epithelium. Its main cell type is the keratinocyte. From deep to superficial, its strata are the basale, spinosum, granulosum, lucidum, and corneum. The stratum lucidum occurs only in thick skin found on palms and soles.

2. The dividing cells of the stratum basale are the source of new keratinocytes for the growth and renewal of the epidermis.

3. In the stratum spinosum, keratinocytes contain strong tonofilaments. In the stratum granulosum, keratinocytes contain keratohyaline granules, which combine with the tonofilaments to form tension-resisting keratin. Stratum granulosum cells also secrete an extracellular waterproofing glycolipid. Keratinized cells of the stratum corneum protect the skin and are shed as skin flakes and dandruff.

4. Scattered among the keratinocytes in the deepest layers of the epidermis are pigment-producing melanocytes, sensory Merkel cells, and the Langerhans cells of the immune system.

SP Exercise: Chapter 5, Layers of the Epidermis.

Dermis (p. 115)

5. The leathery dermis, composed of connective tissue proper, is well supplied with vessels and nerves. Also located in the dermis are skin glands and hair follicles.

6. The superficial papillary layer consists of areolar connective tissue, whereas the deep reticular layer is a dense irregular connective tissue. The papillary layer abuts the undulating undersurface of the epidermis and is responsible for the epidermal friction ridges that produce fingerprints. In the reticular layer, less-dense regions between the collagen bundles produce cleavage lines.

Hypodermis (pp. 115–116)

7. The hypodermis underlies the skin. Composed primarily of adipose tissue, it stores fat, insulates the body against heat

loss, and absorbs and deflects blows. During weight gain, this layer thickens most markedly in the thighs and breasts of females and in the anterior abdominal wall of males.

SP Exercise: Chapter 5, Skin Structure.

Skin Color (pp. 116–117)

8. Skin color reflects the amounts of pigments (melanin and/or carotene) in the skin and the oxygenation level of hemoglobin in the dermal blood vessels.

9. Melanin, made by melanocytes and transferred to keratinocytes, protects keratinocyte nuclei from the damaging effects of UV radiation. The epidermis also produces vitamin D under the stimulus of UV radiation.

APPENDAGES OF THE SKIN (pp. 117–122)

1. Skin appendages, which develop from the epidermis but project into the dermis, include hairs and hair follicles, sebaceous glands, sweat glands, and nails.

Hair and Hair Follicles (pp. 117–120)

2. A hair is produced by a tube-shaped hair follicle and consists of heavily keratinized cells. Each hair has an outer cuticle, a cortex, and usually a central medulla. Hairs have roots and shafts. Hair color reflects the amount and variety of melanin present.

3. A hair follicle consists of an inner, epithelial root sheath that includes the hair matrix and an outer, connective tissue root sheath derived from the dermis. The hair-stem cells are just above the hair bulb. A knot of capillaries in the hair papilla nourishes the hair, and a nerve knot surrounds the hair bulb. Arrector pili muscles pull the hairs erect and produce goose bumps when we are cold or scared.

4. Fine and coarse hairs are called vellus hairs and terminal hairs, respectively. At puberty, under the influence of androgens, terminal hairs appear in the axillae and pubic regions in both sexes.

5. Hairs grow in cycles that consist of growth and resting phases. Hair thinning results from factors that lengthen follicular resting phases, including natural age-related atrophy of hair follicles and expression of the male pattern baldness gene during adulthood.

SP Exercise: Chapter 5, Hair Structure.

Sebaceous Glands (pp. 120–121)

6. Sebaceous glands are simple alveolar glands that usually empty into hair follicles. Their oily holocrine secretion is called sebum.

7. Sebum lubricates the skin and hair, slows water loss across the skin, and acts as a bactericidal agent. Sebaceous glands secrete increased amounts of sebum at puberty under the influence of androgens. Blocked and infected sebaceous glands are the basis of pimples and acne.

Sweat Glands (p. 121)

8. Eccrine sweat glands are simple coiled tubular glands that secrete sweat, a modified filtrate of the blood. Evaporation of sweat cools the skin and the body.

9. Apocrine sweat glands, which may function as scent glands, occur primarily in the axillary, anal, and genital regions. Their secretion contains proteins and fatty acids; bacterial action on this section causes body odor.

Nails (pp. 121–122)

10. A nail is a scale-like modification of the epidermis that covers the dorsal tip of a finger or toe. The actively growing region is the nail matrix.

DISORDERS OF THE INTEGUMENTARY SYSTEM (pp. 122–124)

1. The most common skin disorders result from microbial infections.

Burns (pp. 122–123)

2. In severe burns, the initial threat is loss of body fluids leading to circulatory collapse. The secondary threat is overwhelming infection.

3. The extent of a burn may be estimated by using the rule of nines. The severity of burns is indicated by the terms *first-, second-,* and *third-degree.* Third-degree burns harm the full thickness of the skin and require grafting for successful recovery.

Skin Cancer (pp. 123–124)

4. The major risk factor for skin cancer is exposure to UV sunlight.

5. Basal cell carcinoma and squamous cell carcinoma are the most common types of skin cancer and usually are curable if detected early. Melanoma, a cancer of melanocytes, is less common but dangerous. The ABCD or ABCD(E) rule can be used to determine whether a mole or spot is a melanoma.

THE SKIN THROUGHOUT LIFE (pp. 124–125)

1. Epidermis develops from embryonic ectoderm, dermis and hypodermis from mesoderm, and melanocytes from neural crest cells.

2. At 5–6 months, a fetus has a downy lanugo coat. Fetal sebaceous glands produce vernix caseosa that protects the skin within the amniotic sac.

3. In old age, the skin thins and loses elasticity. Damage from sunlight leads to wrinkles, age spots, and loose and leathery skin; this reflects a loss of collagen and accumulation of elastin-containing material in the dermis.

REVIEW QUESTIONS

Multiple Choice/Matching Questions

1. Which epidermal cell type functions in the immune system? (a) keratinocyte, (b) melanocyte, (c) Langerhans cell, (d) Merkel cell.

2. Match the names of epidermal layers with the following descriptions:

____ **(1)** desmosomes and shrinkage artifacts give its cells "prickly" projections	**(a)** stratum basale
____ **(2)** its cells are flat, dead bags of keratin	**(b)** stratum corneum
____ **(3)** its cells divide, and it is also called the stratum germinativum	**(c)** stratum granulosum
____ **(4)** contains keratohyaline and lamellated granules	**(d)** stratum lucidum
____ **(5)** present only in thick skin	**(e)** stratum spinosum

3. The ability of the epidermis to resist rubbing and abrasion is largely due to the presence of (a) melanin, (b) carotene, (c) collagen, (d) keratin.

4. Skin color is determined by (a) melanin, (b) carotene, (c) oxygenation level of the blood, (d) all of these.

5. What is the major factor accounting for the waterproof nature of the skin? (a) desmosomes in stratum corneum, (b) glycolipid between stratum corneum cells, (c) the thick insulating fat of the hypodermis, (d) the leathery nature of the dermis.

6. Circle the false statement about vitamin D: (a) Dark-skinned people make no vitamin D. (b) If the production of vitamin D is inadequate, one may develop weak bones. (c) If the skin is not exposed to sunlight, one may develop weak bones. (d) Vitamin D is needed for the uptake of calcium from food in the intestine.

7. Use logic to deduce the answer to this question. Given what you know about skin color and skin cancer, the highest rate of skin cancer on earth occurs among (a) blacks in tropical Africa, (b) scientists in research stations in Antarctica, (c) whites in northern Australia, (d) Norwegians in the United States, (e) blacks in the United States.

8. An arrector pili muscle (a) is associated with each sweat gland, (b) causes the hair to stand up straight, (c) enables each hair to be stretched when wet, (d) squeezes hair upward so it can grow out.

9. Sebum (a) lubricates the skin and hair; (b) consists of dead cells and fatty substances; (c) collects dirt, so hair has to be washed; (d) all of these.

10. An eccrine sweat gland is a ____ gland: (a) compound tubular, (b) compound tubuloacinar, (c) simple squamous, (d) simple tubular.

Short Answer Essay Questions

11. Is a bald man really hairless? Explain.

12. Explain why thin skin is also called hairy skin, and why thick skin is also called hairless skin. (Hint: Note their locations.)

13. Distinguish first-, second-, and third-degree burns.

14. Why does skin that is exposed to sunlight age so much more than nonexposed skin?

15. Explain why no skin cancers originate from stratum corneum cells.

16. Explain each of these familiar phenomena in terms of what you learned in this chapter: (a) goose bumps, (b) dandruff, (c) stretch marks from gaining weight, (d) leaving fingerprints, (e) the almost hairless body of humans, (f) getting gray hairs.

17. Eric, an anatomy student, said it was so hot one day that he was "sweating like a pig." His professor overheard him and remarked, "That is a stupid expression, Eric. No pig ever sweats nearly as much as a person does!" Explain her remark.

18. List three functions of the skin.

19. After studying the skin in anatomy class, Toby grabbed the large "love handles" around his own abdomen, and said, "I have a thick hypodermis, but this layer performs some valuable functions!" What are the functions of the hypodermis?

20. What type of tissue(s) form each of the following structures? (a) dermis, (b) epidermis, (c) hypodermis. What is the developmental source of each?

CRITICAL REASONING + CLINICAL APPLICATION QUESTIONS

1. What are the two most important clinical problems encountered by severe burn patients? Explain each in terms of the absence of skin.
2. Dean, a 40-year-old beach boy, complains to you that although his suntan made him popular when he was young, now his face is all wrinkled, and he has several moles that are growing rapidly and are as large as coins. When he shows you the moles, you immediately think "ABCD." What do these letters mean, and what is Dean's disease?
3. Long-term patients confined to hospital beds are turned every 2 hours. Why? Why is this effective?
4. Carmen slipped on some ice and split her chin on the sidewalk. As the emergency room physician was giving her six stitches, he remarked that the split was right along a cleavage line. How cleanly is her wound likely to heal, and is major scarring likely to occur?
5. A man got his finger caught in a machine at the factory. The damage was less serious than expected, but nonetheless the entire nail was torn off his right index finger. The nail parts lost were the body, root, bed, matrix, and eponychium. Explain whether this nail is likely to grow back, and why.
6. On a diagram of the human body, mark off various regions according to the rule of nines. What percent of the total body surface is affected if someone is burned (a) over the entire posterior trunk and buttocks? (b) over an entire lower limb? (c) on the entire front of the left upper limb?

A.D.A.M.® INTERACTIVE ANATOMY QUESTIONS

1. Link: Dissectible anatomy; Female: anterior, Layer 0. Name three structures of the integumentary system that can be viewed with this image.
2. Link: Dissectible anatomy; Female: anterior, Layer 7. Name the specialized sweat glands in this image that secrete milk.

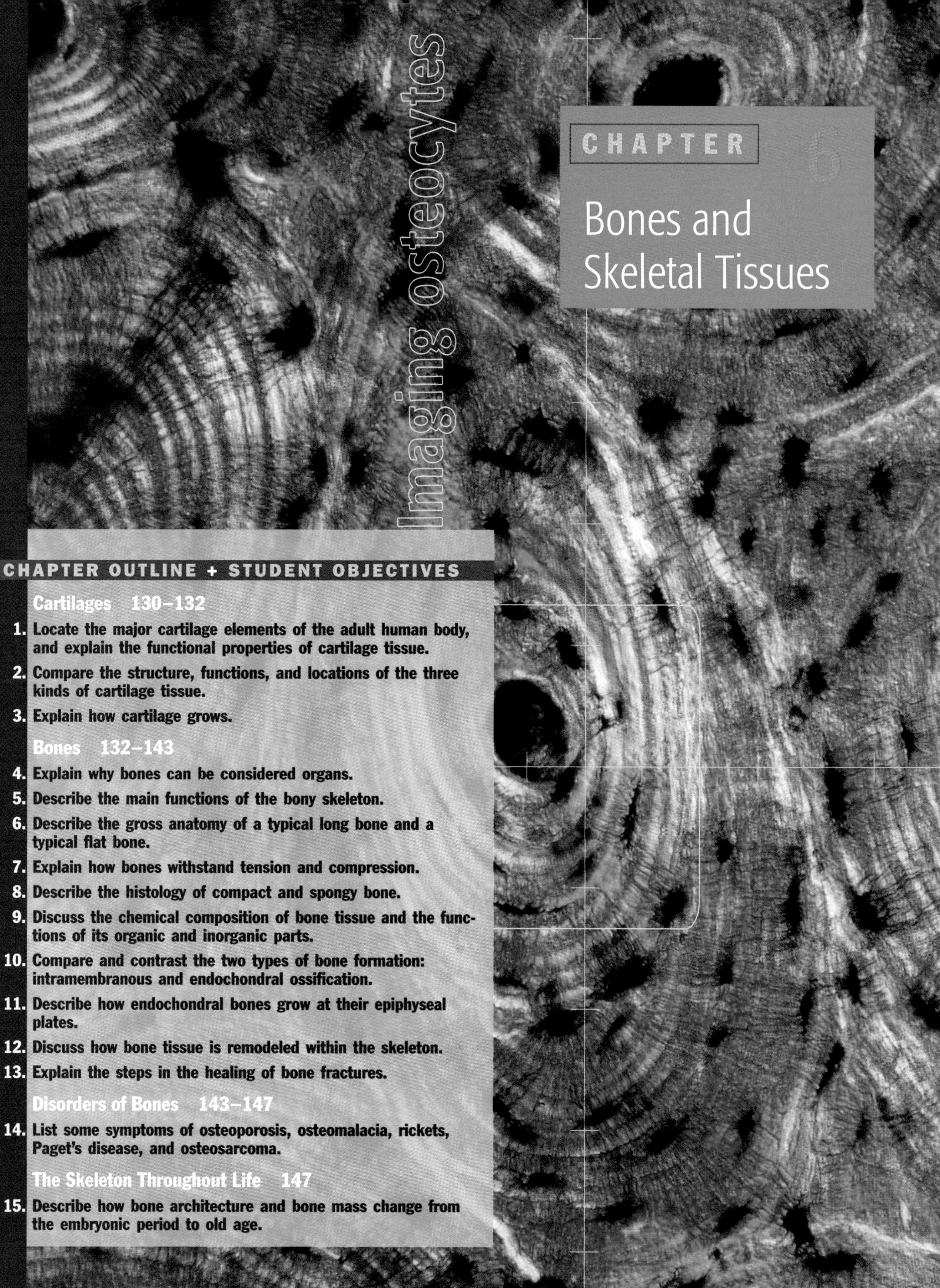

CHAPTER 6

Bones and Skeletal Tissues

CHAPTER OUTLINE + STUDENT OBJECTIVES

Cartilages 130–132

1. Locate the major cartilage elements of the adult human body, and explain the functional properties of cartilage tissue.
2. Compare the structure, functions, and locations of the three kinds of cartilage tissue.
3. Explain how cartilage grows.

Bones 132–143

4. Explain why bones can be considered organs.
5. Describe the main functions of the bony skeleton.
6. Describe the gross anatomy of a typical long bone and a typical flat bone.
7. Explain how bones withstand tension and compression.
8. Describe the histology of compact and spongy bone.
9. Discuss the chemical composition of bone tissue and the functions of its organic and inorganic parts.
10. Compare and contrast the two types of bone formation: intramembranous and endochondral ossification.
11. Describe how endochondral bones grow at their epiphyseal plates.
12. Discuss how bone tissue is remodeled within the skeleton.
13. Explain the steps in the healing of bone fractures.

Disorders of Bones 143–147

14. List some symptoms of osteoporosis, osteomalacia, rickets, Paget's disease, and osteosarcoma.

The Skeleton Throughout Life 147

15. Describe how bone architecture and bone mass change from the embryonic period to old age.

ALL OF US HAVE HEARD EXPRESSIONS LIKE "bone tired," "dry as a bone," and "bag of bones"—pretty unflattering (and inaccurate) images of our main skeletal elements. Our brains, not our bones, convey feelings of fatigue, and living bones are far from dry. As for "bag of bones," bones are indeed more prominent in some of us, but without bones to form our internal skeleton we would creep along like slugs, lacking a defined shape. Bones may conjure up images of graveyards and death, but they are very much alive. As we will see, bone is one of our most dynamic tissues. Along with its bones, the skeleton contains cartilages, which are more flexible and resilient than bones, though not as strong. In this chapter we discuss the structure, function, and growth of cartilage and bone tissue. The individual bones of the skeleton are considered in Chapters 7 and 8.

CARTILAGES

In this section we examine the sites and structure, types, and growth of cartilage tissue.

Location and Basic Structure

Pieces of cartilage are found throughout the adult human body. These cartilages, listed as shown from roughly top to bottom in Figure 6.1, include (1) cartilage in the external ear; (2) cartilages in the nose; (3) **articular cartilages,** which cover the ends of most bones at movable joints; (4) **costal cartilages,** which connect the ribs to the sternum (breastbone); (5) cartilages in the larynx (voice box), including the *epiglottis,* a flap that keeps food from entering the larynx and the lungs; (6) cartilages that hold open the air tubes of the respiratory system; (7) cartilage in the discs between the vertebrae; and (8) cartilage in the pubic symphysis.

Cartilage is far more abundant in the embryo than in the adult: Most of the skeleton is first formed in fast-growing cartilage, which is subsequently replaced by bone tissue in the fetal and childhood periods (see pp. 139–142).

A typical piece of cartilage in the skeleton is composed of cartilage tissue, which contains no nerves or blood vessels. The cartilage is surrounded by a layer of dense irregular connective tissue, the *perichondrium* (per″ĭ-kon′dre-um; "around the cartilage"), which acts like a girdle to resist outward expansion when the cartilage is subjected to pressure. It also functions in the growth and repair of cartilage.

Cartilage tissue consists primarily of water and is very resilient; that is, it has the ability to spring back to its original shape after being compressed. For more information on the unique properties of cartilage tissue, see the box on p. 132.

Types of Cartilage

Three types of cartilage tissue occur within the body: *hyaline cartilage, elastic cartilage,* and *fibrocartilage.* While reading about these, keep in mind that cartilage is a connective tissue that consists of cells called *chondrocytes* and an abundant extracellular matrix containing a jelly-like ground substance and fibers (see Chapter 4, p. 99).

Hyaline Cartilage

Hyaline cartilage (hi′ah-līn; "glass"), which looks like frosted glass when viewed by the unaided eye, is the most abundant kind of cartilage (see Figure 6.1). When viewed under the light microscope, its chondrocytes appear spherical (see Figure 4.15g on p. 95). Each chondrocyte occupies a cavity in the matrix called a *lacuna* (lah-ku′nah), literally a "pit" or "cavity." The only type of fiber in the matrix is a kind of collagen unit fibril, which forms networks that are too thin to be seen with a light microscope. Hyaline cartilage provides support through flexibility and resilience.

Elastic Cartilage

Elastic cartilage is similar to hyaline cartilage, but its matrix contains many elastic fibers along with the delicate collagen fibrils (see Figure 4.15h, p. 95). This cartilage is more elastic than hyaline cartilage and better able to tolerate repeated bending. The epiglottis, which bends down to cover the glottis (opening) of the larynx each time we swallow, is made of elastic cartilage, as is the highly bendable cartilage in the outer ear.

Fibrocartilage

Fibrocartilage is an unusual tissue that resists both strong compression and strong tension forces. It occurs in certain ligaments that are compressed as well as pulled, and in certain cartilages that are pulled as well as compressed. It is a perfect structural intermediate between hyaline cartilage and dense regular connective tissue (Figure 4.15i; p. 96). Microscopically, it consists of rows of thick collagen fibers (as in dense regular connective tissue) alternating with rows of chondrocytes, each of which is surrounded by a layer of cartilage matrix. Two specific locations of fibrocartilage are in the *annulus fibrosus* ligaments in the discs between the vertebrae and in the cartilages of the knee called *menisci.*

Growth of Cartilage

A piece of cartilage grows in two ways. In **appositional** (ap″o-zish′un-al) **growth,** a "growth from outside," cartilage-forming cells (chondroblasts) in the surrounding perichondrium produce the new cartilage tissue. In **interstitial** (in″ter-stish′al) **growth,** a "growth from within," the chondrocytes within the cartilage divide and

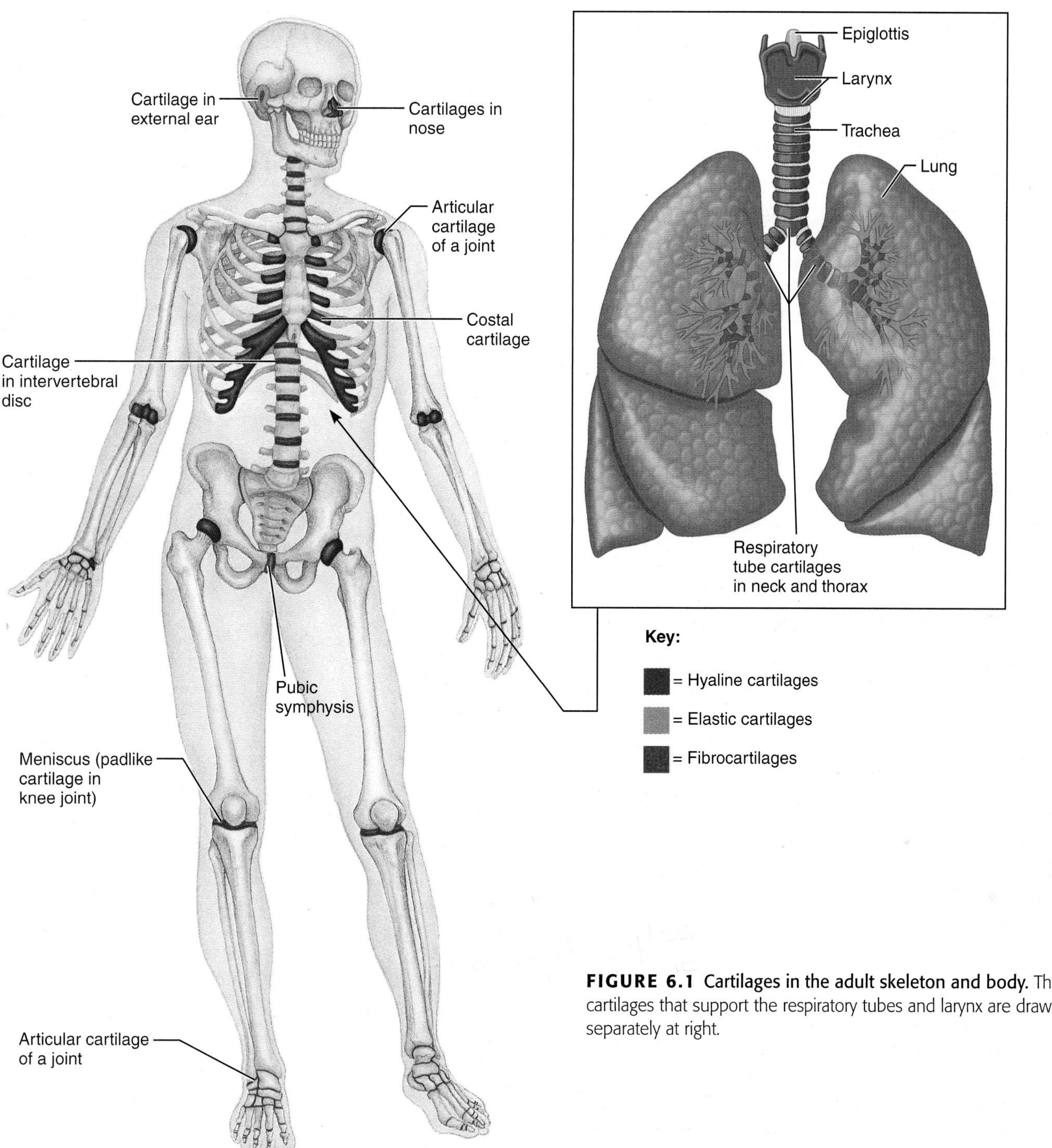

FIGURE 6.1 Cartilages in the adult skeleton and body. The cartilages that support the respiratory tubes and larynx are drawn separately at right.

secrete new matrix. Cartilage stops growing in the late teens when the skeleton itself stops growing, and chondrocytes do not divide again.

Critical Reasoning

Given that chondrocytes do not divide after the teen years, it's not surprising that cartilage regenerates poorly in adults. How does injured cartilage heal and regenerate in adults?

The little healing that occurs within cartilage in adults results from the ability of the surviving chondrocytes to secrete more extracellular matrix.

Under certain conditions, crystals of calcium phosphate precipitate in the matrix of cartilage. Such calcification is both a sign of aging and a normal stage in the growth of most bones (see p. 139). Note, however, that this **calcified cartilage** is not bone: Bone and cartilage are always distinct tissues.

A CLOSER LOOK

The Marvelous Properties of Cartilage

Without healthy cartilage, which among other functions constitutes the spongy cushions of our moveable joints, vigorous activities such as playing tennis are next to impossible.

Cartilage is resilient: If you push on it and then ease the pressure, it bounces back. Such resilience allows the articular cartilages at the ends of our bones to cushion the impact of various activities, including lifting heavy objects, sprinting, and jumping. Once an action is complete, the cartilage tissue springs back to await the next movement.

Another important feature of cartilage tissue is its rapid growth, which enables it to keep pace with the rapid growth of the embryo. Indeed, most of the embryo's "bones" originate as cartilage models that are only gradually replaced by bone tissue. Throughout childhood and adolescence, growth plates of cartilage remain in most bones and bring about most growth of the skeleton. One would expect cartilage to require a rich blood supply for such rapid growth, yet cartilage contains no blood vessels. Instead, the cells (chondrocytes) within cartilage are specially adapted to live under low oxygen levels, and can survive on the little oxygen that diffuses to them from distant vessels in the surrounding perichondrium. In fact, cartilage and blood vessels are strictly incompatible: Cartilage secretes chemicals that prevent blood vessels from growing into it, and it cannot develop in parts of the embryo that are well supplied with blood vessels or oxygen.

The explanation for the resilience of cartilage lies in its ability to hold tremendous amounts of water. In fact, cartilage is 60% to 80% water, largely because the complex sugars in its ground substance have so many water-attracting negative charges. When the cartilage is compressed, the water molecules are forced away from the negative charges. As the pressure increases, the negative charges are pushed closer together until they repel each other, resisting further compression. Then, when the pressure is released, the water molecules rush back to their original sites, causing the cartilage to spring back forcefully to its original shape. These dynamics also play a vital role in the nourishment of cartilage: The flow of fluid during and after compression carries nutrients to the chondrocytes. For this reason, long periods of inactivity can weaken the cartilages of the joints.

The rapid growth of cartilage is another result of its capacity to hold fluid. Because cartilage matrix is mostly water, nutrients from distant capillaries can diffuse through it quickly enough to supply the metabolic needs of the rapidly dividing cartilage-forming cells (chondroblasts). There is no need for a slow, time-consuming process of growing new capillary beds within the cartilage tissue itself. Furthermore, the ground substance secreted by the chondroblasts attracts so much water that the cartilage expands (grows) quickly with little expenditure of materials.

Due to the water-attracting nature of its ground substance, cartilage has a natural tendency to swell with water. To prevent excessive swelling, its matrix contains a network of thin, unstretchable collagen fibrils. This collagen also gives cartilage its ability to resist tension when pulled upon by external forces. However, even though cartilage is strong in resisting tension and compression, it is weak in resisting shear forces (twisting and bending). Because of this weakness, torn cartilages are a common sports injury.

The cartilages of tennis players and other athletes should be well nourished and well hydrated because of the great physical demands vigorous sports place on the joints.

BONES

The bones of the skeleton are *organs* because they contain several different tissues. Although bones are dominated by bone tissue, they also contain nervous tissue in their nerves, blood tissue in blood vessels, cartilage in articular cartilages, and epithelial tissue lining the blood vessels.

Functions of Bones

Besides contributing to body shape, bones perform other important functions:

1. **Support.** Bones provide a hard framework that supports the weight of the body and cradles its soft organs. For example, the bones of the legs are pillars that support the body trunk when we stand, and the bony pelvis acts like a bowl to hold the pelvic organs.

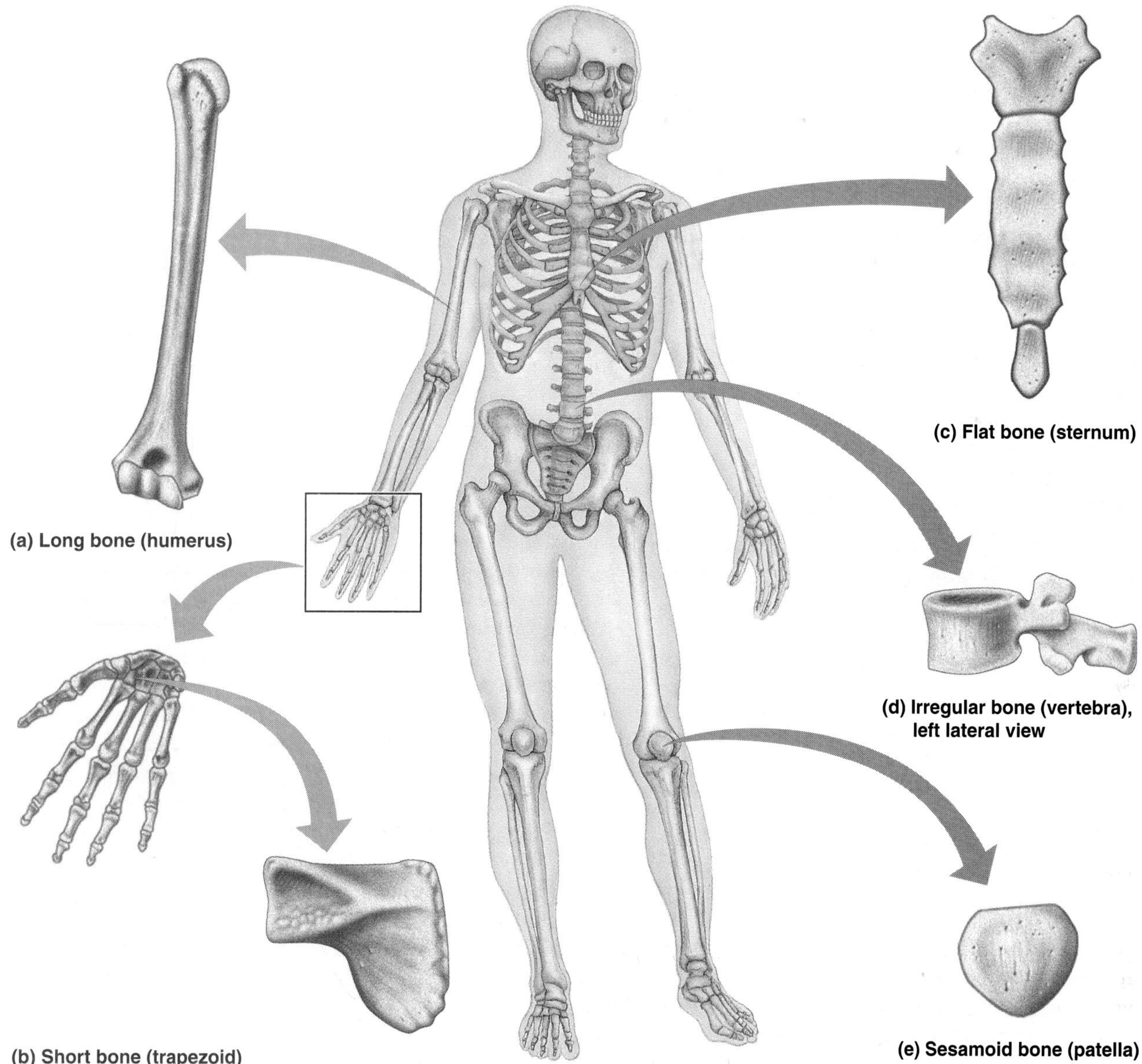

FIGURE 6.2 Classification of bones on the basis of shape.

2. **Protection.** The bones of the skull form a protective case for the brain. The vertebrae surround the spinal cord, and the rib cage helps protect the organs of the thorax.
3. **Movement.** Skeletal muscles, which attach to the bones by tendons, use the bones as levers to move the body and its parts. As a result, we can walk, grasp objects, and move our rib cage to breathe. The arrangement of the bones and the structure of the joints determine the types of movement that are possible.
4. **Mineral storage.** Bone serves as a reservoir for minerals, the most important of which are calcium and phosphate. The stored minerals are released into the bloodstream as ions for distribution to all parts of the body as needed.
5. **Blood-cell formation.** Bones contain red and yellow *bone marrow.* Red marrow makes the blood cells, and yellow marrow is a site of fat storage, with little or no role in blood-cell formation. Bone marrow and the production of blood cells are described in detail in Chapter 17 (p. 512).

Classification of Bones

Bones come in many sizes and shapes. For example, the tiny pisiform bone of the wrist is the size and shape of a pea, whereas the femur (thighbone) is large and elongated. The shape of each bone fills a particular need. The femur, for example, must withstand great weight and pressure, and its hollow cylindrical design provides maximum strength with minimum weight.

Bones are classified by their shape as long, short, flat, or irregular (Figure 6.2, p. 133).

1. **Long bones.** As their name suggests, long bones are considerably longer than they are wide. A long bone has a shaft plus two ends. Most bones in the limbs are long bones. Note that these bones are named for their elongated shape, not their overall size: The bones within each finger and toe are long bones, even though they are very small.
2. **Short bones.** Short bones are roughly cube-shaped. Bones of the wrist are examples (Figure 6.2b).

   Sesamoid bones (ses′ah-moid; "shaped like a sesame seed") are a special type of short bone that form within a tendon. A good example is the kneecap, or patella. Sesamoid bones vary in size and number in different people. Some clearly act to alter the direction of pull of a tendon, but the function of others is unknown.
3. **Flat bones.** Flat bones are thin, flattened, and usually somewhat curved. Most cranial bones of the skull are flat, as are the ribs, sternum (breastbone), and scapula (shoulder blade).
4. **Irregular bones.** Irregular bones have various shapes that do not fit into the previous categories. Examples are the vertebrae and hip bones.

Gross Anatomy of Bones

Compact and Spongy Bone

Almost every bone of the skeleton has a dense outer layer that looks smooth and solid to the naked eye. This external layer is **compact bone** (Figures 6.3 and 6.4). Internal to this is **spongy bone,** also called *cancellous bone,* a honeycomb of small needle-like or flat pieces called *trabeculae* (trah-bek′u-le; "little beams"). In this network, the open spaces between the trabeculae are filled with red or yellow bone marrow.

Structure of a Typical Long Bone

With few exceptions, all the long bones in the body have the same general structure (Figure 6.3).

Diaphysis and Epiphyses The tubular **diaphysis** (di-af′ĭ-sis), or shaft, forms the long axis of a long bone; the **epiphyses** (e-pif′ĭ-sez) are the bone ends (Figure 6.3a). In many cases, the epiphyses are larger in diameter than the diaphysis. The joint surface of each epiphysis is covered with a thin layer of articular cartilage. Between the diaphysis and each epiphysis of an adult long bone is an **epiphyseal line.** This line is a remnant of the epiphyseal plate, a disc of hyaline cartilage that grows during childhood to lengthen the bone (see p. 140).

Blood Vessels Unlike cartilage, bone tissue is well vascularized. In fact, at any given time between 3% and 11% of the blood in the body is in the skeleton. The main vessels serving the diaphysis are the *nutrient artery* (Figure 6.3c) and *nutrient vein.* Together these run through a hole in the wall of the diaphysis, the *nutrient foramen* (fo-ra′men; "opening"). The nutrient artery runs inward to supply the bone marrow and the spongy bone, and it sends branches outward to help supply the compact bone. Several *epiphyseal arteries* and *veins* serve each epiphysis in the same way (not shown in Figure 6.3).

The Medullary Cavity The interior of all bones consists largely of spongy bone. However, the very center of the diaphysis of long bones contains no bone tissue at all and is called the **medullary cavity** (med′u-lar-e; "middle") or *marrow cavity* (Figure 6.3a). As its name implies, this cavity is filled with yellow bone marrow. Recall that the spaces between the trabeculae of spongy bone are also filled with marrow.

Membranes A membrane of connective tissue called the **periosteum** (per″e-os′te-um; "around the bone") covers the entire outer surface of each bone (except on the ends of the epiphyses, where articular cartilage occurs). This periosteal membrane has two sublayers: a superficial layer of dense irregular connnective tissue, which resists tension placed on a bone during bending, and a deep layer that contacts the compact bone. This deep layer contains bone-depositing cells called *osteoblasts* (os′te-o-blasts″; "bone generators") and bone-destroying cells called *osteoclasts* ("bone breakers"). These cells remodel bone surfaces throughout our lives. The periosteum is richly supplied with nerves, which is why broken bones are painful, and with blood vessels, which bleed profusely when a bone is fractured. The vessels that supply the periosteum are branches from the nutrient and epiphyseal vessels. The periosteum is secured to the underlying bone by **Sharpey's fibers** *(perforating fibers),* thick bundles of collagen that run from the periosteum into the bone matrix (see Figure 6.3c). The periosteum also provides insertion points for the tendons and ligaments that attach to a bone. At these points, the Sharpey's fibers are exceptionally dense.

Whereas periosteum covers the external surface of bones, the *internal* bone surfaces are covered by a much thinner membrane of connective tissue called **endosteum** (en-dos′te-um; "within the bone"). Specifically, endosteum covers the trabeculae of spongy bone; it also lines the central canals of osteons (see p. 137). Like periosteum, endosteum contains both osteoblasts and osteoclasts.

Structure of Short, Irregular, and Flat Bones

Short, irregular, and flat bones have much the same composition as long bones: periosteum-covered compact bone externally and endosteum-covered spongy bone internally. These bones are not cylindrical, so they have no diaphysis. They contain bone marrow (between the trabeculae of their spongy bone), but no marrow cavity

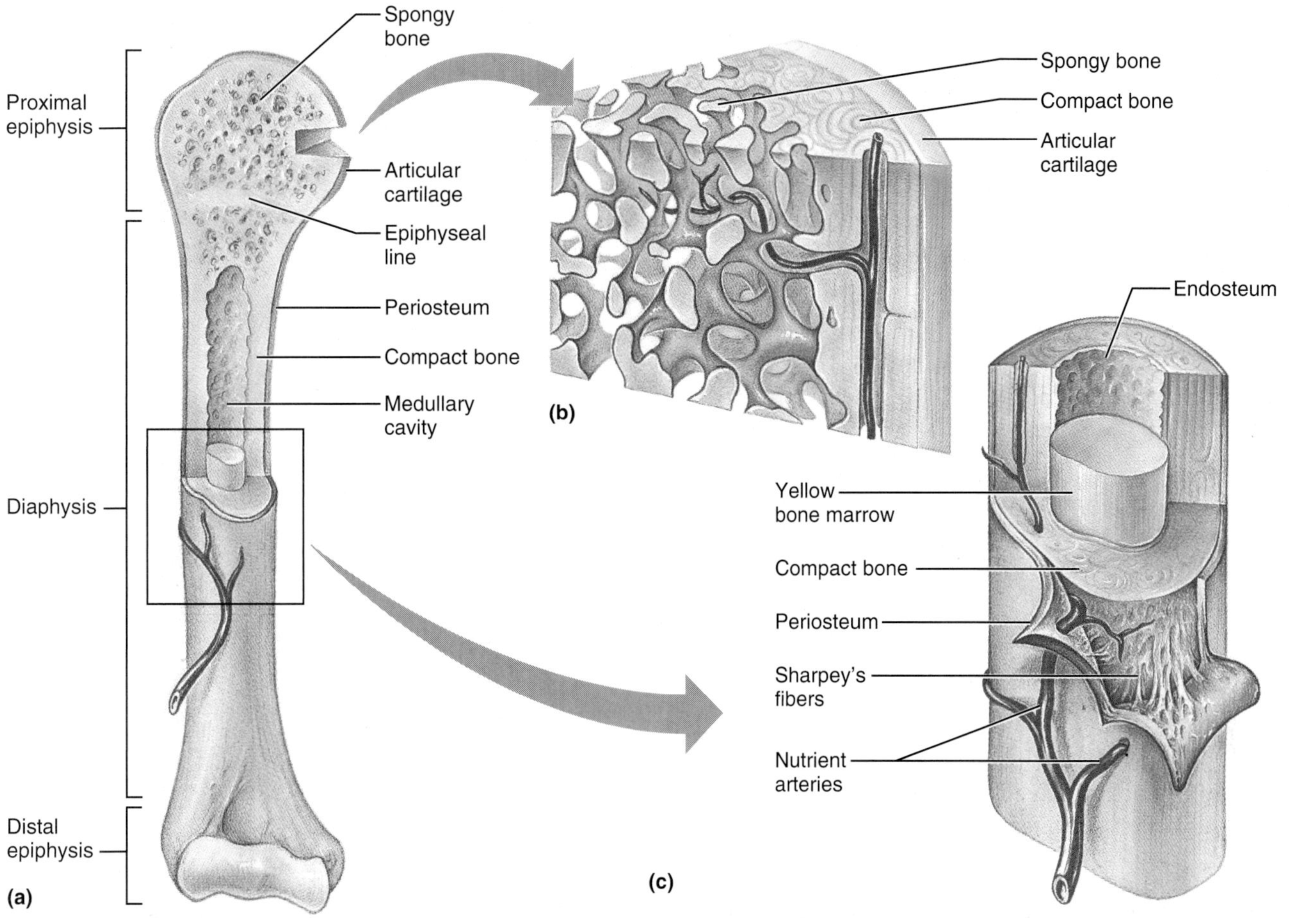

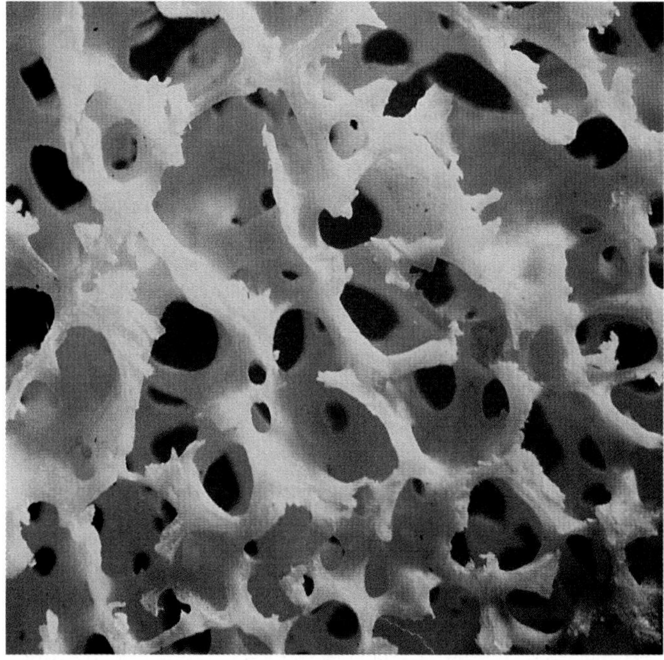

FIGURE 6.3 The structure of a long bone (humerus of arm). **(a)** Anterior view, with the superior half of the bone sectioned frontally to show the interior. **(b)** Enlargement of (a), showing the spongy bone and compact bone of the epiphysis. **(c)** Enlargement of the diaphysis from (a). **(d)** Micrograph of spongy bone, showing a network of bone trabeculae (15×).

is present. Figure 6.4 depicts a typical flat bone of the skull. In flat bones, the internal spongy bone is called **diploë** (dip′lo-e; "folded"). The whole arrangement of a flat bone might be likened to a reinforced sandwich.

Bone Design and Stress

The anatomy of each bone reflects the stresses most commonly placed on it. Bones are subjected to compression as weight bears down on them or as muscles pull on them. The loading usually is applied off center, however, and threatens to *bend* the bone (Figure 6.5). Bending compresses the bone on one side and stretches it (subjects it to tension) on the other. Both compression and tension are greatest at the external bone surfaces. *To resist these maximal stresses, the strong, compact bone tissue occurs in the external region of the bone.* Internal to this region, however, tension and compression forces tend to cancel each other out, resulting in less overall stress. Thus, compact bone does not form in the interiors; spongy bone is sufficient. Because no stress occurs at the bone's center, the lack of bone tissue in the central medullary cavity does not impair the

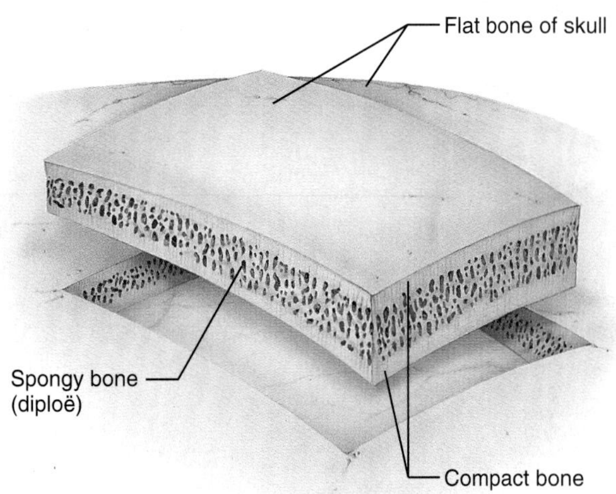

FIGURE 6.4 Structure of a flat bone of the skull. As in other bones, a thick periosteum covers the outer surfaces of the compact bone, and a thin endosteum lines the network of spongy bone. For simplicity, however, these membranes are not illustrated.

strength of long bones. Furthermore, the presence of spongy bone and marrow cavities serves to lighten the heavy skeleton and to provide room for the bone marrow.

Spongy bone is not a random network of bone fragments. Instead, the trabeculae of spongy bone align along stress lines in an organized pattern of tiny struts as crucially positioned as the flying buttresses that support the walls of a Gothic cathedral.

Microscopic Structure of Bone

Like other connective tissues, bone tissue consists of cells separated by an extracellular matrix. In addition to the usual matrix components of collagen fibers and ground substance, however, bone matrix contains mineral crystals. It also contains a small amount of tissue fluid, although bone has less water than other connective tissues do.

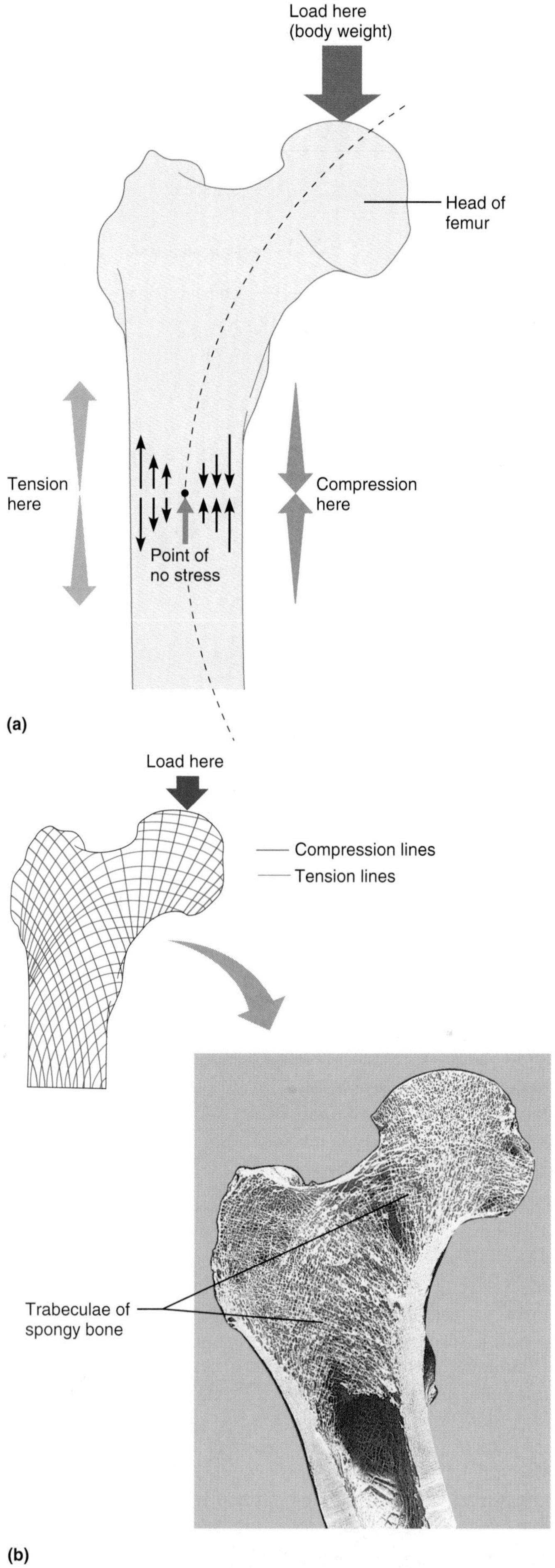

FIGURE 6.5 How bone anatomy relates to stress. (a) Because the load placed on most bones is off center, bones are subjected to bending stress. Body weight is transmitted to the head of the femur and threatens to bend the bone along the arc indicated by the dashed line. This bending compresses the bone at one side (converging red arrows) and places it under tension at the other side (diverging blue arrows). Thus, the strongest forces occur at the periphery, where they are resisted by the strongest, compact bone. Because tension and compression cancel each other internally at the point of no stress, much less bone material is needed internally. **(b)** A diagram of the stress lines (stress trajectories) experienced by the loaded femur (above) and the actual pattern of trabeculae of spongy bone within a femur. The pattern in the diagram matches that in the photo, illustrating that spongy bone provides support along the stress lines.

Compact Bone

Viewed by the unaided eye, compact bone looks solid. However, microscopic examination reveals that it is riddled with passageways for blood vessels, lymphatic vessels, and nerves (Figure 6.6a). An important structural component of compact bone is the **osteon** (os′te-on; "bone"), or **Haversian** (ha-ver′shan) **system.** Osteons are long, cylindrical structures oriented parallel to the long axis of the bone and to the main compression stresses. Functionally, osteons can be viewed as miniature weight-bearing pillars. Structurally, an osteon is a group of concentric tubes resembling the rings in a cross section of a tree trunk (Figures 6.6c and 6.7). Each of the tubes is a **lamella** (lah-mel′ah; "little plate"). More precisely, a lamella is a layer of bone matrix in which all the collagen fibers run in a single direction. However, the fibers of adjacent lamellae always run in roughly opposite directions. This alternating pattern is optimal for withstanding torsion, or twisting, stresses (Figure 6.7). Because the adjacent lamellae in an osteon reinforce one another to resist twisting, you can think of an osteon as a "twister resistor."

Advances in Understanding Collagen fibers are not the only part of bone lamellae that are beautifully ordered. Electron microscopists recently discovered that the tiny *mineral crystals* align with the collagen, and thus also orient in opposite directions in adjacent lamellae. ✱

Through the core of each osteon runs a canal called the **central canal,** or **Haversian canal** (see Figure 6.6a). Like all internal bone cavities, it is lined by endosteum. The central canal contains its own blood vessels, which supply nutrients to the bone cells of the osteon, and its own nerve fibers. **Perforating canals,** also called **Volkmann's** (fōlk′mahnz) **canals,** also occur in compact bone, lying at right angles to the central canals and connecting the blood and nerve supply of the periosteum to that of the central canals and the marrow cavity.

The mature bone cells, or **osteocytes,** are spider-shaped (see Figure 6.6b). Their bodies occupy small cavities in the solid matrix called **lacunae,** and their "spider legs" occupy thin tubes called **canaliculi** (kan″ah-lik′u-li). These "little canals" run through the matrix, connecting neighboring lacunae to one another and to the nearest capillaries, such as those in the central canals. Within the canaliculi, the legs of neighboring osteocytes touch each other and form gap junctions (see p. 86). Nutrients diffusing from capillaries in the central canal pass across these gap junctions, from one osteocyte to the next, throughout the entire osteon. This direct transfer from cell to cell is the only way to supply the osteocytes with the nutrients they need, because the intervening bone matrix is too solid and impermeable to act as a diffusion medium itself.

What is the function of osteocytes? They seem to be essential for maintaining the bone matrix. As evidence of this, if osteocytes die, the surrounding matrix is reabsorbed.

Not all lamellae in compact bone occur within osteons. Lying between the osteons are groups of incomplete lamellae called **interstitial** (in″ter-stish′al) **lamellae** (Figure 6.6c). These are simply the remains of old osteons that have been cut through by bone remodeling. Additionally, **circumferential lamellae** occur in the external and internal surfaces of the layer of compact bone; each of these lamellae extends around the entire circumference of the diaphysis (see Figure 6.6a). Functioning like an osteon but on a much larger scale, the circumferential lamellae effectively resist twisting of the *entire* long bone.

Spongy Bone

The microscopic anatomy of spongy bone is less complex than that of compact bone. Each trabecula contains several layers of lamellae and osteocytes but is too small to contain osteons or vessels of its own. The osteocytes receive their nutrients from capillaries in the endosteum surrounding the trabecula.

Chemical Composition of Bone Tissue

Bone tissue has both organic and inorganic components. The organic components—cells, fibers, and ground substance—account for 35% of the tissue mass. These organic substances, particularly collagen, contribute the flexibility and tensile strength that allow bone to resist stretching and twisting. Collagen is remarkably abundant in bone tissue.

The balance of bone tissue, 65% by mass, consists of inorganic hydroxyapatites (hi-drok″se-ap′ah-tītz), or mineral salts, primarily calcium phosphate. These mineral salts are present as tiny crystals that lie in and around the collagen fibrils in the extracellular matrix. The crystals pack tightly, thereby providing bone with its exceptional hardness, which enables it to resist compression. These mineral salts also explain how bones can endure for centuries, providing information on the sizes, shapes, and even some of the diseases (for example, arthritis) of ancient peoples.

The proper combination of organic and inorganic elements allows bones to be exceedingly durable and strong without being brittle. Amazingly, healthy bone is half as strong as steel in resisting compression and fully as strong as steel in resisting tension.

Critical Reasoning

A long bone that is taken from a fresh cadaver and soaked in weak acid for several weeks will maintain its original shape but will look leathery and can be easily tied into a knot. Explain this transformation.

The acid dissolves away the bone's mineral salts, leaving only the organic component, mainly collagen. The fact that a demineralized bone can be knotted demonstrates the great flexibility provided by the collagen within bone. It also confirms that without its mineral content, bone bends too easily to support weight.

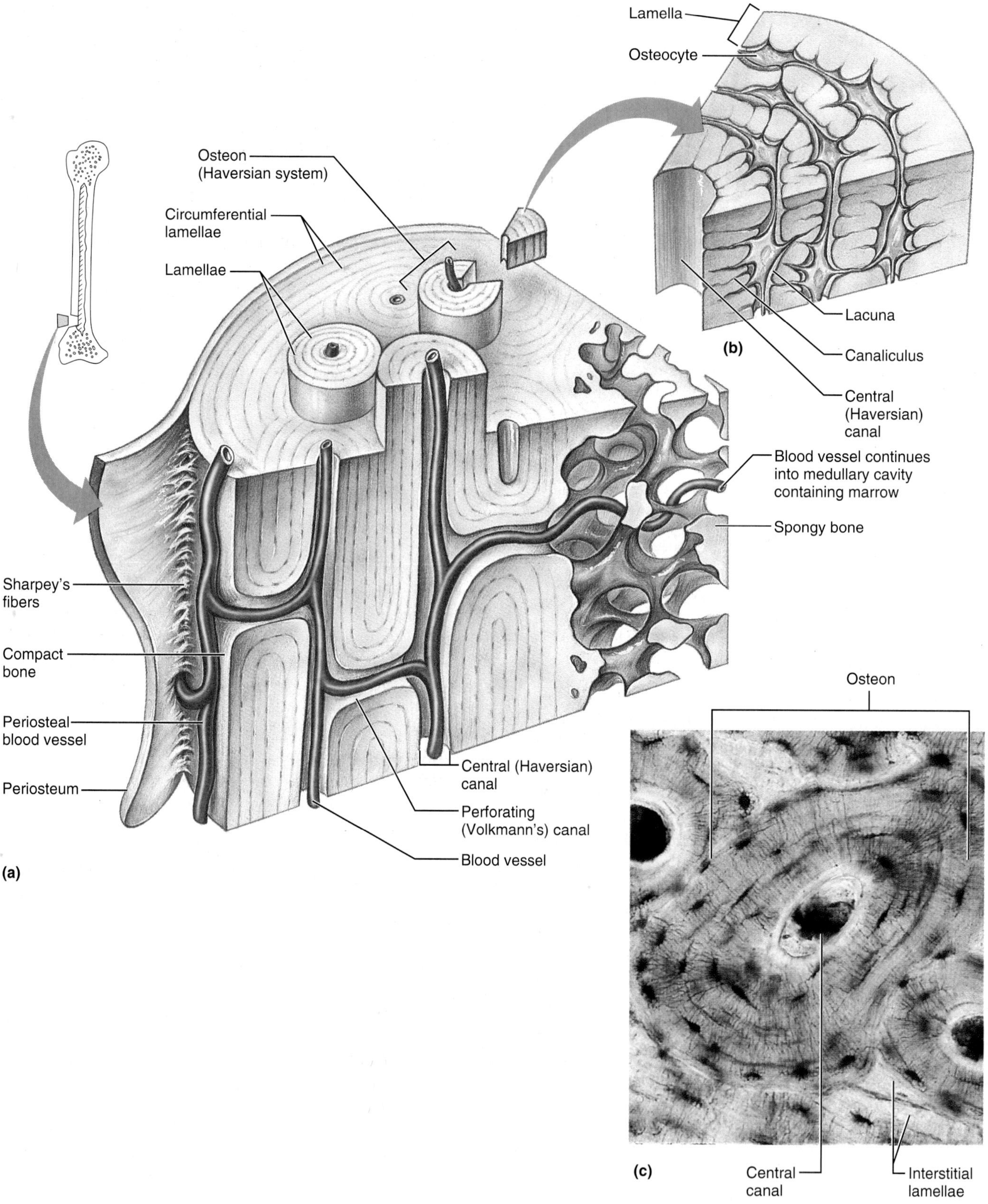

Lamella
Osteocyte
Osteon
(Haversian system)
Circumferential
lamellae
Lamellae
Lacuna
(b)
Canaliculus
Central
(Haversian)
canal
Blood vessel continues
into medullary cavity
containing marrow
Spongy bone
Sharpey's
fibers
Compact
bone
Periosteal
blood vessel
Periosteum
Central (Haversian)
canal
Perforating
(Volkmann's) canal
Blood vessel
(a)
Osteon
(c)
Central
canal
Interstitial
lamellae

FIGURE 6.6 Microscopic structure of compact bone. (a) A pie-shaped segment of compact bone from the wall of the humerus (the bone of the arm). Note the pillar-shaped osteons and the passageways holding blood vessels. At upper right **(b)** is a more highly magnified view of a part of one osteon. **(c)** Photomicrograph of a cross section of one osteon and parts of others (250×). The concentric rings in the osteons are the lamellae.

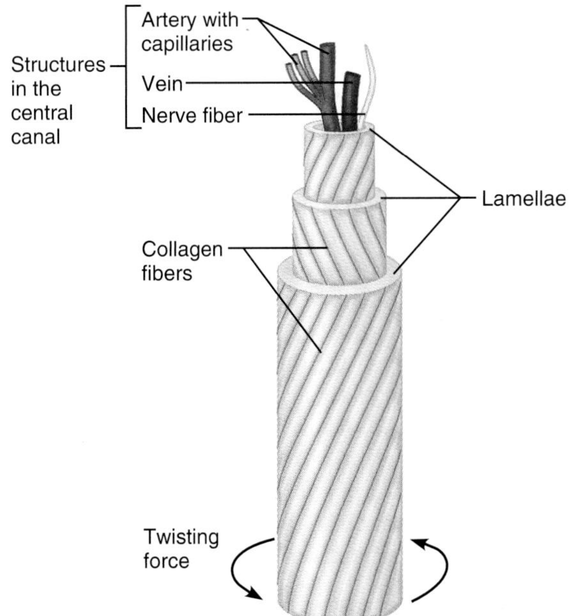

FIGURE 6.7 Diagram of an isolated osteon. The osteon is drawn as if it were pulled out like a telescope to show the lamellae composing it. Note the different orientations of the collagen fibers in the three different lamellae. Such alternating fibers allow the osteon to withstand twisting forces, as indicated by the arrows below. The structures within a central canal are also shown.

Bone Development

Osteogenesis (os″te-o-jen′ĕ-sis) and **ossification** are both names for the process of bone-tissue formation. Osteogenesis begins in the embryo, proceeds through childhood and adolescence as the skeleton grows, and then occurs at a slower rate in the adult as part of a continual remodeling of the full-grown skeleton.

Before week 8, the skeleton of the human embryo consists only of hyaline cartilage and some membranes of mesenchyme. Bone tissue first appears in week 8 and eventually replaces most cartilage and mesenchymal membranes in the skeleton. Some bones, called **membrane bones,** develop from a mesenchymal membrane through a process called **intramembranous ossification.** Other bones develop by replacing a piece of hyaline cartilage through a process called **endochondral ossification** (*endo* = within; *chondro* = cartilage). These bones are called **endochondral bones** or **cartilage bones.**

Intramembranous Ossification

Membrane bones form directly from mesenchyme without first being modeled in cartilage. All bones of the skull, except a few at the base of the skull, are of this category. The clavicles (collarbones) are the only bones formed by intramembranous ossification that are not in the skull.

Intramembranous ossification proceeds in the following way: During week 8 of embryonic development, certain cells cluster within the mesenchymal membrane and become bone-forming **osteoblasts** (Figure 6.8, step 1). These cells begin secreting the organic part of bone matrix, called **osteoid** (os′te-oid; "bone-like"), which then becomes mineralized (Figure 6.8, step 2). Once surrounded by their own matrix, the osteoblasts are called osteocytes. The new bone tissue forms between embryonic blood vessels, which are woven in a random network (Figure 6.8, step 3). The result is **woven bone tissue,** with trabeculae arranged in networks. This embryonic tissue lacks the lamellae that occur in mature spongy bone. During this same stage, more mesenchyme condenses just external to the developing membrane bone and becomes the periosteum.

To complete the development of a membrane bone, the trabeculae at the periphery grow thicker until plates of compact bone are present on both surfaces (Figure 6.8, step 4). In the center of the membrane bone, the trabeculae remain distinct, and spongy bone results. The final pattern is that of the flat bone depicted in Figure 6.4.

Endochondral Ossification

All bones from the base of the skull down, except for the clavicles, are endochondral bones. They are first modeled in hyaline cartilage, which then is gradually replaced by bone tissue. Endochondral ossification begins late in the second month of development and is not completed until the skeleton stops growing in early adulthood. Growing endochondral bones increase both in length and in width, but now we consider only the increase in length, using a large long bone as an example (Figure 6.9).

Stage 1: A bone collar forms around the diaphysis In the late embryo (week 8), the endochondral bone begins as a piece of cartilage called a *cartilage model.* Like all cartilages, it is surrounded by a perichondrium. Then, at the end of week 8 of development, the perichondrium surrounding the diaphysis becomes a bone-forming periosteum. Osteoblasts in this new periosteum lay down a collar of bone tissue around the diaphysis.

Stage 2: Cartilage calcifies in the center of the diaphysis At the same time the bone collar forms, the chondrocytes in the center of the diaphysis enlarge (hypertrophy) and signal the surrounding cartilage matrix to calcify. The matrix of this calcified cartilage is impermeable to diffusing nutrients. Cut off from all nutrients, the chondrocytes die and disintegrate, leaving cavities in the cartilage. No longer kept healthy by chondrocytes, the cartilage matrix starts to deteriorate. This does not seriously weaken the diaphysis, which is well stabilized by the bone collar around it. These changes affect only the center of the diaphysis. Elsewhere, the cartilage remains healthy and continues to grow, causing the entire endochondral bone to elongate.

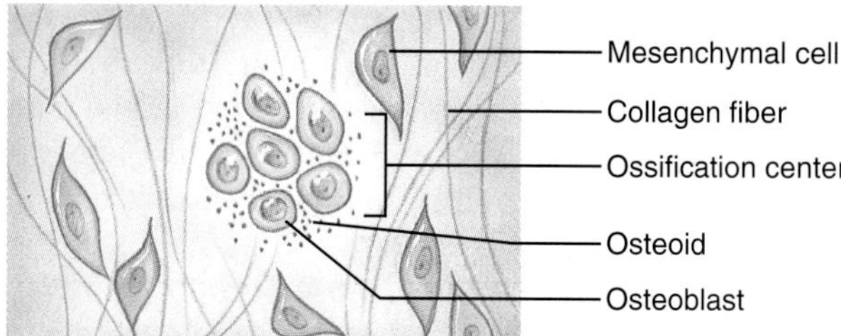

① **An ossification center appears in the fibrous connective tissue membrane.**

- Selected centrally located mesenchymal cells cluster and differentiate into osteoblasts, forming an ossification center.

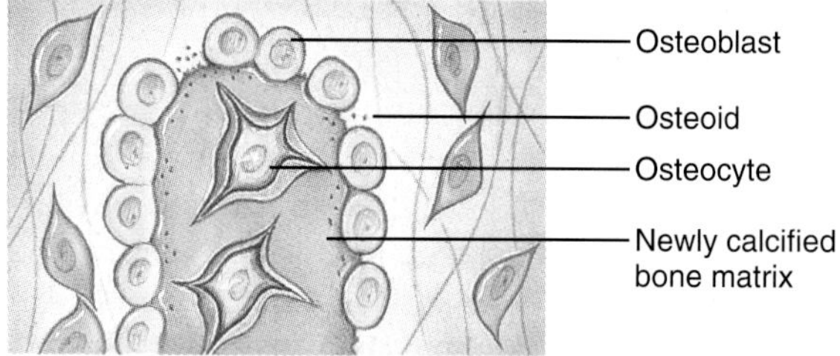

② **Bone matrix (osteoid) is secreted within the fibrous membrane.**

- Osteoblasts begin to secrete osteoid, which is mineralized within a few days.
- Trapped osteoblasts become osteocytes.

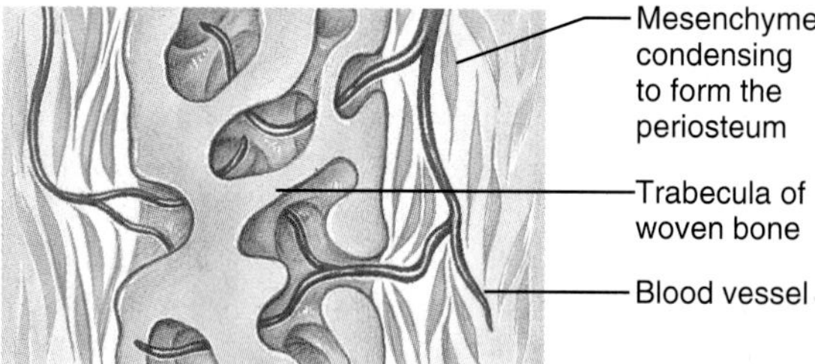

③ **Woven bone and periosteum form.**

- Accumulating osteoid is laid down between embryonic blood vessels, which form a random network. The result is a network (instead of lamellae) of trabeculae (woven bone).
- Vascularized mesenchyme condenses on the external face of the woven bone and becomes the periosteum.

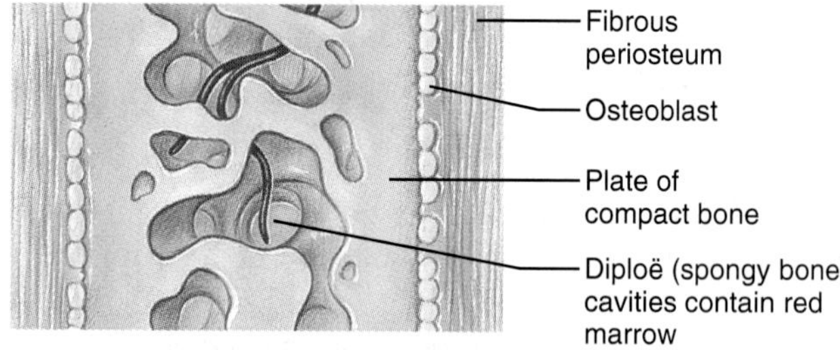

④ **Bone collar of compact bone forms and red marrow appears.**

- Trabeculae just deep to the periosteum thicken, forming a compact bone layer that is later replaced with mature lamellar bone.
- Spongy bone (diploë), consisting of distinct trabeculae, persists internally and its vascular tissue becomes red marrow.

FIGURE 6.8 Intramembranous ossification: Development of a flat bone of the skull in the fetus. The diagrams for steps 3 and 4 are at a much lower magnification than those in the preceding steps.

Stage 3: The periosteal bud invades the diaphysis, and the first bone trabeculae form Immediately following Stage 2 in the third month of development, the cavities within the diaphysis are invaded by a collection of elements called the **periosteal bud.** This bud consists of a nutrient artery and vein (p. 134), along with the cells that will form the bone marrow. Most importantly, the bud contains bone-forming and bone-destroying cells (preosteoblasts and osteoclasts). The entering osteoclasts partly erode the calcified-cartilage matrix, and the entering osteoblasts start to secrete osteoid around the remaining fragments of this matrix, forming bone-covered trabeculae. In this way, the earliest version of spongy bone appears within the diaphysis.

Therefore, by the third month of development, bone tissue has begun to appear both around the diaphysis and in its center. This bone tissue of the diaphysis makes up the **primary ossification center.**

Throughout the rest of the fetal period, the cartilage of the epiphysis continues to grow rapidly, with the part nearest the diaphysis continually calcifying and being replaced by the bone trabeculae of the expanding primary ossification center. Osteoclasts in turn break down the ends of these bone trabeculae to form a central, boneless medullary cavity that grows ever larger (for more details, see "Anatomy of the Epiphyseal Growth Areas" below).

Stage 4: The epiphyses ossify as secondary ossiffication centers Shortly before or after birth, the epiphyses gain bone tissue: First, the cartilage in the center of each epiphysis calcifies and degenerates. Then, a bud containing the epiphyseal vessels invades each epiphysis and bone trabeculae appear, just as they appeared earlier in the primary ossification center. The areas of bone formation in the epiphyses are called **secondary ossification centers.** The larger long bones of the body can have several ossification centers in each epiphysis.

After the secondary ossification centers have appeared, hyaline cartilage remains at only two places (Figure 6.9: Stage 5): (1) on the epiphyseal surfaces, where it forms the articular cartilages; and (2) between the diaphysis and epiphysis, where it forms the **epiphyseal plates.** These are also called *epiphyseal discs* or *growth plates,* and, as we shortly will see, they are responsible for lengthening the bones during the two decades following birth.

Anatomy of the Epiphyseal Growth Areas

In both the epiphyses of the fetus and the epiphyseal plates of the child, the cartilage is organized in a way

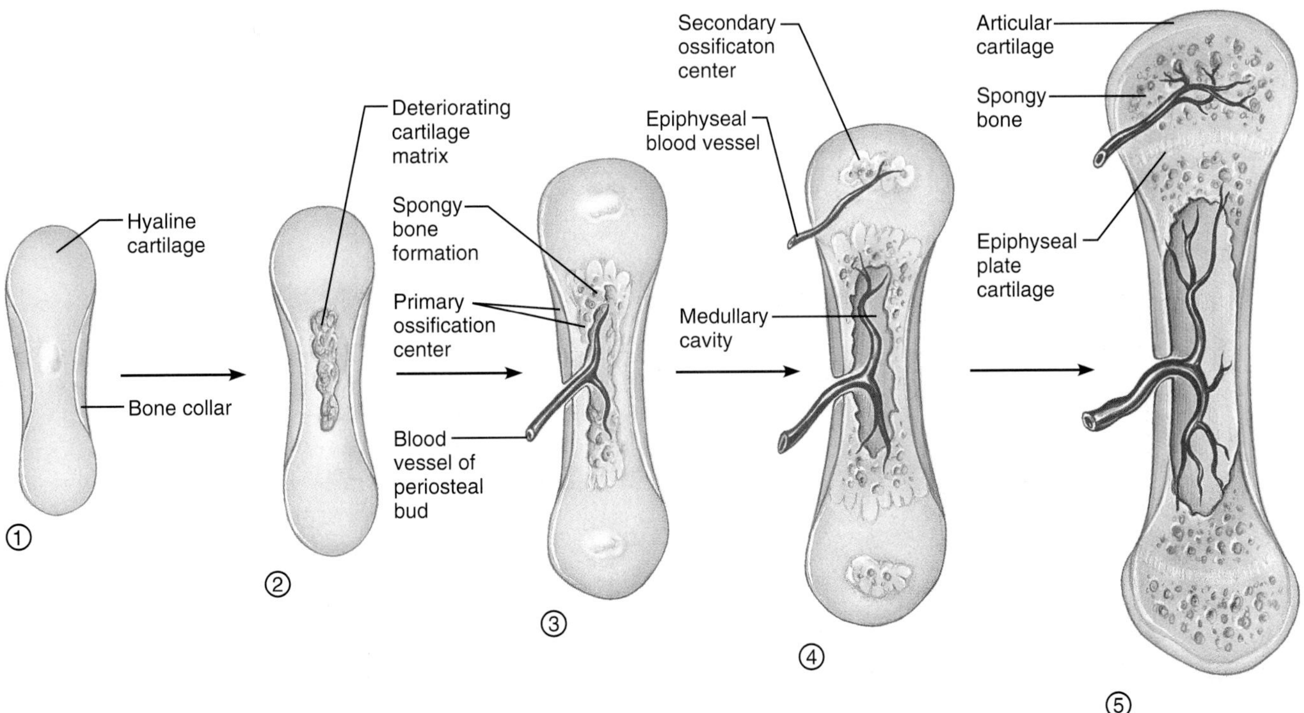

FIGURE 6.9 Stages in endochondral ossification of a long bone. Stage 1: Formation of a bone collar around the diaphysis of the hyaline cartilage model (week 8 of development). Stage 2: Calcification and cavitation of the hyaline cartilage in the center of the diaphysis (weeks 8–9). Stage 3: Invasion of the center of the diaphysis by the periosteal bud and formation of the first trabeculae of spongy bone (week 9). Stage 4: Epiphyses ossify as secondary ossification centers (around the time of birth). Stage 5: Fully ossified epiphyses, growth at the epiphyseal plates (from birth to the start of adulthood).

that allows it to grow exceptionally quickly and efficiently (Figure 6.10). The cartilage cells form tall columns, like coins in a stack. The chondroblasts at the "top" of the stack (zone 1) divide quickly, pushing the epiphysis away from diaphysis, thereby causing the entire long bone to lengthen. Meanwhile, the older chondrocytes deeper in the stack enlarge, signal the surrounding matrix to calcify (zone 2), and then die and disintegrate (zone 3 or 4). This process leaves long spicules (trabeculae) of calcified cartilage on the diaphysis side of the epiphysis-diaphysis junction (zone 4), hanging like stalactites from the roof of a cave. These spicules are partly eroded by osteoclasts, then covered with bone tissue by osteoblasts, and finally are eaten away from their tips by osteoclasts. Because the bone spicules are removed from within the diaphysis at the same rate that they form at the epiphysis, they stay a constant length and the marrow cavity grows longer as the long bone lengthens.

Postnatal Growth of Endochondral Bones

During childhood and adolescence, the endochondral bones lengthen entirely by growth of the epiphyseal plates. Because its cartilage is replaced with bone tissue on the diaphysis side about as quickly as it grows, the epiphyseal plate maintains a constant thickness while the whole bone lengthens. As adolescence draws to an end, the chondroblasts in the epiphyseal plates divide less often, and the plates become thinner. Eventually, they exhaust their supply of mitotically active cartilage cells, so the cartilage stops growing and is replaced by bone tissue. Long bones stop lengthening when the bone of the epiphyses and diaphysis fuses. This process, called *closure of the epiphyseal plates,* occurs at about 18 years of age in females and 21 years of age in males. After our epiphyseal plates close, we can grow no taller.

Critical Reasoning

Our discussion so far has revealed the surprising fact that bone tissue is not responsible for the increase in length of growing endochondral bones. What other tissue is responsible?

Hyaline cartilage does the growing. Bone tissue merely replaces the cartilage.

Growing bones must widen as they lengthen. How do they widen? Simply, osteoblasts in the periosteum add bone tissue to the external face of the diaphysis as osteoclasts in the endosteum remove bone from the internal surface of the diaphysis wall. These two processes occur at about the same rate, so that the circumference of the long bone expands and the bone

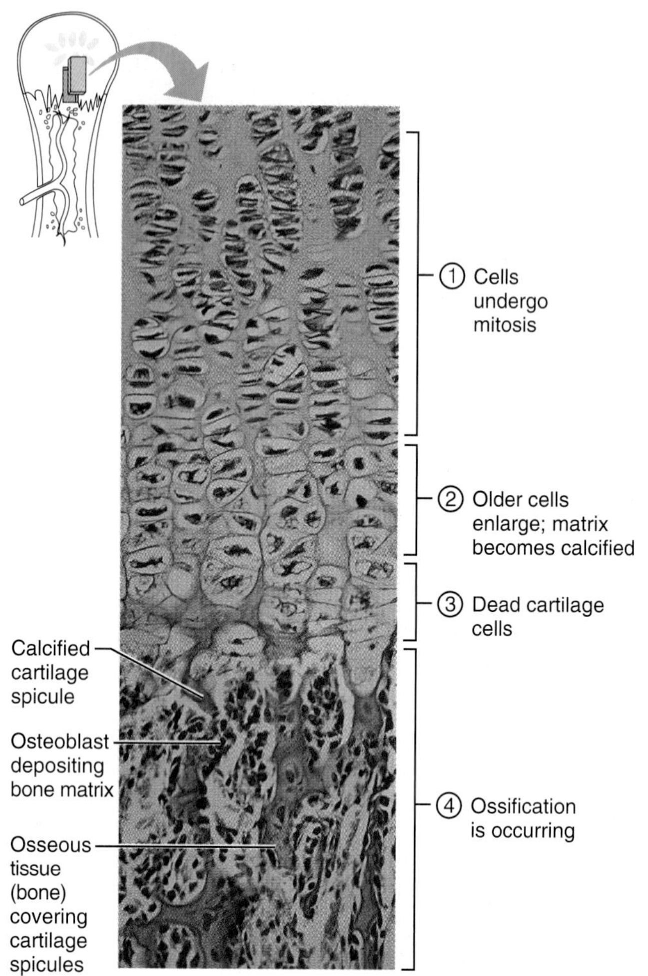

FIGURE 6.10 Organization of the cartilage within the epiphysis of a growing long bone (250×). Cartilage cells are stacked in vertical columns like coins. This neat arrangement first appears in the early fetus, and after birth it occurs in the epiphyseal plate. The following activities take place in the four zones: (1) Chondroblasts undergo mitosis, multiplying vigorously. (2) Chondrocytes enlarge, then the matrix calcifies. (3) Chondrocytes die and disintegrate, leaving ragged pieces (spicules) of cartilage matrix in zone 4. (4) Bone matrix (osseous tissue) covers the cartilage spicules, forming spicules (trabeculae) of bone. Recent advances in microscopic technique suggest the chondrocytes may not die in zone 3, but a bit lower, at the top of zone 4.

widens. Growth of a bone by the addition of bone tissue to its surfaces is called **appositional growth.**

We have been discussing the growth and development of large long bones. The other types of endochondral bones grow in slightly different ways. Short bones, such as those in the wrist, arise from only a single ossification center. Most of the irregular bones, such as the hip bone and vertebrae, develop from several distinct ossification centers. Small long bones, such as those in the palm and fingers, form from a primary ossification center (diaphysis) plus a single secondary center; that is, they have just one epiphysis.

Bone growth is regulated by several hormones, primarily growth hormone (produced by the pituitary gland), which stimulates the epiphyseal plates to grow. Thyroid hormones modulate the effects of growth hormone, ensuring that the skeleton retains its proper proportions as it grows. The sex hormones (androgens and estrogens) first promote bone growth in the growth spurt at adolescence and later induce the epiphyseal plates to close, ending growth.

Bone Remodeling

Bones appear to be the most lifeless of body organs when we see them in the lab, and once they are formed, they seem set for life. Nothing could be farther from the truth. Bone is a very dynamic and active tissue. Large amounts of bone matrix and thousands of osteocytes are continuously being removed and replaced within the skeleton, and the small-scale architecture of our bones constantly changes. As much as half a gram of calcium may enter or leave the adult skeleton each day! The spongy bone in our skeleton is entirely replaced every 3 or 4 years, the compact bone every 10 years.

In the adult skeleton, bone deposit and bone removal occur at the periosteal and endosteal surfaces. Together, the two processes constitute **bone remodeling.** They are coordinated by "packets" or cohorts of adjacent osteoblasts and osteoclasts called remodeling units. In healthy young adults, the total mass of bone in the body stays constant, an indication that the rates of deposit and reabsorption are essentially equal. The remodeling process is not uniform, however. Some bones (or bone parts) are very heavily remodeled, but others are not. For example, the distal region of the femur is fully replaced every 5 to 6 months, whereas the diaphysis of the femur changes much more slowly.

Bone deposition is accomplished by *osteoblasts,* hemisphere-shaped cells with short cellular processes. These cells lay down organic osteoid on bone surfaces, and calcium salts crystallize within this osteoid. This calcification process takes about a week. As stated earlier, the osteoblasts transform into osteocytes when they become surrounded by bone matrix.

Bone reabsorption is accomplished by **osteoclasts** (Figure 6.11). Each of these giant cells has many nuclei. Osteoclasts crawl along bone surfaces, essentially digging pits as they break down the bone tissue. The part of their plasma membrane that touches the bone surface is highly folded, or ruffled. Forming a tight seal against the bone, this expanded membrane secretes concentrated hydrochloric acid, which dissolves the mineral part of the matrix. The liberated calcium ions (Ca^{2+}) and phosphate ions (PO_4^{3-}) enter the tissue fluid and the bloodstream. Lysosomal enzymes are also released across the ruffled membrane and digest the organic part of the bone matrix. Finally, osteoclasts apparently take up collagen and dead osteocytes by phagocytosis.

Bone-forming osteoblasts derive from mesenchyme cells or, in the adult, from mesenchyme-like stem cells in the connective tissues of the nearby bone marrow. Osteoclasts, by contrast, arise from immature blood cells

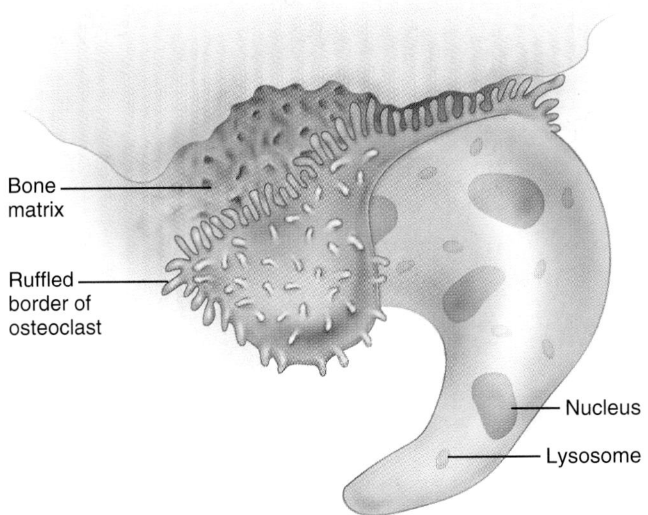

FIGURE 6.11 **An osteoclast, or bone-degrading cell.** Osteoclasts are large and multinucleate and have a ruffled surface border that degrades bone (border pulled back for visibility).

called *hematopoietic stem cells* (p. 513), and they may be related to macrophages. Many of these stem cells fuse together to form each osteoclast.

The bones of the skeleton are continually remodeled for two reasons. First, bone remodeling helps maintain constant concentrations of Ca^{2+} and PO_4^{3-} in body fluids. When the concentration of Ca^{2+} in body fluids starts to fall, a hormone is released by the parathyroid (par″ah-thi′roid) glands of the neck. This *parathyroid hormone* stimulates osteoclasts to reabsorb bone, a process that releases more Ca^{2+} into the blood.

Second, bone is remodeled in response to the mechanical stress it experiences. Accordingly, both the osteons of compact bone and the trabeculae of spongy bone are constantly replaced by new osteons and trabeculae that are more precisely aligned with newly experienced compressive stresses. Furthermore, bone grows thicker in response to the forces experienced during exercise and gains in weight.

Critical Reasoning

Given that bones grow stronger with use, what are the effects of disuse on bones?

In the absence of mechanical stress, bone tissue is lost, which is why the bones of bedridden people atrophy. A loss of bone under near-zero-gravity conditions is the main obstacle to long missions in outer space. To slow bone loss, astronauts perform isometric exercises during space missions.

Repair of Bone Fractures

Despite their strength, bones are susceptible to **fractures,** or breaks. In young people, most fractures result from trauma (sports injuries, falls, or car accidents, for example) that twists or smashes the bones. In old age, bones thin and weaken, and fractures occur more often. A fracture in which the bone breaks clearly but does not penetrate the skin is a **simple fracture.** When broken ends of the bone protrude through the skin, the fracture is **compound.** Other common types of fractures are explained in Table 6.1.

A fracture is treated by **reduction,** the realignment of the broken bone ends. In **closed reduction,** the bone ends are coaxed back into position by the physician's hands. In **open reduction,** the bone ends are joined surgically with pins or wires. After the broken bone is reduced, it is immobilized by a cast or traction to allow the healing process to begin. Healing time is about 6 to 8 weeks for a simple fracture, but it is longer for large, weight-bearing bones and for the bones of elderly people.

The healing of a simple fracture involves several phases (Figure 6.12 on page 146).

1. **Hematoma formation.** The fracture is usually accompanied by hemorrhaging. Blood vessels break both in the periosteum and inside the bone, releasing blood that clots to form a hematoma. The stages of inflammation, described in Chapter 4 (p. 102), are evident in and around the clot.
2. **Fibrocartilaginous callus formation.** Within a few days, new blood vessels grow into the clot. The periosteum and endosteum near the fracture site show a proliferation of bone-forming cells, which then invade the clot, filling it with repair tissue called *soft callus* (kal′us; "hard skin"). Initially, the soft callus is a fibrous granulation tissue (p. 103). This tissue then transforms into a dense connective tissue containing fibrocartilage and hyaline cartilage. At this point, the soft callus is also called a **fibrocartilaginous callus.**
3. **Bony callus formation.** Within a week, trabeculae of new bone begin to form in the callus, mostly by endochondral ossification. These trabeculae span the width of the callus and unite the two fragments of the broken bone. The callus is now called a bony callus, or *hard callus,* and its trabeculae grow thicker and stronger and become firm by about 2 months after the injury.
4. **Bone remodeling.** Over a period of many months, the bony callus is remodeled. The excess bony material is removed from both the exterior of the bone shaft and the interior of the medullary cavity. Compact bone is laid down to reconstruct the shaft walls. The repaired area resembles the original unbroken bone region, since it is responding to the same set of mechanical stresses.

DISORDERS OF BONES

Osteoporosis

Osteoporosis (os″te-o-po-ro′sis; "bone-porous-condition") is characterized by low bone mass and a deterioration of the microscopic architecture of the bony

Table 6.1 Common Types of Fractures

Fracture Type	Illustration	X ray
COMMINUTED Bone fragments into 3 or more pieces Particularly common in the aged, whose bones are more brittle	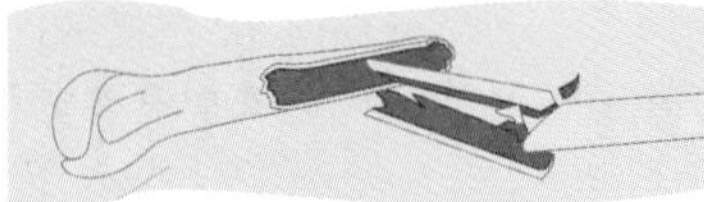	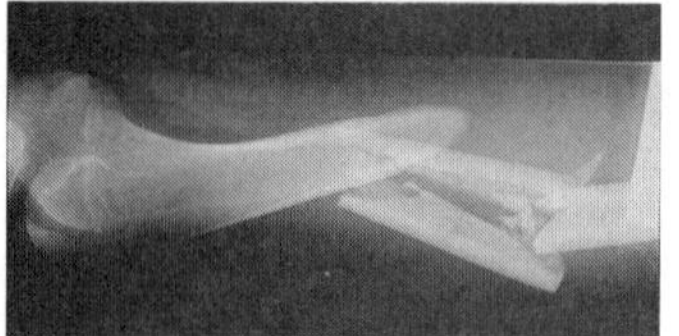
COMPRESSION Bone is crushed Common in porous bones (i.e. osteoporotic bones) subjected to extreme trauma, as in a fall	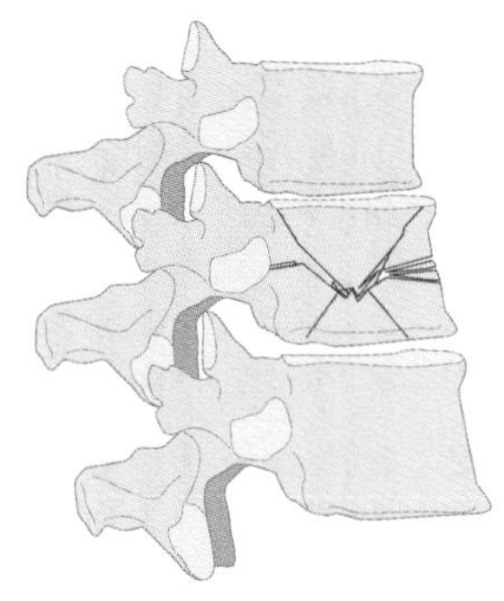	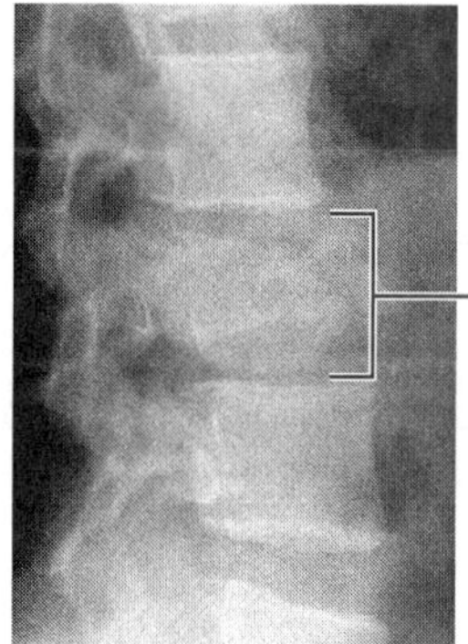
DEPRESSED Broken bone portion is pressed inward Typical of skull fracture	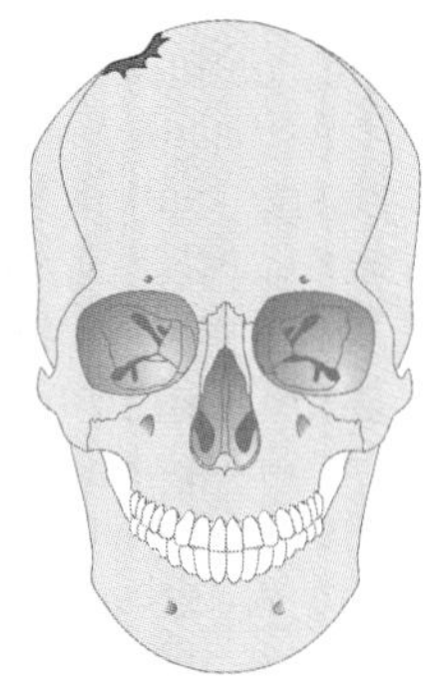	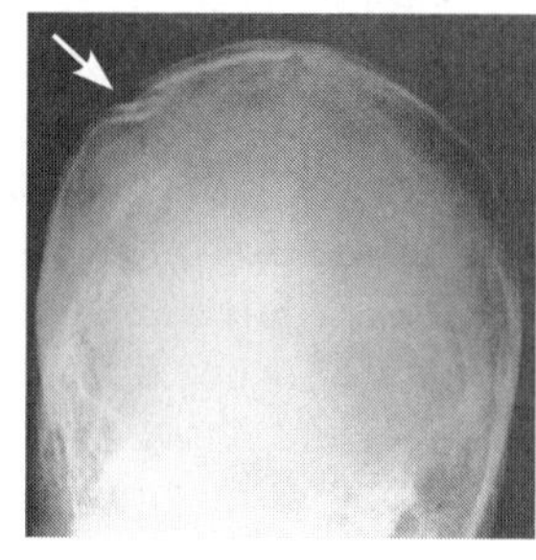

skeleton (Figure 6.13 on page 146). Bone reabsorption outpaces bone deposition, in association with elevated numbers of osteoclasts. Osteoporotic bones become porous and light: The compact bone becomes thinner and less dense than normal, and the spongy bone has fewer trabeculae. The loss of bone mass often leads to fractures. Although the mass of the bone declines, the chemical composition of its matrix remains normal. Even though osteoporosis affects the whole skeleton, the vertebral column is most vulnerable, and compression fractures of the vertebrae are frequent. The femur (thigh bone), especially its neck near the hip joint, is also very susceptible to fracture. A broken hip is a common problem in people with osteoporosis.

Osteoporosis occurs most often in the aged, particularly in women who have gone through menopause. Although men develop it to a lesser degree, 30% of American women have osteoporosis between the ages of 60 to 70; 70% have it by age 80. Moreover, 30% of all Caucasian women (the most susceptible group) will experience a bone fracture due to osteoporosis. Estrogen deficiency is strongly implicated in osteoporosis in older women because the secretion of estrogens, which helps maintain bone density, wanes after menopause. Additional factors that contribute include insufficient exercise to stress the bones and a diet poor in calcium and protein.

Fracture Type	Illustration	X ray
EPIPHYSEAL Epiphyseal plate tears, separating epiphysis from diaphysis Tends to occur where cartilage cells are dying and calcification of the matrix is occuring	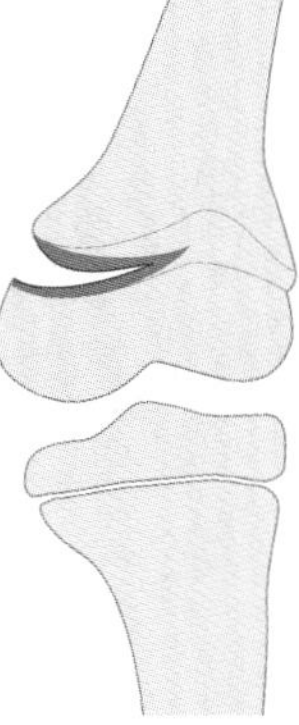	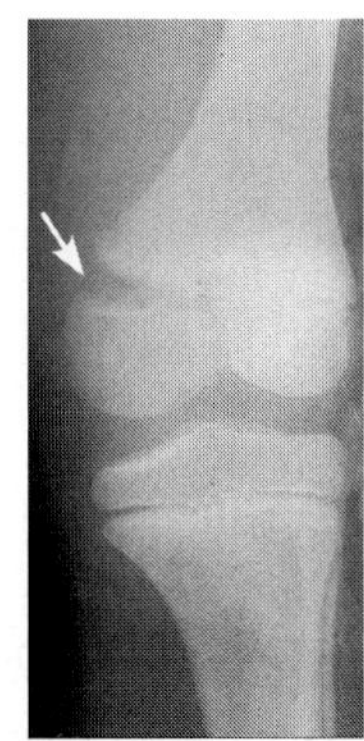
GREENSTICK Bone breaks incompletely, much in the way a green twig breaks. Only one side of the shaft splits, the other side bends. Common in children, whose bones have relatively more organic matrix and are more flexible than those of adults 	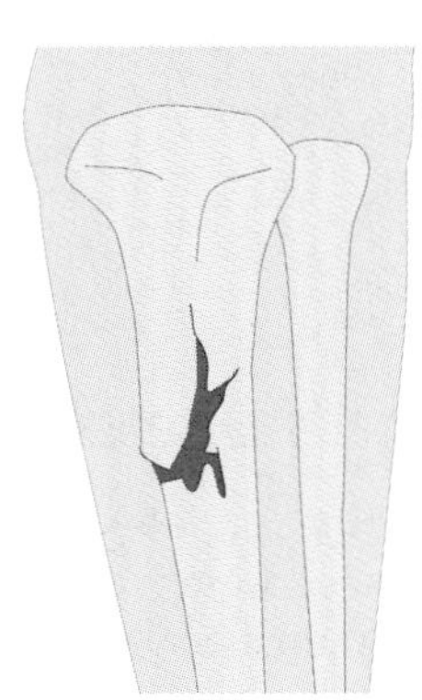	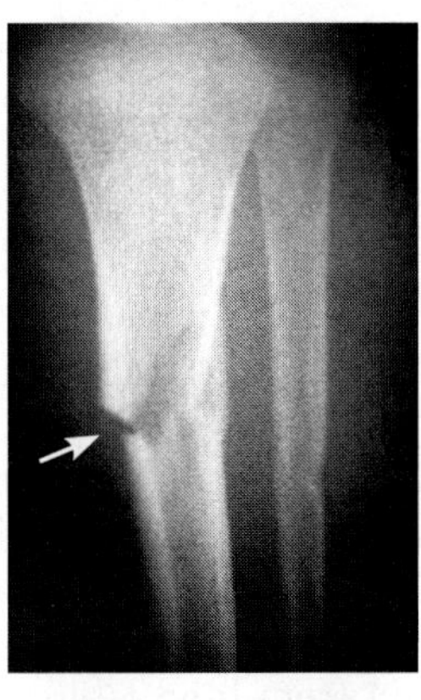
SPIRAL Ragged break occurs when excessive twisting forces are applied to a bone Common sports fracture	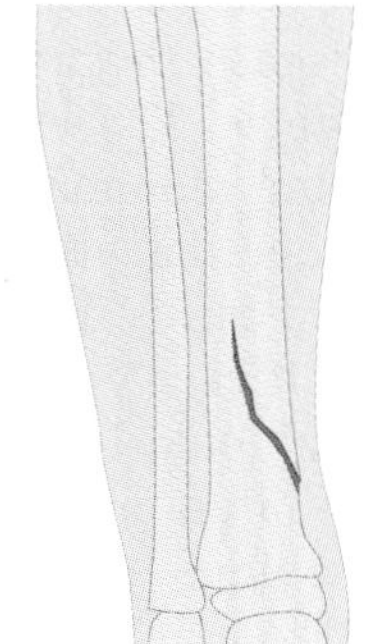	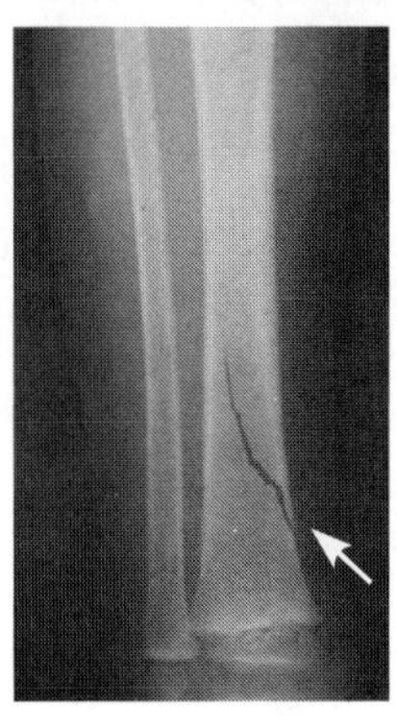

Advances in Understanding Osteoporosis has traditionally been treated by supplemental calcium and vitamin D, increased exercise, and estrogen replacement. Newer treatments also use either bisphosphonate drugs such as alendronate, which suppress osteoclast activity, or hormones that do the same thing (calcitonin, for example). Selective estrogen receptor modulators (SERMs) such as raloxifene and tamoxifen mimic the beneficial effects of estrogen but target bone alone, without any undesired stimulation of tissues of the breast and uterus. These new treatments do not provide a cure, but they do increase bone mass to a moderate degree and lessen the risk of fractures. ✳

Osteomalacia and Rickets

The term **osteomalacia** (os″te-o-mah-la′she-ah; "soft bones") applies to a number of disorders in adults in which the bones are inadequately mineralized. Even though osteoid matrix is produced, calcification does not occur, so the bones soften and weaken. The main symptom is pain when weight is put on the affected bone.

Rickets, the analogous disease in children, is accompanied by many of the same signs and symptoms. Because young bones are still growing rapidly, however, rickets is more severe than osteomalacia. Along with weakened and bowed leg bones, malformities of the child's head and rib cage are common. Because the epiphyseal plates cannot be replaced with calcified bone,

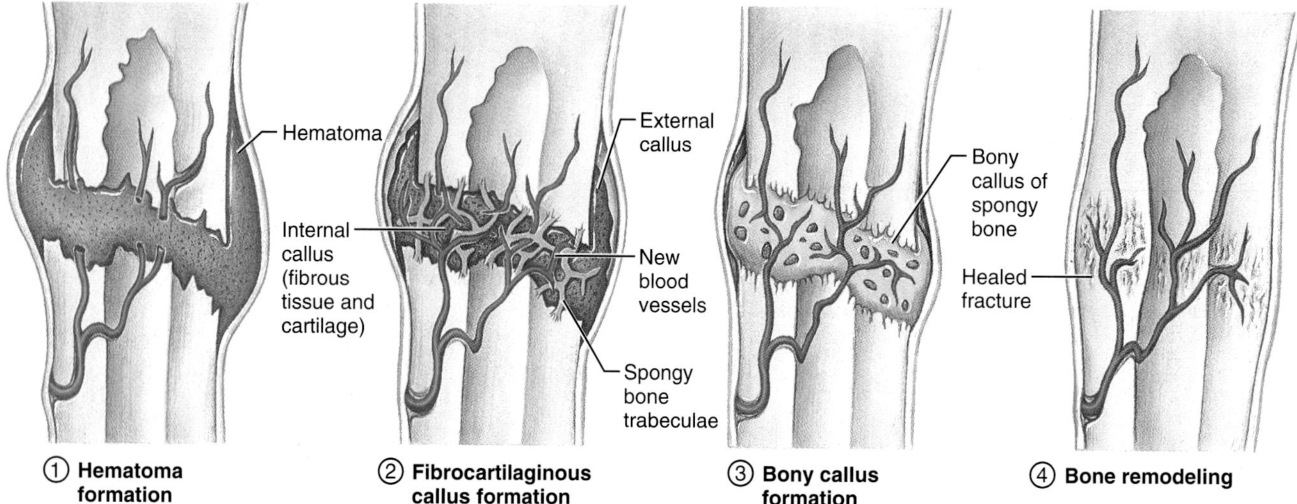

FIGURE 6.12 Stages in the healing of a bone fracture. See page 143 for a detailed explanation.

they grow abnormally thick, and the epiphyses of the long bones become abnormally long.

Osteomalacia and rickets are caused by inadequate amounts of vitamin D or calcium phosphate in the diet. They are cured by drinking vitamin D-fortified milk and exposing the skin to sunlight. (The role of vitamin D in the uptake of calcium from food is discussed on p. 117.

Critical Reasoning

Why may repeated pregnancies cause a woman to develop osteomalacia?

Pregnancy and lactation (making milk) lead to a transfer of large amounts of calcium from a woman's skeleton to the growing skeleton of the fetus. Therefore, it is especially important for women to eat calcium-rich foods during pregnancy, while lactating, and between pregnancies as well.

Paget's Disease

Paget's (paj'ets) **disease** is characterized by excessive rates of bone deposition and bone reabsorption. The newly formed bone, called *Pagetic bone,* has an abnormally high ratio of immature woven bone to mature compact bone. This, along with reduced mineralization, makes the bones soft and weak. Late in the course of the disease the activity of osteoblasts outpaces that of osteoclasts. Therefore, the bones can thicken, but in an irregular manner, and the medullary cavities may fill with bone. Paget's disease may affect many parts of the skeleton but is usually a localized and intermittent condition. It rarely occurs before age 40 and affects about 3% of all elderly people in North America. It progresses slowly, often produces no symptoms, and is seldom life-threatening. Its cause is unknown, but it may be initiated by a virus, such as the virus that causes distemper in dogs. A treatment that involves inhibiting osteoclasts with bisphosphonates and other drugs shows considerable success.

Osteosarcoma

Recall that a *sarcoma* is any cancer arising from a connective tissue cell or muscle cell and that *osteo* means "bone." Clearly, then, **osteosarcoma** is a form of bone cancer.

Osteosarcoma primarily affects young people between 10 and 25 years of age. It usually originates in a long bone of the upper or lower limb, with 50% of

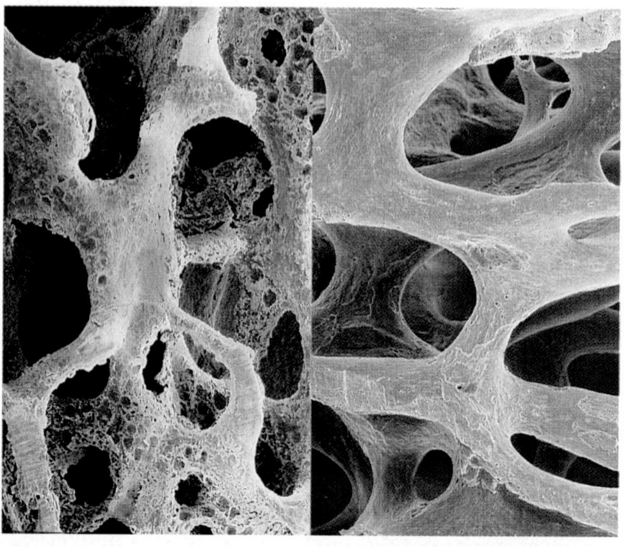

FIGURE 6.13 Osteoporosis. Note the contrast in architecture of osteoporotic (left) and normal (right) spongy bone. Scanning electron micrographs (about 20×); artificially colored.

cases arising near the knee. The cancer cells derive from osteoblast-like cells of mesenchymal origin in the parts of the diaphyses nearest the epiphyses. Secreting osteoid and growing quickly, the tumor alters the affected bone by eroding the medullary cavity internally and the compact bone externally. The tumor metastasizes, and most deaths result from secondary tumors in the lungs. Most osteosarcomas are recognized by the pain and the visible swelling they produce in the affected bone, and the diagnosis is confirmed by X rays or other medical imaging techniques. Treatment begins by removing the cancerous region of the bone and replacing it with bone grafts or prostheses (although a limb amputation is necessary in severe cases). This is followed by chemotherapy and surgical removal of any metastases in the lung. The survival rate is 60% to 70% if the disease is detected early.

THE SKELETON THROUGHOUT LIFE

As previously noted, cartilage grows quickly during youth and then stops growing during early adulthood. In the elderly, it shows fewer chondrocytes and some degradation and calcification of its matrix.

Bones can be said to be on a timetable from the time they form until death. The mesoderm germ layer gives rise to embryonic mesenchyme cells, which in turn produce the membranes and cartilages that form most of the embryonic skeleton. These structures then ossify according to a predictable schedule. Although each bone of the skeleton has its own developmental schedule, most long bones start ossifying by week 8 and have obvious primary ossification centers by week 12 (Figure 6.14). So precise is the ossification timetable that the age of a fetus in the womb can be determined from X-ray images or sonograms of the fetal skeleton.

At birth, all bones are relatively smooth and featureless. However, as the child increasingly uses its muscles, the bones develop projections and other markings. Children's bones are not particularly weak, but the cartilage of their epiphyseal plates is not as strong as bone. Thus, childhood injuries often knock the epiphyses off the diaphysis. To treat such injuries, the bone parts are manipulated back into place, then stabilized with metal pins.

As mentioned earlier, the skeleton keeps growing until the age of 18 to 21 years. In children and adolescents, the rate of bone formation exceeds the rate of bone reabsorption. In young adults, these two processes are in balance. In old age, reabsorption predominates. Beginning in the fourth decade of life, the mass of both compact and spongy bone declines. Among young adults, skeletal mass is generally greater in males than in females, and in blacks than in whites. With age, the rate of bone loss is greater in whites than in blacks, and in females than in males.

As bone mass declines with age, other changes occur. An increasing number of osteons remain incompletely formed, and mineralization is less complete. The amount of nonliving compact bone increases, reflecting a diminished blood supply to the bones in old age.

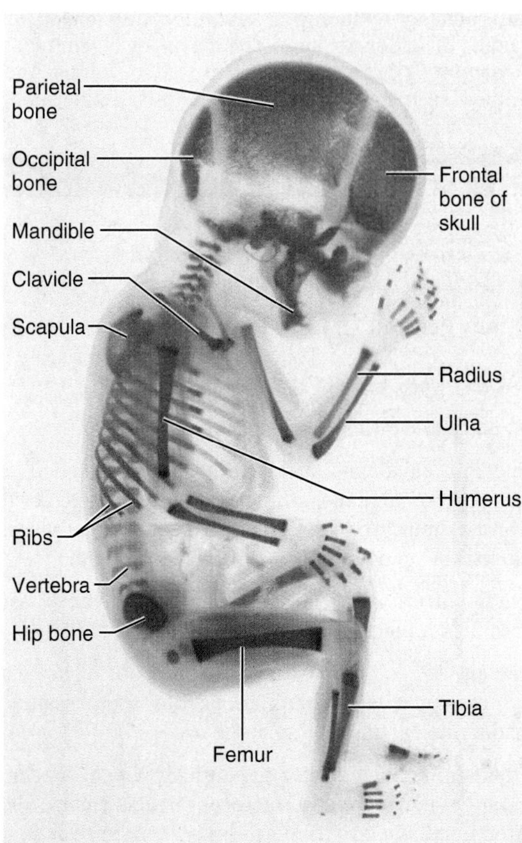

FIGURE 6.14 Primary ossification centers in the skeleton of a 12-week-old fetus.

RELATED CLINICAL TERMS

Achondroplasia (a-kon″dro-pla′ze-ah) (*a* = without; *chondro* = cartilage; *plasia* = shape) A congenital condition involving defective growth of cartilage and defective endochondral ossification. Achondroplasia results in typical dwarfism, in which the limbs are short but the membrane bones are of normal size. Not rare, it affects between 1 in 15,000 and 1 in 40,000 people. The precise genetic mutation responsible for this condition was recently identified.

Bone graft Transplantation of a piece of bone from one part of a person's skeleton to another part where bone has been damaged or lost. The graft, often taken from the crest of the iliac bone of the hip, encourages regrowth of lost bone.

Bony spur An abnormal projection on a bone due to bone overgrowth; is common in aging bones.

Ostealgia (os″te-al′je-ah) (*algea* = pain) Pain in a bone.

Osteomyelitis (os″te-o-mi-ĕ-li′tis; "bone and marrow inflammation"). Bacterial infection of the bone and bone marrow that either enters bones from infections in surrounding tissues or through the bloodstream, or follows a compound bone fracture.

Pathologic fracture Fracture occurring in a diseased bone and involving slight or no physical trauma. An example is a broken hip caused by osteoporosis: The hip breaks first, causing the person to fall.

Traction ("pulling") Placing a sustained tension on a region of the body in order to keep the parts of a fractured bone in the proper alignment. Traction also keeps the bone immobilized as it heals. Without traction of the lower limb, for example, strong spasms of the large muscles of the thigh would separate the fracture in a broken femur. Traction is also used to immobilize fractures of the vertebral column, because any movement there could crush the spinal cord.

CHAPTER SUMMARY

Media study tools that could provide you additional help in reviewing specific key topics of Chapter 6 are referenced below.

SP = Study Partner; AIA = A.D.A.M.® Interactive Anatomy

CARTILAGES (pp. 130–132)

Location and Basic Structure (p.130)

1. Important cartilages in the adult body are articular cartilages, costal cartilages, the epiglottis, cartilages of the respiratory tubes, and cartilage in intervertebral discs. Cartilage makes up most of the embryonic skeleton.
2. Perichondrium is the girdle of dense connective tissue that surrounds a piece of cartilage.
3. Cartilage is resilient: Water squeezed out of its matrix by compression rushes back in as the compression eases, causing the cartilage to spring back.
4. Growing cartilage enlarges quickly because the small amount of matrix it manufactures attracts a large volume of water. Cartilage is avascular and is weak in resisting shearing stresses.

Types of Cartilage (p. 130)

5. Locations of hyaline, elastic, and fibrocartilage are shown in Figure 6.1.
6. Hyaline cartilage is the most common type. Its matrix contains fine collagen fibrils. Elastic cartilage resembles hyaline cartilage, but its matrix also contains elastic fibers that make it pliable. Fibrocartilage contains thick collagen fibers and can resist both compression and extreme tension.

AIA Link: Atlas; Region: Head and Neck; Laryngeal Muscles (anterior).

Growth of Cartilage (pp. 130–131)

7. Cartilages grow from within (interstitial growth) and externally, as the perichondrium adds cartilage tissue at the periphery (appositional growth). In adults, damaged cartilage regenerates poorly. In the growing and aged skeleton, cartilage calcifies.

BONES (pp. 132–143)

1. A bone in the skeleton is an organ.

Functions of Bones (pp. 132–133)

2. In addition to giving the body shape, bones protect and support soft organs, serve as levers for muscles to pull on, store calcium, and contain bone marrow that makes blood cells.

Classification of Bones (pp. 133–134)

3. Bones are classified on the basis of their shape as long, short, flat, or irregular.

SP Exercise: Chapter 7, Classification of Bones.

Gross Anatomy of Bones (pp. 134–136)

4. Bones have an external layer of compact bone and are filled internally with spongy bone, in which trabeculae are arranged in networks.
5. A long bone is composed of a diaphysis or shaft, and epiphyses or ends. Epiphyseal vessels serve each epiphysis, and nutrient vessels serve the diaphysis. Bone marrow occurs within the spongy bone and in a central medullary (marrow) cavity. A periosteum covers the outer surface of each bone, whereas an endosteum covers the inner bone surfaces.

AIA Link: Dissectible Anatomy; Gender: Male; View (anterior); Layer 330. Reduce image size and scroll to lower limbs.

6. Flat bones consist of two plates of compact bone separated by a layer of spongy bone.
7. Both the density of bone material and the magnitude of bending stresses decline from the superficial to the deep regions of the bones. Thus, the strongest forces occur at the periphery, where they are resisted by the strongest, compact bone. The spaces within bones lighten the skeleton and contain bone marrow.
8. The trabeculae of spongy bone are arranged along the dominant lines of stress experienced by the bone.

SP Exercise: Chapter 6, Bone Structure.

SP Exercise: Chapter 6, Synonyms for Bone Parts.

Microscopic Structure of Bone (pp. 136–137)

9. An important structural unit in compact bone is the osteon, essentially a pillar consisting of a central canal surrounded by concentric lamellae. Osteocytes, embedded in lacunae, are connected to each other and to the central canal by canaliculi.
10. Bone lamellae are concentric tubes of bone matrix. The collagen fibers in adjacent lamellae run in roughly opposite directions. This arrangement gives bone tissue great strength in resisting torsion (twisting).

SP Exercise: Chapter 6, Bone Tissue.

Chemical Composition of Bone Tissue (pp. 137–138)

11. Bone consists of cells (osteocytes, osteoblasts, and osteoclasts) plus an extracellular matrix. The matrix contains organic substances secreted by osteoblasts, including collagen, which gives bone tensile strength. Crystals of calcium phosphate salts, which precipitate in this matrix, make bone hard and able to resist compression.

Bone Development (pp. 139–140)

12. Flat bones of the skull form by intramembranous ossification of embryonic mesenchyme. A network of bone tissue woven around capillaries appears first and is then remodeled into a flat bone.

13. Before it mineralizes, the organic part of bone matrix is called osteoid.

14. Most bones develop by endochondral ossification of a hyaline cartilage model, starting in the late embryonic period (week 8). The stages of development of a long bone are (1) formation of a bone collar around the diaphysis; (2) calcification and cavitation in the center of the diaphysis; (3) growth of a periosteal bud into the center of the shaft, and formation of the first bone trabeculae; this is followed by the appearance of the medullary cavity and continued rapid growth throughout the fetal period; (4) ossification of the epiphyses and formation of the epiphyseal plates, around the time of birth.

15. The primary ossification center is the diaphysis. The secondary ossification centers are in the epiphyses.

Anatomy of the Epiphyseal Growth Area (pp. 140–141)

16. The growing cartilage of the fetal epiphyses and the postnatal epiphyseal plates is organized into several zones, which allow rapid growth. These zones are explained in Figure 6.10.

Postnatal Growth of Endochondral Bones (pp. 141–142)

17. Endochondral bones lengthen during youth through the growth of epiphyseal plate cartilages, which close in early adulthood.

18. Bones increase in width through appositional growth.

Bone Remodeling (pp. 142–143)

19. New bone tissue is continuously deposited and reabsorbed in response to hormonal and mechanical stimuli. Together, these processes are called bone remodeling.

20. Osteoid is secreted by osteoblasts at areas of bone deposit. Calcium salt is then deposited in the osteoid.

21. Osteoclasts break down bone tissue by secreting digestive enzymes and acid onto bone surfaces. This process releases Ca^{2+} and PO_4^{3-} into the blood. Parathyroid hormone stimulates this reabsorption of bone.

22. Compressive forces and gravity acting on the skeleton help maintain bone strength, as bones thicken at sites of stress.

Repair of Bone Fractures (p.143)

23. Fractures are treated by open or closed reduction. Healing involves the formation of a hematoma, a fibrocartilaginous callus, and a bony callus, and then a remodeling of the callus into the original bone pattern.

DISORDERS OF BONES (pp. 143–147)

1. Osteoporosis is a condition in which bone breakdown outpaces bone formation, causing bones to weaken. Postmenopausal women are most susceptible.

2. Osteomalacia and rickets occur when bones are inadequately mineralized, making them soft and deformed. The most common cause is inadequate intake of vitamin D.

3. Paget's disease is characterized by excessive and abnormal remodeling of bone.

4. Osteosarcoma is the most common form of bone cancer.

THE SKELETON THROUGHOUT LIFE (p. 147)

1. Bone formation in the fetus occurs in a predictable and precisely timed manner.

2. The mass of the skeleton increases dramatically during puberty and adolescence, when bone formation exceeds reabsorption.

3. Bone mass is constant in young adulthood, but beginning in one's 40s, bone reabsorption exceeds formation.

REVIEW QUESTIONS

Multiple Choice/Matching Questions

1. Which is a function of the skeletal system? (a) support, (b) blood cell formation, (c) mineral storage, (d) providing levers for muscle activity, (e) all of these.

2. Articular cartilages are located (a) at the ends of bones, (b) between the ribs and the sternum (breastbone), (c) between the epiphysis and diaphysis, (d) in the nose.

3. Cartilages relate to their perichondrium in the same way that bones relate to their (a) articular cartilages, (b) spongy bone layer, (c) osteons, (d) marrow, (e) periosteum.

4. Bone versus cartilage tissue. Indicate whether each of the following statements is true (T) or false (F).

____ **(1)** Cartilage is more resilient than bone.

____ **(2)** Cartilage is especially strong in resisting shear (bending and twisting) forces.

____ **(3)** Cartilage can grow faster than bone in the growing skeleton.

____ **(4)** In the adult skeleton, cartilage heals and regenerates faster than bone when damaged.

____ **(5)** Neither bone nor cartilage contains capillaries.

____ **(6)** Bone tissue contains very little water compared to other connective tissues, while cartilage tissue contains a large amount of water.

____ **(7)** Nutrients diffuse quickly through cartilage matrix but very poorly through the solid bone matrix.

5. A bone that has essentially the same width, length, and height is most likely (a) a long bone, (b) a short bone, (c) a flat bone, (d) an irregular bone.

6. The shaft of a long bone is properly called the (a) epiphysis, (b) periosteum, (c) diaphysis, (d) compact bone.

7. The osteon exhibits (a) a central canal containing blood vessels, (b) concentric lamellae of matrix, (c) osteocytes in lacunae, (d) all of these.

8. The flat bones of the skull develop from (a) areolar tissue, (b) hyaline cartilage, (c) mesenchyme membranes, (d) compact bone.

9. The following events apply to the endochondral ossification process as it occurs in the primary ossification center. Put these events in their proper order by assigning each a number (1–6).

____ **(a)** Cartilage in the diaphysis calcifies and contains cavities.

____ **(b)** A collar of bone is laid down around the hyaline cartilage model just deep to the periosteum.

____ **(c)** Periosteal bud invades the center of the diaphysis.

____ **(d)** Perichondrium of shaft becomes more vascularized and becomes a periosteum.

____ **(e)** Osteoblasts first deposit bone tissue around the cartilage spicules within the diaphysis.

____ **(f)** Osteoclasts remove the bone from the center of the diaphysis, leaving a medullary cavity that then houses marrow.

10. The remodeling of bone tissue is a function of which cells? (a) chondrocytes and osteocytes, (b) osteoblasts and osteoclasts, (c) chondroblasts and osteoclasts, (d) osteoblasts and osteocytes.

11. A fracture in which the bone ends penetrate soft tissue and skin is (a) greenstick, (b) compound, (c) simple, (d) comminuted, (e) compression.

12. The disorder in which bones are porous and thin but the chemistry of the bone matrix remains normal is (a) osteomalacia, (b) osteoporosis, (c) osteomyelitis, (d) Paget's disease.

13. Where within an epiphyseal plate is the calcified cartilage located? (a) nearest the diaphysis, (b) in the medullary cavity, (c) farthest from the diaphysis, (d) in the primary ossification center, (e) all of these.

14. Endosteum is in all these places, except: (a) around the exterior of the femur, (b) on the trabeculae of spongy bone, (c) lining the central canal of an osteon, (d) often directly touching bone marrow.

Short Answer Essay Questions

15. Explain (a) why cartilages are resilient and (b) why cartilage can grow so quickly in the developing skeleton.

16. Some anatomy students are joking between classes, imagining what bone organs would look like if they had spongy bone on the outside and compact bone on the inside, instead of the other way around. You tell the students that such imaginary bones would be of poor mechanical design and would break very easily when bent. Explain why this is so.

17. When and why do the epiphyseal plates close?

18. (a) During what period of life does skeletal mass increase dramatically? Begin to decline? (b) Why are fractures most common in elderly people?

19. In a piece of cartilage in the young skeleton, interstitial and appositional growth occur together. Compare and contrast interstitial and appositional growth.

20. Photocopy a picture of a skeleton, then color its membrane bones one color and its endochondral bones another color.

CRITICAL REASONING + CLINICAL APPLICATION QUESTIONS

1. Explain why people confined to wheelchairs due to paralyzed lower limbs have thin, weak bones in their thighs and legs.

2. While walking home from class, 52-year-old Ike broke a bone and damaged a knee cartilage in a fall. Assuming no special tissue grafts are made, which will probably heal faster, the bone or the cartilage? Why?

3. Carlos went to weight-lifting camp in the summer between seventh and eighth grade. He noticed that the trainer put tremendous pressure on participants to improve their strength. After an especially vigorous workout, Carlos's arm felt extremely sore and weak around the elbow. He went to the camp doctor, who took X rays and then told him that the injury was serious—that the "end, or epiphysis, of his upper arm bone was starting to twist off." What had happened? Could the same thing happen to Carlos's 23-year-old sister, Selena, who was also starting a weight-lifting program? Why or why not?

4. Ming posed the following question: "If the epiphyseal growth plates are growing so fast, why do they stay thin? Growing things are supposed to get larger or thicker, but these plates remain the same thickness." How would you answer her?

5. Old Norse stories tell of a famous Viking named Egil, an actual person who lived around 900 A.D. His skull was greatly enlarged and misshapen, and the cranial bones were hardened and thickened (6 cm, or several inches, thick!). After he died, his skull was dug up and it withstood the blow of an ax without damage. In life, he had headaches from the pressure exerted by enlarged vertebrae on his spinal cord. So much blood was diverted to his bones to support their extensive remodeling that his fingers and toes always felt cold and his heart was damaged through overexertion. What bone disorder did Egil probably have?

6. Bernice, a 75-year-old woman, stumbled slightly while walking, then felt a terrible pain in her left hip. At the hospital, X rays revealed that the hip was broken. Furthermore, the spongy bone throughout her spine was very thin. What condition is likely responsible for these findings?

A.D.A.M.® INTERACTIVE ANATOMY QUESTIONS

1. Link: Atlas; Body wall and back: Isolated vertebrae (anterior/lateral). The vertebrae are classified as what kinds of bones according to their shape classification?

2. Link: Atlas; Region: Head and neck; Laryngeal muscles (anterior). Name the two types of cartilage in the larynx.

CHAPTER 7

Bones, Part 1: The Axial Skeleton

X-ray of lumbar spine

CHAPTER OUTLINE + STUDENT OBJECTIVES

1. Define the axial skeleton and contrast it with the appendicular skeleton.
2. Describe the kinds of markings on bones.

The Skull 154–169

3. Name and describe the bones of the skull. Identify their important markings.
4. Compare the functions of the cranial and facial bones.
5. Define the bony boundaries of the orbit, nasal cavity, and paranasal sinuses.

The Vertebral Column 170–177

6. Describe the general structure of the vertebral column, and list its components.
7. Name a function performed by both the spinal curvatures and the intervertebral discs.
8. Discuss the structure of a typical vertebra, and describe the special features of cervical, thoracic, and lumbar vertebrae.

The Bony Thorax 177–179

9. Describe the ribs and sternum.
10. Differentiate true ribs from false ribs, and explain how they relate to floating ribs.

Disorders of the Axial Skeleton 179

11. List three types of abnormal curvatures of the spinal column and explain spinal stenosis.

The Axial Skeleton Throughout Life 179–182

12. Describe how the axial skeleton changes as we grow.

THE WORD SKELETON COMES FROM A GREEK word meaning "dried-up body" or "mummy," a rather disparaging description. Nonetheless, this internal framework is a greater triumph of design and engineering than any skyscraper. The skeleton is strong yet light, wonderfully adapted for the weight-bearing, locomotive, protective, and manipulative functions it performs.

The skeleton consists of **bones, cartilages, joints,** and **ligaments.** Joints, also called **articulations,** are the junctions between skeletal elements. Ligaments connect bones and reinforce most joints. Whereas bones are described in this and next chapter, joints and their ligaments are discussed in detail in Chapter 9.

The 206 named bones of the human skeleton are grouped into the axial and appendicular skeletons (Figure 7.1). The **appendicular** (ap″en-dik′u-lar) **skeleton,** which we examine in Chapter 8, consists of the bones of the upper and lower limbs, including the pectoral (shoulder) and pelvic girdles that attach the limbs to the axial skeleton.

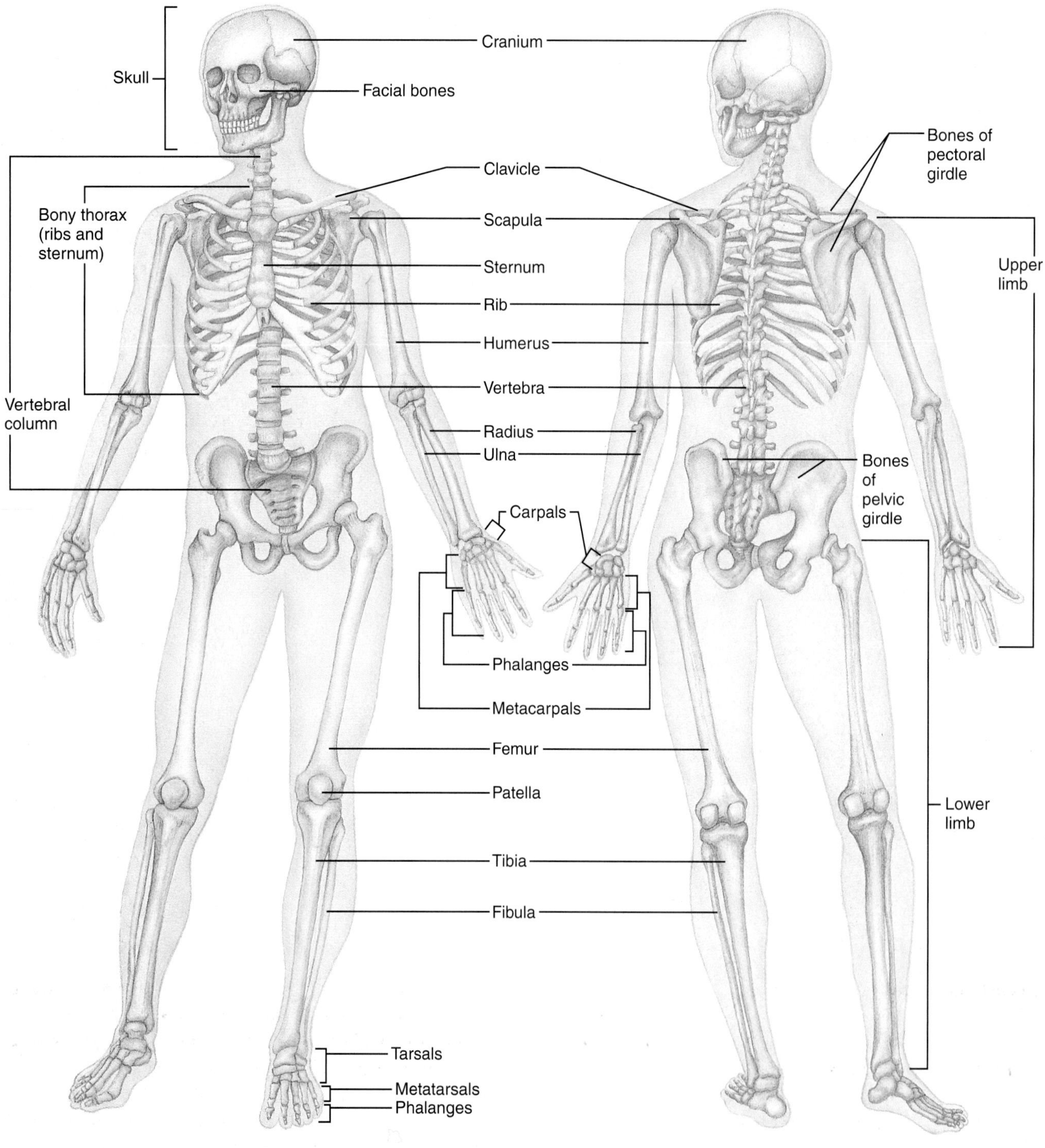

FIGURE 7.1 The human skeleton. Bones of the axial skeleton are colored green. Bones of the appendicular skeleton are gold.

Table 7.1 Bone Markings

Name of Bone Marking	Description
PROJECTIONS THAT ARE SITES OF MUSCLE AND LIGAMENT ATTACHMENT	
Tuberosity (too″be-ros′ĭ-te)	Large rounded projection; may be roughened
Crest	Narrow ridge of bone; usually prominent
Trochanter (tro-kan′ter)	Very large, blunt, irregularly shaped process (the only examples are on the femur)
Line	Narrow ridge of bone; less prominent than a crest
Tubercle (too′ber-kl)	Small rounded projection or process
Epicondyle (ep″ĭ-kon′dīl)	Raised area on or above a condyle
Spine	Sharp, slender, often pointed projection
Process	Any bony prominence

Tuberosity
Anterior crest
Tibia of leg
Trochanter
Intertrochanteric line
Adductor tubercle
Medial epicondyle
Femur of thigh

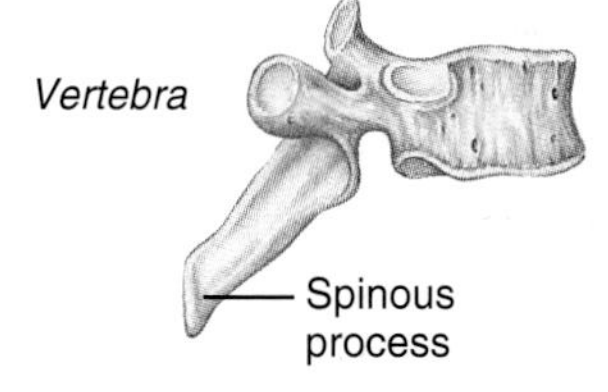

Name of Bone Marking	Description
PROJECTIONS THAT HELP TO FORM JOINTS	
Head	Bony expansion carried on a narrow neck
Facet	Smooth, nearly flat articular surface
Condyle (kon′dīl)	Rounded articular projection
Ramus (ra′mus)	Arm-like bar of bone

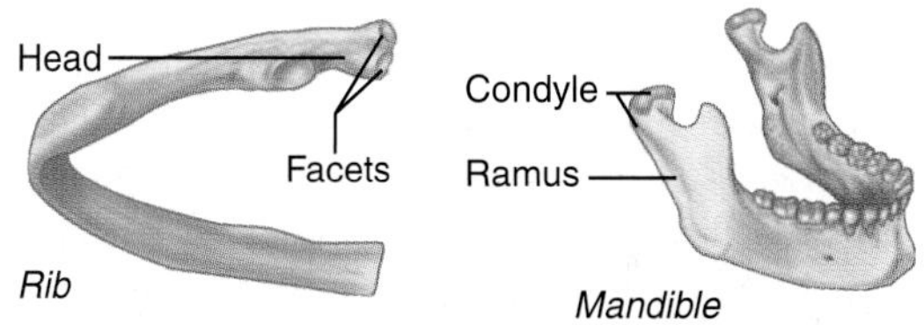

Name of Bone Marking	Description
DEPRESSIONS AND OPENINGS ALLOWING BLOOD VESSELS AND NERVES TO PASS	
Meatus (me-a′tus)	Canal-like passageway
Sinus	Cavity within a bone, filled with air and lined with mucous membrane
Fossa (fos′ah)	Shallow basin-like depression in a bone, often serving as an articular surface
Groove	Furrow
Fissure	Narrow, slit-like opening
Foramen (fo-ra′men)	Round or oval opening through a bone

Meatus
Sinus
Fossa
Groove
Inferior orbital fissure
Foramen
Skull

The **axial skeleton,** which forms the long axis of the body, is the focus of this chapter. Its 80 named bones are arranged into three major regions: the *skull, vertebral column,* and *bony thorax* (see Figure 7.1). This axial division of the skeleton supports the head, neck, and trunk, and protects the brain, spinal cord, and the organs in the thorax.

Before we examine the axial skeleton, we should note that many bones of both the axial and appendicular skeleton display **bone markings** (Table 7.1). These markings include (1) projections that provide attachment sites for muscles and ligaments; (2) projections that help to form joints; and (3) depressions and openings for the passage of vessels and nerves. Bone markings provide a wealth of information about the functions of bone and muscles, and of the relationship of bones to their associated soft structures.

Now we turn our attention to the first major region of the axial skeleton: the skull.

THE SKULL

The **skull** is the body's most complex bony structure (Figures 7.2 to 7.4). It is formed by cranial and facial bones. The *cranial bones,* or **cranium** (kra′ne-um), enclose and protect the brain and provide attachment sites for some head and neck muscles. The *facial bones* (1) form the framework of the face; (2) form cavities for the sense organs of sight, taste, and smell; (3) provide openings for the passage of air and food; (4) hold the teeth; and (5) anchor the muscles of the face.

Most skull bones are flat bones and are firmly united by interlocking, immovable joints called sutures (soo′cherz; "seams"). The suture lines have an irregular, saw-toothed appearance. The longest and most important sutures—the *coronal, sagittal, squamous,* and *lambdoid sutures*—connect the cranial bones. Most other skull sutures connect facial bones and are named according to the specific bones they connect.

Overview of Skull Geography

It is worth surveying basic skull "geography" before describing the individual bones. With the lower jaw removed, the skull resembles a lopsided, hollow, bony sphere. The facial bones form its anterior aspect, and the cranium forms the rest. The cranium can be divided into a vault and a base. The *cranial vault,* also called the *calvaria* (kal-va′re-ah; "bald part of skull") or skullcap, forms the superior, lateral, and posterior aspects of the skull, as well as the forehead region. The *cranial base,* or *floor,* is the inferior part. Internally, prominent bony ridges divide the cranial base into three distinct "steps," or fossae—the anterior, middle, and posterior *cranial fossae* (see Figure 7.4d). The brain sits snugly in these cranial fossae and is completely enclosed by the cranial vault. Overall, the brain is said to occupy the *cranial cavity.*

In addition to its large cranial cavity, the skull contains many smaller cavities, including the *middle ear* and *inner ear cavities* (carved into the lateral aspects of the cranial base), the *nasal cavity,* and the *orbits.* The nasal cavity lies in and posterior to the nose, and the orbits house the eyeballs. Air-filled sinuses that occur in several bones around the nasal cavity are the *paranasal sinuses.*

Moreover, the skull has about 85 named openings (foramina, canals, fissures). The most important of these provide passageways for the spinal cord, the major blood vessels serving the brain, and the 12 pairs of *cranial nerves,* which conduct impulses to and from the brain. Cranial nerves, which are discussed in Chapter 14 (p. 418), are classified by number, using the Roman numerals I–XII.

The skull bones and their markings are illustrated in Figures 7.2 to 7.4 and summarized in Table 7.2 on pp. 168–169. The colored box beside each bone's name in the table corresponds to the color of that bone in the figures.

Cranial Bones

The eight large bones of the cranium are the paired parietal and temporal bones and the unpaired frontal, occipital, sphenoid, and ethmoid bones. Together these bones form the brain's protective "shell." Because its superior aspect is curved, the cranium is self-bracing. This allows the bones to be thin, and, like an eggshell, the cranium is remarkably strong for its weight.

Frontal Bone

The **frontal bone,** shaped like a cockle shell, forms the forehead and roofs of the orbits. Just superior to the orbits, it protrudes slightly to form *superciliary* (soo″per-sil′e-a-re) *arches.* These "brow ridges" lie just deep to our eyebrows, and the word *supercilium* means "eyebrow" in Latin. The **supraorbital margin,** or superior margin of each orbit, is pierced by a **supraorbital foramen,** or *supraorbital notch* (see Figure 7.2a). This opening transmits the supraorbital artery and nerve, which supply the forehead. The smooth part of the frontal bone between the superciliary arches in the midline is the **glabella** (glah-bel′ah; "smooth, without hair"). Just inferior to it, the frontal bone meets the nasal bones at the *frontonasal suture.* The regions of the frontal bone lateral to the glabella contain the air-filled **frontal sinuses** (see Figure 7.3b).

Internally, the frontal bone contributes to the **anterior cranial fossa** (see Figures 7.4c and 7.4d), which holds the large frontal lobes of the brain.

Parietal Bones and the Major Sutures

The two large **parietal bones,** shaped like curved rectangles, make up the bulk of the cranial vault; that is, they form most of the superior part of the skull, as well as its lateral walls (*parietal* = wall). The sites at which the parietal bones articulate (form a joint) with other cranial bones are the four largest sutures:

1. The **coronal suture,** running in the coronal plane, occurs anteriorly where the parietal bones meet the frontal bone (see Figure 7.3).
2. A **squamous suture** occurs where each parietal bone meets a temporal bone inferiorly, on each lateral aspect of the skull (see Figure 7.3).
3. The **sagittal suture** occurs where the right and left parietal bones meet superiorly in the midline of the cranium (see Figure 7.2b).
4. The **lambdoid suture** occurs where the parietal bones meet the occipital bone posteriorly (see Figure 7.2b). This suture is so named because it resembles the Greek letter lambda (Λ).

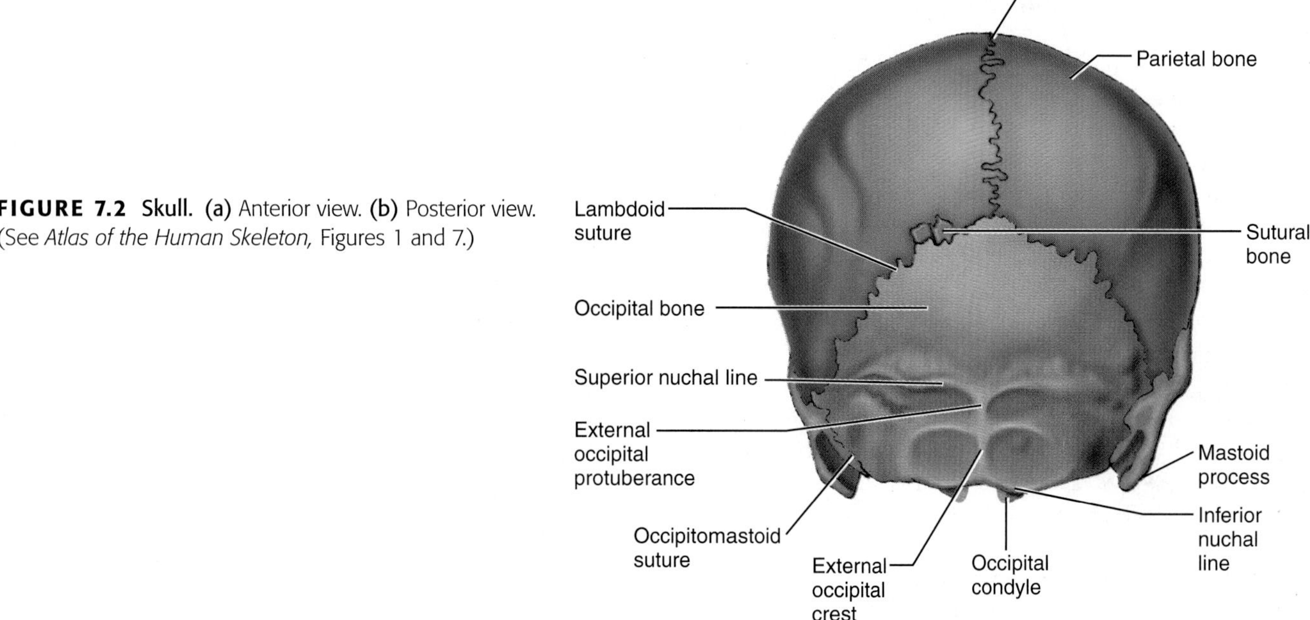

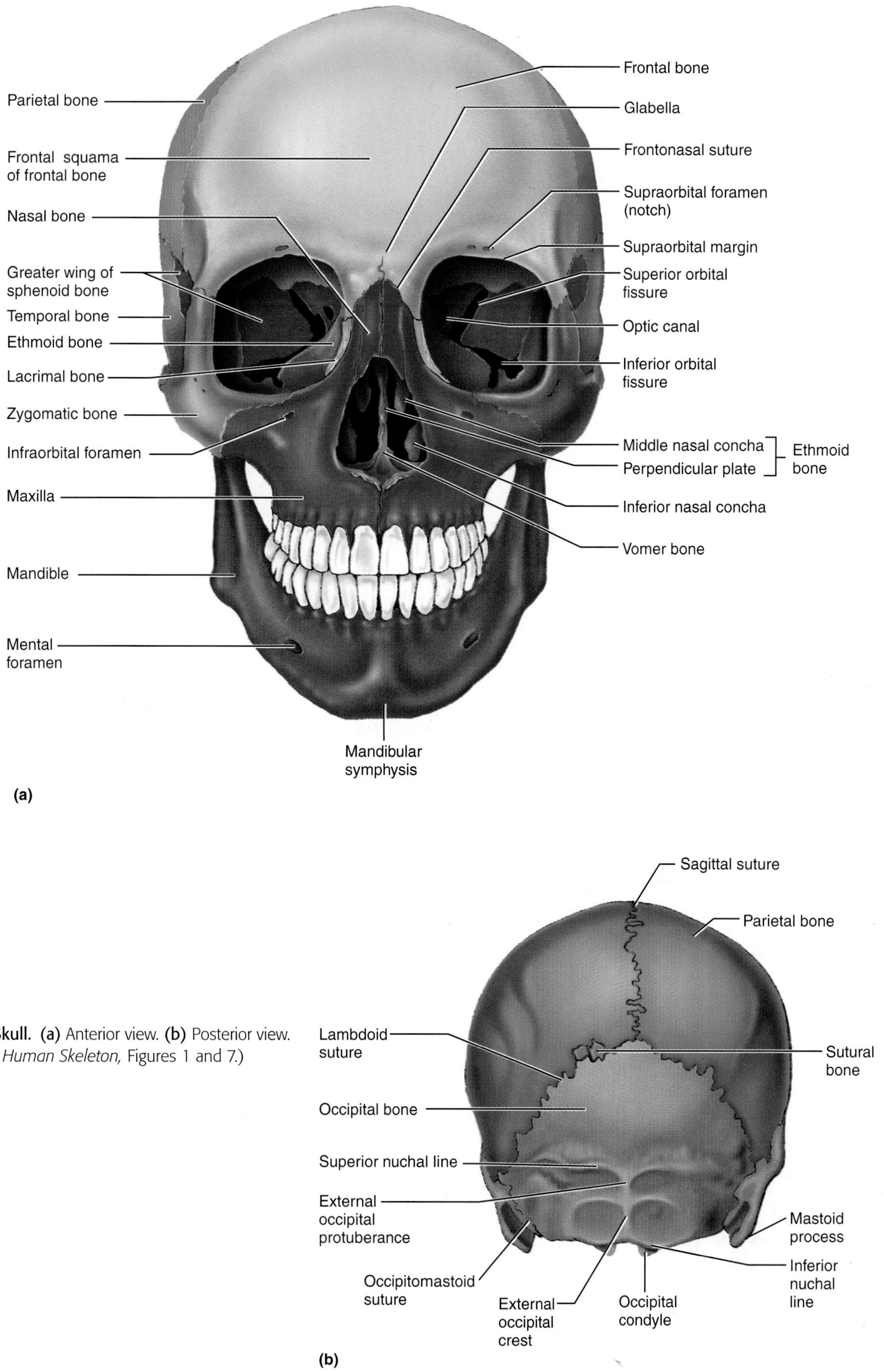

FIGURE 7.2 Skull. (a) Anterior view. (b) Posterior view. (See *Atlas of the Human Skeleton,* Figures 1 and 7.)

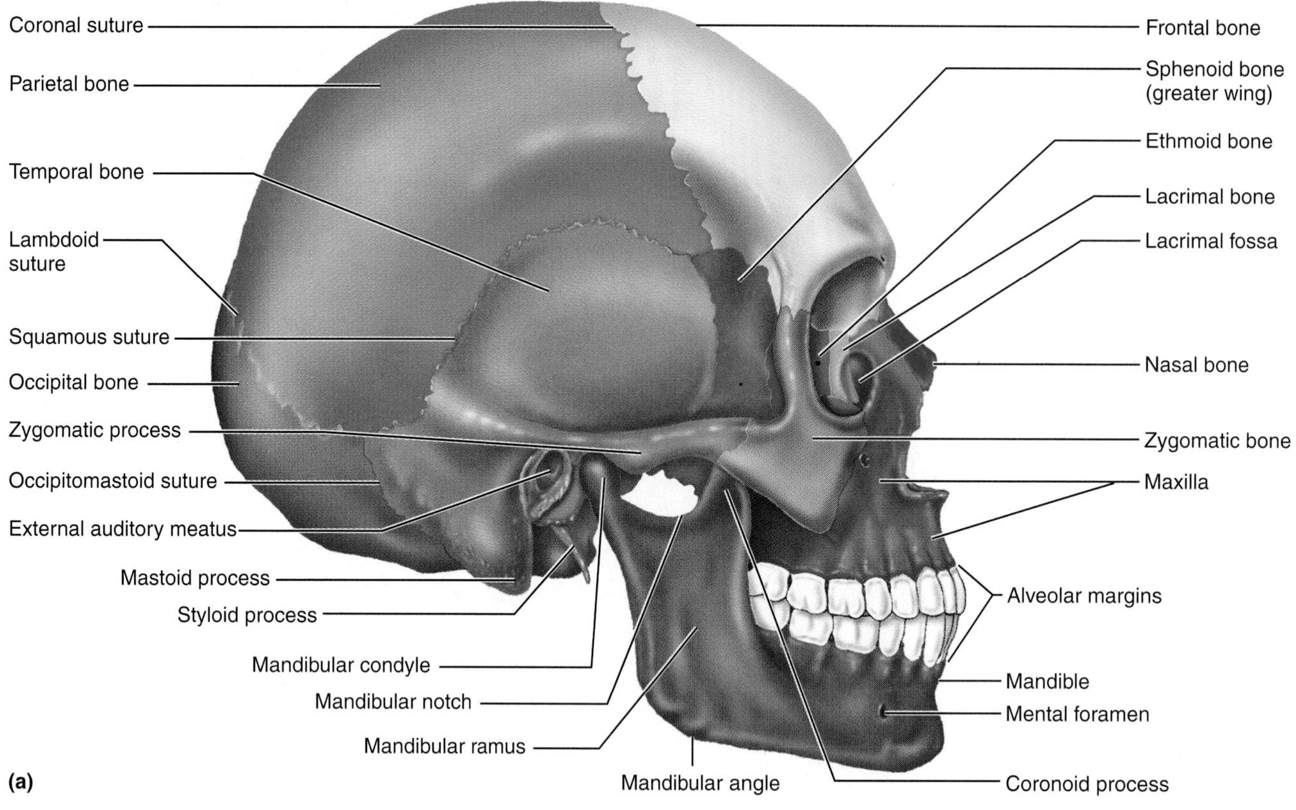

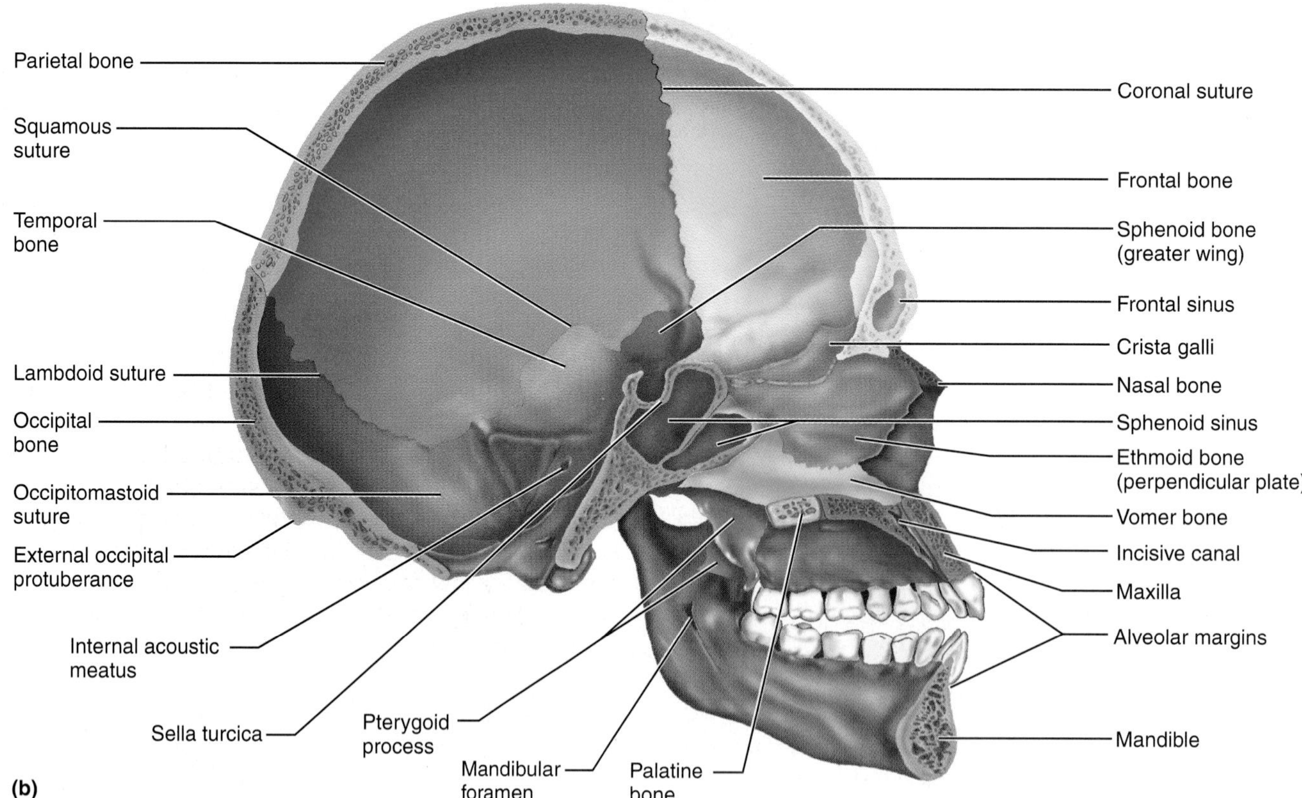

FIGURE 7.3 Lateral aspect of the skull. **(a)** External anatomy of the right side. **(b)** View of a skull cut through the midline, showing the internal anatomy of the left half. **(c)** Photo of skull cut through the midline, same view as in (b). (See *Atlas of the Human Skeleton,* Figures 2 and 3.)

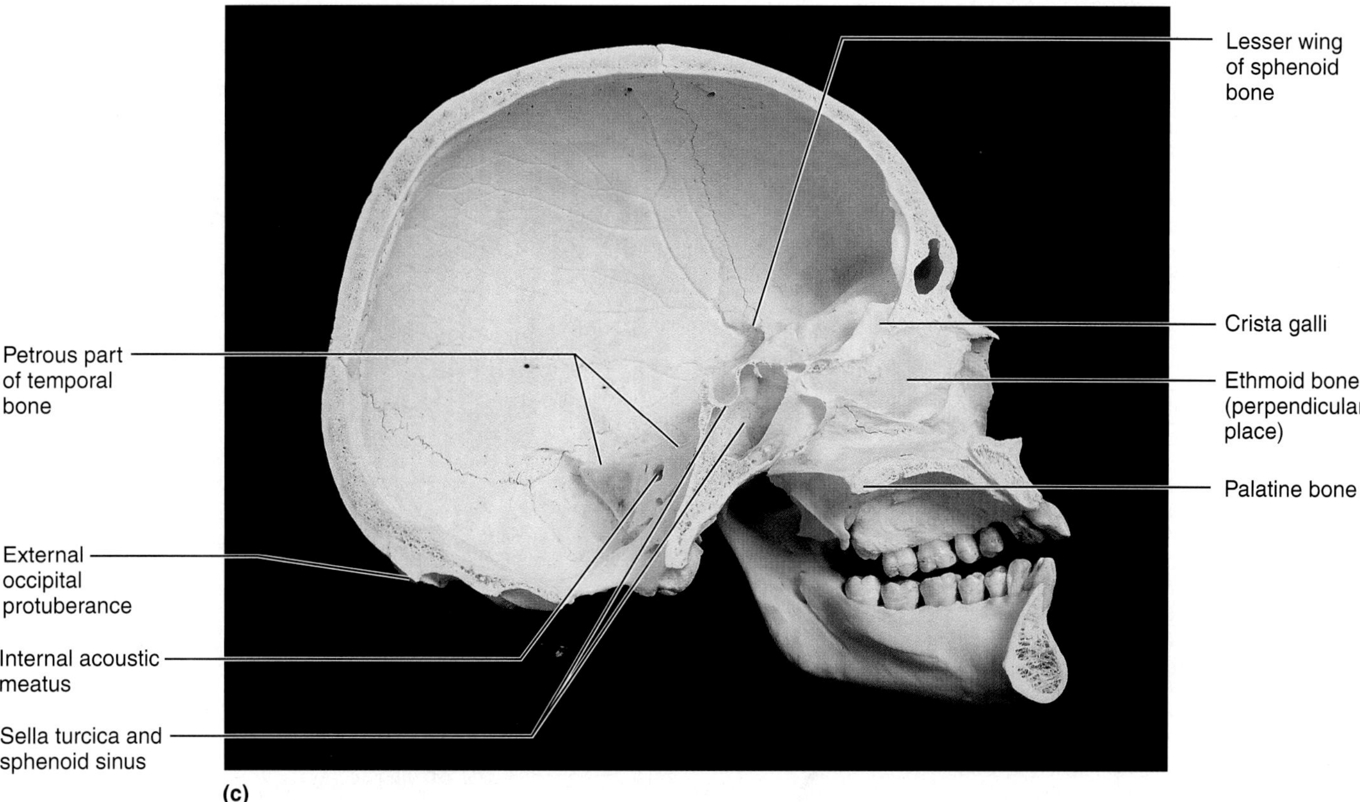

FIGURE 7.3, *continued.* Midsagittally cut skull.

Sutural Bones

Sutural bones are small bones that occur *within* the sutures, especially in the lambdoid suture (see Figure 7.2b). They are irregular in shape, size, and location, and not all people have them. Functionally insignificant, they develop between the major cranial bones during the fetal period and persist throughout life.

Occipital Bone

The **occipital bone** (ok-sip′ĭ-tal; "back of the head") makes up the posterior part of the cranium and cranial base (see Figure 7.2b). It articulates with the parietal bones at the lambdoid suture and with the temporal bones at the *occipitomastoid sutures* (see Figure 7.3). Internally, it forms the walls of the **posterior cranial fossa** (see Figures 7.4c and 7.4d), which holds a part of the brain called the cerebellum. In the base of the occipital bone is the **foramen magnum,** literally "large hole" (Figure 7.4a). Through this opening, the inferior part of the brain connects with the spinal cord. The foramen magnum is flanked laterally by two rocker-like **occipital condyles,** which articulate with the first vertebra of the vertebral column in a way that helps allow the head to nod "yes." Hidden medial and superior to each occipital condyle is a **hypoglossal** (hi″po-glos′al) **canal,** through which runs cranial nerve XII, the hypoglossal nerve. Anterior to the foramen magnum, the occipital bone joins the sphenoid bone via a band of bone called the **basioccipital** (ba″se-ok-sip′ĭ-tal; "base of the occipital").

Several markings occur on the external surface of the occipital bone (see Figures 7.2b and 7.4a). The **external occipital protuberance** is a knob in the midline, at the junction of the base and the posterior wall of the skull. The *external occipital crest* extends anteriorly from the protuberance to the foramen magnum. This crest secures the *ligamentum nuchae* (nu′ke; "of the neck"), an elastic, sheet-shaped ligament that lies in the median plane of the posterior neck and connects the neck vertebrae to the skull. Extending laterally from the occipital protuberance are the **superior nuchal** (nu′kal) **lines,** and running laterally from a point halfway along the occipital crest are the **inferior nuchal lines.** The nuchal lines and the bony regions between them anchor many muscles of the neck and back. The superior nuchal line marks the upper limit of the neck.

Temporal Bones

The temporal bones are best viewed laterally (Figure 7.5, p. 160). They lie inferior to the parietal bones and form the inferolateral region of the skull and parts of the cranial floor. The terms *temporal* and *temple,* from the Latin word for "time," refer to the fact that gray hairs, a sign of time's passage, appear first at the temples.

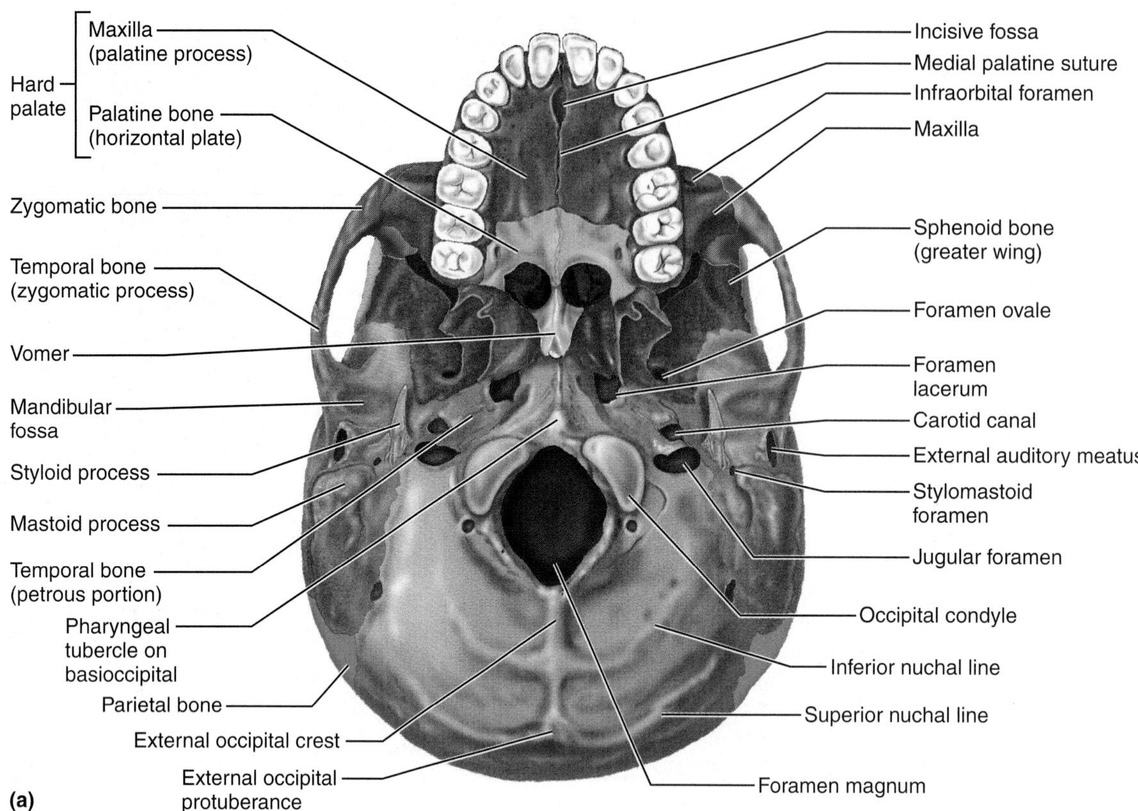

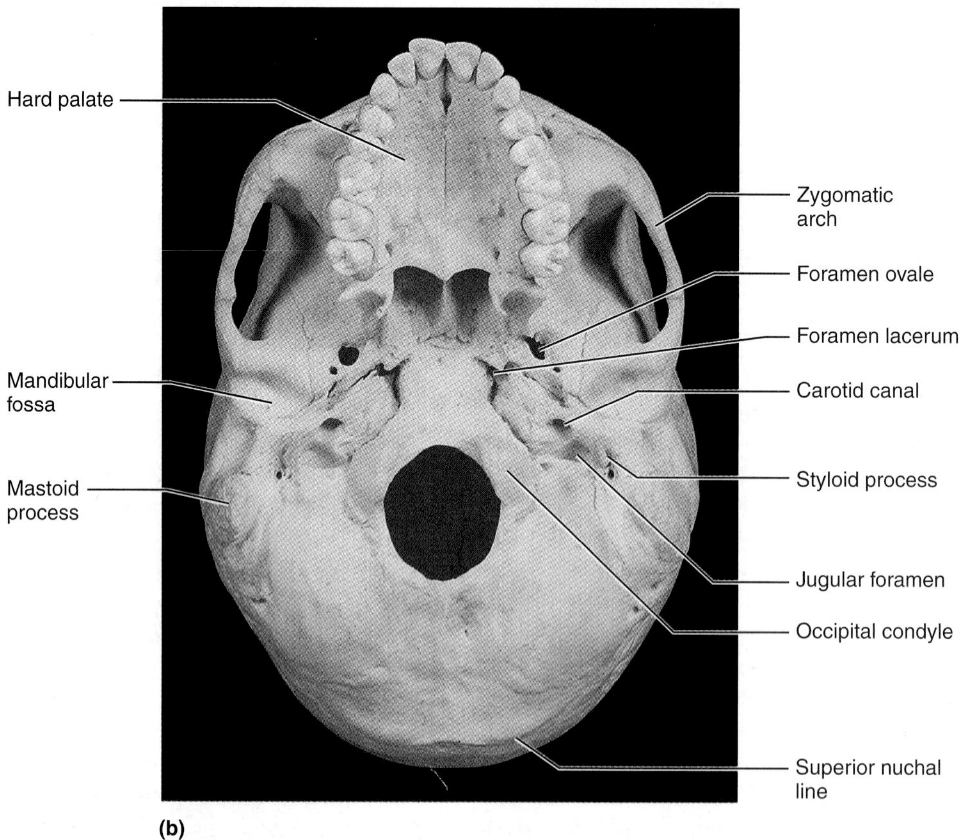

FIGURE 7.4 Inferior aspect of the skull. **(a)** Inferior view of the skull (mandible removed). **(b)** Photo of inferior view of the skull. **(c)** Superior view of the floor of the cranial cavity (calvaria removed). **(d)** The same view as (c), emphasizing the three cranial fossae. The anterior fossa lies superior to the middle fossa, which lies superior to the posterior fossa. (See *Atlas of the Human Skeleton,* Figures 4 and 5.)

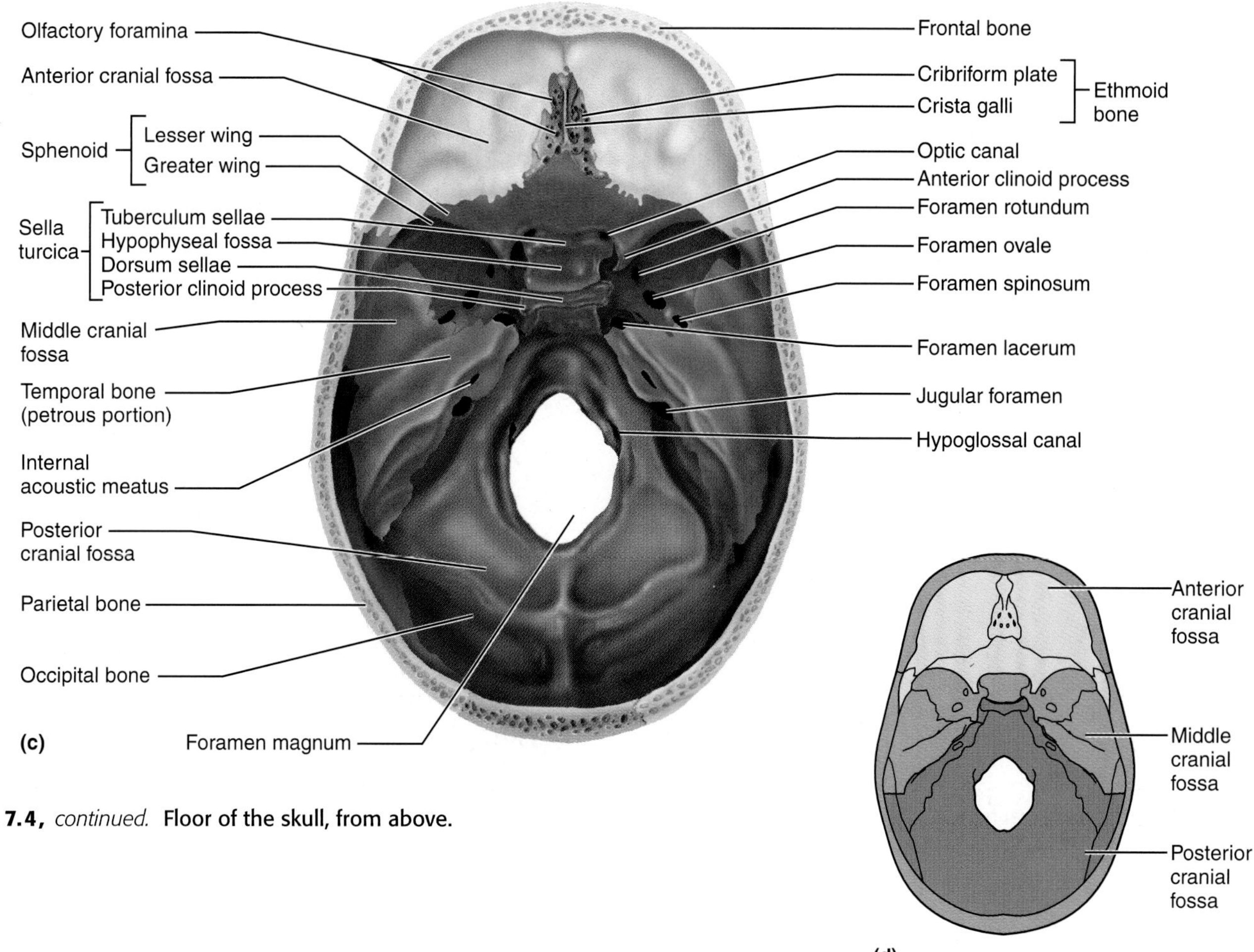

FIGURE 7.4, *continued.* Floor of the skull, from above.

Each temporal bone has an intricate shape and is described in terms of its four major regions: the *squamous, tympanic, mastoid,* and *petrous regions.* The plate-shaped **squamous region** (see Figure 7.5) abuts the squamous suture. It has a bar-like **zygomatic process** (zi″go-mat′ik; "cheek") that projects anteriorly to meet the zygomatic bone of the face. Together, these two bony structures form the *zygomatic arch,* which defines the projection of the cheek. The oval **mandibular** (man-dib′u-lar) **fossa** on the inferior surface of the zygomatic process receives the mandible (lower jawbone), forming the freely movable *temporomandibular joint* (jaw joint).

The **tympanic region** (tim-pan′ik; "eardrum") surrounds the **external acoustic meatus,** also called the **external auditory meatus,** or external ear canal. It is through this canal that sound enters the ear. The external acoustic meatus and the tympanic membrane (eardrum) at its deep end are parts of the external ear. In a dried skull, the tympanic membrane has been removed. Thus, part of the *middle ear cavity* deep to the tympanic region may be visible through the meatus. Projecting inferiorly from the tympanic region is the needle-like **styloid process** (sti′loid; "stake-like"). This process is an attachment point for some muscles of the tongue and pharynx and for a ligament that connects the skull to the hyoid bone of the neck.

The **mastoid region** (mas′toid; "breast-shaped") has a prominent **mastoid process,** an anchoring site for some neck muscles. This process can be felt as a lump just posterior to the ear. The **stylomastoid foramen** (see Figure 7.4a) is located between the styloid and mastoid processes. Cranial nerve VII, the facial nerve, leaves the skull through this foramen.

The mastoid process is full of air sinuses called *mastoid air cells,* which lie just posterior to the middle ear cavity. Infections can spread from the throat to the middle ear to the mastoid cells. Such an infection, called *mastoiditis,* can even spread to the brain, from which the mastoid air cells are separated by only a thin roof of bone. This was a serious problem before the late 1940s, when antibiotics became available.

The **petrous region** (pet′rus; "rocky") of the temporal bone projects medially and contributes to the cranial base. It appears as a bony wedge between the occipital bone posteriorly and the sphenoid bone anteriorly (see Figure 7.4a). From within the cranial cavity this very dense region looks like a mountain ridge (Figure 7.4c).

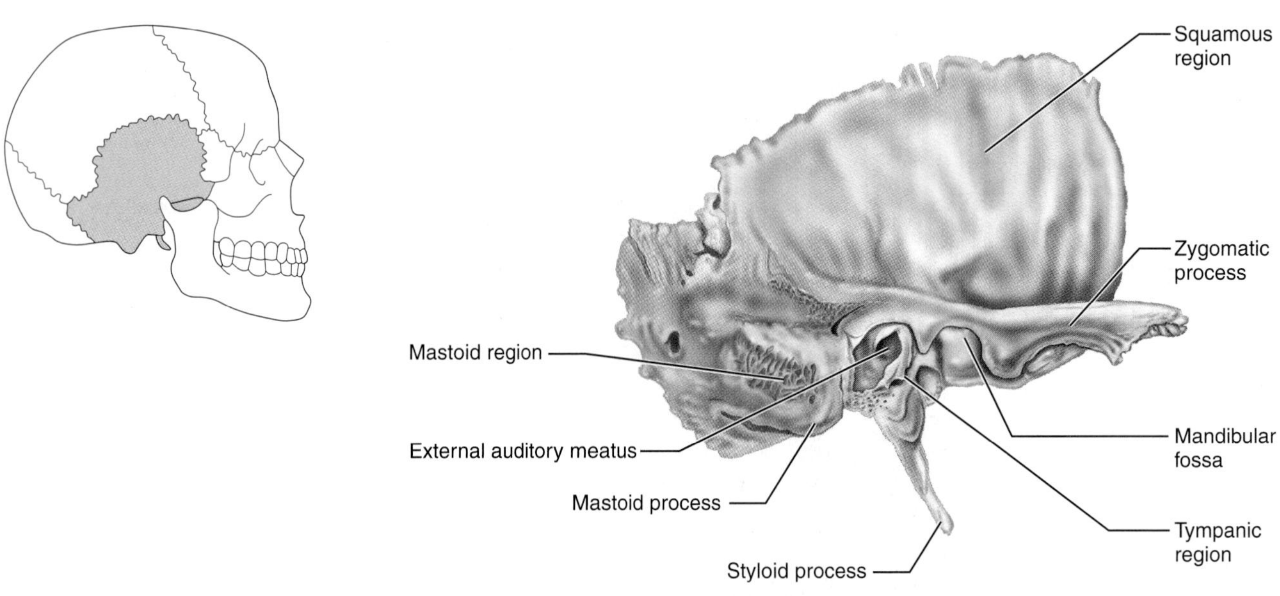

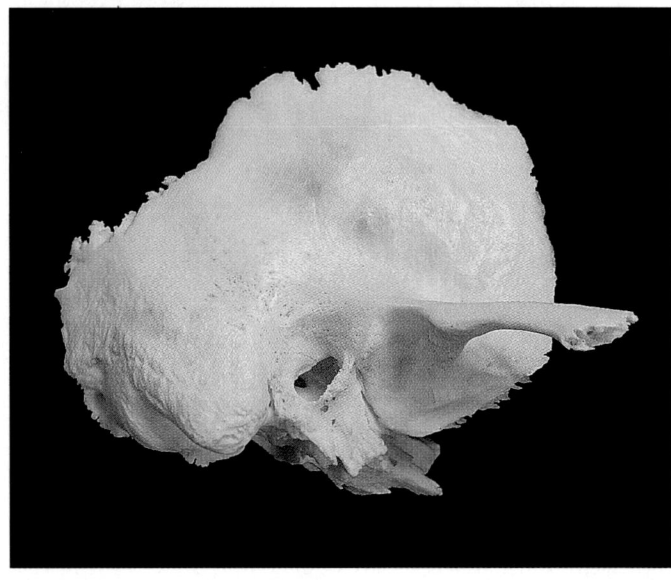

FIGURE 7.5 The temporal bone. Right lateral view. (See *Atlas of the Human Skeleton,* Figures 3 and 8.)

The posterior slope of this ridge lies in the posterior cranial fossa, whereas the anterior slope is in the **middle cranial fossa,** the fossa that holds the temporal lobes of the brain (see Figure 7.4d). Housed inside the petrous region are the cavities of the middle and inner ear, which contain the sensory apparatus for hearing and balance.

Several foramina penetrate the bone of the petrous region (see Figure 7.4a). The large **jugular foramen** is located where the petrous region joins the occipital bone. Through this foramen pass the largest vein of the head, the internal jugular vein, and cranial nerves IX, X, and XI. The **carotid** (ka-rot′id) **canal** opens in the petrous on the skull's inferior aspect, just anterior to the jugular foramen. It transmits the internal carotid artery, the main artery to the brain, into the cranial cavity. The **foramen lacerum** (la′ser-um; "lacerated") is a jagged opening between the medial tip of the petrous temporal bone and the sphenoid bone. This foramen is almost completely closed by cartilage in a living person, but it is so conspicuous in a dried skull that students usually ask its name. The **internal acoustic (auditory) meatus** lies in the cranial cavity on the posterior face of the petrous region (see Figure 7.4c). It transmits cranial nerves VII and VIII.

Sphenoid Bone

The **sphenoid bone** (sfe′noid; "wedge-shaped") spans the width of the cranial floor (see Figure 7.4) and, historically, has been said to resemble a bat with its wings spread out. It is considered the keystone of the cranium because it forms a central wedge that articulates with every other cranial bone. As shown in Figure 7.6, it consists of a central **body** and three pairs of processes: the *greater wings, lesser wings,* and *pterygoid* (ter′ĭ-goid)

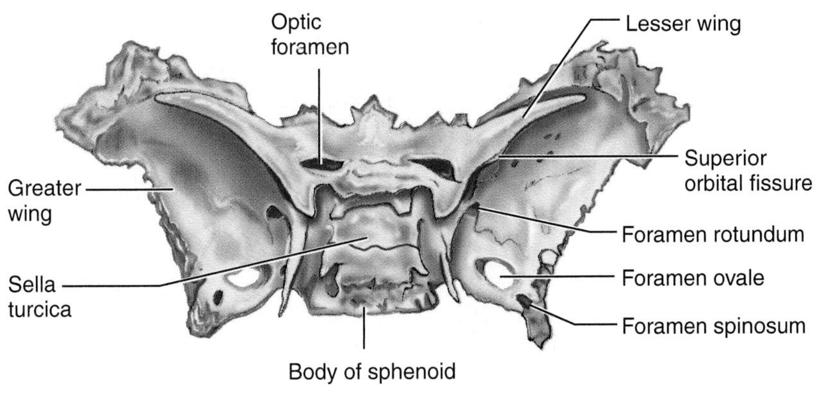

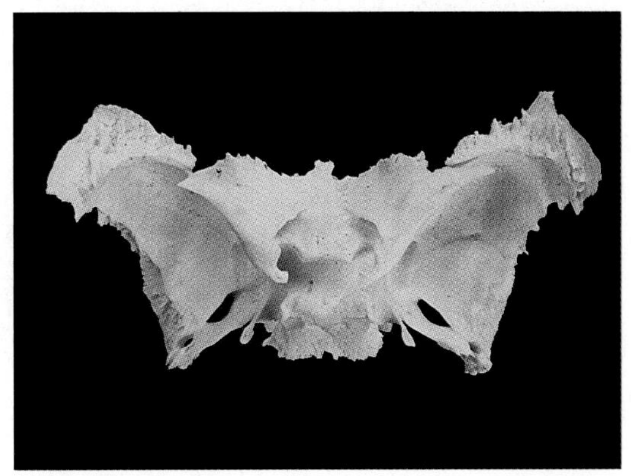

(a) Superior view, as in Figure 7.4c

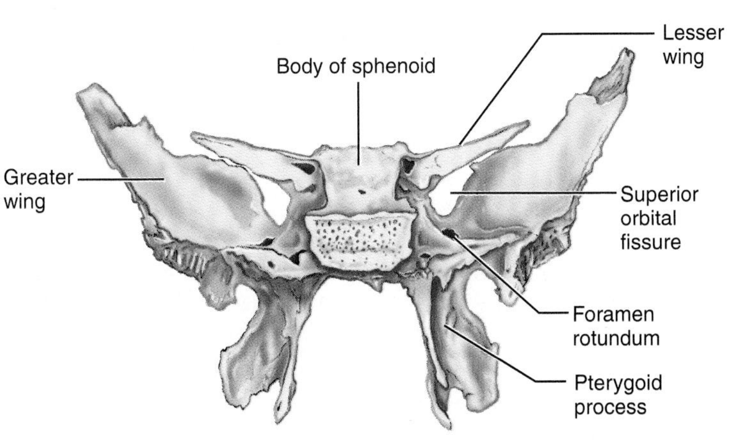

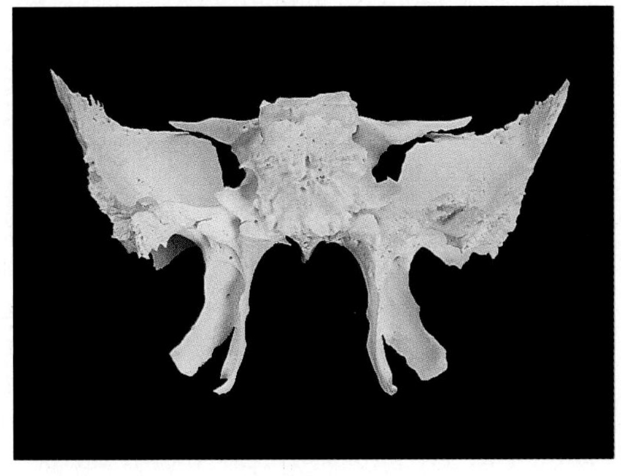

(b) Posterior view

FIGURE 7.6 **The sphenoid bone.** **(a)** Superior view. **(b)** Posterior view. **(c)** Radiograph of posterior view. (See *Atlas of the Human Skeleton,* Figures 2 and 9.)

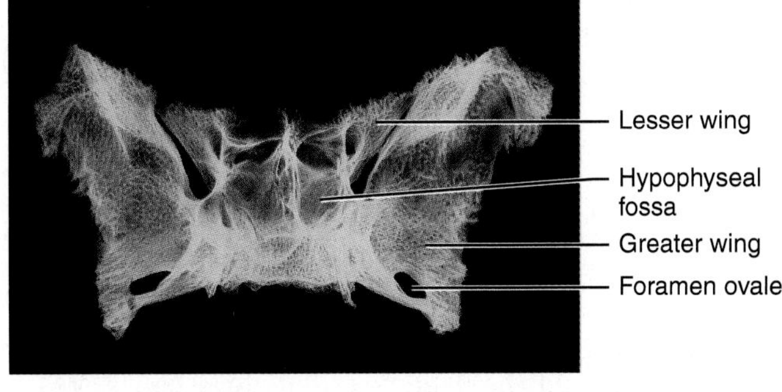

(c) Superior view (x ray)

processes. The superior surface of the body bears a saddle-shaped prominence, the **sella turcica** (sel′ah ter′sik-ah; "Turkish saddle"). The seat of this saddle, called the *hypophyseal* (hi″po-fiz′e-al) *fossa,* holds the pituitary gland, or hypophysis. Within the sphenoid body are the paired **sphenoid sinuses** (see Figure 7.3b). The **greater wings** project laterally from the sphenoid body, forming parts of the middle cranial fossa (see Figure 7.4c) and the orbit. On the lateral wall of the skull, the greater wing appears as a flag-shaped area medial to the zygomatic arch (see Figure 7.3a). The horn-shaped **lesser wings** form part of the floor of the anterior cranial fossa (see Figure 7.4c and d) and a part of the orbit. The trough-shaped **pterygoid** ("wing-like") **processes** project inferiorly from the greater wings (see Figures 7.3b and 7.6b). These processes, which have both medial and lateral plates, are attachment sites for the pterygoid muscles that help close the jaw in chewing.

The sphenoid bone has five important openings (see Figure 7.6) on each side. The **optic foramen** lies just anterior to the sella turcica. Cranial nerve II, the optic nerve, passes through this hole from the orbit into the cranial cavity. The other four openings lie in a crescent-shaped row just lateral to the sphenoid body on each side.

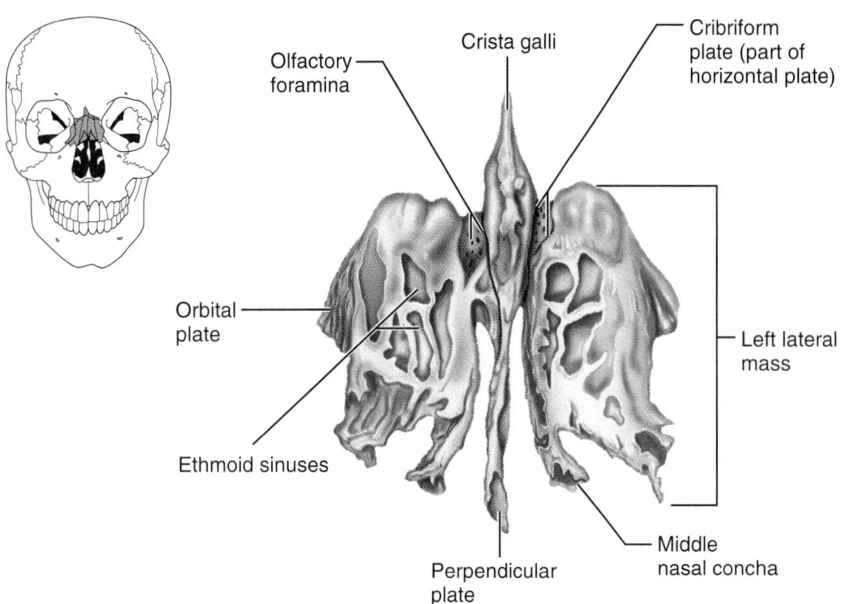

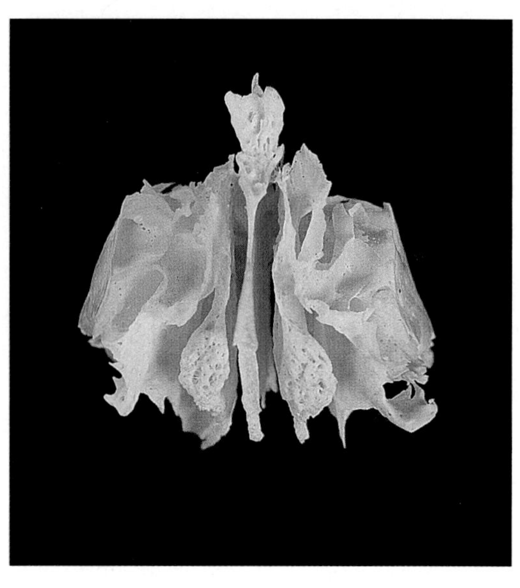

FIGURE 7.7 The ethmoid bone. Anterior view. (See *Atlas of the Human Skeleton,* Figures 2 and 10.)

The most anterior of these openings, the **superior orbital fissure,** is a long slit between the greater and lesser wings. It transmits several structures to and from the orbit, such as the cranial nerves that control eye movements (III, IV, and VI). This fissure is best seen in an anterior view of the orbit (see Figure 7.9, p. 165). The **foramen rotundum** lies in the medial part of the greater wing. It is usually oval, despite its name, which means "round opening." The **foramen ovale** (o-val′e) is an oval hole posterolateral to the foramen rotundum. The foramen rotundum and foramen ovale are passageways through which two large branches of cranial nerve V (namely, the maxillary and mandibular nerves) enter the face. Posterior and lateral to the foramen ovale lies the small **foramen spinosum** (spi-no′sum), named for a short spine that projects from its margin on the inferior aspect of the skull. This foramen transmits the middle meningeal artery, which supplies blood to the broad inner surfaces of the parietal and the squamous temporal bones.

Ethmoid Bone

The **ethmoid bone** (Figure 7.7) is the most deeply situated bone of the skull. Lying between the sphenoid bone posteriorly and the nasal bones anteriorly, it forms most of the medial bony area between the nasal cavity and the orbits. It is remarkably thin-walled and delicate.

Its superior surface is formed by paired, horizontal **cribriform** (krib′rĭ-form) **plates** that contribute to the roof of the nasal cavities and the floor of the anterior cranial fossa (see Figure 7.4c). The cribriform plates are perforated by tiny holes called *olfactory foramina* (*cribriform* = perforated like a sieve). The filaments of cranial nerve I, the olfactory nerve, pass through these holes as they run from the nasal cavity to the brain. Between the two cribriform plates, in the midline, is a superior projection called the **crista galli** (kris′tah gal′li; "rooster's comb"). A fibrous membrane called the falx cerebri (p. 388) attaches to the crista galli and helps to secure the brain within the cranial cavity.

The **perpendicular plate** of the ethmoid bone projects inferiorly in the median plane. It forms the superior part of the nasal septum, the vertical sheet that divides the nasal cavity into right and left halves. Flanking the perpendicular plate on each side is a delicate **lateral mass** riddled with the **ethmoid sinuses.** The ethmoid bone is named for these sinuses, as *ethmos* means "sieve" in Greek. Extending medially from the lateral masses are the thin **superior** and **middle nasal conchae** (kong′ke), which protrude into the nasal cavity (see Figure 7.10a). The conchae are curved like scrolls and are named after the conch shells one finds on warm ocean beaches. The lateral surfaces of the ethmoid's lateral masses are called **orbital plates** because they contribute to the medial walls of the orbits.

Facial Bones

The skeleton of the face consists of 14 bones (see Figures 7.2a and 7.3). These are the unpaired mandible and the vomer, plus the paired maxillae, zygomatics, nasals, lacrimals, palatines, and inferior nasal conchae.

Mandible

The U-shaped **mandible** (man′dĭ-bl), or lower jawbone, is the largest, strongest bone in the face (Figure 7.8a). It has a horizontal *body* that forms the inferior jawline, and two upright *rami* (ra′mi; "branches"). Each **ramus** meets the body posteriorly at a **mandibular angle.** At the superior margin of each ramus are two processes. The anterior **coronoid process** (kor′o-noid; "crown-shaped") is a flat, triangular projection. On it inserts the temporalis muscle, which elevates the lower jaw during chewing. The posterior *condylar process* enlarges superiorly to form the **mandibular condyle,** or *head of the mandible.* It articulates with the temporal bone to form the temporomandibular joint. The coronoid and condylar processes are separated by the **mandibular notch.**

The **body** of the mandible anchors the lower teeth and forms the chin. Its superior border is the **alveolar** (al-ve′o-lar) **margin.** The tooth sockets, called *alveoli,* open onto this margin. Anteriorly, the line of fusion between the two halves of the mandible is called the **mandibular symphysis** or *symphysis menti* (sim′fi-sis men′ti), literally "growing together at the chin" (see Figure 7.2a).

Several openings pierce the mandible. On the medial surface of each ramus is a **mandibular foramen** (see Figure 7.8a), through which a nerve responsible for tooth sensation (inferior alveolar nerve) enters the mandibular body and supplies the roots of the lower teeth. Dentists inject anesthetic into this foramen before working on the lower teeth. The **mental** ("chin") **foramen,** which opens on the anterolateral side of the mandibular body, transmits blood vessels and nerves to the lower lip and the skin of the chin.

CLINICAL APPLICATIONS

FRACTURED MANDIBLE Except for the thin nasal bones of the nose, the mandible is the most frequently fractured bone of the face. Most mandibular fractures result from assault or motor-vehicle accidents, and often involve a vertical split through the mandibular body. Treatment usually involves immobilizing the jaw, which makes eating and talking difficult. ✚

Maxillary Bones

The **maxillary bones,** or **maxillae** (mak-sil′e; "jaws"), form the upper jaw and the central part of the facial skeleton (see Figures 7.2a and 7.8b). They are considered the keystone bones of the face because they articulate with all other facial bones except the mandible.

Like the mandible, the maxillae have an **alveolar margin** that contains teeth in alveoli. The **palatine** (pal′ah-tēn) **processes** project medially from the alveolar margins to form the anterior region of the **hard palate,** or bony roof of the mouth (see Figure 7.4a). The **frontal processes** extend superiorly to reach the frontal bone, forming part of the lateral aspect of the bridge of the nose (see Figure 7.8b). The maxillae lie just lateral to the nasal cavity and contain the **maxillary sinuses** (see Figure 7.11). These sinuses, the largest of the paranasal air sinuses, extend from the orbit down to the roots of the upper teeth. Laterally, the maxillae articulate with the zygomatic bones at the **zygomatic processes** (see Figure 7.8b).

The maxilla, along with several other bones, forms the borders of the **inferior orbital fissure** in the floor of the orbit (Figure 7.9). This fissure transmits several vessels and nerves (see Table 7.2), including the maxillary nerve or its continuation, the infraorbital nerve. The infraorbital nerve proceeds anteriorly to enter the face through the **infraorbital foramen** (Figure 7.9).

Zygomatic Bones

The irregularly shaped **zygomatic bones** are commonly called the cheekbones (*zygoma* = cheekbone). Each joins the zygomatic process of a temporal bone posteriorly, the zygomatic process of the frontal bone superiorly, and the zygomatic process of the maxilla anteriorly (see Figure 7.3a). The zygomatic bones form the prominences of the cheeks and define part of the margin of each orbit.

Nasal Bones

The paired, rectangular **nasal bones** join medially to form the bridge of the nose (see Figure 7.2a). They articulate with the frontal bone superiorly, the maxillae laterally, and the perpendicular plate of the ethmoid bone posteriorly. Inferiorly, they attach to the cartilages that form most of the skeleton of the external nose (see Figure 6.1 on p. 131).

Lacrimal Bones

The delicate, fingernail-shaped **lacrimal** (lak′rĭ-mal) **bones** are depicted in Figures 7.2a, 7.3a, and 7.9. Located in the medial orbital walls, they articulate with the frontal bone superiorly, the ethmoid bone posteriorly, and the maxilla anteriorly (see Figure 7.9). Each lacrimal bone contains a deep groove that contributes to a *lacrimal fossa.* This fossa contains a *lacrimal sac* that gathers tears, allowing the fluid to drain from the eye surface into the nasal cavity (*lacrima* means "tears").

Palatine Bones

The **palatine bones** lie posterior to the maxillae (Figure 7.8c and 7.10a). These paired, L-shaped bones articulate with each other at their inferior **horizontal plates,** which complete the posterior part of the hard palate (see Figure 7.4a). The **perpendicular plates** form the posterior part of the lateral walls of the nasal cavity and a small part of the orbits (see Figure 7.9).

Vomer

The slender, plow-shaped **vomer** (vo′mer; "plowshare") lies in the nasal cavity, where it forms the inferior part of the nasal septum (see Figures 7.2a, 7.3b, and 7.10b).

Inferior Nasal Conchae

The paired **inferior nasal conchae** are thin, curved bones in the nasal cavity (see Figures 7.2a and 7.10a). Projecting medially from the lateral walls of the nasal cavity, just inferior to the middle nasal conchae of the ethmoid bone, they are the largest of the three pairs of conchae.

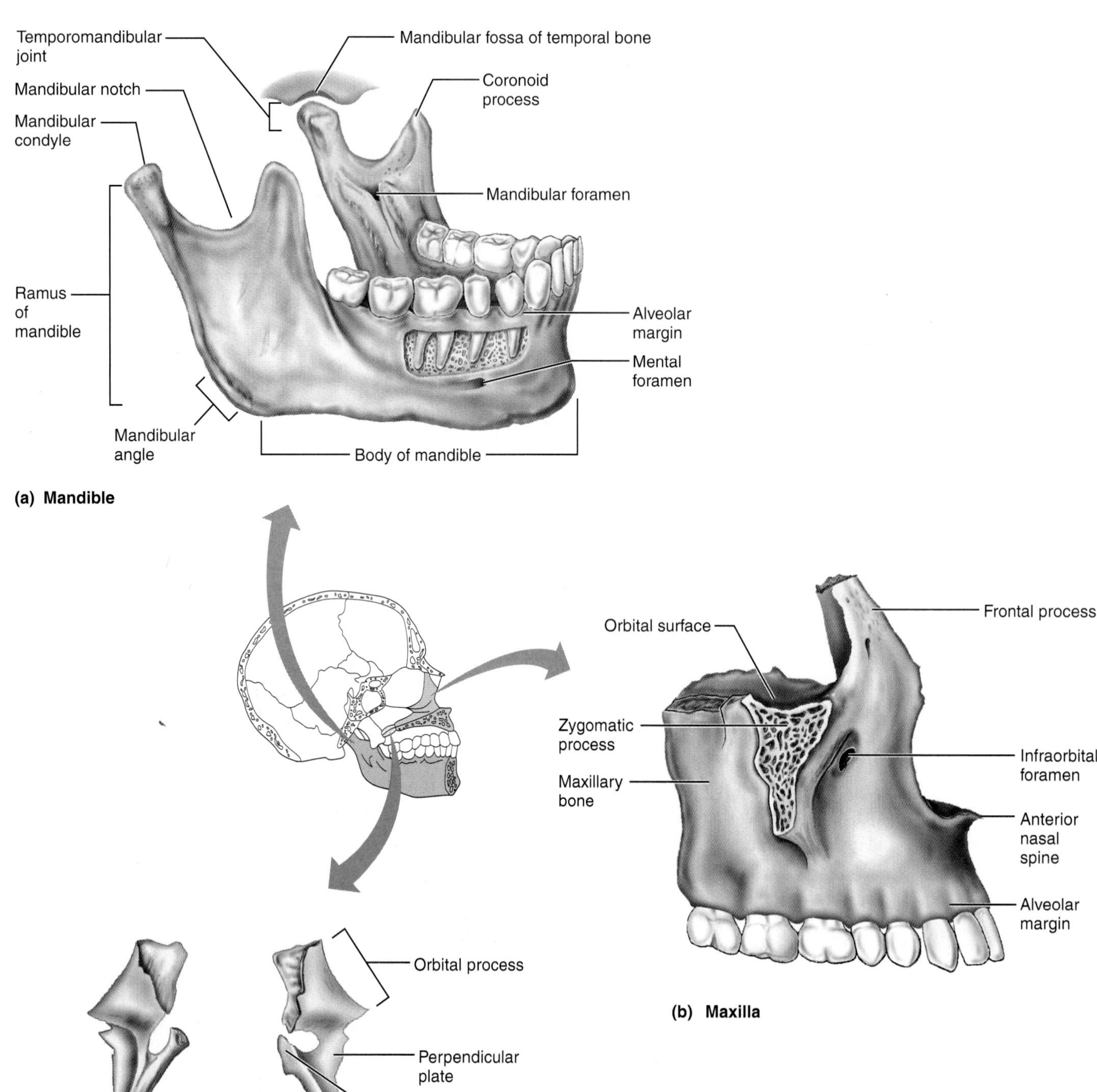

FIGURE 7.8 **Detailed anatomy of some facial bones.** **(a)** The mandible. **(b)** The maxilla. **(c)** Palatine bones. Note that the bones in (c) are enlarged more than those in (a) and (b). (See *Atlas of the Human Skeleton,* Figures 11, 12, and 13.)

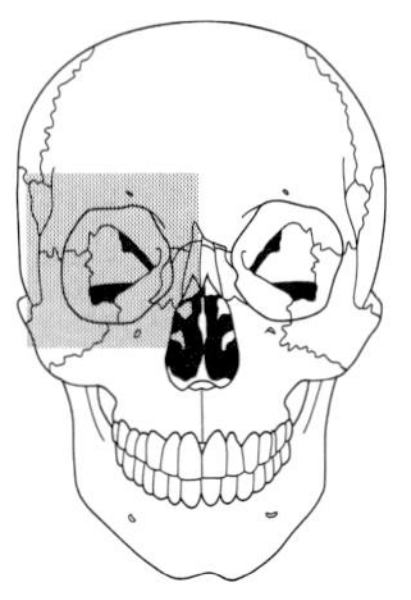

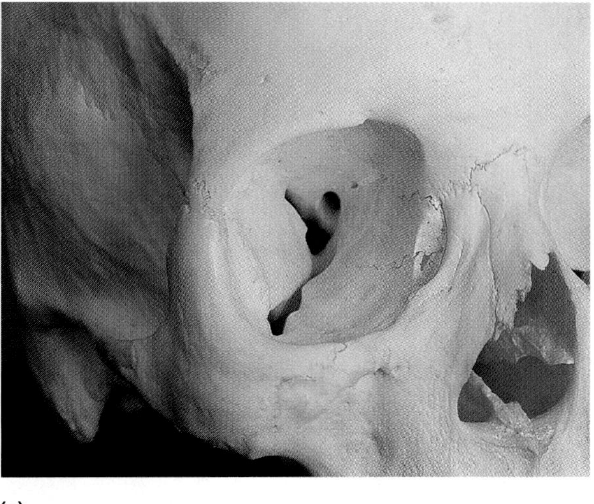

(a)

FIGURE 7.9 Bones of the orbit. **(a)** Photograph. **(b)** Illustration. (See *Atlas of the Human Skeleton,* Figure 14.)

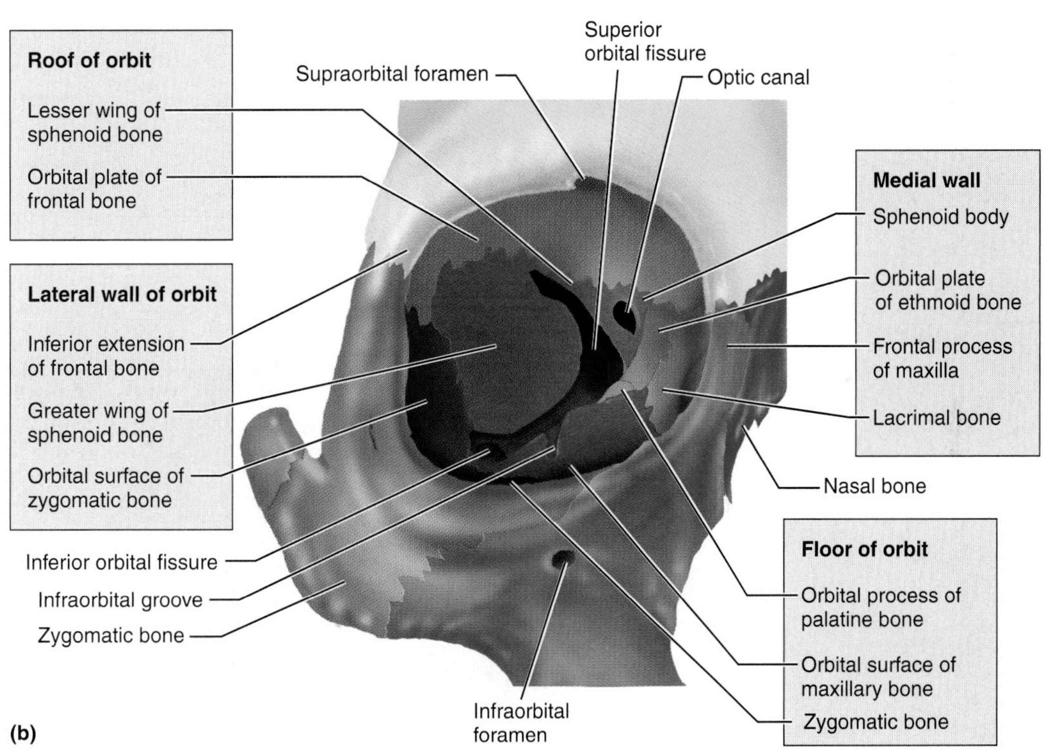

Special Parts of the Skull

Next we discuss four special parts of the skull: the orbits and the nasal cavity, which are restricted regions of the skull formed from many bones; the paranasal sinuses, which are extensions of the nasal cavity; and the hyoid bone.

Orbits

The **orbits** are cone-shaped bony cavities that hold the eyes, as well as the muscles that move the eyes, some fat, and the tear-producing glands. The walls of each orbit are formed by parts of seven bones—the frontal, sphenoid, zygomatic, maxilla, palatine, lacrimal, and ethmoid bones (see Figure 7.9). The superior and inferior orbital fissures, optic foramen, and lacrimal fossa (described earlier) are also in the orbit.

Nasal Cavity

The **nasal cavity** (Figure 7.10) is constructed of bone and cartilage. Its *roof* is the ethmoid's cribriform plates. The *floor* is formed by the palatine processes of the

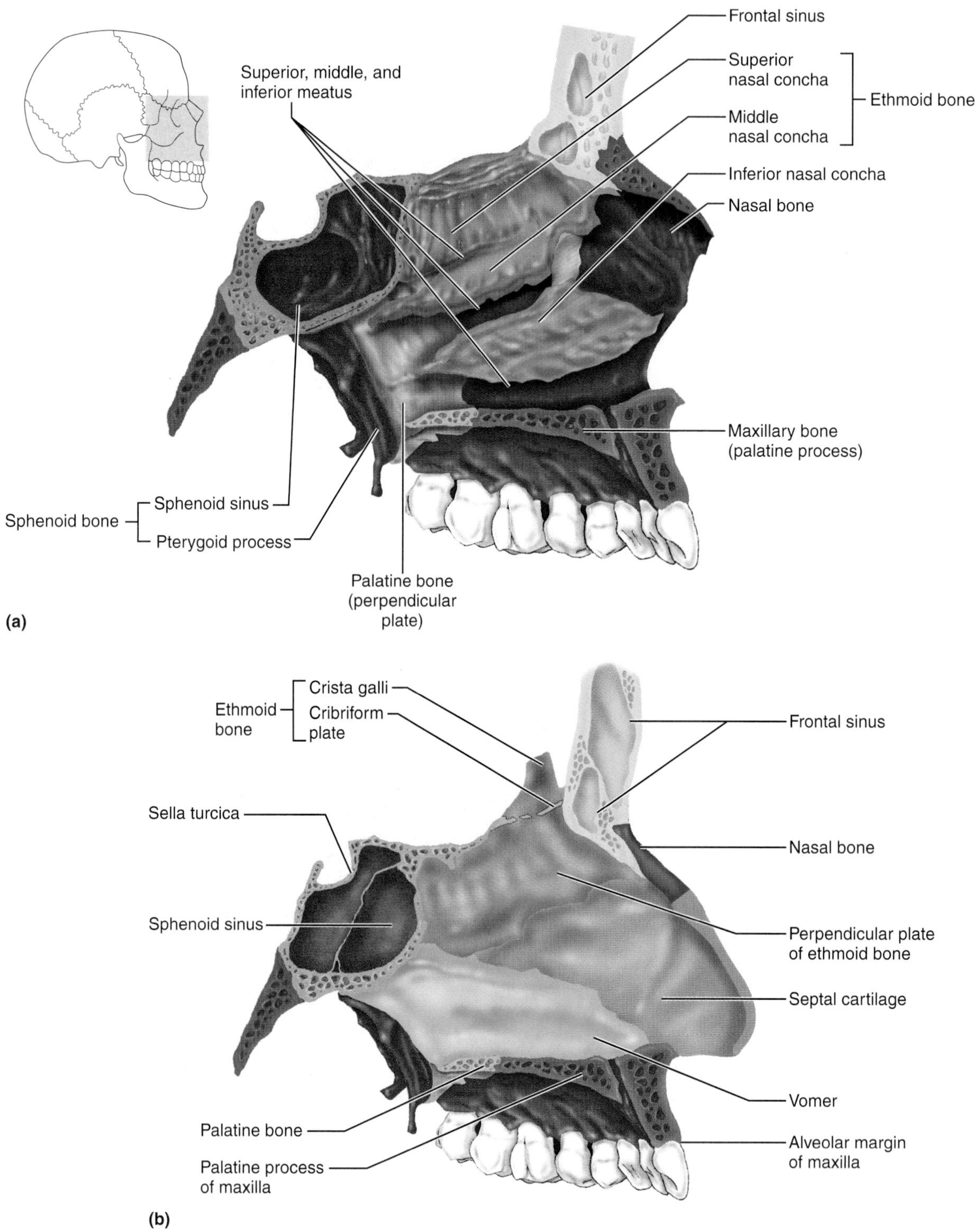

FIGURE 7.10 Bones of the nasal cavity. (a) Bones forming the left lateral wall of the nasal cavity (nasal septum removed). **(b)** Nasal cavity with nasal septum in place, showing how the ethmoid bone, septal cartilage, and vomer make up the septum. (See *Atlas of the Human Skeleton,* Figure 15.)

maxillae and the horizontal plates of the palatine bones. Keep in mind that these same nasal-floor structures also form the roof of the mouth and are collectively called the hard palate. Contributing to the *lateral walls* of the nasal cavity are the nasal bones, the superior and middle conchae of the ethmoid, the inferior nasal conchae, a part of the maxilla, and the perpendicular plates of the palatine bones (Figure 7.10a). On these

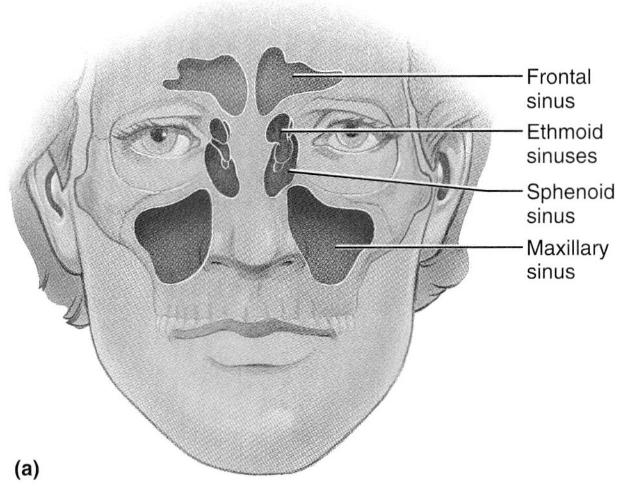

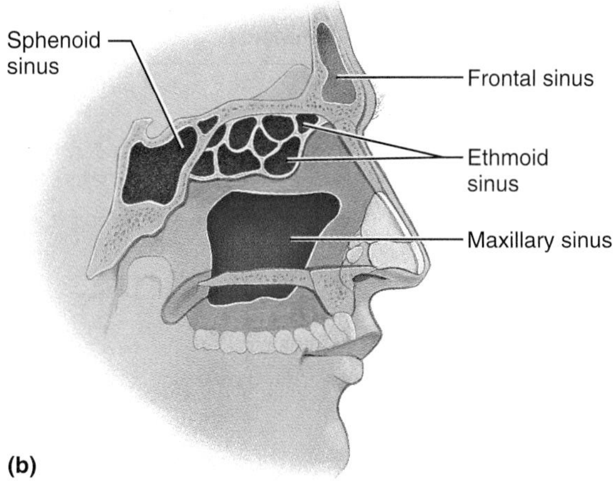

FIGURE 7.11 Paranasal sinuses.

lateral walls, each of the three conchae forms a roof over a groove-shaped air passageway called a *meatus* (me-a′tus; "a passage"). Therefore, there are **superior, middle,** and **inferior meatuses.**

Recall that the nasal cavity is divided into right and left halves by the nasal septum. The bony part of this septum is formed by the vomer inferiorly and by the perpendicular plate of the ethmoid superiorly (Figure 7.10b). A sheet of cartilage, called the *septal cartilage,* completes the septum anteriorly.

The walls of the nasal cavity are covered with a mucosa that moistens and warms inhaled air. This membrane also secretes mucus that traps dust, thereby cleansing the air of debris.

Critical Reasoning

What is the function of the three pairs of nasal conchae that project into the nasal cavity?

These scroll-shaped bones swirl the air flowing through the nasal cavity. This turbulence increases the contact of inhaled air with the mucosa throughout the nasal cavity, thereby increasing the efficiency with which the air is warmed, moistened, and filtered.

Paranasal Sinuses

Several skull bones—the frontal, ethmoid, sphenoid, and both maxillary bones—contain air-filled sinuses that are called **paranasal sinuses** (*para* = near) because they cluster around the nasal cavity (Figure 7.11). In fact, they are extensions of the nasal cavity, lined by the same mucous membrane and probably serving the same function of warming, moistening, and filtering inhaled air. The paranasal sinuses also lighten the skull, causing the bones they occupy to look rather moth-eaten in an X-ray image. These sinuses connect to the nasal cavity through small openings, most of which occur at the meatuses inferior to the conchae. For more on paranasal sinuses, see p. 608.

The Hyoid Bone

Not really part of the skull but associated with it, the **hyoid bone** (hi′oid; "U-shaped") lies just inferior to the mandible in the anterior neck (Figure 7.12). This bone resembles both an archer's bow and a miniature version of the mandible. It has a **body** and two pairs of **horns,** each of which is also called a *cornu,* the Latin word for "horn." The hyoid is the only bone in the skeleton that does not articulate directly with any other bone. Instead, its lesser horns are tethered by thin *stylohyoid ligaments* to the styloid processes of the temporal bones. Other ligaments connect the hyoid to the larynx (voice box) inferior to it. Functionally, the hyoid bone acts as a movable base for the tongue. Furthermore, its body and greater horns are points of attachment for neck muscles that raise and lower the larynx during swallowing.

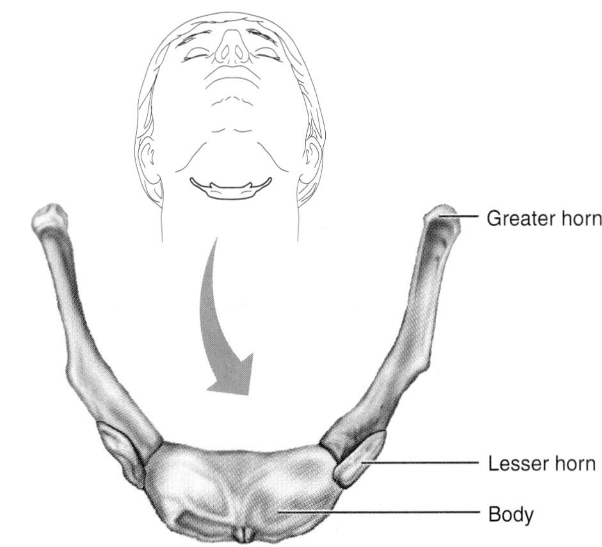

FIGURE 7.12 The hyoid bone.

Table 7.2 Bones of the Skull

Bone*	Comments	Important Markings
CRANIAL BONES		
Frontal (1) (Figures 7.2a, 7.3, and 7.4c)	Forms forehead; superior part of orbits and anterior cranial fossa; contains sinuses	**Supraorbital foramina (notches):** allow the supraorbital arteries and nerves to pass
Parietal (2) (Figures 7.2 and 7.3)	Forms most of the superior and lateral aspects of the skull	
Occipital (1) (Figures 7.2b, 7.3, and 7.4)	Forms posterior aspect and most of the base of the skull	**Foramen magnum:** allows the spinal cord to issue from the brain stem to enter the vertebral canal **Hypoglossal canal:** allows passage of the hypoglossal nerve (cranial nerve XII) **Occipital condyles:** articulate with the atlas (first vertebra) **External occipital protuberance** and **nuchal lines:** sites of muscle attachment **External occipital crest:** attachment site of ligamentum nuchae
Temporal (2) (Figures 7.3, 7.4, and 7.5)	Forms inferolateral aspects of the skull and contributes to the middle cranial fossa; has squamous, mastoid, tympanic, and petrous regions	**Zygomatic process:** helps to form the zygomatic arch, which forms the prominence of the cheek **Mandibular fossa:** articular point of the mandibular condyle **External auditory (acoustic) meatus:** canal leading from the external ear to the eardrum **Styloid process:** attachment site for hyoid bone and several neck muscles **Mastoid process:** attachment site for several neck and tongue muscles **Stylomastoid foramen:** allows cranial nerve VII (facial nerve) to pass **Jugular foramen:** allows passage of the internal jugular vein and cranial nerves IX, X, and XI **Internal acoustic meatus:** allows passage of cranial nerves VII and VIII **Carotid canal:** allows passage of the internal carotid artery
Sphenoid (1) (Figures 7.2a, 7.3, 7.4, and 7.6)	Keystone of the cranium; contributes to the middle cranial fossa and orbits; main parts are the body, greater wings, lesser wings, and pterygoid processes	**Sella turcica:** hypophyseal fossa portion is the seat of the pituitary gland **Optic foramina:** allow passage of cranial nerve II and the opthalmic arteries **Superior orbital fissures:** allow passage of cranial nerves III, IV, VI, part of V (opthalmic division), and opthalmic vein **Foramen rotundum:** allows passage of the maxillary division of cranial nerve V **Foraman ovale:** allows passage of the mandibular division of cranial nerve V **Foramen spinosum:** allows passage of the middle meningeal artery

Bone*	Comments	Important Markings
Ethmoid (1) (Figures 7.2a, 7.3, 7.4c, 7.7, and 7.10)	Helps to form the anterior cranial fossa; forms part of the nasal septum and the lateral walls and roof of the nasal cavity; contributes to the medial wall of the orbit	**Crista galli:** attachment point for the falx cerebri, a dural membrane fold **Cribriform plates:** allow passage of filaments of the olfactory nerves (cranial nerve 1) **Superior** and **middle nasal conchae:** form part of lateral walls of nasal cavity; increase turbulence of air flow
Ear ossicles (malleus, incus, and stapes) (2 each)	Found in middle ear cavity; involved in sound transmission; see Chapter 16	
FACIAL BONES		
Mandible (1) (Figures 7.2a, 7.3, and 7.8a)	The lower jaw	**Coronoid processes:** insertion points for the temporalis muscles **Mandibular condyles:** articulate with the temporal bones in the jaw (temporomandibular) joints **Mandibular symphysis:** medial fusion point of the mandibular bones **Alveoli:** sockets for the teeth **Mandibular foramina:** permit the inferior alveolar nerves to pass **Mental foramina:** allow blood vessels and nerves to pass to the chin and lower lip
Maxilla (2) (Figure 7.2a, 7.3, 7.4, and 7.8b)	Keystone bones of the face; form the upper jaw and parts of the hard palate, orbits, and nasal cavity walls	**Alveoli:** sockets for teeth **Zygomatic processes:** help form the zygomatic arches **Palatine processes:** form the anterior hard palate; meet medially in middle palatine suture **Frontal processes:** form part of lateral aspect of bridge of nose **Incisive fossa:** permits blood vessels and nerves to pass through hard palate (fused palatine processes) **Inferior orbital fissures:** permit maxillary branch of cranial nerve V, the zygomatic nerve, and blood vessels to pass **Infraorbital foramen:** allows passage of infraorbital nerve to skin of face
Zygomatic (2) (Figures 7.2a, 7.3a, 7.4a)	Form the cheek and part of the orbit	
Nasal (2) (Figures 7.2a and 7.3)	Construct the bridge of the nose	
Lacrimal (2) (Figures 7.2a and 7.3a)	Form part of the medial orbit wall	**Lacrimal fossa:** houses the lacrimal sac, which helps to drain tears into the nasal cavity
Palatine (2) (Figures 7.3b, 7.4a, and 7.8c)	Form posterior part of the hard palate and a small part of nasal cavity and orbit walls	
Vomer (1) (Figures 7.2a and 7.10b)	Part of the nasal septum	
Inferior nasal conchae (2) (Figures 7.2a and 7.10a)	Form part of the lateral walls of the nasal cavity	

*The color code beside each bone name corresponds to the bone's color in Figures 7.2 to 7.10. The number in parentheses ( ) following the bone name denotes the total number of such bones in the body.

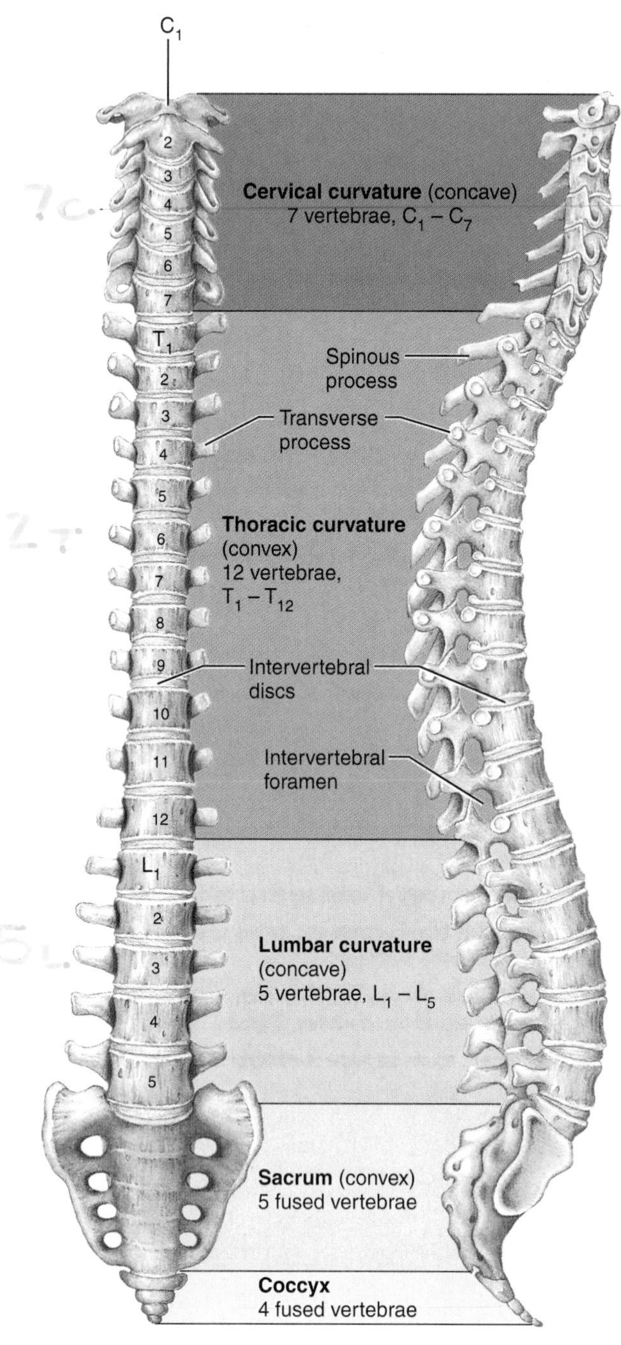

FIGURE 7.13 **The vertebral column.** Note the four curvatures in the lateral view at right. The terms *convex* and *concave* are relative to the *posterior* aspect of the column. (See *Atlas of the Human Skeleton,* Figure 17.)

THE VERTEBRAL COLUMN

The **vertebral column** (Figure 7.13), also called the *spinal column* or *spine,* is formed from 26 bones connected into a flexible, curved structure. The main support of the body axis, the vertebral column extends from the skull to the pelvis, where it transmits the weight of the trunk to the lower limbs. It also surrounds and protects the delicate spinal cord and provides attachment points for the ribs and for muscles of the neck and back. In the fetus and infant, the vertebral column consists of 33 separate bones, or **vertebrae** (ver′te-bre). Inferiorly, nine of these eventually fuse to form two composite bones, the sacrum and the tiny coccyx (tailbone). The remaining 24 bones persist as individual vertebrae separated by *intervertebral discs* (discussed shortly).

Like a tremulous television transmitting tower, the vertebral column cannot stand upright by itself. It must be held in place by an elaborate system of supports. Serving this role are the strap-like ligaments of the back and the muscles of the trunk. (The trunk muscles are discussed in Chapter 11.) The major supporting ligaments are the **anterior** and **posterior longitudinal ligaments** (Figure 7.14), which run vertically along the anterior and posterior surfaces of the bodies of the vertebrae, from the neck to the sacrum. The anterior longitudinal ligament is wide and attaches strongly to both the bony vertebrae and the intervertebral discs. Along with its supporting role, this thick anterior ligament prevents hyperextension of the back (bending too far backward). The posterior longitudinal ligament, which is narrow and relatively weak, attaches only to the intervertebral discs. This ligament helps to prevent hyperflexion (bending the vertebral column too sharply forward).

Several other posterior ligaments connect each vertebra to those immediately superior and inferior (Figure 7.14a). Of these, the **ligamentum flavum** (fla′vum; "yellow"), which contains elastic connective tissue, is especially strong: It stretches as we bend forward, then recoils as we straighten to an erect position.

Intervertebral Discs

Each **intervertebral disc** (see Figure 7.14) is a cushion-like pad composed of an inner sphere, the **nucleus pulposus** (pul-po′sus; "pulp"), and an outer collar of about 12 concentric rings, the **annulus (anulus) fibrosus** (an′u-lus fi-bro′sus; "fibrous ring"). Each nucleus pulposus is gelatinous and acts like a rubber ball, enabling the spine to absorb compressive stress. In the anulus fibrosus, the outer rings consist of ligament and the inner ones consist of fibrocartilage. The main function of these rings is to contain the nucleus pulposus, limiting its expansion when the spine is compressed. However, the rings also bind the successive vertebrae together, resist tension on the spine, and absorb compressive forces themselves. Collagen fibers in adjacent rings in the anulus cross like an X, allowing the spine to withstand twisting. This arrangement creates the same anti-twisting design provided by bone lamellae in osteons (see Figure 6.7 on p. 139).

The intervertebral discs act as shock absorbers during walking, jumping, and running, and they allow the spine to bend. At points of compression, the discs flatten and bulge out a bit between the vertebrae. The discs

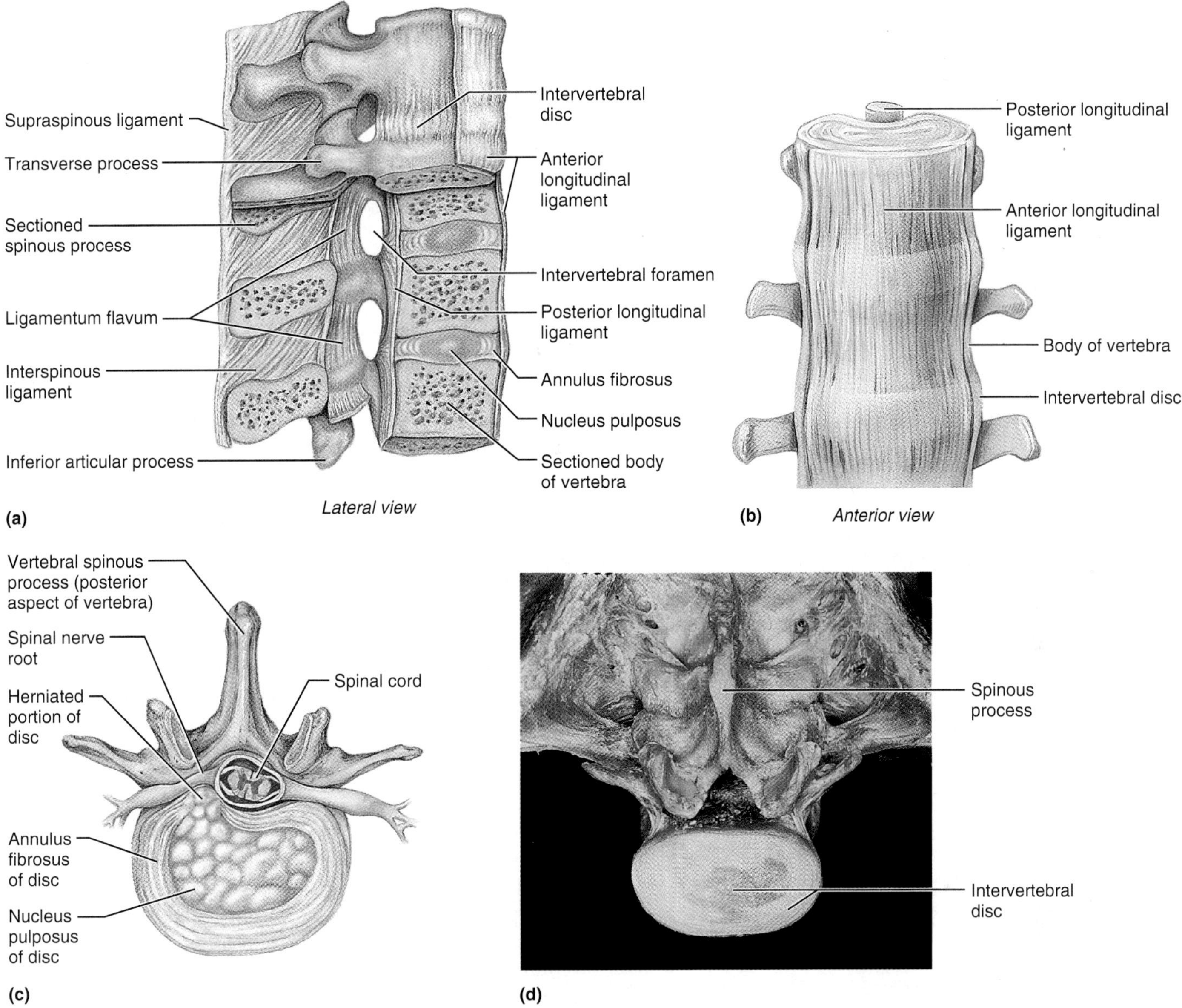

FIGURE 7.14 Ligaments and intervertebral discs of the spine. **(a)** Lateral view of part of the spinal column (anteror is to the right). The lower vertebrae have been cut sagittally to reveal the vertebral canal and the associated ligaments. Some of these ligaments are quite elastic (for example, the ligamentum flavum). Also note the parts of the intervertebral discs. **(b)** Anterior view of part of the spinal column, showing the anterior longitudinal ligament. **(c)** A herniated intervertebral disc. **(d)** Photograph of an intervertebral disc from a cadaver.

are thickest in the lumbar (lower back) and cervical (neck) regions of the vertebral column, a property that enhances the flexibility of these regions. Collectively, the intervertebral discs make up about 25% of the height of the vertebral column. They flatten somewhat by the end of each day, so our height is 1 to 2 centimeters less at night than when we awake in the morning.

Clinical Applications

HERNIATED DISC Severe or sudden physical trauma to the spine—such as results, for example, from lifting a heavy object—may cause one more **herniated discs** (also called a **prolapsed disc** or, in common terms, a slipped disc). This condition usually involves rupture of the anulus fibrosus followed by protrusion of the nucleus pulposus. Aging is a contributing factor, because the nucleus pulposus loses its cushioning properties over time, and the anulus fibrosus weakens and tears. This mechanical fatigue allows the nucleus to rupture out through the anulus. The anulus is thinnest posteriorly, so the nucleus usually herniates in that direction. Actually, the posterior longitudinal ligament prevents the herniation from proceeding directly posteriorly, so the rupture proceeds postero*laterally*—toward the spinal nerve roots exiting from the spinal cord (see Figure 7.14c). The resulting pressure on these nerve roots causes pain

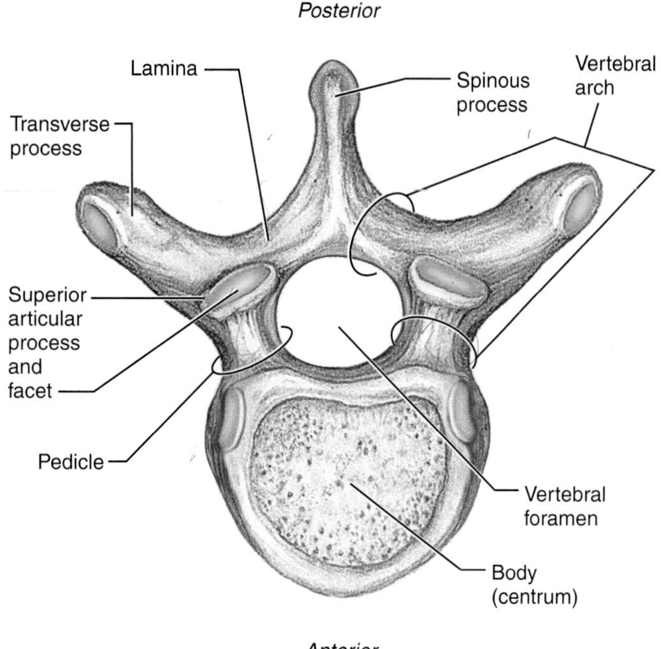

FIGURE 7.15 Structure of a typical vertebra. Superior view.

or numbness. In most cases, the pain eventually resolves itself, so treatment is usually conservative. Bedrest and traction are no longer recommended, but moderate exercise, massage, heat therapy, and painkillers are. If these treatments fail, the herniated disc may be removed surgically, and bone grafts are used to fuse the adjacent vertebrae.

In many of these cases, back pain does not result from the disc herniating, but instead from small nerves that enter tears in the disc on newly invading veins. A new, simple, and low-pain treatment called IDET (intradiscal electrothermal annuloplasty) threads a fine catheter with a heated tip into the disc and burns away the invading nerves at the same time that it seals the tears in the anulus rings. ✚

Regions and Normal Curvatures

The vertebral column, which is about 70 cm (28 inches) long in an average adult, has five major regions (see Figure 7.13). The 7 vertebrae of the neck are the **cervical vertebrae,** the next 12 are the **thoracic vertebrae,** and the 5 that support the lower back are the **lumbar vertebrae.** Remembering the common meal times of 7 A.M., 12 noon, and 5 P.M. can help you recall the number of vertebrae in these three regions. The vertebrae become progressively larger from the cervical to the lumbar region, as they must support progressively more weight. Inferior to the lumbar vertebrae is the **sacrum** (sa′krum), which articulates with the hip bones of the pelvis. The most inferior part of the vertebral column is the tiny **coccyx** (kok′siks).

All people have seven cervical vertebrae. Variations in numbers of vertebrae in the other regions occur in about 5% of people.

Critical Reasoning

Which region of the vertebral column (cervical, thoracic, or lumbar) is most likely to experience a herniated disc?

Most herniated discs occur in the lumbar region, which continually experiences the largest compressive stresses. Lumbar vertebrae must support the weight of the entire upper body, plus whatever heavy objects one lifts with the upper body.

From a lateral view, four curvatures that give the vertebral column an S-shape are visible (see Figure 7.13, right). The **cervical** and **lumbar curvatures** are concave posteriorly, whereas the **thoracic** and **sacral curvatures** are convex posteriorly. These curvatures increase the resilience of the spine, allowing it to function like a spring rather than a straight, rigid rod.

General Structure of Vertebrae

Vertebrae from all regions share a common structural pattern (Figure 7.15). A vertebra consists of a **body,** or **centrum,** anteriorly, and a **vertebral arch** posteriorly. The disc-shaped body is the weight-bearing region. Together, the body and vertebral arch enclose an opening called the **vertebral foramen.** Successive vertebral foramina of the articulated vertebrae form the long *vertebral canal,* through which the spinal cord passes.

The vertebral arch is a composite structure formed by two pedicles and two laminae. The sides of the arch are **pedicles** (ped′ĭ-k′lz; "little feet"), short bony walls that project posteriorly from the vertebral body. The **laminae** (lam′ĭ-ne; "sheets") are flat roof plates that complete the arch posteriorly.

Seven different processes project from each vertebral arch. The **spinous process,** or **vertebral spine,** is a median, posterior projection arising at the junction of the two laminae. A **transverse process** projects laterally from each pedicle-lamina junction. Both the spinous and transverse processes are attachment sites for muscles that move the vertebral column and for ligaments that stabilize it. The paired **superior** and **inferior articular processes** protrude superiorly and inferiorly, respectively, from the pedicle-lamina junctions. The inferior articular processes of each vertebra form movable joints with the superior articular processes of the vertebra immediately inferior. Thus, successive vertebrae are joined by intervertebral discs and by their articular processes. The smooth joint surfaces of these processes are *facets* ("little faces").

The pedicles have notches on their superior and inferior borders, forming lateral openings between adjacent vertebrae called **intervertebral foramina** (see Figure 7.13). The spinal nerves issuing from the spinal cord pass through these foramina.

Regional Vertebral Characteristics

The different regions of the spine perform slightly different functions, so vertebral structure shows regional variation. In general, the types of movements that can occur between vertebrae are (1) flexion and extension (anterior bending and posterior straightening of the spine), (2) lateral flexion (bending the upper body to the right or left), and (3) rotation, in which the vertebrae rotate on one another in the long axis of the vertebral column. Refer to Table 7.3 on p. 174 while reading the following discussions of vertebral characteristics.

Cervical Vertebrae

The seven cervical vertebrae, identified as C_1–C_7, are the smallest, lightest vertebrae. The first two, C_1 and C_2, are unusual, so we will skip them for the moment. The "typical" cervical vertebrae, C_3–C_7, have the following distinguishing features:

1. The body is wider laterally than in the anteroposterior dimension.
2. Except in C_7, the spinous process is short, projects directly posteriorly, and is bifid (bi′fid; "cleaved in two"); that is, split at its tip.
3. The vertebral foramen is large and generally triangular.
4. Each transverse process contains a hole, a **transverse foramen,** through which the vertebral blood vessels pass. These vessels ascend and descend through the neck to help serve the brain.
5. The superior articular facets face superoposteriorly, whereas the inferior articular facets face inferoanteriorly. The orientation of these articulations allows the neck to carry out an extremely wide range of movements: flexion and extension, lateral flexion, and rotation.

The spinous process of C_7 is not bifid and is much larger than those of the other cervical vertebrae (see Figure 7.17a). Because its large spinous process can be seen and felt through the skin, C_7 is called the **vertebra prominens** ("prominent vertebra") and is used as a landmark for counting the vertebrae in living people.

The first two cervical vertebrae are the *atlas* (C_1) and the *axis* (C_2) (Figure 7.16). No intervertebral disc lies between them, and they have unique structural and functional features.

The **atlas** lacks a body and a spinous process. Essentially, it is a ring of bone consisting of **anterior** and **posterior arches,** plus a **lateral mass** on each side. Each lateral mass has articular facets on both its superior and inferior surfaces. The superior articular facets receive the occipital condyles of the skull. Thus, they "carry" the skull, just as the giant Atlas supported the heavens in Greek mythology. These joints participate in flexion and extension of the head on the neck, as when you nod "yes." The inferior articular facets form joints with the axis.

The **axis,** which has a body, a spinous process, and the other typical vertebral processes, is not as specialized as the atlas. In fact, its only unusual feature is the knob-like **dens** ("tooth"), or **odontoid process** (o-don′toid; "tooth-like"), projecting superiorly from its body. The dens is actually the "missing" body of the atlas that fuses with the axis during embryonic development. Cradled within the anterior arch of the atlas (see Figure 7.17a), the dens acts as a pivot for the rotation of the atlas and skull. Hence, this joint participates in rotating the head from side to side to indicate "no." *Axis* is a good name for the second cervical vertebra because its dens allows the head to rotate on the neck's axis.

CLINICAL APPLICATIONS

THE DENS AND FATAL TRAUMA In cases of severe head trauma in which the skull is driven inferiorly toward the spine, the dens may be forced into the brain stem, causing death. Alternatively, if the neck is jerked forward, as in an automobile collision, the dens may be driven posteriorly into the cervical spinal cord. This injury is also fatal. ✚

Thoracic Vertebrae

The 12 **thoracic vertebrae,** T_1–T_{12}, (see Figure 7.17b and Table 7.3), all articulate with ribs. Their other unique characteristics are:

1. From a superior view, the vertebral body is roughly heart-shaped. Laterally, each side of the vertebral body bears two half facets, or **demifacets** (dem′e-fas″ets) one at the superior edge and the other at the inferior edge, for the heads of the ribs. Vertebra T_1 differs from this general pattern in that its body bears a full facet for the first rib and a demifacet for the second rib; furthermore, the bodies of T_{10}–T_{12} have only single facets to receive their respective ribs.
2. The spinous process is long and points inferiorly.
3. The vertebral foramen is circular.
4. With the exception of T_{11} and T_{12}, the transverse processes have facets that articulate with the tubercles of the ribs (see Figure 7.17b).
5. The superior and inferior articular facets lie mainly in the frontal plane; that is, the superior facets face posteriorly, while the inferior facets face anteriorly. Such articulations essentially prohibit flexion and extension, but they allow rotation between successive vertebrae. Much of the ability to rotate the trunk comes from the thoracic region of the vertebral column. Lateral flexion is also possible but is limited by the ribs.

Table 7.3 Regional Characteristics of Cervical, Thoracic, and Lumbar Vertebrae

Characteristic	Cervical[a] (3-7)	Thoracic[b]	Lumbar[c]
Body	Small, wide side to side	Larger than cervical; heart shaped; bears two costal demifacets	Massive; kidney shaped
Spinous process	Short; bifid; projects directly posteriorly	Long; sharp; projects inferiorly	Short; blunt; rectangular; projects directly posteriorly
Vertebral foramen	Triangular	Circular	Triangular
Transverse processes	Contain foramina	Bear facets for ribs (except T_{11} and T_{12})	Thin and tapered
Superior and inferior articulating processes	Superior facets directed superoposteriorly	Superior facets directed posteriorly	Superior facets directed posteromedially (or medially)
	Inferior facets directed inferoanteriorly	Inferior facets directed anteriorly	Inferior facets directed anterolaterally (or laterally)
Movements allowed	Flexion and extension; lateral flexion; rotation; the spine region with the greatest range of movement	Rotation; lateral flexion possible but limited by ribs; flexion and extension prevented	Flexion and extension; some lateral flexion; rotation prevented

SUPERIOR VIEW

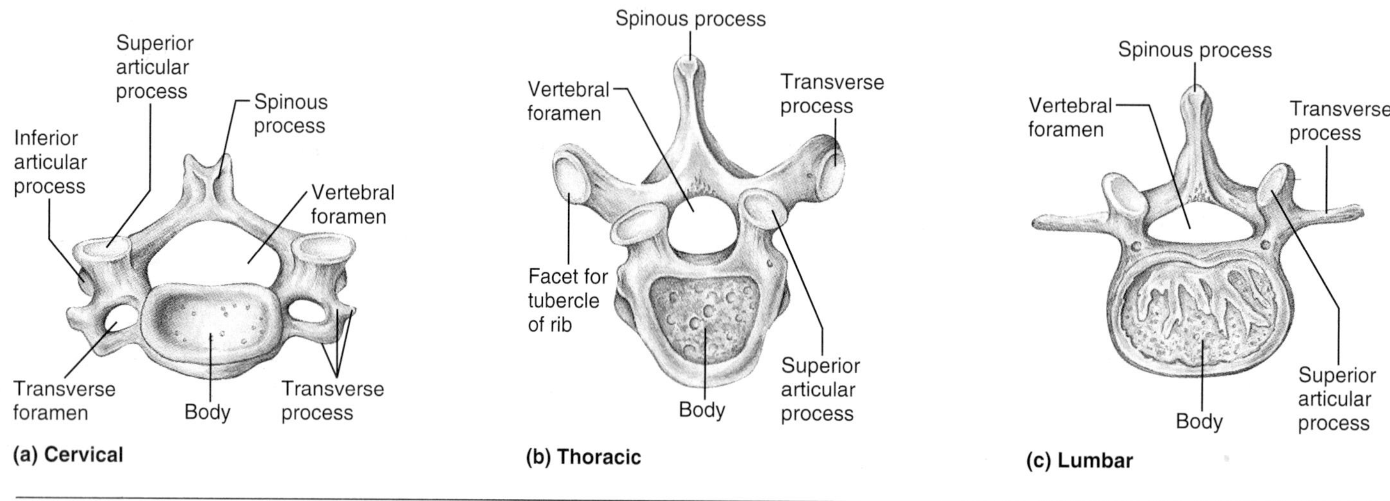

(a) Cervical **(b) Thoracic** **(c) Lumbar**

RIGHT LATERAL VIEW

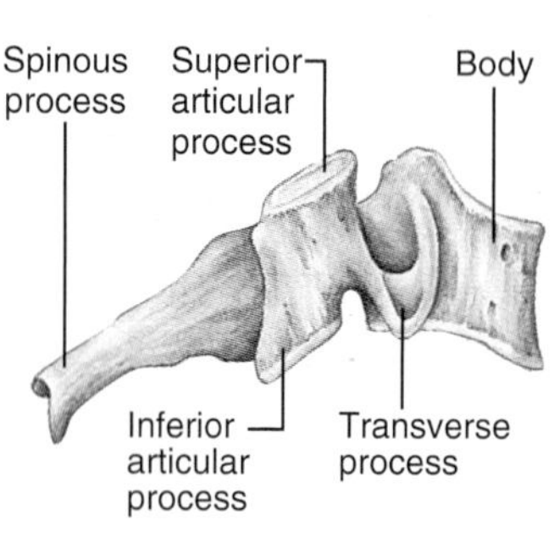

(a) Cervical

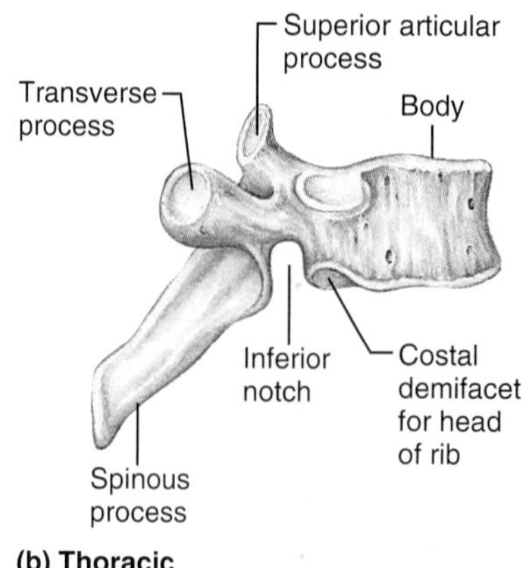

(b) Thoracic

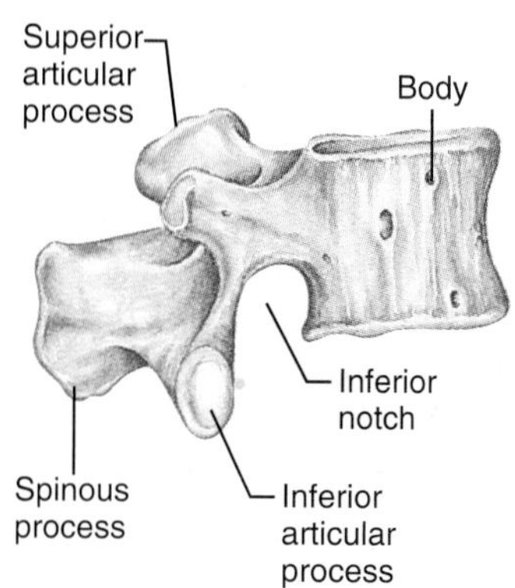

(c) Lumbar

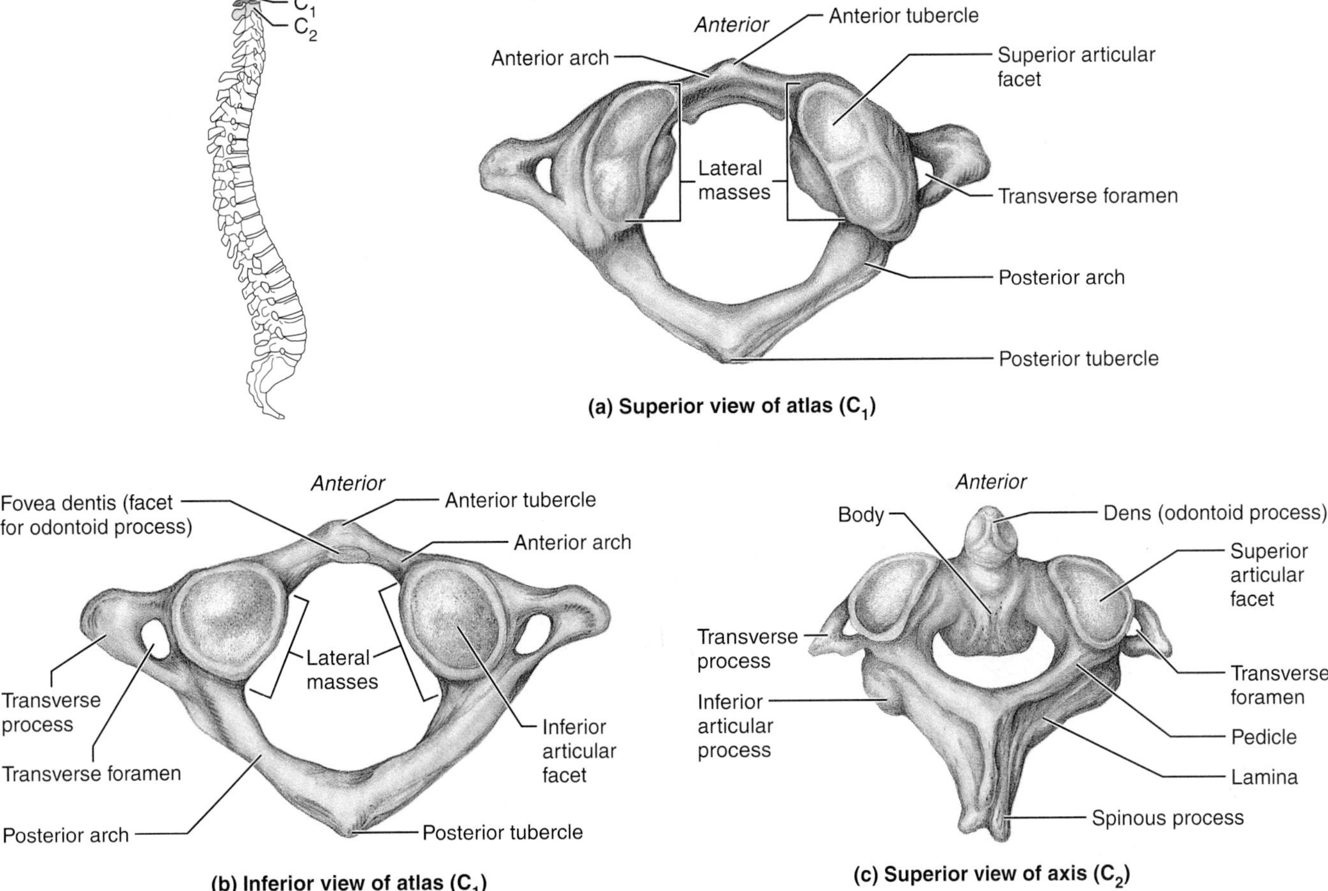

FIGURE 7.16 The first and second cervical vertebrae. In each vertebra shown, the anterior aspect is above. (See *Atlas of the Human Skeleton,* Figure 18.)

Lumbar Vertebrae

The lumbar region of the vertebral column, the area commonly referred to as the small of the back, receives the most stress. The enhanced weight-bearing function of the five **lumbar vertebrae** (L_1–L_5) is reflected in their sturdy structure: Their bodies are massive and appear kidney-shaped from a superior view (see Table 7.3). Their other characteristics are:

1. The pedicles and laminae are shorter and thicker than those of other vertebrae.
2. The spinous processes are short, flat, and hatchet-shaped, and they project straight posteriorly. These processes are robust for the attachment of large back muscles.
3. The vertebral foramen is triangular.
4. The superior articular facets face posteromedially (or medially), whereas the inferior articular facets face anterolaterally (or laterally) (see Figure 7.17c). Such articulations provide stability by preventing rotation between the lumbar vertebrae. Flexion and extension are possible, however. The lumbar region flexes, for example, when you do sit-ups or bend forward to pick a coin up off the ground. Additionally, lateral flexion is allowed by this spinal region.

Critical Reasoning

In your anatomy course, you may be handed an isolated vertebra and asked to determine whether it is cervical, thoracic, or lumbar. Based on what you have just read, devise a reliable scheme for distinguishing these three types of vertebrae.

You are encouraged to create your own scheme, but here is one that works: First, look for a transverse foramen in each transverse process (if one is present, you have a cervical vertebra). Second, examine the vertebral body for evidence of demifacets or facets for ribs (if these are present, you have a thoracic vertebra). Third, if neither of the first two features is present, look for the distinctive features of lumbar vertebrae (massive body, hatchet-shaped spinous process, and so on).

Sacrum

The curved, triangular **sacrum** shapes the posterior wall of the pelvis (Figure 7.18). It is formed by five fused vertebrae (S_1–S_5). Superiorly, it articulates with L_5 through a pair of **superior articular processes** and an intervertebral disc. Inferiorly, it joins the coccyx.

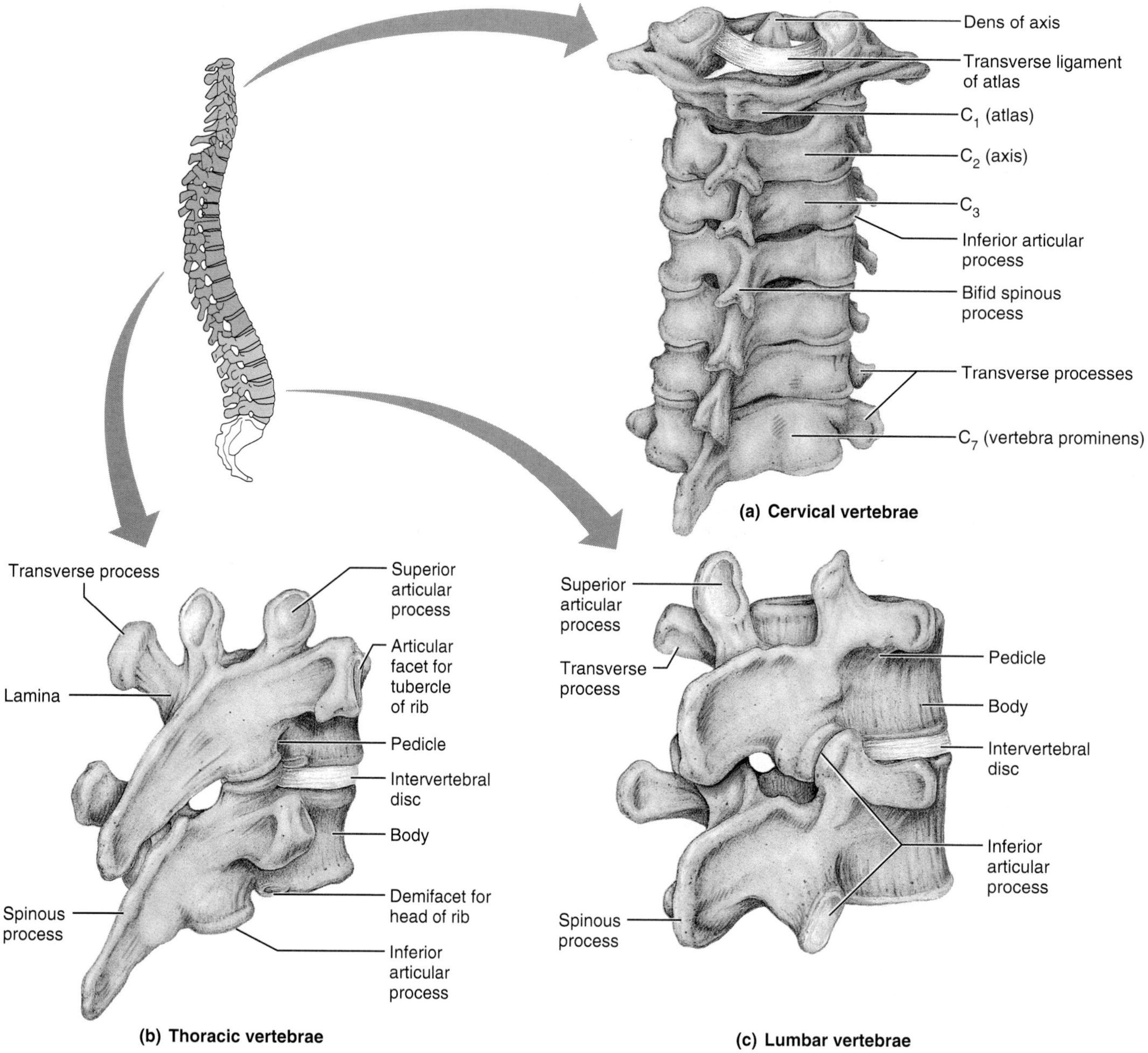

FIGURE 7.17 Posterolateral views of articulated vertebrae. In (a), note the prominent spinous process of C_7, the vertebra prominens. (See *Atlas of the Human Skeleton,* Figures 19, 20, and 21.)

The anterosuperior margin of the first sacral vertebra bulges anteriorly into the pelvic cavity as the **sacral promontory** (prom′on-tor″e; "a high point of land projecting into the sea"). The human body's center of gravity lies about 1 cm posterior to this landmark. Four **transverse lines** cross the anterior surface of the sacrum, marking the lines of fusion of the sacral vertebrae. The **ventral sacral foramina** transmit the ventral divisions (ventral rami) of the sacral spinal nerves. The large region lateral to these foramina is simply called the *lateral part.* This part expands superiorly into the flaring **ala** (a′lah; "wing"), which develops from fused *rib* elements of S_1–S_5. (Embryonic rib elements form in association with *all* vertebrae, although they only become true ribs in the thorax. Elsewhere, they fuse into the ventral surfaces of the transverse processes.) The alae articulate with the two hip bones to form the *sacroiliac* (sa″kro-il′e-ak) *joints* of the pelvis.

On the dorsal surface, in the midline, is the bumpy **median sacral crest,** which represents the fused spinous processes of the sacral vertebrae. Lateral to it are the **dorsal sacral foramina,** which transmit the dorsal rami of the sacral spinal nerves. Just lateral to these is the **lateral sacral crest,** representing the tips of the transverse processes of the sacral vertebrae.

The vertebral canal continues within the sacrum as the **sacral canal.** The laminae of the fifth (and sometimes the fourth) sacral vertebrae fail to fuse medially,

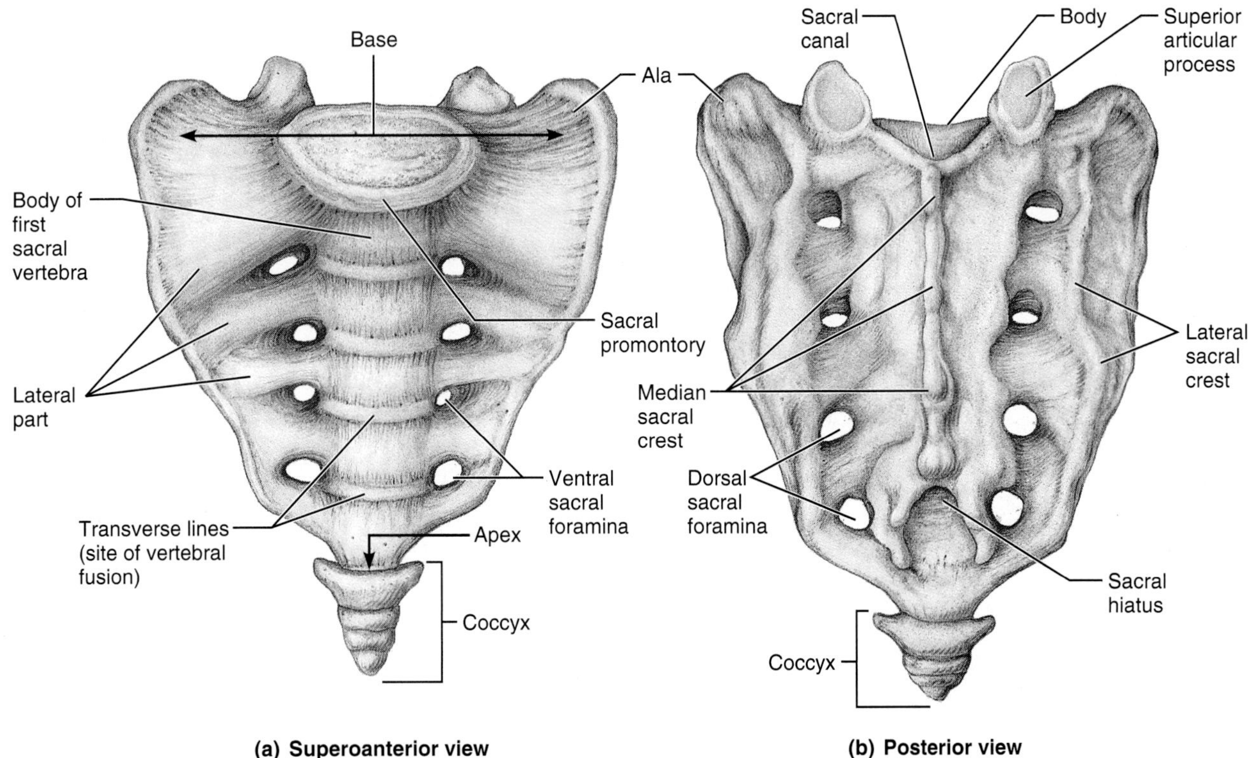

FIGURE 7.18 The sacrum and coccyx. (See *Atlas of the Human Skeleton,* Figure 22.)

leaving an enlarged external opening called the **sacral hiatus** (hi-a′tus; "gap") at the inferior end of the sacral canal.

Coccyx

The **coccyx,** or tailbone, is small and triangular (see Figure 7.18). The name *coccyx* is from a Greek word for "cuckoo," and the bone was so named because of a fancied resemblance to a bird's beak. The coccyx consists of four (or in some cases, three or five) vertebrae fused together. Except for the slight support it affords the pelvic organs, it is an almost useless bone. Injury to the coccyx, such as from an abrupt fall, is extremely painful. Occasionally, a baby is born with an unusually long coccyx. In most such cases, this bony "tail" is discreetly snipped off by a physician.

THE BONY THORAX

The bony framework of the chest (thorax), called the **bony thorax** or **thoracic cage,** is roughly cone-shaped and includes the thoracic vertebrae posteriorly, the ribs laterally, and the sternum and costal cartilages anteriorly (Figure 7.19). The bony thorax forms a protective cage around the heart, lungs, and other thoracic organs. It also supports the shoulder girdles and upper limbs and provides attachment points for many muscles of the back, neck, chest, and shoulders. In addition, the *intercostal spaces* (*inter* = between; *costa* = the ribs) are occupied by the intercostal muscles, which lift and depress the thorax during breathing.

Sternum

The **sternum** (breastbone) lies in the anterior midline of the thorax. Vaguely resembling a dagger, it is a flat bone about 15 cm long consisting of three sections: the manubrium, body, and xiphoid process. The **manubrium** (mah-nu′bre-um; "knife handle"), the superior section, is shaped like the knot in a necktie. Its *clavicular notches* articulate with the clavicles (collarbones) superolaterally. Just below this, the manubrium also articulates with the first and second ribs. The **body,** or midportion, makes up the bulk of the sternum. It is formed from four separate bones, one inferior to the other, that fuse after puberty. The sides of the sternal body are notched where it articulates with the costal cartilages of the second to seventh ribs. The **xiphoid process** (zif′oid; "sword-like") forms the inferior end of the sternum. This tongue-shaped process is a plate of hyaline cartilage in youth, and it does not fully ossify until about age 40. In some people, the xiphoid process projects dorsally. Blows to the chest can push such a xiphoid into the underlying heart or liver, causing massive hemorrhage.

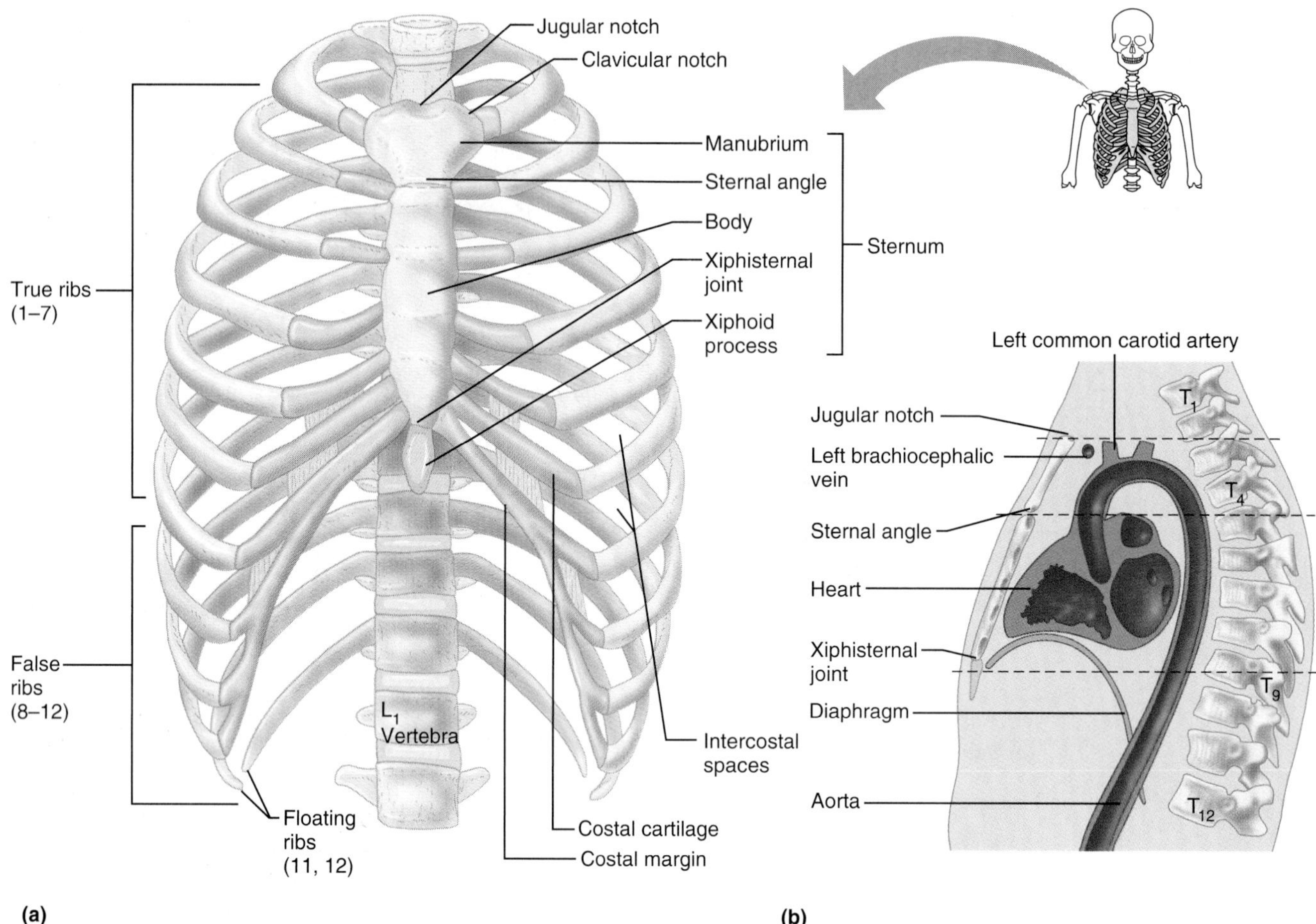

FIGURE 7.19 The bony thorax. (a) Anterior view. (b) Midsagittal section through the thorax, showing the relation of key parts of the sternum (sternal angle, xiphisternal joint) to the vertebral column. (See *Atlas of the Human Skeleton,* Figure 23.)

The sternum has three important anatomical landmarks: the jugular notch, the sternal angle, and the xiphisternal joint (see Figure 7.19a). The **jugular notch,** also called the *suprasternal notch,* is the central indentation in the superior border of the manubrium. It generally lies in the same horizontal plane as the disc between the second and third thoracic vertebrae (see Figure 7.19b). The **sternal angle** is a horizontal ridge across the anterior suface of the sternum, where the manubrium joins the body. This fibrocartilage joint acts like a hinge, allowing the sternal body to swing anteriorly when we inhale. The sternal angle is in line with the disc between the fourth and fifth thoracic vertebrae. Anteriorly, it lies at the level of the second ribs. It is a handy reference point for finding the second rib and thus for providing orientation during a physical examination (Chapter 26, p. 779). The **xiphisternal** (zif″ĭ-ster′nal) **joint** is where the sternal body and xiphoid process fuse. It lies at the level of the ninth thoracic vertebra.

Ribs

Twelve pairs of **ribs** (Latin: *costa)* form the flaring sides of the thoracic cage (see Figure 7.19a). All ribs attach to the thoracic vertebrae posteriorly and run anteroinferiorly to reach the front of the chest. The superior seven pairs, which attach directly to the sternum by their *costal cartilages,* are the **true ribs,** or *vertebrosternal* (ver″tĕ-bro-ster′nal) *ribs.* The inferior five pairs, ribs 8–12, are called **false ribs** because they attach to the sternum either indirectly or not at all. Ribs 8–10 attach to the sternum indirectly, as each joins the costal cartilage above it; these are called *vertebrochondral* (ver″tĕ-bro-kon′dral) *ribs.* Ribs 11 and 12 are called **floating ribs** or *vertebral ribs* because they have no anterior attachments. Instead, their costal cartilages lie embedded in the muscles of the lateral body wall. The ribs increase in length from pair 1 to 7, then decrease in length from pair 8 to 12. The inferior margin of the rib cage, or **costal margin,** is formed by the costal cartilages of ribs 7 to 10. The right and left costal margins diverge from the region of the xiphisternal joint, where they form the *infrasternal angle* (*infra* = below).

A typical rib is a bowed flat bone (Figure 7.20a–c). The bulk of a rib is simply called the **shaft.** Its superior border is smooth, but its inferior border is sharp and thin and has a **costal groove** on its inner face that lodges the intercostal nerves and vessels. In addition to the shaft, each rib has a head, neck, and tubercle. The wedge-shaped **head** articulates with the vertebral bodies by two facets: One facet joins the body of the thoracic vertebra of the same number; the other joins the body of the vertebra immediately superior. The **neck** of a rib is the short, constricted region just lateral to the head. Just lateral to the neck, the knob-like **tubercle** articulates with the transverse process of the thoracic vertebra of the same number. Lateral to the tubercle, the shaft angles sharply anteriorly (at the **angle** of the rib) and extends to the costal cartilage anteriorly. The costal cartilages provide secure but flexible attachments of ribs to the sternum and contribute to the elasticity of the thoracic cage.

The *first rib* is atypical, as it is flattened from superior to inferior and is quite broad (Figure 7.20d). It forms a horizontal table to support the subclavian vessels, the large artery and vein servicing the upper limb, which groove its superior surface. There are other exceptions to the typical rib pattern: Rib 1 and ribs 10–12 articulate with only one vertebral body, and ribs 11 and 12 do not articulate with a vertebral transverse process.

DISORDERS OF THE AXIAL SKELETON

Now we briefly discuss two kinds of disorders of the axial skeleton: abnormal spinal curvatures and a narrowing of the vertebral canal.

Abnormal Spinal Curvatures

There are several types of abnormal spinal curvatures. Some are congenital (present at birth), whereas others result from disease, poor posture, or unequal pull of muscles on the spine.

Scoliosis (sko″le-o′sis; "twisted disease") is an abnormal *lateral* curvature of more than 10 degrees that occurs most often in the thoracic region. One type, of unknown cause, is common in adolescents, particularly girls. Other, more severe cases result from abnormally structured vertebrae, lower limbs of unequal length, or muscle paralysis. If muscles on one side of the back are nonfunctional, those on the opposite side pull unopposed and move the spine out of alignment. To prevent permanent deformity, scoliosis is treated with body braces or surgery before the child stops growing. Severe scoliosis can compress a lung, causing difficulty in breathing.

Kyphosis (ki-fo′sis; "humped disease"), or hunchback, is an exaggerated thoracic curvature that is most common in aged women because it often results from spinal fractures that follow osteoporosis. It may also result from either tuberculosis of the spine or osteomalacia.

Lordosis (lor-do′sis: "bent-backward disease"), or swayback, is an accentuated lumbar curvature. Temporary lordosis is common in people carrying a "large load in front," such as obese men and pregnant women. Like kyphosis, it can result from spinal tuberculosis or osteomalacia.

Stenosis of the Lumbar Spine

The condition known as **stenosis of the lumbar spine** (ste-no′sis; "narrowing") is a narrowing of the vertebral canal in the lumbar region. This disorder can compress the roots of the spinal nerves and cause back pain. The narrowing may result from degenerative or arthritic changes in the vertebral joints and ligaments, or be caused by bone spurs projecting into the canal.

THE AXIAL SKELETON THROUGHOUT LIFE

The membrane bones of the skull begin to ossify late in the second month of development. In these flat bones, bone tissue grows outward from **ossification centers** within the mesenchyme membranes. At birth, the skull bones remain incomplete and are separated by still-unossified remnants of the membranes, called **fontanels** (fon″tah-nelz′) (Figure 7.21). The major fontanels—the **anterior, posterior, mastoid,** and **sphenoidal fontanels**—allow the skull to be compressed slightly as the infant passes through the narrow birth canal, and they accommodate brain growth in the baby. A baby's pulse surging in these "soft spots" feels like gushing water (*fontanel* means "little fountain"). The large, diamond-shaped anterior fontanel can be felt for 1½ to 2 years after birth. The others usually are replaced by bone by the end of the first year.

Whereas the bones of the skullcap and face form directly from mesenchyme by intramembranous ossification, those at the base of the skull develop from cartilage by endochondral ossification. More specifically, the skull's endochondral bones are the occipital (except for its extreme posterosuperior part), the petrous and mastoid parts of the temporal, most of the sphenoid, and the ethmoid.

At birth, the skull bones are thin. The frontal bone and the mandible begin as paired bones that fuse medially during childhood. The tympanic part of the temporal bone is merely a C-shaped ring in the newborn.

Several congenital abnormalities may distort the skull. Most common is **cleft palate** (p. 70), in which the right and left halves of the palate fail to join medially. This defect leaves an opening between the mouth and

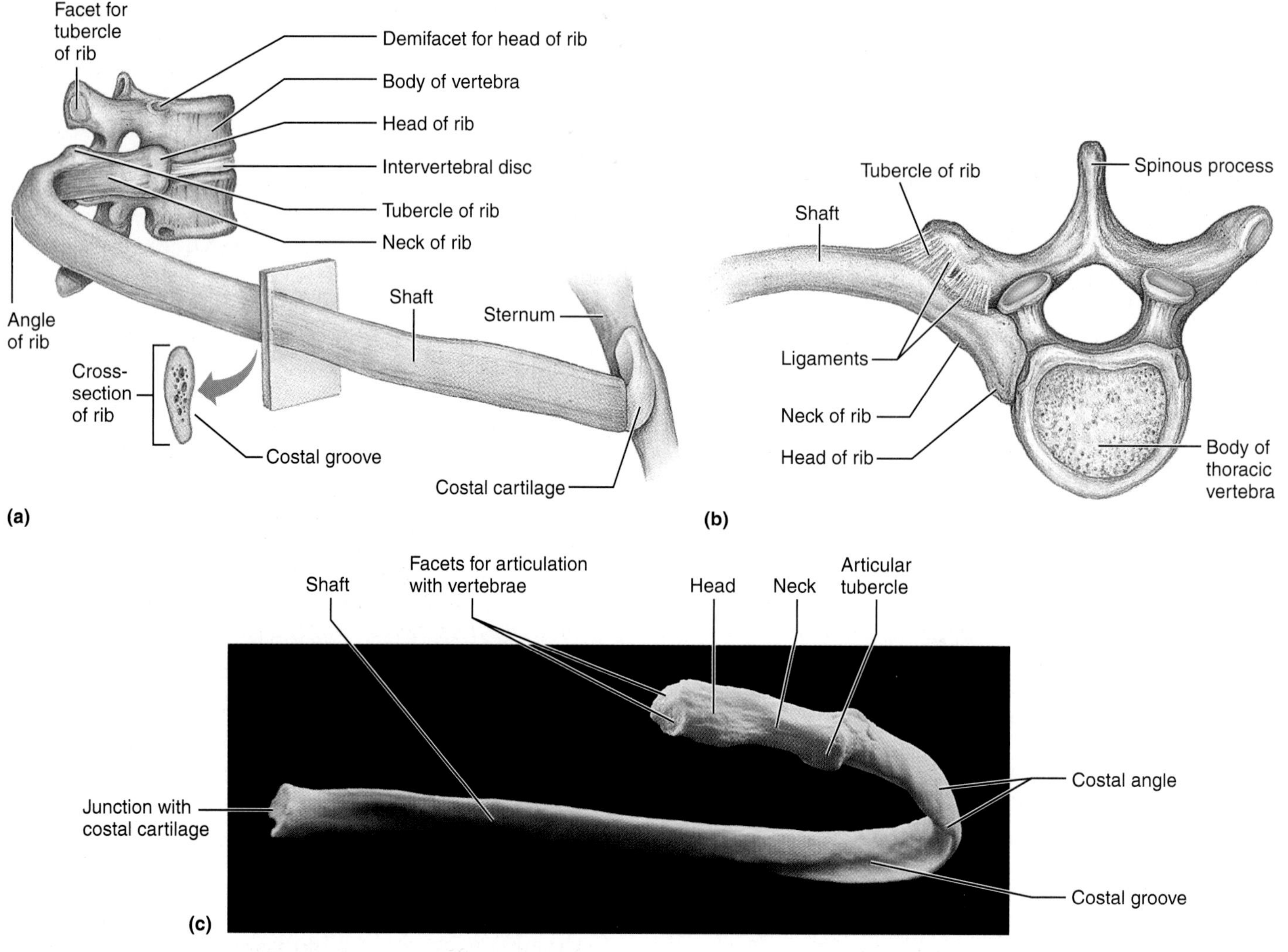

FIGURE 7.20 Ribs. **(a)** Lateral view of a typical true rib (ribs 2 through 7), showing its articulations with the vertebral column and sternum. **(b)** Superior view of a rib articulating with a thoracic vertebra. **(c)** A typical rib (rib 6, right), posterior view. **(d)** Rib 1, superior view. All ribs illustrated in this figure are right ribs. (See *Atlas of the Human Skeleton,* Figure 20f and g.)

the nasal cavity that interferes with sucking and the baby's ability to nurse. Cleft palate can lead to aspiration (inhalation) of food into the nasal cavity and lungs, resulting in pneumonia. Often accompanied by cleft lip (Figure 7.22, p. 182), it is repaired surgically.

Advances in Understanding It has recently been found that the likelihood of cleft palate is minimized if the mother takes multivitamins containing folic acid during early pregnancy. ✷

The skull changes throughout life, but the changes are most dramatic during childhood. At birth, the baby's cranium is huge relative to its small face (Figure 7.21). The maxillae and mandible are comparatively tiny, and the contours of the face are flat. A baby's cranium is big because it fits around the large and rapidly growing brain. By 9 months of age, the skull is already half its adult size. By 2 years, it is three-quarters of its full size, and by 8 or 9 years, the cranium is almost full-sized. However, between the ages of 6 and 13, the head appears to enlarge substantially because the face literally grows out from the skull as the jaws, cheekbones, and nose become more prominent. The enlargement of the face correlates with an expansion of the nose, paranasal sinuses, and chewing muscles, plus the development of the large permanent teeth.

In the vertebral column, only the thoracic and sacral curvatures are well-developed at birth. Both of these *primary curvatures* are convex posteriorly, so that an infant's spine arches (is C-shaped) like that of a four-legged animal. The *secondary curvatures,* the cervical and lumbar curvatures, are concave posteriorly and

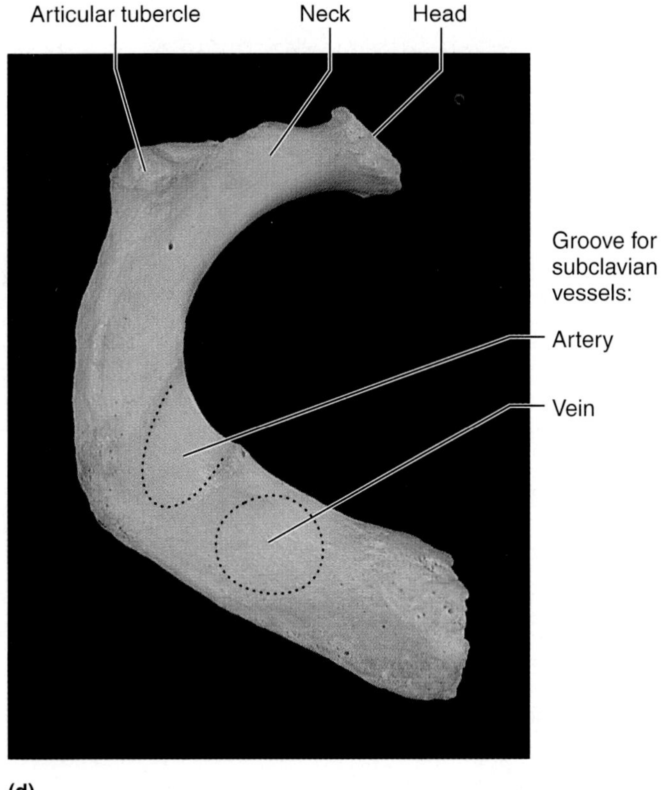

FIGURE 7.20, *continued.*

develop during the first 2 years of childhood. They result from a reshaping of the intervertebral discs rather than from modification of the vertebrae themselves. The cervical curvature is present before birth, but is not pronounced until the baby starts to lift its head at 3 months, and the lumbar curvature develops when the baby begins to walk at about 1 year. The lumbar curvature positions the weight of the upper body over the lower limbs, providing optimal balance during standing.

Problems with the vertebral column, such as lordosis or scoliosis, may appear during the early school years, when rapid growth of the long limb bones stretches many muscles. Children normally have a slight lordosis because holding the abdomen anteriorly helps to counterbalance the weight of their relatively large heads, held slightly posteriorly. During childhood, the thorax grows wider, but true adult posture (head erect, shoulders back, abdomen in, and chest out) does not develop until adolescence.

Aging affects many parts of the skeleton, especially the spine. The water content of the intervertebral discs declines. As the discs become thinner and less elastic, the risk of herniation increases. By 55 years, a loss of several centimeters from a person's height is common. Further shortening of the trunk can be produced by osteoporosis of the spine that leads to kyphosis, which is sometimes called "dowager's hump" in the elderly.

The thorax becomes more rigid with increasing age, largely because the costal cartilages ossify. This loss of elasticity of the rib cage leads to shallow breathing, which in turn leads to less efficient gas exchange in the lungs.

Recall that all bones lose mass with age. Skull bones lose less mass than most, but they lose enough to change the facial contours of the elderly. As the bony tissue of the mandible and maxilla decreases, the jaws come to look small and child-like once again. If an elderly person loses his or her teeth, this loss of bone from the jaws is accelerated because the bone of the alveolar region (tooth sockets) is reabsorbed.

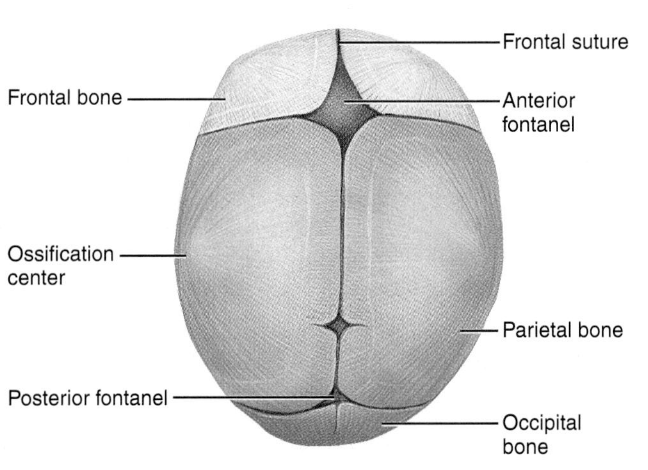

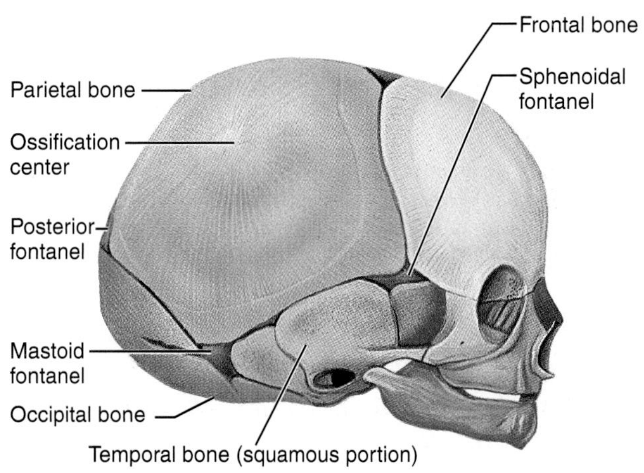

FIGURE 7.21 **Skull of a newborn.** Note the fontanels and ossification centers. (See *Atlas of the Human Skeleton,* Figure 16.)

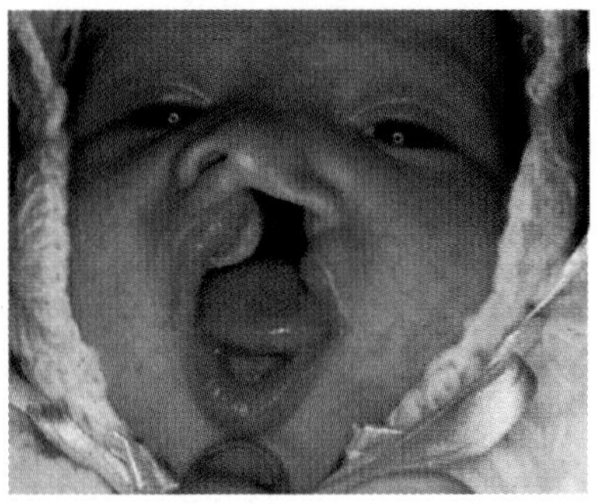

(a)

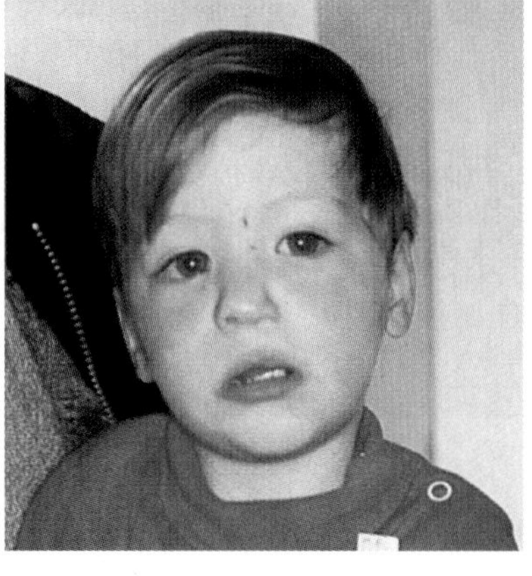

(b)

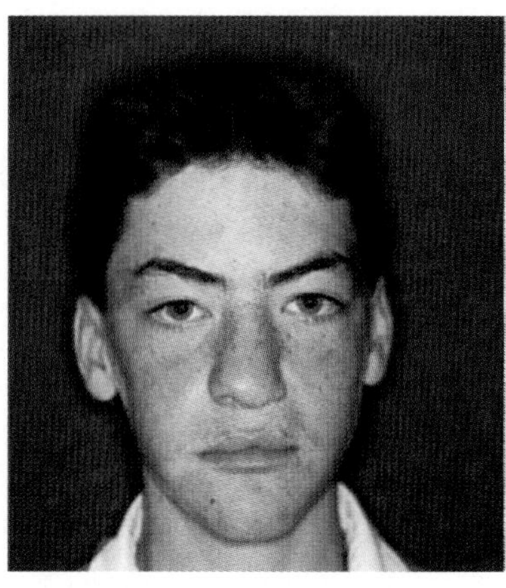

(c)

FIGURE 7.22 Cleft palate. **(a)** A boy born with a cleft palate and lip; **(b)** the boy as a toddler, following surgical repair during infancy; **(c)** by adolescence, there was little evidence of the original defect.

RELATED CLINICAL TERMS

Craniotomy ("cutting the skull") Surgery to remove part of the cranium, usually done to gain access to the brain—for example, to remove a brain tumor, a blood clot, or a sample of brain tissue for a biopsy. The piece of skull is removed by drilling a series of round holes through the bone at regular intervals to outline a square and then cutting between these holes with a string-like saw. At the end of the operation, the square piece of bone is replaced, and it heals normally.

Laminectomy (lam″ĭ-nek′to-me; "lamina-cutting") Surgical removal of a vertebral lamina. Laminectomies are usually done to reach and relieve a herniated disc.

Orthopedist (or″tho-pe′dist) or **orthopedic surgeon** A physician who specializes in restoring lost function or repairing damage to bones and joints.

Spinal fusion Surgical procedure involving insertion of bone chips as a tissue graft to stabilize a portion of the vertebral column, particularly after the fracture of a vertebra or the prolapse of a disc.

Whiplash An injury to the neck region caused by a rapid sequence of hyperextension followed by hyperflexion of the neck. Often results when a car is hit from the rear. Damage and pain usually result from stretching of the ligaments and joints between the neck vertebrae, and torn neck muscles. Less commonly involved are herniated intervertebral discs and cracks in the vertebrae. Treatment typically involves use of a neck brace, painkillers and muscle-relaxing drugs, and physical therapy.

CHAPTER SUMMARY

Media study tools that could provide you additional help in reviewing specific key topics of Chapter 7 are referenced below.

SP = Study Partner; AIA = A.D.A.M.® Interactive Anatomy

1. The axial skeleton forms the long axis of the body. Its parts are the skull, vertebral column, and bony thorax. The appendicular skeleton, by contrast, consists of the pectoral (shoulder) and pelvic girdles and the long bones of the limbs.

2. Bone markings are landmarks that represent sites of muscle attachment, articulation, and passage of blood vessels and nerves (see Table 7.1, p. 153).

AIA Link: Atlas; View: Muscle Attachments; Skull (anterior).

THE SKULL (pp. 154–169)

1. The cranium forms the vault and base of the skull and protects the brain. The facial skeleton provides openings for the eyes, openings for respiratory and digestive passages, and areas of attachment for facial muscles.

AIA Link: Atlas; Region: Body Wall and Back; Muscle Attachments–Trunk (anterior).

2. Almost all bones of the skull are joined by immobile sutures.

3. The eight bones of the cranium are the paired parietal and temporal bones and the unpaired frontal, occipital, ethmoid, and sphenoid bones (see Table 7.2, pp. 168–169).

4. The 14 facial bones are the paired maxillae, zygomatics, nasals, lacrimals, palatines, and inferior nasal conchae, plus the unpaired mandible and vomer (see Table 7.2).

AIA Link: 3D Anatomy; 3D Skull: Maxilla–default.

5. The orbits and nasal cavity are complicated regions of the skull, each formed by many bones (see Figures 7.9 and 7.10).

6. The air-filled paranasal sinuses occur in the frontal, ethmoid, sphenoid, and maxillary bones. They are extensions of the nasal cavity, to which they are connected.
7. The bowed hyoid bone in the anterior neck serves as a movable base for the tongue, among other functions.

SP Exercise: Chapter 7, Anatomy of the Skull.

THE VERTEBRAL COLUMN (pp. 170–177)

1. The vertebral column includes 26 irregular bones: 24 vertebrae plus the sacrum and coccyx.
2. Intervertebral discs, with their nucleus pulposus cores and anulus fibrosus rings, act as shock absorbers and give flexibility to the spine. Herniated discs usually involve rupture of the anulus followed by protrusion of the nucleus.
3. The vertebral column has five major regions: the cervical vertebrae (7), the thoracic vertebrae (12), the lumbar vertebrae (5), the sacrum, and the coccyx. The cervical and lumbar curvatures are concave posteriorly, and the thoracic and sacral curvatures are convex posteriorly.
4. With the exception of C_1, all cervical, thoracic, and lumbar vertebrae have a body, two transverse processes, two superior and two inferior articular processes, a spinous process, and a vertebral arch.
5. Special characteristics distinguish the cervical, thoracic, and lumbar vertebrae from one another (see Table 7.3, p. 174). The orientation of the articular facets determines the movements possible in the different regions of the spinal column. The sacrum and coccyx, groups of fused vertebrae, contribute to the bony pelvis.

AIA Link: Atlas; System: Skeletal; Pelvic Surface of Sacrum.

6. The atlas and axis (C_1 and C_2) are atypical vertebrae. The ring-like atlas supports the skull and helps make nodding movements possible. The axis has a dens that helps the head to rotate.

THE BONY THORAX (pp. 177–179)

1. The bony thorax or thoracic cage includes the 12 pairs of ribs, the sternum, and the thoracic vertebrae.
2. The sternum consists of the manubrium, body, and xiphoid process. The sternal angle marks the joint between the manubrium and body, where the second ribs attach.
3. The first seven pairs of ribs are called true ribs; the rest are false ribs. The eleventh and twelfth are called floating ribs. A typical rib consists of a head with facets, a neck, a tubercle, and a shaft. A costal cartilage occurs on the ventral end of each rib.

AIA Link: Atlas; System: Skeletal; Muscle Attachments–Trunk (anterior).

Disorders of the Axial Skeleton (p. 179)

1. Scoliosis, kyphosis, and lordosis are three types of abnormal spinal curvature. Stenosis of the lumbar spine is a pathological narrowing of the vertebral canal in the lumbar region.

The Axial Skeleton Throughout Life (pp. 179–182)

1. Fontanels are membrane-covered regions between the cranial bones in the infant skull. They allow growth of the brain and ease birth passage.
2. Most skull bones are membrane bones. The skull's only endochondral bones are most of the occipital and sphenoid, the petrous and mastoid parts of the temporal, and the ethmoid.
3. Growth of the cranium early in life is closely related to the rapid growth of the brain. Enlargement of the face in late childhood follows tooth development and expansion of the nose, paranasal sinuses, and chewing muscles.
4. The vertebral column arches (is C-shaped) at birth. The cervical curvature becomes pronounced when a baby starts to lift its head, and the lumbar curvature appears when a toddler starts to walk.
5. In old age, the intervertebral discs thin. This, along with osteoporosis, leads to a gradual decrease in height.

REVIEW QUESTIONS

Multiple Choice/Matching Questions

1. Place the letters in column B next to the appropriate descriptions in column A. (Some blanks will require more than one letter.)

Column A	Column B
____ **(1)** connected by the coronal suture	**(a)** ethmoid
____ **(2)** keystone bone of the cranium	**(b)** frontal
____ **(3)** keystone bone of the face	**(c)** mandible
____ **(4)** form the hard palate	**(d)** maxillary
____ **(5)** foramen magnum	**(e)** occipital
____ **(6)** forms the chin	**(f)** palatine
____ **(7)** contain paranasal sinuses	**(g)** parietal
____ **(8)** contains mastoid sinus	**(h)** sphenoid
	(i) temporal

2. A facet is (a) a sharp, almost pointed projection, (b) a smooth, nearly flat articular surface, (c) a canal-like passageway, (d) the same as a condyle.
3. Herniated intervertebral discs tend to herniate (a) superiorly, (b) anterolaterally, (c) posterolaterally, (d) laterally.
4. The parts of the sternum that articulate at the sternal angle are (a) xiphoid and body, (b) xiphoid and manubrium, (c) manubrium and body, (d) second ribs.
5. The only rib whose shaft is flattened in the horizontal plane, instead of vertically, is the (a) first rib, (b) seventh rib, (c) eleventh rib, (d) twelfth rib.
6. The name of the first cervical vertebra is (a) atlas, (b) axis, (c) occiput, (d) vertebra prominens.

Short Answer Essay Questions

7. In the fetus, how do the relative proportions of the cranium and face compare with those in the skull of a young adult?
8. Name and diagram the four normal vertebral curvatures. Which are primary and which are secondary?
9. List two specific structural characteristics each for cervical, thoracic, and lumbar vertebrae that would enable anyone to identify each type correctly.
10. (a) What is the function of intervertebral discs? (b) Distinguish the anulus fibrosus from the nucleus pulposus of a disc. (c) Which part herniates in the condition called prolapsed disc?
11. Is a floating rib a true rib or a false rib? Explain.
12. Briefly describe the anatomical characteristics and impairment of function seen in cleft palate.
13. Compare the skeleton of a young adult to that of an 85-year-old person, with respect to bone mass in general and the basic bony structure of the skull, thorax, and vertebral column.
14. Identify what types of movement are allowed by the lumbar region of the vertebral column, and compare these with the movements allowed by the thoracic region.
15. List the bones in each of the three cranial fossae.
16. Describe the important features of the sternum.
17. Define these structures in a way that shows you can tell them apart: vertebral arch, vertebral canal, vertebral foramen, and intervertebral foramen.
18. Describe where the four major fontanels are located in relation to the major sutures of the skull.
19. Professor Ron Singer pointed to the foramen magnum of the skull and said, "That is where the food goes. The food passes down through this hole when you swallow." Some of the students believed him, but others said this was a big mistake. Can you correct his statement?

CRITICAL REASONING + CLINICAL APPLICATION QUESTIONS

1. Antonio was hit in the face with a bad-hop grounder during practice. An X-ray film revealed multiple fractures of the bones around an orbit. Name the bones that form the margins of an orbit.
2. Lindsey had polio as a child and was partly paralyzed in one lower limb for over a year. Although she is no longer paralyzed, she now has a severe lateral curvature of the lumbar spine. Explain what has happened, and name her condition.
3. Mr. Chester, a heavy beer drinker with a large potbelly, complained of severe lower back pains. After X rays were taken, he was found to have displacement of his lumbar vertebrae. What would this condition be called, and what would cause it?

A.D.A.M.® INTERACTIVE ANATOMY QUESTIONS

1. Link: Atlas; Region: Head and neck; cervical vertebrae (lateral). List the three sutures visible in this image that join the flat bones of the skull.
2. Link: Atlas; System: Skeletal; Walls of the orbit (lateral). Name the seven bones that form the orbit of the eye.

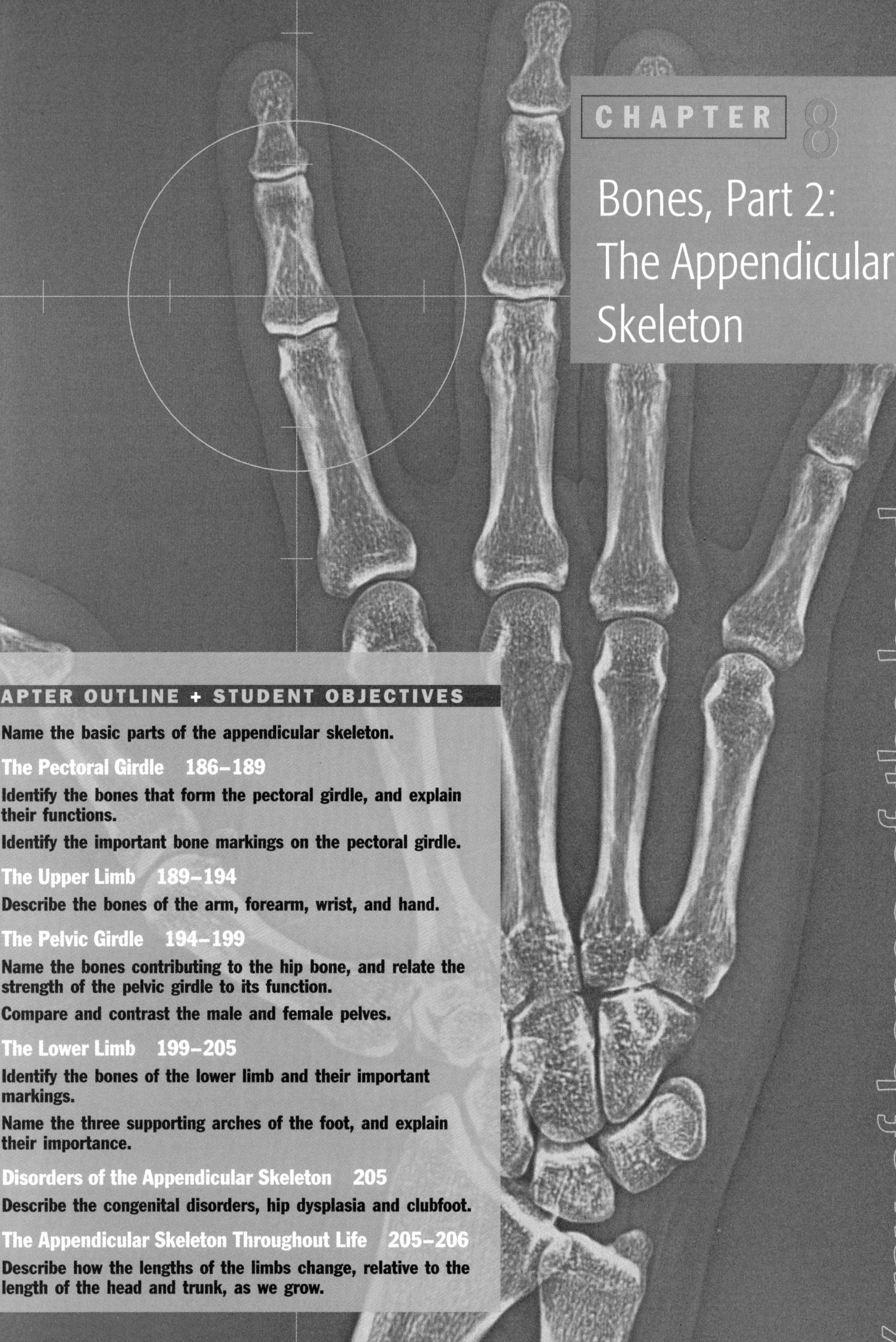

CHAPTER 8

Bones, Part 2: The Appendicular Skeleton

CHAPTER OUTLINE + STUDENT OBJECTIVES

1. Name the basic parts of the appendicular skeleton.

The Pectoral Girdle 186–189

2. Identify the bones that form the pectoral girdle, and explain their functions.
3. Identify the important bone markings on the pectoral girdle.

The Upper Limb 189–194

4. Describe the bones of the arm, forearm, wrist, and hand.

The Pelvic Girdle 194–199

5. Name the bones contributing to the hip bone, and relate the strength of the pelvic girdle to its function.
6. Compare and contrast the male and female pelves.

The Lower Limb 199–205

7. Identify the bones of the lower limb and their important markings.
8. Name the three supporting arches of the foot, and explain their importance.

Disorders of the Appendicular Skeleton 205

9. Describe the congenital disorders, hip dysplasia and clubfoot.

The Appendicular Skeleton Throughout Life 205–206

10. Describe how the lengths of the limbs change, relative to the length of the head and trunk, as we grow.

LIMB BONES AND THEIR GIRDLES ARE ***APPENDED,*** or attached, to the axial skeleton. Thus, they are collectively called the **appendicular skeleton** (see Figure 7.1, p. 152). The pectoral girdles (pek'tor-al; "chest") attach the upper limbs to the trunk, whereas the pelvic girdle secures the lower limbs. Although the bones of the upper and lower limbs differ in their functions, they share the same basic structural plan. That is, each limb is composed of three basic segments: the hand, forearm, and arm in the upper limb; and the foot, leg, and thigh in the lower limb.

The appendicular skeleton enables us to carry out the wide variety of movements typical of our freewheeling and object-manipulating lifestyle. Each time we take a step, throw a ball, or write with a pencil, we use our appendicular skeleton.

As you read about the appendicular skeleton, refer to Table 8.1 on page 194. We begin with the pectoral girdle and upper limb.

THE PECTORAL GIRDLE

The **pectoral girdle,** or *shoulder girdle,* consists of a *clavicle* (klav'ĭ-k'l) anteriorly and a *scapula* (skap'u-lah) posteriorly (Figure 8.1). The paired pectoral girdles and their associated muscles form the shoulders. The term *girdle* implies a belt completely circling the body, but these girdles do not quite satisfy this description: Anteriorly, the medial end of each clavicle joins the sternum and first rib, and the lateral ends of the clavicles join the scapulae at the shoulder. However, the two scapulae fail to complete the ring posteriorly, because their medial borders do not join each other or the axial skeleton.

Besides attaching the upper limb to the trunk, the pectoral girdle provides attachment for many muscles that move the limb. This girdle is light and allows the upper limbs to be quite mobile. This mobility springs from two factors:

1. Because only the clavicle attaches to the axial skeleton, the scapula can move quite freely across the thorax, allowing the arm to move with it.
2. The socket of the shoulder joint—the scapula's glenoid cavity—is shallow, so it does not restrict the movement of the humerus (arm bone). Although this arrangement is good for flexibility, it is bad for stability: Shoulder dislocations are fairly common.

Clavicles

The **clavicles** ("little keys"), or collarbones, are slender, slightly curved long bones that extend horizontally across the superior thorax (Figure 8.1a), just deep to the skin. Each clavicle is cone-shaped at its medial **sternal end,** which attaches to the manubrium, and flattened at its lateral **acromial** (ah-kro'me-al) **end,** which articulates with the scapula (Figure 8.1b and c). The medial two-thirds of the clavicle is convex anteriorly, whereas the lateral third is concave anteriorly. The superior surface is almost smooth, but the inferior surface is ridged and grooved for the ligaments and muscles that attach to it, many of which act to bind the clavicle to the rib cage and scapula. For example, the thick **trapezoid line** and the bold **conoid tubercle** near the acromial end provide attachment for a ligament that runs to the scapula's coracoid process (defined on page 188), and a roughened impression near the sternal end attaches to a ligament running to the first rib (costoclavicular ligament).

The clavicles perform several functions. Besides providing attachment for muscles, they act as braces; that is, they hold the scapulae and arms out laterally from the thorax. This function becomes obvious when a clavicle is fractured: The entire shoulder region collapses medially. The clavicles also transmit compression forces from the upper limbs to the axial skeleton, as when someone puts both arms forward and pushes a car to a gas station.

CLINICAL APPLICATIONS

FRACTURES OF THE CLAVICLE The clavicles are not very strong, and they often fracture. This can occur when a person falls on the lateral border of a shoulder, is hit directly on the clavicle in a contact sport, or uses outstretched arms to break a fall. The curves in the clavicle ensure that it usually fractures anteriorly (outward) at its middle third. If it were to fracture posteriorly (inward), bone splinters would pierce the main blood vessels to the arm, the subclavian vessels, which lie just deep to the clavicle. ✚

Scapulae

The **scapulae,** or shoulder blades, are thin, triangular flat bones (Figure 8.2). They lie on the dorsal surface of the rib cage, between rib 2 superiorly and rib 7 inferiorly. Each scapula has three borders. The **superior border** is the shortest and sharpest. The **medial border,** or *vertebral border,* parallels the vertebral column. The thick **lateral border,** or *axillary border,* abuts the axilla (armpit) and ends superiorly in a shallow fossa, the **glenoid cavity** (gle'noid; "pit-shaped")(Figure 8.2c). This cavity articulates with the humerus of the arm, forming the shoulder joint.

Like all triangles, the scapula has three corners, or *angles.* The glenoid cavity lies at the scapula's **lateral angle.** The **superior angle** is where the superior and medial borders meet, and the **inferior angle** is at the junction of the medial and lateral borders. The inferior angle moves as the arm is raised and lowered, and is an important landmark for studying scapular movements.

The anterior, or costal, surface of the scapula is slightly concave and relatively featureless. The posterior

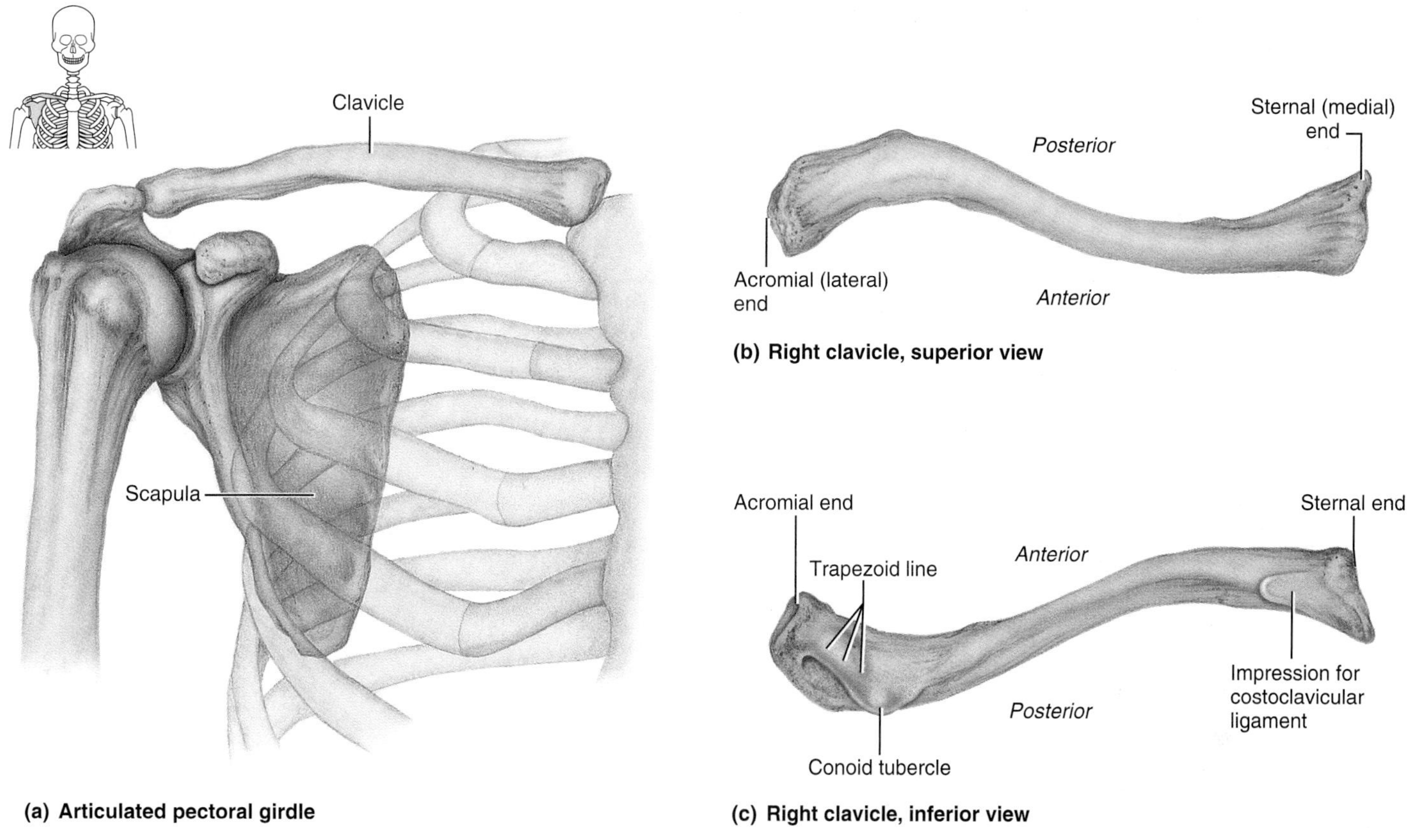

(a) Articulated pectoral girdle

(b) Right clavicle, superior view

(c) Right clavicle, inferior view

Posterior

Acromial end (lateral)

Sternal end (medial)

Anterior

(d)

Trapezoid line

Anterior

Facet for acromion (lateral)

Sternal end (medial)

Conoid tubercle

Posterior

For costoclavicular ligament

(e)

FIGURE 8.1 Overview of the pectoral girdle (a), and the clavicle (b–e). The changes in (b) and (c) match the photos in (d) and (e), respectively. (See *Atlas of the Human Skeleton,* Figure 24d and e.)

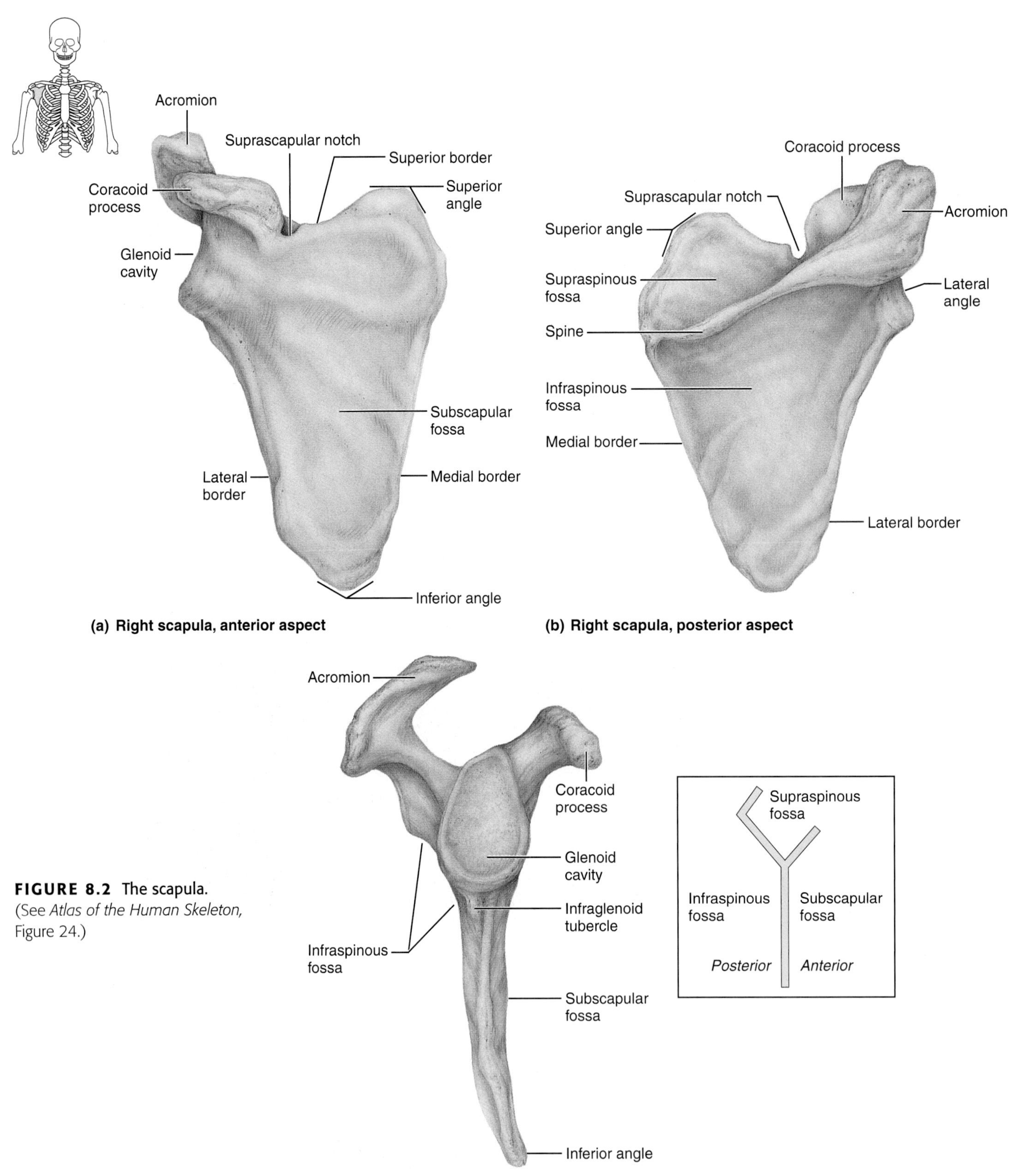

FIGURE 8.2 The scapula. (See *Atlas of the Human Skeleton,* Figure 24.)

surface, by contrast, bears a prominent **spine** that is easily felt through the skin. The spine ends laterally in a flat, triangular projection called the **acromion** (ah-kro′me-on; "apex of shoulder"), which articulates with the acromial end of the clavicle.

The **coracoid** (kor′ah-coid) **process** projects anteriorly from the lateral part of the superior scapular border. The root *corac* means "like a crow's beak," but this process looks more like a bent finger. It is an attachment point for the biceps muscle of the arm. Strong

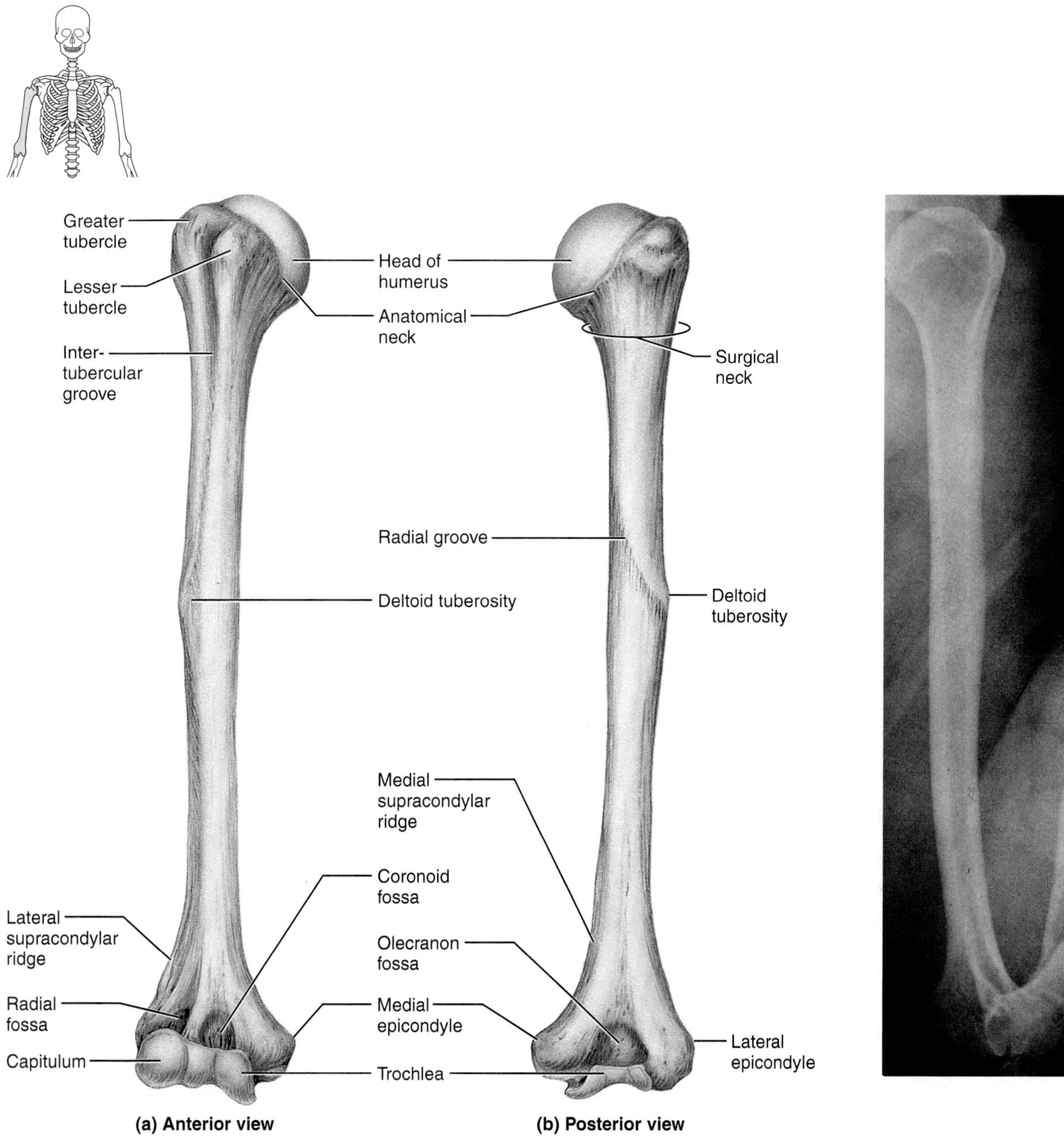

FIGURE 8.3 The humerus of the right arm: illustrations and an X ray. (See *Atlas of the Human Skeleton,* Figure 25.)

ligaments also bind the coracoid process to the clavicle. Just medial to the coracoid process lies the **suprascapular notch** (a nerve passageway), and just lateral to it lies the glenoid cavity.

Several large fossae occur on both surfaces of the scapula and are named according to location. The **infraspinous** and **supraspinous fossae** lie inferior and superior to the spine, respectively. The **subscapular fossa** is the shallow concavity formed by the entire anterior surface of the scapula.

THE UPPER LIMB

Thirty bones form the skeleton of the upper limb (see Figures 8.3 to 8.6). They are grouped into bones of the arm, forearm, and hand.

Arm

Anatomists use the term *arm* to designate the part of the upper limb between the shoulder and elbow only.

The **humerus** (hu'mer-us) is the only bone of the arm (Figure 8.3). The largest and longest bone in the upper limb, it articulates with the scapula at the shoulder and with the radius and ulna (forearm bones) at the elbow.

At the proximal end of the humerus is the hemispherical **head,** which fits into the glenoid cavity of the scapula. Just inferior to the head is a slight constriction, the **anatomical neck.** Just inferior to this are the lateral **greater tubercle** and the more-medial **lesser tubercle.** These tubercles are separated by the **intertubercular groove,** or *bicipital* (bi-sip'ĭ-tal) *groove.* The tubercles are sites where muscles attach. The intertubercular groove guides a tendon of the biceps muscle to its attachment point at the rim of the glenoid cavity. Just inferior to the tubercles is the **surgical neck,** so named because it is the most frequently fractured part of the humerus. About midway down the shaft, on the lateral side, is the **deltoid tuberosity.** This V-shaped, roughened area is an attachment site for the deltoid muscle of the shoulder. Near the deltoid tuberosity, a **radial groove** descends obliquely along the posterior surface of the shaft. This groove marks the course of the radial nerve, an important nerve of the upper limb.

At the distal end of the humerus are two condyles, a medial **trochlea** (trok'le-ah; "pulley") and a lateral **capitulum** (kah-pit'u-lum; "small head"). The trochlea looks like an hourglass turned on its side, and the capitulum is shaped like half a ball. These condyles articulate with the ulna and the radius, respectively. They are flanked by the **medial** and **lateral epicondyles,** which serve as attachment sites for muscles of the forearm. Directly above these epicondyles are **medial** and **lateral supracondylar ridges.**

Directly superior to the trochlea, on the posterior surface of the humerus, is the deep **olecranon** (o-lek'rah-non) **fossa.** In the corresponding position on the anterior surface is a shallower **coronoid** (kor'o-noid) **fossa.** A small **radial fossa** lies lateral to the coronoid fossa.

Forearm

Forming the skeleton of the forearm are two parallel long bones, the radius and ulna (Figures 8.4 and 8.5). Their proximal ends articulate with the humerus, whereas their distal ends reach the wrist. The radius and ulna articulate with each other both proximally and distally at the small *radioulnar* (ra″de-o-ul'nar) *joints.* Furthermore, they are interconnected along their entire length by a flat ligament called the **interosseous membrane** (in″ter-os'e-us; "between the bones"). In the anatomical position, the radius lies laterally (on the thumb side), and the ulna medially. However, when the palm faces posteriorly, the distal end of the radius crosses over the ulna, and the two bones form an X (see Figure 9.6a on p. 223).

Ulna

The ulna (ul'nah; "elbow"), which is slightly longer than the radius, has the main responsibility for forming the elbow joint with the humerus (Figure 8.5). It looks much like a monkey wrench. At its proximal end are two prominent projections, the **olecranon** ("elbow") **process** and **coronoid** ("crown-shaped") **process,** separated by a deep concavity, the **trochlear notch** (see Figure 8.5b). Together, these two processes grip the trochlea of the humerus, forming a hinge joint that allows the forearm to bend upon the arm (flex), then straighten again (extend). When the forearm is fully extended, the olecranon process "locks" into the olecranon fossa of the humerus. When the forearm is flexed, the coronoid process of the ulna fits into the coronoid fossa of the humerus. On the lateral side of the coronoid process is a smooth depression, the **radial notch,** where the ulna articulates with the head of the radius.

Distally, the shaft of the ulna narrows and ends in a knob-like **head** (Figure 8.5c). Medial to this is a **styloid** ("stake-shaped") **process,** from which a ligament runs to the wrist. The head of the ulna is separated from the bones of the wrist by a disc of fibrocartilage and plays little or no role in hand movements.

Radius

The **radius** ("rod") is thin at its proximal end and widened at its distal end—the opposite of the ulna. The proximal **head** of the radius is shaped like the end of a spool of thread (see Figures 8.4 and 8.5a). The superior surface of this head is concave, and it articulates with the capitulum of the humerus. The radial head is fractured more often than any other bony element of the elbow joint, usually when one falls onto an outstretched hand. Medially, the head of the radius articulates with the radial notch of the ulna at the *proximal radioulnar joint.* Just distal to the head is a rough bump, the **radial tuberosity,** a site of attachment of the biceps muscle of the arm. The distal part of the radius is flared (Figure 8.5c). It has a medial **ulnar notch,** which articulates with the ulna at the *distal radioulnar joint,* and a lateral **styloid process,** which anchors a ligament that runs to the wrist. The extreme distal end of the radius is concave and articulates with carpal bones of the wrist. Whereas the ulna contributes heavily to the elbow joint, the radius contributes heavily to the wrist joint. When the radius moves, the hand moves with it.

Hand

The skeleton of the hand includes the bones of the *carpus,* or wrist; the bones of the *metacarpus,* or palm; and the *phalanges,* or bones of the fingers (Figure 8.6).

Carpus

A wristwatch is actually worn on the distal forearm, not on the wrist at all. The true wrist, or **carpus** (kar'pus),

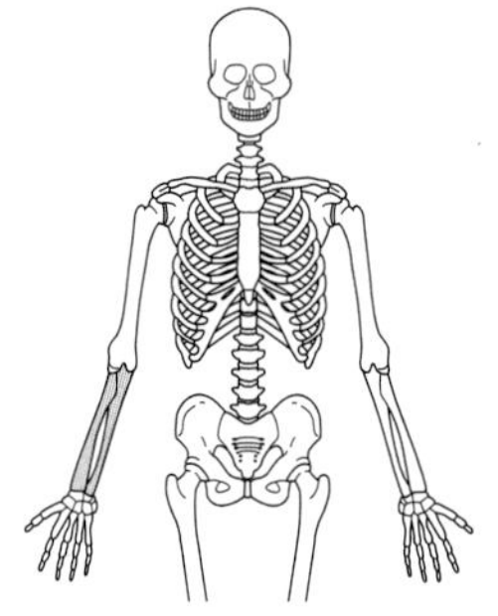

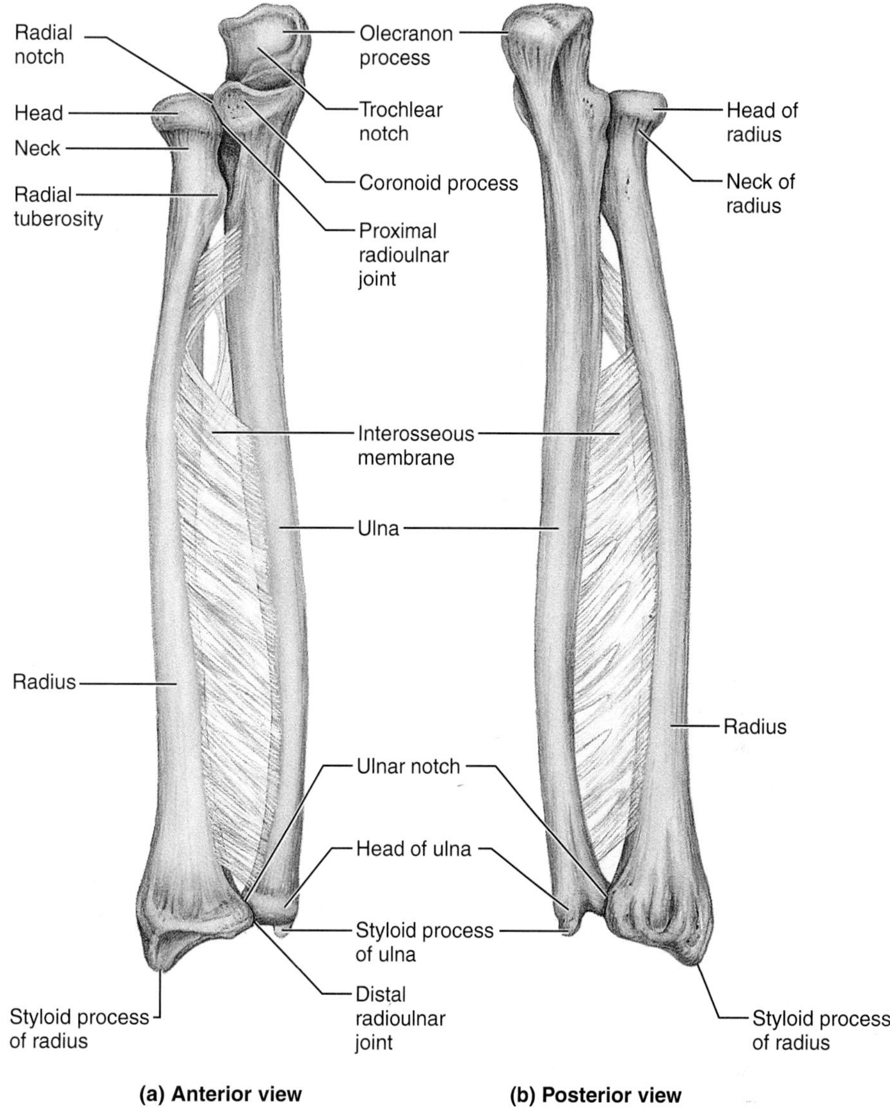

FIGURE 8.4 Radius and ulna of the right forearm. (See *Atlas of the Human Skeleton,* Figure 26.)

is the proximal region of the hand, just distal to the wrist joint. The carpus contains eight marble-sized short bones, or **carpals** (kar′palz), closely united by ligaments. Gliding movements occur between the carpals, making the wrist rather flexible. The carpals are arranged in two irregular rows of four bones each (see Figure 8.6). In the proximal row, from lateral (thumb side) to medial, are the **scaphoid** (skaf′oid; "boat-shaped"), **lunate** (lu′-nāt; "moon-like"), **triquetral** (tri-kwet′ral; "triangular"), and **pisiform** (pi′sĭ-form; "pea-shaped"). Only the scaphoid and lunate bones articulate with the radius to form the wrist joint. The carpals of the distal row, again from lateral to medial, are the **trapezium** (trah-pe′ze-um; "little table"), **trapezoid** ("four-sided"), **capitate** (kap′i-tāt; "head-shaped"), and **hamate** (ham′āt; "hooked").

The scaphoid is the most frequently fractured carpal bone. Such a fracture often results from falling on an outstretched hand. The impact bends the scaphoid, which then breaks at its narrow midregion.

Here is a "memory phrase" to help you recall the carpals in the order given above: **S**ally **l**eft **t**he **p**arty **t**o **t**ake **C**athy **h**ome. As in all such memory aids, the first letter of each word is the first letter of the term you need to remember.

Clinical Applications

CARPAL TUNNEL SYNDROME The arrangement of carpal bones makes the carpus concave anteriorly. A ligamentous band covers this concavity superficially, forming the *carpal tunnel.* Through this narrow tunnel run many long muscle tendons, which extend from the forearm to the fingers. Also crowded in

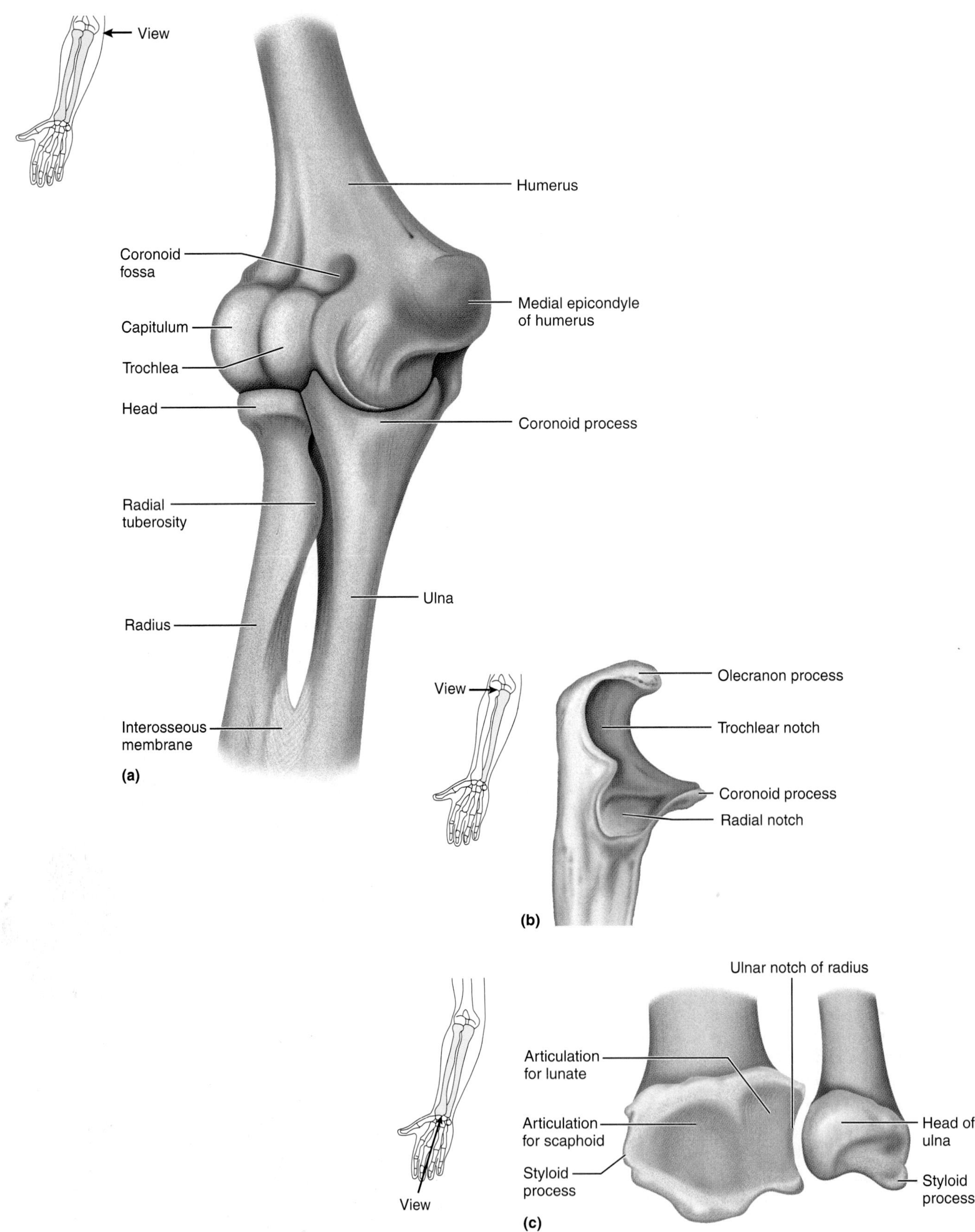

FIGURE 8.5 Details of bones of the forearm and elbow. **(a)** Elbow region in medial view. **(b)** Proximal part of the ulna in lateral view. **(c)** Distal ends of the radius and ulna at the wrist. (See *Atlas of the Human Skeleton,* Figure 26c, d, and e.)

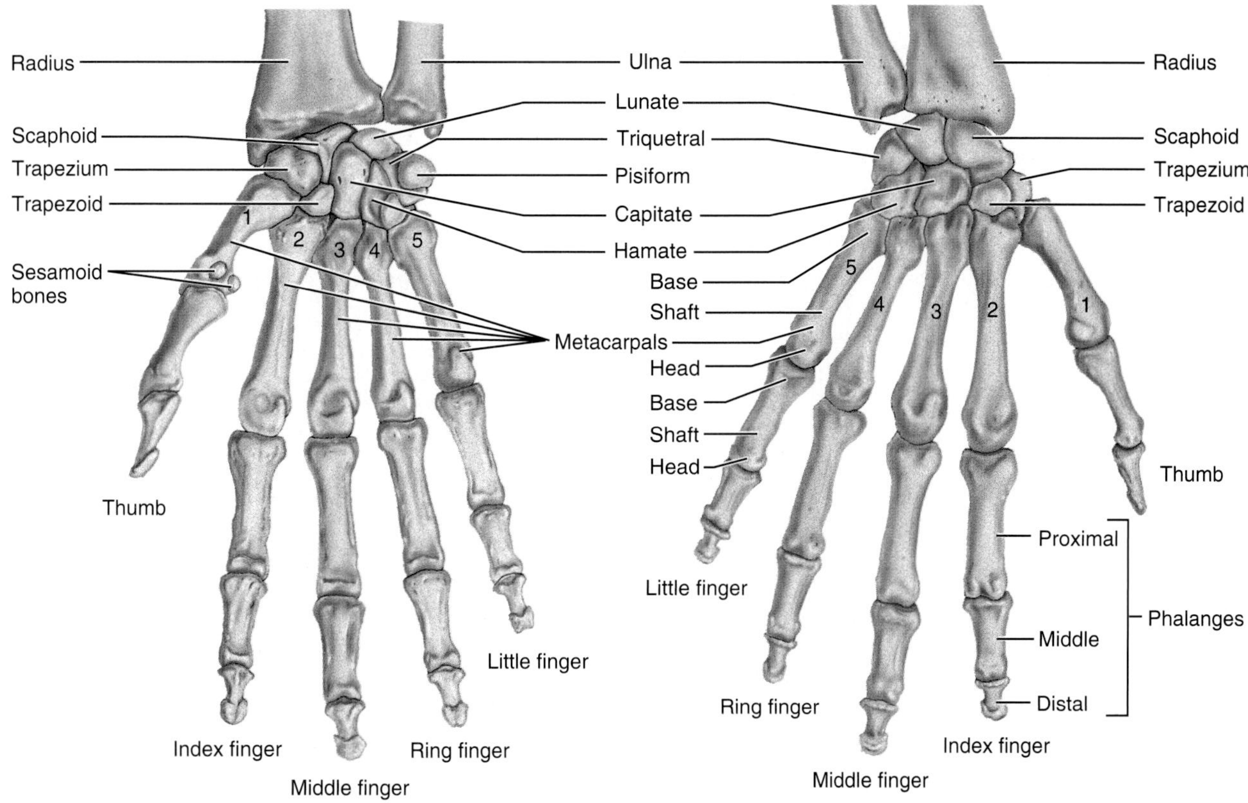

FIGURE 8.6 **Bones of the hand** (a) Anterior (palm) view of the right hand, illustrating the carpals, metacarpals, and phalanges. (b) Posterior view. (c) X-ray image. The white bar on the proximal phalanx of finger 4 indicates where a ring would be worn on the ring finger. (See *Atlas of the Human Skeleton,* Figure 27.)

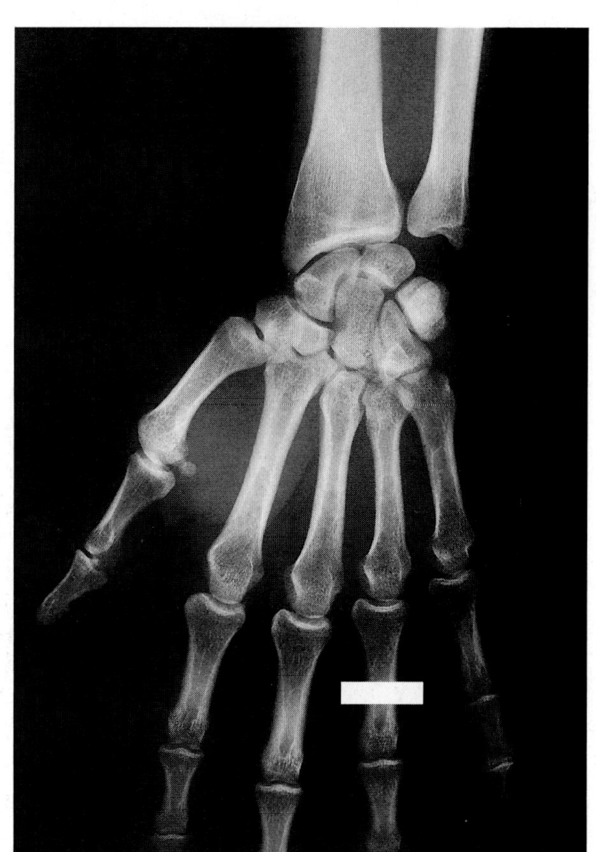
(c)

the tunnel is the median nerve, which (roughly speaking) innervates the lateral half of the hand, including the muscles that move the thumb. Inflammation of any element in the carpal tunnel, such as tendons swollen from overuse, can compress the median nerve. This nerve impairment is called **carpal tunnel syndrome,** and it affects many workers who repeatedly flex their wrists and fingers, such as meat packers and people who type at computers all day. Because the median nerve is impaired, the skin of the lateral part of the hand tingles or becomes numb, and movements of the thumb weaken. Pain is greatest at night. This condition can be treated by resting the hand in a splint during sleep, by anti-inflammatory drugs, or by surgery.

Carpal tunnel syndrome is just one of a series of overuse disorders that can affect the tendons, muscles, and joints of the upper limbs and back. Collectively, these conditions are called **repetitive stress injuries.** ✚

Metacarpus

Five **metacarpals** radiate distally from the wrist to form the **metacarpus,** or palm of the hand (*meta* = beyond). These small long bones are not named individually but instead are numbered 1 to 5, from thumb to little finger (see Figure 8.6). The *bases* of the metacarpals articulate with the carpals proximally, and with each other on

Table 8.1 Bones of the Appendicular Skeleton

Body Region	Bones*	Location	Markings
PART I: BONES OF THE PECTORAL GIRDLE AND UPPER LIMB			
Pectoral girdle (Figures 8.1, 8.2)	Clavicle (2)	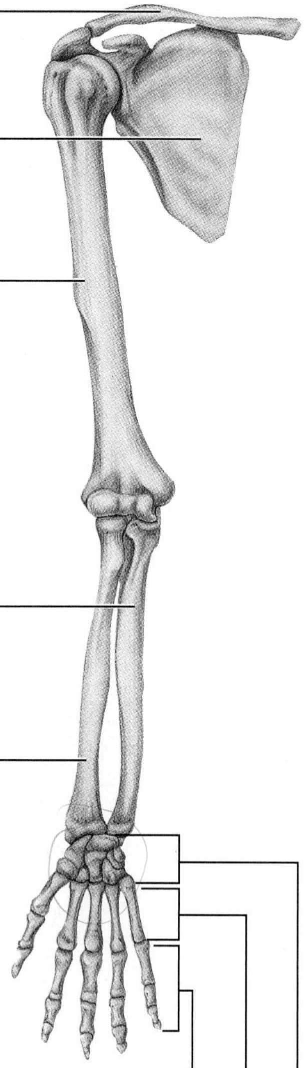 Clavicle is in superoanterior thorax; articulates medially with sternum and laterally with scapula	Acromonial end; sternal end
	Scapula (2)	Scapula is in posterior thorax; forms part of the shoulder; articulates with humerus and clavicle	Glenoid cavity; spine; acromion; coracoid process; infraspinous, supraspinous, and subscapular fossae
Upper Limb Arm (Figure 8.3)	Humerus (2)	Humerus is sole bone of arm; between scapula and elbow	Head; greater and lesser tuberosities; intertubercular groove; deltoid tuberosity; trochlea; capitulum; coronoid and olecranon fossae; radial groove; epicondyles
Forearm (Figures 8.4, 8.5)	Ulna (2)	Ulna is medial bone of forearm between elbow and wrist; forms elbow joint	Coronoid process; olecranon process; radial notch; trochlear notch; styloid process; head
	Radius (2)	Radius is lateral bone of forearm; carries wrist	Radial tuberosity; styloid process; head; ulnar notch
Hand (Figure 8.6)	8 Carpals (16) • scaphoid • lunate • triquetral • pisiform • trapezium • trapezoid • capitate • hamate	Carpals form a bony crescent at the wrist; arranged in two rows of four bones each	
	5 Metacarpals (10)	Metacarpals form the palm; one in line with each digit	
	14 Phalanges (28) • distal • middle • proximal	Phalanges form the fingers; three in digits 2–5; two in digit 1 (the thumb)	

Anterior view of pectoral girdle and upper limb

*The number in parentheses ( ) following the bone name denotes the total number of such bones in the body.

Body Region	Bones*	Illustration	Location	Markings
PART II: BONES OF THE PELVIC GIRDLE AND LOWER LIMB				
Pelvic girdle (Figure 8.7)	Coxal (2) (hip)	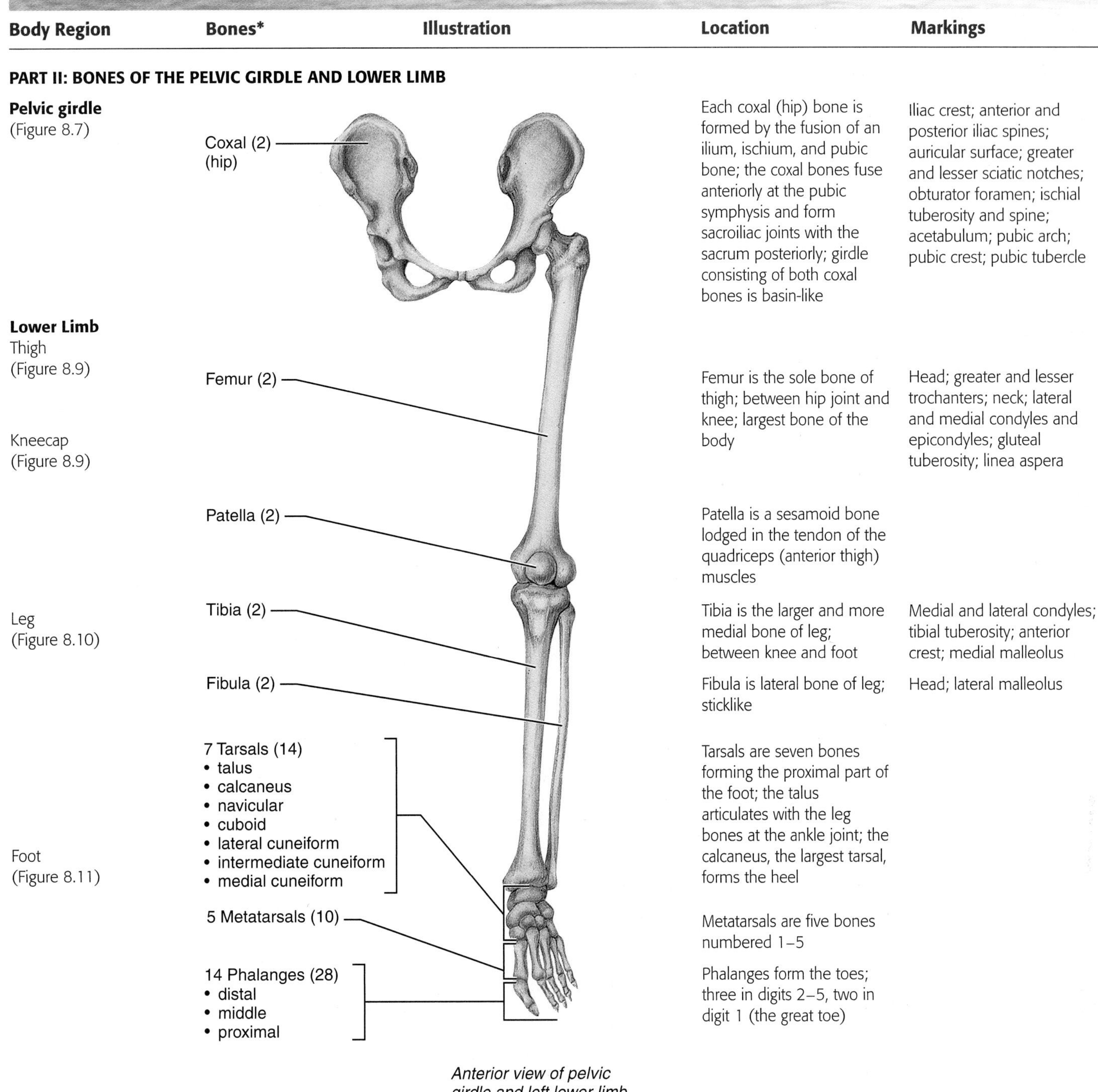	Each coxal (hip) bone is formed by the fusion of an ilium, ischium, and pubic bone; the coxal bones fuse anteriorly at the pubic symphysis and form sacroiliac joints with the sacrum posteriorly; girdle consisting of both coxal bones is basin-like	Iliac crest; anterior and posterior iliac spines; auricular surface; greater and lesser sciatic notches; obturator foramen; ischial tuberosity and spine; acetabulum; pubic arch; pubic crest; pubic tubercle
Lower Limb Thigh (Figure 8.9)	Femur (2)		Femur is the sole bone of thigh; between hip joint and knee; largest bone of the body	Head; greater and lesser trochanters; neck; lateral and medial condyles and epicondyles; gluteal tuberosity; linea aspera
Kneecap (Figure 8.9)	Patella (2)		Patella is a sesamoid bone lodged in the tendon of the quadriceps (anterior thigh) muscles	
Leg (Figure 8.10)	Tibia (2)		Tibia is the larger and more medial bone of leg; between knee and foot	Medial and lateral condyles; tibial tuberosity; anterior crest; medial malleolus
	Fibula (2)		Fibula is lateral bone of leg; sticklike	Head; lateral malleolus
Foot (Figure 8.11)	7 Tarsals (14) • talus • calcaneus • navicular • cuboid • lateral cuneiform • intermediate cuneiform • medial cuneiform		Tarsals are seven bones forming the proximal part of the foot; the talus articulates with the leg bones at the ankle joint; the calcaneus, the largest tarsal, forms the heel	
	5 Metatarsals (10)		Metatarsals are five bones numbered 1–5	
	14 Phalanges (28) • distal • middle • proximal		Phalanges form the toes; three in digits 2–5, two in digit 1 (the great toe)	

Anterior view of pelvic girdle and left lower limb

*The number in parentheses ( ) following the bone name denotes the total number of such bones in the body.

their lateral and medial sides. Distally, the bulbous *heads* of the metacarpals articulate with the proximal phalanges of the fingers to form knuckles. Metacarpal 1, associated with the thumb, is the shortest and most mobile.

Phalanges of the Fingers

The digits, or fingers, are numbered 1 to 5 beginning with the thumb, or *pollex* (pol′eks). The fingers contain miniature long bones called **phalanges** (fah-lan′jēz). The singular of this term is *phalanx* (fa′langks; "a closely knit row of soldiers"). In most people, the third finger is the longest. With the exception of the thumb, each finger has three phalanges: *proximal, middle,* and *distal.* The thumb has no middle phalanx.

Next we turn to the pelvic girdle and lower limb.

THE PELVIC GIRDLE

The **pelvic girdle,** or *hip girdle,* attaches the lower limbs to the spine and supports the visceral organs of the pelvis (see Table 8.1). The full weight of the upper body passes through this girdle to the lower limbs. Whereas the pectoral girdle barely attaches to the thoracic cage, the pelvic girdle attaches to the axial skeleton by some of the strongest ligaments in the body. Furthermore, whereas the glenoid cavity of the scapula is shallow, the corresponding socket in the pelvic girdle is a deep cup that firmly secures the head of the femur (thighbone). Consequently, the lower limbs have less freedom of movement than the upper limbs but are much more stable.

The pelvic girdle consists of the paired **hip bones** (Figure 8.7). A hip bone is also called a **coxal** (kok′ sal) **bone,** an **os coxae** (*os* = bone; *coxa* = hip), or an **innominate** ("no name") **bone.** Each hip bone unites with its partner anteriorly and with the sacrum posteriorly. The deep, basin-like structure formed by the hip bones, sacrum, and coccyx is the **bony pelvis** (Figure 8.7a).

The hip bone is large and irregularly shaped (Figure 8.7b and c). During childhood, it consists of three separate bones: the *ilium, ischium,* and *pubis.* In adults, these bones are fused and their boundaries are indistinguishable. Their names are retained, however, to refer to different regions of the composite hip bone. At the Y-shaped junction of the ilium, ischium, and pubis is a deep hemispherical socket, the **acetabulum** (as″ĕtab′u-lum), on the lateral pelvic surface (see Figure 8.7b). The acetabulum ("vinegar cup") receives the ball-shaped head of the femur at the hip joint.

Ilium

The **ilium** (il′e-um; "flank") is a large, flaring bone that forms the superior region of the hip bone. It consists of an inferior **body** and a superior wing-like **ala** ("wing"). The thickened superior margin of the ala is the **iliac crest.** Many muscles attach to this crest, which is thickest at the *tubercle of the iliac crest* (see Figure 8.7b). Each iliac crest ends anteriorly in a blunt **anterior superior iliac spine** and posteriorly in a sharp **posterior superior iliac spine.** Located inferior to these are the **anterior** and **posterior inferior iliac spines.** The anterior superior iliac spine is an especially prominent anatomical landmark (p. 781) and is easily felt through the skin.

Posteriorly, just inferior to the posterior inferior iliac spine, the ilium is deeply indented to form the **greater sciatic notch** (si-at′ik; "of the hip"). The sciatic nerve, the largest nerve in the body, passes through this notch to enter the thigh. The broad posterolateral surface of the ilium, the *gluteal surface* (glu′te-al; "buttocks"), is crossed by three ridges: the **posterior, anterior,** and **inferior gluteal lines.** These lines define the attachment sites of the gluteal (buttocks) muscles.

The internal surface of the iliac ala is concave. This broad concavity is called the **iliac fossa** (see Figure 8.7c). Posterior to this fossa lies a roughened **auricular surface** (aw-rik′u-lar; "ear-shaped"), which articulates with the sacrum, forming the *sacroiliac joint.* The weight of the body is transmitted from the vertebral column to the pelvis through this joint. Running anteriorly and inferiorly from the auricular surface is a robust ridge called the **arcuate line** (ar′ku-āt; "bowed"), which helps define the superior boundary of the true pelvis (p. 198). The inferior part of the ilium joins with the ischium and pubic bones.

CLINICAL APPLICATIONS

HIP POINTER A **hip pointer,** which is a bruise of the iliac crest, commonly occurs in contact sports as a result of hard contact with other players or the ground. It can also result from violent contraction of the abdominal-wall muscles that attach to the iliac crest, as occurs when an athlete twists the abdomen to try to dodge a blow. ✚

Ischium

The **ischium** (is′ke-um; "hip") forms the posteroinferior region of the hip bone (see Figure 8.7b). Shaped roughly like an L or an arc, it has a thicker, superior **body** and a thinner, inferior **ramus** (*ramus* = branch). Anteriorly, the ischial ramus joins the pubis. Features associated with the ischial body include the *ischial spine, lesser sciatic notch,* and *ischial tuberosity.* The triangular **ischial spine,** which lies posterior to the acetabulum and projects medially, is an attachment point for a ligament from the sacrum and coccyx, the *sacrospinous ligament* (Figure 8.8a). Just inferior to the ischial spine is the **lesser sciatic notch,** through which pass nerves and vessels that serve the perineum (area around the anus and external genitals). The inferior surface of the ischial

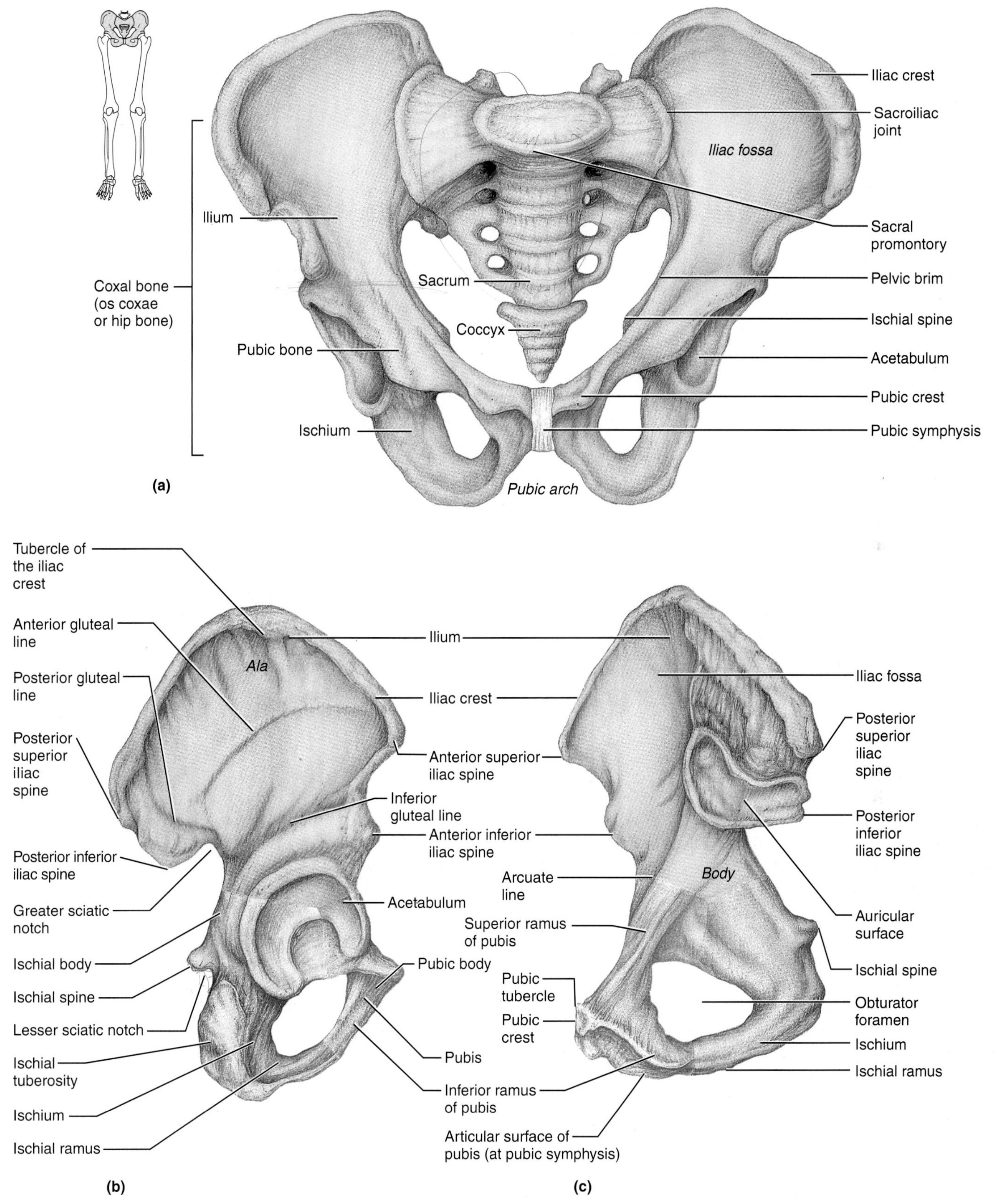

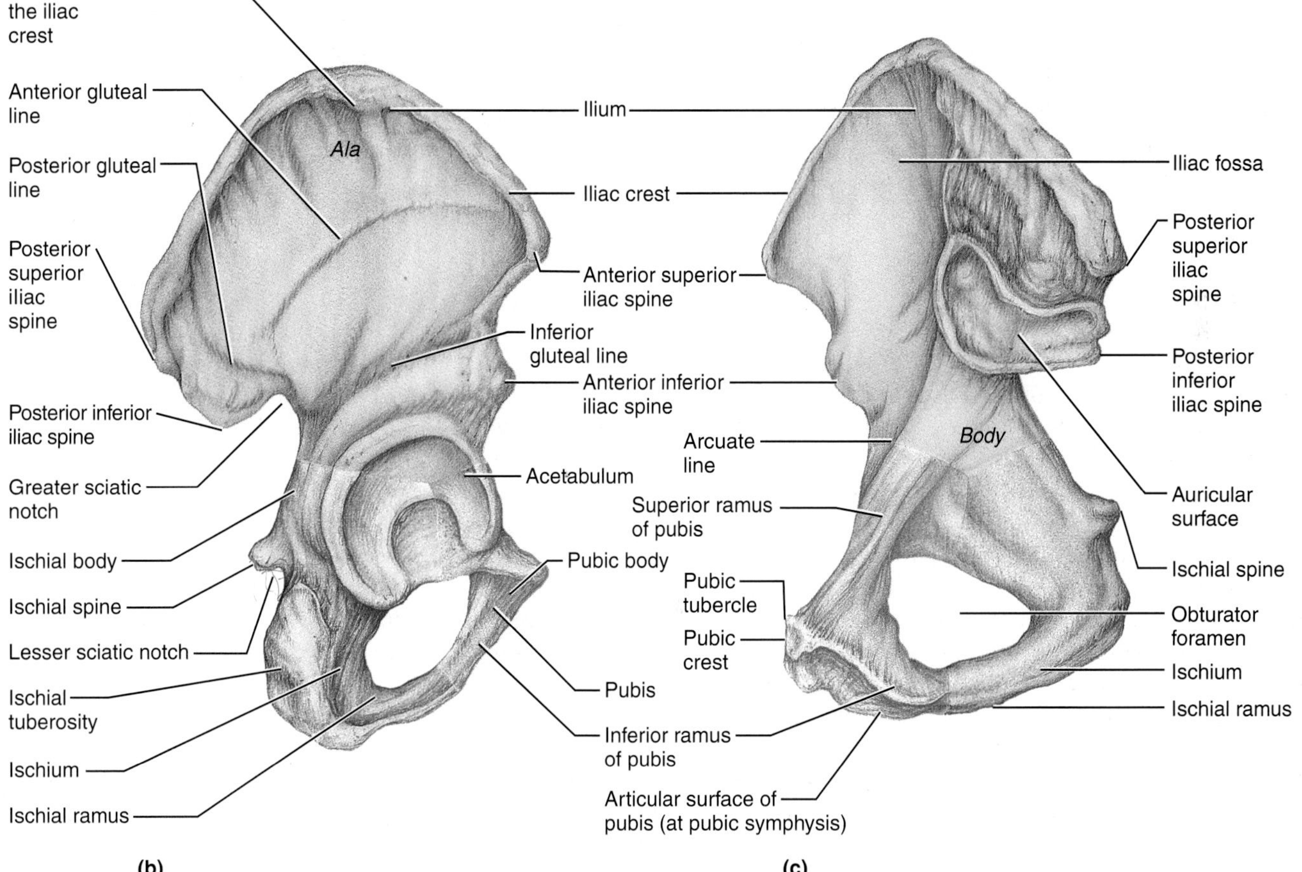

FIGURE 8.7 **Bones of the pelvic girdle.** **(a)** Bony pelvis showing the two hip bones (coxal bones) and the sacrum. **(b)** Lateral view of the right hip bone, showing the ilium, ischium, and pubic bones (anterior is to the right). **(c)** Medial view of the right hip bone (anterior is to the left). (See *Atlas of the Human Skeleton,* Figure 28.)

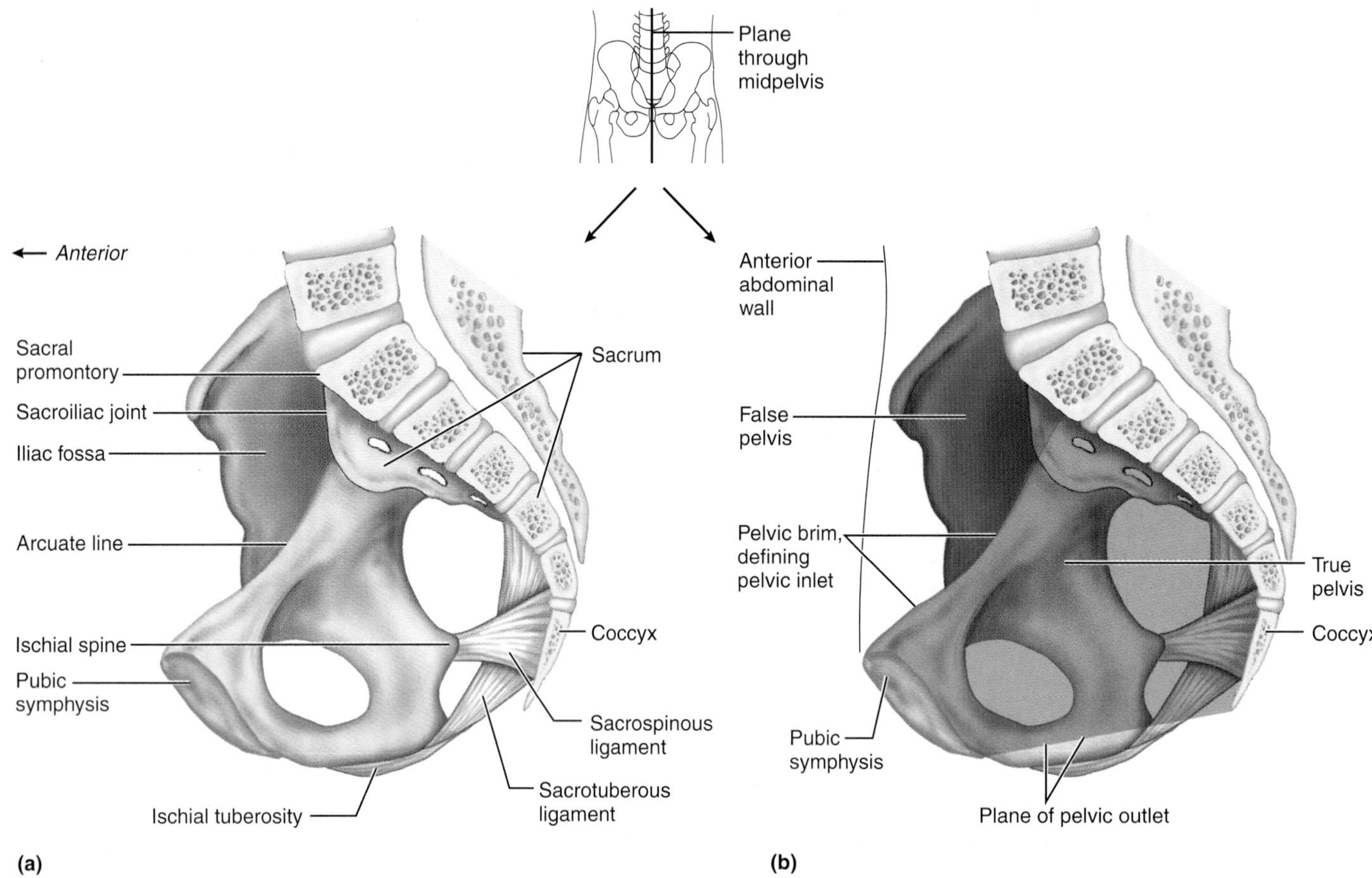

FIGURE 8.8 More features of the pelvis. These are medial views, as in Figure 8.7c (also, see the insert above). **(a)** This part provides orientation, showing how the hip bone joins to the sacrum and coccyx. Notice the sacrospinous and sacrotuberous ligaments. **(b)** Basic pelvic divisions: the true and false pelves.

body is rough and thickened as the **ischial tuberosity.** When we sit, our weight is borne entirely by the ischial tuberosities, which are the strongest parts of the hip bones. A massive *sacrotuberous ligament* (Figure 8.8a) runs from the sacrum to each ischial tuberosity and helps hold the pelvis together.

Pubis

The **pubis** (pu′bis; "sexually mature"), or *pubic bone,* forms the anterior region of the hip bone. In the anatomical position, it lies nearly horizontally, and the bladder rests upon it. Essentially, the pubis is V-shaped, with **superior** and **inferior rami** issuing from a flat **body** (see Figure 8.7b and c). The body of the pubis lies medially, and its anterior border is thickened to form a **pubic crest.** At the lateral end of the pubic crest is the knob-like **pubic tubercle,** an attachment point for an important ligament, the *inguinal ligament* (see Figure 11.11, p. 291). The two rami of the pubic bone extend laterally, where the inferior ramus joins to the ischial ramus, and the superior ramus joins with the bodies of the ischium and ilium. A thin ridge called the *pectineal line* (not illustrated) lies along the superior pubic ramus.

A large hole, the **obturator** (ob″tu-ra′tor) **foramen,** occurs between the pubis and ischium (Figure 8.7c). Students ask the function of this foramen, reasonably assuming something big goes through it. However, that is not the case: Although a few vessels and nerves do pass through it, the obturator foramen is almost completely closed by a fibrous membrane, the obturator membrane. In fact, the word *obturator* literally means "closed up."

In the midline, the bodies of the two pubic bones are joined by a disc of fibrocartilage. This joint is the *pubic symphysis* or *symphysis pubis* (see Figure 8.7a). Inferior to this joint, the inferior pubic rami and the ischial rami form an arch shaped like an inverted V, the **pubic (subpubic) arch.** The angle of this arch helps to distinguish the male pelvis from the female pelvis.

True and False Pelves

The bony pelvis is divided into two parts, the *false (greater) pelvis* and the *true (lesser) pelvis* (Figure 8.8).

These parts are separated by the **pelvic brim,** a continuous oval ridge that runs from the pubic crest through the arcuate line, the rounded inferior edges of the sacral ala, and the sacral promontory (see Figure 8.7a). The **false pelvis,** superior to the pelvic brim, is bounded by the alae of the iliac bones. It is actually part of the abdomen and contains abdominal organs. The **true pelvis** lies inferior to the pelvic brim. It forms a deep "bowl" containing the pelvic organs.

Pelvic Structure and Childbearing

The major differences between "typical" male and female pelves are summarized in Table 8.2. So consistent are these differences that an anatomist can determine the sex of a skeleton with 90% certainty merely by examining the pelvis. The female pelvis is adapted for childbearing: It tends to be wider, shallower, and lighter than that of a male. These features provide more room in the true pelvis, which must be wide enough for an infant's head to pass during birth.

The *pelvic inlet* is delineated by the pelvic brim (Figure 8.8b). Its largest diameter is from side to side (see Table 8.2). As labor begins, the infant's head enters this inlet, its forehead facing one ilium and its occiput facing the other. If the mother's sacral promontory is too large, it can block the entry of the infant into the true pelvis. The *pelvic outlet* is the inferior margin of the true pelvis, as shown in the photos at the bottom of Table 8.2. The outlet's anterior boundary is the pubic arch; its lateral boundaries are the ischial tuberosities, and its posterior boundary is the sacrum and coccyx. Both the coccyx and the ischial spines protrude into the outlet, so a sharply angled coccyx or unusually large ischial spines can interfere with delivery. The largest dimension of the pelvic outlet is the anteroposterior diameter. Generally, after the infant's head passes through the inlet, it rotates so that the forehead faces posteriorly and the occiput anteriorly. This is the usual position of the head as it leaves the mother's body (see Figure 24.27 on p. 732). Thus, during birth, the infant's head makes a quarter turn to follow the widest dimensions of the true pelvis.

THE LOWER LIMB

The lower limbs carry the entire weight of the erect body and experience strong forces when we jump or run. Thus, the bones of the lower limbs are thicker and stronger than the comparable bones of the upper limbs. The three segments of the lower limb are the thigh, the leg, and the foot (see Table 8.1 on p. 195)

Thigh

The **femur** (fe′mur; "thigh") is the single bone of the thigh (Figure 8.9b). It is the largest, longest, strongest bone in the body. Its durable structure reflects the fact that the stress on this bone can reach 280 kg per cm^2, or two tons per square inch! The femur courses medially as it descends toward the knee.

Critical Reasoning

Propose a functional explanation for the medial course of the femur.

Such a medial course places the knee joints closer to the body's center of gravity in the midline and thus provides for better balance. The medial course of the femur is more pronounced in women because of their wider pelvis. Thus, there is a greater angle between the femur and the tibia (shinbone), which is vertical. Some orthopedists think this explains the greater incidence of knee problems in female athletes.

The ball-like **head** of the femur has a small central pit called the **fovea capitis** (fo′ve-ah cap′ĭ-tis; "pit of the head"). A short ligament, the *ligament of the head of the femur,* runs from this pit to the acetabulum of the hip bone. The head of the femur is carried on a **neck,** which does not descend straight vertically but angles laterally to join the shaft. This angled course reflects the fact that the femur articulates with the lateral aspect, rather than the inferior region, of the pelvis. The neck is the weakest part of the femur and is often fractured in a "broken hip." At the junction of the shaft and neck are the lateral **greater trochanter** and posteromedial **lesser trochanter.** Various muscles attach to these robust projections. The two trochanters are interconnected by the *intertrochanteric line* anteriorly and by the prominent *intertrochanteric crest* posteriorly. Inferior to the intertrochanteric crest on the posterior surface of the shaft is the **gluteal tuberosity.** The inferior part of this tuberosity blends into a long vertical ridge, the **linea aspera** (lin′e-ah as′per-ah), which means "rough line." Both of these markings are sites of muscle attachment. Except for the linea aspera, the entire shaft of the femur is smooth and rounded.

Distally, the femur broadens to end in **lateral** and **medial condyles** shaped like wide wheels. The most raised points on the sides of these condyles are the **lateral** and **medial epicondyles,** to which ligaments attach. The **adductor tubercle** is a bump on the upper part of the medial condyle. Anteriorly, the two condyles are separated by a smooth **patellar surface,** which articulates with the kneecap, or patella. Posteriorly, they are separated by a deep **intercondylar notch.** Extending superiorly from the respective condyles to the linea aspera are the **lateral** and **medial supracondylar lines.**

The **patella** (pah-tel′ah; "small pan") is a triangular sesamoid bone enclosed in the tendon that secures the quadriceps muscles of the anterior thigh to the tibia (Figure 8.9a). It protects the knee joint anteriorly and improves the leverage of the thigh muscles acting across the knee.

Table 8.2 Comparison of the Male and Female Pelves

Characteristic	Female	Male
General structure and functional modifications	Titled forward; adapted for childbearing; true pelvis defines the birth canal; cavity of the true pelvis is broad, shallow, and has a greater capacity	Tilted less far forward; adapted for support of a male's heavier build and stronger muscles; cavity of the true pelvis is narrow and deep
Bone thickness	Less; bones lighter, thinner, and smoother	Greater; bones heavier and thicker, and markings are more prominent
Acetabula	Smaller; farther apart	Larger; closer
Pubic angle/arch	Broader (80° to 90°); more rounded	Angle is more acute (50° to 60°)
Anterior view	Pelvic brim	Pubic arch
Sacrum	Wider; shorter; sacral curvature is accentuated	Narrow; longer; sacral promontory more ventral
Coccyx	More moveable; straighter	Less moveable; curves ventrally
Left lateral view	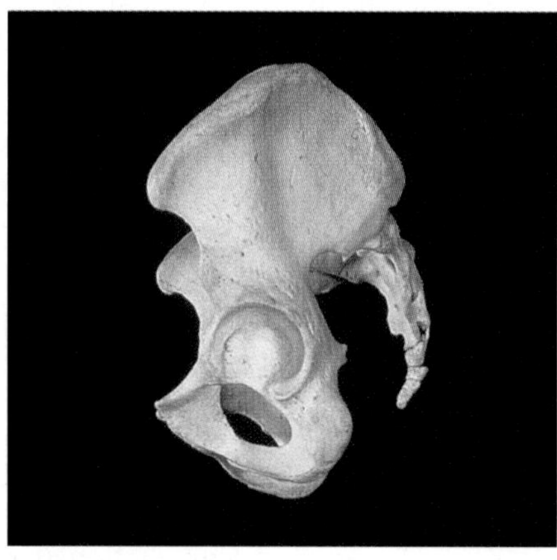	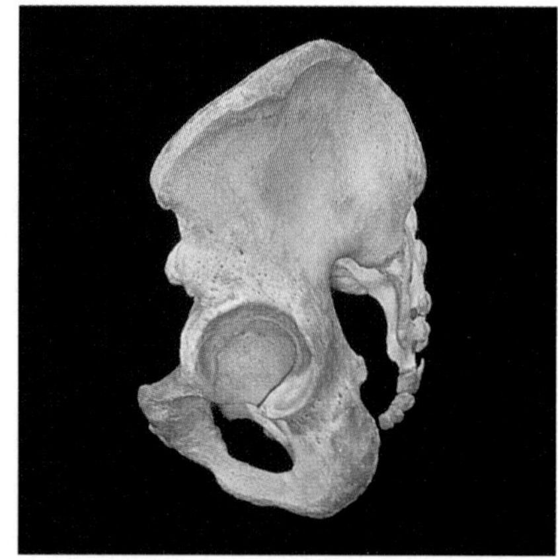
Pelvic inlet (brim)	Wider; oval from side to side	Narrow; basically heart-shaped
Pelvic outlet	Wider; ischial tuberosities shorter, farther apart and everted	Narrower; ischial tuberosities longer, sharper, and point more medially
Posteroinferior view	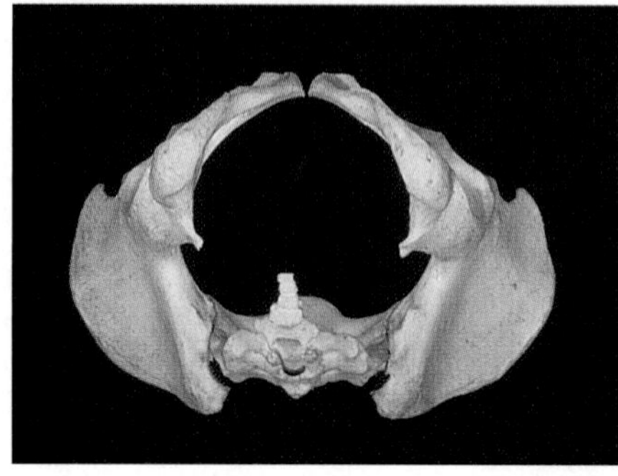	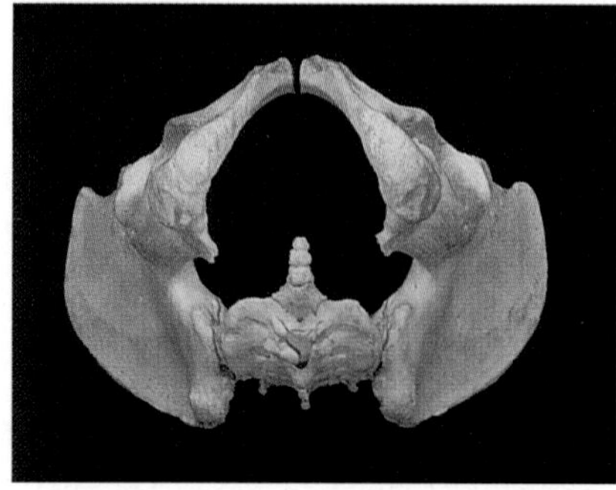

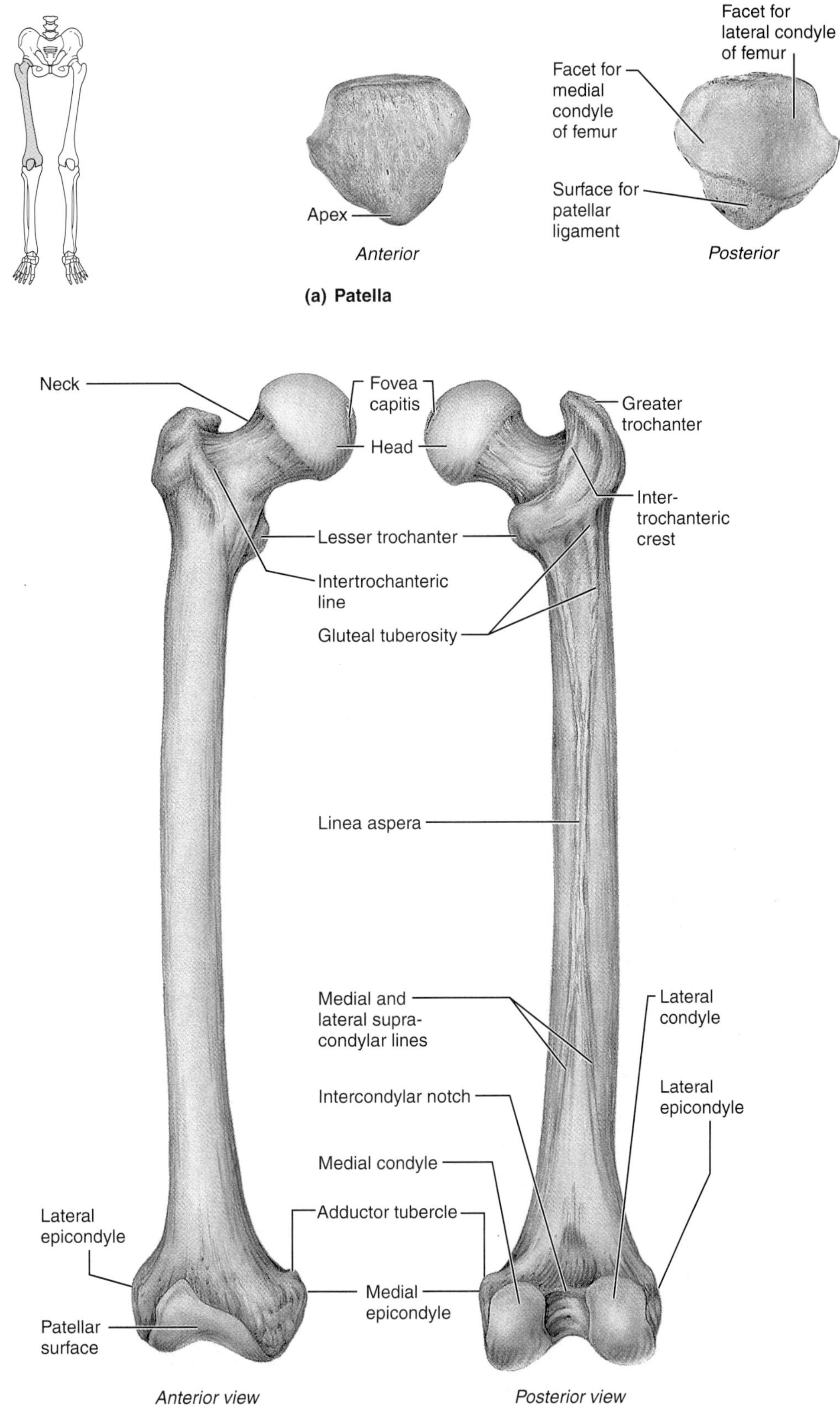

FIGURE 8.9 The right patella (a) and femur (b). (See *Atlas of the Human Skeleton*, Figure 29.)

Leg

Anatomists use the term *leg* to refer to the part of the lower limb between the knee and the ankle. Two parallel bones, the *tibia* and *fibula,* form the skeleton of the leg (Figure 8.10). The tibia is more massive than the stick-like fibula and lies medial to it. These two bones articulate with each other both proximally and distally. However, unlike the joints between the radius and ulna of the forearm, the *tibiofibular* (tib″e-o-fib′u-lar) *joints* allow almost no movement. Thus, the two leg bones do not cross one another when the leg rotates. An **interosseous membrane** connects the tibia and fibula along their entire length. The tibia articulates with the femur to form the knee joint, and with the talus bone of the foot at the ankle joint. The fibula, by contrast, does not contribute to the knee joint and merely helps stabilize the ankle joint.

Tibia

The **tibia** (tib′e-ah; "shinbone") receives the weight of the body from the femur and transmits it to the foot. It is second only to the femur in size and strength. At its broad proximal end are the **medial** and **lateral condyles,** which resemble two thick checkers lying side by side on the top of the shaft. Superiorly, their articular surfaces are slightly concave and are separated by an irregular projection, the **intercondylar eminence.** The tibial condyles articulate with the corresponding condyles of the femur. The inferior part of the lateral tibial condyle bears a facet that joins the fibula at the *proximal tibiofibular joint.* Just inferior to the condyles, the tibia's anterior surface displays a rough **tibial tuberosity.** A strong ligament from the patella, the patellar ligament, attaches to this tuberosity.

The shaft of the tibia is triangular in cross section. Its anterior border is the sharp, subcutaneous **anterior crest.** Distally, the end of the tibia is flat where it articulates with the talus of the foot. Medial to this joint surface, the tibia has an inferior projection called the **medial malleolus** (mah-le′o-lus; "little hammer"), which forms the medial bulge of the ankle. The *fibular notch* occurs on the lateral side of the distal tibia and participates in the *distal tibiofibular joint.*

Fibula

The **fibula** (fib′u-la; "pin") is a thin long bone with two expanded ends. Its superior end is its **head,** and its inferior end is the **lateral malleolus.** This malleolus forms the lateral bulge of the ankle and articulates with the talus bone of the foot. The shaft of the fibula is heavily ridged and appears to have been twisted a quarter turn. The fibula does not bear weight, but several muscles originate from it.

Critical Reasoning

Which parts of the leg bones participating in the ankle joint are fractured most often?

The answer is the malleoli, because these are the least massive parts. They break when the foot is forcefully inverted or everted at the ankle—that is, when one lands either on the lateral side of the foot and twists the sole medially (inversion) or on the medial side and turns the sole laterally (eversion).

Foot

The skeleton of the foot includes the bones of the *tarsus,* the bones of the *metatarsus,* and the *phalanges,* or toe bones (Figure 8.11). The foot has two important functions: It supports our body weight, and it acts as a lever to propel the body forward when we walk or run. A single bone could serve both these purposes but would be poorly adapted to uneven ground. Segmentation makes the foot pliable, avoiding this problem.

Tarsus

The **tarsus** (tar′sus) makes up the posterior half of the foot and contains seven bones called **tarsals.** It is comparable to the carpus of the hand. The weight of the body is borne primarily by the two largest, most posterior tarsal bones: the **talus** (ta′lus; "ankle"), which articulates with the tibia and fibula superiorly, and the strong **calcaneus** (kal-ka′ne-us; "heel bone"), which forms the heel of the foot. The calcaneus carries the talus on its superior surface. The thick tendon of the calf muscles attaches to the posterior surface of the calcaneus. The part of the calcaneus that touches the ground is the **tuber calcanei,** and a medial, shelf-like projection is the **sustentaculum tali** (sus″ten-tak′u-lum ta′le; "supporter of the talus"). The remaining tarsal bones are the lateral **cuboid** (ku′boid; "cube-shaped"), the medial **navicular** (nah-vik′u-lar; "boat-like"), and the anterior **medial, intermediate,** and **lateral cuneiforms** (ku-ne′ĭ-form; "wedge-shaped").

Metatarsus

The **metatarsus** of the foot, which corresponds to the metacarpus of the hand, consists of five small long bones called **metatarsals.** These bones are numbered 1 to 5 beginning on the medial side of the foot. The first metatarsal at the base of the big toe is the largest, and it plays an important role in supporting the weight of the body. The metatarsals are more nearly parallel to one another than are the metacarpals in the palm. Distally, where the metatarsals articulate with the proximal phalanges of the toes, the enlarged head of the first metatarsal forms the "ball" of the foot.

CLINICAL APPLICATIONS

METATARSAL STRESS FRACTURE One of the most common foot injuries, **metatarsal stress fracture,** results from repetitive stress on the foot, typically as a result of increasing one's running mileage too quickly. The second and third metatarsals are most often affected. Treatment generally involves resting the foot and wearing stiff or well-cushioned shoes. ✚

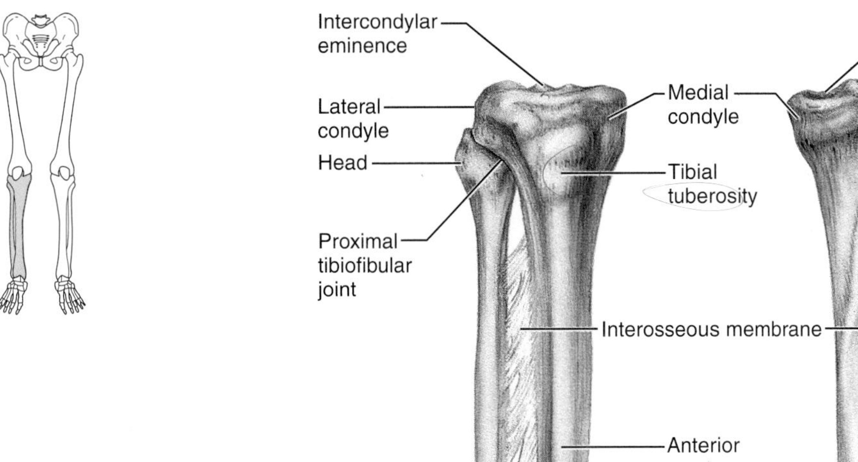

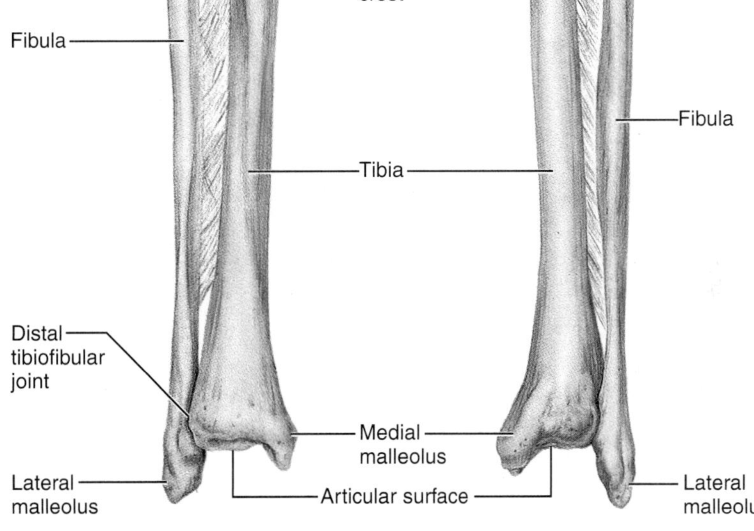

FIGURE 8.10 **The tibia and fibula of the right leg.** Parts **(a)** and **(b)** show the complete bones, while **(c)** and **(d)** are enlargements of the proximal end of the tibia. (See *Atlas of the Human Skeleton,* Figure 30.)

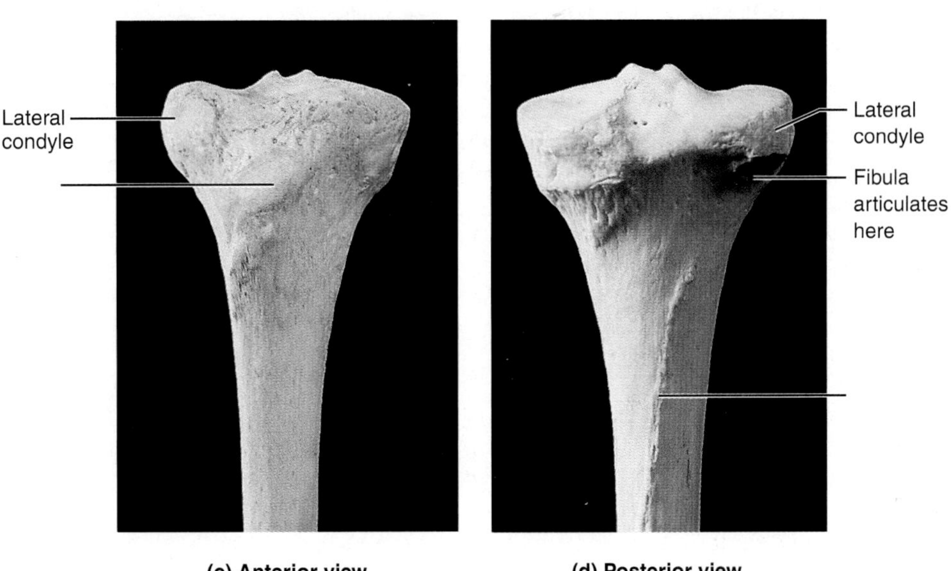

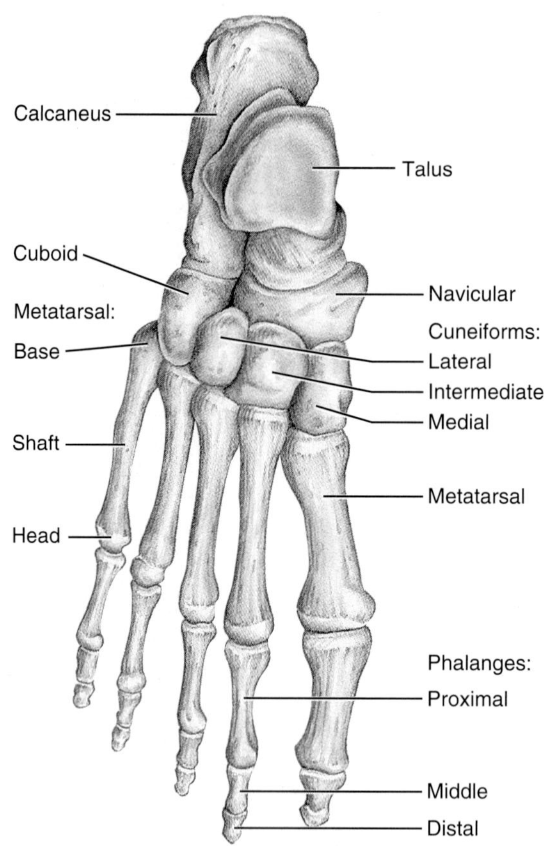

(a) Superior view

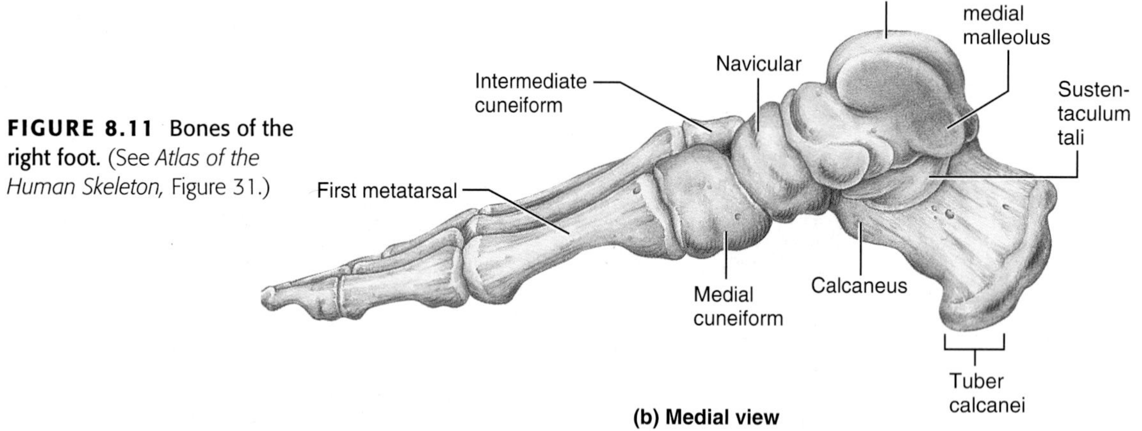

(b) Medial view

FIGURE 8.11 Bones of the right foot. (See *Atlas of the Human Skeleton,* Figure 31.)

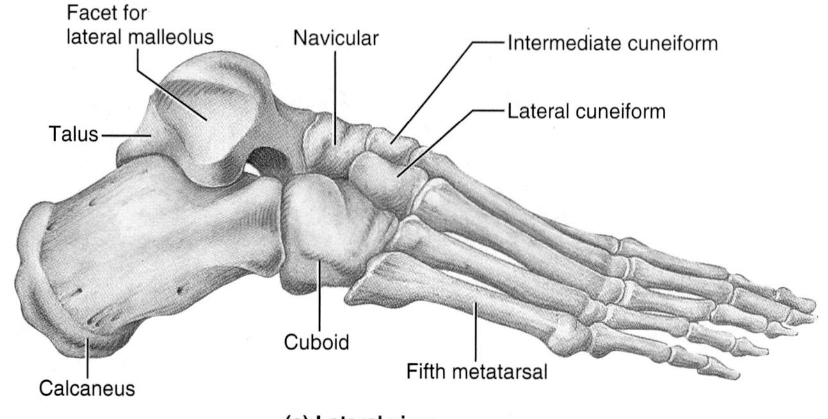

(c) Lateral view

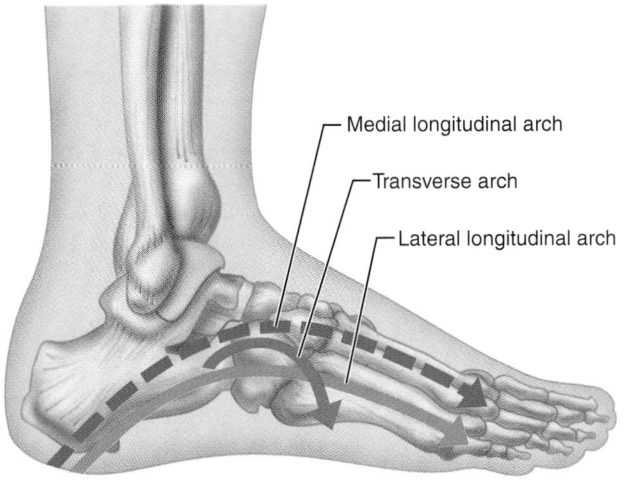

FIGURE 8.12 Arches of the foot.

Phalanges of the Toes

The 14 phalanges of the toes are smaller than those of the fingers, and thus are less nimble. Still, their general structure and arrangement are the same: There are three phalanges in each digit except the great toe (the *hallux*), which has only two phalanges. As in the hand, these toe bones are named *distal, middle,* and *proximal phalanges.*

Arches of the Foot

A segmented structure can support weight only if it is arched. The foot has three arches: the medial and lateral *longitudinal* arches and the *transverse* arch (Figure 8.12). These arches are maintained by the interlocking shapes of the foot bones, by strong ligaments, and by the pull of some tendons during muscle activity; the ligaments and tendons also provide resilience. As a result, the arches "give" when weight is applied to the foot, then spring back when the weight is removed.

If you examine your wet footprints, you will see that the foot's medial margin, from the heel to the distal end of the first metatarsal, leaves no print. This is because the **medial longitudinal arch** curves well above the ground. The talus is the keystone of this arch, which originates at the calcaneus, rises to the talus, and then descends to the three medial metatarsals. The **lateral longitudinal arch** is very low. It elevates the lateral edge of the foot just enough to redistribute some of the body weight to the calcaneus and some to the head of the fifth metatarsal (that is, to the two ends of the arch). The cuboid bone is the keystone of this lateral arch. The two longitudinal arches serve as pillars for the **transverse arch,** which runs obliquely from one side of the foot to the other, following the line of the joints between the tarsals and metatarsals. Together, the three arches form a half dome that distributes approximately half our standing and walking weight to the heel bones and half to the heads of the metatarsals.

As previously mentioned, various tendons run inferior to the foot bones and help support the arches of the foot. The muscles associated with these tendons are less active during standing than walking. Therefore, people who stand all day at their jobs may develop fallen arches or "flat foot." Running on hard surfaces can also cause arches to fall, unless one wears shoes that give proper arch support.

DISORDERS OF THE APPENDICULAR SKELETON

Most disorders of the appendicular skeleton are bone fractures, which were discussed already with the individual bones (also see Table 6.1, p. 144). Here, we consider two other important disorders, both of which are birth defects.

Hip dysplasia (congenital dislocation of the hip) is a relatively common birth defect; in fact, up to 4% of babies are treated for it. In this condition, which affects females more than males, either the acetabulum fails to form completely or the ligaments of the hip joint are loose. In either case the head of the femur tends to slip out of its socket. Early diagnosis and treatment are essential if permanent crippling is to be avoided. Treatment generally involves using a splint or a harness of straps to hold the femur in its proper position, so that the acetabulum can grow properly and the ligaments can tighten on their own. In extreme cases or those that are diagnosed late, surgery may be needed to repair and tighten the ligaments of the hip.

In **clubfoot,** which occurs in about 1 of every 1000 births, the soles of the feet turn medially and the toes point inferiorly. This disorder may be genetically induced, or it may result from the abnormal positioning of the feet (such as being folded against the chest) during fetal development. Clubfoot is treated by applying one cast after another to adjust the position of the growing foot, or in extreme cases, by surgery.

THE APPENDICULAR SKELETON THROUGHOUT LIFE

During youth, the growth of the appendicular skeleton not only increases the body's height but also changes the body's proportions (Figure 8.13). More specifically, the **upper-lower (UL) body ratio** changes with age. In this ratio, the *lower body segment* (L) is the distance from the top of the pelvic girdle to the ground, whereas the *upper body segment* (U) is the difference between the lower body segment's height and the person's total height.

At birth, the UL ratio is about 1.7 to 1. Thus, the head and trunk are more than 1.5 times as long as the lower limbs. The lower limbs grow faster than the trunk

A CLOSER LOOK

Clinical Advances in Bone Repair

Although bones have remarkable self-regenerative powers, some conditions are just too severe for bones to effect repair. Examples include extensive shattering (as in automobile accidents), poor circulation in old bones, osteomyelitis that kills part of a bone, and certain birth defects. Here we examine some recent advances that are allowing medical personnel to deal with healing problems that bones cannot resolve themselves.

Electrical stimulation of fracture sites has increased the speed and completeness of healing in slowly healing fractures. Practitioners have known for many years that bone tissue is deposited in regions of negative electrical charge, and absorbed in regions of positive charge, but the precise manner in which electrical fields promote healing is unknown. One theory is that the fields prevent parathyroid hormone from stimulating the bone-absorbing osteoclast cells at the fracture site. Another theory is that the fields induce production of growth chemicals that stimulate the bone-forming osteoblast cells.

Ultrasound, introduced as the basis of an imaging technique in Chapter 1, can speed the repair of fresh fractures. Daily exposures to low-power ultrasound waves reduce the healing time of broken arm and shin bones by 25% to 35%. It apparently stimulates cartilage cells to make callus.

The most troublesome injuries to bones are non-union fractures or segmental defects, in which the two parts of a split bone fail to join. Such fractures traditionally have been treated with grafts, in which sections of bone are taken from the hip and inserted into the gap. However, this requires several painful grafting sessions, and one-third of the grafts fail to heal. A potential improvement is the **free vascular fibular graft technique,** in which pieces of the fibula are used to replace missing bone. One reason that traditional grafts often fail is that a blood supply cannot reach their interior, but this technique grafts normal blood vessels along with the bone sections. Subsequent remodeling leads to a good replica of the pre-injury bone.

Much research has gone into developing **bone substitutes** to fill the gaps in non-union defects. Crushed bone from human cadavers is mixed with water to form a paste that can be molded into the desired shape or packed into small, difficult-to-reach spaces. However, cadaver bone is a foreign tissue that the immune system may reject, and the body sometimes fails to replace it with new bone, as it must for healing to occur. Furthermore, there is a slight but real risk that the cadaver bone contains disease organisms.

Research has produced several types of artificial bone materials. Some are inorganic ceramics, which are porous and serve as a scaffolding on which new bone can grow, but are not very strong and can cause irritation and rejection. Somewhat more effective is so-called "sea bone," coral that is heat-treated in a way that kills its living cells and converts its calcium carbonate to hydroxyapatite, the mineral in bone. The coral graft is then carved to the desired shape, coated with a natural substance that induces bone formation (bone morphogenic protein), and implanted. Osteoblasts and blood vessels migrate from the adjacent natural bone into the coral implant, gradually replacing it with living bone. Coral derivatives are still too weak and brittle, however, and aluminum and silicate added to increase their strength can cause inflammation.

More recently, a paste made from powdered mother of pearl has shown a remarkable ability to stimulate new bone formation when injected into a defective human maxilla. No inflammation was evident.

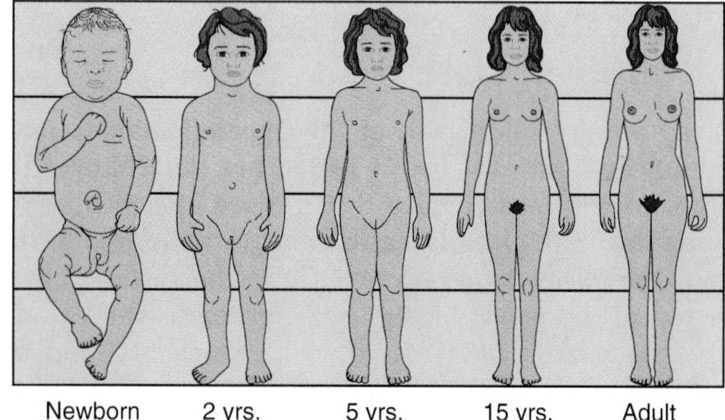

FIGURE 8.13 Changes in body proportions throughout life. During growth, the arms and legs grow faster than the head and trunk, as can be seen in this depiction of different-aged individuals all drawn at the same height.

from this time on, however, and by age 10, the UL ratio is about 1 to 1, and it changes little thereafter. During puberty, the female pelvis broadens in preparation for childbearing, and the entire male skeleton becomes more robust.

Once adult height is reached, a healthy appendicular skeleton changes very little until middle age. Then it loses mass, and osteoporosis and limb fractures become more common.

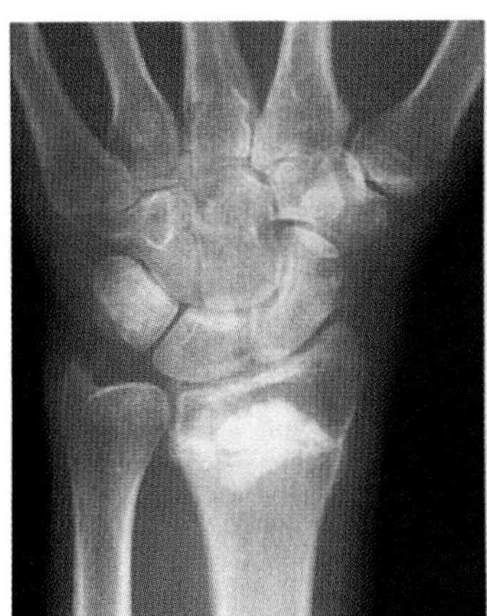
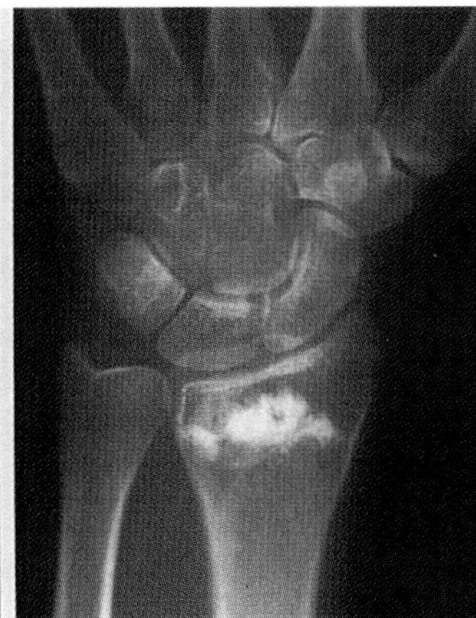

The X ray on the left shows a fracture of the radius repaired with Norian SRS, which appears as the whiter area. The X ray on the right is the same fracture three years later.

Norian SRS, a new bone cement made of calcium phosphate, provides immediate structural support to fractured or osteoporotic sites. Mixed at the time of surgery, Norian SRS is injected into areas of damaged bone to create an internal "cast." The paste hardens in minutes and cures into a substance with greater compressive strength than spongy bone. Because its crystalline structure is the same as that of a natural bone, it continues to provide support as it is gradually remodeled and replaced by host bone. However, Norian SRS, can only be used on bone ends, because it cannot resist the strong compressive and bending stresses occurring at the central shaft.

Perhaps the most promising product being tested is Osteomedica's bioceramic. Made from mildly heated, chemically synthesized hydroxyapatite, bioceramic is stronger than the bone substitutes just described. Better, it can be molded and carved into long pieces that are placed in gaps in the shafts of long bones. It essentially works as an artificial bone graft. In the limbs of animals, it has been shown to bear weight well after several months of initial healing and natural remodeling.

RELATED CLINICAL TERMS

Bunion A deformity of the great toe involving lateral displacement of this digit and medial displacement of metatarsal 1. Also includes a bony swelling and a bursitis (p. 236) on the medial side of the head of the first metatarsal. Caused by tight or ill-fitting shoes (or, more rarely, arthritis or genetic factors).

Knock-knee A deformity in which the two knees rub or knock together during walking. Knock-knee usually occurs in children due to irregular growth of the bones of the lower limb, injury to the ligaments, or injury to the bone ends at the knee.

Lisfranc injury Damage to the joints between the tarsal and metatarsal bones of the foot that results from violently twisting or bending the anterior part of the foot on the posterior part, as can occur in a fall. Generally involves the cuneiforms and the first three metatarsals. The metatarsal bases may be dislocated and the intercuneiform joints damaged, often with fractures of these bones.

Pelvimetry Measurement of the dimensions of the inlet and outlet of the pelvis, usually to determine whether it is of adequate size to allow normal delivery of a baby.

Podiatry The specialized field dealing with the study and care of the foot, including its anatomy, disorders, and medical and surgical treatment.

CHAPTER SUMMARY

Media study tools that could provide you additional help in reviewing specific key topics of Chapter 8 are referenced below.

SP = Study Partner; AIA = A.D.A.M.® Interactive Anatomy

1. The appendicular skeleton consists of the pectoral and pelvic girdles and the bones of the upper and lower limbs.

THE PECTORAL GIRDLE (pp. 186–189)

1. The pectoral girdles are specialized for mobility. Each consists of a clavicle and a scapula and attaches an upper limb to the bony thorax.
2. The clavicles hold the arms laterally away from the thorax and transmit pushing forces from the upper limbs to the thorax.

AIA Link: Atlas; Region: Upper Limb; Bones of the Arm and Shoulder (posterior).

3. Each triangular scapula articulates with a clavicle and a humerus. The borders, angles, and markings of the scapula are summarized in Table 8.1 on pp. 194–195.

THE UPPER LIMB (pp. 189–194)

1. Each upper limb consists of 30 bones and is specialized for mobility.
2. The skeleton of the arm consists solely of the humerus. Bones of the forearm are the radius and ulna; those of the hand are the carpals, metacarpals, and phalanges.

AIA Link: Atlas; Region: Upper limb; Bones of Forearm and Hand (anterior).

SP Exercise: Chapter 8, Ulna and Radius.

THE PELVIC GIRDLE (pp. 194–199)

1. The pelvic girdle, specialized for bearing weight, is composed of two hip bones that connect the lower limbs to the vertebral column. Together with the sacrum, the hip bones form the basin-like bony pelvis.
2. Each hip bone (coxal bone) consists of an ilium, ischium, and a pubis fused together. The cup-like acetabulum is at the Y-shaped region of fusion of these three bones.

AIA Link: Atlas; System: Skeletal; Male bony pelvis (lateral).

3. The ilium is the superior flaring part of the hip bone. Each ilium forms a secure joint with the sacrum. The ischium is a curved bar of bone; during sitting, one's weight is borne by the ischial tuberosities. The V-shaped pubic bones join anteriorly at the pubic symphysis.
4. The pelvic brim, an oval ridge that includes the pubic crest, arcuate line of the ilium, and sacral promontory, separates the superior false pelvis from the inferior true pelvis.
5. The male pelvis is relatively deep and narrow, with larger, heavier bones; the female pelvis, which forms the birth canal, is comparatively shallower and wider (see Table 8.2, p. 200).

THE LOWER LIMB (pp. 199–205)

1. The lower limb consists of the thigh, leg, and foot, and is specialized for weight bearing and locomotion.
2. The long, thick femur is the only bone of the thigh. Its ball-shaped head articulates with the acetabulum.

AIA Link: Atlas; Region; Lower Limb; Bones of Lower Limb (anterior).

SP Exercise: Chapter 8, Bones of the Appendicular Skeleton.

3. The bones of the leg are the tibia (which participates in both the knee and ankle joints) and the slender fibula.
4. The bones of the foot are the tarsals, metatarsals, and phalanges. The most important tarsals are the calcaneus (heel bone) and the talus. The talus articulates with the leg bones at the ankle joint.

AIA Link: Atlas; Region; Lower Limb; Bones of the Leg and Foot (anterior).

5. The foot is supported by three arches that distribute the weight of the body to the heel and ball of the foot.

SP Exercise: Chapter 7, The Skeleton.

DISORDERS OF THE APPENDICULAR SKELETON (p. 205)

1. Hip dysplasia (congenital dislocation of the hip) and clubfoot are common birth defects.

THE APPENDICULAR SKELETON THROUGHOUT LIFE (pp. 205–206)

1. Fast growth of the lower limbs causes the upper-lower (UL) body ratio to change from 1.7:1 at birth to 1:1 at 10 years and beyond.

REVIEW QUESTIONS

Multiple Choice/Matching Questions

1. Match the terms in the key with the bone descriptions that follow:

Key: **(a)** clavicle **(b)** ilium **(c)** ischium **(d)** pubis **(e)** sacrum **(f)** scapula **(g)** sternum

____ **(1)** bone of the axial skeleton to which the pectoral girdle attaches

____ **(2)** its markings include the glenoid cavity and acromion process

____ **(3)** its features include the ala, crest, and greater sciatic notch

____ **(4)** membrane bone that transmits forces from upper limb to bony thorax

____ **(5)** bone of pelvic girdle that articulates with the axial skeleton

____ **(6)** bone that bears weight during sitting

____ **(7)** most anteroinferior bone of the pelvic girdle

____ **(8)** bone of the axial skeleton to which the pelvic girdle attaches

2. Match the choices in the key with the bone descriptions that follow:

Key: **(a)** carpals **(b)** femur **(c)** fibula **(d)** humerus **(e)** radius **(f)** tarsals **(g)** tibia **(h)** ulna

____ **(1)** articulates with the acetabulum and the tibia

____ **(2)** its malleolus forms the lateral aspect of the ankle

____ **(3)** bone that "carries" the hand and wrist

____ **(4)** the wrist bones

____ **(5)** bone shaped much like a monkey wrench

____ **(6)** articulates with the capitulum of the humerus

____ **(7)** largest bone is the calcaneus

3. Which of the following bony features is *not* near or in the shoulder joint? (a) acromion, (b) greater tubercle, (c) glenoid cavity, (d) anatomical neck of humerus, (e) deltoid tuberosity.

4. Which of the following bony features is not in or near the hip joint? (a) acetabulum, (b) sacral promontory, (c) greater trochanter, (d) neck of femur.

Short Answer Essay Questions

5. The major function of the pectoral girdle is mobility. (a) What is the major function of the pelvic girdle? (b) Relate these functional differences to the anatomical differences between the girdles.

6. List three differences between the male and female pelves.

7. Describe the function of the arches of the foot.

8. Briefly describe the anatomical characteristics and impairment of function seen in hip dysplasia.

9. Define and distinguish between the true pelvis and false pelvis.

10. Lance was a bright anatomy student, but he sometimes called the leg bones "fibia" and "tibula." Correct this common mistake.

11. Draw the scapula in posterior view, and label all the borders, angles, fossae, and important markings visible in this view.

12. Identify the associated bone and the location of each of the following bone markings: (a) greater trochanter, (b) linea aspera, (c) trochlea, (d) coronoid process, (e) deltoid tuberosity, (f) greater tubercle, (g) greater sciatic notch.

13. (a) Which body regions do anatomists call the arm and the leg? (b) What is the *medial* side of the hand?

14. Tom Williams, a teaching assistant in anatomy class, picked up a bone and pretended it was a telephone. He put the big hole in this bone up to his ear and said, "Hello, obturator, obturator (operator, operator)." Name the bone and the structure he was helping the students to learn.

CRITICAL REASONING + CLINICAL APPLICATION QUESTIONS

1. Malcolm was rushed to the emergency room after trying to break his fall with outstretched arms. The physician took one look at his shoulder and saw that Malcolm had a broken clavicle (and no other injury). Describe the position of Malcolm's shoulder. What part of the clavicle was most likely broken? Malcolm was worried about injury to the main blood vessels to his arm (the subclavian vessels), but he was told such injury was unlikely. Why could the doctor predict this?

2. Racheal, a hairdresser, developed flat feet. Explain why.

3. Justiniano worked in a poultry-packing plant, where his job was cutting open chickens and stripping out their visceral organs. After work, he typed for long hours on his computer keyboard, because he was writing a novel based on his work in the plant. Soon, his wrist and hand began to hurt whenever he flexed them, and he began to awaken at night with pain and tingling on the thumb half of his hand. What was his probable condition?

A.D.A.M.® INTERACTIVE ANATOMY QUESTIONS

1. Link: Atlas; System: Skeletal; Bones of forearm and hand (anterior). Name the two carpal bones that form the wrist joint with the radius.

2. Link: Atlas; System: Skeletal; Bones of legs and foot (anterior). Name the two largest tarsal bones at the back of the foot, one of which articulates with the fibula and the tibia.

CHAPTER 9 Joints

X-ray of the knee joint

CHAPTER OUTLINE + STUDENT OBJECTIVES

1. Define joint.

Classification of Joints 212

2. Classify the joints by structure and by function.

Fibrous Joints 212–213

3. Describe the general structure of fibrous joints. Provide examples of the three types of fibrous joints.

Cartilaginous Joints 213–214

4. Describe cartilaginous joints, and give examples of the two main types.

Synovial Joints 214–234

5. Describe the structural characteristics shared by all synovial joints.
6. Explain how synovial joints function.
7. List three factors that influence the stability of synovial joints.
8. Define bursa and tendon sheath.
9. Name and describe the common types of body movements.
10. Name six classes of synovial joints based on the shapes of their joint surfaces and the types of movement they allow. Give examples of joints in each class.
11. Describe the key features of the jaw, shoulder, elbow, hip, knee, and ankle joints.

Disorders of Joints 234–238

12. Name the most common injuries to joints, and discuss the problems associated with each.
13. Name and describe the types of arthritis.

The Joints Throughout Life 238–239

14. Describe how joints develop and how their function may be affected by aging.

THE RIGID ELEMENTS OF THE SKELETON MEET at sites called **joints,** or **articulations.** The Greek root *arthro* means "joint," and the scientific study of joints is called arthrology (ar-throl′o-je). Most articulations join bone to bone, but some join bone to cartilage, or teeth to their bony sockets. Joints both hold the skeleton together and allow it to be mobile. The graceful movements of dancers and gymnasts attest to the great variety of motions that joints allow. Even though joints are always the weakest points of any skeleton, their structure enables them to resist crushing, tearing, and the various forces that would drive them out of alignment.

CLASSIFICATION OF JOINTS

Joints can be classified either by function or structure. The *functional classification* focuses on the amount of movement allowed. Accordingly, **synarthroses** (sin″ar-thro′sēz) are immovable joints, **amphiarthroses** (am″fe-ar-thro′sēz) are slightly movable joints, and **diarthroses** (di″ar-thro′sēz) are freely movable joints. Diarthroses predominate in the limbs, whereas synarthroses and amphiarthroses are largely restricted to the axial skeleton.

The ***structural classification*** is based on the material that binds the bones together and on the presence or absence of a joint cavity. Structurally, joints are classified as **fibrous, cartilaginous,** or **synovial joints** (Table 9.1).

In this chapter we organize our discussion of joints according to the structural classification, considering the functional properties where appropriate. As you read along, refer to Table 9.2 on pp. 216–218, which summarizes the structural and functional characteristics of the major joints in the human body.

FIBROUS JOINTS

In **fibrous joints** (Figure 9.1), the bones are connected by fibrous tissue, namely dense regular connective tissue. No joint cavity is present. Most fibrous joints are immovable or only slightly movable. The types of fibrous joints are ***sutures, syndesmoses,*** and ***gomphoses.***

Sutures

In **sutures**, literally "seams," the bones are tightly bound by a minimal amount of fibrous tissue (Figure 9.1a). Sutures occur only between bones of the skull, and their fibrous tissue is continuous with the periosteum around these flat bones. At sutures, the edges of the joining bones are wavy and interlocking. Sutures not only knit the bones together but also allow growth so that the skull can expand with the brain during childhood. During middle age, the fibrous tissue ossifies, and the skull bones fuse together. At this stage, the closed sutures are more precisely called ***synostoses*** (sin″os-to′sēz), literally "bony junctions." Because movement of the cranial bones would damage the brain, the immovable nature of sutures is a protective adaptation.

Syndesmoses

In **syndesmoses** (sin″des-mo′sēz), the bones are connected exclusively by ligaments, bands of fibrous tissue longer than those that occur in sutures (Figure 9.1b). In fact, the name ***syndesmosis*** derives from the Greek word for "ligament."

The amount of movement allowed at a syndesmosis depends on the length of the connecting fibers. For example, the anterior tibiofibular ligament between the

Table 9.1 Summary of Joint Classes

Structural class	Structural characteristics	Types	Mobility
Fibrous	Bone ends/parts united by collagenic fibers	(1) Suture (short fibers)	Immobile (synarthrosis)
		(2) Syndesmosis (longer fibers)	Slightly mobile (amphiarthrosis) and immobile
		(3) Gomphosis (periodontal ligament)	Immobile
Cartilaginous	Bone ends/parts united by cartilage	(1) Synchondrosis (hyaline cartilage)	Immobile
		(2) Symphysis (fibrocartilage)	Slightly movable
Synovial	Bone ends/parts covered with articular cartilage and enclosed within an articular capsule lined with synovial membrane	(1) Plane (4) Condyloid (2) Hinge (5) Saddle (3) Pivot (6) Ball and socket	Freely movable (diarthrosis); movements depend on design of joint

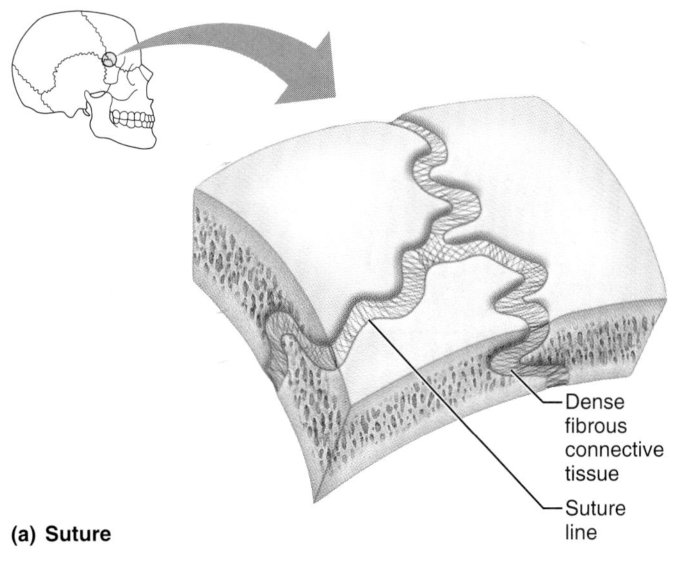

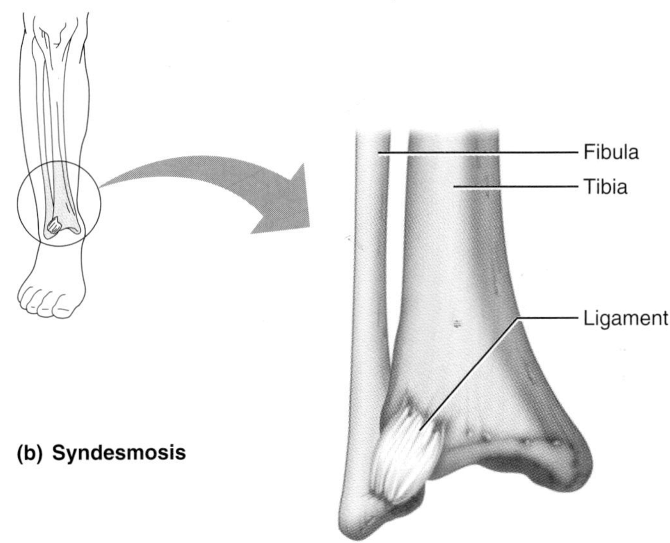

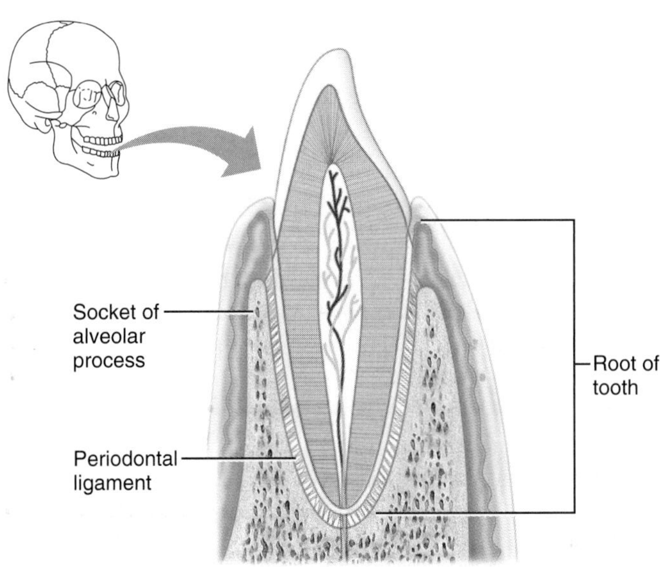

distal ends of the tibia and fibula is short (see Figure 9.1b), allowing only slightly more movement than a suture does. Such slight movement is called "give." True movement is still prevented, so this tibiofibular joint is classified as an immovable synarthrosis. (Note, however, that some authorities prefer to call this joint a slightly movable amphiarthrosis.) The ligament-like *interosseous membrane* that connects the radius and ulna along their length is also a syndesmosis, but one with longer fibers (see Figure 8.4). This membrane forms a freely movable joint (diarthrosis) that permits the rotation of the radius around the ulna.

Gomphoses

A **gomphosis** (gom-fo′sis; "bolt") is a peg-in-socket joint. The only example is the articulation of a tooth with its socket (Figure 9.1c). In this case, the connecting ligament is the short *periodontal ligament*.

CARTILAGINOUS JOINTS

In **cartilaginous** (kar″tĭ-laj′ĭ-nus) **joints,** the articulating bones are united by cartilage (Figure 9.2). Cartilaginous joints, which lack a joint cavity and are not highly movable, are of two types: *synchondroses* and *symphyses*.

Synchondroses

A joint where *hyaline* cartilage unites the bones is a **synchondrosis** (sin″kon-dro′sis; "junction of cartilage"). The epiphyseal plates are synchondroses (Figure 9.2a). Functionally, these plates are classified as immovable synarthroses. Another example is the immovable joint between the first rib's costal cartilage and the manubrium of the sternum (Figure 9.2b).

Symphyses

A joint where *fibrocartilage* unites the bones is a **symphysis** (sim′fĭ-sis; "growing together"). Examples include the intervertebral discs (Figure 9.2c) and the pubic symphysis of the pelvis. Even though fibrocartilage is the main element of a symphysis, hyaline cartilage is also present in the form of articular cartilages on the bony surfaces. We noted in Chapter 6 that fibrocartilage resists both tension and compression stresses and can act as a resilient shock absorber. Symphyses, then, are slightly movable joints (amphiarthroses) that provide strength with flexibility.

FIGURE 9.1 Fibrous joints. **(a)** Sutures of the skull are fibrous joints with very short connecting fibers. **(b)** In a syndesmosis, the fibrous tissue (ligament) is longer than that in sutures. **(c)** A gomphosis is formed by the periodontal ligament that holds a tooth in a socket.

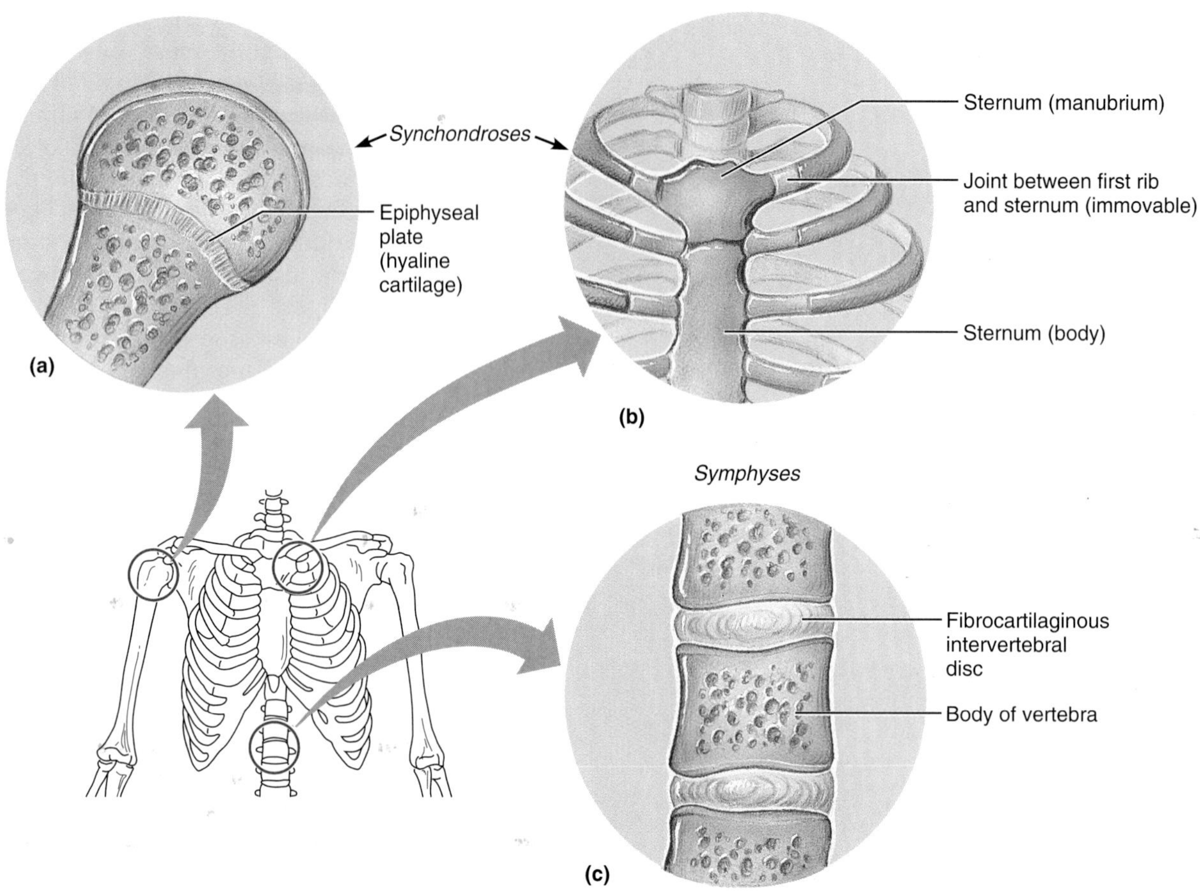

FIGURE 9.2 Cartilaginous joints. (a) The epiphyseal plate in a growing long bone is a synchondrosis (hyaline cartilage joint); recall that epiphyseal plates are only temporary joints that disappear when bones stop growing. (b) Another synchondrosis is the joint between rib 1 and the manubrium of the sternum. (c) The intervertebral discs are symphyses (fibrocartilage joints).

SYNOVIAL JOINTS

Synovial joints (sĭ-no′ve-al; "joint eggs") are the most movable joints of the body, and all are diarthroses. (In fact, many authorities equate the terms *synovial joint* and *diarthrosis*.) Each synovial joint contains a fluid-filled *joint cavity*. Most joints of the body are in this class, especially those in the limbs (see Table 9.2).

Most synovial joints have just two articulating surfaces and are called *simple*. A few, however, have more than two such surfaces and are called *compound*. Examples of compound joints are the elbow, where the two condyles of the humerus articulate with both the ulna and the radial head, and the knee, where the femur and tibia each have two articular condyles.

General Structure of Synovial Joints

The basic features of synovial joints (Figure 9.3a) include the following:

1. **Articular cartilage.** The ends of the opposing bones are covered by articular cartilages composed of hyaline cartilage (p. 130). These spongy cushions absorb compression placed on the joint and thereby keep the bone ends from being crushed.
2. **Joint cavity (synovial cavity).** A feature unique to synovial joints, the joint cavity is really just a potential space that holds a small amount of synovial fluid.

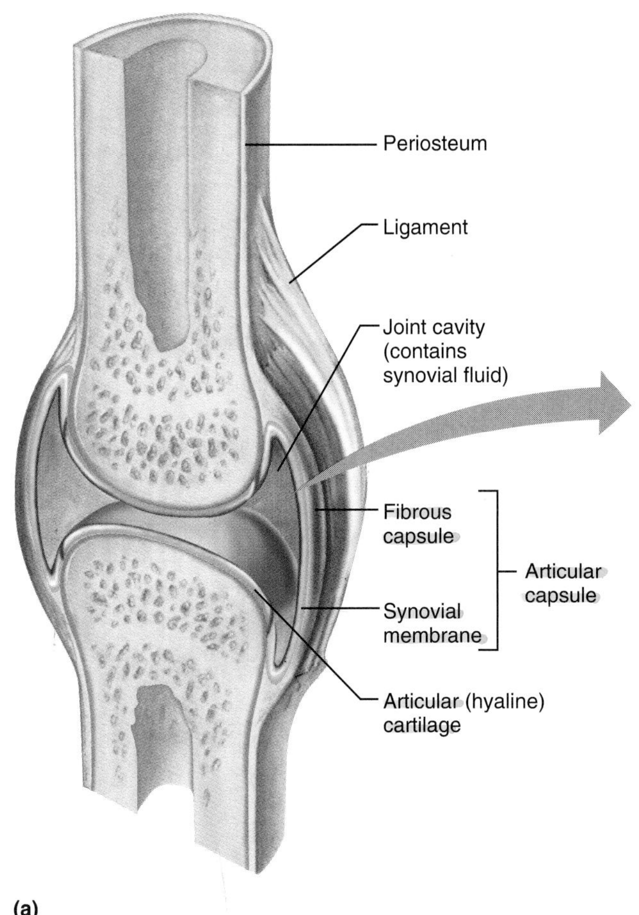

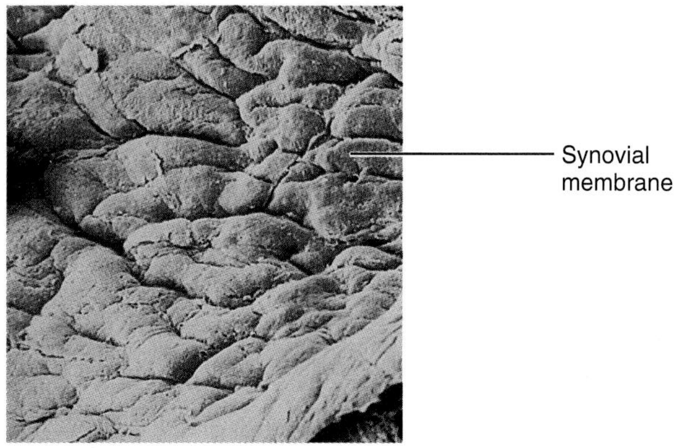

Synovial membrane

(b)

Synovial membrane

Articular cartilage

Joint cavity

Articular disc

Fibrous capsule

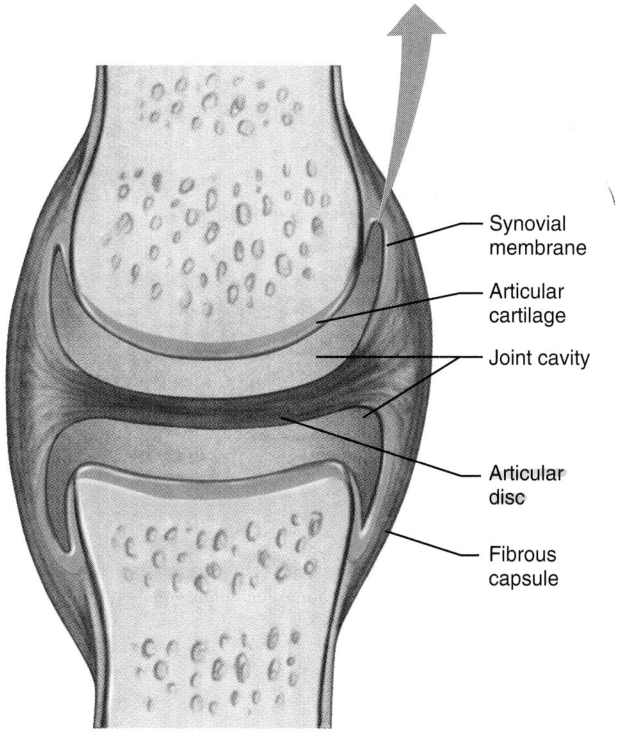

(c)

FIGURE 9.3 Synovial joints. **(a)** A typical synovial joint between the ends of two bones. **(b)** Scanning electron micrograph of the surface of a synovial membrane from the knee joint (13×). **(c)** A synovial joint that contains an articular disc.

3. **Articular capsule.** The joint cavity is enclosed by a two-layered articular capsule, or joint capsule. The outer layer is a **fibrous capsule** of dense irregular connective tissue that is continuous with the periosteum layer of the joining bones. It strengthens the joint so that the bones are not pulled apart. The inner layer of the capsule is a **synovial membrane** composed of loose connective tissue. Besides lining the joint capsule, this membrane covers all the internal joint surfaces not covered by cartilage. Its function is to make synovial fluid.

4. **Synovial fluid.** The viscous liquid inside the joint cavity is called synovial fluid because it resembles raw egg white (*ovum* = egg). Synovial fluid is primarily a filtrate of blood, arising from capillaries in the synovial membrane. It also contains special glycoprotein molecules, secreted by the fibroblasts in the synovial membrane, that make synovial fluid a slippery lubricant that eases the movement at the joint. Synovial fluid not only occupies the joint cavity but also occurs *within* the articular cartilages. The pressure placed on joints during their normal movements squeezes synovial fluid into and out of

Table 9.2 Structural and Functional Characteristics of Body Joints

Illustration	Joint	Articulating bones	Structural type*	Functional type; movements allowed
	Skull	Cranial and facial bones	Fibrous; suture	Synarthrotic; no movement
	Temporo-mandibular	Temporal bone of skull and mandible	Synovial; modified hinge (contains articular disc)	Diarthrotic; gliding and uniaxial rotation; slight lateral movement, elevation, depression, protraction and retraction of mandible
	Atlanto-occipital	Occipital bone of skull and atlas	Synovial; condyloid	Diarthrotic; biaxial; flexion, extension, abduction, adduction, circumduction of head on neck
	Atlantoaxial	Atlas (C_1), and axis (C_2)	Synovial; pivot	Diarthrotic; uniaxial; rotation of the head
	Intervertebral	Between adjacent vertebral bodies	Cartilaginous; symphysis	Amphiarthrotic; slight movement
	Intervertebral	Between articular processes	Synovial; plane	Diarthrotic; gliding
	Vertebrocostal	Vertebrae (transverse processes or bodies) and ribs	Synovial; plane	Diarthrotic; gliding of ribs
	Sternoclavicular	Sternum and clavicle	Synovial; shallow saddle (contains articular disc)	Diarthrotic; multiaxial (allows clavicle to move in all axes)
	Sternocostal	Sternum and rib 1	Cartilaginous; synchondrosis	Synarthrotic; no movement
	Sternocostal	Sternum and ribs 2–7	Synovial; double plane	Diarthrotic; gliding

*Fibrous joints indicated by orange circles, cartilaginous joints by blue circles, and synovial joints by purple circles.

the articular cartilages, nourishing the cells in these cartilages and lubricating their free surfaces.

5. **Reinforcing ligaments.** Some synovial joints are reinforced and strengthened by band-like ligaments. Most often, the ligaments are intrinsic, or capsular; that is, they are thickened parts of the fibrous capsule itself. In other cases they are distinct *extracapsular ligaments* located just outside the capsule, or *intracapsular ligaments* that are internal to the capsule. Intracapsular ligaments are covered with a synovial membrane that separates them from the joint cavity through which they run.
6. **Nerves and vessels.** Synovial joints are richly supplied with sensory nerve fibers that innervate the articular capsule. Some of these fibers detect pain, but most monitor how much the capsule is being stretched. This monitoring of joint stretching is one of several ways by which the nervous system senses our posture and body movements (see pp. 336–337). Synovial joints also have a rich blood supply. Most of the blood vessels supply the synovial membrane, where extensive capillary beds produce the blood filtrate that is the basis of synovial fluid.

Each synovial joint in the body is served by branches from *several* major nerves and blood vessels. These branches come from many different directions and supply overlapping areas of the joint capsule.

Critical Reasoning

Propose a functional explanation for the multiple, overlapping pattern of nerves and blood vessels that supply a joint capsule.

Illustration	Joint	Articulating bones	Structural type*	Functional type; movements allowed
	Acromio-clavicular	Acromion process of scapula and clavicle	Synovial; plane	Diarthrotic; gliding and rotation of scapula on clavicle
	Shoulder (glenohumeral)	Scapula and humerus	Synovial; ball and socket	Diarthrotic; multiaxial; flexion; extension, abduction, adduction, circumduction, rotation of humerus/arm
	Elbow	Ulna (and radius) with humerus	Synovial; hinge	Diarthrotic; uniaxial; flexion; extension of forearm
	Radioulnar (proximal)	Radius and ulna	Synovial; pivot	Diarthrotic; uniaxial; rotation of radius around long axis of forearm to allow pronation and supination
	Radioulnar (distal)	Radius and ulna	Synovial; pivot (contains articular disc)	Diarthrotic; uniaxial; rotation (convex head of ulna rotates in ulnar notch of radius)
	Wrist (radiocarpal)	Radius and proximal carpals	Synovial; condyloid	Diarthrotic; biaxial; flexion, extension, abduction, adduction, circumduction of hand
	Intercarpal	Adjacent carpals	Synovial; plane	Diarthrotic; gliding
	Carpometacarpal of digit 1 (thumb)	Carpal (trapezium) and metacarpal 1	Synovial; saddle	Diarthrotic; biaxial; flexion, extension, abduction, adduction, circumduction, opposition of metacarpal 1
	Carpometacarpal of digits 2–5	Carpal(s) and metacarpal(s)	Synovial; plane	Diarthrotic; gliding of metacarpals
	Knuckle (metacarpo-phalangeal)	Metacarpal and proximal phalanx	Synovial; condyloid	Diathrotic; biaxial; flexion, extension, abduction, adduction, circumduction of fingers
	Finger (interphalangeal)	Adjacent phalanges	Synovial; hinge	Diarthrotic; uniaxial; flexion, extension of fingers

***Fibrous joints** indicated by orange circles, **cartilaginous joints** by blue circles, and **synovial joints** by purple circles.

Such overlapping provides functional redundancy, so that when an injury to a joint destroys some vessels and nerves, others survive and keep the joint functioning. Furthermore, when the normal movements at a joint compress a blood vessel, other vessels stay open and keep the joint nourished.

Certain synovial joints contain a disc of fibrocartilage (Figure 9.3c), alternatively called an **articular disc,** an **intra-articular disc,** or a **meniscus** (mĕ-nis′kus; "crescent"). Articular discs occur in the temporomandibular (jaw) joint, knee joint, and a few others (see Table 9.2). Such a disc extends internally from the capsule and completely or partly divides the joint cavity in two. Articular discs occur in joints whose articulating bone ends have somewhat different shapes. Here is why: When two articulating surfaces fit poorly, they touch each other only at small points, where the loading forces become highly concentrated; this can damage the articular cartilages and lead to arthritis. To avoid this, the disc fills the gaps and improves the fit, thereby distributing the load more evenly and minimizing wear and damage. These discs may also allow two different movements at the same joint—a distinct movement across each face of the disc.

How Synovial Joints Function

Synovial joints are elaborate lubricating devices that allow joining bones to move across one another with a minimum of friction. Without this lubrication, rubbing

Illustration	Joint	Articulating bones	Structural type*	Functional type; movements allowed
	Sacroiliac	Sacrum and coxal bone	Synovial; plane	Diarthrotic; little movement, slight gliding possible (more during pregnancy)
	Pubic symphysis	Pubic bones	Cartilaginous; symphysis	Amphiarthrotic; slight movement (enhanced during pregnancy)
	Hip (coxal)	Coxal bone and femur	Synovial; ball and socket	Diarthrotic; multiaxial; flexion, extension, abduction, adduction, rotation, circumduction of femur/thigh
	Knee (tibiofemoral)	Femur and tibia	Synovial; modified hinge	Diarthrotic; biaxial; flexion, extension of leg, some rotation allowed
	Knee (femoropatellar)	Femur and patella	Synovial; plane	Diarthrotic; gliding of patella
	Tibiofibular	Tibia and fibula (proximally)	Synovial; plane	Diarthrotic; gliding of fibula
	Tibiofibular	Tibia and fibula (distally); both anterior and posterior ligaments exist	Fibrous; syndesmosis	Synarthrotic; slight "give" during dorsiflexion of foot
	Ankle	Tibia and fibula with talus	Synovial; hinge	Diarthrotic; uniaxial; dorsiflexion and plantar flexion of foot
	Intertarsal	Adjacent tarsals	Synovial; plane	Diarthrotic; gliding; inversion and eversion of foot
	Tarsometatarsal	Tarsal(s) and metatarsal(s)	Synovial; plane	Diarthrotic; gliding of metatarsals
	Metatarso-phalangeal	Metatarsal and proximal phalanx	Synovial; condyloid	Diarthrotic; biaxial; flexion, extension, abduction, adduction, circumduction of great toe
	Toe (interpha-langeal)	Adjacent phalanges	Synovial; hinge	Diarthrotic; unaxial; flexion, extension of toes

***Fibrous joints** indicated by orange circles, **cartilaginous joints** by blue circles, and **synovial joints** by purple circles.

would wear away the joint surfaces, and excessive friction could overheat and destroy the joint tissues, essentially "cooking" them.

Synovial joints are routinely subjected to *compressive* forces that occur when the muscles that move bones pull the bone ends together. As the opposing articular cartilages touch, synovial fluid is squeezed out of them, producing a film of lubricant between the cartilage surfaces. The two moving cartilage surfaces ride on this slippery film, not on each other. Synovial fluid is such an effective lubricant that it yields less friction than ice sliding on ice. When the pressure on the joint ceases, the synovial fluid rushes back into the articular cartilages like water into a sponge, ready to be squeezed out again the next time a load is placed on the joint. This mechanism is called **weeping lubrication.**

Critical Reasoning

What is the source of the popping sounds produced when people "crack their knuckles" by pulling on the fingers?

According to one theory, pulling on a synovial joint creates a suction that rapidly pulls water out of the articular cartilages, which then recoil with a snap. A different theory holds that the suction draws respiratory gases out of the capillaries of the synovial membrane. The gases coalesce into bubbles in the synovial fluid that then burst. Incidentally, habitually cracking one's knuckles may cause them to enlarge over time, but there is no evidence that this habit damages the joints.

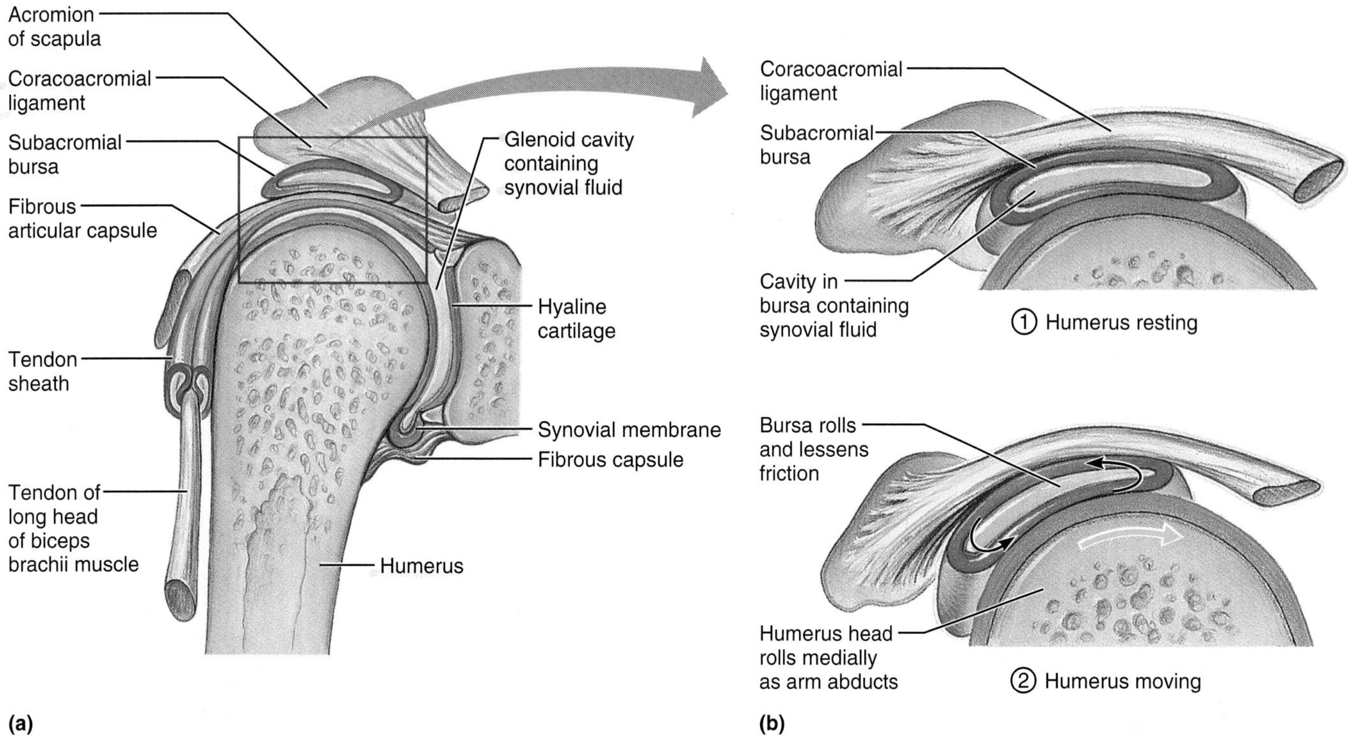

FIGURE 9.4 Bursae and tendon sheaths. Both these structures, which are filled with synovial fluid, are friction-reducing devices. **(a)** Frontal section through the right shoulder joint. **(b)** Enlargement of (a), showing how a bursa eliminates friction where a ligament (or other structure) would rub against a bone.

Bursae and Tendon Sheaths

Bursae and *tendon sheaths* are not synovial joints, but they contain synovial fluid and often are associated with synovial joints. Essentially closed bags of lubricant, these structures act like "ball bearings" to reduce friction between body elements that move over one another (Figure 9.4). A **bursa** (ber′sah), a Latin word meaning "purse," is a flattened fibrous sac lined by a synovial membrane. Bursae occur where ligaments, muscles, skin, tendons, or bones overlie each other and rub together. A **tendon sheath** is essentially an elongated bursa that wraps around a tendon like a bun around a hot dog (see Figure 9.4a). Tendon sheaths occur only on tendons that are subjected to friction, such as those that travel through joint cavities or are crowded together within narrow canals (in the wrist region, for example).

Factors Influencing the Stability of Synovial Joints

Because joints are regularly pushed and pulled, they must be stabilized so that they do not dislocate (come out of alignment). The stability of a synovial joint depends on three factors: the shapes of the articular surfaces, the number and position of stabilizing ligaments, and muscle tone.

Articular Surfaces

The articular surfaces of the bones in a joint fit together in a complementary manner. Although their shapes determine what kinds of movement are possible at the joint, articular surfaces seldom play a major role in joint stability: Most joint sockets are just too shallow. Still, some joint surfaces have deep sockets or grooves that do provide stability. The best example is the ball and deep socket of the hip joint; other examples are the elbow and ankle joints.

Ligaments

The capsules and ligaments of synovial joints help hold the bones together and prevent excessive or undesirable motions. As a rule, the more ligaments a joint has, the stronger it is. When other stabilizing factors are inadequate, however, undue tension is placed on the ligaments, and they fail. Once stretched, ligaments, like taffy, stay stretched. However, a ligament can stretch only about 6% beyond its normal length before it snaps apart.

Muscle Tone

The most important factor in joint stabilization is **muscle tone,** a constant, low level of contractile force generated by a muscle even when it is not causing movement. Muscle tone helps stabilize joints by keeping tension on the muscle tendons that cross over joints just external to the joint capsule. This stabilizing factor is especially important in reinforcing the shoulder and knee joints, and in supporting the joints in the arches of the foot.

Critical Reasoning

People who can bend their thumb back to touch their forearm or place both heels behind their neck are said to be "double-jointed," incorrectly implying that they have more joints than is normal. What is the correct explanation for these abilities?

The joint ligaments and joint capsules of "double-jointed" individuals are simply looser and more stretchable than is the case with the majority of people.

Movements Allowed by Synovial Joints

As muscles contract, they cause bones to move at the synovial joints. The resulting movements are of three basic types: (1) *gliding* of one bone surface across another; (2) *angular movements,* which change the angle between the two bones; and (3) *rotation* about a bone's long axis.

Gliding

In **gliding,** the nearly flat surfaces of two bones slip across each other (Figure 9.5a). Gliding occurs at the joints between the carpals and tarsals and between the flat articular processes of the vertebrae.

Angular Movements

Angular movements (Figure 9.5b–e) increase or decrease the angle between two bones. These movements, which may occur in any plane of the body, include flexion, extension, abduction, adduction, and circumduction.

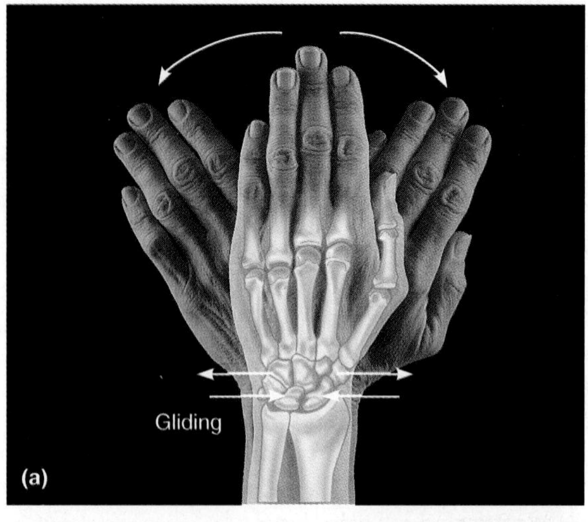

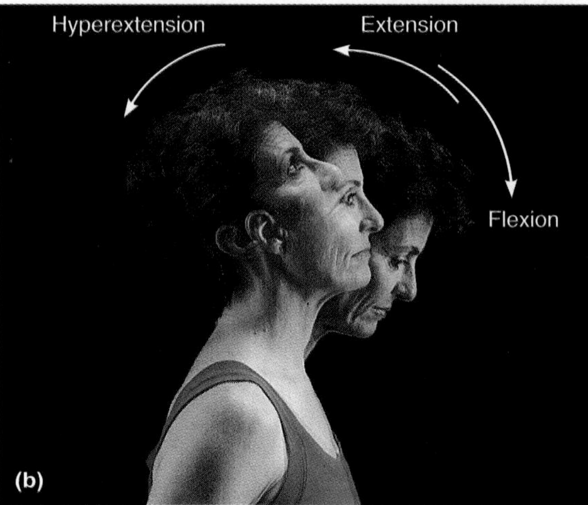

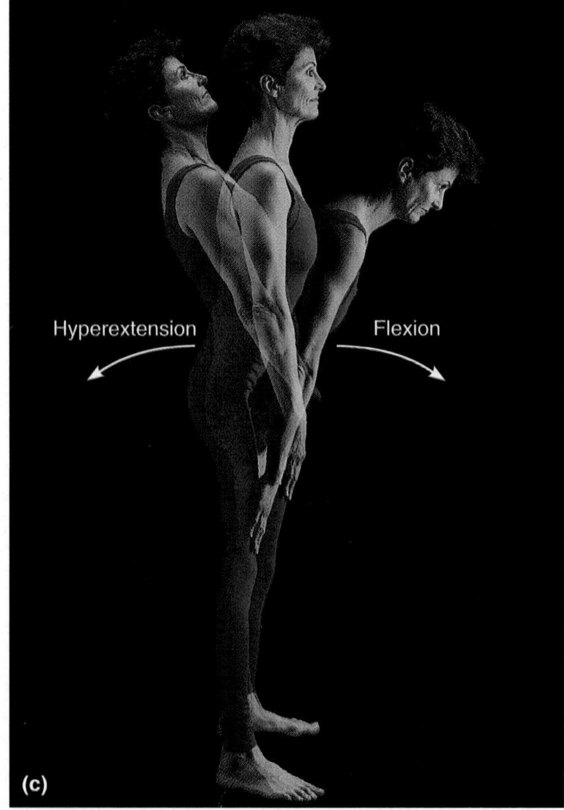

FIGURE 9.5 **Movements allowed by synovial joints.** **(a)** Gliding movements, as the distal carpals glide on the proximal carpals. **(b–e)** Angular movements: **(b)** Flexion and extension of the neck on the trunk. **(c)** Flexion and hyperextension of the vertebral column. **(d)** Flexion and extension of the arm at the shoulder and of the leg at the knee. (Legend continues on p. 222.)

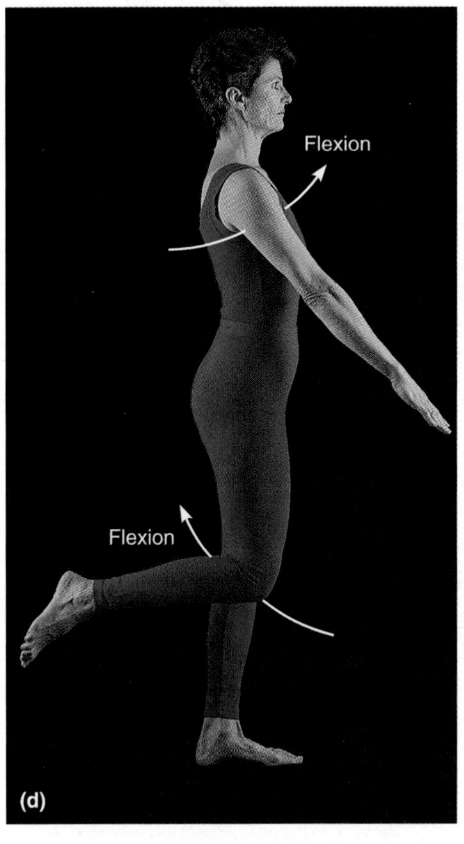

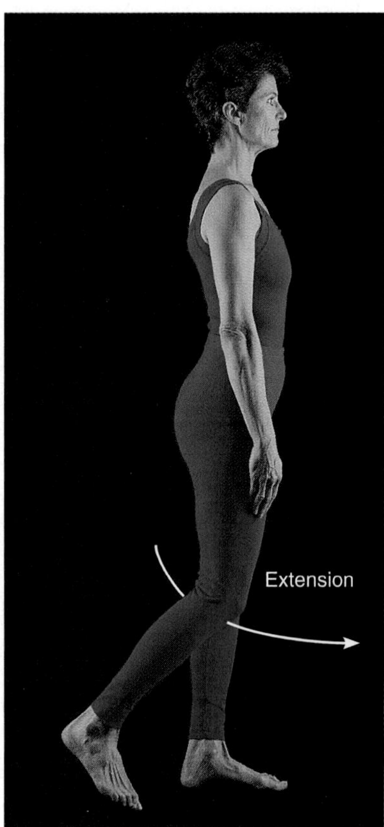

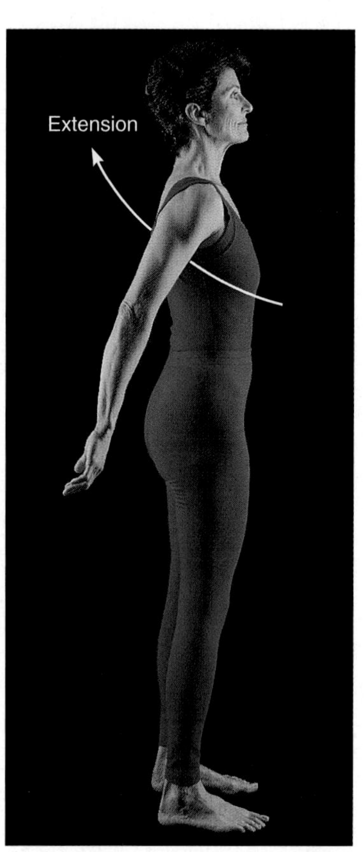

Flexion **Flexion** is bending that *decreases the angle* between the bones, bringing these bones closer together (see Figure 9.5b–d). This movement usually occurs in the sagittal plane of the body. Examples include bending the neck or trunk forward, bending the fingers, and bending the forearm toward the arm at the elbow. In less obvious examples, the arm is flexed at the shoulder when lifted in an anterior direction (see Figure 9.5d), and the thigh is flexed at the hip when lifted anteriorly.

Extension **Extension** is the reverse of flexion and occurs at the same joints (see Figure 9.5b–d). It *increases the angle* between the joining bones and is a straightening action. Straightening the fingers after making a fist is an example of extension. Bending a joint back beyond its straight position is called **hyperextension** (literally, "superextension," see Figure 9.5c). At the shoulder, extension carries the arm posteriorly (see Figure 9.5d).

Abduction **Abduction,** from Latin words meaning "moving away," is movement of a limb *away* from the body midline. Raising the arm or thigh laterally is an example of abduction (see Figure 9.5e). For the fingers or toes, *abduction* means spreading them apart. In this case, the "midline" is the longest digit: the third finger or the second toe. Note that bending the trunk away from the body midline to the right or left is called *lateral flexion* instead of abduction.

Adduction **Adduction** ("moving toward") is the opposite of abduction: It is the movement of a limb *toward* the body midline (see Figure 9.5e) or, in the case of the digits, toward the midline (longest digit) of the hand or foot.

Circumduction **Circumduction** ("moving in a circle") is moving a limb or finger so that it describes a cone in space (see Figure 9.5e). This is a complex movement that combines flexion, abduction, extension, and adduction in succession. Circumduction combines almost all of the movements possible at the shoulder and hip joints. Therefore, circumducting your limbs is the best way to exercise many different limb muscles simultaneously.

Rotation

Rotation is the turning movement of a bone around its own long axis. This is the only movement allowed between the first two cervical vertebrae, and it also occurs at the hip and shoulder joints (Figure 9.5f). Rotation may be directed toward the median plane or away from it. For example, in **medial rotation** of the lower

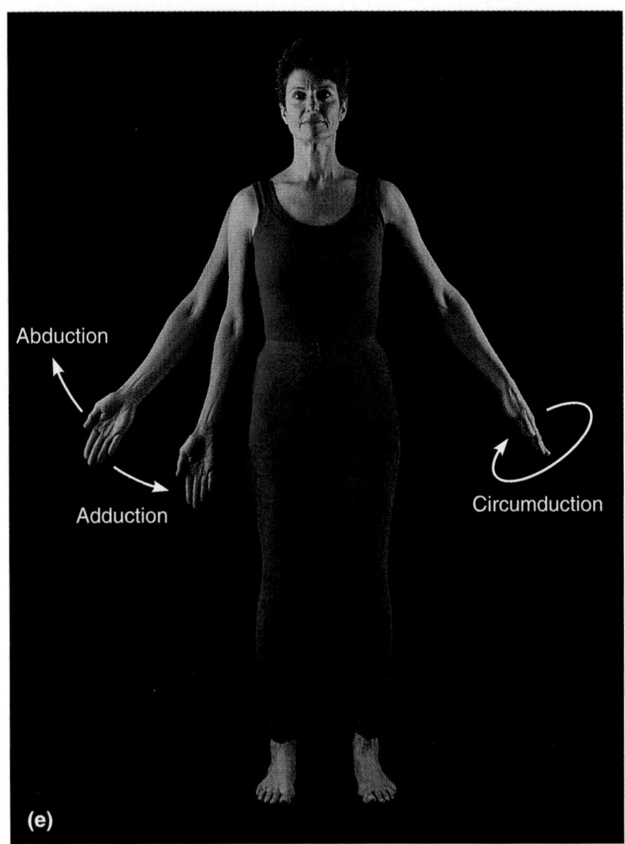

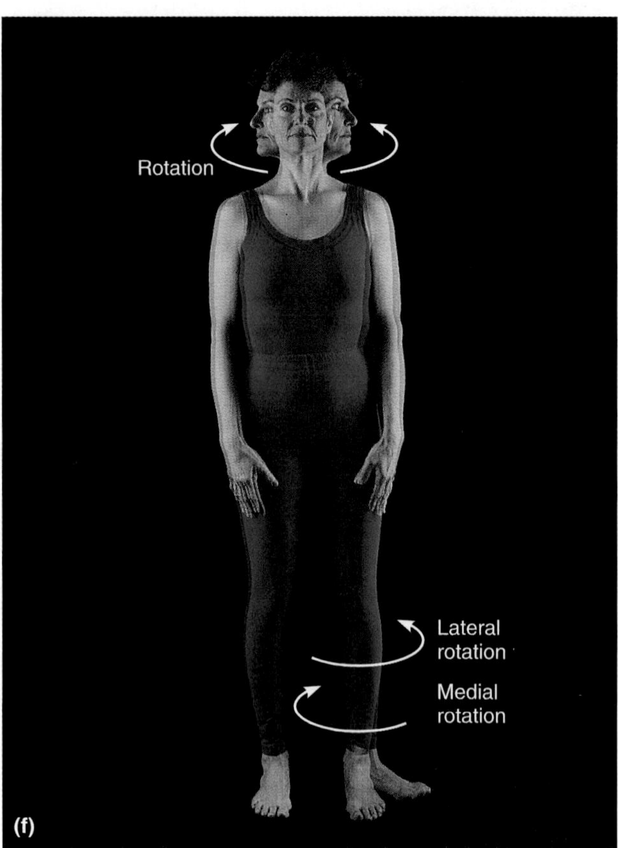

FIGURE 9.5 Movements allowed by synovial joints, *continued.* Angular movements, continued: **(e)** Abduction, adduction, and circumduction of the upper limb at the shoulder. **(f)** Rotation of the head, neck, and lower limb.

limb, the limb's anterior surface turns toward the median plane of the body; **lateral rotation** is the opposite of that movement. The vertebral column also rotates, twisting the whole trunk to the right or left.

Special Movements

Certain movements do not fit into any of the previous categories and occur at only a few joints. Some of these special movements are shown in Figure 9.6.

Supination and Pronation The terms **supination** (soo″pĭ-na′shun; "turning backward") and **pronation** (pro-na′shun; "turning forward") refer to movements of the radius around the ulna (Figure 9.6a). Supination occurs when the forearm rotates laterally so that the palm faces anteriorly, whereas pronation occurs when the forearm rotates medially so that the palm faces posteriorly. Pronation brings the radius across the ulna so that the two bones form an X. A helpful memory trick: If you lift a cup of soup up to your mouth *on your palm,* you are supinating ("soup"-inating).

Dorsiflexion and Plantar Flexion Up and down movements of the foot at the ankle are given special names. Lifting the foot so that its superior surface approaches the shin is called **dorsiflexion,** whereas depressing the foot (pointing the toes) is called **plantar flexion** (Figure 9.6b). Dorsiflexion of the foot corresponds to extension of the hand at the wrist, whereas plantar flexion of the foot corresponds to flexion of the hand.

Inversion and Eversion **Inversion** and **eversion** are also special movements of the foot (Figure 9.6c). To invert the foot, turn the sole medially; to evert the foot, turn the sole laterally.

Protraction and Retraction Nonangular movements in the anterior and posterior directions are called **protraction** and **retraction,** respectively (Figure 9.6d). The mandible is protracted when you jut out your jaw and retracted when you bring it back.

Elevation and Depression **Elevation** means lifting a body part superiorly (Figure 9.6e). Moving the elevated part inferiorly is **depression.** During chewing, the mandible is alternately elevated and depressed.

Opposition In the palm of the hand, the saddle joint between metacarpal 1 and the carpals allows a move-

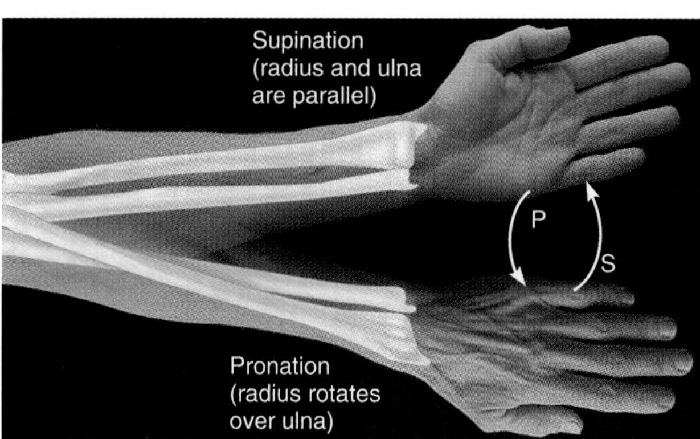

(a) Supination (S) and pronation (P)

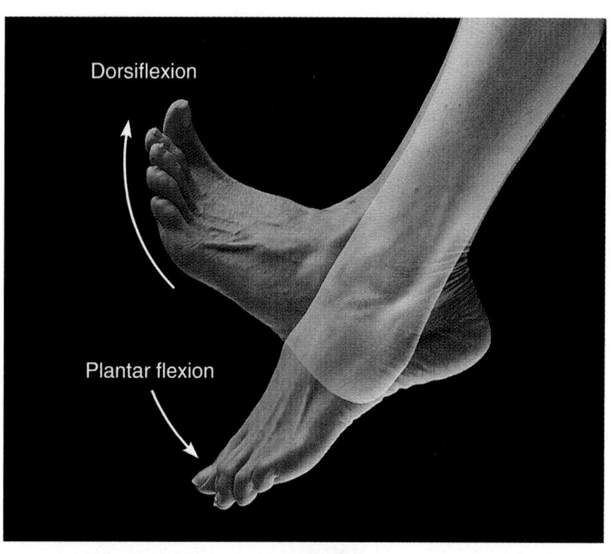

(b) Dorsiflexion and plantar flexion

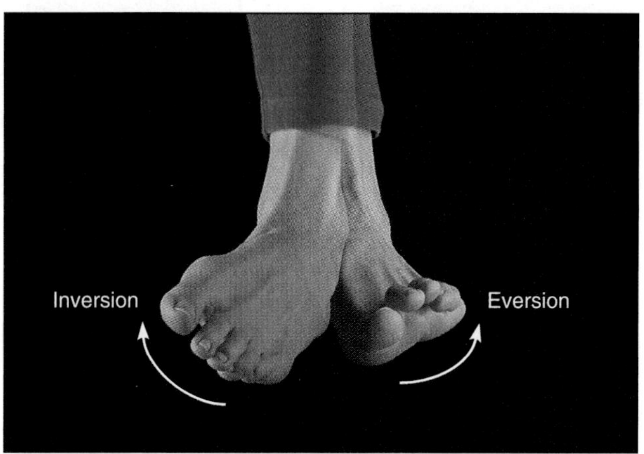

(c) Inversion and eversion

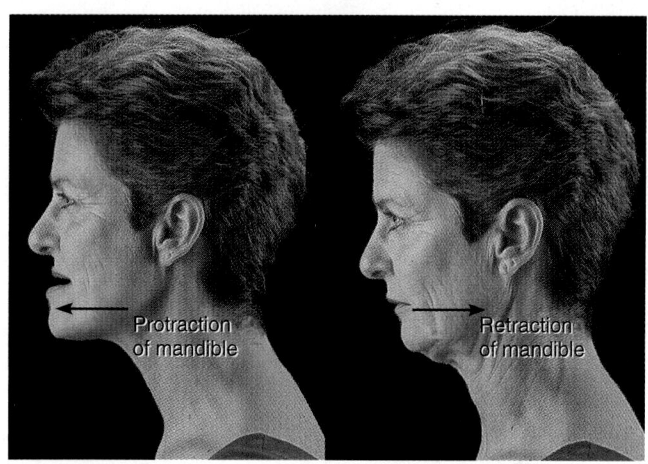

(d) Protraction and retraction

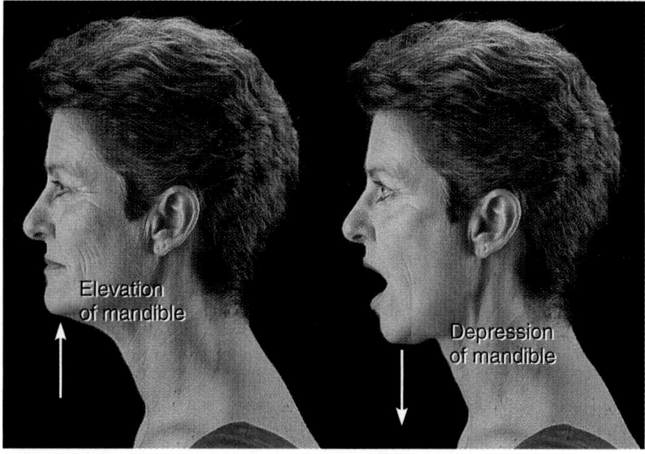

(e) Elevation and depression

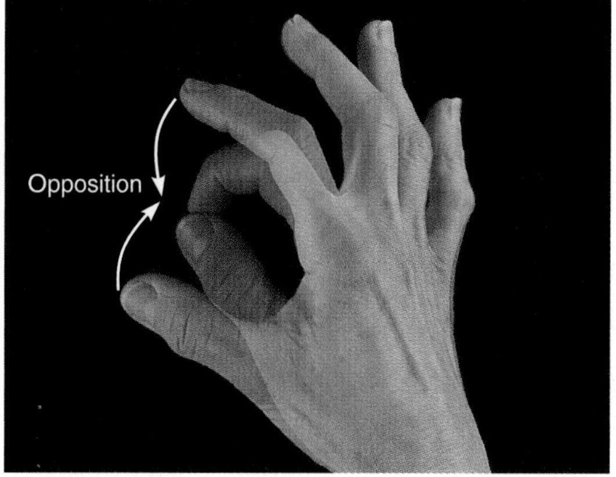

(f) Opposition

FIGURE 9.6 Some special body movements.

FIGURE 9.7 Synovial joints, classified by the shape of their articular surfaces. **(a)** Plane joint. **(b)** Hinge joint. **(c)** Pivot joint. **(d)** Condyloid joint. **(e)** Saddle joint. **(f)** Ball-and-socket joint.

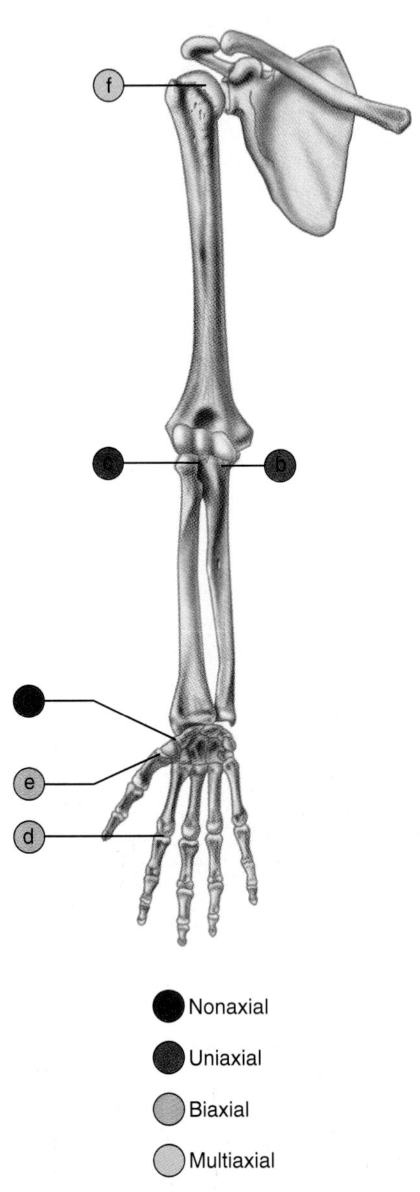

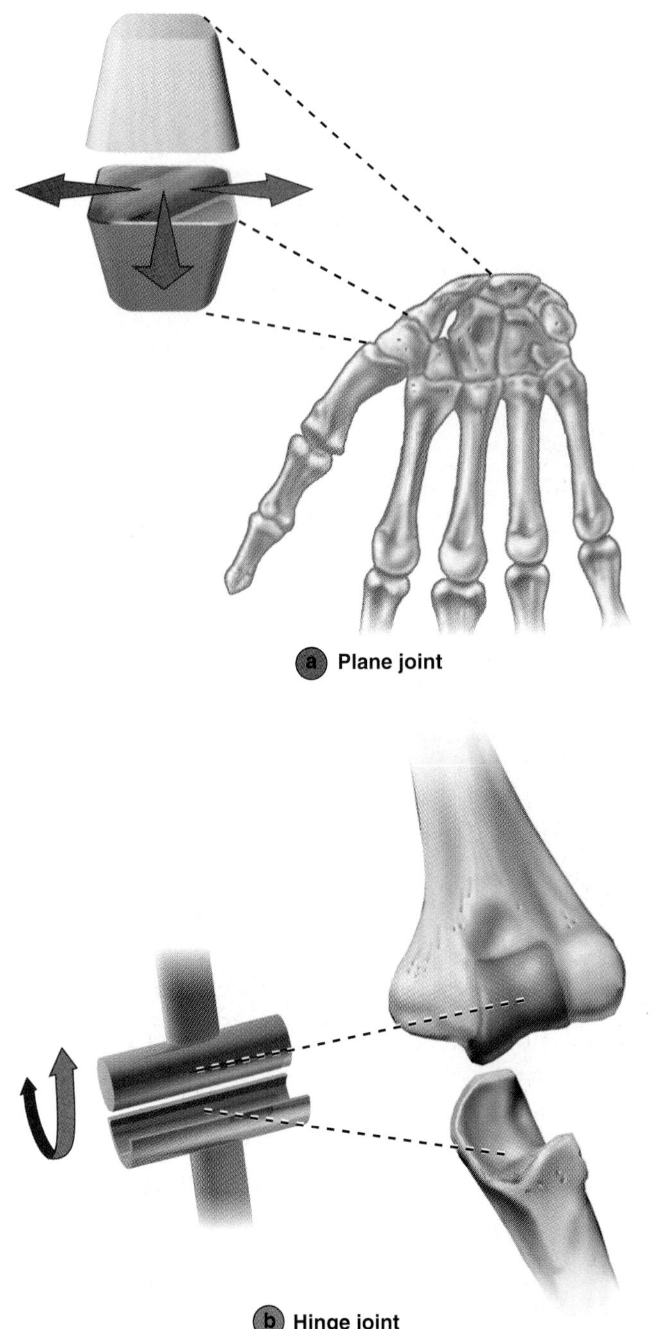

ment called *opposition* of the thumb (Figure 9.6f). This is the action by which you move your thumb to touch the tips of the other fingers on the same hand. This unique action is what makes the human hand such a fine tool for grasping and manipulating objects.

Synovial Joints Classified by Shape

The shapes of the articulating bone surfaces determine the movements allowed at a joint. Based on such shapes, our synovial joints can be classified as *plane, hinge, pivot, condyloid, saddle,* and *ball-and-socket joints* (Figure 9.7).

Plane Joints

In a **plane joint** (Figure 9.7a), the articular surfaces are essentially flat planes, and only short gliding movements are allowed. Plane joints are the gliding joints introduced earlier, such as the intertarsal joints, intercarpal joints, and the joints between the articular processes of the vertebrae. Their movements are **nonaxial;** that is, gliding does not involve rotation around any axis.

Hinge Joints

In a **hinge joint** (Figure 9.7b), the cylindrical end of one bone fits into a trough-shaped surface on another bone. Angular movement is allowed in just one plane, like a

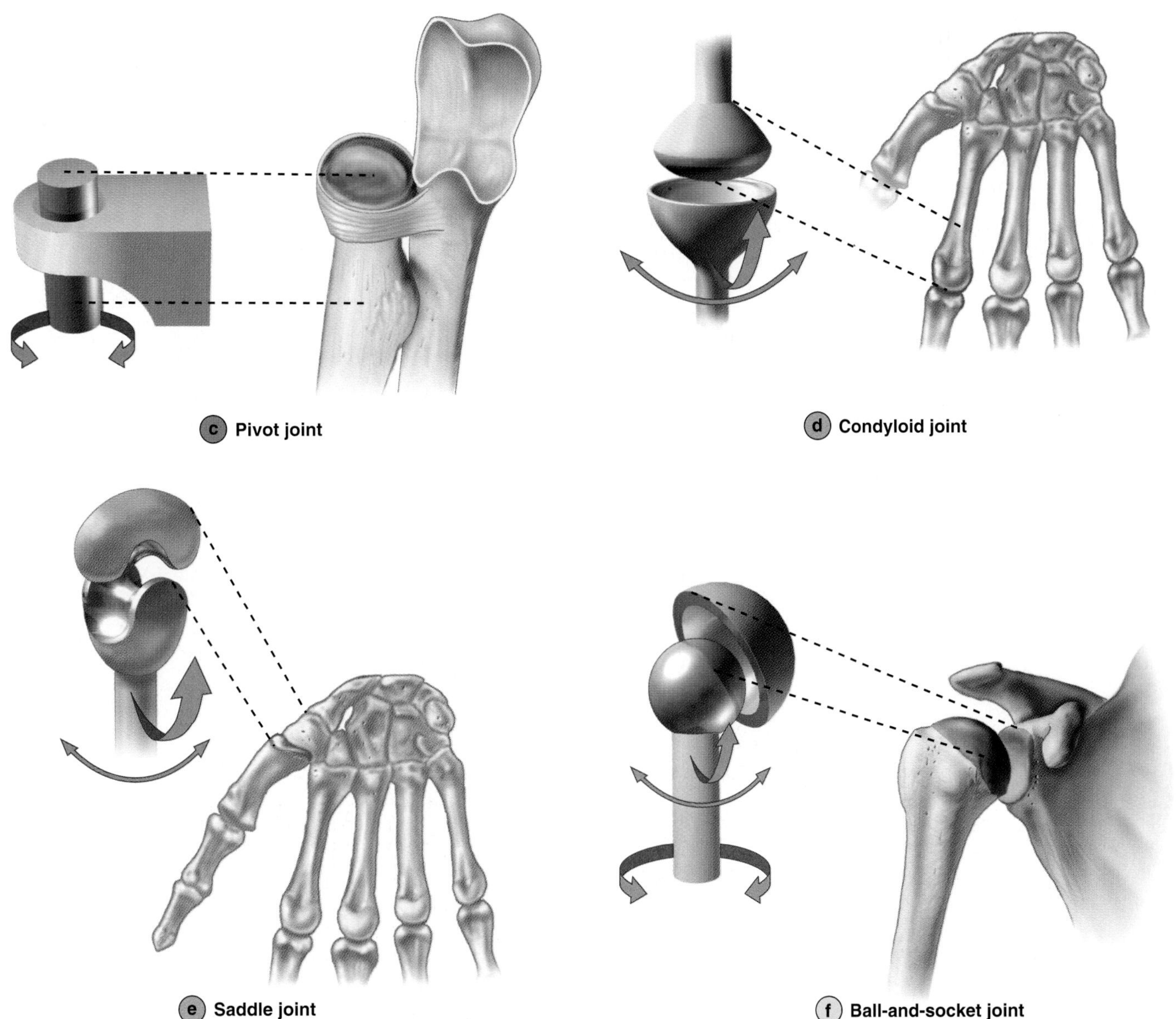

door on a hinge. Examples are the elbow joint, ankle joint, and the joints between the phalanges of the fingers. Hinge joints are classified as **uniaxial** (u″ne-aks′e-al; "one axis"); they allow movement around one axis only, as indicated by the single red arrow in Figure 9.7b.

Pivot Joints

In a **pivot joint** (Figure 9.7c), the rounded end of one bone fits into a ring that is formed by another bone plus an encircling ligament. Because the rotating bone can turn only around its long axis, pivot joints are also uniaxial joints (see the single arrow in Figure 9.7c). One example of a pivot joint is the proximal radioulnar joint, where the head of the radius rotates within a ring-like ligament secured to the ulna. Another example is the joint between the atlas and the dens of the axis (see Figure 7.17a on p.176).

Condyloid Joints

In a **condyloid joint** (kon′dĭ-loid; "knuckle-like"), the egg-shaped articular surface of one bone fits into an oval concavity in another (Figure 9.7d). Both of these articular surfaces are oval. Condyloid joints allow the moving bone to travel (1) from side to side (abduction-adduction) and (2) back and forth (flexion-extension), but the bone cannot rotate around its long axis. Because movement occurs around two axes, indicated by the two

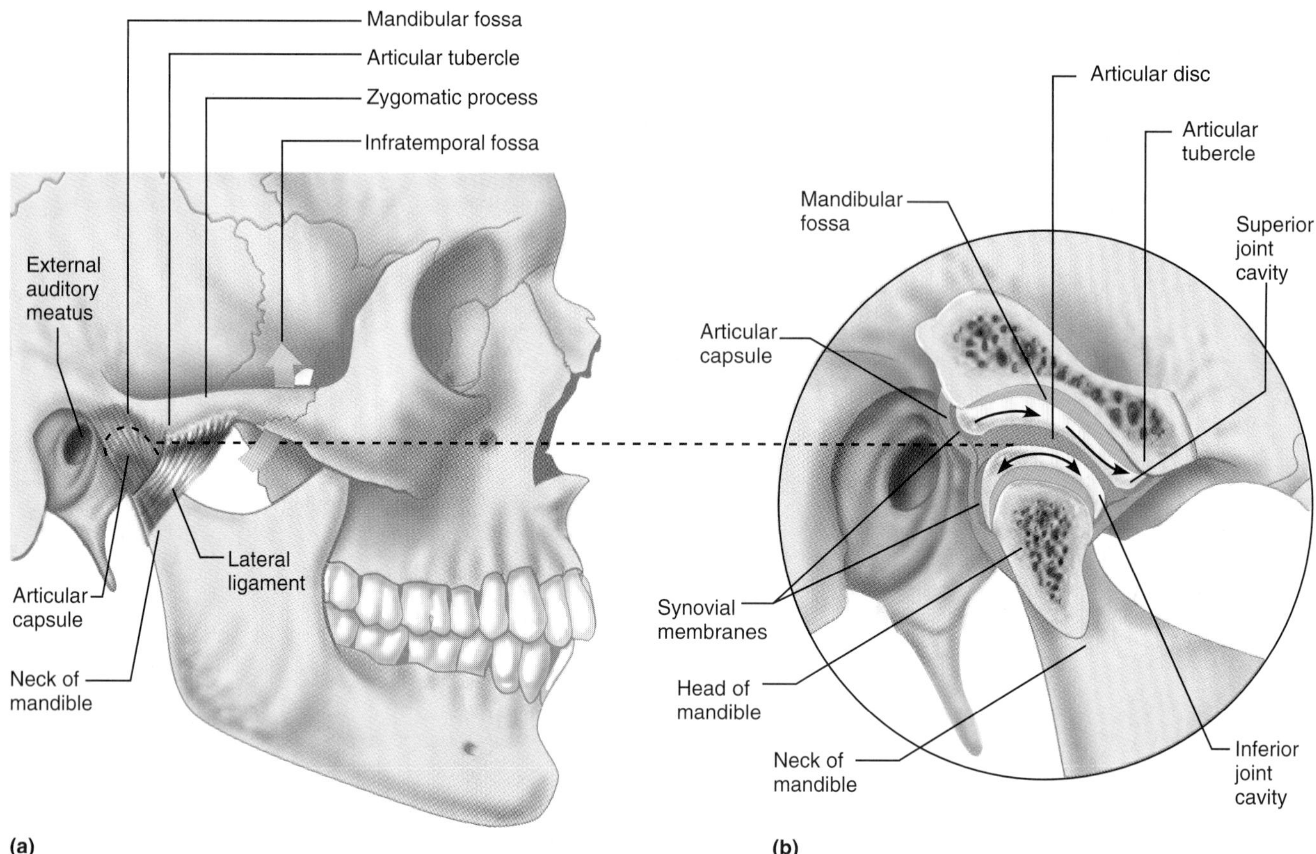

FIGURE 9.8 The temporomandibular (jaw) joint. (a) Location of the joint in the skull, showing the important surrounding structures. (b) Enlargement of a sagittal section through the joint, showing the articular disc, the superior and inferior compartments of the joint cavity, and the two main movements that occur (arrows): The inferior joint cavity lets the head of the mandible rotate in opening and closing the mouth, whereas the superior joint cavity lets the head of the mandible move forward to brace against the articular tubercle when the mouth opens wide.

arrows in Figure 9.7d, these joints are **biaxial** (*bi* = two), as in knuckle (metacarpophalangeal) and wrist joints.

Saddle Joints

In **saddle joints,** each articular surface has both convex and concave areas, just like a saddle. Nonetheless, these biaxial joints allow essentially the same movements as condyloid joints (Figure 9.7e). The best example of a saddle joint is the first carpometacarpal joint, in the ball of the thumb.

Ball-and-Socket Joints

In a **ball-and-socket joint** (Figure 9.7f), the spherical head of one bone fits into a round socket in another. These are **multiaxial** joints that allow movement in all axes, including rotation (see the three arrows in Figure 9.7f). The shoulder and hip are ball-and-socket joints.

Selected Synovial Joints

In this section we consider six important synovial joints in detail: temporomandibular, shoulder, elbow, hip, knee, and ankle joints. While reading about them, keep in mind that they all contain articular cartilages, fibrous capsules, and synovial membranes.

Temporomandibular Joint

The **temporomandibular joint (TMJ),** or jaw joint, lies just anterior to the ear (Figure 9.8a). At this joint, the head of the mandible articulates with the inferior surface of the squamous temporal bone. The head of the mandible is egg-shaped, whereas the articular surface on the temporal bone has a more complex shape: Posteriorly, it forms the concave **mandibular fossa,** while anteriorly it forms a dense knob called the **articular tubercle.** Enclosing the joint is a loose articular capsule, the lateral aspect of which is thickened into a **lateral ligament.** Within the capsule is an articular disc, which divides the synovial cavity into superior and inferior compartments (Figure 9.8b). The two surfaces of the disc allow two distinct kinds of movement at the TMJ. First, the concave inferior surface receives the mandibular head and allows the familiar hinge-like movement of depressing and elevating the mandible while opening and closing the mouth. Second, the superior surface of the disc glides anteriorly with the mandibular head whenever the mouth is opened wide. This anterior movement braces the mandibular head against the dense bone of the articular tubercle, so that the mandible is not forced superiorly through the thin roof of the mandibular fossa when one bites hard foods such as nuts or hard candies. To demonstrate the anterior gliding of your mandible, place a finger on the mandibular head just anterior to your ear opening, and yawn.

CLINICAL APPLICATIONS

TEMPOROMANDIBULAR DISORDERS Because of its shallow socket, the TMJ is the most easily dislocated joint in the body. Even a deep yawn can dislocate it. This joint almost always dislocates anteriorly; that is, the mandibular head glides anteriorly, ending up in a skull region called the infratemporal fossa (see Figure 9.8a). In such cases, the mouth remains wide open and cannot close. To realign a dislocated TMJ, the physician places his or her thumbs in the patient's mouth between the lower molars and the cheeks, and then pushes the mandible inferiorly and posteriorly.

At least 5% of Americans suffer from painful conditions of the TMJ called **temporomandibular disorders.** The most common symptoms are pain in the ear and face, tenderness of the jaw muscles, popping or clicking sounds when the mouth opens, and stiffness of the TMJ. Usually caused by painful spasms of the chewing muscles, temporomandibular disorders affect people who respond to stress by grinding their teeth. However, it can also result from an injury to the TMJ or from poor occlusion of the teeth. Treatment usually focuses on getting the jaw muscles to relax using massage, stretching the muscles, applying moist heat or ice, administering muscle-relaxant drugs, and adopting general stress management techniques. Patients often wear a bite plate while sleeping to stop teeth grinding. In severe cases, surgery on the joint may be necessary. ✚

Shoulder (Glenohumeral) Joint

In the shoulder joint (Figure 9.9), stability has been sacrificed to provide the most freely moving joint of the body. This ball-and-socket joint is formed by the head of the humerus and the shallow glenoid cavity of the scapula. Even though the glenoid cavity is slightly deepened by a rim of fibrocartilage called the **glenoid labrum** (*labrum* = lip) (see Figure 9.9d), this shallow cavity contributes little to joint stability. The articular capsule (see Figure 9.9c) is remarkably thin and loose (qualities that contribute to the joint's freedom of movement) and extends from the margin of the glenoid cavity to the anatomical neck of the humerus. The only strong thickening of the capsule is the superior **coracohumeral ligament,** which helps support the weight of the upper limb. The anterior part of the capsule thickens slightly into three rather weak **glenohumeral ligaments** (see Figure 9.9d).

Muscle tendons that cross the shoulder joint contribute most to the joint's stability. One of these is the tendon of the long head of the biceps brachii muscle (see Figure 9.9a and d). This tendon attaches to the superior margin of the glenoid labrum, travels within the joint cavity, and runs in the intertubercular groove of the humerus, in the process securing the head of the humerus tightly against the glenoid cavity. Four other tendons and the associated muscles make up the **rotator cuff** (see Figure 9.9e), which encircles the shoulder joint and blends with the joint capsule. The rotator cuff muscles include the subscapularis, supraspinatus, infraspinatus, and teres minor (see Chapter 11, p. 297). The rotator cuff can be severely stretched or torn when the arm is moved vigorously and is often injured in baseball pitchers who throw too hard too often.

CLINICAL APPLICATIONS

SHOULDER DISLOCATIONS *Shoulder dislocations* are common injuries. Because the structures reinforcing this joint are weakest anteriorly and inferiorly, the head of the humerus easily dislocates forward and downward. The glenoid cavity provides poor support when the humerus is rotated laterally and abducted, as occurs when a football player makes an arm tackle or a baseball fielder hits the ground diving for a ball. These situations cause many shoulder dislocations, as do blows to the top and back of the shoulder. A *shoulder separation* is a dislocation of the acromioclavicular joint resulting from falling onto an outstretched hand or onto the side of the shoulder. ✚

Elbow Joint

The elbow joint (Figure 9.10) is a hinge that allows only extension and flexion. Even though both the radius and ulna articulate with the condyles of the humerus, it is the close gripping of the humerus by the ulna's trochlear notch that forms the hinge and stabilizes the joint. The articular capsule attaches to the humerus and ulna and to the **annular ligament** (an′u-lar; "ring-like") of the radius, a ring around the head of the radius (see Figure 9.10b). Laterally and medially, the capsule thickens into strong ligaments that prevent lateral and medial movements: the **radial collateral ligament,** a triangular band on the lateral side (see Figure 9.10b), and the **ulnar collateral ligament** on the medial side (see Figure 9.10c). Tendons of several arm muscles, such as the biceps brachii and triceps brachii, cross the elbow joint and provide stability (see Figure 9.10a).

CLINICAL APPLICATIONS

ELBOW TRAUMA Although the elbow is a very stable joint, it often experiences trauma and is dislocated more frequently than any other major joint except the jaw and shoulder. Elbow dislocation is usually caused by falling hand-forward onto an outstretched arm, which pushes the ulna posteriorly. Such a fall may posteriorly dislocate the radius as well, and it often fractures the articulating elements.

Athletes who repeatedly swing a racket or throw a ball subject their forearms to strong outward-bending forces that can weaken and ultimately rupture the ulnar collateral ligament. ✚

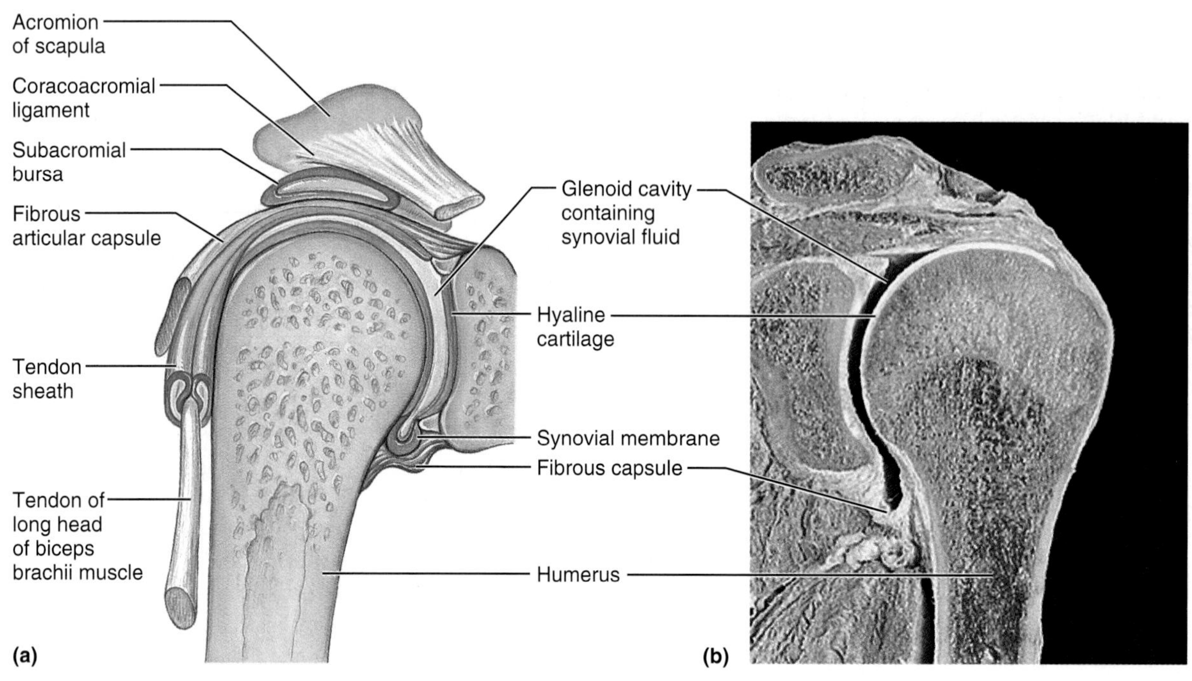

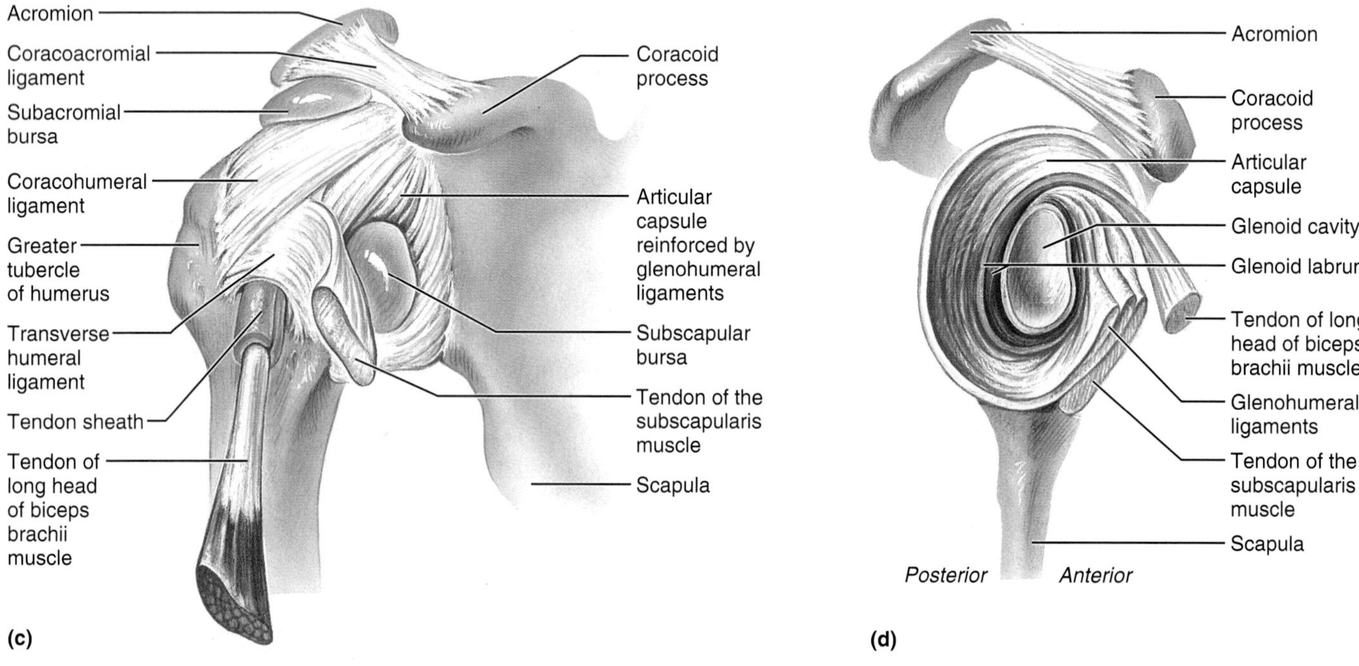

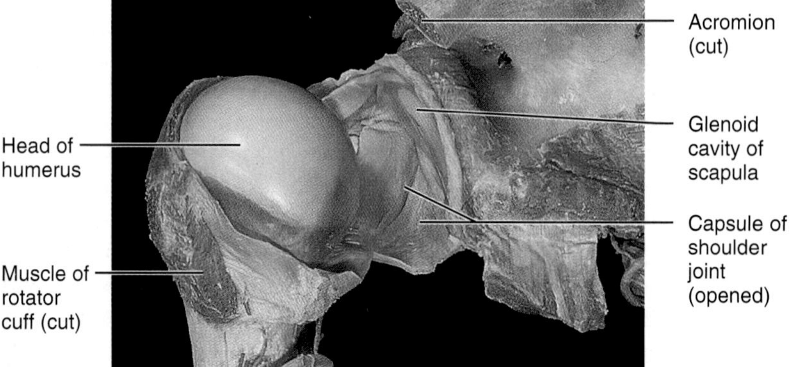

FIGURE 9.9 The shoulder joint. **(a)** Frontal section through the right shoulder joint. **(b)** Photo of a frontally sectioned left shoulder of a cadaver. **(c)** Anterior view of the right shoulder joint with the joint capsule in place. **(d)** Socket of the right shoulder joint, viewed laterally. The humerus has been removed. **(e)** Photo of an opened shoulder joint (anterior view).

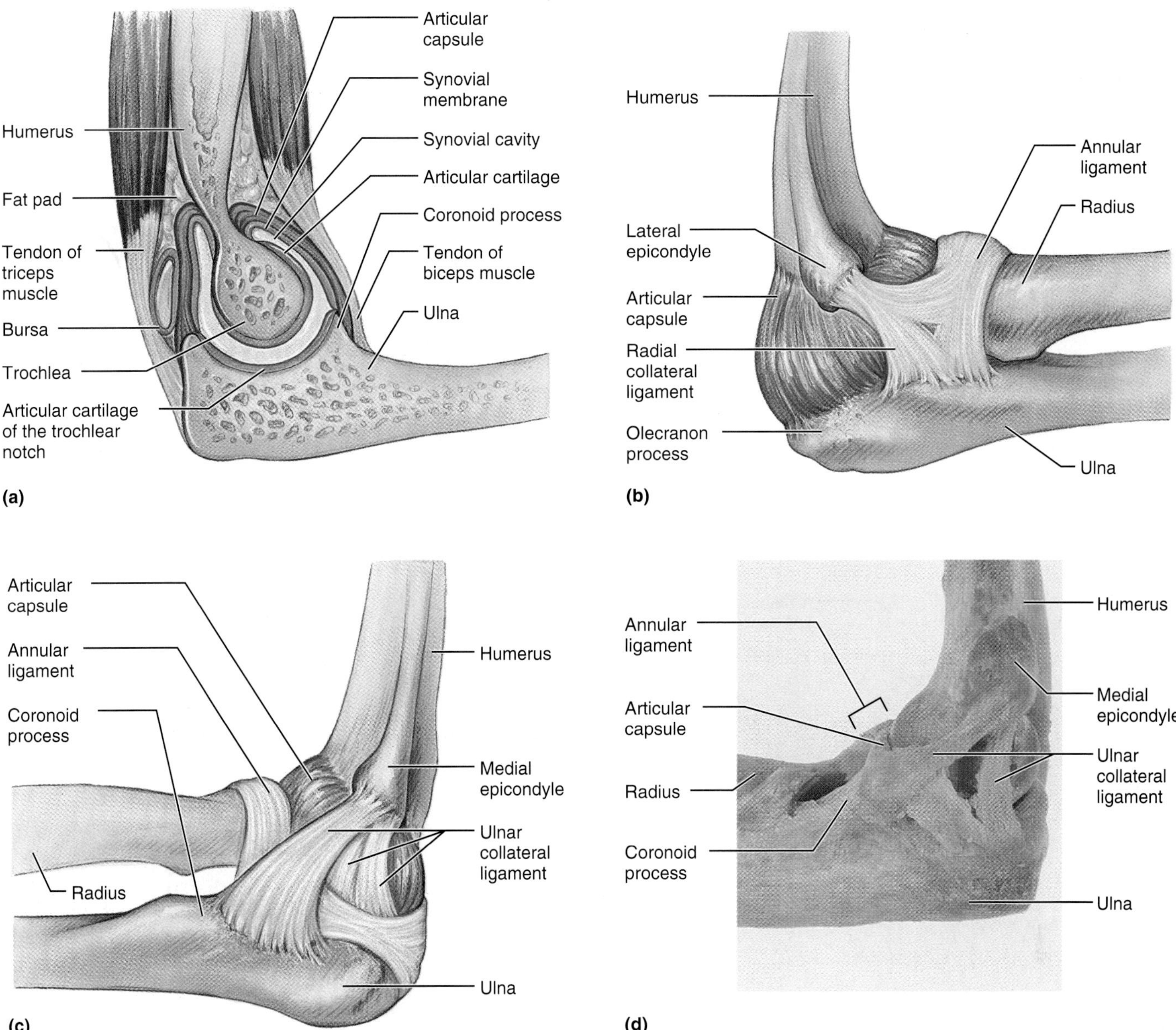

FIGURE 9.10 The elbow joint. **(a)** Sagittal section through the middle of the right elbow joint (lateral view). **(b)** Lateral view of the right elbow joint. **(c)** and **(d)** Medial views of right elbow: diagram (c) and cadaver photo (d).

Hip Joint

The hip (coxal) joint, like the shoulder joint, has a ball-and-socket structure (Figure 9.11). It has a wide range of motion, but not nearly as wide as that of the shoulder joint. Movements occur in all possible axes but are limited by the joint's ligaments and deep socket.

The hip joint is formed by the spherical head of the femur and the deeply cupped acetabulum of the hip bone. The depth of the acetabulum is enhanced by a circular rim of fibrocartilage called the **acetabular labrum** (see Figure 9.11a and d). Because the diameter of this labrum is smaller than that of the head of the femur, the femur cannot easily slip out of the socket, and hip dislocations are rare. The joint capsule runs from the rim of the acetabulum to the neck of the femur (see Figure 9.11b and c).

Three ligamentous thickenings of this capsule reinforce the joint: the **iliofemoral ligament,** a strong, V-shaped, anteriorly located ligament; the **pubofemoral ligament,** a triangular thickening of the capsule's inferior region; and the **ischiofemoral ligament,** a spiraling, posteriorly located ligament. These three ligaments

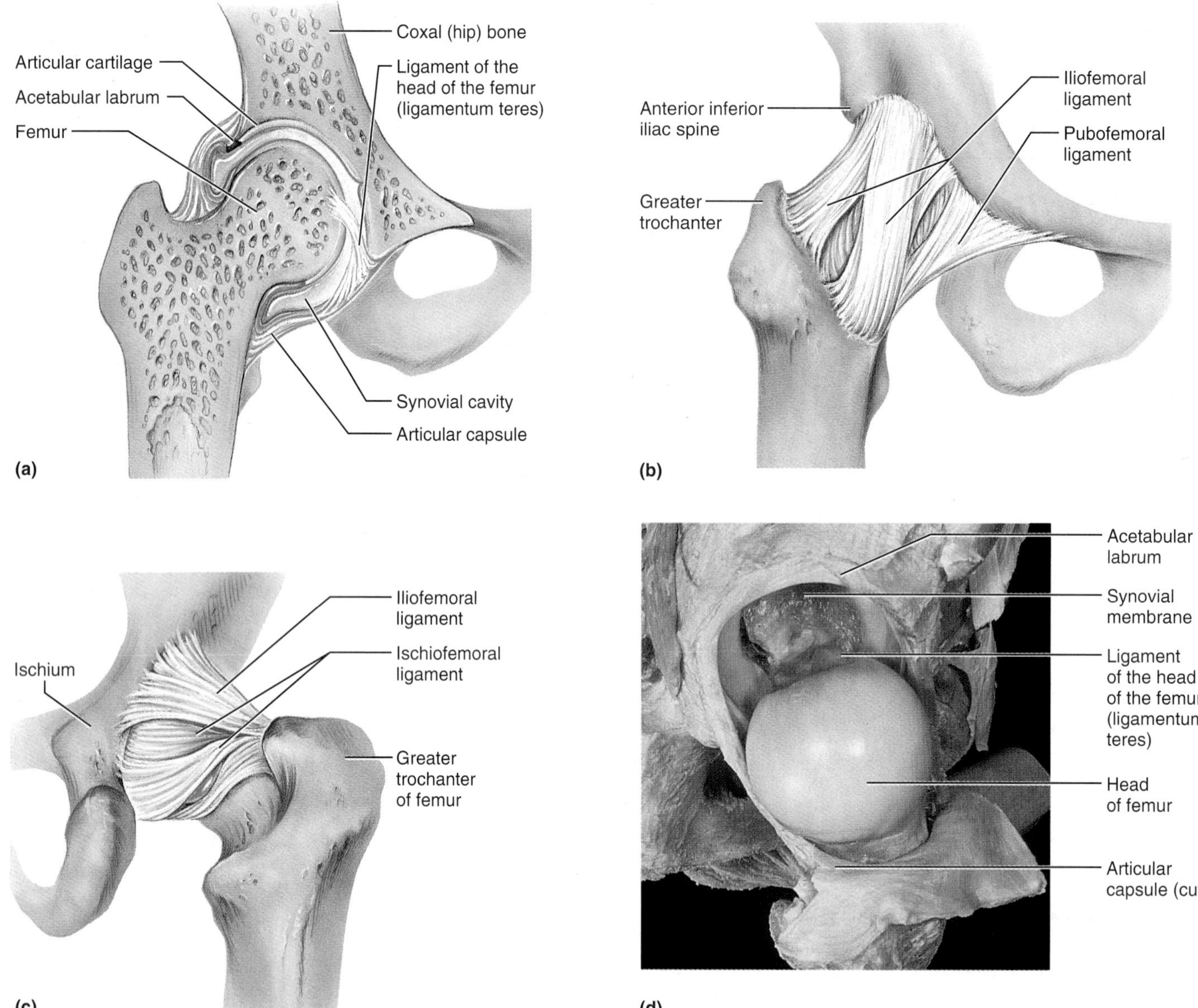

FIGURE 9.11 The hip joint. **(a)** Frontal section through the right hip joint. **(b)** Anterior view of right hip joint, with the capsule in place. **(c)** Posterior view of right hip joint. **(d)** Photograph of the interior of the hip joint, lateral view.

are arranged in such a way that they "screw" the head of the femur into the acetabulum when a person stands erect, thereby increasing the stability of the joint.

The **ligament of the head of the femur** (see Figure 9.11a) is a flat, intracapsular band that runs from the head of the femur to the inferior region of the acetabulum. This ligament remains slack during most hip movements, so it is not important in stabilizing the joint. Its mechanical function is unknown, but it does contain an artery that helps supply the head of the femur. Damage to this artery may lead to arthritis of the hip joint.

Muscle tendons that cross the hip joint contribute to its stability, as do the fleshy parts of many hip and thigh muscles that surround the joint. In this joint, however, stability comes chiefly from the cupped socket and the capsular ligaments.

Clinical Applications

HIP DISLOCATION The main cause of hip dislocation is severe trauma, usually in motor-vehicle accidents, such as when a flexed thigh hits the dashboard. In 30% to 50% of cases the bone tissue of the femoral head slowly starts to die because of damage to the ligament of the head and its artery. ✚

Knee Joint

The knee joint, the largest and most complex joint in the body (Figure 9.12), primarily acts as a hinge. However, it also permits some medial and lateral rotation when in the flexed position and during the act of leg extension. Structurally, it is compound and bicondyloid, because both the femur and tibia have two condylar surfaces. In this joint, the wheel-shaped condyles of the femur roll along the almost-flat condyles of the tibia like tires on a road. Sharing the knee cavity is an articulation between the patella and the inferior end of the femur (see Figure 9.12a); this *femoropatellar joint* is a plane joint that allows the patella to glide across the distal femur as the knee bends.

The synovial cavity of the knee joint has a complex shape (see Figure 9.12a), with several incomplete subdivisions and several extensions that lead to "blind alleys." At least a dozen bursae are associated with this joint, some of which are shown in the figure. The **subcutaneous prepatellar bursa** is often injured when the knee is bumped anteriorly.

Two fibrocartilage menisci occur within the joint cavity, between the femoral and tibial condyles. These C-shaped **lateral** and **medial menisci** attach externally to the condyles of the tibia (see Figure 9.12b). Besides evening the distribution of both compressive load and synovial fluid, the menisci help to guide the condyles during flexion, extension, and rotation movements. They also help prevent side-to-side rocking of the femur on the tibia, thereby stabilizing the joint.

The capsule of the knee joint can be seen on the posterior and lateral aspects of the knee (see Figure 9.12d), where it covers most parts of the femoral and tibial condyles. Anteriorly, however, the capsule is absent. Instead, this anterior area is covered by three broad ligaments that run inferiorly from the patella to the tibia (see Figure 9.12c): the **patellar ligament,** flanked by the **medial** and **lateral patellar retinacula** (ret″ĭ-nak′u-lah; "retainers"). The patellar ligament is actually a continuation of the tendon of the main muscles on the anterior thigh, the quadriceps femoris. Physicians tap the patellar ligament to test the knee-jerk reflex.

The joint capsule of the knee is reinforced by several capsular and extracapsular ligaments, all of which become taut when the knee is extended to prevent hyperextension of the leg at the knee. These ligaments are:

1. The extracapsular **fibular** and **tibial collateral ligaments** (see Figure 9.12c–e). The fibular collateral ligament descends from the lateral epicondyle of the femur to the head of the fibula. The tibial collateral ligament runs from the medial epicondyle of the femur to the medial condyle of the tibia. Besides halting leg extension and preventing hyperextension, these collateral ligaments prevent the leg from moving laterally and medially at the knee.
2. The **oblique popliteal ligament** (pop″lĭ-te′al; "back of the knee") crosses the posterior aspect of the capsule (see Figure 9.12d). Actually a part of the tendon of the semimembranosus muscle that fuses with the joint capsule, it helps stabilize the joint.
3. The **arcuate popliteal ligament** arcs superiorly from the head of the fibula to the posterior aspect of the joint capsule (see Figure 9.12d).

In addition, the knee joint is stabilized by two strong *intracapsular* ligaments called *cruciate ligaments* (kru′she″āt) because they cross each other like an X (*crus* = cross) (see Figure 9.12a, b, and e). Each runs from the tibia to the femur and is named for its site of attachment to the tibia. The **anterior cruciate ligament** attaches to the *anterior* part of the tibia, in the intercondylar area. From there, it passes posteriorly to attach to the head of the femur on the medial side of the lateral condyle. The **posterior cruciate ligament** arises from the *posterior* intercondylar area of the tibia and passes anteriorly to attach to the femur on the lateral side of the medial condyle.

Functionally, the cruciate ligaments prevent undesirable movements at the knee joint (Figure 9.13). The anterior cruciate helps prevent anterior sliding of the tibia. The posterior cruciate, which is even stronger than the anterior cruciate, prevents forward sliding of the femur or backward displacement of the tibia. The two cruciates also function together to lock the knee when one stands (discussed shortly).

The tendons of many muscles reinforce the joint capsule and act as the main stabilizers of the knee joint. Most important are the tendons of the quadriceps femoris and semimembranosus muscles (see Figure 9.12c and d). The greater the strength and tone of these muscles, the less the chance of knee injury.

The knees have a built-in locking device that provides steady support for the body in the standing position. As a person stands up, the flexed leg begins to extend at the knee, and the femoral condyles roll like ball bearings on the tibial condyles. Then, as extension nears completion, the lateral femoral condyle stops rolling before the medial condyle stops. This causes the femur to rotate medially on the tibia until both cruciate and both collateral ligaments stretch tight and halt all movement (see Figure 9.13b). The tension in these ligaments locks the knee into a rigid structure that cannot be flexed again until it is unlocked by a muscle called the popliteus (Muscle Gallery Table 11.15, p. 323), which rotates the femur laterally on the tibia.

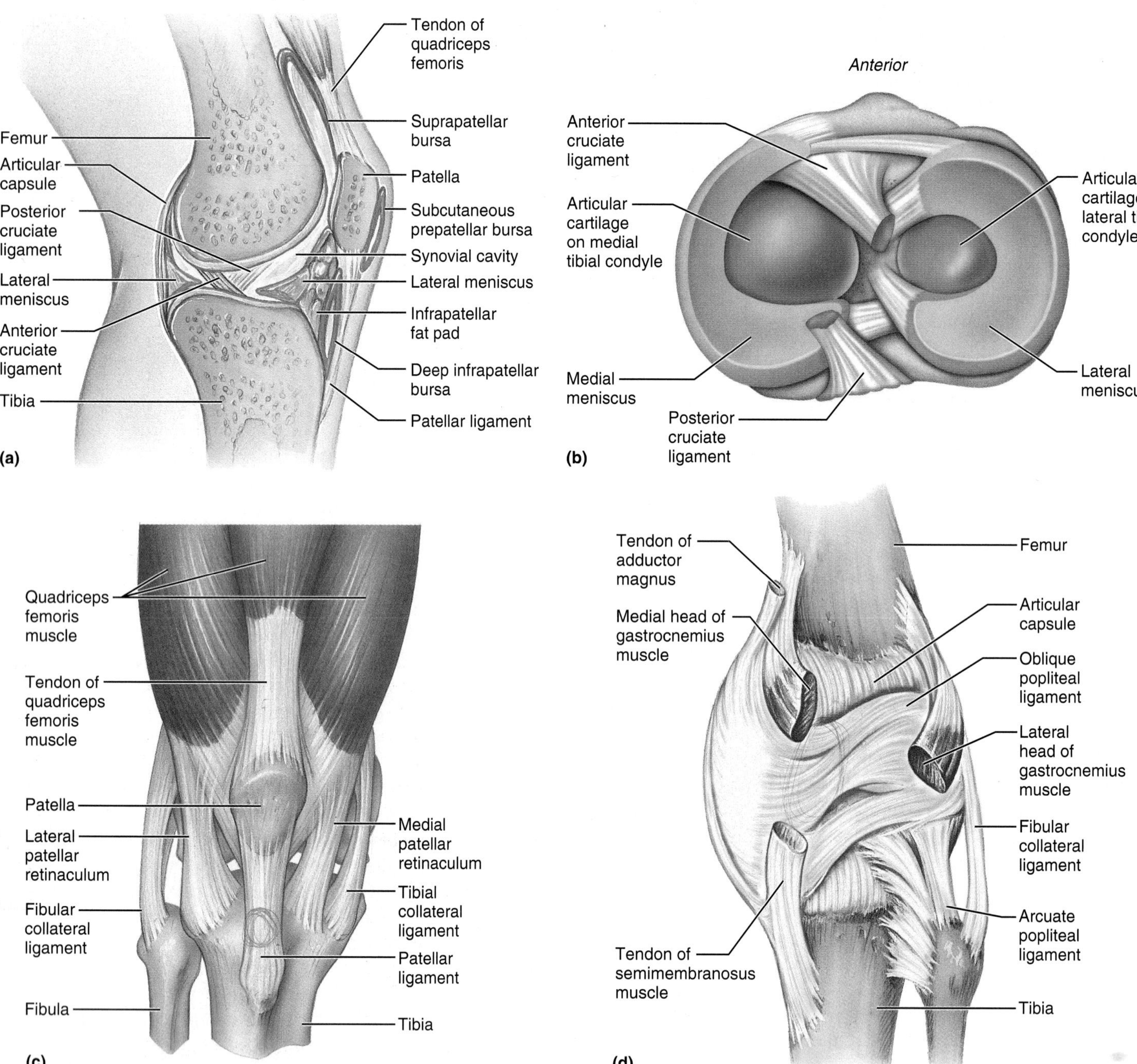

FIGURE 9.12 The knee joint. **(a)** Sagittal section through the right knee joint. **(b)** Superior view of the right tibia in the knee joint, showing the menisci and cruciate ligaments. **(c)** Anterior view of right knee. **(d)** Posterior view of the joint capsule, including ligaments. **(e)** Anterior view of flexed knee, showing the cruciate ligaments. The articular capsule has been removed, and the quadriceps tendon has been cut and reflected distally. **(f)** Photograph of an opened knee joint similar to (e).

Clinical Applications

KNEE INJURIES Knee injuries occur frequently in contact sports because even though the knee is held strongly together by ligaments and muscles, its articular surfaces offer no stability. That is, the nearly flat tibial surface has no socket to secure the femoral condyles. Thus, the knee joint is especially vulnerable to *horizontal* blows, such as occur in tackling and body blocks in football. Most dangerous are *lateral* blows (Figure 9.14), which tear the tibial collateral ligament and the medial meniscus attached to it, as well as the anterior cruciate ligament (ACL).

Injuries to the ACL alone are increasing rapidly, especially in women as women's sports become more vigorous and competitive. Most ACL injuries result when a runner stops and changes direction quickly, twisting a hyperextended leg. A torn ACL heals poorly, so it must be replaced surgically, usually with

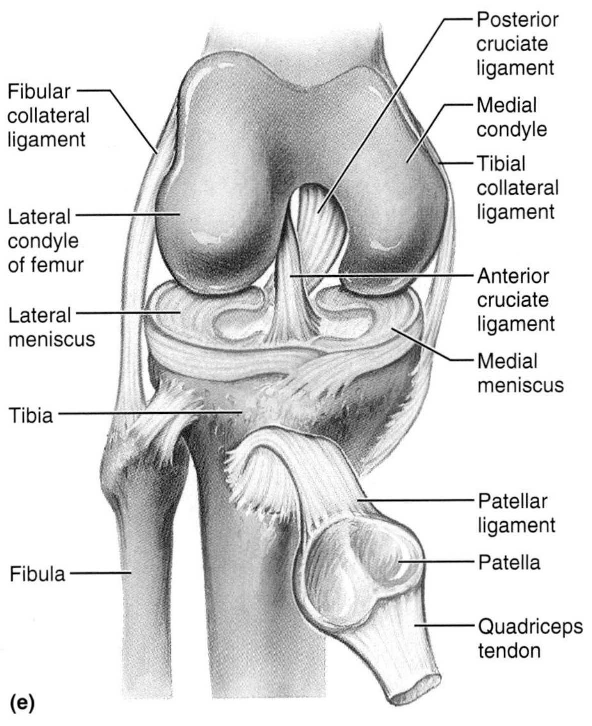

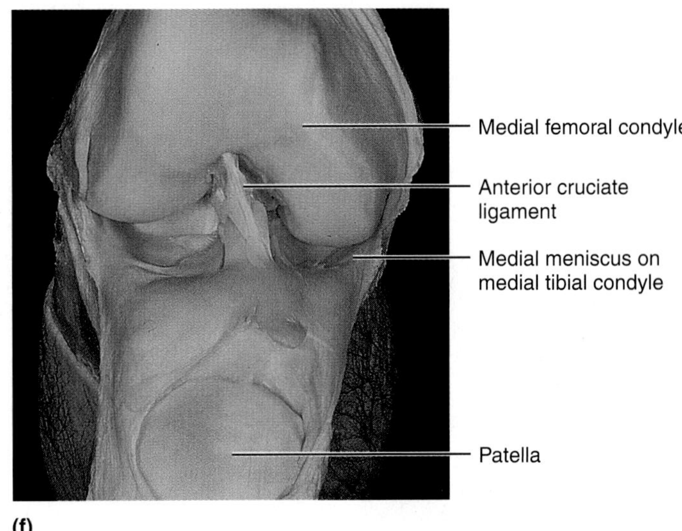

FIGURE 9.12 *continued.*

a graft from either the patellar ligament, the Achilles tendon, or the semitendinosus tendon (see pp. 313, 322, and 318). Injuries to the *posterior* cruciate ligament, caused by posteriorly directed blows to the upper tibia, are less common and more able to heal on their own. ✚

Ankle Joint

The ankle is a hinge joint between (1) the united inferior ends of the tibia and fibula, and (2) the talus of the foot (Figure 9.15). This joint, which allows only dorsiflexion and plantar flexion, has a capsule that is thin anteriorly and posteriorly, but thickened with ligaments

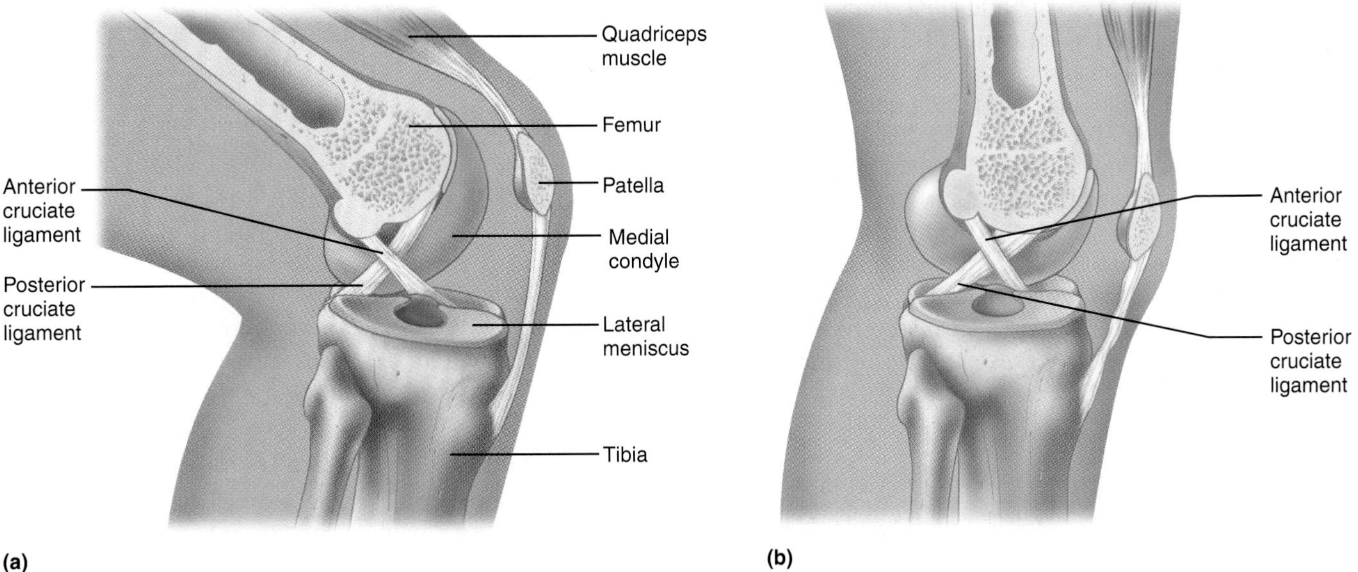

FIGURE 9.13 The cruciate ligaments prevent undesirable movements at the knee joint. **(a)** When the knee is flexed (shown here) or extended, the anterior cruciate prevents anterior slipping movements of the tibia, whereas the posterior cruciate prevents posterior slipping movements. **(b)** When the knee is extended, both cruciates become taut and help lock the knee into a rigid structure.

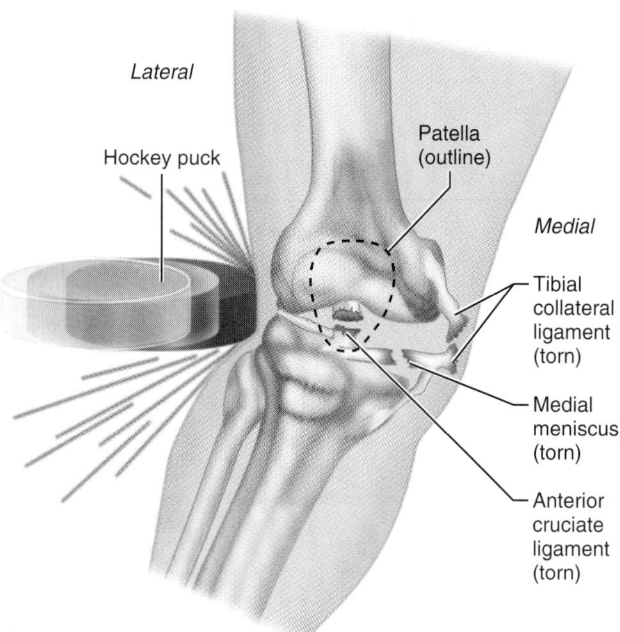

FIGURE 9.14 How a blow to the lateral side of the knee produces a common injury. Anterior view of a knee being hit by a hockey puck. By separating the femur from the tibia medially, such blows tear both the tibial collateral ligament and the medial meniscus because the two are attached. The anterior cruciate ligament also tears.

medially and laterally. The strong **medial (deltoid) ligament** (Figure 9.15a) runs from the medial malleolus of the tibia down to a long line of insertion on the navicular and talus bones, and on the sustentaculum tali of the calcaneus bone. On the other side of the ankle, the **lateral ligament** (Figure 9.15b) consists of three bands that run from the fibula's lateral malleolus to the foot bones: the horizontal **anterior** and **posterior talofibular ligaments,** and the **calcaneofibular ligament,** which runs inferoposteriorly to reach the calcaneus. Functionally, the medial and lateral ligaments prevent anterior and posterior slippage of the talus and the foot.

In forming the deep, U-shaped socket of the ankle joint, the inferior ends of the tibia and fibula are joined by ligaments of their own: the **anterior** and **posterior tibiofibular ligaments** and the lower part of the interosseus membrane (Figure 9.15b and c). These ligaments, called the syndesmosis part of the ankle joint, stabilize the socket so that forces can be transmitted to it from the foot.

CLINICAL APPLICATIONS

ANKLE SPRAINS The ankle is the most frequently injured joint in the lower limb, and ankle sprains are the most common sports injuries. (A sprain is a stretched or torn ligament in a joint; see the next section, on Disorders of Joints.) Most ankle sprains are caused by excessive inversion of the foot and involve the lateral ligament. *Syndesmosis ankle sprains,* by contrast, are caused by extreme dorsiflexion, internal rotation, or external rotation of the foot; all these actions move the talus in ways that wedge the tibia away from the fibula and stretch the tibiofibular ligaments. Ankle sprains are treated with RICE: **r**est, **i**ce, **c**ompression, and **e**levation, followed by ankle-strengthening exercises as recovery begins. ✚

DISORDERS OF JOINTS

The structure and function of joints make them especially vulnerable to a variety of disorders. Because they experience the same strong forces as the hard skeleton, yet consist of soft tissue, joints are prone to *injuries* from traumatic stress. And because their function in movement subjects them to friction and wear, joints can be afflicted by *inflammatory and degenerative processes.*

Joint Injuries

Here we briefly explore three types of joint injuries: sprains, dislocations, and torn cartilage.

Sprains

In a **sprain,** the ligaments reinforcing a joint are stretched or torn. Common sites of sprains are the lumbar region of the spine, the ankle, and the knee. Partly torn ligaments will eventually repair themselves, but they heal slowly because ligaments are poorly vascularized. Sprains tend to be painful and immobilizing. Completely ruptured ligaments, by contrast, require prompt surgical repair because inflammation in the joint will break down the neighboring tissues and turn the injured ligament to "mush." Surgical repair can be a difficult task: A ligament consists of hundreds of fibrous strands, and sewing a ligament back together has been compared to sewing two hairbrushes together. When important ligaments are too severely damaged to be repaired, they must be removed and replaced with grafts or substitute ligaments.

Dislocations

A **dislocation (luxation)** occurs when the bones of a joint are forced out of alignment. This injury is usually accompanied by sprains, inflammation, pain, and difficulty in moving the joint. Dislocations may result from serious falls or blows and are common sports injuries. The jaw, shoulder, finger, and thumb joints are most commonly dislocated. Like fractures, dislocations must be reduced; that is, the bone ends must be returned to their proper positions by a physician. **Subluxation** is a partial or incomplete dislocation of a joint.

Critical Reasoning

Why is it common for a person to experience repeated dislocations at a given joint?

The initial dislocation stretches the joint capsule and ligaments. Once the capsule is loosened, the joint is more likely to dislocate again. Injured ligaments eventually shorten to their original length, but this aspect of healing can take years.

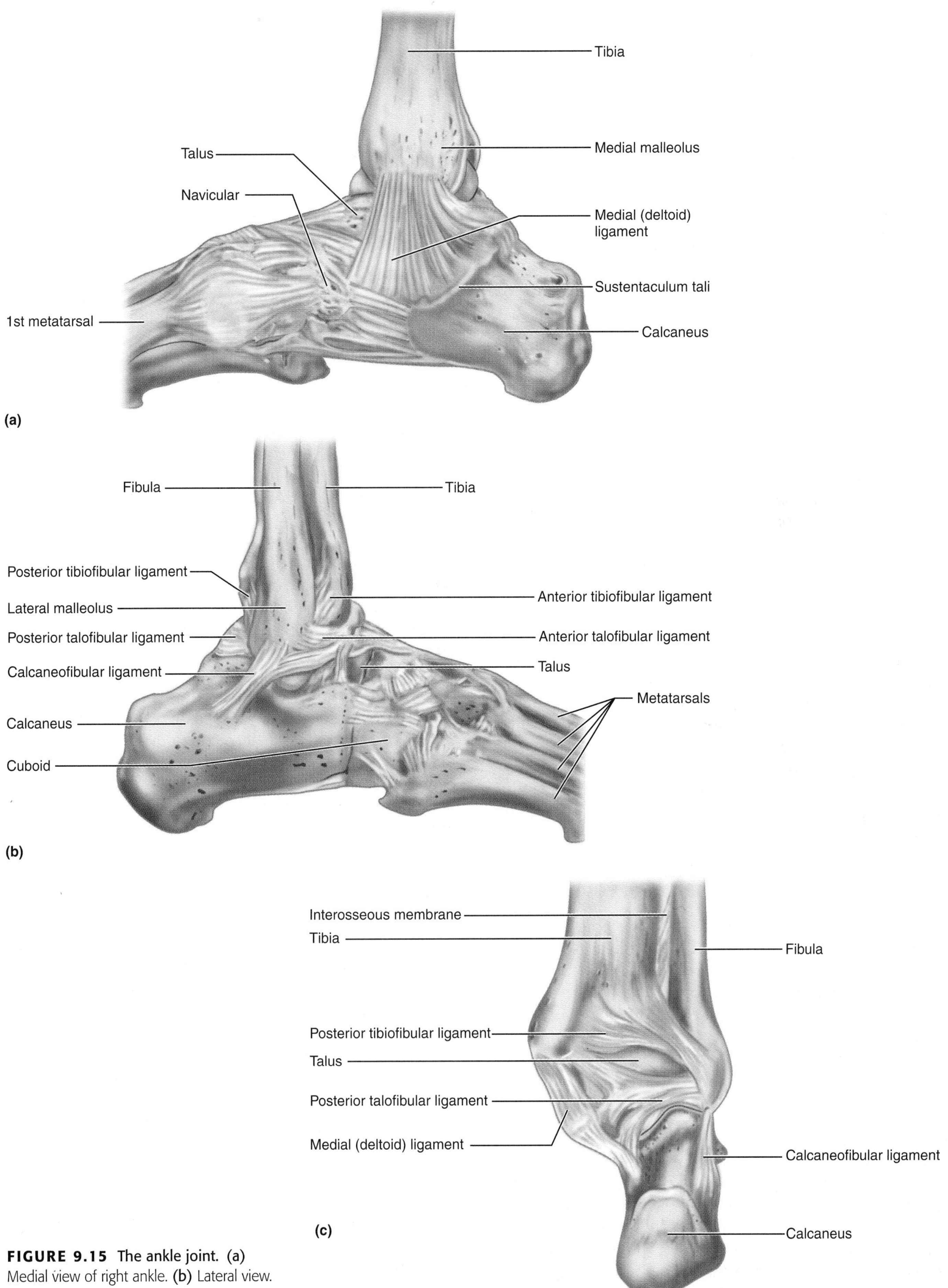

FIGURE 9.15 The ankle joint. **(a)** Medial view of right ankle. **(b)** Lateral view. **(c)** Posterior view.

Torn Cartilage

Even though most cartilage injuries involve tearing of the meniscus in the knee, tears and overuse damage to the cartilages of other joints are becoming increasingly common in highly competitive athletes, particularly gymnasts.

Cartilage tears often happen when a meniscus is simultaneously subjected to both high compression and shear stresses. For example, a tennis player lunging to return a ball can rotate the flexed knee medially with so much force that it tears both the joint capsule and the medial meniscus attached to it. Cartilage is avascular, so it can rarely repair itself; thus, torn cartilage usually stays torn. Because cartilage fragments (also called *loose bodies*) can cause joints to lock or bind, the recommended treatment often is surgical removal of the damaged cartilage.

Repair is achieved using a remarkable outpatient procedure called **arthroscopic surgery** (ar-thro-skop′ik; "looking into joints"). Using miniaturized video and surgical equipment inserted into the joint through a single small incision, the surgeon removes the cartilage fragments and repairs the ligaments, in the process minimizing scarring and tissue damage and speeding healing. Arthroscopic surgery has applications in many joints, not just the knee.

Advances in Understanding Damaged joint cartilages heal poorly (p. 131), and coaxing them to heal better is a major goal of medical research. Although much progress remains to be made, it has been found that a patient's own chondrocytes, when cultured in the lab and then grafted onto joint surfaces, can produce enough new cartilage to fill small holes or breaks in the articular cartilages of the knee. This is called **autologous cartilage implantation,** and its long-term goal is to grow thicker sheets of cartilage to cover entire joint surfaces. ✷

Inflammatory and Degenerative Conditions

Inflammatory conditions that affect joints include inflammations of bursae and tendon sheaths, various forms of arthritis, and Lyme disease.

Bursitis and Tendonitis

Bursitis, inflammation of a bursa, usually results from a physical blow or friction, although it may also be caused by arthritis or bacterial infection. In response, the bursa swells with fluid. Falling on one's knee can cause a painful bursitis of the subcutaneous prepatellar bursa (see Figure 9.12a), known as **housemaid's knee.** Resting and rubbing one's elbows on a desk can lead to **student's elbow,** or **olecranon bursitis,** the swelling of a bursa just deep to the skin of the posterior elbow. Severe cases of bursitis may be treated by injecting inflammation-reducing drugs into the bursa. Excessive fluid accumulation may require fluid removal by needle aspiration.

Tendonitis is inflammation of tendon sheaths. Its causes, symptoms, and treatments mirror those of bursitis.

Arthritis

The term **arthritis** describes over 100 kinds of inflammatory or degenerative diseases that damage the joints. In all its forms, arthritis is the most widespread crippling disease in the United States: One of every seven Americans suffers from its effects. All forms of arthritis have, to a greater or lesser degree, the same initial symptoms: pain, stiffness, and swelling of the joint.

Osteoarthritis (Degenerative Joint Disease) The most common type of arthritis is **osteoarthritis (OA),** a chronic (long-term) degenerative condition that is often called "wear-and-tear arthritis." It is most common in the aged and is probably related to the normal aging process. OA affects women more often than men, but 85% of all Americans develop this condition. OA affects the articular cartilages, causing them to soften, fray, crack, and erode.

The cause of OA is unknown. According to current theory, normal use causes joints to release metalloproteinase enzymes that break down the cartilage matrix (especially the collagen fibrils); meanwhile, the chondrocytes continually fix the damage by secreting more matrix. Whenever the strain on a joint is repeated or excessive, too much of the cartilage-destroying enzyme is thought to be released, causing OA. Because this process occurs most where an uneven orientation of forces causes extensive microdamage, badly aligned or overworked joints are likely to develop OA.

The bone directly below the articular cartilage is also affected, becoming dense and stiff. As the disease progresses, bone spurs tend to grow around the margins of the damaged cartilages, encroaching on the joint cavity and perhaps restricting joint movement. Patients complain of stiffness upon waking in the morning, but this decreases within a half hour. However, there is always joint pain during use. The affected joints may make a crunching noise (called *crepitus*) as their roughened surfaces rub together during movement.

The joints most commonly affected in OA are those of the fingers, knuckles, hips, and knees. The nonsynovial joints between the vertebral bodies are also susceptible, expecially in the cervical and lumbar regions of the spine.

The course of OA is slow and irreversible. It is not usually crippling, except for some severe cases involving the hip and knee joints. Inflammation may or may not accompany the degeneration of the joints, but it is usually not severe. In many cases the symptoms of OA can be controlled with a pain reliever such as aspirin or acetaminophen plus a program of low-impact exercise.

Advances in Understanding Rubbing a hot-pepper-like substance called capsaicin on the skin over the joint also helps lessen the pain of OA. Some claim to obtain relief from nutrient supplements consisting of macromolecules normally present in cartilage called glucosamine and chondroitin sulfate. The effectiveness of these, however, is still being tested in clinical trials. ✷

Rheumatoid Arthritis Another kind of arthritis, **rheumatoid arthritis** (roo′mah-toid) **(RA)** is a chronic inflammatory disorder. It is a complex disease: Along with pain and swelling of the joints, its manifestations include osteoporosis, muscle weakness, and problems with the heart and blood vessels.

The course of RA is variable: It may develop gradually or in spurts that are years apart, and it is marked by flare-ups and remissions (*rheumat,* from the Greek, means "susceptible to change"). Its onset usually occurs between the ages of 30 and 50, but it may arise at any age. Although not as common as osteoarthritis, RA afflicts millions, about 1% of all people. It affects three times as many women as men, and wanes when a woman is pregnant, so apparently it is influenced by female sex hormones. RA tends to affect many joints simultaneously and bilaterally (on both sides of the body), especially the small joints of the fingers, wrists, ankles, and feet.

RA is an **autoimmune disease**—a disorder in which the body's immune system attacks its own tissues. The cause of this reaction is unknown, but RA might follow infection by certain bacteria and viruses that bear surface molecules similar to molecules normally present in the joints, and when the body is stimulated to attack the foreign molecules, it inappropriately destroys its own joint tissues as well.

RA begins with an inflammation of the synovial membrane. Capillaries in this membrane leak tissue fluid and white blood cells into the joint cavity. This excess synovial fluid then causes the joint to swell. With time, the inflamed synovial membrane thickens into a *pannus* ("rag"), a coat of granulation tissue that clings to the articular cartilages. The pannus erodes the cartilage and, often, the underlying bone. As the cartilage is destroyed, the pannus changes into a fibrous scar tissue that interconnects the bone ends. This scar tissue eventually ossifies, and the bone ends fuse together, immobilizing the joint. This end condition, called **ankylosis** (ang″kĭ-lo′sis; "stiff condition"), often produces bent, deformed fingers (Figure 9.16).

Most drugs used to treat RA are either anti-inflammatory or antibiotic, or they block the immune response. These must be given for long periods and are only partly successful; there is no cure. Joint prostheses are the last resort for severely crippled patients. Some RA sufferers have had over a dozen of their joints replaced by artificial joints! (See A Closer Look, p. 238.)

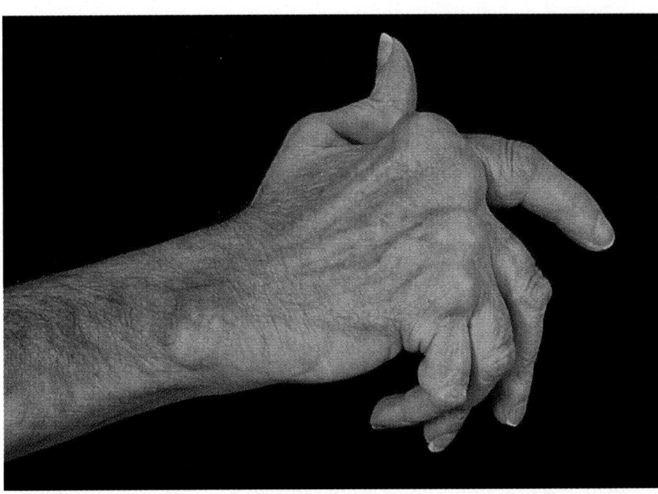

FIGURE 9.16 A hand deformed by rheumatoid arthritis. Note the enlarged joints, a product of inflammation of the synovial membranes.

Gouty Arthritis (Gout) Ordinarily, proper blood levels of uric acid, a normal waste product of nucleic acid metabolism, are maintained by excreting it in the urine. In people who are under-excreters, however, uric acid rises to abnormal levels in the blood and body fluids, and precipitates as solid crystals of urate in the synovial membranes. An inflammatory response follows as the body tries to attack and digest the crystals, producing an agonizingly painful attack of **gouty** (gow′te) **arthritis,** or **gout.** The initial attack involves a single joint, usually in the lower limb, often at the base of the big toe. Other attacks usually follow, months to years later. Gout is far more common in males than in females because men naturally have higher levels of uric acid in their blood (perhaps because estrogens increase the rate of uric-acid excretion).

Untreated gout can cause the ends of articulating bones to fuse, immobilizing the joint. Fortunately, effective treatment is available. For acute attacks, nonsteroidal anti-inflammatory drugs such as ibuprofen are used. For the long term, urate-lowering drugs and control of diet (avoidance of alcohol and red meat) are effective.

Lyme Disease

Lyme disease is an inflammatory disease that often results in joint pain and arthritis, especially in the knee joint. Caused by bacteria called spirochetes and transmitted by the bites of ticks that live on deer and mice, Lyme disease is also characterized by a skin rash, flu-like symptoms, and foggy thinking. The symptoms may persist for years if the disease is untreated, although they eventually decline in most cases. Sometimes, however, they proceed to neurological disorders and irregularities of the heartbeat. Because symptoms vary from person to person, the disease is difficult to diagnose. Lyme disease is treatable with antibiotics, especially if

A Closer Look

The Development of Artificial Joints

Centuries ago, the task of developing artificial joints involved designing suits of armor so that they could allow mobility while protecting the human joints beneath. These days, the task is producing joint prostheses (artificial joints) that can function effectively and safely as replacement joints.

Efforts to create joint prostheses date back to the years following World War II and the Korean War, which produced large numbers of wounded veterans who needed artificial limbs and joints. The initial problems centered on finding materials that are durable, nontoxic, and resistant to rejection and the corrosive effects of organic acids in the blood. In 1963, Sir John Charnley, an English orthopedic surgeon, performed the first total hip replacement using a device consisting of a stainless steel ball on a stem and a cup-shaped polyethylene plastic socket anchored to the pelvis by methyl methacrylate cement.

Charnley's operation revolutionized the therapy of arthritic hips. Next came knee prostheses, although ten years passed before total knee replacements that operate smoothly became available. Now the number of knee replacements equals the number of hip replacements. The field has advanced: Today the metal parts of prostheses consist of strong cobalt and titanium alloys, and replacement joints are also available for the fingers, elbow, and shoulder.

Total hip and knee replacements last 10 to 15 years in elderly patients who do not excessively stress the joint. Most such operations are performed to reduce pain, and most restore about 80% of joint function. For young, active people, however, replacement joints are rarely strong enough or durable enough, but making them so is a major goal of current research in prosthetics.

The problem is that prostheses work loose over time, so researchers are seeking ways of producing better fits between the prosthetic implant and bone. The first-ever robot surgeon, named Robodoc, is being developed to more precisely drill a hole into the top part of the femur's shaft to more securely receive the femoral prosthesis in artificial hips. Researchers are exploring ways to improve the strength of the cement in typical prostheses (simply eliminating air bubbles from the cement increases durability), as well as ways to stimulate bone growth around the implant in cementless prostheses. Two other problems of great interest are the bone resorption that gradually occurs around the prosthesis and inflammation caused by polyethylene particles that wear off the joint surfaces.

Dramatic changes are occurring in the way that artificial joints are made. CAD/CAM (computer-aided design and computer-aided manufacturing) techniques are being used to produce individualized joints. The computer is fed the patient's X rays along with information about the patient's specific problems. It then draws from a database of many normal joints and generates possible designs and modifications for a joint prosthesis. Once the best design is selected, the computer writes a program to direct the machines that shape the joint prosthesis. This computerization saves both time and money.

Clearly, then, the development of artificial joints has come a long way. The advent of new technologies and space age materials is sure to provide exciting new advances in joint replacement.

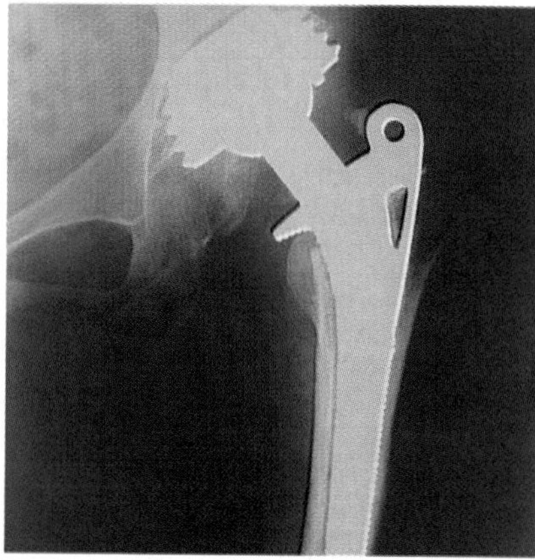

False-color X ray of an artificial hip joint. The metal prosthesis at the top of the femur appears bright yellow.

detected early, but it takes a long time to kill the infecting bacteria. This disease is spreading in the United States, mostly in the northeastern states, northern Wisconsin, Minnesota, and northern California. First-generation vaccines have been approved by the U.S. Food and Drug Administration.

THE JOINTS THROUGHOUT LIFE

Synovial joints develop from mesenchyme that fills the spaces between the cartilaginous "bone models" in the late embryo. The outer region of this intervening mesenchyme condenses to become the fibrous joint capsule, whereas the inner region hollows to become the joint cavity. By week 8, these joints resemble adult joints in form and arrangement; that is, their synovial membranes are developed, and synovial fluid is being secreted into the joint cavities. These basic structural features are genetically determined, but after birth a joint's size, shape, and flexibility are modified by use. Active joints grow larger and have thicker capsules, ligaments, and bony supports than they would if never used.

During youth, many injuries to joints tear or knock the epiphysis off the shaft, a vulnerability that ends after the epiphyseal plates close in early adulthood. From that time on, comparable injuries merely result in sprains.

The main joint problem of advancing age is osteoarthritis. Just as bones must be stressed to maintain their strength, joints must be exercised to keep their health. Exercise squeezes synovial fluid in and out of articular cartilages, providing the cartilage cells with the nourishment they need. And although exercise cannot prevent osteoarthritis, it strengthens joints and slows degeneration of the articular cartilages. It also strengthens the muscles that stabilize the joints. Exercise should not be overdone, however, because too much stress worsens osteoarthritis. Because the buoyancy of water relieves much of the stress on weight-bearing joints, people who exercise in pools often retain good joint function throughout life.

RELATED CLINICAL TERMS

Ankylosing spondylitis (ang″kĭ-lo′sing spon″dĭ-li′tis) (*ankyl* = stiff joint; *spondyl* = vertebra) A distinctive kind of rheumatoid arthritis that mainly affects males. It usually begins in the sacroiliac joints and progresses superiorly along the spine. The vertebrae become interconnected by so much fibrous tissue that the spine becomes rigid ("poker back").

Arthroplasty ("joint reforming") Replacing a diseased joint with an artificial joint (see the box on p. 238).

Chondromalacia patellae ("softening of cartilage of the patella") Damage and softening of the articular cartilages on the posterior surface of the patella and the anterior surface of the distal femur. This condition, seen most often in adolescent athletes, produces a sharp pain in the knee when the leg is extended (in climbing stairs, for example). Chondromalacia may result when the quadriceps femoris, the main group of muscles on the anterior thigh, pulls unevenly on the patella, persistently rubbing it against the femur in the knee joint. Chondromalacia can often be corrected by exercises that strengthen weakened parts of the quadriceps muscles.

Patellofemoral pain syndrome Persistent pain in the region behind the patella; results from rubbing pressure between the femoral condyles and the patella, as may occur with overuse of the quadriceps femoris muscles or an abnormally shaped patella. Is distinguished from chondromalacia patellae (see above) by usual absence of damage to the cartilage and generally younger age at onset. Treatment is exercise that strengthens the different quadriceps muscles evenly and gradually.

Synovitis Inflammation of the synovial membrane of a joint. Can result from a blow, an infection, or arthritis. Synovitis leads to the production of excess fluid, causing the joints to swell. Such accumulation of fluid in the joint cavity is called *effusion*.

Tenosynovitis (ten′o-sin″o-vi″tis) Painful inflammation of a tendon or its sheath. Occurs most often in the hands, wrists, feet, or ankles as a result of repeated, intense use, and it may be temporarily disabling. The condition is common in pianists and typists (see the discussion of carpal tunnel syndrome on p. 193). It can also be caused by arthritis and bacterial infection of the tendon sheath. Therapy involves immobilization of the affected body region or surgery to drain an infected tendon sheath.

Valgus and varus injuries Valgus means bent outward, away from the body midline, such as in abduction of the leg at the knee or the forearm at the elbow. Varus means bent inward, such as in adduction of these elements. Because the knee and elbow are not designed for such movements, strong valgus and varus bending injures them.

CHAPTER SUMMARY

Media study tools that could provide you additional help in reviewing specific key topics of Chapter 9 are referenced below.

SP = Study Partner; AIA = A.D.A.M.® Interactive Anatomy

1. Joints (articulations) are sites where elements of the skeleton meet. They hold bones together and allow various degrees of movement.

CLASSIFICATION OF JOINTS (p. 212)

1. Joints are classified functionally as synarthrotic (no movement), amphiarthrotic (slight movement), or diarthrotic (free movement). They are classified structurally as fibrous, cartilaginous, or synovial.

FIBROUS JOINTS (pp. 212–213)

1. In fibrous joints, the bones are connected by fibrous connective tissue. No joint cavity is present. Nearly all fibrous joints are synarthrotic.
2. The types of fibrous joints are sutures (between skull bones), syndesmoses ("ligament joints"), and gomphoses (articulation of the teeth with their sockets).

AIA Link: Atlas; Type: Cadaver photographs; Mandible (superior).

CARTILAGINOUS JOINTS (p. 213)

1. In cartilaginous joints, the bones are united by cartilage. No joint cavity exists.
2. Synchondroses are immovable joints of hyaline cartilage, such as epiphyseal plates. Symphyses are slightly movable fibrocartilage joints, such as intervertebral discs and the pubic symphysis.

AIA Link: Atlas; View: Lateral; Bones of Trunk (lateral).

SYNOVIAL JOINTS (pp. 214–234)

1. Most joints in the body are synovial. Synovial joints are diarthrotic.

General Structure of Synovial Joints (pp. 214–217)

2. Synovial joints have a fluid-containing cavity and are covered by an articular capsule. The capsule has an outer fibrous region, often reinforced by ligaments, and an inner synovial membrane that produces synovial fluid. The articulating ends of bone are covered with impact-absorbing articular cartilages. Nerves in the capsule provide a sense of "joint stretch." Some joints contain fibrocartilage discs (menisci), which distribute loads evenly and may allow two movements at one joint.

AIA Link: Atlas; Region: Lower Limb; Knee Joint (posterior).

3. Synovial fluid is mainly a filtrate of the blood, but it also contains molecules that make it a friction-reducing lubricant.

How Synovial Joints Function (pp. 217–219)

4. Synovial joints reduce friction. The cartilage-covered bone ends glide on a slippery film of synovial fluid squeezed out of the articular cartilages. This mechanism is called weeping lubrication.

Bursae and Tendon Sheaths (p. 219)

5. A bursa is a fibrous sac lined by a synovial membrane and containing synovial fluid. Tendon sheaths, which are similar to bursae, wrap around certain tendons. Bursae and tendon sheaths are lubricating devices that allow adjacent structures to move smoothly over one another.

Factors Influencing the Stability of Synovial Joints (pp. 219–220)

6. Joints are the weakest part of the skeleton. Factors that stabilize joints are the shapes of the articulating surfaces, ligaments, and the tone of muscles whose tendons cross the joint.

Movements Allowed by Synovial Joints (pp. 220–224)

7. Contracting muscles produce three common kinds of bone movements at synovial joints: gliding, angular movements (flexion, extension, abduction, adduction, and circumduction), and rotation.

8. Special movements include supination and pronation of the forearm, dorsiflexion and plantar flexion of the foot, inversion and eversion of the foot, protraction and retraction, elevation and depression, and opposition of the thumb.

SP Exercise: Chapter 9, Special Body Movements.

Synovial Joints Classified by Shape (pp. 224–226)

9. The shapes of the articular surfaces reflect the kinds of movements allowed at a joint. Based on such shapes, joints are classified as plane (nonaxial), hinge or pivot (uniaxial), condyloid or saddle (biaxial), or ball-and-socket (multiaxial) joints.

Selected Synovial Joints (pp. 226–234)

10. The temporomandibular joint is formed by (1) the head of the mandible and (2) the mandibular fossa and articular tubercle of the temporal bone. This joint allows both a hinge-like opening of the mouth and an anterior gliding of the mandible. It often dislocates anteriorly and is frequently the site of stress-induced temporomandibular disorders.

AIA Link: Atlas; View: Lateral; Temporomandibular Joint (lateral).

11. The shoulder (glenohumeral) joint is a ball-and-socket joint between the glenoid cavity and the head of the humerus. It is the body's most freely movable joint. Its socket is shallow, and its capsule is lax and reinforced by ligaments superiorly and anteriorly. The tendons of the biceps brachii and the rotator cuff help stabilize the joint. The humerus often dislocates anteriorly, then inferiorly, at the shoulder joint.

12. The elbow is a hinge joint in which the ulna and radius articulate with the humerus. It allows flexion and extension. Its articular surfaces are highly complementary and help stabilize it. Radial and ulnar collateral ligaments prevent side-to-side movement of the forearm.

13. The hip joint is a ball-and-socket joint between the acetabulum of the hip bone and the head of the femur. It is adapted for weight bearing. Its articular surfaces are deep and secure, and its capsule is strongly reinforced by three ligamentous thickenings.

14. The knee is a complex and shallow joint formed by the articulation of the tibial and femoral condyles (and anteriorly by the patella with the femur). Extension, flexion, and some rotation are allowed. C-shaped menisci occur on the articular surfaces of the tibia. The capsule, which is absent anteriorly, is reinforced elsewhere by several capsular ligaments. For example, the tibial and fibular collateral ligaments help prevent hyperextension, abduction, and adduction of the leg. The intracapsular cruciate ligaments help prevent anterior-posterior displacement of the joint surfaces and help lock the knee when one stands. The tone of the muscles crossing this joint is important for knee stability. In contact sports, lateral blows are responsible for many knee injuries. Among the knee elements, the anterior cruciate ligament is injured especially often.

15. The ankle joint is a hinge joint between (1) the distal tibia and fibula and (2) the talus of the foot. Its medial and lateral ligaments allow plantar flexion and dorsiflexion, but prevent anterior and posterior displacements of the foot.

SP Exercise: Chapter 9, Structure and Function of Body Joints.

DISORDERS OF JOINTS (pp. 234–238)

Joint Injuries (pp. 234–236)

1. Sprains are the stretching and tearing of joint ligaments. Because ligaments are poorly vascularized, healing is slow.

2. Joint dislocations move the surfaces of articulating bones out of alignment. Such injuries must be reduced.

3. Trauma can tear joint cartilages. Sports injuries to the knee menisci are common and can result from twisting forces as well as from direct blows to the knee. Arthroscopic surgery is used to repair the injured knee, as well as other joints.

Inflammatory and Degenerative Conditions (pp. 236–238)

4. Bursitis and tendonitis are inflammation of a bursa and a tendon sheath, respectively.

5. Arthritis is the inflammation or degeneration of joints accompanied by pain, swelling, and stiffness. Arthritis includes many different diseases and affects millions of people.

6. Osteoarthritis is a degenerative condition that first involves the articular cartilages. It follows wear or excessive loading, and is common in the aged. Weight-bearing joints are most often affected.

7. Rheumatoid arthritis, the most crippling kind of arthritis, is an autoimmune disease involving severe inflammation of the joints, starting with the synovial membranes.

8. Gout is joint inflammation caused by a deposition of urate salts in the synovial membranes.

9. Arthritis is a major symptom of Lyme disease, a condition caused by bacteria transmitted by tick bites.

THE JOINTS THROUGHOUT LIFE (pp. 238–239)

1. Joints develop from mesenchyme between the cartilaginous "bone models" in the embryo.
2. Joints usually function well until late middle age, when osteoarthritis almost always appears.

SP Exercise: Chapter 9, General Structure of a Synovial Joint.

REVIEW QUESTIONS

Multiple Choice/Matching Questions

1. Match the terms in the key to the correct descriptions. (More than one choice might apply.)

   Key: **(a)** fibrous joints **(b)** cartilaginous joints **(c)** synovial joints

____ **(1)** have no joint cavity

____ **(2)** types are sutures and syndesmoses

____ **(3)** dense connective tissue fills the space between the bones

____ **(4)** almost all joints of the skull

____ **(5)** types are synchondroses and symphyses

____ **(6)** all are diarthroses

____ **(7)** the most common type of joint in the body

____ **(8)** nearly all are synarthrotic

____ **(9)** shoulder, hip, knee, and elbow joints

2. Characteristics of a synovial joint include (a) articular cartilage, (b) a joint cavity, (c) a lubricant, (d) articular capsule, (e) all of these.

3. In general, the most important factor determining the stability of synovial joints is (a) interlocking shapes of the articular surfaces, (b) reinforcing ligaments, (c) muscle tone, (d) synovial fluid, which acts like glue, (e) the body's wrapping of skin, which holds the limbs together.

4. Characteristics of a symphysis include (a) presence of fibrocartilage, (b) ability to resist large compression and tension stresses, (c) presence of a joint cavity, (d) very high mobility, (e) both a and b.

5. Synovial joints are most richly innervated by nerve fibers that (a) monitor how much the capsule is stretched, (b) supply the articular cartilages, (c) cause the joint to move, (d) monitor pain when the capsule is injured.

6. Match the parts of a synovial joint listed in the key to their functions below. (More than one choice applies in some cases.)

   Key: **(a)** articular cartilage **(b)** ligaments and fibrous capsule **(c)** synovial fluid **(d)** muscle tendon

____ **(1)** keeps bone ends from crushing when compressed; resilient

____ **(2)** resists tension placed on joints

____ **(3)** lubricant that minimizes friction and abrasion of joint surfaces

____ **(4)** keeps joints from overheating

____ **(5)** helps prevent dislocation

Short Answer Essay Questions

7. Define joint.

8. Where does synovial fluid come from?

9. Explain weeping lubrication of the synovial joint surfaces.

10. Compare and contrast (1) flexion and extension with (2) adduction and abduction.

11. Name two specific examples of each: hinge joint, plane joint, condyloid joint, ball-and-socket joint.

12. What are the functions of the menisci of the knee? Of the anterior and posterior cruciate ligaments?

13. Why are sprains and injuries to joint cartilages particularly troublesome?

14. Name the most common direction in which each of the following joints tends to dislocate: (a) shoulder, (b) elbow, (c) hip.

15. Look at a skeleton, or a picture of one, in posterior view. Classify the various joints you see as containing either hyaline cartilage, fibrocartilage, or fibrous tissue (fibrous joints).

16. What are the technical names of and movements allowed at the following joints? (a) joints in the toes, (b) wrist joint, (c) jaw joint, (d) joint between sacrum and coxal bones, (e) knuckle joint in the hand.

17. List the functions of the following parts of a synovial joint: (a) fibrous part of the capsule, (b) synovial fluid, (c) articular disc.

18. Compare a bursa and a tendon sheath in terms of their structure, function, and locations.

19. Provide the technical names of the joints between (a) the scapula and clavicle, (b) the articular processes of successive vertebrae, (c) the ribs and sternum, (d) the ribs and vertebrae, (e) the various tarsal bones.

20. Name the joint that contains the (a) glenoid cavity and labrum, (b) cruciate ligaments and menisci, (c) annular ligament and head of radius, (d) coronoid process, trochlea, and radial collateral ligament, (e) medial and lateral ligaments and talus.

CRITICAL REASONING + CLINICAL APPLICATION QUESTIONS

1. Harry was jogging down the road when he tripped in a pothole. As he fell, his ankle twisted violently to the side. The diagnosis was severe dislocation and spraining of the ankle. The surgeon stated that she would perform a reduction of the dislocation and then attempt to repair the injured ligament using arthroscopy. (a) What kinds of movements can normally occur at the ankle joint? (b) Was the doctor telling Harry that his bones were broken? (c) What is reduction of a joint? (d) Why is it necessary to repair ligaments surgically? (e) How will the use of arthroscopic surgery minimize Harry's recovery time and his suffering?

2. Dan Park, an exhausted anatomy student, was attending a lecture. After 30 minutes he began to doze. He woke up a while later (the professor's voice was too loud to permit a good nap) and let go with a tremendous yawn. To his great distress, he couldn't close his mouth—his lower jaw was stuck open. What had happened?

3. Mrs. Estevez, who is 37 years old, visited her physician and complained of unbearable pain in several joints of both hands. The joints were very red, swollen, and warm to the touch. When asked if she had ever had such an episode in the past, she said she had had a similar attack 2 years earlier that had disappeared as suddenly as it had come. Her diagnosis was arthritis. (a) What type? (b) What is the precipitating cause of joint inflammation in this particular type of arthritis?

4. At work, a box fell from a shelf onto Ming's acromial region. In the emergency room, the physician could feel that the head of her humerus had moved into the axilla. What had happened to Ming?

5. At his ninety-fourth birthday party, Jim was complimented on how good he looked and was asked about his health. He answered, "I feel good, except some of my joints ache and are stiff, especially my knees and hips and lower back, and especially when I wake up in the morning." A series of X rays and an MRI scan taken a few weeks earlier had revealed that the articular cartilages of these joints were rough and flaking off, and that bone spurs were present at the ends of some of Jim's bones. What is Jim's probable condition?

6. On the evening news, Samantha heard that the deer population in her state had increased greatly in the past few years, and she knew that deer were often seen walking the streets of her suburban community. Suddenly, she exclaimed, "So that's why several children in my son's class at school have gotten Lyme disease this year!" Explain her reasoning.

A.D.A.M.® INTERACTIVE ANATOMY QUESTIONS

1. Link: Atlas; System: Skeletal; Knee Joint (posterior). Name the ligaments that arise between the lateral and medial condyles of the tibia.

2. Link: Atlas; System: Skeletal; Temporomandibular Joint (lateral). (a) Name two markings of the temporal bone that lie superior to the articular disc of the temporomandibular joint. (b) What is the structural classification of this joint?

CHAPTER 10
Muscle Tissue

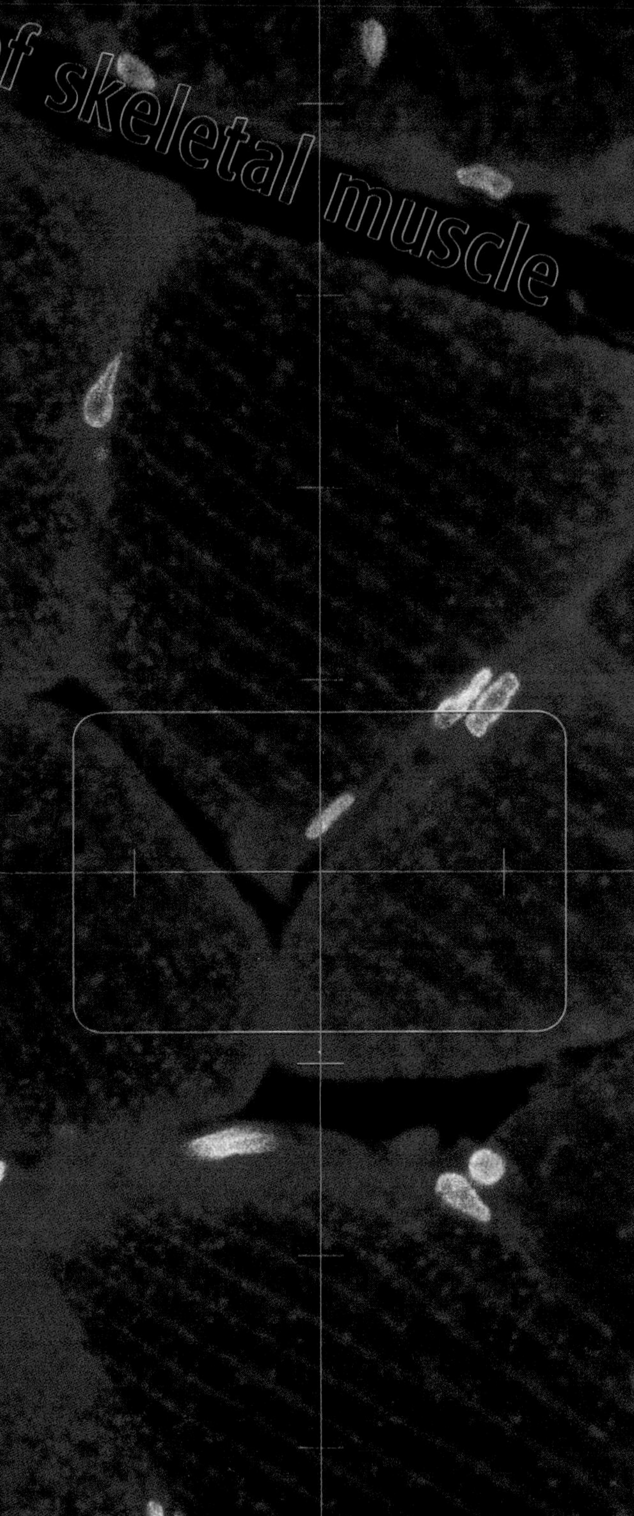

Digital microscopy of skeletal muscle

CHAPTER OUTLINE + STUDENT OBJECTIVES

Overview of Muscle Tissue 244

1. List four functional properties that distinguish muscle tissue from other tissues.
2. Compare and contrast skeletal, cardiac, and smooth muscle tissue.

Skeletal Muscle 244–253

3. Name the layers of connective tissue that occur in and around a skeletal muscle, and briefly describe a muscle's blood and nerve supply.
4. Describe the various ways in which muscles attach to their origins and insertions.
5. Define muscle fascicles.
6. Describe the microscopic structure of a skeletal muscle cell (fiber) and the arrangement of its contractile filaments into sarcomeres and myofibrils.
7. Explain the sliding filament theory of muscle contraction. What is the role of titin?
8. Describe the sarcoplasmic reticulum and T tubules in muscle fibers.
9. Compare and contrast the three kinds of skeletal muscle fibers.

Cardiac Muscle 254

10. Compare cardiac (heart) muscle to skeletal muscle.
11. Describe the structure and function of intercalated discs.

Smooth Muscle 254–257

12. Compare smooth muscle fibers and skeletal muscle fibers in terms of their structure and functions.
13. Describe how smooth muscle cells form a sheet-like tissue.

Disorders of Muscle Tissue 257–259

14. Explain some symptoms of muscular dystrophy, myofascial pain syndrome, and fibromyalgia.

Muscle Tissue Throughout Life 259–261

15. Describe the embryonic development and capacity for regeneration of muscle tissue.
16. Explain the changes that occur in skeletal muscle with age.

MUSCLE IS FROM A LATIN WORD MEANING "little mouse," a name that arose because flexing muscles look like mice scurrying beneath the skin. Indeed, the rippling muscles of weight lifters are often the first thing that comes to mind when one hears the word *muscle.* However, muscle is also the main tissue in the heart and in the walls of other hollow organs. In all its forms, muscle tissue makes up nearly half the body's mass.

OVERVIEW OF MUSCLE TISSUE

We begin our discussion by examining the functions and types of muscle tissue.

Functions

Muscle tissue has the following functions:

1. *Movement.* Most of the musculature attaches to the skeleton and moves the body by moving the bones. The musculature in the walls of visceral organs also produces movement by squeezing fluid and other substances through these hollow organs.
2. *Maintenance of posture.* Certain skeletal muscles contract periodically to maintain posture, enabling the body to remain in a standing or sitting position.
3. *Joint stabilization.* The role of muscle tone in stabilizing and strengthening joints was discussed in Chapter 9 (p. 220).
4. *Heat generation.* Muscle contractions produce heat that plays a role in maintaining normal body temperature at 37 °C (98.6 °F).

Muscle tissue has some special functional features that distinguish it from other tissues:

1. **Contractility.** Most significantly, muscle contracts forcefully. Its long cells shorten and generate a strong pulling force as they contract.
2. **Excitability.** Nerve signals or other factors excite muscle cells, causing an electrical impulse to travel along the muscle cell's plasma membrane. This impulse then stimulates the muscle cell to contract.
3. **Extensibility.** After it contracts and then relaxes, muscle tissue can be stretched back to its original length by the contraction of an opposing muscle.
4. **Elasticity.** After being stretched, muscle tissue can recoil passively and resume its resting length.

Types of Muscle Tissue

There are three types of muscle tissue: *skeletal, cardiac,* and *smooth.*

Skeletal muscle tissue is packaged into **skeletal muscles,** discrete organs that attach to and move the skeleton, such as the biceps brachii muscle of the arm. This tissue makes up a full 40% of body weight. Its cells show obvious stripes or striations, and its contraction is subject to voluntary control: For example, you can "will" your arm to move.

Cardiac muscle tissue occurs only in the wall of the heart. Like skeletal muscle, its cells are striated, but its contraction is not voluntary. As a rule, we have no direct conscious control over how fast our heart beats.

Most **smooth muscle tissue** occupies the walls of our hollow internal organs, such as the stomach, urinary bladder, blood vessels, and respiratory passages. Its cells lack striations. Like cardiac muscle, smooth muscle is not subject to direct voluntary control.

Sometimes, cardiac and smooth muscle are together called *visceral muscle,* a term reflecting the fact that both occur in the visceral organs. The various properties of the types of muscle tissue are summarized in Table 10.1.

Before exploring the differences among the three muscle types, we should point out some similarities. First, the cells of skeletal and smooth muscle (but not cardiac muscle) are called **fibers** because they are highly elongated. Second, muscle contraction depends on **myofilaments** (*myo* = muscle), which fill most of the cytoplasm of each muscle cell. There are two kinds of myofilaments, one containing *actin* and the other containing *myosin.* Recall from Chapter 2 (p. 40) that these two proteins generate contractile force in every cell in the body and that this contractile property is most highly developed in muscle cells. The final similarity among the muscle tissues is one of terminology: The plasma membrane of muscle cells, instead of being called a plasmalemma, is called a **sarcolemma** (sar″ko-lem′ah) (*sarcos* = flesh or muscle; *lemma* = sheath), and the cytoplasm is called **sarcoplasm.** Despite the different terms, the membranes and cytoplasm of muscle cells are not fundamentally different from those of other cell types.

SKELETAL MUSCLE

The many skeletal muscles of the body vary widely in shape, ranging from spindle-shaped cylinders to triangles to broad sheets. Each muscle is an organ made of several kinds of tissue: In addition to consisting mostly of muscle cells (fibers), a muscle also contains connective tissue, blood vessels, and nerves. In the following sections, we examine skeletal muscle anatomy from the gross level to the microscopic level. As the text discusses each level, see it summarized in Table 10.2 on p. 254.

Basic Features of a Skeletal Muscle

Connective Tissue and Fascicles

Several sheaths of connective tissue hold a skeletal muscle and its constituent bundles and fibers together. These

Table 10.1 Comparison of Skeletal, Cardiac, and Smooth Muscle

Characteristic	Skeletal	Cardiac	Smooth
Body location	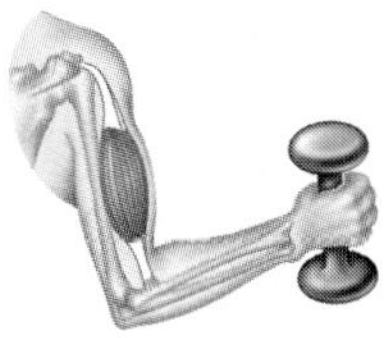 Attached to bones or (some facial muscles) to skin	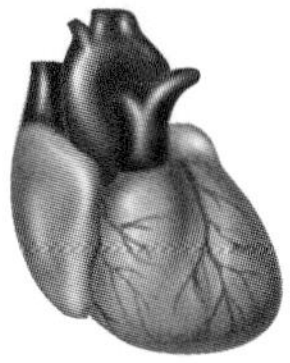 Walls of the heart	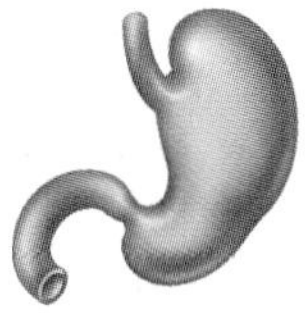 Mostly in walls of hollow organs, such as the stomach, respiratory tubes, bladder, blood vessels, and uterus
Cell shape and appearance	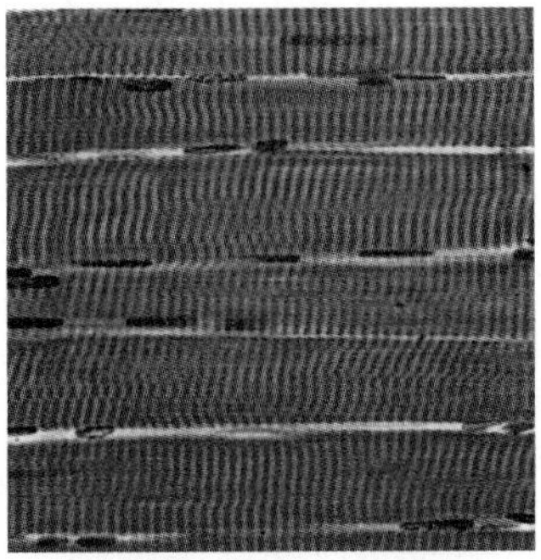  Single, very long, cylindrical, multinucleate cells with very obvious striations	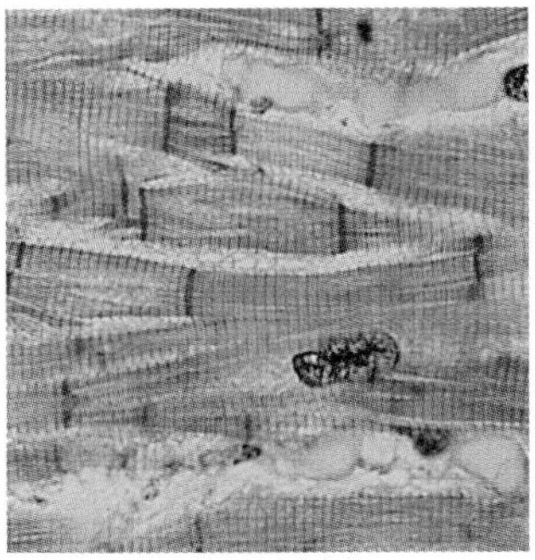 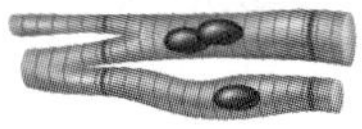 Branching chains of cells; uni- or binucleate; striations	  Single, fusiform, uninucleate; no striations
Regulation of contraction	Voluntary	Involuntary	Involuntary
Connective tissue components	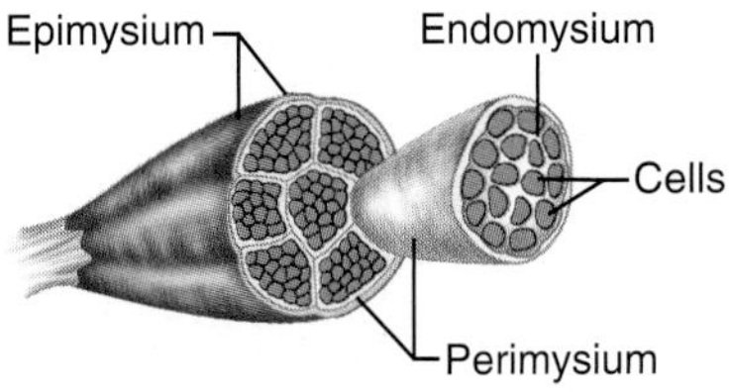 Epimysium, perimysium, and endomysium	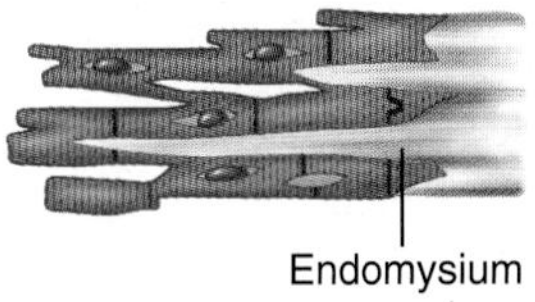 Endomysium attached to fibrous skeleton of heart	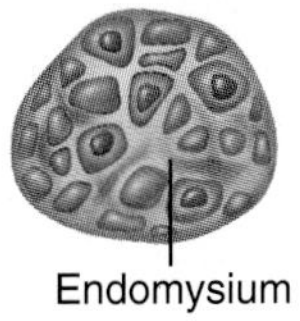 Endomysium
Presence of myofibrils composed of sarcomeres	Yes	Yes, but myofibrils are of irregular thickness	No, but actin and myosin filaments are present throughout
Presence of T tubules and site of invagination	Yes; at A-I junctions	Yes; at Z lines; larger diameter than those of skeletal muscle	No

➤

Characteristic	Skeletal	Cardiac	Smooth
Elaborate sarcoplasmic reticulum	Yes	Less than skeletal muscle; scant terminal cisternae	As well developed as in cardiac muscle
Presence of gap junctions	No	Yes; at intercalated discs	Yes; in single-unit muscle
Cells exhibit individual neuromuscular junctions	Yes	No	Not in single-unit muscle; yes in multiunit muscle
Source of Ca^{2+} for calcium pulse	Sarcoplasmic reticulum (SR)	SR and from extracellular fluid	SR and from extracellular fluid

sheaths, from external to internal, include the following (Figure 10.1):

1. **Epimysium.** An "overcoat" of dense irregular connective tissue surrounds the whole muscle (organ). This coat is the **epimysium** (ep″ĭ-mis′e-um), a name that means "outside the muscle." Sometimes the epimysium blends with the deep fascia that lies between neighboring muscles.
2. **Perimysium.** Within each skeletal muscle, individual muscle fibers (cells) are grouped into **fascicles** (fas′ĭ-klz; "bundles") that resemble bundles of sticks.

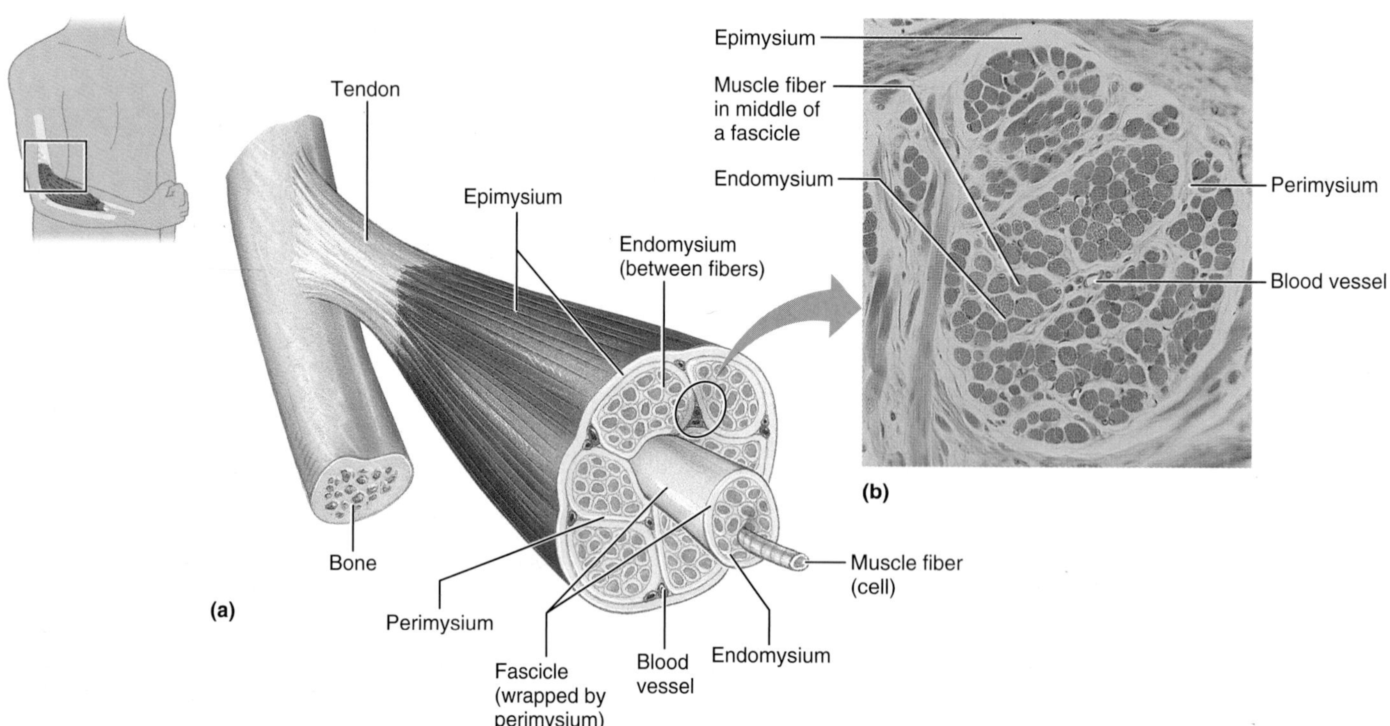

FIGURE 10.1 Connective tissue sheaths in skeletal muscle. **(a)** From external to internal, the connective tissue sheaths are epimysium, perimysium, and endomysium. **(b)** Photomicrograph of a cross section of a skeletal muscle (95 ×).

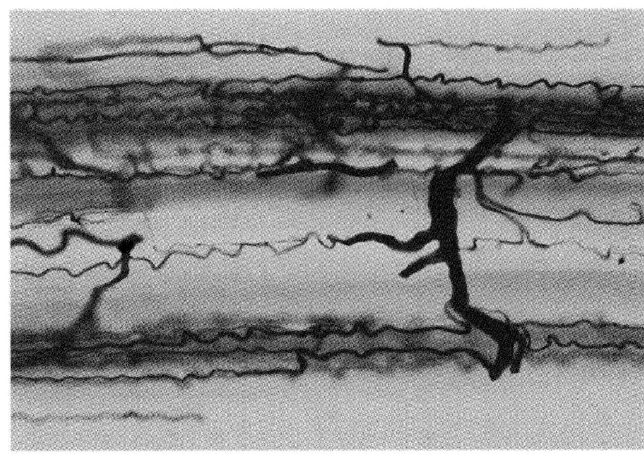

FIGURE 10.2 Photomicrograph of the capillary network surrounding skeletal muscle fibers. The arterial supply was injected with dark red gelatin to demonstrate the capillary bed. The long muscle fibers, which run horizontally across the photo, are stained orange. Note the wavy appearance of the thin red capillaries (170 ×).

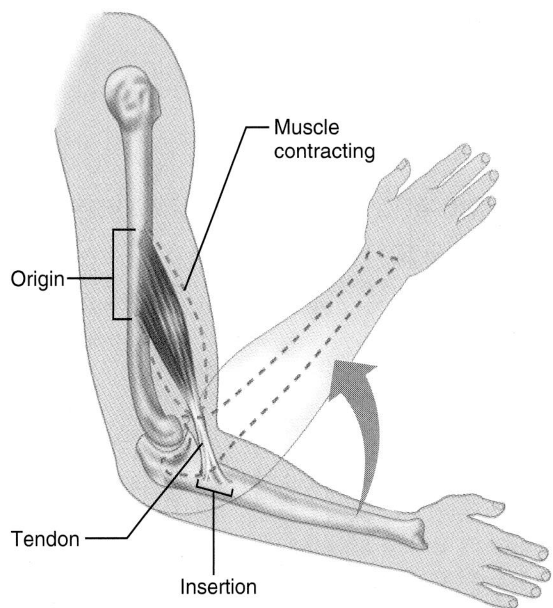

FIGURE 10.3 Muscle attachments (origin and insertion). When a skeletal muscle contracts, its insertion is pulled toward its origin. The muscle illustrated here is the brachialis muscle of the upper limb. In this particular muscle, the origin is by a direct attachment to the humerus, and the insertion is by an indirect attachment (tendon) to the ulna.

Surrounding each fascicle is a layer of fibrous connective tissue called **perimysium** (per″ĭ-mis′e-um; "around the muscle [fascicles]").

3. **Endomysium.** Within a fascicle, each muscle fiber is surrounded by a fine sheath of connective tissue consisting mostly of reticular fibers. This layer is the **endomysium** (en″do-mis′e-um; "within the muscle").

Notice in Figure 10.1 that the epimysium, perimysium, and endomysium are continuous with the tendons that join muscles to bones. Therefore, when muscle fibers contract, they pull on the connective tissue sheaths, which transmit the force to the bone being moved. The connective tissue sheaths also provide muscle tissue with much of its natural elasticity and carry the blood vessels and nerves that serve the muscle fibers.

Nerves and Blood Vessels

In general, each skeletal muscle is supplied by one nerve, one artery, and one or more veins—all of which enter or exit the muscle near the middle of its length. The nerves and vessels branch repeatedly in the intramuscular connective tissue, with the smallest branches serving individual muscle fibers. Each muscle fiber in a skeletal muscle is contacted by one nerve ending, which signals the fiber to contract. Such a contact is called a *neuromuscular junction.* This and other details of the innervation of muscle are considered in Chapter 14 (p. 416).

The rich blood supply to muscle reflects the high demand of contracting muscle fibers for nutrients and oxygen. Capillaries in the endomysium form a network (Figure 10.2). These long capillaries are wavy when the muscle fibers are contracted and are stretched straight when the muscle is extended.

Muscle Attachments

Most skeletal muscles run from one bone to another, crossing at least one movable joint. When such a muscle contracts, it causes one of these bones to move while the other bone usually remains fixed. The less movable attachment of a muscle is called its **origin,** whereas the more movable attachment is the **insertion** (Figure 10.3). Thus, the insertion is pulled toward the origin. In the muscles of the limbs, the origin lies *proximal* to the insertion.

Many muscles, especially in the limbs, span two or more joints. Such muscles are called biarticular ("two-joint") or multi-joint muscles. For example, look at Figure 10.3 and imagine a muscle that originates on the humerus, runs along the forearm without attaching, and inserts on the wrist bones. Contraction of such a muscle would cause movements at two joints: that of the forearm at the elbow, and that of the hand at the wrist.

Muscles attach to their origins and insertions via strong fibrous connective tissue. In *direct,* or *fleshy, attachments,* the attaching strands of connective tissue are so short that the muscle fascicles themselves appear to abut the bone. In *indirect attachments,* by contrast, the connective tissue extends well beyond the muscle to form a rope-like **tendon** or a flat sheet called an *aponeurosis* (ap″o-nu-ro′sis). Indirect attachments are more common than direct attachments, and most muscles have tendons. Raised bone markings are often present where tendons meet bones. These markings include tubercles, trochanters, and crests (see Table 7.1

on p. 153). Although most tendons and aponeuroses attach to bones, a few attach to skin, to cartilage, to sheets of fascia, or to a seam of fibrous tissue called a *raphe* (ra′fe; "seam").

Microscopic and Functional Anatomy of Skeletal Muscle Tissue

In this section we examine the structures and functions of parts of skeletal muscle tissue at and below the level of the cell—the skeletal muscle fiber.

The Skeletal Muscle Fiber

Skeletal muscle fibers are long and cylindrical (Figure 10.4; see also Table 10.2 on p. 254). These fibers are huge cells: Their diameter is 10–100 μm—up to ten times that of an average body cell—and their length is phenomenal: from several centimeters in short muscles to dozens of centimeters in long muscles. It would be inaccurate, however, to call these fibers the biggest cells in the body, because each one is actually formed by the fusion of hundreds of embryonic cells. Because it develops this way, the skeletal muscle fiber contains many nuclei. These nuclei lie at the cell periphery, just deep to the sarcolemma (see Figure 10.4b).

Myofibrils and Sarcomeres

Under the light microscope, light and dark stripes called *striations* are visible in skeletal muscle fibers (see Figure 10.4a and b). These striations result from the internal structure of long rods called **myofibrils** within the sarcoplasm. Note that these are *fibrils,* as distinguished from the larger *fibers* and the smaller *filaments* (discussed shortly). Myofibrils are unbranched cylinders that are present in large numbers, making up over 80% of the sarcoplasm. You might think of the myofibrils as specialized contractile cellular organelles unique to muscle fibers.

The different myofibrils in a fiber are separated and surrounded by narrow regions of sarcoplasm (Figure 10.5) that contain rows of mitochondria and glycosomes that supply energy for muscle contraction. Unfortunately, distinguishing individual myofibrils in histological sections is difficult (see Figure 10.4a) because the striations of adjacent myofibrils line up almost perfectly.

A myofibril is a long row of repeating segments called **sarcomeres** (sar′ko-mērz; "muscle segments") (see Figure 10.4c and d). The sarcomere is the basic unit of contraction in skeletal muscle. The boundaries at the two ends of each sarcomere are called **Z discs (Z lines).** Attached to each Z disc and extending toward the center of the sarcomere are many fine myofilaments called **thin (actin) filaments**, which consist primarily of the protein actin, although they contain other proteins as well. In the center of the sarcomere and overlapping the inner ends of the thin filaments is a cylindrical bundle of **thick (myosin) filaments.** Whereas thick filaments consist largely of myosin molecules, they also contain *ATPase enzymes* that split ATP (energy-storing molecules) to release the energy required for muscle contraction. Both ends of a thick filament are studded with knobs called *myosin heads* or *cross bridges* (see Figure 10.4d).

Knowledge of sarcomere structure enables us to explain the pattern of striations in skeletal muscle fibers. The dark bands, called **A bands,** correspond to the full length of the thick filaments; they also include the inner ends of the thin filaments, which overlap the thick filaments (see Figure 10.4c). The central part of each A band, where no thin filaments occur, is the **H zone.** The **M line** in the center of the H zone contains tiny rods that hold the adjacent thick filaments together (see Figure 10.4d). Alternating with the A bands are the light **I bands,** the regions where only thin filaments occur. Notice that each I band lies within two adjacent sarcomeres and has a Z disc running through its center (see Figure 10.4c). Recall that the striations of adjacent myofibrils align perfectly, allowing us to see the various bands in whole muscle fibers (see Figure 10.4a and b).

Mechanism of Contraction

The mechanism of skeletal muscle contraction is explained by the **sliding filament theory** (Figure 10.6). Contraction results as the myosin heads of the thick filaments attach to the thin filaments at both ends of the sarcomere and pull the thin filaments toward the center of the sarcomere by swiveling inward. After a myosin head pivots at its "hinge," it lets go, "recocks," binds to the thin filament farther along its length, and pivots again. This cycle is repeated many times during a single contraction. It should be emphasized that the thick and thin filaments themselves do not shorten: They merely slide past one another. The sliding filament mechanism is initiated by the binding of calcium ions to the thin filaments and is powered by ATP.

Figure 10.7 shows how contraction affects the striation pattern of skeletal muscle. The action of the thick filaments forcefully pulls the two Z discs closer together, causing each sarcomere to shorten. As the Z discs move closer together, they cause the I bands to narrow, and the H zones disappear completely. However, the A bands stay the same width because the length of the thick filaments does not change. When a muscle is stretched, on the other hand, the I bands and H zones widen as the Z discs move apart. Again, no change in the width of the A bands occurs.

Muscle Extension

We noted that muscle tissue is extensible and that muscle fibers are stretched (extended) back to their original length after they contract. What is responsible for this stretching? Basically, a skeletal muscle—and its

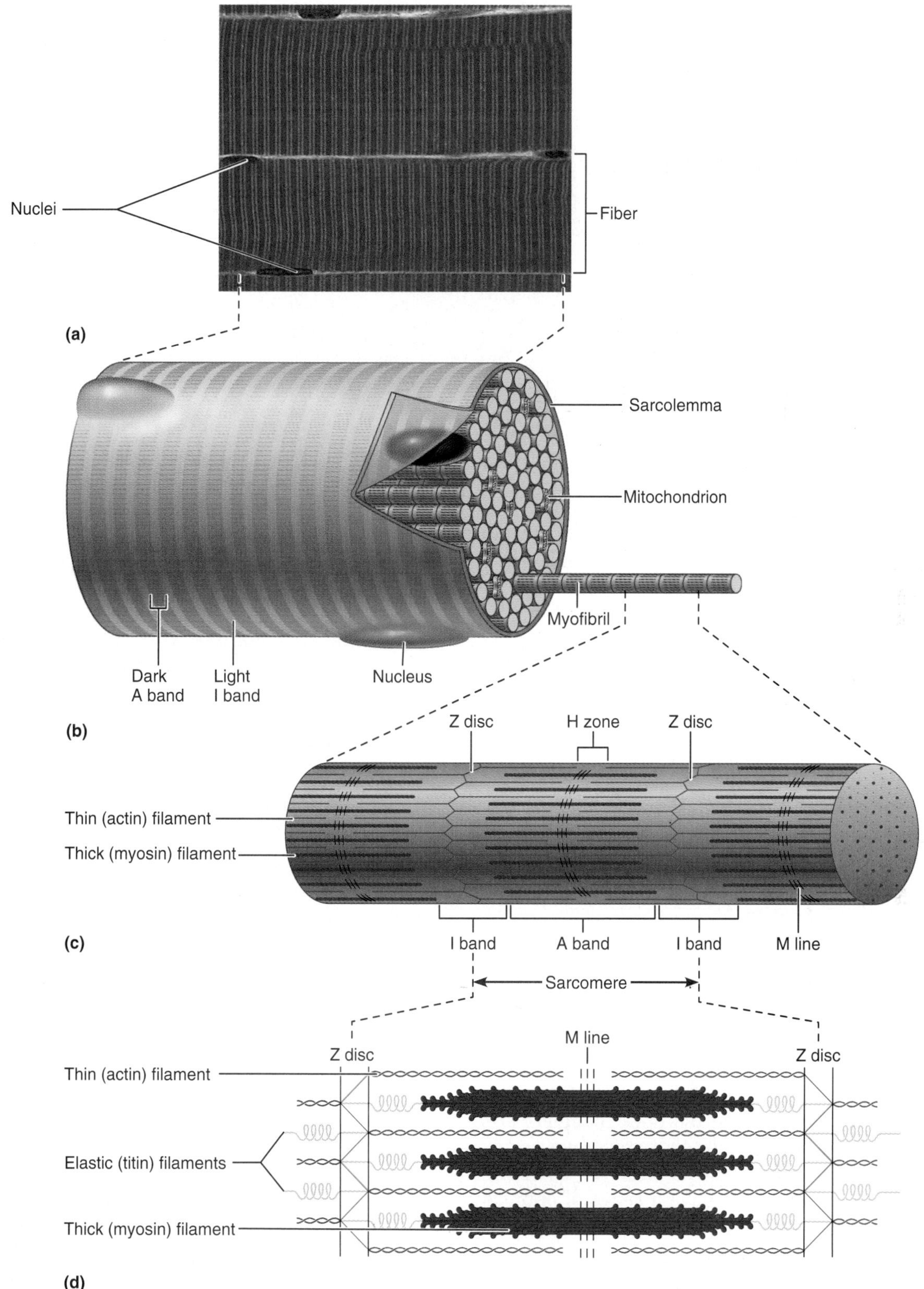

FIGURE 10.4 Microscopic anatomy of the skeletal muscle fiber (cell). **(a)** Photomicrograph of several skeletal muscle fibers. The borders between these fibers are indicated by the thin white horizontal streaks. Note the light and dark striations (400 ×). **(b)** Diagram of part of a muscle fiber (cell), showing the cylindrical myofibrils it contains. **(c)** A part of one myofibril, enlarged to show the sarcomeres and myofilaments responsible for the striations. Each sarcomere, or contractile unit, extends from one Z disc to the next. **(d)** Enlargement of one sarcomere sectioned longitudinally. Notice the interdigitation of the thin and thick filaments and the titin molecules connecting the latter to the Z discs. The many small knobs projecting from the thick filaments are myosin heads.

FIGURE 10.5 Skeletal muscle fiber: Electron micrograph of the sarcoplasm. Parts of two striped myofibrils are shown, separated from one another by strands of sarcoplasm containing mitochondria. As in Figure 10.4, the A and I bands and a Z line are labeled. Magnified 23,000 ×.

contained fibers—is stretched by the movement *opposite* to that which the muscle normally produces. For example, a muscle that normally *abducts* the arm at the shoulder is stretched by *adducting* the arm at this joint.

Muscle Fiber Length and the Force of Contraction

The ideal resting length for skeletal muscle fibers is the length at which they can generate the strongest pulling force. The greatest force is produced when a fiber starts out just slightly stretched, so that its thin and thick filaments overlap to only a moderate extent (see Figure 10.4d). This is because under these conditions, the myosin heads can move and pull along the whole length of the thin filaments. Under other conditions, contraction is suboptimal: If a muscle fiber is stretched so much that the thick and thin filaments do not overlap at all, the myosin heads have nothing to attach to, and no pulling force can be generated. Alternatively, if the sarcomeres are so compressed that the thick filaments are touching the Z discs, little further shortening can occur.

What is true for a muscle fiber is true for an entire muscle. Whole skeletal muscles have a range of optimal operational length of about 80% to 120% of their normal resting length. The sites of muscle attachments tend to keep muscles within that optimal range; that is, the joints normally do not let any bone move so widely that its attached muscles could shorten or stretch beyond their optimal range.

The Role of Titin

Titin (ti′tin) is a spring-like molecule in sarcomeres that resists overstretching. It extends from the Z disc to the thick filament and actually runs within the latter to attach to the M line (see Figure 10.4d). When first

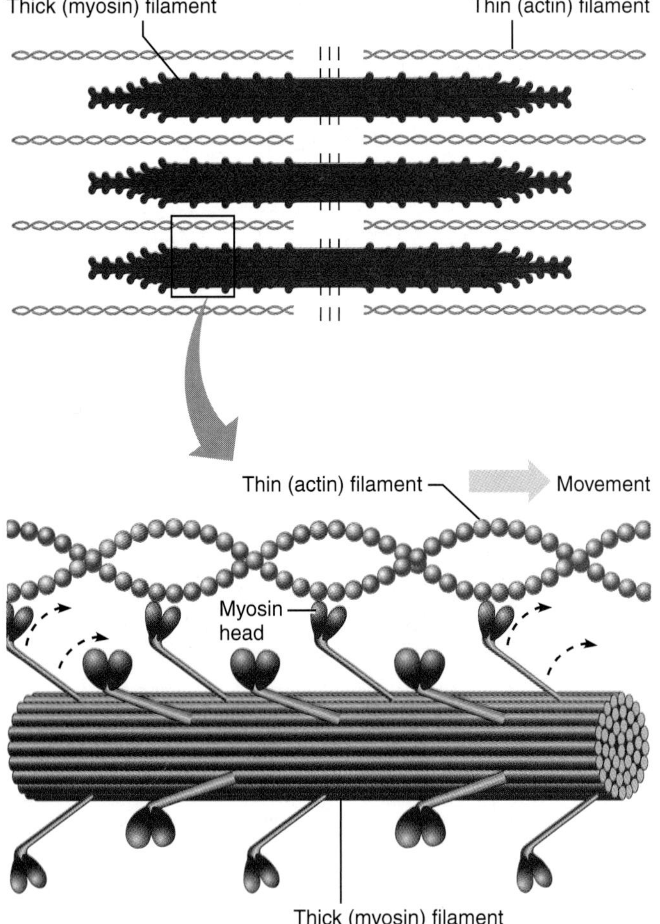

FIGURE 10.6 Sliding filament mechanism of contraction of a skeletal muscle. Myosin heads attach to actin in the thin filaments, then pivot to pull the thin filaments inward toward the center of the sarcomere.

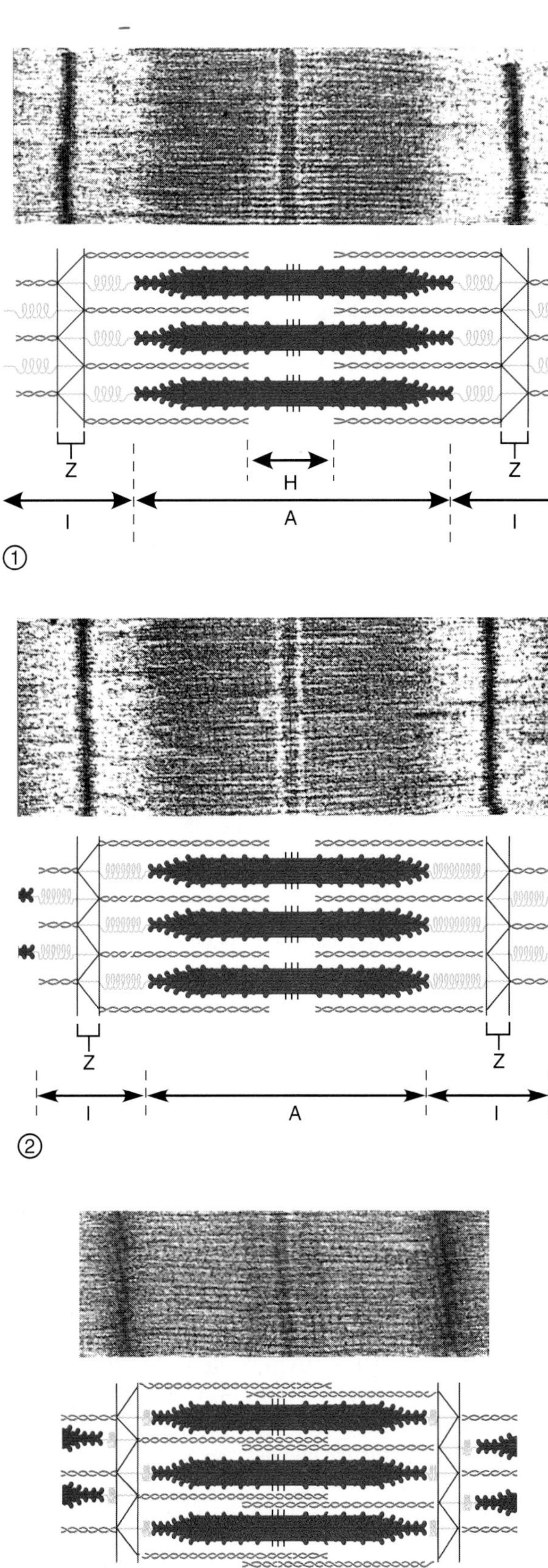

identified, titin generated much excitement because it is the largest protein ever discovered. It has two basic functions: (1) It holds the thick filaments in place in the sarcomere, thereby maintaining the organization of the A band; and (2) it unfolds when the muscle is stretched, then refolds when the stretching force is released, thereby contributing to muscle elasticity. Titin does not resist stretching in the ordinary range of extension, but it becomes stiffer the more it uncoils; therefore, it strongly resists any excessive stretching that tries to pull the sarcomere entirely apart.

Sarcoplasmic Reticulum and T Tubules

Skeletal muscle fibers (cells) contain two sets of intracellular tubules that participate in the regulation of muscle contraction: sarcoplasmic reticulum and T tubules (Figure 10.8). **Sarcoplasmic reticulum** is an elaborate smooth endoplasmic reticulum (see p. 36) whose interconnecting tubules surround each myofibril like the sleeve of a loosely crocheted sweater surrounds your arm. Most of these tubules run longitudinally along the myofibril. Others form larger, perpendicular cross-channels over the A-I junctions called **terminal cisternae** ("end sacs"). These cisternae occur in pairs on either side of a T tubule (discussed shortly).

The sarcoplasmic reticulum stores large quantities of calcium ions (Ca^{2+}), which are released when the muscle is stimulated to contract. These calcium ions then diffuse through the cytoplasm to the thin filaments, where they trigger the sliding filament mechanism of contraction. After the contraction, the ions are pumped back into the sarcoplasmic reticulum for storage.

Muscular contraction is ultimately controlled by nerve-generated impulses that travel along the sarcolemma of the muscle cell. These impulses are further conducted by the **T tubules,** which are deep invaginations of the sarcolemma that run between each pair of terminal cisternae of the sarcoplasmic reticulum. Each impulse signals the release of calcium from the adjacent terminal cisterns. Because T tubules are continuations of the sarcolemma, they conduct each impulse to the deepest regions of the muscle cell, thereby ensuring that the deep-lying myofibrils contract at the same time as the superficial ones, instead of lagging behind.

The complex of a T tubule flanked by two terminal cisternae at each A-I junction is called a **triad** (tri′ad; "group of three").

FIGURE 10.7 Changes in striations during the contraction of skeletal muscle. Electron micrographs of a sarcomere that is relaxed in (1), partly contracted in (2), and fully contracted in (3). A diagram that interprets each micrograph is included below it. Note that the two Z discs are moving closer together and closer to the ends of the thick filaments (closer to the A band). This action causes the I band and H zone to shrink, but the A band stays a constant length. The micrographs are magnified 28,000 ×.

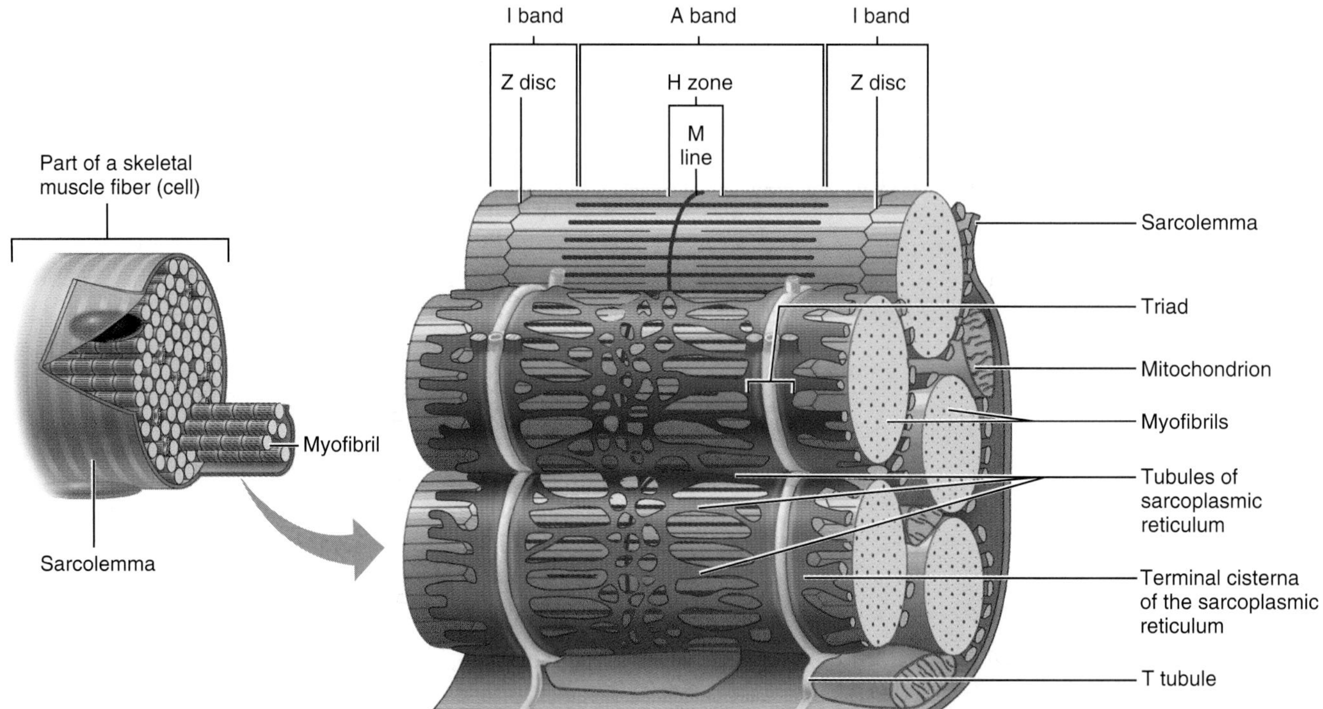

FIGURE 10.8 Sarcoplasmic reticulum and T tubules in the skeletal muscle fiber. The sarcoplasmic reticulum and T tubules are in the sarcoplasm of the muscle cell, between the myofibrils.

Types of Skeletal Muscle Fibers

Not all skeletal muscle fibers are alike. Instead, they can be divided into three main classes based on the strength, speed, and endurance of their contraction: *red slow-twitch, white fast-twitch,* and *intermediate fast-twitch* fibers (Figure 10.9). Most of the muscles in the body contain all three fiber types, but the proportions of each type differ among different muscles.

Red Slow-Twitch Fibers The red color of these relatively thin fibers stems from their abundant content of *myoglobin* (mi″o-glo′bin), an oxygen-binding pigment in their sarcoplasm. Because red fibers obtain their energy from aerobic (oxygen-requiring) metabolic reactions, they have a relatively large number of mitochondria (the sites of aerobic metabolism) and a rich supply of capillaries. Red slow-twitch fibers contract slowly, are extremely resistant to fatigue as long as enough oxygen is present, and deliver prolonged contractions. There are many such fibers in the postural muscles of the lower back, muscles that must contract continuously to keep the spine straight and maintain posture. Because they are thin, red slow-twitch fibers do not generate much power.

CLINICAL APPLICATIONS

RHABDOMYOLYSIS Although oxygen-binding myoglobin helps increase our endurance, it can cause problems if it leaks from muscle. This occurs in **rhabdomyolysis** (rab″do-mi-ol′i-sis; "disintegration of rod-shaped [skeletal] muscle"), a condition in which skeletal muscles are damaged by crushing or by extreme, destructive exercise. It was originally recognized in survivors of bombings and earthquakes who had been pinned under fallen buildings. In rhabdomyolosis, myoglobin pours into the bloodstream and clogs the blood-filtering kidneys, causing kidney failure, which eventually leads to heart failure. ✚

White Fast-Twitch Fibers These fibers are pale because they contain little myoglobin. About twice the diameter of red slow-twitch fibers, they contain more myofilaments and generate much more power. Because white fibers depend on anaerobic pathways (no oxygen is used) to make ATP, they contain few mitochondria or capillaries but many glycosomes containing glycogen as a fuel source. White fast-twitch fibers contract rapidly and tire quickly. They are common in the muscles of the upper limbs, which often lift heavy objects for *brief* periods—in moving furniture across a room, for example.

Intermediate Fast-Twitch Fibers These fibers have a diameter between those of the other two fiber types. Like white fibers, they contract quickly; like slow-twitch fibers, they are oxygen-dependent and have a high myoglobin content and a rich supply of capillaries. Because intermediate fast-twitch fibers depend largely on aerobic metabolism, they are fatigue-resistant, but less so than red slow-twitch fibers. They are more powerful than red fibers, but less so than white fibers. They are abundant in the muscles of the lower limbs, which must move the

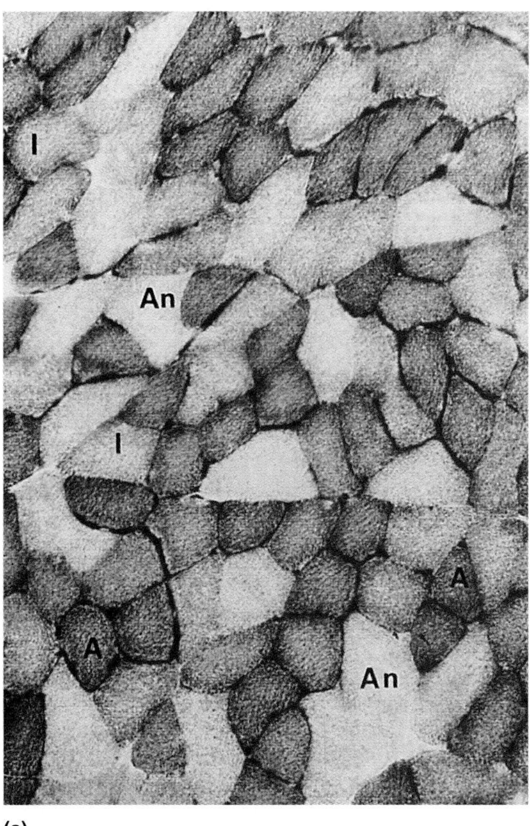

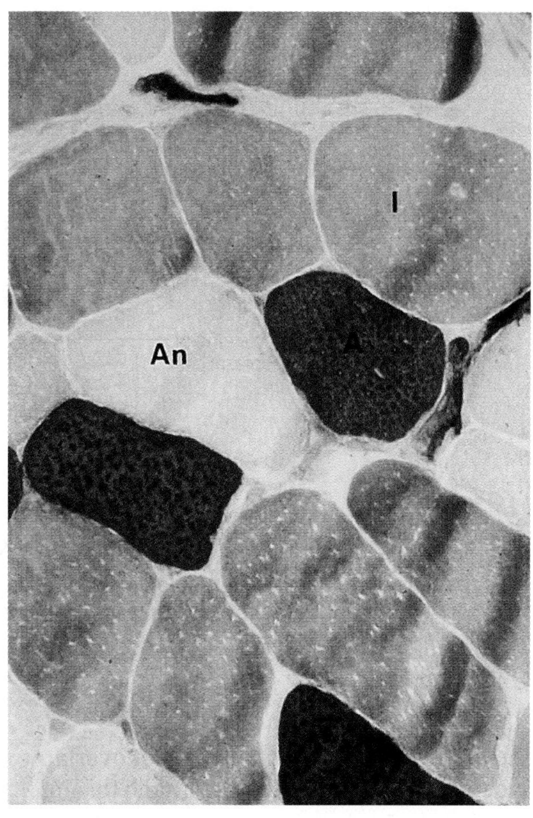

FIGURE 10.9 The three types of fibers in each skeletal muscle. Fibers are shown in cross section. From generally smallest to largest, they are: red slow-twitch fibers (labeled A for aerobic), intermediate fast-twitch fibers (I), and white fast-twitch fibers (labeled An for anaerobic). The staining technique used in this preparation differentiates the fibers by the abundance of their mitochondria, mitochondrial enzymes, and other features. Magnifications: **(a)** 200 ×; **(b)** 600 ×.

body for long periods during walking and jogging.

Because muscles contain a mixture of the three fiber types, each muscle can perform different tasks at different times. A muscle in the calf of the leg, for example, uses its white fibers to propel the body in a short sprint, uses intermediate fast-twitch fibers in long-distance running, and uses red slow-twitch fibers in maintaining a standing posture.

Although everyone's muscles contain mixtures of the three fiber types, some people have relatively more of one type. These differences are genetically controlled and no doubt influence the athletic capabilities of endurance versus strength. Much controversy has raged over whether one fiber type can transform into another when a person changes exercise regimes. Certainly, laboratory tests have produced such interconversions in lab animals by stretching the fibers and using electrical stimulation to change the amount of contractile force the fibers generate. Whether this transformation also occurs in exercising humans is not clear, but it is known that lifting heavy loads increases the diameter and strength of existing white fibers. It is by this process that weight lifters develop large muscles.

Critical Reasoning

How does weight lifting increase the size of one's white muscle fibers?

Weight lifting leads to increased production of contractile proteins, myofilaments, and myofibrils. As the number and size of myofibrils increase, the fibers enlarge. The fibers do not divide mitotically. However, it has been proposed that these enlarged fibers split longitudinally, and that the split cells then continue to enlarge. Most studies designed to investigate this splitting, however, have instead shown that new fibers arise from the proliferation and fusion of immature muscle-forming cells called satellite cells *(see p. 261). In either case, it now seems clear that weight lifting leads to increased numbers as well as increased sizes of muscle cells.*

We have compared the three kinds of skeletal muscle fibers in terms of speed of contraction and resistance to fatigue. But *all* these skeletal fibers contract faster and tire more easily than do smooth muscle and cardiac muscle cells, which are considered next.

Table 10.2 Structure and Organizational Levels of Skeletal Muscle

Structure and organizational level	Description	Connective tissue wrappings
Muscle (organ)	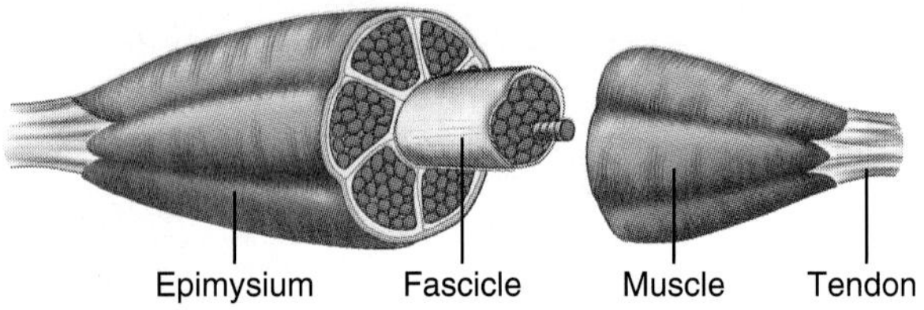 Consists of hundreds to thousands of muscle cells, plus connective tissue wrappings, blood vessels, and nerve fibers	Covered externally by the epimysium
Fascicle (a portion of the muscle)	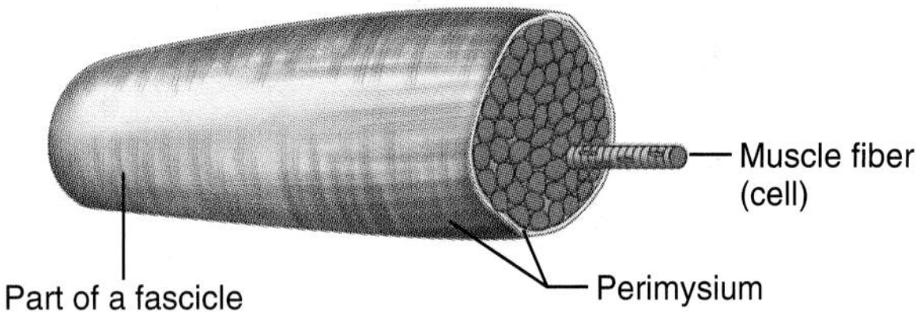 Discrete bundle of muscle cells, segregated from the rest of the muscle by a connective tissue sheath	Surrounded by a perimysium
Muscle fiber (cell)	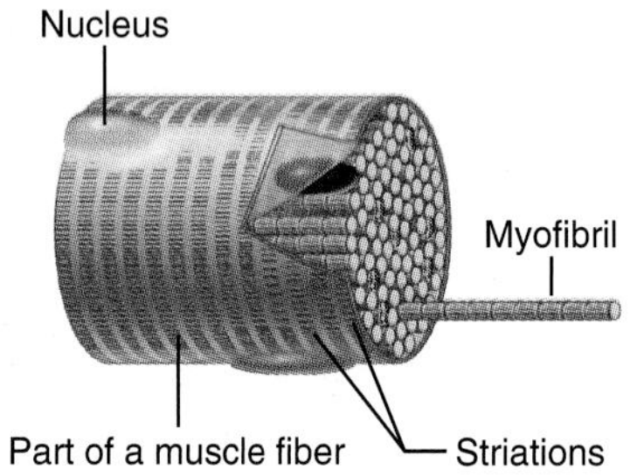 Elongated multinucleate cell; has a banded (striated) appearance	Surrounded by the endomysium
Myofibril (organelle containing myofilaments)	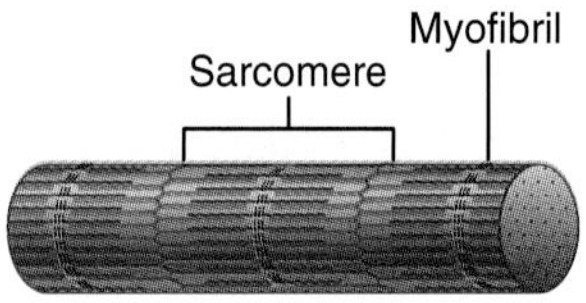 Rodlike contractile element; myofibrils occupy most of the muscle cell volume; appear banded, and bands of adjacent myofibrils are aligned; composed of sarcomeres, end to end	
Sarcomere (segment of a myofibril)	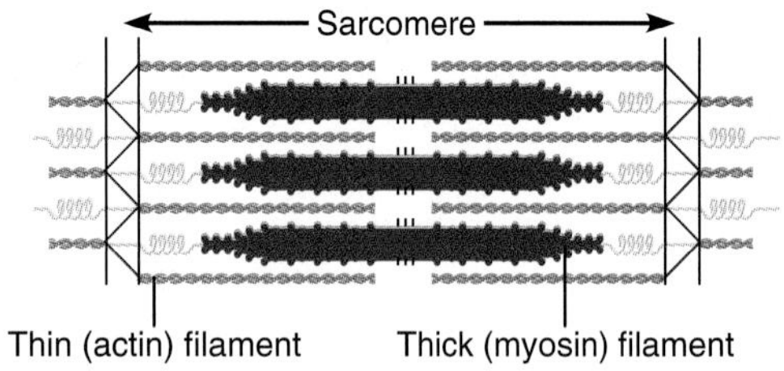 The contractile unit, composed of myofilaments made up of contractile proteins	
Myofilament or filament	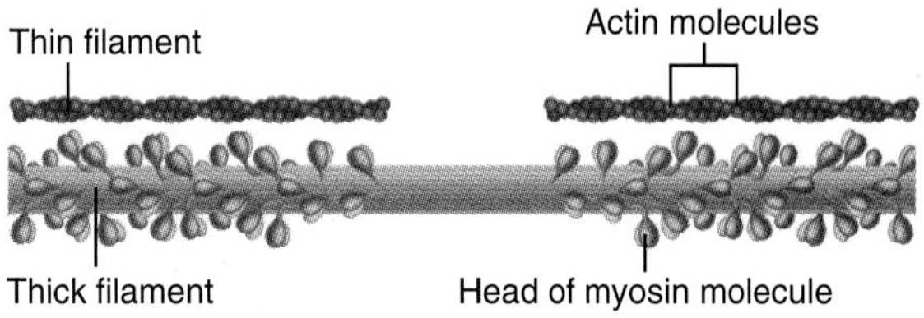 Myofilaments are of two types—thick and thin—composed of contractile proteins; the thick filaments contain bundled myosin molecules; the thin filaments contain actin molecules (plus other proteins); the sliding of the thin filaments past the thick filaments produces muscle shortening	

CARDIAC MUSCLE

Cardiac muscle is the muscle tissue of the heart wall. Arranged in bundles (Figure 10.10a), it forms a thick layer called the myocardium. The contractions of cardiac muscle pump blood through the blood vessels of the circulatory system. Cardiac muscle is striated like skeletal muscle, and it contracts through the sliding filament mechanism.

Cardiac muscle cells are single cells (Figure 10.10b–d), not fused cell colonies like skeletal muscle fibers. They average about 15 μm in width and 100 μm in length, and are separated from one another by a delicate endomysium (connective tissue). Unlike all other muscle cells, cardiac muscle cells branch and join together at complex junctions, called *intercalated discs,* so that they form cellular networks. Each cell contains one or two nuclei in its center, not at the periphery. This is the only class of muscle in which a cell is not called a fiber. Instead, the term *cardiac muscle fiber* refers to a long row of joined cardiac muscle cells.

The striated cardiac muscle cells contain typical sarcomeres with A and I bands, H zones, titin, and Z discs and M lines (Figures 10.10e and 10.11). *Myofibrils*, however, are difficult to discern because instead of being the simple cylinders present in skeletal muscle fibers, they branch extensively and vary in diameter. This irregularity is due to the great abundance of mitochondria between the myofibrils. The mitochondria make large amounts of ATP by aerobic metabolism. Their abundance ensures that cardiac muscle is highly resistant to fatigue. If cardiac muscle cells were to tire easily, even routine exercise would exhaust the heart and result in death.

Like skeletal muscle, cardiac muscle cells are triggered to contract by Ca^{2+} entering the sarcoplasm. However, compared to skeletal muscle, the system for delivering the Ca^{2+} is less elaborate, the sarcoplasmic reticulum is simpler, and the T tubules are less abundant—occurring just *once* per sarcomere at the Z disc instead of *twice* per sarcomere at the A-I junctions (Figure 10.10e). The contraction-triggering mechanism works like this: A small amount of Ca^{2+} enters the cardiac muscle cell from the extracellular tissue fluid and signals the sarcoplasmic reticulum to release its Ca^{2+}; these ions in turn diffuse into the sarcomeres and trigger the sliding of the filaments.

Cardiac muscle cells are joined by complex junctions called **intercalated discs** (in-ter′kah-la″ted; "inserted between"). At these junctions, the adjacent sarcolemmas interlock through meshing "fingers," and the effect is rather like one empty egg carton stacked inside another (Figure 10.10d). Intercalated discs have a combination of three types of cell junctions introduced in Chapter 4: (1) desmosomes, (2) long, desmosome-like junctions called **fasciae adherans** (singular: **fascia adherans**), and (3) gap junctions (Figure 10.10e). Desmosomes and fasciae adherans hold the cells together, whereas the gap junctions allow ions to pass between cells. The free movement of ions between cells allows the direct transmission of an electrical impulse through the entire network of cardiac muscle cells. This impulse in turn signals all the muscle cells in a heart chamber (atria or ventricles) to contract at the same time.

Unlike skeletal muscle tissue, not all cells of cardiac muscle tissue are innervated. In fact, an isolated cardiac muscle cell will contract in a rhythmic manner without any innervation at all. This inherent rhythmicity of cardiac muscle cells is the basis of the rhythmic heartbeat, as we explain further in Chapter 18 (p. 532).

As with skeletal muscle, the amount of force that cardiac muscle cells can generate depends on their length. Significantly, these cells normally remain slightly *shorter* than their optimal length. Therefore, when they are stretched by a greater volume of blood returning to the heart, their contraction force increases and they can pump the additional blood.

SMOOTH MUSCLE

Most smooth muscle of the body is found in the walls of visceral organs like the urinary bladder, uterus, and intestines (Figure 10.12a and b). More specifically, smooth muscle has six major locations: in the walls of the circulatory vessels, respiratory tubes, digestive tubes, urinary organs, and reproductive organs, and inside the eye. Its cells (fibers) contract by the sliding filament mechanism, but the details are not as well understood as in the other muscle types.

Each smooth muscle fiber is a spindle-shaped cell, with one centrally located nucleus (Figure 10.12c and d). Typically, these fibers have a diameter of 2–10 μm and a length of several hundred μm, and are separated from one another by an endomysium. In the walls of hollow viscera, the fibers are grouped into *sheets* of smooth muscle tissue (Figure 10.12b). Often two sheets are present, with their fibers oriented at right angles to each other. In one of the sheets, the **longitudinal layer,** the muscle fibers run parallel to the long axis of the organ. In the other, the **circular layer,** the fibers run around the circumference of the organ. The circular layer constricts the hollow organ, and the longitudinal layer shortens the organ's length. These muscle layers may generate alternate waves of contraction and relaxation that propel substances through the organ, a process called *peristalsis* (per″ĭ-stal′sis; "around contraction"). For example, peristalsis moves food through the digestive tube.

Smooth muscle fibers have no striations when viewed by light microscopy, and electron microscopy confirms that there are no sarcomeres (Figure 10.12c–f). Interdigitating thick and thin filaments are present in these cells, however, and fill much of the sarcoplasm (Figure 10.12e and f). These contractile myofilaments lie nearly parallel to the long axis of the fiber, but at a slightly oblique angle.

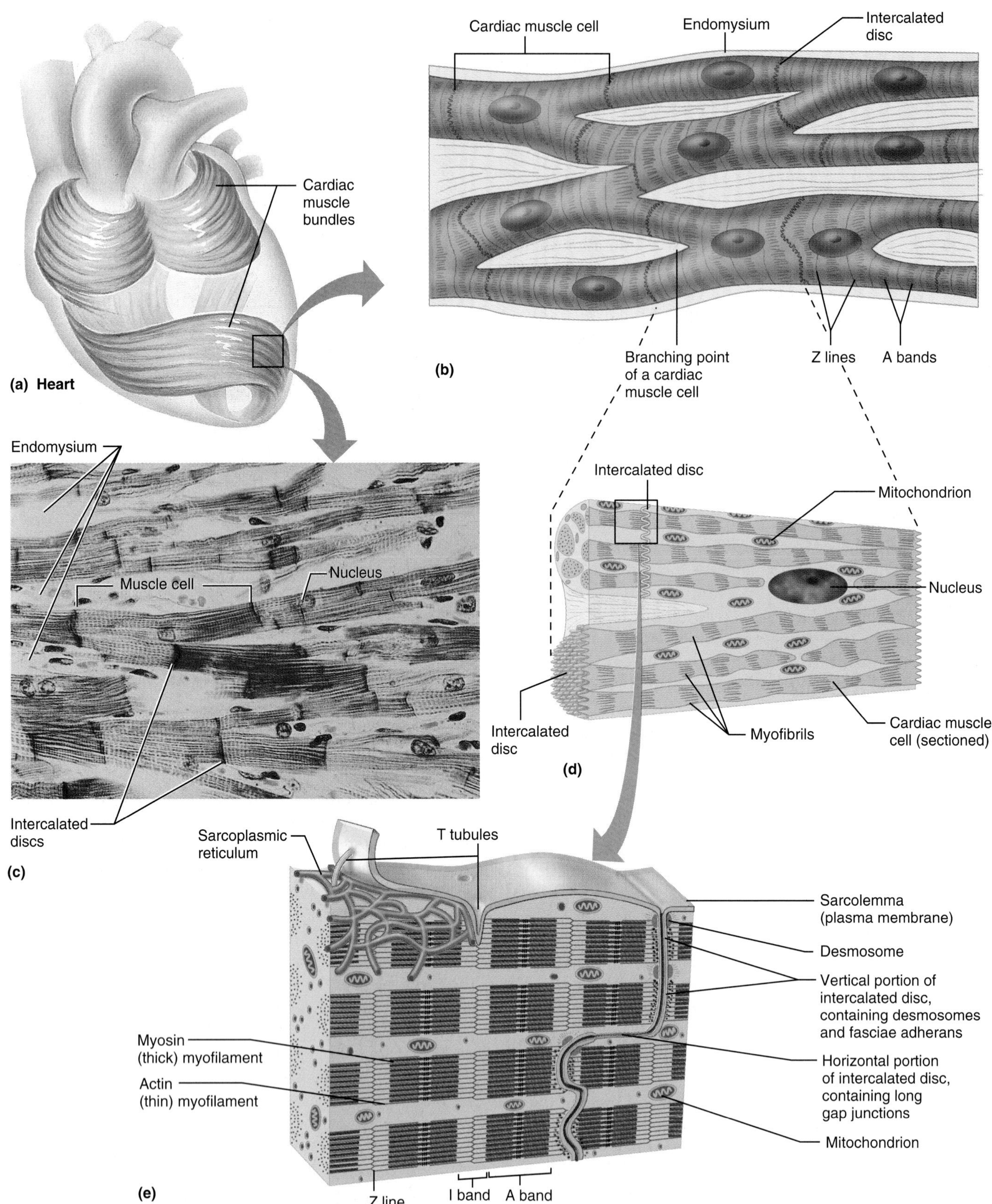

FIGURE 10.10 Cardiac muscle. **(a)** Arrangement of cardiac muscle bundles in the wall of the heart. **(b)** Cardiac muscle tissue: The muscle cells branch and are interconnected by intercalated discs to form a network. **(c)** Photomicrograph of cardiac muscle tissue (440 ×). The cells are striated, but the striations can be difficult to see using light microscopy. **(d)** Diagram of a cardiac muscle cell. At upper left, this cell is joined to part of another cell by an intercalated disc. At lower left, the "pegs" show the exposed surface of another intercalated disc. **(e)** Section showing details of the sarcoplasm and intercalated disc of two joining cardiac muscle cells.

Advances in Understanding The precise arrangement of the myofilaments in the smooth muscle cell has just recently become known, and it has been found that the myofilaments operate by interacting with elements of the cytoskeleton (Figure 10.12e). Through the sarcoplasm run longitudinal bundles of tension-resisting *intermediate filaments,* along which lie **dense bodies** at regular intervals. These bodies anchor the thin filaments, and thus they correspond to the Z discs of skeletal muscle. Through this anchoring attachment, the sliding myofilaments shorten the muscle cell by pulling on the cytoskeleton. Furthermore, just deep to the sarcolemma lie strips of cytoskeletal material called **dense plaques.** Like the dense bodies, these plaques anchor some thin filaments, but they also attach the intermediate filaments to membrane proteins that bind the muscle cell to collagen in the endomysium outside the cell. Therefore, it is through the dense plaques that the contracting cells transmit their pulling force to the surrounding connective tissue. ✷

As in skeletal and cardiac muscle, the entry of Ca^{2+} into the cytoplasm signals the smooth muscle fiber to contract. Some of these calcium ions enter from the extracellular fluid, whereas others are stored and released by an intracellular sarcoplasmic reticulum (not illustrated). Although this reticulum of tubules and sheets is difficult to demonstrate with standard techniques for electron microscopy, it is as large as in cardiac muscle fibers (1% to 8% of cell volume). T tubules and triads are absent from smooth muscle cells, as the sarcoplasmic reticulum directly touches the surface sarcolemma. Tiny spherical infoldings of this surface membrane called *caveolae,* of unknown function, were previously thought to be T tubules. As shown in Figure 10.12e, these caveolae lie beneath the strips of sarcolemma that are not underlain by dense plaques.

The contraction of smooth muscle is slow, sustained, and resistant to fatigue. Whereas smooth muscle takes 30 times longer to contract and relax than does skeletal muscle, it can maintain its contractile force for a long time without tiring, perhaps because the myofilaments latch together. This is a valuable feature, because the smooth muscle in the walls of the small arteries and visceral organs must sustain a moderate degree of contraction without fatiguing, day in and day out. Because smooth muscle's energy requirements are low, mitochondria are not abundant in its fibers.

We do not have direct voluntary control over the contraction of our smooth muscle, because it is innervated by a "nonconscious" part of the nervous system, the autonomic nervous system (see Chapter 15). In general, only a few smooth-muscle fibers in each muscle sheet are innervated, and the impulse spreads through gap junctions between adjacent fibers. Through such intercellular communication, the whole sheet contracts as a single unit. This arrangement is called *single-unit innervation.* There are exceptions: In the iris of the eye and the arrector pili muscles of the skin, every smooth muscle fiber is innervated individually, an arrangement called *multiunit innervation.* Finally, it should be mentioned that smooth-muscle contraction does not always require a nervous signal, as it can be stimulated by stretching the muscle fibers or by hormones.

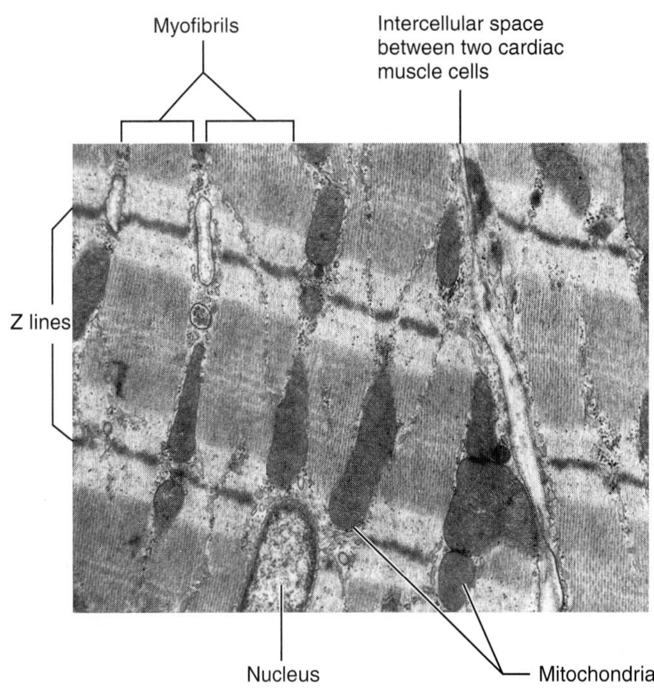

FIGURE 10.11 Cardiac muscle cell: Electron micrograph of the sarcoplasm. Notice the many mitochondria and the various sizes of myofibrils in the sarcoplasm. Magnified 13,000 ×.

DISORDERS OF MUSCLE TISSUE

With the exception of the heart muscle (discussed in Chapter 18), the body's muscle tissues experience remarkably few disorders.

The few clinical problems associated with smooth muscle tissue stem from external irritants. In the stomach and intestines, for example, irritation might result from ingestion of excess alcohol or spicy foods, or from bacterial infections. Under such conditions, contractions of smooth muscle increase in an attempt to rid the body of the irritating agents, and diarrhea and/or vomiting results.

Given good nutrition and sufficient exercise, skeletal muscle is amazingly resistant to infection throughout life. Noninfectious disorders of skeletal muscle, however, include *muscular dystrophy, myofascial pain syndrome,* and *fibromyalgia.*

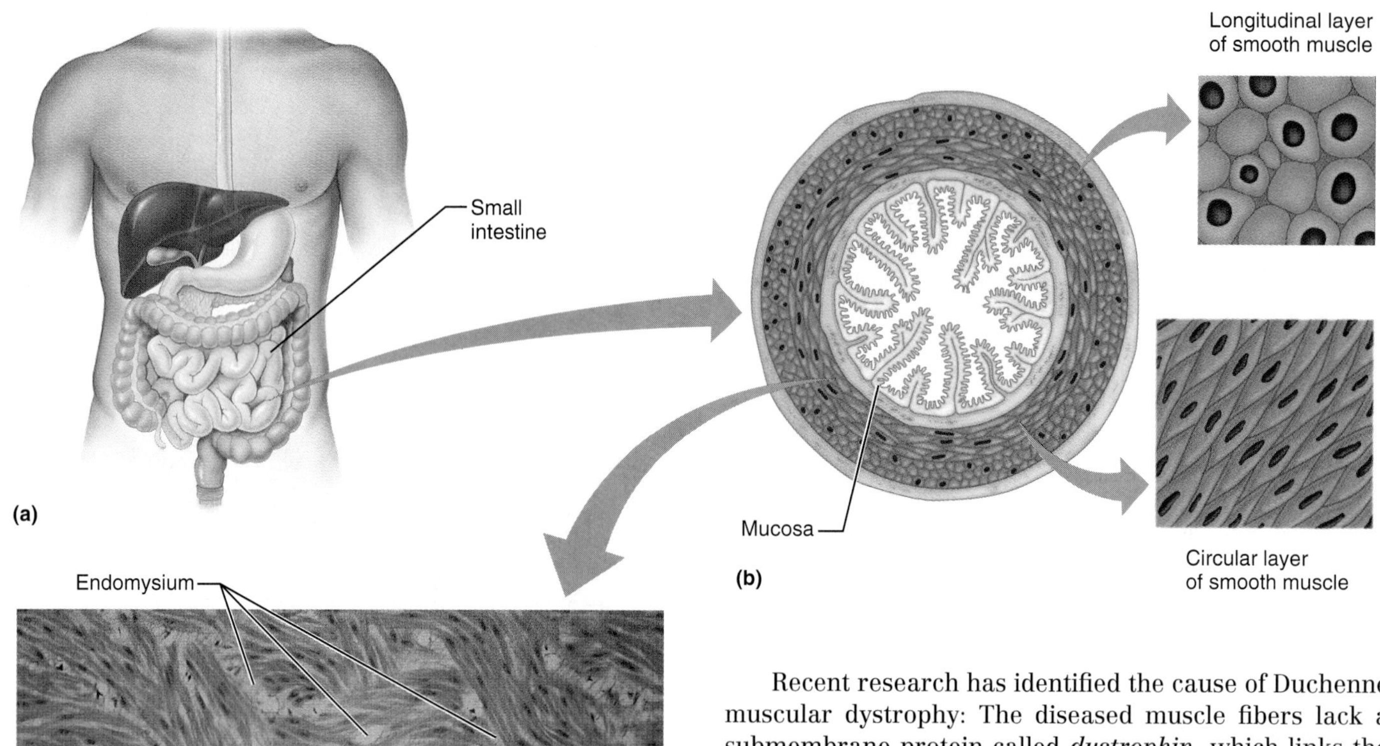

Muscular Dystrophy

Muscular dystrophy is a group of inherited muscle-destroying diseases that generally appear in childhood. The affected muscles enlarge with fat and connective tissue, but the muscle fibers themselves degenerate. The most common and most serious form is **Duchenne muscular dystrophy,** which is inherited as a sex-linked recessive disease; that is, females carry and transmit the abnormal gene, which is expressed almost exclusively in males. It affects 1 of every 3500 boys. This tragic disease is usually diagnosed when the boy is between 2 and 10 years old. Active, apparently normal children become clumsy and start to fall frequently as their muscles weaken. The disease progresses from the pelvic muscles to the shoulder muscles to the head and chest muscles. Victims rarely live past age 20 and usually die of respiratory infections or respiratory failure.

Recent research has identified the cause of Duchenne muscular dystrophy: The diseased muscle fibers lack a submembrane protein called *dystrophin,* which links the cytoskeleton to the extracellular matrix. Without this strengthening protein the sarcolemma weakens, and the leakage of extracellular calcium ions into the muscle fibers can fatally disrupt muscle function.

Scientists have explored possible treatments by injecting mice with embryonic muscle cells called myoblasts (discussed shortly), which then fuse with the unhealthy muscle fibers and induce them to produce dystrophin. Genes that signal the manufacture of dystrophin have also been injected. Similar treatments have not yet been successful in humans. Success may be near, however, because measurements suggest we are now getting about half as many healthy myoblasts into human dystrophic muscles as will be needed to improve their strength. Another treatment is to coax dystrophic muscle to produce more *utrophin,* a protein that is related to dystrophin and can substitute for it functionally.

Another kind of muscular dystrophy, called **myotonic dystrophy,** can appear at any age between birth and age 60. The symptoms of this inherited, slowly progressing disease include skeletal-muscle spasms followed by muscle weakness, and abnormal heart rhythm. The underlying genetic defect and the pattern of inheritance are now well understood.

Myofascial Pain Syndrome

In this very common condition, pain is caused by tightened bands of muscle fibers, which twitch when the skin over them is touched. The sensitive areas of skin are called trigger points. Myofascial pain syndrome is mostly associated with overused or strained *postural* muscles, and the pain is often felt some distance from the trigger point, in predictable places called reference zones. This

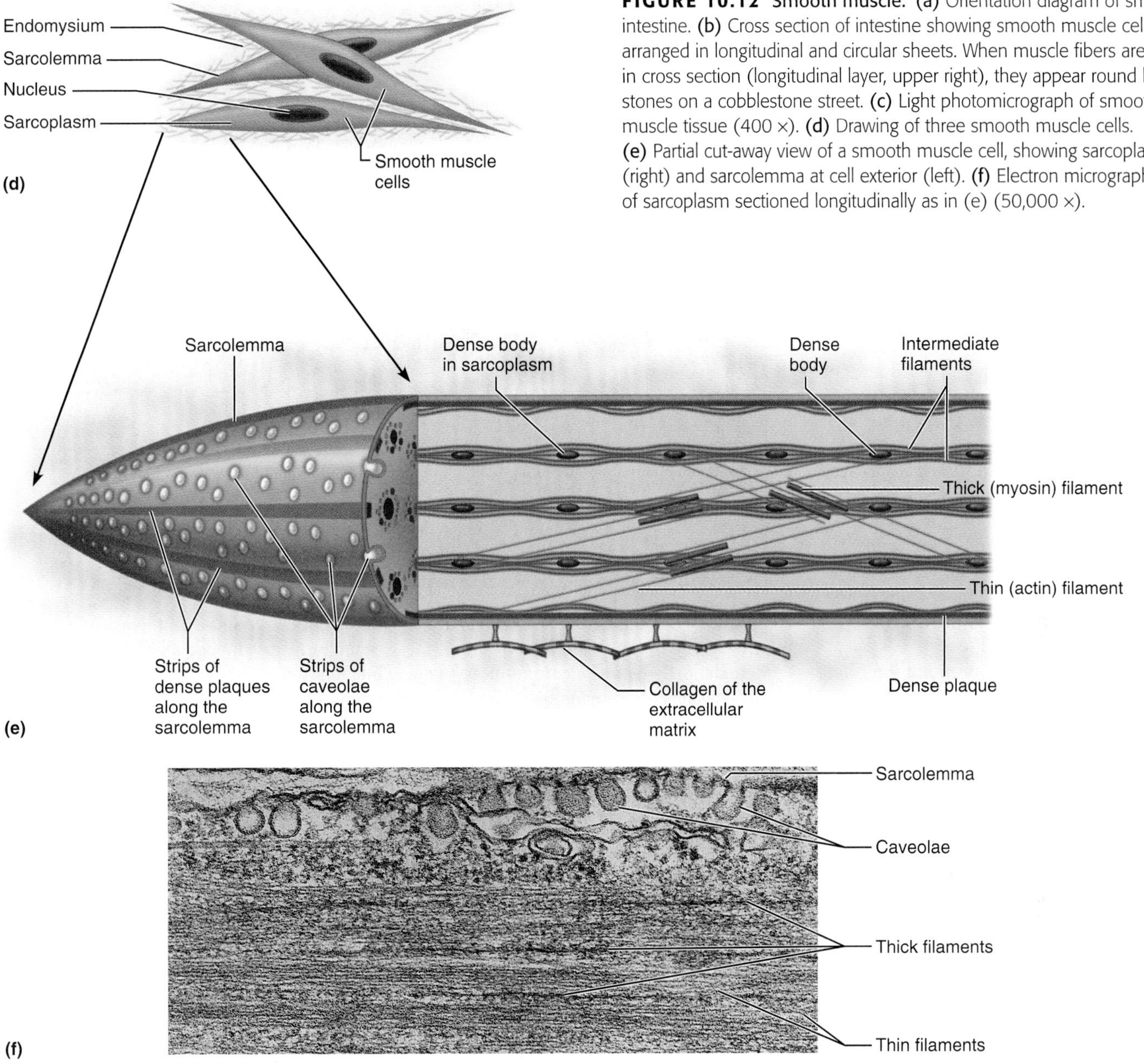

FIGURE 10.12 Smooth muscle. **(a)** Orientation diagram of small intestine. **(b)** Cross section of intestine showing smooth muscle cells arranged in longitudinal and circular sheets. When muscle fibers are cut in cross section (longitudinal layer, upper right), they appear round like stones on a cobblestone street. **(c)** Light photomicrograph of smooth muscle tissue (400 ×). **(d)** Drawing of three smooth muscle cells. **(e)** Partial cut-away view of a smooth muscle cell, showing sarcoplasm (right) and sarcolemma at cell exterior (left). **(f)** Electron micrograph of sarcoplasm sectioned longitudinally as in (e) (50,000 ×).

syndrome affects up to half of all people, mostly those from 30 to 60 years old. The pain is treated with nonsteroidal anti-inflammatory drugs and by stretching the affected muscle. Massage also helps, and exercising the affected muscle can lead to long-term recovery.

Fibromyalgia

Fibromyalgia (fi″bro-mi-al′ge-ah) is a mysterious chronic-pain syndrome of unknown cause (*algia* = pain). Its symptoms include severe musculoskeletal pain, fatigue, sleep abnormalities, and headache. It affects about 2% of all people, mostly women. The most common sites of pain are the lower back or neck, but for a condition to be identified as fibromyalgia, pain must be present in at least 11 of 18 standardized points that are spread widely over the body. Not all of these points are over muscles, and muscle problems do not seem to be the primary cause. However, fibromyalgia is included in this muscle chapter so it can be contrasted with myofascial pain syndrome (see above), with which it may be confused. Fibromyalgia is treated with antidepressants, exercise, and pain relievers.

MUSCLE TISSUE THROUGHOUT LIFE

With rare exceptions, all muscle tissues develop from embryonic mesoderm cells called *myoblasts.* Myoblasts fuse to form skeletal muscle fibers, which then begin to make thick and thin myofilaments and gain the ability

A Closer Look

Anabolic Steroid Abuse

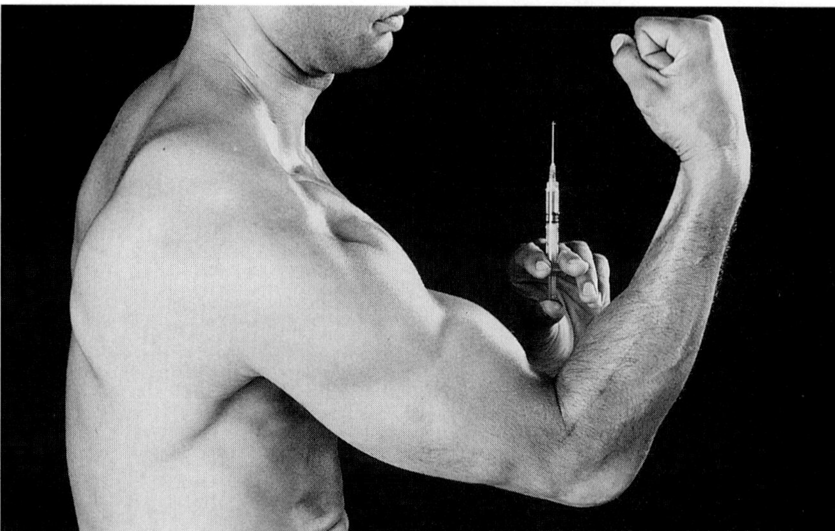

Given the lure of fame and fortune, it comes as no surprise that some athletes will do anything to improve their performance—including using anabolic steroids.

These synthetic substances mimic the actions of natural testosterone, which, among other effects, stimulates the manufacture of muscle proteins, leading to an increase in the number of myofilaments and in the thickness of skeletal muscle fibers. After pharmaceutical companies introduced anabolic steroids in the 1950s to treat victims of certain muscle-wasting diseases and to prevent muscle shrinkage in patients immobilized after surgery, many athletes became convinced that megadoses would produce stronger and bigger muscles. Steroid use—and abuse—continues today.

It is difficult to determine the current incidence of anabolic steroid use: Because most sports federations ban it, few athletes will talk about it. Still, some 1 million U.S. athletes—4% to 12% of all young men, and up to 3% of young women—are thought to inject and/or ingest some form of the drug. The Olympics probably have been tainted with steroid abuse for over 40 years. There is little doubt that many shot putters, discus throwers, and weight lifters are heavy steroid users, and several football players have admitted using it. In 1998, home run king Mark McGwire admitted using a potent but legal-for-baseball variety called androstenedione during his eclipsing of Babe Ruth's single-season home run record. Will trying to emulate their "sports heroes" have dire consequences for countless young athletes? (McGwire has since quit taking androstenedione.)

Competitive body builders typically combine months of intermittent, gradually increasing doses of the drugs with heavy weight-training workouts. Then, shortly before a competition, they stop usage so that urine tests can no longer detect the drug. The "cat-and-mouse game" between users and officials charged with catching them is never-ending.

But do steroids do what their proponents claim? Even though research has demonstrated increases in overall strength and body weight in steroid users, it is not clear whether this translates into athletic performance requiring fine muscle coordination and endurance. Studies by the American College of Sports Medicine indicate that steroids do not increase aerobic capacity.

Steroid use is clearly dangerous. Side effects include a bloated face, shriveled testes, impotence and sterility, acne and baldness, stunted growth in teens (due to premature closure of the epiphyseal plates), liver damage, enlarged breasts in males, and a deep voice and beard in females. Long-term usage predisposes users to coronary heart disease, blood that clots too readily, torn tendons, and liver cancer. Behavioral/psychological changes include increased aggression, irritability, tendency toward violence, depression, and delusions.

Male steroids are not the only muscle-building chemicals being used for athletic and cosmetic enhancement. Creatinine, which helps muscles to replenish their ATP stores, increases muscle mass with exercise but can cause kidney damage. Insulin-like growth factor-1 may build muscle at the expense of fat, but could also stimulate the growth of many other tissues in unpredictable ways. Its action is similar to that of growth hormone (see p. 752), another muscle-increasing substance that some athletes abuse.

The reasons some athletes use these drugs are obvious: In order to win, they'll do anything short of killing themselves. Or are they unwittingly doing that as well?

to contract. Ordinarily, the skeletal muscles are contracting by week 7, when the embryo is only about 2 cm long. Nerves grow into the muscle masses from the spinal cord, bringing the skeletal muscles under the control of the nervous system.

In contrast, the myoblasts that produce cardiac and smooth muscle cells do not fuse. However, the individual cells of these tissues connect through gap junctions at a very early embryonic stage. The cardiac muscle of the heart begins pumping blood 3 weeks after fertilization.

Skeletal muscle fibers never undergo mitosis after they are formed, and cardiac muscle cells stop dividing at about age 9. Both these fiber types, however, do lengthen and thicken during childhood and adolescence to keep up with the growing body. Furthermore, skeletal muscle fibers are surrounded during the late fetal period and thereafter by scattered *satellite cells,* which are immature cells that resemble undifferentiated myoblasts. During youth, satellite cells fuse into the existing muscle fibers to help them grow, and following injury to a young or adult muscle, satellite cells fuse together to produce new skeletal muscle fibers. This capacity for regeneration is not complete, however. Severely damaged skeletal muscle is replaced primarily by scar tissue.

Advances in Understanding Cardiac muscle has no satellite cells and does not regenerate effectively. Even though it was long thought to have no regenerative capacity whatsoever, recent studies suggest that cardiac muscle cells do divide at a modest rate. If encouraged by biotechnology to divide faster, they might one day be able to heal the damaged myocardium after heart attacks. ✷

Smooth muscle cells retain the capacity for division throughout life, and this tissue heals and regenerates well. It has by far the best regenerative ability of all muscle tissues.

Critical Reasoning

Our discussion of the muscular system during adulthood will begin with a commonly asked question. On average, the body strength of adult men is greater than that of adult women. Do you think this difference has a biological basis?

Evidence indicates that it does. Individuals vary, but on average, women's skeletal muscles make up 36% of body mass, compared to 42% in men. The greater muscular development of men is due mostly to the effects of androgen hormones (mainly testosterone) on skeletal muscle, not to effects of exercise. Because men are usually larger than women, the average difference in strength is even greater than the percentage difference in muscle mass would suggest (body strength per unit muscle mass, however, is the same in both sexes). Strenuous muscle exercise causes greater enlargement of muscles in males than in females, again because of the influence of male sex hormones.

Some athletes take large doses of synthetic male sex hormones ("steroids") to increase their muscle mass. This illegal and dangerous practice is discussed in the box A Closer Look on page 260.

As we age, the amount of connective tissue in our skeletal muscles naturally increases, the number of muscle fibers decreases, and the muscles become stringier, or more sinewy. Because skeletal muscles form so much of a person's body mass, body weight declines in many elderly people. The loss of muscle leads to a decrease in muscular strength, usually by 50% by age 80. This condition is called **sarcopenia** (sar″ko-pe′ne-ah), literally "flesh wasting." It can have grave implications for health, because muscle wasting leads to many serious falls in the elderly. The proximate cause of sarcopenia may be a reduction in the rate at which the aging satellite cells can rebuild new muscle. Fortunately, sarcopenia can be reversed by exercise, even in people of very advanced age.

RELATED CLINICAL TERMS

Lower back pain Backache. May be due to a herniated or cracked disc, but is usually due to injured ligaments and muscle strain (see below). The injured back muscles contract in spasms, causing rigidity in the lumbar region and painful movement. Backaches plague 80% of all Americans at some time in their lives, but most cases resolve themselves.

Myalgia (mi-al′je-ah; "muscle pain") Muscle pain resulting from any muscle disorder.

Myopathy (mi-op′ah-the) (*path* = disease) Any disease of muscle.

RICE Acronym for **r**est, **i**ce, **c**ompression, and **e**levation, the standard treatment for a pulled muscle, as well as for excessively stretched tendons or ligaments (also see p. 234).

Spasm A sudden, involuntary twitch of skeletal (or smooth) muscle, ranging in severity from merely irritating to very painful. May be due to chemical imbalances or injury. Spasms of the eyelid or facial muscles, called *tics,* may result from psychological factors. Massaging the affected area may help to end the spasm. A *cramp* is a prolonged spasm that causes a muscle to become taut and painful.

Strain Tearing of a muscle, often due to a sudden movement that excessively stretches the muscle. Also known as muscle "pull." May involve the muscle-tendon junction. Bleeding within the muscle and inflammation lead to pain.

CHAPTER SUMMARY

Media study tools that could provide you additional help in reviewing specific key topics of Chapter 10 are referenced below.

SP = Study Partner; AIA = A.D.A.M.® Interactive Anatomy

OVERVIEW OF MUSCLE TISSUE (p. 244)

1. Muscle tissue produces movement, maintains posture, stabilizes joints, and generates body heat. It has the special properties of contractility, excitability, extensibility, and elasticity.

2. The three types of muscle tissue are skeletal, smooth, and cardiac muscle. A cell in skeletal or smooth muscle (but not cardiac muscle) is called a fiber. The sarcoplasm of muscle cells is packed with myofilaments that generate contractile force.

3. Skeletal muscle attaches to the skeleton, has striated cells, and can be controlled voluntarily. Cardiac muscle occurs in the heart wall, has striated cells, and is controlled involuntarily. Smooth muscle occurs chiefly in the walls of hollow organs, has nonstriated cells, and is controlled involuntarily.

SKELETAL MUSCLE (pp. 244–253)

Basic Features of a Skeletal Muscle (pp. 244–248)

1. Skeletal muscles have various shapes, such as spindle-shaped cylinders, triangles, and sheets. Each skeletal muscle is an organ. From large to small, the levels of organization in a muscle are whole muscle, fascicle, fiber, myofibril, sarcomere, and myofilament.

2. The connective-tissue elements of a skeletal muscle are the epimysium around the whole muscle, the perimysium around a fascicle, and the endomysium around fibers. A fascicle is a bundle of muscle fibers.

3. Every skeletal muscle fiber is stimulated to contract by a nerve cell. Skeletal muscle has a rich blood supply. Fine nerve fibers and capillaries occupy the endomysium.

4. Each muscle extends from an immovable (or less movable) attachment, called the origin, to a more movable attachment, called an insertion.

5. Muscles attach to bones through tendons, aponeuroses, or direct (fleshy) attachments. Some muscles cross two or more joints.

AIA Link: Atlas; System: Muscular; Palmar aponeurosis.

Microscopic and Functional Anatomy of Skeletal Muscle Tissue (pp. 248–253)

6. A skeletal muscle fiber is a long, striated cell formed from the fusion of many embryonic cells. It contains many peripherally located nuclei.

AIA Link: Dissectible anatomy; Male: Anterior; Layer 10.

7. Myofibrils are striped cylinders that fill most of the sarcoplasm of skeletal muscle cells. A myofibril is a row of sarcomeres arranged end to end. A sarcomere extends from one Z disc to the next. Thin (actin) filaments extend centrally from each Z disc. Thick (myosin) filaments occupy the center of each sarcomere and overlap the inner ends of the thin filaments.

8. According to the sliding filament theory of muscle contraction, the thin filaments are pulled toward the center of the sarcomere by a pivoting action of myosin heads on the thick filaments.

9. The myofilaments determine the striation pattern in skeletal (and cardiac) muscle fibers. There are A bands where thick filaments are located, I bands that contain only thin filaments and Z discs, plus M lines and H zones, where only thick filaments occur. During contraction, the Z discs move closer together, and the I bands and H zones narrow.

10. Between contractions, muscles are extended, or stretched by the skeletal movements caused by the contraction of opposing muscles. The huge titin molecules in the sarcomere resist overextension, and, along with the connective-tissue elements, give muscle its elasticity.

11. The muscles attach to the skeleton in a way that keeps them at a near-optimal length for generating maximum contractile forces.

12. The sarcoplasmic reticulum is a specialized smooth endoplasmic reticulum in the muscle fiber. T tubules are deep invaginations of the sarcolemma. When a nerve cell stimulates a muscle fiber, it sets up an impulse in the sarcolemma that signals the sarcoplasmic reticulum to release Ca^{2+}, which then initiates the sliding of the myofilaments (muscle contraction).

13. There are three types of skeletal muscle fibers: (1) red slow-twitch fibers (fatigue-resistant and best for maintaining posture), (2) white fast-twitch fibers (for short bursts of power), and (3) intermediate fast-twitch fibers (for long-term production of fairly strong contraction). Most muscles in the body contain a mixture of these fiber types.

SP Exercise: Chapter 10, Organized Levels of Skeletal Muscles.

CARDIAC MUSCLE (p. 254)

1. Cardiac muscle cells are branching, striated cells with one or two central nuclei. They contain irregularly shaped myofibrils made of typical sarcomeres. Cardiac muscle contracts by the sliding filament mechanism.

2. Cardiac muscle cells have many mitochondria and depend on aerobic respiration to form ATP. They are very resistant to fatigue.

3. Contraction of cardiac muscle is triggered by Ca^{2+}, but some of this calcium comes from the extracellular fluid. The sarcoplasmic reticulum and T tubules are somewhat simpler than those of skeletal muscle.

4. Adjacent cardiac cells are connected by intercalated discs that contain several types of junctions: desmosomes, fasciae adherans, and gap junctions.

5. Networks of cardiac muscle cells contract together because they are electrically coupled by gap junctions. Cardiac muscle cells contract with their own inherent rhythm, the basis of the heartbeat.

SMOOTH MUSCLE (pp. 254–257)

1. Smooth muscle fibers are small, spindle-shaped cells with one central nucleus. They have no striations or sarcomeres, but are filled with myofilaments that contract by the sliding filament mechanism. The cytoskeleton runs in parallel, longitudinal strips through the sarcoplasm and beneath the sarcolemma, containing dense bodies and plaques for the attachment of actin myofilaments.
2. The sarcoplasmic reticulum is as well developed as in cardiac muscle, although more difficult to demonstrate. T tubules are absent from smooth muscle cells.
3. Smooth muscle fibers are most often arranged in circular and longitudinal sheets.
4. Smooth muscle is innervated by involuntary nerve fibers, which usually contact only a few muscle fibers per sheet. The impulse that signals contraction usually spreads from fiber to fiber through gap junctions.
5. Smooth muscle contracts for extended periods at low energy cost and without fatigue.

DISORDERS OF MUSCLE TISSUE (pp. 257–259)

1. The disorders discussed in this section include muscular dystrophy, myofascial pain syndrome, and fibromyalgia.

MUSCLE TISSUE THROUGHOUT LIFE (pp. 259–261)

1. Muscle tissue develops from embryonic mesoderm cells called myoblasts. Skeletal muscle fibers form by the fusion of many myoblasts. Smooth and cardiac muscle cells develop from single myoblasts.
2. Mature skeletal muscle tissue has some ability to regenerate because of its satellite cells, but cardiac tissue does not regenerate well enough to allow functional recovery. Smooth muscle regenerates best.
3. On the average, men have more muscle mass than women. This disparity is due to the effects of male sex hormones.
4. Skeletal muscles are richly vascularized and resistant to infection, but in old age they shrink, become fibrous, and lose strength. This condition, called sarcopenia, is reversible through exercise.

REVIEW QUESTIONS

Multiple Choice/Matching Questions

1. The connective tissue that lies just outside the sarcolemma of an individual muscle cell is called the (a) epimysium, (b) perimysium, (c) endomysium, (d) endosteum.
2. A fascicle is (a) a muscle, (b) a bundle of muscle cells enclosed by a connective tissue sheath, (c) a bundle of myofibrils, (d) a group of myofilaments.
3. Thick and thin myofilaments have different properties. For each phrase below, indicate whether the filament described is thick or thin (write *thick* or *thin* in the blanks).

____ **(1)** contains actin

____ **(2)** contains myosin heads

____ **(3)** contains myosin

____ **(4)** does not lie in the H zone

____ **(5)** does not lie in the I band

____ **(6)** attaches to a Z disc

4. Match the correct level of skeletal muscle organization given in the key with its description:

Key: **(a)** muscle **(b)** fascicle **(c)** fiber **(d)** myofibril **(e)** myofilament

____ **(1)** could be called an elaborate organelle; made of sarcomeres

____ **(2)** an organ

____ **(3)** a bundle of cells

____ **(4)** a group of large molecules

____ **(5)** a cell

5. Write *yes* or *no* in each blank below to indicate whether each of the following narrows when a skeletal or cardiac muscle fiber contracts.

____ **(1)** H band

____ **(2)** A band

____ **(3)** I band

____ **(4)** M line

6. The function of T tubules in muscle contraction is to (a) make and store glycogen, (b) release Ca^{2+} into the cell interior and then pick it up again, (c) transmit an impulse deep into the muscle cell, (d) make proteins.
7. Which fiber type would be the most useful in the leg muscles of a long-distance runner? (a) white fast-twitch, (b) white slow-twitch, (c) intermediate fast-twitch.
8. The ions that first enter a muscle cell when an impulse passes over its sarcolemma and then trigger muscle contraction are (a) calcium, (b) chloride, (c) sodium, (d) potassium.
9. Fill in each blank with the correct single answer from the key.

Key: **(a)** skeletal muscle **(b)** cardiac muscle **(c)** smooth muscle

____ **(1)** striated and involuntary

____ **(2)** striated and voluntary

____ **(3)** not striated and involuntary

____ **(4)** its cells form networks

____ **(5)** is present in wall of bladder

____ **(6)** has intercalated discs

____ **(7)** its fibers are giant, multinucleate cells

____ **(8)** is best at regenerating when injured

____ **(9)** has no A or I bands

____ **(10)** its fibers are in sheets in walls of most viscera

Short Answer Essay Questions

10. Name and explain the four special functional characteristics of muscle tissue.

11. (a) Distinguish a tendon from an aponeurosis and a fleshy attachment. (b) Define origin and insertion, and explain how they differ.

12. Explain the sliding filament theory of contraction by drawing and labeling a relaxed and a contracted sarcomere.

13. List the structural differences between the three distinct types of skeletal muscle fibers.

14. Which is more resistant to fatigue, cardiac muscle or skeletal muscle? What is the anatomical basis for this difference, and why is it important?

15. Describe the structure and function of an intercalated disc.

16. Cindy Wong was a good anatomy student, but she realized she was mixing up the following "sound-alike" structures in skeletal muscle: myofilaments, myofibrils, fibers, and fascicles. Therefore, she compiled a brief table to define and differentiate these four structures. Construct a table like hers.

17. What is the function of the sarcoplasmic reticulum in a skeletal muscle cell?

18. Define sarcolemma and sarcoplasm.

19. Where is titin located, and what are its functions?

CRITICAL REASONING + CLINICAL APPLICATION QUESTIONS

1. A certain anatomy student decided that his physique left much to be desired. He joined a health club and began to "pump iron" three times a week. After 3 months of training, during which he was able to lift increasingly heavy weights, his arm and chest muscles became much larger. Explain what happened to the fibers in these muscles.

2. Diego, who had not kept in shape, went out to play a game of touch football. As he was running, the calf region of his leg began to hurt. He went to a doctor the next day, who told him he had a strain. Diego kept insisting that no joints hurt. Clearly, he was confusing a *strain* with a *sprain*. Explain the difference.

3. Chickens are capable of only brief bursts of flight, and their flying muscles consist of white fibers. The breast muscles of ducks, by contrast, consist of red and intermediate fibers. What can you deduce about the flying abilities of ducks?

4. Takashi, an osteopathic physician, saw that Mrs. and Mr. Rogers were suffering because their son was fighting a long battle with Duchenne muscular dystrophy. To comfort them, Takashi said that one day physicians hope to be able to cure this disease by injecting healthy myoblasts into the weakened muscles. What are myoblasts?

5. Why are muscle infections relatively rare (compared to respiratory or skin infections, for example)?

6. As a sprinter, Lateesha knew that the best way to treat pulled muscles was through "RICE." What does that mean?

A.D.A.M.® INTERACTIVE ANATOMY QUESTIONS

1. Link: Atlas; System: Muscular; Posterior thigh. Name the three muscles known as the hamstrings.

2. Link: Dissectible anatomy; Male (anterior); layer 29; scroll to abdomen. In which direction do the transverse abdominal muscle fibers run?

CHAPTER 11

Muscles of the Body

Imaging of striated muscle

CHAPTER OUTLINE + STUDENT OBJECTIVES

Lever Systems: Bone-Muscle Relationships 266–268

1. Explain the three types of lever systems in which muscles participate, and indicate the arrangement of elements (effort, fulcrum, and load) in each.

Arrangements of Fascicles in Muscles 268

2. Name some muscles whose fascicles have the following arrangements: parallel, convergent, pennate, circular.

Interactions of Skeletal Muscles in the Body 268–270

3. Describe the functions of prime movers (agonists), antagonists, synergists, and fixators.

Naming the Skeletal Muscles 270

4. List the criteria used in naming muscles.

Development-Based Organization of the Muscles 270–272

5. Organize the body's muscles into four functional groups based on their developmental origin.

Major Skeletal Muscles of the Body 272–331

6. Name and identify the muscles described in Tables 11.1 to 11.17. State the origin, insertion, and action of each.

AS WE EXPLAINED IN THE PREVIOUS CHAPTER, muscle tissue includes skeletal, cardiac, and smooth muscle. However, in the study of the muscular system, the **skeletal muscles** take center stage. They make up the flesh of our body, and they also produce many types of movements; the blinking of an eye, standing on tiptoe, swallowing food, and wielding a sledgehammer are just a few examples. Before we examine the individual muscles and their specific actions, we first discuss the general principles of leverage, describe the ways in which muscles act with or against each other to bring about movements, and consider the criteria used in naming muscles.

LEVER SYSTEMS: BONE-MUSCLE RELATIONSHIPS

The operation of most skeletal muscles involves leverage—the use of a lever to aid the movement of some object. A **lever** is a rigid bar that moves on a fixed point, or **fulcrum,** when a force is applied to it. The applied force, or **effort,** is used to move a resistance, or **load.** The bones of the skeleton act as levers, the joints as fulcrums. Muscle contraction provides the effort, applying this force where the muscle attaches to the bone. The load that is moved is the bone itself, along with the overlying tissues and anything else you are trying to move with that particular lever.

A lever allows a given effort to move a heavier load, or to move a load farther, than that effort otherwise could. If the load is close to the fulcrum and the effort is applied far from the fulcrum, a small effort exerted over a large distance can move a large load over a small distance (Figure 11.1a). We say that such a lever operates at a **mechanical advantage.** When a jack lifts a car, for example, the car moves up only a little with each large downward "push" of the jack handle, but not much muscular effort is needed. If, however, the load is farther from the fulcrum than is the effort, the effort applied must be greater than the load to be moved (Figure 11.1b). Such a lever operates at a **mechanical disadvantage.** Nonetheless, such a situation can be advantageous to us, because it allows the load to be moved rapidly over a large distance. Wielding a shovel is an example. Shovel and jack are different kinds of levers, but both follow the same universal **law of levers:** When the effort is farther from the fulcrum than is the load, the lever operates at a mechanical advantage, but when the effort is nearer than the load, the lever operates at a mechanical disadvantage.

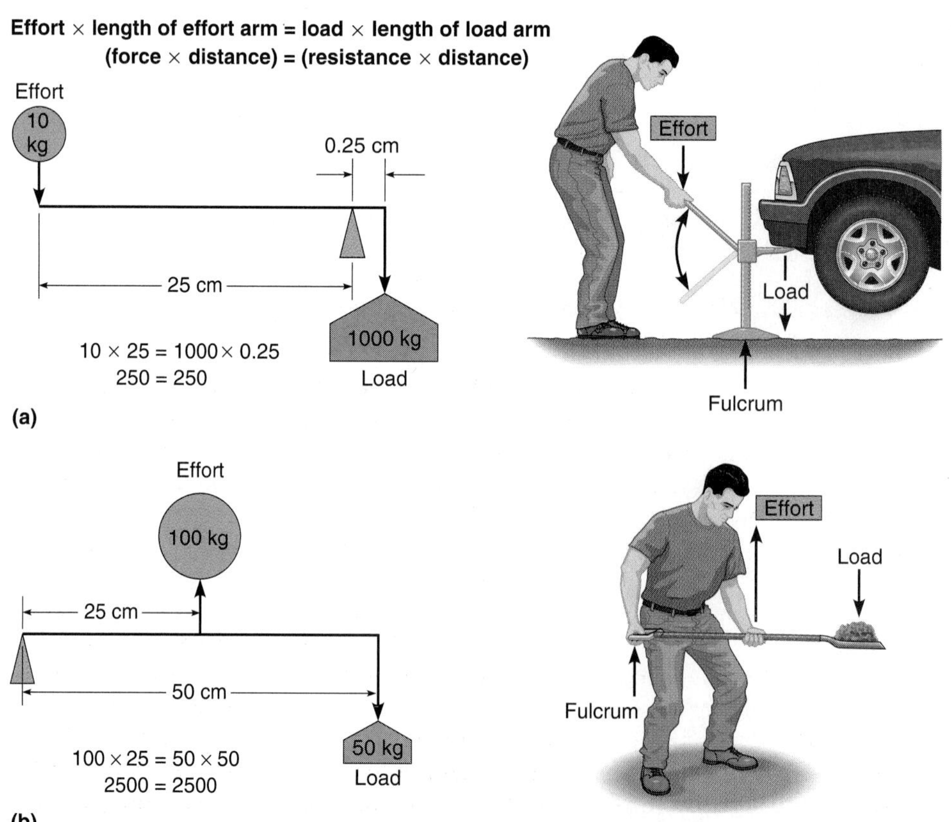

FIGURE 11.1 Lever systems. The equation at the top expresses the relationships among forces and distances for any lever system. **(a)** When the effort is farther from the fulcrum (green triangle) than is the load, a smaller effort (10 kg) can lift a larger load (1000 kg). Such a lever operates at a mechanical advantage. **(b)** When the load is farther from the fulcrum than is the effort, a larger effort (100 kg) is needed to lift a smaller load (50 kg). Such a lever operates at a mechanical disadvantage.

FIGURE 11.2 Classes of lever systems. **(a)** In a first-class lever, the arrangement of elements is load-fulcrum-effort. A scissors is one example. In the body, a first-class lever raises the head off the chest: The posterior neck muscles provide the effort; the fulcrum is the joint between the atlas and occipital condyles (green triangle); and the load is the weight of the head. Some first-class lever systems work at a mechanical advantage, whereas others work at a mechanical disadvantage. **(b)** In a second-class lever, the arrangement is fulcrum-load-effort, as exemplified by a wheelbarrow. This class of lever always works at a mechanical advantage. Second-class leverage is used when the calf muscles lift the weight of the body on the ball of the foot. **(c)** In a third-class lever, the arrangement is load-effort-fulcrum. Lifting or holding something with tweezers is an example. This class of lever always works at a mechanical disadvantage. Third-class leverage is used when the biceps brachii muscle lifts the forearm by flexing it at the elbow.

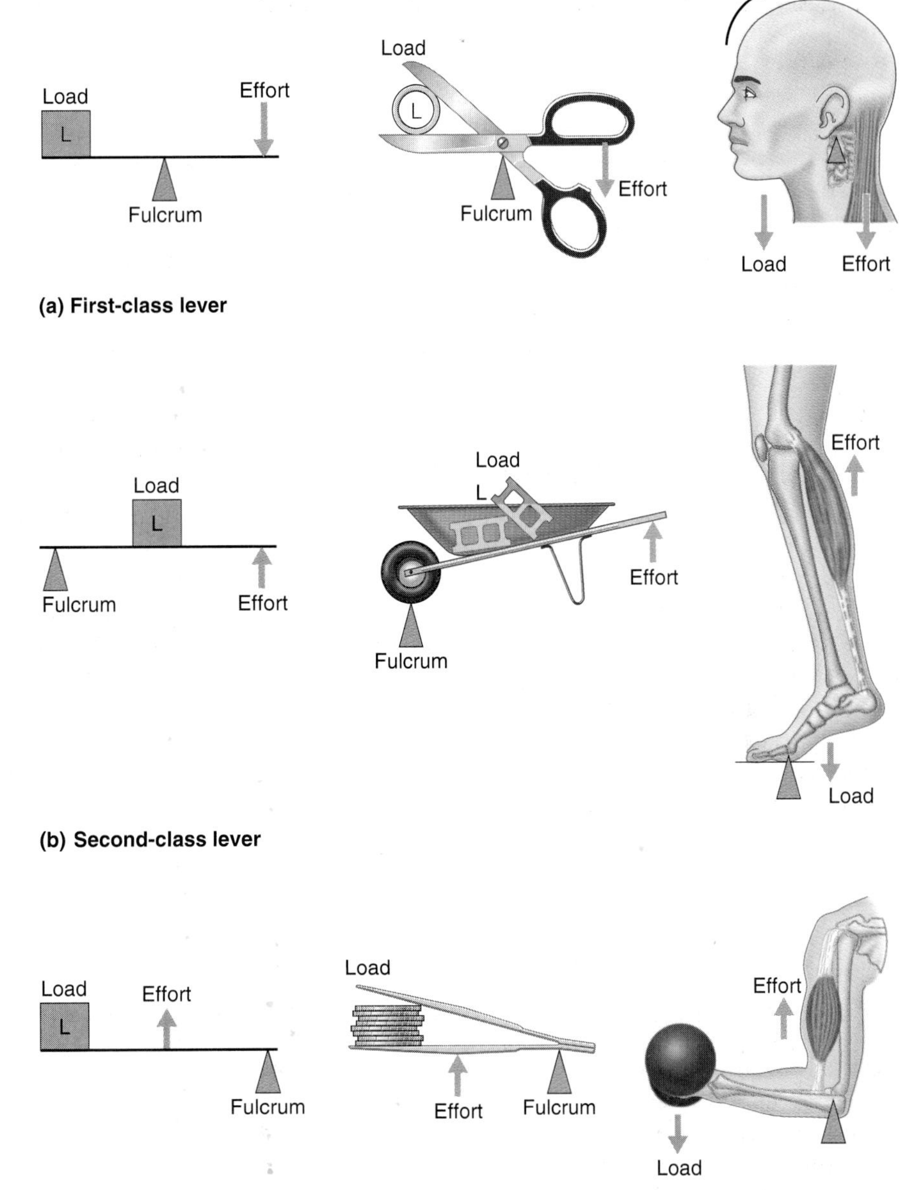

Depending on the relative positions of the three elements—effort, fulcrum, and load—a lever belongs to one of three classes. In **first-class levers,** the effort is applied at one end, and the load is at the other, with the fulcrum somewhere between (Figure 11.2a). Seesaws and scissors are first-class levers. Such leverage also occurs when you lift your head off your chest. Some first-class levers operate at a mechanical advantage (effort is farther from joint than is load), but others operate at a mechanical disadvantage (load is farther from joint than is effort).

In **second-class levers,** the effort is applied at one end, and the fulcrum is at the other, with the load between (Figure 11.2b). A wheelbarrow is this kind of lever. Second-class leverage is not very common in the body. The best example is the act of standing on your toes. In this action, the joints in the ball of the foot are the fulcrum, the load is the whole weight of the body, and the calf muscles exert the effort, pulling the heel superiorly. Second-class levers all work at a mechanical advantage because the muscle insertion (the effort) is farther from the fulcrum than is the load. Second-class levers enable great strength, but speed and range of motion are sacrificed.

In **third-class levers,** effort is applied between the load and the fulcrum (Figure 11.2c). Lifting or holding something with tweezers uses third-class leverage. These levers work speedily, but always at a mechanical disadvantage. Most skeletal muscles of the body are in third-class lever systems, as exemplified by the biceps brachii muscle of the arm. The fulcrum here is the elbow joint; the force is exerted on the proximal region of the radius; and the load is the distal part of the forearm, plus anything carried in the hand. Third-class lever systems permit a muscle to be inserted very close to the joint across which movement occurs, which allows fast, extensive movements with relatively little shortening of the muscle. This in turn permits us to move our limbs quickly, such as when we run or throw.

In summary, differences in the positioning of the three lever-elements modify the activity of muscles with respect to (1) speed of contraction, (2) range of movement, and (3) the weight of the load that can be lifted. In lever systems that operate at a mechanical disadvantage, force is lost, but a greater speed and range of movement is gained. Systems that operate at a mechanical advantage tend to be slower and more stable and are used where strength is a priority.

ARRANGEMENT OF FASCICLES IN MUSCLES

Recall from the previous chapter that skeletal muscles consist of fascicles (bundles of fibers), which are large enough to be seen with the unaided eye (Figure 11.3). In different muscles, the fascicles are arranged in different patterns, including *parallel, convergent, pennate,* and *circular* patterns. One can learn much about the action of a particular muscle from the arrangement of its fascicles.

In a **parallel** arrangement, the long axes of the fascicles run parallel to the long axis of the muscle itself. Muscles with this arrangement are either strap-like, such as the sternocleidomastoid muscle of the neck, or fusiform (spindle-shaped) with an expanded central *belly,* such as the biceps brachii of the arm.

In the **convergent** pattern, the origin of the muscle is broad, and the fascicles *converge* toward the tendon of insertion. Such a muscle can be triangular or fan-shaped. The pectoralis major muscle in the anterior thorax is an example.

In a **pennate** (pen'āt) pattern, the fascicles are short and attach obliquely to a tendon that runs the whole length of the muscle. This pattern makes the muscle look like a feather (*penna* = feather). If the fascicles insert into only one side of the tendon, the muscle is **unipennate** (*uni* = one). The flexus pollicis longus muscle on the anterior forearm is unipennate. If the fascicles insert into the tendon from both sides, the arrangement is **bipennate** (*bi* = two). The rectus femoris muscle of the thigh is bipennate. A **multipennate** arrangement looks like many feathers situated side by side, with all their quills inserting into one large tendon. The deltoid muscle, which forms the roundness of the shoulder, is multipennate.

Fascicles arranged in concentric rings form a **circular** pattern. Muscles with this arrangement surround external body openings, which they close by contracting. The general name for such a circular muscle is ***sphincter*** (sfingk'ter; "squeezer"). Specific examples are the orbicularis oris muscle around the mouth and the orbicularis oculi around the eyes.

The arrangement of its fascicles influences both a muscle's range of motion and its power. Skeletal muscle fibers can shorten by about one-third of their resting length as they contract. The more nearly parallel the fibers are to the muscle's long axis, the more the whole muscle can shorten. Although muscles with parallel fascicles can shorten the most, they usually are not powerful. The power of a muscle depends more on the total number of fibers it contains. The stocky bipennate and multipennate muscles contain the most fibers; thus, they shorten very little but tend to be very powerful.

INTERACTIONS OF SKELETAL MUSCLES IN THE BODY

It is easy to understand that different muscles often work together to bring about a single movement. What is not so obvious, however, is that muscles often work *against* one another in a productive way. No single muscle can reverse the motion it produces, because a muscle cannot "push" after pulling. Therefore, for whatever action one muscle (or muscle group) can do, there must be other muscles that can "undo" the action (or do the opposite action). In general, groups of muscles that produce opposite movements lie on opposite sides of a given joint.

With respect to their actions, muscles are classified into several *functional* types. A muscle that has the major responsibility for producing a specific movement is the **prime mover,** or **agonist** (ag'o-nist; "leader"), of that motion. For example, the biceps brachii is a prime mover for flexing the forearm at the elbow (see Figure 11.2c). Sometimes, two muscles contribute so heavily to the same movement that both are called agonists.

Muscles that oppose or reverse a particular movement act as **antagonists** (an-tag'o-nists; "against the leader"). When a prime mover is active, its antagonists are being stretched and sometimes remain relaxed. Usually, however, the antagonists contract slightly during the movement to keep the movement from overshooting its mark or to slow it near its completion. Antagonists can also be prime movers in their own right; that is, an agonist for one movement can serve as an antagonist for the opposite movement. For example, flexion of the forearm by the biceps brachii is antagonized by the triceps brachii, the prime mover for extending the forearm.

In addition to prime movers and antagonists, most movements also involve one or more **synergists** (sin'er-jist; "together-worker"). Synergists help the prime movers, either by adding a little extra force to the same movement or by reducing undesirable extra movements

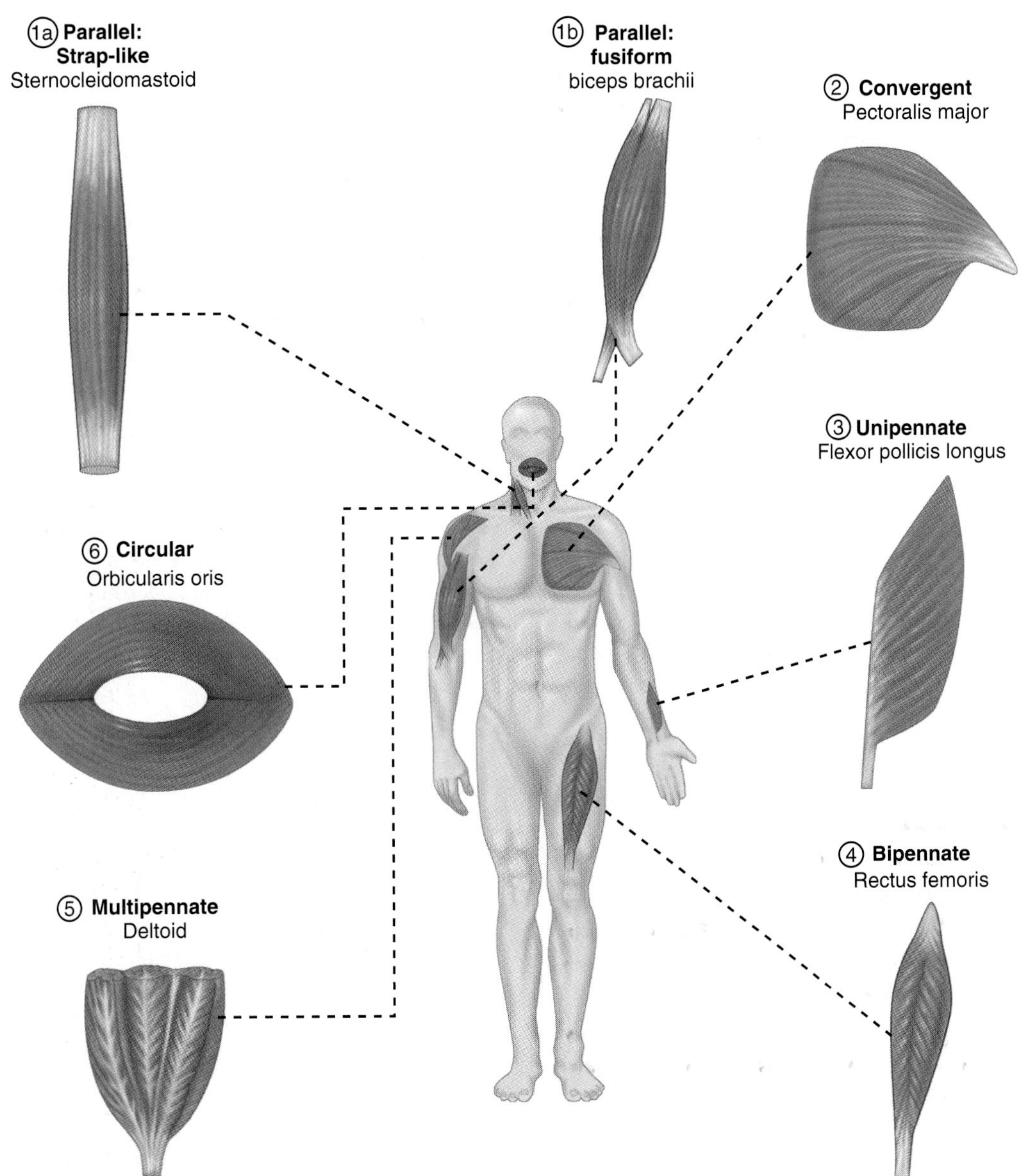

FIGURE 11.3 The different arrangements of fascicles in various muscles. The lines in the muscles define the fascicles.

that the prime mover may produce. This second function deserves more explanation: Some prime movers cross several joints and can cause movements at all these joints, but synergists act to cancel some of these movements. For example, the muscles that flex the fingers cross both the wrist and finger joints, but you can make a fist without flexing your wrist because synergists stabilize the wrist. Additionally, some prime movers can cause several kinds of movement *at the same joint,* and synergists prevent the particular movement that is inappropriate at a given time.

Some synergists hold a bone firmly in place so that a prime mover has a stable base on which to move a body part. Such synergists are called **fixators.** An example is the muscles that fix the scapula when the arm moves. Muscles that maintain posture and stabilize joints also act as fixators.

In summary, although prime movers seem to "get all the credit" for causing movements, the actions of antagonistic and synergistic muscles are also important.

Critical Reasoning

We have seen how important it is that muscles work together. When bodybuilders develop the muscles on one side of a limb more than on the other side, they are said to be "muscle bound." Describe the symptoms of this condition, and explain how they are produced.

The greater muscle tone in the more-developed muscles puts constant tension on the joints, inhibiting the joints' flexibility of movement. An overdeveloped biceps brachii, for example, keeps the forearm partly flexed at the elbow, making it difficult for the triceps brachii to perform its function of extending the forearm. Although muscle-bound athletes may have impressive physiques, their movements are awkward, and they perform poorly in sports. To avoid this problem, modern bodybuilding programs strive for an even development of opposing muscle groups.

NAMING THE SKELETAL MUSCLES

Skeletal muscles are named according to several criteria, each of which describes the muscle in some way. Learning these criteria can simplify the task of learning muscle names.

1. **Location.** Some names indicate where a muscle is located. For example, the brachialis muscle is in the arm (*brachium* = arm), and intercostal muscles lie between the ribs (*costa* = rib).
2. **Shape.** Some muscles are named for their shapes. For example, the deltoid is triangular (*delta* = triangle), and the right and left trapezius muscles together form a trapezoid.
3. **Relative size of the muscle.** The terms *maximus* (largest), *minimus* (smallest), *longus* (long), and *brevis* (short) are part of the names of some muscles—such as the gluteus maximus and gluteus minimus muscles of the buttocks.
4. **Direction of the fascicles and the muscle fibers.** The names of some muscles tell the direction in which their fascicles (and muscle fibers) run. In muscles with the term *rectus* (straight) in their name, the fascicles are parallel to the body midline, whereas *transversus* and *oblique* mean that the fascicles lie at a right angle or at an oblique angle to the midline, respectively. Specific examples include the rectus abdominis, transversus abdominis, and external oblique muscles of the abdomen.
5. **Location of attachments.** The names of some muscles reveal their points of origin and insertion. Recall that the origin is the less movable attachment of a muscle, whereas the insertion is the more movable attachment. The origin is always named first. For instance, the brachioradialis muscle in the forearm originates on the bone of the brachium (arm), or humerus, and inserts on the radius.
6. **Number of origins.** When the term *biceps* ("two heads"), *triceps* ("three heads"), or *quadriceps* ("four heads") is in its name, a muscle has two, three, or four origins, respectively. For example, the biceps brachii has two origins.
7. **Action.** When muscles are named for their action, words such as *flexor, extensor, adductor,* or *abductor* appear in the name. For example, the adductor longus, on the medial thigh, adducts the thigh at the hip. The names of many forearm and leg muscles begin with *extensor* and *flexor,* indicating how they move the hand, foot, and digits.

Often, several different criteria are combined in the naming of a muscle. For instance, the name *extensor carpi radialis longus* tells us the muscle's action (extensor), the joint it acts on (*carp* = wrist), and that it lies along the radius in the forearm (radialis). The name also indicates that the muscle is longer (longus) than some other wrist extensor muscles. Even though such long names can be difficult to pronounce, they are extremely informative.

Critical Reasoning

Deduce some characteristics of the following muscles, based on their names: tibialis anterior; temporalis; erector spinae.

The tibialis anterior lies on the tibia in the anterior region of the leg; the temporalis is located in the temple on the temporal bone; and the erector spinae is a muscle that extends the vertebral column (makes the spine erect).

A DEVELOPMENT-BASED ORGANIZATION OF THE MUSCLES

Before we cover the individual muscles of the body in Tables 11.1–11.17 on pages 276–331, we present an overview of the muscles that groups them according to their embryonic origin and general function. This classification scheme can provide an aid to organizing the formidable array of skeletal muscles into more manageable "chunks" of material.

First, recall from Chapter 3 that all muscles develop from the mesoderm germ layer (see p. 67). Figure 11.4 provides a review of some parts of the mesoderm that are present during the second month of development, especially the myotomes and splanchnic mesoderm (Figure 11.4a and b). Note also that the first seven myotome-like structures in the head collectively are called **somitomeres** (so-mit′o-mērz″; "somite pieces").

In this development-based scheme, muscles are organized into four groups: (1) musculature of the visceral organs, (2) pharyngeal arch muscles, (3) axial muscles, and (4) limb muscles.

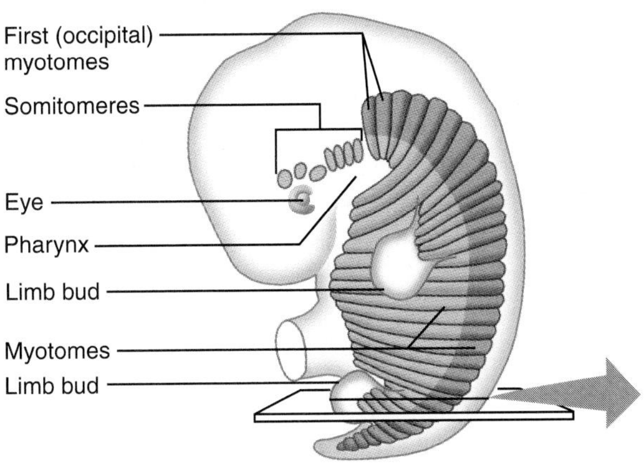

First (occipital) myotomes
Somitomeres
Eye
Pharynx
Limb bud
Myotomes
Limb bud

(a) 6-week embryo

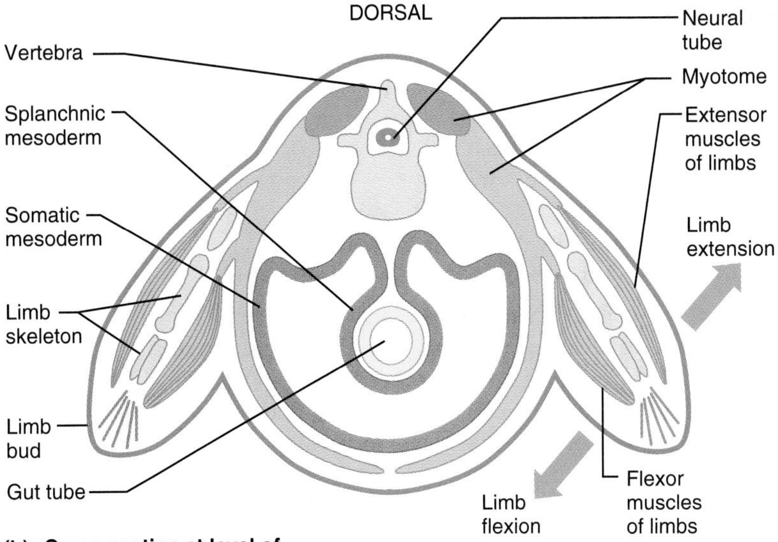

DORSAL
Neural tube
Vertebra
Myotome
Splanchnic mesoderm
Extensor muscles of limbs
Somatic mesoderm
Limb extension
Limb skeleton
Limb bud
Gut tube
Limb flexion
Flexor muscles of limbs

(b) Cross section at level of lower limb buds

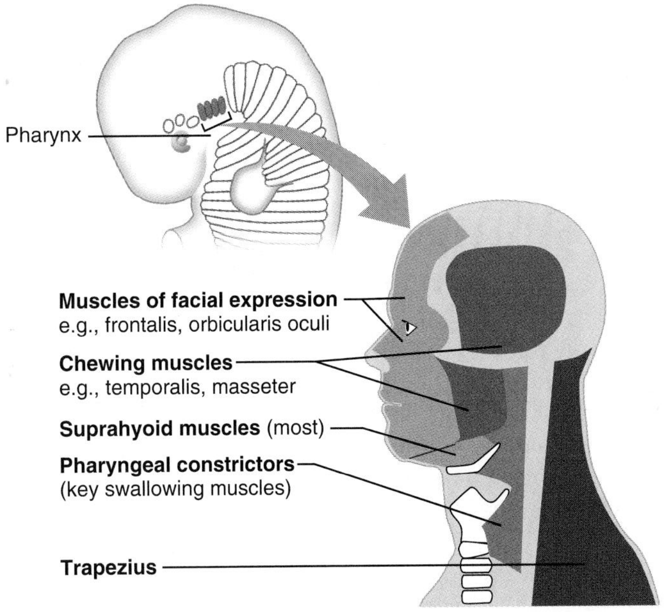

Pharynx
Muscles of facial expression
e.g., frontalis, orbicularis oculi
Chewing muscles
e.g., temporalis, masseter
Suprahyoid muscles (most)
Pharyngeal constrictors
(key swallowing muscles)
Trapezius

(c) Pharyngeal arch (branchiomeric) muscles: 4th-7th somitomeres

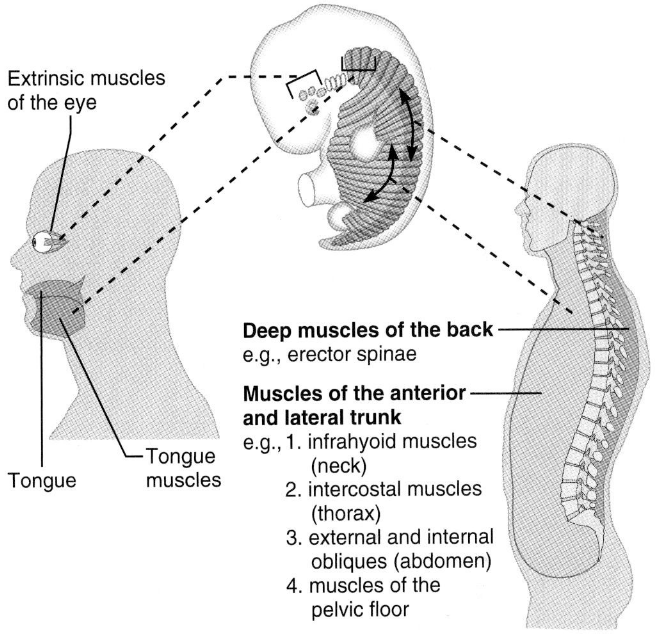

Extrinsic muscles of the eye
Deep muscles of the back
e.g., erector spinae
Muscles of the anterior and lateral trunk
e.g., 1. infrahyoid muscles (neck)
2. intercostal muscles (thorax)
3. external and internal obliques (abdomen)
4. muscles of the pelvic floor
Tongue
Tongue muscles

(d) Axial muscles: 1st-3rd somitomeres and from myotomes

Extensors:
e.g., triceps brachii, extensor digitorum longus
Flexors:
e.g., biceps brachii, flexor carpi ulnaris
Extensors
Flexors
Extensors:
e.g., quadriceps femoris, tibialis anterior
Flexors:
e.g., hamstrings, gastrocnemius

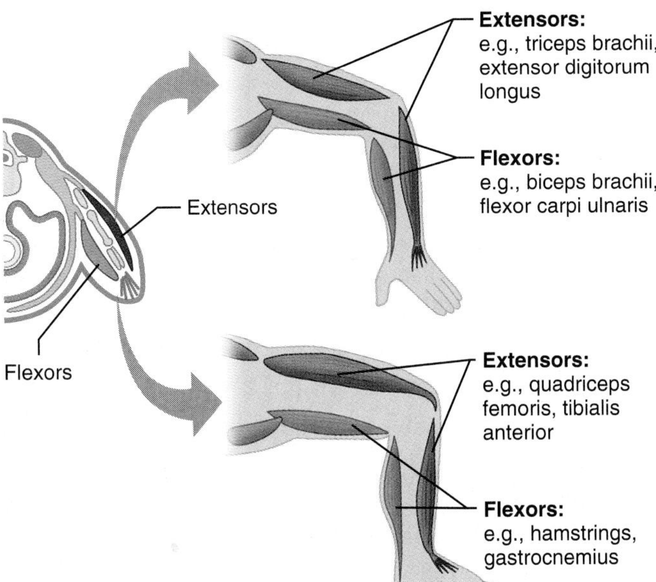

(e) Limb muscles: From the limb muscle mass, ultimately from myotomes.

FIGURE 11.4 Development and basic organization of the muscles. **(a)** A six-week embryo, showing the myotomes and somitomeres that give rise to muscles. **(b)** Cross section through the trunk and lower limbs of the six-week embryo, showing divisions of the muscle-forming mesoderm. The plane of section is shown in (a). The remaining portion of the figure, (c)–(e), shows three of the main functional classes of muscles, based on developmental origin. **(c)** Pharyngeal arch muscles. **(d)** Axial muscles. **(e)** Limb muscles.

1. **Musculature of the visceral organs.** Recall that the visceral musculature includes both smooth muscle and cardiac muscle. This musculature develops from the splanchnic mesoderm around the early gut (Figure 11.4b). Because the visceral musculature does not form skeletal muscles, it is not considered further in this chapter (but see Chapter 10).
2. **Pharyngeal arch muscles.** This group includes the skeletal muscles of the pharynx (throat region of the digestive tract), plus some other muscles in the head and neck (Figure 11.4c). The unifying feature of these muscles is that they all develop around the embryonic pharynx, from the fourth to seventh somitomeres. Their original function was to squeeze things through the pharynx, as in swallowing, but they have diversified to do more than that. Another name for the pharyngeal arch muscles is **branchiomeric muscles** (brang″ke-o-mer′ik; "gill segment").

   The major muscles in this group include the muscles of facial expression (Table 11.1), chewing muscles (Table 11.2), and two sets of muscles involved in swallowing: namely, suprahyoid muscles and the pharyngeal constrictors (Table 11.3). A neck muscle called the sternocleidomastoid (Table 11.4) and a back muscle called the trapezius (Table 11.8) are also placed in this group, although these two muscles develop only partly from somitomeres (and partly from nearby myotomes).
3. **Axial muscles.** These are the skeletal muscles of the thorax, abdomen, and pelvis, plus many muscles of the neck and a few in the head (Figure 11.4d). They are called axial muscles because they lie anterior and posterior to the body axis (vertebral column), and their main functions are moving the trunk and maintaining posture. They develop from the myotomes and some somitomeres.

   The *dorsal* regions of the myotomes become the deep muscles of the back (Table 11.4), which extend the spine as we stand erect.

   The *ventral* regions of the myotomes become the muscles of the anterior and lateral parts of the trunk and neck, which flex the spine (among other functions). These muscles include muscles of the anterior neck (infrahyoid muscles; Table 11.3), respiratory muscles of the thorax (Table 11.5), muscles of the anterior abdominal wall (Table 11.6), and muscles of the pelvic floor (Table 11.7).

   The first three somitomeres become the extrinsic eye muscles, which move the eyes (see p. 472), whereas the occipital myotomes become the muscles that move the tongue (Table 11.2).
4. **Limb muscles.** The upper and lower limbs arise as limb buds, and limb muscles develop from the lateral parts of the nearby myotomes (Figure 11.4e). The muscle mass that lies *dorsal* to the limb bones becomes the *extensor* muscles of that limb, whereas the *ventral* mass becomes the limb's *flexor* muscles. These muscles function in locomotion and the manipulation of objects. Because they develop in the same way, the muscles of the upper and lower limbs are similar and directly comparable.

   In the adult *upper* limbs (Tables 11.10–11.13), the extensor muscles lie on the limb's posterior side and extend its parts (forearm, hand, and fingers). The flexor muscles lie on the anterior side of the limb and flex these parts.

   In the adult *lower* limbs (Tables 11.14–11.17), the extensor muscles occupy the anterior side of the limb and extend the leg at the knee, dorsiflex the foot at the ankle, and extend the toes. The flexor muscles occupy the posterior side of the lower limb and flex the leg at the knee, plantar flex the foot at the ankle, and flex the toes.

   Also part of this group are the muscles that attach the limbs to their girdles and the muscles that attach the girdles to the trunk (Tables 11.8, 11.9, and 11.14).

MAJOR SKELETAL MUSCLES OF THE BODY

There are over 600 muscles in the body, and learning them can be difficult. The first requirement is to make sure you understand all the body movements shown in Figures 9.5 and 9.6 (pp. 220–223). Some instructors have their students learn only a few major groups of muscles, whereas others require a more detailed study. The information in Tables 11.1 through 11.17 should provide enough depth to satisfy even the latter approach.

In these tables, the muscles are grouped by function and by location, roughly from head to foot. Every table starts by introducing a muscle group and then describes each muscle's shape, location, attachments, actions, and innervation. As you study each muscle, look at its attachments and the direction of its fascicles, and try to understand how these features determine its specific action. Figure 11.5 summarizes the major superficial muscles in the body and will help you tie together the information in the tables. A good way to learn muscle actions is to act out the movements yourself and then feel the contracting muscles bulge beneath your skin.

The following list summarizes the sequence of the tables in this chapter:

- **Table 11.1** Muscles of the Head, Part I: Facial Expression (Figure 11.6); pp. 276–277
- **Table 11.2** Muscles of the Head, Part II: Mastication and Tongue Movement (Figure 11.7); pp. 278–279
- **Table 11.3** Muscles of the Anterior Neck and Throat: Swallowing (Figure 11.8); pp. 280–282
- **Table 11.4** Muscles of the Neck and Vertebral Column: Head Movements and Trunk Extension (Figure 11.9); pp. 283–286

- **Table 11.5** Deep Muscles of the Thorax: Breathing (Figure 11.10); pp. 287–288
- **Table 11.6** Muscles of the Abdominal Wall: Trunk Movements and Compression of Abdominal Viscera (Figure 11.11); pp. 289–291
- **Table 11.7** Muscles of the Pelvic Floor and Perineum: Support of Abdominopelvic Organs (Figure 11.12); pp. 292–293
- **Table 11.8** Superficial Muscles of the Anterior and Posterior Thorax: Movements of the Scapula (Figure 11.13); pp. 294–296
- **Table 11.9** Muscles Crossing the Shoulder Joint: Movements of the Arm (Humerus) (Figure 11.14); pp. 297–299
- **Table 11.10** Muscles Crossing the Elbow Joint: Flexion and Extension of the Forearm (Figure 11.14); p. 300
- **Table 11.11** Muscles of the Forearm: Movements of the Wrist, Hand, and Fingers (Figures 11.15 and 11.16); pp. 301–306
- **Table 11.12** Summary of Actions of Muscles Acting on the Arm, Forearm, and Hand (Figure 11.17); pp. 307–308
- **Table 11.13** Intrinsic Muscles of the Hand: Fine Movements of the Fingers (Figure 11.18); pp. 309–311
- **Table 11.14** Muscles Crossing the Hip and Knee Joints: Movements of the Thigh and Leg (Figures 11.19 and 11.20); pp. 312–318
- **Table 11.15** Muscles of the Leg: Movements of the Ankle and Toes (Figures 11.21 to 11.23); pp. 319–325
- **Table 11.16** Summary of Actions of Muscles Acting on the Thigh, Leg, and Foot (Figure 11.24); pp. 326–328
- **Table 11.17** Intrinsic Muscles of the Foot: Toe Movement and Foot Support (Figure 11.25); pp. 329–331

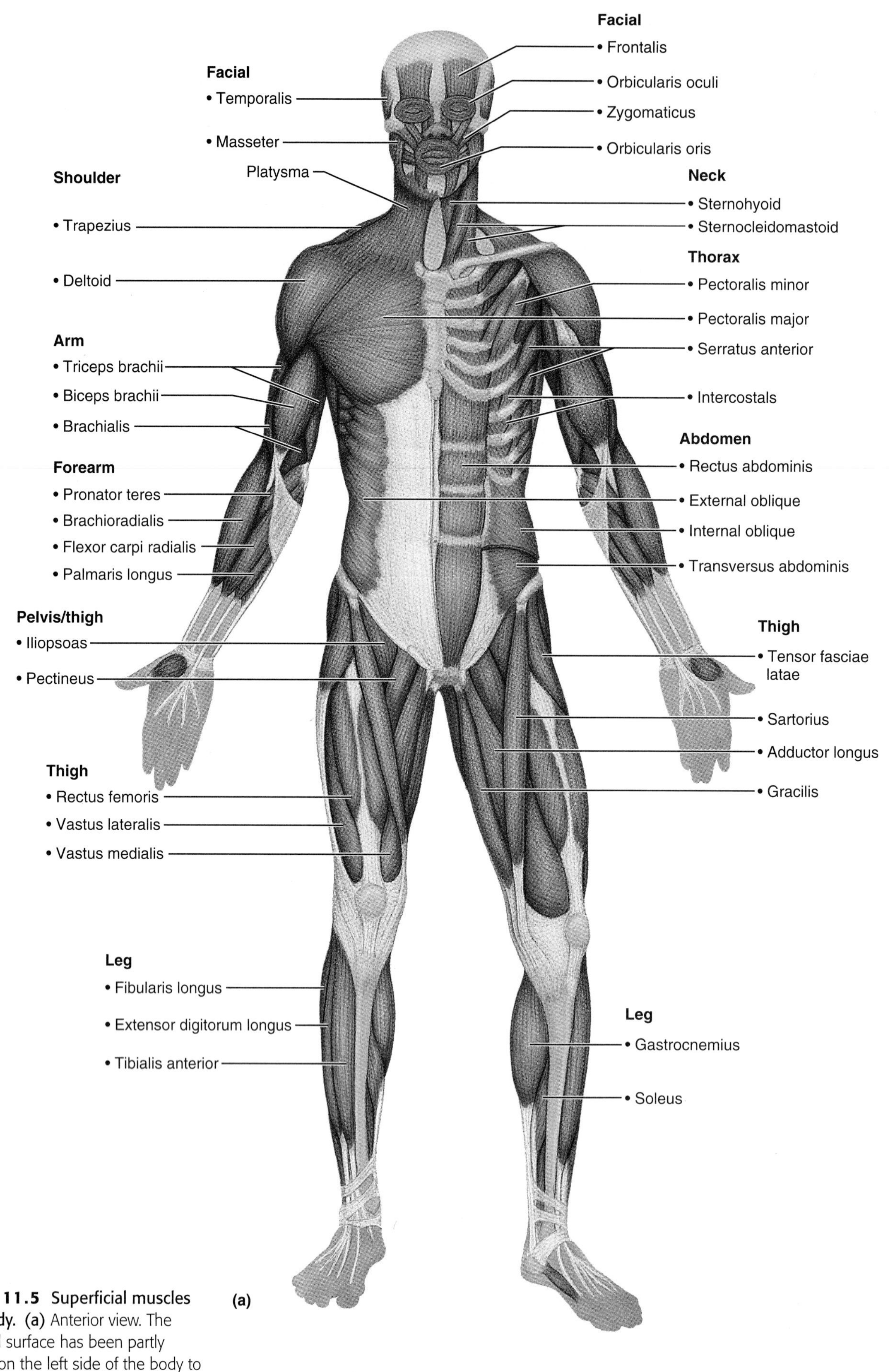

FIGURE 11.5 Superficial muscles of the body. **(a)** Anterior view. The abdominal surface has been partly dissected on the left side of the body to show deeper muscles. **(b)** Posterior view.

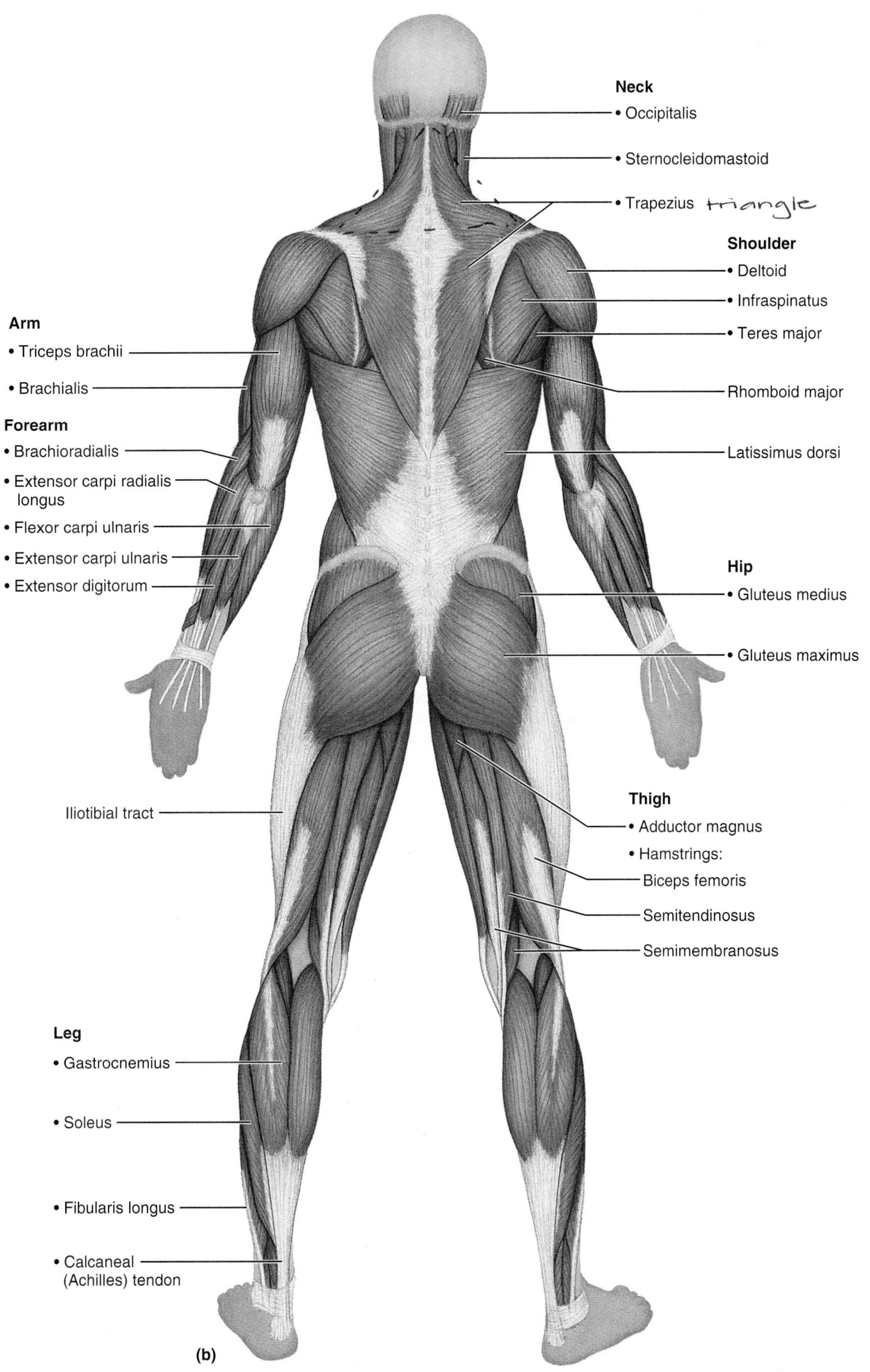
Neck
• Occipitalis
• Sternocleidomastoid
• Trapezius
Shoulder
• Deltoid
• Infraspinatus
• Teres major
Rhomboid major
Latissimus dorsi
Arm
• Triceps brachii
• Brachialis
Forearm
• Brachioradialis
• Extensor carpi radialis longus
• Flexor carpi ulnaris
• Extensor carpi ulnaris
• Extensor digitorum
Hip
• Gluteus medius
• Gluteus maximus
Iliotibial tract
Thigh
• Adductor magnus
• Hamstrings:
Biceps femoris
Semitendinosus
Semimembranosus
Leg
• Gastrocnemius
• Soleus
• Fibularis longus
• Calcaneal (Achilles) tendon

(b)

Muscle Gallery: Table 11.1 Muscles of the Head, Part I: Facial Expression (Figure 11.6)

The muscles that promote facial expression lie in the face and scalp, just deep to the skin. They are thin and variable in shape, and adjacent muscles in this group tend to be fused. Unlike other skeletal muscles, facial muscles insert on the skin, not on the bones. In the scalp, the main muscle is the **epicranius,** which has distinct anterior and posterior parts. In the face, the muscles clothing the facial bones lift the eyebrows and flare the nostrils, close the eyes and lips, and provide one of the best tools for influencing others—the smile. The great importance of the facial muscles in nonverbal communication becomes clear when they are paralyzed, as in some stroke victims. All muscles listed in this table are innervated by *cranial nerve VII, the facial nerve.* The muscles that move our eyeballs as we look in various directions are described in Chapter 16 (p. 472).

Muscle	Description	Origin (O) and Insertion (I)	Action	Nerve Supply
MUSCLES OF THE SCALP				
Epicranius (Occipitofrontalis) (ep″ĭ-kra′ne-us; ok-sip″ĭ-to-fron-ta′lis) (*epi* = over; *cran* = skull)	Bipartite muscle consisting of the frontalis and occipitalis muscles connected by a cranial aponeurosis, the galea aponeurotica; the alternate actions of these two muscles pull scalp forward and backward			
• **Frontalis** (fron-ta′lis) (*front* = forehead)	Covers forehead and dome of skull; no bony attachments	O—galea aponeurotica I—skin of eyebrows and root of nose	With aponeurosis fixed, raises the eyebrows (as in surprise); wrinkles forehead skin horizontally	Facial nerve (cranial VII)
• **Occipitalis** (ok-sip″ĭ-tal′is) (*occipito* = base of skull)	Overlies posterior occiput; by pulling on the galea, fixes origin of frontalis	O—occipital and temporal (mastoid) bones I—galea aponeurotica	Fixes aponeurosis and pulls scalp posteriorly	Facial nerve
MUSCLES OF THE FACE				
Corrugator supercilii (kor′ah-ga-ter soo″per-sĭ′le-i) (*corrugo* = wrinkle; *supercilium* = eyebrow)	Small muscle; activity associated with that of orbicularis oculi	O—arch of frontal bone above nasal bone I—skin of eyebrow	Draws eyebrows together and inferiorly; wrinkles skin of forehead vertically (as in frowning)	Facial nerve
Orbicularis oculi (or-bik′u-lar-is ok′u-li) (*orb* = circular; *ocul* = eye)	Thin, flat sphincter muscle of eyelid; surrounds rim of the orbit	O—frontal and maxillary bones and ligaments around orbit I—tissue of eyelid	Closes eye; protects eye from intense light and injury; various parts can be activated individually; produces blinking, squinting, and draws eyebrows inferiorly	Facial nerve
Zygomaticus (zi-go-mat′ĭ-kus), major and minor (*zygomatic* = cheekbone)	Muscle pair extending diagonally from cheekbone to corner of mouth	O—zygomatic bone I—skin and muscle at corner of mouth	Raises lateral corners of mouth upward (smiling muscle)	Facial nerve
Risorius (ri-zor′e-us) (*risor* = laughter)	Slender muscle inferior and lateral to zygomaticus	O—lateral fascia associated with masseter muscle I—skin at angle of mouth	Draws corner of lip laterally; tenses lips; synergist of zygomaticus	Facial nerve
Levator labii superioris (lĕ-va′tor la′be-i soo-per″e-or′is) (*leva* = raise; *labi* = lip; *superior* = above, over)	Thin muscle between orbicularis oris and inferior eye margin	O—zygomatic bone and infraorbital margin of maxilla I—skin and muscle of upper lip	Opens lips; raises and furrows the upper lip	Facial nerve
Depressor labii inferioris (de-pres′or la′be-i in-fer″e-or′is) (*depressor* = depresses; *infer* = below)	Small muscle running from mandible to lower lip	O—body of mandible lateral to its midline I—skin and muscle of lower lip	Draws lower lip inferiorly (as in a pout)	Facial nerve
Depressor anguli oris (ang′gu-li or′-is) (*angul* = angle, corner; *or* = mouth)	Small muscle lateral to depressor labii inferioris	O—body of mandible below incisors I— skin and muscle at angle of mouth below insertion of zygomaticus	Zygomaticus antagonist; draws corners of mouth downward and laterally (as in a "tragedy mask" grimace)	Facial nerve

Muscle	Description	Origin (O) and Insertion (I)	Action	N S
Orbicularis oris (*or* = mouth)	Complicated, multilayered muscle of the lips with fibers that run in many different directions; most run circularly	O—arises indirectly from maxilla and mandible; fibers blended with fibers of other facial muscles associated with the lips I—encircles mouth; inserts into muscle and skin at angles of mouth	Closes lips; purses and protrudes lips; kissing and whistling muscle	Facial nerve
Mentalis (men-ta′lis) (*ment* = chin)	One of the muscle pair forming a V-shaped muscle mass on chin	O—mandible below incisors I—skin of chin	Protrudes lower lip; wrinkles chin	Facial nerve
Buccinator (bu′si-na″ter) (*bucc* = cheek or "trumpeter")	Thin, horizontal cheek muscle; principal muscle of cheek; deep to masseter (see also Figure 11.7)	O—molar region of maxilla and mandible I—orbicularis oris	Draws corner of mouth laterally; compresses cheek (as in whistling and sucking); holds food between teeth during chewing; well developed in nursing infants	Facial nerve
Platysma (plah-tiz′mah) (*platy* = broad, flat)	Unpaired, thin, sheetlike superficial neck muscle; not strictly a head muscle, but plays a role in facial expression	O—fascia of chest (over pectoral muscles and deltoid) I—lower margin of mandible, and skin and muscle at corner of mouth	Helps depress mandible; pulls lower lip back and down, i.e., produces downward sag of mouth; tenses skin of neck (e.g., during shaving)	Facial nerve

FIGURE 11.6 Muscles of the scalp, face, and neck, lateral view.

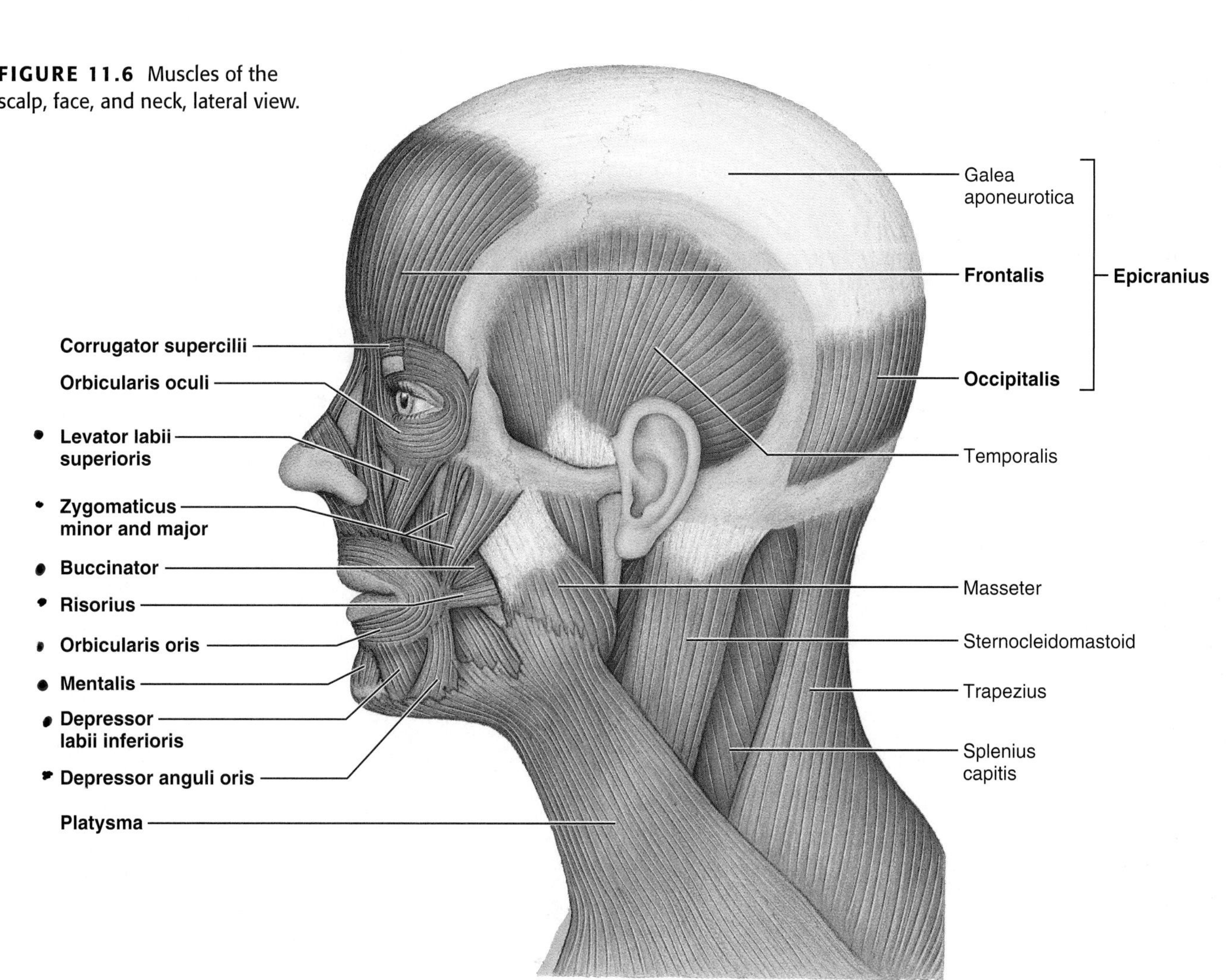

Muscle Gallery: Table 11.2 Muscles of the Head, Part II: Mastication and Tongue Movement (Figure 11.7)

Four main pairs of muscles are involved in mastication (chewing), and all are innervated by the mandibular division of *cranial nerve V, the trigeminal nerve.* The prime movers of jaw closure and biting are the powerful **masseter** and **temporalis** muscles, which can be felt bulging through the skin when the teeth are clenched. Side-to-side grinding movements are brought about by the **pterygoid** muscles. The **buccinator** muscles in the cheeks (see Table 11.1) also play a role in chewing. For lowering the mandible, gravity is usually sufficient, but if there is resistance to jaw opening, the jaw-opening muscles are activated (**digastric** and **mylohyoid** muscles; see Table 11.3).

The tongue consists of muscle fibers that curl, squeeze, and fold the tongue during speaking and chewing. These **intrinsic tongue muscles,** which change the tongue's shape but do not really move it, are considered in Chapter 22 (p. 641) with the digestive system. Only the **extrinsic tongue muscles** are covered in this table. These move the tongue laterally, anteriorly, and posteriorly. The tongue muscles are all innervated by *cranial nerve XII, the hypoglossal nerve.*

Muscle	Description	Origin (O) and Insertion (I)	Action	Nerve Supply
MUSCLES OF MASTICATION				
Masseter (mah-se'ter) (*maseter* = chewer)	Powerful muscle that covers lateral aspect of mandibular ramus	O—zygomatic arch and zygomatic bone I—angle and ramus of mandible	Prime mover of jaw closure; elevates mandible	Trigeminal nerve (cranial V)
Temporalis (tem"por-ă'lis) (*tempora* = time; pertaining to the temporal bone)	Fan-shaped muscle that covers parts of the temporal, frontal, and parietal bones	O—temporal fossa I—coronoid process of mandible via a tendon that passes deep to zygomatic arch	Closes jaw; elevates and retracts mandible; maintains position of the mandible at rest; deep anterior part may help protract mandible	Trigeminal nerve
Medial pterygoid (me'de-ul ter'ĭ-goid) (*medial* = toward median plane; *pterygoid* = wing-like)	Deep two-headed muscle that runs along internal surface of mandible and is largely concealed by that bone	O—medial surface of lateral pterygoid plate of sphenoid bone, maxilla, and palatine bone I—medial surface of mandible near its angle	Synergist of temporalis and masseter muscles in elevation of the mandible; acts with lateral pterygoid muscle to protract mandible and to promote side-to-side (grinding) movements	Trigeminal nerve
Lateral pterygoid (*lateral* = away from median plane)	Deep two-headed muscle; lies superior to medial pterygoid muscle	O—greater wing and lateral pterygoid plate of sphenoid bone I—condyle of mandible and capsule of temporomandibular joint	Protracts mandible (pulls it anteriorly); provides forward sliding and side-to-side grinding movements of the lower teeth	Trigeminal nerve
Buccinator	See Table 11.1	See Table 11.1	Trampoline-like action of buccinator muscles helps keep food between grinding surfaces of teeth during chewing	Facial nerve (cranial VII)
MUSCLES PROMOTING TONGUE MOVEMENTS (EXTRINSIC MUSCLES)				
Genioglossus (je"ne-o-glah'sus) (*geni* = chin; *glossus* = tongue)	Fan-shaped muscle; forms bulk of inferior part of tongue; its attachment to mandible prevents tongue from falling backward and obstructing respiration	O—internal surface of mandible near symphysis I—inferior aspect of the tongue and body of hyoid bone	Primarily protracts tongue, but can depress or act in concert with other extrinsic muscles to retract tongue	Hypoglossal nerve (cranial XII)
Hyoglossus (hi'o-glos"us) (*hyo* = pertaining to hyoid bone)	Flat, quadrilateral muscle	O—body and greater horn of hyoid bone I—inferolateral tongue	Depresses tongue and draws its sides downward	Hypoglossal nerve
Styloglossus (sti-lo-glah'sus) (*stylo* = pertaining to styloid process)	Slender muscle running superiorly to and at right angles to hyoglossus	O—styloid process of temporal bone I—lateral inferior aspect of tongue	Retracts (and elevates) tongue	Hypoglossal nerve

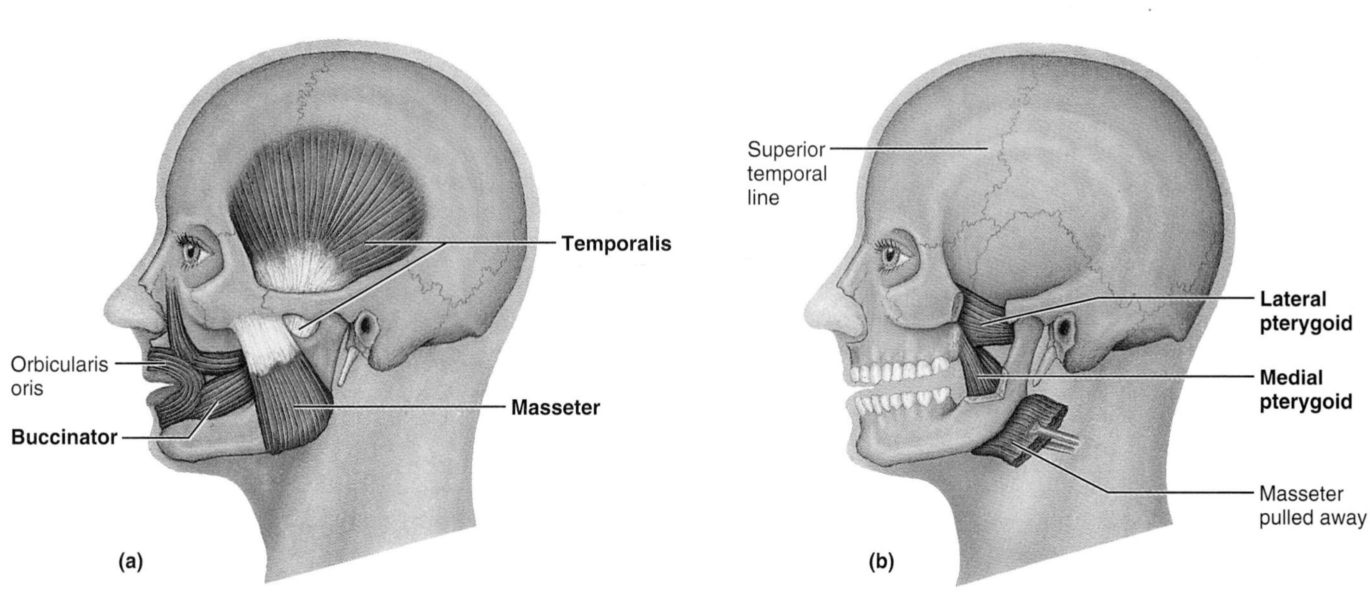

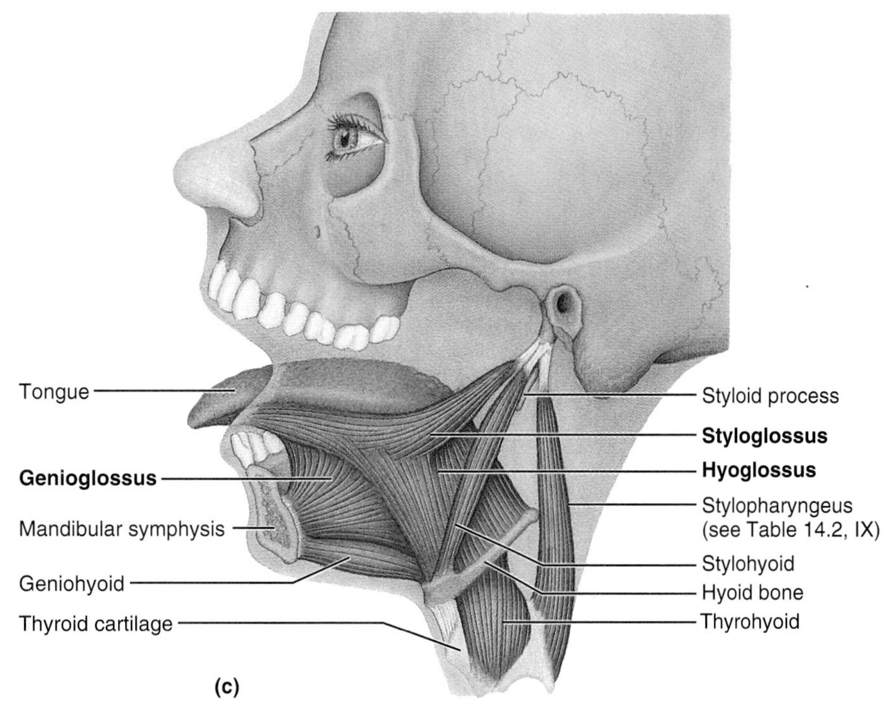

FIGURE 11.7 Muscles promoting mastication and tongue movements. **(a)** Lateral view of the temporalis, masseter, and buccinator muscles. **(b)** Lateral view of the deep chewing muscles, the medial and lateral pterygoid muscles. **(c)** Extrinsic muscles of the tongue. Some suprahyoid muscles of the throat are also illustrated.

Muscle Gallery: Table 11.3 Muscles of the Anterior Neck and Throat: Swallowing (Figure 11.8)

The neck is divided into two triangles (anterior and posterior) by the sternocleidomastoid muscle (Figures 11.8 and 26.9 on p. 777). This table considers the muscles of the *anterior* triangle of the neck. These muscles are divided into **suprahyoid** and **infrahyoid** groups, which lie superior and inferior to the hyoid bone, respectively. Most of these muscles participate in swallowing.

Swallowing begins when the tongue and buccinator muscles squeeze food posteriorly along the roof of the mouth toward the pharynx. Then, many muscles in the pharynx and neck contract in sequence to complete the swallowing process: (1) The suprahyoid muscles aid in closing the air passageway of the larynx—so that food is not inhaled into the lungs—by lifting the larynx superiorly and anteriorly, under the cover of a protective flap (the epiglottis). By pulling anteriorly on the hyoid bone, suprahyoid muscles also widen the pharynx to receive the food. (2) The muscles of the pharynx wall, the **pharyngeal constrictors,** squeeze the food inferiorly into the esophagus. (3) As swallowing ends, the infrahyoid muscles pull the hyoid bone and larynx inferiorly to their original position.

Swallowing also includes mechanisms that prevent food from being squeezed superiorly into the nasal cavity, but these mechanisms will be considered later (Chapter 21, p. 608).

Muscle	Description	Origin (O) and Insertion (I)	Action	Nerve Supply
SUPRAHYOID MUSCLES (soo″prah-hi′oid)	Muscles that help form floor of oral cavity, anchor tongue, elevate hyoid, and move larynx superiorly during swallowing; lie superior to hyoid bone			
Digastric (di-gas′trik)(*di* = two; *gaster* = belly)	Consists of two bellies united by an intermediate tendon, forming a V shape under the chin	O—lower margin of mandible (anterior belly) and mastoid process of the temporal bone (posterior belly) I—by a connective tissue loop to hyoid bone	Acting in concert, the digastric muscles elevate hyoid bone and steady it during swallowing and speech; acting from behind, they open mouth and depress mandible	Mandibular branch of trigeminal nerve (cranial V) for anterior belly; facial nerve (cranial VII) for posterior belly
Stylohyoid (sti″lo-hi′oid) (also see Figure 11.7)	Slender muscle below angle of jaw; parallels posterior belly of digastric muscle	O—styloid process of temporal bone I—hyoid bone	Elevates and retracts hyoid, thereby elongating floor of mouth during swallowing	Facial nerve
Mylohyoid (mi″lo-hi′oid) (*myle* = molar)	Flat, triangular muscle just deep to digastric muscle; this muscle pair forms a sling that forms the floor of the anterior mouth	O—medial surface of mandible I—hyoid bone and median raphe	Elevates hyoid bone and floor of mouth, enabling tongue to exert backward and upward pressure that forces food bolus into pharynx	Mandibular branch of trigeminal nerve

Muscle	Description	Origin (O) and Insertion (I)	Action	Nerve Supply
SUPRAHYOID MUSCLES, continued				
Geniohyoid (je'ne-o-hy"oid) (also see Figure 11.7) (*geni* = chin)	Narrow muscle in contact with its partner medially; runs from chin to hyoid bone	O—inner surface of mandibular symphysis I—hyoid bone	Pulls hyoid bone superiorly and anteriorly, shortening floor of mouth and widening pharynx for receiving food during swallowing	First cervical spinal nerve via hypoglossal nerve (cranial XII)
INFRAHYOID MUSCLES	Strap-like muscles that depress the hyoid bone and larynx during swallowing and speaking (see also Figure 11.9c)			
Sternohyoid (ster"no-hi'oid) (*sterno* = sternum)	Most medial muscle of the neck; thin; superficial except inferiorly, where covered by sternocleidomastoid	O—manubrium and medial end of clavicle I—lower margin of hyoid bone	Depresses larynx and hyoid bone if mandible is fixed; may also flex skull	C_1–C_3 through ansa cervicalis (slender nerve root in cervical plexus)
Sternothyroid (ster"no-thi'roid) (*thyro* = thyroid cartilage)	Lateral and deep to sternohyoid	O—posterior surface of manubrium of sternum I—thyroid cartilage	Pulls thyroid cartilage (plus larynx and hyoid bone) inferiorly	As for sternohyoid
Omohyoid (o"mo-hi'oid) (*omo* = shoulder)	Strap-like muscle with two bellies united by an intermediate tendon; lateral to sternohyoid	O—superior surface of scapula I—hyoid bone, lower border	Depresses and retracts hyoid bone	As for sternohyoid
Thyrohyoid (thi"ro-hi'oid) (also see Figure 11.7)	Appears as a superior continuation of sternothyroid muscle	O—thyroid cartilage I—hyoid bone	Depresses hyoid bone or elevates larynx if hyoid is fixed	First cervical nerve via hypoglossal
PHARYNGEAL CONSTRICTOR MUSCLES				
Superior, middle, and inferior pharyngeal constrictors (far-in'je-al)	Composite of three paired muscles whose fibers run circularly in pharynx wall; arranged so that the superior muscle is innermost and inferior one is outermost; substantial overlap	O—attached anteriorly to mandible and medial pterygoid plate (superior), hyoid bone (middle), and laryngeal cartilages (inferior) I—posterior median raphe of pharynx	Working as a group and in sequence, all constrict pharynx during swallowing, which propels a food bolus to esophagus	Pharyngeal plexus [branches of vagus (X) nerve]

➤

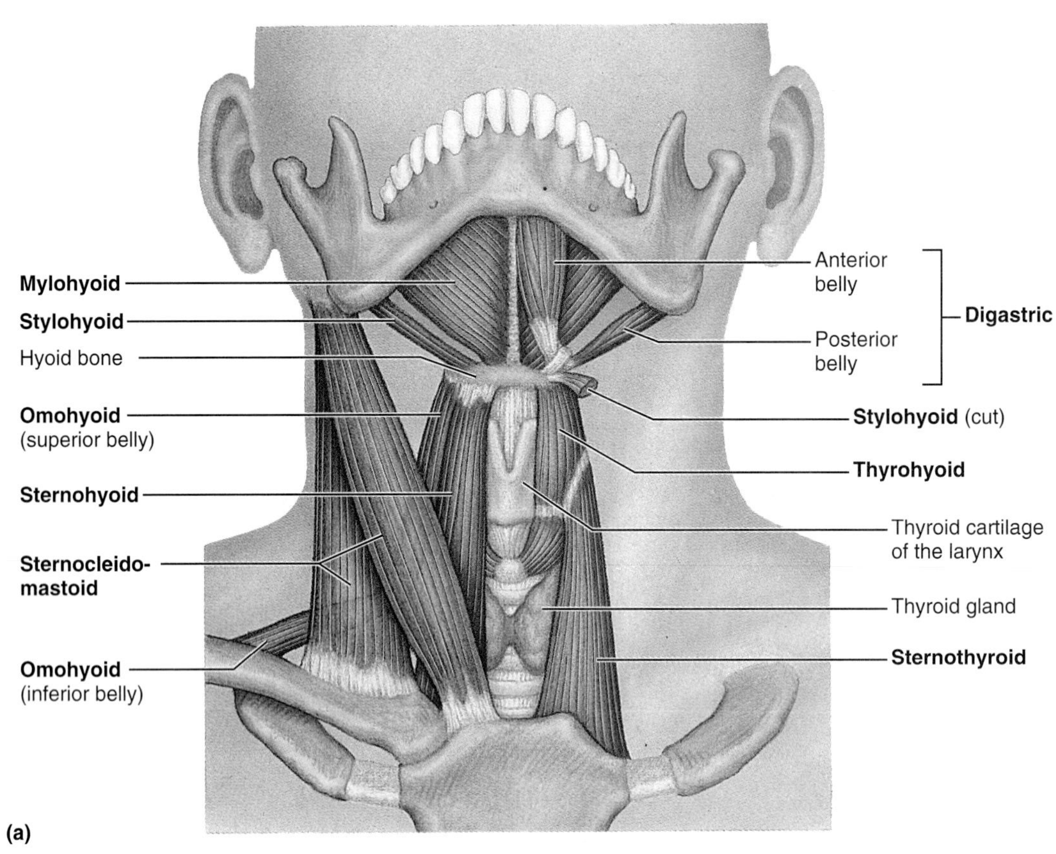

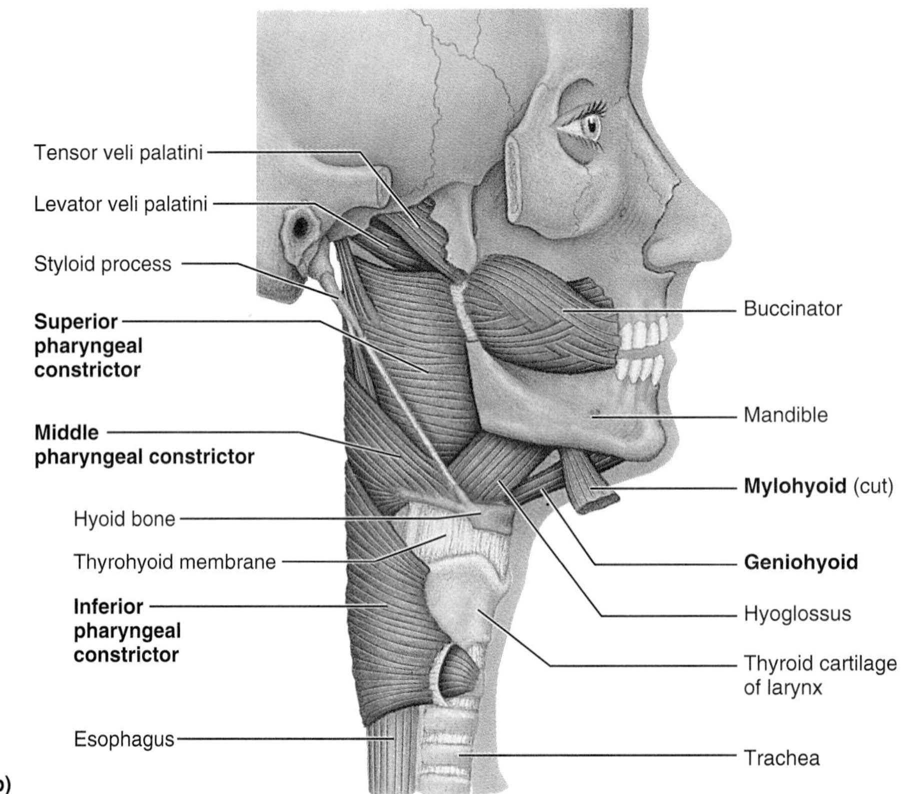

FIGURE 11.8 Muscles of the anterior neck and throat that promote swallowing. (a) Anterior view of the suprahyoid and infrahyoid muscles. The sternocleidomastoid muscle (not involved in swallowing) is shown at the left to provide an anatomical landmark. The picture's right half shows deeper muscles than those in the left half. (b) Lateral view of the constrictor muscles of the pharynx. These muscles are shown in their proper anatomical relationship to the buccinator (a chewing muscle) and to the hyoglossus muscle (which promotes tongue movements).

Muscle Gallery: Table 11.4 Muscles of the Neck and Vertebral Column: Head Movements and Trunk Extension (Figure 11.9)

Head Movements. The head is moved by muscles that originate on the axial skeleton inferiorly. *Flexion* of the head is mainly brought about by the **sternocleidomastoid** muscles (Figure 11.9a), with some help from the suprahyoid and infrahyoid muscles (see Table 11.3). The head can be *tilted* or *turned* from side to side as the neck is laterally flexed or rotated. These actions, which result when muscles contract on one side of the neck only, are performed by the sternocleidomastoids and the deeper neck muscles considered in this table. *Extension* of the head is aided by the trapezius muscle of the back (see Figure 11.5a), but the main extensors of the head are the **splenius** muscles deep to the trapezius (Figure 11.9b).

Trunk Extension. Trunk extension is brought about by the *deep,* or *intrinsic, muscles of the back* (Figure 11.9d and e). These muscles also maintain the normal curvatures of the spine, acting as postural muscles. As we consider these back muscles, keep in mind that they are deep; the superficial back muscles that cover them (see Figure 11.5b) primarily run to the bones of the upper limb and are considered in Tables 11.8 and 11.9.

The deep muscles of the back form a broad, thick column running from the sacrum to the skull (see Figure 11.9d). Many muscles of varying length contribute to this mass. It helps to regard each of these muscles as a string, which when pulled causes one or many vertebrae to extend or rotate on the vertebrae inferior to it. The largest of the deep back muscles in the **erector spinae** group. Many of the deep back muscles are long, so that large regions of the spine can be extended at once. When these muscles contract on just one side of the body, they help bend the spine laterally. Such lateral flexion is automatically accompanied by some rotation of the vertebral column. During vertebral movements, the articular facets of the vertebrae glide on each other.

There are also many short muscles that extend from one vertebra to the next (see Figure 11.9e). These small muscles act primarily as synergists in the extension and rotation of the spine and as stabilizers of the spine. They are not described in this table, but you can deduce their actions by examining their origins and insertions in Figure 11.9e.

The trunk muscles considered in this table are extensors. *Flexion* of the trunk is brought about by muscles that lie anterior to the vertebral column (see Table 11.6).

Clinical Applications

TORTICOLLIS Torticollis (tor″tĭ-kol′is; "turned neck") is a condition in which the neck stays rotated to one side, keeping the head tilted in that direction. All of the various forms of this condition result from problems in a sternocleidomastoid muscle on one side of the neck only. The adult-onset form, or spasmotic torticollis, is caused by sustained spasms in the muscle. In congenital forms, on the other hand, either the developing fetus twists into an awkward position because of a lack of amniotic fluid, or excessive stretching of the muscle occurs during birth. The resulting damage prevents the muscle from lengthening as the child grows. ✚

Muscle	Description	Origin (O) and Insertion (I)	Action	Nerve Supply
ANTEROLATERAL NECK MUSCLES (FIGURE 11.9a AND c)				
Sternocleidomastoid (ster″no-kli″do-mas′toid) (*sterno* = breastbone; *cleido* = clavicle; *mastoid* = mastoid process)	Two-headed muscle located deep to platysma on anterolateral surface of neck; fleshy parts on either side of neck delineate limits of anterior and posterior triangles; key muscular landmark in neck; spasms of one of these muscles may cause torticollis	O—manubrium of sternum and medial portion of clavicle I—mastoid process of temporal bone and superior nucal line of occipital bone	Prime mover of active head flexion; simultaneous contraction of both muscles causes neck flexion, generally against resistance as when one raises head when lying on back; acting alone, each muscle rotates head toward shoulder on opposite side and tilts or laterally flexes head to its own side	Accessory nerve (cranial nerve XI); cervical spinal nerves C_2 and C_3 (ventral rami)
Scalenes (ska′lēnz)—anterior, middle, and posterior (*scalene* = uneven)	Located more laterally than anteriorly on neck; deep to platysma and sternocleidomastoid	O—transverse processes of cervical vertebrae I—anterolaterally on first two ribs	Elevate first two ribs (aid in inspiration); flex and rotate neck	Cervical spinal nerves

➤

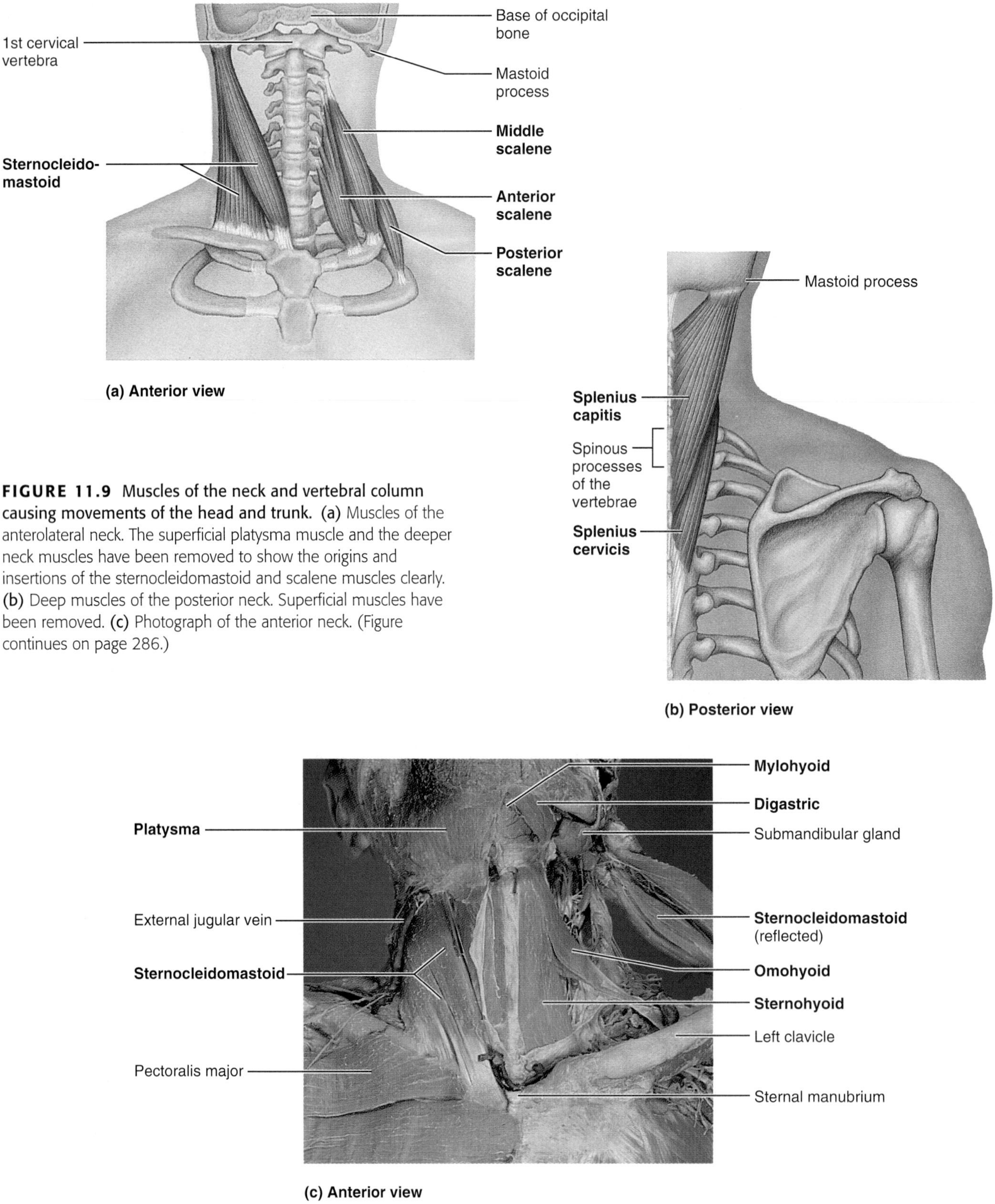

1st cervical vertebra
Base of occipital bone
Mastoid process
Middle scalene
Sternocleido-mastoid
Anterior scalene
Posterior scalene

(a) Anterior view

Mastoid process
Splenius capitis
Spinous processes of the vertebrae
Splenius cervicis

(b) Posterior view

FIGURE 11.9 Muscles of the neck and vertebral column causing movements of the head and trunk. **(a)** Muscles of the anterolateral neck. The superficial platysma muscle and the deeper neck muscles have been removed to show the origins and insertions of the sternocleidomastoid and scalene muscles clearly. **(b)** Deep muscles of the posterior neck. Superficial muscles have been removed. **(c)** Photograph of the anterior neck. (Figure continues on page 286.)

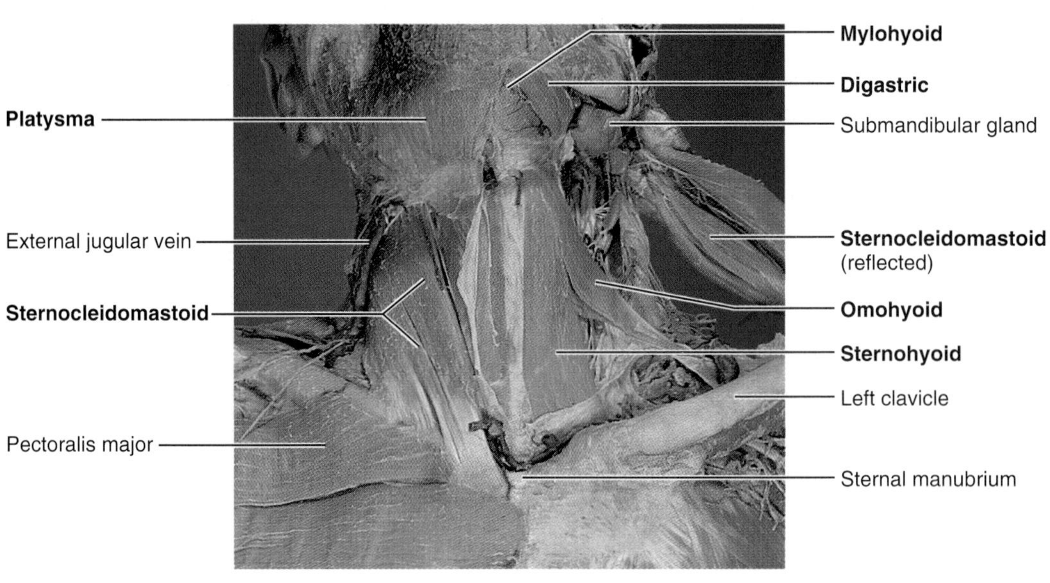

(c) Anterior view

Muscle	Description	Origin (O) and Insertion (I)	Action	Nerve Supply
INTRINSIC MUSCLES OF THE BACK (FIGURE 11.9b–e)				
Splenius (sple′ne-us)—capitis and cervicis portions (kah-pit′us; ser-vis′us) (*splenion* = bandage; *caput* = head; *cervi* = neck) (Figures 11.9b and 11.6)	Broad bipartite superficial muscle (capitis and cervicis parts) extending from upper thoracic vertebrae to skull; capitis portion known as "bandage muscle" because it covers and holds down deeper neck muscles	O—ligamentum nuchae,* spinous processes of vertebrae C_7–T_6 I—mastoid process of temporal bone and occipital bone (capitis); transverse processes of C_2–C_4 vertebrae (cervicis)	Act as a group to extend or hyperextend head; when splenius muscles on one side are activated, head is rotated and bent laterally toward same side	Cervical spinal nerves (dorsal rami)
Erector spinae (e-rek′tor spi′ne) Also called **sacrospinalis** (Figure 11.9d, left side)	Prime mover of back extension; erector spinae muscles, each consisting of three columns— the iliocostalis, longissimus, and spinalis muscles—form intermediate layer of intrinsic back muscles; erector spinae provide resistance that helps control action of bending forward at the waist, and act as powerful extensors to promote return to erect position; during full flexion (i.e., when touching fingertips to floor), erector spinae are relaxed and strain is borne entirely by ligaments of back; on reversal of the movement, these muscles are initially inactive, and extension is initiated by hamstring muscles of thighs and gluteus maximus muscles of buttocks. As a result of this peculiarity, lifting a load or moving suddenly from a bent-over position is potentially injurious to muscles and ligaments of back and intervertebral discs; erector spinae muscles readily go into painful spasms following injury to back structures.			
• **Iliocostalis** (il″e-o-kos-tă′lis)—lumborum, thoracis, and cervicis portions (lum′bor-um; tho-ra′sis) (*ilio* = ilium; *cost* = rib)	Most lateral muscle group of erector spinae muscles; extend from pelvis to neck	O—iliac crests (lumborum); inferior 6 ribs (thoracis); ribs 3 to 6 (cervicis) I—angles of ribs (lumborum and thoracis); transverse processes of cervical vertebrae C_6–C_4 (cervicis)	Extend vertebral column, maintain erect posture; acting on one side, bend vertebral column to same side	Spinal nerves (dorsal rami)
• **Longissimus** (lon-jis′ĭ-mus)—thoracis, cervicis, and capitis parts (*longissimus* = longest)	Intermediate tripartite muscle group of erector spinae; extend by many muscle slips from lumbar region to skull; mainly pass between transverse processes of the vertebrae	O—transverse processes of lumbar through cervical vertebrae I—transverse processes of thoracic or cervical vertebrae and to ribs superior to origin as indicated by name; capitis inserts into mastoid process of temporal bone	Thoracis and cervicis act together to extend vertebral column, and acting on one side, bend it laterally; capitis extends head and turns the face toward same side	Spinal nerves (dorsal rami)
• **Spinalis** (spi-nă′lis)—thoracis and cervicis parts (*spin* = vertebral column, spine)	Most medial muscle column of erector spinae; cervicis usually rudimentary and poorly defined	O—spines of upper lumbar and lower thoracic vertebrae I—spines of upper thoracic and cervical vertebrae	Extends vertebral column	Spinal nerves (dorsal rami)
Semispinalis (sem′e-spĭ-nă′lis)—thoracis, cervicis, and capitis regions (*semi* = half; *thorac* = thorax) (Figure 11.9d, right side)	Composite muscle forming part of deep layer of intrinsic back muscles; extends from thoracic region to head	O—transverse processes of C_7–T_{12} I—occipital bone (capitis) and spinous processes of cervical (cervicis) and thoracic vertebrae T_1 to T_4 (thoracis)	Extends vertebral column and head and rotates them to opposite side; acts synergistically with sternocleidomastoid muscle of opposite side	Spinal nerves (dorsal rami)
Quadratus lumborum (kwod-ra′tus lum-bor′um) (*quad* = four-sided; *lumb* = lumbar region) (See also Figure 11.19a)	Fleshy muscle forming part of posterior abdominal wall	O—iliac crest and lumbar fascia I—transverse processes of upper lumbar vertebrae and lower margin of 12th rib	Flexes vertebral column laterally when acting separately; when pair acts jointly, lumbar spine is extended and 12th rib is fixed; maintains upright posture; assists in forced inspiration	T_{12} and upper lumbar spinal nerves (ventral rami)

*The ligamentum nuchae (lig″ah-men′tum noo′ke) is a strong, elastic ligament in the midsagittal plane extending from the occipital bone of the skull along the tips of the spinous processes of the cervical vertebrae. It binds the cervical vertebrae together and inhibits excessive head and neck flexion, thus preventing damage to the spinal cord in the vertebral canal.

➤

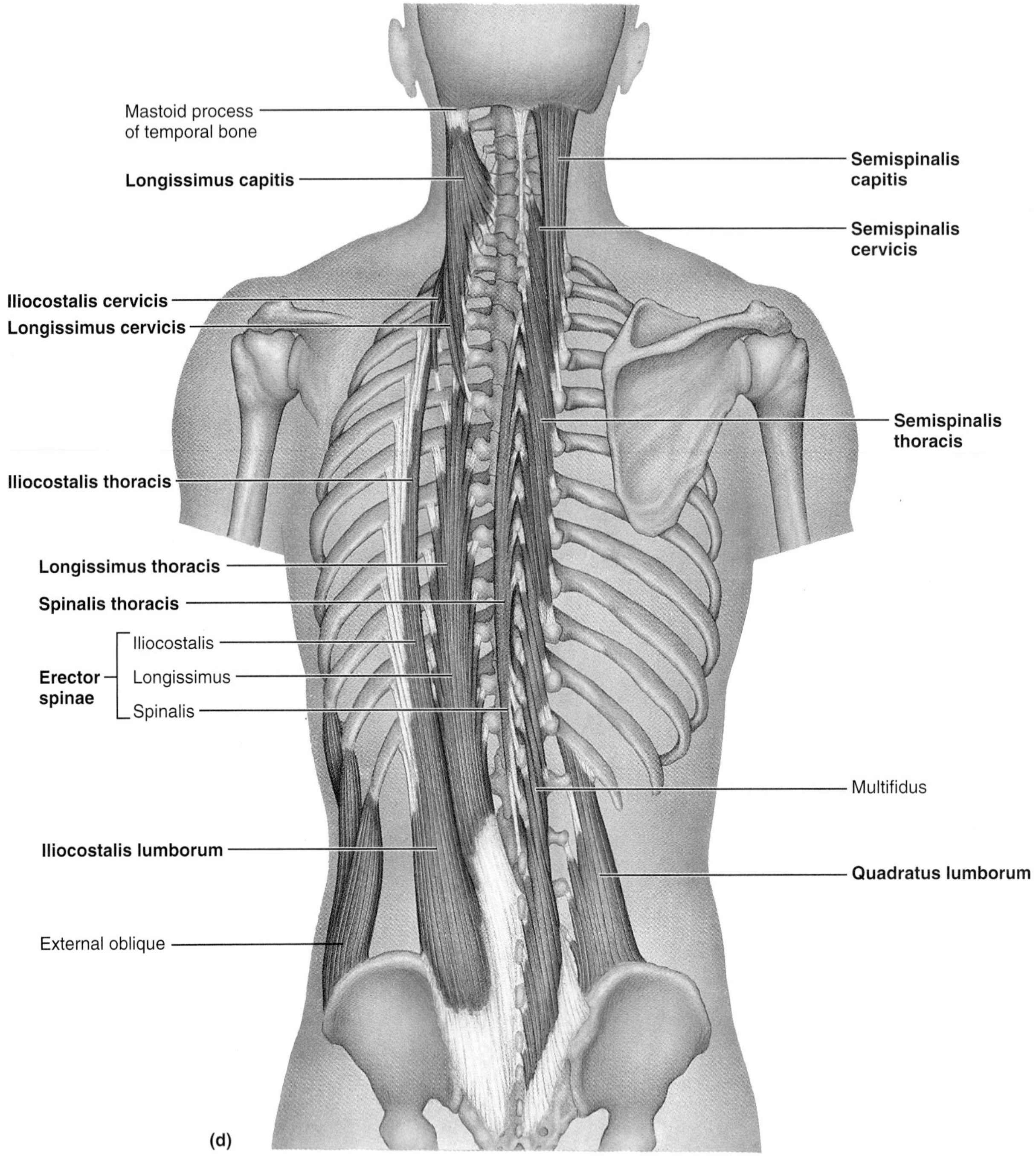

Mastoid process of temporal bone
Longissimus capitis
Semispinalis capitis
Semispinalis cervicis
Iliocostalis cervicis
Longissimus cervicis
Semispinalis thoracis
Iliocostalis thoracis
Longissimus thoracis
Spinalis thoracis
Erector spinae: Iliocostalis, Longissimus, Spinalis
Multifidus
Iliocostalis lumborum
Quadratus lumborum
External oblique
(d)

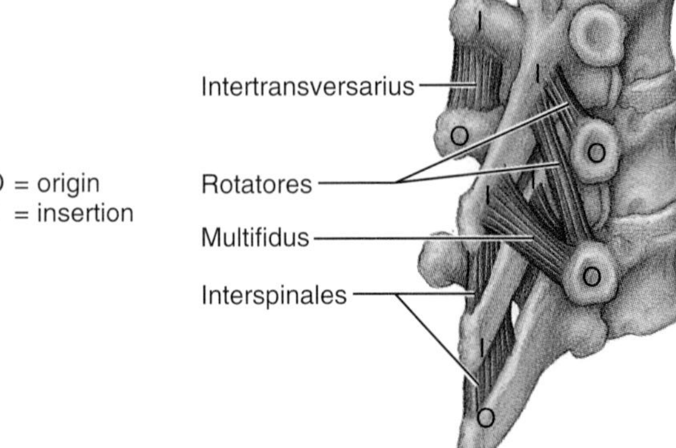

FIGURE 11.9 **Muscles of the neck and vertebral column causing movements of the head and trunk,** *continued.* **(d)** Deep muscles of the back. The superficial muscles and the splenius muscles have been removed. The three-muscle columns (iliocostalis, longissimus, and spinalis) composing the erector spinae are shown at the left; the deeper semispinalis muscles are shown at the right. **(e)** The deepest muscles of the back (rotatores, multifidus, interspinales, and intertransversarius muscles), associated with the vertebral column.

Muscle Gallery: Table 11.5 Deep Muscles of the Thorax: Breathing (Figure 11.10)

An important function of the deep muscles of the thorax is to provide the movements necessary for ventilation, or breathing. Breathing has two phases—inspiration, or inhaling, and expiration, or exhaling—caused by cyclical changes in the volume of the thoracic cavity.

The thoracic muscles are very short: Most run only from one rib to the next. They form three layers in the wall of the thorax. The 11 **external intercostal muscles** form the most superficial layer. Their function is controversial, but they seem to lift the rib cage, which increases its anterior-posterior and lateral dimensions. This enlargement reduces intra-thoracic pressure and draws air into the lungs. Thus, the external intercostals seem to be inspiratory muscles. The 11 **internal intercostals** form the intermediate muscle layer. They may aid expiration during heavy breathing by depressing the rib cage, which decreases thoracic volume and helps expel air. The internal intercostals do not operate in normal quiet expiration, however. Quiet expiration is a passive phenomenon, resulting only from elastic recoil of the lungs. The third and deepest muscle layer of the thoracic wall attaches to the internal surfaces of the ribs. It has three discontinuous parts (from posterior to anterior): the *subcostals, innermost intercostals,* and *transversus thoracis*. These are small, and their function is unknown, so they are not listed in this table.

The **diaphragm,** the most important muscle of respiration (Figures 11.10b, 11.10c, and 1.11), forms a complete partition between the thoracic and abdominopelvic cavities. In the relaxed state, the diaphragm is dome-shaped, but it flattens as it contracts, increasing the volume of the thoracic cavity. Thus, the diaphragm is a powerful muscle of inspiration. It contracts rhythmically during respiration, but one can also contract it voluntarily to push down on the abdominal viscera and increase the pressure in the abdominopelvic cavity. This pressure helps to evacuate the contents of the pelvic organs (feces, urine, or a baby). It also helps in lifting heavy weights: When one takes a deep breath to fix the diaphragm, the abdomen becomes a firm pillar that will not buckle under the weight being lifted. The muscles of the anterior abdominal wall also help to increase the intra-abdominal pressure (see Table 11.6).

With the exception of the diaphragm, which is innervated by *phrenic nerves* from the neck, all muscles described in this table are served by nerves running between the ribs, called *intercostal nerves*.

When breathing is forced and heavy, as during exercise, additional muscles become active in ventilation. For example, in forced inspiration, the scalene and sternocleidomastoid muscles of the neck help lift the ribs. Forced expiration is aided by muscles that pull inferiorly on the ribs (quadratus lumborum) or push the diaphragm superiorly by compressing the abdominal organs (abdominal wall muscles).

Muscle	Description	Origin (O) and Insertion (I)	Action	Nerve Supply
External intercostals (in″ter-kos′talz) (*external* = toward the outside; *inter* = between; *cost* = rib)	11 pairs lie between ribs; fibers run obliquely (down and forward) from each rib to rib below; in lower intercostal spaces, fibers are continuous with external oblique muscle forming part of abdominal wall	O—inferior border of rib above I—superior border of rib below	With first ribs fixed by scalene muscles, pull ribs toward one another to elevate rib cage; aids in inspiration; synergists of diaphragm	Intercostal nerves
Internal intercostals (*internal* = toward the inside, deep)	11 pairs lie between ribs; fibers run deep to and at right angles to those of external intercostals (i.e., run downward and posteriorly); lower internal intercostal muscles are continuous with fibers of internal oblique muscle of abdominal wall	O—superior border of rib below I—inferior border (costal groove) of rib above	With 12th ribs fixed by quadratus lumborum, muscles of posterior abdominal wall, and oblique muscles of the abdominal wall, the internal intercostals draw ribs together and depress rib cage; aid in forced expiration; antagonistic to external intercostals	Intercostal nerves
Diaphragm (di′ah-fram) (*dia* = across; *phragm* = partition)	Broad muscle pierced by the aorta, inferior vena cava, and esophagus; forms floor of thoracic cavity; in relaxed state is dome shaped; fibers converge from margins of thoracic cage toward a boomerang-shaped central tendon	O—inferior internal surface of rib cage and sternum, costal cartilages of last six ribs and lumbar vertebrae I—central tendon	Prime mover of inspiration; flattens on contraction, increasing vertical dimensions of thorax; when strongly contracted, dramatically increases intra-abdominal pressure	Phrenic nerves

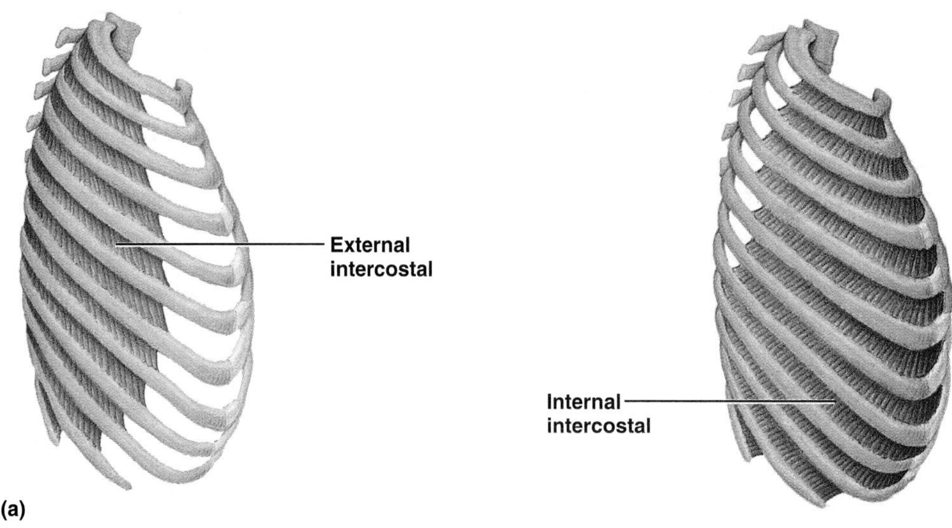

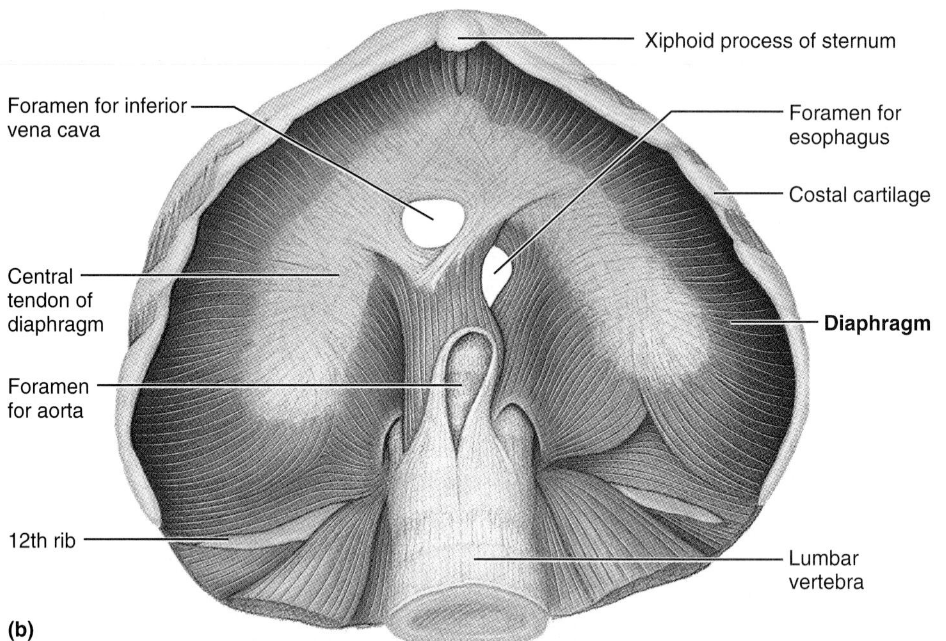

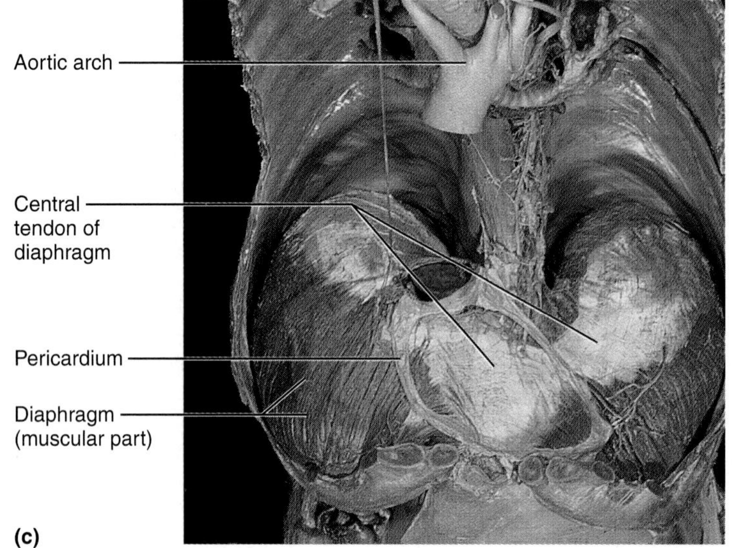

FIGURE 11.10 Muscles of quiet respiration. **(a)** Deep muscles of the thorax. The external intercostals (inspiratory muscles) are shown at the left and the internal intercostals (expiratory muscles) at the right. These two muscle layers run obliquely and at right angles to each other. **(b)** Inferior view of the diaphragm, the prime mover of inspiration. Notice that its muscle fibers converge toward a central tendon, an arrangement that causes the diaphragm to flatten and move inferiorly as it contracts. The diaphragm and its tendon are pierced by the great vessels (aorta and inferior vena cava) and the esophagus. **(c)** Photograph of the diaphragm, superior view.

Muscle Gallery: Table 11.6 Muscles of the Abdominal Wall: Trunk Movements and Compression of Abdominal Viscera (Figure 11.11)

The walls of the abdomen have no bony reinforcements (no ribs). Instead, the abdominal wall is a composite of sheet-like muscles. Forming the anterior and lateral parts of the abdominal wall are three broad, flat muscle sheets, layered one on the next: the **external oblique, internal oblique,** and **transversus abdominis.** These are direct continuations of the three intercostal muscles in the thorax. The fascicles of the external oblique run inferomedially, at right angles to fascicles of the internal oblique muscle (which primarily run superomedially), whereas the fascicles of the deep transversus abdominis are strictly horizontal. This arrangement of different fascicle directions lends great strength to the abdominal wall, just as plywood gets its strength from sheets of wood whose grains run in different directions. These three muscles end anteriorly in broad, white aponeuroses (tendon-like sheets). The aponeuroses enclose a fourth muscle pair, the strap-like **rectus abdominis,** and extend medially to insert on the **linea alba** ("white line"), a tendinous raphe (seam) that runs vertically from the sternum to the pubic symphysis. The tight enclosure of the rectus abdominis within aponeuroses ensures that this vertical muscle cannot snap forward like a bowstring when it contracts to flex the trunk.

The four muscles of the abdominal wall perform many functions. They help contain the abdominal organs. They also flex the trunk; therefore, performing sit-ups causes the rectus abdominis to bulge beneath the skin. Other functions include lateral flexion and rotation of the trunk. In addition, these muscles help produce forced, heavy breathing: When they contract simultaneously, they pull the ribs inferiorly and squeeze the abdominal contents. This action in turn pushes the diaphragm superiorly, aiding forced expiration. When the abdominal muscles contract with the diaphragm, the increased intra-abdominal pressure helps promote micturition (voiding urine), defecation, vomiting, childbirth, sneezing, coughing, laughing, burping, screaming, and nose blowing. Next time you perform one of these activities, feel the abdominal wall muscles contract under your skin. These muscles also contract during heavy lifting—sometimes so forcefully that hernias result.

Clinical Applications

HERNIA An abdominal hernia is an abnormal protrusion of abdominal contents out of the abdominal cavity through a weak point in the muscles of the abdominal wall. The most commonly herniated elements are coils of small intestine and parts of the greater omentum (a large membrane, or mesentery, in the abdominal cavity). A hernia can result from increased intra-abdominal pressure during lifting and straining. An element that herniates from the abdominal cavity usually lies deep to the skin or superficial fascia, where it can form a visible bulge on the body surface. The most common types of hernias occur in the groin in the region of the aponeurosis of the external oblique muscle (inguinal hernia: p. 710), in the superior thigh (femoral hernia: p. 791), through the foramen for the esophagus in the diaphragm (hiatal hernia: p. 646), and in or near the navel (umbilical hernia: p. 781). ✚

Muscle	Description	Origin (O) and Insertion (I)	Action	Nerve Supply
MUSCLES OF THE ANTERIOR AND LATERAL ABDOMINAL WALL				
	Four paired flat muscles; very important in supporting and protecting abdominal viscera; important in promoting movement of vertebral column (flexion and lateral bending)			
Rectus abdominis (rek'tus ab-dom'ĭ-nis) (*rectus* = straight; *abdom*=abdomen)	Medial muscle pair; extend from pubis to rib cage; ensheathed by aponeuroses of lateral muscles; segmented by 3 tendinous intersections	O—pubic crest and symphysis I—xiphoid process and costal cartilages of ribs 5–7	Flex and rotate lumbar region of vertebral column; fix and depress ribs, stabilize pelvis during walking, increase intra-abdominal pressure	Intercostal nerves (T_6 or T_7–T_{12})
External oblique (o-blēk') (*external* = toward outside; *oblique* = running at an angle)	Largest and most superficial of the three lateral muscles; fibers run downward and medially (same direction outstretched fingers take when hands put into pants pockets); aponeurosis turns under inferiorly, forming inguinal ligament	O—by fleshy strips from outer surfaces of lower eight ribs I—most fibers insert into linea alba via a broad aponeurosis; some insert into pubic crest and tubercle and iliac crest	When pair contracts simultaneously, aid rectus abdominis muscles in flexing vertebral column and in compressing abdominal wall and increasing intra-abdominal pressure; acting individually, aid muscles of back in trunk rotation and lateral flexion	Intercostal nerves (T_7–T_{12})

➤

Muscle	Description	Origin (O) and Insertion (I)	Action	Nerve Supply
Internal oblique (*internal* = toward the inside; deep)	Most fibers run superiorly and medially; the muscle fans such that its inferior fibers run inferiorly and medially	O—lumbar fascia, iliac crest, and inguinal ligament I—linea alba, pubic crest, last three or four ribs and costal margin	As for external oblique	Intercostal nerves (T_7–T_{12}) and L_1
Transversus abdominis (trans-ver′sus) (*transverse* = running straight across or transversely)	Deepest (innermost) muscle of abdominal wall; fibers run horizontally	O—inguinal ligament, lumbar fascia, cartilages of last six ribs; iliac crest I—linea alba, pubic crest	Compresses abdominal contents	Intercostal nerves (T_7–T_{12}) and L_1

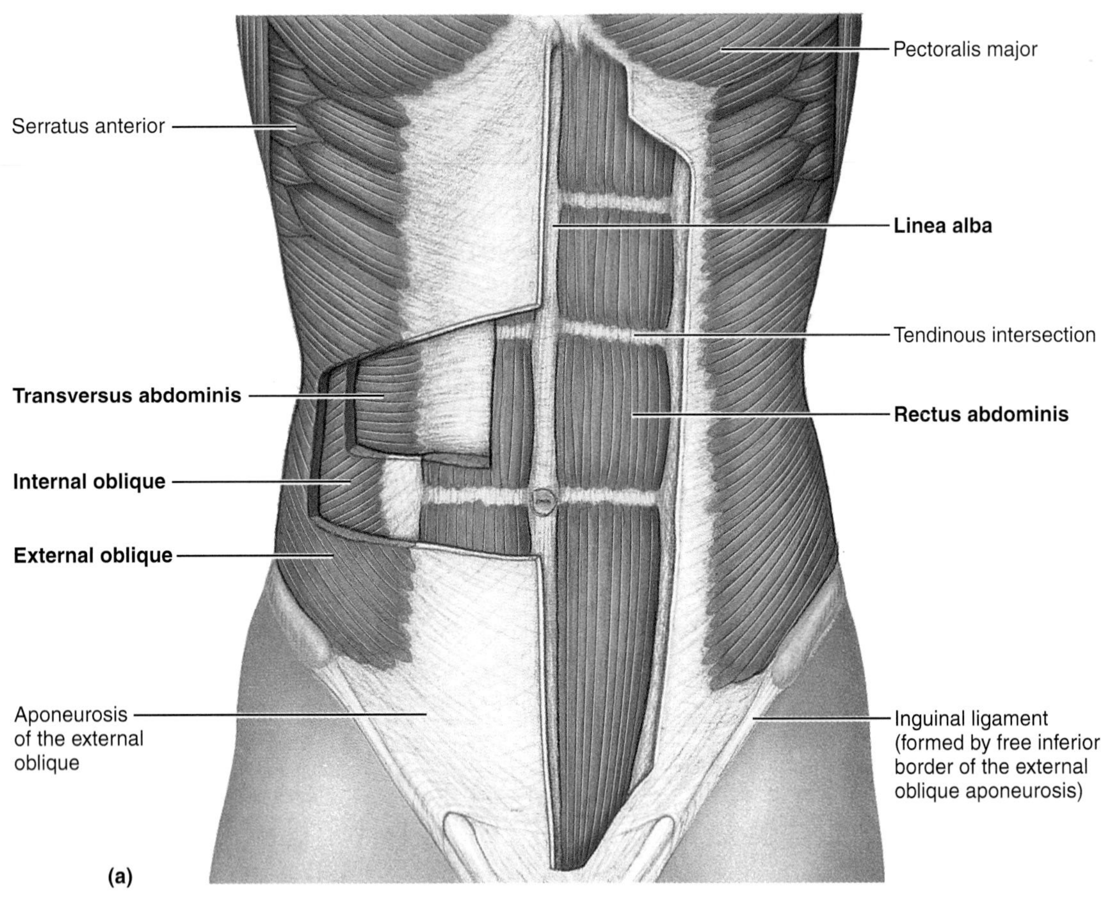

(a)

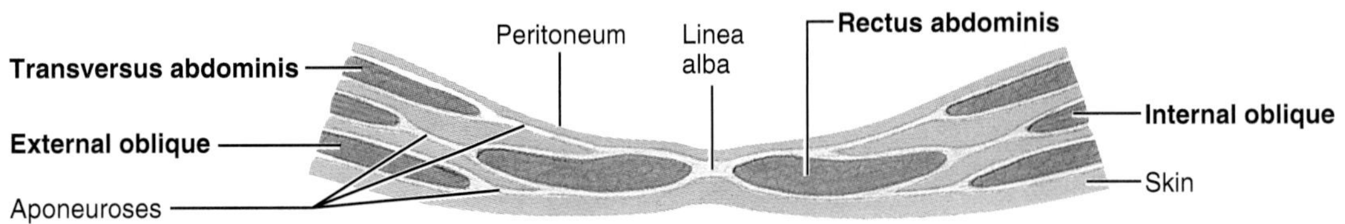

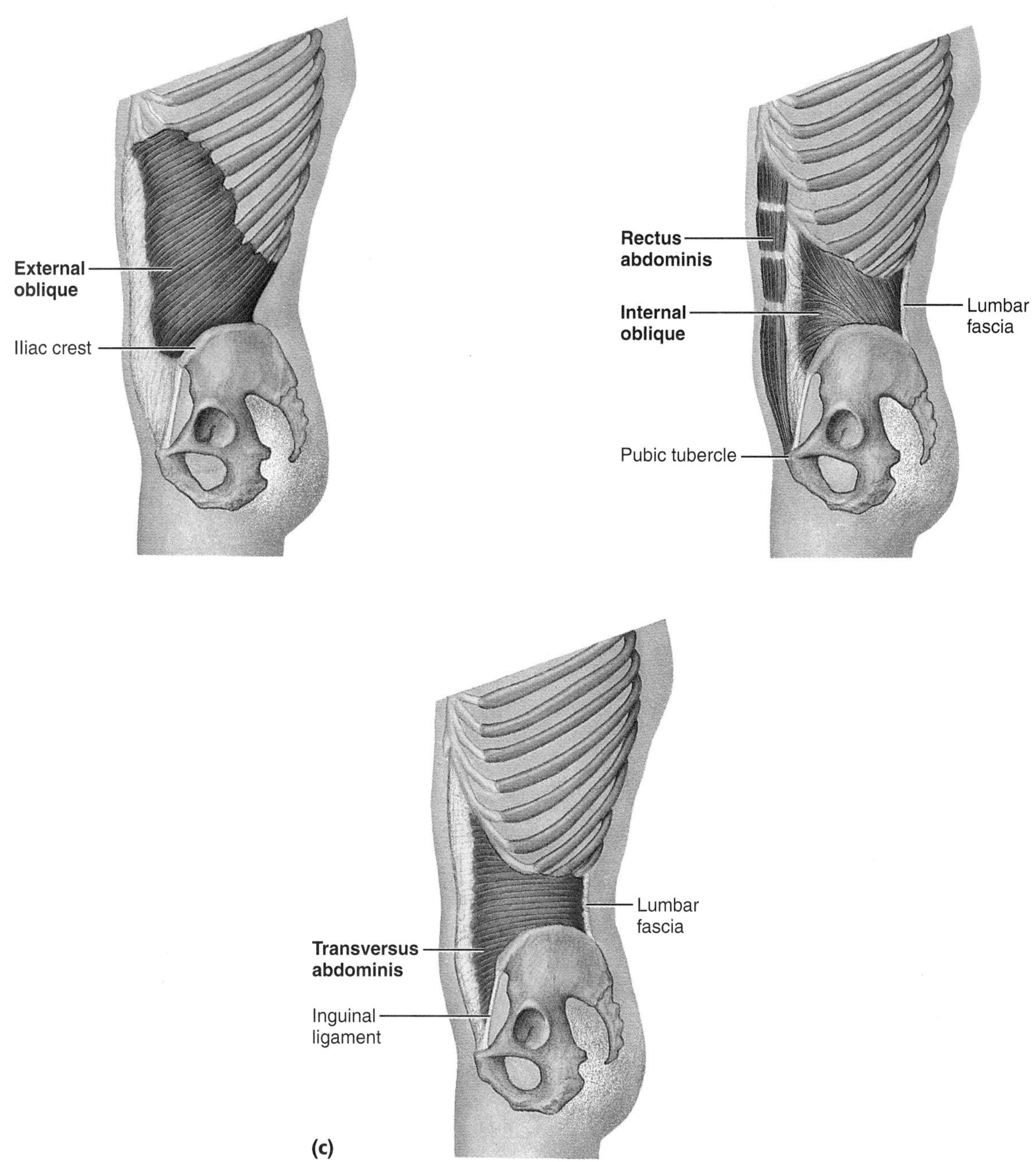

FIGURE 11.11 **Muscles of the abdominal wall.** **(a)** Anterior view of the muscles forming the anterior and lateral abdominal wall. The superficial muscles have been partially cut away on the left side of the illustration to reveal the deeper internal oblique and transversus abdominis muscles. **(b)** Transverse section through the anterior abdominal wall (midregion), showing how the aponeuroses of the abdominal muscles contribute to the rectus abdominis sheath. **(c)** Lateral views of the trunk, illustrating the fiber directions and attachments of the external oblique, internal oblique, and transversus abdominis muscles.

Muscle Gallery: Table 11.7 Muscles of the Pelvic Floor and Perineum: Support of Abdominopelvic Organs (Figure 11.12)

The pelvic floor, or **pelvic diaphragm** (Figure 11.12a) is a sheet consisting of two muscles, the **levator ani** and the small **coccygeus.** This diaphragm supports the pelvic organs, seals the inferior opening of the bony pelvis, and lifts superiorly to help release feces during defecation. The pelvic diaphragm is pierced by the rectum and urethra (the tube for urine) and (in females) by the vagina. The body region inferior to the pelvic diaphragm is the *perineum* (Figure 11.12b and c). In the anterior half of the perineum is a triangular sheet of muscle called the **urogenital diaphragm.** It contains the **sphincter urethrae** muscle, which surrounds the urethra. One uses this muscle voluntarily to prevent urination. Just inferior to the urogenital diaphragm is the superficial perineal space, which contains muscles **(bulbospongiosus, ischiocavernosus)** that help maintain erection of the penis and clitoris. In the posterior half of the perineum, circling the anus, lies the **external anal sphincter** (Figure 11.12b and p. 657). This muscle is used voluntarily to prevent defecation. Just anterior to this sphincter, at the exact midpoint of the perineum, is the **central tendon.** Many perineal muscles insert on this strong tendon and, in so doing, are able to support the heavy organs in the pelvis.

Muscle	Description	Origin (O) and Insertion (I)	Action	Nerve Supply
MUSCLES OF THE PELVIC DIAPHRAGM (FIGURE 11.12a)				
Levator ani (lĕ-va′tor a′ne) (*levator* = raises; *ani* = anus)	Broad, thin, tripartite muscle (pubococcygeus, puborectalis, and iliococcygeus parts); its fibers extend inferomedially, forming a muscular "sling" around male prostate (or female vagina), urethra, and anorectal junction before meeting in the median plane	O—extensive linear origin inside pelvis from pubis to ischial spine I—inner surface of coccyx, levator ani of opposite side, and (in part) into the structures that penetrate it	Supports and maintains position of pelvic viscera; resists downward thrusts that accompany rises in intrapelvic pressure during coughing, vomiting, and expulsive efforts of abdominal muscles; forms sphincters at anorectal junction and vagina; lifts anal canal during defecation	S_3, S_4, and inferior rectal nerve (branch of pudendal nerve)
Coccygeus (kok-sij′e-us) (*coccy* = coccyx)	Small triangular muscle lying posterior to levator ani; forms posterior part of pelvic diaphragm	O—spine of ischium I—sacrum and coccyx	Supports pelvic viscera; supports coccyx and pulls it forward after it has been reflected posteriorly by defecation and childbirth	S_4 and S_5
MUSCLES OF THE UROGENITAL DIAPHRAGM (FIGURE 11.12b)				
Deep transverse perineus (per″ĭ-ne′us) (*deep* = far from surface; *transverse* = across; *perine* = near anus)	Together the pair spans distance between ischial rami; in females, lies posterior to vagina	O—ischial rami I—midline central tendon of perineum; some fibers into vaginal wall in females	Supports pelvic organs; steadies central tendon	Pudendal nerve
Sphincter urethrae (*sphin* = squeeze)	Muscle encircling urethra and vagina (female)	O—ischiopubic rami I—midline raphe	Constricts urethra; allows voluntary inhibition of urination; helps support pelvic organs	Pudendal nerve
MUSCLES OF THE SUPERFICIAL PERINEAL SPACE (FIGURE 11.12c)				
Ischiocavernosus (is′ke-o-kav′ern-o′sus) (*ischi* = hip; *caverna* = hollow chamber)	Runs from pelvis to base of penis or clitoris	O—ischial tuberosities I—crus of corpus cavernosa of male penis or female clitoris	Retards venous drainage and maintains erection of penis or clitoris	Pudendal nerve
Bulbospongiosus (bul″bo-spun″je-o′sus) (*bulbon* = bulb; *spongio* = sponge)	Encloses base of penis (bulb) in males and lies deep to labia in females	O—central tendon of perineum and midline raphe of male penis I—anteriorly into corpus cavernosa of penis or clitoris	Empties male urethra; assists in erection of penis in males and of clitoris in females	Pudendal nerve
Superficial transverse perineus (*superficial* = closer to surface)	Paired muscle bands posterior to urethral (and in females, vaginal) opening; variable; sometimes absent	O—ischial tuberosity I—central tendon of perineum	Stabilizes and strengthens central tendon of perineum	Pudendal nerve

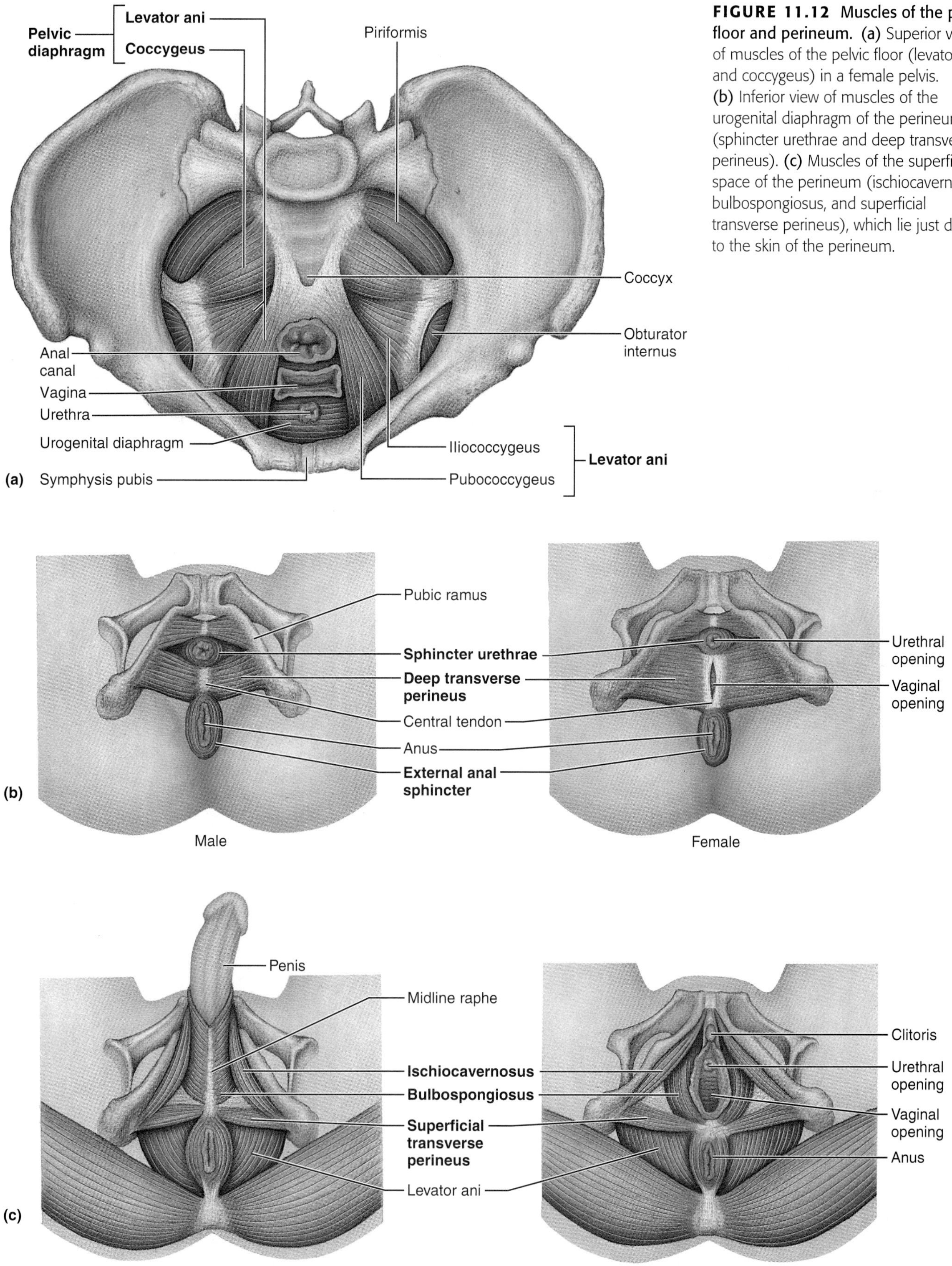

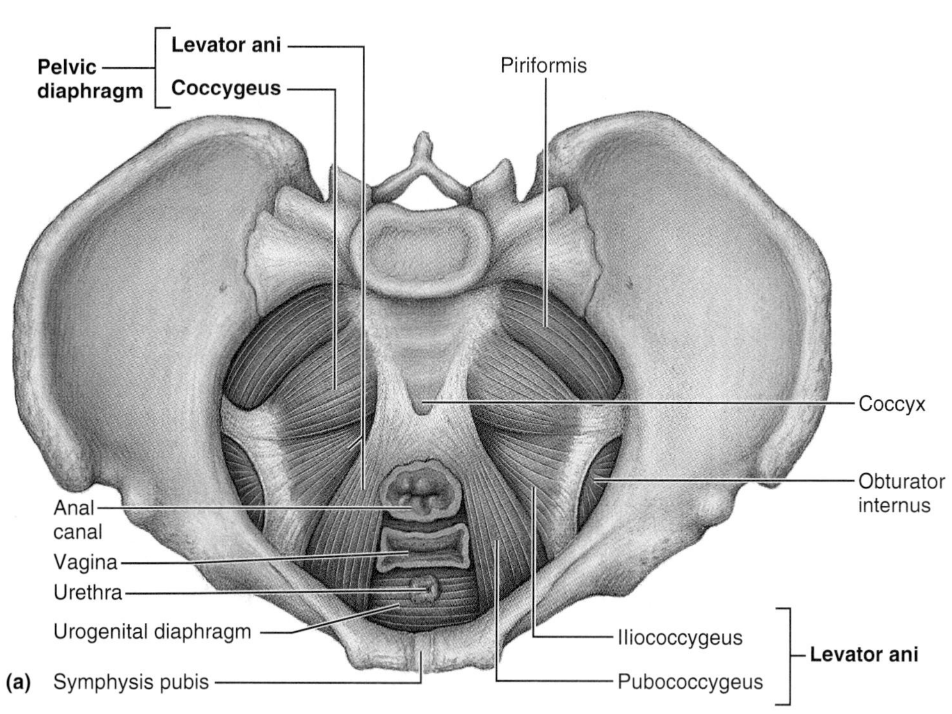

Pubic ramus
Sphincter urethrae
Deep transverse perineus
Central tendon
Anus
External anal sphincter
Urethral opening
Vaginal opening
(b)
Male
Female

Penis
Midline raphe
Ischiocavernosus
Bulbospongiosus
Superficial transverse perineus
Levator ani
Clitoris
Urethral opening
Vaginal opening
Anus
(c)

FIGURE 11.12 Muscles of the pelvic floor and perineum. (a) Superior view of muscles of the pelvic floor (levator ani and coccygeus) in a female pelvis. (b) Inferior view of muscles of the urogenital diaphragm of the perineum (sphincter urethrae and deep transverse perineus). (c) Muscles of the superficial space of the perineum (ischiocavernosus, bulbospongiosus, and superficial transverse perineus), which lie just deep to the skin of the perineum.

Muscle Gallery: Table 11.8 Superficial Muscles of the Anterior and Posterior Thorax: Movements of the Scapula (Figure 11.13)

Most superficial muscles on the thorax are *extrinsic shoulder muscles*, which run from the ribs and vertebral column to the shoulder girdle. They both fix the scapula in place and move it to increase the arm's range of movements. The *anterior* muscles of this group are **pectoralis major, pectoralis minor, serratus anterior,** and **subclavius** (Figure 11.13a). The *posterior* muscles are **trapezius, latissimus dorsi, levator scapulae,** and the **rhomboids** (Figure 11.13b). The pectoralis major and latissimus dorsi move the scapula indirectly by moving the arm, so they are considered in Table 11.9.

The scapula undergoes a variety of movements on the posterior rib cage. In *elevation* (superior movement) of the scapula, the prime movers are the levator scapulae and the superior fascicles of the trapezius; the rhomboids are synergists. Note that the fascicles of all these muscles run inferolaterally (see Figure 11.13b); thus, they are ideal elevators of the scapula. *Depression* (inferior movement) of the scapula is mostly due to gravity (weight of the arm). When the scapula must be depressed against resistance, however, this action is done by the serratus anterior, pectoralis minor, the inferior part of the trapezius, and—especially—the latissimus dorsi. The fascicles of all these muscles run superiorly to insert on the scapula or upper humerus, so it is logical that they depress the scapula. *Protraction* or *abduction* of the scapula moves it laterally and anteriorly, as in punching. This action is mainly carried out by the serratus anterior, whose horizontal fibers pull the scapula anterolaterally. *Retraction* or *adduction* of the scapula moves it medially and posteriorly. This is brought about by the mid part of the trapezius and the superior part of the latissimus dorsi, whose fibers run horizontally from the vertebral column to pull the scapula and humerus medially. *Upward rotation* of the scapula (in which the scapula's inferior angle moves laterally) allows one to lift the arm above the head: The serratus anterior swings the inferior angle laterally while the superior part of the trapezius, gripping the scapular spine, pulls the top of the scapula medially. In *downward rotation* of the scapula (where the inferior angle swings medially, as in paddling a canoe), the rhomboids pull the inferior part of the scapula medially while the levator scapulae steadies the superior part.

Muscle	Description	Origin (O) and Insertion (I)	Action	Nerve Supply
MUSCLES OF THE ANTERIOR THORAX (FIGURE 11.13a)				
Pectoralis minor (pek"to-ra'lis mi'nor) (*pectus* = chest, breast; *minor* = lesser)	Flat, thin muscle directly beneath and obscured by pectoralis major	O—anterior surfaces of ribs 3–5 (or 2–4) I—coracoid process of scapula	With ribs fixed, draws scapula forward and downward; with scapula fixed, draws rib cage superiorly	Medial and lateral pectoral nerves (C_6–C_8)
Serratus anterior (ser-a'tus) (*serratus* = saw)	Lies deep to scapula, deep and inferior to pectoral muscles on lateral rib cage; forms medial wall of axilla; origins have serrated, or sawtooth, appearance	O—by a series of muscle slips from ribs 1–8 (or 9) I—entire anterior surface of vertebral border of scapula	Prime mover to protract and hold scapula against chest wall; rotates scapula so that its inferior angle moves laterally and upward; raises point of shoulder; important role in abduction and raising of arm and in horizontal arm movements (pushing, punching); called "boxer's muscle"	Long thoracic nerve (C_5–C_7)

Muscle	Description	Origin (O) and Insertion (I)	Action	Nerve Supply
MUSCLES OF THE ANTERIOR THORAX (FIGURE 11.13a), continued				
Subclavius (sub-kla've-us) (*sub* = under, beneath; *clav* = clavicle)	Small cylindrical muscle extending from rib I to clavicle	O—costal cartilage of rib I I—groove on inferior surface of clavicle	Helps stabilize and depress pectoral girdle	Nerve to subclavius (C_5 and C_6)
MUSCLES OF THE POSTERIOR THORAX (FIGURE 11.13b)				
Trapezius (trah-pe'ze-us) (*trapezion* = irregular four-sided figure)	Most superficial muscle of posterior thorax; flat, and triangular in shape; upper fibers run inferiorly to scapula; middle fibers run horizontally to scapula; lower fibers run superiorly to scapula	O—occipital bone, ligamentum nuchae, and spines of C_7 and all thoracic vertebrae I—a continuous insertion along acromion and spine of scapula and lateral third of clavicle	Stabilizes, raises, retracts, and rotates scapula; middle fibers retract (adduct) scapula; superior fibers elevate scapula or can help extend head when scapula is fixed; inferior fibers depress scapula (and shoulder)	Accessory nerve (cranial nerve XI); C_3 and C_4
Levator scapulae (skap'u-le) (*levator* = raises)	Located at back and side of neck, deep to trapezius; thick, straplike muscle	O—transverse processes of C_1–C_4 I—medial border of the scapula, superior to the spine	Elevates/adducts scapula in concert with superior fibers of trapezius; tilts glenoid cavity downward; when scapula is fixed, flexes neck to same side	Cervical spinal nerves and dorsal scapular nerve (C_3–C_5)
Rhomboids (rom'boidz)—major and minor (*rhomboid* = diamond shaped)	Two rectangular muscles lying deep to trapezius and inferior to levator scapulae; rhomboid minor is the more superior muscle	O—spinous processes of C_7 and T_1 (minor) and spinous processes of T_2–T_5 (major) I—medial border of scapula	Act together (and with middle trapezius fibers) to retract scapula, thus "squaring shoulders"; rotate scapulae so that glenoid cavity rotates downward (as when arm is lowered against resistance; e.g., paddling a canoe); stabilize scapula	Dorsal scapular nerve (C_4 and especially C_5)

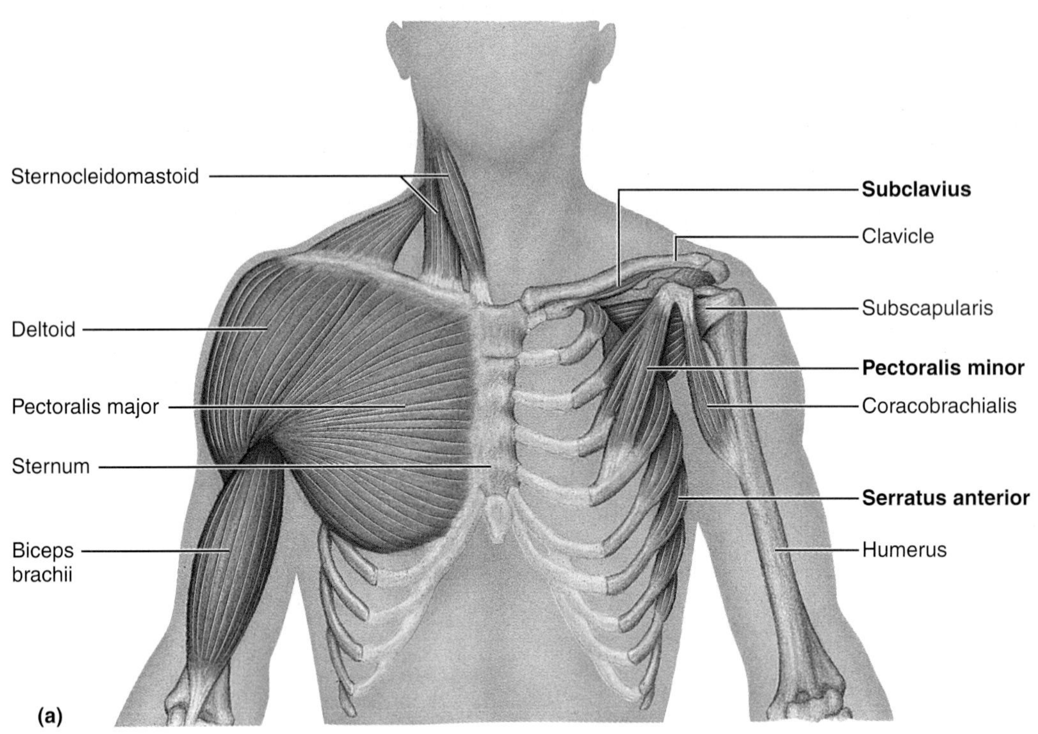

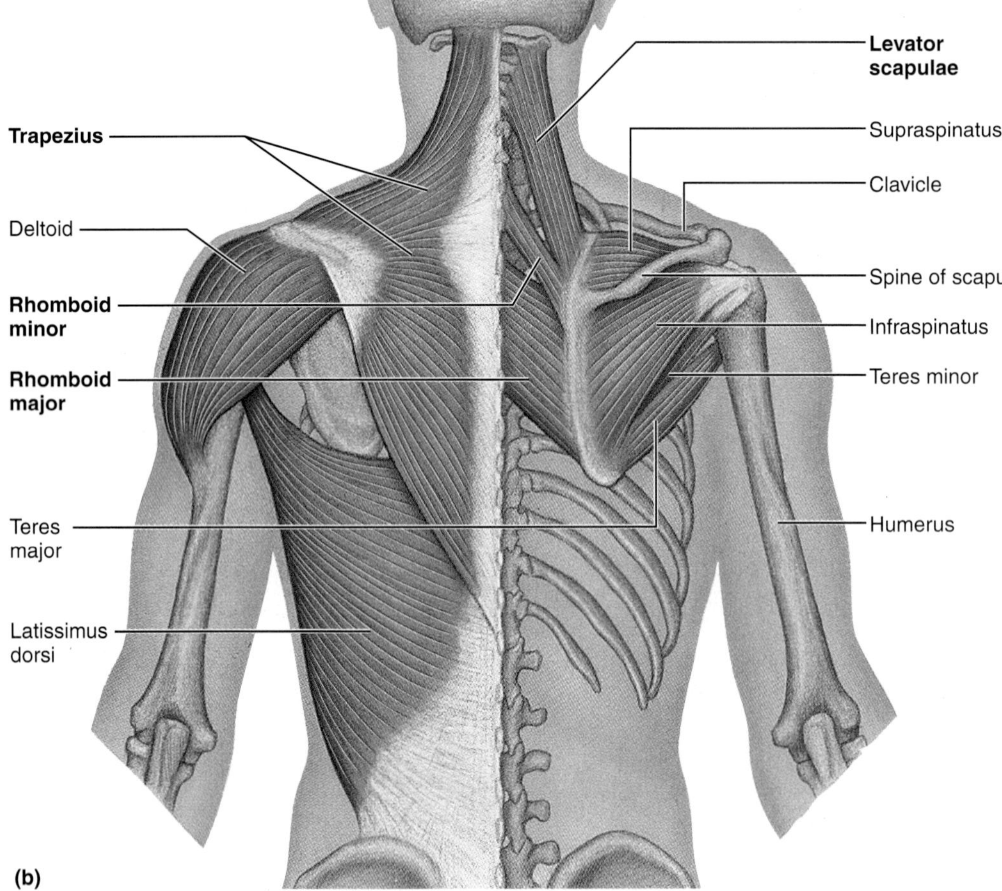

FIGURE 11.13 Superficial muscles of the anterior and posterior thorax and shoulder acting on the scapula and arm. **(a)** Anterior view. The superficial muscles, which cause arm movements, are shown at the left. These muscles have been removed at the right to show the muscles that stabilize or move the pectoral girdle. **(b)** Posterior view. The superficial muscles of the back are shown for the left side of the body, but are removed at the right to reveal the deeper muscles acting on the scapula and the rotator cuff muscles that help stabilize the shoulder joint.

Muscle Gallery: Table 11.9 Muscles Crossing the Shoulder Joint: Movements of the Arm (Humerus) (Figure 11.14)

Recall that the shoulder joint is the most flexible joint in the body but pays the price of instability. Many muscles cross each shoulder joint to insert on the humerus. All muscles that act on the humerus originate from the pectoral girdle (although two of these muscles, the latissimus dorsi and pectoralis major, primarily originate from the axial skeleton).

As we consider the arm movements, remember that the term *arm* refers to the *upper* arm. Of the nine muscles covered here, the three largest ones are powerful prime movers: the **pectoralis major, latissimus dorsi,** and **deltoid** (Figure 11.14a and b). The other six are synergists and fixators. Four of these were introduced on p. 227 as muscles of the *rotator cuff:* **supraspinatus, infraspinatus, subscapularis,** and **teres minor.** This cuff reinforces the capsule of the shoulder joint to prevent dislocation of the humerus. The remaining two muscles are the **teres major** and **coracobrachialis.**

Generally speaking, muscles that originate *anterior* to the shoulder joint *flex* the arm (lift it anteriorly). These flexors include the pectoralis major, the anterior fibers of the deltoid, and the coracobrachialis. Muscles that originate *posterior* to the shoulder joint, by contrast, tend to *extend* the arm. These extensors include the latissimus dorsi, the posterior fibers of deltoid, and the teres major. The middle region of the deltoid muscle is the prime *abductor* of the arm and extends over the superior and lateral side of the humerus to pull it laterally. The main arm *adductors* are the pectoralis major anteriorly and the latissimus dorsi posteriorly. *Lateral* and *medial rotation* of the arm are primarily brought about by the small muscles.

The interactions among these nine muscles are complex, and each contributes to several movements. A summary of their actions is given on page 307 in Table 11.12 (Part 1).

Muscle	Description	Origin (O) and Insertion (I)	Action	Nerve Supply
Pectoralis major (pek″to-ra′lis ma′jer) (*pect* = breast, chest; *major* = larger)	Large, fan-shaped muscle covering upper portion of chest; forms anterior axillary fold; divided into clavicular and sternal parts	O—sternal end of clavicle, sternum, cartilage of ribs 1–6 (or 7), and aponeurosis of external oblique muscle I—fibers converge to insert by a short tendon into greater tubercle of humerus	Prime mover of arm flexion; rotates arm medially; adducts arm against resistance; with scapula (and arm) fixed, pulls rib cage upward, and thus can help in climbing, throwing, pushing, and in forced inspiration	Lateral and medial pectoral nerves (C_5–C_8, and T_1)
Latissimus dorsi (lah-tis′ĭ-mus dor′si) (*latissimus* = widest; *dorsi* = back)	Broad, flat, triangular muscle of lower back (lumbar region); extensive superficial origins; covered by trapezius superiorly; contributes to the posterior wall of axilla	O—indirect attachment via thoracolumbar fascia into spines of lower six thoracic vertebrae, lumbar vertebrae, lower 3 to 4 ribs, and iliac crest; also from scapula's inferior angle I—spirals around teres major to insert in floor of intertubercular groove of humerus	Prime mover of arm extension; powerful arm adductor; medially rotates arm at shoulder; depresses scapula; because of its power in these movements, it plays an important role in bringing the arm down in a power stroke, as in striking a blow, hammering, swimming, and rowing; with arms reaching overhead, it pulls the rest of the body upward and forward	Thoracodorsal nerve (C_6–C_8)

➤

Muscle	Description	Origin (O) and Insertion (I)	Action	Nerve Supply
Deltoid (del'toid) (*delta* = triangular)	Thick, multipennate muscle forming rounded shoulder muscle mass; responsible for roundness of shoulder; a site commonly used for intramuscular injection, particularly in males, where it tends to be quite fleshy	O—embraces insertion of the trapezius; lateral third of clavicle; acromion and spine of scapula I—deltoid tuberosity of humerus	Prime mover of arm abduction when all its fibers contract simultaneously; antagonist of pectoralis major and latissimus dorsi, which adduct the arm; if only anterior fibers are active, can act powerfully in flexion and medial rotation of humerus, and is therefore synergist of pectoralis major; if only posterior fibers are active, causes extension and lateral rotation of arm; active during rhythmic arm swinging movements during walking	Axillary nerve (C_5 and C_6)
Subscapularis (sub-scap"u-lar'is) (*sub* = under; *scapular* = scapula)	Forms part of posterior wall of axilla; tendon of insertion passes in front of shoulder joint; a rotator cuff muscle	O—subscapular fossa of scapula I—lesser tubercle of humerus	Chief medial rotator of humerus; assisted by pectoralis major; helps to hold head of humerus in glenoid cavity, thereby stabilizing shoulder joint	Subscapular nerves (C_5–C_7)
Supraspinatus (soo"prah-spi-nah'tus) (*supra* = above, over; *spin* = spine)	Named for its location on posterior aspect of scapula; deep to trapezius; a rotator cuff muscle	O—supraspinous fossa of scapula I—superior part of greater tubercle of humerus	Stabilizes shoulder joint; helps to prevent downward dislocation of humerus, as when carrying a heavy suitcase; assists in abduction	Suprascapular nerve
Infraspinatus (in"frah-spi-nah'tus) (*infra* = below)	Partially covered by deltoid and trapezius; named for its scapular location; a rotator cuff muscle	O—infraspinous fossa of scapula I—greater tubercle of humerus posterior to insertion of supraspinatus	Helps to hold head of humerus in glenoid cavity, stabilizing the shoulder joint; rotates humerus laterally	Suprascapular nerve
Teres minor (te'rēz) (*teres* = round; *minor* = lesser)	Small, elongated muscle; lies inferior to infraspinatus and may be inseparable from that muscle; a rotator cuff muscle	O—lateral border of dorsal scapular surface I—greater tubercle of humerus inferior to infraspinatus insertion	Same action(s) as infraspinatus muscle	Axillary nerve
Teres major	Thick, rounded muscle; located inferior to teres minor; helps to form posterior wall of axilla (along with latissimus dorsi and subscapularis)	O—posterior surface of scapula at inferior angle I—crest of lesser tubercle on anterior humerus; insertion tendon fused with that of latissimus dorsi	Extends, medially rotates, and adducts humerus; synergist of latissimus dorsi	Lower subscapular nerve
Coracobrachialis (kor"ah-ko-bra"ke-al'is) (*coraco* = coracoid; *brachi* = arm)	Small, cylindrical muscle	O—coracoid process of scapula I—medial surface of humerus shaft	Flexion and adduction of the humerus; synergist of pectoralis major	Musculocutaneous nerve (C_5–C_7)

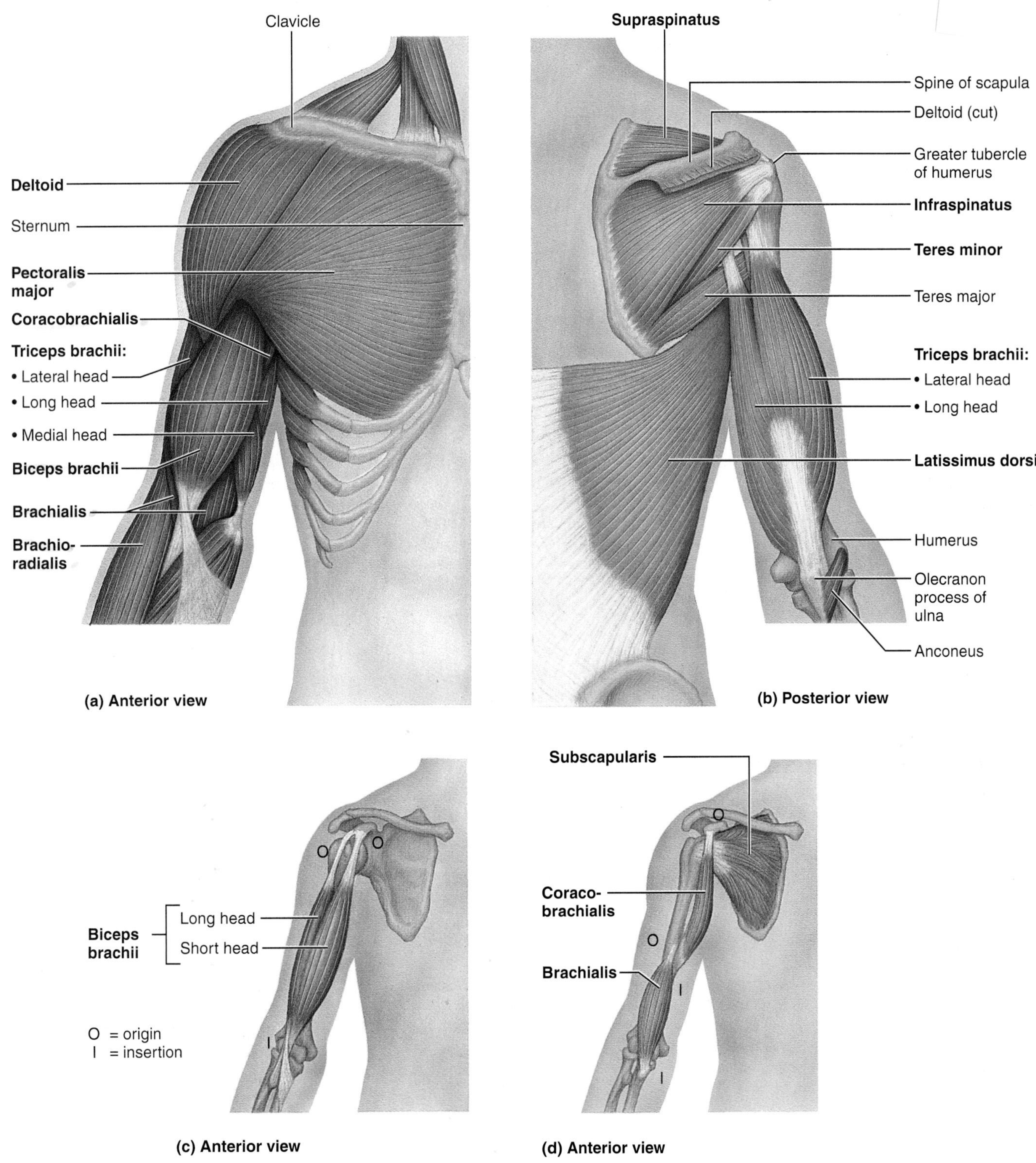

FIGURE 11.14 Muscles crossing the shoulder and elbow joint, which cause movements of the arm and forearm, respectively. **(a)** Superficial muscles of the anterior thorax, shoulder, and arm, anterior view. **(b)** The triceps brachii muscle of the posterior arm, shown in relation to the deep scapular muscles. The deltoid muscle of the shoulder has been removed. **(c)** The biceps brachii muscle of the anterior arm, shown in isolation. **(d)** The brachialis muscle arising from the humerus, and the coracobrachialis and subscapularis muscles arising from the scapula, shown in isolation.

Muscle Gallery: Table 11.10 Muscles Crossing the Elbow Joint Flexion and Extension of the Forearm (Figure 11.14)

This table focuses on muscles that lie in the arm but move the forearm. These muscles cross the elbow joint to insert on the forearm bones, which they flex and extend. Walls of fascia divide the arm into two muscle compartments: the *posterior extensors* and *anterior flexors*. The main *extensor* of the forearm is the **triceps brachii.**

The anterior muscles of the arm *flex* the forearm. In order of decreasing strength, these are the **brachialis, biceps brachii,** and **brachioradialis.** The biceps brachii and brachialis insert on the radius and ulna, respectively, and contract together. The brachioradialis (Figure 11.15a) is a weak flexor of the forearm and actually lies in the forearm more than in the arm.

The biceps brachii not only flexes the forearm but also *supinates* it by rotating the radius at the proximal radioulnar joint. This muscle cannot flex the forearm without also supinating it, so it is ineffective when one lifts a heavy object with a pronated hand that must stay pronated. (This is why doing chin-ups with palms facing anteriorly is harder than with palms facing posteriorly.)

The biceps and triceps also cross the shoulder joint, but they cause only weak movements at the shoulder.

The actions of the muscles in this table are summarized on page 307 in Table 11.12 (Part II).

Muscle	Description	Origin (O) and Insertion (I)	Action	Nerve Supply
POSTERIOR MUSCLES				
Triceps brachii (tri'seps bra'ke-i) (*triceps* = three heads; *brachi* = arm)	Large fleshy muscle; the only muscle of posterior compartment of arm; three-headed origin; long and lateral heads lie superficial to medial head	O—long head: infraglenoid tubercle of scapula; lateral head: posterior shaft of humerus; medial head: posterior humeral shaft distal to radial groove I—by common tendon into olecranon process of ulna	Powerful forearm extensor (prime mover, particularly medial head); antagonist of forearm flexors; long and lateral heads mainly active in extension against resistance; long head tendon may help stabilize shoulder joint and assist in arm adduction	Radial nerve (C_6–C_8)
Anconeus (an-ko'ne-us) (*ancon* = elbow) (also see Figure 11.16)	Short triangular muscle; closely associated (blended) with distal end of triceps on posterior humerus	O—lateral epicondyle of humerus I—lateral aspect of olecranon process of ulna	Abducts ulna during forearm pronation; synergist of triceps brachii in elbow extension	Radial nerve
ANTERIOR MUSCLES				
Biceps brachii (bi'seps) (*biceps* = two heads)	Two-headed fusiform muscle; bellies unite as insertion point is approached; tendon of long head helps stabilize shoulder joint	O—short head: coracoid process; long head: tubercle above glenoid cavity and lip of glenoid cavity; tendon of long head runs within capsule and descends into intertubercular groove of humerus I—by common tendon into radial tuberosity	Flexes ebow joint and supinates forearm; these actions usually occur at same time (e.g., when you open a bottle of wine it turns the corkscrew and pulls the cork); weak flexor of arm at shoulder	Musculo-cutaneous nerve (C_5 and C_6)
Brachialis (bra'ke-al-is)	Strong muscle that is immediately deep to biceps brachii on distal humerus	O—front of distal humerus; embraces insertion of deltoid muscle I—coronoid process of ulna and capsule of elbow joint	A major forearm flexor (lifts ulna as biceps lifts the radius)	Musculo-cutaneous nerve
Brachioradialis (bra"ke-o-ra"de-a'lis) (*radi* = radius, ray) (also see Figure 11.15a)	Superficial muscle of lateral forearm; forms lateral boundary of cubital fossa; extends from distal humerus to distal forearm	O—lateral supracondylar ridge at distal end of humerus I—base of styloid process of radius	Synergist in forearm flexion; acts to best advantage when forearm is partially flexed and semi-pronated; stabilizes the elbow during rapid flexion *and* extension	Radial nerve (an important exception: the radial nerve typically serves extensor muscles)

Muscle Gallery: Table 11.11 Muscles of the Forearm: Movements of the Wrist, Hand, and Fingers (Figure 11.15 and 11.16)

The many muscles in the forearm perform several basic functions: Some move the hand at the wrist, some move the fingers, and a few help supinate and pronate the forearm. Most of these muscles are fleshy proximally and have long tendons distally, almost all of which insert in the hand. At the wrist, these tendons are anchored firmly by band-like thickenings of deep fascia called **flexor** and **extensor retinacula** ("retainers"). These "wrist bands" keep the tendons from jumping outward when tensed. Crowded together in the wrist and palm, the muscle tendons are surrounded by slippery tendon sheaths that minimize friction as they slide against one another.

Many forearm muscles arise from the distal humerus; thus, they cross the elbow joint as well as the wrist and finger joints. However, their actions on the elbow are slight. At the *wrist* joint, the forearm muscles bring about flexion, extension, abduction, and adduction of the hand, but at the *finger* joints these muscles mostly just flex and extend the fingers. (The other movements of fingers are brought about by small muscles in the hand itself.)

Sheaths of fascia divide the forearm muscles into two main compartments, an *anterior flexor compartment* and a *posterior extensor compartment*. Both of these compartments, in turn, contain a superficial and a deep muscle layer. Most flexors in the anterior compartment originate from a common tendon on the medial epicondyle of the humerus. The flexors are innervated by two nerves that descend through the anterior forearm, the median and ulnar nerves—especially by the median nerve. Two of the anterior compartment muscles are not flexors but pronators, the **pronator teres** and **pronator quadratus.**

In general, the posterior compartment muscles extend the hand and fingers. One exception is the **supinator,** which helps supinate the forearm. Another exception is the brachioradialis (see Table 11.10), which flexes the forearm at the elbow. Many of the posterior compartment muscles arise by a common tendon from the lateral epicondyle of the humerus. All are innervated by the radial nerve, the main nerve on the posterior aspect of the forearm.

Most muscles that move the palm and fingers are located in the forearm, not in the hand itself. The hand must perform so many different movements that it could not possibly contain all of its own muscles: If it did, it would be as large and clumsy as a baseball glove. Instead, the fingers are largely "operated" by tendons originating in the forearm, like a wooden puppet operated by strings. Still, small muscles do occur inside the hand, where they control our most delicate finger movements. These are described in Table 11.13.

The actions of the forearm muscles are summarized in Table 11.12 (Part III).

Clinical Applications

TENNIS ELBOW Tennis elbow is a tenderness due to trauma or overuse of the tendon of origin of the forearm extensors at the lateral epicondyle of the humerus. This condition is caused (and subsequently aggravated) when these muscles contract forcefully to extend the hand at the wrist—as in executing a tennis backhand or lifting snow with a shovel. Despite its name, tennis elbow involves neither the elbow joint nor the olecranon process, and most cases result from work activities rather than sports.

Muscle	Description	Origin (O) and Insertion (I)	Action	Nerve Supply
PART I: ANTERIOR MUSCLES (FIGURE 11.15)	These eight muscles of the anterior fascial compartment are listed from the lateral to the medial aspect. Most arise from a common flexor tendon attached to the medial epicondyle of the humerus and have additional origins as well. Most of the tendons of insertion of these flexors are held in place at the wrist by a thickening of deep fascia called the *flexor retinaculum.*			
SUPERFICIAL MUSCLES				
Pronator teres (pro′na′tor te′rēz) (*pronation* = turning palm posteriorly, or down; *teres* = round)	Two-headed muscle; seen in superficial view between proximal margins of brachioradialis and flexor carpi radialis; forms medial boundary of cubital fossa	O—medial epicondyle of humerus; coronoid process of ulna I—by common tendon into lateral radius, midshaft	Pronates forearm; weak flexor of elbow	Median nerve
Flexor carpi radialis (flek′sor kar′pe ra″de-a′lis) (*flex* = decrease angle between two bones; *carpi* = wrist; *radi* = radius)	Runs diagonally across forearm; midway, its fleshy belly is replaced by a flat tendon that becomes cord-like at wrist	O—medial epicondyle of humerus I—base of second and third metacarpals; insertion tendon easily seen and provides guide to position of radial artery (used for pulse taking) at wrist	Powerful flexor of wrist; abducts hand; weak synergist of elbow flexion	Median nerve
Palmaris longus (pahl-ma′ris lon′gus) (*palma* = palm; *longus* = long)	Small fleshy muscle with a long insertion tendon; often absent; may be used as guide to find median nerve that lies lateral to it at wrist	O—medial epicondyle of humerus I—fascia of palm (palmar aponeurosis)	Weak wrist flexor; tenses skin and fascia of palm during hand movements; weak synergist for elbow flexion	Median nerve

Muscle	Description	Origin (O) and Insertion (I)	Action	Nerve Supply
Flexor carpi ulnaris (ul-na′ris) (*ulnar* = ulna)	Most medial muscle of this group; two-headed; ulnar nerve lies lateral to its tendon	O—medial epicondyle of humerus; olecranon process and posterior surface of ulna I—pisiform and hamate bones and base of fifth metacarpal	Powerful flexor of wrist; also adducts hand in concert with extensor carpi ulnaris (posterior muscle); stabilizes wrist during finger extension	Ulnar nerve (C_7 and C_8)
Flexor digitorum superficialis (dĭ″ji-tor′um soo″per-fish″e-a′lis) (*digit* = finger, toe; *superficial* = close to surface)	Two-headed muscle; more deeply placed (therefore, actually forms an intermediate layer); overlain by muscles above but visible at distal end of forearm	O—medial epicondyle of humerus, coronoid process of ulna; shaft of radius I—by four tendons into middle phalanges of fingers 2–5	Flexes wrist and middle phalanges of fingers 2–5; the important finger flexor when speed and flexion against resistance are required	Median nerve (C_7, C_8, and T_1)

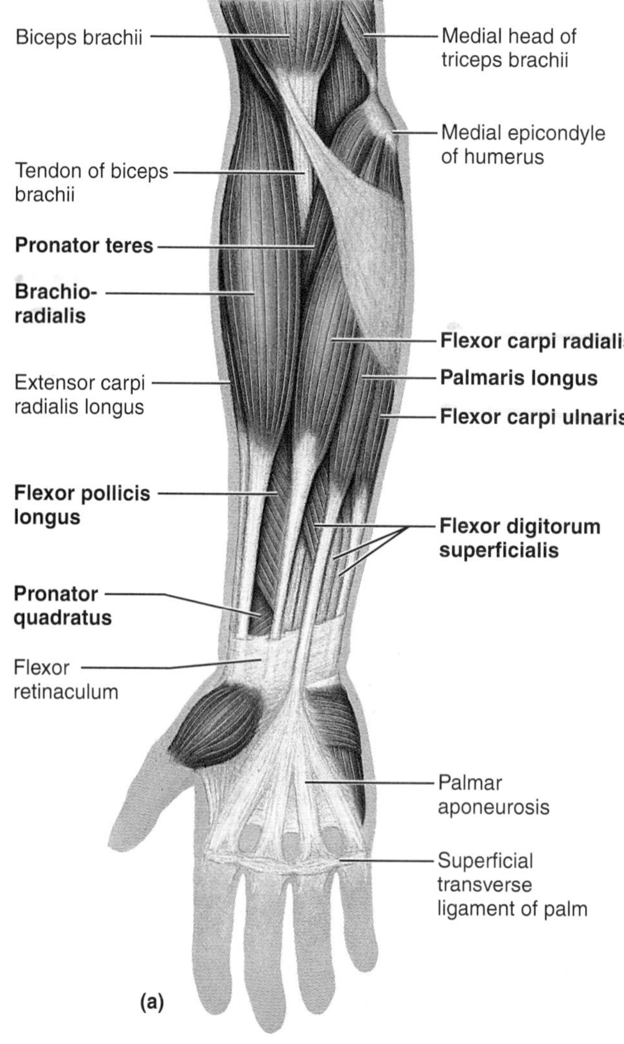

Muscle	Description	Origin (O) and Insertion (I)	Action	Nerve Supply
DEEP MUSCLES				
Flexor pollicis longus (pah'lĭ-kis) (*pollix* = thumb)	Partly covered by flexor digitorum superficialis; lies lateral and parallel to flexor digitorum profundus	O—anterior surface of radius and interosseous membrane I—distal phalanx of thumb	Flexes distal phalanx of thumb	Branch of median nerve (C_8 and T_1)
Flexor digitorum profundus (pro-fun'dus) (*profund* = deep)	Extensive origin; overlain entirely by flexor digitorum superficialis	O—anteromedial surface of ulna and interosseous membrane I—by four tendons into distal phalanges of fingers 2–5	Slow-acting flexor of any or all fingers; assists in flexing wrist; the only muscle that can flex distal interphalangeal joints	Medial half by ulnar nerve; lateral half by median nerve

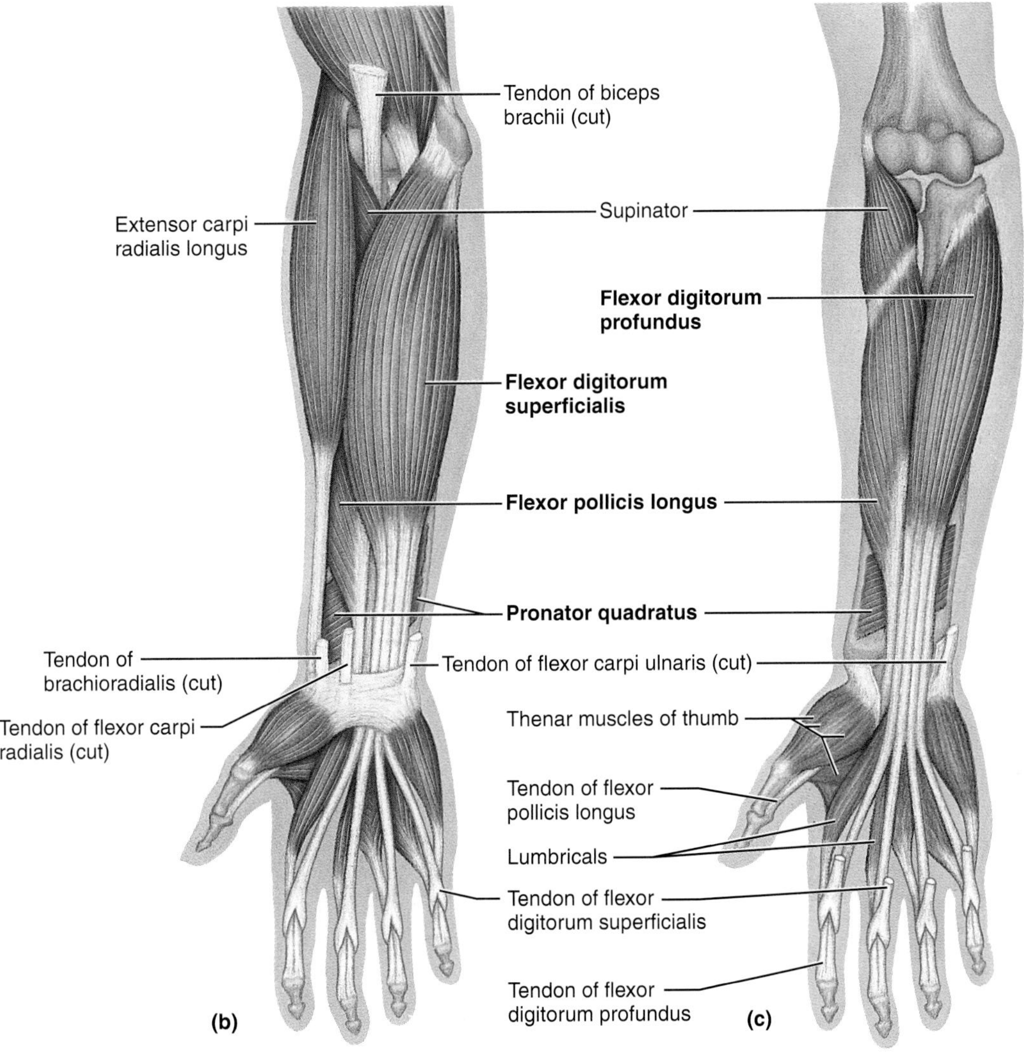

FIGURE 11.15 Muscles of the anterior fascial compartment of the forearm acting on the wrist and fingers. **(a)** Superficial view of the muscles of the right forearm and hand. **(b)** The brachioradialis, flexors carpi radialis and ulnaris, and palmaris longus muscles have been removed to reveal the somewhat deeper flexor digitorum superficialis. **(c)** Deep muscles of the anterior compartment. Superficial muscles have been removed. The lumbricals and thenar muscles of the hand (intrinsic hand muscles) are also illustrated.

Muscle	Description	Origin (O) and Insertion (I)	Action	Nerve Supply
ANTERIOR MUSCLES (Figure 11.15), continued				
Pronator quadratus (kwod-ra'tus) (*quad* = square, four-sided)	Deepest muscle of distal forearm; passes downward and laterally; only muscle that arises solely from ulna and inserts solely into radius	O—distal portion of anterior ulnar shaft I—distal surface of anterior radius	Prime mover of forearm pronation; acts with pronator teres; also helps hold ulna and radius together	Median nerve (C_8 and T_1)
PART II: POSTERIOR MUSCLES (FIGURE 11.16)	These muscles of the posterior fascial compartment are listed from the lateral to the medial aspect. They are all innervated by the radial nerve or its branches. More than half of the posterior compartment muscles arise from a common extensor origin tendon attached to the posterior surface of the lateral epicondyle of the humerus and adjacent fascia. The extensor tendons are held in place at the posterior aspect of the wrist by the *extensor retinaculum* (not illustrated), which prevents "bowstringing" of these tendons when the wrist is hyperextended. The extensor muscles of the fingers end in a broad hood over the dorsal side of the digits, the *extensor expansion.*			
SUPERFICIAL MUSCLES				
Brachioradialis (Table 11.10)	This most-anterior muscle of the posterior fascial compartment is seen on the front of the forearm. Because it originates well up in the arm, it is covered in Table 11.10 (see Figure 11.15a)	See Table 11.10	See Table 11.10	See Table 11.10
Extensor carpi radialis longus (ek-sten'sor) (*extend* = increase angle between two bones)	Parallels brachioradialis on lateral forearm, and may blend with it	O—lateral supracondylar ridge of humerus I—base of second metacarpal	Extends wrist in conjunction with the extensor carpi ulnaris and abducts wrist in conjunction with the flexor carpi radialis	Radial nerve (C_6 and C_7)
Extensor carpi radialis brevis (brĕ'vis) (*brevis* = short)	Somewhat shorter than extensor carpi radialis longus and lies deep to it	O—lateral epicondyle of humerus I—base of third metacarpal	Extends and abducts wrist; acts synergistically with extensor carpi radialis longus to steady wrist during finger flexion	Deep branch of radial nerve
Extensor digitorum	Lies medial to extensor carpi radialis brevis; a detached portion of this muscle, called *extensor digiti minimi,* extends little finger	O—lateral epicondyle of humerus I—by four tendons into extensor expansions and distal phalanges of fingers 2–5	Prime mover of finger extension; extends wrist; can abduct (flare) fingers	Posterior interosseous nerve, a branch of radial nerve (C_5 and C_6)

Muscle	Description	Origin (O) and Insertion (I)	Action	Nerve Supply
SUPERFICIAL MUSCLES, continued				
Extensor carpi ulnaris	Most medial of superficial posterior muscles; long, slender muscle	O—lateral epicondyle of humerus and posterior border of ulna I—base of fifth metacarpal	Extends wrist in conjuction with the extensor carpi radialis and adducts wrist in conjunction with flexor carpi ulnaris	Posterior interosseous nerve
DEEP MUSCLES				
Supinator (soo″pĭ-na′tor) (*supination* = turning palm anteriorly or upward)	Deep muscle at posterior aspect of elbow; largely concealed by superficial muscles	O—lateral epicondyle of humerus; proximal ulna I—proximal end of radius	Assists biceps brachii to forcibly supinate forearm; works alone in slow supi- ination; antagonist of pronator muscles	Posterior interosseous nerve
Abductor pollicis longus (ab-duk′tor) (*abduct* = movement away from median plane)	Lateral and parallel to extensor pollicis longus; just distal to supinator	O—posterior surface of radius and ulna; interosseous membrane I—base of first metacarpal and trapezium	Abducts and extends thumb	Posterior interosseous nerve
Extensor pollicis brevis and **longus**	Deep muscle pair with a common origin and action; overlain by extensor carpi ulnaris	O—dorsal shaft of radius and ulna; interosseous membrane I—base of proximal (brevis) and distal (longus) phalanx of thumb	Extends thumb	Posterior interosseous nerve
Extensor indicis (in′dĭ-kis) (*indicis* = index finger)	Tiny muscle arising close to wrist	O—posterior surface of distal ulna; interosseous membrane I—extensor expansion of index finger; joins tendon of extensor digitorum	Extends index finger and assists in wrist extension	Posterior interosseous nerve

➤

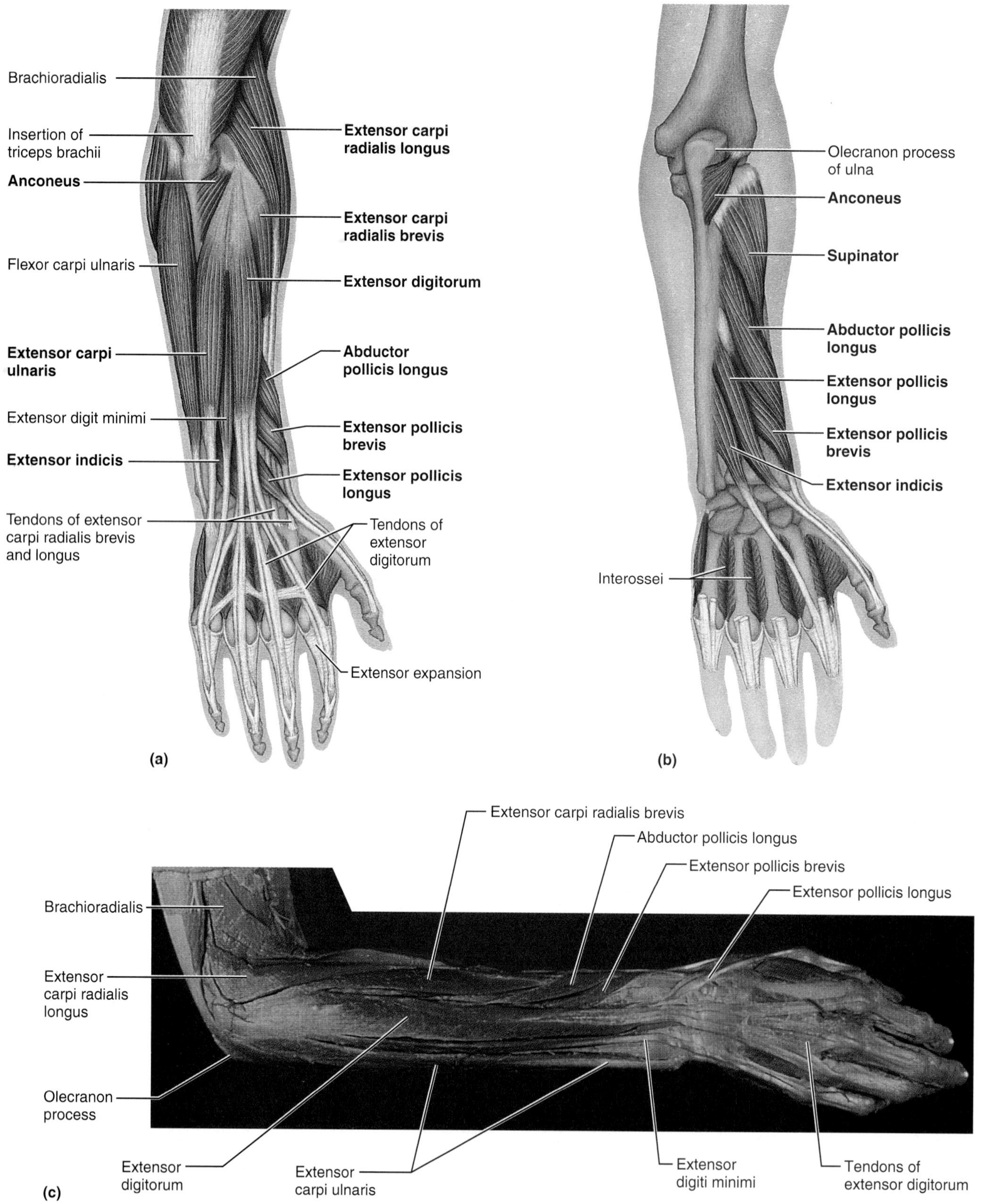

FIGURE 11.16 **Muscles of the posterior fascial compartment of the forearm acting on the wrist and fingers.** **(a)** Superficial muscles of the right forearm, posterior view. **(b)** Deep posterior muscles of the right forearm; superficial muscles have been removed. The interossei, the deepest layer of intrinsic hand muscles, are also illustrated. **(c)** Photograph of the posterior muscles of the right forearm.

Muscle Gallery: Table 11.12 Summary of Actions of Muscles Acting on the Arm, Forearm, and Hand (Figure 11.17)

PART I: MUSCLES ACTING ON THE ARM (HUMERUS) (PM = PRIME MOVER)	Actions at the Shoulder					
	Flexion	**Extension**	**Abduction**	**Adduction**	**Medial Rotation**	**Lateral Rotation**
Pectoralis major	X (PM)			X (PM)	X	
Latissimus dorsi		X (PM)		X (PM)	X	
Deltoid	X (PM) (anterior fibers)	X (PM) (posterior fibers)	X (PM)		X (anterior fibers)	X (posterior fibers)
Subscapularis					X (PM)	
Supraspinatus			X			
Infraspinatus						X (PM)
Teres minor				X (weak)		X (PM)
Teres major		X		X	X	
Coracobrachialis	X			X		
Biceps brachii	X					
Triceps brachii				X		

PART II: MUSCLES ACTING ON THE FOREARM	Actions			
	Elbow Flexion	**Elbow Extension**	**Pronation**	**Supination**
Biceps brachii	X (PM)			X
Triceps brachii		X (PM)		
Anconeus		X		
Brachialis	X (PM)			
Brachioradialis	X			
Pronator teres	X (weak)		X	
Pronator quadratus			X (PM)	
Supinator				X

PART III: MUSCLES ACTING ON THE WRIST AND FINGERS	Actions on the Wrist				Actions on the Fingers	
	Flexion	**Extension**	**Abduction**	**Adduction**	**Flexion**	**Extension**
Anterior Compartment:						
Flexor carpi radialis	X (PM)		X			
Palmaris longus	X (weak)					
Flexor carpi ulnaris	X (PM)			X		
Flexor digitorum superficialis	X (PM)				X	
Flexor pollicis longus					X (thumb)	
Flexor digitorum profundus	X				X	
Posterior Compartment:						
Extensor carpi radialis longus and brevis		X	X			
Extensor digitorum		X (PM)				X (and abducts)
Extensor carpi ulnaris		X		X		
Abductor pollicis longus			X		(abducts thumb)	
Extensor pollicis longus and brevis						X (thumb)
Extensor indicis						X (index finger)

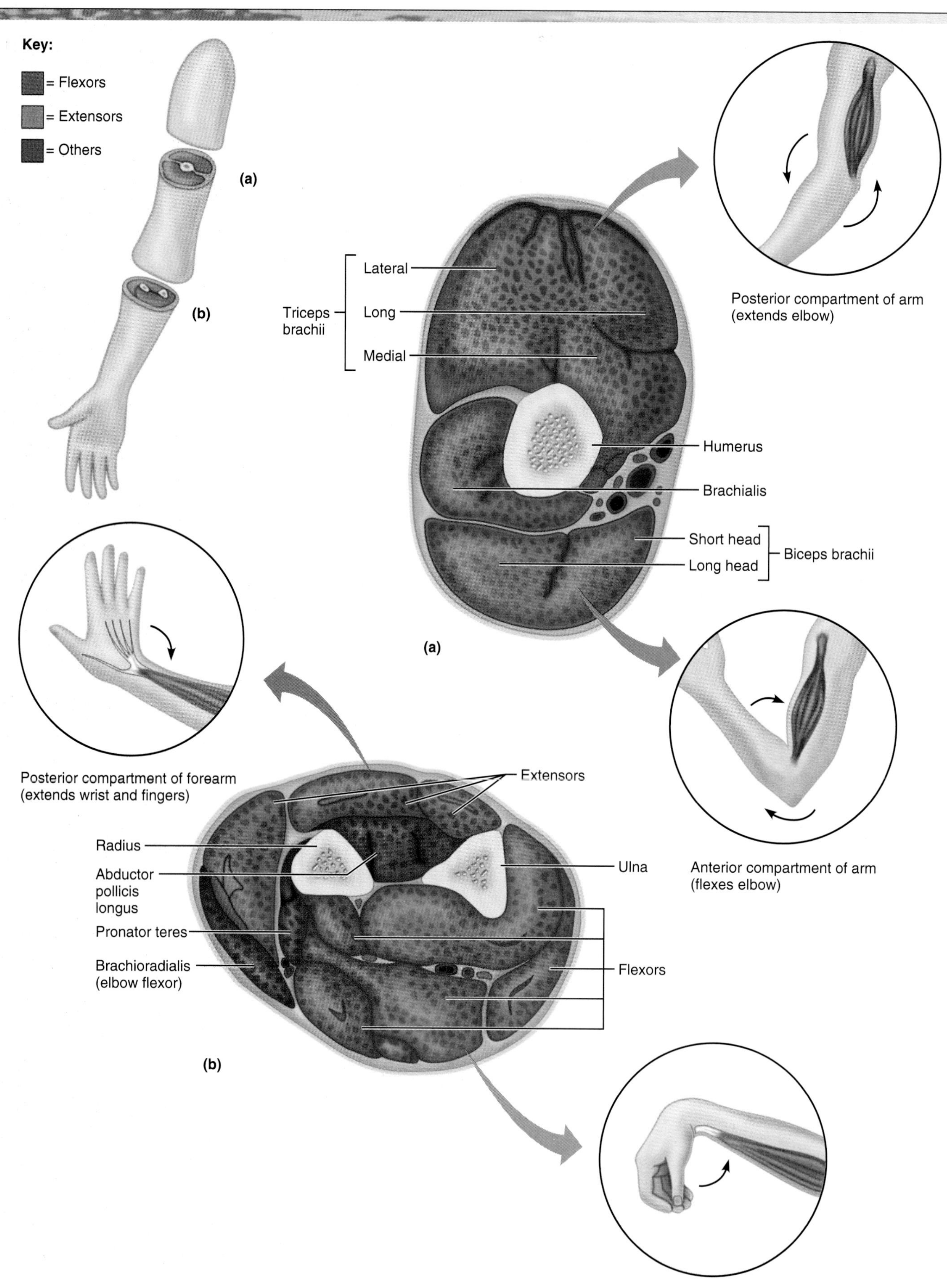

FIGURE 11.17 Summary of actions of muscles of the arm and forearm. (a) Muscles of the arm. (b) Muscles of the forearm.

Muscle Gallery: Table 11.13 Intrinsic Muscles of the Hand: Fine Movements of the Fingers (Figure 11.18)

This table considers the small muscles that lie entirely in the hand. All are in the palm, none on the hand's dorsal side. All move the metacarpals and fingers, but because they are small, they are weak. Therefore, they mostly control precise movements (such as threading a needle), leaving the powerful movements of the fingers ("power grip") to the forearm muscles. The intrinsic muscles include the main abductors and adductors of the fingers, as well as the muscles that produce the movement of opposition—moving the thumb toward the little finger—that enables one to grip objects in the palm (the handle of a hammer, for example). Many muscles in the palm are specialized in movement of the thumb, and surprisingly many move the little finger.

Movements of the thumb are defined differently from movements of other fingers, because the thumb lies at a right angle to the rest of the hand. The thumb flexes by bending medially along the palm, not by bending anteriorly, as do the other fingers. (Be sure to start with your hand in the anatomical position or this will not be clear!) The thumb straightens, or extends, by pointing laterally (as in hitchhiking), not posteriorly, as do the other fingers. To abduct the fingers is to splay them laterally, but to abduct the thumb is to point it anteriorly. Adduction of the thumb brings it back posteriorly.

The intrinsic muscles of the palm are divided into three groups: (1) those in the *thenar eminence* (ball of the thumb); (2) those in the *hypothenar eminence* (the ball of the little finger); and (3) muscles in the mid palm. The thenar and hypothenar muscles are almost mirror images of each other, each containing a small flexor, an abductor, and an opponens muscle. The midpalmar muscles are called **lumbricals** and **interossei.** We rely on these muscles to extend our fingers at the interphalangeal joints, a movement that the great finger extensors from the forearm cannot perform. The interossei are also the main abductors and adductors of the fingers.

Muscle	Description	Origin (O) and Insertion (I)	Action	Nerve Supply
THENAR MUSCLES IN BALL OF THUMB (the′nar) (*thenar* = palm)				
Abductor pollicis brevis (*pollex* = thumb)	Lateral muscle of thenar group; superficial	O—flexor retinaculum and nearby carpals I—lateral base of thumb's proximal phalanx	Abducts thumb (at carpometacarpal joint)	Median nerve (C_8 and T_1)
Flexor pollicis brevis	Medial and deep muscle of thenar group	O—flexor retinaculum and nearby carpals I—lateral side of base of proximal phalanx of thumb	Flexes thumb (at carpometacarpal and metacarpophalangeal joints)	Median (or occasionally ulnar) nerve (C_8 and T_1)
Opponens pollicis (o-pōn′enz) (*opponens* = opposition)	Deep to abductor pollicis brevis, on metacarpal I	O—flexor retinaculum and trapezium I—whole anterior side of metacarpal I	Opposition: moves thumb to touch tip of little finger	Median (or occasionally ulnar) nerve
Adductor pollicis	Fan-shaped with horizontal fibers; distal to other thenar muscles; oblique and transverse heads	O—capitate bone and bases of metacarpals 2–4; front of metacarpal 3 I—medial side of base of proximal phalanx of thumb	Adducts and helps to oppose thumb	Ulnar nerve (C_8 and T_1)
HYPOTHENAR MUSCLES IN BALL OF LITTLE FINGER				
Abductor digiti minimi (dĭ′jĭ-ti min′ĭ-mi) (*digiti minimi* = little finger)	Medial muscle of hypothenar group; superficial	O—pisiform bone I—medial side of proximal phalanx of little finger	Abducts little finger at metacarpophalangeal joint	Ulnar nerve
Flexor digiti minimi brevis	Lateral deep muscle of hypothenar group	O—hamate bone and flexor retinaculum I—same as abductor digiti minimi	Flexes little finger at metacarpophalangeal joint	Ulnar nerve
Opponens digiti minimi	Deep to abductor digiti minimi	O—same as flexor digiti minimi brevis I—most of length of medial side of metacarpal 5	Helps in opposition: brings metacarpal 5 toward thumb to cup the hand	Ulnar nerve

➤

Muscle	Description	Origin (O) and Insertion (I)	Action	Nerve Supply
MIDPALMAR MUSCLES				
Lumbricals (lum′brĭ-k′lz) (*lumbric* = earthworm)	Four worm-shaped muscles in palm, one to each finger (except thumb); odd because they originate from the tendons of another muscle	O—lateral side of each tendon of flexor digitorum profundus in palm I—lateral edge of extensor expansion on first phalanx of fingers 2–5	By pulling on the extensor expansion over the first phalanx, they flex fingers at metacarpophalangeal joints but extend fingers at interphalangeal joints	Median nerve (lateral two) and ulnar nerve (medial two)
Palmar interossei (interossei = between bones)	Four long, cone-shaped muscles in the spaces between the metacarpals; lie ventral to the dorsal interossei	O—the side of each metacarpal that faces the midaxis of the hand (metacarpal 3); but absent from metacarpal 3 I—extensor expansion on first phalanx of each finger (except finger 3), on side facing midaxis of hand	Adductors of fingers: pull fingers in toward third digit; act with lumbricals to extend fingers at interphalangeal joints and flex them at metacarpophalangeal joints	Ulnar nerve
Dorsal interossei	Four bipennate muscles filling spaces between the metacarpals; deepest palm muscles, also visible on dorsal side of hand (see Figure 11.16b)	O—sides of metacarpals I—extensor expansion over first phalanx of fingers 2–4 on side opposite midaxis of hand (finger 3), but on *both* sides of finger 3	Abduct (diverge) fingers; extend fingers at interphalangeal joints and flex them at metacarpophalangeal joints	Ulnar nerve

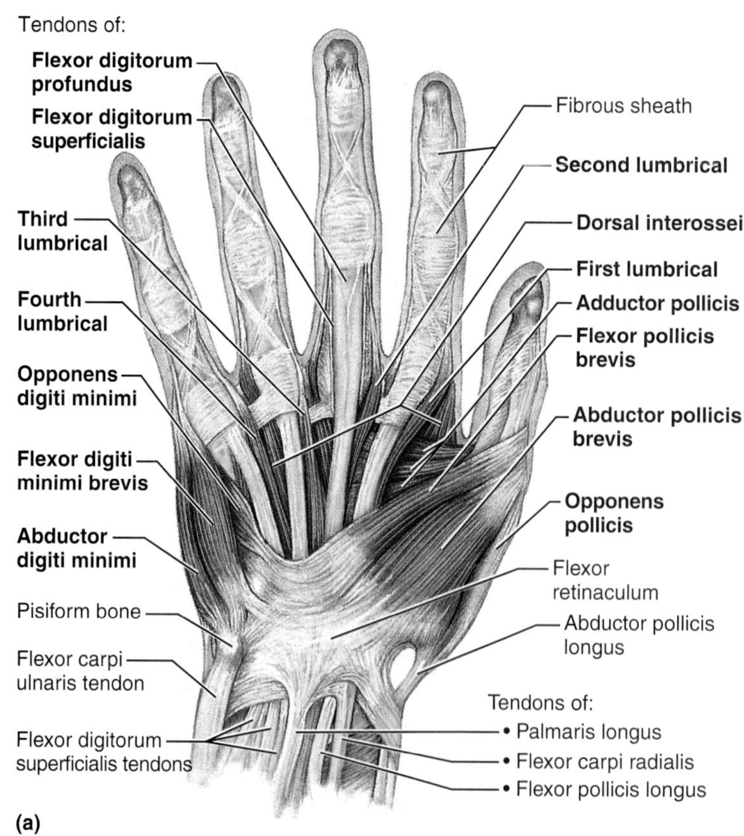

(a)

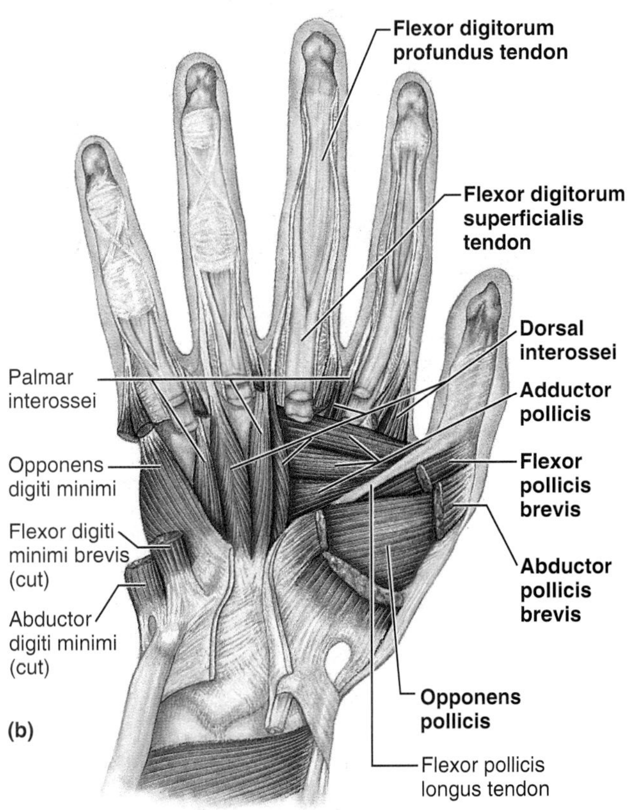

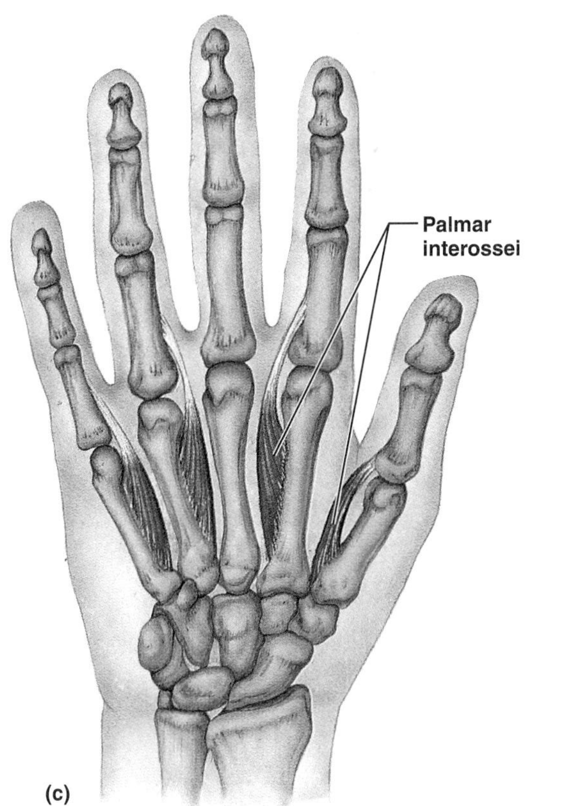

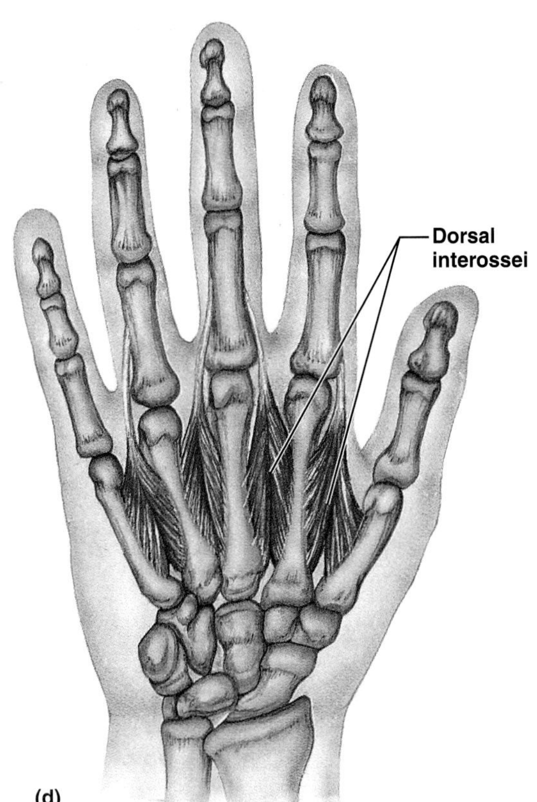

FIGURE 11.18 Hand muscles, ventral views of right hand.

Muscle Gallery: Table 11.14 Muscles Crossing the Hip and Knee Joints: Movements of the Thigh and Leg (Figures 11.19 and 11.20)

In this table we consider the muscles that span both the hip and knee joints and produce movements at both joints.

The *anterior muscles* of the hip and thigh tend to flex the femur at the hip and to extend the leg at the knee—producing the foreswing phase of walking. The *posterior muscles* of the hip and thigh, by contrast, mostly extend the thigh and flex the leg—the backswing phase of walking. A third group of muscles, the *medial*, or *adductor, muscles* on the medial aspect of the thigh, move the thigh only, not the leg. In the thigh, the anterior, posterior, and adductor muscles are separated by walls of fascia into *anterior, posterior,* and *medial compartments*. The deep fascia of the thigh (*fascia lata*) surrounds and encloses all three groups of muscles like a support stocking.

Movements at the hip joint are summarized first, then movements at the knee joint. The muscles that *flex* the thigh at the hip originate from the vertebral column and pelvis and pass anterior to the hip joint. These muscles include the **iliopsoas, tensor fasciae latae, rectus femoris,** and **pectineus** (Figure 11.19a).

The thigh *extensors,* by contrast, arise posterior to the hip joint and include the **gluteus maximus** and **hamstrings** (Figure 11.20a). As mentioned previously, the thigh **adductors** originate medial to the hip joint (Figure 11.19b). *Abduction* of the thigh is brought about mainly by the **gluteus medius** and **gluteus minimus,** buttocks muscles that lie lateral to the hip joint (Figure 11.20b). The adductors and abductors function during walking—not for moving the lower limb, but for shifting the trunk from side to side so that the body's center of gravity is always balanced directly over the limb that is on the ground. *Medial* and *lateral rotation* of the femur are accomplished by many different muscles.

At the knee joint, flexion and extension are the main movements. The only *extensor* of the leg at the knee is the four-part **quadriceps femoris** muscle from the anterior thigh (Figure 11.19a). The quadriceps is antagonized by the hamstrings of the posterior compartment (Figure 11.20a), which are the prime movers of knee *flexion*.

The actions of these muscles are summarized in Table 11.16.

Muscle	Description	Origin (O) and Insertion (I)	Action	Nerve Supply
PART I: ANTERIOR AND MEDIAL MUSCLES (FIGURE 11.19) ORIGIN ON THE PELVIS OR BACKBONE				
Iliopsoas (il″e-o-so′us)	Iliopsoas is a composite of two closely related muscles (iliacus and psoas major) whose fibers pass under the inguinal ligament (see Figure 11.11) to insert via a common tendon on the femur.			
• **Iliacus** (il-e-ak′us) (*iliac* = ilium)	Large fan-shaped, more lateral muscle	O—iliac fossa, ala of sacrum I—lesser trochanter of femur via iliopsoas tendon	Iliopsoas is the prime mover in thigh flexion and in flexing trunk (as during bowing)	Femoral nerve (L_2 and L_3)
• **Psoas** (so′us) **major** (*psoa* = loin muscle; *major* = larger)	Longer, thicker, more medial muscle of the pair. (Butchers refer to this muscle, in beef or pork, as the tenderloin)	O—by fleshy slips from transverse processes, bodies, and discs of lumbar vertebrae and T_{12} I—lesser trochanter of femur via iliopsoas tendon	As above; also causes lateral flexion of vertebral column; important postural muscle	Ventral rami L_1–L_3
Sartorius (sar-tor′e-us) (*sartor* = tailor)	Straplike superficial muscle running obliquely across anterior surface of thigh to knee; longest muscle in body; crosses both hip and knee joints	O—anterior superior iliac spine I—winds around medial aspect of knee and inserts into medial aspect of proximal tibia	Flexes, abducts, and laterally rotates thigh; flexes knee (weak); helps produce the cross-legged position in which tailors are often depicted	Femoral nerve
MUSCLES OF THE MEDIAL COMPARTMENT OF THE THIGH				
Adductors (ah-duk′torz)	Large muscle mass consisting of three muscles (longus, brevis, and magus) forming medial aspect of thigh; arise from inferior part of pelvis and insert at various levels on femur; all used in movements that press thighs together, as when astride a horse; important in pelvic tilting movements that occur during walking and in fixing the hip when the knee is flexed and the foot is off the ground; entire group innervated by obturator nerve. Strain or stretching of this muscle group is called a "pulled groin."			
• **Adductor longus** (*longus* = long)	Overlies middle aspect of adductor magnus; most anterior of adductor muscles	O—pubis near pubic symphysis I—linea aspera	Adducts, flexes, and medially rotates thigh	Obturator nerve (L_2–L_4)

Muscle	Description	Origin (O) and Insertion (I)	Action	Nerve Supply
MUSCLES OF THE MEDIAL COMPARTMENT OF THE THIGH, continued				
• **Adductor brevis** (*brevis* = short)	In contact with obturator externus muscle; largely concealed by adductor longus and pectineus	O—body and inferior ramus of pubis I—linea aspera above adductor longus	Adducts and medially rotates thigh	Obturator nerve
• **Adductor magnus** (mag'nus) (*adduct* = move toward midline; *magnus* = large)	A triangular muscle with a broad insertion; is a composite muscle that is part adductor and part hamstring in action	O—ischial and pubic rami and ischial tuberosity I—linea aspera, medial supracondylar line, and adductor tubercle of femur	Anterior part adducts and medially rotates and flexes thigh; posterior part is a synergist of hamstrings in thigh extension	Obturator nerve and sciatic nerve (L_2–L_4)
Pectineus (pek-tin'e-us) (*pecten* = comb)	Short, flat muscle; overlies adductor brevis on proximal thigh; abuts adductor longus medially	O—pectineal line of pubis (and superior ramus) I—a line from lesser trochanter to the linea aspera on posterior aspect of femur	Adducts, flexes, and medially rotates thigh	Femoral and sometimes obturator nerve
Gracilis (grah-sĭ'is) (*gracilis* = slender)	Long, thin, superficial muscle of medial thigh	O—inferior ramus and body of pubis and adjacent ischial ramus I—medial surface of tibia just inferior to its medial condyle	Adducts thigh, flexes and medially rotates leg, especially during walking	Obturator nerve
MUSCLES OF THE ANTERIOR COMPARTMENT OF THE THIGH				
Quadriceps femoris (kwod'rĭ-seps fem'o-lis)	Has four separate heads (*quadriceps* = four heads) that form the flesh of front and sides of thigh; these heads (rectus femoris, and lateral, medial, and intermediate vastus muscles) have a common insertion tendon, the **quadriceps tendon,** which inserts into the patella and then via the **patellar ligament** into tibial tuberosity. The quadriceps is a powerful knee extensor used in climbing, jumping, running, and rising from seated position; group is innervated by femoral nerve; the tone of quadriceps plays an important role in strengthening knee joint			
• **Rectus femoris** (rek'tus) (*rectus* = straight; *femoris* = femur)	Superficial muscle of anterior thigh; runs straight down thigh; longest head and only muscle of group to cross hip joint	O—anterior inferior iliac spine and superior margin of acetabulum I—patella and tibial tuberosity via patellar ligament	Extends knee and flexes thigh at hip	Femoral nerve (L_2–L_4)
• **Vastus lateralis** (vas'tus lat"er-a'lis) (*vastus* = large; *lateralis* = lateral)	Largest head of the group, forms lateral aspect of thigh; a common intramuscular injection site	O—greater trochanter, intertrochanteric line, linea aspera I—as for rectus femoris	Extends and stabilizes knee	Femoral nerve
• **Vastus medialis** (me"de-a'lis) (*medialis* = medial)	Forms inferomedial aspect of thigh	O—linea aspera, medial supracondylar line, intertrochanteric line I—as for rectus femoris	Extends knee; inferior fibers stabilize patella	Femoral nerve
• **Vastus intermedius** (in"ter-me'de-us) (*intermedius* = intermediate)	Obscured by rectus femoris; lies between vastus lateralis and vastus medialis on anterior thigh	O—anterior and lateral surfaces of proximal femur shaft I—as for rectus femoris	Extends knee	Femoral nerve
Tensor fasciae latae (ten'sor fă'she-e la'te) (*tensor* = to make tense; *fascia* = band; *lata* = wide)	Enclosed between fascia layers of anterolateral aspect of thigh; functionally associated with medial rotators and flexors of thigh	O—anterior aspect of iliac crest and anterior superior iliac spine I—iliotibial tract*	Flexes and abducts thigh; rotates thigh medially; steadies trunk on thigh by making iliotibial tract taut	Superior gluteal nerve (L_4 and L_5)

*The iliotibial tract is a thickened lateral portion of the *fascia lata* (the fascia that ensheathes all the muscles of the thigh). It extends as a tendinous band from the iliac crest to the knee (see Figure 11.20).

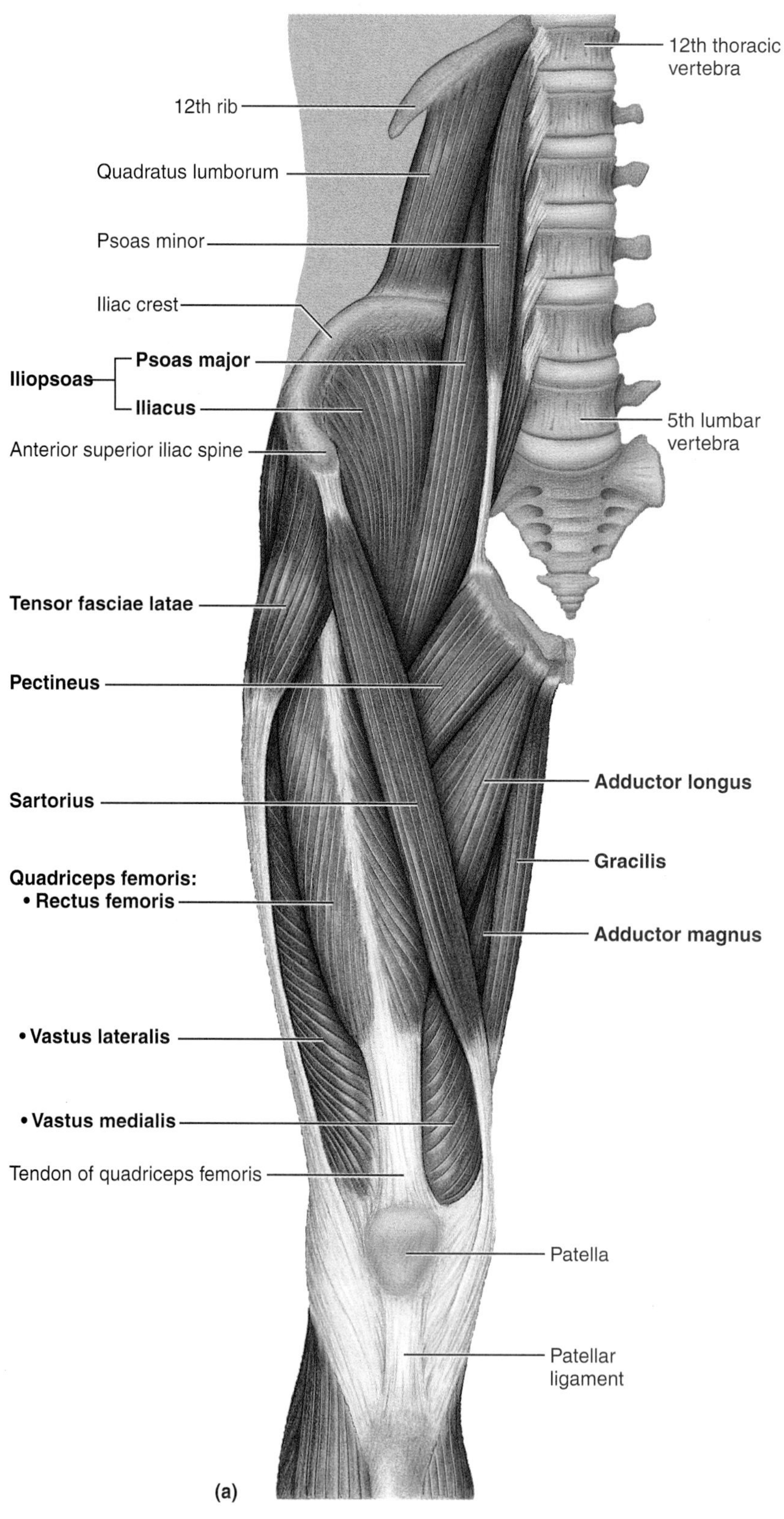
12th thoracic vertebra
12th rib
Quadratus lumborum
Psoas minor
Iliac crest
Iliopsoas
Psoas major
Iliacus
5th lumbar vertebra
Anterior superior iliac spine
Tensor fasciae latae
Pectineus
Adductor longus
Sartorius
Gracilis
Quadriceps femoris:
• Rectus femoris
Adductor magnus
• Vastus lateralis
• Vastus medialis
Tendon of quadriceps femoris
Patella
Patellar ligament
(a)

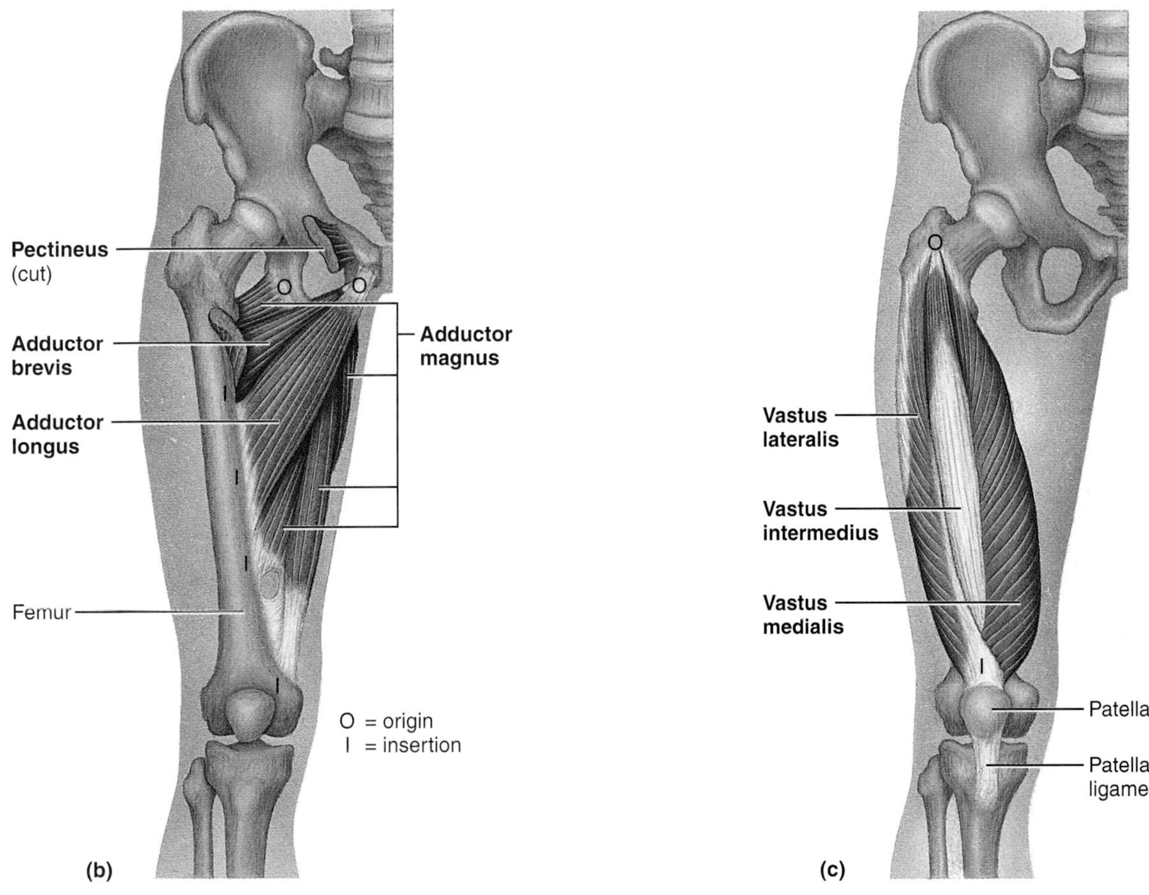

FIGURE 11.19 Anterior and medial muscles promoting movements of the thigh and leg. **(a)** Anterior view of the deep muscles of the pelvis and superficial muscles of the right thigh. **(b)** Adductor muscles of the medial compartment of the thigh. Other muscles have been removed so that the origins and insertions of the adductor muscles can be seen. **(c)** The vastus muscles of the quadriceps group. The rectus femoris muscle of the quadriceps group and surrounding muscles have been removed to reveal the attachments and extent of the vastus muscles.

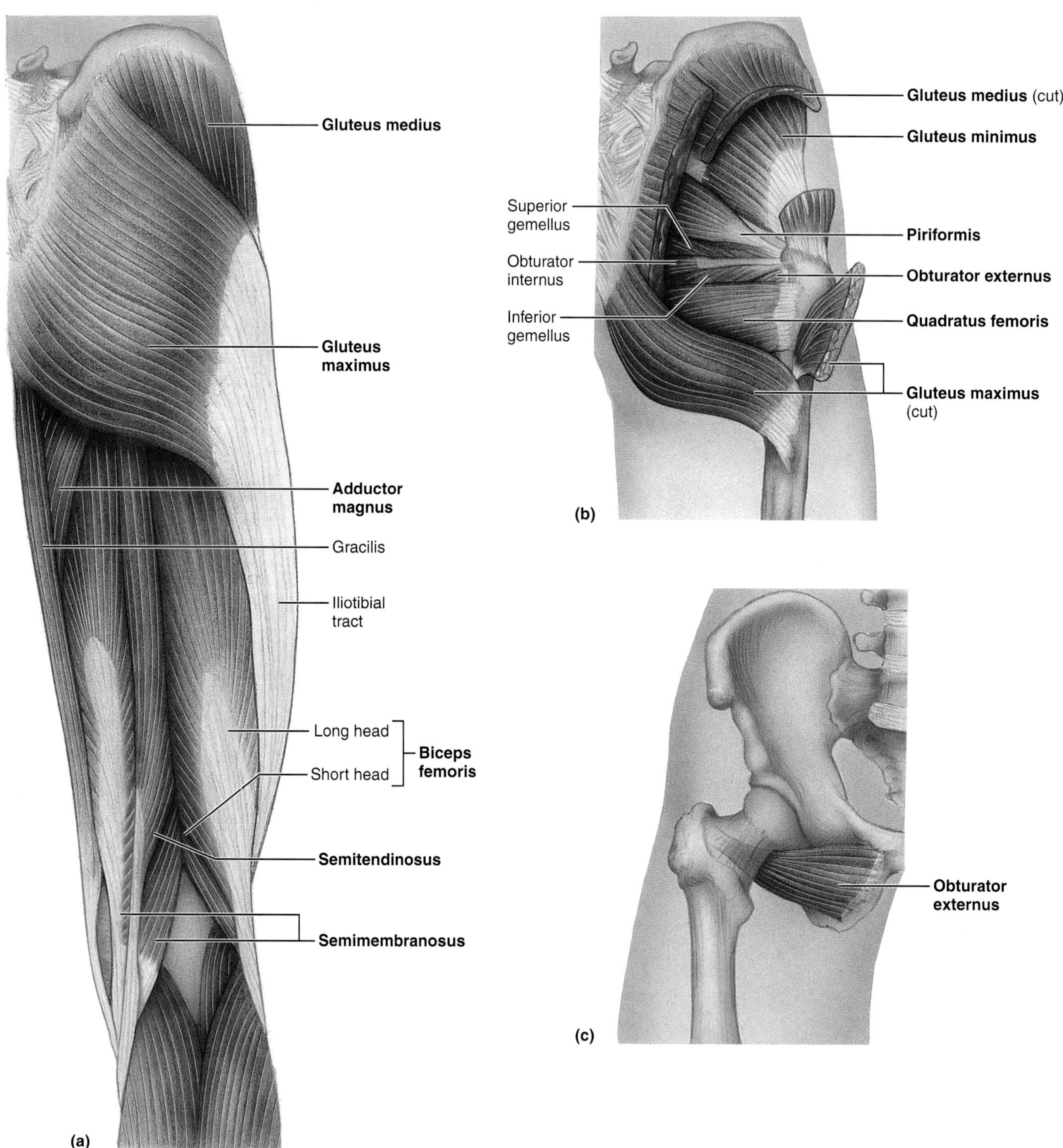

FIGURE 11.20 Posterior muscles of the right hip and thigh. (a) Superficial view showing the gluteus muscles of the buttock and hamstring muscles of the thigh. (b) Deep muscles of the gluteal region, which act primarily to rotate the thigh laterally. The superficial gluteus maximus and medius have been removed. (c) Anterior view of the isolated obturator externus muscle, showing its course as it travels from its origin on the anterior pelvis to the posterior aspect of the femur.

Muscle	Description	Origin (O) and Insertion (I)	Action	Nerve Supply
PART II: POSTERIOR MUSCLES (FIGURE 11.20) **GLUTEAL MUSCLES—ORIGIN ON PELVIS OR SACRUM**				
Gluteus maximus (gloo'te-us mak'sĭ-mus) (*glutos* = buttock; *maximus* = largest)	Largest and most superficial of gluteus muscles; forms bulk of buttock mass; fascicles are thick and coarse; a site of intramuscular injection (dorsal gluteal site); overlies large sciatic nerve; covers ischial tuberosity only when standing; when sitting, moves superiorly, leaving ischial tuberosity exposed in the subcutaneous position	O—dorsal ilium, sacrum, and coccyx I—gluteal tuberosity of femur; iliotibial tract	Major extensor of thigh; complex, powerful, and most effective when thigh is flexed and force is necessary, as in rising from a forward flexed position and in thrusting the thigh posteriorly in climbing stairs and running; generally inactive during standing and walking; laterally rotates and abducts thigh	Inferior gluteal nerve (L_5, S_1, and S_2)
Gluteus medius (me'de-us) (*medius* = middle)	Thick muscle largely covered by gluteus maximus; important site for intramuscular injections (ventral gluteal site); considered safer than dorsal gluteal site because there is less chance of injuring sciatic nerve	O—between anterior and posterior gluteal lines on lateral surface of ilium I—by short tendon into lateral aspect of greater trochanter of femur	Abducts and medially rotates thigh; steadies pelvis; its action is extremely important in walking; e.g., muscle of limb planted on ground tilts or holds pelvis in abduction so that pelvis on side of swinging limb does not sag; the foot of swinging limb can thus clear the ground	Superior gluteal nerve (L_5 and S_1)
Gluteus minimus (mĭ'nĭ-mus) (*minimus* = smallest)	Smallest and deepest of gluteal muscles	O—between anterior and inferior gluteal lines on external surface of ilium I—anterior border of greater trochanter of femur	As for gluteus medius	Superior gluteal nerve (L_5 and S_1)
LATERAL ROTATORS				
Piriformis (pir'ĭ-form-is) (*piri* = pear; *forma* = shape)	Pyramidal muscle located on posterior aspect of hip joint; inferior to gluteus minimus; issues from pelvis via greater sciatic notch	O—anterolateral surface of sacrum (opposite greater sciatic notch) I—superior border of greater trochanter of femur	Rotates extended thigh laterally; because inserted above head of femur, can also assist in abduction of thigh when hip is flexed; stabilizes hip joint	S_1 and S_2, L_5
Obturator externus (ob"tu-ra'tor ek-ster'nus) (*obturator* = obturator foramen; *externus* = outside)	Flat, triangular muscle deep in upper medial aspect of thigh	O—outer surface of obturator membrane, external surface of pubis and ischium, and margins of obturator foramen I—by a tendon into trochanteric fossa of posterior femur	As for piriformis	Obturator nerve

➤

Muscle	Description	Origin (O) and Insertion (I)	Action	Nerve Supply
Obturator internus (in-ter'nus) (*internus* = inside)	Surrounds obturator foramen within pelvis; leaves pelvis via lesser sciatic notch and turns acutely forward to insert in femur	O—inner surface of obturator membrane, greater sciatic notch, and margins of obturator foramen I—greater trochanter in front of piriformis	As for piriformis	L_5 and S_1
Gemellus (jĕ-mĕ'lis)—superior and inferior (*gemin* = twin, double; *superior* = above; *inferior* = below)	Two small muscles with common insertions and actions; considered extra-pelvic portions of obturator internus	O—ischial spine (superior); ischial tuberosity (inferior) I—greater trochanter of femur	As for piriformis	L_5 and S_1
Quadratus femoris (*quad* = four-sided square)	Short, thick muscle; most inferior of lateral rotator muscles; extends laterally from pelvis	O—ischial tuberosity I—intertrochanteric crest of femur	Rotates thigh laterally and stabilizes hip joint	L_5 and S_1
MUSCLES OF THE POSTERIOR COMPARTMENT OF THE THIGH				
Hamstrings	The hamstrings are fleshy muscles of the posterior thigh (biceps femoris, semitendinosus, and semimembranosus); they cross both the hip and knee joints and are prime movers of thigh extension and knee flexion; group has a common origin site and is innervated by sciatic nerve; ability of hamstrings to act on one of the two joints spanned depends on which joint is fixed; i.e., if knee is fixed (extended), they promote hip extension; if hip is extended, they promote knee flexion; however, when hamstrings are stretched, they tend to restrict full accomplishment of antagonistic movements; e.g., if knees are fully extended, it is difficult to flex the hip fully (and touch your toes), and when the thigh is fully flexed as in kicking a football, it is almost impossible to extend the knee fully at the same time (without considerable practice); name of this muscle group comes from old butchers' practice of using the tendons to hang hams for smoking; "pulled hamstrings" are common sports injuries in those who run very hard, e.g., football halfbacks			
• **Biceps femoris** (*biceps* = two heads)	Most lateral muscle of the group; arises from two heads	O—ischial tuberosity (long head); linea aspera, lateral supracondylar line, and distal femur (short head) I—common tendon passes downward and laterally (forming lateral border of popliteal fossa) to insert into head of fibula and lateral condyle of tibia	Extends thigh and flexes knee; laterally rotates leg, especially when knee is flexed	Sciatic nerve (L_5–S_2)
• **Semitendinosus** (sem"e-ten"dĭ-no'sus) (*semi* = half; *tendinosus* = tendon)	Lies medial to biceps femoris; although its name suggests that this muscle is largely tendinous, it is quite fleshy; its long, slender tendon begins about two-thirds of the way down thigh	O—ischial tuberosity in common with long head of biceps femoris I—medial aspect of upper tibial shaft	Extends thigh at hip; flexes knee; with semimembranosus, medially rotates leg	Sciatic nerve
• **Semimembranosus** (sem"e-mem"brah-no'sus) (*membranosus* = membrane)	Deep to semitendinosus	O—ischial tuberosity I—medial condyle of tibia; through oblique ligament to lateral condyle of femur	Extends thigh and flexes knee; medially rotates leg	Sciatic nerve

Muscle Gallery: Table 11.15 Muscles of the Leg: Movements of the Ankle and Toes (Figures 11.21 to 11.23)

The deep fascia of the leg is continuous with the fascia lata that surrounds the thigh. Like a "knee sock" deep to the skin, this leg fascia surrounds the leg muscles and binds them tightly, preventing excess swelling of these muscles during exercise and also aiding venous return. Inward extensions from the leg fascia divide the leg muscles into *anterior, lateral,* and *posterior compartments* (Figure 11.24b), each with its own nerve and blood supply. Distally, the leg fascia thickens to form the **extensor, fibular,** and **flexor retinacula,** "ankle bracelets" that hold the tendons in place where they run to the foot. As in the wrist and hand, the distal tendons are covered with slippery tendon sheaths.

The various muscles of the leg cause movements at the ankle joint (dorsiflexion and plantar flexion), at the intertarsal joints (inversion and eversion of the foot), or at the toes (flexion, extension). The muscles in the *anterior extensor compartment* of the leg (Figure 11.21) are directly comparable to the extensor muscle group of the forearm. Mostly, they extend the toes and dorsiflex the foot. Although dorsiflexion is not a powerful movement, it keeps the toes from dragging during walking. The *lateral compartment muscles* (Figure 11.22) are called *fibularis* (formerly *peroneal*; *peron* = the fibula) muscles. They evert and plantar flex the foot and do not correspond to any muscles in the forearm. Muscles of the *posterior flexor compartment* (Figure 11.23) are comparable to the flexor muscle group of the forearm. They flex the toes and plantar flex the foot. Plantar flexion is the most powerful movement at the ankle: It lifts the weight of the entire body. Plantar flexion is necessary for standing on tiptoe and provides the forward thrust in walking and running.

The actions of the muscles in this table are summarized in Table 11.16 (Part II).

Muscle	Description	Origin (O) and Insertion (I)	Action	Nerve Supply
PART I: MUSCLES OF THE ANTERIOR COMPARTMENT (FIGURES 11.21 AND 11.22)				
	All muscles of the anterior compartment are dorsiflexors of the ankle and have a common innervation, the deep fibular nerve. Paralysis of the anterior muscle group causes *foot drop*, which requires that the leg be lifted unusually high during walking to prevent tripping over one's toes. "Shin splints" (p. 332) is a painful inflammatory condition of the muscles of the anterior compartment.			
Tibialis anterior (tib″e-a′lis) (*tibial* = tibia; *anterior* = toward the front)	Superficial muscle of anterior leg; laterally parallels sharp anterior margin of tibia	O—lateral condyle and upper 2/3 of tibial shaft; interosseous membrane I—by tendon into inferior surface of medial cuneiform and first metatarsal bone	Prime mover of dorsiflexion; inverts foot; assists in supporting medial longitudinal arch of foot	Deep fibular nerve (L_4 and L_5)
Extensor digitorum longus (*extensor* = increases angle at a joint; *digit* = finger or toe; *longus* = long)	Unipennate muscle on anterolateral surface of leg; lateral to tibialis anterior muscle	O—lateral condyle of tibia; proximal 3/4 of fibula; interosseous membrane I—middle and distal phalanges of toes 2–5 via extensor expansion	Prime mover of toe extension (acts mainly at metatarsophalangeal joints); dorsiflexes foot	Deep fibular nerve (L_5 and S_1)
Fibularis (peroneus) tertius (fib-u-lah′ris ter′shus) (*perone* = fibula; *tertius* = third)	Small muscle; usually continuous and fused with distal part of extensor digitorum longus; not always present	O—distal anterior surface of fibula and interosseous membrane I—tendon inserts on dorsum of fifth metatarsal	Dorsiflexes and everts foot	Deep fibular nerve (L_5 and S_1)
Extensor hallucis (hal′u-sis) **longus** (*hallux* = great toe)	Deep to extensor digitorum longus and tibialis anterior; narrow origin	O—anteromedial fibula shaft and interosseous membrane I—tendon inserts on distal phalanx of great toe	Extends great toe; dorsiflexes foot	Deep fibular nerve (L_5 and S_1)

➤

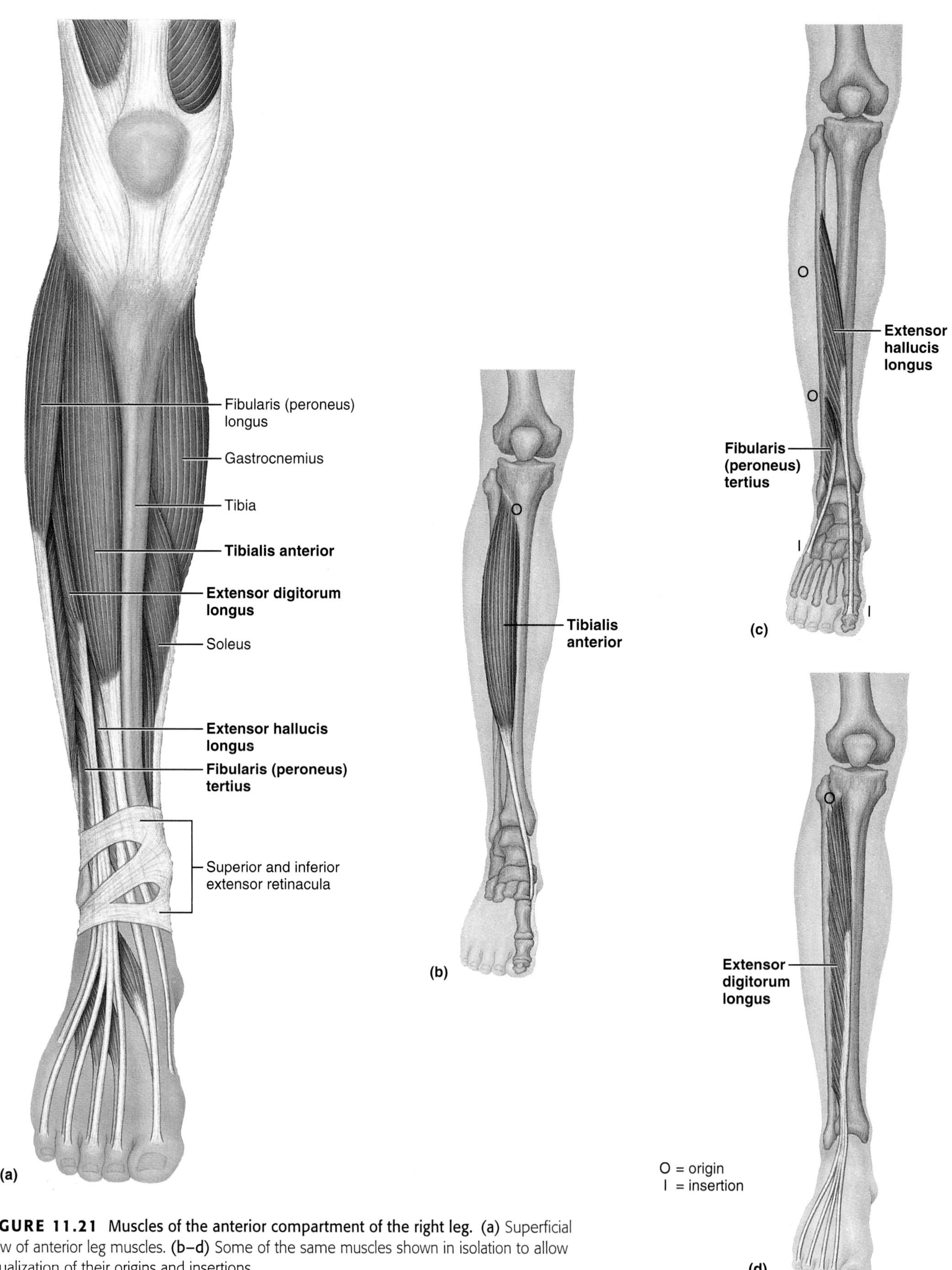

FIGURE 11.21 Muscles of the anterior compartment of the right leg. **(a)** Superficial view of anterior leg muscles. **(b–d)** Some of the same muscles shown in isolation to allow visualization of their origins and insertions.

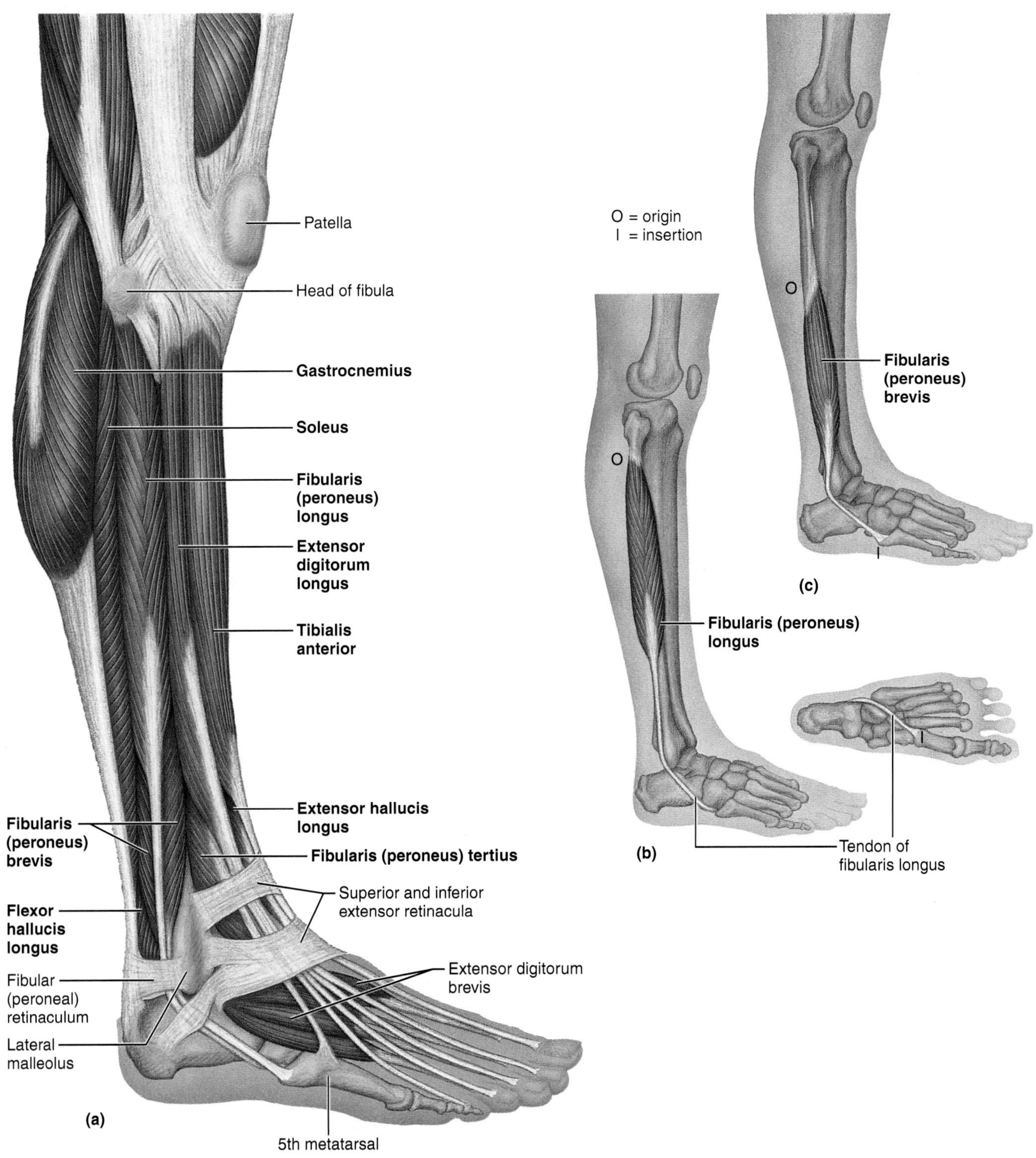

FIGURE 11.22 **Muscles of lateral compartment of the right leg.** **(a)** Superficial view of lateral aspect of the leg, illustrating the position of the lateral compartment muscles (fibularis longus and brevis) relative to anterior and posterior leg muscles. **(b)** View of fibularis longus in isolation; inset illustrates the insertion of the fibularis longus on the plantar surface of the foot. **(c)** Isolated view of the fibularis brevis muscle.

Muscle	Description	Origin (O) and Insertion (I)	Action	Nerve Supply
PART II: MUSCLES OF THE LATERAL COMPARTMENT (FIGURES 11.22 AND 11.23)				
	These muscles have a common innervation, the superficial fibular nerve. Besides plantar flexion and foot eversion, these muscles stabilize the lateral ankle and lateral longitudinal arch of the foot.			
Fibularis (peroneus) longus (See also Figure 11.21)	Superficial lateral muscle; overlies fibula	O—head and upper portion of lateral side of fibula I—by long tendon that curves under foot to first metatarsal and medial cuneiform	Plantar flexes and everts foot; may help keep foot flat on ground	Superficial fibular nerve (L_5–S_2)
Fibularis (peroneus) brevis (*brevis* = short)	Smaller muscle; deep to fibularis longus; enclosed in a common sheath	O—distal fibula shaft I—by tendon running behind lateral malleolus to insert on proximal end of fifth metatarsal	Plantar flexes and everts foot	Superficial fibular nerve
PART III: MUSCLES OF THE POSTERIOR COMPARTMENT (FIGURE 11.23)				
	The muscles of the posterior compartment have a common innervation, the tibial nerve. They act in concert to plantar flex the ankle.			
SUPERFICIAL MUSCLES				
Triceps surae (tri″seps sur′e) (See also Figure 11.22)	Refers to muscle pair (gastrocnemius and soleus) that shapes the posterior calf and inserts via a common tendon into the calcaneus of the heel; this **calcaneal** or **Achilles tendon** is the largest tendon in the body; prime movers of ankle plantar flexion			
• **Gastrocnemius** (gas″truk-ne′me-us) (*gaster* = belly; *kneme* = leg)	Superficial muscle of pair; two prominent bellies that form proximal curve of calf	O—by two heads from medial and lateral condyles of femur I—posterior calcaneus via calcaneal tendon	Plantar flexes foot when knee is extended; because it also crosses knee joint, it can flex knee when foot is dorsiflexed	Tibial nerve (S_1 and S_2)
• **Soleus** (so′le-us) (*soleus* = fish)	Broad, flat muscle, deep to gastrocnemius on posterior surface of calf	O—extensive cone-shaped origin from superior tibia, fibula, and interosseous membrane I—as for gastrocnemius	Plantar flexes foot; important locomotor and postural muscle during walking, running, and dancing	Tibial nerve

Muscle	Description	Origin (O) and Insertion (I)	Action	Nerve Supply
SUPERFICIAL MUSCLES, continued				
Plantaris (plan-tar'is) (*planta* = sole of foot)	Generally a small feeble muscle, but varies in size and extent; may be absent	O—posterior femur above lateral condyle I—via a long, thin tendon into calcaneus or calcaneal tendon	Assists in knee flexion and plantar flexion of foot	Tibial nerve
DEEP MUSCLES				
Popliteus (pop-lit'e-us) (*poplit* = back of knee)	Thin, triangular muscle at posterior knee; passes downward and medially to tibial surface	O—lateral condyle of femur and lateral meniscus of knee I—proximal tibia	Flexes and rotates leg medially to unlock knee from full extension when flexion begins; with tibia fixed, rotates thigh laterally	Tibial nerve (L_4–S_1)
Flexor digitorum longus (*flexor* = decreases angle at a joint)	Long, narrow muscle; runs medial to and partially overlies tibialis posterior	O—extensive origin on the posterior tibia I—tendon runs behind medial malleolus and splits to insert into distal phalanges of toes 2–5	Plantar flexes and inverts foot; flexes toes; helps foot "grip" ground	Tibial nerve (S_2 and S_3)
Flexor hallucis longus (See also Figure 11.22)	Bipennate muscle; lies lateral to inferior aspect of tibialis posterior	O—middle part of shaft of fibula; interosseous membrane I—tendon runs under foot to distal phalanx of great toe	Plantar flexes and inverts foot; flexes great toe at all joints; "push off" muscle during walking	Tibial nerve (S_2 and S_3)
Tibialis posterior (*posterior* = toward the back)	Thick, flat muscle deep to soleus; placed between posterior flexors	O—extensive origin from superior tibia and fibula and interosseous membrane I—tendon passes behind medial malleolus and under arch of foot; inserts into several tarsals and metatarsals 2–4	Prime mover of foot inversion; plantar flexes foot; stabilizes medial longitudinal arch of foot (as during ice skating)	Tibial nerve (L_4 and L_5)

➤

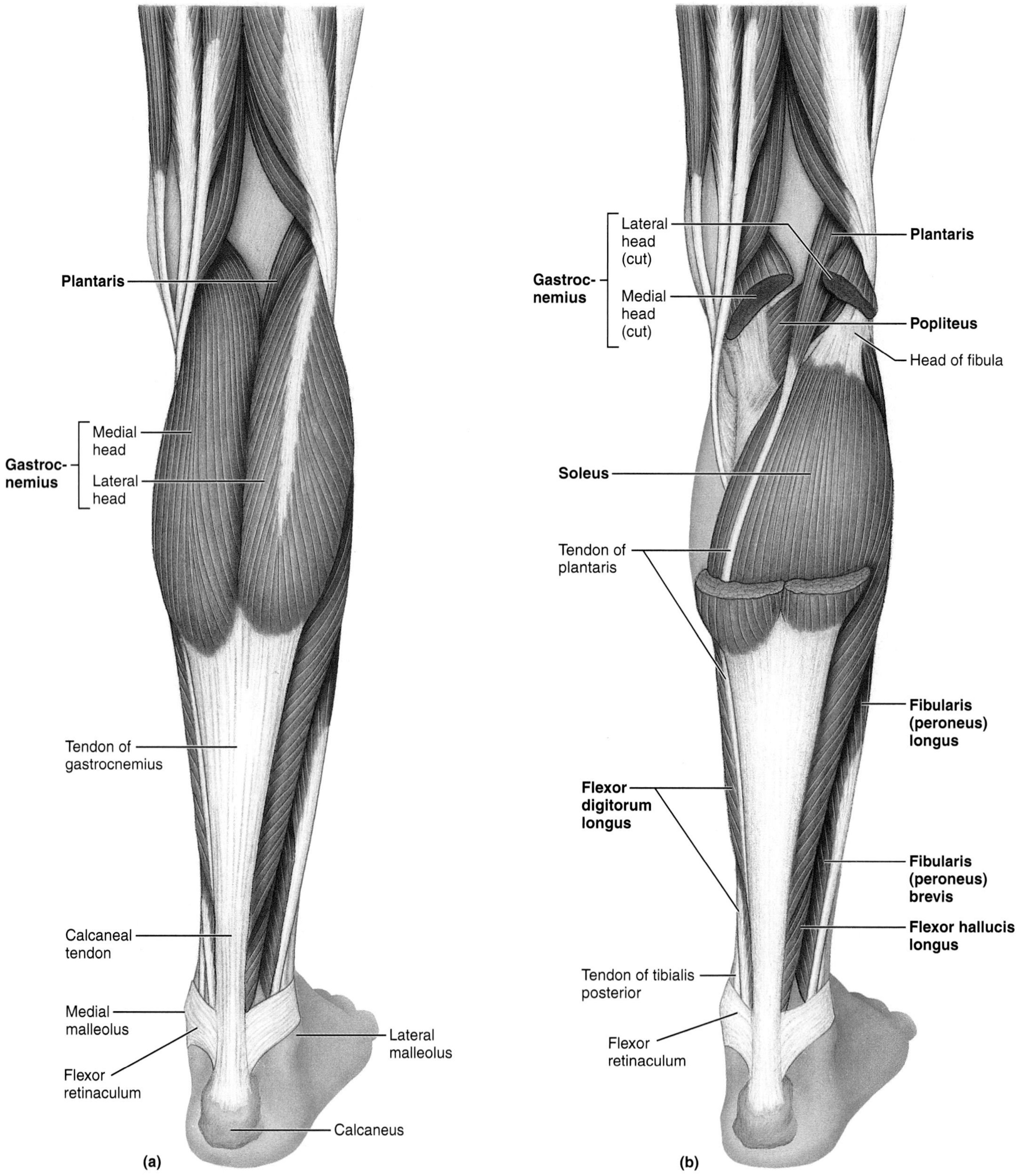

FIGURE 11.23 Muscles of the posterior compartment of the right leg.
(a) Superficial view of the posterior leg. **(b)** The fleshy gastrocnemius has been removed to show the soleus immediately deep to it. **(c)** The triceps surae has been removed to show the deep muscles of the posterior compartment. **(d–f)** Individual deep muscles are shown in isolation so that their origins and insertions may be visualized.

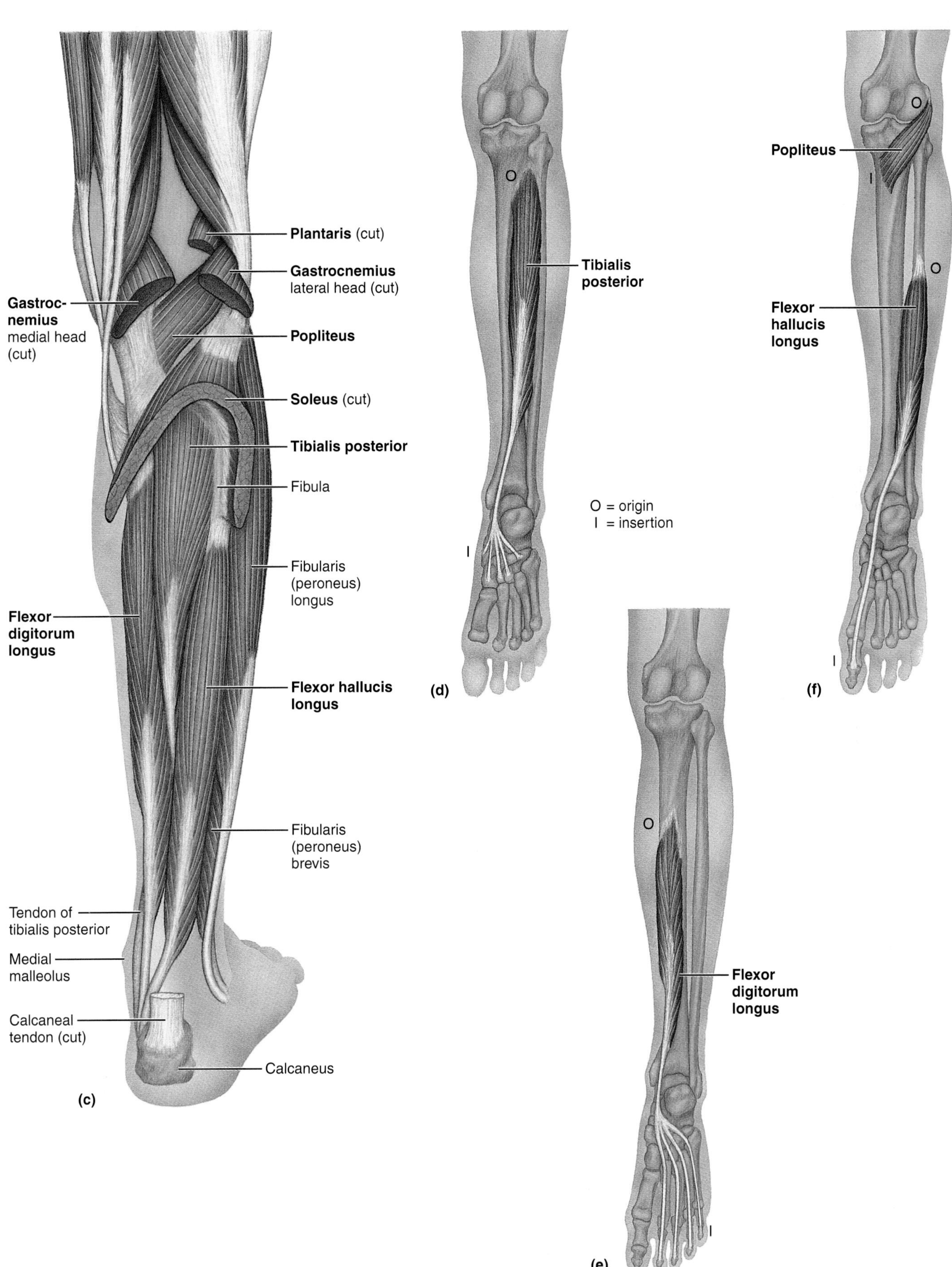

Plantaris (cut)
Gastrocnemius lateral head (cut)
Gastroc-nemius medial head (cut)
Popliteus
Soleus (cut)
Tibialis posterior
Fibula
Fibularis (peroneus) longus
Flexor digitorum longus
Flexor hallucis longus
Fibularis (peroneus) brevis
Tendon of tibialis posterior
Medial malleolus
Calcaneal tendon (cut)
Calcaneus
(c)
O
Tibialis posterior
I
(d)
O = origin
I = insertion
Popliteus
O
I
O
Flexor hallucis longus
I
(f)
O
Flexor digitorum longus
I
(e)

Muscle Gallery: Table 11.16 Summary of Actions of Muscles Acting on the Thigh, Leg, and Foot (Figure 11.24)

PART I: MUSCLES ACTING ON THE THIGH AND LEG (PM = prime mover)	Actions at the Hip Joint						Actions at the Knee	
	Flexion	Extension	Abduction	Adduction	Medial Rotation	Lateral Rotation	Flexion	Extension
Anterior and Medial Muscles:								
Iliopsoas	X (PM)							
Sartorius	X		X			X	X	
Adductor magnus		X		X	X			
Adductor longus	X			X	X			
Adductor brevis	X			X	X			
Pectineus	X			X	X			
Gracilis				X	X		X	
Rectus femoris	X							X (PM)
Vastus muscles								X (PM)
Tensor fasciae latae	X		X		X			X
Posterior Muscles:								
Gluteus maximus		X (PM)	X			X		
Gluteus medius			X (PM)		X			
Gluteus minimus			X		X			
Piriformis			X			X		
Obturator internus						X		
Obturator externus						X		
Gemelli						X		
Quadratus femoris						X		
Biceps femoris		X (PM)					X (PM)	
Semitendinosus		X					X (PM)	
Semimembranosus		X					X (PM)	
Gastrocnemius							X	
Plantaris							X	
Popliteus							X (and rotates leg medially)	

PART II: MUSCLES ACTING ON THE ANKLE AND TOES	**Actions at the Ankle Joint**				**Actions at the Toes**	
	Plantar Flexion	**Dorsiflexion**	**Inversion**	**Eversion**	**Flexion**	**Extension**
Anterior Compartment:						
Tibialis anterior		X (PM)	X			
Extensor digitorum longus		X				X (PM)
Fibularis tertius		X		X		
Extensor hallucis longus		X	X (weak)			X (great toe)
Lateral Compartment:						
Fibularis longus and brevis	X			X		
Posterior Compartment:						
Gastrocnemius	X (PM)					
Soleus	X (PM)					
Plantaris	X					
Flexor digitorum longus	X		X		X (PM)	
Flexor hallucis longus	X		X		X (great toe)	
Tibialis posterior	X		X (PM)			

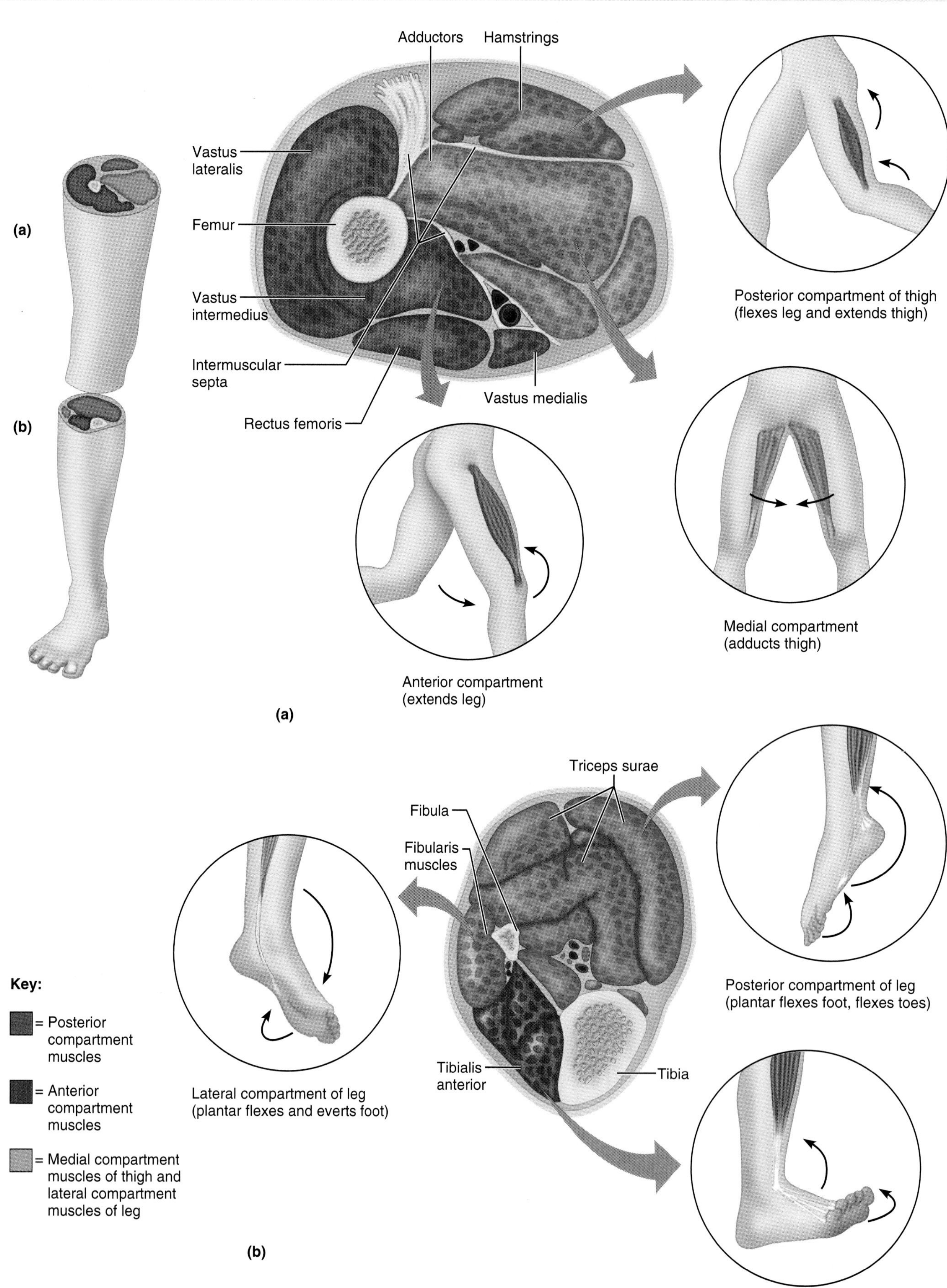

FIGURE 11.24 Summary of actions of muscles of the thigh and leg. (a) Muscles of the thigh. (b) Muscles of the leg.

Muscle Gallery: Table 11.17 Intrinsic Muscles of the Foot: Toe Movement and Foot Support (Figure 11.25)

The intrinsic muscles of the foot help to flex, extend, abduct, and adduct the toes. Furthermore, along with the tendons of some leg muscles that enter the sole, the foot muscles support the arches of the foot. There is a single muscle on the foot's dorsum (superior aspect), and many muscles on the plantar aspect (the sole). The plantar muscles occur in four layers, from superficial to deep. Overall, the foot muscles are remarkably similar to those in the palm of the hand.

Muscle	Description	Origin (O) and Insertion (I)	Action	Nerve Supply
MUSCLE ON DORSUM OF FOOT				
Extensor digitorum brevis (Figure 11.22a)	Small, four-part muscle on dorsum of foot; deep to the tendons of extensor digitorum longus; corresponds to the extensor indicis and extensor pollicis muscles of forearm	O—anterior part of calcaneus bone; extensor retinaculum I—base of proximal phalanx of big toe; extensor expansions on toes 2–4	Helps extend toes at metatarso-phalangeal joints	Deep fibular nerve (S_1 and S_2)
MUSCLES ON SOLE OF FOOT (FIGURE 11.25)				
First Layer (most superficial)				
• **Flexor digitorum brevis**	Bandlike muscle in middle of sole; corresponds to flexor digitorum superficialis of forearm and inserts into digits in the same way	O—tuber calcanei I—middle phalanx of toes 2–4	Helps flex toes	Medial plantar nerve (S_2 and S_3)
• **Abductor hallucis** (hal'yu-kis)(*hallux* = the big toe)	Lies medial to flexor digitorum brevis (recall the similar thumb muscle, abductor pollicis brevis)	O—tuber calcanei and flexor retinaculum I—proximal phalanx of big toe, medial side, through a tendon shared with flexor hallucis brevis (see below)	Abducts big toe	Medial plantar nerve
• **Abductor digiti minimi**	Most lateral of the three superficial sole muscles; (recall similar abductor muscle in palm)	O—tuber calcanei I—lateral side of base of little toe's proximal phalanx	Abducts and flexes little toe	Lateral plantar nerve (S_2 and S_3)
Second Layer				
• **Flexor accessorius (quadratus plantae)**	Rectangular muscle just deep to flexor digitorum brevis in posterior half of sole; two heads (see Figure 11.25b and c)	O—medial and lateral sides of calcaneus I—tendon of flexor digitorum longus in mid-sole	Straightens out the oblique pull of flexor digitorum longus	Lateral plantar nerve
• **Lumbricals**	Four little "worms" (like lumbricals in hand)	O—from each tendon of flexor digitorum longus I—extensor expansion on proximal phalanx of toes 2–5, medial side	By pulling on extensor expansion, flex toes at metatarsophalangeal joints and extend toes at interphalangeal joints	Medial plantar nerve (first lumbrical) and lateral plantar nerve (second to fourth lumbricals)
Third Layer				
• **Flexor hallucis brevis**	Covers metatarsal I; splits into two bellies—recall flexor pollicis brevis of thumb (see Figure 11.25c)	O—lateral cuneiform and cuboid bones I—via two tendons onto both sides of the base of the proximal phalanx of big toe; each tendon has a sesamoid bone in it	Flexes big toe's metatarsophalangeal joint	Medial plantar nerve

➤

Muscle	Description	Origin (O) and Insertion (I)	Action	Nerve Supply
Third Layer, continued				
• **Adductor hallucis**	Oblique and transverse heads; deep to lumbricals (recall adductor pollicis in thumb)	O—from bases of metatarsals 2–4 and from sheath of fibularis longus tendon (oblique head); from a ligament across metatarsophalangeal joints (transverse head) I—base of proximal phalanx of big toe, lateral side	Helps maintain the transverse arch of foot; weak adductor of big toe	Lateral plantar nerve (S_2 and S_3)
• **Flexor digiti minimi brevis**	Covers metatarsal 5 (recall same muscle in hand)	O—base of metatarsal 5 and sheath of fibularis longus tendon I—base of proximal phalanx of toe 5	Flexes little toe at metatarsophalangeal joint	Lateral plantar nerve
Fourth Layer (deepest)				
• **Plantar and dorsal interossei**	Three plantar and four dorsal; similar to palmar and dorsal interossei of hand in locations, attachments, and actions; however, the long axis of foot around which these muscles orient is the second digit, not third	See palmar and dorsal interossei (Table 11.13)	See palmar and dorsal interossei (Table 11.13)	Lateral plantar nerve

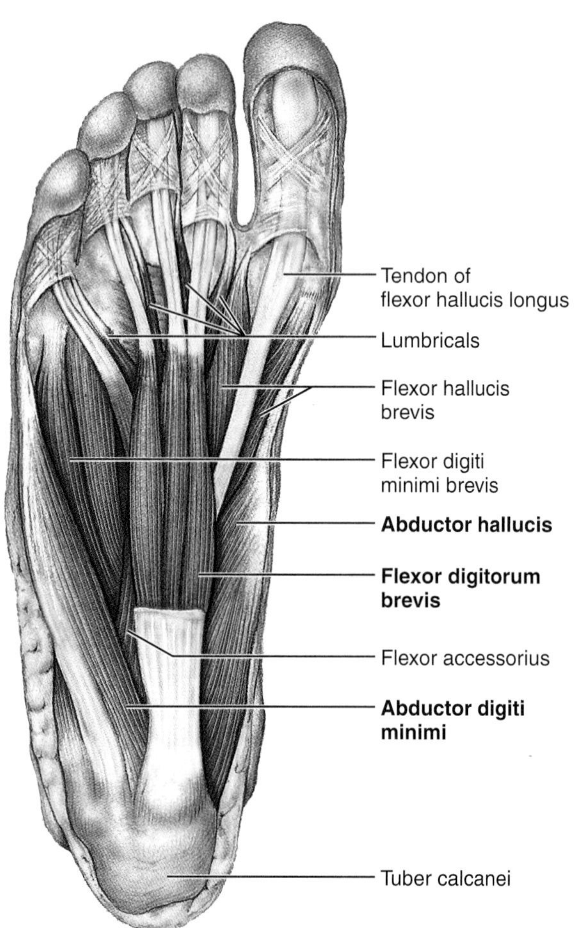

(a) First layer

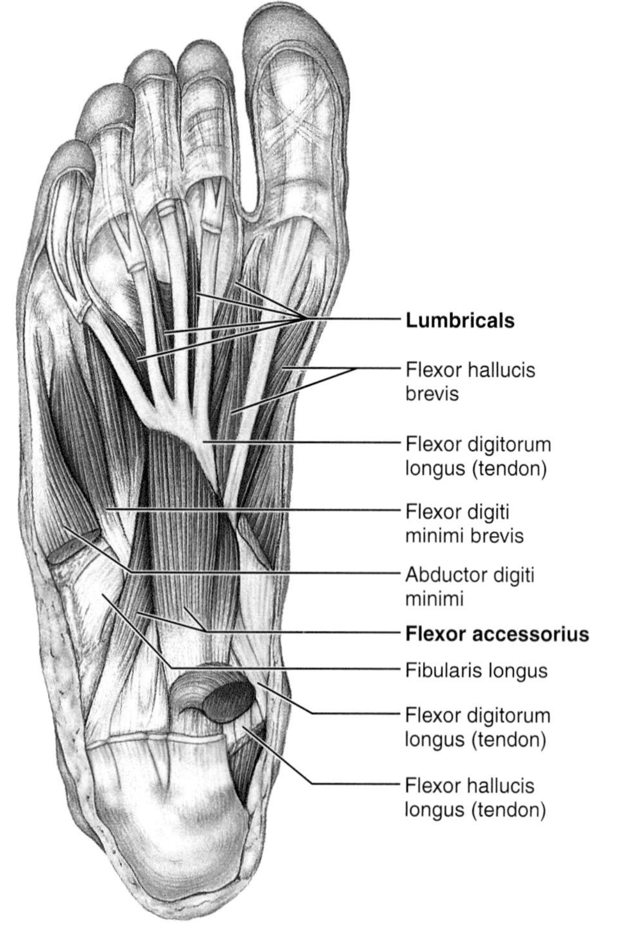

(b) Second layer

FIGURE 11.25 Muscles of the right foot, plantar aspect (part (e) is a dorsal view).

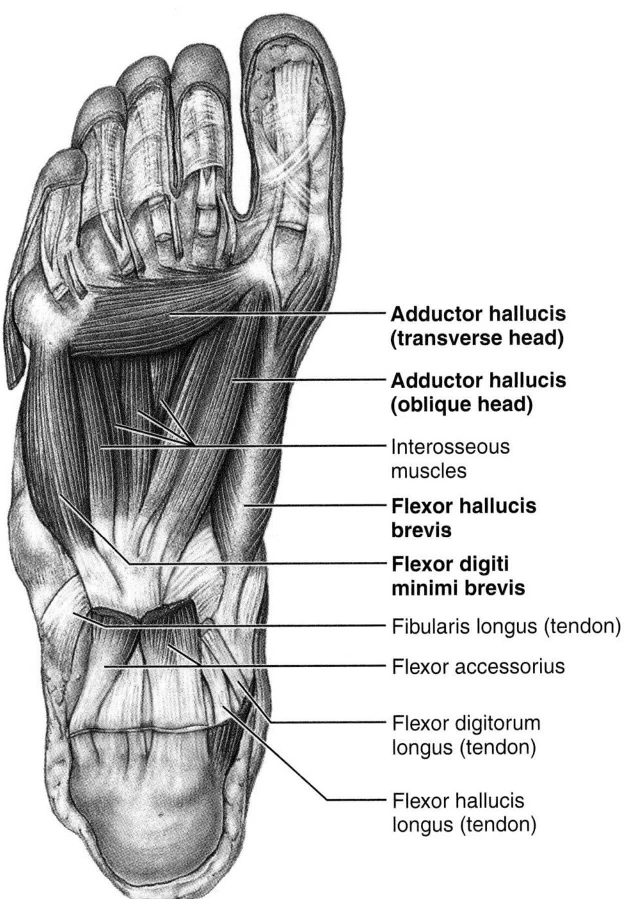

(c) Third layer

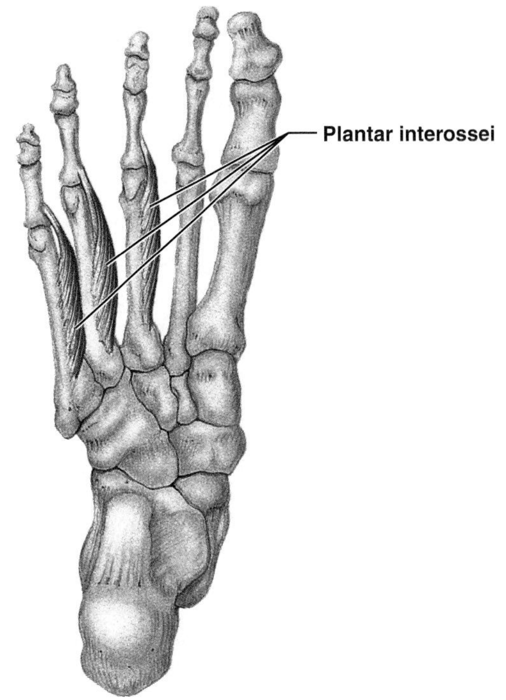

(d) Fourth layer: plantar interossei

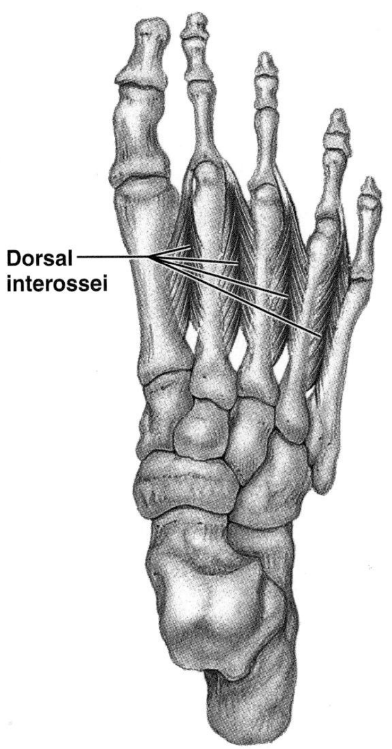

(e) Fourth layer: dorsal interossei

RELATED CLINICAL TERMS

Adductor strain Pulled adductor muscle in the thigh, usually the adductor longus. Common sports injury, caused by extreme (or forced) thigh abduction when the thigh is adducting (as when the planted limb slips sideways when a soccer player is kicking a ball). Can also occur if one's hip joint is unstable, and the hip-stabilizing adductors strain from continually compensating for the instability. Imaging or other techniques are used to distinguish this from other injuries to the groin region, such as hernias, cracked neck of femur, and so on.

Charley horse Tearing of a muscle, followed by bleeding into the tissues (hematoma) and severe pain; a common sports injury. Charley horse of the quadriceps femoris occurs frequently in football players.

Electromyography The recording, study, and interpretation of graphic records of the electrical activity of contracting muscles. Electrodes inserted into the muscles detect the impulses that cross muscle-cell membranes and stimulate contraction; related equipment provides a visual display of the impulses. This is the best and most important technique for determining the functions of the muscles and muscle groups of body.

Hallux valgus (hal′uks val′gus) (*valgus* = bent away from body midline) Permanent displacement of the big toe toward the other toes, often caused by tight shoes. In this condition, the sesamoid bones that underlie the head of the first metatarsal have been pushed laterally, away from the big toe. Thus, the abductor hallucis muscle of the sole, whose insertion tendon contains one of the sesamoids, cannot abduct the displaced toe back to its proper position.

Mallet finger Extreme, persistent flexion of the distal phalanx of a finger, caused by forceful bending, as when a baseball hits the finger end-on. Part of the bony insertion of the extensor digitorum muscles is stripped off, so the finger flexors flex the joint unopposed.

Rupture of calcaneal tendon The calcaneal (Achilles) tendon is the strongest tendon in the body, but its rupture is surprisingly common, particularly in older people as a result of stumbling and in sprinters who tense the tendon too severely while starting a race. The rupture leads to a gap just superior to the heel, and the calf bulges as the triceps surae muscles are released from their insertion. As a result, plantar flexion is very weak, but dorsiflexion is exaggerated. The tendon must be repaired surgically.

Shin splints Common term for pain in the anterior leg caused by swelling of the tibialis anterior muscle. This muscle swells after heavy exercise in the absence of adequate prior conditioning. Because it is tightly wrapped by fascia, the tibialis anterior cuts off its own circulation as it swells and presses painfully on its own nerves.

CHAPTER SUMMARY

Media study tools that could provide you additional help in reviewing specific key topics of Chapter 11 are referenced below.

SP = Study Partner; AIA = A.D.A.M.® Interactive Anatomy

LEVER SYSTEMS: BONE-MUSCLE RELATIONSHIPS (pp. 266–268)

1. A lever is a rigid bar that moves on a fulcrum. When an effort is applied to the lever, a load is moved. In the body, bones are the levers, joints are the fulcrums, and the effort is exerted by skeletal muscles pulling on their insertions.
2. When the effort is farther from the fulcrum than is the load, the lever works at a mechanical advantage (is slow and strong). When the effort is closer to the fulcrum than is the load, the lever works at a mechanical disadvantage (moves quickly and far but takes extra effort).
3. First-class levers (effort-fulcrum-load) may operate at a mechanical advantage or disadvantage. Second-class levers (fulcrum-load-effort) all work at a mechanical advantage. Third-class levers (fulcrum-effort-load) all work at a mechanical disadvantage. Most muscles of the body are in third-class lever systems to provide speed of movement.

SP Exercise: Chapter 11, Lever Systems.

ARRANGEMENTS OF FASCICLES IN MUSCLES (p. 268)

1. For a summary of the different fascicle arrangements, see Figure 11.3.

INTERACTIONS OF SKELETAL MUSCLES IN THE BODY (pp. 268–270)

1. Skeletal muscles are arranged in opposing groups across movable joints such that one group can reverse or modify the action of the other.
2. Prime movers (agonists) bear the main responsibility for a particular movement. Synergists help the prime movers, prevent undesirable movements, or stabilize joints (as fixators). Antagonists produce the opposite movement.

NAMING THE SKELETAL MUSCLES (p. 270)

1. Muscles are named for their location, shape, relative size, fascicle direction, location of attachments, number of origins, and action.

A DEVELOPMENT-BASED ORGANIZATION OF THE MUSCLES (pp. 270–272)

1. Based on embryonic origin and general functions, the musculature can be divided into four main groups: visceral musculature, pharyngeal arch (branchiomeric) muscles, axial muscles, and limb muscles.
2. *Pharyngeal arch muscles* are skeletal muscles that develop from somitomeres and surround the pharynx, but some migrate into the head and back during development. The *axial muscles* of the neck and trunk derive from myotomes. They include dorsal muscles that extend the spine and ventral muscles that flex the spine. Of the axial muscles in the

head, the eye muscles and the tongue muscles develop from somitomeres and occipital myotomes, respectively. *Limb muscles* also derive from myotomes and exhibit the same arrangement in the upper and lower limbs: extensors on one side, flexors on the other.

MAJOR SKELETAL MUSCLES OF THE BODY (pp. 272–331)

1. The muscles are divided into functional groups and presented in detail in Tables 11.1–11.17. The muscles of facial expression (see Table 11.1) are thin and primarily insert on the facial skin. They close the eyes and lips, compress the cheeks, and allow us to smile and show our feelings.
2. Chewing muscles include the masseter and temporalis that elevate the mandible and the two pterygoids that grind food. Extrinsic muscles of the tongue anchor the tongue and control its movements (see Table 11.2).
3. Deep muscles in the anterior neck help us swallow by lifting and then depressing the larynx (suprahyoid and infrahyoid groups). Pharyngeal constrictors squeeze food into the esophagus (see Table 11.3).

AIA Link: Atlas; System: Muscular; Pharyngeal constrictor muscles.

4. Muscles moving the head and muscles extending the spine are considered in Table 11.4. Flexion and rotation of the head and neck are brought about by the sternocleidomastoid and scalenes in the anterior and lateral neck. Extension of the spine and head is brought about by deep muscles of the back and the posterior neck (for example, the erector spinae and splenius muscles).
5. Movements of breathing are promoted by the diaphragm and by the intercostal muscles of the thorax (see Table 11.5). Contraction of the diaphragm increases abdominal pressure during straining, an action that is aided by the muscles of the abdominal wall.
6. The sheet-like muscles of the anterior and lateral abdominal wall (see Table 11.6) are layered like plywood. This layering forms a strong wall that protects, supports, and compresses the abdominal contents. These muscles can also flex the trunk.

AIA Link: Atlas; System: Muscular; Deep muscles of the trunk (anterior).

7. Muscles of the pelvic floor and perineum (see Table 11.7) support our pelvic organs, inhibit urination and defecation, and aid erection.
8. The superficial muscles of the thorax attach to the pectoral girdle and fix or move the scapula during arm movements (see Table 11.8).
9. All muscles that move the arm at the shoulder joint originate at least partly from the shoulder girdle (see Table 11.9). Four muscles form the musculotendinous rotator cuff, which stabilizes the shoulder joint. As a rule, muscles located on the anterior thorax flex the arm (pectoralis major), whereas the posterior muscles extend it (latissimus dorsi). Abduction is mostly brought about by the deltoid. Adduction and rotation are brought about by many muscles on both the anterior and posterior sides.

AIA Link: Atlas; Region: Upper limb; Deep muscles of anterior arm.

10. Most muscles that lie within the arm move the forearm (see Table 11.10). Anterior arm muscles are flexors of the forearm (biceps, brachialis), whereas posterior arm muscles are extensors of the forearm (triceps).
11. Forearm muscles primarily move the hand (see Table 11.11). Anterior forearm muscles flex the hand and fingers (but include two pronators), whereas the posterior muscles are mostly extensors (but include the strong supinator).
12. The intrinsic muscles of the hand aid in fine movements of the fingers (see Table 11.13) and in opposition, which helps us grip things in our palms. These small muscles are divided into thenar, hypothenar, and midpalmar groups.
13. Muscles that bridge the hip and knee joints move the thigh and the leg (see Table 11.14). Anterior muscles mostly flex the thigh and extend the leg (quadriceps femoris), medial muscles adduct the thigh (adductors), and posterior muscles (gluteals, hamstrings) primarily abduct and extend the thigh and flex the leg.
14. Muscles in the leg act on the ankle and toes (see Table 11.15). Leg muscles in the anterior compartment are mostly dorsiflexors of the foot (extensor group). Muscles in the lateral compartment are everters and plantar flexors of the foot (fibular group), and posterior leg muscles are strong plantar flexors (gastrocnemius and soleus, flexor group).

AIA Link: Atlas; Region: Lower limb; Dissection of medial leg and foot.

15. The intrinsic muscles of the foot (see Table 11.17) support the foot arches and help move the toes. Most occur in the sole, arranged in four layers. They resemble the small muscles in the palm of the hand.

SP Exercise: Chapter 11, Anterior View of Superficial Muscles of the Body.

SP Exercise: Chapter 11, Superficial Muscles (Posterior View).

REVIEW QUESTIONS

Multiple Choice/Matching Questions

1. A muscle whose fascicles are arranged at an angle to a central longitudinal tendon has this arrangement: (a) circular, (b) longitudinal, (c) parallel, (d) pennate, (e) convergent.
2. A muscle that helps an agonist by causing a similar movement or by stabilizing a joint over which the agonist acts is (a) an antagonist, (b) a prime mover, (c) a synergist, (d) a fulcrum.

3. Match the muscles in Column B to their embryonic origins in Column A.

Column A	Column B
____ **(1)** from head (occipital) myotomes	**(a)** biceps brachii
____ **(2)** from trunk (nonlimb) myotomes	**(b)** erector spinae
____ **(3)** limb extensor group	**(c)** tongue muscles
____ **(4)** limb flexor group	**(d)** chewing muscles
____ **(5)** pharyngeal arch muscles	**(e)** triceps brachii

4. The muscle that closes the eyes is (a) the occipitalis, (b) the zygomaticus, (c) the corrugator supercilii, (d) the orbicularis oris, (e) none of these.
5. The arm muscle that both flexes the forearm at the elbow and supinates it is the (a) brachialis, (b) brachioradialis, (c) biceps brachii, (d) triceps brachii.
6. The chewing muscles that protrude the mandible and grind the teeth from side to side are (a) the buccinators, (b) the masseters, (c) the temporalis, (d) the pterygoids, (e) none of these.
7. Muscles that depress the hyoid bone and larynx include all of the following *except* the (a) sternohyoid, (b) omohyoid, (c) geniohyoid, (d) sternothyroid.
8. The quadriceps femoris muscles include all of the following *except* the (a) vastus lateralis, (b) biceps femoris, (c) vastus intermedius, (d) rectus femoris.
9. A prime mover of thigh flexion at the hip is the (a) rectus femoris, (b) iliopsoas, (c) vastus intermedius, (d) gluteus maximus.
10. Someone who sticks out a thumb to hitch a ride is ____ the thumb. (a) extending, (b) abducting, (c) adducting, (d) opposing.
11. The major muscles used in doing a chin-up are the (a) triceps brachii and pectoralis major, (b) infraspinatus and biceps brachii, (c) serratus anterior and external oblique, (d) latissimus dorsi and brachialis.

Short Answer Essay Questions

12. Define and distinguish between first-, second-, and third-class levers. Which classes operate at a mechanical advantage?
13. (a) Name the four pairs of muscles that act together to compress the abdominal contents. (b) How do their fiber (fascicle) directions contribute to the strength of the abdominal wall? (c) Of these four muscles, which is the strongest flexor of the vertebral column? (d) Name the prime mover of inhalation (inspiration) in breathing.
14. List all six possible movements of the humerus that can occur at the shoulder joint, and name the prime mover(s) of each movement.
15. (a) Name two forearm muscles that are both powerful extensors and abductors of the hand at the wrist. (b) Name the only forearm muscle that can flex the distal interphalangeal joints. (c) Name the hand muscles responsible for opposition.
16. (a) Name the lateral rotators of the hip. (b) Explain the main function of each of the three gluteal muscles.
17. What is the functional reason why the muscle group on the dorsal leg (calf) is so much larger than the muscle group in the ventral region of the leg?
18. Define (a) a fascia compartment in a limb and (b) a retinaculum in a limb.
19. Name two muscles in each of the following compartments or regions: (a) thenar eminence (ball of thumb), (b) posterior compartment of forearm, (c) anterior compartment of forearm—deep muscle group, (d) anterior muscle group in the arm, (e) muscles of mastication, (f) third muscle layer of the foot, (g) posterior compartment of leg, (h) medial compartment of thigh, (i) posterior compartment of thigh.
20. (a) What muscles develop from the embryonic structures called somitomeres? (b) What are the major groups of pharyngeal arch (branchiomeric) muscles? (c) What three constrictor muscles squeeze swallowed food through the pharynx to the esophagus?
21. Describe the location and function of the levator ani muscle.

CRITICAL REASONING + CLINICAL APPLICATION QUESTIONS

1. Rosario, a student nurse, was giving Mr. Graves a back rub. What two broad superficial muscles of his back were receiving most of her attention?
2. When Mrs. O'Brian returned to her doctor for a follow-up visit after childbirth, she complained that she was having problems controlling urine flow (was incontinent) when she sneezed. The physician gave Mrs. O'Brian instructions in performing exercises to strengthen the muscles of the pelvic floor. What are these specific muscles?
3. Assume you are trying to lift a heavy weight off the ground with your right hand. Explain why it will be easier to flex your forearm at the elbow when your forearm is supinated than when it is pronated.
4. Chao-Jung abducted her thigh very strongly while riding a horse and pulled some groin muscles in her medial upper thigh. Precisely what muscles were most likely strained, and what is the medical name for her condition? (Hint: See this chapter's Related Clinical Terms.)

A.D.A.M.® INTERACTIVE ANATOMY QUESTIONS

1. Link: Atlas; System: Muscular; Deep muscles of trunk (anterior). What are the insertions of the rectus abdominis?

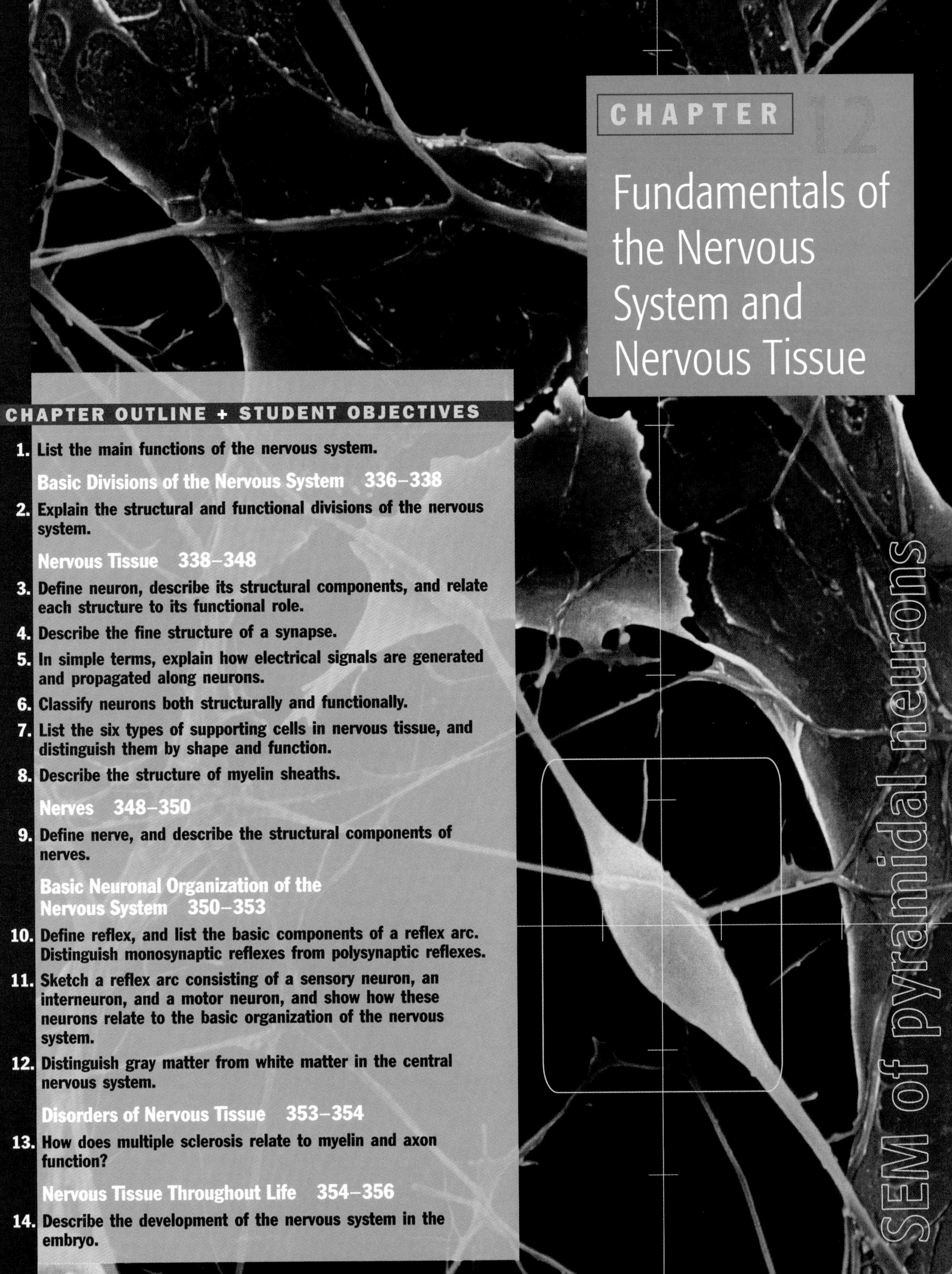

CHAPTER 12

Fundamentals of the Nervous System and Nervous Tissue

CHAPTER OUTLINE + STUDENT OBJECTIVES

1. List the main functions of the nervous system.

Basic Divisions of the Nervous System 336–338

2. Explain the structural and functional divisions of the nervous system.

Nervous Tissue 338–348

3. Define neuron, describe its structural components, and relate each structure to its functional role.
4. Describe the fine structure of a synapse.
5. In simple terms, explain how electrical signals are generated and propagated along neurons.
6. Classify neurons both structurally and functionally.
7. List the six types of supporting cells in nervous tissue, and distinguish them by shape and function.
8. Describe the structure of myelin sheaths.

Nerves 348–350

9. Define nerve, and describe the structural components of nerves.

Basic Neuronal Organization of the Nervous System 350–353

10. Define reflex, and list the basic components of a reflex arc. Distinguish monosynaptic reflexes from polysynaptic reflexes.
11. Sketch a reflex arc consisting of a sensory neuron, an interneuron, and a motor neuron, and show how these neurons relate to the basic organization of the nervous system.
12. Distinguish gray matter from white matter in the central nervous system.

Disorders of Nervous Tissue 353–354

13. How does multiple sclerosis relate to myelin and axon function?

Nervous Tissue Throughout Life 354–356

14. Describe the development of the nervous system in the embryo.

AS YOU ARE DRIVING DOWN THE FREEWAY, A horn blares on your right, and immediately you swerve to your left. Charlie's note on the kitchen table reads, "See you later—I will have the stuff ready at 6"; you know that the "stuff" is chili with taco chips, and your mouth waters in anticipation. While you are dozing, your infant son cries out softly, and instantly you awaken. What do these events have in common? All are everyday examples of the function of your nervous system, which has the cells in your body humming with activity nearly all of the time.

The nervous system is the master control and communications system of the body. Every thought, action, instinct, and emotion reflects its activity. Its cells communicate via electrical signals, which are rapid and specific and usually produce almost immediate responses.

The nervous system has three overlapping functions (Figure 12.1): (1) It uses its millions of *sensory receptors* to monitor changes occurring both inside and outside the body. Each of these changes is called a *stimulus,* and the gathered information is called **sensory input.** (2) It processes and interprets the sensory input and makes decisions about what should be done at each moment—a process called **integration.** (3) It dictates a response by activating the *effector organs,* our muscles or glands; the response is called **motor output.** Some examples illustrate how these functions work together. When you are driving and hear a horn to your right, your nervous system integrates this information (a horn means "danger"), and your arm muscles contract to turn the wheel to the left (motor output). As another example, when you taste food your nervous system integrates this sensory information and signals your salivary glands to secrete more saliva into your mouth.

This chapter begins with an overview of the fundamental divisions of the nervous system. We then focus on the functional anatomy of nervous tissue, especially of the nerve cells, or *neurons,* which are the key to the efficient system of neural communication. Finally, we discuss how the arrangement of neurons determines the structural organization of the nervous system.

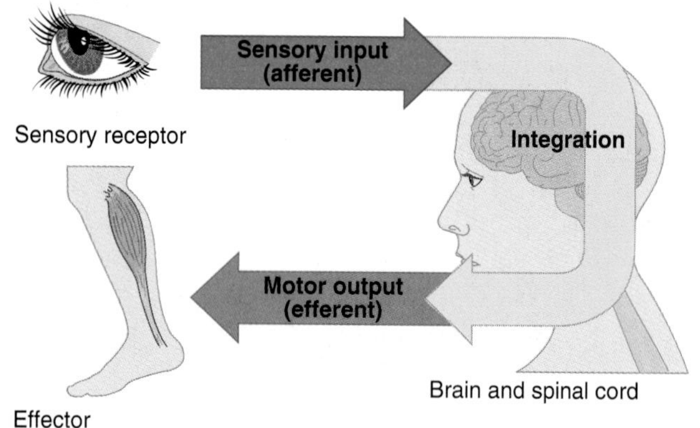

FIGURE 12.1 Simplified functions of the nervous system.

BASIC DIVISIONS OF THE NERVOUS SYSTEM

We have only one, highly integrated nervous system. However, for the sake of convenience we can divide it into two anatomical parts: the central nervous system and the peripheral nervous system (Figure 12.2). The **central nervous system (CNS)** consists of the *brain* and the *spinal cord,* which occupy the cranium and the vertebral canal, respectively. The CNS is the integrating and command center of the nervous system: It interprets incoming sensory signals and dictates motor responses based on past experiences, reflexes, and current conditions. The **peripheral nervous system (PNS),** the part of the nervous system outside the CNS, consists mainly of the *nerves* that extend from the brain and spinal cord. *Cranial nerves* carry signals to and from the brain, whereas *spinal nerves* carry signals to and from the spinal cord. These peripheral nerves serve as communication lines that link all regions of the body to the central nervous system.

As mentioned, the nervous system receives sensory inputs and dictates motor outputs (Figure 12.3). **Sensory,** or **afferent** (af′er-ent), signals are picked up by sensory receptors throughout the body and carried by nerve fibers of the PNS into the CNS (*afferent* means "carrying toward"). **Motor,** or **efferent** (ef′er-ent) signals are carried away from the CNS by nerve fibers in the PNS to innervate the muscles and glands, causing these organs to contract or secrete (*efferent* means "carrying away"). Both the sensory inputs and motor outputs are further divided according to the body regions they serve: The *somatic body region* consists of the structures external to the ventral body cavity (skin, skeletal musculature, bones), whereas the *visceral body region* mostly contains the viscera within the ventral body cavity (digestive tube, lungs, heart, bladder, and so on). This scheme results in the four main subdivisions: (1) *somatic sensory* (the sensory innervation of the outer part of the body); (2) *visceral sensory* (the sensory innervation of the viscera); (3) *somatic motor* (the motor innervation of skeletal musculature); and (4) *visceral motor* (the motor innervation of muscle in the viscera and of glands). Because they are essential to an understanding of the nervous system, we next describe each subdivision in greater detail (see Figure 12.3).

1. **Somatic sensory:** The **general somatic senses** are the senses whose receptors are spread widely throughout the outer part of the body (in this context the term *general* means "widespread"). These include the many senses experienced on the skin and in the body wall, such as touch, pain, pressure, vibration, and temperature. Another group of general somatic senses is the **proprioceptive senses,** or **proprioception** (pro″pre-o-sep′shun; "sensing one's own body"). These senses, of which most people have not heard, detect the amount of stretch in our

muscles, tendons, and joint capsules. They inform us of the position and movement of our bodies in space, giving us a "body sense." To demonstrate proprioception, flex and extend your fingers without looking at them—you will be able to feel exactly which joints are moving.

The **special somatic senses** are the somatic senses whose receptors are confined to relatively small areas rather than spread widely throughout the body (in this context, the term *special* means "localized"). Most special senses are confined to the head, including hearing and balance (with receptors in the inner ear), vision (with receptors in the eye), and smell (with receptors in the nose). Note that the technical name for the sense of balance is **equilibrium.**

2. **Visceral sensory:** The **general visceral senses** include stretch, pain, and temperature, which can be felt widely in the digestive and urinary tracts, reproductive organs, and other viscera. Nausea and hunger are also general visceral senses. Taste can be considered a **special visceral sense,** because the taste receptors are localized to the tongue and some surrounding structures.
3. **Somatic motor:** The **general somatic motor** part of the nervous system signals the contraction of the skeletal muscles in the body, which develop from myotomes and somitomeres (see Chapter 11, p. 270). We have voluntary control over the contraction of our skeletal muscles, so the somatic motor system is often called the *voluntary nervous system.* Because skeletal muscles are widely distributed throughout the body, there is no *special* somatic motor category.

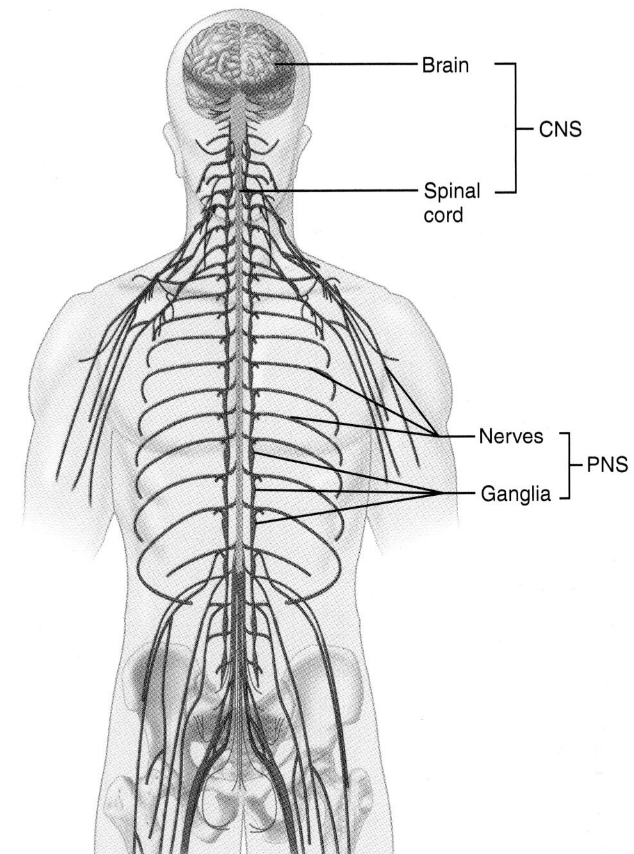

FIGURE 12.2 Basic divisions of the nervous system: The CNS and the PNS.

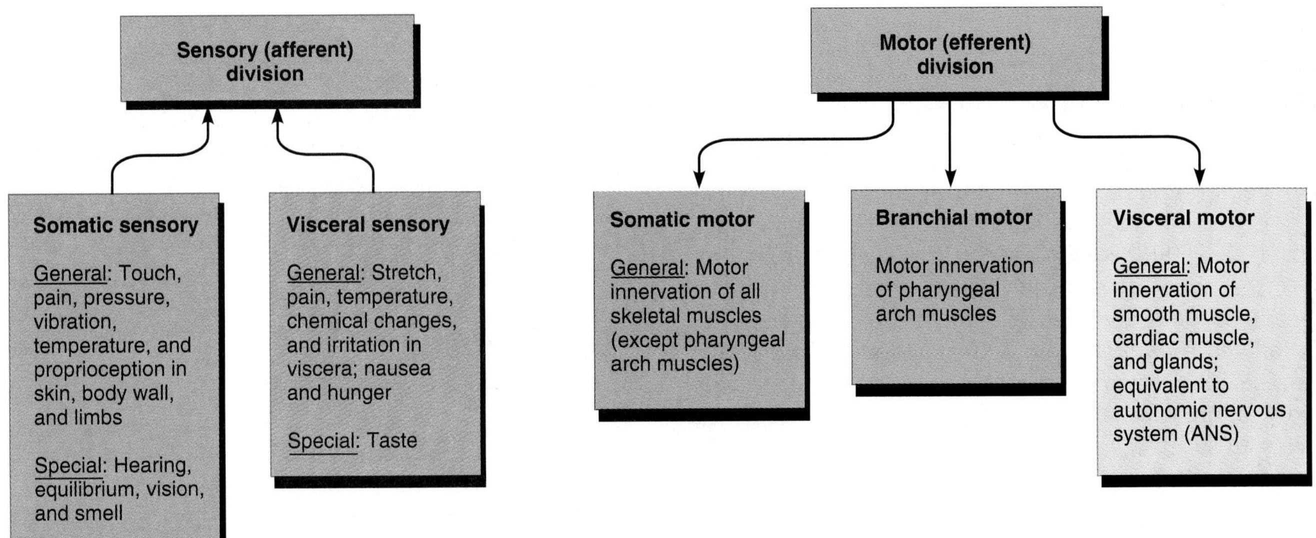

FIGURE 12.3 Types of sensory and motor information carried by the nervous system. Note that "general" means widespread, and "special" means localized when applied to these senses and motor innervations.

We do, however, consider the motor innervation of the pharyngeal arch (branchial) muscles to be a separate category: **branchial motor.** (Pharyngeal arch muscles were explained in Figure 11.4c and on page 272.) Their innervation was once classified as special visceral motor because these muscles originate around the embryonic pharynx (a visceral structure that is part of the digestive tube), but their innervation is now considered more akin to somatic motor because branchial muscles are typical skeletal muscles derived from somitomeres. We also have voluntary control over their contraction.

4. **Visceral motor:** The **general visceral motor** part of the nervous system regulates the contraction of the smooth muscle and cardiac muscle of the body and secretion by the body's many glands. General visceral motor neurons make up the *autonomic* (aw-to-nom'ik) *nervous system (ANS),* which controls the function of the visceral organs (see Chapter 15). Because we generally have no voluntary control over such activities as the pumping of the heart and movement of food through the digestive tract, the ANS is also called the *involuntary nervous system.*

We will return to these basic functional components of the nervous system throughout the next few chapters.

NERVOUS TISSUE

The nervous system consists mostly of **nervous tissue,** whose cells are densely packed and tightly intertwined. Although exceedingly complex, nervous tissue is made up of just two main types of cells: (1) *neurons,* the excitable nerve cells that transmit electrical signals, and (2) *supporting cells,* nonexcitable cells that surround and wrap the neurons. Both of these cell types develop from the same embryonic tissues: neural tube and neural crest (see p. 62). First we examine neurons.

The Neuron

The human body contains many billions of **neurons,** or **nerve cells** (Figure 12.4), which are the basic structural units of the nervous system. Neurons are highly specialized cells that conduct electrical signals from one part of the body to another. These signals, which are transmitted along the plasma membrane, often take the form of *nerve impulses,* or *action potentials.* Basically, an impulse is a reversal of electrical charge that travels rapidly along the neuronal membrane (see p. 342). Besides their ability to conduct electrical signals, neurons have other special characteristics:

1. They have *extreme longevity.* Neurons can live and function for a lifetime, over 100 years.
2. They *do not divide.* As the fetal neurons assume their roles as communication links in the nervous system, they lose their ability to undergo mitosis. There can be a high price for this characteristic of neurons, for they cannot replace themselves if destroyed.
3. They have an exceptionally *high metabolic rate,* requiring continuous and abundant supplies of oxygen and glucose. Neurons cannot survive for more than a few minutes without oxygen.

Neurons typically are large, complex cells. Although they vary in structure, they all have a *cell body* from which one or more *processes* project (see Figure 12.4).

The Cell Body

The **cell body** is also called a *soma* (= body) or a *perikaryon* (per″ĭ-kar'e-on; "around the nucleus"). Although the cell bodies of different neurons vary widely in size (from 5 to 140 μM in diameter), all consist of a single nucleus surrounded by cytoplasm. In all but the smallest neurons, the nucleus is spherical and clear and contains a dark nucleolus near its center (Figure 12.5). Such a nucleus resembles an owl's eye. The cytoplasm contains all the usual cellular organelles as well as distinctive **chromatophilic (Nissl) bodies** (kro-mah'to-fil-ic). These bodies, whose name means "color-loving" or "easily stained," are large clusters of rough endoplasmic reticulum and free ribosomes that stain darkly with basic dyes. These cellular organelles continually renew the membranes of the cell and the protein part of the cytosol. **Neurofibrils** are bundles of intermediate filaments *(neurofilaments)* that run in a network between the chromatophilic bodies (see Figure 12.4). Like all intermediate filaments (see p. 40), neurofilaments keep the cell from being pulled apart when it is subjected to tensile forces. *Lipofuscin* (lip″o-fu'sin), a yellow-brown pigment, is a harmless by-product of lysosomal activity and accumulates with age. The cell body is the focal point for the outgrowth of the neuron processes during embryonic development. In most neurons, the plasma membrane of the cell body acts as a receptive surface that receives signals from other neurons.

Most neuron cell bodies are located within the CNS, where they are protected by the bones of the skull and vertebral column. However, clusters of cell bodies called **ganglia** lie along the nerves in the PNS (see Figure 12.2). The singular of *ganglia* is **ganglion** (gang'le-on), literally a "knot in a string."

Neuron Processes

Arm-like **processes** extend from the cell bodies of all neurons. These processes are of two types, *dendrites* and *axons* (see Figure 12.4), which differ from each other both in structure and in the functional properties of their plasma membranes. The convention is to

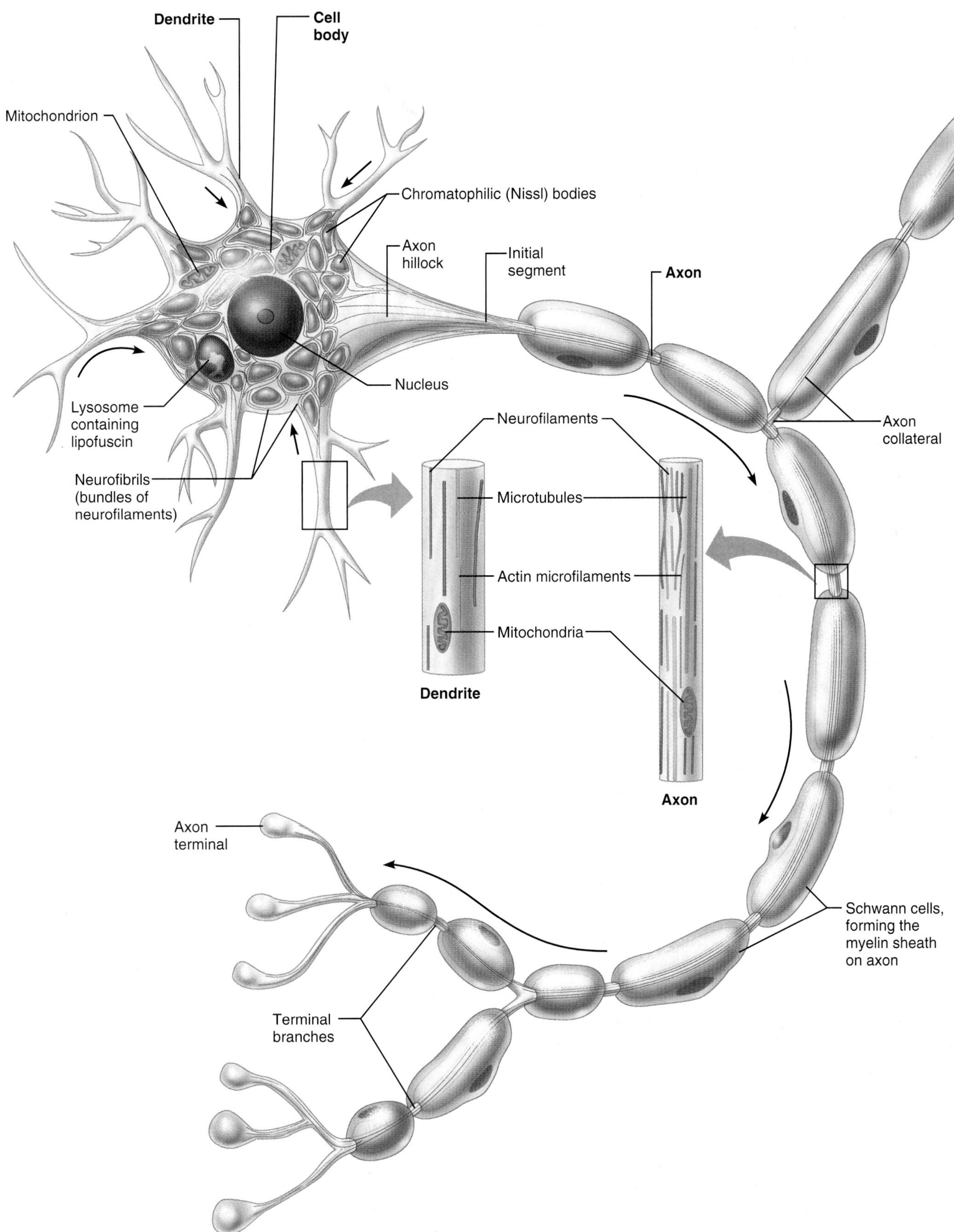

FIGURE 12.4 Structure of a typical large neuron (a motor neuron). The arrows indicate directions in which signals travel along the dendrites and the axon. The insets at center are enlargements of a dendrite and the axon. The axon of this particular neuron is covered by a myelin sheath (myelin is discussed on page 347).

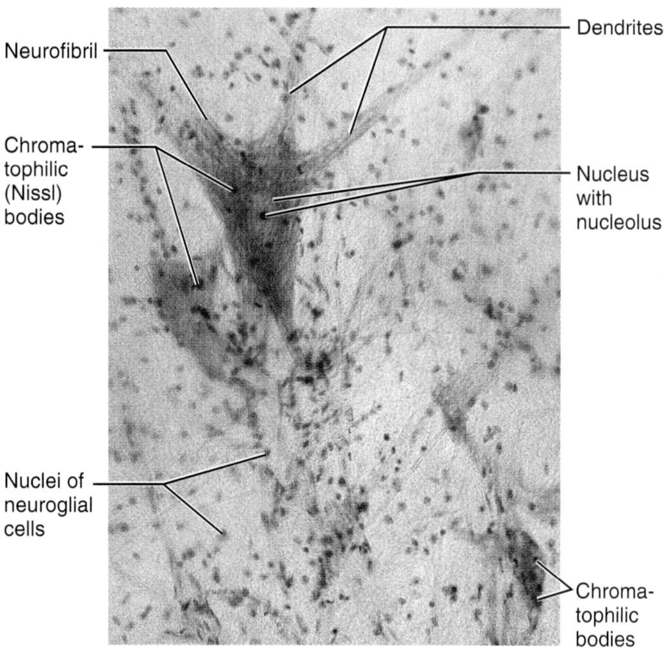

FIGURE 12.5 Neuron cell bodies in the CNS. For this micrograph, a section through the spinal cord was stained with a dye that reveals the large, round nuclei and the chromatophilic bodies of neurons, but does not stain the cell processes beyond the first parts of the dendrites (100 ×).

describe the processes using a motor neuron as an example of a typical neuron. Motor neurons do indeed resemble most neurons in the arrangement of their processes, but sensory neurons and some tiny neurons differ from the "typical" pattern presented here.

Dendrites Most neurons have numerous **dendrites,** processes that branch like the limbs on a tree (*dendro* = tree). Virtually all organelles that occur in the cell body also occur in dendrites, and chromatophilic bodies extend into the basal part of each dendrite. Dendrites function as *receptive sites,* providing an enlarged surface area for receiving signals from other neurons. By definition, dendrites conduct electrical signals *toward* the cell body. As will be explained shortly, these signals are not true impulses (action potentials) but another type of electrical signal called *graded potentials,* which are also carried by the membrane of the cell body itself (see p. 342).

Axons A neuron has only one **axon** (ak'son; "axis, axle"), whose **initial segment** arises from a cone-shaped region of the cell body called the **axon hillock** ("little hill"). Axons are thin processes of uniform diameter throughout their length. By definition, axons are impulse generators and conductors that transmit nerve impulses *away* fom their cell body.

Chromatophilic bodies and the Golgi apparatus are absent from the axon and the axon hillock alike. In fact, axons lack ribosomes and all organelles involved in protein synthesis, so they must receive their proteins from the cell body. Neurofilaments, actin microfilaments, and microtubules are especially evident in axons, where they provide structural strength. These cytoskeletal elements also aid in the transport of substances to and from the cell body as the axonal cytoplasm is continually recycled and renewed. This movement of substances along axons is called **axonal transport.**

The axon of some neurons is short, but in others it can be extremely long. For example, the axons of the motor neurons that control muscles in the foot extend from the lumbar region of the spine to the sole, a distance of a meter or more (3–4 feet). Any long axon is called a **nerve fiber.**

Critical Reasoning

Neurons with the longest axons have the largest cell bodies, whereas neurons with short axons tend to have small cell bodies. Suggest an explanation for this relationship.

Because the cell body must continuously maintain and renew the components of the axon, a longer axon requires a larger cell body to house more cellular "machinery" to produce the axon-sustaining materials.

Although axons branch far less frequently than dendrites, occasional branches do occur along their length. These branches, called **axon collaterals,** extend from the axon at more or less right angles (see Figure 12.4). Whether an axon remains undivided or has collaterals, it usually branches profusely at its terminus (end): Ten thousand of these **terminal branches** per neuron is not unusual. They end in knobs that are variously called **axon terminals,** *end bulbs,* or *boutons* (boo-tonz'; "buttons"). These knobs contact other neurons to form specialized cell junctions called *synapses* (discussed shortly).

A nerve impulse is typically generated at the axon's initial segment and is conducted along the axon to the axon terminals, where it causes the release of chemicals called *neurotransmitters* into the extracellular space. The neurotransmitters excite or inhibit the neurons with which the axon is in close contact. Because each neuron typically both receives signals from and sends signals to scores of other neurons, it "carries on conversations" with many different neurons at the same time.

Axon diameter varies considerably among the different neurons of the body. Axons with larger diameters conduct impulses faster than those with smaller diameters because of a basic law of physics: The resistance to the passage of an electrical current decreases as the diameter of any "cable" increases.

Synapses

The site at which neurons communicate is called a **synapse** (sin'aps; "union"), a cell junction that mediates the transfer of information from one neuron to the next (Figure 12.6). Because signals pass across most synapses

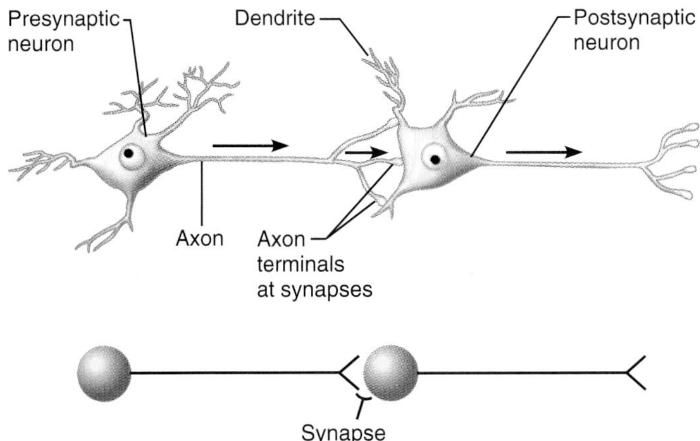

FIGURE 12.6 Two neurons communicating at synapses. Synapses occur where the axon terminals of a presynaptic neuron contact the membrane of a postsynaptic neuron. Arrows indicate the direction of flow of electrical signals. The lower drawing is a highly simplified version of the drawing above. Such simplified drawings are used to represent synapsing neurons in many figures in this text.

in one direction only, synapses determine the direction of information flow through the nervous system.

The neuron that conducts signals toward a synapse is called the **presynaptic neuron;** the neuron that transmits electrical activity away from the synapse is called the **postsynaptic neuron** (see Figure 12.6). As you might anticipate, most neurons function as both presynaptic (information-sending) and postsynaptic (information-receiving) neurons, getting information from some neurons and dispatching it to others.

There are several types of synapses (Figure 12.7). Most synapses occur between the axon terminals of one neuron and the dendrites of another neuron; these are called **axodendritic synapses.** However, many synapses also occur between axons and neuron cell bodies; these are called **axosomatic synapses.** Less common, and far less understood, are synapses between two axons *(axoaxonic),* between two dendrites *(dendrodendritic),* or between a dendrite and a cell body *(dendrosomatic).*

Structurally, synapses are elaborate cell junctions. We focus here on the axodendritic synapse (Figure 12.8) because its structure is representative of most synapse types. On the presynaptic side, the axon terminal contains **synaptic vesicles.** These are membrane-bound sacs filled with *neurotransmitters,* the molecules that transmit signals across the synapse. Mitochondria are abundant in the axon terminal, as the secretion of neurotransmitters requires a great deal of energy. At the synapse, the plasma membranes of the two neurons are separated by a **synaptic cleft.** On the undersurfaces of the opposing cell membranes are regions of dense material, called **presynaptic** and **postsynaptic densities.** In three dimensions, the former is actually a honeycombed grid and the latter is a perforated plate.

How does such a synapse function? When an impulse travels along the axon of the presynaptic neuron, it signals the synaptic vesicles to fuse with the presynaptic membrane at the presynaptic density. The released neurotransmitter molecules diffuse across the synaptic cleft and bind to the postsynaptic membrane at the postsynaptic density. This binding changes the membrane charge on the postsynaptic neuron, influencing the generation of a nerve impulse or action potential in that neuron (discussed further in the next section).

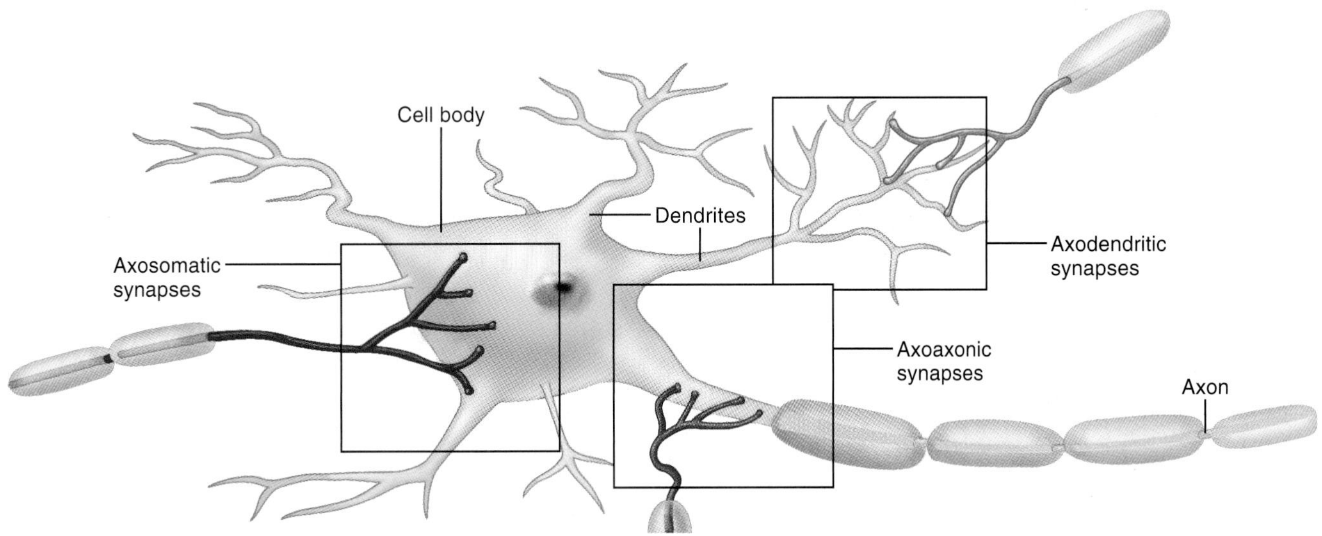

FIGURE 12.7 Some important types of synapses. Several axons are synapsing on the large neuron in the center of this figure. Axosomatic, axoaxonic, and axodendritic synapses are shown. Axon terminals of presynaptic neurons are indicated in different colors for the different types of synapses.

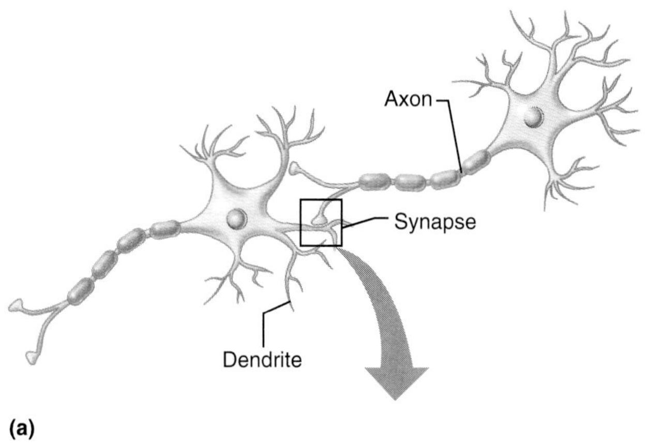

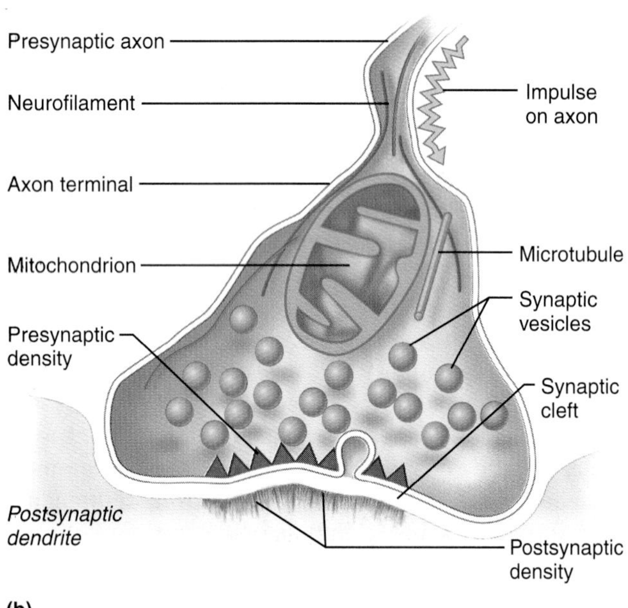

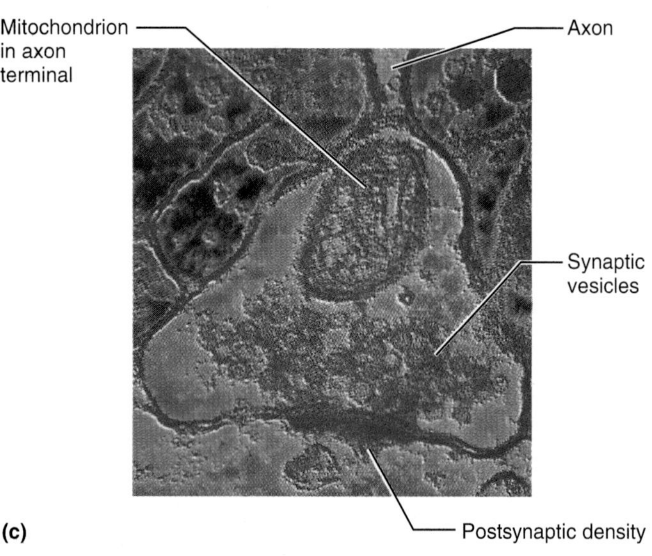

The Signals Carried by Neurons

We have said that the plasma membranes of neurons conduct electrical signals, and that synapses relay the signals from neuron to neuron. Now we will explain these signals, in a simplified way. In a resting (unstimulated) neuron, the membrane is *polarized;* that is, its inner, cytoplasmic side is negatively charged with respect to its outer, extracellular side (Figure 12.9a). When a neuron is stimulated experimentally (pinched or shocked, for example), the permeability of the plasma membrane changes at the site of the stimulus, allowing positive ions to rush in (Figure 12.9b). As a result, the inner face of the membrane becomes less negative, an event called *depolarization*.

Action Potentials on Axons Any part of the neuron depolarizes if stimulated, but at the *axon* alone something else can happen. If the stimulus applied to an axon is strong enough, it will trigger a **nerve impulse,** or **action potential,** in which the membrane is not only depolarized, but its polarity is completely reversed so it becomes negative externally and positive internally (Figure 12.9c). Once begun, this nerve impulse travels rapidly down the entire length of the axon without decreasing in strength (Figure 12.9d). After the impulse has passed, the membrane repolarizes itself (Figure 12.9e).

Graded Potentials on Dendrites and the Cell Body In the human body, natural stimuli are not applied directly to axons, but to the dendrites and the cell body, which constitute the receptive zone of the neuron. When the membrane of this receptive zone is stimulated, it does not undergo a polarity reversal. Instead, it undergoes a local depolarization in which the inner surface of the membrane merely becomes less negative (but not positive). This local depolarization, called a **graded potential,** spreads from the receptive zone to the axon hillock (trigger zone), decreasing in strength as it travels. If this depolarizing signal is strong enough when it reaches the initial segment of the axon, it acts as the trigger that initiates an action potential in the axon. Thus, signals from the receptive zone determine if the axon will fire an impulse.

Synaptic potentials Most neurons in the body do not receive stimuli directly from the environment but are "stimulated" only by signals received at synapses from

FIGURE 12.8 Structure of a synapse between an axon terminal and a dendrite. (a) Orientation diagram, showing the location of the synapse. **(b)** Enlarged diagram of the elements of the synapse. The actual synaptic junction consists of the pre- and postsynaptic densities and synaptic cleft. **(c)** Transmission electron micrograph from which the diagram in part (b) was drawn (60,000 ×).

other neurons. Synaptic input influences impulse generation in the following way. First, neurotransmitters released by presynaptic neurons alter the permeability of the postsynaptic membrane to certain ions. This may lead to an inflow of positive ions, which, as previously explained, depolarizes the postsynaptic membrane and drives the postsynaptic neuron toward impulse generation. Synapses that behave in this way are **excitatory synapses.** By contrast, **inhibitory synapses** have the opposite effect: They cause the external surface of the postsynaptic membrane to become even more positive, thereby reducing the ability of the postsynaptic neuron to generate an action potential. Thousands of excitatory and inhibitory synapses act on every neuron, competing to determine whether or not that neuron will generate an impulse.

Classification of Neurons

Neurons may be classified both by structure and by function.

Structural Classification of Neurons Neurons are grouped structurally according to the number of processes that extend from the cell body. By this classification, neurons are *multipolar, bipolar,* or *unipolar* (*polar* = ends, poles).

Most neurons are **multipolar neurons;** that is, they have more than two processes (Figure 12.10a). Usually, multipolar neurons have numerous dendrites and an axon. However, some small multipolar neurons have no axon and rely strictly on their dendrites for conducting signals. Well over 99% of neurons in the body belong to the multipolar class.

Bipolar neurons have two processes that extend from opposite sides of the cell body (Figure 12.10b). These very rare neurons occur in some of the special sensory organs (inner ear, smell region of the nose, retina of the eye), where they mostly serve as sensory neurons.

Unipolar neurons have a short, single process that emerges from the cell body and divides like a T into two long branches (Figure 12.10c). Most unipolar neurons start out as bipolar neurons whose two processes fuse together near the cell body during development. Therefore, they are more properly called *pseudounipolar neurons* (*pseudo* = false). Unipolar neurons make up the typical sensory neurons, which are discussed shortly.

Functional Classification of Neurons The functional classification scheme groups neurons according to the direction the nerve impulse travels relative to the CNS. Based on this criterion, there are *sensory neurons, motor neurons,* and *interneurons* (Figure 12.11).

Sensory neurons, or *afferent neurons,* transmit impulses *toward* the CNS from sensory receptors in the PNS. Virtually all sensory neurons are unipolar, and their cell bodies are in ganglia outside the CNS. As previously mentioned, the short, single process near the neuron cell body divides into two longer branches. One of these branches runs centrally into the CNS and is called the **central process,** whereas the other branch extends peripherally to the receptors and is called the **peripheral process.** Both of these processes function as one, carrying impulses directly from the peripheral receptors to the CNS.

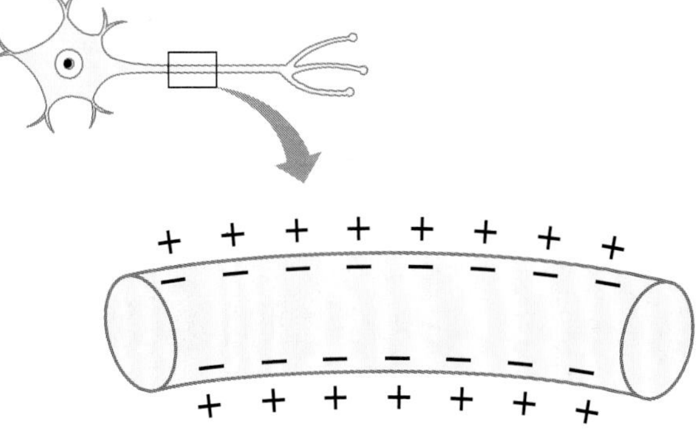

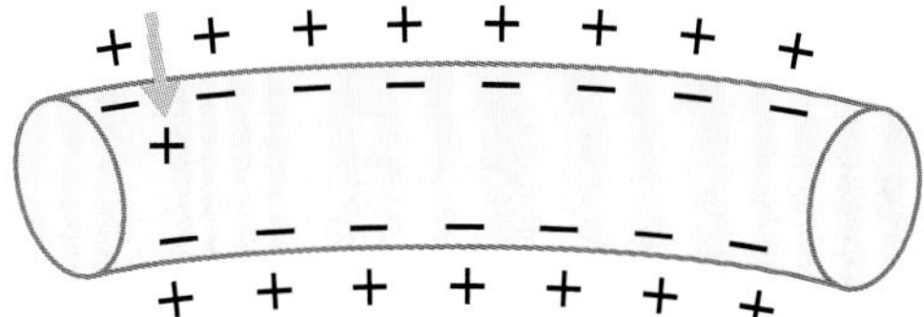

(a) Resting membrane

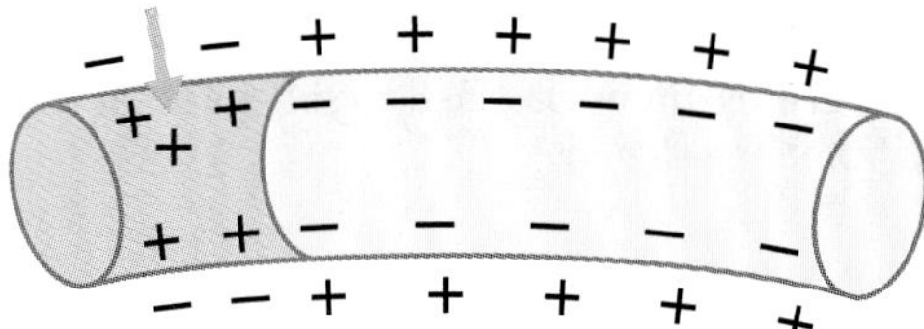

(b) Stimulus causes depolarization

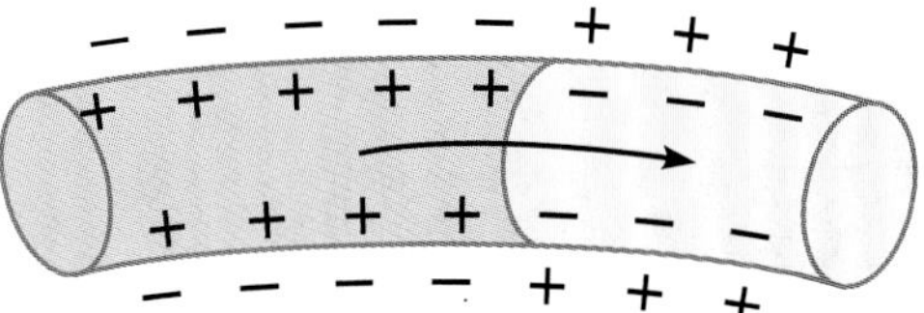

(c) Depolarization and generation of the action potential

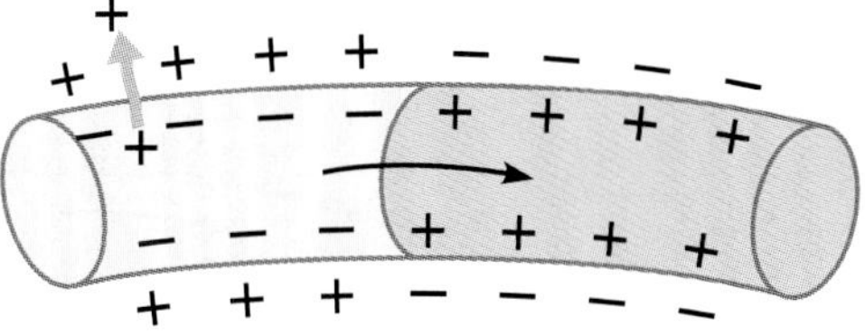

(d) Propagation of the action potential

(e) Repolarization

FIGURE 12.9 A segment of an axon, illustrating the generation and propagation of an action potential (nerve impulse). See the text for an explanation of stages (a) through (e).

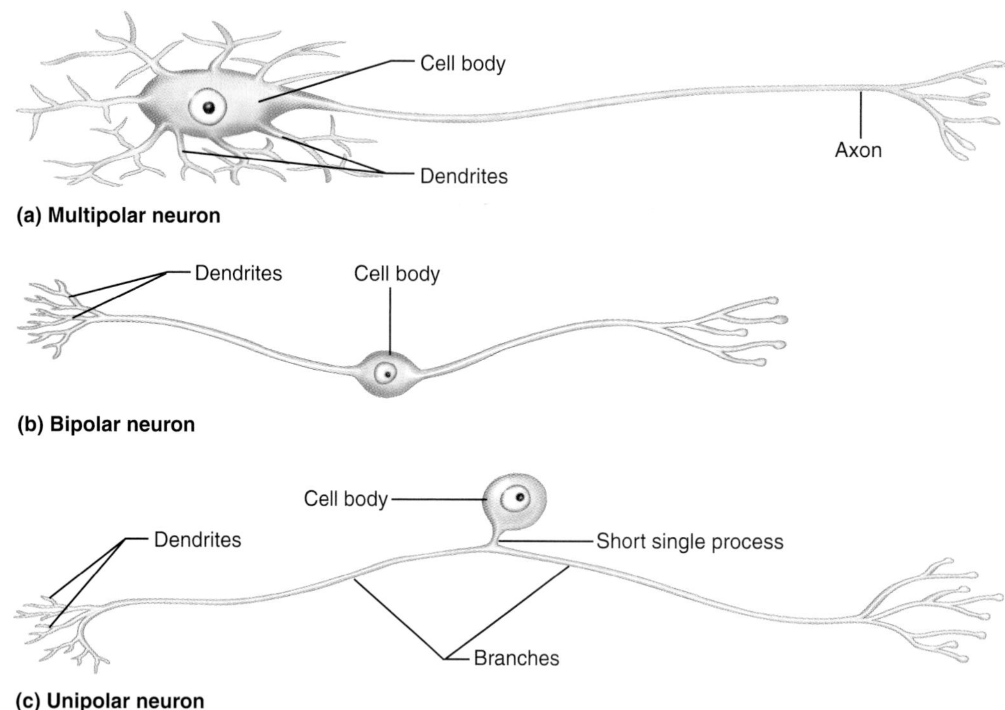

FIGURE 12.10 Neurons classified by structure (the number of processes projecting from the cell body).

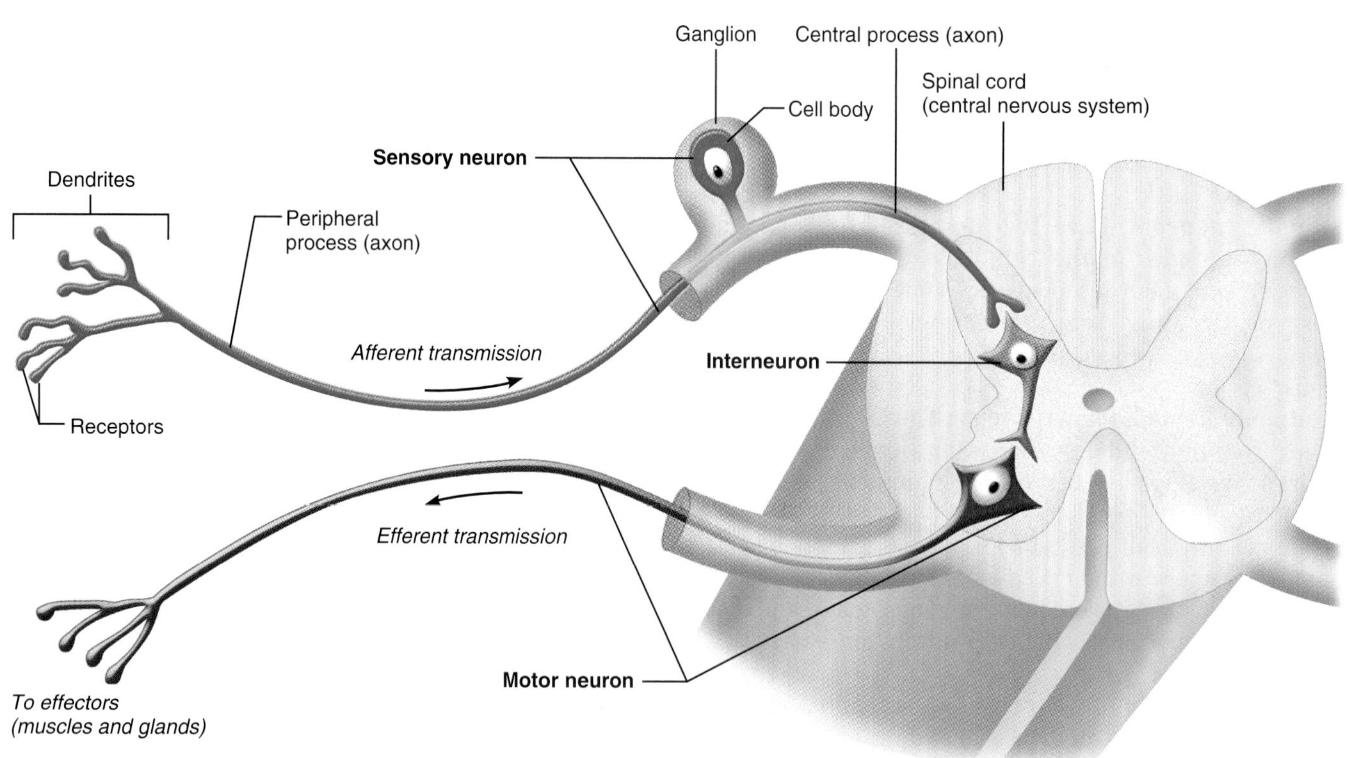

FIGURE 12.11 Neurons classified by function, as sensory neurons (blue), motor neurons (red), and interneurons (purple). Sensory neurons (top) transmit impulses from sensory receptors (in skin, viscera, muscles, and tendons) to the central nervous system. Most sensory neurons are unipolar neurons, with cell bodies in ganglia in the PNS. Motor neurons (below) transmit impulses from the CNS to the effectors. Interneurons (association neurons), which are confined to the CNS, complete the communication line between sensory and motor neurons.

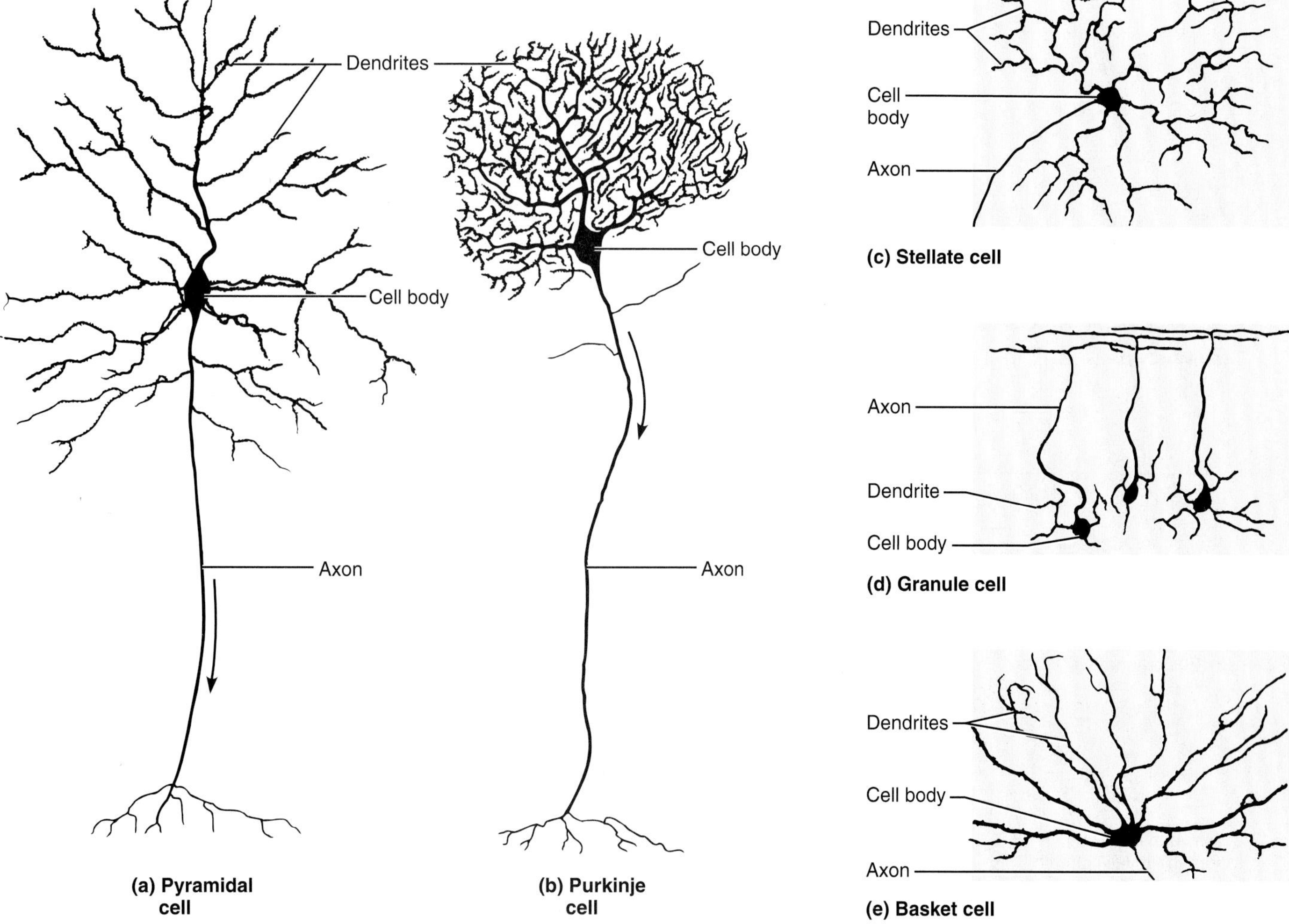

FIGURE 12.12 Variety of interneurons (a–e). All neurons shown are from an extremely cell-rich part of the brain called the cerebellum, except (a), which is from the largest brain region, the cerebrum. Each neuron's distinct appearance depends partly on the length of its axon but mostly on the number and branching pattern of its dendrites. Neurons with more complex dendrites have greater synaptic input; that is, more axon terminals synapse on such neurons.

Are these processes of sensory neurons dendrites, or are they axons? The *central process* is clearly an axon because it (1) carries a nerve impulse and (2) carries that impulse away from the cell body—the two criteria that define an axon (p. 340). The *peripheral process,* by contrast, is ambiguous: It carries nerve impulses, as does an axon, but these signals travel *toward* the cell body, a fundamental feature of dendrites. We choose to call the peripheral process an *axon* and a nerve fiber, because its fine structure is identical to that of true axons. For sensory neurons, we use the term *dendrite* to mean only the small branches in the receptor region, at the very distal end of the peripheral process (see Figure 12.11).

Motor neurons, or *efferent neurons,* carry impulses *away* from the CNS to effector organs (muscles and glands). Motor neurons are multipolar, and their cell bodies are located in the CNS (except for some neurons of the autonomic nervous system). Motor neurons form junctions with effector cells, signaling muscles to contract or glands to secrete.

Interneurons, or *association neurons,* lie between motor and sensory neurons and are confined entirely to the CNS (see Figure 12.11). Interneurons link together into chains that form complex neural pathways. The fact that interneurons make up 99.98% of the neurons of the body reflects the vast amount of information processed in the human CNS. Almost all interneurons are multipolar, but they show great diversity in size and in the branching patterns of their processes (Figure 12.12).

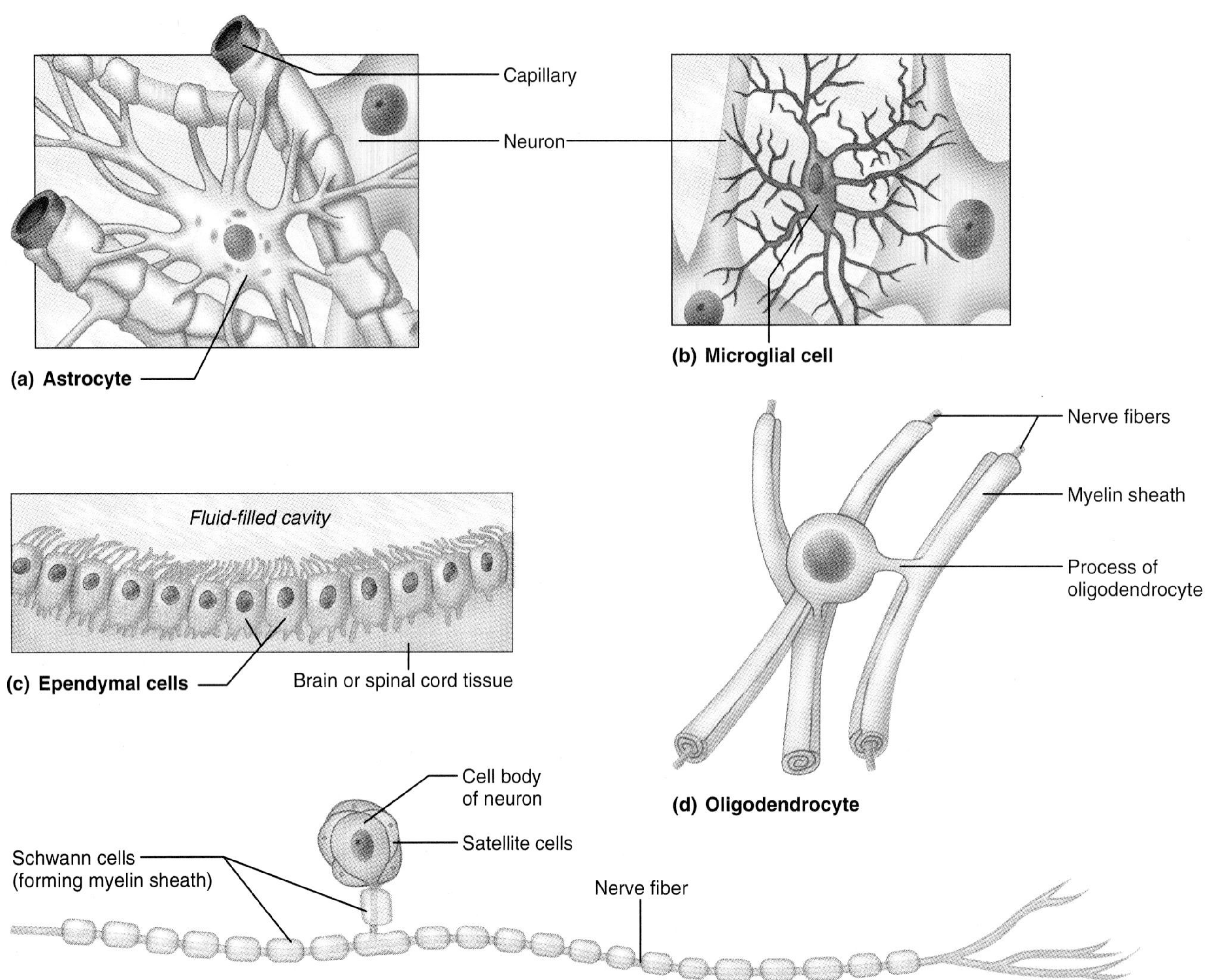

FIGURE 12.13 Supporting cells of nervous tissue. (a–d) The four types of neuroglial cells in the CNS. Note in (c) that ependymal cells line the fluid-filled central cavity of the spinal cord and brain. (e) The two types of supporting cells in the PNS: Schwann cells and satellite cells around a sensory neuron.

Supporting Cells

All neurons associate closely with non-nervous **supporting cells,** of which there are six types—four in the CNS and two in the PNS (Figure 12.13). Whereas each type has a unique specific function, in general these cells provide a supportive scaffolding for neurons. Furthermore, they cover all nonsynaptic parts of the neurons, thereby insulating the neurons and keeping the electrical activities of adjacent neurons from interfering with each other.

The importance of supporting cells in insulating nerve fibers from one another is illustrated in the painful disorder called **tic douloureux** (tik doo″loo-roo′; "wincing in pain"). In this condition, the supporting cells around the sensory nerve fibers in the main nerve of the face, the trigeminal nerve, degenerate and are lost. As a result, impulses in nerve fibers that carry *touch* sensations proceed to influence and stimulate the uninsulated *pain* fibers in the same nerve, leading to a perception of pain by the brain. Because of this crossover, the softest touch to the face can produce agonizing pain. (For more information on tic douloureux, see Table 14.3, page 421.)

Supporting Cells in the CNS

The supporting cells in the CNS are collectively called **neuroglia** (nu-rog′le-ah; "nerve glue") or **glial** (gle′al) **cells.** Most authorities restrict usage of the term *neuroglia* to the supporting cells in the CNS, but others consider all supporting cells neuroglia, including those in the PNS.

Like neurons, most glial cells have branching processes and a central cell body (Figure 12.13a–d). Neuroglia can be distinguished from neurons, however, by their much smaller size and by their darker-staining nuclei (see Figure 12.5). They outnumber neurons in the CNS by about 10 to 1, and they make up about half the mass of the brain. Unlike neurons, glial cells can divide throughout life.

Critical Reasoning

Which of the two cell types in neural tissue—neurons or neuroglia—gives rise to more brain tumors?

The glial cells do. Because glial cells can divide, they accumulate the "mistakes" in DNA replication that may transform them into neoplastic cells. Such an accumulation does not occur in neurons, which do not divide. Therefore, most tumors that originate in the brain (60%) are gliomas, *tumors formed by uncontrolled proliferation of glial cells. Two percent of all cancers are gliomas, and their incidence is increasing. These are difficult cancers to treat, and the one-year survival rate is under 50%.*

Star-shaped **astrocytes** (as'tro-sītz; "star cells") are the most abundant glial cells (Figure 12.13a). They have many radiating processes with bulbous ends. Some of these bulbs cling to neurons (including the axon terminals), whereas others cling to capillaries. Functionally, astrocytes take up and release ions to control the ionic environment around neurons. This is valuable because the concentrations of various ions outside the axons must be kept within narrow limits for nerve impulses to be generated and conducted. Astrocytes also function to recapture and recycle released neurotransmitters.

Advances in Understanding Recent findings have revealed why astrocytes contact both neurons and capillaries. The contacts with neurons allow astrocytes to sense when the neurons are highly active and releasing large amounts of the common neurotransmitter, glutamate, at their synapses. In response to such activity, the astrocytes increase their rate of glutamate uptake from the synapses. At the same time, the astrocytes extract blood sugar from the capillaries they contact, thereby obtaining the energy they need to fuel the process of glutamate uptake. ✷

Microglia, the smallest and least abundant of the neuroglia, have elongated cell bodies and cell processes with many pointed projections, like a thorny bush (see Figure 12.13b). They are phagocytes, the macrophages of the CNS. They migrate to, and then engulf, invading microorganisms and injured or dead neurons. Unlike other neuroglial cells, microglia do not originate in nervous tissue. Instead, like the other macrophages of the body, they derive from blood cells called monocytes. The monocytes that become microglia migrate to the CNS during the embryonic and fetal periods.

Recall from Chapter 3 (p. 62) that the CNS originates in the embryo as a hollow neural tube and retains a central cavity throughout life. **Ependymal cells** (ĕ-pen'dĭ-mal; "wrapping garment") form a simple epithelium that lines the central cavity of the spinal cord and brain (see Figure 12.13c). Here these cells provide a fairly permeable layer between the *cerebrospinal fluid* that fills this cavity and the tissue fluid that bathes the cells of the CNS. Ependymal cells bear cilia that help circulate the cerebrospinal fluid.

Oligodendrocytes (ol″ĭ-go-den'dro-sītz) (see Figure 12.13d) have fewer branches than astrocytes; indeed, their name means "few-branch cells." They line up in small groups and wrap their cell processes around the thicker axons in the CNS, producing insulating coverings called *myelin sheaths* (discussed in detail shortly).

Supporting Cells in the PNS

The two kinds of supporting cells in the PNS are *satellite cells* and *Schwann cells* (see Figure 12.13e), very similar cell types that differ mainly in location. **Satellite cells** surround neuron cell bodies within ganglia. (Their name comes from a fancied resemblance to the moons or satellites around a planet.) **Schwann cells,** also called *neurolemmocytes,* surround all axons in the PNS and form myelin sheaths around many of these axons.

Now that we've considered the six kinds of neuroglia, we turn our attention to an important structure produced by oligodendrocytes and Schwann cells: the myelin sheath.

Myelin Sheaths

Myelin sheaths are segmented structures, each composed of the lipoprotein **myelin,** that surround the thicker axons of the body (Figure 12.14). Each segment of myelin consists of the plasma membrane of a supporting cell rolled in concentric layers around the axon (see Figure 12.15 and Figure 12.16 on p. 350). Myelin sheaths form an insulating layer that prevents the leakage of electrical current from the axon, increasing the speed of impulse conduction along the axon and making impulse propagation more energy efficient.

We next examine myelin sheaths as they occur in the PNS and in the CNS.

Myelin Sheaths in the PNS

As previously stated, the myelin sheaths in the PNS are formed by Schwann cells (see Figure 12.14a). Myelin sheaths develop during the fetal period and the first year or so of postnatal life. To form the myelin sheath, the Schwann cells first indent to receive the axon and then wrap themselves around the axon repeatedly, in a jellyroll fashion (Figure 12.15b). Initially the wrapping is loose, but the cytoplasm of the Schwann cell is gradually squeezed outward from between the membrane layers. When the wrapping process is finished, many concentric layers of Schwann cell plasma membrane ensheathe the axon in a tightly packed coil of membranes that is the true myelin sheath. The nucleus and most of the cytoplasm of the Schwann cell end up just external to the myelin layers. This external material is called the **neurilemma** ("neuron sheath") (see Figure 12.15b).

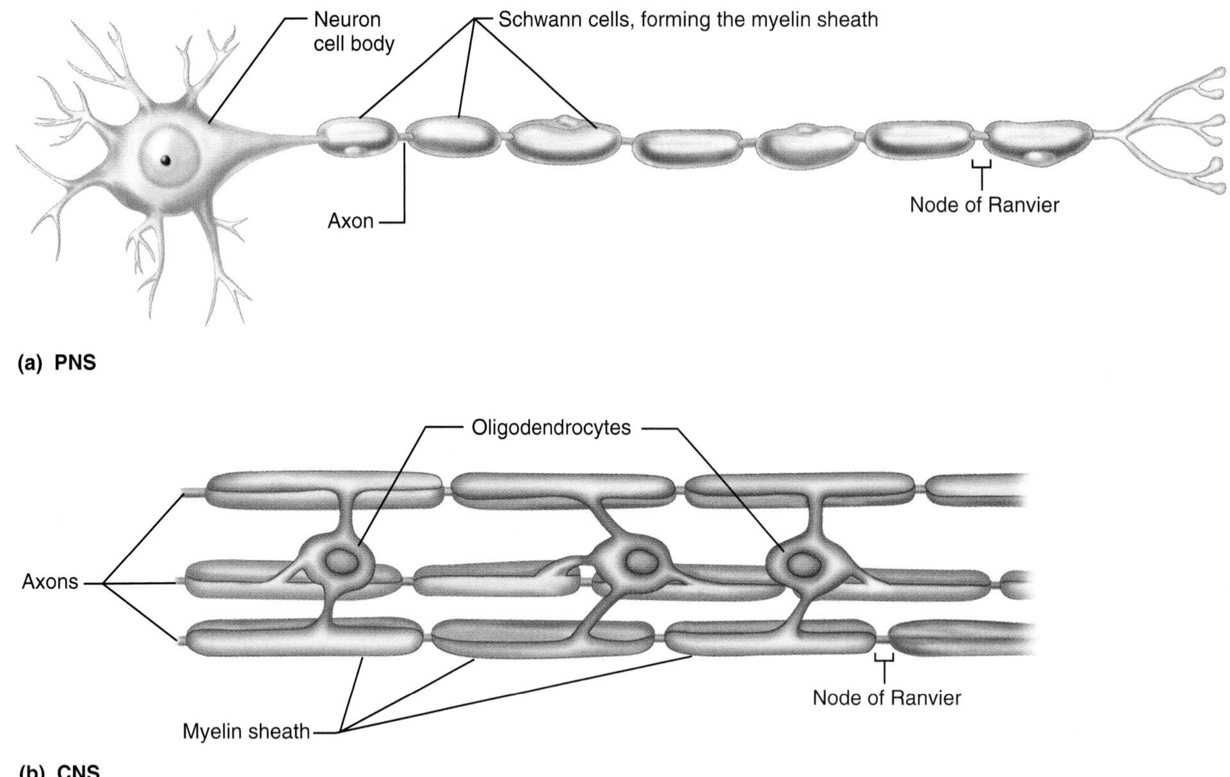

FIGURE 12.14 **Myelin sheaths in the PNS and CNS.** **(a)** In the PNS, Schwann cells form the myelin sheath around axons. **(b)** In the CNS, oligodendrocytes form the myelin sheath around axons. Note that each oligodendrocyte contributes myelin to several axons, whereas each Schwann cell is associated with one axon only.

Because the adjacent Schwann cells along a myelinated axon do not touch one another, there are gaps in the myelin sheath. These gaps, called **nodes of Ranvier,** or *neurofibral nodes,* occur at regular intervals about 1 mm apart (see Figures 12.14 and 12.15a). In myelinated axons, nerve impulses do not travel along the myelin-covered regions of the axonal membrane, but instead jump from the membrane of one node of Ranvier to the next, in a way that greatly speeds impulse conduction.

Only the thick, rapidly conducting axons are sheathed with myelin. In contrast, thin, slowly conducting axons lack a myelin sheath and are called **unmyelinated axons** (see Figure 12.15c). Schwann cells surround such axons but they do not wrap around them in concentric layers of membrane. A single Schwann cell can partly enclose 15 or more unmyelinated axons, each of which occupies a separate tubular recess in the surface of the Schwann cell.

Myelin Sheaths in the CNS

Oligodendrocytes form the myelin sheaths in the brain and spinal cord (see Figure 12.14b). In contrast to Schwann cells, each oligodendrocyte has multiple processes that coil around several different axons. Nodes of Ranvier are present, although they are more widely spaced than those in the PNS.

As in the PNS, the thinnest axons in the CNS are unmyelinated. These unmyelinated axons are covered by the many long processes of glial cells that are so abundant in the CNS.

NERVES

A **nerve** is a cord-like organ in the peripheral nervous system (see Figure 12.2). Each nerve consists of many axons (nerve fibers) arranged in parallel bundles and enclosed by successive wrappings of connective tissue (Figure 12.17). Almost all nerves contain both myelinated and unmyelinated sensory and motor fibers.

Within a nerve, each axon is surrounded by Schwann cells, then a delicate layer of loose connective tissue called **endoneurium** (en″do-nu′re-um). Groups of axons are bound into bundles called **nerve fascicles** by a wrapping of connective tissue called the **perineurium.** Finally, the whole nerve is surrounded by a tough fibrous sheath, the **epineurium.** The three layers of

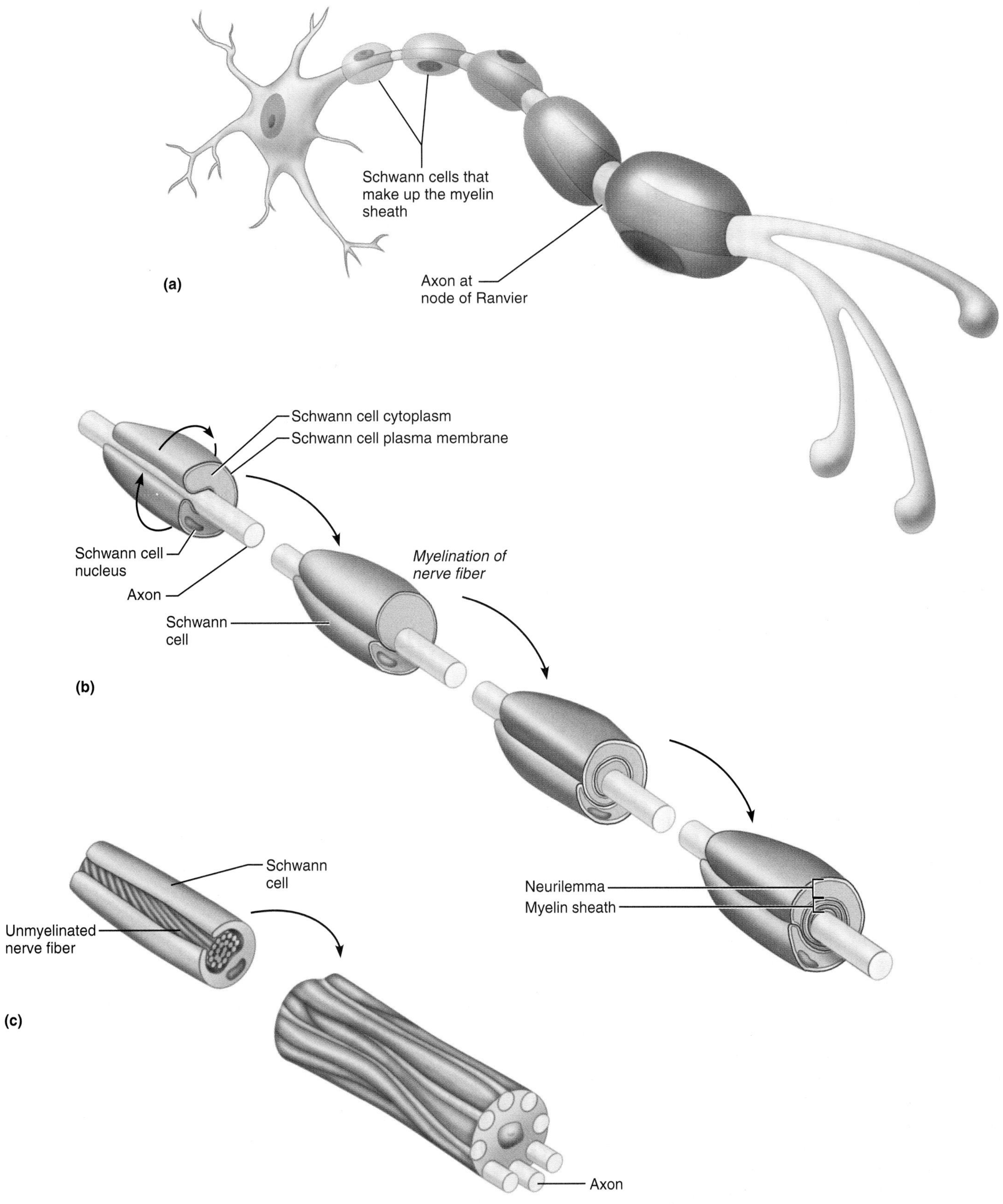

FIGURE 12.15 Schwann cells on myelinated and unmyelinated axons in the PNS. **(a)** Orientation diagram, showing Schwann cells comprising the myelin sheath of a fully myelinated axon. **(b)** Diagram showing how myelin forms: In the fetus and baby, each Schwann cell forms a segment of the myelin sheath by coiling around a peripheral axon. The resulting concentric layers of membrane are the myelin. **(c)** Unmyelinated axons in the PNS. These axons first are engulfed by Schwann cells, and then lie in tunnels in the latter, but no coiling of the membrane ever occurs. Thus, there is no myelin.

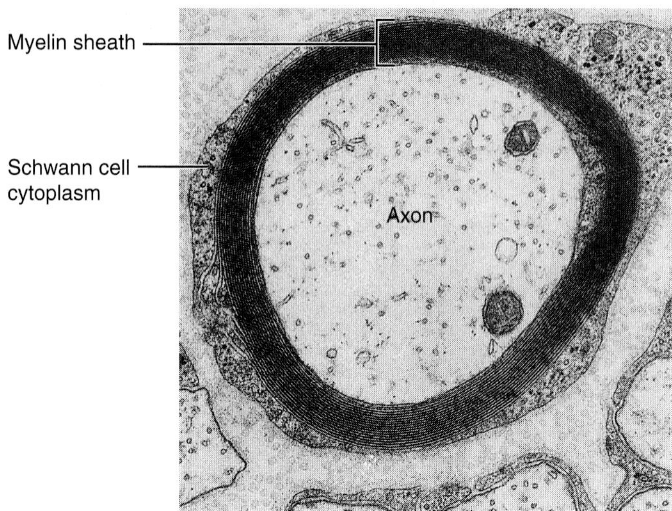

FIGURE 12.16 Electron micrograph of a myelinated axon, cross-sectional view (40,000 ×). Note the many dark layers of Schwann cell membrane that comprise the myelin sheath.

connective tissue in nerves correspond exactly to those in skeletal muscle, the endomysium, perimysium, and epimysium (see p. 246). The connective tissue in a nerve also contains the blood vessels that nourish the axons and Schwann cells.

In this chapter we have defined *neurons, nerve fibers,* and *nerves.* Special effort must be made to avoid confusing these sound-alike terms. A neuron is a nerve cell, a nerve fiber is a long axon, and a nerve is a collection of nerve fibers in the PNS.

BASIC NEURONAL ORGANIZATION OF THE NERVOUS SYSTEM

Reflex Arcs

Reflex arcs are simple chains of neurons that both explain our simplest, reflexive behaviors and determine the basic structural plan of the nervous system. Reflex arcs are responsible for **reflexes,** which are defined as rapid, automatic *motor* responses to stimuli. Reflexes are unlearned, unpremeditated, and involuntary. Examples are jerking away your hand after it accidentally touches a hot stove and vomiting in response to some food that irritates your stomach. As you can see from these examples, reflexes are either *somatic reflexes* resulting in the contraction of skeletal muscles or *visceral reflexes* activating smooth muscle, cardiac muscle, or glands.

Every reflex arc has five essential components, each of which activates the next (Figure 12.18a):

1. The *receptor* is the site where the stimulus acts—for example, the dendritic endings of a sensory neuron.
2. The *sensory neuron* transmits the afferent impulses to the CNS.
3. The *integration center* consists of one or more synapses in the CNS (represented by the spinal cord in Figure 12.18). In the simplest reflex arcs, the integration center is a single synapse between a sensory neuron and a motor neuron. In more complex reflexes it involves multiple synapses that send signals through long chains of interneurons.
4. The *motor neuron* conducts efferent impulses from the integration center to an effector.
5. The *effector* is the muscle or gland cell that responds to the efferent impulses by contracting or secreting.

The left side of Figure 12.18b illustrates the simplest of all reflexes, the **monosynaptic reflex** (mon′o-sĭ-nap″tik; "one synapse"). The example shown is the familiar "knee-jerk" reflex: The impact of a hammer on the patellar ligament stretches the quadriceps muscles of the thigh. This stretching initiates an impulse in a sensory neuron that directly activates a motor neuron in the spinal cord, which then signals the quadriceps muscle to contract. This contraction counteracts the original stretching caused by the hammer.

Many skeletal muscles of the body participate in such monosynaptic *stretch reflexes,* which help maintain equilibrium and upright posture. In these reflexes, the sensory neurons sense the stretching of muscles that occurs when the body starts to sway, and then the motor neurons activate muscles that adjust the body's position to prevent a fall. Because they contain just one synapse, stretch reflexes are the fastest of all reflexes—for speed is essential to keep one from falling to the ground.

Far more common are **polysynaptic reflexes,** in which one or more interneurons are part of the reflex pathway between the sensory and motor neurons. Most of the simple reflex arcs in the body contain a single interneuron and therefore have a total of three neurons. *Withdrawal reflexes,* by which we pull away from danger, are three-neuron reflexes. Such a reflex is shown in the right side of Figure 12.18b: Pricking a finger with a tack initiates an impulse in the sensory neuron, which activates the interneuron in the CNS. The interneuron then signals the motor neuron to contract the muscle that withdraws the hand.

The common three-neuron reflex arcs have a special importance in the science of neuroanatomy because they reveal the fundamental design of the entire nervous system, which is the topic of the next section (p. 351).

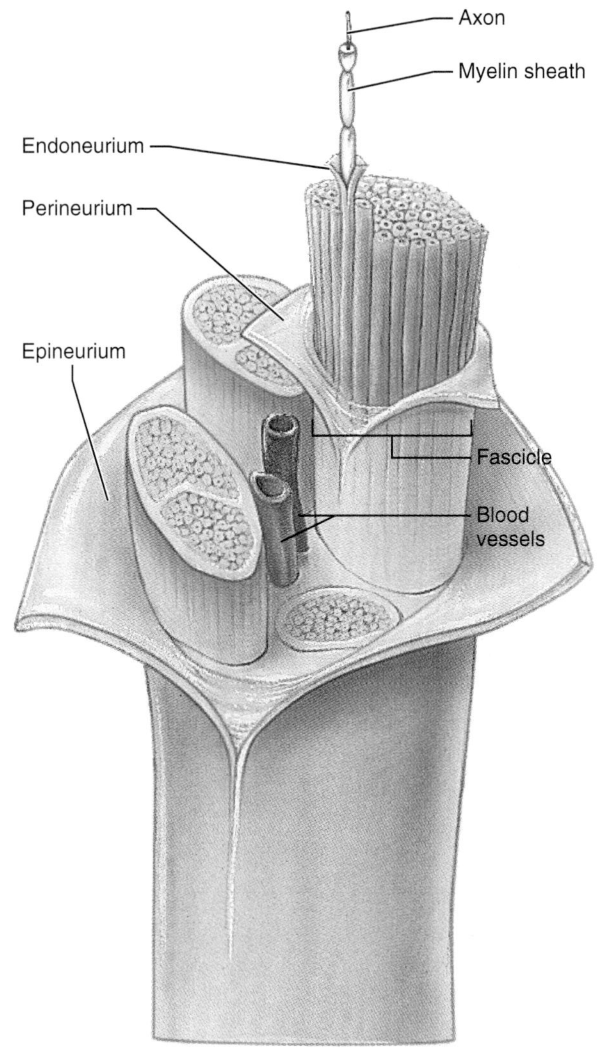

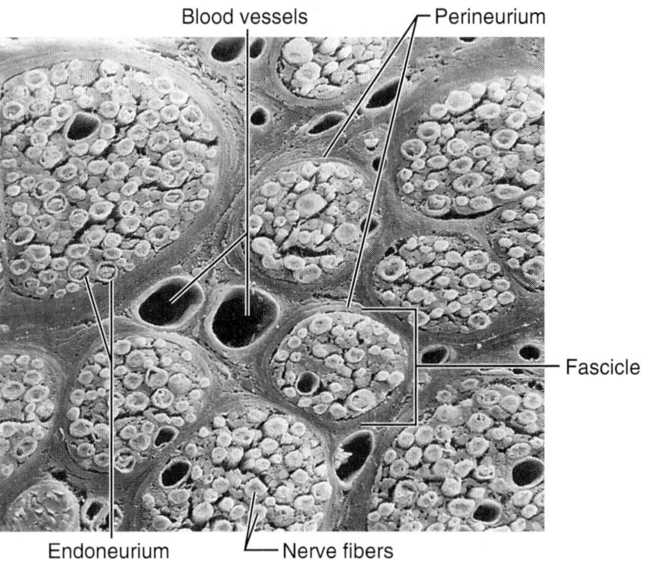

Simplified Design of the Nervous System

Figure 12.19 illustrates how three-neuron reflex arcs form the basis of the structural plan of the nervous system. Even though this figure emphasizes the reflex arcs that are associated with the spinal cord, similar reflex arcs are also associated with the brain. In the lower half of the figure, note that the locations of the sensory neurons (dorsally), motor neurons (ventrally), and interneurons (centrally) mirror their locations in reflex arcs. Note as well that the cell bodies of the sensory neurons lie outside the CNS in **sensory ganglia,** and that their central processes enter the *dorsal* aspect of the spinal cord to synapse with interneurons there. In the CNS, the cell bodies of most interneurons lie dorsal to those of the motor neurons. The long axons of motor neurons exit the *ventral* aspect of the spinal cord and lie in the

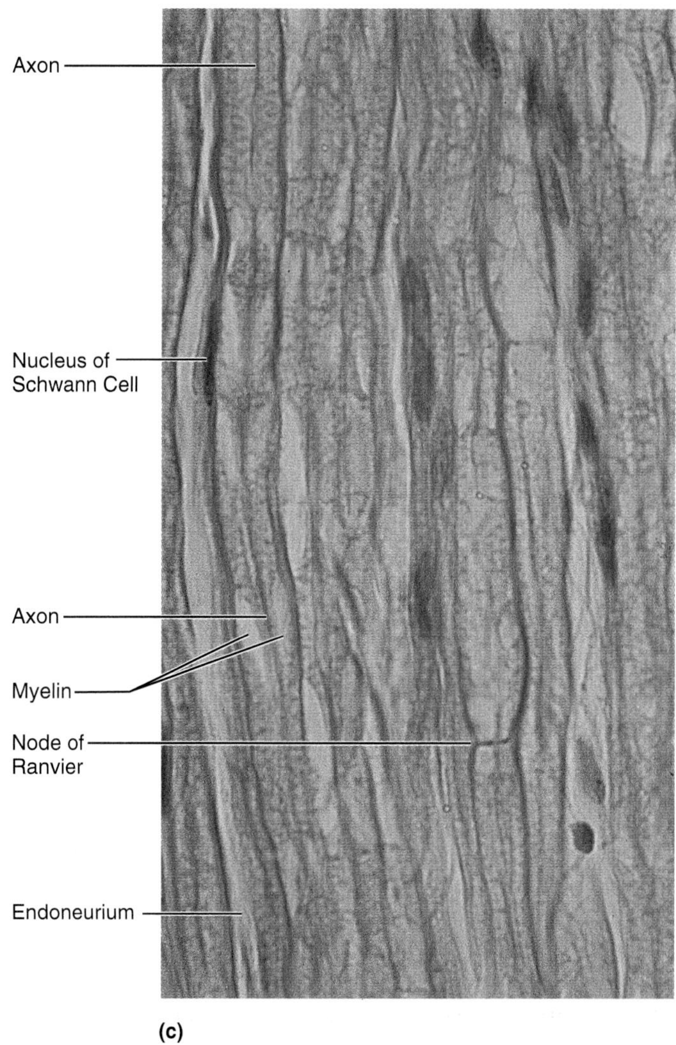

FIGURE 12.17 Structure of nerve. (a) Three-dimensional diagrammatic view of a nerve dissected to reveal its nerve fibers and its connective-tissue wrappings. (b) Scanning electron micrograph of a cross section of part of a nerve (400 ×). ©R. Kessel and R. Kardon. (c) Longitudinal section of a nerve, as viewed by light microscopy (800 ×).

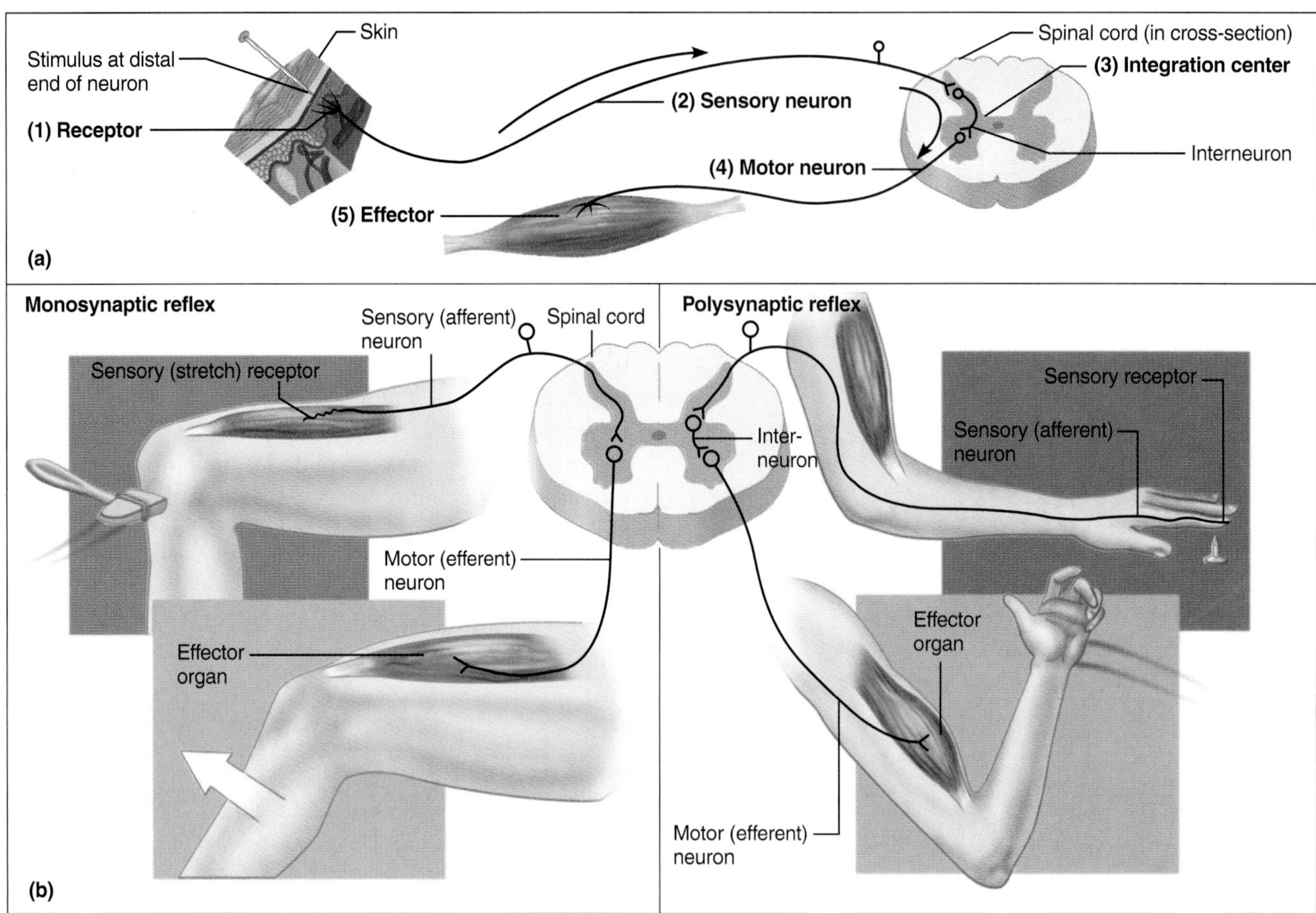

FIGURE 12.18 Simple reflex arcs. **(a)** The five basic components of all reflex arcs, as illustrated in a three-neuron reflex arc: At left, a pin stimulates a *receptor,* which sends an impulse through a *sensory neuron.* The signal then crosses the *integration center* (one or more synapses in the CNS) to activate the *motor neuron,* which activates the *effector.* **(b)** A monosynaptic stretch reflex (left) and a polysynaptic withdrawal reflex (right). A monosynaptic reflex arc has two neurons, whereas a polysynaptic reflex arc has more than two neurons (in this case, three).

PNS. The *nerves* of the PNS consist of the motor axons plus the long peripheral processes of the sensory neurons. These motor and sensory nerve fibers extend throughout the body to reach the peripheral effectors and receptors.

Even though reflex arcs determine its basic organization, the human nervous system is obviously more than just a series of simple reflex arcs. To appreciate its complexity, we must expand our conception of interneurons. Interneurons include not only the intermediate neurons of reflex arcs, but also all the neurons that are confined entirely within the CNS. The complexity of the CNS arises from the organization of the vast number of interneurons in the spinal cord and brain into complex neural circuits that process information; that is, the complexity of the CNS results from long chains of interneurons (top of Figure 12.19) that are interposed between each sensory and motor neuron. Although tremendously oversimplified, the information contained in Figure 12.19 is a useful way to conceptualize the organization of neurons in the CNS.

The CNS has distinct regions of gray and white matter that reflect the arrangement of its neurons. The **gray matter** is a gray-colored zone that surrounds the hollow central cavity of the CNS. It is H-shaped in the spinal cord, where its dorsal half contains cell bodies of interneurons and its ventral half contains the cell bodies of motor neurons (see Figure 12.19). Thus, *gray matter is the site where neuron cell bodies are clustered.* More specifically, gray matter is a mixture of neuron cell bodies, dendrites, and short, unmyelinated axons. External to the gray matter is **white matter,** which contains no neuron cell bodies but millions of axons. Its white color comes from the myelin sheaths around many of the axons. Most of these axons ascend from the spinal cord to the brain or descend from the brain to the spinal cord, allowing these two regions of the CNS to communicate with each other. Thus, *white matter consists of*

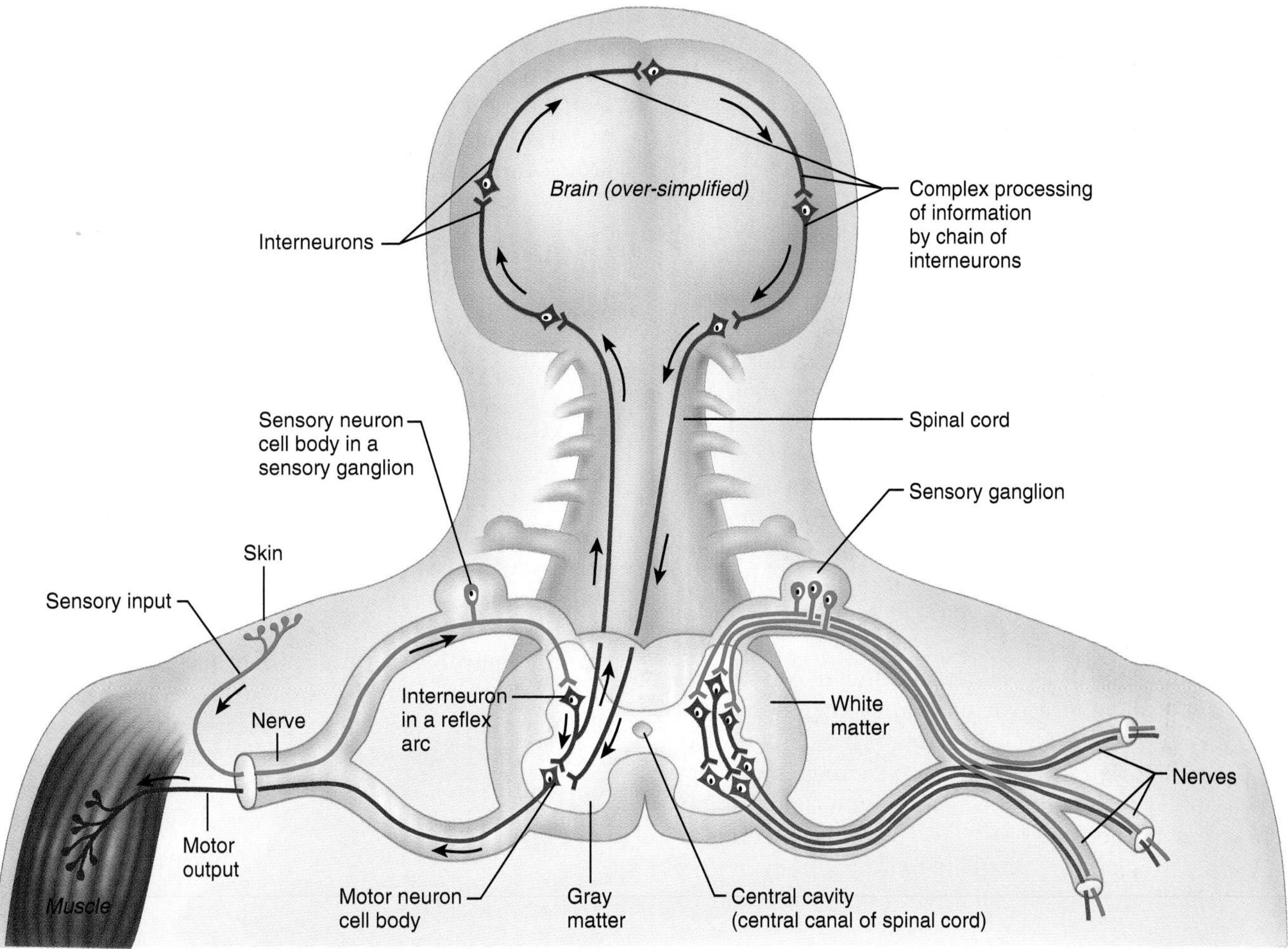

FIGURE 12.19 Simplified diagram of the human nervous system, based on the locations of sensory neurons, motor neurons, and interneurons. In this posterior view, the spinal cord is cut in cross section, with the brain shown above. Most importantly, note the reflex arcs formed by the sensory neurons, interneurons, and motor neurons. Also shown are the arrangement of gray and white matter and the processing of information by circuits of interneurons in the spinal cord and brain. The arrangement of neurons in the spinal cord is realistically shown, but those in the brain are greatly oversimplified. The exact arrangement of gray and white matter in the brain is considered in Chapter 13 (p. 362).

axons running between different parts of the CNS. Within the white matter, axons traveling to similar destinations form axon bundles called **tracts.**

In summary, the CNS has white matter external to gray matter, which surrounds the hollow central cavity. The anatomy of the CNS is described more completely in Chapter 13.

DISORDERS OF NERVOUS TISSUE

Multiple sclerosis (MS), the most common cause of neural disability in young adults, is a progressive disease that destroys patches of myelin in the brain and spinal cord, leading to sensory disorders and weakened musculature. This disease, which varies greatly in intensity among individuals who have it, is characterized by periods of relapse (disability) and remission (recovery). Common signs and symptoms, which vary according to which part of the CNS is affected, include numbness or pain on the skin of some part of the body, disturbances of vision, muscle weakness or paralysis, difficulty in maintaining balance, slurred speech, bladder incontinence, fatigue, and depression.

The cause of MS is incompletely understood. It is known to be an autoimmune disease in which the immune system attacks the myelin around axons in the CNS, thereby interfering with the conduction of signals along the axons. The most important immune cells involved, called lymphocytes, break down the myelin, and macrophages then phagocytize the remains. The damage is also accompanied by inflammation, which can destroy the axons themselves. Although the immune attack is probably directed against several different proteins in the myelin membranes, these proteins have not been identified with certainty. Perhaps the myelin proteins resemble proteins produced by certain viruses to which susceptible people were exposed early in life, and against which the body now directs a misguided immune response. Indeed, viral infection can trigger a relapse of MS in affected individuals—but so can almost any type of stress or injury.

In the United States, MS affects about 1 of every 1000 people, for a total of over 250,000 individuals, and its incidence seems to be increasing. Both genetic and environmental factors seem to influence who gets MS. Supporting the contribution of genetics is the fact that a person's chances of developing MS are increased about thirtyfold if an immediate relative has it. An environmental contribution is suggested by an incidence that is five times higher in temperate Europe and the United States than in tropical countries. Where people live for the first 15 years of life provides a good predictor of future risk.

MS usually appears in early adulthood and affects two or three times more women than men. The reason for this sex-based difference is unknown, but it may relate to the fact that in general, females have more vigorous immune responses than males. Sex hormones may also play a role, because MS is likely to go into remission during pregnancy and then flare up 3–8 months after a woman gives birth. Men tend to get MS later in life, but they decline toward paralysis faster.

MS affects different people in different ways. Some have only a single attack and recover without a recurrence, many have over a decade of alternating remissions and relapses, and a few deteriorate rapidly and continuously within the first year. Remission seems to result from an unexplained halting of the immune response, which gives axons time to repair themselves and to recover their functions. However, after 10–15 years of periodic relapses, the accumulated damage to the axons is often too great, and 50% of MS sufferers decline rapidly in their second decade of the disease.

MS is not always easy to recognize. Initial diagnosis involves ruling out other conditions that have similar symptoms. A definitive but invasive test for MS identifies MS-specific antibodies in cerebrospinal fluid taken from the center of the spinal cord. A less invasive approach tests for a delay in the time it takes visual signals to travel from the eye through the brain, as measured by monitoring brain waves detected on the outside of the skull. Finally, magnetic resonance imaging (p. 22) can reveal the disease's characteristic lesions localized in the vicinity of blood vessels in the brain.

The search for a cure is hampered by our incomplete understanding of the cause of MS. Treatment is directed toward relieving the symptoms. The anti-inflammatory steroid drug methylprednisolone lessens the severity and duration of the relapses, but it does not reduce the frequency of the attacks. Other drugs, such as interferon beta-1a and copolymer-1 antigen, are designed to decrease the frequency of both relapses and the appearance of new lesions, but trials indicate only modest success. Future treatments will likely be directed at targeting the lymphocytes that mediate the immune attack on myelin.

NERVOUS TISSUE THROUGHOUT LIFE

Now that we have introduced the basic organization of the nervous system (Figure 12.19), we can consider how this organization develops in the embryo and fetus. Other developmental aspects of the nervous system are considered in later chapters.

Recall from Chapter 3 that the nervous system develops from the dorsal ectoderm, which invaginates to form the neural tube and the neural crest (Figure 12.20; see also Figure 3.7 on p. 65). The neural tube, whose walls begin as a layer of **neuroepithelial cells,** becomes the CNS. These cells divide, migrate externally, and become neuroblasts (future neurons), which never again divide (Figure 12.20b). Just external to the neuroepithelium, the neuroblasts cluster into **alar** and **basal plates,** the future gray matter (Figure 12.20c). Dorsally, the neuroblasts of the alar plate become interneurons, which remain in the CNS. Ventrally, the neuroblasts of the basal plate become motor neurons and sprout axons that grow out to the effector organs. (Some interneurons also form from the basal plate and remain ventrally among the motor neurons.) Axons that sprout from the young interneurons form the white matter by growing outward along the length of the CNS. These events occur in both the spinal cord and the brain.

Most of the events described so far take place in the second month of development, but neurons continue to form rapidly until about the sixth month. At that time, neuron formation slows markedly, although it may continue at a reduced rate into childhood. Just before neuron formation slows, the neuroepithelium begins to produce astrocytes and oligodendrocytes. The earliest of these glial cells extend outward from the neuroepithelium and provide pathways along which young neurons migrate to reach their final destinations. As the division of its cells slows, the neuroepithelium becomes the ependymal layer.

Sensory neurons arise not from the neural tube but from the neural crest (see Figure 12.20c). This explains why the cell bodies of sensory neurons lie *outside* the CNS. Like motor neurons and interneurons, sensory neurons stop dividing during the fetal period.

Neuroscientists are actively investigating how forming neurons "hook up" with each other. As the growing axons elongate at *growth cones*, they are attracted by chemical signals from other neurons called *neurotrophins* (nu-ro-tro'finz). At the same time, the receiving dendrites send out thin, wiggling extensions to reach the approaching axons and form synapses. Which

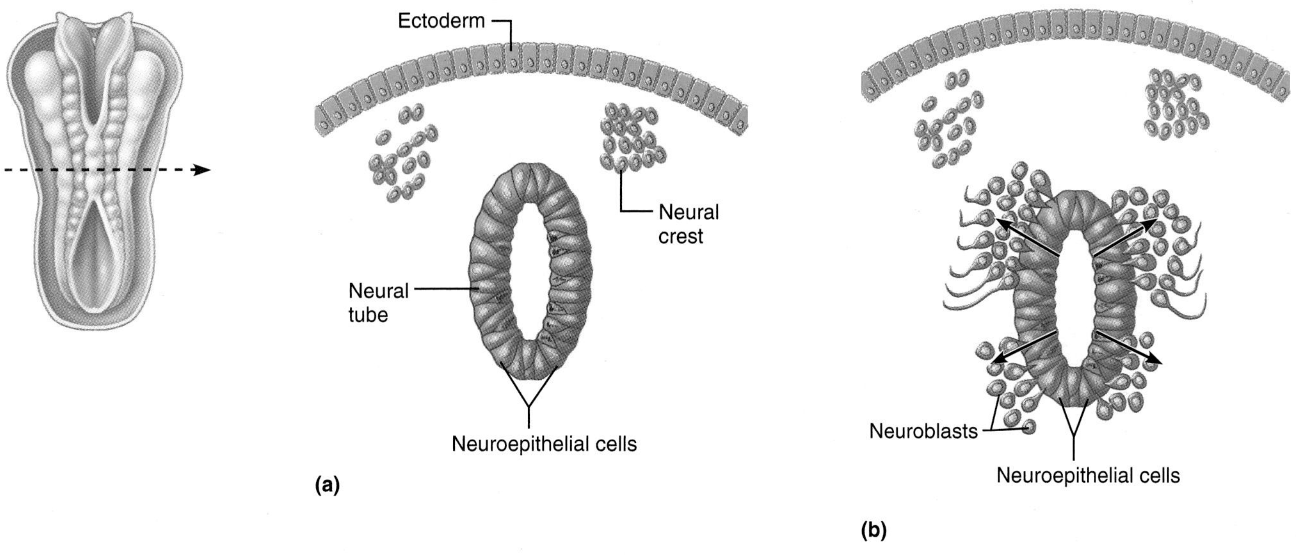

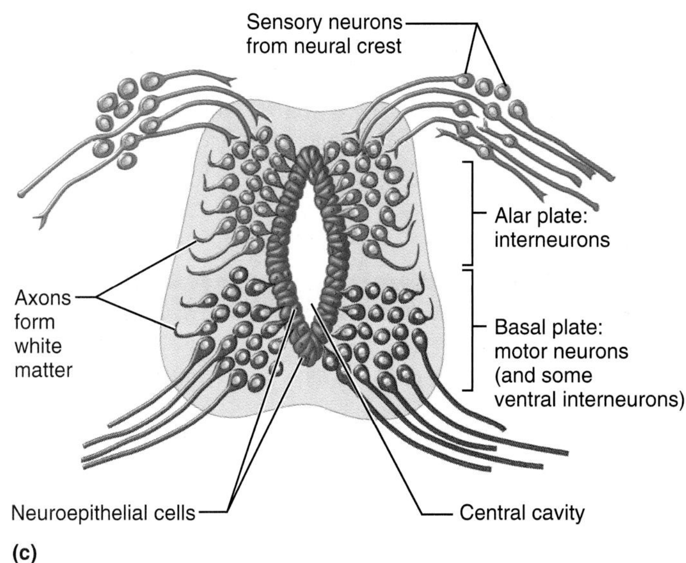

FIGURE 12.20 Development of the nervous system in weeks 4 and 5 of the embryonic period. (a) Cross section of the neural tube, the future spinal cord and brain. For orientation, a 22-day embryo is shown at left in dorsal view. **(b)** Neuroblasts (future neurons) arise through division of neuroepithelial cells and migrate externally. **(c)** Neuroblasts form the alar plate (future interneurons) and basal plate (future motor neurons). Neural crest cells form the sensory neurons.

synaptic connections are made, and which persist, are determined by two factors: (1) the amount of neurotrophin initially received, and (2) the degree to which a synapse is used after being established. Neurons that make "bad" connections are signaled to die via apoptosis (see p. 53). Of the many neurons formed during the embryonic period, about two-thirds die before birth. This initial overproduction of neurons ensures that all necessary neural connections will be made and that mistaken connections will be eliminated.

As mentioned previously, fully differentiated neurons do not divide, probably because these cells are too large and complex to break and then reestablish their many synaptic connections. Furthermore, in the postnatal period there is no obvious replacement of dead neurons by any neural stem cells, so traditionally such stem cells were thought not to exist. This is called the no-new-neurons doctrine.

Advances in Understanding Recently, the no-new-neurons doctrine was overturned by the discovery of neural stem cells in adult humans. That is, some dividing ependymal cells have been shown to form new neurons in two regions of the adult brain, namely the hippocampus and the olfactory bulb (see Chapter 13). Perhaps such ependymal stem cells can some day be induced to form new neurons and new neural networks in people with brain injuries or degenerative brain diseases. ✱

Because there is no effective replacement of dead or damaged neurons after the fetal period, neural injuries tend to cause permanent dysfunction in children and adults. If axons alone are destroyed, however, the cell bodies often survive and the axons may regenerate. If a nerve is severed in the PNS, the sprouting axons can grow peripherally along solid bands formed by the surviving Schwann cells that surrounded the original axons in the region beyond the injury. Thus, an eventual reinnervation of the target organ, with partial recovery of function, is sometimes possible in the PNS. In the CNS, however, the neuroglia never form bands to guide the regrowing axons and even hinder such axons by secreting growth-inhibiting chemicals. Therefore, there is no effective regeneration after injury to the spinal cord or brain.

Clinical Applications

REGENERATION AND SPINAL CORD INJURIES Trauma to the spine has left 200,000 living Americans paralyzed with spinal cord injury. To help these people, extensive medical research is being conducted on spinal cord regeneration. The main goals of this research are to (1) identify and block the chemicals that inhibit axonal growth in the injured CNS; (2) add neurotrophins to induce axonal sprouting and remyelinization; (3) make the damaged region of the spinal cord more like the PNS by implanting Schwann cells or segments of peripheral nerves, along which the CNS axons may grow; (4) isolate neuronal stem cells and add them to the damaged spinal cord. Currently, the only helpful treatment for spinal injury relies on the fact that much of the axonal damage is secondary—caused by inflammation, free-radical production, and apoptosis that develop in the cord after the injury. Administering the anti-inflammatory drug methylprednisolone within 8 hours after injury minimizes this secondary damage and improves the prognosis of many patients. (For more on spinal cord injuries, see p. 403.)

RELATED CLINICAL TERMS

Neuroblastoma (nu″ro-blas-to′mah) (*oma* = tumor) A malignant tumor in children arising from cells that have retained a neuroblast-like structure. These blastomas sometimes originate in the brain, but most are of neural crest origin in the PNS.

Neurologist (nu-rol′o-jist) A medical specialist in the study of the nervous system and its disorders.

Neuropathy (nu-rop′ah-the) Any disease of the nervous tissue, but particularly a degenerative disease of nerves.

Neurotoxin (nu″ro-tok′sin) Substance that is poisonous or destructive to nervous tissue; examples are botulism and tetanus toxins.

Rabies (*rabies* = madness) A viral infection of the nervous system transferred to humans by the bites of—or other contact with—infected mammals such as dogs, bats, racoons, foxes, and skunks. Once it enters the body, the virus is transported through peripheral nerve axons to the CNS, where it causes inflammation of the brain, delirium, and death. Because of extensive vaccination of dogs and careful medical treatment of animal bites, human rabies is now very rare in the United States, but it kills about 40,000 people a year worldwide. A vaccine- and antibody-based treatment is effective if given before symptoms appear.

CHAPTER SUMMARY

Media study tools that could provide you additional help in reviewing specific key topics of Chapter 12 are referenced below.

SP = Study Partner; AIA = A.D.A.M.® Interactive Anatomy

1. The nervous system controls most of the other organ systems in the body. Its chief functions are to monitor, integrate, and respond to information in the environment.

BASIC DIVISIONS OF THE NERVOUS SYSTEM (pp. 336–338)

1. The central nervous system consists of the spinal cord and brain. The peripheral nervous system is external to the CNS and consists of nerves and ganglia.
2. The nervous system receives sensory inputs and dictates motor outputs. Sensory (afferent) signals are carried from sensory receptors through the PNS into the CNS. Motor (efferent) signals are carried away from the CNS and through the PNS to the effectors, which are muscles and glands. As shown in Figure 12.3, the types of sensory inputs and motor outputs are further categorized as somatic (outer body) and visceral (mainly inner body), and general (widespread) and special (localized). Branchial innervation refers to the motor innervation of the pharyngeal (branchial) muscles.
3. Proprioception refers to a series of senses that monitor the degree of stretch in muscles, tendons, and joint capsules. Proprioception thus senses the positions and movements of our body parts.

NERVOUS TISSUE (pp. 338–348)

The Neuron (pp. 338–345)

1. Neurons are long-lived, nondividing cells. Each has a cell body and cell processes. The processes are axons and dendrites.
2. The neuron cell body contains a nucleus resembling an owl's eye. Its cytoplasm contains supportive neurofibrils and chromatophilic (Nissl) bodies, which are concentrations of rough endoplasmic reticulum and free ribosomes. The chromatophilic bodies in the cell body manufacture proteins and membranes for the entire neuron, thereby continuously maintaining and renewing the contents of the cell, including the processes. Except for those found in ganglia, all neuron cell bodies are in the CNS.
3. Most neurons have a number of branched dendrites, receptive sites that conduct signals from other neurons toward the neuron cell body.
4. Most neurons have one axon, which generates and conducts nerve impulses away from the neuron cell body. Impulses begin at the initial segment of the axon, arising from the cone-shaped axon hillock. Impulses end at the knob-like axon terminals, which participate in synapses and release neurotransmitter molecules.
5. A synapse is a functional junction between neurons. The major categories of synapses are axodendritic, axosomatic, and axoaxonic.

6. At synapses, information is transferred from a presynaptic neuron to a postsynaptic neuron. Synaptic vesicles in the presynaptic cell fuse with the presynaptic membrane and empty neurotransmitter molecules into the synaptic cleft. Taken up by the postsynaptic membrane, this neurotransmitter drives the postsynaptic neuron toward or away from firing.
7. Electrical transmission occurs along neuronal membranes, as follows: Signals received at dendrites and the cell body set up graded potentials (depolarizations) that spread to the axon, where they trigger action potentials (reversals of membrane charge that race along the axon).

SP Exercise: Chapter 12, Structure of a Motor Nerve.

8. Anatomically, neurons are classified by the number of processes issuing from their cell body as multipolar, bipolar, or unipolar.
9. Functionally, neurons are classified according to the direction in which they conduct impulses. Sensory neurons conduct impulses toward the CNS, motor neurons conduct away from the CNS, and interneurons lie in the CNS between sensory and motor neurons.

Supporting Cells (pp. 346–348)

10. Non-nervous supporting cells act to support, protect, nourish, and insulate neurons.
11. Neuroglial cells are the supporting cells in the CNS. They include star-shaped astrocytes, phagocytic microglia, ependymal cells that line the central cavity, and myelin-forming oligodendrocytes. Schwann cells (neurolemmocytes) and satellite cells are the supporting cells in the PNS.
12. Thick axons are myelinated. Myelin speeds impulse conduction along these axons.
13. The myelin sheath is a coat of supporting-cell membranes wrapped in layers around the axon. This sheath is formed by Schwann cells in the PNS and by oligodendrocytes in the CNS. The sheath has gaps called nodes of Ranvier or neurofibral nodes. Unmyelinated axons are surrounded by supporting cells, but they are not wrapped by layers of membrane.

NERVES (pp. 348–350)

1. A nerve is a bundle of axons in the PNS. Each axon is enclosed by an endoneurium; each fascicle of axons is wrapped by a perineurium; and the whole nerve is surrounded by the epineurium. Because they contain more than one kind of tissue, nerves are organs.

BASIC NEURONAL ORGANIZATION OF THE NERVOUS SYSTEM (pp. 350–353)

Reflex Arcs (p. 350)

1. Reflexes are rapid, automatic responses to stimuli. They can be either somatic or visceral.
2. Reflexes are mediated by chains of neurons called reflex arcs. The minimum number of elements in a reflex arc is *five:* a receptor, a sensory neuron, an integration center, a motor neuron, and an effector.
3. A few fast reflexes for maintaining balance have only two neurons (sensory and motor). These are monosynaptic stretch reflexes.
4. Most reflexes in humans are polysynaptic. The simplest of these, such as withdrawal reflexes, have three neurons: a sensory neuron, an interneuron, and a motor neuron.

SP Exercise: Chapter 12, Components of a Reflex Arc.

Simplified Design of the Nervous System (pp. 351–353)

5. Simple three-neuron reflex arcs can be said to form the basis of the structural plan of the entire nervous system. Sensory neurons enter the spinal cord dorsally; motor axons exit it ventrally; and the intervening interneurons are confined to the CNS. The nerves in the PNS consist of the peripheral axons of the sensory and motor neurons. The cell bodies of motor neurons and interneurons make up the internal gray matter of the CNS, whereas the cell bodies of sensory neurons lie external to the CNS in sensory ganglia.
6. Throughout most of the CNS, the inner gray matter (in which neuron cell bodies are located) is surrounded by outer white matter (which consists of fiber tracts). The extreme center of the spinal cord and brain is a hollow central cavity.

DISORDERS OF NERVOUS TISSUE (pp. 353–354)

1. Multiple sclerosis is an autoimmune disease in which destruction of myelin in the CNS leads to neuronal dysfunction. It is characterized by periods of relapses and remissions. Common symptoms include visual disturbances, muscle weakness, fatigue, and depression.

NERVOUS TISSUE THROUGHOUT LIFE (pp. 354–356)

1. The brain and spinal cord develop from the embryonic neural tube, which begins as a layer of dividing neuroepithelial cells. These cells migrate externally to become the neuroblasts of the dorsal *alar plate* (future interneurons) and ventral *basal plate* (future motor neurons and some interneurons). Neuroblasts of the neural crest, external to the neural tube, become the sensory neurons. Neuroblasts sprout axons, which grow toward their targets.
2. After injury, effective axonal regeneration may occur in the PNS but not in the CNS.

REVIEW QUESTIONS

Multiple Choice/Matching Questions

1. Which of the following structures is not part of the central nervous system? (a) the brain, (b) a nerve, (c) the spinal cord, (d) a tract.

2. Match the names of the supporting cells in column B with their correct descriptions in column A.

Column A	Column B
____ **(1)** form myelin in the CNS	**(a)** astrocytes
____ **(2)** line the central cavity of the brain	**(b)** ependymal cells
	(c) microglia
____ **(3)** form myelin in the PNS	**(d)** oligodendrocytes
____ **(4)** CNS phagocytes	**(e)** satellite cells
____ **(5)** regulate ionic composition of the fluid around neurons in the CNS	**(f)** Schwann cells
____ **(6)** remove neurotransmitters in the CNS	

3. Which of the following structures is in the somatic part of the human body? (a) bladder, (b) biceps muscle, (c) lung, (d) stomach.

4. Classify the following inputs and outputs as either somatic sensory (SS), visceral sensory (VS), somatic motor (SM), visceral motor (VM), or branchial motor (BM).

____ **(1)** pain from skin

____ **(2)** taste

____ **(3)** efferent innervation of a gland

____ **(4)** efferent innervation of the gluteus maximus

____ **(5)** a stomachache

____ **(6)** a sound one hears

____ **(7)** efferent innervation of the masseter (a pharyngeal arch muscle)

5. An example of an effector is (a) the eye, (b) a gland, (c) a sensory neuron, (d) a motor neuron.

6. Which of the following parts of a neuron occupies the gray matter in the spinal cord? (a) tracts of long axons, (b) motor neuron cell bodies, (c) sensory neuron cell bodies, (d) nerves.

7. A ganglion is a collection of (a) neuron cell bodies, (b) axons of motor neurons, (c) interneuron cell bodies, (d) axons of sensory neurons.

8. A synapse between an axon terminal and a neuron cell body is classified as (a) axodendritic, (b) axoaxonic, (c) axosomatic, (d) axoneuronic.

9. Myelin is most like which of the following cell parts introduced in Chapter 2? (a) the cell nucleus, (b) smooth endoplasmic reticulum, (c) ribosomes, (d) the plasma membrane.

10. Match the following parts of the adult central nervous system with the embryonic cells that give rise to them: (a) alar plate cells, (b) basal plate cells, (c) neural crest cells.

____ **(1)** sensory neurons

____ **(2)** motor neurons

____ **(3)** dorsal interneurons

Short Answer Essay Questions

11. Define proprioception.

12. Sin Young incorrectly classified proprioception as general somatic *motor* because it refers to the innervation of muscles. Actually, proprioception is general somatic *sensory*. Explain why.

13. Define interneuron.

14. Distinguish gray matter from white matter of the CNS in terms of location and composition.

15. Describe the appearance of a cell nucleus of a neuron.

16. Reexamine the different neurons depicted in Figure 12.12, and indicate which one probably has the most axon terminals synapsing on it. Explain your reasoning.

17. Distinguish a nerve from a nerve fiber and a neuron.

18. Explain why damage to peripheral nerve fibers is often reversible, whereas damage to CNS fibers rarely is.

19. Draw a reflex arc in place in the nervous system (recall Figure 12.19). Would the reflex still function if the neurons in the sensory ganglia were destroyed?

20. Define axon and dendrite.

CRITICAL REASONING + CLINICAL APPLICATION QUESTIONS

1. Two anatomists were arguing about a sensory neuron. One anatomist said its peripheral process is an axon and gave two good reasons, but the other called the peripheral process a dendrite and also gave a good reason. Cite all three reasons given, and state your own opinion.

2. A CT scan and other diagnostic tests indicated that Laressa had developed an oligodendroglioma. Can you deduce from its name what an oligodendroglioma is?

3. The following event received worldwide attention in 1962: A boy playing in a train yard fell under a train, and his right arm was cut off cleanly by the wheels. Surgeons reattached the arm, sewing nerves and vessels back together. The surgery proceeded very well. The arm immediately regained its blood supply, yet the boy could not move the limb or feel anything in it for months. Explain why it took longer to reestablish innervation than circulation.

4. Rochelle developed multiple sclerosis when she was 27. After 8 years she had lost a good portion of her ability to control her skeletal muscles. Explain how this happened.

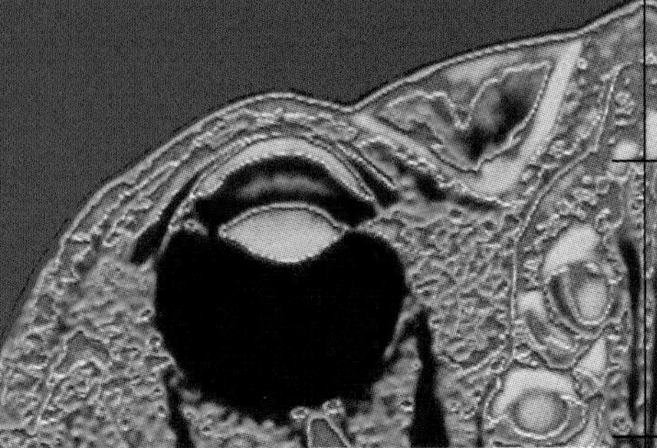

CHAPTER 13

The Central Nervous System

CHAPTER OUTLINE + STUDENT OBJECTIVES

The Brain 360–391

1. Describe the development of the five embryonic divisions of the brain.
2. Name the major parts of the adult brain.
3. Name and describe the locations of the ventricles of the brain.
4. List the major lobes, fissures, and functional areas of the cerebral cortex.
5. Name three classes of fiber tracts in the white matter of the cerebrum.
6. Describe the form and function of the basal nuclei (basal ganglia).
7. Name the divisions of the diencephalon and their functions.
8. Identify the three basic subdivisions of the brain stem, and list the major nuclei in each.
9. Describe the structure and function of the cerebellum.
10. Describe the locations and functions of the reticular formation and the limbic system.
11. Explain how meninges, cerebrospinal fluid, and the blood-brain barrier protect the CNS.
12. Explain the formation of cerebrospinal fluid, and describe its pattern of circulation.

The Spinal Cord 391–401

13. Describe the gross structure of the spinal cord, including the arrangement of its gray matter and white matter.
14. List the largest tracts in the spinal cord, and explain their positions in the major neuronal pathways to and from the brain.

Disorders of the Central Nervous System 401–403

15. Describe the signs and symptoms of concussions, brain contusions, strokes, and Alzheimer's disease.
16. Explain the effects of severe injuries to the spinal cord.

The Central Nervous System Throughout Life 403–404

17. Describe these congenital disorders: anencephaly, cerebral palsy, and spina bifida.
18. Explain the effects of aging on brain structure.

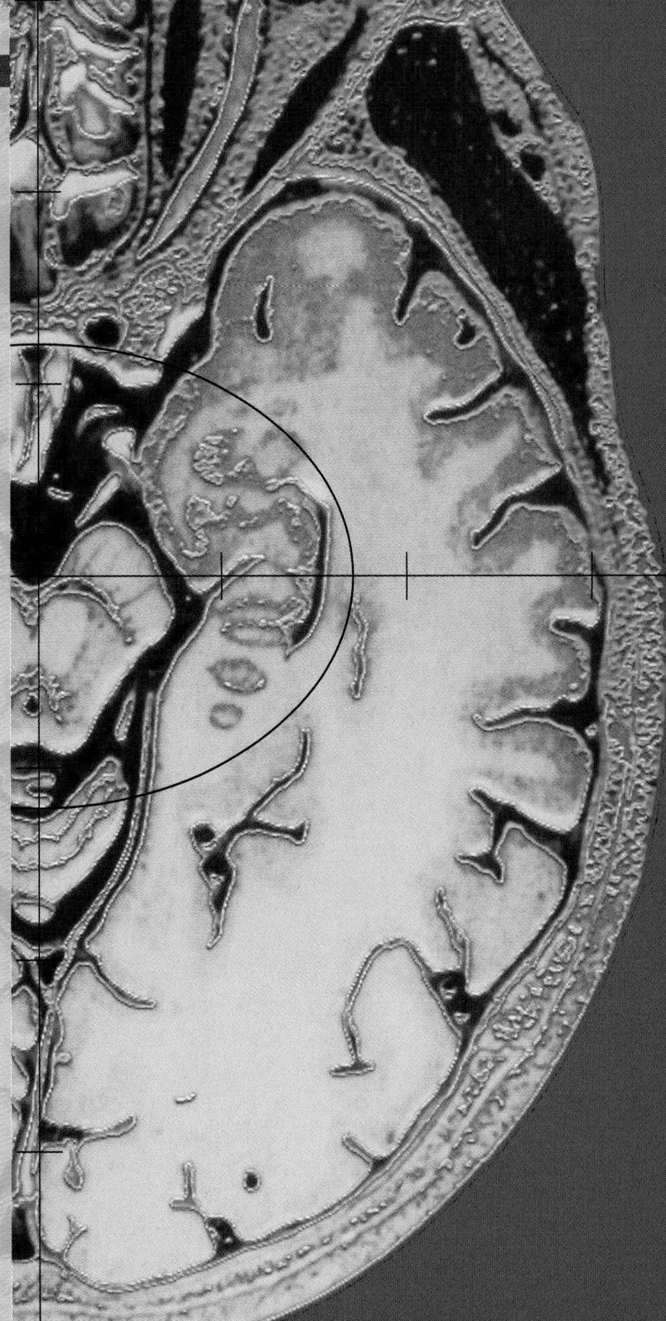

MRI of the brain

THIS CHAPTER COVERS THE BRAIN AND SPINAL cord, which make up the central nervous system (CNS). Historically, the CNS has been likened to the central switchboard of a telephone system that interconnects and routes a dizzying number of incoming and outgoing calls. Nowadays, many compare it to a supercomputer. These analogies may be apt for some of the workings of the spinal cord, but neither really does justice to the fantastic complexity of the human brain. As the basis of each person's unique behavior, the brain is one of the most amazing things known.

In this chapter we use some directional terms that are unique to the CNS (Figure 13.1): The higher brain regions are said to lie **rostrally** (literally, "toward the snout"), whereas the inferior parts of the CNS are said to lie **caudally** ("toward the tail").

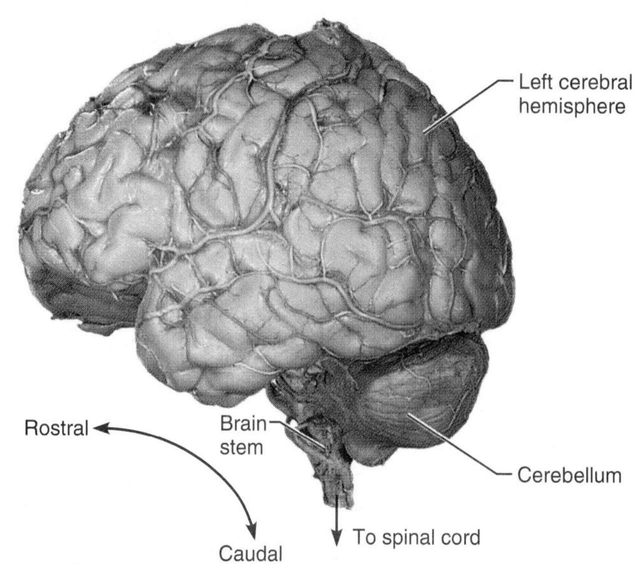

FIGURE 13.1 The human brain in lateral view. Also shown are the meanings of the directional terms rostral and caudal.

THE BRAIN

The **brain** performs our most complex neural functions—those associated with intelligence, consciousness, instincts, and so on. Furthermore, through the cranial nerves that attach to it, the brain is involved in the sensory and motor innervation of the head. Shown in Figure 13.1, it is approximately two large handfuls of pinkish gray tissue, somewhat the consistency of cold oatmeal. The average adult human brain weighs about 1500 g, or 3.3 pounds.

Embryonic Development of the Brain

An examination of development can be helpful in understanding the structures of the adult brain. As mentioned previously, the brain arises as the rostral part of the neural tube in the 4-week embryo (Figure 13.2a). It immediately starts to expand, and constrictions that define three **primary brain vesicles** appear (Figure 13.2b). These three vesicles are the **prosencephalon** (pros″en-sef′ah-lon), or **forebrain;** the **mesencephalon** (mes″en-sef′ah-lon), or **midbrain;** and the **rhombencephalon** (romb″en-sef′ah-lon), or **hindbrain.** (Note that the word stem *encephalos* means "brain.") The caudal portion of the neural tube becomes the spinal cord.

In week 5, the three primary vesicles give rise to five **secondary brain vesicles** (Figure 13.2c). The forebrain divides into the **telencephalon** ("endbrain") and the **diencephalon** ("interbrain"). The **mesencephalon** remains undivided, but the hindbrain divides into the **metencephalon** ("afterbrain") and the **myelencephalon** ("brain most like the spinal cord").

Each of the five secondary brain vesicles then develops rapidly to produce the major structures of the adult brain (Figure 13.2d). The greatest change occurs in the telencephalon, which has two lateral swellings that look like Mickey Mouse's ears and becomes the large **cerebral hemispheres,** together called the **cerebrum** (ser-e′brum). The diencephalon develops three main divisions: the thalamus, the hypothalamus, and the epithalamus. Farther caudally, the ventral part of the metencephalon becomes the **pons** while the **cerebellum** (ser″e-bel′um) develops from the metencephalon's dorsal roof. The myelencephalon is now called the **medulla oblongata** (mĕ-dul′ah ob″long-gah′tah). The midbrain, pons, and medulla together constitute the **brain stem.** Finally, the central cavity of the neural tube enlarges in certain regions to form the hollow **ventricles** (ven′trĭ-klz; "little bellies") of the brain (Figure 13.2e).

During the late embryonic and the fetal periods, the brain continues to grow rapidly, and changes occur in the relative positions of its parts (Figure 13.3). Because the brain grows faster than the membranous skull that contains it, it develops two major bends or flexures—a *midbrain flexure* and a *cervical flexure* (Figure 13.3a). Space restrictions also force the cerebral hemispheres to grow posteriorly over the rest of the brain (Figure 13.3b), and soon these hemispheres completely envelop the diencephalon and midbrain (Figure 13.3c). As each cerebral hemisphere grows, it bends into a horseshoe shape, as indicated by the three arrows in Figure 13.3c. By week 26, the continued growth of the cerebral hemispheres causes their surfaces to crease and fold (Figure 13.3c), until the hemispheres are wrinkled like a walnut (Figure 13.3d). This wrinkling allows more neurons to fit within the limited space, and it may result from tension on the young axons as they become arranged in a way that minimizes the lengths of the interconnections they form among the various parts of the cerebrum.

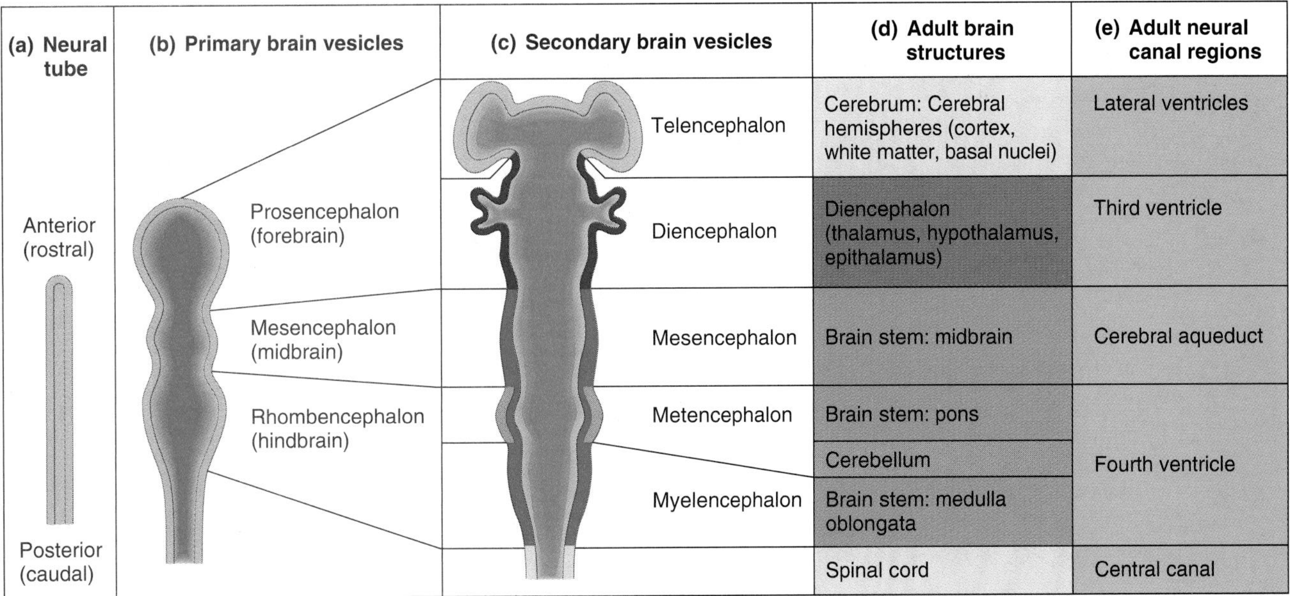

FIGURE 13.2 Embryonic development of the brain in simplified frontal sections. (a) The neural tube subdivides into (b) three primary brain vesicles (week 4), which divide into (c) five secondary brain vesicles (week 5), which differentiate into (d) the adult brain structures. (e) Adult derivatives of the embryonic neural canal (the central cavity of the CNS).

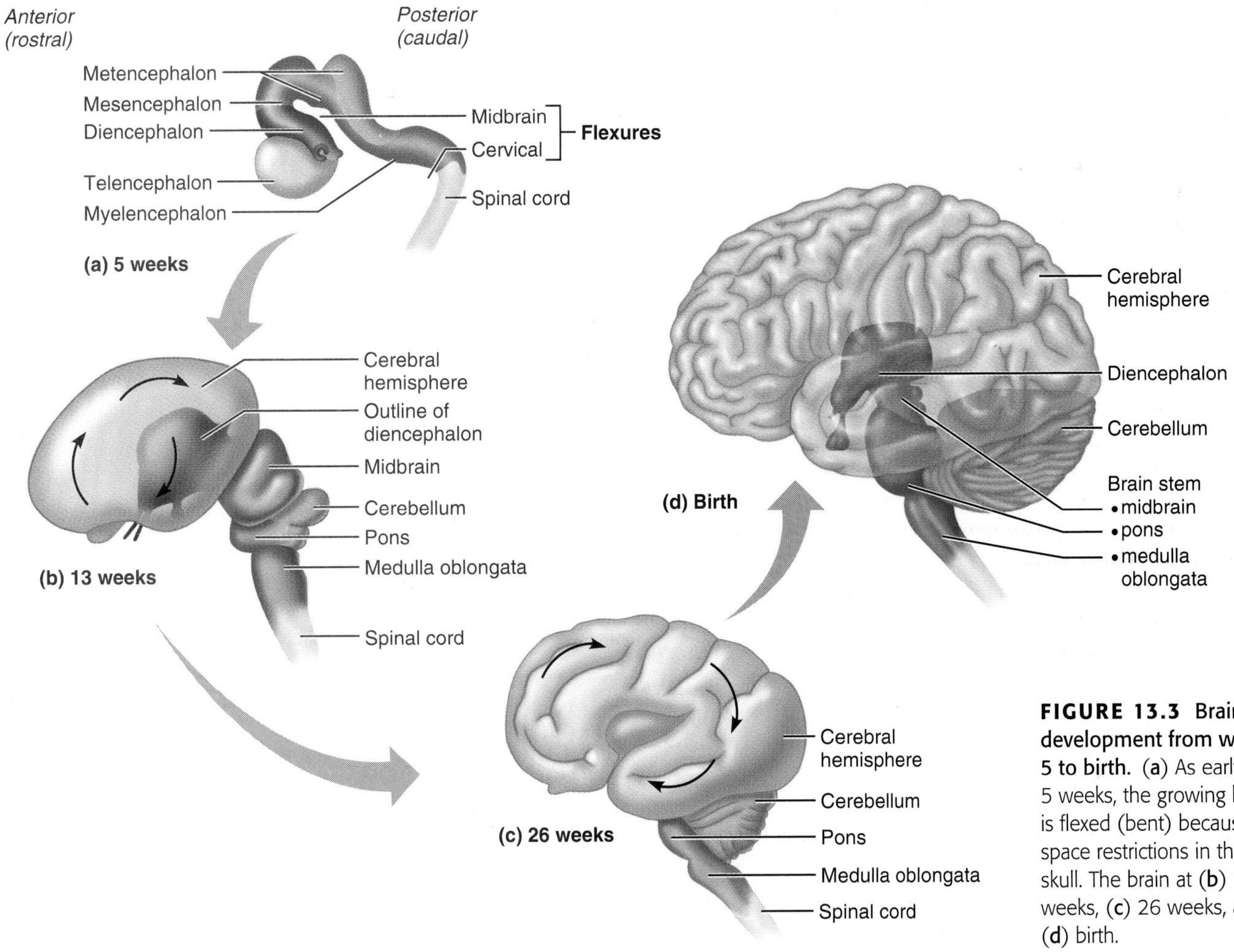

FIGURE 13.3 Brain development from week 5 to birth. (a) As early as 5 weeks, the growing brain is flexed (bent) because of space restrictions in the skull. The brain at (b) 13 weeks, (c) 26 weeks, and (d) birth.

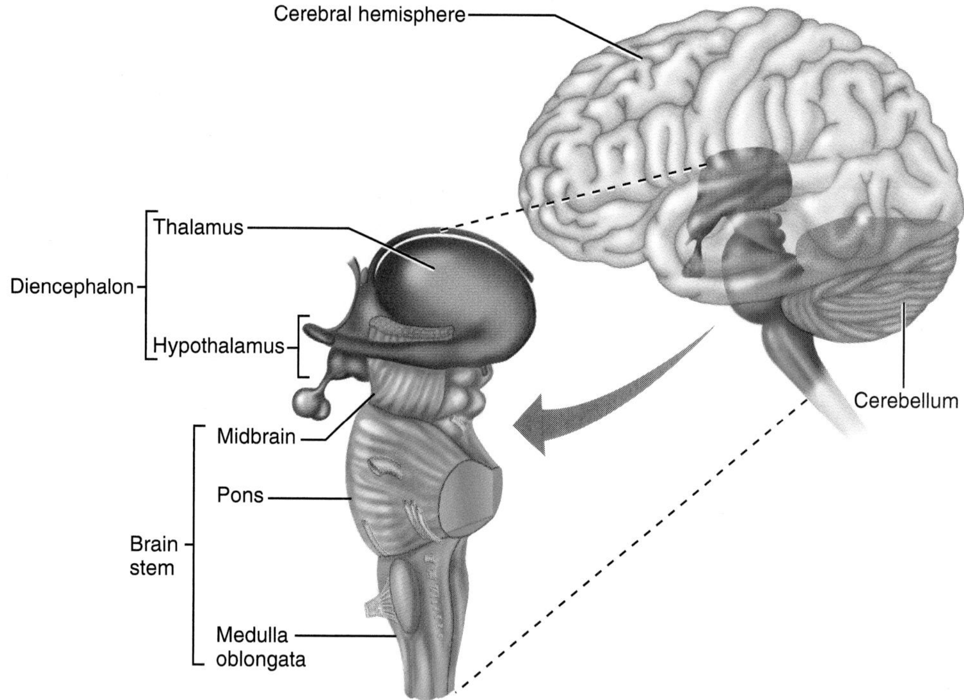

FIGURE 13.4 Parts of the brain. The entire brain, with an enlargement of its diencephalon and brain stem shown below and to the left. The four main parts of the brain are the cerebral hemispheres, diencephalon, brain stem, and cerebellum.

Basic Parts and Organization of the Brain

The brain is often divided into the four parts shown in Figure 13.4: the (1) *cerebral hemispheres,* (2) *diencephalon,* (3) *brain stem* (midbrain, pons, and medulla), and (4) *cerebellum.* (Although most anatomists favor this scheme, some classify the diencephalon as either a portion of the cerebrum or a portion of the brain stem.)

Next, we consider the organization of gray and white matter in the brain. Recall from Chapter 12 (p. 352) that the CNS has an inner region of gray matter, external to which lies white matter. The brain has this basic design, but it also has additional regions of gray matter that are not present in the spinal cord (Figure 13.5). This difference occurs because, during brain development, certain groups of neurons migrate externally to form collections of gray matter in regions that otherwise consist of white matter. The most extreme examples of this outward migration occur in the cerebellum and cerebrum, where sheets of gray matter lie at the surface of the brain. Each of these external sheets of gray matter is called a *cortex* ("bark on a tree"). All other gray matter in the brain is in the form of spherical or irregularly shaped clusters of neuron cell bodies called *brain nuclei* (not to be confused with cell nuclei). Functionally, the large amount of gray matter in the brain allows this part of the CNS to perform very complex neural functions—because gray matter contains many small interneurons that process information.

To help you visualize the spatial relationships among the basic parts of the brain, we first consider the hollow ventricles that lie deep within the brain. Then we explore each of the four brain parts, from rostral to caudal. Finally, a summary of the brain regions is provided in Table 13.1 on page 390.

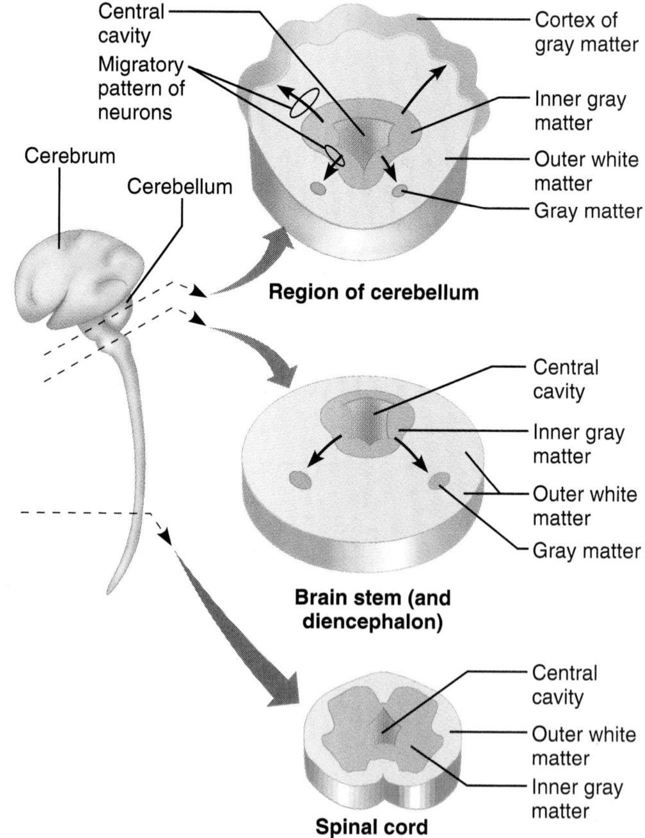

FIGURE 13.5 Gray and white matter in the CNS. Sections through brain and spinal cord of a four-month fetus, with dorsal aspect above. In the spinal cord (at bottom), the gray matter is entirely internal to the white matter. In the brain, however, some central gray migrates outward during development, either into the region of the white matter (brain stem) or onto the surface as a cortex (cerebellum). This migration is indicated by black arrows. The cerebrum resembles the cerebellum in having an outer cortex of gray matter.

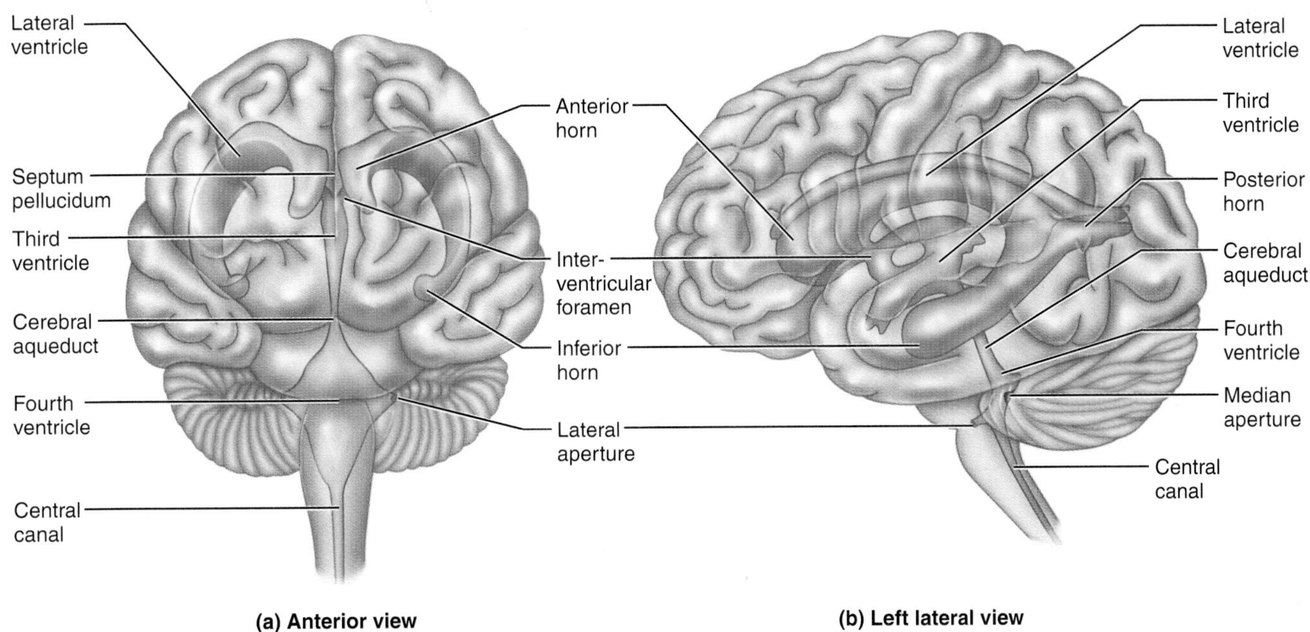

FIGURE 13.6 The ventricles of the brain. (a) Anterior view. (b) Lateral view. The ventricles are drawn as if the brain were transparent. Note that each lateral ventricle has three parts called the anterior, inferior, and posterior horns.

Ventricles of the Brain

The ventricles are expansions of the brain's central cavity, filled with cerebrospinal fluid and lined by ependymal cells (Figure 13.6). They are continuous with each other and with the central canal of the spinal cord. The paired **lateral ventricles,** once called the *first and second ventricles,* lie in the cerebral hemispheres. Their horseshoe shape reflects the bending of the cerebral hemispheres during development. Anteriorly, the two lateral ventricles lie close together, separated only by a thin median membrane called the *septum pellucidum* (sep′tum pĕ-lu′si-dum; "transparent wall") (Figures 13.6a and 13.15).

The **third ventricle** lies in the diencephalon. Anteriorly, it connects to each lateral ventricle through an **interventricular foramen.** In the mesencephalon, the thin tube-like central cavity is called the **cerebral aqueduct,** which connects the third and fourth ventricle. The **fourth ventricle** lies in the hindbrain, dorsal to the pons and the superior half of the medulla. Caudally, the fourth ventricle connects to the central canal of the inferior medulla and spinal cord.

Three openings occur in the walls of the fourth ventricle: the paired **lateral apertures** in its side walls and the **median aperture** in its roof. These holes connect the ventricles with the *subarachnoid space,* which surrounds the whole CNS (see p. 389). This connection allows cerebrospinal fluid to fill both the ventricles and the subarachnoid space.

The Cerebral Hemispheres

The cerebral hemispheres, which make up the most superior region of the brain (see Figure 13.4), account for 83% of total brain mass. They so dominate the brain that many people mistakenly use the words *brain* and *cerebral hemispheres* interchangeably. The cerebral hemispheres cover the diencephalon and the top of the brain stem in much the same way a mushroom cap covers the top of its stalk.

Grooves are evident on and around the cerebral hemispheres (Figure 13.7). The deepest of these grooves are **fissures,** which separate major portions of the brain. The **transverse fissure** separates the cerebral hemispheres from the cerebellum inferiorly (Figure 13.7a), whereas the median **longitudinal fissure** separates the right and left cerebral hemispheres (Figure 13.7c). The many grooves on the surface of the cerebral hemispheres are called **sulci** (sul′ki; singular: **sulcus,** "furrow"). Between these sulci are twisted ridges of brain tissue called **gyri** (ji′ri; singular: **gyrus,** "twister"). The more prominent gyri and sulci are similar in all people and are important anatomical landmarks.

Some of the deeper sulci are used to divide each cerebral hemisphere into five major lobes: the *frontal, parietal, occipital,* and *temporal lobes,* and the *insula* (see Figure 13.7). Most of these lobes are named for the skull bones overlying them. The **frontal lobe** is separated from the **parietal lobe** by the **central sulcus,** which lies in the frontal plane. Bordering the central sulcus are two important gyri, the **precentral gyrus**

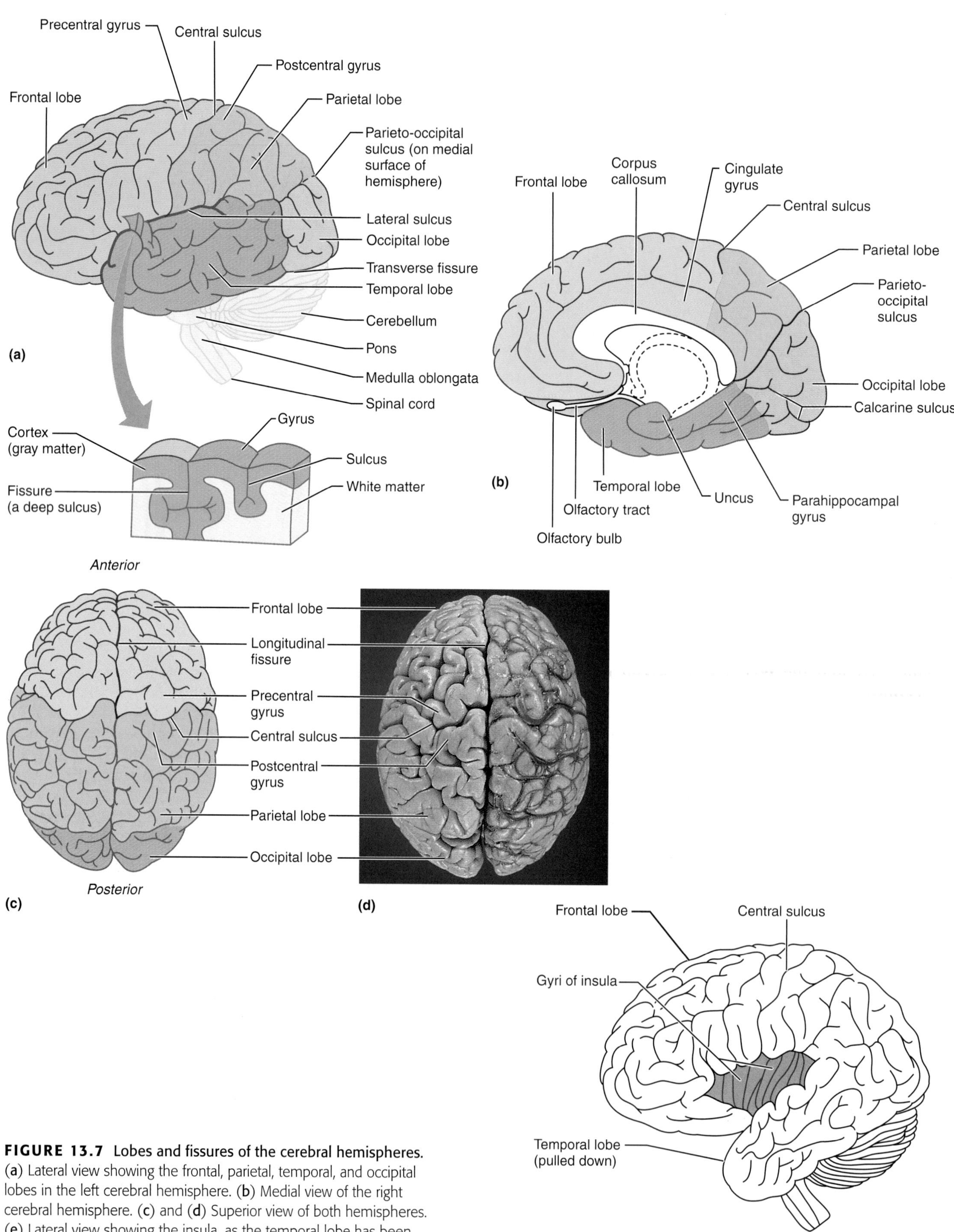

FIGURE 13.7 Lobes and fissures of the cerebral hemispheres. (a) Lateral view showing the frontal, parietal, temporal, and occipital lobes in the left cerebral hemisphere. (b) Medial view of the right cerebral hemisphere. (c) and (d) Superior view of both hemispheres. (e) Lateral view showing the insula, as the temporal lobe has been pulled down to widen the lateral sulcus.

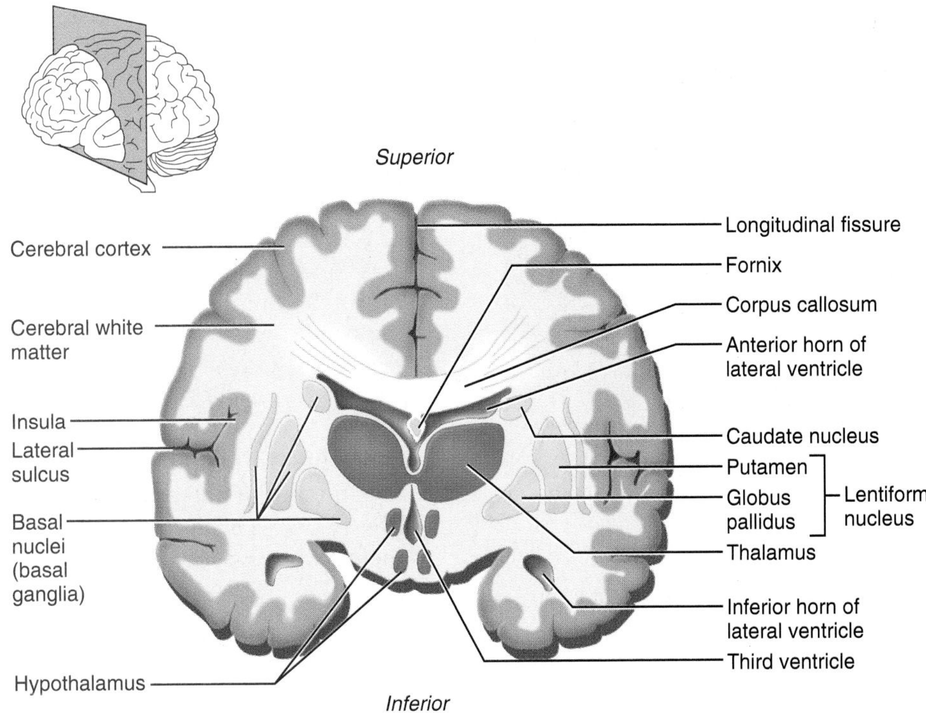

FIGURE 13.8 Internal structure of the forebrain. The cerebrum is sectioned frontally to show the positions of the superficial cerebral cortex, the intermediate white matter, and the deep basal nuclei. The diencephalon, enclosed by the cerebral hemispheres, is also evident.

anteriorly and the **postcentral gyrus** posteriorly. The **occipital lobe,** which lies farthest posteriorly, is separated from the parietal lobe by several landmarks, the most conspicuous of which is the **parieto-occipital sulcus** on the medial surface of the hemisphere (Figure 13.7b). On the lateral side of the hemisphere, the flaplike **temporal lobe** is separated from the overlying parietal and frontal lobes by the deep **lateral sulcus,** which is so deep that, despite its name, it is actually a fissure. A fifth lobe of the cerebral hemisphere is the **insula** (in′su-lah; "island") which is buried deep within the lateral sulcus and forms part of its floor (see Figure 13.7e). The insula is covered by parts of the temporal, parietal, and frontal lobes.

The cerebral hemispheres fit snugly into the skull. The frontal lobes lie in the anterior cranial fossa, whereas the anterior parts of the temporal lobes fill the middle cranial fossa. The posterior cranial fossa houses the brain stem and cerebellum. The occipital lobes are located superior to the cerebellum (see Figure 13.7a) and do not occupy a cranial fossa.

A frontal section through the forebrain reveals the three largest regions within the cerebrum: a superficial **cerebral cortex** of gray matter, the **cerebral white matter** internal to it, and the **basal nuclei (basal ganglia)** deep in the white matter (Figure 13.8). We next examine each of these regions in turn.

Cerebral Cortex

The cerebral cortex is the "executive suite" of the nervous system, the home of our "conscious mind." It enables us to be aware of ourselves and our sensations, to initiate and control voluntary movements, and to communicate, remember, and understand.

Because it is composed of gray matter, the cerebral cortex contains neuron cell bodies, dendrites, and very short unmyelinated axons, but no fiber tracts. Even though it is only 2–4 mm thick (about 1/8 inch), its many folds triple its surface area to the size of a large pizza, and it accounts for about 40% of the total mass of the brain. The cortex contains billions of neurons arranged in six layers.

In the late 1800s, anatomists studying the cerebral cortex were able to map subtle variations in the thickness of the six layers and in the structure of the contained neurons over the entire surface of the cerebral hemispheres. The most successful of these efforts was that of Korbinian Brodmann, who in 1909 divided the cortex into 52 *structurally* different areas. Some of the most important **Brodmann areas** are numbered in Figure 13.10 on page 367.

With a structural map emerging, early neurologists were anxious to localize *functional* regions of the cortex. Traditionally, the methods for determining the functions of brain regions were quite limited. Most involved either studying the mental defects of people and animals

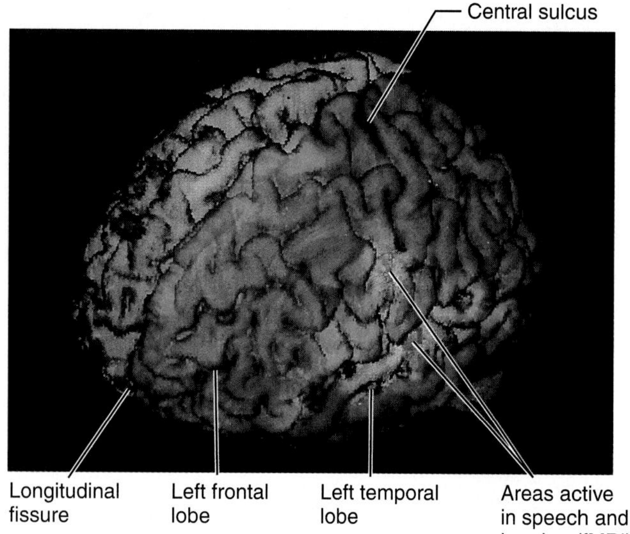

FIGURE 13.9 Functional neuroimaging of the cerebral cortex. The rostral direction is to the left. Functional magnetic resonance image (fMRI) of the cerebral cortex of a person during speaking and hearing reveals activity (blood flow) in the posterior frontal and superior temporal lobes, respectively.

that had sustained localized brain injuries or recording neuronal activity in the brains of uninjured animals. Rarely, a person undergoing brain surgery would let scientists mildly stimulate and record electrical activity at the surface of the exposed cerebrum. Since about 1990, however, two **functional neuroimaging techniques—PET** and **fMRI** (see pp. 21–23) have yielded a flood of new discoveries and revolutionized brain science. Recall that these techniques work by revealing areas of maximum blood flow to the brain, which are presumed to be areas of maximum mental activity as well. A sample fMRI scan is shown in Figure 13.9.

Advances in Understanding Two other techniques are just starting to be used for localizing brain functions. In **transcranial magnetic stimulation,** a pulse of magnetism temporarily prevents a localized area of the brain from functioning, then the resulting functional deficits are noted, and the function of the area is deduced. **Electroencephalography (EEG)** is the familiar technique of recording electromagnetic "brain waves" through electrodes placed on the scalp. Because the skull blocks most brain waves, EEG was previously limited to coarse measurement of the activity of the entire brain, but it is now so sensitive that it can locate the brain's most active regions. If its spatial localization can be improved further, EEG will become extremely valuable because it can detect and map much *faster* changes in brain activity than can fMRI or PET. ✷

Over the years, neurologists have established that structurally distinct areas of the cerebral cortex perform distinct motor and sensory functions. However, some higher mental functions, such as memory and language, are spread over very large cortical areas.

We now recognize three kinds of functional areas: *motor areas,* which control voluntary motor functions; *sensory areas,* which provide for conscious awareness of sensation; and *association areas,* which mainly integrate diverse information to enable purposeful action.

We discuss these three kinds of areas next. As you read, do not confuse motor and sensory *areas* with motor and sensory *neurons*—all the neurons in the cerebral cortex are interneurons, which by definition are confined entirely to the CNS.

Motor Areas The cortical areas that control motor functions lie in the posterior part of the frontal lobe (see Figure 13.10). These areas are the *primary motor cortex,* the *premotor cortex,* the *frontal eye field,* and *Broca's area.*

1. **Primary motor cortex.** The primary motor area, or *somatic motor area,* is located along the precentral gyrus of the frontal lobe. It corresponds to Brodmann area 4. In this area are large neurons called *pyramidal cells* (see Figure 12.12a, p. 345), the axons of which form the massive *pyramidal* or *corticospinal tract* (see p. 397), which descends through the brain stem and spinal cord. Here, the pyramidal axons signal the motor neurons to bring about precise or skilled voluntary movements of the body, especially of our forearms, fingers, and facial muscles. The projection of the pyramidal axons is **contralateral,** meaning they cross over to the opposite side of the brain and spinal cord (*contra* = opposite). Therefore, the *left* and *right* primary motor cortexes control muscles on the *right* and *left* sides of the body, respectively.

   The human body is represented spatially in the primary motor cortex of each hemisphere. In other words, the pyramidal cells that control hand movement are in one place, those that control foot movement are in another, and so on, as mapped on the left side of Figure 13.11. This map was first constructed by mildly stimulating parts of the precentral gyrus with an electrode and watching which part of the patient's body moved. Note that the body is represented upside down, with the head on the inferolateral part of the precentral gyrus and the toes at the superomedial end. Also note that the face and hand representations are disproportionately large: The face and hand muscles can perform very delicate and skilled movements because so many pyramidal cells control them. The body map on the motor cortex is called a **motor homunculus** (ho-mung'ku-lus; "little man"). As we will see, the body (and the outside world) are represented spatially in many parts of the CNS, a general principle called **somatotopy** (so-mat'o-tōp'e), which literally means "body mapping."

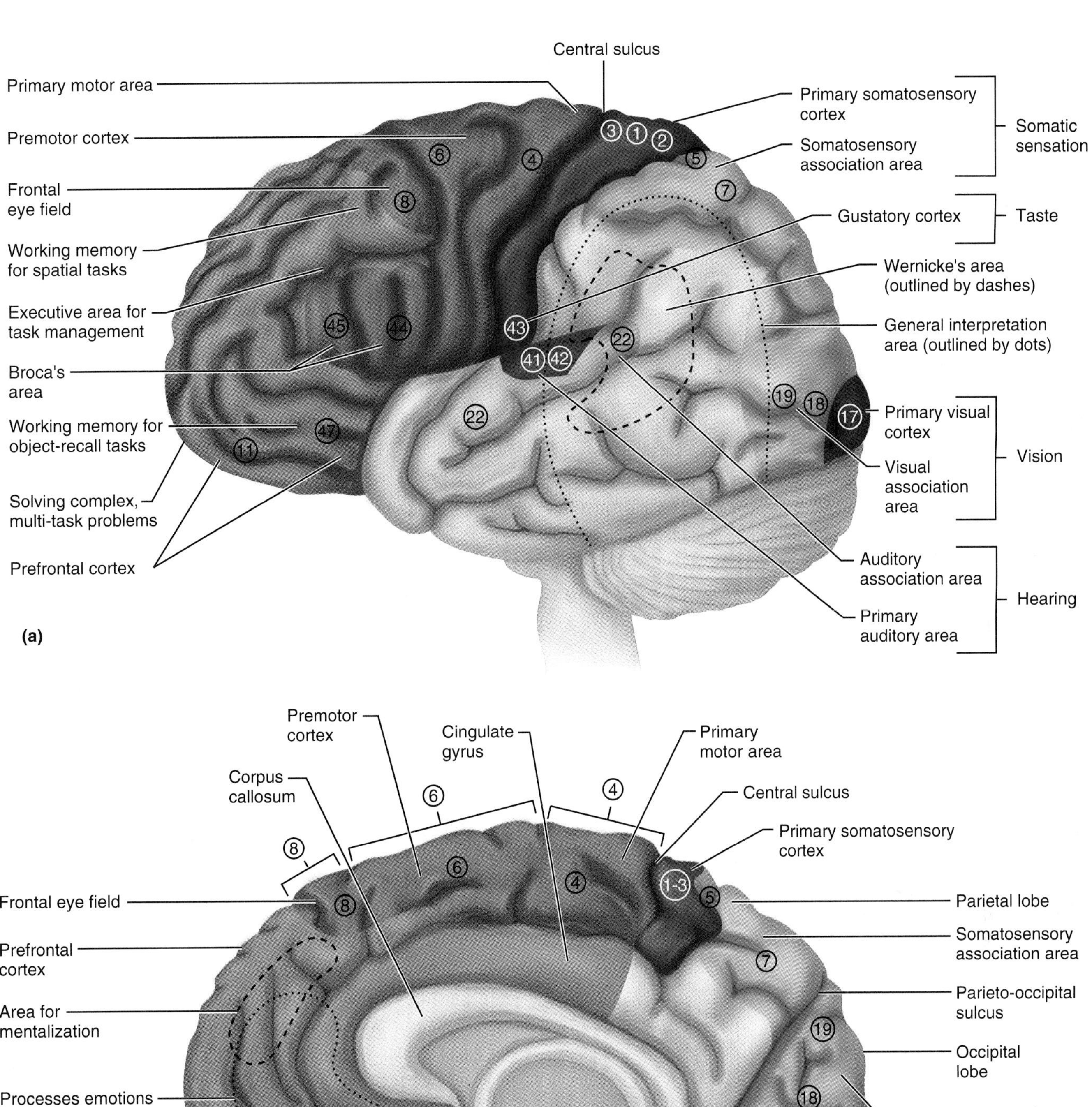

FIGURE 13.10 Functional and structural areas of the cerebral cortex. (a) Lateral view, left cerebral hemisphere. (b) Medial view, right hemisphere. Numbers indicate Brodmann's structural areas, whereas colors and outlines represent functional regions of the cortex.

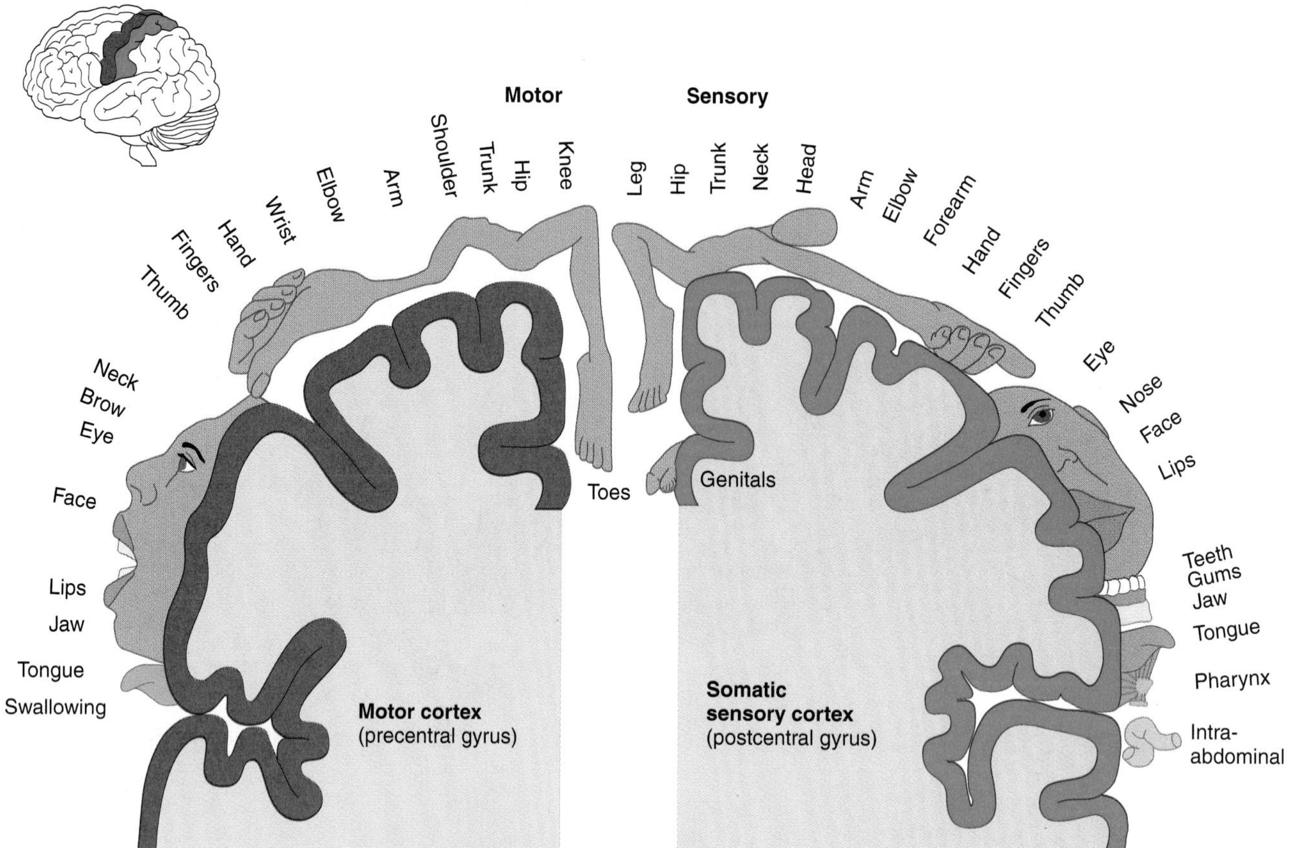

FIGURE 13.11 Body maps in the primary motor cortex and somatosensory cortex of the cerebrum. (a) Frontal sections through the precentral motor gyrus (left) and the postcentral sensory gyrus (right). The relative amount of cortical tissue devoted to the function of each body region is indicated by the size of that region on the body maps.

Critical Reasoning

Strokes, tumors, or wounds can destroy limited areas of the brain. What are the effects of localized lesions to the primary motor cortex?

Damage to an area of the primary motor cortex paralyzes the body muscles controlled by that area. Only voluntary *control is lost, however, because the muscles can still contract reflexively. Because the projection of the pyramidal axons is contralateral, lesions in the right cerebral hemisphere affect the left side of the body, and those in the left hemisphere affect the right side of the body.*

Recent studies indicate that the concept of a motor map works at the relatively crude level shown in Figure 13.11, but it cannot provide a map of separate body muscles in the precentral gyrus. Instead, each pyramidal cell stimulates the contraction of several related muscles that work together to produce an often-executed movement.

2. **Premotor cortex.** Just anterior to the precentral gyrus is the premotor cortex, which coincides with Brodmann area 6 (see Figure 13.10). This region controls more complex movements than does the primary motor cortex. It receives highly processed sensory information (visual, auditory, general somatic sensory) from other parts of the cortex, especially from the visual regions. Using this information, it controls voluntary actions that are dependent on sensory feedback about spatial relations, such as moving an arm through a maze of obstacles to grasp a hidden object. The premotor cortex also seems to be involved in the *planning* of movements.
3. **Frontal eye field.** The frontal eye field lies anterior to the premotor cortex, in the inferior part of Brodmann area 8 (see Figure 13.10). It controls voluntary movements of the eyes, especially when we look quickly at something, as in moving our eyes to follow a moving target.
4. **Broca's** (Bro′kahz) **area.** This area lies anterior to the inferior part of the premotor cortex in the left or "language-dominant" cerebral hemisphere, where it overlaps Brodmann areas 44 and 45 (see Figure 13.10a). It manages speech production, controlling the movements necessary for speaking. The corresponding region in the right or "intuitive-emotional" hemisphere controls the emotional overtones given to spoken words. People with damage to Broca's area cannot speak but can understand most aspects

of the speech of others. Functional neuroimaging studies suggest that Broca's area functions with the nearby premotor cortex to store working memories (i.e., short-term memories) of the words one is about to speak. It also breaks down words to form a mental image of the sounds to be spoken.

Sensory Areas The cortical areas involved with conscious awareness of sensation occur in the parietal, temporal, and occipital lobes. There is a distinct cortical area for each of the major senses: the general somatic senses, vision, hearing, balance, and taste (recall Figure 12.3). Some association areas are so closely related to adjacent sensory areas that they are included in this section, whereas the other association areas are considered in the next section (p. 370).

1. **Primary somatosensory cortex.** The primary somatosensory cortex is located along the postcentral gyrus of the parietal lobe, just posterior to the primary motor cortex. It corresponds to Brodmann areas 1–3 (see Figure 13.10) and is involved with conscious awareness of the general somatic senses (skin senses and proprioception; see Figure 12.3). This sensory information is picked up by sensory receptors in the periphery of the body and relayed through the spinal cord and brain stem to the somatosensory cortex. There, cortical neurons process the sensory information and identify the precise area of the body being stimulated. This ability to localize (assign a location to) a stimulus precisely is called **spatial discrimination.** The projection is contralateral; that is, the right cerebral hemisphere receives its sensory input from the left side of the body. As in the primary motor cortex, somatotopy is exhibited: A **sensory homunculus** can be constructed for the postcentral gyrus (see the right half of Figure 13.11). This map was first demonstrated by stroking patients on various areas of their skin and then recording exactly where electrical activity was detected in the cerebral cortex. The amount of somatosensory cortex devoted to a body region is related to the sensitivity of that region—that is, to the number of its sensory receptors. The lips and fingertips are our most sensitive body parts, and hence the largest parts of the sensory homunculus.

   Damage to the primary somatosensory cortex destroys the conscious ability to feel and localize touch, pressure, and vibrations on the skin. Most ability to feel pain and temperature is also lost, although these can still be felt in a vague, poorly localized way.

2. **Somatosensory association area.** The somatosensory association cortex lies posterior to the primary somatosensory cortex and corresponds roughly to Brodmann areas 5 and 7 (see Figure 13.10). With many connections to the primary somatosensory cortex, this association area integrates different sensory inputs (touch, pressure, and others) into a comprehensive understanding of what is being felt. For example, when you reach into your pocket, the somatosensory association area draws upon stored memories of past sensory experiences and perceives the objects you feel as, say, coins or keys.

3. **Visual areas.** The **primary visual (striate) cortex,** which corresponds to Brodmann area 17, is on the posterior and medial part of the occipital lobe (see Figure 13.10). More precisely, most of it is located on the medial aspect of the occipital lobe, buried within the deep **calcarine sulcus** (kal′kar-in; "spur-shaped") (see Figure 13.10b). The largest of all cortical sensory areas, the primary visual cortex receives visual information that originates on the retina of the eye. If this cortical area is damaged, the person has no conscious awareness of what is being viewed and is functionally blind. There is a map of visual space on the primary visual cortex that is analogous to the body map on the somatosensory cortex. The right half of visual space is represented on the left visual cortex, and the left half is on the right cortex. The primary visual area is the first of a series of cortical areas that process visual input, and the processing here is at a comparatively low level—noting the orientation of objects being viewed and putting the inputs from the two eyes together.

   The **visual association area** surrounds the primary visual area and covers much of the occipital lobe. It coincides with Brodmann areas 18 and 19 (see Figure 13.10). Communicating with the primary visual area, the visual association area continues the processing of visual information by analyzing color, form, and movement.

   Elegant experiments on monkeys, coupled with functional neuroimaging of humans, have revealed that complex visual processing extends far forward from the occipital lobe into the temporal and parietal lobes. Overall, about 30 cortical areas for visual processing have been identified, each more sophisticated than the last. Visual information proceeds anteriorly through these visual areas in two streams (Figure 13.12): (1) The *ventral stream* extends through the inferior part of the entire temporal lobe and is responsible for recognizing objects, words during reading, and faces (facial recognition involves the right hemisphere only); (2) the *dorsal stream* extends through the posterior parietal cortex to the postcentral gyrus and perceives spatial relationships among different objects. These ventral and dorsal streams have been dubbed the "what" and "where" pathways, respectively.

Advances in Understanding Recent findings demonstrate the importance of the dorsal stream (in the parietal lobe) for spatial perception. For example, it was found that the superior part of the parietal lobe calculates how we should move our limbs through space, then

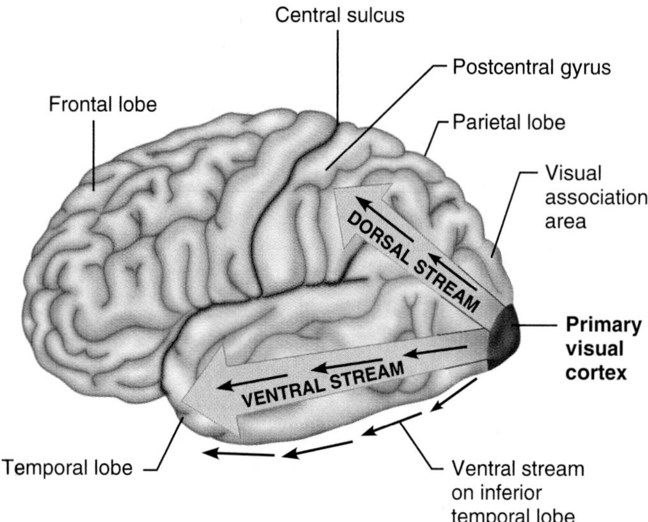

FIGURE 13.12 The ventral ("what") stream and dorsal ("where") stream for processing visual information, in the posterior part of the cerebral cortex. The ventral stream also occupies the inferior surface of the temporal lobe, as indicated by the line of arrows below the brain.

sends this information forward to the motor cortex, which dictates these movements. Furthermore, the parietal lobe is important for abstract mathematical abilities, which are highly visual-spatial in nature. Recent examination of the preserved brain of Albert Einstein, the mathematical genius who was voted the greatest physicist of the last millennium, showed that parts of his parietal lobe were unusually large and well developed. ✷

4. **Auditory areas.** The **primary auditory cortex,** which functions in conscious awareness of sound, is on the superior edge of the temporal lobe, primarily inside the lateral sulcus. It corresponds to Brodmann areas 41 and 42 (see Figure 13.10a). When sound waves excite the sound receptors of the inner ear, impulses are transmitted to the primary auditory cortex, where this information is related to loudness, rhythm, and especially to pitch (high and low notes). A "pitch map" has been compiled on the auditory cortex.

   The **auditory association area** lies just posterior to the primary auditory area, in the posterior part of Brodmann area 22 (see Figure 13.10a). This area permits the evaluation of a sound as, say, a screech, thunder, music, and so on. Memories of past sounds seem to be stored here. In one hemisphere (usually the left), the auditory association area lies in the center of **Wernicke's** (Ver′nĭ-kēz) **area,** a functional brain region involved in recognizing and understanding spoken words. Damage to Wernicke's area interferes with the ability to comprehend speech.

5. **Gustatory (taste) cortex.** The gustatory (gus′tah-to″re) cortex is involved in the conscious awareness of taste stimuli. It corresponds to Brodmann area 43 and lies on the roof of the lateral sulcus (see Figure 13.10a). As expected, this taste area occurs on the tongue in the somatosensory homunculus.

6. **Vestibular (equilibrium) cortex.** Researchers have had difficulty locating the part of the cortex that is responsible for conscious awareness of the sense of balance; that is, of the position of the head in space. Both electrical stimulation and neuroimaging studies now place this region in the posterior part of the insula, deep to the lateral sulcus (see Figure 13.7e).

7. **Olfactory (smell) cortex.** The primary olfactory cortex lies on the medial aspect of the cerebrum in a small region called the *piriform lobe* (pir′ĭ-form; "pear-shaped"), which is dominated by the hook-like **uncus** (see Figure 13.10b). The olfactory nerves (cranial nerve I) from the nasal cavity transmit impulses that ultimately are relayed to the olfactory cortex. The outcome is conscious awareness of smells.

   The olfactory cortex is part of a brain area called the *rhinencephalon* (ri″nen-sef′ah-lon; "nose brain"), which includes all parts of the cerebrum that directly receive olfactory signals: the piriform lobe, the **olfactory tract,** the **olfactory bulb** (see Figure 13.10b), and some nearby structures. The rhinencephalon connects to the brain area that is involved in emotions, the *limbic system,* which explains why smells often trigger emotions. The limbic system is discussed further on page 385.

   The part of the frontal lobe directly above the orbit—namely, the orbitofrontal cortex—is involved in higher-order processing of smells: consciously identifying and recalling specific odors, and telling different smells apart.

Association Areas The association areas include all cortical regions other than the primary sensory areas and motor areas. Their name reflects the fact that some of these areas tie together, or make *associations* between, the different kinds of sensory information. They also seem to *associate* new sensory inputs with memories of past experiences. They do so much more than this, however, that the name "association areas" is fading from use and will probably be replaced by "higher-order processing areas." We have already explored the association areas that directly border the primary sensory regions (the somatosensory, visual, and auditory association areas). The remaining association areas are considered next.

1. **Prefrontal cortex.** The prefrontal cortex is the large region of the frontal lobe that lies anterior to the motor areas (see Figure 13.10). The most complicated cortical region of all, it performs many *cognitive functions.* **Cognition** is all aspects of thinking, perceiving, and of intentionally remembering and recalling information. The prefrontal cortex is nec-

essary for abstract ideas, reasoning and judgment, impulse control, persistence, long-term planning, complex problem solving, mental flexibility, social skills, appreciating humor, empathy, and conscience. It also seems to be related to mood and has close links to the emotional (limbic) part of the forebrain. The tremendous elaboration of this prefrontal region distinguishes humans from other animals.

Critical Reasoning

Predict some effects of injury to the prefrontal cortex.

Tumors or other lesions of the prefrontal cortex cause mental and personality disorders. Wide swings in mood may occur, and there may be a loss of attentiveness, of inhibitions, and of judgment. The affected person may be oblivious to social restraints, perhaps becoming careless about personal appearance, or irrational, rashly attacking a 7-foot-tall assailant instead of fleeing.

Functional neuroimaging techniques have begun to reveal the functions of specific parts of the prefrontal cortex. Completion of multistep problem-solving tasks requires the temporary storage of information in working memory. The working memories of *spatial* relations for such tasks are stored in the dorsolateral prefrontal cortex just anterior to the frontal eye field, whereas working memories of objects and faces are stored farther ventrally, below Broca's area (see Figure 13.10a). More significant is the discovery of a region that runs and manages cognitive tasks by directing our attention to the relevant information in working memory. This "executive area" lies between the working-memory sites, just above the anterior part of Broca's area.

In other new studies, the extreme anterior pole of the frontal cortex (anterior to "11" in Figure 13.10a) was found to be active in solving the most complex problems—problems in which many subproblems had to be completed before a solution could be obtained. Intriguingly, this finding seems to support a general "rule" of neuroscience that says the farther rostrally one goes in the CNS, the more complex are the neural functions performed.

Functional cortical areas are also being identified on the *medial* side of the frontal lobe (see Figure 13.10b). The area just anterior to the corpus callosum may process emotions involved in complex personal and social interactions; lesions here cause severe depression. Also, prefrontal regions superior and anterior to the corpus callosum are involved in "mentalization," the ability to understand and manipulate other people's thoughts and emotions. It is not surprising that all these medial-prefrontal areas border the emotional or limbic part of the brain (see p. 385).

2. **General interpretation area.** The function of the area traditionally called the *general interpretation area* is currently under investigation. Because this area, located in the posterolateral cerebral cortex (see Figure 13.10a), is at the interface of the visual, auditory, and somatosensory association areas, it was thought to integrate all these types of sensory information. The main evidence for its existence was that people with damage in this area of their dominant hemisphere cannot recognize things or understand experiences that require multisensory coordination—a deficit called **agnosia** (ag-no′ze-ah: "not knowing"). Although agnosia certainly exists, newer studies show that most of the so-called general interpretation area is actually involved in the visual processing of spatial relationships. Therefore, the confusion experienced by people with agnosia may not result from any multisensory deficit, but instead could stem from not knowing where any objects—even their own body—are located in space. Today, many neuroscientists either abandon the concept of a multisensory general interpretation area or confine it to a much smaller region superior and posterior to Wernicke's area.

3. **Language area.** A large area surrounding the lateral sulcus in the left cerebral hemisphere is involved in various functions related to language. Five parts of this area have been identified: (1) Broca's area (speech production), (2) Wernicke's area (speech comprehension), (3) the lateral prefrontal cortex just anteroinferior to Broca's area (deep conceptual analysis of spoken words), (4) most of the lateral and inferior temporal lobe (coordination of auditory and visual aspects of language, as when naming viewed objects and reading words), and (5) parts of the insula, deep to the lateral sulcus, between Broca's and Wernicke's areas (initiation of word articulation and recognition of rhymes and sound sequences).

   The corresponding areas on the right hemisphere, although not involved in the mechanics of language, act in the creative interpretation of words and in controlling the emotional overtones of speech.

4. **Insula.** The insula is large and the functions of its cortex are not well understood. As previously mentioned, some parts function in language and in the sense of balance. Other parts seem to have visceral functions, including the conscious perception of visceral sensations (upset stomach, full bladder, and some aspects of smell) and behavioral influences on cardiovascular activity (as when a startled person's heart stops beating for a short time).

Lateralization of Cortical Functioning We have already seen that some division of labor exists between the right and left cerebral hemispheres. We have already noted, for example, that the two hemispheres control opposite sides of the body. Furthermore, the hemispheres are specialized for different cognitive functions: In most people (90 to 95%), the left hemisphere has more control over language abilities, math, and logic, whereas the right hemisphere is more involved with visual-spatial skills, reading facial expressions, intuition, emotion, and artistic and musical skills. Whereas the right hemisphere deals with the big picture, the left deals with the details, which it then interprets logically. In the remaining 5 to 10% of the population, either these roles of the hemispheres are reversed, or the two hemispheres share their cognitive functions equally. Functional differences between the hemispheres may be greater in males than in females, as described in this chapter's A Closer Look box on page 373.

Despite their differences, the two cerebral hemispheres are nearly identical in structure and share most functions and memories. Such sharing is possible because the hemispheres communicate through interconnecting fiber tracts called *commissures* (discussed shortly).

Cerebral White Matter

Now that we have discussed the gray matter of the cerebral cortex, we turn next to the underlying **cerebral white matter.** Different areas of the cerebral cortex communicate extensively with one another, and with the brain stem and spinal cord via the many axons that form the cerebral white matter. Most of these fibers are myelinated and bundled into large tracts. These fibers and tracts are classified as (1) *commissural,* (2) *association,* and (3) *projection,* according to where they run (Figure 13.13).

1. **Commissures,** composed of **commissural fibers,** run between the two hemispheres. Specifically, they interconnect corresponding gray areas of the right and left hemispheres, allowing the two hemispheres to function together as a coordinated whole. The largest commissure is the **corpus callosum** (kor'pus kah-lo'sum, "thickened body"), a broad band that lies superior to the lateral ventricles, deep within the longitudinal fissure (see Figure 13.13a).
2. **Association fibers** connect different parts of the *same* hemisphere. Short association fibers connect neighboring cortical areas, whereas long association fibers connect widely separated cortical lobes. For example, many association fibers run between two portions of the language area, Broca's and Wernicke's areas.
3. **Projection fibers** either descend from the cerebral cortex to more caudal parts of the CNS or ascend to the cortex from lower regions. It is through projection fibers that sensory information reaches the cortex, and motor instructions leave it. These fibers run vertically, whereas most commissural and association fibers run horizontally.

   Deep to the cerebral white matter, the projection fibers form a compact bundle called the **internal capsule,** which passes between the thalamus and some of the basal nuclei (see Figures 13.13b and 13.14a). Superior to the internal capsule, the projection fibers running to and from the cerebral cortex fan out to form the **corona radiata** (kŏ-ro'nah ra-de-ah'tah; "radiating crown").

Basal Nuclei (Basal Ganglia)

Deep within the cerebral white matter lies a group of nuclei collectively called the **basal nuclei** (or **basal ganglia***): the **caudate** ("tail-like"), the **lentiform** ("lens-shaped"), and **amygdala** ("almond-shaped") **nuclei** (Figure 13.14). The caudate nucleus arches superiorly over the thalamus and lies medial to the internal capsule. Together, the lentiform and caudate nuclei are called the **corpus striatum** (kor'pus stri-a'tum; "striated body") because some fibers of the internal capsule passing through them create a striated appearance. The lentiform nucleus is divided into a medial part, the **globus pallidus** ("pale globe"), and a lateral part, the **putamen** (pu-ta'men; "pod") (see Figures 13.14b, 13.13b, and 13.8).

The amygdala (ah-mig'dah-lah) sits on the tip of the tail of the caudate nucleus (see Figure 13.14a). Traditionally, it has been grouped with the basal nuclei, but functionally it belongs to the limbic system (see p. 385).

Functionally, the basal nuclei can be viewed as complex neural calculators that cooperate with the cerebral cortex in controlling movements. Indeed, they communicate extensively with the cerebral cortex, receiving inputs from many cortical areas and sending almost all their output back to the motor cortex (through a relay in the thalamus). Despite extensive research, the precise roles of the basal nuclei have proven elusive. There is some evidence that they *start*, *stop*, and *regulate the intensity* of the voluntary movements that are ordered and executed by the cerebral cortex, and they may *select* the appropriate muscles or movements for a task and inhibit the relevant antagonistic muscles. They may also control rhythmic, repetitive tasks and participate in the learning of habits. Finally, in a nonmotor role, the basal nuclei in some way estimate the passage of time, like an hourglass.

Clinical Applications

DYSKINESIA Degenerative conditions of the various basal nuclei produce *dyskinesia* (dis-ki-ne'ze-ah), literally, "bad movements." Two examples of dyskinetic conditions are Parkinson's disease and Huntington's disease.

Parkinson's disease is characterized by slow and jerky movements; tremor of the face, limbs, or hands; muscle rigidity; and great difficulty in starting voluntary movements. In this common disorder, an overactive globus pallidus has the effect of a

*These are true nuclei—collections of neuron cell bodies within the CNS—so the term *basal nuclei* is correct. The more commonly used term *basal ganglia* is technically incorrect because ganglia are PNS structures.

A Closer Look

Sex-Related Differences in Brain Structure

Although behavioral differences between the sexes are recognized in every society, upon examination many prove to be the result of cultural influences, not genes. However, some of the differences do appear to be grounded in biology.

Evidence for differences in behavior between the sexes can be found from infancy onward. Newborn boys, for example, show greater motor strength and muscle tone, while infant girls show greater sensitivity to touch, taste, and light, and they smile more often. Studies of children and adults have shown that females tend to be more nurturing, do better at some verbal tasks, at recalling landmarks, and at precisely manipulating objects, whereas males tend to be more aggressive, and are better in spatial-visual tasks and at throwing things to hit targets. Females seem to have better memories, especially in recalling words and associating names with faces. Males seem to have a greater degree of lateralization of skills to the separate cerebral hemispheres. For example, men are more likely than women to have trouble understanding words following injury to the left (language) cerebral hemisphere, despite a recent finding that language shows an equally strong left-lateralization in both sexes.

How can the sex differences in behaviors and skills be explained? They must relate to sex hormones and the structure of the brain. When males enter puberty, their aggression increases and there is little doubt that the rise in male sex hormones surging through their blood contributes to this change. Radioisotope techniques have shown that both male and female sex hormones (testosterone and estrogens, respectively) concentrate selectively in certain brain regions that play an important role in aggression, courtship, and mating behaviors—behaviors in which the sexes differ most. In 1973, it was demonstrated for the first time that male and female brains differ structurally in some ways. In rats, a nucleus in the anterior part of the hypothalamus (now called the sexually dimorphic nucleus) was shown to have a distinctive synaptic pattern in each sex.

A girl plays with a sleek toy ambulance or fire truck, perhaps to rescue her sick doll, while a boy plays with a toy version of a powerful garbage truck.

Castration of male animals shortly after birth produced the female hypothalamic pattern, whereas injection of testosterone into females triggered the development of the male pattern. This discovery rocked the neuroscience community because it was the first evidence of the structural brain differences between the sexes and the first demonstration that sex hormones circulating before, at, or after birth can change the brain. Many more sexual differences are now documented in the rat hypothalamus, and some in the human hypothalamus as well. Based on these and later observations, scientists concluded that the basic plan of the mammalian brain is female and it stays that way unless "told to do otherwise" by masculinizing (male) hormones. This conclusion has proven to be a bit too extreme, because estrogens are now known to influence the development of a few parts of the brain in females and to improve the mood and mental state of human females throughout life. Still, it does appear that fetal testosterone is responsible for development of the brain pattern seen in males.

Other structural differences between the central nervous systems of men and women have been found, and include the following:

1. Females tend to have a longer left temporal lobe and a larger visual cortex.
2. Sex-related differences exist in the corpus callosum, the band of axons connecting the two cerebral hemispheres. In males it is more cylindrical and uniform in diameter, whereas in females it is wider and more bulbous posteriorly.
3. In males only, the gray matter of the right cerebral cortex is thicker than that of the left. Overall, the density of neurons in the cerebral cortex of females may exceed that in males.
4. In the spinal cord, clusters of motor neurons that serve the external genitals are much larger in males than in females, and they contain receptors for testosterone but not estrogens. (These neurons innervate muscles that maintain erection, such as the bulbospongiosus and ischiocavernosus.)

The newest findings and the claims made from them are quite controversial. Quantitative measurements—which are very hard to perform accurately—suggest that men have bigger brains (by 16%) and more neurons in their cerebral cortex, even after correcting for the smaller average body size of women. On the other hand, men's brains shrink much faster in old age (again, by 16%). Functional neuroimaging studies suggest that sad feelings activate larger brain areas in women, and that primitive and aggressive parts of the emotional brain are more active in men than in women. In neither of these imaging studies, however, were the experimental subjects undergoing genuine emotional experiences, so one cannot conclude they prove women are more emotional or men are more aggressive.

What should we make of these findings? Are these anatomical and behavioral differences between the sexes of only minor importance, or do they underlie most manifestations of maleness or femaleness? We do not know yet, for studies of sexual differences in the brain are still in an early stage.

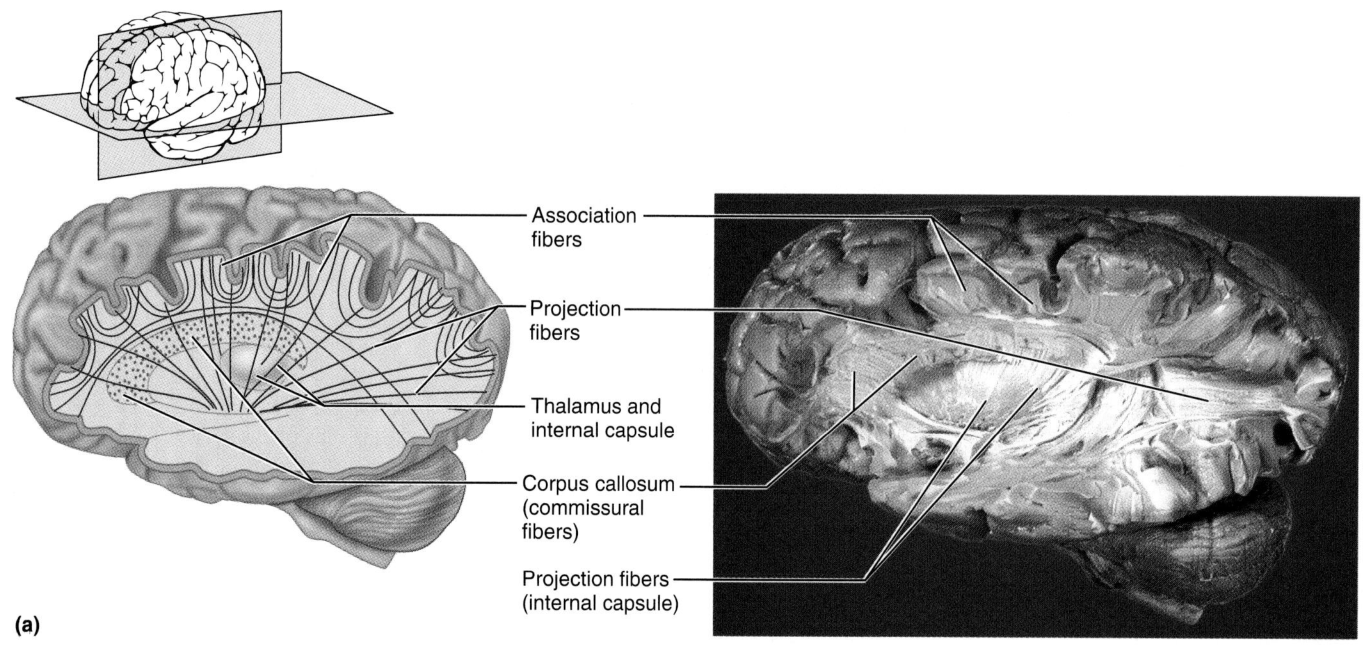

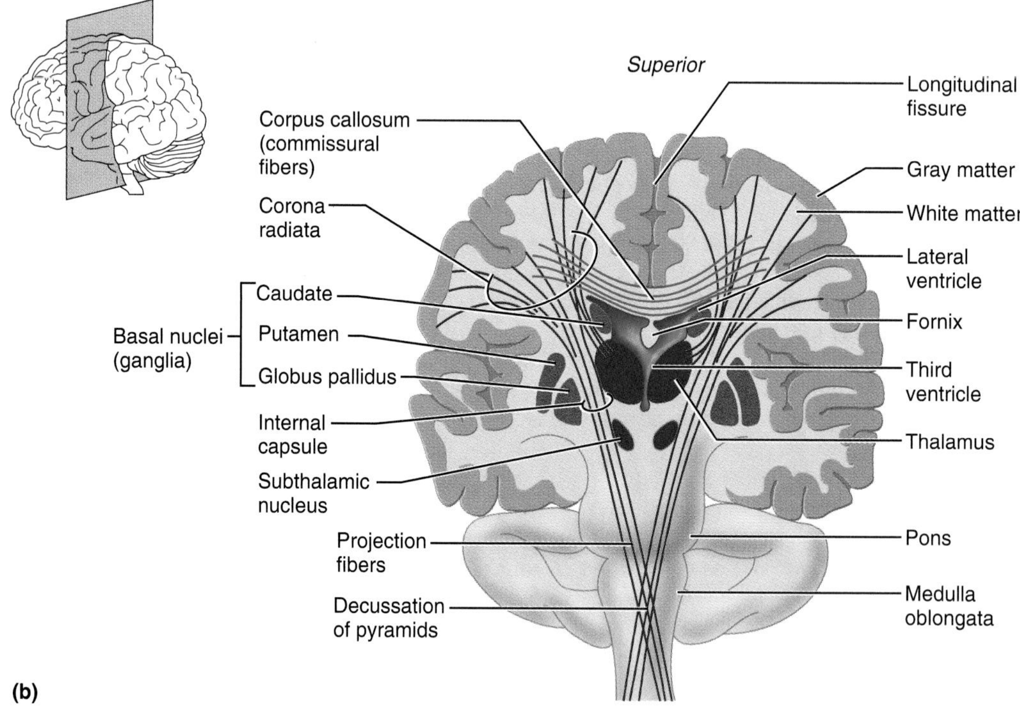

FIGURE 13.13 Types of fiber tracts in the white matter of the cerebral hemispheres. (a) Cut-open view of right hemisphere, with photo. Note the association fibers, projection fibers, and the corpus callosum, a large commissure that connects the right and left hemispheres. (b) Frontal section of the brain. Note that the internal capsule and corona radiata consist of projection fibers.

stuck brake, continuously inhibiting the motor cortex. Parkinson's disease results from degeneration of a midbrain nucleus called substantia nigra (see p. 380), but its ultimate cause is unknown. It may be caused by accumulations of environmental toxins or the buildup of iron or free radicals in the brain.

Huntington's disease involves an overstimulation of motor activities, such that the limbs jerk uncontrollably. The signs of the disease are caused by degeneration of the corpus striatum, and the cerebral cortex eventually degenerates as well. The condition is inherited, and the precise nature of the genetic defect in the DNA has been discovered. ✚

The Diencephalon

The **diencephalon,** the second of the four main parts of the brain, forms the central core of the forebrain and is surrounded by the cerebral hemispheres. Colored blue in Figures 13.15 and 13.16, the diencephalon consists largely of three paired structures—the thalamus, the hypothalamus, and the epithalamus. These border the third ventricle and consist primarily of gray matter.

FIGURE 13.14 Basal nuclei (basal ganglia). (a) Three-dimensional view of the basal nuclei (pink), deep in the cerebrum. (The fibers of the internal capsule and corona radiata are also shown.) (b) Transverse section through the cerebrum and diencephalon, with photo, showing the relationship of the basal nuclei to the thalamus and ventricles.

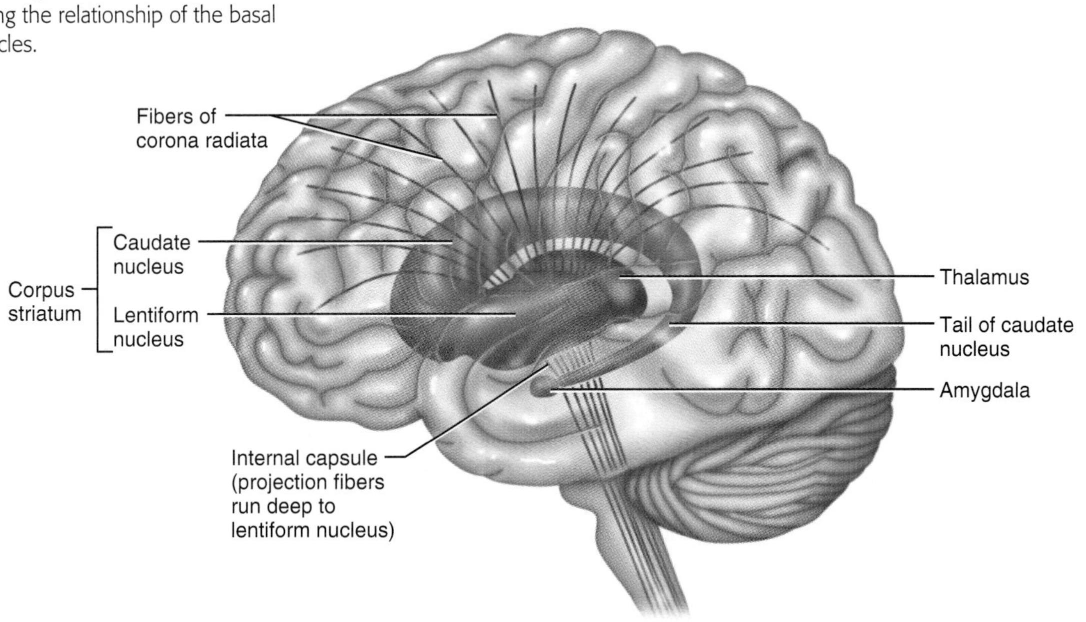

Fibers of corona radiata

Corpus striatum: Caudate nucleus; Lentiform nucleus

Thalamus

Tail of caudate nucleus

Amygdala

Internal capsule (projection fibers run deep to lentiform nucleus)

(a)

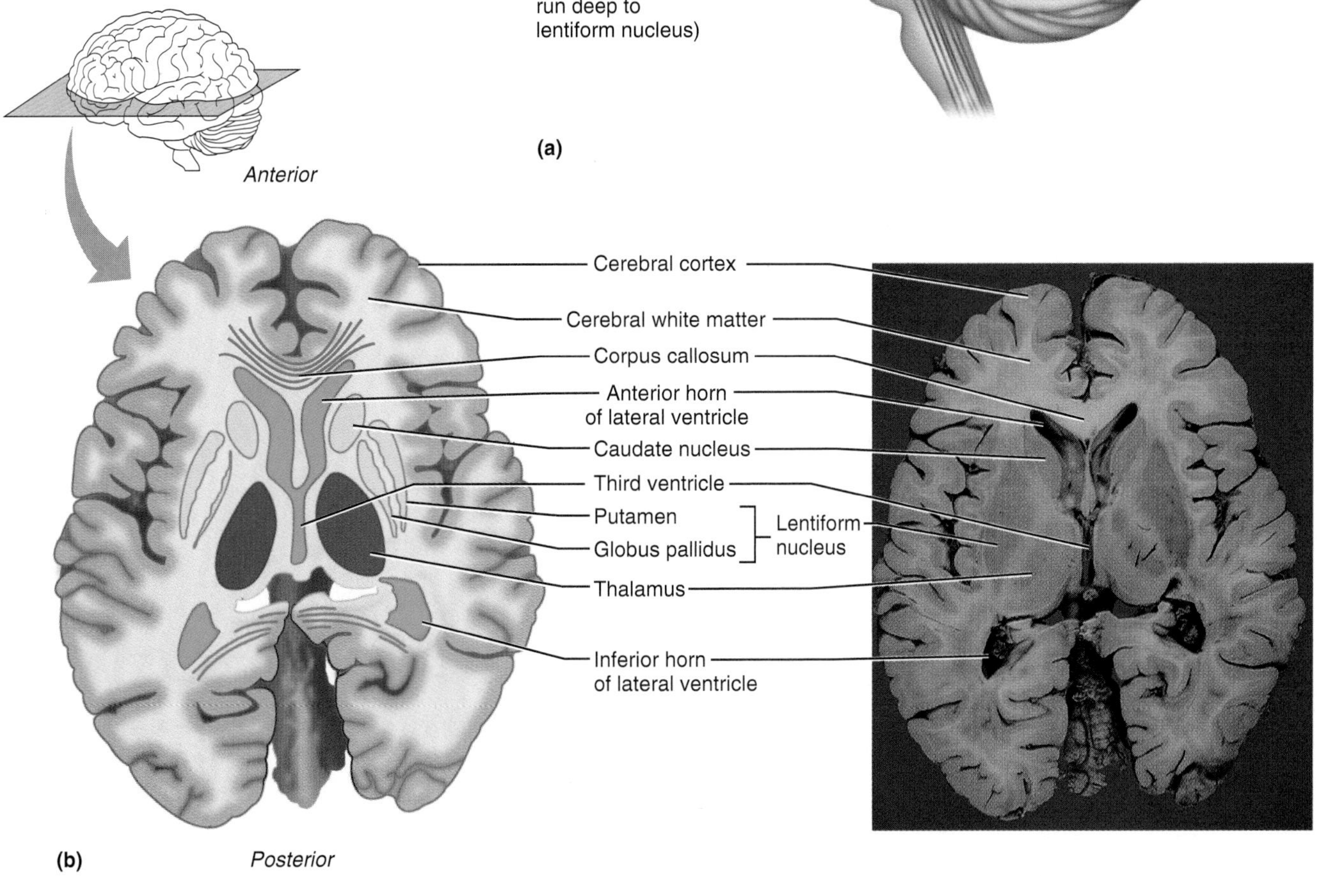

Anterior

Cerebral cortex

Cerebral white matter

Corpus callosum

Anterior horn of lateral ventricle

Caudate nucleus

Third ventricle

Putamen

Globus pallidus

Lentiform nucleus

Thalamus

Inferior horn of lateral ventricle

(b)

Posterior

The Thalamus

The egg-shaped **thalamus** (thal′ah-mus) makes up 80% of the diencephalon and forms the superolateral walls of the third ventricle (Figure 13.15). *Thalamus* is a Greek word meaning "inner room," which well describes this deep brain region. In some people, the right and left thalamus are joined by a small midline connection, the *interthalamic adhesion* (*intermediate mass).*

The thalamus contains about a dozen major nuclei, each of which sends axons to particular portions of the cerebral cortex (Figure 13.17). Some thalamic nuclei act as relay stations for the sensory information ascending to the primary sensory areas of the cortex. Afferent impulses from all the conscious senses converge on the thalamus and synapse in at least one of its nuclei. For example, the **ventral posterior lateral nucleus**

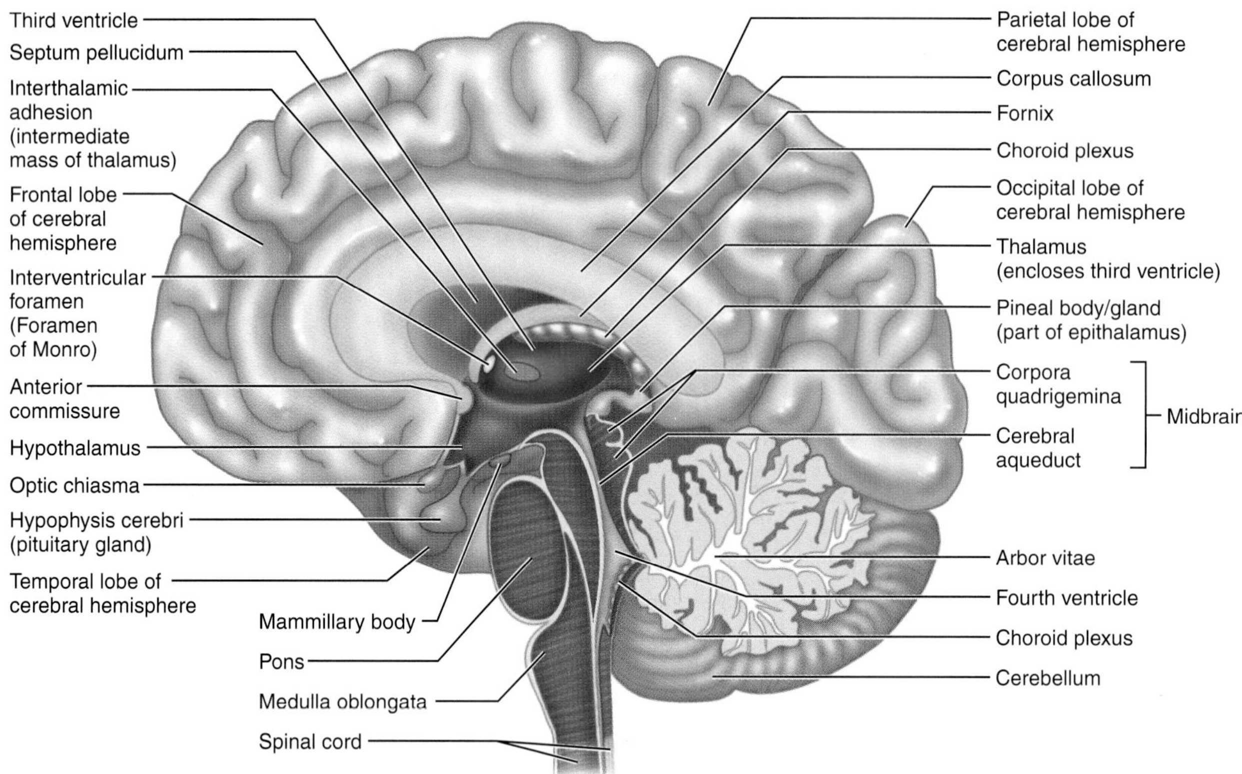

FIGURE 13.15 Brain sectioned in the midsagittal plane, highlighting the structures of the diencephalon and brain stem.

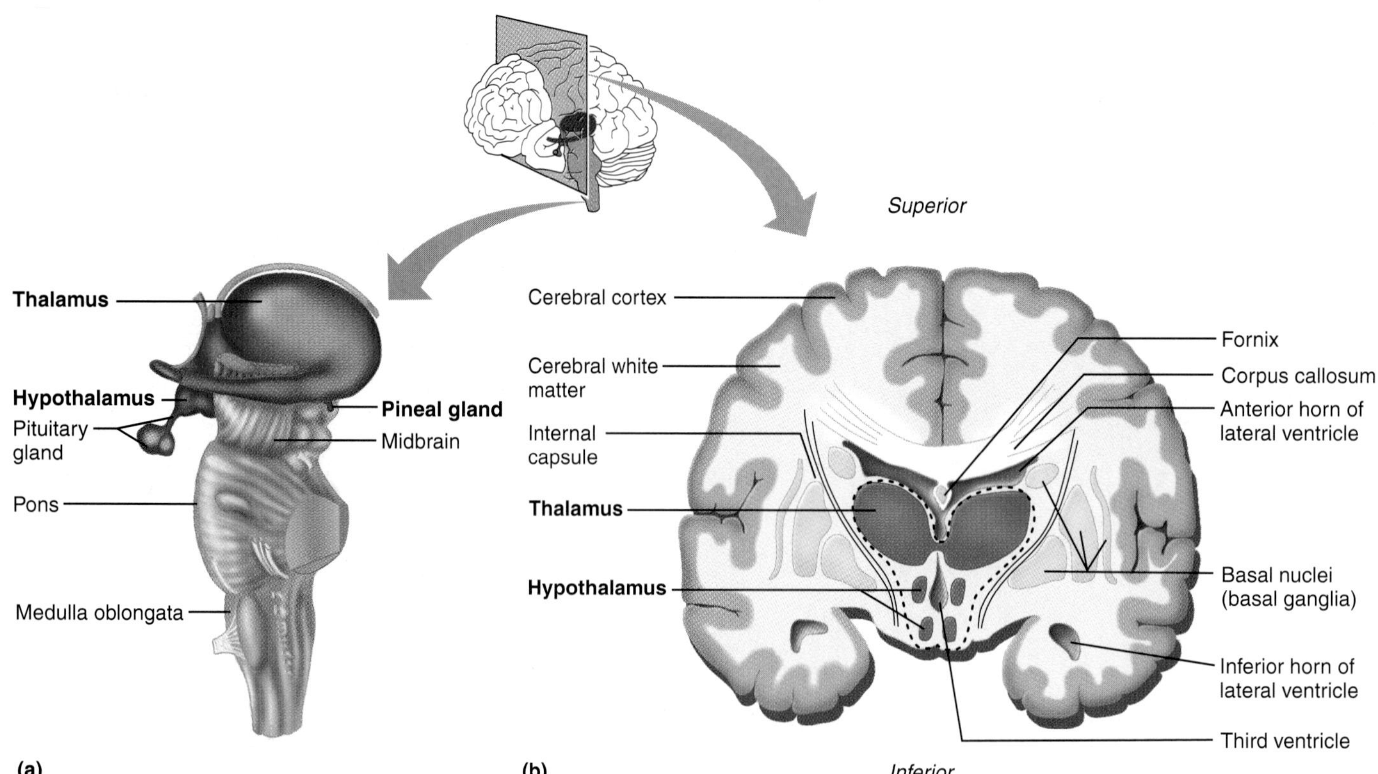

FIGURE 13.16 The diencephalon. (a) Lateral view of the diencephalon in relation to the brain stem. (b) Frontal section through the forebrain, in which the diencephalon is outlined by a dotted line.

receives general somatic sensory information, and the **lateral** and **medial geniculate nuclei** (jĕ-nik′u-lāt; "knee shaped") receive visual and auditory information, respectively.

Sensory inputs are not the only type of information relayed through the thalamus. *Every part of the brain that communicates with the cerebral cortex must relay its signals through a nucleus of the thalamus.* The thalamus can therefore be thought of as the "gateway" to the cerebral cortex.

The thalamus not only relays information to the cortex but also actively processes the information as it passes through. The thalamic nuclei organize and amplify or "tone down" the signals headed for the cerebral cortex. They are just one example of the many *relay nuclei* in the brain, all of which process and "edit" information before sending it along.

The **subthalamic nucleus** lies in the caudal part of the diencephalon (see Figure 13.13b). Functionally, it belongs to the basal nuclei, and bilateral surgical destruction of the subthalamic nuclei is used to relieve many of the symptoms of Parkinson's disease.

The Hypothalamus

The **hypothalamus** (hi″po-thal′ah-mus; "below the thalamus") is the inferior portion of the diencephalon (see Figures 13.15 and 13.18). It forms the inferolateral walls of the third ventricle. On the underside of the brain, it lies between the **optic chiasma** (Figure 13.19) (point of crossover of cranial nerves II, the optic nerves) and the posterior border of the **mammillary bodies,** rounded bumps that bulge from the hypothalamic floor (*mammillary* = "little breast"). Also projecting inferiorly from the hypothalamus is the *pituitary gland,* which secretes many hormones (see Chapter 25, p. 745).

The hypothalamus, like the thalamus, contains about a dozen nuclei of gray matter (Figure 13.18). Functionally, the hypothalamus is the main visceral control center of the body, regulating many activities of the visceral organs. Its functions include the following:

1. **Control of the autonomic nervous system.** Recall that the autonomic nervous system is composed of the peripheral motor neurons that regulate contraction of smooth and cardiac muscle and the secretion of glands (see p. 338). The hypothalamus is one of the major brain regions involved in directing the autonomic neurons. In doing so, it regulates heart rate and blood pressure, movement of the digestive tube, the secretion of sweat glands and salivary glands, and many other visceral activities.
2. **Control of emotional responses.** The hypothalamus lies at the center of the emotional part of the brain, the limbic system. Regions involved in pleasure, rage, sex drive, and fear are located in the hypothalamus.
3. **Regulation of body temperature.** The body's thermostat is in the hypothalamus. Some hypothalamic neurons sense blood temperature and then initiate the body's cooling or heating mechanisms as needed (sweating or shivering, respectively). Hypothalamic centers also induce fever.
4. **Regulation of hunger and thirst sensations.** By sensing the concentrations of nutrients and salts in the blood, certain hypothalamic neurons mediate feelings of hunger and thirst.
5. **Control of behavior.** The hypothalamus controls our motivation for feeding, thereby determining how much we eat. It also influences sexual behavior.
6. **Regulation of sleep-wake cycles.** Acting with other brain regions, the hypothalamus helps regulate the complex phenomenon of sleep. It is responsible for the timing of the sleep cycle. The nuclei involved are the **suprachiasmatic nucleus** (soo″prah-ki-az-mat′ik) above the optic chiasma and the nearby **preoptic nucleus** (see Figure 13.18). The suprachiasmatic nucleus is the body's biological clock, which regulates the timing of many daily rhythms **(circadian rhythms).** It receives information on daylight-darkness cycles from the eye through the optic nerve, and then sends signals to the preoptic nucleus. In response to such signals, the preoptic nucleus induces sleep. Other hypothalamic nuclei, near the mammillary body, mediate arousal from sleep.
7. **Control of the endocrine system.** The hypothalamus controls the secretion of hormones by the pituitary gland, which in turn regulates many functions of the visceral organs (see Chapter 25).
8. **Formation of memory.** The nucleus in the mammillary body receives many inputs from the major memory-processing structure of the cerebrum, the hippocampal formation (see p. 387) and therefore may relate to memory formation.

Through experiments that stimulate or remove parts of the hypothalamus, researchers have localized each of these functions to general regions of the hypothalamus called *functional centers.* For the most part, functional centers can be only roughly matched with specific structural nuclei. For example, feeding-initiation centers are in the lateral part of the hypothalamus, whereas feeding-inhibition centers are in the ventromedial part.

Critical Reasoning

Predict some of the functional disorders that can result from injuries to the hypothalamus.

In general, hypothalamic lesions cause disorders in visceral functions and in emotions. Thus, injuries to the hypothalamus can result in severe weight loss or obesity, sleep disturbances, dehydration, and a broad range of emotional disorders.

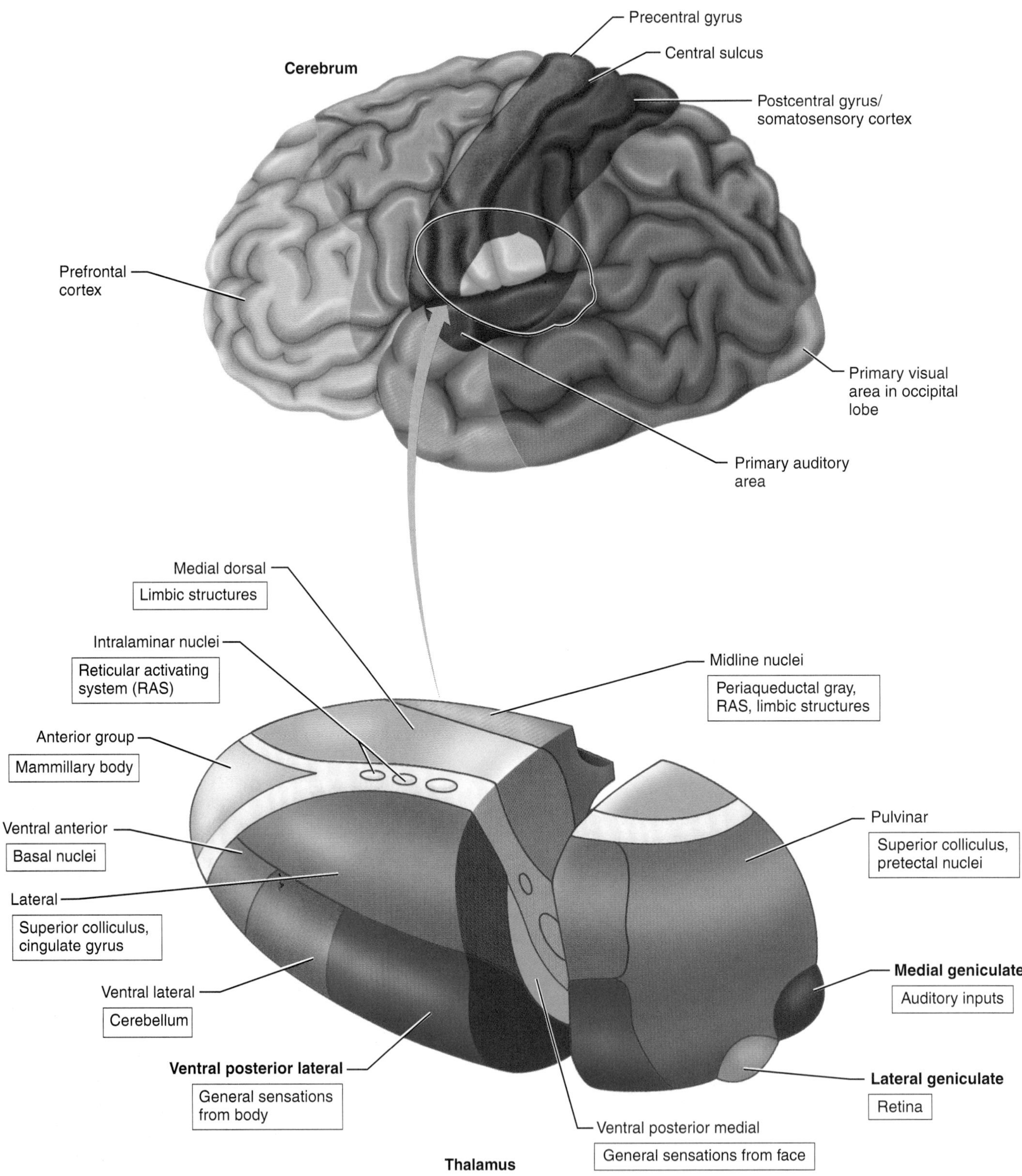

FIGURE 13.17 The thalamus and its relationship to the cerebral cortex. Below, the thalamic nuclei are indicated, as are the brain regions that send input to each thalamic nucleus (named in the boxes). Each nucleus of the thalamus sends its axons to a particular portion of the cerebral cortex (shown above), as indicated by the matching colors.

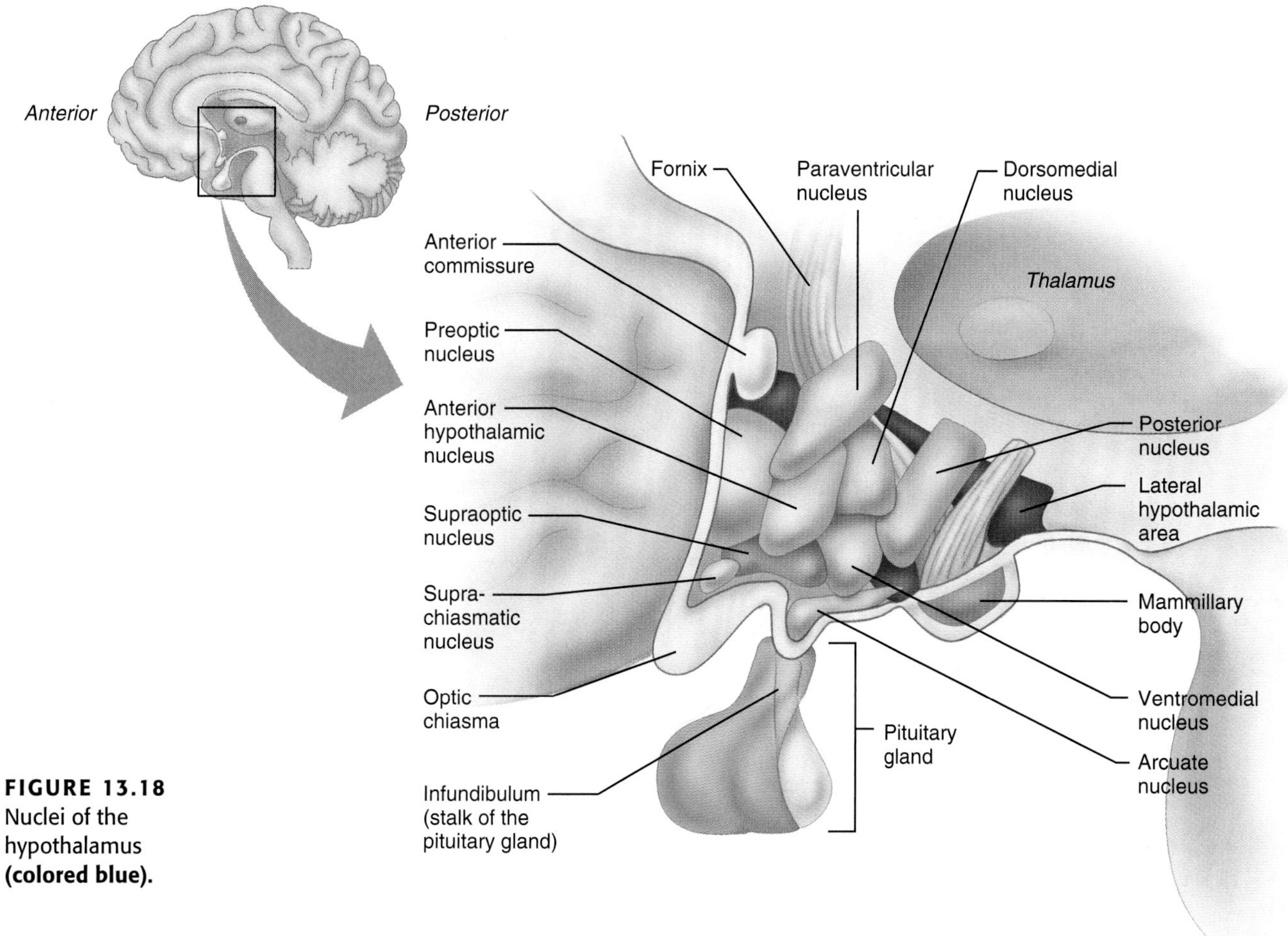

FIGURE 13.18 Nuclei of the hypothalamus **(colored blue).**

The Epithalamus

The **epithalamus,** the third and most dorsal part of the diencephalon (see Figure 13.15) forms part of the roof of the third ventricle. It consists of one tiny group of nuclei and a small, unpaired knob called the **pineal gland** or **pineal body** (pin′e-al; "pinecone-shaped"). This gland, which derives from ependymal glial cells, is a hormone-secreting organ. Under the influence of the hypothalamus, the pineal gland secretes the hormone **melatonin** (mel″ah-to′nin), which signals the body to prepare for the nighttime stage of the sleep-wake cycle.

The Brain Stem

The third of the four major parts of the brain is the **brain stem.** From rostral to caudal, the three regions of the brain stem are the *midbrain, pons,* and *medulla oblongata* (Figures 13.19, 13.20, and 13.21). Each of these regions is roughly an inch long, and together they make up only 2.5% of total brain mass. Lying in the posterior cranial fossa of the skull, on the basiocciput, the brain stem has several general functions: (1) It produces the rigidly programmed, automatic behaviors necessary for our survival; (2) it acts as a passageway for all the fiber tracts running between the cerebrum and the spinal cord; and (3) it is heavily involved with the innervation of the face and head, as 10 of the 12 pairs of cranial nerves attach to it. The brain stem has the same structural plan as the spinal cord, with outer white matter surrounding an inner region of gray matter. However, there are also nuclei of gray matter located within the white matter (see Figure 13.5).

The Midbrain

The first of the three regions of the brain stem, the midbrain, lies between the diencephalon rostrally and the pons caudally (see Figure 13.20). Its central cavity is the cerebral aqueduct, which divides it into a *tectum* ("roof") dorsally and paired *cerebral peduncles* ventrally. On the ventral surface of the brain, the cerebral peduncles form vertical pillars that appear to hold up the forebrain (Figure 13.20a, b); hence their name, which means "the little feet of the cerebrum." These peduncles contain the pyramidal (corticospinal) motor tracts descending toward the spinal cord. The ventral part of each peduncle that contains this tract is called the *crus cerebri* ("leg of the cerebrum"). Dorsally, the midbrain has another pair of bands, the **superior cerebellar peduncles** (Figure 13.20b, c), which connect the midbrain to the cerebellum.

A cross section through the midbrain reveals its internal structure (Figure 13.21a). Around the cerebral aqueduct is the **periaqueductal gray matter,** which has

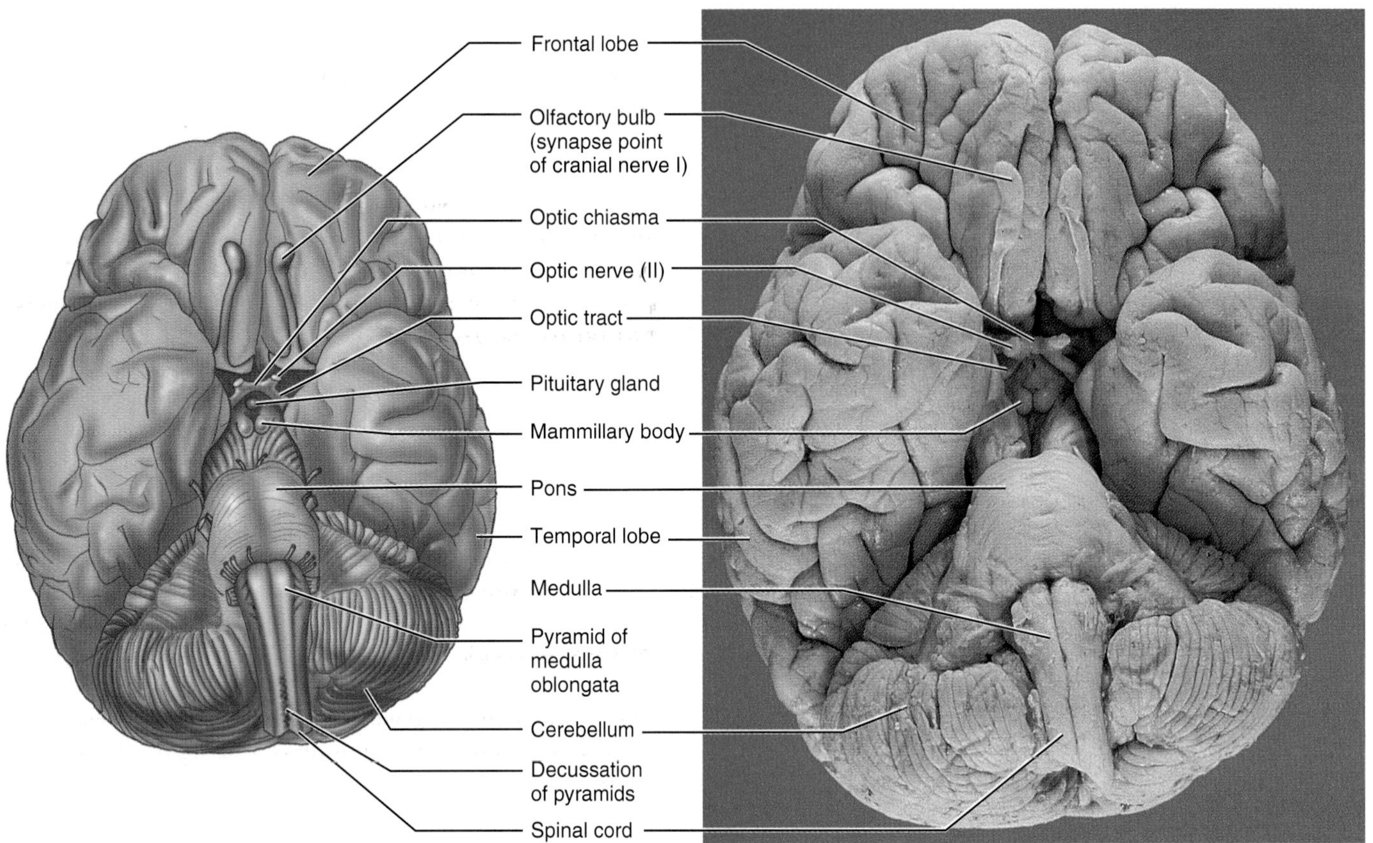

FIGURE 13.19 Ventral view of the brain, showing the diencephalon and brain stem. Note that very little of the midbrain—lying between the pons and mammillary bodies—is visible in this view.

two somewhat related functions. First, it is involved in the "fright-and-flight" (sympathetic) reaction, constituting the midbrain link between the amygdala of the forebrain (which perceives fear) and the autonomic pathway (which directly signals the physiological reactions associated with fear). More specifically, the periaqueductal gray matter elicits a terror-induced increase in heart rate and skyrocketing blood pressure, wild fleeing or defensive freezing, the flexing of the spine (as in rolling into a ball for protection), and the suppression of pain upon injury. Second, the periaqueductal gray matter also seems to mediate our response to visceral pain (as when we feel nauseous), during which it decreases heart rate and blood pressure, produces a cold sweat, and discourages movement.

The most ventral part of the periaqueductal gray matter contains the cell bodies of motor neurons that contribute to two cranial nerves, the oculomotor (III) and the trochlear (IV) nuclei. These cranial nerves control most of the muscles that move the eyes (see Chapter 14).

Many nuclei lie external to the periaqueductal gray matter, scattered within the surrounding white matter of the midbrain. The largest of these nuclei form the **corpora quadrigemina** (kwod″rĭ-jem′ĭ-nah; "quadruplets"), which make up the tectum (see Figure 13.20c). These nuclei, which form four bumps on the dorsal surface of the midbrain, are more specifically named the superior and inferior colliculi. The **superior colliculi** (ko-lik′u-li; "little hills") are nuclei that act in visual reflexes, as, for example, when our eyes track and follow moving objects even if we are not consciously looking at the objects. The **inferior colliculi,** by contrast, belong to the auditory system. Among other functions, they act in reflexive responses to sounds, as occurs when one is startled by a loud noise; one's head and eyes reflexively turn toward the sound because of the function of the inferior colliculi.

Also embedded in the white matter of the midbrain are two pigmented nuclei: *substantia nigra* (*nigra* = black) and the *red nucleus* (see Figure 13.21a). The band-like **substantia nigra** (sub-stan′she-ah ni′grah), whose neuronal cell bodies contain dark melanin pigment, is in the cerebral peduncle deep to the corticospinal tract. This nucleus is functionally linked to the basal nuclei, with axons that project to the globus pallidus. Degeneration of these neurons in the substantia nigra is the cause of Parkinson's disease.

The oval **red nucleus** lies deep to the substantia nigra. Its reddish hue is due to a rich blood supply and to the presence of iron pigment in the cell bodies of its neurons. It has a minor motor function: helping to bring about flexion movements of the limbs. The red nucleus is the largest nucleus of the *reticular formation* (see pp. 387–388), a system of otherwise small nuclei that are scattered throughout the core of the brain stem.

The Pons

The second region of the brain stem, the **pons,** is a bulging region wedged between the midbrain and the medulla oblongata (see Figure 13.20). Dorsally, it is separated from the cerebellum by the fourth ventricle. As its name suggests, the pons ("bridge") forms a ventral bridge between the right and left halves of the cerebellum (see Figure 13.20a).

Several cranial nerves attach to the pons: the trigeminal (V), which innervates the skin of the face and the chewing muscles; the abducens (VI), which innervates an eye-moving muscle; and the facial (VII), which supplies the muscles of facial expression, among other functions. These cranial nerves and their functions are discussed in more depth in Chapter 14.

Near the fourth ventricle, the pons contains the nuclei of the cranial nerves that attach to it (see Figure 13.20d). *Cranial nerve nuclei* comprise two general groups: *motor nuclei* consisting of cell bodies of the motor neurons that contribute to cranial nerves, and *sensory nuclei* consisting of interneurons that receive direct input from the sensory neurons in cranial nerves. The details of these complex nuclei are beyond the scope of this book and the nuclei are illustrated primarily for reference purposes.

More parts can be seen in a cross section through the pons (Figure 13.21b). Ventral to the cranial nerve nuclei lies a part of the reticular formation, and ventral to that is the thick pyramidal motor tract descending from the cerebral cortex. Interspersed among the fibers of this tract are numerous **pontine nuclei,** relay nuclei in a path by which the cerebral motor cortex communicates with the cerebellum concerning the coordination of voluntary movements (see p. 385). The pontine nuclei send axons to the cerebellum in the thick **middle cerebellar peduncles** (see Figure 13.20).

The Medulla Oblongata

The conical **medulla oblongata,** or simply **medulla,** is the third and most caudal part of the brain stem. It is continuous with the spinal cord at the level of the foramen magnum of the skull. As mentioned previously, a part of the fourth ventricle lies dorsal to the rostral half of the medulla. Here, the ventricle has a thin, sheet-like roof that contains a capillary-rich membrane called a *choroid plexus* (see Figures 13.21c and 13.15).

The medulla has several externally visible landmarks. Flanking the ventral midline are two longitudinal ridges called **pyramids** (see Figure 13.20a) formed by the pyramidal tracts. In the caudal part of the medulla, most of these pyramidal fibers cross over to the opposite side of the brain at a point called the **decussation of the pyramids** (de″kus-sa′shun; "a crossing"). As previously noted, the result of this crossover is that each cerebral hemisphere controls the voluntary movements of the opposite side of the body.

Other external landmarks on the medulla are the inferior cerebellar peduncles and the olives (Figure 13.20b). The **inferior cerebellar peduncles** are fiber tracts that connect the medulla to the cerebellum dorsally. Each **olive,** which does indeed resemble an olive, lies just lateral to a pyramid and contains an **inferior olivary nucleus,** a large wavy fold of gray matter (see Figure 13.21c). This nucleus is a relay station for sensory information traveling to the cerebellum, especially for proprioceptive information ascending from the spinal cord.

The five most inferior pairs of cranial nerves attach to the medulla (see Figure 13.20a): The vestibulocochlear (VIII), which is the sensory nerve of hearing and equilibrium; the glossopharyngeal (IX), which innervates part of the tongue and pharynx; the vagus (X), which innervates many visceral organs in the thorax and abdomen; the accessory nerve (XI), which innervates some muscles of the neck; and the hypoglossal (XII), which innervates tongue muscles (see Chapter 14). The sensory and motor nuclei of these cranial nerves lie near the fourth ventricle (Figure 13.21c). Only two groups of these nuclei are considered here: The **cochlear** (kok′le-ar) **nuclei** receive auditory information, and the **vestibular nuclei** receive information on equilibrium (also see Figure 13.20d). The vestibular nuclei send axons to the spinal cord and midbrain, where they help to (1) bring about subconscious movements by which the body maintains its equilibrium and (2) stabilize the gaze of the eyes as the head moves.

The most caudal part of the medulla contains some large nuclei in its dorsal roof, the **nucleus cuneatus** (ku-ne-a′tus) and the **nucleus gracilis** (gras′ĭ-lis). As shown in Figure 13.33a on p. 398, these nuclei serve as relays in the pathway by which general somatic sensory information ascends through the spinal cord to the somatosensory cerebral cortex.

The core of the medulla contains much of the reticular formation (see Figure 13.21c), some nuclei of which influence our autonomic (visceral motor) functions. The most important visceral centers in the medulla's reticular formation are:

1. The *cardiac center* adjusts the force and rate of the heartbeat.

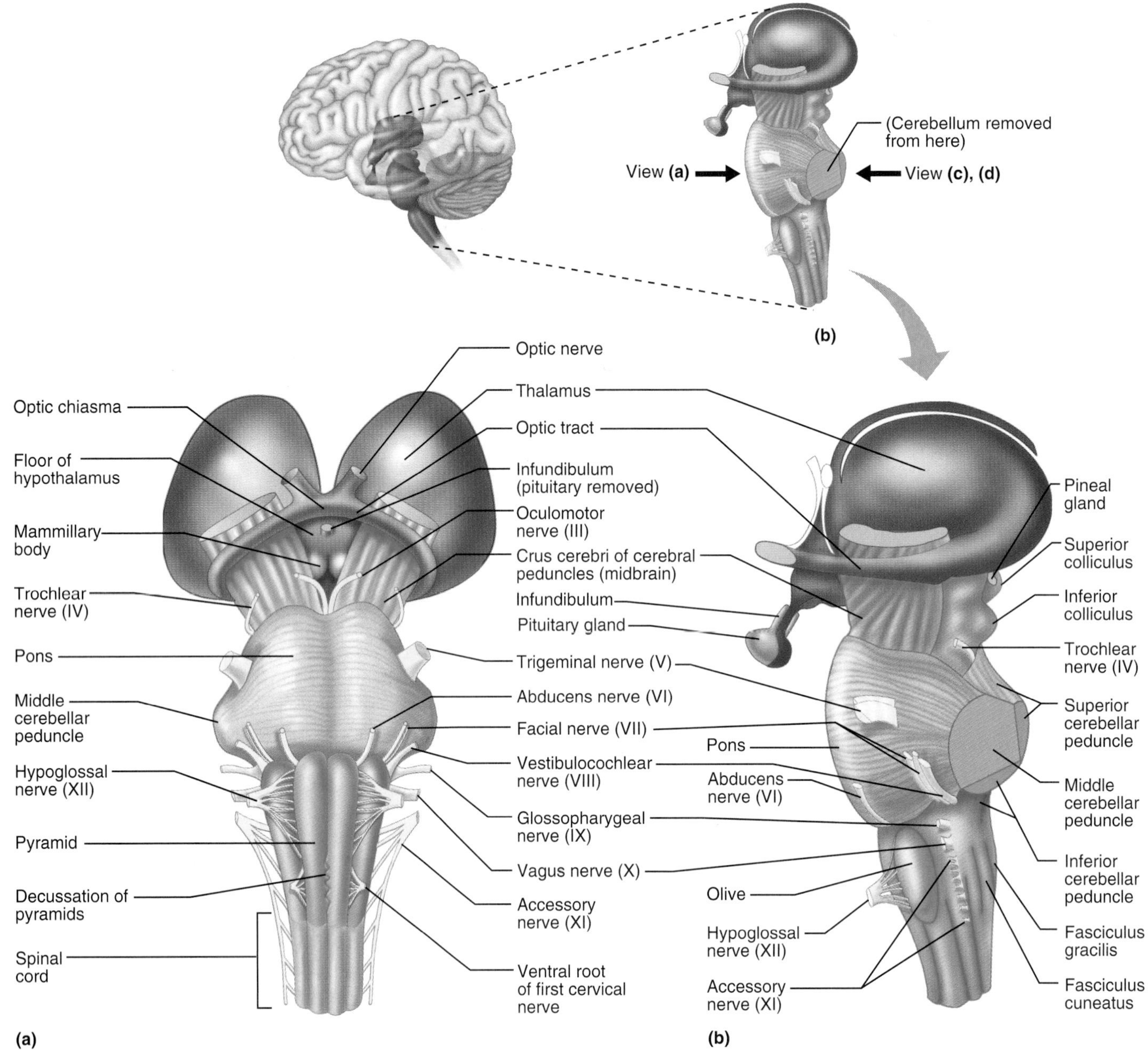

2. The *vasomotor center* regulates blood pressure by stimulating or inhibiting the contraction of smooth muscle in the walls of blood vessels, thereby constricting or dilating the vessels. Constriction of arteries throughout the body causes blood pressure to rise, whereas dilation reduces blood pressure.
3. The *medullary respiratory center* controls the basic rhythm and rate of breathing. This center is in the ventrolateral part of the medulla's reticular formation, a region called the *pre-Bötzinger complex*.
4. Additional centers regulate hiccuping, swallowing, coughing, and sneezing.

This list contains functions also attributed to the hypothalamus and periaqueductal gray matter (see pp. 377 and 380). The overlap is easily explained: The two higher regions exert their control over the visceral functions by relaying their instructions through the medulla's reticular centers, which then carry out those instructions.

The Cerebellum

The cauliflower-like **cerebellum,** the fourth of the brain's major parts, makes up 11% of the mass of the brain and is exceeded in size only by the cerebrum. It

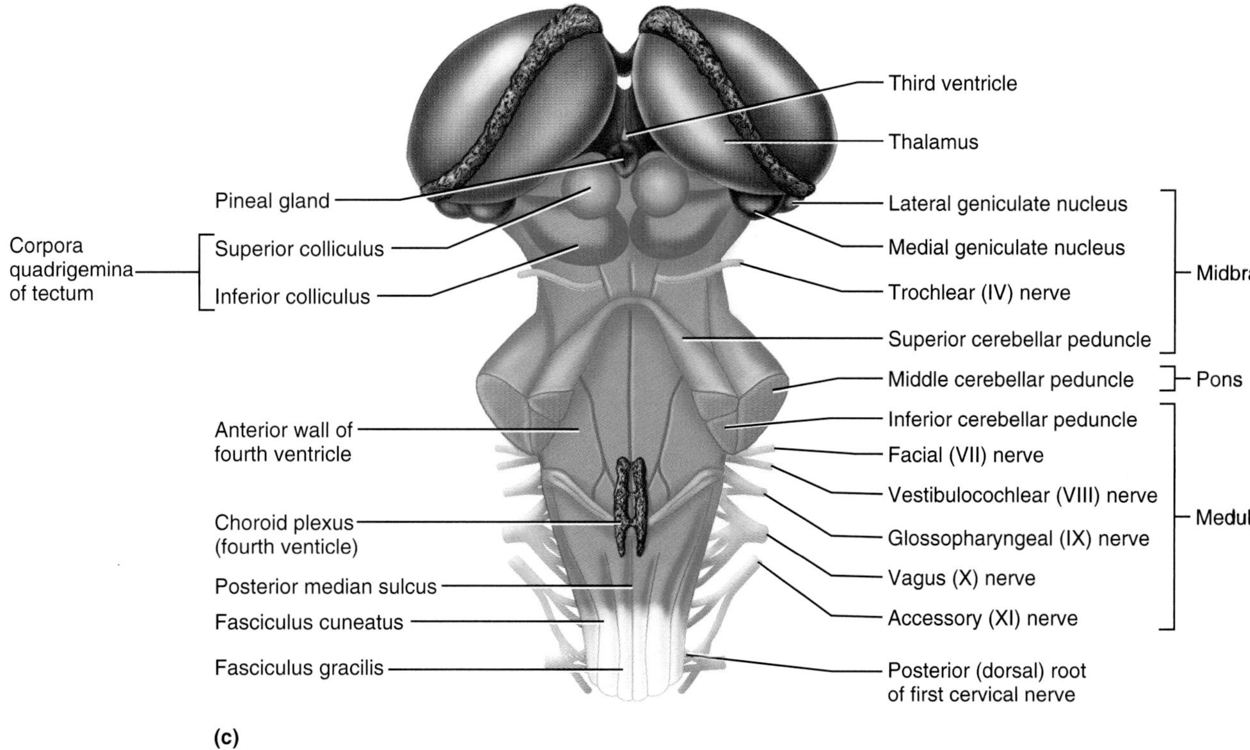

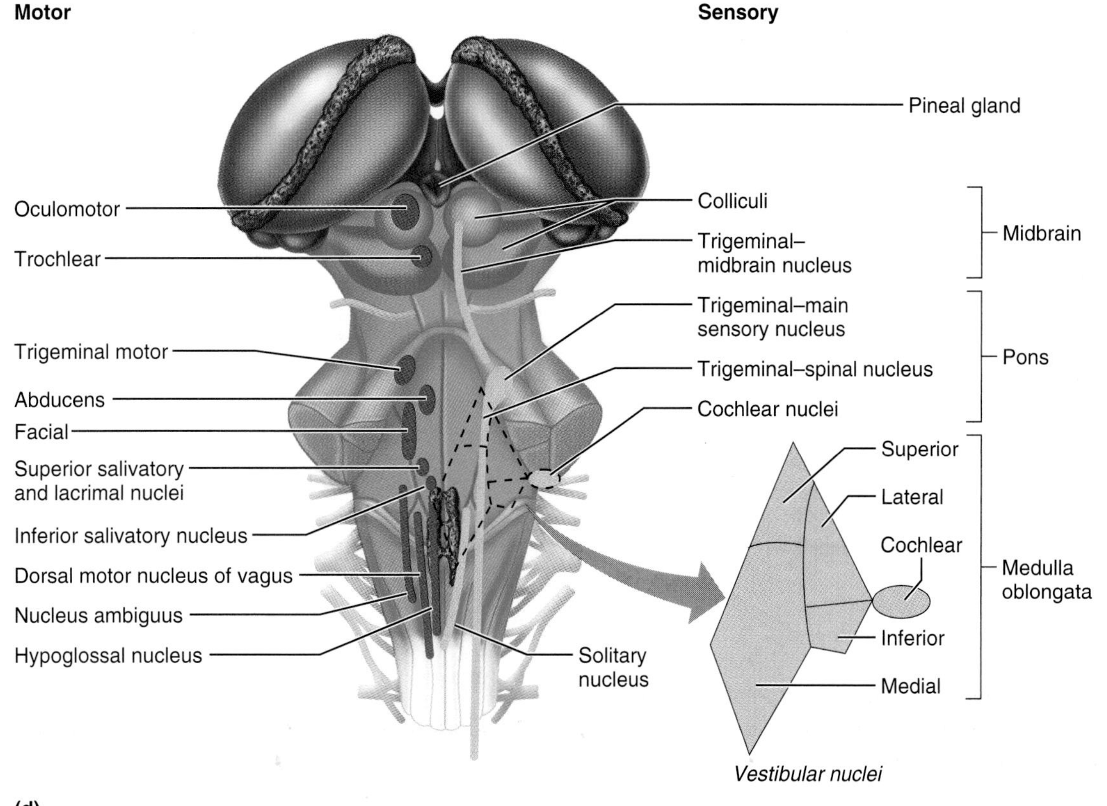

FIGURE 13.20 **Brain stem and diencephalon.** (**a**) Ventral view. (**b**) Lateral view. (**c**) Dorsal view. (**d**) Transparent brain stem in dorsal view, revealing the cranial nerve nuclei within. The motor nuclei are shown on the left in red, the sensory nuclei on the right in blue.

Posterior
Superior colliculus
Periaqueductal gray matter
Oculomotor nucleus
Medial lemniscus
Red nucleus
Substantia nigra
Pyramidal (corticospinal) fibers, forming crus cerebri
Cerebral aqueduct
Reticular formation
Cerebral peduncle
Anterior

(a) Midbrain

Superior cerebellar peduncle
Trigeminal-main sensory nucleus
Trigeminal motor nucleus
Middle cerebellar peduncle
Trigeminal nerve
Medial lemniscus
Fourth ventricle
Reticular formation
Pontine nuclei
Fibers of pyramidal tract

(b) Pons

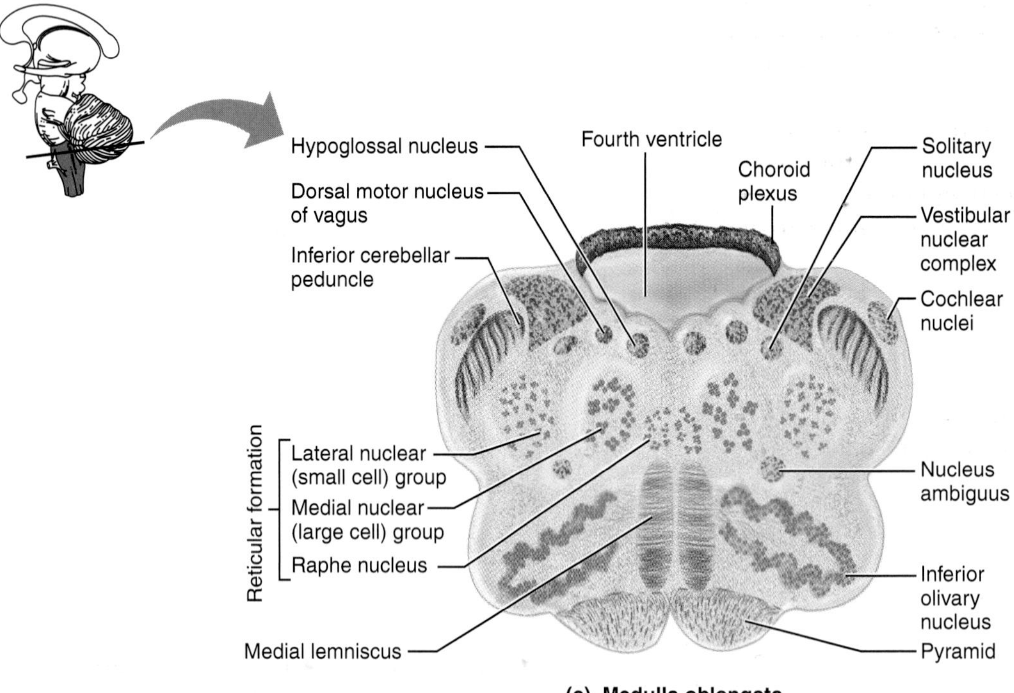

(c) Medulla oblongata

FIGURE 13.21 Cross sections through the brain stem. (a) Midbrain, (b) pons, (c) medulla oblongata.

is located dorsal to the pons and medulla, from which it is separated by the fourth ventricle (see Figure 13.22a). Functionally, the cerebellum smooths and coordinates body movements that are directed by other brain regions, and it helps maintain posture and equilibrium.

The cerebellum consists of two expanded **cerebellar hemispheres** connected medially by the worm-like **vermis** (Figure 13.22b). Its surface is folded into many plate-like ridges called **folia** (fo′le-ah; "leaves"), which are separated by grooves called **fissures.** Each hemisphere is subdivided into three lobes: the large **anterior** and **posterior lobes,** and the small **flocculonodular** (flok″u-lo-nod′u-lar) **lobe.** The flocculonodular lobes, altogether shaped like a moose head with antlers, are hidden ventral to the posterior lobe (see Figure 13.22a). Functionally, the flocculonodular lobes adjust posture to maintain equilibrium, whereas the anterior and posterior lobes coordinate body movements.

Like the cerebrum, the cerebellum has three regions: an outer *cortex* of gray matter, internal *white matter,* and deeply situated gray matter called *deep cerebellar nuclei* (see Figure 13.22c). The cortex is a neuron-rich calculator whose function is to smooth our body movements. The white matter consists of axons that carry information to and from the cortex. The deep cerebellar nuclei give rise to axons that relay the instructions from the cerebellar cortex to other parts of the brain.

To calculate how to coordinate body movements, the cerebellar cortex must continuously receive three types of information about how the body is moving: (1) *information on equilibrium,* relayed from receptors in the inner ear through the vestibular nuclei to the cerebellar cortex; (2) *information on the current movements of the limbs, neck, and trunk,* which travels from proprioceptors in muscles, tendons, and joints up the spinal cord to the cerebellar cortex; and (3) *information from the cerebral cortex,* which passes from the cerebral cortex through the pyramidal tract and pontine nuclei to the cerebellar cortex.

In functioning, the cerebellum first receives information on the movements being planned (and later ordered) by the motor cerebral cortex, then it compares these intended movements to the body movements that are actually occurring, and finally it sends instructions to the cerebral cortex on how to resolve any differences between the intended and actual movements. Using this feedback from the cerebellum, the motor cerebral cortex continuously readjusts the motor commands it sends to the spinal cord, fine-tuning movements so that they are well coordinated.

Cerebellar Peduncles

The superior, middle, and inferior cerebellar peduncles are thick tracts of nerve fibers that connect the cerebellum to the brain stem (see Figure 13.22a and 13.20). Their fibers carry the information that travels from and to the cerebellum. Namely, the **superior cerebellar peduncles,** connecting the cerebellum to the midbrain, carry instructions from the cerebellum toward the cerebral cortex. The **middle cerebellar peduncles,** connecting the pons to the cerebellum, carry the information from the cerebral cortex and the pontine nuclei. The **inferior cerebellar peduncles,** arising from the medulla, carry information on equilibrium from the vestibular nuclei and information on proprioception from the spinal cord.

Unlike the contralateral distribution of fibers to and from the cerebral cortex, virtually all fibers that enter and leave the cerebellum are **ipsilateral** (ip″sĭ-lat′er-al; *ipsi* = same)—run to and from the *same* side of the body.

Critical Reasoning

Describe some effects of damage to various parts of the cerebellum.

Damage to the anterior and posterior lobes leads to disorders of coordination, characterized by slow or jerky movements that tend to overreach their targets. People with such damage are unable to touch their finger to their nose with their eyes closed. Speech may be slurred, slow, and singsongy. Damage to the flocculonodular nodes, by contrast, leads to disorders of equilibrium. Affected individuals have a wide stance and an unsteady "drunken sailor" gait that predisposes them to falling. In all cases, damage in one half of the cerebellum (left or right) affects that same side of the body.

Advances in Understanding Recent findings have revealed that the cerebellum participates in cognition. Even though such a "thinking" role surprised many neurobiologists, perhaps it should not have, given that the cerebellar cortex contains half of all the neurons in the brain and receives some of its input from the cognitive, prefrontal lobes of the cerebrum. Functional neuroimaging studies and studies of individuals with cerebellar damage indicate that the cerebellum plays some role in language, problem solving, and the planning of tasks. Overall, its cognitive function may be to recognize, use, and predict *sequences of events* that we experience or perceive. ✷

Functional Brain Systems

Functional brain systems are networks of neurons that function together despite spanning large distances within the brain. Here we discuss two important functional systems: the *limbic system,* spread widely throughout the forebrain, and the *reticular formation,* which spans the brain stem.

The Limbic System

The **limbic system** is a group of structures on the medial aspect of each cerebral hemisphere and diencephalon (Figure 13.23). In the cerebrum, the limbic structures form a broad ring (*limbus* = headband), that

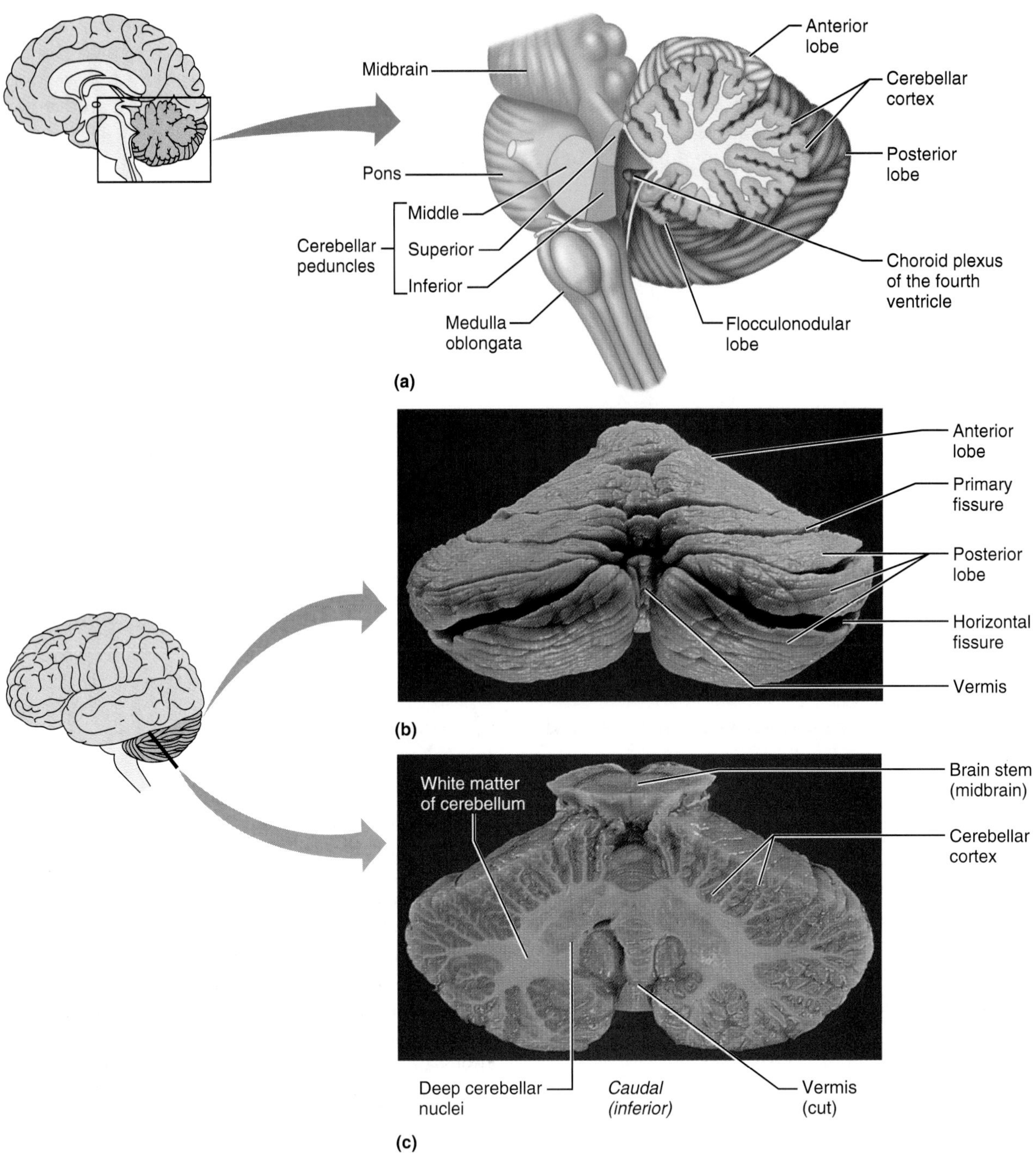

FIGURE 13.22 Cerebellum. (a) Lateral view of the brain stem, from which the left half of the cerebellum has been removed to reveal its cerebellar peduncles. A midsagittal section through the cerebellum, and the cerebellum's right half, are evident. (b) Posterior (dorsal) view of the cerebellum, in which superior is above. (c) The cerebellum sectioned to reveal its three regions; like (b), this is a posterior view but the midbrain is sectioned horizontally and the cerebellum is sectioned frontally.

includes the *septal nuclei, cingulate gyrus, hippocampal formation,* and part of the amygdala. It also includes a small nucleus (basal nucleus of Meynert) just posterior to the septal nuclei and inferior to the corpus striatum. In the diencephalon, the main limbic structures are the hypothalamus and the anterior thalamic nuclei. The **fornix** ("arch") and other fiber tracts link the limbic system together. The limbic system also overlaps the rhinencephalon ("nose brain"; see p. 370) in several places.

The limbic system is the "emotional brain." Two parts of the cerebral limbic structures seem especially important in emotions. The first structure is the *amyg-*

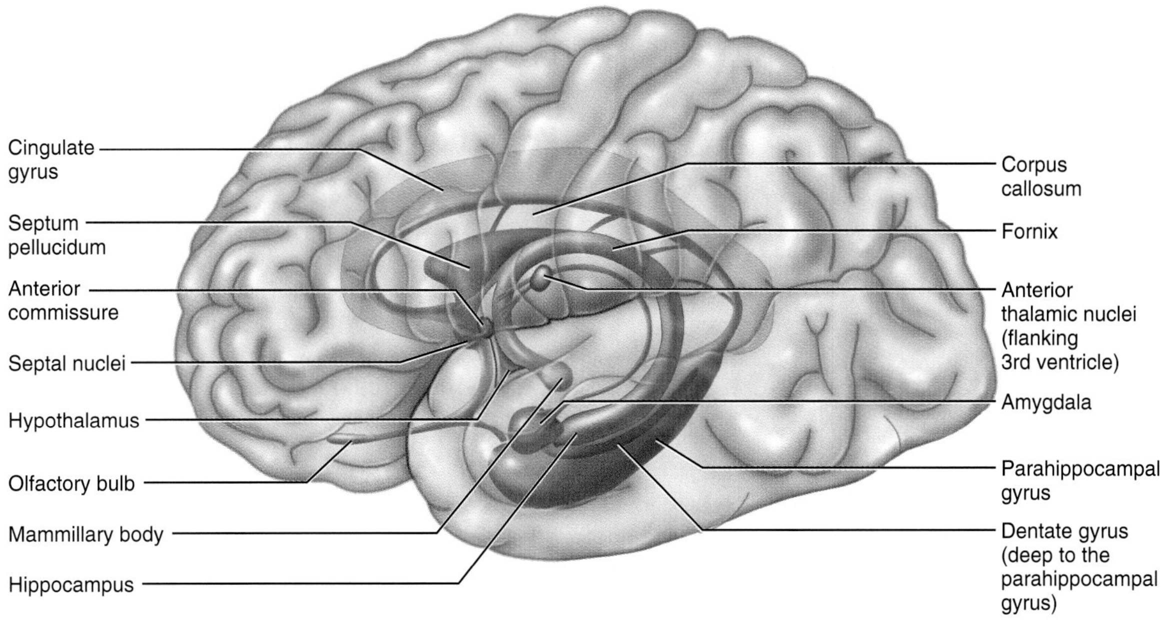

FIGURE 13.23 The limbic system (orange). Although here seen as projected through a lateral view of the cerebrum, the limbic structures lie on the medial side of the cerebral hemisphere. The brain stem is not illustrated. Even though authorities disagree about whether the hypothalamus and anterior thalamic nuclei are parts of the limbic system, all agree that they are very closely related to this system.

dala, which contains the key nuclei for processing fear and then stimulating the appropriate sympathetic response to fear. The amygdala also enables us to recognize menacing facial expressions in others and to detect the precise direction of the gaze of someone who is looking at us. The second important structure is the **cingulate gyrus** (sing′gu-lāt; "belt-shaped") (see Figure 13.23), which allows us to shift between thoughts and to express our emotions through gestures. The anterior part of this gyrus interprets pain as unpleasant and resolves mental conflict during frustrating tasks.

The limbic system also functions in consolidating and retrieving memories. The structures involved, both of which are in the medial aspect of the temporal lobe, are the amygdala and the hippocampal formation, which consists of the **hippocampus** (hip″o-kam′pus; "sea horse") and the **parahippocampal gyrus.** The hippocampal formation encodes, consolidates, and later retrieves memories of facts and events. It first receives information to be remembered from the rest of the cerebral cortex; then it processes these data and returns them to the cortex, where they are stored as long-term memories. The amygdala, by contrast, forms memories of experiences based entirely on their emotional impact, especially if related to fear. If we later are reminded of these experiences, the amygdala retrieves the memories and causes us to reexperience the original emotion. This is beneficial because it lets us make difficult and risky decisions correctly, based on memories of all our past emotional experiences.

The limbic system communicates with many other regions of the brain. Most output from the limbic system is relayed through the hypothalamus and the reticular formation (discussed next), the portions of the brain that control our visceral responses. This fact explains why people under emotional stress experience visceral illnesses such as high blood pressure and heartburn. The limbic system also interacts heavily with the prefrontal lobes of the cerebral cortex. Thus, our feelings (mediated by the emotional brain) and our thoughts (mediated by the thinking brain) interact closely.

The Reticular Formation

The **reticular formation,** which runs through the central core of the medulla, pons, and midbrain (Figure 13.24), consists of loosely clustered neurons in what is otherwise white matter. These reticular neurons form three columns that extend the length of the brain stem: (1) the midline **raphe** (ra′fe) **nuclei,** which are flanked laterally by (2) the **medial nuclear group** and then (3) the **lateral nuclear group** (see Figure 13.21c).

Because they have long, branching axons, single reticular neurons project to such widely separated regions as the thalamus, cerebellum, and spinal cord. Such widespread connections make reticular neurons ideal for governing the arousal of the brain as a whole.

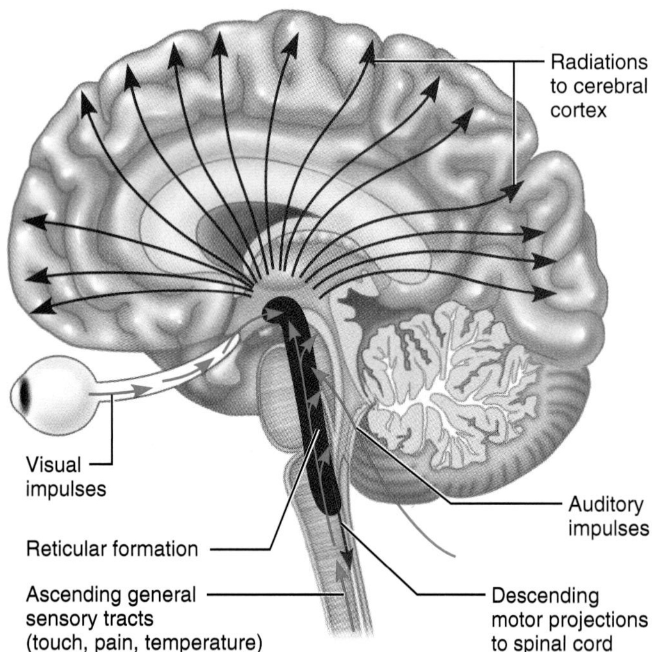

FIGURE 13.24 The reticular formation. The reticular formation runs the length of the brain stem. A part of this formation, the reticular activating system (RAS), keeps the cerebral cortex alert and conscious. Ascending arrows represent sensory inputs to the RAS, which stimulate outputs from the RAS to the cerebrum. Although the entire reticular formation is shown in this figure, the RAS is concentrated at the junction of the pons and medulla. The reticular formation also provides output that controls some muscle activity throughout the body (indicated by the descending arrow).

For example, certain reticular neurons send a continuous stream of impulses to the cerebrum (through relays in the thalamus), thereby maintaining the cerebral cortex in an alert, conscious state. This arm of the reticular formation, which maintains consciousness and alertness, is called the *reticular activating system (RAS).*

Axons from all the great ascending sensory tracts synapse on RAS neurons, keeping these reticular neurons active and enhancing their arousing effect on the cerebrum. The fact that visual, auditory, and touch stimuli keep us awake and mentally alert explains why many students like to study in a crowded room such as a cafeteria; they are stimulated by the bustle of such an environment. The RAS also functions in sleep and in arousal from sleep. Conversely, general anesthesia, alcohol, tranquilizers, and sleep-inducing drugs depress the RAS and lead to a loss of alertness or consciousness. Knockout blows produce the same effect because they twist the brain stem. Severe injury to the RAS is the cause of coma.

The reticular formation also has a *motor* arm that sends axons to the spinal cord. Some of these axons control the motor neurons to skeletal muscles during coarse movements of the limbs (see *reticulospinal tract* in Table 13.3, p. 401), whereas others influence autonomic neurons to regulate visceral functions such as heart rate, blood pressure, and respiration (see pp. 381–382).

The functions of the major parts of the brain are summarized in Table 13.1 on page 390.

Protection of the Brain

Nervous tissue is soft and delicate, and the irreplaceable neurons can be injured by even slight pressure. However, the brain is protected from injury by the skull (see Chapter 7), by surrounding membranes called *meninges,* and by a watery cushion of *cerebrospinal fluid.* Furthermore, the brain is protected from harmful substances in the blood by the *blood-brain barrier.*

Meninges

The **meninges** (mě-nin′jēz; "membranes") are three membranes of connective tissue that lie just external to the brain and spinal cord. Their functions are to (1) cover and protect the CNS, (2) enclose and protect the blood vessels that supply the CNS, and (3) contain the cerebrospinal fluid. From external to internal, the meninges are the *dura mater, arachnoid mater,* and *pia mater* (Figure 13.25). Here we discuss the cranial meninges; the meninges of the spinal cord are discussed on p. 391.

The Dura Mater The leathery **dura mater** (du′rah ma′ter; "tough mother") is the strongest of the meninges. Where it surrounds the brain, it is a two-layered sheet of fibrous connective tissue (Figure 13.25a). The more superficial **periosteal layer** attaches to the internal surface of the skull bones (it is the periosteum); the deeper **meningeal layer** forms the true external covering of the brain. These two layers of dura mater are fused together, except where they separate to enclose the blood-filled *dural sinuses* (see p. 566). These sinuses act as veins, collecting blood from the brain and conducting it to the large internal jugular veins of the neck. The largest dural sinus is the *superior sagittal sinus* in the superior midline.

In several places, the meningeal dura mater extends inward to form flat partitions that subdivide the cranial cavity (Figure 13.26) and limit movement of the brain within the cranium. These partitions include the following:

1. The **falx cerebri** (falks ser′ě-bri). This large vertical sheet lies in the median plane in the longitudinal fissure between the cerebral hemispheres. A sickle-shaped (*falx* = sickle) partition, it attaches anteriorly to the crista galli of the ethmoid bone.
2. The **falx cerebelli** (ser″ě-bel′li). Continuing inferiorly from the posterior part of the falx cerebri, this vertical partition runs along the vermis of the cerebellum in the posterior cranial fossa.
3. The **tentorium** (ten-to′re-um) **cerebelli.** Resembling a tent over the cerebellum (*tentorium* means "tent"), this almost-horizontal sheet lies in the transverse fissure between the cerebrum and cerebellum.

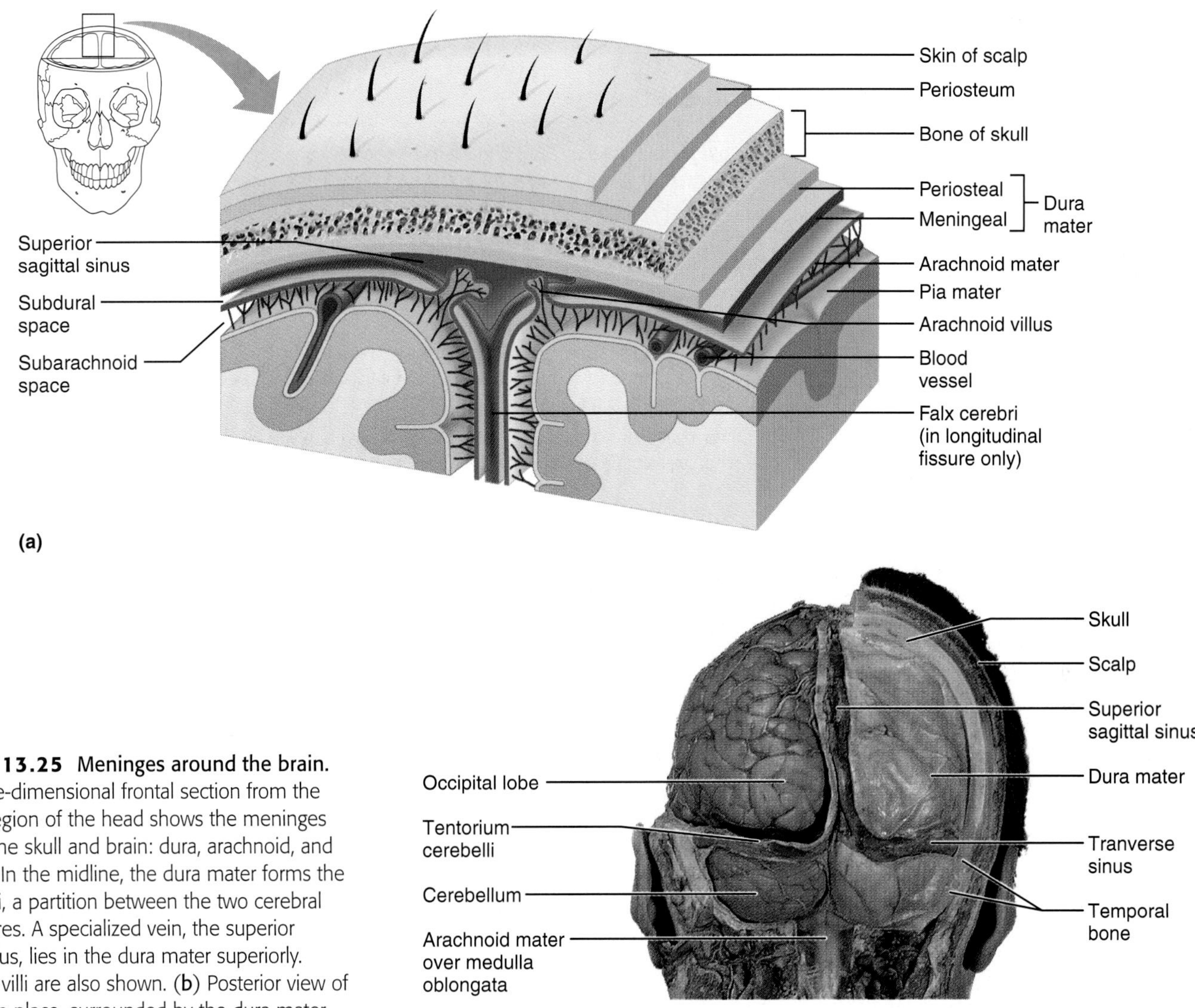

FIGURE 13.25 Meninges around the brain. (**a**) A three-dimensional frontal section from the superior region of the head shows the meninges between the skull and brain: dura, arachnoid, and pia mater. In the midline, the dura mater forms the falx cerebri, a partition between the two cerebral hemispheres. A specialized vein, the superior sagittal sinus, lies in the dura mater superiorly. Arachnoid villi are also shown. (**b**) Posterior view of the brain in place, surrounded by the dura mater and dural sinuses.

The Arachnoid Mater The **arachnoid** (ah-rak′noid) **mater** lies just deep to the dura mater (see Figure 13.25a). Between these two layers is a thin potential space called the **subdural space,** which contains only a film of fluid. The dura and arachnoid surround the brain loosely, never dipping into the sulci on the cerebral surface. Deep to the arachnoid membrane is the wide **subarachnoid space,** which is spanned by web-like threads that hold the arachnoid mater to the underlying pia mater. (This web is the basis of the name *arachnoid,* which means "spider-like.") The subarachnoid space is filled with cerebrospinal fluid, and it also contains the largest blood vessels that supply the brain. Because the arachnoid is fine and elastic, these vessels are rather poorly protected.

Over the superior part of the brain, the arachnoid forms knob-like projections called **arachnoid villi** or *arachnoid granulations.* These villi project superiorly through the dura mater into the superior sagittal sinus (see Figure 13.25a), and into some other dural sinuses as well. They act as valves that allow cerebrospinal fluid to pass from the subarachnoid space into the dural blood sinuses (see p. 391).

The Pia Mater The **pia mater** (pi′ah; "gentle mother") is a layer of delicate connective tissue richly vascularized with fine blood vessels. Unlike the other meninges, it clings tightly to the brain surface, following every convolution. As its arteries enter the brain tissue, they carry ragged sheaths of pia mater internally for short distances.

Meningitis, or inflammation of the meninges, is caused by a bacterial or viral infection that can spread to the underlying nervous tissue and cause brain inflammation, or **encephalitis** (en″sef-ah-li′tis). Meningitis is usually diagnosed by taking a sample of cerebrospinal fluid from the subarachnoid space, and examining it for the presence of microbes.

Table 13.1 Functions of Major Parts of the Brain

Part	Function
CEREBRAL HEMISPHERES (pp. 363–374) 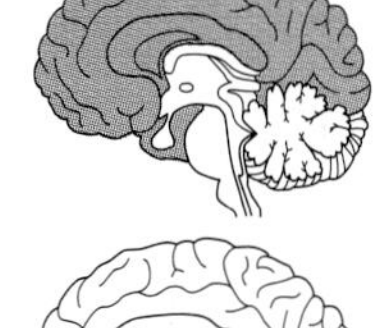	Cortical gray matter localizes and interprets sensory inputs, controls voluntary and skilled skeletal muscle activity, and functions in cognitive and emotional processing; basal nuclei (basal ganglia) are subcortical motor centers that help control movements
DIENCEPHALON (pp. 374–379) 	Thalamic nuclei are relay stations in conduction of (1) sensory impulses to cerebral cortex for interpretation, and (2) impulses from all other regions of the brain that communicate with the cerebral cortex; thalamus is also involved in neural processing
	Hypothalamus is chief integration center of autonomic (involuntary) nervous system; it functions in regulation of body temperature, food intake, water balance, thirst, and biological rhythms and drives; regulates hormonal output of anterior pituitary gland and is an endocrine organ in its own right (produces ADH and oxytocin); part of limbic system
Limbic system (pp. 385–387)	Emotional brain; a functional system involving cerebral and diencephalon structures that mediates emotional repsonse; it also forms and retrieves memories
BRAIN STEM	
Midbrain (pp. 379–381) 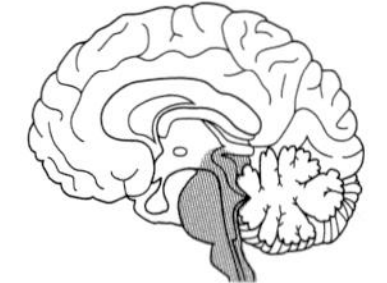	Conduction pathway between higher and lower brain centers (e.g., cerebral peduncles contain the fibers of the pyramidal tracts); its superior and inferior colliculi are visual and auditory reflex centers; substantia nigra and red nuclei are subcortical motor centers; contains nuclei for cranial nerves III and IV
Pons (p. 381)	Conduction pathway between higher and lower brain centers; pontine nuclei relay information from the cerebrum to the cerebellum; houses nuclei of cranial nerves V–VII
Medulla oblongata (pp. 381–382)	Conduction pathway between higher brain centers and spinal cord; site of decussation of the pyramidal tracts; houses nuclei of cranial nerves VIII–XII; contains nuclei cuneatus and gracilis (synapse points of ascending sensory pathways transmitting sensory impulses from skin and proprioceptors), and visceral nuclei controlling heart rate, blood vessel diameter, respiratory rate, vomiting, coughing, etc.; its inferior olivary nuclei provide the sensory relay to the cerebellum
Reticular formation (pp. 387–388)	A functional brain stem system that maintains cerebral cortical alertness (reticular activating system) and filters out repetitive stimuli; its motor nuclei help regulate skeletal and visceral muscle activity
CEREBELLUM (pp. 382–385) 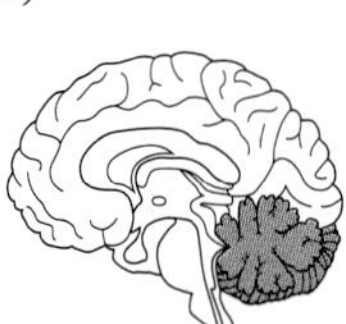	Processes information received from cerebral cortex and from proprioceptors and visual and equilibrium pathways, and provides "instructions" to cerebral motor cortex and subcortical motor centers that result in proper balance and posture and smooth, coordinated skeletal muscle movements; also has some cognitive functions

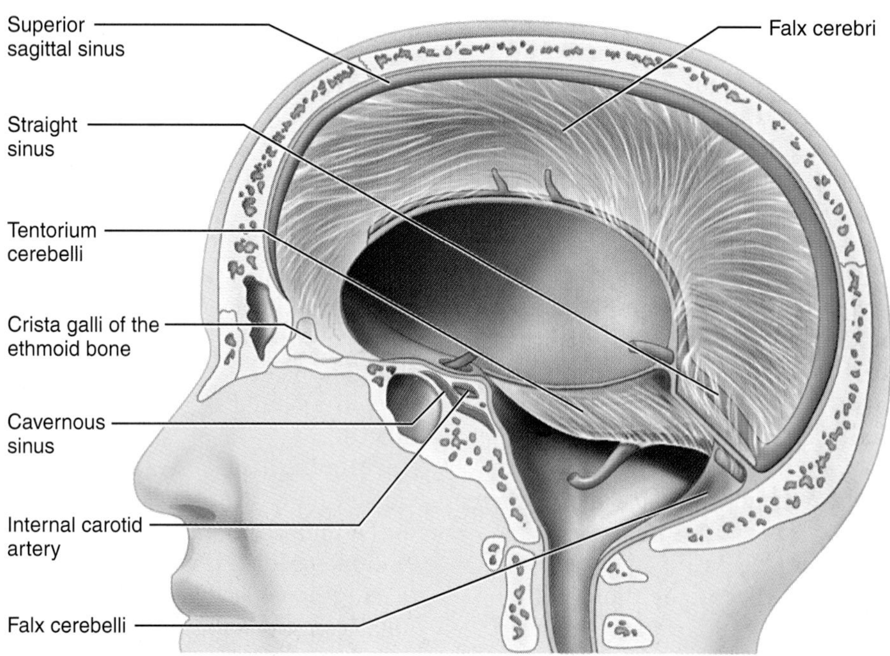

FIGURE 13.26 Partitions of dura mater in the cranial cavity. Note the falx cerebri, falx cerebelli, and tentorium cerebelli. Some vein-like dural sinuses are also shown in blue.

Cerebrospinal Fluid

Cerebrospinal fluid (CSF), a watery broth located in and around the brain and spinal cord, provides a liquid cushion that gives buoyancy to the central nervous system. The brain actually floats in the CSF, which reduces its weight by 97% and prevents this delicate organ from crushing under its own weight. CSF also cushions the brain and spinal cord from blows and jolts. Moreover, even though the brain has a rich blood supply, CSF helps to nourish the brain, to remove wastes produced by neurons, and to carry chemical signals between different parts of the central nervous system. Amazingly, very little CSF is needed to perform all these functions: Only 100–160 ml, about half a cup, is present at any time. Although CSF resembles the blood plasma from which it arises, it has more sodium and chloride ions and less protein.

Figure 13.27 depicts the sites of CSF production and its circulation. Most CSF is made in the **choroid plexuses,** membranes in the roofs of the four brain ventricles that consist of a layer of ependymal cells covered externally by a capillary-rich pia mater. CSF continuously forms from blood plasma by filtration from the capillaries and then passes through the ependymal cells into the ventricles. After entering the ventricles, the CSF moves freely through these chambers. Some CSF runs into the central canal of the spinal cord, but most enters the subarachnoid space through the lateral and median apertures in the walls of the fourth ventricle. In the subarachnoid space, CSF bathes the outer surfaces of the brain and spinal cord. CSF then passes through the arachnoid villi and enters the blood of the dural sinuses. Cerebrospinal fluid arises from the blood and returns to it at a rate of about 500 ml a day.

CLINICAL APPLICATIONS

HYDROCEPHALUS In the condition called **hydrocephalus** (hi″dro-sef′ah-lus; "water on the brain"), excessive accumulation of CSF in the ventricles or subarachnoid space can exert a crushing pressure on the brain. Its causes include: a tumor or inflammatory swelling that closes off the cerebral aqueduct or the fourth ventricle; meningitis, which causes scarring that closes the subarachnoid space and arachnoid villi; and, in infants, an overdeveloped choroid plexus that secretes too much CSF. Hydrocephalus in newborns results in an enlarged cranium because the young skull bones have not yet fused (Figure 13.28), but in adults, fluid pressure quickly builds up and damages the brain because the cranium is unyielding. Hydrocephalus is treated surgically by running a shunt (plastic tube) from the ventricles into a vein in the neck. ✚

Blood-Brain Barrier

The brain has a rich supply of capillaries that provide its nervous tissue with nutrients, oxygen, and all other vital molecules. However, some blood-borne molecules that can cross other capillaries of the body cannot cross the brain capillaries. Blood-borne toxins, such as urea, mild toxins from food, and bacterial toxins, are prevented from entering brain tissue by the **blood-brain barrier,** which protects the neurons of the CNS.

The blood-brain barrier results primarily from special features of the endothelium (see p. 78) that makes up the walls of the brain capillaries. The endothelial cells in brain capillaries are joined together around their entire perimeters by tight junctions, making those capillaries the least permeable capillaries in the body (see p. 547). Even so, the blood-brain barrier is not an absolute barrier. All nutrients (including oxygen) and ions needed by the neurons pass through, some by special transport mechanisms in the plasma membranes of the endothelial cells. Furthermore, because the barrier is ineffective against fat-soluble molecules, which easily diffuse through all cell membranes, the barrier allows alcohol, nicotine, and anesthetics to reach brain neurons.

More information on the blood-brain barrier is provided on p. 549.

THE SPINAL CORD

The **spinal cord** runs through the vertebral canal of the vertebral column, from the foramen magnum superiorly to the level of vertebra L_1 or L_2 inferiorly (Figure 13.29). Its function can be described in several ways: (1) Through the spinal nerves that attach to it, the spinal cord is involved in the *sensory and motor innervation* of the entire body inferior to the head; (2) it provides a *two-way conduction pathway* for signals between the body and the brain; and (3) it is a *major center for reflexes* (see Figure 12.18).

Like the brain, the spinal cord is protected by bone, meninges, and cerebrospinal fluid (Figure 13.30). The tough dura mater, called the **spinal dural sheath,** differs somewhat from the dura around the brain: It does not attach to the surrounding bone and corresponds only to the meningeal layer of the brain's dura mater. Just external to this spinal dura is a rather large **epidural space** filled with cushioning fat and a network of veins. Anesthetics are often injected into the epidural space to numb the spinal cord and thereby relieve pain in body regions inferior to the injection site. The subdural space, arachnoid mater, subarachnoid space, and pia mater around the spinal cord resemble the corresponding elements around the brain. Inferiorly, the dura and arachnoid extend to the level of S_2, well beyond the end of the spinal cord.

The spinal cord does not extend the full length of the vertebral canal but ends in the superior lumbar region because of prenatal events. Until the third month of development, the spinal cord does run all the way to the coccyx, but thereafter it grows slower than the caudal vertebral column. As the vertebral column grows caudally, the spinal cord assumes a progressively more rostral position. By the time of birth, the spinal cord

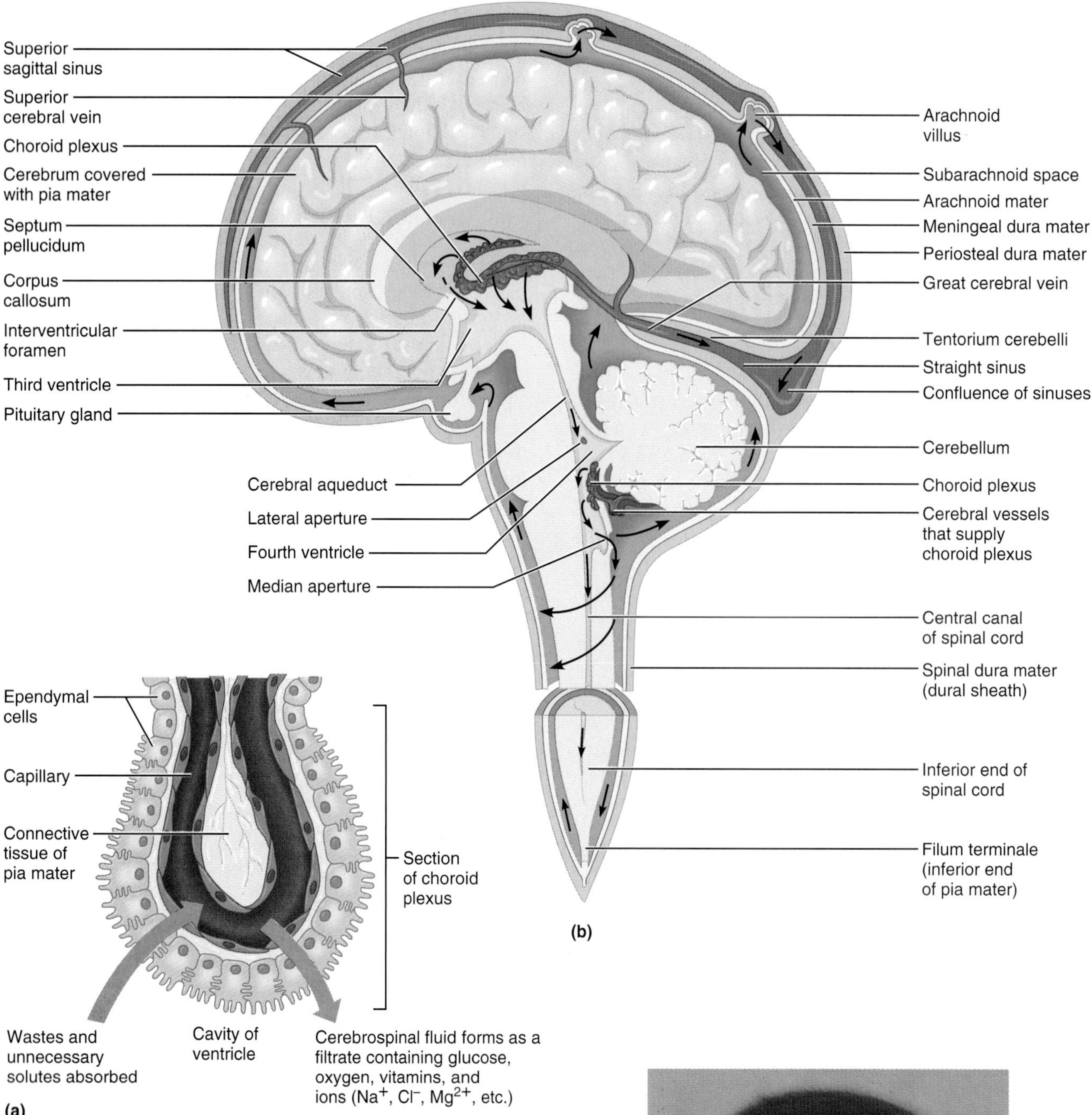

FIGURE 13.27 The sites of formation and circulation pattern of cerebrospinal fluid. (a) Formation: Cerebrospinal fluid, a filtrate of the blood, forms at the choroid plexuses, each of which consists of a knot of capillaries surrounded by a layer of ependymal cells. The filtrate moves out of the capillaries, is processed by the ependymal cells, and enters the ventricles as cerebrospinal fluid. (b) Circulation: Once formed, cerebrospinal fluid flows through the ventricles, median and lateral apertures, subarachnoid space, arachnoid villi, and into the blood of the superior sagittal sinus.

FIGURE 13.28 Hydrocephalus in the newborn. Due to inadequate drainage of cerebrospinal fluid the head enlarges, resulting in abnormal growth of the skull.

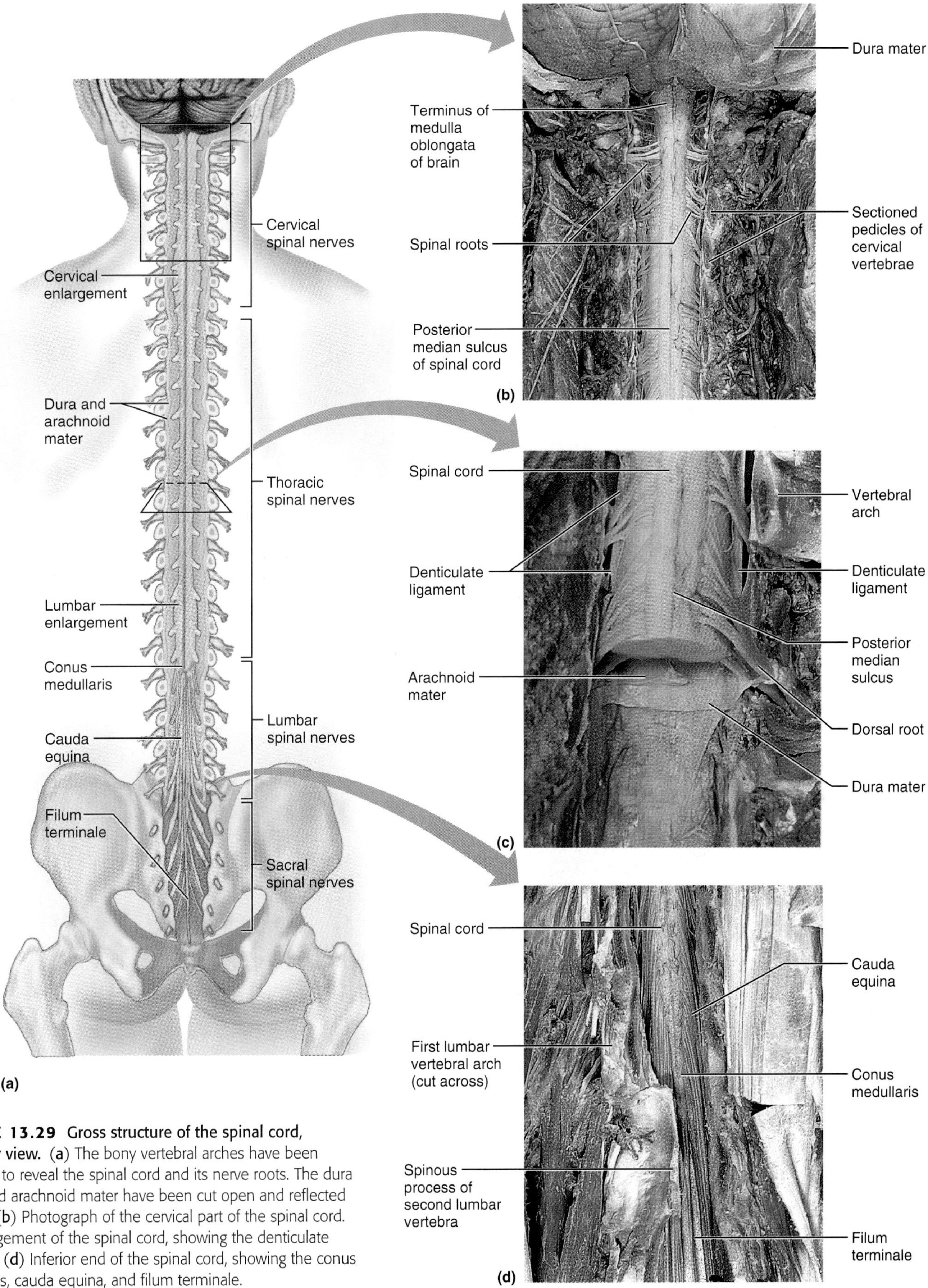

FIGURE 13.29 Gross structure of the spinal cord, posterior view. (a) The bony vertebral arches have been removed to reveal the spinal cord and its nerve roots. The dura mater and arachnoid mater have been cut open and reflected laterally. (b) Photograph of the cervical part of the spinal cord. (c) Enlargement of the spinal cord, showing the denticulate ligament. (d) Inferior end of the spinal cord, showing the conus medullaris, cauda equina, and filum terminale.

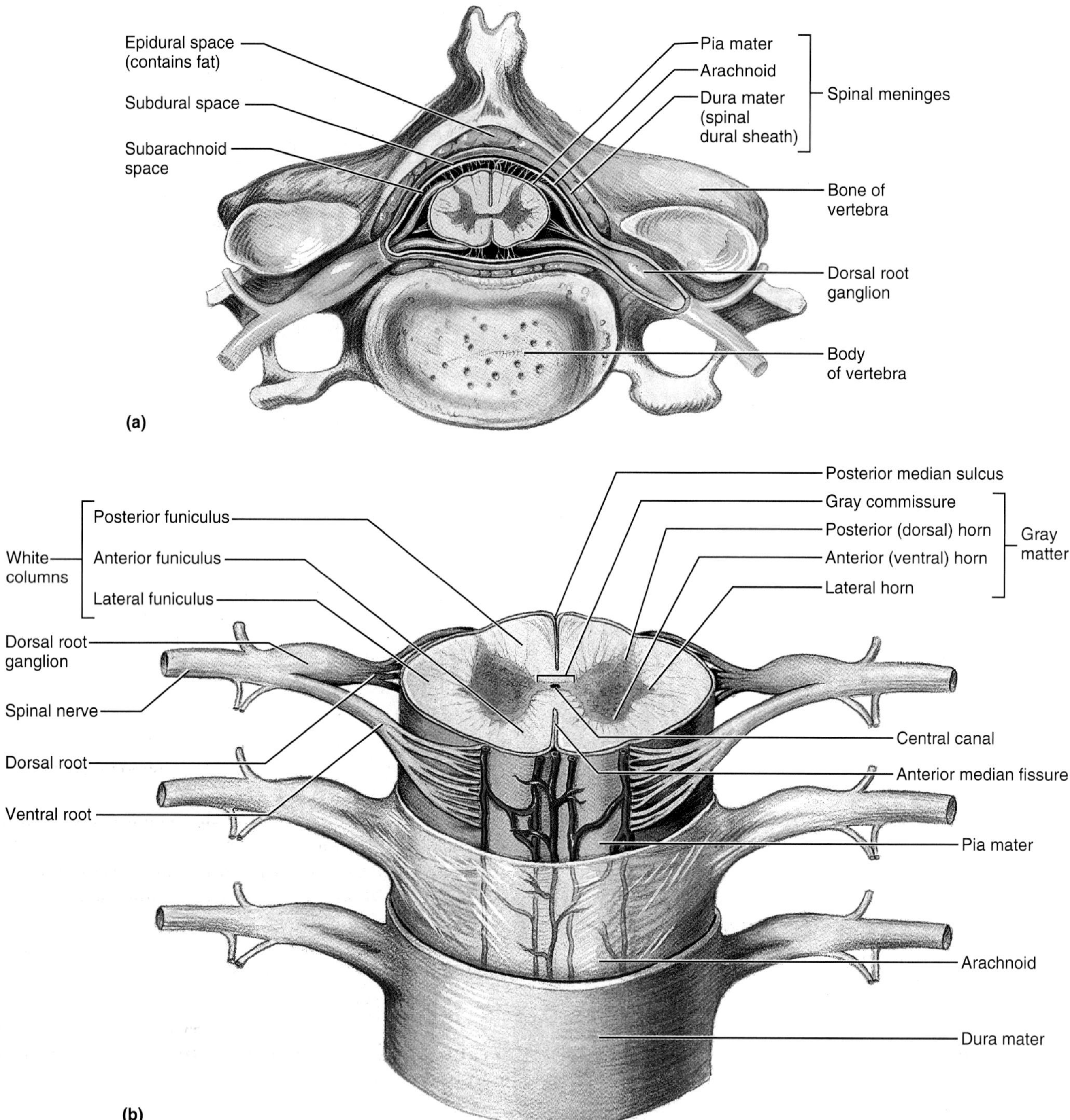

FIGURE 13.30 Anatomy of the spinal cord. (**a**) Cross section through the spinal cord in the cervical region illustrating its relationship to a surrounding cervical vertebra. (**b**) Three-dimensional ventral view of the spinal cord and its meningeal coverings. The dorsal direction is at the top in both these figures.

ends at L_3. During childhood, it attains its adult position, terminating at the level of the intervertebral disc between L_1 and L_2. (This is merely the average level; it varies among different people, from the inferior margin of T_{12} to the superior margin of L_3.)

Critical Reasoning

Referring to Figure 13.29, what is the safest place along the vertebral canal to insert a needle to withdraw a sample of cerebrospinal fluid from the subarachnoid space in an adult?

*Because the adult spinal cord ends superior to the midlumbar region, a needle can safely enter the subarachnoid space inferior to that location, a procedure that is called a **lumbar puncture (spinal tap).** The patient bends forward, and the needle is inserted between the spines and laminae of successive vertebrae, either between L_3 and L_4 or between L_4 and L_5. Lumbar punctures are used to sample the CSF for microbial or chemical analysis, to measure the pressure of the CSF, and to inject antibiotics, anesthetics, or X-ray contrast media into the subarachnoid space.*

At its inferior end, the spinal cord tapers into the **conus medullaris** (ko′nus med′u-lar″is; "cone of the spinal cord") (see Figure 13.29a and d). This cone in turn tapers into a long filament of connective tissue covered with pia mater, the **filum terminale** ("end filament"), which attaches to the coccyx inferiorly, anchoring the spinal cord in place so it is not jostled by body movements. Furthermore, the spinal cord is anchored to the bony walls of the vertebral canal throughout its length by saw-toothed lateral shelves of pia mater called the **denticulate ligaments** (den-tik′u-lāt; "toothed") (see Figure 13.29c).

Thirty-one pairs of *spinal nerves* (PNS structures) attach to the spinal cord through dorsal and ventral nerve roots (see Figure 13.30b). The spinal nerves lie in the intervertebral foramina, from which they send lateral branches throughout the body. Based on the vertebral locations in which they lie, the 31 spinal nerves are divided into cervical (8), thoracic (12), lumbar (5), sacral (5), and coccygeal (1) groups. In its cervical and lumbar regions, where the nerves to the upper and lower limbs arise, the spinal cord shows obvious enlargements called the **cervical** and **lumbar enlargements** (see Figure 13.29a). Because the cord does not reach the inferior end of the vertebral column, the lumbar and sacral nerve roots must descend for some distance before reaching their intervertebral foramina. This collection of nerve roots at the inferior end of the vertebral canal is the **cauda equina** (kaw′dah e-kwi′nah; "horse's tail").

Just as the vertebral column is segmented, so too is the spinal cord. These segments of the spinal cord, which are not visible externally, are designated according to the numbering of the spinal nerves that issue from them—for example, sixth cervical, fifth thoracic, second lumbar, and third sacral. The segments of the spinal cord all lie *superior* to their corresponding vertebrae because of the rostral shift of the spinal cord during development. For example, spinal cord segment T_1 is at the level of vertebra C_7; cord segment T_{10} is at vertebra T_9; and cord segment L_3 is at vertebra T_{12}.

In a cross-sectional view, the spinal cord is wider laterally than anteroposteriorly (see Figure 13.30a). Two deep grooves—the **posterior median sulcus** and the wider **anterior median fissure**—run the length of the cord and partly divide it into right and left halves (see Figure 13.30b).

Gray Matter of the Spinal Cord and Spinal Roots

As previously mentioned, the spinal cord consists of an outer region of white matter and an inner region of gray matter. As in other parts of the CNS, the gray matter of the cord consists of a mixture of neuron cell bodies, short unmyelinated axons and dendrites, and neuroglia. In cross section, the gray matter is shaped like the letter H (see Figure 13.30). The crossbar of the H, the **gray commissure,** contains the narrow central cavity of the spinal cord, the **central canal.** The two posterior arms of the H are the **posterior horns,** whereas the two anterior arms are the **anterior horns.** In three dimensions, these arms run the entire length of the spinal cord and are called *dorsal* and *ventral columns,* respectively. Additionally, small **lateral columns**—called **lateral horns** in section—are present in the thoracic and superior lumbar segments of the spinal cord.

Figure 13.31 depicts the basic organization of the spinal gray matter (also see Figure 12.19). The posterior horns consist entirely of interneurons. These interneurons receive information from sensory neurons whose cell bodies lie outside the spinal cord in *dorsal root ganglia,* and whose axons reach the cord in *dorsal roots.* The anterior (and the lateral) horns contain cell bodies of motor neurons that send their axons out of the cord in *ventral roots* to supply muscles and glands. (Interneurons also occur in the anterior horns, but they are not emphasized in the figure.) The size of the anterior motor horns varies along the length of the spinal cord, reflecting the amount of skeletal musculature innervated at each level. Accordingly, the anterior horns are largest in the cervical and lumbar regions of the cord, which innervate the upper and lower limbs, respectively.

The spinal gray matter can be further divided according to the innervation of the somatic and visceral regions of the body. This scheme recognizes four zones of spinal cord gray matter: **somatic sensory (SS), visceral sensory (VS), visceral motor (VM),** and **somatic motor (SM)** (see Figure 13.31 for more details).

White Matter of the Spinal Cord

The white matter of the spinal cord is composed of myelinated and unmyelinated axons that allow communication between different parts of the spinal cord and between the cord and brain. These fibers are of three types according to the direction they run:

1. **Ascending.** Most of the ascending fibers in the spinal cord carry sensory information from the sensory neurons of the body up to the brain.
2. **Descending.** Most descending fibers carry motor instructions from the brain to the spinal cord, to stimulate contraction of the body's muscles and secretion of its glands.

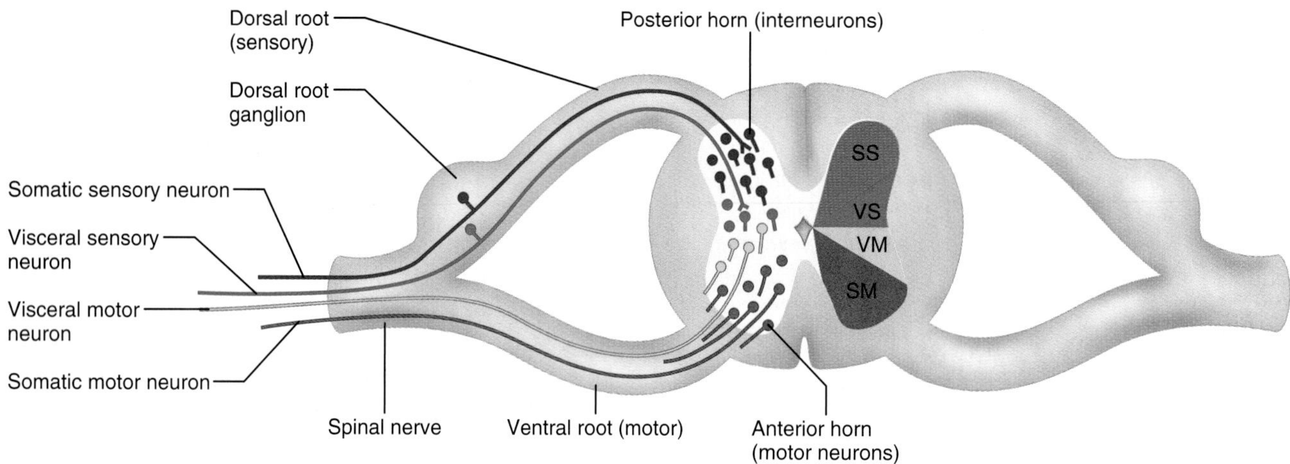

FIGURE 13.31 **General organization of the gray matter of the spinal cord.** The H-shaped gray matter of the spinal cord can be divided into a dorsal sensory half and a ventral motor half. SS: interneurons receiving input from somatic sensory neurons. VS: interneurons receiving input from visceral sensory neurons. VM: visceral motor cell bodies. SM: somatic motor cell bodies. Note that the dorsal roots, ventral roots, and spinal nerve are part of the PNS, not of the spinal cord.

3. **Commissural.** Some fibers cross from one side of the cord to the other.

Of these three types, the ascending and descending fibers make up most of the white matter.

The white matter on each side of the spinal cord is divided into three **white columns,** or **funiculi** (fu-nik′u-li; "long ropes"), named according to their positions—the **posterior, anterior,** and **lateral funiculi** (see Figure 13.30b). The anterior and lateral funiculi are continuous with one another and are divided by an imaginary line that extends out from the anterior gray horn. The posterior funiculus, also called the *dorsal white column,* is subdivided into a medial **fasciculus gracilis** ("slender bundle") and a lateral **fasciculus cuneatus** ("wedge-shaped bundle") (Figure 13.32).

The three funiculi contain many fiber tracts, each of which consists of axons with similar destinations and functions. The principal ascending and descending

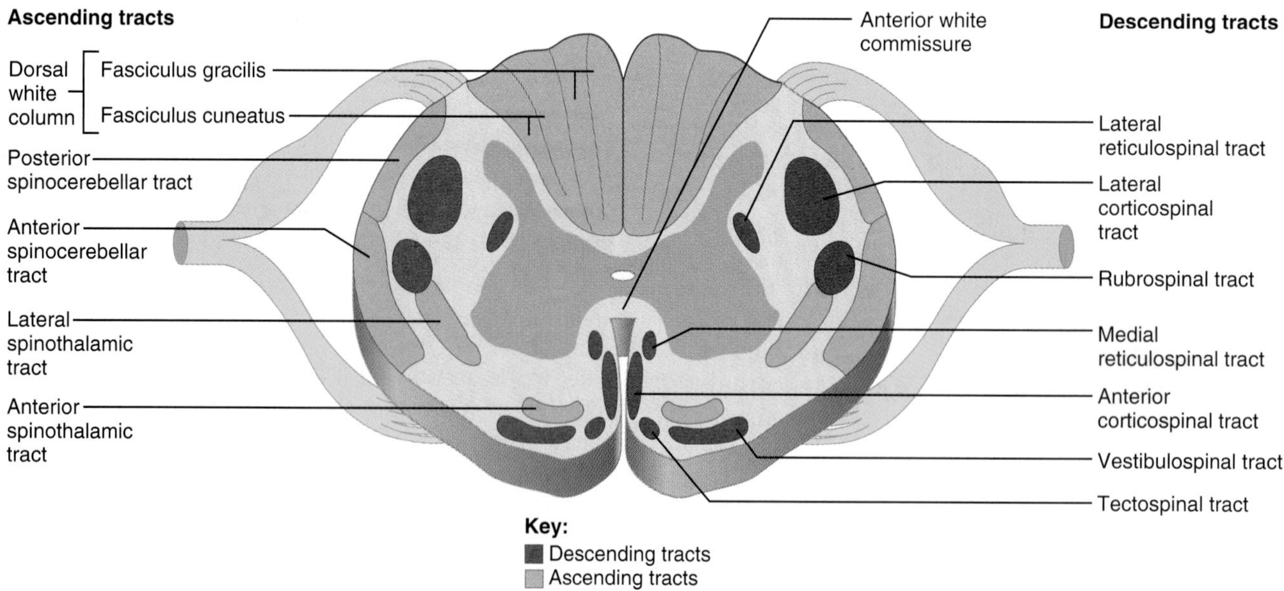

FIGURE 13.32 **Major fiber tracts in the white matter of the spinal cord.** Dorsal is above in this section. Ascending (sensory) tracts are blue, descending (motor) tracts are red. These tracts are not as distinct as the figure indicates, and they overlap one another considerably. The white matter that lies directly ventral to the central canal is the anterior white commissure, the area where axons cross from one side of the cord to the other.

tracts of the spinal cord are shown in cross section in Figure 13.32. For the most part, these spinal tracts are named according to their origin and destination. Next we consider the major spinal tracts more closely.

Sensory and Motor Pathways

All major spinal tracts are segments of multineuron *pathways* that connect the brain to the body periphery, carrying sensory information to the brain and motor instructions to the effectors of the body (Figures 13.33 and 13.34). Before discussing them, we can make several generalizations about the great ascending and descending pathways:

1. Most pathways cross from one side of the CNS to the other, or decussate, at some point along their course.
2. Most pathways consist of a chain of two or three neurons that contribute to successive tracts along the pathway.
3. Most pathways exhibit somatotopy. In this context, such "body mapping" means that axons in the tracts are spatially arranged in a specific way, according to the body region they supply. For example, in one ascending tract, the axons transmitting impulses from the superior parts of the body lie lateral to the axons carrying impulses from the inferior body parts.
4. All pathways are paired right and left; one member of the pair is present on each side of the spinal cord or brain.

Ascending (Sensory) Pathways

The ascending pathways (Figure 13.33) conduct general somatic sensory impulses superiorly through chains of two or three neurons (composed of *first-, second-,* and *third-order neurons*) to various regions of the brain. There are four main ascending pathways: The *dorsal column* and *spinothalamic* pathways transmit sensory impulses from the body to the primary somatosensory cortex for conscious interpretation. The *posterior* and *anterior spinocerebellar* pathways convey information on proprioception to the cerebellum, which uses this information to coordinate body movements. These pathways are summarized in Table 13.2 on p. 399, and each is described below.

The **dorsal column pathway** (right side of Figure 13.33a) carries information on fine touch, pressure, and conscious aspects of proprioception. These are discriminative senses—senses we can localize very precisely on our body surface. In this pathway, the axons of the sensory neurons enter the spinal cord and send branches up the dorsal white column in the fasciculus gracilis or fasciculus cuneatus. These axons ascend to the medulla oblongata and synapse with second-order neurons in the nucleus gracilis or nucleus cuneatus. Axons from these nuclei form a tract called the **medial lemniscus** (lem-nis′kus; "ribbon"), which decussates and then ascends to the thalamus. From there, third-order neurons send axons to the primary somatosensory cortex on the postcentral gyrus. Here, the sensory information is processed and reaches our consciousness as precisely localized sensations.

The **spinothalamic pathway** (Figure 13.33b) carries information on pain, temperature, deep pressure, and the coarser aspects of touch—senses of which we are aware but cannot localize with exact precision on the body surface. In this pathway, the axons of sensory neurons enter the spinal cord, where they synapse on interneurons in the posterior gray horn. Axons of these second-order neurons then decussate, enter the lateral and anterior funiculi as the **lateral** and **anterior spinothalamic tracts,** and ascend to the thalamus. From there, third-order axons project to the primary somatosensory cortex on the postcentral gyrus, where the information is processed into conscious sensations. Note that the brain usually interprets the sensory information carried by the spinothalamic pathway as unpleasant—pain, burns, cold, and so on.

The **posterior** and **anterior spinocerebellar pathways** carry information on the subconscious aspects of proprioception to the cerebellum (left side of Figure 13.33a). Instead of crossing over once, these fibers either do not decussate or cross twice, undoing the decussation.

Descending (Motor) Pathways

The descending tracts (Figure 13.34) that deliver motor instructions from the brain to the spinal cord can be divided into two groups: (1) pyramidal tracts and (2) all others. The following is an overview; additional information is provided in Table 13.3 on page 401.

The **pyramidal,** or **corticospinal, tracts** (Figure 13.34a) control precise and skilled voluntary movements, such as writing and threading a needle. In these tracts, the axons of pyramidal neurons descend from the cerebral motor cortex to the spinal gray matter—mostly to the ventral horns. There, the axons synapse with short interneurons that activate spinal motor neurons. In this way, the corticospinal tracts exert influence over the limb muscles, especially muscles that move the hand and fingers. The axons of the corticospinal tracts decussate along their course, most in the medulla within the decussation of the pyramids but some in the spinal cord.

The pyramidal tracts were considered so important by earlier anatomists that they named all other descending tracts "extrapyramidal tracts." These descending tracts include the **tectospinal tract** from the superior colliculus (the tectum of the midbrain), the **vestibulospinal tract** from the vestibular nuclei, the **rubrospinal tract** from the red nucleus (*rubro* = red) (Figure 13.34b), and the **reticulospinal tract** from the reticular formation (Table 13.3). These tracts seem to

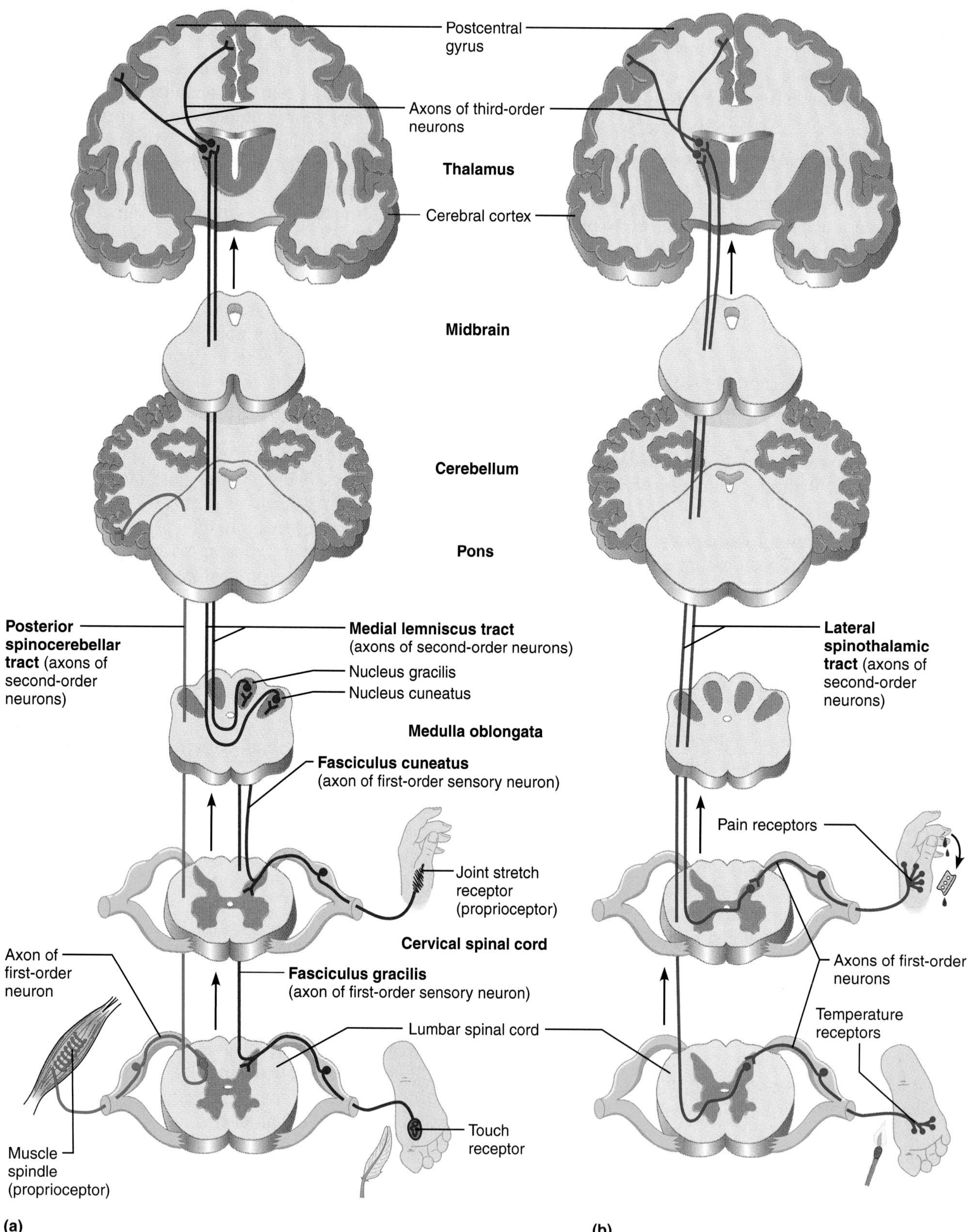

FIGURE 13.33 **Major ascending pathways for the general somatic senses.** These pathways relay sensory information from the body surface (of the neck and inferiorly) to the brain. Every pathway contains thousands of fibers, not just the few depicted. (**a**) Posterior spinocerebellar pathway (left) and dorsal column pathway (right). (**b**) Spinothalamic pathway.

Table 13.2 Major Ascending (Sensory) Pathways and Spinal Cord Tracts

Spinal cord tract	Locaton (funiculus)	Origin	Termination	Function
DORSAL COLUMN PATHWAYS				
Fasciculus cuneatus and fasciculus gracilis	Posterior	Central axons of sensory (first-order) neurons enter dorsal root of the spinal cord and branch; branches enter posterior white column on same side without synapsing	By synapse with second-order neurons in nucleus cuneatus and nucleus gracilis in medulla; fibers of medullary neurons cross over and ascend in medial lemniscus tracts to thalamus, where they synapse with third-order neurons; thalamic neurons then transmit impulses to somatosensory cortex	Both tracts transmit sensory impulses from general sensory receptors of skin and proprioceptors, which are interpreted as discriminative touch, pressure, and "body sense" (limb and joint position) in opposite somatosensory cortex; cuneatus transmits afferent impulses from upper limbs, upper trunk, and neck; it is not present in spinal cord below level of T_6; gracilis carries impulses from lower limbs and inferior body trunk
SPINOTHALAMIC PATHWAYS				
Lateral spinothalamic	Lateral	Association (second-order) neurons of posterior horn; fibers cross to opposite side before ascending	By synapse with third-order neurons in thalamus; impulses then conveyed to somatosensory cortex by thalamic neurons	Transmits impulses concerned with pain and temperature to opposite side of brain; eventually interpreted in somatosensory cortex
Anterior spinothalamic	Anterior	Association neurons in posterior horns; fibers cross to opposite side before ascending	By synapse with third-order neurons in thalamus; impulses eventually conveyed to somatosensory cortex by thalamic neurons	Transmits impulses concerned with crude touch and pressure to opposite side of brain for interpretation by somatosensory cortex
SPINOCEREBELLAR PATHWAYS				
Posterior spinocerebellar*	Lateral (posterior part)	Association (second-order) neurons in posterior horn on same side of cord; fibers ascend without crossing	By synapse in cerebellum	Transmits impulses from trunk and lower limb proprioceptors on one side of body to same side of cerebellum; subconscious proprioception
Anterior spinocerebellar*	Lateral (anterior part)	Association (second-order) neurons of posterior horn; contains crossed fibers that cross back to the opposite side in the pons	By synapse in cerebellum	Transmits impulses from the trunk and lower limb on the same side of body to cerebellum; subconscious proprioception

*These spinocerebellar tracts carry information from the lower limb and trunk only. The corresponding tracts for the upper limb and neck (called *rostral spinocerebellar* and *cuneocerebellar tracts*) are beyond the scope of this book.

bring about those body movements that are subconscious, coarse, or postural.

The cerebellum smooths and coordinates all movements dictated by the subcortical motor nuclei, just as it coordinates all skilled movements controlled by the pyramidal system. Therefore, axons from the cerebellum project to and influence the red nucleus, vestibular nuclei, and reticular nuclei (axons not illustrated).

Furthermore, it is now well known that pyramidal tract neurons project to and influence the activity of

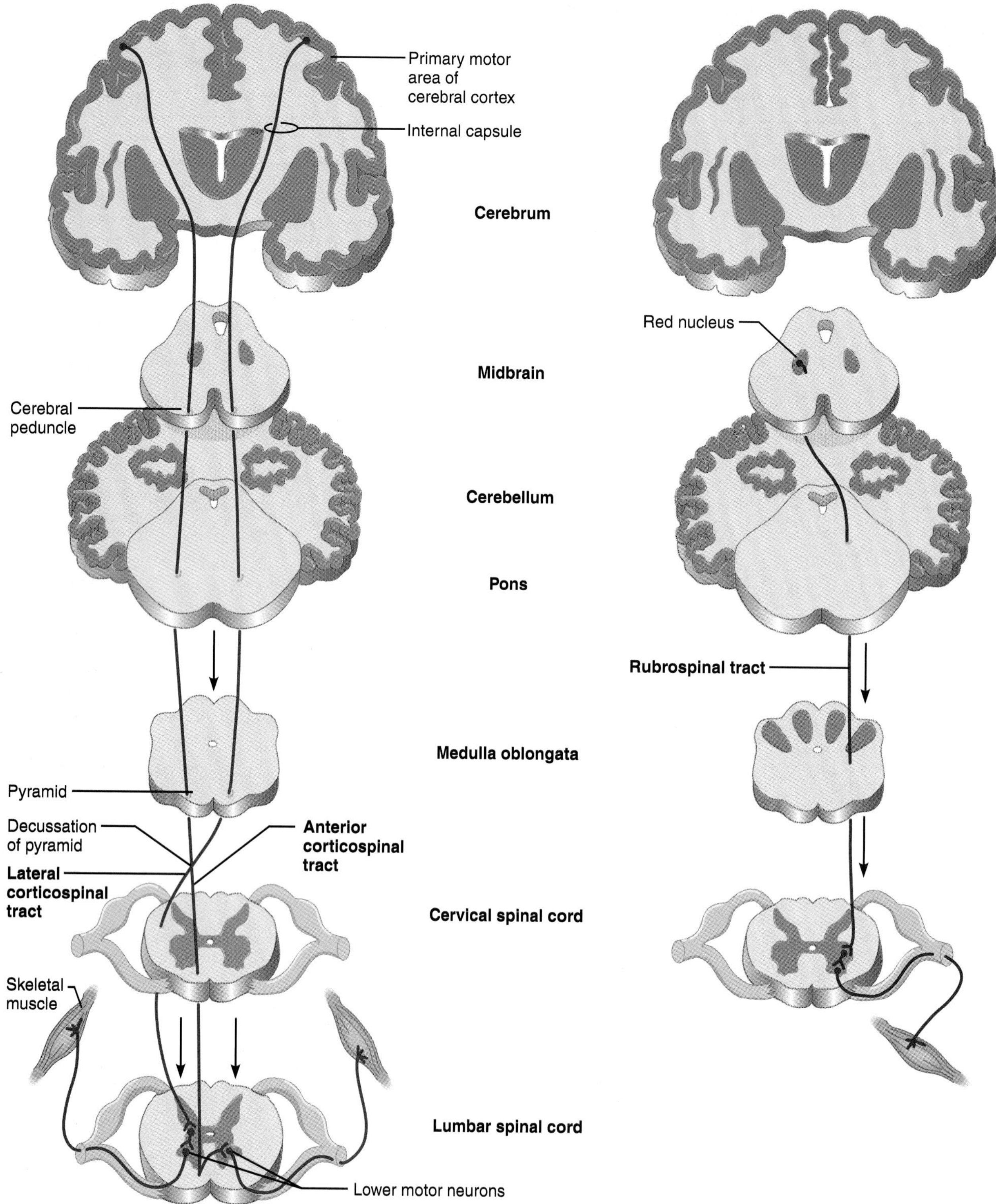

FIGURE 13.34 Descending motor pathways by which the brain influences movements. (a) Pathways of the pyramidal (corticospinal) tracts. (b) Pathway of the rubrospinal tract, an extrapyramidal pathway from the red nucleus.

Table 13.3 Major Descending (Motor) Pathways and Spinal Cord Tracts

Spinal cord tract	Locaton (funiculus)	Origin	Termination	Function
PYRAMIDAL				
Lateral corticospinal	Lateral	Pyramidal neurons of motor cortex of the cerebrum; decussate in pyramids of medulla	Anterior horn interneurons that influence motor neurons and occasionally directly to anterior horn motor neurons	Transmits motor impulses from cerebrum to spinal cord motor neurons (which activate skeletal muscles on opposite side of body); voluntary motor tract
Anterior corticospinal	Anterior	Pyramidal neurons of motor cortex; fibers cross over at the spinal cord level	Anterior horn (as above)	Same as lateral corticospinal tract
OTHER MOTOR PATHWAYS (formerly called extrapyramidal)				
Tectospinal	Anterior	Superior colliculus of midbrain of brain stem (fibers cross to opposite side of cord)	Anterior horn (as above)	Turns neck so eyes can follow a moving object
Vestibulospinal	Anterior	Vestibular nuclei in medulla of brain stem (fibers descend without crossing)	Anterior horn (as above)	Transmits motor impulses that maintain muscle tone and activate limb and trunk extensors and muscles that move head; in this way, helps maintain balance during standing and moving
Rubrospinal	Lateral	Red nucleus of midbrain of brain stem (fibers cross to opposite side just inferior to the red nucleus)	Anterior horn (as above)	Transmits motor impulses concerned with muscle tone of distal limb muscles (mostly flexors) on opposite side of body
Reticulospinal (anterior, medial, and lateral)	Anterior and lateral	Reticular formation of brain stem (medial nuclear group of pons and medulla); both crossed and uncrossed fibers	Anterior horn (as above)	Transmits impulses concerned with muscle tone and many visceral motor functions; may control most unskilled movements

most of the "extrapyramidal" nuclei. Because *extrapyramidal* literally means "independent of the pyramidal (tracts)," modern anatomists have rejected that term, preferring instead to use the names of the individual motor pathways. However, clinicians still classify many motor disorders as being pyramidal or extrapyramidal in origin.

Our discussion of sensory and motor pathways has focused on the innervation of body regions inferior to the head. In general, the pathways to and from the head are similar to those for the trunk and limbs, except that the axons are located in cranial nerves and the brain stem, not in the spinal cord.

DISORDERS OF THE CENTRAL NERVOUS SYSTEM

In our consideration of the many disorders that can affect the central nervous system, we proceed from brain to spinal cord.

Brain Dysfunction

Brain dysfunctions are varied and extensive. Here, we focus on two types: traumatic brain injuries and degenerative brain diseases.

Traumatic Brain Injuries

Blows to the head are the leading cause of *accidental* death in the United States. Such blows send destructive shock waves through the brain. A million new cases of traumatic brain injury occur in the United States each year, mainly from traffic accidents (50%), sports injuries (30%), and violence (20%). A **concussion** occurs when brain injury is slight and the symptoms are mild and transient. The result of marked destruction of brain tissue, by contrast, is a brain **contusion.** A person suffering a contusion of the cerebral cortex may remain conscious, but a contusion of the brain stem usually produces coma.

Following blows to the head, fatal consequences may result from subdural or subarachnoid hemorrhaging (bleeding from ruptured vessels into these spaces). People who initially are lucid after head trauma and then start to deteriorate neurologically are, in all probability, hemorrhaging intracranially. Accumulating blood within the skull increases intracranial pressure. If this pressure forces the brain stem through the foramen magnum, the control of heart rate, blood pressure, and respiration is lost. Surgeons treat intracranial hemorrhages by removing the blood mass and repairing the ruptured vessels.

Another consequence of serious head injury is swelling of the brain, which leads to brain-damaging pressure within the cranium. The swelling results both from the edema of inflammation and from water uptake by brain tissue. Current treatments seek to limit the swelling and minimize secondary damage to the neurons.

Degenerative Brain Diseases

The most important degenerative diseases of the brain are cerebrovascular accidents and Alzheimer's disease.

Cerebrovascular Accidents Cerebrovascular accidents (CVAs), commonly called *strokes,* are the most common disorders of the nervous system and the third leading cause of death in the United States. Strokes occur when blockage or interruption of the flow of blood to a brain area causes brain tissue to die from lack of oxygen. Such a deprivation of blood to a tissue is called **ischemia** (is-ke′me-ah; "to hold back blood").

Twenty percent of strokes are caused by burst or torn cranial vessels; the other 80% are caused by a blood clot blocking these vessels. In half of these latter cases the clot comes from outside the head (from the heart, for example), and in the other half the clot forms in place on the roughened walls of brain arteries that are narrowed by atherosclerosis. Whatever the cause, the nature of the neurological deficits that ensue depends on exactly which artery is affected and which area of the brain it supplies. People who survive a CVA may be paralyzed on one side of the body or have sensory or language deficits.

Fewer than 35% of the individuals who survive an initial CVA are alive three years later, because those who suffer CVAs from blood clots are likely to have recurrent clotting problems—and more CVAs. Some of the survivors partly recover their lost faculties, especially if given rehabilitation therapy, because undamaged neurons sprout new branches that spread into the damaged brain area and take over lost functions. This is an example of **neural plasticity,** the regenerative process by which the damaged CNS rewires itself. However, because neuronal spread is limited to only about 5 mm, large lesions always result in less-than-complete recovery.

Strategies for treating CVAs focus on: (1) earlier detection of strokes in progress using faster neuroimaging techniques; (2) early administration of a clot dissolver, such as *tissue-type plasminogen activator,* to strokes caused by clots; and (3) development of neuroprotective drugs that keep damaged neurons from dying.

Alzheimer's Disease **Alzheimer's** (Altz′hi-merz) **disease** is a progressive degenerative disease of the brain that ultimately results in dementia (mental deterioration). Between 5% and 15% of people over age 65 develop this condition, and up to half of all people over 85 die from it. Although usually confined to the elderly, Alzheimer's disease may begin in middle age. Victims exhibit a wide variety of mental defects, including loss of memory (particularly for recent events), shortened attention span, depression, and disorientation. The disease worsens over a period of years until, eventually, sufferers experience hallucinations.

Alzheimer's disease is associated with structural changes in the hippocampus and the association areas of the cerebral cortex, areas involved with memory and thought. The cerebral gyri shrink and many neurons ultimately are lost, especially those that produce the neurotransmitter acetylcholine. Additionally, functional neuroimaging has documented a loss of neuronal activity in the affected regions (Figure 13.35). Microscopically, brain tissue of Alzheimer's patients contains *senile plaques* between the neurons and *neurofibrillar tangles* in the neurons. The cause of the disease is unclear. The main hypothesis focuses on *amyloid precursor protein* (APP), which builds up in the senile plaques and produces a neurotoxic derivative called *amyloid-beta peptide* (Aß). Recently the enzyme that cleaves Aß from APP was identified, and it is hoped that inhibiting this enzyme will prove an effective treatment for Alzheimer's disease.

Spinal Cord Damage

Any localized damage to the spinal cord or spinal roots leads to some functional loss, either *paralysis* (loss of

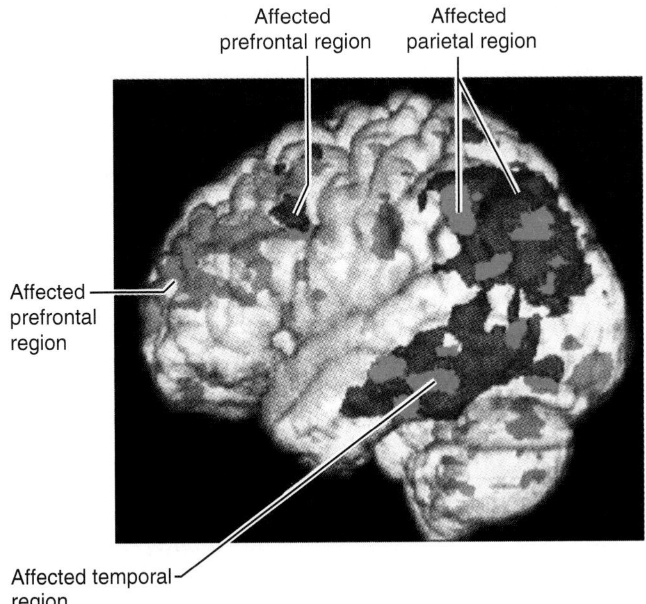

FIGURE 13.35 Areas of depressed neural activity (colored blue and purple) in the cerebrum of patients with Alzheimer's disease, based on functional neuroimaging (PET).

motor function) or *paresthesia* (abnormal or lost sensation; see p. 439). Severe damage to the anterior horn or ventral motor roots results in a complete or **flaccid paralysis** of the skeletal muscles served. No longer stimulated by neurons, the muscles shrink and waste away. By contrast, damage to the cerebral motor cortex leaves the spinal motor neurons and spinal reflexes intact such that the muscles remain healthy, but their movements are no longer under voluntary control. This condition is termed **spastic paralysis.**

Spinal injuries—resulting from traffic accidents, skiing or diving mishaps, falls, and gunshot wounds—can crush, tear, or cut through (transect) the spinal cord. Transection leads to a total loss of voluntary movements and conscious sensation in body regions inferior to the cut. If the damage occurs between the T_1 and L_2 segments of the cord, the lower limbs are affected, but not the upper limbs. This condition is called **paraplegia** (par″ah-ple′je-ah). If the damage is in the cervical region, all four limbs are affected, a condition known as **quadriplegia** (*quadri* = four; *plegia* = a blow). Spinal transection superior to the mid-neck destroys the ability to breathe because C_3–C_5 is where the motor neurons to the diaphragm are located. Such injuries are fatal unless the victim is kept alive by artificial resuscitation at the accident site and by a respirator thereafter. In evaluating spinal cord injuries, it is important to remember that the spinal cord segments are slightly higher than the corresponding vertebral levels (see p. 395).

THE CENTRAL NERVOUS SYSTEM THROUGHOUT LIFE

Established during the first month of development, the brain and spinal cord continue to grow and mature throughout the prenatal period. A number of congenital malformations originate during this period, the most serious of which are congenital hydrocephalus (see p. 391) and neural tube defects.

Neural tube defects result from a delay in the closure of the neural groove into a neural tube (see Figure 3.7). **Anencephaly** (an″en-sef′ah-le; "without a brain") is caused by the failure of the rostral part of the groove to close and form a complete brain, so the cerebrum and cerebellum never develop; only the brain stem and traces of the basal nuclei form. The anencephalic newborn is totally vegetative, and mental life as we know it is absent. Death occurs soon after birth. About 2 in 1000 births are anencephalic newborns.

Spina bifida (spi′nah bif′ĭ-dah; "forked spine") refers to a variety of neural tube defects involving incomplete formation of the bony vertebral arches (absence of vertebral laminae). In these cases, the delay in formation of the spinal cord and meninges does not leave time for the bony vertebral laminae to form. In extremely delayed cases, the neural groove remains unclosed and functionless, but in most cases a complete spinal cord forms but then becomes damaged. Spina bifida involves about 3 of every 2000 births and most often affects the lumbosacral region.

In **spina bifida cystica,** the most common variety, the meninges around the cord are expanded into a baglike cyst called a **meningocele** (mĕ-ning′go-sēl) that protrudes dorsally from the infant's spine. The spinal cord usually moves back into the dorsal part of the cyst, in which case it is called a **myelomeningocele** (mi″ĕ-lo-mĕ-ning′go-sēl), a "spinal cord in a sac of meninges" (Figure 13.36). Because the wall of the cyst is thin, irritants

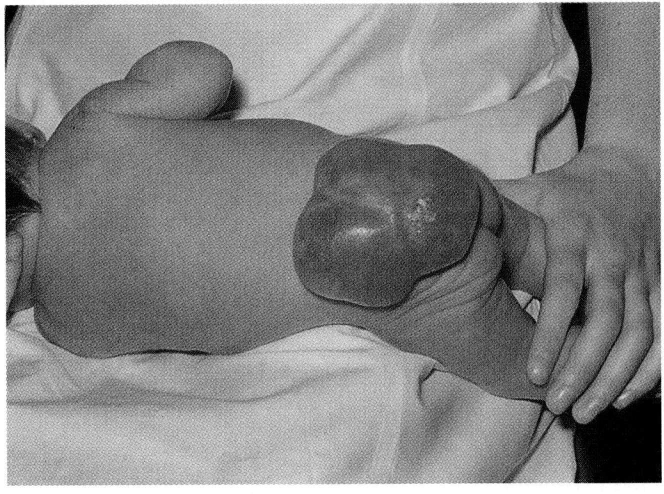

FIGURE 13.36 Newborn with a lumbar myelomeningocele.

such as urea in the surrounding amniotic fluid can pass through it and harm the spinal cord during the fetal period, as can physical jostling and infections after birth. Worse, the movement of the spinal cord into the cyst pulls the medulla oblongata down through the foramen magnum of the skull, constricting the brain ventricles so that hydrocephalus results. Immediate surgical intervention can relieve the hydrocephalus and may sometimes prevent paralysis.

Up to 70% of cases of spina bifida are caused by inadequate amounts of the B vitamin folate in the maternal diet, so folate is now added to all bread, flour, and pasta products sold in the United States. Additionally, experiments indicate that another B vitamin, inositol, further reduces the incidence of spina bifida in mice, a result that may apply to humans as well.

Critical Reasoning

Why should spina bifida babies be delivered by cesarean section, not vaginally?

Passage of the infant through the narrow birth canal is likely to compress the meningocele and damage the spinal cord. Of course, planning for a cesarean delivery requires that the spina bifida be recognized prenatally—by a fetal ultrasound exam or by a test called maternal serum feroprotein screening before midpregnancy.

We now turn to another CNS-related congenital condition. Infections of the placenta and other factors that interfere with the blood supply to the fetal brain can lead to **cerebral palsy,** a lifelong neural disorder in which the voluntary muscles are poorly controlled or paralyzed. This condition reflects damage to the cerebral motor cortex or, less often, to the cerebellum or basal nuclei. In addition to spasticity, speech difficulties, and other motor impairments, the symptoms can also include seizures, mental retardation, deafness, and visual impairments. Cerebral palsy is the largest single cause of crippling in children. It affects 2 of every 1000 full-term births, but it occurs 70 times more frequently among premature infants.

After birth, different parts of the normal brain complete their development at different times, and the neural connections made in the brain of the growing baby are strongly molded by experiences. PET scans have revealed that the thalamus and the somatosensory cortex are active in a 5-day-old baby, but that the visual cortex is not, a finding that supports the observation that infants of this age respond to touch but have poor vision. By 11 weeks, more of the cortex is active, and the baby can reach for objects. By 8 months, the cortex is highly active; the child can think about and remember what she or he sees, and is starting to understand the sounds of language. At 10 to 18 months, the prefrontal lobe starts to communicate with the limbic regions, so the child can begin to control emotional urges. Maturation of the nervous system continues through childhood and reflects the progressive myelination and thickening of its axons.

Despite the great plasticity of young brains, there are critical periods after which learning is no longer possible. For example, no one has ever learned a first language after about age 5, and childhood abuse can so greatly retard the maturation of the frontal lobes that normal cognitive functioning is not attained. Recent studies suggest these critical periods are not as rigid as once was believed, providing more hope for deprived children.

Advances in Understanding Formerly it was thought that the brain's circuitry is fully formed by late childhood, but new findings show that such maturation is not complete until the early 20s. In teenagers, the "rational" frontal lobe is still organizing itself, at the same time that "emotional" limbic structures are in developmental overdrive from the effects of sex hormones. This may explain why some teenagers seem moody or abandon good judgment for risky, thrill-seeking behaviors. ✷

The brain reaches its maximum weight in the young adult, but with age it shrinks and its cognitive abilities slowly decline. Traditionally, this decline has been attributed to neuronal death, but improved techniques now suggest that the number of brain neurons does not decrease with age—at least in the cerebral cortex and hippocampus. The reason for declining brain function with age may instead lie in changes in neural circuitry or in lower amounts of neurotransmitters. Studies show that the elderly compensate for this decline by using more regions of their cerebral cortex, and that keeping mentally active can slow their rate of brain shrinkage. Finally, the choroid plexus calcifies in the elderly, drastically reducing the production of cerebrospinal fluid late in life.

RELATED CLINICAL TERMS

Amyotrophic lateral sclerosis (ALS) (ah-mi″o-trof′ik = muscle degeneration or muscle atrophy; skle-ro′sis = hardening) Disease involving degeneration of the pyramidal tracts or brain-stem motor nuclei, with the resultant wasting of the skeletal muscles. As the pyramidal tracts deteriorate, scar tissue hardens the lateral parts of the spinal cord where these tracts lie. Initial symptoms include weakness in using hands or difficulty in swallowing or speaking. ALS appears after age 40, and its progress is relentless and always fatal; currently, half of those affected die within 3 years. However, new intensive-care programs may extend survivorship to 15 years. The cause of ALS is unclear, but it may result from an excess of the neurotransmitter glutamate in the primary motor cerebral cortex, where the glutamate overstimulates, then kills, the neurons.

Encephalopathy (en-sef″ah-lop′ah-the) Any disease or disorder of the brain.

Microcephaly (mi″kro-sef′ah-le) Congenital condition involving the formation of a small brain and skull. May result from damage to the brain before birth or from premature fusion of the sutures of the skull. Most microcephalic children are severely retarded.

Myelitis (mi″ĕ-li-tis) (*myel* = spinal cord) Inflammation of the spinal cord.

Myelogram (*gram* = recording) X ray of the spinal cord after injection of a contrast medium.

Phantom limb pain A common phenomenon in people with amputated limbs in which the perception of pain occurs in the "phantom" body part that is no longer present. Although it can be spontaneous, the pain is often induced by touching the skin of the limb stump or the face. The cause of this pain is now understood. First, the amputation removes all axonal input to the limb representation on the primary somatosensory cerebral cortex. Next, this denervated cortical area is reinnervated by growing axons from the nearby cortical regions that contain face and stump representations, such that touches on the face and stump are interpreted as coming from the now-missing limb.

Tourette's syndrome A brain disorder that affects about 1 of every 3000 Americans, mostly males. Characterized by tics (sudden, fast, stereotyped movements or sounds) such as blinks, grimaces, barks, and yelps. May involve obscene vocalizations, repeating the words and actions of others, and obsessive/compulsive behavior. In this condition, the basal nuclei do not properly regulate a neurotransmitter called dopamine, and thus are unable to interact with the motor cerebral cortex to inhibit stereotypical behaviors. Treated with psychotropic drugs. After the pre-teen years, the severity of the symptoms usually declines with age.

CHAPTER SUMMARY

Media study tools that could provide you additional help in reviewing specific key topics of Chapter 13 are referenced below.

SP = Study Partner; AIA = A.D.A.M.® Interactive Anatomy

THE BRAIN (pp. 360–391)

1. The brain provides for voluntary and involuntary movements, interpretation and integration of sensation, consciousness, and cognitive function. It is also involved with the innervation of the head through the cranial nerves.

Embryonic Development of the Brain (pp. 360–361)

2. The brain develops from the rostral part of the embryonic neural tube. By week 5, the early forebrain, midbrain, and hindbrain have become the telencephalon (cerebrum) and diencephalon, mesencephalon (midbrain), metencephalon (pons, cerebellum), and myelencephalon (medulla oblongata).

3. The cerebral hemispheres grow rapidly, enveloping the diencephalon and rostral brain stem. Their expanding surfaces fold into a complex pattern of sulci and gyri.

Basic Parts and Organization of the Brain (p. 362)

4. A widely used classification divides the brain into the cerebral hemispheres, diencephalon, brain stem (midbrain, pons, medulla), and cerebellum.

5. The brain stem consists of white matter external to central gray matter, whereas the cerebrum and cerebellum also have an additional, external cortex of gray matter.

Ventricles of the Brain (p. 363)

6. The two lateral ventricles are in the cerebral hemispheres; the third ventricle lies in the diencephalon; the cerebral aqueduct is in the midbrain; and the fourth ventricle lies in the hindbrain.

AIA Link: Atlas; System: Nervous; Ventricles of brain (lateral).

SP Exercise: Chapter 13, Ventricles of the Brain.

The Cerebral Hemispheres (pp. 363–374)

7. The large cerebral hemispheres exhibit gyri, sulci, and fissures. The five lobes of each hemisphere are the frontal, parietal, occipital, and temporal lobes and the insula.

8. The cerebral cortex is the site of conscious sensory perception, voluntary initiation of movements, and higher thought functions. Structural regions of the six-layered cerebral cortex were identified by Brodmann and others.

9. Functional areas of the cerebral cortex include the following: (a) *motor areas in the frontal lobe:* primary motor cortex, premotor cortex, frontal eye field, and Broca's area for speech production; (b) *sensory areas in the parietal, occipital, and temporal lobes:* primary somatosensory cortex, primary visual cortex, primary auditory cortex, gustatory cortex, vestibular cortex, and olfactory cortex; and (c) *association areas:* various sensory association areas, prefrontal cortex for the highest cognitive processes, and language areas.

10. Areas for higher-order visual processing occupy much of the temporal lobe (ventral processing stream for recognizing objects) and the posterior parietal lobe (dorsal stream for identifying spatial relationships). The language areas occupy a large part of the left cerebral cortex around the lateral sulcus.

11. The cortex of each cerebral hemisphere receives sensory information from, and sends motor instructions to, the opposite (contralateral) side of the body. The body is mapped in an upside-down "homunculus" on the somatosensory and primary motor areas, thereby providing examples of somatotopy.

12. The pyramidal, or corticospinal, tracts descend from the primary motor and premotor areas of the cortex and run through the brain stem to the spinal cord, where they signal motor neurons to perform skilled voluntary movements.

13. In most people, the left cerebral hemisphere is specialized for language and math skills, and the right hemisphere is more concerned with visual-spatial and creative abilities.

14. In the white matter deep to the cerebral cortex, the types of fibers include commissural fibers between hemispheres, association fibers between cortical areas within a hemisphere, and projection fibers to and from the brain stem and spinal cord.

15. The basal nuclei (basal ganglia), embedded deep within the cerebral white matter, include the caudate, lentiform, and amygdala nuclei. Functionally, this group of nuclei works with the cerebral motor cortex to control complex movements, among other functions.

AIA Link: Atlas; System: Nervous; Basal ganglia (superior).

SP Exercise: Chapter 13, Structures of the Brain.

The Diencephalon (pp. 374–379)

16. The diencephalon, which consists of the thalamus, hypothalamus, and epithalamus, encloses the third ventricle.

17. The thalamus, an egg-shaped group of brain nuclei, is the gateway to the cerebral cortex. It is a major relay station for sensory impulses ascending to the sensory cortex and for impulses from all brain regions that communicate with the cortex.

18. The hypothalamus, a series of brain nuclei, is the brain's most important visceral control center. It regulates sleep cycles, hunger, thirst, body temperature, secretion of the pituitary gland, the autonomic nervous system, and some emotions and behaviors.

19. The small epithalamus contains the pineal gland, which secretes a hormone called melatonin that is involved in the nighttime stage of the sleep-wake cycle.

AIA Link: Atlas; System: Nervous; Hippocampus and fornix (superior).

The Brain Stem **(pp. 379–382)**

20. The midbrain is divided into a tectum and paired cerebral peduncles, with the latter containing the pyramidal motor tracts in the crus cerebri. In the tectum, the superior and inferior colliculi mediate visual and auditory reflexes. In the peduncles, the red nucleus and substantia nigra participate in motor functions. The periaqueductal gray matter elicits the fear response, but it also contains motor nuclei of cranial nerves III and IV (eye movers).

21. In the pons, nuclei of cranial nerves V–VII lie near the fourth ventricle. The ventral region of the pons contains the pyramidal tracts plus the pontine nuclei that project to the cerebellum.

22. The medulla oblongata contains the pyramids and their decussation, all formed by the pyramidal tracts. The olives contain relay nuclei to the cerebellum. Nuclei of cranial nerves VIII–XII lie near the fourth ventricle. Centers in the medullary reticular formation regulate respiration, heart rate, blood pressure, and other visceral functions.

AIA Link: Atlas; System: Nervous; Sagittal section of head and neck.

The Cerebellum **(pp. 382–385)**

23. The cerebellum smooths and coordinates body movements and helps maintain posture and equilibrium. Its gross parts, the paired cerebellar hemispheres and median vermis, are divided transversely into three lobes: anterior, posterior, and flocculonodular. Its surface is covered with folia (ridges) and fissures.

24. From superficial to deep, the main regions of the cerebellum are the cortex, white matter, and the deep cerebellar nuclei.

25. The cerebellar cortex receives sensory information on current body movements, then compares this information to how the body should be moving. In this way, it in effect carries out a dialogue with the cerebral cortex on how to coordinate precise voluntary movements. The cerebellum also performs cognitive functions, which are not yet well understood but may involve predicting sequential events.

26. The cerebellum connects to the brain stem by the superior, middle, and inferior cerebellar peduncles, thick fiber tracts that carry information to and from the cerebellum.

Functional Brain Systems **(pp. 385–388)**

27. The limbic system consists of many cerebral and diencephalic structures on the medial aspect of the forebrain. It is the "emotional-visceral" part of the brain, and it also forms and retrieves memories. The hippocampus and amygdala consolidate memories of facts and emotions, respectively.

28. The reticular formation includes diffuse nuclei that span the length of the brain stem. Its reticular activating system maintains the conscious state of the cerebral cortex, and the reticulospinal tracts signal many nonskilled body movements.

Protection of the Brain **(pp. 388–391)**

29. The brain is protected by skull bones, meninges, cerebrospinal fluid, and the blood-brain barrier.

30. The three meninges, which enclose both brain and spinal cord, are the outer dura mater, the arachnoid, and the inner pia mater. Internal projections of the dura form septa that secure the brain in the skull (falx cerebri, falx cerebelli, tentorium cerebelli).

31. Cerebrospinal fluid both floats and cushions the CNS. It derives from blood plasma at the choroid plexuses in the roof of the ventricles, circulates through the ventricles and into the subarachnoid space, and returns to the blood through arachnoid villi at the superior sagittal sinus.

32. The blood-brain barrier derives from the relative impermeability of brain capillaries (complete tight junctions between endothelial cells). It lets water, oxygen, nutrients, and fat-soluble molecules enter the neural tissue but prevents entry of most harmful substances.

SP Exercise: Chapter 13, Meninges of the Brain.

THE SPINAL CORD **(pp. 391–401)**

1. The spinal cord attaches to the spinal nerves that innervate the neck, limbs, and trunk. It also acts as a reflex center and contains axon tracts running to and from the brain. It extends from the foramen magnum to the level of the first or second lumbar vertebra.

2. Thirty-one pairs of spinal nerve roots issue from the spinal cord. The most inferior bundle of roots resembles a horse's tail (cauda equina). The spinal cord is enlarged in its cervical and lumbar regions, reflecting the innervation of the limbs.

AIA Link: Dissectible anatomy; Male: Posterior; Layer 180.

Gray Matter of the Spinal Cord and Spinal Roots **(p. 395)**

3. The H-shaped gray matter of the spinal cord has two anterior horns containing motor neurons and two posterior horns containing interneurons. The posterior horns can be subdivided into somatic and visceral sensory regions, the anterior horns into visceral and somatic motor regions.

4. The roots of the spinal nerves—dorsal sensory roots and ventral motor roots—are PNS structures that attach to the spinal cord.

White Matter of the Spinal Cord **(pp. 395–397)**

5. The white matter of the cord is divided into posterior, lateral, and anterior funiculi containing ascending and descending fiber tracts.

Sensory and Motor Pathways (pp. 397–400)

6. The ascending tracts in the funiculi of the spinal cord belong to sensory pathways that run from the body periphery to the brain. These include the spinocerebellar pathways (for proprioception) to the cerebellum and two pathways to the somatosensory cortex: the dorsal column pathway (for discriminative touch and proprioception) and the spinothalamic pathway (for pain, temperature, and coarse touch).
7. The descending tracts in the funiculi of the spinal cord belong to motor pathways that connect the brain to the body muscles. These include the pyramidal (corticospinal) tracts that control skilled movements and other fiber tracts that control subconscious or coarse movements.

DISORDERS OF THE CENTRAL NERVOUS SYSTEM (pp. 401–403)

Brain Dysfunction (pp. 401–402)

1. Blows to the head may cause concussions or brain contusions. These injuries can be aggravated by intracranial bleeding and swelling of the brain.
2. Cerebrovascular accidents (strokes) result when the blood vessels to the brain tissue either are blocked or hemorrhage, causing the tissue to die.
3. Alzheimer's disease is a degenerative brain disease affecting the cerebral cortex and hippocampus. Mental functions diminish progressively.

Spinal Cord Damage (pp. 402–403)

4. Transection or crushing of the spinal cord permanently eliminates voluntary movements and sensation inferior to the damage site.

THE CENTRAL NERVOUS SYSTEM THROUGHOUT LIFE (pp. 403–404)

1. Birth defects that involve the brain include anencephaly, spina bifida, and cerebral palsy. Spina bifida is usually associated with myelomeningocele and hydrocephalus.
2. The brain forms many neuronal connections during childhood, all based on early life experiences. Growth of the brain stops in young adulthood, and cognitive functions decline slowly with age.

REVIEW QUESTIONS

Multiple Choice/Matching Questions

1. Choose the proper response from the key for each of the following descriptions of various brain areas. Some letters will be used more than once.

Key: **(a)** cerebellum; **(b)** corpora quadrigemina; **(c)** corpus striatum; **(d)** corpus callosum; **(e)** hypothalamus; **(f)** medulla; **(g)** midbrain; **(h)** pons; **(i)** thalamus

____ **(1)** basal nuclei (basal ganglia) involved in motor activities; related to Huntington's disease

____ **(2)** region where there is a crossover of fibers of pyramidal tracts

____ **(3)** control of temperature, autonomic nervous system, hunger, and water balance

____ **(4)** houses the substantia nigra and red nucleus

____ **(5)** involved in visual and auditory reflexes; found in midbrain

____ **(6)** part of diencephalon with vital centers controlling heart rate, some aspects of emotion, and blood pressure

____ **(7)** all inputs to cerebral cortex must first synapse in one of its nuclei

____ **(8)** brain area that has folia and coordinates movements

____ **(9)** brain region that contains the cerebral aqueduct

____ **(10)** associated with fourth ventricle and contains nuclei of cranial nerves V–VII

____ **(11)** thick tract between the two cerebral hemispheres

2. A patient suffered a cerebral hemorrhage that damaged the postcentral gyrus in her right cerebral hemisphere. As a result, she (a) cannot voluntarily move her left arm or leg, (b) feels no sensation on the left side of her body, (c) feels no sensation on the right side of her body, (d) can no longer play the violin.
3. Destruction of the anterior horn cells of the spinal cord results in loss of (a) intelligence, (b) all sensation, (c) pain, (d) motor control.
4. For each of the following brain structures, write G (for *gray matter*) or W (for *white matter*) as appropriate.

____ **(1)** cortex of cerebellum

____ **(2)** pyramids

____ **(3)** internal capsule and corona radiata

____ **(4)** red nucleus

____ **(5)** medial lemniscus

____ **(6)** cranial nerve nuclei

____ **(7)** cerebellar peduncle

5. Which of the following areas is most likely to store visual memories? (a) visual association area, (b) Wernicke's area, (c) premotor cortex, (d) primary somatosensory cortex.

Short Answer Essay Questions

6. Make a diagram showing the five secondary brain vesicles of an embryo, and then list the basic adult brain regions derived from each.
7. (a) Make a rough sketch of a lateral view of the left cerebral hemisphere. (b) You may be thinking, "But I just can't draw!" So, name the hemisphere involved in most people's ability to draw. (c) On your drawing, locate the following areas, and identify their major functions: primary motor cortex, premotor cortex, somatosensory association area, primary visual area, prefrontal cortex, Broca's area.

8. (a) What is the basic function of the basal nuclei (basal ganglia)? (b) Which of these nuclei form the lentiform nuclei? (c) Which one arches over the thalamus?

9. Describe the role of the cerebellum in maintaining smooth, coordinated skeletal muscle activity.

10. (a) Where is the limbic system located? (b) What structures make up this system? (c) How is the limbic system important in behavior?

11. (a) Describe the location of the reticular formation in the brain. (b) What does RAS mean, and what is its function?

12. (a) List four ways in which the CNS is protected. (b) How is cerebrospinal fluid formed and drained? Describe its pathway within and around the brain.

13. (a) What are the superior and inferior boundaries of the spinal cord in the vertebral canal? (b) A correctly executed lumbar spinal tap could poke the cauda equina or filum terminale, but not the conus medullaris. Explain.

14. (a) Which of the ascending tracts in the spinal cord carry sensory impulses concerned with discriminative touch and pressure? (b) Which are concerned with proprioception only?

15. (a) In the spinothalamic pathway, where are the cell bodies of the first-order neurons located? Of the third-order neurons? (b) Trace the descending pathway that is concerned with skilled voluntary movements.

16. List a few changes in brain structure that occur with aging.

17. Define somatotopy, and give one example of it in the brain and one example in the spinal cord.

18. A brain surgeon removed a piece of a woman's skull and cut through all the meninges to reach the brain itself. Name all the layers that were cut, from skin to brain.

19. Gloria and Betty disagree with the claims about behavior in the A Closer Look box in this chapter. Still, they concede that structural differences exist between the brains of men and women, although they consider these differences to be minor and insignificant. What are some of these structural differences?

20. Sketch the spinal cord in cross section, and label the following structures (alternatively, you could cover up the labels in Figures 13.30b and 13.32, and then just find the structures): (a) lateral funiculus of white matter, (b) ventral root, (c) anterior (ventral) horn of gray matter, (d) central canal, (e) dorsal root ganglion, (f) dorsal column tracts: fasciculus gracilis and fasciculus cuneatus, (g) anterior corticospinal tract, (h) lateral corticospinal tract.

21. Describe the location and function of the periaqueductal gray matter.

22. Describe the location and function of the ventral stream of visual-processing areas in the cerebral cortex. Then explain why the dorsal and ventral visual streams are called the "where" and "what" streams, respectively.

CRITICAL REASONING + CLINICAL APPLICATION QUESTIONS

1. Kimberly learned that the basic design of the CNS is gray matter on the inside and white matter on the outside. When her friend L'Shawn said that there is gray matter on the outside of the cerebrum of the brain, Kim became confused. Clear up her confusion.

2. When Ralph had brain surgery to remove a small intracranial hematoma (blood mass), the operation was done under local anesthesia because he was allergic to general anesthesia. The surgeon removed a small part of the skull, and Ralph remained conscious. The operation went well, and when Ralph asked the surgeon to stimulate his (unharmed) postcentral gyrus mildly with an electrode, the surgeon did so. Indicate which of the following happened, and explain why: (a) Ralph was seized with uncontrollable rage. (b) He saw things that were not there. (c) He asked to see what was touching his hand, but nothing was. (d) He started to kick. (e) He heard his mother's voice from 30 years ago.

3. When their second child was born, Kiko and Taka were told their baby had spina bifida and that he would be kept in intensive care for a week, after which time a "shunt" would be put in. Also, immediately after the birth, an operation was performed on the infant's lower back. The parents were told that this operation went well but that their son would always be a "little weak in the ankles." Explain the statements in quotation marks in more informative and precise language.

4. Cesar, a brilliant computer analyst, was hit on the forehead by a falling rock while mountain climbing. It was soon obvious to his coworkers that his behavior had changed dramatically. He had been a smart dresser, but now he was unkempt. One morning, he was discovered defecating into a wastebasket. When he started revealing the most secret office gossip to everyone, his supervisor ordered Cesar to report to the company doctor. What region of Cesar's brain was affected by the cranial blow?

5. One war veteran was quadriplegic, and another was paraplegic. Explain the relative locations of their injuries.

6. Every time Spike went to a boxing match, he screamed for a knockout. Then his friend Rudy explained what happens when someone gets knocked out. What is the explanation?

A.D.A.M.® INTERACTIVE ANATOMY QUESTIONS

1. Link: Atlas; System: Nervous; Spinal cord vessels and meninges. (a) Name the three meninges that protect the spinal cord and brain. (b) At their medial end, spinal nerves attach to the spinal cord through which structures?

2. Link: Atlas; Region: Head and Neck; Sagittal Section of Brain. (a) Name the gland that sits just inferior to the splenium of the corpus callosum. (b) Commisures run between the two cerebral hemispheres, allowing them to function as a whole. Name the three parts of the largest commisure, the corpus callosum.

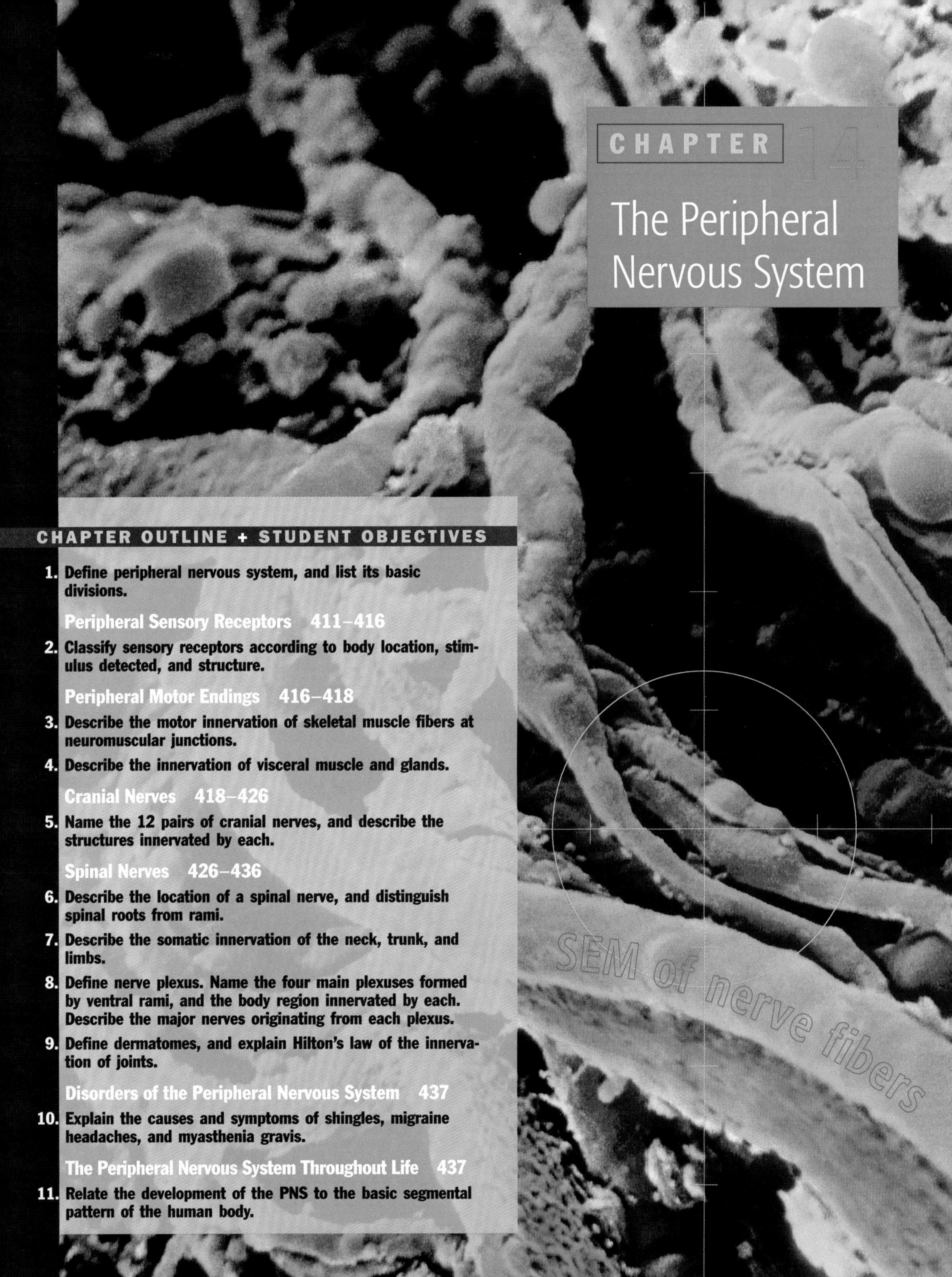

CHAPTER 14

The Peripheral Nervous System

CHAPTER OUTLINE + STUDENT OBJECTIVES

1. Define peripheral nervous system, and list its basic divisions.

Peripheral Sensory Receptors 411–416

2. Classify sensory receptors according to body location, stimulus detected, and structure.

Peripheral Motor Endings 416–418

3. Describe the motor innervation of skeletal muscle fibers at neuromuscular junctions.
4. Describe the innervation of visceral muscle and glands.

Cranial Nerves 418–426

5. Name the 12 pairs of cranial nerves, and describe the structures innervated by each.

Spinal Nerves 426–436

6. Describe the location of a spinal nerve, and distinguish spinal roots from rami.
7. Describe the somatic innervation of the neck, trunk, and limbs.
8. Define nerve plexus. Name the four main plexuses formed by ventral rami, and the body region innervated by each. Describe the major nerves originating from each plexus.
9. Define dermatomes, and explain Hilton's law of the innervation of joints.

Disorders of the Peripheral Nervous System 437

10. Explain the causes and symptoms of shingles, migraine headaches, and myasthenia gravis.

The Peripheral Nervous System Throughout Life 437

11. Relate the development of the PNS to the basic segmental pattern of the human body.

THE HUMAN BRAIN, FOR ALL ITS SOPHISTICATION, would be useless without its sensory and motor connections to the outside world. Our very sanity depends on a continual flow of sensory information from the environment. When blindfolded volunteers were suspended in a tank of warm water (a sensory-deprivation tank), they began to hallucinate: One saw pink and purple elephants, another heard a singing chorus, and others had taste hallucinations. Our sense of well-being also depends on our ability to carry out motor instructions sent from the CNS. For example, many victims of spinal cord injuries experience despair at being unable to move or take care of their own needs.

The **peripheral nervous system (PNS)**—the nervous system structures outside the brain and spinal cord—provides these vital links to the real world. Its nerves thread through almost every part of the body, allowing the CNS to receive information and to take action. Figure 14.1 reviews the *functional* components of the PNS as presented in Chapter 12 (pp. 336–338). Recall that the sensory inputs and motor outputs carried by the PNS are categorized as either *somatic* (outer body) or *visceral* (visceral organs), and that sensory inputs are also classified as either *general* (widespread) or *special* (localized). The general visceral motor part of the PNS is the autonomic nervous system, which has *parasympathetic* and *sympathetic* divisions (see the right side of Figure 14.1 and Chapter 15). Additionally, note that "branchial motor" refers to the motor innervation of the pharyngeal arch muscles.

Figure 14.2 shows the basic *structural* components of the PNS, which are the following:

1. The **sensory receptors.** Sensory receptors pick up stimuli (environmental changes) from inside and outside the body, and then initiate impulses in sensory axons.
2. The **motor endings.** The motor endings are the axon terminals of motor neurons that innervate the effectors (muscle fibers and glands).
3. The **nerves** and **ganglia.** As defined in Chapter 12, nerves are bundles of peripheral axons, and ganglia are clusters of peripheral cell bodies, such as the sensory cell bodies depicted in Figure 14.2. Even though most nerves contain both sensory and motor axons and are called *mixed nerves,* a few nerves are purely sensory or purely motor in function.

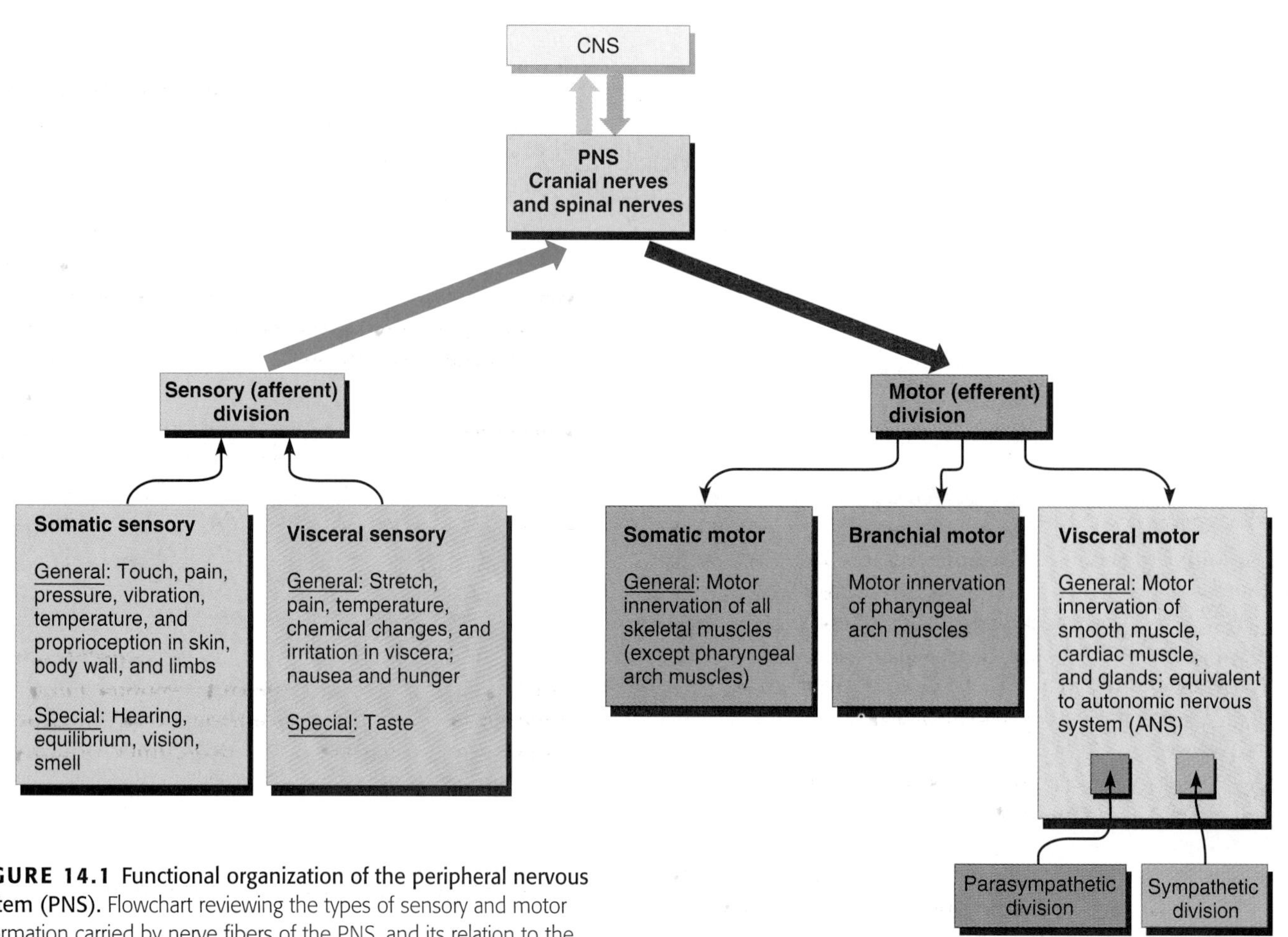

FIGURE 14.1 Functional organization of the peripheral nervous system (PNS). Flowchart reviewing the types of sensory and motor information carried by nerve fibers of the PNS, and its relation to the central nervous system (CNS). Arrows indicate the direction in which the sensory and motor information travel.

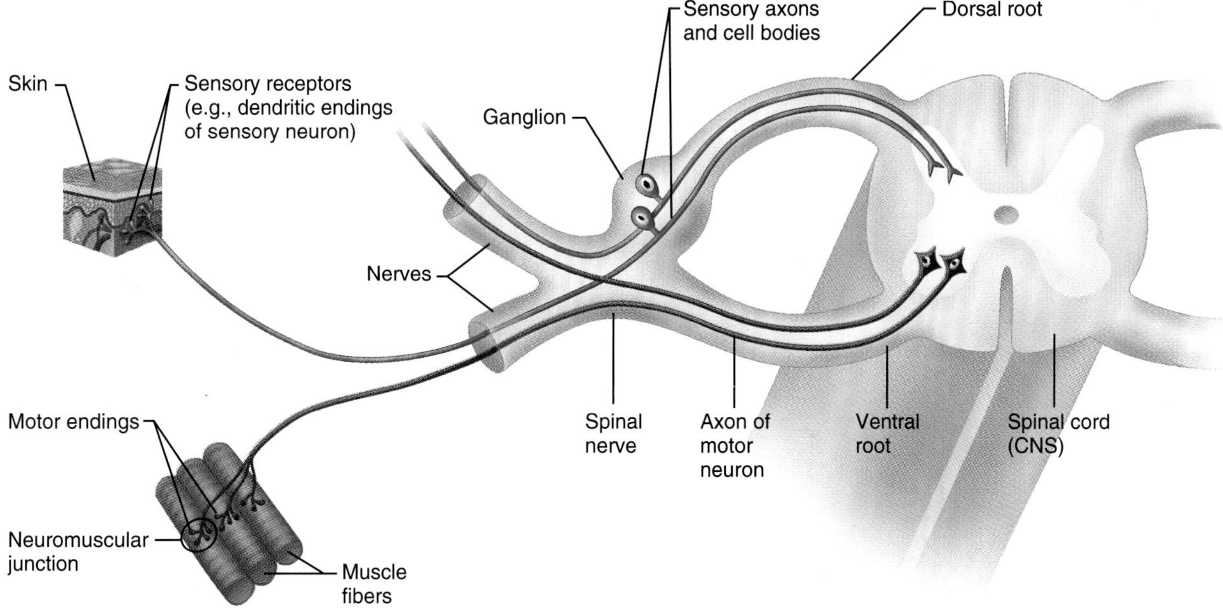

FIGURE 14.2 Basic anatomical scheme of the PNS in the region of a spinal nerve. Note the sensory receptors and motor endings at left, the ganglion, and the sensory and motor axons that constitute nerves.

We begin this chapter by describing in turn the peripheral sensory receptors and the peripheral motor endings of the PNS. We then discuss the nerves of the body, starting with the *cranial nerves* (the nerves attached to the brain) and ending with the *spinal nerves* (the nerves attached to the spinal cord). Although the visceral part of the PNS is mentioned, this chapter focuses on somatic functions. A more complete consideration of the visceral nervous system is deferred to Chapter 15.

PERIPHERAL SENSORY RECEPTORS

Peripheral sensory receptors are structures that pick up sensory stimuli and then initiate signals in the sensory axons. Most receptors fit into two main categories: (1) *dendritic endings of sensory neurons,* such as the one depicted in Figure 14.2; and (2) complete *receptor cells,* which are specialized epithelial cells or small neurons that transfer sensory information to sensory neurons. Dendritic endings of sensory neurons monitor most types of general sensory information (such as touch, pain, pressure, temperature, and proprioception), whereas specialized receptor cells monitor most types of special sensory information (taste, vision, hearing, and equilibrium).

Sensory receptors may also be classified according to (a) location, (b) the type of stimulus they detect, and (c) their structure. We next examine these different classification schemes in turn.

Classification by Location

Sensory receptors are divided into three classes based either on their location in the body or the location of the stimuli to which they respond:

1. **Exteroceptors** (ek″ster-o-sep′torz) are sensitive to stimuli arising outside the body. Accordingly, most exteroceptors are located at or near the body surface and include receptors for touch, pressure, pain, and temperature in the skin and most receptors of the special sense organs.
2. **Interoceptors** (in″ter-o-sep′torz), also called *visceroceptors,* receive stimuli from the internal viscera, such as the digestive tube, bladder, and lungs. Different interoceptors monitor a variety of stimuli, including changes in chemical concentration, taste stimuli, the stretching of tissues, and temperature. Their activation causes us to feel visceral pain, nausea, hunger, or fullness.
3. **Proprioceptors** are located in the musculoskeletal organs such as skeletal muscles, tendons, joints, and ligaments. Recall that proprioceptors monitor the degree of stretch of these locomotory organs and send input on body movements to the CNS (see p. 336).

Classification by Stimulus Detected

A second way to classify sensory receptors is by the kinds of stimuli that most readily activate them. **Mechanoreceptors** respond to mechanical forces such

Table 14.1 General Sensory Receptors Classified by Structure and Function

Anatomical Class (structure)	Illustration	Functional Class According to Location (L) and Stimulus Type (S)	Body Location
UNENCAPSULATED Free dendritic nerve endings of sensory neurons	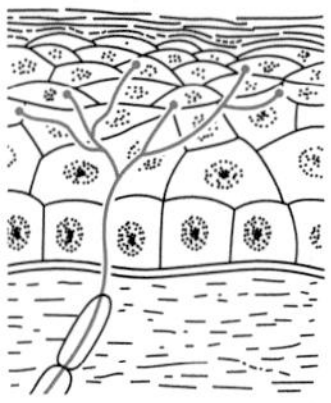	L: Exteroceptors, interoceptors, and proprioceptors S: Nociceptors (pain), thermoreceptors (heat and cold), possibly mechanoreceptors (pressure)	Most body tissues; densest in connective tissues (ligaments, tendons, dermis, joint capsules, periostea) and epithelia (epidermis, cornea, mucosae, and glands)
Modified free dendritic endings: Merkel discs	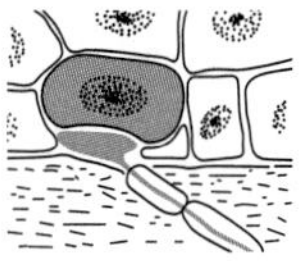	L: Exteroceptors S: Mechanoreceptors (light pressure)	Basal layer of epidermis
Root hair plexuses	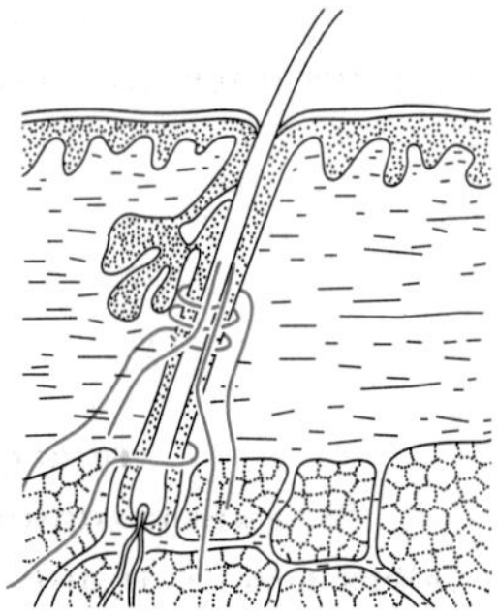	L: Exteroceptors S: Mechanoreceptors (hair deflection)	In and surrounding hair follicles
ENCAPSULATED Meissner's corpuscles	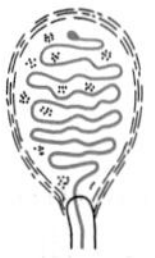	L: Exteroceptors S: Mechanoreceptors (light pressure, discriminative touch, vibration of low frequency)	Dermal papillae of hairless skin, particularly nipples, external genitalia, fingertips, eyelids
Krause's end bulbs		L: Exteroceptors S: Mechanoreceptors (probably modified Meissner's corpuscles)	Connective tissue of mucosae (mouth, conjunctiva of eye) and of hairless skin near body openings (lips)
Pacinian corpuscles		L: Exteroceptors, interoceptors, and some proprioceptors S: Mechanoreceptors (deep pressure, stretch, vibration of high frequency); rapidly adapting	Subcutaneous tissue; periostea, mesentery, tendons, ligaments, joint capsules, most abundant on fingers, soles of feet, external genitalia, nipples
Ruffini's corpuscles	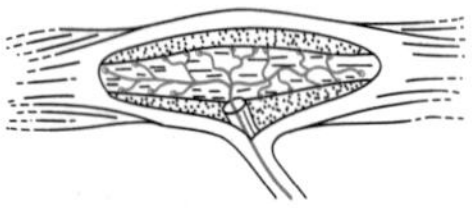	L: Exteroceptors and proprioceptors S: Mechanoreceptors (deep pressure and stretch); slowly adapting	Deep in dermis, hypodermis, and joint capsules

Anatomical Class (structure)	Illustration	Functional Class According to Location (L) and Stimulus Type (S)	Body Location
PROPRIOCEPTORS: Muscle spindles	Intrafusal fibers	L: Proprioceptors S: Mechanoreceptors (muscle stretch)	Skeletal muscles, particularly those of the extremities
Golgi tendon organs		L: Proprioceptors S: Mechanoreceptors (tendon stetch)	Tendons
Joint kinesthetic receptors (Pacinian and Ruffini corpuscles, bare dendritic endings, and receptors resembling Golgi tendon organs)		L: Proprioceptors S: Mechanoreceptors and nociceptors	Joint capsules of synovial joints

as touch, pressure, stretch, vibrations, and itch; **thermoreceptors** respond to temperature changes; **chemoreceptors** respond to chemicals in solution (such as molecules tasted or smelled) and to changes in blood chemistry; **photoreceptors** in the eye respond to light; and **nociceptors** (no″se-sep′torz) respond to harmful stimuli that result in pain (*noci* = harm).

Classification by Structure

The third way to classify sensory receptors is by their structure. The special senses are discussed in Chapter 16, so here we consider only the **general sensory receptors.** All these widely distributed receptors are dendritic endings of sensory neurons that monitor touch, pressure, vibration, stretch, pain, temperature and proprioception. Structurally, sensory receptors are divided into two broad groups: (1) *free (naked)* dendritic endings and (2) *encapsulated* dendritic endings surrounded by a capsule of connective tissue. These general sensory receptors are summarized in Table 14.1.

Before we consider these two groups of receptors, it is important to note that there is no perfect "one receptor–one function" relationship. Instead, one receptor type can respond to several different kinds of stimuli, and different receptor types can respond to similar stimuli.

Free Dendritic Endings

Free dendritic endings *(naked dendritic endings)* of sensory fibers invade almost all tissues of the body but are particularly abundant in epithelia and in the connective tissue that underlies epithelia (Figure 14.3). These receptors respond chiefly to pain and temperature (though some respond to tissue movements caused by pressure). One way to characterize free endings functionally is to say that they monitor the *affective* senses, those to which we have an emotional response—and we certainly respond emotionally to pain!

Certain free dendritic endings contribute to **Merkel discs** *(tactile menisci),* which lie in the epidermis of the skin. Each such disc consists of a disc-shaped epithelial cell (a Merkel cell; see p. 114) innervated by a dendrite. Merkel discs seem to be *slowly adapting* receptors for light touch; that is, they continue to respond and send out action potentials even after a long period of continual stimulation. **Root hair plexuses,** which are free dendritic endings that wrap around hair follicles, are receptors for light touch that monitor the bending of hairs. Unlike Merkel discs, they are *rapidly adapting,* meaning the sensation disappears quickly even if the stimulus is maintained. The tickle of a mosquito landing on your forearm is mediated by root hair plexuses.

Advances in Understanding The skin's *itch receptor* is a newly discovered receptor consisting of free dendritic endings. Located in the dermis, this receptor escaped detection until 1997 because of its thin diameter. Its discovery startled those scientists who had not realized that itch is a distinct sense, but had believed it to be a mild form of pain. ✳

Encapsulated Dendritic Endings

All **encapsulated dendritic endings** consist of one or more end fibers of sensory neurons enclosed in a capsule of connective tissue. All seem to be mechanoreceptors, and their capsules serve either to amplify the stimulus or to filter out the wrong types of stimuli. Encapsulated receptors vary widely in shape, size, and distribution in the body. The main types are *Meissner's corpuscles, Krause's end bulbs, Pacinian* and *Ruffini's corpuscles* (see Figure 14.3), and *proprioceptors.*

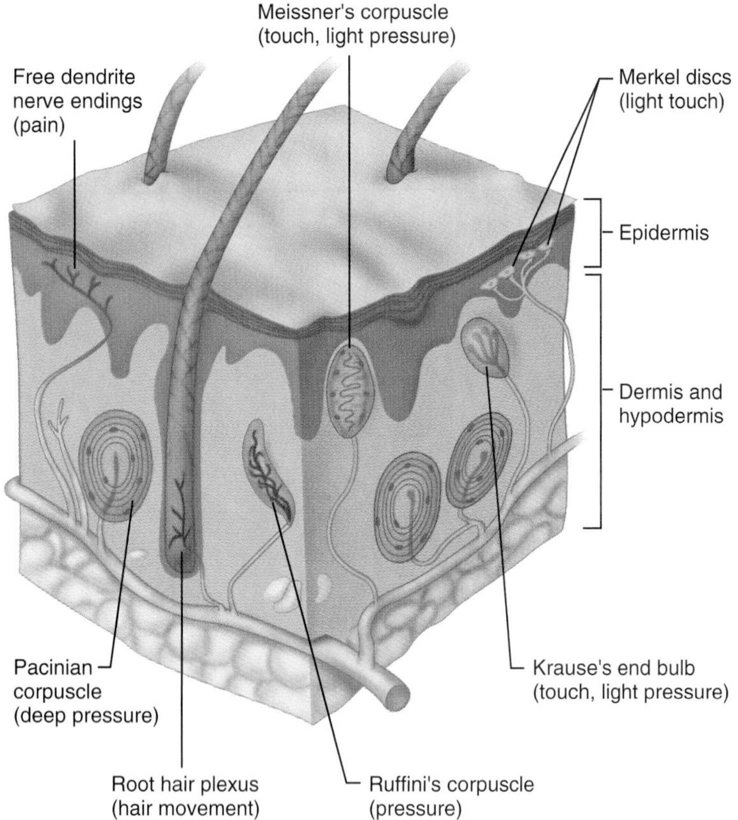

FIGURE 14.3 Structure of free and encapsulated general sensory receptors. Receptors of both structural groups are shown in the skin, the organ with the greatest variety of receptor types. Krause's bulbs and Meissner's corpuscles are not common in hairy skin, but are included here for illustrative purposes.

Meissner's Corpuscles In a **Meissner's corpuscle** *(tactile corpuscle),* a few spiraling dendrites are surrounded by Schwann cells, which in turn are surrounded by an egg-shaped capsule of connective tissue. These corpuscles, which occur in the dermal papillae beneath the epidermis, are rapidly adapting receptors for fine, discriminative touch. They mainly occur in sensitive and hairless areas of the skin, such as the soles, palms, fingertips, nipples, and lips. Apparently, Meissner's corpuscles perform the same "light touch" function in hairless skin that root hair plexuses perform in hairy skin.

Krause's End Bulbs Once thought to be thermoreceptors for cold, **Krause's end bulbs** are now known to be a type of Meissner's corpuscle for fine touch. Whereas true Meissner's corpuscles occupy the skin, Krause's end bulbs occur mostly in mucous membranes—in the conjunctiva of the eye and the lining of the mouth, for example.

Pacinian Corpuscles Scattered throughout the deep connective tissues of the body are **Pacinian corpuscles** *(large lamellated corpuscles).* They occur, for example, in the hypodermis deep to the skin. Although they are sensitive to deep pressure, they respond only to the initial application of that pressure before they tire and stop firing. Therefore, Pacinian corpuscles are rapidly adapting receptors that are best suited to monitor *vibration,* an on-off pressure stimulus. These corpuscles are large enough to be visible to the unaided eye—about 0.5–1 mm wide and 1–2 mm long. In section, a Pacinian corpuscle resembles a cut onion: Its single dendrite is surrounded by up to 60 layers of flattened Schwann cells, which in turn are covered by a capsule of connective tissue.

Ruffini's Corpuscles Located in the dermis and elsewhere are **Ruffini's corpuscles,** which contain an array of dendritic endings enclosed in a thin, flattened capsule. Like Pacinian corpuscles, they respond to pressure and touch. However, they adapt slowly and thus can monitor *continuous pressure* placed on the skin.

Proprioceptors Virtually all proprioceptors are encapsulated dendritic endings that monitor stretch in the locomotory organs. Proprioceptors include *muscle spindles, Golgi tendon organs,* and *joint kinesthetic receptors.*

Muscle spindles *(neuromuscular spindles)* measure the changing length of a muscle as that muscle contracts and is stretched back to its original length (Figure 14.4). An average muscle contains some 50 to 100 muscle spindles, which are embedded in the perimysium between the fascicles (see pp. 246–247). Structurally, each spindle contains several modified skeletal muscle fibers called **intrafusal muscle fibers** (*intra* = within; *fusal* = the spindle) surrounded by a connective tissue capsule. Intrafusal muscle fibers have fewer striations than do the ordinary

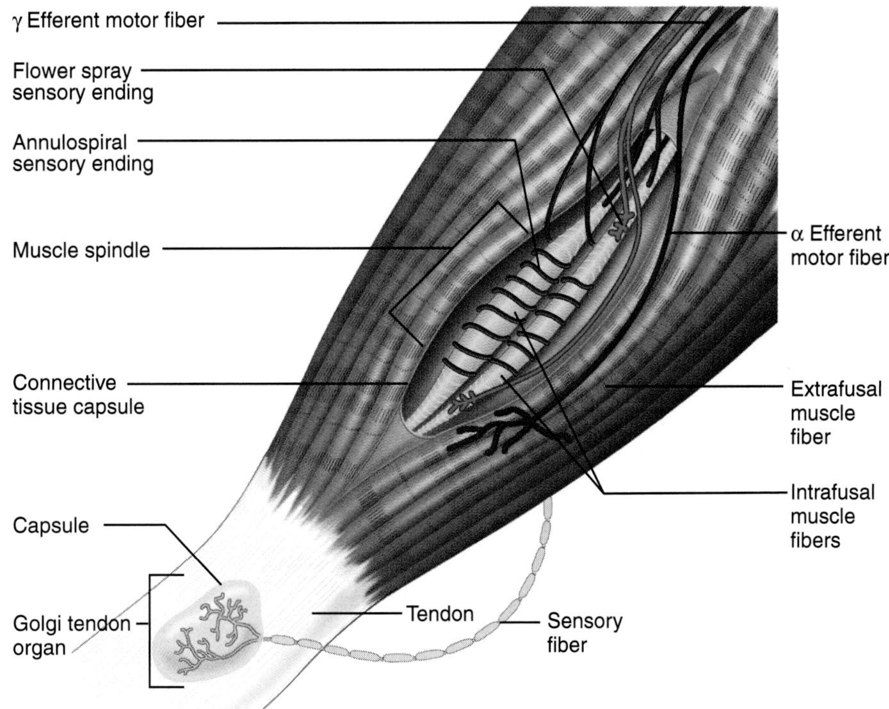

FIGURE 14.4 Structure of proprioceptors. Two types of these encapsulated receptors are depicted: muscle spindles (above) and Golgi tendon organs (below).

muscle cells outside the spindles, which in this context are called **extrafusal muscle fibers** ("outside the spindle"). The intrafusal fibers are innervated by the dendrites of several sensory neurons. Some of these sensory dendrites twirl around the middle of the intrafusal fibers as **annulospiral** ("ring-spiral") **sensory endings,** whereas others supply the ends of the intrafusal fibers in **flower spray sensory endings** that resemble tiny bouquets.

Recall that muscles are stretched by the contraction of antagonist muscles and also by the movements that occur when we begin to lose our balance (see pp. 248–250). The muscle spindles sense this lengthening in the following way: When a whole muscle is stretched, its intrafusal fibers are also stretched. This stretching activates the sensory neurons that innervate the spindle, causing them to signal the spinal cord and brain. The CNS then activates spinal motor neurons called α **(alpha) efferent neurons** (see Figure 14.4) that cause the entire muscle (extrafusal fibers) to generate contractile force and resist further stretching. This response can take the form of a monosynaptic spinal reflex that rapidly prevents a fall; alternatively, the response can be controlled by the cerebellum, in which case it is involved in the regulation of *muscle tone,* the steady force generated by noncontracting muscles to resist stretching.

Also innervating the intrafusal fibers of the muscle spindle are the axons of spinal *motor* neurons called γ **(gamma) efferent fibers** (see Figure 14.4). They let the brain preset the sensitivity of the spindle to stretch; that is, when the brain signals the gamma motor neurons to fire, the intrafusal muscle fibers contract and become tense so that very little stretch is needed to stimulate the sensory dendrites, making the spindles highly sensitive to applied stretch. Gamma motor neurons are most active when balance reflexes must be razor sharp, as required when walking a tightrope.

Golgi (gol′je) **tendon organs** are proprioceptors located near the muscle-tendon junction, where they monitor tension within tendons (see Figure 14.4). Each consists of an encapsulated bundle of tendon fibers (collagen fibers) within which sensory dendrites are intertwined. When a contracting muscle pulls on its tendon, Golgi tendon organs are stimulated, and their sensory neurons send this information to the cerebellum. These receptors also induce a spinal reflex that both relaxes the contracting muscle and activates its antagonist. This relaxation reflex is important in motor activities that involve rapid alternation between flexion and extension, such as running.

Joint kinesthetic receptors (kin″es-thet′ik; "movement feeling") are proprioceptors that monitor stretch in the synovial joints. Specifically, they are sensory dendritic endings within the joint capsules. Four types of joint kinesthetic receptors are present within each joint capsule:

1. *Pacinian corpuscles:* These rapidly adapting stretch receptors are ideal for measuring acceleration and rapid movement of the joints.
2. *Ruffini's corpuscles:* These slowly adapting stretch receptors are ideal for measuring the positions of

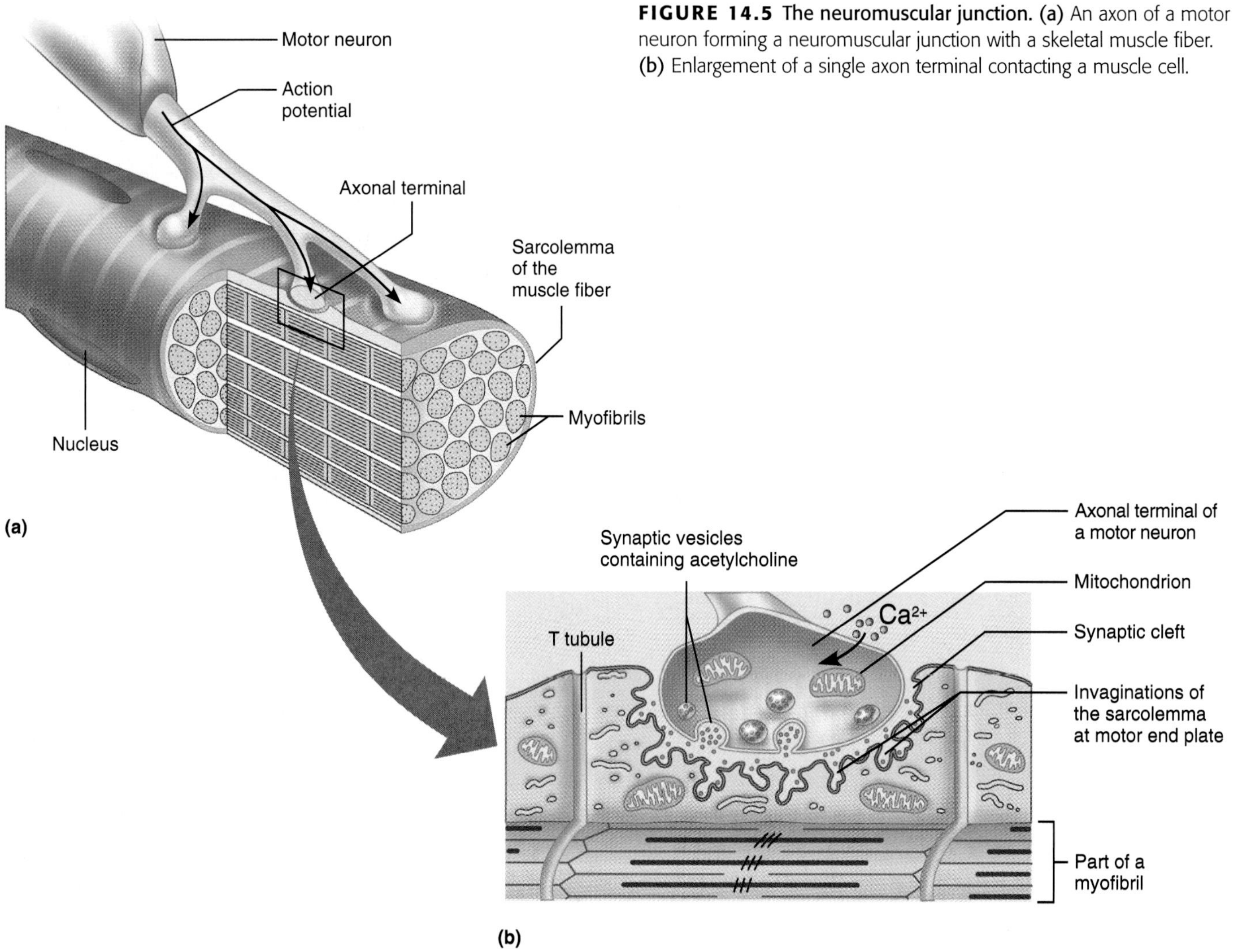

FIGURE 14.5 The neuromuscular junction. (a) An axon of a motor neuron forming a neuromuscular junction with a skeletal muscle fiber. **(b)** Enlargement of a single axon terminal contacting a muscle cell.

nonmoving joints and the stretch of joints that undergo slow, sustained movements.

3. *Free dendritic endings:* May be pain receptors.
4. Receptors resembling *Golgi tendon organs:* Their function in joints is not known.

Joint receptors, like the other two classes of proprioceptors, send information on body movements to the cerebellum and cerebrum, as well as to spinal reflex arcs.

PERIPHERAL MOTOR ENDINGS

Now that we have discussed the sensory nerve endings that receive stimuli, we turn to the motor endings that activate the effectors (muscles and glands) of the body. We first examine the innervation of skeletal muscle, then the innervation of visceral muscle and glands.

Innervation of Skeletal Muscle

Motor axons innervate skeletal muscle fibers at junctions called **neuromuscular junctions,** or *motor end plates* (Figures 14.2 and 14.5). A single neuromuscular junction is associated with each muscle fiber. These junctions are similar to the synapses between neurons. The neural part of the junction is a cluster of typical axon terminals separated from the plasma membrane (sarcolemma) of the underlying muscle cell by a synaptic cleft. As in typical synapses, the axon terminals contain synaptic vesicles that release neurotransmitter when a nerve impulse reaches the terminals. The neurotransmitter—acetylcholine—diffuses across the synaptic cleft and binds to receptor molecules on the sarcolemma, where it induces an impulse that signals the muscle cell to contract.

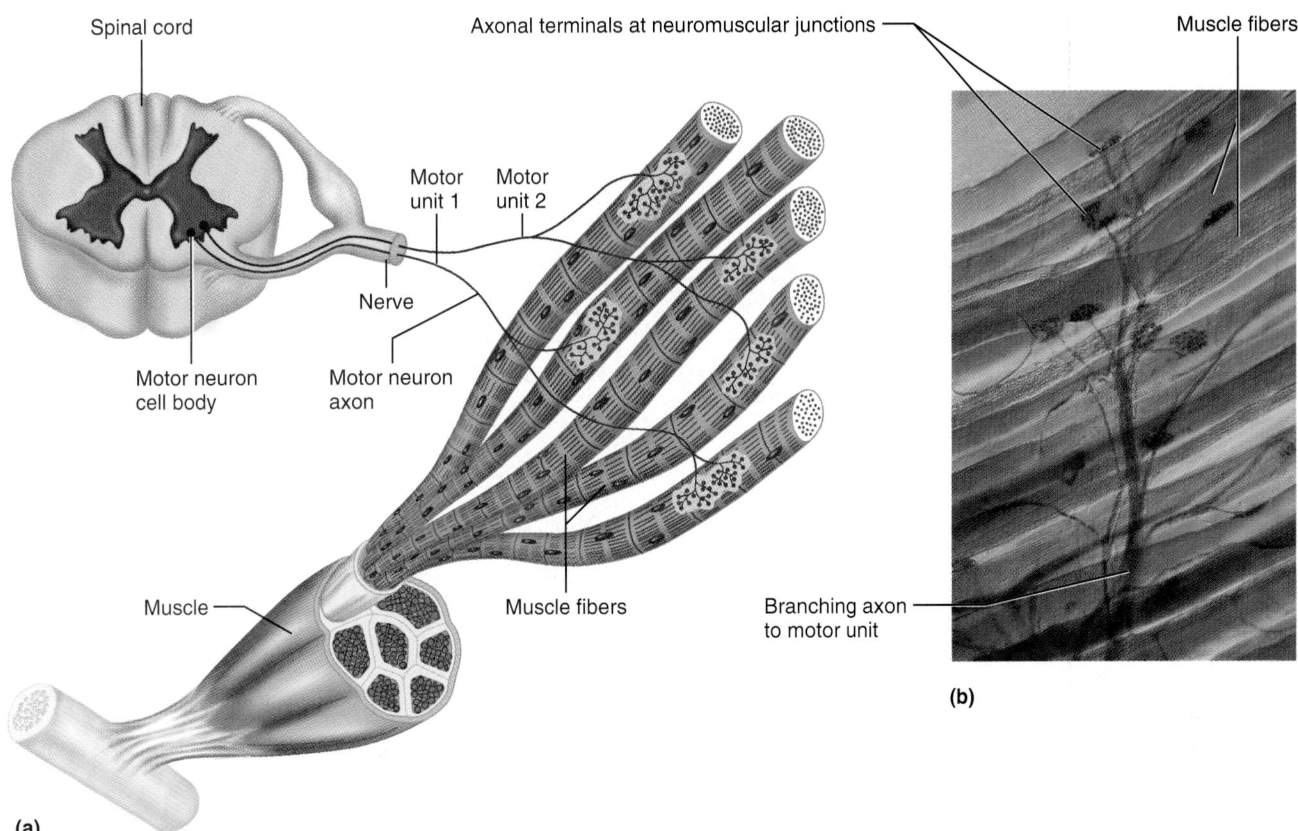

FIGURE 14.6 Motor units. Each motor unit consists of one motor neuron and all the muscle fibers it innervates. **(a)** Drawing of two motor units. The muscle fibers of a motor unit are distributed throughout the entire skeletal muscle and are scattered more widely than is depicted. **(b)** Photomicrograph of a part of a motor unit (80 ×). Note the branching axon giving rise to clusters of axon terminals at neuromuscular junctions.

Although they resemble synapses, neuromuscular junctions have several unique features. Each axon terminal lies in a trough-like depression of the sarcolemma, which in turn shows groove-like invaginations (Figure 14.5b). The invaginations and the synaptic cleft contain a basal lamina (not illustrated), a structure never seen in the synapses between neurons. This basal lamina contains the enzyme acetylcholinesterase (as″ĕ-til-ko″lin-es′ter-ās), which breaks down acetylcholine immediately after the neurotransmitter signals a single contraction. This assures that each nerve impulse in the motor axon produces just one twitch of the muscle cell, preventing any undesirable additional twitches that would result if acetylcholine were to linger in the synaptic cleft.

Each motor axon branches to innervate a number of muscle fibers within a skeletal muscle. A motor neuron and all the muscle fibers it innervates are called a **motor unit** (Figure 14.6). When a motor neuron fires, all the skeletal muscle cells in the motor unit contract together. Although the average number of muscle fibers in a motor unit is 150, a motor unit may contain as many as several hundred or as few as four muscle fibers. Muscles that require very fine control (such as the muscles moving the fingers and eyes) have few muscle fibers per motor unit, whereas bulky, weight-bearing muscles, whose movements are less precise (such as the hip muscles), have many muscle fibers per motor unit. The muscle fibers of a single motor unit are not clustered together but spread throughout the muscle. As a result, stimulation of a single motor unit causes a weak contraction of the entire muscle.

Innervation of Visceral Muscle and Glands

The contacts between visceral motor endings and the visceral effectors are much simpler than the elaborate neuromuscular junctions present on skeletal muscle. Near the smooth muscle or gland cells it innervates, a visceral motor axon swells into a row of knobs (varicosities) resembling beads on a necklace. These varicosities are the presynaptic terminals, which contain synaptic vesicles filled with neurotransmitter. Some of these axon terminals form shallow indentations on the membrane of the effector cell, but many axon terminals remain a considerable distance from any cell. Because it takes time for neurotransmitters to diffuse across these wide synaptic clefts, visceral motor responses tend to be slower than somatic motor reflexes.

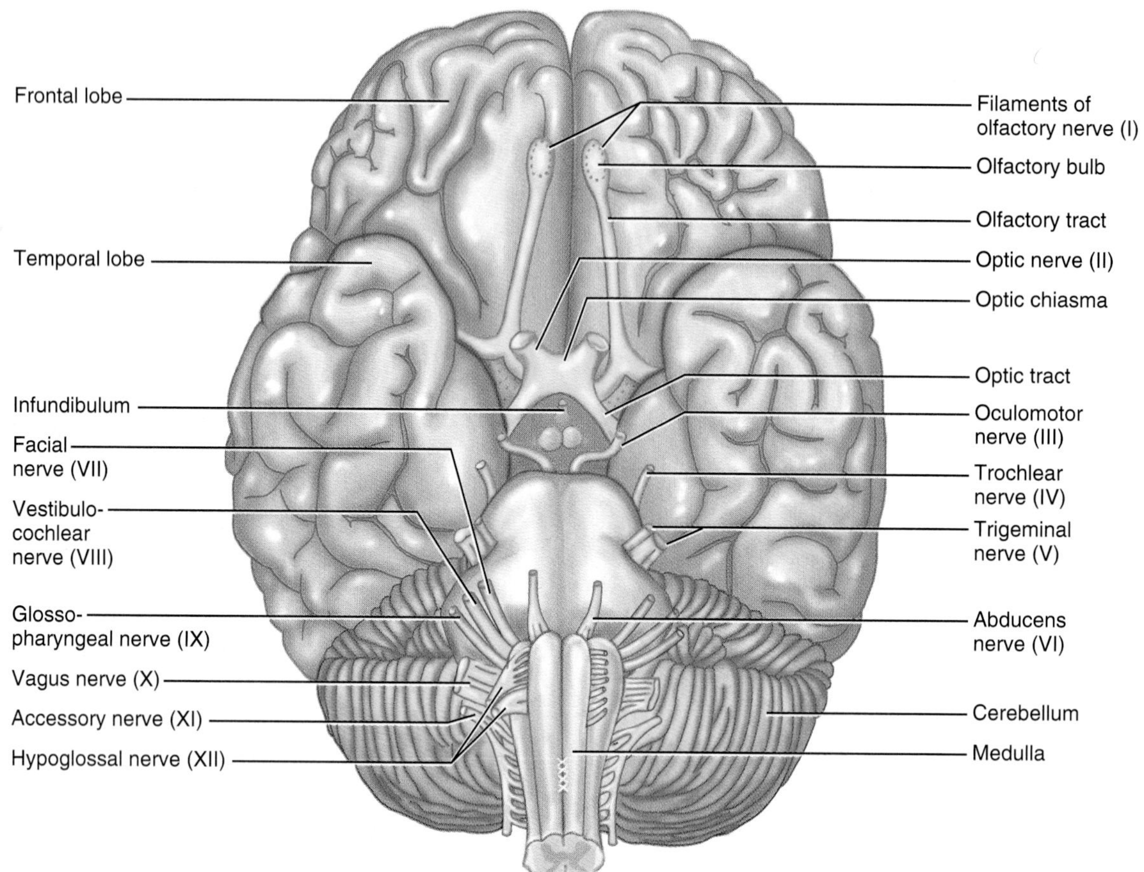

FIGURE 14.7 The 12 pairs of cranial nerves. Note that almost all of the cranial nerves attach to the *ventral* side of the brain.

The motor innervation of cardiac muscle cells resembles that of smooth muscle fibers and glands. However, the axon terminals are of a uniform diameter and do not include varicosities at the sites where they release their neurotransmitters.

CRANIAL NERVES

The rest of this chapter is devoted to the specific nerves of the body, beginning with the 12 pairs of **cranial nerves** that attach to the brain and pass through various foramina in the skull (Figure 14.7). These nerves are numbered from I through XII in a rostral to caudal direction. The first two pairs attach to the forebrain, the rest to the brain stem. Except for the vagus nerve (X), which extends into the abdomen, the cranial nerves serve only head and neck structures.

The following discussion introduces the cranial nerves, and is summarized in Table 14.2. For a more detailed presentation of the individual cranial nerves, see Table 14.3.

- **I. Olfactory.** These are the sensory nerves of smell.
- **II. Optic.** Because it develops as an outgrowth of the brain, this sensory nerve of vision is not a true nerve at all. It is more correctly called a brain tract.
- **III. Oculomotor.** The name *oculomotor* means "eye mover." This nerve innervates four of the *extrinsic eye muscles*—muscles that move the eyeball in the orbit.
- **IV. Trochlear.** The name *trochlear* means "pulley." This nerve innervates an extrinsic eye muscle that hooks through a pulley-shaped ligament in the orbit.
- **V. Trigeminal.** The name *trigeminal* means "threefold," which refers to this nerve's three major branches. The trigeminal nerve provides sensory innervation to the face and motor innervation to the chewing muscles.
- **VI. Abducens.** This nerve was so named because it innervates the muscle that *abducts* the eyeball (turns the eye laterally).
- **VII. Facial.** This nerve innervates the muscles of facial expression and other structures.
- **VIII. Vestibulocochlear.** This sensory nerve of hearing and equilibrium was once called the *auditory nerve*.

Table 14.2 Cranial Nerves Summarized According to Functional Groups

Cranial Nerve	Name	Sensory Function	Motor Function	Parasympathetic Autonomic Fibers
PURELY SENSORY:				
I	Olfactory	Smell	—	–
II	Optic	Vision	—	–
VIII	Vestibulocochlear	Hearing and equilibrium	—	–
PRIMARILY OR EXCLUSIVELY MOTOR:				
III	Oculomotor	—[1]	Somatic motor	+
IV	Trochlear	—[1]	Somatic motor	–
VI	Abducens	—[1]	Somatic motor	–
XII	Hypoglossal	—	Somatic motor	–
MIXED (MOTOR AND SENSORY):				
V	Trigeminal	General sensation	Branchial motor	–
VII	Facial	General sensation and taste	Branchial motor	+
IX	Glossopharyngeal	General sensation and taste	Branchial motor	+
X	Vagus	General sensation and taste	Branchial motor	+
XI	Accessory[2]	—	Branchial motor	–

[1]These motor nerves do contain proprioceptive (sensory) fibers from the eye muscles they innervate.
[2] The accessory nerve, although exclusively motor, is related to the vagus (X), so it is placed with the vagus in this table.

IX. Glossopharyngeal. The name *glossopharyngeal* means "tongue and pharynx," structures that this nerve helps to innervate.

X. Vagus. The name *vagus* means "vagabond" or "wanderer." This nerve "wanders" beyond the head into the thorax and abdomen.

XI. Accessory. This nerve can be considered an accessory part of the vagus nerve, and we will include it with the vagus in this introductory discussion. It was once called the *spinal accessory nerve.*

XII. Hypoglossal. The name *hypoglossal* means "below the tongue." This nerve runs inferior to the tongue and innervates the tongue muscles.

The following saying can help you remember the first letters of the names of the 12 cranial nerves in their proper order: "**Oh**, **Oh**, **Oh**, **to** **t**ouch **and** **f**eel **v**ery **g**ood **v**elvet, **ah**!"

The cranial nerves contain the sensory and motor nerve fibers that innervate the head. The cell bodies of the *sensory* neurons lie either in receptor organs (e.g., the nose for smell, or the eye for vision) or within **cranial sensory ganglia,** which lie along some cranial nerves (V, VII–X) just external to the brain. The cranial sensory ganglia are directly comparable to the dorsal root ganglia on the spinal nerves (see p. 395). The cell bodies of most cranial *motor* neurons occur in cranial nerve nuclei in the ventral gray matter of the brain stem—just as cell bodies of spinal motor neurons occur in the ventral gray matter of the spinal cord.

Based on the types of sensory motor fibers they contain, the 12 cranial nerves can be divided into three functional groups (Table 14.2):

1. **Purely sensory nerves** (I, II, VIII) that contain *special somatic sensory* fibers for smell (I), vision (II), and hearing and equilibrium (VIII).
2. **Primarily or exclusively motor nerves** (III, IV, VI, XII) that contain the *general somatic motor* fibers to skeletal muscles of the eye and tongue. (Cranial nerve XI, the accessory nerve, is also exclusively motor, but contains *branchial* motor fibers and is combined with the vagus nerve in the next group.)
3. **Mixed (motor and sensory) nerves** (V, VII, IX, and the X/XI combination). These mixed nerves supply sensory innervation to the face (through *general somatic sensory* fibers) and to the mouth and viscera *(general visceral sensory),* including the taste buds for the sense of taste *(special visceral sensory).* These nerves also innervate all pharyngeal arch muscles *(branchial motor),* such as chewing muscles and the muscles of facial expression.

Overview continues on p. 426

Table 14.3 Cranial Nerves

I The Olfactory (ol-fak′to-re) Nerves

Origin and course: Olfactory nerve fibers arise from olfactory receptor cells located in olfactory epithelium of nasal cavity, then group into bundles called filaments and pass through cribriform plate of ethmoid bone to synapse in olfactory bulb; fibers of olfactory bulb neurons extend posteriorly as olfactory tract, which runs beneath frontal lobe to enter cerebral hemispheres and terminates in primary olfactory cortex; see also Figure 16.3

Function: Purely sensory; carry afferent impulses for sense of smell

CLINICAL APPLICATIONS

ANOSMIA Fracture of ethmoid bone or lesions of olfactory fibers may result in partial or total loss of smell, a condition known as *anosmia* (an-oz′me-ah).

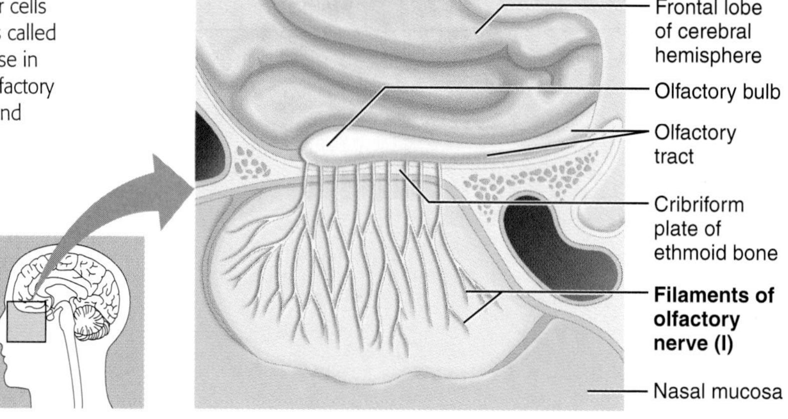

II The Optic Nerves

Origin and course: Fibers arise from retina of eye to form optic nerve, which passes through optic foramen of orbit; the optic nerves converge to form the optic chiasma (ki-az′mah) where partial crossover of fibers occur, continue on as optic tracts, enter thalamus, and synapse there; thalamic fibers run (as the optic radiation) to occipital (visual) cortex, where visual interpretation occurs; see also Figure 16.15

Function: Purely sensory; carry afferent impulses for vision

CLINICAL APPLICATIONS

OPTIC NERVE DAMAGE Damage to an optic nerve results in blindness in the eye served by the nerve; damage to visual pathway distal to optic chiasma results in partial visual losses; visual defects are called *anopsias* (an-op′se-as).

III The Oculomotor (ok″u-lo-mo′ter) Nerves

Origin and course: Fibers extend from ventral midbrain (near its junction with pons) and pass through bony orbit, via *superior orbital fissure,* to eye

Function: Primarily motor; although oculomotor nerves contain a few proprioceptive afferents, they are chiefly motor nerves (as implied by their name, meaning "eye mover"); each nerve includes:

- Somatic motor fibers to four of the six extrinsic eye muscles (inferior oblique and superior, inferior, and medial rectus muscles) that help direct eyeball, and to levator palpebrae superioris muscle, which raises upper eyelid (see p. 471)
- Parasympathetic (autonomic) motor fibers to constrictor muscles of iris, which cause pupil to constrict, and to ciliary muscle, which controls lens shape for visual focusing
- Sensory (proprioceptor) afferents, which run from same four extrinsic eye muscles to midbrain

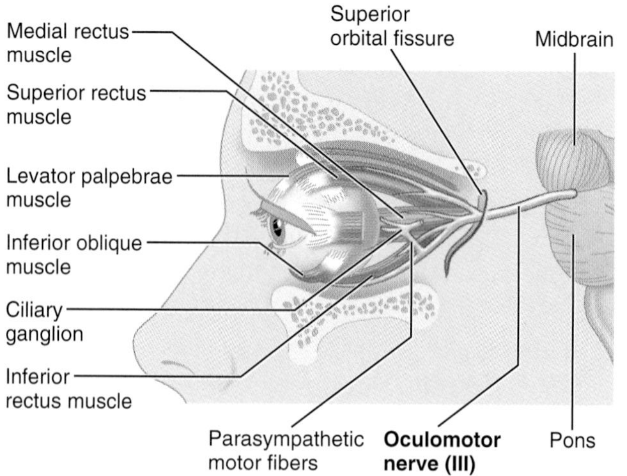

CLINICAL APPLICATIONS

OCULOMOTOR NERVE PARALYSIS In oculomotor nerve paralysis, the eye cannot be moved up or inward, and at rest, eye turns laterally (*external strabismus* [strah-biz′mus]) because the actions of the two extrinsic eye muscles not served by cranial nerve III are unopposed; upper eyelid droops *(ptosis)* and the person has double vision and trouble focusing on close objects.

IV The Trochlear (trok′le-ar) Nerves

Origin and course: Fibers emerge from dorsal midbrain and course ventrally around midbrain to enter orbits through *superior orbital fissures* along with oculomotor nerves

Function: Primarily motor; supply somatic motor fibers to, and carry proprioceptor fibers from, one of the extrinsic eye muscles, the superior oblique muscle

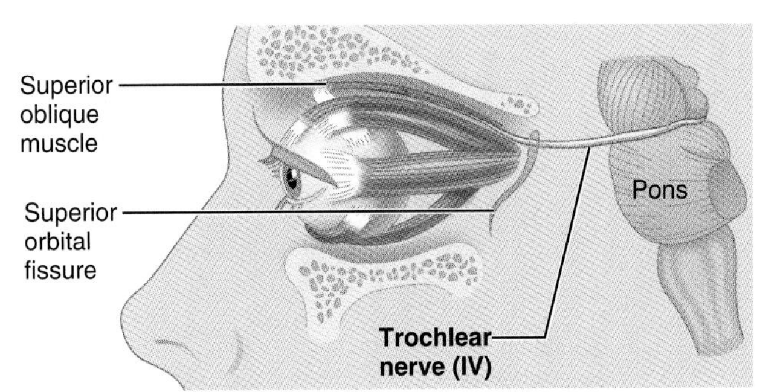

CLINICAL APPLICATIONS

TROCHLEAR NERVE DAMAGE Trauma to, or paralysis of, a trochlear nerve results in double vision and reduced ability to rotate eye inferolaterally. ✚

V The Trigeminal Nerves

Largest of cranial nerves; fibers extend from pons to face, and form three divisions (*trigemina* = threefold): ophthalmic, maxillary, and mandibular divisions; major general sensory nerves of face; transmit afferent impulses from touch, temperature, and pain receptors; cell bodies of sensory neurons of all three divisions are located in large *trigeminal* (also called *semilunar* or *gasserian*) *ganglion;* the mandibular division also contains some branchial motor fibers that innervate chewing muscles.

Dentists desensitize upper and lower jaws by injecting local anesthetic (such as novocain) into alveolar branches of maxillary and mandibular divisions, respectively; because this blocks pain-transmitting fibers of teeth, the surrounding tissues become numb.

	OPHTHALMIC DIVISION (V_1)	MAXILLARY DIVISION (V_2)	MANDIBULAR DIVISION (V_3)
Origin and course:	Fibers run from face to pons via superior orbital fissure	Fibers run from face to pons via foramen rotundum	Fibers pass through skull via foramen ovale
Function:	Conveys sensory impulses from skin of anterior scalp, upper eyelid, and nose, and from nasal cavity mucosa, cornea, and lacrimal gland	Conveys sensory impulses from nasal cavity mucosa, palate, upper teeth, skin of cheek, upper lip, lower eyelid	Conveys sensory impulses from anterior tongue (except taste buds), lower teeth, skin of chin, temporal region of scalp; supplies motor fibers to, and carries proprioceptor fibers from, muscles of mastication

CLINICAL APPLICATIONS

TIC DOULOUREUX *Tic douloureux* (tik du″lu-ru′), or *trigeminal neuralgia* (nu-ral′je-ah), caused by inflammation of trigeminal nerve, is widely considered to produce the most excruciating pain known; the stabbing pain lasts for a few seconds to a minute, but it can be relentless, occurring a hundred times a day (the term *tic* refers to victim's wincing during pain); usually triggered by some sensory stimulus, such as brushing teeth or even a puff of air hitting the face; cause is unknown but may reflect pressure on the trigeminal nerve root; analgesics only partially effective; in severe cases, nerve is cauterized, poisoned, or cut proximal to trigeminal ganglion; this relieves the agony, but also results in loss of sensation on that side of face; many cases respond well to removal of small vessels that compress the nerve near pons. ✚

➤

V The Trigeminal Nerves

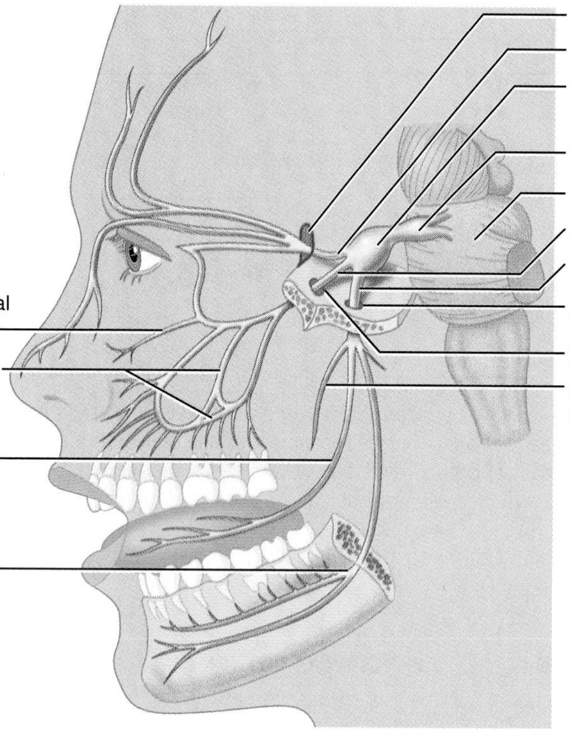

Superior orbital fissure
Ophthalmic division (V_1)
Trigeminal (semilunar or gasserian) ganglion
Trigeminal nerve (V)
Pons
Maxillary division (V_2)
Mandibular division (V_3)
Foramen ovale
Foramen rotundum
Anterior trunk to chewing muscles
Infraorbital nerve
Superior alveolar nerves
Lingual nerve
Inferior alveolar nerve

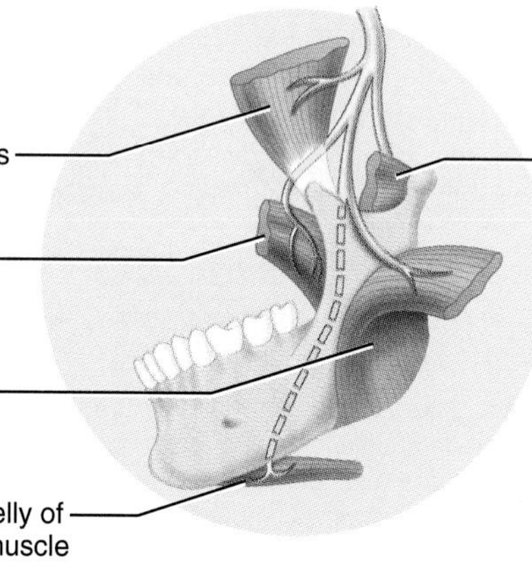

Temporalis muscle
Lateral pterygoid muscle
Medial pterygoid muscle
Masseter muscle
Anterior belly of digastric muscle

Inset shows motor branches of the mandibular division (V_3)

V_1
V_2
V_3

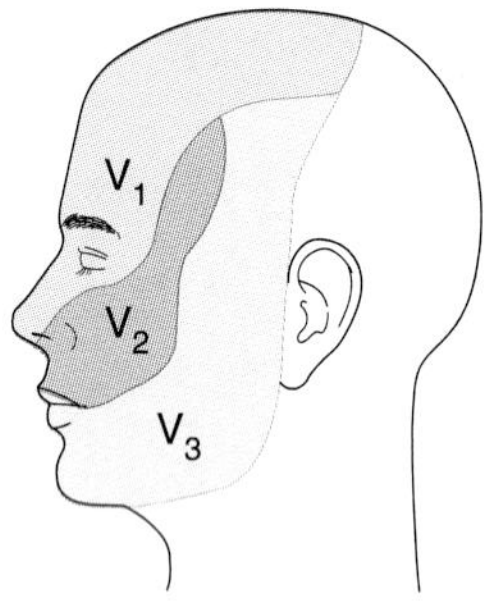

Distribution of sensory fibers of each division

VI The Abducens (ab-du′senz) Nerves

Origin and course: Fibers leave inferior pons and enter orbit via superior orbital fissure to run to eye

Function: Primarily motor; supplies somatic motor fibers to lateral rectus muscle, an extrinsic muscle of the eye; conveys proprioceptor impulses from same muscle to brain.

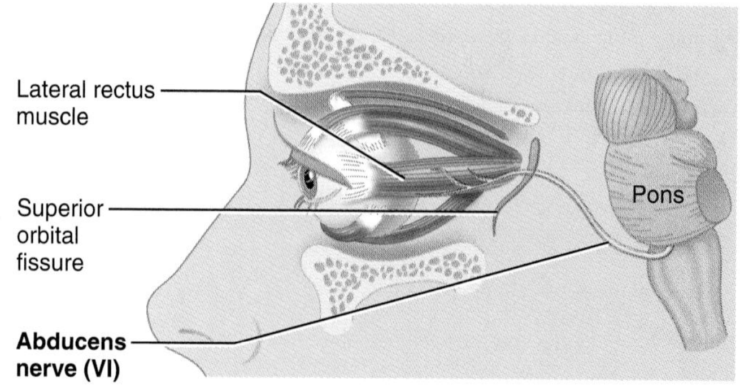

Clinical Applications

ABDUCENS NERVE PARALYSIS In abducens nerve paralysis, the eye cannot be moved laterally; at rest, affected eyeball turns medially *(internal strabismus).* ✚

Lateral rectus muscle
Superior orbital fissure
Abducens nerve (VI)
Pons

VII The Facial Nerves

Origin and course: Fibers issue from pons, just lateral to abducens nerves (see Figure 14.7), enter temporal bone via *internal acoustic meatus,* and run within that bone before emerging through *stylomastoid foramen;* nerve then courses to lateral aspect of face

Function: Mixed nerves that are the chief motor nerves of face; have five major branches on face: temporal, zygomatic, buccal, mandibular, and cervical; see **(a)**

- Convey branchial motor impulses to skeletal muscles of face (muscles of facial expression), except for chewing muscles served by trigeminal nerves, and transmit proprioceptor impulses from same muscles to pons; see **(b)**
- Transmit parasympathetic (autonomic) motor impulses to lacrimal (tear) glands, nasal and palatine glands, and submandibular and sublingual salivary glands. Some of the cell bodies of these parasympathetic motor neurons are in *pterygopalatine* (ter″ĭ-go-pal′ah-tīn) and *submandibular ganglia* on the trigeminal nerve; see **(b)**
- Convey sensory impulses from taste buds of anterior two-thirds of tongue and a tiny patch of skin on the ear; cell bodies of these sensory neurons are in *geniculate ganglion*; see **(c)**

Clinical Applications

BELL'S PALSY *Bell's palsy,* characterized by paralysis of facial muscles on affected side and partial loss of taste sensation, may develop rapidly (often overnight); caused by herpes simplex (viral) infection, which produces inflammation and swelling of the facial nerve; lower eyelid droops, corner of mouth sags (making it difficult to eat or speak normally), eye constantly drips tears and cannot be completely closed; condition may disappear spontaneously without treatment. ✚

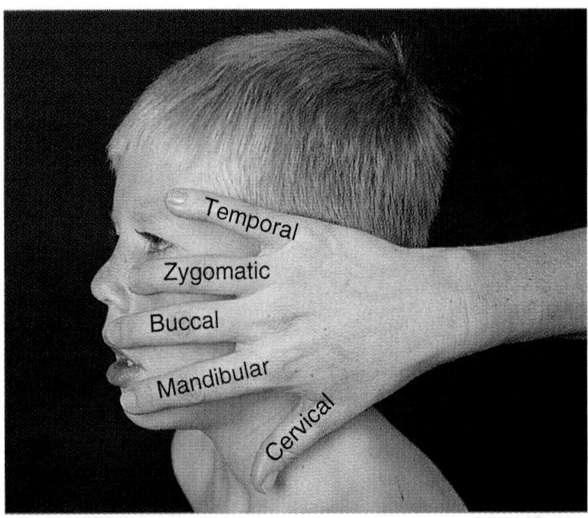

(a) A simple method of remembering the five major motor branches of facial nerve

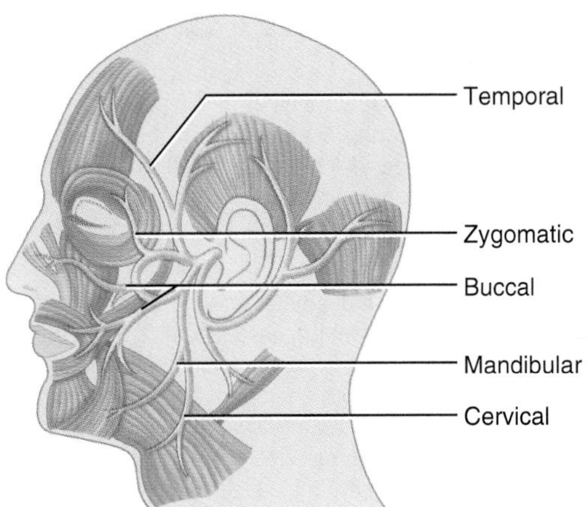

(b) Motor branches to muscles of facial expression and scalp muscles

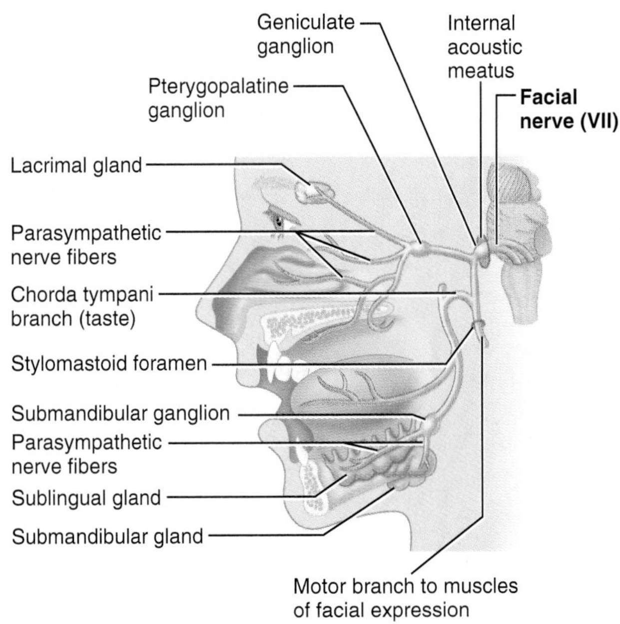

(c) Parasympathetic efferents and sensory afferents

VIII The Vestibulocochlear (ves-tib″u-lo-kok′le-ar) Nerves

Origin and course: Fibers arise from hearing and equilibrium apparatus located within inner ear of temporal bone and pass through internal acoustic meatus to enter brain stem at pons-medulla border; afferent fibers from hearing receptors (in cochlea) form *cochlear division;* those from equilibrium receptors (in semicircular canals and vestibule) form *vestibular division;* the two divisions merge to form vestibulocochlear nerve; see also Figure 16.20

Function: Purely sensory; vestibular branch transmits afferent impulses for sense of equilibrium, and sensory nerve cell bodies are located in *vestibular ganglia;* cochlear branch transmits afferent impulses for sense of hearing, and sensory nerve cell bodies are located in *spiral ganglion* within cochlea

CLINICAL APPLICATIONS

VESTIBULOCOCHLEAR NERVE DAMAGE **Lesions of cochlear nerve or cochlear receptors result in *central* or *nerve deafness,* whereas damage to vestibular division produces dizziness, rapid involuntary eye movements, loss of balance, nausea and vomiting.** ✚

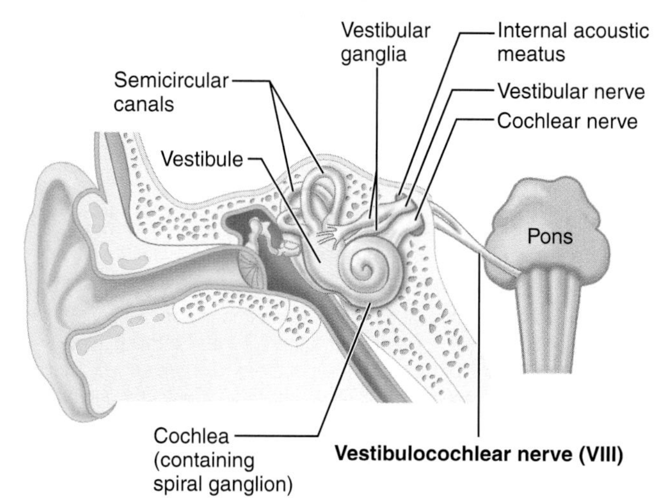

IX The Glossopharyngeal (glos″o-fah-rin′je-al) Nerves

Origin and course: Fibers emerge from medulla and leave skull via *jugular foramen* to run to throat

Function: Mixed nerves that innervate part of tongue and pharynx; provide branchial motor fibers to, and carry proprioceptor fibers from, a superior pharyngeal muscle called the stylopharyngeus, which elevates the pharynx during swallowing; provide parasympathetic motor fibers to parotid salivary gland; (some of the nerve cell bodies of these parasympathetic motor neurons are located in the *otic ganglion* on the trigeminal nerve)

Sensory fibers conduct taste and general sensory (touch, pressure, pain) impulses from pharynx and posterior tongue, from chemoreceptors in the carotid body (which monitor O_2 and CO_2 tension in the blood and help regulate respiratory rate and depth), and from pressure receptors of carotid sinus (which help to regulate blood pressure by providing feedback information); also innervates a small area of skin on the external ear and some of the membrane lining the middle ear cavity; sensory neuron cell bodies are located in *superior* and *inferior ganglia*

CLINICAL APPLICATIONS

GLOSSOPHARYNGEAL NERVE DAMAGE **Injury or inflammation of glossopharyngeal nerves impairs swallowing and taste on the posterior third of the tongue, particularly for sour and bitter substances.** ✚

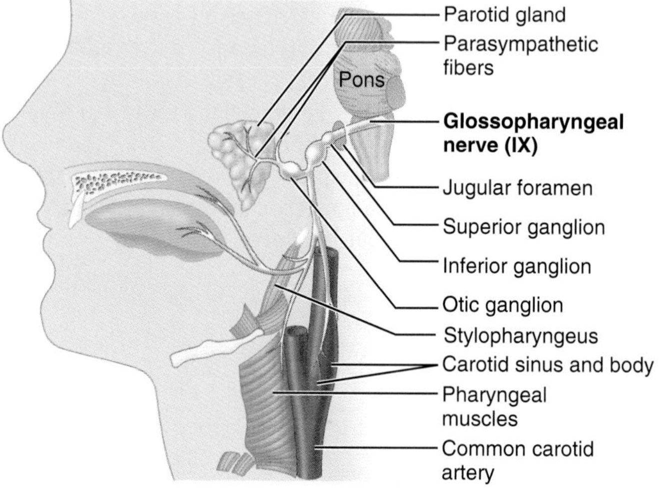

X The Vagus (va′gus) Nerves

Origin and course: The only cranial nerves to extend beyond head and neck region; fibers emerge from medulla, pass through skull via jugular foramen, and descend through neck region into thorax and abdomen; see also Figure 15.6

Function: Mixed nerves; branchial motor fibers serve skeletal muscles of pharynx and larynx (involved in swallowing); parasympathetic motor fibers supply heart, lungs, and abdominal viscera and are involved in regulation of heart rate, breathing, and digestive system activity; transmit sensory impulses from thoracic and abdominal viscera, from the carotid sinus (pressoreceptor for blood pressure) and the carotid and aortic bodies (chemoreceptors for respiration), and taste buds of posterior tongue (epiglottis) and pharynx; also innervate a tiny area of skin on the external ear and some of the membrane lining the middle ear cavity; carry proprioceptor fibers from muscles of larynx and pharynx

Clinical Applications

VAGUS NERVE DAMAGE Because most muscles of the larynx ("voice box") are innervated by laryngeal branches of the vagus, vagal nerve paralysis can lead to hoarseness or loss of voice; other symptoms are difficulty swallowing and impaired digestive system mobility. Total destruction of both vagus nerves is incompatible with life, because these parasympathetic nerves are crucial in maintaining the normal state of visceral organ activity; without their influence, the activity of the sympathetic nerves, which mobilize and accelerate vital body processes (and shut down digestion), would be unopposed. ✚

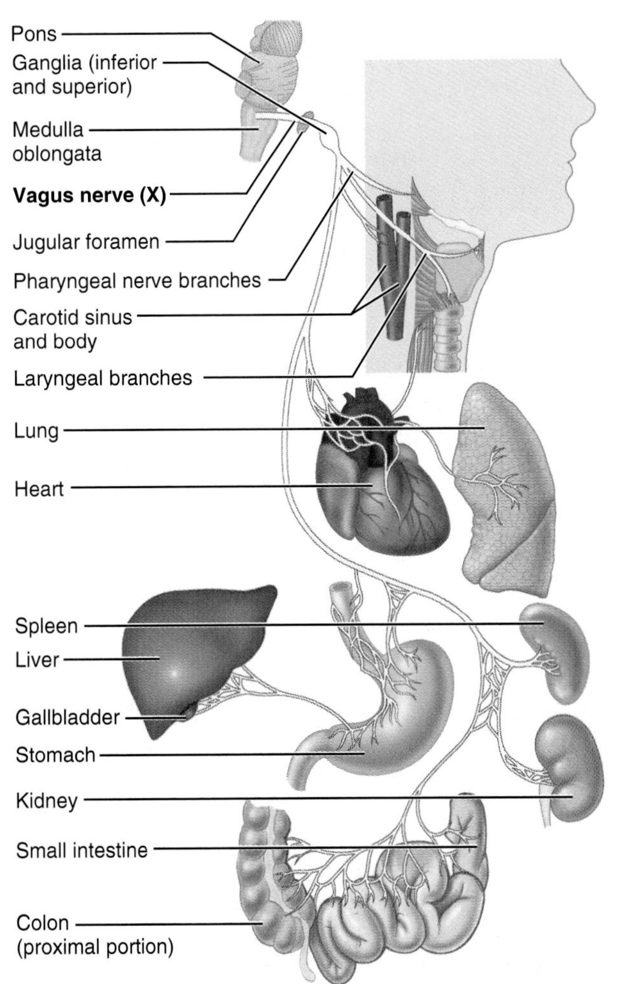

XI The Accessory Nerves

Origin and course: Unique in that they are formed by union of a *cranial root* and a *spinal root;* cranial root emerges from lateral aspect of medulla of brain stem; spinal root arises from superior region (C_1–C_5) of spinal cord. Spinal portion passes upward along spinal cord, enters skull via foramen magnum, and temporarily joins cranial root; the resulting accessory nerve exits from skull through *jugular foramen;* then cranial and spinal fibers diverge; cranial root fibers join vagus, whereas spinal root runs to the sternocleidomastoid and trapezius muscles

Function: Strictly motor; cranial division joins with fibers of vagus nerve (X) to supply branchial motor fibers to larynx, pharynx, and soft palate; spinal root supplies branchial motor fibers to trapezius and sternocleidomastoid muscles, which together move head and neck

Clinical Applications

DAMAGE TO ACCESSORY NERVES Injury to the spinal root of one accessory nerve causes head to turn toward injury side as result of sternocleidomastoid muscle paralysis; shrugging of that shoulder (role of trapezius muscle) becomes difficult. ✚

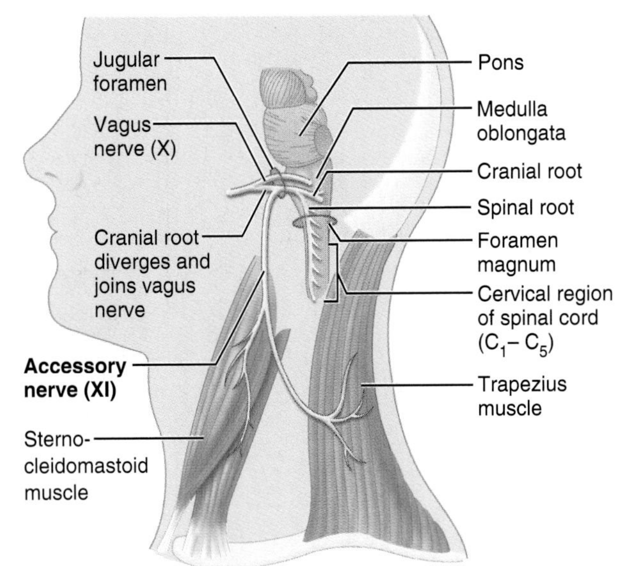

➤

XII The Hypoglossal (hi″po-glos′al) **Nerves**

Origin and course: As their name implies (*hypo* = below; *glossal* = tongue), hypoglossal nerves mainly serve the tongue; fibers arise by a series of roots from medulla and exit from skull via *hypoglossal canal* to travel to tongue; see also Figure 14.7

Function: Strictly motor in function; carry somatic motor fibers to intrinsic and extrinsic muscles of tongue; hypoglossal nerve control allows not only food mixing and manipulation by tongue during chewing, but also tongue movements that contribute to swallowing and speech.

CLINICAL APPLICATIONS

HYPOGLOSSAL NERVE DAMAGE Damage to hypoglossal nerves causes difficulties in speech and swallowing; if both nerves are impaired, the person cannot protrude the tongue; if only one side is affected, the tongue deviates (leans) toward affected side; eventually paralyzed side begins to atrophy. ✚

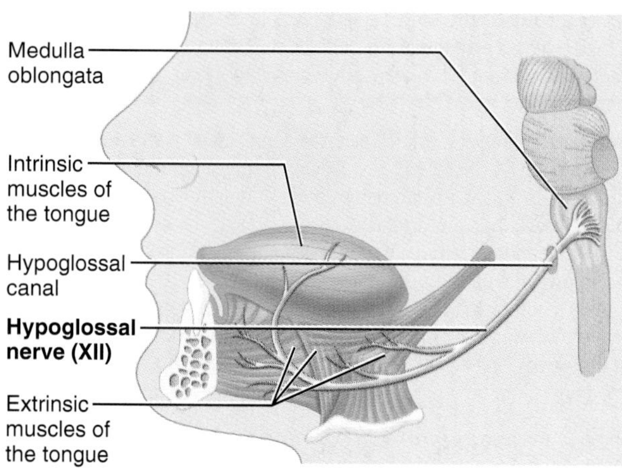

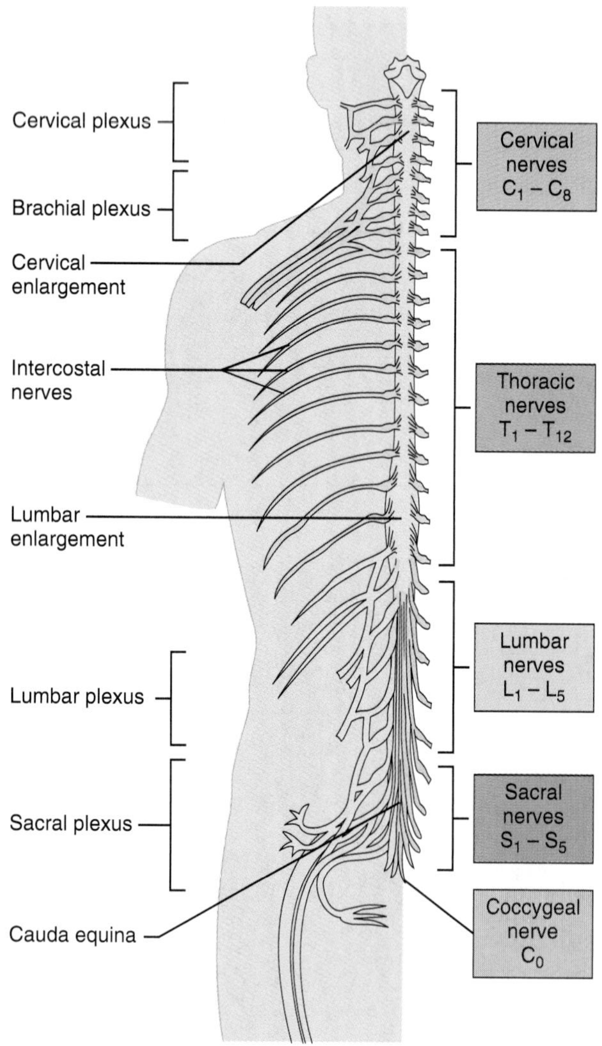

FIGURE 14.8 Spinal nerves, posterior view. The short spinal nerves are shown at right; their ventral rami are shown at left. Most ventral rami form nerve plexuses (cervical, brachial, lumbar, and sacral). The long, horizontal nerves in the region of the ribs are the intercostal nerves.

Additionally, four of the cranial nerves (III, VII, IX, X) contain *general visceral motor* fibers that regulate visceral muscle and glands throughout much of the body. Because these motor fibers belong to the parasympathetic division of the autonomic nervous system (ANS), these four cranial nerves have parasympathetic fibers (see Table 14.2). The ANS is considered in Chapter 15; here we note only that the ANS innervates body structures through chains of two motor neurons, and that the cell bodies of the second neurons occupy *autonomic motor ganglia* in the PNS.

Now that you have read this overview, study the individual cranial nerves discussed in detail in Table 14.3.

SPINAL NERVES

Thirty-one pairs of **spinal nerves,** each containing thousands of nerve fibers, attach to the spinal cord (Figure 14.8). These short nerves, which have long branches that supply most of the body inferior to the head, are named according to their point of issue from the vertebral column: There are 8 pairs of cervical nerves (C_1–C_8), 12 pairs of thoracic nerves (T_1–T_{12}), 5 pairs of lumbar nerves (L_1–L_5), 5 pairs of sacral nerves (S_1–S_5), and 1 pair of coccygeal nerves (designated C_0). Notice that there are eight pairs of cervical nerves but only seven cervical vertebrae. This discrepancy is easily explained: The first cervical spinal nerve lies *superior* to the first vertebra, whereas the last cervical nerve exits *inferior* to the seventh cervical vertebra, leaving six nerves in between. Below the cervical region, every spinal nerve exits *inferior* to the vertebra of the same number.

As mentioned in Chapter 13, each spinal nerve connects to the spinal cord by a **dorsal root** and a **ventral root** (Figures 14.9 and 13.30b). Each root forms from a series of **rootlets** that attach along the whole length of

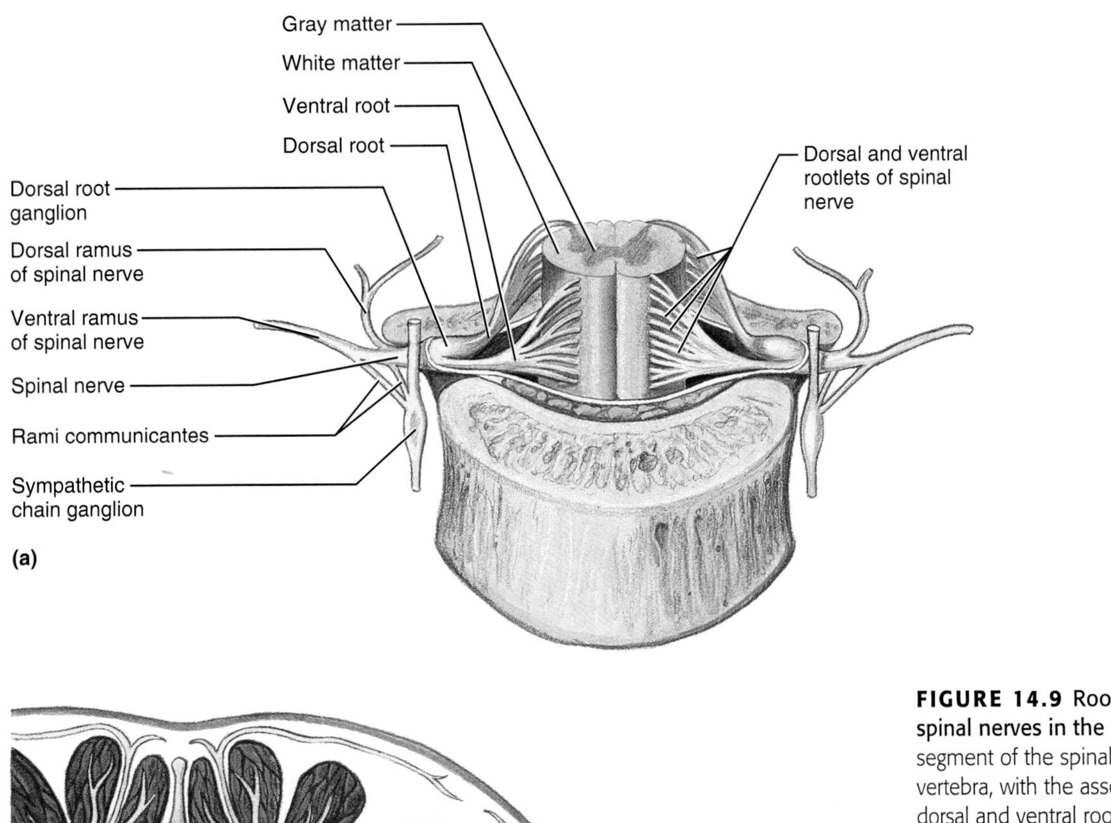

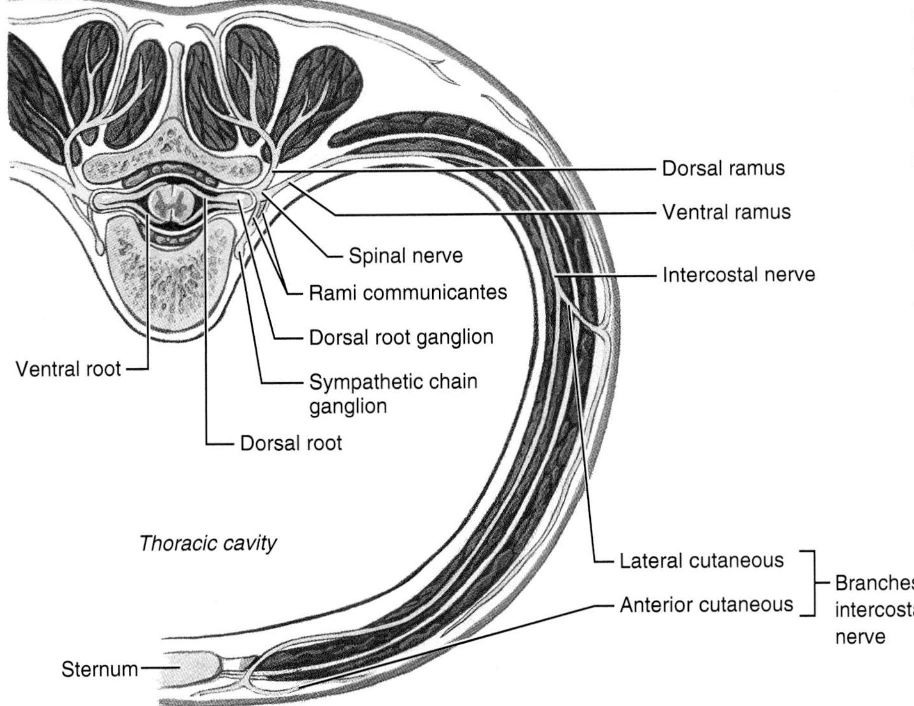

FIGURE 14.9 Roots and branches of the spinal nerves in the thorax. (a) A single segment of the spinal cord within one vertebra, with the associated nerves. The dorsal and ventral roots arise medially as rootlets and converge laterally to form the spinal nerve in the intervertebral foramen. **(b)** Cross section of thorax showing the main roots and branches of a spinal nerve. Note the dorsal and ventral roots and rami, and the rami communicantes. In the thorax, each ventral ramus continues as an intercostal nerve.

the corresponding spinal cord segment (see Figure 14.9a). The dorsal root contains *sensory* fibers arising from cell bodies in the *dorsal root ganglion,* whereas the ventral root contains *motor* fibers arising from the anterior gray column of the spinal cord. The spinal nerve lies at the junction of the dorsal and ventral roots, just lateral to the dorsal root ganglion. The spinal nerves and dorsal root ganglia both lie within the intervertebral foramina, against the bony pedicles of the vertebral arches.

Directly lateral to its intervertebral foramen, each spinal nerve branches into a **dorsal ramus** (ra′mus; "branch") and a **ventral ramus.** Connecting to the base of the ventral ramus are *rami communicantes* leading to *sympathetic chain ganglia*—visceral structures discussed in Chapter 15. Each of the branches of the spinal nerve, like the spinal nerve itself, contains both motor and sensory fibers.

The rest of this chapter focuses on the dorsal and ventral rami and their branches (see Figure 14.9b). These rami supply the entire *somatic* region of the body

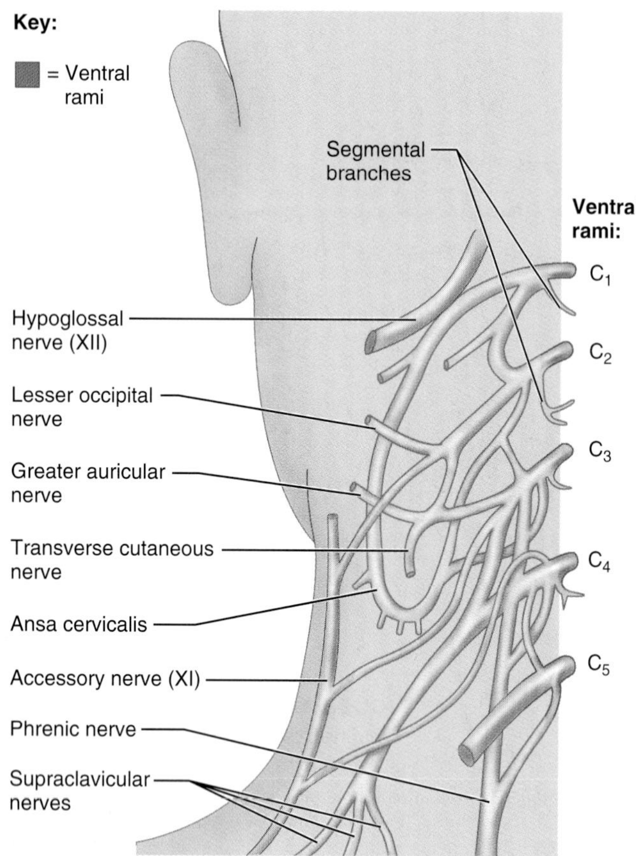

FIGURE 14.10 The cervical plexus. The nerves that are depicted in gray connect to the cervical plexus but do not belong to it. (See also Table 14.4.)

(skeletal musculature and skin) from the neck inferiorly. Whereas the dorsal rami supply the dorsum of the neck and trunk (that is, our "back"), the ventral rami are much thicker and supply a much larger area: the anterior and lateral regions of the neck and trunk, and all regions of the limbs.

Before we proceed, it is important to review the difference between the *roots* and *rami.* The roots lie medial to the spinal nerves and are either strictly sensory (dorsal root) or strictly motor (ventral root). The rami, by contrast, are lateral branches of the spinal nerves, and each contains both sensory fibers and motor fibers.

In the pages that follow we explore how the rami and their branches innervate various regions of the body. First we consider the innervation of the back and the anterior thoracic and abdominal wall. Then we discuss the anatomy and roles of the structures called *nerve plexuses* before we examine the innervation of the neck, the upper limbs, the lower limbs, the joints, and finally the skin. The text throughout these discussions refers to the innervation of major muscle groups; for more specific information on muscle innervation, see Tables 11.1–11.17 on pages 276–331.

Innervation of the Back

The innervation of the back (posterior part of the trunk and neck) by the dorsal rami follows a neat, segmented pattern. Via several branches, each dorsal ramus innervates a horizontal strip of muscle and skin in line with its emergence point from the vertebral column (Figure 14.9b). This pattern of innervation is far simpler than that of the rest of the body by the ventral rami.

Innervation of the Anterior Thoracic and Abdominal Wall

Only in the thorax are the ventral rami arranged in a simple and segmented pattern. The thoracic ventral rami run anteriorly, one deep to each rib, as the neatly segmented **intercostal nerves** (see Figures 14.8 and 14.9b). These nerves supply the intercostal muscles, the skin of the anterior and lateral thorax, and most of the abdominal wall inferior to the rib cage. Along its course, each intercostal nerve gives off **lateral** and **anterior cutaneous branches** to the adjacent skin (see Figure 14.9b).

Two nerves in this thoracic series are unusual. The last (T_{12}) lies inferior to the twelfth rib, and is thus a *subcostal* ("below the ribs") *nerve* rather than an intercostal nerve. The first (most superior) intercostal nerve is exceptionally small because most fibers of T_1 enter the brachial plexus (discussed shortly).

Introduction to Nerve Plexuses

A **nerve plexus** is a network of nerves. The ventral rami of all spinal nerves except T_2–T_{12} branch and join one another lateral to the vertebral column, forming nerve plexuses (see Figure 14.8). These interlacing networks occur in the cervical, brachial, lumbar, and sacral regions and primarily serve the limbs. Note that these plexuses are formed by *ventral* rami, not by dorsal rami. Within the plexuses, fibers from the different rami crisscross each other and redistribute so that (1) each end branch of the plexus contains fibers from several different spinal nerves, and (2) fibers from each ventral ramus travel to the body periphery via several different routes or branches. As a result, each muscle in a limb receives its nerve supply from more than one spinal nerve. This arrangement of fibers ensures that the destruction of a single spinal nerve cannot completely paralyze any limb muscle.

The Cervical Plexus and Innervation of the Neck

The **cervical plexus** is buried deep in the neck, under the sternocleidomastoid muscle, and is formed by the ventral rami of the first four cervical nerves (Figure 14.10). The plexus forms an irregular series of interconnecting loops from which branches arise. The structures these branches serve are listed in Table 14.4. Most

Table 14.4 Branches of the Cervical Plexus (see also Figure 14.10)

Nerves	Ventral Rami	Structures Served
CUTANEOUS BRANCHES (SUPERFICIAL)		
Lesser occipital	C_2, C_3	Skin on posterolateral aspect of neck
Greater auricular	C_2, C_3	Skin of ear, skin over parotid gland
Transverse cervical	C_2, C_3	Skin on anterior and lateral aspect of neck
Supraclavicular (anterior, middle, and posterior)	C_3, C_4	Skin of shoulder and clavicular region
MOTOR BRANCHES (DEEP)		
Ansa cervicalis (superior and inferior roots)	C_1–C_3	Infrahyoid muscles of neck (omohyoid, sternohyoid, and sternothyroid)
Segmental and other muscular branches	C_1–C_5	Deep muscles of neck (geniohyoid and thyrohyoid) and portions of scalenes, levator scapulae, trapezius, and sternocleidomastoid muscles
Phrenic	C_3–C_5	Diaphragm (sole motor nerve supply)

branches of the cervical plexus are **cutaneous nerves** (nerves that supply only skin) that innervate the neck, the back of the head, and the most superior part of the shoulder. Other branches innervate muscles in the anterior neck region. The most important nerve from this plexus is the **phrenic** (fren′ik) **nerve,** which receives fibers from C_5 in addition to its main input from C_3 and C_4. The phrenic nerve courses inferiorly through the thorax and innervates the diaphragm (*phren* = the diaphragm), providing both motor and sensory innervation to this most vital respiratory muscle.

Critical Reasoning

Name the condition in which the phrenic nerve induces abrupt, rhythmic contractions of the diaphragm. What would happen if the phrenic nerves were cut?

The condition is hiccups, which often originate as a reflexive response to sensory irritation of the diaphragm or stomach. Swallowing spicy food or acidic soda pop can lead to hiccups via this reflex. If both phrenic nerves are cut (or if the spinal cord is damaged superior to C_3–C_5), respiratory arrest will occur.

The Brachial Plexus and Innervation of the Upper Limb

The **brachial plexus** (Figure 14.11) lies partly in the neck and partly in the axilla (armpit) and gives rise to almost all of the nerves that supply the upper limb. The brachial plexus can sometimes be felt just superior to the clavicle at the lateral border of the sternocleidomastoid muscle. (Note that the name is "brachial," referring to the arm, not "branchial" or "branchiomeric" ["gill"] referring to the pharynx.)

The brachial plexus is formed by the intermixing of the ventral rami of the inferior four cervical nerves (C_5–C_8) and most of the T_1 ramus as well. Additionally, it may receive small branches from C_4 or T_2.

Because the brachial plexus is very complex, some consider it to be the anatomy student's nightmare. It is composed of four groups of stems and branches. From medial to lateral, these four components are the ventral rami (misleadingly called *roots**), which form *trunks,* which form *divisions,* which then form *cords* (see Figure 14.11a and c). You can remember the sequence of roots, trunks, divisions, and cords by using the saying "**R**eally **t**ired? **D**rink **c**offee." We now examine each of these components.

The five **roots** of the brachial plexus, ventral rami C_5–T_1, lie deep to the sternocleidomastoid. At the lateral border of this muscle, these rami unite to form the **upper, middle,** and **lower trunks,** each of which branches into an **anterior** and **posterior division.** The anterior division mostly carries fibers that serve the limb's anterior part, whereas the posterior division's fibers mostly serve the limb's posterior part. The divisions pass deep to the clavicle and enter the axilla, where they give rise to the **lateral, medial,** and **posterior cords,** named for how they relate to the axillary artery. Along the brachial plexus, small nerves branch off to supply muscles of the superior thorax and shoulder (and some skin of the shoulder as well).

*Do not confuse these roots of the brachial plexus with the dorsal and ventral roots of spinal nerves.

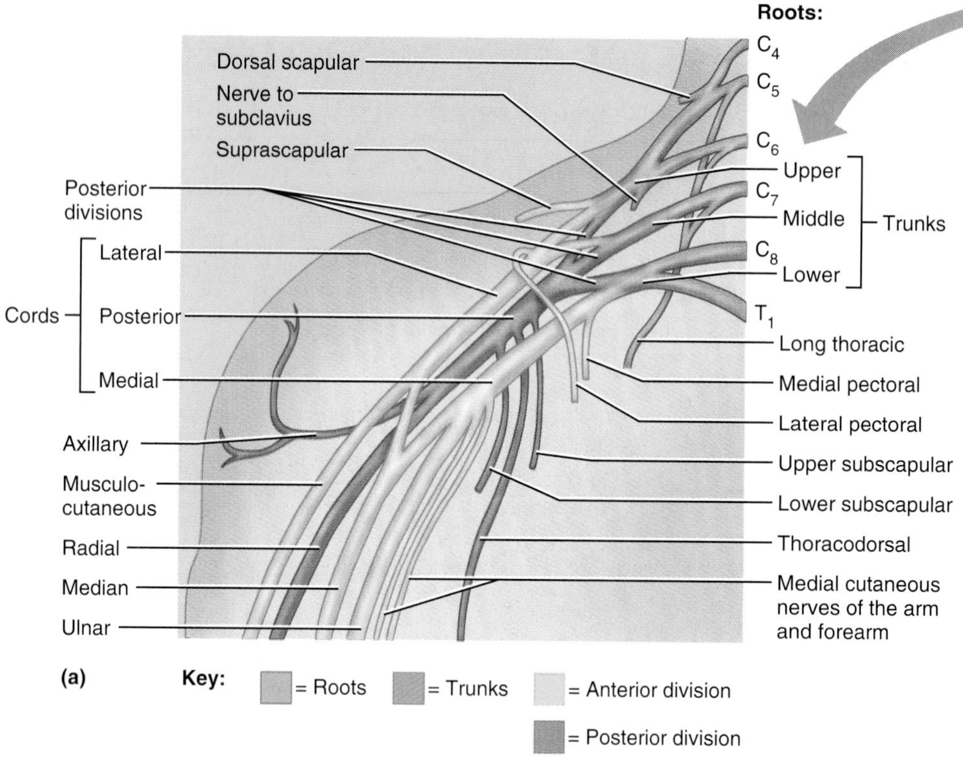

Roots:
C4
C5
C6
C7
C8
T1
Dorsal scapular
Nerve to subclavius
Suprascapular
Posterior divisions
Cords
Lateral
Posterior
Medial
Axillary
Musculo-cutaneous
Radial
Median
Ulnar
Upper
Middle
Lower
Trunks
Long thoracic
Medial pectoral
Lateral pectoral
Upper subscapular
Lower subscapular
Thoracodorsal
Medial cutaneous nerves of the arm and forearm
(a)
Key:
= Roots
= Trunks
= Anterior division
= Posterior division

Humerus
Radial nerve
Musculo-cutaneous nerve
Ulna
Radius
Ulnar nerve
Median nerve
Radial nerve (superficial branch)
Dorsal branch of ulnar nerve
Superficial branch of ulnar nerve
Digital branch of ulnar nerve
Muscular branch
Digital branch
Median nerve
(d)

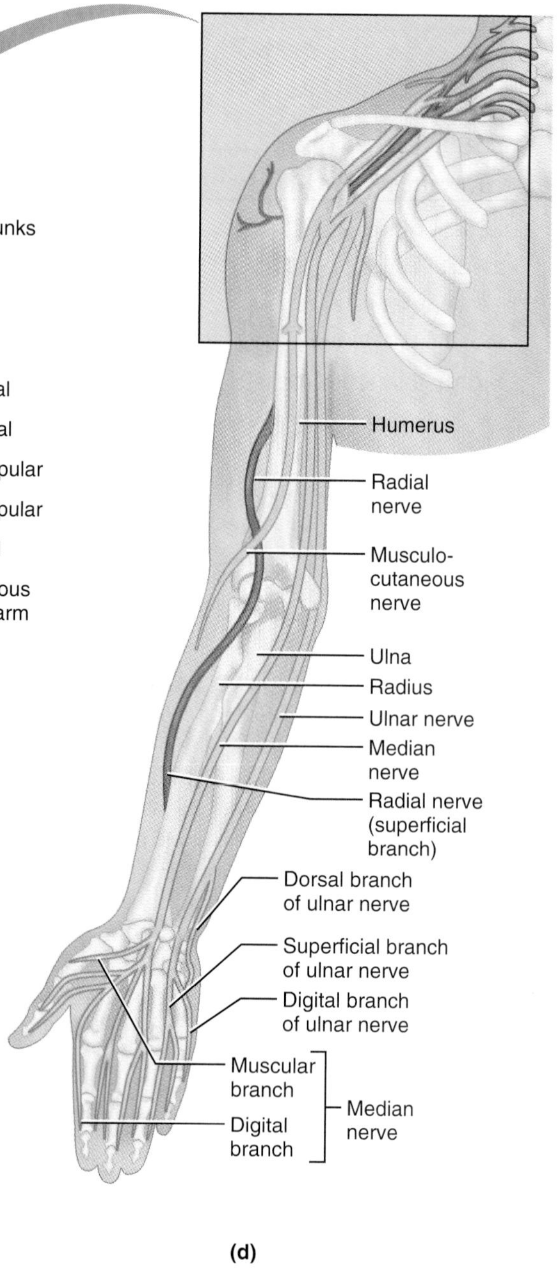

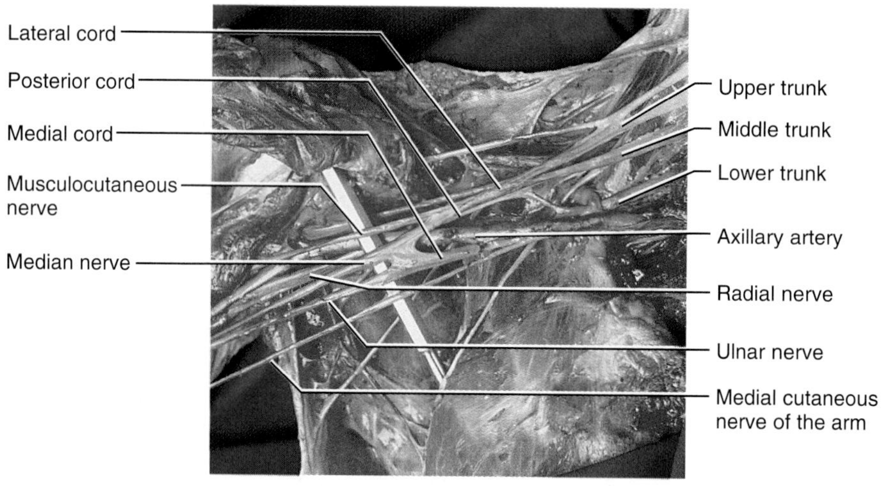

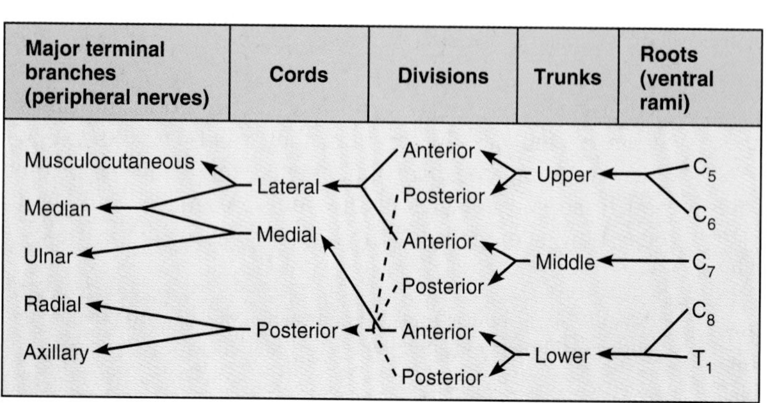

Lateral cord
Posterior cord
Medial cord
Musculocutaneous nerve
Median nerve
Upper trunk
Middle trunk
Lower trunk
Axillary artery
Radial nerve
Ulnar nerve
Medial cutaneous nerve of the arm
(b)

Major terminal branches (peripheral nerves)
Cords
Divisions
Trunks
Roots (ventral rami)
Musculocutaneous
Median
Ulnar
Radial
Axillary
Lateral
Medial
Posterior
Anterior
Posterior
Anterior
Posterior
Anterior
Posterior
Upper
Middle
Lower
C5
C6
C7
C8
T1
(c)

FIGURE 14.11 The brachial plexus. (a) Roots (rami C_5–T_1), trunks, divisions, and cords of the brachial plexus. (b) Photograph of the brachial plexus from a cadaver. (c) Flowchart summarizing relationships within the brachial plexus; the dashed lines to the posterior cord merely indicate that the posterior divisions lie posterior to the anterior divisions. (d) The major nerves of the upper limb. (See also Table 14.5.)

Table 14.5 Branches of the Brachial Plexus (see also Figure 14.11)

Nerves	Cord and Spinal Roots (Ventral Rami)	Structures Served
Musculocutaneous	Lateral cord (C_5–C_7)	Muscular branches: flexor muscles in anterior arm (biceps brachii, brachialis, coracobrachialis) Cutaneous branches: skin on lateral forearm (extremely variable)
Median	By two branches, one from medial cord (C_8, T_1) and one from the lateral cord (C_5–C_7)	Muscular branches to flexor group of anterior forearm (palmaris longus, flexor carpi radialis, flexor digitorum superficialis, flexor pollicis longus, lateral half of flexor digitorum profundus, and pronator muscles); intrinsic muscles of lateral palm (most thenar muscles and two lumbricals) Cutaneous branches: skin of lateral two-thirds of hand on palm side, and dorsum of fingers 2 and 3
Ulnar	Medial cord (C_8 and T_1)	Muscular branches: flexor muscles in anterior forearm (flexor carpi ulnaris and medial half of flexor digitorum profundus); most intrinsic muscles of hand (hypothenar, all interossei, medial two lumbricals, adductor pollicis) Cutaneous branches: skin of medial third of hand, both anterior and posterior aspects
Radial	Posterior cord (C_5–C_8, T_1)	Muscular branches: posterior muscles of arm and forearm (triceps brachii, anconeus, supinator, brachioradialis, extensors carpi radialis longus and brevis, extensor carpi ulnaris, and several muscles that extend the fingers) Cutaneous branches: skin of posterolateral surface of entire limb (except dorsum of fingers 2 and 3)
Axillary	Posterior cord (C_5, C_6)	Muscular branches: deltoid and teres minor muscles Cutaneous branches: shoulder joint and skin covering inferior half of deltoid muscle
Dorsal scapular	Branches of C_5 rami	Rhomboid muscles and levator scapulae
Long thoracic	Branches of C_5–C_7 rami	Serratus anterior muscle
Subscapular	Posterior cord; branches of C_5 and C_6 rami	Teres major and subscapularis muscles
Suprascapular	Upper trunk (C_5, C_6)	Shoulder joint; supraspinatus and infraspinatus muscles
Pectoral (lateral and medial)	Branches of lateral and medial cords (C_5–T_1)	Pectoralis major and minor muscles

Clinical Applications

BRACHIAL PLEXUS INJURIES Injuries to the brachial plexus are common and can be serious. These injuries usually result from stretching the plexus, as can occur in football when a tackler yanks the arm of a ball carrier, or when a blow depresses a shoulder and laterally flexes the head away from that shoulder. Overstretching or crushing the brachial plexus can lead to weakness or even paralysis of the upper limb; transient injuries to this plexus that eventually heal are sometimes called "burners." ✚

The brachial plexus ends in the axilla, where its three cords give rise to the main nerves of the upper limb (see Figure 14.11a and d). Next we discuss the distribution and targets of five of the most important of these peripheral nerves—the *axillary, musculocutaneous, median, ulnar,* and *radial nerves.* For more details, see Table 14.5.

The **axillary nerve,** a branch of the posterior cord, runs posterior to the surgical neck of the humerus and innervates the deltoid and teres minor muscles. Its sensory fibers supply the capsule of the shoulder joint and the skin covering the inferior half of the deltoid muscle.

The **musculocutaneous nerve,** the main terminal branch of the lateral cord, courses within the anterior arm. Descending between the biceps brachii and brachialis muscles, it innervates these arm flexors. Distal to the elbow, it becomes cutaneous and enables skin sensation on the lateral forearm.

The **median nerve** innervates most muscles of the anterior forearm and the lateral palm. After originating from the lateral and medial cords, it descends through the arm without branching, lying medial and posterior to the biceps brachii muscle. Just distal to the elbow region, the median nerve gives off branches to most muscles in the flexor compartment of the forearm (but not the flexor carpi ulnaris and the medial part of the flexor digitorum profundus). Passing through the carpal tunnel and reaching the hand, it innervates five intrinsic muscles in the lateral part of the palm, including the thenar muscles used to oppose the thumb.

Critical Reasoning

Because the median nerve descends through the exact middle of the forearm and wrist, it is often severed during wrist-slashing suicide attempts. Describe some effects of transecting this nerve near the wrist.

Destruction of the median nerve makes it difficult to oppose the thumb toward the little finger; thus, the ability to pick up small objects is diminished.

The **ulnar nerve** branches off the medial cord of the brachial plexus, then descends along the medial side of the arm. At the elbow it passes posterior to the medial epicondyle of the humerus; then it follows the ulna along the forearm, where it supplies the flexor carpi ulnaris and the medial (ulnar) part of the flexor digitorum profundus. The ulnar nerve then continues into the hand, where it innervates most of the intrinsic hand muscles and the skin on the medial side of the hand. A dorsal branch supplies the skin of the dorsal medial hand.

CLINICAL APPLICATIONS

ULNAR NERVE INJURIES Because of its exposed position at the elbow, the ulnar nerve is vulnerable to injury. Striking the "funny bone"—the spot where this nerve rests against the medial epicondyle—causes tingling of the little finger. Repeated throwing can also damage the nerve here because it stresses the medial part of the elbow region. Even more commonly, the ulnar nerve is injured in the carpal region of the palm, another place where it is superficially located. Individuals with damage to this nerve can neither adduct and abduct their fingers (because the interossei and medial lumbrical muscles are paralyzed) nor form a tight grip (because the hypothenar muscles are useless). Without the interossei muscles to extend the fingers at the interphalangeal joints, the fingers flex at these joints, and the hand contorts into a *clawhand.* ✚

The **radial nerve,** a continuation of the posterior cord, is the largest branch of the brachial plexus, innervating almost the entire posterior side of the upper limb, including the limb extensor muscles. As it descends through the arm, it wraps around the humerus in the radial groove, sending branches to the triceps brachii. It then curves anteriorly around the lateral epicondyle at the elbow, where it divides into a superficial and deep branch. The superficial branch (see Figure 14.11d) descends along the lateral edge of the radius under cover of the brachioradialis muscle; distally, it supplies the skin on the dorsolateral surface of the hand. The deep branch of the radial nerve (not illustrated) runs posteriorly to supply most of the extensor muscles on the forearm.

CLINICAL APPLICATIONS

RADIAL NERVE INJURIES Trauma to the radial nerve results in **wristdrop,** an inability to extend the hand at the wrist. If the lesion occurs far enough superiorly, the triceps muscle is paralyzed, so that the forearm cannot be actively extended at the elbow. Many fractures of the humerus follow the radial groove and harm the radial nerve there. This nerve can also be crushed in the axilla by improper use of a crutch or if a person falls asleep with an arm draped over the back of a chair. ✚

The Lumbar Plexus and Innervation of the Lower Limb

The **lumbar plexus** (Figure 14.12) arises from the first four lumbar spinal nerves (L_1–L_4) and lies within the psoas major muscle in the posterior abdominal wall. Its smaller branches innervate parts of the abdominal wall and the psoas muscle itself, but the main branches descend to innervate the anterior thigh. The **femoral nerve,** the largest terminal branch, runs deep to the inguinal ligament to enter the thigh. From there it divides into several large branches. Motor branches of the femoral nerve innervate the anterior thigh muscles, including the important quadriceps femoris. Cutaneous branches of the femoral nerve serve the skin of the anterior thigh and the medial surface of the leg from the knee to the foot. The **obturator nerve** passes through the large obturator foramen of the pelvis, enters the medial thigh, and innervates the adductor muscle group plus some skin on the superomedial thigh. These and other branches of the lumbar plexus are summarized in Table 14.6.

Compression of the spinal roots of the lumbar plexus by a herniated disc causes a major disturbance in gait. This is because the femoral nerve innervates muscles that flex the thigh at the hip (rectus femoris and iliacus), and muscles that extend the leg at the knee (the entire quadriceps femoris group). Other symptoms are pain or anesthesia of the anterior thigh and of the medial thigh if the obturator nerve is impaired.

The Sacral Plexus and Innervation of the Lower Limb

The **sacral plexus** (Figure 14.13) arises from spinal nerves L_4–S_4 and lies immediately caudal to the lumbar plexus. Because some fibers from the lumbar plexus contribute to the sacral plexus via the **lumbosacral trunk,** the two plexuses are often considered together as the *lumbosacral plexus.* Of the dozen named branches of the sacral plexus, about half serve the buttock and lower limb, whereas the rest innervate parts of the pelvis and perineum. The most important branches are described next and summarized in Table 14.7.

The largest branch is the **sciatic** (si-at′ik) **nerve,** the thickest and longest nerve in the body. It supplies all of the lower limb except the anterior and medial regions of the thigh. Actually, the sciatic is two nerves—the *tibial* and *common fibular nerves*—wrapped in a common sheath. It leaves the pelvis by passing through the greater sciatic notch, then courses deep to the broad gluteus maximus muscle and enters the thigh just medial to the hip joint (*sciatic* literally means "of the hip"). From there it descends through the posterior thigh deep to the hamstrings, which it innervates. Superior to the knee region, it branches into the tibial nerve and the common fibular nerve.

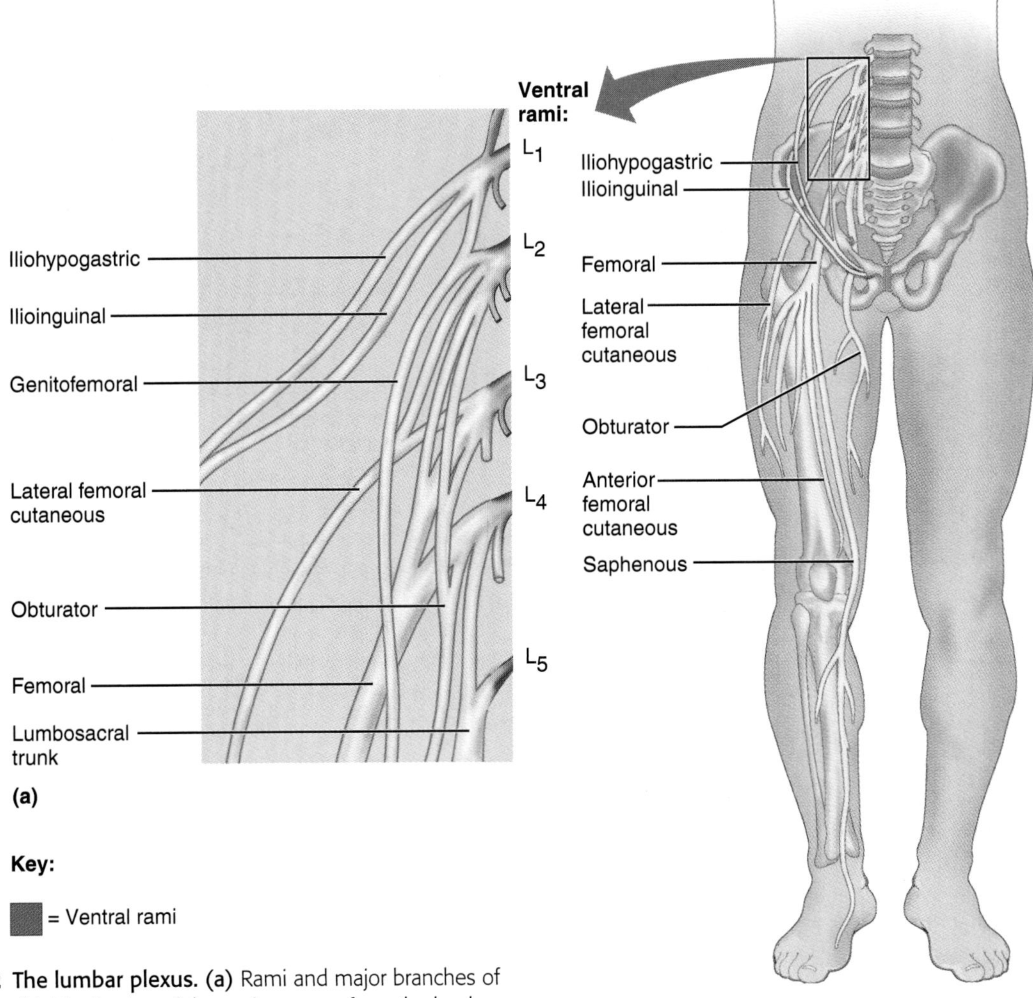

FIGURE 14.12 **The lumbar plexus.** **(a)** Rami and major branches of the lumbar plexus. **(b)** Distribution of the major nerves from the lumbar plexus to the lower limb. (See also Table 14.6.)

Table 14.6 **Branches of the Lumbar Plexus (see also Figure 14.12)**

Nerves	Ventral Rami	Structures Served
Femoral	L_2–L_4	Skin of anterior and medial thigh via *anterior femoral cutaneous* branch; skin of medial leg and foot, hip; and knee joints via *saphenous branch;* motor to anterior muscles of thigh (quadriceps and sartorius); pectineus, iliacus
Obturator	L_2–L_4	Motor to adductor magnus (part), longus and brevis muscles, gracilis muscle of medial thigh, obturator externus; sensory for skin of medial thigh and for hip and knee joints
Lateral femoral cutaneous	L_2, L_3	Skin of lateral thigh; some sensory branches to peritoneum
Iliohypogastric	L_1	Skin on side of buttock and skin on pubis; proprioceptor (and motor?) to the most inferior parts of the oblique and transversus abdominis muscles
Ilioinguinal	L_1	Skin of external genitalia and proximal medial aspect of the thigh
Genitofemoral	L_1, L_2	Skin of scrotum in males, of labia majora in females, and of anterior thigh inferior to middle portion of inguinal region; cremaster muscle in males

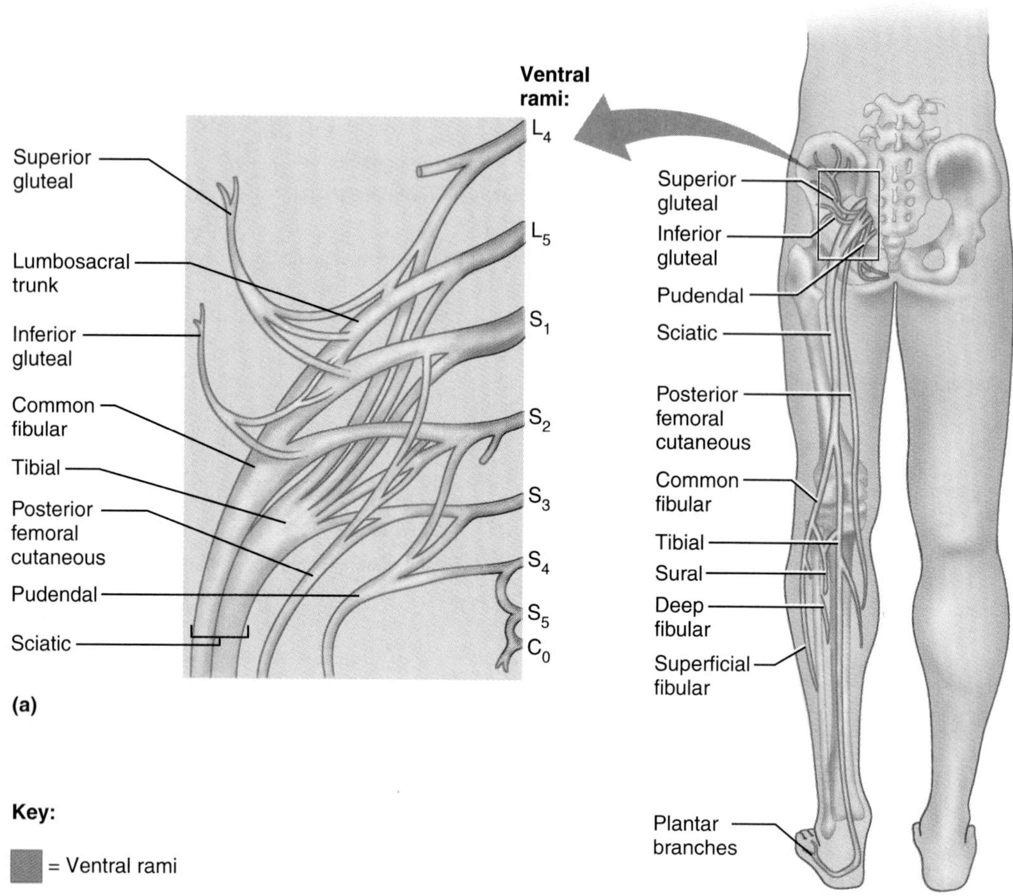

FIGURE 14.13 The sacral plexus. (a) Rami and major branches of the sacral plexus. (b) Distribution of the major nerves from the sacral plexus to the lower limb. (See also Table 14.7.)

The **tibial nerve** courses through the popliteal fossa (the region just posterior to the knee joint) and then descends through the calf deep to the soleus muscle. At the ankle, it passes posterior to the medial malleolus and divides into the two main nerves of the sole of the foot, the medial and lateral **plantar nerves.** The tibial nerve and its branches supply almost all muscles in the posterior region of the lower limb. As for the cutaneous innervation, the tibial nerve supplies (1) the skin of the sole of the foot, through the plantar nerves, and (2) a vertical strip of skin along the posterior leg through a **sural nerve,** to which the common fibular nerve also contributes.

The **common fibular nerve,** or *common peroneal nerve* (*perone* = fibula), supplies most structures on the anterolateral aspect of the leg. It descends laterally from its point of origin in the popliteal fossa and enters the superior part of the leg, where it wraps around the neck of the fibula and divides into deep and superficial branches. The **superficial fibular (peroneal) nerve** supplies the fibular muscles in the lateral compartment of the leg and most of the skin on the superior surface of the foot. The **deep fibular (peroneal) nerve** serves the muscles of the anterior compartment of the leg—the extensors that dorsiflex the foot.

Except for the sciatic nerve, the largest branches of the sacral plexus are the **superior** and **inferior gluteal nerves,** which innervate the gluteal muscles. Other branches of the sacral plexus supply the lateral rotators of the thigh and the muscles of the pelvic floor and perineum. The **pudendal nerve** (pu-den'dal; "shameful") innervates the muscles and skin of the perineum, stimulates erection, and is responsible for voluntary inhibition of defecation and urination (Table 11.7, p. 292). As it passes anteriorly to enter the perineum, the pudendal nerve lies just medial to the ischial tuberosity. The pudendal nerve may be injected with anesthetic, either to help block pain in the perineum during childbirth, or as preparation for surgery on the anal or genital regions. In such a **pudendal nerve block,** the injection needle is inserted into the nerve just medial to the ischial tuberosity.

Clinical Applications

SCIATIC NERVE INJURIES The sciatic nerve can be injured in many ways. **Sciatica** (si-at"ĭ-kah), characterized by a stabbing pain over the course of this nerve, often occurs when a herniated lumbar disc presses on the sacral dorsal roots within the vertebral canal. Furthermore, wounds to the buttocks or posterior thigh can sever the sciatic nerve, in which case the leg can-

Table 14.7 Branches of the Sacral Plexus (see also Figure 14.13)

Nerves	Ventral Rami	Structures Served
Sciatic nerve	L_4, L_5, S_1–S_3	Composed of two nerves (tibial and common fibular) in a common sheath that usually diverge just proximal to the knee
• Tibial (including medial and lateral plantar branches)	L_4–S_3	Cutaneous branches: to skin of posterior surface of leg and sole of foot Motor branches: to muscles of back of thigh, leg, and foot (to hamstrings except short head of biceps femoris, to posterior part of adductor magnus, to triceps surae, tibialis posterior, popliteus, flexor digitorum longus, flexor hallucis longus, and intrinsic muscles of foot)
• Common fibular (superficial and deep branches)	L_4–S_2	Cutaneous branches: to skin of anterior and lateral surface of leg and dorsum of foot Motor branches: to short head of biceps femoris of thigh, fibular muscles of lateral compartment of leg, tibialis anterior, and extensor muscles of toes (extensor hallucis longus, extensors digitorum longus and brevis)
Superior gluteal	L_4, L_5, S_1	Motor branches: to gluteus medius and minimus and tensor fasciae latae
Inferior gluteal	L_5–S_2	Motor branches: to gluteus maximus
Posterior femoral cutaneous	S_1–S_3	Skin of most inferior part of buttock, posterior thigh, and popliteal region; length variable; may also innervate part of skin of calf and heel
Pudendal	S_2–S_4	Supplies most of skin and muscles of perineum (region encompassing external genitalia and anus and including clitoris, labia, and vaginal mucosa in female, and scrotum and penis in males); external anal sphincter

not be flexed at the knee (because the hamstrings are paralyzed) and the foot cannot be moved at the ankle. When only the tibial nerve is injured, the paralyzed muscles in the calf cannot plantar flex the foot, and a shuffling gait results. When the common fibular nerve is damaged, dorsiflexion is lost and the foot drops into plantar flexion, a condition called **footdrop.** The common fibular nerve is susceptible to injury in the superolateral leg, where it is superficially located and easily crushed against the neck of the fibula. ✚

Innervation of Joints of the Body

Because injuries to joints are common, health professionals should know which nerves serve which synovial joints. Memorizing so much information is almost impossible, however, because every joint capsule receives sensory branches from several different nerves (see pp. 216–217).

The most helpful guideline for deducing which nerves supply a given joint is provided by **Hilton's law:** *Any nerve that innervates a muscle producing movement at a joint also innervates the joint itself (and the skin over it).* Applied to the knee joint, for example, the process of deduction would be as follows: Because the knee is crossed by the quadriceps, hamstring, and gracilis muscles (see Table 11.14), and because the nerves to these muscles are the femoral, obturator, and branches of the sciatic nerves (see Tables 14.6 and 14.7), then all these nerves innervate the knee joint as well.

Critical Reasoning

Which four nerves send branches that innervate the capsule of the elbow joint?

Because the elbow joint is crossed by the biceps brachii, the triceps brachii, and many flexor muscles on the forearm (including the flexor carpi ulnaris), the nerves that innervate the elbow joint are the musculocutaneous, radial, median, and ulnar nerves.

Innervation of the Skin: Dermatomes

The area of skin innervated by the cutaneous branches from a single spinal nerve is called a **dermatome,** literally a "skin segment." All spinal nerves except C_1 participate in dermatomes. The map of dermatomes illustrated in Figure 14.14 was constructed by recording areas of numbness in patients who had experienced injuries to specific spinal roots. In the *trunk* region, the dermatomes are almost horizontal, relatively uniform in width, and in direct line with their spinal nerves. In the *limbs,* however, the pattern of dermatomes is less straightforward. In the upper limb, the skin is supplied by the nerves participating in the brachial plexus: the ventral rami of C_5 to T_1 or T_2. In the lower limb, lumbar nerves supply most of the anterior surface, whereas sacral nerves supply much of the posterior surface. This distribution basically reflects the areas supplied by the lumbar and sacral plexuses, respectively.

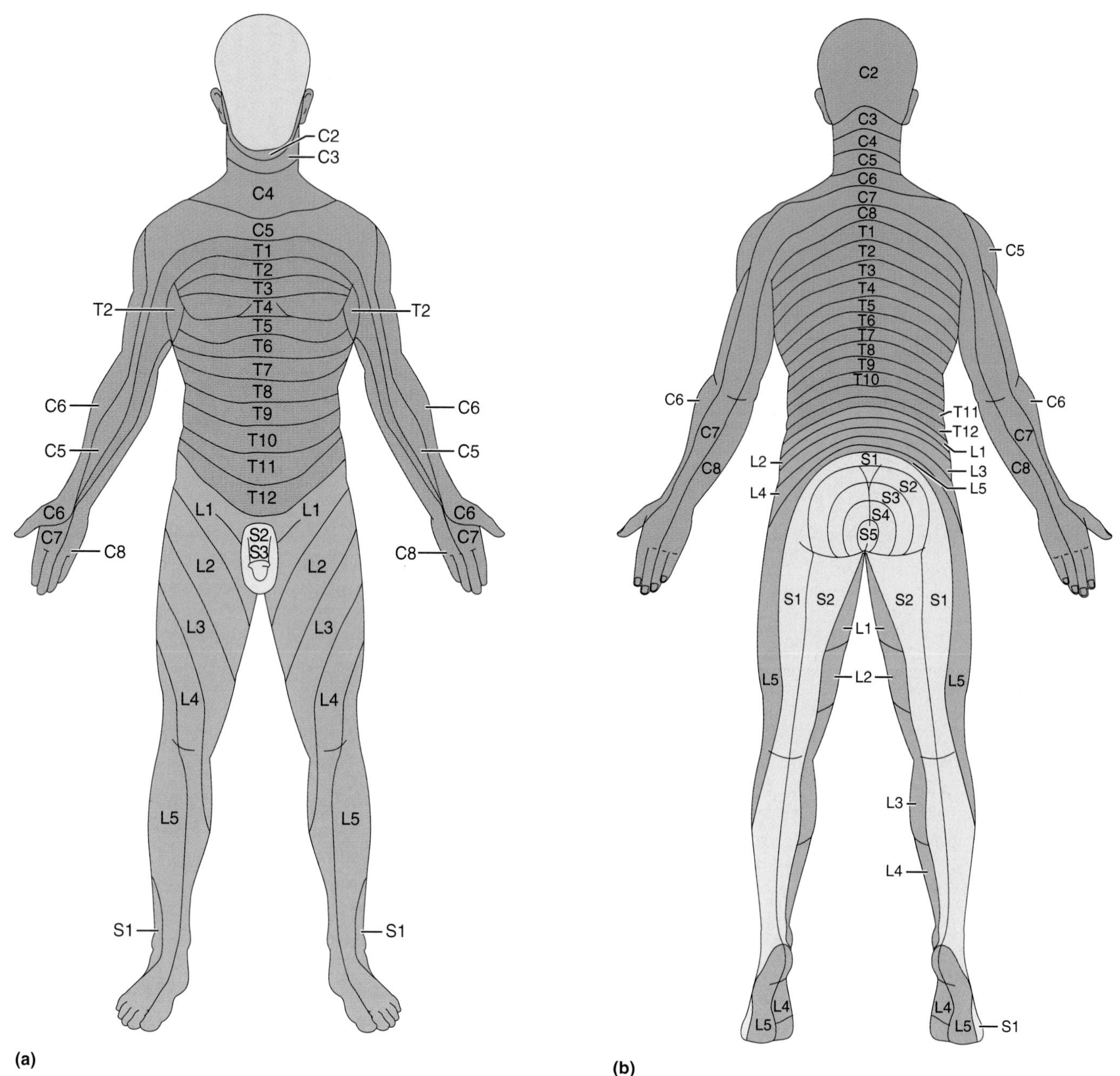

FIGURE 14.14 Map of dermatomes (a) anterior view; (b) posterior view. A dermatome is an area of skin innervated by the sensory fibers of a single spinal nerve (both ventral and dorsal rami). The pattern of dermatomes in the limbs is controversial; studies using different mapping techniques have produced slightly different maps of limb dermatomes.

Adjacent dermatomes are not as clearly demarcated from one another as a dermatome map indicates. On the trunk, neighboring dermatomes overlap each other by a full 50%; as a result, destruction of a single spinal nerve does not produce complete numbness anywhere. On the limbs, by contrast, the overlap is not complete: Some skin patches are innervated by one spinal nerve only.

CLINICAL APPLICATIONS

CLINICAL IMPORTANCE OF DERMATOMES Clinicians use dermatomes in at least two ways. First, dermatomes can help pinpoint the level of spinal injuries. Numbness in all dermatomes below T_1, for example, indicates a spinal cord injury at T_1. Additionally, because the varicella-zoster virus that produces shingles (see next section) resides in a single dorsal root ganglion, a flare-up of this infection may produce skin blisters along the entire course of the associated dermatome. Finally, dermatomes may also have surgical applications, as anesthetics may be injected into specific spinal nerves or roots to desensitize specific skin regions during surgery.

DISORDERS OF THE PERIPHERAL NERVOUS SYSTEM

This section considers two disorders that involve sensory neurons (shingles and migraine headaches) and one disorder of motor innervation (myasthenia gravis). For information on another group of motor disorders—polio and postpolio syndrome—see A Closer Look on p. 438.

Shingles (herpes zoster) is a viral infection of sensory neurons to the skin. It is characterized by a rash of scaly, painful blisters usually confined to a narrow strip of skin on one side of the trunk (that is, to one or several adjacent dermatomes). Shingles is caused by the varicella-zoster virus (herpesvirus), which also produces chicken pox. The disease stems from a childhood infection of chicken pox, during which the viruses are transported from the skin lesions, through the peripheral processes of the sensory neurons, to the cell bodies in a sensory ganglion. There, in the nuclei of the nerve cells, the viruses remain dormant for many years, held in check by the immune system. When the immune system is weakened, the viruses multiply and travel back through the sensory axons to the skin, producing the inflammatory rash of shingles. Often caused by stress, shingles attacks last for several weeks. In a minority of those with shingles, the attacks recur, separated by periods of healing and remission. Shingles is experienced by 1–2 of every 1000 people, mostly those over 50 years of age. Treatment is with pain-relieving drugs and, under certain conditions, anti-viral medicines.

Migraine headaches are extremely painful, episodic headaches that affect 25 million Americans. The cause of migraines was long debated, but they are now known to relate to the sensory innervation of the brain's cerebral arteries by the trigeminal nerve: A signal from the brain stem causes the sensory dendrites of the ophthalmic division to release chemicals onto the cerebral arteries that they innervate, signaling these arteries to dilate (widen). This dilation then compresses and irritates these sensory dendrites, causing the headache. New drugs can relieve migraines by blocking pain receptors on the trigeminal endings.

Myasthenia gravis (mi″es-the′ne-ah gră′vis; "severe muscle weakness") is a disorder of the neuromuscular junctions. This crippling disease is characterized by a progressive weakening of muscles, especially those of the head, neck, and limbs. In this autoimmune disorder, the body's own defense molecules (antibodies) destroy acetylcholine receptors in the sarcolemma of the motor end plates. Once inactivated, these receptors cannot bind acetylcholine, so the muscle fibers cannot contract. Treatments involve drugs that inhibit the enzyme acetylcholinesterase and suppress the immune response.

THE PERIPHERAL NERVOUS SYSTEM THROUGHOUT LIFE

The spinal nerves start to form late in the fourth week of development. Motor axons grow outward from the early spinal cord, whereas sensory peripheral axons grow out from the neural crest of the dorsal root ganglia (see Figure 12.20, p. 355). The bundles of these motor and sensory axons exit between successive vertebrae, where they converge to form the early spinal nerves. Each of the 31 spinal nerves sends motor fibers into a single myotome, and each sends its sensory fibers to the overlying band of skin, which accounts for the segmented pattern of dermatomes. Cranial nerves grow to innervate the head in a roughly comparable manner; however, some of their sensory neurons—especially those for the special senses—develop from plate-shaped thickenings of the head ectoderm called **ectodermal placodes** (plak′ōdz; "plate-like"), rather than from neural crest.

During week 5 of development, nerves reach the organs they innervate. Shortly thereafter, some of the embryonic muscles migrate to new locations, and some skin dermatomes become displaced as they are pulled along the elongating limbs. Even though they may migrate for considerable distances, these muscles and skin areas always retain their original nerve supply.

Aging of the nervous system has been discussed in Chapters 12 and 13. In the PNS, sensory receptors atrophy to some degree with age, and tone in the muscles of the face and neck decreases. Reflexes slow a bit with age, but this change seems to reflect a slowing of central processing more than changes in peripheral nerve fibers.

A Closer Look

Postpolio Syndrome: The Plight of Some "Recovered" Polio Victims

Poliomyelitis, or infantile paralysis, was the most feared infectious disease in America in the first half of the twentieth century. In 1949, a public service message in a nationwide magazine warned parents to take precautions during the annual summer polio epidemic—to ensure that their children avoid swimming in water not declared safe by health departments, avoid crowds and "new contacts," and keep their hands and food scrupulously clean. Tragically, such precautions were of little use.

Polio is caused by a virus spread via fecal contamination (in swimming pools or on dirty hands, for example) and by coughs. It enters the mouth, multiplies in the gut, and then travels to various preferred cell types in the body—mainly motor neurons. The initial symptoms resemble the flu, including fever, headache, and stiffness of the neck and back, but about 10% of cases progress further, destroying some motor neurons of the spinal cord or brain. The result is the elimination of the motor functions of certain nerves and paralysis of muscles (including the respiratory muscles; see the photo).

Fortunately, only about 3% of those who contracted polio were paralyzed permanently; the rest recovered either full or partial use of their muscles. Still, during the peak of the polio epidemic in the early 1950s, over 22 cases were reported for every 100,000 Americans. The epidemic was halted in the mid-1950s by the development of the Salk and Sabin polio vaccines. Through aggressive vaccination, polio was eliminated from the entire Western Hemisphere in 1991, and since then from Europe and China as well. It persists in parts of Asia and Africa, leading to the current goal of complete eradication by the year 2005.

Even though the devastating epidemics appear to be a thing of the past, many of the "recovered" survivors are now experiencing a surprising "postpolio" deterioration. People now in their 50s, 60s, and 70s who struggled to regain health and to live active lives have developed new symptoms: lethargy; sensitivity to cold; sharp, burning pains in muscles and joints; and weakness and loss of mass in muscles that had earlier recovered. When these symptoms first appeared in the 1980s, they were either dismissed as the flu or a psychosomatic illness, or misdiagnosed as multiple sclerosis, arthritis, lupus, or amyotrophic lateral sclerosis (see p. 404). But as the number of complaints increased, more attention was given to this group of symptoms, now named *postpolio syndrome (PPS)*. About one-third of polio survivors, approximately 500,000 Americans, have developed PPS. Certain survivors seem to be particularly vulnerable: those who contracted polio after age 10, those who needed ventilatory support, and those in whom all four limbs were affected.

The cause of PPS is not known. It is unlikely that the virus has persisted or been reactivated, because all efforts to find live poliovirus in the body fluids of PPS patients have failed. A more favored explanation is that polio survivors, like all of us, lose neurons throughout life. Whereas unimpaired nervous systems can enlist nearby neurons to compensate for the losses, polio survivors (who have already lost many motor neurons) may have too few left to take over. What is particularly ironic is that PPS seems to affect those who worked hardest to overcome polio's effects—perhaps the grueling hours of physical therapy to regain their strength ultimately burned out motor neurons. A newer, related hypothesis is that the virus originally weakened many of the neurons that it did not kill, and that these weakened neurons are now losing connections to muscles. An entirely different theory holds that the virus left fragments of its genetic material inside surviving neurons, and although once harmless, these fragments have initiated an immune response that is slowly destroying the neurons. The fact that both viral genes and antibodies against them were recently found in PPS victims seems to support this idea.

Fortunately, PPS progresses slowly and is not life threatening. Its victims are advised to conserve energy by resting more and by using canes, walkers, and braces to support their wasting muscles. Still, many PPS sufferers are emotionally devastated—cruelly surprised by the return of an old enemy and angry because no remedy exists. The medical community has recognized their plight and established research funding and postpolio assessment clinics across the nation, but this is little comfort to those experiencing the limbo of PPS.

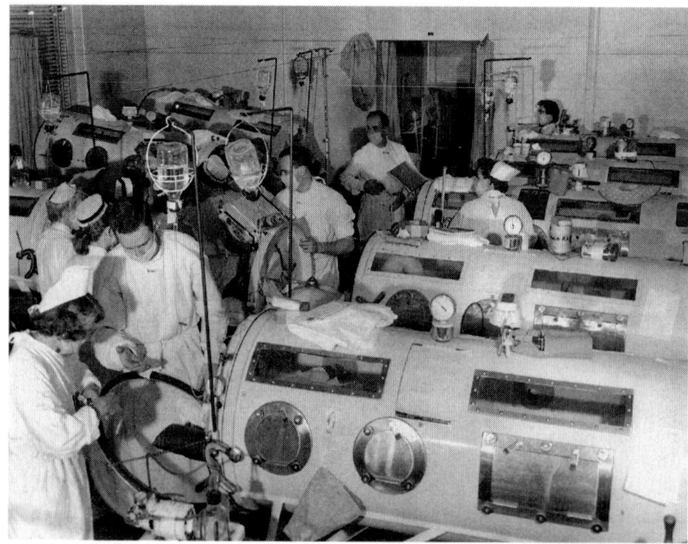

Large numbers of iron lungs in a March of Dimes Respiratory Center in the 1950s. These devices, developed in the late 1920s, provided breathing assistance for polio patients whose respiratory muscles were paralyzed. After several weeks in an iron lung, many patients recovered the ability to breathe on their own.

RELATED CLINICAL TERMS

Neuralgia (nu-ral′je-ah) (*algia* = pain) Sharp, spasm-like pain along the course of one or more nerves, usually caused by inflammation or injury to the nerves.

Neuritis Inflammation of a nerve. There are many different forms of neuritis with different effects, including increased or decreased nerve sensitivity, paralysis of the structure served, and pain.

Nerve injuries The following terms describe various injuries, listed here in order of increasing severity. (1) *Neurapraxia* is a temporary and incomplete loss of nerve function, resulting from pressure and the ensuing ischemia; only the myelin is seriously harmed, and it regenerates so that function returns in a few weeks. (2) *Axonotmesis* is an injury that causes breaks in the axons, but the surrounding epineurium remains intact; axonal regrowth proceeds at the rate of about a centimeter a week. (3) In *neurotmesis*, the whole nerve is severed. Surgical reconnection is necessary for regeneration to occur.

Paresthesia (par″es-the′ze-ah; "faulty sensation") An abnormal sensation (numbness, burning, or tingling) resulting from a disorder of a sensory nerve.

Peripheral neuropathy Any disease of a nerve or its neurons; causes include diabetes, alcoholism, and leprosy.

Scapular winging Condition in which the medial border of the scapula projects posteriorly, due to the paralysis of the serratus anterior muscle. The ultimate cause is damage to the long thoracic nerve supplying the serratus anterior.

CHAPTER SUMMARY

Media study tools that could provide you additional help in reviewing specific key topics of Chapter 14 are referenced below.

SP = Study Partner; AIA = A.D.A.M.® Interactive Anatomy

1. The PNS consists of sensory receptors, motor endings that innervate effectors, nerves, and ganglia. Most nerves contain both sensory and motor fibers (mixed nerves), but some are purely sensory or purely motor.

PERIPHERAL SENSORY RECEPTORS (pp. 411–416)

1. Sensory receptors monitor stimuli (environmental changes) from both inside and outside the body. The widespread receptors for the general senses tend to be dendrites of sensory neurons, whereas the localized receptors for the special senses tend to be distinct receptor cells.

Classification by Location (p. 411)

2. Receptors are classified by location as exteroceptors for external stimuli, interoceptors for internal stimuli in the viscera, and proprioceptors for measuring stretch in muscles and the skeleton.

Classification by Stimulus Detected (pp. 411–413)

3. Receptors are classified by stimuli detected as mechanoreceptors, thermoreceptors, chemoreceptors, photoreceptors, and nociceptors.

Classification by Structure (pp. 413–416)

4. Dendritic endings of sensory neurons monitor general sensory information. Structurally, they are either (1) free dendritic endings or (2) encapsulated dendritic endings. The free endings are mainly receptors for pain and temperature, although two are for light touch: Merkel discs and root hair plexuses. The encapsulated endings are mechanoreceptors: They include Meissner's corpuscles and Krause's end bulbs (discriminative touch), Pacinian corpuscles and Ruffini's corpuscles (deep pressure), and most proprioceptors.

5. The proprioceptors include muscle spindles, Golgi tendon organs, and joint kinesthetic receptors. Muscle spindles are small bags of muscle fibers (intrafusal fibers) innervated by sensory dendrites within the skeletal muscles of the body. They monitor muscle length by measuring muscle stretch.

6. Golgi tendon organs are encapsulated dendritic endings embedded in muscle tendons. They monitor the tension generated in tendons during muscle contraction. Joint kinesthetic receptors occur in joint capsules. Some joint receptors measure tension placed on the joint (Pacinian corpuscles and Ruffini's corpuscles). Free nerve endings in joints are receptors for pain.

PERIPHERAL MOTOR ENDINGS (pp. 416–418)

Innervation of Skeletal Muscle (pp. 416–417)

1. Motor axons innervate skeletal muscle fibers at synapse-like neuromuscular junctions (motor end plates). The motor axon terminal releases acetylcholine, which signals the muscle cell to contract. The synaptic cleft is a grooved gap filled with acetylcholinesterase, an enzyme that rapidly breaks down acetylcholine.

2. A motor unit consists of one motor neuron and all the skeletal muscle fibers it innervates. Motor units contain different numbers of muscle fibers distributed widely within a muscle. All muscle fibers in the unit contract simultaneously.

Innervation of Visceral Muscle and Glands (pp. 417–418)

3. Visceral motor neurons do not form elaborate neuromuscular junctions with the visceral muscle and glands they innervate. Their axon terminals (varicosities) may even end some distance from the effector cells.

CRANIAL NERVES (pp. 418–426)

1. Twelve pairs of cranial nerves originate from the brain and issue through the skull to innervate the head and neck. Only the vagus nerves extend into the thoracic and abdominal cavities.

2. The cranial nerves can be grouped according to function. Some are purely *sensory,* containing special somatic sensory fibers (I, II, VIII); some are primarily or exclusively *motor,* containing general somatic motor fibers (III, IV, VI, and XII); and some are *mixed,* containing both sensory and motor fibers (V, VII, IX, X/XI). Note that in this grouping scheme, the purely motor accessory nerve (XI) is combined with the vagus (X), to which it is related.

3. The following are the cranial nerves:

- The olfactory nerves (I): purely sensory; concerned with smell; attach to the olfactory bulb of the telencephalon.

AIA Link: Atlas; System: Nervous; Olfactory nerve in nasal cavity.

- The optic nerves (II): purely sensory; transmit visual impulses from the retina to the thalamus.
- The oculomotor nerves (III): primarily motor; emerge from the midbrain and serve four extrinsic muscles that move the eye and some smooth muscle within the eye; also carry proprioceptive impulses from the skeletal muscles served.
- The trochlear nerves (IV): primarily motor; emerge from the dorsal midbrain and carry motor and proprioceptive impulses to and from superior oblique muscles of the eyes.
- The trigeminal nerves (V): contain both sensory and motor fibers; emerge from the pons; the main general sensory nerves of the face, nasal cavity, and mouth; three divisions: ophthalmic, maxillary, and mandibular; the mandibular branch contains motor fibers that innervate the chewing muscles.
- The abducens nerves (VI): primarily motor; emerge from the pons and supply motor and proprioceptive innervation to the lateral rectus muscles of the eyes.

AIA Link: Atlas; System: Nervous; Abducent nerve in orbit.

- The facial nerves (VII): contain both sensory and motor fibers; emerge from the pons; motor nerves to muscles of facial expression and many glands in head; also carry sensory impulses from the taste buds of anterior two-thirds of the tongue.
- The vestibulocochlear nerves (VIII): purely sensory; enter brain at pons/medulla border; transmit impulses from the hearing and equilibrium receptors of the inner ears.
- The glossopharyngeal nerves (IX): contain both sensory and motor fibers; issue from the medulla; transmit sensory impulses from the posterior tongue (including taste) and the pharynx; innervate the stylopharyngeus muscle and parotid gland.
- The vagus nerves (X): contain both sensory and motor fibers; arise from the medulla; most of its motor fibers are autonomic; carries motor fibers to, and sensory fibers from, the pharynx, larynx, and visceral organs of the thorax and abdomen.

AIA Link: Atlas; System: Nervous; Accessory Nerve (lateral).

- The accessory nerves (XI): motor only; have a cranial root arising from the medulla and a spinal root arising from the cervical spinal cord; cranial root runs in vagus to supply muscles of larynx and pharynx; spinal root supplies trapezius and sternocleidomastoid.
- The hypoglossal nerves (XII): motor only; issue from the medulla; supply motor innervation to the tongue muscles.

SP Exercise: Chapter 14, Location of the Cranial Nerves.

SPINAL NERVES (pp. 426–436)

1. The 31 pairs of spinal nerves are numbered according to the region of the vertebral column from which they issue: cervical, thoracic, lumbar, sacral, and coccygeal. There are eight cervical spinal nerves but only seven cervical vertebrae.

SP Exercise: Chapter 14, Distribution of Spinal Nerves.

2. Each spinal nerve forms from the union of a dorsal sensory root and a ventral motor root in an intervertebral foramen. On each dorsal root is a dorsal root ganglion containing the cell bodies of sensory neurons. The branches of each spinal nerve are the dorsal and ventral rami (somatic branches) and rami communicantes (visceral branches).
3. Dorsal rami serve the muscles and skin of the posterior trunk, whereas ventral rami serve the muscles and skin of the lateral and anterior trunk. Ventral rami also serve the limbs.
4. The back, from neck to sacrum, is innervated by the dorsal rami in a neatly segmented pattern.
5. The anterior and lateral wall of the thorax (and abdomen) are innervated by thoracic ventral rami—the segmented intercostal and subcostal nerves. Thoracic ventral rami do not form nerve plexuses.

AIA Link: Atlas; System: Nervous; Spinal nerves of trunk (anterior).

6. The major nerve plexuses are networks of successive ventral rami that exchange fibers. They primarily innervate the limbs.

SP Exercise: Chapter 14, Formation and Branches of Spinal Nerves.

7. The cervical plexus (C_1–C_4) innervates the muscles and skin of the neck and shoulder. Its phrenic nerve (C_3–C_5) serves the diaphragm.
8. The brachial plexus serves the upper limb, including the muscles of the shoulder girdle. It arises primarily from C_5–T_1. From proximal to distal, the brachial plexus consists of "roots" (rami), trunks, divisions, and cords.
9. The main nerves arising from the brachial plexus are the axillary (to the deltoid muscle), musculocutaneous (to flexors on the arm), median (to anterior forearm muscles and lateral palm), ulnar (to anteromedial muscles of the forearm and the medial hand), and radial (to the posterior part of the limb).
10. The lumbar plexus (L_1–L_4) innervates the anterior and medial muscles of the thigh, through the femoral and obturator nerves, respectively. The femoral nerve also innervates the skin on the anterior thigh and medial leg.
11. The sacral plexus (L_4–S_4) supplies the muscles and skin of the posterior thigh and almost all of the leg. Its main branch is the large sciatic nerve, which consists of the tibial nerve (to most of the hamstrings and all muscles of the calf and sole) and the common fibular nerve (to muscles of the anterior and lateral leg and the overlying skin). Other branches of the sacral plexus supply the posterior and lateral muscles of the pelvic girdle (such as the gluteal muscles) and the perineum (pudendal nerve).
12. The sensory nerves to a joint are branches of the nerves innervating the muscles that cross that joint (Hilton's law).
13. A dermatome is a segment of skin innervated by the sensory fibers of a single spinal nerve. Adjacent dermatomes overlap to some degree, more so on the trunk than on the limbs. Loss of sensation in specific dermatomes reveals sites of damage to spinal nerves or the spinal cord.

DISORDERS OF THE PERIPHERAL NERVOUS SYSTEM (p. 437)

1. Shingles and migraine headaches are painful sensory disorders of dermatomes and the brain arteries, respectively. Myasthenia gravis is a muscle-weakening disorder of the neuromuscular junctions.

THE PERIPHERAL NERVOUS SYSTEM THROUGHOUT LIFE (p. 437)

1. During embryonic development, each spinal nerve grows out between newly formed vertebrae to provide the motor innervation of an adjacent myotome (future trunk muscle) and the sensory innervation of the adjacent skin region (dermatome).

REVIEW QUESTIONS

Multiple Choice/Matching Questions

1. Proprioceptors include all of the following except (a) muscle spindles, (b) Golgi tendon organs, (c) Merkel discs, (d) Pacinian corpuscles in joint capsules.
2. The large, onion-shaped pressure receptors in deep connective tissues are (a) Merkel discs, (b) Pacinian corpuscles, (c) free nerve endings, (d) Krause's end bulbs.
3. Match the receptor type from column B to the correct description in column A.

Column A	Column B
____ **(1)** pain, itch, and temperature receptors	**(a)** Ruffini's corpuscles
____ **(2)** look and function like Meissner's corpuscles but occur in mucosae	**(b)** Golgi tendon organs
____ **(3)** contain intrafusal fibers and flower spray endings	**(c)** muscle spindles
____ **(4)** discriminative touch receptors in hairless skin (fingertips)	**(d)** Krause's end bulbs
____ **(5)** contain dendrites wrapped around thick collagen bundles in a tendon	**(e)** free dendritic endings
____ **(6)** rapidly adapting deep-pressure receptors	**(f)** Pacinian corpuscles
____ **(7)** slowly adapting deep-pressure receptors	**(g)** Meissner's corpuscles

4. Match the names of the cranial nerves in column B to the appropriate description in column A.

Column A	Column B
____ **(1)** moves four extrinsic eye muscles	**(a)** abducens
____ **(2)** is the major sensory nerve of the face	**(b)** accessory
____ **(3)** serves the sternocleidomastoid and trapezius	**(c)** facial
____ **(4)** are purely sensory (three nerves)	**(d)** glossopharyngeal
____ **(5)** serves the tongue muscles	**(e)** hypoglossal
____ **(6)** allows chewing of food	**(f)** oculomotor
____ **(7)** is impaired in tic douloureux	**(g)** olfactory
____ **(8)** helps to regulate heart activity	**(h)** optic
____ **(9)** helps hearing and equilibrium	**(i)** trigeminal
____ **(10)** contain parasympathetic motor fibers (four nerves)	**(j)** vestibulocochlear
____ **(11)** serves muscles of facial expression	**(k)** vagus
	(l) trochlear

5. For each of the following muscles or body regions, identify the plexus and the peripheral nerve (or branch of one) involved. Use one choice from Key A followed by one choice from Key B.

Key A: Plexuses	Key B: Nerves
(a) brachial	**(1)** common fibular
(b) cervical	**(2)** femoral
(c) lumbar	**(3)** median
(d) sacral	**(4)** musculocutaneous
	(5) obturator
	(6) phrenic
	(7) radial
	(8) tibial
	(9) ulnar

Structure innervated

____, ____ **(1)** the diaphragm

____, ____ **(2)** muscles of the posterior thigh and posterior leg

____, ____ **(3)** anterior thigh muscles

____, ____ **(4)** medial thigh muscles

____, ____ **(5)** anterior arm muscles that flex the forearm

____, ____ **(6)** muscles that flex the wrist and fingers (two nerves)

____, ____ **(7)** muscles that extend the wrist and fingers

____, ____ **(8)** skin and extensor muscles of the posterior arm

____, ____ **(9)** fibular muscles, tibialis anterior, and toe extensors

6. Which one of the following contains only motor fibers? (a) dorsal root, (b) dorsal ramus, (c) ventral root, (d) ventral ramus.

7. The trigeminal nerve contains which class(es) of nerve fibers? (a) somatic sensory only, (b) somatic motor and proprioceptor, (c) somatic sensory, visceral sensory, and branchial motor.

8. Whereas *branchial* refers to the pharynx, *brachial* refers to the (a) back, (b) gills, (c) arm, (d) thigh.

9. Which cranial nerves contain branchial motor axons? (a) I, II, VII; (b) V, VII, IX, X, XI; (c) III, IV, VI, XI, XII.

10. Which of the following components occur in all of the mixed cranial nerves V, VII, IX, and X? (choose as many as apply): (a) general somatic sensory; (b) special somatic sensory; (c) general visceral sensory; (d) special visceral sensory (taste), (e) general somatic motor, (f) branchial motor.

Short Answer Essay Questions

11. In the sensory receptors called "encapsulated dendritic endings," what is the "capsule" made of?

12. Define motor unit.

13. Name two ways in which a neuromuscular junction differs from a typical synapse between two neurons.

14. (a) Describe the roots to and the composition of a spinal nerve. (b) Are dorsal and ventral roots in the CNS or the PNS? (c) Name the branches of a spinal nerve (other than the rami communicantes), and tell what basic region of the body each branch supplies.

15. (a) Define nerve plexus. (b) List the spinal nerves of origin of the four major nerve plexuses, and name the general body regions served by each plexus.

16. In the brachial plexus, what specific rami make up each of the three trunks?

17. Use Hilton's law to identify the nerves that supply the hip joint.

18. Adrian and Abdul, two anatomy students, were arguing about the facial nerve. Adrian said that it innervates all of the skin of the face, and that it is called the facial nerve for this reason. Abdul claimed the facial nerve does not innervate facial skin at all. Who was more correct?

19. Choose the correct answer, and explain why it is correct. Through which pair of intervertebral foramina do spinal nerves L_5 leave the vertebral canal? (a) holes in the sacrum, (b) just superior to vertebra L_5, (c) just superior to S_1, (d) just inferior to S_4.

20. There are 40 roots in the cauda equina. To what spinal nerves do they belong? (Hint: see pp. 393 and 395.)

CRITICAL REASONING + CLINICAL APPLICATION QUESTIONS

1. As Harry was falling off a ladder, he reached out and grabbed a tree branch with his hand. His overstretched upper limb became weak and numb. What major nervous structure had he injured?

2. Frita, in her early 70s, had problems chewing. When she was asked to stick out her tongue, it deviated to the right, and its right half was quite wasted. What nerve was injured?

3. Ted is a war veteran who was hit in the back with small pieces of shrapnel. His skin is numb in the center of his buttocks and along the entire posterior side of a lower limb, but there is no motor problem. Indicate which of the following choices is the most likely site of his nerve injury, and explain your choice: (a) a few dorsal roots of the cauda equina, (b) spinal cord transection at C_6, (c) spinal cord transection at L_5, (d) femoral nerve transected in the lumbar region.

4. The quarterback for Wattsamatta U. suffered torn menisci in his right knee joint when he was tackled from the side. The same collision crushed his common fibular nerve against the neck of his right fibula. What locomotory problems could he expect to experience from this nerve injury?

5. In a patient that developed carpal tunnel syndrome, the nerve injury began as a neurapraxia and then became an axonotmesis. (1) Recall what nerve is affected in carpal tunnel syndrome. (2) Explain the terms *neurapraxia* and *axonotmesis*. (3) Estimate how long full recovery from this nerve injury will take. (Hints: See p. 193 and this chapter's Related Clinical Terms.)

A.D.A.M.® INTERACTIVE ANATOMY QUESTIONS

1. Link: Atlas; System: Nervous; external, middle and inner ear. (a) Branches of which cranial nerve innervate the inner ear? (b) The vestibular branch innervates which structure of the inner ear?

2. Link: Atlas; System: Nervous; Brachial plexus. (a) The brachial plexus serves which part of the body? (b) Name two nerves that branch from the lateral cord of the brachial plexus.

3. Link: Atlas; Region: Head and neck: Lateral rectus muscle (lateral). (a) Which cranial nerve innervates the lateral rectus muscle? (b) Through which opening does this nerve enter the orbit?

CHAPTER 15
The Autonomic Nervous System and Visceral Sensory Neurons

CHAPTER OUTLINE + STUDENT OBJECTIVES

1. Define the autonomic nervous system (ANS), and explain its relationship to the peripheral nervous system as a whole.

Introduction to the Autonomic Nervous System 444–447

2. Compare autonomic neurons to somatic motor neurons.
3. Describe the basic differences between the parasympathetic and sympathetic divisions of the autonomic nervous system.

The Parasympathetic Division 447–451

4. Describe the anatomy of the parasympathetic division, and explain how it relates to the brain, the cranial nerves, and the sacral spinal cord.

The Sympathetic Division 452–457

5. Describe the anatomy of the sympathetic division, and explain how it relates to the spinal cord and spinal nerves.
6. Explain the sympathetic function of the adrenal medulla.

Central Control of the Autonomic Nervous System 457–458

7. Explain how various regions of the CNS help to regulate the autonomic nervous system.

Visceral Sensory Neurons 458–459

8. Describe the role and location of visceral sensory neurons relative to autonomic neurons.
9. Explain the concept of referred pain.

Visceral Reflexes 459

10. Explain how spinal and peripheral reflexes regulate some functions of visceral organs.

Disorders of the Autonomic Nervous System 459–460

11. Briefly describe some diseases of the autonomic nervous system.

The Autonomic Nervous System Throughout Life 460

12. Describe the embryonic development of the autonomic nervous system.
13. List some effects of aging on autonomic functions.

CONSIDER THE FOLLOWING SITUATIONS: YOU wake up at night after having eaten at a restaurant where the food did not taste quite right, and you find yourself waiting helplessly for your stomach to "decide" whether it can hold the food down. A few days later, you are driving to school after drinking too much coffee and wish in vain that your full bladder would stop its uncomfortable contractions. Later that day, your professor asks you a hard question in front of the class, and you try not to let them see you sweat—but the sweat runs down your face anyway. All of these are examples of visceral motor functions that are not easily controlled by the conscious will and that sometimes seem to "have a mind of their own." These functions are performed by the **autonomic nervous system (ANS),** a motor system that does indeed operate with a certain amount of independence (*autonomic* = "self-governing").

The ANS is the system of motor neurons that innervate the smooth muscle, cardiac muscle, and glands of the body. By controlling these effectors, the ANS regulates such visceral functions as heart rate, blood pressure, digestion, and urination, which are essential for maintaining the stability of the body's internal environment. The ANS is the *general visceral motor* division of the peripheral nervous system and is distinct from the *general somatic motor* and *branchial motor* divisions, which innervate the skeletal muscles (Figure 15.1).

Although we will focus on autonomic (general visceral motor) functions, we will also consider the **general visceral senses.** The general visceral sensory system continuously monitors the activities of the visceral organs so that the autonomic motor neurons can make adjustments as necessary to ensure optimal performance of visceral functions.

INTRODUCTION TO THE AUTONOMIC NERVOUS SYSTEM

In this section we first compare the autonomic nervous system to the somatic motor system, and then we examine the two divisions of the ANS.

Comparison of the Autonomic and Somatic Motor Systems

Our previous discussions have focused largely on the *somatic* motor system, which innervates skeletal muscles. Recall that each somatic motor neuron runs from

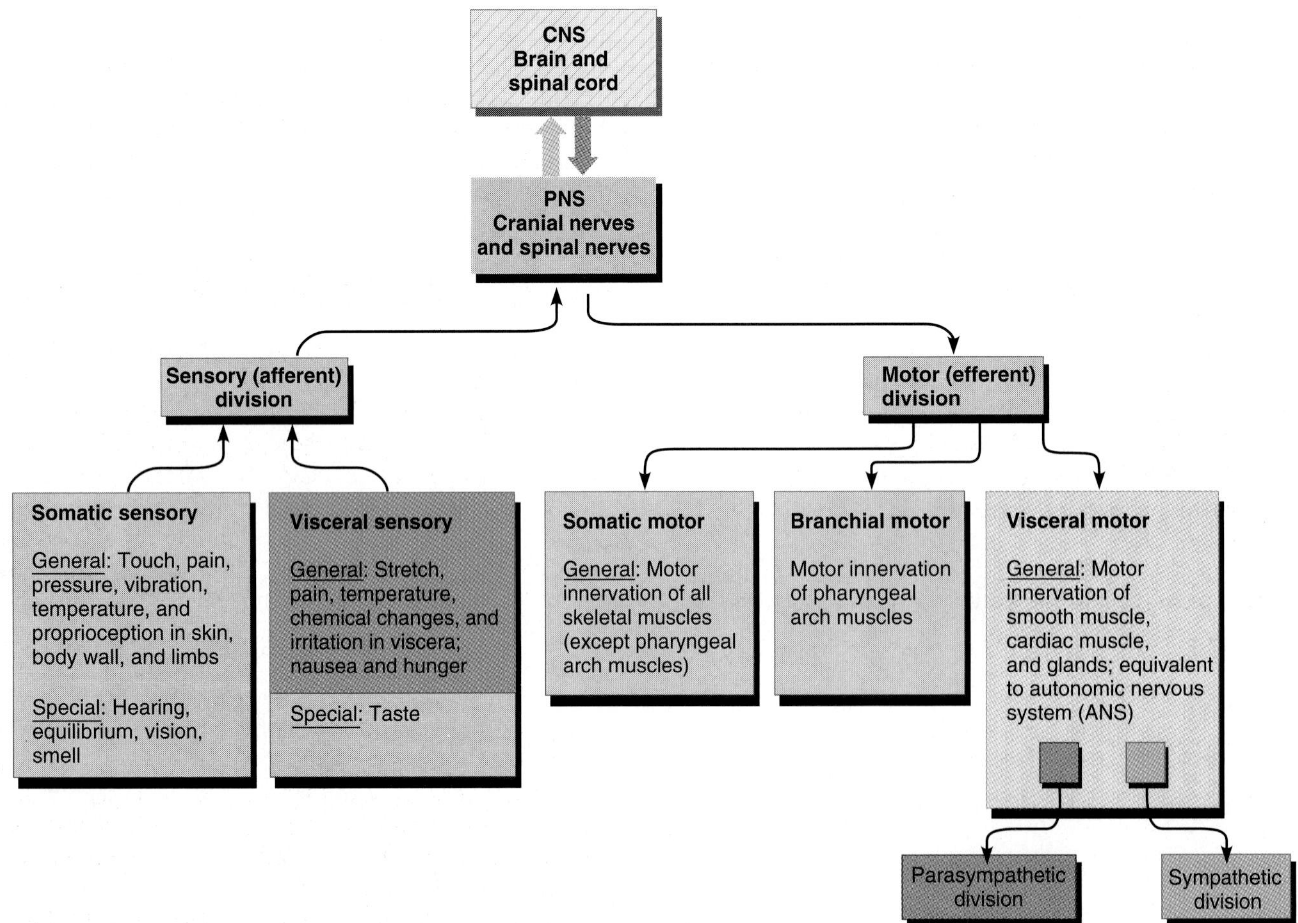

FIGURE 15.1 Positions of the general visceral motor system (autonomic nervous system), and the general visceral sensory system in the nervous system hierarchy.

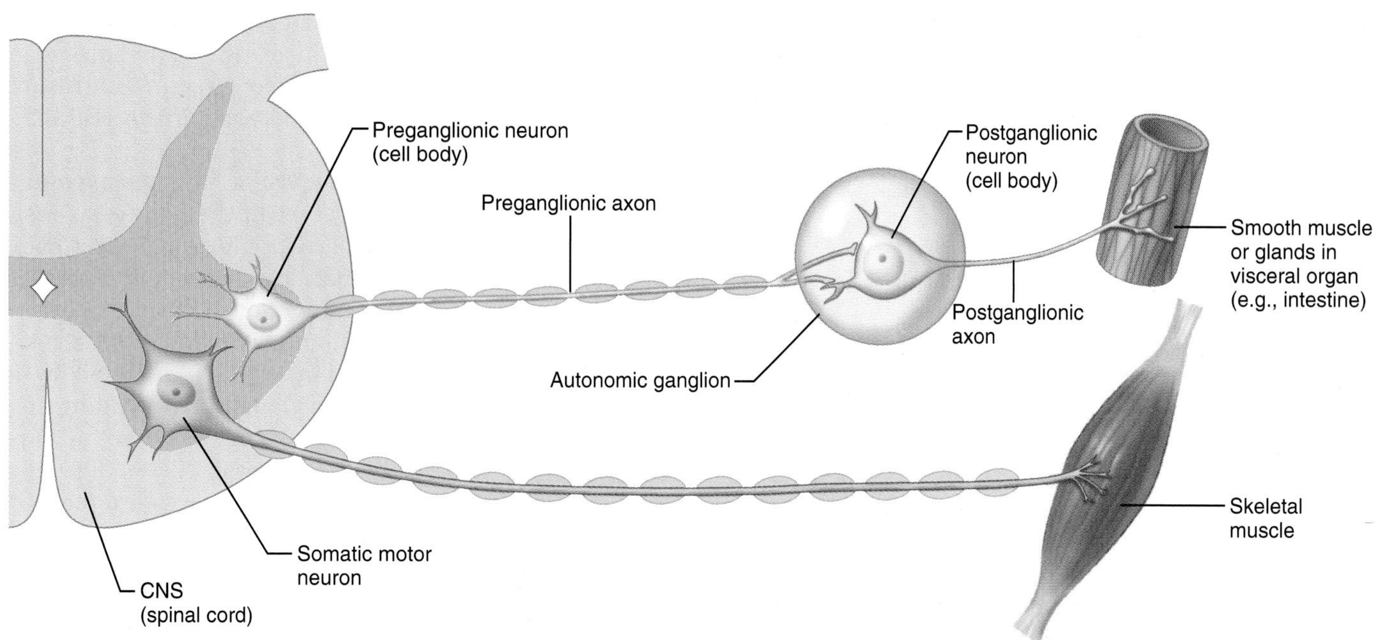

FIGURE 15.2 Autonomic and somatic motor nervous systems. In the somatic motor system (pink neuron below), single neurons extend from the CNS to skeletal muscle cells. In the autonomic nervous system (yellow neurons above), a chain of two motor neurons—preganglionic and postganglionic neurons—runs between the CNS and the visceral effector cells.

the central nervous system all the way to the muscle being innervated, and that each motor unit consists of a single neuron plus the skeletal muscle cells it innervates. Typical somatic motor axons are thick, heavily myelinated fibers that conduct nerve impulses rapidly.

By contrast, the comparable motor unit in the ANS includes a chain of *two* motor neurons (Figure 15.2). The first of these is called a **preganglionic neuron,** the cell body of which lies within the spinal cord or brain. Its axon, the **preganglionic axon,** synapses with the second motor neuron, the **postganglionic neuron,** in a peripheral **autonomic ganglion.** The **postganglionic axon** then extends to the visceral organs. Functionally, the preganglionic neuron signals the postganglionic neuron, which then stimulates muscle contraction or gland secretion in the effector organ. Preganglionic axons are thin, lightly myelinated fibers, whereas postganglionic axons are even thinner and are unmyelinated. Consequently, the conduction of impulses through the autonomic nervous system is slower than conduction through the somatic motor system.

It is important to emphasize that the autonomic ganglia are *motor* ganglia containing the cell bodies of motor neurons. They are *not sensory* ganglia, like the dorsal root ganglia considered in previous chapters.

Critical Reasoning

Given that two-neuron motor units are unique to the ANS, is the motor innervation of the pharyngeal arch muscles *(chewing and swallowing muscles and so on) through two-neuron chains or via single motor neurons?*

Because the pharyngeal arch muscles are typical skeletal muscles in structure and developmental origin, they are innervated by single motor neurons—just like the general skeletal muscles of the body (see Figure 15.1). Thus, branchial motor and somatic motor innervation resemble each other and differ from autonomic innervation.

Divisions of the Autonomic Nervous System

The ANS has two divisions, the *sympathetic* and *parasympathetic* (par″ah-sim″pah-thet′ik) *divisions,* which are shown issuing from the brain and spinal cord in Figure 15.3. Both divisions have chains of two motor neurons that mostly innervate the same visceral organs, but they cause opposite effects: Whereas one division stimulates some smooth muscle to contract or a gland to secrete, the other division inhibits that action. Additionally, whereas the sympathetic division mobilizes the body during extreme situations such as fear, exercise, or rage, the parasympathetic division enables us to unwind and relax and works to conserve body energy. In other words, the parasympathetic division controls routine maintenance functions, and the sympathetic division becomes active when extra effort is needed. The balance between the two divisions keeps our body systems running smoothly. Next we consider these functional differences in greater detail.

The **sympathetic division** is responsible for the "fight, flight, or fright" response. Its activity is evident during vigorous exercise, excitement, or emergencies. A

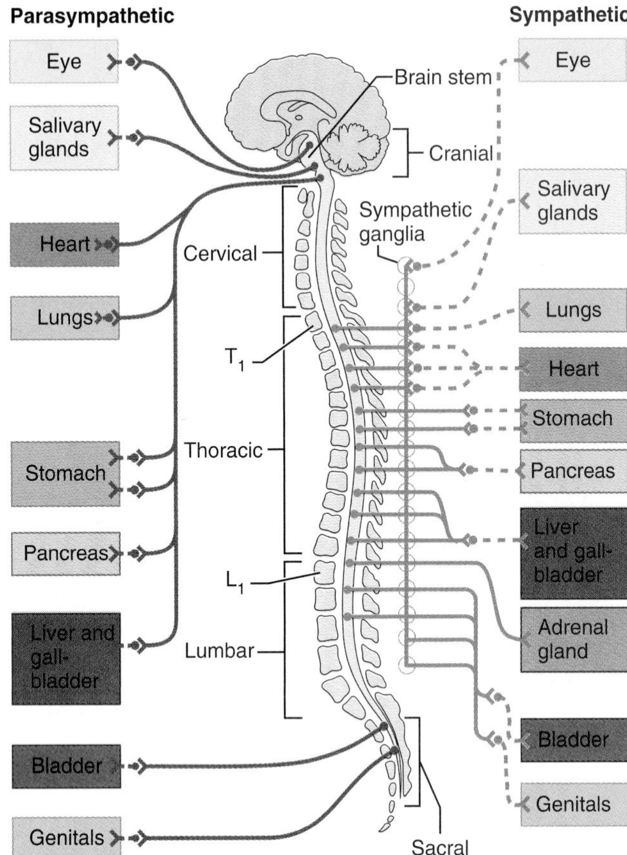

FIGURE 15.3 Overview of the sympathetic and parasympathetic divisions of the ANS, issuing from the brain and spinal cord. The two-neuron pathways are shown for both divisions: Solid lines indicate preganglionic axons, whereas broken lines indicate postganglionic axons.

pounding heart, fast and deep breathing, dilated (widened) eye pupils, and cold, sweaty skin are signs of mobilization of the sympathetic division, and all of them help us survive danger: The increased heart rate and breathing rate deliver more blood and oxygen to the skeletal muscles used for fighting or running; widened pupils let in more light for clearer vision; and cold skin indicates that blood is being diverted from the skin to more vital organs, such as the brain. Furthermore, the small air tubes in the lungs (bronchioles) dilate, increasing the uptake of oxygen; oxygen consumption by the body's cells increases; and the liver releases more sugar into the blood to provide for the increased energy needs of the cells. In this way, our "motors are revved up" for vigorous activity. Temporarily nonessential functions, such as digestion and motility of the urinary tract, are inhibited: When you are running from a mugger, digesting lunch can wait.

The sympathetic system also innervates blood vessels, sending signals to the smooth muscle in their walls. Even though sympathetic input causes the smooth muscle in *some* vessels (those in skeletal muscles, for example) to relax so that the vessel dilates, the bulk of sympathetic input signals smooth muscle in blood vessels to contract, producing **vasoconstriction.** This narrowing of vessel diameter forces the heart to work harder in pumping blood around the vascular circuit. As a result, sympathetic activity causes blood pressure to rise during excitement and stress.

Unlike the sympathetic division, the **parasympathetic division** is most active when the body is at rest. This division is chiefly concerned with conserving body energy even as it directs vital "housekeeping" activities such as digestion and the elimination of feces and urine. The "buzzwords" to remember are "resting and digesting." Parasympathetic function is best illustrated by a person who is relaxing after dinner and reading the newspaper. Heart rate and respiratory rates are at low-normal levels, and the gastrointestinal tract is digesting food. The pupils are constricted as the eyes are focused for close vision.

As we explore the sympathetic and parasympathetic divisions in detail, you will find their effects on individual visceral organs are easy to learn if you just remember "fight, flight, or fright" (sympathetic) versus "resting and digesting" (parasympathetic). Furthermore, a dynamic counteraction exists between the two divisions, such that they balance each other during times when we are neither highly excited nor completely at rest.

In addition to these functional differences, there are anatomical and biochemical differences between the sympathetic and parasympathetic divisions. First, the two divisions issue from different regions of the CNS (see Figure 15.3). The sympathetic division can also be called the **thoracolumbar division** because its fibers emerge from the thoracic and superior lumbar parts of the spinal cord. The parasympathetic division can also be termed the **craniosacral division** because its axons emerge from the brain *(cranial part)* and the sacral spinal cord *(sacral part).*

A second anatomical difference between the two divisions is that sympathetic pathways have short preganglionic fibers and long postganglionic fibers, whereas the opposite is true for parasympathetic pathways (Figure 15.4). Therefore, all sympathetic ganglia lie near the spinal cord and vertebral column, whereas all parasympathetic ganglia lie far from the CNS, in or near the organs innervated.

A third anatomical difference between the two divisions is that sympathetic axons branch profusely, whereas parasympathetic fibers do not (see Figure 15.4). Such extensive branching allows each sympathetic neuron to influence a number of different visceral organs, enabling many organs to mobilize simultaneously during the "fight, flight, or fright" response. Indeed, the literal translation of *sympathetic,* "experienced together," reflects the bodywide mobilization it provides. Parasympathetic effects, by contrast, are more localized and discrete.

The main biochemical difference between the two divisions of the ANS involves the neurotransmitter released by the *post*ganglionic axons (see Figure 15.4).

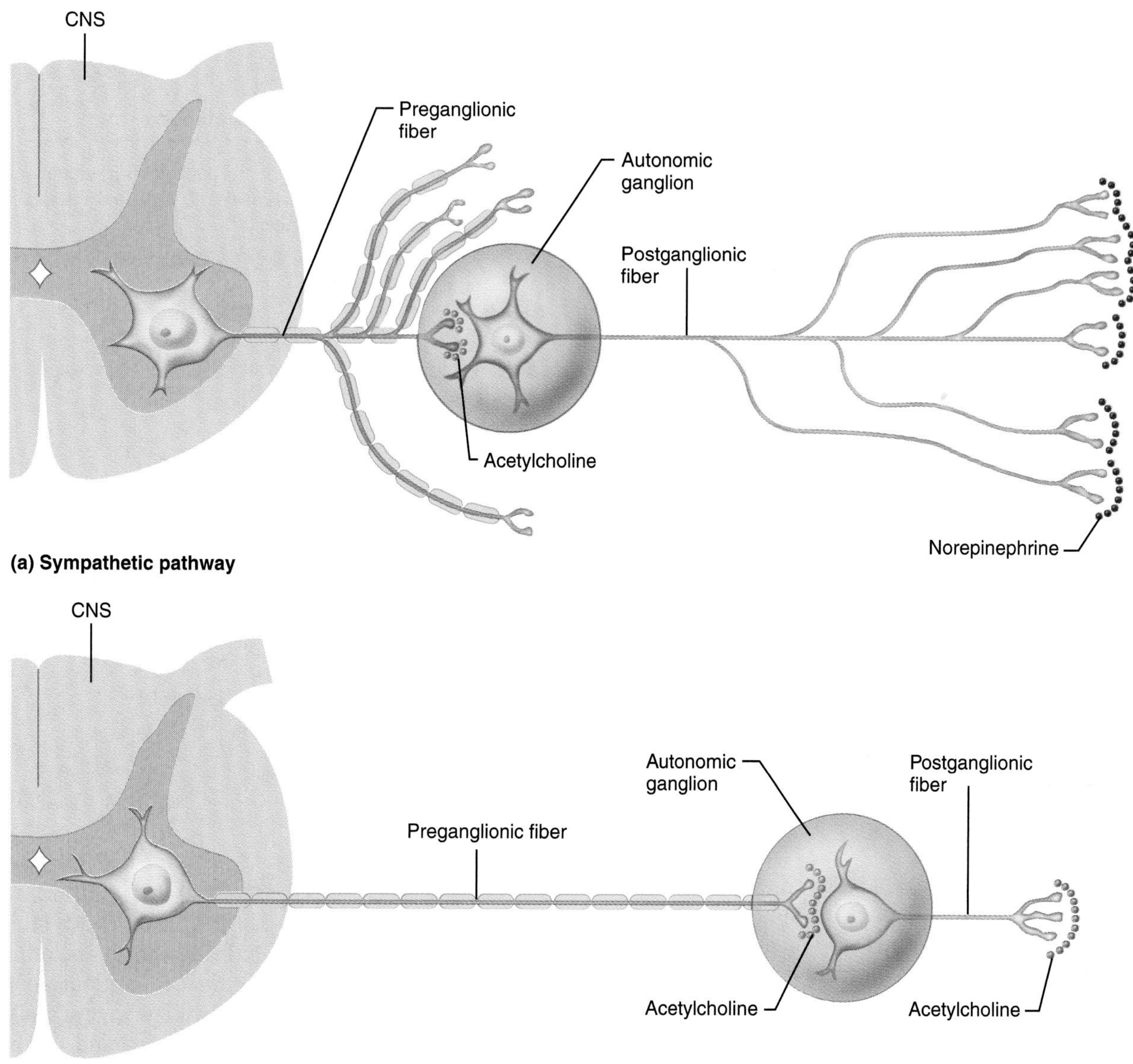

FIGURE 15.4 Some basic differences between the sympathetic and parasympathetic divisions. (a) The sympathetic division has shorter preganglionic than postganglionic neurons, its axons branch more, and its postganglionic axons typically secrete norepinephrine. **(b)** The parasympathetic division has longer preganglionic than postganglionic neurons, its axons branch less, and its postganglionic axons secrete acetylcholine.

In the sympathetic division, most postganglionic axons release *norepinephrine* (also called *noradrenaline*); these fibers are termed *adrenergic.** The postganglionic neurotransmitter in the parasympathetic division is *acetylcholine (ACh);* these fibers are termed *cholinergic.* The *pre*ganglionic axon terminals of both divisions are always cholinergic (release ACh).

*Not all postganglionic fibers in the sympathetic division are adrenergic: Those that innervate the sweat glands and blood vessels in skeletal muscle are cholinergic.

The main anatomical and physiological differences between the parasympathetic and sympathetic divisions are summarized in Table 15.1. Next we turn to a more detailed discussion of the anatomical organization of each division, starting with the somewhat simpler parasympathetic division.

THE PARASYMPATHETIC DIVISION

The *cranial outflow* of the parasympathetic division comes from the brain and innervates organs in the head,

Table 15.1 Anatomical and Physiological Differences Between the Parasympathetic and Sympathetic Divisions

Characteristic	Sympathetic	Parasympathetic
Origin	Thoracolumbar outflow; lateral horn of gray matter of spinal cord segments T_1–L_2	Craniosacral outflow: brain stem nuclei of cranial nerves III, VII, IX, and X; spinal cord segments S_2–S_4
Location of ganglia	Ganglia close to CNS: alongside vertebral column (sympathetic trunk ganglia) and anterior to vertebral column (prevertebral ganglia)	Ganglia in or close to visceral organ served
Relative length of pre- and postganglionic fibers	Short preganglionic; long postganglionic	Long preganglionic; short postganglionic
Rami communicantes (see p. 452)	Gray and white rami communicantes; white contain myelinated preganglionic fibers; gray contain unmyelinated postganglionic fibers	None
Degree of branching of preganglionic fibers	Extensive	Minimal
Functional role	Prepares body to cope with emergencies and intense muscular activity	Maintenance functions; conserves and stores energy
Neurotransmitters	All preganglionic fibers release ACh; most postganglionic fibers release norepinephrine (adrenergic fibers); some postganglionic fibers (e.g., those serving sweat glands and blood vessels of skeletal muscles) release ACh; neurotransmitter activity augmented by release of adrenal medullary hormones (epinephrine and norepinephrine)	All fibers release ACh (cholinergic fibers)

neck, thorax, and most of the abdomen, whereas the *sacral outflow* comes from the sacral spinal cord and supplies the rest of the abdominal organs and the pelvic organs (Figure 15.5). We begin with the cranial outflow.

Cranial Outflow

The cranial parasympathetic outflow is contained in several cranial nerves. More specifically, the preganglionic fibers run via the oculomotor (III), facial (VII), glossopharyngeal (IX), and vagus (X) nerves (see the top part of Figure 15.5). The cell bodies of these preganglionic neurons are located in motor cranial nerve nuclei in the gray matter of the brain stem (see Figure 13.20d, p. 383). The precise locations of both the preganglionic and postganglionic neurons of each cranial parasympathetic pathway are next examined in turn.

Outflow via the Oculomotor Nerve (III)

The parasympathetic fibers of the oculomotor nerve innervate smooth muscles in the eye that cause the pupil to constrict and the lens of the eye to bulge—actions that allow focusing on close objects in the field of vision. In this two-neuron pathway, the preganglionic axons found in the oculomotor nerve issue from cell bodies in the accessory *oculomotor nucleus* in the midbrain (see Figure 13.20d); the postganglionic cell bodies lie in the **ciliary ganglion,** in the posterior part of the orbit just lateral to the optic nerve (see Table 14.3, p. 420).

Outflow via the Facial Nerve (VII)

The parasympathetic fibers of the facial nerve stimulate the secretion of many glands in the head, including the lacrimal (tear) gland above the eye; mucus-secreting glands in the nasal cavity; and two salivary glands inferior to the mouth (the submandibular and sublingual glands). In the pathway leading to the lacrimal and nasal glands, the preganglionic neurons originate in the *lacrimal nucleus* in the pons and synapse with postganglionic neurons in the **pterygopalatine ganglion,** just posterior to the maxilla (see Table 14.3, p. 423). In the pathway leading to the submandibular and sublingual glands, the preganglionic neurons originate in the *superior salivatory nucleus* in the pons and synapse with postganglionic neurons in the **submandibular ganglion,** deep to the mandibular angle.

Outflow via the Glossopharyngeal Nerve (IX)

The parasympathetic fibers of the glossopharyngeal nerve stimulate secretion of a large salivary gland called the parotid gland, which lies anterior to the ear. The preganglionic neurons originate in the *inferior salivatory nucleus* in the medulla and synapse with postganglionic neurons in the **otic ganglion** inferior to the foramen ovale of the skull (see Table 14.3, p. 424).

The three cranial nerves considered so far (III, VII, IX) supply the entire parasympathetic innervation of the head. Note, however, that only the *pre*ganglionic fibers

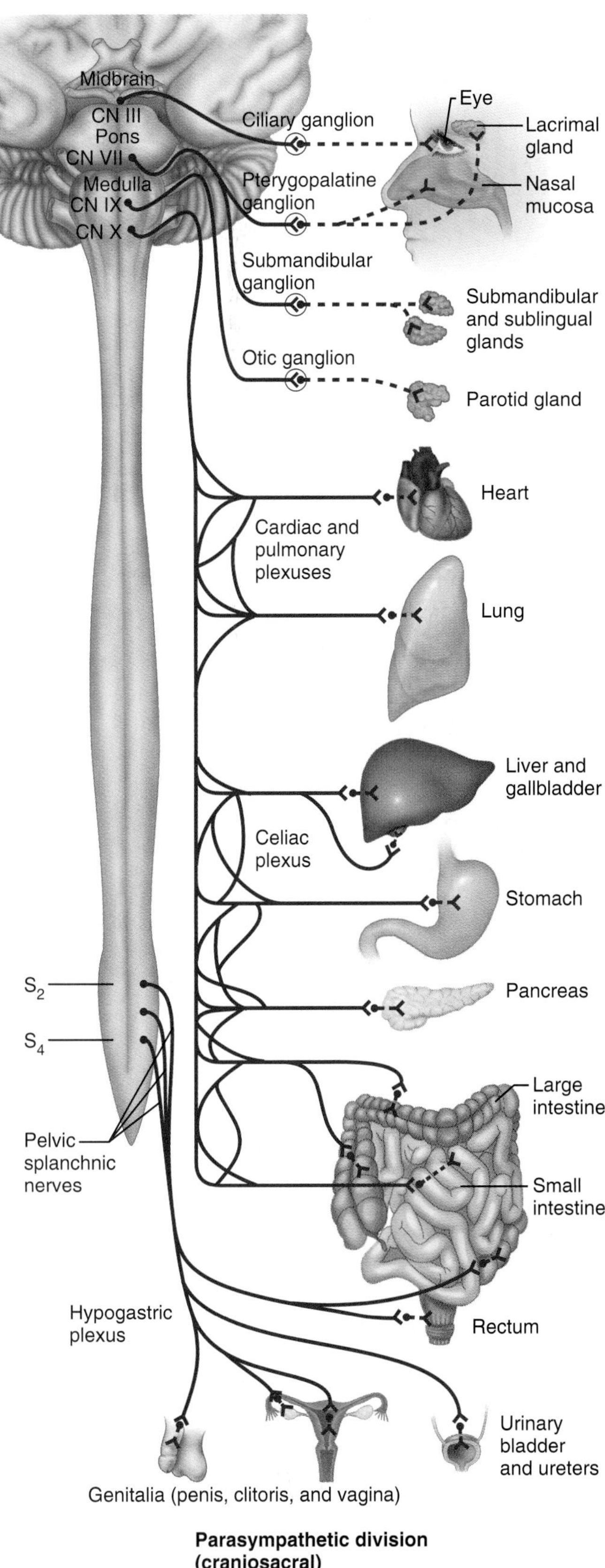

FIGURE 15.5 Parasympathetic division of the ANS. Solid lines indicate preganglionic nerve fibers, broken lines indicate postganglionic fibers. Note: CN = cranial nerve.

run within these three nerves. The *post*ganglionic fibers do not because the distal ends of the preganglionic fibers join branches of the trigeminal nerve (V) and then the postganglionic fibers travel in the trigeminal to their final destinations. This routing to the trigeminal nerve probably occurs because the trigeminal has the widest distribution within the face and thus is best suited for reaching the widely separated glands and smooth muscles in the head.

Outflow via the Vagus Nerve (X)

Parasympathetic fibers from one additional cranial nerve, the vagus, innervate the visceral organs of the thorax and most of the abdomen (see Figure 15.5). Note that this does not include innervation of the pelvic organs, and that the vagal innervation of the digestive tube ends halfway along the large intestine. The vagus is an extremely important part of the ANS, containing nearly 90% of the preganglionic parasympathetic fibers in the body. Functionally, the parasympathetic fibers in the vagus nerve bring about typical "resting and digesting" activities in visceral muscle and glands—stimulation of digestion, reduction in the heart rate, and constriction of the bronchi in the lungs, for example.

The preganglionic cell bodies are mostly in the *dorsal motor nucleus* in the medulla, and the preganglionic axons run through the entire length of the vagus nerve. Most postganglionic neurons are confined within the walls of the organs being innervated, and their cell bodies form *intramural ganglia* (in″trah-mu′ral; "within the walls").

Because the vagus nerve is essential to the functioning of many organs, we will briefly follow its path through the body (Figure 15.6). As the vagus descends through the neck and trunk, it sends branches through many *autonomic nerve plexuses* to the organs being innervated. Specifically, it sends branches through the **cardiac plexus** to the heart, through the **pulmonary plexus** to the lungs, through the **esophageal plexus** to the esophagus, into the stomach wall, and through the **celiac plexus** and the **superior mesenteric plexus** to the other abdominal organs (intestines, liver, pancreas, and so on).

Sacral Outflow

The sacral part of the parasympathetic outflow emerges from segments S_2–S_4 of the spinal cord (see the bottom part of Figure 15.5). Continuing where the vagus ends, it innervates the organs in the pelvis, including the distal half of the large intestine, the bladder, and reproductive organs such as the uterus, and the erectile tissues of the external genitalia. Parasympathetic effects on these organs include stimulation of defecation, voiding of urine, and erection.

The preganglionic cell bodies of the sacral parasympathetics lie in the visceral motor region of the spinal gray matter (see Figure 13.31, p. 396). The axons of

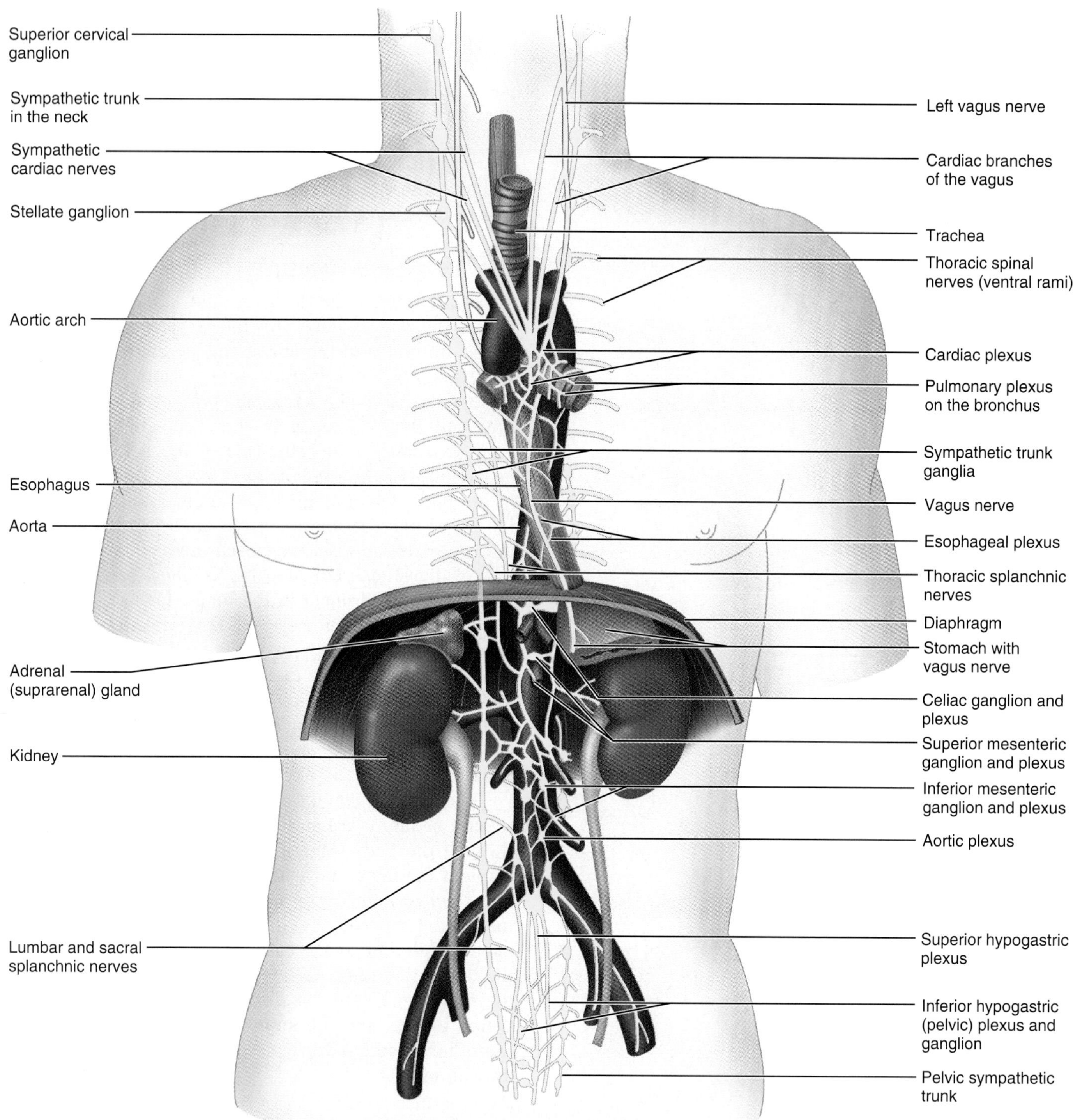

FIGURE 15.6 The vagus nerve, autonomic nerve plexuses, and autonomic ganglia throughout the body. All autonomic plexuses are shared by both parasympathetic and sympathetic fibers, but the ganglia in these plexuses are almost exclusively sympathetic. Note also the sympathetic trunk and its ganglia.

these preganglionic neurons run in the ventral roots to the ventral rami, from which they branch to form **pelvic splanchnic nerves** (see Figure 15.5). These nerves then run through an autonomic plexus in the pelvic floor, the **inferior hypogastric plexus** (or *pelvic plexus;* see Figure 15.6) to reach the pelvic organs. Some preganglionic fibers synapse in ganglia in this plexus, but most synapse in intramural ganglia in the organs.

The various effects of the parasympathetic division are summarized in Table 15.2.

Table 15.2 Effects of the Parasympathetic and Sympathetic Divisions on Various Organs

Target Organ/System/Activity	Parasympathetic Effects	Sympathetic Effects
Eye (iris)	Stimulates constrictor muscles; constricts eye pupils	Stimulates dilator muscles, dilates eye pupils
Eye (ciliary muscle)	Stimulates ciliary muscles, which results in bulging of the lens for accommodation and close vision	No innervation
Glands (nasal, lacrimal, salivary, gastric, pancreas)	Stimulates secretory activity	Inhibits secretory activity; causes vasoconstriction of blood vessels supplying the glands
Sweat glands	No innervation	Stimulates copious sweating (cholinergic fibers)
Adrenal medulla	No innervation	Stimulates medulla cells to secrete epinephrine and norepinephrine into bloodstream
Arrector pili muscles attached to hair follicles	No innervation	Stimulates to contract (erects hairs and produces goose bumps)
Heart muscle	Decreases rate; slows and steadies heart	Increases rate and force of heartbeat
Heart: coronary blood vessels	Causes vasoconstriction	Causes vasodilation
Bladder/urethra	Causes contraction of smooth muscle of bladder wall; relaxes urethral sphincter; promotes voiding	Causes relaxation of smooth muscle of bladder wall; constricts urethral sphincter; inhibits voiding
Lungs	Constricts bronchioles	Dilates bronchioles and mildly constricts blood vessels
Digestive tract organs	Increases motility (peristalsis) and amount of secretion by digestive organs; relaxes sphincters to allow movement of foodstuffs along tract	Decreases activity of glands and muscles of digestive system and constricts sphincters (e.g., anal sphincter); causes vasoconstriction
Liver	No effect	Epinephrine stimulates liver to release glucose to blood
Gallbladder	Stimulates activity (gallbladder contracts to expel bile)	Inhibits activity (gallbladder is relaxed)
Kidney	No effect	Causes vasoconstriction; decreases urine output
Penis	Causes erection (vasodilation)	Causes ejaculation
Vagina/clitoris	Causes erection (vasodilation) of clitoris	Causes contraction of vagina
Blood vessels	Little or no effect	Constricts most vessels and increases blood pressure; constricts vessels of abdominal viscera and skin to divert blood to muscles, brain, and heart when necessary; dilates vessels of the skeletal muscles (cholinergic fibers) during exercise
Blood coagulation	No innervation	Increases coagulation
Cellular metabolism	No innervation	Increases metabolic rate
Adipose tissue	No innervation	Stimulates lipolysis (fat breakdown)
Mental activity	No innervation	Increases alertness

THE SYMPATHETIC DIVISION

Our discussion of the sympathetic division has three parts. First we examine the basic organization of this division, then we describe the sympathetic innervation of specific body regions, and finally we explore the role of the adrenal gland in the sympathetic division.

Basic Organization

The sympathetic division issues from the thoracic and superior lumbar part of the spinal cord, from segments T_1 to L_2 (Figure 15.7). Its preganglionic cell bodies lie in the visceral motor region of the spinal gray matter, where they form the lateral gray horn (see Figure 13.30b, p. 394).

The sympathetic division is more complex than the parasympathetic, in part because it innervates more organs. It not only supplies all the visceral organs in the internal body cavities, but also all visceral structures in the superficial regions of the body: the sweat glands, the hair-raising arrector pili muscles of the skin, and the smooth musculature in the walls of all arteries and veins. It is easy to remember that the sympathetic system alone innervates these structures: Everyone knows that we sweat when under stress, our hair stands up when we are terrified, and our blood pressure skyrockets (because of widespread vasoconstriction) when we get excited. The *parasympathetic* division does *not* innervate sweat glands, arrector pili, or (with minor exceptions) blood vessels.

Another reason the sympathetic division is more complex is that it has more ganglia, which fall into two classes: (1) *sympathetic trunk ganglia* and (2) *prevertebral,* or *collateral, ganglia*. Recall that all autonomic ganglia consist of postganglionic cell bodies.

Sympathetic Trunk Ganglia

Each of the many **sympathetic trunk ganglia,** which are located along both sides of the vertebral column from the neck to the pelvis, are linked by short nerves into long **sympathetic trunks** *(sympathetic chains)* that resemble strings of beads (Figure 15.8). The sympathetic trunk ganglia are also called *chain ganglia* and *paravertebral* ("near the vertebrae") *ganglia*. These ganglia are joined to the ventral rami of nearby spinal nerves by **white** and **gray rami communicantes** (singular: **ramus communicans**; "communicating arm"). As shown in Figure 15.8, white rami communicantes lie lateral to gray rami communicantes.

There is approximately one sympathetic trunk ganglion for each spinal nerve. Note, however, that the number of sympathetic trunk ganglia and spinal nerves is not identical, because some adjacent ganglia fuse during development. Such fusion is most evident in the neck region, where there are eight spinal nerves but only three sympathetic trunk ganglia: the **superior, middle,** and **inferior cervical ganglia** (see Figure 15.7). Furthermore, the inferior cervical ganglion usually fuses with the first thoracic ganglion to form the **stellate** ("star-shaped") **ganglion** in the superior thorax (see Figure 15.6).

Overall, there are 22–24 sympathetic trunk ganglia per side, and a typical person may have 3 cervical, 11 thoracic, 4 lumbar, 4 sacral, and 1 coccygeal ganglia. The cervical ganglia lie just anterior to the transverse processes of the cervical vertebrae; the thoracic ganglia lie on the heads of the ribs; the lumbar ganglia lie on the anterolateral sides of the vertebral bodies; the sacral ganglia lie medial to the sacral foramina; and the coccygeal ganglion is anterior to the coccyx.

Take care not to confuse the sympathetic trunk ganglia with the *dorsal root ganglia.* Recall that dorsal root ganglia are sensory and lie along the dorsal roots in the intervertebral foramina, whereas sympathetic trunk ganglia are motor and lie anterior to the ventral rami.

Prevertebral Ganglia

The **prevertebral ganglia** differ from the sympathetic trunk ganglia in at least three ways: (1) They are not paired and are not segmentally arranged; (2) they occur only in the abdomen and pelvis; and (3) they all lie anterior to the vertebral column (hence the name *prevertebral*), mostly on the large artery called the abdominal aorta. The main prevertebral ganglia are the *celiac, superior mesenteric, inferior mesenteric,* and *inferior hypogastric ganglia* (see Figures 15.6 and 15.7), which lie within the autonomic nerve plexuses of the same names.

Sympathetic Pathways

Now that the gross parts of the sympathetic division have been introduced, we can discuss how the preganglionic and postganglionic neurons fit into these parts to innervate the different body regions: the body periphery, the head, thorax, abdomen, and pelvis (Figures 15.9 through 15.13). Let us begin with an overview of these two-neuron pathways: In the sympathetic pathways to all body regions (Figures 15.9 through 15.13), preganglionic neurons in the thoracolumbar spinal cord invariably send their motor axons through the adjacent ventral root into the spinal nerve, white ramus communicans, and associated sympathetic trunk ganglion. From there, these preganglionic axons synapse with the postganglionic neurons, either within a sympathetic trunk ganglion (Figures 15.9 through 15.11) or in a prevertebral ganglion (Figure 15.12), and the postganglionic axon extends to the visceral organ. In many cases, the preganglionic fiber ascends or descends in the sympathetic trunk before synapsing; this allows sympathetic outputs, which come only from the thoracolumbar (mid-body) region, to supply the superior and inferior body regions, such as the head and pelvis.

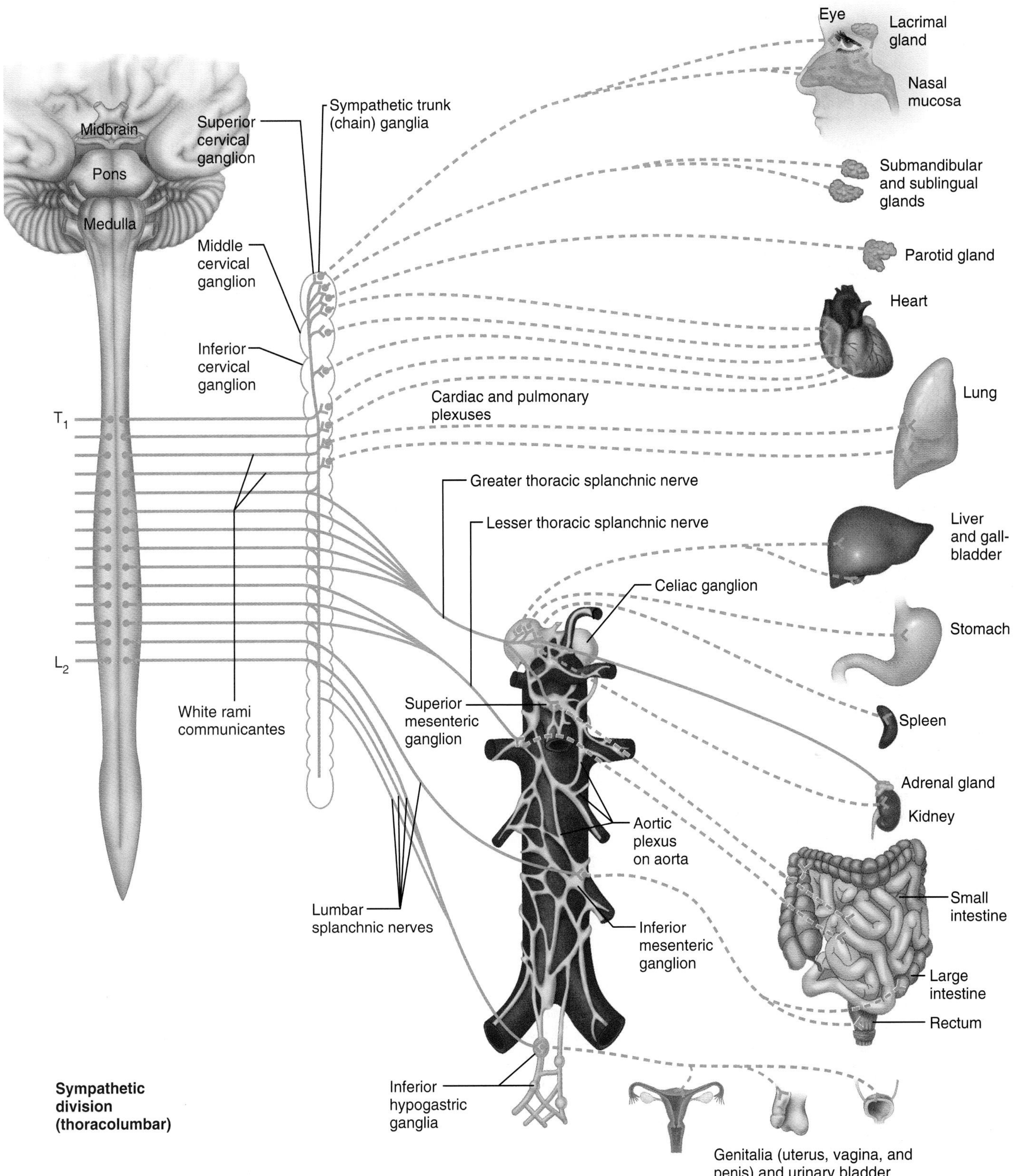

FIGURE 15.7 **Sympathetic division of the ANS.** Solid lines indicate preganglionic nerve fibers, broken lines indicate postganglionic fibers.

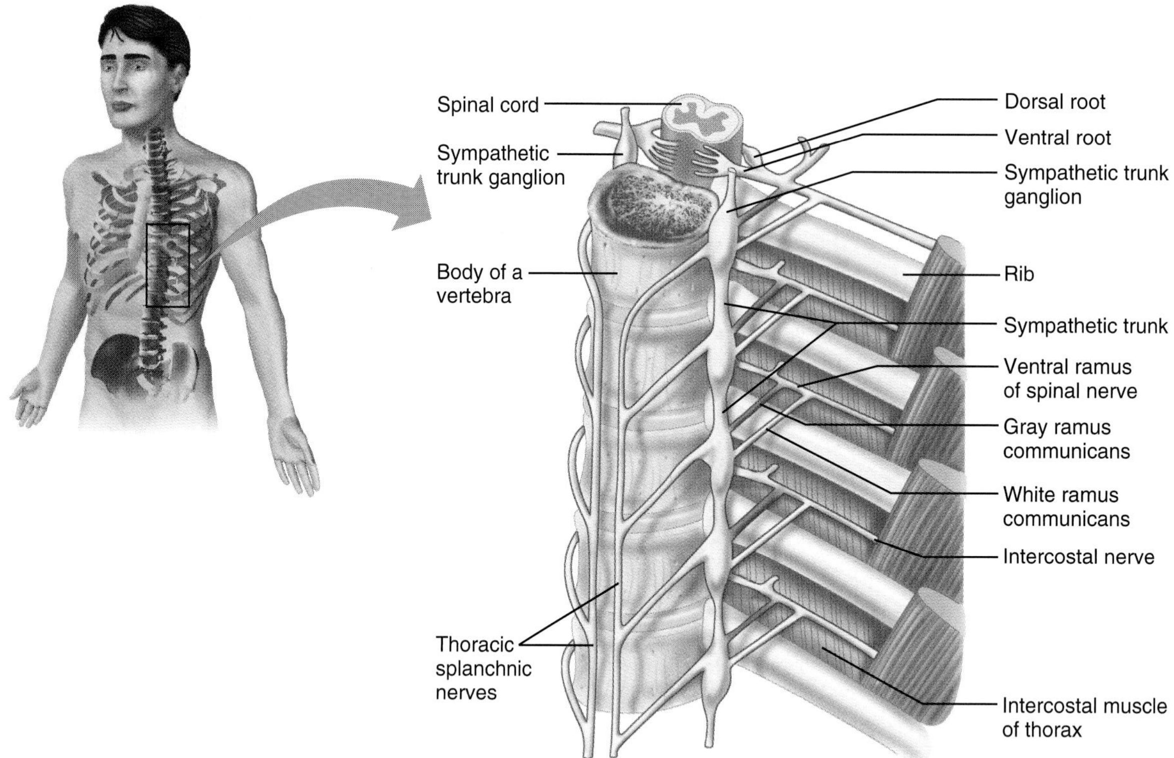

FIGURE 15.8 The sympathetic trunks and trunk ganglia in the thorax. Diagram of the left sympathetic trunk in the posterior thorax, along the side of the vertebral column.

With this overview in mind, we can now discuss the pathways to the specific body regions in more detail.

Sympathetic Pathways to the Body Periphery

Sympathetic pathways to the body periphery (Figure 15.9) innervate the sweat glands, the arrector pili muscles in the skin, and peripheral blood vessels. In these pathways, the preganglionic fibers enter the sympathetic trunk ganglia and synapse there with postganglionic cells bodies. From the sympathetic trunk ganglion, the postganglionic axons travel in a *gray* ramus communicans to the ventral ramus. The axons then follow branches of the dorsal and ventral rami to the skin, where they supply the arrector pili and sweat glands. Anywhere along this path, the postganglionic axons can "jump onto" nearby blood vessels—and then follow and innervate those vessels.

At this point, the differences between the two types of rami communicantes become clear. The *gray* rami carry only the *post*ganglionic fibers headed for peripheral structures, whereas the *white* rami contain all the *pre*ganglionic fibers traveling to the sympathetic trunk ganglion. Gray rami are gray because postganglionic fibers are unmyelinated, and white rami are white because preganglionic fibers are myelinated. Because the white rami carry the sympathetic outflow from the spinal cord, they occur only in the region of such outflow—on sympathetic trunk ganglia between T_1 and L_2. Gray rami, by contrast, occur on all the sympathetic trunk ganglia because sweat glands, arrector pili, and blood vessels must be innervated in *all* body segments.

Sympathetic Pathways to the Head

In the sympathetic innervation of the head (Figure 15.10; also see Figure 15.7), the preganglionic fibers originate in the first four thoracic segments of the spinal cord (T_1–T_4). From there, these fibers ascend in the sympathetic trunk to synapse in the superior cervical ganglion. From this ganglion, the postganglionic fibers associate with large arteries that carry them to glands, smooth muscle, and vessels throughout the head. Functionally, these sympathetic fibers (1) inhibit the lacrimal, nasal, and salivary glands (the reason fear causes dry mouth), and (2) stimulate the muscle in the iris that dilates the pupil in the eye. The sympathetic fibers also supply the *superior tarsal muscle,* a smooth muscle in the upper eyelid that prevents drooping of the eyelid whenever the eyes are open. This muscle also lifts the eyelid to open the eyes wide when one is frightened.

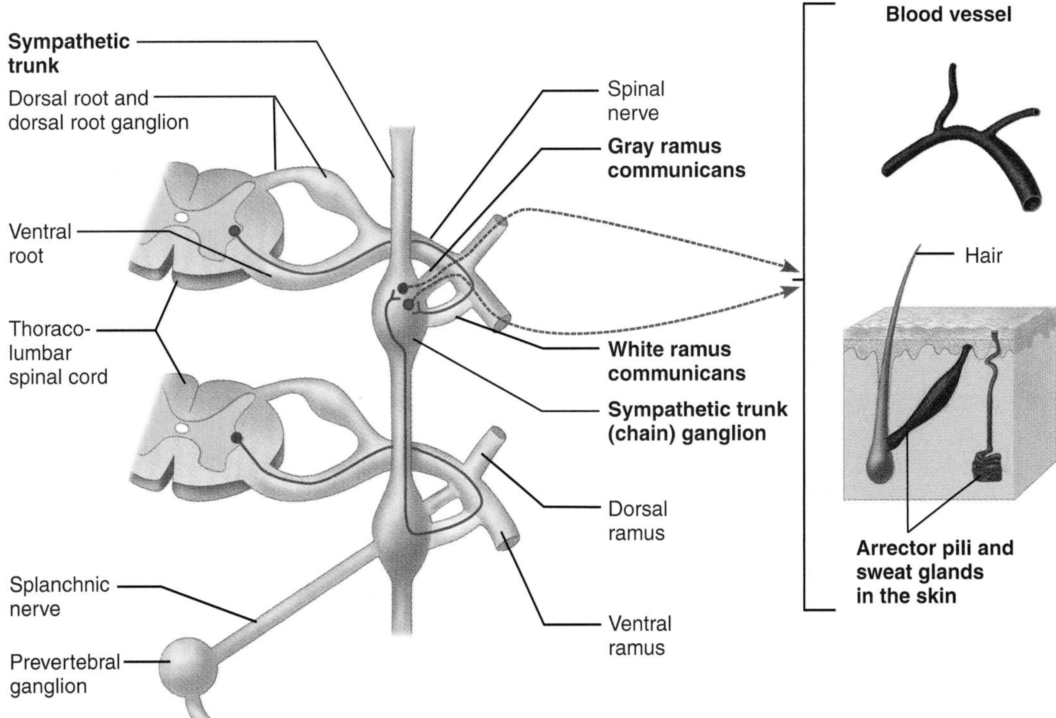

FIGURE 15.9 Sympathetic pathways to the body periphery (sweat glands and arrector pili in the skin, peripheral blood vessels). Again, preganglionic fibers are indicated by solid green lines, postganglionic fibers by dashed lines. Note that the synapses between preganglionic and postganglionic neurons are invariably in the sympathetic trunk ganglia and that the postganglionic fibers run in the gray rami communicans.

Sympathetic Pathways to the Thoracic Organs

In the sympathetic innervation of thoracic organs (Figure 15.11 and 15.7), the preganglionic fibers originate at spinal levels T_1–T_6. Some of these fibers synapse in the nearest sympathetic trunk ganglion, and the postganglionic fibers run directly to the organ being supplied. Fibers to the lungs and esophagus take this direct route, as do some fibers to the heart. Along the way, the postganglionic fibers pass through the pulmonary, esophageal, and cardiac plexuses (see Figure 15.6).

Many of the sympathetic fibers to the *heart,* however, take a less direct path. The preganglionic fibers ascend in the sympathetic trunk to synapse in the cervical ganglia of the sympathetic trunk. From there, the postganglionic fibers descend through the cardiac plexus and into the heart wall. Many nerves to the heart (a thoracic organ) arise in the neck because the heart develops in the neck region of the embryo (see Chapter 18, p. 537).

Functionally, the thoracic sympathetic nerves speed the heart rate and dilate the blood vessels that supply the heart wall (so that the heart muscle itself receives more blood). They also dilate the respiratory air tubes and inhibit the muscle and glands in the esophagus,

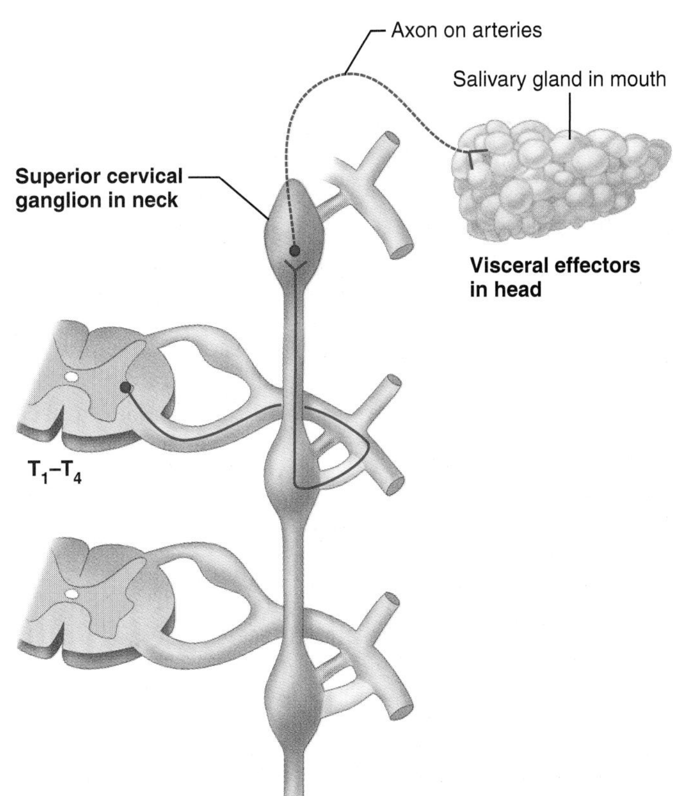

FIGURE 15.10 Sympathetic pathways to the head. Note that the synapses between preganglionic and postganglionic neurons are in the superior cervical ganglion of the sympathetic trunk.

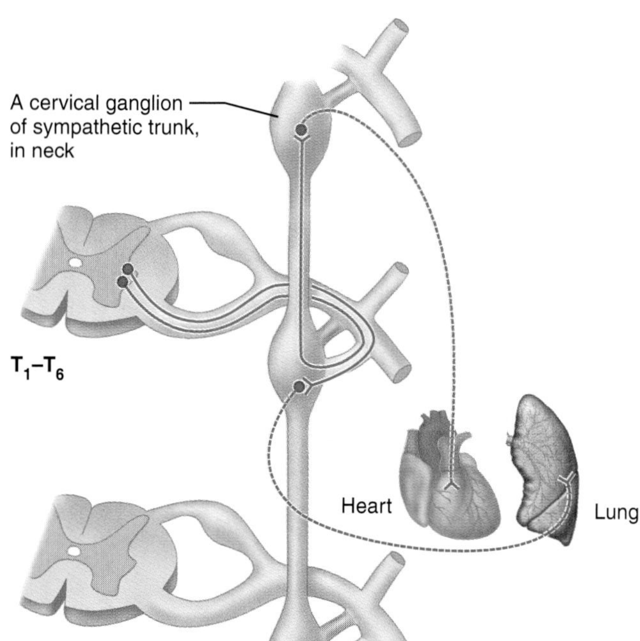

FIGURE 15.11 Sympathetic pathways to the thoracic viscera. The synapses between preganglionic and postganglionic neurons are in the thoracic and cervical ganglia of the sympathetic trunk.

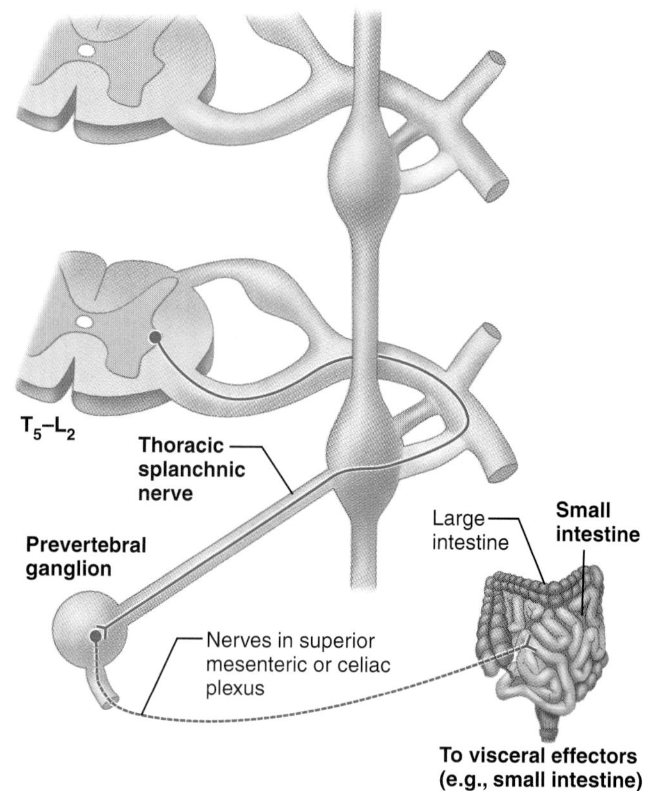

FIGURE 15.12 Sympathetic pathways to the abdominal viscera. The synapses between preganglionic and postganglionic neurons are in prevertebral ganglia.

effects that are integral to the fight, flight, or fright response.

Sympathetic Pathways to the Abdominal Organs

In the sympathetic innervation of abdominal organs (Figure 15.12), the preganglionic fibers originate in the inferior half of the thoracolumbar spinal cord (T_5–L_2). From there, these fibers pass through the adjacent sympathetic trunk ganglia and travel in **thoracic splanchnic nerves** (greater, lesser, and least) to synapse in prevertebral ganglia in the large plexuses on the abdominal aorta. These ganglia include the **celiac** and **superior mesenteric ganglia,** along with some smaller ones (see Figure 15.7). Postganglionic fibers from these ganglia then follow the main branches of the aorta to the stomach, liver, kidney, spleen, and intestines (through the proximal half of the large intestine). For the most part, the sympathetic fibers *inhibit* the activity of the muscles and glands in these visceral organs.

Sympathetic Pathways to the Pelvic Organs

In the sympathetic innervation of pelvic organs (Figure 15.13), the preganglionic fibers originate in the most inferior part of the thoracolumbar spinal cord (T_{10}–L_2), then descend in the sympathetic trunk to the lumbar and sacral ganglia of the sympathetic trunk. Some fibers synapse there, and the postganglionic fibers run in *lumbar* and *sacral splanchnic nerves* to plexuses on the lower aorta and in the pelvis—namely, the **inferior mesenteric plexus,** the **aortic plexus,** and the **hypogastric plexuses.** Other preganglionic fibers, by contrast, pass directly to these autonomic plexuses and synapse in prevertebral ganglia there—the **inferior mesenteric ganglia** and **inferior hypogastric ganglia** (see Figure 15.7). Postganglionic fibers proceed from these plexuses to the pelvic organs, including the bladder, the reproductive organs, and the distal half of the large intestine. These sympathetic fibers inhibit urination and defecation and promote ejaculation.

The Role of the Adrenal Medulla in the Sympathetic Division

On the superior aspect of each kidney lies an **adrenal (suprarenal) gland** (Figure 15.14). The internal portion of this gland—the **adrenal medulla**—is a major organ of the sympathetic nervous system. The adrenal medulla constitutes the largest and most specialized of all the sympathetic ganglia, a collection of modified postganglionic neurons that completely lack nerve processes

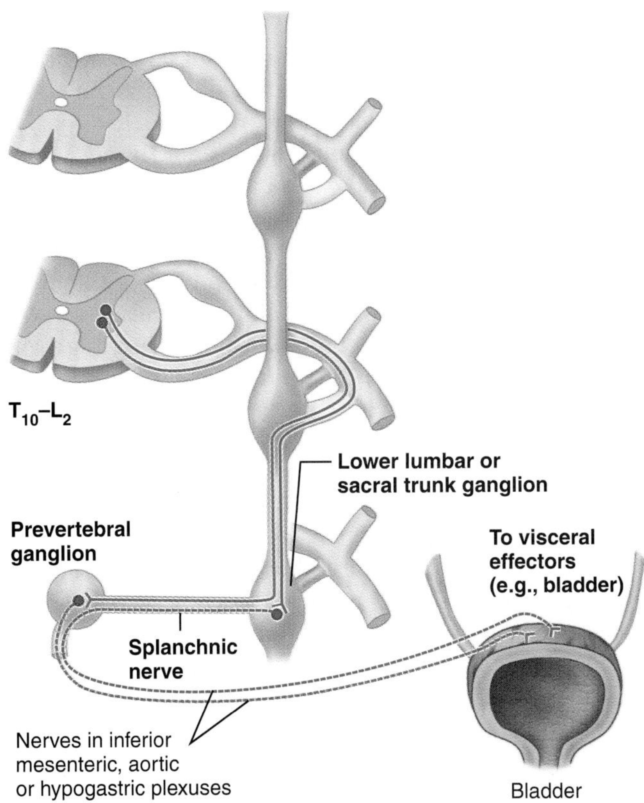

FIGURE 15.13 Sympathetic pathways to the pelvic viscera. The synapses between preganglionic and postganglionic neurons are in both prevertebral and sympathetic trunk ganglia.

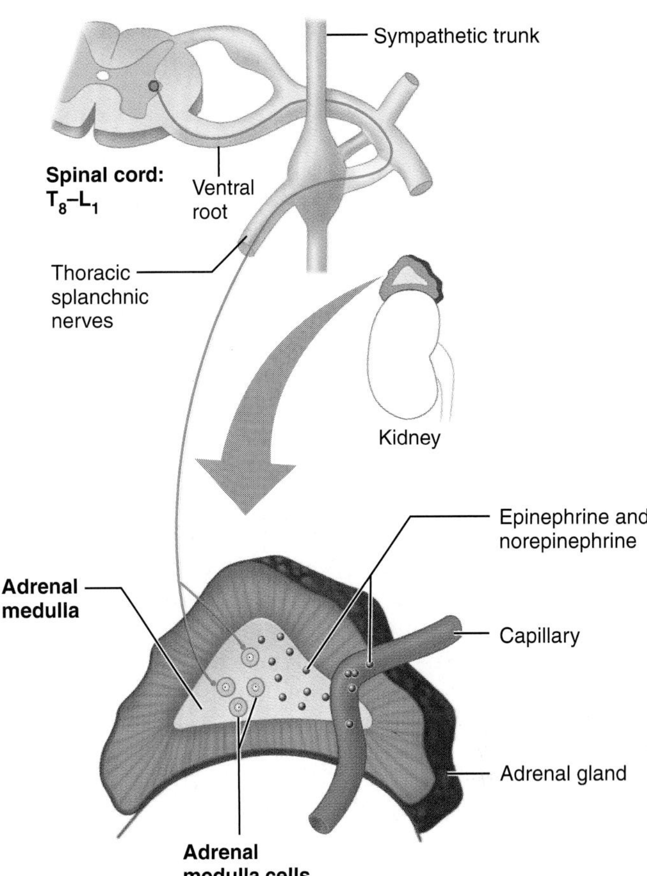

FIGURE 15.14 The adrenal medulla. Sympathetic innervation of the adrenal medulla cells induces them to secrete the excitatory hormones epinephrine and norepinephrine into the blood capillaries, producing excitatory effects throughout the body.

(see Figure 15.14). These neuron-like cells do not innervate any single structure; instead, they secrete great quantities of two excitatory hormones into the blood of nearby capillaries during the fight, flight, or fright response. The hormones secreted are norepinephrine (the chemical secreted by other postganglionic sympathetic neurons as a neurotransmitter) and greater amounts of a related excitatory molecule called **epinephrine (adrenaline).** Once released, these hormones travel throughout the body in the bloodstream, producing the widespread excitatory effects that we have all experienced as a "surge of adrenaline."

The cells of the adrenal medulla are stimulated to secrete by preganglionic sympathetic fibers that arise from cell bodies in the T_8–L_1 region of the spinal cord. From there, they run in the thoracic splanchnic nerves and pass through the celiac plexus before reaching the adrenal medulla (see Figure 15.7). Not surprisingly, the adrenal medulla has a more concentrated sympathetic innervation than any other organ in the body.

The various effects of the sympathetic division are summarized in Table 15.2 on page 451.

CENTRAL CONTROL OF THE AUTONOMIC NERVOUS SYSTEM

Even though the ANS is not considered to be under direct voluntary control, its activities are regulated by the central nervous system. Several sources of central control exist—the brain stem and spinal cord, the hypothalamus and amygdala, and the cerebral cortex.

Control by the Brain Stem and Spinal Cord

The reticular formation of the brain stem appears to exert the most direct influence over autonomic functions. Centers in the medulla oblongata (see pp. 381–382) regulate heart rate *(cardiac centers),* the diameter of blood vessels *(vasomotor center),* and many digestive activities. Also, the periaqueductal gray matter of the midbrain controls many autonomic functions, especially the sympathetic fear response during attack by predators (p. 380).

Control of autonomic functions at the level of the spinal cord involves the spinal visceral reflexes (see p. 350). Note, however, that even though the defecation and urination reflexes are integrated by the spinal cord, they are subject to conscious inhibition from the brain.

Control by the Hypothalamus and Amygdala

The main *integration center* of the autonomic nervous system is the hypothalamus. In general, the medial and anterior parts of the hypothalamus direct parasympathetic functions, whereas the lateral and posterior parts direct sympathetic functions. These hypothalamic centers influence the preganglionic autonomic neurons in the spinal cord and brain, both through direct connections and through relays in the reticular formation and periaqueductal gray matter. It is through the ANS that the hypothalamus controls heart activity, blood pressure, body temperature, and digestive functions.

Recall that the amygdala is the main limbic region for emotions, including fear (see p. 387). Through communications with the hypothalamus and periaqueductal gray matter, the amygdala stimulates sympathetic activity, especially previously learned fear-related behavior.

Control by the Cerebral Cortex

Even though it was once thought that the autonomic nervous system was not subject to voluntary control by the cerebral cortex, people can exert some conscious control over some autonomic functions by developing control over their thoughts and emotions. For example, feelings of extreme calm achieved during meditation are associated with parasympathetic activation in which the cerebral cortex influences the parasympathetic centers in the hypothalamus via various limbic structures. Voluntary sympathetic activation can occur when we decide to recall a frightful experience; in this case the cerebral cortex acts through the amygdala.

VISCERAL SENSORY NEURONS

The visceral division of the PNS contains sensory as well as motor (autonomic) neurons (see Figure 15.1). General visceral sensory neurons monitor stretch, temperature, chemical changes, and irritation within the visceral organs. The brain interprets this visceral information as feelings of hunger, fullness, pain, nausea, or well-being. Almost all of the receptors for these visceral senses are free (unencapsulated) dendritic endings widely scattered throughout the visceral organs. Visceral sensations tend to be difficult to localize with precision: For example, we usually cannot distinguish whether gas pains originate in the stomach or in the intestine, or whether a pain in the lower abdomen originates from the uterus or the appendix.

Like somatic sensory neurons, the cell bodies of visceral sensory neurons are located in dorsal root ganglia and in the sensory ganglia of cranial nerves, from which their central processes extend to the CNS through the dorsal roots. The long peripheral fibers of these sensory neurons accompany the autonomic motor fibers to the visceral organs. For example, many visceral sensory fibers accompany the *parasympathetic* fibers in the vagus nerve and monitor visceral sensations in the many organs served by this nerve. Other visceral sensory fibers accompany the *sympathetic* fibers, running from the visceral organs into the autonomic plexuses and then through the splanchnic nerves, sympathetic trunk, rami communicantes, spinal nerves, and dorsal roots to enter the spinal cord. *Most visceral pain fibers follow this sympathetic route to the CNS.*

The pathways by which visceral sensory information is relayed through the spinal cord to the cerebral cortex are not yet fully understood. Basically, most visceral inputs travel along the spinothalamic (and other) pathways to the thalamus. Additonally, axons carrying pain sensations from internal organs such as the esophagus and intestine course through the most superficial part of the dorsal columns. Neurons in the thalamus relay visceral sensory information to the cerebral cortex for conscious perception. Visceral sensory information also reaches and influences the visceral control centers in the hypothalamus and medulla oblongata.

Visceral *pain* exhibits an unusual feature that is worth noting: Surprisingly, one often feels no pain when a visceral organ is cut or scraped. When pieces of mucous membrane are snipped from the uterus or the intestines to be examined for cancer, for example, most patients report little discomfort. Visceral pain does result, however, from chemical irritation or inflammation of visceral organs, from spasms of the smooth muscle in these organs (cramping), and from excessive stretching of the organ. In these cases, visceral pain can be severe.

People suffering from visceral pain often perceive this pain to be somatic in origin—that is, as if it originated from the skin or outer body. This phenomenon is called **referred pain.** For example, heart attacks can produce referred pain in cutaneous areas of the superior thoracic wall and the medial aspect of the left arm (Figure 15.15). The cause of referred pain is not fully understood. However, it is known that both the affected organ and the region of the body wall to which the pain is referred are innervated by the same spinal segments. (For example, both the heart and the skin area to which heart pain projects are innervated by sensory neurons from T_1–T_5.) One possible explanation is that damage to the visceral organ causes painful, reflexive vasoconstriction in the vessels supplying the corresponding somatic segments.

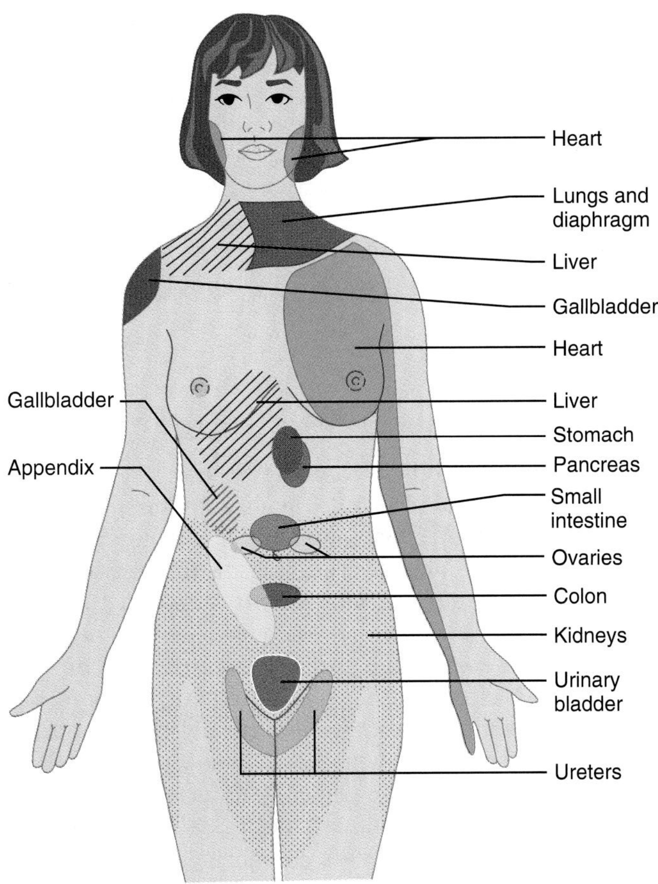

FIGURE 15.15 A map of referred pain. This map shows the anterior skin areas to which pain is referred from certain visceral organs.

VISCERAL REFLEXES

Visceral sensory and autonomic neurons participate in **visceral reflex arcs,** including the *defecation reflex,* in which the rectum is stretched by feces and the smooth muscle of the large intestine responds by contracting, and the *micturition reflex,* in which the smooth muscle in the urine-filled bladder contracts. Many visceral reflexes are simple spinal reflexes in which sensory neurons activate spinal interneurons, which in turn activate preganglionic autonomic neurons. Other visceral reflex arcs, however, do not involve the central nervous system at all—they are strictly *peripheral reflexes.* In some of these peripheral reflexes, branches from visceral sensory fibers synapse with postganglionic motor neurons *within sympathetic ganglia.* Furthermore, complete three-neuron reflex arcs (with small sensory, motor, and intrinsic neurons) exist *entirely within the wall of the digestive tube;* these neurons constitute the *enteric nervous system* (see pp. 639–640). The peripheral reflexes carry out highly localized autonomic responses involving either small segments of an organ or a few different visceral organs. These reflexes allow the peripheral part of the visceral nervous system to control some of its own activity, making it partly independent of the brain and spinal cord. This fact further illustrates the general concept that the autonomic nervous system operates partly on its own.

DISORDERS OF THE AUTONOMIC NERVOUS SYSTEM

Because the ANS is involved in nearly every important process that occurs within the body, it is not surprising that abnormalities of autonomic functioning can have far-reaching effects. Such abnormalities can impair blood delivery and elimination processes and can even threaten life itself.

Raynaud's disease is characterized by intermittent attacks during which the skin of the fingers and toes becomes pale, then blue and painful. When such an attack ends and the vessels dilate, the fingers refill with blood and become red. Commonly provoked by exposure to cold or by emotional stress, this disease is thought to be an exaggerated, sympathetic vasoconstriction response in the affected body regions. The severity of Raynaud's disease ranges from mere discomfort to such extreme constriction of vessels that gangrene (tissue death) results. Treatment typically involves administering drugs that inhibit vasoconstriction, but in severe cases it may be necessary to cut the preganglionic sympathetic fibers serving the affected region. Raynaud's disease is rather common among the elderly, affecting 9% of all elderly women and 3–5% of elderly men.

Hypertension, or high blood pressure, ostensibly a circulatory disorder, can result from overactive sympathetic vasoconstriction promoted by continual stress. Hypertension is always serious because (1) it increases the work load on the heart, possibly precipitating heart disease, and (2) it increases the wear and tear on artery walls. Stress-induced hypertension is treated with drugs that prevent smooth muscle cells in the walls of blood vessels from binding norepinephrine and epinephrine.

The **mass reflex reaction,** a massive uncontrolled activation of both autonomic and somatic motor neurons, affects quadriplegics and paraplegics with spinal cord injuries above the level of T_6. The cord injury is followed by a temporary loss of all reflexes inferior to the level of the injury. When reflex activity returns, it is exaggerated because of the lack of inhibitory input from higher (brain) centers. The ensuing episodes of mass reflex reaction involve surges of both visceral and somatic motor output from large regions of the spinal cord. The usual trigger for such an episode is a strong

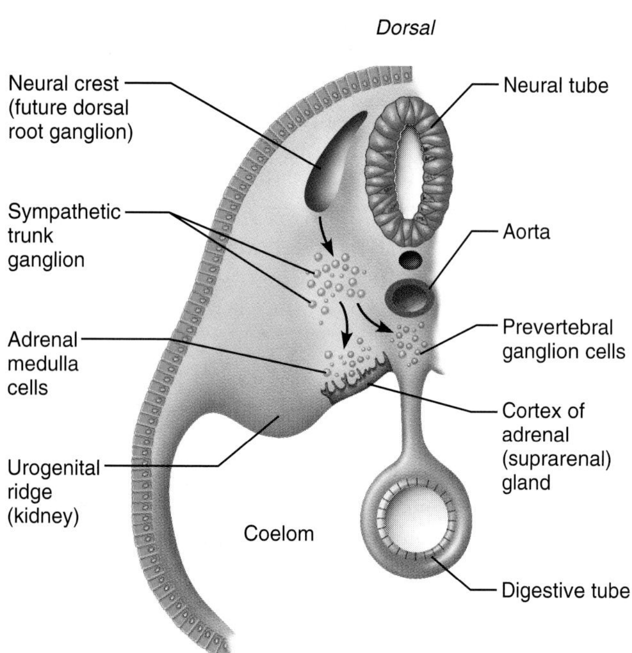

FIGURE 15.16 Embryonic development of some sympathetic structures, including the adrenal medulla. Transverse section through a 5-week embryo. As shown by the arrows, neural crest gives rise to sympathetic trunk ganglia, from which some cells migrate to form the adrenal medulla and prevertebral ganglia.

stimulus to the skin or the overfilling of a visceral organ, such as the bladder. During the mass reflex episode, the body goes into flexor spasms, the limbs move wildly, the colon and bladder empty, and profuse sweating occurs. Most seriously, sympathetic activity raises blood pressure to life-threatening levels. The precise mechanism of the mass reflex reaction is unknown, but it has been likened to a type of epilepsy of the spinal cord.

Achalasia (ak″ah-la′ze-ah) **of the cardia** is a condition in which some defect in the autonomic innervation of the esophagus results in a loss of that organ's ability to propel swallowed food inferiorly. Additionally, the smooth muscle surrounding the inferior end of the esophagus (cardiac sphincter) remains contracted, preventing the passage of food into the stomach. (*Achalasia* means "failure to relax.") The accumulation of food stretches the esophagus to enormous width, and meals cannot be kept down. The cause of this condition, which usually appears in young adults, is not precisely understood. The preferred treatment is a longitudinal surgical incision through the muscle at the inferior end of the esophagus.

Congenital megacolon, or **Hirschsprung's disease,** is a birth defect in which the parasympathetic innervation of the distal region of the large intestine fails to develop normally because migrating neural crest cells fail to reach this region (see the next section). Feces accumulates proximal to the immobile bowel segment, greatly distending this area (*megacolon* = enlarged large intestine). The condition is corrected surgically by removing the inactive part of the infant's intestine.

THE AUTONOMIC NERVOUS SYSTEM THROUGHOUT LIFE

Preganglionic neurons of the autonomic nervous system develop from the neural tube, as do somatic motor neurons (see p. 354). Postganglionic neurons, however, develop from the neural crest—as do the visceral sensory neurons. Note that *all neurons with cell bodies in the PNS derive from the neural crest, and all neurons with cell bodies in the CNS derive from the neural tube.*

In the development of the sympathetic division (Figure 15.16), some cells migrate ventrally from the neural crest to form the sympathetic trunk ganglia. From there, other cells migrate ventrally to form both the prevertebral ganglia on the aorta and the adrenal medulla. Next, the sympathetic trunk ganglia and prevertebral ganglia receive axons from spinal preganglionic neurons, and they in turn send postganglionic axons to the visceral organs.

In the development of the parasympathetic division, the postganglionic neurons also derive from the neural crest and reach the visceral organs by migrating along the growing preganglionic axons.

During youth, impairments of visceral nervous function are usually due to injuries to either the spinal cord or autonomic nerves. With advancing age, the efficiency of the ANS begins to decline. Elderly people are often constipated because the autonomically controlled mobility of their gastrointestinal tract is reduced. In the eyes, frequent infections can result from diminished formation of tears, which contain bacteriocidal enzymes, and the pupils cannot dilate as widely or as quickly. Whenever a young and healthy person rises to a standing position, the sympathetic division induces bodywide vasoconstriction raising blood pressure so that blood can be pumped to the head and brain. The response becomes sluggish with age, so elderly people may faint if they stand up too quickly. Although these age-related problems are distressing, they usually are not life-threatening and can be easily alleviated. For example, standing up slowly gives the sympathetic nervous system time to adjust blood pressure, eye drops are available for dry eyes, and drinking ample fluid helps alleviate constipation.

RELATED CLINICAL TERMS

Atonic bladder (a-tahn'ik; "without tone") A condition in which the bladder becomes flaccid and overfills, allowing urine to dribble out. Atonic bladder results from the temporary loss of the micturition reflex following injury to the spinal cord.

Chagas' disease An incurable parasitic infection, in which the protozoan *Trypanosoma cruzi* seriously damages many internal organs, especially the heart and the autonomic innervation of the digestive tract. This infection, which affects 20 million people in Latin America and kills 45,000 per year, is transmitted by the kissing bug that lives in the thatched walls and roofs of traditional housing and comes out at night to bite people and suck their blood: The bug defecates and people become infected by scratching the itchy bite wound, thereby introducing the infected bug feces into the wound. After about two decades of infection, the parasite may destroy the heart wall or the parasympathetic neurons in the colon, so feces are not propelled through and defecation does not occur. Death can result from heart failure or from blood poisoning by fecal compounds.

Horner's syndrome A condition that follows damage to the sympathetic trunk in the inferior neck region on one side of the body. The symptoms, which occur on the head on the affected side only, include drooping of the upper eyelid (ptosis), pupil constriction, flushing of the face, and inability to sweat. Horner's syndrome can indicate the presence of disease or infection in the neck.

Vagotomy (va-got'o-me) Cutting or severing of a vagus nerve, often to decrease the secretion of stomach acid and other caustic digestive juices that aggravate ulcers.

CHAPTER SUMMARY

Media study tools that could provide you additional help in reviewing specific key topics of Chapter 15 are referenced below.

SP = Study Partner; AIA = A.D.A.M.® Interactive Anatomy

1. The ANS is the system of motor neurons that innervates smooth muscle, cardiac muscle, and glands. It is the general visceral motor division of the PNS. The ANS largely operates below the level of consciousness.

INTRODUCTION TO THE AUTONOMIC NERVOUS SYSTEM (pp. 444–447)

Comparison of the Autonomic and Somatic Motor Systems (pp. 444–445)

1. In the somatic and branchial motor divisions of the nervous system, a single motor neuron forms the pathway from the CNS to the effectors. Autonomic motor pathways, by contrast, consist of chains of two neurons: the preganglionic neuron, whose cell body is in the CNS, and the postganglionic neuron, whose cell body is in an autonomic ganglion.

Divisions of the Autonomic Nervous System (pp. 445–447)

2. The ANS has parasympathetic and sympathetic divisions. Both innervate the same organs but produce opposite effects. The sympathetic division prepares us for "fight, flight, or fright," whereas the parasympathetic division is active during "resting and digesting." Details of the effects of both divisions are listed in Table 15.2 (p. 451).

AIA Link: Atlas; System: Nervous; Autonomic nervous system–viscera 2.

3. The parasympathetic division is craniosacral and has relatively long preganglionic axons. The sympathetic division is thoracolumbar and has relatively long postganglionic axons.
4. The two divisions differ in the neurotransmitters they release at the effector organ. Acetylcholine is released by parasympathetic postganglionic fibers, whereas norepinephrine is released by most sympathetic postganglionic fibers.

SP Exercise: The Somatic and Autonomic Nervous Systems.

THE PARASYMPATHETIC DIVISION (pp. 447–451)

Cranial Outflow (pp. 448–449)

1. Cranial parasympathetic fibers arise in the brain stem nuclei of cranial nerves III, VII, IX, and X and synapse in ganglia in the head, thorax, and abdomen. Fibers in III serve smooth musculature in the eye via a relay in the ciliary ganglion. Fibers in VII serve the submandibular, sublingual, lacrimal, and nasal glands via relays in the submandibular and pterygopalatine ganglia. Fibers in IX serve the parotid gland via a relay in the otic ganglion.
2. Parasympathetic fibers in the vagus nerve (X) innervate organs in the thorax and most of the abdomen, including the heart, lungs, esophagus, stomach, liver, and most of the intestines. The fibers in the vagus are preganglionic. Almost all postganglionic neurons are located in intramural ganglia within the organ walls.

Sacral Outflow (pp. 449–451)

3. Sacral parasympathetic pathways innervate the pelvic viscera. The preganglionic fibers issue from the visceral motor region of the gray matter of the spinal cord (S_2–S_4) and form the pelvic splanchnic nerves. Most of these fibers synapse in intramural ganglia in the pelvic organs.

AIA Link: Atlas; System: Nervous; Autonomic nervous system—female organs 2.

THE SYMPATHETIC DIVISION (pp. 452–457)

Basic Organization (p. 452)

1. The preganglionic sympathetic cell bodies are in the lateral horn of the spinal gray matter from the level of T_1 to L_2.
2. The sympathetic division supplies some peripheral structures that the parasympathetic division does not: arrector pili, sweat glands, and the smooth muscle of blood vessels.
3. Sympathetic ganglia include 22–24 pairs of sympathetic trunk ganglia (also called chain ganglia and paravertebral ganglia), which are linked together to form sympathetic trunks on both sides of the vertebral column, and unpaired prevertebral ganglia (collateral ganglia), most of which lie on the aorta in the abdomen.

Sympathetic Pathways (pp. 452–456)

4. Every preganglionic sympathetic fiber leaves the lateral gray horn of the thoracolumbar spinal cord through a ventral root and spinal nerve. From there, it runs in a white ramus communicans to a sympathetic trunk ganglion or prevertebral ganglion, where it synapses with the postganglionic neuron that extends to the visceral effector. Many preganglionic axons ascend or descend in the sympathetic trunk to synapse in a ganglion at another body level.
5. In the sympathetic pathway to the body periphery (to innervate arrector pili, sweat glands, and peripheral blood vessels), the preganglionic fibers synapse in the sympathetic trunk ganglia and the postganglionic fibers run in gray rami communicantes to the dorsal and ventral rami of the spinal nerves for peripheral distribution.
6. In the sympathetic pathway to the *head,* the preganglionic fibers synapse in the superior cervical ganglion. From there, most postganglionic fibers associate with a large artery that distributes them to the glands and smooth musculature of the head.
7. In the sympathetic pathway to *thoracic organs,* most preganglionic fibers synapse in the nearest sympathetic trunk ganglion, and the postganglionic fibers run directly to the organs (lungs, esophagus). Many postganglionic fibers to the heart, however, descend from the cervical ganglia in the neck.
8. In the sympathetic pathway to *abdominal organs,* the preganglionic fibers run in splanchnic nerves to synapse in prevertebral ganglia on the aorta. From these ganglia, the postganglionic fibers follow large arteries to the abdominal viscera (stomach, liver, kidney, and most of the large intestine).
9. In the sympathetic pathway to *pelvic organs,* the preganglionic fibers synapse in sympathetic trunk ganglia or in prevertebral ganglia on the aorta, sacrum, and pelvic floor. Postganglionic fibers travel through the most inferior autonomic plexuses to the pelvic organs.

The Role of the Adrenal Medulla in the Sympathetic Division (pp. 456–457)

10. The adrenal glands, one superior to each kidney, contain a medulla of modified postganglionic sympathetic neurons that secrete the hormones epinephrine (adrenaline) and norepinephrine into the blood. The result is the "surge of adrenaline" felt during excitement.
11. The cells of the adrenal medulla are innervated by preganglionic sympathetic neurons, which signal them to secrete the hormones.

SP Exercise: Chapter 15, Parasympathetic and Sympathetic Divisions of the ANS.

CENTRAL CONTROL OF THE AUTONOMIC NERVOUS SYSTEM (pp. 457–458)

1. Visceral motor functions are influenced by the medulla oblongata, the periaqueductal gray matter, spinal visceral reflexes, the hypothalamus, the amygdala, and the cerebral cortex. We can voluntarily regulate some autonomic activities by gaining extraordinary control over our emotions.

VISCERAL SENSORY NEURONS (pp. 458–459)

1. General visceral sensory neurons monitor temperature, pain, irritation, chemical changes, and stretch in the visceral organs.
2. Visceral sensory fibers run within the autonomic nerves, especially within the vagus and the sympathetic nerves. The sympathetic nerves carry most pain fibers from the visceral organs of the body trunk.
3. A simplified description of most visceral sensory pathways to the brain is the following: sensory neurons to spinothalamic tract to thalamus to cerebral cortex.
4. Visceral pain is induced by stretching, infection, and cramping of internal organs but seldom by cutting or scraping these organs.
5. Pain in visceral organs is referred to somatic regions of the body that receive innervation from the same spinal cord segments.

VISCERAL REFLEXES (p. 459)

1. Many visceral reflexes are spinal reflexes, such as defecation reflexes. Some visceral reflexes, however, involve only peripheral neurons.

DISORDERS OF THE AUTONOMIC NERVOUS SYSTEM (pp. 459–460)

1. Most autonomic disorders reflect problems with the control of smooth muscle. Abnormalities in vascular control, which occur in Raynaud's disease, hypertension, and the mass reflex reaction, are most devastating.

THE AUTONOMIC NERVOUS SYSTEM THROUGHOUT LIFE (p. 460)

1. Preganglionic neurons develop from the neural tube. Postganglionic neurons and visceral sensory neurons develop from the neural crest.
2. The efficiency of the ANS declines in old age: Gastrointestinal motility and production of tears decrease, and the vasomotor response to standing slows.

REVIEW QUESTIONS

Multiple Choice/Matching Questions

1. Which of the following does *not* characterize the ANS? (a) two-neuron motor chains, (b) preganglionic cell bodies in the CNS, (c) presence of postganglionic cell bodies in ganglia, (d) innervation of skeletal muscle.

2. For each of the following terms or phrases, write either S (for sympathetic) or P (for parasympathetic) division of the autonomic nervous system.

____ **(1)** short preganglionic fibers, long postganglionic fibers

____ **(2)** intramural ganglia

____ **(3)** craniosacral outflow

____ **(4)** adrenergic fibers

____ **(5)** cervical ganglia of the sympathetic trunk

____ **(6)** otic and ciliary ganglia

____ **(7)** more widespread response

____ **(8)** increases heart rate, respiratory rate, and blood pressure

____ **(9)** increases motility of stomach and secretion of lacrimal and salivary glands

____ **(10)** innervates blood vessels

____ **(11)** most active when you are lolling in a hammock

____ **(12)** most active when you are running in the Boston Marathon

____ **(13)** gray rami communicantes

____ **(14)** synapse in celiac ganglion

____ **(15)** relates to fear response induced by the amygdala

3. The thoracic splanchnic nerves contain which kind of fibers? (a) preganglionic parasympathetic, (b) postganglionic parasympathetic, (c) preganglionic sympathetic, (d) postganglionic sympathetic.

4. The prevertebral ganglia contain which kind of cell bodies? (a) preganglionic parasympathetic, (b) postganglionic parasympathetic, (c) preganglionic sympathetic, (d) postganglionic sympathetic.

5. Preganglionic sympathetic neurons develop from (a) neural crest, (b) neural tube, (c) alar plate, (d) endoderm.

6. Which of the following is the best way to describe how the ANS is controlled? (a) completely under control of voluntary cerebral cortex, (b) entirely controls itself, (c) completely under control of brain stem, (d) little control by cerebrum, major control by hypothalamus and amygdala, and major control by spinal and peripheral reflexes.

7. The white rami communicantes contain what kind of fibers? (a) preganglionic parasympathetic, (b) postganglionic parasympathetic, (c) preganglionic sympathetic, (d) postganglionic sympathetic.

8. Prevertebral sympathetic ganglia are involved with the innervation of the (a) abdominal organs, (b) thoracic organs, (c) head, (d) arrector pili, (e) all of these.

Short Answer Essay Questions

9. Is the visceral sensory nervous system part of the autonomic nervous system? Explain your answer.

10. (a) Describe the anatomical relationship of the white and gray rami communicantes to a spinal nerve, and to the dorsal and ventral rami. (b) Why are gray rami communicantes gray?

11. What effect does sympathetic activation have on each of the following structures? Sweat glands, pupils of the eye, adrenal medulla, heart, lungs, blood vessels of skeletal muscles, blood vessels of digestive viscera, salivary glands.

12. Which of the effects listed in Question 11 would be reversed by parasympathetic activity?

13. Suppose that a mad scientist seeking to invent a death ray produces a beam that destroys a person's ciliary, pterygopalatine, and submandibular ganglia (and nothing else). List all the signs that would be apparent in the victim. Would the victim die, or would the scientist have to go back to the laboratory to try again?

14. A friend asks you how the parasympathetic division, which comes only from cranial and sacral regions, is able to innervate organs in the thorax and abdomen. How would you answer that question?

15. What manifestations of decreased ANS efficiency are seen in elderly people?

16. The students in anatomy class were having difficulty answering Question 4, which asks what kind of cell bodies are located in the autonomic ganglia. Frustrated by being asked about this same question repeatedly, the exasperated TA stood on a desk and yelled, "Autonomic ganglia always contain the *post*ganglionic cell bodies!" Describe how that information enabled the students to work out which kinds of cell bodies are present in the (a) sympathetic trunk ganglia, (b) intramural ganglia, and (c) head ganglia (ciliary, pterygopalatine, submandibular, and otic).

17. Describe the sympathetic pathways to the (a) submandibular salivary gland, (b) bladder, (c) sweat glands in the skin, (d) adrenal medulla, (e) heart, (f) stomach.

18. Describe the parasympathetic pathways to the (a) smooth muscles in the eye, (b) bladder, (c) small intestine.

19. Describe the path of the vagus through the neck, thorax, and abdomen.

20. What structures are innervated by the *sacral* part of the parasympathetic division?

CRITICAL REASONING + CLINICAL APPLICATION QUESTIONS

1. A one-year-old infant has a swollen abdomen and is continually constipated. On examination, a mass is felt over the distal colon, and X-ray findings show the colon to be greatly distended in that region. What is likely wrong with this infant?
2. Roweena, a high-powered marketing executive, develops a stomach ulcer. She complains of a deep abdominal pain that she cannot quite locate, plus superficial pain on her abdominal wall. Exactly where on the abdominal wall is the superficial pain most likely to be located? (Hint: See Figure 15.15.)
3. Constipated people often have hardened feces that can tear the inner lining of the rectum and anus during defecation. Those with tears in the skin of the anus complain of sharp pain, whereas those with tears higher up in the intestinal mucosa do not complain of such pain. How do you explain this difference?
4. A woman with a history of kidney infections was feeling pain in the skin of the lower part of her abdomen and even in her thighs. Does this make sense? Why or why not?
5. Which clinical condition has the classical symptoms of blue fingertips that later turn bright red?

A.D.A.M.® INTERACTIVE ANATOMY QUESTIONS

1. Link: Atlas; System: Nervous; Ciliary ganglion pathways #1. Which cranial nerve serves the smooth musculature in the eye via the parasympathetic ciliary ganglion?
2. Link: Atlas; System: Nervous; Pterygopalatine ganglion #2. Which cranial nerve serves the lacrimal gland by way of the pterygopalatine ganglion?

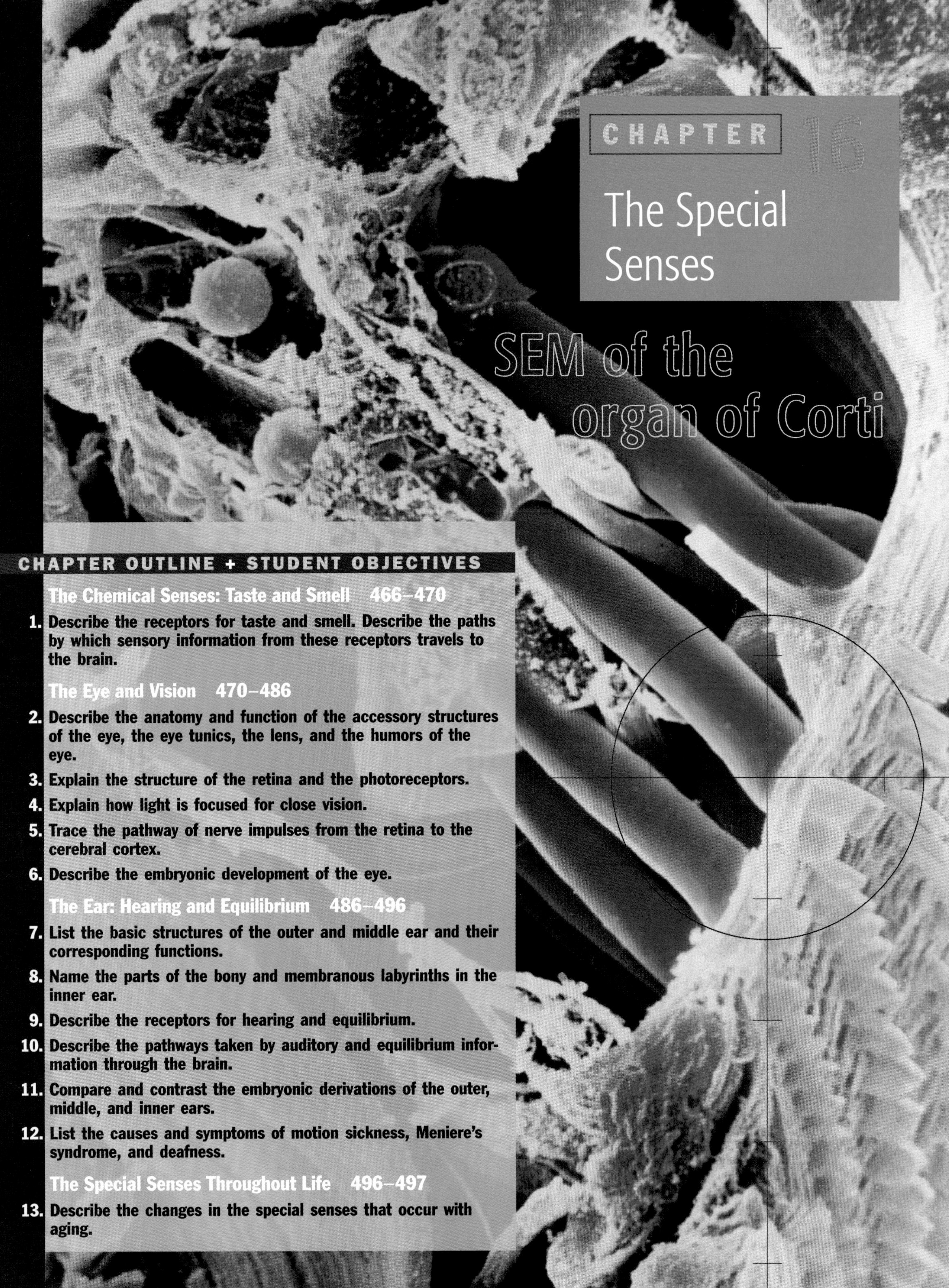

CHAPTER 16

The Special Senses

SEM of the organ of Corti

CHAPTER OUTLINE + STUDENT OBJECTIVES

The Chemical Senses: Taste and Smell 466–470

1. Describe the receptors for taste and smell. Describe the paths by which sensory information from these receptors travels to the brain.

The Eye and Vision 470–486

2. Describe the anatomy and function of the accessory structures of the eye, the eye tunics, the lens, and the humors of the eye.
3. Explain the structure of the retina and the photoreceptors.
4. Explain how light is focused for close vision.
5. Trace the pathway of nerve impulses from the retina to the cerebral cortex.
6. Describe the embryonic development of the eye.

The Ear: Hearing and Equilibrium 486–496

7. List the basic structures of the outer and middle ear and their corresponding functions.
8. Name the parts of the bony and membranous labyrinths in the inner ear.
9. Describe the receptors for hearing and equilibrium.
10. Describe the pathways taken by auditory and equilibrium information through the brain.
11. Compare and contrast the embryonic derivations of the outer, middle, and inner ears.
12. List the causes and symptoms of motion sickness, Meniere's syndrome, and deafness.

The Special Senses Throughout Life 496–497

13. Describe the changes in the special senses that occur with aging.

PEOPLE ARE RESPONSIVE CREATURES. THE aroma of baking bread makes our mouths water; lightning makes us blink and thunder makes us recoil. Many such sensory stimuli reach us each day and are processed by our nervous systems.

We are usually told that we have five senses: touch, taste, smell, sight, and hearing. Actually, touch is a large group of general senses that we considered in Chapter 14. The other four traditional senses—*smell, taste, sight,* and *hearing*—are called special senses. Receptors for a fifth special sense—*equilibrium,* or the sense of balance—are also located in the ear.

In contrast to the widely distributed receptors for the general senses, the **special sensory receptors** are localized and confined to the head region. The receptors for the special senses are not dendritic endings of sensory neurons but distinct **receptor cells.** These are neuron-like epithelial cells or small peripheral neurons that transfer sensory information to other neurons in afferent pathways to the brain. The special sensory receptors are either housed in complex sensory organs (eye or ear) or in distinctive epithelial structures (taste buds or olfactory epithelium). Their sensory information travels via cranial nerves.

This chapter explores the functional anatomy of the five special senses: the chemical senses of taste and smell, vision in the eye, and hearing and equilibrium in the ear.

THE CHEMICAL SENSES: TASTE AND SMELL

The receptors for taste (gustation) and smell (olfaction) are classified as **chemoreceptors** because they respond to chemical substances—to food chemicals dissolved in saliva and to airborne chemicals that dissolve in fluids on the nasal membranes, respectively. We turn first to the sense of taste.

Taste (Gustation)

In our exploration of the sense of taste we first examine the anatomy of the taste receptors in taste buds, and then we discuss the types of taste sensations and their pathway through the nervous system.

Taste Buds

The taste receptors occur in **taste buds** in the mucosa of the mouth and pharynx. The majority of the 10,000 or so taste buds are on the surface of the tongue; a few others occur on the posterior region of the palate (roof of the mouth), on the inner surface of the cheeks, on the posterior wall of the pharynx, and on the epiglottis (a leaf-shaped flap behind the tongue).

Most taste buds occur in two types of peg-like projections of the tongue mucosa called *papillae* (pah-pil′e) (Figure 16.1a). Specifically, they occur in small *fungiform papillae* scattered over the entire surface of the tongue, and in large *vallate (circumvallate) papillae* arranged in an inverted V near the back of the tongue. Taste buds occur within the epithelium that covers the papillae, on the apical surface of fungiform papillae and in the side walls of vallate papillae (Figure 16.1b).

Each taste bud is a globular collection of 50–100 epithelial cells that resemble a bud on a tree or a closed tulip flower (Figure 16.1c and d). Under the light microscope, each contains three major cell types: **supporting cells, gustatory cells** (receptor cells), and **basal cells.** Supporting cells, the most abundant type, insulate the gustatory (receptor) cells from each other and from the surrounding epithelium of the tongue. Basal cells are immature and regularly replace the other two cell types. Sensory nerve fibers enter the taste buds and synapse with the receptor cells. More recently, electron microscopy has revealed *four* types of cells in taste buds: basal cells, plus dark, light, and intermediate cells. Preliminary evidence suggests that the intermediate and light cells are the gustatory receptors, and that the dark cells are supporting cells.

Long microvilli project from receptor cells and supporting cells and extend through a **taste pore** to the surface of the epithelium. There, these microvilli are bathed in saliva containing the dissolved molecules that stimulate taste. Such molecules bind to the plasma membrane of the microvilli, inducing the receptor cells to generate impulses in the sensory nerve fibers that innervate them.

The cells in taste buds are replenished every 7–10 days by the division of the basal cells, replacing the taste cells that are scraped and burned off during eating. If an entire taste bud is destroyed, a new one will form after its nerve ending grows back into the regenerating epithelium.

Taste Sensations and the Gustatory Pathway

Taste sensations can be described in terms of four basic qualities: sweet, sour, salty, and bitter.

Advances in Understanding A fifth distinct taste sensation, called *umami* (Japanese: "deliciousness"), has recently been identified. It is elicited by a substance called glutamate and is responsible for the "beef steak" taste of meat and for enhancing the other taste qualities. It is the taste of the flavor-enhancing food additive monosodium glutamate. ✷

Although there are no *structural* differences among the taste buds in different areas, the tongue's anterior area is most sensitive to sweet and salty substances, its lateral areas to sour, its posterior area to bitter (Figure 16.1a), and the posterior pharynx to umami.

Taste information reaches the brain stem and cerebral cortex through the *gustatory pathway* (Figure 16.2). Sensory fibers carrying taste information from the tongue occur primarily in two cranial nerves: The *facial nerve* (VII) transmits impulses from taste receptors in

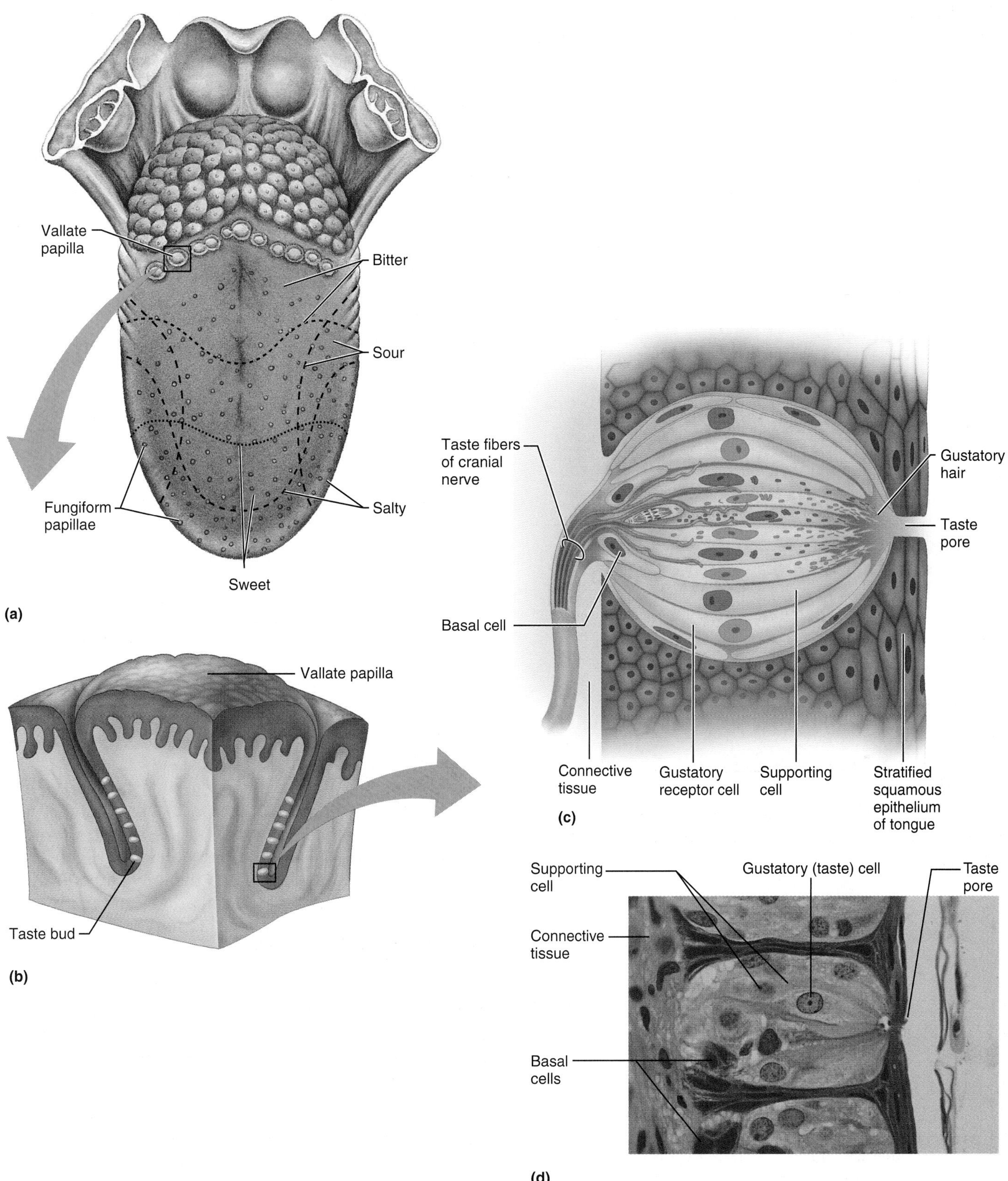

FIGURE 16.1 Taste buds. **(a)** Taste buds on the tongue occur in fungiform and vallate papillae, peg-like projections of the tongue mucosa. Tongue areas most sensitive to sweet, salty, sour, and bitter tastes are also shown. **(b)** A vallate papilla sectioned to show the taste buds in the epithelium of its lateral walls. **(c)** Enlarged view of a taste bud showing the gustatory cells (the taste receptor cells), supporting cells, and basal cells. **(d)** Photomicrograph of a taste bud (600 ×).

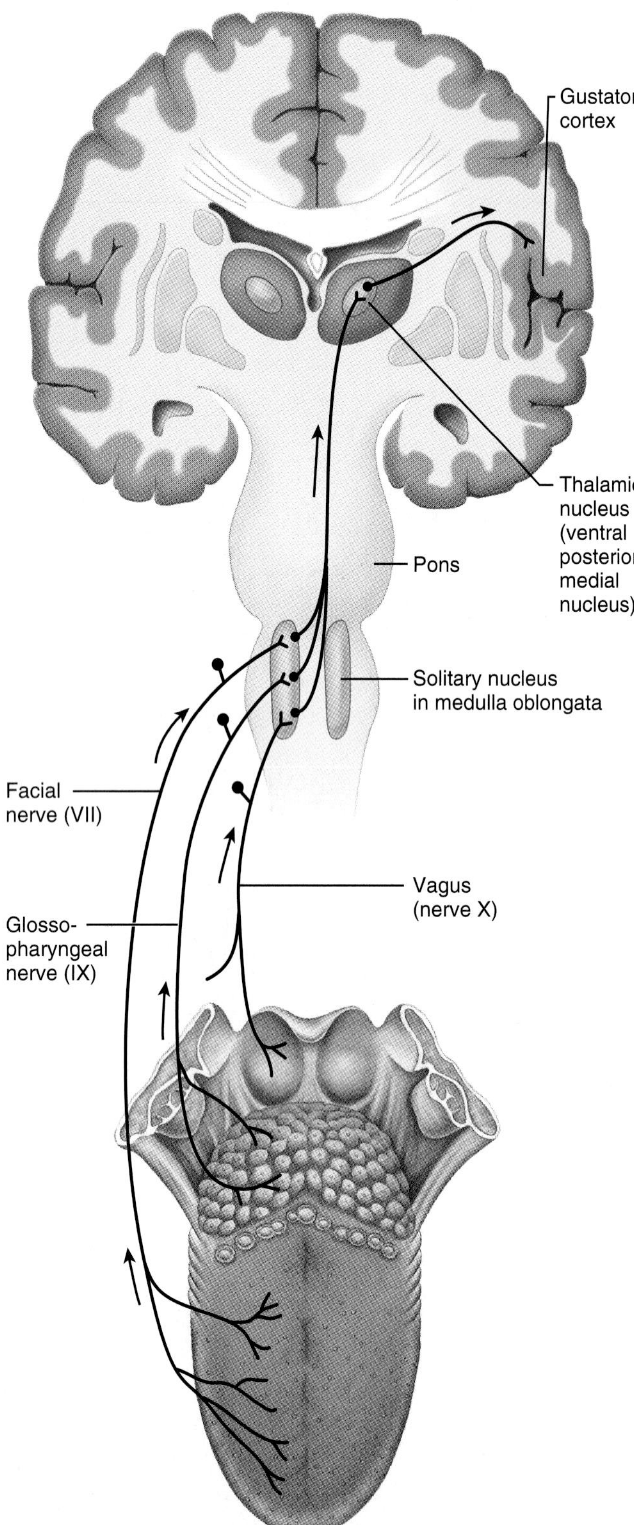

FIGURE 16.2 The gustatory pathway from taste buds on the tongue (below) to the gustatory area of the cerebral cortex (above).

the anterior two-thirds of the tongue, whereas the ***glossopharyngeal nerve*** (IX) carries sensations from the tongue's posterior third, as well as from the few buds in the pharynx. Additional taste impulses from the few taste buds on the epiglottis and lower pharynx are conducted by the ***vagus nerve*** (X). All the sensory neurons that carry taste information synapse in a nucleus in the medulla called the ***solitary nucleus.*** From there, impulses are transmitted to the thalamus and ultimately to the gustatory area of the cerebral cortex in the parietal lobe.

Critical Reasoning

Which one of these taste sensations—sweet, sour, salty, or bitter—is diminished by injuries to the glossopharyngeal nerve?

Damage to the glossopharyngeal nerve reduces one's ability to taste bitter substances because this nerve supplies the tongue's posterior, bitter-sensitive region. Damage to the facial nerve, by contrast, diminishes one's ability to taste sweet, sour, and salty substances on the anterior two-thirds of the tongue.

Next we turn our attention to the sense of smell.

Smell (Olfaction)

The receptors for smell lie in a yellow-tinged patch of epithelium on the roof of the nasal cavity (Figure 16.3a). Specifically, the receptors are a part of the **olfactory epithelium** that covers the superior nasal concha and the superior part of the nasal septum and is bathed by swirling air that has been inhaled into the nasal cavity. Sniffing draws more air across the olfactory epithelium and thus intensifies the sense of smell.

The olfactory epithelium is a pseudostratified columnar epithelium (Figure 16.3b and c) that contains millions of bowling-pin-shaped neurons called **olfactory receptor cells.** These are surrounded by columnar **supporting cells.** At the base of the epithelium lie short **basal cells,** undifferentiated cells that continually form new olfactory receptor cells. This makes the latter the only neurons in the body that undergo replacement throughout adult life.

The cell bodies of olfactory receptor cells are located in the olfactory epithelium. Each receptor cell has an apical dendrite that projects to the epithelial surface and ends in a knob from which long **olfactory cilia** "hairs" radiate (see Figure 16.3b). At the surface these cilia act as the receptive structures for smell by binding odor molecules to receptor proteins located in the plasma membrane of the cilia. The surface of the epithelium is also coated with a layer of mucus secreted by the nearby supporting cells and by glands in the underlying connective tissue (lamina propria). This mucus, which "captures" and dissolves odor molecules from the air, is renewed continuously, flushing away "old" odor molecules so that new odors always have access to the olfactory cilia. Unlike other cilia in the body, olfactory cilia are largely immotile.

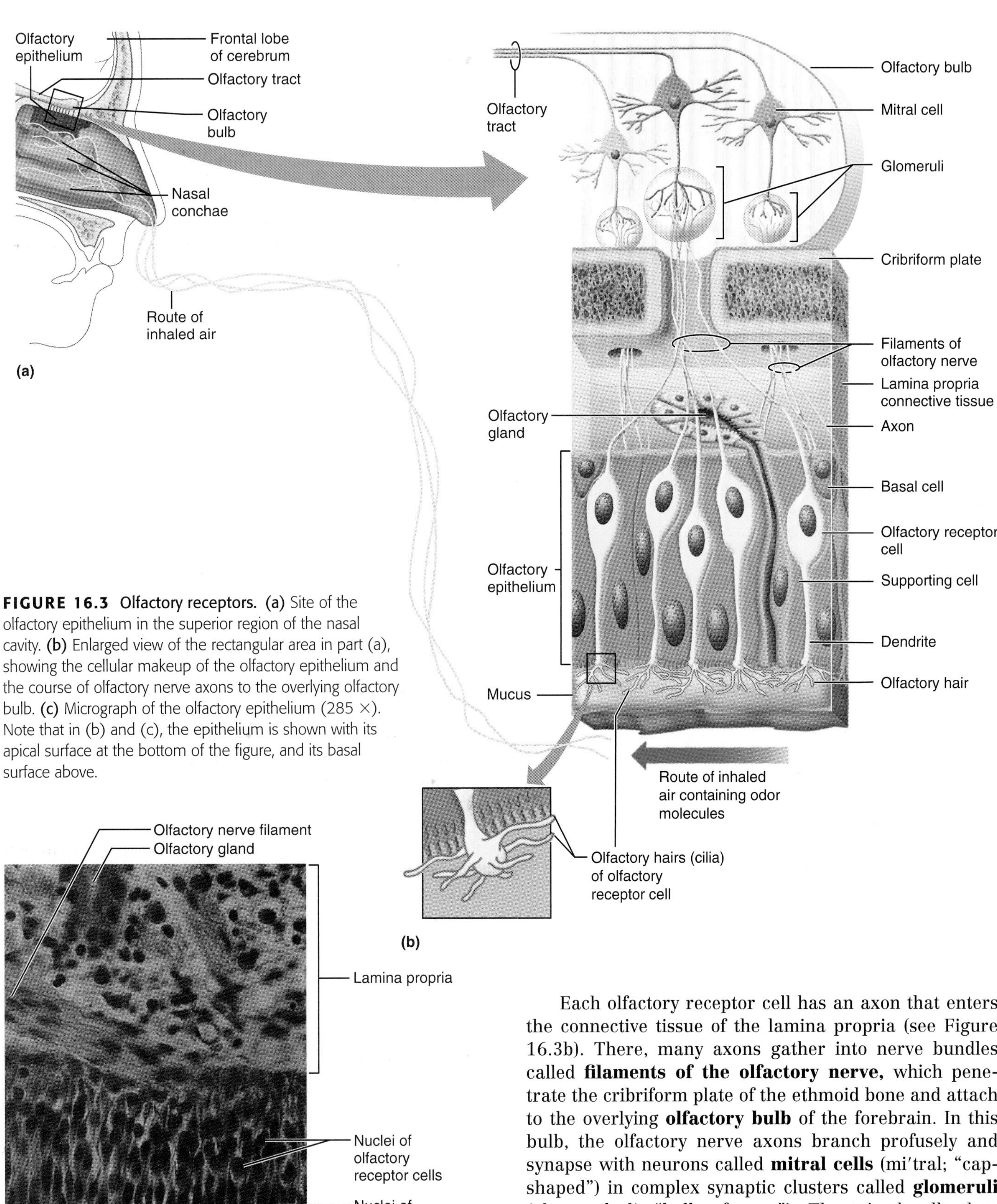

FIGURE 16.3 Olfactory receptors. **(a)** Site of the olfactory epithelium in the superior region of the nasal cavity. **(b)** Enlarged view of the rectangular area in part (a), showing the cellular makeup of the olfactory epithelium and the course of olfactory nerve axons to the overlying olfactory bulb. **(c)** Micrograph of the olfactory epithelium (285 ×). Note that in (b) and (c), the epithelium is shown with its apical surface at the bottom of the figure, and its basal surface above.

Each olfactory receptor cell has an axon that enters the connective tissue of the lamina propria (see Figure 16.3b). There, many axons gather into nerve bundles called **filaments of the olfactory nerve,** which penetrate the cribriform plate of the ethmoid bone and attach to the overlying **olfactory bulb** of the forebrain. In this bulb, the olfactory nerve axons branch profusely and synapse with neurons called **mitral cells** (mi′tral; "cap-shaped") in complex synaptic clusters called **glomeruli** (glo-mer′lu-li; "balls of yarn"). The mitral cells then relay the olfactory information to other parts of the brain.

After synapsing with the receptor neurons, the mitral cells send their axons through the olfactory tract to (1) the limbic region, by which smells elicit emotions, and (2) the piriform lobe of the cerebral cortex, represented by the yellow and orange area around the uncus in Figure 13.10b. The piriform lobe is thought to process olfactory information into a conscious perception of odor; it also sends this information through a thalamic relay to the orbitofrontal cortex (also seen in Figure 13.10b), where the smells are analyzed and compared to other smells.

Advances in Understanding Much is now known about the functional organization of receptor cells and glomeruli. Each receptor cell contains just one of the approximately 1000 types of smell receptors, each of which binds to some part of an odor molecule, much as a key fits a lock. Thus, one receptor can bind to a variety of odor molecules that share a common structural part, and each odor molecule can bind to different receptors that fit its various parts. Many copies of each receptor-cell type are scattered throughout the olfactory epithelium, but all receptor cells of the same type send their axons to the same glomerulus in the olfactory bulb.

Through this arrangement, each odor molecule activates several types of receptor cells, thereby activating its own distinct pattern of several glomeruli. This pattern, in turn, serves as a "signature" that identifies the odor when it arrives at higher brain centers. Moreover, the glomeruli also amplify faint smells: The convergence onto one glomerulus of input from many receptor cells greatly amplifies the signal transmitted to the mitral cell. ✷

Disorders of the Chemical Senses

It is possible to have disorders of either taste or olfaction, but olfactory disorders are more common. Absence of the sense of smell, called **anosmia** (an-oz′me-ah; "without smell"), typically results from blows to the head that tear the olfactory nerves, from colds or allergies, or from physical blockage of the nasal cavity (by polyps, for example). The culprit in fully one-third of all cases of loss of chemical senses is zinc deficiency, and the cure is rapid once a dietary zinc supplement is prescribed.

Brain disorders can distort the sense of smell. Some people have **uncinate** (un′sĭ-nāt) **fits,** olfactory hallucinations in which they perceive some imaginary odor, like that of gasoline or rotting meat. Uncinate fits are so called because the primary olfactory cortex is in the uncinate region, or uncus, of the cerebrum (see p. 370). These fits may result from irritation of the olfactory pathway by brain surgery or head trauma. Olfactory auras—smells imagined by some epileptics just before they go into a seizure—are brief uncinate fits.

Embryonic Development of the Special Senses

The development of the olfactory epithelium and taste buds is straightforward. Olfactory epithelium derives from paired olfactory placodes, which are plate-like thickenings of the surface ectoderm on the embryonic snout region (see Figure 21.18). Taste buds develop, upon stimulation by gustatory nerves, from the epithelium lining the embryonic mouth and pharynx, an epithelium that ultimately comes from ectoderm and endoderm.

THE EYE AND VISION

Vision is our dominant sense: Approximately 70% of the sensory receptors in the body are in the eyes, and 40% of the cerebral cortex is involved in processing visual information (see p. 369). The visual receptor cells (photoreceptors) sense and encode the patterns of light that enter the eye; subsequently, the brain invests these signals with meaning, fashioning visual images of the world around us.

The visual organ is the **eye,** or **eyeball,** a spherical structure with a diameter of about 2.5 cm (1 inch). Only the anterior one-sixth of the eye's surface is visible; the rest of the eye lies in the cone-shaped bony *orbit,* where it is surrounded by a protective cushion of fat. Behind the eye, the posterior half of the orbit contains the optic nerve, the arteries and veins to the eye, and the extrinsic eye muscles.

Before we examine the structure of the eyeball itself, we discuss the variety of structures surrounding the eye.

Accessory Structures of the Eye

The accessory structures of the eye include the eyebrows, eyelids, conjunctiva, lacrimal apparatus, and extrinsic eye muscles (Figures 16.4 through 16.6).

Eyebrows

The **eyebrows** consist of coarse hairs in the skin on the superciliary arches (brow ridges of the skull). They shade the eyes from sunlight and prevent perspiration running down the forehead from reaching the eyes.

Eyelids

Anteriorly, the eyes are protected by the mobile **eyelids,** or **palpebrae** (pal′pĕ-bre). The upper and lower lids are separated by the **palpebral fissure** (eye slit) and meet each other at the medial and lateral **angles (canthi),** or eye corners (see Figure 16.4). The medial canthus contains a reddish elevation called the **lacrimal caruncle** (kar′ung-k′l; "a bit of flesh"). Glands here produce the gritty "eye sand" reputedly left by the legendary Sandman at night. In most Asian people, a vertical fold of skin called the *epicanthic fold* occurs on both sides of the nose and sometimes covers the medial canthus.

The eyelids themselves are thin, skin-covered folds supported internally by connective tissue structures called **tarsal plates** (see Figure 16.5a). These stiff plates give the eyelids their curved shape and serve as attach-

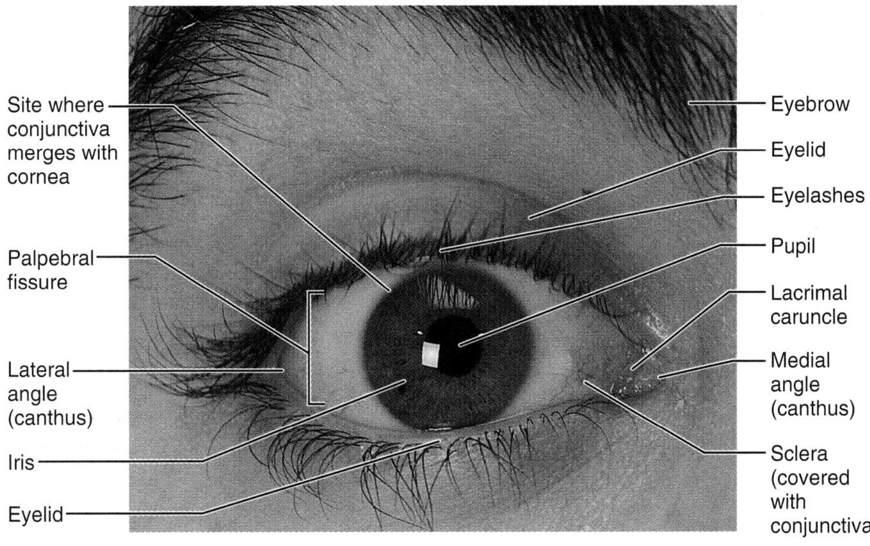

FIGURE 16.4 Surface anatomy of the right eye.

ment sites for the eye-closing muscle, the *orbicularis oculi* (see p. 276).

The **levator palpebrae superioris** ("lifter of the upper eyelid") is the skeletal muscle that voluntarily opens the eye. It runs anteriorly from the posterior roof of the orbit, enters the upper eyelid, and inserts on the tarsal plate. The inferior part of the aponeurosis of this muscle contains fibers of smooth muscle, called the *superior tarsal muscle,* an involuntary muscle that prevents the upper eyelid from drooping (see p. 454).

Projecting from the free margin of each eyelid are the **eyelashes.** Because the follicles of these hairs are richly innervated by nerve endings (root hair plexuses), even slight pressure on the eyelashes will trigger reflexive blinking.

Several types of glands occur in the eyelids. **Tarsal glands** (*Meibomian glands*) are modified sebaceous glands embedded in the tarsal plates (Figure 16.5a). About 25 of these vertical glands line up side-by-side in the upper eyelid; fewer than this line up in the lower lid. The tarsal gland ducts, which open along the edge of the eyelids, release an oil that spreads over the surface of the eye, slowing the evaporation of water and perhaps keeping the eyelids from sticking together. Other glands are associated with the hair follicles of the eyelashes: These *ciliary glands* (*cilium* = eyelash) include typical sebaceous glands whose ducts open into the hair follicles, and modified sweat glands that lie between the follicles. Infection of a tarsal gland results in an unsightly cyst called a **chalazion** (kah-la′ze-on; "swelling"), whereas infection of the ciliary glands is called a **sty.**

Conjunctiva

The **conjunctiva** (con″junk-ti′vah; "joined together") is a transparent mucous membrane that covers the inner surfaces of the eyelids as the **palpebral conjunctiva** and folds back over the anterior surface of the eye as the **ocular (bulbar) conjunctiva** (see Figure 16.5a). The ocular conjunctiva, which covers the white of the eye but not the cornea (the transparent tissue over the iris and pupil), is a very thin membrane, and blood vessels are clearly visible beneath it. (These vessels are responsible for "bloodshot" eyes.) When an eye is closed, the slit-like space that forms between the eye surface and the eyelids is the **conjunctival sac,** which is where a contact lens would lie.

Microscopically, the conjunctiva consists of a stratified columnar epithelium underlain by a thin lamina propria of loose connective tissue. Its epithelium contains scattered goblet cells that perform the main function of the conjunctiva: secreting a lubricating mucus that prevents the eyes from drying. A deficiency of vitamin A, which is required for the maintenance of epithelia throughout the body, prevents mucus secretion by the conjunctiva. As a result, the conjunctiva dries up and becomes scaly, impairing vision.

Inflammation of the conjunctiva, called **conjunctivitis,** is a relatively common condition in which irritation of the eyes makes them red. A highly contagious form of conjunctivitis caused by bacteria or viruses is called **pinkeye.**

Lacrimal Apparatus

The **lacrimal apparatus** (lak′rĭ-mal; "tear"), which keeps the surface of the eye moist with lacrimal fluid (tears), consists of a gland and ducts that drain the lacrimal fluid into the nasal cavity (see Figure 16.5b). The **lacrimal gland,** lying in the orbit superolateral to the eye, produces lacrimal fluid, which enters the superior part of the conjunctival sac through several small excretory ducts. Blinking of the eye then spreads this fluid inferiorly across the eyeball to the medial canthus, where it passes through tiny openings called **lacrimal**

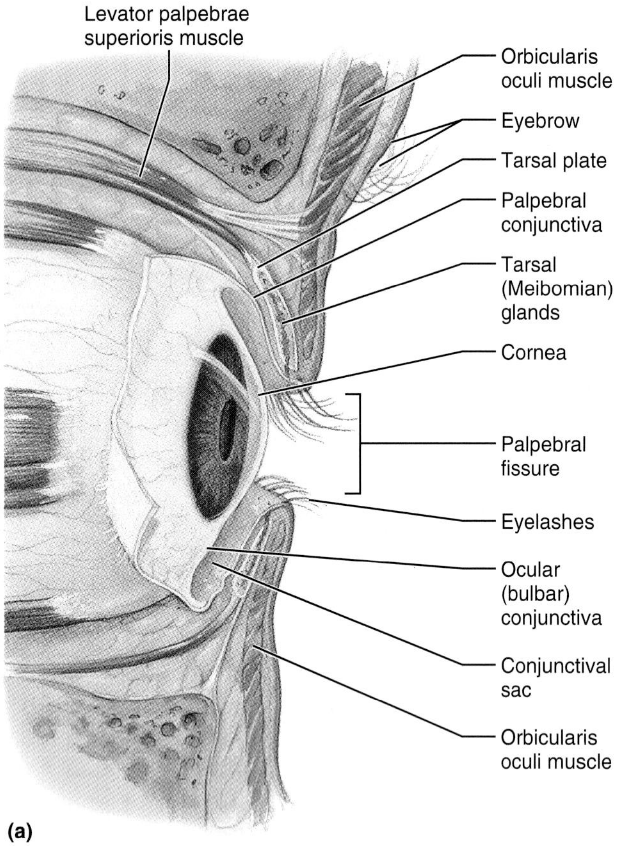

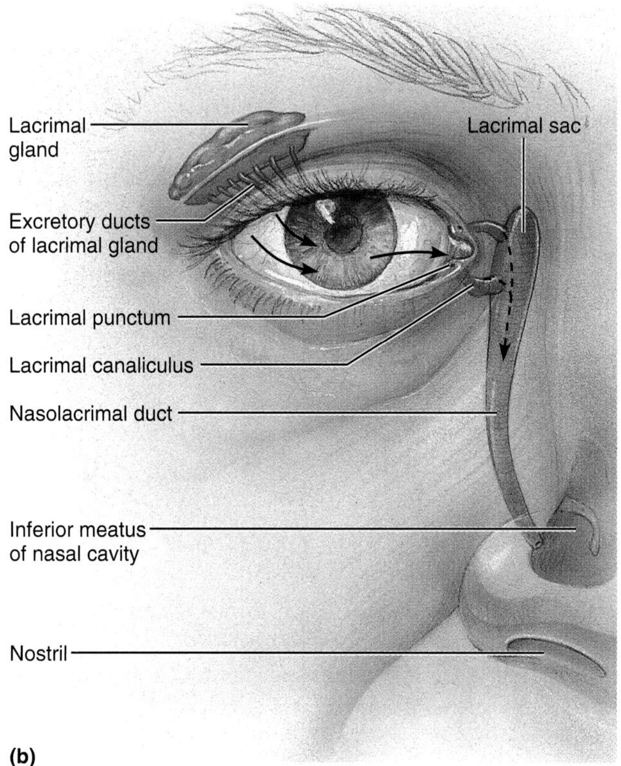

FIGURE 16.5 Accessory structures of the eye. **(a)** Lateral view; some structures are shown in sagittal section. **(b)** The lacrimal (tear) apparatus. Arrows indicate the direction of flow of lacrimal fluid (tears) after its secretion by the lacrimal gland.

puncta ("points") into two small tubes called **lacrimal canaliculi.** From these canals, the fluid drains into the **lacrimal sac** in the medial orbital wall. Finally, the fluid enters the **nasolacrimal duct,** which empties into the nasal cavity at the inferior nasal meatus. Because lacrimal fluid ultimately empties into the nasal cavity, we get the sniffles when we cry.

Lacrimal fluid contains mucus, antibodies, and **lysozyme,** an enzyme that destroys bacteria. When the eye surface is irritated by dust or fumes (from an onion, for example), lacrimal secretion increases to wash away the irritant.

Critical Reasoning

Explain why people with colds often have watery eyes.

During a cold, the viral infection in the nasal passages can spread into the nasolacrimal duct and lacrimal sac, producing inflammation that causes these passages to swell shut. The resulting blockage prevents the drainage of lacrimal fluid from the eye surface; the fluid accumulates, and the eyes water.

Extrinsic Eye Muscles

The movement of each eyeball is controlled by six strap-like **extrinsic** (outer) **eye muscles** (Figure 16.6). These muscles, which originate from the walls of the orbit and insert onto the outer surface of the eyeball, direct the gaze and hold the eyes in the orbits.

Four of the extrinsic eye muscles are *rectus* muscles (*rectus* = straight). These originate from a common tendinous ring, the **annular ring** (see Figure 16.6b), at the posterior point of the orbit. From there, they run straight to their insertions on the anterior half of the eyeball. The **lateral rectus muscle** turns the eye laterally (outward), whereas the **medial rectus muscle** turns it medially (inward). The **superior** and **inferior rectus muscles** turn the eye superiorly (elevate it) and inferiorly (depress it), respectively. It is easy to deduce the actions of the rectus muscles from their names and locations.

The actions of the two *oblique* muscles are not so easily deduced, however, because they take indirect paths through the orbit. The **superior oblique muscle** (see Figure 16.6b) originates posteriorly near the annular ring, runs anteriorly along the medial orbit wall, and then loops through a ligamentous sling, the **trochlea** ("pulley"), which is suspended from the frontal bone in the anteromedial part of the orbit roof. From there, its tendon runs posteriorly and inserts on the eye's posterolateral surface (see Figure 16.6). Because its tendon approaches from an anterior direction, the superior oblique depresses the eye, and because its tendon also approaches from a medial direction, it also turns the eye laterally. Because the **inferior oblique muscle** (see Figure 16.6a) originates on the anteromedial part of the orbit floor and angles back to insert on the posterolateral part of the eye, the inferior oblique elevates the eye and turns it somewhat laterally.

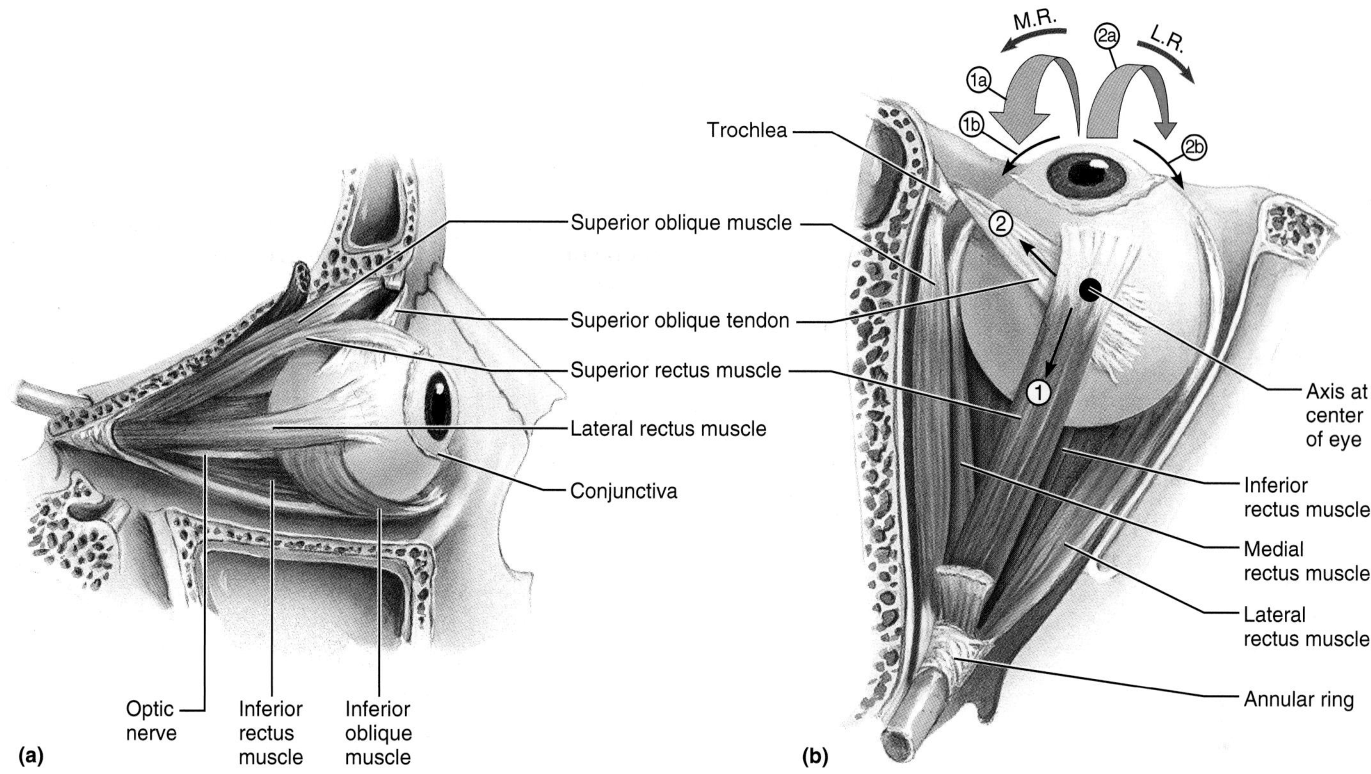

Name	Action	Controlling cranial nerve
Lateral rectus	Turns eye laterally	VI (abducens)
Medial rectus	Turns eye medially	III (oculomotor)
Superior rectus	Elevates eye	III (oculomotor)
Inferior rectus	Depresses eye	III (oculomotor)
Inferior oblique	Elevates eye and turns it laterally	III (oculomotor)
Superior oblique	Depresses eye and turns it laterally	IV (trochlear)

(c)

FIGURE 16.6 Extrinsic muscles of the eye. **(a)** Lateral view of the right eye in the orbit. **(b)** Superior view of the right eye, showing the functions of the eye muscles. The large dot is the center of the eye, defining the axis around which the eye turns. The *medial rectus* is the main muscle for turning the eye medially, and the *lateral rectus* for turning the eye laterally, as indicated by the arrows labelled M.R. and L.R., respectively. Contraction of the *superior rectus* muscle (arrow ①) elevates the eye to look upward (arrow 1a); however, because this muscle inserts medial to the center point of the eye, it also turns the eye medially (arrow 1b). Contraction of the *superior oblique* (arrow ②) depresses the eye to look downward (arrow 2a); however, because its tendon runs medial to the center of the eye, this muscle also turns the eye laterally (arrow 2b). The inferior rectus and inferior oblique muscles run in exactly the same directions as the superior rectus muscle and the superior oblique tendon, respectively, but on the *inferior* aspect of the eye. Therefore, the *inferior rectus* depresses the eye and turns it medially, and the *inferior oblique* elevates the eye and turns it laterally. **(c)** Summary of names, actions, and innervations of the extrinsic eye muscles.

Why, then, are the two oblique muscles needed, when the four rectus muscles seem to provide all the eye movements—medial, lateral, elevation, and depression—required? The simplest way to answer this question is to point out that the superior and inferior recti cannot elevate or depress the eye *without also turning it medially* because they approach the eye from a medial as well as from a posterior direction (see Figure 16.6b). For an eye to be *strictly* elevated or depressed, therefore, the lateral pull of the oblique muscles must cancel the medial pull of the superior and inferior recti. Also, the superior and inferior recti can neither elevate nor depress an eye that has been turned far medially to look inward, so the two oblique muscles must do that.

Clinical Applications

STRABISMUS The extrinsic muscles of both eyes are closely controlled by centers in the midbrain, so that the eyes move together in unison when we look at objects. If this coordination is disrupted, double vision results, because the two eyes do not look at the same point in the visual field. This misalignment of the eyes is called **strabismus** (strah-biz′mus), meaning "cross-eyed" or "squint-eyed." In strabismus, the affected eye is turned either medially or laterally with respect to the normal eye. Strabismus results from weakness or paralysis of extrinsic eye muscles, from damage to the oculomotor nerve, or from other causes. Immediate surgical correction is recommended because over time the brain comes to disregard the image from the affected eye; as a result, the entire visual pathway from that eye degenerates, making the eye functionally blind.

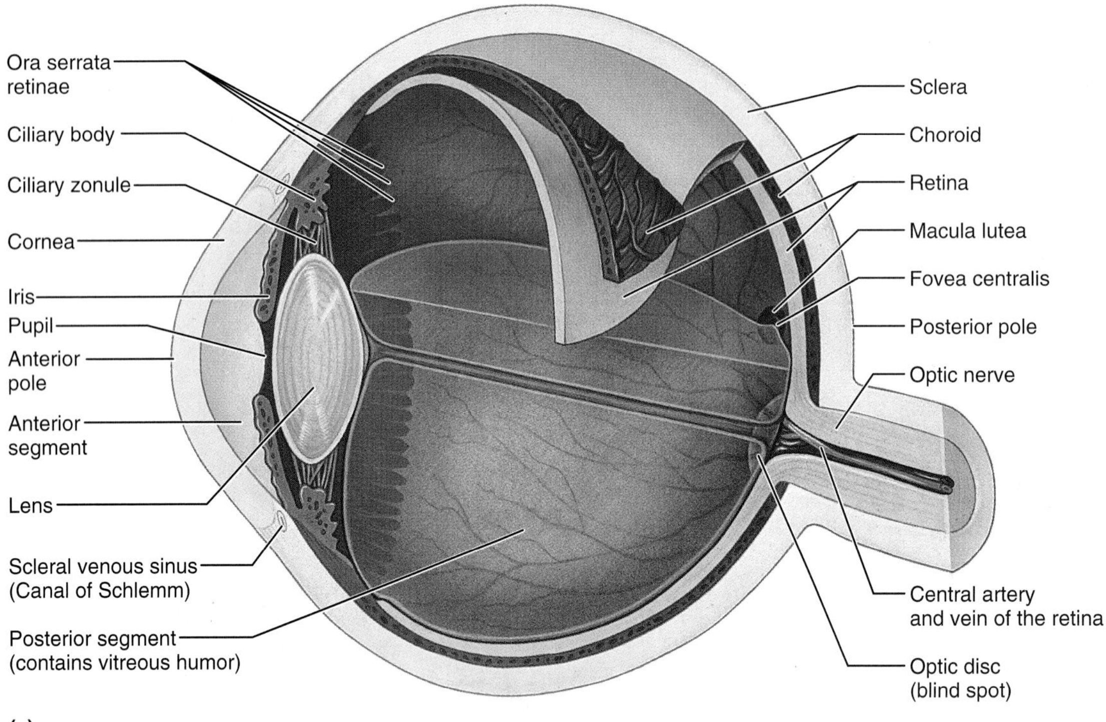

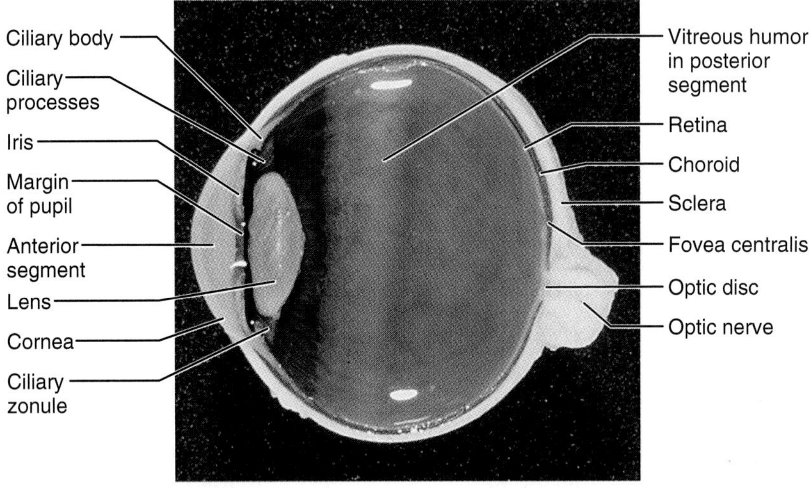

FIGURE 16.7 Medial view of the lateral half of the right eye. (a) Diagrammatic view. Above, a part of the choroid and retina have been reflected for easier viewing of the eye's three tunics. The vitreous humor is illustrated only in the bottom half of the eyeball. **(b)** Photograph of a human eye.

Anatomy of the Eyeball

The eye is a complex organ whose many components not only protect and support the delicate photoreceptor cells but also gather, focus, and process light into precise images. We begin with a brief overview of eye anatomy (Figure 16.7). Because the eyeball is shaped roughly like a globe of the earth, it is said to have poles. Its most anterior point is the **anterior pole,** and its most posterior point is the **posterior pole.** Its external wall consists of three layers, or *tunics,* and its internal cavity contains fluids called *humors.* The *lens,* a structure that helps to focus light, is supported vertically within the internal cavity, dividing it into anterior and posterior segments. The anterior segment is filled with the liquid *aqueous humor,* whereas the posterior segment is filled with the jelly-like *vitreous humor.*

Next we examine in turn the three tunics that form the external wall of the eye: the fibrous tunic, the vascular tunic, and the sensory tunic (retina).

The Fibrous Tunic

The **fibrous tunic,** the most external layer, consists of dense connective tissue arranged into two different regions: sclera and cornea (see Figure 16.7). The white, opaque tough **sclera** (skle′rah; "hard") forms the posterior five-sixths of the fibrous tunic. Seen anteriorly as the "white of the eye," the sclera protects the eyeball

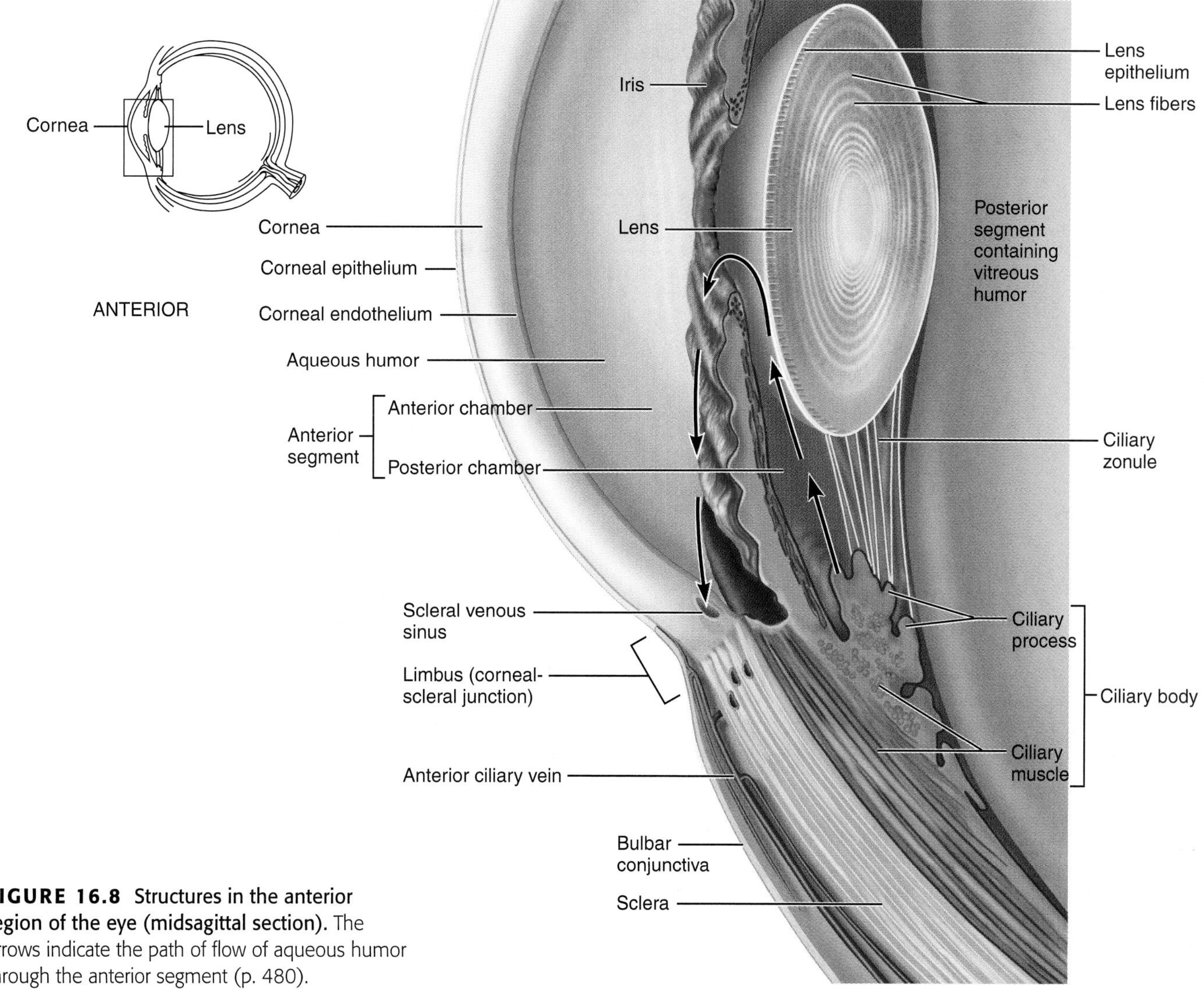

FIGURE 16.8 Structures in the anterior region of the eye (midsagittal section). The arrows indicate the path of flow of aqueous humor through the anterior segment (p. 480).

and provides shape and a sturdy anchoring site for the extrinsic eye muscles. The sclera corresponds to the dura mater that covers the brain.

The anterior sixth of the fibrous tunic is the transparent **cornea,** through which light enters the eye. This round window bulges anteriorly from its junction with the sclera. The cornea consists of a thick layer of dense connective tissue sandwiched between a superficial *corneal epithelium* and a deep *corneal endothelium* (Figure 16.8). The junction between the cornea and sclera is the **limbus.** Here, between the corneal epithelium and the conjunctiva, are epithelial stem cells that continually renew the corneal epithelium. The cornea's connective-tissue layer contains hundreds of sheets of collagen fibers stacked like the pages in a book; the transparency of the cornea is due to this regular alignment of collagen fibers. The cornea not only lets light into the eye but also forms part of the light-bending apparatus of the eye (see p. 481).

The cornea is avascular—it receives oxygen from the air in front of it, and oxygen and nutrients from the aqueous humor that lies posterior to it. Near the cornea, however, the sclera part of the limbus contains a larger vessel called the **scleral venous sinus** (see Figure 16.8). The inner wall of this sinus, adjacent to the aqueous humor, is a permeable net that allows the sinus to drain aqueous humor out of the eye (see p. 480).

The cornea is richly supplied with nerve endings, most of which are pain receptors. (This is why some people can never adjust to wearing contact lenses.) Touching the cornea causes reflexive blinking and an increased secretion of tears. Even with these protective responses, the cornea is vulnerable to damage by dust, slivers, and other objects. Fortunately, its capacity for regeneration and healing is extraordinary.

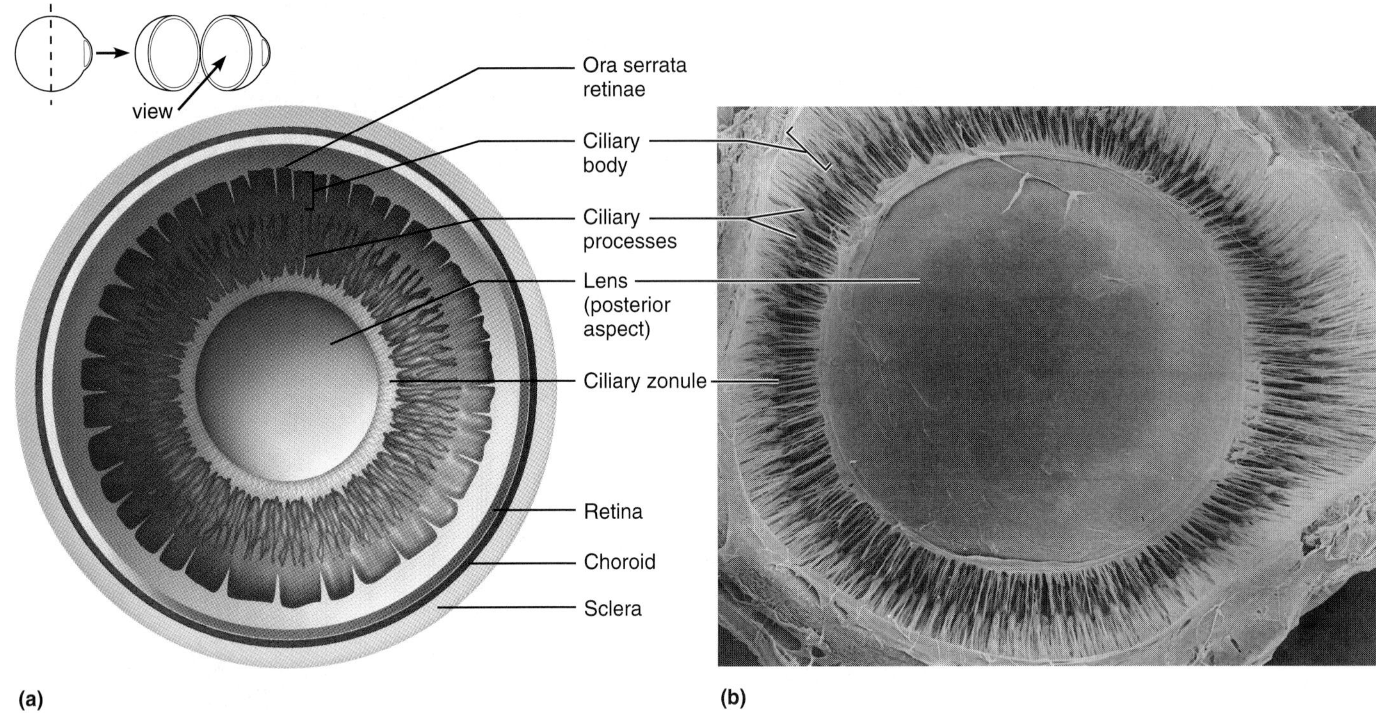

FIGURE 16.9 Posterior view of the anterior half of the eye. **(a)** Note the ciliary processes on the ciliary body, the ciliary zonule, and the ora serrata retinae. **(b)** Scanning electron micrograph of the posterior aspect of the lens, ciliary zonule, and ciliary body. Note the thin fibrils of the ciliary zonule, which holds the lens in the eye (28 ×).

CLINICAL APPLICATIONS

CORNEAL TRANSPLANTS "Eye banks" are institutions that receive and store corneas for use in the surgical replacement of severely damaged corneas, a procedure called a **corneal transplant.** The cornea is one of the few structures in the body that can be transplanted from one person to another with only minimal risk of rejection: Because it has no blood vessels, it is beyond the reach of the body's immune system. In the 10% of transplanted corneas that are rejected, the damage to the recipient's original cornea was so extensive that it destroyed the limbus. Without the limbus, the corneal epithelium cannot regenerate, so *conjunctival* epithelium overgrows the new cornea, in the process attracting new blood vessels carrying the immune cells that cause rejection. This problem can be alleviated by grafting corneal epithelium from the patient's healthy eye onto the newly transplanted cornea. ✚

The Vascular Tunic

The **vascular tunic,** the middle coat of the eyeball, has three parts: the *choroid,* the *ciliary body,* and the *iris* (see Figure 16.7).

The **choroid** (ko'roid; "membrane") is a highly vascular, darkly pigmented membrane that forms the posterior five-sixths of the vascular tunic. Its many blood vessels nourish the other tunics. The brown color of the choroid is produced by melanocytes, whose pigment—melanin—helps absorb light, thereby preventing light from scattering within the eye and creating visual confusion. Overall, the choroid layer of the eye corresponds to the arachnoid and pia mater around the brain.

Anteriorly, the choroid is continuous with the **ciliary body,** a thickened ring of tissue that encircles the lens (see Figures 16.7 and 16.8). The ciliary body consists chiefly of smooth muscle called the **ciliary muscle** (see Figure 16.8), which acts to focus the lens. Nearest the lens, the posterior surface of the ciliary body is thrown into radiating folds called **ciliary processes** (Figure 16.9). The halo of fine fibrils that extends from these processes to attach around the entire circumference of the lens is called the **ciliary zonule** or *suspensory ligament of the lens* (see Figures 16.8 and 16.9).

The **iris** ("rainbow") is the visible, colored part of the eye. It lies between the cornea and lens, and its base attaches to the ciliary body (see Figure 16.8). Its round central opening, the *pupil,* allows light to enter the eye. The iris contains both circularly arranged and radiating smooth muscle fibers, the *sphincter* and *dilator pupillae* muscles, that act to vary the size of the pupil. In bright light and for close vision, the sphincter pupillae contracts to constrict the pupil. In dim light and for distant vision, the dilator pupillae contracts to widen the pupil, allowing more light to enter the eye. (Constriction and dilation of the pupil are controlled by parasympathetic and sympathetic fibers, respectively; see Chapter 15.) The constriction of the pupils that occurs in intense, potentially damaging light is a protective response called the **pupillary light reflex.**

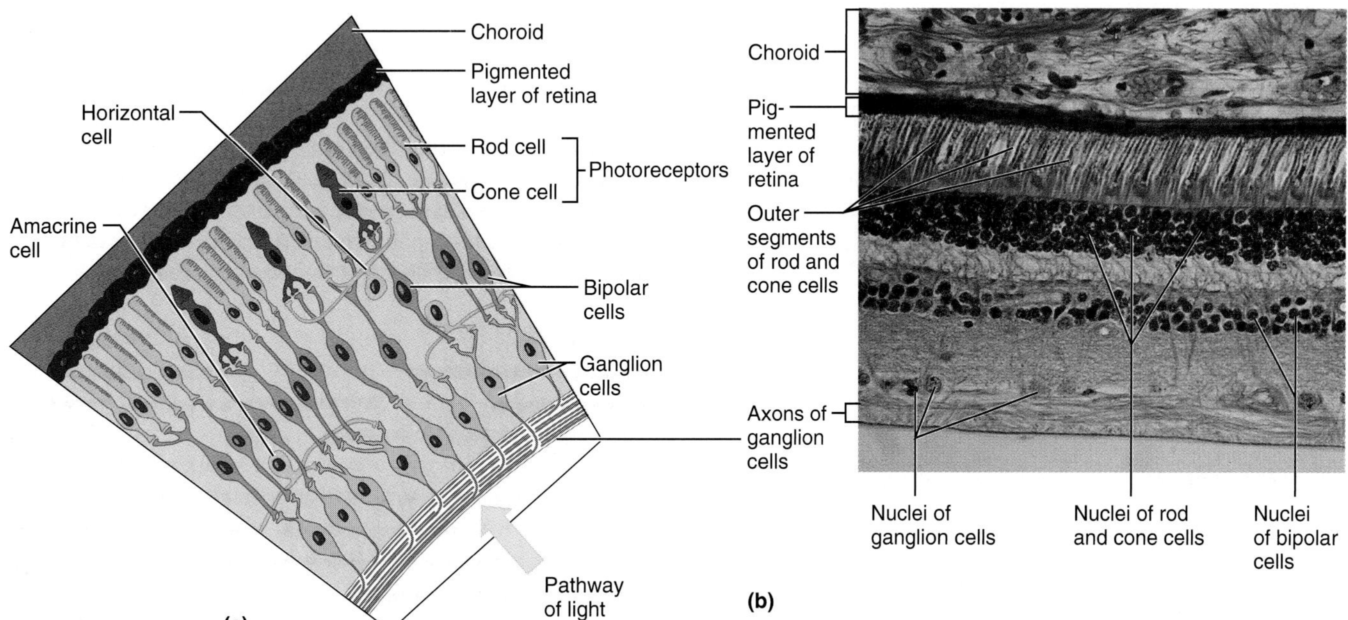

FIGURE 16.10 Microscopic anatomy of the retina. **(a)** At upper left, this is a schematic view of the major types of neurons in the neural layer of the retina. Light passes outward through the entire thickness of the retina before exciting the photoreceptor cells; then, electrical signals flow inward from neuron to neuron. **(b)** Photomicrograph of the retina (160 ×). **(c)** Schematic view of the posterior part of the eyeball illustrating how the axons of the ganglion cells form the optic nerve, which leaves the back of the eye at the optic disc.

Although irises come in many colors, they contain only brown pigment. All people except albinos have a layer of pigmented cells on the posterior surface of the iris, and brown-eyed people have many pigment cells in the body of the iris as well. Blue-eyed people, by contrast, have no pigment in the body of the iris. The blue color represents the posteriorly located pigment as it appears through the colorless body of the iris. Hazel-eyed people have some pigment in the body of the iris but less than do brown-eyed people.

The Sensory Tunic (Retina)

The sensory tunic or **retina,** the deepest tunic of the eye, consists of two layers: a thin pigmented layer and a far thicker neural layer (Figure 16.10). The outer **pigmented layer,** which lies against the choroid, is a single layer of flat-to-columnar melanocytes that, like those in the choroid, absorb light to prevent it from scattering within the eye. The much thicker inner **neural layer** is a sheet of nervous tissue that contains the light-sensitive photoreceptor cells. The neural and pigmented layers of the retina are held together by a thin film of extracellular matrix, but they are not tightly fused. Only the neural layer plays a direct role in vision.

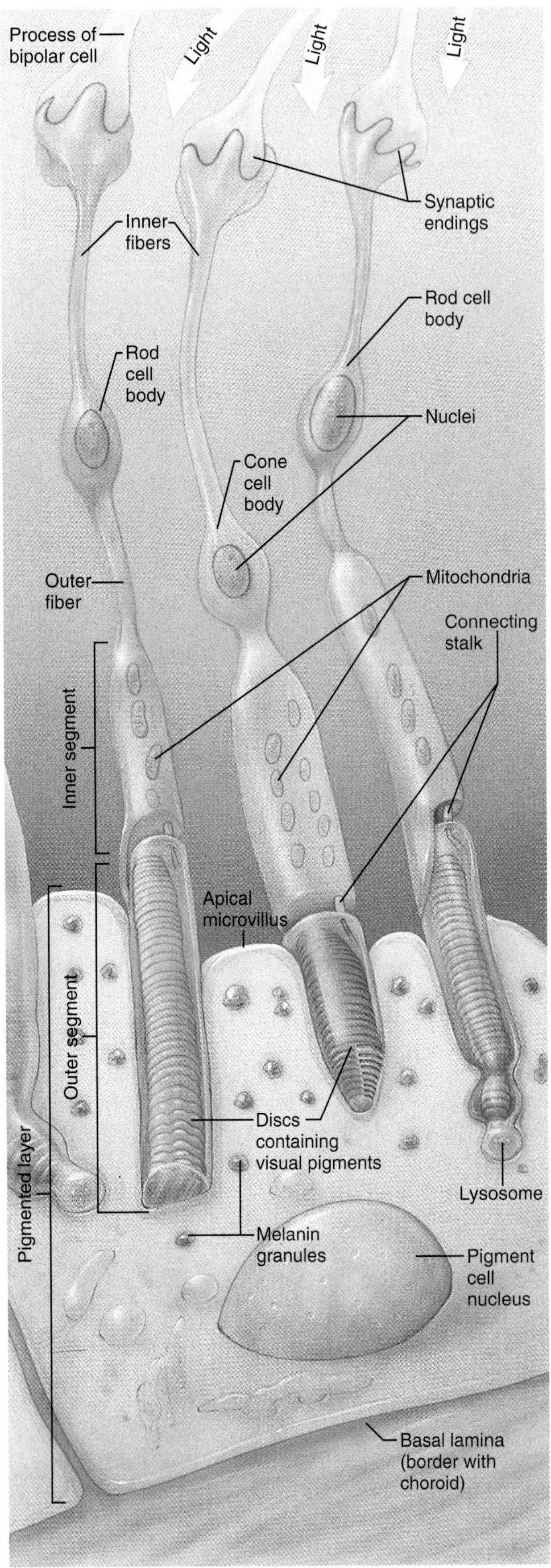

FIGURE 16.11 Three photoreceptors (rod cells and cone cells) in the retina. Note the various parts of these cells. The outer segments of the photoreceptors (below) indent the pigmented cells of the retina like pegs in sockets.

The neural layer contains three main types of neurons. From external to internal, these are the **photoreceptor cells, bipolar cells,** and **ganglion cells** (see Figure 16.10a). When exposed to light, the photoreceptor neurons signal the bipolar cells, which in turn signal the ganglion cells to generate action potentials. Axons from the ganglion cells run along the internal surface of the retina and converge posteriorly to form the *optic nerve,* which runs from the eye to the brain (see Figure 16.10c).

Along with its three main cell types, the retina also contains interneurons—including amacrine cells and horizontal cells—whose processes run laterally (parallel to the surface of the retina). It makes sense that the retina contains interneurons (and glial cells as well) given that the retina develops as a part of the brain (see p. 485). The retinal interneurons process and organize visual information before it is sent to higher brain centers for further processing.

Photoreceptors The photoreceptor cells are of two types: *rod cells* and *cone cells.* The more numerous **rod cells** are more sensitive to light and permit vision in dim light. Because rod cells provide neither sharp images nor color vision, things look gray and fuzzy when viewed in dim light. **Cone cells,** by contrast, operate best in bright light and enable high-acuity color vision. Three subtypes of cone cells are sensitive to blue, red, and green light, respectively.

Photoreceptors are considered neurons, but they also resemble tall epithelial cells turned upside down, with their "tips" immersed in the pigmented layer (Figure 16.11). Both rod cells and cone cells have an *outer segment* joined to an *inner segment* by a connecting stalk containing a cilium. In each rod cell, the inner and outer segments together form a rod-shaped structure (hence the name *rod*), which connects to the nucleus-containing *cell body* by an *outer fiber.* In each cone cell, by contrast, the inner and outer segments form a cone-shaped structure (hence the name *cone*), one end of which joins to the cell body directly. In both cell types, the cell body is continuous with an *inner fiber* that bears synaptic endings.

The outer segments are the receptor regions of the rod and cone cells. Each outer segment is a modified cilium whose plasma membrane has folded inward to form hundreds of membrane-covered discs. The presence of light-absorbing ***visual pigments*** within the vast membrane of these discs greatly magnifies the surface area available for trapping light. When light particles (photons) hit the visual pigment, it splits in two, changing the membrane charge of the photoreceptors and causing them to signal the bipolar neurons with which they synapse. This reaction initiates the flow of visual information to the brain.

The photoreceptors are highly vulnerable to damage by intense light or heat. These cells cannot regenerate if destroyed, but they continually renew and replace their outer segments through the addition of new discs. In this normal recycling process, as new discs are added to one end of the stack, old discs are removed at the other end by retinal pigment cells phagocytizing the tips of the rods and cones (see the lysosome in Figure 16.11).

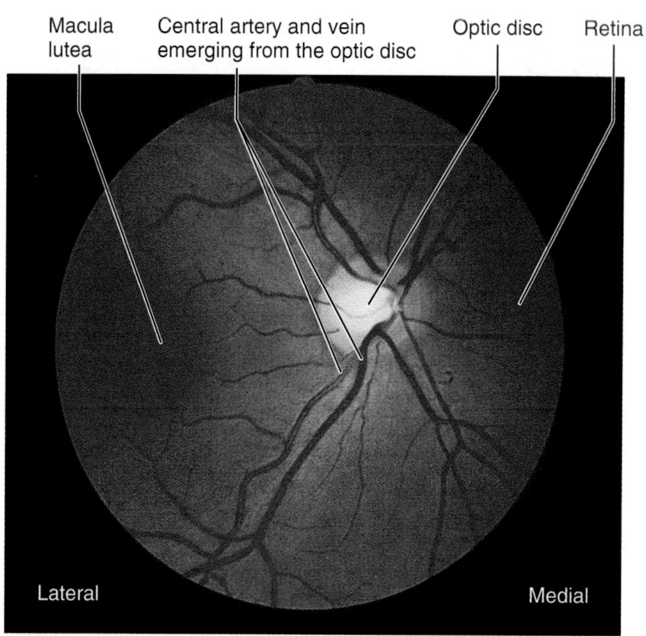

FIGURE 16.12 The posteromedial wall of the retina as seen with an ophthalmoscope (right eye). Note the optic disc, from which radiate the central vessels of the retina.

Regional Specializations of the Retina In certain regions of the eye, the retina differs from its "typical" structure just described. Here, we consider these specialized regions.

In the anterior part of the eye, the retina is incomplete. Its neural layer ends at the posterior margin of the ciliary body, called the **ora serrata retinae** (o′rah se-rah′tah ret′ĭ-ne; "saw-toothed mouth of the retina") (see Figure 16.9a). The retina's pigmented layer, by contrast, extends anteriorly beyond the ora serrata to cover the ciliary body and the posterior face of the iris.

The posterior part of the eye contains several special areas of retina (see Figure 16.7). Lying precisely at the eye's posterior pole is the **macula lutea** (mak′u-lah lu′te-ah; "yellow spot"), the center of which is a tiny pit called the **fovea centralis** (fo′ve-ah sen-trah′lis; "central pit"). The fovea contains only cones; the macula contains mostly cones; and the density of cones declines progressively with increasing distance from the macula. Because of its high concentration of cones, the fovea provides maximal visual acuity. Because the fovea lies directly in the anterior-posterior axis of the eye, we see things most clearly when we look straight at them—which is why peripheral vision is not as sharp as central vision. A few millimeters medial to the fovea is the **optic disc** *(optic papilla),* a circular elevation where the axons of ganglion cells converge to exit the eye as the optic nerve. The optic disc is called the *blind spot* because it lacks photoreceptors, and light focused on it cannot be seen.

Blood Supply of the Retina The retina receives its blood from two different sources. The outer third of the retina, containing the photoreceptors, is supplied by capillaries in the choroid, whereas its inner two-thirds is supplied by the **central artery** and **vein of the retina,** which enter and leave the eye by running through the center of the optic nerve (see Figure 16.7). These vessels radiate from the optic disc, giving rise to a rich network of tiny vessels that weave among the axons on the retina's inner face. This vascular network is clearly visible through an ophthalmoscope, a hand-held instrument that shines light through the pupil and illuminates the retina (Figure 16.12). Physicians observe the tiny retinal vessels for signs of hypertension, diabetes, and other diseases that damage our smallest blood vessels.

CLINICAL APPLICATIONS

DETACHED RETINA The retina's pattern of vascularization contributes to a potentially blinding condition called **retinal detachment,** in which the loosely joined neural and pigmented layers of the retina separate ("detach") from one another. The detachment begins with a tear in the retina, which may result from a small hemorrhage, a blow to the eye, or age-related degeneration. The tear allows the jelly-like vitreous humor from the eye's interior (see the next section) to seep between the two retinal layers. With the neural layer detached, the photoreceptors, now separated from their blood supply in the choroid, are kept alive temporarily by nutrients obtained from the inner retinal capillaries. However, because these capillaries are too distant to supply them permanently, the photoreceptors soon die. Early symptoms of retinal detachment include seeing flashing lights or spots that float across the field of vision, and having objects appear as if seen through a veil. If the detachment is diagnosed early, blindness may be prevented by reattaching the retina with laser surgery or the application of extreme cold or heat. Alternately, a gas bubble may be injected into the vitreous humor to push the retina back into place. ✚

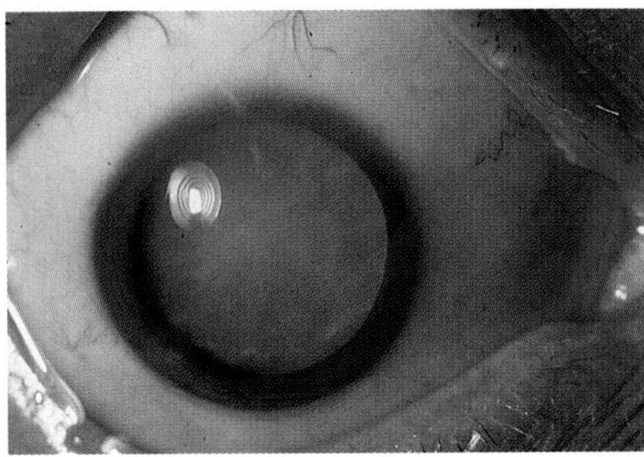

FIGURE 16.13 Photograph of a cataract. The *lens* is milky and opaque, not the cornea.

Internal Chambers and Fluids

As noted previously, the lens and its halo-like ciliary zonule divide the eye into posterior and anterior segments (see Figures 16.7a and 16.8). The **posterior segment** is filled with the clear **vitreous humor** (*vitreus* = glassy), a jelly-like substance that contains fine fibrils of collagen and a ground substance that binds tremendous amounts of water. Indeed, water constitutes over 98% of its volume. The functions of vitreous humor are to (1) transmit light, (2) support the posterior surface of the lens and hold the neural retina firmly against the pigmented layer, and (3) help maintain *intraocular pressure* (the normal pressure within the eye), thereby counteracting the pulling forces of the extrinsic eye muscles.

The **anterior segment** of the eye (see Figure 16.8) is divided into an **anterior chamber** between the cornea and iris, and a **posterior chamber** between the iris and lens. The entire anterior segment is filled with **aqueous humor,** a clear fluid similar to blood plasma (*aqueous* = watery). Unlike the vitreous humor, which forms in the embryo and lasts a lifetime, aqueous humor is renewed continuously and is in constant motion. This process is indicated by the arrows in Figure 16.8. After being formed as a filtrate of the blood from capillaries in the ciliary processes, the aqueous humor enters the posterior chamber, flows through the pupil into the anterior chamber, and drains into the scleral venous sinus, which returns it to the blood. An equilibrium in the rates at which the aqueous humor forms and drains results in a constant intraocular pressure, which supports the eyeball internally. Furthermore, the aqueous humor supplies nutrients and oxygen to the avascular lens and cornea.

CLINICAL APPLICATIONS

GLAUCOMA When the aqueous humor drains more slowly than it forms, the result is **glaucoma** (glaw-ko′mah), a disease in which intraocular pressure increases to dangerous levels and ultimately damages the head of the optic nerve. Glaucoma results from an obstruction of outflow, usually from a clogging of the permeable net through which aqueous humor drains into the scleral venous sinus. The resulting destruction of the optic nerve eventually causes blindness. Even though vision can be saved if the condition is detected early, glaucoma often steals sight so slowly and painlessly that most people do not realize they have a problem until the damage is done. Late signs include blurred vision, seeing halos around lights, and headaches. The examination for glaucoma involves testing for high intraocular pressure by measuring the force necessary to deform the sclera. A glaucoma exam should be done yearly after age 40, because glaucoma affects fully 2% of people over that age. Early glaucoma is treated with eye drops that lower the pressure in the eye. ✚

The Lens

The **lens** is a thick, transparent, biconvex disc that changes shape to allow precise focusing of light on the retina (see Figures 16.7 and 16.8). It is enclosed in a thin elastic capsule and is held in place posterior to the iris by its ciliary zonule. Like the cornea, it lacks blood vessels, for vessels interfere with transparency.

The lens has two components: the lens epithelium and the lens fibers. The **lens epithelium,** confined to the anterior surface, consists of cuboidal cells. The subset of epithelial cells around the edge of the lens disc transforms continuously into the elongated **lens fibers** that form the bulk of the lens. These fibers, which are packed together like the layers in an onion, contain no nuclei and few organelles. They do, however, contain precisely folded proteins which make them transparent. New lens fibers are added continuously, so the lens enlarges throughout life. It becomes denser, more convex, and less elastic with age. As a result, its ability to focus light is gradually impaired.

CLINICAL APPLICATIONS

CATARACT A cataract (kat′ah-rakt; "waterfall") is a clouding of the lens (Figure 16.13) that causes the world to appear distorted, as if seen through frosted glass or through a waterfall. Some cataracts are congenital, but most result from age-related changes in the lens. Recent evidence indicates that excessive exposure to sunlight, heavy smoking, and severe bouts of diarrhea cause cataracts. No matter the cause, cataracts seem to result from an inadequate delivery of nutrients to the deeper lens fibers. Fortunately, surgical removal of the damaged lens and its replacement by an artificial lens can save a cataract patient's sight. ✚

The Eye as an Optical Device

Here we consider how light is focused on the retina. From each point on an object being viewed, light rays radiate in every direction. Some of these light rays enter the eye of the viewer (Figure 16.14). Rays from a distant point are parallel to one another as they reach the eye, whereas rays from a nearby point are diverging markedly as they enter the eye. If we are to see clearly, the eye must be able to bend all these light rays so that they converge on the retina at a single *focal point.* The light-bending parts of the eye, called **refractory media,** are the cornea, the lens, and the humors. The cornea does most of the light-bending, the lens does some of it, and the humors do a minimal amount.

Although the lens is not as powerful as the cornea in bending light, its curvature is adjustable. This adjustability allows the eye to focus on nearby objects—a process called **accommodation** (Figure 16.14c). A resting eye, with its lens stretched along its long axis by tension in the ciliary zonule, is "set" to focus the almost-parallel rays from distant points. Therefore, distance vision is the natural state. The diverging rays from *nearby* points must be bent more sharply if they are to focus on the retina. To accomplish this the lens is made rounder: The ciliary muscle contracts in a complex way that releases most of the tension on the ciliary zonule. No longer stretched, the lens becomes rounder as a result of its own elastic recoil. Accommodation is controlled by the parasympathetic fibers that signal the ciliary muscle to contract (see p. 448).

Focusing on nearby objects is always accompanied by pupillary constriction, which prevents the most divergent light rays from entering the eye and passing through the extreme edges of the lens. Such rays would not be focused properly and would cause blurred vision.

For simplicity, our discussion of eye focusing has been confined to single-point images; how the eye focuses the "multiple-point" images of large objects is beyond the scope of this text. However, it must be noted that the convex lens of the eye, just like the convex lens of a camera, produces images that are upside down and reversed from right to left. Therefore, an inverted and reversed image of the visual field is projected onto each retina. The cerebral cortex then "flips" the image back, so that we see things as they are actually oriented.

Common focusing disorders of the eye are discussed in this chapter's A Closer Look box on pages 482–483.

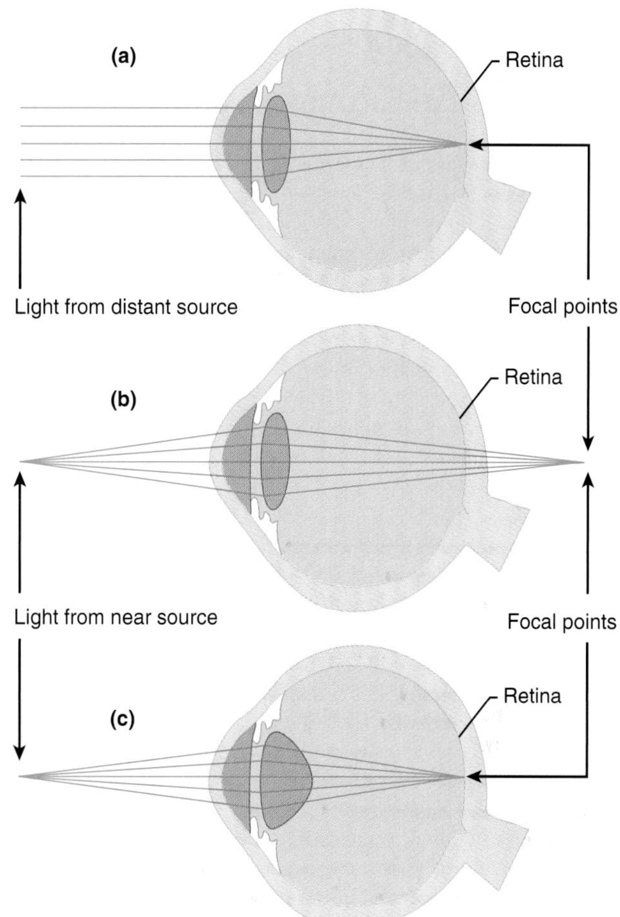

FIGURE 16.14 The eye as an optical device. (a) The resting eye is set for distance vision, such that parallel light rays from a distant point are focused directly on the retina. **(b)** The resting eye is not set for viewing nearby points—because divergent light rays are focused far behind the retina. **(c)** Focusing on a nearby point involves accommodation: The lens rounds, bending the divergent rays more sharply so that they converge on the retina. (For simplicity, the substantial light-bending effect of the cornea is ignored in these diagrams.)

Visual Pathways

Having considered the eye, we now will discuss the paths by which visual information leaves the eye and travels to the brain for complex processing. Most of this visual information goes to the cerebral cortex, which is responsible for conscious "seeing" (recall Chapter 13), but some goes to nuclei in the midbrain and diencephalon, which control reflexes and subconscious behaviors that require visual input.

Visual Pathway to the Cerebral Cortex

Visual information travels to the cerebral cortex through the main **visual pathway** (Figure 16.15 on p. 484). This pathway begins in the retina as light activates the photoreceptors, which signal bipolar cells, which signal ganglion cells. Axons of the ganglion cells then exit the eye in the **optic nerve.** At the X-shaped **optic chiasma** (ki-as′mah; "cross"), which lies anterior to the hypothalamus, the axons from the medial half of each eye decussate and then continue in an **optic tract.** The paired optic tracts sweep posteriorly around the hypothalamus and send most of their axons to the **lateral geniculate nucleus of the thalamus,** where they synapse with thalamic neurons. Axons of those neurons then project through the internal capsule to form the **optic radiation** of fibers in the cerebral white matter. These fibers reach the primary **visual cortex** in the occipital lobe, which is essential for the conscious perception of visual images.

A CLOSER LOOK

Eye-Focusing Disorders

It seems that whenever people who wear glasses or contact lenses discuss their vision, one of them says something like, "I need my glasses to see faraway objects clearly, so am I nearsighted or farsighted?" Someone else might say, "Without my glasses, nearby objects appear blurry; does that mean I'm nearsighted or farsighted?" Here we clarify the meaning of nearsightedness and farsightedness as we explore the anatomical basis of eye-focusing disorders.

The eye that focuses images correctly on the retina is said to have **emmetropia** (em″ĕ-tro′pe-ah; "harmonious vision"). Such an eye is depicted in part (a) of the figure.

Nearsightedness—a condition formally called **myopia** (mi″o′pe-ah; "short vision")—occurs when the parallel light rays from distant objects are not focused on the retina, but in front of it, as shown in part (b) in the figure. Therefore, *distant* objects appear blurry to myopic people, but nearby objects are in focus because as an object nears the eye, its focal plane naturally moves posteriorly and comes to lie on the retina. Myopia results from an eyeball that is too long, a lens that bends light too much, or a cornea that is too highly curved. Correction requires the use of *concave* corrective lenses that diverge the light rays before they enter the eye, so that the focal point occurs more posteriorly. The person who asked the first question is nearsighted, for *near*sighted people see *near* objects clearly and need corrective lenses to focus on distant objects.

Farsightedness—formally called **hyperopia** (hi″per-o′pe-ah; "far vision")—occurs when the parallel light rays from distant objects are focused *behind* the retina—at least in the resting eye, in which the lens is relatively flat and the ciliary muscle is relaxed, as in part (c) in the figure. Hyperopia usually results from an eyeball that is too short. People with hyperopia can see distant objects clearly because their ciliary muscles contract continuously to increase the light-bending power of the lens, which moves the focal plane forward onto the retina. However, the diverging rays from *nearby* objects are focused so far behind the retina that even the full accommodative power of the lens cannot bring the focal point onto the retina. Therefore, nearby objects appear blurry. Furthermore, hyperopic people are subject to eyestrain as their endlessly contracting ciliary muscles tire from overwork. Correction of hyperopia requires *convex* corrective lenses that converge the light rays before they enter the eye. The person who asked the second question is farsighted, for *far*sighted people can see *far*away objects clearly and require corrective lenses to focus on nearby objects.

Another eye-focusing disorder is **presbyopia** (pres″be-o′pe-ah; "old-person's vision"), which develops between ages 40 and 65 in essentially all people. Presbyopia is the loss of elasticity and accommodative power of the lens. A presbyopic eye that is otherwise normal can see distant objects clearly (recall that the normal eye of young and old alike is set for distance vision), but it cannot accommodate to focus on nearby objects. To allow near vision, corrective glasses with convex lenses are prescribed, in effect supplementing the action of the eye's convex lens. However, many people with presbyopia do not have perfect *distance* vision either, so they wear bifocal glasses, in which the upper part aids in distance vision, the lower part in closer vision.

Unequal curvatures in different parts of the cornea or lens cause **astigmatism** (ah-stig′mah-tizm). In this condition, blurry images occur because points of light are focused not as points on the retina but as lines (*astigma* = not a spot). Special cylinder-shaped lenses or contacts are used to correct this problem. Eyes that are myopic or hyperopic as well as astigmatic require more complex correction.

Modern surgical techniques offer freedom from the bother of wearing glasses or contact lenses, especially for individuals with nearsightedness, which affects 20 million people in the United States. Most of these techniques change the curvature of the cornea, the eye structure most responsible for bending light. In *radial keratectomy,* incisions are made in the cornea in a starburst or "pie-shaped" pattern, causing the cornea to collapse to a flatter shape that does not bend light as sharply. Newer techniques include *photorefractive keratectomy,* which resculpts the cornea's shape with lasers, and *intrastromal corneal implants,* which are plastic arcs placed in the periphery of the cornea to lessen its curvature.

The partial decussation of axons in the optic chiasma relates to depth perception, which is also called stereoscopic or three-dimensional vision. To explain this, we first divide the retina of each eye into a medial and lateral half (see the dashed lines bisecting the eyes in Figure 16.15a). Next recall that the lens system of each eye reverses all images, as indicated by the solid arrows inside the eyes in the figure. Because of this reversal, the medial half of each retina receives light rays from the lateral (peripheral) part of the visual field; that is, from objects that lie either far to the left or to the right rather than straight ahead. Correspondingly, the lateral half of each retina receives an image of the central part of the visual field. Only those axons from the *medial* halves of the two retinas cross over at the optic chiasma. The result is that all information from the left half of the visual field is directed through the right optic tract to be perceived by the right cerebral cortex. Likewise, the right half of visual space is perceived by the left visual cortex. Each cerebral cortex receives an image of half the visual field, as viewed by the two different eyes from slightly different angles. The cortex then compares these

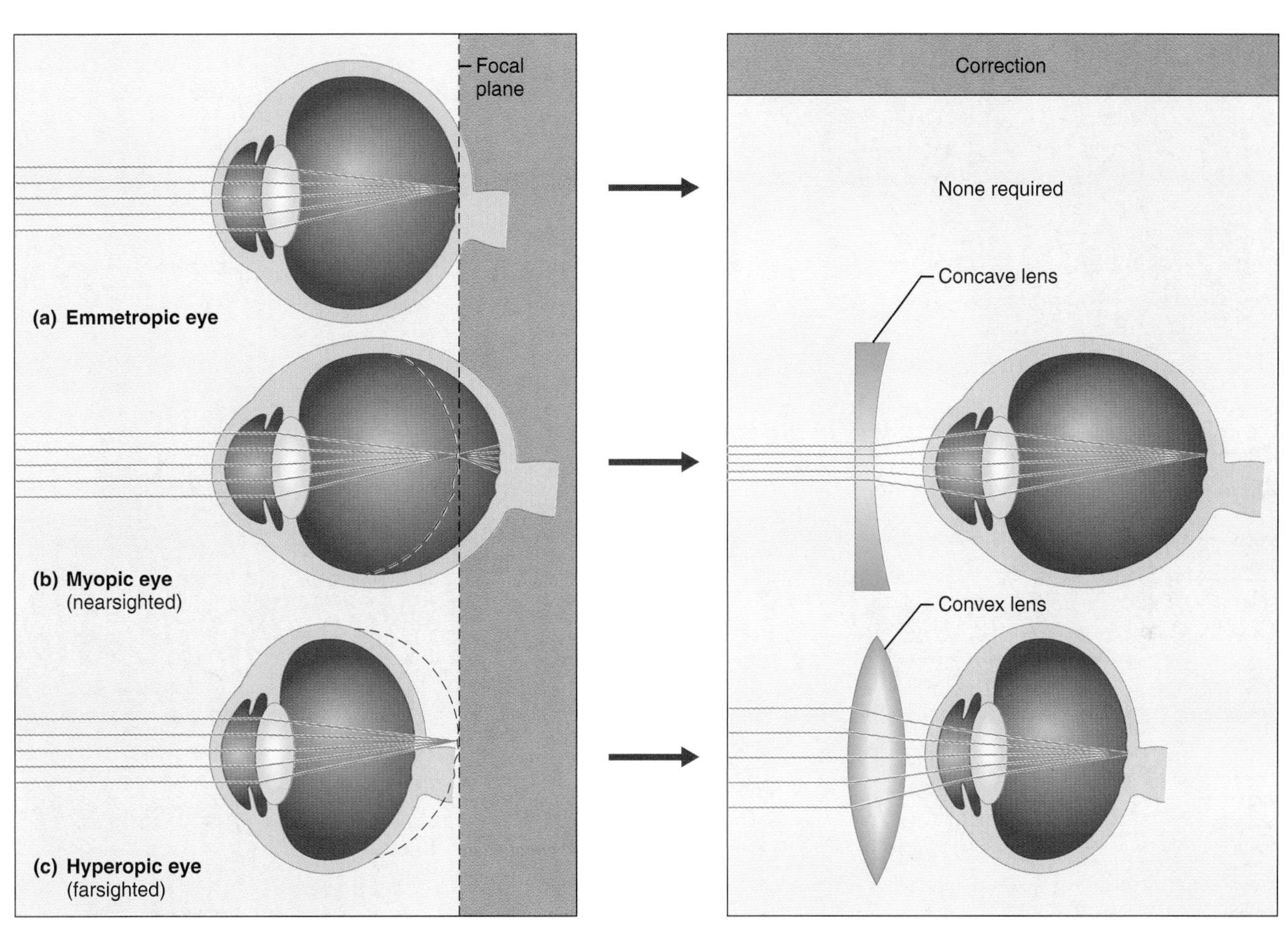

Eye-focusing disorders. (a) In the emmetropic (normal) eye, light is focused properly on the retina. **(b)** In a myopic eye, light is focused at a point anterior to the retina and then diverges again. **(c)** In the hyperopic eye, light is focused at a point behind the retina. The appropriate kinds of corrective lenses are shown at right.

two similar but different images and, in doing so, assembles the perception of depth.

These relationships explain patterns of blindness that follow damage to different visual structures. Destruction of one eye or one optic nerve eliminates true depth perception and causes a loss of peripheral vision on the side of the damaged eye. Thus, if the "Left eye" in Figure 16.15a were lost, nothing could be seen in the visual area colored yellow in that figure. However, if damage occurs beyond the optic chiasma—in an optic tract, the thalamus, or the visual cortex—then the entire opposite half of the visual field is lost. For example, a stroke affecting the left visual cortex leads to blindness (blackness) throughout the right half of the visual field.

Critical Reasoning

Describe the effect on vision of a tumor in the hypothalamus or pituitary gland that compresses the optic chiasma, destroying the axons crossing through this structure.

Peripheral vision would be lost on both the right and left sides of the visual field. Central vision would remain intact, so the person could only see straight ahead (tunnel vision).

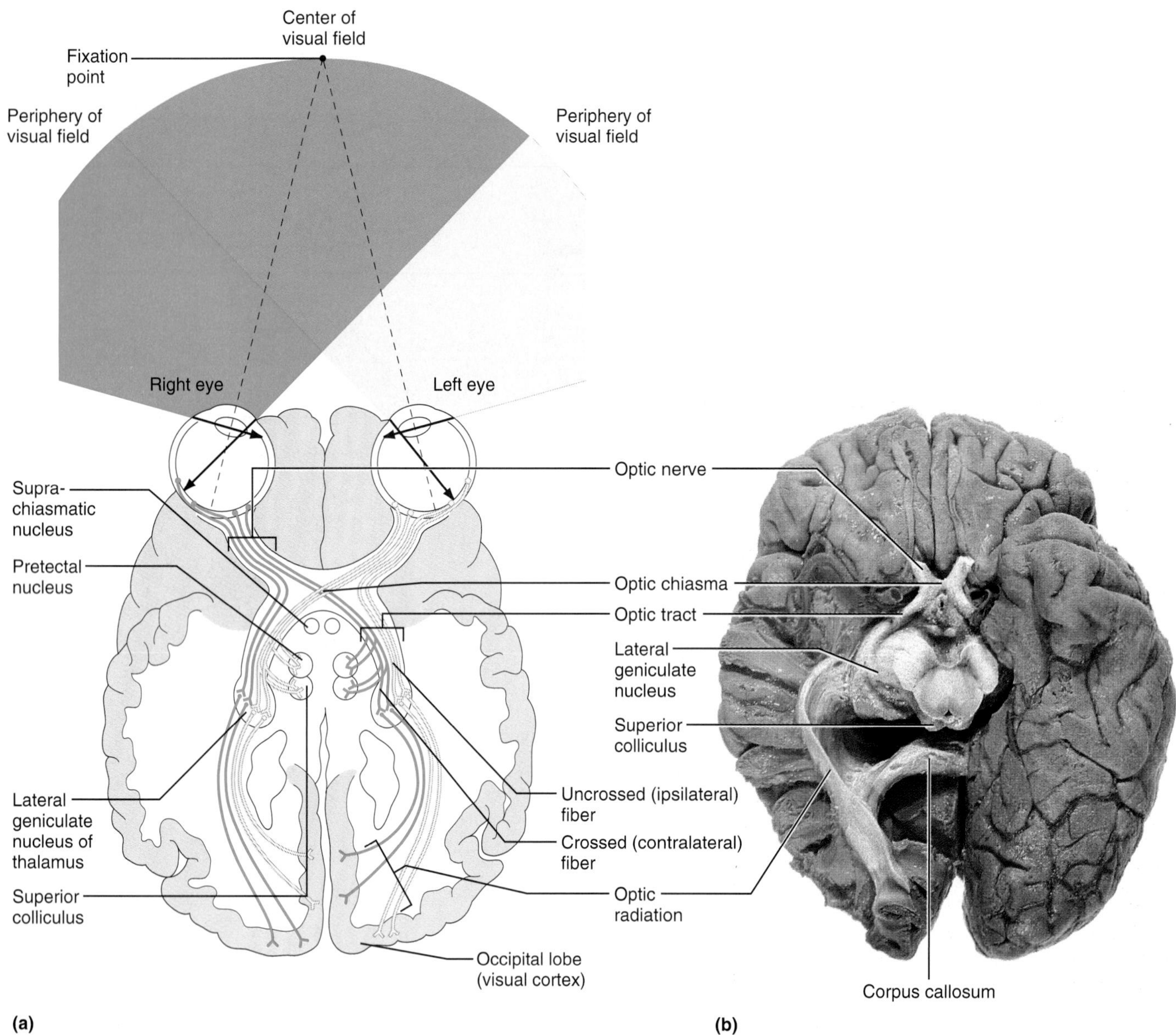

FIGURE 16.15 Visual pathway to the brain, and visual fields of the eyes. Note that the brain and eyes are shown in inferior view, so the right eye is on the left side of this figure, the left eye on the right. **(a)** In the visual pathway, ganglion cells from the retina form the optic nerves, optic chiasma, and optic tract to the thalamus. From the thalamus, the visual information projects to the visual cortex in the occipital lobe of the cerebrum. The blue visual field is that seen by the right eye, whereas the yellow visual field is seen by the left eye; the green area is seen by both eyes and is the area of stereoscopic vision. **(b)** Photograph of the structures involved in the visual pathway.

Visual Pathways to Other Parts of the Brain

Some axons from the optic tracts send branches to the midbrain (see Figure 16.15a). These branches go to the **superior colliculi,** reflex nuclei controlling the extrinsic eye muscles (see p. 380), and to the **pretectal nuclei,** which mediate the pupillary light reflexes. Other branches from the optic tracts run to the *suprachiasmatic nucleus* of the hypothalamus, which is the "timer" that runs our daily biorhythms (see p. 377) and requires visual input to keep it in synchrony with the daylight/darkness cycle.

Disorders of the Eye and Vision

The most common of the visual disorders—glaucoma, cataracts, and disorders of accommodation—have already been discussed. Here, some other common and important eye disorders are considered, namely *age-related macular degeneration, retinopathy of prematurity,* and *trachoma*.

Age-related macular degeneration (AMD) is a progressive deterioration of the retina that affects the macula lutea and leads to loss of central vision. It is the main cause of vision loss in those over age 65, and

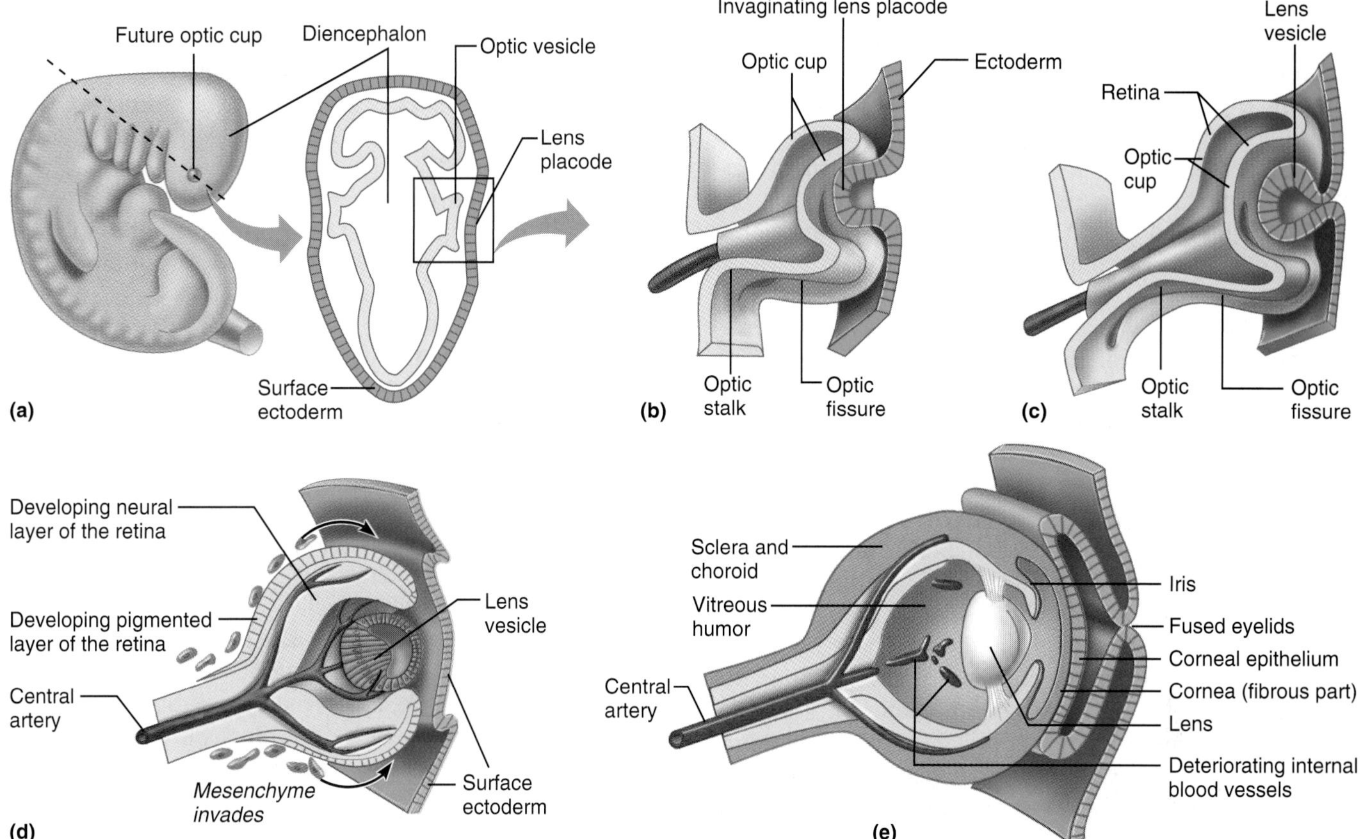

FIGURE 16.16 Embryonic development of the eye. **(a)** Section through the brain and optic vesicles in a 4- to 5-week embryo, at the time of formation of the lens placode. (The broken line in the diagram at left shows the plane of the section.) **(b)** Contact between the optic cup and the lens placode signals the lens placode to invaginate. **(c)** The lens placode forms a lens vesicle (about day 40). **(d)** The central artery in the optic fissure reaches the interior of the eye, and the optic fissure closes. The optic cup forms both layers of the retina. **(e)** Mesenchyme, shown in pink, surrounds and invades the optic cup to form the vascular and fibrous tunics and the vitreous humor (done by day 50). Lens vesicle becomes the lens. Surface ectoderm becomes the epithelium of the cornea and conjunctiva.

200,000 Americans develop it each year. Early stages or mild forms of AMD involve the buildup of visual pigments in the macula, perhaps caused by damage to the pigment-containing cones or by a malfunction of the phagocytic cells that normally remove the damaged pigment. Continued accumulation of the pigment is associated with the "dry" form of AMD, in which many of the macular photoreceptors die. Far less common is the "wet" form, in which new blood vessels grow into the retina from the choroid layer, then bleed and cause scarring and detachment of the retina. The ultimate cause is unknown and AMD is largely untreatable, although laser treatments can kill some of the growing vessels in the wet form.

Retinopathy of prematurity is a visual impairment that affects many of those born so prematurely that they need to receive oxygen in an oxygen tent. When the infant is weaned off the high concentrations of oxygen, new blood vessels start to grow widely within the eyes. These vessels have weak walls and they hemorrhage, leading to retinal detachment and then blindness. Cryosurgery—applying cold needles to a circle of points around the eyeball just external to the ora serrata to kill many of the growing vessels at their source—has been only slightly successful in preventing loss of vision in this disorder.

Trachoma (trah-ko′mah; "rough growth") is a highly contagious infection of the conjunctiva and cornea, caused by the bacterium *Chlamydia trachomatis*. It is transmitted by hand-to-eye contact, by flies that go from eye to eye, or by placing contaminated objects in or near the eye (towels, eye liner, etc.). Symptoms begin with an inflammation of the conjunctiva of the upper eyelid; then the conjunctiva and cornea become highly vascularized, and finally, scarred. The corneal scarring reduces vision and causes blindness. Common worldwide, trachoma blinds millions of people in third-world countries. It is effectively treated with eye ointments containing antibiotic drugs.

Embryonic Development of the Eye

The eyes develop as outpocketings of the brain. By week 4, paired lateral outgrowths called **optic vesicles** protrude from the diencephalon (Figure 16.16a). Soon, these hollow vesicles indent to form double-layered

optic cups (Figure 16.16b). The proximal parts of the outgrowths, called the **optic stalks,** form the basis of the optic nerves.

Once a growing optic vesicle reaches the overlying surface ectoderm, it signals the ectoderm to thicken and form a **lens placode.** By week 5, this placode has invaginated to form a **lens vesicle** (Figure 16.16c). Shortly thereafter, the lens vesicle pinches off into the optic cup, where it becomes the lens.

The internal layer of the optic cup differentiates into the neural retina, whereas the external layer becomes the pigmented layer of the retina (Figure 16.16d). The **optic fissure,** a groove on the underside of each optic stalk and cup, serves as a direct pathway for blood vessels to reach and supply the interior of the developing eye. When this fissure closes, the optic stalk becomes a tube through which the optic nerve fibers, originating in the retina, grow centrally to reach the diencephalon. The blood vessels that were originally within the optic fissure now lie in the center of the optic nerve.

The fibrous tunic, vascular tunic, and vitreous humor form from head mesenchyme that surrounds the early optic cup and invades the cup's interior. The central interior of the eyeball has a rich blood supply during development, but these blood vessels degenerate, leaving only those in the vascular tunic and retina (Figure 16.16e).

THE EAR: HEARING AND EQUILIBRIUM

The **ear,** the receptor organ for both hearing and equilibrium, has three main regions: the *outer ear,* the *middle ear,* and the *inner ear* (Figure 16.17). The outer and middle ears participate in hearing only, whereas the inner ear functions in both hearing and equilibrium.

The Outer (External) Ear

The **outer (external) ear** consists of the auricle and the external auditory canal. The **auricle,** or **pinna,** is what most people call the ear—the shell-shaped projection that surrounds the opening of the external auditory canal. Most of the auricle, including the **helix** (rim), consists of elastic cartilage covered with skin. Its fleshy, dangling **lobule** ("earlobe"), however, lacks supporting cartilage. The function of the auricle is to gather and funnel (and thereby amplify) sound waves coming into the external auditory canal. Moreover, the way that sound bounces off the ridges and cavities of the auricle provides the brain with clues about whether the sound came from above or below.

The **external auditory canal,** or external acoustic meatus (see p. 159), is a short tube (about 2.5 cm long) running medially, from the auricle to the eardrum. Near the auricle, its wall consists of elastic cartilage, but its medial two-thirds tunnels through the temporal bone. The entire canal is lined with skin that contains hairs, as well as sebaceous glands and modified apocrine sweat glands called *ceruminous* (sĕ-roo′mĭ-nus) *glands.* The ceruminous and sebaceous glands secrete yellow-brown *cerumen,* or earwax (*cere* = wax). Earwax traps dust and repels insects, keeping these substances out of the auditory canal.

Sound waves entering the external auditory canal hit the thin, translucent **tympanic membrane,** or eardrum (*tympanum* = drum), which forms the boundary between the outer and middle ears. It is shaped like a flattened cone, the apex of which points medially into the middle ear cavity. Sound waves that travel through the air set the eardrum vibrating, and the eardrum in turn transfers the vibrations to tiny bones in the middle ear (discussed next).

CLINICAL APPLICATIONS

PERFORATED EARDRUM Undue pressure from a cotton swab or sharp object in the external auditory canal can tear the tympanic membrane, a condition called a **perforated eardrum.** A more common cause, however, is a middle ear infection, in which the accumulation of pus medial to the eardrum exerts pressure that bursts the thin membrane. Permanent deafness rarely results because perforated eardrums heal well, but even small amounts of scarring can permanently diminish hearing acuity. ✚

The Middle Ear

The **middle ear,** or *tympanic cavity,* is a small, air-filled space inside the petrous part of the temporal bone. It is lined by a thin mucous membrane and is shaped like a rounded hat box standing on its side (Figure 16.17). Its *lateral boundary* is the tympanic membrane, whereas its *medial boundary* is a wall of bone that separates it from the inner ear. Two small holes penetrate this medial wall: a superior **oval window** *(vestibular window)* and an inferior **round window** *(cochlear window).* Superiorly, the middle ear arches upward as the **epitympanic recess** (*epi* = over); its *superior boundary* is the roof of the petrous bone, which is so thin here that middle ear infections can spread to the overlying meninges and brain. The *posterior wall* of the middle ear opens into the **mastoid antrum,** a canal leading to the *mastoid air cells* in the mastoid process. As discussed in the section on the skull (p. 159), infections can spread from the middle ear to the mastoid air cells, using the antrum as a passageway. The *anterior wall* of the middle ear lies just behind the internal carotid artery (main artery to the brain), and it also contains the opening of the pharyngotympanic tube (discussed shortly). The *inferior boundary* of the middle ear is a thin, bony floor, under which lies the important internal jugular vein. Middle ear infections may burst through this floor and clot the blood in this vein.

The **pharyngotympanic tube,** formerly named the *auditory tube* or *eustachian tube,* links the middle ear to the pharynx (Figures 16.17 and 16.18). About 4 cm

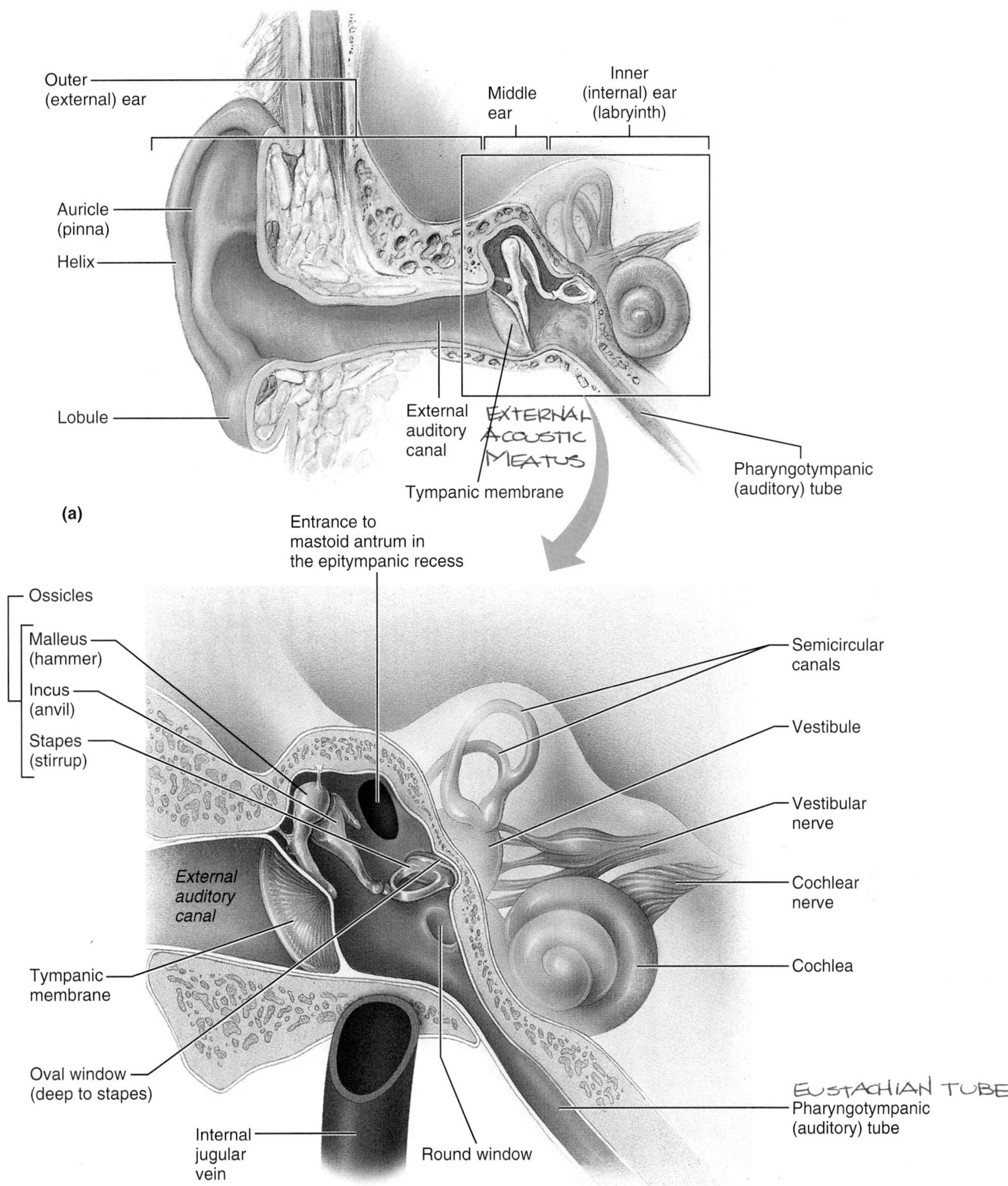

FIGURE 16.17 **Structure of the ear.** **(a)** The three regions of the ear: outer, middle, and inner. **(b)** Enlarged view of the middle and inner ear. In the inner ear, the bony labyrinth is shown, but the membranous labyrinth is not.

(1.5 inches) long, it runs medially, anteriorly, and inferiorly. Its lateral third consists of bone and occupies a groove on the inferior surface of the skull; its medial two-thirds is cartilage and opens into the side wall of the superior pharynx behind the nasal cavity. This normally flattened and closed tube can be opened briefly by swallowing or yawning so that the air pressure in the middle ear equalizes with the outside air pressure. This is important because the eardrum does not vibrate freely unless the pressure on both its surfaces is the

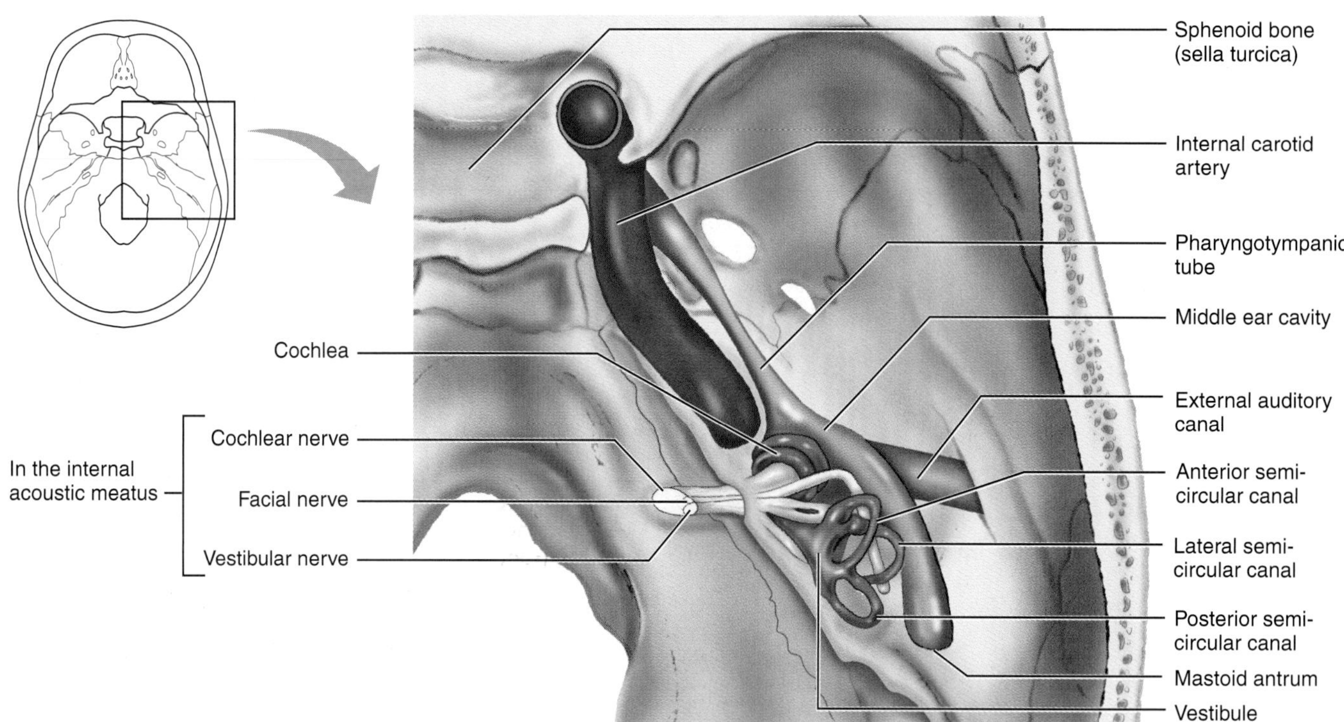

FIGURE 16.18 Superior view of ear structures in the floor of the cranial cavity of the skull (right ear). The structures in this figure are shown as though the surrounding temporal bone (petrous part) were transparent.

same. Differences in air pressure build up across the eardrum during rapid changes in altitude (as in an airplane). The next time your ears "pop" in such a situation, yawning is the best way to open your pharyngotympanic tubes and equalize the air pressures.

CLINICAL APPLICATIONS

MIDDLE EAR INFECTIONS Otitis media (o-ti′tis; "ear inflammation"), an infection and inflammation of the middle ear, usually starts as a throat infection that spreads to the middle ear through the pharyngotympanic tube. Fluid and pus can build up in the middle ear cavity and exert painful pressure within this enclosed space. More children than adults develop otitis media because the child's pharyngotympanic tube is shorter and enters the pharynx at a less acute angle. Extremely common, otitis media accounts for one-third of all visits to pediatricians in the United States, and is frequently cured by antibiotics. However, the overuse of antibiotics has lead to bacterial resistance, making persistent and recurrent cases increasingly difficult to treat.

Children with persistent otitis media sometimes have their eardrums lanced and have ear tubes inserted through the eardrum. Such a **myringotomy** (mir″ing-got′o-me; "lancing the eardrum") allows the middle ear to drain and relieves the pressure. The tiny tube that is inserted through the eardrum during myringotomy permits the pus to drain into the external ear. This tube is left in the eardrum and falls out by itself within a year. ✚

The tympanic cavity is spanned by the three smallest bones in the body, the **ossicles** (Figures 16.17 and 16.19), which transmit the vibrations of the eardrum across the cavity to a fluid in the inner ear. From lateral to medial, the ossicles are the **malleus** (mal′e-us), or hammer, which looks like a club with a knob on top; the **incus** (ing′kus), or anvil, which resembles a tooth with two roots (see Figure 16.19); and the **stapes** (sta′pēz), which looks like the stirrup of a saddle. The "handle" of the malleus attaches to the eardrum, and the base of the stapes vibrates against the *oval* window. Most people have trouble remembering whether the stapes fits into the oval window or into the round window inferior to it: To avoid this problem, remember that the footplate of a saddle stirrup is usually oval, not round.

Critical Reasoning

What would be the result of a condition in which the footplate of the stapes fuses to the oval window?

The stapes cannot move, and deafness results. This condition, called ***otosclerosis*** *(o″to-sklĕ-ro′sis; "hardening of the ear"), is caused by an excessive growth of bone tissue in the walls of the middle ear cavity. This common age-related problem, which affects 1 in every 200 people, can be treated by a delicate and difficult surgery that removes the stapes and replaces it with a prosthetic (artificial) one.*

Tiny ligaments suspend the ossicles in the middle ear, and tiny synovial joints link the ossicles into a chain. By concentrating the vibrations of the eardrum onto the much smaller oval window, the ossicles amplify the pressure of the sound vibrations about 20-fold. Without the ossicles, we could hear only loud sounds.

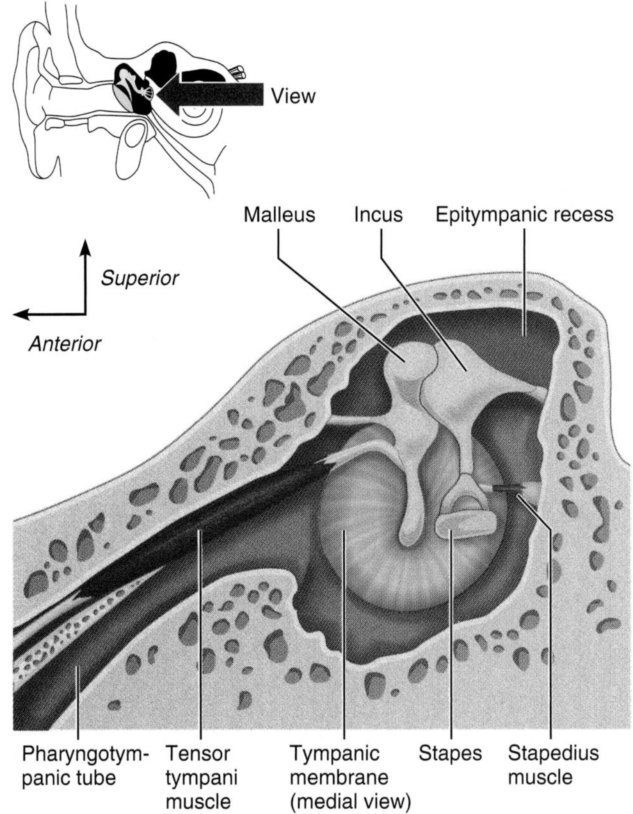

FIGURE 16.19 The three ossicles in the right middle ear (medial view).

Two tiny skeletal muscles occur in the middle ear cavity (see Figure 16.19). The **tensor tympani** (ten′sor tim′pah-ni) originates on the cartilage part of the pharyngotympanic tube and inserts on the malleus, whereas the **stapedius** (stah-pe′de-us) runs from the posterior wall of the middle ear to the stapes. When the ears are assaulted by very loud sounds, these muscles contract reflexively to limit the vibration of the ossicles and thus prevent damage to the hearing receptors (discussed shortly).

The Inner (Internal) Ear

The **inner (internal) ear,** also called the **labyrinth** ("maze") because of its maze-like, complex shape (see Figures 16.17 and 16.18), lies within the thick, protective walls of the petrous part of the temporal bone. The inner ear consists of two main divisions: the bony labyrinth and the membranous labyrinth (Figure 16.20).

The **bony labyrinth** is a *cavity* in the petrous bone consisting of a system of twisting channels that has three parts. From posterolateral to anteromedial, these parts are the *semicircular canals,* the *vestibule,* and the *cochlea* (see Figure 16.17). Textbooks often picture the bony labyrinth as if it were a solid object, but it is actually a cavity.

The **membranous labyrinth** is a continuous series of membrane-walled sacs and ducts that fit loosely within the bony labyrinth and more or less follow its contours (see Figure 16.20). The main parts of the membranous labyrinth are (1) the *semicircular ducts,* one inside each semicircular canal; (2) the *utricle* and *saccule,* both in the vestibule; and (3) the *cochlear duct* in the cochlea. The wall of the membranous labyrinth—its "membrane"—is a thin layer of connective tissue lined by a simple squamous epithelium. Parts of this epithelium are thickened and contain the receptors for equilibrium and hearing. The parts of the bony and membranous labyrinths are summarized in Table 16.1.

The membranous labyrinth is filled with a clear fluid called **endolymph** (en′do-limf; "internal water"). External to the membranous labyrinth, the bony labyrinth is filled with another clear fluid called **perilymph** (per′ĭ-limf; "surrounding water"). Nowhere are the perilymph and endolymph continuous with one another. Whereas the endolymph is confined to the membranous labyrinth, the perilymph is continuous with the cerebrospinal fluid that fills the subarachnoid space.

We now will explore the basic parts of the bony and membranous labryinth. We begin with the vestibule because of its central location in the labyrinth and because its contained receptors measure the simplest aspects of the sense of equilibrium.

The Vestibule

The **vestibule** is the central, egg-shaped cavity of the bony labyrinth (see Figure 16.20). It lies just medial to the middle ear, and the oval window is in its lateral bony wall. Suspended within its perilymph are the two egg-shaped parts of the membranous labyrinth, the **utricle** (u′tri-k′l; "leather bag") and the **saccule** ("little sac"). The utricle is continuous with the semicircular ducts, the saccule with the cochlear duct.

The utricle and saccule each house a spot of sensory epithelium called a **macula** (mak′u-lah; "spot") (Figure 16.21a). Both the macula of the utricle and the macula of the saccule contain receptor cells that monitor the position of the head when the head is held still. This aspect of the sense of balance is called *static equilibrium.* Furthermore, these receptor cells also monitor straight-line changes in the speed and the direction of head movements—that is, *linear acceleration*—but not rotational movements of the head.

Each macula is a patch of epithelium containing columnar **supporting cells** and scattered receptors called **hair cells** (see Figure 16.21b). The hair cells synapse with sensory fibers of the **vestibular nerve,** which is the vestibular division of the vestibulocochlear nerve. Named for the hairy look of its free surface, each hair cell has many stereocilia (long microvilli) and a single kinocilium (a true cilium) protruding from its apex. The tips of these stiff hairs are embedded in an overlying **otolithic** (o″to-lith′ik) **membrane,** which is actually

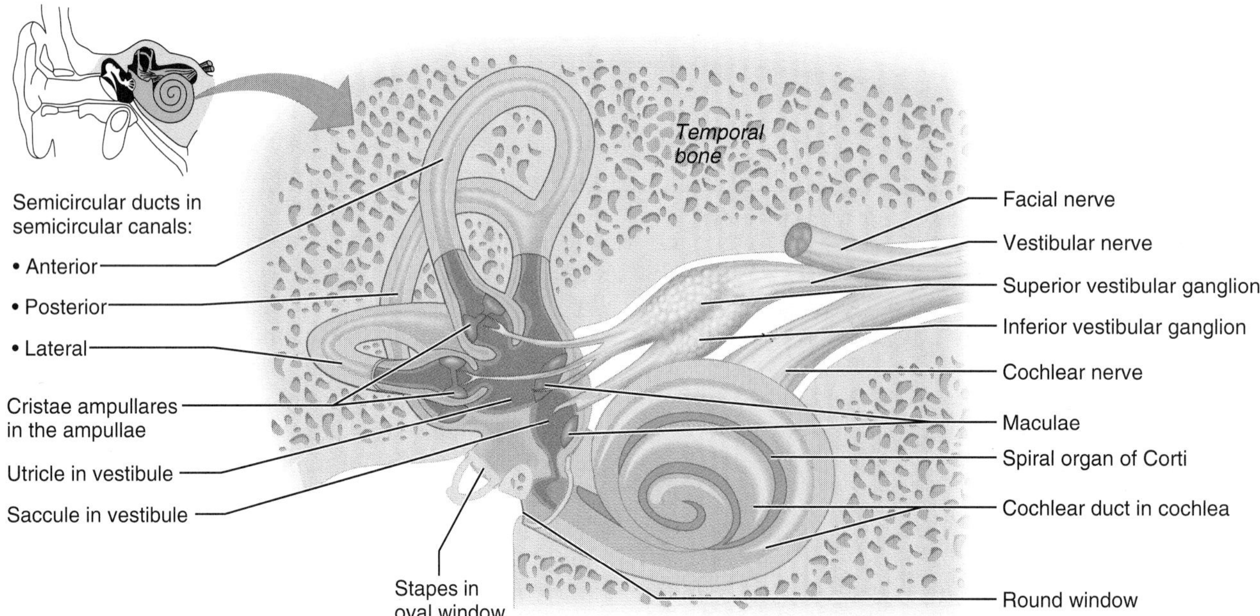

FIGURE 16.20 Membranous labyrinth (orange) in the bony labyrinth (blue and gray) of the inner ear.

a jelly-like disc that contains heavy crystals of calcium carbonate called **otoliths** ("ear stones").

It is easy to understand how the maculae and otoliths contribute to the sense of static equilibrium. The macula of the *utricle* has a *horizontal* orientation within the ear (Figure 16.21a). When one holds the head in a tilted position (Figure 16.21c), the heavy otolithic membrane pulls downward, bending the receptor hairs and signaling the vestibular nerve to tell the brain that the head is tilted. The macula of the *saccule,* on the other hand, has a *vertical* orientation within the ear (Figure 16.21a), so its heavy otoliths pull downward on the hairs whenever our head is upright, signaling the brain that the head is in this untilted position. It is also easy to understand how both maculae monitor the movements of linear acceleration: Whenever our body jolts forward, upward, or sideways in a straight line, the heavy otolithic membrane lags behind, again bending the hairs and signaling the brain.

The maculae are innervated by two branches of the vestibular nerve (see Figure 16.20). The sensory neurons in this nerve are bipolar neurons, with cell bodies located in the *superior* and *inferior vestibular ganglia.* These ganglia lie in the internal acoustic meatus of the petrous temporal bone. The further path of vestibular nerve fibers within the brain is discussed shortly (see p. 494).

Having considered the receptors for static equilibrium and linear acceleration, we now turn to the balance receptors for *rotational* acceleration of the head.

Table 16.1 The Inner Ear: Basic Structures of the Bony and Membranous Labyrinths

Bony Labyrinth	Membranous Labyrinth (within bony labyrinth)	Functions of the Membranous Labyrinth
1. Semicircular canals	Semicircular ducts	Equilibrium: rotational (angular) acceleration of the head
2. Vestibule	Utricle and saccule	Equilibrium: static equilibrium and linear acceleration of the head
3. Cochlea	Cochlear duct	Hearing

The Semicircular Canals

The three **semicircular canals** of the bony labyrinth lie posterior and lateral to the vestibule (see Figure 16.20). Each of these canals actually describes about two-thirds of a circle and has an expansion at one end called an **ampulla** ("flask"). Each canal lies in one of the three planes of space: The **anterior** and **posterior semicircular canals** lie in vertical planes at right angles to each other, whereas the **lateral semicircular canal** *(horizontal canal)* lies almost horizontally. Snaking through each semicircular canal is a membranous **semicircular duct,** each of which has a swelling called a **membranous ampulla** within the corresponding bony ampulla.

Each membranous ampulla houses a small crest called a **crista ampullaris,** or "crest of the ampulla" (Figure 16.22). The cristae contain the receptor cells that measure *rotational (angular) acceleration* of the head, as occurs when a figure skater spins or a gymnast does a flip. Each crista has an epithelium on its top that, like the maculae, contains *supporting cells* and receptor *hair cells.* The "hairs" of these hair cells project into a tall, jelly-like mass that resembles a pointed cap, the **cupula** (ku′pu-lah; "little barrel"); the basal parts of the hair cells synapse with fibers of the vestibular nerve.

Because the three semicircular ducts lie in three different planes, each crista responds to head rotation in a different plane of space. When the head starts to rotate, the endolymph in the semicircular duct lags behind at first, pushing on the cupula and bending the hairs (see Figure 16.22b). As their hairs bend, the hair cells depolarize and change the pattern of impulses carried by vestibular nerve fibers to the brain.

The Cochlea

The **cochlea** (kok′le-ah; "snail shell") is a spiraling chamber in the bony labyrinth (see Figures 16.17 and 16.20). It is about half the size of a split pea. From its attachment to the vestibule at its base, it coils for about two and a half turns around a pillar of bone called the **modiolus** (mo-di′o-lus) (Figure 16.23a). The modiolus is shaped like a screw whose tip lies at the *apex of the cochlea,* pointing anterolaterally. Just as screws have threads, the modiolus has a spiraling thread of bone called the **spiral lamina.** Running through the bony core of the modiolus is the **cochlear nerve,** which is the cochlear division of the vestibulocochlear nerve.

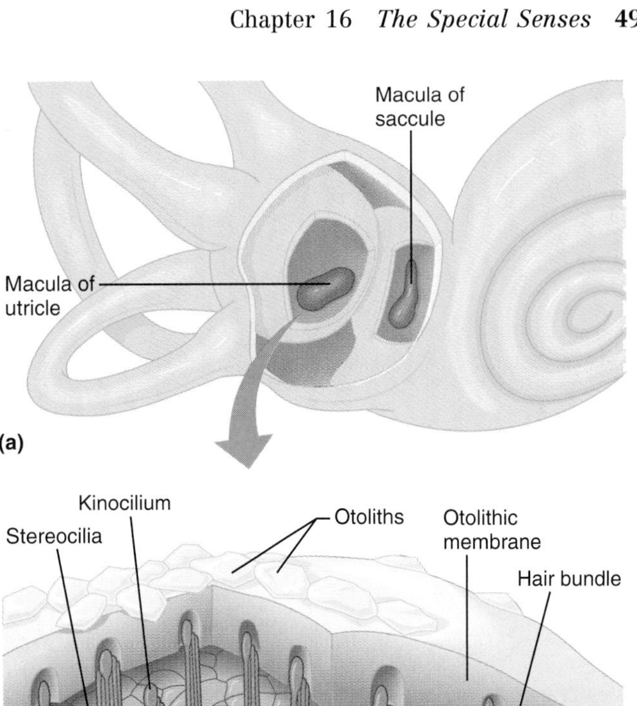

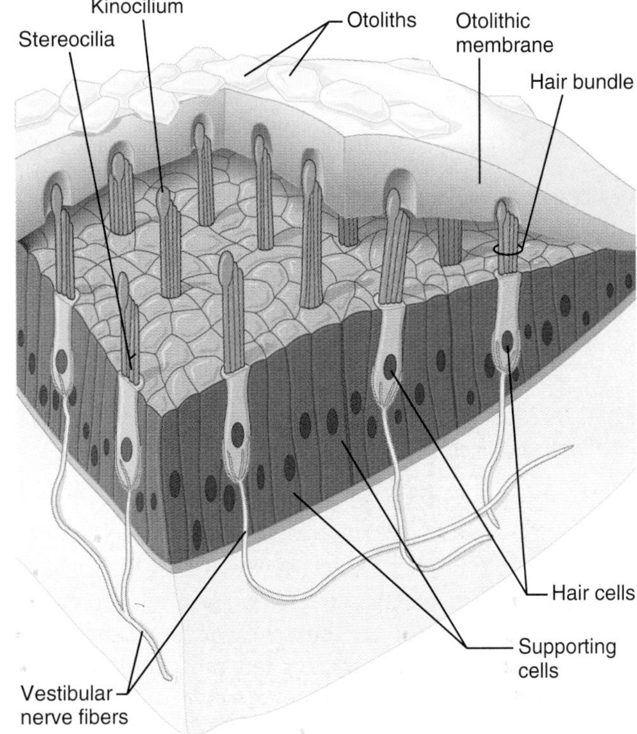

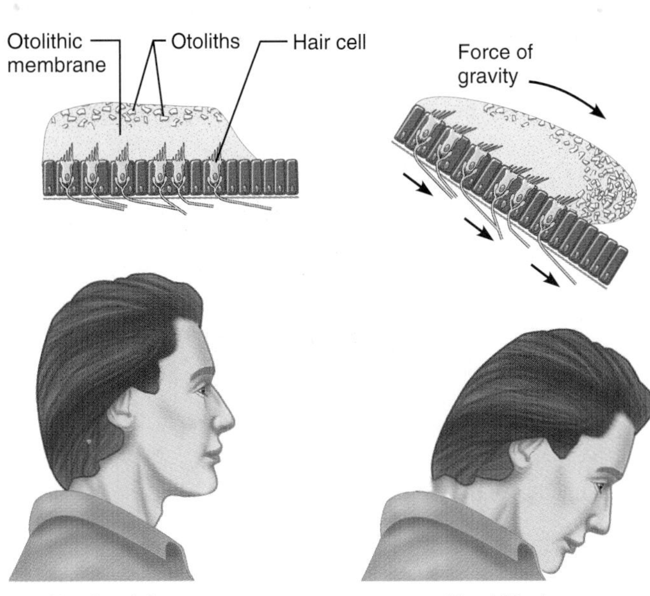

FIGURE 16.21 Anatomy and function of the maculae in the inner ear. **(a)** This diagram shows the horizontal orientation of the macula in the utricle and the vertical orientation of the macula in the saccule. **(b)** Enlargement showing that a macula is a spot of epithelium containing hair cells. Note the overlying otolithic membrane. **(c)** In the macula of the utricle, tilting the head causes the heavy otolithic membrane to drop under the force of gravity, bending the hairs of the hair cells and generating signals in the axons of the vestibular nerve.

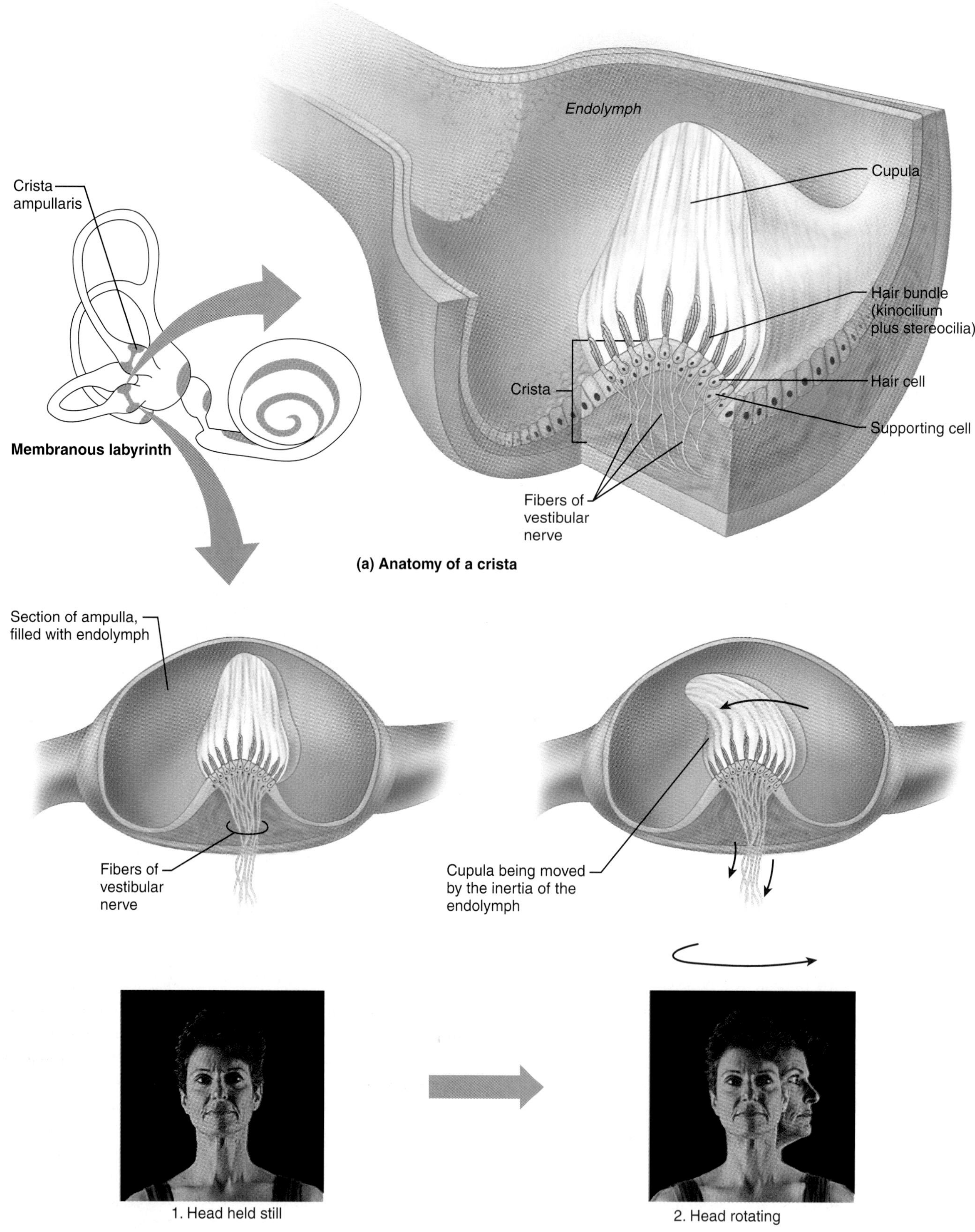

FIGURE 16.22 Structure and function of the crista ampullaris in the inner ear. **(a)** Microscopic anatomy of the crista and cupula in the lateral semicircular duct, which has been cut in half. **(b)** Function of the crista: During rotational acceleration of the head, the endolymph in the semicircular duct lags, bending the cupula away from the direction of head movement and stimulating the hair cells.

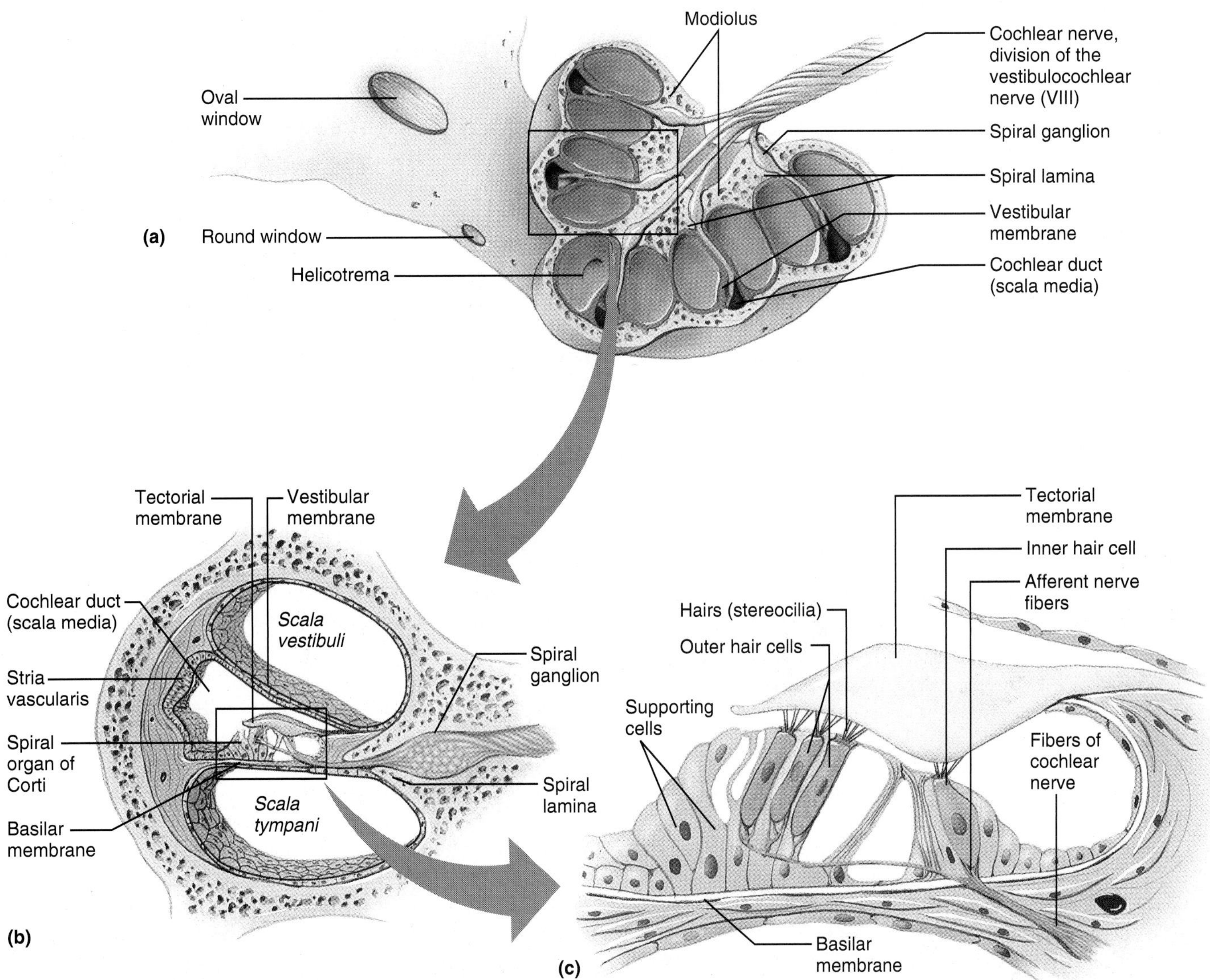

FIGURE 16.23 Anatomy of the cochlea. **(a)** A section through the right cochlea (superior half removed). The area containing the helicotrema is the apex of the cochlea. **(b)** Magnified cross-sectional view of one turn of the cochlea, showing the three scalas: scala vestibuli, scala tympani, and cochlear duct (scala media). **(c)** Detailed structure of the spiral organ of Corti.

The coiled part of the membranous labyrinth within the cochlea, called the **cochlear duct** (Figure 16.23b), contains the receptors for hearing. This duct winds through the cochlea and ends blindly in the cochlear apex. Within the cochlea, the endolymph-filled cochlear duct (or **scala media**) lies between two perilymph-filled chambers, the **scala vestibuli** and **scala tympani** (*scala* = ladder) (Figure 16.23b). The scala vestibuli is continuous with the vestibule near the base of the cochlea, where it abuts the oval window. The scala tympani, on the other hand, ends at the round window at the base of the cochlea. The scala vestibuli and scala tympani are continuous with each other at the apex of the cochlea, in a region called the **helicotrema** (hel″ĭ-ko-tre′mah; "the hole in the spiral") (see Figure 16.23a).

The cochlear duct contains the receptors for hearing (see Figure 16.23b). The "roof" of the cochlear duct, separating it from the scala vestibuli, is the **vestibular membrane.** The external wall of this duct is the *stria vascularis* ("vascularized streak"), an unusual epithelium that contains capillaries and secretes the endolymph of the inner ear. The floor of the cochlear duct consists of the spiral lamina of bone plus an attached sheet of fibers called the **basilar membrane,** which supports the **spiral organ of Corti,** the receptor epithelium for hearing. This tall epithelium (see Figure 16.23c) is very similar to the receptor epithelia for equilibrium, consisting of columnar *supporting cells* and several long rows of **inner** and **outer hair cells,** which are the receptor cells. At the cell's apex, the tips of the hairs

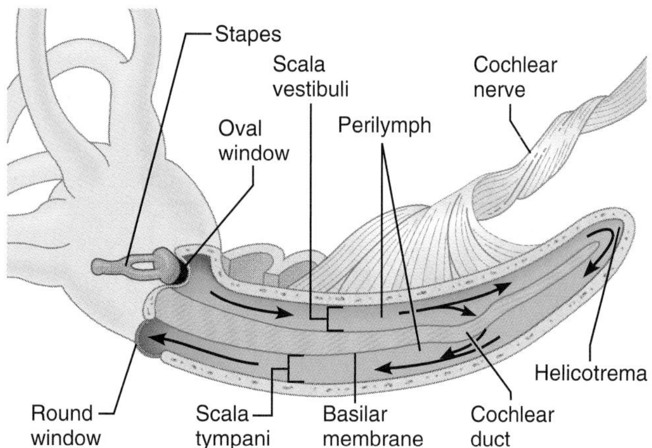

FIGURE 16.24 Role of the cochlea in hearing. The cochlea (in blue) is drawn as if uncoiled to make the events that occur in it easier to follow. Vibrations of the stapes on the oval window transmit vibrations to the perilymph of the scala vestibuli, which are in turn transmitted to the endolymph of the cochlear duct (in green). These vibrations cause the basilar membrane to vibrate, stimulating the cochlear nerve. The structure of the basilar membrane is such that it segregates sound according to frequency—its basal part (left) vibrates in response to high-pitched sounds, whereas its apical part (right) vibrates in response to low-pitched sounds. The vibrations of the basilar membrane set the perilymph vibrating in the underlying scala tympani (see the arrows in that scala). These vibrations then travel to the round window (at left), where they push on the membrane that covers that window, thereby dissipating their remaining energy into the air of the middle ear cavity. Without this release mechanism, echoes would reverberate within the rigid cochlear box, disrupting sound reception.

(microvilli) are embedded in a gel-like **tectorial membrane** ("roofing membrane"); at their base the hair cells synapse with sensory fibers of the cochlear nerve. These nerve fibers belong to bipolar neurons, whose cell bodies occupy a **spiral ganglion** in the spiral lamina and modiolus (see Figure 16.23a) and whose central fibers project to the brain.

How do sound waves stimulate the hair cells in the spiral organ of Corti? A simplified explanation is presented here, and some additional details are provided in the legend for Figure 16.24. First, sound vibrations travel from the eardrum through the ossicles, causing the stapes to oscillate back and forth against the oval window. This oscillation sets up pressure waves in the perilymph of the scala vestibuli and in the endolymph of the cochlear duct. These waves cause the basilar membrane to vibrate up and down. The hair cells in the spiral organ move along with the basilar membrane (see Figure 16.23c), but the overlying tectorial membrane (in which the hairs are anchored) does not move. Therefore, the movements of the hair cells cause their hairs to bend. Each time such bending occurs, the hair cells release neurotransmitters that excite the cochlear-nerve fibers, which carry the vibratory (sound) information to the brain.

Advances in Understanding Anatomists have only recently explained why there are both inner and outer hair cells in the spiral organ. The *inner* hair cells have been identified as the true receptors that transmit the vibrations of the basilar membrane to the cochlear nerve, whereas the *outer* hair cells are sound amplifiers. In response to the slightest vibration of the basilar membrane, the outer hair cells actively contract and then relax, thereby amplifying the membrane's movements by lifting and lowering it. The enhanced movements of the basilar membrane are then sensed by the inner-hair-cell receptors. Overall, this mechanism amplifies sounds some 100 times, so that we can hear the faintest sounds. ✷

Equilibrium and Auditory Pathways

As is the case for all sensory information, information on equilibrium and hearing travels to the brain for processing and integration.

The **equilibrium pathway** transmits information on the position and movements of the head via the vestibular nerve to the brain stem. Equilibrium is the only special sense for which most information goes to the *lower* brain centers—which are primarily reflex centers—rather than to the "thinking" cerebral cortex. This pathway reflects the fact that our responses to a loss of balance, such as stumbling, must be rapid and reflexive: By the time we "thought about" correcting our fall, we would already be on the ground. The vestibular nuclei in the medulla (see p. 381) and the cerebellum (see p. 382) are the major brain centers for processing information on equilibrium. Furthermore, a *minor* pathway to the cerebral cortex provides conscious awareness of the position and movements of the head. In this minor pathway, vestibular nerve fibers project to the vestibular nuclei, then to the thalamus, and then to the posterior insula of the cerebrum.

The ascending **auditory pathway** transmits auditory information from the cochlear receptors to the cerebral cortex (Figure 16.25). First, impulses pass through the cochlear nerve to the **cochlear nuclei** in the medulla. From there, some neurons project to the **superior olivary nuclei,** which lie at the junction of the medulla and pons. Beyond this, the axons ascend in the **lateral lemniscus** (a fiber tract) to the **inferior colliculus** (the auditory reflex center in the midbrain), which projects to the **medial geniculate nucleus (body)** of the thalamus. Axons of the thalamic neurons then project to the primary auditory cortex, which provides conscious awareness of sound. The auditory pathway is unusual in that not all of its fibers cross over to the other side of the brain. Therefore, each primary auditory cortex receives impulses from both ears. Clinically, this phenomenon makes identifying damage to the primary auditory cortex on one side difficult, because such damage produces only a minimal loss of hearing.

The superior olivary nuclei and inferior colliculus are not merely relay stations along the auditory pathway, but perform important functions of their own. For example, both these structures participate in the localization of sounds. For more on the function of the inferior colliculus, see p. 380.

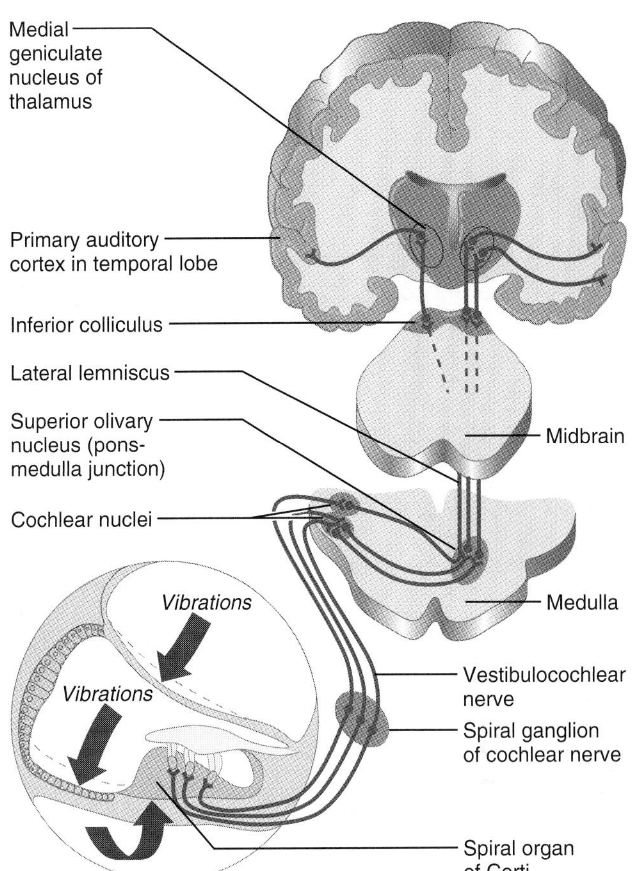

FIGURE 16.25 The auditory pathway, from the spiral organ of Corti to the primary auditory cortex in the temporal lobe. For clarity, only the pathway from the right ear is shown. This diagram is highly simplified; there are over 20 different auditory nuclei in the pathway, most of which are very small.

Disorders of Equilibrium and Hearing

Here we consider two disorders of equilibrium, motion sickness and Meniere's syndrome, and then we briefly discuss the two types of deafness.

Motion Sickness

Motion sickness is a common disorder of equilibrium in which particular motions (such as riding in a car or aboard a ship) lead to nausea and vomiting. The cause of this condition has been difficult to determine. The most popular theory says it is due to a mismatch of sensory inputs. For example, if you are in a rocking ship during a storm, visual inputs indicate that your body is fixed with reference to a stationary environment (your cabin), but your vestibular apparatus detects movement. The brain's resulting confusion somehow leads to motion sickness. Another theory points out that the vestibular nuclei lie near, and may project to, the centers in the medulla that control vomiting. Antimotion drugs can relieve the symptoms of motion sickness.

Meniere's Syndrome

In a disorder called **Meniere's syndrome,** the membranous labyrinth is apparently distorted by excessive amounts of endolymph. Affected individuals experience a variety of symptoms: equilibrium so disturbed that standing is nearly impossible; transient but repeated attacks of vertigo, nausea, and vomiting; and "howling" in the ears, such that hearing is impaired and perhaps ultimately lost. Although less-severe cases can be managed by antimotion drugs, more debilitating attacks may require diuretics (drugs that increase urine output) and restriction of dietary intake of salt—both of which decrease the volume of extracellular fluid, and consequently that of endolymph. Severe cases may require surgery either to drain excess endolymph from the inner ear or to cut the vestibular nerve to relieve the vertigo. A last resort, usually deferred until all hearing is lost, is removal of the entire labyrinth.

Deafness

Any hearing loss, no matter how slight, is considered deafness. The two types of deafness, *conduction deafness* and *sensorineural deafness,* have different causes.

Conduction deafness occurs when sound vibrations cannot be conducted to the inner ear. It can be caused by earwax blocking the external auditory canal, a ruptured eardrum, otitis media, or otosclerosis.

Sensorineural deafness results from damage to the hair cells or to any part of the auditory pathway to the brain. Most often it results from the normal, gradual loss of hearing receptor cells that proceeds throughout life. In other cases, hair cells can be destroyed at an earlier age by a single, explosively loud noise or by repeated exposure to loud music, factory noise, or airport noise. Strokes and tumors that damage the auditory cortex can also cause sensorineural deafness. When deafness reflects damage to hair cells, hearing aids can help. Traditional hearing aids simply amplify sounds, an effective strategy if the loss of hair cells is not too great. For complete sensorineural deafness, **cochlear implants** are available. Placed in the temporal bone, these devices convert sound energy into electrical signals and deliver these signals directly to the cochlear nerve fibers. Modern models have 20 electrodes, each responding to a different frequency, and are so effective that even children who were born deaf can hear well enough to learn to speak.

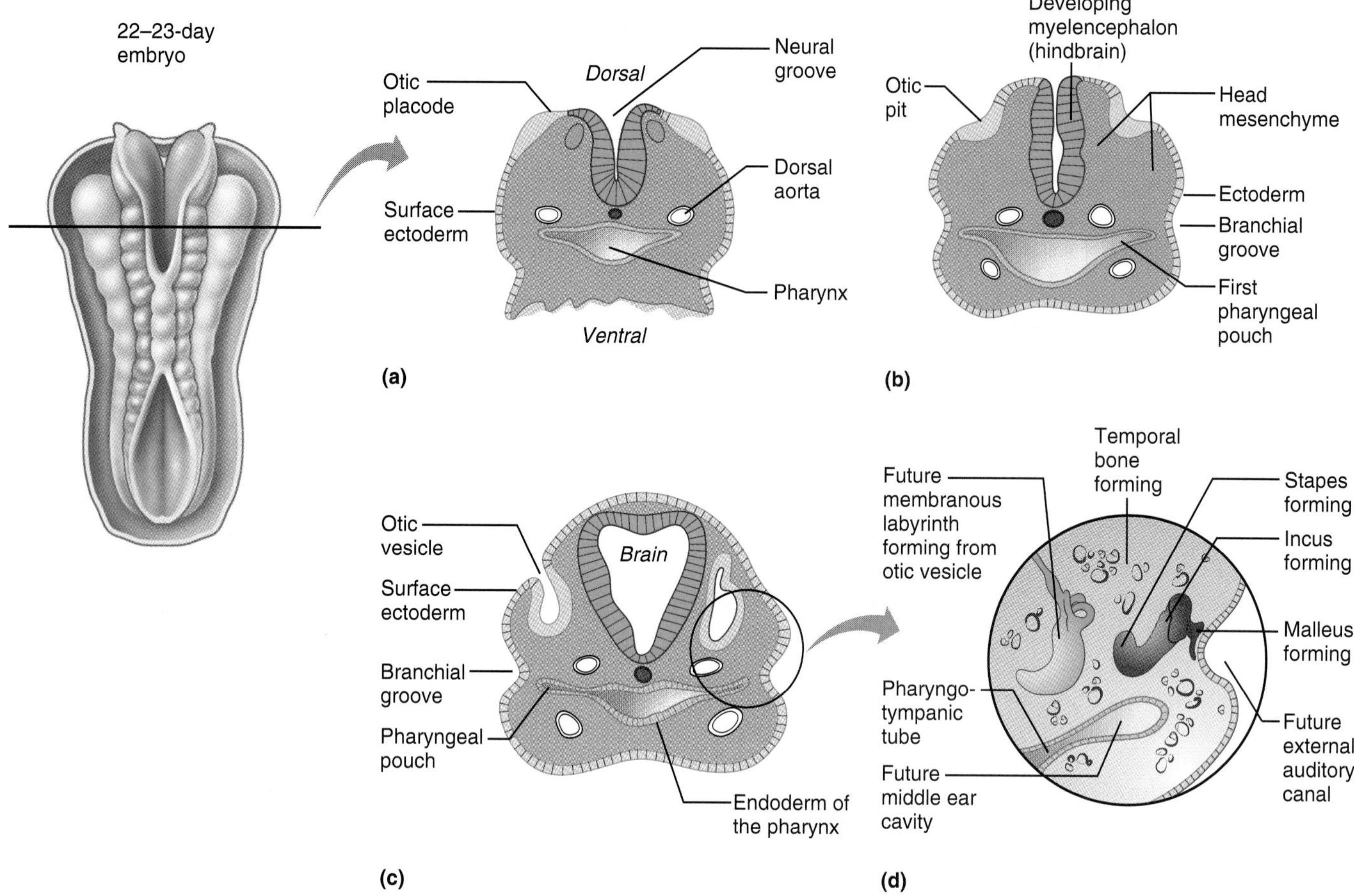

FIGURE 16.26 Embryonic development of the ear. At left, a surface view of the embryo shows the plane of the sections. **(a)** By about day 22, otic placodes have formed as thickenings of the ectoderm. Endoderm (yellow) lines the pharynx. **(b)** The otic placodes invaginate to become otic pits, and branchial grooves start to push inward from the surface ectoderm. **(c)** At about 28 days, otic vesicles are forming from the otic pits. Also note the pharyngeal pouch. **(d)** Between weeks 5 and 8, the membranous labyrinth forms from the otic vesicle while the first pharyngeal pouch gives rise to the pharyngotympanic tube and middle ear cavity, and the branchial groove develops into the external auditory canal. Head mesenchyme forms the surrounding bony structures.

Embryonic Development of the Ear

Development of the ear begins in the fourth week after conception (Figure 16.26). First the inner ear begins to form from a thickening of the surface ectoderm called the **otic placode,** which lies lateral to the hindbrain on each side of the head. This placode invaginates to form the **otic pit.** Then its edges fuse to form the **otic vesicle,** which detaches from the surface epithelium. The otic vesicle takes on a complex shape and becomes the membranous labyrinth. The mesenchyme tissue around the otic vesicle becomes the petrous temporal bone—that is, the walls of the bony labyrinth.

As the inner ear develops, the middle ear starts to form. Lateral outpocketings called *pharyngeal pouches* form from the endoderm-lined pharynx (see p. 12). The middle ear cavity and the pharyngotympanic tube develop from the first of the pharyngeal pouches. The ossicles, which will bridge the middle ear cavity, develop from cartilage bars associated with the first and second pharyngeal pouches.

Turning to the external ear, the external auditory canal differentiates from the first **branchial groove,** an indentation of the surface ectoderm. The auricle of the outer ear grows from a series of bulges around this branchial groove.

THE SPECIAL SENSES THROUGHOUT LIFE

All special senses are functional, to a greater or lesser degree, at birth. Smell and taste are sharp in newborns, and the likely reason that infants relish food that adults consider bland is that children have more taste buds than do adults. Most people experience no difficulties with their chemical senses throughout childhood and young adulthood, but starting in the fourth decade of life, the ability to taste and smell declines. This decline reflects a gradual loss of the chemoreceptors, which are replaced more slowly than in younger years.

Even though the fetus cannot see in the darkness of the uterus, the photoreceptors are fully formed in the posterior retina by 25 weeks after conception, and the neuronal connections of the visual pathway have developed even earlier. Visual experiences during the first 8 months after birth fine-tune these synaptic connections.

Congenital problems of the eyes are relatively rare, although maternal rubella during the first trimester of pregnancy may cause congenital blindness or cataracts. In newborns, the eyeballs are foreshortened, so all infants are hyperopic; only gray tones are perceived, eye movements are uncoordinated, and often only one eye is used at a time. By 3 months, however, an image can be focused on the retina's fovea centralis for sharp vision, and babies can follow moving objects with their eyes. By the age of 6 months, depth perception is present, color vision is well developed, and the earlier hyperopia is almost gone.

As a person ages, the lens loses its clarity and becomes discolored, and it begins to scatter light (the resulting glare is why older people should look slightly away from oncoming headlights while driving at night). The dilator muscles of the iris become less efficient, so the pupils stay partly constricted. All these changes decrease the amount of light reaching the retina, such that visual acuity is dramatically lower in people over 70. In addition, elderly individuals are susceptible to conditions that cause blindness, such as glaucoma, retinal detachment, diabetes mellitus, and age-related macular degeneration. Age-related presbyopia is discussed in the A Closer Look box on page 482.

Newborn infants can hear, but their early responses to sounds are mostly reflexive—for example, crying and clenching the eyelids in response to a startling noise. Low-pitched and middle-pitched sounds can be heard at birth, but the ability to hear high-pitched sounds is a postnatal development. By the third or fourth month, infants can localize sounds and will turn to the voices of family members. By 12 months, babies know all the sounds of their native language, and critical listening begins in toddlers as they learn to speak.

Congenital structural abnormalities of the external ear, including partly or fully missing pinnae and closed or absent auditory canals, are fairly common; less common is sensorineural deafness due to maternal rubella infections during pregnancy. Except for the common ear infections of childhood, few problems usually affect the ears until old age. By about age 60, however, deterioration of the spiral organ of Corti becomes noticeable. We are born with about 20,000 hair cells in each ear, but they are gradually lost. The ability to hear high-pitched sounds fades first. This gradual loss of hearing with age is called **presbycusis** (pres″bĭ-ku′sis), literally "old hearing." It is the most common type of sensorineural deafness. Although it is considered a disability of old age, it is becoming more common in younger people as the modern world grows noisier.

Advances in Understanding Much excitement was generated in the mid-1990s when it was documented that hair cells in the cochlea of adult mammals, long thought to be irreplaceable, are actually replaced at a very slow rate. This raised hope that one day scientists will find chemical compounds that accelerate this hair-cell replacement and reverse age-related hearing loss. ✷

RELATED CLINICAL TERMS

Ageusia (ah-gu′ze-ah) Loss or impairment of the taste sense.

Blepharitis (blef′ah-ri′tis) (*blephar* = eyelash; *itis* = inflammation of) Inflammation of the margins of the eyelids.

Congenital deafness One of every 500–1000 children is born deaf, due to such physical factors as an immobile stapes, overproduction of perilymph, and various malformations of the middle and inner ear. The ultimate causes of congenital deafness are less well understood, though it can result from maternal mumps, syphilis, or rubella during pregnancy. Many congenital cases are genetically based, and at least four different genes have been identified that when mutated, cause deafness.

Ophthalmology (of″thal-mol′o-je; "eye study") The science that studies the eye and eye diseases. An *ophthalmologist* is a medical doctor whose specialty is treating eye disorders. By contrast, an *optometrist* is a licensed nonphysician who measures vision and prescribes corrective lenses.

Otitis externa Inflammation and infection of the external auditory canal, caused by bacteria or fungi that enter the canal from outside, especially when the canal is moist (such as after swimming).

Otorhinolaryngology (o″to-ri″no-lar″ing-gol′o-je) The science that studies the ear, nose, and larynx and the diseases of these body regions.

Scotoma (sko-to′mah) (*scoto* = darkness) A blind spot in the visual field other than the normal blind spot caused by the optic disc; often reflects the presence of a brain tumor pressing on nerve fibers along the visual pathway.

Tinnitus (tĭ-ni′-tus) Persistent noise—a ringing, whistling, humming, buzzing, or screeching—that seems to come from the ears in 10% to 20% of all elderly people, causing great distress and annoyance. It often first appears after a loud noise or an injury to the head or cochlea. Recent evidence suggests that tinnitus is analagous to phantom limb pain (see p. 405)—it is a "phantom cochlear noise" caused by destruction of some neurons along the auditory pathway. With the other neurons of this pathway now deprived of their normal input, nearby axons grow in and reinnervate these neurons, and the CNS interprets background signals from the new axons as noise. Treatments include masking the noise with soothing sounds, counseling, and biofeedback; drugs are largely unsuccessful.

CHAPTER SUMMARY

Media study tools that could provide you additional help in reviewing specific key topics of Chapter 16 are referenced below.

SP = Study Partner; AIA = A.D.A.M.® Interactive Anatomy

THE CHEMICAL SENSES: TASTE AND SMELL (pp. 466–470)

Taste (pp. 466–468)

1. Most taste buds are on the tongue, in the epithelium of fungiform and vallate (circumvallate) papillae.

AIA Link: Atlas; View: Superior; Surface of tongue (dorsal).

2. Taste buds contain gustatory receptor cells, supporting cells, and basal cells that give rise to the other two cell types. The receptor cells are excited when taste-stimulating chemicals bind to their microvilli.
3. The four basic qualities of taste—sweet, sour, salty, and bitter—are sensed best on different regions of the tongue.
4. The sense of taste is served by cranial nerves VII, IX, and X, which send impulses to the medulla. From there, impulses travel to the thalamus and the taste area of the cerebral cortex.

Smell (Olfaction) (pp. 468–470)

5. The olfactory epithelium is located in the roof of the nasal cavity. This epithelium contains receptor, supporting, and basal cells.
6. The olfactory receptor cells are ciliated bipolar neurons. Odor molecules bind to the cilia, exciting the neurons. Axons of these receptor neurons form the filaments of the olfactory nerve (cranial nerve I).
7. Olfactory nerve axons transmit impulses to the olfactory bulb. Here, these axons synapse with mitral cells in structures called glomeruli.
8. After receiving input from the olfactory receptor neurons, the mitral cells send this olfactory information through the olfactory tract to the olfactory cortex and limbic system.

Disorders of the Chemical Senses (p. 470)

9. Disorders of smell include anosmia (inability to smell) and uncinate fits (smell hallucinations).

Embryonic Development of the Chemical Senses (p. 470)

10. The olfactory epithelium and taste buds develop from the epithelia on the face and mouth/pharynx, respectively.

THE EYE AND VISION (pp. 470–486)

1. The eye is located in the bony orbit and is cushioned by fat. The cone-shaped orbit also contains nerves, vessels, and extrinsic muscles of the eye.

Accessory Structures of the Eye (pp. 470–473)

2. Eyebrows shade and protect the eyes.
3. Eyelids protect and lubricate the eyes by reflexive blinking. Each eyelid contains a supporting tarsal plate, the roots of the eyelashes, and tarsal, ciliary, and sebaceous glands.
4. Muscles in the eyelids include the levator palpebrae superioris, which opens the eye, and the orbicularis oculi, which closes the eye.

AIA Link: Atlas; View: Anterior; Orbicularis oculi muscle (anterior).

5. The conjunctiva is a mucosa that covers the inner surface of the eyelids (palpebral conjunctiva) and the white of the eye (ocular conjunctiva). Its mucus lubricates the eye surface.
6. The lacrimal gland secretes lacrimal fluid (tears), which is blinked medially across the eye surface and drained into the nasal cavity through the lacrimal canaliculi, lacrimal sac, and nasolacrimal duct.
7. The six extrinsic eye muscles are the lateral and medial rectus (which turn the eye laterally and medially, respectively); the superior and inferior rectus (which elevate and depress the eye, respectively, but also turn it medially); and the superior and inferior obliques (which depress and elevate the eye, respectively, but also turn it laterally).

Anatomy of the Eyeball (pp. 474–480)

8. The wall of the eye has three tunics. The most external, fibrous tunic consists of the posterior sclera and the anterior cornea. The tough sclera protects the eye and gives it shape. The cornea is the clear window through which light enters the eye.

AIA Link: Atlas; View: Medial; Sagittal section of the eyeball.

9. The middle, pigmented vascular tunic consists of the choroid, the ciliary body, and the iris. The choroid provides nutrients to the retina's photoreceptors and prevents the scattering of light within the eye. The ciliary body contains smooth ciliary muscles that control the shape of the lens and ciliary processes that secrete aqueous humor. The iris contains smooth muscle that changes the size of the pupil.
10. The sensory tunic, or retina, consists of an outer pigmented layer and an inner neural layer. The neural layer contains photoreceptors (rod and cone cells), other types of neurons, and neuroglia. Light influences the photoreceptors, which signal bipolar neurons, which signal ganglion neurons. The axons of ganglion neurons run along the inner retinal surface toward the optic disc.
11. The outer segments of the rods and cones contain light-absorbing pigment in membrane-covered discs. Light splits this pigment to initiate the flow of signals through the visual pathway.
12. Two important spots on the posterior retinal wall are (1) the macula lutea with its central fovea (area of highest visual acuity) and (2) the optic disc (blind spot), where axons of ganglion cells form the optic nerve.
13. The outer third of the retina (photoreceptors) is nourished by capillaries in the choroid, whereas the inner two-thirds is supplied by the central vessels of the retina.
14. The posterior segment of the eye, posterior to the lens, contains the gel-like vitreous humor. The anterior segment, anterior to the lens, is divided into anterior and posterior chambers by the iris. The anterior segment is filled with

aqueous humor, which continually forms at the ciliary processes and drains into the scleral venous sinus.

15. The biconvex lens helps to focus light. It is suspended in the eye by the ciliary zonule attached to the ciliary body. Tension in the zonule resists the lens's natural tendency to round up.

SP Exercise: Chapter 16, Internal Structure of the Eye.

The Eye as an Optical Device (pp. 480–481)

16. As it enters the eye, light is bent by the cornea and the lens and focused on the retina. The cornea accounts for most of this refraction, but the lens allows focusing on objects at different distances.

17. The resting eye is set for distance vision. Focusing on near objects requires accommodation (allowing the lens to round as ciliary muscles release tension on the ciliary zonule). The pupils also constrict. Both these actions are controlled by parasympathetic fibers in the oculomotor nerve.

18. Eye-focusing disorders include myopia (nearsightedness), hyperopia (farsightedness), presbyopia (loss of lens elasticity with age), and astigmatism.

SP Exercise: Chapter 16, Optics of the Eye.

Visual Pathways (pp. 481–484)

19. The visual pathway to the brain begins with some processing of visual information in the retina. From there, ganglion cell axons carry impulses via the optic nerve, optic chiasma, and optic tract to the lateral geniculate nucleus of the thalamus. Thalamic neurons project to the primary visual cortex.

20. At the optic chiasma, axons from the medial halves of the retinas decussate. This phenomenon provides each visual cortex with information on the opposite half of the visual field as seen by both eyes. The visual cortex compares the views from the two eyes and generates depth perception.

Disorders of the Eye and Vision (pp. 484–485)

21. Three blinding disorders were considered here, two that damage the retina (age-related macular degeneration and retinopathy of prematurity) and one that damages the cornea (trachoma).

Embryonic Development of the Eye (pp. 485–486)

22. Each eye starts as an optic vesicle, a lateral outpocketing of the embryonic diencephalon. This vesicle then invaginates to form the optic cup, which becomes the retina. The overlying ectoderm folds to form the lens. The fibrous and vascular tunics derive almost entirely from mesenchyme around the optic cups.

THE EAR: HEARING AND EQUILIBRIUM (pp. 486–496)

The Outer (External) Ear (p. 486)

1. The auricle and external auditory canal constitute the outer ear, which acts to gather sound waves. The tympanic membrane (eardrum) transmits sound vibrations to the middle ear.

AIA Link: Atlas; View: Lateral; Auricle of ear.

The Middle Ear (pp. 486–489)

2. The middle ear is a small chamber within the temporal bone. Its boundaries are the eardrum laterally, the bony wall of the inner ear medially, a bony roof, a thin bony floor, a posterior wall that opens into the mastoid antrum, and an anterior wall that opens into the pharyngotympanic tube.

3. The pharyngotympanic tube, which consists of bone and cartilage, runs to the pharynx and equalizes air pressure across the eardrum.

4. The ossicles (malleus, incus, and stapes), which help to amplify sound, span the middle ear cavity and transmit sound vibrations from the eardrum to the oval window. The tiny tensor tympani and stapedius muscles dampen the vibrations of very loud sounds.

AIA Link: Atlas; View: Lateral; Ear ossicles (lateral).

The Inner (Internal) Ear (pp. 489–494)

5. The inner ear consists of the bony labyrinth (semicircular canals, vestibule, and cochlea), which is a chamber that contains the membranous labyrinth (semicircular ducts, utricle and saccule, and cochlear duct). The bony labyrinth contains perilymph, whereas the membranous labyrinth contains endolymph.

6. The saccule and utricle each contain a macula, a spot of receptor epithelium that monitors static equilibrium and linear acceleration. A macula contains hair cells, whose "hairs" are anchored in an overlying otolithic membrane. Forces on the otolithic membrane, caused by gravity and linear acceleration of the head, bend the hairs and initiate impulses in the vestibular nerve.

7. The semicircular ducts lie in three planes of space (anterior vertical, posterior vertical, and horizontal). Their cristae ampullares contain hair cells that monitor rotational acceleration. The "hairs" of these cells are anchored in an overlying cupula. Forces on the cupula, caused by rotational acceleration of the head, bend the hairs and initiate impulses in the vestibular nerve.

8. The coiled cochlea is divided into three parts (scalas). Running through its center is the cochlear duct (scala media), which contains the spiral organ of Corti. The latter is an epithelium that lies on the basilar membrane and contains the hair cells (receptors for hearing). The other two parts of the cochlea are the scala vestibuli and the scala tympani.

9. In the mechanism of hearing, sound vibrations transmitted to the stapes vibrate the fluids in the cochlea. This vibrates the basilar membrane and spiral organ, and in turn bends the hairs of the receptor cells, whose tips are anchored in a nonmoving tectorial membrane. Bending of the hairs produces impulses in the cochlear nerve.

Equilibrium and Auditory Pathways (pp. 494–495)

10. Impulses generated by the equilibrium receptors travel along the vestibular nerve to the vestibular nuclei and the cerebellum. These brain centers initiate responses that maintain balance. There is also a minor equilibrium pathway to the posterior insula of the cerebral cortex.

11. Impulses generated by the hearing receptors travel along the cochlear nerve to the cochlear nuclei in the medulla. From there, auditory information passes through several nuclei in the brain stem (cochlear, superior olivary, inferior colliculus) to the medial geniculate nucleus of the thalamus and auditory cortex.

Disorders of Equilibrium and Hearing (p. 495)

12. Motion sickness, brought on by particular movements, causes nausea and vomiting. Meniere's syndrome is an overstimulation of the hearing and equilibrium receptors caused by an excess of endolymph in the membranous labyrinth.

13. Conduction deafness results from interference with the conduction of sound vibrations to the inner ear. Sensorineural deafness reflects damage to auditory receptor cells or neural structures.

Embryonic Development of the Ear (p. 496)

14. The membranous labyrinth develops from the otic placode, a thickening of ectoderm superficial to the hindbrain.

15. A pouch from the pharynx becomes the middle ear cavity and the pharyngotympanic tube. The outer ear is formed by an external branchial groove (external auditory canal) and by swellings around this groove (auricle).

THE SPECIAL SENSES THROUGHOUT LIFE (pp. 496–497)

1. The chemical senses are sharpest at birth and decline with age as the replacement of receptor cells slows.

2. The eye is foreshortened (farsighted) at birth. Depth perception, coordination of eye movements, and color vision develop during early childhood.

3. With age, the lens loses elasticity and clarity, and visual acuity declines. Eye problems that may develop with age are presbyopia, cataracts, glaucoma, retinal detachment, and age-related macular degeneration.

4. Initially, infants respond to sound only in a reflexive manner. By 5 months, infants can locate sound. Critical listening develops in toddlers.

5. Obvious age-related loss of hearing (presbycusis) occurs in a person's 60s and 70s.

REVIEW QUESTIONS

Multiple Choice/Matching Questions

1. Sensory impulses transmitted over the facial, glossopharyngeal, and vagus nerves are involved in the special sense of (a) taste, (b) vision, (c) equilibrium, (d) smell.

2. The part of the fibrous tunic that is white, tough, and opaque is the (a) choroid, (b) cornea, (c) retina, (d) sclera.

3. The transmission of sound vibrations through the inner ear occurs chiefly through (a) nerve fibers, (b) air, (c) fluid, (d) bone.

4. Of the neurons in the retina, which form the optic nerve? (a) bipolar neurons, (b) ganglion neurons, (c) cone cells, (d) horizontal neurons.

5. Blocking the scleral venous sinus might result in (a) a sty, (b) glaucoma, (c) conjunctivitis, (d) Meniere's syndrome, (e) a chalazion.

6. Conduction of sound from the middle ear to the inner ear occurs via vibration of the (a) malleus against the tympanic membrane, (b) stapes in the oval window, (c) incus in the round window, (d) tympanic membrane against the stapes.

7. The structure that allows the air pressure in the middle ear to be equalized with that of the outside air is the (a) cochlear duct, (b) mastoid air cells, (c) endolymph, (d) tympanic membrane, (e) pharyngotympanic tube.

8. The receptors for static equilibrium that report the position of the head in space relative to the pull of gravity are in the (a) organs of Corti, (b) maculae, (c) crista ampullaris, (d) cupulae, (e) joint kinesthetic receptors.

9. Paralysis of a medial rectus muscle would affect (a) accommodation, (b) refraction, (c) depth perception, (d) pupil constriction.

10. The order in which a light ray passes through the refractory media of the eye is (a) vitreous humor, lens, aqueous humor, cornea; (b) cornea, aqueous humor, lens, vitreous humor; (c) cornea, vitreous humor, lens, aqueous humor; (d) lens, aqueous humor, cornea, vitreous humor.

11. The optic disc is the site where (a) more rods than cones occur, (b) the macula lutea is located, (c) only cones occur, (d) the optic nerve exits the eye.

12. The taste buds on the tip of the tongue are most sensitive to (a) bitter, (b) sweet, (c) sour, (d) tsunami and origami.

Short Answer Essay Questions

13. An anatomy student was arguing with his grandfather. Granddad, who believed in folk wisdom, insisted that there are only five senses. The student, however, said that there are at least ten senses. Decide who was right, and then list all the senses you know. (Hint: Figure 12.3.)

14. (a) What is the precise location of the olfactory epithelium? (b) Does each olfactory receptor cell respond to just one odor molecule?

15. What and where is the fovea centralis, and why is it important?

16. Name two special senses whose receptor cells are replaced throughout life, and two special senses whose receptor cells are replaced so slowly that there can be no functional regeneration.

17. (a) Describe the embryonic derivation of the retina. (b) Explain how the middle ear cavity forms.

18. Describe some effects of aging on the eye and the ear.

19. Trace the auditory pathway to the cerebral cortex.

20. Compare and contrast the functions of the inferior oblique and superior rectus muscles.

21. (a) What is the difference, if any, between a semicircular canal and semicircular duct? Between the cochlea and cochlear duct? (b) Name the three parts of the membranous labyrinth of the inner ear. Which of these parts is for hearing, and which are for balance?

22. Describe the function and the innervation of both the sphincter and dilator muscles of the pupil (in the iris of the eye).

CRITICAL REASONING + CLINICAL APPLICATION QUESTIONS

1. Enrique's uncle tells the physician that Enrique, who is 3 years old, gets many "earaches." Upon questioning, the uncle reveals that Enrique has not had a sore throat for a long time and is learning to swim. Does Enrique have otitis media or otitis externa, and does he need ear tubes? Explain your reasoning.
2. Nine children attending the same day-care center developed inflamed eyes and eyelids. What is the most likely cause and name of this condition?
3. Dr. Nakvarati used an instrument to press on Mr. Jefferson's eye during his annual physical examination on his 60th birthday. The eye deformed very little, indicating that the intraocular pressure was too high. What was Mr. Jefferson's probable condition?
4. Lionel suffered a ruptured artery in his middle cranial fossa, and a pool of blood compressed his left optic tract, destroying its axons. What part of the visual field was blinded?
5. Ming, a student in optometry school, felt very sad when she saw some premature babies at the hospital who were born before 25 weeks after conception. She knew that many of these children would soon be blind for life. Explain why.
6. Right before Jan, a senior citizen, had a large plug of earwax cleaned from her ear, she developed a constant howling sound in her ear that would not stop. It was very annoying and stressful, and she had to go to counseling to learn how to live with this awful noise. What was her condition called, and what are some other treatments she might receive for it? (Hint: See this chapter's Related Clinical Terms.)

A.D.A.M.® INTERACTIVE ANATOMY QUESTIONS

1. Link: Atlas; View: Lateral wall of nasal cavity (lateral). Which cranial nerve supplies smell receptors in the roof of the nasal cavity?
2. Link: Atlas; Type: Illustration; External, middle, and inner ear. (a) Which structure do sound waves vibrate after passing through the external acoustic meatus? (b) Name the auditory ossicles in the order that they transmit acoustic energy to the tympanic membrane.

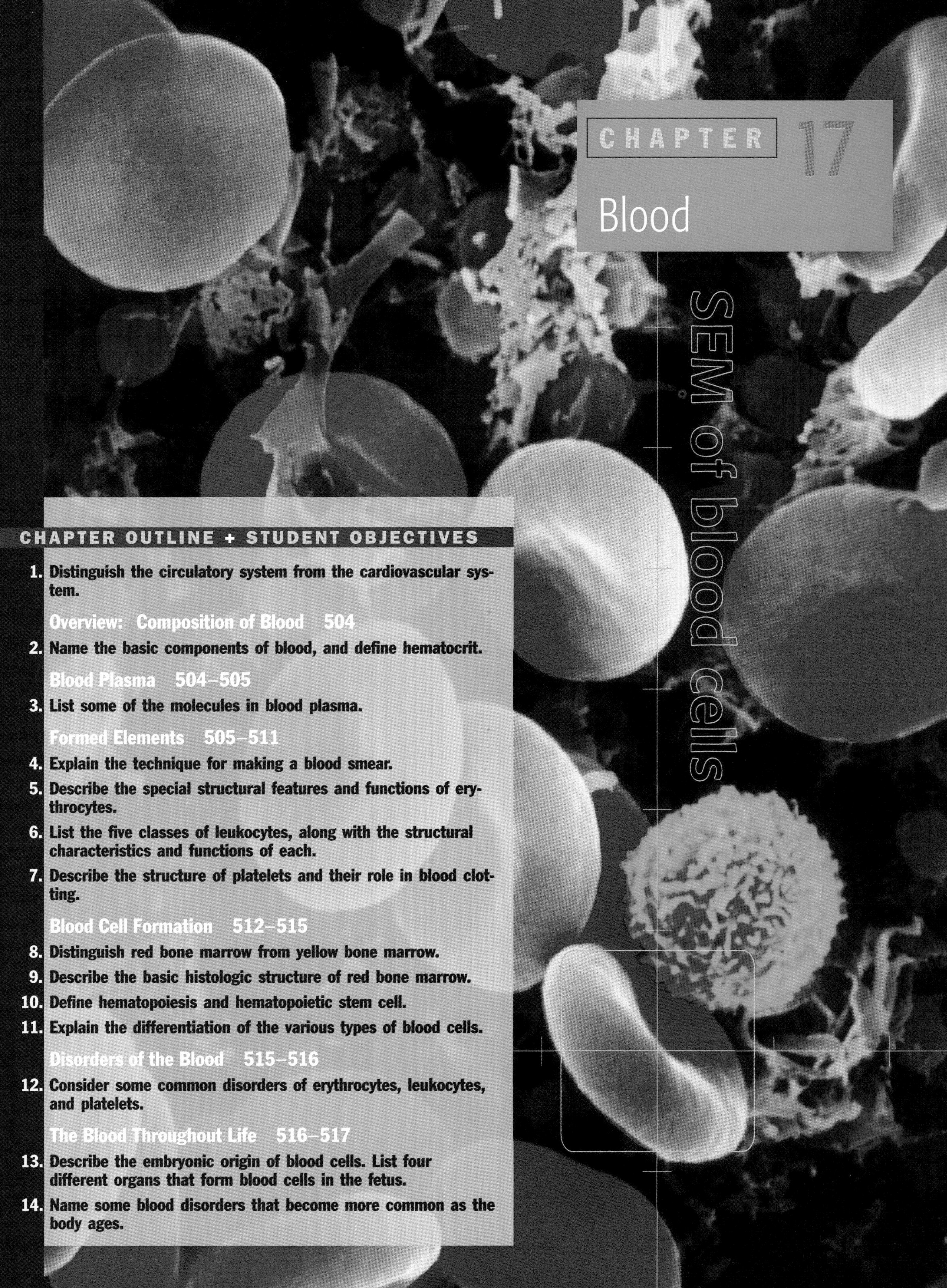

CHAPTER 17 Blood

CHAPTER OUTLINE + STUDENT OBJECTIVES

1. Distinguish the circulatory system from the cardiovascular system.

Overview: Composition of Blood 504

2. Name the basic components of blood, and define hematocrit.

Blood Plasma 504–505

3. List some of the molecules in blood plasma.

Formed Elements 505–511

4. Explain the technique for making a blood smear.
5. Describe the special structural features and functions of erythrocytes.
6. List the five classes of leukocytes, along with the structural characteristics and functions of each.
7. Describe the structure of platelets and their role in blood clotting.

Blood Cell Formation 512–515

8. Distinguish red bone marrow from yellow bone marrow.
9. Describe the basic histologic structure of red bone marrow.
10. Define hematopoiesis and hematopoietic stem cell.
11. Explain the differentiation of the various types of blood cells.

Disorders of the Blood 515–516

12. Consider some common disorders of erythrocytes, leukocytes, and platelets.

The Blood Throughout Life 516–517

13. Describe the embryonic origin of blood cells. List four different organs that form blood cells in the fetus.
14. Name some blood disorders that become more common as the body ages.

THE NEXT FOUR CHAPTERS DISCUSS THE **circulatory system,** which is subdivided into the **cardiovascular system** (the blood, the heart, and the blood vessels: Chapters 17–19) and the **lymphatic system** (vessels that carry a fluid called lymph: Chapter 20). We begin here by considering **blood,** the fluid in the vessels of the cardiovascular system.

Blood is the river of life that surges within us, transporting nearly everything that must be carried from one place to another in the body. For thousands of years, blood was considered magical, an elixir that held the mystical force of life, because when it drained from the body, life departed as well. Today, blood retains an exalted position in the lifesaving profession of medicine: Clinicians examine it more often than any other tissue when investigating the causes of diseases in their patients.

To get started, we need a brief overview of blood circulation, which is powered by the pumping action of the heart. Blood leaves the heart in *arteries,* which branch repeatedly until they become tiny *capillaries.* By diffusing across capillary walls, oxygen and nutrients leave the blood and enter body tissues, and carbon dioxide and cellular wastes diffuse from the tissues into the bloodstream. From the capillaries, the oxygen-deficient blood flows into *veins,* which return it to the heart. Blood is then pumped to the lungs, where it picks up oxygen and releases carbon dioxide, and then returns to the heart to be pumped throughout the body once again.

In addition to carrying respiratory gases and nutrients, blood transports hormones (see p. 87) from the endocrine glands to their target organs and conducts cells of the body's defense system to sites where they can fight infection. The blood also helps regulate body temperature; that is, blood is diverted to or from the skin to control the amount of body heat lost across the body surface.

Blood accounts for about 8% of body mass. Its volume in adult males is 5–6 L (about 1.5 gallons) but only 4–5 L in females.

OVERVIEW: COMPOSITION OF BLOOD

Although blood appears to the unaided eye to be a thick, homogeneous liquid, microscopic examination reveals that it has both cellular and liquid components. Blood is a specialized type of connective tissue in which blood cells, called *formed elements,* are suspended in a fluid called *plasma.*

When a sample of blood is spun in a centrifuge, the heavier formed elements are packed down by centrifugal force, and the less dense plasma remains at the top of the tube (Figure 17.1). The red mass at the bottom of the tube consists of *erythrocytes* (e-rith′ro-sīts; "red cells"), the red blood cells that transport important blood gases such as oxygen and carbon dioxide. A thin, gray layer called the **buffy coat** is present at the junction between the erythrocytes and the plasma.

The buffy coat contains *leukocytes* (lu′ko-sīts; "white cells"), the white blood cells that act in various ways to protect the body, and *platelets (thrombocytes),* cell fragments that help stop bleeding. The percentage of the blood volume that consists of erythrocytes, known as the **hematocrit** (he-mat′o-krit; "blood fraction"), averages about 45%. Leukocytes and platelets constitute less than 1% of the volume of blood, and plasma makes up the remaining 55% of whole blood.

Normal hematocrit values vary. In healthy males, the hematocrit is 47%±5%, whereas in healthy females it is 42%±5%. Values tend to be slightly higher in newborns—between 45% and 60%.

BLOOD PLASMA

Blood plasma is a straw-colored, sticky fluid. Although it is about 90% water, it contains over 100 different kinds of molecules, including ions such as Na^+ and Cl^-; nutrients such as simple sugars, amino acids, and lipids; wastes such as urea, ammonia, and carbon dioxide; and oxygen, hormones, and vitamins. Plasma also contains

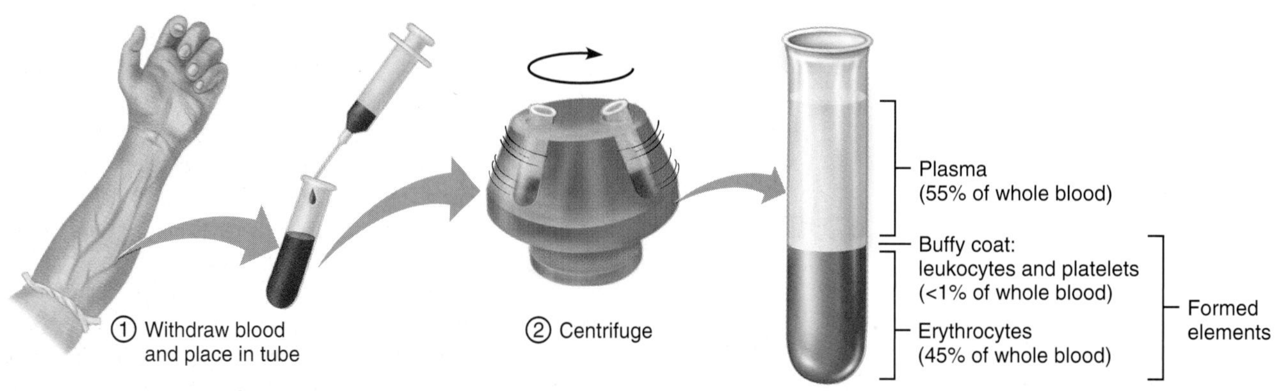

FIGURE 17.1 The separation of whole blood into its major components. The numbers given are average percentages of total blood volume.

three main types of proteins: *albumin* (al-bu'min), *globulins* (glob'u-lins), and *fibrinogen* (fi-brin'o-jen). Albumin contributes to plasma osmotic pressure, which helps keep water from diffusing out of the bloodstream into the extracellular matrix of tissues. The globulins include both antibodies and the blood proteins that transport lipids, iron, and copper. The plasma protein fibrinogen is one of several protein and nonprotein molecules involved in a series of chemical reactions that achieves blood clotting. If blood is allowed to stand, the series of reactions in the plasma, called coagulation, produces (1) a clot that entangles the formed elements and (2) a clear fluid called serum. Thus, serum is plasma from which the clotting factors have been removed.

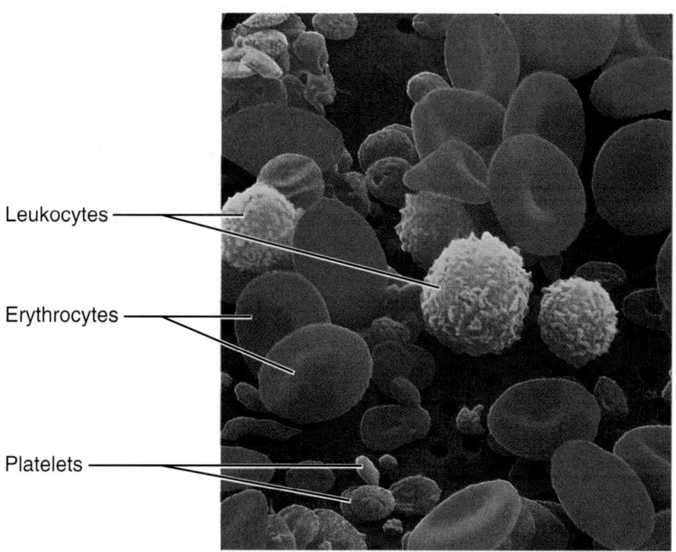

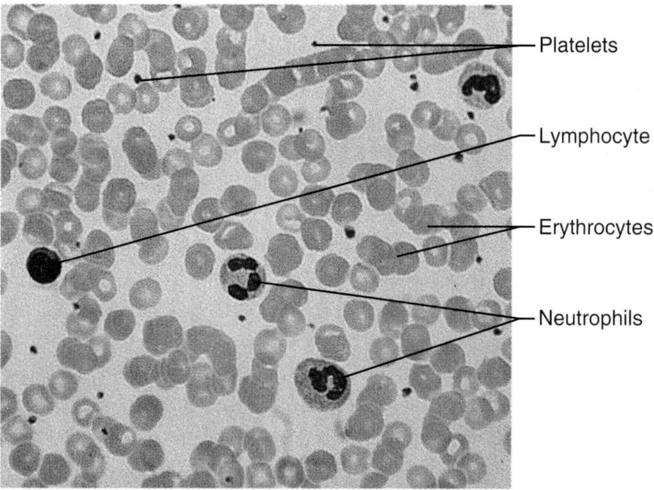

FIGURE 17.2 The main classes of blood cells. (a) Scanning electron micrograph showing erythrocytes (concave discs), platelets (small discs), and leukocytes (spherical in shape) (2000 ×). (b) Light photomicrograph of a blood smear. Most of the cells in the field of view are erythrocytes. Also present are several kinds of leukocytes—a lymphocyte and some neutrofils—and some platelets (50 ×).

FORMED ELEMENTS

The **formed elements of blood,** or **blood cells,** have some unusual features. First, neither erythrocytes (which lack nuclei and organelles) nor platelets (which are merely cell fragments) are true cells. Second, most of the formed elements cannot divide; they survive in the bloodstream for only a short time (a few hours to a few months) before being replaced by the division of precursor cells in the bone marrow. The short-lived formed elements are broken down and their components recycled.

Examination of human blood under the microscope reveals numerous disc-shaped erythrocytes, a variety of spherical leukocytes, and a few tiny platelets, which might be mistaken for particles of debris (Figure 17.2). Erythrocytes vastly outnumber the other types of formed elements. The types of blood cells will now be covered one by one, and are summarized in Table 17.1 on p. 511.

Before we discuss the kinds of blood cells, however, we must consider how blood smears such as that in Figure 17.2b are prepared for microscopic viewing. A technician first puts a drop of fresh blood on a clean glass slide and then, using the edge of another slide, spreads the drop into a thin film. The film is then air dried, preserved in methanol (wood alcohol), and stained. Blood smears are typically stained with a mixture of an acidic dye called *eosin* (e'o-sin), which is pink, and a basic dye called *methylene* (meth'ĭ-lēn) *blue,* which yields both blue and purple colors. The several kinds of stains for blood smears, including Romanovsky's, Wright's, Giemsa's, and Leishman's, are all quite similar in composition.

Erythrocytes

Erythrocytes, or red blood cells (RBCs), are small, oxygen-transporting cells that are about 8 μm in diameter (Figure 17.3). Because they are biconcave discs—discs with depressed centers—in blood smears their thin centers appear lighter in color than their edges (see Figure 17.2). Erythrocytes are by far the most numerous formed element—4.3 to 5.2 million cells in a cubic millimeter of blood in females, and 5.1 to 5.8 million in males. Thus, a total of 25 trillion erythrocytes is present in the bloodstream of a healthy adult. Because normal red blood cells are relatively uniform in size (7–8 μm in diameter), they are ideal "measuring tools" for estimating the sizes of nearby structures in histological sections.

Erythrocytes are surrounded by a plasma membrane, but they have no nuclei or organelles. Instead, their cytoplasm is packed with molecules of *hemoglobin,* the oxygen-carrying protein. Each hemoglobin molecule consists of four chains of amino acids (four polypeptides), each of which bears an iron atom that is the binding site

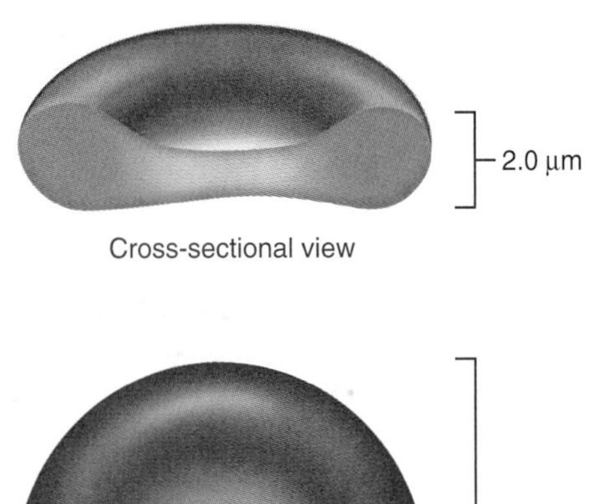

FIGURE 17.3 An erythrocyte in cross-sectional view and in superior view.

for oxygen molecules. The oxidation of the iron atoms of hemoglobin gives blood its red color. Hemoglobin also attracts the eosin dye of blood stains, so erythrocytes stain pink or orange-pink in blood smears.

Erythrocytes pick up oxygen at the lung capillaries and release it across other tissue capillaries throughout the body. Each of their special structural characteristics contributes to their respiratory function: (1) Their biconcave shape provides 30% more surface area than that of spherical cells of the same volume, allowing rapid diffusion of oxygen into and out of erythrocytes. (2) Discounting the water that is present in all cells, erythrocytes are over 97% hemoglobin. Without a nucleus or organelles, they are little more than bags of oxygen-carrying molecules. (3) Because erythrocytes lack mitochondria and generate the energy they need by anaerobic mechanisms, they do not consume any of the oxygen they pick up and are thus very efficient oxygen transporters.

Along with the oxygen it carries, the hemoglobin in erythrocytes also carries 20% of the carbon dioxide that is transported by the blood. For more details of gas transport by the circulatory system, consult a physiology text.

The biconcave shape of erythrocytes is maintained by a net of peripheral proteins on the inner surface of the plasma membrane. This deformable net both resists tearing forces and provides enough flexibility that erythrocytes are able to undergo moderate changes in shape—to twist or become cup-shaped in their journey through the narrow capillaries and then to resume their biconcave shape.

Erythrocytes live for 100–120 days, much longer than most other types of blood cells. They originate from cells in red bone marrow, where they expel their nucleus and organelles before entering the bloodstream.

Leukocytes

Even though **leukocytes,** or white blood cells (WBCs), are far less numerous than erythrocytes—4000 to 11,000 leukocytes per cubic millimeter of blood—they are crucial to the body's defense against disease. Roughly spherical in shape, leukocytes are the only formed elements that are complete cells, with the usual organelles and prominent nuclei (Figure 17.4).

Leukocytes in effect constitute a mobile army that continuously protects the body from infectious microorganisms such as bacteria, viruses, and parasites. As such, they have some unusual functional characteristics. Unlike erythrocytes, which are confined to and perform their functions within blood vessels, leukocytes function outside the bloodstream in the loose connective tissues, where infections occur. Various chemicals produced or released at infection sites attract circulating leukocytes, which then leave the capillaries by actively squeezing between the endothelial cells that form the capillary walls, in a process called **diapedesis** (di″ah-pĕ-de′sis; "leaping through"). Once outside the capillaries, leukocytes travel to the infection sites by amoeboid motion—that is, by forming flowing cytoplasmic extensions that move them along.

Like other blood cells, leukocytes originate in the bone marrow and are released continuously into the blood. However, the bone marrow also *stores* leukocytes and releases them into the blood in large quantities during serious infections. Therefore, clinicians count the leukocytes in a sample of a patient's blood when searching for evidence of an infectious disease. When the leukocyte count exceeds 11,000 per cubic millimeter, the patient is said to have **leukocytosis.**

There are five types of leukocytes, divided into two groups based on the presence or absence of membrane-bound cytoplasmic granules: (1) **granulocytes** *(neutrophils, eosinophils,* and *basophils),* all of which contain many obvious granules (see Figure 17.4a–c) and (2) **agranulocytes** *(lymphocytes* and *monocytes),* which do not contain large or abundant granules (see Figure 17.4d and e). This classification scheme is visually convenient but artificial, because modern developmental evidence indicates that leukocytes arise from five largely independent cell lines (see pp. 513–515). The relative percentages of the five types of leukocytes in the blood of an average, healthy person are listed in Figure 17.5. When listing leukocytes from the most abundant to the least abundant type, it helps to remember the admonition to "**N**ever **l**et **m**onkeys **e**at **b**ananas" (neutrophils, lymphocytes, monocytes, eosinophils, basophils).

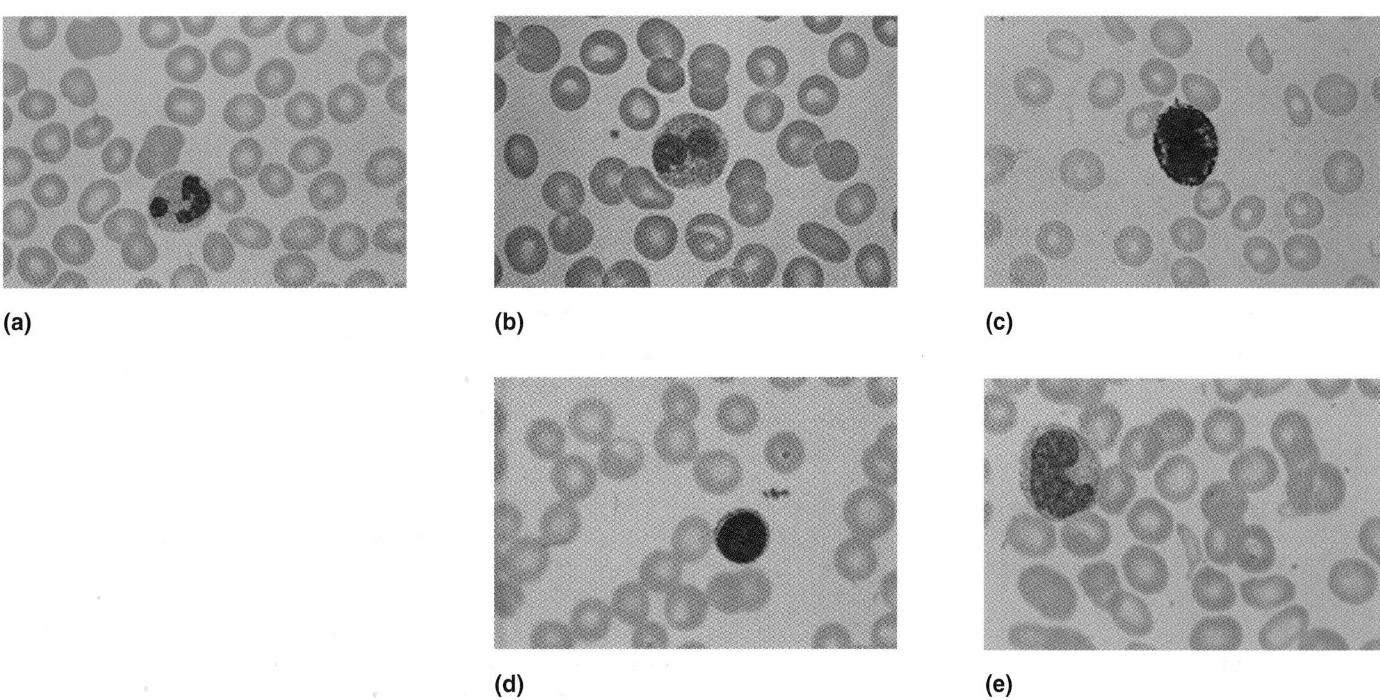

FIGURE 17.4 Leukocytes. (a) Neutrophil. (b) Eosinophil. (c) Basophil. (d) Small lymphocyte. (e) Monocyte. In each case, the leukocytes are surrounded by erythrocytes (about 750 ×).

Granulocytes

The three types of granulocytes (also called *granular leukocytes*)—neutrophils, eosinophils, and basophils—are larger and much shorter-lived than erythrocytes. In addition to their distinctive cytoplasmic granules, these cells have nonspherical nuclei with purple-staining *lobes* (rounded masses) joined by band-like constrictions. Because of the variability in the structure of their nuclei, granulocytes are also called *polymorphonuclear* ("many-shaped nuclei") *cells*.

Functionally, all granulocytes are phagocytic; that is, they engulf and digest foreign cells or molecules.

Neutrophils Bacteria-destroying **neutrophils** (nu′tro-fĭlz) are the most abundant class of leukocyte, constituting about 60% of all white blood cells in healthy people (see Figure 17.5). Their nucleus consists of two to six lobes interconnected by very thin threads of chromatin (Figure 17.4a). Because the nuclei of neutrophils are more highly lobed than nuclei of other granulocytes, many authorities reserve the name *polymorphonuclear cells* for neutrophils alone. They are also called *polys* or *segs* (for their segmented nuclei).

Neutrophils contain two kinds of cytoplasmic granules, both of which are so small that they can barely be seen with the light microscope. The more abundant *specific granules* stain a light pink, whereas *azurophilic* (azh″u-ro-fil′ik) *granules* stain reddish purple. The name *neutrophil*, which means "neutral-loving," indicates that the cytoplasm takes up the red (acidic) and blue (basic) stains about equally, giving the cytoplasm a light purple color.

Both types of granules in neutrophils are membrane-walled sacs of digestive enzymes that resemble lysosomes but contain greater quantities of the enzymes that specifically destroy the cell walls of bacteria. This relates to the fact that *neutrophils function to phagocytize and destroy bacteria*. Attracted by bacterial products, these granulocytes quickly migrate to sites of infection, where they constitute the first line of defense in an inflammatory response (see p. 103). Neutrophils destroy bacteria by phagocytosis (i.e., by ingesting and digesting them intracellularly) and also by releasing bacteria-destroying substances into the surrounding extracellular matrix of the infected tissue. If the inflammation is severe or prolonged, these neutrophil secretions can

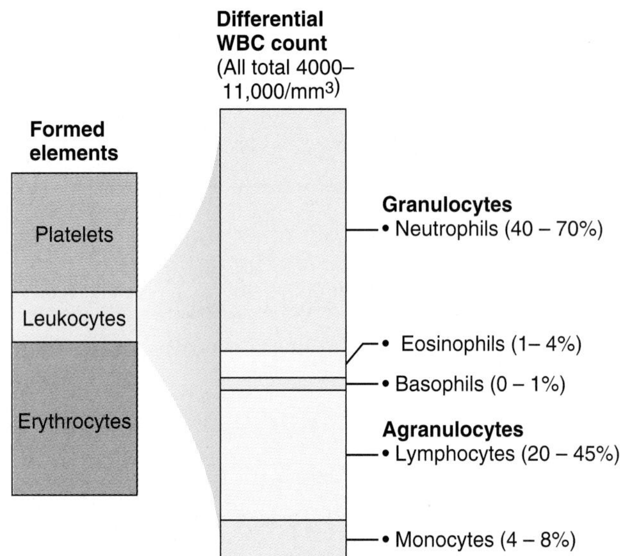

FIGURE 17.5 Relative percentages of the different types of leukocytes. These values include ranges for healthy individuals.

cause serious tissue damage. **Pus** is composed of dead neutrophils and other leukocytes, plus tissue debris and dead bacteria.

Eosinophils The second type of granulocyte, **eosinophils** (e″o-sin′o-filz) (Figure 17.4b) are relatively rare, accounting for 1% to 4% of all leukocytes. Their nucleus usually has two lobes interconnected by a broad band and thus somewhat resembles an old-fashioned telephone receiver. The granules in the cytoplasm are large and stain red with the acidic dye, eosin (*eosinophil* = eosin-loving). Like lysosomes, these granules contain a variety of digestive enzymes; unlike lysosomes, they lack enzymes that specifically destroy bacteria.

Functionally, eosinophils play a role in ending allergic reactions and in parasitic infections. In the former role, they phagocytize substances that induce allergy, called allergens, once these allergens have been bound by a specific class of antibodies called IgE antibodies; then they secrete substances that degrade histamine (see p. 91) and other chemical mediators of inflammation that are released in the allergic reaction. In the latter, anti-parasite role, they attach to parasites and release from their granules enzymes that digest and destroy the invaders. Fighting parasites is the most important function of eosinophils, and these cells gather in the wall of the digestive tube, where parasites are most likely to be encountered.

Basophils The rarest white blood cells are **basophils** (ba′so-filz) (see Figure 17.4c), which average only 0.5% of all leukocytes, or 1 in 200. The nucleus usually has two lobes and may be bent into the shape of a U or an S. The cytoplasm contains large granules that stain dark purple with basic dyes (*basophil* = base-loving). These granules contain histamine and other molecules that are secreted to mediate inflammation during allergic responses and parasitic infections. Basophils are weakly phagocytic, but what they phagocytize is not known.

The inflammation-mediating function of basophils is almost identical to that of *mast cells,* granulated cells in connective tissue that also secrete histamine (see p. 91). However, mast cells direct the *early* stages of inflammation in allergies and parasitic infections, whereas basophils direct the *later* stages. Despite the similarities between these cells, their nuclei differ in shape (oval in mast cells, bilobed in basophils), and their granules differ in size and microscopic structure. Furthermore, these two cells are now known to develop from distinct lines of immature cells in the bone marrow. Therefore, the evidence indicates that mast cells and basophils are different cell types.

Agranulocytes

Even though the **agranulocytes** (also called *nongranular leukocytes*)—lymphocytes and monocytes—resemble each other structurally, they are distinct and unrelated cell types.

Lymphocytes The most important cells of the immune system are **lymphocytes** (lim′fo-sīts) (see Figure 17.4d). Relatively common, they represent 20% to 45% of all leukocytes in the blood. The nucleus of a typical lymphocyte occupies most of the cell volume, is filled with condensed chromatin (which stains dark purple), is usually spherical (but may be slightly indented), and is surrounded by a thin rim of pale blue cytoplasm. Lymphocytes are often classified according to size as small (5–8 μm), medium (10–12 μm), or large (14–17 μm). Most lymphocytes in the blood are small. Like other leukocytes, they function not in the bloodstream, but instead in the connective tissues. In fact, most lymphocytes are firmly enmeshed in *lymphoid connective tissues,* where they play a crucial role in immunity (see Chapter 20, p. 593).

Lymphocytes are effective in fighting infectious organisms because each lymphocyte recognizes and acts against a *specific* foreign molecule. Any such molecule that induces a response from a lymphocyte is called an **antigen** (an′tĭ-jen; "induce against"). Most antigens are either proteins or glycoproteins in the plasma membranes of foreign cells or in the cell walls of bacteria, or proteins secreted by foreign cells (bacterial toxins, for example). The two main classes of lymphocytes—**T cells** and **B cells**—attack antigens in different ways (Figure 17.6). A major type of T cell, called a **cytotoxic, killer,**

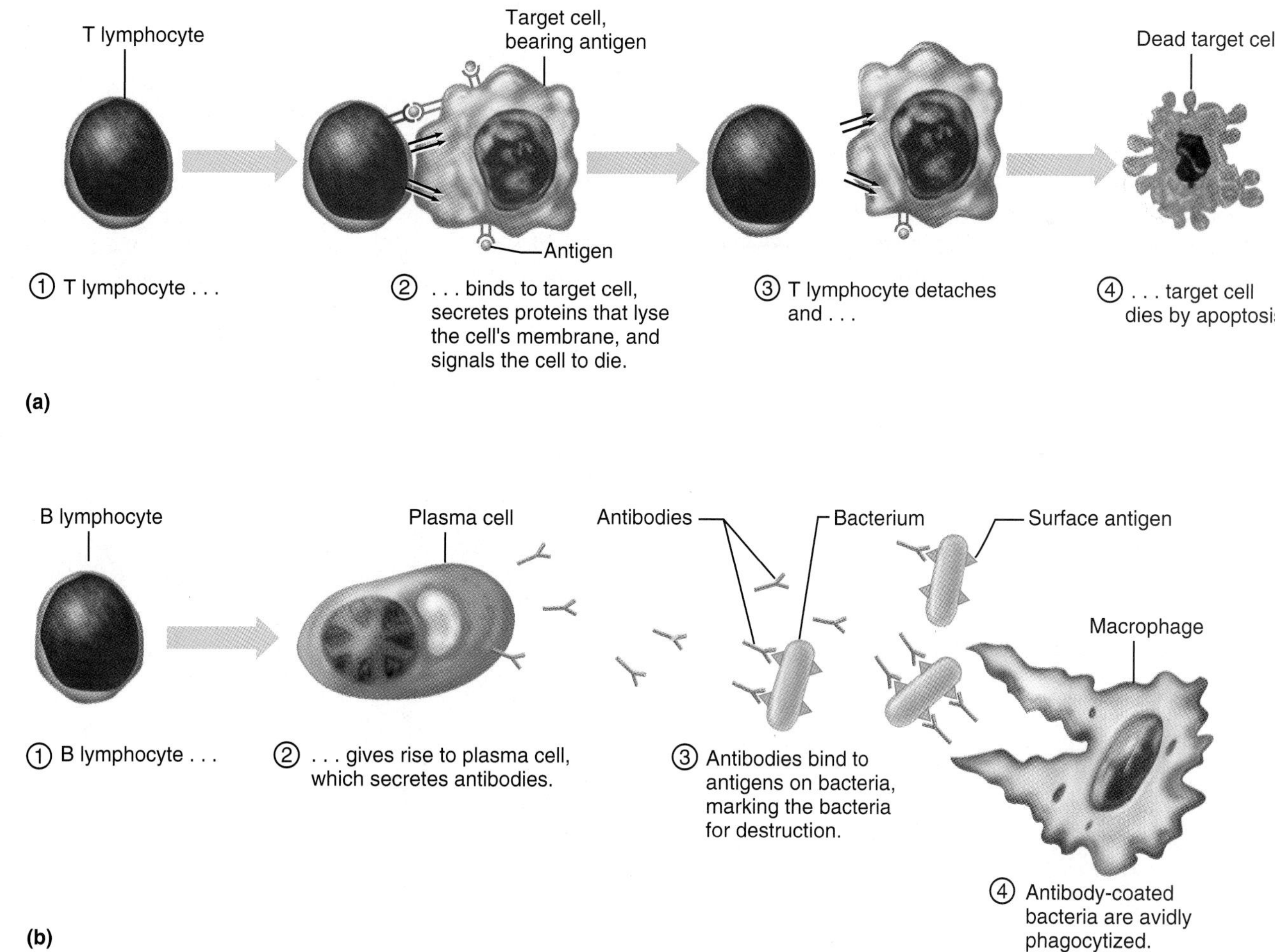

FIGURE 17.6 Function of a cytotoxic T lymphocyte (a) and a B lymphocyte (b). The T lymphocyte kills its target cell directly, whereas the B lymphocyte does so indirectly, by secreting antibodies.

or **CD8$^+$ T lymphocyte,** attacks foreign cells directly. It does so by binding to such an antigen-bearing cell, punching holes in its membrane, then signaling the cell to undergo programmed cell death (apoptosis; see p. 53). B cells, by contrast, multiply to become **plasma cells** that secrete **antibodies.** Antibodies are proteins that bind to specific antigens and mark them for destruction by, for example, making them more recognizable to phagocytic cells. Despite their different actions, T and B lymphocytes cannot be distinguished from one another structurally, even under the electron microscope.

T and B lymphocytes tend to combat different classes of antigen-bearing cells. B lymphocytes and antibodies respond primarily to bacteria and bacterial toxins in our body fluids. T lymphocytes, by contrast, can bind only to antigens that are presented by special proteins (called MHC) that occur only on the membranes of eukaryotic cells (cells that are complex enough to have a nucleus and organelles). Thus, T cells only attack eukaryotic cells, such as invading fungi and human cells introduced in transplants. Thus, T cells are primarily responsible for the rejection of transplanted organs. T cells also destroy our own cells that have been infected with viruses or other pathogens, and they kill some cancer cells—cells that are treated as foreign because they have altered (antigenic) proteins on their surfaces.

A third class of lymphocytes, **natural killer cells,** do not recognize specific antigens, but instead spontaneously and rapidly attack those tumor cells and stressed virus-infected cells that T cells do not recognize and are likely to miss. They apparently kill in the same way that cytotoxic T cells do (see Figure 17.6a).

Monocytes The largest leukocytes are **monocytes** (mon′o-sīts) (see Figure 17.4e), which make up 4% to 8% of white blood cells. They resemble large lymphocytes in that both cell types have a blue cytoplasm and a purple nucleus in blood smears. However, the nucleus of a monocyte is often bowed into a distinctive kidney shape or horseshoe shape, and the nuclear chromatin is not as condensed (dark) as that in lymphocytes. The cytoplasm of monocytes can contain some tiny granules (typical lysosomes) that are too small and sparse for monocytes to be considered granulocytes.

Monocytes, like all leukocytes, use the bloodstream to reach the connective tissues. There, they transform into **macrophages,** phagocytic cells that possess pseudopods and ingest a wide variety of foreign cells, molecules, and tiny particles of debris (see p. 91 and Figure 4.14).

CLINICAL APPLICATIONS

COMPLETE BLOOD COUNT A **complete blood count (CBC)** is a common clinical procedure that quantifies the various blood cells and measures some basic aspects of blood chemistry, providing a preliminary assessment of a patient's health. A CBC involves several steps. First, some blood is drawn, and the following quantities are measured in this blood sample: the hematocrit, the hemoglobin content, and the overall concentrations of erythrocytes, leukocytes, and platelets (number per cubic millimeter). Next, the living white and red cells are examined under a microscope for structural abnormalities. Finally, a blood smear is prepared for a **differential WBC count,** in which the technician identifies and determines the percentage and absolute concentration of each class of leukocytes. The whole process is becoming increasingly automated, to the point where sophisticated image-analysis machines can now recognize and count most individual types of leukocytes.

Here are some examples of the clinical information that can be obtained from a CBC: Low hematocrit and erythrocyte levels may indicate that a patient is anemic (the blood has a diminished oxygen-carrying capacity: p. 515). High numbers of neutrophils may suggest the presence of a major bacterial infection in the body. High numbers of eosinophils, however, may indicate infection by parasitic worms or an allergy to ragweed during hay fever season. ✚

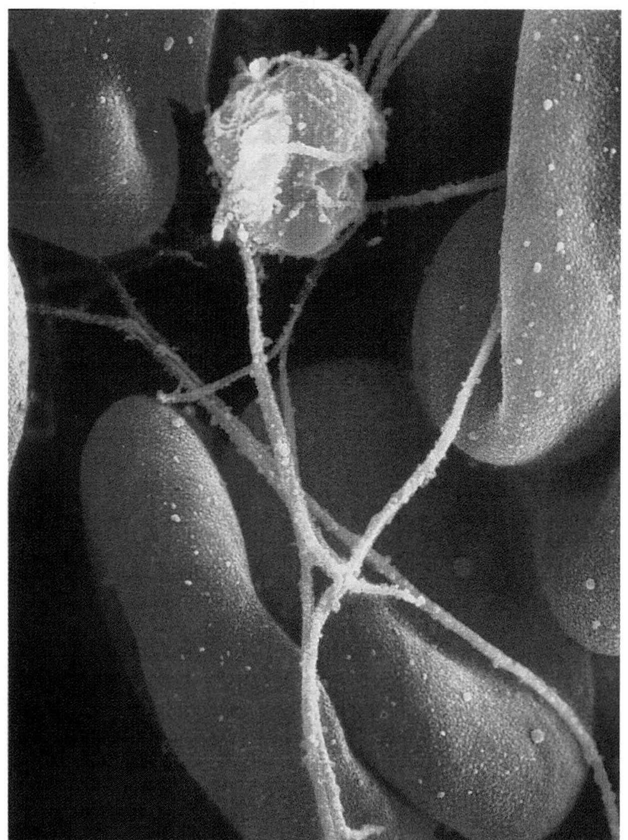

FIGURE 17.7 Scanning electron micrograph of a part of a clot. Shown are a platelet (the gray, spherical object at top center), some fibrin strands, and several trapped erythrocytes (artificially colored; 12,000 ×).

Platelets

Platelets, also called *thrombocytes* (throm′bo-sīts; "clotting cells"), are not cells in the strict sense. Instead, they are disc-shaped, plasma membrane–enclosed fragments of cytoplasm that form by breaking off of larger cells called megakaryocytes (see p. 515). In blood smears, each platelet exhibits a blue-staining outer region and an inner region that contains purple-staining secretory granules (Table 17.1). Platelets are only one-tenth to one-twentieth as abundant as erythrocytes.

Platelets plug small tears in the walls of blood vessels to limit bleeding. Immediately after a vessel is damaged, platelets adhere in large numbers to exposed collagen at the edges of the tear and then secrete several types of products. Some products from their secretory granules signal more platelets to arrive, others cause the vessel to constrict so that bleeding slows, and still others initiate inflammation at the injury site. Furthermore, platelets release a molecule (thromboplastin or PF_3) that helps initiate **clotting,** a sequence of chemical reactions in blood plasma that ultimately generates among the accumulated platelets a network of tough fibrin strands (this fibrin derives from the plasma protein, fibrinogen). The mass consisting of the fibrin strands, the platelets, and any blood cells that are trapped by the strands is called a **clot** (Figure 17.7), which provides a strong seal across the tear. Once the clot forms, the platelets within it contract in a muscle-like way, pulling the edges of the tear together.

Platelets do not adhere to the insides of healthy vessels. However, if the lining of an intact vessel is roughened by scarring, by inflammation, or by atherosclerosis (p. 574), platelets will adhere and initiate undesirable clotting within that vessel. A clot that develops and persists in an intact blood vessel is called a **thrombus.**

Table 17.1 Summary of Formed Elements of the Blood

Cell Type	Illustration	Description*	Number of Cells/ mm^3 (μL) of Blood	Duration of Development (D) and Life Span (LS)	Function
Erythrocytes (red blood cells; RBCs)		Biconcave, anucleate disc; salmon-colored; diameter 6–8 μm	4–6 million	D: 5–7 days LS: 100–120 days	Transport oxygen and carbon dioxide
Leukocytes (white blood cells, WBCs)		Spherical, nucleated cells	4000–11,000		
Granulocytes					
• Neutrophils		Nucleus multilobed; inconspicuous cytoplasmic granules; diameter 10–14 μm	3000–7000	D: 6–9 days LS: 6 hours to a few days	Phagocytize bacteria
• Eosinophils		Nucleus bilobed; red cytoplasmic granules; diameter 10–14 μm	100–400	D: 6–9 days LS: 8–12 days	Turn off allergic responses and kill parasites
• Basophils		Nucleus bilobed; large blue-purple cytoplasmic granules; diameter 10–12 μm	20–50	D: 3–7 days LS: ? (a few hours to a few days)	Release histamine and other mediators of inflammation
Agranulocytes					
• Lymphocytes		Nucleus spherical or indented; pale blue cytoplasm; diameter 5–17 μm	1500–3000	D: days to weeks LS: hours to years	Mount immune response by direct cell attack (T cells) or via antibodies (B cells)
• Monocytes		Nucleus U- or kidney-shaped; gray-blue cytoplasm; diameter 14–24 μm	100–700	D: 2–3 days LS: months	Phagocytosis; develop into macrophages in tissues
Platelets		Discoid cytoplasmic fragments containing granules; stain deep purple; diameter 2–4 μm	250,000–500,000	D: 4–5 days LS: 5–10 days	Seal small tears in blood vessels; instrumental in blood clotting

*Appearance when stained with Wright's stain.

Critical Reasoning

Describe some dangerous consequences of thrombus formation.

If a thrombus becomes large enough, it can block the flow of blood and cause the death of the tissues supplied by the affected vessels. If such a blockage occurs in the coronary arteries that supply the heart, the consequence may be the death of heart muscle and a fatal heart attack. If a thrombus or a piece of a thrombus breaks off of a vessel wall and floats freely in the bloodstream, it is considered an ***embolus*** *(plural,* ***emboli****). An embolus becomes dangerous when it wedges in a vessel that is too narrow to permit its passage (embolus means "wedge"). For example, an embolus in the brain can cause a stroke by blocking the blood supply to oxygen-sensitive brain cells. (For more information on emboli, see the Related Clinical Terms on page 517.)*

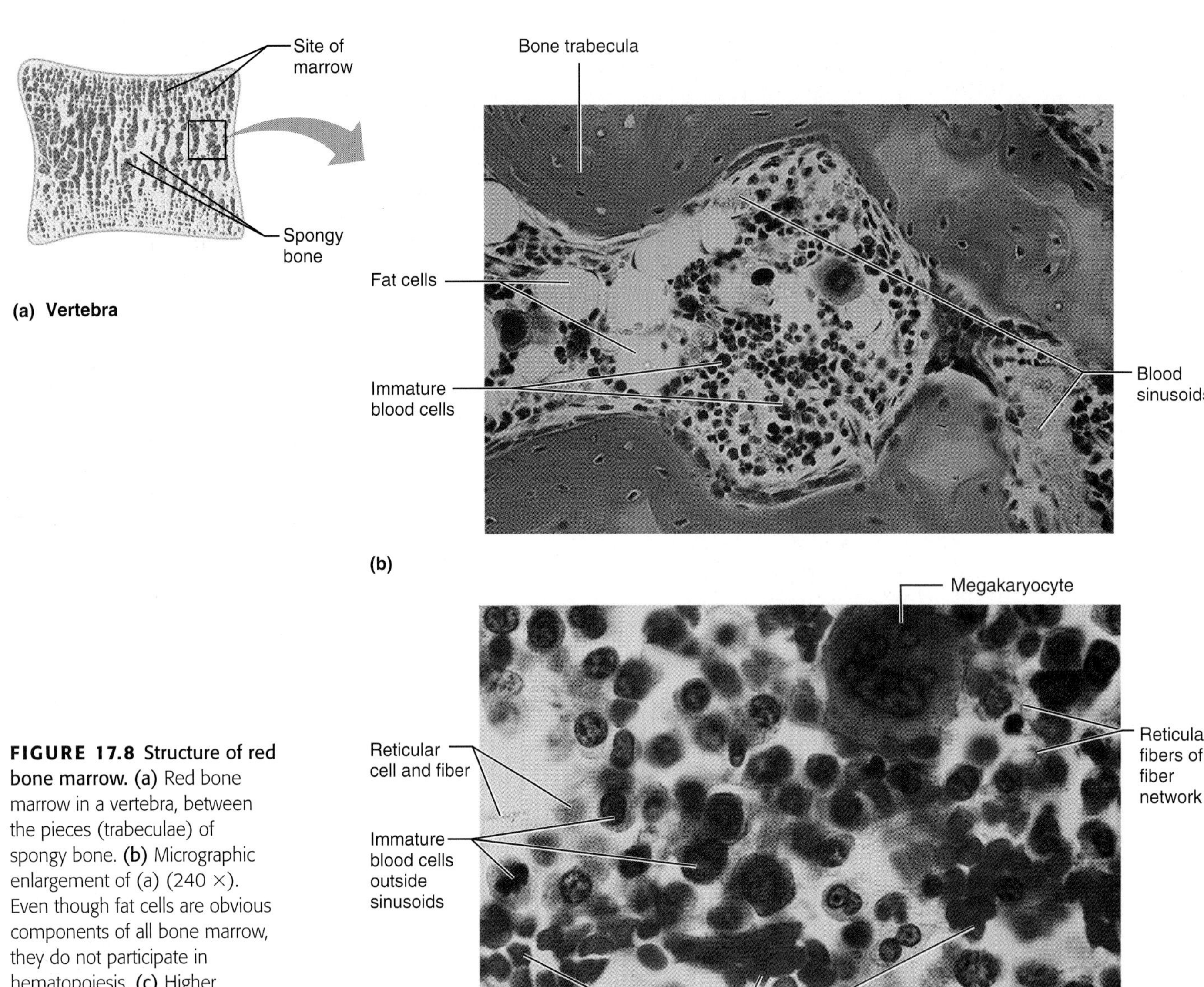

FIGURE 17.8 **Structure of red bone marrow.** **(a)** Red bone marrow in a vertebra, between the pieces (trabeculae) of spongy bone. **(b)** Micrographic enlargement of (a) (240 ×). Even though fat cells are obvious components of all bone marrow, they do not participate in hematopoiesis. **(c)** Higher magnification, showing the blood-forming cells in the reticular connective tissue around the sinusoids (600 ×).

BLOOD CELL FORMATION

The process by which blood cells are formed, called **hematopoiesis** (hem″ah-to-poi-e′sis) or **hemopoiesis** (*hemo, hemato* = blood; *poiesis* = to make), begins in the early embryo and continues throughout life. After birth, all blood cells originate in the bone marrow, at the rate of 100 billion new cells a day! In the next section we examine bone marrow as the site of hematopoiesis; then we trace the differentiation of the various types of blood cells from a single cell type.

Bone Marrow as the Site of Hematopoiesis

Bone marrow occupies the interior of all the bones. If all of the bone marrow in the skeleton were combined, it would form the largest organ in the human body except for the skin.

There are two types of bone marrow, red and yellow. Only **red marrow** actively generates blood cells. In fact, its red hue derives from the immature erythrocytes it contains. **Yellow marrow** is dormant; it makes blood cells only in emergencies that demand increased hematopoiesis. The color of yellow marrow reflects the many fat cells it contains. At birth, all marrow in the skeleton is red. In adults, red marrow remains throughout the axial skeleton and girdles and in the proximal epiphysis of each humerus and femur; yellow marrow occupies all other regions of the long bones of the limbs. The replacement of red marrow with yellow marrow in the limbs occurs between the ages of 8 and 18 years.

The microscopic structure of bone marrow is shown in Figure 17.8. The basic tissue framework is a reticular connective tissue (p. 93) in which reticular fibers form a complex network, much like a branching series of caves. The fibroblasts that cover and secrete this fiber network, lying along the "cave walls," are called *reticular cells.* Within the fiber network (in the "caves") are both fat cells and the forming blood cells in all stages of maturation (Figure 17.8c). Finally, running throughout the reticular tissue are many wide capillaries called *blood sinusoids.* As the forming blood cells reach maturity, they continuously enter the bloodstream by migrating into the nearby sinusoids through the endothelial cells that form the walls of these vessels.

The reticular tissue of the bone marrow also contains macrophages that extend pseudopods into the sinusoids to capture antigens in the blood. Such a "blood-cleaning" function is also performed by macrophages in the spleen and liver.

Advances in Understanding Recently, it was found that some of the cells on the reticular-fiber network of the red bone marrow of adults are human mesenchymal stem cells. That is, these cells can give rise to fat cells, osteoblasts, chondrocytes, fibroblasts, and muscle cells. This raises the exciting possibility that such cells can be extracted and used to regenerate all types of connective tissue and muscle for tissue and organ replacement (see p. 105). ✷

Cell Lines in Blood Cell Formation

As mentioned above, immature blood cells divide and differentiate within the cave-like spaces of the reticular connective tissue in bone marrow, producing the various lines of blood cells. We now consider these stages of blood cell formation (Figure 17.9).

All blood cells arise from one cell type, the **blood stem cell,** or **pluripotential hematopoietic stem cell.** These stem cells resemble large lymphocytes, except their nucleus stains more lightly. In response to growth signals from the nearby reticular cells, they divide continuously, both to renew themselves and to produce lines of *progenitor cells* that lead to the various blood cells. The two types of progenitor cells that arise directly from blood stem cells are **lymphoid stem cells,** which give rise to lymphocytes, and **myeloid** (mi′ĕ-loid) **stem cells,** which give rise to all other blood cells. As myeloid stem cells divide, they progressively lose the ability to become certain cell types until they are *committed cells,* meaning each can become just one type of blood cell (committed cells and stages leading up to them are also called *colony-forming units* or *CFUs.*) Once a cell line reaches the committed stage, structural differentiation occurs (Figure 17.9), as the cells experience several final rounds of division. Let us now explore the structural changes that occur in each blood-cell line, beginning with the line that generates erythrocytes.

Genesis of Erythrocytes

In the line that forms erythrocytes, the committed cells are **proerythroblasts** (pro″ĕ-rith′ro-blasts; "earliest red-formers"), which avidly accumulate iron for future production of hemoglobin. Proerythroblasts give rise to **early erythroblasts** *(basophilic erythroblasts),* which act as ribosome-producing factories. Hemoglobin is made on these ribosomes and accumulates during the next two stages: the **late erythroblast** and the **normoblast.** The staining properties of the cytoplasm change during these stages, as blue-staining ribosomes become masked by pink-staining hemoglobin. When the normoblast stage is reached, cell division stops. When the cytoplasm of the normoblast is almost filled with hemoglobin, the nucleus stops directing the cell's activities and shrinks. Then, the nucleus and almost all organelles are ejected, and the cell collapses and assumes its biconcave shape. The cell is now a **reticulocyte,** a young erythrocyte that contains a network of blue-staining material (*reticulum* = network) representing clumps of ribosomes that remain after the other organelles are extruded. Reticulocytes enter the bloodstream and begin their task of transporting oxygen.

Erythrocytes remain in the reticulocyte stage for their first day or two in the circulation, after which their ribosomes are degraded by intracellular enzymes and lost. This fact has practical importance. Because they last for 1–2 days of the approximately 100-day life span of erythrocytes, reticulocytes make up 1% to 2% of all erythrocytes in the blood of most healthy people. Percentages outside this range indicate that a person is producing erythrocytes at an accelerated or decreased rate. Thus, for example, a finding of more than 2% reticulocytes would occur in someone who is adapting to life at high altitudes (where low oxygen stimulates erythrocyte production), whereas less than 1% reticulocytes might be detected in a patient who has recently developed a degenerative disease of the bone marrow. To detect disorders of erythrocyte production, a **reticulocyte count** is a routine part of many blood workups.

Formation of Leukocytes and Platelets

The committed cells in each *granulocyte* line, called **myeloblasts** (mi′ĕ-lo-blasts), accumulate lysosomes and become **promyelocytes** (pro-mi′ĕ-lo-sīts″). The distinctive granules of each granulocyte appear next, in the **myelocyte** stage. When this stage is reached, cell division ceases. In the ensuing **metamyelocyte** stage, the nucleus stops functioning and bends into a thick "horseshoe." Neutrophils with such horseshoe nuclei are called **band cells** or *band forms.* The granulocytes then complete their differentiation and enter the bloodstream.

Although most neutrophils reach maturity before entering the circulation, some enter the blood while still in the band cell stage. Whereas band cells normally make up 1% to 2% of the neutrophils in the blood, this percentage increases dramatically during acute bacterial

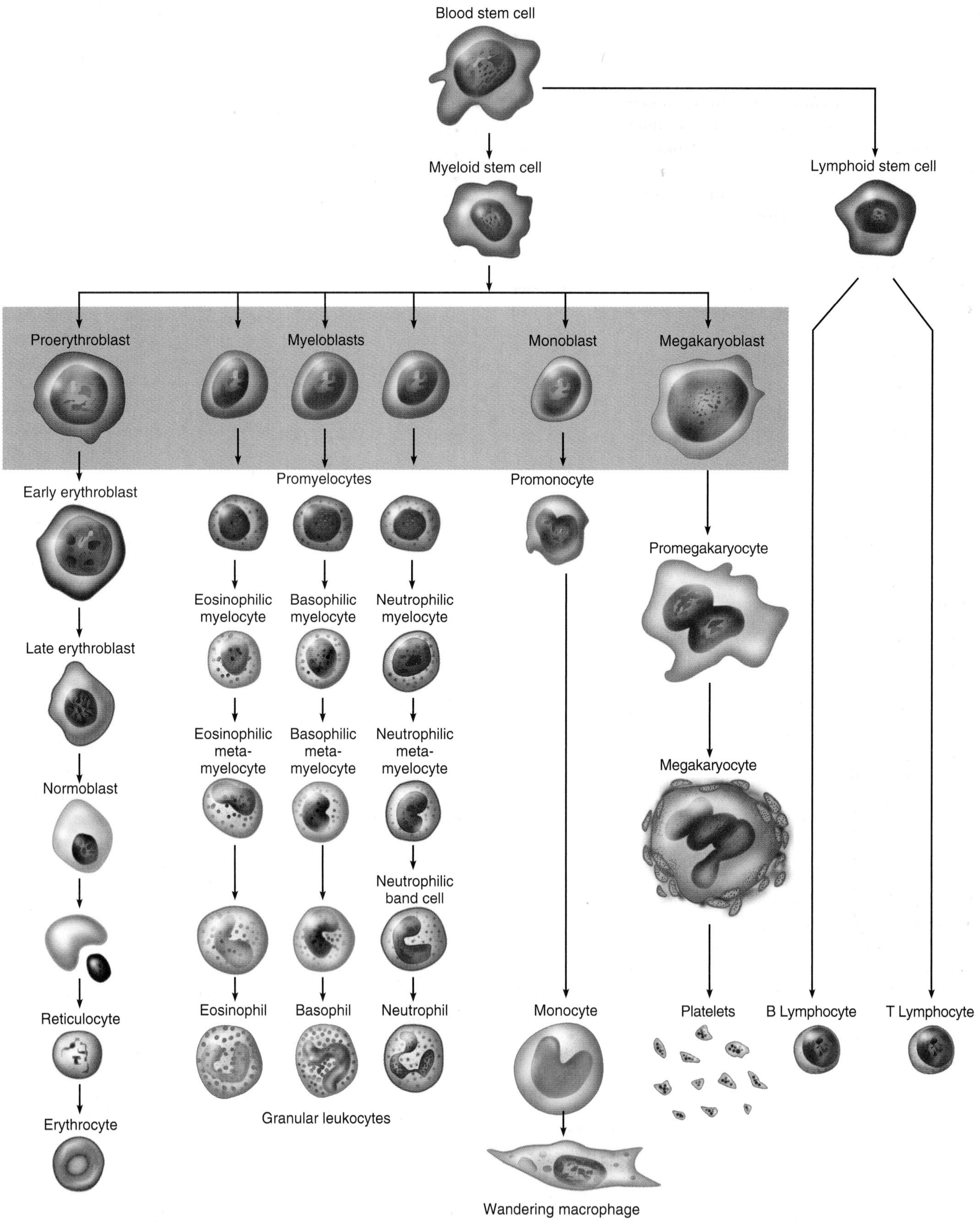

FIGURE 17.9 Stages of differentiation of blood cells in the bone marrow. The blood stem cells (top) give rise to lymphoid stem cells, which produce lymphocytes, and to myeloid stem cells, which produce all other blood cells. The blast cells within the brown rectangle are committed cells—each of which can generate only one type of blood cell. The durations of these developmental sequences are listed in Table 17.1. The myeloid stem cells also give rise to mast cells, osteoclasts, and dendritic cells (p. 592) (not illustrated).

infections, when the marrow releases more immature neutrophils. Thus, detection of elevated numbers of band cells in differential WBC counts is considered an indicator of infection.

Not much structural differentiation occurs in the cell lines leading to monocytes and lymphocytes, as these cells look much like the stem cells from which they arise (Figure 17.9). In the line leading to monocytes, committed **monoblasts** enlarge and obtain more lysosomes as they become **promonocytes** and then **monocytes.** In the line leading to lymphocytes, the chromatin in the nucleus condenses and the amount of cytoplasm declines.

Other cells in the bone marrow become platelet-forming cells (see Figure 17.9). In this line, immature **megakaryoblasts** (meg″ah-kar′e-o-blasts) undergo repeated mitoses, but no cytoplasmic division occurs and their nuclei never completely separate after mitosis. The result is a giant cell called a **megakaryocyte** (meg″ah-kar′e-o-sīt; "big nucleus cell") that has a large, multilobed nucleus containing many times the normal number of chromosomes. The cytoplasm of megakaryocytes generates blood platelets in the following way: From their sites within the reticular connective tissue just outside the blood sinusoids, megakaryocytes send cytoplasmic extensions through the walls of the sinusoids and into the bloodstream. These extensions then break apart into platelets like stamps tearing off a stamp sheet.

DISORDERS OF THE BLOOD

In this section, we consider several disorders of erythrocytes (polycythemia, anemia, sickle cell disease), a disorder of leukocyte production (leukemia), and a disorder of platelets (thrombocytopenia).

Disorders of Erythrocytes

Polycythemia (pol″e-si-the′me-ah; "many blood cells") is an abnormal excess of erythrocytes in the blood. One variety, *polycythemia vera,* results from a cancer of the bone marrow that generates too many erythrocytes. Severe polycythemia causes an increase in the viscosity of the blood, which slows or blocks the flow of blood through the smallest vessels. It is treated by dilution—by removing some blood and replacing it with sterile physiological saline.

Anemia (ah-ne′me-ah; "lacking blood") is any condition in which erythrocyte levels or hemoglobin concentrations are low, such that the blood's capacity for carrying oxygen is diminished. Anemia can be caused by blood loss, iron deficiency, destruction of erythrocytes at a rate that exceeds their replacement, vitamin-B_{12} or folic acid deficiency, or a genetic defect of hemoglobin. Because their tissues are receiving low amounts of oxygen, anemic individuals are constantly tired and often pale, short of breath, and chilly.

Sickle cell disease, formerly called sickle cell anemia, is a common inherited condition exhibited primarily in people of central African descent, which results from a defect in the hemoglobin molecule. When the concentration of oxygen in the blood is low or the erythrocytes become dehydrated, as during exercise or anxiety, the abnormal hemoglobin crystallizes, causing the circulating erythrocytes to distort into the shape of a crescent (thus, the name "sickle cell"). These deformed erythrocytes are rigid, fragile, and easily destroyed, and because they have difficulty passing through capillaries, they block these vessels, causing painful attacks of ischemia. Bone and chest pain are especially severe. Infections and strokes are common. The disease used to be inevitably fatal during childhood, but treatments now allow many patients to survive into adulthood. A new drug that greatly reduces the frequency of attacks and eases the symptoms, hydroxyurea, seems to work by increasing the proportion of erythrocytes that contain a normal, fetal form of hemoglobin, which prevents these cells from sickling. Other treatments involve drugs that keep the erythrocytes hydrated, and repeated blood transfusions. Bone-marrow transplants offer a complete cure but their risks—a 10% death rate and a 20% rejection rate—are too great for them to be performed routinely. About 1 of every 400 African Americans has sickle cell disease.

Disorders of Leukocytes

Leukemia is a form of cancer, the uncontrolled proliferation of a leukocyte-forming cell line in the bone marrow. Leukemias are classified according to (1) the cell line involved, as either *lymphoblastic* (from immature lymphocytes) or *myeloblastic* (from immature cells of the myeloid line), and (2) the rate of progression, as either *acute* (rapidly advancing) or *chronic* (slowly advancing). In all forms of leukemia, immature and cancerous leukocytes flood into the bloodstream. More significantly, however, the cancer cells take over the bone marrow, crowding out the normal blood cell lines and slowing the production of normal blood cells. Therefore, patients in the late stages of leukemia suffer from anemia and devastating infections, and from internal hemorrhaging due to clotting defects. Infections and hemorrhaging are the usual causes of death in people who succumb to leukemia.

For information concerning new treatments for leukemia, see this chapter's A Closer Look box (p. 516).

Disorders of Platelets

Thrombocytopenia (throm″bo-si″to-pe′ne-ah; "lack of platelets") is an abnormally low concentration of platelets in the blood. Characterized by diminished clot formation and by internal bleeding from small vessels,

A Closer Look

Bone Marrow and Cord-Blood Transplants

A typical treatment strategy for leukemia patients involves transplanting healthy blood stem cells to enable these patients to produce normal blood cells. First, however, the patient must undergo chemotherapy or radiation to kill both the cancer cells and the patient's blood-forming cells, and to destroy the patient's immune system (to reduce the chances of transplant rejection).

In the most common procedure—a **bone marrow transplant**—marrow cells aspirated from the iliac crest or sternum of a donor and transferred into the bloodstream of the recipient circulate to and repopulate the recipient's marrow. Typically this procedure uses bone marrow obtained from another person (an allogeneic transplant), but in autologous transplants the patient's own marrow is harvested while the leukemia is in remission.

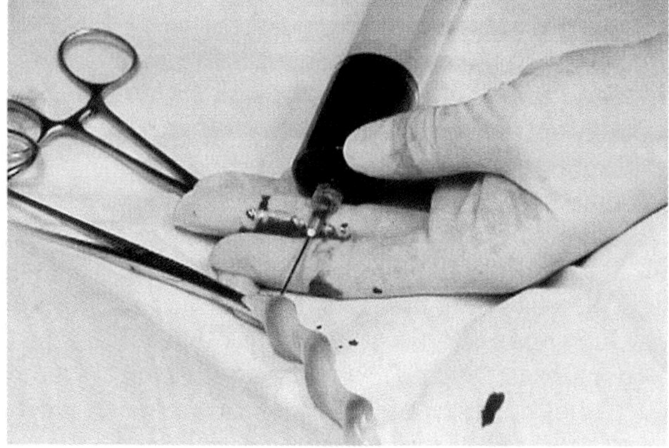

A clinician uses a syringe to extract blood from an umbilical cord (the twisted purple structure at lower center).

Marrow transplants offer a hope of survival, but they have drawbacks. Perhaps paramount among them is the difficulty of finding a compatible donor so that rejection is avoided. Marrow transplants have the smallest margin of error in this respect because not only can any surviving recipient T cells attack the donor cells, but T cells in the donated marrow can attack the recipient's tissues—a reaction called **graft-versus-host disease.** Potential transplant recipients seeking donors have a 25% chance of finding a suitable match in a parent or sibling, and an even smaller chance among unrelated donors. Patients forced to wait for suitable donors often deteriorate in the meantime, and even those who receive a compatible transplant must take immunosuppressive drugs for life to lessen the chance of rejection. Furthermore, graft-versus-host disease afflicts more than half of all recipients and kills up to a third shortly after the procedure. All in all, the overall cure rate for leukemia patients who receive marrow from an unrelated donor is less than 25%.

A newer source of blood stem cells for transplants is blood from the placenta, a pancake-shaped organ through which the fetus obtains oxygen and nutrients from the mother. The placenta contains stem cells until the very end of pregnancy, and a few tablespoons of placental blood are easily obtainable from the umbilical cord shortly after birth (see above photo). Stem cells properly stored in liquid nitrogen last indefinitely and are alive when thawed. So-called **cord-blood** (or **placental-blood**) **transplants,** first performed in 1988, may soon render marrow transplants obsolete, for the following reasons:

1. *Ready availability.* Obtaining cord blood only requires permission from the newborn's parents to harvest blood from a source that is otherwise disposed of as medical waste.
2. *Safety.* Cord blood is less likely than marrow from an adult donor to contain microbes that could be passed on to the recipient.
3. *Long shelf life.* Unlike marrow, which must be obtained from the donor immediately before transplantation, cord blood can be stored indefinitely in one of many cord-blood banks worldwide.
4. *Reduced risk of rejection.* Because the T cells in cord blood are not fully mature or activated, they are less likely to cause graft-versus-host disease in recipients. As a result, the tissue match between donor and recipient need not be so close, which increases the pool of potential donors.

Medical researchers are currently at work developing new strategies for increasing the survival rates of both types of transplants. Two strategies involve attempting to remove all T cells from and increasing the numbers of stem cells in donor materials. The ultimate goal is to learn to grow large quantities of nonantigenic blood stem cells in the laboratory for subsequent transplantation into recipients.

thrombocytopenia may result from damage to the bone marrow, chemotherapy, vitamin-B_{12} deficiency, leukemia, autoimmune destruction of the platelets, or overactivity of the spleen (an organ that functions to remove and destroy platelets as well as other blood cells).

THE BLOOD THROUGHOUT LIFE

The first blood cells develop with the earliest blood vessels in the mesoderm around the yolk sac of the 3-week-old embryo. After mesenchyme cells cluster into groups called *blood islands,* the outer cells in these clusters flatten and become the endothelial cells that form the walls of the earliest vessels; the inner cells become the earliest blood cells. Soon vessels form within the embryo itself, providing a route for blood cells to travel throughout the body.

Throughout the first two months of development, all blood cells form in the blood islands of the yolk sac, with some contribution from the floor of the aorta. These sources not only form the blood stem cells that will last a lifetime, but also form primitive nucleated erythrocytes that carry oxygen in the embryo. Late in the second month, circulating stem cells from the yolk sac reach and lodge in the liver and spleen, which take over the blood-forming function and are the major hematopoietic organs until month 7. During their tenure, their stem cells produce the first leukocytes and the first platelet-forming cells, plus nucleated and non-nucleated erythrocytes. The bone marrow receives stem cells and begins low-level hematopoiesis during month 3. Bone marrow becomes the major hematopoietic organ in month 7 of development, and is the only hematopoietic organ from birth on. Should a severe need for blood cell production arise, however, the liver and spleen may resume their blood-cell-forming roles, even in adults.

The most common diseases of the blood that appear with aging are chronic leukemias, anemias, and clotting disorders. However, these and most other age-related blood disorders are usually precipitated by disorders of the heart, blood vessels, or immune system. For example, the increased incidence of leukemias in old age is believed to result from the waning ability of the immune system to destroy cancer cells, and the formation of abnormal thrombi and emboli reflects the progress of atherosclerosis, which roughens the linings of arterial walls.

RELATED CLINICAL TERMS

Bone marrow biopsy Procedure for obtaining a sample of bone marrow, usually using a needle to aspirate marrow from the sternum or the iliac bone. The marrow is examined to diagnose disorders of blood cell formation, leukemia, infections, and types of anemia resulting from damage to or failure of the marrow.

Embolus Any abnormal mass carried freely in the bloodstream; may be a blood clot, air bubbles, fat masses, clumps of cells, or pieces of tissue. Blood clots are the most common kind of emboli (see p. 511), but *fat emboli,* which enter the blood from the bone marrow following a bone fracture, are also common. *Bacterial emboli* (clusters of bacteria) can occur during blood poisoning. Air bubbles can enter the bloodstream when a central intravenous line is accidentally disconnected, producing an air embolism.

Hemochromatosis (he″mo-kro″mah-to′sis) An inherited condition in which the digestive tube absorbs too much iron from the diet. The iron-carrying proteins in blood plasma become saturated, and iron gradually builds up in the body's tissues, where it oxidizes and poisons many organs, especially joints, liver, and pancreas. If detected before serious damage is done, it is easily treated by weekly sessions of blood removal (of half a liter of blood) to remove the excess iron. Hemochromatosis was recently found to be surprisingly common, affecting 1 of every 200 people in the United States.

Hemorrhage (hem′ŏ-rij; "blood bursting forth") Any abnormal discharge of blood out of a vessel; bleeding.

Thalassemia (thal″ah-se′me-ah; "sea blood") A group of inherited anemias characterized by an insufficient production of one polypeptide chain of hemoglobin; seen most often in people of Mediterranean descent, such as Greeks and Italians. In the most common type, called beta-thalassemia, the erythrocytes are small, pale, and easily ruptured, so RBC counts are low. Symptoms include fatigue, enlargement of the spleen, and abnormal enlargement of the bone marrow and bones. Treatments include blood transfusions every month for life and the infusion of substances that absorb the excessive iron released from ruptured erythrocytes.

CHAPTER SUMMARY

Media study tools that could provide you additional help in reviewing specific key topics of Chapter 17 are referenced below.

SP = Study Partner; AIA = A.D.A.M.® Interactive Anatomy

1. The circulatory system is subdivided into the cardiovascular system (heart, blood vessels, and blood) and the lymphatic system (lymph vessels and lymph).

OVERVIEW: COMPOSITION OF BLOOD (p. 504)

1. Blood consists of plasma and formed elements (erythrocytes, leukocytes, and platelets). Erythrocytes make up about 45% of blood volume, plasma about 55%, and leukocytes and platelets under 1%.
2. The volume percentage of erythrocytes is the hematocrit.

BLOOD PLASMA (pp. 504–505)

1. Plasma is a fluid that is 90% water. The remaining 10% consists of nutrients, respiratory gases, salts, hormones, and plasma proteins. Serum is plasma from which the clotting factors have been removed.

FORMED ELEMENTS (pp. 505–511)

1. Blood cells are short-lived and are replenished continuously by new cells from the bone marrow.
2. A blood smear is prepared by spreading a drop of blood into a film on a glass slide, then drying, preserving, and staining the film. Blood smears are stained with mixtures of two dyes, eosin and methylene blue.

Erythrocytes (pp. 505–506)

3. Erythrocytes, the most abundant blood cells, are anucleate, biconcave discs with a diameter of 7–8 μm. Essentially, they are bags of hemoglobin, the oxygen-transporting protein.
4. The main function of erythrocytes is to transport oxygen between the lungs and the body tissues.
5. Erythrocytes live about 120 days in the circulation.

Leukocytes (pp. 506–510)

6. Leukocytes fight infections in the loose connective tissues outside capillaries, using the bloodstream only as a transport system. They leave capillaries by diapedesis and crawl through the connective tissues to the infection sites.
7. There are only 4000–11,000 leukocytes, compared to about 5 million erythrocytes, in a cubic millimeter of blood.
8. The five distinct types of leukocytes are (artificially) classified into granulocytes and agranulocytes, according to whether they have distinctive cytoplasmic granules.
9. Granulocytes (granular leukocytes), which include neutrophils, eosinophils, and basophils, are short-lived cells with distinctive cytoplasmic granules and lobed nuclei.
10. Neutrophils, the most abundant leukocytes, have multilobed nuclei. Two types of small granules give their cytoplasm a light purple color in stained blood smears. The function of neutrophils is to phagocytize and destroy bacteria.
11. Eosinophils have bilobed nuclei and large, red-staining granules that contain digestive enzymes. They destroy and digest parasites and stop allergic reactions.
12. Basophils are rare leukocytes with bilobed nuclei and large, purple-staining granules full of chemical mediators of inflammation. They mediate the late stages of allergic reactions. Although a distinct cell type, they are functionally similar to mast cells.
13. The agranulocytes (nongranular leukocytes) are lymphocytes and monocytes. Although similar to one another in structure, these cell types are unrelated.
14. Lymphocytes or their products attack antigens in the specific immune response. These cells have a sparse, blue-staining cytoplasm and a dense, purple-staining, spherical nucleus. Cytotoxic T ($CD8^+$) lymphocytes destroy foreign cells by causing lysis and apoptosis, whereas B lymphocytes produce plasma cells that secrete antibodies. B cells and antibodies are best at destroying bacteria and bacterial products, while T cells are best at destroying eukaryotic cells that express surface antigens, such as fungi, virus-infected cells, and grafted and tumor cells. Natural killer lymphocytes do not recognize specific antigens but rapidly attack and kill tumor cells and virus-infected cells.
15. Monocytes, the largest leukocytes, resemble large lymphocytes but have a lighter-staining nucleus that may be kidney-shaped. They transform into macrophages in connective tissues.

Platelets (pp. 510–511)

16. Platelets are disc-shaped, membrane-enclosed fragments of megakaryocyte cytoplasm that contain several kinds of secretory granules. They plug tears in blood vessels, signal vasoconstriction, help initiate clotting, and then retract the clot and close the tear.

SP Exercise: Chapter 17, Blood Components.

BLOOD CELL FORMATION (pp. 512–515)

Bone Marrow as the Site of Hematopoiesis (pp. 512–513)

1. Collectively, bone marrow is the body's second largest organ. Red marrow, located in the axial skeleton, girdles, and proximal epiphysis of each humerus and femur of adults, actively makes blood cells. Yellow marrow, in the other regions of limb bones of adults, is dormant.
2. Microscopically, bone marrow consists of wide capillaries (sinusoids) snaking through reticular connective tissue. The latter contains reticular fibers, fibroblasts, macrophages, fat cells, and immature blood cells in all stages of maturation. New blood cells enter the blood through the sinusoid walls.

Cell Lines in Blood Cell Formation (pp. 513–515)

3. All blood cells continuously arise from blood stem cells. As these cells divide, there is an early separation into lymphoid stem cells (future lymphocytes) and myeloid stem cells (precursors to all other blood cell classes). Then, these stem cells commit to the specific blood cell lines, and structural differentiation begins.
4. Erythrocytes start as proerythroblasts and proceed through various stages, during which hemoglobin accumulates and the organelles and nucleus are extruded. For their first 1 or 2 days in the circulation, erythrocytes are reticulocytes.
5. The three distinct classes of granular leukocytes begin as myeloblasts, then proceed through various stages in which they gain specific granules and their nucleus shuts down, distorting into a horseshoe shape. The cells then mature and enter the bloodstream. Some neutrophils enter the circulation as immature band cells.
6. Monocytes and lymphocytes look somewhat like stem cells. Structural changes are minimal in the developmental pathways of these cells.
7. In the platelet line, immature megakaryoblasts become giant megakaryocytes with multilobed nuclei. The cytoplasm of megakaryocytes breaks up to form platelets in the blood.

DISORDERS OF THE BLOOD (pp. 515–516)

1. Polycythemia is an excess of erythrocytes in the blood, and anemia is any condition in which the blood's capacity for carrying oxygen is diminished. Sickle cell disease is an inherited condition in which the erythrocytes distort into a sickle shape and block the capillaries. Leukemia is cancer of leukocyte-forming cells in bone marrow, and thrombocytopenia is abnormally low concentration of platelets in the blood.

THE BLOOD THROUGHOUT LIFE (pp. 516–517)

1. The blood stem cells develop in blood islands on the yolk sac in the 3-week-old embryo and then travel to the hematopoietic organs.
2. The hematopoietic organs in the fetus are the liver, spleen, and bone marrow.
3. Blood disorders that may develop with age include increased incidences of leukemia, anemia, and clotting disorders.

REVIEW QUESTIONS

Multiple Choice/Matching Questions

1. Which of the following descriptions of erythrocytes is *false?* (a) are shaped like biconcave discs, (b) have a life span of about 120 days, (c) contain hemoglobin, (d) have lobed nuclei.

2. Rank the following leukocytes in order of their relative abundance in the blood of a normal, healthy person, from 1 (most abundant) to 5 (least abundant).

____ **(a)** lymphocytes

____ **(b)** basophils

____ **(c)** neutrophils

____ **(d)** eosinophils

____ **(e)** monocytes

3. The white blood cell that releases histamine and other mediators of inflammation is the (a) basophil, (b) neutrophil, (c) monocyte, (d) eosinophil.

4. Which of the following blood cells are phagocytic? (a) lymphocytes, (b) erythrocytes, (c) neutrophils, (d) lymphocytes and neutrophils.

5. The blood cell that can attack a specific antigen is (a) a lymphocyte, (b) a monocyte, (c) a neutrophil, (d) a basophil, (e) an erythrocyte.

6. Match the descriptions of the blood cells at left with their names at right. Some names will be used more than once.

Column A	Column B
____ **(1)** destroys parasites	**(a)** erythrocyte
____ **(2)** has two types of granules	**(b)** neutrophil
	(c) eosinophil
____ **(3)** does not use diapedesis	**(d)** basophil
	(e) lymphocyte
____ **(4)** stops the allergic response	**(f)** monocyte
____ **(5)** the only cell that is not spherical	
____ **(6)** granulocyte that phagocytizes bacteria	
____ **(7)** the largest blood cell	
____ **(8)** granulocyte with the smallest granules	
____ **(9)** a young one is a reticulocyte	
____ **(10)** lives for about 4 months	
____ **(11)** most abundant blood cell	
____ **(12)** includes T cells and B cells	

7. In typically stained monocytes, the blue cytoplasm and the purple nuclei are colored by this dye: (a) eosin, (b) brilliant cresyl blue, (c) basin (as in basophilic), (d) methylene blue, (e) none of the above—you need two dyes for two colors.

Short Answer Essay Questions

8. Which class or classes of formed elements form the buffy coat in a hematocrit tube?

9. (a) What is the basic functional difference between red bone marrow and yellow bone marrow? (b) Where is yellow marrow located?

10. (a) Describe the steps of erythrocyte formation in the bone marrow. (b) What name is given to the immature type of red blood cell that is newly released into the circulation?

11. Describe the structure of platelets, and explain their functions.

12. What is the difference between a blood stem cell and a committed cell in the bone marrow?

13. What is the relationship between megakaryocytes and platelets?

14. Looking at a blood smear under the microscope, Tina became confused by the unusual nuclei of granulocytes. She kept asking why each eosinophil had two nuclei and why neutrophils had four or five nuclei. How would you answer her?

15. Compare and contrast each of the following pairs of terms: (a) circulatory system and cardiovascular system, (b) complete blood count and differential WBC count.

16. Emilio argued that a polymorphonuclear cell is a neutrophil, but Moshe insisted that a polymorphonuclear cell is any granulocyte. Who is correct?

17. On her test, Floris wrote that the function of platelets was "clotting." The instructor, however, demanded a more complete answer. Indicate what the instructor had in mind.

18. When examining leukocytes in a blood count, is it possible to tell B lymphocytes from T lymphocytes? Explain your answer.

19. Compare an eosinophil to a basophil, both in structure and in function.

20. What are the advantages of placental-blood transplants over bone-marrow transplants?

CRITICAL REASONING + CLINICAL APPLICATION QUESTIONS

1. After young Janie was diagnosed as having acute lymphocytic leukemia, her parents could not understand why infection was a major problem for Janie when her WBC count was so high. Provide an explanation for Janie's parents.
2. Freddy could tell that a neuron cell body on his slide was as wide as ten erythrocytes. How wide was the neuron cell body (in micrometers)?
3. The Jones family, who raise sheep on their ranch, let their sheep dog Rooter lick their faces, and sometimes kiss him on the mouth. The same day that the veterinarian diagnosed Rooter as having tapeworms, a blood test indicated that both the family's daughters had blood eosinophil levels of over 3000 per cubic millimeter. What is the connection?
4. A reticulocyte count indicated that 5% of Tyler's red blood cells were reticulocytes. His blood test also indicated he had polycythemia and a hematocrit of 65%. Explain the connections among these three facts.
5. Cancer patients being treated with chemotherapy drugs, which are designed to destroy rapidly mitotic cells, are monitored closely for changes in their RBC and WBC counts. Why?

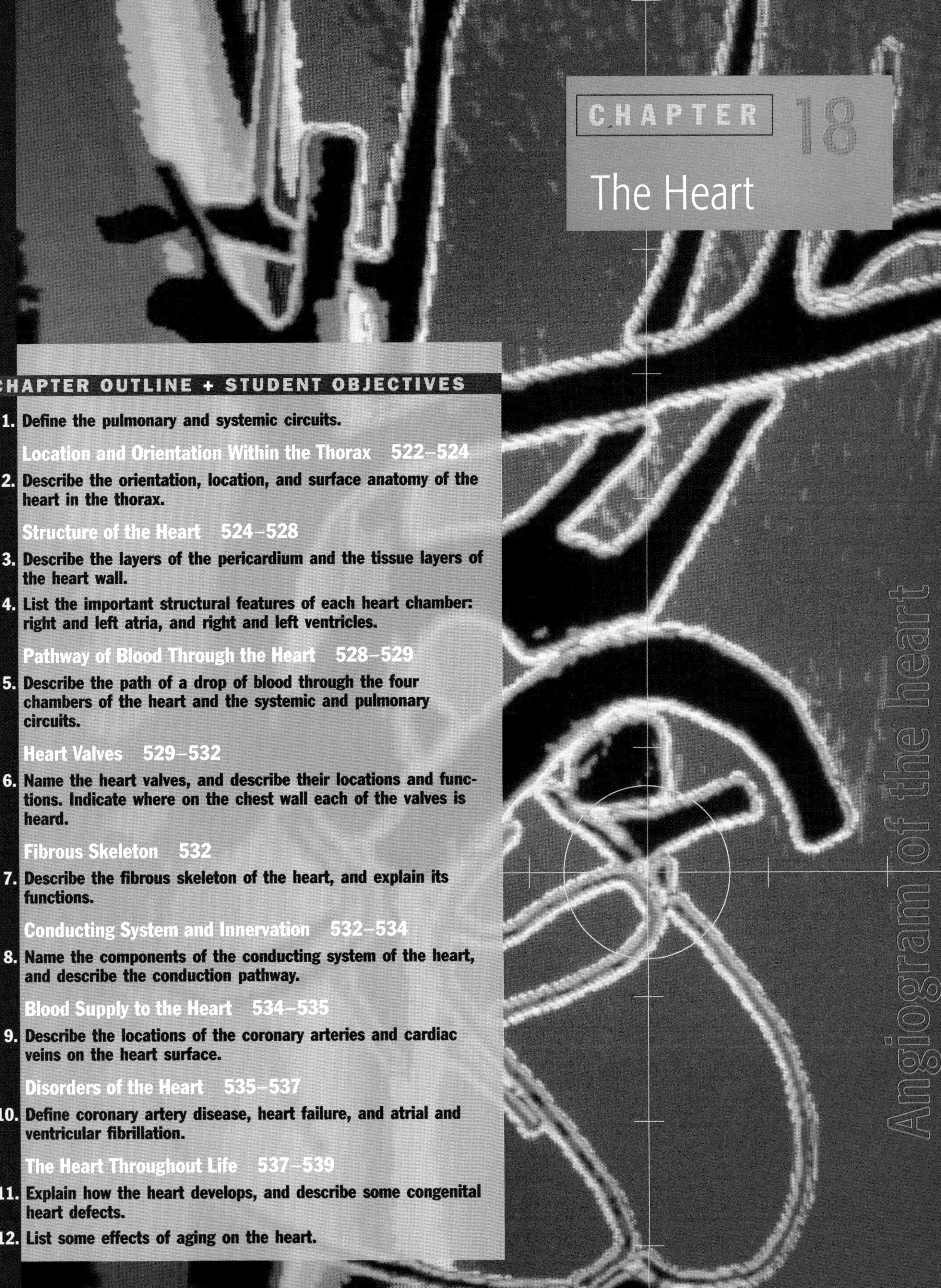

CHAPTER 18

The Heart

CHAPTER OUTLINE + STUDENT OBJECTIVES

1. Define the pulmonary and systemic circuits.

Location and Orientation Within the Thorax 522–524

2. Describe the orientation, location, and surface anatomy of the heart in the thorax.

Structure of the Heart 524–528

3. Describe the layers of the pericardium and the tissue layers of the heart wall.
4. List the important structural features of each heart chamber: right and left atria, and right and left ventricles.

Pathway of Blood Through the Heart 528–529

5. Describe the path of a drop of blood through the four chambers of the heart and the systemic and pulmonary circuits.

Heart Valves 529–532

6. Name the heart valves, and describe their locations and functions. Indicate where on the chest wall each of the valves is heard.

Fibrous Skeleton 532

7. Describe the fibrous skeleton of the heart, and explain its functions.

Conducting System and Innervation 532–534

8. Name the components of the conducting system of the heart, and describe the conduction pathway.

Blood Supply to the Heart 534–535

9. Describe the locations of the coronary arteries and cardiac veins on the heart surface.

Disorders of the Heart 535–537

10. Define coronary artery disease, heart failure, and atrial and ventricular fibrillation.

The Heart Throughout Life 537–539

11. Explain how the heart develops, and describe some congenital heart defects.
12. List some effects of aging on the heart.

THE CEASELESSLY BEATING HEART IN THE THORAX has intrigued people for thousands of years. The ancient Greeks believed the heart to be the seat of intelligence. Others thought it was the source of emotions. Although these theories have proved false, we do know that emotions affect heart rate. Only when your heart pounds or occasionally skips a beat do you become acutely aware of how much you depend on this dynamic organ for your very life.

The heart is a muscular double pump with two functions: (1) Its right side receives oxygen-poor blood from the body tissues and then pumps this blood to the lungs to pick up oxygen and dispel carbon dioxide, and (2) its left side receives the oxygenated blood returning from the lungs and pumps this blood throughout the body to supply oxygen and nutrients to the body tissues (Figure 18.1). The blood vessels that carry blood to and from the lungs form the **pulmonary circuit** (*pulmonos* = lung), whereas the vessels that transport blood to and from all body tissues form the **systemic circuit.** The heart has two receiving chambers called **atria** (*atrium* = entranceway) that receive blood returning from the pulmonary and systemic circuits. The heart also has two main pumping chambers called **ventricles** ("hollow bellies") that pump blood around the two circuits. Next we look at the heart as it is situated within the thorax.

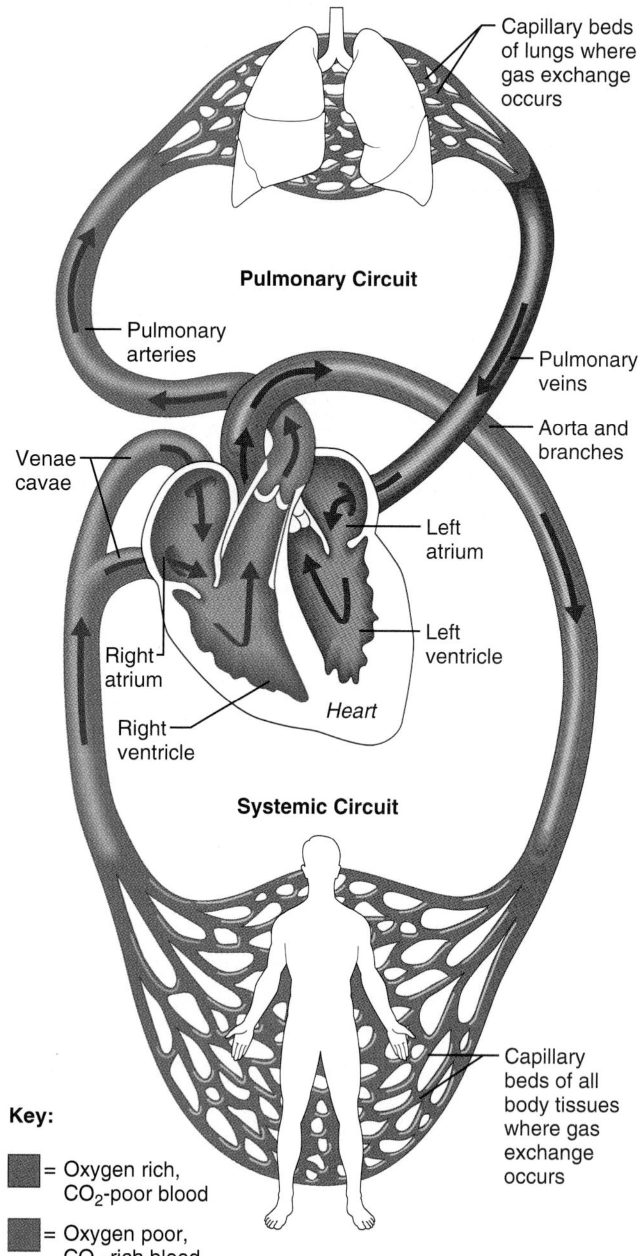

FIGURE 18.1 The route by which the heart (center) pumps blood around the pulmonary and systemic circuits. The right side of the heart is the pulmonary pump that propels oxygen-poor blood to the lungs to pick up oxygen. The left side of the heart is the systemic pump that pumps oxygen-rich blood to tissues throughout the body.

LOCATION AND ORIENTATION WITHIN THE THORAX

The heart's modest size belies its incredible strength and durability. Only about the size of a fist, this hollow, cone-shaped organ looks enough like the popular valentine image to satisfy the sentimentalists among us. Typically it weighs between 250 and 350 grams—less than a pound.

The heart lies in the thorax posterior to the sternum and costal cartilages and rests on the superior surface of the diaphragm (Figure 18.2a). It is the largest organ in the mediastinum, which is the thick partition between the two lungs (and pleural cavities) (Figure 18.2b and c). The heart assumes an oblique position in the thorax, with its pointed **apex** lying to the left of the midline and anterior to the rest of the heart (Figure 18.2c). If you press your fingers between the fifth and sixth ribs just inferior to the left nipple, you may feel the beating of your heart where the apex contacts the thoracic wall. Cone-shaped objects have a base as well as an apex, and the heart's **base** is its broad posterior surface.

The heart is said to have four corners defined by four points projected onto the anterior thoracic wall, and you may sketch these points on Figure 18.2a. The *superior right* point lies where the costal cartilage of the third rib joins the sternum. The *superior left* point lies at the costal cartilage of the second rib, a finger's breadth lateral to the sternum. The *inferior right* point lies at the costal cartilage of the sixth rib, a finger's breadth lateral to the sternum. Finally, the *inferior left* point (the apex point) lies in the fifth intercostal space at the midclavicular line—that is, at a line extending inferiorly from the midpoint of the left clavicle. The imaginary lines that connect these four corner points delineate the normal size and location of the heart. Clinicians must know these normal landmarks, because an enlarged or displaced heart can indicate heart disease or other disease conditions.

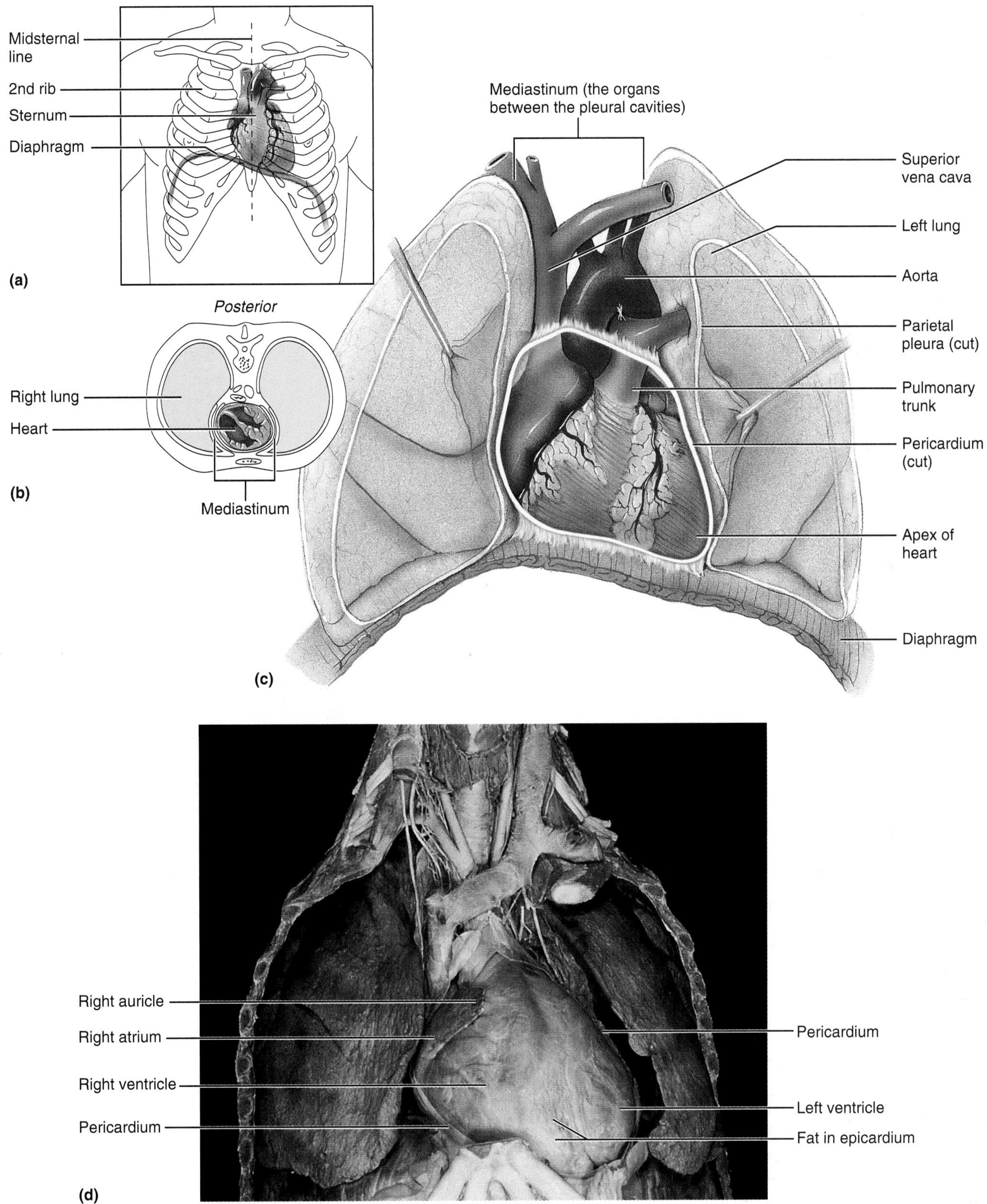

FIGURE 18.2 Location of the heart in the thorax. (a) Relation of the heart to the sternum and ribs in a person who is lying down. (In a standing person, the heart lies slightly inferior to this position.) (b) Cross section through the mid thorax showing the heart in the mediastinum between the lungs. (c) Relation of the heart and great vessels to the lungs. (d) Photograph of the heart in the mediastinum. Note the fat in the outer layer of the heart (epicardium).

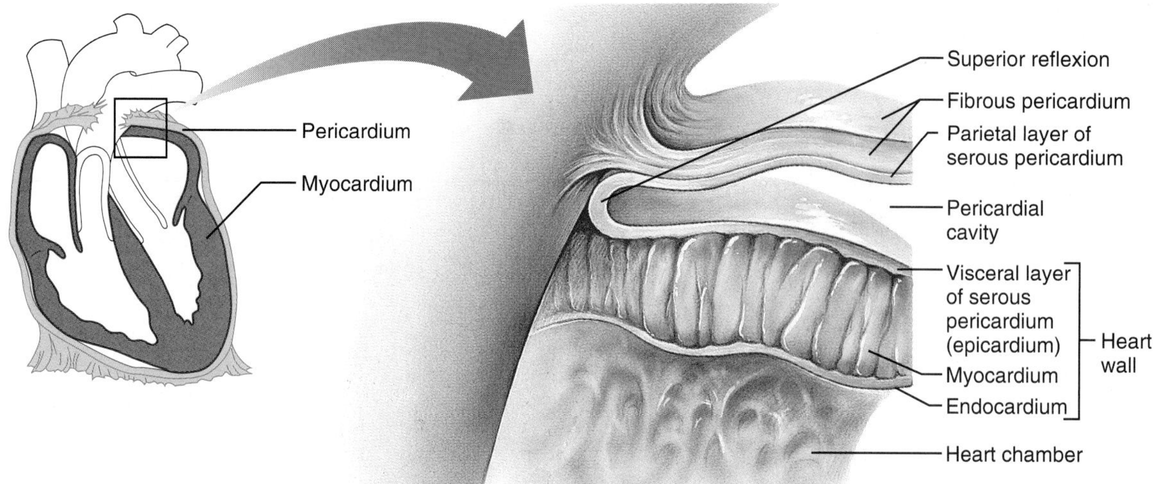

FIGURE 18.3 The layers constituting the pericardium and the heart wall.

Critical Reasoning

Assume that the four corner points of a patient's heart, as revealed on an X-ray film, relate to the rib cage as follows: second right rib at sternum, tip of xiphoid process, seventh intercostal space lateral to the left nipple, and second left rib at sternum. Are the size and position of this patient's heart normal?

By sketching and interconnecting these four points on Figure 18.2a, you can see that this patient's heart is enlarged and perhaps displaced slightly to the left.

STRUCTURE OF THE HEART

Our discussion of heart anatomy begins with superficial structures and proceeds inward. In turn we will consider the heart's covering, its wall, and its chambers.

Coverings

The **pericardium** (per″ĭ-kar′de-um; "around the heart") is a triple-layered sac that encloses the heart (Figure 18.3). The outer layer of this sac, called the **fibrous pericardium,** is a strong layer of dense connective tissue. It adheres to the diaphragm inferiorly, and superiorly it is fused to the roots of the great vessels that leave and enter the heart. The fibrous pericardium acts as a tough outer coat that holds the heart in place and keeps it from overfilling with blood.

Deep to the fibrous pericardium is the double-layered **serous pericardium,** a closed sac sandwiched between the fibrous pericardium and the heart. The outer, **parietal layer of the serous pericardium** adheres to the inner surface of the fibrous pericardium. At a superior reflection, the parietal layer is continuous with the **visceral layer of the serous pericardium** or **epicardium,** which lies on the heart and is considered a part of the heart wall (discussed shortly). Between the parietal and visceral layers of serous pericardium is a slit-like space, called the **pericardial cavity,** which contains a lubricating film of serous fluid that reduces friction between the beating heart and the outer wall of the pericardial sac.

CLINICAL APPLICATIONS

PERICARDITIS AND CARDIAC TAMPONADE Infection and inflammation of the pericardium, or **pericarditis,** can lead to a roughening of the serous lining of the pericardial cavity. As a consequence, the beating heart produces a creaking sound called pericardial friction rub, which can be heard with a stethoscope. Pericarditis is characterized by pain behind the sternum. Over time, it can lead to adhesions of the heart to the outer pericardial wall, or the pericardium can scar and thicken, then contract and inhibit the heart's movements.

In severe, acute cases of pericarditis, large amounts of fluid resulting from the inflammatory response exude into the pericardial cavity. In this condition, called **cardiac tamponade** (tam″po-nād′; "a heart plug"), the excess fluid compresses the heart, limiting the expansion of the heart between beats and diminishing its ability to pump blood. Physicians treat cardiac tamponade by inserting a hypodermic needle into the pericardial cavity to drain the excess fluid. Cardiac tamponade also results if ***blood*** accumulates inside the pericardial cavity, as occurs when a penetrating wound to the heart (such as a stab wound) causes blood to leak out of the heart and into the pericardial cavity. ✚

Layers of the Heart Wall

The wall of the heart has three layers: a superficial *epicardium,* a middle *myocardium,* and a deep *endocardium* (Figure 18.3). All three layers are richly supplied with blood vessels.

The **epicardium** ("upon the heart") is the visceral layer of the serous pericardium, as previously explained. This serous membrane is often infiltrated with fat, especially in older people (see Figure 18.2d). The **myocardium** ("muscle heart"), which forms the bulk of the heart, consists mainly of cardiac muscle and is the layer that actually contracts. The myocardium's elongated, circularly and spirally arranged networks of cardiac muscle cells, called *bundles* (Figure 18.4), squeeze blood through the heart in the proper directions: inferiorly through the atria and superiorly through the ventricles. The **endocardium** ("inside the heart") is a sheet of endothelium resting on a thin layer of connective tissue. Located deep to the myocardium, it lines the heart chambers and makes up the heart valves (see p. 529).

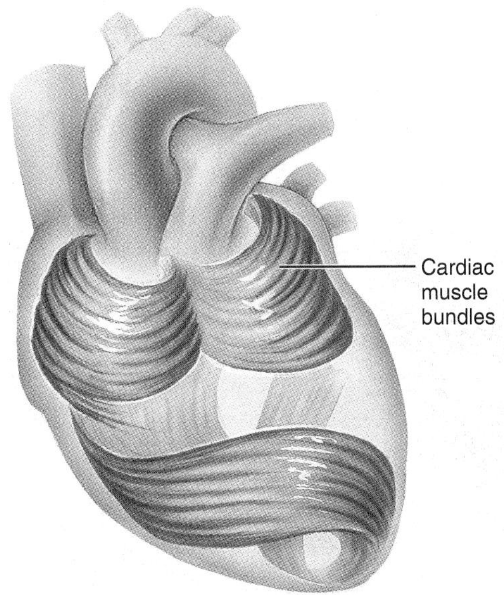

FIGURE 18.4 The circular and spiral arrangement of cardiac muscle bundles in the myocardium of the heart.

Heart Chambers

The four heart chambers are the *right* and *left atria* superiorly, and the *right* and *left ventricles* inferiorly (Figures 18.5 and 18.6). Internally, the heart chambers are divided longitudinally by a partition, called either the **interatrial septum** (see Figure 18.5b) or the **interventricular septum** (see Figure 18.5c), depending on which chambers it separates. Externally, the boundaries of the four chambers are marked by two grooves. The first, the **coronary sulcus** (see Figure 18.5a), forms a "crown" by circling the boundary between the atria and the ventricles (*corona* = crown). The second groove consists of (1) the **anterior interventricular sulcus** (see Figure 18.5a), which marks the anterior position of the interventricular septum between the two ventricles, and (2) the **posterior interventricular sulcus** (see Figure 18.5d), which separates the two ventricles on the heart's inferior surface.

Right Atrium

The **right atrium,** which forms the entire right border of the human heart, is the receiving chamber for oxygen-poor blood returning from the systemic circuit. The right atrium receives blood via three veins: the *superior vena cava* and *inferior vena cava* (see Figure 18.5b) and the *coronary sinus* (see Figure 18.5b and d).

Externally, the right atrium exhibits no outstanding feature except the **right auricle,** a small flap shaped like a dog's ear (*auricle* = little ear) that projects to the left from the superior corner of the atrium (see Figure 18.2d). Internally, the right atrium has two parts (see Figure 18.5b): a smooth-walled posterior part and an anterior part lined by horizontal ridges called the **pectinate muscles** (*pectinate* = like the teeth of a comb). These two parts of the atrium are separated by a large, C-shaped ridge called the **crista terminalis** ("terminal crest"). The crista is an important landmark in locating the sites where veins enter the right atrium: The superior vena cava opens into the atrium just posterior to the superior bend of the crista; the inferior vena cava opens into the atrium just posterior to the inferior bend of the crista; and the coronary sinus opens into the atrium just anterior to the inferior end of the crista. Additionally, just posterior to this end of the crista is the **fossa ovalis,** a depression in the interatrial septum that marks the spot where an opening existed in the fetal heart (the *foramen ovale;* see p. 538).

Inferiorly and anteriorly, the right atrium opens into the right ventricle through the *tricuspid valve (right atrioventricular valve)* (see Figure 18.5b and c).

Right Ventricle

The **right ventricle** receives blood from the right atrium and pumps it into the pulmonary circuit via an artery called the **pulmonary trunk** (see Figure 18.5c). Externally, the right ventricle forms most of the anterior surface of the heart. Internally, its walls are marked by irregular ridges of muscle, called **trabeculae carneae** (trah-bek′u-le kar′ne-e; "crossbars of flesh"). Additionally, cone-shaped **papillary muscles** project from the walls into the ventricular cavity (*papilla* = nipple).

Thin, strong bands called **chordae tendineae** (kor′de ten′dĭ-ne-e; "tendinous cords") project superiorly from the papillary muscles to the flaps (cusps) of the tricuspid (right atrioventricular) valve. The popular expression "tugging on my heartstrings" refers to these bands. Superiorly, the opening between the right ventricle and the pulmonary trunk contains the *pulmonary semilunar valve* (or simply, *pulmonary valve*).

Left Atrium

The **left atrium,** which makes up most of the heart's posterior surface, or base, receives oxygen-rich blood returning from the lungs through two right and two left **pulmonary veins** (see Figures 18.5d). The only part of

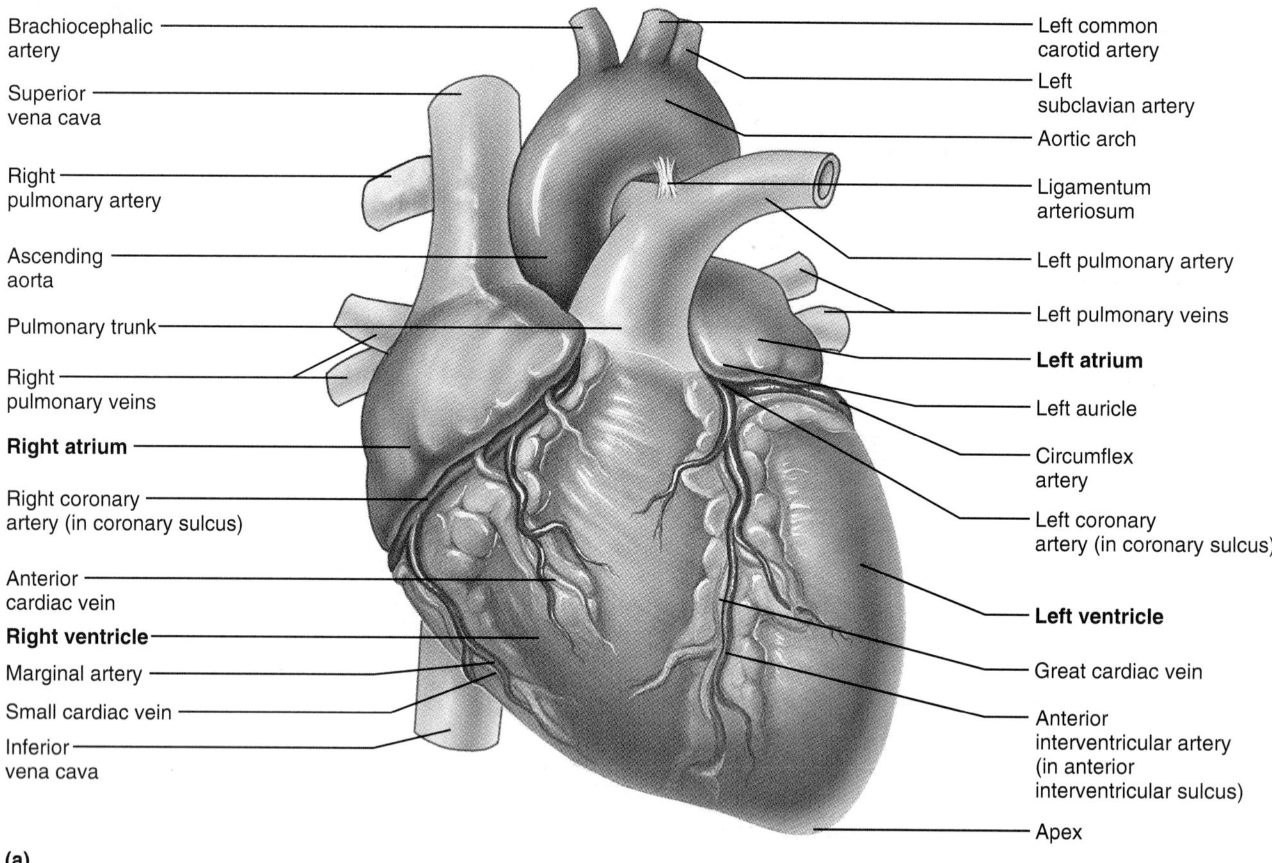

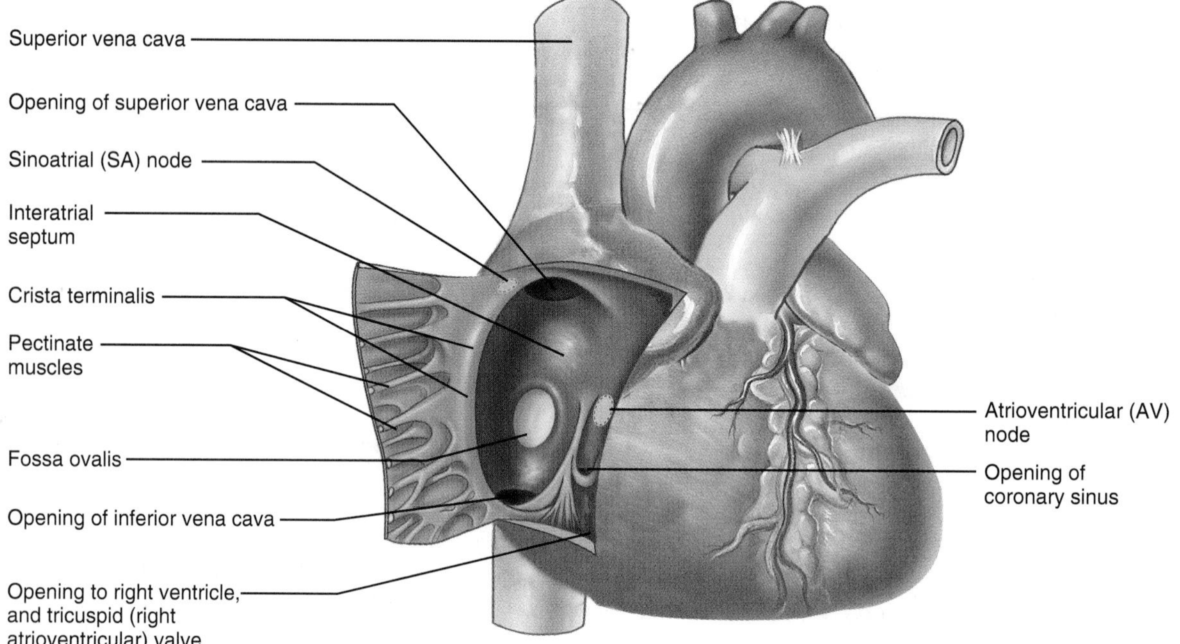

FIGURE 18.5 Gross anatomy of the heart. (a) Anterior view. (b) Anterior view emphasizing the right atrium, which has been opened; the anterior wall of the atrium has been reflected to the side. (c) Frontal section showing the interior chambers and valves. (d) Inferior view; the surface shown rests on the diaphragm (dorsal aspect is at the top of this figure).

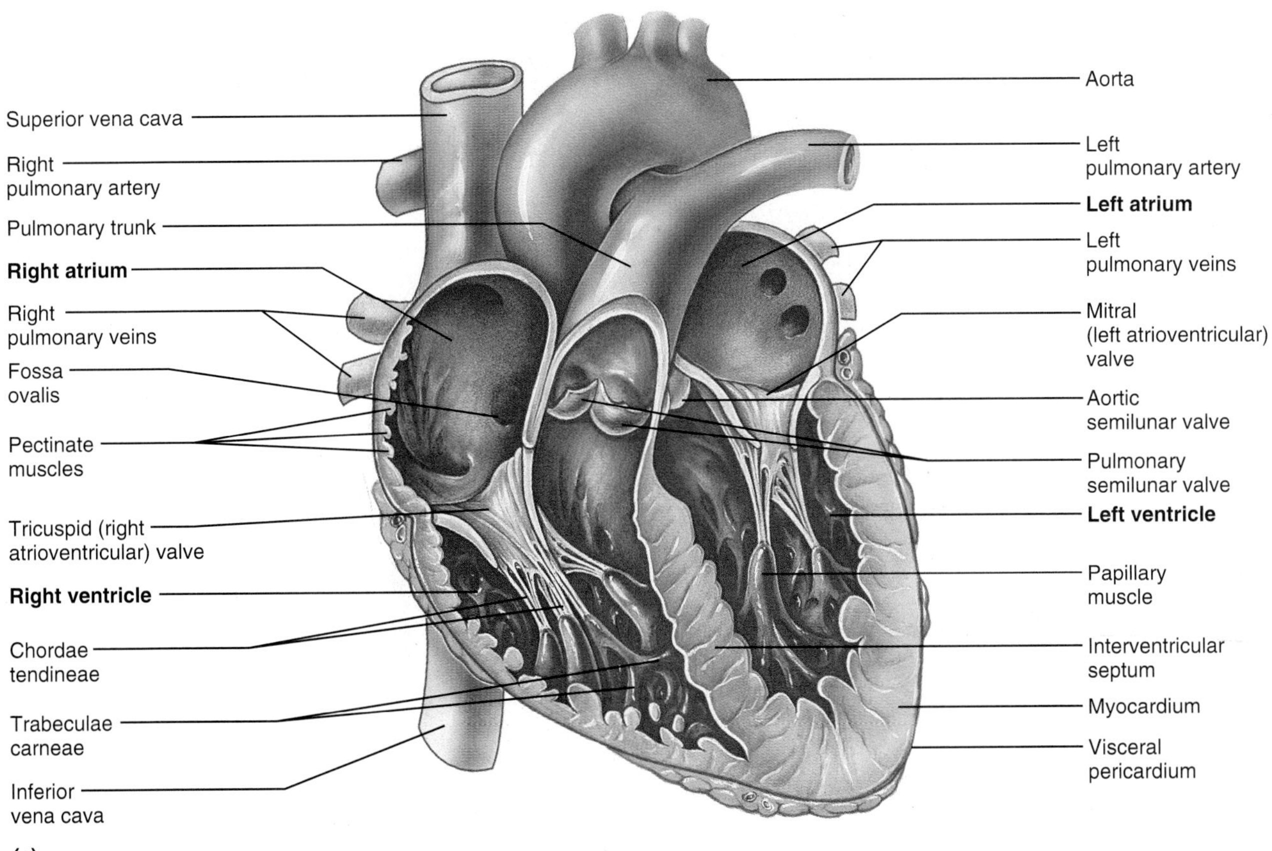
Superior vena cava
Right pulmonary artery
Pulmonary trunk
Right atrium
Right pulmonary veins
Fossa ovalis
Pectinate muscles
Tricuspid (right atrioventricular) valve
Right ventricle
Chordae tendineae
Trabeculae carneae
Inferior vena cava
Aorta
Left pulmonary artery
Left atrium
Left pulmonary veins
Mitral (left atrioventricular) valve
Aortic semilunar valve
Pulmonary semilunar valve
Left ventricle
Papillary muscle
Interventricular septum
Myocardium
Visceral pericardium

(c)

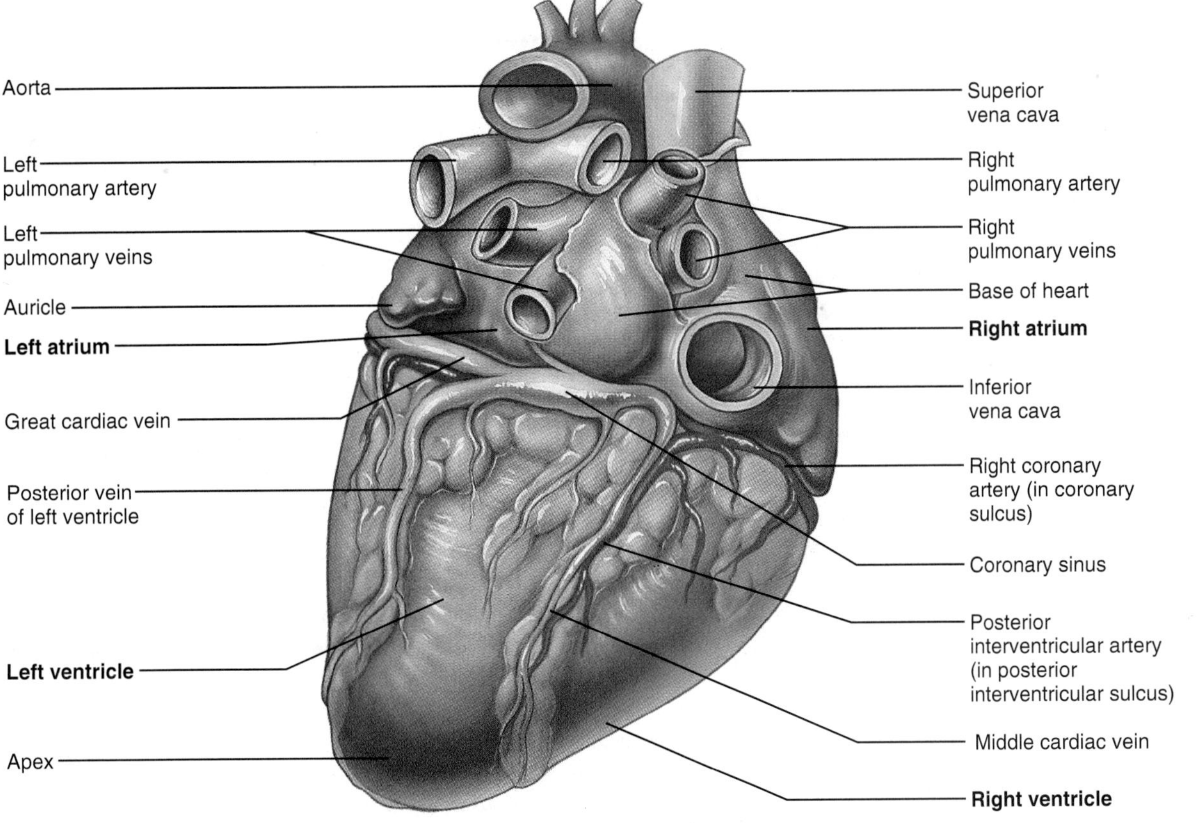
Aorta
Left pulmonary artery
Left pulmonary veins
Auricle
Left atrium
Great cardiac vein
Posterior vein of left ventricle
Left ventricle
Apex
Superior vena cava
Right pulmonary artery
Right pulmonary veins
Base of heart
Right atrium
Inferior vena cava
Right coronary artery (in coronary sulcus)
Coronary sinus
Posterior interventricular artery (in posterior interventricular sulcus)
Middle cardiac vein
Right ventricle

(d)

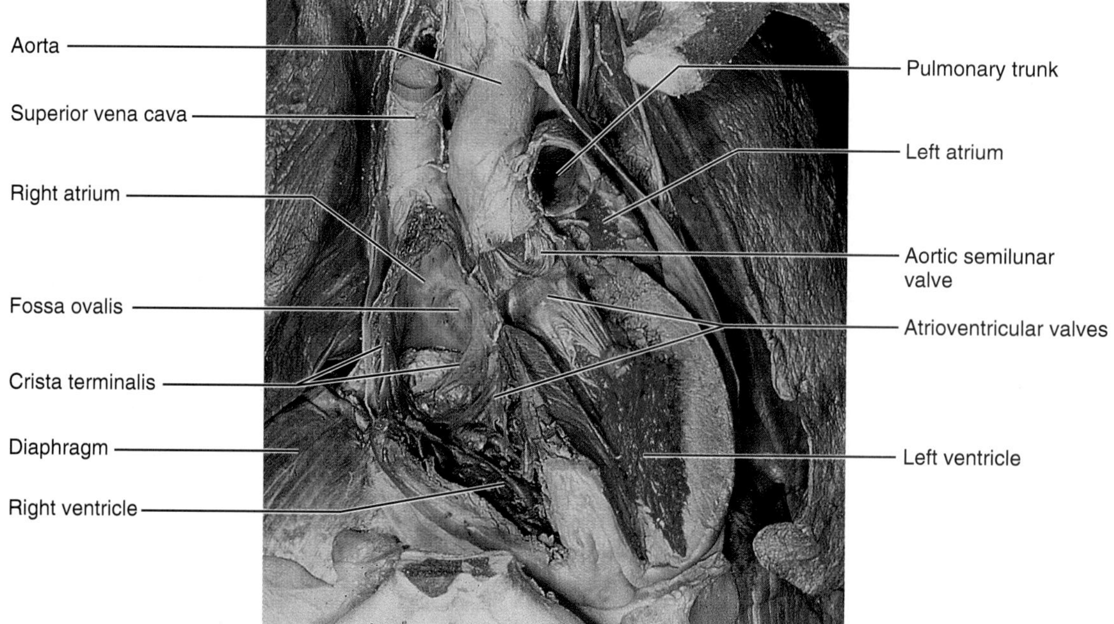

FIGURE 18.6 Photograph of a heart from a cadaver, sectioned in the frontal plane. This view is comparable to that in Figure 18.5c, except that the base of the aorta and much of the pulmonary trunk have been removed.

the left atrium visible anteriorly is its triangular left auricle (see Figure 18.5a). Internally, most of the atrial wall is smooth, with pectinate muscles lining the auricle only. The left atrium opens into the left ventricle through the *mitral valve (left atrioventricular valve)* (see Figure 18.5c).

Left Ventricle

The **left ventricle,** which forms the apex of the heart and dominates the heart's inferior surface (see Figure 18.5c and d), pumps blood into the systemic circuit. Like the right ventricle, it contains trabeculae carneae, papillary muscles, chordae tendineae, and the cusps of an atrioventricular (mitral) valve. Superiorly, the left ventricle opens into the stem artery of the systemic circulation (the aorta) through the *aortic semilunar valve* (or simply, *aortic valve*).

PATHWAY OF BLOOD THROUGH THE HEART

Here we summarize the route of blood through the four chambers of the heart by following the path of a single drop of blood around the pulmonary and systemic circuits (see Figures 18.1 and 18.5). We begin with oxygen-poor systemic blood as it arrives at the right side of the heart. Blood coming from body regions superior to the diaphragm (excluding the heart wall) enters the right atrium via the superior vena cava; blood returning from body regions inferior to the diaphragm enters via the inferior vena cava; and blood draining from the heart wall itself is collected by and enters through the coronary sinus. The blood passes from the right atrium through the tricuspid valve to the right ventricle, propelled by gravity and the contraction of the right atrium. Then, the right ventricle contracts, propelling the blood through the pulmonary semilunar valve into the pulmonary trunk and around the pulmonary circuit for oxygenation. The freshly oxygenated blood returns via the four pulmonary veins to the left atrium and passes through the mitral valve to the left ventricle, propelled by gravity and the contraction of the left atrium. The left ventricle then contracts and propels the blood through the aortic semilunar valve into the aorta and its branches. After delivering oxygen and nutrients to the body tissues through the systemic capillaries, the oxygen-poor blood returns through the systemic veins to the right atrium—and the whole cycle repeats itself continually.

The fact that a given drop of blood passes through the heart chambers sequentially (one after another), does not mean that the four chambers contract in that order. Rather, the two atria always contract *together,* followed by the simultaneous contraction of the two ventricles.

A single sequence of atrial contraction followed by ventricular contraction is called a **heartbeat,** and the heart of an average person at rest beats 70–80 times per minute. The term that describes the contraction of a heart chamber is **systole** (sis′to-le; "contraction"); the time during which a heart chamber is relaxing and filling with blood is termed **diastole** (di-as′to-le; "expansion"). Thus, both atria and ventricles experience systole and diastole. Beware, however, that "systole" and "diastole" can also be used in a different way, to refer to the stages of the heartbeat when the *ventricles*—the dominant heart chambers—are contracting and filling, respectively.

Now that we have considered how the heart pumps blood, we can readily explain why the muscular walls of its different chambers differ in thickness. The walls of the atria are much thinner than those of the ventricles (see Figure 18.5c) because the atria need to exert little effort to propel blood inferiorly into the ventricles (much of ventricular filling is done by gravity). Furthermore, the wall of the left ventricle (the systemic pump) is at least three times as thick as that of the right ventricle (the pulmonary pump) (Figure 18.7). Consequently, the left ventricle can generate much more force than the right and is a stronger pump, reflecting the fact that the systemic circuit is much longer than the pulmonary circuit and offers greater resistance to blood flow. The thick wall of the left ventricle gives this chamber a circular shape and flattens the cavity of the adjacent right ventricle into the shape of a crescent (see Figure 18.7).

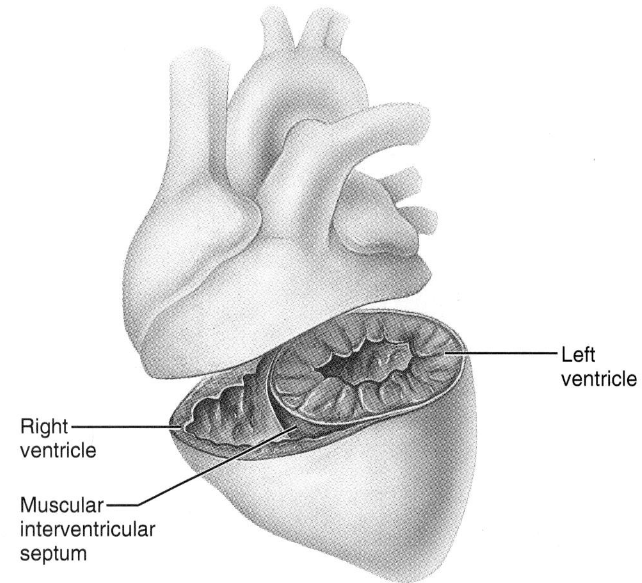

FIGURE 18.7 Anatomical differences between the right and left ventricles. The thicker, circular wall of the left ventricle encroaches on the cavity of the thinner-walled right ventricle, making the cavity of the right ventricle crescent-shaped.

HEART VALVES

The heart valves—the paired atrioventricular (AV) and semilunar valves—enforce the one-way flow of blood through the heart, from the atria to the ventricles and into the great arteries that leave the superior part of the heart. Next we examine the structure and function of heart valves, and how they produce heart sounds.

Valve Structure

Each heart valve consists of two or three *cusps,* which are flaps of endocardium reinforced by cores of dense connective tissue (Figure 18.8). Located at the junctions of the atria and their respective ventricles are the atrioventricular valves: the **right atrioventricular (tricuspid) valve,** which has three cusps, and the **left atrioventricular valve,** which has only two cusps. The latter is also called the **mitral** (mi′tral) **valve** because its cusps resemble the two sides of a bishop's hat, or miter. Located at the junctions of the ventricles and the great arteries are the **aortic** and **pulmonary (semilunar) valves,** each of which has three pocket-like cusps shaped roughly like crescent moons (*semilunar* = half moon).

Valve Function

Heart valves are pushed open (to allow blood flow) and shut (to prevent the backflow of blood) in response to differences in blood pressure on their two sides. The two *atrioventricular valves* prevent the backflow of blood into the atria during contraction of the ventricles (Figure 18.9). When the ventricles are relaxed (in diastole), the cusps of the AV valves hang limply into the ventricular chambers while blood flows into the atria and down through the open AV valves into the ventricles. When the ventricles start to contract, the pressure within them rises and forces the blood superiorly against the valve cusps, pushing the edges of the cusps together and closing the AV valves. The chordae tendineae and papillary muscles that attach to these valves serve as guy wires that anchor the cusps in their closed position to prevent reflux of ventricular blood into the atria. That is, if the cusps were not anchored in this manner, they would be forced superiorly into the atria and would evert, in much the same way that an umbrella is blown inside out by a gusty wind.

Critical Reasoning

State the precise function of the papillary muscles.

The papillary muscles contract when the rest of the ventricle contracts (during ventricular systole), in the process pulling on the chordae tendineae and preventing the AV valves from everting. The papillary muscles do not *open these valves.*

The two *semilunar valves* prevent backflow from the great arteries into the ventricles (Figure 18.10). When the ventricles contract and raise the intraventricular pressure, the semilunar valves are forced open, and their cusps are flattened against the arterial walls as the blood rushes past them. When the ventricles relax, blood that tends to flow back toward the heart fills the cusps of the semilunar valves and forces them shut.

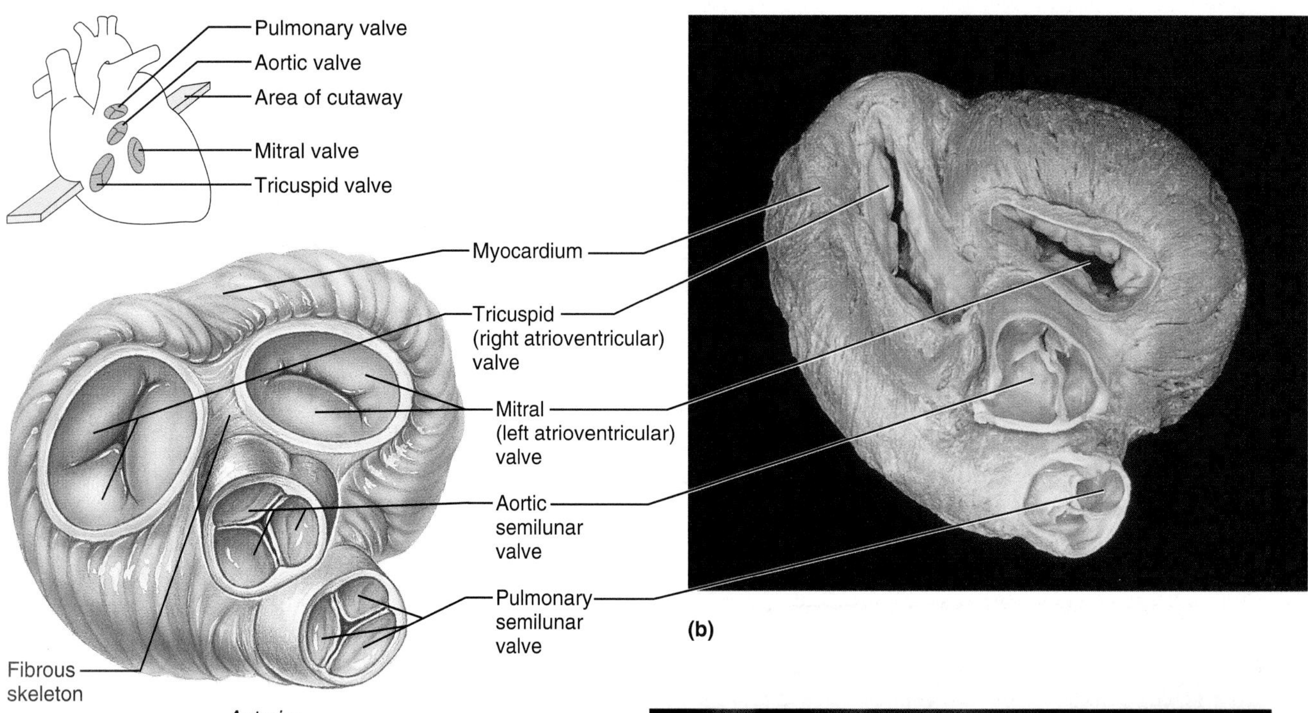

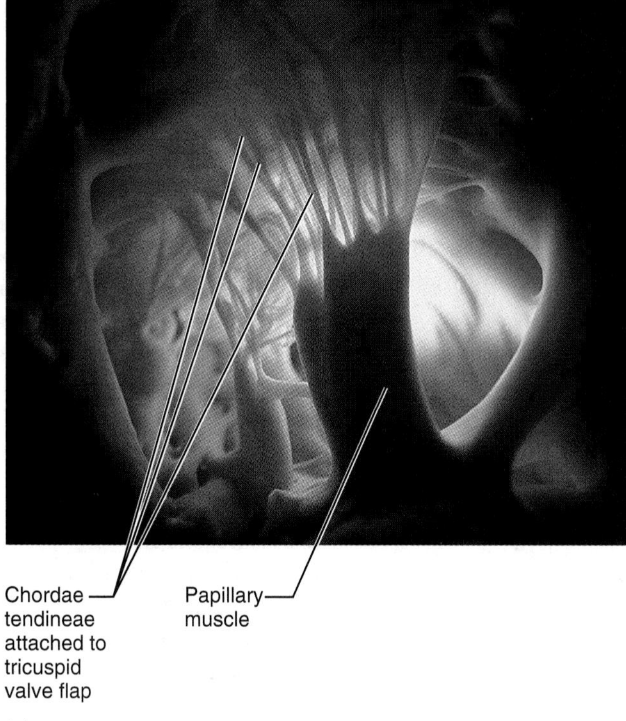

FIGURE 18.8 Structure and location of the heart valves. **(a)** Superior view of the four heart valves (atria removed). The inset shows the plane of section through the heart. **(b)** Photograph providing a view similar to that in part (a). **(c)** Photograph of the tricuspid valve. This inferior-to-superior view shows both the right ventricle (foreground) and the right atrium (background).

Heart Sounds

The closing of the valves causes vibrations in the adjacent blood and heart walls that account for the familiar "lub-dup" sounds of each heartbeat: The "lub" sound is produced by the closing of the AV valves at the start of ventricular systole, whereas the "dup" is produced by the closing semilunar valves at the end of ventricular systole. Because the mitral valve closes slightly before the tricuspid closes and the aortic valve generally closes just before the pulmonary valve closes, all four valve sounds are discernable by placing a stethoscope on the anterior chest wall.

Even though all four heart valves lie in roughly the same plane (see Figure 18.8a)—the plane of the coronary sulcus (see Figure 18.5a)—the clinician does not listen directly over the respective valves, because the sounds take oblique paths through the heart chambers to reach the chest wall. Conveniently, each valve is best heard near a different heart corner: the pulmonary valve near the superior left point, the aortic valve near the superior right point, the mitral valve at the apex point, and the tricuspid near the inferior right point

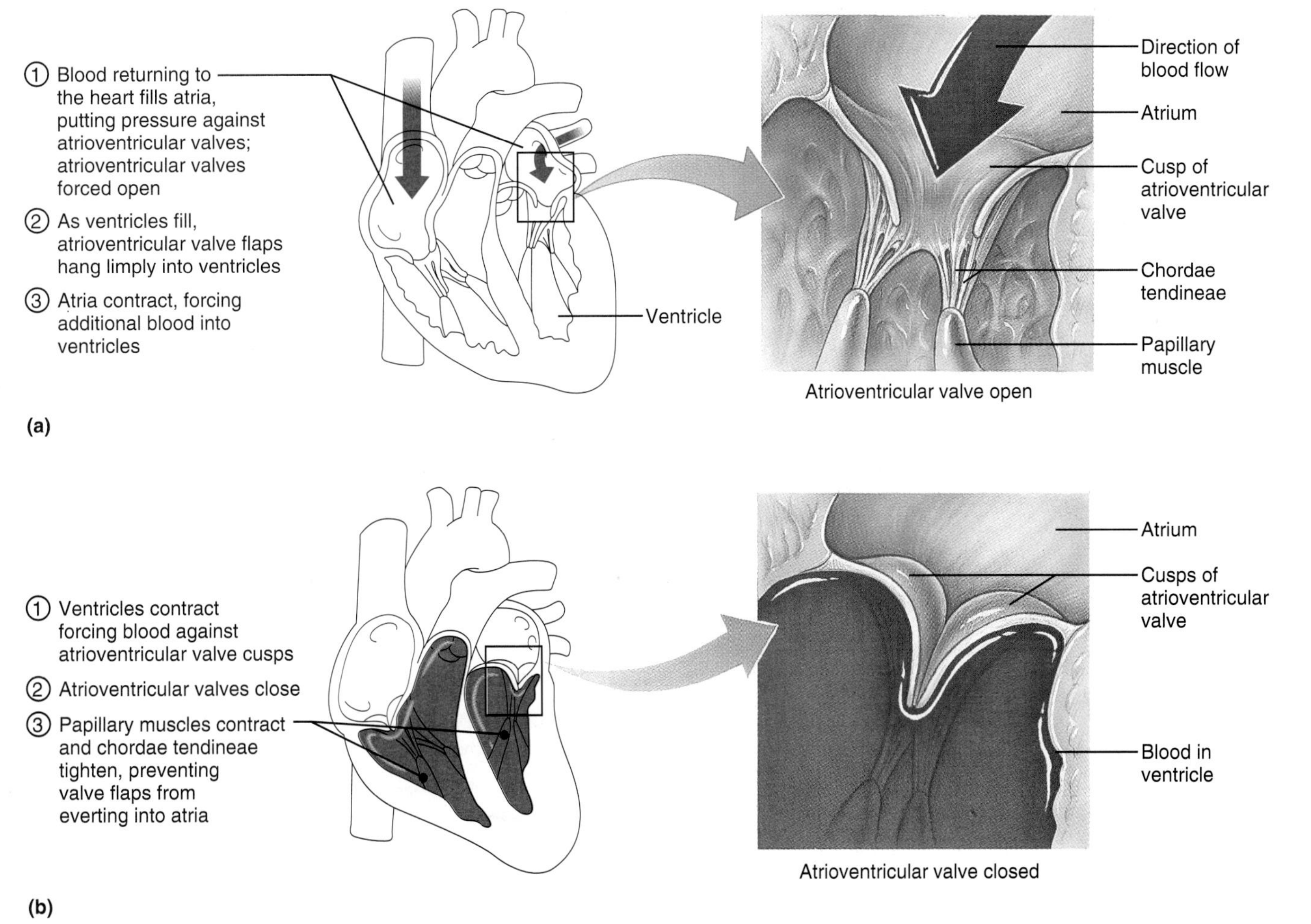

① Blood returning to the heart fills atria, putting pressure against atrioventricular valves; atrioventricular valves forced open

② As ventricles fill, atrioventricular valve flaps hang limply into ventricles

③ Atria contract, forcing additional blood into ventricles

Ventricle

Direction of blood flow

Atrium

Cusp of atrioventricular valve

Chordae tendineae

Papillary muscle

Atrioventricular valve open

(a)

① Ventricles contract forcing blood against atrioventricular valve cusps

② Atrioventricular valves close

③ Papillary muscles contract and chordae tendineae tighten, preventing valve flaps from everting into atria

Atrium

Cusps of atrioventricular valve

Blood in ventricle

Atrioventricular valve closed

(b)

FIGURE 18.9 Function of the atrioventricular valves. (a) When the ventricles are relaxed, the valves are forced open by the blood pressure exerted on their atrial side. **(b)** When the ventricles contract, forcing the contained blood superiorly, the valves are pushed shut.

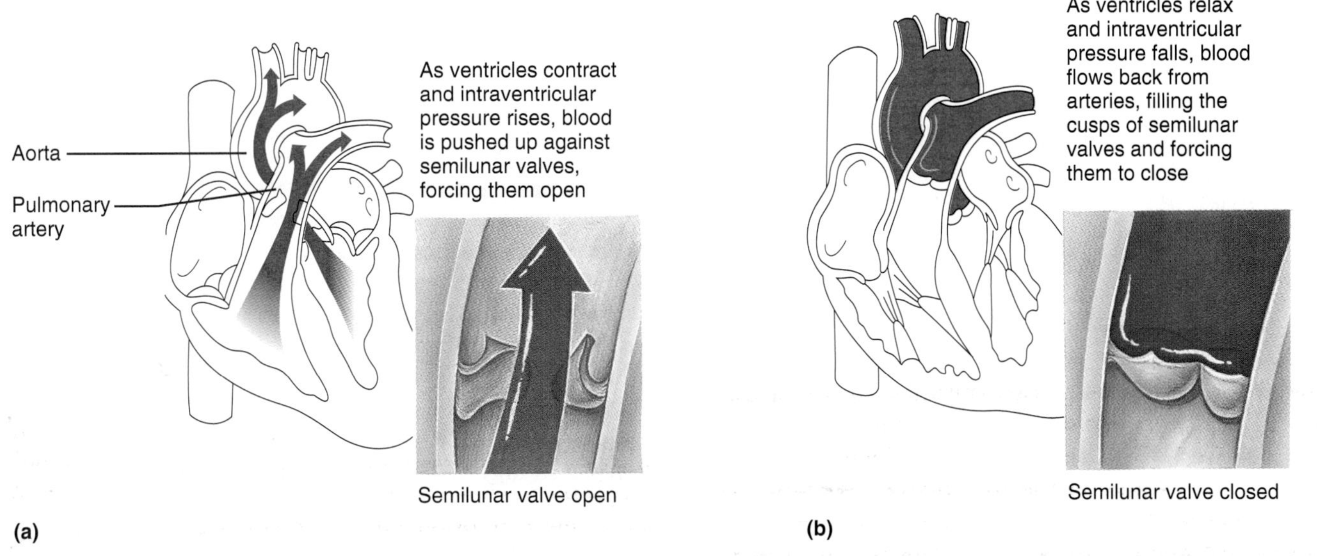

FIGURE 18.10 Function of the semilunar valves. (a) During ventricular contraction, the valves are pushed open, and their cusps are flattened against the artery walls. **(b)** When the ventricles relax, the backflowing blood closes the valves.

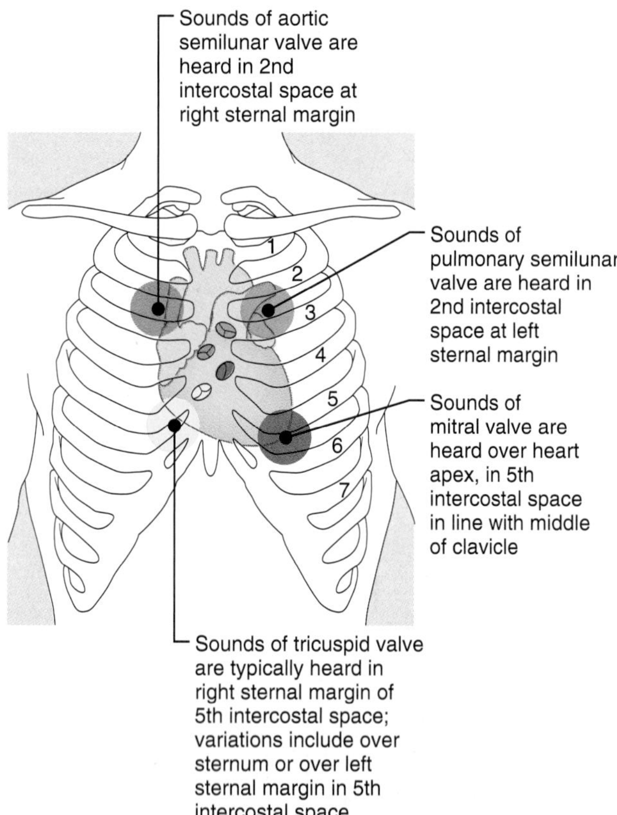

FIGURE 18.11 Points on the anterior thoracic surface where sounds of the heart valves are heard most clearly. These four points are almost identical to the four heart corners described on page 522, displaced only slightly to avoid sound-blocking ribs.

(Figure 18.11). Abnormal heart sounds often arise from disorders of the heart valves.

CLINICAL APPLICATIONS

VALVE DISORDERS Defects of the heart valves are common and have a number of causes, ranging from congenital and genetic abnormalities, to an inadequate blood supply to the valves (resulting from a heart attack), to bacterial infection of the endocardium. Valvular defects lead to valvular disorders, most of which are classified into one of two types. Valves that leak because they fail to close properly are considered **incompetent** (or are said to exhibit **insufficiency**). An incompetent valve produces a distinct blowing sound after the valve closes. By contrast, valves with narrowed openings, such as occur when cusps have fused or become stiffened by calcium deposits, are termed **stenotic.** Stiffened valves cannot open properly. Stenosis of the aortic valve produces a distinctive "click" sound during ventricular systole as blood passing through the constricted opening becomes turbulent and vibrates. Both incompetent and stenotic valves lessen the heart's efficiency and thus increase its work load; ultimately the heart may weaken severely.

The mitral valve, and the aortic valve to a lesser extent, are most often involved in valve disorders because they are subjected to the great forces resulting from contraction of the powerful left ventricle. In **mitral valve prolapse,** an inherited weakness of the collagen in the valve and chordae tendineae allows one or more cusps of this valve to "flop" into the left atrium during ventricular systole. This condition, which is the most common heart valve disorder (affecting 2.5% to 5% of the population), is characterized by a distinctive heart sound—a click followed by a swish (backflow of blood). Most cases are mild and harmless, but severe cases may lead to heart failure or a disruption of heart rhythm.

Treatment of valve disorders typically involves surgical repair of damaged valves, but when this approach is unsuccessful, replacement with either synthetic or pig valves is performed. Replacement techniques are neither permanent nor problem-free. +

FIBROUS SKELETON

The **fibrous skeleton** of the heart lies in the plane between the atria and the ventricles and surrounds all four heart valves rather like handcuffs (see Figure 18.8a). Composed of dense connective tissue, it has four functions:

1. It anchors the valve cusps.
2. It prevents overdilation of the valve openings as blood pulses through them.
3. It is the point of insertion for the bundles of cardiac muscle in the atria and ventricles (see Figure 18.4).
4. It blocks the direct spread of electrical impulses from the atria to the ventricles (discussed shortly).

CONDUCTING SYSTEM AND INNERVATION

We begin this section by examining the means by which heart muscle generates and conducts electrical impulses. Then we discuss the pathways through which this intrinsic production of impulses can be altered by extrinsic neural controls.

Conducting System

Cardiac muscle cells have an intrinsic ability to generate and conduct impulses that signal these same cells to contract rhythmically. These properties are intrinsic to the heart muscle itself and do not depend on extrinsic nerve impulses. Even if all nerve connections to the heart are severed, the heart continues to beat rhythmically, a fact amply demonstrated by transplanted hearts.

The **conducting system** of the heart is a series of specialized cardiac muscle cells that carries impulses throughout the heart musculature, signaling the heart chambers to contract in the proper sequence. It also initiates each contraction sequence, thereby setting basic heart rate. The components of the conducting system are the ***sinoatrial node, internodal fibers, atrioventricular node, atrioventricular bundle,*** right and left ***bundle branches,*** and ***Purkinje fibers*** (Figure 18.12).

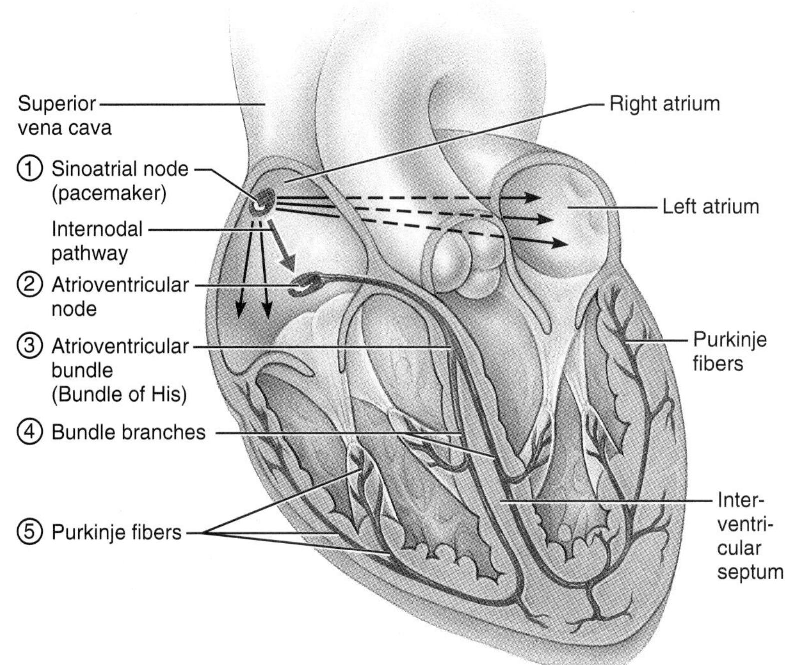

FIGURE 18.12 The conducting system of the heart. This system consists of specialized cardiac muscle cells, not nerves. The impulse that signals the heart to contract is initiated by the sinoatrial node and then passes successively through the atrial myocardium and the internodal pathway to the atrioventricular node, the atrioventricular bundle, the right and left bundle branches, the Purkinje fibers, and the ventricular myocardium.

The impulse that signals each heartbeat begins at the **sinoatrial (SA) node,** a crescent-shaped mass of muscle cells that lies in the wall of the right atrium, just inferior to the entrance of the superior vena cava. The SA node, which sets the basic heart rate by generating 70–80 impulses per minute, is the heart's pacemaker.

The sequence that controls each heartbeat—atrial contraction followed by ventricular contraction—is as follows: From the SA node, impulses spread in a wave along the cardiac muscle fibers of the atria, signaling the atria to contract. Some of these impulses travel along an **internodal pathway** to the **atrioventricular (AV) node** in the inferior part of the interatrial septum, where they are delayed for a fraction of a second. After this delay, the impulses race through the **atrioventricular bundle** (formerly, *bundle of His*), which enters the interventricular septum and divides into right and left **bundle branches,** or *crura* ("legs"). About halfway down the septum, the crura become bundles of **Purkinje** (Pur-kin′je) **fibers** (also called *conduction myofibers*), which approach the apex of the heart, then turn superiorly into the ventricular walls. This arrangement ensures that the contraction of the ventricles begins at the apex of the heart and travels superiorly, so that ventricular blood is ejected superiorly into the great arteries. The brief delay of the contraction-signaling impulses at the AV node enables the ventricles to fill completely before they start to contract. Because the fibrous skeleton between the atria and ventricles is nonconducting, it prevents impulses in the atrial wall from proceeding directly on to the ventricular wall. As a result, only those signals that go through the AV node can continue on.

Examination of the microscopic anatomy of the heart's conducting system reveals that the cells of the nodes and AV bundle are small, but otherwise typical, cardiac muscle cells. Each Purkinje fiber, by contrast, is a long row of special, large-diameter, barrel-shaped cells called *Purkinje myocytes*. These muscle cells contain relatively few myofilaments because they are adapted more for conduction than for contraction. Their large diameter maximizes the speed of impulse conduction. Purkinje fibers are located in the deepest part of the ventricular endocardium, between the endocardium and myocardium layers.

CLINICAL APPLICATIONS

DAMAGE TO THE CONDUCTING SYSTEM Because the atria and ventricles are insulated from each other by the electrically inert fibrous skeleton, the only route for impulse transmission from the atria to the ventricles is through the AV node and AV bundle. Therefore, damage to either of these, called a **heart block,** interferes with the ability of the ventricles to receive the pacing impulses. Without these signals, the ventricles beat at an intrinsic rate that is slower than that of the atria—and too slow to maintain adequate circulation. In such cases, an artificial pacemaker set to discharge at the appropriate rate is usually implanted. For other conditions involving damage to the conducting system, see this chapter's Disorders section (p. 537). ✚

Innervation

Although the heart's inherent rate of contraction is set by the SA node, this rate can be altered by extrinsic neural controls (Figure 18.13). The nerves to the heart consist of *visceral sensory* fibers, *parasympathetic* fibers that slow heart rate, and *sympathetic* fibers that increase the rate and force of heart contractions. The parasympathetic nerves arise as branches of the vagus nerve in the neck and thorax, whereas the sympathetic nerves travel to the heart from the cervical and upper

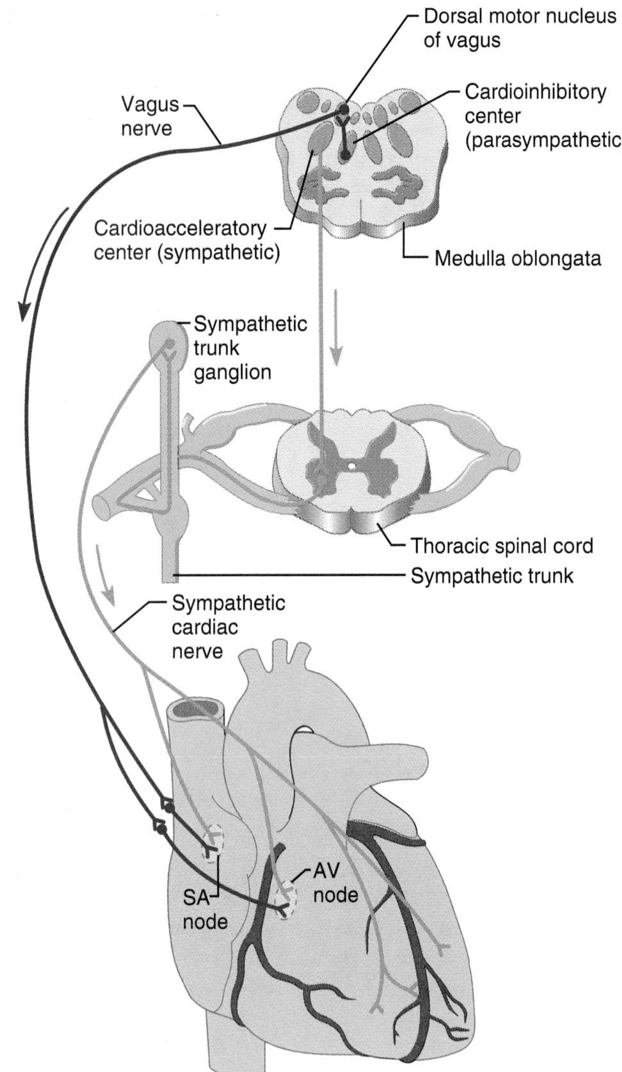

FIGURE 18.13 Autonomic innervation of the heart by the sympathetic (green) and parasympathetic (purple) systems.

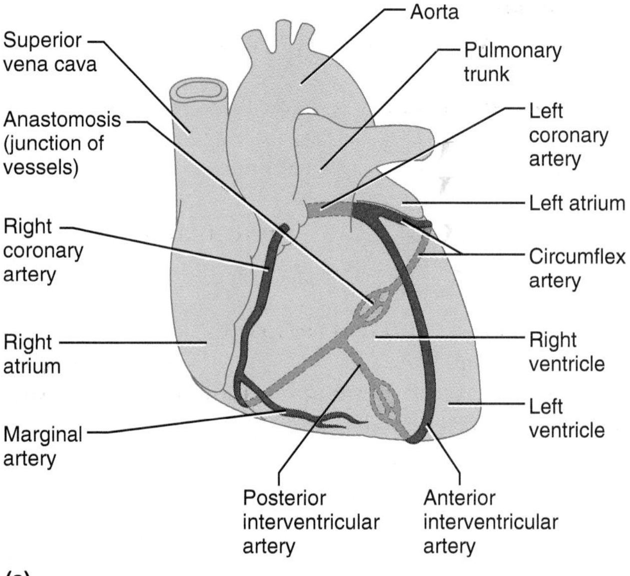

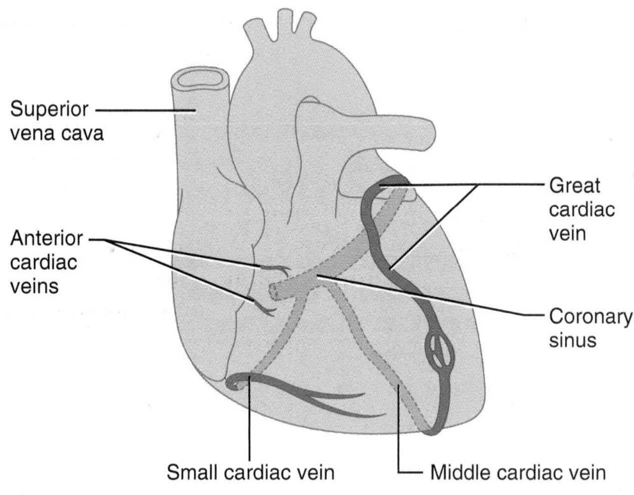

FIGURE 18.14 **Coronary blood vessels.** **(a)** The main coronary arteries. **(b)** The major cardiac veins. (The lighter-colored vessels in each case lie more posteriorly on the heart.)

thoracic chain ganglia (see Figure 15.7, p. 453). All nerves serving the heart pass through the cardiac plexus on the trachea before entering the heart. Although these autonomic nerve fibers project to cardiac musculature throughout the heart, they project most heavily to the SA and AV nodes and the coronary arteries.

As mentioned previously (see p. 381), the autonomic input to the heart is controlled by *cardiac centers* in the reticular formation of the medulla of the brain. In the medulla, the **cardioinhibitory center** influences parasympathetic neurons, whereas the **cardioacceleratory center** influences sympathetic neurons. These medullary cardiac centers, in turn, are influenced by such higher brain regions as the hypothalamus, periaqueductal gray matter, amygdala, and insular cortex (see Chapter 13).

BLOOD SUPPLY TO THE HEART

Although the heart is filled with blood, this contained blood provides little nourishment to the heart walls, which are too thick to obtain much nutrition by diffusion alone. Instead, the functional blood supply of the heart is delivered by the right and left *coronary arteries* (Figure 18.14a). These systemic arteries arise from the base of the aorta and run in the coronary sulcus.

The **left coronary artery** arises from the left side of the aorta, passes posterior to the pulmonary trunk, then divides into two branches: the *anterior interventricular* and *circumflex arteries*. The **anterior interventricular artery** descends in the anterior interventricular sulcus toward the apex of the heart, sending branches into the interventricular septum and onto the anterior walls of both ventricles on the heart's left side.

The **circumflex artery** follows the coronary sulcus posteriorly and supplies the left atrium and the posterior part of the left ventricle.

The **right coronary artery** emerges from the right side of the aorta and descends in the coronary sulcus on the anterior surface of the heart, between the right atrium and right ventricle. At the inferior border of the heart, it branches to form the **marginal artery.** Continuing into the posterior part of the coronary sulcus, the right coronary artery gives off a large branch called the **posterior interventricular artery** in the posterior interventricular sulcus. Overall, the branches of the right coronary artery supply the right atrium and almost all of the right ventricle.

The arrangement of the coronary arteries varies considerably among individuals. For example, in about 15% of people, the left coronary artery gives rise to *both* interventricular arteries. In other people (4%), a single coronary artery emerges from the aorta and supplies the entire heart.

Cardiac veins, which carry deoxygenated blood from the heart wall into the right atrium, also occupy the sulci on the heart surface (Figure 18.14b). The largest of these veins, the **coronary sinus,** occupies the posterior part of the coronary sulcus and returns almost all the venous blood from the heart to the right atrium. Draining into the coronary sinus are three large tributaries: the **great cardiac vein** in the anterior interventricular sulcus, the **middle cardiac vein** in the posterior interventricular sulcus, and the **small cardiac vein** running along the heart's inferior right margin. The anterior surface of the right ventricle contains several horizontal **anterior cardiac veins** that empty directly into the right atrium.

DISORDERS OF THE HEART

In the following discussion of disorders of the heart we examine some conditions of three broad types: coronary artery disease, heart failure, and conducting system disorders.

Coronary Artery Disease

Atherosclerosis is an accumulation of fatty deposits in the inner lining of the body's arteries that can block blood flow through these arteries (see p. 574 for details). When atherosclerosis affects the coronary arteries, it leads to **coronary artery disease,** in which the arteries supplying the heart wall are narrowed or blocked. A common symptom of this disease is **angina pectoris** (an-ji′nah pek′tor-us; "choked chest"), thoracic pain caused by inadequate oxygenation of heart muscle cells, which weaken but do not die. Although the pain of angina usually results directly from tissue hypoxia, it can also result from stress-induced spasms of the atherosclerotic coronary arteries. Angina attacks occur most often during exercise, when the vigorously contracting heart may demand more oxygen than the narrowed coronary arteries can deliver. The onset of angina should be considered a warning sign of other, more serious conditions.

If the blockage of a coronary artery is more complete or prolonged, the oxygen-starved cardiac muscle cells die—a condition called **myocardial infarction,** or a heart attack. Few experiences are more frightening than a heart attack: A sharp pain strikes with lightning speed through the chest (and sometimes the left arm and left side of the neck) and does not subside. Death from cardiac arrest occurs almost immediately in about one-third of cases. Heart attacks kill either directly (due to severe weakening of the heart) or indirectly (due to heart-rhythm disruptions caused by damage to the conducting system).

Ironically, angina sufferers and heart-attack survivors are lucky, in that they have received warnings that allow them an opportunity to make dietary and behavioral changes, to take medications, or even to have coronary bypass surgery (see p. 574) to restore the blood supply to their oxygen-deprived heart. Unfortunately, however, for many people a painless but fatal heart attack is the first and only clear symptom of coronary artery disease. Such individuals are victims of **silent ischemia,** a condition in which blood flow to the heart is interrupted often, exactly as in angina, but without any pain to provide warning. Silent ischemia, which can also occur in some heart-attack survivors, can be detected by measuring heart rhythm with an electrocardiograph (ECG) during exercise, a procedure called ambulatory ECG monitoring.

Heart Failure

Heart failure is a progressive weakening of the heart as it fails to keep pace with the demands of pumping blood and thus cannot meet the body's need for oxygenated blood. This condition may be due to weakened ventricles (damaged, for example, by a heart attack), to failure of the ventricles to fill completely during diastole (for example, as a result of mitral valve stenosis), to overfilling of the ventricles (for example, because of stenosis or insufficiency of the aortic valve), or, most commonly, to congestive heart failure.

In **congestive heart failure** (also called *dilated cardiomyopathy*), the heart enlarges greatly while its pumping efficiency progressively declines. The cause of this condition, which affects 5 million Americans and is increasing in frequency, is unknown. One hypothesis is that it may involve a destructive positive feedback loop in which an initially weakened heart signals the sympathetic nervous system to signal the heart to pump harder, which further weakens the heart, which again signals the sympathetic system, and so on.

A condition called **cor pulmonale** (*cor* = heart; *pulmonale* = lung) is the enlargement and sometimes ultimate failure of the right ventricle resulting from elevated blood pressure in the pulmonary circuit. It follows

A CLOSER LOOK

Treating Heart Disease

The HeartMate LVAD

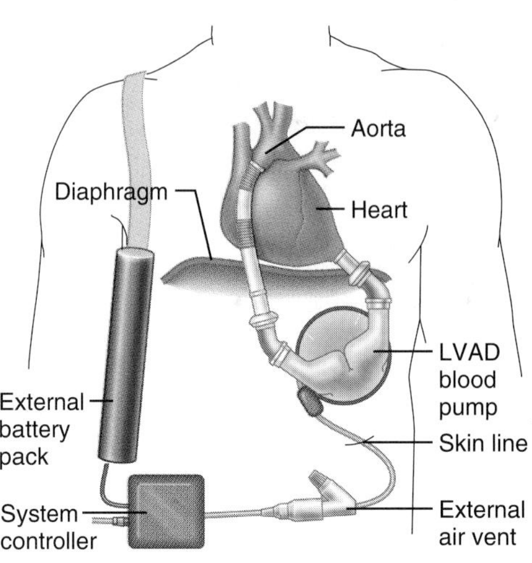

Even though medical treatment receives a lot of attention, practitioners in many fields of medicine advocate prevention over treatment. In the case of heart disease, the approach to preventing heart attacks involves a variety of strategies: encouraging exercise and a healthy low-fat diet; assessing patients' risk factors such as family history and blood cholesterol levels; and viewing the coronary arteries with medical imaging techniques, including angiograms (see p. 578), ultrasonography, and CT scans, some of which can image the beating heart to detect calcium deposits that indicate atherosclerosis in coronary arteries.

For some individuals—those at risk for heart disease and those already diagnosed with heart problems—***drug therapies*** can help. Aspirin reduces the clotting that can plug coronary vessels, and nitrate drugs such as nitroglycerine dilate the coronary vessels. Beta blockers reduce the rate and force of heart contractions by blocking the sympathetic response of the heart to stress, thereby reducing strain on the heart. Calcium channel blockers prevent the coronary arterial muscle spasms that often trigger heart attacks.

But what are the options for patients whose hearts are so devastated by heart disease that drugs are "too little, too late"? Currently, ***heart transplants*** are the primary option, but this procedure is complicated and not without risk. After a compatible donor is located and a difficult and traumatic surgery performed, the recipient must undergo lifelong treatment with anti-T cell immunosuppressive drugs to ward off rejection of the transplanted organ. After about one year, antibodies may attack the inner layer of the coronary arteries; this type of rejection, which occurs in about 40% of the transplanted hearts by the third year, does not respond to any immunosuppressive treatment and may lead to fatal heart attacks. Despite these and other complications, heart transplants are probably the best option available—over 75% of heart-transplant patients now survive for 2 years, and about half of heart recipients survive for more than 12 years.

The demand for donor hearts for transplantation greatly exceeds the supply. Fully one-third of patients awaiting heart transplants in the United States die before receiving one. The shortage of donated hearts has fueled considerable research into developing alternatives to human hearts. Some approaches, such as using pig hearts (xenotransplantation) and wrapping skeletal muscle around the failing ventricles (cardiomyoplasty) have proven prone to rejection and proven ineffective, respectively, and the ultimate goal of growing a new heart from cultures of cardiac muscle cells seems decades away.

Two new approaches aimed at increasing vascularization have shown some promise. Administering a natural chemical called ***vascular endothelial growth factor (VEGF)*** seems to aid the healing of damaged hearts by stimulating the growth of new coronary vessels in the heart wall. A treatment called ***transmyocardial revascularization*** uses a laser inserted into the heart cavity to form 15–40 small channels in the left ventricular wall; this technique apparently stimulates the coronary vessels to send more branches into the myocardium.

Still, at present the most successful alternative to transplantation employs mechanical cardiac-assist devices that are significant improvements over the unsuccessful artificial hearts of the 1980s. Among these devices are ***left ventricular assist devices (LVADs),*** including one made by HeartMate. These mechanical pumps, which are approved for temporary use in patients awaiting heart transplants, are implanted into the abdomen, connected to the left ventricle, and powered by an external battery that in portable models is carried in a backpack or shoulder holster (see the illustration). The results have been sufficiently promising that some 20% of patients with failing hearts (particularly those too old to be eligible for transplants) might be candidates for permanently implanted LVADs.

Finally, an experimental surgery called ***ventricular reduction*** seems to work well in strengthening the left ventricle of hearts enlarged and weakened by congestive heart failure. This technique, developed in a hospital in the jungles of Brazil by Dr. Randas Batista in 1994, removes a triangular piece of the left ventricle, thereby restoring the enlarged heart to its proper size and enabling it to beat more efficiently and with reduced strain on the ventricular wall.

a blockage or constriction of the vessels in the lungs, which increases resistance to blood flow and forces the right ventricle to work harder. Whereas acute cases of cor pulmonale may develop suddenly from an embolism in the pulmonary vessels, chronic cases are usually associated with chronic lung diseases such as emphysema.

For more information concerning treatments for victims of heart attack and heart failure, see the Closer Look box on this page.

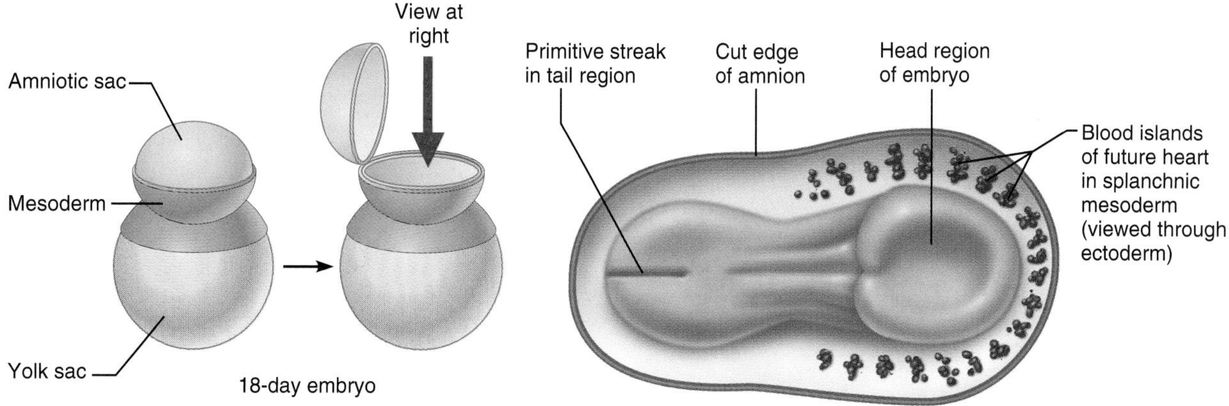

FIGURE 18.15 Appearance of the heart in the embryonic disc (days 18–20). As shown in the sequence at left, the top of the amnion has been removed to allow a clear view of the dorsal surface of the embryo. For further orientation, see Figure 3.7b on p. 65. The heart originates from a crescent of blood islands around the future head and neck (far right side of the figure).

Disorders of the Conduction System

Here we discuss a type of *arrhythmia*—a variation from the normal rhythm of the heartbeat—called fibrillation.

In **ventricular fibrillation,** the ventricles are unable to pump blood into the arteries because the rapid, random firing of electrical impulses within ventricular cardiac muscle prevents coordinated contraction of the ventricle. The fibrillating ventricles can be likened to a quivering bag of worms. Ventricular fibrillation results from a crippled conducting system and is the most common cause of cardiac arrest and sudden death in patients with hearts damaged by coronary artery disease.

In **atrial fibrillation,** multiple waves of impulses circle within the atrial myocardium, randomly stimulating the AV node, which then signals the ventricles to contract quickly and irregularly. The resulting lack of smooth movement of blood through the heart can enable the formation of clots, parts of which can break off, reach the brain, and cause strokes. The fibrillations are usually discontinuous, occurring in episodes characterized by palpitations (sensations of an unduly quick and irregular heartbeat), anxiety, fatigue, and shortness of breath. The cause of this condition, which affects 5% of individuals over age 65 and 10% over 75, is unknown, but it is often associated with coronary artery or heart-valve disease. Treatment typically involves drug therapy and administering anticoagulants.

THE HEART THROUGHOUT LIFE

In this section we first examine the events in the development of the heart and consider some common congenital heart defects; then we discuss age-related changes in the heart.

Development of the Heart

An understanding of heart development is clinically important because congenital abnormalities of the heart account for nearly half of all deaths from birth defects. One of every 150 newborns has some congenital heart defect.

As explained in Chapter 17 (see p. 516), all blood vessels begin as condensations of mesodermal mesenchyme called blood islands. Subsequently, the blood islands destined to become the heart form in the splanchnic mesoderm around the future head and neck of the embryonic disc (Figure 18.15). The heart folds neatly into the thorax region when the flat embryonic disc lifts up off the yolk sac to assume its three-dimensional body shape around day 20 or 21 (see p. 66).

When the embryonic heart first reaches the thorax, it is a pair of tubes in the body midline that fuse into a single tube on about day 21 (Figure 18.16a). The heart starts pumping about day 22, by which time four bulges have developed along the heart tube (Figure 18.16b). These bulges are the earliest heart chambers and are unpaired. From tail to head, following the direction of blood flow, the four chambers are the *sinus venosus, atrium, ventricle,* and *bulbus cordis*:

1. **Sinus venosus** (ven-o′sus; "of the vein"). This chamber, which initially receives all blood from the veins of the embryo, will become the smooth-walled part of the right atrium and the coronary sinus; it also gives rise to the sinoatrial node. Furthermore, recent evidence indicates that the sinus venosus also contributes to the back wall of the *left* atrium (which mostly derives from the bases of the developing pulmonary veins).

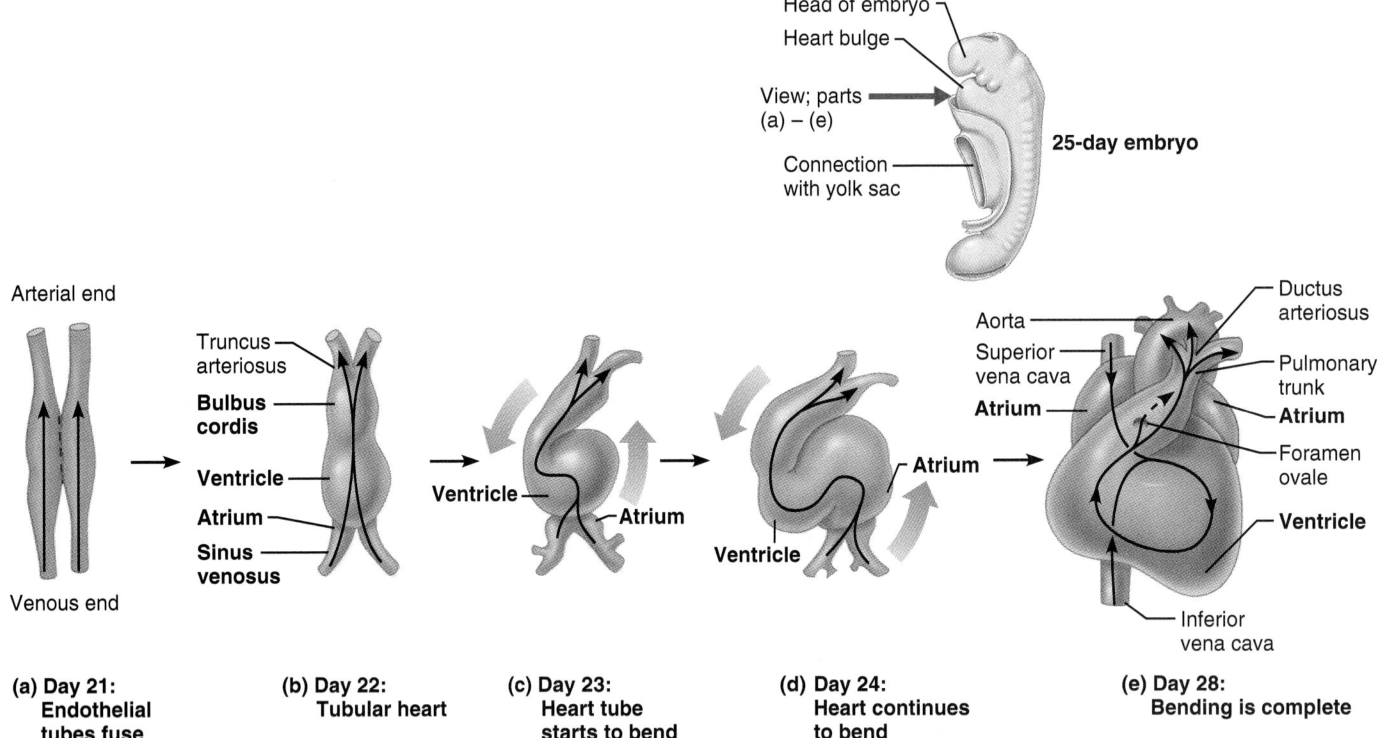

FIGURE 18.16 **Stages in heart development during week 4 (a–e).** As shown above in the orientation diagram of the 25-day embryo, parts (a) through (e) are all ventral views of the heart, with the cranial direction toward the top of the figures.

2. **Atrium.** This embryonic chamber eventually becomes the ridged parts of the right and left atria—namely, the parts lined by pectinate muscles.
3. **Ventricle.** The strongest pumping chamber of the early heart, the embryonic ventricle gives rise to the *left* ventricle.
4. **Bulbus cordis.** This chamber, and its most cranial extension, the *truncus arteriosus,* give rise to the pulmonary trunk and first part of the aorta. It also gives rise to the *right* ventricle.

At the time these four chambers appear, the heart starts bending into an S shape (Figure 18.16c and d). The ventricle moves caudally and the atrium cranially, assuming their adult positions. This bending occurs because the ventricle and bulbus cordis grow quickly, and the heart is unable to accommodate elongation within the confines of the pericardial sac.

During month 2 of development, the heart divides into its four definitive chambers by the formation of its midline septum and valves. These structures develop from cardiac cushions, which are regional thickenings of the inner lining of the heart wall. For example, most of the interatrial septum forms by growing caudally from the heart's roof, and most of the interventricular septum forms by growing cranially from the heart's apex. Additionally, neural crest cells (see pp. 62 and 66) migrate into the area where atrium meets ventricle. Here, these cells contribute to the developing heart valves and to the bases of the pulmonary trunk and ascending aorta, the great arteries that form by the splitting of the bulbus cordis. These month-2 events are so complex that perfect orchestration does not always occur and many developmental defects result.

Most other details of heart development are beyond the scope of this book, but one thing should be mentioned: The two atria remain interconnected by a hole in the interatrial septum—the foramen ovale—until birth, at which time this hole closes to become the fossa ovalis (see Figure 18.5c). The foramen ovale plays an important role in blood circulation before birth, and is discussed in the section on fetal circulation in the next chapter (see p. 576).

Figure 18.17 illustrates and explains the common congenital heart defects, almost all of which can be traced back to month 2 of development. The most common of these is a **ventricular septal defect** (Figure 18.17a), in which the superior (cranial) region of the interventricular septum fails to form, leaving a hole between the two ventricles. As you study this and the other defects in the figure, note that they produce two basic kinds of effects in newborns: Either inadequately oxygenated blood reaches the body tissues (because oxygen-poor systemic blood mixes with oxygenated pulmonary blood), or the ventricles labor under an increased work load (due to narrowed valves and vessels), or both effects occur. Modern surgical techniques can usually correct these congenital defects.

Aorta
Pulmonary valve
Right ventricle
Inter-ventricular septum
Aortic valve
Left ventricle

(a) Normal heart. Arrows indicate the path of blood flow through the heart. Red = oxygen-rich blood; blue = oxygen-poor blood.

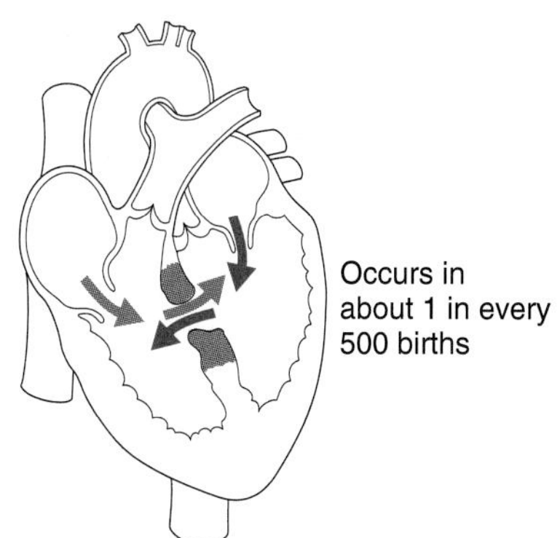

(b) Ventricular septal defect. The superior part of the inter-ventricular septum fails to form; thus, blood mixes between the two ventricles.

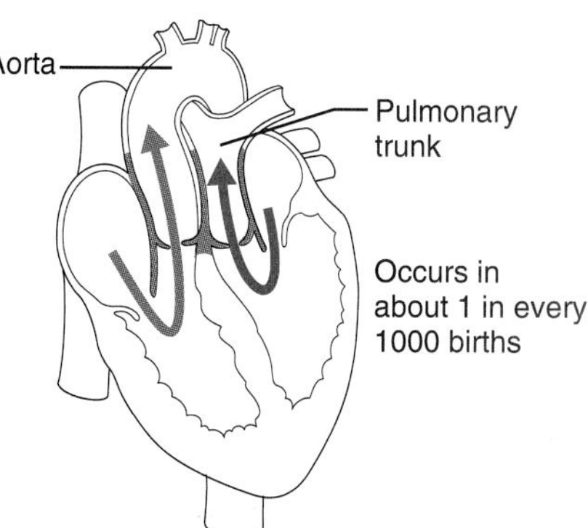

(c) Transposition of the great vessels. Aorta comes from right ventricle, pulmonary trunk from left. Results when the bulbus cordis does not divide properly. Unoxygenated blood passes repeatedly around systemic circuit, while oxygenated blood recycles around the pulmonary circuit.

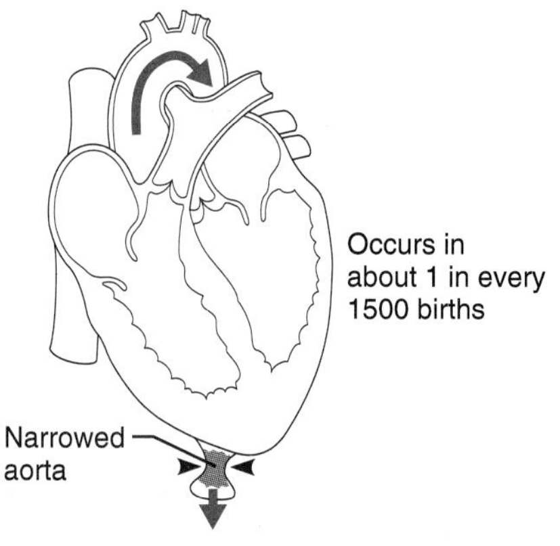

(d) Coarctation of the aorta. A part of the aorta is narrowed, increasing the work load on the left ventricle.

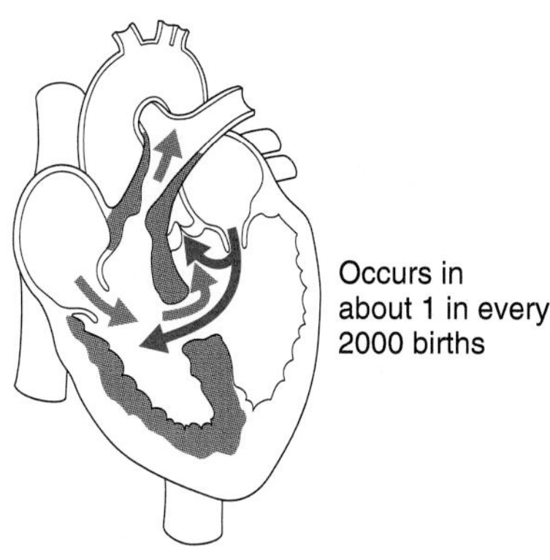

(e) Tetralogy of Fallot. Multiple defects (tetra= four): Pulmonary trunk too narrow and pulmonary valve stenosed; ventricular septal defect; aorta opens from both ventricles; wall of right ventricle thickened from overwork.

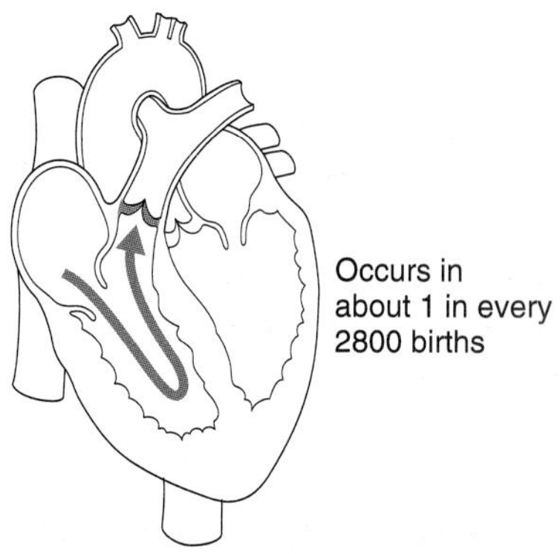

(f) Pulmonary stenosis. The pulmonary semilunar valve is narrowed, lessening the flow of blood to the lungs.

FIGURE 18.17 Congenital heart defects. The defects are arranged according to relative frequency of occurrence. Purple areas indicate the locations of the defects.

The Heart in Adulthood and Old Age

In the absence of congenital heart problems, the resilient heart usually functions well throughout life. In individuals who exercise regularly and vigorously, the heart gradually adapts to the increased demand by increasing in strength and size. Aerobic exercise also helps clear fatty deposits from the walls of the coronary vessels, thereby retarding the process of atherosclerosis. Barring some chronic illness, this beneficial response to exercise persists into old age.

Age-related changes that affect the heart include the following:

1. **Hardening and thickening of the cusps of the heart valves.** These changes occur particularly in those valves subjected to the highest blood pressures (mitral and aortic valves). Thus, abnormal heart sounds are more common in elderly people.
2. **Decline in cardiac reserve.** Although the passage of years seems to cause few changes in the resting heart rate, the aged heart is slower at increasing its output to pump more blood. Sympathetic control of the heart becomes less efficient, and heart rate gradually becomes more variable. Maximum heart rate declines, although this problem is much less severe in seniors who are physically active.
3. **Fibrosis of cardiac muscle.** As one ages, more and more cardiac muscle cells die and are replaced by fibrous scar tissue. This fibrosis, which is much more extensive in men than in women, lowers the maximum amount of blood that the heart can pump per unit time. Also, in conjunction with the aging of muscle-cell membranes, fibrosis hinders the initiation and transmission of contraction-signaling impulses, leading to abnormal heart rhythms and other conduction problems.

RELATED CLINICAL TERMS

Asystole (a-sis'to-le) (*a* = without) Failure of the heart to contract.

Cardiac catheterization Diagnostic procedure involving the passage of a fine catheter (tube) from a blood vessel at the body surface to the heart. Blood-oxygen content, blood pressure, and blood flow are measured, and heart structures can be visualized. Findings help detect problems with the heart valves, heart deformities, and other cardiac malfunctions.

Cardiomyopathy (kar'de-o-mi-op"ah-the) Any disease of the myocardium that weakens the heart's ability to pump.

Echocardiography (ek"o-kar"de-og'rah-fe) Ultrasound imaging of the heart; used not only for imaging but also to measure blood flow through the heart.

Endocarditis (en"do-kar-di'tis) Inflammation of the endocardium, usually confined to the endocardium of the heart valves. Endocarditis often results from infection by bacteria that have entered the bloodstream, but may result from fungal infections or an autoimmune response. Drug addicts may develop endocarditis by injecting themselves with contaminated needles. Additionally, the bacteria can enter during routine dentistry and ear-piercing procedures.

Hypertrophic cardiomyopathy Inherited condition in which the wall of the left ventricle (especially the interventricular septum) is abnormally thick and composed of disorganized muscle cells. Affects 1 of every 500 people. Because the lumen of the left ventricle is too small to hold all the atrial blood returning to it, blood pressure in the pulmonary circuit is elevated. Additionally, the grossly thickened myocardium becomes ischemic because it cannot get enough oxygen from the coronary vessels that supply it. Exercising can lead to fainting, chest pain, or even sudden death.

Myocarditis (mi'o-kar-di'tis) Inflammation of the heart's myocardium. Sometimes follows an untreated streptococcus infection in children; may be extremely serious because it can weaken the heart and impair its ability to pump blood.

Percussion Tapping the thorax or abdomen wall with the fingertips and using the nature of the resulting sounds to estimate the location, density, and size of the underlying organs. Percussion of the thoracic wall can be used to estimate the size of a patient's heart.

CHAPTER SUMMARY

Media study tools that could provide you additional help in reviewing specific key topics of Chapter 18 are referenced below.

SP = Study Partner; AIA = A.D.A.M.® Interactive Anatomy

1. The heart is a double pump whose right side pumps blood to the lungs for oxygenation and whose left side pumps blood throughout the body to nourish body tissues; that is, the right side is the pump of the pulmonary circuit, and the left side is the pump of the systemic circuit.
2. The four chambers of the heart are right and left atria (receiving chambers) and right and left ventricles (main pumping chambers).

AIA Link: Atlas; System: Cardiovascular; Heart and Great Vessels (posterior).

LOCATION AND ORIENTATION WITHIN THE THORAX (pp. 522–524)

1. The cone-shaped human heart, about the size of a fist, lies obliquely within the mediastinum. Its apex points anteriorly and to the left. Its base is its posterior surface.

AIA Link: Atlas; System: Cardiovascular; Thoracic Viscera (anterior).

2. From an anterior view, the heart has a trapezoidal shape. The locations of its four corner points are described on page 522.

STRUCTURE OF THE HEART (pp. 524–528)

Coverings (p. 524)

1. The pericardium encloses the heart. It consists of a superficial layer (fibrous pericardium and parietal layer of the serous pericardium), and a deeper layer that covers the heart surface (visceral layer of the serous pericardium, or epicardium). The pericardial cavity, between the two layers, contains lubricating serous fluid.

AIA Link: Dissectible Anatomy: Male; Anterior Layer 169. Scroll to Thorax.

Layers of the Heart Wall (pp. 524–525)

2. The layers of the heart wall, from external to internal, are the epicardium, the myocardium (which consists of cardiac muscle), and the endocardium (endothelium and connective tissue).

AIA Link: Atlas; System: Cardiovascular; Right Atrium and Ventricle (anterior).

Heart Chambers (pp. 525–528)

3. On the external surface of the heart, several sulci separate the four heart chambers: the coronary sulcus and the anterior and posterior interventricular sulci. Internally, the right and left sides of the heart are separated by the interatrial and interventricular septa.
4. The right atrium has the following features: right auricle and pectinate muscles anteriorly; a smooth-walled posterior part; crista terminalis; openings of the coronary sinus and the superior and inferior venae cavae; SA and AV nodes; and fossa ovalis.

AIA Link: Atlas; System: Cardiovascular; Right Atrium (lateral).

5. The right ventricle contains the following features: trabeculae carneae, papillary muscles, chordae tendineae, and right atrioventricular (tricuspid) valve. The pulmonary semilunar valve lies at the base of the pulmonary trunk.

AIA Link: Atlas; System: Cardiovascular; Right Ventricle (anterior).

6. The left atrium has a large, smooth-walled, posterior region into which open the four pulmonary veins. Anteriorly, its auricle is lined by pectinate muscles.
7. The left ventricle, like the right ventricle, contains papillary muscles, chordae tendineae, trabeculae carneae, and an atrioventricular valve (mitral). The aortic semilunar valve lies at the base of the aorta.

AIA Link: Atlas; System: Cardiovascular; Left Atrium and Ventricle (lateral).

PATHWAY OF BLOOD THROUGH THE HEART (pp. 528–529)

1. A drop of blood circulates along the following path: right atrium to right ventricle to pulmonary circuit to left atrium to left ventricle to systemic circuit to right atrium.
2. In each heartbeat, both atria contract together, followed by simultaneous contraction of both ventricles.

HEART VALVES (pp. 529–532)

Valve Structure (p. 529)

1. The four heart valves are the atrioventricular valves (tricuspid and mitral) and the semilunar valves (aortic and pulmonary). All except the mitral have three cusps.

AIA Link: Atlas; System: Cardiovascular; Cardiac Veins (superior).

Valve Functions (p. 529)

2. The atrioventricular valves prevent backflow of blood into the atria during contraction of the ventricles. The pulmonary and aortic semilunar valves prevent backflow into the ventricles during relaxation of the ventricles.

Heart Sounds (pp. 530–532)

3. Using a stethoscope, physicians can listen to the sounds produced by the closing atrioventricular and semilunar valves. Each valve is best heard near a different heart corner on the anterior chest wall (see Figure 18.11).

AIA Link: Atlas; System: Cardiovascular; Heart Valve Auscultation Sites.

FIBROUS SKELETON (p. 532)

1. The fibrous skeleton of the heart surrounds the valves between the atria and ventricles. It anchors the valve cusps, is the point of insertion of the heart musculature, and blocks the direct spread of electrical impulses from the atria to the ventricles.

CONDUCTING SYSTEM AND INNERVATION (pp. 532–534)

Conducting System (pp. 532–533)

1. The conducting system of the heart is an interconnected series of cardiac muscle cells that initiates each heartbeat, sets the basic rate of the heartbeat, and coordinates the contraction of the heart chambers. The impulse that signals heart contraction travels from the SA node (the pacemaker) through the atrial myocardium and internodal pathway to the AV node, to the atrioventricular bundle, bundle branches, Purkinje fibers, and the ventricular musculature.

SP Exercise: Chapter 18, Intrinsic Conduction System of the Heart.

Innervation (pp. 533–534)

2. Heart innervation consists of visceral sensory fibers, heart-slowing vagal parasympathetic fibers, and sympathetic fibers that accelerate heart rate.

BLOOD SUPPLY TO THE HEART (pp. 534–535)

1. The coronary vessels supplying the heart wall occupy the coronary sulcus and interventricular sulci. The right and left coronary arteries branch from the aorta to supply the heart wall. Venous blood, collected by the cardiac veins (great, middle, small, and anterior), empties into the coronary sinus and right atrium.

AIA Link: Atlas; System: Cardiovascular; Coronary Arteries (anterior).

DISORDERS OF THE HEART (pp. 535–537)

1. Coronary artery disease is caused by atherosclerotic blockage of the coronary arteries, and can lead to angina pectoris and myocardial infarction (heart attack). Heart failure is a weakening of the heart, which fails to pump blood fast enough to meet the body's needs. Atrial and ventricular fibrillations are rapid, irregular, and uncoordinated contractions of the respective heart chambers, resulting from damage to the conducting system.

THE HEART THROUGHOUT LIFE (pp. 537–539)

Development of the Heart (pp. 537–538)

1. The heart develops from splanchnic mesoderm around the head and neck of the embryonic disc. When it folds into the thorax, it is a double tube that soon fuses into one and starts pumping blood (day 22). It soon bends into an S shape. Its four earliest chambers are the sinus venosus, atrium, ventricle, and bulbus cordis.
2. The four final heart chambers are defined during month 2 through the formation of valves and dividing walls. Failures in this complex process account for most congenital defects of the heart (see Figure 18.17).

The Heart in Adulthood and Old Age (p. 539)

3. Age-related changes in the heart include hardening and thickening of the valve cusps, decline in cardiac reserve, fibrosis of cardiac muscle, and atherosclerosis.

REVIEW QUESTIONS

Multiple Choice/Matching Questions

1. The most external part of the pericardium is the (a) parietal layer of serous pericardium, (b) fibrous pericardium, (c) visceral layer of serous pericardium, (d) pericardial cavity.
2. Which heart chamber forms most of the heart's inferior surface? (a) right atrium, (b) right ventricle, (c) left atrium, (d) left ventricle.
3. How many cusps does the right atrioventricular valve have? (a) two, (b) three, (c) four.
4. The sequence of contraction of the heart chambers is (a) random, (b) left chambers followed by right chambers, (c) both atria followed by both ventricles, (d) right atrium, right ventricle, left atrium, left ventricle.

5. The middle cardiac vein runs with which artery? (a) marginal artery, (b) aorta, (c) coronary sinus, (d) anterior interventricular artery, (e) posterior interventricular artery.

6. The base of the heart (a) is its posterior surface, (b) lies on the diaphragm, (c) is the same as its apex, (d) is its superior border.

7. Which of the following is an *in*correct statement about the crista terminalis of the right atrium? (a) It separates the smooth-walled part from the part with pectinate muscles. (b) It is shaped like the letter C. (c) The coronary sinus and inferior vena cava open near its inferior part. (d) It lies mostly in the interatrial septum.

8. The aortic semilunar valve closes (a) at the same time the mitral valve closes, (b) just after the atria contract, (c) just before the ventricles contract, (d) just after the ventricles contract.

9. The ventricle of the embryonic heart gives rise to what adult structure(s)? (a) bulbus cordis, (b) both ventricles, (c) left ventricle, (d) the aorta, (e) none of these.

10. Which layer of the heart wall is the thickest? (a) endocardium, (b) myocardium, (c) epicardium, (d) endothelium.

11. The inferior left corner of the heart is located at the (a) second rib slightly lateral to the sternum, (b) third rib at the sternum, (c) sixth rib slightly lateral to the sternum, (d) fifth intercostal space at the midclavicular line.

Short Answer Essay Questions

12. Ben Garber was annoyed when the teaching assistant who ran the discussion section of his anatomy class said that the atria pump blood to the lungs, and the ventricles pump blood throughout the body. Correct this error.

13. Describe the location of the heart within the thorax.

14. Trace a drop of blood through all the heart chambers and heart valves, and through the basic vascular circuits, from the time it enters the left atrium until it enters the left atrium again.

15. (a) Name the elements of the heart's conducting system in order, beginning with the pacemaker. (b) Is the conducting system made of nerves? Explain. (c) What are the functions of this conducting system?

16. During a lethal heart attack, a blood clot lodges in the first part of the circumflex branch of the left coronary artery, blocking blood flow through this vessel. What regions of the heart will become ischemic and die?

17. To expand on the previous question, sketch the heart and draw all the coronary vessels in their correct locations. (Alternatively, you could locate these vessels on a realistic diagram of the heart.)

18. On a diagram of a frontally sectioned heart, indicate the location of the fibrous skeleton.

19. When you view a heart's *anterior* surface, which of its four chambers appears the largest?

20. Make a drawing of the adult heart and the associated large vessels. Then color and label the adult regions that derive from each *embryonic* heart chamber: (a) sinus venosus, (b) embryonic atrium, (c) embryonic ventricle, (d) bulbus cordis.

CRITICAL REASONING + CLINICAL APPLICATION QUESTIONS

1. After studying Figure 18.17, classify the five congenital heart defects depicted according to whether they produce (1) mixing of oxygenated and unoxygenated blood, (2) increased work load for the ventricles, or (3) both of these problems.

2. Ms. Hamad, who is 73 years old, is admitted to the coronary care unit of a hospital with a diagnosis of left ventricular failure resulting from a myocardial infarction. Her heart rhythm is abnormal. Explain what a myocardial infarction is, how it is likely to have been caused, and why the heart rhythm is affected.

3. You have been called on to demonstrate where to listen for heart sounds. Explain where on the chest wall you would place a stethoscope to listen for (a) incompetence of the aortic semilunar valve and (b) stenosis of the mitral valve.

4. After a man was stabbed in the chest, his face became blue and he lost consciousness from lack of oxygenated blood to the brain. The diagnosis was cardiac tamponade rather than severe blood loss through internal bleeding. What is cardiac tamponade, how did it cause the observed symptoms, and how is it treated?

5. A heroin addict felt tired, weak, and feverish, with vague pains and aches. Finally, he went to a doctor, terrified that he had AIDS. Instead, the doctor found an abnormal heart sound, and the final diagnosis was not AIDS but endocarditis. What is the most likely way this patient contracted endocarditis?

6. Another patient had an abnormal heart sound that indicated a stenotic valve. Define this condition and contrast it with an incompetent heart valve.

A.D.A.M.® INTERACTIVE ANATOMY QUESTIONS

1. Link: Dissectible anatomy; Male (anterior): layer 176; scroll to thorax. (a) Name the three major arteries from the arch of the aorta. (b) Name the two major veins that converge to form the superior vena cava.

CHAPTER 19 Blood Vessels

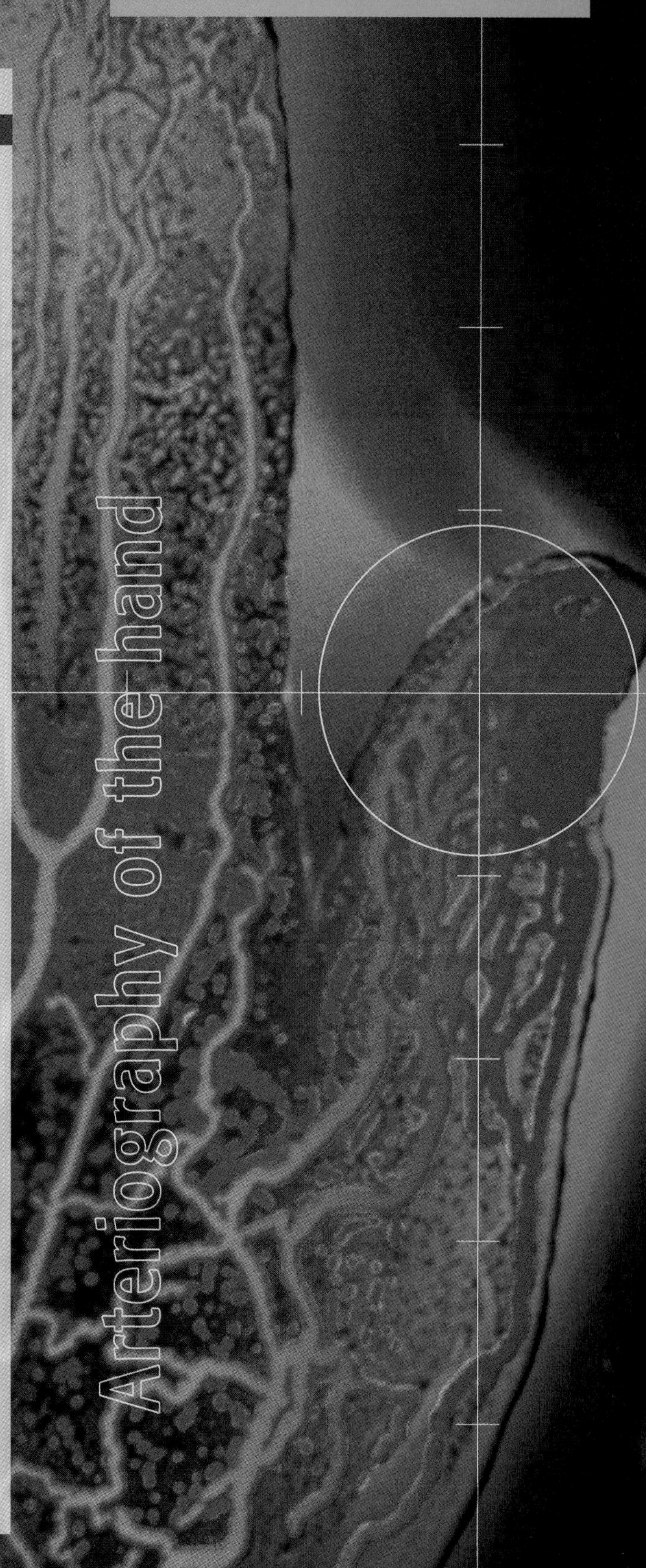

CHAPTER OUTLINE + STUDENT OBJECTIVES

PART ONE: GENERAL CHARACTERISTICS OF BLOOD VESSELS 544–551

Structure of Blood Vessel Walls 544

1. Describe the three tunics that form the wall of an artery or vein.

Types of Blood Vessels 544–550

2. Compare and contrast the structure and functions of elastic arteries, muscular arteries, and arterioles.
3. Describe the structure and function of capillaries, sinusoids, and capillary beds, and explain the structural basis of capillary permeability.
4. Compare postcapillary venules to capillaries.
5. Explain how to distinguish a vein from an artery in histological sections.
6. Explain the function of valves in veins.

Vascular Anastomoses 550–551

7. Define vascular anastomoses and explain their functions.

Vasa Vasorum 551

8. Define vasa vasorum.

PART TWO: BLOOD VESSELS OF THE BODY 552–578

The Pulmonary Circulation 552–553

9. Name the major vessels of the pulmonary circuit.

The Systemic Circulation 553–573

10. List the major arteries and veins of the systemic circuit. Describe their locations and the body regions they supply.
11. Describe the structure and special function of the hepatic portal system, and explain the significance of portal-systemic anastomoses.

Disorders of the Blood Vessels 573–576

12. Define atherosclerosis, aneurysm, deep vein thrombosis and venous disease of the lower limb, microangiopathy of diabetes, and arteriovenous malformation.

Blood Vessels Throughout Life 576–578

13. Trace the cardiovascular circuit in the fetus, and explain how it changes at birth.
14. List some effects of aging on the blood vessels.

THE BLOOD VESSELS OF THE BODY FORM A closed delivery system powered by the pumping heart. The realization that blood is pumped in a circle dates back to the 1620s and is based on the careful experiments of William Harvey, an English physician. Before that time, it was taught—as proposed by the ancient Greek physician Galen—that blood moved through the body like an ocean tide, first moving out from the heart, then ebbing back into the heart via the same vessels.

Blood vessels are not rigid tubes but dynamic structures that pulsate, constrict and relax, and even proliferate, according to the changing needs of the body. In this chapter we examine the structure and function of these important circulatory pathways through the body.

The three major types of blood vessels are *arteries, capillaries,* and *veins.* When the heart contracts, it forces blood into the large arteries that leave the ventricles. The blood then moves into successively smaller branches of arteries, finally reaching the smallest branches, the *arterioles* (ar-te′re-ōlz; "little arteries"), which feed into the *capillaries* of the organs. Blood leaving the capillaries is collected by *venules,* small veins that merge to form larger veins that ultimately empty into the heart. This pattern of vessels applies to both the pulmonary and systemic circuits. Altogether, the blood vessels in an adult human body stretch for 100,000 km or 60,000 miles, a distance equivalent to almost 2½ times around the world!

Notice that arteries are said to "branch," "diverge," or "fork" as they carry blood *away* from the heart. Veins, by contrast, are said to "join," "merge," "converge," or "serve as tributaries" as they carry blood *toward* the heart.

This chapter is divided into two parts: Part One considers the general characteristics of blood vessels, and Part Two discusses the specific vessels in the pulmonary and systemic circulations.

PART ONE

GENERAL CHARACTERISTICS OF BLOOD VESSELS

In the first part of this chapter we begin by examining the structure of the blood vessel walls, and then we describe the structure and function of each type of blood vessel.

STRUCTURE OF BLOOD VESSEL WALLS

The walls of all blood vessels, except the very smallest, are composed of three distinct layers, or *tunics* ("cloaks")—the tunica intima, tunica media, and tunica externa—that surround the central blood-filled space, the **lumen** (Figure 19.1). As we describe these tunics, we will be covering features that are common to both arteries and veins.

The innermost tunic of a vessel wall is the **tunica intima** (too′nĭ-kah in′tĭ-mah), which might be considered to be in "intimate" contact with the blood in the lumen. This tunic contains the *endothelium,* the simple squamous epithelium that lines the lumen of all vessels (see p. 78). The flat endothelial cells form a smooth surface that minimizes the friction of blood moving across them. In vessels larger than about 1 mm in diameter, a thin layer of loose connective tissue, the **subendothelial layer,** lies just external to the endothelium.

The middle tunic, or **tunica media,** consists primarily of *circularly* arranged sheets of smooth muscle fibers, between which lie circular sheets of elastin and collagen fibrils. Contraction of the smooth muscle cells decreases the diameter of the vessel, a process called *vasoconstriction,* whereas their relaxation increases the vessel's diameter, a process called *vasodilation;* both activities are regulated by *vasomotor nerve fibers* of the sympathetic division of the autonomic nervous system. The elastin and collagen convey elasticity and strength for resisting the blood pressure that is placed on the vessel wall by each heartbeat. In arteries, which bear most of the responsibility for maintaining and resisting blood pressure, the tunica media is the thickest layer, and it is thicker in arteries than in veins.

The outermost layer of the vessel wall is the **tunica externa** (formerly *tunica adventitia:* ad″ven-tish′e-ah). This tunic is a layer of connective tissue that contains many collagen and elastic fibers. Its cells and fibers run *longitudinally.* Functionally, the tunica adventitia protects the vessel, further strengthens its wall, and anchors the vessel to surrounding structures.

Critical Reasoning

The heart can be viewed as an enlarged blood vessel; accordingly, match each layer of the heart wall (epicardium, endocardium, and myocardium: see p. 524) with its corresponding vascular tunic.

The endocardium corresponds to the tunica intima, the myocardium corresponds to the tunica media, and the connective-tissue part of the epicardium corresponds to the tunica externa.

TYPES OF BLOOD VESSELS

Our examination of the three types of blood vessels—arteries, capillaries, and veins—discusses these vessels in the order in which a drop of blood would encounter them on its journey through the systemic or pulmonary circuit. Accordingly, we begin with arteries.

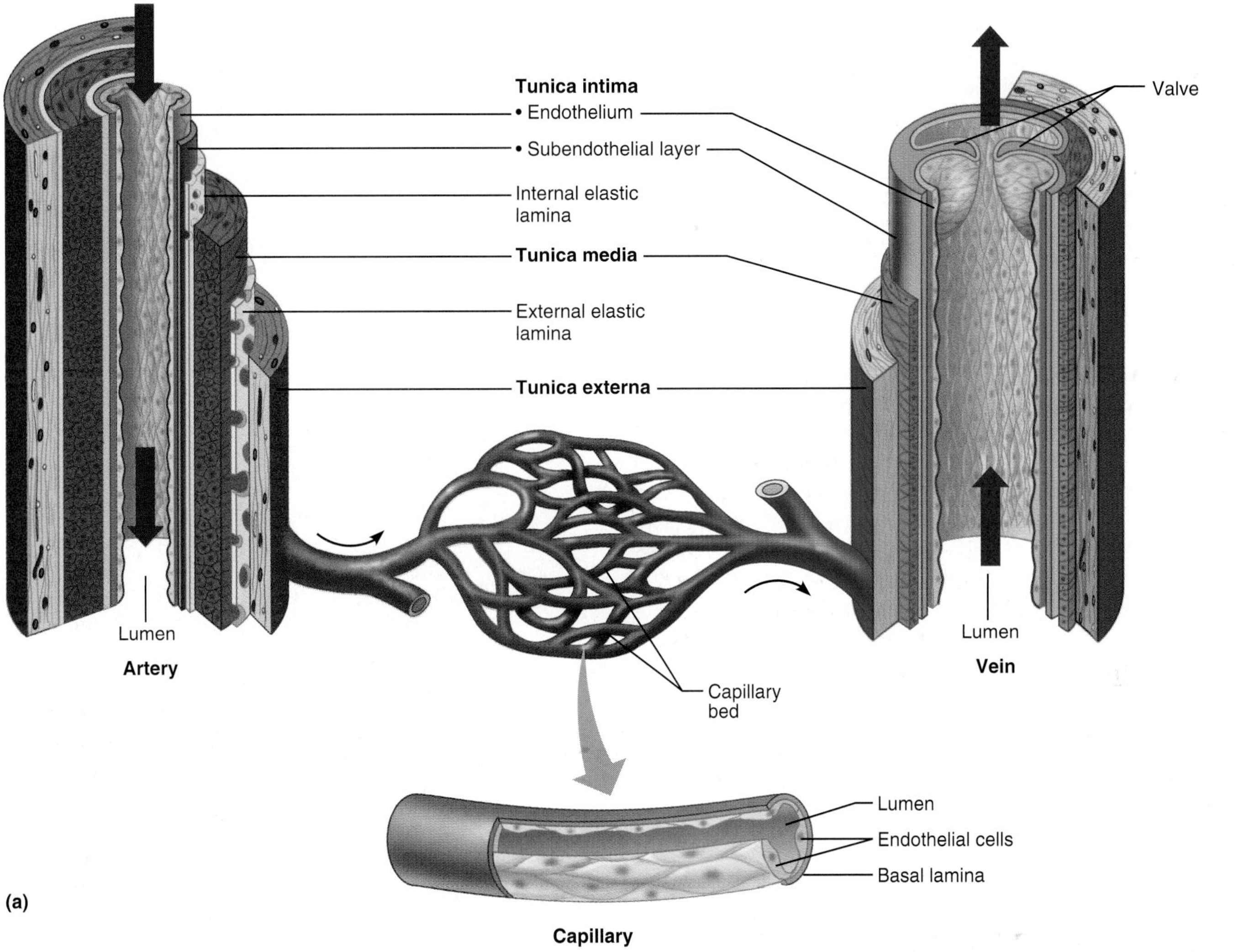

FIGURE 19.1 Structure of arteries, veins, and capillaries. (a) This drawing emphasizes the three tunics in the walls of a muscular artery (left) and a medium-sized vein (right). The artery has a thicker wall, a thicker tunica media, a narrower lumen, and thickened elastic laminae not present in the vein. The vein, by contrast, has a thicker tunica externa, a wider lumen, and valves. Note the capillary bed between the artery and vein, and the diagram of the capillary below this bed. (b) Photomicrograph of a muscular artery and a vein, in cross section (40 ×).

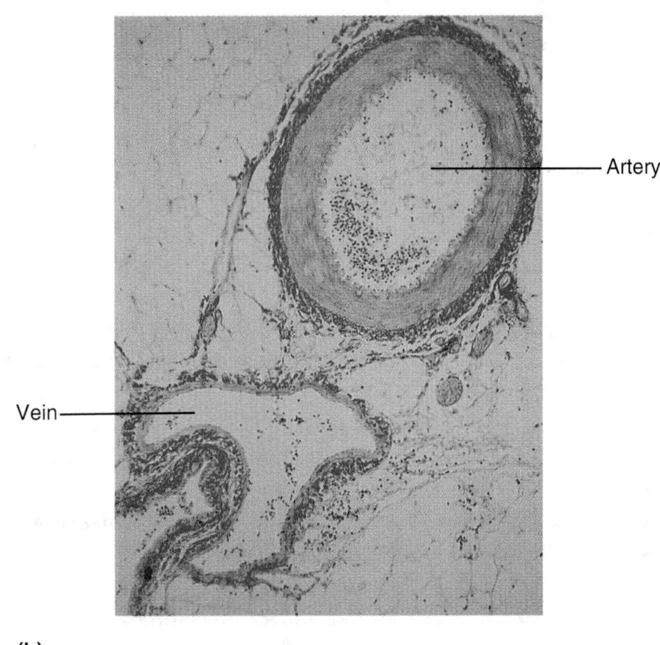

Arteries

Arteries are vessels that carry blood away from the heart. It is a common misconception that *all* arteries carry oxygen-rich blood, whereas *all* veins carry oxygen-poor blood. This notion is correct for the systemic circuit, but it is not correct for the pulmonary circuit, whose arteries carry oxygen-poor blood to the lungs for oxygenation (see Figure 18.1).

Our discussion of arteries will proceed from elastic arteries, to muscular arteries, to arterioles.

Elastic Arteries

Elastic arteries (Figure 19.2) are the largest arteries near the heart—the aorta and its major branches—with diameters ranging from 2.5 cm (about the width of the thumb) to 1 cm (slightly less than the width of the little finger). Their large lumen allows them to serve as low-resistance conduits for conducting blood between the heart and the medium-sized muscular arteries. For this reason, elastic arteries are sometimes called *conducting arteries*. More elastin occurs in the walls of these arteries than in any other type of vessel, and the sheets of elastin in the tunica media are remarkably thick.

The high elastin content of conducting arteries dampens the surges of blood pressure resulting from the rhythmic contractions of the heart. When the heart forces blood into the arteries, the elastic elements in these vessels expand in response to increased blood pressure, in effect storing some of the energy of the flowing fluid; then, when the heart relaxes, the elastic elements recoil, propelling the blood onward. As the blood proceeds through smaller arteries, there is a marked decline both in its absolute pressure (due to resistance imparted by the arterial walls) and in the size of the pressure vacillations (due to the elastic recoil of the arteries just described). By the time the blood reaches the thin-walled capillaries, which are too weak to withstand strong surges in blood pressure, the pressure of the blood is completely steady.

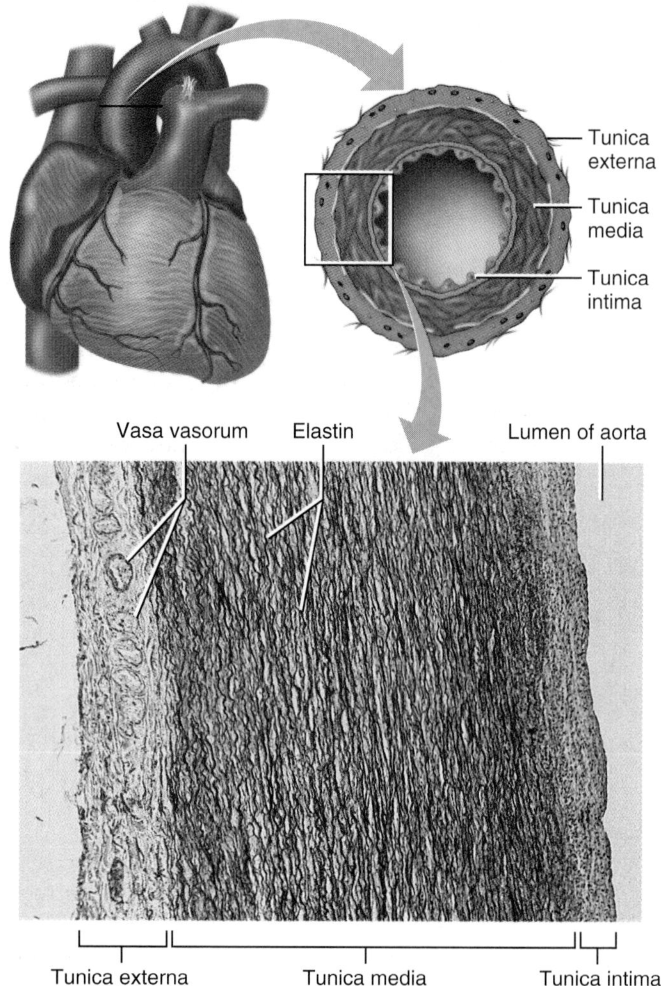

FIGURE 19.2 **Structure of an elastic artery.** Section through the wall of the aorta, showing its layers, its abundant elastin (stained brown), and the vasa vasorum in the tunica externa (30 ×).

Critical Reasoning

Describe some consequences of the loss of elasticity in the conducting arteries, as occurs in arteriosclerosis ("hardening of the arteries").

In the absence of the pressure-dampening effect of the elastic arteries, the walls of arteries throughout the body must withstand pulses of higher maximum pressure. Battered by higher-than-normal pressures, the arteries may eventually rupture, as occurs in some strokes, or may weaken and form a sac-like dilatation, a condition called an aneurysm *(see p. 573)**. Atherosclerosis, a common type of arteriosclerosis, is described in detail in the A Closer Look box on pp. 574–575.*

Muscular Arteries

Muscular (distributing) arteries lie distal to the elastic arteries and supply groups of organs, individual organs, and parts of organs. These "middle-sized" arteries, which range in diameter from about 1 cm to 0.3 mm (as thick as a sharpened pencil lead), constitute most of the named arteries seen in the anatomy laboratory. They are called "muscular" because their tunica media is thicker relative to the size of the lumen than that of any other type of vessel (see Figure 19.1a). By actively changing the diameter of the artery, this muscular layer regulates the amount of blood flowing to the organ supplied according to the specific needs of that organ.

As in all vessels, concentric sheets of elastin occur within the tunica media of muscular arteries—although these sheets are not as thick or abundant as those of elastic arteries. Additionally, as a feature unique to muscular arteries, especially thick sheets of elastin lie on each side of the tunica media: A wavy *internal elastic lamina* lies between the tunica media and the tunica intima, and an *external elastic lamina* lies between the tunica media and tunica externa (Figure 19.1a). The elastin in muscular arteries, like that in elastic arteries, helps dampen the pulsatile pressure produced by the heartbeat.

Arterioles

Arterioles are the smallest arteries, with diameters ranging from about 0.3 mm to 10 μm. Their tunica media contains only one or two layers of smooth muscle cells. Larger arterioles have all three tunics plus an internal elastic lamina. Smaller arterioles, which lead into the capillary beds, are little more than a single layer of smooth muscle cells spiraling around an underlying endothelium.

The diameter of each arteriole is regulated in two ways: (1) Local factors in the tissues signal the smooth musculature to contract or relax, thus regulating the amount of blood sent downstream to each capillary bed;

and (2) the sympathetic nervous system adjusts the diameter of arterioles throughout the body to regulate systemic blood pressure. For example, a widespread sympathetic vasoconstriction raises blood pressure during fight-or-flight responses (see Chapter 15, p. 446).

Capillaries

Capillaries are the smallest blood vessels, with a diameter of 8–10 μm, just large enough to enable erythrocytes to pass through in single file. They are composed of only a single layer of endothelial cells surrounded by a basal lamina (see Figure 19.1a). They are the body's most important blood vessels because they renew and refresh the surrounding tissue fluid (interstitial fluid; p. 91) with which all body cells are in contact, delivering to this fluid the oxygen and nutrients cells need, and removing carbon dioxide and nitrogenous wastes that cells deposit into the fluid. Along with these universal functions, some capillaries also perform site-specific functions. In the lungs, oxygen enters the blood (and carbon dioxide leaves it) through capillaries. Capillaries in the small intestine receive digested nutrients, those in the endocrine glands pick up hormones, and those in the kidneys function in the removal of nitrogenous wastes from the body.

In the following sections we examine the structures by which capillaries supply cells in tissues—capillary beds—and then we examine the anatomical basis of how they deliver and pick up substances—capillary permeability.

Capillary Beds

A *capillary bed* is a network of the body's smallest vessels that run throughout almost all tissues, especially the loose connective tissues (Figure 19.3). A terminal arteriole leads to a **metarteriole**—a vessel that is structurally intermediate between an arteriole and a capillary—off of which branch true capillaries. The metarteriole then continues into a **thoroughfare channel,** a vessel structurally intermediate between a capillary and a venule. The thoroughfare channel then proceeds to join a venule, receiving true capillaries along the way. Smooth muscle cells called **precapillary sphincters** wrap around the root of each true capillary where it leaves the metarteriole.

The precapillary sphincters regulate the flow of blood to the tissue according to that tissue's needs for oxygen and nutrients. When the tissue is functionally active, the sphincters are relaxed, enabling blood to flow through the wide-open capillaries and supply the surrounding tissue cells. When the tissue has lower demands (such as when nearby tissue cells already have adequate oxygen), then the precapillary sphincters contract, closing off the true capillaries and forcing blood to flow straight from the metarteriole into the thoroughfare channel and venule—thereby bypassing the

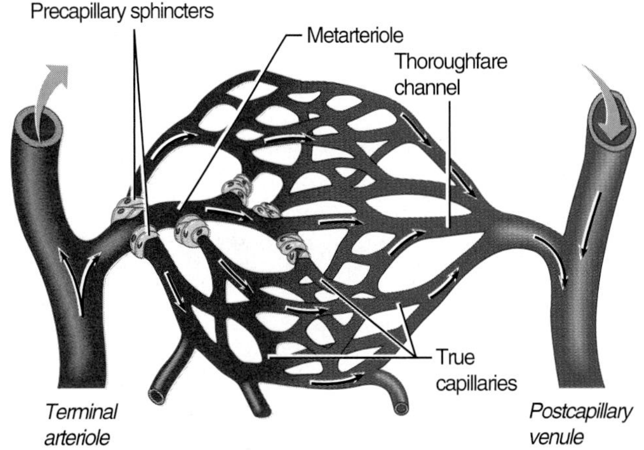

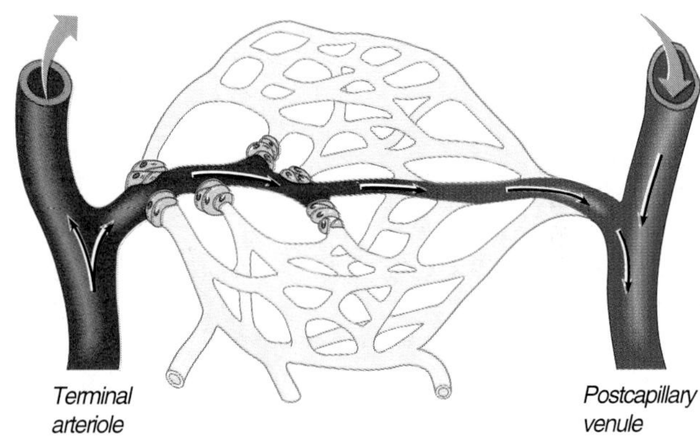

FIGURE 19.3 Generalized anatomy of a capillary bed, as seen in a mesentery. **(a)** When precapillary sphincters relax, blood fills the true capillaries. **(b)** When these sphincters contract, they force most blood to flow straight from metarterioles to thoroughfare channels, bypassing the true capillaries.

true capillaries. In this way, capillary beds precisely control the amount of blood supplying a tissue at any time.

Even though most tissues and organs have a rich capillary supply, not all do. Tendons and ligaments are poorly vascularized. Epithelia and cartilage contain no capillaries; instead, they receive nutrients indirectly from nearby vascularized connective tissues. The cornea and the lens have no capillary supply at all and are nourished by the aqueous humor and other sources (see pp. 475 and 480).

Capillary Permeability

The structure of capillaries is well suited for their function in the exchange of nutrients and wastes between the blood and the tissues through the tissue fluid. As mentioned, a capillary is a tube consisting of a layer of thin endothelial cells surrounded by a basal lamina (Figure 19.4). The endothelial cells are held together by tight junctions and occasional desmosomes. Tight junctions

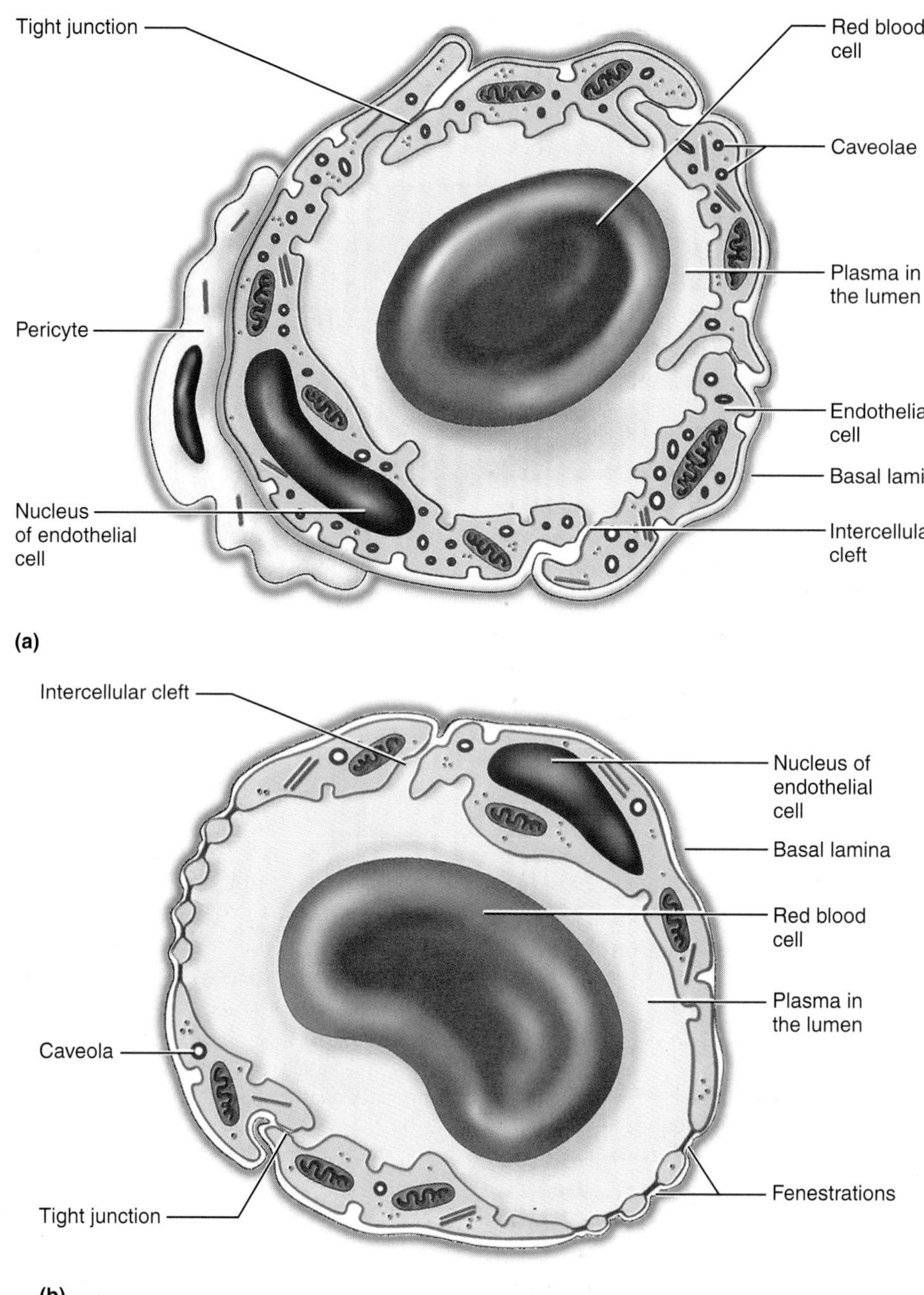

FIGURE 19.4 **Structure of capillaries, cut in cross section.** (a) Continuous capillary. (b) Fenestrated capillary. (c) Sinusoid.

block the passage of small molecules (p. 85), but such junctions do not surround the whole perimeter of the endothelial cells. Instead, they leave gaps of unjoined membrane, called **intercellular clefts,** through which small molecules exit and enter the capillary. External to the endothelial cells, the delicate capillary is strengthened and stabilized by scattered **pericytes** (per′ĭ-sīts; "surrounding cells"), spider-shaped cells whose thin processes form a network that is widely spaced so as not to interfere with capillary permeability.

Some capillaries are *continuous* and others are *fenestrated* (fen′is-tra-ted). The only structural difference between these two types is that fenestrated capillaries have pores (fenestrations) spanning the endothelial cells and continuous capillaries lack such pores. Continuous capillaries (Figure 19.4a) are the more common type, occurring in most organs of the body, such as skeletal muscles, skin, and the central nervous system. Fenestrated capillaries (Figure 19.4b), on the other hand, occur only where there are exceptionally high rates of

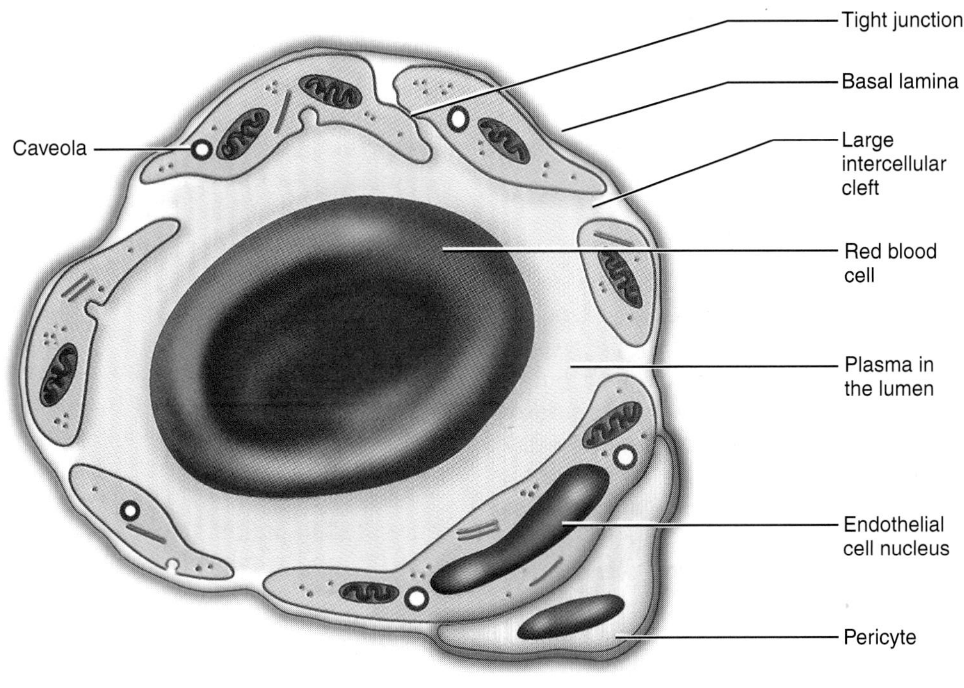

exchange of small molecules between the blood and the surrounding tissue fluid. For example, capillaries in the small intestine, which receive the digested nutrients from food, are fenestrated. So are capillaries in the synovial membranes of joints, where many water molecules exit the blood to contribute to the synovial fluid.

Routes of Capillary Permeability Molecules pass into and out of capillaries via four routes: (1) direct diffusion through the endothelial cell membranes; (2) through the intercellular clefts; (3) through cytoplasmic vesicles, or caveolae (see p. 33 and Figure 19.4), that invaginate from the plasma membrane and migrate across the endothelial cell; and (4) through the fenestrations in fenestrated capillaries. Most exchange of small molecules is thought to occur through the intercellular clefts; caveolae apparently transport a few larger molecules, such as small proteins. Carbon dioxide and oxygen seem to be the only important molecules that diffuse directly through endothelial cells, because these uncharged molecules easily diffuse through the lipid-containing membranes of cells.

Low-Permeability Capillaries: The Blood-Brain Barrier The *blood-brain barrier,* which prevents all but the most vital molecules (and normally even leukocytes) from leaving the blood and entering brain tissue (see p. 391), derives from the structure of capillaries in the brain. More accurately stated, the blood-brain barrier actually derives from the *lack* of the structural features that account for capillary permeability: Brain capillaries have complete tight junctions, so intercellular clefts are absent; they are continuous, not fenestrated; and they lack caveolae. Instead, the vital molecules that must cross brain capillaries are "ushered through" by highly selective transport mechanisms in the plasma membranes of the endothelial cells. (It should be noted that the blood-brain barrier is *not* a barrier against uncharged and lipid-soluble molecules such as oxygen, carbon dioxide, and some anesthetics, which diffuse unhindered through the endothelial cells and freely enter brain tissue.)

Advances in Understanding It was recently discovered that during prolonged emotional stress, the tight junctions between the endothelial cells of brain capillaries are opened, so that the blood-brain barrier fails and toxic substances in the blood can enter brain tissue. Further study of this phenomenon could one day help medical scientists who are seeking ways to deliver beneficial drugs—antibiotics and chemicals to kill brain tumors—across the blood-brain barrier into the brain. ✷

Sinusoids

Some organs contain wide, leaky capillaries called **sinusoids** or **sinusoidal capillaries** (Figure 19.4c). Each sinusoid follows a twisted path and has both expanded and narrowed regions. Sinusoids are usually fenestrated, and their endothelial cells have fewer cell junctions than do ordinary capillaries. In some sinusoids, in fact, the intercellular clefts are wide open. Sinusoids occur wherever there is an extensive exchange of *large* materials, such as proteins or cells, between the blood and surrounding tissue. For example, they occur in the bone marrow and spleen, where many blood cells move through their walls. The large diameter and twisted course of sinusoids ensure that blood slows when flowing through these vessels, allowing time for the many exchanges that occur across their walls.

Veins

Veins are the blood vessels that conduct blood from the capillaries toward the heart. Those veins in the systemic circuit carry blood that is relatively oxygen-poor, but the pulmonary veins carry oxygen-rich blood returning from the lungs. Because blood pressure declines substantially while passing through the high-resistance arterioles and capillary beds, blood pressure in the venous part of the circulation is much lower than in the arterial part. Because they need not withstand as much pressure, the walls of veins are thinner than those of comparable arteries.

The smallest veins are called **venules,** which are 8–100 μm in diameter. The smallest venules, called **postcapillary venules,** consist of an endothelium on which lie pericytes, and these venules function very much like capillaries. In fact, during inflammatory responses more fluid and leukocytes leave the circulation through postcapillary venules than through capillaries. Larger venules have one or two layers of smooth muscle cells (a tunica media) and a thin tunica externa as well.

Venules join to form **veins.** The lumens of veins are larger than those of arteries of comparable size; at any given time veins hold fully 65% of the body's blood. In veins, the tunica externa is thicker than the tunica media, a relationship that is the opposite of that in arteries. In the body's largest veins—the venae cavae, which return systemic blood to the heart—the tunica externa is further thickened by longitudinal bands of smooth muscle. Veins have less elastin in their walls than do arteries, because veins need not dampen any pulsations (all of which are smoothed out by arteries before the blood reaches the veins).

Several mechanisms counteract the low venous blood pressure by helping to move the blood along its course back to the heart. One structural feature of some veins is valves that prevent the backflow of blood away from the heart (Figure 19.5a). Each of these valves has several cusps formed from the tunica intima. The flow of blood toward the heart pushes the cusps apart so the valve opens, and any backflow pushes the cusps together so the valve closes. Valves are most abundant in veins of the limbs, where the superior direction of venous flow is most directly opposed by gravity. A few valves occur in the veins of the head and neck, but none are located in veins of the thoracic and abdominal cavities.

One functional mechanism that aids the return of venous blood to the heart is the normal movement of our body and limbs, as when we walk: Swinging a limb moves the blood in this limb, and the venous valves ensure this blood moves only in the proper direction. Another mechanism aiding venous return is the so-called skeletal muscular pump, in which contracting skeletal muscles press against the thin-walled veins, forcing valves proximal to the area of contraction open and propelling blood toward the heart (Figure 19.5b). Valves distal to the contracting muscles are closed by backflowing blood.

The effectiveness of venous valves in preventing the backflow of blood is easily demonstrated. Hang one hand by your side until the veins on its dorsal surface become distended with blood. Next, place two fingertips against one of the distended veins, and, pressing firmly, move the superior finger proximally along the vein, and then release that finger. The vein stays flat and collapsed despite the pull of gravity. Finally, remove the distal fingertip, and watch the vein refill rapidly with blood.

Clinical Applications

VARICOSE VEINS When the valves in veins weaken and fail, the result is **varicose veins,** in which veins are twisted and swell with large amounts of pooled blood and the venous drainage is slowed considerably. Fully 15% of all adults suffer from varicose veins, usually in the lower limbs. Females are affected more often than males. Varicose veins can be hereditary but also occur in people whose jobs require prolonged standing in one position, such as store clerks, hairdressers, dentists, and nurses. (The causal sequence here is that in nonmoving legs the venous blood drains so slowly that it accumulates, stretches the venous walls and valves, and causes the valves to fail.) Obesity and pregnancy can cause or worsen the problem, because excessive weight constricts the leg-draining veins in the superior thigh. Another cause of varicose veins—elevated venous pressure—results when straining to deliver a baby or to have a bowel movement raises the intra-abdominal pressure, preventing drainage of blood from the veins of the anal canal at the inferior end of the large intestine. The resulting varicosities in these anal veins are called **hemorrhoids.**

A severe case of varicose veins so effectively slows the circulation through a body region that the tissues in that region die of oxygen starvation. To prevent this, physicians either remove the affected veins or inject them with an irritating solution that causes them to scar and fuse shut. Drainage of the body region then proceeds normally, through alternate vascular pathways called anastomoses, which are the topic of the next section. ✚

VASCULAR ANASTOMOSES

Where vessels unite or interconnect, they form *vascular anastomoses* (ah-nas″to-mo′sēz; "coming together"). Most organs receive blood from more than one arterial branch, and neighboring arteries often communicate with one another to form *arterial anastomoses.* Arterial anastomoses provide alternate pathways, or *collateral channels,* for blood to reach a given body region. If one arterial branch is blocked or cut, the collateral channels can often provide the region with an adequate blood

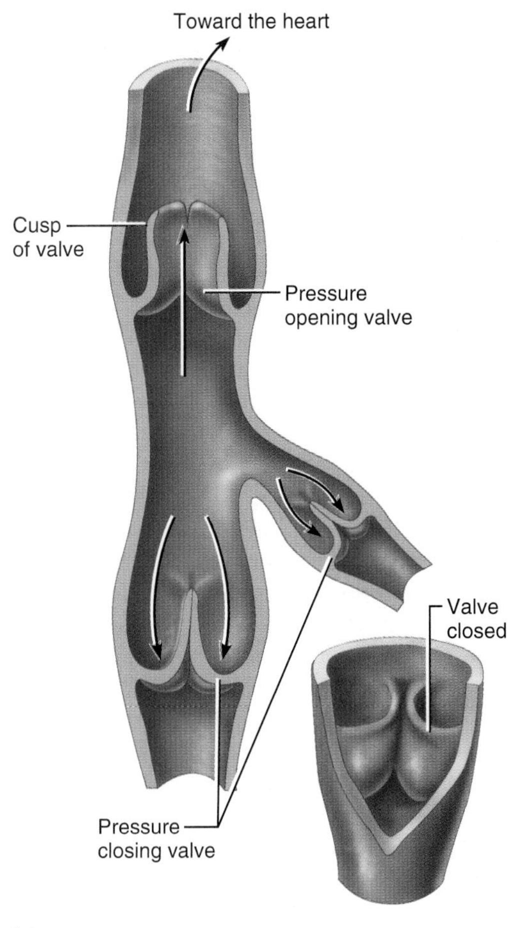

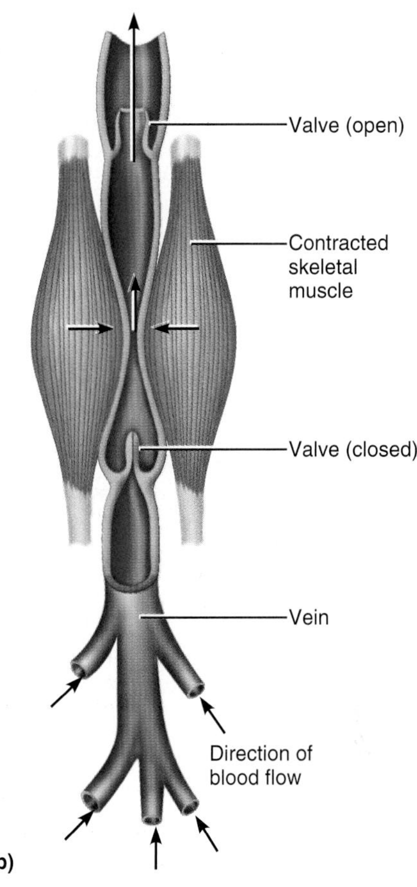

FIGURE 19.5 Valves in veins, and the muscular pump. **(a)** General function of the valves: The valves are opened by blood flowing toward the heart and are closed by backflow. **(b)** How the skeletal muscular pump aids venous return: Contracting skeletal muscles press against a vein and propel blood toward the heart, forcing valves proximal to the muscles to open and valves distal to close.

supply. Arterial anastomoses occur around joints, where active body movements may hinder blood flow through one channel, as well as in the abdominal organs, brain, and heart. Because of the many anastomoses among the smaller branches of the coronary artery in the heart wall, a coronary artery can be 90% occluded by atherosclerosis before a myocardial infarction occurs. By contrast, because arterial anastomoses are poorly developed in the kidneys, in the spleen, in the parts of bone diaphyses nearest the epiphyses, and in the central artery of the retina, blockage of such arteries causes severe tissue damage.

Veins anastomose much more freely than arteries. You may be able to see *venous anastomoses* through the skin on the dorsum of your hand. Because of the abundant anastomoses, occlusion of a vein rarely blocks blood flow or leads to tissue death.

VASA VASORUM

The walls of blood vessels contain living cells and therefore require a blood supply of their own. For this purpose, the larger arteries and veins have tiny arteries, capillaries, and veins in their tunica externa (see Figure 19.2). These little vessels, the **vasa vasorum** (va′sah va-sor′um; "vessels of the vessels"), nourish the outer half of the wall of the larger vessel. The inner half, by contrast, gets its nutrients from the blood in the vessel's own lumen. Small blood vessels need no vasa vasorum because their walls are entirely supplied by luminal blood. The vasa vasorum to a large vessel may arise either as tiny branches from the same vessel or as small branches from other, nearby arteries and veins.

PART TWO

BLOOD VESSELS OF THE BODY

Recall from Chapter 18 that the complex system of blood vessels in the body called the *vascular system* has two basic circuits: The *pulmonary circuit* carries blood to and from the lungs for the uptake of oxygen and the removal of carbon dioxide, whereas the *systemic circuit* carries oxygenated blood throughout the body and picks up carbon dioxide from body tissues (see Figure 18.1). Blood vessels in the systemic circuit also (1) pick up nutrients from the digestive tract and deliver them to cells throughout the body, and (2) receive nitrogenous wastes from body cells and transport them to the kidneys for elimination in the urine.

As you read about the vessels, note that arteries and veins tend to run together, side by side (see Figure 19.1b). In many places, these vessels also run with nerves. Note also that by convention, vessels carrying oxygen-rich blood are depicted as red, whereas those transporting oxygen-poor blood are depicted as blue, regardless of vessel type.

THE PULMONARY CIRCULATION

The pulmonary circulation begins as oxygen-poor blood leaves the right ventricle of the heart via the **pulmonary trunk** (Figure 19.6). This large artery exits the ventricle anterior to the aorta, ascends to the aorta's left, and reaches the concavity of the aortic arch, where it branches at a T-shaped divergence into the **right** and **left pulmonary arteries.** Each pulmonary artery penetrates the medial surface of a lung and then divides into several **lobar arteries**, of which there are three in the right lung and two in the left lung. Within the lung, the arteries branch along with the lung's air passageways (bronchi). As the branching arteries decrease in size, they become arterioles and finally the pulmonary capillaries that surround the delicate air sacs (lung alveoli). Gas exchange occurs across these capillaries, and the newly oxygenated blood enters venules and then progressively larger veins. The largest venous tributaries form the superior and inferior **pulmonary veins,** which exit the medial aspect of each lung. In the mediastinum posterior to the heart, the four pulmonary veins run horizontally, just inferior to the pulmonary arteries, and empty into the left atrium.

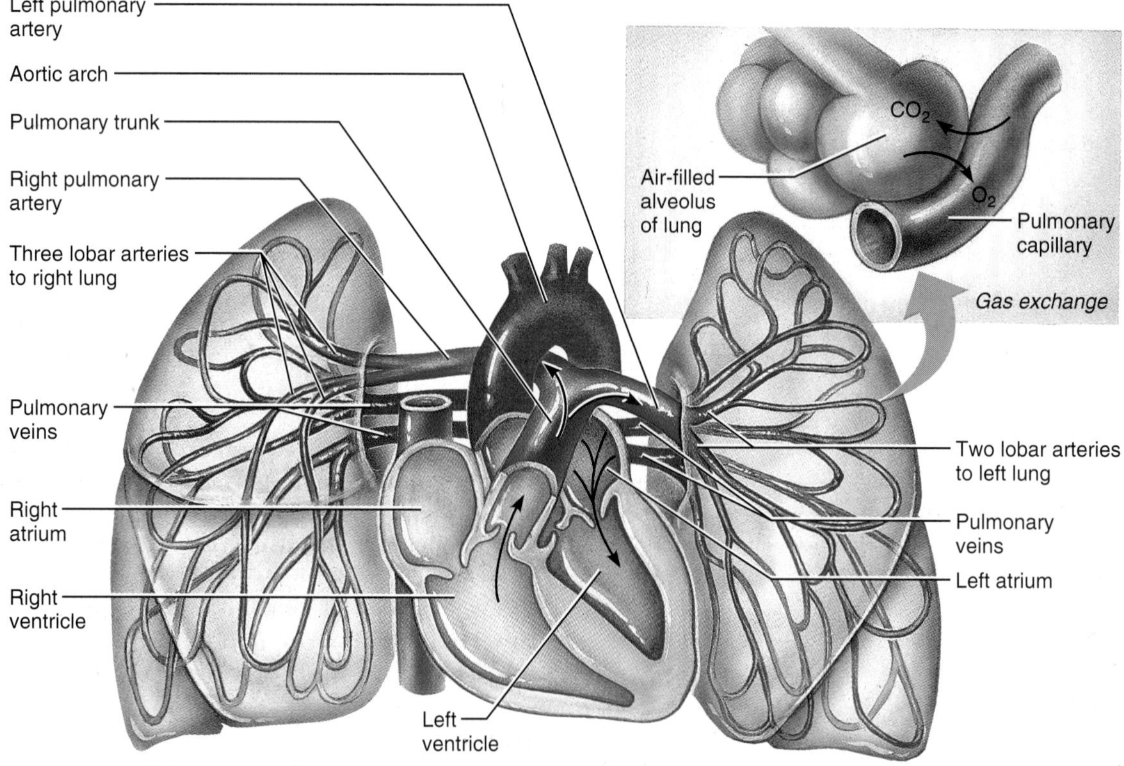

FIGURE 19.6 **Pulmonary circulation.** The pulmonary trunk divides into right and left pulmonary arteries, which subdivide into the lobar arteries serving the lobes (main divisions) of the lungs. After passing through the pulmonary capillaries (inset at upper right), where gas exchange occurs, blood drains to the pulmonary veins (two per lung), which return the blood to the heart. The pulmonary arteries are colored blue to indicate that their blood is oxygen-poor, while pulmonary veins are colored red to indicate that their blood is oxygen-rich.

The arteries and veins of the pulmonary circuit have thinner walls than do systemic vessels of comparable diameter, reflecting the fact that the maximum arterial pressure here is only one-sixth that in the systemic circuit.

THE SYSTEMIC CIRCULATION

Before we examine the systemic vessels, it is worth noting that the vessels on the right and left sides of the body are not always mirror images of each other. That is, some of the large, deep vessels of the trunk region are asymmetrical (their initial symmetry is lost during embryonic development). In the head and limbs, by contrast, almost all vessels are bilaterally symmetrical. We begin the discussion by examining the systemic arteries.

Systemic Arteries

The systemic arteries (Figure 19.7) carry oxygenated blood from the heart to the capillaries of organs throughout the body. We will examine the aorta first and then consider the systemic arteries in a generally superior-to-inferior direction.

Aorta

The **aorta,** the largest artery in the body, leaves the heart, arcs superiorly, and then descends along the bodies of the vertebrae to the inferior part of the abdomen (see Figure 19.7). Along this course, the aorta is divided into the following three parts:

Ascending Aorta The ascending aorta, one of the great vessels leaving the heart, arises from the left ventricle and ascends for only about 5 cm (Figure 19.8). It begins posterior to the pulmonary trunk, passes to the right of that vessel, and then curves left to become the aortic arch. The only branches of the ascending aorta are the two *coronary arteries* that supply the wall of the heart (see pp. 534–535).

Aortic Arch Arching posteriorly and to the left, the aortic arch lies posterior to the manubrium of the sternum. The **ligamentum arteriosum,** a fibrous remnant of a fetal artery called the *ductus arteriosus,* interconnects the aortic arch and the pulmonary trunk (see Figure 19.8 and p. 576).

Three arteries branch from the aortic arch and run superiorly (see Figure 19.8). The first and largest branch is the **brachiocephalic trunk** (bra″ke-o-sĕ-fal′ik; "arm-head"), which ascends to the base of the neck, where it divides into the *right common carotid* and *right subclavian* arteries. The second and third branches of the aortic arch are the **left common carotid** and **left subclavian arteries,** respectively. These three branches of the aorta supply the head and neck, upper limbs, and the superior part of the thoracic wall. Note that the brachiocephalic trunk on the right has no corresponding artery on the left because the left common carotid and subclavian arteries arise directly from the aorta.

Descending Aorta Continuing from the aortic arch, the **descending aorta** runs inferiorly on the bodies of the thoracic and lumbar vertebrae. It has two parts, the *thoracic aorta* and the *abdominal aorta.*

The **thoracic aorta** (see Figures 19.7 and 19.8) descends on the bodies of the thoracic vertebrae (T_5–T_{12}) just to the left of the midline. Along the way, it sends many small branches to the thoracic organs and body wall (see pp. 557–559).

The thoracic aorta pierces the diaphragm at the level of vertebra T_{12} and enters the abdominal cavity as the **abdominal aorta** (see Figure 19.7), which lies on the lumbar vertebral bodies in the midline. The abdominal aorta ends at the level of vertebra L_4, where it divides into the right and left *common iliac arteries,* which supply the pelvis and lower limbs.

Arteries of the Head and Neck

Four pairs of arteries supply the head and neck: the *common carotid arteries* plus three branches from each subclavian artery—the *vertebral artery,* the *thyrocervical trunk,* and the *costocervical trunk* (Figure 19.9a).

Common Carotid Arteries Most parts of the head and neck receive their blood from the common carotid arteries, which ascend through the anterior neck just lateral to the trachea. These vessels, which are covered by the relatively thin sternocleidomastoid and infrahyoid muscles, are more vulnerable than most other arteries in the body because their relatively superficial location makes them vulnerable to slashing wounds. If a common carotid artery is cut, the victim can bleed to death in minutes. At the superior border of the larynx—the level of the "Adam's apple"—each common carotid ends by dividing into an external and internal carotid artery.

The **external carotid arteries** supply most tissues of the head external to the brain and orbit. As each external carotid ascends, it sends a branch to the thyroid gland and larynx **(superior thyroid artery),** to the tongue **(lingual artery),** to the skin and muscles of the anterior face **(facial artery),** and to the posterior part of the scalp **(occipital artery).** Near the temporomandibular joint, each external carotid ends by splitting into the superficial temporal and maxillary arteries. The **superficial temporal artery** ascends just anterior to the ear and supplies most of the scalp. Branches of this vessel bleed profusely in scalp wounds. The **maxillary artery** runs anteriorly into the maxillary bone by passing through the chewing muscles. Along the way, it sends branches to the upper and lower teeth, the cheeks, nasal cavity, and muscles of mastication.

A clinically important branch of the maxillary artery is the *middle meningeal artery* (not illustrated), which

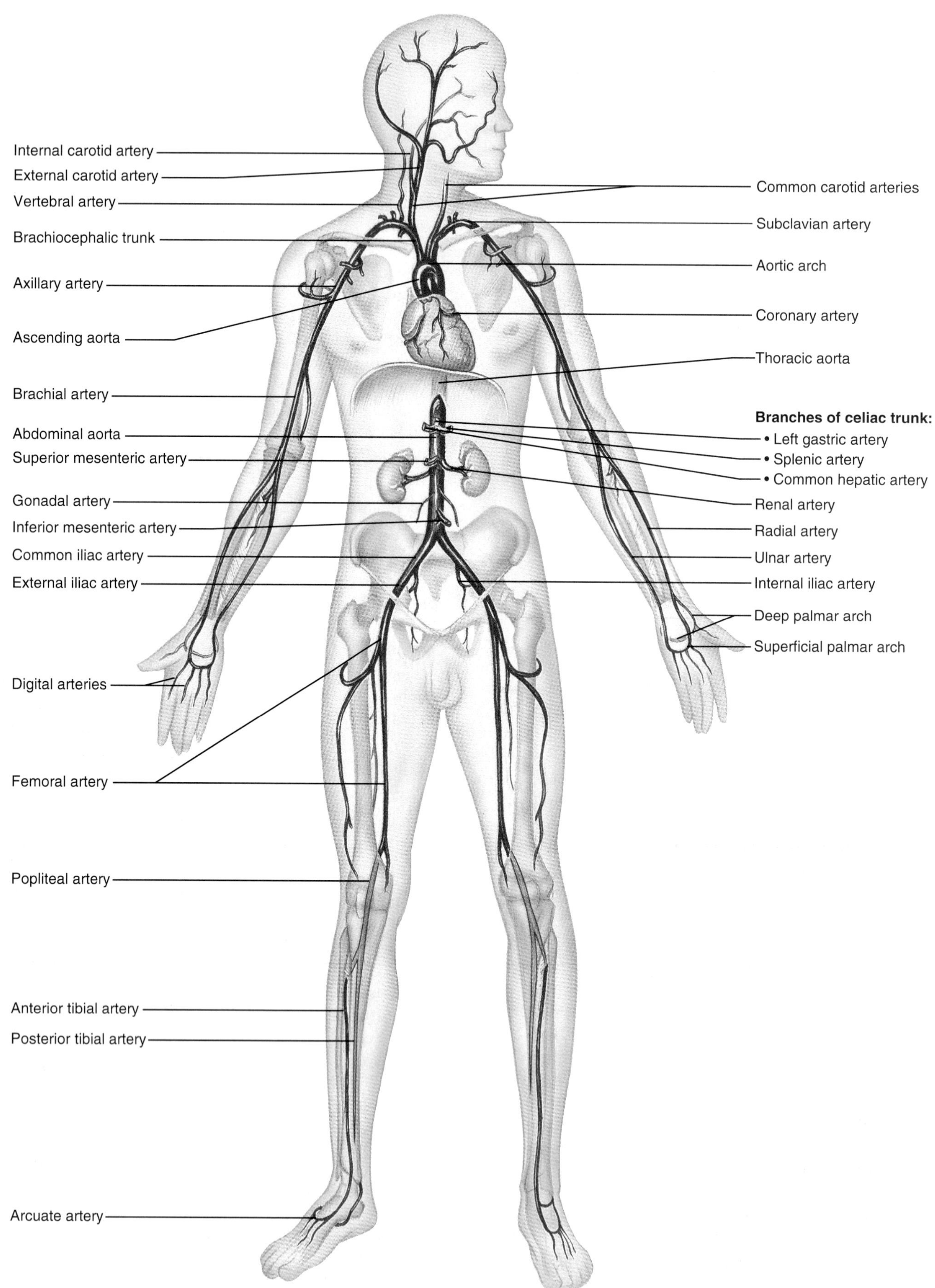

FIGURE 19.7 Major arteries of the systemic circulation.

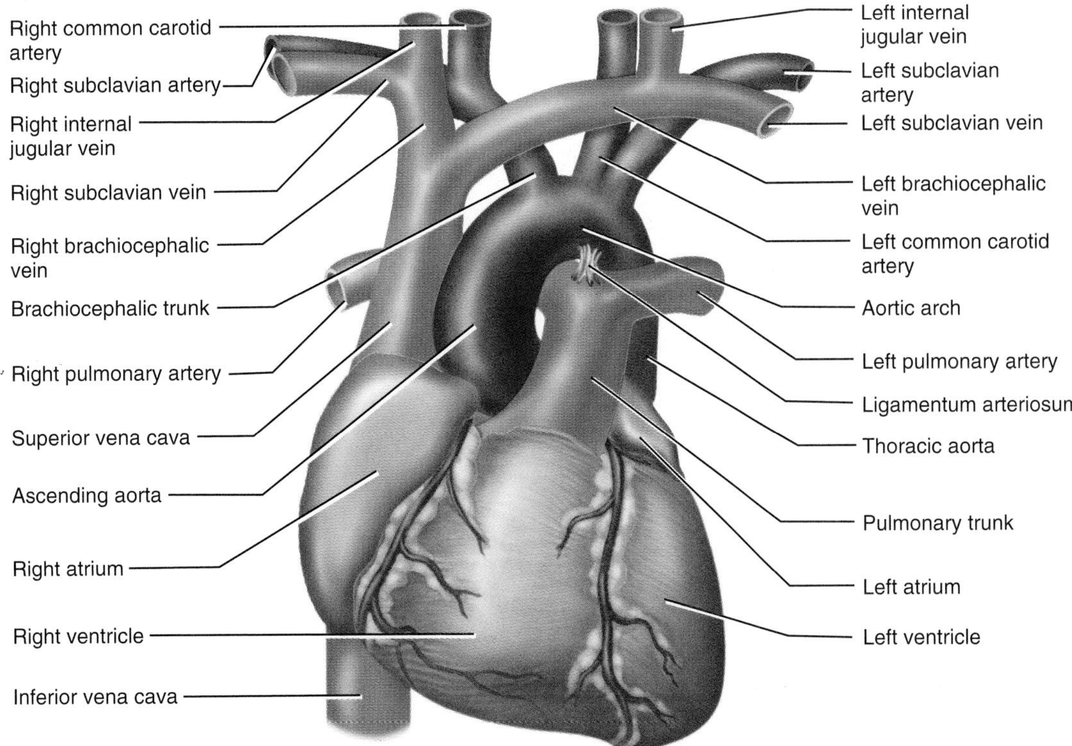

FIGURE 19.8 The great vessels that exit and enter the heart.

enters the skull through the foramen spinosum and supplies the broad inner surfaces of the parietal bone and squamous region of the temporal bone, as well as the underlying dura mater. Blows to the side of the head often tear this artery, producing an intracranial hematoma that can compress the cerebrum and disrupt brain function.

The **internal carotid arteries** supply the orbits and most of the cerebrum. Each internal carotid ascends through the superior neck directly lateral to the pharynx and enters the skull through the carotid canal in the temporal bone. From there, it runs medially through the petrous region of the temporal bone, runs forward along the body of the sphenoid bone, and bends superiorly to enter the sella turcica just posterior to the optic foramen (see Figure 16.18, p. 488). Here, it gives off the **ophthalmic artery** to the eye and orbit (see Figure 19.9a) and divides into the *anterior* and *middle cerebral arteries* (Figure 19.9c).

Each **anterior cerebral artery** anastomoses with its partner on the opposite side via a short **anterior communicating artery** (see Figure 19.9c) and supplies the medial surfaces of the frontal and parietal lobes. Each **middle cerebral artery** runs through the lateral fissure of a cerebral hemisphere and supplies the lateral parts of the temporal and parietal lobes. Together, the anterior and middle cerebral arteries supply over 80% of the cerebrum; the rest of the cerebrum is supplied by the posterior cerebral artery (described shortly).

Vertebral Arteries The blood supply to the posterior brain comes from the right and left **vertebral arteries,** which arise from the subclavian arteries at the root of the neck (see Figure 19.9a). The vertebral arteries ascend through the foramina in the transverse processes of cervical vertebrae C_6 to C_1 and enter the skull through the foramen magnum. Along the way, they send branches to the vertebrae and cervical spinal cord. Within the cranium, the right and left vertebral arteries join to form the unpaired **basilar** (bas′ĭ-lar) **artery** (see Figure 19.9c), which ascends along the ventral midline of the brain stem, sending branches to the cerebellum, pons, and inner ear. At the border of the pons and midbrain, it divides into a pair of **posterior cerebral arteries,** which supply the occipital lobes plus the inferior and medial parts of the temporal lobes of the cerebral hemispheres.

Short **posterior communicating arteries** connect the posterior cerebral arteries to the middle cerebral arteries anteriorly. The two posterior communicating arteries and the single anterior communicating artery complete the formation of an arterial anastomosis called the **cerebral arterial circle** (formerly the **circle of Willis**). This circle forms a loop around the pituitary gland and optic chiasma, and it unites the brain's anterior and posterior blood supplies provided by the internal carotid and vertebral arteries.

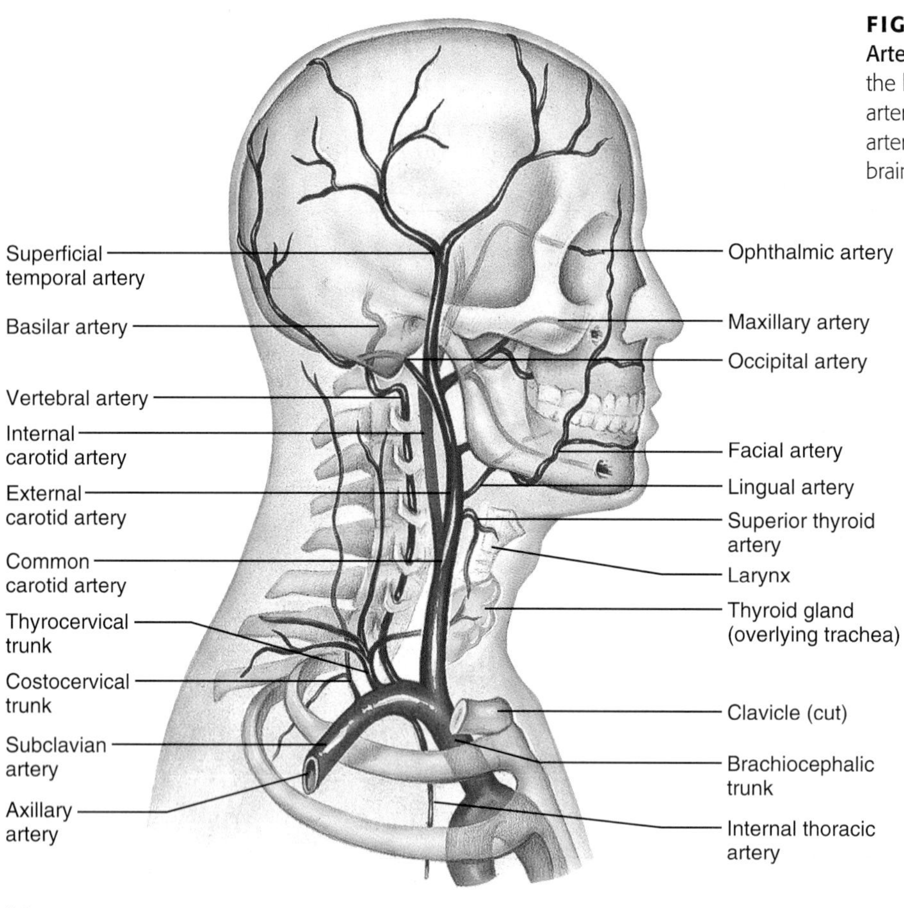

FIGURE 19.9
Arteries of the head, neck, and brain. **(a)** Arteries of the head and neck, right aspect. **(b)** Colorized arteriograph (X ray of injected vessels) showing the arterial supply to the brain. (c) Major arteries serving the brain (ventral view).

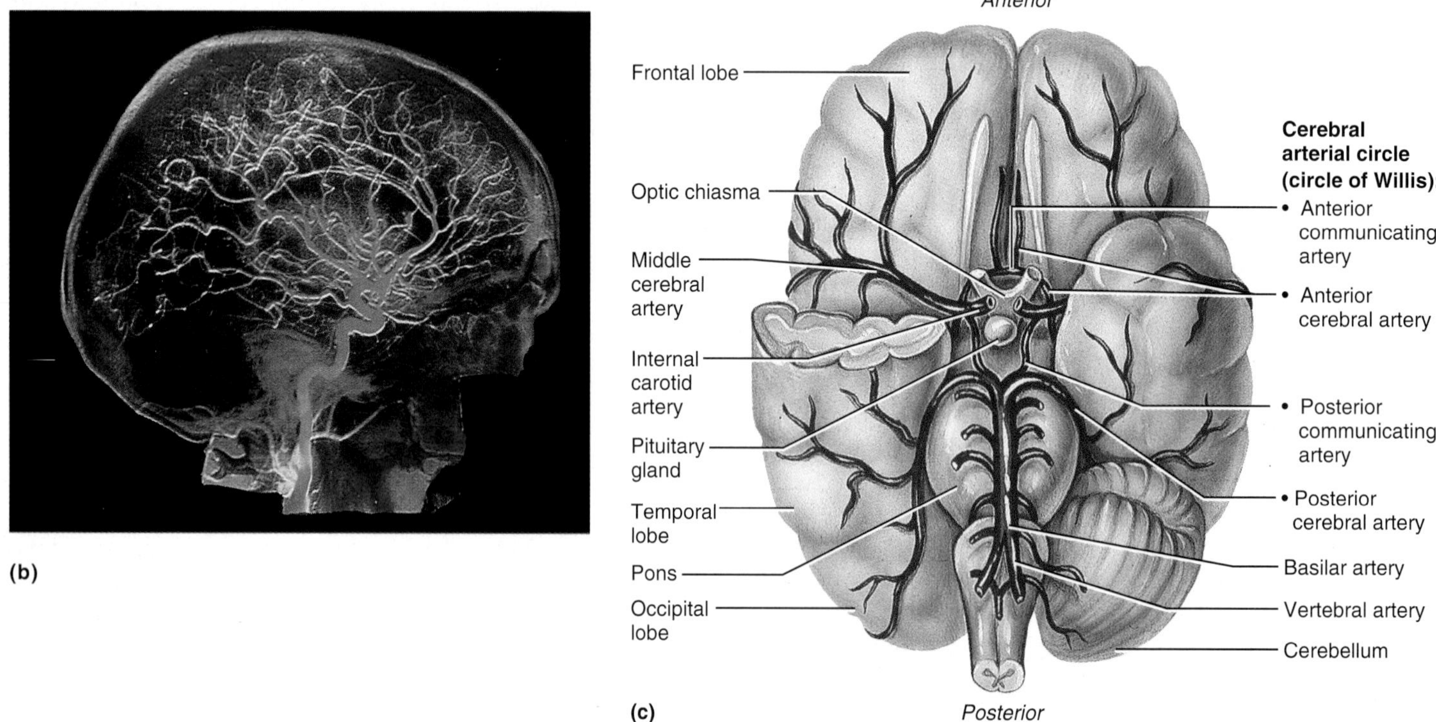

Critical Reasoning

Describe the adaptive function performed by the cerebral arterial circle.

By interconnecting the arteries that supply the anterior, posterior, left, and right parts of the brain, this anastomosis provides alternate routes for blood to reach brain areas that are affected if either a carotid or vertebral artery becomes occluded.

Thyrocervical and Costocervical Trunks The rest of the neck receives its blood from two smaller branches of the subclavian arteries, the *thyrocervical* and *costocervical trunks* (see Figure 19.9a).

The **thyrocervical** (thi″ro-ser′vĭ-kal) **trunk,** which arises first, sends two branches posteriorly over the scapula to help supply the scapular muscles, and one branch anteriorly to the inferior part of the thyroid gland *(inferior thyroid artery).* The ascending neck branch illustrated in Figure 19.9a comes off the base of the inferior thyroid artery and helps to supply the cervical vertebrae and spinal cord.

The **costocervical trunk** sends a branch superiorly into the deep muscles of the neck and a branch inferiorly to supply the two most superior intercostal spaces.

Arteries of the Upper Limb

The upper limb is supplied by arteries that arise from the subclavian artery (Figure 19.10). After giving off its branches to the neck, each subclavian artery runs laterally onto the first rib, where it underlies the clavicle. From here, the subclavian artery enters the axilla as the *axillary artery.*

Axillary Artery The axillary artery descends through the axilla, giving off the following branches: (a) the **thoracoacromial** (tho″rah-ko-ah-kro′me-al) **artery,** which arises just inferior to the clavicle and branches to supply much of the pectoralis and deltoid muscles; (b) the **lateral thoracic artery,** which descends along the lateral edge of pectoralis minor and sends important branches to the breast; (c) the **subscapular artery,** which serves the dorsal and ventral scapular regions and the latissimus dorsi muscle; and (d) **anterior** and **posterior circumflex humeral arteries,** which wrap around the surgical neck of the humerus and help supply the deltoid muscle and shoulder joint. The axillary artery continues into the arm as the *brachial artery.*

Brachial Artery The brachial artery descends along the medial side of the humerus deep to the biceps muscle and supplies the anterior arm muscles. One major branch, the **deep artery of the arm** (also called *profunda brachii = deep brachial*), wraps around the posterior surface of the humerus with the radial nerve and serves the triceps muscle. As the brachial artery nears the elbow, it sends several small branches inferiorly that anastomose with branches ascending from arteries in the forearm to supply the elbow joint. The brachial artery crosses the anterior aspect of the elbow joint in the midline of the arm, a site at which its pulse is easily felt, and where one listens when measuring blood pressure. Immediately beyond the elbow joint, the brachial artery splits into the *radial* and *ulnar arteries,* which descend through the anterior aspect of the forearm.

Radial Artery The radial artery descends along the medial margin of the brachioradialis muscle, supplying muscles of the lateral anterior forearm, the lateral part of the wrist, and the thumb and index finger. At the root of the thumb, it lies very near the surface and provides a convenient site for taking the pulse (see p. 789).

Ulnar Artery The ulnar artery, which descends along the medial side of the anterior forearm, lies between the superficial and deep flexor muscles and sends branches to the muscles that cover the ulna. Proximally, it gives off a major branch called the **common interosseous artery,** which splits immediately into *anterior* and *posterior interosseous arteries.* These vessels descend along the respective surfaces of the interosseous membrane between the radius and the ulna. The anterior interosseous artery supplies the deep flexor muscles, whereas the posterior interosseous artery and its branches supply all the extensors on the posterior forearm.

Palmar Arches In the palm, branches of the radial and ulnar arteries join to form two horizontal arches, the **superficial** and **deep palmar arches.** The superficial arch underlies the skin and fascia of the hand, whereas the deep arch lies against the metacarpal bones. The **digital arteries,** which supply the fingers, branch from these arches. (The radial and ulnar arteries also form a *carpal arch* on the dorsum of the wrist. Branches from this dorsal arch run distally along the metacarpal bones.)

Arteries of the Thorax

We first consider the vessels supplying the thoracic wall; then we examine those supplying the thoracic viscera.

Arteries of the Thoracic Wall The anterior thoracic wall receives its blood from the **internal thoracic artery** (formerly the *internal mammary artery*) (see Figure 19.10). This vessel, which branches from the subclavian artery superiorly, descends just lateral to the sternum and just deep to the costal cartilages. **Anterior intercostal arteries** branch off at regular intervals and run horizontally to supply the ribs and the structures in the intercostal spaces. The internal thoracic artery also sends branches superficially to supply the skin and mammary gland. It ends inferiorly at the costal margin, where it divides into a branch to the anterior abdominal wall and a branch to the anterior part of the diaphragm.

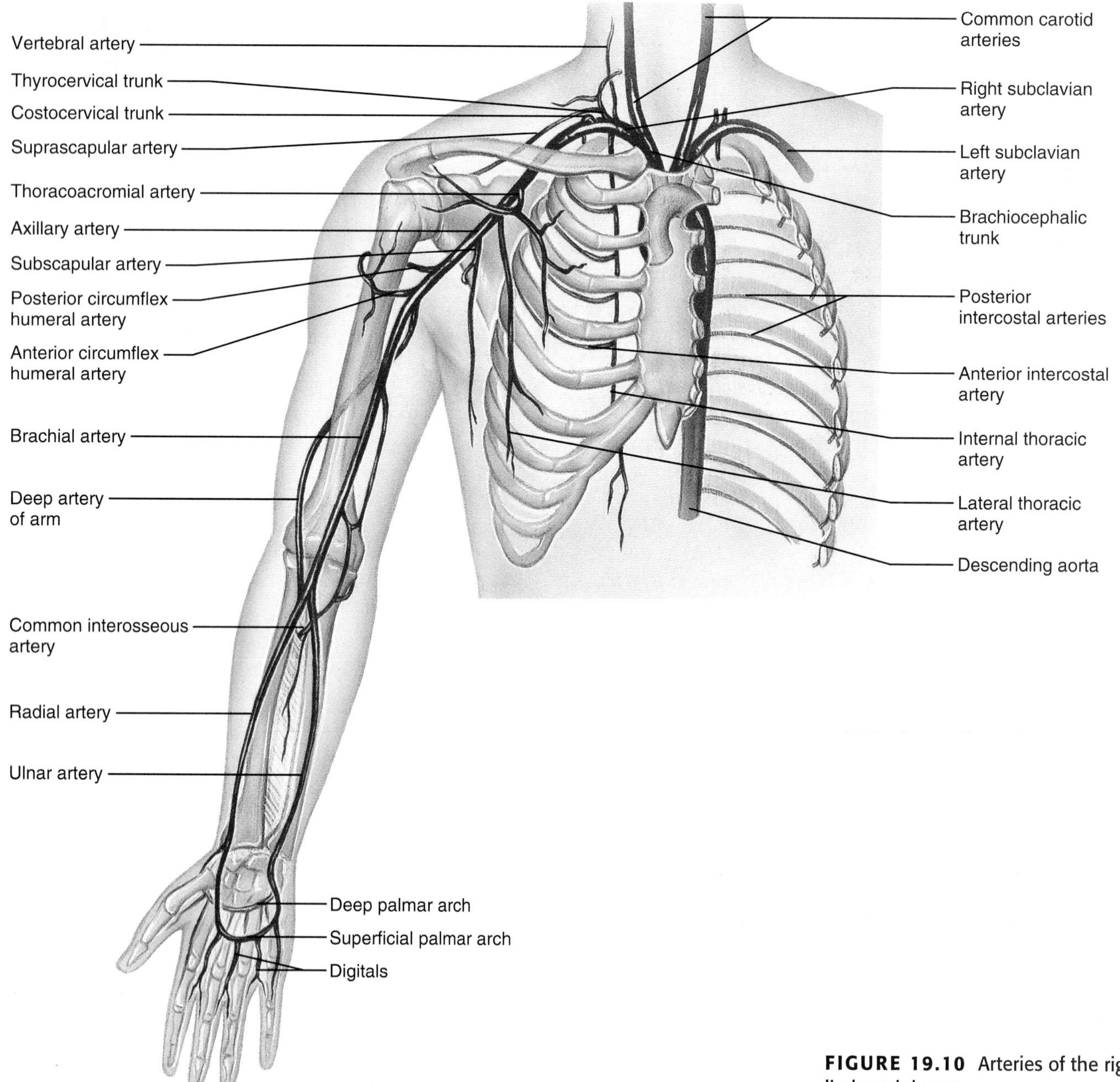

FIGURE 19.10 Arteries of the right upper limb and thorax.

Critical Reasoning

In coronary bypass surgery (see the box on pp. 574–575), the internal thoracic artery can be used as the bypass vessel by suturing its distal end onto the obstructed coronary artery in the heart wall. Explain why using this vessel is superior to using the great saphenous vein of the leg.

Because it is an artery, the internal thoracic artery has a thicker wall than the saphenous vein, so it is less likely either to be damaged by the pressure of the heartbeat or to develop artherosclerosis. As a result, the arterial graft lasts longer and leads to higher long-term rates of survival than the venous graft. Another advantage of using the internal thoracic artery is that it lies near the heart, so it need not be transplanted from some distant site in order to be grafted.

The posterior thoracic wall receives its blood from the **posterior intercostal arteries.** The superior two pairs arise from the costocervical trunk, whereas the inferior nine pairs issue from the thoracic aorta. All of the posterior intercostal arteries run anteriorly in the costal grooves of their respective ribs. In the lateral thoracic wall, they anastomose with the anterior intercostal arteries. Inferior to the twelfth rib, one pair of *subcostal arteries* (not illustrated) branches from the thoracic aorta. Finally, a pair of *superior phrenic arteries* (not illustrated) leaves the most inferior part of the thoracic aorta to supply the posterior, superior part of the diaphragm.

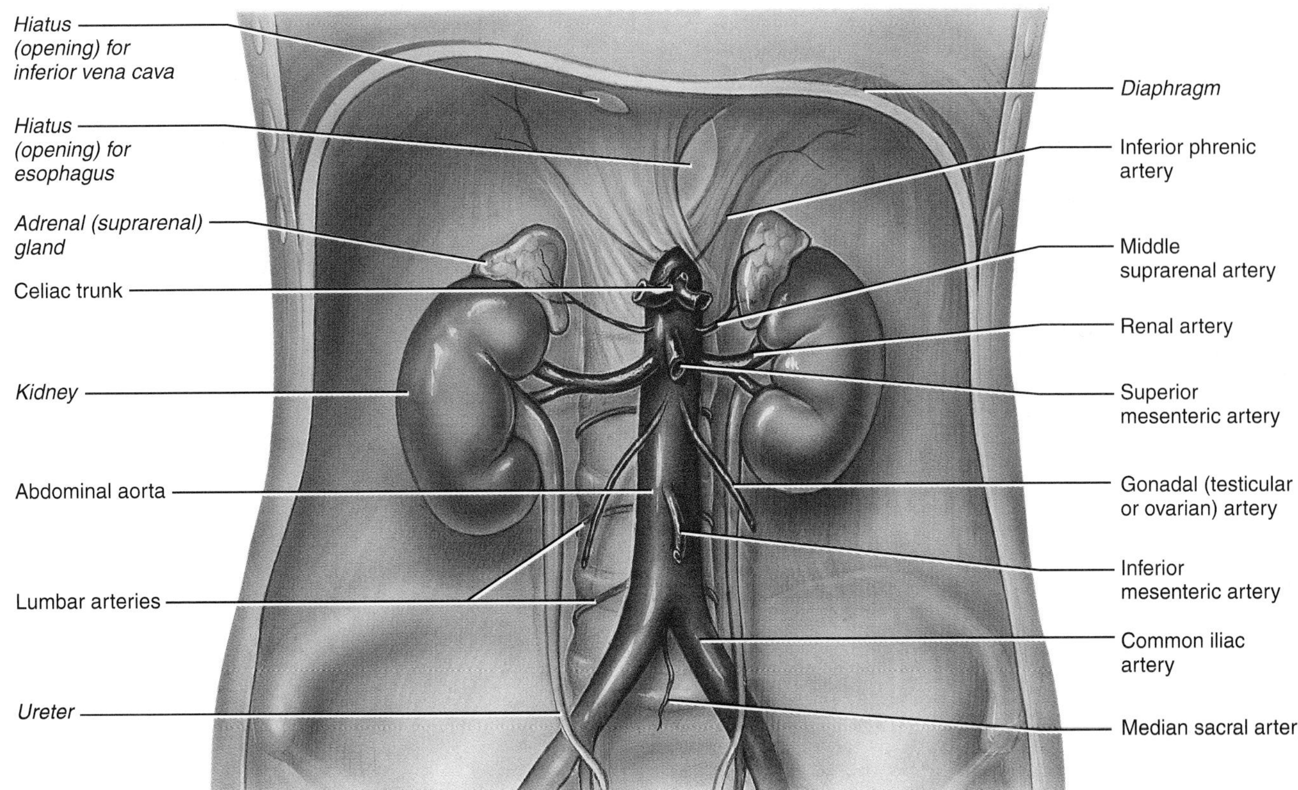

FIGURE 19.11 Major branches of the abdominal aorta.

Arteries of the Thoracic Visceral Organs Many thoracic viscera receive their functional blood supply from small branches off the thoracic aorta. Because these vessels are so small, they are not illustrated. The *bronchial arteries* supply systemic (oxygenated) blood to the lung structures. Usually, two bronchial arteries serve the left lung, and one serves the right lung; in some people they arise from posterior intercostal arteries instead of the aorta. Bronchial arteries enter the lung's medial surface along with the large pulmonary vessels.

The thoracic aorta also sends several small branches to the esophagus directly anterior to it, as well as to the posterior part of the mediastinum and pericardium.

Arteries of the Abdomen

The arteries to the abdominal organs arise from the abdominal aorta (see Figures 19.7 and 19.11). In a person at rest, about half of the entire arterial flow is present in these vessels. We consider these arteries in their order of issue from the aorta, indicating the vertebral level at which each artery arises.

Inferior Phrenic Arteries The paired *inferior phrenic arteries* branch from the abdominal aorta at the level of vertebra T_{12}, just inferior to the aortic opening (hiatus) in the diaphragm (Figure 19.11). These arteries supply the inferior surface of the diaphragm.

Celiac Trunk The short, wide, unpaired *celiac* (se′le-ak) *trunk* (Figure 19.12) supplies the viscera in the superior part of the abdominal cavity (*coelia* = abdominal cavity). Specifically, it sends branches to the stomach, liver, pancreas, spleen, and a part of the small intestine (duodenum). It emerges midventrally from the aorta at the level of T_{12} and divides almost immediately into three branches: the *left gastric, splenic,* and *common hepatic arteries.*

The **left gastric artery** (*gaster* = stomach) runs superiorly and to the left, to the junction of the stomach with the esophagus, where it gives off several esophageal branches and descends along the right (lesser) curvature of the J-shaped stomach.

The **splenic artery** runs horizontally and to the left, posterior to the stomach, to enter the spleen. Along the way, it passes superior to the pancreas, sending branches to this organ. Near the spleen, it sends several short branches superiorly to the stomach's dome *(short gastric arteries)* and sends a major branch along the stomach's left (greater) curvature—the **left gastroepiploic** (gas″tro-ep″ĭ-plo′ik) **artery** (also called the *left gastro-omental artery*).

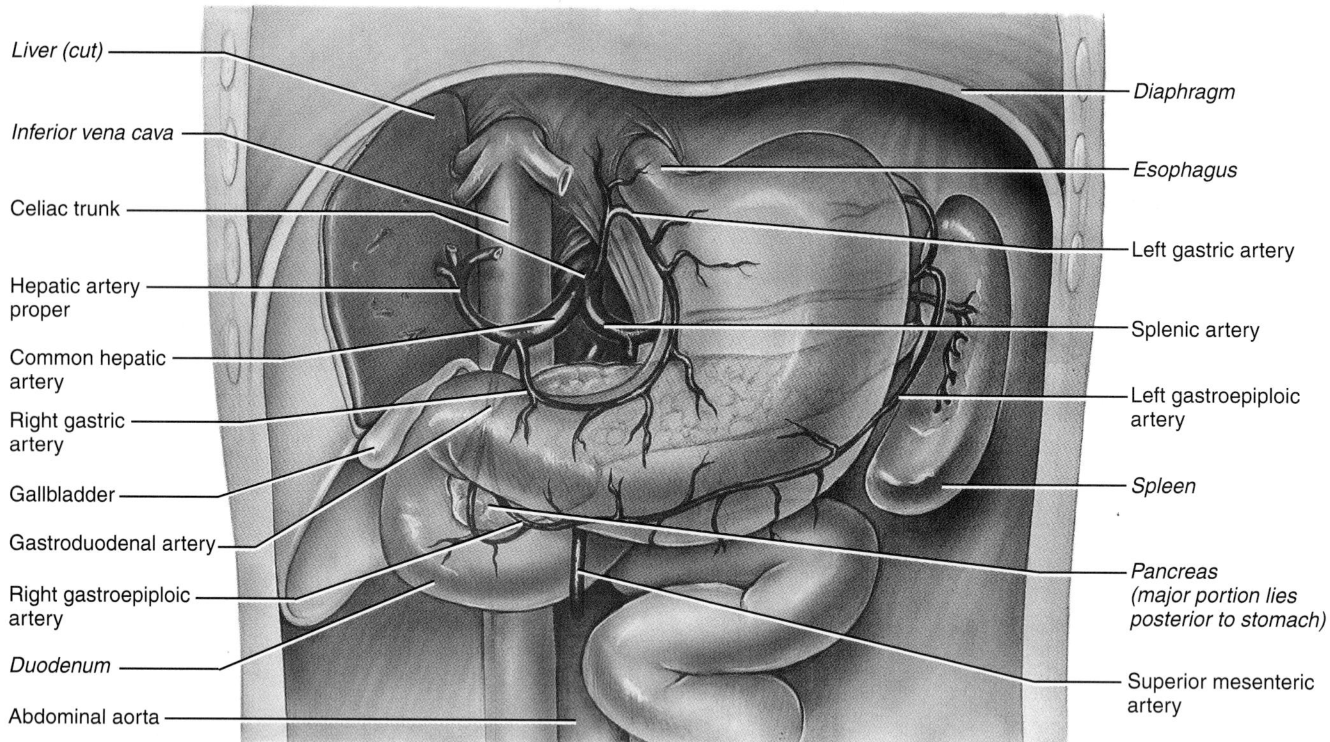

FIGURE 19.12 The celiac trunk and its main branches. To make this artery visible, the left half of the liver has been removed.

The **common hepatic artery** (*hepar, hepat* = liver) is the only branch of the celiac trunk that runs to the right. At the junction of the stomach with the small intestine (duodenum), this artery divides into an ascending branch, the *hepatic artery proper,* and a descending branch, the *gastroduodenal artery.* The **hepatic artery proper** divides into *right* and *left branches* just before entering the liver; the *cystic artery* to the gallbladder usually arises from the right branch of the hepatic artery. The **right gastric artery,** which can arise either from the hepatic artery proper or from the common hepatic artery, runs along the stomach's lesser curvature from the right. The **gastroduodenal artery** (gas″tro-du″o-de′nal), the descending branch of the common hepatic artery, runs inferiorly on the head of the pancreas. It helps supply the pancreas, plus the nearby duodenum. Its major branch, the **right gastroepiploic artery** (= *right gastro-omental artery*), runs along the stomach's greater curvature from the right.

Superior Mesenteric Artery The large, unpaired *superior mesenteric* (mes″en-ter′ik) artery serves most of the intestines (Figure 19.13). It arises midventrally from the aorta, posterior to the pancreas at the level of L_1. From there, it runs inferiorly and anteriorly to enter the mesentery, a drape-like membrane that supports the long, coiled parts of the small intestine (the jejunum and the ileum). The superior mesenteric artery angles gradually to the right as it descends through the mesentery. From its left side arise many **intestinal arteries** to the jejunum and ileum. From its right side emerge branches that supply the proximal half of the large intestine: the ascending colon, cecum, and appendix via the **ileocolic artery;** and part of the transverse colon, via the **right** and **middle colic arteries.**

Suprarenal Arteries The paired **middle suprarenal arteries** (see Figure 19.11), which emerge from the sides of the aorta, at L_1, supply blood to the adrenal (suprarenal) glands on the superior poles of the kidneys. The adrenal glands also receive *superior suprarenal* branches from the nearby inferior phrenic arteries, and *inferior suprarenal* branches (not illustrated) from the nearby renal arteries.

Renal Arteries The paired *renal arteries* to the kidneys (*ren* = kidney) stem from the sides of the aorta, between vertebrae L_1 and L_2 (Figure 19.11). The kidneys remove nitrogenous wastes from the blood delivered via the renal arteries. As mentioned previously, the transportation of cellular wastes to the kidney is an important function of the vascular system, so the renal circulation is a major functional subdivision of the systemic circuit.

Gonadal Arteries The paired arteries to the gonads are more specifically called **testicular arteries** in males and **ovarian arteries** in females (see Figure 19.11). They branch from the aorta at L_2, the level where the gonads

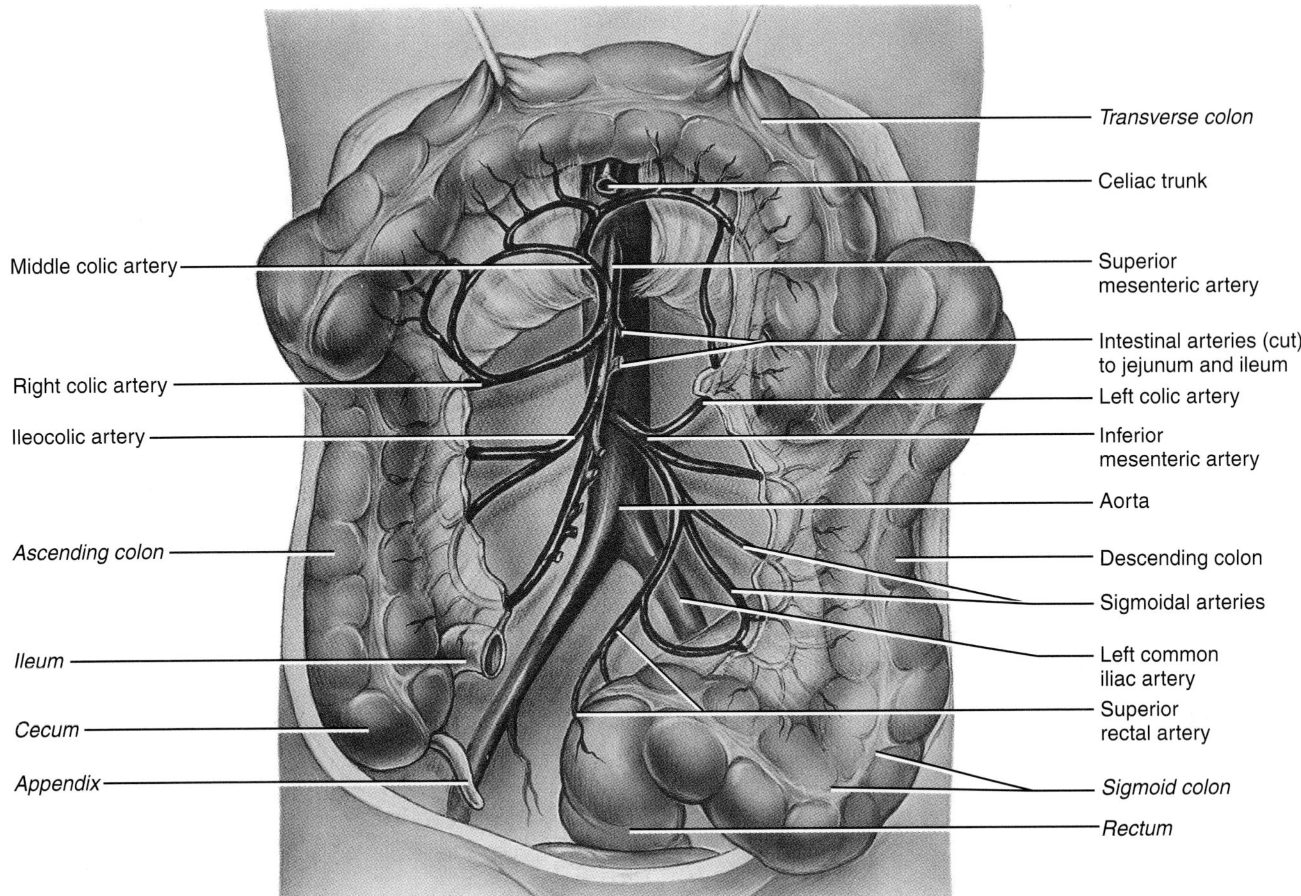

FIGURE 19.13 **Distribution of the superior and inferior mesenteric arteries.** The transverse colon has been pulled superiorly, and much of the mesentery (the yellow membrane) has been removed. In the body, the arteries in this figure are covered by the coils of the small intestine (jejunum and ileum, which also were removed).

first develop in the embryo. The ovarian arteries extend inferiorly into the pelvis to serve the ovaries and part of the uterine tubes. The longer testicular arteries extend through the pelvis to the scrotum, where they serve the testes.

Inferior Mesenteric Artery The unpaired *inferior mesenteric artery* (see Figures 19.11 and 19.13) is the final major branch of the abdominal aorta, arising midventrally at the level of L_3. It serves the distal half of the large intestine—from the last part of the transverse colon to the middle part of the rectum. Its branches are the **left colic, sigmoidal,** and **superior rectal arteries.**

Lumbar Arteries Four pairs of *lumbar arteries* arise from the posterolateral surface of the aorta in the lumbar region (see Figure 19.11). These segmental arteries run horizontally to supply the posterior abdominal wall.

Median Sacral Artery The unpaired *median sacral artery* issues from the most inferior part of the aorta. As it descends, this thin artery supplies the sacrum and coccyx along the midline.

Common Iliac Arteries At the level of L_4, the aorta splits into the right and left common iliac arteries (see Figure 19.11), which supply the inferior part of the anterior abdominal wall, as well as the pelvic organs and the lower limbs (described next).

Arteries of the Pelvis and Lower Limbs

At the level of the sacroiliac joint on the pelvic brim, each common iliac artery forks into two branches: the *internal iliac artery,* which mainly supplies the pelvic organs, and the *external iliac artery,* which supplies the lower limb (Figure 19.14).

Internal Iliac Arteries The branches of *internal iliac arteries* (Figure 19.15) supply blood to the pelvic walls, pelvic viscera, buttocks, medial thighs, and perineum. Among the most important branches are the superior and inferior **gluteal arteries,** which run posteriorly to supply the gluteal muscles; the **internal pudendal artery,** which leaves the pelvic cavity to supply the perineum and external genitalia; and the **obturator artery,** which

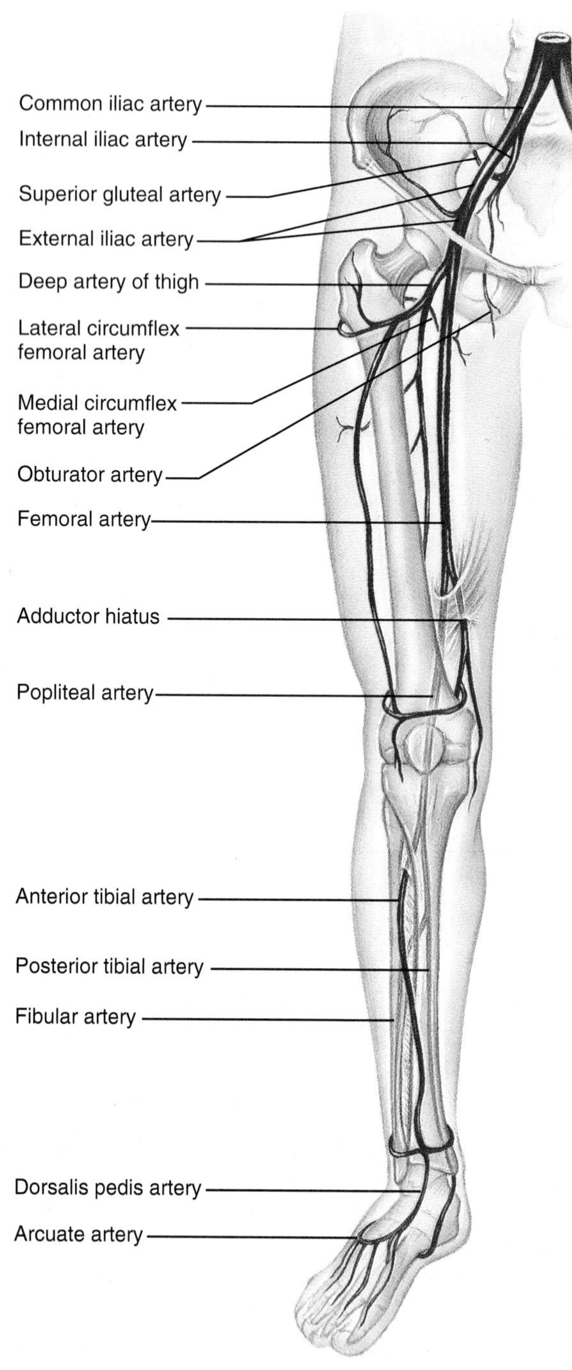

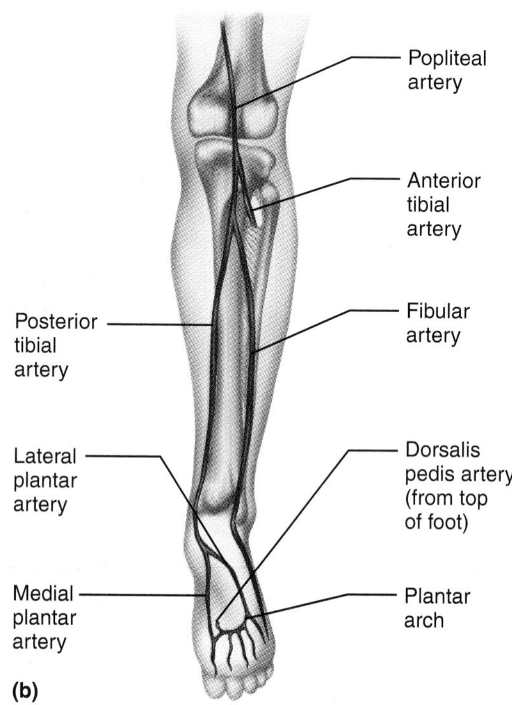

FIGURE 19.14 Arteries of the right pelvis and lower limb. **(a)** Anterior view. **(b)** Posterior view of the leg.

descends through the obturator foramen into the thigh adductor muscles. Other branches run to the bladder, the rectum, the uterus and vagina (in females), and the pelvic reproductive glands (in males).

External Iliac Artery The right and left *external iliac arteries* carry blood to the lower limbs (see Figure 19.14). Originating from the common iliac arteries in the pelvis, each external iliac artery descends along the arcuate line of the ilium bone, sends some small branches to the anterior abdominal wall, and enters the thigh by passing deep to the midpoint of the inguinal ligament. At this point, the external iliac artery is called the femoral artery.

Femoral Artery The *femoral artery* descends vertically through the thigh medial to the femur and along the anterior surface of the adductor muscles. Inferiorly, the femoral artery passes through a gap in the adductor magnus muscle (the adductor hiatus) and emerges posterior to the distal femur as the *popliteal artery.*

Even though the superior part of the femoral artery is enclosed in a tube of dense fascia, it is relatively superficial and not protected by any overlying musculature. This lack of protection makes the proximal femoral artery a convenient place to take a pulse, but it also makes it susceptible to injury.

Several arteries arise from the femoral artery in the thigh (see Figure 19.14a). The largest branch, which arises superiorly and is called the **deep artery of the thigh** (or *profunda femoris* = *deep femoral*), is the main supplier of the thigh muscles: adductors, hamstrings, and quadriceps. Proximal branches of the deep femoral artery are the **medial** and **lateral circumflex femoral arteries,** which circle the neck and upper shaft of the femur. The medial circumflex artery supplies the head of the femur, and if a fracture of the hip tears this artery, the bone tissue of the head of the femur dies. A long, descending branch of the lateral circumflex artery (see Figure 19.14a) runs along the anterior aspect of the vastus lateralis muscle, which it supplies.

Popliteal Artery The *popliteal artery* (pop″lĭ-te′al), the inferior continuation of the femoral artery, lies within the popliteal fossa (the region posterior to the knee), a deep location that offers protection from injury. The popliteal artery gives off several *genicular arteries* (je-nik′u-lar; "knee") that circle the knee joint like horizontal hoops. Just inferior to the head of the fibula, the popliteal artery splits into the *anterior* and *posterior tibial arteries.*

Anterior Tibial Artery The *anterior tibial artery* runs through the anterior muscular compartment of the leg, descending along the interosseous membrane lateral to the tibia and sending branches to the extensor muscles along the way. At the ankle, it becomes the **dorsalis pedis artery** ("artery of the dorsum of the foot"). At the base of the metatarsal bones, the **arcuate artery** branches from dorsalis pedis and sends smaller branches distally along the metatarsals. The end part of dorsalis pedis penetrates into the sole, where it forms the medial end of the plantar arch (described shortly).

The dorsalis pedis artery is superficial and provides a place to feel the pulse (the "pedal pulse point"). The absence of this pulse can indicate that the blood supply to the leg is inadequate. Routine checking of the pedal pulse is indicated for patients known to have impaired circulation to the legs and for those recovering from surgery to the leg or groin.

Posterior Tibial Artery The *posterior tibial artery* (Figure 19.14b), which descends through the posteromedial part of the leg, lies directly deep to the soleus muscle. Proximally, it gives off a large branch, the **fibular (peroneal) artery,** which descends along the medial aspect of the fibula. Together, the posterior tibial and fibular arteries supply the flexor muscles in the leg, and the fibular arteries send branches to the fibularis muscles.

Inferiorly, the posterior tibial artery passes behind the medial malleolus of the tibia. On the medial side of the foot, it divides into **medial** and **lateral plantar arteries** (Figure 19.14b). These serve the sole, and the lateral plantar artery forms the lateral end of the **plantar arch.** Metatarsal and digital arteries to the toes arise from the plantar arch.

Figure 19.16 summarizes the main systemic arteries in the form of a flow chart.

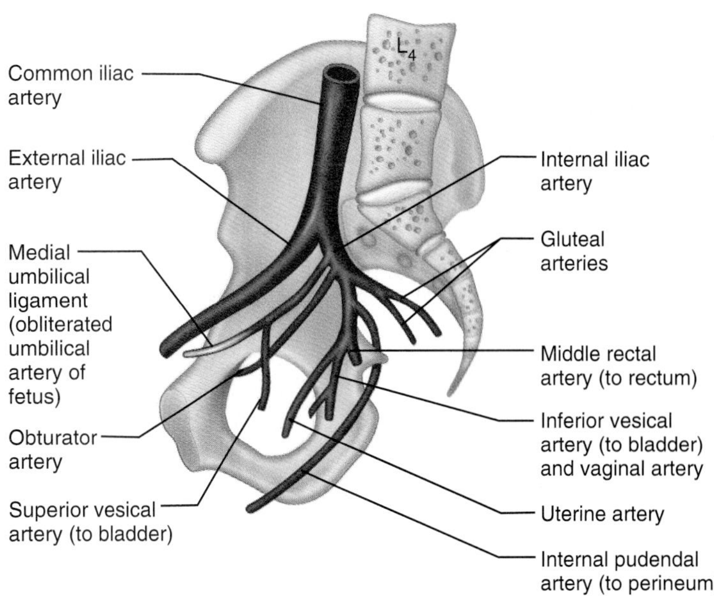

FIGURE 19.15 Internal iliac artery. The right half of the female pelvis, in medial view. The branches of the internal iliac artery are variable, arising in somewhat different places in different individuals.

Systemic Veins

Having considered the arteries of the body, we now turn to the veins (Figure 19.17). Although most veins run with corresponding arteries, there are some important differences in the distributions of arteries and veins:

1. Whereas just one systemic artery leaves the left ventricle (the aorta), three major veins enter the right atrium: the superior and inferior venae cavae and the coronary sinus.
2. Whereas all large and medium-sized arteries have deep locations for protection, many veins, called *superficial veins,* lie just beneath the skin, unattended by any arteries.
3. Several parallel veins often take the place of a single larger vein. Such multivein bundles, including networks of anastomosing veins, are called *venous plexuses.*
4. Two important body areas have unusual patterns of venous drainage. First, veins from the brain drain into *dural sinuses,* which are not typical veins but endothelium-lined channels supported by walls of dura mater. Second, venous blood draining from the digestive organs enters a special subcirculation, the *hepatic portal system,* and passes through capillaries in the liver before the blood reenters the general systemic circulation. The dural sinuses and hepatic portal system are considered shortly (see pp. 566, 570).

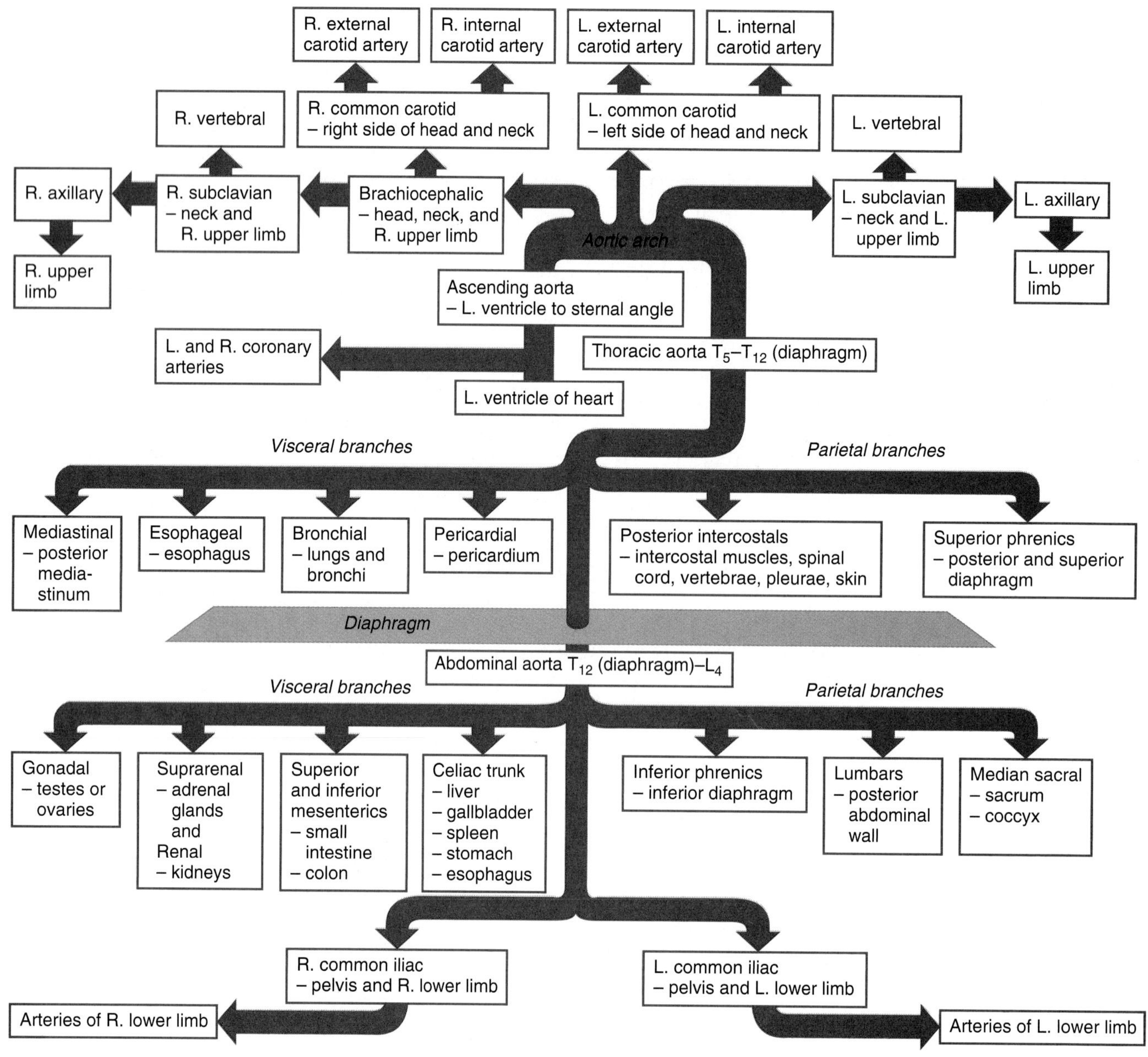

FIGURE 19.16 Flow chart summarizing the main arteries of the systemic circulation.

Venae Cavae and Their Major Tributaries

The unpaired *superior* and *inferior venae cavae,* the body's two largest veins, empty directly into the right atrium of the heart (see Figure 19.8). The name *vena cava* means "cave-like vein."

Superior Vena Cava The superior vena cava (see Figure 19.17) receives the systemic blood from all body regions superior to the diaphragm excluding the heart wall. This vein arises from the union of the **left** and **right brachiocephalic veins** posterior to the manubrium and descends to join the right atrium. Of the two brachiocephalic veins, the left is longer and nearly horizontal, whereas the right is vertical (see Figure 19.8). Each brachiocephalic vein is formed by the union of an *internal jugular vein* and *a subclavian vein.*

Inferior Vena Cava The inferior vena cava, which ascends along the posterior wall of the abdominal cavity and is the widest blood vessel in the body (Figure 19.18), returns blood to the heart from all body regions inferior to the diaphragm (see Figure 19.17). It begins inferiorly at the union of the two common iliac veins on the body of vertebra L_5, and ascends on the right side of the vertebral bodies to the right of the abdominal aorta. Upon penetrating the diaphragm at T_8, the inferior vena cava joins the right atrium.

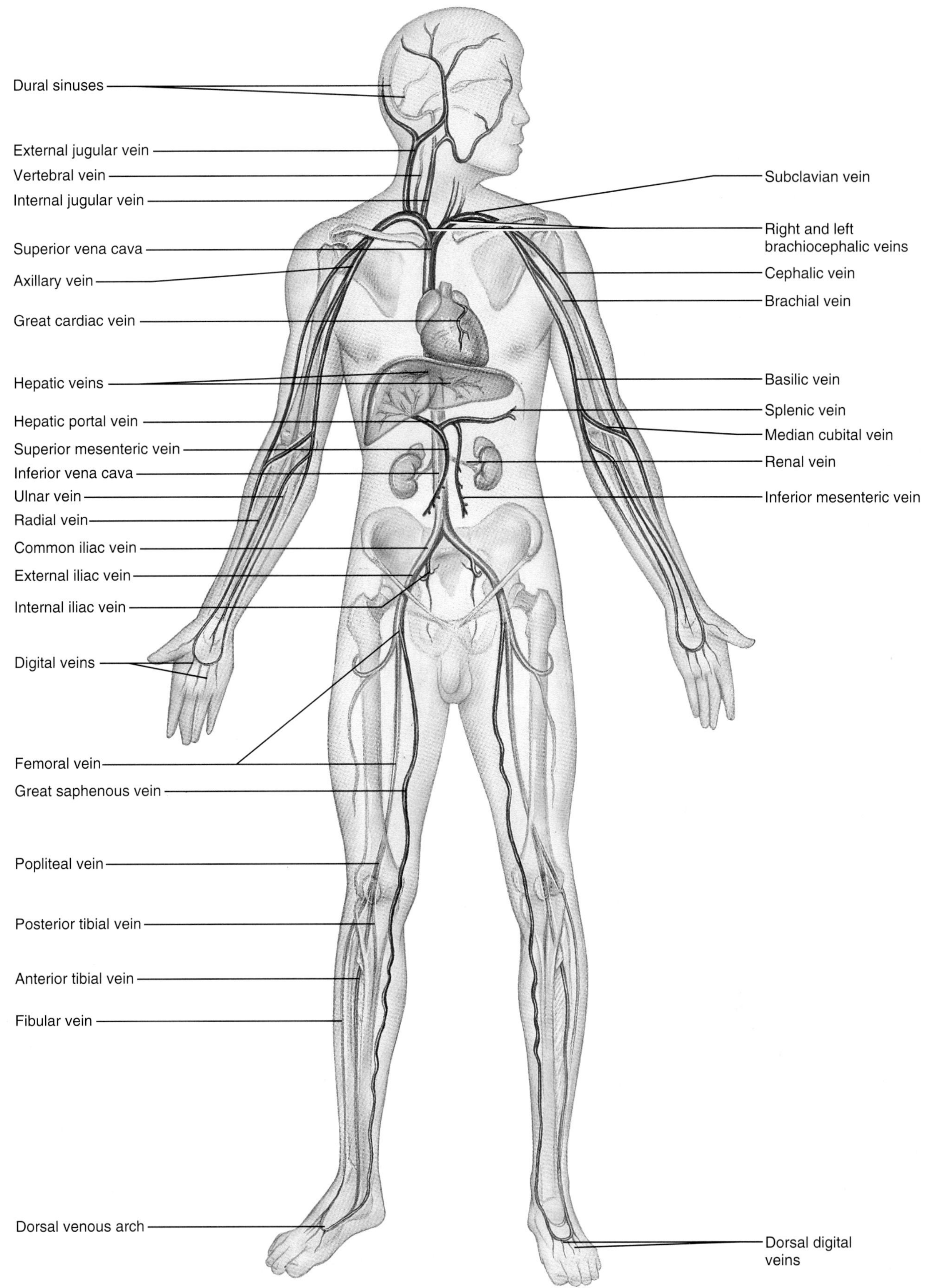

FIGURE 19.17 Major veins of the systemic circulation.

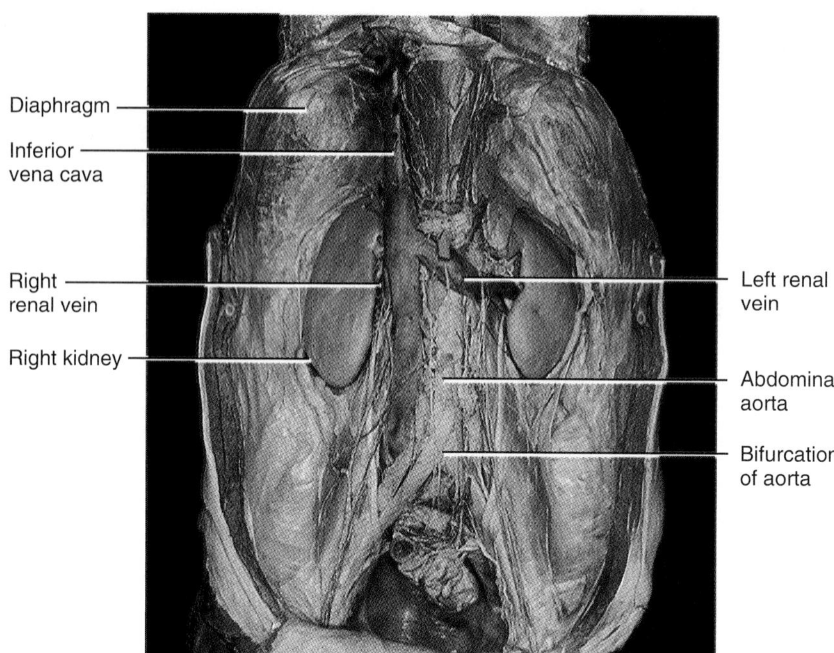

FIGURE 19.18 Photograph of the posterior abdominal wall, revealing the position of the inferior vena cava. Note that this vein lies just to the right of the abdominal aorta.

Veins of the Head and Neck

Most blood draining from the head and neck enters three pairs of veins: (1) the internal jugular veins from the dural sinuses, (2) the external jugular veins, and (3) the vertebral veins (Figure 19.19a). Even though most of the extracranial veins have the same names as the extracranial arteries (facial, ophthalmic, occipital, and superficial temporal), their courses and interconnections differ substantially.

Dural Sinuses Most veins of the brain drain into the intracranial dural sinuses (Figure 19.19b), which form an interconnected series of channels in the skull and lie between the two layers of cranial dura mater (see p. 388). The **superior** and **inferior sagittal sinuses** lie in the falx cerebri between the cerebral hemispheres. The inferior sagittal sinus drains posteriorly into the **straight sinus.** The superior sagittal and straight sinuses then drain posteriorly into the **transverse sinuses,** which run in shallow grooves on the internal surface of the occipital bone. Each transverse sinus in turn drains into an S-shaped sinus **(sigmoid sinus),** which becomes the **internal jugular vein** as it leaves the skull through the jugular foramen.

The paired **cavernous sinuses** flank the body of the sphenoid bone laterally, and each has an internal carotid artery *running within it.* The following cranial nerves also run within the cavernous sinus (or in its wall of dura mater) on their way to the orbit and the face: the oculomotor, trochlear, abducens, and the maxillary and ophthalmic divisions of the trigeminal nerve. The cavernous sinus communicates with the **ophthalmic vein** of the orbit (see Figure 19.19a), which in turn communicates with the facial vein, which drains the nose and upper lip.

Several other dural sinuses exist, but they are small and will not be considered here.

Clinical Applications

MEDICAL IMPORTANCE OF THE CAVERNOUS SINUS Squeezing pimples on the nose or upper lip can spread infection through the facial vein into the cavernous sinus and, from there, through the other dural sinuses in the skull. For this reason, the nose and upper lip are called the *danger triangle of the face.*

Blows to the head can rupture the internal carotid artery within the confines of the cavernous sinus. Leaked blood then accumulates within this sinus and exerts crushing pressure on the contained cranial nerves, disrupting their functions. For example, the damage to the oculomotor, trochlear, and abducens nerves leads to a loss of ability to move the eyes. ✚

Internal Jugular Veins The large internal jugular veins drain almost all of the blood from the brain (see Figure 19.19a). From its origin at the base of the skull, each such vein descends through the neck, first lying lateral to the internal carotid artery and then to the common carotid artery. Along the way, the internal jugular vein receives blood from some deep veins of the face and neck—branches of the *facial* and *superficial temporal veins.* At the base of the neck, the internal jugular vein joins the subclavian vein to form the brachiocephalic vein. The jugular veins are named for their end point, as *jugulum* means "the throat just above the clavicle."

Just as wounds to the neck can cut the carotid arteries, they can also sever the internal jugular veins. However, cut veins do not bleed as quickly as arteries of comparable size, because the blood pressure in them is lower than in arteries. For this reason, the chances of surviving are higher if a neck wound affects the vein and not the artery.

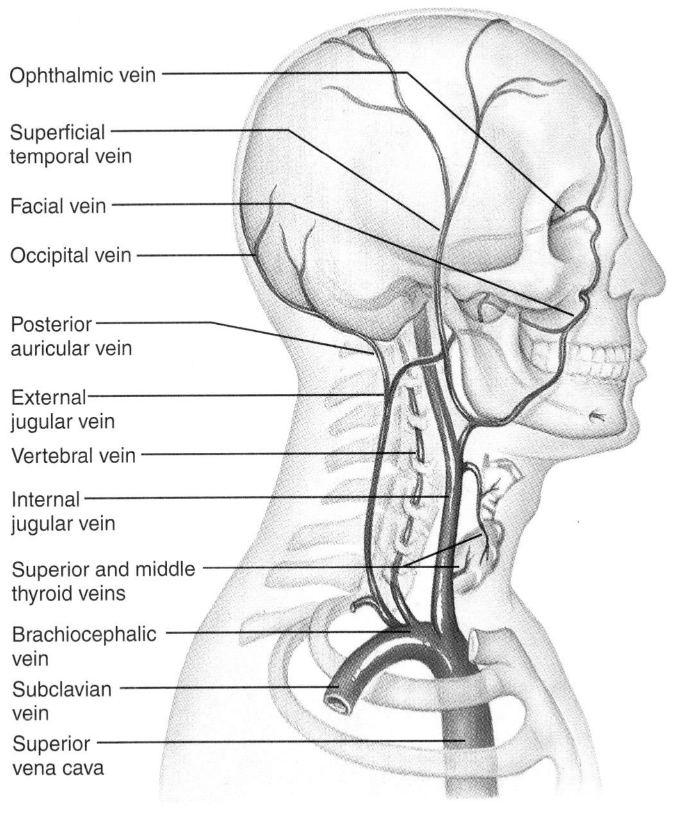

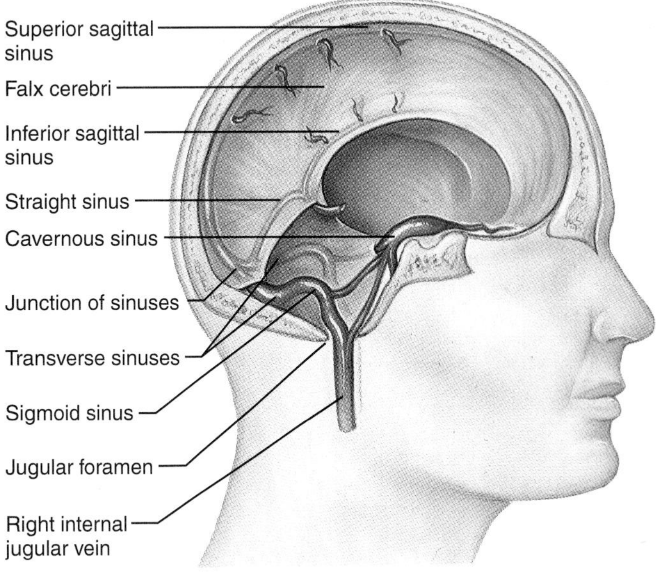

FIGURE 19.19 Venous drainage of the head, neck, and brain. **(a)** Veins of the head and neck, right aspect. The clavicle has been removed and the external jugular vein has been pulled slightly posteriorly so that the vertebral and internal jugular veins can be seen. **(b)** Dural sinuses in the cranium, lateral view.

External Jugular Veins The external jugular vein is a superficial vein that descends vertically through the neck on the surface of the sternocleidomastoid muscle. Superiorly, its tributaries drain the posterior scalp, lateral scalp, and some of the face (see Figure 19.19a). The external jugular vein, which is not accompanied by any corresponding artery, empties inferiorly into the subclavian vein.

Vertebral Veins Unlike the vertebral arteries, the vertebral veins do not serve much of the brain, instead draining only the cervical vertebrae, cervical spinal cord, and small muscles in the superior neck. Originating inferior to the occipital condyle, each vertebral vein descends through the transverse foramina of vertebrae C_1–C_6 in the form of a venous plexus. Emerging from C_6 as a single vein, the vertebral vein continues inferiorly to join the brachiocephalic vein in the root of the neck.

Veins of the Upper Limbs

We consider the veins of the upper limbs (Figure 19.20a) in two groups: deep veins and superficial veins.

Deep Veins The deep veins of the upper limbs (see Figure 19.20a) follow the paths of their companion arteries and have the same names. All except the largest, however, are actually two parallel veins that flank their artery on both sides. The **deep** and **superficial palmar venous arches** of the hand empty into the **radial** and **ulnar veins** of the forearm, which unite just inferior to the elbow joint to form the **brachial vein** of the arm. As the brachial vein enters the axilla, it empties into the **axillary** vein, which becomes the **subclavian vein** at the first rib.

Superficial Veins The superficial veins of the upper limb (Figure 19.20a) are larger than the deep veins and are visible beneath the skin. They anastomose frequently along their course. They begin with the *dorsal venous network* (not illustrated) on the dorsum of the hand. The **cephalic vein** starts at the lateral side of this network, then bends around the distal radius to enter the anterior forearm. From there, this vein ascends through the anterolateral side of the entire limb and ends inferior to the clavicle, where it joins the axillary vein. The **basilic vein** arises from the medial aspect of the hand's dorsal venous network, then ascends along the posteromedial forearm and the anteromedial surface of the arm. In the axilla, the basilic vein joins the brachial vein to become the axillary vein. On the anterior aspect of the elbow joint, in the region called the cubital fossa, the **median cubital vein** connects the basilic and cephalic veins. The median cubital vein is easy to find in most people and is used to obtain blood or to administer substances intravenously (see p. 787). The **median vein of the forearm** ascends in the center of the forearm; its termination point at the elbow is highly variable.

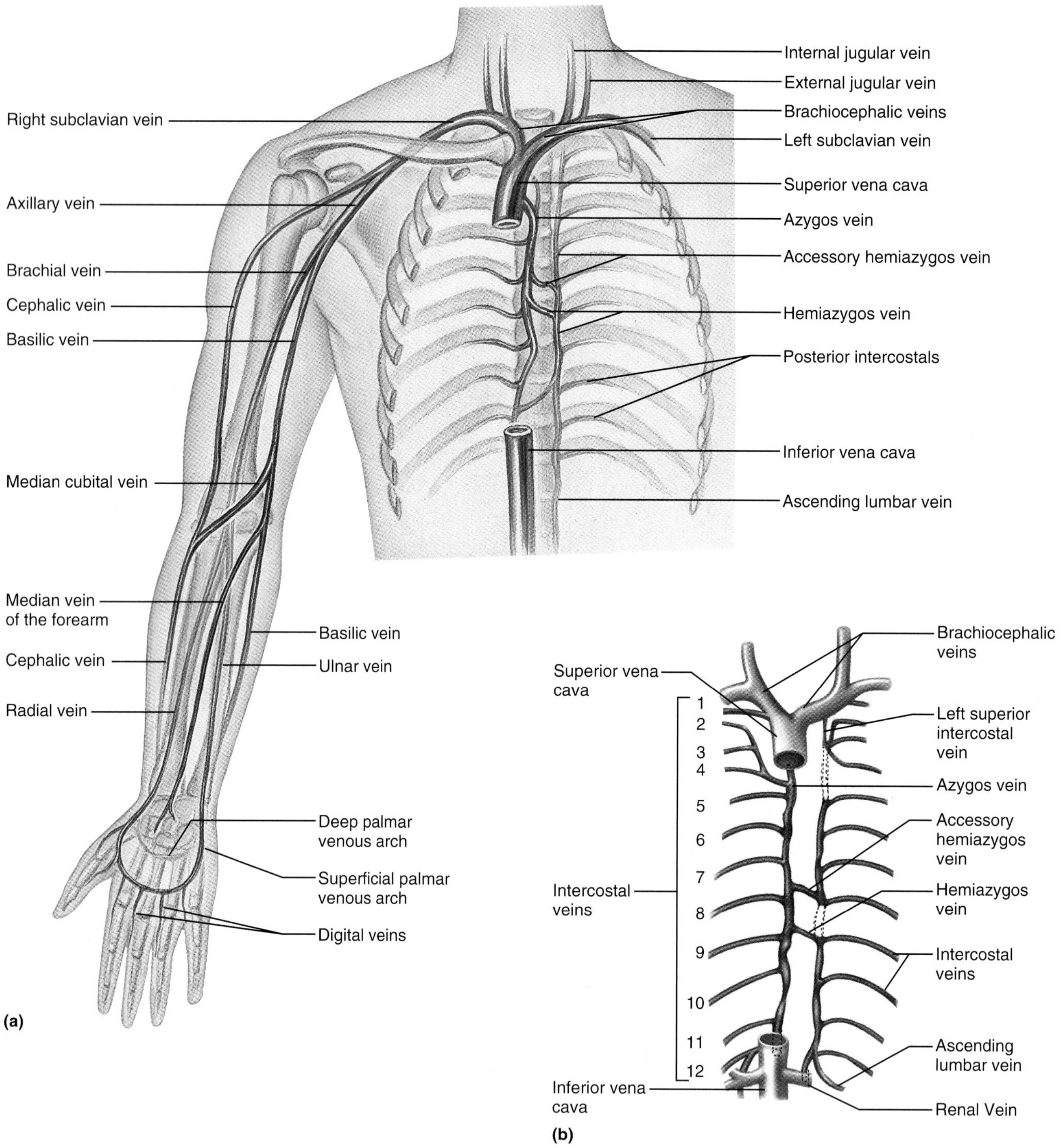

FIGURE 19.20 **Veins of the right upper limb and thorax wall.** **(a)** Drawing of the superficial and deep veins. Each of the deep veins is actually two parallel veins running side by side, but they are drawn as single veins for clarity. Also for clarity, the many anastomoses of the superficial veins are not shown. **(b)** Diagram of the azygos system of veins; these veins are highly variable among different individuals.

Veins of the Thorax

Whereas blood draining from the first few intercostal spaces enters the brachiocephalic veins, blood from the other intercostal spaces, as well as from some of the thoracic viscera, drains into a group of veins called the *azygos* (āz′ĭ-gos, or a-zi′gus) *system* (see Figure 19.20b). This group of veins, which flank the vertebral column and ultimately empty into the superior vena cava, consists of the *azygos vein,* the *hemiazygos vein,* and the *accessory hemiazygos vein.*

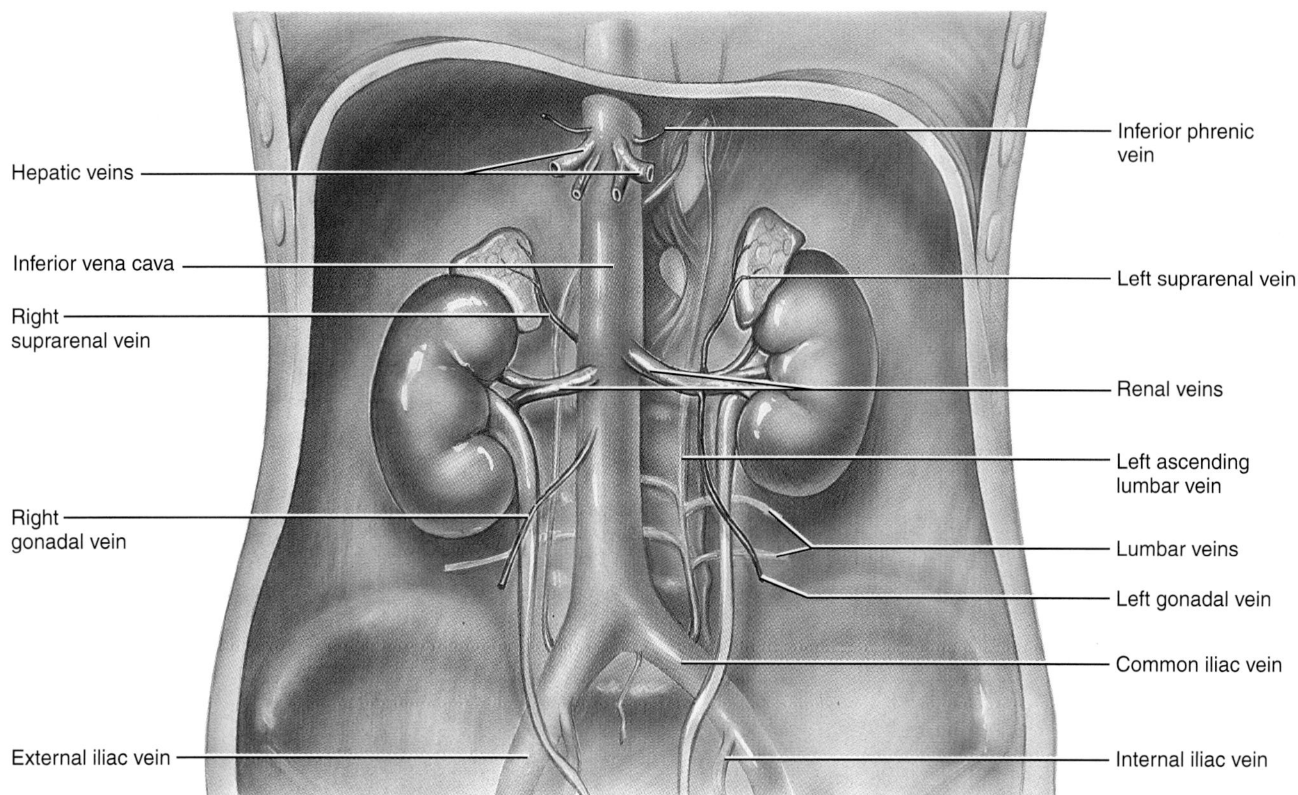

FIGURE 19.21 Tributaries of the inferior vena cava.

Azygos Vein The azygos vein, whose name means "unpaired," ascends along the right or the center of the thoracic vertebral bodies. It receives all of the right **posterior intercostal veins** (except the first), plus the subcostal vein. Superiorly, at the level of T_4, the azygos arches over the great vessels that run to the root of the right lung and joins the superior vena cava.

Hemiazygos Vein The hemiazygos vein (hem″ĭ-āz′ĭ-gos), which ascends on the left side of the vertebral column (see Figure 19.20b), corresponds to the inferior half of the azygos on the right (*hemiazygos* = half the azygos). It receives the ninth through eleventh left posterior intercostal veins and the subcostal vein. At about midthorax, the hemiazygos runs roughly horizontally across the vertebrae and joins the azygos vein.

Accessory Hemiazygos Vein The accessory hemiazygos vein can be thought of as a superior continuation of the hemiazygos, receiving the fourth (or fifth) through the eighth left posterior intercostal veins; it also courses to the right to join the azygos.

The superior parts of the azygos and accessory hemiazygos veins also receive the small *bronchial veins,* which drain unoxygenated systemic blood from the bronchi in the lungs. Veins from the thoracic part of the esophagus also enter the azygos system.

Veins of the Abdomen

Blood returning from the abdominopelvic viscera and the abdominal wall reaches the heart via the inferior vena cava (Figure 19.21). Most venous tributaries of this great vein share the names of the corresponding arteries.

Lumbar Veins Several pairs of lumbar veins drain the posterior abdominal wall, running horizontally with the corresponding lumbar arteries.

Gonadal (Testicular or Ovarian) Veins The right and left gonadal veins ascend along the posterior abdominal wall with the gonadal arteries. Whereas the right vein drains into the anterior surface of the inferior vena cava at L_2, the left gonadal vein drains into the left renal vein.

Renal Veins The right and left renal veins drain the kidneys; each lies just anterior to the corresponding renal artery.

Suprarenal Veins Although each adrenal gland has several main arteries, it has just one vein. Whereas the right suprarenal vein empties into the nearby inferior vena cava, the left suprarenal vein drains into the left renal vein.

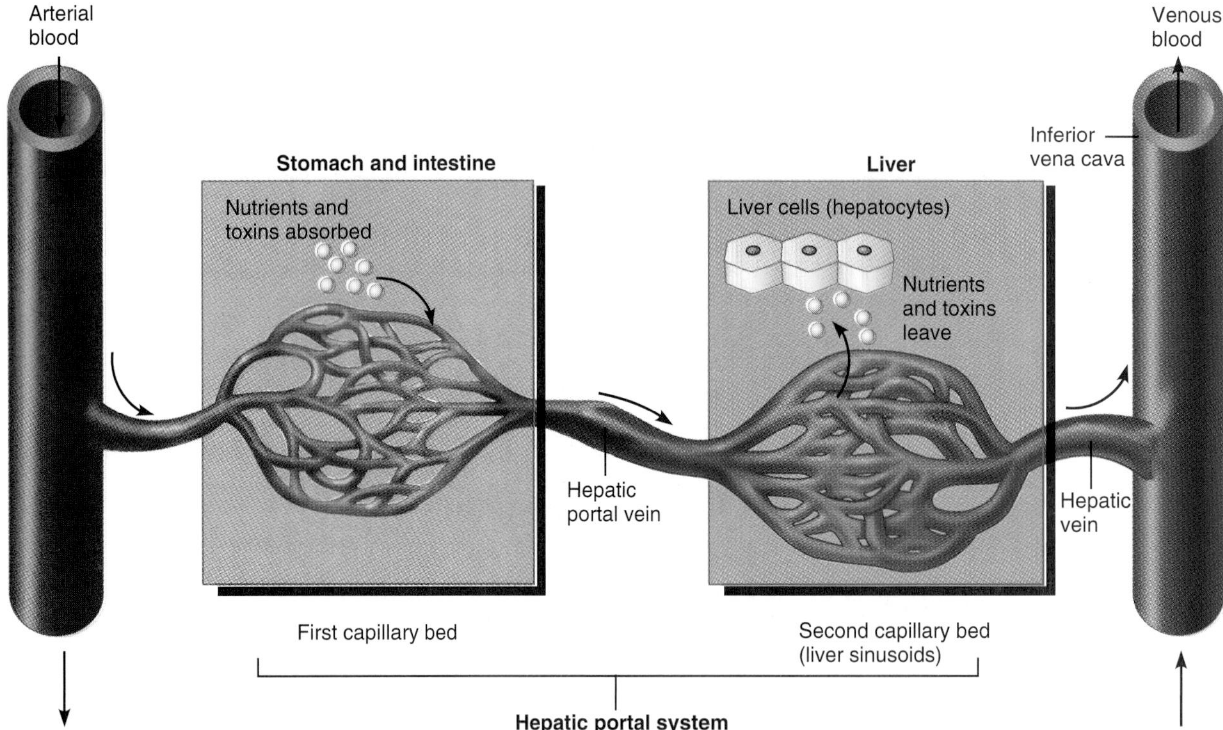

FIGURE 19.22 The basic scheme of the hepatic portal system and associated vessels. Note the presence of two capillary beds within the portal system. Nutrients and toxins picked up from capillaries in the stomach and intestine are transported to the liver cells for processing. From the liver sinusoids, the blood continues onward into the hepatic veins and inferior vena cava.

Hepatic Veins The right and left hepatic veins exit the liver superiorly and empty into the most superior part of the inferior vena cava. These robust veins carry all the blood that originated in the digestive organs in the abdominal and pelvic cavities and arrived via the hepatic portal system (discussed next).

Hepatic Portal System The hepatic portal system is a specialized part of the vascular circuit that serves a function unique to digestion: It picks up digested nutrients from the stomach and intestines and delivers these nutrients to the liver for processing and storage.

Like all portal systems, the hepatic portal system is a series of vessels in which two separate capillary beds lie between the arterial supply and the final venous drainage (Figure 19.22). In this case, capillaries in the stomach and intestines receive the digested nutrients and then drain into the tributaries of the **hepatic portal vein.** This vein then delivers the nutrient-rich blood to a second capillary bed—the *liver sinusoids*—through which nutrients reach liver cells for processing. (The liver cells also break down any toxins that entered the blood through the digestive tract.) After passing through the liver sinusoids, the blood enters the hepatic veins and inferior vena cava, thereby reentering the general systemic circulation. (For a more complete discussion of liver functions, see p. 658.)

As you study the hepatic portal system, be careful not to confuse the *hepatic veins* with the *hepatic portal vein* (also called the *portal vein*). The following veins of the hepatic portal system serve as tributaries of the hepatic portal vein (Figure 19.23):

1. **Superior mesenteric vein.** This large vein ascends just to the right of the superior mesenteric artery. It drains the entire small intestine, the first half of the large intestine (ascending and transverse colon), and some of the stomach. Its superior part lies posterior to the stomach and pancreas.
2. **Splenic vein.** Even though the spleen is not a digestive organ, venous blood leaving it drains through the hepatic portal system. As a result, any microbes that escape the spleen's infection-fighting activities (see p. 595) are carried to the liver for destruction. The splenic vein runs horizontally, posterior to the stomach and pancreas, and joins the superior mesenteric vein to form the hepatic portal vein. Its tributaries correspond to the branches of the splenic artery.
3. **Inferior mesenteric vein.** This vein ascends along the posterior abdominal wall, well to the left of the inferior mesenteric artery. Its tributaries drain the organs that are supplied by that artery—namely, the

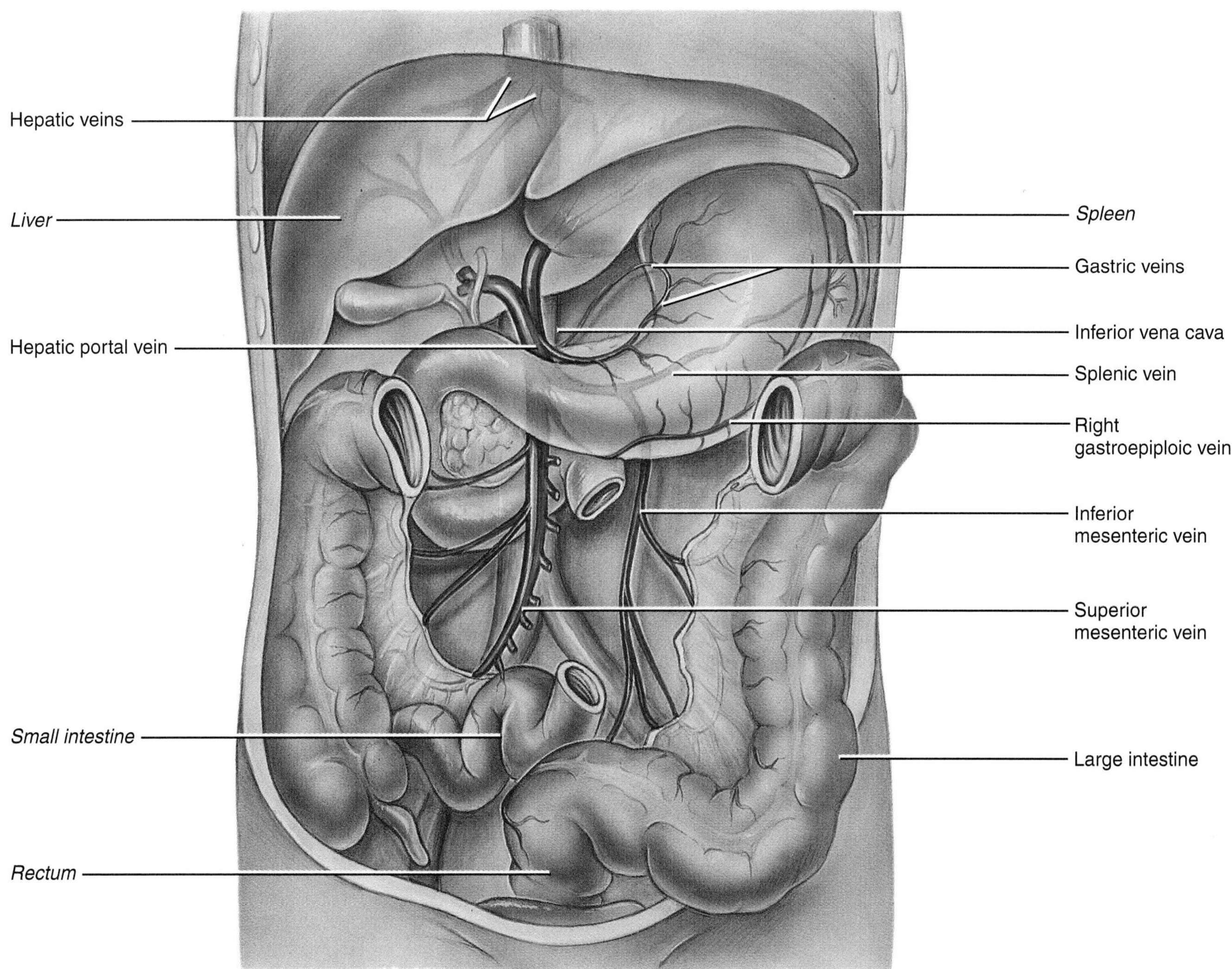

FIGURE 19.23 Veins of the hepatic portal system.

distal region of the colon and the superior rectum. The inferior mesenteric vein empties into the splenic vein posterior to the stomach and pancreas.

The *hepatic portal vein* is a short, vertical vessel that lies directly inferior to the liver and anterior to the inferior vena cava (see Figure 19.23). Inferiorly, it begins posterior to the pancreas as the union of the superior mesenteric and splenic veins. Superiorly, it enters the underside of the liver and divides into right and left branches, whose smaller branches reach the liver sinusoids.

On its way to the liver, the hepatic portal vein receives the right and left gastric veins from the stomach.

Veins of the Pelvis and Lower Limbs

Veins draining the pelvis and lower limbs (Figure 19.24) are either deep or superficial.

Deep Veins Like those in the upper limb, most deep veins in the lower limbs share the names of the arteries they accompany, and all but the largest are actually two parallel veins. Arising on the sole of the foot from the union of the medial and lateral **plantar veins,** the **posterior tibial vein** ascends deep within the calf muscles and receives the **fibular (peroneal) vein.** The **anterior tibial vein,** which is the superior continuation of the **dorsalis pedis vein** of the foot, ascends to the superior part of the leg, where it unites with the posterior tibial vein to form the **popliteal vein.** This in turn ascends to become the **femoral vein,** which drains the thigh and becomes the **external iliac vein** upon entering the pelvis. In the pelvis, the external iliac vein unites with the **internal iliac vein** to form the **common iliac vein.**

Superficial Veins Two large superficial veins, the *great* and *small saphenous veins* (sah-fe′nus; "obvious"), issue from the **dorsal venous arch** of the foot. The saphenous

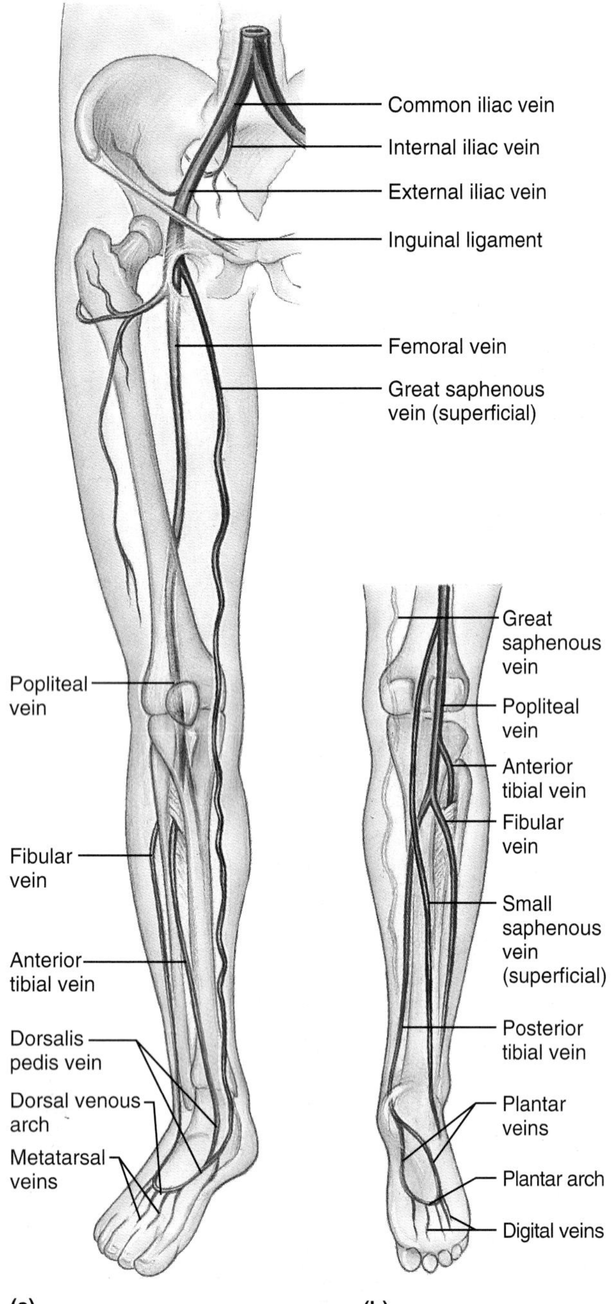

FIGURE 19.24 Veins of the right lower limb and pelvis. (a) Anterior view. (b) Posterior view of the leg and foot. Most of the deep veins are actually two veins running side by side, but they are drawn as single veins for clarity.

veins anastomose frequently with each other and with deep veins along their course. The **great saphenous vein,** the longest vein in the body, ascends along the medial aspect of the entire limb to empty into the femoral vein just distal to the inguinal ligament. The **small saphenous vein** runs along the lateral side of the foot and then along the posterior calf. Posterior to the knee, it empties into the popliteal vein.

CLINICAL APPLICATIONS

MEDICAL IMPORTANCE OF THE SAPHENOUS VEINS The great saphenous vein is the vessel most commonly used in coronary artery bypass operations; this is considered in detail in the A Closer Look box on p. 574. Here, it should be noted that when suturing this vein onto a coronary artery, the surgeon must orient it so that blood flow will open, rather than close, its valves.

The saphenous veins are more likely to weaken and become varicose than any other veins in the lower limb, because they are poorly supported by surrounding tissue. Furthermore, when valves begin to fail in veins throughout a lower limb, the normal contractions of the leg muscles can squeeze blood out of the deep veins into the superficial veins through the anastomoses between these two groups of veins. This influx of blood engorges and weakens the saphenous veins even further. Even when the saphenous veins do not fail, they are often involved in *venous disease of the lower limb* (see the Disorders section on p. 574).

Figure 19.25 summarizes the systemic veins of the body in flow-chart form.

Portal-Systemic Anastomoses

In conditions that lead to scarring and degeneration (cirrhosis) of the liver, especially chronic alcoholism, blood flow through the liver sinusoids is blocked. This blockage raises the blood pressure throughout the hepatic portal system and results in **portal hypertension.** Fortunately, some veins of the portal system anastomose with veins that drain into the venae cavae, providing emergency pathways through which the "backed up" portal blood can return to the heart. These pathways are the *portal-systemic (portal-caval) anastomoses.* The main ones are (1) veins in the inferior esophagus, (2) the hemorrhoidal veins in the wall of the anal canal, and (3) superficial veins in the anterior abdominal wall around the navel. These connecting veins are small, however, and they swell and burst when forced to carry large volumes of portal blood.

Critical Reasoning

Deduce some symptoms exhibited by victims of cirrhosis when these anastomosing veins start to fail.

Alcoholics with cirrhosis of the liver may vomit blood from torn esophageal veins, develop hemorrhoids from swollen hemorrhoidal veins, and exhibit a snake-like network of distended veins through the skin around the navel. This network is called a caput medusae *(kap'ut me-du'se)—the Medusa head—after a monster in Greek mythology whose hair was made of writhing snakes.*

Of all the symptoms of cirrhosis, swelling and bursting of veins in the inferior esophagus is the most serious. Bleeding of esophageal veins is associated with a 50% mortality rate, and if the bleeding recurs, another 30% die. Varicose esophageal veins are treated by injecting a hardening agent into them or by tying them off with bands. Also, an implanted metal tube can be run

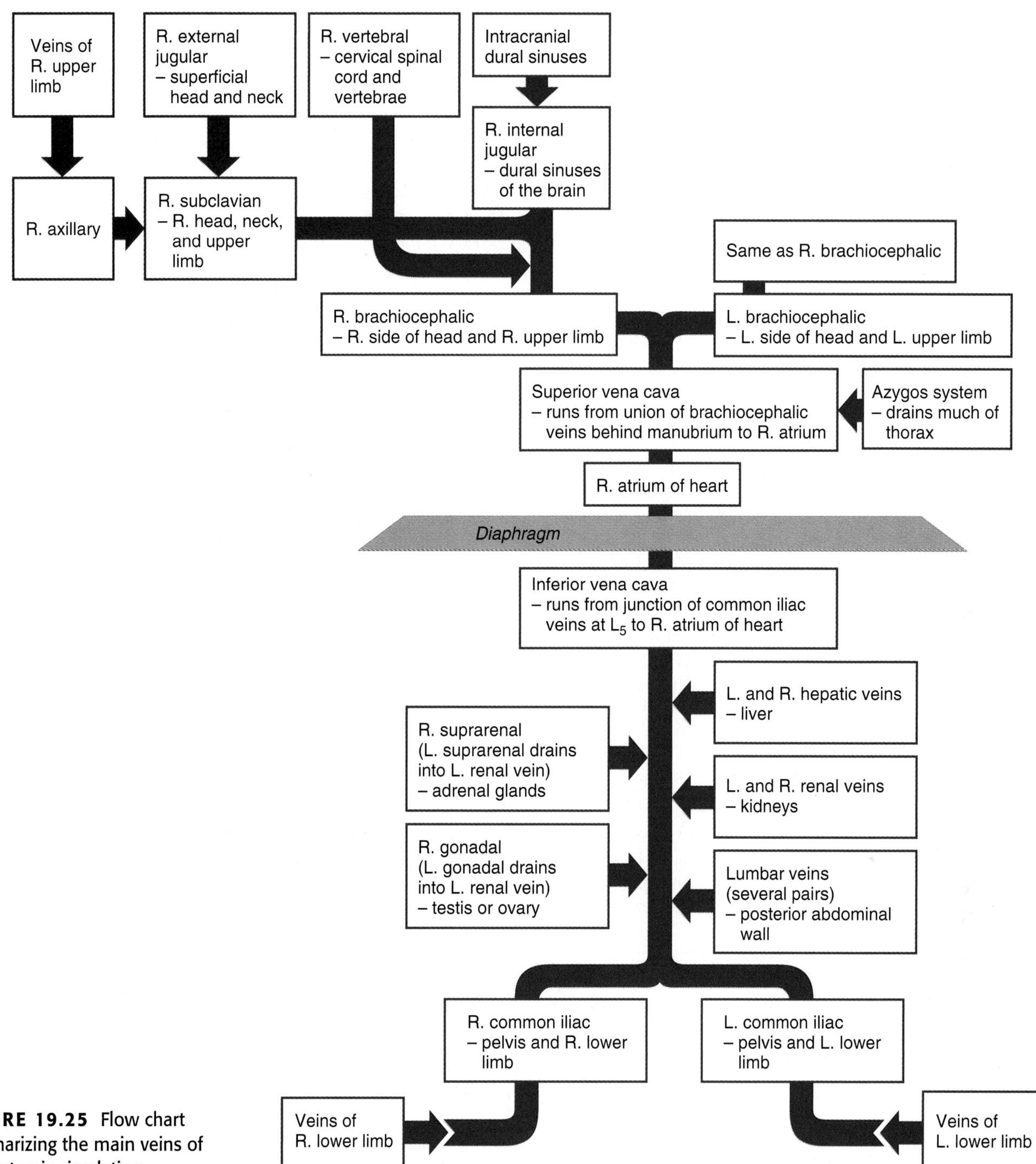

FIGURE 19.25 Flow chart summarizing the main veins of the systemic circulation.

from the inferior vena cava behind the liver into the portal vein, creating a new portal-caval shunt that relieves portal hypertension entirely.

DISORDERS OF THE BLOOD VESSELS

The primary disorder of blood vessels, atherosclerosis, is covered in this chapter's A Closer Look on p. 574. The rest of this section considers some other common and serious disorders of arteries (aneurysms), of veins (deep vein thrombosis of the lower limb, venous disease), and of capillaries (microangiopathy of diabetes, arteriovenous malformation).

An **aneurysm** (an′u-rizm; "widening") is a sac-like widening or outpocketing of an artery (or vein) that places the vessel at risk of rupturing. The aneurysm may result from a congenital weakness of an artery wall or, more often, from a gradual weakening of the vessel by hypertension or arteriosclerosis. The most common sites of aneurysms are the abdominal aorta and the arteries to the brain and kidneys. Aortic aneurysms are present in 1% of women and 8% of men over age 65, and their rupture causes 10,000 deaths per year in the United

A CLOSER LOOK

Atherosclerosis: Controlling a Silent Killer

Arterial disease is the primary cause of death worldwide. **Arteriosclerosis** (ar-te″re-o-sklĕ-ro′sis; "hardening of the arteries") is a general term for the pathological thickening and loss of elasticity of arterial walls. It includes three distinct diseases: (1) the rare *Mönckeberg's arteriosclerosis,* characterized by scattered deposits of calcium in the tunica media of small arteries; (2) *medial arteriosclerosis,* in which the smooth muscle and elastin in the tunica media degenerate with age and are replaced by fibrous tissue; and (3) **atherosclerosis** (ath″er-o″skle-ro′sis), by far the most common and significant type of arterial disease.

In atherosclerosis, the arterial tunica intima thickens with deposits called **atheromas** ("growths that resemble oatmeal") or **atherosclerotic plaques** that narrow the arterial lumen, as seen in the histological section in photo a. Once this narrowing occurs, a circulating blood clot or an arterial spasm can block the vessel completely.

Atherosclerosis primarily affects the coronary arteries to the heart wall, the internal carotid arteries to the head, and the abdominal aorta. Narrowing and blockage of the brain's arteries lead to strokes, and blockage of coronary arteries causes heart attacks (myocardial infarctions). Strokes and heart attacks account for almost half of all deaths in America—heart attacks alone account for one-third.

What triggers this scourge of the arteries? According to the **response to injury** hypothesis, the initial event is injury to the endothelium, which can be caused by high blood pressure (hypertension), by a blow to the vessel, by viral (cytomegalovirus) or bacterial *(Chlamydia pneumoniae)* infection, or by toxins in the blood such as carbon monoxide and free radicals from cigarette smoke. Most often, however, the injury is from accumulations of high concentrations of low-density lipoproteins (LDLs), the molecules in blood that transport cholesterol from a fatty diet. Within the tunica intima the accumulating LDLs become irritants that damage the endothelial cells and stimulate an inflammatory response, attracting immune system cells that accumulate and intensify the inflammation. Plaque formation has thus begun, and the incipient plaques, called fatty streaks, can appear as early as the pre-teen years. The inflammatory response persists and progresses, until smooth muscle cells and fibrous elements become incorporated into the growing plaque. Eventually the inflammation causes cell death (necrosis) and produces advanced plaques, called complicated plaques, that start to calcify, become even more fibrous, and narrow the lumen of the vessel. If the thin fibrous cap over the necrotic core of the plaque breaks off, the exposed core stimulates formation of a massive clot, causing a thrombus that blocks the artery and may give rise to an embolus. Such events in the coronary arteries are the main causes of lethal heart attacks. Even if the cap remains in place, dangerous clots can form on the rough surfaces of the plaques.

It has become popular in recent years to view atherosclerosis as a chronic inflammatory disease. Indeed, measuring the concentration of inflammatory chemicals in the blood is one of the best ways to assess one's risk of a heart attack, and some studies on mice suggest that anti-inflammatory drugs can slow the progression of atherosclerosis. Antioxidants such as vitamin E also seem promising in this respect.

The most common treatment for atherosclerotic coronary arteries, used when several coronary arteries are occluded in several places, is **coronary bypass surgery.** In this procedure, a vessel such as the saphenous vein of the leg or the internal thoracic artery is removed and implanted on the heart wall. This bypass vessel runs from the aorta to the distal, unclogged part of the damaged coronary artery. Cultured arteries, grown in the lab from endothelial and smooth muscle cells, may one day be used in coronary bypass grafts.

Another common treatment, used when a coronary artery is obstructed in just one place, is **balloon angioplasty.** This procedure uses a catheter with a balloon tightly packed into its tip. When the catheter reaches the obstruction, the balloon is inflated, stretching the arterial wall and widening the lumen (see photos b and c). Because it is minimally invasive, balloon angioplasty is being used increasingly. Its use is sure to increase even more because it has

States. If detected before rupture—by palpation, ultrasound, or CT scans—they are treated by replacing the affected section of the vessel with a synthetic graft or by placing a strong-walled tube inside the part of the vessel where the aneurysm is present.

Deep vein thrombosis of the lower limb is the formation of clots in the veins of the lower extremity (usually in the thigh). In half the patients with this condition, the clot detaches, travels to the heart and pulmonary trunk, then blocks a branch of the pulmonary artery (pulmonary embolism). Extremely common, deep vein thrombosis affects 1% to 2% of hospital patients or 2 million Americans per year. It is usually caused by the sluggish flow of blood in veins of inactive and bedridden patients, so that thrombi form on the cusps of the venous valves. Alternatively, it can result from abnormal clotting chemistry or from inflammatory damage to the venous endothelium (in this latter case it is called *thrombophlebitis*). Deep vein thrombosis can be diagnosed by ultrasonagraphy of the veins in the lower limb and is treated with anticlotting drugs. Another treatment involves inserting a filter into the inferior vena cava to prevent emboli from reaching the lungs.

Venous disease is another common venous disorder of the lower limb, affecting 600,000 mostly elderly people in the United States. This disease is characterized by inadequate drainage of venous blood from the limb, whose tissues become ischemic and vulnerable to dam-

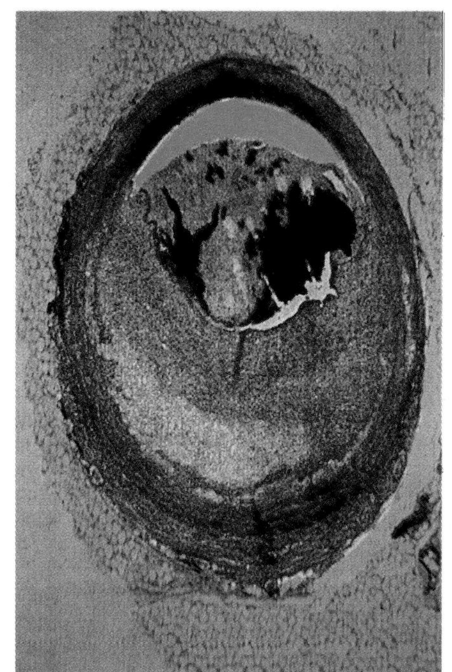

(a)

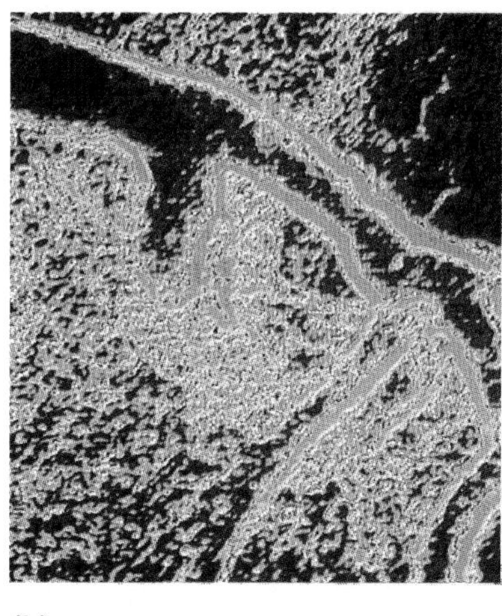

(b)

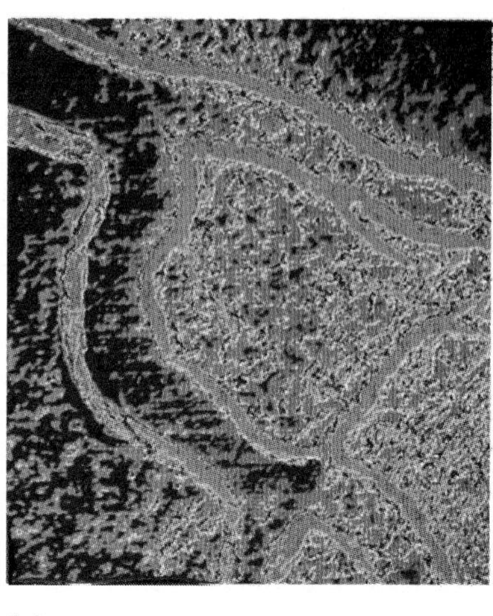

(c)

(a) In this cross sectional image of a human artery, a yellow and pink atherosclerotic plaque thickens the vessel wall and narrows the lumen. **(b)** X-ray images of an occluded artery and **(c)** the same vessel after being cleared by balloon angioplasty. The x-ray views in (b) and (c) were prepared by digital subtraction angiography, a computer imaging technique.

recently been improved so that it can be used on patients in whom more than one coronary vessel is occluded.

Although balloon angioplasty is faster, cheaper, and much less risky than coronary bypass surgery, it has a major shortcoming: New blockages occur after a few months in 30% to 50% of cases. When such a ***restenosis*** ("re-narrowing") follows angioplasty, it is a response to the damage caused by the balloon, and involves a proliferation of smooth muscle cells rather than a new atherosclerotic plaque. The insertion of a stent—a short tube made of metal mesh—into the artery during angioplasty keeps the lumen open and lessens the risk of restenosis by about half.

The aggressive use of cholesterol-lowering drugs such as lovastatin has proven effective in lowering the risk of heart attacks. Other drugs, which dissolve dangerous clots that block atherosclerotic arteries, include **streptokinase,** an enzyme extracted from bacteria, and **tissue plasminogen activator (t-PA),** a natural human product made by genetic engineering techniques. Injecting streptokinase or t-PA directly into the heart through a catheter restores blood flow quickly and puts an early end to many heart attacks in progress.

Factors contributing to atherosclerosis include a high-cholesterol diet, hypertension, emotional stress, smoking, obesity, and lack of exercise. Better management of these risk factors—especially of hypertension, smoking, and diet—has contributed to a 35% decrease in deaths from cardiovascular disease in the United States since 1970. Actually, such preventative measures account for half of this improvement; the advances in technology and better treatment following heart attacks account for the other half. Much more must be done, however, before atherosclerosis is under control.

age and ulceration. It is caused by the failure of valves in interconnecting veins that normally function to prevent blood in the limb's deep veins from flowing outward into the superficial saphenous veins. When these valves fail, the skeletal muscle pump propels the deep venous blood into the saphenous veins, which cannot drain so much blood quickly enough. Due to the backup of blood in the saphenous veins, so much fluid pours out of the capillaries serving these veins that the leg tissue develops edema and becomes ischemic. The slightest trauma to this oxygen-starved tissue leads to tissue damage, so ulcers form in the lower limb. Venous disease can lead to varicose saphenous veins (pp. 550, 572) when the overloaded saphenous veins fail and tear. Treatment of this disease involves elevating and compressing the lower limb to speed the drainage of blood.

Microangiopathy of diabetes (mi″kro-an″je-op′ah-the; "small vessel disease") is a common complication of long-term diabetes mellitus. The elevated blood sugar levels of diabetes lead to deposition of glycoproteins in the basal lamina of the body's capillaries. This results in thickened but leaky capillary walls, and a slowed rate of turnover of the tissue fluid upon which tissue cells rely for oxygen and nutrients. The organs most affected and damaged by this microangiopathy are the kidneys, retina, peripheral nerves, and feet (that is, foot ulcers commonly develop).

Arteriovenous malformation is a congenital condition in which capillaries fail to develop in a certain location, so that an artery continues directly into a vein. Relatively common among birth defects, it affects 1 in 700 people and usually occurs in the cerebrum of the brain. Without an intervening bed of capillaries to lower the high pressure of arterial blood before it reaches the vein, the vein weakens and forms a bulging aneurysm, which can compress nearby structures or burst to cause a stroke. Treatment is by surgically closing or removing the damaged segment of vein.

BLOOD VESSELS THROUGHOUT LIFE

As described in Chapter 17, the earliest blood vessels develop from blood islands around the yolk sac in the 2-week embryo. By the end of week 3, mesoderm *within* the embryo itself has begun to form networks of blood vessels that grow throughout the body, both by sprouting extensions and by splitting into more branches. At first, the vessels consist only of endothelium, but adjacent mesenchymal cells soon surround these tubes, forming the muscular and fibrous tunics of the vessel walls. Ultimately, the networks are remodeled into treelike arrangements of larger and smaller vessels.

Once the basic pattern of vessels is established in the embryo, vessel formation slows, but it still continues throughout life: to support the growth of a fetus, to facilitate wound healing, and to rebuild the vessels lost each month after a woman's menstrual period. In adults, as in embryos, new capillaries sprout from existing vessels as tubular outgrowths of the endothelium, in direct relation to tissue demands for oxygen.

Fetal Circulation

All major vessels are in place by the third month of development, and blood is flowing through these vessels in the same direction as in adults (Figure 19.26). However, there are two major differences between fetal and postnatal circulation: First, the fetus must supply blood to the *placenta* (see p. 729), a pancake-shaped organ at the end of the umbilical cord through which oxygen and nutrients are obtained from the mother's uterus. Second, because the fetal respiratory organ is the placenta and the fetus does not breathe, its lungs do not need much blood. Therefore, the fetus sends very little blood around the pulmonary circuit. Despite these special features and needs, the fetus must be able to convert rapidly at birth to the postnatal circulatory pattern. We now explore the two special features of fetal circulation in more detail.

Vessels to and from the Placenta

The fetal vessels that carry blood to and from the placenta are called *umbilical vessels* because they run in the umbilical cord (see Figure 19.26a). The paired **umbilical arteries** branch from the internal iliac arteries in the pelvis and carry blood to the placenta to pick up oxygen and nutrients. The unpaired **umbilical vein** returns this blood to the fetus, delivering some of it to the portal vein so its nutrients can be processed by the liver cells. However, there is too much returning blood for the liver to process, so most of the blood is diverted through a shunt called the **ductus venosus** ("venous duct"). Whether it goes through the portal system or the ductus venosus, all returning blood proceeds into the hepatic veins, inferior vena cana, and right atrium of the heart. In the inferior vena cava, this oxygen-rich placental blood mixes with the oxygen-poor blood returning to the heart from the caudal parts of the fetal body.

At birth, most parts of these umbilical vessels are discarded when the umbilical cord is cut, but the parts remaining in the baby's body constrict rapidly and then gradually degenerate into fibrous bands called *ligaments* (see Figure 19.26b). Throughout postnatal life, the remnant of the umbilical vein is the **ligamentum teres** ("round ligament"). The ductus venosus becomes the **ligamentum venosum** on the liver's inferior surface, and the umbilical arteries become **medial umbilical ligaments** in the anterior abdominal wall inferior to the navel.

Shunts Away from the Pulmonary Circuit

As mentioned, the fetal lungs and pulmonary circuit need very little blood. However, as in the adult, the right heart receives all blood returning from the systemic circuit and pumps blood into the pulmonary trunk, thereby threatening to overload the lungs. To avoid this problem, blood is diverted through two shunts: the *foramen ovale* and the *ductus arteriosus*.

1. **Foramen ovale.** Slightly less than half the blood entering the fetal heart is diverted from the right atrium to the left atrium through a hole in the interatrial septum, the foramen ovale (see p. 538). That is, each time the atria contract, the full right atrium squeezes some of its blood through this hole into the almost-empty left atrium. The foramen ovale is actually a valve with two flaps that prevent any blood from flowing in the opposite direction.
2. **Ductus arteriosus.** Despite the shunt through the foramen ovale, most of the blood that enters the fetal right atrium continues into the right ventricle, which pumps it into the pulmonary trunk, where it threatens to overload the immature lungs. This is avoided by the existence of another bypass: A wide arterial shunt called the ductus arteriosus carries blood from the most cranial part of the pulmonary trunk to the adjacent aortic arch, so that only a trickle reaches the lungs. The blood in the aorta goes to nourish the tissues throughout the body, and some proceeds to the placenta to pick up more oxygen and nutrients.

What happens to the foramen ovale and the ductus arteriosus at birth (Figure 19.26b)? When the newborn takes its first breaths, the lungs receive more blood, and the ductus arteriosus constricts and closes. For the first

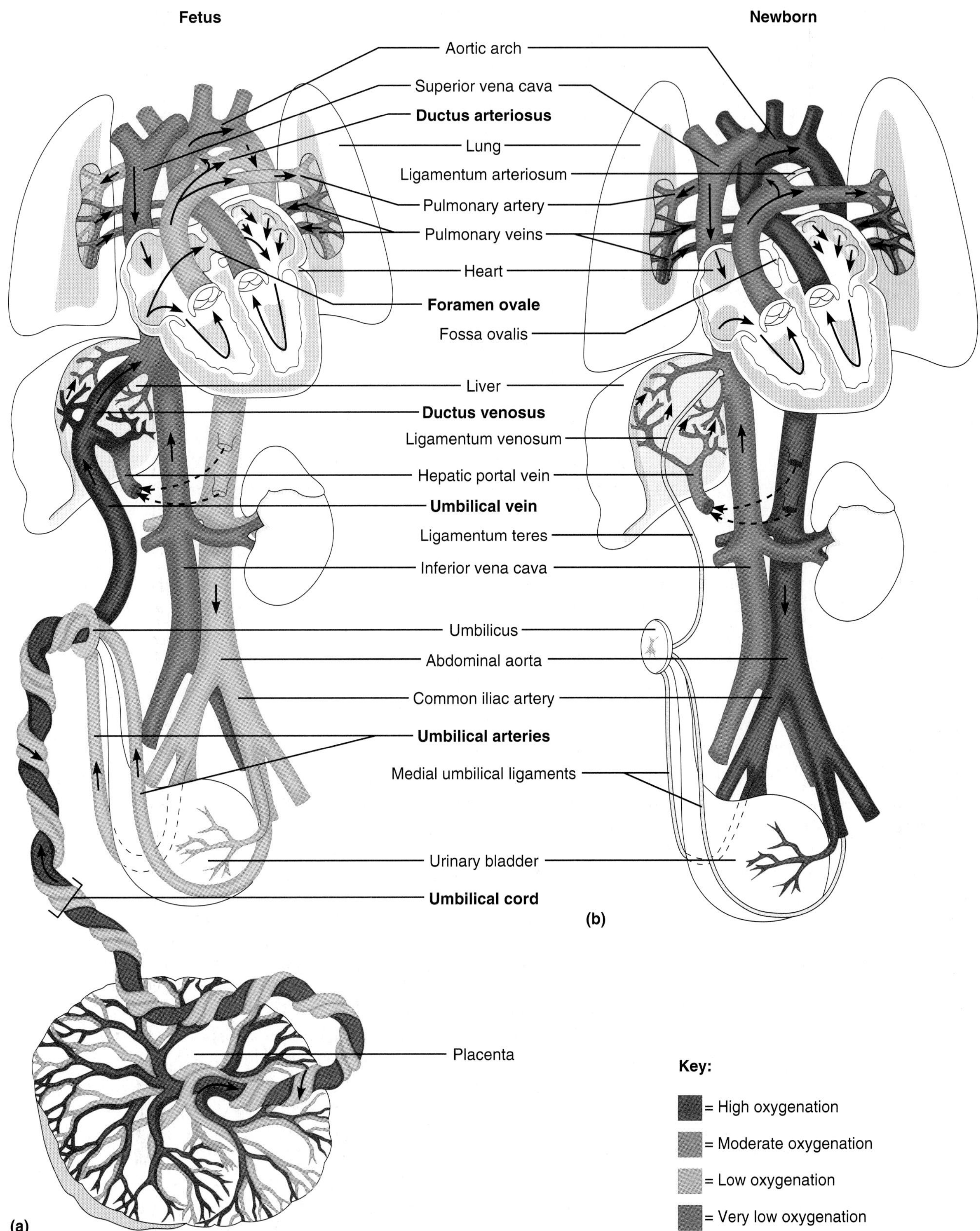

FIGURE 19.26 Fetal and newborn circulation compared. Arrows indicate the direction of blood flow. **(a)** Special adaptations for fetal life. The umbilical arteries carry waste-laden, low-oxygen blood from the fetus to the placenta, whereas the umbilical vein carries oxygen-rich and nutrient-rich blood from the placenta to the fetus. The ductus arteriosus and foramen ovale shunt blood away from the nonfunctional lungs, and the ductus venosus enables much of the blood to bypass the liver circulation. The ductus venosus lies in a deep cleft between the two major lobes of the liver and is therefore a shunt *around* the liver. **(b)** Changes in the cardiovascular system at birth. The umbilical cord and placenta are lost, and the parts of the umbilical vessels remaining in the body close, as do the ductus arteriosus and foramen ovale.

time, oxygenated pulmonary blood begins pouring into the left atrium, raising the pressure within this chamber. This pressure pushes the two valve flaps of the foramen ovale together, closing it. Both the foramen ovale and ductus arteriosus are now closed, and the postnatal circulatory plan is established.

Although the foramen ovale and ductus arteriosus close shortly after birth, they do not immediately fuse shut. It takes about 3 months for the ductus arteriosus to become the solid ligamentum arteriosum, and about a year for the flaps of the foramen ovale to fuse together.

Critical Reasoning

Occasionally, either the ductus arteriosus or the foramen ovale stays patent (open) after birth. Patent ductus arteriosus and patent foramen ovale each characterize about 8% of all congenital heart defects, or 1 in every 1850 births. What functional deficit do both these conditions produce?

Because they both lead to a mixing of oxygen-rich pulmonary blood with oxygen-poor systemic blood after birth, the blood reaching the tissues is not fully oxygenated. In such cases, the baby is blue from cyanosis. Still, the affected babies can survive, and surgeons usually wait until the child is a few years old (and can better withstand the risks of thoracic surgery) before repairing the defects.

Blood Vessels in Adulthood

The most important aspect of the aging of the vascular system is the progression of atherosclerosis (see p. 574). Because of the high fat content in the American diet, almost everyone in the United States develops atherosclerotic plaques in the major arteries before the age of 40. Although the degenerative process of atherosclerosis begins in youth, its consequences are rarely apparent until middle to old age, when it may lead to myocardial infarction or stroke.

Sex has an effect on how atherosclerosis develops with age. Until puberty, the blood vessels of boys and girls look alike, but from puberty to about the age of 45 years, women have strikingly less atherosclerosis than men, probably because of protective effects of estrogen. This "gap between the sexes" closes between the ages of 45 and 65, and after age 65 the incidence of heart disease is slightly higher in women. Furthermore, when a woman does experience a heart attack, she is more likely to die from it than is a man. Some of the reasons for this are unpreventable (the average woman having a heart attack is 10 years older, and thus weaker, than the average male victim, and the vessel-damaging aspects of diabetes seem to affect women more), but others are preventable. Many people wrongly believe that women do not get atherosclerosis, so some women may not recognize the symptoms of angina or a heart attack in time to seek help. Although cardiovascular disease is the number one killer of women in the United States, with an age-adjusted mortality rate that is four to six times greater than that caused by breast cancer, estrogen-replacement therapy seems to be a good preventative measure.

RELATED CLINICAL TERMS

Angiogram (an'je-o-gram") (*angio* = vessel; *gram* = writing) Diagnostic technique involving the infusion of a radiopaque substance into the bloodstream for X-ray examination of specific blood vessels. Angiograms are the major way of diagnosing occlusion of coronary arteries and risk of heart attack.

Angiosarcoma Cancer originating from the endothelium of a blood vessel; may develop in liver vessels following exposure to chemical carcinogens.

Blue baby A baby with cyanosis (skin appears blue) due to relatively low levels of oxygen in the blood. This condition is caused by any congenital defect that leads to low oxygenation of the systemic blood, including patent foramen ovale and patent ductus arteriosus (see above), failure of the lungs to inflate at birth, and other conditions (see pp. 538–539).

Carotid endarterectomy Surgical procedure that scrapes away atherosclerotic plaques that block the base of the internal carotid artery; done to decrease the risk of a stroke.

Phlebitis (fle-bi'tis) (*phleb* = vein; *itis* = inflammation) Inflammation of a vein, accompanied by painful throbbing and redness of the skin over the inflamed vein; most often caused by bacterial infection or local physical trauma.

Phlebotomy (fle-bot'o-me) (*tomy* = cut) An incision made in a vein for withdrawing blood.

CHAPTER SUMMARY

Media study tools that could provide you additional help in reviewing specific key topics of Chapter 19 are referenced below.

SP = Study Partner; AIA = A.D.A.M.® Interactive Anatomy

PART ONE: GENERAL CHARACTERISTICS OF BLOOD VESSELS (pp. 544–551)

1. The main types of blood vessels are arteries, capillaries, and veins. Arteries carry blood away from the heart and toward the capillaries; veins carry blood away from the capillaries and toward the heart. Arteries "branch," whereas veins "serve as tributaries."

STRUCTURE OF BLOOD VESSEL WALLS (p. 544)

1. All but the smallest blood vessels have three tunics: tunica intima (with an endothelium), tunica media (mostly smooth muscle), and tunica externa (external connective tissue). The tunica media, which changes the diameter and strengthens the wall of the vessel, is thicker in arteries than in other types of blood vessels.

TYPES OF BLOOD VESSELS (pp. 544–550)

1. The three classes of arteries, from large to small, are elastic arteries, muscular arteries, and arterioles. Elastic (conducting) arteries, the largest arteries near the heart, contain more elastin than does any other vessel type. The elastin in all arteries helps dampen the pressure pulses produced by the heartbeat.
2. Muscular (distributing) arteries carry blood to specific organs and organ regions. These arteries have a thick tunica media and are most active in vasoconstriction.
3. Arterioles, the smallest arteries, have one or two layers of smooth muscle cells in their tunica media. They regulate the flow of blood into capillary beds.
4. Capillaries, which have diameters slightly larger than erythrocytes, are endothelium-walled tubes arranged in networks called capillary beds. The most permeable capillaries have fenestrations (pores) in their endothelial cells; other capillaries are continuous (lack pores). Spider-shaped pericytes help support the capillary wall.
5. The four routes involved in capillary permeability are (1) through intercellular clefts between endothelial cells, (2) through the pores of fenestrated capillaries, (3) direct diffusion of respiratory gases across the endothelium, and (4) through endothelial cytoplasmic vesicles (caveolae).
6. Sometimes, blood is shunted straight through a capillary bed (from arteriole to metarteriole to thoroughfare channel to venule) and does not enter the true capillaries to the surrounding tissue. Precapillary sphincters determine how much blood enters the true capillaries.
7. Sinusoids are wide, twisted, leaky capillaries.
8. The smallest veins, postcapillary venules, function like capillaries.
9. Because venous blood is under less pressure than arterial blood, the walls of veins are thinner than those of arteries of comparable size. Veins also have a wider lumen, a thinner tunica media, and a thicker tunica externa.
10. Many veins have valves that prevent the backflow of blood. Varicose veins are veins whose valves have failed.

VASCULAR ANASTOMOSES (pp. 550–551)

11. The joining together of arteries serving a common organ, called an arterial anastomosis, provides collateral channels for blood to reach the same organ. Anastomoses between veins are more common than anastomoses between arteries.

VASA VASORUM (p. 551)

1. Vasa vasorum are small arteries, capillaries, and veins that supply the outer part of the wall of larger blood vessels.

PART TWO: BLOOD VESSELS OF THE BODY (pp. 552–578)

THE PULMONARY CIRCULATION (pp. 552–553)

1. The pulmonary trunk splits inferior to the aortic arch into the right and left pulmonary arteries. These arteries divide into lobar arteries, then branch repeatedly in the lung. Arising from the capillary beds, pulmonary venous tributaries empty newly oxygenated blood into the superior and inferior pulmonary veins of each lung. The four pulmonary veins extend from the lungs to the left atrium.

AIA Link: Atlas; System: Cardiovascular; Heart and Great Vessels (posterior).

THE SYSTEMIC CIRCULATION (pp. 553–573)

1. Figures 19.16 and 19.25 summarize the main arteries and veins of the systemic circulation. The systemic circuit begins with the aorta (ascending aorta, aortic arch, descending aorta) and ends with the two large venae cavae and the coronary sinus.

AIA Link: Dissectible Anatomy; Male: Anterior; Layer 238. Scroll to Thorax.

2. A portal system is a set of vessels in which *two* capillary beds, interconnected by a vein, lie between the initial artery and final vein. The hepatic portal system is a special subcirculation that drains the digestive organs of the abdomen and pelvis (see Figure 19.22).

AIA Link: Atlas; System: Cardiovascular; Hepatic Portal Vein.

3. Cirrhosis of the liver leads to portal hypertension, in which blood backs up through the portal-systemic anastomoses, overloading these delicate veins. This overload leads to esophageal bleeding, hemorrhoids, and caput medusae.

SP Exercise: Chapter 19, Making Connections: The Cardiovascular System.

DISORDERS OF THE BLOOD VESSELS (pp. 573–576)

1. *Atherosclerosis,* the most important vascular disorder, is discussed on pp. 574–575. An *aneurysm* is a widening or outpocketing of an artery (or vein). *Deep vein thrombosis of the lower limb* is the formation of clots in veins of the lower extremity. In *venous disease,* the failure of venous valves in some lower-limb veins results in inadequate drainage of venous blood from the limb, whose tissues consequently become ischemic and vulnerable to ulceration. In *microangiopathy of diabetes,* the capillaries are leaky due to sugar deposits in their basal lamina. In *arteriovenous malformation,* the capillaries fail to develop between some important artery and the vein that drains it, often leading to a rupture or an aneurysm of the vein.

BLOOD VESSELS THROUGHOUT LIFE (pp. 576–578)

1. The first blood vessels develop from blood islands on the yolk sac in the 3-week embryo. Soon, vessels begin forming from mesoderm inside the embryo. The formation of new vessels, whether in the embryo or adult, occurs mainly through the sprouting of endothelial buds from existing vessels.

2. The pattern of blood vessels is the same in both the fetus and the newborn, and blood flows through these vessels in the same directions. However, the fetus also has vessels to and from the placenta (umbilical arteries, umbilical vein, and ductus venosus) and two shunts that bypass the nearly nonfunctional pulmonary circuit (the foramen ovale in the interatrial septum, and the ductus arteriosus between the pulmonary trunk and aortic arch).

3. Fetal shunts and vessels close shortly after birth. Failure of the foramen ovale or ductus arteriosus to close leads to a mixing of oxygenated and unoxygenated blood in the newborn.

4. The most important age-related vascular disorder is the progression of atherosclerosis.

REVIEW QUESTIONS

Multiple Choice/Matching Questions

1. Which of the following statements does *not* correctly describe veins? (a) They have less elastic tissue and smooth muscle than arteries. (b) They are subject to lower blood pressures than arteries. (c) They have larger lumens than arteries. (d) They always carry deoxygenated blood.

2. In atherosclerosis, which of the following layers of the artery wall thickens most? (a) tunica media, (b) tunica intima, (c) tunica externa.

3. Blood flow through the capillaries is steady despite the rhythmic pumping action of the heart because of the (a) elasticity of the large arteries only, (b) elasticity of all the arteries, (c) ligamentum arteriosum, (d) venous valves.

4. Fill in the blanks with the name of the appropriate vascular tunic (intima, media, or externa).

______ **(1)** contains endothelium

______ **(2)** the tunic with the largest vasa vasorum

______ **(3)** the thickest tunic in veins

______ **(4)** the outer tunic; is mostly connective tissue

______ **(5)** is mostly smooth muscle

______ **(6)** forms the valves of veins

5. Which of the following vessels is bilaterally symmetrical (that is, has an identical member of a pair on either side of the body)? (a) internal carotid artery, (b) brachiocephalic trunk, (c) azygos vein, (d) superior mesenteric vein.

6. Tell which artery is *missing* from the following sequence, which traces the flow of arterial blood to the right hand: Blood leaves the heart and passes through the aorta, the right subclavian artery, the axillary and brachial arteries, and through either the radial or ulnar artery to a palmar arch. (a) left coronary, (b) brachiocephalic, (c) cephalic, (d) right common carotid.

7. Which of the following veins do not drain directly into the inferior vena cava? (a) lumbar veins, (b) hepatic veins, (c) inferior mesenteric vein, (d) renal veins.

8. A stroke that blocks a posterior cerebral artery will most likely affect (a) hearing, (b) sight, (c) smell, (d) higher thought processes.

9. Tell which artery is *missing* from the following sequence, which traces the flow of arterial blood to the left parietal and temporal lobes of the brain: Blood leaves the heart and passes through the aorta, the left common carotid artery, and the middle cerebral artery. (a) vertebral, (b) brachiocephalic, (c) internal carotid, (d) basilar.

10. Tell which two veins are *missing* from the following sequence: Tracing the drainage of *superficial* venous blood from the leg, blood enters the great saphenous vein, femoral vein, inferior vena cava, and right atrium. (a) coronary sinus and superior vena cava, (b) posterior tibial and popliteal, (c) fibular (peroneal) and popliteal, (d) external and common iliacs.

11. The inferior mesenteric artery supplies the (a) rectum and part of the colon, (b) liver, (c) small intestine, (d) spleen.

12. Tell which two vessels are *missing* from the following sequence: Tracing the drainage of venous blood from the small intestine, we find that blood enters the superior mesenteric vein, hepatic vein, inferior vena cava, and right atrium. (a) coronary sinus and left atrium, (b) celiac and common hepatic veins, (c) internal and common iliac veins, (d) hepatic portal vein and liver sinusoids.

13. Tell which vessel is *missing* from the following vascular circuit: Tracing the circulation to and from the placenta, a drop of blood travels in the fetus from left ventricle to aorta to a common iliac artery to an internal iliac artery to an umbilical artery to capillaries in the placenta to ductus venosus to a hepatic vein to inferior vena cava to right atrium to foramen ovale to left atrium to left ventricle. (a) umbilical vein, (b) ligamentum arteriosum, (c) internal carotid artery, (d) capillaries in the small intestine.

Short Answer Essay Questions

14. (a) What structural features are responsible for the permeability of capillary walls? (b) Which of these features are absent from brain capillaries in the blood-brain barrier?

15. Distinguish among elastic arteries, muscular arteries, and arterioles relative to their location, histology, and functions.

16. (a) Name an organ containing fenestrated capillaries. (b) What is a sinusoid?

17. (a) Explain in your own words why varicose veins are more common in the lower limbs than elsewhere in the body. (b) Give a functional reason why valves are more abundant in veins of the upper and lower limbs than in veins of the neck.

18. Sketch the arterial circle at the base of the brain, and label the important arteries that branch off this circle.

19. (a) What are the two large tributaries that form the hepatic portal vein? (b) What is the function of the hepatic portal circulation?

20. State the location and basic body regions drained by the azygos vein.

21. Name all four pulmonary veins.

22. What are the ductus venosus and ductus arteriosus, where are they located, what are their functions, and how are they different?

CRITICAL REASONING + CLINICAL APPLICATION QUESTIONS

1. In an eighth-grade health class, the teacher warned the students not to squeeze pimples or pluck hairs on their nose and upper lip. The students made fun of this warning, but they got the message when the teacher explained the danger triangle of the face. What is the danger of infections in this area?
2. Logically deduce which is the more difficult and dangerous surgery on a child: closing a patent ductus arteriosus or closing a patent foramen ovale. Explain your reasoning.
3. Samantha accidentally received a small but deep puncture wound from broken glass in the exact midline of the anterior side of her distal forearm. She worried during the ten-minute drive to the hospital that she would bleed to death because she had heard stories about people dying from slashing their wrist. Look at Figure 19.10, and determine whether Sam's fear of bleeding to death is justified. Explain your reasoning.
4. Your friend, who knows little about science, is reading a magazine article about a patient who had an "aneurysm at the base of his brain that suddenly grew much larger." The surgeons' first goal was to "keep it from rupturing," and their second goal was to "relieve the pressure on the brain stem and cranial nerves." The surgeons were able to "replace the aneurysm with a section of plastic tubing," so the patient recovered. Your friend asks you what all this means. Explain. (Hint: Start by checking this chapter's Disorders section.)

A.D.A.M.® INTERACTIVE ANATOMY QUESTIONS

1. Link: Dissectible anatomy; Male (anterior): layer 240; scroll to abdomen. (a) Name the arteries that supply blood to the kidneys. (b) Name the two arteries that branch from the inferior end of the abdominal aorta.
2. Link: Dissectible anatomy: Male (anterior): Layer 87; scroll to upper limb. The left subclavian artery supplies blood to the shoulder and arm. As it enters the shoulder, the artery name changes, and as it enters the arm, the name changes again. (a) Give the name of the artery as it passes through the shoulder. (b) Give the name of the artery as it enters the arm.

CHAPTER 20

The Lymphatic and Immune Systems

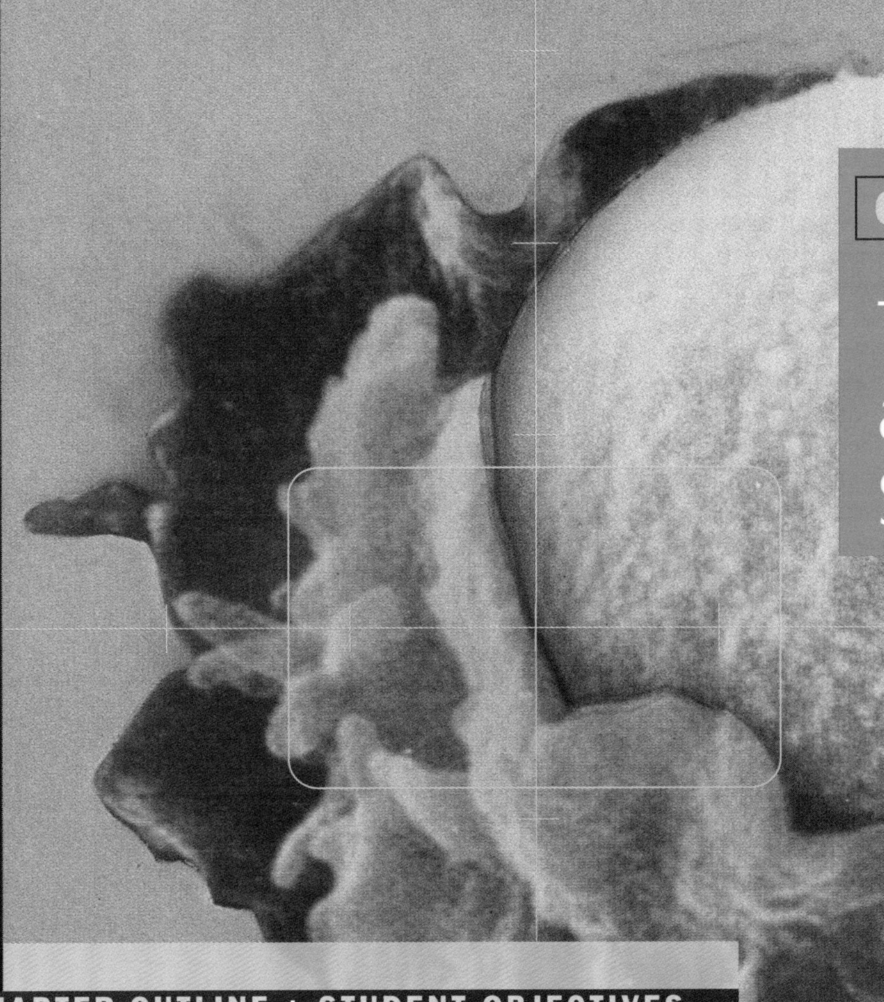

SEM of lymphocyte and yeast cell

CHAPTER OUTLINE + STUDENT OBJECTIVES

The Lymphatic System 584–589

1. Describe the structure and distribution of lymphatic vessels.
2. Explain how lymph forms and the mechanisms by which it is transported.
3. List and explain the important functions of the lymphatic vessels.
4. Describe how lymph nodes function as lymphatic organs.

The Immune System 589–598

5. Describe the function, recirculation, and activation of lymphocytes.
6. Relate the structure of lymphoid tissue to its infection-fighting function.
7. Describe the locations, histological structure, and immune functions of the following lymphoid organs: lymph nodes, spleen, thymus, tonsils, aggregated lymphoid follicles in the intestine, and appendix.

Disorders of the Lymphatic and Immune Systems 598–599

8. Describe the basic characteristics of chylothorax, lymphangitis, mononucleosis, Hodgkin's disease, and non-Hodgkin's lymphoma.

The Lymphatic and Immune Systems Throughout Life 599–600

9. Outline the development of the lymphatic vessels and lymphoid organs.

NOW THAT WE HAVE CONSIDERED THE cardiovascular system in the previous three chapters, we turn to the closely related lymphatic and immune systems. The main structures of the lymphatic system are the **lymphatic vessels,** which transport fluid that has escaped from the blood vessels back to the bloodstream. The main components of the immune system are **lymphocytes, lymphoid tissue,** and **lymphoid organs** (such as the spleen, lymph nodes, and thymus), which fight infections and confer immunity to disease.

Both the lymphatic and immune structures are of utmost importance to students entering the health professions: As we will see, the lymphatic vessels provide a route by which disease organisms travel throughout the body, whereas the lymphoid tissues and organs function to contain and destroy these organisms.

THE LYMPHATIC SYSTEM

An elaborate system of lymphatic vessels runs throughout the body. These vessels collect a fluid called **lymph** (*lympha* = clear water) from the loose connective tissue around blood capillaries and carry this fluid to the great veins at the root of the neck. Because lymph flows only *toward* the heart, the lymphatic vessels form a one-way system rather than a full circuit (Figure 20.1).

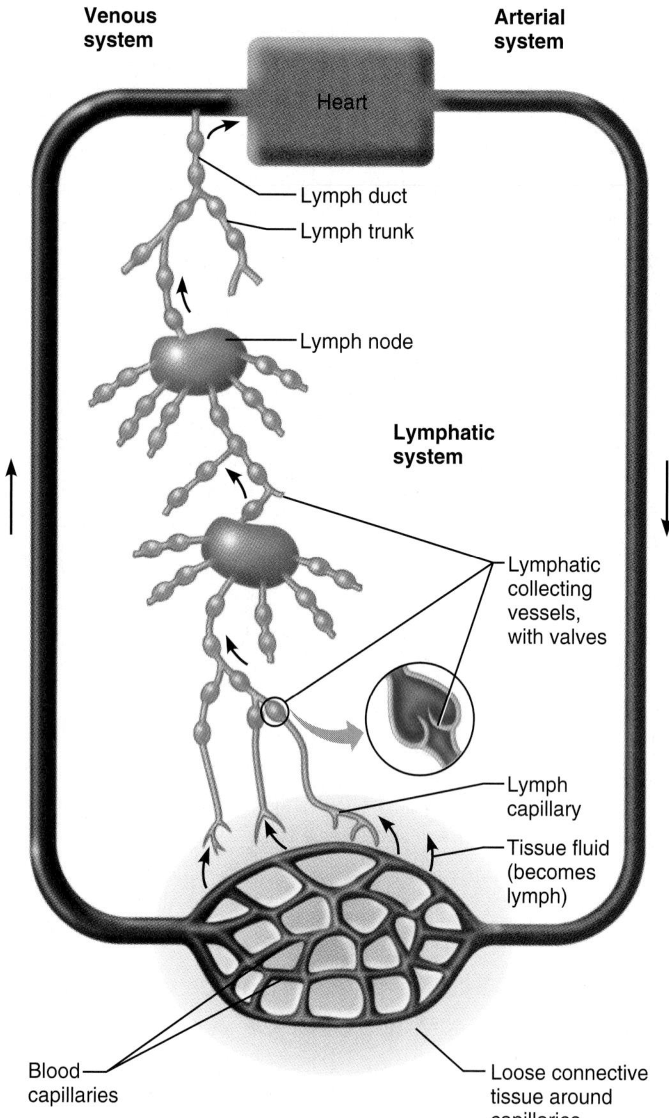

FIGURE 20.1 Simplified scheme of the lymphatic vessels as they relate to the blood vessels of the cardiovascular circuit. Proceeding from the bottom to the top of this figure, we see that lymph, which begins as tissue fluid derived from blood capillaries in the loose connective tissue, enters the lymph capillaries, travels through the lymphatic collecting vessels and lymph nodes, continues through the lymph trunks and lymph ducts, and enters the bloodstream in the great veins at the root of the neck.

There are several orders of lymphatic vessels. The smallest vessels, those that first receive lymph, are the *lymph capillaries.* These vessels drain into larger *lymphatic collecting vessels,* along which are scattered *lymph nodes.* The collecting vessels then drain into *lymph trunks,* which unite to form *lymph ducts,* which empty into the veins of the neck. We will consider each order of lymphatic vessel shortly, after we explore the basic functions of the lymphatic system.

Recall from Chapter 4 that all blood capillaries are surrounded by a loose connective tissue that contains *tissue fluid* (see Figure 20.1), which arises from blood filtered through the capillary walls and consists of the small molecules of blood plasma, including water, various ions, nutrient molecules, and respiratory gases. Even though tissue fluid is continuously leaving and reentering the blood capillaries, for complex reasons slightly *more* tissue fluid arises from the arteriole end ("upstream" end) of each capillary bed than reenters the blood at the venule end ("downstream" end). The lymphatic vessels function to collect this excess tissue fluid and return it to the bloodstream. Indeed, any blockage of the lymphatic vessels causes the affected body region to swell with excess tissue fluid, a condition called *edema* (see p. 103).

The lymphatic vessels also perform another, related function. Blood proteins leak slowly but steadily from blood capillaries into the surrounding tissue fluid, and *the lymph vessels return these leaked proteins to the bloodstream.* Recall that the proteins in blood generate osmotic forces that are essential for keeping water in the bloodstream (see p. 505). If proteins were allowed to leak from the capillaries without being returned to the bloodstream, a massive outflow of water from the blood to the tissues would soon follow, and the entire cardiovascular system would collapse from insufficient volume.

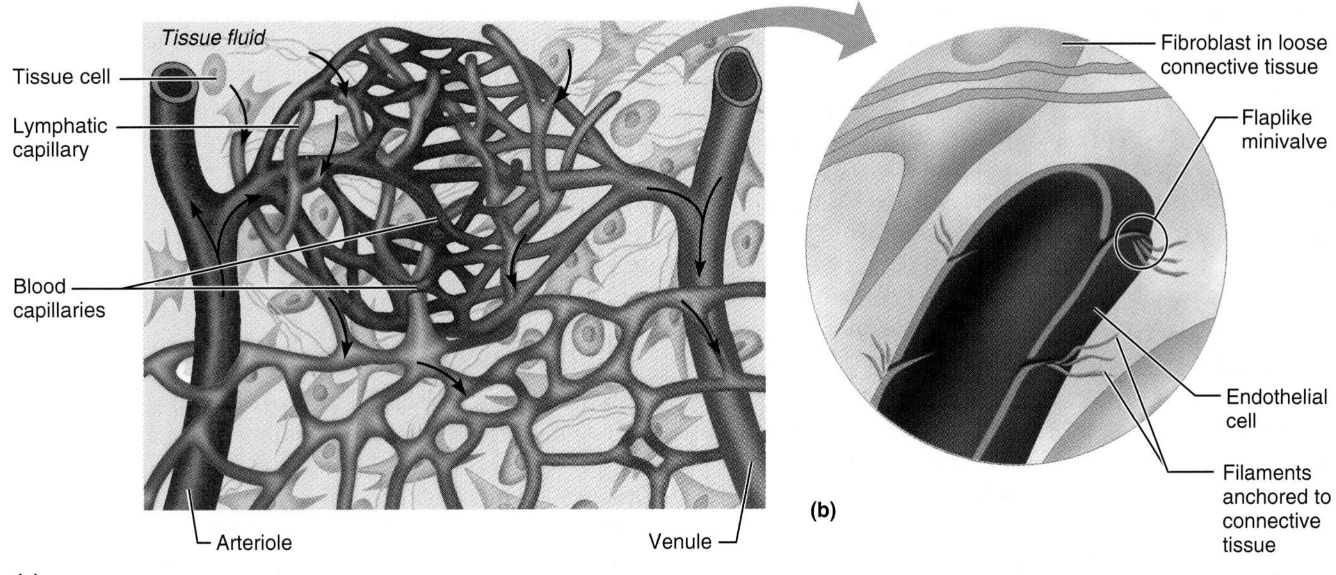

FIGURE 20.2 Location and structure of lymph capillaries. **(a)** Structural relationship between blood capillaries and the lymph capillaries (shown in green). Tissue fluid is in the loose connective tissue around the capillaries. **(b)** Lymphatic capillaries begin as closed-end tubes. The endothelial cells of their walls loosely overlap one another, forming flap-like minivalves.

Lymph Capillaries

Lymph capillaries, the permeable vessels that receive the tissue fluid, are located near blood capillaries in the loose connective tissue (Figure 20.2a). Like blood capillaries, their wall consists of a single layer of endothelial cells. Now known to be closed-end vessels (Figure 20.2b), lymph capillaries are so remarkably permeable that they were once thought to be open-ended. We now know that their permeability results from the structure and arrangement of the endothelial cells: They have few intercellular junctions, and the edges of adjacent cells overlap, forming easily opened minivalves. Bundles of fine collagen filaments anchor the endothelial cells to the surrounding connective tissue. As a result, any increase in the volume of the tissue fluid separates the minivalve flaps, opening gaps in the wall and allowing the fluid to enter. Once this lymph enters the lymphatic capillaries, it cannot leak out, because backflow forces the minivalve flaps together.

Although the high permeability of lymph capillaries allows the uptake of large quantities of tissue fluid and large protein molecules, it also allows any bacteria, viruses, or cancer cells in the loose connective tissue to enter these capillaries with ease. These pathogenic agents can then travel throughout the body via the lymphatic vessels. However, this threat is averted in part by the lymph nodes, which destroy most pathogens in the lymph (see pp. 586 and 595).

Lymph capillaries are widespread, occurring almost everywhere blood capillaries occur. However, lymph capillaries are absent from bone and teeth, from bone marrow, and from the entire central nervous system, where excess tissue fluid drains through the nervous tissue into the cerebrospinal fluid. The cerebrospinal fluid then returns this tissue fluid to the blood at the superior sagittal sinus (see pp. 391–392).

One set of lymph capillaries, called *lacteals* (lak′te-alz), has a unique function. Located in the finger-like villi of the mucosa of the small intestine, lacteals receive digested fats from the intestine, which causes the lymph draining from the digestive viscera to become milky white (*lacte* means "milk"). This fatty lymph is called *chyle* (kīl; "juice"), and, like all lymph, it is carried to the bloodstream.

Lymphatic Collecting Vessels

From the lymph capillaries, lymph enters **lymphatic collecting vessels,** which accompany blood vessels: In general, the superficial lymphatic collecting vessels in the skin travel with superficial *veins*, whereas the deep lymphatic collecting vessels of the trunk and digestive viscera travel with the deep *arteries*.

Lymphatic collecting vessels are narrow and delicate, so they usually are not seen in the dissecting laboratory. They have the same tunics as blood vessels (tunica intima, tunica media, and tunica externa), but their walls are always much thinner. This thinness reflects the fact that lymph flows under very low pressure, because lymphatic vessels are not connected to the pumping heart. To direct the flow of lymph, lymphatic collecting vessels contain more valves than do veins (see Figure 20.1). At the base of each valve, the vessel bulges, forming a pocket in which lymph collects and forces the valve shut. Because of these bulges, each lymph collecting vessel resembles a string of beads. This distinctive appearance, which characterizes the larger lymph trunks and lymph ducts as well, allows physicians to recognize lymph vessels in X-ray films taken after these vessels are injected with radiopaque dye. This radiographic procedure is called **lymphangiography** (lim′fan″je-og′rah-fe; "lymph vessel picturing").

Unaided by the force of the heartbeat, lymph is propelled through lymph vessels by a series of weaker mechanisms. Both the bulging of contracting skeletal muscles and the pulsations of nearby arteries push on the lymph vessels, squeezing lymph through them. Additionally, the muscular tunica media of the lymph vessels contracts to help propel the lymph, and the normal movements of the limbs and trunk keep the lymph flowing. Despite these propulsion mechanisms, the transport of lymph is sporadic and slow, which explains why people who stand for long times at work may develop severe edema around the ankles by the end of the workday. The edema usually disappears if the legs are exercised, as, for example, by walking home. The seemingly useless nervous habit of people who bounce and wiggle their legs while sitting actually performs the important function of moving lymph up the legs.

CLINICAL APPLICATIONS

THE LYMPH VESSELS AND EDEMA A body region whose lymph collecting vessels have been blocked or removed will become swollen and puffy with edema. One cause is the tropical parasitic disease **elephantiasis** (el″ĕ-fan-ti′ah-sis), in which worms block the lymph vessels that drain the lower limb or external genitalia. The resulting edema, accompanied by a massive proliferation of fibrous (scar) tissue, produces a grossly enlarged lower limb that resembles an elephant's leg or, in males, a hugely swollen scrotum. Elephantiasis, which currently affects 120 million people in 73 countries, is transmitted by mosquitos, which carry blood-borne immature worms from person to person. Drugs are available to kill these immature worms, but the treatment takes five years. Because this cure is available, the World Health Organization has launched a worldwide campaign to eradicate elephantiasis by 2020.

Edema of the arm often follows a mastectomy (removal of a cancerous breast), in which the lymph vessels and nodes that drain the arm are removed at the axilla. Such women often suffer severe edema for several months, until the lymph vessels grow back. (Lymph vessels regenerate quite well and can even grow through scar tissue.) Because this condition, called *lymphedema of the arm,* is very uncomfortable and difficult to treat, surgeons who perform mastectomies are now encouraged to abandon the standard, aggressive procedure of removing all the axillary lymph nodes, and instead to remove only those nodes to which the cancer is likely to have spread. ✚

Lymph Nodes

Lymph nodes, which cleanse the lymph of pathogens, are bean-shaped organs situated along lymphatic collecting vessels (see Figure 20.1). The popular term "lymph glands" is not correct, because they are not glands at all. There are about 500 lymph nodes in the human body, ranging in diameter from 1 to 25 mm (up to 1 inch). Large clusters of superficial lymph nodes in the cervical, axillary, and inguinal regions, plus some important groups of deep nodes, are depicted in Figure 20.3.

The superficial *cervical nodes* along the jugular veins and carotid arteries receive lymph from the head and neck. *Axillary nodes* in the armpit and the *inguinal nodes* in the superior thigh filter lymph from the upper and lower limbs, respectively. Nodes in the mediastinum, such as the deep *tracheobronchial nodes,* receive lymph from the thoracic viscera. Deep nodes along the abdominal aorta, called *aortic nodes,* filter lymph from the posterior abdominal wall. Finally, deep nodes along the iliac arteries, called *iliac nodes,* filter lymph from pelvic organs and the lower limbs.

The microscopic anatomy of a lymph node is shown in Figure 20.4. The node is surrounded by a fibrous **capsule** of dense connective tissue, from which fibrous strands called **trabeculae** (trah-bek′u-le; "beams") extend inward to divide the node into compartments. Lymph enters the convex aspect of the node through several **afferent lymphatic vessels** and exits from the indented region on the other side, the **hilus** (hi′lus), through **efferent lymphatic vessels.** Within the node, between the afferent and efferent vessels, lymph percolates through **lymph sinuses.** These large lymph capillaries are spanned internally by a crisscrossing network of reticular fibers covered by star-shaped endothelial cells. Many macrophages live on this fiber network, phagocytizing pathogens and foreign particles in the lymph that flows through the sinuses. Most lymph passes through several nodes, so it is usually free of pathogens by the time it leaves its last node and enters the lymph trunks on its way to the great veins of the neck.

Along with its lymph sinuses, a lymph node also contains tadpole-shaped masses of lymphoid tissue and is divided into outer (cortex) and inner (medulla) regions. These structures are explained on p. 595, which considers the role of lymph nodes in the immune system.

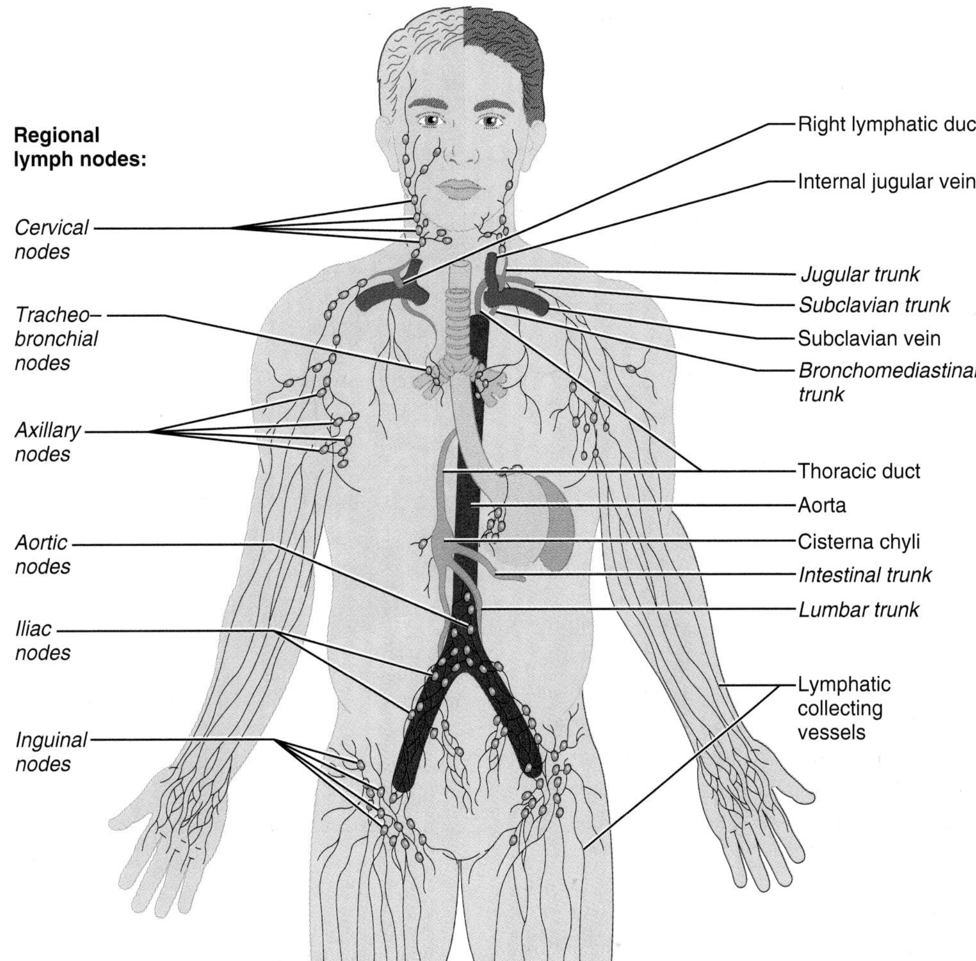

FIGURE 20.3 Overview of the lymph nodes (left) and lymph trunks and ducts (right). The cervical, axillary, and inguinal nodes are superficial; the tracheobronchial, aortic, and iliac nodes are deep. The superior right quarter of the body (shaded green) is drained by the right lymphatic duct (and its associated trunks), whereas the remainder of the body is drained by the thoracic duct (and its associated trunks).

CLINICAL APPLICATIONS

SWOLLEN LYMPH NODES Sometimes, lymph nodes are overwhelmed by the very agents they are trying to destroy. In one instance, when large numbers of undefeatable bacteria or viruses are trapped by the nodes but are not destroyed, the nodes become enlarged, inflamed, and very tender to the touch. Such infected lymph nodes are called *buboes* (bu'bōz), the most obvious symptom of bubonic plague. In another case, metastasizing cancer cells that enter lymph vessels and are trapped in the local lymph nodes continue to multiply there. The fact that cancer-infiltrated lymph nodes are swollen but not painful helps to distinguish cancerous nodes from those infected by microorganisms. (Pain results from inflammation, and cancer cells do not induce the inflammatory response.) Potentially cancerous lymph nodes can be located by palpation, as when a physician examining a patient for breast cancer feels for swollen axillary lymph nodes. Physicians can also locate enlarged, cancerous lymph nodes by using CT and MRI scans. ✚

Lymph Trunks

After leaving the lymph nodes, the largest lymphatic collecting vessels converge to form **lymph trunks** (see Figure 20.3). These trunks drain large areas of the body and are large enough to be found by a skilled dissector. The five major lymph trunks, from inferior to superior, are:

1. **Lumbar trunks.** These paired trunks, which lie along the sides of the aorta in the inferior abdomen, receive all lymph draining from the lower limbs, the pelvic organs, and from some of the anterior abdominal wall.
2. **Intestinal trunk.** This *unpaired* trunk, which lies near the posterior abdominal wall in the midline, receives fatty lymph (chyle) from the stomach, intestines, and other digestive organs.

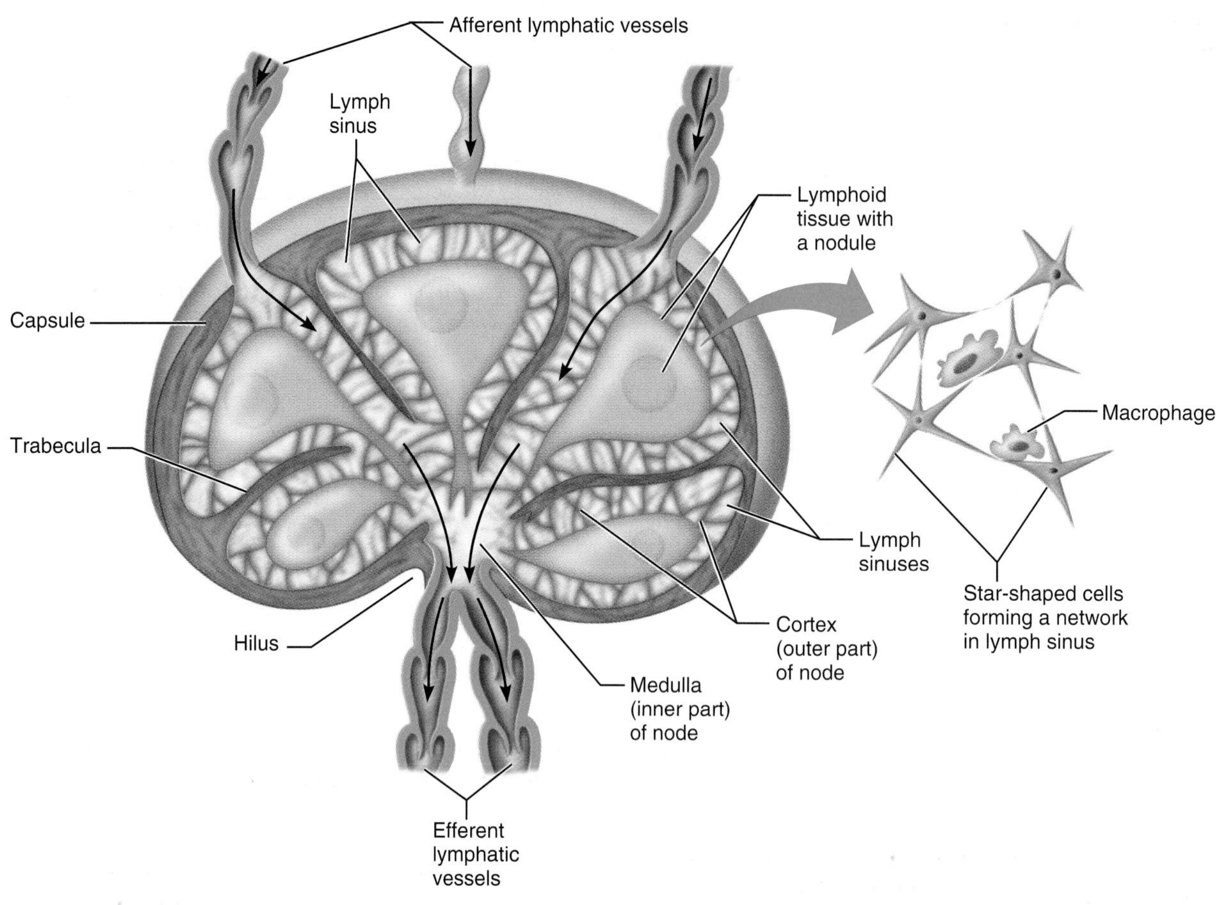

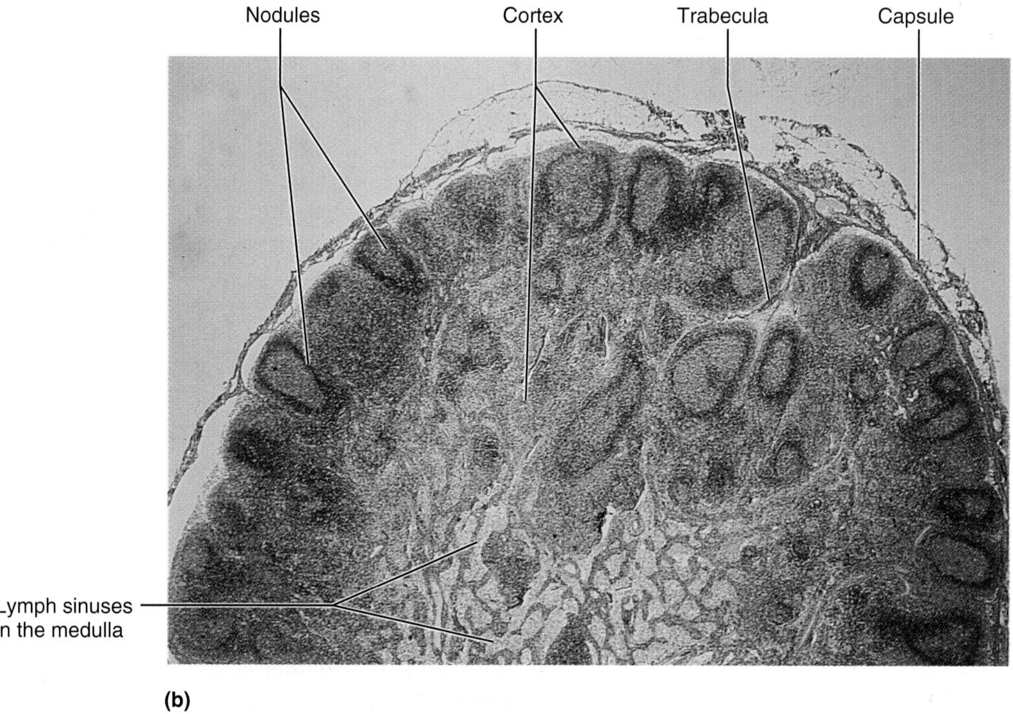

FIGURE 20.4 Structure of a lymph node (also see Figure 20.7b). **(a)** Diagram of a lymph node cut in longitudinal section, and the associated lymphatic vessels. Arrows indicate the direction of the lymph flow into, through, and out of the node. At right is an enlargement of the cells in a lymph sinus; notice the network of star-shaped cells, upon which reside macrophages that remove pathogens from the lymph. **(b)** Photomicrograph of a part of a lymph node (28 ×).

3. **Bronchomediastinal** (brong″ko-me″de-ah-sti′nal) **trunks.** Ascending near the sides of the trachea, these paired trunks collect lymph from the thoracic viscera and thoracic wall.
4. **Subclavian trunks.** Located near the base of the neck, these paired trunks receive lymph from the upper limbs; they also drain the inferior neck and the superior thoracic wall.
5. **Jugular trunks.** Located at the base of each internal jugular vein, these paired trunks drain lymph from the head and neck.

Lymph Ducts

The lymph trunks drain into the largest lymphatic vessels, the **lymph ducts** (Figures 20.3 and 20.5). Whereas some individuals have two lymph ducts, others have just one.

1. **Thoracic duct.** The *thoracic duct* is present in all individuals. Its most inferior part, located at the union of the lumbar and intestinal trunks, is the **cisterna chyli** (sis-ter′nah ki′li; "sac of chyle"), which lies on the bodies of vertebrae L_1 and L_2 (see Figure 20.5a). From there, the thoracic duct ascends along the vertebral bodies. In the superior thorax, it turns left and empties into the venous circulation at the junction of the left internal jugular and left subclavian veins. The thoracic duct is often joined near its end by the left jugular, subclavian, and/or bronchomediastinal trunks. Alternately, any or all of these three lymph trunks can empty separately into the nearby veins. When it is joined by the three trunks, the thoracic duct drains three quarters of the body: the left side of the head, neck, and thorax; the left upper limb; and the body's entire lower half (see Figure 20.3).
2. **Right lymphatic duct.** Some people have a short *right lymphatic duct,* formed by the union of the right jugular, subclavian, and bronchomediastinal trunks. When present, this duct empties into the neck veins at or near the junction of the right internal jugular and subclavian veins. The right lymphatic duct and its trunks drain the upper right quarter of the body (see Figure 20.3). In individuals without a right lymphatic duct, the three trunks empty separately into the neck veins.

This completes the description of the lymphatic vessels. To summarize their functions, they (1) return excess tissue fluid to the bloodstream, (2) return leaked proteins to the blood, and (3) carry absorbed fat from the intestine to the blood (through lacteals). Moreover, in addition to their roles as lymph filters, the lymph nodes along lymph collecting vessels fight disease in their roles as lymphoid organs of the immune system. The components of the immune system are the topic of the next section.

THE IMMUNE SYSTEM

The **immune system** is central to the body's fight against disease. Unlike the body's other defense systems, it recognizes and attacks *specific* foreign molecules and it destroys pathogens more and more effectively with each new exposure. The immune system centers around the key defense cells called *lymphocytes,* but it also includes *lymphoid tissue* and the *lymphoid organs (lymph nodes, spleen, thymus, tonsils, aggregated lymphoid nodules in the small intestine,* and *appendix).* We now consider these immune components one by one, from the cellular level to the organ level.

Lymphocytes

As you will recall from Chapter 4 (p. 102), infectious microorganisms that penetrate the epithelial barriers of the body enter the underlying loose connective tissues, where they are attacked by the inflammatory response, by macrophages, and finally, by **lymphocytes** of the immune system. Recall from Chapter 17 (p. 508) that lymphocytes are white blood cells and that each lymphocyte recognizes and attacks its own type of foreign molecule, called an *antigen.* Also recall that *B lymphocytes* multiply to become plasma cells that secrete antibodies, whereas *cytotoxic ($CD8^+$) T lymphocytes* destroy antigen-bearing cells by penetrating their membranes and inducing programmed cell death (see Figure 17.6, p. 509).

B and T cells continuously travel in the blood and lymph streams to reach infected connective tissues throughout the body, where they fight infection. They repeatedly enter and exit these connective tissues, including the often-infected lymphoid tissue, by squeezing through the walls of capillaries and venules. This repeated movement of activated lymphocytes between the circulatory vessels and the connective tissues, called *recirculation,* ensures that lymphocytes reach all infection sites quickly.

Lymphocyte Activation

Immature lymphocytes go through several stages before they are able to attack antigens. Most lymphocytes pass through these stages during infancy and childhood, but many do so in adulthood as well.

Figure 20.6 provides an overview of lymphocyte activation. Lymphocytes originate in the bone marrow from lymphoid stem cells, some of which travel in the bloodstream to the thymus in the thorax and become T lymphocytes *(T* stands for *thymus),* while others stay in the bone marrow and become B lymphocytes. These new T and B lymphocytes divide rapidly and generate many lymphocyte families (clones), each of which is able to recognize one unique type of antigen (this is called gaining *immunocompetence*). Next, the young T or B lymphocyte travels through the bloodstream to an infected connective tissue, where it binds to its specific

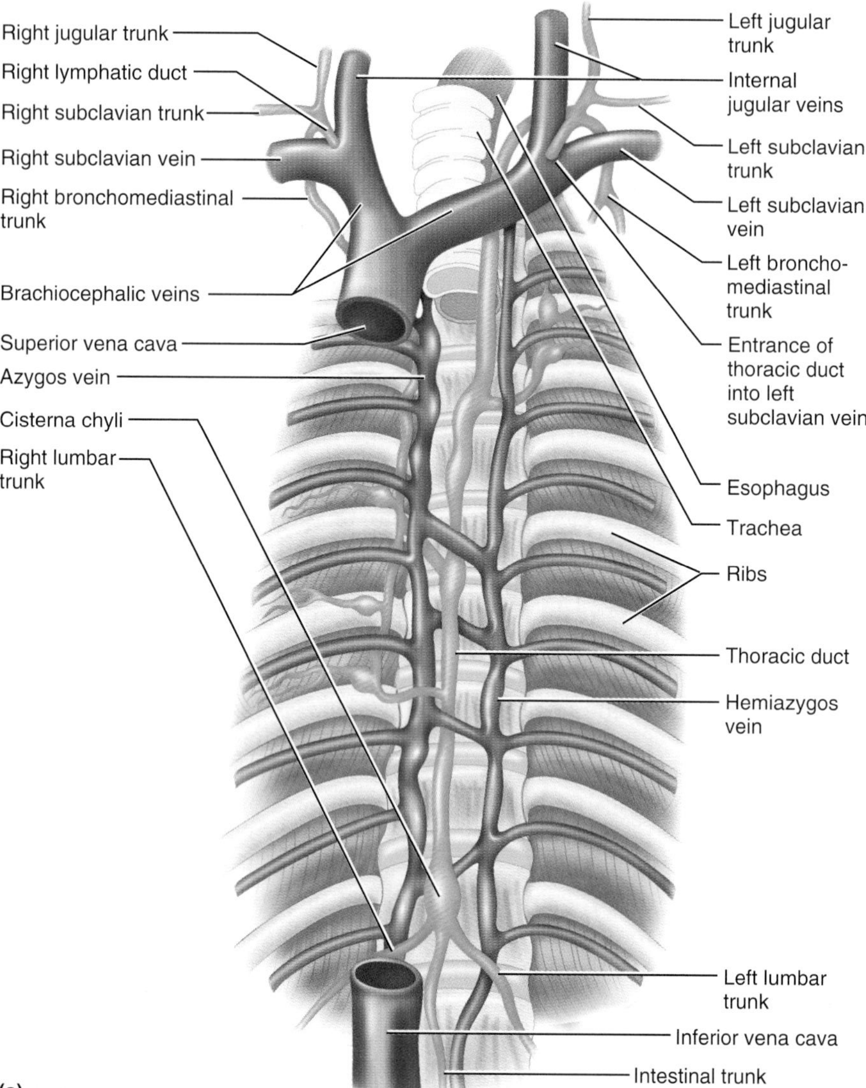

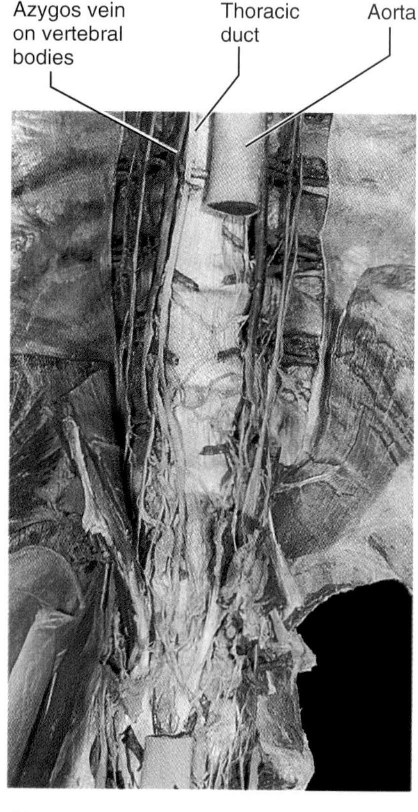

FIGURE 20.5 The lymphatic trunks and ducts in relation to surrounding structures. (a) Enlarged anterior view of the posterior thoracic and abdominal walls. In this specimen, the thoracic duct drains into the left subclavian vein, but in other individuals it drains into the left brachiocephalic or the left internal jugular vein. (b) Photograph of the thoracic duct on the vertebral column.

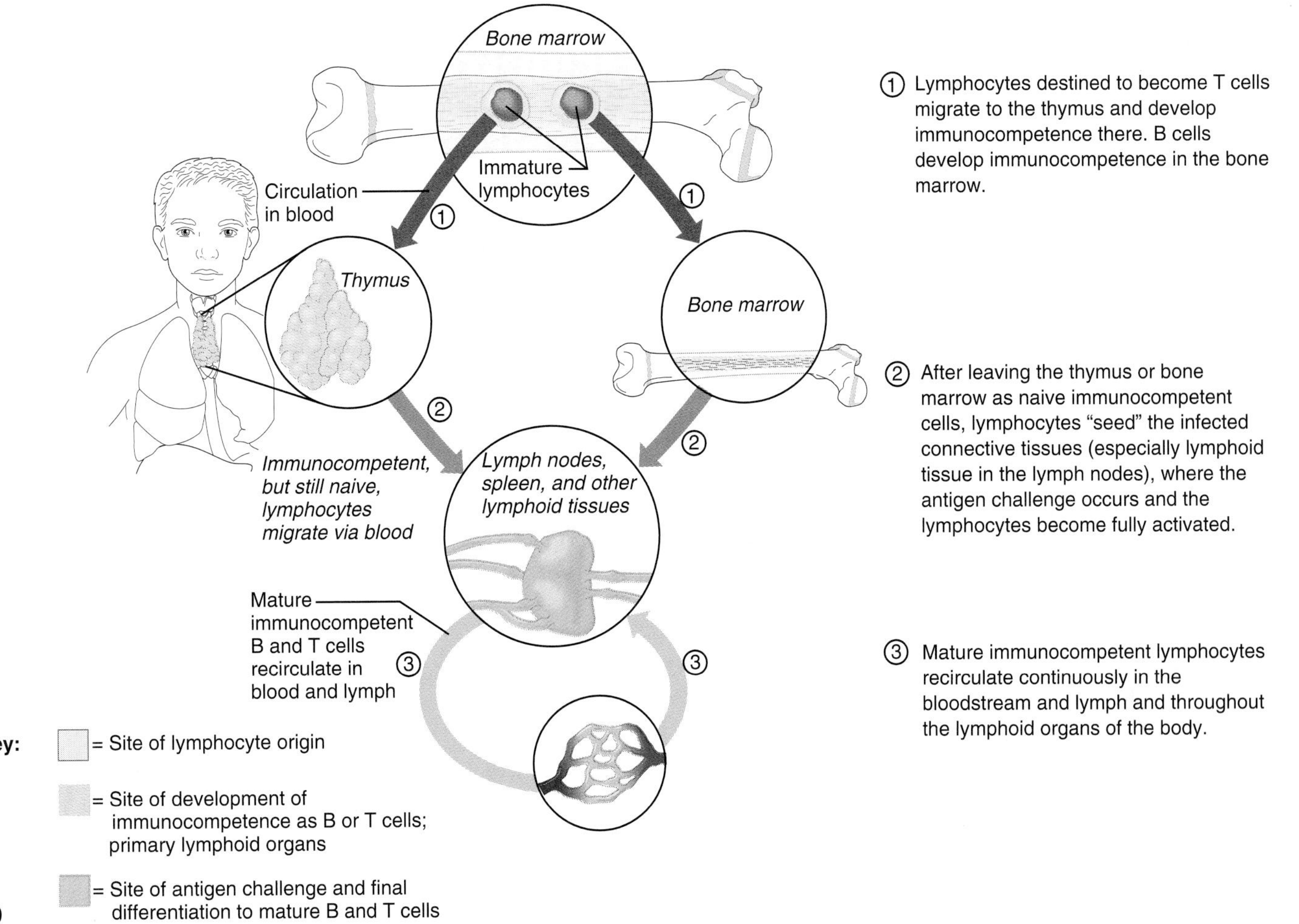

Bone marrow

Immature lymphocytes

Circulation in blood

Thymus

Bone marrow

Immunocompetent, but still naive, lymphocytes migrate via blood

Lymph nodes, spleen, and other lymphoid tissues

Mature immunocompetent B and T cells recirculate in blood and lymph

① Lymphocytes destined to become T cells migrate to the thymus and develop immunocompetence there. B cells develop immunocompetence in the bone marrow.

② After leaving the thymus or bone marrow as naive immunocompetent cells, lymphocytes "seed" the infected connective tissues (especially lymphoid tissue in the lymph nodes), where the antigen challenge occurs and the lymphocytes become fully activated.

③ Mature immunocompetent lymphocytes recirculate continuously in the bloodstream and lymph and throughout the lymphoid organs of the body.

Key:

= Site of lymphocyte origin

= Site of development of immunocompetence as B or T cells; primary lymphoid organs

= Site of antigen challenge and final differentiation to mature B and T cells

(a)

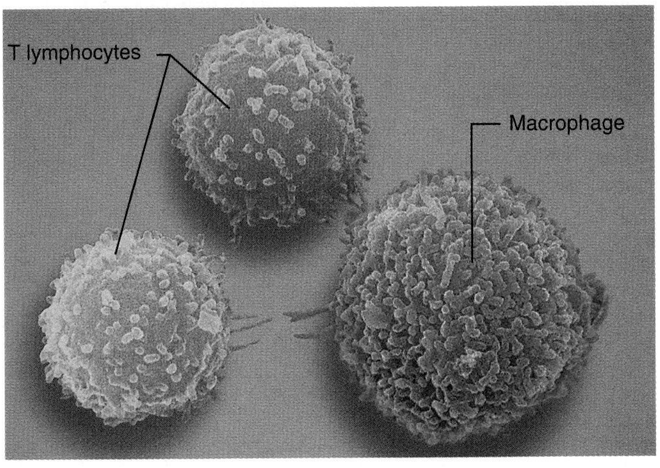

(b)

FIGURE 20.6 **Lymphocyte activation.** **(a)** Flow chart of the activation process: Lymphocytes originate in the bone marrow, develop immunocompetence (the ability to recognize a specific antigen) in the thymus (T cells) or bone marrow (B cells), and become fully activated in infected connective tissues when presented with their antigen (antigen challenge); such antigen challenges are especially frequent in lymph nodes, as shown. **(b)** During the antigen challenge, a macrophage (or dendritic cell) presents antigens to lymphocytes, as shown in this artificially colored scanning electron micrograph (3700 ×).

A Closer Look

AIDS: The Modern-Day Plague?

In 1347, bubonic plague first hit Europe, and by 1400, 70% of all Europeans had died from the "Black Death." In 1987, the U.S. Secretary of Health and Human Services warned that *acquired immune deficiency syndrome (AIDS)* might be the modern-day plague. These are strong words, but are they true?

Even though AIDS was first identified in the United States in 1981 among homosexual men and intravenous drug users of both sexes, it had already afflicted the heterosexual population of Africa for some years. In fact, tissue preserved in 1959 from a central African man was recently identified as the earliest known case. AIDS may have spread from the blood of hunted chimpanzees to their butchers in the years after World War II.

AIDS is a viral disease that progresses through three stages: (1) An acute stage, which develops within several weeks of infection and lasts about two weeks, is characterized by flu-like symptoms such as fever, fatigue, rash, headache, sore throat, swollen lymph nodes, muscle and joint pain, night sweats, and diarrhea. (2) Next comes a long period without symptoms that lasts an average of 10 years. (3) Finally, full-blown AIDS develops, characterized by collapse of the immune system that results in increasingly frequent opportunistic infections, including tuberculosis, an otherwise-rare pneumonia called *Pneumocystis carinii* pneumonia, a fungal infection called *candidiasis* in the esophagus, an eye infection called *cytomegalovirus retinitis*, and the purple skin lesions of *Kaposi's sarcoma* (see p. 600). Furthermore, there is wasting (weight loss) accompanied by diarrhea. More than one-third of AIDS victims develop dementia. Untreated AIDS ends in death from wasting or overwhelming infection. This final stage lasts a few months to several years.

The agent that causes AIDS is the human immunodeficiency virus (HIV), which is transmitted solely through body secretions—blood, semen, and possibly vaginal secretions. It is not transmitted by casual contact, because the virus dries and dies when exposed to air. Most commonly, HIV enters the body during sexual contact in which the mucosa is torn and bleeds, or where open lesions induced by other sexually transmitted diseases give the virus access to the blood. It can also enter through blood-contaminated needles, and it was transmitted early in the epidemic—before blood supplies were made 99.9% safe—through blood transfusions. Although HIV has also been detected in saliva and tears, it is not believed to be spread through those secretions.

After entering the body, HIV travels to the lymphoid tissues, where it infects and destroys helper ($CD4^+$) T cells, resulting in severe depression of immunity. (Helper T cells are the cornerstone of the immune system, because they regulate the populations of B cells and cytotoxic T cells, and without them these lymphocytes cannot be activated or maintained.) HIV also enters dendritic cells, macrophages, and microglia cells, all of which share a surface protein called CD4, to which HIV attaches. The virus invades the brain, probably in infected microglia, and somehow induces destruction of neurons, accounting for the dementia of some AIDS patients.

Although it is not known exactly how HIV kills or diminishes helper T cells, here is the current interpretation of the relation between HIV and the stages of the disease: In the initial, acute stage, rapid virus multiplication stimulates the immune system. Antibody levels rise and cytotoxic ($CD8^+$) T cells fight the infection. The long, asymptomatic stage is not a time of viral dormancy, but a silent war between closely matched combatants. Most of the viruses are destroyed and helper T cells multiply mightily, but HIV replicates and mutates so rapidly that it produces deadly strains that evade the immune system by hiding in the $CD4^+$ cells of the lymphoid tissues of the body, where the viruses reproduce furiously and kill helper cells and dendritic cells. (Some recent studies suggest that, contrary to current view, helper T cells do not reproduce rapidly during the asymptomatic stage, and that their reproduction is actually depressed by the virus.) Finally, the immune system is so weakened that the exhausted $CD8^+$ cells can no longer contain the virus. The number of helper T cells declines, and the terminal, AIDS stage begins.

The worldwide AIDS epidemic is entering its third decade, and AIDS is still spreading fast. The number of people with HIV infection worldwide jumped from 10 million in 1994 to 30

---

antigen, an encounter called *antigen challenge*. Upon this encounter, the lymphocyte becomes fully activated (gains the ability to attack its antigen), proliferates rapidly, and produces mature lymphocytes that recirculate throughout the body seeking pathogens to attack.

During the antigen challenge, an activating lymphocyte interacts with several other cell types. The lymphocyte receives its antigen from an *antigen presenting cell,* such as a macrophage that has recently phagocytized the antigen, or a star-shaped **dendritic cell,** a "professional" antigen gatherer that patrols the body seeking antigens and carries them to places where lymphocytes gather. A distinct type of lymphocyte, called a **helper ($CD4^+$) T lymphocyte,** secretes chemical signals that greatly stimulate the proliferation of activating B and cytotoxic T lymphocytes. Helper T cells are important because their signals amplify and fine-tune the immune response. Their importance is illustrated by *acquired immune deficiency syndrome (AIDS),* a viral disease in which a drastic decline in the body's helper T cells greatly weakens the immune system. AIDS is considered in more detail in the box on this page.

As an activating T or B cell proliferates within infected connective tissue, it produces two types of million in 1999, and 6 million new cases occur each year. That means about 1 of every 200 people on the planet is now infected. Although the rate of spread has slowed in Europe (500,000 cases in 1999) and North America (900,000 cases), it is increasing faster in Southeast Asia (6 million cases) and is a terrible health problem in sub-Saharan Africa (with 21 million cases). Even though predictions that AIDS would decimate the U.S. population have not proven true, AIDS is now the world's fourth biggest killer (after coronary artery disease, strokes, and respiratory disorders), and the number-one cause of death from infectious disease.

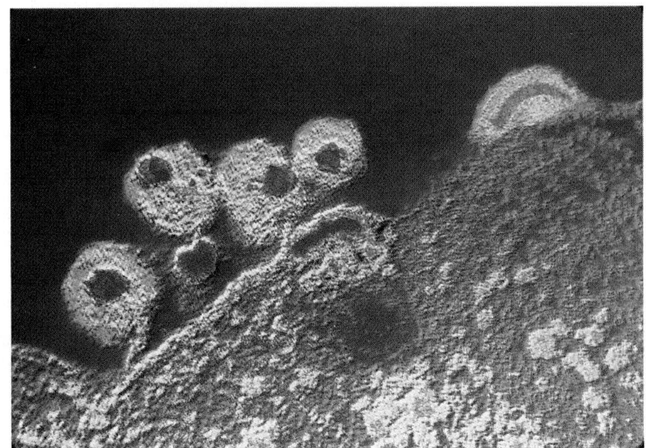

New HIV viruses emerge (yellow and red dots) from an infected human cell.

Whereas AIDS was first detected in the United States among homosexual men, this group currently represents only about half of all cases; intravenous drug users make up some 40%. In about 20% of untreated cases, HIV is transmitted from infected mother to her fetus or newborn child.

After years of research, scientists announced in June 1996 that a three-drug "cocktail" can diminish viral loads to undetectable levels, allowing recovery in many patients. The antiviral drugs involved are two ***reverse transcriptase inhibitors,*** which prevent replication of the virus's genetic material, and a ***protease inhibitor,*** which prevents HIV from assembling functional proteins. This triple cocktail has been miraculous for many AIDS patients, despite the many pills that must be taken daily and the high cost ($10,000–$12,000 per year). Some 60% of patients have no detectable virus in their blood after a year or more of treatment; their immune system partially recovers, and opportunistic infections tend to clear up. Increasingly, patients are receiving the cocktail during the early, symptom-free stage, and even immediately following infection, in hopes that it will prevent progression to AIDS.

Unfortunately, the triple cocktail does not entirely eliminate HIV from the body, so even short breaks in treatment allow viral levels to soar. Thus patients must take the drugs for life—or until the drugs stop working. Because it mutates so quickly within the body, HIV has developed resistance to every antiviral drug yet tested, including the triple cocktail. Moreover, continued use of protease inhibitors can have serious side effects, including increased risk of heart attacks. Now the initial excitement for the triple cocktail has given way to the realization that treated patients are only in remission, and that more permanent solutions are needed—perhaps an anti-HIV vaccine.

First-generation vaccines, which were directed against a viral surface protein, failed. Although vaccinating with whole, weakened HIV viruses may hold promise, some evidence suggests that such viruses may revert to virulent strains. Only a new vaccine called AIDSVAX, containing two components of the viral envelope, has been approved for clinical trials in the United States. Preliminary results concerning its effectiveness are expected by 2003.

Until a cure is found, the only certain way to avoid AIDS is to avoid contact with HIV-infected body fluids and to practice abstinence. Short of that, the best defense is to practice "safe sex" by using latex condoms, and to know one's sexual partner well.

mature lymphocytes, effector and memory. The short-lived *effector lymphocytes* attack the pathogen immediately and then die. *Memory lymphocytes,* by contrast, wait until the body encounters their antigen again—maybe decades later. When a memory lymphocyte finally encounters its antigen, its proliferative response and its attack are most vigorous and rapid. Memory lymphocytes are the basis of acquired immunity; that is, they guard against subsequent infections and prevent us from getting many diseases more than once. Both effector and memory lymphocytes come in T and B varieties.

Lymphoid Tissue

Lymphoid tissue, the most important tissue of the immune system, is an often-infected connective tissue in which vast quantities of lymphocytes gather to fight invading microorganisms. This tissue has two general locations (Figure 20.7): (1) in the frequently infected mucous membranes (p. 99) of the digestive, respiratory, urinary, and reproductive tracts, where it is called **mucosa-associated lymphoid tissue** or **MALT;** and (2) in all lymphoid organs but the thymus. Besides serving

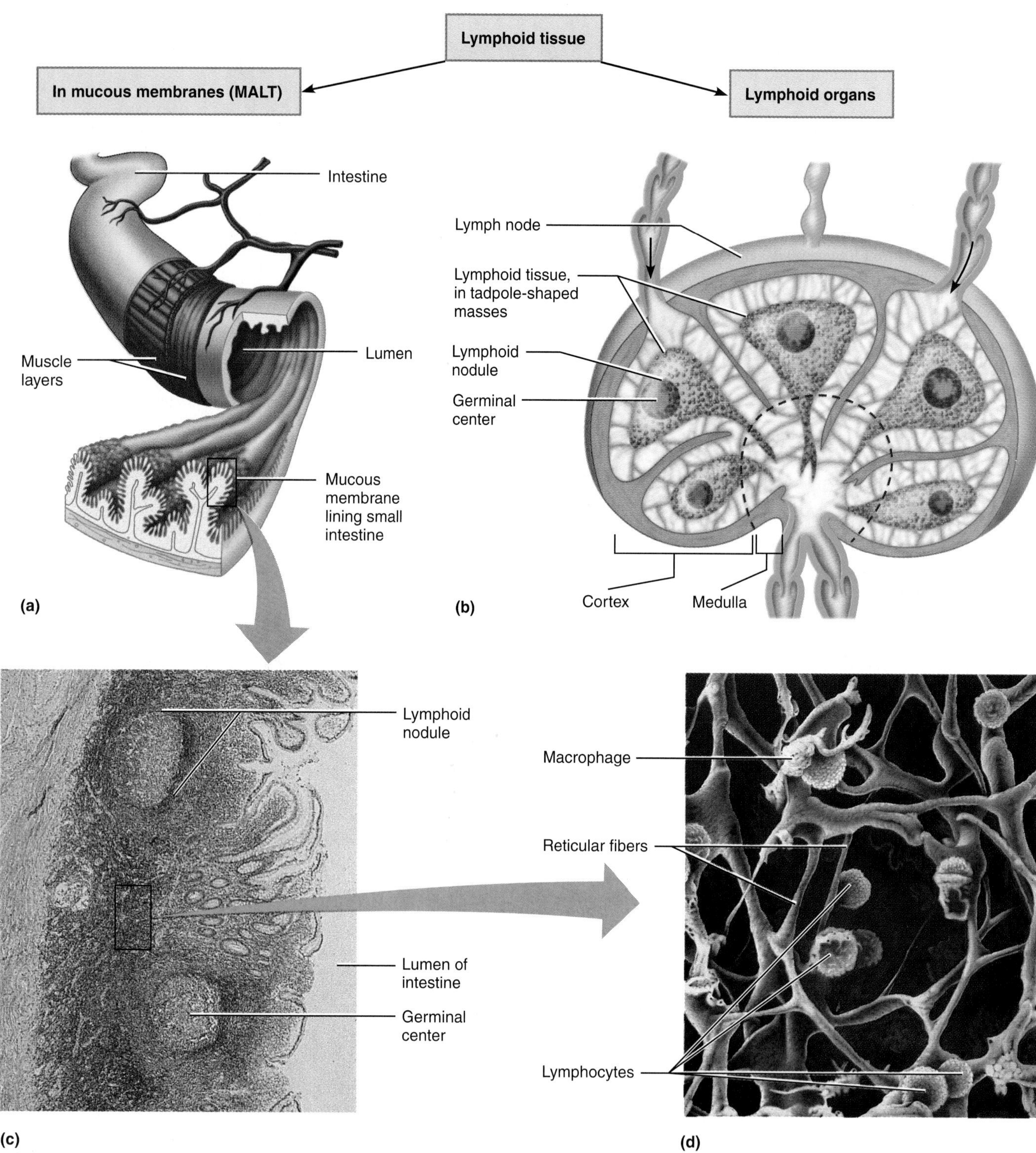

FIGURE 20.7 Lymphoid tissue. This tissue has two general locations: (1) in the mucous membranes of the body, where it is called mucosa-associated lymphoid tissue, or MALT **(a)**; and (2) in all lymphoid organs but the thymus **(b)**. Part b emphasizes the arrangement of lymphoid tissue in a lymph node (also see Figure 20.4). **(c)** Micrograph showing the microscopic anatomy of lymphoid tissue (from the mucosa of the small intestine; 30 ×). As indicated by the many small purple dots, lymphoid tissue is packed with large numbers of lymphocytes, and it also contains round lymphoid nodules with germinal centers. **(d)** Higher magnification of a small region of lymphoid tissue, as viewed by scanning electron microscopy (850 ×). Here, most of the lymphocytes have been removed so that the other components of the tissue—the reticular fiber network and macrophages—are evident.

as the main battleground in the fight against infection, lymphoid tissue is also where most lymphocytes become activated and most effector and memory lymphocytes are generated.

The structural features of lymphoid tissue (Figure 20.7c and d) serve its infection-fighting role. It is a reticular connective tissue whose basic framework is a network of reticular fibers secreted by reticular cells (fibroblasts). Within the spaces of this network reside the many T and B lymphocytes that arrive continuously from venules coursing through this tissue. Macrophages on the fiber network kill invading microorganisms by phagocytosis and, along with dendritic cells, they activate nearby lymphocytes by presenting them with antigens.

Evident within lymphoid tissue are scattered, spherical clusters of densely packed lymphocytes, called **lymphoid nodules** or **follicles** (Figure 20.7c). Note the name is *nodules* and is not to be confused with lymph *nodes*. These nodules often exhibit lighter-staining centers, called **germinal centers** of dividing lymphocytes. Each nodule derives from the activation of a single B cell, whose rapid proliferation generates the thousands of lymphocytes in the nodule. Newly produced B cells migrate away from the nodule to become plasma cells.

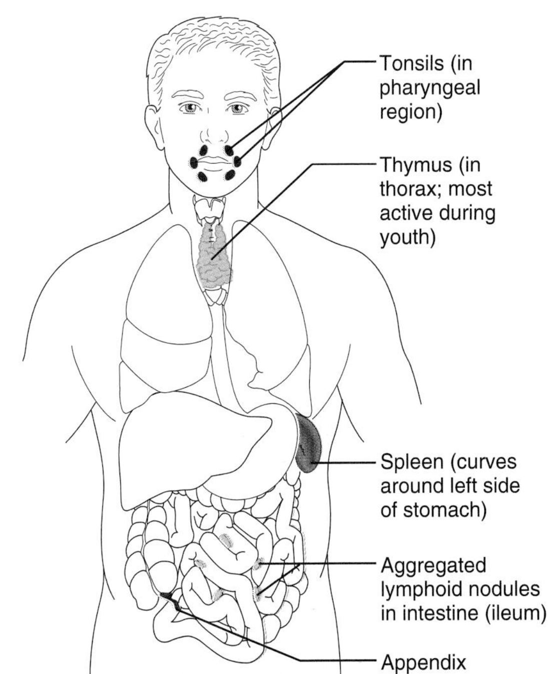

FIGURE 20.8 **Lymphoid organs.** Locations of the tonsils, spleen, thymus, aggregated lymphoid nodules, and appendix.

Lymphoid Organs

The **lymphoid organs** (Figure 20.8) are the lymph nodes, spleen, thymus, tonsils, aggregated lymphoid nodules in the small intestine, and appendix. The thymus functions to produce T lymphocytes, and the other lymphoid organs are elaborately designed to gather and destroy infectious microorganisms within their lymphoid tissue. We begin our discussion with the organs at which the lymphatic and immune systems intersect—the lymph nodes.

Lymph Nodes

Although the lymph-filtering components of lymph nodes have already been considered (p. 586), lymph nodes are more than just lymph filters. The regions of the node between the lymph sinuses are tadpole-shaped masses of lymphoid tissue (Figure 20.7b). As the lymph percolates through the lymph sinuses, some of the contained antigens leak out through the sinus walls into this lymphoid tissue. Most antigen challenges in the human body occur here in the lymph nodes, where the antigens are not only destroyed, but they also activate B and T lymphocytes, adding to the body's valuable supply of memory lymphocytes that offer long-term immunity.

Lymph nodes have two histologically distinct regions, an external **cortex** ("outer bark") and a **medulla** ("middle") near the hilus. As seen in Figure 20.7b, all the lymphoid nodules and most B cells occupy the lymphoid tissue of the most superficial part of the cortex (the tadpoles' "heads"). Deeper in the cortex, the lymphocytes are primarily T cells, especially helper T cells that increase the activity of B cells in the nearby nodules. Thin, inward extensions from the cortical lymphoid tissue (the tadpoles' "tails") help define the medulla. These cord-like medullary extensions contain both T and B lymphocytes, plus plasma cells.

Spleen

The soft, blood-rich **spleen** (Figure 20.9) is the largest lymphoid organ. Its size varies greatly among individuals, but on average it is the size of a fist. This unpaired organ, which lies in the superior left part of the abdominal cavity just posterior to the stomach (see Figure 20.8), is roughly egg-shaped, although its anterior surface is concave rather than convex. The large *splenic vessels* enter and exit the anterior surface along a line called the **hilus** (see Figure 20.9a).

The spleen has two main blood-cleansing functions: (1) the removal of blood-borne antigens (its immune function), and (2) the removal and destruction of aged or defective blood cells. Additionally, the spleen is a site of hematopoiesis in the fetus and stores blood platelets throughout life.

The spleen, like lymph nodes, is surrounded by a fibrous capsule from which trabeculae extend inward (see Figure 20.9b). The larger branches of the splenic artery run in the trabeculae and send smaller arterial branches into the substance of the spleen. These branches are called **central arteries** because they are

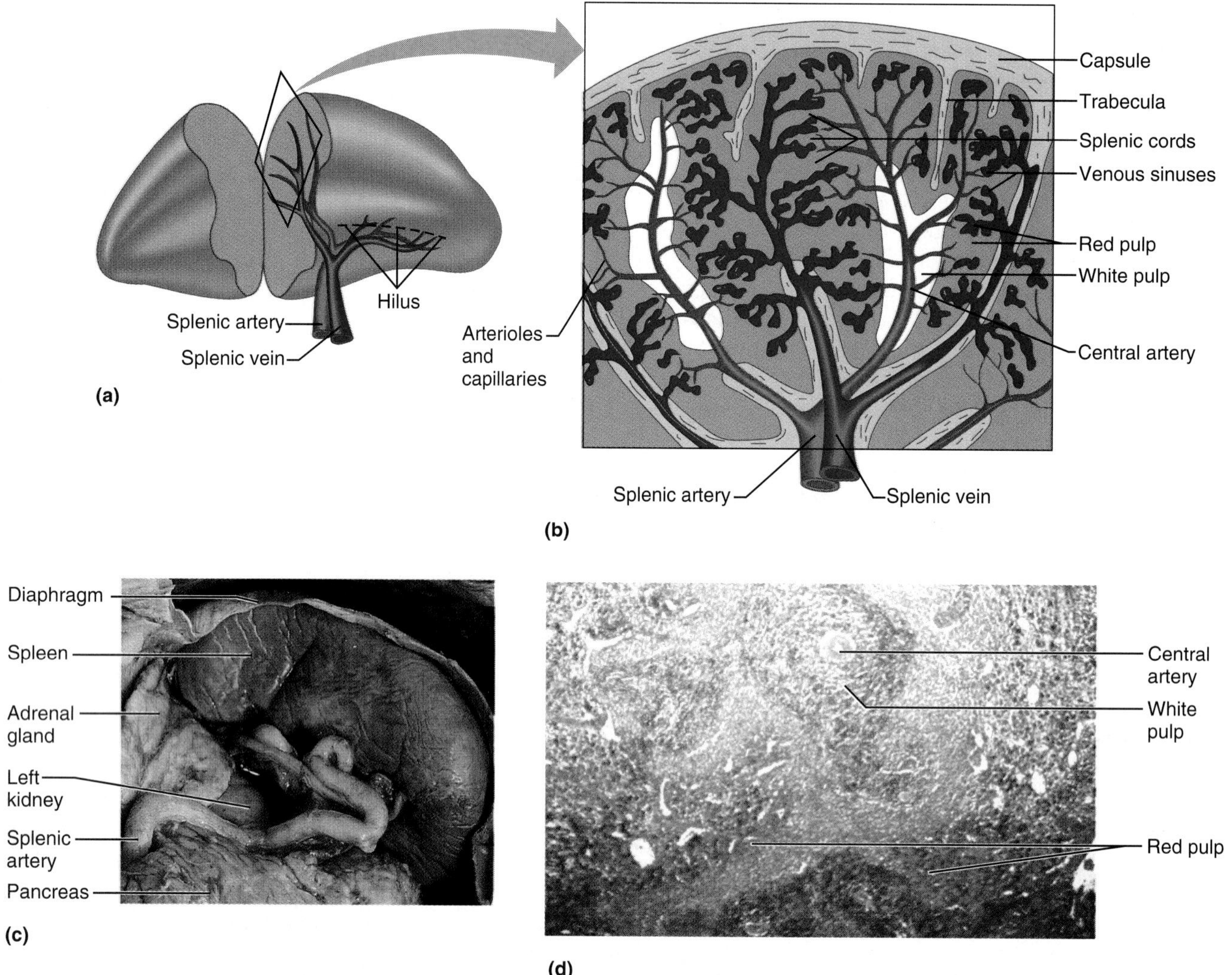

FIGURE 20.9 Structure of the spleen. **(a)** Diagram of the spleen, anterior view. **(b)** Diagram of the histological structure of the spleen. **(c)** Photo of a spleen from a cadaver, anterior view. **(d)** Photomicrograph of spleen tissue (110 ×). The white pulp, a lymphoid tissue with many lymphocytes stained purple here, is surrounded by red pulp containing abundant erythrocytes, stained red here.

enclosed by (lie near the *center* of) thick sleeves of lymphoid tissue that collectively constitute the **white pulp** of the spleen. Blood-borne antigens enter this lymphoid tissue and are destroyed as they activate the immune response. Surrounding the white pulp is **red pulp,** which has two parts: (1) **venous sinuses,** blood sinusoids that arise from the distal branches of the central arteries outside of the white pulp; and (2) **splenic cords,** which consist of a reticular connective tissue that is exceptionally rich in macrophages. Whole blood leaks from the sinuses into this connective tissue, where macrophages then phagocytize any defective blood cells. Hence, red pulp is responsible for the spleen's ability to dispose of worn-out blood cells, whereas white pulp provides the immune function of the spleen.

In histological sections (see Figure 20.9d), the white pulp appears as islands in a sea of red pulp. Note that the naming of the pulp regions reflects their appearances in fresh spleen tissue rather than in stained histological sections; with many stains, the white pulp actually appears darker than the red pulp.

Clinical Applications

SPLENECTOMY Because the capsule of the spleen is relatively thin, physical injury or a serious infection may cause the spleen to rupture, leading to severe loss of blood due to hemorrhage into the peritoneal cavity. In such cases, the spleen must be removed quickly and its artery tied off, a surgical procedure called a **splenectomy.** A person can live a relatively healthy life without a spleen, because macrophages in the bone marrow and liver can take over most of the spleen's functions. Such a person will be more susceptible to infections, however, so surgeons performing splenectomies now leave some of the spleen in place whenever possible. Alternatively, healthy fragments of the spleen are surgically reattached immediately after the operation. In some such cases, the spleen can regenerate. ✚

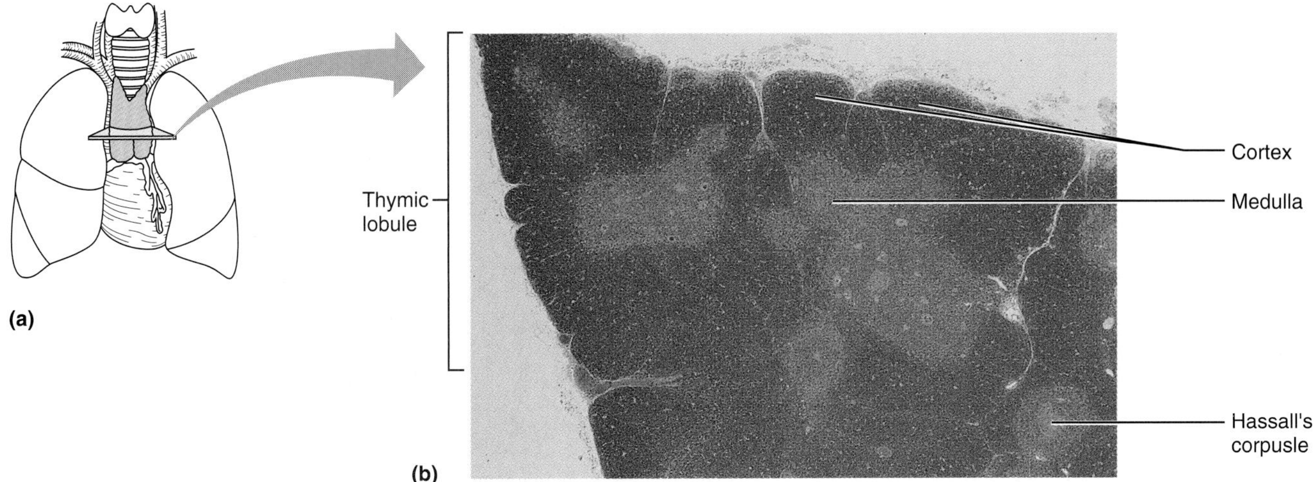

FIGURE 20.10 Structure of the thymus. (a) Location of the thymus. Here, the thymus lies in the superior thorax, but it may extend superior or inferior to this. (b) Photomicrograph of a part of the thymus, showing several lobules with cortex and medulla regions, and thymic (Hassall's) corpuscles, appearing as light dots, in the thymic medulla (30 ×).

Thymus

The two-lobed **thymus** lies in the superior thorax and inferior neck, just posterior to the sternum (Figure 20.10a). As noted earlier, the thymus is the site at which immature lymphocytes develop into T lymphocytes. Specifically, it secretes thymic hormones such as thymosin and thymopoietin, which cause T lymphocytes to gain immunocompetence. Prominent in newborns, the thymus continues to increase in size during childhood, when it is most active. During late adolescence, it begins to atrophy gradually, as its functional tissue is slowly replaced with fibrous and fatty tissue. At age 20 it still has about 80% of its functional tissue but at age 40 it typically retains only 5%. By old age only 2% remains, and the thymus is a fatty mass that is difficult to distinguish from the surrounding connective tissue. Even as it atrophies over time, the thymus continues to produce immunocompetent cells throughout adulthood (but at a reduced rate).

The thymus contains numerous **lobules** arranged like the fronds in a head of cauliflower. Each lobule in turn contains an outer cortex and an inner medulla (see Figure 20.10b). The **cortex** stains dark because it is packed with rapidly dividing T lymphocytes gaining immunocompetence; the **medulla** contains fewer lymphocytes and stains lighter. The medulla contains **thymic (Hassall's) corpuscles,** which seem to be collections of degenerating epithelial reticular cells (described shortly). The number and size of these corpuscles increase with age.

Advances in Understanding The functional differences between the thymic cortex and medulla have recently been elucidated. The cortex certainly generates millions of families of antigen-specific T lymphocytes, but it also selects for those families that will best recognize one's own cells when antigens are bound to these "self" cells (recall that attacking antigen-bearing self cells, when virus-infected or cancerous, is a major function of mature T cells; see p. 509). This is the test of *positive selection*, and families of lymphocytes that fail this test die. The thymic medulla, by contrast, is the site of *negative selection,* the destruction of those T-cell families that bind too strongly to one's own proteins and would attack healthy self tissues in harmful autoimmune reactions if allowed to live. ✳

The thymus differs from the other lymphoid organs in two basic ways: First, it functions strictly in lymphocyte maturation and thus is the only lymphoid organ that does not directly fight antigens. In fact, the *blood-thymus barrier,* analogous to the blood-brain barrier, keeps blood-borne antigens from leaking out of thymic capillaries and prematurely activating the immature thymic lymphocytes. Second, the tissue framework of the thymus is not a true lymphoid connective tissue. Because the thymus arises like a gland from the epithelium lining the embryonic pharynx, its basic tissue framework consists of star-shaped epithelial cells rather than reticular fibers. These *epithelial reticular cells* secrete the thymic hormones that stimulate T cells to become immunocompetent. Note as well that the thymus has no lymphoid nodules because it lacks B cells.

Tonsils

The **tonsils,** perhaps the simplest lymphoid organs, are mere swellings of the mucosa lining the pharynx. There are four groups of tonsils, whose precise locations are indicated in Figure 21.3 on p. 606. The **palatine tonsils** lie directly posterior to the mouth and palate on the lateral sides of the pharyngeal wall. These are the largest tonsils and the ones most often infected and removed

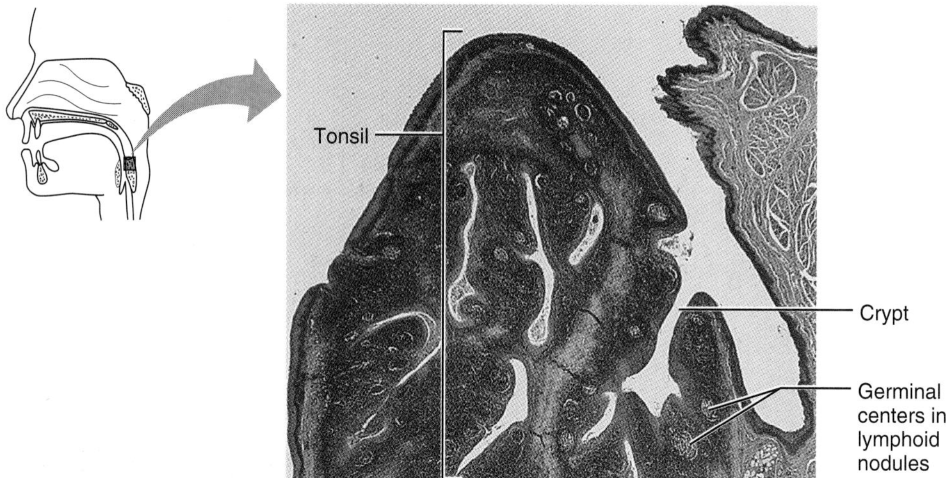

FIGURE 20.11 **Histological structure of the palatine tonsil.** The luminal surface of the tonsil is covered by epithelium, which invaginates deeply to form crypts. The purple area is lymphoid tissue, containing scattered spherical structures, the lymphoid nodules (8 ×).

during childhood, in a surgical procedure called *tonsillectomy*. The **lingual tonsil** lies on the posterior surface of the tongue, the **pharyngeal tonsil** (adenoids) lies on the pharyngeal roof, and the **tubal tonsils** are just behind the openings of the pharyngotympanic tubes into the pharynx. The four tonsils are arranged in a ring around the entrance to the pharynx to gather and remove many pathogens that enter the pharynx in inspired air and swallowed food. The tonsils process the antigens, then set up immune responses.

The histological structure of tonsils is shown in Figure 20.11. These swellings of mucosa, like all mucosae, consist of an epithelium underlain by a connective tissue lamina propria (see p. 99). In the tonsils, the underlying lamina propria consists of abundant mucosa-associated lymphoid tissue (MALT) packed with lymphocytes and scattered lymphoid nodules. The overlying epithelium invaginates deep into the interior of the tonsil, forming blind-ended **crypts** that trap bacteria and particulate matter. The trapped bacteria work their way through the epithelium to the underlying lymphoid tissue, causing the activation of lymphocytes. Such trapping of bacteria leads to many tonsil infections during childhood, but it also generates a great variety of memory lymphocytes for long-term immunity.

Aggregated Lymphoid Nodules and the Appendix

Many bacteria permanently inhabit the hollow interior of the intestines and are constantly infecting the intestinal walls. To fight these invaders, MALT is especially abundant in the intestine. In fact, in two parts of the intestine, MALT is so large, permanent, and densely packed with lymphocytes that it is said to form lymphoid organs: the aggregated lymphoid nodules and the appendix.

Aggregated lymphoid nodules (formerly *Peyer's patches*) are clusters of lymphoid nodules in the walls of the distal part (ileum) of the small intestine (Figure 20.12). About 40 of these patches are present, averaging about a centimeter long and a centimeter wide.

Lymphoid tissue is also heavily concentrated in the wall of the **appendix,** a tubular offshoot of the first part (cecum) of the large intestine (see Figure 20.8). Histological sections reveal that dense lymphoid tissue uniformly occupies over half the thickness of the appendix wall.

Besides destroying the microorganisms that invade them, the aggregated lymphoid nodules and appendix sample many different antigens from within the digestive tube and generate a wide variety of memory lymphocytes to protect the body.

DISORDERS OF THE LYMPHATIC AND IMMUNE SYSTEMS

This section discusses two disorders of the lymph vessels *(chylothorax* and *lymphangitis)* and three disorders of lymphocytes and the lymphoid organs *(mononucleosis, Hodgkin's disease,* and *non-Hodgkin's lymphoma).*

Chylothorax ("chyle in the thorax") is the leakage of the fatty lymph, chyle, from the thoracic duct into a pleural cavity in the thorax. It is caused by tearing or a blockage of the thoracic duct, due to chest trauma or to compression by a nearby tumor. Complications can result from large amounts of lymph in the pleural cavity compressing and collapsing the lungs, from metabolic problems caused by the loss of fatty nutrients from the circulation, or from diminished blood volume following the loss of fluid from the circulation. Treatment involves draining the lymph out of the pleural cavity and surgically repairing any damage to the thoracic duct.

Lymphangitis (lim′fan-ji′tis; "lymph vessel inflammation") is inflammation of a lymph vessel. Like the

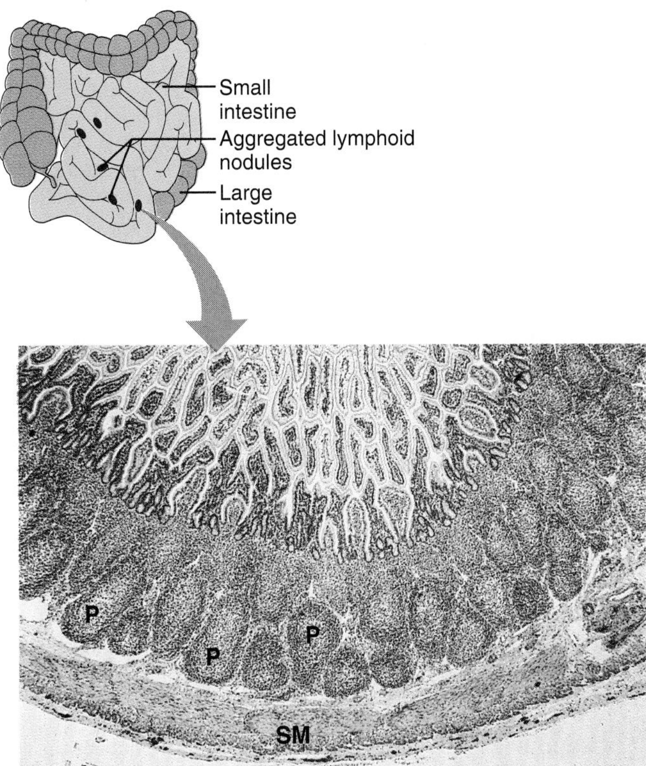

FIGURE 20.12. **Histological structure of an aggregated lymphoid nodule (Peyer's patch) in the wall of the ileum of the small intestine (20 ×).** The lumen of the intestine is seen at the top of the photomicrograph. P = one nodule of the Peyer's patch. SM = smooth muscle in the external part of the intestinal wall.

large blood vessels, the lymph vessels are supplied with blood by vasa vasorum. When lymph vessels are infected and inflamed, their vasa vasorum become congested with blood. The superficial lymphatic vessels then become visible through the skin as red lines that are tender to the touch.

Mononucleosis is a viral disease common in adolescents and young adults. Its symptoms include fatigue, fever, sore throat, and swollen lymph nodes. It is caused by the Epstein-Barr virus, which specifically attacks B lymphocytes. This attack leads to a massive activation of T lymphocytes, which in turn attack the virus-infected B cells. The large numbers of oversized T lymphocytes that circulate in the bloodstream were originally misidentified as monocytes (*mononucleosis* = condition of monocytes). Mononucleosis is transmitted in saliva ("kissing disease") and usually lasts 4 to 6 weeks.

Hodgkin's disease is a malignancy of the lymph nodes characterized by swollen, nonpainful nodes, fatigue, and, often, persistent fever and night sweats. It is characterized by distinctive giant cells in the nodes called Reed-Sternberg cells, which are of uncertain origin but may be modified lymphocytes or macrophages. Hodgkin's disease is treated with radiation therapy, and the cure rate is high relative to other cancers.

Non-Hodgkin's lymphoma includes all cancers of lymphoid tissues except Hodgkin's disease. It involves the uncontrolled multiplication and metastasis of undifferentiated lymphocytes: usually B cells but sometimes T cells. Lymph nodes become swollen, the spleen may enlarge, gut-lymphoid tissue may be affected, and many organs may eventually be involved. Non-Hodgkin's lymphoma is increasing in frequency and is now the fifth most common type of cancer. A low-grade type, which affects the elderly, grows slowly but because it is resistant to chemotherapy drugs, it is often fatal. An intermediate or high-grade type, which affects young people, grows quickly but can respond to chemotherapy with a 40% to 50% remission rate.

THE LYMPHATIC AND IMMUNE SYSTEMS THROUGHOUT LIFE

The lymphatic system develops from a number of sources. Both lymphatic vessels and the main clusters of lymph nodes grow from *lymphatic sacs,* which are projections from the large veins in the embryo. The thymus originates as an outgrowth of the endoderm lining the embryonic pharynx (see Figure 25.14), detaches from the pharynx, and migrates caudally into the thorax. Starting early in the fetal period, the thymus receives the first progenitors of T lymphocytes from the blood-forming organs (yolk sac, liver, then bone marrow), which also produce the first B lymphocytes at this time. All the nonthymic lymphoid organs and tissues (spleen, lymph nodes, and MALT) arise from mesodermal mesenchyme. Except for the spleen and tonsils, these organs are poorly developed before birth. However, shortly after birth, they become heavily populated by circulating lymphocytes and start to gain their functional properties.

Advances in Understanding The immune system of newborns was long thought to be too immature to attack invading pathogens. However, it was found recently that the original experiments were flawed. New, redesigned experiments have shown that newborns respond to new antigens just as vigorously as do adults, with both T cells and antibodies. ✷

In childhood, many memory lymphocytes are formed, and a wide range of immunity is attained. After childhood, some immune organs become less active and begin to shrink. The tonsils are regressing by age 14 (except the lingual tonsil, which does not shrink until about age 30). As previously mentioned, the thymus regresses after late adolescence but keeps producing T cells throughout life.

The lymphoid organs and immune system normally serve us well until late in life, when their efficiency begins to wane and the ability to fight infection declines. This reduced effectiveness seems mostly due to a decrease in the production and responsiveness of T cells, which in turn causes similar declines in B cells. Old age is accompanied by an increased susceptibility to disease. The greater incidence of cancer in the elderly is assumed to be another example of the declining ability of the immune system to destroy harmful cells. Just why this decline occurs is not understood, but "genetic aging" and its consequences are probably at fault.

RELATED CLINICAL TERMS

Kaposi's sarcoma Tumor-like lesions of the skin and some internal organs, seen in some AIDS patients. The lesions are vascularized and arise from proliferating capillary endothelial cells. The capillaries are so permeable that they leak erythrocytes, which gives the lesions a purple color. Kaposi's sarcoma is now known to be caused by a previously unrecognized type of herpes virus that infects the endothelium. Confusion exists as to whether the lesions are merely due to cellular hyperplasia, or are true tumors. Treatment is with chemotherapy.

Lymphadenopathy (lim-fad′ĕ-nop-ah-the) (*adeno* = gland; *pathy* = disease) Any disease of the lymph nodes.

Lymphedema Swelling due to accumulation of lymphatic fluid in the loose connective tissues.

Lymphoma Any neoplasm (tumor) of the lymphoid tissue, whether benign or malignant.

Sentinel lymph node in cancer The first node that receives lymph draining from a body area suspected of having a tumor. When examined for the presence of cancer cells, this node gives the best indication of whether metastasis (spread) through the lymph vessels has occurred.

Splenomegaly (sple″no-meg′ah-le; "spleen big") Enlargement of the spleen, usually resulting from blood diseases such as mononucleosis, malaria, leukemia, or polycythemia vera. A way that physicians diagnose spleen enlargement is by feeling for the spleen's notched superior border through the skin of the abdominal wall just anterior to the costal margin. A healthy spleen will never reach this far anteriorly.

Tonsillitis Congestion of the tonsils, typically with infecting bacteria, which causes them to become red, swollen, and sore.

CHAPTER SUMMARY

Media study tools that could provide you additional help in reviewing specific key topics of Chapter 20 are referenced below.

SP = Study Partner; AIA = A.D.A.M.® Interactive Anatomy

1. The lymphatic system consists of lymphatic circulatory vessels that carry lymph; and the immune system contains the lymphocytes, lymphoid tissue, and lymphoid organs, which are involved in the body's fight against disease.

THE LYMPHATIC SYSTEM (pp. 584–589)

1. The vessels of the lymphatic system, from smallest to largest, are lymph capillaries, lymphatic collecting vessels (with lymph nodes), trunks, and ducts.
2. Lymph is excess tissue fluid, which originates because slightly more fluid leaves blood capillaries than returns there. Lymph vessels pick up this excess fluid and return it to the great veins at the root of the neck.

AIA Link: Atlas; System: Lymphatic; Lymph Flow of the Neck (Anterior).

3. Lymph vessels also retrieve blood proteins that leak from capillaries and return these proteins to the bloodstream.

Lymph Capillaries (p. 585)

4. Lymph capillaries weave through the loose connective tissues of the body. These closed-end tubes are highly permeable to entering tissue fluid and proteins because their endothelial cells are loosely joined. Disease-causing microorganisms and cancer cells also enter the permeable lymph capillaries and spread widely via the lymph vessels.
5. Lymph capillaries called lacteals absorb digested fat from the small intestine.

Lymphatic Collecting Vessels (pp. 585–586)

6. Lymphatic collecting vessels run alongside arteries and veins but have thinner walls and many more valves than do veins, and resemble a string of beads.
7. Lymph flows very slowly through lymphatic collecting vessels. Flow is maintained by normal body movements, contractions of skeletal muscles, arterial pulsations, and contraction of smooth muscle in the wall of the lymph vessel. Lymphatic valves prevent backflow.

Lymph Nodes (pp. 586–587)

8. Clustered along the lymphatic collecting vessels, bean-shaped lymph nodes cleanse the lymph stream of infectious agents and cancer cells. Lymph enters the node via afferent lymphatic vessels and exits via efferent vessels at the hilus. In between, the lymph percolates through lymph sinuses, where macrophages remove lymph-borne pathogens. For a picture of the body's main groups of lymph nodes, see Figure 20.3.

AIA Link: Atlas; System: Lymphatic; Lymph Nodes of the Thorax (Anterior).

Lymph Trunks (pp. 587–589)

9. The lymph trunks (lumbar, intestinal, bronchomediastinal, subclavian, and jugular) each drain a large body region. All except the intestinal trunk are paired.

Lymph Ducts (p. 589)

10. The right lymphatic duct (and/or the nearby trunks) drains lymph from the superior right quarter of the body. The thoracic duct (and/or the nearby trunks) drains lymph from the rest of the body. These two ducts empty into the junction of the internal jugular and subclavian veins. The thoracic duct starts at the cisterna chyli at L_1–L_2 and ascends along the thoracic vertebral bodies.

SP Exercise: Chapter 20, The Lymphatic System.

THE IMMUNE SYSTEM (pp. 589–598)

1. Lymphoid organs and lymphoid tissues house millions of lymphocytes, important cells of the immune system that recognize specific antigens.

Lymphocytes (pp. 589–593)

2. B and T lymphocytes fight infectious microorganisms in the loose and lymphoid connective tissues of the body—B cells by producing antibody-secreting plasma cells, and cytotoxic ($CD8^+$) T cells by directly killing antigen-bearing cells.

3. Mature lymphocytes patrol connective tissues throughout the body by passing in and out of the circulatory vessels (recirculation).

Lymphocyte Activation (pp. 589–593)

4. Lymphocytes arise from stem cells in the bone marrow. T cells develop immunocompetence in the thymus, while B cells develop immunocompetence in the bone marrow. Immunocompetent lymphocytes then circulate to the loose and lymphoid connective tissues, where antigen binding (the antigen challenge) leads to lymphocyte activation.

5. The antigen challenge involves an interaction among the activating lymphocyte, an antigen-presenting cell (dendritic cell or macrophage), and a helper ($CD4^+$) T lymphocyte. A newly activated T or B cell divides quickly to produce many short-lived effector lymphocytes and some long-lived memory lymphocytes. Recirculating memory lymphocytes provide long-term immunity.

Lymphoid Tissue (pp. 593–595)

6. Lymphoid tissue is an often-infected reticular connective tissue in which many B and T lymphocytes gather to fight pathogens or become activated. It is located in the mucous membranes (as MALT) and in the lymphoid organs (except the thymus).

7. Lymphoid tissue contains lymphoid nodules (follicles) with germinal centers. Each nodule contains thousands of B lymphocytes, all derived from one activated B cell.

Lymphoid Organs (pp. 595–598)

8. Within a lymph node, tadpole-shaped masses of lymphoid tissue lie between the sinuses. This lymphoid tissue receives some of the antigens that pass through the node, leading to lymphocyte activation and memory-cell production. The masses help divide each node into an outer cortex and an inner medulla.

SP Exercise: Chapter 20, Structure of a Lymph Node.

9. The spleen lies in the superior left part of the abdominal cavity. The splenic vessels enter and exit the hilus on the anterior surface.

AIA Link: Atlas; View: Anterior; Blood Supply to Pancreas/Spleen.

10. The spleen has two main functions: (1) cleansing antigens from the blood and (2) destroying worn-out blood cells. The first function is performed by the white pulp, the second by the red pulp. White pulp consists of sleeves of lymphoid tissue, each surrounding a central artery. Red pulp consists of venous sinuses and strips of blood-filled reticular connective tissue called splenic cords, whose macrophages remove worn-out blood cells.

11. The thymus, located in the superoanterior thorax and neck, is most active during youth. Its hormones, secreted by epithelial reticular cells, signal the contained T lymphocytes to gain immunocompetence.

AIA Link: Atlas; Dissectible Anatomy; Male; Anterior; Layer 164. Scroll to Thorax.

12. The thymus has lobules, each with an outer cortex packed with maturing T cells and an inner medulla containing fewer T cells and degenerative thymic (Hassall's) corpuscles.

13. The thymus neither directly fights antigens nor contains true lymphoid tissue.

14. The tonsils in the pharynx, aggregated lymphoid nodules in the small intestine, and the wall of the appendix are parts of MALT in which the lymphoid tissue contains an exceptionally high concentration of lymphocytes and nodules.

SP Exercise: Chapter 20, Making Connections: The Lymphatic System.

DISORDERS OF THE LYMPHATIC AND IMMUNE SYSTEMS (pp. 598–599)

1. Chylothorax is the leakage of chyle from the thoracic duct into the pleural cavity. Lymphangitis is inflammation of a lymph vessel. Mononucleosis is a viral infection of B lymphocytes, which are attacked by T lymphocytes, leading to flu-like symptoms. Hodgkin's disease is a type of lymph-node cancer, and non-Hodgkin's lymphoma includes all other cancers of lymphoid tissue.

THE LYMPHATIC AND IMMUNE SYSTEMS THROUGHOUT LIFE (pp. 599–600)

1. Lymphatic vessels develop from lymphatic sacs attached to the embryonic veins. The thymus develops from pharyngeal endoderm, and the other lymphoid organs derive from mesenchyme.

2. Lymphoid organs become populated by lymphocytes, which arise from hematopoietic tissue.

3. With aging, the immune system becomes less responsive. Thus, the elderly suffer more often from infections and cancer.

REVIEW QUESTIONS

Multiple Choice/Matching Questions

1. Lymph capillaries (a) are open-ended like drinking straws, (b) have continuous tight junctions like those of the capillaries of the blood-brain barrier, (c) have endothelial cells separated by flap-like minivalves that open wide, (d) have special barriers that stop cancer cells from entering, (e) all of the above.

2. The basic structural framework of most lymphoid organs consists of (a) areolar connective tissue, (b) hematopoietic tissue, (c) reticular connective tissue, (d) adipose tissue.

3. Lymph nodes cluster in all the following body areas except the (a) brain, (b) axillae, (c) groin, (d) neck.

4. The germinal centers in lymph nodes are sites of (a) the lymph sinuses, (b) proliferating B lymphocytes, (c) T lymphocytes, (d) a and c, (e) all of the above.

5. The red pulp of the spleen (a) contains venous sinuses and a macrophage-rich connective tissue, (b) is another name for

the fibrous capsule, (c) is lymphoid tissue, (d) is the part of the spleen that destroys worn-out erythrocytes, (e) a and d.

6. Lymphocytes that develop immunocompetence in the thymus are (a) B lymphocytes, (b) T lymphocytes.

7. Which of the following lymphoid organs have a cortex and a medulla? (More than one choice is correct.) (a) lymph nodes, (b) spleen, (c) thymus, (d) Peyer's patches, (e) tonsils.

8. Which one of the following lymphoid organs has no lymphoid tissue and therefore does not contain lymphoid nodules or germinal centers? (a) lymph nodes, (b) spleen, (c) thymus, (d) Peyer's patches, (e) tonsils.

9. Developmentally, the embryonic lymphatic vessels are most closely associated with the (a) veins, (b) arteries, (c) nerves, (d) thymus.

10. It sometimes is difficult to distinguish the different lymphoid organs from one another in histological sections. How would you tell the thymus from a lymph node? (a) Only the thymus has a cortex and medulla; (b) lymphocytes are far less densely packed in the thymus than in the lymph node; (c) the thymus contains no blood vessels; (d) only the thymus has distinct lobules and thymic (Hassall's) corpuscles.

11. In some people, the thoracic duct does not receive any of the lymph trunks at the base of the neck (jugular, subclavian, or bronchomediastinal). In such people, what part of the body does the thoracic duct drain? (a) upper right quarter, (b) upper left quarter, (c) upper half, (d) lower half.

Short Answer Essay Questions

12. Compare the basic functions of a lymph node to those of the spleen.

13. If you saw a blood vessel and a lymphatic collecting vessel running side by side, how could you tell them apart?

14. Trace the entire course of the thoracic duct. Then tell the locations of all the lymph trunks.

15. List and briefly explain three important functions of the lymphatic vessels.

16. Which three of the six groups of lymph nodes shown in Figure 20.3 are easily felt (palpated) through the skin during a physical examination?

17. As Olivia began reading through this chapter, she suddenly stopped and said, "Oops, I think I have to go back to Chapter 17 and review the basic functional differences between B and T lymphocytes." Explain these differences.

18. As Germaine was reading this chapter for the second time, she suddenly had an insight, exclaiming, "Lymph comes from the blood, and then it returns to the blood!" Is this insight correct? Explain.

19. Billie Joe is looking at histological slides of the spleen under a microscope. She says that she sees lymphoid nodules in the white pulp. Is she correct? Explain.

20. Aparna, a gross anatomist who has been teaching human anatomy for 35 years, told her class that in adults, the thymus is degenerate and has no function. Is she correct? Explain.

CRITICAL REASONING + CLINICAL APPLICATION QUESTIONS

1. A friend tells you that she has tender, swollen "glands" along the left side of the front of her neck. You notice that she has a bandage on her left cheek that is not fully hiding a large infected cut there. The glands are not really glands. Exactly what are her swollen "glands," and how did they become swollen?

2. When young Joe went sledding, the runner of a friend's sled hit him in the left side and ruptured his spleen. Joe almost died before he got to a hospital. What is the immediate danger of a ruptured spleen?

3. The man in the hospital bed next to Joe is an alcoholic with cirrhosis of the liver and portal hypertension. His spleen is seriously enlarged. Based on what you learned in Chapter 19 (pp. 570–572), how could portal hypertension lead to splenomegaly?

4. Mrs. Roselli has undergone a left radical mastectomy. Her left arm is severely swollen and painful, and she is unable to raise it higher than her shoulder. (a) Explain the origin of her signs and symptoms. (b) Is she likely to have relief from these symptoms in time? Explain.

5. Traci arrives at the clinic complaining of pain and redness of her right ring finger. The finger and the dorsum of her hand have edema, and red streaks are apparent on her right forearm. Antibiotics are prescribed, and the nurse applies a sling to the affected arm. Why is it important that Traci not move the affected arm excessively?

6. Simi realized she was very ill. During a recent trip to the tropics, she had contracted both malaria and tuberculosis, diseases whose microorganisms travel throughout the bloodstream. Back home for only a week, she then contracted mononucleosis. When she went to the doctor, he was able to feel the notched superior border of her spleen projecting far anterior to the left costal margin in her abdominal wall. Was something wrong with her spleen? Explain.

A.D.A.M.® INTERACTIVE ANATOMY QUESTIONS

1. Link: Atlas; System: Lymphatic; Lymph Flow of Head (Lateral). (a) Name the lymph node that drains the posterior area of the head. (b) The lymph system forms a one-way circuit in the body. In which direction does the lymph flow?

2. Link: Atlas; System: Lymphatic; Lymph Flow of Tongue (Dorsal). (a) Lymph flows from the palatine tonsil to which lymph node? (b) Lymph flowing from the lateral areas of the tongue enters which lymph node?

CHAPTER 21

The Respiratory System

SEM of lung tissue

CHAPTER OUTLINE + STUDENT OBJECTIVES

Functional Anatomy of the Respiratory System 604–621

1. Identify the respiratory tubes and passageways in order, from the nose to the alveoli in the lungs. Distinguish the structures of the conducting zone from those of the respiratory zone.
2. List and describe several protective mechanisms of the respiratory system.
3. Describe the structure and functions of the larynx.
4. Explain how the walls of the upper respiratory passages differ histologically from those in the lower parts of the respiratory tree.
5. Describe the structure of a lung alveolus and of the respiratory membrane.
6. Describe the gross structure of the lungs and the pleurae.

Ventilation 621–623

7. Explain the relative roles of the respiratory muscles and lung elasticity in the act of ventilation.
8. Define surfactant, and explain its function in ventilation.
9. Explain how the brain and peripheral chemoreceptors control the ventilation rate.

Disorders of the Respiratory System 623–628

10. Consider the causes and consequences for respiration of asthma, chronic bronchitis, emphysema, lung cancer, pneumonia, tuberculosis, cystic fibrosis, and nosebleeds.

The Respiratory System Throughout Life 628–629

11. Trace the development of the respiratory system in the embryo and fetus.
12. Describe the normal changes that occur in the respiratory system from infancy to old age.

IN A CLASSIC ROCK SONG FROM THE 1970S, the singer woos his lover by claiming "All I need is the air that I breathe, and to love you." The admission in a love song that respiration comes before romance is powerful testimony to the fact that breathing is our most urgent need. The trillions of cells in the body need a continuous supply of oxygen to carry out their vital functions. We can live without water for days, and without food for weeks, but we cannot live without oxygen for even a few minutes. As our cells use oxygen, furthermore, they produce carbon dioxide, a waste product the body must eliminate. The major function of the **respiratory system** is to fulfill these needs—that is, to supply the body with oxygen and dispose of carbon dioxide. To accomplish this, the following four processes, collectively called **respiration,** must occur:

1. **Pulmonary ventilation.** Air must be moved in and out of the lungs so that the gases in the air sacs (alveoli) of the lungs are continuously replaced. This movement is commonly called **ventilation** or breathing.
2. **External respiration.** Gas exchange (oxygen loading and carbon dioxide unloading) must occur between the blood and air at the lung alveoli.
3. **Transport of respiratory gases.** Oxygen and carbon dioxide must be transported between the lungs and the cells of the body. This is accomplished by the cardiovascular system, with blood serving as the transporting fluid.
4. **Internal respiration.** At the systemic capillaries, gases must be exchanged between the blood and the tissue cells.*

In this chapter we focus on pulmonary ventilation and external respiration, because they alone are the special responsibility of the respiratory system. However, unless gas transport and internal respiration also occur, the respiratory system cannot accomplish its main goal of taking in oxygen and eliminating carbon dioxide. Thus, the respiratory and cardiovascular systems are closely coupled, and if either system fails, the body's cells begin to die from oxygen starvation.

Because it moves air into and out of the body, the respiratory system is also involved in the sense of smell and with the vocalizations of speech, which we will briefly discuss along with respiration.

FUNCTIONAL ANATOMY OF THE RESPIRATORY SYSTEM

The organs of the respiratory system include the *nose, nasal cavity,* and *paranasal sinuses;* the *pharynx;* the *larynx;* the *trachea;* the *bronchi* and their smaller branches; and the *lungs,* which contain the terminal air sacs, or *alveoli* (Figure 21.1). Functionally, these respiratory structures are divided into respiratory and conducting zones. The *respiratory zone,* the actual site of gas exchange in the lungs, is composed of tiny structures that contain alveoli—namely, the respiratory bronchioles, alveolar ducts, and alveolar sacs (see p. 614). The *conducting zone* includes all other respiratory passageways, which serve as fairly rigid conduits that carry air to the sites of gas exchange. The structures of the conducting zone also filter, humidify, and warm the incoming air. Thus, the air reaching the lungs contains much less dust than it did when it entered the nose and is warm and damp. The functions of the major organs of the respiratory system are summarized in Table 21.1 on page 621.

The Nose and the Paranasal Sinuses

In this section we examine the nose, including the nasal cavity, and the paranasal sinuses.

The Nose

The **nose** is the only externally visible part of the respiratory system. Unlike the often poetic references to other facial features, such as the eyes or lips, the nose is usually the target of irreverence: We are urged to keep our nose to the grindstone, and to keep it out of other people's business. However, considering its many important functions, it deserves to be held in higher esteem. The nose (1) provides an airway for respiration, (2) moistens and warms entering air, (3) filters inhaled air to cleanse it of foreign particles, (4) serves as a resonating chamber for speech, and (5) houses the olfactory (smell) receptors.

The structures of the nose are divided into the *external nose* and the internal *nasal cavity.* The skeletal framework of the **external nose,** shown in Figure 21.2, consists of the frontal and nasal bones superiorly (forming the root and bridge, respectively), the maxillary bones laterally, and flexible plates of hyaline cartilage inferiorly (the lateral, septal, and alar cartilages). The great variation in nose size and shape is largely due to differences in the nasal cartilages. The skin covering the nose's anterior and lateral surfaces is thin and contains many sebaceous glands that open into some of the largest skin pores on the face.

The Nasal Cavity

The **nasal cavity** (see Figures 21.1 and 21.3) lies in and posterior to the external nose. During breathing, air enters this cavity by passing through the external **nares**

*The use of oxygen and the production of carbon dioxide by tissue cells, termed *cellular respiration,* is the cornerstone of all energy-producing chemical reactions in the body. For a discussion of cellular respiration, consult a physiology or cell biology text.

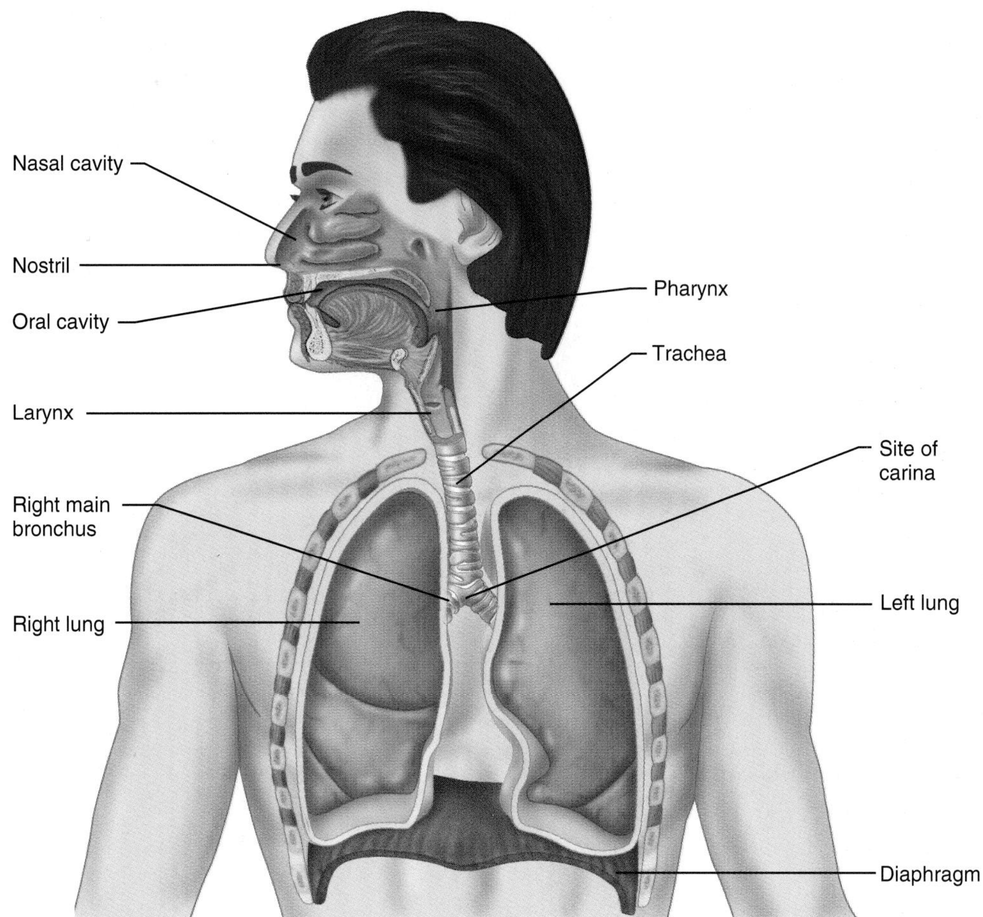

FIGURE 21.1 Organs of the respiratory system. All structures shown here are in the conducting zone. (Structures of the respiratory zone are small and lie deep within the lung.) Organs superior to the larynx are called *upper respiratory structures,* whereas the larynx and all structures inferior to it are the *lower respiratory structures.*

(na′rēz), or *nostrils*. The nasal cavity is divided into right and left halves by the *nasal septum* in the midline; this septum is formed by the perpendicular plate of the ethmoid bone, the vomer and a septal cartilage (see Figure 7.10 on p. 166), all covered by a pink mucous membrane. Posteriorly, the nasal cavity is continuous with the nasal part of the pharynx (nasopharynx) through the **posterior nasal apertures,** also called the *choanae* (ko-a′ne; "funnels") or *internal nares.*

To review the bony boundaries of the nasal cavity (see p. 165), its roof is formed by the ethmoid and sphenoid bones, whereas its floor is formed by the *palate* (pal′at), which separates the nasal cavity from the mouth inferiorly and keeps food out of the airways. Anteriorly, where the palate contains the horizontal processes of the palatine bones and the palatine process of the maxillary bone, it is called the **hard palate;** the posterior part is the muscular **soft palate** (see Figure 21.3).

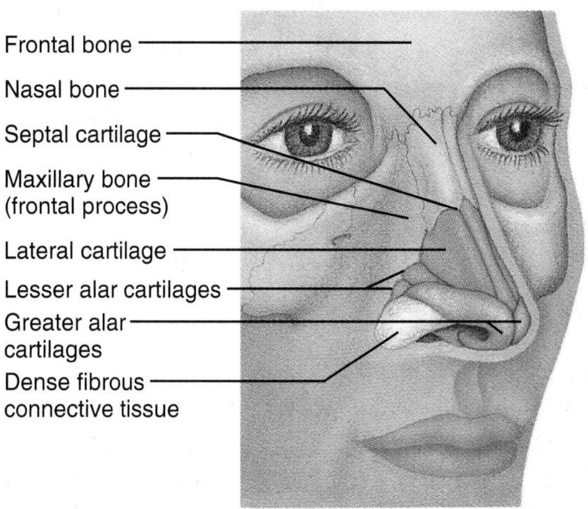

FIGURE 21.2 Skeletal framework of the external nose.

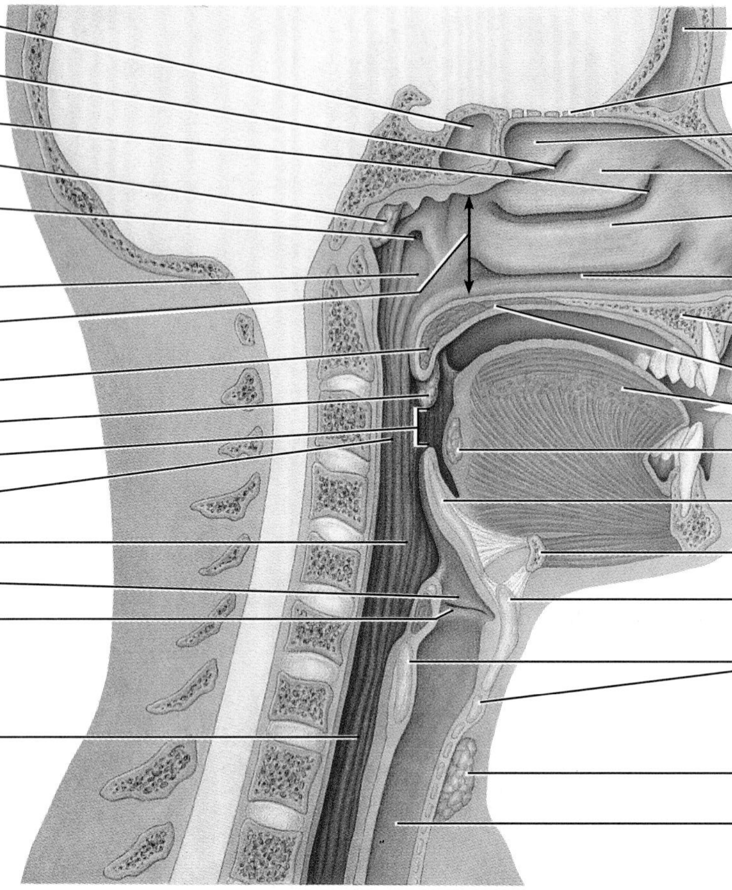

FIGURE 21.3 Basic anatomy of the upper respiratory tract. (a) Midsagittal section of the head and neck. (b) Photograph of a midsagittal section of a cadaver's head in the region of the nasal cavity and mouth.

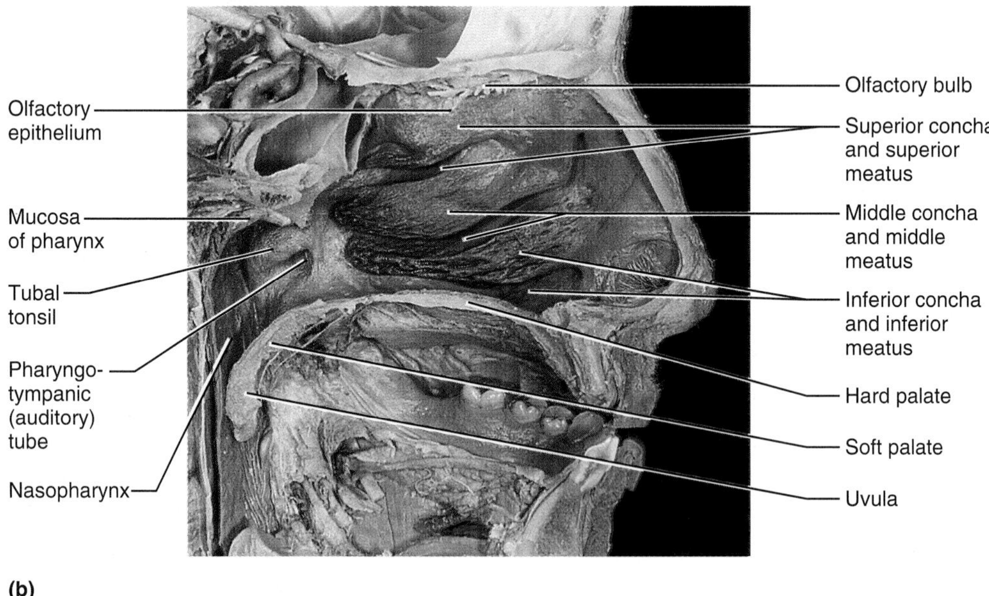

The part of the nasal cavity that lies just superior to the nostrils, within the flared wings of the external nose, is the **vestibule** ("porch, entranceway"). This is lined with skin containing sebaceous and sweat glands and numerous hair follicles. The nose hairs, or *vibrissae* (vi-bris′e) (*vibro* = to quiver), filter large particles, such as insects and lint, from the inspired air. The rest of the nasal cavity is lined with two types of mucous membrane: (1) the small patch of *olfactory mucosa* near the roof of the nasal cavity, which houses the receptors for smell (see pp. 468–470), and (2) the *respiratory mucosa,* which lines the vast majority of the nasal cavity.

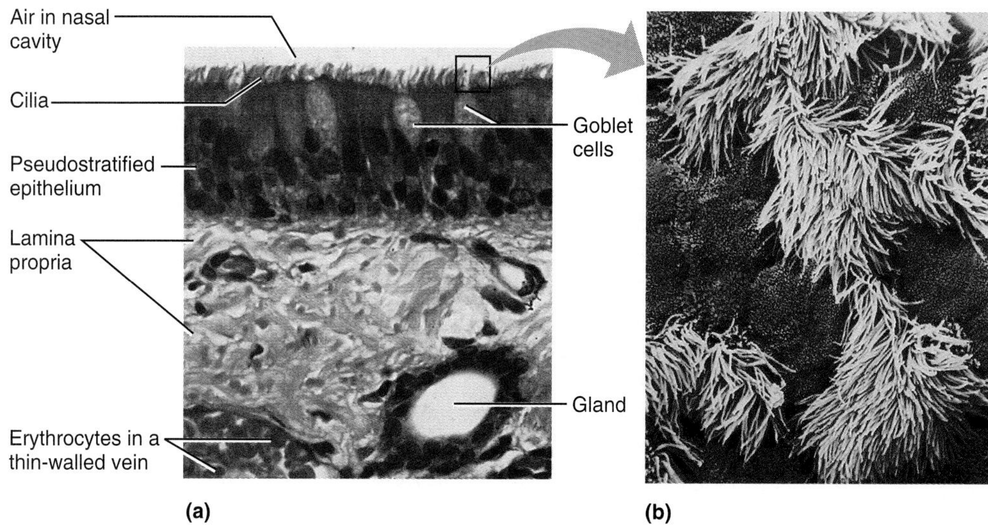

FIGURE 21.4 Respiratory mucosa. (a) Light photomicrograph (400 ×). **(b)** Surface view of the ciliated epithelium (scanning electron micrograph, 2700 ×).

The Respiratory Mucosa The respiratory mucosa (Figure 21.4) consists of a pseudostratified ciliated columnar epithelium containing scattered goblet cells, and the underlying connective tissue lamina propria. This lamina propria is richly supplied with compound tubuloalveolar glands (see p. 88) that contain mucous cells and serous cells. (*Mucous cells* secrete mucus, whereas *serous cells* in glands secrete a watery fluid containing digestive enzymes.) Each day, the nasal glands and the epithelial goblet cells secrete about a quart of mucus containing lysozyme, an enzyme that digests and destroys bacteria. The sticky mucus forms a sheet that covers the surface of the mucosa and traps inhaled dust, bacteria, pollen, viruses, and other debris from the air. Thus, an important function of the respiratory mucosa is to filter the inhaled air. The ciliated cells in the epithelial lining create a gentle current that moves the sheet of contaminated mucus posteriorly to the pharynx, where it is swallowed. In this way, particles filtered from the air are ultimately destroyed by digestive juices in the stomach. In addition, the sheet of mucus is a wet film that moistens the inhaled air.

CLINICAL APPLICATIONS

RHINITIS Rhinitis (ri-ni′tis; "nose inflammation") is inflammation of the nasal mucosa, caused by such things as cold viruses, streptococcal bacteria, and various allergens. This inflammation is accompanied by an excessive production of mucus, which results in nasal congestion, a runny nose, and postnasal drip. ✚

The nasal mucosa is richly supplied with sensory nerve endings. A sneeze reflex is stimulated when irritating particles (dust, pollen, and so on) contact this sensitive mucosa. The resulting sneeze propels air outward in a violent burst, expelling the irritant from the nose.

Rich plexuses of capillaries and thin-walled veins occupy the lamina propria of the nasal mucosa and warm the incoming air that flows across the mucosal surface. When the temperature of inhaled air drops, as when we step outside on a cold day, the vascular plexuses respond by engorging with warm blood, thereby intensifying the air-heating process. Because of the abundance and the superficial location of these vessels, nosebleeds are common and often profuse.

The Nasal Conchae Projecting medially from each lateral wall of the nasal cavity are three mucosa-covered, scroll-like structures: the *superior* and *middle conchae* of the ethmoid bone, and the *inferior concha,* which is a separate bone (see Figure 21.3). The groove inferior to each concha is a *meatus*. As inhaled air rushes over the curved conchae, the resulting turbulence greatly increases the amount of contact between the nasal mucosa and this inspired air. The gases in the inhaled air swirl through the twists and turns of the conchae, but the air's particulate matter is deflected onto the mucus-coated surfaces, where it becomes trapped. As a result, few particles larger than 4 μm get past the nasal cavities.

The conchae and nasal mucosa not only function during inhalation to filter, heat, and moisten the air, but also act during exhalation to reclaim this heat and moisture. That is, the inhaled air cools the conchae, then during exhalation these cooled conchae precipitate moisture and extract heat from the humid air flowing over them. This reclamation mechanism minimizes the amount of moisture and heat lost from the body through breathing, helping us to survive in dry and cold climates.

Advances in Understanding Another fascinating function of the nasal conchae has recently come to light. It has long been known that the mucosa covering the conchae swells periodically as the mucosal arteries vasodilate and vasoconstrict—and every few hours the swelling switches from the right side of the nasal cavity to the left and back. Because swollen conchae slow the flow of air through the nasal cavity, air always flows more slowly through one nostril than through the other. (Try to sense it: through which one of your own nostrils is the air now flowing faster?) Recently, a function was proposed for this phenomenon, based on the discovery that the intensity with which we smell airborne scents depends on how fast these scents flow across the olfactory epithelium (p. 468) in the nose. This means that the olfactory epithelium in the half of the nasal cavity with the slower airflow senses some different odors than in the half with the faster airflow. This in turn improves the effectiveness of our sense of smell by increasing the number of different scents detected in each sniff. ✷

The Paranasal Sinuses

The nasal cavity is surrounded by a ring of air-filled cavities called *paranasal sinuses* located in the frontal, sphenoid, ethmoid, and maxillary bones (see Figure 7.11, p. 167). Recall from pages 165–167 that these sinuses open into the nasal cavity, are lined by the same mucosa, and perform the same air-processing functions as does that cavity. Their mucus drains into the nasal cavities, and the suctioning effect caused by nose blowing helps to drain them.

CLINICAL APPLICATIONS

SINUSITIS Inflammation of the paranasal sinuses, a condition called **sinusitis,** is caused by viral, bacterial, or fungal infections. When the passages that connect the paranasal sinuses to the nasal cavity become blocked by swelling of the inflamed nasal mucosa, air in the sinus cavities is absorbed into the blood vessels of the mucosal lining, resulting in a partial vacuum and a sinus headache localized over the inflamed areas. Later, if the infection persists, inflammatory fluid oozes from the mucosa, fills the obstructed sinus, and exerts painful positive pressure. Most cases of sinusitis originate just inferior and lateral to the middle concha, where the openings of the frontal, maxillary, and one of the ethmoid sinuses lie in close proximity and can be blocked simultaneously. Serious cases of sinusitis are treated by promoting drainage and with antibiotics. ✚

The Pharynx

The **pharynx** (far'ingks) is the funnel-shaped passageway that connects the nasal cavity and mouth superiorly to the larynx and esophagus inferiorly (see Figure 21.1). It descends from the base of the skull to the level of the sixth cervical vertebra and serves as a common passageway for both food and air. In the context of the digestive tract, the pharynx is commonly called the *throat.*

On the basis of location and function, the pharynx is divided into (from superior to inferior) the *nasopharynx, oropharynx,* and *laryngopharynx* (lah-ring"go-far'ingks) (see Figure 21.3a). The muscular wall of the pharynx consists of skeletal muscle throughout its length (see pp. 645–646), but the nature of the mucosal lining varies among the three pharyngeal regions.

The Nasopharynx

The **nasopharynx** lies directly posterior to the nasal cavity, inferior to the sphenoid bone and superior to the level of the soft palate (see Figure 21.3a). Because it is superior to the point where food enters the body, the nasopharynx serves only as an air passageway. During swallowing, the soft palate and its pendulous **uvula** (u'vu-lah; "little grape") reflect superiorly, an action that closes off the nasopharynx and prevents food from entering the nasal cavity. During giggling, this sealing action fails, and swallowed fluids can spray from the nose.

The nasopharynx is continuous with the nasal cavity through the posterior nasal apertures, and its ciliated pseudostratified epithelium takes over the job of propelling mucus where the nasal mucosa leaves off, such that dusty mucus is moved downward through the nasopharynx. High on the posterior nasopharyngeal wall is the **pharyngeal tonsil,** or *adenoids,* which destroys pathogens entering the nasopharynx in the air (see pp. 597–598).

Critical Reasoning

Infection and inflammation of the adenoids is common, especially in children. Deduce how such a condition affects breathing.

Infected and enlarged adenoids obstruct the flow of air through the nasopharynx, so mouth breathing becomes necessary.

A *pharyngotympanic (auditory) tube,* which drains the middle ear, opens into each lateral wall of the nasopharynx. A ridge of pharyngeal mucosa posterior to this opening constitutes the **tubal tonsil** (see Figure 21.3b), whose location provides the middle ear some protection against infections that may spread from the pharynx.

The Oropharynx

The **oropharynx** lies posterior to the oral cavity (mouth); its arch-like entranceway, directly behind the mouth, is the **fauces** (faw'sēz; "throat") (see Figure 21.3a). The oropharynx extends inferiorly from the level of the soft palate to the level of the epiglottis (a flap posterior to the tongue). Both swallowed food and inhaled air pass through the oropharynx.

As the nasopharynx blends into the oropharynx, the lining epithelium changes from pseudostratified columnar to a thick, protective *stratified squamous epithelium.* This structural adaptation reflects the increased friction and greater chemical trauma accompanying the passage of swallowed food through the oropharynx.

Two kinds of tonsils are embedded in the mucosa of the oropharynx: The paired **palatine tonsils** lie in the lateral walls of the fauces, and the **lingual tonsil** covers the posterior surface of the tongue.

The Laryngopharynx

Like the oropharynx superior to it, the **laryngopharynx** serves as a common passageway for food and air and is

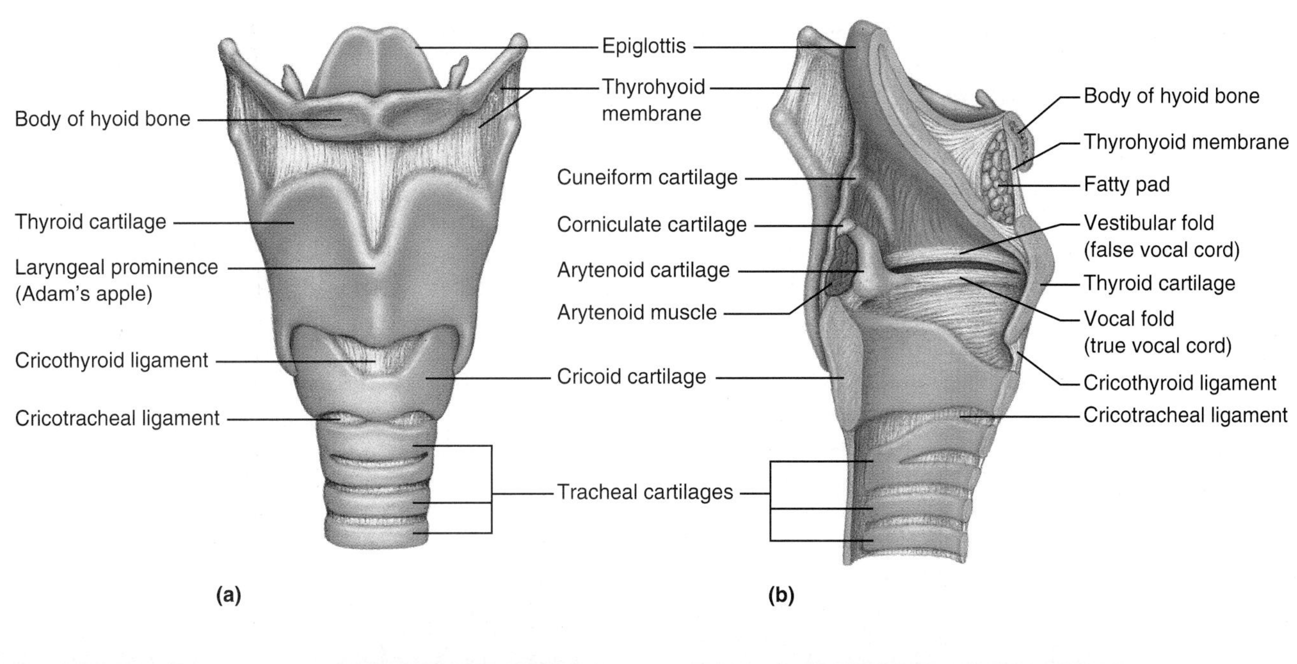

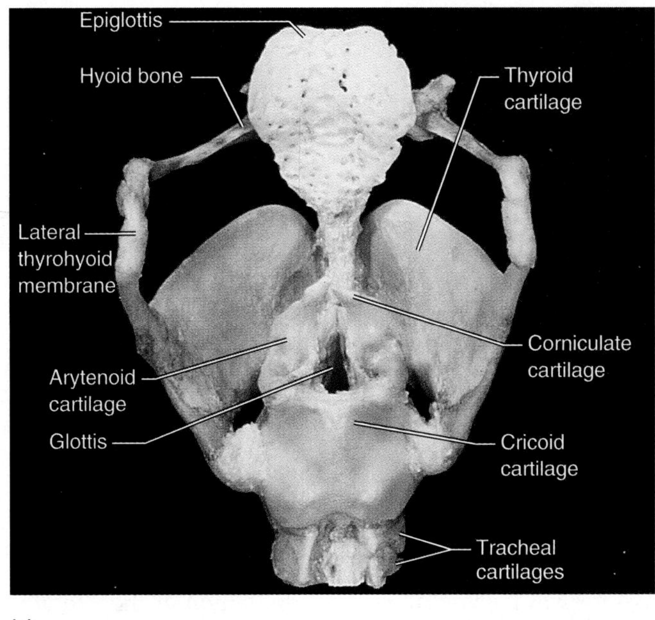

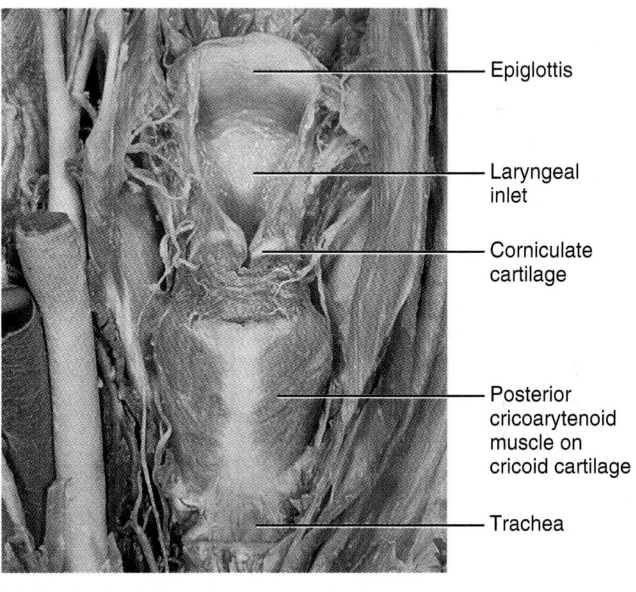

FIGURE 21.5 **Anatomy of the larynx.** **(a)** Anterior view of the skeleton of the larynx. **(b)** Lateral view of the larynx cut in half by a midsagittal section. The anterior aspect is to the right. **(c)** Photograph of the cartilages of the larynx, posterior view. **(d)** Posterior view of the larynx in the neck; this view is from within the pharynx. Much of the mucosal layer has been removed.

lined with a stratified squamous epithelium. The laryngopharynx lies directly posterior to the larynx and is continuous with both the esophagus, which conducts food and fluids to the stomach, and the larynx, which conducts air to the respiratory tract.

The Larynx

The **larynx** (lar′ingks), or voice box, extends from the level of the fourth to the sixth cervical vertebra. Superiorly, it attaches to the hyoid bone and opens into the laryngopharynx (see Figure 21.3a); inferiorly, it is continuous with the trachea (windpipe). The larynx has three functions: In addition to (1) voice production, it (2) provides an open airway and (3) acts as a switching mechanism to route air and food into the proper channels. For the latter purposes, the inlet (superior opening) to the larynx is closed during swallowing and open during breathing.

The framework of the larynx is an intricate arrangement of nine cartilages connected by membranes and ligaments (Figure 21.5). The large, shield-shaped **thyroid cartilage,** which is formed by two cartilage plates, resembles an upright open book, with the book's "spine" lying in the anterior midline of the neck. This "book spine" is the ridge-like **laryngeal** (lah-rin′je-al) **prominence,** which is obvious externally as the **Adam's apple.** The thyroid cartilage is larger in males than in females because male sex hormones stimulate its growth during puberty. Inferior to the thyroid cartilage is the **cricoid cartilage** (kri′koid; "circle"), which is shaped like a signet ring and is perched on top of the trachea.

As shown in Figure 21.5b and c, three pairs of small cartilages lie just superior to the cricoid cartilage in the posterior part of the larynx: the **arytenoid cartilages** (ar″ĭ-te′noid; "ladle-like"), the **corniculate cartilages** (kor-nik′u-lāt; "little horn"), and the **cuneiform cartilages** (ku-ne′ĭ-form; "wedge-shaped"). The most important of these cartilages are the pyramid-shaped arytenoids, which anchor the vocal cords.

The ninth cartilage of the larynx, the spoon-shaped **epiglottis** (ep″ĭ-glot′is), is composed of elastic cartilage and is almost entirely covered by a mucosa. Its stalk attaches anteriorly to the internal aspect of the angle of the thyroid cartilage. From there, the epiglottis projects superoposteriorly and attaches to the posterior aspect of the tongue (*epiglottis* means "upon the back of the tongue"). During swallowing, the entire larynx is pulled superiorly, and the epiglottis tips inferiorly to cover and seal the laryngeal inlet. Because this action keeps food out of the lower respiratory tubes, the epiglottis has been called the "guardian of the airways." The entry into the larynx of anything other than air initiates the cough reflex, which expels the substance and prevents it from continuing into the lungs. Because this protective reflex functions in conscious individuals only, it is never a good idea to give fluids when trying to revive someone who has lost consciousness.

Because the larynx lies so far inferiorly in the neck and must ascend so far during swallowing that it sometimes cannot reach the protective cover of the epiglottal lid before food enters the laryngeal inlet, the larynx seems poorly adapted for survival. Indeed, humans are the only animals that routinely choke to death. However, the low position of the larynx is essential to humans' ability to talk: Its inferior location allows for greater movement of the tongue in shaping sounds and for an exceptionally long pharynx, which acts as a resonating chamber for speech. This latter feature greatly improves the quality of the vowel sounds we produce.

Within the larynx, paired *vocal ligaments* run anteriorly from the arytenoid cartilages to the thyroid cartilage. These ligaments, composed largely of elastic fibers, form the core of a pair of mucosal folds called the **vocal folds** or **(true) vocal cords** (see Figures 21.5b and 21.6). Because the mucosa covering them is avascular, the vocal folds appear pearly white. Air exhaled from the lungs causes these folds to vibrate in a wave motion and to clap together, producing the basic sounds of speech. The medial opening between the vocal folds through which air passes is called the **rima glottidis,** and the vocal folds together with this rima compose the **glottis.** Another pair of horizontal mucosal folds that lies directly superior to the vocal folds, the **vestibular folds,** or **false vocal cords,** play no part in sound production. However, they define a slit-like cavity between themselves and the true vocal cords (see Figure 21.5b) that enhances high-frequency sounds, functioning like the tweeter on stereo speakers.

The epithelium lining the superior part of the larynx, an area subject to food contact, is stratified squamous. But inferior to the vocal folds, the epithelium is pseudostratified ciliated columnar and entraps dust. In this epithelium, the power stroke of the cilia is directed upward, toward the pharynx, so that dust-trapping mucus is continuously moved superiorly from the lungs. Sometimes, we help move this mucus up and out of the larynx by "clearing the throat."

Voice Production

Speech involves the intermittent release of exhaled air and the opening and closing of the glottis. In this process, the length of the vocal folds and the size of the rima glottidis are varied by intrinsic laryngeal muscles, most of which move the arytenoid cartilages (see Figure 21.6). As the length and tension of the vocal folds change, the pitch of the produced sound changes. Generally, the tenser the vocal folds, the faster the exhaled air causes them to vibrate, and the higher the pitch.

As a boy's larynx enlarges during puberty, his laryngeal prominence grows anteriorly into a large Adam's apple, lengthening the vocal folds. Because longer vocal folds vibrate more slowly than short folds do, the voice becomes deeper. For this reason, most men have lower voices than females or young boys. The voices of adolescent boys frequently "crack," alternating between high-pitched and low-pitched sounds, because the boys have not yet learned to control the action of their longer vocal folds.

Loudness of the voice depends on the force with which air rushes across the vocal folds. The greater the force, the stronger the vibrations and the louder the sound. The vocal folds do not move at all when we whisper, but they vibrate vigorously when we yell.

Although the vocal folds produce the basic sounds of speech, the quality of the voice depends on many other structures. As previously mentioned, the entire length of the pharynx acts as a resonating chamber to amplify the quality of sound; the oral cavity, nasal cavity, and paranasal sinuses also contribute to vocal resonance. In addition, normal speech and good enunciation depend on the "shaping" of sounds into recognizable consonants and vowels by the pharynx, tongue, soft palate, and lips.

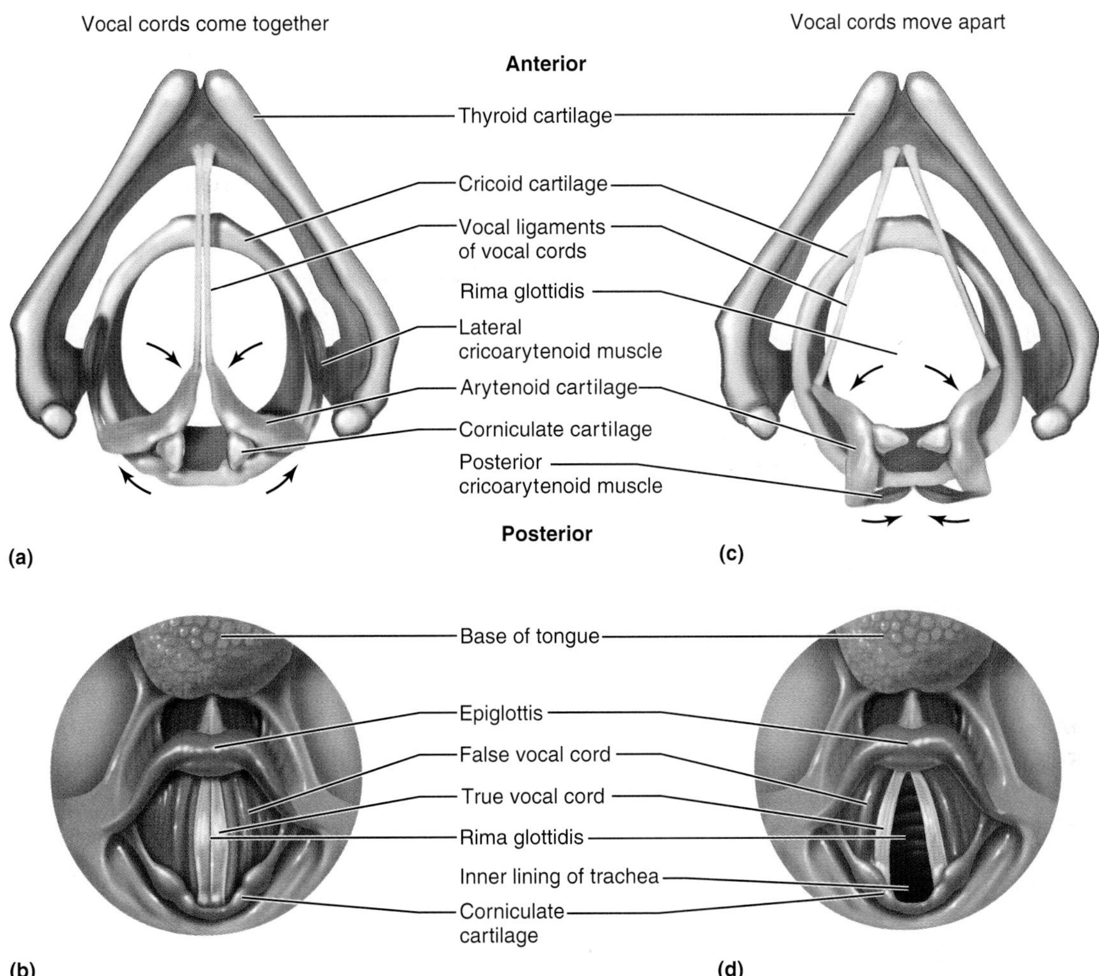

FIGURE 21.6 Movements of the vocal cords. These are superior views of the larynx, with the vocal cords together (in **a** and **b** at left), and with the vocal cords apart (in **c** and **d** at right). Parts (a) and (c) show the skeletal elements and two of the many pairs of laryngeal muscles: the lateral and posterior cricoarytenoid muscles, which rotate the arytenoid cartilages. Parts (b) and (d), by contrast, are corresponding views of the living larynx, with all mucous membranes in place. The epiglottis has been left out of (a) and (c) because it would obscure the view of other laryngeal cartilages.

Critical Reasoning

Describe how the voices of individuals suffering from a bad cold become hoarse.

The infection stimulates inflammation of the larynx, or ***laryngitis,*** *causing the vocal folds to swell and interfering with their ability to vibrate. These events alter the tone of the voice, producing hoarseness, and in severe cases may result in an inability to speak above a whisper. Hoarseness is also caused by overuse of the voice, growths on the vocal cords, inhalation of irritating chemicals (as in tobacco smoking), or paralysis of some laryngeal muscles.*

Sphincter Functions of the Larynx

Under certain conditions, the vocal folds act as a sphincter that prevents the passage of air. During abdominal straining, such as occurs when one strains to defecate, the abdominal muscles contract and the glottis closes to prevent exhalation, raising intrathoracic and intra-abdominal pressure. These events, collectively called **Valsalva's maneuver,** help to evacuate the rectum; they also stabilize the body trunk when one lifts a heavy load.

Innervation of the Larynx

The larynx receives its sensory and motor innervation through a superior laryngeal branch of each vagus nerve and from the *recurrent laryngeal nerves,* which branch off the vagus in the superior thorax and loop superiorly to reascend through the neck. The left recurrent laryngeal nerve loops under the aortic arch, whereas the right recurrent laryngeal nerve loops under the right subclavian artery. The backtracking course of these nerves is so unusual that the ancient Greeks mistook them for slings supporting the great arteries. Damage to these nerves disrupts speech. Transection of one recurrent laryngeal nerve immobilizes one vocal fold, producing a degree of hoarseness. In such cases the other vocal fold can compensate, and speech remains almost normal. However, if both recurrent laryngeal nerves are transected, speech (except for whispering) is lost entirely.

Because each recurrent laryngeal nerve is related to the apex of a lung in the thorax, cancers in the apical region of the lung can compress this nerve and cause hoarseness.

The Trachea

The flexible **trachea** (tra′ke-ah), or windpipe, descends from the larynx through the neck and into the mediastinum; it ends by dividing into the two main bronchi (primary bronchi) in the mid thorax (see Figure 21.1). Early anatomists mistook the trachea for a rough-walled artery (*trachea* = rough).

The tracheal wall contains 16 to 20 C-shaped rings of hyaline cartilage (Figure 21.7a) joined to one another by intervening membranes of fibroelastic connective tissue. Consequently, the trachea is flexible enough to permit bending and elongation, but the cartilage rings prevent it from collapsing and keep the airway open despite the pressure changes that occur during breathing. The open posterior parts of the cartilage rings, which abut the esophagus, contain smooth muscle fibers of the **trachealis muscle** and soft connective tissue. Because the posterior wall of the trachea is not rigid, the esophagus can expand anteriorly as swallowed food passes through it. Contraction of the trachealis muscle decreases the diameter of the trachea: During coughing and sneezing, this action helps to expel irritants from the trachea by accelerating the exhaled air to a speed of 165 kph (100 mph)! A ridge on the internal aspect of the last tracheal cartilage, called the **carina** (kah-ri′nah; "keel"), marks the point where the trachea branches into the *two main (primary) bronchi* (see Figure 21.1). The mucosa that lines the carina is highly sensitive to irritants, and the cough reflex often originates here.

The microscopic structure of the wall of the trachea (Figure 21.7b) consists of several layers common to many tubular organs of the body: the *mucous membrane, submucosa,* and *adventitia.* The **mucous membrane,** as usual, consists of an inner epithelium and a lamina propria. The epithelium is the same air-filtering pseudostratified epithelium that occurs throughout most of the respiratory tract; its cilia continuously propel dust-laden sheets of mucus superiorly toward the pharynx. The lamina propria contains many elastic fibers and is separated from the submucosa by a distinct sheet of elastin (not illustrated). This elastin, which occurs in all smaller air tubes as well, enables the trachea to stretch during inhalation and recoil during exhalation. The **submucosa** ("below the mucosa"), another layer of connective tissue, contains glands with both serous and mucous cells, called *seromucous glands,* which help produce the sheets of mucus within the trachea. The **adventitia** (ad″ven-tish′e-ah), the most external layer, is a connective tissue that contains the tracheal cartilages.

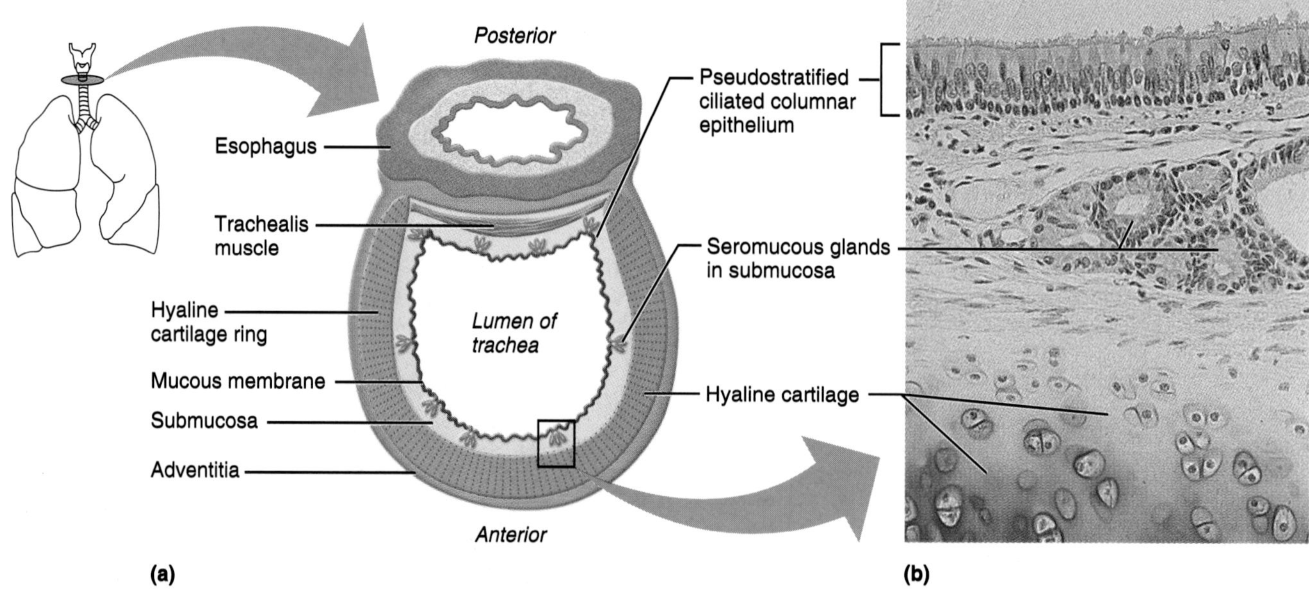

FIGURE 21.7 The trachea. **(a)** Gross view of the trachea at left, and cross-sectional view at center. The cross section shows the relationship of the trachea to the esophagus, the position of the cartilage rings in the tracheal wall, and the trachealis muscle connecting the free ends of the cartilage rings. **(b)** Photomicrograph of a portion of the tracheal wall (150 ×). The lamina propria lies between the epithelium and the gland-filled submucosa.

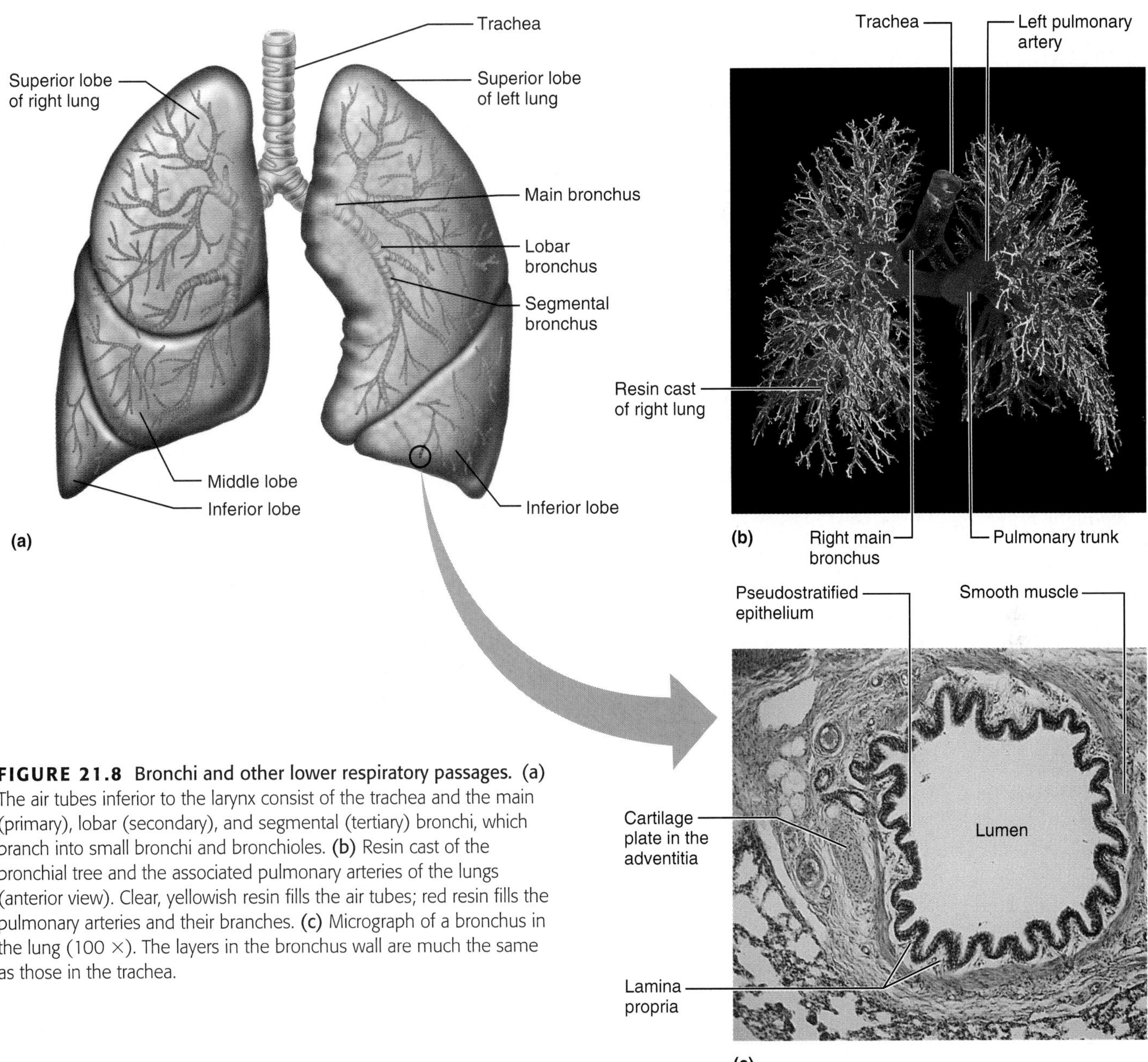

FIGURE 21.8 Bronchi and other lower respiratory passages. **(a)** The air tubes inferior to the larynx consist of the trachea and the main (primary), lobar (secondary), and segmental (tertiary) bronchi, which branch into small bronchi and bronchioles. **(b)** Resin cast of the bronchial tree and the associated pulmonary arteries of the lungs (anterior view). Clear, yellowish resin fills the air tubes; red resin fills the pulmonary arteries and their branches. **(c)** Micrograph of a bronchus in the lung (100 ×). The layers in the bronchus wall are much the same as those in the trachea.

The Bronchi and Subdivisions: The Bronchial Tree

In this section we discuss the final structures in the conduction zone before we examine the anatomy of the respiratory zone.

Bronchi in the Conducting Zone

The right and left **main bronchi** (brong′ki), also called *primary bronchi,* are the largest conduits in the *bronchial tree,* a system of respiratory passages that branches extensively within the lungs (Figure 21.8). The two main bronchi are branches of the trachea in the mediastinum. This bifurcation occurs at the level of the sternal angle (T_4) in the cadavers studied in the anatomy laboratory, but in living, standing individuals, it typically occurs at T_7. Each main bronchus runs obliquely through the mediastinum before plunging into the medial depression (hilus) of a lung. The main bronchi lie directly posterior to the large pulmonary vessels that supply the lungs (see Figure 21.8b). Because the right main bronchus is wider, shorter, and more vertical than the left, an accidentally inhaled object, such as a button or marble, is more likely to lodge in the right main bronchus.

As they approach and enter the lungs, the main bronchi divide into secondary or **lobar bronchi**—three on the right and two on the left—each of which supplies one lung lobe. The lobar bronchi branch into tertiary or **segmental bronchi,** which in turn divide repeatedly into smaller bronchi: fourth-order, fifth-order, and so on. Overall, there are about 23 orders of air tubes in the lungs, the tiniest almost too small to be seen without a microscope. The tubes smaller than 1 mm in diameter are called **bronchioles** ("little bronchi"), and the smallest of these, the **terminal bronchioles,** are less than 0.5 mm in diameter.

The tissue composition of the wall of each main bronchus mimics that of the trachea, but as the conducting tubes become smaller, the following changes occur:

1. **The cartilage changes.** The cartilage rings are replaced by irregular *plates* of cartilage as the main bronchi enter the lungs (see Figure 21.8c). By the level of the bronchioles, supportive cartilage is no longer present in the tube walls. By contrast, elastin, which occurs in the walls throughout the bronchial tree, does not diminish.
2. **The epithelium changes.** The mucosal epithelium thins as it changes from pseudostratified columnar to simple columnar and then to simple cuboidal in the smallest bronchioles. Neither cilia nor mucus-producing cells are present in the smallest bronchioles, where the sheets of air-filtering mucus end. Any inhaled dust particles that travel beyond the bronchioles are not trapped in mucus but instead are removed by macrophages in the alveoli (discussed shortly).
3. **Smooth muscle becomes important.** A complete layer of smooth muscle, which first appears in the walls of the large bronchi and is present throughout the smaller bronchi and bronchioles, regulates the amount of air entering the alveoli. The musculature relaxes to widen the air tubes during sympathetic excitement, when respiratory needs are great, and it constricts the air tubes under parasympathetic direction when respiratory needs are low (see Table 15.2, p. 451). Strong contractions of the bronchial smooth muscles narrow the air tubes during asthma attacks (see p. 624).

The Respiratory Zone

The respiratory zone is the end part of the respiratory tree in the lungs, consisting of structures that contain air-exchange chambers called **alveoli** (Figure 21.9). The first respiratory-zone structures, which branch from the terminal bronchioles of the conducting zone, are **respiratory bronchioles.** These can be recognized by the scattered alveoli protruding from their walls. The respiratory bronchioles lead into **alveolar ducts,** straight ducts whose walls consist almost entirely of alveoli. The alveolar ducts then lead into terminal clusters of alveoli called **alveolar sacs.** Note that alveoli and alveolar sacs are not the same things: An alveolar sac is analogous to a bunch of grapes, in which the individual grapes are the alveoli. The opening from an alveolar duct into an alveolar sac is called an **atrium,** meaning an "entrance chamber."

About 300 million air-filled alveoli crowd together within the lungs, accounting for most of lung volume and providing a tremendous surface area for gas exchange. The total area of all alveoli in an average pair of lungs is 140 square meters, or 1500 square feet, which is 40 times greater than the surface area of the skin!

The wall of each alveolus consists of a single layer of squamous epithelial cells called **type I cells** surrounded by a delicate basal lamina (Figure 21.10). The extreme thinness of this wall—0.5 μm—is hard to imagine. It is fifteen times thinner than a sheet of tissue paper! The external surfaces of the alveoli are densely covered with a "cobweb" of pulmonary capillaries (see Figure 21.10a and b), each of which is surrounded by a thin sleeve of the finest areolar connective tissue. Together, the alveolar and capillary walls and their fused basal laminas form the **respiratory membrane** (also called the *air-blood barrier*), where oxygen and carbon dioxide are exchanged between the alveolus and the blood. Air is present on one side of the membrane, and blood flows past on the other (see Figure 21.10c and d). Gases pass easily through this thin membrane: Oxygen diffuses from the alveolus into the blood, and carbon dioxide diffuses from the blood to enter the air-filled alveolus.

Scattered among the type I squamous cells in the alveolar walls are cuboidal epithelial cells called **type II cells** (see Figure 21.10c), which secrete a fluid that coats the internal alveolar surfaces. This fluid contains a detergent-like substance called *surfactant* (see p. 623), without which the alveolar walls would stick together during exhalation.

Lung alveoli also have the following significant features:

1. Alveoli are surrounded by fine *elastic fibers* of the same type that surround structures along the entire respiratory tree (Figure 21.10b).
2. Adjacent alveoli interconnect via **alveolar pores** (see Figure 21.10c), which allow the equalization of air pressure throughout the lung and provide alternate routes for air to reach alveoli whose bronchi have collapsed due to disease.
3. Internal alveolar surfaces provide a site for the free movement of **alveolar macrophages** *(dust cells),* which actually live in the air space and remove the tiniest inhaled particles that were not trapped by mucus. Most dust-filled macrophages migrate from the "dead-end" alveoli superiorly into the bronchioles, where ciliary action carries them into the pharynx to be swallowed. By this mechanism over 2 million dust cells are cleared each hour!

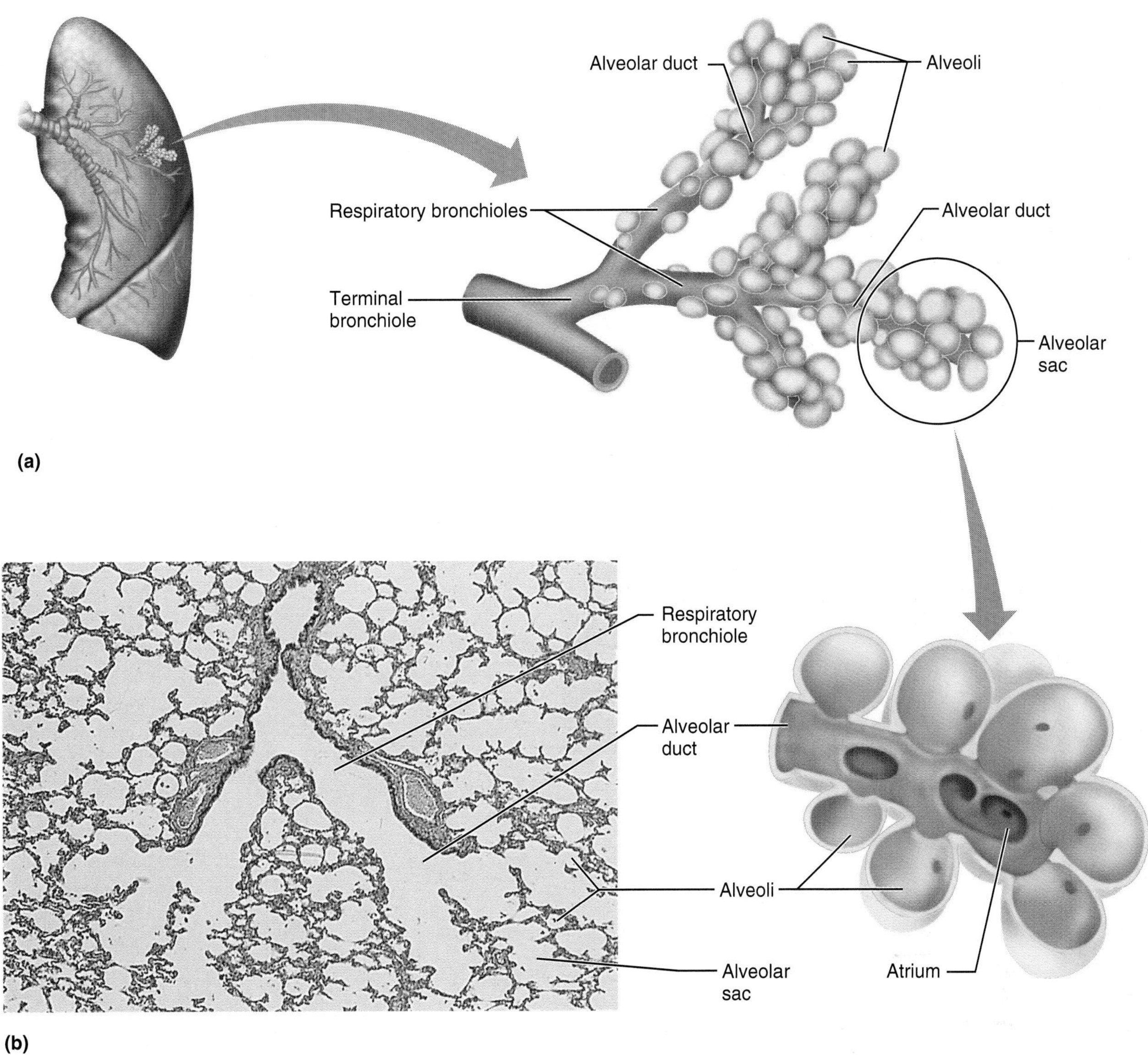

FIGURE 21.9 **Structures of the respiratory zone.** **(a)** Diagram of a respiratory bronchiole, alveolar ducts, alveolar sacs, and alveoli. **(b)** Photomicrograph of a part of the lung (30 ×).

The Lungs and Pleurae

We begin this section by discussing the pleurae, the membranes that cover the lungs and line the pleural cavity. Then in turn we examine the gross anatomy of the lungs and their blood supply and innervation.

The Pleurae

Around each lung is a flattened sac whose walls consist of a serous membrane called **pleura** (ploo′rah; "the side"). As shown in Figure 21.11, the outer layer of this sac is the *parietal pleura,* whereas the inner layer, directly on the lung, is *visceral pleura.* Let us now describe these layers of pleura more thoroughly. The **parietal pleura** covers the internal surface of the thoracic wall, the superior surface of the diaphragm, and the lateral surfaces of the mediastinum. From the mediastinum, it reflects laterally to enclose the great vessels running to the lung (lung root). In the area where these vessels enter the lung, the parietal pleura becomes the **visceral pleura,** which covers the external lung surface.

The **pleural cavity,** the space between the parietal and visceral pleurae, is actually a slit-like potential space filled with a layer of *pleural fluid.* Secreted by the pleurae, this lubricating fluid allows the lungs to glide without friction over the thoracic wall during breathing movements. The fluid also holds the parietal and visceral pleurae together, just as a film of oil or water would hold two glass plates together. The two pleurae can easily slide from side to side across each other, but

Smooth muscle
Alveolus
Capillaries
Elastic fibers
(b)

(a)

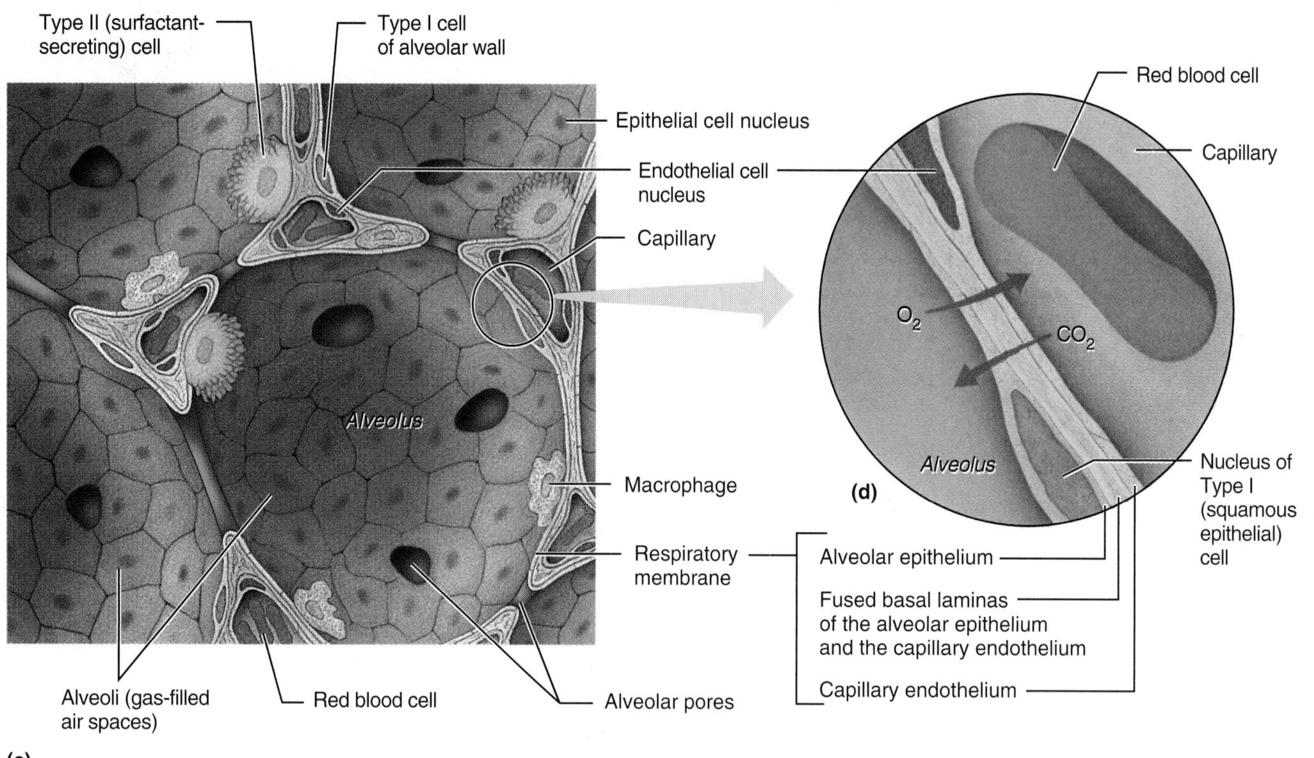

FIGURE 21.10 Anatomy of alveoli and the respiratory membrane. **(a)** Scanning electron micrograph of casts of several alveoli and the surrounding pulmonary capillaries (255 ×). © R. Kessel and R. Kardon. **(b)** External view of alveoli in an alveolar duct and sac, covered with capillaries and elastic fibers. **(c)** Detailed anatomy of alveoli in cut-open view. **(d)** The respiratory membrane, which is composed of a squamous alveolar type I cell, a capillary endothelial cell, and the fused basal laminas of these cells. Oxygen (O_2) diffuses from the alveolar air into the pulmonary capillary blood while carbon dioxide (CO_2) diffuses from the pulmonary blood into the alveolus.

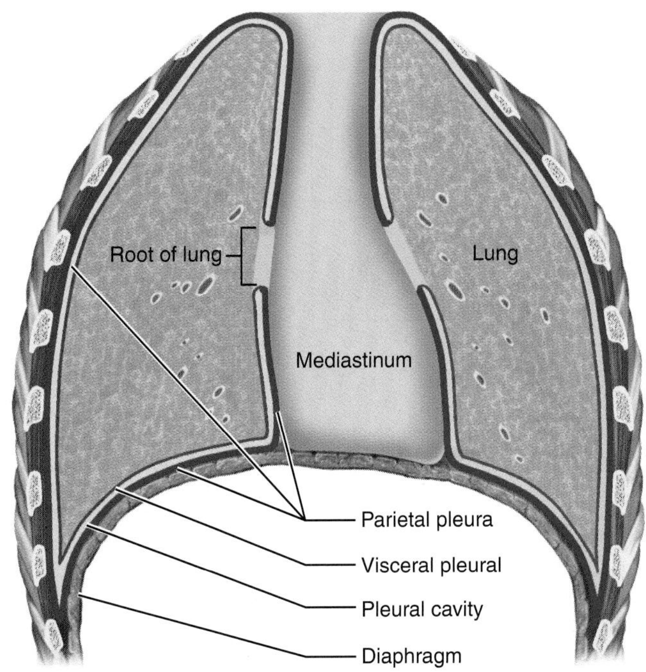

FIGURE 21.11 Diagram of the pleurae and pleural cavities.

their separation is strongly resisted. Consequently, the lungs cling tightly to the thoracic wall and are forced to expand and recoil as the volume of the thoracic cavity increases and decreases during breathing (see pp. 621–623).

The pleurae also help divide the thoracic cavity into three separate compartments—the central mediastinum and two lateral pleural compartments, each containing a lung. This compartmentalization helps to prevent the moving lungs or heart from interfering with one another. Compartmentalization also limits the spread of local infections and the extent of traumatic injury.

CLINICAL APPLICATIONS

PLEURISY AND PLEURAL EFFUSION Lung infections such as pneumonia produce inflammation of the pleura, called **pleurisy (pleuritis).** The rubbing together (friction) of the two inflamed pleural membranes produces a stabbing chest pain with each breath. Because the visceral pleura is relatively insensitive (see the discussion of visceral pain on p. 458), the pain of pleurisy actually originates from the parietal pleura only. If pleurisy persists, the inflamed pleurae may secrete excess pleural fluid, which then overfills the pleural cavity and exerts pressure on the lungs. Although this type of pleurisy hinders breathing movements, it is less painful than the type resulting from friction.

Pleural effusion is a general term for the accumulation of one of various kinds of fluid in the pleural cavity. Although it characterizes many cases of pleurisy, it can also result from (1) hemorrhage of a damaged lung or lung vessel or (2) leakage of a serous fluid from the lung capillaries when either right ventricular failure or heart failure causes blood to back up in the pulmonary circuit. ✚

Gross Anatomy of the Lungs

The paired **lungs** and their pleural sacs occupy all the thoracic cavity lateral to the mediastinum, which houses the heart, great blood vessels, trachea, main bronchi, esophagus, and other organs (Figure 21.12). Each lung is roughly cone-shaped. The anterior, lateral, and posterior surfaces of a lung contact the ribs and form a continuously curving *costal surface.* Just deep to the clavicle is the **apex,** the rounded, superior tip of the lung. The concave inferior surface that rests on the diaphragm is the **base.** On the *medial (mediastinal) surface* of each lung is an indentation, the **hilus,** through which blood vessels, bronchi, lymph vessels, and nerves enter and exit the lung. Collectively, these structures attach the lung to the mediastinum and are called the **root** of the lung. The largest components of this root are the pulmonary artery and veins and the main (primary) bronchus. Figure 21.13b and c shows the precise arrangement of the root structures at the hilus of each lung.

Because the heart is tilted slightly to the left of the median plane of the thorax, the left and right lungs differ slightly in shape and size. The left lung is somewhat smaller than the right and has a **cardiac notch,** a deviation in its anterior border that accommodates the heart (see Figure 21.12a). Several deep fissures divide the two lungs into different patterns of **lobes.** The left lung is divided into two lobes, the **upper lobe** and the **lower lobe,** by the **oblique fissure** (see Figure 21.13a). The right lung is partitioned into three lobes, the **upper, middle,** and **lower lobes,** by the **oblique** and **horizontal fissures.** (The upper lobes are also called superior lobes; the lower lobes, inferior lobes.) As previously mentioned, each lung lobe is served by a lobar (secondary) bronchus and its branches.

Each of the lobes, in turn, contains a number of pyramid-shaped **bronchopulmonary segments** (Figure 21.14 on p. 620) separated from one another by thin partitions of dense connective tissue. Each segment receives air from an individual segmental (tertiary) bronchus. Each lung has ten bronchopulmonary segments arranged in similar, but not identical, patterns in the two lungs.

The bronchopulmonary segments have clinical significance in that they limit the spread of some diseases within the lung, because infections do not easily cross the connective-tissue partitions between them. Furthermore, because only small veins span these partitions, surgeons can neatly remove segments without cutting any major blood vessels.

The smallest subdivision of the lung that can be seen with the naked eye is the **lobule.** Appearing on the lung surface as hexagons ranging from the size of a pencil eraser to the size of a penny (see Figure 21.13b and c), each lobule is served by a large bronchiole and its branches. In most city dwellers and in smokers, the connective tissue that separates the individual lobules is blackened with carbon.

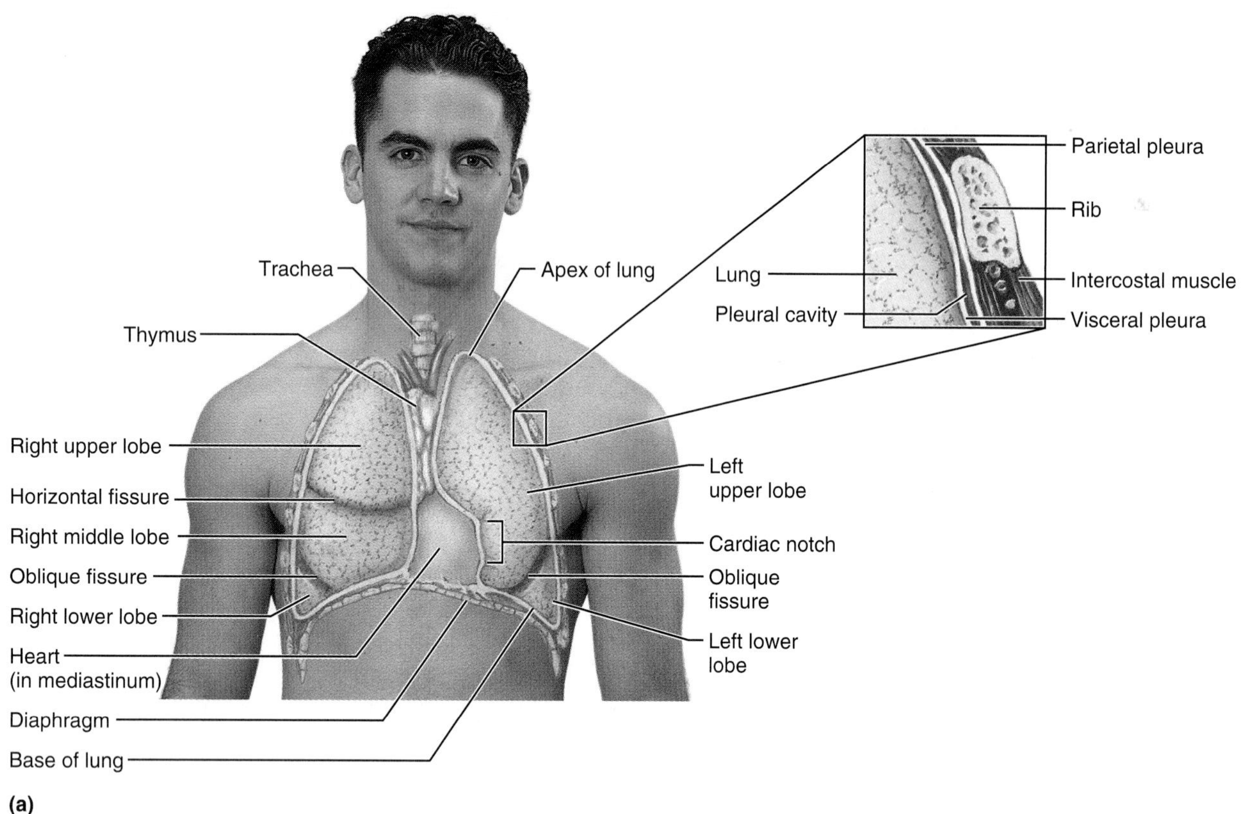

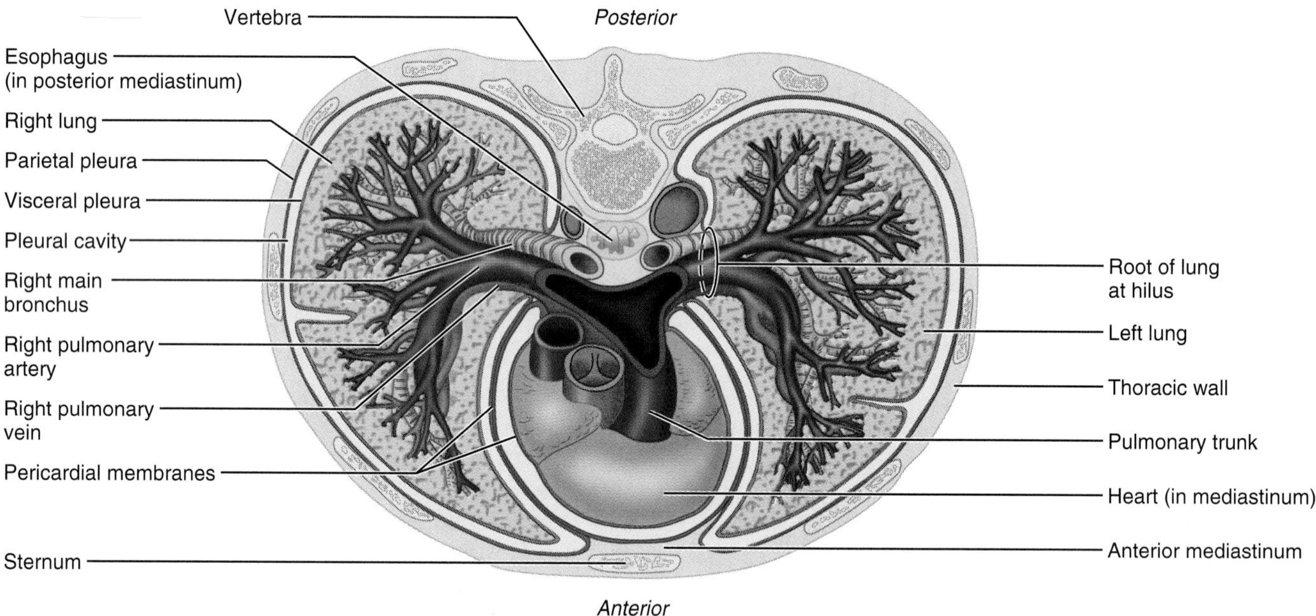

FIGURE 21.12 Anatomical relationships of organs in the thoracic cavity. **(a)** Anterior view of thoracic structures. The lungs flank the central mediastinum. The inset at upper right shows details of the pleurae and pleural cavity. **(b)** Transverse cut through the superior part of the thorax, showing the lungs and the main organs in the mediastinum. The plane of section is through the lung roots just superior to the heart, looking down on the heart.

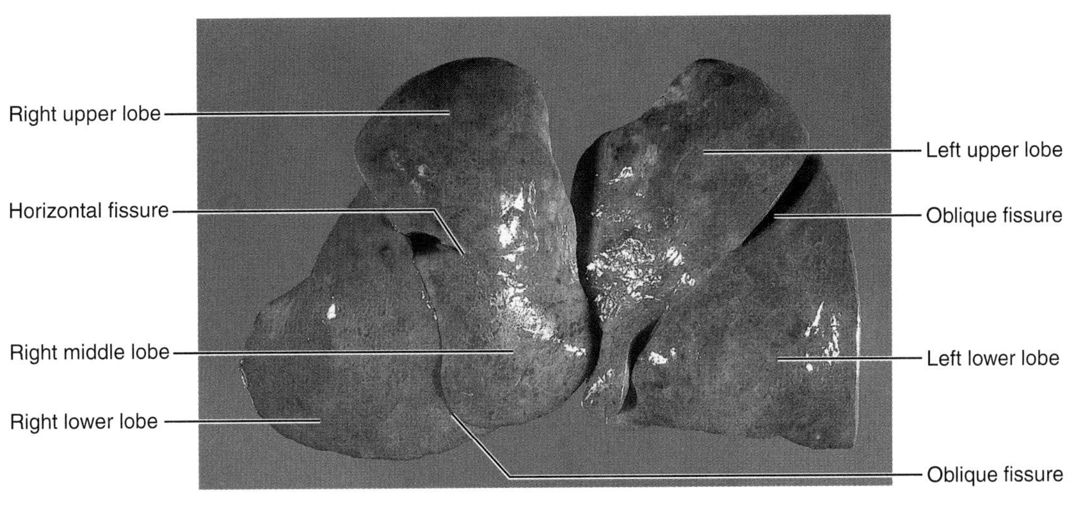

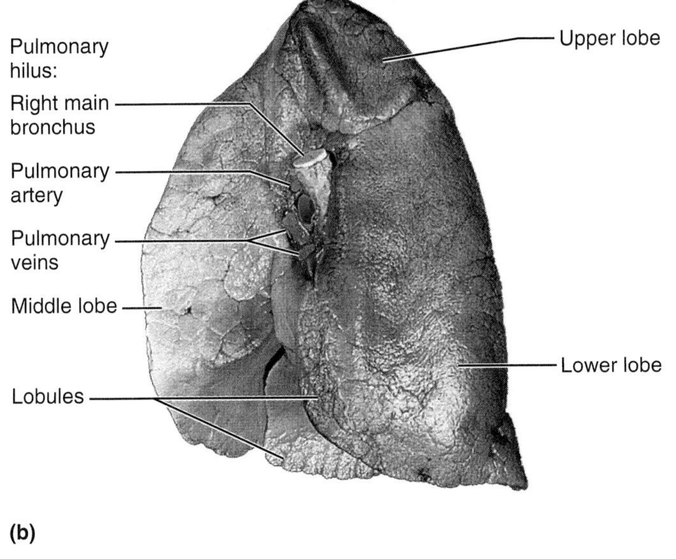

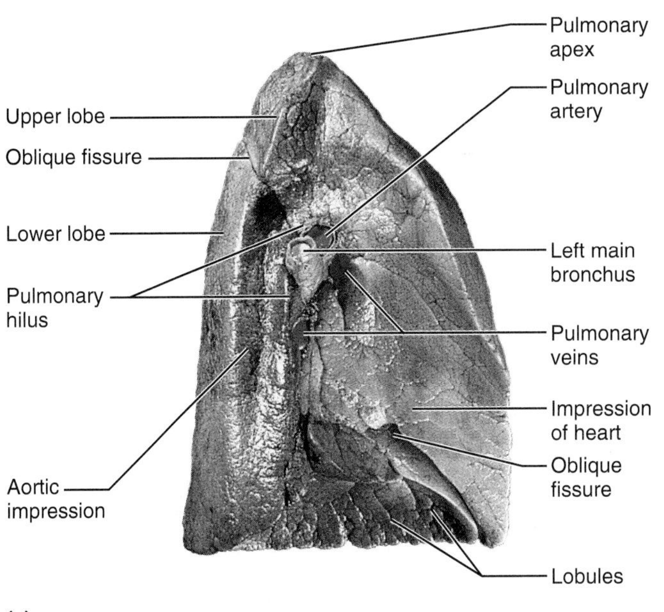

FIGURE 21.13
Photographs of the lungs. **(a)** Right and left lungs, anterior view. **(b)** Medial view of right lung. **(c)** Medial view of left lung.

As previously mentioned, the lungs consist largely of air tubes and spaces. The balance of the lung tissue, its *stroma* (stro′mah; "supporting mattress"), is a framework of connective tissue containing many elastic fibers. As a result, the lungs are light, soft, spongy, elastic organs that each weigh only about 0.6 kg (1.25 pounds). The elasticity of healthy lungs helps to reduce the effort of breathing, as described shortly.

Blood Supply and Innervation of the Lungs

The **pulmonary arteries** (see p. 552) deliver oxygen-poor blood to the lungs for oxygenation (see Figure 21.12b). In the lung, these arteries branch along with the bronchial tree, generally lying *posterior* to the corresponding bronchi. The smallest arteries feed into the **pulmonary capillary networks** around the alveoli (see Figure 21.10b). Oxygenated blood is carried from the alveoli of the lungs to the heart by the **pulmonary veins,** whose tributaries generally lie *anterior* to the corresponding bronchi within the lungs. However, some venous tributaries run in the connective-tissue partitions between the lung lobules and between the bronchopulmonary segments.

As explained in Chapter 19 (pp. 559, 569), the *bronchial arteries* and *veins* provide and drain systemic blood to and from the lung tissues. These small vessels enter and exit the lungs at the hilus, and within the lung they lie on the branching bronchi.

The lungs are innervated by sympathetic, parasympathetic, and visceral sensory fibers that enter each lung through the *pulmonary plexus* on the lung root. From

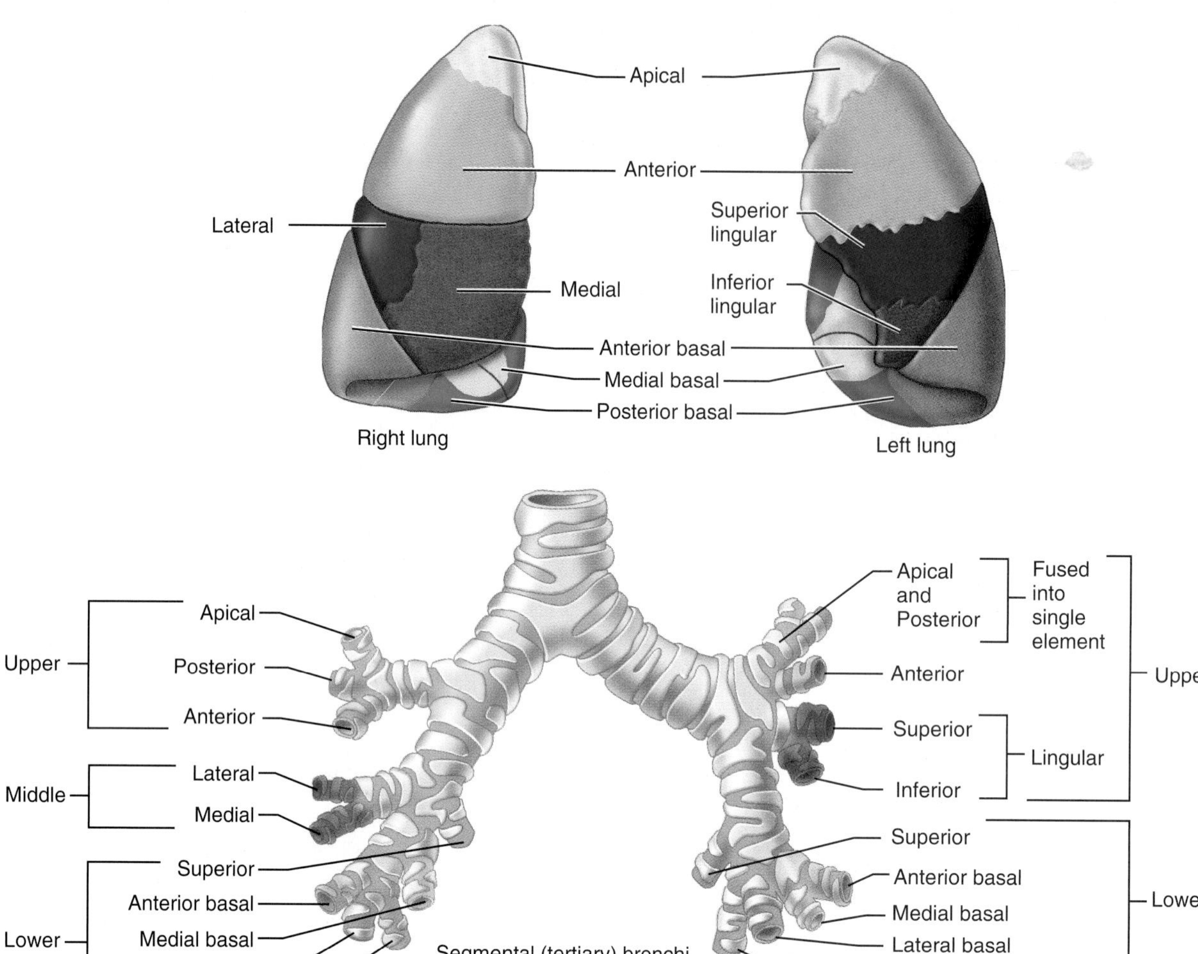

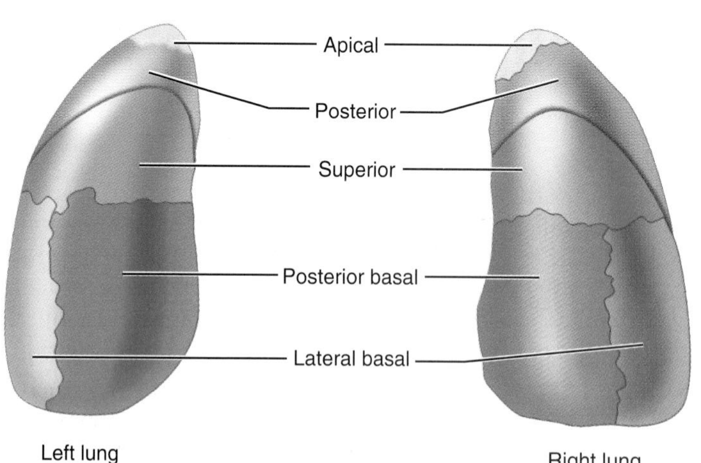

FIGURE 21.14 Bronchopulmonary segments. Also shown are the segmental (tertiary) bronchi that supply each bronchopulmonary segment (center diagram). Because most introductory anatomy courses do not require their students to learn the individual bronchopulmonary segments, this figure is provided for reference only.

Table 21.1 Principal Organs of the Respiratory System

Structure	Description, general and distinctive features	Function
Nose	Jutting external portion supported by bone and cartilage; internal nasal cavity divided in half by midline nasal septum and lined with mucosa	Produces mucus; filters, warms, and moistens incoming air; resonance chamber for speech
	Roof of nasal cavity contains olfactory epithelium	Receptors for sense of smell
	Nasal cavity surrounded by paranasal sinuses	Sinuses function the same as nasal cavity; also lighten skull
Pharynx	Passageway connecting nasal cavity to larynx and oral cavity to esophagus; three subdivisions: nasopharynx, oropharynx, and laryngopharynx	Passageway for air and food
	Houses tonsils	Tonsils respond to inhaled antigens
Larynx	Connects pharynx to trachea; framework of cartilage and dense connective tissue; opening (rima glottidis) can be closed by epiglottis or vocal folds	Air passageway; prevents food from entering lower respiratory tract
	Houses true vocal cords	Voice production
Trachea	Flexible tube running from larynx and dividing inferiorly into two main (primary) bronchi; walls contain C-shaped cartilages that are incomplete posteriorly where trachealis muscle occurs	Air passageway; filters, warms, and moistens incoming air
Bronchial tree	Consists of right and left main bronchi, which subdivide within the lungs to form lobar (secondary) and segmental (tertiary) bronchi, smaller bronchi, and bronchioles; bronchiolar walls contain complete layer of smooth muscle; constriction of this muscle impedes expiration	Air passageways connecting trachea with alveoli; warms and moistens incoming air
Alveoli	Microscopic chambers at termini of bronchial tree; walls of simple squamous epithelium underlain by thin basement membrane; external surfaces intimately associated with pulmonary capillaries	Main sites of gas exchange
	Special alveolar cells produce surfactant	Reduces surface tension; helps prevent lung collapse
Lungs	Paired composite organs located within pleural cavities of thorax; composed primarily of alveoli and respiratory passageways; stroma is fibrous elastic connective tissue, allowing lungs to recoil passively during expiration	House passageways smaller than main bronchi
Pleurae	Serous membranes; parietal pleura lines thoracic cavity; visceral pleura covers external lung surfaces	Produce lubricating fluid and compartmentalize lungs

there, these nerve fibers lie along the bronchial tubes and blood vessels within the lungs. As previously mentioned, parasympathetic fibers constrict the air tubes, whereas sympathetic fibers dilate them (p. 451).

VENTILATION

In our examination of ventilation, we first discuss the mechanical aspects of breathing and then discuss the control of breathing by the nervous system.

The Mechanism of Ventilation

Breathing, or **pulmonary ventilation,** consists of two phases: *inspiration* (inhalation), the period when air flows into the lungs, and *expiration* (exhalation), the period when gases exit the lungs. Here we discuss the mechanical factors that promote this movement of gases; for more information on the *muscles* of ventilation, see Table 11.5 on p. 287.

Inspiration

The process of **inspiration** is easy to understand if you think of the thoracic cavity as an expandable container with a single entrance at the top: the tube-like trachea. The volume of this container is increased by enlarging all of its dimensions, which increases the volume and thus decreases the pressure within it. This decrease in internal gas pressure in turn causes air to enter the container from the atmosphere, because gases always flow from areas of high pressure to areas of low pressure. During normal quiet inspiration, the inspiratory muscles—the diaphragm and external intercostal muscles—are activated to produce those actions, as follows:

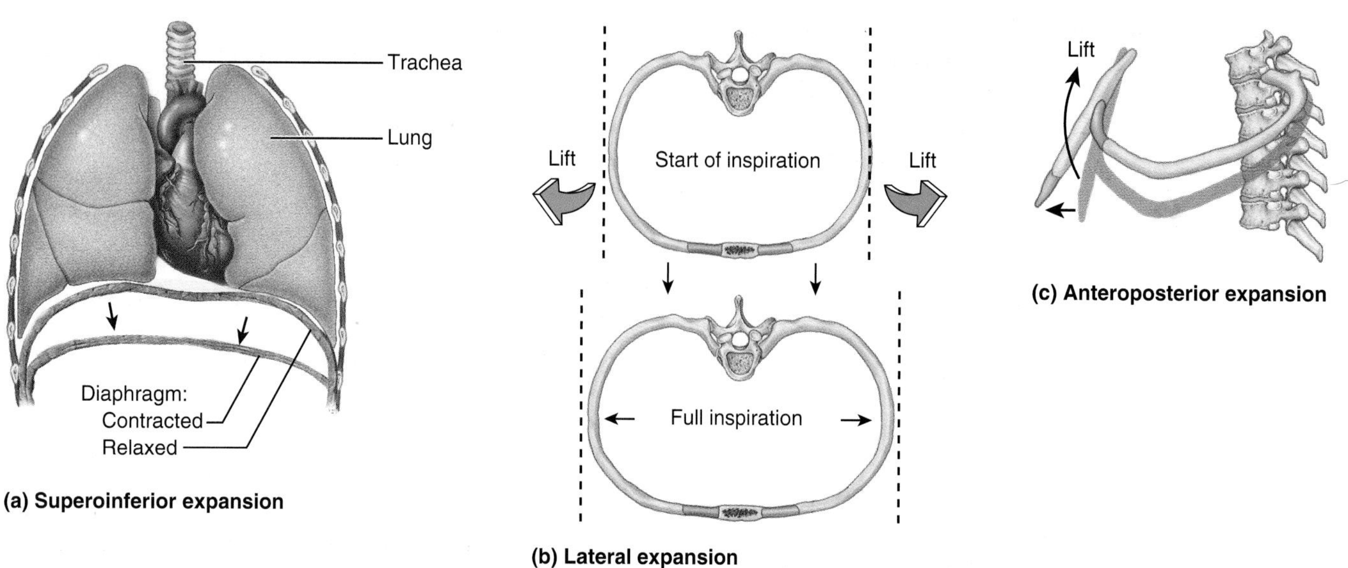

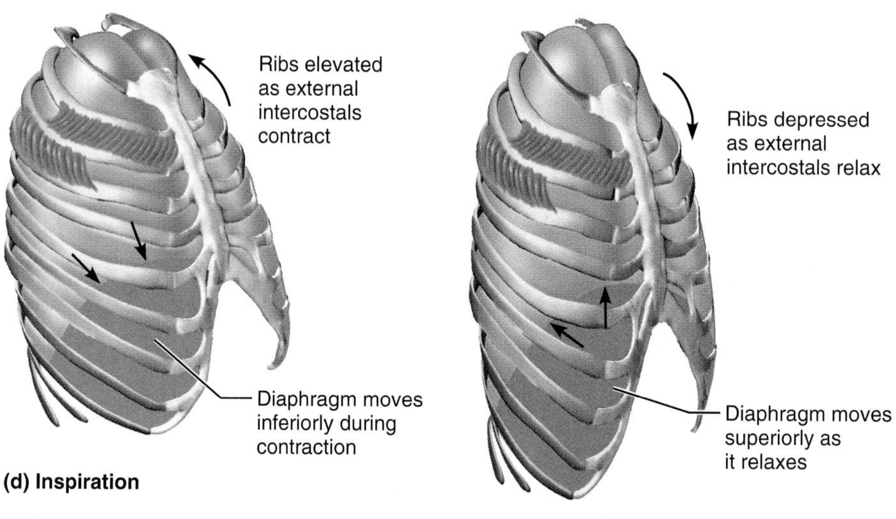

FIGURE 21.15 Changes in thoracic volume during breathing. (a)–(c) Ways in which the volume of the thorax increases during inspiration. **(a)** The diaphragm descends as it contracts, increasing the superior-to-inferior dimension of the thorax. **(b)** The external intercostal muscles raise the ribs, causing the thorax to expand laterally. **(c)** As the ribs are raised, the anteroposterior dimension of the thorax also increases. **(d)** Lateral view of the thorax during inspiration (enlarged thoracic volume) and during expiration (decreased thoracic volume).

1. **Action of the diaphragm.** When the dome-shaped diaphragm contracts, it moves inferiorly and flattens (called "superoinferior expansion" in Figure 21.15a). As a result, the vertical dimension (height) of the thoracic cavity increases.
2. **Action of the intercostal muscles.** The external intercostal muscles contract to raise the ribs. Because the ribs normally extend anteroinferiorly from the vertebral column, lifting them enlarges both the right-to-left dimension of the thoracic cavity ("lateral expansion" in Figure 21.15b) and the anterior-to-posterior diameter of the thorax ("anteroposterior expansion" in Figure 21.15c). Although these actions expand the thoracic dimensions by only a few millimeters along each plane, this movement is sufficient to increase the volume of the thoracic cavity by almost half a liter (a pint)—the amount of air that typically enters the lungs during a normal quiet inspiration.

During deep or forced inspiration, additional muscles contract and further increase thoracic volume. The rib cage is elevated by the scalenes and sternocleidomastoid muscle of the neck, and by the pectoralis minor of the chest. Additionally, the back extends as the thoracic curvature is straightened by the erector spinae muscles.

Expiration

Quiet **expiration** in healthy people is chiefly a passive process. As the inspiratory muscles relax, the rib cage drops under the force of gravity and the relaxing diaphragm moves superiorly. At the same time, the many elastic fibers within the lungs recoil. The result is that the volumes of the thorax and lungs decrease simultaneously, which increases the pressure within the lungs and pushes the air out.

By contrast, *forced* expiration is an active process produced by contraction of muscles in the abdominal wall, primarily the oblique and transversus abdominis muscles. These contractions (1) increase the intra-abdominal pressure, which forces the diaphragm superiorly, and (2) sharply depress the rib cage and so decrease thoracic volume. The internal intercostal muscles, quadratus lumborum, and the latissimus dorsi also help to depress the rib cage.

CLINICAL APPLICATIONS

COLLAPSED LUNG A lung will collapse if air enters the pleural cavity, a condition called **pneumothorax** (nu″mo-tho′raks; "air thorax"). The air breaks the seal of pleural fluid that holds the lung to the thoracic wall, allowing the elastic lung to collapse like a deflating balloon. Pneumothorax usually results from chest trauma or overexertion that raises intrathoracic pressure such that a lung pops like one pops a blown-up paper bag; it can also result from a wound that penetrates the thoracic wall or from a disease process that erodes a hole through the external surface of the lung. Pneumothorax is reversed surgically by closing the "hole" through which air enters the pleural cavity and then gradually withdrawing the air from this cavity using chest tubes. This treatment allows the lung to reinflate and resume its normal function.

Obstruction of a bronchus by a plug of mucus, an inhaled object, a tumor, or enlarged lymph nodes may also cause a lung to collapse; the collapse occurs as the air beyond the point of blockage is gradually reabsorbed into the pulmonary capillaries. Finally, pleural effusion (see p. 617) can cause collapse as the accumulating fluid compresses the lung.

Because the lungs are in completely separate pleural cavities, one lung can collapse without affecting the function of the other. ✚

In a healthy lung, the alveoli remain open at all times and do not collapse during exhalation. At first glance this seems to contradict the laws of physics, because a watery film coats the internal surfaces of the alveoli, and water molecules have a high attraction for one another (called *surface tension*) that should collapse the alveoli after each breath. Collapse does not occur because the alveolar film also contains **surfactant** (ser-fak′tant), detergent-like molecules, secreted by the type II alveolar cells, that interfere with the attraction between water molecules, thereby reducing surface tension and enabling the alveoli to remain open.

CLINICAL APPLICATIONS

RESPIRATORY DISTRESS SYNDROME Because pulmonary surfactant is not produced until the end of fetal life, its absence can have dire consequences for infants born prematurely. In such infants, the alveoli collapse during exhalation and must be completely reinflated during each inspiration, an effort that requires a tremendous expenditure of energy that can lead to exhaustion and respiratory failure. This condition, called **respiratory distress syndrome (RDS)**, is responsible for one-third of all infant deaths. It is treated by using positive-pressure respirators to force air into the alveoli and keep them inflated between breaths, and by administering natural or artificial surfactants. Even though such *surfactant therapy* has saved many lives since its introduction in 1990, many survivors still suffer from *chronic lung disease (bronchopulmonary dysplasia)* throughout childhood and beyond. This condition results from inflammatory injury to the respiratory zone, possibly caused by the high pressures the respirator must place on the delicate lungs in order to distribute surfactant evenly. *Liquid ventilation*—having the respirator deliver an oxygen-carrying fluid called perfluorocarbon instead of air—could solve all these problems in the near future. ✚

Neural Control of Ventilation

The brain's most important respiratory center is in the reticular formation of the medulla oblongata. This center, called RVLM (rostral ventrolateral medulla) or the pre-Bötzinger complex, is a pacemaker whose neurons generate the basic ventilatory rhythm and rate (with input, perhaps, from centers in the pons and dorsal medulla). This basic pattern can be modified by higher centers of the brain, such as the limbic system and hypothalamus (by which emotions influence breathing—as when we gasp), and the cerebral cortex (through which we have conscious control over the rate and depth of breathing).

Although the medulla's respiratory center sets a baseline ventilatory rate, this rate is also modified by input from receptors that sense the chemistry of blood. These chemoreceptors respond to falling concentrations of oxygen, rising levels of carbon dioxide, or increased acidity of the blood by signaling the respiratory center to increase the rate and depth of breathing, which returns the blood gases to their normal concentrations. The chemoreceptors are of two types: *central chemoreceptors,* which are mainly in the medulla, and *peripheral chemoreceptors,* which are the **aortic bodies** on the aortic arch and the **carotid bodies** in the fork of the common carotid artery (Figure 21.16). The aortic bodies send their sensory information to the medulla through the vagus nerves, whereas the carotid bodies send theirs through the glossopharyngeal (and perhaps the vagus) nerves. For more information on the peripheral chemoreceptors, see Table 14.3 on pp. 424–425.

DISORDERS OF THE RESPIRATORY SYSTEM

Because the respiratory system is open to the outside world and exposed to large volumes of air, it is highly susceptible to airborne microorganisms, pollutants, and irritants, leading to a remarkable number of common respiratory disorders. One of the most lethal respiratory diseases, lung cancer, is given special consideration in this chapter's A Closer Look box on p. 626. In this section, we consider some other common respiratory disorders, beginning with the more-serious group that primarily involves the *lower* respiratory structures *(bronchial asthma, chronic obstructive pulmonary disease, pneumonia, tuberculosis,* and *cystic fibrosis)* and

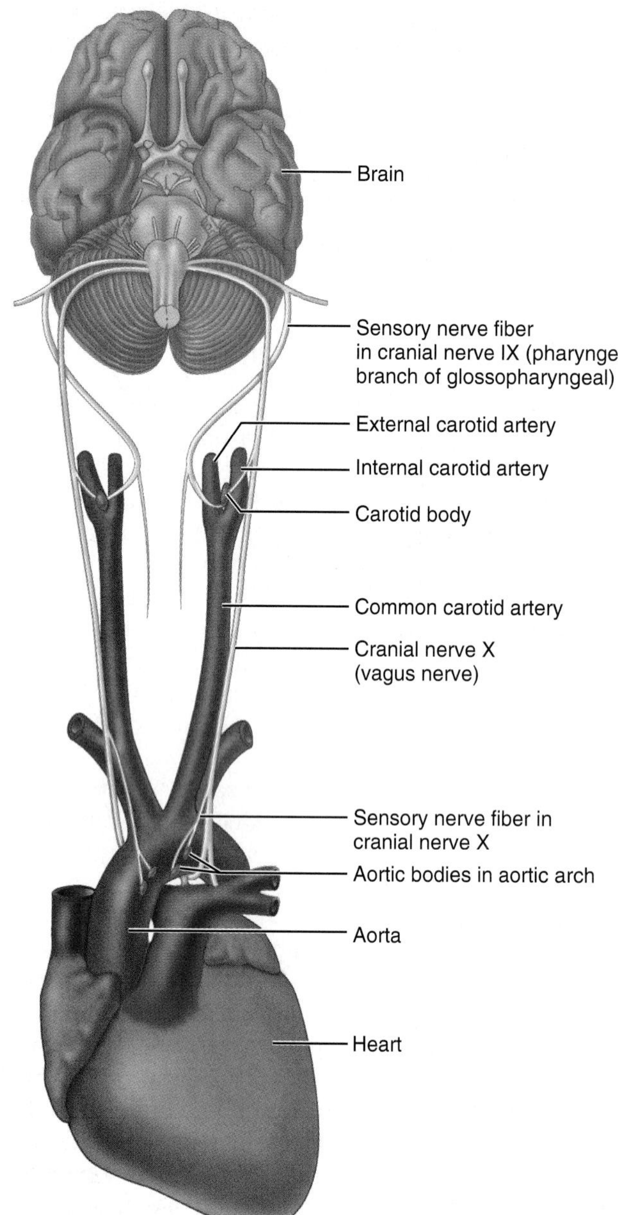

FIGURE 21.16 Location of the peripheral chemoreceptors in the carotid and aortic bodies. Also shown are the sensory pathways from these receptors through cranial nerves IX and X to the respiratory center in the medulla.

then proceeding to disorders of the *upper* respiratory structures *(epistaxis* and *epiglottitis).*

Disorders of Lower Respiratory Structures

Bronchial Asthma

Bronchial asthma, a respiratory condition that affects about 7% of adults and 10% of children and is increasing in frequency, is a type of allergic inflammation in people who are hypersensitive to irritants in the air or to stress. Attacks may be triggered by inhaling substances to which the sufferer is allergic (dust mites, pollen, molds, bits of cockroach), by inhaling dust or smoke, by respiratory infections, by emotional upset, or by the mild shock of breathing cold air. Symptoms of asthma attacks include coughing, wheezing, and shortness of breath.

An asthma attack starts with an early phase, in which mast cells (p. 91) stimulate both contraction of the bronchial smooth musculature (bronchoconstriction) and secretion of mucus in these airways. Within several hours, a late phase develops as eosinophils, neutrophils, a certain type of helper T lymphocytes, and basophils accumulate in the bronchi and bronchioles, where these leukocytes secrete additional inflammatory chemicals that damage the bronchial mucosa and increase the severity of bronchoconstriction and mucous secretion.

Until about 1990, bronchoconstriction was considered the primary symptom of asthma, and quick relief is still provided by drugs that either inhibit smooth-muscle contraction (bronchodilators) or inhibit the parasympathetic stimulation of such contraction (anticholinesterases). Recently, the realization that asthma is primarily an inflammatory disease has led to new treatments using anti-inflammatory drugs (glucocorticoids [p. 756] and nonsteroidal anti-inflammatories). These drugs allow much better long-term management, leading to fewer asthma attacks and less damage to the airways.

Chronic Obstructive Pulmonary Disease

Chronic obstructive pulmonary disease (COPD) is a category of disorders in which airflow in and out of the lungs is difficult, or obstructed. It mostly refers to *obstructive emphysema* or *chronic bronchitis* (or both of these occurring together), and is a major cause of death and disability in the United States. These two diseases share certain features: The patients almost always have a history of smoking; they experience difficult or labored breathing called **dyspnea** (disp-ne′ah; "bad breathing"); coughing and pulmonary infections occur frequently; and most COPD victims ultimately develop respiratory failure.

Obstructive emphysema (em″fĭ-se′mah; "to inflate") is characterized by a permanent enlargement of the alveoli accompanied by a deterioration of the alveolar walls (Figure 21.17). The disease is most often associated with a smoking-related chronic inflammation of the lungs and increased activity of lung macrophages, whose lysosomal enzymes seem responsible for destroying the alveolar walls and breaking down elastin. Chronic inflammation also leads to fibrosis (formation of scar tissue), and the lungs become progressively less elastic, making expiration difficult and exhausting. For complex reasons, the bronchioles open during inspiration but collapse during expiration, trapping huge volumes of air in the alveoli. This enlarges the lung, leads to the development of an expanded "barrel chest," and flattens the

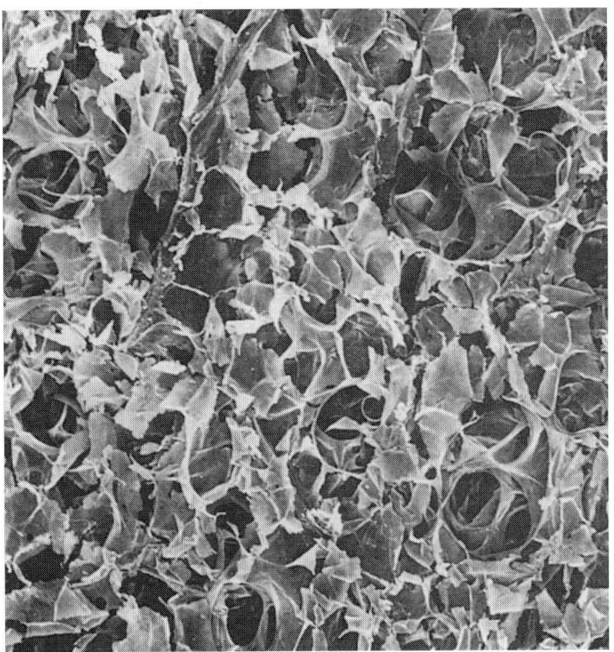

(a)

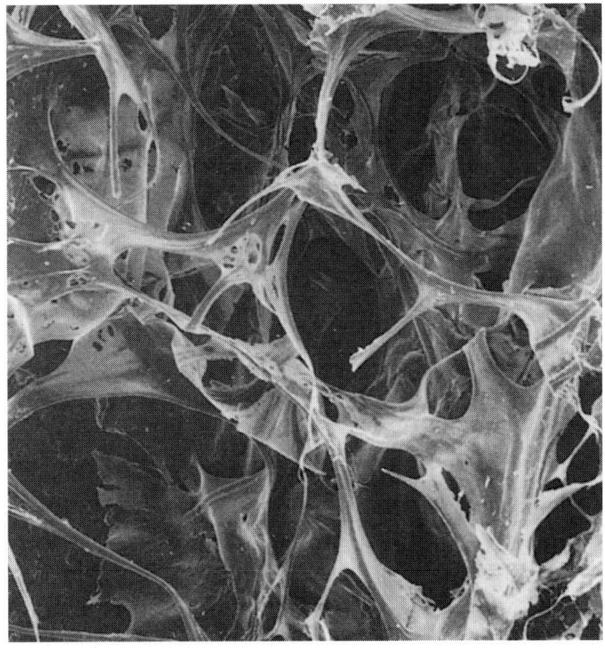

(b)

FIGURE 21.17 Scanning electron micrograph of the alveoli (a) in a normal lung and (b) in a lung with emphysema. Note that the alveoli are very much larger in the emphysematic lung (25 ×).

diaphragm, which decreases ventilatory efficiency. Damage to the lung capillaries increases the resistance in the pulmonary vascular circuit, forcing the heart's right ventricle to enlarge through overwork.

In **chronic bronchitis,** inhaled irritants lead to a prolonged secretion of excess mucus by the mucosa of the lower respiratory passages, and to inflammation and fibrosis of this mucosa. These responses obstruct the airways, severely impairing ventilation and gas exchange. Infections frequently ensue because bacteria and viruses thrive in the stagnant pools of mucus. Coughing is especially persistent and productive. Patients with chronic bronchitis are sometimes called "blue bloaters" because lowered blood oxygenation occurs early in the disease, often resulting in cyanosis. However, the degree of dyspnea is usually moderate compared to that of emphysema sufferers. So significant is cigarette smoking as a causative factor of chronic bronchitis that this disease would be an insignificant health problem if cigarettes were unavailable. Air pollution is another, although minor, causative factor.

COPD is routinely treated with bronchodilators and anti-inflammatory drugs (including glucocorticoids) in aerosol form. Severe dyspnea and lowered uptake of oxygen nearly mandates supplemental oxygen use. A new, controversial treatment introduced in 1994, called *lung volume reduction surgery,* has been performed on some emphysema patients. In this surgery, part of the grossly enlarged lung is removed to give the remaining lung tissue room to expand. At least in the short term, enough breathing capacity is restored (55% of lung function, on average) to allow these patients to live something close to a normal life.

Pneumonia

Pneumonia is infectious inflammation of the lungs in which fluid accumulates in the alveoli. Of the over 50 known varieties, most are caused by viruses or bacteria (however, the type of pneumonia associated with AIDS, *Pneumocystis carinii* pneumonia, is caused by a fungus). Extremely common, pneumonia is the sixth most frequent cause of death in the United States because almost any severely ill person can develop it.

Tuberculosis

Tuberculosis (TB) is a lung disease caused by the bacterium *Mycobacterium tuberculosis,* which is spread by coughing and primarily enters the body in inhaled air. TB typically affects the lungs but can spread through lymph vessels to other organs. A massive inflammatory and immune response contains the primary infection

A CLOSER LOOK

Lung Cancer: The Facts Behind the Smoke Screen

Lung cancer accounts for fully one-third of all cancer deaths in the United States, and its incidence is increasing daily. It is the most prevalent type of malignancy in both sexes, and it has a notoriously low survival rate: The overall 5-year survival rate is just 12%, and the average person survives only about 9 months after diagnosis. The survival rate is low because most types of lung cancer are tremendously aggressive and metastasize rapidly and widely, and because most cases cannot be diagnosed until they are too advanced. (Only recently have new, whole-lung CT scans been shown to detect early-stage tumors well enough to make screening for this cancer worthwhile.)

For years, Americans were remarkably unaware of the link between lung cancer and cigarette smoking, despite the fact that over 90% of lung cancer patients are smokers. Professional athletes still promoted cigarettes in the 1950s, and cigarettes were advertised as a harmless way to keep one's weight down ("Reach for a cigarette rather than a sweet"). As late as the 1960s, smoking was considered socially desirable, even romantic, even though it gives smokers bad breath and yellow teeth, wrinkles their skin, and leaves an unpleasant residue on the clothes and hair of everyone exposed to the smoke. Smoking even a single cigarette increases one's heart rate, constricts peripheral blood vessels throughout the body, disrupts the flow of air in the lungs, and affects one's brain and mood. Long-term smoking contributes to atherosclerosis and heart disease, strokes, cataracts, and early onset of osteoporosis. Furthermore, each year secondhand tobacco smoke causes 3000 lung-cancer deaths among nonsmokers in the United States, and it appears to be linked to heart disease. People who work in bars and restaurants are especially vulnerable.

Ordinarily, sticky mucus and the action of cilia do a fine job of protecting the lungs from chemical and biological irritants, but smoking overwhelms these cleansing devices until they eventually become nonfunctional. Even though continuous irritation prompts the production of more mucus, smoking slows the movements of cilia that clear this mucus, and it also depresses the activity of lung macrophages. One result is a pooling of mucus in the lower respiratory tree and an increased frequency of pulmonary infections, including pneumonia and COPD. However, it is the 15 or so carcinogens in tobacco smoke that ultimately lead to lung cancer. The worst of these is a chemical called nitrosamine, but the tars in tobacco also contain carcinogens that eventually cause the epithelial cells lining the bronchial tree to proliferate wildly and lose their characteristic structure.

The three most common types of lung cancer are (1) **squamous cell carcinoma** (20% to 40% of cases), which arises in the epithelium of the larger bronchi and tends to form masses that cavitate and bleed; (2) **adenocarcinoma** (25% to 35%), which originates in the peripheral areas of the lung as solitary nodules that develop from bronchial mucous glands and alveolar epithelial cells; and (3) **small cell carcinoma** (20% of cases, but rapidly increasing), which contains lymphocyte-like epithelial cells that originate in the main bronchi and grow aggressively in cords or small grapelike clusters within the mediastinum, from which site metastasis is especially rapid.

The most effective treatment for lung cancer is complete removal of the diseased lung in an attempt to halt metastasis. However, removal is an option open to very few lung cancer patients because metastasis has usually occurred by the time of diagnosis, and most patients' chances of survival are too poor to justify the surgery. In most cases, radiation therapy and chemotherapy are the only options, but most lung cancers are resistant to these treatments. Only small cell carcinoma

within fibrous or calcified nodules in the lungs (tubercles), but the bacteria often survive, break out, and cause repeated infections. Symptoms of TB are coughing, weight loss, mild fever, and chest pain. People with AIDS are at increased risk for contracting tuberculosis. Some strains of TB are now resistant to antibiotics, and TB cases are increasing in the United States. Current vaccines protect more than half of the children who receive them, but are not effective in adults. Fully one-third of the world's population is infected by the TB bacterium, but 90% to 95% of those infected never show symptoms and remain healthy. Still, about 16 million people in the world do have symptomatic TB, and 2 million die from it each year.

Cystic Fibrosis

Cystic fibrosis (CF) is an inherited disease in which the functions of exocrine glands are disrupted throughout the body. Occurring mainly in people of European descent, it kills 1 of every 2400 Americans. CF affects the respiratory system by causing an oversecretion of a viscous mucus by the bronchial glands. This mucus clogs the respiratory passageways and acts as a feeding ground for bacteria, predisposing the child to early death through respiratory infections. Modern antibiotics allow the average CF patient to survive until age 30, an increase in life span of 16 years since 1969. For information on the non-respiratory effects of CF, see p. 667.

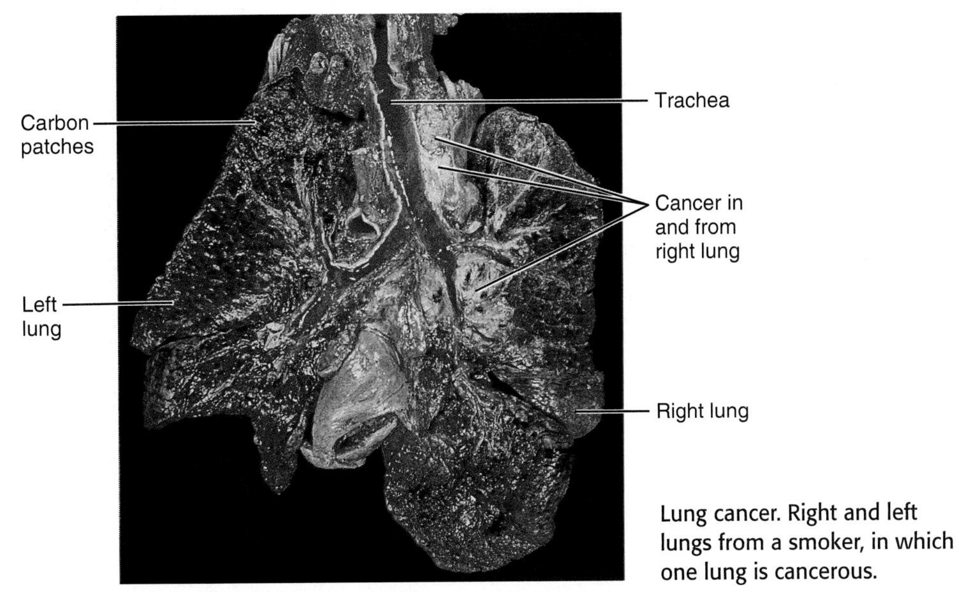

Lung cancer. Right and left lungs from a smoker, in which one lung is cancerous.

responds to chemotherapy, but frequently it returns quickly and spawns brain tumors.

The extreme addictiveness of nicotine in tobacco, recognized by the American Medical Association, the World Health Organization, and the American Psychiatric Association, may explain why over 50 million Americans smoke cigarettes, even though smoking is estimated to contribute to one-fifth of all deaths in the United States. Surprisingly, a recent survey showed that most smokers do not think they are at increased risk for lung cancer or heart disease. To the contrary, a 40-year study of British men, the most complete lung cancer survey ever done, indicates that death rates are three times higher during middle age (35 to 69 years) for long-term smokers than for nonsmokers. Half of all regular smokers were killed by their habit. Only 43% of those who smoked over 25 cigarettes per day lived to age 70, as compared to 79% of non-smokers. These statistics are alarming in light of the fact that more young people are taking up smoking in Great Britain and the United States; the number of American high-school students who smoke increased by 32% between 1991 and 1997.

Fortunately, quitting helps: Whereas the incidence of lung cancer is 20 times greater in smokers than in nonsmokers, this ratio drops to 2 to 1 for ex-smokers who have not smoked in 15 years. In the previously noted British study, those who stopped smoking before age 35 lived full life spans, and quitting at any age prolonged survival. Using such treatments as nicotine patches, nasal sprays, and inhalants has been shown to double the success rate of quitting smoking (from 10% to 20%).

Advances in Understanding In the 1980s, researchers isolated the gene responsible for cystic fibrosis (the *cftr* gene) and showed that a defect in this gene prevents the formation of a membrane channel that carries chloride ions into epithelial cells, such that the fluid covering the respiratory epithelium becomes a salty brine. Recently, it was demonstrated that a natural antibiotic produced by the epithelium is ineffective under such salty conditions, which is why bacteria thrive, leading to inflammation and the excessive mucous secretion. Despite our understanding of the cause of CF, attempts to cure it by delivering nondefective copies of the *cftr* gene to the affected respiratory epithelial cells have not succeeded. ✱

Disorders of Upper Respiratory Structures

Epistaxis

Epistaxis (ep″ĭ-stak′sis; "the act of bleeding at the nose") is a nosebleed, or nasal hemorrhage. It commonly follows trauma to the nose or excessive nose blowing or nose picking. Most nasal bleeding is from the highly vascularized *anterior* part of the nasal septum and can be stopped by pinching the nostrils together or packing them with cotton. Bleeding from the *posterior* part of the nasal cavity, on the other hand, is less common but more serious because it usually involves more bleeding that is harder to stop. Posterior nasal bleeding usually indicates a cardiovascular disease such as high blood pressure or a clotting disorder.

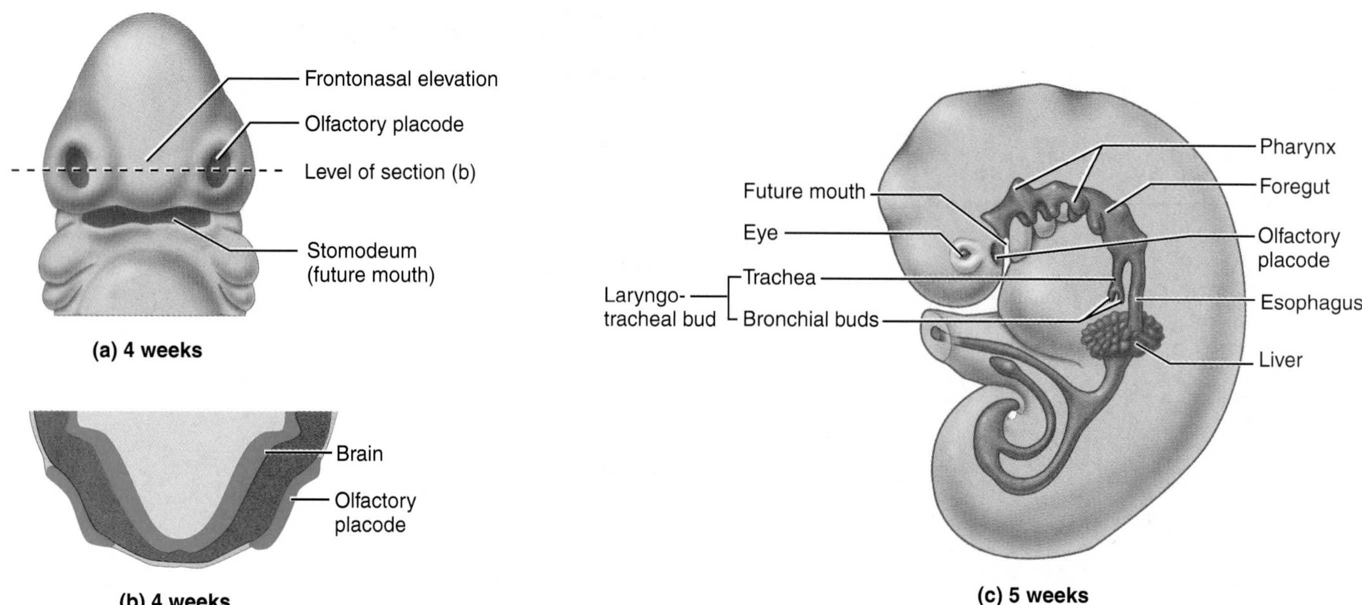

FIGURE 21.18 **Embryonic development of the respiratory system.** **(a)** Ventral view of an embryo's face and head, showing the ectodermal olfactory placodes. **(b)** Transverse section through the region of the dashed line in part (a). Ventral aspect is below. **(c)** Development of the lower respiratory structures. The laryngotracheal bud is shown as an outpocketing from the foregut.

Epiglottitis

Epiglottitis is a rapid inflammation and swelling of the epiglottis, usually occurring in children ages 2 through 7. Resulting from an upper-airway infection by the bacterium *Hemophilus influenzae* and associated with fever, sore throat, and difficulty swallowing, it is a dangerous condition because the swollen larynx can block the laryngopharynx and obstruct breathing. When epiglottitis affects the elderly, it involves some pharyngeal structures along with the epiglottis. Treatments include keeping the airway clear, giving antibiotics, and vaccinating the patient against *H. influenzae*.

THE RESPIRATORY SYSTEM THROUGHOUT LIFE

Because embryos develop in a craniocaudal (head-to-tail) direction, the upper respiratory structures appear before the lower ones. By week 4, a thickened plate of ectoderm called the **olfactory placode** (plak′ōd; "plate") has appeared on each side of the future face (Figure 21.18a and b). These placodes quickly invaginate to form *olfactory pits* that form the nasal cavity, including the olfactory epithelium in its roof. The nasal cavity then connects with the future pharynx of the developing foregut, which forms at the same time.

The lower respiratory organs develop from a tubular outpocketing off the pharyngeal foregut called the **laryngotracheal bud** (Figure 21.18c). The proximal part of this bud forms the trachea, and its distal part branches repeatedly to form the bronchi and their subdivisions, including (eventually) the lung alveoli. The respiratory tubes, like the gut tube from which they arise, are lined by endoderm and covered by splanchnic mesoderm. The endoderm becomes the lining epithelium (and glands) of the trachea, bronchial tree, and alveoli. The splanchnic mesoderm, by contrast, gives rise to all other layers of the tracheal and bronchial walls (including cartilage, smooth muscle, and lamina propria) and to the stroma of the lungs.

The respiratory system reaches functional maturity relatively late in development. No alveoli appear until month 7 (28 weeks), and the alveolar type I cells do not attain their extreme thinness until the time of birth. It is not until 26–30 weeks that a prematurely born baby can survive and breathe on its own. Infants born before this time are most severely threatened by respiratory distress syndrome resulting from inadequate production of surfactant (see p. 623).

During fetal life, the lungs are filled with fluid, and all respiratory exchange occurs across the placenta. At birth, the first breaths bring air into the lungs, and the alveoli inflate and begin to function in gas exchange. During the nearly 2 weeks it takes for the lungs to become fully inflated, the fluid that originally filled the lungs is absorbed into the alveolar capillaries.

At birth, only one-sixth of the final number of lung alveoli are present. The lungs continue to mature throughout childhood, and more alveoli are formed until young adulthood. Research has revealed that in individuals who begin smoking in their early teens, the lungs never completely mature, and those additional alveoli never form.

Under normal conditions, the respiratory system works so efficiently that we are not even aware of it. Most problems that occur are the result of external factors, such as viral or bacterial infections, irritants that trigger asthma in susceptible individuals, or obstruction of the trachea by a piece of food. For many years, tuberculosis and bacterial pneumonia were the worst killers in the United States and Europe; even though antibiotics have decreased their threat to a large extent, they are still dangerous diseases. Influenza, a respiratory flu virus, killed at least 20 million people worldwide in 1918, and each year government health agencies keep a watchful eye on outbreaks caused by new viral strains. By far the most troublesome respiratory disorders, however, are COPD and lung cancer.

As we age, the number of glands in the nasal mucosa declines, as does the flow of blood in this mucosa. Thus, the nose dries and crusts and produces a thickened mucus that makes one want to "clear the throat." In the elderly, the thoracic wall grows more rigid and the lungs gradually lose their elasticity, resulting in a slowly declining ability to ventilate the lungs. The levels of oxygen in the blood may fall slightly while levels of carbon dioxide rise. Additionally, just as the overall efficiency of the immune system declines with age, many of the respiratory system's protective mechanisms become less effective—the activity of the cilia in its epithelial lining decreases, and the macrophages in the lungs become sluggish. The net result is that the elderly are at greater risk for respiratory infections, particularly pneumonia and influenza.

RELATED CLINICAL TERMS

Adenoidectomy (adenotonsillectomy) Surgical removal of an infected pharyngeal tonsil (adenoids).

Atelectasis (at″ĕ-lek′tah-sis) Collapse of the lung, either from airway obstruction or from the compression of pleural effusion (see p. 617). Can also refer to a failure of the lungs to inflate, as in premature infants.

Bronchoscopy (*scopy* = viewing) Use of a viewing tube to examine the internal surface of the main bronchi in the lung. The tube is inserted through the nose or mouth and guided inferiorly through the larynx and trachea. Forceps may be attached to the tip of the tube to remove trapped objects, take biopsy samples, or retrieve samples of mucus for examination.

Croup Disease in children in which viral-induced inflammation causes the air passageways to narrow; is characterized by coughing that sounds like the bark of a dog, hoarseness, and wheezing or grunting sounds during inspiration. Most cases resolve after a few days, but severe cases may require anti-inflammatory aerosols or a tracheotomy (see below) to bypass the obstructed upper respiratory tubes.

Deviated septum Condition in which the nasal septum takes a more lateral course than normal and may obstruct breathing. Deviated septum is often manifest in old age but may also be congenital or result from a blow to the nose.

Sudden infant death syndrome (SIDS) Unexpected death of an apparently healthy infant during sleep. Commonly called crib death, SIDS is the most frequent cause of death in infants under 1 year old. The cause is unknown, but it may reflect immaturity of the brain's respiratory control centers. Since 1992, a campaign to have babies sleep on their backs instead of on their bellies has led to a decline of 40% or more in the incidence of SIDS in the United States.

Tracheotomy (tracheostomy) A surgical technique for opening an airway when the upper respiratory tract is obstructed. In a tracheotomy, the surgeon makes a vertical incision between the second and third tracheal rings in the anterior neck; a tube is then inserted to keep the airway open. In the emergency procedure called cricothyroid trachotomy, a breathing tube is inserted through an incision made through the cricothyroid ligament of the larynx (see Figure 21.5a).

CHAPTER SUMMARY

Media study tools that could provide you additional help in reviewing specific key topics of Chapter 21 are referenced below.

SP = Study Partner; AIA = A.D.A.M.® Interactive Anatomy

1. Respiration involves four processes: pulmonary ventilation, external respiration, cardiovascular transport of respiratory gases, and internal respiration. Both the respiratory system and the cardiovascular system participate in respiration.

FUNCTIONAL ANATOMY OF THE RESPIRATORY SYSTEM (pp. 604–621)

1. The organs of the respiratory system include a conducting zone (nose to terminal bronchioles), which warms, moistens, and filters the inhaled air, and a respiratory zone (respiratory bronchioles to alveoli), where gas exchange occurs.

The Nose and the Paranasal Sinuses (pp. 604–608)

2. The nose and nasal cavity provide an airway for respiration and house the olfactory receptors.
3. The external nose is shaped by bone and by cartilage plates. The nasal cavity begins at the external nares and ends posteriorly at the posterior nasal apertures (choanae). Air-swirling conchae occupy its lateral walls. The paranasal sinuses drain into the nasal cavity.

AIA Link: Atlas; Region: Head & Neck; Paranasal Sinuses (anterior/inferior).

4. The mucosa that lines the nasal cavity and paranasal sinuses processes inspired air. Its epithelium, a pseudostratified ciliated columnar epithelium with goblet cells, is covered with a sheet of mucus that filters dust particles and moistens inhaled air. Its lamina propria contains glands that contribute to the mucus sheet and blood vessels that warm the air.

The Pharynx (pp. 608–609)

5. The pharynx has three regions: The nasopharynx (behind the nasal cavity) is an air passageway, whereas the oropharynx (behind the mouth) and the laryngopharynx (behind the larynx) are passageways for both food and air.

AIA Link: Dissectible anatomy; Male (medial); layer 0; scroll to Head and Neck Region.

6. The soft palate flips superiorly to seal off the nasopharynx during swallowing. The tubal and pharyngeal tonsils occupy the mucosa that lines the nasopharynx. The oropharynx contains the lingual and palatine tonsils. The laryngopharynx opens into the laryngeal inlet anteriorly and into the esophagus inferiorly.

The Larynx (pp. 609–612)

7. The larynx, or voice box, is the entryway to the trachea and lower respiratory tubes. The larynx has a skeleton of nine cartilages: the thyroid; the cricoid; the paired arytenoid, corniculate, and cuneiform cartilages; and the epiglottis.

8. The leaf-shaped epiglottis acts as a lid that prevents food or liquids from entering the lower respiratory channels. The larynx moves superiorly under this protective flap during swallowing.

AIA Link: Atlas; System: Respiratory; Laryngeal Muscles (anterior).

9. The larynx contains the vocal cords (vocal folds). Exhaled air causes these folds to vibrate, producing the sounds of speech. The vocal folds extend anteriorly from the arytenoids to the thyroid cartilage, and muscles that move the arytenoids change the tension on the folds to vary the pitch of the voice. The pharynx, nasal cavity, lips, and tongue also aid in vocal articulation.

AIA Link: Atlas; System: Respiratory; Dissection of Larynx (superior).

10. The main sensory and motor nerves of the larynx (and of speech) are the left and right recurrent laryngeal nerves, which are branches of the vagus nerves.

The Trachea (p. 612)

11. The trachea, which extends from the larynx in the neck to the main bronchi in the thorax, is a tube reinforced by C-shaped cartilage rings, which keep it open. The trachealis muscle narrows the trachea, increasing the speed of airflow during coughing and sneezing.

AIA Link: 3D Anatomy; 3D Lungs: Tracheal Ring Cartilage.

12. The wall of the trachea contains several layers: a mucous membrane, a submucosa, and an outer adventitia. The mucous membrane consists of an air-filtering pseudostratified ciliated columnar epithelium and a lamina propria rich in elastic fibers.

The Bronchi and Subdivisions: The Bronchial Tree (pp. 613–615)

13. The right and left main (primary) bronchi supply the lungs. Within the lungs, they subdivide into lobar (secondary) bronchi to the lobes, segmental (tertiary) bronchi to the bronchopulmonary segments, and finally to bronchioles and terminal bronchioles.

AIA Link: Atlas; System: Respiratory; Bronchial tree (anterior).

14. As the respiratory tubes become smaller, the cartilage in their walls is reduced and finally lost (in bronchioles); the epithelium thins and loses its air-filtering function (in the smallest bronchioles); smooth muscle becomes increasingly important; and elastic fibers continue to surround all of the tubes.

15. The terminal bronchioles lead into the respiratory zone, which consists of respiratory bronchioles, alveolar ducts, and alveolar sacs, all of which contain tiny chambers called alveoli. Gas exchange occurs in the alveoli, across the thin respiratory membrane. This air-blood barrier consists of alveolar type I cells, fused basal laminae, and capillary endothelial cells.

16. Alveoli also contain alveolar type II cells (which secrete surfactant) and freely wandering alveolar macrophages (dust cells) that remove dust particles that reach the alveoli.

SP Exercise: Chapter 21, Anatomy of the Respiratory Membrane.

The Lungs and Pleurae (pp. 615–621)

17. Pleurae are serous membranes. The parietal pleura lines the thoracic wall, diaphragm, and mediastinum; the visceral pleura covers the external surfaces of the lungs. The slit-like pleural cavity between the parietal and visceral pleura is filled with a serous fluid that holds the lungs to the thorax wall and reduces friction during breathing movements.

18. The lungs flank the mediastinum in the thoracic cavity. Each lung is suspended in a pleural cavity by its root (the vessels supplying it) and has a base, an apex, and medial and costal surfaces. The root structures enter the lung hilus.

AIA Link: 3D Anatomy; 3D Lungs: Oblique Fissure of Right Lung.

19. The right lung has three lobes (upper, middle, and lower lobes, as defined by the oblique and horizontal fissures), whereas the left lung has two lobes (upper and lower lobes, as defined by the oblique fissure). The lobes are divided into bronchopulmonary segments, which are supplied by segmental bronchi.

20. The lungs consist primarily of air tubes and alveoli, but they also contain a stroma of elastic connective tissue.

21. The pulmonary arteries carry deoxygenated blood and generally lie posterior to their corresponding bronchi within the lungs. The pulmonary veins, which carry oxygenated blood, tend to lie anterior to their corresponding bronchi. The lungs are innervated by the pulmonary plexus.

VENTILATION (pp. 621–623)

The Mechanism of Ventilation (pp. 621–623)

1. Ventilation consists of inspiration and expiration. Inspiration occurs when the diaphragm and intercostal muscles contract, increasing the volume of the thorax. As the intrathoracic pressure drops, air rushes into the lungs.

2. Expiration is largely passive, occurring as the inspiratory muscles relax. The volume of the thorax decreases and the lungs recoil elastically, raising pressure and forcing air out of the lungs.

3. Surface tension of the alveolar fluid threatens to collapse the alveoli after each breath. This tendency is resisted by surfactant secreted by alveolar type II cells.

Neural Control of Ventilation (p.623)

4. The main respiratory-control center is in the reticular formation of the medulla oblongata, although this center is influenced by input from the limbic system, hypothalamus, and cerebral cortex.

5. Chemoreceptors monitor concentrations of respiratory gases and acid in the blood. The central chemoreceptors are neurons in the medulla, and the peripheral chemoreceptors are the carotid and aortic bodies.

DISORDERS OF THE RESPIRATORY SYSTEM (pp. 623–628)

1. Some major respiratory disorders are bronchial asthma, COPD (obstructive emphysema and chronic bronchitis), lung cancer, and the respiratory effects of cystic fibrosis.

2. Bronchial asthma is a chronic, allergic inflammatory condition.

3. Obstructive emphysema is characterized by permanent enlargement and destruction of alveoli. The lungs lose elasticity, and expiration becomes an active, exhaustive process.

4. Chronic bronchitis is characterized by an excessive production of mucus in the lower respiratory passages, which impairs ventilation and leads to infection. Patients may become cyanotic as a result of low oxygen levels in the blood.

5. Most lung cancers are extremely aggressive and metastasize rapidly. The 5-year survival rate is less than 10%.

6. Cystic fibrosis is an inherited disease in which the functions of the body's exocrine glands are disrupted. It involves an oversecretion of mucus in the bronchi, in which bacteria grow, leading to early death through respiratory infection.

7. Epistaxis (nosebleed) and epiglottitis (inflammation of the epiglottis) are also discussed in this section.

THE RESPIRATORY SYSTEM THROUGHOUT LIFE (pp. 628–629)

1. The nasal cavity develops from the olfactory placodes; the pharynx forms as part of the foregut; and the lower respiratory tubes grow from an outpocketing of the embryonic pharynx (laryngotracheal bud). The epithelium lining the lower respiratory tubes derives from endoderm. Mesoderm forms all other parts of these tubes and the lung stroma as well.

2. The respiratory system completes its development very late in the prenatal period: No alveoli or surfactant appear until month 7, and only one-sixth of the final number of alveoli are present at birth. Respiratory immaturity, especially a lack of surfactant, is the main cause of death of premature infants.

3. With age, the nose dries, the thorax becomes more rigid, the lungs become less elastic, and ventilation capacity declines. Respiratory infections become more common in old age.

REVIEW QUESTIONS

Multiple Choice/Matching Questions

1. When the inspiratory muscles contract, (a) only the lateral dimension of the thoracic cavity increases, (b) only the anteroposterior dimension of the thoracic cavity increases, (c) the volume of the thoracic cavity decreases, (d) both the lateral and the anteroposterior dimensions of the thoracic cavity increase, (e) the diaphragm bulges superiorly.

2. The part of the respiratory mucosa that warms the inhaled air is the (a) pseudostratified epithelium, (b) vessels in the lamina propria, (c) alveolar type I cells, (d) cartilage and bone.

3. Which of the following statements about the vocal cords is false? (a) They are the same as the vocal folds. (b) They attach to the arytenoid cartilages. (c) Exhaled air flowing through the glottis vibrates them to produce sound. (d) They are also called the vestibular folds.

4. In both lungs, the surface that is the largest is the (a) costal, (b) mediastinal, (c) inferior (base), (d) superior (apex).

5. Match the proper type of lining epithelium from the key with each of the following respiratory structures.

Key: **(a)** stratified squamous
(b) pseudostratified columnar
(c) simple squamous
(d) stratified columnar

____ **(1)** nasal cavity

____ **(2)** nasopharynx

____ **(3)** laryngopharynx

____ **(4)** trachea and bronchi

____ **(5)** alveoli (type I cells)

6. Match the proper type of air tube (at right) with the lung region (at left) supplied by that air tube and its branches.

Lung region	Air tube
____ **(1)** bronchopulmonary segment	____ **(a)** main bronchus
____ **(2)** lobule	____ **(b)** lobar bronchus
____ **(3)** alveolar ducts and sacs	____ **(c)** segmental bronchus
____ **(4)** whole lung	____ **(d)** large bronchiole
____ **(5)** lobe	____ **(e)** respiratory bronchiole

7. The respiratory membrane (air-blood barrier) consists of (a) alveolar type I cell, basal laminae, endothelial cell; (b) air, connective tissue, lung; (c) type II cell, dust cell, type I cell; (d) pseudostratified epithelium, lamina propria, capillaries.

8. The trachealis muscle and the smooth muscle around the bronchi develop from which of the following embryonic layers? (a) ectoderm, (b) endoderm, (c) mesoderm, (d) neural crest.

9. A serous cell of a gland secretes (a) the slippery serous fluid in the body cavities, (b) mucus, (c) a watery fluid containing enzymes, (d) tissue fluid.

10. The function of alveolar type I cells is to (a) produce surfactant, (b) propel sheets of mucus, (c) phagocytize dust particles, (d) allow rapid diffusion of respiratory gases.

Short Answer Essay Questions

11. Trace the route of exhaled air from an alveolus to the external nares, naming all the structures through which the air passes. Then say which of these structures are in the respiratory zone and which are in the conducting zone.

12. The cilia lining the upper respiratory passages (superior to the larynx) beat inferiorly, whereas the cilia lining the lower respiratory passages (larynx and inferior) beat superiorly. What is the functional "reason" for this difference?

13. What is the function of the abundant elastin fibers that occur in the stroma of the lung and around all respiratory tubes from the trachea through the respiratory tree?

14. (a) Which tonsils lie in the fauces of the oropharynx? (b) Does the lingual tonsil lie near the superior part of the epiglottis? Explain.

15. (a) It is easy to confuse the hilus of the lung with the root of the lung. Define both structures, and contrast hilus and root. (b) Define the parietal pleura and the visceral pleura.

16. The three terms *choanae, conchae,* and *carina* are easily confused. Define each of them, clarifying the differences.

17. Briefly explain the anatomical "reason" why most men have deeper voices than boys or women.

18. Sketch a picture of the right and left lungs in anterior view, showing all the fissures and lung lobes, as well as the cardiac notch.

19. Explain the functions of the pseudostratified epithelium of the respiratory mucosa.

20. What is the function of the dust cells in the lung alveoli?

CRITICAL REASONING + CLINICAL APPLICATION QUESTIONS

1. (a) Two girls in a high school cafeteria were giggling over lunch, and milk accidentally sprayed out of the nostrils of one of them. Explain in anatomical terms why swallowed fluids can sometimes come out the nose. (b) A boy in the same cafeteria then stood on his head to show he could drink milk upside down without any of it entering his nasal cavity or nose. What prevented the milk from flowing downward into his nose?

2. A surgeon had to remove three adjacent bronchopulmonary segments from the left lung of a patient with tuberculosis. Even though almost half of the lung was removed, there was no severe bleeding, and relatively few blood vessels had to be cauterized (closed off). Why was this surgery so easy to perform?

3. While diapering his 1-year-old boy (who puts almost everything in his mouth), Mr. Gregoire could not find one of the small safety pins he had just removed. Two days later, his son developed a cough and became feverish. What probably had happened to the safety pin, and where (anatomically) would you expect to find it?

4. A taxi driver was carried into an emergency room after being knifed once in the left side of the thorax. The diagnosis was pneumothorax and a collapsed lung. (a) Explain exactly why the lung collapsed. (b) Explain why only one lung (not both) collapsed.

5. A man was driving to work when his car was hit broadside by a car that ran a red light. When freed from the wreckage, the man was deeply cyanotic, and his breathing had stopped. His heart was still beating, but his pulse was fast and thready. His head was cocked at a peculiar angle, and the emergency personnel said he seemed to have a fracture at the level of vertebra C_2. (a) How might these findings account for the cessation of breathing? (b) Define cyanosis (see Chapter 5), and indicate why the man was cyanotic.

6. After a four-year-old child had surgery that successfully closed off a patent ductus arteriosus (see p. 578), the child's voice was hoarse, and an examination revealed that many muscles in the left half of the larynx were paralyzed. What had gone wrong during surgery to cause the paralysis?

A.D.A.M.® INTERACTIVE ANATOMY QUESTIONS

1. Link: Atlas; System: Respiratory; Left lung (medial). (a) How many lobes make up the left lung? (b) Name the structure that fits in the depression on the inferior surface of this lung.

2. Link: Atlas; System: Respiratory; Bronchial tree (anterior). (a) Name the two branches that arise from the trachea and enter the right and left lungs. (b) How many lobar bronchi are there in the right lung?

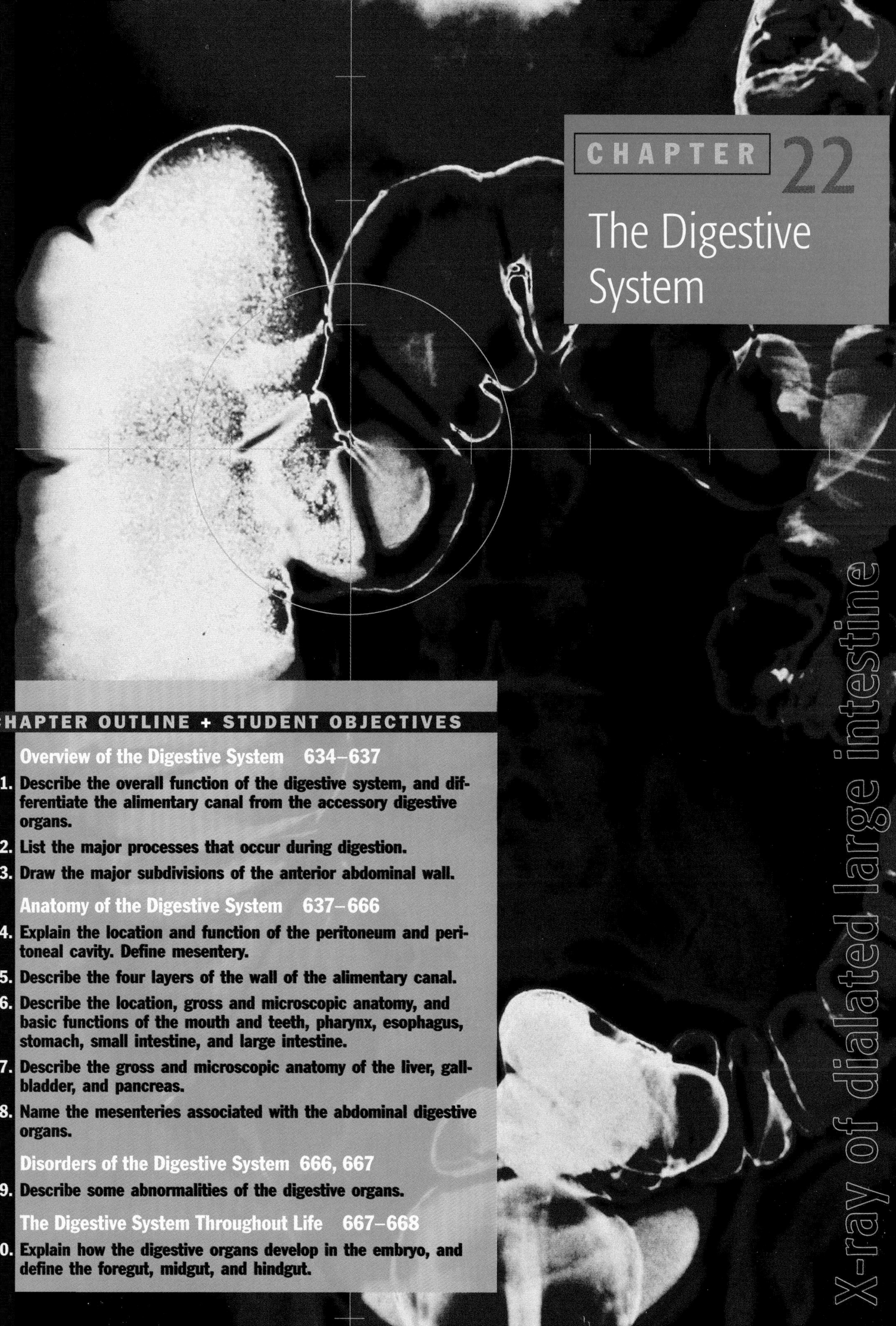

CHAPTER 22

The Digestive System

CHAPTER OUTLINE + STUDENT OBJECTIVES

Overview of the Digestive System 634–637

1. Describe the overall function of the digestive system, and differentiate the alimentary canal from the accessory digestive organs.
2. List the major processes that occur during digestion.
3. Draw the major subdivisions of the anterior abdominal wall.

Anatomy of the Digestive System 637–666

4. Explain the location and function of the peritoneum and peritoneal cavity. Define mesentery.
5. Describe the four layers of the wall of the alimentary canal.
6. Describe the location, gross and microscopic anatomy, and basic functions of the mouth and teeth, pharynx, esophagus, stomach, small intestine, and large intestine.
7. Describe the gross and microscopic anatomy of the liver, gallbladder, and pancreas.
8. Name the mesenteries associated with the abdominal digestive organs.

Disorders of the Digestive System 666, 667

9. Describe some abnormalities of the digestive organs.

The Digestive System Throughout Life 667–668

10. Explain how the digestive organs develop in the embryo, and define the foregut, midgut, and hindgut.

CHILDREN TYPICALLY FEEL A SPECIAL FASCINATION with the workings of the digestive system: They delight in listening to their stomach growl, and some of their earliest questions reveal a tremendous curiosity about bodily wastes. As adults, when we know that a healthy digestive system is essential to the maintenance of life, many of us pay a lot of attention to how the foods we eat affect how we look and feel. In this chapter we explore the structures with which our bodies take in food, break it into nutrient molecules, absorb these molecules into the bloodstream, and then eliminate the indigestible wastes.

OVERVIEW OF THE DIGESTIVE SYSTEM

The various organs of the digestive system (Figure 22.1) can be divided into two main groups; the *alimentary canal* (*aliment* = nourishment) and the *accessory digestive organs.*

The **alimentary canal,** also called the *gastrointestinal (GI) tract,* is the muscular digestive tube that winds through the body. The organs of the alimentary canal are the *mouth, pharynx, esophagus, stomach, small intestine* (small bowel), and *large intestine* (large bowel), the last of which leads to the terminal opening,

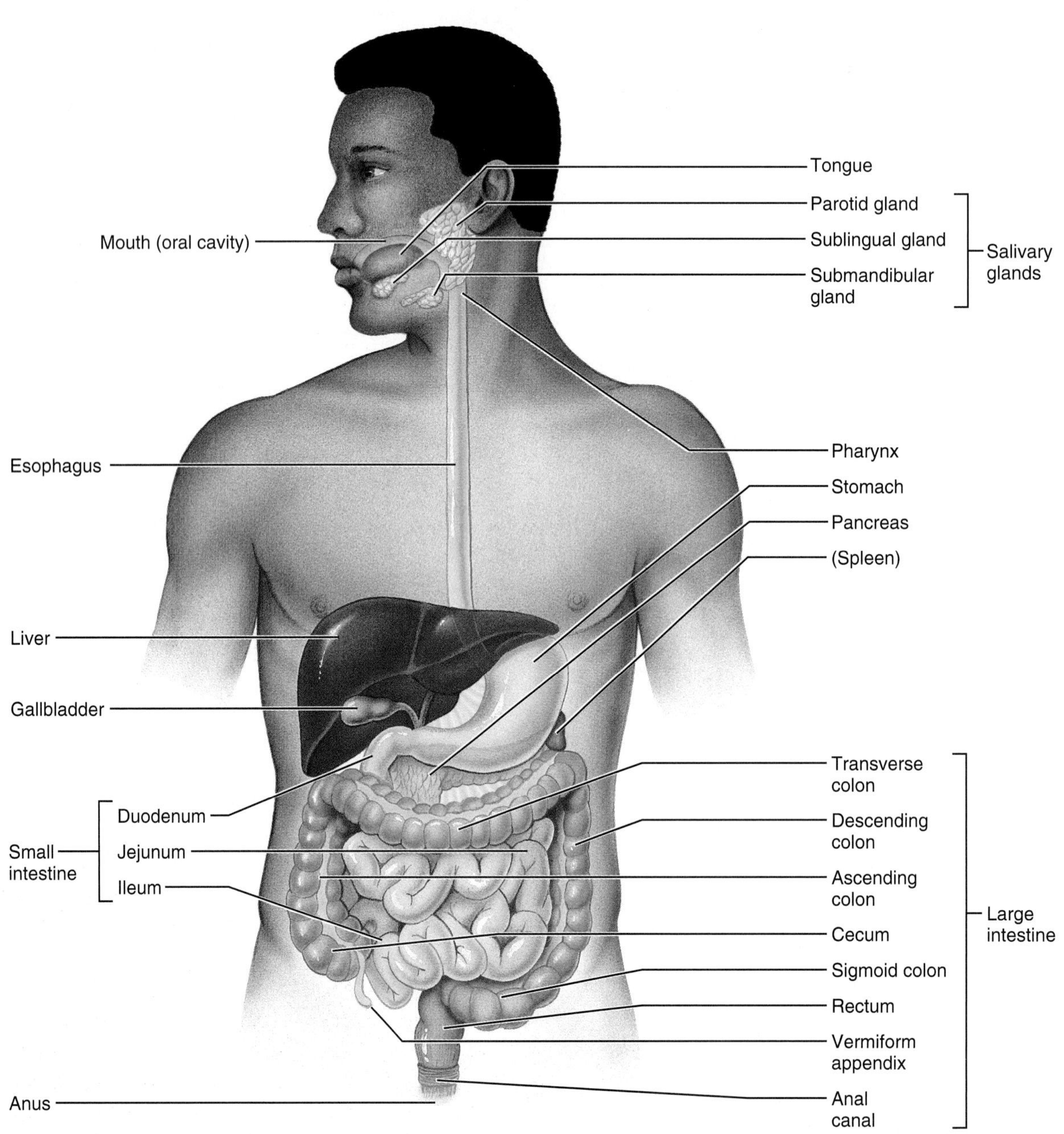

FIGURE 22.1 The alimentary canal and the accessory digestive organs.

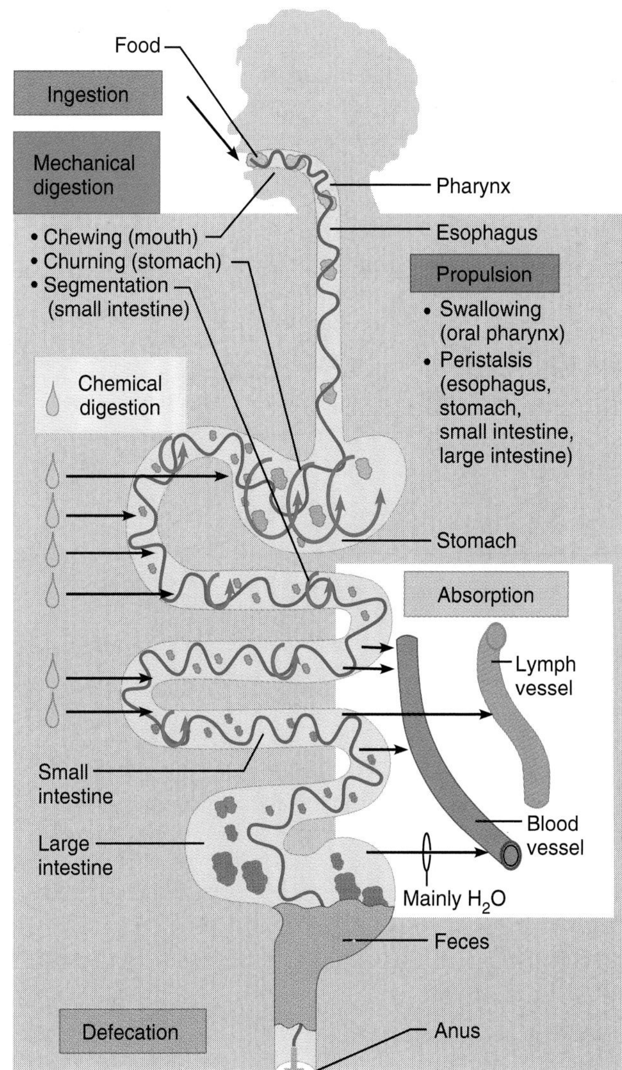

FIGURE 22.2 Schematic summary of digestive processes. The sites at which ingestion, mechanical digestion, propulsion, chemical (enzymatic) digestion, absorption, and defecation occur are indicated. The mucosa of almost the entire alimentary canal secretes mucus, which protects the canal and lubricates its contents.

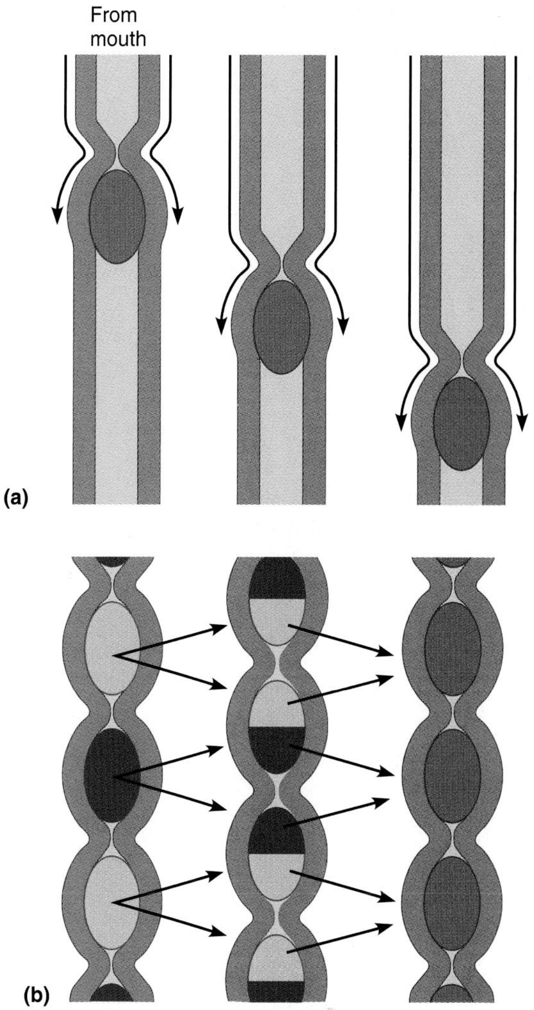

FIGURE 22.3 Peristalsis and segmentation, each shown in a time sequence from left to right. (a) In peristalsis, the major means of propulsion, adjacent segments of the alimentary canal alternately contract and relax, moving food distally along the canal. **(b)** In segmentation, a food-mixing process of mechanical digestion, nonadjacent segments of the intestine alternately contract and relax. Because the active segments are separated by inactive regions, segmentation moves the food onward and then backward, mixing the food rather than propelling it forward.

or *anus.* In a cadaver, the alimentary canal is about 9 m (30 feet) long, but in a living person it is considerably shorter because of its muscle tone. Food material in the alimentary canal is technically considered to be *outside* the body because the canal is open to the external environment at both ends.

The **accessory digestive organs** are the *teeth* and *tongue,* plus the *gallbladder* and various large digestive glands—the *salivary glands, liver,* and *pancreas*—that lie external to and are connected to the alimentary canal by ducts. The accessory digestive glands secrete saliva, bile, and digestive enzymes, all of which contribute to the breakdown of foodstuffs.

Digestive Processes

The organs of the digestive system perform the following six essential food-processing activities: ingestion, propulsion, mechanical digestion, chemical digestion, absorption, and defecation (Figure 22.2).

1. **Ingestion** is the taking of food into the mouth.
2. **Propulsion** is the movement of food through the alimentary canal. It includes swallowing, which is initiated voluntarily, and peristalsis, an involuntary process. *Peristalsis,* the major means of propulsion, involves alternate waves of contraction and relaxation of musculature in the organ walls (Figure 22.3a). Its net effect is to squeeze food from one organ to the next, but some mixing occurs as well.

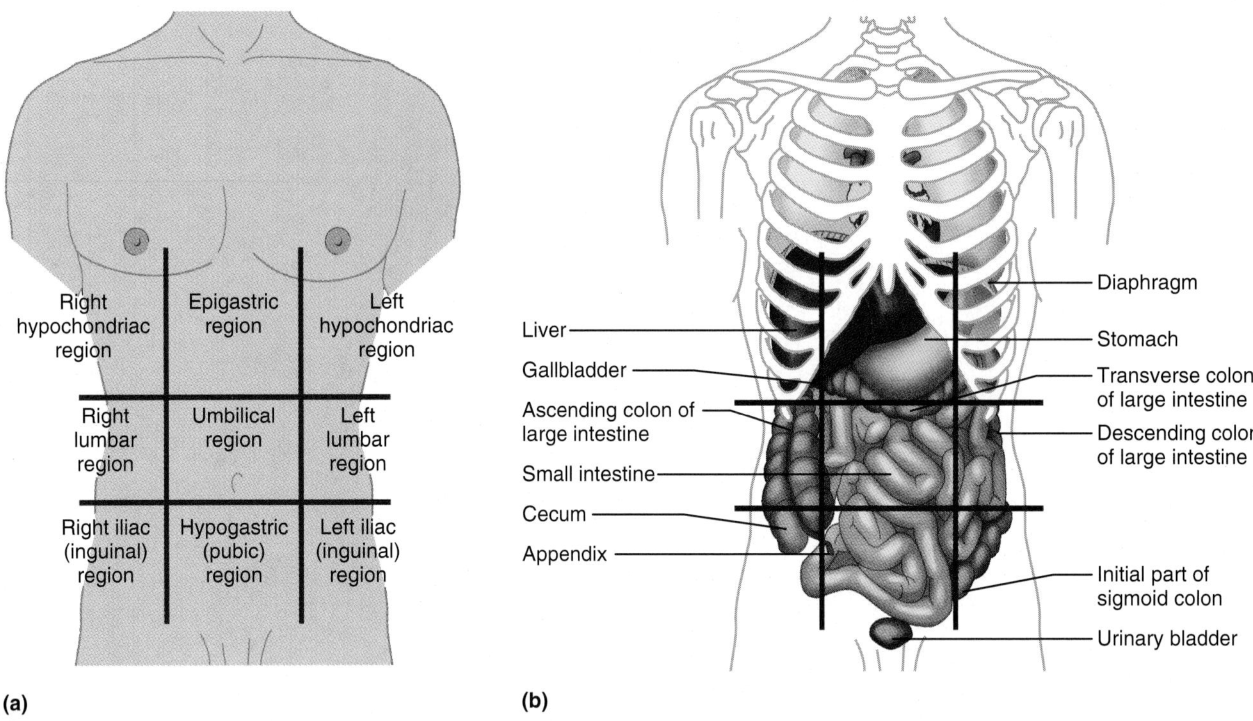

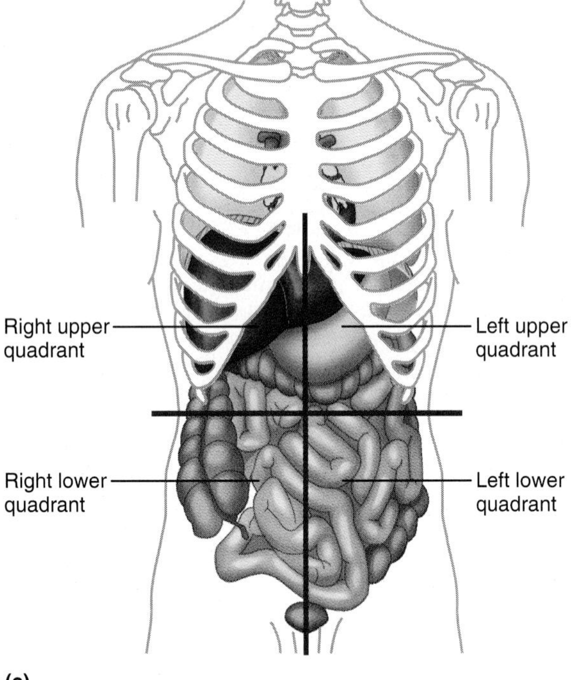

FIGURE 22.4 **Divisions of the anterior abdominal wall for mapping the digestive organs in the abdominopelvic cavity.** **(a)** The nine surface regions of the anterior abdominal wall. **(b)** The abdominal viscera as they relate to the nine surface regions. **(c)** A second, simpler scheme of four quadrants centered at the navel.

3. **Mechanical digestion** physically prepares food for chemical digestion by enzymes. Mechanical processes include chewing, the churning of food in the stomach, and **segmentation,** the rhythmic local constrictions of the intestine (Figure 22.3b). Segmentation mixes food with digestive juices and increases the efficiency of nutrient absorption by repeatedly moving different parts of the food mass over the intestinal wall.
4. **Chemical digestion** is a series of steps in which complex food molecules (carbohydrates, proteins, and lipids) are broken down to their chemical building blocks (simple sugars, amino acids, and fatty acids and glycerol). This process is carried out by enzymes secreted by digestive glands into the lumen of the alimentary canal.
5. **Absorption** is the transport of digested end products from the lumen of the alimentary canal into the blood and lymph capillaries located in the wall of the canal.
6. **Defecation** is the elimination of indigestible substances from the body as feces.

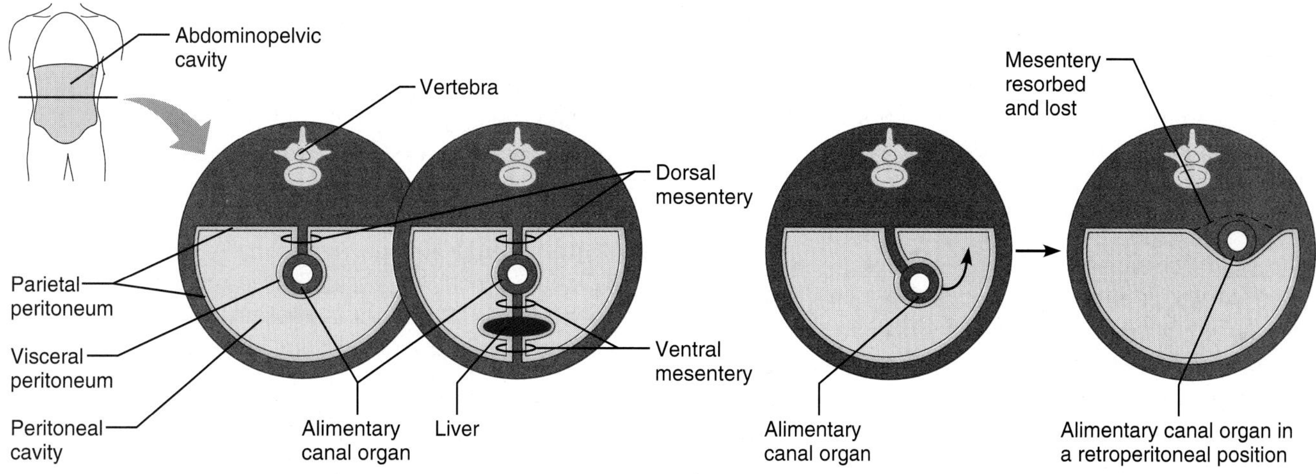

FIGURE 22.5 The peritoneum and peritoneal cavity. **(a)** Simplified schematic cross sections through the abdominopelvic cavity. The peritoneum is depicted as a distinct line around the peritoneal cavity. Note that the peritoneal cavity is actually very much thinner than indicated here—merely a slit between the organs and the body wall. Most regions of the abdominopelvic cavity contain a dorsal mesentery (cross section at left), whereas the superior part of the abdominal cavity also contains a ventral mesentery (cross section at right). **(b)** Some parts of the alimentary canal lose their mesentery during development and become retroperitoneal.

Abdominal Regions and Quadrants

Most digestive organs are contained in the abdominopelvic cavity, the largest division of the ventral body cavity. As an aid to locating the positions of these abdominopelvic organs, clinicians typically use two schemes to divide the anterior abdominal wall into a number of regions.

In one scheme, four lines forming a pattern similar to a tic-tac-toe grid divide the abdominal wall into nine regions (Figure 22.4a and b). The two vertical lines of the grid extend inferiorly from the midpoint of each clavicle. The superior horizontal line is in the *subcostal* ("below the ribs") *plane* and connects the inferior points of the costal margins, whereas the inferior horizontal line is in the *transtubercular plane* and connects the tubercles (widest points) of the iliac crests. (All of these bony landmarks can be felt on the body's surface.) The superior three of the nine regions are the **right** and **left hypochondriac regions** (hi″po-kon′dre-ak; "deep to the ribs") and the central **epigastric region** (ep″ĭ-gas′trik; "superior to the belly"). The middle three regions are the **right** and **left lumbar regions** (or *lateral regions*) and the central **umbilical region.** The inferior three regions are the **right** and **left iliac regions,** or **inguinal regions,** and the central **hypogastric region** ("inferior to the belly") (also called the **pubic region**). Because many abdominal organs move, their positions within the abdominal grid are only approximate.

In a simpler scheme, a vertical and a horizontal line intersecting at the navel (Figure 22.4c) define four regions: the **right** and **left upper** and **lower quadrants.**

ANATOMY OF THE DIGESTIVE SYSTEM

Before we discuss the individual organs of the digestive system, we consider the peritoneal cavity, with which many of these organs associate, and we explore the basic histology of the alimentary canal.

The Peritoneal Cavity and Peritoneum

The digestive organs in the abdominopelvic cavity relate to the *peritoneum* and *peritoneal cavity* (Figure 22.5). Recall from Chapter 1 (see p. 14) that all divisions of the ventral body cavity contain slippery *serous membranes,* the most extensive of which is the **peritoneum** (per″ĭ-to-ne′um) of the abdominopelvic cavity. The **visceral peritoneum,** which covers the external surfaces of most digestive organs, is continuous with the **parietal peritoneum,** which lines the body wall. Between the visceral and parietal peritoneum is the **peritoneal cavity,** a slit-like potential space containing a lubricating serous fluid that is secreted by the peritoneum and allows the organs to glide easily along one another and along the body wall as they move during digestion.

CLINICAL APPLICATIONS

PERITONITIS **Peritonitis** is inflammation and infection of the peritoneum. It can arise from a piercing wound to the abdomen, from a perforating ulcer that leaks stomach juices into the peritoneal cavity, or from poor sterile technique during abdominal surgery. Most commonly, however, it results from a burst appendix that leaks feces into the peritoneal cavity. If it is *generalized* (widespread within the peritoneal cavity), peritonitis is a dangerous condition that can lead to death. Treatment involves rinsing the peritoneum with large amounts of sterile saline solution and giving antibiotics intravenously. ✚

A **mesentery** (mes′en-ter″e) is a double layer of peritoneum—a sheet of two serous membranes fused back to back—that extends to the digestive organs from the body wall (see Figure 22.5a). Mesenteries hold the organs in place, are sites of fat storage, and provide a route by which circulatory vessels and nerves reach the organs in the peritoneal cavity. Mesenteries are described more completely later in the chapter (see p. 663).

Not all digestive organs have a mesentery or are surrounded by the peritoneal cavity on all sides. During embryonic development, some organs (some parts of the intestine, for example) have a mesentery at first but then fuse to the dorsal abdominal wall (see Figure 22.5b), in the process losing their mesentery and lodging behind the peritoneum. Such organs are called **retroperitoneal** (*retro* = behind). By contrast, the digestive organs that keep their mesentery and remain surrounded by the peritoneal cavity are called **intraperitoneal** or **peritoneal** organs. The stomach is an example of such an organ. The specific intraperitoneal and retroperitoneal organs are listed in Table 22.1 on page 666.

It is not completely understood why some digestive organs are located retroperitoneally while others are suspended freely. One explanation is that the fusing of various segments of the alimentary canal to the posterior abdominal wall minimizes the chance that this long tube will knot or kink during peristaltic movements.

Histology of the Alimentary Canal Wall

The walls of the alimentary canal, from the esophagus to the anal canal, have the same four tissue layers (Figure 22.6). In fact, most such layers occur in the hollow organs of the respiratory, urinary, and reproductive systems as well. From the lumen outward, these layers are the *mucosa, submucosa, muscularis externa,* and *serosa.*

The Mucosa

The innermost layer is the **mucosa,** or *mucous membrane.* More complex than other mucous membranes in the body (see p. 99), the typical digestive mucosa contains three sublayers: (1) a lining epithelium, (2) a lamina propria, and (3) a muscularis mucosae.

The lining **epithelium** abuts the lumen of the alimentary canal and performs many functions related to digestion, such as absorbing nutrients and secreting mucus. This epithelium is continuous with the ducts and secretory cells of the various digestive glands, most of which lie fully within the wall and are called *intrinsic glands.* Larger glands, such as the liver and pancreas, extend beyond the walls of the alimentary canal and are called *extrinsic (accessory) glands.*

The **lamina propria** is a loose areolar or reticular connective tissue whose capillaries nourish the lining epithelium and absorb digested nutrients. The lamina propria contains most of the mucosa-associated lymphoid tissue (MALT), which defends against invasion by bacteria and other microorganisms in the alimentary canal (see p. 593).

External to the lamina propria is the **muscularis mucosae,** a thin layer of smooth muscle that produces local movements of the mucosa. For example, the twitching of this muscle layer dislodges sharp food particles that become embedded in the mucosa.

The Submucosa

Just external to the mucosa is the **submucosa,** a layer of connective tissue containing major blood and lymphatic vessels and nerve fibers. Its rich vascular network sends branches to all other layers of the wall. Its connective tissue is a type intermediate between loose areolar and dense irregular—a "moderately dense" connective tissue. The many elastic fibers in the submucosa enable the alimentary canal to return to its shape after food material passes through it.

The Muscularis Externa

External to the submucosa is the **muscularis externa,** also simply called the *muscularis.* Throughout most of the alimentary canal, this tunic consists of two layers of smooth muscle, an inner *circular layer* whose fibers orient around the circumference of the canal, and an outer *longitudinal layer* whose fibers orient along the length of the canal. Functionally, the circular layer squeezes the gut tube, and the longitudinal layer shortens it. Together, these layers are responsible for peristalsis and segmentation. In some places, the circular layer thickens to form sphincters that act as valves to prevent the backflow of food from one organ to the next.

The Serosa

The **serosa,** which is in fact the visceral peritoneum, is the outermost layer of the intraperitoneal organs of

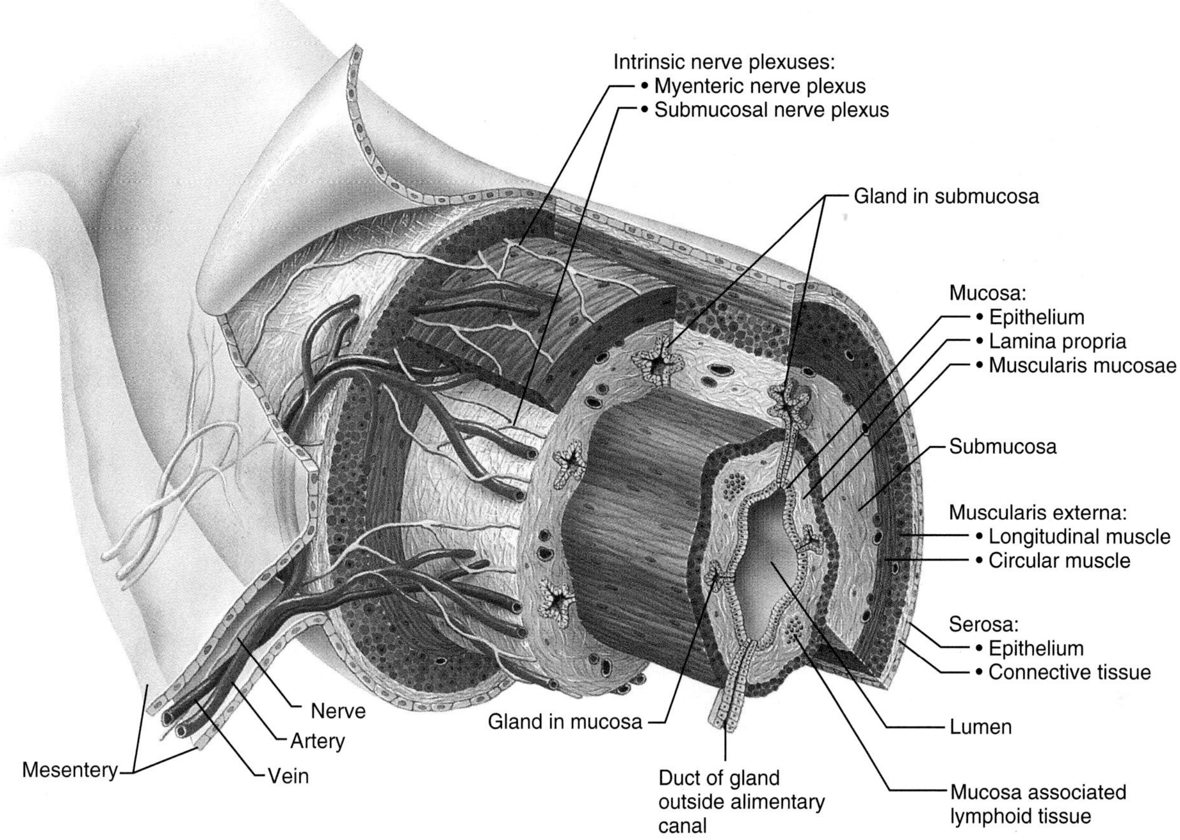

FIGURE 22.6 Histological layers of the alimentary canal (from esophagus through large intestine). The four basic layers in the wall, from the lumen outward, are: 1) mucosa, consisting of an epithelium, lamina propria, and muscularis mucosae; 2) submucosa, 3) muscularis externa, consisting of inner circular and outer longitudinal layers; and 4) serosa (or adventitia). Also note the glands, blood vessels, and the myenteric and submucosal plexus of visceral nerves.

the alimentary canal. Like all serous membranes (see p. 99), it is formed of a simple squamous epithelium (mesothelium) underlain by a thin layer of areolar connective tissue.

Parts of the alimentary canal that are not associated with the peritoneal cavity lack a serosa and have an *adventitia,* an ordinary fibrous connective tissue, as their outer layer. For example, the esophagus in the thorax has an adventitia that binds it to surrounding structures. Retroperitoneal organs have both a serosa and an adventitia—a serosa on the anterior side facing the peritoneal cavity and an adventitia on the posterior side embedded in the posterior abdominal wall.

Nerve Plexuses

Visceral nerve plexuses also occur within the wall of the alimentary canal (see Figure 22.6). The **myenteric nerve plexus** (mi-en-ter′ik; "intestinal muscle") is in the muscularis externa between the circular and longitudinal layers, where it innervates the muscularis externa to control peristalsis and segmentation. The **submucosal nerve plexus** lies within the submucosa, extends inward, and signals the glands in the mucosa to secrete and the muscularis mucosae to contract. Both plexuses contain parasympathetic and sympathetic motor components and visceral sensory fibers, all of which link the alimentary canal to the brain and bring digestion under the influence of the central nervous system.

Despite this external influence from the brain, digestive activity is largely automatic, controlled by an internal nervous system of *enteric neurons* (*enteric* = gut) in both the myenteric and submucosal plexuses. Within the wall of the alimentary canal, enteric neurons form independent reflex arcs of sensory, intrinsic, and motor neurons that control the normal movements of peristalsis and segmentation, as well as glandular secretion by the

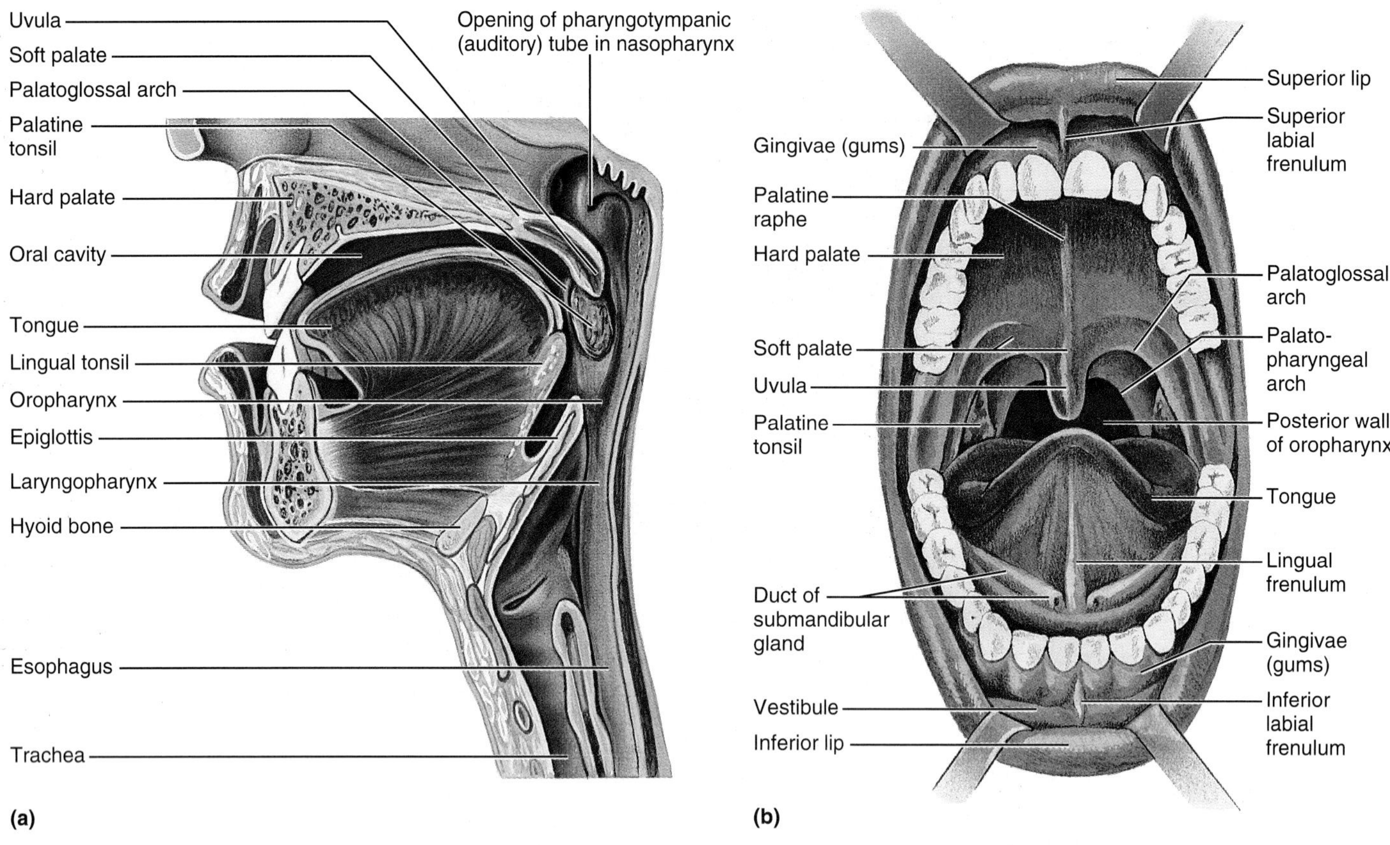

FIGURE 22.7 Anatomy of the mouth. (a) Sagittal section of the oral cavity and pharynx. (b) Anterior view of mouth.

mucosa. The classical autonomic nervous system (parasympathetic and sympathetic) merely speeds or slows this inherent activity, and allows the central nervous system to influence it. Even though few people seem to know that our gut has its own **enteric nervous system,** this system is not trivial: It consists of 100 million neurons, as many as in the entire spinal cord!

The parasympathetic and sympathetic inputs to the digestive organs are postganglionic sympathetic fibers, preganglionic parasympathetic fibers, and postganglionic parasympathetic neurons (see Chapter 15). Recall that parasympathetic input stimulates digestive functions, whereas sympathetic input inhibits digestion.

The Mouth and Associated Organs

In addition to discussing the structures that constitute the mouth itself, this section examines the tongue, the salivary glands, and the teeth.

The Mouth

Food enters the alimentary canal through the mouth, where it is chewed, manipulated by the tongue, and moistened with saliva. The **mouth,** or **oral cavity** (Figure 22.7), is a mucosa-lined cavity whose boundaries are the lips anteriorly, the cheeks laterally, the palate superiorly, and the tongue inferiorly. Its anterior opening is the **oral orifice.** Posteriorly, the mouth borders the fauces of the oropharynx (see p. 608).

The mouth is divided into the vestibule and the oral cavity proper. The **vestibule** (ves′tĭ-būl; "porch") is the slit between the teeth and the cheeks (or lips). When you brush the outer surface of your teeth, your toothbrush is in the vestibule. The **oral cavity proper** is the region of the mouth that lies internal to the teeth.

Histology of the Mouth The walls of the oral cavity consist of just a few layers of tissue: an internal mucosa made of an epithelium and lamina propria only, a thin submucosa in some areas, and an external layer of muscle or bone. The lining of the mouth, a thick stratified squamous epithelium, protects it from abrasion by sharp pieces of food during chewing. The lining epithelium of the tongue, palate, lips, and gums may show slight keratinization, which provides extra protection against abrasion.

The Lips and Cheeks The **lips** (or **labia**) and the **cheeks,** which help keep food inside the mouth during chewing, are composed of a core of skeletal muscle covered by skin. Whereas the cheeks are formed largely by the buccinator muscles, the orbicularis oris muscle (p. 277) forms the bulk of the lips.

The lips are thick flaps extending from the inferior boundary of the nose to the superior boundary of the chin. The region of the lip where one applies lipstick or lands a kiss is called the *red margin,* a transition zone where the highly keratinized skin meets the oral mucosa. This margin is poorly keratinized and translucent, so it derives its color from blood in the underlying capillaries. Because the red margin lacks sweat and sebaceous glands, it must be moistened with saliva periodically to prevent drying and cracking. The **labial frenulum** (fren′u-lum; "little bridle of the lip") is a median fold that connects the internal aspect of each lip to the gum (see Figure 22.7b).

The Palate The **palate,** which forms the roof of the mouth, has two distinct parts: the *hard palate* anteriorly and the *soft palate* posteriorly (see Figure 22.7 and p. 605). The bony hard palate forms a rigid surface against which the tongue forces food during chewing. The muscular soft palate is a mobile flap that rises to close off the nasopharynx during swallowing. (To demonstrate this action, try to breathe and swallow at the same time.) Dipping inferiorly from the free edge of the soft palate is the finger-like uvula. Laterally, the soft palate is anchored to the tongue by the **palatoglossal arches** and to the wall of the oropharynx by the **palatopharyngeal arches** (see Figures 22.7 and 22.9). These two folds form the boundaries of the fauces, the arched area of the oropharynx that contains the palatine tonsils.

The Tongue

The **tongue,** which occupies the floor of the mouth (see Figure 22.7a), is predominantly a muscle constructed of interlacing fascicles of skeletal muscle fibers. During chewing, the tongue grips food and constantly repositions it between the teeth. Tongue movements also mix the food with saliva and form it into a compact mass called a *bolus* (bo′lus; "lump"); then, during swallowing, the tongue moves posteriorly to push the bolus into the pharynx. In speech, the tongue helps to form some consonants (k, d, t, and l, for example). Finally, it houses most of the taste buds.

The tongue has both intrinsic and extrinsic muscle fibers. The *intrinsic muscles,* which are confined within the tongue and are not attached to bone, have fibers that run in several different planes (Figure 22.8). These intrinsic muscles change the shape of the tongue but do not change its position. The *extrinsic muscles,* including the genioglossus muscle shown in Figure 22.8, extend to the tongue from bones of the skull or the soft palate (see p. 278). These extrinsic muscles alter the position of the tongue: They protrude it, retract it, and move it laterally. The tongue is divided by a median septum of connective tissue, and both halves contain identical groups of muscles.

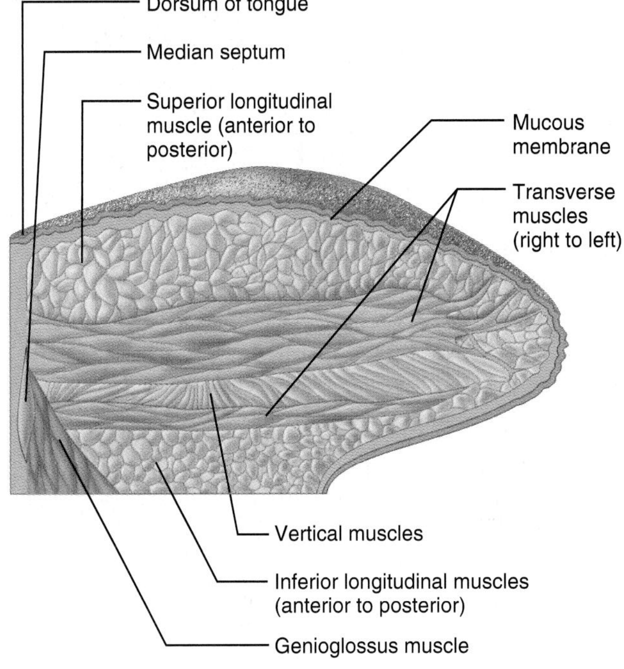

FIGURE 22.8 Frontal section through the left half of the tongue, showing its intrinsic skeletal muscles and the genioglossus, an extrinsic muscle.

A fold of mucosa on the undersurface of the tongue, the **lingual frenulum** (see Figure 22.7b), both secures the tongue to the floor of the mouth and limits its posterior movements. Individuals in which the lingual frenulum is abnormally short or extends exceptionally far anteriorly are said to be "tongue-tied," and their speech is distorted because movement of the tongue is restricted. This congenital condition, called **ankyloglossia** (ang″kĭ-lo-glos′e-ah; "fused tongue"), is corrected surgically by snipping the frenulum.

The superior surface of the tongue (Figure 22.9) bears three major types of peg-like projections of the mucosa: the filiform, fungiform, and vallate (circumvallate) *papillae.* The terms *papillae* and *taste buds* are not synonymous; instead, the fungiform and vallate papillae *contain* taste buds.

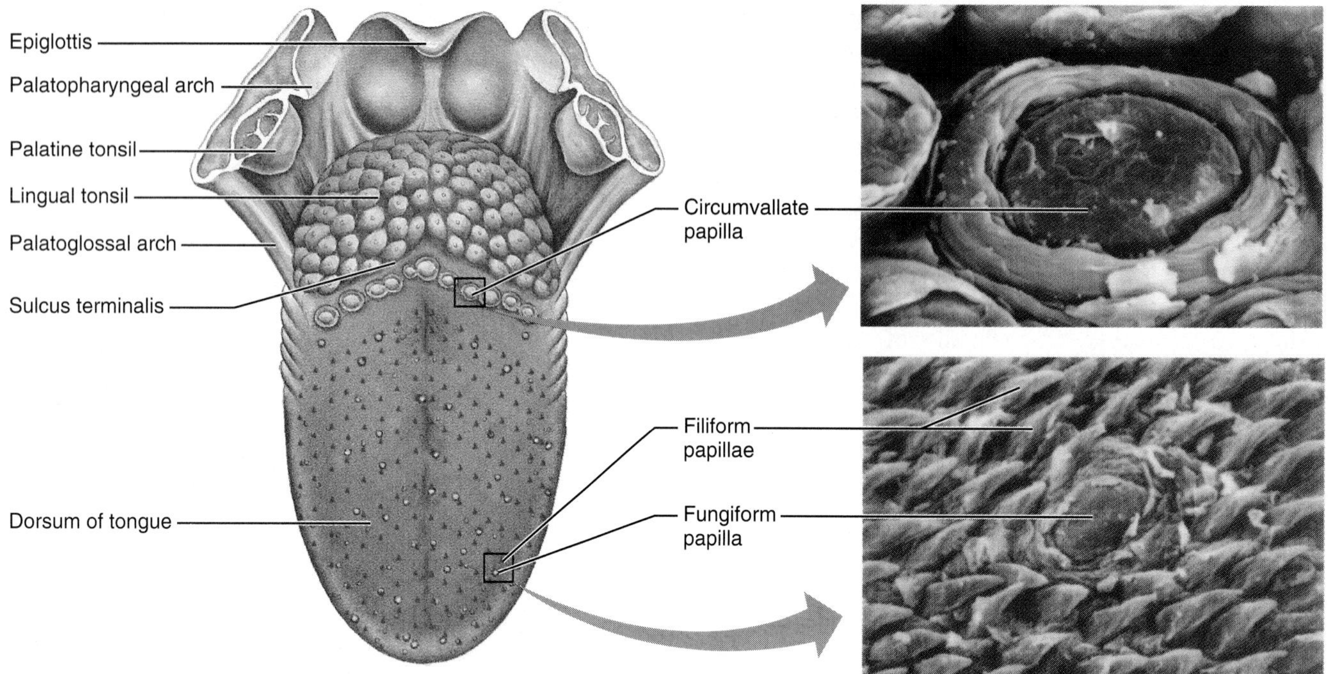

FIGURE 22.9 The superior surface of the tongue (left). At right are scanning electron micrographs of a circumvallate papilla (top; 20 ×), and filiform and fungiform papillae (bottom; 30 ×). (Micrographs © R. Kessel and C. Y. Shih.)

The conical, pointed, and keratinized **filiform papillae** (fil′ĭ-form; "thread-shaped") roughen the tongue, enabling it to grasp and manipulate food during chewing. These smallest and most numerous papillae line up in parallel rows. They give the tongue's surface its whitish appearance.

The **fungiform papillae** (fun′jĭ-form; "mushroom-shaped"), which resemble tiny mushrooms, have a vascular core that gives them a red color. Although less abundant than filiform papillae, they are scattered widely over the tongue surface. Taste buds occur in the epithelium on the *tops* of these papillae.

Ten to twelve large **vallate** or **circumvallate papillae** (ser″cum-val′āt) line up in a V-shaped row that is two-thirds of the way posteriorly on the tongue surface and directly anterior to a groove called the **sulcus terminalis,** which marks the border between the mouth and pharynx. Each circumvallate papilla is surrounded by a circular ridge (*circumvallate* means "surrounding wall"), from which it is separated by a deep furrow. Taste buds occupy the epithelium on the *sides* of these papillae (see Figure 16.1).

The posterior third of the tongue, which lies in the oropharynx, not in the mouth, is covered not with papillae but with the bumpy *lingual tonsil* (see Figure 22.9).

The Salivary Glands

The salivary glands produce *saliva,* a complex mixture of water, ions, mucus, and enzymes that performs many functions: It moistens the mouth, dissolves food chemicals so that they can be tasted, wets food, binds the food together into a bolus, and enzymes within it begin the digestion of starches. Saliva contains a bicarbonate buffer that neutralizes the acids that are produced by oral bacteria and that initiate tooth decay. Additionally, it contains bactericidal enzymes, antiviral substances, antibodies, and a cyanide compound, all of which kill harmful oral microorganisms. Saliva also contains proteins that stimulate the growth of beneficial bacteria to outcompete harmful bacteria in the mouth.

All **salivary glands** are compound tubuloalveolar glands. Small *intrinsic salivary glands* are scattered within the mucosa of the tongue, palate, lips, and cheeks. Saliva from these glands keeps the mouth moist at all times. By contrast, large *extrinsic salivary glands,* which lie external to the mouth but connect to it through their ducts (Figure 22.10), secrete saliva only when we eat or anticipate a meal such that our mouth waters. These paired extrinsic glands are the *parotid, submandibular,* and *sublingual glands.*

The largest extrinsic gland is the **parotid** (pah-rot'id) **gland.** True to its name (*par* = near; *otid* = the ear), it lies anterior to the ear, between the masseter muscle and the skin (see Figure 22.10a). Its **parotid duct** runs parallel to the zygomatic arch, penetrates the muscle of the cheek, and opens into the mouth lateral to the second upper molar. Because the branches of the facial nerve run through the parotid gland on their way to the muscles of facial expression, surgery on this gland can lead to facial paralysis.

Critical Reasoning

Mumps *is caused by a virus that spreads from one person to another in the saliva. Its dominant symptom is inflammation and swelling of the parotid gland. Deduce why people with mumps say it hurts to open their mouth or chew.*

Movements of the mandible pull on the irritated parotid glands and the bulging masseter muscles compress these glands, causing pain.

The **submandibular gland** *(submaxillary gland),* which is about the size of a walnut, lies along the medial surface of the mandibular body, just anterior to the angle of the mandible. Its duct raises the mucosa of the floor of the mouth and opens directly lateral to the tongue's lingual frenulum (see Figure 22.7b).

The **sublingual gland** lies in the floor of the oral cavity, inferior to the tongue. Its 10 to 12 ducts open into the mouth, directly superior to the gland (see Figure 22.10a).

The secretory cells of the salivary glands are *serous cells,* which produce a watery secretion containing the enzymes and ions of saliva, and *mucous cells,* which produce mucus. (Note, however, that because the serous cells in human salivary glands are now known to secrete a small amount of mucus as well, some scientists refer to these cells as *seromucous cells.*) The parotid glands contain only serous cells; the submandibular and intrinsic glands contain approximately equal numbers of serous and mucous cells; and the sublingual glands contain mostly mucous cells. The smallest ducts of all salivary glands are formed by a simple cuboidal epithelium.

The Teeth

The **teeth** lie in sockets (alveoli) in the gum-covered margins of the mandible and maxilla. We masticate, or chew, by raising and lowering the mandible and by moving it from side to side while using the tongue to position food between the teeth. In the process, the teeth tear and grind the food, breaking it into smaller fragments.

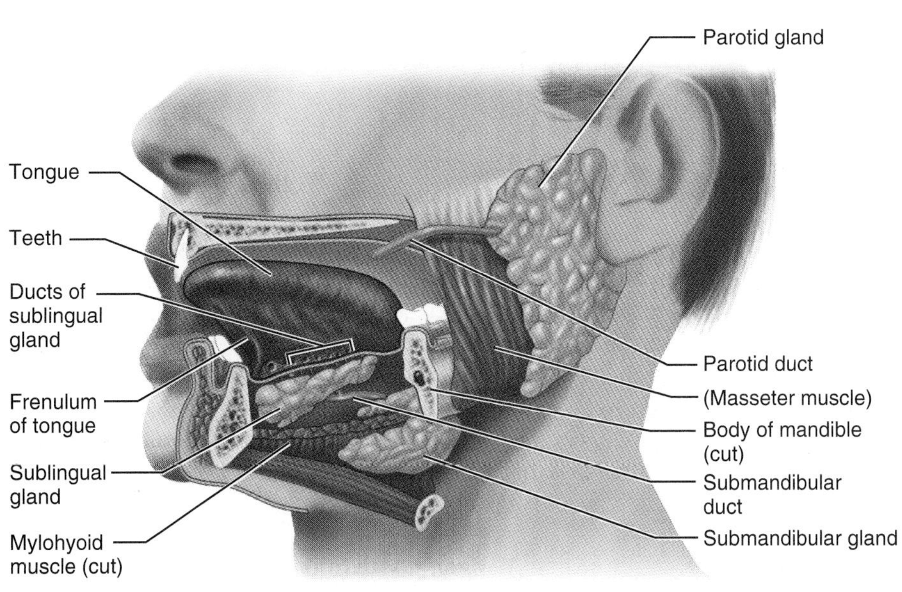

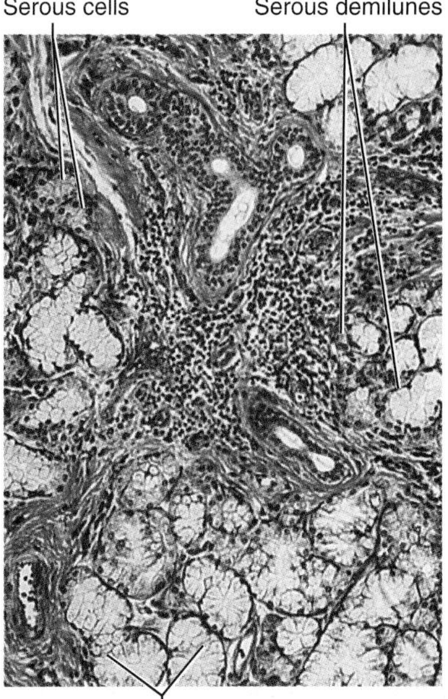

FIGURE 22.10 The extrinsic salivary glands. (a) Locations of the parotid, submandibular, and sublingual glands and their ducts. **(b)** Photomicrograph of the sublingual gland (85 ×), which consists of a series of compound tubuloalveolar glands exactly like the one shown in Figure 4.11g on page 88. The sublingual gland contains mostly mucous cells (white) with a few serous cells (pink). A serous demilune is a crescent-shaped cap of serous cells at the end of a tubule of mucous cells. The red y-shaped structure at upper center is a duct.

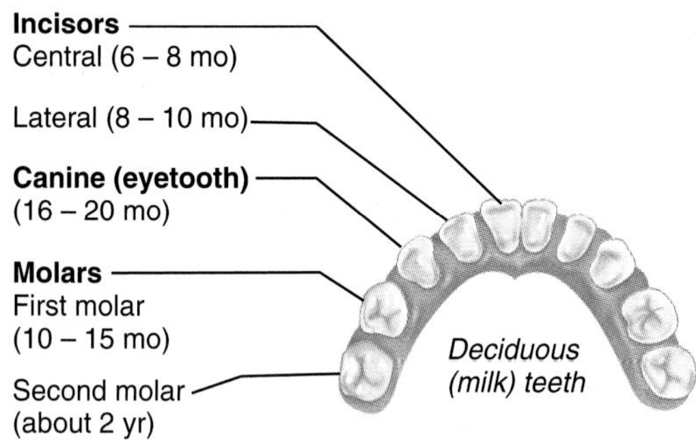

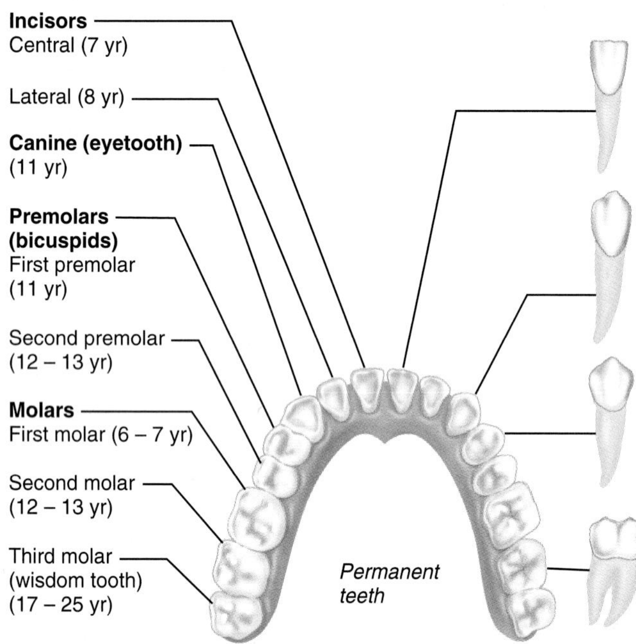

FIGURE 22.11 Human deciduous and permanent teeth. Approximate ages at which the teeth erupt are included within parentheses. Only the lower jaw is shown (the upper teeth are similar). Shapes of the individual teeth are shown at right. Note that all incisors, canines, and premolars have one root, and that lower molars have three roots; upper molars (not illustrated) have two roots.

Dentition and the Dental Formula During their lifetime, humans have two sets of teeth, or *dentitions*. By age 21, the primary dentition, called the **deciduous teeth** (de-sid′u-us; "falling off"), has been replaced by the permanent dentition (Figure 22.11).

At about 6 months after birth, the lower central incisors become the first of the deciduous teeth to appear. Additional pairs of teeth erupt at varying intervals until all 20 deciduous teeth have emerged, by about 2 years of age. As the deep-lying **permanent teeth** enlarge and develop, the roots of the deciduous teeth are resorbed until these teeth loosen and fall out, typically between the ages of 6 and 12 years. Generally, by the end of adolescence, all permanent teeth have erupted except for the third molars (also called *wisdom teeth*), which emerge between the ages of 17 and 25 years. There are 32 permanent teeth in a full set, but in some people the wisdom teeth are either completely absent or fail to erupt.

Instead of emerging normally, teeth may remain embedded deep in the jawbone and push on the roots of the other teeth. Such *impacted teeth* cause pressure and pain and must be removed by a dentist or oral surgeon. Wisdom teeth are the most commonly impacted teeth.

Teeth are classified according to their shape and function as incisors, canines, premolars, or molars (see Figure 22.11). The chisel-shaped **incisors** are adapted for nipping off pieces of food, and the cone-shaped **canines** (cuspids, eyeteeth) tear and pierce. The **premolars** (bicuspids) and **molars** have broad crowns with rounded *cusps* (surface bumps) for grinding food. The molars (literally, "millstones") have four or five cusps and are the best grinders. During chewing, the upper and lower molars lock repeatedly together, the cusps of the uppers fitting into the valleys in the lowers, and vice versa. This action generates tremendous crushing forces.

The *dental formula* is a shorthand way of indicating the numbers and relative positions of the different classes of teeth in the mouth. This formula is written as a ratio of uppers over lowers, for just half of the mouth (because right and left halves are the same). The total number of teeth is calculated by multiplying the dental formula by 2. The formula for the permanent dentition (two incisors, one canine, two premolars, and three molars) is written as:

$$\frac{\text{2I, 1C, 2P, 3M}}{\text{2I, 1C, 2P, 3M}} \times 2 \quad \text{(equals 32 teeth)}$$

The dental formula for the deciduous teeth is:

$$\frac{\text{2I, 1C, 2M}}{\text{2I, 1C, 2M}} \times 2 \quad \text{(equals 20 teeth)}$$

The upper teeth are served by the superior alveolar nerves, which are branches of the maxillary division of the trigeminal nerve, whereas the lower teeth are served by the inferior alveolar nerves, which are branches of the mandibular division of the trigeminal nerve (see p. 422). The arterial supply for the teeth comes via the superior and inferior alveolar arteries, which are branches of the maxillary artery from the external carotid (see p. 553).

Tooth Structure Each tooth has two main regions, the exposed **crown** and the **root(s)** in the socket. These regions meet at the **neck** near the gum line (Figure 22.12). The surface of the crown, which bears the forces of chewing, is covered by a 2.5-mm-thick layer of **enamel,** the hardest substance in the body. Enamel lacks cells and vessels, and 99% of its mass consists of densely packed hydroxyapatite (a calcium salt) crystals arranged in force-resisting rods or prisms oriented perpendicular to the tooth's surface.

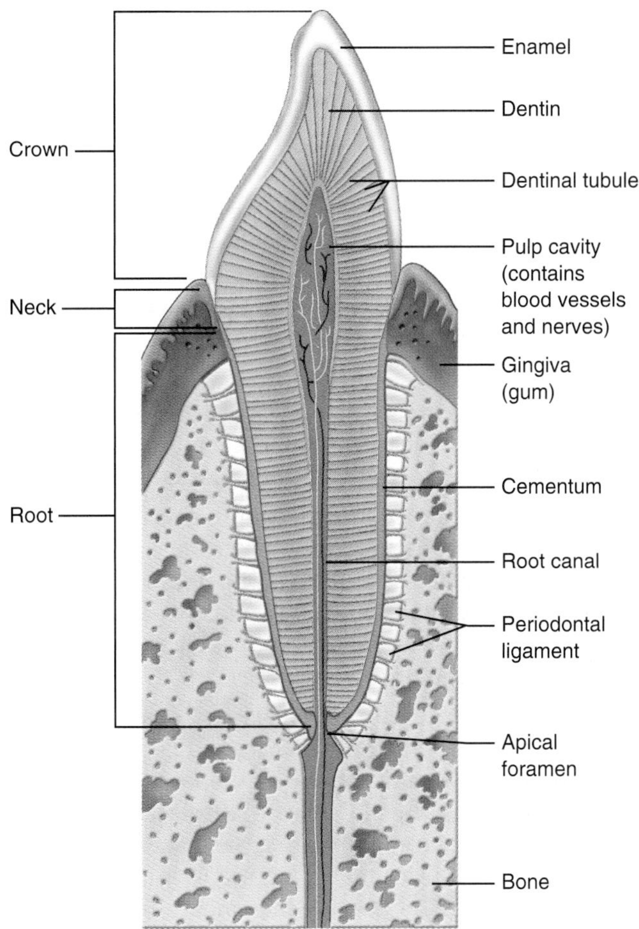

FIGURE 22.12 Longitudinal section of a canine tooth within its bony alveolus (socket).

Dentin, or **dentine,** underlies the enamel cap and forms the bulk of the tooth. This is a bone-like tissue with mineral and collagen components, but it is harder than bone and lacks internal blood vessels. Dentin contains unique radial striations called **dentinal tubules,** each of which is filled with an elongated cellular process of an *odontoblast* (o-don′to-blast; "tooth former"), the cell type that secretes and maintains the dentin. The spherical cell bodies of odontoblasts lie just deep to the dentin, within the underlying pulp cavity.

The **pulp cavity,** in the center of the tooth, is filled with dental **pulp,** a loose connective tissue containing the tooth's vessels and nerves. Pulp supplies nutrients for the tooth's hard tissues and provides for tooth sensation. The part of the pulp cavity in the root is the **root canal.** The opening into the root canal at the tip of each root is the **apical foramen.** When a tooth is damaged by a blow or by a deep cavity, the pulp may die and become infected. In such cases, **root canal therapy** must be performed. In this procedure, all of the pulp is drilled out, and the pulp cavity is sterilized and filled with an artificial, inert material before the tooth is capped.

The external surface of the tooth root is covered by a calcified connective tissue called **cementum.** Essentially a bone layer, cementum attaches the tooth to the **periodontal ligament** (per″e-o-don′tal; "around the tooth") or **periodontium.** This ligament anchors the tooth in the bony socket of the jaw. The periodontal ligament is continuous with the gum, or **gingiva** (jin-jĭ′vah), at the neck of the tooth.

Dental **cavities,** or **caries** (kar′ēz; "rottenness"), result from a gradual demineralization of the enamel and dentin by bacterial action. The decay process begins with the accumulation of **dental plaque,** a film of sugar, bacteria, and other debris that adheres to the teeth. Metabolism of the trapped sugars by the bacteria produces acids, which dissolve the calcium salts from the teeth. Once the salts have leached out, the remaining organic matrix is broken down by protein-digesting bacterial enzymes. Frequent brushing helps prevent tooth decay by removing plaque.

CLINICAL APPLICATIONS

GUM DISEASE Even more serious than tooth decay is the effect of plaque on the *gums.* As plaque accumulates around the necks of the teeth, the contained bacteria release toxins that irritate the gums and cause them to pull away from the teeth. The plaque calcifies into a layer called **calculus** (kal′ku-lus; "stone"), on which more plaque accumulates, further inflaming the gums. This condition, **gingivitis** (jin″jĭ-vi′tis), can be reversed if the calculus is removed, but if it is neglected the bacteria invade the periodontal tissues, forming pockets of infection that destroy the periodontal ligament and dissolve away the bone around the tooth. Called **periodontitis,** this condition begins around age 35 and eventually affects 75% of people. (Periodontitis often results in the loss of teeth and is the reason many people wear dentures.) Still, tooth loss is not inevitable. Even advanced cases of periodontitis can be treated by cleaning the infected pockets around the roots, then cutting and stitching the gums to shrink the pockets. Newly available gels, which are injected into the pockets, contain either antibiotics or substances that stimulate regeneration of the gum, cementum, and periodontal ligament.

Dentists recommend that people floss between their teeth every day, beginning at a young age, because flossing removes plaque, thereby minimizing one's chances of developing periodontitis and losing one's teeth later in life. ✚

The Pharynx

From the mouth, swallowed food passes posteriorly into the **oropharynx** and then the **laryngopharynx** (see Figure 22.7a and p. 608), both of which are passageways for food, fluids, and inhaled air.

The histology of the pharyngeal wall resembles that of the mouth in that the oropharynx and laryngopharynx are lined by a stratified squamous epithelium, which protects them against abrasion. (Swallowed food often contains rough particles, even after mastication.) The external muscle layer consists primarily of three (superior, middle, and inferior) *pharyngeal constrictor* muscles (see Figure 11.8b, p. 282), muscles of swallowing that encircle the pharynx and partially overlap one another. Like three stacked, clutching fists, they contract in sequence, from superior to inferior, to squeeze the bolus into the esophagus. The pharyngeal constrictors are *skeletal* muscles, as swallowing is a voluntary action.

The Esophagus

In this section we examine the gross and microscopic structure of this alimentary-canal organ.

Gross Anatomy

The **esophagus** is a muscular tube that propels swallowed food to the stomach. Its lumen is collapsed when it is empty. The esophagus begins as a continuation of the pharynx in the mid neck, descends through the thorax on the anterior surface of the vertebral column (see Figure 22.1), and it passes through the *esophageal hiatus* (hi-a′tus; opening) in the diaphram to enter the abdomen (see Figure 19.11). Its abdominal part, which is only about 2 cm long, joins the stomach at the **cardiac orifice** (see Figure 22.14a), where a *cardiac sphincter* acts to close off the lumen and prevent regurgitation of acidic stomach juices into the esophagus. The only anatomical evidence of this sphincter is a minimal thickening of the smooth muscle in the wall. The edges of the esophageal hiatus also help prevent regurgitation.

CLINICAL APPLICATIONS

HIATAL HERNIA AND GASTROESOPHAGEAL REFLUX DISEASE

In a **hiatal hernia,** the superior part of the stomach pushes through an enlarged esophageal hiatus into the thorax following a weakening of the diaphragmatic muscle fibers around the hiatus. Because the diaphragm no longer reinforces the action of the cardiac sphincter, the acidic stomach juices are persistently regurgitated, eroding the wall of the esophagus and causing a burning pain.

The regurgitation associated with hiatal hernia is just one form of **gastroesophageal reflux disease (GERD),** a condition that affects at least 4% of Americans. Most cases of GERD are due to abnormal relaxation or weakness of the cardiac sphincter (and probably, of the sphincter mechanism of the esophageal hiatus as well). Symptoms include heartburn behind the sternum, regurgitation of stomach contents, and belching. The patient may aspirate the acids that are burped up, leading to hoarseness, coughing, and bronchial asthma. After persistent exposure to the acidic stomach contents, the lower esophagus develops ulcers, and the epithelium there becomes abnormal and precancerous. Treatment, which is usually successful, involves administering antacids and drugs that decrease the secretion of stomach acids. In severe cases, surgery is used to reconstruct a valve in the lower esophagus. ✚

Microscopic Anatomy

Unlike the mouth and pharynx, the esophagus wall (Figure 22.13a) contains all four layers of the alimentary canal described on page 638. The following histological features are of interest:

1. The lining epithelium is a nonkeratinized stratified squamous epithelium. At the junction of the esophagus and stomach, this thick, abrasion-resistant layer changes abruptly to the thin simple columnar epithelium of the stomach, which is specialized for secretion (Figure 22.13b).
2. When the esophagus is empty, its mucosa and submucosa are thrown into longitudinal folds, but during passage of a bolus, these folds flatten out.
3. The wall of the esophagus contains mucous glands, primarily compound tubuloalveolar glands that extend from the submucosa to the lumen. As a bolus passes, it compresses these glands, causing them to secrete a lubricating mucus, which smoothes the further passage of the bolus through the esophagus.
4. The muscularis externa consists of skeletal muscle in the superior third of the esophagus, a mixture of skeletal and smooth muscle in the middle third, and smooth muscle in the inferior third. This arrangement is easy to remember if the esophagus is viewed as the zone where the skeletal muscle of the mouth and pharynx gives way to the smooth muscle of the stomach and intestines.
5. The most external esophageal layer is an adventitia, not a serosa, because the thoracic segment of the esophagus is not suspended in the peritoneal cavity.

The Stomach

The J-shaped **stomach** (Figure 22.14), the widest part of the alimentary canal, is a temporary storage tank in which food is churned and turned into a paste called **chyme** (kīm; "juice"). The stomach also starts the breakdown of food proteins by secreting **pepsin,** a protein-digesting enzyme that can function only under acidic conditions, and hydrochloric acid, a strong acid that destroys many harmful bacteria in the food. Although most nutrients are absorbed in the small intestine, some substances are absorbed through the stomach, including water, electrolytes, and some drugs (aspirin and alcohol). Food remains in the stomach for roughly 4 hours.

Gross Anatomy

The stomach, which extends from the esophagus to the small intestine, lies in the superior left part of the peritoneal cavity, in the left hypochondriac, epigastric, and umbilical regions of the abdomen (see Figure 22.4b). It is directly inferior to the diaphragm and anterior to the spleen and pancreas, and its upper part is hidden behind the left side of the liver. Although the stomach is

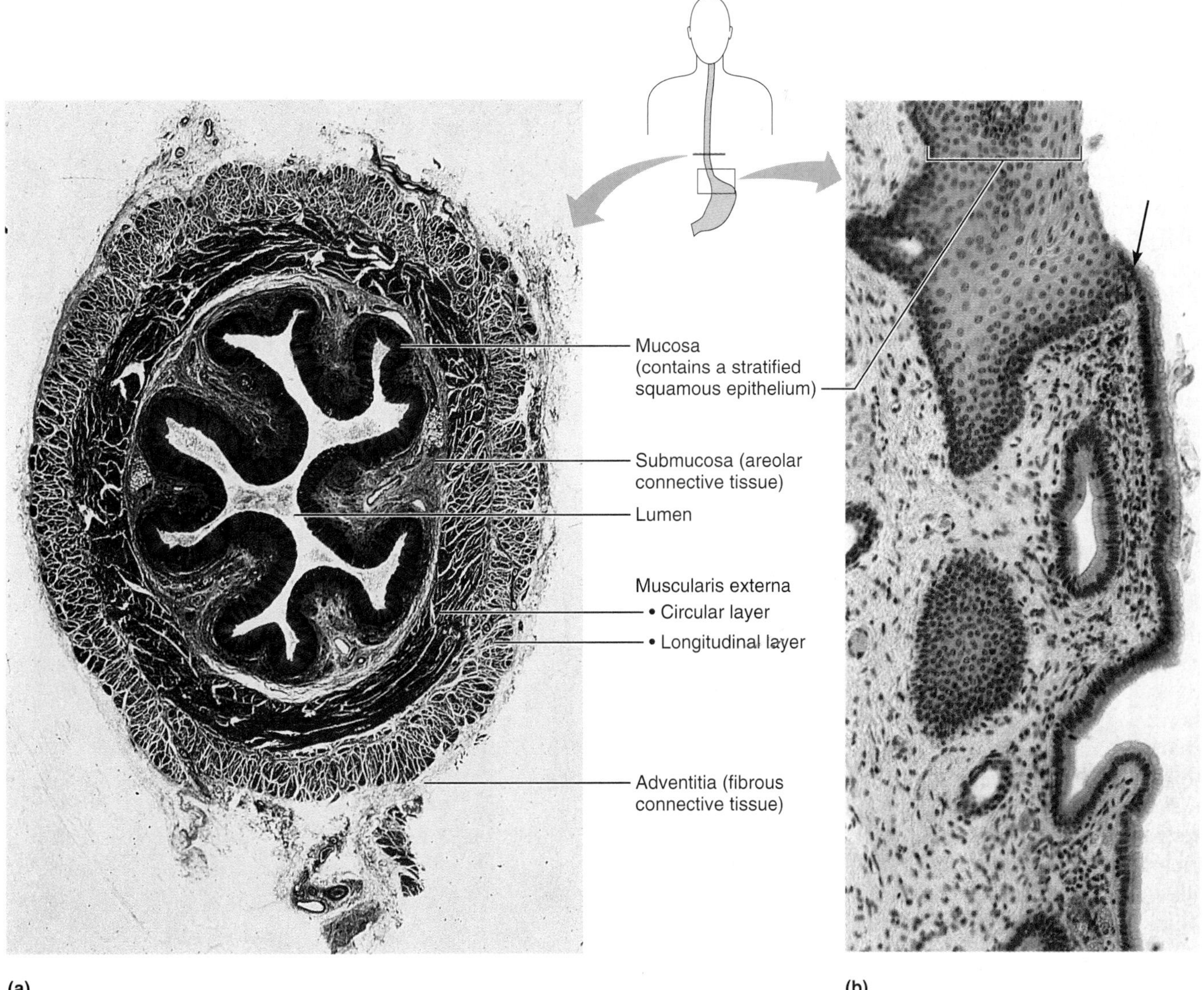

FIGURE 22.13 **Microscopic structure of the esophagus.** **(a)** Cross-sectional view of the esophagus (5×), showing its four layers. Folds of the mucosa and submucosa give the central lumen an irregular shape. **(b)** Longitudinal section at the junction of the esophagus and stomach; the arrow marks the abrupt change between the stratified squamous epithelium lining the esophagus and the simple columnar epithelium lining the stomach (130 ×).

anchored at both ends by esophageal and intestinal attachments, it is quite mobile in between. It tends to lie high and run horizontally in short, stout people (steer-horn stomach) and is elongated vertically in many tall, thin people (J-shaped stomach). A full, J-shaped stomach may extend low enough to reach the pelvis!

The main regions of the stomach are shown in Figure 22.14a. The **cardiac region,** or **cardia** ("near the heart"), is a ring-shaped zone encircling the cardiac orifice at the junction with the esophagus. The **fundus,** the stomach's dome, is tucked under the diaphragm. The large midportion of the stomach, the **body,** ends at the funnel-shaped **pyloric region,** composed of the wider **pyloric antrum** ("cave") and the narrower **pyloric canal.** The pyloric region ends at the terminus of the stomach, the **pylorus** ("gatekeeper") containing the **pyloric sphincter** (Figure 22.14b), which controls the entry of chyme into the intestine. The convex left surface of the stomach is its **greater curvature,** and the concave right margin is the **lesser curvature.**

The stomach's structure accounts for its great distensibility—it easily holds 1.5 L of food and has a maximum capacity of about 4 L (1 gallon). The internal surface of the empty stomach contains numerous longitudinal folds

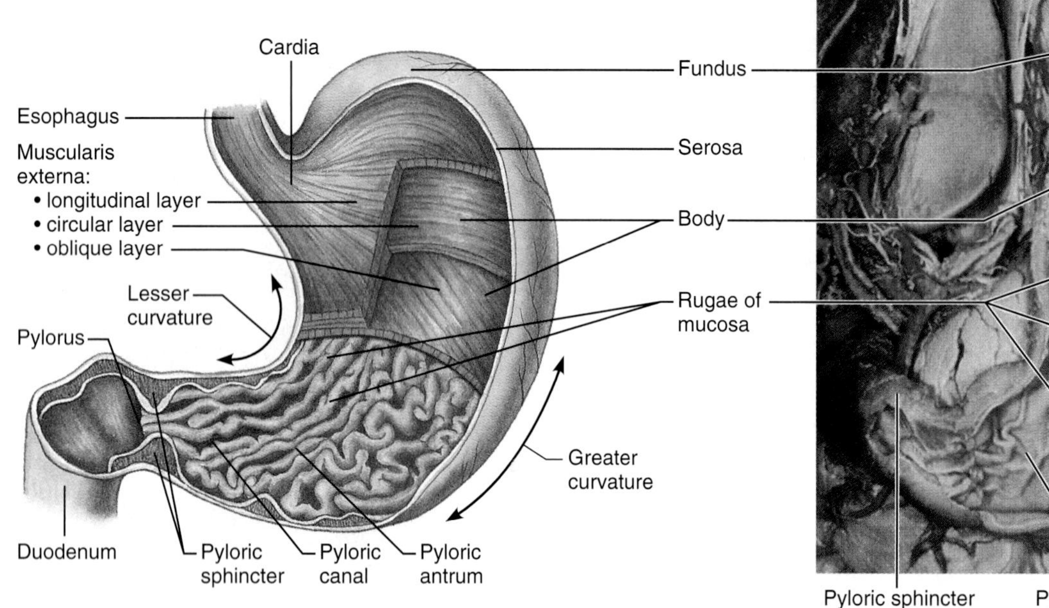

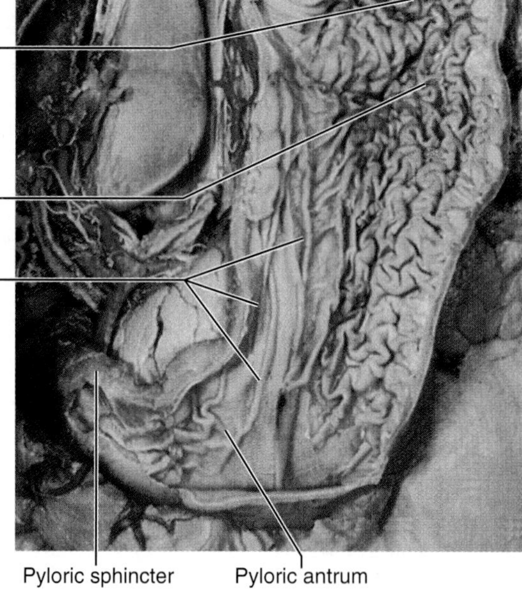

FIGURE 22.14 Gross anatomy of the stomach. **(a)** Basic regions. **(b)** Layers of the muscularis externa (upper half), and the internal lining with rugae (lower half). **(c)** Photograph of internal aspect of stomach.

of mucosa called **rugae** (roo′ge; "wrinkles") (see Figure 22.14b and c), which flatten as the stomach fills; the resulting expansion in volume accommodates the increasing quantity of food within the stomach.

The stomach is innervated by sympathetic fibers that derive from the thoracic splanchnic nerves by way of the celiac plexus, and by parasympathetic fibers that derive from the vagus (see Chapter 15). The stomach contains no submucosal nerve plexus, so the myenteric plexus innervates its mucosa as well as its muscularis externa. The arteries to the stomach arise from the celiac trunk and include the right and left gastrics, short gastrics, and right and left gastroepiploics; the corresponding veins drain into the portal, splenic, and superior mesenteric veins (see Chapter 19).

Microscopic Anatomy

The wall of the stomach has the typical layers of the alimentary canal (see p. 638), but its muscularis externa shows special features. Along with circular and longitudinal layers of smooth muscle, the muscularis externa also has another (innermost) layer that runs ***obliquely*** (see Figure 22.14b). The circular and longitudinal layers churn and pummel food into smaller fragments, whereas the oblique layer jack-knifes the stomach into a V shape to ram the chyme into the small intestine. The pyloric sphincter is a thickening of the circular layer.

The histology of the stomach is illustrated in Figure 22.15. The lining epithelium ("surface epithelium" in the figure) is simple columnar and consists entirely of goblet cells, which secrete a coat of bicarbonate-buffered mucus that protects the stomach wall from the destructive effects of acid and pepsin in the lumen. This mucous

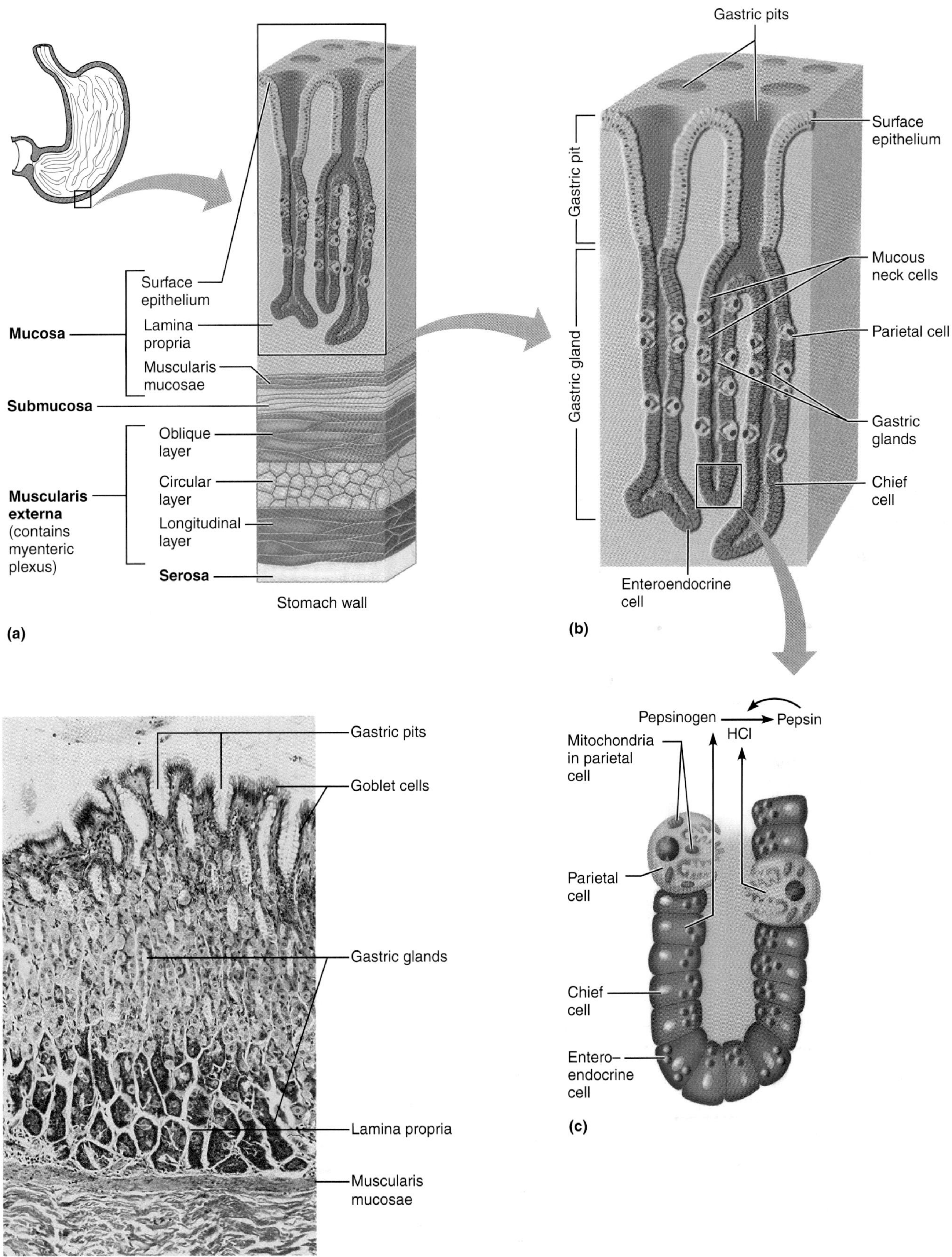

FIGURE 22.15 Microscopic anatomy of the stomach. **(a)** Layers of the stomach wall. **(b)** Enlarged view of the mucosa, with gastric pits and gastric glands. **(c)** Locations of parietal, chief, and enteroendocrine cells in a gastric gland. **(d)** Micrograph of the stomach mucosa, a view similar to part (b) (120 ×).

A Closer Look

Peptic Ulcers

Archie, a 53-year-old factory worker, began to experience a burning pain in his upper abdomen an hour or two after each meal. At first, he blamed the quality of his home cooking, but he also experienced the same symptoms after eating at the factory cafeteria or in restaurants. Archie always responded to stress by drinking and smoking heavily, and his abdominal pain became worse during a hectic week when he worked 15 hours overtime on the assembly line. When after 2 months of increasingly severe pain he finally consulted a physician, he learned he had a peptic ulcer—defined as a crater-like erosion of the mucosa of any part of the alimentary canal that is exposed to stomach secretions.

Peptic ulcers are very common, affecting 1 of every 8 Americans of both sexes. Although they may appear at any age, they develop most frequently in individuals between the ages of 50 and 70 years. Left untreated, they tend to recur—to flare up again after a period of recovery—for the rest of one's life.

Even though a few peptic ulcers occur in the lower esophagus following the regurgitation of stomach contents (see the discussion of gastroesophageal reflux disease on p. 646), about 98% occur either in the pyloric region of the stomach (gastric ulcers) or the first part of the duodenum of the small intestine (duodenal ulcers). Duodenal ulcers are about three times more common than gastric ulcers, and both types may produce a gnawing or burning sensation in the epigastric region of the abdomen. Such pain often begins 1–3 hours after a meal (or awakens one at night) and is typically relieved by eating. Other symptoms include a loss of appetite, burping, nausea, and vomiting, but not all individuals experience these symptoms, and some experience no symptoms at all.

Ulcers are typically round, sharply defined craters about 1–4 mm in diameter (see the photo) in which the mucosa has eroded to the depth of the muscularis mucosae. The base of the ulcer contains necrotic tissue, granulation tissue, scar tissue, and possibly eroded blood vessels.

Peptic ulcers can cause serious complications. In about 20% of cases, eroded blood vessels bleed into the alimentary canal, causing vomiting of blood and blood in the feces. In 5% to 10% of patients, scarring within the stomach obstructs the pyloric opening, blocking digestion (see "Pyloric stenosis" on p. 669). About 5% of ulcers *perforate,* allowing duodenal or stomach contents to leak into the peritoneal cavity and causing either peritonitis or the digestion and destruction of the nearby pancreas. A perforated ulcer is a life-threatening condition.

The traditional "common knowledge" was that stress causes ulcers, and the stereotypical ulcer patient was the overworked business executive. Even though research has not demonstrated a causal link between stress and ulcers, a stressful lifestyle does seem to aggravate existing ulcers. In response to the traditional view that ulcers resulted from various factors that caused either an oversecretion of stomach acid or a failure of the mucous lining to protect against this acid, patients were correctly told to stop smoking and drinking alcohol (especially wine), and at all costs to avoid taking aspirin and ibuprofen—all of which aggravate ulcers. Additionally, antacids (which neutralize stomach acids) and liquids that coat the stomach mucosa were administered. These simple measures allowed about two-thirds of ulcers to heal.

Then, in the 1970s, research produced drugs that suppress the secretion of stomach acid by impeding the stimulatory action of histamine on parietal cells. Considered miraculous at the time, these drugs "cured" up to 90%

---

secretion is necessary because the stomach mucosa is exposed to some of the harshest conditions in the entire alimentary canal. The oversecretion of stomach acid can result in peptic ulcers, the topic of this chapter's A Closer Look box.

The surface of the stomach mucosa is dotted with millions of cup-shaped **gastric pits,** which open into tubular **gastric glands** (see Figure 22.15b). Goblet cells invariably line the gastric pits, but the cells lining the gastric glands vary among the different regions of the stomach. In the pyloric and cardiac regions (not illustrated), the cells of the glands are primarily mucous cells. In the fundus and body, by contrast, the gastric glands contain three types of secretory cells: mucous neck cells, parietal (oxyntic) cells, and chief (zymogenic) cells (see Figure 22.15b and c).

1. **Mucous neck cells,** which occur in the upper ends, or necks, of the gastric glands, secrete a different type of mucus from that secreted by the surface goblet cells. The specific function of this mucus is not known.
2. **Parietal (oxyntic) cells,** which occur mainly in the middle regions of the glands, produce the stomach's hydrochloric acid (HCl) by pumping hydrogen and chloride ions into the lumen of the gland. Although parietal cells appear spherical when viewed by light microscopy, they actually have three thick prongs like those of a pitchfork. Many long microvilli cover each prong, providing a large surface area to enable rapid movement of H^+ and Cl^- out of the cells. The cytoplasm contains many mitochondria that supply the large amount of energy expended in pumping

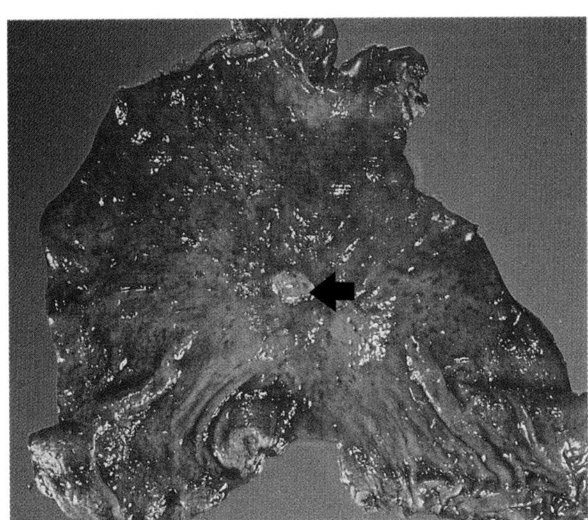

A peptic ulcer in the mucosa of the stomach is visible at the arrow.

of ulcers in a few weeks, eliminating the routine surgical removal of affected portions of the stomach. One of these drugs, Zantac (ranitidine), became the best-selling drug ever.

Still, in 1994 an influential panel assembled by the National Institutes of Health recommended that physicians stop using acid-suppressing drugs as the primary treatment for ulcers. One reason cited was that neither these drugs nor the older treatments truly cured ulcers; 95% returned within 2 years of treatment. More importantly, peptic ulcers were finally recognized as an infectious disease that could be treated effectively with antibiotics.

Since the early 1980s the accumulating evidence suggested that ulcers are caused by a spiral-shaped, acid-resistant bacterium named *Helicobacter pylori,* which inhabits the stomach (and may also infect the duodenum) and seems to spread via fecally contaminated food or water. Over one-third of Americans harbor this bacterium, but many are either resistant or have harmless strains.

Helicobacter binds specifically to gastric epithelium and induces the over-secretion of acid and the inflammation that lead to ulcers—and to many cases of chronic dyspepsia (long-term indigestion) and some stomach cancers as well. A simple week-long regimen of antibiotics permanently cures peptic ulcers in 95% of all patients. Now, a "triple therapy" consisting of the antibiotics metronidazole and tetracycline plus bismuth subcitrate routinely destroys even resistant strains of *H. pylori.* Proton-pump inhibitors, which lessen the production of stomach acid by parietal cells, can be given in conjunction with the triple therapy to speed recovery. (Note, however, that because ***esophageal*** peptic ulcers associated with gastroesophageal reflux disease are not caused by *H. pylori,* these esophageal ulcers are still treated using acid-suppressing drugs and not antibiotics.)

Now, based on knowledge of closely related bacteria and remarkably successful vaccination trials in mice, development of a vaccine against *H. pylori* is expected. With preventive vaccination coupled with antibiotic cure, the number of ulcer sufferers should plummet, possibly making the goal of eradicating peptic ulcers attainable within 25 years!

these ions. Parietal cells also secrete **gastric intrinsic factor,** a protein necessary for the absorption of vitamin B_{12} by the small intestine. The body uses this vitamin in the manufacture of red blood cells.

3. **Chief (zymogenic) cells,** which occur mainly in the basal parts of the glands, make and secrete the enzymatic protein **pepsinogen** (pep-sin′o-jen), which is activated to pepsin when it encounters acid in the apical region of the gland (Figure 22.15c). Chief cells have features typical of protein-secreting cells: a well-developed rough endoplasmic reticulum (rough ER) and Golgi apparatus, plus secretory granules in the apical cytoplasm.

At least two other epithelial cell types occur in the gastric glands but also extend beyond these glands:

1. *Enteroendocrine cells* (en″ter-o-en′do-krin; "gut endocrine") are hormone-secreting cells scattered throughout the lining epithelium and glands of the alimentary canal. These cells (Figure 22.15c) release their hormones into the capillaries of the underlying lamina propria. One of these hormones, gastrin, signals parietal cells to secrete HCl when food enters the stomach. Most enteroendocrine cells that produce gastrin are in the stomach's pyloric region.
2. *Undifferentiated stem cells* (not illustrated) are located throughout the stomach, at the junction of the gastric glands and gastric pits. These cells divide continuously, replacing the entire lining epithelium of goblet cells every 3–7 days. Such rapid replacement is vital because individual goblet cells can survive for only a few days in the harsh environment of the stomach.

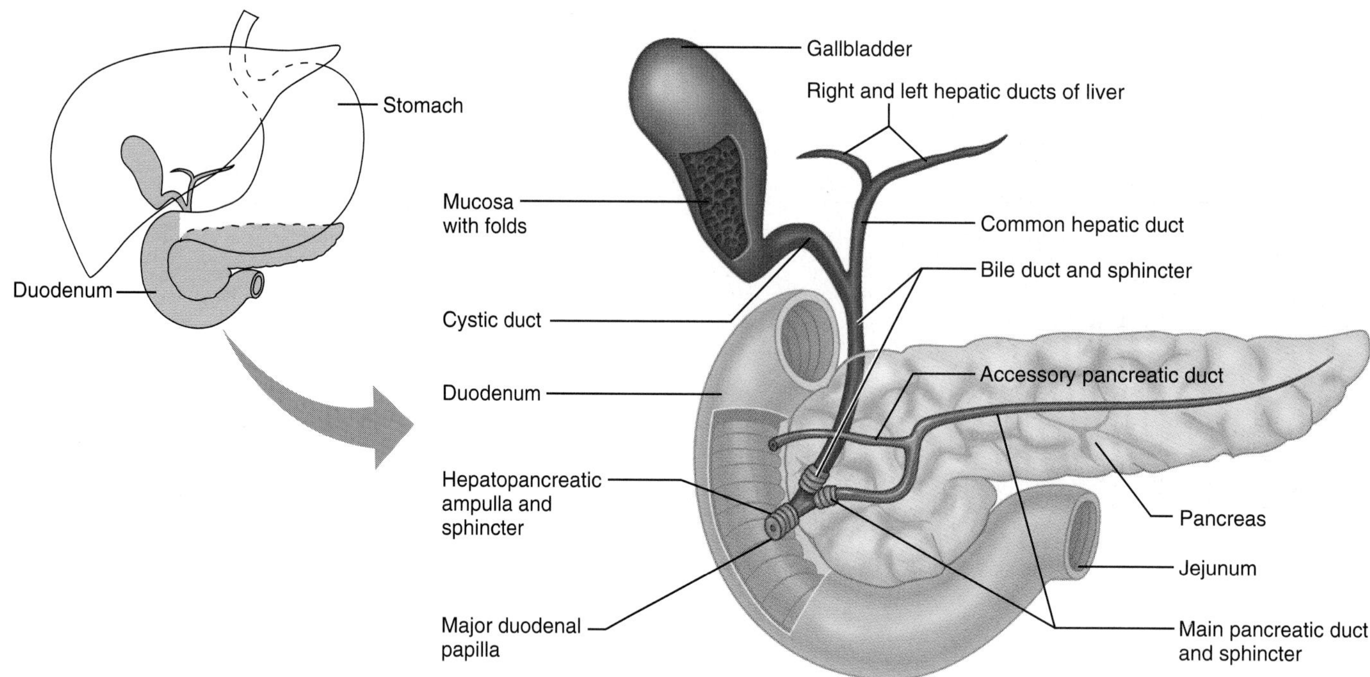

FIGURE 22.16 The duodenum of the small intestine, and related organs. For orientation, see the inset above, where these organs are colored purple. Note the various ducts opening into the duodenum from the pancreas, gallbladder, and liver. (In this figure, the gallbladder has been reflected upward.)

The Small Intestine

The **small intestine** is the longest part of the alimentary canal (see Figure 22.1) and the site of most enzymatic digestion and virtually all absorption of nutrients. Most digestive enzymes that operate within the small intestine are secreted not by the intestine, but by the pancreas. During digestion, the small intestine undergoes active segmentation movements, shuffling the chyme back and forth and thereby maximizing its contact with the nutrient-absorbing mucosa. Peristalsis propels chyme through the small intestine in about 3–6 hours.

Gross Anatomy

The small intestine is a convoluted tube that runs from the pyloric sphincter, in the epigastric region of the abdomen, to the first part of the large intestine, in the lower right quadrant. It is shorter in living people (2.7–5 meters) than in preserved cadavers (6–7 meters), where loss of muscle tone and the effects of preservatives have caused it to lengthen.

The small intestine has three subdivisions (see Figure 22.1): the **duodenum** (du″o-de′num; "twelve finger-widths long"), the **jejunum** (je-joo′num; "empty"), and the **ileum** (il′e-um; "twisted intestine"), which contribute 5%, almost 40%, and almost 60% of the length of the small intestine, respectively. Whereas most of the C-shaped duodenum lies retroperitoneally, the jejunum and ileum form sausage-like coils that hang from the posterior abdomen by a mesentery and are framed by the large intestine. The jejunum makes up the superior left part of this coiled intestinal mass, whereas the ileum makes up the inferior right part.

Even though the *duodenum* is the shortest subdivision of the small intestine, it has the most features of interest (Figure 22.16). It receives digestive enzymes from the pancreas and bile from the liver and gallbladder via the *main pancreatic duct* and the *bile duct,* which enter the wall of the duodenum where they form a bulb called the **hepatopancreatic ampulla** (hep″ah-to-pan″kre-ah′tik am-pul′ah; "flask from the liver and pancreas"). This ampulla opens into the duodenum via a volcano-shaped mound called the **major duodenal papilla.** Entry of bile and pancreatic juice into the duodenum is controlled by sphincters of smooth muscle that surround the hepatopancreatic ampulla and the ends of the pancreatic and bile ducts.

The small intestine is innervated by parasympathetic fibers from the vagus and by sympathetic fibers from the thoracic splanchnic nerves, both relayed through the superior mesenteric (and celiac) plexus. Its arterial supply comes primarily via the superior mesenteric artery (see p. 560). The veins run parallel to the arteries and typically drain into the superior mesenteric vein; from there, the nutrient-rich venous blood drains into the hepatic portal vein (see p. 570), which carries it to the liver.

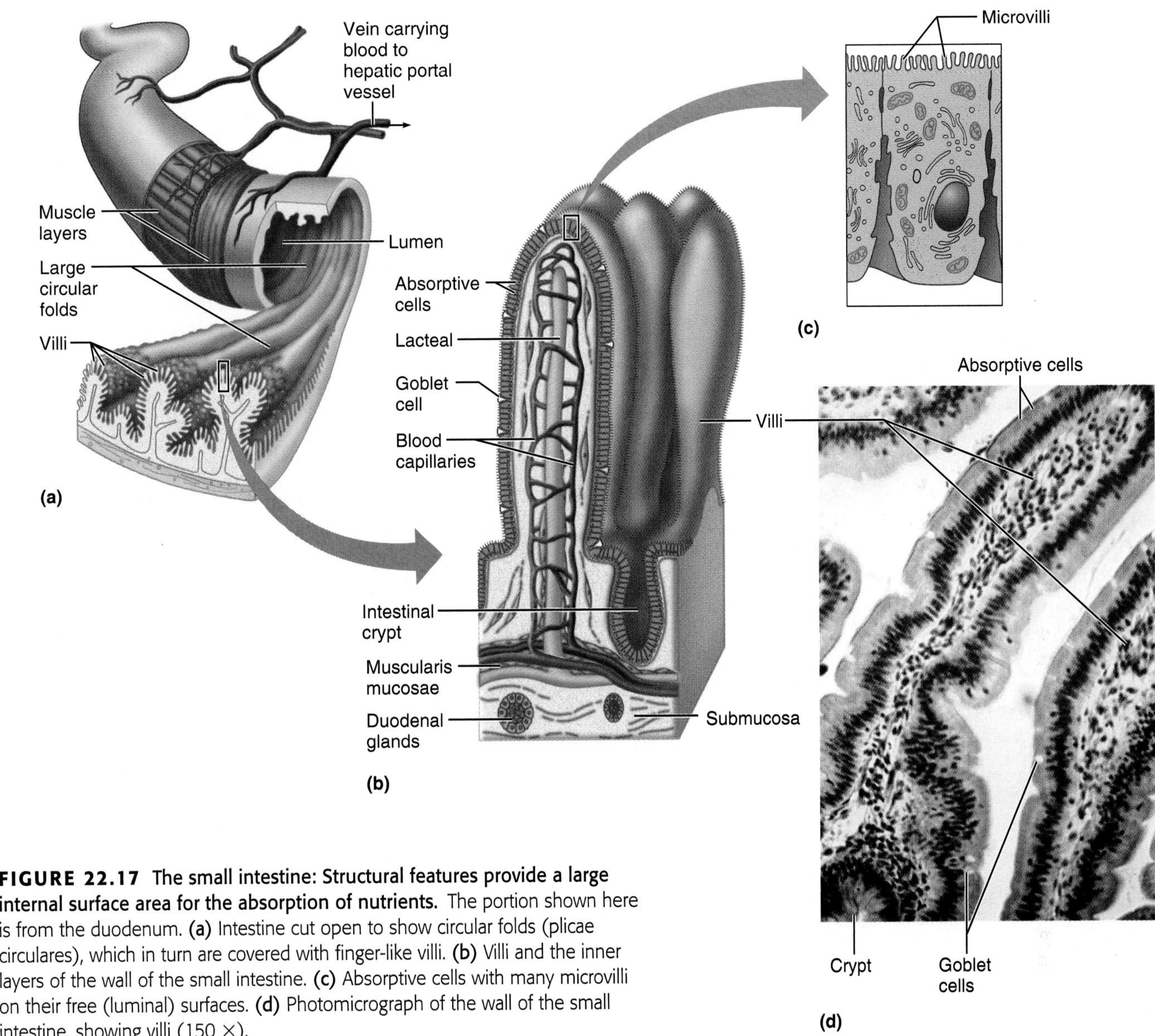

FIGURE 22.17 The small intestine: Structural features provide a large internal surface area for the absorption of nutrients. The portion shown here is from the duodenum. **(a)** Intestine cut open to show circular folds (plicae circulares), which in turn are covered with finger-like villi. **(b)** Villi and the inner layers of the wall of the small intestine. **(c)** Absorptive cells with many microvilli on their free (luminal) surfaces. **(d)** Photomicrograph of the wall of the small intestine, showing villi (150 ×).

Microscopic Anatomy

As previously noted, nearly all nutrient absorption occurs in the small intestine, which is highly adapted for this function. In addition to the huge surface area for absorption provided by the small intestine's great length, several other structural features provide even more absorptive surface area, and serve other functions related to digestion as well.

Modifications for Absorption The wall of the small intestine has three structural modifications that amplify its absorptive surface enormously: *circular folds, villi,* and extensive *microvilli* (Figure 22.17). Because most absorption occurs in the proximal region of the small intestine, these specializations decrease in number toward the distal end.

The **circular folds** or **plicae circulares** (pli′ke) are permanent, transverse ridges of the mucosa and submucosa. They are nearly 1 cm tall. Besides increasing the absorptive surface area, these folds force the chyme to spiral through the intestinal lumen, slowing its movement and allowing time for complete absorption of nutrients.

Villi are finger-like projections of the mucosa that give it a velvety texture, much like the soft nap of a towel. Over 1 mm tall, and thus large enough to be seen with the unaided eye, villi are covered by a simple columnar epithelium made up primarily of **absorptive cells** specialized for absorbing digested nutrients (see Figure 22.17b and c). (These absorptive cells are also called *enterocytes,* literally "intestinal cells.") Within the lamina-propria core of each villus is a network of blood capillaries and a wide lymph capillary called a **lacteal** (see p. 585). With the exception of absorbed fats, which enter the lacteals, all end products of nutrient digestion enter the blood capillaries.

The villi are able to move during digestion because each contains a slip of smooth muscle that extends into the core of the villus from the muscularis mucosae. The movements enhance absorption efficiency by increasing the amount of contact between the villi and the nutrients in the intestinal lumen, and they also squeeze lymph through the lacteals.

The apical surfaces of the absorptive cells have many **microvilli** (see Figure 22.17c). Whereas such projections occur on most epithelial surfaces in the body, those in the small intestine are exceptionally long and densely packed. Besides amplifying the absorptive surface, the plasma membrane of these microvilli contains enzymes that complete the final stages of the breakdown of nutrient molecules.

The amount of absorptive surface in the small intestine is remarkable. Together, the circular folds, villi, and microvilli increase the intestinal surface area to about 200 square meters, equivalent to the floor area of an average two-story house!

Histology of the Wall All typical layers of the alimentary canal occur in the small intestine. The lining epithelium, which occurs not only on the villi but on the intestinal surface between villi, contains the previously mentioned absorptive cells plus some scattered goblet cells and enteroendocrine cells:

1. *Absorptive cells* (see Figure 22.17c). These cells contain many mitochondria because the uptake of digested nutrients is an energy-demanding process. They also contain an abundant endoplasmic reticulum, which assembles the newly absorbed lipid molecules into lipid-protein complexes called *chylomicrons* (ki″lo-mi′kronz). (Once made, the chylomicrons enter the lacteal capillaries, so it is in this form that absorbed fat enters the circulation.)
2. *Goblet cells* (see Figure 22.17b). These cells secrete onto the internal surface of the intestine a coat of mucus that lubricates the chyme and forms a protective barrier that prevents enzymatic digestion of the intestinal wall.
3. *Enteroendocrine cells.* The enteroendocrine cells of the duodenum (not illustrated) secrete several hormones, including *cholecystokinin* (ko″le-sis″to-ki′nin; "gallbladder activator"), which signals the gallbladder to release stored bile and the pancreas to secrete its digestive enzymes, and *secretin,* which signals the pancreatic ducts to secrete a bicarbonate-rich juice to neutralize the acidic chyme entering the duodenoum.

Between the villi, the mucosa contains tubes called **intestinal crypts,** *intestinal glands,* or *crypts of Lieberkühn* (Lēb′er-kun) (see Figure 22.17b). The epithelial cells that line these crypts secrete *intestinal juice,* a watery liquid that mixes with chyme in the intestinal lumen. Furthermore, undifferentiated epithelial cells lining the intestinal crypts renew the mucosal epithelium by dividing rapidly and moving continuously onto the villi. These are among the most quickly dividing cells of the body, completely renewing the inner epithelium of the small intestine every 3–6 days. Such rapid replacement is necessary because individual epithelial cells cannot long withstand the destructive effects of the digestive enzymes in the intestinal lumen.

Critical Reasoning

Explain why cancer chemotherapies that stop the replication of cellular DNA throughout the body cause nausea, diarrhea, and vomiting.

Because these treatments preferentially destroy all of the body's most rapidly dividing cells, chemotherapy usually obliterates the dividing cells of the intestinal epithelium in addition to cancer cells. The resulting destruction of the lining of the alimentary tract disrupts digestion and causes these signs and symptoms.

Although the intestinal crypts contain mostly undifferentiated cells, they also contain mature *Paneth cells* (not illustrated). These epithelial cells secrete enzymes that destroy certain bacteria and may help determine which kinds of bacteria live in the intestinal lumen. The permanent bacterial residents of the intestinal lumen, called the *intestinal flora,* manufacture some essential vitamins, which the intestines absorb. Vitamin K is one substance produced by the intestinal bacteria.

The lamina propria and submucosa of the small intestine contain many areas of lymphoid tissue, including *aggregated lymphoid nodules (Peyer's patches)* in the ileum (see p. 598).

The submucosa of the small intestine is a typical connective tissue. In the duodenum only, it contains a set of compound tubular **duodenal glands** (also called *Brunner's glands*), whose ducts open into the intestinal crypts (see Figure 22.17b). These glands secrete an alkaline, bicarbonate-rich mucus that helps neutralize the acidity of the chyme from the stomach and contributes to the protective layer of mucus on the inner surface of the small intestine.

The outer layers of the small intestine (muscularis externa and serosa) have no unusual features.

The Large Intestine

The **large intestine** is the last major organ in the alimentary canal (Figure 22.18). The material that reaches it is a largely digested residue that contains few nutrients. During the 12–24 hours that this residue remains in the large intestine, little additional breakdown of food occurs, except for the small amount of digestion performed by the many bacteria living there. Even though the large intestine absorbs these few remaining nutri-

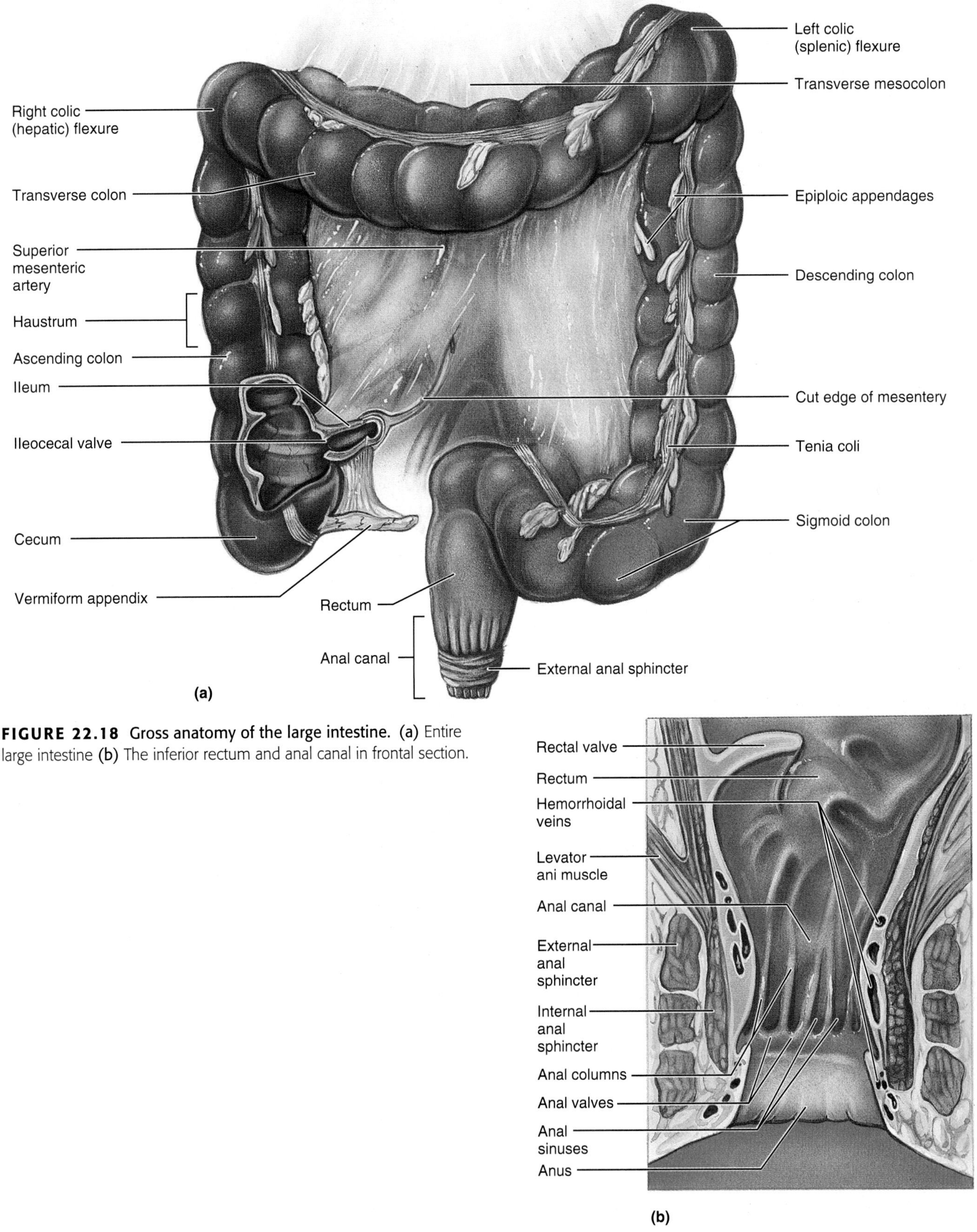

Left colic (splenic) flexure
Transverse mesocolon
Right colic (hepatic) flexure
Transverse colon
Epiploic appendages
Superior mesenteric artery
Descending colon
Haustrum
Ascending colon
Ileum
Cut edge of mesentery
Ileocecal valve
Tenia coli
Sigmoid colon
Cecum
Vermiform appendix
Rectum
Anal canal
External anal sphincter
(a)

FIGURE 22.18 **Gross anatomy of the large intestine.** (a) Entire large intestine (b) The inferior rectum and anal canal in frontal section.

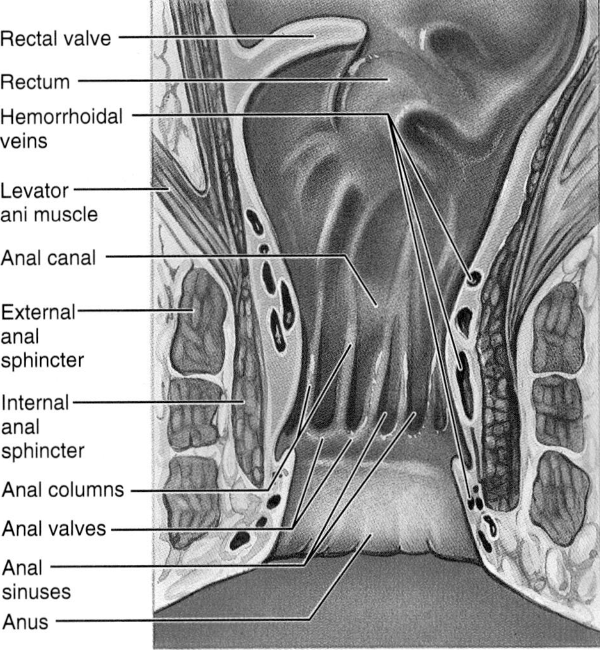

ents, its main function is to absorb water and electrolytes from the digested mass, resulting in semisolid feces. Propulsion through the large intestine is sluggish and weak, except for **mass peristaltic movements,** which pass over the colon a few times a day to force the feces powerfully toward the rectum.

Gross Anatomy

The large intestine frames the small intestine on 3½ sides, forming an open rectangle (see Figure 22.1). This organ, which is wider than the small intestine but less than half as long (1.5 meters), has the following subdivisions: *cecum, vermiform appendix, colon, rectum,* and *anal canal* (Figure 22.18a).

Over most of its length, the large intestine exhibits three special features: teniae coli, haustra, and epiploic appendages. **Teniae (taeniae) coli** (te′ne-e ko′li; "ribbons of the colon") are three longitudinal strips, spaced at equal intervals around the circumference of the cecum and colon. They are thickenings of the longitudinal layer of the muscularis externa, which is thin except at these sites. Because the teniae maintain muscle tone, they cause the large intestine to pucker into sacs, or **haustra** (haw′strah; "to draw up"). **Epiploic appendages** (ep″ĭ-plo′ik; "membrane-covered"), also called *omental appendices,* are fat-filled pouches of visceral peritoneum that hang from the intestine. Their significance is unknown.

The Cecum and Vermiform Appendix The large intestine begins with the sac-like **cecum** (se′kum; "blind pouch") in the right iliac fossa. The opening of the ileum of the small intestine into the cecum's medial wall is surrounded internally by the **ileocecal valve** (Figure 22.18a), which is formed by two raised edges of the mucosa. A sphincter in the distal ileum keeps the valve closed until there is food in the stomach, at which time the sphincter reflexively relaxes, opening the valve. As the cecum fills, its walls stretch, pulling the edges of the ileocecal valve together and closing the opening. This action prevents reflux of feces from the cecum back into the ileum.

The **vermiform appendix** (ver′mĭ-form; "worm-shaped") is a blind tube that opens into the posteromedial wall of the cecum. Although almost always illustrated as hanging inferiorly, it more often lies "tucked up" posterior to the cecum in the right iliac fossa. The appendix has large masses of lymphoid tissue in its wall (p. 598) and probably functions like the tonsils, gathering antigens, neutralizing harmful pathogens, and generating memory lymphocytes for long-term immunity. This function predisposes the appendix, like the tonsils, to serious infections.

CLINICAL APPLICATIONS

APPENDICITIS Appendicitis, acute inflammation of the appendix, results from a blockage that traps infectious bacteria within its lumen. The blockage often is caused by a lump of feces or by a virus-induced swelling of the lymphoid tissue of the appendix wall. Unable to empty its contents, the blocked appendix swells with the mucus it secretes, squeezing off its venous drainage and leading to ischemic necrosis and infection. If the appendix ruptures, bacteria and feces are released onto the abdominal organs, causing peritonitis. Because the symptoms of appendicitis vary greatly, this condition is notoriously difficult to diagnose. Often, however, the first symptom is pain in the umbilical region, followed by loss of appetite, fever, nausea, vomiting, and relocalization of pain to the lower right quadrant of the abdominal surface. Appendicitis occurs most often between the ages of 15 and 25, and it affects 7% of people in western countries. Immediate surgical removal of the appendix, called **appendectomy** (ap″en-dek′to-me), is the usual treatment. A new, high-resolution imaging technique called *limited computed tomography* is generating much excitement because it greatly improves the chances of diagnosing appendicitis noninvasively. ✚

The Colon The **colon** (ko′lon) has several distinct segments (see Figure 22.18a). From the cecum, the **ascending colon** ascends along the right side of the posterior abdominal wall and reaches the level of the right kidney, where it makes a right-angle turn, the **right colic flexure** (also called the **hepatic flexure** because the liver lies directly superior to it). From this flexure, the **transverse colon** extends to the left across the peritoneal cavity. Directly anterior to the spleen, it bends acutely downward at the **left colic (splenic) flexure** and descends along the left side of the posterior abdominal wall as the **descending colon.** Inferiorly, the colon enters the true pelvis as the S-shaped **sigmoid colon** (*sigma* = the Greek letter corresponding to the letter *s).*

CLINICAL APPLICATIONS

DIVERTICULOSIS AND DIVERTICULITIS When the diet lacks fiber, the contents of the colon are reduced in volume, and the contractions of the circular muscle in the colon exert greater pressures on its wall. When this pressure promotes the formation of multiple sacs called *diverticula* (di″ver-tik′u-lah), which are small outward herniations of the mucosa through the colon wall, the resulting condition is termed **diverticulosis.** This condition arises most frequently in the sigmoid colon. Occurring in 30% to 40% of all Americans over age 50, and in half of those over 70, diverticulosis generally leads to nothing more than dull pain, although it may rupture an artery in the colon and produce bleeding from the anus. Increasing the amount of fiber in the diet generally relieves the symptoms.

In about 20% of diverticulosis cases, however, patients develop the more serious condition called **diverticulitis,** in which the inflamed diverticula become infected and may perforate, leaking feces into the peritoneal cavity. In serious cases, the affected region of the colon is removed by surgery and antibiotics are given to fight the peritonitis. ✚

Rectum In the pelvis, the sigmoid colon joins the **rectum** (see Figure 22.18a), which descends along the inferior half of the sacrum. The rectum has no teniae coli;

instead, its longitudinal muscle layer is complete and well-developed, so that it can generate strong contractions for defecation. Even though the word *rectum* means "straight," the rectum actually has several tight bends. Internally, these bends are represented as three *transverse folds of the rectum,* or **rectal valves** (see Figure 22.18b), which prevent feces from being passed along with flatus (gas).

Anal Canal The last subdivision of the large intestine is the **anal canal** (see Figure 22.18a). About 3 cm long, it begins where the rectum passes through the levator ani, the muscle that forms the pelvic floor. Thus, the anal canal lies entirely external to the abdominopelvic cavity.

Internally, the superior half of the anal canal contains longitudinal folds of mucosa, the **anal columns** (see Figure 22.18b). Neighboring anal columns join each other inferiorly at crescent-shaped transverse folds called **anal valves.** The pockets just superior to these valves are **anal sinuses,** which release mucus when they are compressed by feces, providing lubrication that eases fecal passage during defecation.

The horizontal line along which the anal valves lie is called the *pectinate* ("comb-shaped") or *dentate* ("teeth-shaped") *line.* Because the mucosa superior to this line is innervated by visceral sensory fibers, it is relatively insensitive to pain. Inferior to the pectinate line, however, the mucosa is sensitive to pain because it is innervated by somatic nerves.

CLINICAL APPLICATIONS

HEMORRHOIDS Hemorrhoids are varicosities of the hemorrhoidal veins in the anal canal (see Figure 22.18b) that often result from straining to deliver a baby or to defecate. Because they are stretched and inflamed, the swollen veins throb and bulge into the lumen of the anal canal. Internal hemorrhoids and external hemorrhoids occur above and below the pectinate line, respectively. External hemorrhoids are itchier and more painful, but only internal hemorrhoids tend to bleed. About 75% of Americans develop hemorrhoids at some time in their lives.

Severe hemorrhoids are often treated by tying them at their base with small rubber bands, whereupon they wither and fall away. They can also be injected with a hardening agent or exposed to electricity or strong infrared light to coagulate the blood within them. These unpleasant-sounding treatments are simple and effective and have largely eliminated the difficult surgical removals that were formerly performed. ✚

The wall of the anal canal contains two sphincter muscles: an **internal anal sphincter** of smooth muscle and an **external anal sphincter** of skeletal muscle (see Figure 22.18b). The former is a thickening of the circular layer of the muscularis, whereas the latter is a distinct muscle (also see Figure 11.12b on p. 293). The external sphincter contracts voluntarily to inhibit defecation, whereas the internal sphincter contracts involuntarily, both to prevent feces from leaking from the anus between defecations and to inhibit defecation during emotional stress. During toilet training, children learn to control the external anal sphincter.

Vessels and Nerves The first half of the large intestine—to a point two-thirds of the way along the transverse colon—is supplied by the superior mesenteric vessels. Its sympathetic innervation is from the superior mesenteric and celiac ganglia and plexuses, and its parasympathetic innervation is from the vagus nerve.

The distal half of the large intestine is largely supplied by the inferior mesenteric vessels, although the lower rectum and the anal canal are served by rectal branches of the internal iliac vessels. The sympathetic innervation of the distal half of the large intestine is via the inferior mesenteric and hypogastric plexuses, and the parasympathetic innervation is from the pelvic splanchnic nerves. The final part of the anal canal is innervated by somatic nerves, such as the pudendal nerve, rather than by visceral nerves.

Defecation The rectum is usually empty, but when feces are squeezed into it by mass peristaltic movements, the stretching of the rectal wall initiates the defecation reflex. Mediated by the spinal cord, this parasympathetic reflex signals the walls of the sigmoid colon and rectum to contract and the anal sphincters to relax. If one decides to delay defecation, the reflexive contractions end, and the rectum relaxes. Another mass movement occurs a few minutes later, initiating the defecation reflex again—and so on, until one chooses to defecate or the urge to defecate becomes unavoidable.

During defecation, the musculature of the rectum contracts to expel the feces. This process is supplemented by the voluntary contraction of the diaphragm and the abdominal wall muscles, which increases intra-abdominal pressure, and of the levator ani muscle (see p. 292), which lifts the anal canal superiorly, leaving the feces inferior to the anus and thus outside the body.

Microscopic Anatomy

The wall of the large intestine (Figure 22.19) resembles that of the small intestine in some ways and differs from it in others. Villi are absent, which reflects the fact that fewer nutrients are absorbed in the large intestine. Intestinal crypts are present, however, as simple tubular glands. The internal surface of the colon (including the intestinal crypts) is lined by a simple columnar epithelium containing the same cell types as in the small intestine. Goblet cells are more abundant here, however, for they secrete large amounts of lubricating mucus that eases the passage of feces toward the end of the alimentary canal. The *absorptive cells,* labeled "columnar cells" in Figure 22.19a, take in water and electrolytes. Finally, undifferentiated stem cells occur at the bases of the intestinal crypts, and full epithelial replacement occurs every week or so.

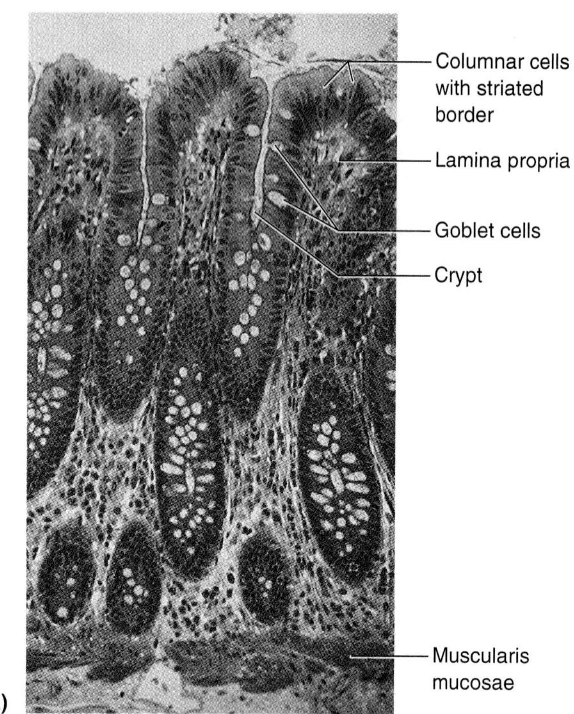

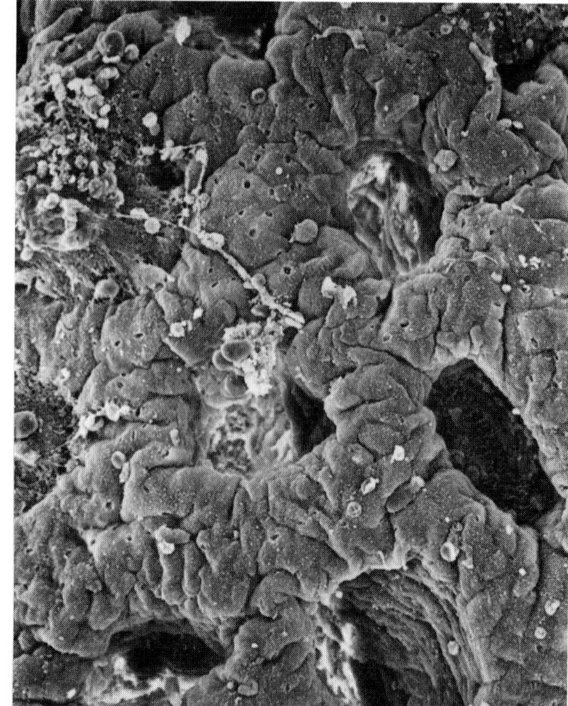

FIGURE 22.19 The mucosa of the large intestine. (a) Photomicrograph. Note the abundance of goblet cells (120 ×). (b) Scanning electron micrograph of the mucosal surface of the colon; the dark areas are entrances to the intestinal crypts (575 ×).

The other layers of the wall are rather typical. The lamina propria and submucosa contain more lymphoid tissue than occurs elsewhere in the alimentary canal, but this is not surprising, considering the extensive bacterial flora of the large intestine. The specializations of the muscularis externa and serosa, namely the teniae coli and epiploic appendages, were discussed previously.

The *anal canal* is a zone of epithelial transition in which the simple columnar epithelium of the intestine abruptly changes to stratified squamous epithelium near the level of the pectinate line. At the extreme inferior end of the anal canal, the mucosa merges with the true skin that surrounds the anus.

The Liver

The ruddy **liver** is the largest gland in the body, weighing about 1.4 kg (3 pounds) in an average adult. Amazingly versatile, it performs over 500 functions! Its digestive function is to produce **bile,** a green alkaline liquid that is stored in the gallbladder and secreted into the duodenum. Bile salts emulsify fats in the small intestine; that is, they break up fatty nutrients into tiny particles that are more accessible to digestive enzymes from the pancreas. The liver also performs many metabolic functions: It picks up glucose from nutrient-rich blood returning from the alimentary canal and stores this carbohydrate as glycogen for subsequent use by the body; it processes fats and amino acids and stores certain vitamins; it detoxifies many poisons and drugs in the blood; and it makes the blood proteins. Almost all of these functions are carried out by a type of cell called a **hepatocyte** (hep′ah-to-sīt″), or simply a *liver cell.*

Gross Anatomy

The liver lies inferior to the diaphragm in the right superior part of the abdominal cavity (see Figures 22.1 and 22.20c), filling much of the right hypochondriac and epigastric regions and extending into the left hypochondriac region. It lies almost entirely within the rib cage, which protects this highly vascular organ from blows that could rupture it. The liver is shaped like a wedge, the wide base of which faces right and the narrow apex of which lies just inferior to the level of the left nipple.

The liver has two surfaces: the **diaphragmatic** and **visceral** surfaces (see Figures 22.20 and 22.21). The **diaphragmatic surface** faces anteriorly and superiorly, whereas the **visceral surface** faces posteroinferiorly. Even though most of the liver is covered with a layer of visceral peritoneum, the superior part, called the **bare area,** is fused to the diaphragm and is therefore devoid of peritoneum.

The liver has a **right lobe** and a **left lobe,** which traditionally were considered to be divided by the **falciform ligament** on the anterior part of the diaphragmatic surface (see Figure 22.20a) and the **fissure** on the visceral surface (see Figure 22.21). The falciform liga-

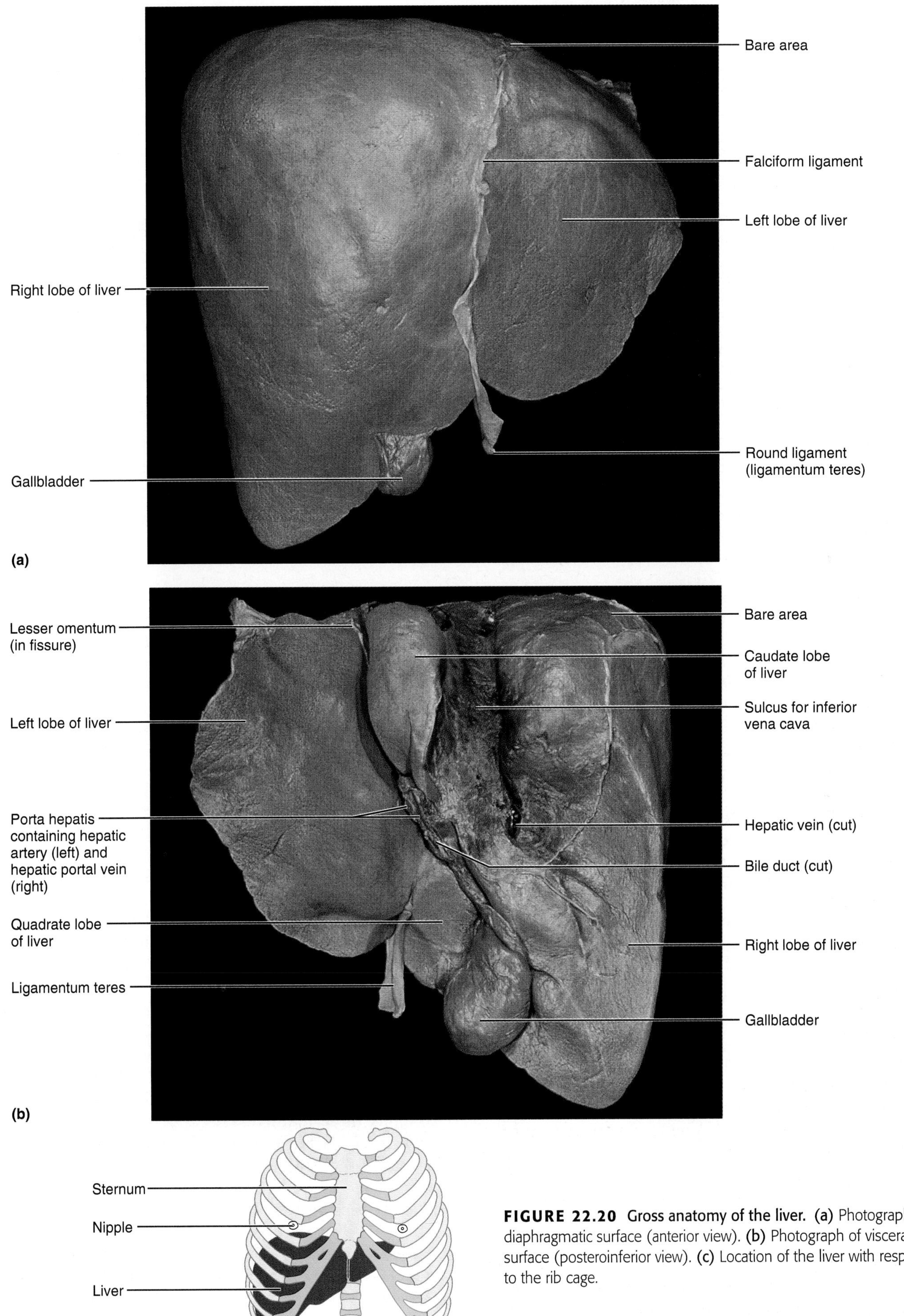

FIGURE 22.20 **Gross anatomy of the liver.** **(a)** Photograph of diaphragmatic surface (anterior view). **(b)** Photograph of visceral surface (posteroinferior view). **(c)** Location of the liver with respect to the rib cage.

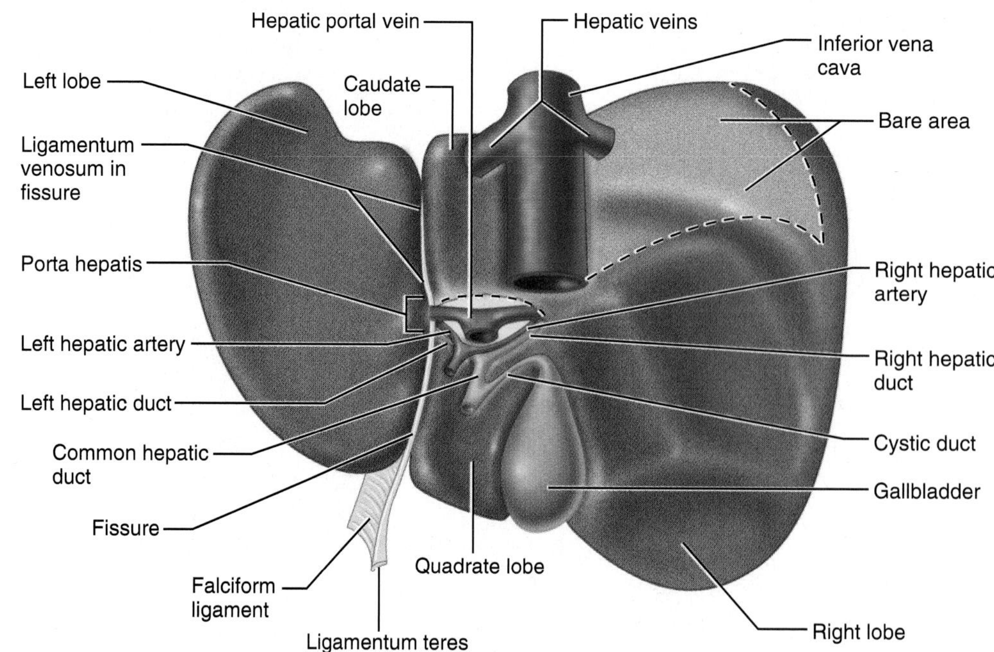

FIGURE 22.21 Visceral surface of the liver (posteroinferior view). Note the vessels and ducts that enter and leave the liver.

ment is a vertical mesentery that binds the liver to the anterior abdominal wall, and the fissure is a deep groove in the same sagittal plane as the falciform ligament. Two other lobes, the **quadrate lobe** and the **caudate lobe,** are visible on the visceral surface just to the right of the fissure (see Figures 22.20b and 22.21). Long considered part of the right lobe, these lobes are now considered parts of the left lobe, with which they share nerves and vessels.

An important area near the center of the visceral surface is the **porta hepatis** (por'tah hep-ah'tis; "gateway to the liver"), where most of the major vessels and nerves enter and leave the liver (see Figure 22.21). The right and left branches of the *hepatic portal vein,* which carry nutrient-rich blood from the stomach and intestines, enter the porta hepatis, as do the right and left branches of the *hepatic artery* carrying oxygen-rich blood to the liver. The **right** and **left hepatic ducts,** which carry bile from the respective liver lobes, exit from the porta hepatis and fuse to form the **common hepatic duct,** which extends inferiorly toward the duodenum. Autonomic nerves reach the liver from the celiac plexus and consist of both sympathetic and parasympathetic (vagal) fibers. Other important structures on the liver's visceral surface are the *gallbladder* and the *inferior vena cava,* which lie to the right of the quadrate and caudate lobes, respectively (see Figure 22.21). The inferior vena cava receives the *hepatic veins* carrying blood out of the liver.

Several structures relate to the liver's fissure. Lying in the fissure's inferior half is the **ligamentum teres** (*teres* = round), or **round ligament.** This cord-like ligament, the remnant of the umbilical vein in the fetus (see p. 576), ascends to the liver from the navel, within the inferior margin of the falciform ligament. Additionally, the superior half of the liver's fissure contains the **ligamentum venosum** (see Figure 22.21), a cord-like remnant of the ductus venosus of the fetus (see p. 576).

Microscopic Anatomy

The liver contains over a million classical **liver lobules** (Figure 22.22), each about the size of a sesame seed. Each lobule is shaped like a hexagonal (six-sided) solid and consists of plates of liver cells, or hepatocytes, radiating out from a **central vein** (see Figure 22.22c). If you were to look at the top of a thick paperback book opened so wide that its two covers touched each other, you would have a rough model of the liver lobule, with the spreading pages representing the plates of hepatocytes, and the hollow cylinder formed by the rolled spine representing the central vein. The hepatocytes in each plate are organized like bricks in a wall.

At almost every corner of the lobule is a **portal triad** (tri'ad; "three"), also called a **portal tract** or *portal canal* (see Figure 22.22c and d). The portal triad contains three main vessels: a portal arteriole that is a branch of the hepatic artery, a portal venule that is a branch of the hepatic portal vein, and a **bile duct** (which carries bile away from the liver lobules). Note that the blood vessels bring both arterial and venous blood to the lobule. The arterial blood supplies the hepatocytes with oxygen, and blood from the portal vein delivers substances from the intestines for processing by the hepatocytes.

Between the plates of hepatocytes are large capillaries, the **liver sinusoids.** Near the portal triads, these

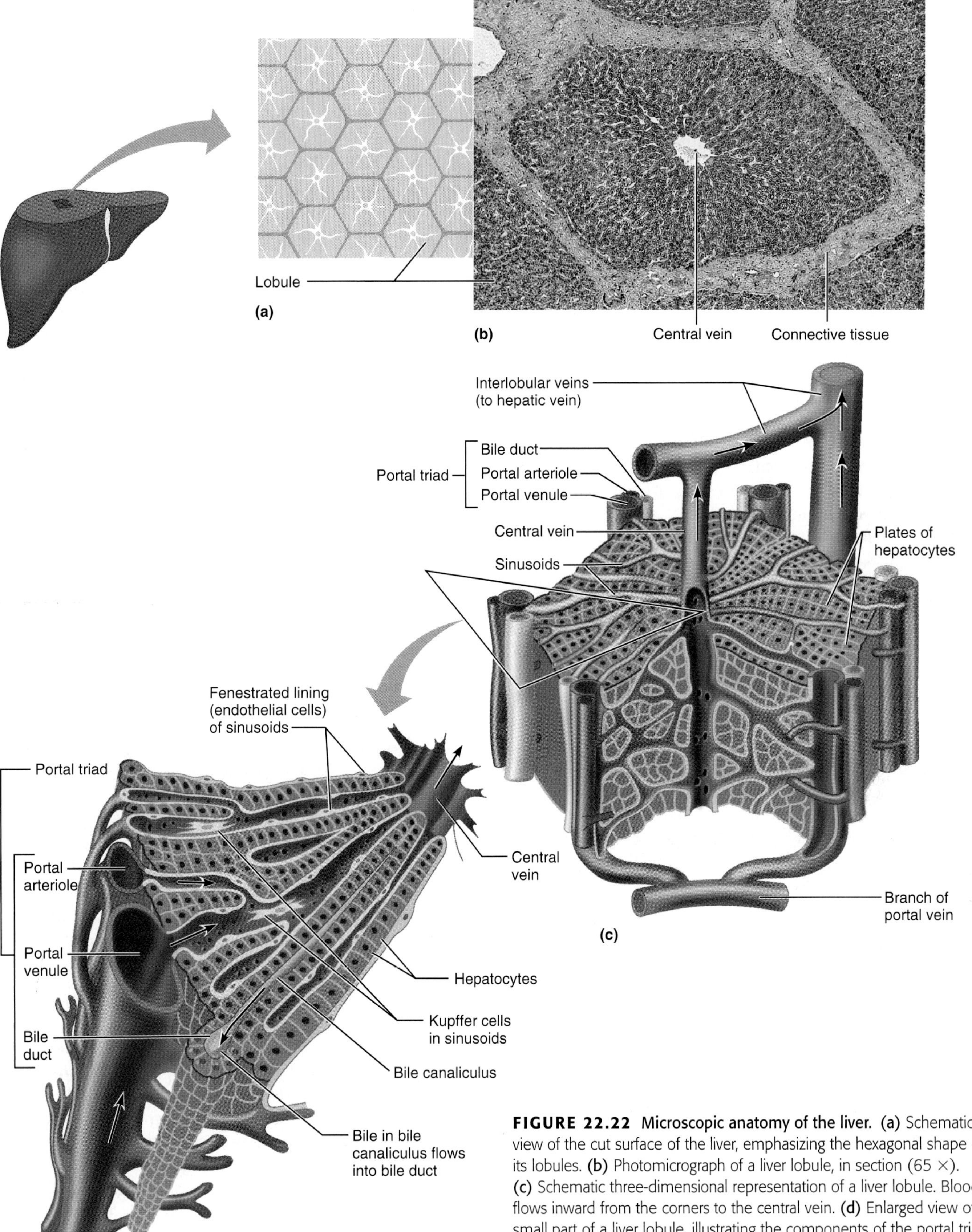

FIGURE 22.22 **Microscopic anatomy of the liver.** **(a)** Schematic view of the cut surface of the liver, emphasizing the hexagonal shape of its lobules. **(b)** Photomicrograph of a liver lobule, in section (65 ×). **(c)** Schematic three-dimensional representation of a liver lobule. Blood flows inward from the corners to the central vein. **(d)** Enlarged view of a small part of a liver lobule, illustrating the components of the portal triad (portal tract) region, the positions of the bile canaliculi, Kupffer cells in the sinusoids, and hepatocytes (the brown square cells). The directions of blood flow and bile flow are indicated by arrows.

sinusoids receive blood from both the portal arteriole and venule, and carry this blood inward to reach the central vein (see Figure 22.22d). From there, the central veins form tributaries (interlobular veins) that ultimately lead to the hepatic veins and then to the inferior vena cava outside the liver.

Inside the sinusoids are star-shaped **Kupffer cells,** or *hepatic macrophages,* which destroy bacteria and other foreign particles in the blood flowing past them. Thus, even though microorganisms in the intestine may enter the intestinal capillaries, few of them make it past the liver. Besides cleansing the blood of microorganisms, Kupffer cells also destroy worn-out blood cells—as do macrophages of the spleen and bone marrow.

The liver sinusoids are lined by an exceptionally leaky, fenestrated endothelium. Vast quantities of blood plasma pour out of the sinusoids, bathing the hepatocytes with fluid. Hepatocytes require proximity to such a large blood supply because so many of their functions depend on interactions with the fluid portion of blood.

Hepatocytes possess large numbers of many different organelles that enable them to carry out their many functions. For example, the abundant rough ER manufactures the blood proteins; the well-developed smooth ER helps produce bile salts and detoxifies blood-borne poisons; abundant peroxisomes detoxify other poisons (including alcohol; see p. 39); and the large Golgi apparatus packages the abundant secretory products from the endoplasmic reticulum. Energy for these functions is provided by an exceptional abundance of mitochondria. Additionally, hepatocytes are loaded with glycosomes, a feature that reflects their role in storing sugar for the regulation of blood glucose levels.

Finally, hepatocytes have a great capacity for cell division and regeneration: Judging from experiments on laboratory animals, if half of a person's liver were removed, it would all regenerate in a few weeks! Cellular replacement occurs through the division of mature hepatocytes and of *liver stem cells (oval cells),* which are located near the bile ducts at the portal triads.

Collectively, hepatocytes produce about 500–1000 ml of bile each day. The secreted bile enters tiny intercellular spaces or channels, called **bile canaliculi** ("little canals"), that lie between adjacent hepatocytes (see Figure 22.22d). These canaliculi carry bile outward through each lobule, emptying into the bile ducts in the portal triads. From there, the bile flows into progressively larger ducts, exiting the liver through the hepatic ducts at the porta hepatis. Beyond this, additional bile-carrying ducts lead to the duodenum, as described shortly.

CLINICAL APPLICATIONS

CIRRHOSIS Cirrhosis (sĭ-ro′sis; "orange-colored") is a progressive inflammation of the liver that usually results from chronic alcoholism. Even though the alcohol-poisoned hepatocytes are continuously replaced, the liver's connective tissue regenerates faster, so the liver becomes fibrous and fatty, and its function declines. The scar tissue impedes the flow of blood through the liver, causing portal hypertension (see p. 572). The patient may grow confused or comatose as toxins accumulate in the blood and depress brain functions. Besides alcoholism, other causes of cirrhosis include hepatitis (see the Disorders section on p. 667) and autoimmune attack on the bile ducts. ✚

The Gallbladder

The **gallbladder** is a muscular sac, resting in a shallow depression on the visceral surface of the liver (see Figure 22.21). It functions to store and concentrate bile produced by the liver. Its rounded head, or *fundus,* protrudes from the liver's inferior margin (Figure 22.20a). Internally, a honeycomb pattern of mucosal foldings (see Figure 22.16) enables the mucosa to expand as the gallbladder fills.

The gallbladder's duct, the **cystic duct** (*cyst* = bladder; see Figure 22.16), joins the common hepatic duct from the liver to form the **bile duct** *(common bile duct),* which empties into the duodenum. The liver secretes bile continuously, but sphincters at the end of the bile duct and at the hepatopancreatic ampulla are closed when bile is not needed for digestion. At these times, bile backs up through the cystic duct into the gallbladder for storage. Later, when fatty chyme from a meal enters the duodenum, the gallbladder's muscular wall contracts and the sphincters at the end of the duct system relax, expelling bile into the duodenum.

By way of a summary, here is the complete path by which bile flows from the liver to the duodenum: from right and left hepatic ducts to common hepatic duct to cystic duct to gallbladder for storage; then, from gallbladder through cystic duct to bile duct to duodenum.

Histologically, the wall of the gallbladder has fewer layers than the wall of the alimentary canal: (1) a mucosa consisting of a simple columnar epithelium and a lamina propria, (2) one layer of smooth muscle, and (3) a thick outer layer of connective tissue that is covered by a serosa wherever it is not in direct contact with the liver. The columnar cells of the lining epithelium concentrate the bile by absorbing some of its water and ions.

CLINICAL APPLICATIONS

GALLSTONES Bile is the normal vehicle in which cholesterol is excreted from the body, and bile salts keep the cholesterol dissolved within bile. Either too much cholesterol or too few bile salts can lead to the crystallization of cholesterol in the gallbladder, producing **gallstones** that can plug the cystic duct and cause agonizing pain when the gallbladder or its duct contracts. Gallstones are easy to diagnose because they show up well with ultrasound imaging (see p. 21). Treatments include administering drugs that might dissolve the stones, and *laparoscopic cholecystectomy* (ko″le-sis-tek′to-me), a minimally invasive surgical technique for removing the gallbladder. In this procedure, a viewing scope and surgical tools threaded through small holes in the anterior abdominal wall are used to excise the gallbladder. Any gallstones remaining in the common bile duct are vaporized using a laser. ✚

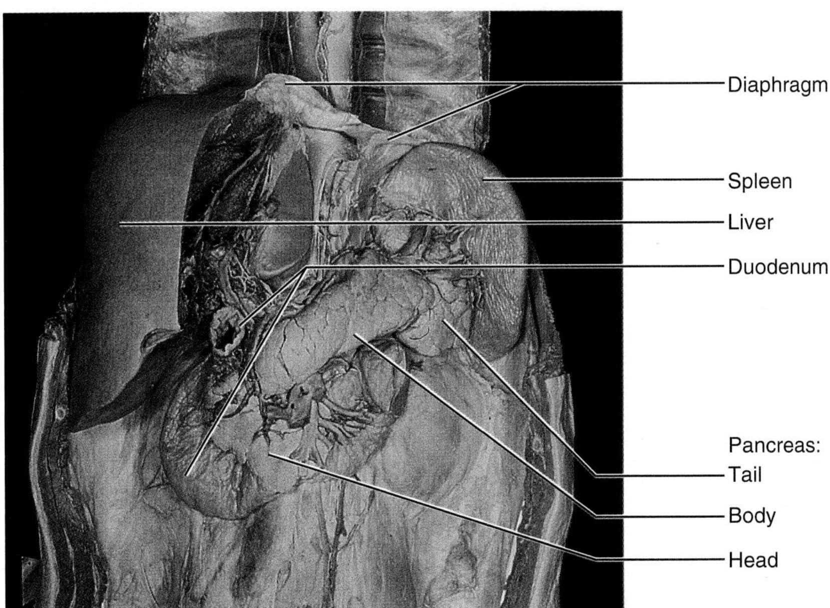

FIGURE 22.23 Photograph of the pancreas and some adjacent organs in the posterior abdominal wall. The stomach and much of the liver have been removed.

The Pancreas

The **pancreas** ("all meat") is both an exocrine gland and an endocrine gland. Its exocrine function is to produce most of the enzymes that digest foodstuffs in the small intestine, whereas its main endocrine function is to produce hormones that regulate the levels of sugar in the blood. The pancreas, which lies retroperitoneally in the epigastric and left hypochondriac regions of the abdomen (Figure 22.23), is shaped like a tadpole, with head, body, and tail regions; its head lies in the C-shaped curvature of the duodenum, and its tail extends to the left to touch the spleen.

The pancreas receives blood through branches of the hepatic, splenic, and superior mesenteric vessels. Its autonomic nerves are from the celiac plexus. Sympathetic input derives from the thoracic splanchnic nerves, whereas parasympathetic input is from the vagus nerve.

The **main pancreatic duct** extends through the length of the pancreas (see Figure 22.16). As previously mentioned, this duct joins the bile duct to form the hepatopancreatic ampulla and empties into the duodenum at the major duodenal papilla. An **accessory pancreatic duct** lies in the head of the pancreas and drains into the main duct.

Microscopically, the pancreas consists of many exocrine glands intermixed with fewer clusters of endocrine cells. These exocrine glands—compound acinar glands that open into the two large ducts like clusters of grapes attached to two main vines—will be considered first. The acini of these glands consist of serous **acinar cells** (Figure 22.24a), which make, store, and secrete at least 22 kinds of pancreatic enzymes capable of digesting the various categories of foodstuffs. The enzymes are stored in inactive form in intracellular secretory granules called **zymogen granules** (zi′mo-jen; "fermenting"). The acinar cells also contain an elaborate rough ER and Golgi apparatus, typical features of cells that secrete proteins. From each acinus, the secreted product travels through the pancreatic duct system to the duodenum, where the enzymes are activated. Furthermore, the epithelial cells that line the smallest pancreatic ducts secrete a bicarbonate-rich fluid that helps neutralize acidic chyme in the duodenum. Also among the duct cells are stem cells that form new acini and endocrine cells.

In its endocrine function, the pancreas secretes two major hormones, insulin and glucagon, which lower and raise blood sugar levels, respectively. The hormone-secreting cells are clustered into spherical bodies called **pancreatic islets (islets of Langerhans)** (Figure 22.25). As can be seen in the figure, the islets lie directly adjacent to the acini of the exocrine glands, although they do not connect to these acini or to any of the pancreatic ducts. The endocrine pancreas is discussed in detail on page 758.

CLINICAL APPLICATIONS

PANCREATITIS Pancreatitis (pan″kre-ah-ti′tis)—inflammation of the pancreas, usually accompanied by necrosis of pancreatic tissue—is a painful condition that is usually caused by the blockage of the pancreatic duct, either by gallstones or by an alcoholism-induced precipitation of protein. The blockage results in activation of the pancreatic enzymes in the pancreas instead of the small intestine. Pancreatitis can lead to nutritional deficiencies, diabetes, pancreatic infections, and, finally, to death through circulatory shock as the mediators of inflammation pour into the bloodstream from the damaged pancreas. ✚

Mesenteries

We now consider the mesenteries that support the abdominal digestive organs. Recall that mesenteries are double-layered sheets of peritoneum that connect the

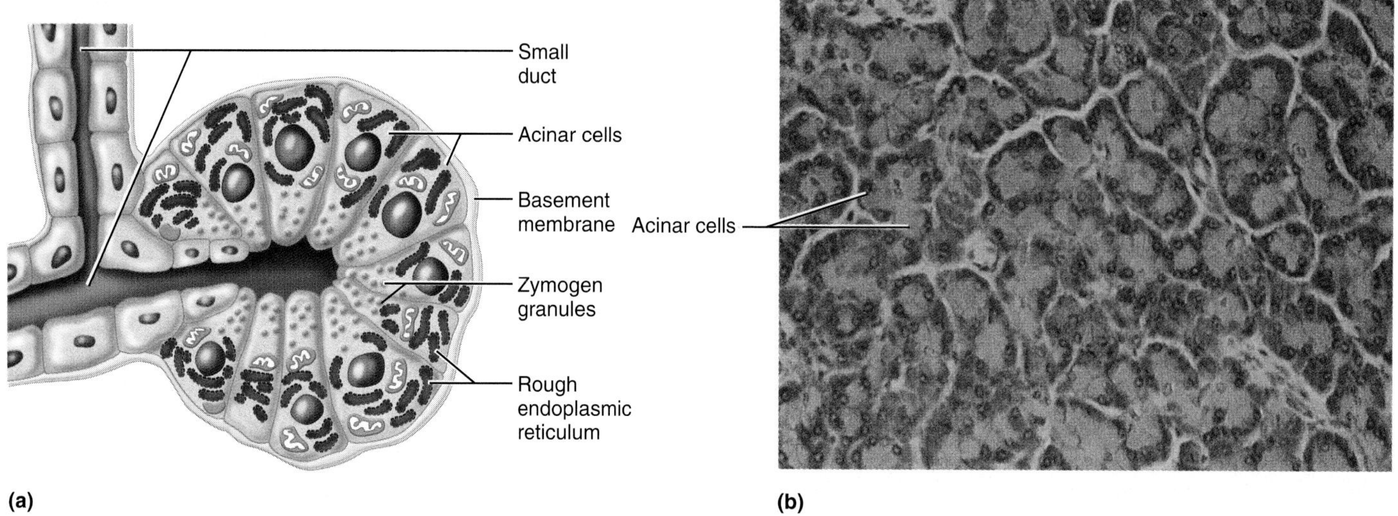

FIGURE 22.24 **The exocrine pancreas: Histology.** **(a)** One acinus. Note the individual acinar cells, which secrete the pancreatic digestive enzymes. **(b)** Photomicrograph of pancreatic acinar tissue (220 ×).

peritoneal organs to the dorsal and ventral body wall (see Figure 22.5a). Most mesenteries are *dorsal* mesenteries, extending dorsally from the alimentary canal to the posterior abdominal wall. In the superior abdomen, however, a *ventral* mesentery extends ventrally from the stomach and liver to the anterior abdominal wall. As you read about the different parts of the dorsal and ventral mesenteries, note that some mesenteries are called "ligaments," even though these peritoneal sheets are not the same as the fibrous ligaments that interconnect bones. For a summary of the mesenteries described next, see Table 22.1 on p. 666.

The two ventral mesenteries are the falciform ligament and the lesser omentum. The **falciform ligament** (fal′sĭ-form; "sickle-shaped") binds the anterior aspect of the liver to the anterior abdominal wall and diaphragm (Figure 22.26a). The **lesser omentum** (o-men′tum; "fatty skin") runs from the fissure of the liver and the

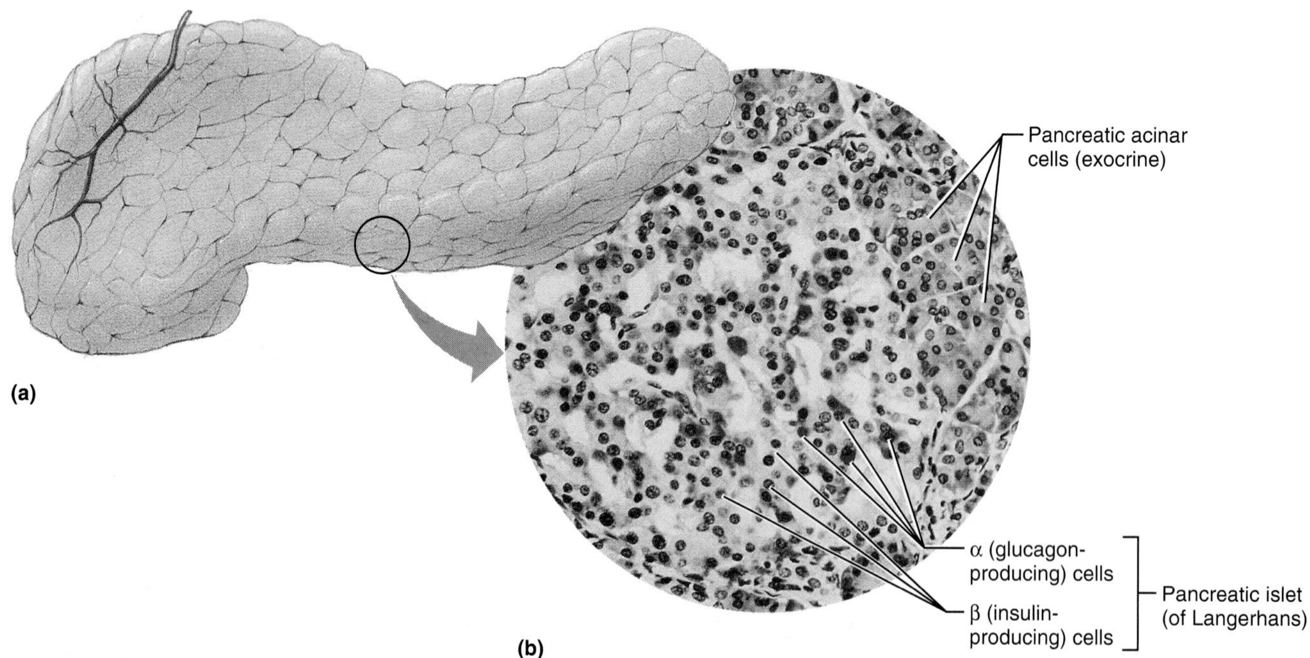

FIGURE 22.25 **The endocrine pancreas.** **(a)** The entire pancreas.
(b) Photomicrograph of pancreas tissue (300 ×) showing a lighter-staining pancreatic islet (islet of Langerhans) surrounded by pancreatic acini.

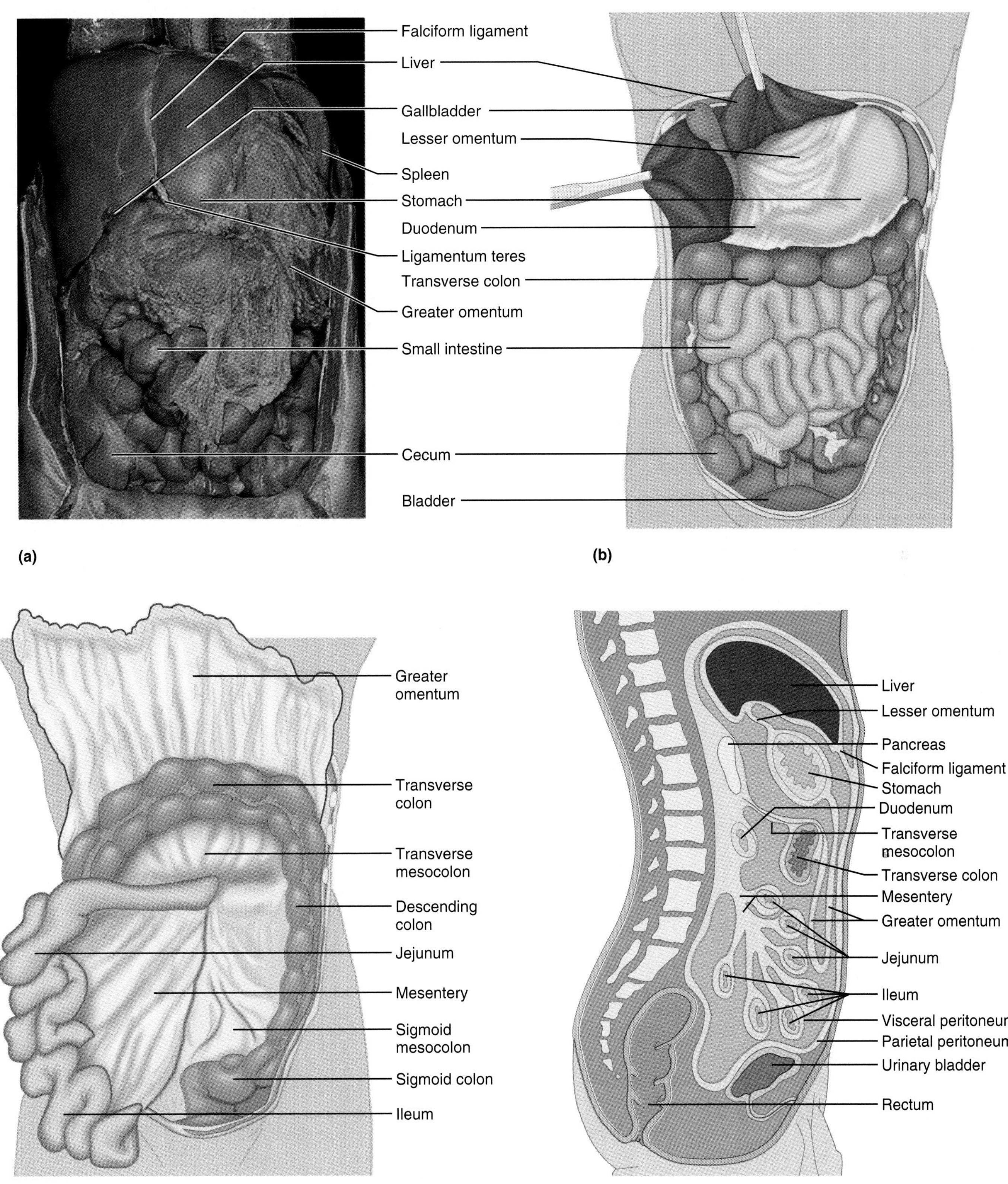

FIGURE 22.26 The mesenteries. **(a)** Superficial view of the abdominal organs, with only the anterior abdominal wall removed. The greater omentum is in its normal position covering the abdominal viscera. **(b)** The lesser omentum attaches the lesser curvature of the stomach to the posterior half of the fissure of the liver (the liver is lifted out of the way, and the greater omentum has been removed). **(c)** The greater omentum and transverse colon have been reflected superiorly to reveal the mesentery proper ("Mesentery"), the transverse mesocolon, and the sigmoid mesocolon. **(d)** Sagittal section through the abdominopelvic cavity, showing the attachments of the various mesenteries to the posterior and anterior body wall.

porta hepatis to the lesser curvature of the stomach (Figure 22.26b).

All remaining mesenteries are dorsal mesenteries. The **greater omentum** (see Figure 22.26a and d) connects the greater curvature of the stomach to the posterior abdominal wall, but in a very roundabout way: It is tremendously elongated and extends inferiorly to cover the transverse colon and coils of small intestine like a butterfly net. In addition to containing a great deal of fat, the greater omentum has a remarkable ability to limit the spread of infections within the peritoneal cavity; for example, it can wrap and enclose an inflamed appendix.

The duodenum of the small intestine is retroperitoneal, but the long coils of the jejunum and ileum are supported by the **mesentery,** or *mesentery proper* (see Figure 22.26c and d). This sheet fans inferiorly from the posterior abdominal wall like long, pleated curtains.

Most parts of the large intestine are retroperitoneal and thus lack a mesentery. The transverse and sigmoid colons, however, are exceptions: The transverse colon is held to the posterior abdominal wall by the **transverse mesocolon** (mez″o-ko′lon; "mesentery of the colon"), a nearly horizontal sheet that is fused to the underside of the greater omentum, so it can be viewed only inferiorly (see Figure 22.26c). The **sigmoid mesocolon** is the mesentery that connects the sigmoid colon to the posterior pelvic wall.

DISORDERS OF THE DIGESTIVE SYSTEM

This chapter discusses major disorders of most of the digestive organs. We have already covered GERD and the esophagus on p. 646 and ulcers of the stomach and duodenum on p. 650. Here, we consider *intestinal obstruction* and *inflammatory bowel disease* of the small and large intestines, *viral hepatitis* of the liver, and the effects of *cystic fibrosis* on the pancreas.

Intestinal Obstruction

Any hindrance to the movement of chyme or feces through the intestine is called **intestinal obstruction.** Most obstructions are *mechanical*—due either to hernias of the bowel or twists that pinch the bowel shut, to intestinal tumors or adhesions, or to foreign objects lodged in the bowel. *Nonmechanical* obstruction, by contrast, is due to a halt in peristalsis, often when movement of the intestine is inhibited by trauma or when the intestine is touched during surgery. About 85% of all obstructions occur in the small intestine; the remainder affect the large intestine. Common symptoms include cramps, vomiting, nausea, and failure to pass gas and feces.

Inflammatory Bowel Disease

Up to 2 of every 1000 people are affected by **inflammatory bowel disease,** a noncontagious, periodic inflammation of the intestinal wall characterized by chronic leukocyte infiltration of this wall. Symptoms include cramping, diarrhea, weight loss, and intestinal bleeding. Of the two subtypes of this condition, the form called *Crohn's disease* is the more serious, with deep ulcers and fissures developing along the entire intestine, but primarily in the terminal ileum. The other form, *ulcerative colitis,* is characterized by a shallow inflammation of the large-intestinal mucosa, mainly in the rectum. Although formerly thought to be a nervous condition, inflammatory bowel disease is now understood to be an abnormal immune and inflammatory response to bacterial antigens that normally occur in the intestine. Treatment involves adopting a special diet that is low in fiber and in dairy products, reducing stress, taking antibiotics, and most effectively, administering anti-inflammatory and immunosuppressant drugs.

Table 22.1 Summary of Intraperitoneal and Retroperitoneal Digestive Organs in the Abdomen and Pelvis

Intraperitoneal Organs (and their mesenteries)	Retroperitoneal Organs (lack mesenteries)
Liver (falciform ligament and lesser omentum)	Duodenum (almost all of it)
Stomach (greater and lesser omentum)	Ascending colon
Ileum and jejunum (mesentery proper)	Descending colon
Transverse colon (transverse mesocolon)	Rectum
Sigmoid colon (sigmoid mesocolon)	Pancreas

Viral Hepatitis

Hepatitis, the general term for any inflammation of the liver, is largely of viral origin. Upon infection, most types of **viral hepatitis** lead to flu-like symptoms and jaundice (yellow skin and mucous membranes, an indication that the liver is not removing bile pigments from the blood to make bile). The major types of hepatitis are A, B, C, and G.

Hepatitis A, spread by the fecal-oral route, often in contaminated food or water, is characterized by an acute infection without long-term damage followed by recovery and lifelong immunity. Treatments include administration of antibodies, and effective preventive vaccines are available.

Hepatitis B is transmitted via infected blood or body fluids, or from mothers to newborns at birth. Most infected individuals recover and gain immunity, but some develop chronic liver disease and eventually cirrhosis, with an increased likelihood of developing liver cancer. Many hepatitis B patients can be helped with interferon (a substance that enhances the immune response against viruses) plus a combination of drugs that stops viral replication; an effective vaccine is also available.

Hepatitis C, like hepatitis B, is transmitted via body fluids and can lead to cirrhosis and liver cancer, but it has raised greater concern because it usually produces no short-term symptoms. As a result, it is difficult to diagnose, and many individuals do not know they are infected until they have spread the virus or develop serious symptoms, sometimes 20 years after infection. Some 4 million Americans have hepatitis C, and its spread is a serious health concern. No vaccine yet exists, but interferon and a drug that inhibits viral replication can help many patients.

Hepatitis G is as widespread as type C but seems to cause little liver damage.

Cystic Fibrosis and the Pancreas

Cystic fibrosis (described in more detail on p. 626) primarily disrupts secretions in the respiratory system, but most of the body's other secretory epithelia are affected as well. Although the pancreas and intestinal glands, submandibular glands, and bile ducts in the liver all become blocked with thick secretions, most serious is the effect on the pancreas, in which the clogged ducts prevent the pancreatic juices from reaching the small intestine. As a result, fats and other nutrients are not digested or absorbed and the feces are bulky and fat-laden, a problem that can be treated by administering pancreatic enzymes with meals. Eventually, the pancreas may become a mass of cystically dilated ducts surrounded by dense fibrous tissue and no acini.

THE DIGESTIVE SYSTEM THROUGHOUT LIFE

Embryonic Development

Recall from Chapter 3 (see p. 66) that the alimentary canal originates when the flat embryo folds into the shape of a cylinder, enclosing a tubular part of the yolk sac within its body. This folding produces the *primitive gut,* a tube of endoderm that is covered by splanchnic mesoderm. The endoderm gives rise to the lining epithelium of the alimentary canal and to the secretory cells of the digestive glands; the splanchnic mesoderm gives rise to all other layers in the wall of the alimentary canal.

Initially, the middle region of the primitive gut is open to the yolk sac through the **vitelline duct** (vi-tel′in; "yolk"), a key landmark that divides the embryonic gut into three basic regions: foregut, midgut, and hindgut (Figure 22.27a). The embryonic **foregut** becomes the first segment of the adult digestive system, from the pharynx to the point in the duodenum where the bile duct enters. The embryonic **midgut** becomes the adult segment beginning at the duodenum and extending to a point two-thirds of the way along the transverse colon. The **hindgut** forms the rest of the large intestine. The abdominal foregut, midgut, and hindgut—and their adult derivatives—are supplied by the celiac, superior mesenteric, and inferior mesenteric arteries, respectively.

The caudal part of the early hindgut joins a tube-like outpocketing called the **allantois** (ah-lan′to-is; "sausage"). The expanded junction between the hindgut and the allantois is the **cloaca** (klo-a′kah; "sewer"), which gives rise to the rectum and most of the anal canal, among other structures (see pp. 694, and 734–736).

In the mouth region of the embryo, the endoderm-lined gut touches the surface ectoderm to form an **oral membrane,** which lies in a depression called the **stomodeum** (sto″mo-de′um; "on the way to becoming the mouth"). Similarly, at the end of the hindgut, endoderm meets ectoderm to form the **cloacal membrane** in a pit called the **proctodeum** (prok″to-de′um; "on the way to becoming the anus"). The oral membrane lies at the future mouth/pharynx boundary (the fauces), and the cloacal membrane lies in the future anal canal, roughly where the pectinate line will occur. The oral and cloacal membranes are reabsorbed during month 2, thereby opening the alimentary canal to the outside.

During weeks 4 and 5, the embryonic gut starts to elongate, bend, and form outpocketings (Figure 22.27b). Salivary glands arise as outpocketings from the mouth; the pharynx develops four or five pairs of lateral *pharyngeal pouches* (see Figure 1.7b); and the future lungs and trachea bud off the distal pharynx (see p. 628). A spindle-shaped enlargement of the abdominal foregut is the first sign of the stomach, the liver and pancreas arise as buds from the last part of the foregut, and the midgut elongates into the **primitive intestinal loop.** In

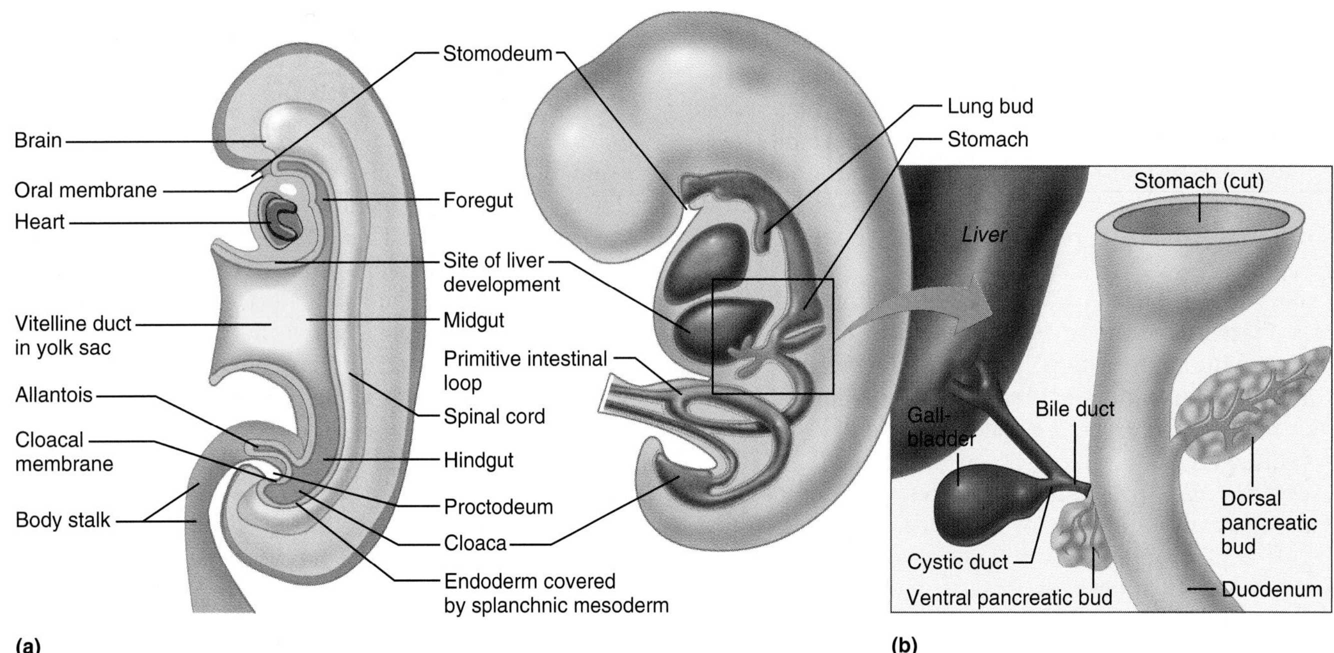

FIGURE 22.27 Embryonic development of the digestive system. (a) Embryo at 3 weeks (23 days). The primitive gut (composed of foregut, midgut, and hindgut) has formed. The midgut is still open and continuous with the yolk sac. The oral and cloacal membranes will be reabsorbed in a few weeks to form the oral and anal openings. **(b)** By 5 weeks of development, the liver and pancreas are budding off of the distal foregut. The pancreas forms from two pancreatic buds (ventral and dorsal) that later join.

months 2 and 3, this loop rotates and elongates to bring the intestines into their final positions.

The Digestive System in Later Life

Unless abnormal interferences occur, the digestive system operates through childhood and adulthood with relatively few problems. However, contaminated or spicy foods sometimes cause an inflammation of the alimentary canal called **gastroenteritis** (gas″tro-en″tĕ-ri′tis), the symptoms of which include nausea, vomiting, cramps, loss of appetite, or diarrhea. As previously mentioned, appendicitis is common in teenagers and young adults. Gallstones and ulcers are problems of middle age.

During old age, the activity of the digestive organs declines: Fewer digestive juices are produced, the absorption of nutrients becomes less efficient, and peristalsis slows. So much water is reabsorbed from the slow-moving fecal mass in the large intestine that the feces become hard and compacted. The result is a decrease in the frequency of bowel movements and, often, constipation.

Diverticulosis and cancer of the digestive organs are other common problems of the aged. Cancer of the stomach, colon, liver, or pancreas rarely exhibits early signs, and the cancer has often metastasized before the person seeks medical attention. These forms of cancer are deadly—colon cancer (also see p. 51) is the second leading cause of cancer deaths in the United States. However, if detected early, cancers of the digestive viscera are sometimes curable, colon and liver cancer more so than pancreatic or gallbladder cancer. Perhaps the best advice is to have regular medical checkups. Half of all rectal cancers can be felt digitally during rectal exams, and nearly 80% of colon cancers can be seen during a colonoscopy (see "Endoscopy" in the Related Clinical Terms). Evidence suggests that diets high in plant fiber decrease the incidence of colon cancer.

RELATED CLINICAL TERMS

Anal fissure A longitudinal tear in the mucosa of the anal canal, often caused by the passage of hard, dry feces. Usually heals naturally, but most fissures that do not heal are in the posterior midline, which is poorly vascularized. Symptoms include pain and bleeding during defecation. Treatment includes using laxatives to soften the feces, glycerin suppositories, or in persistent cases, surgery.

Ascites (ah-si′tēz) (*asci* = bag, bladder) Abnormal accumulation of serous fluid that has leaked out of peritoneal capillaries into the peritoneal cavity; may be caused by portal hypertension following liver cirrhosis or by heart or kidney disease. Excessive ascites causes visible bloating of the abdomen.

Bruxism (bruk'sizm) Grinding or clenching of teeth, usually at night during sleep in response to a stressful day; can wear down and crack teeth and lead to temporomandibular disorder (see p. 227).

Endoscopy (en-dos'ko-pe) (*endo* = inside; *scopy* = viewing) The viewing of the lining of a ventral body cavity or tubular organ with a flexible, tube-like device called an endoscope, which contains a lens and a light radiating from its tip. Endoscopes are used to view the internal surfaces of various parts of the alimentary canal, including the stomach (gastroscopy), the colon (colonoscopy), and the sigmoid colon (sigmoidoscopy). **Laparoscopy** (lap″ah-ros'ko-pe; "flank viewing") is the use of an endoscope inserted into the peritoneal cavity through the anterior abdominal wall, typically to assess the condition of the digestive organs and the pelvic reproductive organs in females.

Enteritis (*enteron* = intestine) Inflammation of the intestine, especially the small intestine.

Liver biopsy (bi'op-se) (*bio* = living; *opsis* = vision, viewing) Removal from the liver of a small piece of living tissue, which is then examined for signs of disease. The puncturing needle is inserted through the seventh, eighth, or ninth intercostal space, in the right midaxillary line (straight inferiorly from the axilla) after the patient has exhaled as much air as possible. (Exhalation minimizes the chances that the needle will pierce the lung.)

Pyloric stenosis (*stenosis* = narrowing, constriction) Congenital condition in about 1 in 400 newborns, in which the pyloric sphincter of the stomach is abnormally constricted; the condition's characteristic sign, projectile vomiting, usually does not appear until the baby begins to eat solid food. This condition can usually be repaired surgically. Can also occur in adults through scarring caused by an ulcer or by a tumor that blocks the pyloric opening.

Rectocele (rek'to-sēl) (*recto* = rectum; *cele* = sac) Condition in females in which the rectum pushes on the vagina and bulges into the posterior vaginal wall. Usually results from tearing of the supportive muscles of the pelvic floor during childbirth, which then allows the unsupported pelvic viscera to sink inferiorly. May also be associated with *rectal prolapse,* in which the rectal mucosa protrudes from the anus.

CHAPTER SUMMARY

Media study tools that could provide you additional help in reviewing specific key topics of Chapter 22 are referenced below.

SP = Study Partner; AIA = A.D.A.M.® Interactive Anatomy

OVERVIEW OF THE DIGESTIVE SYSTEM (pp. 634–637)

1. The digestive system includes the alimentary canal (mouth, pharynx, esophagus, stomach, and small and large intestines) and accessory digestive organs (teeth, tongue, salivary glands, liver, gallbladder, and pancreas).

Digestive Processes (pp. 635–636)

2. The digestive system carries out the processes of ingestion, propulsion, mechanical digestion, chemical digestion, absorption, and defecation.

Abdominal Regions and Quadrants (p. 637)

3. The nine regions of the anterior abdominal wall are defined by two horizontal planes (subcostal and transtubercular planes) and by the vertical midclavicular planes. These regions are two hypochondriacs and an epigastric, two lumbars (laterals) and an umbilical, and two iliacs (inguinals) and a hypogastric (pelvic). The anterior abdominal wall can also be divided into four quadrants.

AIA Link: Atlas; Region: Abdomen; Abdominopelvic Region.

ANATOMY OF THE DIGESTIVE SYSTEM (pp. 637–665)

The Peritoneal Cavity and Peritoneum (pp. 637–638)

1. The serous membrane in the abdominopelvic cavity is the peritoneum, which has a parietal layer (on the internal surface of the body wall) and a visceral layer (on the viscera). The slit between the visceral and parietal peritoneum, the peritoneal cavity, contains slippery serous fluid, which decreases friction as the organs move.

2. A mesentery is a double layer of peritoneum that tethers the movable digestive organs to the body wall. Mesenteries also store fat and carry blood vessels and nerves. Visceral organs that lack a mesentery and are fused to the posterior body wall are called retroperitoneal organs.

AIA Link: Atlas; Region: Abdomen; Intestines & Mesentery.

Histology of the Alimentary Canal Wall (pp. 638–640)

3. The esophagus, stomach, and intestine share the same tissue layers: an inner mucosa, a fibrous submucosa, a muscularis externa, and an outer serosa (visceral peritoneum) or adventitia. The mucosa consists of a lining epithelium, lamina propria, and muscularis mucosae.

4. Visceral nerve plexuses (myenteric and submucosal) occur in the wall of the alimentary canal. These plexuses contain parasympathetic, sympathetic, and visceral sensory fibers, as well as enteric neurons.

The Mouth and Associated Organs (pp. 640–645)

5. Food enters the alimentary canal through the mouth (oral cavity), which consists of an external vestibule and an internal oral cavity proper.

6. The mouth is lined by a stratified squamous epithelium, which resists abrasion by food fragments.

7. The lips and cheeks keep food inside the mouth during chewing. The red margin is the red part of the lips that borders the oral orifice.

8. The tongue is predominantly a mucosa-covered skeletal muscle. Its intrinsic muscles change its shape, and its extrinsic muscles change its position. Three classes of papillae occur on the tongue's superior surface: filiform papillae, which grip food during chewing, and fungiform and vallate papillae, which contain taste buds.

AIA Link: Atlas; System: Digestive; Surface of Tongue (dorsal).

9. Saliva is produced by intrinsic salivary glands in the oral mucosa and by three pairs of large extrinsic salivary glands—the parotid, submandibular, and sublingual glands. The salivary glands are compound tubuloalveolar glands containing varying amounts of serous and mucous cells.

AIA Link: Atlas; System: Digestive; Glands of Head & Neck (lateral).

10. Teeth tear and grind food in the chewing process (mastication). The 20 deciduous teeth begin to fall out at age 6 and are gradually replaced during childhood and youth by the 32 permanent teeth.

11. Teeth are classified as incisors, canines, premolars, and molars. Each tooth has an enamel-covered crown and a cementum-covered root. The bulk of the tooth is dentin (dentine), which surrounds the central pulp cavity. A periodontal ligament (periodontium) secures the tooth to the bony alveolus.

The Pharynx (pp. 645–646)

12. During swallowing, food passes from the mouth through the oropharynx and laryngopharynx, which are lined by a stratified squamous epithelium. The pharyngeal constrictor muscles squeeze food into the esophagus during swallowing.

The Esophagus (p. 646)

13. The esophagus descends from the pharynx through the posterior mediastinum and into the abdomen. There it joins the stomach at the cardiac orifice, where a functional sphincter prevents superior regurgitation of stomach contents.

AIA Link: Atlas; System: Digestive; Sagittal Section of Neck.

14. The esophageal mucosa contains a stratified squamous epithelium. The muscularis consists of skeletal muscle superiorly and smooth muscle inferiorly. The esophagus has an adventitia rather than a serosa.

The Stomach (pp. 646–651)

15. The J-shaped stomach churns food into chyme and secretes HCl and pepsin. Lying in the superior left part of the abdomen, its major regions are the cardia, fundus, body, pyloric region, and pylorus (containing the pyloric sphincter). Its right and left borders are the lesser and greater curvatures. When the stomach is empty, its internal surface exhibits rugae.

AIA Link: Atlas; System: Digestive; Stomach & Spleen (anterior).

16. The internal surface of the stomach is lined by goblet cells and dotted with gastric pits that lead into tubular gastric glands. Secretory cells in the gastric glands include pepsinogen-producing chief cells, parietal cells that secrete HCl, mucous neck cells, and enteroendocrine cells that secrete hormones (including gastrin).

17. The stomach wall is protected against self-digestion and acid by an internal coat of mucus and by the rapid regeneration of its lining epithelium.

The Small Intestine (pp. 652–654)

18. The small intestine is the main site of digestion and nutrient absorption. Its segments are the duodenum, jejunum, and ileum.

AIA Link: Atlas; System: Digestive; Barium in Small Bowel.

19. The duodenum lies retroperitoneally in the superior abdomen. The bile duct and main pancreatic duct join to form the hepatopancreatic ampulla and empty into the duodenum through the major duodenal papilla.

AIA Link: Atlas; System: Digestive; Pancreatic & Bile Ducts.

20. The small intestine has four features that increase its surface area and allow rapid absorption of nutrients: its great length, circular folds, villi, and abundant microvilli on the absorptive cells. Absorbed nutrients enter capillaries in the core of the villi.

21. The epithelium lining the internal intestinal surface contains absorptive, goblet, and enteroendocrine cells. Between the villi are the intestinal crypts (intestinal glands), which secrete intestinal juice and continuously renew the lining epithelium.

22. Other special features of the small intestine are (1) aggregated lymphoid nodules in the ileum and (2) the duodenal mucous glands.

The Large Intestine (pp. 654–658)

23. The large intestine, which concentrates feces by absorbing water, forms an open rectangle around the small intestine. Its subdivisions are the cecum and vermiform appendix, colon (ascending, transverse, descending, and sigmoid), rectum, and anal canal. Special features on the external surface of the colon and cecum are teniae coli, haustra, and epiploic appendages.

AIA Link: Atlas; System: Digestive; Large Intestine in situ.

24. The cecum lies in the right iliac fossa and contains the ileocecal valve. Attached to the cecum is the vermiform appendix, which contains abundant lymphoid tissue.

25. Near the end of the large intestine, the sigmoid colon enters the pelvis and joins the rectum. The rectum pierces the pelvic floor and joins the anal canal, which ends at the anus. Internally, the rectum contains the transverse rectal folds, and the anal canal contains anal columns, anal valves, the pectinate line, and anal sinuses.

AIA Link: Atlas; System: Digestive; Frontal Section of Anal Canal.

26. The defecation reflex produces a contraction of the rectal walls when feces enter the rectum. During defecation, a person strains to increase the intra-abdominal pressure and lifts the anal canal. Defecation can be inhibited by the anal sphincters.

27. Like the small intestine, the large intestine is lined by a simple columnar epithelium with absorptive and goblet cells and contains intestinal crypts. It lacks villi, secretes more mucus, and contains more lymphoid tissue than the small intestine.

SP Exercise: Chapter 22, Gastrointestinal Tract Activities.

The Liver (pp. 658–662)

28. The liver performs many functions, including processing of absorbed nutrients and secreting fat-emulsifying bile.

29. The liver lies in the superior right region of the abdomen. It has diaphragmatic and visceral surfaces and four lobes (right, left, quadrate, and caudate). The quadrate and caudate are now considered parts of the left lobe.

AIA Link: Atlas; System: Digestive; Hepatic Portal Vein.

30. Most vessels enter or leave the liver through the porta hepatis. Other structures on the visceral surface are the inferior vena cava, hepatic veins, gallbladder, fissure, round ligament, and ligamentum venosum.

AIA Link: Atlas; System: Digestive; Liver (inferior).

31. Several structures attach to or run in the liver's fissure, including the round ligament and lesser omentum. The liver's bare area lies against the diaphragm.

32. Classical liver lobules are hexagonal structures consisting of plates of hepatocytes that are separated by blood sinusoids and converge on a central vein. Portal triads (venule branch of portal vein, arteriole branch of hepatic artery, bile duct) occur at most corners of the lobule.

33. Blood flows through the liver in the following sequence: through the branches of the hepatic artery and portal vein, through the sinusoids to supply the hepatocytes, and then to the central veins, hepatic veins, inferior vena cava, and heart. In the sinusoids, Kupffer cells (macrophages) remove debris from the blood.

34. Hepatocytes perform almost all liver functions. These liver cells secrete bile into the bile canaliculi, which proceeds to the bile ducts, to the hepatic ducts, and through the common hepatic duct to the gallbladder and duodenum.

SP Exercise: Chapter 22, Structure of Liver Lobules.

The Gallbladder (p. 662)

35. The gallbladder, a green muscular sac, stores and concentrates bile. Its duct is the cystic duct. When fatty chyme enters the small intestine, the gallbladder squeezes bile into the bile duct and duodenum.

The Pancreas (p. 663)

36. The tadpole-shaped pancreas runs horizontally across the posterior abdominal wall, between the duodenum and the spleen.

AIA Link: Atlas; System: Digestive; Blood Supply to Pancreas/Spleen.

37. The pancreas is an exocrine gland containing many compound acinar glands that empty into the main and accessory pancreatic ducts. The serous acinar cells secrete digestive enzymes, and the smallest duct cells secrete an alkaline fluid. Both products are emptied into the duodenum.

38. The pancreas is also an endocrine gland, with hormone-secreting cells contained in spherical pancreatic islets.

Mesenteries (pp. 663–666)

39. The ventral mesenteries, associated with the abdominal digestive organs, are the falciform ligament and lesser omentum. The dorsal mesenteries are the greater omentum, mesentery proper, transverse mesocolon, and sigmoid mesocolon.

SP Exercise: Chapter 22, The Digestive System.

DISORDERS OF THE DIGESTIVE SYSTEM (pp. 666–667)

1. Disorders considered in this chapter include hiatal hernia of the esophagus, gastroesophageal reflux disease (p. 646), peptic ulcers (p. 650), intestinal obstruction, inflammatory bowel disease, viral hepatitis, and the obstructive effect of cystic fibrosis on the pancreatic ducts.

THE DIGESTIVE SYSTEM THROUGHOUT LIFE (pp. 667–668)

Embryonic Development (pp. 667–668)

1. As the 3-week-old embryo assumes its cylindrical body shape, it encloses the primitive gut, a tube of endoderm covered by splanchnic mesoderm. The endoderm becomes the lining epithelium (and gland cells), and the splanchnic mesoderm gives rise to all other layers of the wall of the alimentary canal.

2. The embryonic gut is divided into a foregut, midgut, and hindgut, which form distinct regions of the adult digestive system.

3. The primitive embryonic gut tube, straight at first, soon grows outpocketings (liver, pancreas, pharyngeal pouches), shows swellings (stomach, cloaca), and lengthens into a primitive intestinal loop. This loop rotates and elongates to bring the intestines into their final positions.

The Digestive System in Later Life (p. 668)

4. Various diseases may plague the digestive organs throughout life. Appendicitis is common in young adults; gastroenteritis and food poisoning can occur at any time; and ulcers and gallbladder problems increase in middle age.

5. The efficiency of digestive processes declines in the elderly, and constipation becomes common. Diverticulosis and cancers (such as colon and stomach cancer) appear with increasing frequency among older individuals.

REVIEW QUESTIONS

Multiple Choice/Matching Questions

1. Which of the following organs is retroperitoneal? (a) pharynx, (b) stomach, (c) ascending colon, (d) ileum.

2. The submucosal nerve plexus of the intestine (a) innervates the mucosa layer, (b) lies in the mucosa layer, (c) controls peristalsis, (d) contains only motor neurons.

3. Match each of the mesenteries at right with the appropriate description at left.

Descriptions	Mesenteries
____ **(1)** connects the ileum and jejunum to the posterior abdominal wall	**(a)** greater omentum
____ **(2)** connects anterior surface of the liver to the anterior abdominal wall	**(b)** lesser omentum
____ **(3)** connects the large intestine to the pelvic wall	**(c)** falciform ligament
____ **(4)** attaches to the greater curvature of the stomach; has the most fat	**(d)** mesentery proper
____ **(5)** runs from the stomach's lesser curvature to the fissure of the liver	**(e)** transverse mesocolon
____ **(6)** a mesentery of the large intestine that is fused to the underside of the greater omentum	**(f)** sigmoid mesocolon

4. Match each of the organs at left with the type of epithelium that lines its lumen, at right.

Organ	Epithelium types
____ **(1)** oral cavity	**(a)** simple squamous
____ **(2)** oropharynx	**(b)** simple cuboidal
____ **(3)** esophagus	**(c)** simple columnar
____ **(4)** stomach	**(d)** stratified squamous
____ **(5)** small intestine	**(e)** stratified cuboidal
____ **(6)** colon	

5. The pointed type of tongue papilla that is not involved with taste reception is (a) filiform, (b) fungiform, (c) circumvallate, (d) dermal.

6. Which of the following statements about the gallbladder is *false*? (a) It makes bile. (b) Its duct is called the cystic duct. (c) It has a fundus lying inferior to the liver. (d) It has mucosal folds similar to the stomach rugae.

7. Which of the following correctly describe the flow of blood through the classical liver lobule and beyond? (More than one choice is correct.) (a) portal venule to sinusoids to central vein to hepatic vein to inferior vena cava, (b) porta hepatis to hepatic vein to portal venule, (c) portal venule to central vein to hepatic vein to sinusoids, (d) portal arteriole to sinusoids to central vein to hepatic vein.

8. The exocrine glands in the pancreas are (a) simple tubular, (b) simple acinar, (c) compound tubular, (d) compound acinar, (e) compound tubuloalveolar.

9. Which cell type occurs in the stomach mucosa, has three prongs, contains many mitochondria and many microvilli, and pumps hydrogen ions? (a) absorptive cell, (b) parietal cell, (c) goblet cell, (d) muscularis externa cell, (e) mucous neck cell.

10. Which one of the following features is shared by both the small and large intestines? (a) intestinal crypts, (b) Peyer's patches (aggregated nodules), (c) teniae coli, (d) haustra, (e) circular folds, (f) intestinal villi.

11. A digestive organ that has a head, neck, body, and tail is the (a) pancreas, (b) gallbladder, (c) greater omentum, (d) stomach.

Short Answer Essay Questions

12. Make a simple drawing of the organs of the alimentary canal and label each organ. Also label the gross subparts of the stomach and intestines. Then add three more labels to your drawing—*salivary glands, liver,* and *pancreas*—and use arrows to show where each of these three glandular organs empties its secretion into the alimentary canal.

13. Name the layers of the wall of the alimentary canal. Describe the tissue composition and the major function of each layer. Then, tell the embryonic source of each layer.

14. (a) Write the dental formulas for both the deciduous and the permanent teeth. (b) Define dental pulp and describe its location.

15. What are the effects of aging on the activity of the digestive system?

16. Bianca went on a trip to the Bahamas during spring vacation and did not study for her anatomy test scheduled early the following week. On the test, she mixed up the following pairs of structures: (a) the rectal valves and the anal valves, (b) pyloric region and pylorus of the stomach, (c) anal canal and anus, (d) villi and microvilli (in the small intestine), (e) hepatic vein and hepatic portal vein, (f) gastric pits and gastric glands. Help her by defining and differentiating between these sound-alike structures.

17. (a) Make a rough sketch of the visceral surface of the liver, and label the following five structures: fissure, porta hepatis, inferior vena cava, gallbladder, caudate lobe. (b) To follow up on part (a), it is easy to learn the structure of the visceral surface of the liver if you can find and mark the big "H" there. To mark the left vertical limb of the H, draw a line through the fissure. To mark the crossbar of the H, draw a horizontal line through the porta hepatis. Then, to mark the right limb of the H, draw a continuous vertical line through the inferior vena cava and gallbladder. Which two lobes are inside the H? Which limb of the H separates the liver's right and left lobes?

18. Name four structural features that increase the absorptive surface area of the small intestine.

19. (a) List the three major vessels of the portal triad, and describe the location of this triad with respect to the liver lobule. (b) Name three organelles that are abundant within hepatocytes, and explain how each of these organelles contributes to liver functions.

20. Of the three big extrinsic salivary glands, which is the largest? Which one underlies the tongue?

21. Trace the entire duct systems of the liver and pancreas. (Hint: Study these ducts in Figure 22.16, and then try to draw them from memory.)

22. On a diagram of the adult digestive tract (such as Figure 22.1), mark the boundaries between the derivatives of the embryonic foregut, midgut, and hindgut.

CRITICAL REASONING + CLINICAL APPLICATION QUESTIONS

1. This chapter described schemes that divided the anterior abdominal wall into nine regions or four quadrants. List the regions that lie either entirely or partially within (a) the upper right quadrant and (b) the lower left quadrant.

2. A 21-year-old man with severe appendicitis did not seek treatment in time and died a week after his abdominal pain and fever began. Explain why appendicitis can quickly lead to death.

3. Duncan, an inquisitive 8-year-old, saw his grandfather's dentures soaking overnight in a glass of water. He asked his grandfather how his real teeth had fallen out. Assuming the grandfather remembered the events correctly but was not a medical expert and used layperson's terms, reconstruct the kind of story the man is likely to have told.

4. Eva, a middle-aged attorney, complains of a burning pain in the "pit of her stomach," usually beginning about 2 hours after eating and lessening after she drinks a glass of milk. When asked to indicate the site of pain, she points to her epigastric region. When her GI tract is examined by endoscopy, a gastric ulcer is visualized. What are the possible consequences of nontreatment?

5. A doctor used an endoscope and located some polyps (precancerous tumors) in the wall of the large intestine of an important government official in his seventies. What is an endoscope?

6. The janitor who cleaned the anatomy lab had a protruding abdomen that looked like the biggest "beer-belly" the students had ever seen, even though the rest of his body did not look fat at all. Another janitor on the floor told some students that the man was a recovering alcoholic and that he had "over a hundred pounds of fluid in his belly." What was the man's probable condition? (Hint: See this chapter's Related Clinical Terms.)

A.D.A.M.® INTERACTIVE ANATOMY QUESTIONS

1. Link: Atlas; System: Digestive; Stomach & Spleen (anterior) (a) Give the name of the folds in the stomach that expand to accommodate food. (b) What part of the small intestine receives the digested food from the stomach?

2. Link: Atlas; System: Digestive; Liver (posterior) (a) Which structure stores bile produced by the liver? (b) Give the name of the gallbladder's duct.

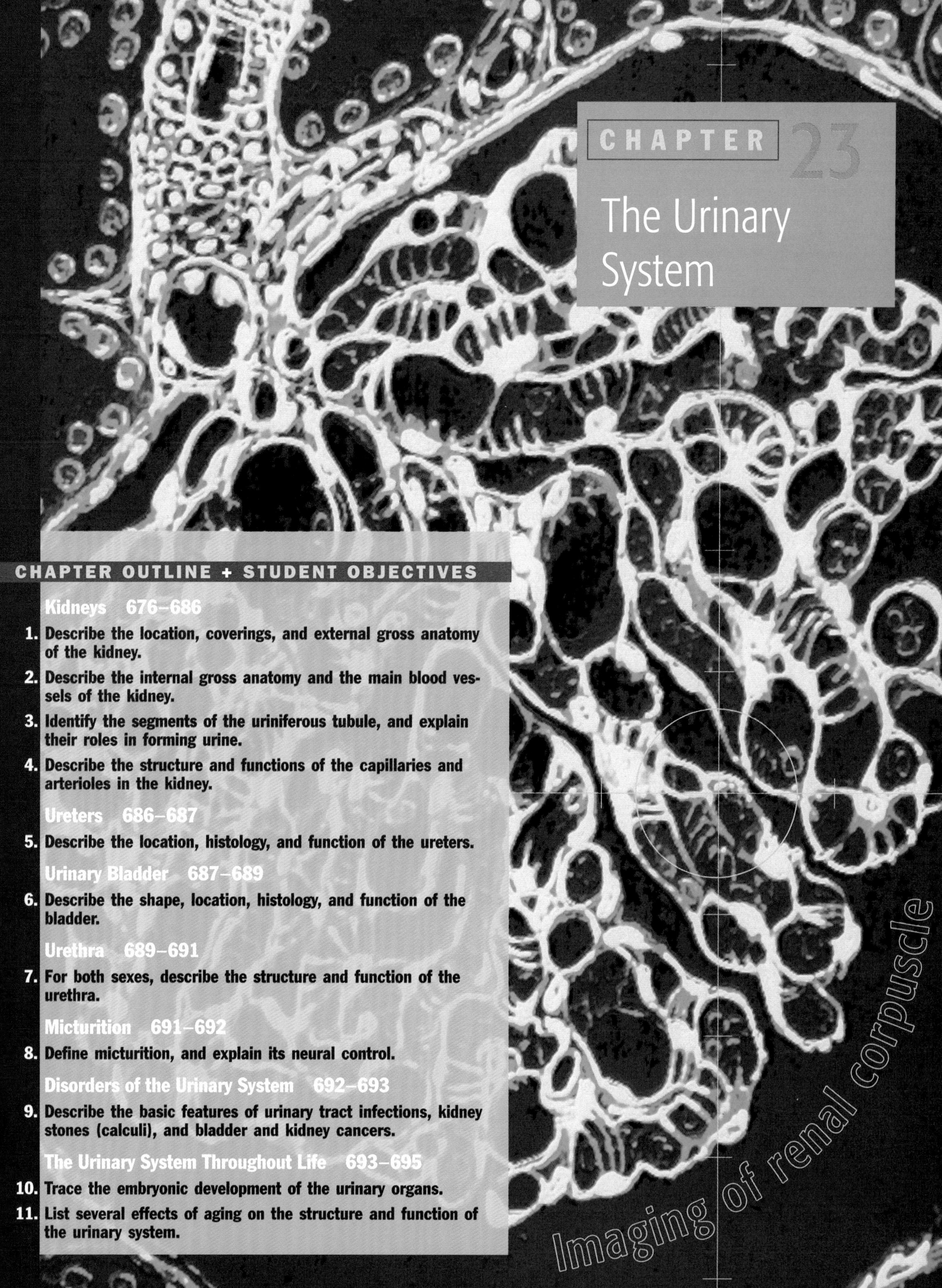

CHAPTER 23 The Urinary System

CHAPTER OUTLINE + STUDENT OBJECTIVES

Kidneys 676–686

1. Describe the location, coverings, and external gross anatomy of the kidney.
2. Describe the internal gross anatomy and the main blood vessels of the kidney.
3. Identify the segments of the uriniferous tubule, and explain their roles in forming urine.
4. Describe the structure and functions of the capillaries and arterioles in the kidney.

Ureters 686–687

5. Describe the location, histology, and function of the ureters.

Urinary Bladder 687–689

6. Describe the shape, location, histology, and function of the bladder.

Urethra 689–691

7. For both sexes, describe the structure and function of the urethra.

Micturition 691–692

8. Define micturition, and explain its neural control.

Disorders of the Urinary System 692–693

9. Describe the basic features of urinary tract infections, kidney stones (calculi), and bladder and kidney cancers.

The Urinary System Throughout Life 693–695

10. Trace the embryonic development of the urinary organs.
11. List several effects of aging on the structure and function of the urinary system.

THE KIDNEYS MAINTAIN THE PURITY AND CHEMical constancy of the blood and the other extracellular body fluids. Much like a water purification plant that keeps a city's water drinkable and disposes of its wastes, the kidneys (Figure 23.1) are usually unappreciated until they malfunction and body fluids become contaminated. Every day, the kidneys filter many liters of fluid from the blood, sending toxins, metabolic wastes, excess water, and excess ions out of the body in urine while returning needed substances from the filtrate to the blood. The main waste products excreted in urine are three nitrogenous compounds: (1) *urea,* derived from the breakdown of amino acids during normal recycling of the body's proteins; (2) *uric acid,* which results from the turnover of nucleic acids; and (3) *creatinine* (kre-at′ĭ-nin), formed by the breakdown of creatine phosphate, a molecule in muscle that stores energy for the manufacture of adenosine triphosphate (ATP). Although the lungs, liver, and skin also participate in excretion, the kidneys are the major excretory organs.

Disposing of wastes and excess ions is only one aspect of kidney function; the kidneys also regulate the volume and chemical makeup of the blood, maintaining the proper balance between water and salts and between acids and bases. The kidneys perform what would be tricky work for a chemical engineer, and perform it efficiently most of the time.

Besides the urine-forming kidneys, the other organs of the urinary system (Figure 23.1b) are the paired *ureters* (u-re′terz; "pertaining to urine"), the *urinary bladder,* and the *urethra* (u-re′thrah). The ureters are tubes that carry urine from the kidney to the bladder, which is a temporary storage sac for urine, and the urethra is a tube that carries urine to the body exterior.

KIDNEYS

Our discussion of the kidneys begins with gross anatomy and then proceeds to microscopic anatomy.

Gross Anatomy of the Kidneys

We examine the gross anatomy of the kidneys by discussing their location and external appearance, then their internal structure, and finally their gross vasculature and nerve supply.

Location and External Anatomy

The red-brown, bean-shaped **kidneys** lie in the superior lumbar region of the posterior abdominal wall (see Figures 23.1 and 23.2). They extend from the level of the 11th or 12th thoracic vertebra superiorly to the third

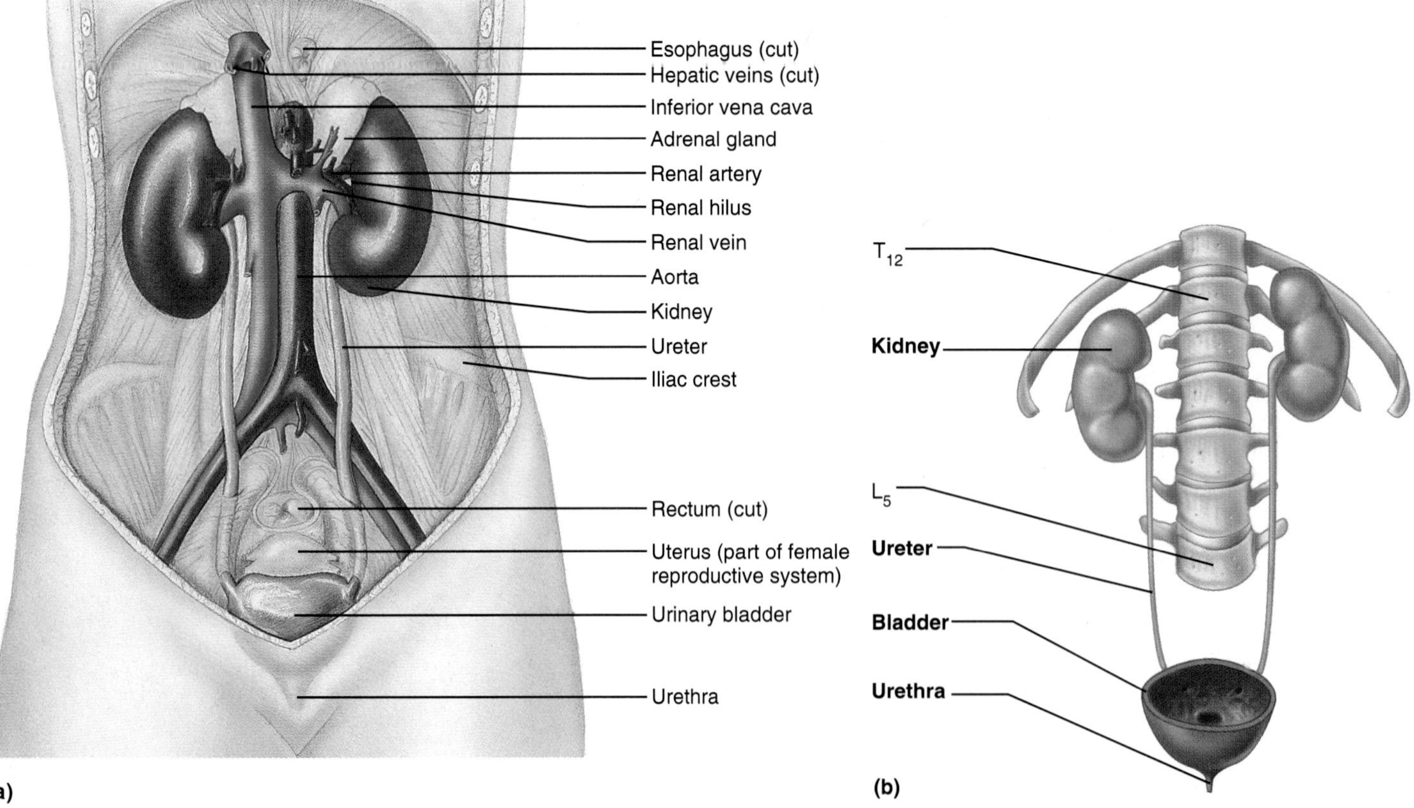

FIGURE 23.1 Organs of the urinary system. **(a)** Kidneys and other urinary organs in the posterior abdomen and the pelvis of a female. **(b)** Relationship of the kidneys to the vertebrae and lower ribs.

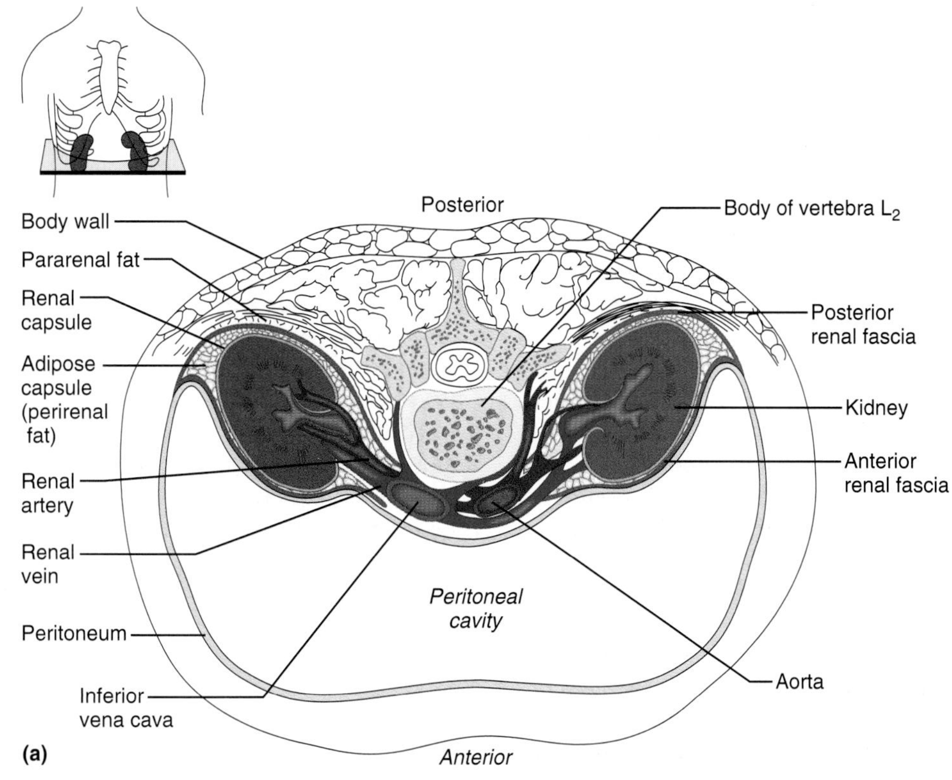

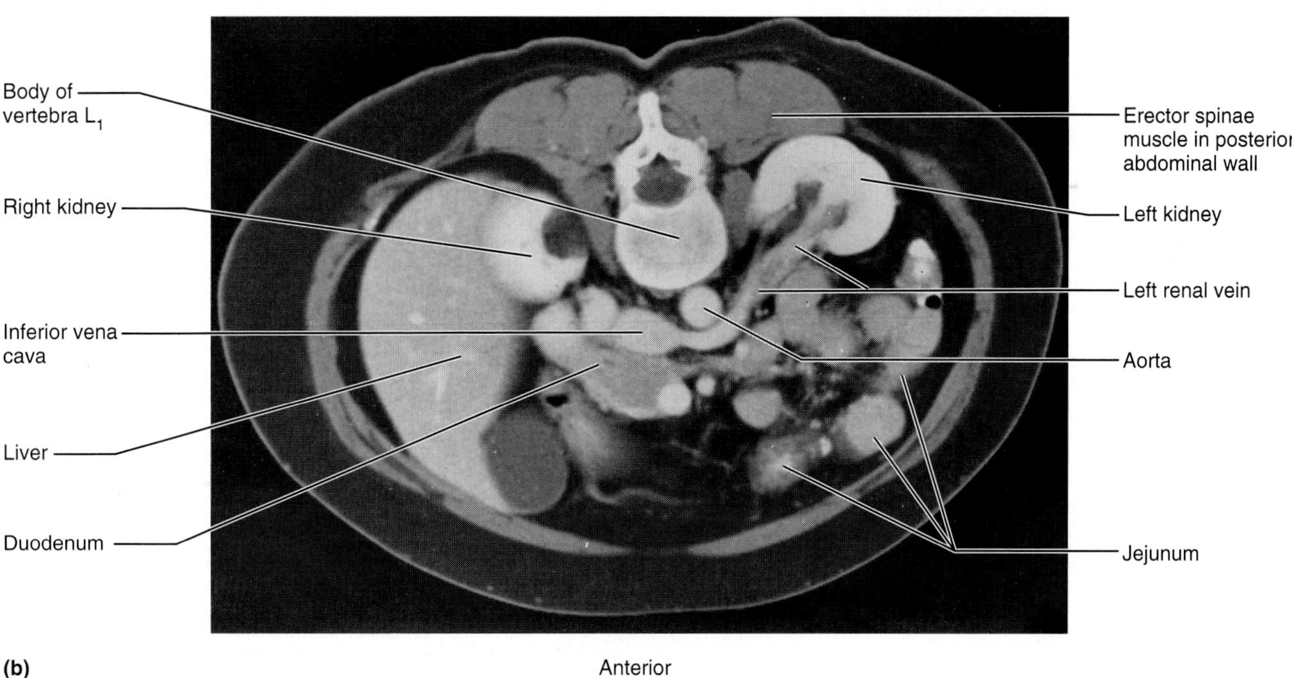

FIGURE 23.2 Position of the kidneys within the posterior abdominal wall (transverse section). (a) The kidneys are retroperitoneal and surrounded by layers of fascia and fat: the renal capsule, adipose capsule of perirenal fat, the renal fascia, and the pararenal fat (posteriorly). (b) CT scan through the abdomen, showing the relationship of the kidneys to the peritoneal organs of the digestive system.

lumbar vertebra inferiorly, and thus they receive some protection from the lower two ribs (see Figure 23.1b). The right kidney is crowded by the liver and lies slightly inferior to the left kidney. An average adult kidney is about 12 cm tall, 6 cm wide, and 3 cm thick—the size of a large bar of soap. The lateral surface of each kidney is convex; the medial surface is concave and has a vertical cleft called the renal **hilus,** where vessels and nerves enter and leave the kidney. On the superior part of each kidney lies an adrenal (suprarenal) gland, an endocrine gland that is functionally unrelated to the kidney (see p. 754).

Several layers of supportive tissue surround each kidney (see Figure 23.2a). A thin, tough layer of dense connective tissue called the **renal capsule** (*ren* = kidney) adheres directly to the kidney's surface, maintaining its shape and forming a barrier that can inhibit the spread of infection from the surrounding regions. Just external to the renal capsule is an **adipose capsule** consisting of **perirenal fat** (per′e-re″nal; "around the kidney"), and external to that is an envelope of **renal fascia.** Finally, the **pararenal fat** (par′ah-re″nal; "near the kidney") lies external and mostly posterior to the renal fascia. The perirenal and pararenal fat layers cushion the kidney against blows and help hold the kidneys in place within the body. A loss of this fat, as during rapid weight loss, leaves the kidneys unsupported, and they drop to a lower position, kinking the ureter and blocking the drainage of urine. The resulting accumulation of urine in the kidney, called **hydronephrosis** (hi″dro-nĕ-fro′sis; "water in the kidney"), produces pressure and may lead to necrosis of the kidney and renal failure.

CLINICAL APPLICATIONS

SURGICAL APPROACH TO THE KIDNEY Surgery is performed on kidneys for a variety of reasons, such as tumor removal and removal of a kidney for a kidney transplant. To reach a kidney, the surgeon usually cuts through the posterolateral abdominal wall, where the kidney lies closest to the body surface (Figure 23.2a). With this approach, very few muscles, vessels, or nerves need to be cut. However, the incision must be made inferior to the level of T_{12} to avoid puncturing the pleural cavity, which lies posterior to the superior third of each kidney. Recall that puncturing the pleural cavity leads to pneumothorax and collapse of the lung (p. 623). ✚

Internal Gross Anatomy

A frontal section through a kidney (Figure 23.3) reveals two distinct regions of kidney tissue: *cortex* and *medulla.* The more superficial region, the **renal cortex,** is light in color and has a granular appearance. Deep to the cortex is the darker **renal medulla,** which consists of cone-shaped masses called **medullary pyramids** or **renal pyramids.** The broad base of each pyramid abuts the cortex, whereas the pyramid's apex, or **papilla,** points internally. The medullary pyramids exhibit striations because they contain roughly parallel bundles of tiny urine-collecting tubules. The **renal columns,** inward extensions of the renal cortex, separate adjacent pyramids.

The human kidney is said to have *lobes,* each of which is a single medullary pyramid plus the cortical tissue that surrounds that pyramid. There are 5 to 11 lobes and pyramids in each kidney.

The *renal sinus* is a large space within the medial part of the kidney opening to the exterior through the renal hilus. This sinus is actually a "filled space" in that it contains the renal vessels and nerves, some fat, and the urine-carrying tubes called the *renal pelvis* and *calices.* The **renal pelvis** (*pelvis* = basin), a flat, funnel-shaped tube, is simply the expanded superior part of the ureter. Branching extensions of the renal pelvis form two or three **major calices** (ca′lih-sēs; also spelled **calyces;** singular: **calyx**), each of which divides to form several **minor calices,** cup-shaped tubes that enclose the papillae of the pyramids (*calyx* = cup). The calices collect urine draining from the papillae and empty it into the renal pelvis; the urine then flows through the renal pelvis and into the ureter, which transports it to the bladder for storage.

CLINICAL APPLICATIONS

PYELONEPHRITIS When an infection of the renal pelvis and calices, called *pyelitis* (pi″ĕ-li′tis; "inflammation of the renal pelvis"), spreads to involve the rest of the kidney as well, the result is **pyelonephritis** (pi″ĕ-lo-nĕ-fri′tis; *nephros* = kidney). Although it usually results from the spread of the fecal bacterium *Escherichia coli* from the anal region superiorly through the urinary tract, pyelonephritis also occurs when blood-borne bacteria lodge in the kidneys and proliferate there. In severe cases, the kidney swells and scars, abscesses form, and the renal pelvis fills with pus. Left untreated, the infected kidneys may be severely damaged, but timely administration of antibiotics usually achieves a total cure. ✚

Gross Vasculature and Nerve Supply

Given that the kidneys continuously cleanse and reconstitute the blood, it is not surprising that they have a rich blood supply (see Figure 23.3b). Under normal resting conditions, about one-fourth of the heart's systemic output reaches the kidneys via the large **renal arteries.** These arteries issue at right angles from the abdominal aorta, between the first and second lumbar vertebrae (see Figure 23.1a and Figure 19.11 on p. 559). Because the aorta lies slightly to the left of the body midline, the right renal artery is longer than the left. As each renal artery approaches a kidney, it divides into five **segmental arteries** that enter the hilus (see Figure 23.3b). Within the renal sinus, each segmental artery branches into several **lobar arteries** that then divide into **interlobar arteries,** which lie in the renal columns between the medullary pyramids.

At the medulla-cortex junction, the interlobar arteries branch into **arcuate arteries** (ar′ku-āt; "shaped like a bow"), which arch over the bases of the medullary pyramids. Radiating outward from the arcuate arteries and supplying the cortical tissue are small **interlobular arteries,** which, as their name implies, divide the cortical tissue into *lobules.* (Note that *interlobar* and

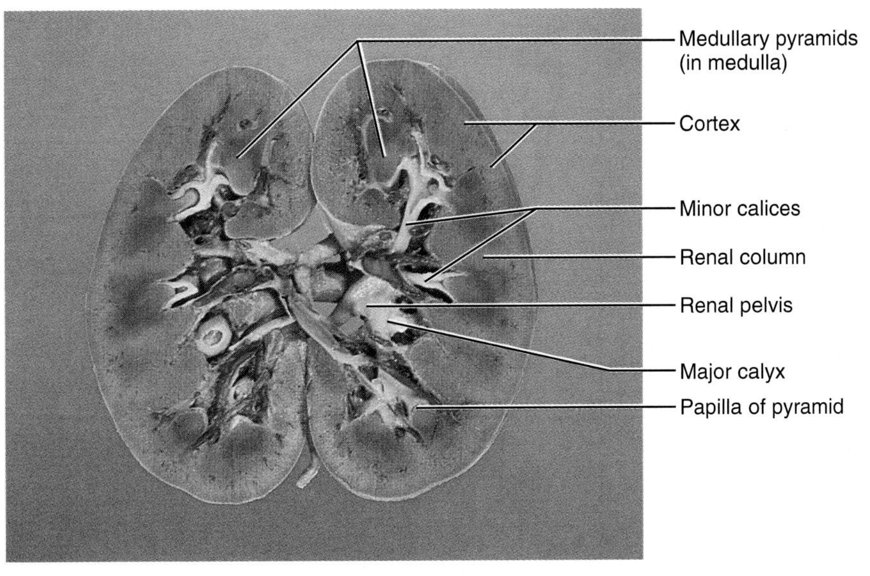

(a)

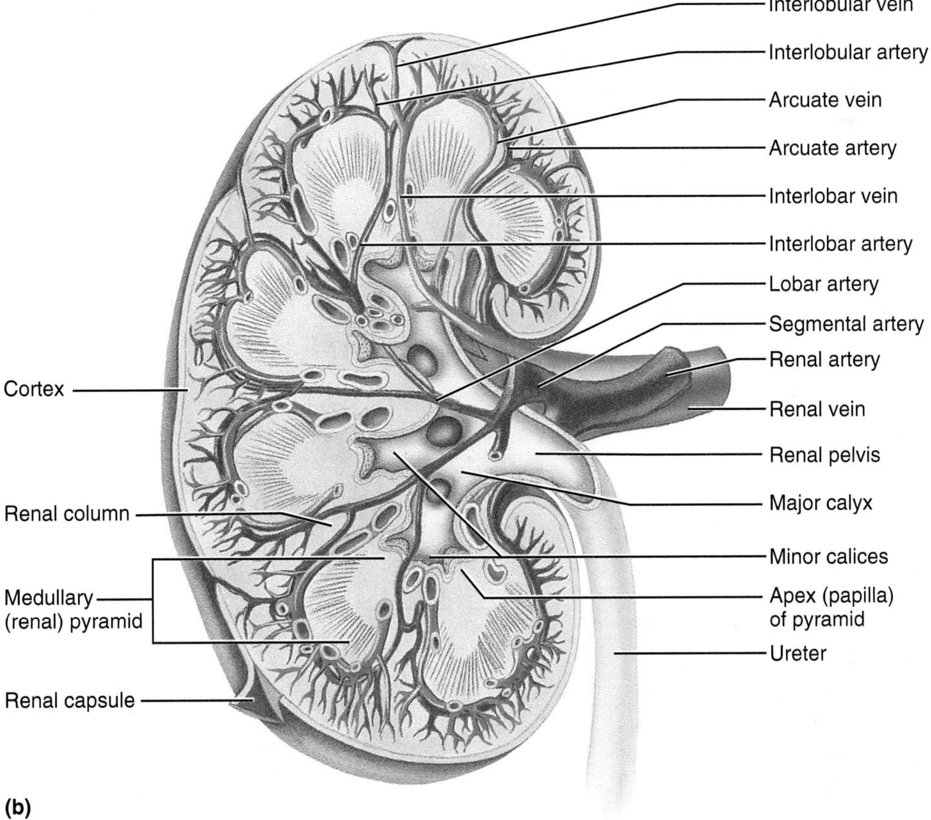

(b)

FIGURE 23.3 Internal anatomy of the kidney. (a) Photograph of a frontally sectioned kidney. **(b)** Diagram of a posterior view of the anterior half of a frontally sectioned kidney, emphasizing major blood vessels. The renal sinus is the kidney's medial space that contains the renal pelvis, calices, lobar vessels, nerves, and fat.

interlobular vessels are distinctly different.) More than 90% of the blood entering the kidney perfuses the cortex.

The *veins* of the kidney essentially trace the pathway of the arteries in reverse: Blood leaving the renal cortex drains sequentially into the **interlobular, arcuate, interlobar,** and **renal veins** (there are no lobar or segmental veins). The renal vein issues from the kidney at the hilus and empties into the inferior vena cava (see Figure 23.1a). Because the inferior vena cava lies on the right side of the vertebral column, the left renal vein is about twice as long as the right. Along its course, each renal vein lies anterior to the corresponding renal artery, and both blood vessels lie anterior to the renal pelvis at the hilus of the kidney.

The nerve supply of the kidney is provided by the *renal plexus,* a network of autonomic fibers and autonomic ganglia on the renal arteries. This plexus is an offshoot of the celiac plexus (see Figure 15.12, p. 456). The renal plexus is supplied by sympathetic fibers from the most inferior thoracic splanchnic nerve, the first lumbar splanchnic nerve, and other sources. These sympathetic fibers both control the diameters of the kidney

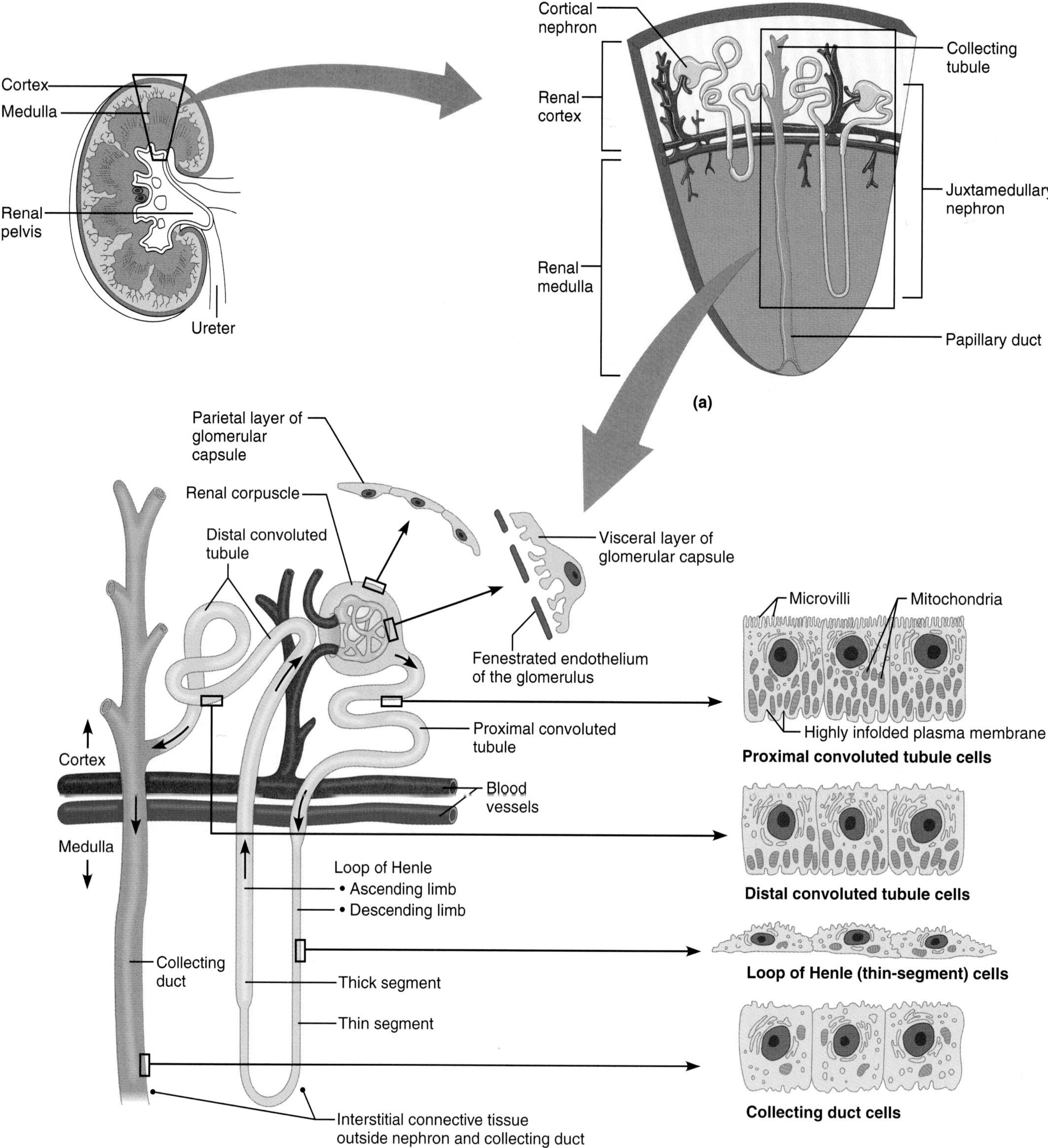

FIGURE 23.4 The uriniferous tubule (nephron plus collecting tubule). **(a)** A wedge-shaped section (lobe) of kidney tissue, showing the location of nephrons and collecting tubules in the kidney. Although thousands of collecting tubules and nephrons occupy each lobe of the kidney, only a few are shown here. **(b)** Schematic view of a uriniferous tubule, its various portions, and the ultrastructure of the epithelial cells that form these portions.

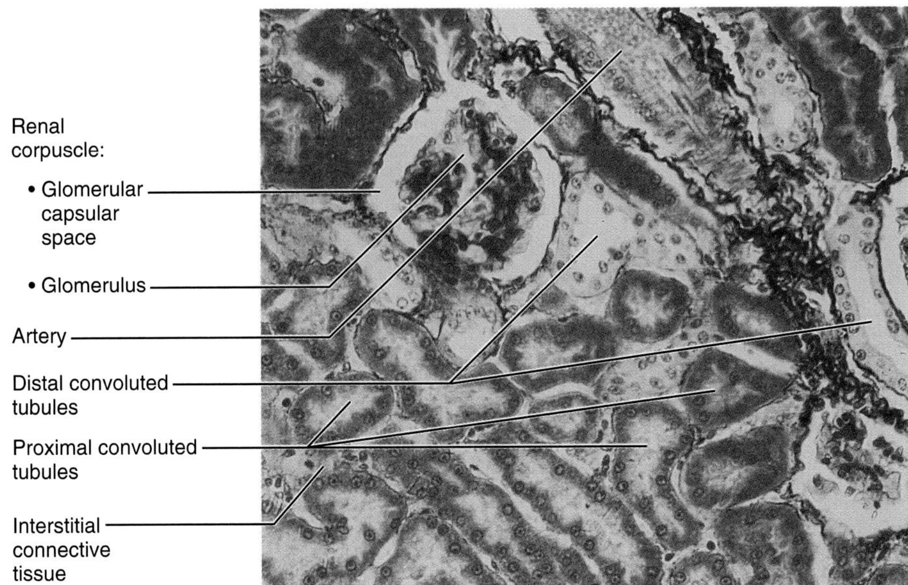

FIGURE 23.5 Photomicrograph of the renal cortex, showing parts of the nephron. Note the renal corpuscles and the sections through the various tubules. In the proximal convoluted tubules, the lumens appear "fuzzy" because they are filled with the long microvilli of the epithelial cells. In the distal tubules, by contrast, the lumens are clear (220 ×).

arteries and influence the urine-forming functions of the uriniferous tubules (discussed next).

Microscopic Anatomy of the Kidneys

As shown in Figures 23.4 and 23.5, the main structural and functional unit of the kidney is the **uriniferous tubule** (u″rĭ-nif′er-us; "urine-carrying"). More than a million of these tubules are crowded together within each kidney, separated from one another by small amounts of a loose areolar connective tissue called **interstitial connective tissue.** Each uriniferous tubule (Figure 23.4b) has two major parts: (1) a urine-forming *nephron,* composed of a renal corpuscle, a proximal convoluted tubule, a loop of Henle, and a distal convoluted tubule, and (2) a *collecting duct (collecting tubule),* which is involved in concentrating urine by removing water from it. Throughout its length the uriniferous tubule is lined by a simple epithelium (one cell thick) that is adapted for various aspects of the production of urine.

Before we examine each of the structures of the uriniferous tubule in detail, we first briefly discuss the mechanisms by which it produces urine.

Mechanisms of Urine Production

The uriniferous tubule produces urine through three interacting mechanisms: filtration, reabsorption, and secretion. In **filtration,** a filtrate of the blood leaves the kidney capillaries and enters the nephron, as shown by arrow [a] in Figure 23.6. This filtrate resembles tissue fluid in that it contains all the small molecules of blood plasma. As it proceeds through the uriniferous tubule, the filtrate is processed into urine by the mechanisms of reabsorption and secretion. During **reabsorption,** most

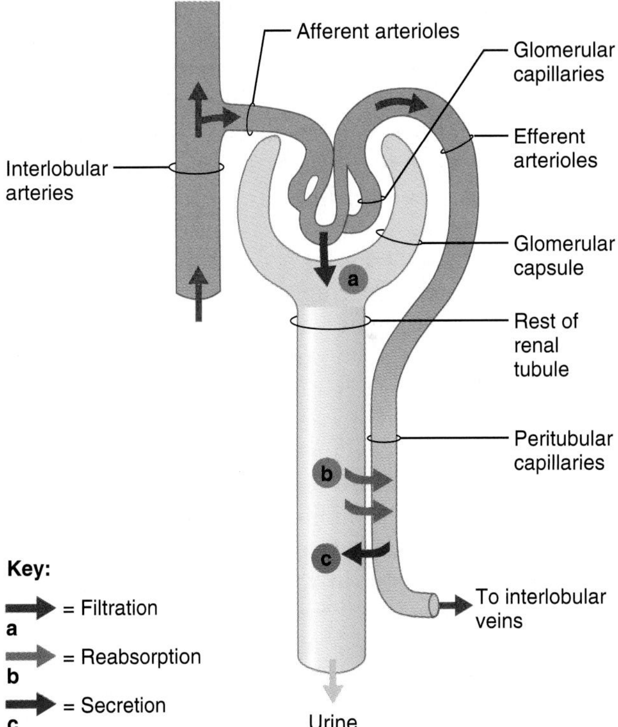

FIGURE 23.6 Basic kidney functions: The urine-forming part of the kidney, depicted as a single, generalized uriniferous tubule. Millions of these tubules act together in a kidney. The three major mechanisms by which the kidneys form urine are **(a)** glomerular filtration, **(b)** tubular reabsorption, and **(c)** tubular secretion.

of the nutrients, water, and essential ions are reclaimed from the filtrate, and returned to the blood of capillaries in the surrounding connective tissue (arrow [b] in Figure 23.6). In fact, 99% of the volume of the renal filtrate is reabsorbed in this manner. As the essential molecules are reclaimed from the filtrate, the remaining wastes and unneeded substances contribute to the urine that ultimately leaves the body. Supplementing this passive method of waste disposal is the active process of **secretion,** which moves additional undesirable molecules into the tubule from the blood of surrounding capillaries (arrow [c] in Figure 23.6). Next we explore the basic portions of the uriniferous tubule and see how each contributes to the processes of filtration, reabsorption, and secretion.

The Nephron

Recall that the parts of the nephron are the renal corpuscle and a tubular section consisting of the proximal convoluted tubule, loop of Henle, and distal convoluted tubule (Figure 23.4b).

Renal Corpuscle The first part of the nephron, where filtration occurs, is the spherical **renal corpuscle** (Figure 23.7). Renal corpuscles, which occur strictly in the cortex, consist of a tuft of capillaries called a **glomerulus** (glo-mer′u-lus; "ball of yarn") surrounded by a cup-shaped, hollow **glomerular capsule** *(Bowman's capsule).* The glomerulus lies in its glomerular capsule like a fist pushed deeply into an underinflated balloon (see Figure 23.7a). This tuft of capillaries is supplied by an afferent arteriole and drained by an efferent arteriole (see p. 686). The endothelium of the glomerulus is fenestrated (has pores), and thus these capillaries are highly porous, allowing large quantities of fluid and small molecules to pass from the capillary blood into the hollow interior of the glomerular capsule, the **capsular space.** This fluid is the filtrate that is ultimately processed into urine. Only about 20% of the fluid leaves the glomerulus and enters the capsular space; 80% remains in the blood within this capillary.

The external **parietal layer** of the glomerular capsule, which is a simple squamous epithelium (see Figure 23.7a), simply contributes to the structure of the capsule and plays no part in the formation of filtrate. By contrast, the capsule's **visceral layer** clings to the glomerulus and consists of unusual, branching epithelial cells called **podocytes** (pod′o-sītz; "foot cells"). The branches of the octopus-like podocytes (see Figure 23.7b) end in **foot processes,** or **pedicels** ("little feet"), which interdigitate with one another as they surround the glomerular capillaries. The filtrate passes into the capsular space through thin clefts between the foot processes, called **filtration slits** or *slit pores* (see Figure 23.7b).

The **filtration membrane (filtration barrier),** the actual filter that lies between the blood in the glomerulus and the capsular space, consists of three layers (see Figure 23.7c): (1) the fenestrated endothelium of the capillary; (2) the filtration slits between the foot processes of podocytes, each of which is covered by a thin **slit diaphragm;** and (3) an intervening basement membrane consisting of the fused basal laminae of the endothelium and the podocyte epithelium. The capillary pores (fenestrations) restrict the passage of the largest elements such as blood cells, whereas the basement membrane and slit diaphragm hold back all but the smallest proteins while letting through small molecules such as water, ions, glucose, amino acids, and urea. The relative roles of the latter two layers in this filter are under debate: Despite the traditional view that the basement membrane is the main or only molecular filter, when podocytes and slit diaphragms are stripped away by disease, the basement membrane leaks many proteins, indicating that the slit diaphragms also contribute extensively to filtration in the normal kidney.

Tubular Section of the Nephron After forming in the renal corpuscle, the filtrate proceeds into the long tubular section of the nephron (see Figure 23.4b), which begins as the elaborately coiled *proximal convoluted tubule,* makes a hairpin loop called the *loop of Henle,* winds and twists again as the *distal convoluted tubule,* and ends by joining a collecting tubule. This meandering nature of the nephron increases its length and enhances its capabilities in processing the filtrate that flows through it.

Each part of the nephron's tubular section has a unique cellular anatomy that reflects its filtrate-processing function. The **proximal convoluted tubule,** confined entirely to the renal cortex, is most active in reabsorption and secretion. Its walls are formed by cuboidal epithelial cells whose luminal (exposed) surfaces have long microvilli that seem to fill the tubule lumen with a "fuzz" in photomicrographs (Figures 23.4b and 23.5). These microvilli increase the surface area of these cells tremendously, maximizing their capacity for reabsorbing water, ions, and solutes from the filtrate. Furthermore, the cells of the proximal tubule contain many mitochondria, which provide the energy for reabsorption, and a highly infolded basolateral membrane (plasma membrane on their basal and lateral cell surfaces) that contains many ion-pumping enzymes responsible for reabsorbing molecules from the filtrate.

The U-shaped **loop of Henle,** or *loop of the nephron,* consists of a descending limb and an ascending limb. The first part of the **descending limb** is continuous with the proximal tubule and has a similar structure; the rest of the descending limb is the **thin segment,** the narrowest part of the nephron, with walls consisting of a permeable simple squamous epithelium. The thin segment may continue into the ascending limb, and inevitably it joins the **thick segment** of the ascending limb (or thick ascending limb, for short). The cell structure of this thick segment resembles that of the distal convoluted tubule (discussed next).

The **distal convoluted tubule,** like the proximal convoluted tubule, is confined to the renal cortex, has walls of simple cuboidal epithelium, and is specialized for the selective secretion and reabsorption of ions. It is

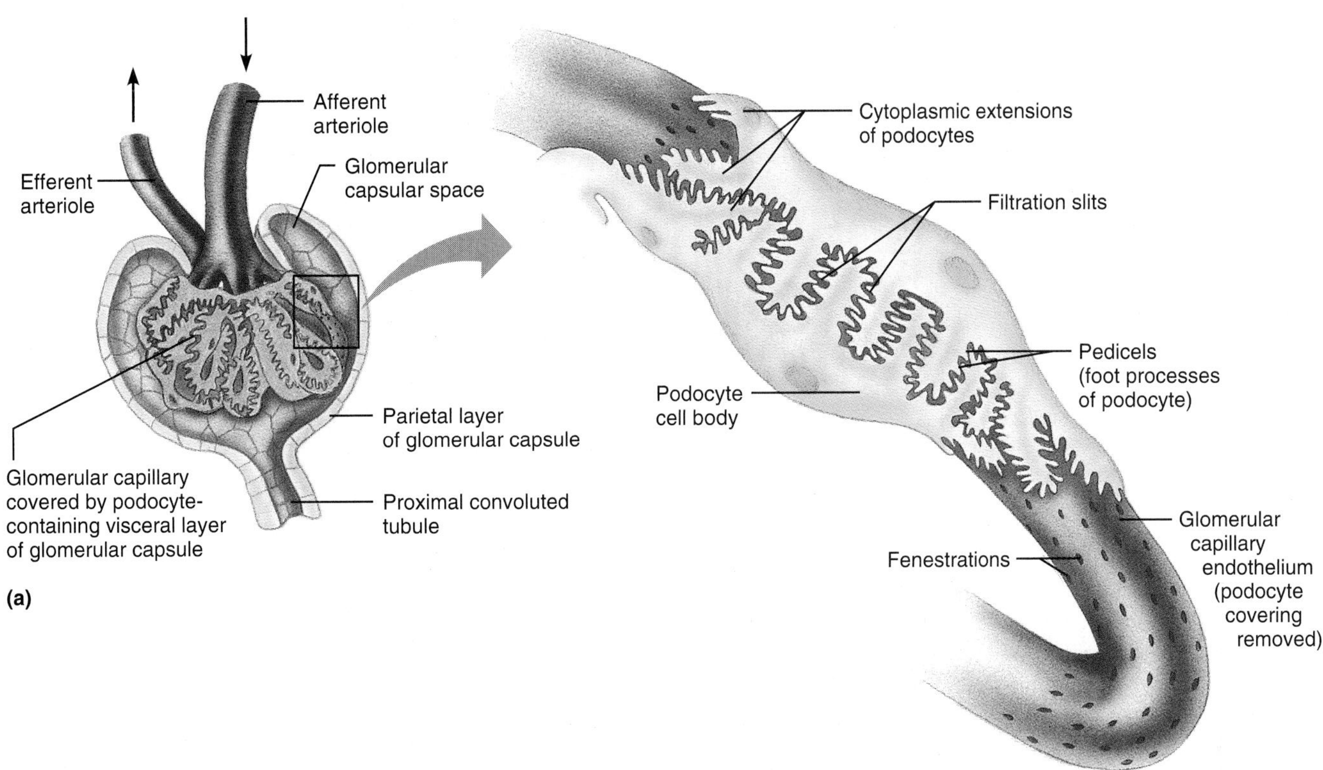

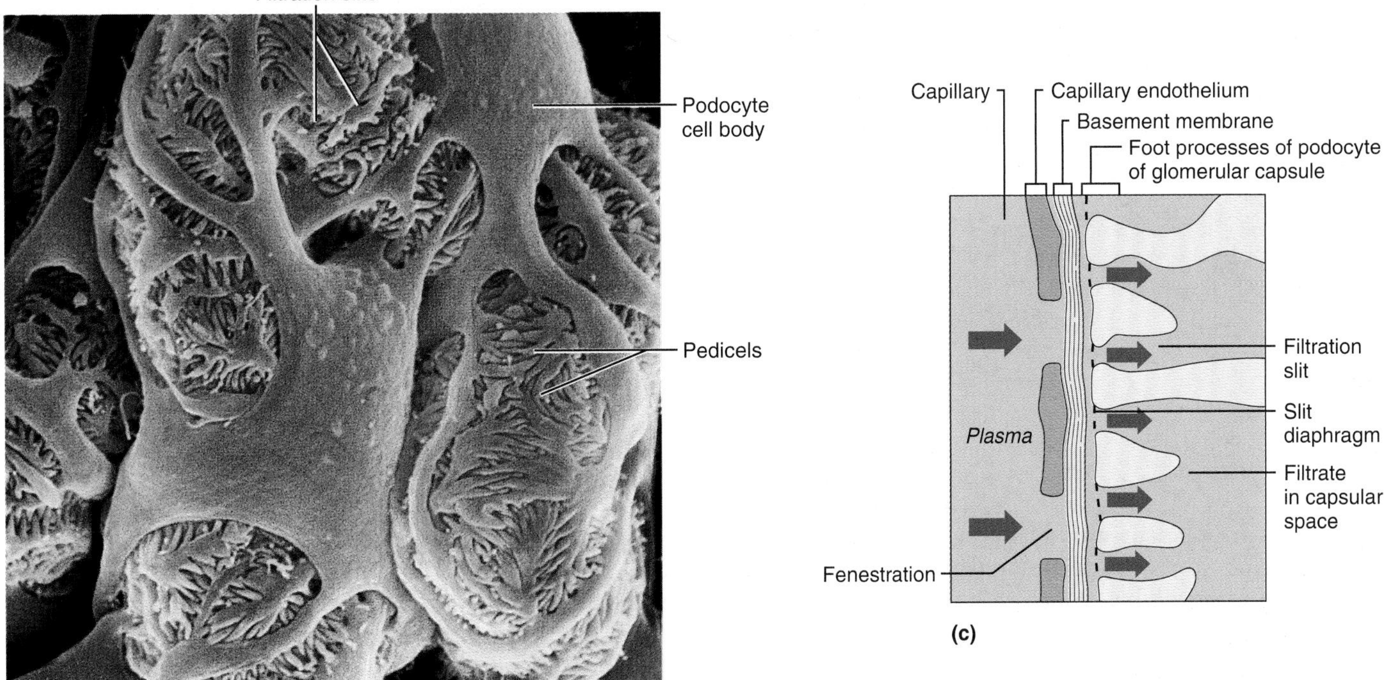

FIGURE 23.7 Renal corpuscle and the filtration membrane. **(a)** The renal corpuscle, at left, consists of a glomerulus (a tuft of capillaries) surrounded by a glomerular capsule, like a fist pushed deep within a balloon. The enlargement at right shows how a glomerular capillary is covered by the visceral (inner) layer of the glomerular capsule, or podocyte epithelium. Some podocytes have been removed to show the fenestrations (pores) in the underlying capillary wall. **(b)** Scanning electron micrograph of podocytes clinging to the glomerular capillaries (artificially colored; 3900 ×). This view is similar to that in the right half of part (a). **(c)** The filtration membrane: diagram of a section through this membrane, showing its three layers. The interior of the glomerular capillary is at left, the glomerular capsular space is at right.

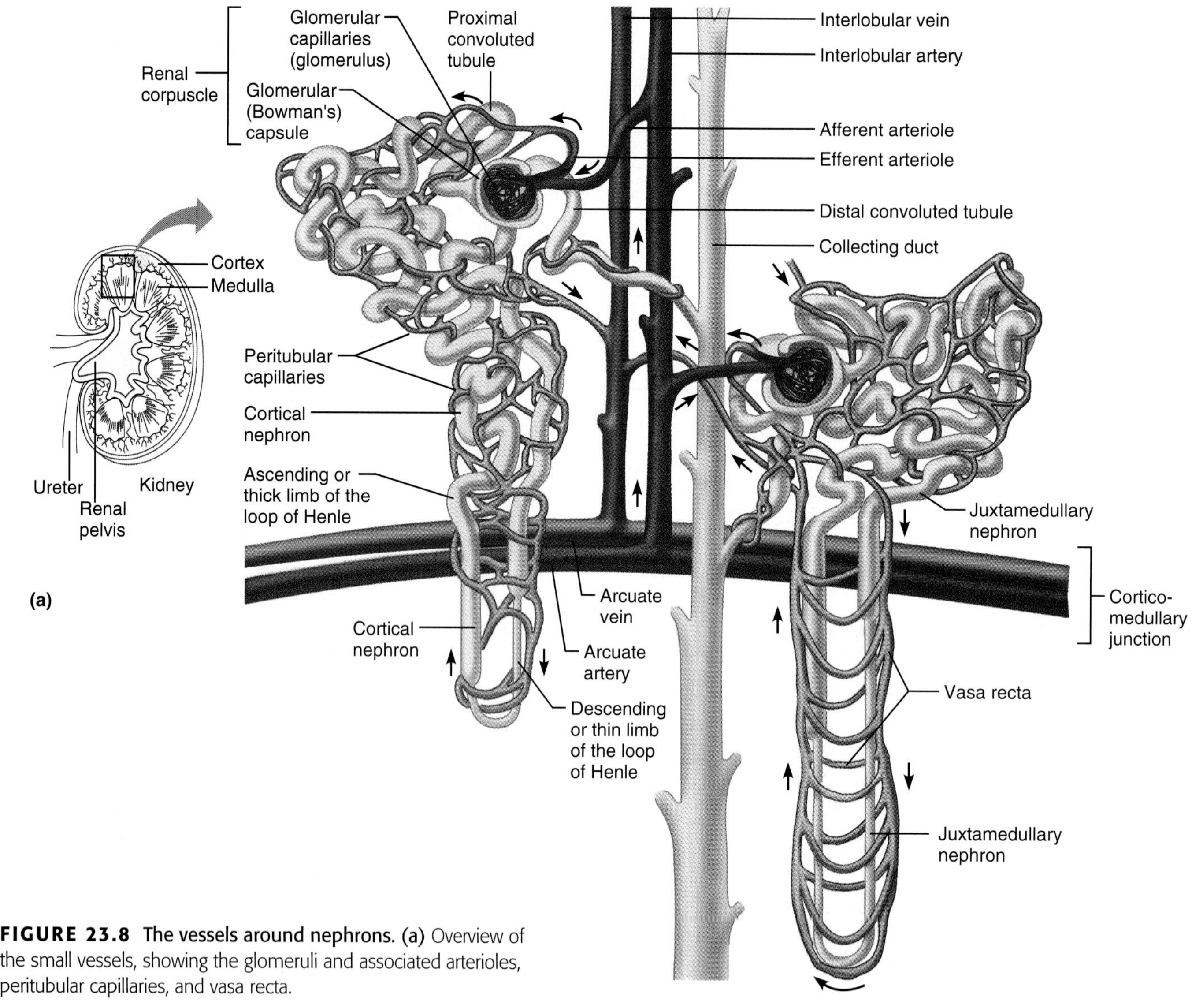

FIGURE 23.8 The vessels around nephrons. **(a)** Overview of the small vessels, showing the glomeruli and associated arterioles, peritubular capillaries, and vasa recta.

less active in reabsorption than the proximal tubule, however, and its cells do not have an abundance of absorptive microvilli (see Figure 23.4b). Still, cells of the distal tubule have many mitochondria and infoldings of the basolateral membrane, features that are typical of the ion-pumping cells in the body.

Classes of Nephrons Although all nephrons have the structures just described, they are divided into two categories according to location (see Figure 23.4a). **Cortical nephrons,** which represent 85% of all nephrons, are located almost entirely within the cortex, with their loops of Henle dipping only a short distance into the medulla. The remaining 15%, called **juxtamedullary** ("near the medulla") **nephrons** because their renal corpuscles lie near the cortex-medulla junction, have loops of Henle that deeply invade the medulla and thin segments that are much longer than those of cortical nephrons. These long loops of Henle, in conjunction with nearby collecting tubules, contribute to the kidney's ability to produce a concentrated urine.

Collecting Tubules

Urine passes from the distal tubules of the nephrons into the **collecting tubules,** each of which receives urine from several nephrons and runs straight through the cortex into the deep medulla (see Figure 23.4a). There, adjacent collecting tubules join to form larger **papillary ducts,** which empty into the minor calices through the renal papillae. Histologically, the walls of the collecting

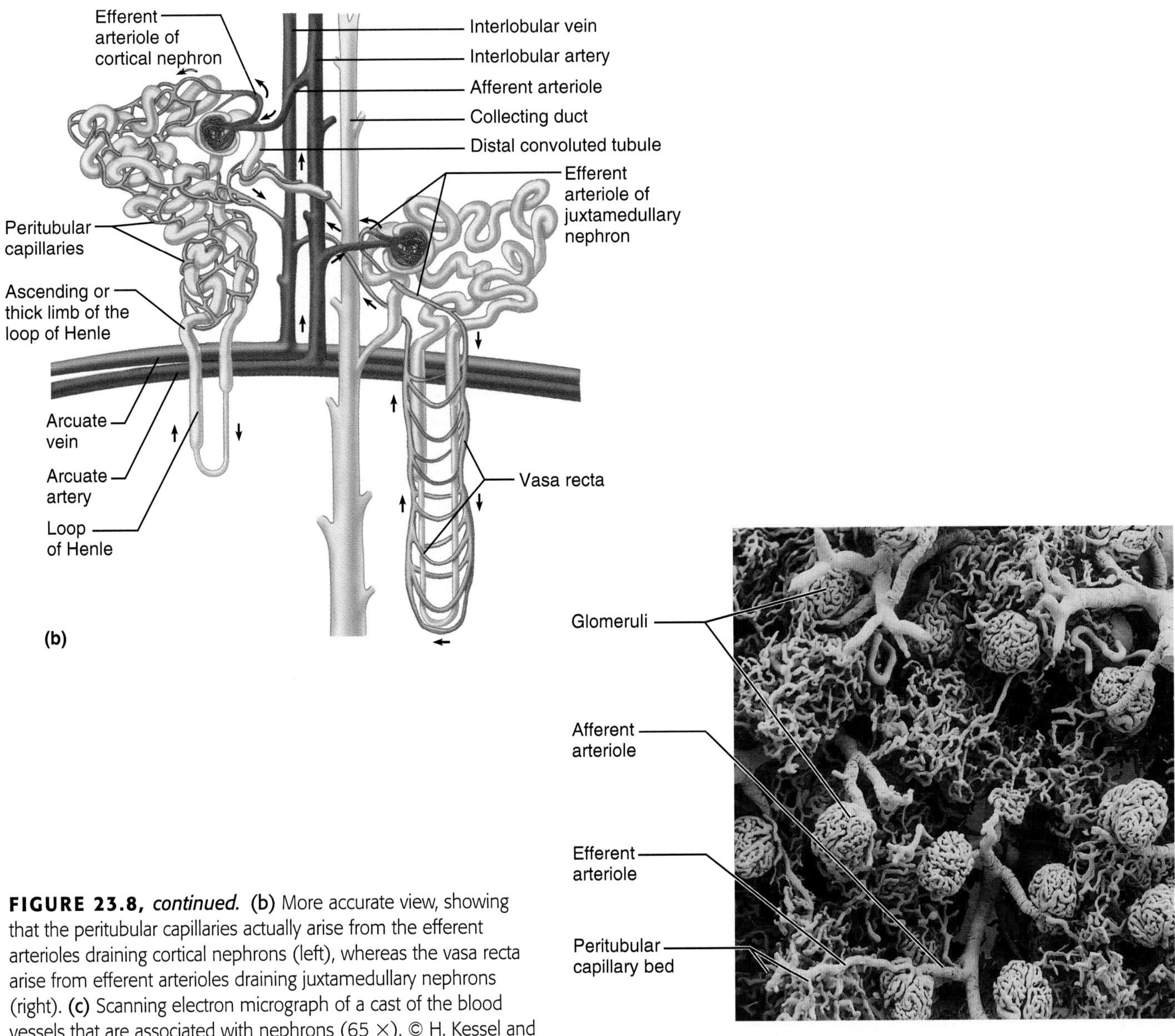

FIGURE 23.8, *continued.* (b) More accurate view, showing that the peritubular capillaries actually arise from the efferent arterioles draining cortical nephrons (left), whereas the vasa recta arise from efferent arterioles draining juxtamedullary nephrons (right). (c) Scanning electron micrograph of a cast of the blood vessels that are associated with nephrons (65 ×). © H. Kessel and R. Kardon.

tubules consist of a simple cuboidal epithelium (see Figure 23.4b), which thickens to become simple columnar in the papillary ducts. Most of these cells have few organelles, but a few are rich in mitochondria and participate in reabsorption and secretion of ions.

The most important role of the collecting tubules, however, is to conserve body fluids, a function they share with the distal tubules of the nephron. When the body must conserve water, the posterior part of the pituitary gland secretes antidiuretic hormone (see Chapter 25, p. 752), which increases the permeability of the collecting tubules and distal tubules to water. As a result, water leaves the urine that is in these tubules and is reabsorbed into the surrounding blood vessels, decreasing the total volume of urine produced.

Microscopic Blood Vessels Associated with Uriniferous Tubules

Nephrons associate closely with two kinds of capillary beds, the *glomerulus* and the *peritubular capillaries,* and also with the capillary-like *vasa recta* (Figure 23.8a).

Glomeruli: As previously discussed, the capillaries of the glomerulus produce the filtrate that moves through the rest of the uriniferous tubule and becomes

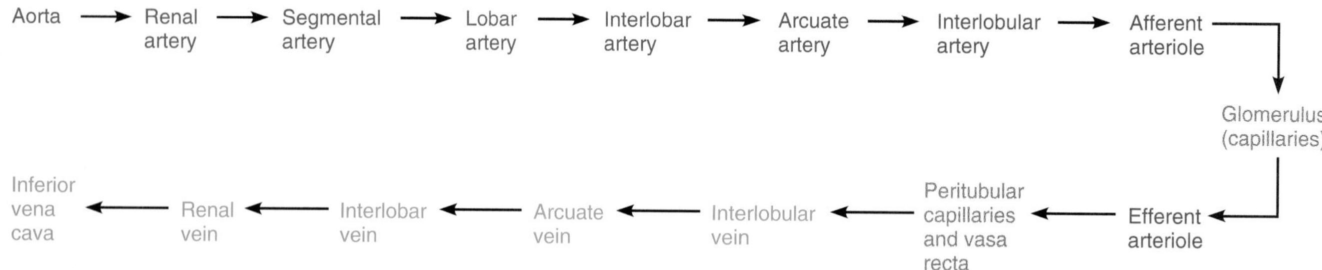

FIGURE 23.9 Summary of blood vessels supplying the kidney.

urine. The glomerulus differs from all other capillary beds in the body: It is both fed and drained by arterioles—an **afferent arteriole** and an **efferent arteriole,** respectively. The *afferent arterioles* arise from the interlobular arteries that run through the renal cortex. Because arterioles are high-resistance vessels and the efferent arteriole is narrower than the afferent arteriole, the blood pressure in the glomerulus is extraordinarily high for a capillary bed and easily forces the filtrate out of the blood and into the glomerular capsule. The kidneys generate 1 L (about 1 quart) of this filtrate every 8 minutes, but, as previously mentioned, only 1% ends up as urine; the other 99% is reabsorbed by the uriniferous tubule and returned to the blood in the peritubular capillary beds.

Peritubular capillaries: The **peritubular capillaries** *(intertubular capillaries)* arise from the efferent arterioles draining the cortical glomeruli (Figure 23.8b). These capillaries, which lie in the interstitial connective tissue of the renal cortex, cling closely to the convoluted tubules and empty into nearby venules of the renal venous system. The peritubular capillaries are adapted for absorption: They are low-pressure, porous capillaries that readily absorb solutes and water from the tubule cells after these substances are reabsorbed from the filtrate. Furthermore, all molecules that are *secreted* by the nephrons into the urine are derived from the blood of nearby peritubular capillaries.

Vasa recta: In the deepest part of the renal cortex, efferent arterioles from the juxtamedullary glomeruli continue into thin-walled looping vessels called **vasa recta** (va′sah rek′tah; "straight vessels") (Figure 22.8b). These hairpin loops descend into the medulla, running alongside the loops of Henle. The vasa recta are part of the kidney's urine-concentrating mechanism.

Figure 23.9 provides a complete summary of the microscopic and gross blood vessels of the kidney.

Juxtaglomerular Apparatus

The **juxtaglomerular apparatus** (juks″tah-glo-mer′u-lar; "near the glomerulus"), a structure that functions in the regulation of blood pressure, is an area of specialized contact between the first part of the distal convoluted tubule and the afferent arteriole (Figure 23.10). Within the apparatus, the structures of both the tubule and the arteriole are modified.

The wall of the afferent arteriole contains **juxtaglomerular cells,** modified smooth muscle cells with secretory granules containing a hormone called **renin** (re′nin; "kidney hormone"). The juxtaglomerular cells seem to be mechanoreceptors that secrete renin in response to falling blood pressure in the afferent arteriole. (As shown in the figure, a few juxtaglomerular cells may exist in the *efferent* arteriole as well.)

The **macula densa** (mak′u-lah den′sah; "dense spot"), which is the portion of the distal tubule adjacent to the juxtaglomerular cells, consists of tall, closely packed epithelial cells that act as chemoreceptors for monitoring solute concentrations in the filtrate. Whenever such solute concentrations fall below a certain level, the cells of the macula densa signal the juxtaglomerular cells to secrete renin, which, among other effects, increases blood-solute concentration, blood volume, and most importantly, blood pressure.

Interstitial Connective Tissue

As mentioned, a thin layer of interstitial connective tissue surrounds the uriniferous tubules and contains the small blood vessels of the kidney (see Figures 23.4b and 23.5). This interstitium contains the typical elements of loose connective tissue—ground substance, collagen, fibroblasts, and macrophages—as well as some endocrine cells whose hormones help to lower blood pressure and raise the hematocrit (see p. 759).

URETERS

Leaving the kidneys, we now discuss the next structures along the urinary tract, the ureters, beginning with their gross structure and proceeding to their microscopic anatomy.

Gross Anatomy

The **ureters** are slender tubes, about 25 cm (10 inches) long, that carry urine from the kidneys to the bladder (see Figure 23.1). Each ureter begins superiorly, at the level of L_2, as a continuation of the renal pelvis. From there, it descends retroperitoneally through the

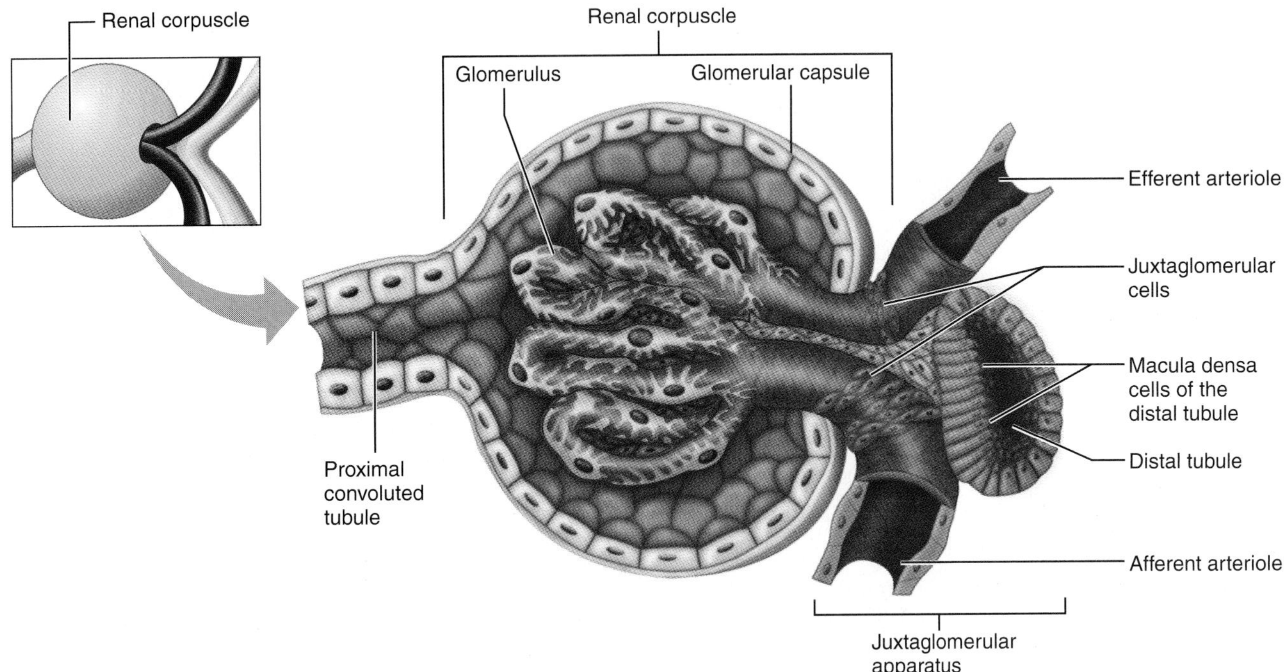

FIGURE 23.10 Juxtaglomerular apparatus. This apparatus, which lies near the glomerulus, consists of the macula densa of the distal tubule and juxtaglomerular cells of the afferent (and efferent) arterioles.

abdomen, enters the true pelvis by crossing the pelvic brim at the sacroiliac joint, enters the posterolateral corner of the bladder, and then runs medially within the posterior bladder wall before opening into the bladder's interior. This oblique entry into the bladder prevents backflow of urine from the bladder into the ureters—because any increase of pressure within the bladder compresses the bladder wall, thereby closing the distal ends of the ureters.

Clinical Applications

PYELOGRAPHY The radiographic procedure for examining the ureters and renal calices is called pyelography (pi″ĕ-log′rah-fe; "recording the renal pelvis"); the resulting image is called a **pyelogram** (Figure 23.11). Radiologists introduce the X-ray contrast medium via one of two routes; either by injecting it into the ureters through a catheter in the bladder *(retrograde pyelography)* or by injecting it through a vein so that it reaches the ureters when excreted by the kidneys *(intravenous pyelography)*. Pyelography enables the clinician to examine the ureters along their course from the kidneys to the bladder as these tubes extend along an imaginary line defined by the transverse processes of the five lumbar vertebrae (Figure 23.11). ✚

Microscopic Anatomy

The histological structure of the tubular ureters is the same as that of the renal calices and renal pelvis, with walls having three basic layers: a mucosa, a muscularis, and an adventitia (Figure 23.12). The lining **mucosa** is composed of a *transitional epithelium* that stretches when the ureters fill with urine (see p. 82) and a lamina propria composed of a stretchy, fibroelastic connective tissue containing rare patches of lymphoid tissue. The middle **muscularis** consists of two layers: an inner longitudinal layer and an outer circular layer of smooth muscle. A third layer of muscularis, an external longitudinal layer, appears in the inferior third of the ureter. The external **adventitia** of the ureter wall is a typical connective tissue.

The ureters play an active role in transporting urine. Distension of the ureter by entering urine stimulates its muscularis to contract, setting up peristaltic waves that propel urine to the bladder. This means that urine does *not* reach the bladder due to gravity alone. Although the ureters are innervated by both sympathetic and parasympathetic nerve fibers, neural control of their peristalsis appears to be insignificant compared to the local stretch response of ureteric smooth muscle.

URINARY BLADDER

The **urinary bladder,** a collapsible, muscular sac that stores and expels urine, lies inferior to the peritoneal cavity on the pelvic floor, just posterior to the pubic

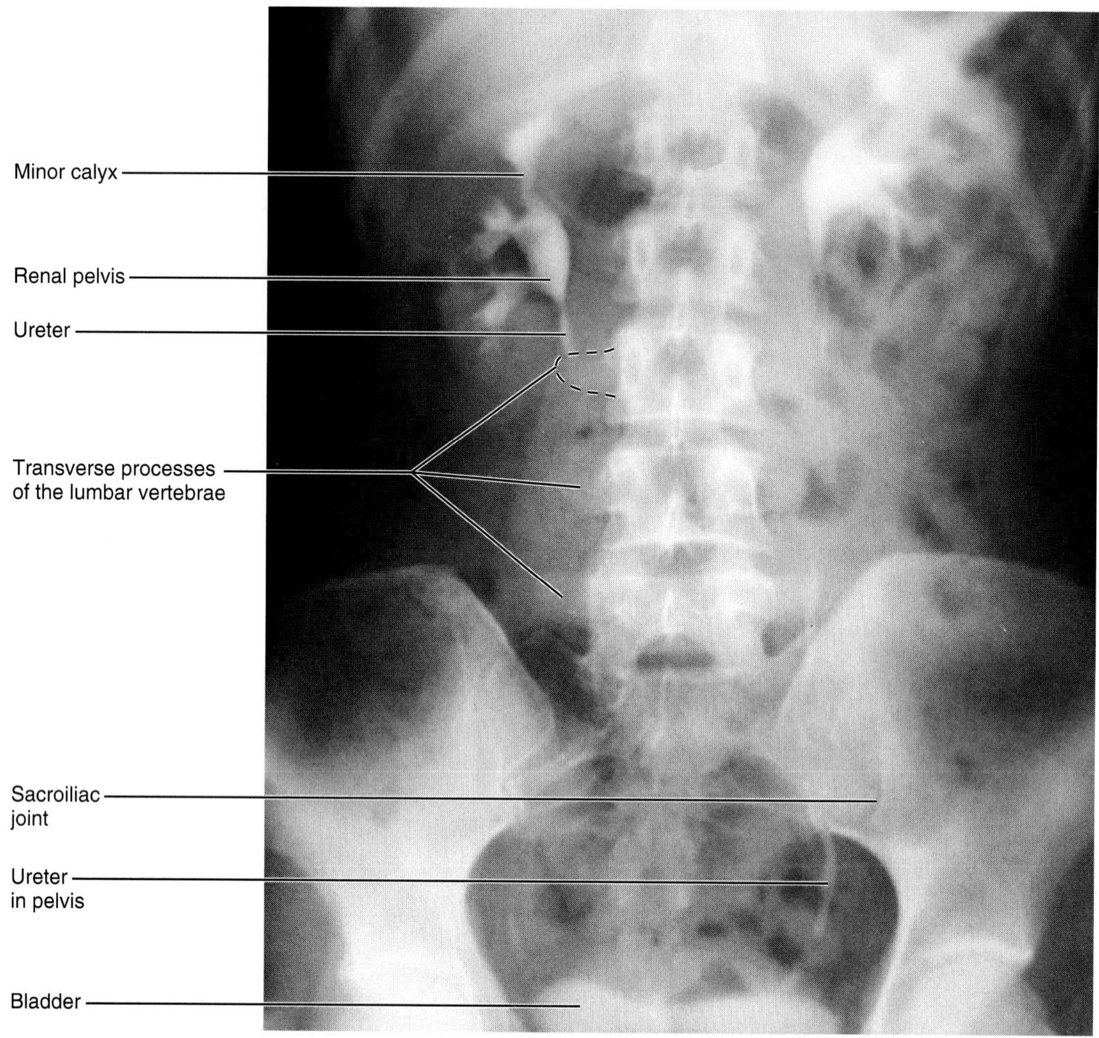

FIGURE 23.11 X-ray image of the ureters (pyelogram) in the posterior abdominal wall and the pelvis. Also note the minor calices and the renal pelvis (above).

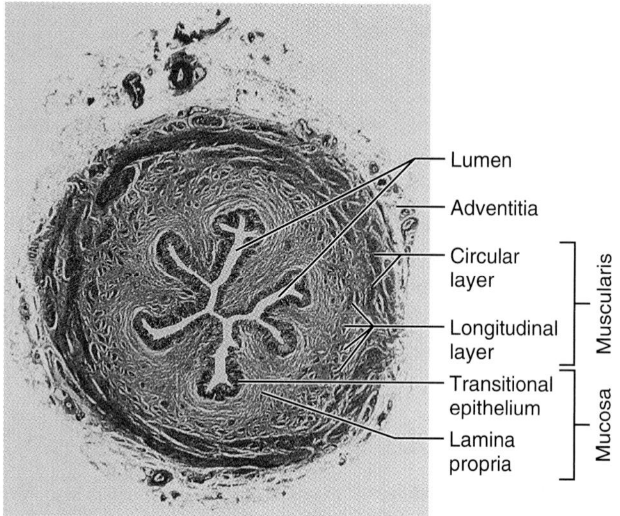

FIGURE 23.12 Microscopic structure of the ureter (cross-sectional view, 10 ×). Note the basic layers. In the empty ureter, the mucosa folds into longitudinal ridges, giving the lumen an irregular, branched appearance; these mucosal folds stretch and flatten to accommodate large pulses of urine.

symphysis (Figure 23.13). In males, the bladder lies anterior to the rectum; in females, the bladder lies just anterior to the vagina and uterus.

A full bladder is roughly spherical and expands superiorly into the abdominal cavity (see Figure 23.13); an empty bladder, by contrast, lies entirely within the pelvis and has the shape of an upside-down pyramid with four triangular surfaces and four corners, or angles

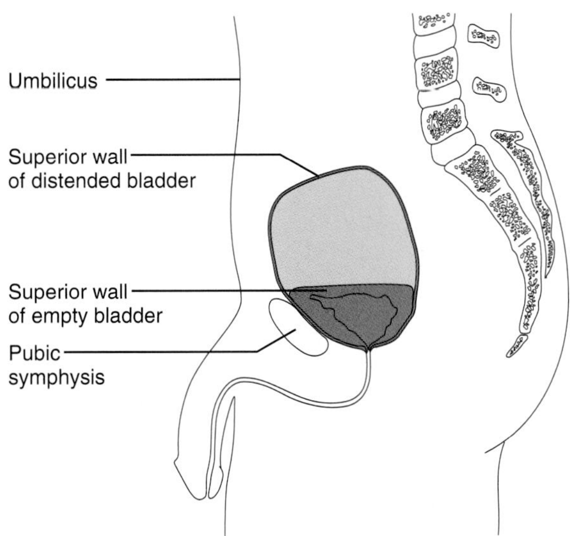

FIGURE 23.13 The urinary bladder: Positions and shapes of a full and an empty bladder.

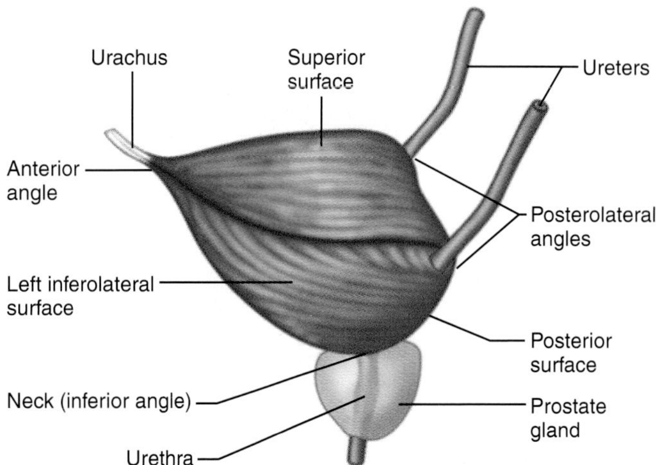

FIGURE 23.14 The empty bladder: An inverted pyramid with four surfaces and four angles. This bladder, from a male, is oriented like the bladder in Figure 23.13. Its four surfaces are the superior, posterior, and left and right inferolateral surfaces. Note that either a tube (ureter, urethra) or a band (the urachus) attaches to each angle of the bladder.

(Figure 23.14). The two *posterolateral angles* receive the ureters. At the bladder's *anterior angle* (or apex) is a fibrous band called the **urachus** (u′rah-kus; "urinary canal of the fetus"), the closed remnant of an embryonic tube called the allantois (see p. 667). The *inferior angle (neck)* drains into the urethra. In males, the prostate gland, a reproductive gland, lies directly inferior to the bladder, where it surrounds the urethra (see p. 711).

In the interior of the bladder, openings for both ureters and the urethra define a triangular region on the posterior wall called the **trigone** (tri′gon; "triangle") (Figure 23.15). The trigone is of special clinical importance because infections tend to persist in this region.

The arteries supplying the bladder are branches of the internal iliac arteries, primarily the *superior* and *inferior vesical arteries* (*vesical* = sac = the bladder) (see Figure 19.15 on p. 563). Veins draining the bladder form a plexus on the bladder's inferior and posterior surfaces that empties into the internal iliac veins. Nerves extending to the bladder from the hypogastric plexus consist of parasympathetic fibers (ultimately from the pelvic splanchnic nerves), a few sympathetic fibers (ultimately from the lower thoracic and upper lumbar splanchnic nerves), and visceral sensory fibers.

The wall of the bladder has three layers (Figure 23.16): (1) a mucosa with a distensible transitional epithelium (p. 82) and a lamina propria; (2) a thick muscular layer; and (3) a fibrous adventitia (except on the superior surface of the bladder, which is covered by parietal peritoneum). The muscular layer, called the **detrusor muscle** (de-tru′sor; "to thrust out"), consists of highly intermingled smooth muscle fibers arranged in inner and outer longitudinal layers and a middle circular layer.

Critical Reasoning

What is the function of the detrusor muscle?

Contraction of this muscle squeezes urine from the bladder during urination.

The bladder's great distensibility makes it uniquely suited for its function of storing urine. When there is little urine in it, the bladder collapses into its basic pyramidal shape, its walls thick and its mucosa thrown into folds, or rugae (see Figure 23.15). But as urine accumulates, the rugae flatten, and the wall of the bladder thins as it stretches, allowing the bladder to store larger amounts of urine without a significant rise in internal pressure (at least until 300 mL has accumulated). A full adult bladder holds about 500 mL (or 1 pint) of urine, 15 times its empty volume.

Voiding of the bladder, or micturition, is discussed on page 691.

URETHRA

The **urethra** is a thin-walled tube that drains urine from the bladder and conveys it out of the body (see Figure 23.15). Whereas in both sexes this tube consists of smooth muscle and an inner mucosa, in males the muscle layer becomes very thin toward the distal end of the urethra. The lining epithelium changes from a transitional epithelium near the bladder to a stratified and pseudostratified columnar epithelium in mid urethra (sparse in females), and then to a stratified squamous epithelium near the end of the urethra.

At the bladder-urethra junction, a thickening of the detrusor muscle forms the **internal urethral sphincter.**

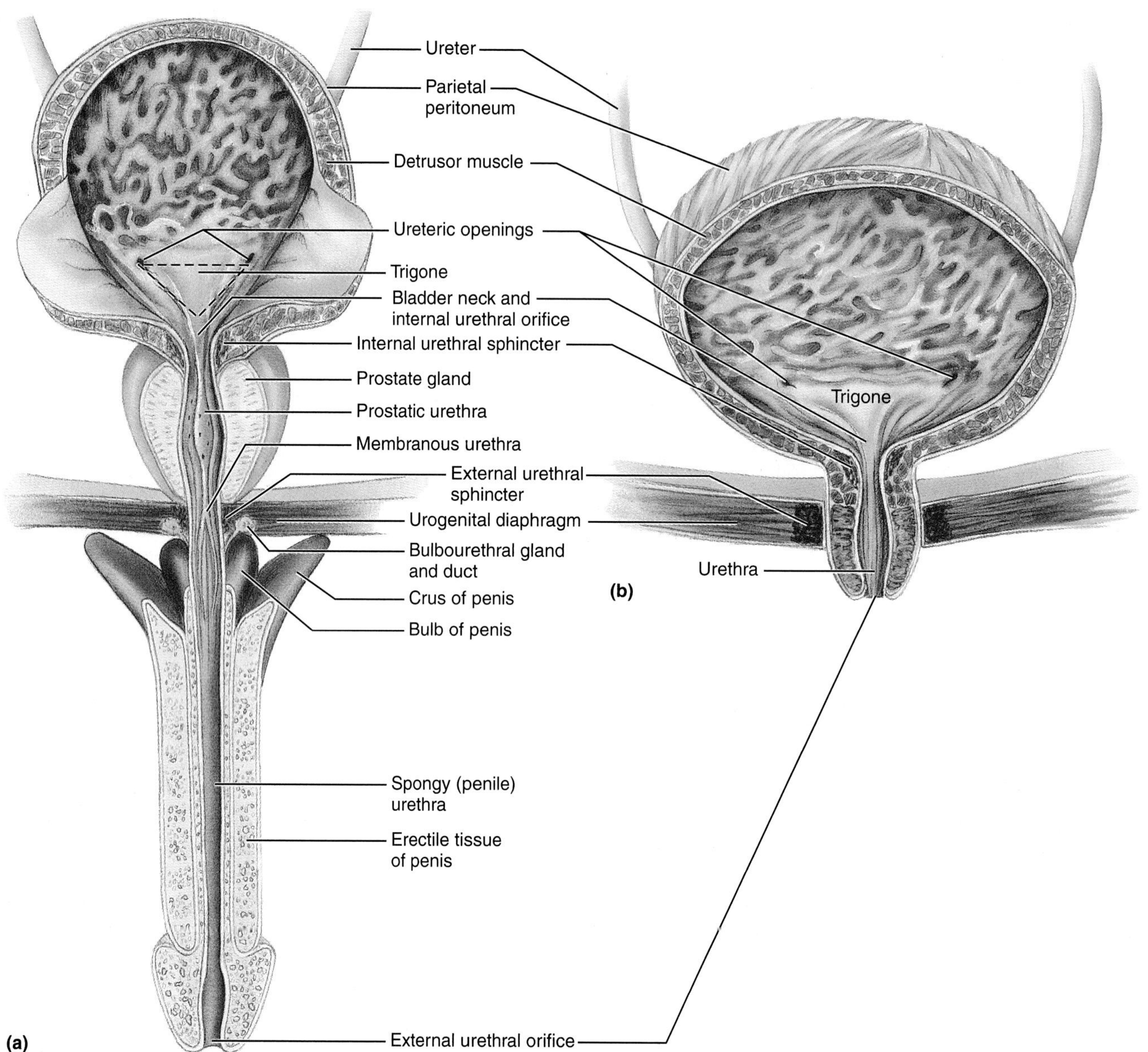

FIGURE 23.15 **Structure of the urinary bladder and urethra.** In these views from **(a)** a male and **(b)** a female, the anterior wall of the bladder has been cut away to reveal the trigone inside. The network-like pattern on the internal surface of the bladder is formed by mucosal folds, or rugae, which flatten as the bladder fills with urine. Note that the long urethra of the male has three regions: prostatic, membranous, and spongy (penile).

This is an involuntary sphincter of smooth muscle that keeps the urethra closed when urine is not being passed and prevents dribbling of urine between voidings. A second sphincter, the **external urethral sphincter** *(sphincter urethrae),* surrounds the urethra within the sheet of muscle called the *urogenital diaphragm.* This external sphincter is a skeletal muscle that we use voluntarily to inhibit urination until the proper time. The levator ani muscle of the pelvic floor also serves as a voluntary constrictor of the urethra. (For more information on the sphincter urethrae and levator ani muscles, see Table 11.7 on p. 292).

The length and functions of the urethra differ in the two sexes. In females, the urethra is just 3–4 cm (1.5 inches) long and is bound to the anterior wall of the vagina by connective tissue. It opens to the outside at the **external urethral orifice** *(external urethral meatus),* a small, often difficult-to-locate opening that lies

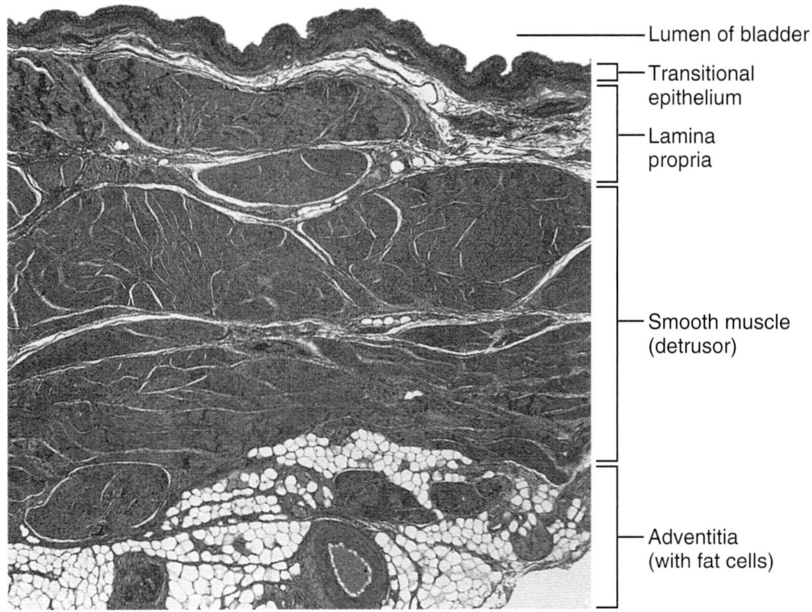

FIGURE 23.16 Histological section through the bladder wall (30 ×). Although the wall of the bladder is thicker than that of the ureter, the layers in these two organs are essentially identical.

anterior to the vaginal opening and posterior to the clitoris (see Figure 24.20, p. 724).

In males, the urethra is about 20 cm long (8 inches) and has three named regions: the **prostatic urethra,** which is about 2.5 cm long and runs in the prostate gland; the **membranous urethra,** which runs for about 2.5 cm through the membrane-like urogenital diaphragm; and the **spongy,** or **penile, urethra,** which is about 15 cm long, passes through the entire penis, and opens at the tip of the penis via the **external urethral orifice.** The male urethra carries ejaculating semen as well as urine (although not simultaneously) from the body. The reproductive function of the male urethra is covered in Chapter 24 (p. 710).

Critical Reasoning

Given these facts about the male and female urethra, why are urinary tract infections more common in females than in males?

Because the urethra in females is short and its external orifice lies close to the anus, fecal bacteria can easily enter the urethra. To minimize this risk, females should always wipe from front to back after defecation. For more on urinary tract infections, see the Disorders section on p. 692.

MICTURITION

Micturition (mik″tu-rish′un), also called voiding or urination, is the act of emptying the bladder. It is brought about by the contraction of the bladder's detrusor muscle, assisted by the muscles of the abdominal wall, which contract to raise intra-abdominal pressure.

Advances in Understanding Previously, micturition was thought to be controlled by a spinal reflex arc, which the brain influenced but did not control. It has been found, however, that micturition is indeed controlled by the brain. As urine accumulates in the bladder (Figure 23.17), distension of the bladder wall activates stretch receptors, which send sensory impulses through visceral sensory neurons to the sacral region of the spinal cord, and then up to a micturition center in the dorsal part of the pons. Acting as an on-off switch, the micturition center in the pons signals the parasympathetic neurons that stimulate contraction of the detrusor muscle, thereby emptying the bladder. At the same time, the sympathetic pathways to the bladder—which would prevent micturition—are inhibited. ✱

The micturition center in the pons is heavily influenced by rostral brain regions, such as the inferior-frontal region of the cerebral cortex (which enables a conscious decision that it is safe to micturate) and the anterior cingulate gyrus (involved in emotional evaluation of the urge to micturate).

The pons, cerebral cortex, and other parts of the CNS also *inhibit* micturition. They do so by (1) stimulating the somatic motor neurons to the external urethral sphincter, and (2) activating sympathetic pathways that relax the detrusor and stimulate the internal urethral sphincter to contract.

CLINICAL APPLICATIONS

MICTURITION DYSFUNCTIONS Urinary incontinence, the inability to control micturition, is normal in babies who have not learned voluntarily to close their external urethral sphincter. Reflexive voiding occurs each time a baby's bladder fills enough to activate its stretch receptors, but the internal sphincter prevents dribbling of urine between voidings just as it does in

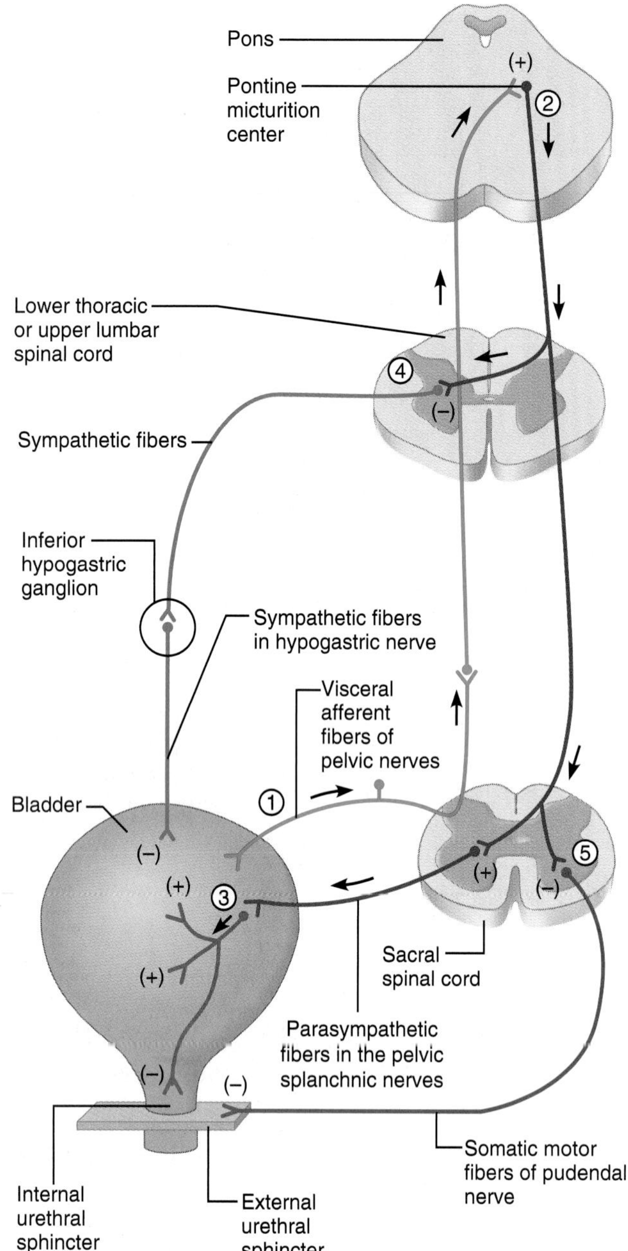

FIGURE 23.17 Micturition. In ①, stretching of the bladder wall by accumulating urine sends visceral afferent impulses to the pontine micturition center in the brain. In ② and ③, this center stimulates (+) the parasympathetic pathway to the detrusor muscle of the bladder, which contracts to expel urine from the bladder. During micturition, in ④ and ⑤, the pontine center inhibits (−) the *sympathetic* pathway to the bladder (which would otherwise relax the detrusor muscle and contract the internal urethral sphincter), and also inhibits the somatic motor neurons to the external urethral sphincter.

adults. Incontinence is very common in the elderly, affecting half of those who are in nursing homes or are homebound. Its many causes can be divided into three main types: (1) **urge incontinence,** in which the detrusor has uncontrolled contractions; (2) **stress incontinence,** in which the urethral sphincter mechanisms malfunction, such that coughing and sneezing can force urine through these sphincters; and (3) **overflow incontinence,** in which the bladder overfills and urine dribbles from the urethra. Incontinence should not be viewed as an inevitable result of aging, because up to 80% of the elderly with this condition can be helped by behavioral training, medications, or surgery.

Urinary retention, in which the bladder is unable to expel its contained urine, is common in patients recovering from surgeries involving general anesthesia, after which the detrusor muscle apparently needs time to recover before resuming activity. It is also common in older men, because enlargement of the prostate squeezes the urethra (see p. 711). If urinary retention is prolonged, a slender rubber drainage tube called a **catheter** (kath′ĕ-ter; "thrust in") must be inserted through the urethra and into the bladder to drain the urine and prevent overfilling of the bladder. The straight, short urethra of females is much easier to catheterize than the long, curved urethra of males. ✚

DISORDERS OF THE URINARY SYSTEM

This section first considers major infectious and noninfectious disorders that affect the urinary tract (*urinary tract infections* and *renal calculi,* respectively), and then discusses cancers of the major urinary organs *(kidney cancer* and *bladder cancer).*

Urinary Tract Infections

Most **urinary tract infections** occur in sexually active young women, because intercourse drives bacteria from the vagina and the external genital region (and from the anus as well) through the nearby opening of the short urethra and toward the bladder. The use of spermicides magnifies the problem, because they tend to kill normal resident bacteria and allow pathogenic fecal bacteria to colonize the vagina.

Overall, urinary tract infections occur in about 40% of women. Infection of the bladder, called **cystitis,** can spread superiorly to infect the ureters and kidneys, causing **pyelitis** and **pyelonephritis** (see p. 678). Symptoms of urinary tract infections include a burning sensation during micturition, increased urgency and frequency of micturition, fever, and sometimes cloudy or blood-tinged urine. When the kidneys are involved, back pain and severe headache often occur as well. Most urinary tract infections are easily cured by antibiotics, the most important of which is trimethoprim-sulfamethoxazole.

In men, urinary tract infections usually result from long-term catheterization of the penile urethra because it is difficult to keep indwelling catheters sterile.

Renal Calculi

In 12% of men and 5% of women in North America, calcium, magnesium, or uric acid salts in the urine crystallize and precipitate in the calices or renal pelvis, forming kidney stones, or **renal calculi** (kal′ku-li; "little stones"). Most calculi are smaller than 5 mm in diameter and thus can pass through the urinary tract without causing serious problems. The calculi, however, cause pain when they obstruct a ureter, thereby blocking the drainage of urine and increasing intrarenal pressure. The most severe pain results when the contracting walls of the ureter contact the sharp calculi during their periodic peristaltic contractions. Pain from kidney stones radiates from the flank to the anterior abdominal wall, and then perhaps to the groin. Calculi tend to lodge in three especially narrow regions of the ureters: (1) at the level of L_2, where the renal pelvis first narrows into the ureter; (2) at the sacroiliac joint, where the ureter enters the true pelvis; and (3) where the ureters enter the bladder. The clinician should be aware of these three points when searching for kidney stones on X-ray, CT, and ultrasound images.

Predisposing factors for calculi are (1) oversaturation of the renal filtrate with calcium ions, uric acid, or a substance called oxalate, which leads to the precipitation of crystals from the urine; (2) abnormal acidity or alkalinity of the urine; (3) dehydration; (4) stasis (blockage of flow) of urine in the urinary tract; and (5) bacterial infection, because at least some calculi precipitate around bacteria. Individuals with a history of kidney stones are encouraged to drink large volumes of water because this leads to a urine so dilute that salts will not precipitate and form calculi.

Surgical removal of calculi has been the traditional treatment, but because kidney stones often recur, physicians have long sought alternatives to the trauma of repeated surgeries. Now, surgery has been largely replaced by *extracorporeal shock wave therapy,* in which ultrasonic shock waves delivered from outside the patient's body break up the calculi.

Cancer of Urinary Organs

Bladder cancer, which accounts for about 3% of cancer deaths and is five times more common in men than in women, typically involves neoplasms of the bladder's lining epithelium. Blood in the urine is a common warning sign. This form of cancer may be induced by organic carcinogens that are deposited in the urine after being absorbed from the environment. Substances that have been linked to bladder cancer include tars from tobacco smoke, certain industrial chemicals, and some artificial sweeteners. Bladder cancer is usually lethal if it metastasizes, but even then surgical removal of the bladder and chemotherapy increase survival time significantly, to an average of 5 years after detection.

Kidney cancer—a cancer arising from the epithelial cells of the uriniferous tubules or renal pelvis/calices—accounts for 2% of cancer deaths in the United States, but its incidence is rising. Risk factors for this type of cancer, which is twice as common in men as in women, include obesity, high blood pressure, and, perhaps, a high-protein diet. Most tumors are over 3 cm in diameter and have metastasized when detected (typically during a kidney examination using ultrasound imaging, CT, or MRI). In such cases the prognosis is poor, and the average survival time is only 12–18 months. Because renal cancers are resistant to radiation and chemotherapy, standard treatment is the surgical removal of the entire kidney and the associated adrenal gland and lymph nodes.

CLINICAL APPLICATIONS

KIDNEY TRANSPLANT A kidney transplant is the transfer of a functioning kidney from a donor to a recipient whose kidneys are failing. About 30% of transplanted kidneys come from living donors (usually a relative or spouse), and 70% come from cadavers (whose kidneys can be maintained for about 36 hours after death). Kidney transplants, which are more common than transplants of any other major organ, have a high rate of success (80% to 90% survival after 3 years if from a living donor, 70% if from a cadaver) and are comparatively easy to perform (only a few vessels need to be cut and rejoined). A single kidney is transplanted, because one is sufficient to carry out excretory functions. The failing kidney is usually left in place, and the new kidney is transplanted into the right iliac fossa of the pelvis, which has more room than the crowded lumbar region. As with other organ transplants, the recipient must take immunosuppressive drugs. Current efforts are directed toward improving long-term survival rates, which are diminished by chronic rejection and other factors.

THE URINARY SYSTEM THROUGHOUT LIFE

The embryo develops not just one pair of kidneys, but three pairs, one after another, starting cranially and proceeding caudally. The three pairs are the pronephros, mesonephros, and metanephros. These different kidneys develop from the *urogenital ridges,* paired elevations of intermediate mesoderm on the dorsal abdominal wall (Figure 23.18). Of the three pairs of kidneys that form, only the last pair persists to become the adult kidneys. Initially, during week 4, the first kidney, or **pronephros** (pro-nef′ros), forms as a set of nephrons and then quickly degenerates. Although the pronephros is never functional and is gone by week 6, it sends a **pronephric duct** *(primary excretory duct)* to the cloaca, and this duct is used by the kidneys that develop later. As the second nephron system, called the **mesonephros** ("middle kidney"), claims the pronephric duct, this duct becomes the **mesonephric duct.** The nephrons of the mesonephros, in turn, degenerate once the third kidney, the **metanephros** (met″ah-nef′ros), becomes functional.

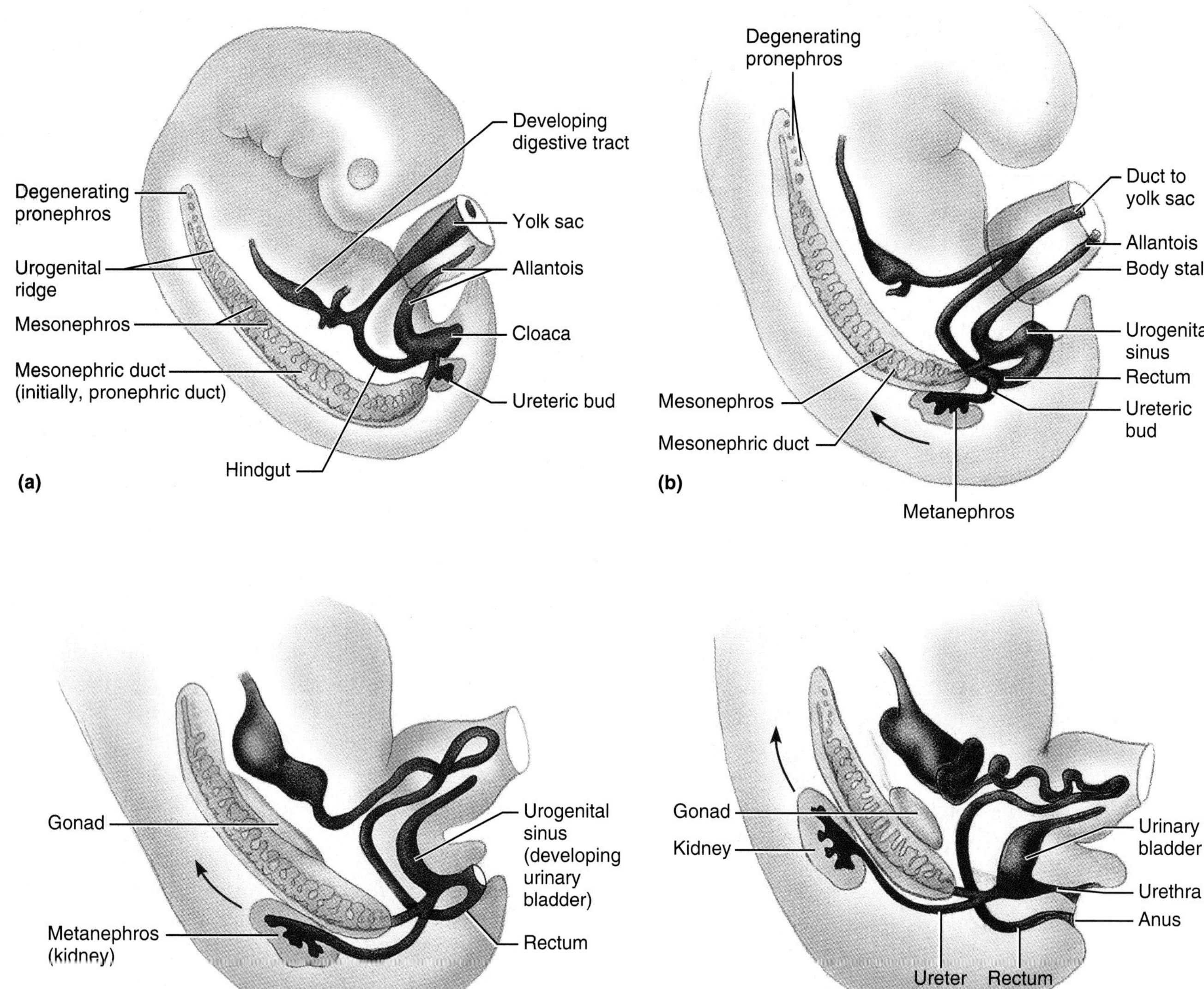

FIGURE 23.18 Development of the urinary organs in the embryo. **(a)** Week 5. **(b)** Week 6. **(c)** Week 7. **(d)** Week 8.

The metanephros is the definitive kidney (*metanephros* means "ultimate kidney"). It develops in the pelvic region in the following way: Starting in week 5, a hollow **ureteric bud** grows from the mesonephric duct into the urogenital ridge, inducing the mesoderm there to form the nephrons. The ureteric bud, in turn, develops into the renal pelvis, calices, and collecting tubules; its unexpanded proximal part becomes the ureter.

After forming in the pelvis, the metanephric kidneys ascend to their final position in the abdomen, receiving their blood supply from successively higher sources as they ascend. Whereas the lower renal vessels usually degenerate as the upper ones appear, these early vessels often fail to degenerate. As a result, 30% of people have multiple renal arteries.

The metanephric kidneys are actively forming urine by month 3 of fetal life. Even though the placenta (see p. 729) performs all excretory functions before birth, the production of urine by the fetal kidneys is an important function because the urine is voided into the amniotic sac, where it helps maintain a sufficient volume of amniotic fluid.

As the metanephric kidneys develop, the *cloaca* (see p. 667) divides into two parts: (1) the future rectum and anal canal, and (2) the **urogenital sinus,** into which the urinary and genital ducts empty (see Figure 23.18b). The urogenital sinus becomes the urinary bladder and the urethra. The *allantois* (see p. 667), an extension of the urogenital sinus into the umbilical cord, becomes the urachus of the bladder.

As the embryonic kidneys ascend through the narrow pelvic brim, they face a tight squeeze during which the right and left kidneys come close together, and in about 1 out of every 600 people they fuse into one U-shaped **horseshoe kidney.** This condition is usually harmless, but it can be associated with obstructed drainage and hydronephrosis. In **pelvic kidney,** one of the two kidneys stays in the bony pelvis throughout life. Because a pelvic kidney often blocks the birth canal, a woman with such a kidney may have difficulty while giving birth.

In **childhood polycystic kidney,** a rare, inherited condition, the infant's kidneys contain many urine-filled sacs (cysts) that are dilations along the collecting ducts in the renal cortex and medulla. Because the cysts impede the drainage of urine and crush the surrounding renal tissue, most children with this disease die of kidney failure early in life. **Autosomal dominant polycystic kidney disease,** another inherited condition, is considered in the A Closer Look box on page 696.

Because an infant's bladder is small and its kidneys cannot concentrate urine for the first 2 months, a newborn baby voids 5 to 40 times a day, depending on the amount of fluid intake. As the child grows, she or he voids less often but produces progressively more urine. By adolescence, the adult rate of urine output—an average of 1500 mL/day—is achieved.

Control of the voluntary urethral sphincter is attained as the nervous system matures. By 15 months, most toddlers are aware of having voided. By 18 months, most children can hold urine in the bladder for about 2 hours, the first sign that toilet training for micturition can begin. Daytime control usually is achieved well before nighttime control. As a rule, it is unrealistic to expect complete nighttime control before the age of 4 years.

From childhood through late middle age, most problems that affect the urinary system are infections. *Escherichia coli,* normal resident bacteria of the lower digestive tract that seldom cause problems there, produce 80% of all urinary tract infections. Sexually transmitted diseases, which are chiefly infections of the reproductive tract, can also inflame the urinary tract, clogging some urinary ducts. Childhood streptococcal infections, such as a severe strep throat or scarlet fever, can lead to long-term inflammatory damage of the kidneys.

Only about 3% of elderly people have histologically normal kidneys, and kidney function declines with advancing age. The kidneys shrink, the nephrons decrease in size and number, and the tubules become less efficient at secretion and reabsorption. By age 70, the average rate of filtration is only half that of middle-aged adults, a functional decline believed to result from narrowing of the renal arteries by atherosclerosis. Diabetics are particularly at risk for renal disease, and over 50% of individuals who have had diabetes mellitus for 20 years (regardless of their age) are in renal failure owing to the vascular ravages of this disease.

The bladder of an aged person is shrunken, and the desire to urinate is often delayed. Loss of muscle tone in the bladder causes an annoying increased frequency of urination. Other common age-related problems are incontinence and urethral constriction by an enlarged prostate.

RELATED CLINICAL TERMS

Cystocele (sis-to′-sēl) (*cyst* = the bladder; *cele* = sac, hernia) Condition in women in which the urinary bladder pushes on the vagina and bulges into the anterior superior vaginal wall; may result from tearing of the supportive muscles of the pelvic floor during childbirth, which then causes the unsupported pelvic viscera to sink inferiorly. This condition changes the position of the upper urethra in ways that can lead either to urinary retention or incontinence; a tearing or weakening of the external urethral sphincter also promotes stress incontinence.

Cystoscopy (sis-tos′ko-pe) The threading of a thin viewing tube through the urethra into the bladder to examine the surface of the bladder mucosa. Can detect bladder tumors, kidney stones, and infections.

Dialysis (di-al′ĭ-sis) A technique for removing wastes from the blood and replenishing acid-base buffers in individuals with failing kidneys; involves the diffusion of solutes across a membrane situated between the blood and a dialysis solution. The membrane can either be in an artificial kidney machine, through which the patient's blood is run (a process called *hemodialysis*), or the patient's own peritoneum, in a process called *peritoneal dialysis.* In the latter case, the dialysis fluid is injected into the patient's peritoneal cavity and removed when it has accumulated wastes.

Renal infarct Area of dead (necrotic) renal tissue; may result from infection, hydronephrosis, or blockage of the blood supply to the kidney; commonly caused by blockage of an interlobar artery, which lacks anastomoses.

Urologist (u-rol′o-jist) Physician who specializes in diseases of the urinary structures in both sexes, and in diseases of the *reproductive* tract of males.

Vesicoureteric reflux (ves″ĭ-ko-u-re′ter-ik; "bladder to ureter") Abnormal backflow of urine from the bladder into the ureter during micturition. In infants, it often reflects a congenital abnormality of the valve mechanism of the ureters in the bladder wall (see p. 687). When it appears in adults, it can result from increased pressures during voiding or from an obstruction of the urethra. Reflux can lead to pressure damage in the kidneys, kidney stones, and infection of the entire urinary tract as bacteria are flushed upward from the urethra. Treatment is with antibiotics, surgical reconstruction of the vesicoureteric valves or, in severe cases, by moving the end of the ureter so that it empties into the intestine.

A CLOSER LOOK

ADPKD: The Most Interesting Disease You've Never Heard Of?

Imagine a lethal "birth defect" that does not affect children. Imagine an inherited condition that is 10 times more common than sickle-cell disease, 15 times more common than cystic fibrosis, and 20 times more common than Huntington's disease—yet most people have never heard of it. Imagine a condition that swells the kidneys to many times their normal size, then destroys kidney function—yet can be treated effectively enough to prolong life into the sixth or seventh decade. All these features, plus the fact that it poses prickly ethical questions, place this condition—**autosomal dominant polycystic kidney disease (ADPKD),** also called *adult polycystic kidney disease*—among the most fascinating of diseases.

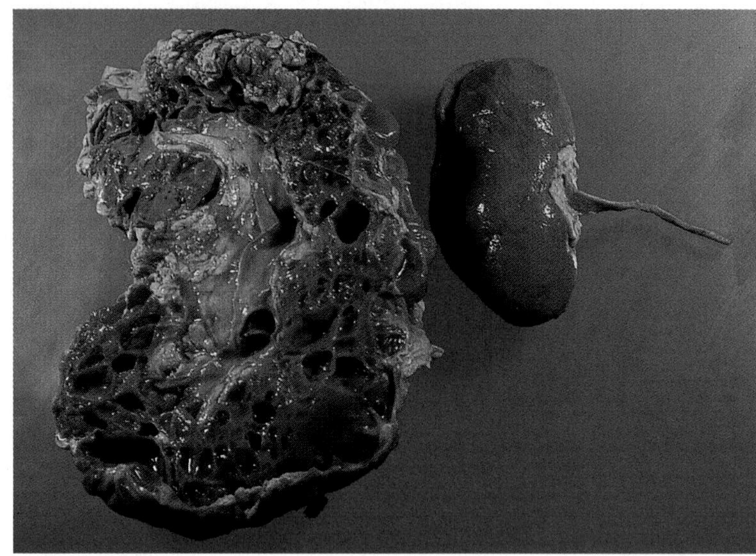

At left is an enlarged kidney affected by autosomal dominant polycystic kidney disease (ADPKD). The large holes in this frontally-sectioned kidney are the cysts. At right is a normal kidney for comparison.

ADPKD is characterized by the slow enlargement within both kidneys of fluid-filled sacs (cysts) that can arise in any part of the uriniferous tubule (see the photo). The cysts, which form in only a small proportion of the millions of tubules, begin within the fetus as epithelial proliferations of the tubular wall, from which they disconnect as they secrete and accumulate more and more fluid. Cysts are tiny and inconsequential at birth, are generally too small to detect in ultrasound images until age 20, and usually cause no obvious symptoms until one's 30s or 40s, by which time they are large enough to cause compression-related disruptions of kidney function. Within an average of ten more years, the kidneys become "knobby" and grossly enlarged, reaching weights up to 14 kg (30 lb) each. Most victims die from kidney failure or the effects of high blood pressure.

ADPKD is a relatively common inherited condition affecting 1 in 500 individuals (or some 500,000 Americans).

CHAPTER SUMMARY

Media study tools that could provide you additional help in reviewing specific key topics of Chapter 23 are referenced below.

SP = Study Partner; AIA = A.D.A.M.® Interactive Anatomy

1. In forming urine, the kidneys cleanse the blood of nitrogenous wastes, toxins, excess ions and water, and other unnecessary or undesirable substances. Kidneys also maintain the proper chemical composition of the blood and other body fluids. The organs for transporting and storing urine are the ureters, the bladder, and the urethra.

KIDNEYS (pp. 676–686)

Gross Anatomy of the Kidneys (pp. 676–681)

1. The bean-shaped kidneys lie retroperitoneally in the posterior abdominal wall, extending from the level of T_{11} or T_{12} to L_3.
2. The supportive and protective layers around the kidney are the renal capsule, the perirenal fat, the renal fascia, and the pararenal fat.

AIA Link: Atlas; System: Urinary; Dissection of Kidneys (anterior).

3. A kidney has an external cortex, a deeper medulla consisting of medullary pyramids, and a medial renal sinus that contains the main vessels and nerves of the kidney as well as the renal pelvis (the wide upper part of the ureter). Tributaries of the renal pelvis, the major and minor calices, collect urine draining from the papillae of the medullary pyramids. The medial, slit-like opening of the renal sinus is the hilus.

AIA Link: Atlas; System: Urinary; Renal Arteries.

4. The vascular pathway through a kidney is as follows: renal artery → segmental arteries → lobar arteries → interlobar arteries → arcuate arteries → interlobular arteries → afferent arterioles → glomeruli → efferent arterioles → peritubular capillary beds (or vasa recta) → interlobular veins → arcuate veins → interlobar veins → renal veins.

The words *autosomal dominant* in its name mean that if one parent has it, each of his or her offspring has a 50% chance of inheriting it. Genetic screening tests are able to identify the presence of ADPKD at any stage of life, beginning with amniocentesis and chorionic villus sampling of the early fetus. The two genes that, when defective, are responsible for ADPKD encode for giant proteins called polycystins, which seem to play a role in the proliferation, interactions, and differentiation of epithelial cells. Thus, the cells that form the renal cysts may remain undifferentiated and proliferative—and undifferentiated epithelial cells in general are known to secrete a fluid very much like that found in the cysts in ADPKD.

Defects in one gene produce one subtype of the disease, called ADPKD1, which has more severe symptoms, is associated with an average life span of just over 50 years, and represents 85% of cases. Defects in the other gene produce the less severe subtype, ADPKD2, which is associated with an average life span of almost 70 years. (Individuals without ADPKD on average live almost 80 years.)

Of the many signs and symptoms of ADPKD—which vary greatly in severity, even among ADPKD1 sufferers—the most important is hypertension (high blood pressure). Hypertension appears early, affecting 30% of all children with ADPKD, 60% of otherwise asymptomatic young adults, and 80% of older, symptomatic patients. Although the cause of this hypertension remains unclear, the presence of cysts may somehow disrupt the kidney's secretion of renin, a hormone that raises blood pressure (see p. 686). Other common signs and symptoms include pain in the flank, blood in the urine, serious urinary tract infections, kidney stones, and aneurysms in the arteries of the brain.

Many people with ADPKD develop cysts in the *liver* as well as the kidney, causing the liver to enlarge. The liver cysts originate in the bile ducts and grow to several centimeters in size, as in the kidney. However, the liver's structure and function are usually not disrupted unless the cysts hemorrhage or get infected.

Various treatments can prolong the life and health of people with ADPKD. Management of the urinary infections and hypertension is of primary importance while the kidneys still function. Cysts that cause pain or are infected can be drained or removed by minimally invasive surgery. When the kidneys begin to fail, dialysis and kidney transplants become options. Kidney transplants are usually successful and, inexplicably, the presence of the donated, functioning kidney slows or reverses the growth of cysts in the original kidneys, which are usually left in place after the transplant.

ADPKD poses tremendously difficult ethical dilemmas for its victims and their families. Should an affected adult have any children at all, knowing the 50–50 odds of transmission? If fetal screening reveals that a fetus has ADPKD1, is it ethical to end the pregnancy given that the child will probably live over 50 years? If a young person of unknown status has a parent with ADPKD, should that person be tested to determine if she or he has inherited it? Whereas the answer to this last question would seem to be yes, because early detection leads to better control of the hypertension and therefore prolongs life, many young people do not want the emotional trauma of knowing they have a lethal disease—and will probably have difficulty getting health insurance and life insurance because of it. Better treatments in the future may eliminate some of these ethical dilemmas, but for now, counseling and a willingness to make tough choices are essential.

5. The nerve supply of the kidneys, consisting largely of sympathetic fibers, is from the renal plexus.

Microscopic Anatomy of the Kidneys (pp. 681–686)

6. The main structural and functional unit of the kidney is the uriniferous tubule, which consists of a nephron and collecting tubule. The nephron, which forms the urine, consists of a renal corpuscle, a proximal convoluted tubule, a loop of Henle (loop of the nephron), and a distal convoluted tubule.

AIA Link: Atlas; System: Urinary; Diagram of Nephron.

7. The production of urine by the nephron involves three processes: filtration (which produces a filtrate of the blood that is modified into urine); reabsorption (which reclaims the desirable molecules from the filtrate and returns them to the blood); and secretion (in which undesirable molecules are pumped from the peritubular capillaries into the nephron).

8. Filtration occurs at the renal corpuscles in the renal cortex. Each renal corpuscle consists of a glomerulus (a capillary tuft) surrounded by a glomerular capsule. The visceral (inner) layer of this capsule consists of octopus-shaped podocytes, which surround the capillaries of the glomerulus.

9. In passing from the glomerulus to the capsular space, the filtrate passes through the filtration membrane, which consists of the fenestrated glomerular endothelium, the basement membrane, and the filtration slits between podocytes. The main filters are the basement membrane and slit diaphragms over the filtration slits; these structures retain molecules that are larger than small proteins.

10. From the glomerular capsule, the filtrate enters the proximal convoluted tubule (in the renal cortex), whose simple cuboidal epithelial cells show extreme specializations for reabsorption and secretion: many long microvilli, many mitochondria, and infoldings of the basolateral plasmalemma.

11. The loop of Henle consists of descending and ascending limbs. The first part of the descending limb, which resembles the proximal tubule, narrows to join the thin segment, whose walls consist of a simple squamous epithelium. The

thin segment joins the thick segment of the ascending limb, which is structurally similar to the distal convoluted tubule.

12. The distal convoluted tubule (in the renal cortex) is active in reabsorption and secretion, but less so than the proximal tubule. Its simple cuboidal epithelial cells exhibit abundant mitochondria and infoldings of the basolateral membrane, but no elaboration of the surface microvilli.
13. There are two kinds of nephrons: cortical and juxtamedullary. The loops of the juxtamedullary nephrons project deeply into the medulla and contribute to the mechanism that concentrates urine.
14. Collecting tubules play a minor role in reabsorbing and secreting ions. More importantly, collecting tubules (along with distal tubules) reabsorb water to concentrate urine.
15. Each nephron is associated with a glomerulus (the capillary in the renal corpuscle from which the filtrate arises). Also associated with the nephrons are peritubular capillaries (which surround the convoluted tubules and are involved in reabsorption and secretion), and vasa recta (capillary-like vessels around the loop of Henle, which function in concentrating urine). Glomeruli are supplied by afferent arterioles and drained by efferent arterioles, which in turn are continuous with the peritubular capillaries or the vasa recta.
16. The juxtaglomerular apparatus, which occurs at the point of contact between the afferent arteriole and the distal convoluted tubule, consists of juxtaglomerular cells and a macula densa. It functions in the hormonal correction of low blood pressure and low blood volume.
17. The renal interstitium is the loose areolar connective tissue between the uriniferous tubules.

SP Exercise: Chapter 23, Structure of Nephrons.

SP Exercise: Chapter 23, Nephron Activity.

SP Exercise: Chapter 23, Cell Types in Nephrons.

URETERS (pp. 686–687)

Gross Anatomy (pp. 686–687)

1. The ureters are slender tubes descending retroperitoneally from the kidneys to the bladder. Each ureter enters the true pelvis by passing over the sacroiliac joint.

Microscopic Anatomy (p. 687)

2. From external to internal, the layers of the ureter wall are (1) an adventitia of connective tissue, (2) a muscularis of smooth muscle, and (3) a highly distensible mucosa consisting of a lamina propria and a transitional epithelium. The muscularis squeezes urine inferiorly through the ureters.

URINARY BLADDER (pp. 687–689)

1. The urinary bladder, a distensible muscular sac for storing urine, lies in the pelvis just posterior to the pubic symphysis.
2. Whereas a full bladder expands superiorly into the abdominal cavity, an empty bladder is an inverted pyramid with four angles (neck with urethra, anterior angle with urachus, and two posterolateral angles where ureters enter). The trigone is an elevated triangle on the inner surface of the posterior wall that is defined by the ureteric and urethral openings.
3. The bladder wall, from internal to external, consists of a mucosa with a transitional epithelium, the detrusor muscle, and an adventitia (or, on the superior surface of the bladder, a serosa).

URETHRA (pp. 689–691)

1. The urethra is the tube that conveys urine from the bladder out of the body. Various types of stratified epithelium line its lumen.
2. Where the urethra leaves the bladder, it is surrounded by the internal urethral sphincter, an involuntary sphincter of smooth muscle. Inferior to this, where the urethra passes through the urogenital diaphragm, it is surrounded by the external urethral sphincter, a voluntary sphincter of skeletal muscle.
3. In females, the urethra is 3–4 cm long and conducts only urine. In males, the urethra is about 20 cm long and conducts both urine and semen.

MICTURITION (pp. 691–692)

1. Micturition is the act of emptying the bladder.
2. As accumulating urine stretches the bladder, sensory neurons carry this information up the spinal cord to the micturition center in the pons. This center initiates micturition by signalling the parasympathetic neurons to the detrusor muscle.
3. Because the external urethral sphincter is controlled voluntarily, micturition can be delayed temporarily.

DISORDERS OF THE URINARY SYSTEM (pp. 692–693)

1. This section discusses urinary tract infections (in which fecal bacteria spread up the urinary tract), calculi that precipitate from the urine (and can painfully block the ureter), bladder cancer, and kidney cancer.

THE URINARY SYSTEM THROUGHOUT LIFE (pp. 693–695)

1. Three sets of kidneys (pronephric, mesonephric, and metanephric) develop in sequence from the intermediate mesoderm of the embryo. The metanephros—the definitive kidney—forms in the pelvis as the ureteric bud signals nephrons to form in the intermediate mesoderm. The metanephric kidney moves cranially from the pelvis to the lumbar region.
2. The embryonic cloaca forms the urogenital sinus, which becomes both the bladder and the urethra.
3. The kidneys of newborns cannot concentrate urine. Their bladders are small, and voiding is frequent. Neuromuscular maturation generally allows toilet training for micturition to begin by 18 months of age.
4. From youth to middle age, the most common urinary problems are bacterial infections.
5. With age, nephrons are lost, the filtration rate decreases, and the kidneys become less efficient at concentrating urine. The capacity and tone of the bladder decrease with age, leading to frequent voiding and (often) incontinence. Urinary retention is a common problem of elderly men.

REVIEW QUESTIONS

Multiple Choice/Matching Questions

1. The inferior border of the right kidney is at the level of which vertebra? (a) T_{12}, (b) L_1, (c) L_2, (d) L_3.

2. The capillaries of the glomerulus differ from other capillary networks in the body in that they (a) form an anastomosing network, (b) drain into arterioles instead of venules, (c) contain no endothelium, (d) are the only capillaries from which a fluid leaves the blood.

3. Which of the following structures occurs exclusively in the renal medulla? (a) renal corpuscles, (b) distal convoluted tubules, (c) vasa recta, (d) proximal convoluted tubules.

4. Which of the following is not always part of (or in) the renal cortex? (a) glomerulus, (b) distal convoluted tubules, (c) loops of Henle, (d) proximal convoluted tubules, (e) renal columns.

5. The arrangement of the major blood vessels and urine-carrying vessels at the hilus of the kidney (from anterior to posterior) is: (a) renal artery branches, renal pelvis, renal vein; (b) renal vein, renal pelvis, renal artery branches; (c) renal pelvis, renal vein, renal artery branches; (d) renal vein, renal artery branches, renal pelvis.

6. The part of the uriniferous tubule whose epithelial cells contain the longest microvilli and the most mitochondria is the (a) glomerular capsule (podocytes), (b) proximal tubule, (c) thin segment, (d) distal tubule.

7. The only part of the uriniferous tubule that originates from the embryonic ureter (ureteric bud) and is not part of the nephron is the (a) glomerular capsule, (b) proximal convoluted tubule, (c) loop of Henle, (d) distal convoluted tubule, (e) collecting tubule.

8. The main function of the transitional epithelium in the ureter is (a) not the same as that of the transitional epithelium in the bladder, (b) protection against kidney stones, (c) secretion of mucus, (d) reabsorption, (e) stretching.

9. Jim was standing at a urinal in a crowded public rest room, and a long line was forming behind him. He became anxious (a sympathetic response) and suddenly found he could not micturate no matter how hard he tried. Jim's problem was that (a) his internal urethral sphincter was constricted and would not relax, (b) he had formed a kidney stone on the spot, and it was blocking his urethra, (c) his detrusor muscle was contracting too hard, (d) he almost certainly had a burst bladder.

10. A major function of the collecting tubules is (a) secretion, (b) filtration, (c) concentrating urine, (d) lubrication with mucus, (e) stretching.

11. Urine passes downward through the ureters by which mechanism(s)? (a) ciliary action, (b) peristalsis and gravity, (c) gravity alone, (d) suction exerted on the ureters and renal pelvis as the bladder expands with urine.

Short Answer Essay Questions

12. Name the layers of fat and fascia around the kidney, then tell the functions of the fat layers.

13. Pairs of urinary structures that are often confused are the ureters and the urethra; the perirenal and pararenal fat; the interlobar and interlobular arteries; and the renal sinus and renal pelvis. For each pair, differentiate one structure from the other.

14. Trace the path taken by the renal filtrate (and urine) from the glomerulus to the urethra. Name every microscopic and gross tube and structure that it passes through on its journey.

15. (a) Describe the basic process and purpose of tubular reabsorption. (b) Explain the location and function of the peritubular capillaries.

16. List all the layers of the filtration membrane in a renal corpuscle, and describe the function of each layer.

17. Name (a) the four angles of the empty bladder and (b) the tube or band that attaches to each angle. (c) What three openings define the trigone of the bladder?

18. Define micturition, and describe its neural control.

19. Describe the changes that occur in the anatomy and function of the kidney with age.

20. How does the particular path of the ureters through the bladder wall minimize the chances of vesicoureteric reflux and hydronephrosis occurring?

CRITICAL THINKING + CLINICAL APPLICATION QUESTIONS

1. Recently, people at risk for developing bladder cancer have been encouraged to drink more fluids. Describe why this lowers one's chances of getting bladder cancer.

2. What is cystitis? Why do women suffer from cystitis more often than men?

3. Hattie, aged 55, is awakened by excruciating pain that radiates from her right abdomen to the loin and groin regions on the same side. The pain does not occur continuously but recurs at intervals of 3–4 minutes. Diagnose her problem, and cite factors that might favor its occurrence. Also, explain why Hattie's pain comes in "waves."

4. Felicia, a medical student, arrived late to a surgery in which she was to observe the removal of an extensive abscess from the fat around a patient's kidney. Felicia was startled to find the surgical team working to reinflate the patient's lung and to remove air from the pleural cavity. How could renal surgery lead to such events?

5. Maliki, a radiologist, was examining a pyelogram for renal calculi in a patient's right ureter. Which three regions of the ureter did he scrutinize first?

6. Why should parents teach their young daughters to wipe from front to back after defecation? Why does using a spermicide for birth control increase a woman's chance of getting a urinary tract infection?

A.D.A.M.® INTERACTIVE ANATOMY QUESTIONS

1. Link: Atlas; System: Urinary; Section of Female Bladder (anterior) (a) The ureters carry urine from the kidneys to which structure in the pelvic area? (b) Which structure conveys urine out of the body?

2. Link: Atlas; System: Urinary; Renal Arteries (a) The renal pyramids drain directly into which structures of the kidney? (b) Name the tough covering that gives the kidney its shape.

CHAPTER 24

The Reproductive System

x-ray of human sperm in fallopian tube

CHAPTER OUTLINE + STUDENT OBJECTIVES

The Male Reproductive System 702–715

1. Describe the location, structure, and function of the testes.
2. Describe the histology of the testes, and outline the events of spermatogenesis.
3. Describe the location, structure, and functions of the accessory organs of the male reproductive system.
4. Describe the structure of the penis, and explain the mechanism of erection.

The Female Reproductive System 715–726

5. Describe the location, structure, and function of the ovaries.
6. Explain the phases of the ovarian cycle and the stages of oogenesis.
7. Describe the anatomy of the uterine tubes in terms of their function.
8. Explain the location, regions, supportive structures, and layers of the uterus.
9. Outline the three phases of the uterine cycle.
10. Explain the structure of the vagina in terms of its functions.
11. Describe the anatomy of the female external genitalia.
12. Describe the mammary glands, and explain their clinical importance.

Pregnancy and Childbirth 726–732

13. Describe the processes of implantation and placenta formation.
14. Define a placental (chorionic) villus, and explain its functions.
15. List the three phases of labor.

Disorders of the Reproductive System 732–734

16. Consider various cancers of the reproductive organs, especially prostate cancer and breast cancer.

The Reproductive System Throughout Life 734–738

17. Compare and contrast the prenatal development of the male and female sex organs.
18. Note the anatomical changes that occur during puberty and menopause.

MOST ORGAN SYSTEMS OF THE BODY FUNCTION almost continuously to maintain the well-being of the individual. The reproductive system, however, appears to "slumber" until puberty, after which it plays an important role in our adult lives. Although male and female reproductive organs are quite different, they share a common purpose—to produce offspring.

In both females and males, the sex organs, or **genitalia** (jen″ĭ-ta′le-ah), are divided into *primary* and *accessory* organs. The **primary sex organs,** or **gonads** (go′nadz; "seeds"), are the testes in males and the ovaries in females. Gonads produce the sex cells or **gametes** (gam′ēts)—the sperm in males and the ovum (egg) in females—that fuse to form a fertilized egg. All other genitalia in both sexes are **accessory sex organs,** including the internal glands and ducts that nourish the gametes and transport them toward the outside of the body, and the external genitalia as well.

Besides producing sex cells, the gonads also secrete *sex hormones* and therefore function as endocrine glands. Recall that hormones are messenger molecules that travel through the bloodstream to signal various physiological responses in the cells that take them up (see p. 87). Sex hormones play vital roles in the development, maintenance, and function of all sex organs. Other hormones, secreted by the pituitary gland at the base of the brain, also influence reproductive functions.

We begin by considering the somewhat simpler male reproductive system and then progress to the more complex female system.

THE MALE REPRODUCTIVE SYSTEM

In the male reproductive system (Figure 24.1), the gonads are the sperm-producing *testes* (singular: *testis*), which lie in the scrotum. From the testes, sperm travel to the outside of the body through a system of ducts, including (in order): the duct of the *epididymis,* the *ductus deferens,* the *ejaculatory duct,* and finally the *urethra,* which opens at the tip of the *penis.* The accessory sex glands, which empty their secretions into the sex ducts during ejaculation, are the *seminal vesicles, prostate gland,* and *bulbourethral glands.*

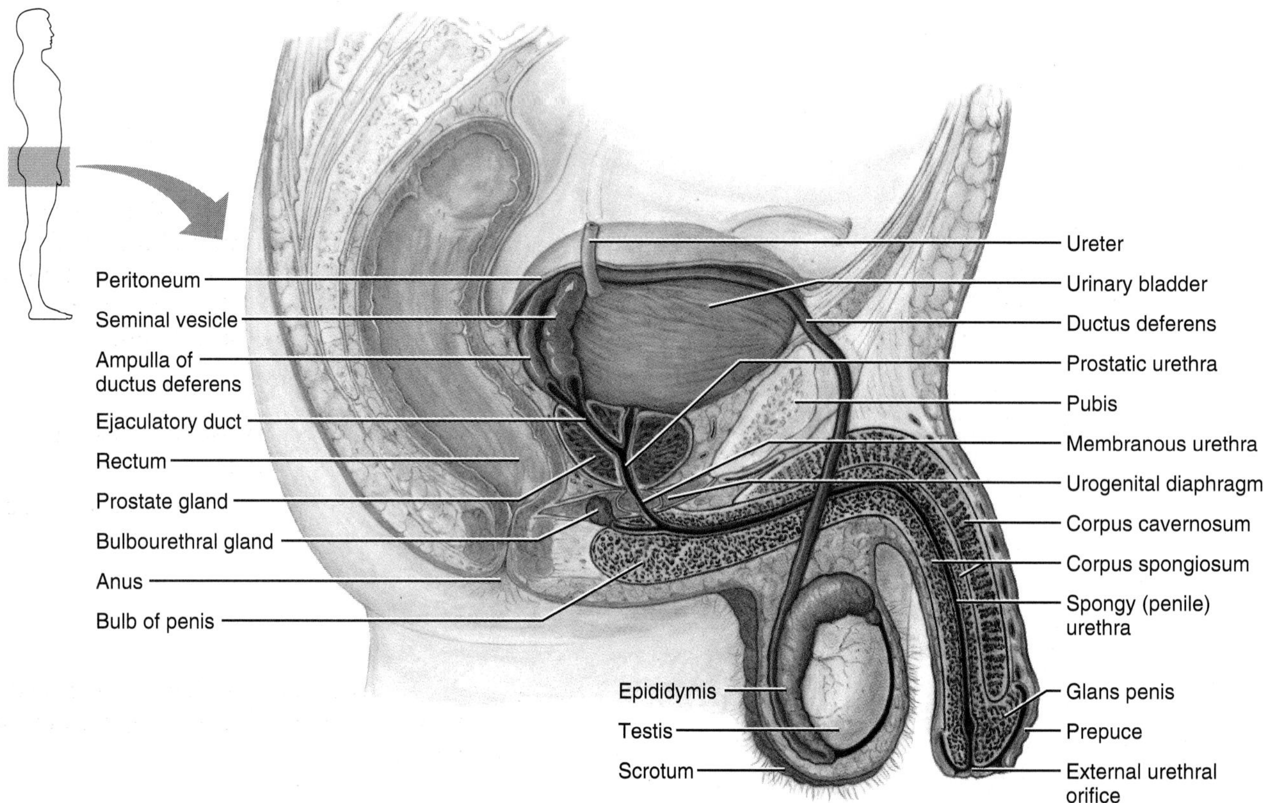

FIGURE 24.1 Reproductive organs of the male, sagittal view. At right, a portion of the pubic bone (pubis) is shown in three-dimensional view to indicate how the ductus deferens enters the pelvic cavity.

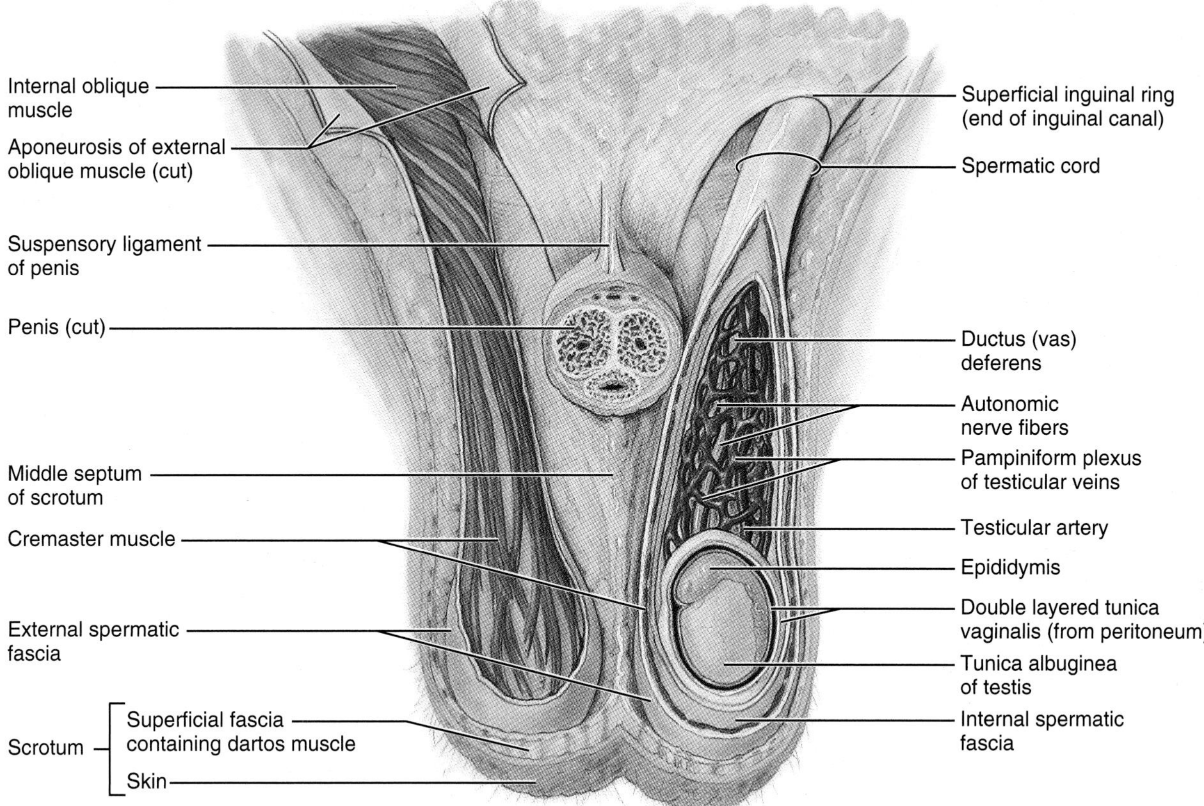

FIGURE 24.2 The scrotum, containing the testis and spermatic cord (anterior view). The scrotum has been opened by the removal of its anterior wall.

The Scrotum

The **scrotum** (skro′tum; "pouch") is a sac of skin and superficial fascia that hangs inferiorly external to the abdominopelvic cavity at the root of the penis (Figure 24.2). It is covered with sparse hairs and contains the paired, oval **testes** (tes′tēz; "witnesses"), or testicles. A septum in the midline divides the scrotum into right and left halves, providing one compartment for each testis.

As we will see (p. 737), the testes first develop deep in the posterior abdominal wall of the embryo and then migrate down into the scrotum, which is much closer to the body surface. Such a superficial location would seem to place a male's entire genetic heritage in a vulnerable position. However, because viable sperm cannot be produced at the core body temperature of 37 °C, the scrotum's superficial position provides an environment that is about 3 °C cooler, an essential adaptation.

Furthermore, the scrotum responds to changes in external temperature. Under cold conditions, the testes are pulled up toward the warm body wall, and the scrotal skin wrinkles to increase its thickness and reduce heat loss. These actions are performed by two muscles in the scrotum: The **dartos muscle** (dar′tos; "skinned"), a layer of smooth muscle in the superficial fascia, is responsible for wrinkling the scrotal skin, whereas the **cremaster muscles** (kre-mas′ter; "a suspender"), consisting of bands of skeletal muscle that extend inferiorly from the internal oblique muscles of the trunk, are responsible for elevating the testes. Under hot conditions, these muscles relax, so the scrotal skin is flaccid and loose, and the testes hang low to increase the skin surface available for cooling (sweating); this also moves the testis farther away from the warm trunk.

Critical Reasoning

Why do doctors often recommend that men not wear tight underwear or tight jeans if they want to father children?

Tight clothing holds the testes close to the warmth of the body, where higher temperatures inhibit sperm production.

The Testes

We begin our discussion with an overview of testis structure and then proceed to the production of sperm in the seminiferous tubules.

Gross Anatomy

Each testis (Figure 24.3) averages about 2.5 cm (1 inch) in width and 4 cm in height. Within the scrotum, each testis is partially enclosed by a serous sac called the

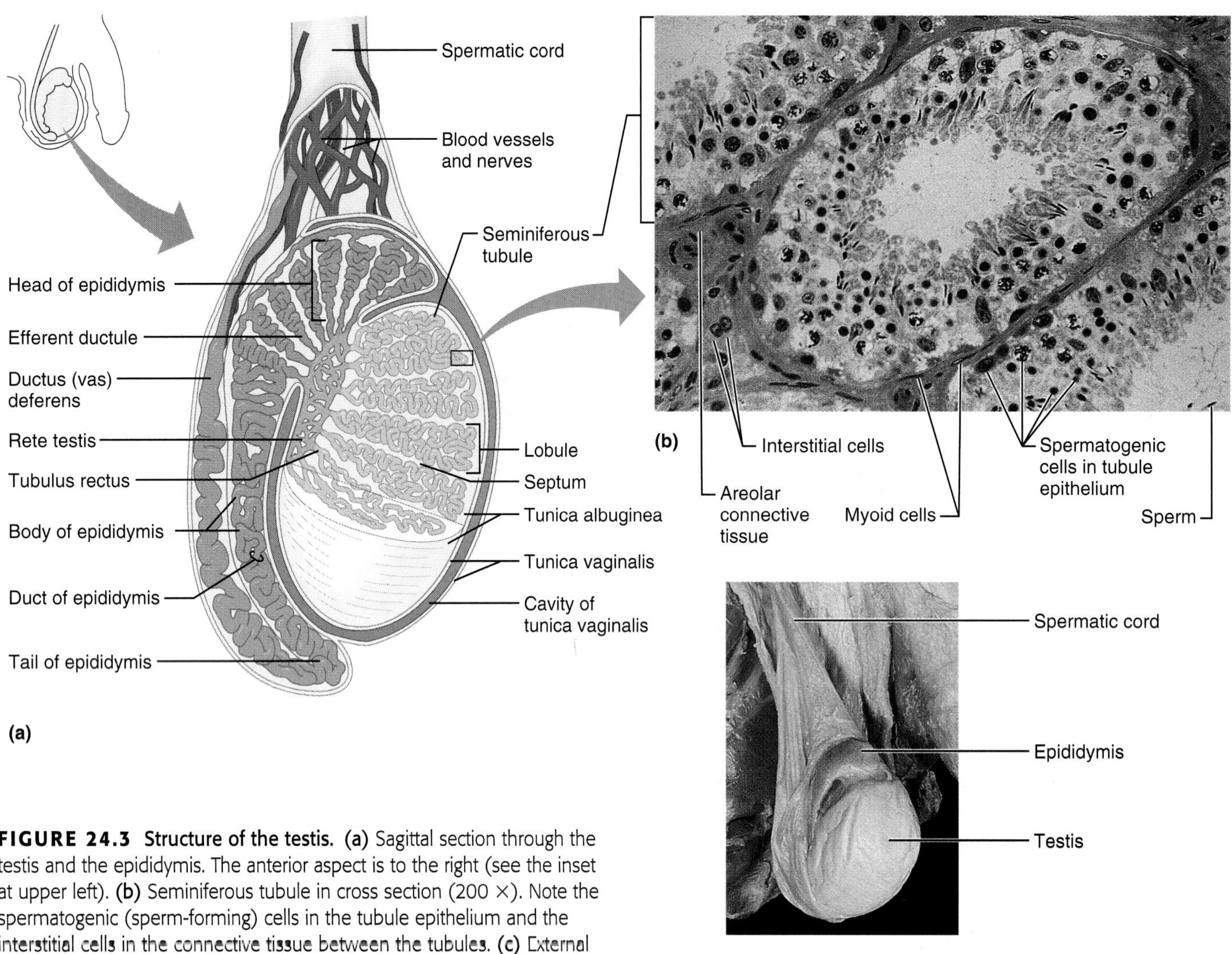

FIGURE 24.3 Structure of the testis. **(a)** Sagittal section through the testis and the epididymis. The anterior aspect is to the right (see the inset at upper left). **(b)** Seminiferous tubule in cross section (200 ×). Note the spermatogenic (sperm-forming) cells in the tubule epithelium and the interstitial cells in the connective tissue between the tubules. **(c)** External view of a testis from a cadaver; same orientation as in part (a).

tunica vaginalis (vaj-in-al′is; "ensheathing coat") (Figure 24.3a). This sac develops as an outpocketing of the abdominal peritoneal cavity that precedes the descending testes into the scrotum (see p. 737). It consists of a superficial parietal layer, an intermediate cavity containing serous fluid, and a deeper visceral layer that hugs the surface of the testis. Just deep to the visceral layer of the tunica vaginalis lies the **tunica albuginea** (al″bu-jin′e-ah; "white coat"), the fibrous capsule of the testis. Septal extensions of the tunica albuginea project inward to divide the testis into 250 to 300 wedge-shaped compartments called **lobules,** each of which contains one to four coiled **seminiferous tubules** (sem″ĭ-nif′er-us; "sperm-carrying"), the actual "sperm factories." Most of the convoluted seminiferous tubules are looped like hairpins.

Posteriorly, the seminiferous tubules of each lobule converge to form a **tubulus rectus** (plural: *tubuli recti*), a straight tubule that conveys sperm into the **rete testis** (re′te; "network of the testis"), a complex network of tiny branching tubes. The rete testis lies in the *mediastinum testis,* a region of dense connective tissue in the posterior part of the testis. From the rete testis, sperm leave the testis through about a dozen *efferent ductules* that enter the *epididymis,* a comma-shaped structure that hugs the outer surface of the testis (see Figures 24.1 and 24.3c).

The testes receive their arterial blood from the long testicular arteries, which branch from the aorta in the superior abdomen (see Figure 19.11, p. 559). The testicular veins, which roughly parallel the testicular arteries in the posterior abdominal wall, arise from a venous network in the scrotum called the **pampiniform plexus** (pam-pin′ĭ-form; "tendril-shaped") (see Figure 24.2). The veins of this plexus, which surround the testicular arteries like climbing vines, absorb heat from the arterial blood, cooling it before it enters the testes and thereby keeping the testes cool. The testes are innervated by both divisions of the autonomic nervous system. The very abundant visceral sensory nerves transmit impulses that result in agonizing pain and nausea when the testes are hit forcefully.

It is worth emphasizing that the testicular vessels and nerves do not stem from the nearby pelvis (as one might expect), but extend to the testis from a distant, superior location—generally from the level of L_2 on the posterior abdominal wall (see Figures 19.11 and 19.21). This level is the embryonic location of the testes, where the testes receive their vessels before migrating down into the scrotum.

The Seminiferous Tubules and Spermatogenesis

A histological section through a lobule of the mature testis (see Figure 24.3b) reveals numerous *seminiferous tubules* separated from each other by an areolar connective tissue. The sperm-forming tubules consist of a thick stratified epithelium surrounding a hollow central lumen. The epithelium consists of spherical *spermatogenic* ("sperm-forming") *cells* embedded in columnar *sustentacular* ("supporting") *cells* (Figure 24.4).

Spermatogenic Cells

The spermatogenic cells are stages in the process of sperm formation or **spermatogenesis** (sper″mah-to-jen′ĕ-sis), which begins at puberty and forms about 400 million sperm per day in an adult male. The least differentiated cells in the sequence are **spermatogonia** (sper″mah-to-go′ne-ah; "sperm seed"), which lie peripherally on the epithelial basal lamina. As these cells differentiate, they move inward toward the lumen to produce *primary* and *secondary spermatocytes, spermatids,* and *spermatozoa.* Spermatogenesis may be divided into the following three successive stages (see Figure 24.4b and c, proceeding from top to bottom):

Stage 1: Formation of spermatocytes. The spermatogonia divide vigorously and continuously by mitosis, with each division producing two distinctive daughter cells. **Type A** daughter cells remain at the basal lamina to maintain the germ cell line, whereas **type B** cells move toward the lumen to become **primary spermatocytes** (sper″mah′to-sītz).

Stage 2: Meiosis. The spermatocytes undergo **meiosis** (mi-o′sis; "a lessening"), a process that reduces the number of chromosomes found in typical body cells (denoted $2n$ and termed *diploid*) to half that number (n, termed *haploid*). In this process, which is greatly simplified here and in Figure 24.4b, two successive but distinctly different nuclear divisions (termed meiosis I and meiosis II) achieve the reduction of chromosome number from the diploid complement ($2n = 46 = 23$ pairs) of chromosomes to the haploid complement ($n = 23$ chromosomes).

Meiosis, which is an essential part of gamete formation in both sexes, ensures that the diploid complement of chromosomes is reestablished at fertilization, when the genetic material of the two haploid gametes joins to make a diploid zygote. (It also prevents the chromosomes from doubling in number with each successive generation.)

Within the seminiferous tubules, the cells undergoing meiosis I are by definition the primary spermatocytes; these cells each produce two **secondary spermatocytes,** each of which then undergoes meiosis II and produces two small cells called **spermatids.** Thus, four haploid spermatids result from the meiotic divisions of each original diploid primary spermatocyte.

Stage 3: Spermiogenesis. In the third and final stage of spermatogenesis, called **spermiogenesis** (sper″me-o-jen′ĕ-sis), spermatids differentiate into sperm (Figure 24.5). Each spermatid undergoes a streamlining process as it fashions a tail and sheds superfluous cytoplasm. The resulting sperm cell, or **spermatozoon** (sper″mah-to-zo′on; "animal seed"), has a *head,* a *tail,* and a *midpiece,* which is the thickened first part of the tail. The **head** of the sperm consists of a flat, sunflower-seed-shaped nucleus, which contains highly condensed chromatin, and a helmet-like **acrosome** (ak′ro-sōm; "tip piece"), a lysosome-like vesicle made by the Golgi apparatus and containing digestive enzymes that enable the sperm to penetrate and enter an egg. The **midpiece** of the sperm contains mitochondria spiraled tightly around the core of the tail. The long **tail** is an elaborate flagellum whose motile cytoskeleton has grown out from a centriole near the nucleus. The mitochondria provide the energy needed for the whip-like movements of the tail that will propel the sperm through the reproductive tract of the female.

The newly formed sperm detach from the epithelium of the seminiferous tubule and enter the lumen of the tubule (see Figure 24.4c). Whereas these young sperm look mature, they are nearly immotile and do not gain the ability to swim until after they have left the testis (see p. 708).

Overall, the process of spermatogenesis is controlled by the stimulating action of two hormones: **follicle-stimulating hormone (FSH)** from the anterior part of the pituitary gland, and **testosterone,** the primary male sex hormone produced by the testes.

Sustentacular Cells

The spermatogenic cells are surrounded by **sustentacular cells,** or *Sertoli* (ser-to′le) *cells,* which extend from the basal lamina to the lumen of the seminiferous tubule (see Figure 24.4c). The sustentacular cells, bound to each other by tight junctions on their lateral membranes, divide the seminiferous tubule into two compartments. The **basal compartment** extends from the basal lamina to the tight junctions and contains the spermatogonia and earliest primary spermatocytes; the **adluminal** ("near the lumen") **compartment** lies internal to the tight junctions and includes the more-advanced spermatogenic cells and the lumen of the tubule.

The tight junctions between the sustentacular cells constitute the **blood-testis barrier** (Figure 24.4c). This barrier prevents the escape of membrane antigens of differentiating sperm through the basal lamina and into the bloodstream, where they would activate the immune system (sperm and late-stage spermatogenic cells produce unique proteins to which the immune system is not

FIGURE 24.4 Seminiferous tubule and spermatogenesis (sperm formation). **(a)** Scanning electron micrograph of a cross-sectioned seminiferous tubule (225 ×). Copyright by R. Kessel and R. Kardon. **(b)** Flow chart of the events of spermatogenesis, showing the relative positions of various spermatogenic cells in the epithelium of the seminiferous tubule. These cells are held together by intercellular "bridges" until they are released as spermatozoa. The entire sequence of sperm formation illustrated here takes about 75 days. **(c)** Enlarged view of the wall of the seminiferous tubule, showing the spermatogenic cells and two large sustentacular cells (green). The epithelium is oriented with its apical surface below. The largest spermatogenic cells, the primary spermatocytes, are readily identified by the thick, condensed chromosomes in their nuclei.

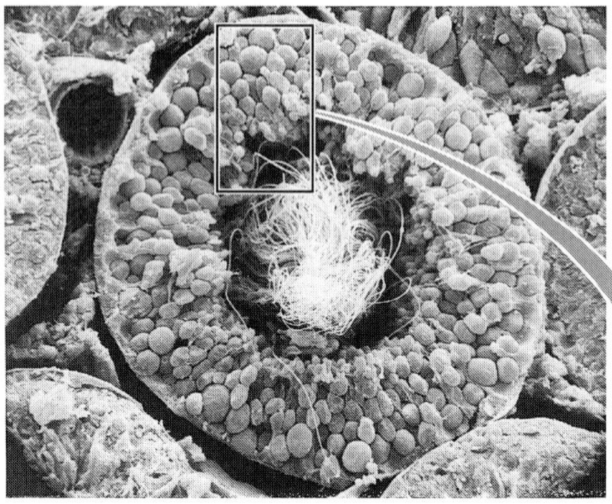

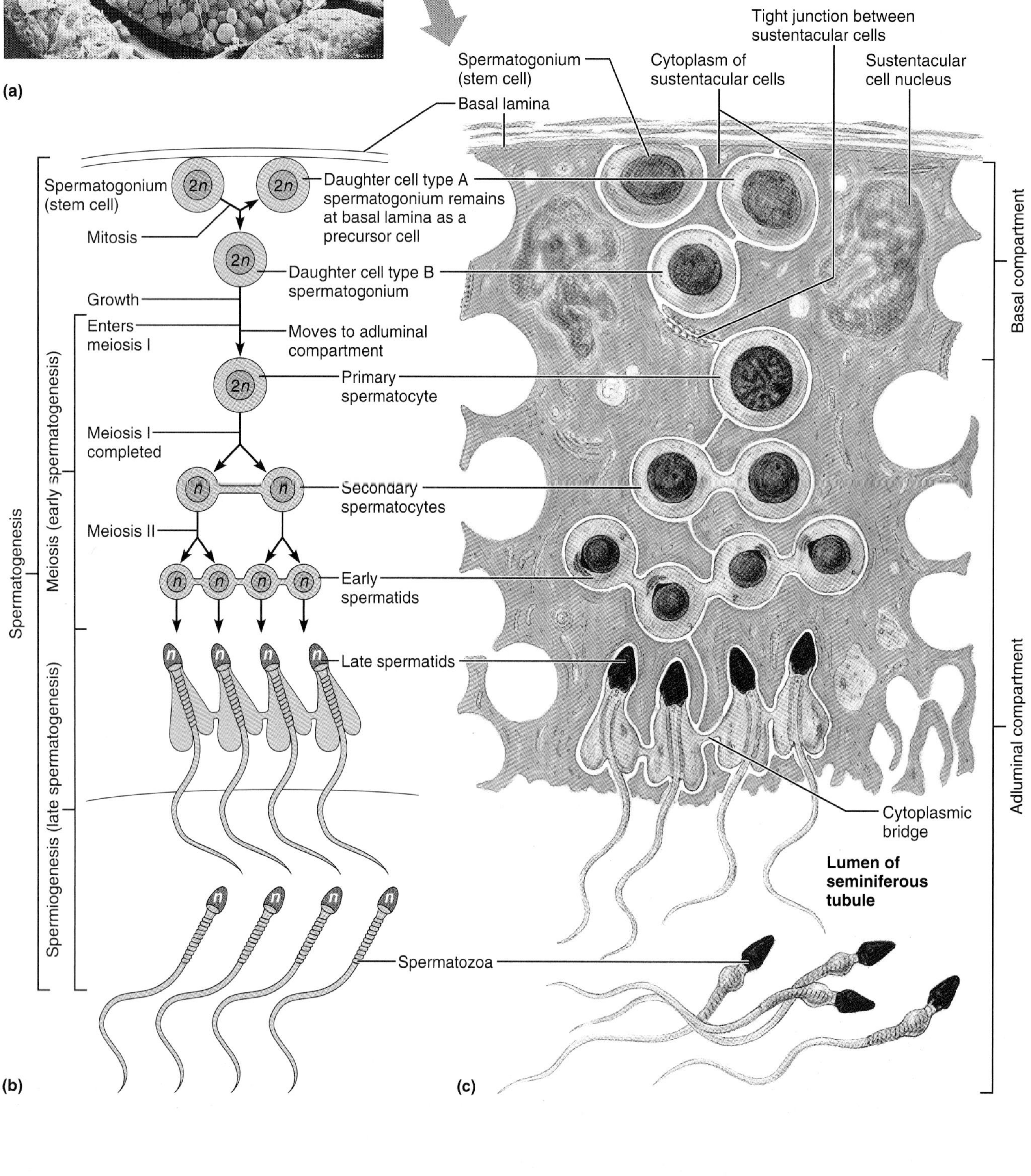

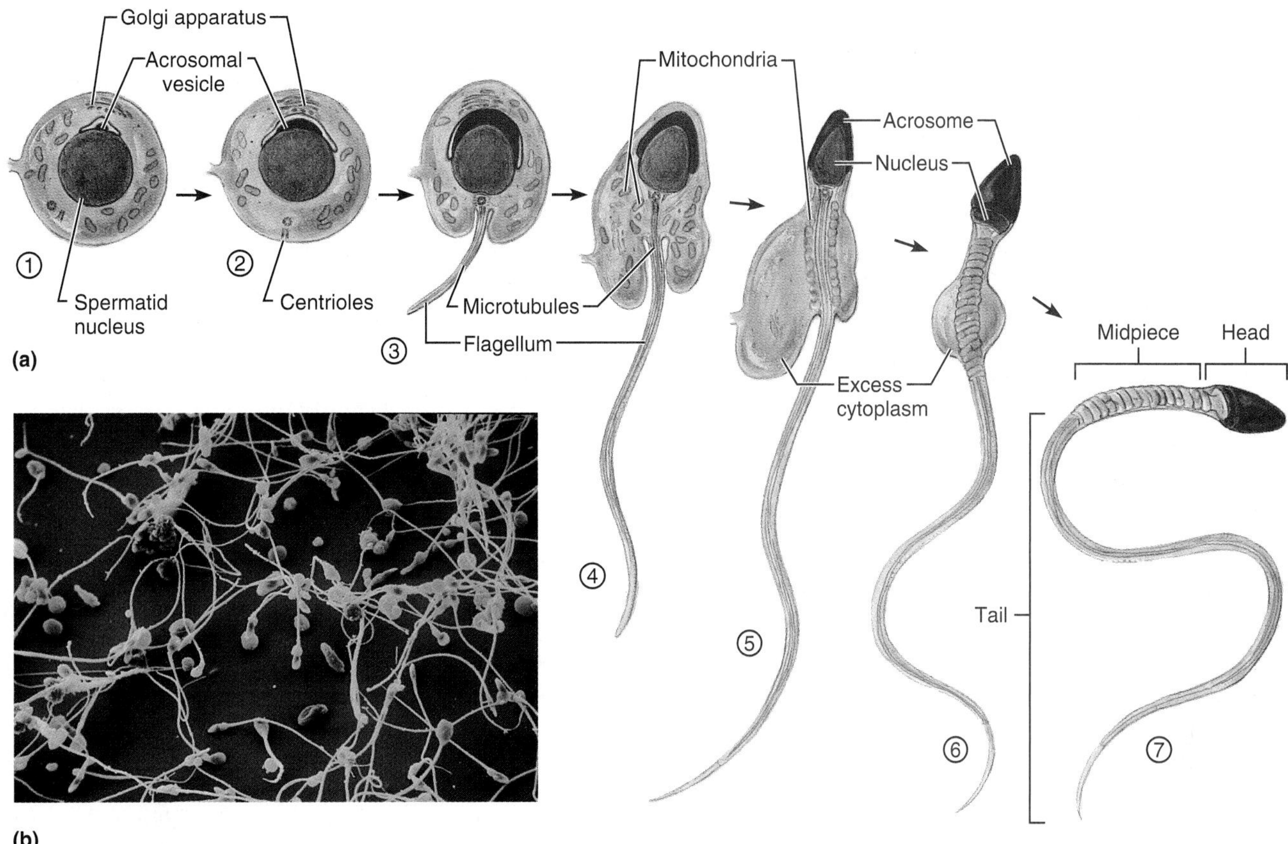

FIGURE 24.5 Spermiogenesis: Transformation of a spermatid into a sperm. (a) The steps in spermiogenesis are as follows: (1) The spermatid's Golgi apparatus produces enzyme-filled vesicles that fuse together to form the acrosome. (2) The acrosome positions itself at the anterior end of the nucleus, and the centrioles move to the opposite end. (3) Microtubules assemble from a centriole and grow to form the flagellum that is the sperm tail. (4) Mitochondria multiply in the cytoplasm. (5) The mitochondria position themselves around the proximal core of the flagellum, the chromatin condenses in the nucleus, the nucleus flattens, and excess cytoplasm is shed from the cell. (6) Structure of an immature sperm that has just been released from a sustentacular cell into the lumen of the seminiferous tubule. (7) A structurally mature sperm with a streamlined shape that allows active swimming. **(b)** Scanning electron micrograph of mature sperm (550 ×).

tolerant). The resulting autoimmune response directed at one's own sperm would destroy the gametes and produce sterility. Because developing spermatogenic cells must cross the blood-testis barrier on their way to the lumen, the tight junctions between sustentacular cells allow passage of these cells into the adluminal compartment in a way reminiscent of the opening of canal locks to permit the passage of ships.

Sustentacular cells assist sperm production in many ways. Besides conveying nutrients to the spermatogenic cells, they actively move these cells toward the tubule lumen and phagocytize the cytoplasm that is shed as spermatids become sperm. Sustentacular cells also secrete **testicular fluid** into the tubule lumen, and the pressure of this fluid pushes sperm through the tubule and out of the testes. Furthermore, sustentacular cells secrete **androgen-binding protein,** which concentrates male sex hormone (testosterone) near the spermatogenic cells; the testosterone then stimulates spermatogenesis, either directly or by signaling the sustentacular cells to do so. Finally, sustentacular cells secrete the hormone **inhibin,** which slows the rate of sperm production (when necessary) by inhibiting FSH secretion from the pituitary gland.

Myoid Cells Human seminiferous tubules are surrounded by several layers of smooth-muscle–like **myoid cells** (Figure 24.3b). Contracting rhythmically, these may help to squeeze sperm and testicular fluid through the tubules and out of the testis.

Interstitial Cells The loose connective tissue between the seminiferous tubules contains clusters of **interstitial cells,** or *Leydig* (li′dig) *cells* (see Figure 24.3b), spherical or polygon-shaped cells that make and secrete the male sex hormones, or **androgens,** the main type of which is testosterone. After it is secreted into the nearby blood and lymph capillaries, testosterone circulates throughout the body and maintains all male sex characteristics and sex organs. In fact, all male genitalia atrophy if the testes (and testosterone) are removed. Secretion of testosterone by the interstitial cells is controlled by **luteinizing** (loo′te-in-īz″ing) **hormone (LH),** a hormone from the anterior part of the pituitary gland. By stimulating testosterone secretion, LH controls testosterone's effects on the entire male reproductive system.

The Reproductive Duct System in Males

As previously noted, sperm leaving the seminiferous tubules travel through the *tubuli recti* and *rete testis* (see Figure 24.3a), which are lined by a simple cuboidal epithelium. Sperm then leave the testis through the *efferent ductules,* which are lined by a simple columnar epithelium; here, both ciliated epithelial cells and smooth musculature in the wall help move the sperm along while nonciliated cells absorb most of the testicular fluid. From the efferent ductules the sperm enter the duct of the epididymis.

The Epididymis

The **epididymis** (ep″ĭ-did′ĭ-mis; "beside the testis") is a comma-shaped organ that arches over the posterior and lateral side of the testis (see Figure 24.3a and c). The *head* of the epididymis contains the efferent ductules, which empty into the **duct of the epididymis,** a highly coiled duct that completes the head and forms all of the *body* and *tail* of this organ. With an uncoiled length of over 6 m (20 feet), the duct of the epididymis is longer than the entire intestine!

Histologically, the duct of the epididymis is dominated by a tall, pseudostratified columnar epithelium (Figure 24.6a). The luminal surface of this epithelium bears tufts of long microvilli called **stereocilia** (ster″e-o-sil′e-ah), which are not cilia and do not move. Instead, they provide the tall epithelial cells with a vast surface area for reabsorbing testicular fluid and for transferring nutrients and secretions to the many sperm that are stored in the lumen of the epididymis. External to the epithelium lies a layer of smooth muscle.

The immature, nearly immotile sperm that leave the testes are moved slowly through the duct of the epididymis. During this journey, which takes about 20 days, the sperm gain the ability to swim and also the ability to fertilize the egg through the acrosome reaction (the events of fertilization occur in the female's uterine tube: see p. 727). These maturation processes are stimulated by proteins secreted by the epididymis epithelium. Sperm are ejaculated from the epididymis, *not* directly from the testes. At the beginning of ejaculation, the smooth muscle in the walls of the epididymis contracts, expelling sperm from the tail of the epididymis into the next segment of the duct system, the ductus deferens.

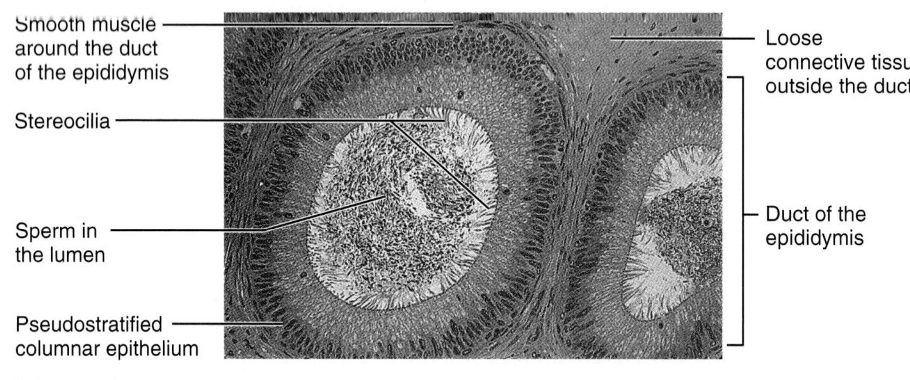

FIGURE 24.6 The epididymis and ductus deferens in histological cross sections. **(a)** The duct of the epididymis. What appear to be several ducts are instead multiple coils of a single duct in section. Note the thick epithelium and stereocilia (120 ×). **(b)** Ductus deferens. Note the thick layer of smooth muscle (30 ×).

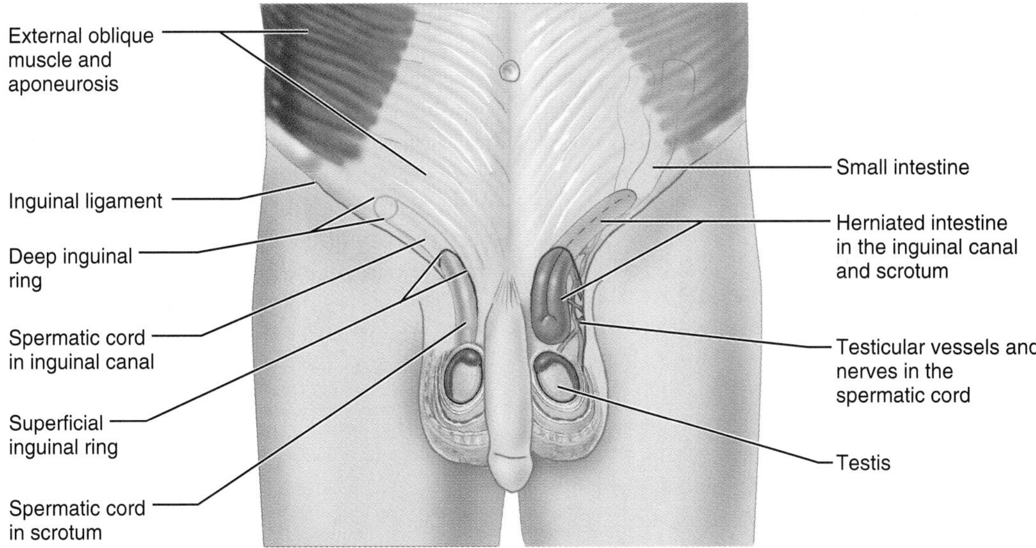

FIGURE 24.7 Spermatic cord in the inguinal canal (at left) and an inguinal hernia (at right). The inguinal canal lies just deep to the aponeurosis of the external oblique muscle. Although a portion of the intestine is shown passing through the inguinal canal toward the scrotum, the greater omentum may also herniate from the abdominal cavity by this path. In either case, an inguinal hernia forms a bulge below the skin in the groin.

Sperm can be stored in the epididymis for several months, after which time they are phagocytized by the epithelial cells of the duct of the epididymis.

The Ductus Deferens

The **ductus deferens** (def′er-ens; "carrying away"), or *vas deferens,* is about 45 cm (18 inches) long. It runs superiorly through the scrotum, pierces the anterior abdominal wall, and enters the pelvic cavity (see Figure 24.1); from there it runs posteriorly along the lateral wall of the true pelvis, arches medially over the ureter, and descends along the posterior wall of the bladder. Its distal end expands as the **ampulla** ("flask") **of the ductus deferens** and then joins with the duct of the seminal vesicle (a gland) to form the short **ejaculatory duct.** Each ejaculatory duct runs within the prostate gland, where it empties into the prostatic urethra.

The wall of the ductus deferens (see Figure 24.6b) consists of (1) an inner mucosa with the same pseudostratified epithelium as that of the epididymis, plus a lamina propria, (2) an extremely thick muscularis, and (3) an outer adventitia. During ejaculation, the smooth muscle in the muscularis creates strong peristaltic waves that rapidly propel sperm through the ductus deferens to the urethra.

Clinical Applications

VASECTOMY Some men take responsibility for birth control by having a **vasectomy** (vah-sek′to-me; "cutting the vas"). In this minor surgery, the physician makes a small incision into both sides of the scrotum, transects both ductus deferens, and then closes the cut ends, either by tying them off or by fusing them shut by cauterization. Although sperm continue to be produced, they can no longer exit the body and are phagocytized in the epididymis.

Some men who have had vasectomies wish to reverse them (after divorcing and remarrying, for example). New microsurgical techniques are making it easier to reopen and reattach the closed ends of the vas, and the rate of successful reversal (as judged by subsequent impregnation) is now over 50%. However, in many cases, a vasectomy leads to true, irreversible sterility. This occurs when sperm leak out of the closed end of the vas, or when the sperm backing up in the epididymis creates so much pressure that the epididymis bursts. Either way, the immune system gains access to the leaked sperm, and the resulting autoimmune destruction of sperm at the leakage site and in the testes produces sterility.

Physicians performing a vasectomy are able to distinguish the ductus deferens from the testicular vessels that surround it because its muscular layer is so thick that it feels like a hard wire when squeezed between the fingertips. ✚

The Spermatic Cord

The ductus deferens is the largest component of the **spermatic cord,** a tube of fascia that also contains the testicular vessels and nerves (see Figure 24.2). The inferior part of the spermatic cord lies in the scrotum, and its superior part runs through the **inguinal canal,** an obliquely oriented trough in the anterior abdominal wall (Figure 24.7). The inguinal canal is formed by the inguinal ligament, which is the free inferior margin of the aponeurosis of the external oblique muscle (p. 289).

This canal begins medially at the **superficial inguinal ring** (a V-shaped opening in the external oblique aponeurosis) and runs laterally to the **deep inguinal ring** (a weak point in the fascia of the anterior body wall where the ductus deferens and testicular vessels enter the pelvic cavity).

CLINICAL APPLICATIONS

INGUINAL HERNIA Because the deep inguinal ring and an area of fascia just medial to it (the inguinal triangle) are weak areas in the abdominal wall, coils of intestine or the greater omentum can be forced anteriorly through these weak areas, into the inguinal canal, sometimes pushing all the way into the scrotum. The resulting condition, called an **inguinal hernia** (see Figure 24.7), is often precipitated by lifting or straining, which raises intra-abdominal pressure and forces the herniated elements through the body wall. Although most herniated elements pop back into the abdominal cavity whenever the pressure decreases, a few remain squeezed in the narrow inguinal canal, which can strangle their blood supply and lead to potentially fatal gangrene.

At least 2% of adult males develop inguinal hernias, which are surgically repaired by closing off the enlarged opening with surgical thread or a strong prosthetic patch. Increasingly popular is a minimally invasive procedure that uses a laparoscope to guide placement of a patch external to the parietal peritoneum on the inner side of the anterior abdominal wall.

Although composed of the same elements, the inguinal canals of *females* are considerably narrower than in males because they contain only a thin band called the round ligament of the uterus (see p. 721). As a result, inguinal hernias are only one-ninth as common in females as in males. ✚

The Urethra

The urethra in males carries sperm from the ejaculatory ducts to the outside of the body (see Chapter 23, p. 691). Figure 24.8a depicts its three parts: the *prostatic urethra* in the prostate gland, the *membranous urethra* in the urogenital diaphragm, and the *spongy (penile) urethra* in the penis. The mucosa of the spongy urethra contains scattered outpocketings called *urethral glands* (not shown) that secrete a mucus that helps lubricate the urethra just before ejaculation.

Accessory Glands

The accessory glands in males include the paired *seminal vesicles*, the single *prostate gland*, and paired *bulbourethral glands* (see Figure 24.1). These glands produce the bulk of the **semen**, which is defined as sperm plus the secretions of the accessory glands and accessory ducts.

The Seminal Vesicles

The **seminal vesicles** (or **seminal glands**), which lie on the posterior surface of the bladder (see Figure 24.1), are hollow glands about the shape and length of a finger (5 to 7 cm). However, because a seminal vesicle is pouched, coiled, and folded back on itself, its true (uncoiled) length is about 15 cm. Internally, the mucosa is folded into a honeycomb pattern of crypts and blind chambers, and the lining epithelium is a secretory pseudostratified columnar epithelium. The external wall is composed of a fibrous capsule of dense connective tissue surrounding a thick layer of smooth muscle. This muscle contracts during ejaculation to empty the gland.

The secretion of the seminal vesicles, which constitutes about 60% of the volume of semen, is a viscous

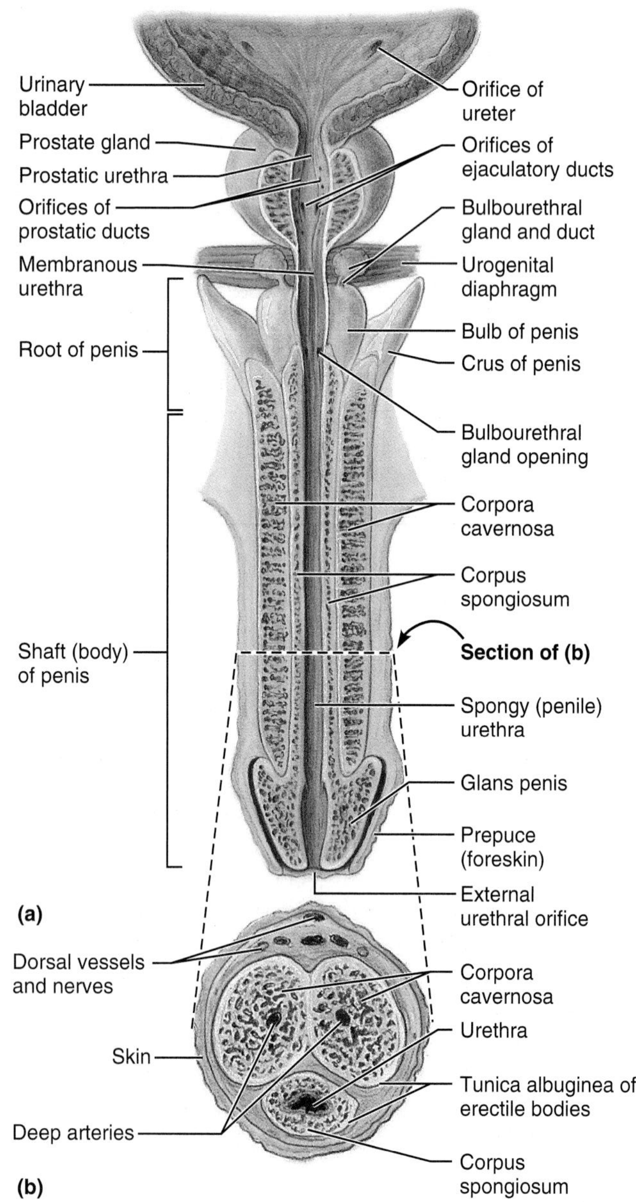

FIGURE 24.8 **Male urethra and penis.** **(a)** Longitudinal (frontal) section through the penis. **(b)** Enlarged cross section through the penis, with dorsal aspect at top (the *dorsal* aspect is the *posterior* aspect of the *erect* penis).

fluid that contains (1) a sugar called fructose and other nutrients that nourish the sperm on their journey; (2) *prostaglandins* (pros″tah-glan′dinz), which stimulate contraction of the uterus to help move sperm through the female reproductive tract; (3) substances that suppress the immune response against semen in females; (4) substances that enhance sperm motility; and (5) enzymes that clot the ejaculated semen in the vagina and then liquefy it so that the sperm can swim out. Finally, the presence in seminal fluid of a yellow pigment that fluoresces under ultraviolet light enables the recognition of semen residues in criminal investigations of sexual assault.

As previously noted, the duct of each seminal vesicle joins the ductus deferens on the same side of the body to form an ejaculatory duct within the prostate gland. Sperm and seminal fluid mix in the ejaculatory duct and enter the prostatic urethra together during ejaculation.

The Prostate Gland

The **prostate** (pros′tāt) **gland,** which is the size and shape of a chestnut, encircles the first part of the urethra just inferior to the bladder (Figure 24.9a). The prostate consists of 20–30 compound tubuloalveolar glands of three classes—*main, submucosal,* and *mucosal glands*—that are embedded in a mass of dense connective tissue and smooth muscle called the **fibromuscular stroma** and surrounded by a connective tissue capsule (Figure 24.9b). The muscle of the stroma contracts during ejaculation to squeeze the prostatic secretion into the urethra.

The prostatic secretion constitutes about one-third of the volume of semen. It is a milky fluid that, like the secretion of the seminal vesicles, contains various substances that enhance sperm motility and enzymes that clot and liquefy ejaculated semen. One of the enzymes that liquefy semen is *prostate-specific antigen (PSA);* measuring the levels of PSA in a man's blood is the most important method of screening for prostate cancer (discussed on p. 733).

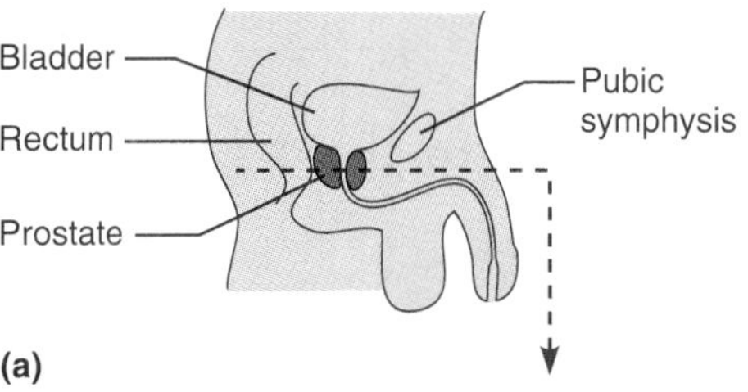

(a)

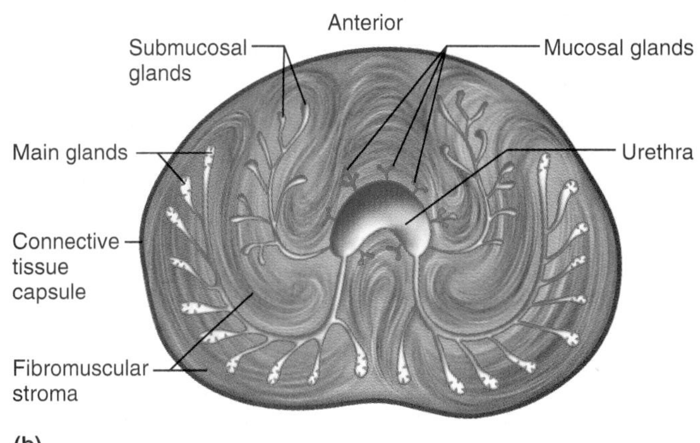

(b)

Anterior

Fibromuscular stroma

Compound glands

Urethra

(c)

FIGURE 24.9 The prostate gland. **(a)** The location of the prostate gland in the pelvis. **(b)** Diagram of a transverse section through the prostate gland. Note the three classes of glands: main, submucosal, and mucosal. **(c)** Histological section through the prostate; same orientation as the diagram in part b (5 ×).

Clinical Applications

BENIGN PROSTATIC HYPERPLASIA A common, noncancerous tumor called **benign prostatic hyperplasia (BPH)** is characterized by uncontrolled growth in the deep, mucosal glands in the prostate, and by proliferation of some of the nearby stroma cells. About 50% of men at age 50 and 80% of men at age 70 develop BPH. Enlargement of the prostate and contraction of its smooth musculature constrict the prostatic urethra, making micturition difficult, and it can lead to urinary retention, continual dribbling of urine, urinary tract infections, and the formation of kidney stones.

Diagnosis of BPH begins with inquiries about the patient's urinary symptoms—whether he is hesitant to urinate, has a weak stream, must strain to urinate, feels that his bladder does not empty fully, has increased urinary frequency and urgency, and must urinate often at night. Next, to determine whether the prostate is enlarged, the physician performs a **digital rectal exam;** a finger inserted into the anal canal can feel the prostate because it lies just anterior to the rectum. The physician may also order a blood test to check the levels of PSA (which are only mildly elevated in BPH; generally less so than in prostate cancer). Treatment for BPH begins with drugs that inhibit the

A Closer Look

Sexually Transmitted Diseases

Sexually transmitted diseases (STDs), also known as venereal diseases, are infectious diseases spread via sexual contact. Over 12 million Americans, a quarter of them adolescents, get STDs each year.

Until recently, gonorrhea and syphilis, caused by bacterial infections, were the most common STDs, but viral diseases have now taken center stage. AIDS, the most notorious viral STD, is discussed in the A Closer Look box in Chapter 20. Here we consider the other prominent STDs—the spread of which, it is important to note, is significantly reduced by the conscientious use of latex condoms (see the photo).

Gonorrhea

The bacterial disease **gonorrhea** (gon″o-re′ah) is caused by invasion of the reproductive and urinary tract mucosae by *Neisseria gonorrhoeae,* a bacterium spread by contact with genital, anal, and pharyngeal surfaces. Commonly called "the clap," its most common symptom in males is urethritis, which is accompanied by painful urination and the discharge of pus from the penis. In females, signs and symptoms vary from none (about 20% of cases) to abdominal discomfort, vaginal discharge, abnormal uterine bleeding, and, occasionally, urethral symptoms similar to those seen in males.

In men, untreated gonorrhea can cause urethral constriction and inflammation of the entire duct system; in women it causes pelvic inflammatory disease and sterility. Although these consequences were lessened by the advent in the 1950s of antibiotics, strains resistant to antibiotic treatment are becoming increasingly prevalent. Currently, the antibiotic most frequently used to treat gonorrhea is ceftriaxone.

Syphilis

Although **syphilis** (sif′ĭ-lis), caused by the corkscrew-shaped bacterium *Treponema pallidum,* is usually transmitted sexually, it can be contracted congenitally from an infected mother. Fetuses infected with syphilis are usually stillborn or die shortly after birth.

In adults, the bacterium easily penetrates intact mucosae and abraded skin. After a 2- to 3-week incubation period, a painless red primary lesion called a **chancre** (shang′ker) appears at the site of the bacterial invasion—in males, typically the penis, but in females the lesion often goes undetected in the vagina or on the cervix. The chancre ulcerates and becomes crusty before healing spontaneously and disappearing within a few weeks.

production or action of testosterone, upon which the tumor cells depend, or drugs that relax the prostate's smooth muscle. If such treatment does not increase ease of urination, the prostate is either removed in a surgical procedure called *transurethral prostatectomy* or destroyed by heat from a microwave device or an electrode inserted into the urethra. ✚

In addition to its susceptibility to tumors, the prostate is also subject to infection in sexually transmitted diseases (STDs). (STDs are the subject of this chapter's A Closer Look box.) **Prostatitis** (pros″tah-ti′tis), inflammation of the prostate, is the single most common reason that men consult a urologist.

The Bulbourethral Glands

The **bulbourethral** (bul″bo-u-re′thral) **glands** are pea-sized glands situated inferior to the prostate gland, within the urogenital diaphragm (see Figures 24.1 and 24.8). These compound tubuloalveolar glands produce a mucus, some of which enters the spongy urethra when a male becomes sexually excited prior to ejaculation. This mucus apparently neutralizes traces of acidic urine in the urethra and lubricates the urethra to smooth the passage of semen during ejaculation.

The Penis

The **penis** ("tail"), the male organ of sexual intercourse, delivers sperm into the female reproductive tract (see Figure 24.8). Together, it and the scrotum make up the external reproductive structures, or external genitalia, of the male. The penis consists of an attached **root** and a free **shaft** or **body** that ends in an enlarged tip called the **glans penis.** The skin covering the penis is loose and slides distally around the glans to form a cuff, called the **prepuce** (pre′pūs), or **foreskin.**

Clinical Applications

CIRCUMCISION Over 60% of male babies in the United States undergo **circumcision** (ser″kum-sizh′un; "cutting around"), the surgical removal of the foreskin that is typically performed shortly after birth. Circumcision is a controversial procedure. Whereas opponents charge that it is a medically unnecessary surgery that causes too much pain, proponents contend that it

Left untreated, syphilis can progress through a symptomatic *secondary phase* lasting 3–12 weeks, and an asymptomatic *latent period* that may either last the patient's lifetime or progress to ***tertiary syphilis,*** characterized by destructive lesions of the CNS, blood vessels, bones, and skin. Penicillin is the treatment of choice for all stages of syphilis.

Chlamydia

Chlamydia (klah-mid′e-ah; *chlamys* = cloak) is a largely unrecognized, silent epidemic that afflicts some 4–5 million Americans annually, making it the most common STD in the United States. Chlamydia is responsible for 25% to 50% of all diagnosed cases of pelvic inflammatory disease, and each year more than 150,000 infants are born to infected mothers. About 20% of men and 30% of women with gonorrhea are also infected by *Chlamydia trachomatis,* the causative agent of chlamydia.

Chlamydia, a bacterium with a virus-like dependence on host cells, produces often-unrecognized symptoms that include urethritis (involving painful, frequent urination, and a thick penile discharge); vaginal discharge; abdominal, rectal, or testicular pain; painful intercourse; and irregular menstruation. In affected women—80% of whom have asymptomatic infections—its most severe consequence is sterility. Newborns infected in the birth canal tend to develop conjunctivitis and respiratory tract inflammations, including pneumonia. Chlamydia is easily treated with tetracycline.

Vaginitis

Vaginitis is typically a manifestation of a local infection with either the flagellated protozoan *Trichomonas vaginalis* or the bacterium *Gardnerella vaginalis,* but several microorganisms, including *Gonococcus* and *Candida,* can cause it. The form caused by *T. vaginalis* is as common in sexually active women as gonorrhea. Among the many symptoms, including painful intercourse, the most common are severe genital itching and a copious, green, foul-smelling vaginal discharge. The drug of choice for treating vaginitis is a single oral dose of metronidazole given simultaneously to both sexual partners to prevent continual reinfection.

Genital Warts

Each year about 1 million Americans develop **genital warts,** an STD caused by the ***human papillomavirus,*** which is actually a group of some 60 viruses. Treatment is difficult and controversial. Some clinicians prefer to leave the warts untreated unless they become very widespread, whereas others recommend their removal by cryosurgery, laser therapy, and/or treatment with alpha interferon.

Genital Herpes

An estimated 25% to 50% of all adult Americans harbor herpes simplex type 2, the common cause of **genital herpes** (her′pēz), and most people who have it do not know it. Direct transmission of this herpesvirus—via infectious secretions—results in infections that remain silent for weeks and then suddenly flare up, causing painful blister-like lesions on the reproductive organs that are usually more of a nuisance than a threat to life. However, congenital herpes infections can cause severe malformations of a fetus. The antiviral agent acyclovir, which speeds healing of lesions and reduces the frequency of flare-ups, is the treatment of choice. Inter Vir-A, an antiviral ointment, provides some relief from the itching and pain that accompany the lesions.

reduces infections of the glans, urethral and urinary-tract infections, and penile cancer. ✚

Internally, the penis contains the spongy urethra and three long cylindrical bodies (corpora) of erectile tissue (Figure 24.8a). Each of these three **erectile bodies** is a thick tube covered by a sheath of dense connective tissue and filled with a network of partitions that consist of smooth muscle and connective tissue. This sponge-like network, in turn, is riddled with vascular spaces. The midventral erectile body surrounding the spongy urethra, the **corpus spongiosum** (kor′pus spun″je-o′sum; "spongy body"), is enlarged *distally* where it forms the glans penis, and *proximally* where it forms a part of the root called the **bulb of the penis.** This bulb is secured to the urogenital diaphragm and is covered externally by the sheet-like bulbospongiosus muscle (see Figure 11.12, p. 293). The paired, dorsal erectile bodies, the **corpora cavernosa** (kav″er-no′sah; "cavernous bodies"), make up most of the mass of the penis. Their proximal ends in the root are the **crura** ("legs") **of the penis** (singular, **crus**); each crus is anchored to the pubic arch of the bony pelvis and is covered by an ischiocavernosus muscle (see Figure 11.12).

Most of the main vessels and nerves of the penis lie near the dorsal midline (see Figure 24.8b). The sensory *dorsal nerves* are branches of the pudendal nerve from the sacral plexus, and the *dorsal arteries* are branches of the internal pudendal arteries from the internal iliac arteries. Two *dorsal veins* (superficial and deep) lie precisely in the dorsal midline and drain all blood from the penis. Finally, a *deep artery* runs within each corpus cavernosum. The autonomic nerves to the penis, which follow the arteries and supply the erectile bodies, arise from the inferior hypogastric plexus in the pelvis.

The chief phases of the male sexual response are (1) erection of the penis, which allows it to penetrate into the female vagina, and (2) ejaculation, which expels semen into the vagina. *Erection* results from engorgement of the erectile bodies with blood. During sexual excitement, the arteries supplying the erectile bodies dilate, increasing the flow of blood to the vascular spaces within, and the smooth muscle in the partitions in these bodies relaxes, allowing the bodies to expand

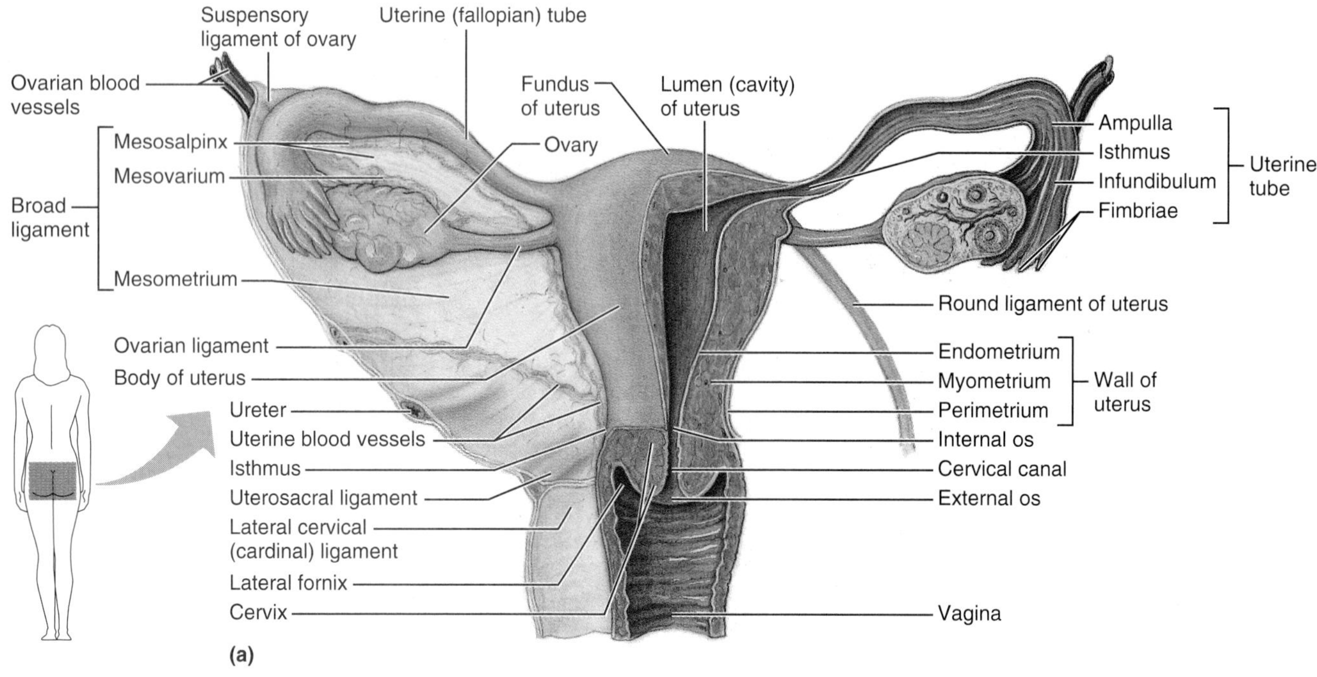

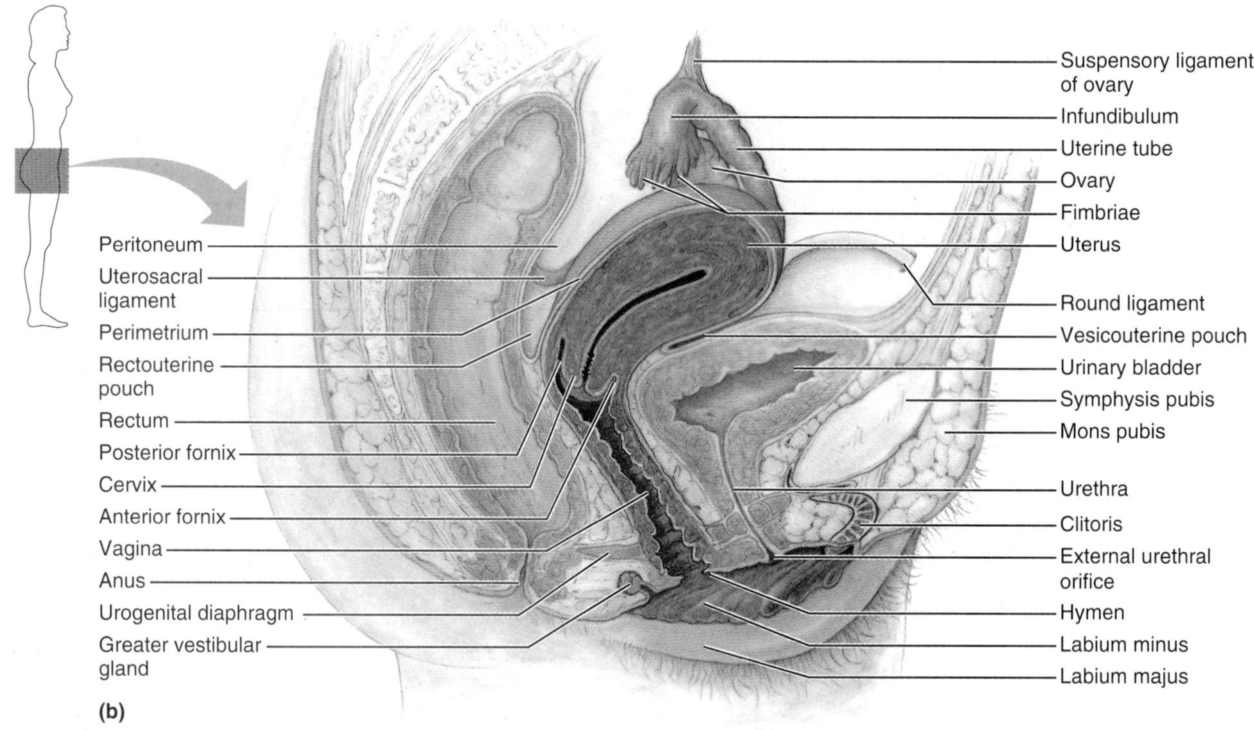

FIGURE 24.10 Female internal reproductive organs. **(a)** Posterior view of the reproductive organs in the pelvis. At right, the posterior walls of the various organs have been removed, as has the broad ligament. **(b)** Midsagittal section of the female pelvis.

as the blood enters them. As the erectile bodies begin to swell, they press on the small veins that normally drain them, slowing venous drainage and maintaining engorgement. Although complex, the control of erection is mainly through parasympathetic autonomic nervous input that dilates the vessels.

Whereas erection is largely under parasympathetic control, *ejaculation* is under sympathetic control. Ejaculation begins with a strong, sympathetically induced contraction of the smooth musculature throughout the reproductive ducts and glands, which squeezes the semen toward and into the urethra. Simultaneously, the

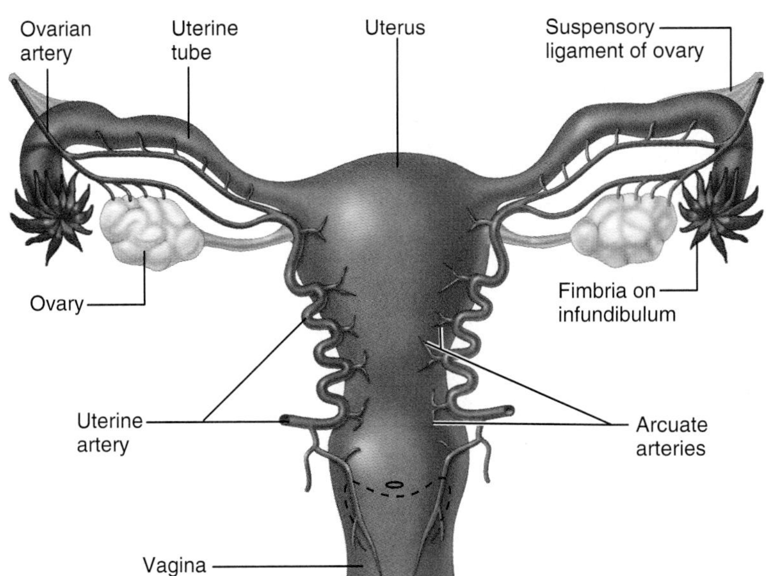

FIGURE 24.11 **Arteries of the internal female genitalia (posterior view).** Corresponding veins run with these arteries.

bulbospongiosus muscle of the penis contracts rapidly, squeezing the semen onward through the penile urethra and out of the body.

Advances in Understanding Recent experiments have revealed that the collagen fibers in the dense connective tissue outside the erectile bodies have a special arrangement that strengthens the penis during erection. Longitudinal fibers that run the length of the penis lie at right angles to circular fibers that form rings around the penile shaft. This is an optimal design for resisting bending forces, so the erect penis does not buckle or kink sharply during intercourse. ✷

The Male Perineum

The male **perineum** (per″ĭ-ne′um; "around the anus") contains the scrotum, the root of the penis, and the anus (see Figure 26.14a). More specifically, it is defined as the diamond-shaped area between the pubic symphysis anteriorly, the coccyx posteriorly, and the ischial tuberosities laterally. The floor of the perineum is formed by muscles described in Chapter 11 (see p. 293).

THE FEMALE REPRODUCTIVE SYSTEM

The female reproductive organs differ from those of males in several important ways. In addition to producing gametes (ova, or "eggs"), they also prepare to support a developing embryo during pregnancy. Furthermore, the female organs undergo changes according to a reproductive cycle, the **menstrual cycle,** which averages 28 days in length. By definition, the menstrual cycle is the female's monthly cycle as it affects *all* her reproductive organs.

Figure 24.10 provides an overview of the female's internal reproductive organs. The gonads are the ovum-producing *ovaries.* The accessory ducts include the *uterine tubes,* where fertilization typically occurs; the *uterus,* where the embryo develops; and the *vagina,* which acts as a birth canal and receives the penis during sexual intercourse. The female's external genitalia are referred to as the *vulva.* Even though the milk-producing *mammary glands* are actually part of the integumentary system, they are considered here because of their reproductive function of nourishing the infant.

The Ovaries

The paired, almond-shaped **ovaries,** which flank the uterus on each side, measure about 3 cm by 1.5 cm by 1 cm. Each ovary lies against the bony lateral wall of the true pelvis, in the fork of the iliac vessels. The surfaces of the ovaries are smooth in young girls but after puberty become scarred and pitted from the monthly release of ova.

The ovaries are surrounded by the peritoneal cavity and held in place by mesenteries and ligaments. The mesentery of the ovary, the horizontal **mesovarium** (mez″o-va′re-um), is part of the **broad ligament,** a large fold of peritoneum that hangs from the uterus and the uterine tubes like a tent (see Figure 24.10a). The **suspensory ligament of the ovary,** a lateral continuation of the broad ligament, attaches the ovary to the lateral pelvic wall. Finally, the ovary is anchored to the uterus medially by the **ovarian ligament,** a distinct fibrous band enclosed within the broad ligament.

The ovaries are served by the *ovarian arteries,* which are branches from the abdominal aorta, and by the *ovarian branches of the uterine arteries,* which arise from the internal iliac artery (Figure 24.11). The ovarian artery, veins, and nerves reach the ovary by traveling within the suspensory ligament and then through the

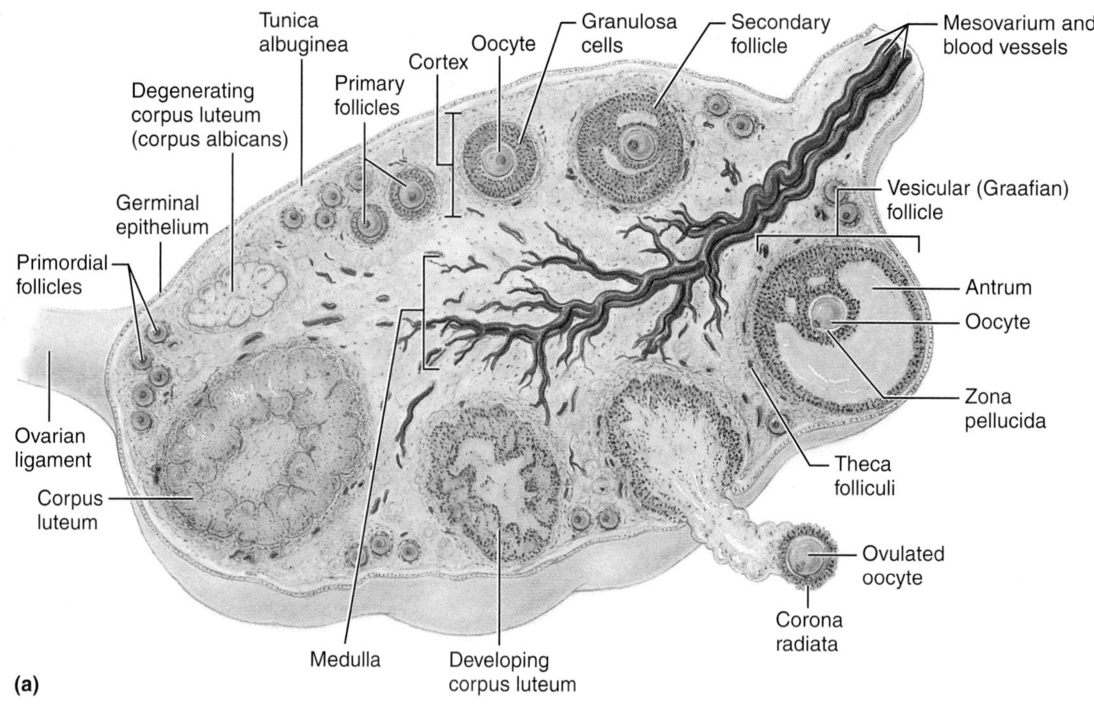

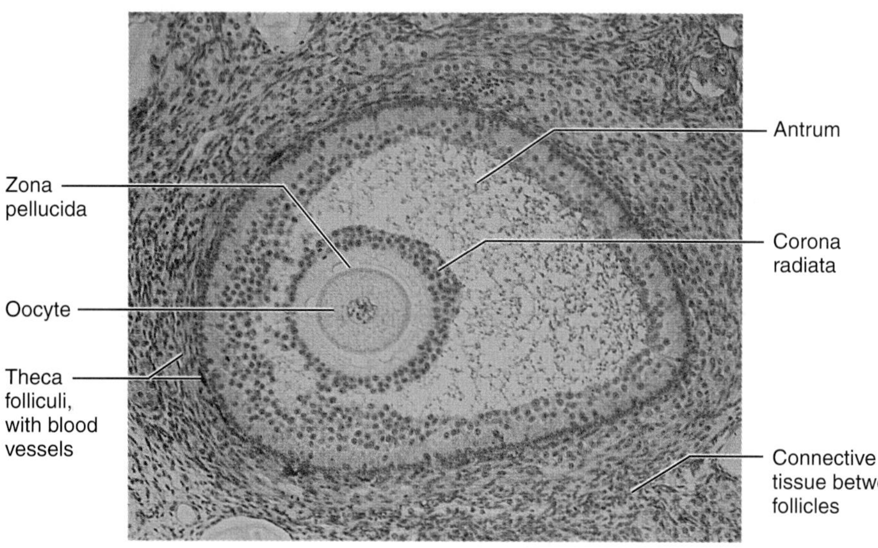

FIGURE 24.12 Structure of the ovary. (a) Frontal section through the ovary; the lateral aspect is to the right. Follicles (the red spheres) are shown in all stages of maturation. The maturing follicles do not actually travel in a circle around the ovary, as illustrated; they are drawn in this way so that one can easily follow the stages of development. **(b)** A micrograph of a nearly mature follicle (130 ×).

mesovarium. The ovaries are innervated by both divisions of the autonomic nervous system.

Figure 24.12 shows the internal structure of the ovary. The ovary is surrounded by a fibrous capsule called the **tunica albuginea,** which is much thinner than the tunica albuginea of the testis. The tunica albuginea is covered by a simple cuboidal epithelium called the **germinal epithelium,** which is a misnomer because it is simply a mesothelium that does *not* germinate the ova. The main substance of the ovary is divided into an outer *cortex* and an inner *medulla.* The **ovarian cortex** houses the developing gametes, which are called **oocytes** (o′o-sīts; "egg cells") while in the ovary. All oocytes occur within sac-like multicellular structures called **follicles** ("little bags"), which enlarge substantially as they mature. The part of the cortex between the follicles is a cell-rich connective tissue. The deep **ovarian medulla** is a loose connective tissue containing the largest blood vessels, nerves, and lymph vessels of the ovary; these vessels enter the ovary through the *hilus,* a horizontal slit in the anterior ovarian surface where the mesovarium attaches.

The Ovarian Cycle

The monthly **ovarian cycle** is the menstrual cycle as it relates to the ovary. It has three successive phases: the

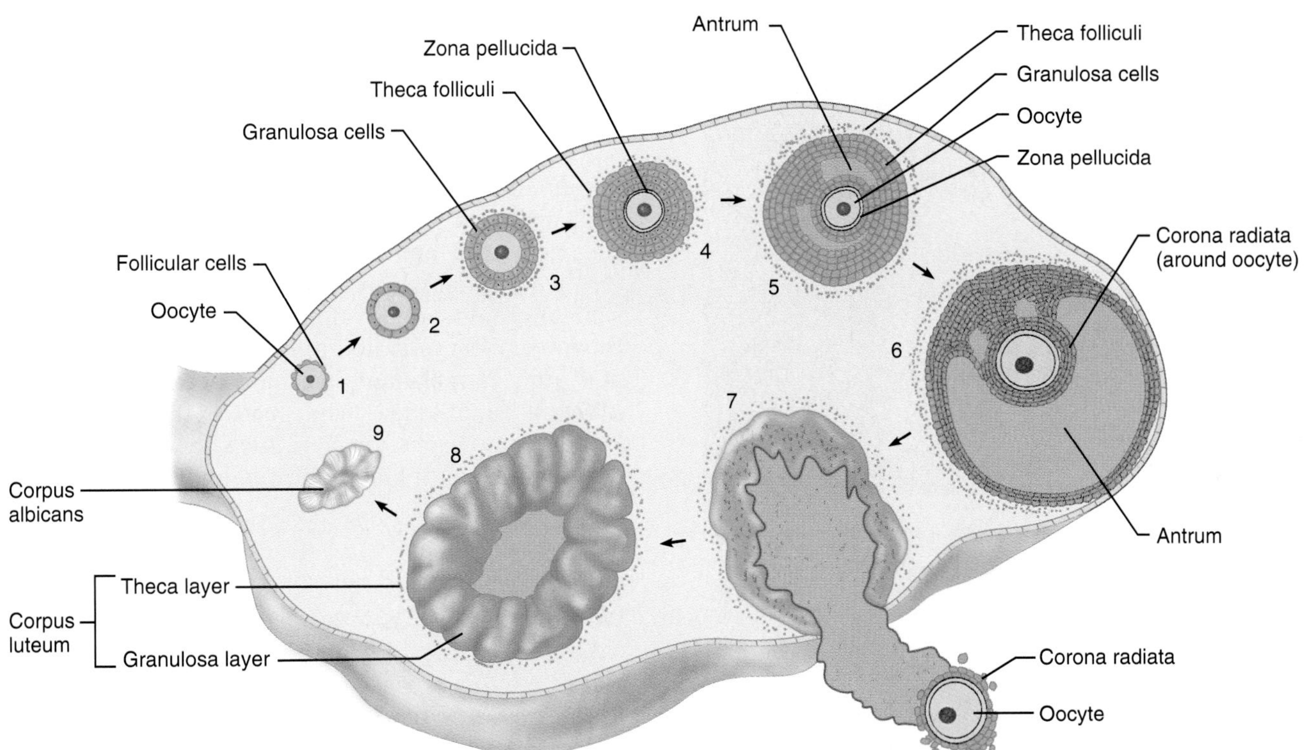

FIGURE 24.13 Schematic diagram of the ovarian cycle: Development and fate of the ovarian follicles. The three phases of this monthly cycle are the follicular phase (follicles 1–6), ovulation (follicle 7), and the luteal phase (follicle 8). (1) A primordial follicle, which is an oocyte surrounded by flat follicular cells. (2) An early primary follicle, as evidenced by its cuboidal follicular cells. (3 and 4) The primary follicle grows as its follicular cells divide to become several layers of granulosa cells. The zona pellucida and theca folliculi appear. (5) Secondary (antral) follicle, as evidenced by the appearance of a fluid-filled antrum. (6) The mature vesicular (Graafian) follicle has a huge antrum and is ready to be ovulated at mid cycle. (7) Ovulation: the follicle ruptures and expels the oocyte surrounded by its corona radiata of granulosa cells. (8) The corpus luteum, formed from the remains of the follicle, consists of theca and granulosa layers. (9) The corpus albicans (scar).

follicular phase, ovulation, and the luteal phase. While reading about this cycle, refer to Figures 24.13 and 24.19.

Follicular Phase (First Half of the Ovarian Cycle) From before birth to the end of a female's reproductive years, the ovarian cortex contains many thousands of follicles, most of which are **primordial follicles** (Figure 24.13, follicle 1). Each of these primordial follicles, from which all subsequent follicle stages arise, consists of an oocyte surrounded by a single layer of flat supportive cells called **follicular cells.** At the start of each ovarian cycle, 6–12 primordial follicles start to grow, initiating the **follicular phase** (follicles 1–6 in Figure 24.13), which lasts 2 weeks. Beyond the smallest stages, growth is stimulated by follicle-stimulating hormone (FSH) from the anterior part of the pituitary gland. Of the many follicles that grow in the follicular phase, most die and degenerate along the way, so that only one follicle each month completes the maturation process and expels its oocyte from an ovary for potential fertilization. Such expulsion, called *ovulation,* occurs just after the follicular phase ends, on about day 14 of the ovarian cycle.

We will now describe the stages of follicle growth in more detail. When a primordial follicle starts to grow, its flat follicular cells become cuboidal, and the oocyte grows larger (follicle 1 becomes follicle 2 in Figure 24.13). Now the follicle is called a **primary follicle.** Next, the follicular cells multiply to form a stratified epithelium around the oocyte (follicle 3 in the figure); from this point on, the follicular cells are called **granulosa cells.**

In the next stage (follicle 4), the oocyte develops a glycoprotein coat called the **zona pellucida** (zo′nah pĕ-lu′sid-ah; "transparent belt"), a protective shell that a sperm must ultimately penetrate to fertilize the oocyte. As the granulosa cells continue to divide, a layer of connective tissue called the **theca folliculi** (the′kah fo-lik′u-li; "box around the follicle") condenses around the exterior of the primary follicle. The external cells of the theca are spindle-shaped and resemble smooth muscle cells; the internal cells of the theca are tear-drop shaped and secrete hormones. That is, these theca cells are stimulated by luteinizing hormone (from the anterior pituitary) to secrete androgens (male sex hormones), which the nearby granulosa cells, under the influence of follicle-stimulating hormone, convert into the female sex

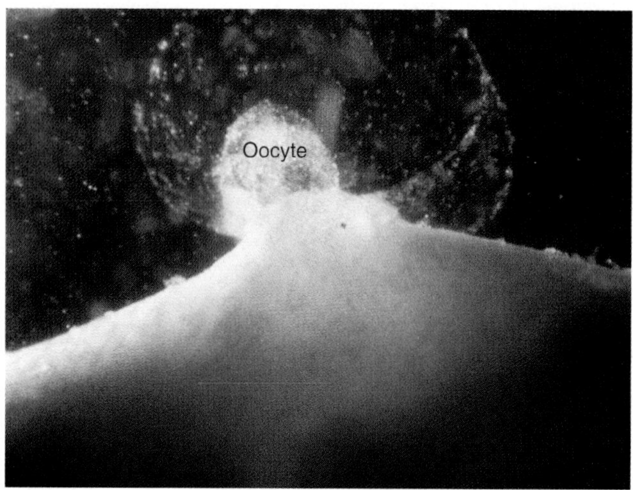

FIGURE 24.14 Ovulation. An oocyte is released from a follicle at the surface of the ovary. (The orange mass below the ejected oocyte is a part of the ovary.)

hormones, **estrogens** (es'tro-jens). Once secreted into the bloodstream, the estrogens stimulate the growth and activity of all female sex organs. Most importantly, these estrogens signal the uterine mucosa to repair itself after each menstrual period (see p. 723).

In the next stage of follicle development (follicle 5), a clear liquid gathers between the granulosa cells and coalesces to form a fluid-filled cavity called the **antrum** ("cave"); the follicle is now a **secondary (antral) follicle.** The antrum expands with fluid until it isolates the oocyte, along with a surrounding coat of granulosa cells called a **corona radiata** (ko-ro'nah ra-de-ah'tah; "radiating crown"), on a stalk at the periphery of the follicle (see Figure 24.12b). Finally, one follicle attains full size (follicle 6 in Figure 24.13), reaching a diameter of 2 cm (almost 1 inch)! This mature follicle, the **vesicular (Graafian) follicle,** is ready to be ovulated.

Ovulation (Midpoint of the Ovarian Cycle) Ovulation occurs about halfway through each ovarian cycle (see follicle 7 in Figure 24.13, and Figure 24.14). In this process, one oocyte exits from one of the woman's two ovaries, enters the peritoneal cavity, and is swept into a uterine tube. The signal for ovulation is the sudden release of a large quantity of luteinizing hormone (LH) from the pituitary gland, just before day 14. In the process of ovulation, the ovarian wall over the follicle bulges, thins, and oozes fluid; this wall then ruptures and the oocyte exits, surrounded by its corona radiata. The forces responsible for this process are not fully understood, but they probably involve an enzymatic breakdown of the follicle wall, followed by a contraction of the muscle-like cells of the external layer of the theca that retracts the follicle and leaves the oocyte outside the ovary.

Luteal Phase (Second Half of the Ovarian Cycle) After ovulation, the part of the follicle that stays in the ovary collapses, and its wall is thrown into wavy folds (follicle 8 in Figure 24.13). This structure, now called the **corpus luteum** ("yellow body"), consists of the remaining granulosa and theca layers.

The corpus luteum is not a degenerative structure but an endocrine gland that persists through the second half, or **luteal phase,** of each ovarian cycle. It secretes estrogens and another hormone called **progesterone** (pro-jes'tĕ-rōn), which acts on the mucosa of the uterus, signaling it to prepare for implantation of an embryo. However, if there is no implantation, the corpus luteum dies after 2 weeks and becomes a scar called a **corpus albicans** (al'bĭ-kans; "white body") (follicle 9 in Figure 24.13). The corpus albicans stays in the ovary for several months, shrinking until it is finally phagocytized by macrophages.

Oogenesis

We now consider how ova are produced in a process called **oogenesis** (o″o-jen'ĕ-sis; "egg generation"), and will relate this process both to a female's life stages and to the events of the ovarian cycle (Figure 24.15). Oogenesis, like spermatogenesis, includes the chromosome-reduction divisions of meiosis ($2n$ to n: see p. 705). Oogenesis takes many years to complete. First, in the fetal period, stem cells called **oogonia** give rise to the female's lifelong supply of oocytes, which are arrested around the time of birth in an early stage of meiosis I. Now called **primary oocytes,** they remain "stalled" in meiosis I for decades, until they are ovulated from their follicle. Only in response to the influence of the LH surge that signals ovulation does a primary oocyte finish meiosis I and enter meiosis II as a **secondary oocyte**—but then it arrests again and does not finish meiosis II until a sperm penetrates its plasma membrane. Only after the completion of meiosis II is the egg technically called an **ovum.**

As you study Figure 24.15, note that oogenesis typically produces four daughter cells: the large ovum and three smaller cells called **polar bodies.** The polar bodies degenerate quickly, without being fertilized or contributing to the developing embryo.

In the next three sections we discuss the accessory ducts of the female reproductive system: the uterine tubes, the uterus, and the vagina. First we examine the uterine tubes.

The Uterine Tubes

The **uterine** (u'ter-in) **tubes,** also called **fallopian tubes** or **oviducts** (see Figure 24.10a), receive the ovulated oocyte and provide a site for fertilization. Each uterine tube begins laterally near an ovary and ends medially, where it empties into the superior part of the uterus. The lateral region of the uterine tube is an open funnel called the **infundibulum** (in″fun-dib'u-lum; "funnel"), the margin of which is surrounded by ciliated, finger-like projections called **fimbriae** (fim'bre-e; "fibers, a

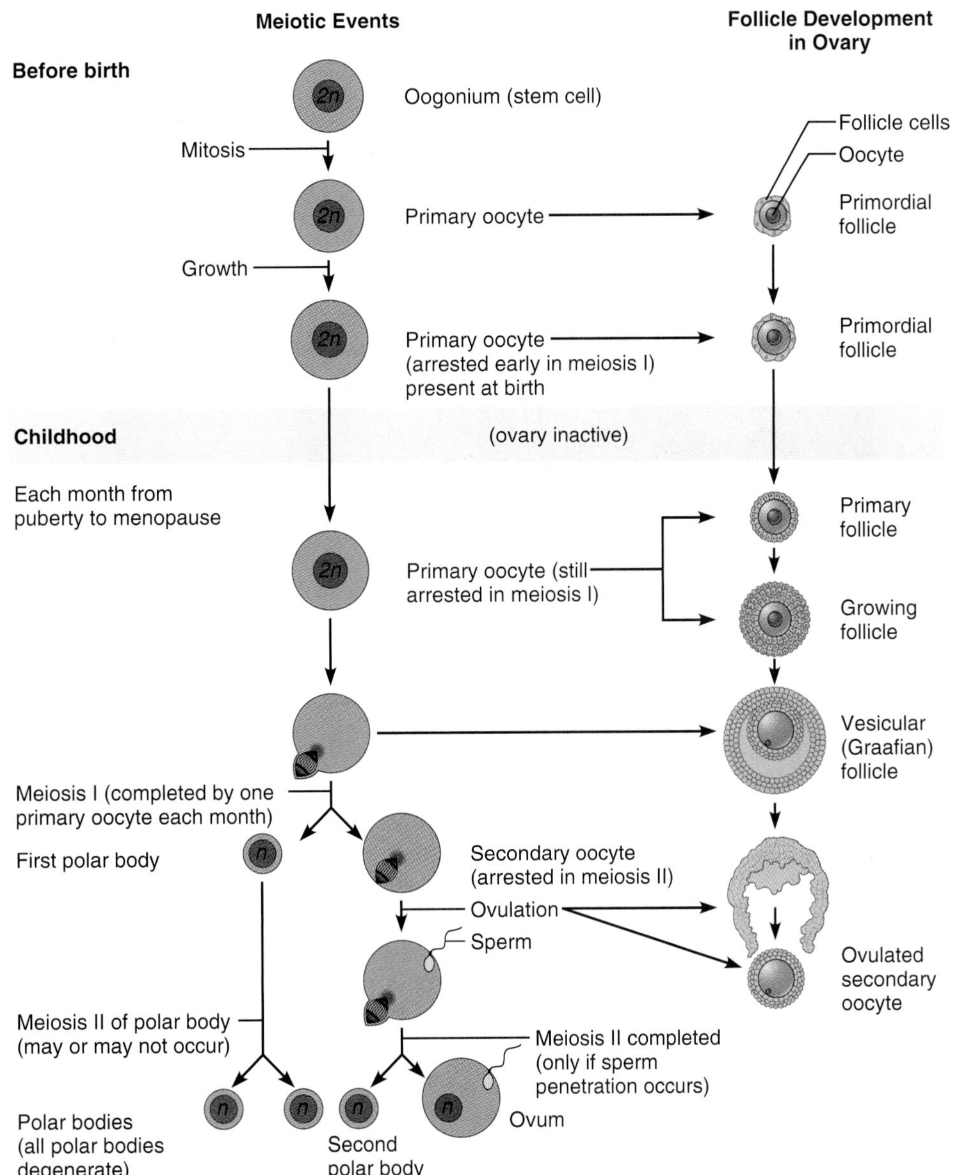

FIGURE 24.15 Oogenesis. Left, flowchart of the events of meiosis. Right, correlation with follicle development and ovulation in the ovary. As in Figure 24.4, $2n$ signifies the diploid state, n the haploid state.

fringe") that drape over the ovary. Medial to the infundibulum is the expanded **ampulla** ("flask"), which forms half the length of the uterine tube and is the site where fertilization usually occurs. Finally, the medial third of the uterine tube is the **isthmus** (is′mus; "a narrow passage").

Unlike the reproductive ducts of males, which are directly continuous with the tubules of the testes, the uterine tubes have little or no direct contact with the ovaries. Instead, each ovulated oocyte is released into the peritoneal cavity of the pelvis (where many oocytes are lost), and the uterine tube performs a complex series of movements to capture the oocyte: The infundibulum bends to cover the ovary while the fimbriae stiffen and sweep the ovarian surface; then the beating cilia on the fimbriae generate currents in the peritoneal fluid that carry the oocyte into the uterine tube, where it begins its journey toward the uterus.

The oocyte's journey is aided by peristaltic waves generated by sheets of smooth muscle in the uterine-tube wall, and by the beating cilia of the uterine tube's simple columnar ciliated epithelium. Alternating with the ciliated cells in this lining epithelium are tall, nonciliated cells that secrete substances that nourish the oocyte (or embryo) and facilitate fertilization.

Externally, the uterine tube is covered by visceral peritoneum and supported by a short mesentery called the **mesosalpinx** (mez″o-sal′pinks), a part of the broad ligament (see Figure 24.10a). (*Mesosalpinx* means "mesentery of the trumpet," a reference to the uterine

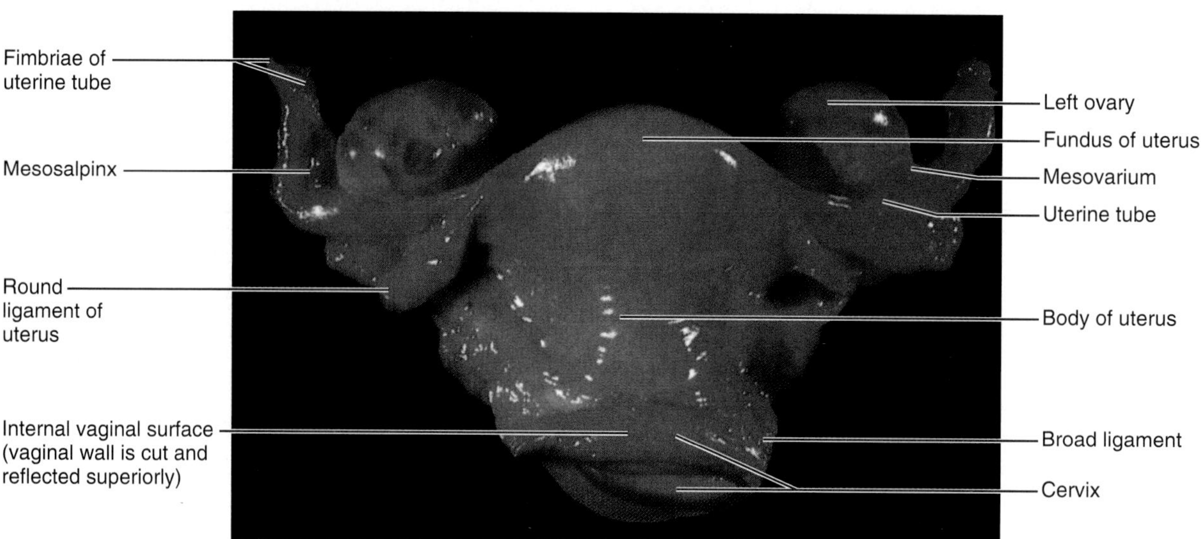

FIGURE 24.16 Anterior view of the internal reproductive organs from a female cadaver. Compare this to the posterior view of the same organs in Figure 24.10a.

tube's trumpet shape; the root word *salpinx* is the basis for the terms **salpingectomy,** the surgical removal of a uterine tube, and **salpingitis,** inflammation of a uterine tube.)

CLINICAL APPLICATIONS

PELVIC INFLAMMATORY DISEASE Widespread infection that originates in the vagina and uterus and spreads to the uterine tubes, ovaries, and ultimately the pelvic peritoneum is called **pelvic inflammatory disease (PID).** This condition, which occurs in about 10% of women in the United States, is usually caused by chlamydia or gonorrhea infection (see this chapter's A Closer Look box on p. 712), but without exception other bacteria infecting the vagina are involved as well. Signs and symptoms include tenderness of the lower abdomen, fever, and a vaginal discharge. Even a single episode of PID causes infertility (due to scarring that blocks the uterine tubes) in 10% to 25% of cases unless halted within a few days. Therefore, patients are immediately given broad-spectrum antibiotics whenever PID is suspected. ✚

The uterine tube is the site of 90% of all *ectopic pregnancies.* As explained in Chapter 3 (see p. 72), ectopic pregnancy is the implantation of an embryo outside the uterus. Implantations in the uterine tube often result in rupture and hemorrhaging that threaten a woman's life.

The Uterus

The **uterus** (womb) lies in the pelvic cavity, anterior to the rectum and posterosuperior to the bladder (see Figure 24.10b). It is a hollow, thick-walled organ whose functions are to receive, retain, and nourish a fertilized egg throughout pregnancy. In a woman who has never been pregnant, the uterus is about the size and shape of a small, inverted pear, but it is somewhat larger in women who have had children. Normally, the uterus is tilted anteriorly, or **anteverted,** at the superior part of the vagina. However, in older women it is often inclined posteriorly, or **retroverted.** The major portion of the uterus is called the **body** (see Figures 24.10a and 24.16). The rounded region superior to the entrance of the uterine tubes is the **fundus,** and the slightly narrowed region inferior to the body is the **isthmus.** Below this, the narrow neck of the uterus is the **cervix,** the inferior tip of which projects into the vagina. Containing much collagen, the cervix forms a tough, fibrous ring that keeps the uterus closed and the fetus within it during pregnancy.

The central lumen of the uterus is quite small (except in pregnancy). It is divided into the *cavity of the body* and the **cervical canal,** which communicates with the vagina inferiorly via the **external os** (*os* = mouth) and with the cavity of the body superiorly via the **internal os.** The mucosal lining of the cervical canal contains *cervical glands* that secrete a mucus that fills the cervical canal and covers the external os, presumably to block the spread of bacteria from the vagina into the uterus. Cervical mucus also blocks the entry of sperm—except at mid cycle, when it becomes less viscous and allows the sperm to pass through.

Supports of the Uterus

Several ligaments and mesenteries help hold the uterus in place (see Figure 24.10a). The uterus is anchored to the lateral pelvic walls by the **mesometrium** (mez″o-me′tre-um; "mesentery of the uterus"), which is the largest division of the broad ligament. Inferiorly, the **lateral cervical (cardinal) ligaments** run horizontally from the uterine cervix and superior vagina to the lat-

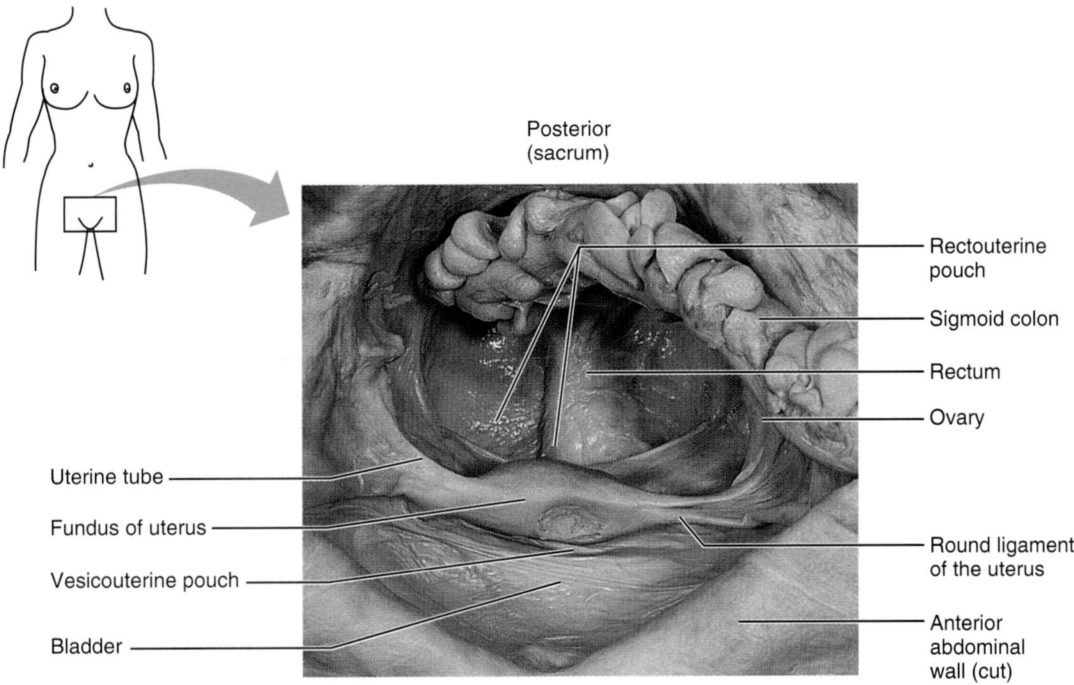

FIGURE 24.17 The female pelvic organs in anterior view. Note the uterus and the vesicouterine and rectouterine pouches of the peritoneal cavity. *Males* have a pouch equivalent to the rectouterine pouch, but because males have no uterus, their corresponding pouch is called the rectovesical pouch (*vesical* = bladder).

eral pelvic walls. These "ligaments" are thickenings of the fascia of the pelvis. The uterus is bound to the anterior body wall by the paired **round ligaments of the uterus,** each of which starts as a continuation of the ovarian ligament on the superolateral aspect of the uterus, descends through the mesometrium and inguinal canal, and anchors in one of the labia majora of the external genitalia.

Despite the presence of these ligaments, most support of the uterus is provided by the muscles of the pelvic floor—namely, the muscles of the urogenital and pelvic diaphragms (see pp. 292–293). If these muscles are torn during childbirth, the unsupported uterus may sink inferiorly, until the tip of the cervix protrudes through the external vaginal opening—a condition called **prolapse of the uterus.**

The undulating course of the peritoneum around and over the various pelvic organs produces several cul-de-sacs, or blind-ended peritoneal pouches (see Figures 24.10b and 24.17). The *vesicouterine pouch* (ves″ĭ-ko-u′ter-in; "bladder to uterus") lies between the uterus and bladder, and the *rectouterine pouch* lies between the uterus and rectum. Because the rectouterine pouch forms the most inferior part of the peritoneal cavity, pus from intraperitoneal infections and blood from internal wounds accumulate there.

The Uterine Wall

The wall of the uterus is composed of three basic layers (see Figure 24.10a): an outer *perimetrium* (per″ĭ-me′tre-um; "around the uterus"), a middle *myometrium* (mi″o-me′tre-um; "muscle of the uterus"), and an inner *endometrium* ("within the uterus"). The **perimetrium,** the outer serous membrane, is the visceral peritoneum. The **myometrium,** the bulky middle layer, consists of interlacing bundles of smooth muscle that contract during childbirth to expel the baby from the mother's body. The **endometrium,** the mucosal lining of the uterine cavity (Figure 24.18), consists of a simple columnar epithelium containing secretory and ciliated cells, underlain by a lamina propria connective tissue. If fertilization occurs, the embryo burrows into the endometrium and resides there for the rest of its development. The endometrium has two chief layers: *stratum functionalis* and *stratum basalis* (see Figure 24.18). The thick, inner **stratum functionalis,** or **functional layer,** undergoes cyclic changes in response to varying levels

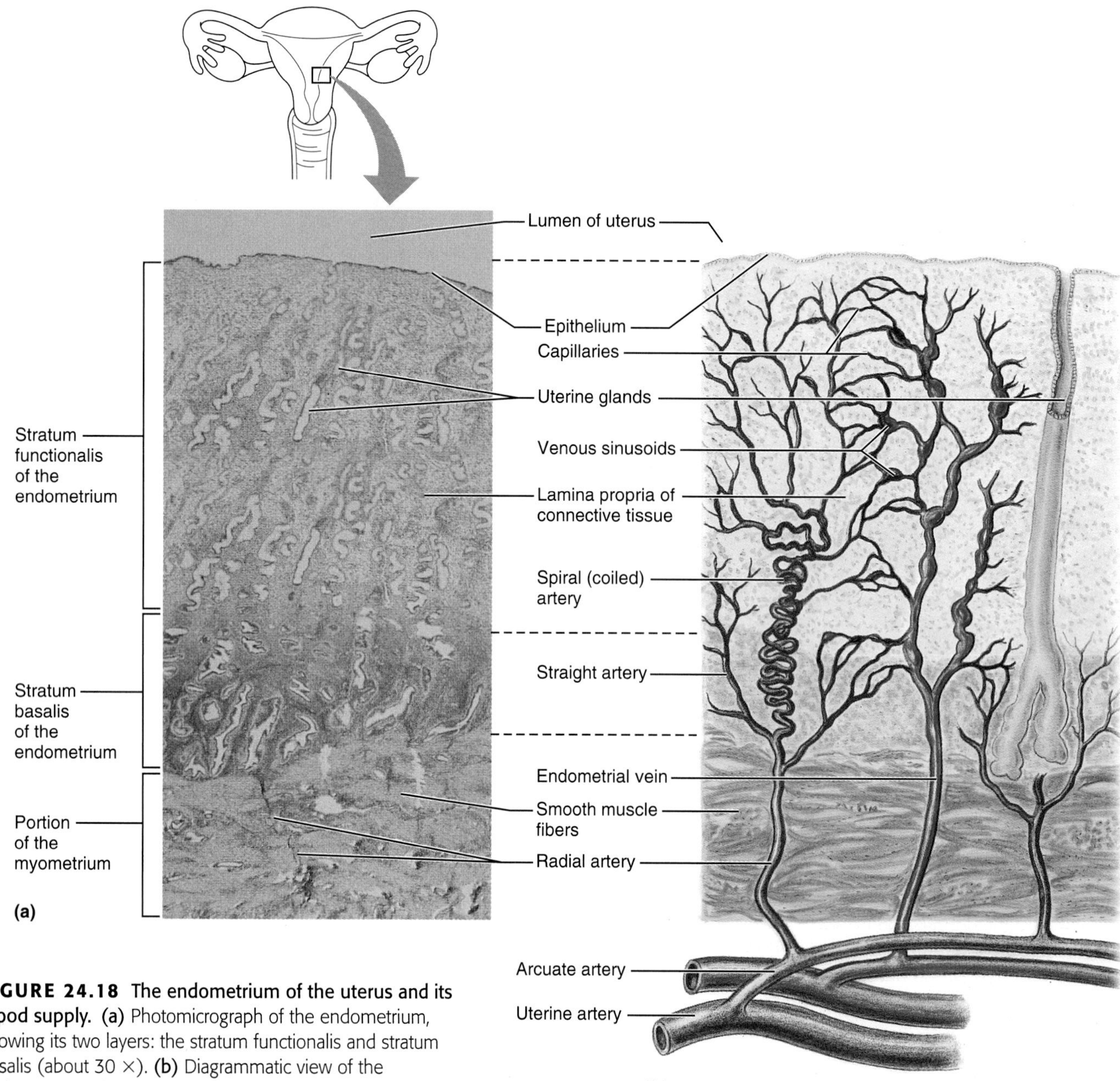

FIGURE 24.18 The endometrium of the uterus and its blood supply. **(a)** Photomicrograph of the endometrium, showing its two layers: the stratum functionalis and stratum basalis (about 30 ×). **(b)** Diagrammatic view of the endometrium, showing the major blood vessels in the uterus.

of ovarian hormones in the blood and is shed during menstruation (about every 28 days). The thin **stratum basalis,** or **basal layer,** is not shed and is responsible for forming a new functional layer after menstruation ends. The endometrium contains straight tubular **uterine glands** that change in length as the endometrium thickens and thins.

To understand the changes in the endometrium during the uterine cycle (discussed next), it is essential to understand the blood supply of the uterus. The **uterine arteries,** which arise from the internal iliac arteries in the pelvis, ascend along the sides of the uterine body, send branches into the uterine wall, and divide into **arcuate arteries** that course through the myometrium (see Figures 24.11 and 24.18b). **Radial arteries** then reach the endometrium, where they give off **straight arteries** *(basal arteries)* to the stratum basalis and **spiral (coiled) arteries** to the stratum functionalis. The spiral arteries undergo degeneration and regeneration during each successive menstrual cycle, and they undergo spasms that cause the functional layer to shed during menstruation. Veins in the endometrium are thin-walled and form an extensive network, with occasional sinusoidal enlargements.

The Uterine Cycle

The **uterine cycle** is the menstrual cycle as it involves the endometrium (see Figure 24.19). Specifically, it is a series of cyclic phases that the endometrium undergoes

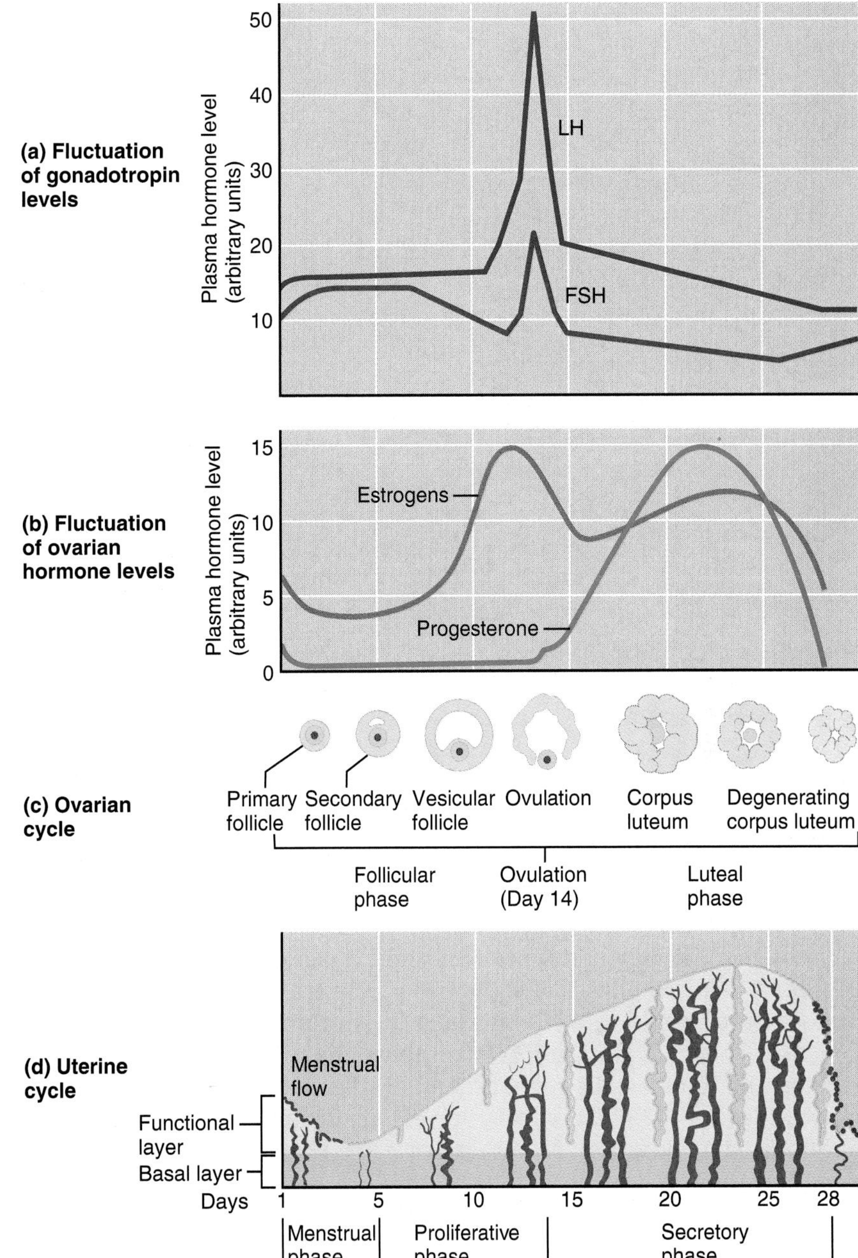

FIGURE 24.19 The menstrual cycle: Structural and hormonal changes. The time bar near the bottom of the figure, reading "Days 1 to 28," applies to all four parts of this figure. **(a)** The fluctuating levels of the pituitary gonadotropin hormones FSH and LH in the blood (FSH = follicle-stimulating hormone; LH = luteinizing hormone). These hormones regulate the events of the ovarian cycle. **(b)** The fluctuating blood levels of the ovarian hormones, estrogens and progesterone, which regulate the endometrial changes of the uterine cycle. **(c)** Changes in the ovarian follicles during the ovarian cycle. **(d)** Changes in the endometrium during the uterine cycle.

The phases of the *ovarian cycle,* depicted in (c), are as follows:

1. Days 1–14: Follicular phase. The ovarian follicles grow under the influence of FSH from the pituitary; see (a). Also during this phase, LH and FSH signal the follicles to secrete increasing amounts of estrogens into the blood; see (b). Circulating estrogens signal the pituitary to secrete more LH, until LH secretion surges; see (a).

2. Day 14: Ovulatory phase. Ovulation is signaled by the surge of LH from the pituitary, and then the corpus luteum forms.

3. Days 14–28: Luteal phase. The corpus luteum secretes progesterone and estrogens into the blood. High levels of progesterone inhibit the secretion of LH and FSH by the pituitary; see (a). Without LH to sustain it, the corpus luteum dies at the end of the luteal phase, and a new ovarian cycle begins (unless an embryo implants in the uterus).

The phases of the *uterine cycle,* depicted in (d), are as follows:

1. Days 1–5: Menstrual phase. The uterus sheds all but the deepest part of its endometrium: The thick functional layer detaches, a process that is accompanied by bleeding for 3–5 days. The detached tissue and blood pass out the vagina as the menstrual flow. By day 5, the growing ovarian follicles are starting to produce more estrogens: see (b).

2. Days 6–14: Proliferative phase. The endometrium rebuilds itself. Under the influence of rising blood levels of estrogens, the basal layer of the endometrium generates a new functional layer. This new layer thickens, its glands enlarge, and its spiral arteries increase in number. Consequently, the endometrium once again becomes thick and well vascularized. All the endometrial changes in the proliferative phase are signaled by estrogens. At the end of this phase (day 14), ovulation occurs in the ovary, forming the corpus luteum; see (c).

3. Days 15–28: Secretory phase. The endometrium prepares for implantation of an embryo. Rising levels of progesterone from the corpus luteum act on the estrogen-primed endometrium, causing the spiral arteries to elongate and coil more tightly and converting the functional layer to a secretory mucosa. The uterine glands enlarge, coil, and secrete glycoproteins into the uterine cavity. These nutrients sustain the embryo until it has implanted in the blood-rich endometrium. All events of the secretory phase are signaled by progesterone.

If fertilization and implantation occur, the endometrium stays in this secretory phase throughout pregnancy. However, if no fertilization occurs, the degenerating corpus luteum stops secreting progesterone. Blood progesterone levels fall, depriving the endometrium of hormonal support. Cells in the functional layer leak lysosomal enzymes (which initiate tissue necrosis) and release prostaglandins (which signal the spiral arteries to kink and constrict spasmodically). Denied blood-borne oxygen, the endometrial cells die and release more lysosomal enzymes, until the functional layer begins to "self-digest," setting the stage for menstruation to begin on day 28. The spiral arteries constrict one final time then suddenly open wide. As blood gushes into the weakened capillaries, they fragment, causing the functional layer to slough off. Thus the menstrual cycle starts over again on this first day of menstrual flow.

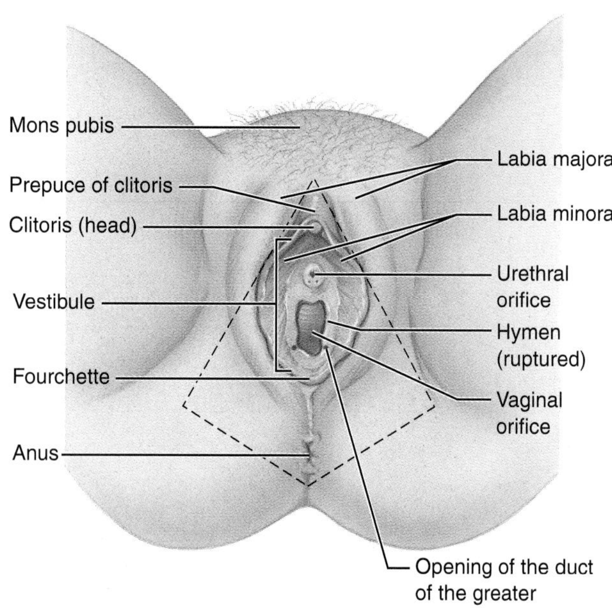

FIGURE 24.20 The external genitalia (vulva) of the female: superficial structures. The diamond-shaped region enclosed by the dashed lines is the perineum.

month after month as it responds to changing levels of ovarian hormones in the blood (see Figure 24.19b). These endometrial phases are closely coordinated with the phases of the ovarian cycle, which in turn are dictated by the pituitary hormones FSH and LH (see Figure 24.19a). The three phases of the uterine cycle are (1) the **menstrual phase** (days 1–5), in which the stratum functionalis is shed; (2) the **proliferative phase** (days 6–14), in which the functionalis rebuilds; and (3) the **secretory phase** (days 15–28), in which the endometrium prepares for implantation of an embryo. Figure 24.19 describes these phases in detail.

CLINICAL APPLICATIONS

ENDOMETRIOSIS Endometriosis (en″do-me″tre-o′sis) is a condition in which fragments of endometrial tissue are present in the uterine tubes, on the ovary, and in the peritoneum of the pelvic cavity. The most widely favored explanation for this out-of-place endometrium is that a reflux of menstrual fluid spreads endometrial cells from the uterus through the uterine tubes. In endometriosis, the pain associated with menstruation is extreme because the endometrial fragments also bleed, and this blood accumulates in the pelvic cavity, forms cysts, and exerts pressure. Present in up to 10% of women of reproductive age, endometriosis causes one-third of all cases of infertility in females, as ectopic endometrial tissue covers the ovaries or blocks the uterine tubes. Treatment involves drugs that halt the secretion of estrogens and suppress menstruation, use of lasers to vaporize patches of ectopic endometrium, and hysterectomy. The environmental pollutants dioxin and PCBs may be implicated in the condition's increasing incidence.

The Vagina

The vagina ("sheath") is a thin-walled tube that lies inferior to the uterus, anterior to the rectum, and posterior to the urethra and bladder (see Figure 24.10b). The vagina is often called the *birth canal,* as it provides a passageway for delivery of an infant. It also receives the penis and semen during sexual intercourse.

The highly distensible wall of the vagina consists of three coats: an outer *adventitia* of fibrous connective tissue, a *muscularis* of smooth muscle, and an inner *mucosa,* which is marked by transverse folds (rugae) (see Figure 24.10a). These ridges stimulate the penis during intercourse, and they flatten as the vagina expands during childbirth. The mucosa consists of a lamina propria, which contains elastic fibers that help the vagina return to its original shape after expanding, and a stratified squamous epithelium that can withstand the friction of intercourse and resist bacterial infection. Additionally, glycogen from epithelial cells shed into the lumen is fermented by beneficial resident bacteria into lactic acid; the resulting acidity discourages the growth of harmful bacteria, but it is also hostile to sperm.

Near the vagina's external opening, the **vaginal orifice,** the mucosa elaborates to form an incomplete diaphragm called the **hymen** (hi′men; "membrane") (Figure 24.20). The hymen is vascular and tends to bleed when ruptured during the first sexual intercourse. However, its durability varies: In some females it is delicate and ruptures during a sports activity, the insertion of a tampon, or a pelvic examination; occasionally the hymen is so tough that it must be breached surgically if intercourse is to occur.

The recess formed where the widened superior part of the vagina encircles the tip of the cervix is called the **fornix** (see Figure 24.10a). The posterior part of this recess, the posterior fornix, is much deeper than either the lateral or the anterior fornices (see Figure 24.10b).

Critical Reasoning

Using Figure 24.10b for reference, deduce how a physician could remove fluid (such as blood or pus) that has accumulated in the peritoneal cavity without subjecting the patient to the trauma of surgically opening the abdominal wall.

To drain such fluids from the peritoneal cavity, a physician could insert a hypodermic needle superiorly through the vagina and the posterior fornix and into the rectouterine pouch.

The External Genitalia and Female Perineum

The female reproductive structures that lie external to the vagina are the **external genitalia,** also called the **vulva** ("covering") or **pudendum** (pyoo-den′-dum) (see Figure 24.20). These structures include the mons pubis, the labia, the clitoris, and structures associated with the vestibule.

The **mons pubis** (mons pu'bis; "mountain on the pubis") is a fatty, rounded pad overlying the pubic symphysis. After puberty, pubic hair covers the skin of this area. Extending posteriorly from the mons are two long, hair-covered, fatty skin folds, the **labia majora** (la'be-ah mah-jor'ah; "larger lips"), which are the female counterpart, or *homologue,* of the male scrotum; that is, they derive from the same embryonic structure. The labia majora enclose two thin, hairless folds of skin called the **labia minora** (mi-nor'ah; "smaller lips"), which enclose a recess called the **vestibule** ("entrance hall") housing the external openings of the urethra and vagina. In the vestibule, the vaginal orifice lies posterior to the urethral orifice. Flanking the vaginal orifice are the paired, pea-sized *greater vestibular glands* (Figure 24.21), which lie deep to the posterior part of the labia on each side and secrete lubricating mucus into the vaginal orifice during sexual arousal, facilitating entry of the penis. At the extreme posterior point of the vestibule, the right and left labia minora come together to form a ridge called the *frenulum of the labia* ("little bridle of the lips") or the **fourchette** (foor-shet').

Just anterior to the vestibule is the **clitoris** (klit'o-ris; "hill"), a protruding structure composed largely of erectile tissue that is sensitive to touch and swells with blood during sexual stimulation. The clitoris is homologous to the penis, having both a glans and a body (although there is no urethra within it). It is hooded by a fold of skin, the **prepuce of the clitoris,** formed by the anterior junction of the two labia minora (Figure 24.20). As in the penis, the body of the clitoris contains paired **corpora cavernosa,** which continue posteriorly into **crura** that extend along the bony pubic arch (Figure 24.21). Unlike the penis, however, the clitoris contains no corpus spongiosum. During sexual stimulation, the homologous **bulbs of the vestibule,** which lie along each side of the vaginal orifice and directly deep to the bulbospongiosus muscle, engorge with blood and may help grip the penis within the vagina.

Advances in Understanding Recent dissections reveal that the bulbs of the vestibule and the base of the clitoris are larger than was previously recognized, large enough to squeeze the urethral orifice shut during intercourse. This closes off the urethra, perhaps to prevent bacteria from entering it and traveling up to infect the bladder. ✷

The female **perineum** is a diamond-shaped region between the pubic arch anteriorly, the coccyx posteriorly, and the ischial tuberosities laterally (see Figure 24.20). In the exact center of the perineum, just posterior to the fourchette, lies the *central tendon* or *perineal body* (see Figure 24.21). This knob is the insertion tendon of most muscles that support the pelvic floor.

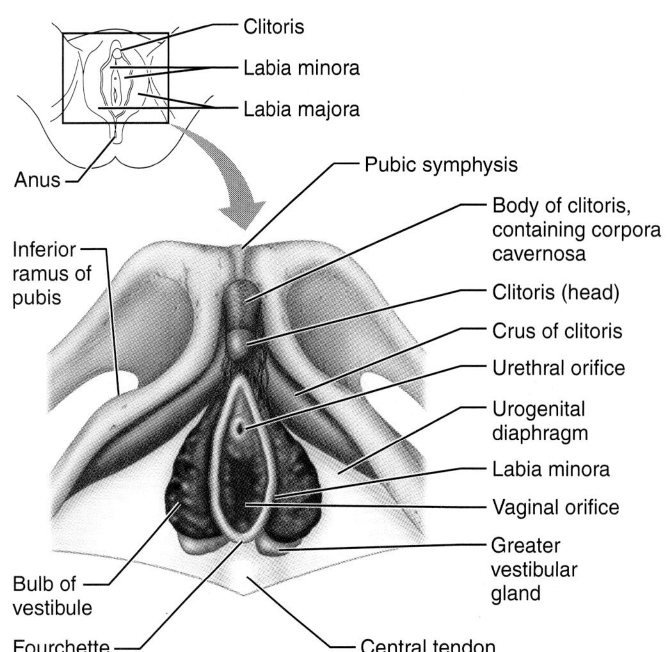

FIGURE 24.21 Deep structures of the external genitalia and perineum of the female. The labia majora and associated skin have been removed to show the underlying erectile bodies. For the muscles that cover each crus of the clitoris and bulb of the vestibule, see Figure 11.12.

CLINICAL APPLICATIONS

EPISIOTOMY Due to its proximity to the vagina, the central tendon is sometimes torn by an infant's head during childbirth. The resulting jagged tear heals poorly, allowing the pelvic organs to sink inferiorly and the uterus to prolapse. To avoid these problems, an **episiotomy** (e-piz"e-ot'o-me; "cutting of the vulva") is performed in 50% to 80% of deliveries in the United States. In this procedure, the vaginal orifice is widened by a posterior cut through the fourchette at the time the baby's head appears at the vestibule. The posterior incision either passes straight through the central tendon or just lateral to it. After the birth, this clean incision is stitched together and allowed to heal. Episiotomies have generated much controversy, as opponents say they are too often performed unnecessarily, they are painful while healing, and that rigorous studies are needed to determine if they are as effective as claimed. ✚

The Mammary Glands

Mammary glands (breasts) are modified sweat glands that are present in both sexes but function only in lactating females (Figure 24.22) when they produce milk to nourish an infant.

Embryonically, the mammary glands form as part of the skin, arising along bilateral lines that run between the axilla and groin on the lateral trunk of the embryo. These embryonic *milk lines* persists only in the mid thorax, where the definitive breasts form. About 1 in 500 people develops extra (accessory) nipples or breasts, which can occur anywhere along the milk line.

The base of the cone-shaped breast in females extends from the second rib superiorly to the sixth rib inferiorly. Its medial border is the sternum, and its lateral boundary is the midaxillary line, a line extended

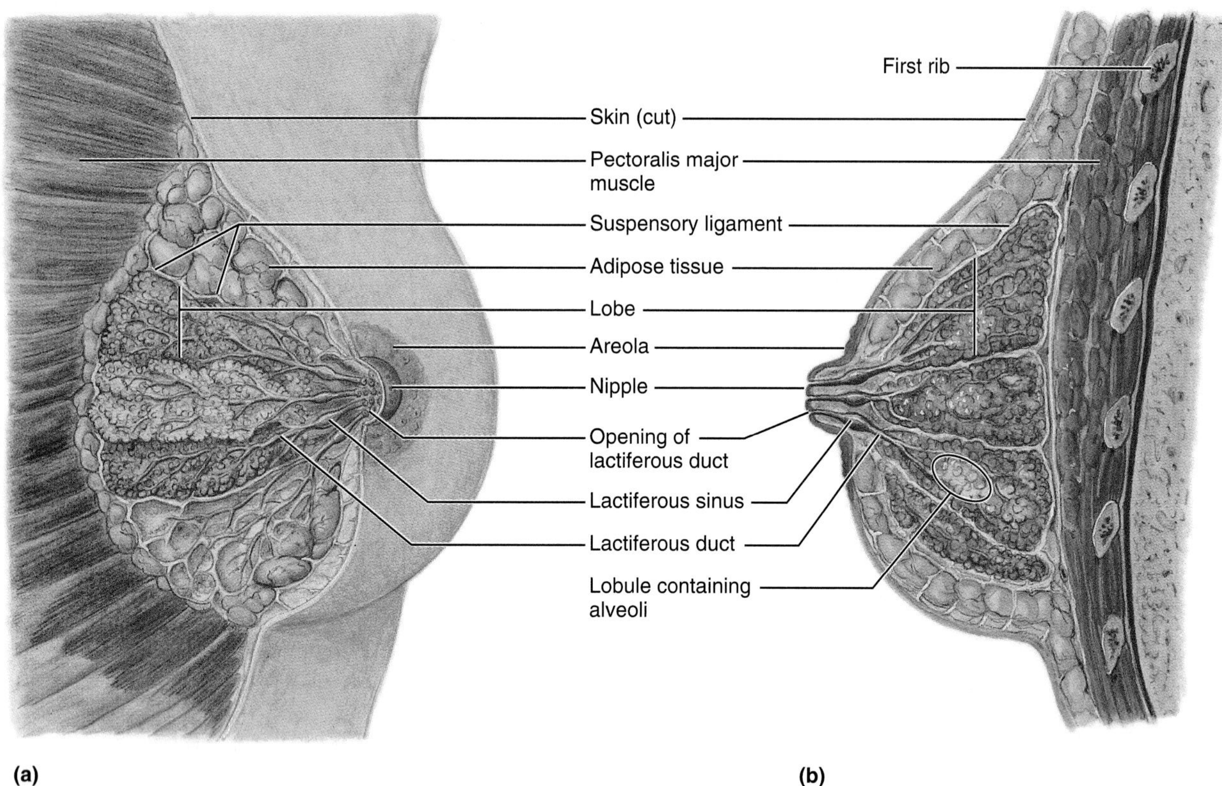

FIGURE 24.22 Structure of a lactating (milk-secreting) mammary gland. **(a)** Anterior view of a partly dissected breast. **(b)** Sagittal section of a breast.

straight inferiorly from the middle of the axilla. The muscles deep to the breast are the pectoralis major and minor and parts of the serratus anterior and external oblique. The main arteries to the breast are the lateral thoracic artery and cutaneous branches of the internal thoracic and posterior intercostals (see Figure 19.10 on p. 558). The lymph vessels draining the breast drain into *parasternal lymph nodes* (nodes along the internal thoracic arteries) and into the *axillary lymph nodes* in the armpit.

The **nipple,** the central protruding area from which an infant sucks milk, is surrounded by a ring of pigmented skin, the **areola** (ah-re′o-lah; "a small area"). During nursing, large sebaceous glands in the areola produce an oily sebum that minimizes chapping and cracking of the skin of the nipple.

Internally, the mammary gland consists of 15 to 25 **lobes** (see Figure 24.22), each of which is a distinct compound alveolar gland opening at the nipple. The lobes are separated from one another by a large amount of adipose tissue and strips of interlobar connective tissue that form **suspensory ligaments of the breasts,** which run from the underlying skeletal muscles to the overlying dermis and provide, as their name implies, support for the breasts. The lobes of the breast consist of smaller units called **lobules,** which are clusters of tiny alveoli or acini. (A bunch of grapes is a good analogy for a lobule.) The walls of the alveoli consist of a simple cuboidal epithelium of milk-secreting cells. From the alveoli, the milk passes through progressively larger ducts until it reaches the largest ducts, called **lactiferous ducts** (*lactiferous* = milk-carrying), which lie within and deep to the nipple. Just deep to the areola, each lactiferous duct has a dilated region called a **lactiferous sinus,** where milk accumulates during nursing.

The glandular structure of the breast is undeveloped in nonpregnant women. During childhood, a girl's breasts consist only of rudimentary ducts, as do the breasts of males throughout life. When a girl reaches puberty, these ducts grow and branch, but lobules and alveoli are still absent. (Breast enlargement at puberty is due to fat deposition.) The glandular alveoli finally form—from the smallest ducts—about halfway through pregnancy; and milk production starts a few days after childbirth.

Breast cancer, which usually arises from the duct system and is a leading cause of cancer deaths in women, is discussed in depth on p. 734.

PREGNANCY AND CHILDBIRTH

In this section we consider the events that occur in the female reproductive tract during pregnancy. Here we focus more on the mother than on her embryo and fetus, as embryology is discussed in Chapter 3.

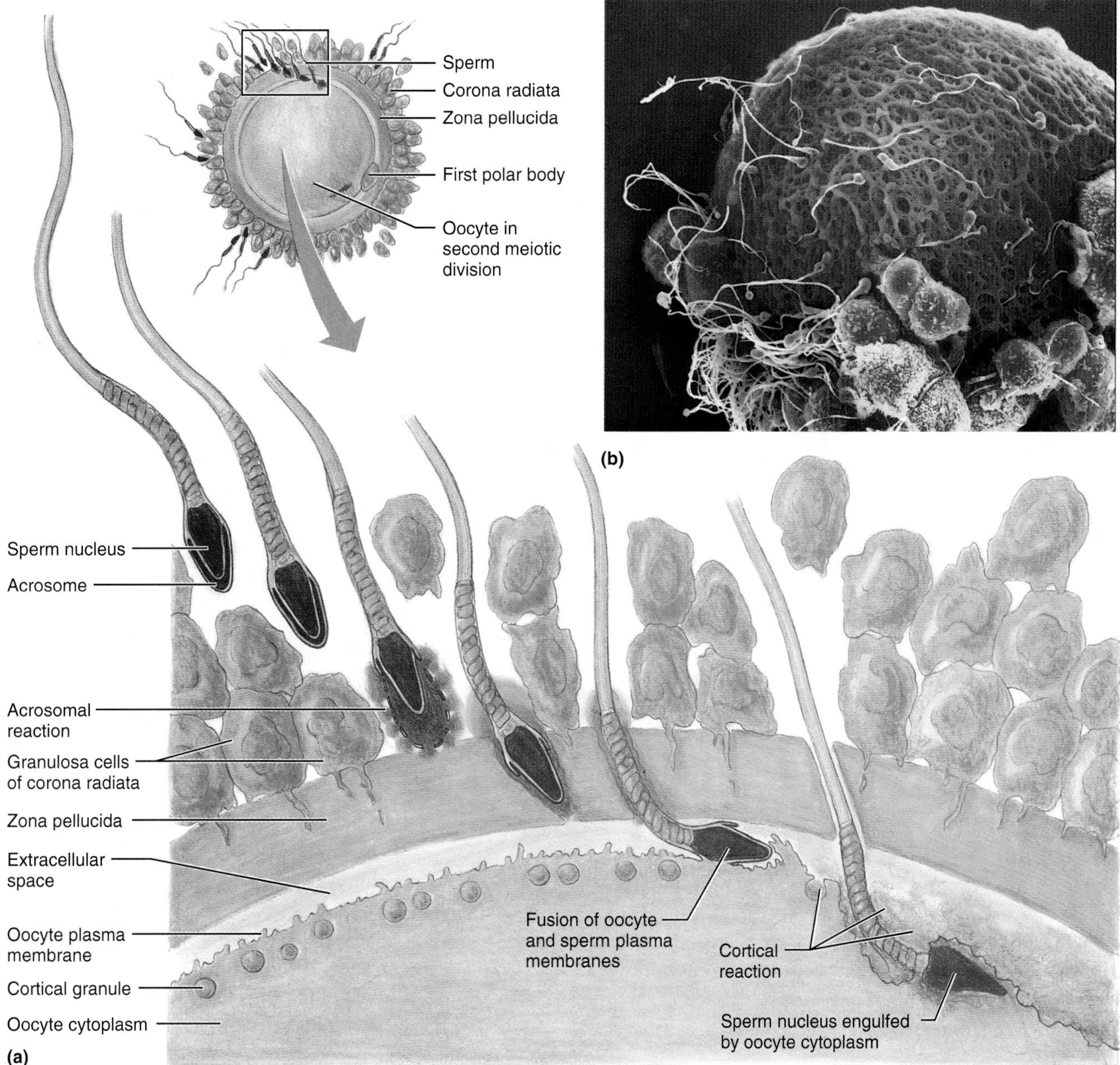

FIGURE 24.23 Events leading to fertilization: The acrosomal and cortical reactions. **(a)** The steps by which a sperm penetrates an oocyte are depicted from left to right. A sperm passes between the cells of the corona radiata, binds to the zona pellucida, releases its acrosomal enzymes, and passes through the zona. Fusion of the plasma membrane of a single sperm with the oocyte membrane induces the cortical reaction. In this reaction, cortical granules in the oocyte release into the extracellular space enzymes that destroy the sperm receptors in the zona pellucida, thereby preventing entry of more than one sperm. **(b)** Scanning electron micrograph of human sperm surrounding an oocyte (600 ×). The smaller, spherical cells are granulosa cells of the corona radiata.

Pregnancy

Events Leading to Fertilization

Before fertilization can occur, sperm deposited in the vagina must swim through the uterus to reach the ovulated oocyte in the uterine tube. Once a sperm binds to the zona pellucida around the oocyte, it undergoes the **acrosomal reaction,** the release of digestive enzymes from its acrosome (Figure 24.23). These enzymes digest a slit in the zona, through which the sperm wriggles and reaches the oocyte. The plasma membranes of the sperm and oocyte then fuse, and the sperm nucleus is engulfed by the oocyte's cytoplasm. This fusion induces the **cortical reaction**, wherein granules in the oocyte secrete enzymes into the extracellular space beneath the zona pellucida. These enzymes destroy the sperm receptors on the zona pellucida, preventing any other sperm from binding to and entering the egg.

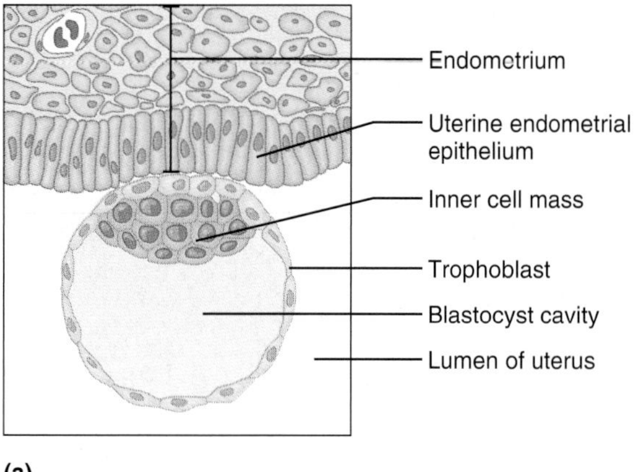

FIGURE 24.24 Implantation of the blastocyst. **(a)** Diagram of a blastocyst that has just adhered to the uterine endometrium (day 6). **(b)** Implanting embryo at a slightly later stage (day 7), showing the cytotrophoblast and syncytiotrophoblast. **(c)** Light micrograph of an implanting embryo (day 10).

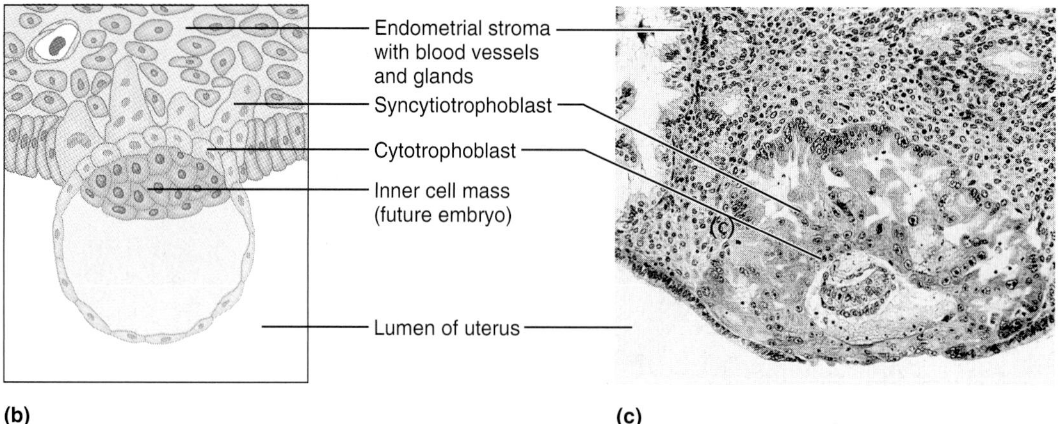

Fertilization occurs at the moment the chromosomes from the male and female gametes come together within the ovum. The fertilized egg (zygote) then initiates cleavage (cell division). By the fourth day after fertilization, the embryo is in the multicellular *blastocyst* stage and has entered the uterus (see p. 60).

Implantation

After reaching the uterus, the blastocyst floats freely in the uterine cavity for about 2 days, receiving nourishment from uterine secretions. During this time, it "hatches" by digesting a large opening in the zona pellucida and squeezing out. Then, about 6 days after fertilization, the blastocyst begins **implantation,** the act of burrowing into the endometrium (Figure 24.24a). At this time, the blastocyst consists of an *inner cell mass* (the future embryo) and an outer **trophoblast** ("nourishment generator"), a layer that will soon provide the embryo with nourishment from the mother's uterus. In the first step of implantation, trophoblast cells proliferate and form two distinct layers (see Figure 24.24b): an inner layer, or **cytotrophoblast** (si″to-trof′o-blast; "cellular part of the trophoblast"), where cell proliferation occurs; and an outer layer, or **syncytiotrophoblast** (sin-sit″e-o-trof′o-blast; "the part of the trophoblast with fused cells"), where these cells lose their plasma membranes and fuse into a multinuclear mass of cytoplasm.

The syncytiotrophoblast projects invasively into the endometrium and digests the uterine cells it contacts. As the endometrium is eroded, the blastocyst becomes deeply embedded within this thick uterine lining (Figure 24.25a on p. 730). Cleft-like spaces called **lacunae** then open within the syncytiotrophoblast and quickly fill with maternal blood that leaks from degraded endometrial blood vessels (see Figure 24.25b and c). Now, 10–12 days after fertilization, embryo-derived tissues have made their first contact with maternal blood, their ultimate source of nourishment. Events now proceed toward the formation of the placenta, the organ that nourishes the developing fetus.

Formation of the Placenta

Both embryonic (trophoblastic) and maternal (endometrial) tissues contribute to the placenta. First, the proliferating trophoblast gives rise to a layer of **extraembryonic mesoderm** on its internal surface (see Figure 24.25b and c). Together, the extraembryonic mesoderm and trophoblast layers are now called the **chorion** (ko′re-on; "membrane"), which folds into finger-like **chorionic villi,** which contact the lacunae containing maternal blood (see Figure 24.25c). At this stage, the embryo's body connects to the chorion outside of it through a **body stalk** *(connecting stalk)* made of extraembryonic mesoderm. This stalk forms the core of the future umbilical cord, as the umbilical arteries and umbilical vein grow into it from the embryo's body. Over the next 2 weeks, the mesodermal cores of the chorionic villi develop capillaries that connect to the umbilical blood vessels growing from the embryo (see Figure 24.25d). With the establishment of this vascular connection, the embryo's own blood now courses within the chorionic villi, very close to (but not mixing with) the maternal blood just external to these villi. Nutrients, wastes, and other substances begin diffusing between the two separate bloodstreams, across the villi. This exchange is occurring by the end of the first month of pregnancy.

Formation of the placenta is completed during months 2 and 3 (see Figure 24.25d and e). The chorionic villi that lie nearest the umbilical cord grow in complexity, whereas those around the rest of the embryo regress and ultimately disappear. The part of the mother's endometrium adjacent to the complex chorionic villi and umbilical cord is called the **decidua basalis** (de-sid′u-ah), whereas the endometrium opposite this, on the uterine-luminal side of the implanted embryo, is the **decidua capsularis** (see Figure 24.25d). The decidua capsularis expands to accommodate the growing embryo, which is called a fetus after month 2 (pp. 58–59), and completely fills the uterine lumen by the start of month 4. At this latter time, the decidua basalis and chorionic villi together make up a thick, pancake-shaped disc at the end of the umbilical cord—the **placenta** ("cake") (Figure 24.25e)—which continues to nourish the fetus for six more months, until birth. Because it detaches and sloughs off from the uterus after birth, the name of its maternal portion—*decidua* ("that which falls off")—is very appropriate.

It is worth emphasizing that the placenta is not given this name until it achieves its pancake shape at the start of month 4 (week 13: Figure 24.25e), even though its chorionic villi are established and functioning long before this time (Figures 24.25c and 24.26).

Anatomy of the Placenta

The exchanges that occur across the chorionic villi between maternal and fetal blood are absolutely essential to prenatal life, for these exchanges provide the fetus with nutrients and oxygen, dispose of its wastes, and allow it to send hormonal signals to the mother. To travel between maternal and fetal blood, substances must pass through the *placental barrier,* which consists of all three layers of each chorionic villus (syncytiotrophoblast, cytotrophoblast, extraembryonic mesoderm) plus the endothelium of the fetal capillaries in the villi (see Figure 24.25f). The cytotrophoblast and mesoderm layers are soon reduced to mere traces, thereby thinning the barrier and allowing a very efficient exchange of molecules.

Many kinds of molecules pass across the chorionic villi of the placenta, by several mechanisms. Sugars, fats, and oxygen diffuse from mother to fetus, whereas urea and carbon dioxide diffuse from fetus to mother. Maternal antibodies are actively transported across the placenta and give the fetus some resistance to disease. The placenta is effective at blocking passage of most bacteria from mother to fetus, but many viruses (German measles virus, chicken pox virus, mononucleosis virus, and sometimes HIV) pass through, as do many drugs and toxins (alcohol, heroin, mercury). Because the placenta is not an effective barrier to drugs, the mother must minimize her intake of harmful substances during pregnancy to minimize the chances of birth defects (see Chapter 3).

The syncytiotrophoblast secretes many substances that reach the mother's blood and regulate the events of pregnancy. One of these substances is an enzyme that "turns off" the mother's T lymphocytes, which prevents her immune system from rejecting the fetus. Most of these substances, however, are hormones—some of which (progesterone and human chorionic gonadotropin, or HCG) maintain the uterus in its pregnant state, and others of which (estrogens and corticotropin-releasing hormone, or CRH; see p. 750) instead promote labor (see p. 731). The production of HCG begins shortly after the syncytiotrophoblast forms, a mere 1 week after fertilization; its presence in the urine of a pregnant woman is the basis of most home pregnancy tests.

Clinical Appplications

DISPLACED PLACENTA Placenta previa (pre′ve-ah; "appearing in front of") is a condition in about 1 in 200 pregnancies, in which the embryo implants in the inferior part—rather than in the usual superior part—of the uterine wall. The placenta may cover the internal os of the cervix. Placenta previa is usually associated with bleeding during the last 3 months of pregnancy as some of the placenta pulls away from the uterus. To limit the bleeding, bed rest is prescribed. Placenta previa should not be confused with **placental abruption,** in which the placenta is in a normal position but becomes partly separated from the uterine wall before birth. This disorder, which is up to eight times more common than placenta previa, also produces vaginal bleeding during pregnancy. Placental abruption can interfere with fetal development by reducing the delivery of nutrients and oxygen to the fetus. ✚

Childbirth

Parturition (par″tu-rish′un; "bring forth young"), the act of giving birth, occurs an average of 266 days after fertilization and 280 days after the last menstrual period. The events that expel the infant from the uterus are referred to as **labor,** the initiation of which is complex.

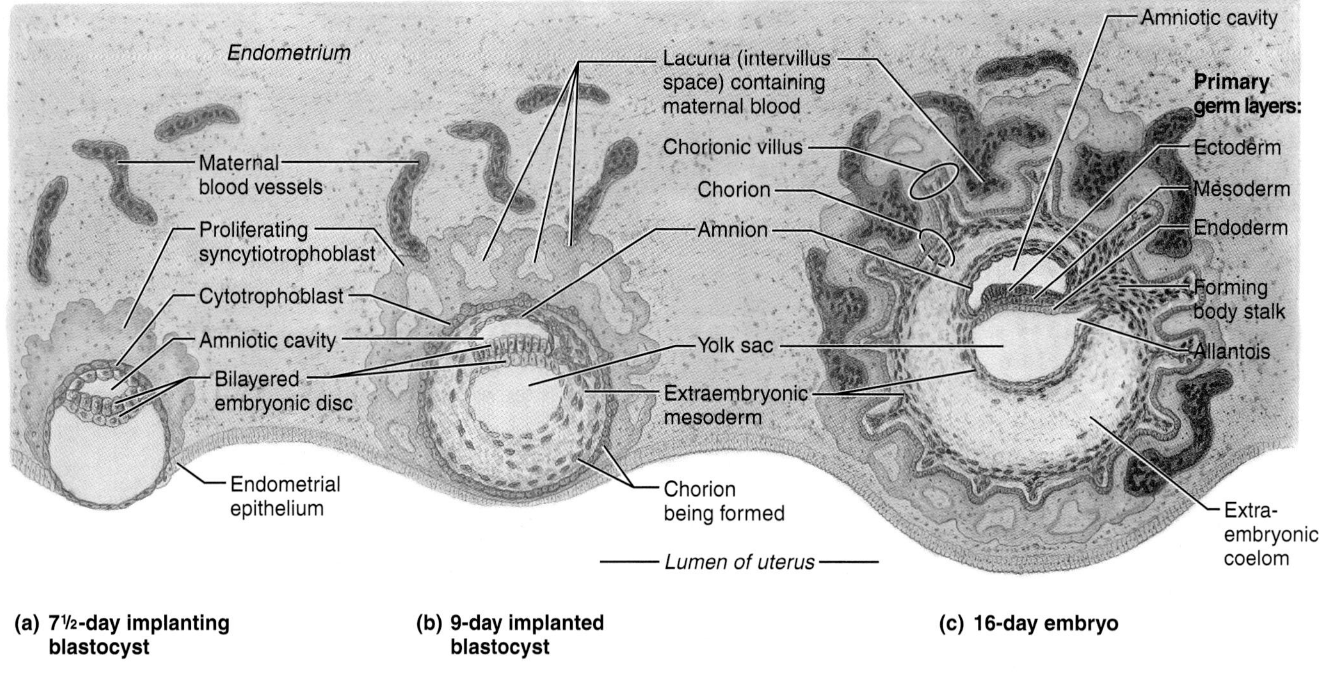

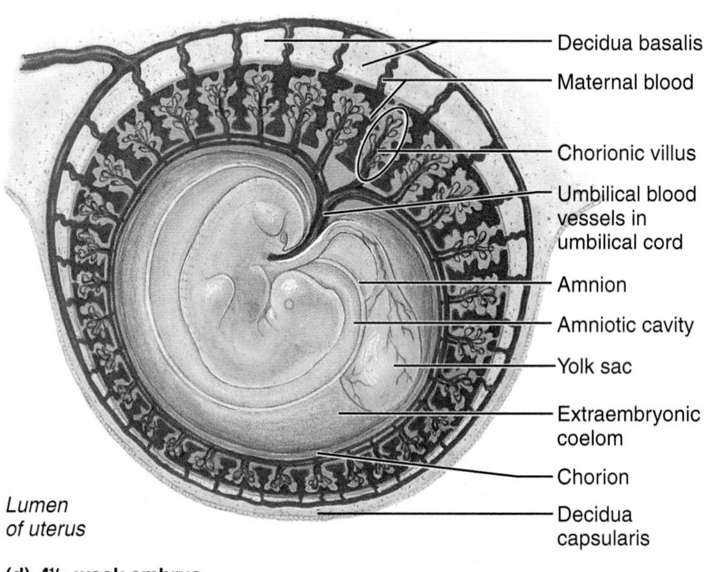

FIGURE 24.25 **Implantation and stages of placenta formation.** **(a)** Implanting blastocyst at 7½ days after fertilization. The syncytiotrophoblast is eroding the mother's endometrium. **(b)** Implantation is completed by day 12, and extraembryonic mesoderm forms deep to the cytotrophoblast. Spaces called lacunae appear in the syncytiotrophoblast and will soon fill with maternal blood. **(c)** By 16 days, the chorionic villi are forming, and the body stalk is present (future umbilical cord). **(d)** At 4½ weeks, the embryo's umbilical vessels are in place, and exchanges between maternal and embryonic blood are occurring across the chorionic villi. **(e)** At 13 weeks, the placenta is fully formed from the chorionic villi and the endometrium (decidua basalis), as a pancake around the end of the umbilical cord. **(f)** Detailed anatomy of the vascular relationships in the placenta, centering on the chorionic villi. Nutrients and other substances pass across the walls of the chorionic villi from maternal blood (in lacunae) to fetal blood (in capillaries in the villi).

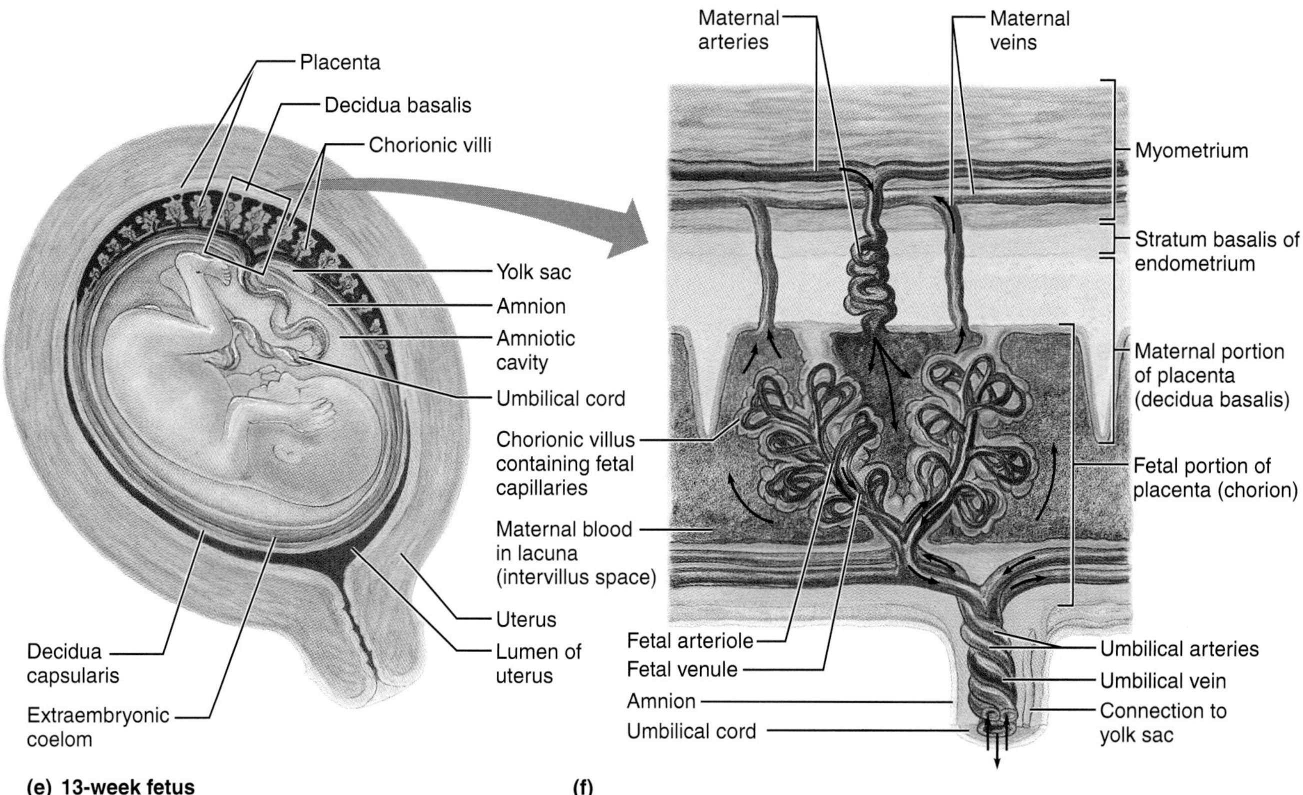

Advances in Understanding Whereas the mother was once thought to initiate labor, it now appears that the fetus does so. When the fetus grows so large that the placenta can no longer fulfill the fetus's nutritional needs, the fetus produces stress-related hormones (such as placental CRH; see p. 729) that stimulate an increased secretion of estrogens from the syncytiotrophoblast. This in turn promotes the secretion of (1) the hormone *oxytocin* (see p. 752) from both the mother's pituitary gland and her uterine decidua, and (2) prostaglandins from the decidua. Together, oxytocin and the prostaglandins signal the uterine musculature to contract and expel the infant. ✷

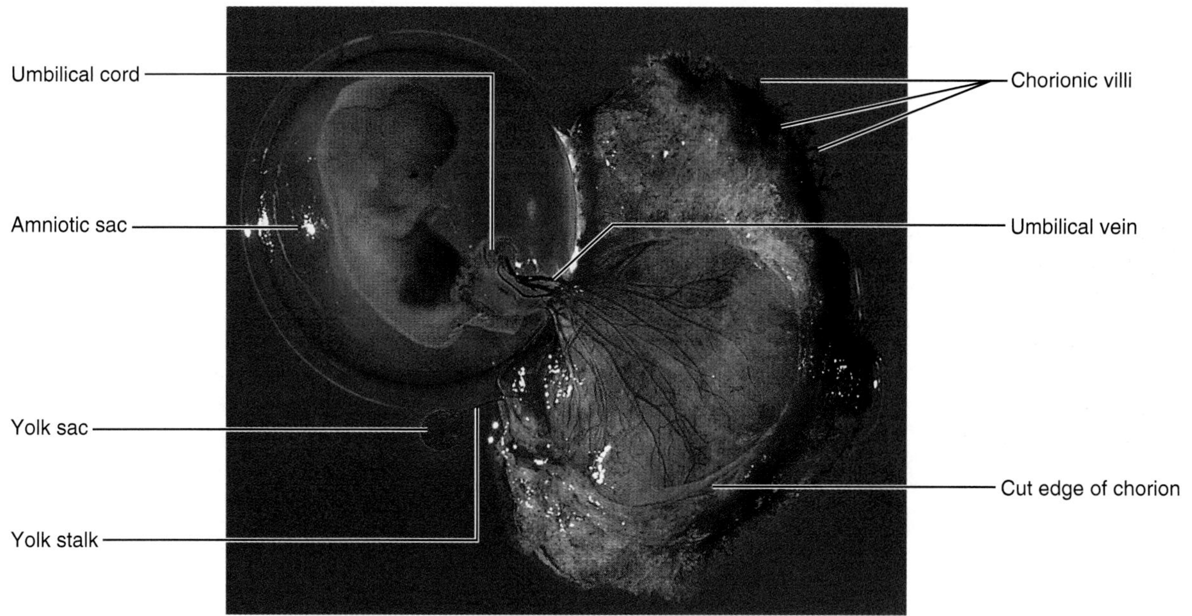

FIGURE 24.26 A late embryo at about 50 days after conception, and the embryonic membranes (at right) that will contribute to the placenta.

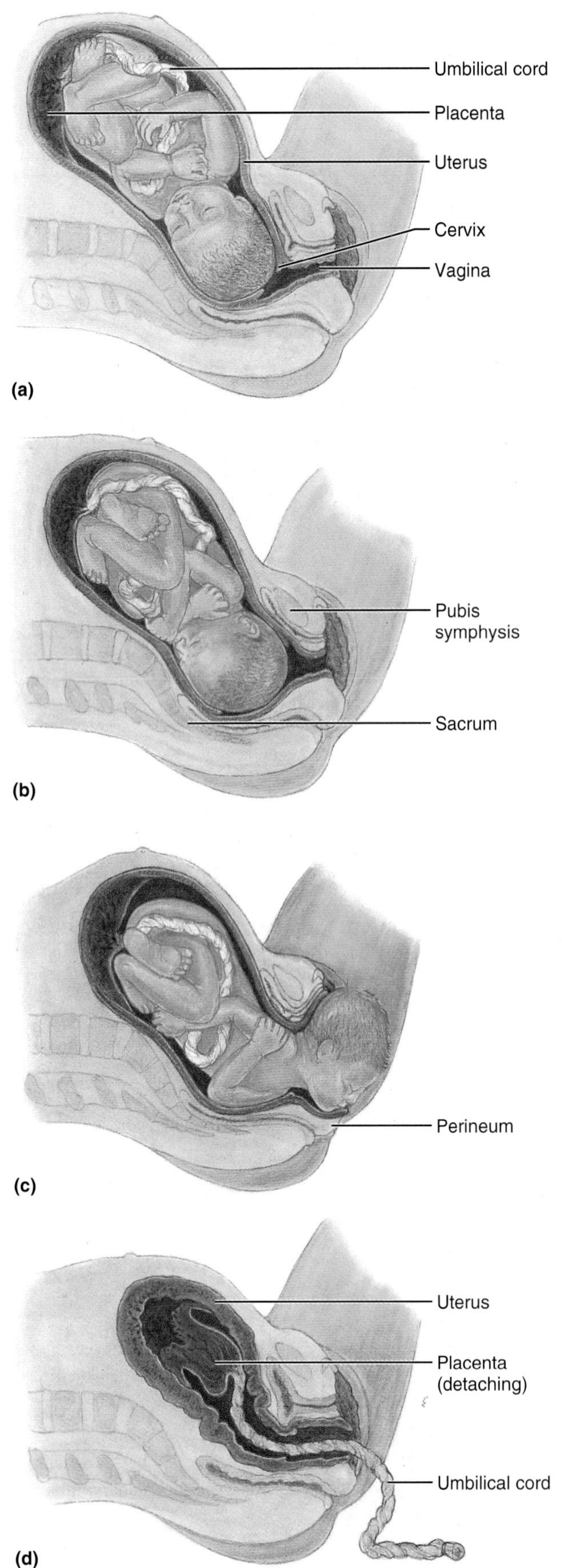

Labor involves three successive stages: the dilation, expulsion, and placental stages (Figure 24.27). The **dilation (first) stage** (see Figure 24.27a and b) begins with the first regular contractions of the uterus and ends when the cervix is fully dilated by the baby's head (about 10 cm in diameter). The dilation stage is the longest part of labor, lasting 6–12 hours or more.

The **expulsion (second) stage** (Figure 24.27c) lasts from full dilation to delivery, or actual childbirth. The uterine contractions become stronger during this stage, and the mother has an increasing urge to bear down with her abdominal muscles. Although this phase can take up to 2 hours, it typically lasts 50 minutes in a first birth and 20 minutes in subsequent births.

The **placental (third) stage** (see Figure 24.27d), or the delivery of the placenta, is accomplished within 15 minutes after birth of the infant. Forceful uterine contractions that continue after birth compress the uterine blood vessels, which both limits bleeding and causes the placenta to detach from the uterine wall. The placenta (now referred to as afterbirth) is easily removed by a slight tug on the umbilical cord. The blood vessels in the umbilical cord are then counted to verify the presence of two umbilical arteries and one umbilical vein (see p. 576) because the absence of one umbilical artery is often associated with cardiovascular disorders in the infant.

DISORDERS OF THE REPRODUCTIVE SYSTEM

In this section we examine a major group of diseases that affect the reproductive system organs of both sexes: cancer. (This chapter's A Closer Look box on page 712 considers another major group of reproductive system disorders, sexually transmitted diseases.) We begin with reproductive system cancers in males and then proceed to cancers in females.

FIGURE 24.27 Stages of labor. **(a)** Dilation stage (early). The baby's head has entered the true pelvis. The widest dimension of the head is along the left-right axis of the pelvis. **(b)** Late dilation stage. The baby's head rotates so that its greatest dimension is in the anteroposterior pelvic axis, as the head moves through the pelvic outlet. Dilation of the cervix is nearly complete. **(c)** Expulsion stage. The baby's head extends at the neck as it reaches the perineum and is delivered. The rest of the body is narrower than the head and follows more easily. **(d)** Placental stage. After the baby has been delivered, the placenta detaches and is also delivered.

Reproductive System Cancers in Males

Here we discuss testicular cancer and prostate cancer.

Testicular Cancer

Testicular cancers, which affect approximately 1 of every 50,000 males, most often between ages 15 and 35, arise most commonly from the rapidly dividing, early-stage spermatogenic cells. The resulting tumors are firm, usually painless lumps, although diffuse pain may be associated with a swelling of the entire testis. Because most testicular cancers are curable if detected early enough, men are advised to examine their testicles regularly for lumps, which are readily felt through the skin of the scrotum. Although most lumps discovered during such self-examinations are not cancers but relatively harmless pockets of fluid called varicoceles and hydroceles (see Related Clinical Terms on p. 738), proper diagnosis includes ruling out testicular cancers.

For unknown reasons, testicular cancer is becoming increasingly common (its incidence in the United States increased by 50% between 1974 and 1990). Despite its increased incidence, this cancer is cured in 95% of all cases, and death rates are falling. Treatment involves removal of the affected testis, radiotherapy, and chemotherapy based on the anticancer drug cisplatin.

Prostate Cancer

Prostate cancer is a slow-growing cancer that arises from the main (peripheral) glands in that organ. Although usually symptomless in the early stages, eventually it can grow sufficiently to impinge on the urethra, and urination may be blocked. After reaching a certain size, prostate cancer metastasizes to the bony pelvis, pelvic lymph nodes, and axial skeleton, producing pain in the pelvis, the lower back, and the bones. Metastasis is most rapid in the relatively few men who develop it before age 60.

Prostate cancer is the second most common cause of cancer death in men (after lung cancer), killing 3% of all men in the United States. Risk factors include a fatty diet and genetic predisposition; indeed, prostate cancer has a more prominent hereditary component than most other cancers. This kind of cancer is twice as common in blacks as in whites, and half as common in Asians.

Digital rectal examination and ultrasound imaging with a device inserted into the rectum can detect prostate cancer by revealing nodules, asymmetry, or hard areas on the prostate, but these techniques detect only the later stages of the disease. More effective is measuring blood levels of prostate-specific antigen (PSA: see p. 711). Although not perfect, PSA tests successfully detect prostate cancer in 70–80% of all cases, and are recommended for all men over age 50. PSA concentration in the blood of normal males is less than 0.4 ng/ml, and when PSA is found to be above this level, cancer is proven by biopsy of the prostate. The PSA test detects prostate cancer an average of 11 years earlier than does a rectal exam. Metastases are detected by bone scans or MRI and confirmed by ultrasound imaging or biopsy.

If the cancer is detected before metastasis, radical prostatectomy (complete removal of the prostate) or radiation therapy (including inserting rice-sized grains of radioactive material directly into the gland) is highly effective—85% of patients are cancer-free after 10 years. (Chemotherapy is relatively ineffective against prostate cancer and is seldom used.) For those in whom metastasis has occurred, however, no effective control measures are available, and the cancer spreads continually. However, it usually grows so slowly that it kills only 20% of these patients (the rest live so long that they die of other diseases or conditions related to old age). Late in the course of metastasis, administration of drugs that either stop the production of or block the action of androgens prolongs remission but does not seem to prolong survival. Overall, prostate cancer kills about 10% of the men it affects.

Reproductive System Cancers in Females

In this section we examine ovarian cancer, endometrial cancer, cervical cancer, and breast cancer.

Ovarian Cancer

Ovarian cancer, which typically arises from cells in the germinal epithelium covering the ovary, affects 1.4% of women and is the fifth most common cause of cancer death in women. This cancer produces few symptoms until it enlarges sufficiently to produce either a feeling of pressure in the pelvis or changes in bowel or bladder habits. Metastasis occurs either via lymph and blood vessels or by direct spread to nearby structures.

Women who have the most ovulations over their lifetime (that is, women who have neither had children nor used oral contraceptives) have the highest risk for ovarian cancer, perhaps because the extensive cell division involved in repairing the site of a ruptured (ovulated) follicle each month predisposes the germinal epithelium cells to cancer.

Diagnosis may involve feeling a mass during a pelvic exam, visualizing it by using an ultrasound probe placed in the vagina, or using a blood test for a protein associated with ovarian cancer (CA-125). Treatment is by surgical removal of the ovary, uterine tubes, and uterus, followed by radiotherapy and especially by chemotherapy, which is improving rapidly. Nonetheless, because early detection is difficult and the prognosis is poor if metastasis has occurred, the overall 5-year survival rate is under 40%.

Endometrial Cancer

Endometrial cancer, which arises from the endometrium of the uterus (usually from the uterine glands), is the fourth most common cancer of women (after lung, breast and colorectal cancer); about 2% of all women develop it. The most important sign of endometrial cancer, bleeding from the vagina, allows early detection. Risk factors include obesity and post-menopausal estrogen-replacement therapy (because estrogens stimulate the growth and division of endometrial cells). Diagnosis typically involves the use of an ultrasound device inserted into the vagina to detect endometrial thickening, followed by an endometrial biopsy. Treatment involves removal of the uterus, commonly followed by pelvic irradiation. Because it is often detected early, endometrial cancer has a relatively high cure rate of 40% to 95%.

Cervical Cancer

Cervical cancer, which usually appears between the ages of 30 and 50 and occurs in about 1% of U.S. women, is a slow-growing cancer that arises from the epithelium covering the tip of the cervix. Ninety percent of cases are caused by sexually transmitted human papillomavirus. In a **Papanicolaou (Pap) smear** or *cervical smear test,* which is the most effective way to detect this cancer in its earliest (precancerous) stage, some cervical epithelial cells are scraped off and examined microscopically for abnormalities. If cancerous or precancerous cells are confined to the cervix, they are excised or removed by freezing or a laser; in such cases survival rates are very high (over 95%). When the cancer has spread throughout the pelvic organs, radiation plus chemotherapy appears to produce 5-year survival rates of over 70%. Because of the importance of early detection, women are advised to have a Pap smear performed every 1–3 years.

Breast Cancer

Breast cancer, the second most common cause of cancer deaths in women and the killer of 3% of U.S. women, typically arises from the smallest ducts in the lobules of the breast. About 97% of cases occur in women over age 55, and one of every eight women will develop breast cancer in her lifetime. Women are advised to perform monthly breast self-examinations to detect lumps in the breast, a change in breast shape, scaling on the nipple or areola, discharge from the nipple, or dimpling or an orange-peel texture on the breast.

Risk factors for breast cancer include a family history of the disease, late menopause, early onset of puberty, first live birth after age 30, and post-menopausal estrogen-replacement therapy. With the possible exception of family history, all these risk factors reflect an increased lifelong exposure to estrogens, which stimulate division of the duct cells and promote cancer.

Breast cancer is a rapidly spreading cancer that metastasizes from the breast through lymph vessels to axillary and parasternal lymph nodes, and then to the chest wall, lungs, liver, brain, and bones. If metastases reach the lungs or liver, average survival time is less than 6 months.

Screening techniques to enable early detection effectively reduce mortality from breast cancer. Women over age 50 are advised to have an annual **mammogram** (X ray of the breast). Suspicious masses are biopsied and microscopically examined to determine whether they are cancerous. If one or more are, the axillary lymph nodes are removed and examined for cancer cells; the percent of these axillary nodes that are affected is both the best indicator of metastasis and the best predictor of survival. One painful result of removing all these lymph nodes, lymphedema of the arm (see p. 586), may soon be prevented by **sentinel node mapping,** in which only the one or two nodes that directly drain the tumor site are examined; if these nodes are cancer-free, the other axillary nodes are left in place.

The once-standard treatment for breast cancer—**radical mastectomy,** the removal of the entire affected breast plus all underlying muscles, fascia, and associated lymph nodes—is now performed only in the most advanced cases. Current standard treatment consists of a **lumpectomy**—the removal of the cancerous mass plus a small rim of surrounding tissue—followed by radiation therapy and then administration of hormones (so-called "selective estrogen replacement modulators" such as tamoxifen) and chemotherapy.

The importance of early detection via regular mammograms is apparent in the survival statistics. Among patients in which breast cancer has metastasized to almost all the axillary lymph nodes, the 5-year survival rate is about 30%, but when the axillary nodes are not involved, the 5-year survival rate is 78%.

THE REPRODUCTIVE SYSTEM THROUGHOUT LIFE

We divide our discussion of the development of the reproductive system into four parts: embryonic development of the sex organs, descent of the gonads, puberty, and menopause.

Embryonic Development of the Sex Organs

The gonads of both males and females begin their development during week 5 as masses of intermediate mesoderm called the **gonadal ridges** (Figure 24.28), which form bulges on the dorsal abdominal wall in the lumbar region, just medial to the mesonephros (transient kidney; see p. 693). The **mesonephric (Wolffian) ducts,** or future male ducts, develop medial to the **paramesonephric (Müllerian) ducts,** or future female ducts, and both sets of ducts empty into the cloaca (future bladder and urethra). At this time, the embryo is said to be in the **sexually indifferent stage** because the gonadal ridge and ducts are structurally identical in both sexes.

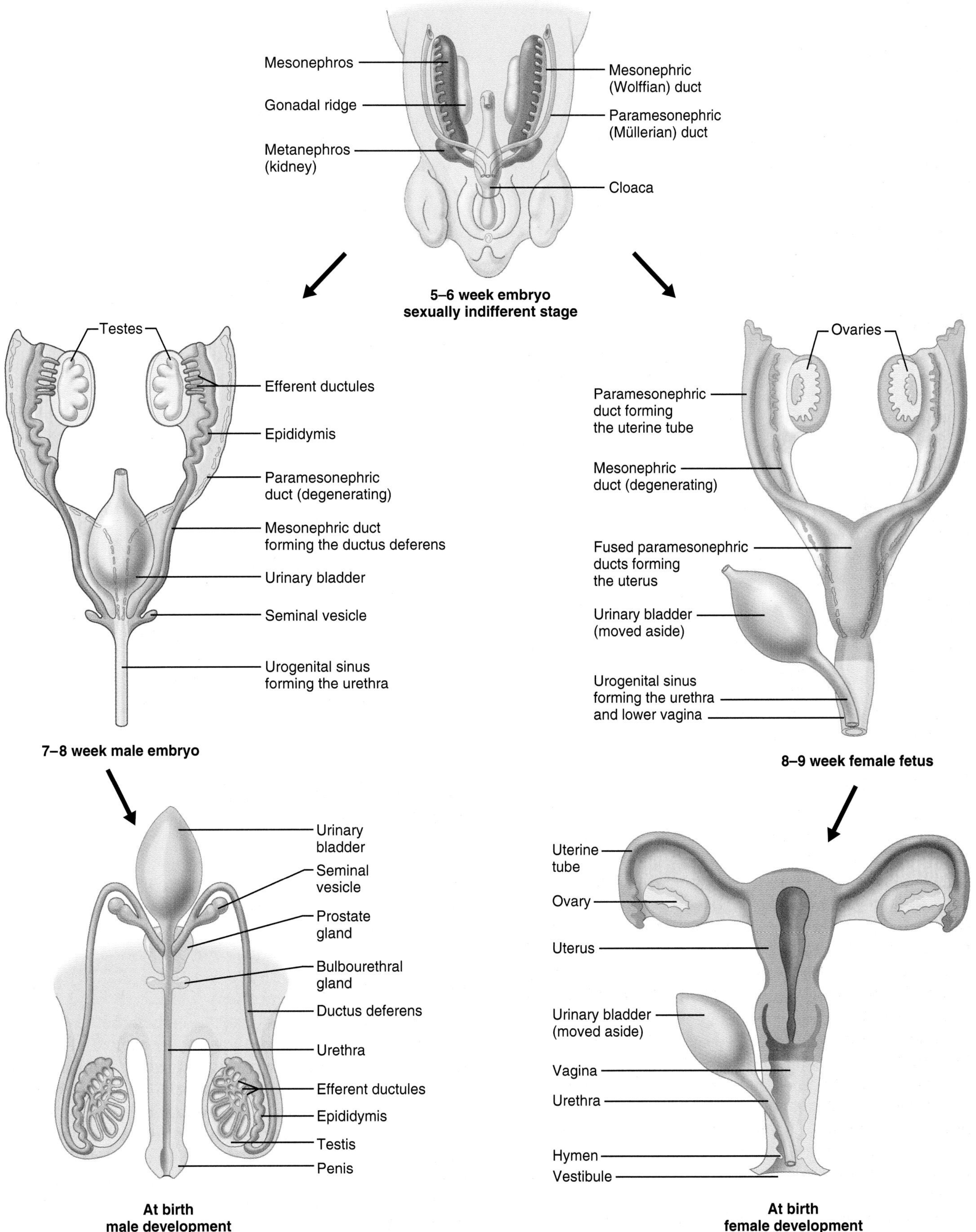

FIGURE 24.28 Development of the internal reproductive organs in both sexes.

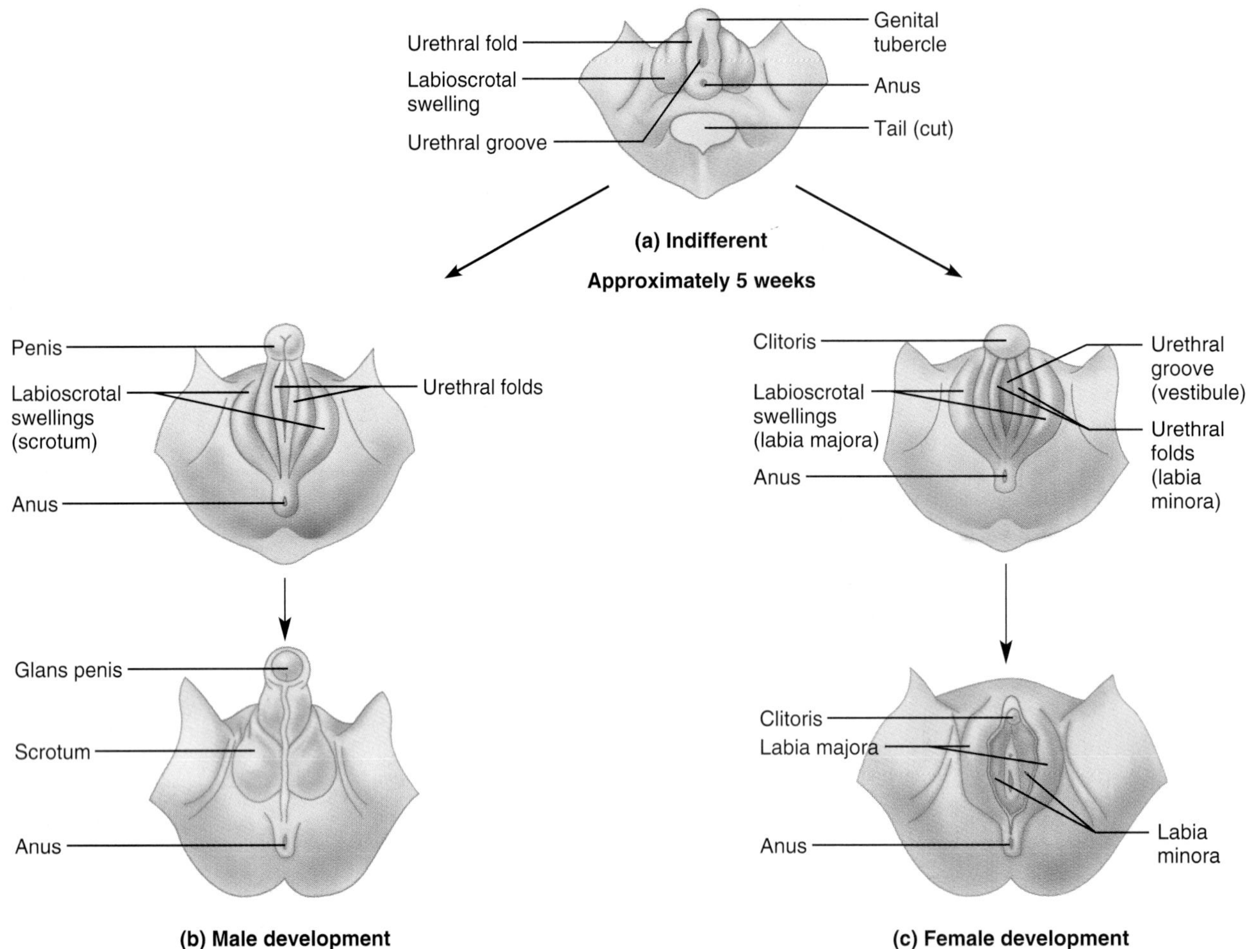

FIGURE 24.29 **Development of the external genitalia in both sexes.** The two pictures at the bottom of the figure show the fully developed perineal region.

In the week after the gonadal ridges appear, *primordial germ cells* migrate to them from the yolk sac and seed the developing gonads with germ cells destined to become male spermatogonia or female oogonia. Once these germ cells are in place, the gonadal ridges differentiate into testes or ovaries, based on the presence or absence of a male-inducing signal, called SRY protein, that comes from the epithelium covering the ridge.

In *male* embryos, this differentiation begins in week 7. Testes cords (the future seminiferous tubules) grow from the gonadal surface into the inner part of the gonad, where they connect to the mesonephric duct via nephrons that become the efferent ductules. With further development, the mesonephric duct becomes the major male sex ducts: epididymis, ductus deferens, and ejaculatory duct. The paramesonephric ducts play no part in male development and soon degenerate.

In female embryos, the process begins slightly later, in about week 8. The outer, or cortical, part of the immature ovaries forms the ovarian follicles. Shortly thereafter, the paramesonephric ducts differentiate into most structures of the female duct system—the uterine tubes, the uterus, and the superior part of the vagina—and the mesonephric ducts degenerate.

Like the gonads, the *external* genitalia develop from identical structures in both sexes (Figure 24.29). During the sexually indifferent stage, both male and female embryos have a small projection called the **genital tubercle** on their external perineal surface. The urogenital sinus of the cloaca (future urethra and bladder; see p. 667) lies deep to the tubercle, and the **urethral groove,** which serves as the external opening of the urogenital sinus, runs between the genital tubercle and the anus. The urethral groove is flanked laterally by the **urethral folds** *(genital folds)* and the **labioscrotal swellings.**

During week 8, the external genitalia begin to develop rapidly. In males, the genital tubercle enlarges, forming most of the penis (the glans and the erectile bodies), and the urethral folds fuse in the midline to form the penile urethra. The two labioscrotal swellings also join in the midline to become the scrotum. In females, the genital tubercle becomes the clitoris, the unfused urethral folds become the labia minora, and the unfused labioscrotal swellings become the labia majora. The urethral groove persists in females as the vestibule.

CLINICAL APPLICATIONS

HYPOSPADIAS A condition called **hypospadias** (hi″po-spa′de-as) is the most common congenital abnormality of the male urethra. Occuring in nearly 1% of male births, it results from a failure of the two urethral folds to fuse completely, producing urethral openings on the undersurface of the penis. As a result, more urine exits from the underside of the penis than from its tip. This condition, which is becoming more common, is typically corrected surgically at about 1 year of age. ✚

Descent of the Gonads

In both sexes, the gonads originate in the dorsal body wall of the lumbar region and then descend caudally during the fetal period. In the male fetus, the testes descend toward the scrotum and are followed by their blood vessels and nerves (Figure 24.30a). The testes reach the pelvis near the inguinal region in month 3, when a finger-like outpocketing of the peritoneal cavity, the **vaginal process,** pushes through the muscles of the anterior abdominal wall to form the inguinal canal. The testes descend no farther until month 7, when suddenly they pass quickly through the inguinal canal and enter the scrotum (Figure 24.30b). After the testes reach the scrotum, the proximal part of each vaginal process closes off, whereas its caudal part becomes the sac-like tunica vaginalis (Figure 24.30c). In some boys, however, the vaginal process does not close at all, and this open path to the scrotum constitutes the route for an inguinal hernia into the vaginal process. Indeed, such **congenital inguinal hernias** are the most common hernias.

The causative mechanism of the descent of the testis is unknown, but most theories focus on a fibrous cord called the **gubernaculum** (gu″ber-nak′u-lum; "governor"), which extends caudally from the testis to the floor of the scrotal sac and is known to shorten (see Figure 24.30). The final descent through the inguinal canal may be aided by an increase in intra-abdominal pressure that pushes the testis into the scrotum. In any case, this final descent through the inguinal canal is stimulated by testosterone.

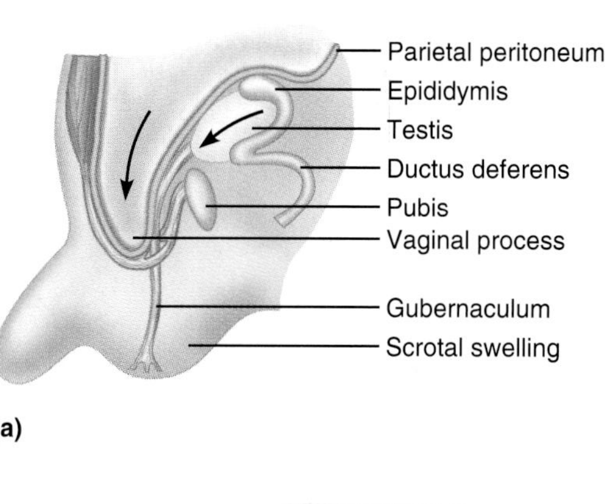

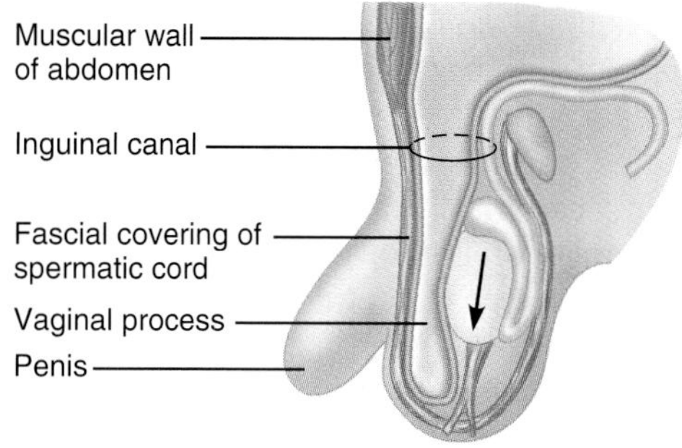

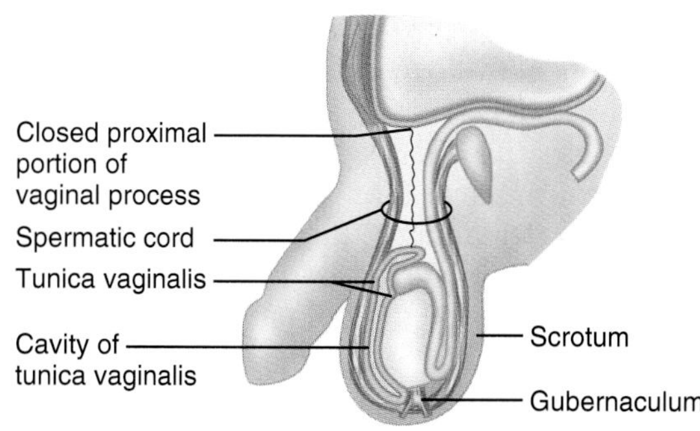

FIGURE 24.30 Descent of the testes. Sagittal sections through the pelvis of the male fetus. **(a)** At 3 months, the testes descend dorsal to the parietal peritoneum, reaching the inguinal region. At this time, the vaginal process (an extension of the peritoneal cavity) pushes caudally into the future scrotum. **(b)** In month 7, the testes descend rapidly into the scrotum via the inguinal canal. **(c)** Finally, by the time of birth, the caudal part of the vaginal process has become the tunica vaginalis ventral to the testis, whereas the proximal portion of the vaginal process is obliterated.

CLINICAL APPLICATIONS

CRYPTORCHIDISM Cryptorchidism (krip-tor′kĭ-dizm; *crypt* = hidden; *orcho* = testis) is a congenital condition occurring in about 5% of newborn males in which one or both testes fail to descend into the scrotum. An undescended testis, which may be located in either the inguinal canal (typically) or in the pelvic cavity, is sterile because viable sperm cannot be produced at the higher temperature there. The interstitial cells of these testes, however, do secrete testosterone. Surgical correction within 6–18 months after birth is recommended to avoid damage to sperm-forming cells. Left uncorrected, undescended testes have an increased likelihood of developing testicular cancer. ✚

Like the testes, the female ovaries descend during fetal development—but only into the pelvis, where the tent-like broad ligament blocks any further descent. Each ovary is guided in its descent by a gubernaculum that is anchored in the labium majus and is homologous to the gubernaculum in the male (see Figure 24.30a). In the female, the caudal part of the gubernaculum becomes the round ligament of the uterus, and its cranial part becomes the ovarian ligament.

Puberty

Puberty is the period of life, generally between ages 10 and 15, when the reproductive organs grow to their adult size and reproduction becomes possible. These changes occur under the influence of rising levels of gonadal hormones, testosterone in males and estrogens in females.

The events of puberty occur in the same sequence in all individuals, but the age when they occur varies widely. In males, the first change is an enlargement of the testes and scrotum around age 13. This is followed by the appearance of pubic, axillary, and facial hair, which are **secondary sex characteristics**—features induced in the *nonreproductive* organs by sex hormones. Other secondary sex characteristics include the enlargement of the larynx that deepens the male voice and an increased oiliness of the skin that can lead to acne. The musculoskeletal system also increases in mass at puberty in a growth spurt that lasts up to 6 years. Sexual maturation is evidenced by the presence of mature sperm in the semen.

In females, the first sign of puberty is budding breasts, often apparent by age 11. *Menarche* (me-nar′ke; "month beginning"), the first menstruation, usually occurs 1–2 years later. Dependable ovulation and fertility await the maturation of pituitary hormonal controls, which takes nearly 2 more years. In the meantime, the estrogen-induced secondary sexual characteristics appear at around age 13, including an increase in subcutaneous fat, especially in the hips and breasts, and widening and lightening of the bones of the pelvic girdle, which are adaptations that facilitate childbirth. Furthermore, the ovaries secrete some androgens that signal the development of pubic and axillary hair, apocrine sweat glands, and oiliness of the skin. The estrogen-induced female growth spurt lasts from about age 12 until age 15 to 17.

Menopause

Most women reach their reproductive peak in their late 20s. Around age 35, the rate at which follicles degenerate in the ovaries accelerates. Later, when few follicles remain in the ovaries, ovulation and menstruation cease. This event, called **menopause,** normally occurs between ages 46 and 54. The cause of menopause is unclear: It may be induced by the exhaustion of follicles from the ovaries, or alternatively, it may be signaled by the brain.

After menopause, the ovaries stop secreting estrogens and testosterone. Deprived of estrogens, the reproductive organs and breasts start to atrophy. As the vagina becomes dry, intercourse may become painful, and vaginal infections become more common. The skin thins gradually, and bone mass declines, sometimes resulting in osteoporosis. Slowly rising blood cholesterol levels place postmenopausal women at risk for cardiovascular disorders (see p. 578).

There is no equivalent of menopause in men. Instead, starting at age 40, testosterone production begins a slow decline of about 1% per year. As a result, bone and muscle mass decrease gradually, as does sperm formation—although not enough to prevent some elderly men from fathering children. Erection becomes less efficient as the amount of connective tissue in the penile erectile bodies increases with age.

RELATED CLINICAL TERMS

Fibroids Slow-growing, benign tumors made of smooth muscle cells and fibrous connective tissue in the wall of the uterus; occur in about 20% of all women over 30. Their precise cause is unknown but may relate to an atypical response to estrogens. Small fibroids produce no symptoms, but large ones can cause pressure and pain in the pelvis, heavy menstrual periods, and infertility. A conservative treatment is myomectomy, removal of the fibroid from its capsule while preserving the rest of the uterus. However, if the fibroids are abundant, the whole uterus may require removal.

Gynecology (gi′ne-kol′o-je) (*gyneco* = woman; *logy* = study of) Specialized branch of medicine that deals with the diagnosis and treatment of reproductive disorders in females.

Hydrocele (hi′dro-sēl) (*hydro* = water; *cele* = sac) A swelling in the scrotum caused by an excessive accumulation of fluid in the cavity of the tunica vaginalis; may result from an infection or injury to the testis that causes the layers of the tunica vaginalis to secrete excess serous fluid. Whereas most hydroceles are small and do not require treatment, large hydroceles can be treated by first aspirating the fluid with a needle inserted into the cavity of the tunica vaginalis and then removing the serous lining of the tunica vaginalis.

Hysterectomy (his″tĕ-rek′to-me) (*hyster* = uterus; *ectomy* = cut out) Surgical removal of the uterus; a very common operation, typically performed to remove fibroids (25–30%), to stop abnormal uterine bleeding (20%), as a treatment for endometriosis (under 20%), and to relieve uterine prolapse (4–17%) and chronic pelvic pain (12–18%). The uterus can be removed either through the anterior abdominal wall or the vagina. In response to evidence that hysterectomies are performed too often, physicians may soon adopt standardized criteria for deciding when they are appropriate.

Orchitis (or-ki′tis) (*orcho* = testis) Inflammation of the testes, sometimes caused by the mumps virus. If testicular infection spreads to the epididymis, it is called epididymo-orchitis.

Testicular torsion Twisting of the spermatic cord and testis within the scrotum, which can starve the testis of its blood supply and produce sudden urgent pain, swelling, and reddening of the scrotum; can result from trauma or a spasm of the cremaster muscle; is most common in those males in whom the testis, instead of attaching to the posterior scrotal wall, hangs freely. Can be diagnosed by various imaging techniques, but surgery to undo the twist is often performed without waiting for imaging.

Tubal ligation Having one's uterine tubes "tied" surgically as a form of birth control. Usually involves threading a cutting instrument and a viewing tube (laparoscope) through a small incision in the anterior abdominal wall and then tying and/or cutting the uterine tubes. Tubal ligation usually is irreversible, although in some cases delicate microsurgery can successfully reopen or reattach the uterine tubes.

Varicocele (var′ĭ-ko-sēl″) Varicose veins in the pampiniform plexus of the spermatic cord that occurs in about 10% to 15% of men; feels like a bag of worms beneath the skin in the scrotum. Although usually harmless, it causes problems often enough to be the most common cause of infertility in men. In these cases, drainage of the pampiniform veins is sufficiently impaired to reduce these vessels' ability to cool the testes, resulting in a declining sperm count or sterility. Treated by surgically removing the varicose veins.

CHAPTER SUMMARY

Media study tools that could provide you additional help in reviewing specific key topics of Chapter 24 are referenced below.

SP = Study Partner; AIA = A.D.A.M.® Interactive Anatomy

1. The function of the reproductive system is to produce offspring. The gonads produce gametes (sperm or ova) and sex hormones. All other reproductive organs are accessory organs.

THE MALE REPRODUCTIVE SYSTEM (pp. 702–715)

The Scrotum (p. 703)

1. The scrotum is the sac of skin and subcutaneous tissue that contains the testes. Its temperature, which is slightly lower than that of the body, is required for viable sperm production. The dartos muscle, cremaster muscle, and pampiniform plexus help regulate the temperature of the testes in the scrotum.

The Testes (pp. 703–708)

2. Each testis, which is partly surrounded by a serous sac called the tunica vaginalis, is divided into many lobules that contain sperm-producing seminiferous tubules separated by areolar connective tissue.

AIA Link: Atlas; System: Reproductive; Fascia in Male Pelvis (medial).

3. The thick epithelium of the seminiferous tubule consists of spermatogenic cells (spermatogonia, primary and secondary spermatocytes, and spermatids), embedded in columnar sustentacular cells. Spermatogenic cells move toward the tubule lumen as they differentiate into sperm by a process called spermatogenesis.

4. The three stages of spermatogenesis are (1) formation of spermatocytes, (2) meiosis, and (3) spermiogenesis.

5. In spermiogenesis, round spermatids become tadpole-shaped sperm cells as an acrosome forms, a flagellum grows, the nucleus flattens as its chromatin condenses, and the superfluous cytoplasm is shed.

6. Sustentacular (Sertoli) cells form the blood-testis barrier, nourish the spermatogenic cells, and move them toward the lumen. They also secrete testicular fluid (for sperm transport), androgen-binding protein, and the hormone inhibin.

7. Myoid cells are smooth-muscle–like cells that surround the seminiferous tubules and help to squeeze sperm through the tubules.

8. Interstitial (Leydig) cells, which are oval cells in the connective tissue between seminiferous tubules, secrete androgens (mostly testosterone) under the influence of luteinizing hormone from the pituitary gland.

SP Exercise: Chapter 24, Internal Structure of the Testes.

The Reproductive Duct System in Males (pp. 708–710)

9. From the seminiferous tubules, sperm travel through the tubuli recti and rete testes, then out of the testes through the efferent ductules into the epididymis.

10. The comma-shaped epididymis hugs the posterolateral surface of the testis. The duct of the epididymis, lined with a pseudostratified columnar epithelium with stereocilia, is where sperm gain the ability to swim and fertilize. Ejaculation begins with contraction of smooth muscle in the duct of the epididymis.

11. The ductus (vas) deferens extends from the epididymis in the scrotum to the ejaculatory duct in the pelvic cavity. During ejaculation, the thick layers of smooth muscle in its wall propel sperm into the urethra by peristalsis.

AIA Link: Atlas; System: Reproductive; Male Pelvic Organs (lateral).

12. The fascia-covered spermatic cord contains the ductus deferens and the testicular vessels. It runs through the anterior abdominal wall from the scrotum to the superficial inguinal ring and then continues through the inguinal canal to the deep inguinal ring.

13. The urethra, which runs from the bladder to the tip of the penis, conducts semen and urine to the body exterior.

Accessory Glands (pp. 710–712)

14. The seminal vesicles (seminal glands)—long, pouched glands posterior to the bladder—secrete a sugar-rich fluid that contributes 60% of the ejaculate.

15. The prostate gland, surrounding the prostatic urethra, is a group of compound glands embedded in a fibromuscular stroma. Its secretion helps to clot, then to liquefy, the semen. Benign prostatic hyperplasia, which affects many men, arises from the deep glands in the prostate.

16. The bulbourethral glands in the urogenital diaphragm secrete mucus into the urethra before ejaculation to lubricate this tube for the passage of semen.

The Penis (pp. 712–715)

17. The penis has a root, a shaft, and a glans covered by foreskin. Its main nerves and vessels lie dorsally. Most of its mass consists of vascular erectile bodies—namely, a corpus

spongiosum and two corpora cavernosa. Engorgement of these bodies with blood causes erection.

The Male Perineum (p. 715)

18. In the male, the diamond-shaped perineum contains the scrotum, the root of the penis, and the anus.

AIA Link: Atlas; System: Reproductive; Triangles of Perineum (Male).

THE FEMALE REPRODUCTIVE SYSTEM (pp. 715–726)

1. The female reproductive system produces ova and sex hormones and houses a developing infant until birth. It undergoes a menstrual cycle of about 28 days.

The Ovaries (pp. 715–718)

2. The almond-shaped ovaries lie on the lateral walls of the pelvic cavity and are suspended by various mesenteries and ligaments.

AIA Link: Atlas; System: Reproductive; Female Pelvic Organs (anterior).

3. The mesovarium (mesentery of the ovary) is part of a fold of peritoneum called the broad ligament. (Other parts of the broad ligament are the mesosalpinx and mesometrium.) Contained within the broad ligament are the suspensory ligament of the ovary and its continuation, the round ligament of the uterus.
4. Each ovary is divided into an outer cortex (follicles separated by connective tissue) and an inner medulla (connective tissue containing the main ovarian vessels and nerves).
5. The ovaries of a newborn female contain many thousands of primordial follicles, each of which consists of an oocyte surrounded by a layer of flat follicular cells.
6. During the follicular phase of each monthly ovarian cycle (days 1–14), 6–12 follicles start maturing. The follicles progress from primordial follicles to early primary follicles, to growing primary follicles, to secondary (antral) follicles, to mature vesicular follicles that are ready for ovulation. Beyond the smallest stages, this progression is stimulated by follicle-stimulating hormone (FSH) from the anterior part of the pituitary gland. Generally, only one follicle per month completes the maturation process.
7. The theca folliculi around a growing follicle cooperates with the follicle itself to produce estrogens from androgens under the stimulatory influence of luteinizing hormone (LH) and FSH. Estrogens stimulate the growth and activity of all female sex organs and signal the proliferative stage of the uterine cycle.
8. Upon stimulation by LH, ovulation occurs at mid cycle (day 14), releasing the oocyte from the ovary into the peritoneal cavity. Ovulation seems to involve a weakening and rupture of the follicle wall followed by violent muscular contraction of the external theca cells.
9. In the luteal phase of the ovarian cycle (days 15–28), the ruptured follicle remaining in the ovary becomes a wavy corpus luteum that secretes progesterone and estrogens. Progesterone functions to maintain the secretory phase of the uterine cycle. If fertilization does not occur, the corpus luteum degenerates in about 2 weeks into a corpus albicans (scar).
10. Oogenesis, the production of female gametes, starts before birth and takes decades to complete. The stem cells (oogonia) appear in the ovarian follicles of the fetus. Primary oocytes stay in meiosis I until ovulation occurs years later, and each secondary oocyte then stays in meiosis II until it is penetrated by a sperm.

The Uterine Tubes (pp. 718–720)

11. Each uterine tube extends from an ovary to the uterus. Its regions, from lateral to medial, are infundibulum, ampulla, and isthmus. The fimbriated distal end creates currents that help draw an ovulated oocyte into the uterine tube.

AIA Link: Atlas; System: Reproductive; Female Pelvic Organs (lateral).

12. The wall of the uterine tube contains a muscularis and a folded inner mucosa with a simple columnar epithelium. Both the smooth muscle and ciliated epithelial cells help propel the oocyte toward the uterus.

The Uterus (pp. 720–724)

13. The hollow uterus, shaped like an inverted pear, has a fundus, a body, an isthmus, and a cervix. Cervical glands fill the cervical canal with a bacteria-blocking mucus.

AIA Link: Atlas; System: Reproductive; Female Pelvic Organs (medial) 1.

14. The uterus is supported by the broad, lateral cervical, and round ligaments. Most support, however, comes from the muscles of the pelvic floor.
15. The uterine wall consists of the outer perimetrium, the muscular myometrium, and a thick mucosa called the endometrium. The myometrium functions to squeeze the baby out during childbirth. The endometrium consists of a stratum functionalis, which sloughs off each month (except in pregnancy), and an underlying stratum basalis, which replenishes the functionalis.
16. The uterine cycle has three phases: the menstrual, proliferative, and secretory phases. The first two phases are a shedding and then a rebuilding of endometrium in the 2 weeks before ovulation. The third phase prepares the endometrium to receive an embryo in the 2 weeks after ovulation.

SP Exercise: Chapter 24, The Female Menstrual Cycle.

The Vagina (p. 724)

17. The vagina, a highly distensible tube that runs from the uterus to the body's exterior, receives the penis and semen during intercourse and acts as the birth canal.
18. The layers of the vagina wall are an outer adventitia, a middle muscularis, and an inner mucosa consisting of an elastic lamina propria and a stratified squamous epithelium. The lumen is acidic.
19. The vaginal fornix is a ring-like recess around the tip of the cervix in the superior vagina.

The External Genitalia and Female Perineum (pp. 724–725)

20. The female external genitalia (vulva) include the mons pubis, the labia majora and minora, the clitoris with its erectile bodies, and the vestibule containing the vaginal and urethral orifices. The mucus-secreting greater vestibular glands and the bulbs of the vestibule (erectile bodies) lie just deep to the labia.

21. The central tendon of the perineum lies just posterior to the vaginal orifice and the fourchette.

AIA Link: Atlas; System: Reproductive; Triangles of Perineum (Female).

The Mammary Glands (pp. 725–726)

22. The mammary glands develop from the skin of the embryonic milk lines. Internally, each breast consists of 15–25 lobes (compound alveolar glands) that secrete milk. The lobes, which are subdivided into lobules and alveoli, are separated by adipose tissue and by supportive suspensory ligaments. The full glandular structure of the breast does not develop until the second half of pregnancy. Breast cancers usually arise from the duct system.

SP Exercise: Chapter 24, Female Reproductive System.

PREGNANCY AND CHILDBIRTH (pp. 726–732)

Pregnancy (pp. 727–729, 730–731)

1. Fertilization usually occurs in the ampulla of a uterine tube. Many sperm must release enzymes from their acrosomes to penetrate the oocyte's zona pellucida. Entry of one sperm into the oocyte prevents the entry of others.

2. Once fertilized, the zygote divides to form a multicellular embryo, which travels to the uterus and implants in the endometrium. As it implants, its trophoblast divides into a cytotrophoblast and an invasive syncytiotrophoblast. Lacunae open in the syncytiotrophoblast and fill with blood from endometrial vessels, forming the first contact between maternal blood and embryonic tissue.

3. The placenta acts as the respiratory, nutritive, and excretory organ of the fetus and produces the hormones of pregnancy. It is formed both from embryonic tissue (chorionic villi) and from maternal tissue (endometrial decidua).

4. The placental exchange barrier consists of the layers of the chorionic villi plus the endothelium of the fetal capillaries inside those villi. Of these layers, the syncytiotrophoblast secretes the placental hormones. No direct mixing of maternal and fetal blood occurs across a healthy placenta.

Childbirth (pp. 729, 731–732)

5. The three stages of labor are the dilation, expulsion, and placental stages.

DISORDERS OF THE REPRODUCTIVE SYSTEM (pp. 732–734)

1. This section focuses on cancers of the reproductive organs, the most serious of which are prostate cancer and breast cancer.

THE REPRODUCTIVE SYSTEM THROUGHOUT LIFE (pp. 734–738)

Embryonic Development of the Sex Organs (pp. 734–737)

1. The gonads of both sexes develop in the dorsal lumbar region from the gonadal ridges of intermediate mesoderm. The primordial germ cells migrate to the gonads from the endoderm of the yolk sac. The mesonephric ducts form most of the ducts and glands in males, whereas the paramesonephric ducts form the uterine tubes and uterus of the female.

2. The external genitalia arise from the genital tubercle (penis and clitoris), urethral folds (penile urethra and labia minora), and labioscrotal swellings (scrotum and labia majora).

Descent of the Gonads (pp. 737–738)

3. The testes form in the dorsal abdomen and descend into the scrotum. The ovaries also descend, but only into the pelvis.

Puberty (p. 738)

4. Puberty is the interval when the reproductive organs mature and become functional. It starts in males with penile and scrotal growth and in females with breast development.

Menopause (p. 738)

5. During menopause, ovarian function declines, and ovulation and menstruation cease.

REVIEW QUESTIONS

Multiple Choice/Matching Questions

1. Which of the following is correct about the female vestibule? (a) The vaginal orifice is the most anterior of the two orifices in the vestibule. (b) The urethra is between the vaginal orifice and the anus. (c) The anus is between the vaginal orifice and the urethra. (d) The urethral orifice is anterior to the vaginal orifice.

2. Match each of the following hollow organs with its lining epithelium. (Some choices may be used more than once.)

Organ	Lining epithelium
____ (1) duct of the epididymis	(a) simple squamous
____ (2) uterine tube	(b) simple cuboidal
____ (3) ductus deferens	(c) simple columnar
____ (4) uterus	(d) pseudostratified columnar
____ (5) vagina	(e) stratified squamous

3. Which of the following structures produces the male sex hormones? (a) seminal vesicles, (b) seminiferous tubules, (c) developing follicles of the testes, (d) interstitial cells.

4. The uterine cycle can be divided into three continuous phases. Starting from the first day of the cycle, their order is (a) menstrual, proliferative, secretory; (b) menstrual, secretory, proliferative; (c) secretory, menstrual, proliferative; (d) proliferative, menstrual, secretory; (e) proliferative, secretory, menstrual.

5. All of the following are true of luteinizing hormone and follicle-stimulating hormone *except* that they are (a) secreted by the pituitary gland, (b) abbreviated as LH and FSH, (c) hormones with important functions in both males and females, (d) the sex hormones secreted by the gonads.

6. The hormone-secreting layer of the placental barrier is (a) the extraembryonic mesoderm, (b) the lacunae, (c) endothelium, (d) cytotrophoblast, (e) none of the above.

7. The function of the stereocilia in the duct of the epididymis and in the ductus deferens is (a) absorption and secretion, (b) beating forward to move sperm through, (c) initiating ejaculation, (d) beating backward to keep the sperm in these ducts for 20 days.

8. The broad ligament of the uterus in the female pelvis (a) is really a mesentery, (b) consists of the mesovarium and mesosalpinx but not the mesometrium, (c) is the same as the round ligament, (d) is the same as the lateral cervical ligament, (e) b and c.

9. The embryonic paramesonephric duct gives rise to the (a) uterus and uterine tubes, (b) ovary, (c) epididymis and vas deferens, (d) urethra.

10. Which of the following statements about the uterine cervix is *false*? (a) It dilates at the end of the first stage of labor. (b) All of it projects into and lies within the vagina. (c) The cervical glands secrete mucus. (d) It contains the cervical canal.

11. If the uterine tube is a trumpet ("salpinx"), what part of it represents the wide, open end of the trumpet? (a) isthmus, (b) ampulla, (c) infundibulum, (d) flagellum.

12. The myometrium is the muscular layer of the uterus, and the endometrium is the (a) serosa, (b) adventitia, (c) submucosa, (d) mucosa.

Short Answer Essay Questions

13. Describe the major structural and functional regions of a sperm cell.

14. (a) For a question about the path of the ductus deferens, Ryan's response—that it runs from the scrotum directly up to the penile urethra—was a common error, because the path of this duct is longer and more complex than he thought. Trace the entire path of the ductus deferens. (b) To expand on part (a), trace the pathway of an ejaculated sperm from the epididymis all the way into the uterine tube in the female.

15. In menstruation, the stratum functionalis is shed from the endometrium. Explain the hormonal and physical factors responsible for this shedding. (Hint: See Figure 24.19.)

16. (a) Outline the structural changes that a maturing follicle undergoes in the follicular stage of the ovarian cycle. (b) At what time in the ovarian cycle does ovulation occur?

17. Name three sets of arteries that must be cauterized during a simple mastectomy.

18. Some anatomy students were saying that a man's bulbourethral glands (and the urethral glands) act like city workers who come around before a parade and clear the street of parked cars. What did they mean by this analogy?

19. A man swam in a cold lake for an hour and then noticed that his scrotum was shrunken and wrinkled. His first thought was that he had lost his testicles. What had really happened?

20. Indicate whether the following statement is true or false, and explain your answer: Complete continuation and a perfect mixing of fetal and maternal blood occur across the placenta.

21. What is the blood-testis barrier, and what is its function?

22. Tell the difference (if any) between each of the following pairs of sound-alike terms: (a) sperm (spermatozoan) and spermatocyte, (b) epididymis and duct of epididymis, (c) ovarian ligament and suspensory ligament of the ovary, (d) vulva and vestibule, (e) spermatogenesis and spermiogenesis.

CRITICAL REASONING + CLINICAL APPLICATION QUESTIONS

1. A person who has had which of the following types of mastectomy will have great difficulty adducting the arms, as in doing push-ups: (a) radical mastectomy, (b) lumpectomy. Explain your answer. (Hint: Try to remember the main muscle used in doing push-ups.)

2. Based on what you know about the sources of semen, explain why a vasectomy either diminishes or does not diminish the volume of the ejaculate very much.

3. A 68-year-old man who has trouble urinating is given a rectal exam. What is his most probable condition, and what is the purpose of the rectal exam?

4. Gina gave birth to six children. During an examination, the doctor sees her cervix protruding from the vaginal orifice. What is Gina's condition, and what has caused it?

A.D.A.M.® INTERACTIVE ANATOMY QUESTIONS

1. Link: Atlas; System: Reproductive; Male Pelvic Organs (anterior) (a) The ampulla of the ductus deferens and the seminal vesicle join to form which duct? (b) The ejaculatory duct empties into which structure?

2. Link: Atlas; System: Reproductive; Viscera of Female Pelvis (superior) (a) The uterine tubes move the ovulated oocyte from the ovary to which structure in the pelvis? (b) Name the finger-like projections that hover over the ovaries and capture the oocyte.

CHAPTER 25

The Endocrine System

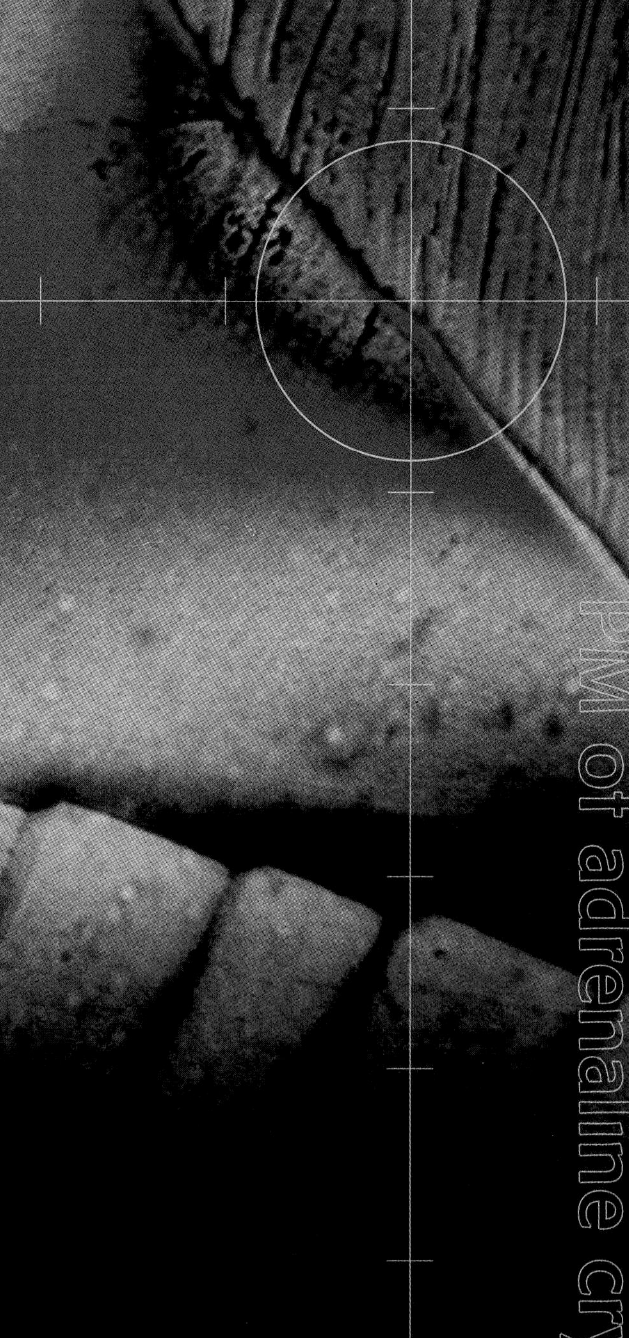

CHAPTER OUTLINE + STUDENT OBJECTIVES

The Endocrine System: An Overview 744–745

1. List the major endocrine organs, and describe their locations.
2. Describe how hormones are classified chemically.
3. Describe the basic interaction between hormones and their target cells. Describe three mechanisms that control hormone secretion.

The Major Endocrine Organs 745–759

4. Name the basic divisions of the pituitary gland.
5. List the cell types in the adenohypophysis, the hormones secreted by each cell, and the basic functions of each hormone.
6. Explain how the hypothalamus controls secretion of adenohypophyseal hormones. Define releasing hormones, and trace their path through the pituitary gland.
7. Describe the structure of the neurohypophysis and the functions of the hormones it releases.
8. Describe the anatomy of the thyroid gland. Define TH and calcitonin, and explain how TH is secreted.
9. Describe the anatomy of the parathyroid gland and the function of parathyroid hormone.
10. Name the two divisions of the adrenal gland. Compare and contrast these divisions in terms of their structure and the hormones they secrete.
11. Describe the ultrastructure of a cell that secretes steroid hormones.
12. Describe the endocrine functions of the pineal, pancreas, thymus, and gonads.

Other Endocrine Structures 759–760

13. Name a hormone produced by the heart, and define DNES.
14. Briefly explain the endocrine functions of the placenta, kidney, and skin.

Disorders of the Endocrine System 760–761

15. Describe the effects of excessive and inadequate hormonal secretion by the pituitary, thyroid, and adrenal glands; define diabetes mellitus.

The Endocrine System Throughout Life 762–763

16. Describe the development of the major endocrine glands.
17. Describe the effect of aging on some endocrine organs.

THE ENDOCRINE SYSTEM IS A SERIES OF DUCTless glands that secrete messenger molecules called **hormones** into the circulation. The circulating hormones travel to distant body cells and signal characteristic physiological responses in those cells. Through its hormonal signals, the endocrine system controls and integrates the functions of other organ systems in the body. In its general integrative role, the endocrine system resembles the nervous system, with which it closely interacts. However, because hormones travel more slowly than nerve impulses, the endocrine system tends to regulate slow processes, such as growth and metabolism, rather than processes that demand rapid responses, such as the contraction of skeletal muscle. Some major processes controlled by the endocrine system are growth of the body and of the reproductive organs, the mobilization of body defenses against stress, maintenance of proper blood chemistry, and control of the rate of oxygen use by the body's cells. The scientific study of hormones and the endocrine glands is called **endocrinology.**

THE ENDOCRINE SYSTEM: AN OVERVIEW

Our overview of the endocrine system is divided into two parts: a brief introduction to the endocrine organs and a discussion of the nature of hormones.

Endocrine Organs

Compared to most other organs of the body, the endocrine organs (Figure 25.1) are small and unimpressive. Indeed, to collect 1 kg (2.2 pounds) of hormone-producing tissue, you would need to collect *all* of the endocrine glands from *nine* adults! In addition, unlike the anatomical continuity typical of most organ systems, the endocrine organs are scattered about in widely separated parts of the body.

The endocrine cells of the body are partly contained within "pure" endocrine organs and partly within organs of other body systems. The purely endocrine organs are the *pituitary gland* at the base of the brain; the *pineal gland* in the roof of the diencephalon; the *thyroid* and *parathyroid glands* in the neck; and the *adrenal glands* on the kidneys (each adrenal gland is actually two glands, an *adrenal cortex* and an *adrenal medulla*). Organs that belong to other organ systems but contain endocrine cells include the *pancreas,* the *thymus,* the *gonads,* and the *hypothalamus* of the brain. Because in addition to its typical nervous functions the hypothalamus produces hormones, it is considered a *neuroendocrine* organ.

Most endocrine cells—like most gland cells in the body—are of *epithelial* origin. However, the endocrine system is so diverse that it also includes hormone-secreting neurons, muscle cells, and fibroblast-like cells.

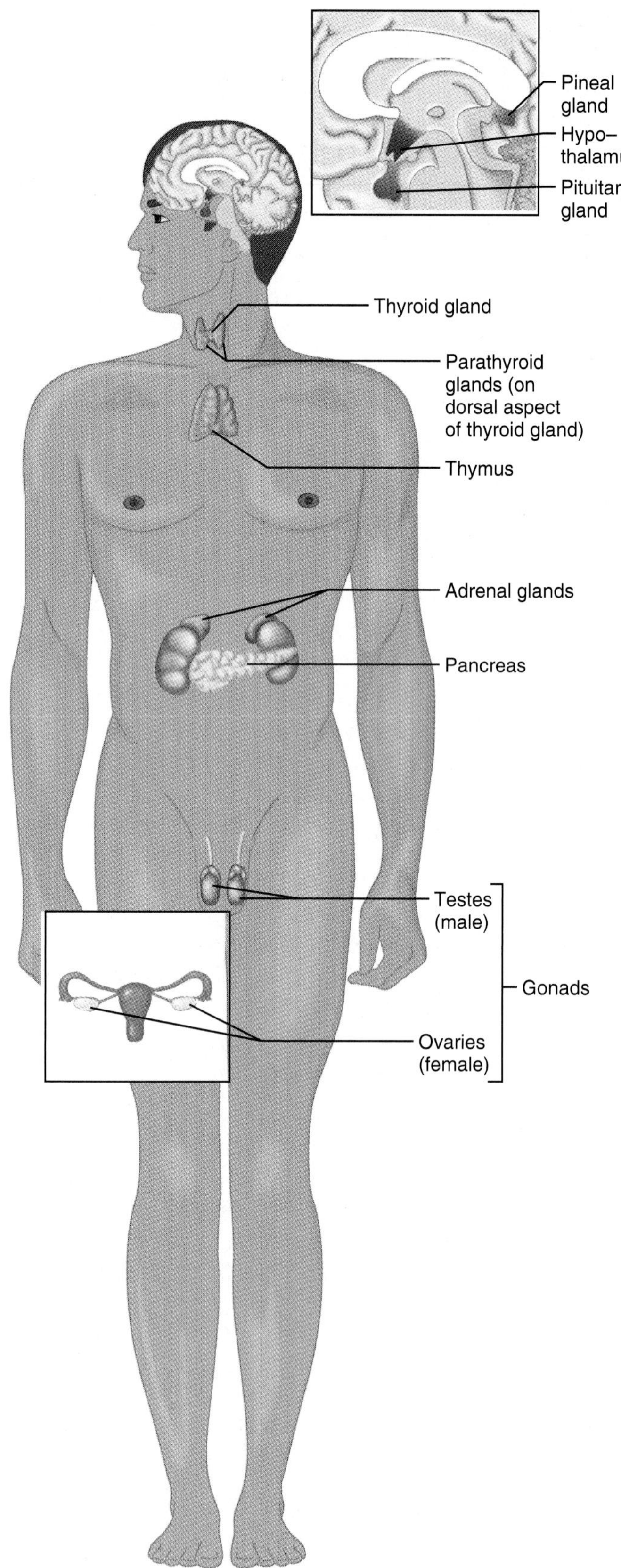

FIGURE 25.1 Location of the major endocrine organs of the body. Endocrine cells also occur in the heart, alimentary canal, kidney, skin, placenta, and elsewhere. The box at the top of this figure is an enlargement of the sectioned brain to its left.

Endocrine glands are richly supplied with blood and lymph vessels. Typically, endocrine cells are arranged in small clusters, cords, or branching networks, an arrangement that maximizes their contact with large numbers of capillaries. After endocrine cells release their hormones into the surrounding extracellular space, the hormones immediately enter the adjacent capillaries.

Hormones

In the following discussion of the nature of hormones, we examine the classes of hormones, basic hormone actions, and the control of hormone secretion.

Classes of Hormones

The body produces many different kinds of hormones, all with distinct chemical structures. Most hormones, however, belong to one of two broad molecular categories: amino acid-based molecules and steroid molecules. **Amino acid-based hormones** include modified amino acids (or *amines*), peptides (short chains of amino acids), and proteins (long chains of amino acids). **Steroids,** by contrast, are lipid molecules derived from cholesterol.

Basic Hormone Action

All major hormones circulate throughout the entire body, then leave the bloodstream at the capillaries and encounter virtually all tissues. Nevertheless, a given hormone influences only specific tissue cells, called its **target cells.** The ability of a target cell to respond to a hormone depends on the presence, on the cell's plasma membrane or in its interior, of specific receptor molecules to which that particular hormone can bind. Once binding has occurred, the target cell reacts in a preprogrammed way.

Each kind of hormone produces its own characteristic effects within the body. These effects, however, depend on the preprogrammed responses of its target cells, not on any information contained in the hormone molecule itself. To illustrate this point, hormones with similar molecular structures often have very dissimilar functions and the same hormone can have different effects on different target cells. Hormones are just molecular triggers—they do not carry any coded information.

Control of Hormone Secretion

The various endocrine cells of the body are stimulated to make and secrete their hormones by three major types of stimuli (Figure 25.2): *humoral, neural,* and *hormonal* stimuli.

Endocrine glands that secrete their hormones in direct response to changing levels of ions or nutrients in the blood are said to be controlled by **humoral stimuli** (*humoral* = relating to the blood and other body fluids). This is the simplest of all endocrine control mechanisms. For example, the cells of the parathyroid gland (see Figure 25.2a) directly monitor the concentration of calcium ions (Ca^{2+}) in the blood and then respond to any decline in this concentration by secreting a hormone that acts to reverse the decline.

The secretion of a few endocrine glands is controlled by **neural stimuli** (see Figure 25.2b). For example, sympathetic nerve fibers stimulate cells in the adrenal medulla to release epinephrine and norepinephrine during "fight, flight, or fright" situations (see p. 755).

Finally, many endocrine glands secrete their hormones in response to **hormonal stimuli** received from other endocrine glands; that is, certain hormones have the sole purpose of signaling the secretion of other hormones (see Figure 25.2c). For example, the hypothalamus of the brain secretes some hormones that stimulate the anterior part of the pituitary gland to secrete its hormones, which then proceed to stimulate hormonal secretion by the body's largest endocrine glands: the thyroid, the adrenal cortex, and the gonads. As we will see, the hypothalamus controls many functions of the endocrine system, through hormonal and other mechanisms (pp. 377 and 750).

No matter how it is stimulated, hormone secretion is always controlled by *feedback loops:* If blood concentrations of a hormone decline below a minimum set point, more hormone is secreted; if the maximum set point is exceeded, then hormone production is halted. This ensures that hormone concentrations stay within a narrow, "desirable" range in the blood.

THE MAJOR ENDOCRINE ORGANS

Now we consider in turn the body's major endocrine organs: the pituitary gland, thyroid gland, parathyroid glands, adrenal (suprarenal) glands, pineal gland, pancreas, thymus, and gonads.

The Pituitary Gland

The **pituitary gland,** or **hypophysis** (hi-pof′ĭ-sis; "undergrowth [from the brain]"), is an important endocrine organ that secretes at least nine major hormones. It sits in the hypophyseal fossa, which is a depression in the sella turcica of the sphenoid bone, just inferior to the brain (Figure 25.3). Usually said to be the size and shape of a pea, the pituitary gland is more accurately described as a pea and a stalk, or perhaps better as a golf club. The stalk, which connects superiorly to the hypothalamus, is the **infundibulum** ("funnel"). The infundibulum projects inferiorly from a part of the hypothalamus called the tuber cinereum, which lies between the optic chiasma anteriorly and the mammillary bodies posteriorly.

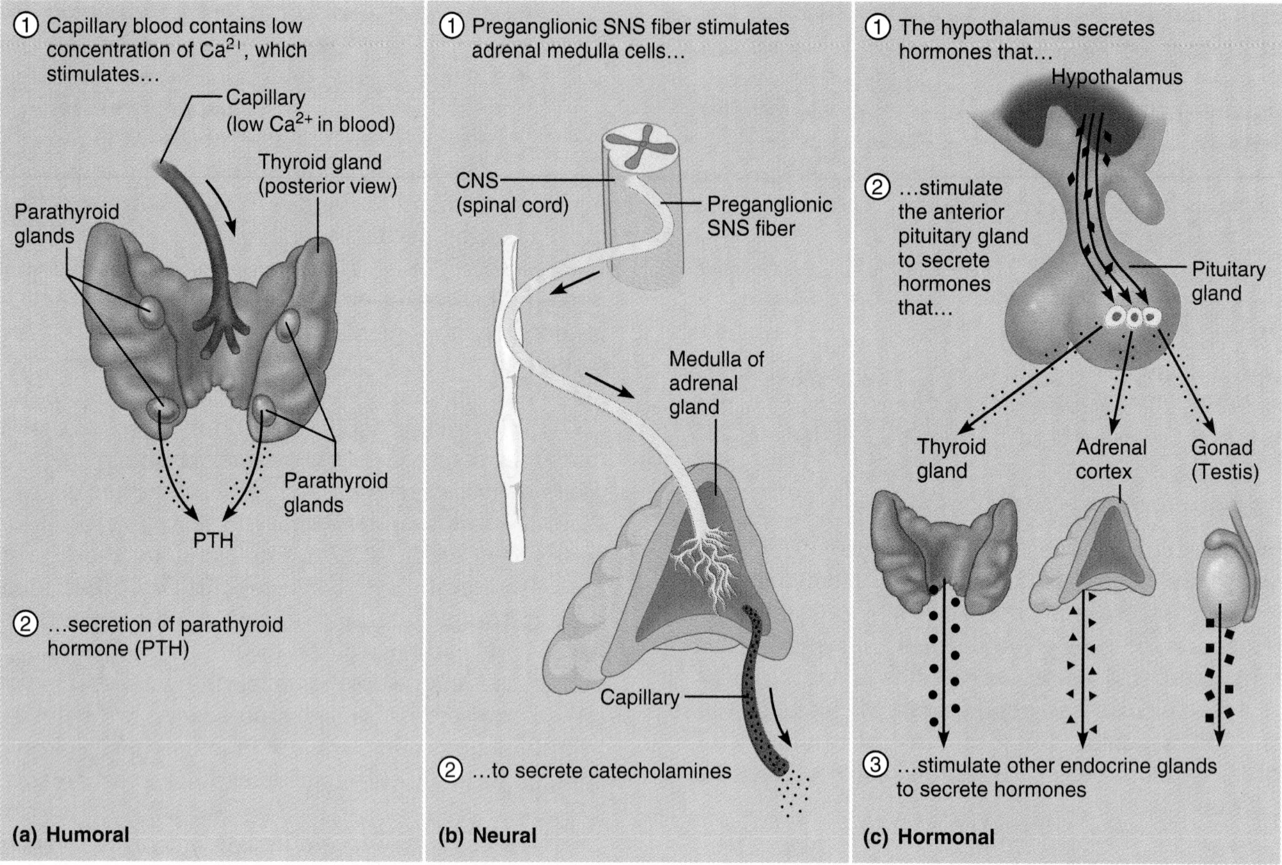

FIGURE 25.2 Control of hormone release: Three different mechanisms. (a) Control by humoral stimuli: Falling concentrations of calcium in the blood cause the cells in the parathyroid gland to secrete their hormone. **(b)** Control by neural stimuli: Sympathetic neurons (preganglionic SNS fibers) stimulate cells of the adrenal medulla to release their hormones into the bloodstream. **(c)** Control by hormonal stimuli: The hypothalamus of the brain releases hormones that stimulate the anterior pituitary gland to secrete its own hormones. Some of the pituitary hormones then proceed to signal the release of hormones by the thyroid gland, the adrenal cortex, and the gonads. In this way, the hypothalamus regulates hormone secretion by the body's major endocrine organs and thereby controls much of the endocrine system.

Critical Reasoning

Judging from Figure 25.3, deduce why tumors of the pituitary gland can lead to blindness.

As it enlarges, a tumorous pituitary gland can expand superiorly to compress the visual axons in the optic chiasma. Additionally, as the tumor continues to enlarge, it can compress the overlying hypothalamus, leading to emotional disturbances, such as uncontrolled rage.

The pituitary gland has two basic divisions, an anterior **adenohypophysis** (ad″ĕ-no-hi-pof′ĭ-sis; "glandular hypophysis") composed of glandular tissue, and a posterior **neurohypophysis** (nu″ro-hi-pof′ĭ-sis; "neural hypophysis"), which is a part of the brain. Each of these divisions has three subdivisions (Figure 25.3c). In the adenohypophysis, the largest subdivision is the anteriormost **pars distalis (anterior lobe).** Just posterior to this lies the **pars intermedia,** and just superior to it lies the **pars tuberalis,** which wraps around the infundibulum like a tube. In the neurohypophysis, the three subdivisions are, from inferior to superior, the **pars nervosa (posterior lobe),** the **infundibular stalk,** and the cone-shaped **median eminence** of the hypothalamus.

Arterial blood reaches the hypophysis through two branches of the internal carotid artery (Figure 25.4). The **superior hypophyseal artery** supplies the entire adenohypophysis and the infundibulum, whereas the **inferior hypophyseal artery** supplies the pars nervosa. The veins from the extensive capillary beds in the pituitary gland drain into the cavernous sinus (see p. 566) and other nearby dural sinuses.

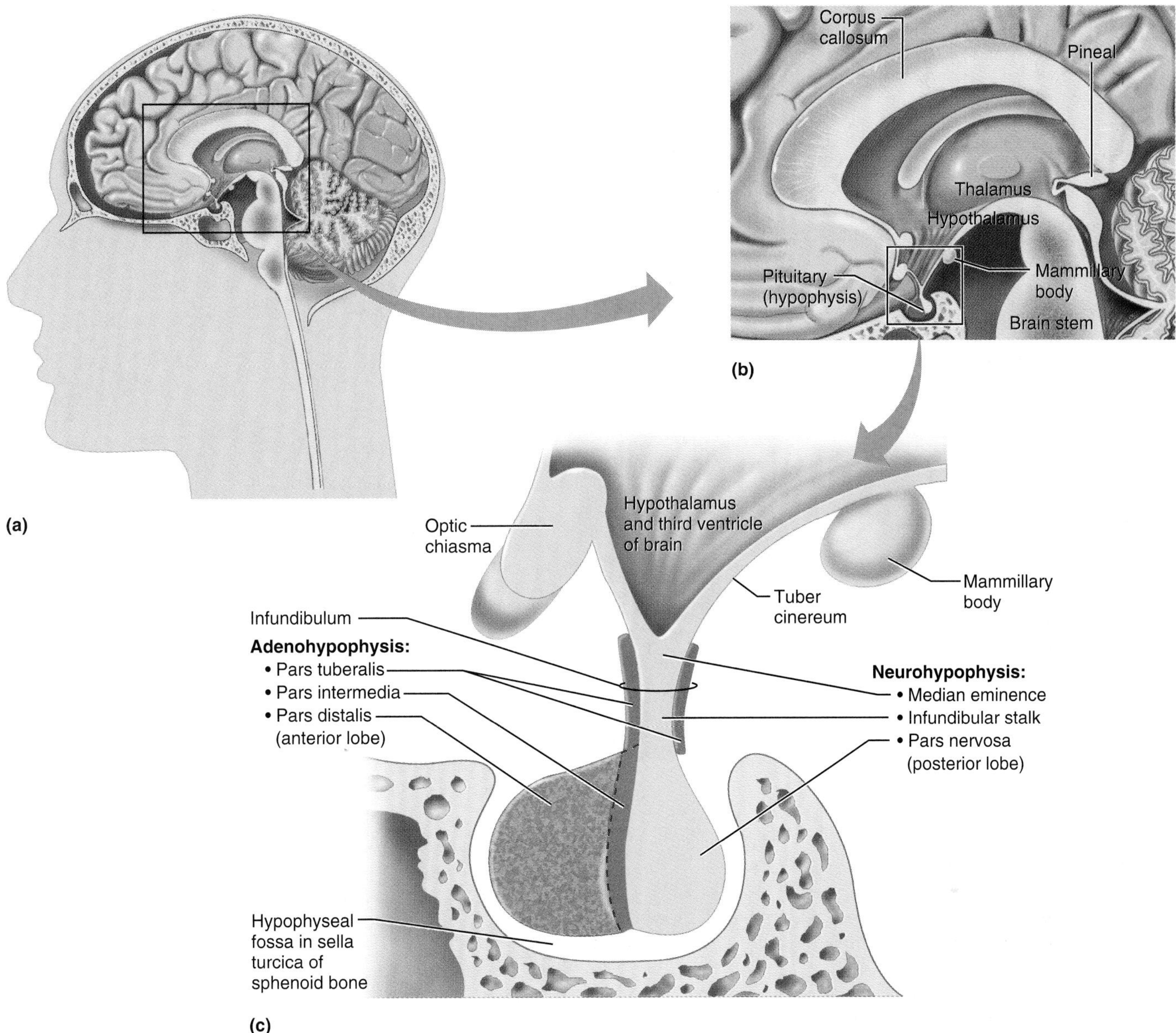

FIGURE 25.3 The pituitary gland (hypophysis). **(a)** Orientation diagram showing the brain cut in half midsagittally. **(b)** Enlargement of the brain region that centers around the diencephalon, showing the location of the pituitary gland. **(c)** Basic regions of the pituitary gland. Note that the anterior aspect is to the left. The adenohypophysis is colored pink and the neurohypophysis is colored peach.

The Adenohypophysis

Our exploration of the adenohypophysis concentrates on its largest division, the *pars distalis,* which contains five different types of endocrine cells that make and secrete at least seven different hormones (Table 25.1 on p. 749). Because these cells secrete protein hormones, they have the cytoplasmic features of protein-secreting cells, such as secretory granules and a well-developed rough endoplasmic reticulum (rough ER) and Golgi apparatus. The five cell classes in the pars distalis are the following:

1. **Somatotropic cells,** the most abundant cells in the pars distalis, secrete **growth hormone (GH),** which is also called **somatotropic hormone (SH)** (so″mah-to-tro′pik; "body changing") or **somatotropin.** This hormone stimulates growth of the entire body by signaling body cells to increase their production of proteins and by stimulating growth of the epiphyseal plates of the skeleton (see p. 142). It stimulates these actions directly, and also indirectly by signaling the liver to secrete *insulin-like growth factor-1* (*somatomedin*), which acts together with GH. The current and potential uses of GH are explored in this chapter's A Closer Look box (p. 752).

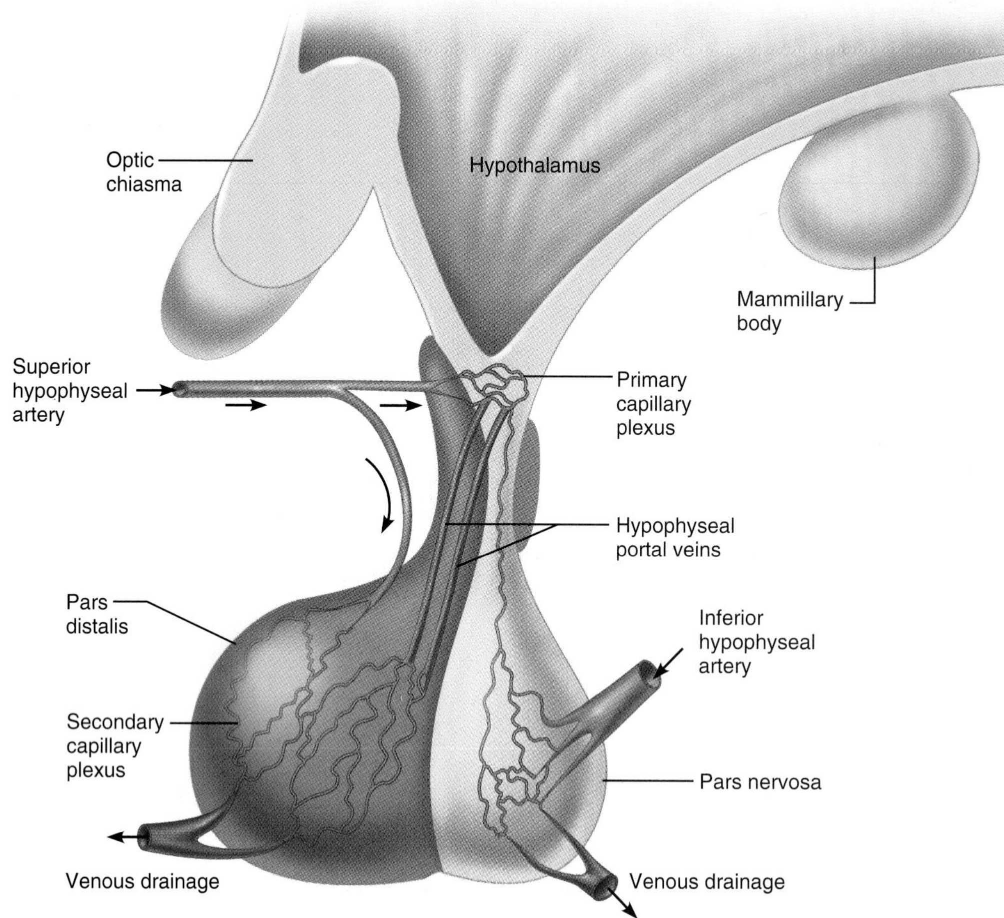

FIGURE 25.4 The major vessels of the pituitary gland. For orientation, see Figure 25.3.

2. **Mammotropic cells** (*mamm* = breast) secrete **prolactin (PRL)** (pro-lak′tin), a hormone that stimulates manufacture of milk by the breasts.
3. **Thyrotropic cells** secrete **thyroid-stimulating hormone (TSH),** which signals the thyroid gland to secrete its own hormone (thyroid hormone; see p. 754), ultimately controlling our metabolic rate.
4. **Corticotropic cells** secrete at least two hormones that are split from a common parent molecule: **adrenocorticotropic hormone (ACTH)** (ad-re″no-kor″tĭ-ko-trōp′ik; "adrenal cortex-changing") and **melanocyte-stimulating hormone (MSH).** ACTH signals the adrenal cortex to secrete hormones (glucocorticoids; see p. 756) that help us cope with stress, whereas MSH apparently darkens skin pigmentation by stimulating melanocytes in the epidermis.
5. **Gonadotropic cells** secrete **follicle-stimulating hormone (FSH)** and **luteinizing hormone (LH),** which together are called **gonadotropins** (go-nad″o-

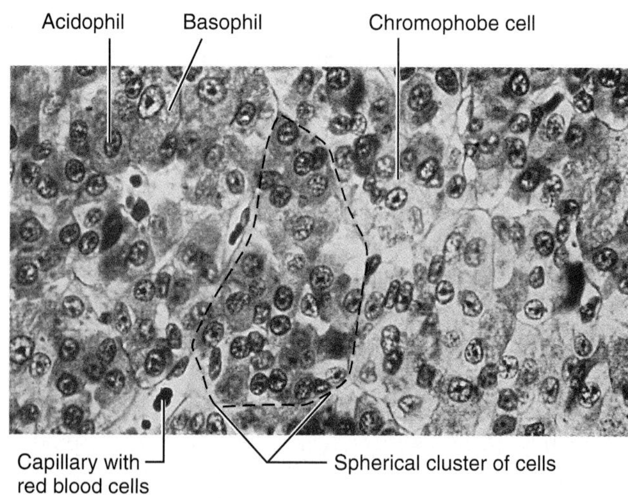

FIGURE 25.5 Histology of the pars distalis of the adenohypophysis (350 ×). Note the acidophil, basophil, and chromophobe cells.

Table 25.1 Cells and Hormones of the Pituitary Gland

Cell Type	Hormone	Target/Effects
ADENOHYPOPHYSIS (PARS DISTALIS)		
1. Somatotropic cell	Growth hormone (GH) (= somatotropic hormone (SH))	Stimulates growth of skeleton at epiphyseal plates; stimulates body cells to synthesize protein
2. Mammotropic cell	Prolactin (PRL)	Signals mammary gland to make milk
3. Thyrotropic cell	Thyroid-stimulating hormone (TSH)	Signals thyroid gland to release thyroid hormone
4. Corticotropic cell	Adrenocorticotropic hormone (ACTH)	Signals adrenal cortex to secrete glucocorticoid hormones
	Melanocyte-stimulating hormone (MSH)	Apparently signals melanocytes in skin to darken skin pigmentation; may increase mental alertness
5. Gonadotropic cell	Follicle-stimulating hormone (FSH)	In females, stimulates maturation of follicles in the ovaries and helps signal production of estrogens by these follicles; in males, stimulates sperm production and the production of androgen-binding protein by sustentacular cells in the testes
	Luteinizing hormone (LH)	In females, triggers ovulation and signals the ovarian follicles to produce androgens (which are converted to estrogens) and progesterone; in males, signals the secretion of androgens by interstitial cells in the testes
NEUROHYPOPHYSIS		
1. Neurons from supraoptic nucleus of hypothalamus	Antidiuretic hormone (ADH) (= vasopressin)	Stimulates kidneys (distal tubules and collecting tubules) to reclaim more water from the urine; raises blood pressure; makes us want to cuddle, groom, and pair bond
2. Neurons from paraventricular nucleus of hypothalamus	Oxytocin	Signals contraction of smooth muscle of male and female reproductive tracts; initiates labor and ejection of milk from breast; makes us want to cuddle, groom, and pair bond

trōp′inz). These hormones act on the gonads, stimulating maturation of the sex cells and inducing the secretion of sex hormones (see Chapter 24). In females, FSH and LH stimulate the maturation of the egg-containing ovarian follicles and the secretion of androgens, estrogens, and progesterone from theca and granulosa cells in the ovary (see pp. 716–718). Furthermore, a large amount of LH is secreted in the middle of the female menstrual cycle to induce ovulation (see p. 718). In males, LH signals the secretion of androgens (mainly testosterone) by interstitial cells in the testes (see pp. 707–708), and FSH stimulates the maturation of sperm cells and the production of androgen-binding protein by sustentacular cells (see p. 707).

Note that four of the seven hormones secreted by the pars distalis (TSH, ACTH, FSH, LH) regulate the secretion of hormones by other endocrine glands; these hormones are called **tropic hormones** (trō′pik; "changing"). The remaining three adenohypophyseal hormones (GH, PRL, and MSH) act directly on nonendocrine target tissues.

The endocrine cells of the pars distalis are clustered into spheres and branching cords separated by capillaries. When tissue of the pars distalis is stained with typical histological dyes and viewed by light microscopy, its five cell types group into three categories (Figure 25.5): (1) **acidophils,** which stain with acidic stains and include the somatotropic and mammotropic cells; (2) **basophils,** which stain with basic stains and include the

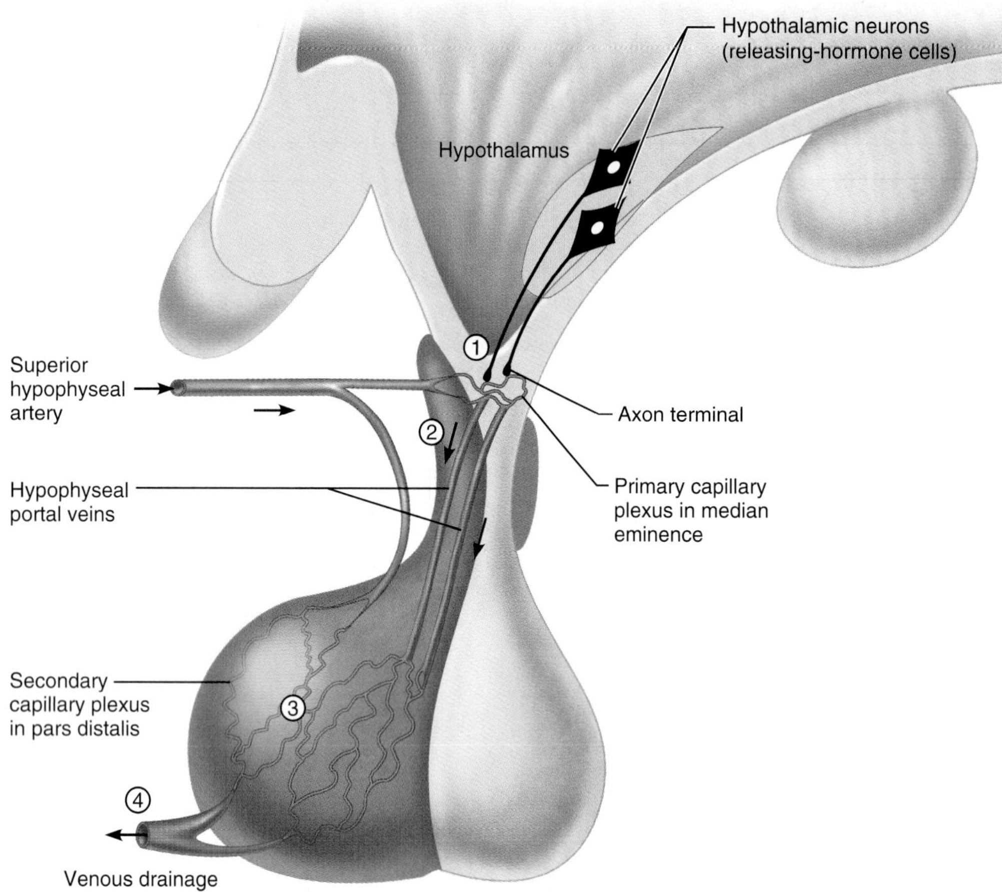

FIGURE 25.6 Structural and functional relationships between the adenohypophysis and the hypothalamus. Some hypothalamic neurons (upper right) secrete releasing hormones **(1)**, which travel through the hypophyseal portal veins **(2)** to the pars distalis **(3)**. There, the releasing hormones signal the adenohypophyseal cells to release their hormones into the capillaries, and these hormones leave the pars distalis in veins **(4)**.

thyrotropic, corticotropic, and gonadotropic cells; and (3) **chromophobes** ("color avoiders"), which stain poorly. Chromophobes are either immature cells or cells whose supply of hormone has been depleted.

Whereas the pars distalis is by far the most important division of the adenohypophysis, the other divisions—the *pars intermedia* and *pars tuberalis*—are not as well understood. The pars intermedia contains corticotropic cells that secrete more MSH than ACTH. The pars tuberalis contains gonadotropic cells, thyrotropic cells, and unique, *pars tuberalis-specific cells* that have receptors for melatonin, a hormone that is secreted by the pineal gland and regulates our daily (circadian) rhythms according to light/dark cycles (photoperiod). Therefore, the pars tuberalis may be where sexual functions and metabolic rate are brought under the influence of circadian rhythms and photoperiod. For more information on the function of the pineal and melatonin, see p. 379.

Hypothalamic Control of Hormone Secretion from the Adenohypophysis

The secretion of hormones by the adenohypophysis is controlled by the hypothalamus of the brain. The hypothalamus exerts its control by secreting peptide hormones called **releasing hormones (releasing factors),** which then signal the cells in the adenohypophysis to release their hormones. The hypothalamus also secretes **inhibiting hormones,** which *turn off* the secretion of hormones by the adenohypophysis when necessary. There are distinct releasing and inhibiting hormones for almost every adenohypophyseal hormone, including *growth hormone–releasing hormone, prolactin-inhibiting hormone, gonadotropin-releasing hormone,* and *corticotropin-releasing hormone.*

Here is how the releasing hormones signal the secretion of adenohypophyseal hormones (Figure 25.6). First, releasing hormones made in hypothalamic neurons

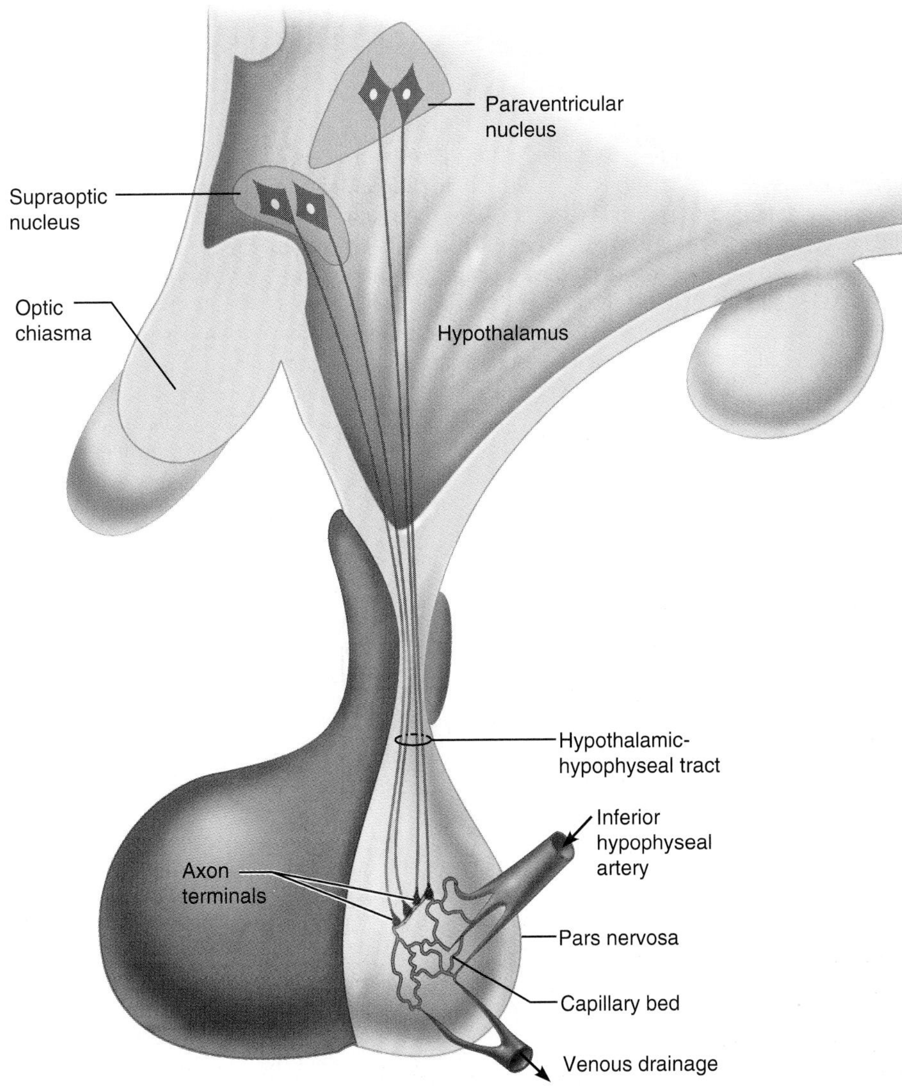

FIGURE 25.7 **Neurohypophysis.** Hypothalamic neurons, whose cell bodies lie in the paraventricular and supraoptic nuclei, send axons down through the neurohypophysis and secrete the neurohypophyseal hormones from their axon terminals in the pars nervosa.

are secreted like neurotransmitters from the axon terminals of these neurons. (Note that in this case *neurons* are serving as endocrine cells.) Upon secretion, the releasing hormones enter a **primary capillary plexus** in the median eminence and then travel inferiorly in **hypophyseal portal veins** to a **secondary capillary plexus** in the pars distalis. Leaving this plexus, the releasing hormones attach to the adenohypophyseal cells and stimulate these cells to secrete hormones (GH, LH, TSH, PRL, and so on) back into the secondary plexus. From there the newly secreted adenohypophyseal hormones proceed into the general circulation and travel to their target organs throughout the body.

The primary and secondary capillary plexuses in the pituitary gland, plus the intervening hypophyseal portal veins, constitute the *hypophyseal portal system*—which, like all portal systems (p. 570), consists of two capillary beds separated by veins, with the first bed receiving molecules that are destined for tissues supplied by the second capillary bed.

In summary, the hypothalamus controls the secretion of hormones by the adenohypophysis, which in turn controls the secretion of hormones by the thyroid gland, the adrenal cortex, and the gonads. In this way, the brain controls the body's largest and most important endocrine glands (see Figure 25.2c).

The Neurohypophysis

The neurohypophysis, which secretes two hormones, is structurally a part of the brain (Figure 25.7). It consists of nervous tissue that contains unmyelinated axons and neuroglial cells. Its axons make up the **hypothalamic-hypophyseal tract,** which arises from neuron cell bodies in the supraoptic and paraventricular nuclei in the hypothalamus and ends in axon terminals in the pars

A CLOSER LOOK

Potential Uses for Growth Hormone

Growth hormone (GH) has been used for pharmaceutical purposes since its discovery in the 1950s. Originally obtained from the pituitary glands of cadavers, it is now made biosynthetically and administered by injection. Although widely used in clinical trials, its use as a prescription drug is restricted until research can fully document its helpful and harmful effects—many of which are very intriguing.

GH is administered legally to children who do not produce it naturally or who have chronic kidney failure, to allow these children to grow to near-normal heights (p. 760). Unfortunately, some physicians succumb to pressure to prescribe GH to children who do produce it but are extremely short.

Researchers have discovered many non-growth effects of GH, mainly by administering this hormone to *adults* with a growth-hormone deficiency. Such studies found that GH decreases body fat and increases lean body mass, bone density, and muscle mass. It also increases the performance and muscle of the heart, decreases blood cholesterol, boosts the immune system, and perhaps improves one's psychological outlook. Such effects have led to abuse of GH by body builders and athletes, which is one reason why this substance remains restricted.

GH may also reverse some effects of aging. Many people naturally stop producing GH after age 60, and this may explain why their ratio of lean-to-fat mass declines and their skin thins. Administration of GH to elderly patients reverses these declines. However, research reveals the administered GH does not increase strength or exercise tolerance in such patients, and a careful study of very sick patients in intensive care units (where GH is routinely given to restore nitrogen balance) found that high doses of GH are associated with increased mortality. For these reasons, earlier media claims that GH is a "youth potion" have proven to be dangerously misleading, and GH should not be administered to the very old or the critically ill.

Can growth hormone help aged patients?

GH may help AIDS patients. Because of improved antibiotics, fewer AIDS patients are dying from opportunistic infections, so more die from the weight loss called "wasting." It has been shown that injections of GH can actually reverse wasting during AIDS, leading to weight gain—a gain of lean muscle. In 1996, the FDA approved the use of GH to treat such wasting.

GH is not a wonder drug, even in cases where it is clearly beneficial. GH treatment is expensive and has side effects: It can lead to fluid retention, carpal tunnel syndrome, joint and muscle pain, high blood sugar, and gynecomastia (breast enlargement in males). Hypertension, heart enlargement, diabetes, and cancer of the colon may result from high doses of GH, and edema and headaches accompany even the lowest doses. Carefully tailored dosages can avoid most of these side effects, however.

Intensive research into the potential benefits of GH should keep this hormone in the public eye for years to come.

nervosa. The neurohypophyseal hormones are made in the neuron cell bodies, transported along the axons, and stored in the axon terminals. When the neurons fire, they release the stored hormones into a capillary bed in the pars nervosa for distribution throughout the body. Therefore, the neurohypophysis does not make hormones but only stores and releases hormones produced in the hypothalamus. The neurohypophysis releases two peptide hormones: **antidiuretic hormone (ADH)** (an″tĭ-di″u-ret′ik; "inhibiting urination"), also called *vasopressin,* and **oxytocin** (ok″sĭ-to′sin; "childbirth hormone") (see Table 25.1).

Made in the neurons of the supraoptic nucleus, ADH targets the collecting tubules and distal tubules in the kidney, which respond by reabsorbing more water from the urine and returning it to the bloodstream (see p. 685). In this way, ADH helps the body retain as much fluid as possible when thirst (dehydration) or fluid loss (severe bleeding) occurs. Also, when this fluid loss lowers blood pressure, ADH raises blood pressure to normal by signaling the peripheral arterioles to constrict (*vasopressin* = vessel constrictor).

Oxytocin, produced in the paraventricular nucleus, induces contraction of the smooth musculature of reproductive organs in both males and females. Most importantly, it signals the myometrium of the uterus to contract, expelling the infant during childbirth (see p. 731); it also induces contraction of muscle-like cells (myoepithelial cells) around the secretory alveoli in the breast to eject milk during breast-feeding. Finally, both oxytocin and ADH cause a desire to cuddle, groom, and bond with a mate.

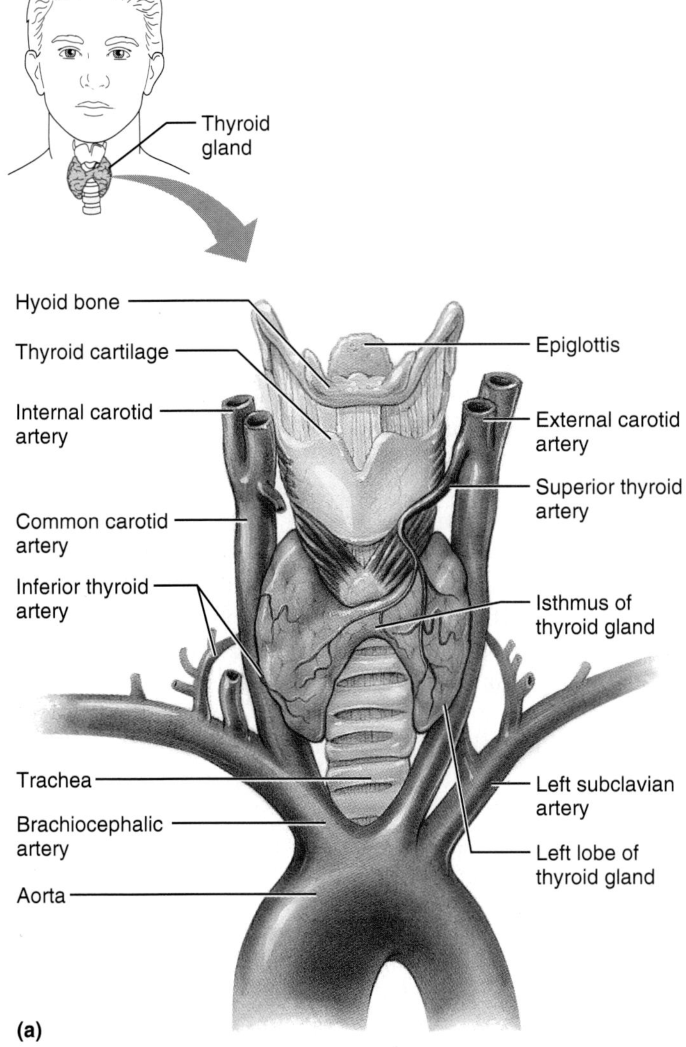

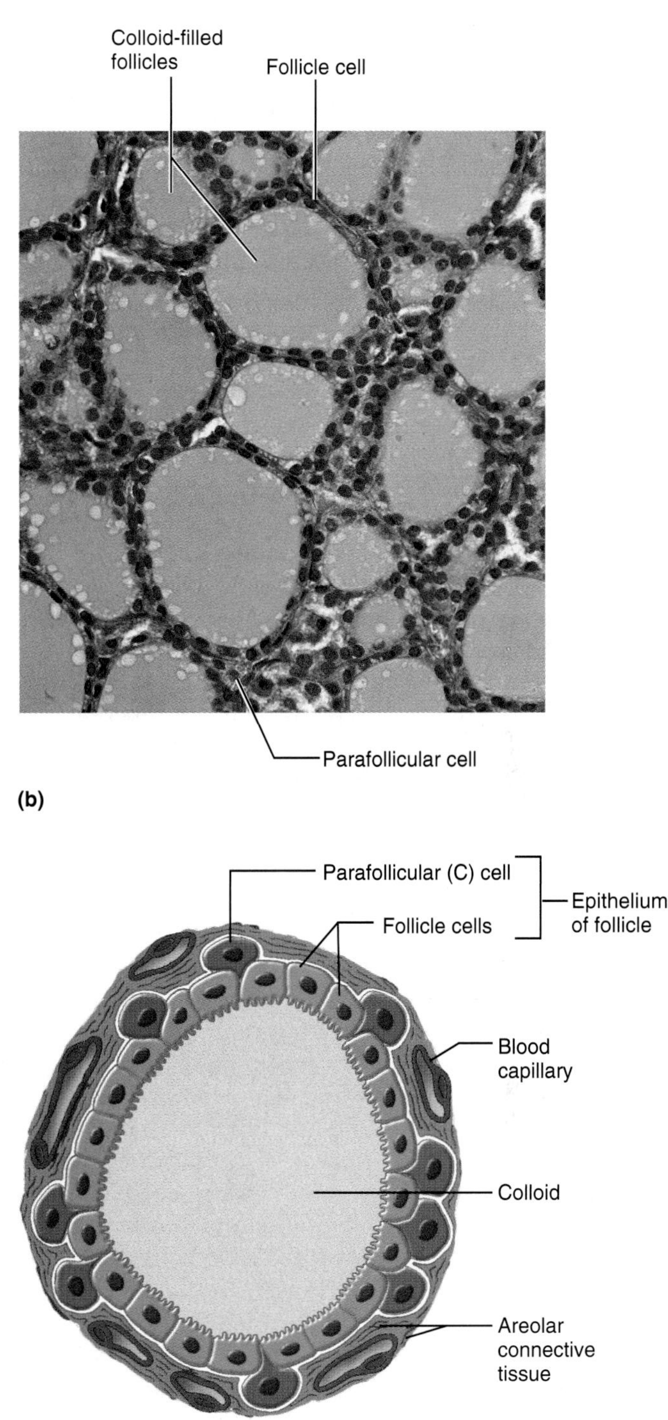

FIGURE 25.8 **The thyroid gland.** **(a)** Location and arterial supply of the thyroid gland, anterior view. **(b)** Micrograph of the thyroid gland, showing the round follicles (250 ×). **(c)** Diagram of a thyroid follicle and the surrounding connective tissue.

CLINICAL APPLICATIONS

THERAPEUTIC USES OF OXYTOCIN Sometimes, such as when a pregnancy has lasted well beyond the estimated due date, a physician will deem it necessary to induce labor. The most effective way of inducing labor is to inject the mother with natural or synthetic oxytocin, which initiates uterine contractions. Oxytocics are also used to stop bleeding following the delivery (by causing constriction of the ruptured blood vessels at the site of placental detachment) or to stimulate the ejection of milk by the mammary glands. ✚

The Thyroid Gland

The butterfly-shaped **thyroid gland** is located in the anterior neck, on the trachea just inferior to the larynx. It has two lateral **lobes** (the butterfly's "wings") connected by a median bridge called the **isthmus** (Figure 25.8a). The thyroid gland is the largest pure endocrine gland in the body and has a prodigious blood supply from the superior and inferior thyroid arteries.

Internally, the thyroid gland is composed of hollow, approximately spherical **follicles** separated by an areolar connective tissue rich in capillaries (Figure 25.8b). The walls of each follicle are formed by a layer of cuboidal or squamous epithelial cells called **follicle (follicular) cells,** and the central lumen is filled with a jelly-like substance called **colloid** (kol′oid; "glue-like") consisting of **thyroglobulin** (thi″ro-glob′u-lin), a protein from which thyroid hormone is ultimately derived. Lying within the follicular epithelium are **parafollicular (C) cells,** which appear to project into the surrounding connective tissue (Figure 25.8c).

The thyroid produces two hormones: thyroid hormone and calcitonin. The follicle cells of the thyroid gland secrete **thyroid hormone (TH),** a name that actually applies to two similar molecules called thyroxine (thi-rok′sin) or T_4, and triiodothyronine (tri″i-o″do-thi′ro-nēn) or T_3. Each of these hormone molecules is constructed from a pair of amino acids and contains the element iodine, which is essential to the function of the hormone. Thyroid hormone affects many target cells throughout the body in its main function of increasing the basal metabolic rate (the rate at which the body uses oxygen to transform nutrients into energy). Individuals who secrete an excess of TH are "hyperactive," nervous, and continually feel warm, whereas those who do not produce enough TH feel sluggish and cold. These and other disorders of the endocrine glands are covered in more detail in the Disorders section on pages 760–761.

TH is made and secreted by the thyroid follicles in the following way. The follicle cells continuously synthesize the precursor protein thyroglobulin and secrete it into the center of the follicle for iodination and storage, making the thyroid gland the only endocrine gland that stores its hormone extracellularly and in large quantities—its supply of TH is enough to last several months. To initiate secretion of the stored TH into the blood, the *pituitary gland* releases thyroid-stimulating hormone (TSH, *not* TH), which signals the follicle cells to reclaim the thyroglobulin by endocytosis. Next, after TH is cleaved off the thyroglobulin molecules by lysosomal enzymes in the cytoplasm of the follicle cell, TH diffuses out of the follicle cells into the capillaries around the follicle, thereby entering the circulation.

The parafollicular cells of the thyroid secrete the protein hormone **calcitonin** (kal″sĭ-to′nin), which depresses excessive levels of Ca^{2+} in the blood by both slowing the calcium-releasing activity of osteoclasts in bone and increasing calcium secretion by the kidney. Calcitonin has no demonstrable function in adults; it seems to act mostly during childhood, when the skeleton grows quickly and osteoclast activity must be slowed to allow bone deposition and bone growth.

The Parathyroid Glands

The small, yellow-brown **parathyroid glands** lie on the posterior surface of the thyroid gland (Figure 25.9). Even though they may be embedded in the substance of the thyroid, the parathyroids always remain distinct organs surrounded by their own connective tissue capsules. Most people have two pairs of parathyroid glands, but the precise number varies among individuals. As many as eight glands have been reported, and some may be located in other regions of the neck or even in the thorax.

Histologically, the parathyroid gland contains thick, branching cords composed of two types of endocrine cells (see Figure 25.9b and c): small, abundant **chief cells** and rare, larger **oxyphil cells** (ok′sĭ-fil; "acid-loving" = "stains with acidic dyes").

Whereas the function of oxyphil cells is unknown, the chief cells produce a small protein hormone called **parathyroid hormone (PTH,** or **parathormone)**, which increases the blood concentration of Ca^{2+} whenever it falls below some threshold value. PTH raises blood calcium by (1) stimulating osteoclasts to release more Ca^{2+} from bone, (2) decreasing the secretion of Ca^{2+} by the kidney, and (3) activating vitamin D, which stimulates the uptake of Ca^{2+} by the intestine. PTH is essential to life because low Ca^{2+} levels lead to lethal neuromuscular disorders.

Note that parathyroid hormone and calcitonin have *opposite,* or antagonistic, effects: PTH raises blood calcium, whereas calcitonin lowers it.

Critical Reasoning

Years ago, before PTH was discovered, it was observed that some patients recovered uneventfully after partial (or even total) removal of the thyroid gland, but others exhibited uncontrolled muscle spasms and severe pain and soon died. Explain why thyroid surgery sometimes proved lethal.

The lethal neuromuscular disorders resulted when surgeons unwittingly removed the parathyroid glands along with the thyroid.

The Adrenal (Suprarenal) Glands

The paired **adrenal (suprarenal) glands** are pyramidal or crescent-shaped organs perched on the superior surface of the kidneys (see Figure 25.10). Each adrenal gland is supplied by up to 60 small *suprarenal arteries,* which form three groups: the superior suprarenal arteries from the inferior phrenic artery, the middle suprarenal arteries from the aorta (see p. 560), and the inferior suprarenal arteries from the renal artery. The left *suprarenal vein* drains into the renal vein, whereas the *right suprarenal vein* drains into the inferior vena cava (see p. 569). The nerve supply, consisting almost exclusively of sympathetic fibers to the adrenal medulla, is richer than the nerve supply of any other visceral organ (see p. 456).

Each adrenal gland is two endocrine glands in one (Figure 25.10a). The internal *adrenal medulla,* more like a "knot" of nervous tissue than a gland, derives from the neural crest and acts as part of the sympa-

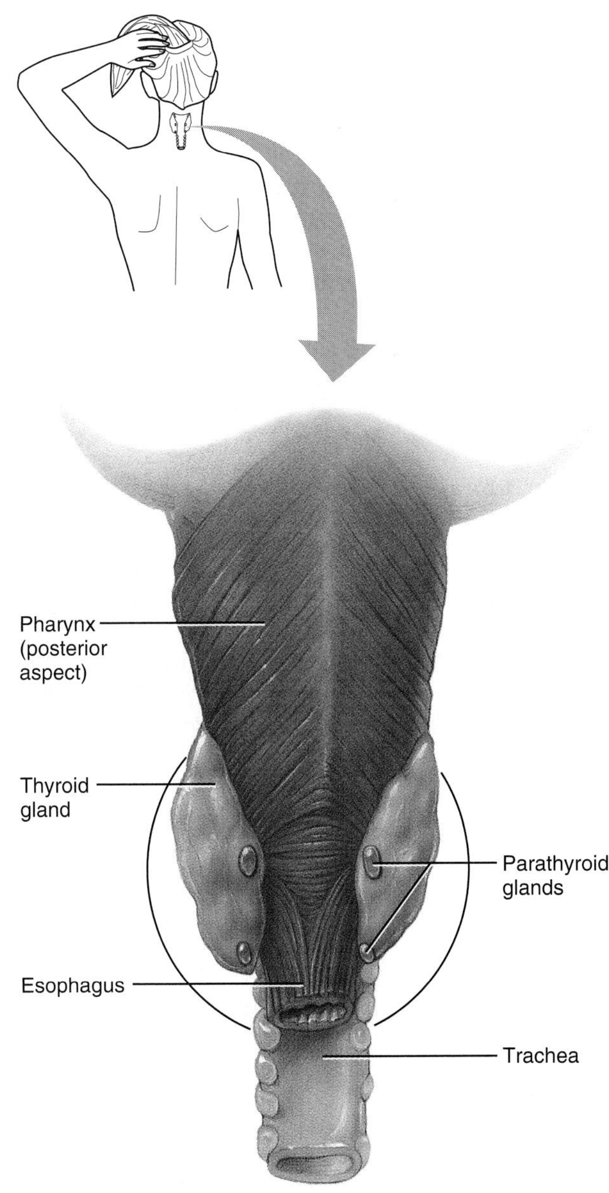

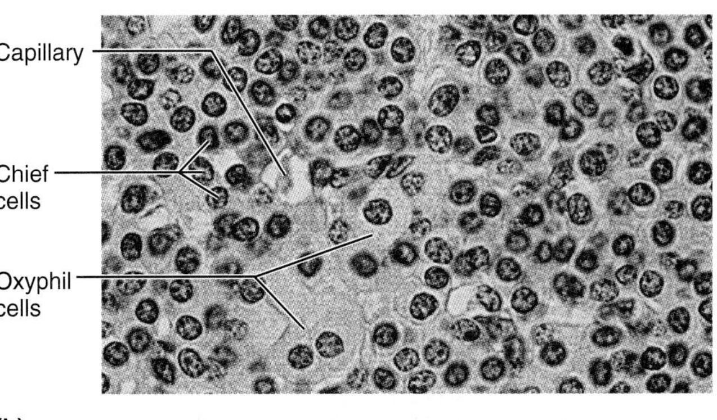

Chief cells

Oxyphil cell

Blood capillary

Fibrous capsule of the parathyroid gland

(c)

FIGURE 25.9 The parathyroid glands. **(a)** Posterior view of the pharynx and trachea showing the location of the parathyroid glands on the posterior aspect of the thyroid gland. **(b)** Micrograph of a section through a parathyroid gland (540 ×). **(c)** Diagram of the histology of the parathyroid gland.

thetic nervous system. The external *adrenal cortex,* surrounding the medulla and forming the bulk of the gland, derives from mesoderm. The cortex and medulla secrete hormones of entirely different chemical types, but all adrenal hormones help us cope with extreme situations—situations associated with danger, terror, or stress.

The Adrenal Medulla

The centrally located **adrenal medulla** is discussed in Chapter 15 as a part of the autonomic nervous system (pp. 456–457), so it is covered only briefly here. Its spherical **chromaffin** (kro-maf′in) **cells** are modified postganglionic sympathetic neurons that secrete catecholamines (the amine hormones epinephrine and norepinephrine) into the blood to enhance the "fight, flight, and fright" response. These cells store their catecholamines in secretory vesicles that can be stained with salts containing chromium metal. (Indeed, *chromaffin* literally means "with an affinity for chromium.") Within the adrenal medulla, the chromaffin cells are arranged in spherical clusters (with some branching cords).

The Adrenal Cortex

The thick **adrenal cortex** secretes a variety of hormones, all of which are *steroids.* Microscopically, this cortex exhibits three distinct layers, or zones (Figure 25.10). From external to internal these are as follows:

1. **Zona glomerulosa** (glo-mer-u-lōs′ah), which contains cells arranged in spherical clusters (*glomerulus* = ball of yarn)
2. **Zona fasciculata** (fah-sik″u-lah′tah), whose cells are arranged in parallel cords (*fascicle* = bundle of parallel sticks) and contain an abundance of lipid droplets

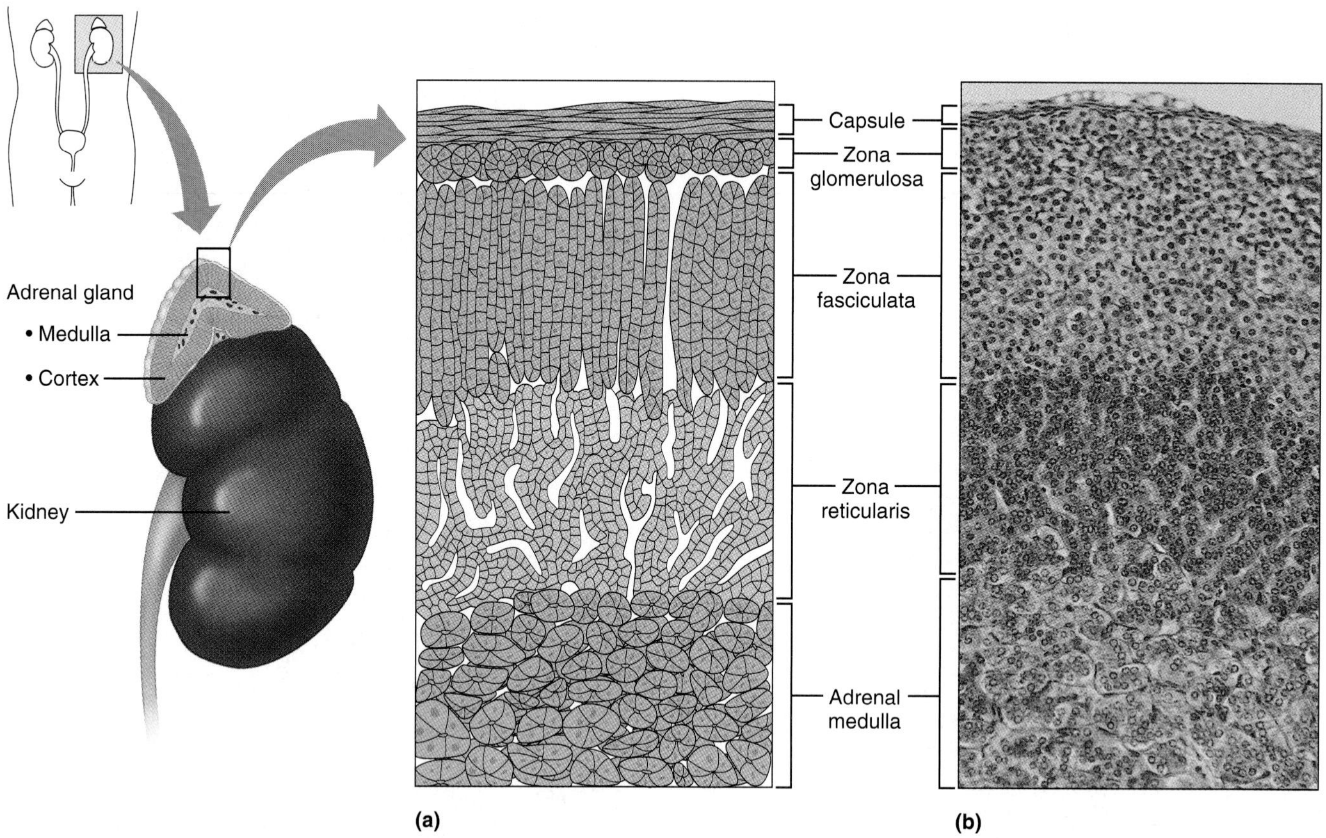

FIGURE 25.10 The adrenal gland. The diagrams at left show the position of an adrenal gland atop a kidney, with the gland cut open to reveal its cortex and medulla. **(a)** Drawing of the histology of the adrenal cortex and a portion of the adrenal medulla. **(b)** Photomicrograph (125 ×) of the tissue drawn in (a).

3. **Zona reticularis** (rĕ-tik′u-lar″is), whose cells are arranged in a branching network (*reticulum* = network) and stain intensely with the pink dye eosin

The hormones secreted by the adrenal cortex are **corticosteroids** (kor″tĭ-ko-ste′roids), which, along with the sex hormones, are the body's major steroid hormones. Adrenal corticosteroids are of two main classes: *mineralocorticoids* (min″er-al-o-kor′tĭ-koids) and *glucocorticoids* (gloo″ko-kor′tĭ-koids).

The main **mineralocorticoid,** called **aldosterone** (al-dos′ter-ōn), is secreted by the zona glomerulosa in response to a decline in either blood volume or blood pressure, as occurs in severe hemorrhage. To compensate for either decline, aldosterone signals the distal and collecting tubules in the kidney to reabsorb more water and sodium into the blood.

Glucocorticoids, of which **cortisol** is the main type, are secreted by the zona fasciculata and zona reticularis to help the body deal with stressful situations such as fasting, anxiety, trauma, crowding, and infection. In essence, glucocorticoids keep blood glucose levels high enough to support the brain's activities while forcing most other body cells to switch to fats and amino acids as energy sources. Glucocorticoids also redirect circulating lymphocytes (p. 508) to lymphoid and peripheral tissues, where most pathogens are. When present in large quantities, however, glucocorticoids depress the inflammatory response and inhibit the immune system. Indeed, glucocorticoids are administered as anti-inflammatory drugs to treat rheumatoid arthritis, severe allergies, tendonitis, joints injuries, and other inflammatory disorders.

The pathway by which stress leads to glucocorticoid secretion begins when the brain perceives a stressful situation. The hypothalamus then sends corticotropin-releasing hormone (CRH) to the adenohypophysis (see p. 750), which secretes ACTH, which in turn travels to the adrenal cortex, where it signals glucocorticoid secretion. In addition to this hormonal pathway, the sympathetic nervous system can also stimulate glucocorticoid secretion.

The zona reticularis also secretes large quantities of an androgen hormone called dehydroepiandrosterone (DHEA), the function of which remains unclear. After its secretion (via the stress-CRH-ACTH pathway) and its release from the adrenal cortex, DHEA is converted to

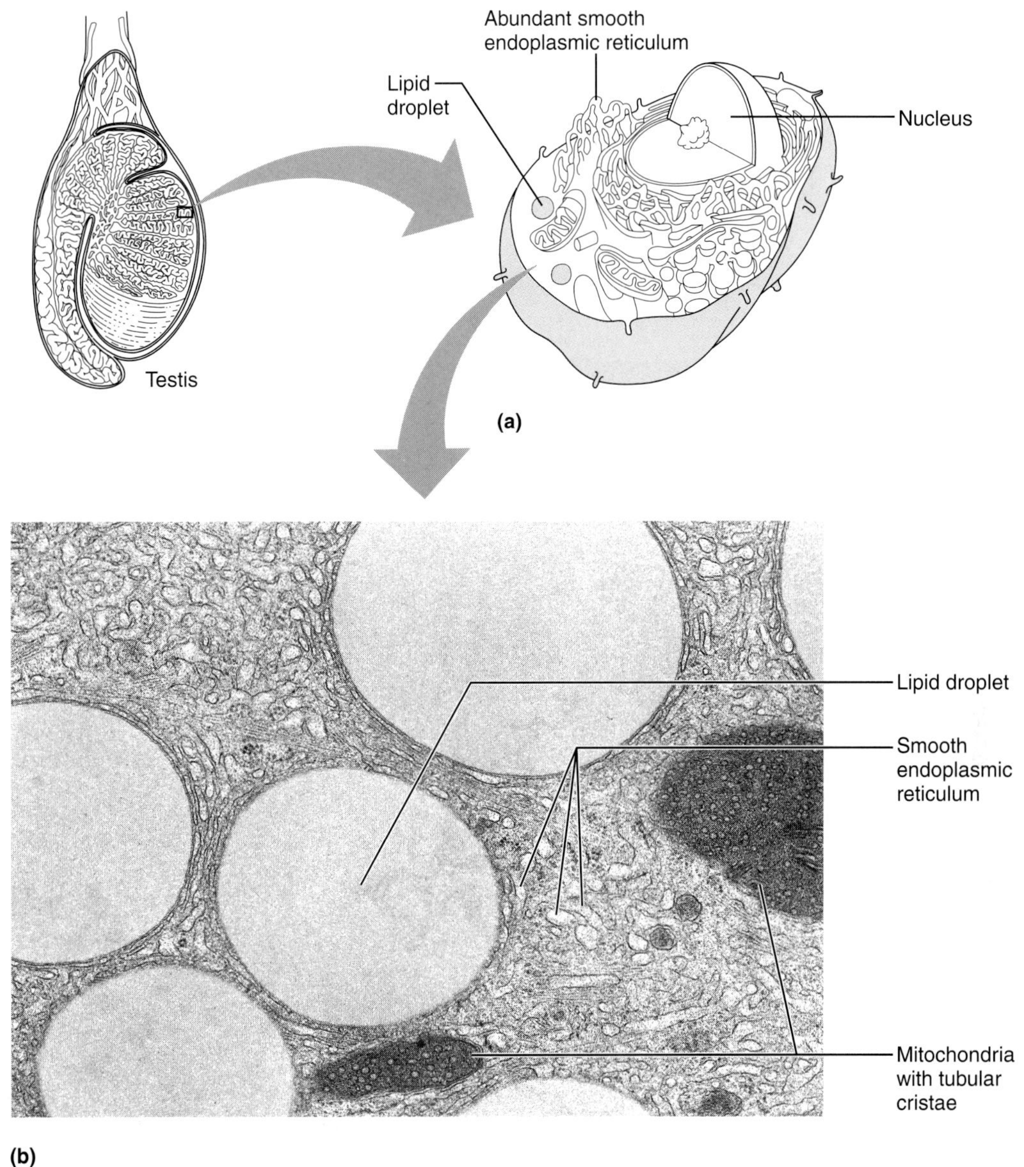

FIGURE 25.11 Ultrastructure of a cell that secretes steroid hormones, an interstitial cell in the testis. Part (a) is an orientation diagram; part (b) is an electron micrograph of a representative part of the cytoplasm (55,000 ×). The special features are labeled.

testosterone and estrogens in the peripheral tissues. Proposed beneficial effects of DHEA include counteracting stress, boosting immunity and muscle mass, and improving mood.

Structure of Steroid-Secreting Cells

Cells that secrete steroid hormones have many distinctive ultrastructural features (Figure 25.11). These features characterize not only the cells of the adrenal cortex but also the testicular and ovarian cells that secrete the steroid sex hormones: the interstitial cells of Leydig, theca folliculi cells, and cells of the corpus luteum (see Chapter 24). Whereas *protein*-secreting gland cells have an elaborate rough ER and abundant secretory granules, steroid-secreting cells have an abundant *smooth* ER and no secretory granules at all. Smooth ER is abundant because this organelle is the site of most stages in the manufacture of lipid-based steroid molecules; secretory granules are lacking because steroids are secreted not by exocytosis (fusion of secretory granules to the plasma membrane), but through a direct outward diffusion of these hormones across the plasma membrane. The mitochondria of steroid-secreting cells have unusual swollen cristae, shaped like tubes instead of shelves, that carry out some of the steps of steroid synthesis. Finally, lipid droplets are abundant in the cytoplasm of steroid-secreting cells, because lipid provides the raw material from which steroids are made.

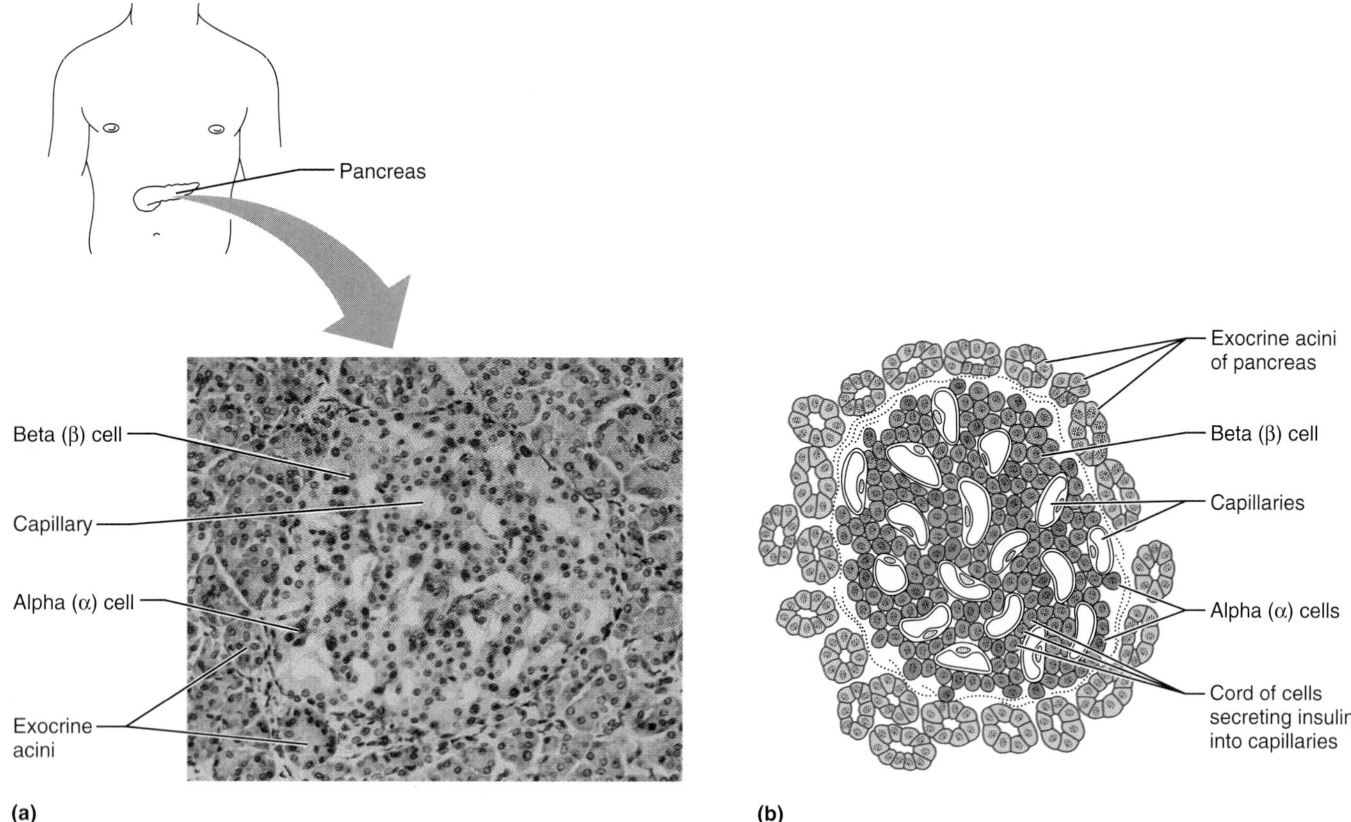

FIGURE 25.12 A pancreatic islet. **(a)** Micrograph (150 ×). In the spherical islet, the alpha and beta cells are stained differentially. **(b)** Diagram of an islet and the surrounding acini, in a view similar to that shown in part (a).

The Pineal Gland

The **pineal gland** (pineal body) is a small, pine cone–shaped structure at the end of a short stalk on the roof of the diencephalon (see Figures 25.1 and Figure 13.15 on p. 376). Its endocrine cells, called *pinealocytes,* are arranged in both spherical clusters and branching cords. Pinealocytes are star-shaped cells with long, branching cell processes. Within the adult pineal gland, dense particles of calcium lie between the cell clusters, forming the "pineal sand." Pinealocytes secrete the hormone **melatonin,** which helps regulate our circadian rhythms (p. 379).

Via a remarkably indirect route, the brain signals the pineal gland to secrete melatonin. During the darkness of night, the "clock" that controls daily rhythms and melatonin secretion—the suprachiasmatic nucleus of the hypothalamus (SCN; see p. 377)—responds to a lack of visual input from the retina by sending signals to the preganglionic sympathetic neurons in the upper thoracic spinal cord; the signals then go to postganglionic neurons in the superior cervical ganglion, whose axons run on the internal carotid artery to stimulate the pineal gland.

Critical Reasoning

What enables radiologists to easily locate the pineal gland and use it as a landmark for identifying structures in X-ray images of the brain?

The dense calcium minerals in the pineal sand are radiopaque (X rays cannot penetrate them) and thus stand out in a radiograph.

The Pancreas

Located in the posterior wall of the abdominal cavity, the tadpole-shaped *pancreas* contains both exocrine and endocrine cells. The exocrine *acinar cells,* forming most of the gland, secrete digestive enzymes into the small intestine during digestion of food (see p. 663).

The endocrine cells of the pancreas are contained in spherical bodies called **pancreatic islets** or **islets of Langerhans** (Figure 25.12), about a million of which are scattered among the exocrine acini. In each islet, the endocrine cells are arranged in twisted, branching cords separated by capillaries. The main cell types in the islets are alpha and beta cells. **Alpha (α) cells (A cells)** secrete **glucagon** (gloo′kah-gon), a hormone that signals *liver* cells to release glucose from their glycogen stores, thus raising blood sugar levels whenever these levels have fallen too low. **Beta (β) cells (B cells)** secrete **insulin** ("hormone from the islets"), a hormone that signals most cells of the body to take up glucose from the blood, thus lowering excessive blood sugar levels (after we digest a sugary snack, for example). Most alpha cells lie at the periphery of the pancreatic islets, whereas the more abundant beta cells occupy the central part.

The pancreatic islets also contain two rare cell types (not illustrated): **Delta (D) cells** secrete **somatostatin** (so″mah-to-stat′in), a hormone that inhibits the secretion of glucagon and insulin by the nearby alpha and beta cells; **F (PP) cells** secrete **pancreatic polypeptide,** a hormone that may inhibit the exocrine activity of the pancreas.

Because the pancreatic hormones are either small proteins (insulin, glucagon, pancreatic polypeptide) or peptides (somatostatin), the islet cells contain the typical microscopic features of cells engaged in protein secretion.

The Thymus

Located in the lower neck and anterior thorax is the lobulated *thymus* (see Figure 25.1). Recall that the thymus is an important immune organ, the site at which T lymphocytes arise from lymphocyte-precursor cells (see p. 597). This transformation seems to be stimulated by **thymic hormones,** which are secreted by the structural cells of the thymus, the epithelial reticular cells. Thymic hormones are a family of peptide molecules, including thymopoietin (thi″mo-poi′e-tin) and thymosin (thi′mo-sin). The thymus is described in detail on p. 597.

The Gonads

The gonads—testes and ovaries—are the main source of the steroid sex hormones. In the male testes, interstitial cells between the sperm-forming tubules secrete **androgens** (mainly **testosterone**), which maintain the reproductive organs and the secondary sex characteristics of males and help promote the formation of sperm. In the female ovaries, androgens secreted by the theca folliculi are directly converted into **estrogens** by the follicular granulosa cells, which also produce **progesterone.** (Estrogens and progesterone are also secreted by the corpus luteum.) Estrogens maintain the reproductive organs and secondary sex characteristics of females, whereas progesterone signals the uterus to prepare for pregnancy. Full details are presented on pp. 708, 718, and 723.

OTHER ENDOCRINE STRUCTURES

Other endocrine cells occur within various organs of the body, including the following:

1. **The heart.** The atria of the heart contain some specialized cardiac muscle cells that secrete **atrial natriuretic peptide (ANP)** (na″tre-u-ret′ik; "producing salty urine"), a hormone that decreases excess blood volume, high blood pressure, and high blood sodium concentration, primarily by signaling the kidney to increase its production of salty urine.

Critical Reasoning

For what therapeutic purposes would pharmaceutical companies seek to design drugs that either mimic ANP or slow the rate at which ANP is broken down in the body?

Because ANP lowers blood pressure, substances that mimic it or increase its longevity in the body might offer great clinical benefits in treating vascular hypertension (high blood pressure). Moreover, control of hypertension reduces the risk of strokes, heart failure, and aneurysms.

2. **The gastrointestinal tract and its derivatives.** *Enteroendocrine cells* are hormone-secreting cells scattered within the epithelial lining of the alimentary canal (p. 651), and related endocrine cells occur within organs that derive from the embryonic gut, such as the respiratory tubes, pancreas, prostate, and thyroid gland. Collectively, all these scattered epithelial cells make up the **diffuse neuroendocrine system (DNES).** To give some concrete examples from this chapter, the parafollicular cells in the thyroid belong to this system, as do the cells of the pancreatic islets. Over 35 kinds of DNES cells secrete amine and peptide hormones that perform such functions as regulating digestion, controlling aspects of blood chemistry, and adjusting local blood flow. Many of these hormones are chemically identical to neurotransmitter molecules, and some of them signal nearby target cells without first entering the bloodstream. These are neuron-like characteristics, explaining why the DNES epithelial cells are called neuroendocrine cells.
3. **The placenta.** Besides sustaining the fetus during pregnancy, the placenta secretes several steroid and protein hormones that influence the course of pregnancy, including estrogens, progesterone, corticotropin-releasing hormone, and human chorionic gonadotropin (see p. 729).
4. **The kidneys.** Various cells in the kidneys produce hormones. The muscle cells of the juxtaglomerular apparatus (see p. 686) secrete the protein hormone *renin,* which indirectly signals the adrenal cortex to secrete aldosterone. Other kidney cells—either those in the interstitial connective tissue between the tubules or the endothelial cells of the peritubular capillaries—secrete *erythropoietin* (e-rith″ro-poi′e-tin), which signals the bone marrow to increase the production of red blood cells.
5. **The skin.** When modified cholesterol molecules present in epidermal cells are exposed to ultraviolet radiation in sunlight, the cells convert these molecules to a precursor of *vitamin D,* a steroid hormone

that is essential for calcium metabolism (see p. 117). The precursor molecules then enter the bloodstream through the dermal capillaries, undergo chemical modification in the liver, and become fully activated vitamin D in the proximal tubules of the kidney. Vitamin D signals the intestine to absorb Ca^{2+} from the diet. Without this vitamin, the bones weaken from insufficient calcium content.

DISORDERS OF THE ENDOCRINE SYSTEM

Most disorders of endocrine glands involve either a hypersecretion (oversecretion) or a hyposecretion (undersecretion) of a given hormone. Hypersecretion often results from a tumor in an endocrine gland, in which the rapidly proliferating tumor cells secrete hormone at an uncontrolled rate. Hyposecretion, by contrast, typically results from damage to the endocrine gland by infection, autoimmune attack, or physical trauma. Here we examine several noteworthy endocrine disorders.

Pituitary Disorders

Some disorders of the adenohypophysis affect the secretion of growth hormone (GH). A tumor that causes hypersecretion of GH in children causes **gigantism,** in which the child grows exceptionally fast and becomes extremely tall, often reaching 2.4 meters (8 feet). If excessive amounts of GH are secreted after the bones' epiphyseal plates have closed, the result is **acromegaly** (ak-ro-meg′ah-le; "enlargement of the extremities"), characterized by enlargement of bony areas that are still responsive to GH—the hands, feet, and face.

Hyposecretion of GH in children produces **pituitary dwarfs,** whose bodies are normally proportioned but rarely reach 4 feet in height. As discussed in this chapter's A Closer Look box (see p. 752), many children with this condition can reach nearly normal stature if given GH injections.

In **diabetes insipidus** (di″ah-be′tēz in-sĭ′pĭ-dus; "passing of dilute [urine]"), the pars nervosa does not make or secrete sufficient antidiuretic hormone (or, more rarely, the kidney does not respond to this hormone). Because individuals with this condition produce large volumes of dilute urine, they compensate by drinking large quantities of water. Diabetes insipidus can be caused by either a blow to the head that damages the posterior pituitary, a tumor that compresses the pars nervosa, or kidney damage.

Disorders of the Thyroid Gland

The most common form of hyperthyroidism is **Graves' disease,** apparently an autoimmune disease in which the immune system makes abnormal antibodies that mimic TSH and stimulate the oversecretion of TH by follicle cells of the thyroid. Typical signs and symptoms of Graves' disease include elevated metabolic rate, rapid heart rate, sweating, nervousness, and weight loss despite normal food intake. Additionally, the eyeballs may protrude, perhaps because of edema in the orbital tissue behind the eyes or because the abnormal antibodies affect the extrinsic eye muscles. Graves' disease develops most often in middle-aged women and affects 1 of every 20 women overall.

Insufficient secretion of TH produces different effects at different stages of life. In adults, such hyposecretion of TH results in **adult hypothyroidism** or **myxedema** (mik″se-de′mah; "mucous swelling"), typically an autoimmune disease in which antibodies attack and destroy thyroid tissue. Signs and symptoms of this condition, which occurs in 7% of women and 3% of men, include a low metabolic rate, weight gain, lethargy, constant chilliness, puffy eyes, edema, and mental sluggishness (but not retardation).

Hypothyroidism can also result from an insufficient amount of iodine in the diet. In such cases, the thyroid gland enlarges, producing in the anterior neck a large visible lump called an **endemic goiter.** Because the cells of the thyroid follicles produce colloid but cannot iodinate it or make functional hormones, the pituitary gland secretes increasing amounts of TSH in a futile attempt to stimulate the thyroid to produce TH. This action causes the follicles to accumulate ever more colloid, and the thyroid gland swells. The commercial marketing of iodized salts has significantly reduced the incidence of goiters in regions that have iodine-poor soils or lack access to iodine-rich shellfish.

In children, hypothyroidism leads to **cretinism** (cre′tĭ-nizm), a condition characterized by a short, disproportionate body, a thick tongue and neck, and mental retardation.

Disorders of the Adrenal Cortex

Hypersecretion of glucocorticoid hormones leads to **Cushing's disease (Cushing's syndrome),** caused either by an ACTH-secreting pituitary tumor or (rarely) by a tumor of the adrenal cortex. This condition is characterized by high blood glucose levels, loss of protein from muscles, and lethargy. The so-called "cushingoid signs" include a swollen face and the redistribution of fat to the posterior neck, causing a "buffalo hump" (Figure 25.13), as well as depression of the immune and inflammatory responses. Mild cases can result from large doses of glucocorticoids prescribed as drugs to suppress inflammation.

Addison's disease, the major hyposecretory disorder of the adrenal cortex, usually involves deficiencies of both glucocorticoids and mineralocorticoids. Blood levels of glucose and sodium drop, and severe dehydration and low blood pressure are common. Other symptoms include fatigue, loss of appetite, and abdominal pain.

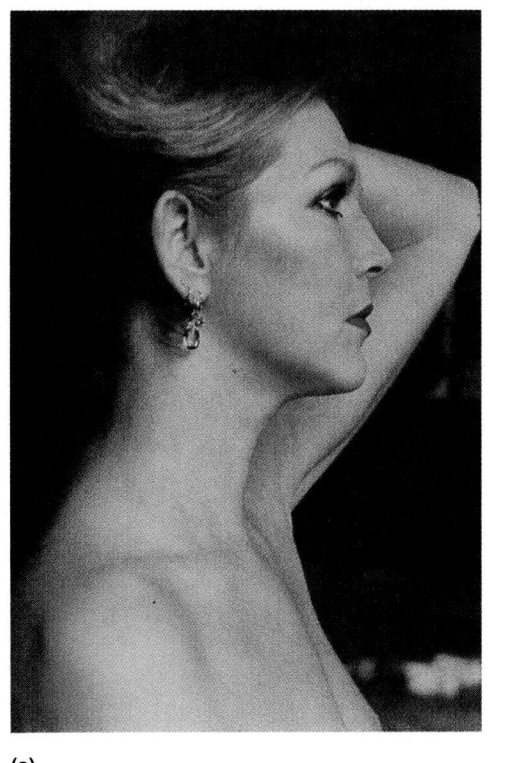

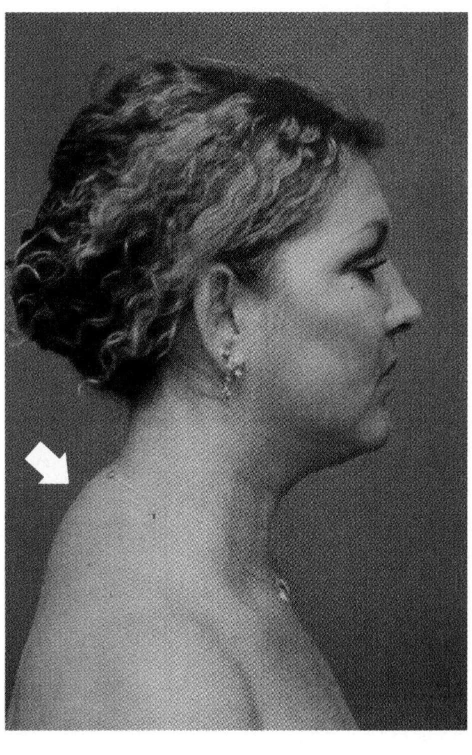

(a) (b)

FIGURE 25.13 A woman before (a) and after the onset (b) of Cushing's disease. In (b), note the swollen face and the "buffalo hump" of fat on the upper back (white arrow).

A Disorder of the Pancreas: Diabetes Mellitus

Diabetes mellitus, which affects about 7% of Americans and has a strong hereditary component, is caused either by insufficient secretion of insulin or resistance of body cells to the effects of insulin. As a result, glucose cannot enter most cells, so blood sugar remains high and glucose appears in an abundant urine. Because glucose is unavailable as fuel, the body's cells metabolize fats, whose acidic breakdown products, ketones, accumulate in the blood. Left untreated, the increased urination depletes the body of water and electrolytes, and the ketone acidosis depresses almost all physiological functions and leads to coma.

Of the two types of diabetes mellitus, the more serious is **type I diabetes** (formerly called **insulin-dependent diabetes**), which develops suddenly, usually before age 15. Because a T cell-mediated autoimmune response destroys the insulin-secreting beta cells in the pancreas, insulin must be administered to type I diabetics several times daily to control blood glucose levels. After 20–30 years with the disease, most type I diabetics develop health-threatening complications: The lipid in their blood predisposes them to atherosclerosis (see p. 574), and the excessive sugar in their body fluids disrupts capillary functions (see *microangiopathy of diabetes* on p. 575).

Advances in Understanding In recent years, research on type I diabetes has (1) demonstrated that regular exercise and careful management of diet and blood sugar level can delay the onset of complications; (2) identified a probable link between drinking cow's milk and developing diabetes in genetically susceptible children; (3) identified two possible auto-antigens responsible for the disorder: a protein on pancreatic beta cells called glutamic acid decarboxylase and a fragment of the insulin protein itself. ✷

Type II (non-insulin-dependent) diabetes is less serious, develops more slowly (usually appearing after age 40), and accounts for over 90% of all cases of diabetes. Most type II diabetics produce some insulin, but their cells have a lowered sensitivity to the effects of insulin. More easily managed than type I, type II diabetes can usually be controlled by dietary modification (such as losing weight and avoiding high-calorie, sugar-rich foods) and regular exercise. If such measures fail, the condition is treated with insulin injections or oral medications that raise blood insulin levels or lower blood glucose levels.

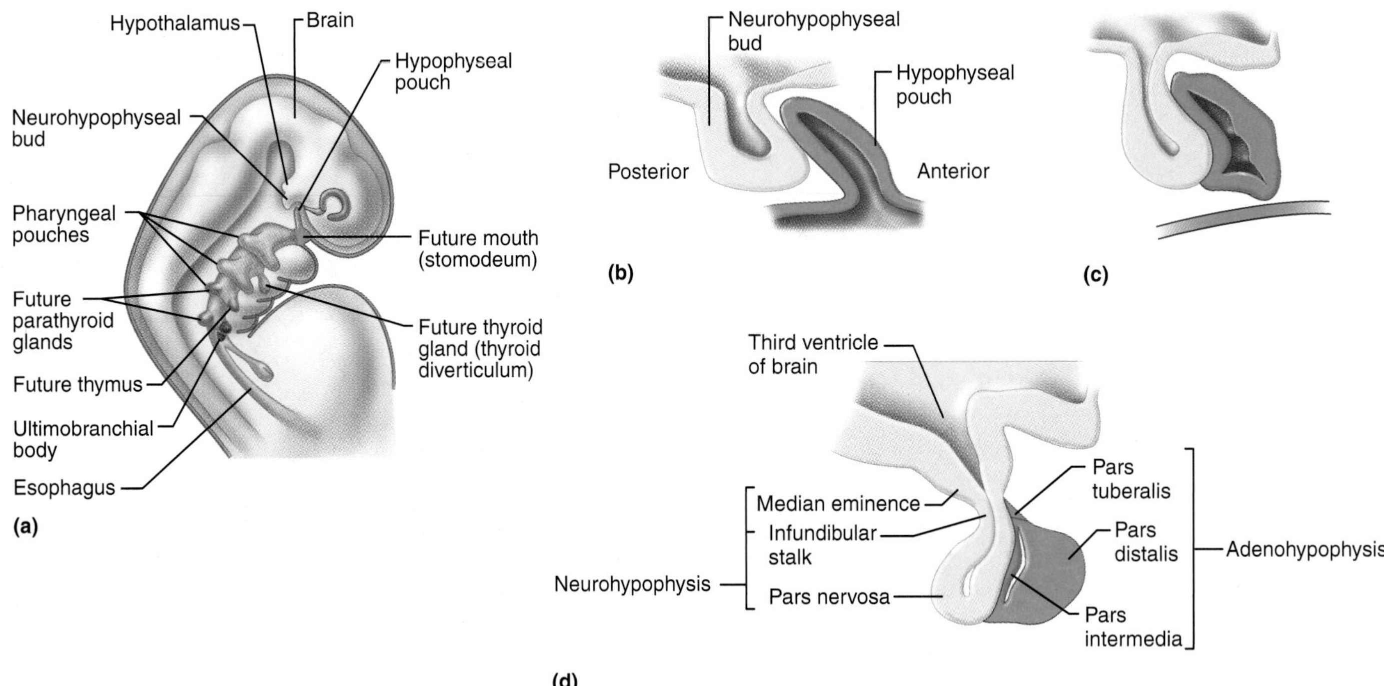

FIGURE 25.14 Embryonic development of some major endocrine organs. **(a)** An embryo in weeks 5 to 6. Four pharyngeal pouches are present on each side of the pharynx, and the hypophyseal pouch is pushing superiorly toward the brain from the roof of the mouth. Also note the future thyroid gland (thyroid diverticulum) on the pharyngeal floor. **(b)–(d)** Stages in the formation of the pituitary gland during weeks 6 to 8 (anterior is to the right). The hypophyseal pouch (adenohypophysis) meets the neurohypophyseal bud (neurohypophysis) that is growing inferiorly from the floor of the brain.

THE ENDOCRINE SYSTEM THROUGHOUT LIFE

The diverse and widely distributed endocrine organs arise from all three germ layers.

The *pituitary gland* has a dual origin (Figure 25.14a–d). The adenohypophysis arises from the roof of the mouth as a pouch of ectoderm—the **hypophyseal pouch,** or *Rathke's pouch.* This pouch contacts the future neurohypophysis, which grows inferiorly from the floor of the brain as the **neurohypophyseal bud.**

The *thyroid gland* forms from a thickening of endoderm on the floor of the pharynx (see Figure 25.14a) that first appears on the posterior part of the future tongue and then migrates caudally into the neck. Both the *parathyroid glands* and the *thymus* arise from the endoderm lining the pharyngeal pouches and then migrate to their final positions in the neck and thorax. The parafollicular cells arise as *ultimobranchial* ("end-gill") *bodies* in the wall of the last pharyngeal pouch and then migrate caudally, into the thyroid gland. The *pineal gland* arises from ependymal cells that cover the roof of the embryonic diencephalon.

The adrenal gland has a dual embryonic origin (Figure 15.16). Whereas its medulla originates from the neural crest cells of nearby sympathetic trunk ganglia, the cortex develops from mesoderm lining the coelom on the dorsal abdominal wall.

Barring hypersecretory and hyposecretory disorders of the endocrine glands, most endocrine organs operate smoothly throughout life until old age. Then, a number of changes become evident. In the adenohypophysis, the amounts of connective tissue and aging pigment (lipofuscin) increase, vascularization decreases, and the number of hormone-secreting cells declines. These changes may or may not affect hormone production: For example, blood levels of TSH decline slightly with normal aging, ACTH remains constant, and gonadotropins actually *increase* with age.

The adrenal cortex also shows structural changes with age, but normal rates of glucocorticoid secretion appear to persist as long as an individual is healthy. No age-related changes in the release of catecholamines by the adrenal medulla have been found.

The synthesis and release of thyroid hormones diminish somewhat with normal aging. Typically, the thyroid follicles are loaded with colloid in the elderly, and fibrosis of the gland occurs. Autoimmune diseases of the thyroid become common, affecting about 10% of older women (see p. 760).

The parathyroid glands change little with age, but a tendency for the concentration of PTH in the blood to increase compensates for a lowered intake of calcium and vitamin D as the quality of the diet declines. Furthermore, the decline in estrogen secretion after menopause may sensitize women to the bone-demineralizing effects of PTH.

Only GH, DHEA, and the sex hormones (estrogens and testosterone) show marked drops in secretion with age. The deterioration of the musculoskeletal, cardiovascular, immune, and other organs that accompany these hormonal declines has been discussed elsewhere in this book (see pp. 261, 578, and 599). So closely do the effects of withdrawing these hormones parallel the symptoms of aging that all three have been "hyped" as "anti-aging drugs" in recent years.

RELATED CLINICAL TERMS

Hypophysectomy Surgical removal of the pituitary gland.

Hypercalcemia Elevated concentrations of calcium ions in the blood caused by either primary hyperparathyroidism or malignancies. In primary hyperparathyroidism, a parathyroid gland spontaneously (usually as a result of a tumor) secretes large amounts of calcium-raising parathyroid hormone. Many different kinds of malignancies also raise blood calcium levels by secreting chemicals that either stimulate osteoclasts to dissolve bone or stimulate calcium reabsorption from the kidney.

Pancreatic transplantation A procedure performed on people with type I diabetes whose lives are threatened by gradually failing kidneys, as *microangiopathy of diabetes* (p. 575) destroys the kidney capillaries. The transplanted pancreas secretes insulin and halts the diabetes, and a healthy kidney may be transplanted at the same time to take over for the recipient's failing kidneys. The donated pancreas is placed in the pelvis and sutured to the bladder so the digestive enzymes it produces can drain into the bladder and flush from the body in the urine, instead of irritating the pelvic peritoneum and causing peritonitis. As with other organ transplants, immunosuppressive drugs must be taken for life. The success rate for pancreatic transplants is about 70%. Techniques are now being developed to transplant only the pancreatic islets.

Pheochromocytoma (fe″o-kro″mo-si″to′mah; "dusky color tumor") Tumor of the chromaffin cells of the adrenal medulla; results in excessive secretion of catecholamine hormones, which produces the symptoms of prolonged sympathetic response (especially hypertension).

Prolactinoma (*oma* = tumor) The most common type of pituitary gland tumor, characterized by a hypersecretion of prolactin, excessive secretion of milk, and menstrual disturbances in women. In men, it causes milk production, a loss of libido, and impotency.

Psychosocial dwarfism Dwarfism (and failure to thrive) resulting from childhood stress and emotional disorders that suppress the hypothalamic release of growth hormone–releasing hormone and thus adenohypophyseal secretion of GH.

Thyroid cancer The most common malignancy of the endocrine glands, affecting about 1% of all people, typically between the ages of 25 and 65; characterized by firm, fixed nodules in the thyroid. Incidence is increasing, especially among individuals who had external irradiation of the head and neck when young. Treatment of this highly curable cancer involves removing some or all of the thyroid, then administering a radioactive form of iodine, which is actively taken up by the remaining thyroid follicular cells (normal and cancerous). The sequestered radio-iodine then kills the cancer cells with a minimum of harm to nonthyroidal tissues.

Thyroid storm (thyroid crisis) A sudden and dangerous increase in the effects of thyroid hormone resulting from excessive amounts of TH in the circulation; metabolic rate is greatly increased, as indicated by fever, rapid heart rate, high blood pressure, nervousness, and tremors. Precipitating factors include stressful situations, excessive intake of TH supplements, and trauma to the thyroid gland.

CHAPTER SUMMARY

Media study tools that could provide you additional help in reviewing specific key topics of Chapter 25 are referenced below.

SP = Study Partner; AIA = A.D.A.M.® Interactive Anatomy

1. Endocrine organs are ductless glands that release hormones into the blood or lymph.
2. Hormones are messenger molecules that travel in the circulatory vessels and signal physiological changes in target cells.
3. Hormonally regulated processes include reproduction, growth, mobilization of body defenses against stress, maintenance of the proper chemistry of the blood and body fluids, and regulation of cellular metabolism.

THE ENDOCRINE SYSTEM: AN OVERVIEW (pp. 744–745)

Endocrine Organs (pp. 744–745)

1. The endocrine organs are small and widely separated from one another within the body. The "pure" endocrine organs are the pituitary, thyroid, parathyroid, adrenal, and pineal glands. Other organs that contain endocrine cells are the gonads, pancreas, kidney, alimentary canal, heart, thymus, and skin. The hypothalamus of the brain is a neuroendocrine organ.
2. Endocrine organs are richly vascularized.
3. Although most endocrine cells are modified epithelial cells, others are neurons, muscle cells, or fibroblast-like cells.

Hormones (p. 745)

4. Most hormones are either amino acid derivatives (amines, peptides, proteins) or steroids (lipid-based molecules derived from cholesterol).
5. Hormones produce their effects by leaving the capillaries and binding to specific receptor molecules in or on their target cells. Such binding triggers a preprogrammed response in the target cell.
6. Endocrine organs are stimulated to release their hormones by humoral, neural, or hormonal stimuli.

7. The hypothalamus of the brain regulates many functions of the endocrine system through the hormones it secretes.

THE MAJOR ENDOCRINE ORGANS (pp. 745–759)

The Pituitary Gland (pp. 745–753)

1. The golf club-shaped pituitary gland is suspended from the diencephalon of the brain by its stalk (infundibulum) and lies in the hypophyseal fossa of the sella turcica of the sphenoid bone. It consists of an anterior adenohypophysis and a posterior neurohypophysis.

AIA Link: Atlas; System: Endocrine; Base of Brain (inferior).

2. The adenohypophysis has three parts: pars distalis (anterior lobe), pars intermedia, and pars tuberalis. The neurohypophysis also has three parts: pars nervosa (posterior lobe), infundibular stalk, and median eminence.

3. The pituitary gland receives its rich blood supply from the superior and inferior hypophyseal arteries.

4. The largest part of the adenohypophysis is the pars distalis. Its five cell types secrete seven protein hormones: (1) somatotropic cells (growth hormone: GH); (2) mammotropic cells (prolactin: PRL); (3) thyrotropic cells (thyroid-stimulating hormone: TSH); (4) corticotropic cells (adrenocorticotropic hormone and melanocyte-stimulating hormone: ACTH and MSH); and (5) gonadotropic cells (follicle-stimulating hormone and luteinizing hormone: FSH and LH). Cells in the pars distalis cluster into spheres (and branching cords).

5. The basic functions of each adenohypophyseal hormone are these: GH stimulates growth of the body and skeleton; PRL signals milk production; TSH signals the thyroid gland to secrete thyroid hormone; ACTH signals the adrenal cortex to secrete glucocorticoids; MSH causes the skin to darken; FSH and LH signal the maturation of sex cells and the secretion of sex hormones. Four of these seven hormones—FSH, LH, ACTH, and TSH—stimulate other endocrine glands to secrete and are called tropic hormones.

6. The hypothalamus of the brain controls the secretion of hormones from the adenohypophysis in the following way: First, certain hypothalamic neurons make releasing hormones and inhibiting hormones, which they secrete into a primary capillary plexus in the median eminence. These hormones then travel through hypophyseal portal veins to a secondary capillary plexus in the pars distalis. They leave this plexus to signal the adenohypophyseal cells to secrete their hormones, which then enter the secondary capillary plexus and travel to their target cells throughout the body.

7. The neurohypophysis, which consists of nervous tissue, contains the hypothalamic-hypophyseal axon tract. The cell bodies of the neurons that form this tract are located in the paraventricular and supraoptic nuclei of the hypothalamus. These neurons synthesize oxytocin and ADH (vasopressin), respectively, and store them in their axon terminals in the pars nervosa. Here, the stored hormones are released into capillaries when the neurons fire.

8. The neurohypophyseal hormones have the following functions: ADH increases reabsorption of water from the urine and raises blood pressure, and oxytocin induces labor and ejection of milk from the breasts. Both these hormones promote cuddling and pair bonding.

The Thyroid Gland (pp. 753–754)

9. The thyroid gland, which lies on the superior trachea, consists of spherical follicles covered by epithelial follicle cells and separated by a capillary-rich connective tissue. The follicles are filled with a colloid of thyroglobulin, a storage protein containing thyroid hormone.

AIA Link: Atlas; System: Endocrine; Glands of Head & Neck (lateral).

10. Thyroid hormone (TH), which contains iodine and increases basal metabolic rate, is made continuously by follicle cells and stored within the follicles until TSH from the pituitary gland signals the follicle cells to reclaim the TH and secrete it into the extrafollicular capillaries.

11. Parafollicular cells protrude from the thyroid follicles and secrete the hormone calcitonin, which can lower blood calcium concentrations in children.

The Parathyroid Glands (p. 754)

12. Several pairs of parathyroid glands lie on the dorsal aspect of the thyroid gland. Their chief cells are arranged in thick, branching cords and secrete parathyroid hormone, which raises low blood calcium levels.

The Adrenal (Suprarenal) Glands (pp. 754–757)

13. The paired adrenal glands lie on the superior surface of each kidney. Each adrenal gland has two distinct parts, an outer cortex and an inner medulla.

AIA Link: Atlas; System: Endocrine; Posterior Abdominal Wall (anterior).

14. The adrenal medulla consists of spherical clusters (and some branching cords) of chromaffin cells. Upon sympathetic stimulation, these cells secrete excitatory catecholamines into the blood—the surge of adrenaline that is experienced during fight, flight, or fright situations.

15. The adrenal cortex has three layers: outer zona glomerulosa, middle zona fasciculata, and inner zona reticularis; the name of each zone describes its histological structure.

16. The steroid hormones secreted by the adrenal cortex (corticosteroids) include mineralocorticoids (mostly from the zona glomerulosa), glucocorticoids (mostly from the zona fasciculata and reticularis), and the androgen DHEA (from the zona reticularis). Mineralocorticoids (mainly aldosterone) conserve water and sodium by increasing reabsorption of these substances by the kidney. Glucocorticoids (mainly cortisol) help the body cope with stress by stabilizing blood glucose levels; in large quantities they also inhibit inflammation and the immune system. The functions of DHEA are unclear but probably beneficial.

17. Steroid-secreting cells, including the cells in the gonads that secrete sex hormones, have an abundant smooth ER, tubular cristae in their mitochondria, abundant lipid droplets, and no secretory granules.

The Pineal Gland (p. 758)

18. The pineal gland, on the roof of the diencephalon, contains pinealocytes, which cluster into spherical clumps and cords separated by dense particles of calcium called pineal sand.

19. Pinealocytes secrete the hormone melatonin, which helps regulate circadian rhythms. This secretion is signaled by the suprachiasmatic nucleus of the hypothalamus through a sympathetic pathway.

The Pancreas (pp. 758–759)

20. The endocrine structures in the pancreas are the spherical pancreatic islets. These islets consist of alpha (A), beta (B), delta (D) and F (PP) cells arranged in twisting cords.

21. Alpha cells secrete glucagon, which raises blood sugar levels, whereas beta cells secrete insulin, which lowers blood sugar levels.

The Thymus (p. 759)

22. The thymus, an important organ of the immune system, secretes thymic hormones that are essential for the production of T lymphocytes.

The Gonads (p. 759)

23. Various cells in the ovaries and testes secrete steroid sex hormones, estrogens and androgens.

OTHER ENDOCRINE STRUCTURES (pp. 759–760)

1. Some muscle cells in the atria of the heart secrete atrial natriuretic peptide (ANP), which stimulates loss of body fluids and salts through the production of a sodium-rich urine.

2. Endocrine cells are scattered within the epithelium of the digestive tube and other gut-derived organs (respiratory tubes and so on). These epithelial cells, which have some neuron-like properties, make up the diffuse neuroendocrine system (DNES). There are many classes of DNES cells, some of which secrete hormones that regulate digestion.

3. The placenta secretes hormones of pregnancy, the kidney secretes renin and erythropoietin, and the skin produces vitamin D.

SP Exercise: Chapter 25, Hormones and Their Targets.

DISORDERS OF THE ENDOCRINE SYSTEM (pp. 760–761)

1. Most disorders of endocrine glands involve either hypersecretion or hyposecretion of a hormone. *Hyper*secretion of GH leads to gigantism, of TH leads to Graves' disease, and of ACTH or glucocorticoids leads to Cushing's disease. *Hypo*secretion of GH leads to pituitary dwarfism, of TH leads to adult hypothyroidism or cretinism, of glucocorticoids and mineralocorticoids leads to Addison's disease, and of insulin leads to type I diabetes mellitus.

THE ENDOCRINE SYSTEM THROUGHOUT LIFE (pp. 762–763)

1. The endocrine glands have diverse developmental origins from all three germ layers. The adenohypophysis arises from ectoderm on the roof of the mouth, and the neurohypophysis arises from the floor of the brain. The endocrine organs in the neck (thyroid, parathyroids, thymus) derive from the endoderm of the pharynx. The pineal gland forms from the ependyma of the roof of the brain, the adrenal medulla develops from sympathetic trunk ganglia, and the adrenal cortex arises from the mesoderm of the dorsal abdominal wall.

2. The efficiency of some endocrine organs gradually declines as the body ages. The hormones whose secretions decline most are GH, DHEA, estrogens, and testosterone.

REVIEW QUESTIONS

Multiple Choice/Matching Questions

1. The major stimulus for the release of estrogens is (a) hormonal, (b) humoral, (c) nervous.

2. Match the hormones in Column A (below) with the descriptions in Column B (at right).

Column A

(a) melatonin

(b) antidiuretic hormone

(c) somatotropin

(d) luteinizing hormone

(e) TH

(f) TSH

(g) prolactin

(h) oxytocin

(i) cortisol

(j) parathyroid hormone

Column B

____ **(1)** stimulates cell division in the epiphyseal plates of growing bones

____ **(2)** involved in water balance; causes the kidneys to conserve water

____ **(3)** stimulates milk production

____ **(4)** stimulates milk ejection

____ **(5)** tropic hormone that signals the gonads to secrete sex hormones

____ **(6)** increases basal metabolic rate

____ **(7)** tropic hormone that stimulates the thyroid gland to secrete thyroid hormone

____ **(8)** adjusts blood sugar levels and helps the body cope with stress

____ **(9)** is secreted by the pineal gland

____ **(10)** increases blood calcium levels

____ **(11)** is secreted by the neurohypophysis (two possible choices)

____ **(12)** the only steroid hormone in the list

3. The pars distalis of the adenohypophysis does not secrete (a) antidiuretic hormone, (b) growth hormone, (c) gonadotropins, (d) TSH.

4. The pure endocrine organs of the body (a) tend to be very large organs, (b) are closely connected with each other, (c) all contribute directly to digestion, (d) tend to lie near the midline of the body.

5. Structure of various endocrine cells. Write (a) or (b) in each blank as appropriate.

Key: **(a)** This cell secretes protein hormones and has a well-developed rough ER and secretory granules.

(b) This cell secretes steroid hormones: has a well-developed smooth ER, no secretory granules, unusual mitochondria,and lipid droplets.

____ **(1)** any endocrine cell in the pars distalis

____ **(2)** interstitial (Leydig) cell in the testis

____ **(3)** chief cell in the parathyroid gland

____ **(4)** zona fasciculata cell

____ **(5)** theca cell in the ovary that secretes sex hormones

____ **(6)** parafollicular cells in the thyroid gland

6. Match each of the following glands with the best description of its histological structure from the key.

Key: **(a)** spherical clusters of cells

(b) parallel cords of cells

(c) branching cords of cells

(d) follicles

(e) nervous tissue

____ **(1)** pars nervosa of pituitary gland

____ **(2)** zona glomerulosa of adrenal gland

____ **(3)** pars distalis of pituitary gland

____ **(4)** thyroid gland

____ **(5)** zona fasciculata of adrenal gland

7. The divisions of the neurohypophysis are the (a) anterior lobe and posterior lobe, (b) pars emphasis, metropolis, and hypothesis, (c) pars distalis, tuberalis, and intermedia, (d) pars nervosa, infundibular stalk, and median eminence, (e) pars glomerulosa, fasciculata, and reticularis.

8. Which of the following types of cell secretes releasing hormones? (a) neurons, (b) chromaffin cells, (c) cells in the pars distalis, (d) parafollicular cells.

9. Chromaffin cells occur in the (a) parathyroid gland, (b) pars distalis, (c) pituitary gland, (d) adrenal gland, (e) pineal gland.

10. The anterior lobe of the pituitary gland is the same as the (a) neurohypophysis, (b) pars nervosa, (c) pars distalis, (d) hypothalamus.

Short Answer Essay Questions

11. (a) Describe the body location of each of the following endocrine glands: anterior and posterior lobe of the pituitary, pineal, thyroid, parathyroids, and adrenals. (b) List the hormones secreted by each of these glands.

12. The adenohypophysis secretes so many hormones that it is often called the master endocrine organ, but it too has a "master." What structure controls the release of anterior pituitary hormones?

13. When Joshua explained to his classmate Jennifer that the thyroid gland contains parathyroid cells in its follicles, and that the parathyroid cells secrete parathyroid hormone and calcitonin, Jennifer told him he was all mixed up again. Correct Josh's mistakes.

14. (a) Define hormone. (b) Name a hormone secreted by a muscle cell and a hormone secreted by a neuron.

15. On a realistic drawing of the endocrine glands in the body, such as a photocopy of Figure 25.1, indicate the gland associated with (a) cretinism, (b) diabetes mellitus, (c) acromegaly, (d) secreting thyroid-stimulating hormone, (e) secreting a hormone that regulates the nightly activities of our circadian rhythms, (f) secreting estrogens, (g) secreting DHEA.

16. On a realistic drawing of the endocrine glands in the whole body, trace the following hormones from their glands of origin all the way to their target organs: (a) vitamin D, (b) glucagon, (c) erythropoietin, (d) oxytocin.

17. On a realistic drawing of the adult body, mark and label the endocrine organs that develop from the (a) roof of the embryonic mouth, (b) floor of the diencephalon, (c) endoderm on the caudal part of the future tongue, (d) endoderm of the third and fourth pharyngeal pouches (two answers here), (e) neural crest of early sympathetic trunk ganglia.

18. On a realistic drawing of the endocrine glands in the body, mark the endocrine gland that is shaped like (a) a pyramid on a kidney, (b) a bow tie, (c) a golf club, (d) spherical islets in a tadpole-shaped digestive organ, (e) an egg or almond in the pelvic cavity, (f) a butterfly.

19. Compare and contrast the functions of both sets of capillaries in the hypophyseal portal system.

20. List the hormones secreted by each of the three zonas of the adrenal cortex.

CRITICAL REASONING + CLINICAL APPLICATION QUESTIONS

1. The brain senses when we are in a stressful situation, and the hypothalamus responds by secreting a releasing hormone called corticotropin-releasing hormone, which, through a sequence of events, helps the body to deal with the stressful situation. Outline this entire sequence, starting with corticotropin-releasing hormone and ending with the release of cortisol. (As you do this, be sure to trace the hormones through the hypophyseal portal system and out of the pituitary gland.)
2. Jeremy, a 5-year-old boy, has been growing by leaps and bounds such that his height is 70% above normal for his age group. A CT scan reveals a pituitary tumor. (a) What hormone is being secreted in excess? (b) What condition will Jeremy exhibit if corrective measures are not taken?
3. An accident victim who had not been wearing a seat belt received trauma to his forehead when he was thrown against the windshield. The physicians in the emergency room worried that his brain stem may have been driven inferiorly through the foramen magnum. To help assess this, they quickly took a standard X-ray film of his head and searched for the position of the pineal gland. How could anyone expect to find this tiny, boneless gland in a radiograph?
4. Mrs. Giardino had an abnormally high concentration of calcium in her blood, and her physicians were certain she had a tumor of the parathyroid gland. However, when surgery was performed on her neck, the surgeon could not find the parathyroid glands at all. Where should the surgeon look next to find the tumorous parathyroid gland?
5. A carnival sideshow came to a town. The fringe group called Political Correctness First said that such "freak" shows were cruel and exploitative, and asked the town council to enforce truth-in-advertising laws. The council agreed, requiring that the fat man be billed as a "man with hypothyroidism," the dwarf and giant be called "people with pituitary disorders," the bearded lady be called a "woman with a tumor of the adrenal's zona reticularis," and the woman whose eyes protruded from the orbits be called "a person with Graves' disease." Explain how each of these endocrine disorders produced the characteristic features of these five people.

A.D.A.M.® INTERACTIVE ANATOMY QUESTIONS

1. Link: Dissectible anatomy; Male: View; lateral: layer 205. (a) The pituitary gland sits in which bone of the skull? (b) Identify the two basic divisions of this gland.
2. Link: Dissectible anatomy; Female: View; Anterior: layer 230; Scroll to pelvic region. The gonads are a major source of steroid sex hormones in both males and females. Identify the gonad in the female.

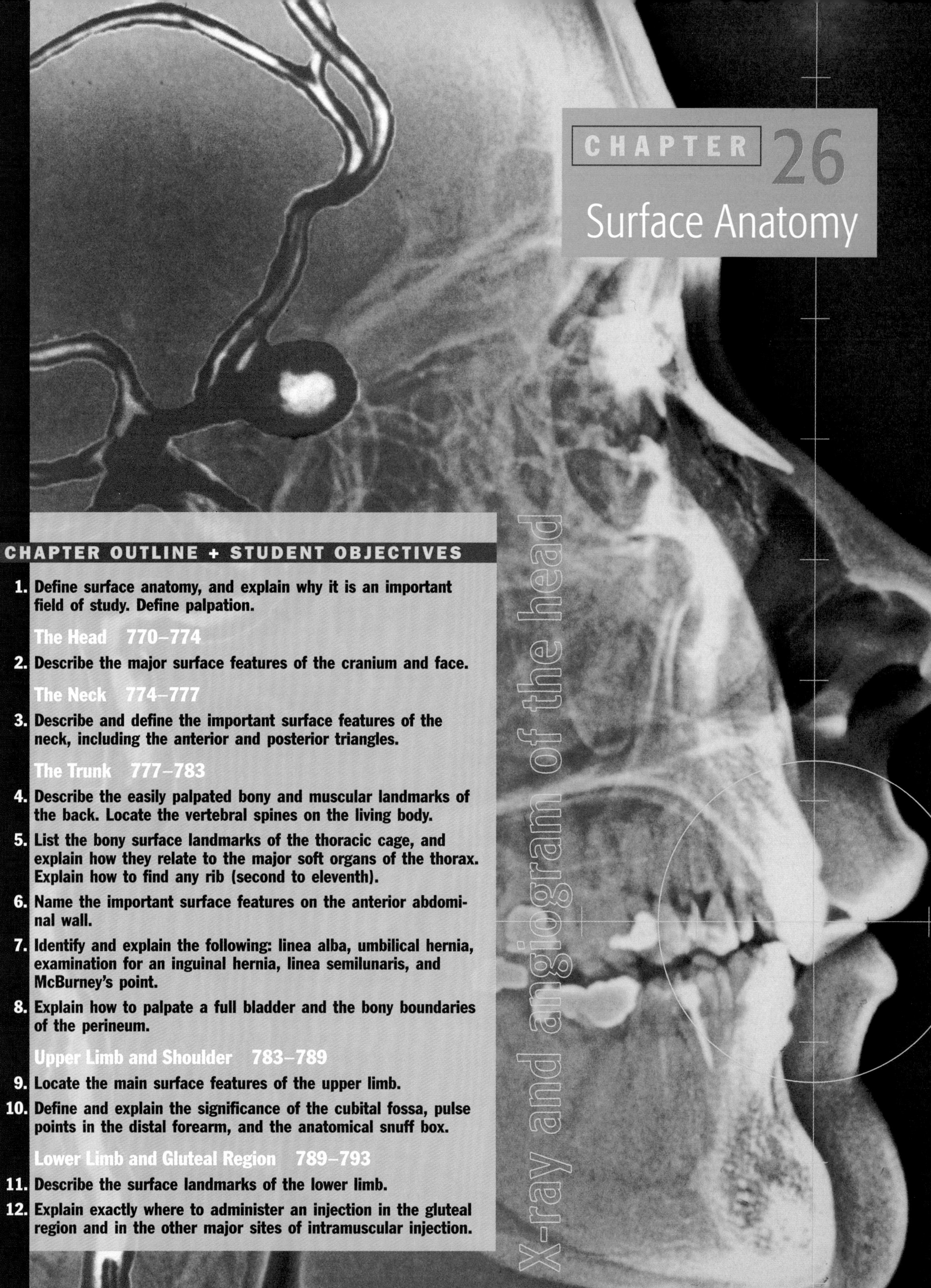

CHAPTER 26 Surface Anatomy

CHAPTER OUTLINE + STUDENT OBJECTIVES

1. Define surface anatomy, and explain why it is an important field of study. Define palpation.

The Head 770–774

2. Describe the major surface features of the cranium and face.

The Neck 774–777

3. Describe and define the important surface features of the neck, including the anterior and posterior triangles.

The Trunk 777–783

4. Describe the easily palpated bony and muscular landmarks of the back. Locate the vertebral spines on the living body.
5. List the bony surface landmarks of the thoracic cage, and explain how they relate to the major soft organs of the thorax. Explain how to find any rib (second to eleventh).
6. Name the important surface features on the anterior abdominal wall.
7. Identify and explain the following: linea alba, umbilical hernia, examination for an inguinal hernia, linea semilunaris, and McBurney's point.
8. Explain how to palpate a full bladder and the bony boundaries of the perineum.

Upper Limb and Shoulder 783–789

9. Locate the main surface features of the upper limb.
10. Define and explain the significance of the cubital fossa, pulse points in the distal forearm, and the anatomical snuff box.

Lower Limb and Gluteal Region 789–793

11. Describe the surface landmarks of the lower limb.
12. Explain exactly where to administer an injection in the gluteal region and in the other major sites of intramuscular injection.

THERE IS AN OLD JOKE ABOUT A STUDENT WHO was kicked out of anatomy class for cheating on an exam: Instead of knowing how many ribs there are, he counted his own. Actually, that student was practicing the study of **surface anatomy,** an extremely valuable branch of anatomical and medical science. Even though surface anatomy does indeed study the external surface of the body, it also studies *internal* organs as they relate to surface landmarks and as they are seen and felt through the skin. Feeling internal structures through the skin with the fingers is called **palpation** (pal-pa′shun; "touching").

Surface anatomy is living anatomy, better studied in live people than in cadavers. It can provide a great deal of information about the living skeleton (almost all bones can be palpated) and about the muscles and vessels that lie near the body surface. Furthermore, because a skilled examiner can learn much about the heart, lungs, and other deep organs by performing a surface assessment, surface anatomy serves as the basis of a standard physical examination. If you are planning a career in the health sciences or physical education, your study of surface anatomy will show you where to take pulses, where to insert tubes and needles, where to make surgical incisions, where to locate broken bones and inflamed muscles, and where to listen for the sounds of the lungs, heart, and intestines.

The study of surface anatomy is influenced by the thickness of the fatty subcutaneous layer deep to the skin, which tends to be greater in small children and women than in men and can be very thick in obese people. Obviously, surface anatomy is more difficult to study on individuals with a thick layer of subcutaneous fat. The examiner can often compensate, however, by pressing harder than usual to find the bony and muscular landmarks being sought, an action called *deep palpation.*

Along with presenting new information, this chapter reviews some of what was presented in the previous twenty-five chapters of this book. Whereas those chapters were organized by organ *systems,* here we take a *regional* approach to surface anatomy, exploring the head first and proceeding to the trunk and the limbs. We ask you to observe and palpate your own body as you work through the chapter, because your body is the best learning tool of all. Even better, whenever possible, get together with a study partner so that you can make your observations and palpations as if they were on a patient. To aid your exploration of living anatomy, Figures 26.1, 26.2, and 26.3 review the bones and muscles you will encounter, and Figure 26.4 previews various points where the arterial pulse can be palpated.

THE HEAD

The head (Figures 26.5 and 26.6) is divided into the cranium and the face.

Cranium

Run your fingers over the superior surface of your head. Notice that the underlying cranial bones lie very near the surface. Proceed to your forehead and palpate the *superciliary arches* ("brow ridges") directly superior to your orbits (see Figure 26.5a). Then move your hand to the posterior surface of the skull, where you can feel the knob-like *external occipital protuberance.* By running your finger directly laterally from this protuberance, feel the ridge-like *superior nuchal line* on the occipital bone. This line marks the superior extent of the muscles of the posterior neck and is a boundary between the head and the neck. Now feel the prominent *mastoid process* on each side of the cranium just posterior to your ear.

Place a hand on your temple, and clench your teeth together in a biting action. You should be able to feel the *temporalis muscle* bulge as it contracts. Next, raise your eyebrows, and feel your forehead wrinkle due to contraction of the subcutaneous *frontalis muscle,* a muscle of facial expression (see Figure 26.6). The frontalis inserts superiorly onto a broad aponeurosis called the *galea aponeurotica* (see Table 11.1, p. 276), which covers the superior surface of the cranium and binds tightly to the overlying subcutaneous tissue and skin to form the true scalp. Push on your scalp to confirm that it slides freely over the underlying cranial bones. Because the scalp is only loosely bound to the skull, people can easily be "scalped" (in industrial accidents, for example). Most arteries of the body constrict and close after they are cut or torn, but the vessels in the richly vascularized scalp are unable to do so because they are held open by the dense connective tissue surrounding them. As a result, scalp wounds bleed profusely. However, because the scalp is so well vascularized, these wounds also heal quickly.

Face

The surface of the face is divided into many different regions, including the orbital, nasal, oral (mouth), and auricular (ear) areas. In the region of the eyes (see Figure 26.6), trace a finger around the entire bony margin of an orbit. The *lacrimal fossa,* which contains the tear-gathering lacrimal sac, may be felt on the medial side of the eye socket. (You can review the surface anatomy of the eye and eyelids in Figure 16.4 on page 471.) Next, touch the most superior part of your nose, its *root* between the eyebrows (see Figure 26.6). Just inferior to this, between your eyes, is the *bridge* of the nose formed by the nasal bones. Continue your finger inferiorly along the nose's anterior margin, the **dorsum nasi,** to its tip, the **apex.** Place one finger in a *nostril* and another finger on the flared wing, the **ala,** that defines the nostril's lateral border. Then feel the **philtrum** (fil′trum; "love"), the shallow vertical groove on your upper lip below the nose.

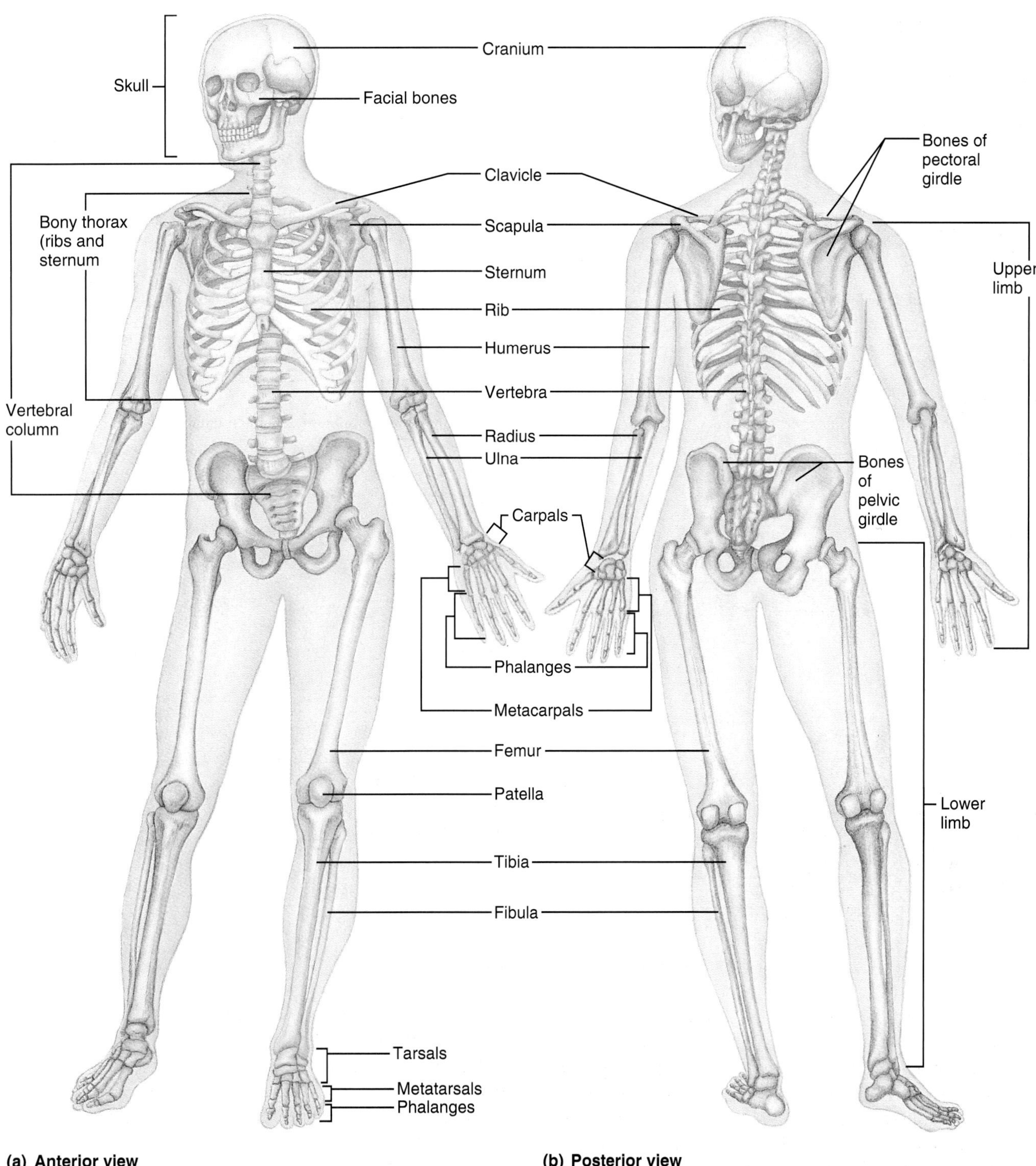

(a) Anterior view **(b) Posterior view**

FIGURE 26.1 Bones of the human skeleton.

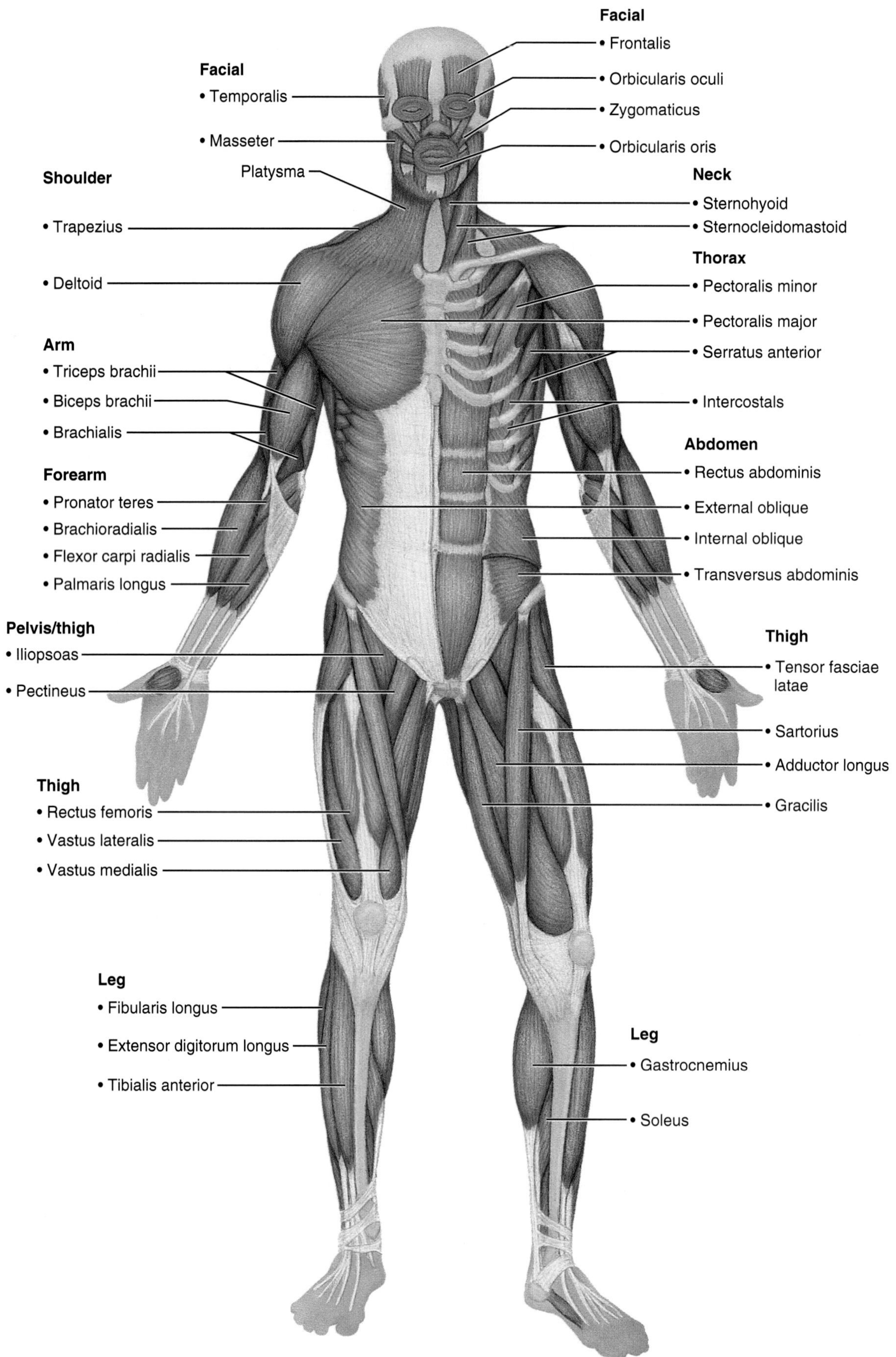

FIGURE 26.2 **Anterior view of the superficial muscles of the body.** On the left side of the trunk, some deeper muscles are shown.

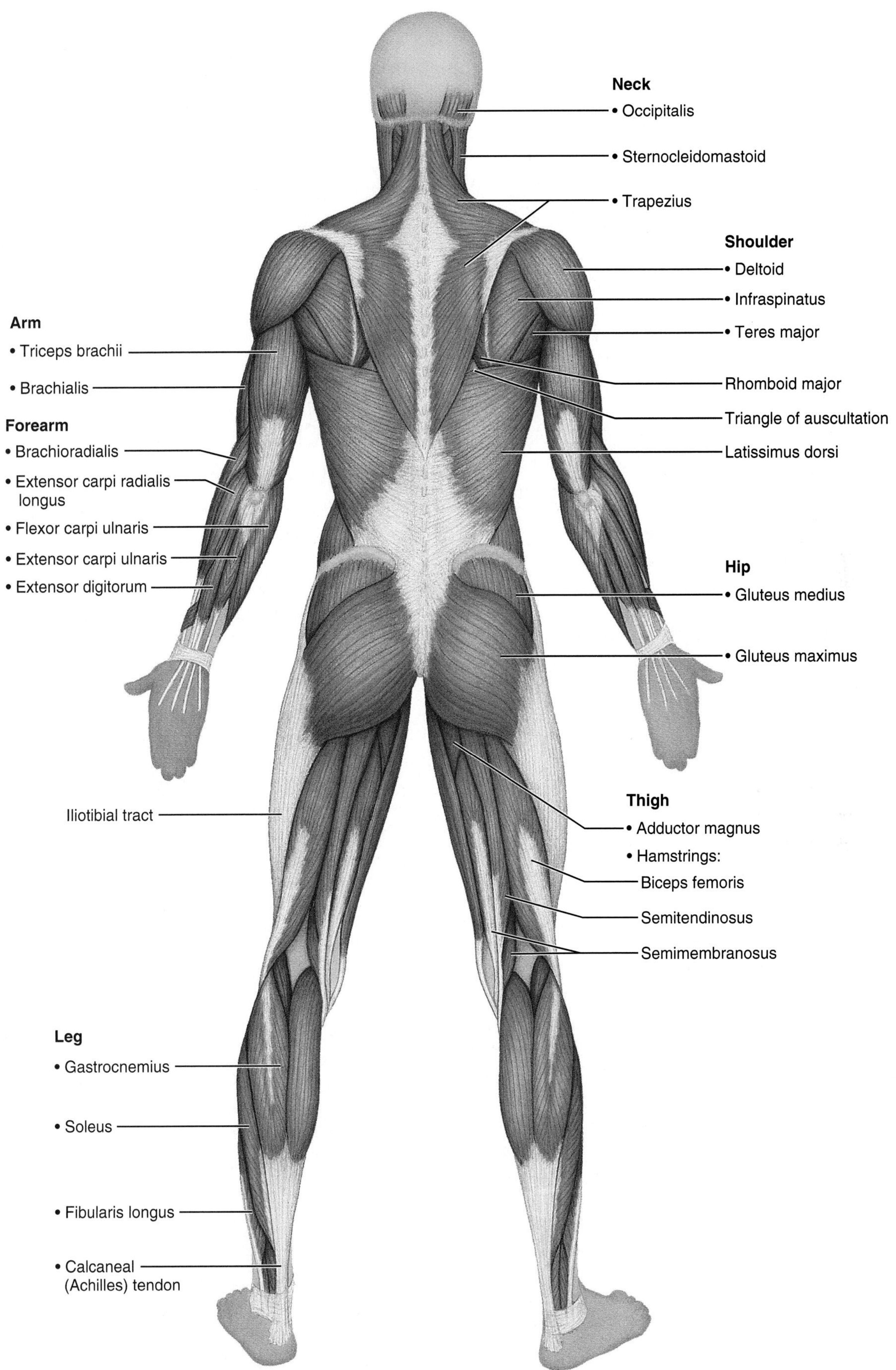

FIGURE 26.3 Posterior view of the superficial muscles of the body.

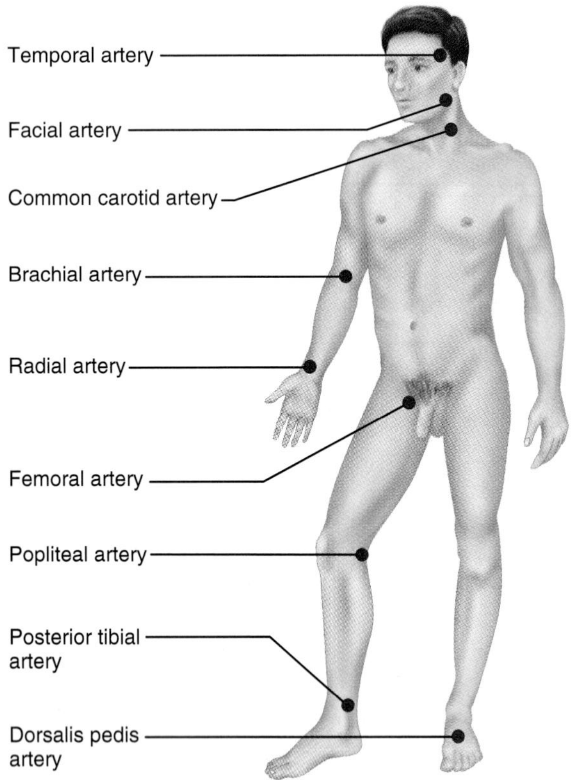

FIGURE 26.4 Points on the body surface where the arterial pulse is most easily felt. The arteries indicated are discussed throughout the chapter, as are some additional pulse points that are not pictured here.

Grasp your *auricle,* the shell-like part of the external ear that surrounds the opening of the *external auditory canal* (see Figure 26.5b). Just posterior to this canal you can feel the deep ear well, the **concha.** Directly anterior to the canal is a stiff projection called the **tragus** (tra′gus; "goat"), which helps protect the opening to the canal. Now trace the ear's outer rim, or *helix,* to the *lobule* (ear lobe) inferiorly. The lobule is easily pierced, and because it is not highly sensitive to pain, it provides a convenient place to obtain a drop of blood for clinical analysis. Additionally, a *pulse oxymeter,* a device that measures blood oxygen levels across the surface of the skin, may be clamped to the ear lobe. The auricle has several other regions, as shown in Figure 26.5b.

Next, place a finger on your temple just anterior to the auricle to feel the pulsations of the *superficial temporal artery* (see Figure 26.5a), which ascends to supply the scalp.

Run your hand anteriorly from the ear's external auditory canal toward the orbit, and feel the *zygomatic arch* just deep to the skin. This "cheek bone" can easily be broken by blows to the face. Next, place your fingers on the skin of your face, and feel it bunch and stretch as you contort your face into a series of smiles, frowns, and grimaces; you are now monitoring the action of the subcutaneous *muscles of facial expression* (see Table 11.1, p. 276).

In the region of the lower jaw, palpate the parts of the bony *mandible:* its anterior *body* and its posterior ascending *ramus* (see Figure 26.5a). Press on the skin over the mandibular ramus, and feel the *masseter muscle* bulge when you bite down. Feel for the anterior border of the masseter, and trace it to the mandible's inferior margin; at this site you will be able to detect the pulse of your *facial artery.* Finally, to feel the *temporomandibular joint,* place a finger directly anterior to the tragus of your ear, and open and close your mouth several times. The bony structure you feel moving is the *head of the mandible.*

THE NECK

Skeletal Landmarks

Begin by running your fingers inferiorly along the back of your neck, in the posterior midline, where you can feel the *spinous processes* of the cervical vertebrae. The spine of C_7 is especially prominent. Now, after placing a finger on your chin, run it inferiorly along the neck's anterior midline. The first hard structure you encounter will be the U-shaped *hyoid bone* (Figure 26.7), which lies in the angle between the floor of the mouth and the vertical part of the neck (Figure 26.8). Directly inferior to this, you will feel the *laryngeal prominence* (Adam's apple) of the thyroid cartilage; this prominence begins with a V-shaped notch superiorly and continues inferiorly as a sharp vertical ridge. Just inferior to the prominence, your finger will sink into a soft depression (formed by the *cricothyroid ligament*) and then will proceed onto the rounded surface of the *cricoid cartilage.* Now swallow several times, and feel the whole larynx move up and down. Continuing inferiorly onto the upper trachea, you might be able to feel the *isthmus of the thyroid gland* as a spongy cushion over the second to fourth tracheal rings (see also Figure 25.8a on p. 753). Then, try to palpate the two soft lateral *lobes* of your thyroid gland along the sides of the trachea. Next, move your finger all the way down to the root of the neck, and rest it in the *jugular notch,* the depression in the superior part of the sternum between the two clavicles. By pushing deeply at this point, you can feel the cartilage rings of the trachea.

Muscles of the Neck

The *sternocleidomastoid,* the most prominent muscle in the neck and the neck's most important surface landmark, can best be seen and felt when your head is turned to the side. If you stand in front of a mirror and turn your face sharply from right to left several times, you will be able to see both heads of this muscle, the

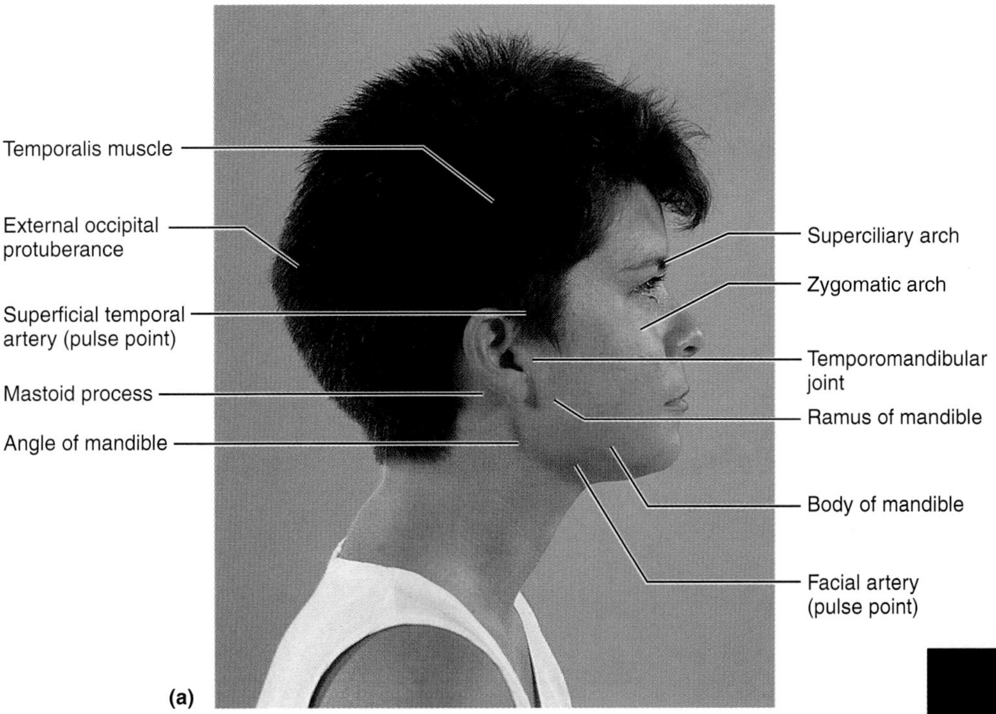

Temporalis muscle
External occipital protuberance
Superficial temporal artery (pulse point)
Mastoid process
Angle of mandible
Superciliary arch
Zygomatic arch
Temporomandibular joint
Ramus of mandible
Body of mandible
Facial artery (pulse point)
(a)

FIGURE 26.5 Surface anatomy of the lateral aspect of the head. (a) Overview. (b) Close-up of an auricle.

Concha
Antihelix
Helix
External auditory canal
Tragus
Antitragus
Lobule
(b)

Frontalis muscle
Lacrimal fossa
Ala of nose
Root and bridge of nose
Dorsum nasi
Apex of nose
Philtrum

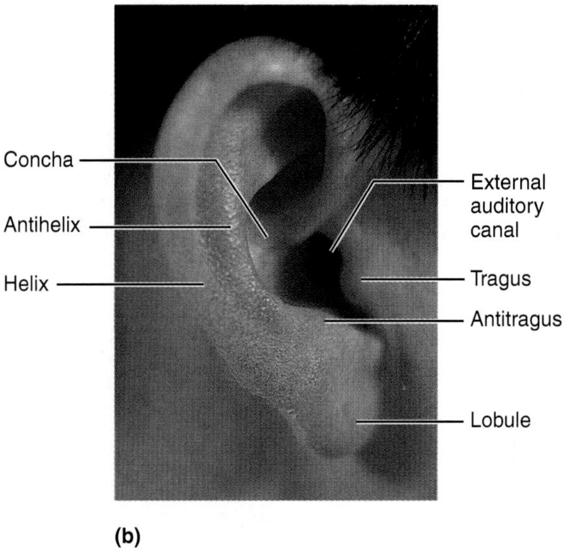

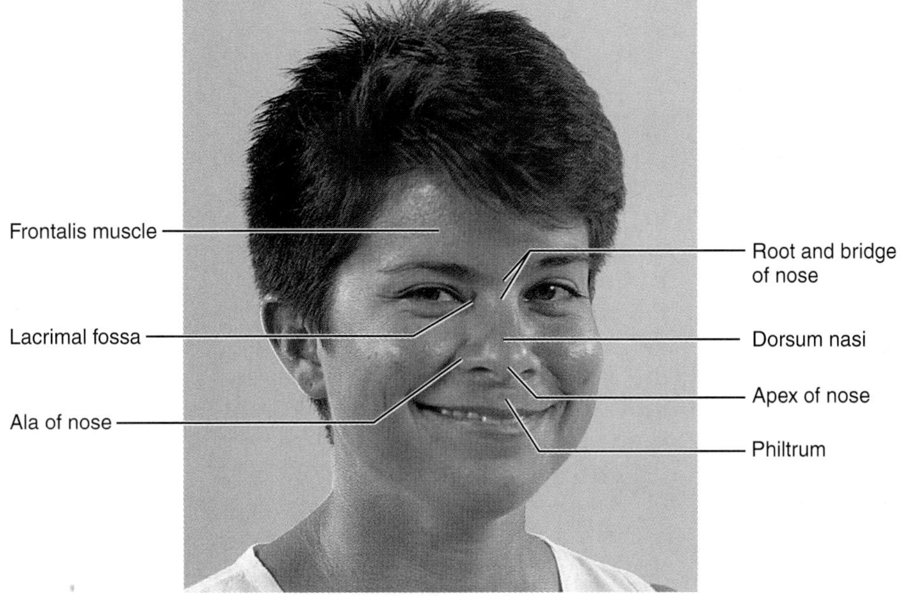

FIGURE 26.6 Surface structures of the face.

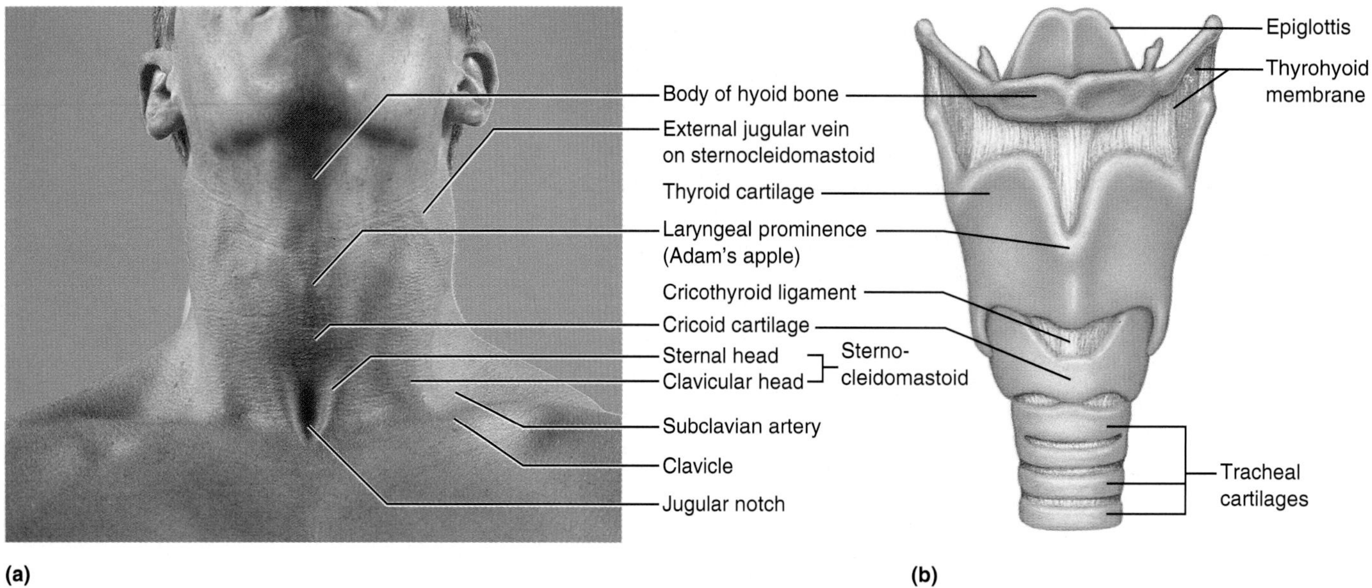

FIGURE 26.7 Anterior surface of the neck. (a) Photograph. (b) Drawing of the underlying skeleton of the larynx.

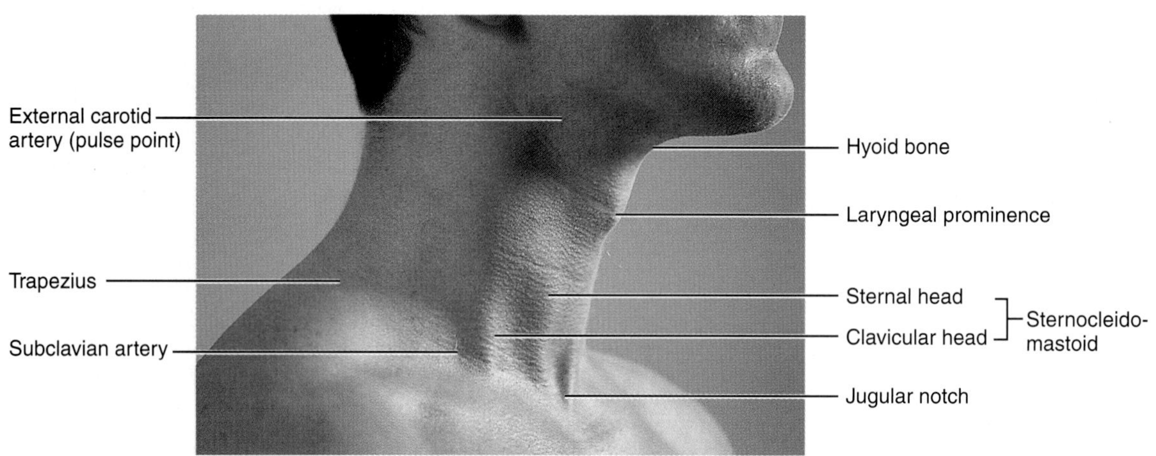

FIGURE 26.8 Lateral surface of the neck.

sternal head medially and the *clavicular head* laterally (see Figures 26.7a, 26.8, and 26.9). Physicians palpate the sternocleidomastoid as they search for any swelling of the *cervical lymph nodes* that lie both superficial and deep to this muscle. The *common carotid artery* and *internal jugular vein* lie just deep to the sternocleidomastoid, a relatively superficial location that exposes these vessels to danger in slashing wounds to the neck. Just lateral to the inferior part of the sternocleidomastoid lies the large *subclavian artery* (see Figure 26.7a) on its way to supply the upper limb. By pushing on the subclavian artery at this point, one can stop the bleeding from a wound anywhere in this limb. Just anterior to the sternocleidomastoid, superior to the level of your larynx, you can feel a carotid pulse—the pulsations of the *external carotid artery* (see Figure 26.8). The *external jugular vein* descends vertically, just superficial to the sternocleidomastoid and deep to the skin (see Figures 26.7a and 26.9b). To make this vein "appear" on your neck, stand before a mirror and gently compress the skin superior to your clavicle with your fingers.

Another large muscle in the neck, on the posterior aspect, is the *trapezius* (see Figure 26.8). You can feel this muscle contract just deep to the skin as you shrug your shoulders.

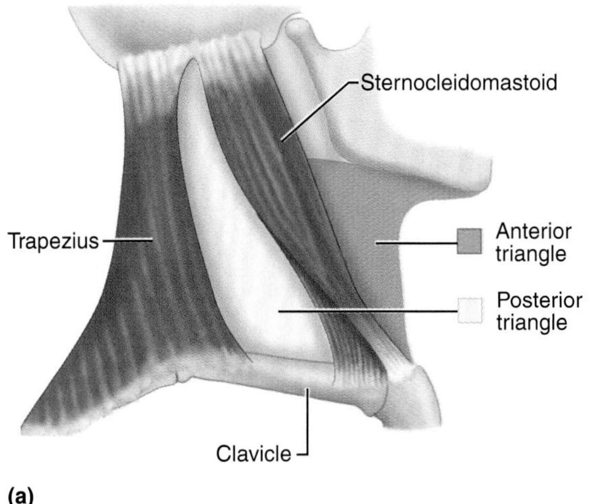

(a)

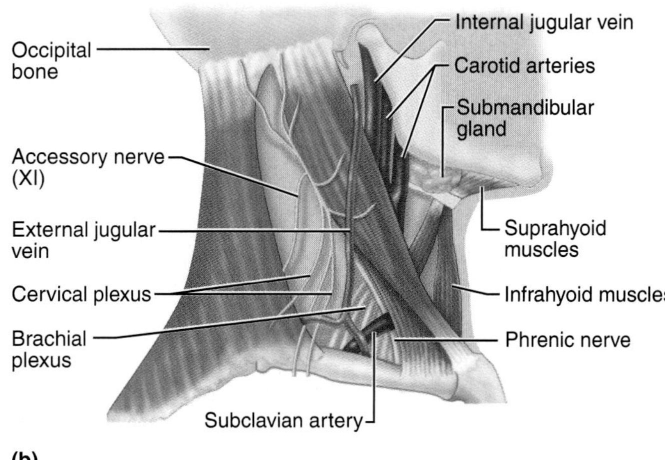

(b)

FIGURE 26.9 Anterior and posterior triangles of the neck. **(a)** Locations of the triangles. **(b)** Some structures in the triangles.

Triangles of the Neck

The sternocleidomastoid muscles divide each side of the neck into a posterior and an anterior triangle (Figure 26.9a). The **posterior triangle** is defined by the sternocleidomastoid anteriorly, the trapezius posteriorly, and the clavicle inferiorly. The **anterior triangle** is defined by the inferior margin of the mandible superiorly, the midline of the neck anteriorly, and the sternocleidomastoid posteriorly.

The contents of these two triangles are shown in Figure 26.9b. The posterior triangle contains many important nerves and blood vessels, including the *accessory nerve* (cranial nerve XI), most of the *cervical plexus,* and the *phrenic nerve.* In the inferior part of the triangle are the external jugular vein, the trunks of the *brachial plexus,* and the subclavian artery—all of which are relatively superficial and easily cut by lacerations or slashing wounds to the neck. In the neck's anterior triangle, the important structures include the *submandibular gland,* the *suprahyoid* and *infrahyoid muscles,* and the parts of the carotid arteries and jugular veins that lie superior to the sternocleidomastoid.

Critical Reasoning

Explain how a wound to the posterior triangle of the neck can lead to long-term loss of sensation in the skin of the neck and shoulder, as well as partial paralysis of the sternocleidomastoid and trapezius muscles.

These effects result from damage to the cervical plexus and accessory nerve, which supply these skin areas and muscles.

THE TRUNK

The trunk consists of the thorax, abdomen, pelvis, and perineum. Because what is commonly called the *back* includes parts of all of these regions, for convenience it is treated separately.

The Back

Bones of the Back

The vertical groove in the center of the back is called the **posterior median furrow** (Figure 26.10a). The *spinous processes* of the vertebrae are visible in this furrow when the spinal column is flexed. Palpate a few of these processes on your back (C_7 and T_1 are the most prominent and the easiest to find). Although locating structures on your own back is difficult, you should be able to palpate the posterior parts of some ribs, as well as the prominent *spine of the scapula* and the scapula's long *medial border.* The scapula lies superficial to ribs 2 through 7, its *inferior angle* is at the level of the spinous process of vertebra T_7, and the medial end of the scapular spine lies opposite the spinous process of T_3.

Now feel the *iliac crests* (superior margins of the iliac bones) in your lower back; you can find these crests easily by resting your hands on your hips. Locate the most superior point of each crest, a point roughly halfway between the posterior median furrow and the lateral side of the body (see Figure 26.10a). A horizontal line through the right and left superior points, the **supracristal line,** intersects L_4, providing a simple way to locate that vertebra. The ability to locate L_4 is

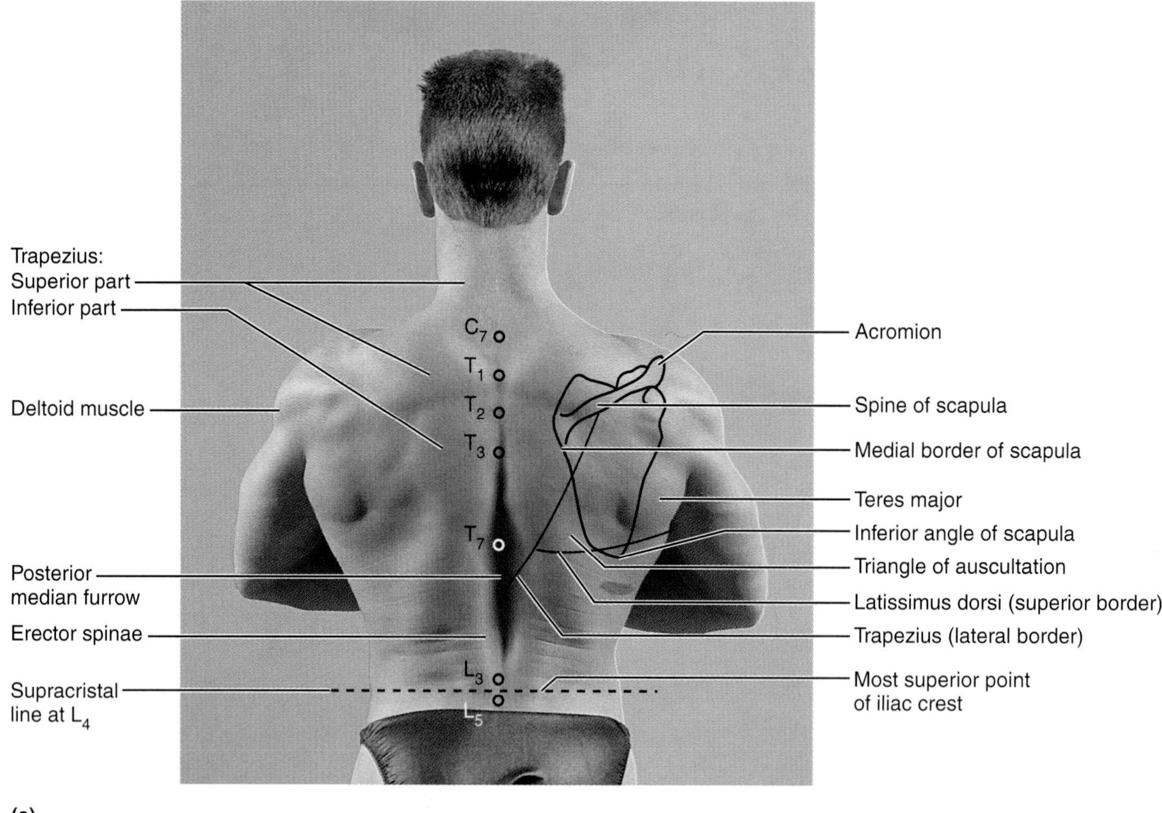

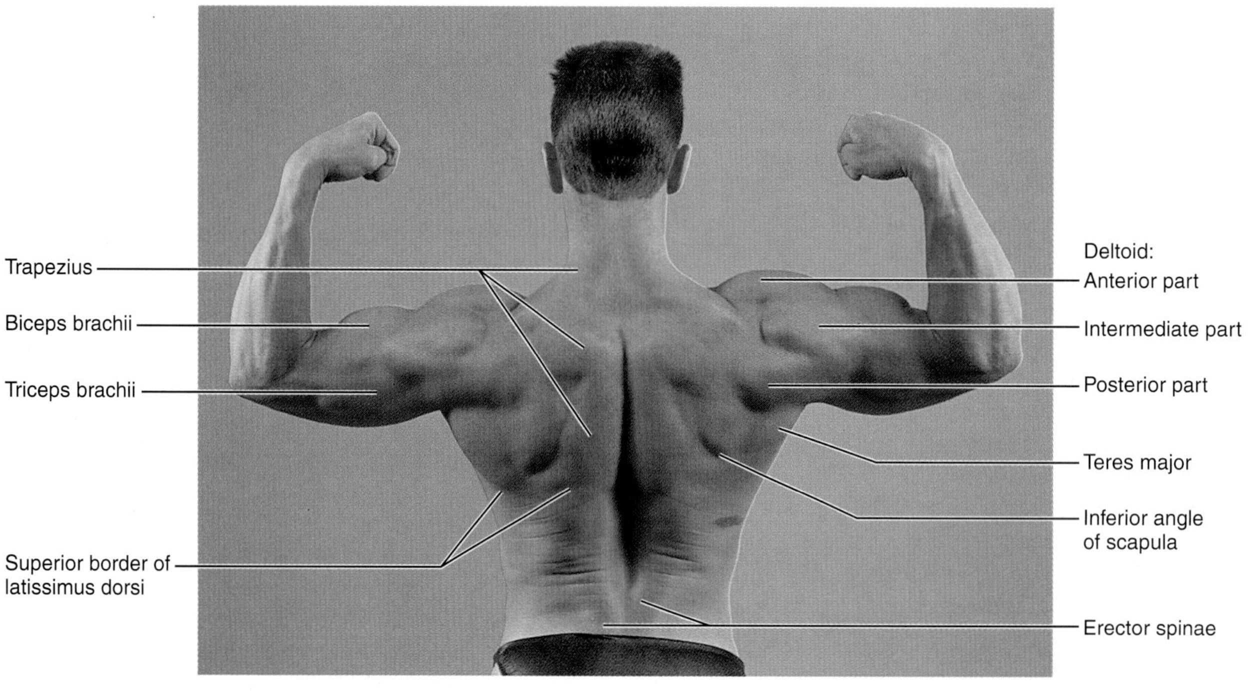

FIGURE 26.10 Surface anatomy of the back, two different poses. In (a), C_7, T_1, T_2, and so on are spinous processes of the indicated vertebrae.

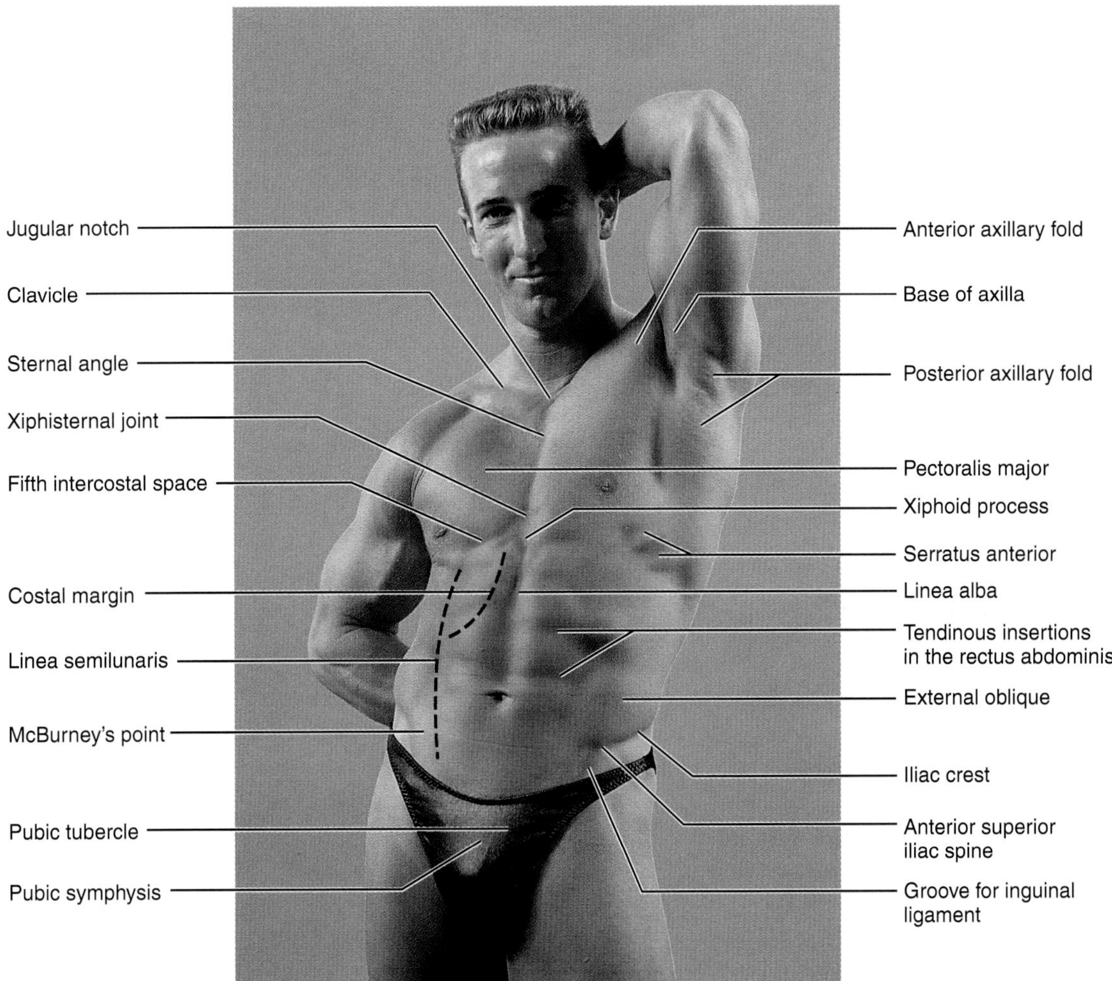

FIGURE 26.11 The anterior thorax and abdomen.

essential for performing a *lumbar puncture,* a procedure in which the clinician inserts a needle into the vertebral canal of the spinal column directly superior or inferior to L_4. Lumbar punctures for withdrawing cerebrospinal fluid are described on page 395.

The *sacrum* is easy to palpate just superior to the cleft in the buttocks, and one can feel the *coccyx* in the extreme inferior part of that cleft, just posterior to the anus.

Muscles of the Back

The largest superficial muscles of the back are the *trapezius* superiorly and *latissimus dorsi* inferiorly (see Figures 26.3 and 26.10b). Furthermore, the deeper *erector spinae* muscles are very evident in the lower back, flanking the vertebral column like thick vertical cords. Feel your erector spinae muscles contract and bulge as you extend your spine.

CLINICAL APPLICATIONS

THE TRIANGLE OF AUSCULTATION The superficial muscles of the back do not cover a small triangular area of the rib cage just medial to the inferior part of the scapula. This so-called **triangle of auscultation** (aw″skul-ta′shun; "listening"), which is bounded by the trapezius medially, the latissimus dorsi inferiorly, and the medial border of the scapula laterally (see Figure 26.3), is the site at which a physician places a stethoscope to listen to lung sounds. To hear the lungs clearly, the physician first directs the patient to fold the arms together in front of the chest and flex the trunk, an action that draws the scapula anteriorly and maximizes the size of the triangle (see Figure 26.10a). ✚

The Thorax

Bones of the Thorax

Start exploring the surface of the anterior thorax (Figure 26.11) by defining the *sternum* (see Figure 26.1a). First use a finger to trace your sternum's triangular *manubrium,* the flat sternal *body,* and the tongue-shaped *xiphoid process* (see Figure 26.11), and then palpate the ridge-like *sternal angle,* where the manubrium meets the body of the sternum. Locating the sternal angle—an easily found, highly reliable landmark—is important

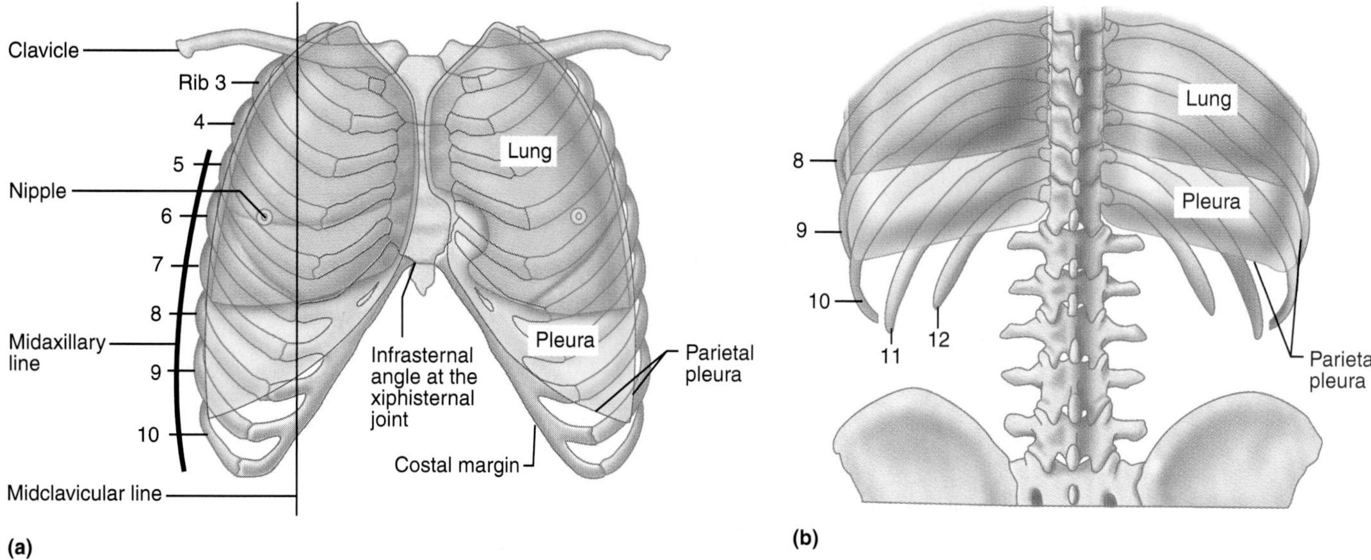

FIGURE 26.12 The bony thoracic cage as it relates to the underlying lungs and pleural cavity. Both the pleural cavities (blue) and the lungs (pink) are indicated. **(a)** Anterior view. **(b)** Posterior view.

because it directs you to the second ribs, and once you find them you can count down to identify all the other ribs (except the first and sometimes the twelfth, which are too deep to be palpated).

By locating the individual ribs, you attain a series of horizontal lines of "latitude" by which to map and locate the underlying visceral organs of the thoracic cavity. Such mapping also requires lines of "longitude," so we define some imaginary vertical lines on the wall of the trunk. Lifting an arm straight up in the air and extending a line inferiorly from the center of the axilla onto the lateral thoracic wall creates the **midaxillary line** (Figure 26.12a); running a vertical line inferiorly from the midpoint of your *clavicle* to the groin defines the **midclavicular line,** which passes about 1 cm medial to the nipple.

Next, feel along the inverted V-shaped inferior edge of your rib cage, the *costal margin.* The superior angle of this margin is termed the **infrasternal angle,** and here lies the *xiphisternal joint* (Figure 26.12a). Deep to this joint, the heart lies on the diaphragm.

The thoracic cage provides many valuable landmarks for locating organs both superficial and deep to it. On the anterior thoracic wall, ribs 2–6 define the superior-to-inferior extent of the breast in females, and the fourth intercostal space defines the location of the *nipple* in men, children, and small-breasted women. The right costal margin runs across the anterior surface of the liver and gallbladder (see Figure 22.4 on p. 636).

Surgeons must be aware of the inferior margin of the *pleural cavities* (see Figure 26.12), because if they accidentally cut into one of these cavities, a lung collapses (see p. 623). The inferior pleural margin lies adjacent to vertebra T_{12} near the posterior midline (see Figure 26.12b) and runs horizontally across the back to reach rib 10 at the midaxillary line. From there, this pleural margin ascends to rib 8 in the midclavicular line (see Figure 26.12a) and to the level of the xiphisternal joint near the anterior midline. The *lungs* do not fill the inferior region of the pleural cavity; instead, their inferior borders run at a level that is two ribs superior to the pleural margin, until they meet that margin near the xiphisternal joint.

The relation of the *heart* to the thoracic cage is considered in Chapter 18 (p. 522). In essence, the superior right corner of the heart lies at the junction of the third rib and the sternum; the superior left corner lies at the second rib, near the sternum; the inferior left corner lies in the fifth intercostal space in the midclavicular line; and the inferior right corner lies at the sixth rib, near the sternum. You may wish to find and interconnect these four corner points on a friend's chest in order to define the heart's size and location. Recall that clinicians listen for the sounds of the four heart valves near each of the heart corners (see Figure 18.11 and p. 530).

Muscles of the Thorax

The main superficial muscles of the anterior thoracic wall are the *pectoralis major* and the anterior slips of the *serratus anterior* (see Figure 26.11). Using this figure as a guide, try to palpate these two muscles on your chest. They both contract during push-ups, and you can confirm this by pushing yourself up from your desk with one arm while palpating the muscles with your opposite hand.

The Abdomen

Bony Landmarks

The anterior abdominal wall (see Figure 26.11) extends from the costal margin down to an inferior boundary that is defined by the following landmarks:

1. **Iliac crest.** Recall that you can locate the iliac crests, the superior margins of the iliac bones, by resting your hands on your hips.
2. **Anterior superior iliac spine.** Representing the most anterior point of the iliac crest, this spine is a prominent landmark that can be palpated in everyone, even those who are overweight.
3. **Inguinal ligament.** This ligament, indicated by a groove on the skin of the groin, runs medially from the anterior superior iliac spine to the pubic tubercle of the pubic bone.
4. **Pubic crest.** You will have to press deeply to feel this crest on the pubic bone near the median *pubic symphysis.* The *pubic tubercle,* the most lateral point of the pubic crest, is easier to palpate, but you must still push deeply.

CLINICAL APPLICATIONS

DIAGNOSING INGUINAL HERNIAS Recall that *inguinal hernias* lie directly superior to the inguinal ligament and may exit from a medial opening called the *superficial inguinal ring* (see p. 710). To locate this ring, palpate the pubic tubercle (Figure 26.13). In examining a male for an inguinal hernia, a physician pushes on the skin that overlies the pubic tubercle, forces a finger into the superficial inguinal ring, and asks the patient to cough. If the patient has an inguinal hernia, the cough will push an intestinal coil through the inguinal canal to touch the physician's fingertip. ✚

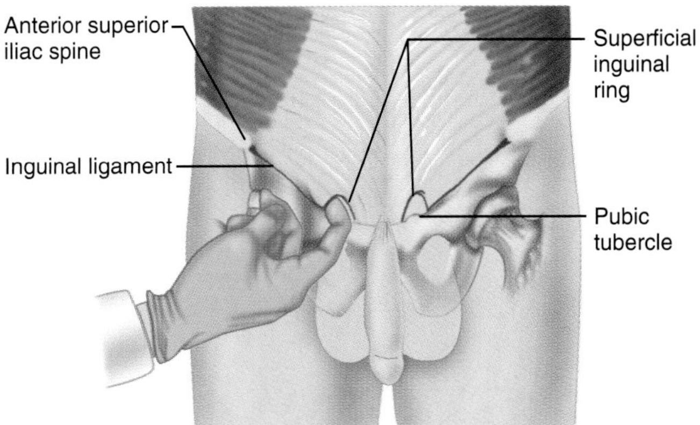

FIGURE 26.13 Clinical examination for an inguinal hernia in a male. The examiner palpates the patient's pubic tubercle, pushes superiorly to invaginate the scrotal skin into the superficial inguinal ring, and asks the patient to cough. If an inguinal hernia exists, it will push inferiorly, and the examiner can feel it touch his or her fingertip.

Muscles and Other Abdominal Surface Features

The central landmark of the anterior abdominal wall is the *umbilicus* (navel). Running superiorly and inferiorly from the umbilicus is the *linea alba* ("white line"), represented in the skin of lean people by a vertical groove (see Figure 26.11). This is a tendinous line that extends from the xiphoid process to the pubic symphysis, just medial to the rectus abdominis muscles (see Table 11.6, p. 289). The linea alba is a favored site for surgical entry into the abdominal cavity because a long cut through it produces no muscle damage and a minimum of bleeding.

CLINICAL APPLICATIONS

UMBILICAL HERNIAS AND MCBURNEY'S POINT At least two kinds of hernias involve the umbilicus and the linea alba. In **acquired umbilical hernias,** which are most common in people over age 40, the linea alba weakens (due to stretching associated with obesity or pregnancy), until coils of intestine push through it just superior to the navel. The herniated coils can be palpated as bulges just deep to the skin. Another type of umbilical hernia, a **congenital umbilical hernia,** occurs in some infants. Normally, the opening of the umbilicus into the stub of the umbilical cord closes a few days after birth through the constriction of a ring of fascia that surrounds the umbilicus. However, in some infants this ring does not constrict, and loops of intestine protrude through the opening, producing a cherry-sized bulge deep to the skin of the navel that enlarges whenever the infant cries. Most congenital umbilical hernias are harmless and correct themselves before the child's second birthday.

McBurney's point is a landmark related to appendicitis. It is a spot on the anterior abdominal skin that lies two-thirds of the way along a line between the umbilicus and the right anterior superior iliac spine (see Figure 26.11). It lies directly superficial to the base of the appendix, is the traditional incision site for appendectomies, and is often the place where the pain of appendicitis is felt most acutely. Palpation of this point that causes strong pain after the pressure is removed (so-called rebound tenderness) can indicate appendicitis. Because this indication cannot provide a precise diagnosis, contact a physician whenever appendicitis is suspected. ✚

Flanking the linea alba on the abdominal wall are the vertical, strap-like *rectus abdominis* muscles (see Figures 26.2 and 26.11). Feel these muscles contract just deep to your skin as you do a bent-knee sit-up or as you bend forward after leaning back in your chair. In the skin of lean people, the lateral margin of each rectus muscle makes a groove known as the **linea semilunaris** (lin′e-ah sem″ĭ-lu-nar′is; "half-moon line"). On your right side, estimate where the linea semilunaris crosses the costal margin of the rib cage; the fundus of the *gallbladder* lies just deep to this spot, so this is the standard point of incision for gallbladder surgery. In muscular people, three horizontal grooves seen in the skin covering the rectus abdominis represent the *tendinous insertions* (or *intersections*), fibrous bands that subdivide the rectus muscle. Because of these subdivisions, each vertical rectus abdominis muscle presents four distinct horizontal bulges.

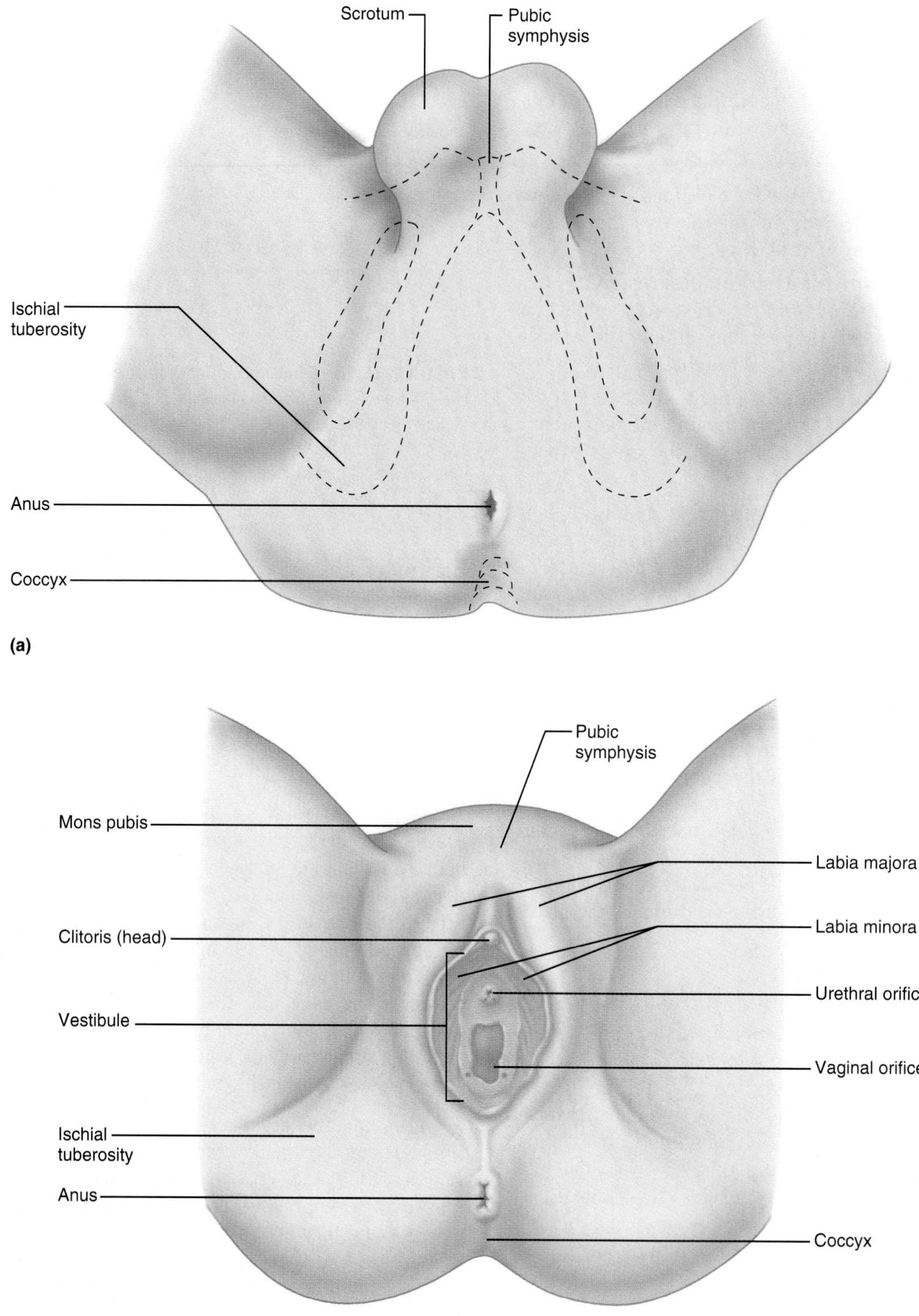

FIGURE 26.14 Surface features of the perineum (a) male, (b) female. In both sexes, the bony boundaries of the diamond-shaped perineum are the pubic symphysis anteriorly, the two ischial tuberosities laterally, and the coccyx posteriorly.

The only other major muscles that can be seen through the anterior abdominal wall are the lateral *external obliques.* Feel these muscles contract as you cough, strain, or in some other way raise your intra-abdominal pressure.

CLINICAL APPLICATIONS

BOWEL SOUNDS To acquire the valuable information conveyed by **bowel sounds,** physicians place a stethoscope in each of the four quadrants of the anterior abdominal wall (see Figure 22.4c, p. 636). Normal bowel sounds, which result from the movement of gas and intestinal contents by peristalsis, are high-pitched gurgles that occur every 5–15 seconds. Less frequent bowel sounds can indicate a halt in intestinal activity, whereas loud sounds may indicate increased activity, associated with inflammation, diarrhea, or other bowel disorders. ✚

The Pelvis and Perineum

Because most of the bony surface features of the *pelvis* are best studied with the abdomen or the lower limb, these features are considered elsewhere in this chapter. Most *internal* pelvic organs can be felt by a rectal exam (see p. 711), but are not palpable through the skin. A full *bladder,* however, becomes firm and can be palpated through the anterior abdominal wall just superior to the pubic symphysis. A bladder that can be palpated more than a few centimeters above this symphysis is dangerously full and retaining urine, and requires drainage by catheterization. The surface anatomy of the *perineum,* which contains the anus and external genitalia, is reviewed in Figure 26.14. Four palpable bony structures define the corners of the diamond-shaped perineum: the pubic symphysis, the ischial tuberosities, and the coccyx.

UPPER LIMB AND SHOULDER

Axilla

The **base of the axilla** is the groove in which the underarm hair grows (see Figure 26.11). Deep to this base lie the *axillary lymph nodes,* the large *axillary vessels* serving the upper limb, and much of the brachial plexus. The base of the axilla forms a "valley" between two thick, rounded ridges of muscle called the **axillary folds.** Just anterior to the base, clutch your **anterior axillary fold,** a fold formed by the pectoralis major muscle. Then grasp your **posterior axillary fold,** which is formed by the latissimus dorsi and teres major muscles of the back as they course toward their insertions on the humerus.

Shoulder

After you again locate the prominent spine of the scapula posteriorly (Figure 26.15), follow this spine to its lateral end, the flattened *acromion* on the shoulder's summit. Then, palpate your *clavicle* anteriorly, tracing this bone from the sternum to the shoulder and noting its curved shape. Now locate the junction between the clavicle and the acromion on the superolateral surface of your shoulder, at the *acromioclavicular joint.* To find this joint, repeatedly thrust your arm anteriorly until you can palpate the precise point of pivoting action. Next, place your fingers on the *greater tubercle* of the humerus, the most lateral bony landmark on the superior surface of the shoulder. It is covered by the thick *deltoid muscle,* which forms the rounded superior part of the shoulder. Intramuscular injections are often given into the deltoid, about 5 cm (2 inches) inferior to the greater tubercle (Figure 26.16a).

Arm

Recall that anatomists define the arm as the part of the upper limb between the shoulder and the elbow. In the arm, the *humerus* can be palpated along its entire length, especially along its medial and lateral sides. Feel the *biceps brachii* muscle contract on your anterior arm when you flex your forearm against resistance. The medial boundary of the biceps is represented by the **medial bicipital furrow (groove)** (see Figure 26.15b). This groove contains the large *brachial artery,* and by pressing on it with your fingertips you can feel your *brachial pulse.* By pressing harder on this artery, one can stop bleeding from a hemorrhage in more distal parts of the limb. The brachial artery is used in measuring blood pressure with a sphygmomanometer (sfig″-mo-mah-nom′ĕ-ter), a device whose cuff is wrapped around the arm superior to the elbow. Next, extend your forearm against resistance, and feel your *triceps brachii* muscle bulge in the posterior arm. All three heads of the triceps (lateral, long, and medial) are visible through the skin of a muscular person (Figure 26.17).

Elbow Region

In the distal part of your arm, near the elbow, palpate the two projections of the humerus, the *lateral* and *medial epicondyles* (see Figure 26.15a and b). Midway between the epicondyles, on the posterior side, feel the *olecranon process* of the ulna, which forms the point of the elbow. Confirm that the two epicondyles and the olecranon all lie in the same horizontal line when the elbow is extended. If these three bony processes do not line up, the elbow is dislocated. Now feel the posterior surface of the medial epicondyle: You are palpating your ulnar nerve. As we explained earlier (p. 432), banging the ulnar nerve at the "funny bone" sends a sharp twinge of pain along the medial side of the forearm and hand.

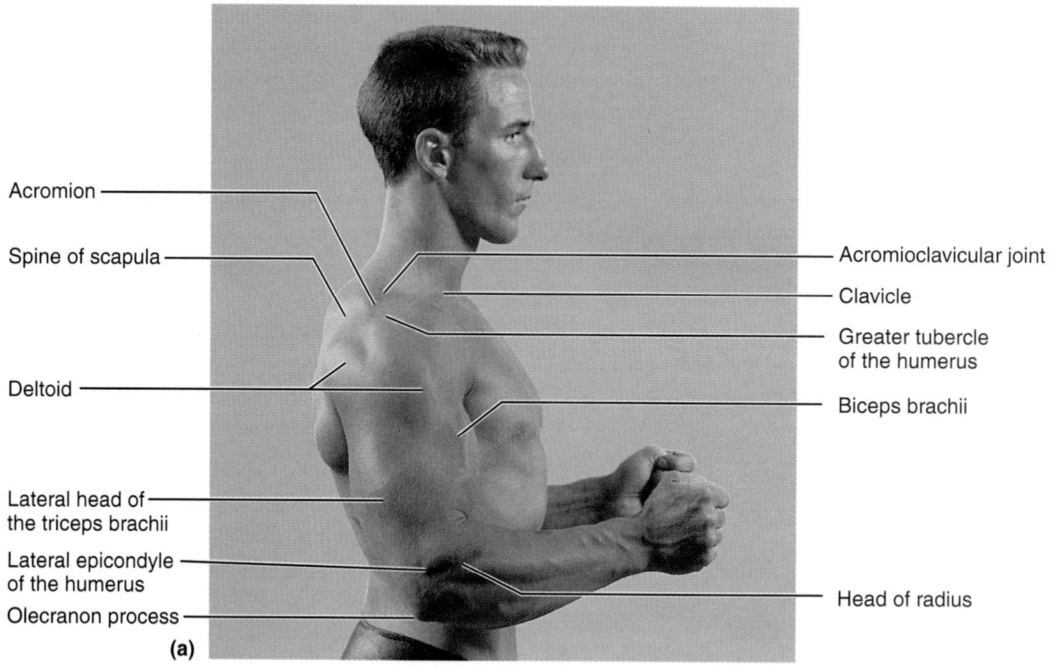

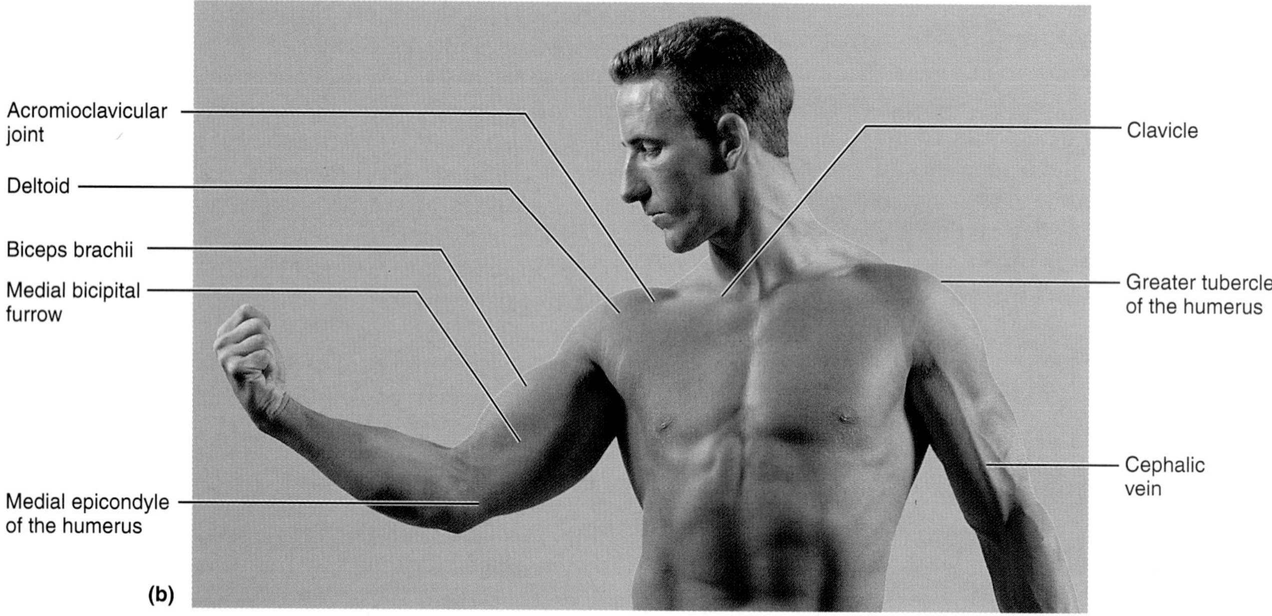

FIGURE 26.15 Shoulder and arm. (a) Lateral view. (b) Anterior and medial view.

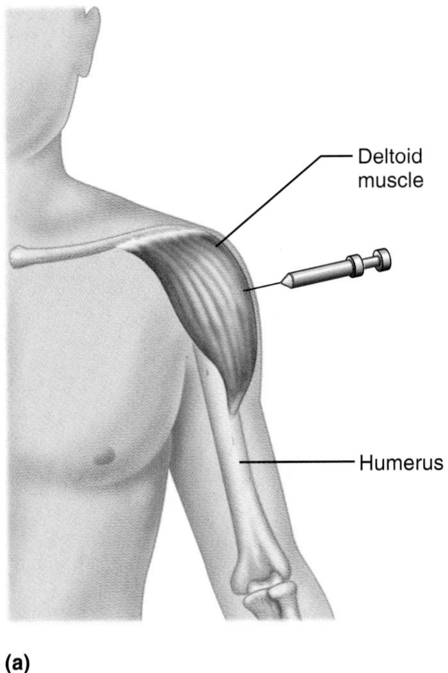

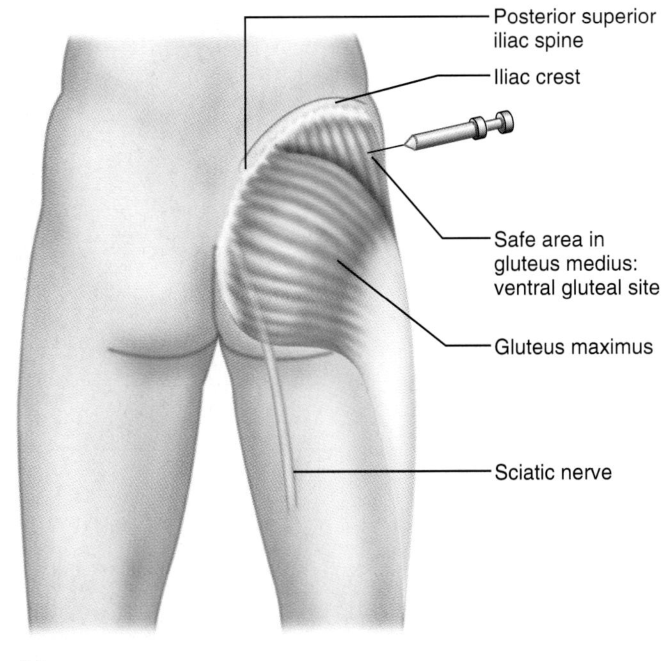

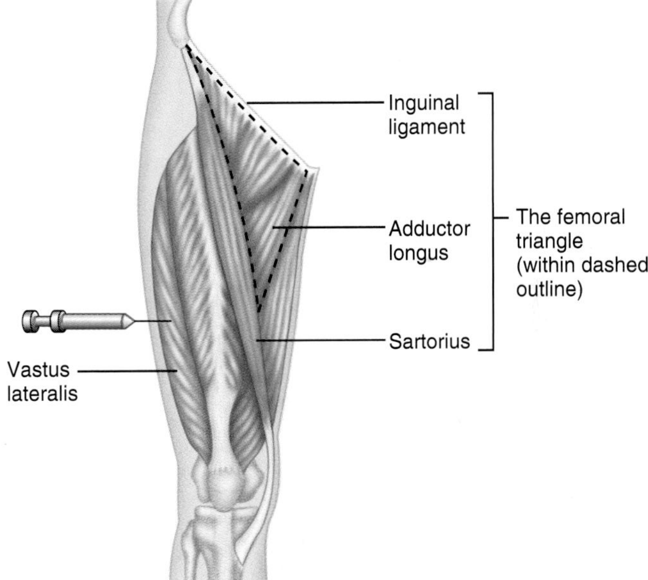

FIGURE 26.16 Three major sites of intramuscular injections. **(a)** Deltoid muscle on arm (for injection volumes of less than 1 ml). **(b)** Ventral gluteal site: gluteus medius (safest and least-painful site). **(c)** Vastus lateralis in lateral thigh (best site in small children). Part (c) also shows the femoral triangle.

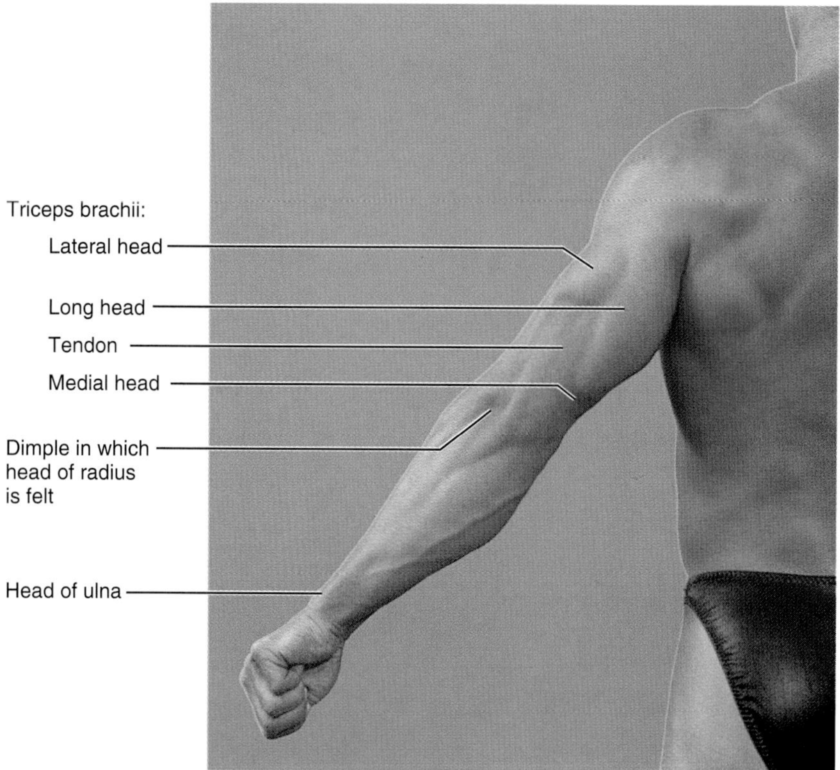

FIGURE 26.17 The three heads of the triceps brachii muscle, which insert on a large tendon.

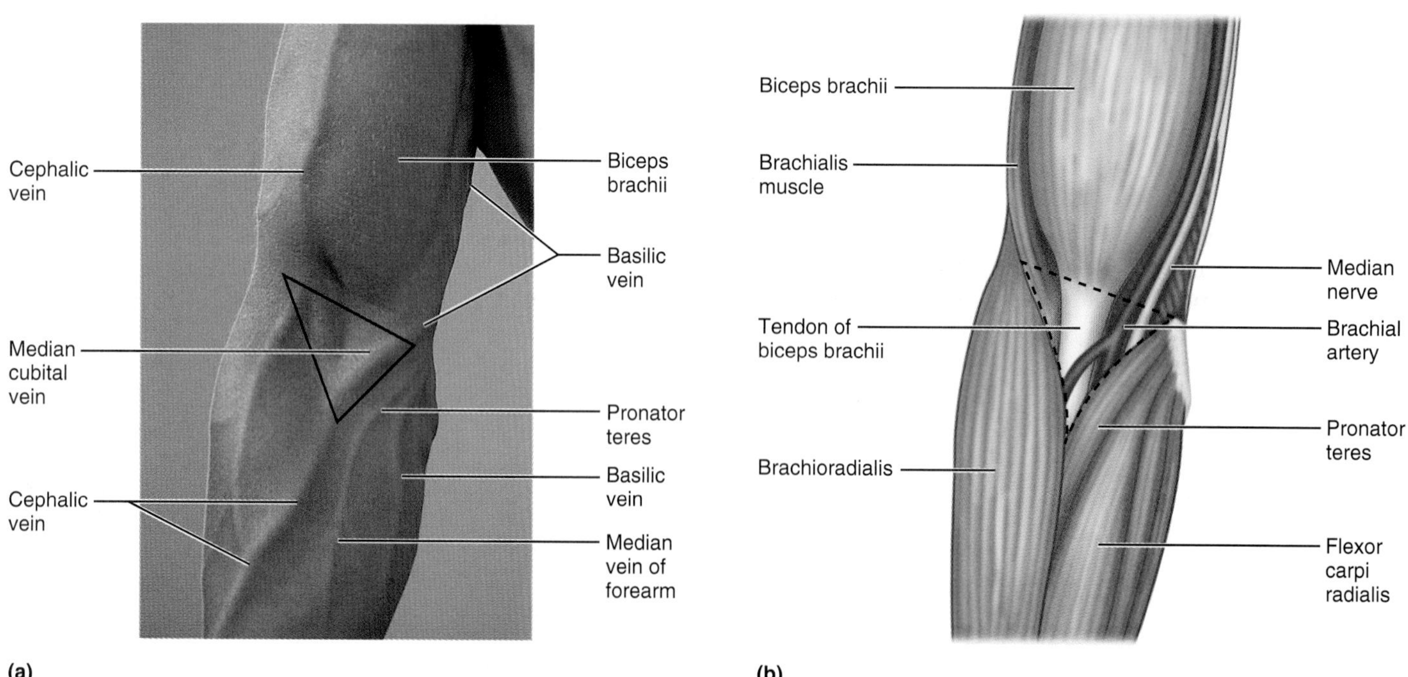

FIGURE 26.18 The cubital fossa (outlined by the triangle) on the anterior surface of the right elbow. (a) Photograph. (b) Drawing of deeper structures in the fossa.

On the anterior surface of the elbow region is a triangular depression called the **cubital fossa** or *antecubital fossa* (*ante* = in front of; *cubit* = elbow) (Figure 26.18). The triangle's superior base is formed by a horizontal line between the humeral epicondyles, and its two inferior sides are defined by the *brachioradialis* and *pronator teres* muscles (see Figure 26.18b). Try to define these boundaries on your own limb: To find the brachioradialis muscle, flex your forearm against resistance, and watch this muscle bulge through the skin of your lateral forearm; to feel your pronator teres contract, palpate the cubital fossa as you pronate your forearm against resistance.

The cubital fossa contains the superficial *median cubital vein* (see Figure 26.18a), a site at which clinicians often draw blood and insert intravenous (IV) catheters to administer medications, transfused blood, and nutrient fluids. Clinicians must be aware that because the large *brachial artery* lies just deep to the median cubital vein (see Figure 26.18b), a needle must be inserted into the vein at a shallow angle, almost parallel to the skin, to avoid puncturing the artery. Other structures that lie deep in the cubital fossa are also shown in Figure 26.18b.

The median cubital vein interconnects the larger *cephalic vein,* which ascends along the lateral side of the forearm and arm, and the *basilic vein,* which ascends through the limb's medial side (see p. 567). Both of these veins are visible through the skin of lean individuals (see Figure 26.18a).

Forearm and Hand

The two parallel bones of the forearm are the medial *ulna* and the lateral *radius.* You can feel the ulna along its entire length as a sharp ridge on the posterior forearm. Confirm that this ridge runs inferiorly from the olecranon process. As for the radius, you can feel its distal half, but most of its proximal half is covered by muscle. You can, however, feel the rotating *head* of the radius. To do this, extend your forearm and note that a dimple forms on the posterior lateral surface of the elbow region (see Figure 26.17). Then, while pressing three fingers into this dimple, rotate your free hand as if you were turning a doorknob. You will feel the head of the radius rotate as you perform this action.

Figure 26.19 shows a way to locate the distal, knob-like *styloid processes* of the radius and ulna. (Do not confuse the ulna's styloid process with the conspicuous *head of the ulna,* from which the styloid process stems.) Confirm that the styloid process of the radius lies about 1 cm (0.4 inch) distal to that of the ulna.

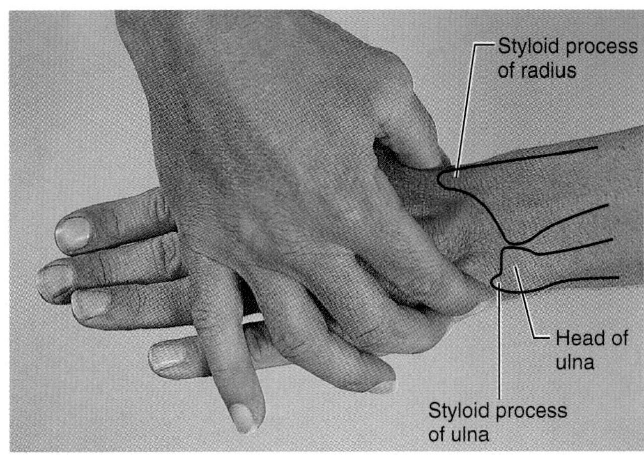

FIGURE 26.19 A way to locate the styloid processes of the ulna and radius. In this picture, the right hand is palpating the left forearm, feeling for the knobs. The styloid process of the radius lies about 1 cm distal to the styloid process of the ulna.

CLINICAL APPLICATIONS

PALPATION OF COLLES' FRACTURE When presented with a patient who has fallen on an outstretched hand and whose wrist has curves that resemble those on a fork, a physician uses the method depicted in Figure 26.19. If palpation reveals that the styloid process of the radius has moved proximally from its normal position, the diagnosis is **Colles' fracture,** an impacted fracture in which the distal end of the radius is forced proximally into the shaft of the radius. ✚

Next, feel the major groups of muscles within your forearm. Flex your hand and fingers against resistance, and feel the anterior *flexor muscles* contract; then extend your hand at the wrist, and feel the tightening of the posterior *extensor muscles.*

The anterior surface of the forearm near the wrist possesses many significant features (Figure 26.20). Flex your fist against resistance, and the tendons of the main wrist flexors will bulge the skin of the distal forearm. Most obvious will be the tendons of the *flexor carpi radialis* and *palmaris longus* muscles (see Figure 26.20a). (However, because the palmaris longus is absent from at least one arm in 30% of all people, your forearm may exhibit just one prominent tendon instead of two.) The

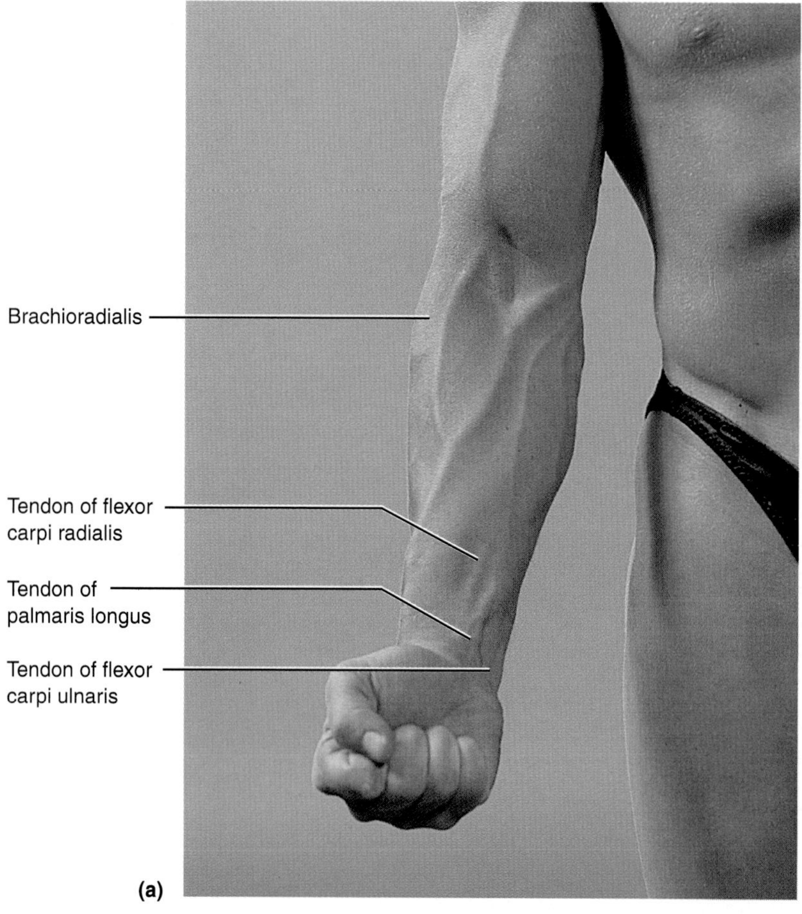

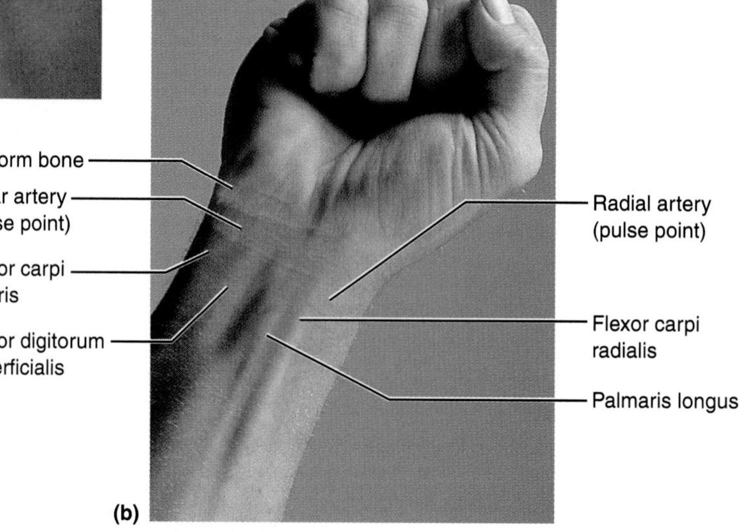

FIGURE 26.20 The anterior surface of the forearm and fist. (a) The entire forearm. (b) Enlarged view of the distal forearm and fist. The tendons of the flexor muscles guide the clinician to two sites for taking pulses.

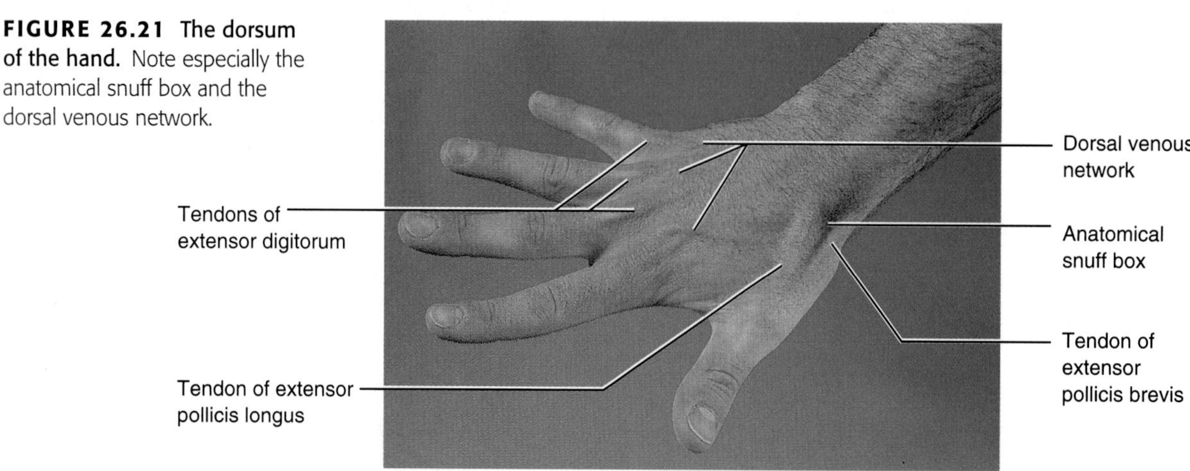

FIGURE 26.21 The dorsum of the hand. Note especially the anatomical snuff box and the dorsal venous network.

radial artery lies just lateral to (on the thumb side of) the flexor carpi radialis tendon, where the pulse is easily detected (see Figure 26.20b). Feel your radial pulse there. The *median nerve* (which innervates the thumb) lies deep to the palmaris longus tendon. Finally, the *ulnar artery* lies on the medial side of the forearm, just lateral to the tendon of the *flexor carpi ulnaris.* Using Figure 26.20b as a guide, locate and feel your ulnar arterial pulse.

By extending your thumb and pointing it posteriorly, you will form a triangular depression in the base of the thumb on the back of your hand—the **anatomical snuff box** (Figure 26.21), so named because people once put snuff (tobacco for sniffing) in this hollow before lifting it up to the nose. Its two elevated borders are defined by the tendons of the thumb extensor muscles, *extensor pollicis brevis* and *extensor pollicis longus.* Because the radial artery runs within the snuff box, this is another site for taking an arterial pulse. The main bone on the floor of the snuff box is the scaphoid bone of the wrist, but the styloid process of the radius is also present here. (If displaced by a bone fracture, the radial styloid process will be felt outside of the snuff box instead of within it.)

On the dorsum of your hand, observe the superficial veins constituting the *dorsal venous network,* which drains superiorly into the cephalic vein. This network provides a site for drawing blood and inserting intravenous catheters and is preferred over the median cubital vein for these purposes. Next, extend your hand and fingers, and observe the tendons of the *extensor digitorum.*

The palmar surface of the hand also contains some features of interest (Figure 26.22), including the *epidermal ridges* ("fingerprints") and many *flexion creases* in the skin. Grasp your *thenar eminence* (the bulge on the palm that contains the thumb muscles) and your *hypothenar eminence* (the bulge on the medial palm that contains muscles of the little finger).

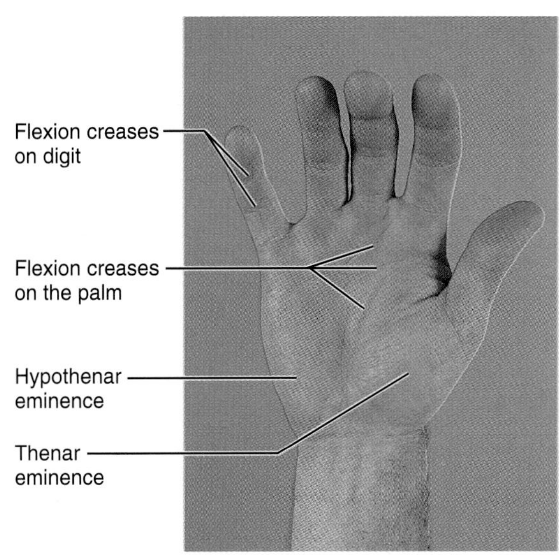

FIGURE 26.22 The palmar surface of the hand.

LOWER LIMB AND GLUTEAL REGION

Gluteal Region

Locate your iliac crests again, and trace each to its most posterior point—the small but sharp *posterior superior iliac spine,* indicated by a distinct, easily found dimple in the skin that lies two to three fingers' widths lateral to the midline of the back (Figure 26.23). The dimple also indicates the position of the *sacroiliac joint,* where

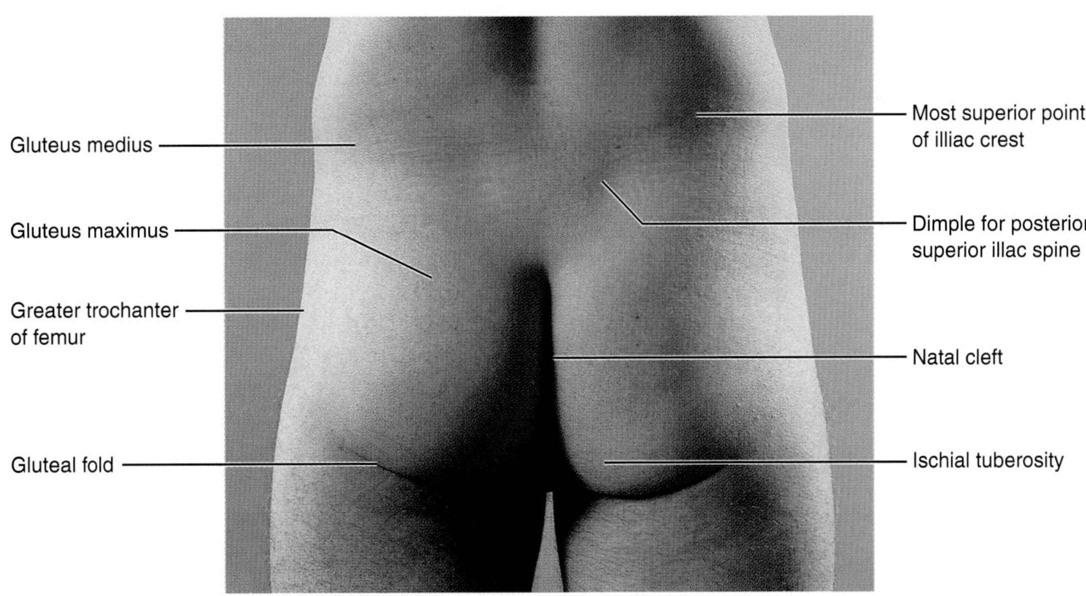

FIGURE 26.23 The gluteal region. Because this region extends from the iliac crests superiorly to the gluteal folds inferiorly, it includes more than just the prominences ("cheeks") of the buttock.

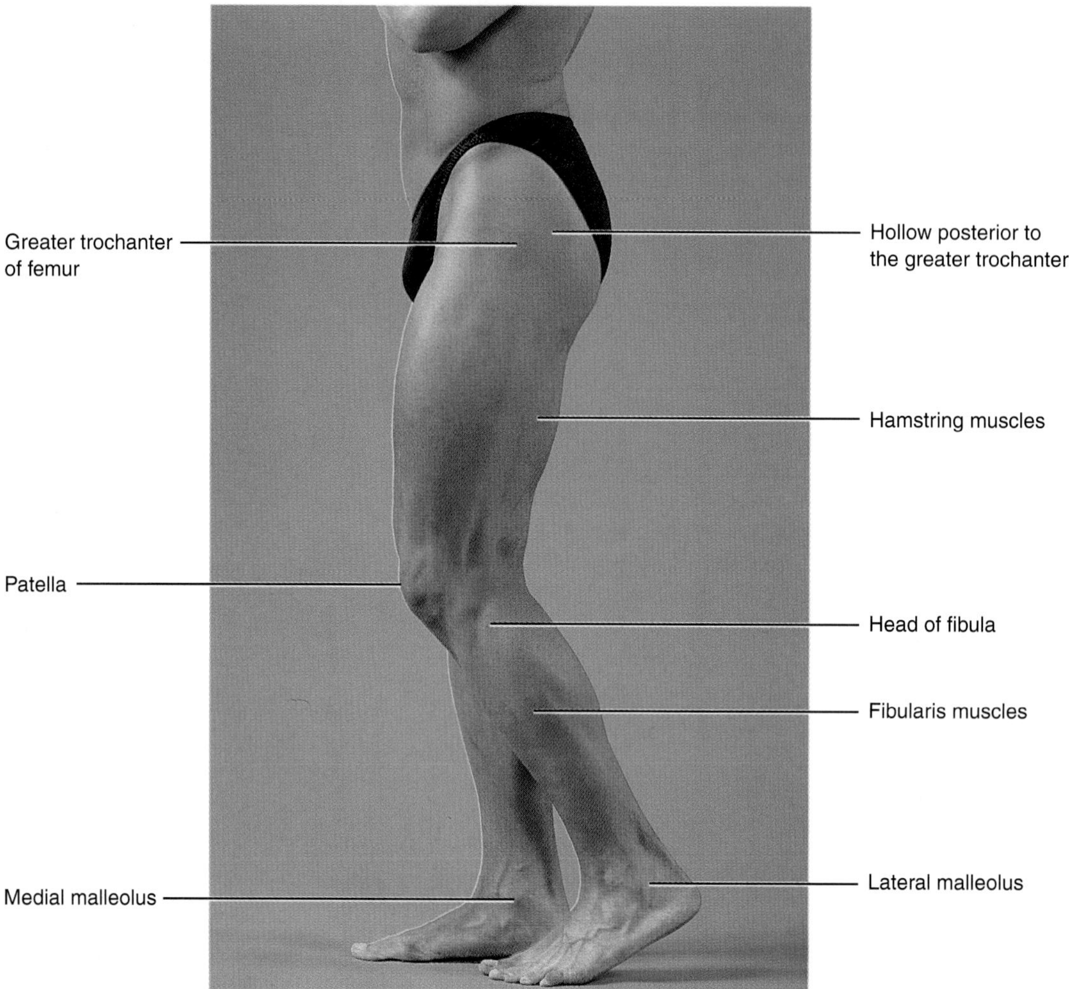

FIGURE 26.24 Lateral surface of the lower limb.

the hip bone attaches to the sacrum of the spinal column. In *bone marrow biopsies,* a physician inserts a needle into the iliac bone 1 cm (0.4 inch) inferolateral to this dimple to aspirate a sample of marrow that is then examined for evidence of blood disorders or leukemia.

Dominating the gluteal region are the two *prominences* ("cheeks") of the buttocks formed by subcutaneous fat and by the thick *gluteus maximus* muscles. The vertical, midline groove between the two prominences is called the **natal cleft** (na′tal; "rump") or *gluteal cleft.* The inferior margin of each prominence is the roughly horizontal **gluteal fold,** which roughly corresponds to the inferior margin of the gluteus maximus. As you sit down or flex your thigh, try to palpate the *ischial tuberosity* just above the medial side of each gluteal fold. The ischial tuberosities are the robust inferior parts of the ischial bones that support the body's weight during sitting. Next, palpate the *greater trochanter* of the femur (Figure 26.24) by placing a finger on the lateral side of the hip, just anterior to a hollow and about a hand's breadth inferior to the iliac crest. Because this trochanter is the most superior point on the lateral femur, you can feel it move with the femur as you alternately flex and extend your thigh.

Clinical Applications

GLUTEAL INTRAMUSCULAR INJECTIONS The gluteal region is a major site for administering intramuscular injections. When inserting the hypodermic needle, extreme care must be taken to avoid piercing the thick *sciatic nerve* (which innervates much of the lower limb) and the gluteal nerves and blood vessels, all of which lie just deep to the gluteus maximus muscle. To avoid these structures, injections are applied to the *gluteus medius* muscle just superior to the cheeks of the buttocks, in a "safe area" called the **ventral gluteal site** (see Figure 26.16b).

A good way to find this site is to draw a line laterally from the posterior superior iliac spine (dimple) to the greater trochanter and then proceed 5 cm (2 inches) superiorly from the midpoint of that line. Alternatively, approach the lateral side of the patient's left hip with your extended right hand (or the right hip with your left hand), and then place your thumb on the anterior superior iliac spine and your index finger as far posteriorly on the iliac crest as it can reach. In so doing the heel of your hand comes to lie on the greater trochanter, and the needle is then inserted in the angle of the V formed between your thumb and index finger. Note that the needle must enter at least 4 cm (1.5 inches) inferior to the iliac crest, or it will pierce bone instead of muscle.

Gluteal injections are not given to small children because their "safe areas" are too small to locate with certainty and their gluteal muscles are too thin. Instead, infants and toddlers receive shots midway down the length of the prominent *vastus lateralis* muscle of the thigh (see Figure 26.16c). ✚

Thigh

The thigh is pictured in Figures 26.24, 26.25, and 26.26. Because much of the femur is enveloped by thick muscles, the thigh has few palpable bony landmarks. Distally, you can feel the *medial* and *lateral condyles of the femur,* and the *patella* anterior to the condyles (see Figure 26.25c). Next, palpate the three groups of thigh muscles: the *quadriceps femoris muscles* anteriorly and the *adductor muscles* medially (see Figure 26.25a), and the *hamstrings* posteriorly (see Figures 26.24 and 26.26). The *vastus lateralis,* already mentioned as a site for intramuscular injections, is the lateral muscle of the quadriceps femoris group (see Figure 26.25b).

The anterosuperior surface of the thigh exhibits a three-sided depression called the **femoral triangle** (see Figures 26.25a and 26.16c). As shown in Figure 26.16c, the superior border of this triangle is formed by the inguinal ligament, and its two inferior borders are defined by the *sartorius* and *adductor longus* muscles. The large *femoral artery* and *vein* (the main vessels to the lower limb) descend vertically through the center of the femoral triangle. To feel the pulse of the femoral artery, press hard inward just inferior to the midinguinal point (halfway between the anterior superior iliac spine and the pubic tubercle). Very hard pressure on this point can stop the bleeding from a hemorrhage in the distal limb. The femoral triangle also contains most of the *inguinal lymph nodes* (which are easily palpated if swollen) and is the site of femoral hernias. In a **femoral hernia,** coils of the intestine push inferiorly from the abdominal cavity, descend deep to the inguinal ligament, and enter the superior thigh, where they form a distinct bulge deep to the skin of the femoral triangle. (Note that a femoral hernia, which lies inferior to the inguinal ligament, is entirely different from an *inguinal* hernia, which lies superior to the inguinal ligament.)

On the posterior of the knee is a diamond-shaped hollow called the **popliteal fossa** (see Figure 26.26), the four borders of which are defined by large muscles. The *biceps femoris* forms the superolateral border, the *semitendinosus* and *semimembranosus* define the superomedial border, and the two heads of the *gastrocnemius* form the two inferior borders. The *popliteal artery* and *vein* (main vessels to the leg) lie deep within this fossa, and you may be able to feel a popliteal pulse if you flex your leg at the knee and push your fingers firmly into the popliteal fossa. If a physician is unable to feel a patient's popliteal pulse, the femoral artery may be narrowed by atherosclerosis. Because the popliteal fossa is covered superficially by a roof of strong fascia that prevents any expansion of this space, anything that causes the fossa to swell—an infected lymph node in the fossa, varicosity of the popliteal vein, or an aneurysm in the popliteal artery—is accompanied by pressure and pain.

Leg and Foot

Locate your patella again, and then follow the thick *patellar ligament* inferiorly from the patella to its insertion on the superior tibia where you can feel a rough projection, the *tibial tuberosity* (see Figure 26.25c). Continue running your fingers inferiorly along the tibia's sharp *anterior border* and its flat *medial surface*—bony landmarks that lie very near the surface throughout their length. It is the exposed medial surface of the tibia that is bruised when someone receives a "bang on the shin."

Now, return to the superior part of your leg, and palpate the expanded *lateral* and *medial condyles of the tibia* just inferior to the knee. You can distinguish the tibial condyles from the *femoral* condyles because you can feel the tibial condyles move with the tibia during knee flexion. Feel the bulbous *head of the fibula* in the superolateral region of the leg (see Figures 26.24 and 26.25c). Try to feel the *common fibular nerve* (nerve to the anterior leg and foot; see p. 434) where it wraps around the fibula's *neck* just inferior to its head. This nerve is often bumped against the bone here and damaged (p. 435).

In the most distal part of the leg, you can feel the *lateral malleolus* of the fibula as the lateral prominence of the ankle (see Figure 26.25d). Notice that this lies slightly lower than the *medial malleolus* of the tibia, which forms the ankle's medial prominence. Place your finger just posterior to the medial malleolus to feel the pulse of your *posterior tibial artery.*

Next, palpate the main muscle groups of your leg, starting with the calf muscles posteriorly (see Figure 26.26). Standing on tiptoes will help you feel the *lateral* and *medial heads of the gastrocnemius* and, inferior to these, the broad *soleus* muscle. Also feel the tension in your *calcaneal (Achilles) tendon* and the point of insertion of this tendon onto the calcaneus bone of the foot. Return to the anterior surface of the leg, and palpate the *anterior muscle compartment* (see Figure 26.25c) while alternately dorsiflexing and plantar flexing your foot. You will feel the tibialis anterior and extensor digitorum muscles contracting and then relaxing. Then, palpate the *fibularis muscles* that cover most of the fibula laterally (see Figure 26.24). The tendons of these

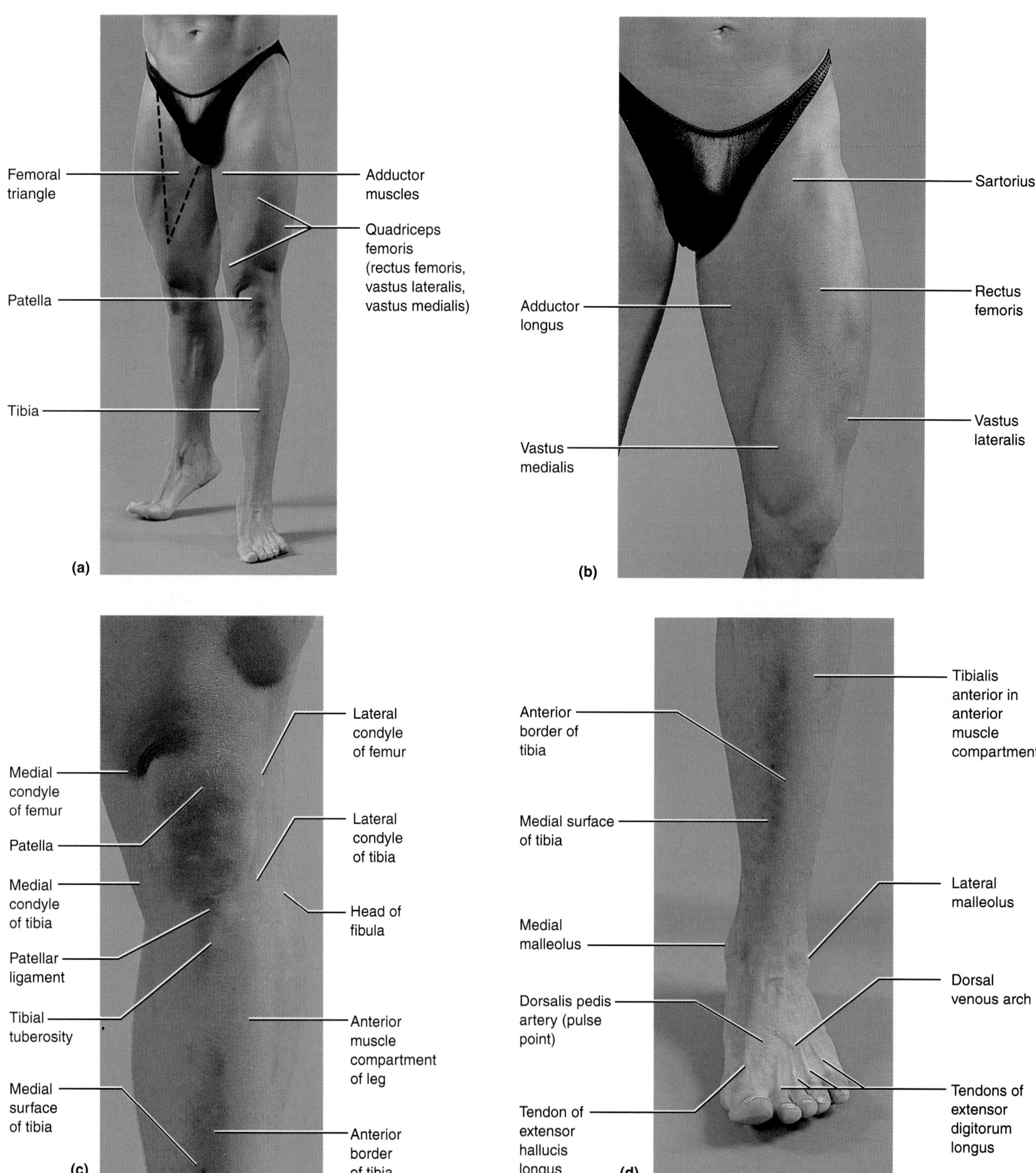

FIGURE 26.25 Anterior surface of the lower limb. **(a)** Both limbs, with the right limb revealing its medial surface. The femoral triangle is outlined on the right limb. **(b)** Closer view of the left thigh. **(c)** The left knee region. **(d)** The dorsum of the left foot.

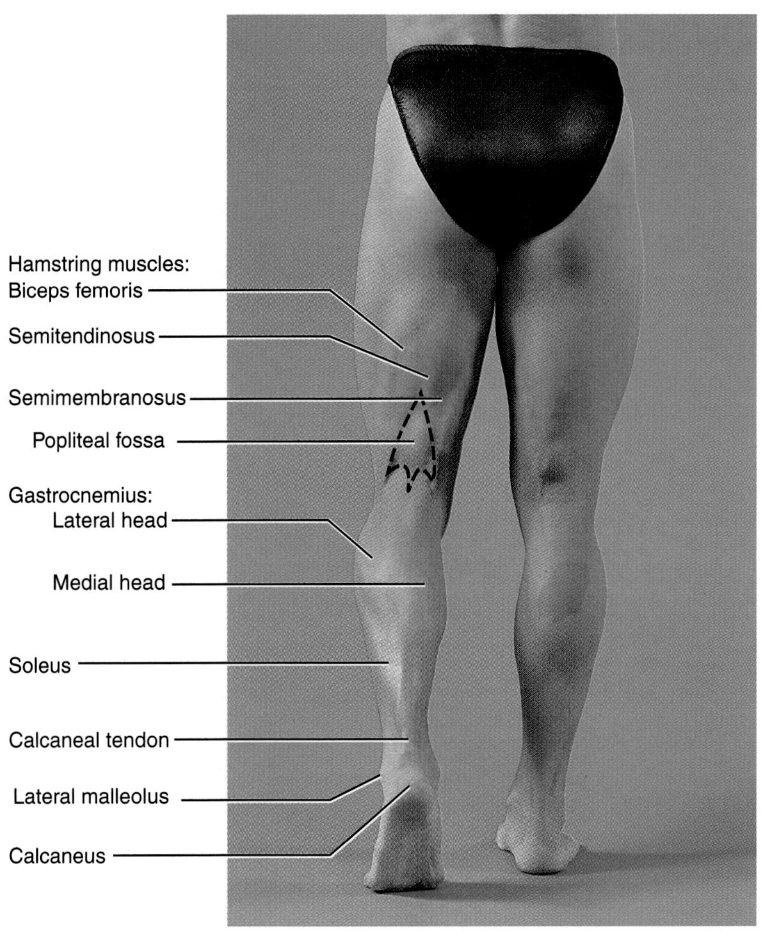

FIGURE 26.26 Posterior surface of the lower limb. Notice the diamond-shaped popliteal fossa posterior to the knee.

muscles pass posterior to the lateral malleolus and can be felt at a point posterior and slightly superior to that malleolus.

Now, observe the dorsum (superior surface) of your foot. You may see the superficial *dorsal venous arch* overlying the proximal part of the metatarsal bones (see Figure 26.25d); this arch gives rise to both of the saphenous veins, the main superficial veins of the lower limb. Visible in lean people, the *great saphenous vein* ascends along the medial side of the entire limb, whereas the *small saphenous vein* ascends through the center of the calf (see Figure 19.24 on p. 572).

As you extend your toes, observe the tendons of the *extensor digitorum longus* and *extensor hallucis longus* muscles on the dorsum of the foot (see Figure 26.25d). Finally, place a finger on the extreme proximal part of the space between the first and second metatarsal bones, where you should be able to feel the pulse of the *dorsalis pedis artery.*

We hope that this chapter has reinforced some of the information you learned in earlier chapters and has helped you appreciate the human body as a living structure. A study of one's own surface anatomy can be a profound and humbling experience—it enables us to see ourselves as mortal, physical beings, and to realize that the structures we see in anatomy books and in the anatomy laboratory really do exist in our own bodies. If you are planning a career in the health-related fields or in physical education, your study of living anatomy will help ease the transition from working with dead specimens in your anatomy course to working with the bodies of active, responsive, sensitive people in a clinical situation.

CHAPTER SUMMARY

Media study tools that could provide you additional help in reviewing specific key topics of Chapter 26 are referenced below.

SP = Study Partner; AIA = A.D.A.M.® Interactive Anatomy

1. Surface anatomy is the study of the external features of the body and of those internal features that can be observed or palpated through the body surface. Surface anatomy reveals more about the skeleton and muscles than about other organs, but it also provides information on many of our deep viscera.
2. Surface anatomy provides the simplest and least invasive method of assessing the health of the living body, so it forms the basis of a routine physical examination. Surface landmarks also show the clinician where to take pulses, make injections, listen to internal organs, make surgical incisions, feel for broken bones, and more.

THE HEAD (pp. 770–774)

Cranium (p. 770)

1. In the cranium, the bony and muscular structures that can be felt through the skin include the superciliary arches, external occipital protuberance, mastoid process, and temporalis muscles.

AIA Link: Atlas; Region: Head & Neck; Surface Anatomy of Head (posterior).

2. The scalp covers the superior surface of the cranium and consists of skin, a highly vascular subcutaneous layer, and the galea aponeurotica.

Face (pp. 770–774)

3. The parts of the nose include the root, bridge, dorsum nasi, apex, and alae. Parts of the auricle include the concha, tragus, helix, and lobule.
4. Palpable features of the face include the zygomatic arch, mandibular body and ramus, masseter muscle, temporomandibular joint, and the pulse points of the superficial temporal and facial arteries.

THE NECK (pp. 774–777)

Skeletal Landmarks (p. 774)

1. By running a finger inferiorly along the anterior midline of the neck, one can palpate the hyoid bone, laryngeal prominence, cricoid cartilage, thyroid gland, trachea, and jugular notch.

Muscles of the Neck (pp. 774–776)

2. The sternocleidomastoid muscle is an important surface landmark of the neck. Near or on this muscle, one can locate swollen cervical lymph nodes, the external carotid pulse, the subclavian artery, and the external jugular vein.

Triangles of the Neck (p. 777)

3. The posterior triangle of the neck is defined by the sternocleidomastoid anteriorly, the trapezius posteriorly, and the clavicle inferiorly. The anterior triangle is defined by the inferior margin of the mandible superiorly, the midline of the neck anteriorly, and the sternocleidomastoid posteriorly.
4. Many nerves and arteries are located relatively close to the surface in the posterior triangle of the neck: the accessory nerve, the brachial and cervical plexuses, the subclavian artery, and the external jugular vein.

THE TRUNK (pp. 777–783)

The Back (pp. 777–779)

1. In the posterior median furrow of the back, one can count and identify the vertebral spinous processes. The most easily located vertebral spines are: C_7 and T_1 (the most prominent spines), T_3 (at the level of the medial end of the spine of the scapula), T_7 (level of the inferior scapular angle), and L_4 (level of the supracristal line).

AIA Link: Atlas; Region: View: Posterior; Vertebral Column (posterior).

2. Superficial muscles of the back include the trapezius and latissimus dorsi, which, along with the medial border of the scapula, define the triangle of auscultation. Physicians place a stethoscope on this triangle to listen for lung sounds.

The Thorax (pp. 779–780)

3. The sternal angle is an important surface landmark on the anterior thorax. By finding the second rib at the sternal angle, one can count inferiorly to identify the other ribs. The midaxillary and midclavicular lines are imaginary vertical lines that extend inferiorly from the center of the axilla and clavicle, respectively. The apex of the inverted V-shaped costal margin (infrasternal angle) is at the xiphisternal joint.

AIA Link: Atlas; Region: Thorax; Landmarks of Thorax (lateral).

4. The thoracic cage provides many reliable bony landmarks for locating the soft organs in the thoracic cavity. By orienting to specific ribs, one can map the location of the pleural cavity, the lungs, and the heart.

The Abdomen (pp. 781–783)

5. Surface landmarks on the anterior abdominal wall include the iliac crest, anterior superior iliac spine, groove for the inguinal ligament, pubic tubercle, pubic crest, pubic symphysis, umbilicus, groove for linea alba, rectus abdominis muscle, and linea semilunaris. A physician's finger placed in the superficial inguinal ring can detect an inguinal hernia.
6. The umbilicus and linea alba are sites of umbilical hernias, and the linea alba is a common incision site for surgical entry into the abdominal cavity.
7. McBurney's point, one-third of the way between the right anterior superior iliac spine and the umbilicus, is usually the site of maximum rebound tenderness during appendicitis and is the traditional site of incision for appendectomy.

8. The fundus of the gallbladder underlies the right costal margin, where the linea semilunaris intersects this margin.
9. Physicians listen for bowel sounds over all four quadrants of the anterior abdominal wall.

AIA Link: Atlas; Region: Abdomen; Abdominopelvic Quadrants.

The Pelvis and Perineum (p. 783)

10. A full urinary bladder can be palpated through the anterior abdominal wall superior to the pubic symphysis. Four palpable structures define the corners of the diamond-shaped perineum: pubic symphysis, coccyx, and the two ischial tuberosities.

UPPER LIMB AND SHOULDER (pp. 783–789)

Axilla (p. 783)

1. Important superficial structures associated with the axilla include the base of the axilla, axillary lymph nodes, and axillary vessels. The anterior and posterior walls of the axilla are called axillary folds.

Shoulder (p. 783)

2. Skeletal landmarks that can be palpated in the shoulder region include the acromion of the scapula, the acromioclavicular joint, and the greater tubercle of the humerus.
3. Intramuscular injections are often given into the deltoid muscle, about 5 cm (2 inches) inferior to the greater tubercle.

Arm (p. 783)

4. On the arm, one can palpate the humerus and the biceps and triceps brachii muscles. The medial bicipital furrow (groove), on the medial side of the arm, indicates the position of the brachial artery.

Elbow Region (pp. 783–787)

5. Bony landmarks in the elbow region include the two epicondyles of the humerus and the olecranon process of the ulna, all lying in the same horizontal plane.
6. The cubital fossa (antecubital fossa) is the triangular depression on the anterior surface of the elbow region; the median cubital vein runs through it. Clinicians often insert needles into this superficial vein to draw blood or to administer IV medications and fluids.

SP Exercise: Chapter 26, Surface Anatomy of the Shoulder and Arm.

Forearm and Hand (pp. 787–789)

7. In the forearm, the palpable bony structures include the entire posterior margin of the ulna and the distal half and head of the radius. At the wrist, the styloid process of the radius lies 1 cm distal to the styloid process of the ulna.
8. One can palpate the flexor and extensor muscle groups on the anterior and posterior forearm, respectively.
9. On the anterior forearm near the wrist, one sees the tendons of the flexor carpi radialis, palmaris longus (absent in some people), and flexor carpi ulnaris. A radial arterial pulse can be felt just lateral to the flexor carpi radialis tendon, and an ulnar arterial pulse just lateral to the flexor carpi ulnaris tendon.
10. The tendons of the thumb extensor muscles define the anatomical snuff box on the posterior base of the thumb. A radial pulse and the styloid process of the radius are felt here.
11. The dorsal venous network on the posterior surface of the hand is an important site for drawing blood and administering IV medications. On the palmar surface of the hand, features include flexion creases and the thenar and hypothenar eminences.

LOWER LIMB AND GLUTEAL REGION (pp. 789–793)

Gluteal Region (pp. 789–790)

1. The posterior point of the iliac crest is the posterior superior iliac spine, represented by a dimple in the skin. Other bony landmarks in the gluteal region are the ischial tuberosities and the greater trochanter of the femur.
2. The right and left gluteal prominences, separated by the natal cleft, consist of gluteus maximus muscles and subcutaneous fatty tissue.
3. The gluteal region is a major site for administering intramuscular injections. For safety, the needle is inserted into the gluteus medius muscle at the ventral gluteal site. Gluteal injections are not administered to small children.

Thigh (p. 791)

4. Bony landmarks that can be palpated in the thigh include the patella and the condyles of the femur. The main groups of thigh muscles are the quadriceps femoris muscles, adductor muscles, and hamstrings. The vastus lateralis is a common site for intramuscular injections, especially in small children.
5. The femoral triangle on the anterior superior thigh contains the femoral artery and vein and most of the inguinal lymph nodes. A femoral hernia bulges the skin of the femoral triangle. One can feel a femoral pulse just inferior to the midinguinal point.
6. The popliteal fossa is a diamond-shaped hollow on the posterior aspect of the knee, defined by four muscular boundaries. It contains the deeply located popliteal artery and vein and is covered superficially by a layer of dense fascia.

Leg and Foot (pp. 791–793)

7. Bony landmarks that can be palpated in the leg include the tibial tuberosity and tibial condyles, the anterior border and medial surface of the tibia, the head of the fibula, and the lateral and medial malleolus. The pulse of the posterior tibial artery is felt just posterior to the medial malleolus.
8. The main muscles of the leg, which can be palpated as they contract, include the gastrocnemius and soleus on the calf, the muscles in the anterior compartment, and the fibularis muscles on the fibula.
9. On the dorsum of the foot, one can observe the dorsal venous arch and feel the pulse of the dorsalis pedis artery.

REVIEW QUESTIONS

Multiple Choice/Matching Questions

1. Rebound tenderness (a) occurs in appendicitis, (b) is whiplash of the neck, (c) is a sore foot from playing basketball, (d) occurs when the larynx falls back into place after swallowing.

2. A pressure point is a place where one presses on an artery through the body surface to stop bleeding farther distally. Which of the following sites is not a pressure point for any major artery? (a) the neck just superior to the clavicle and lateral to the sternocleidomastoid, (b) the sciatic artery in middle of gluteus maximus, (c) the medial bicipital furrow on arm, (d) the midinguinal point in the femoral triangle.

3. The anatomical snuff box (a) is in the nose, (b) contains the styloid process of the radius, (c) is defined by tendons of the flexor carpi radialis and palmaris longus, (d) cannot really hold snuff.

4. Some landmarks on the body surface can be seen or felt, but others are abstractions that you must construct by drawing imaginary lines. Which of the following pairs of structures is abstract and invisible? (a) umbilicus and costal margin, (b) anterior superior iliac spine and natal cleft, (c) linea alba and linea semilunaris, (d) McBurney's point and midaxillary line, (e) philtrum and sternocleidomastoid.

5. A muscle that contributes to the posterior axillary fold is the (a) pectoralis major, (b) latissimus dorsi, (c) trapezius, (d) infraspinatus, (e) pectoralis minor, (f) both a and e are correct.

6. Which of the following is not a pulse point? (a) anatomical snuff box, (b) inferior margin of mandible anterior to masseter muscle, (c) center of distal forearm at palmaris longus tendon, (d) medial bicipital furrow on arm, (e) dorsum of foot between the first two metatarsals.

7. A nurse missed a patient's median cubital vein while trying to draw blood and then inserted the needle far too deeply into the cubital fossa. This error could cause any of the following problems, *except* (a) paralysis of the ulnar nerve, (b) paralysis of the median nerve, (c) bruising the insertion tendon of the biceps brachii muscle, (d) blood spurting from the brachial artery.

8. Which of the following organs is almost impossible to study with the techniques of surface anatomy? (a) heart, (b) lungs, (c) brain, (d) nose.

9. One listens for bowel sounds with a stethoscope that is placed (a) on the four quadrants of the abdominal wall; (b) in the triangle of auscultation; (c) in the right and left midaxillary line, just superior to the iliac crests; (d) inside the patient's bowels (intestines), on the tip of an endoscope.

Short Answer Essay Questions

10. (a) Define palpation. (b) Why is a knowledge of surface anatomy valuable in clinical situations and for evaluating athletic injuries?

11. Explain how one locates the proper site for intramuscular injections into (a) the ventral gluteal site and (b) the deltoid muscle.

12. Esperanza, a pre-physical therapy student, was trying to locate the vertebral spinous processes on the flexed back of her friend Chao, but she kept losing count. Chao told her to check her count against several reliable "guideposts" along the way: the spinous processes of C_7, T_3, T_7, and L_4. Describe how to find each of these four particular vertebrae without having to count any vertebrae.

13. During a physical examination, how would a physician distinguish between an inguinal hernia and a femoral hernia?

14. Where is a standard site for inserting the needle into the iliac bone for a bone marrow biopsy?

15. If necessary, review the surface anatomy of the heart from Chapter 18 to answer this question: Where is the apex point of the heart on the thoracic wall?

16. Describe how to locate the standard points of surgical incision for reaching both the appendix and the gallbladder.

17. Sketch the following neck structures as they relate to the sternocleidomastoid muscle: (a) cervical lymph nodes, (b) pulse point of the external carotid artery, (c) subclavian artery, (d) external jugular vein, (e) brachial plexus.

18. Name the borders of the triangle of auscultation.

19. Tell where to locate the following thigh structures: (a) greater trochanter of the femur, (b) adductor muscles, (c) vastus lateralis, (d) bony origin of the hamstring muscles.

20. Figure 26.4 shows where to feel the pulses of various arteries. Name the artery from which each listed artery branches. Here is an example to get you started: The superficial temporal artery *is a branch of the external carotid artery.*

CRITICAL REASONING + CLINICAL APPLICATION QUESTIONS

1. Walking to her car after her 65th birthday party, Mrs. Schultz tripped on ice and fell forward on her outstretched palms. When she arrived at the emergency room, her right wrist and hand were bent like a fork. Dr. Jefferson felt that the styloid process of her radius was outside of the anatomical snuff box and slightly proximal to the styloid process of the ulna. When he checked her elbow, he found that the olecranon process lay 2 cm proximal to the two epicondyles of the right humerus. Explain all these observations, and describe what had happened to Mrs. Schultz's limb.

2. One of a group of rabbit hunters was accidentally sprayed with buckshot in both of his gluteal prominences. When his companions saw that he would survive, they laughed and joked about where he had been shot. Later, however, they

were horrified and ashamed to find that their friend would be permanently paralyzed and without sensation in both legs from the knee down, as well as on the back of his thighs. What had happened?

3. An athletic trainer was examining a college basketball player to find the site of a pulled muscle. The trainer asked the athlete to extend her thigh forcefully at the hip, but she found this action too painful to perform. Then the trainer palpated her posterior thigh and felt some swelling of the muscles there. In simplest terms, which basic muscle group was injured?

4. The drug fight was over, and the gang member felt exhilarated because his gang had won, and the only wound he had received was a relatively shallow knife slash across the inferior left side of his neck. Then his nightmare began: He suddenly realized his left upper limb was paralyzed; parts of the skin on his neck were numb; a considerable amount of blood was flowing (but not spurting) from the wound; and he could not turn his head to the right or shrug his left shoulder. Worst of all, he was gasping for breath. Based on your knowledge of surface anatomy, explain as many of his symptoms as you can.

5. A two-year-old fell against the edge of a table and opened a 7-cm gash in her scalp. A doctor stitched her up, and the wound healed. For the next year her parents told their friends that they had been frightened by how much the wound bled and were amazed that it had healed in only 2 weeks. Are these observations unusual for a scalp wound? Explain.

6. Gabrielle, a talented college athlete, told her physician that she was experiencing great pain in the hollow behind her left knee. Clinical tests indicated that the pain was caused by a popliteal lymph node swollen with cancer cells that had spread from bone cancer in the tibia. Gabrielle was told that her prognosis was good because the cancer had been detected very early, and that she would not need to have her leg amputated. The following semester, Gabrielle learned in her anatomy course that cancerous lymph nodes do not hurt (see Chapter 20, p. 587). If that is so, why had her swollen popliteal node caused so much pain?

A.D.A.M.® INTERACTIVE ANATOMY QUESTIONS

1. Link: Atlas; Region: Head & Neck; Triangles of the Neck (anterior) (a) Which muscle in the neck provides an important surface landmark? (b) Give the name of the bone that can be palpated at the midline of the neck.

2. Link: Atlas; Region: Upper Limb; Bones of forearm and hand (posterior) (a) Name the bony landmarks in the elbow region. (b) These bony landmarks are located on which bones in the elbow region?

APPENDIX A
The Metric System

Measurement	Unit and Abbreviation	Metric Equivalent	Metric to English Conversion Factor	English to Metric Conversion Factor
Length	1 kilometer (km)	= 1000 (10^3) meters	1 km = 0.62 mile	1 mile = 1.61 km
	1 meter (m)	= 100 (10^2) centimeters = 1000 millimeters	1 m = 1.09 yards 1 m = 3.28 feet 1 m = 39.37 inches	1 yard = 0.914 m 1 foot = 0.305 m
	1 centimeter (cm)	= 0.01 (10^{-2}) meter	1 cm = 0.394 inch	1 foot = 30.5 cm 1 inch = 2.54 cm
	1 millimeter (mm)	= 0.001 (10^{-3}) meter	1 mm = 0.039 inch	
	1 micrometer (μm) [formerly micron (μ)]	= 0.000001 (10^{-6}) meter		
	1 nanometer (nm) [formerly millimicron (mμ)]	= 0.000000001 (10^{-9}) meter		
	1 angstrom (Å)	= 0.0000000001 (10^{-10}) meter		
Area	1 square meter (m^2)	= 10,000 square centimeters	1 m^2 = 1.1960 square yards 1 m^2 = 10.764 square feet	1 square yard = 0.8361 m^2 1 square foot = 0.0929 m^2
	1 square centimeter (cm^2)	= 100 square millimeters	1 cm^2 = 0.155 square inch	1 square inch = 6.4516 cm^2
Mass	1 metric ton (t)	= 1000 kilograms	1 t = 1.103 ton	1 ton = 0.907t
	1 kilogram (kg)	= 1000 grams	1 kg = 2.205 pounds	1 pound = 0.4536 kg
	1 gram (g)	= 1000 milligrams	1 g = 0.0353 ounce 1 g = 15.432 grains	1 ounce = 28.35 g
	1 milligram (mg)	= 0.001 gram	1 mg = approx. 0.015 grain	
	1 microgram (μg)	= 0.000001 gram		
Volume (solids)	1 cubic meter (m^3)	= 1,000,000 cubic centimeters	1 m^3 = 1.3080 cubic yards 1 m^3 = 35.315 cubic feet	1 cubic yard = 0.7646 m^3 1 cubic foot = 0.0283 m^3
	1 cubic centimeter (cm^3 or cc)	= 0.000001 cubic meter = 1 millimeter	1 cm^3 = 0.0610 cubic inch	1 cubic inch = 16.387 cm^3
	1 cubic millimeter (mm^3)	= 0.000000001 cubic meter		
Volume (liquids and gases)	1 kiloliter (kl or kL)	= 1000 liters	1 kL = 264.17 gallons	1 gallon = 3.785 L
	1 liter (l or L)	= 1000 milliliters	1 L = 0.264 gallon 1 L = 1.057 quarts	1 quart = 0.946 L
	1 milliliter (ml or mL)	= 0.001 liter = 1 cubic centimeter	1 ml = 0.034 fluid ounce 1 ml = approx. $\frac{1}{4}$ teaspoon 1 ml = approx. 15–16 drops (gtt.)	1 quart = 946 ml 1 pint = 473 ml 1 fluid ounce = 29.57 ml 1 teaspoon = approx. 5 ml
	1 microliter (μl or μL)	= 0.000001 liter		
Time	1 second (s)	= $\frac{1}{60}$ minute		
	1 millisecond (ms)	= 0.001 second		
Temperature	Degrees Celsius (°C)		°F = $\frac{9}{5}$°C + 32	°C = $\frac{5}{9}$(°F − 32)

APPENDIX B

Answers to Multiple Choice and Matching Questions

Ansv

Chapter 1

1. c
2. (1) d; (2) b; (3) a
3. (a) wrist; (b) hipbone; (c) nose; (d) toes; (e) scalp
4. b
5. (a) D; (b) V; (c) D; (d) V; (e) V
6. (a) 2; (b) 3; (c) 1; (d) 4
7. d
8. c
9. b
10. c
11. e

Chapter 2

1. a
2. a
3. b
4. a
5. (a) Golgi apparatus (b) rough ER (c) condensed (d) smooth (e) nucleosome (f) microtubules (g) mitochondrion (h) mitochondrion
6. d
7. b
8. a
9. (a) metaphase (b) prophase (c) telophase (d) prophase (e) anaphase
10. (a) microtubules (b) intermediate filaments (c) microtubules (d) actin microfilaments (e) intermediate filaments (f) microtubules (g) actin microfilaments
11. (1) b; (2) g; (3) c or b; (4) d; (5) e; (6) f; (7) a
12. c
13. b

Chapter 3

1. (1) a; (2) b; (3) b; (4) a
2. b
3. b
4. c
5. (1) e; (2) g; (3) a; (4) f; (5) i; (6) b; (7) h; (8) d; (9) c
6. (1) f; (2) g; (3) a; (4) c; (5) d; (6) e; (7) b
7. c
8. d
9. a, d
10. c
11. e
12. a
13. a

Chapter 4

1. (1) a; (2) c; (3) d; (4) b; (5) b
2. c and e
3. (1) b; (2) f; (3) a; (4) d; (5) g
4. b
5. a
6. (1) cilia; (2) microvilli; (3) desmosome; (4) tight junction; (5) basement membrane

Chapter 5

1. c
2. (1) e; (2) b or d; (3) a; (4) c; (5) d
3. d
4. d
5. b
6. a
7. c
8. b
9. d
10. d

Chapter 6

1. e
2. a
3. e
4. (1) T; (2) F; (3) T; (4) F; (5) F; (6) T; (7) T
5. b
6. c
7. d
8. c
9. (a) 3; (b) 2; (c) 4; (d) 1; (e) 5; (f) 6
10. b
11. b
12. b
13. a
14. a

Chapter 7

1. (1) b,g; (2) h; (3) d; (4) d,f; (5) e; (6) c; (7) a,b,d,h; (8) i
2. b
3. c
4. c
5. a
6. a

Chapter 8

1. (1) g; (2) f; (3) b; (4) a; (5) b; (6) c; (7) d; (8) e
2. (1) b; (2) c; (3) e; (4) a; (5) h; (6) e; (7) f
3. e
4. b

Chapter 9

1. (1) a,b; (2) a; (3) a; (4) a; (5) b; (6) c; (7) c; (8) a,b; (9) c
2. e
3. c
4. e
5. a
6. (1) a; (2) b,d; (3) c; (4) c; (5) b,d

Chapter 10

1. c
2. b
3. (1) thin; (2) thick; (3) thick; (4) thin; (5) thick; (6) thin
4. (1) d; (2) a; (3) b; (4) e; (5) c
5. (1) yes; (2) no; (3) yes; (4) no
6. c
7. c
8. a
9. (1) b; (2) a; (3) c; (4) b; (5) c; (6) b; (7) a; (8) c; (9) c; (10) c

Chapter 11

1. d
2. c
3. (1) c; (2) b; (3) e; (4) a; (5) d
4. e
5. c
6. d
7. c
8. b
9. b
10. a
11. d

Chapter 12

1. b
2. (1) d; (2) b; (3) f; (4) c; (5) a; (6) a
3. b
4. (1) SS; (2) VS; (3) VM; (4) SM; (5) VS; (6) SS; (7) BM
5. b
6. b
7. a

8. c
9. d
10. (1) c; (2) b; (3) a

Chapter 13

1. (1) c; (2) f; (3) e; (4) g; (5) b; (6) e; (7) i; (8) a; (9) g; (10) h; (11) d
2. b
3. d
4. (1) G; (2) W; (3) W; (4) G; (5) W; (6) G; (7) W
5. a

Chapter 14

1. c
2. b
3. (1) e; (2) d; (3) c; (4) g; (5) b; (6) f; (7) a
4. (1) f; (2) i; (3) b; (4) g,h,j; (5) e; (6) i; (7) i; (8) k; (9) j; (10) c,d,f,k; (11) c
5. (1) b,6; (2) d,8; (3) c,2; (4) c,5; (5) a,4; (6) a,3&9; (7) a,7; (8) a,7; (9) d,1
6. c
7. c
8. c
9. b
10. f (and a)

Chapter 15

1. d
2. (1) S; (2) P; (3) P; (4) S; (5) S; (6) P; (7) S; (8) S; (9) P; (10) S; (11) P; (12) S; (13) S; (14) S; (15) S
3. c
4. d
5. b
6. d
7. c
8. a

Chapter 16

1. a
2. d
3. c
4. b
5. b
6. b
7. e
8. b
9. c
10. b
11. d
12. b

Chapter 17

1. d
2. (a) 2; (b) 5; (c) 1; (d) 4; (e) 3
3. a
4. c
5. a
6. (1) c; (2) b; (3) a; (4) c; (5) a; (6) b; (7) f; (8) b; (9) a; (10) a; (11) a; (12) e
7. d

Chapter 18

1. b
2. d
3. b
4. c
5. e
6. a
7. d
8. d
9. c
10. b
11. d

Chapter 19

1. d
2. b
3. b
4. (1) intima; (2) externa; (3) externa; (4) externa; (5) media; (6) intima
5. a
6. b
7. c
8. b
9. c
10. d
11. a
12. d
13. a

Chapter 20

1. c
2. c
3. a
4. b
5. e
6. b
7. a,c
8. c
9. a
10. d
11. d

Chapter 21

1. d
2. b
3. d
4. a
5. (1) b, (2) b, (3) a, (4) b, (5) c
6. (1) c, (2) d, (3) e, (4) a, (5) b
7. a
8. c
9. c
10. d

Chapter 22

1. c
2. a
3. (1) d; (2) c; (3) f; (4) a; (5) b; (6) e
4. (1) d; (2) d; (3) d; (4) c; (5) c; (6) c
5. a
6. a
7. a,d
8. d
9. b
10. a
11. a

Chapter 23

1. d
2. b
3. c
4. c
5. d
6. b
7. e
8. e
9. a
10. c
11. b

Chapter 24

1. d
2. (1) d; (2) c; (3) d; (4) c; (5) e
3. d
4. a
5. d
6. e
7. a
8. a
9. a
10. b
11. c
12. d

Chapter 25

1. a
2. (1) c; (2) b; (3) g; (4) h; (5) d; (6) e; (7) f; (8) i; (9) a; (10) j; (11) b,h; (12) i
3. a
4. d
5. (1) a; (2) b; (3) a; (4) b; (5) b; (6) a
6. (1) e; (2) a; (3) a; (4) d; (5) b
7. d
8. a
9. d
10. c

Chapter 26

1. a
2. b
3. b
4. d
5. b
6. c
7. a
8. c
9. a

GLOSSARY

Pronunciation Key

The pronunciation key that is used throughout this book takes advantage of the fact that, in scientific terms, long vowels usually occur at the end of syllables, whereas short vowels usually occur at the beginning or in the middle of syllables. Therefore:

1. When vowels are *unmarked,* they have a long sound when they occur at the end of a syllable, and a short sound when they are at the beginning or in the middle of a syllable. For example, in the word *kidney* (kid′ne), you automatically know that the middle 'i' is short and the terminal 'e' sound is long. In the word *renal* (re′nal), the 'e' is long, and the 'a' is short.
2. When a vowel sound violates the above rule, it is marked with a short or long symbol. Only those short vowels that come at the end of a syllable are marked with the short symbol, called a breve (˘); and only those long vowels that occur at the start or middle of a syllable are marked with a long symbol (¯). For example, in the word *pelvirectal* (pel″vĭ-rek′tal), all four vowels are short, but only the 'i' is marked as short, because it is the only vowel that falls at the end of a syllable. In *methane* (meth′ān), the long 'a' is marked as such because it does not fall at the end of a syllable.
3. Short 'a' is *never* marked with a breve. Instead, short 'a' at the end of a syllable is indicated by 'ah.' For example, in the word *papilla* (pah-pil′ah), each 'a' is short and pronounced as 'ah.'
4. In words that have more than one syllable, the syllable with the strongest accent is followed by a prime (′) mark, and syllables with the second strongest accent, where present, are indicated with a double prime (″). The unaccented syllables are followed by dashes. An example of these principles is the word *anesthetic* (an″es-thet′ik), in which *thet′* is emphasized most strongly, *an″* has a secondary emphasis, and the other two syllables are not spoken with any emphasis.

Abdomen (ab-do′men) Region of the body between the diaphragm and the pelvis.
Abduct (ab-dukt′) To move away from the midline of the body.
Absorption Process by which the products of digestion pass through the lining of the alimentary canal into the blood or lymph.
Accessory digestive organs Organs that contribute to the digestive process but are not part of the alimentary canal, including the tongue, teeth, salivary glands, pancreas, liver, and gallbladder.
Acetabulum (as″ĕ-tab′u-lum) Cup-like cavity on lateral surface of hip bone that receives the femur.
Acetylcholine (as″ĕ-til-ko′lēn) Chemical neurotransmitter substance released by some nerve endings.
Actin (ak′tin) A contractile protein in cells, especially abundant in muscle cells.
Action potential A large, transient depolarization event, including polarity reversal, that is conducted along the plasma membrane of a nerve axon or muscle cell without diminishing in intensity.
Acute Producing severe symptoms in the short term; rapidly developing.
Adduct (ah-dukt′) To move toward the midline of the body.
Adenohypophysis (ad″ĕ-no-hi-pof′ĭ-sis) One of the main divisions of the pituitary gland, the other being the neurohypophysis; the glandular part of the pituitary.
Adenoids (ad′ĕ-noids) The pharyngeal tonsil on the roof of the pharynx.
Adenosine triphosphate (ATP) (ah-den′o-sēn) Molecule in cells that stores and releases chemical energy for use in body cells.
Adipose (ad′ĭ-pōs) Fatty.
Adrenal gland (ah-dre′nal) Hormone-secreting gland located superior to the kidney; consists of medulla and cortex areas; also called suprarenal gland.
Adrenaline (ah-dren′ah-lin) *See* Epinephrine.
Adrenergic fibers (ad″ren-er′jik) Nerve fibers that release norepinephrine.
Adrenocorticotropic hormone (ACTH) (ah-dre″no-kor″tĭ-ko-trop′ik) Hormone from the anterior pituitary that influences the activity of the adrenal cortex.
Adventitia (ad″ven-tish′e-ah) Outermost layer or covering of an organ; consists of connective tissue.
Aerobic (a′er-ōb-ik) Oxygen-requiring.
Afferent (af′er-ent) Carrying to or toward a center; in the nervous system, *afferent* means "sensory."
Afferent neuron Nerve cell that carries impulses toward the central nervous system; sensory neuron.
Agonist (ag′o-nist) Muscle that bears primary responsibility for causing a certain movement; also called a *prime mover.*
AIDS (acquired immune deficiency syndrome) Disease caused by the human immunodeficiency virus (HIV): symptoms include severe weight loss, swollen lymph nodes, and many opportunistic infections that lead to death.
Aldosterone (al-dos′ter-ōn) Hormone secreted by the adrenal cortex that stimulates reabsorption of sodium ions and water from the kidney.
Alimentary canal (al″ĭ-men′tar-e) The digestive tube, extending from the mouth to the anus; its basic regions are the oral cavity, pharynx, esophagus, stomach, and small and large intestines.
Allantois (ah-lan′to-is) A tubular extension of the embryonic hindgut and cloaca; becomes the urachus, a fibrous cord attached to the adult bladder.
Allergy (al′er-je) Overzealous immune response to an otherwise harmless substance.
Alveolus (al-ve′o-lus) (1) One of the microscopic air sacs of the lungs; (2) a spherical sac formed by the secretory cells in a gland; also called an *acinus;* (3) the socket of a tooth.
Amino acid (ah-me′no) Organic compound containing nitrogen, carbon, hydrogen, and oxygen; building block of proteins.
Amnion (am′ne-on) Membrane that forms a fluid-filled sac around the embryo and fetus.
Amphiarthrosis (am″fe-ar-thro′sis) A slightly movable joint.
Anaerobic (an″a-er-o′bik) Not requiring oxygen.
Anastomosis (ah-nas″to-mo′sis) A union or joining of blood vessels or other tubular structures.

Androgen (an'dro-jen) A male sex hormone, the main example of which is testosterone.

Anemia (ah-ne'me-ah) Reduced oxygen-carrying capacity of the blood; results from too few erythrocytes or from abnormal hemoglobin.

Aneurysm (an'u-rizm) Blood-filled dilation of a blood vessel, caused by a weakening of the vessel wall.

Angina pectoris (an-ji'nah) Severe, suffocating chest pain caused by temporary lack of oxygen supply to heart muscle.

Antagonist (an-tag'o-nist) Muscle that reverses, or opposes, the action of another muscle.

Anterior The front of an organism, organ, or body part; the ventral direction.

Anterior pituitary *See* Pars distalis.

Antibody A protein molecule that is secreted by a plasma cell (a cell derived from an activated B lymphocyte) and that binds to an antigen in immune responses.

Antidiuretic hormone (ADH) (an"tĭ-di"u-ret'ik) Hormone produced by the hypothalamus and released by the posterior part of the pituitary gland (pars nervosa); stimulates the kidney to reabsorb more water.

Antigen (an'tĭ-jen) A molecule that is recognized as foreign by the immune system, activates the immune system, and reacts with immune cells or antibodies.

Anus (a'nus) The opening at the distal end of the alimentary canal.

Aorta (a-or'tah) Major systemic artery; arises from the left ventricle of the heart.

Apocrine gland (ap'o-krin) A type of sweat gland in the armpit and anal-genital regions; produces a secretion containing water, salts, proteins, and lipids.

Aponeurosis (ap"o-nu-ro'sis) Fibrous sheet connecting a muscle to the body part it moves.

Appendicular skeleton (ap"en-dik'u-lar) Bones of the limbs and limb girdles that are attached to the axial skeleton.

Apoptosis (ap"o-to-sis) Programmed cell death. Controlled cellular suicide that eliminates cells that are stressed, unneeded, excessive, or aged.

Aqueous humor (a'kwe-us) Watery fluid in the anterior segment of the eye.

Arachnoid mater (ah-rak'noid ma'ter) The web-like middle layer of the three meninges.

Areola (ah-re'o-lah) Circular, pigmented area of skin surrounding the nipple.

Arrector pili (ah-rek'tor) Tiny band of smooth muscle attached to each hair follicle; its contraction causes the hair to stand upright.

Arterial circle (of Willis) or **cerebral arterial circle** A union of arteries at the base of the brain.

Arteriole (ar-te're-ōl) A minute artery.

Arteriosclerosis (ar-te"re-o-sklĕ-ro'sis) Hardening of the arteries: any of a number of degenerative changes in the walls of arteries leading to a decrease in their elasticity. (*See* atherosclerosis.)

Artery Vessel that carries blood away from the heart.

Arthritis (ar-thri'tis) Inflammation of joints.

Articular capsule (ar-tik'u-lar) The capsule of a synovial joint; consists of an outer layer of fibrous connective tissue and an inner synovial membrane.

Articulation Joint; point where two elements of the skeleton meet.

Atherosclerosis (ath"er-o"skle-ro'sis) Changes in the walls of large arteries involving the deposit of lipid plaques; the most common variety of arteriosclerosis.

Atlas First cervical vertebra.

Atria (a'tre-ah) Paired, superiorly located heart chambers that receive blood returning to the heart.

Atrioventricular bundle (AV bundle) (a"tre-o-ven-trik'u-lar) Bundle of cardiac muscle cells that conducts impulses from the AV node to the walls of the right and left ventricles; located in the septum (wall) between the two ventricles of the heart; also called *bundle of His.*

Atrioventricular node (AV node) Specialized mass of conducting cells located in the interatrial septum of the heart.

Atrophy (at'ro-fe) Reduction in size or wasting away of an organ or tissue.

Autonomic nervous system (aw"to-nom'ik) General visceral motor division of the peripheral nervous system; innervates smooth and cardiac muscle, and glands.

Avascular (a-vas'ku-lar) Having no blood supply; containing no blood vessels.

Axial skeleton Portion of the skeleton that forms the central (longitudinal) axis of the body; includes the bones of the skull, the vertebral column, and the bony thorax.

Axilla (ak-sil'ah) Armpit.

Axis (1) Second cervical vertebra; (2) imaginary line about which a joint or structure rotates.

Axon Neuron process that carries impulses away from the cell body.

B cells Lymphocytes that oversee humoral immunity; they divide to generate plasma cells, which secrete antibodies; also called *B lymphocytes.*

Basal ganglia *See* Basal nuclei.

Basal lamina A thin sheet of protein that underlies an epithelium.

Basal nuclei Areas of gray matter located deep within the white matter of the cerebral hemispheres; regulate certain aspects of movement; also called *basal ganglia.*

Basement membrane A layer between an epithelium and the underlying connective tissue; consists of both a basal lamina and a network of reticular fibers.

Basophil (ba'so-fil) (1) Type of white blood cell whose cytoplasmic granules stain purple with the basic dyes in blood stains; (2) a gland cell in the anterior pituitary containing cytoplasmic granules that stain with basic dyes.

Benign (be-nīn) Not malignant; not life-threatening.

Biceps (bi'seps) Two-headed, especially applied to certain muscles.

Bile Greenish fluid secreted by the liver, stored in the gallbladder, and released into the small intestine; helps start the breakdown of fats.

Biopsy (bi'op-se) Removing a piece of living tissue to examine it under a microscope. Usually done to diagnose a suspected disease condition.

Bipolar neuron Neuron with just two processes, which extend from opposite sides of the cell body.

Blastocyst (blas'to-sist) Stage of early embryonic development; a hollow ball of cells; the product of cleavage.

Blood-brain barrier The feature that inhibits passage of harmful materials from the blood into brain tissues; reflects relative impermeability of brain capillaries.

Blood pressure Force exerted by blood against a unit area of the blood vessel walls; differences in blood pressure between different areas of the circulation provide the driving force for blood circulation.

Blood stem cell The cell type, present throughout life, from which all blood cells (erythrocytes, leukocytes, platelets) arise. Present in the bone marrow. It not only gives rise to blood cells, but also to mast cells, osteoclasts, and dendritic cells of the immune system. Also called *pluripotential hematopoietic stem cell.*

Bolus (bo'lus) A rounded mass of food prepared by the mouth for swallowing.

Bone remodeling Process involving bone formation and bone destruction in response to mechanical and hormonal factors.

Bony thorax Bones that form the framework of the thorax; includes sternum, ribs, and thoracic vertebrae.

Brain stem Collectively, the midbrain, pons, and medulla of the brain.

Bronchus Any of the air tubes of the respiratory tree between the trachea and bronchioles; bronchi enter and branch within the lungs.

Bursa A fibrous sac lined with synovial membrane and containing synovial fluid; occurs between bones and tendons (or other structures), where it acts to decrease friction during movement.

Calcaneal (Achilles) tendon Tendon that attaches the calf muscles to the heel bone.

Calcitonin (kal″sĭ-to′nin) Hormone released by the thyroid gland that promotes a decrease in calcium levels in the blood.

Calyx (ka′liks) A cup-like tributary of the pelvis of the kidney.

Cancer A malignant, invasive cellular tumor that has the capability of spreading throughout the body or body parts.

Capillary The smallest of blood vessels and the site of exchanges of molecules between the blood and the tissue fluid.

Carcinogen (kar-sin′o-jen) Cancer-causing agent.

Cardiac muscle Muscle tissue of the heart wall.

Cardiac sphincter The circular layer of smooth muscle at the junction of the esophagus and stomach; contracts to prevent reflux of stomach contents into the esophagus.

Cartilage (kar′tĭ-lij) White, semiopaque, resilient connective tissue; gristle.

Catecholamines (kat″ĕ-kol′ah-mēns″) Epinephrine, norepinephrine, and related molecules.

Caudal (kaw′dal) Literally, toward the tail; in humans, toward the inferior portion of the trunk.

Cecum (se′kum) The blind-ended pouch at the beginning of the large intestine.

Cell membrane *See* Plasma membrane.

Central nervous system (CNS) Brain and spinal cord.

Centriole (sen′tre-ōl) Barrel-shaped organelle formed of microtubules and located near the nucleus of the cell; active in cell division.

Cerebellum (ser″ĕ-bel′um) Brain region that is attached to the pons and smoothes and coordinates body movements.

Cerebral aqueduct (sĕ-re′bral ak′wĕ-dukt″) The narrow cavity of the midbrain that connects the third and fourth ventricles.

Cerebral cortex The external, gray matter region of the cerebral hemispheres.

Cerebrospinal fluid (ser″ĕ-bro-spi′nal) Clear fluid that fills the cavities of the central nervous system and surrounds the CNS externally; it floats and cushions the brain and spinal cord.

Cerebrum (ser′ĕ-brum) The cerebral hemispheres (some authorities also include the diencephalon).

Cervix (ser′viks) The inferior, neck-like part of the uterus (*cervix* = neck).

Chemoreceptor (ke″mo-re-sep′tor) Receptor sensitive to chemicals in solution.

Cholesterol (ko-les′ter-ol) A steroid lipid found in animal fats as well as in the plasma membranes of cells.

Cholinergic fiber (ko″lin-er′jik) An axon whose axon terminals release the neurotransmitter acetylcholine.

Chondroblast (kon′dro-blast) Actively mitotic form of a cartilage cell.

Chondrocyte (kon′dro-sīt) Mature form of a cartilage cell.

Chorion (ko′re-on) The outermost fetal membrane; helps form the placenta; technically, it consists of the trophoblast and the extraembryonic mesoderm.

Choroid plexus (ko′roid) A capillary-rich membrane on the roof of the brain that forms the cerebrospinal fluid; technically, it consists of pia mater and ependymal cells.

Chromatin (kro′mah-tin) Strands in the cell nucleus that consist of deoxyribonucleic acid (DNA) and histone proteins.

Chromosome (kro′mo-sōm) Bar-like body of tightly coiled chromatin, visible during cell division; typical human cells have 46 chromosomes.

Chronic (kron′ik) Long-term; prolonged; not acute.

Chyme (kīm) Semifluid, creamy mass consisting of partially digested food and stomach juices.

Cilium (sil′e-um) Motile, hair-like projection from the apical surface of certain epithelial cells.

Circumduction (ser″kum-duk′shun) Movement of a body part so that it outlines a cone in space.

Cirrhosis (si-ro′sis) A chronic disease, particularly of the liver, characterized by an overgrowth of connective tissue, or fibrosis.

Cleavage An early embryonic stage consisting of rapid cell divisions without intervening growth periods; begins with a fertilized ovum and produces a blastocyst.

Cochlea (kok′le-ah) Snail-shaped chamber of the bony labyrinth in the inner ear; houses the receptor for hearing (spiral organ, or organ of Corti).

Cognition (kog-nish′un) All aspects of thinking, perceiving, and intentionally remembering and recalling information; the mental processes involved in obtaining knowledge.

Condyle (kon′dīl) A rounded projection at the end of a bone that articulates with another bone.

Cone cell One of the two types of photoreceptor cells in the retina of the eye; provides for color vision and sharp vision.

Congenital (kon-jen′ĭ-tal) Existing at birth.

Conjunctiva (kon″junk-ti′vah) Thin, protective mucous membrane that covers the white of the eye and the internal surface of eyelids.

Connective tissue A primary tissue; form and function vary widely, but all connective tissues contain a large amount of extracellular matrix; functions include support, holding tissue fluid, and protection from disease.

Constriction Narrowing of a blood or lymph vessel, or of an opening like the pupil. Often caused by the squeezing action of circular musculature. *See* dilation.

Contraction The generation of a pulling force while shortening; this ability is highly developed in muscle cells.

Contralateral (kon″trah-lat′er-al) Concerning the opposite half of the body; when nerve fibers project contralaterally, they cross over to the opposite side of the body (right to left, or vice versa).

Cornea (kor′ne-ah) Transparent anterior portion of the eyeball.

Corona radiata (kŏ-ro′nah ra-de-ah′tah) (1) Crown-like arrangement of granulosa cells around an oocyte in an ovarian follicle after the appearance of an antrum; (2) crown-like arrangement of nerve fibers in the white matter of the cerebrum, radiating to and from every part of the cerebral cortex.

Coronal plane (kŏ-ro′nal) *See* Frontal plane.

Cortex (kor′teks) Outer region of an organ.

Corticosteroids (kor″tĭ-ko-ste′roids) Steroid hormones secreted by the adrenal cortex. Examples are cortisol, aldosterone, and some sex hormones.

Cortisol (kor′tĭ-sol) A glucocorticoid hormone produced by the adrenal cortex.

Cranial nerves The 12 pairs of nerves that attach to the brain.

Cutaneous (ku-ta′ne-us) Pertaining to the skin.

Cytokinesis (si″to-ki-ne′sis) Division of the cytoplasm that occurs after the cell nucleus has divided.

Cytoplasm (si′to-plazm) The part of a cell between the plasma membrane and the nucleus; contains many organelles.

Cytotoxic (CD 8^+) T cell T lymphocyte that directly kills eukaryotic foreign cells, cancer cells, or virus-infected body cells; also called *killer T cell.*

Decussation A crossing of structures in the form of an "x." Often applied to axons that cross the body midline from the left to the right side (or vice versa) of the central nervous system.
Deep Toward the inside; inner; internal.
Defecation (def″ĕ-ka-shun) Elimination of the contents (feces) of the bowel.
Dendrite (den′drīt) Neuron process that transmits signals toward the cell body and serves as receptive region of the neuron; most dendrites branch extensively.
Deoxyribonucleic acid (DNA) (de-ok″sĭ-ri″bo-nu-kle′ik) A nucleic acid found in all living cells; carries the organism's hereditary information.
Depolarization (de-po″lar-ĭ-za″shun) Loss of a state of polarity across a cellular membrane; loss of negative charge inside the cell.
Dermis The leathery layer of skin, deep to the epidermis; composed largely of dense irregular connective tissue.
Desmosome (des′mo-sōm) A cell junction composed of two disc-shaped plaques connected across the intercellular space; the most important junction for holding epithelial cells together.
Diabetes insipidus (di″ah-be′tēz) Disease characterized by passage of a large quantity of dilute urine plus intense thirst and dehydration; caused by inadequate release of antidiuretic hormone.
Diabetes mellitus Disease caused by deficient release of, or deficient use of, insulin; characterized by an inability of the body cells to use sugars at a normal rate and by high blood sugar levels.
Diapedesis (di″ah-pĕ-de″sis) Active movement of white blood cells through the walls of capillaries and venules into the surrounding tissue.
Diaphragm (di′ah-fram) (1) Any partition or wall separating one area from another; (2) the muscular sheet that separates the thoracic cavity from the abdominopelvic cavity.
Diaphysis (di-af′ĭ-sis) Elongated shaft of a long bone.
Diarthrosis (di″ar-thro′sis) Freely movable joint; all synovial joints are diarthroses.
Diastole (di-as′to-le) Period during which the ventricles or atria of the heart relax.
Diencephalon (di″en-sef′ah-lon) The part of the forebrain between the cerebral hemispheres and the midbrain; includes the thalamus, hypothalamus, and third ventricle.
Diffusion (dĭ-fu′zhun) The spreading of particles in a gas or solution from regions of high particle concentration to regions of low concentration, with movement toward a uniform distribution of the particles.
Digestion Chemical and mechanical process of breaking down foodstuffs into molecules that can be absorbed.
Dilation Expansion or widening of a vessel, organ, or opening. (*See* constriction.)
Distal (dis′tal) Away from the attached end of a structure, especially a limb.
Diverticulum (di″ver-tik′u-lum) A pouch or a sac in the walls of a hollow organ or structure.
Dorsal (dor′sal) Pertaining to the back; posterior.
Duct Canal or passageway, usually tubular.
Duodenum (du″o-de′num) First part of the small intestine.
Dura mater (du′rah ma′ter) Most external and toughest of the three membranes (meninges) covering the brain and spinal cord.
Ectoderm (ek′to-derm) Embryonic germ layer that forms both the outer layer of the skin (epidermis) and nervous tissues.
Edema (ĕ-de′mah) Abnormal accumulation of tissue fluid in the loose connective tissue; causes the affected body region to swell.
Effector (ef-fek′tor) Muscle or gland capable of being activated by motor nerve endings.
Efferent (ef′er-ent) Carrying away or away from, especially a nerve fiber that carries impulses away from the central nervous system; efferent neurons are motor neurons.
Elastin (e-las′tin) Main protein in elastic fibers of connective tissues; stretchable and resilient.
Embolus (embolism) (em′bo-lus) Any abnormal mass carried freely in the bloodstream; maybe a blood clot, bubbles of air, mass of fat, or clumps of cells.
Embryo (em′bre-o) The developing human from week 2 through week 8 after fertilization.
Endocarditis (en″do-kar-di′tis) Inflammation of the inner lining of the heart.
Endocardium (en″do-kar′de-um) The layer that lines the inner surface of the heart wall; consists of endothelium and areolar connective tissue.
Endocrine glands (en′do-krin) Ductless glands that secrete hormones into the blood.
Endocrine system Body system consisting of glands that secrete hormones.
Endocytosis (en″do-si-to′sis) Processes by which large molecules and particles enter cells; types are phagocytosis, pinocytosis, and receptor-mediated endocytosis.
Endoderm (en′do-derm) An embryonic germ layer that forms the lining and glands of the digestive and respiratory tubes.
Endometrium (en″do-me′tre-um) Mucous membrane lining the uterus.
Endomysium (en″do-mis′e-um) Thin connective tissue surrounding each muscle cell.
Endoplasmic reticulum (ER) (en″do-plaz′mik rĕ-tik′u-lum) A system of membranous envelopes and tubes in the cytoplasm of a cell; there are smooth and rough varieties.
Endothelium (en″do-the′le-um) The simple squamous epithelium that lines the walls of the heart, blood vessels, and lymphatic vessels.
Enzyme (en′zīm) A protein that acts as a biological catalyst to speed up chemical reactions.
Eosinophil (e″o-sin′o-fil) Granular white blood cell whose granules readily take up a pink dye called eosin.
Epidermis (ep″ĭ-der′mis) Superficial layer of the skin, composed of a keratinized stratified squamous epithelium.
Epididymis (ep″ĭ-did′ĭ-mis) Comma-shaped structure in the scrotum adjacent to the testis; contains a duct in which the sperm mature.
Epiglottis (ep″ĭ-glot′is) A leaf-shaped piece of elastic cartilage that extends from the posterior surface of the tongue to the larynx; covers the opening of the larynx during swallowing.
Epimysium (ep″ĭ-mis′e-um) Sheath of fibrous connective tissue surrounding a muscle.
Epinephrine (ep″ĭ-nef′rin) Chief hormone produced by the adrenal medulla; also called *adrenaline.*
Epiphyseal plate (ep″ĭ-fiz′e-al) Plate of hyaline cartilage at the junction of the diaphysis (shaft) and epiphysis (end) of most bones in the growing skeleton; provides growth in the length of the bone.
Epiphysis (e-pif″ĭ-sis) The end of a long bone, attached to the shaft.
Epithelium (ep″ĭ-the′le-um) A primary tissue that covers body surfaces and lines body cavities; its cells are arranged in sheets; also forms glands.
Equilibrium The sense of balance; measures both the position and the movement of the head.
Erythrocyte (ĕ-rith′ro-sīt) Red blood cell; when mature, an erythrocyte is literally a sac of hemoglobin (oxygen-carrying protein) covered by a plasma membrane.
Estrogens (es′tro-jens) Female sex hormones.
Exocrine glands (ek′so-krin) Glands that secrete onto body surfaces or into body cavities; except for the one-celled goblet cells, all exocrine glands have ducts.

Exocytosis (ek″so-si-to′sis) Mechanism by which substances are moved from the cell interior to the extracellular space; the main mechanism of secretion.
Expiration Act of expelling air from the lungs; exhalation.
External Outside of; superficial to.
Exteroceptor (ek″ster-o-sep′tor) Sensory end organ that responds to stimuli from the external world.
Extracellular Outside a cell.
Extracellular matrix (ma′triks) The material that lies between the cells in connective tissues; consists of fibers, ground substance, and tissue fluid.
Extrinsic (ek-strin′sik) Originating outside an organ or part.
Facet (fas′et) A smooth, nearly flat surface on a bone for articulation.
Fallopian tube (fah-lo′pe-an) *See* Uterine tube.
Fascia (fash′e-ah) Layers of fibrous connective tissue that cover and separate muscles and other structures. Superficial fascia is the fatty hypodermis of the skin.
Fascicle (fas′ĭ-k′l) Bundle of nerve or muscle fibers bound together by connective tissue.
Fenestrated (fen′es-tra-ted) Pierced with one or more small openings or pores.
Fertilization Fusion of the sperm and egg nuclei.
Fetus (fe′tus) Developmental stage lasting from week 9 of development to birth.
Fibrin (fi′brin) Fibrous insoluble protein formed during blood clotting; takes the form of a fiber network.
Fibroblast (fi′bro-blast) Young, actively mitotic cell that secretes the fibers and ground substance of connective tissue proper.
Fibrocyte (fi′bro-sīt) Mature fibroblast; maintains the matrix of connective tissue proper.
Filtration Passage of a solution or suspension through a membrane or filter, with the purpose of holding back the larger particles.
Fissure (fish′er) (1) A groove or cleft; (2) the deepest depressions or inward folds on the brain.
Fixator (fik-sa′tor) Muscle that immobilizes one or more bones, allowing other muscles to act from a stable base.
Flagellum (flah-jel′um) Long, whip-like extension of the plasma membrane of some bacteria and sperm cells; propels the cell.
Follicle (fol′ĭ-k′l) (1) Spherical structure in the ovary consisting of a developing egg cell surrounded by one or more layers of follicle cells; (2) colloid-containing structure in the thyroid gland.
Follicle-stimulating hormone (FSH) Hormone secreted by the anterior pituitary that stimulates the maturation of ovarian follicles in females and the production of sperm in males.
Foramen (fo-ra′men) Hole or opening in a bone or between body cavities.
Forebrain Rostral portion of the brain, consisting of the telencephalon and diencephalon.
Formed elements Blood cells (red and white cells and platelets).
Fossa (fos′ah) A depression, often a joint surface.
Fovea (fo′ve-ah) A pit.
Frontal (coronal) plane Vertical plane that divides the body into anterior and posterior parts.
Fundus (fun′dus) The base of an organ; that part farthest from the opening of an organ.
Funiculus (fu-nik′u-lus) (1) A chord-like structure; (2) a division of the white matter in the spinal cord.
Gallbladder Sac inferior to the right part of the liver; it stores and concentrates bile.
Gamete (gam′ēt) Sex cell; sperm or oocyte.
Gametogenesis (gam″ĕ-to-jen′ĕ-sis) Formation of gametes.
Ganglion (gang′gle-on) Collection of neuron cell bodies outside the central nervous system.
Gap junction A passageway between two adjacent cells; formed by transmembrane proteins called connexons.
Gene One of the biological units of heredity located in chromatin; transmits hereditary information; roughly speaking, one gene codes for the manufacture of one protein.
General Pertaining to sensory inputs or motor outputs that are *widely distributed* through the body rather than localized; opposite of *special.*
Germ layers Three cellular layers (ectoderm, mesoderm, and endoderm) that represent the early specialization of cells in the embryonic body and from which all body tissues arise.
Gestation (jes-ta′shun) The period of pregnancy; averages 280 days in humans.
Gland A structure whose cells are specialized for secretion.
Glial cells (gle′al) *See* Neuroglia.
Glomerular capsule (glo-mer′u-lar) Double-walled cup forming the initial portion of the nephron in the kidney; also called *Bowman's capsule.*
Glomerulus (glo-mer′u-lus) A ball of capillaries forming part of the nephron in the kidney; forms a filtrate that will be modified into urine.
Glottis (glot′is) The opening between the two vocal cords in the larynx.
Glucagon (gloo′kah-gon) Hormone secreted by alpha cells of the pancreatic islets; raises the glucose level of blood.
Glucocorticoids (gloo″ko-kor′tĭ-koids) Hormones secreted by the cortex of the adrenal gland; they increase the concentration of glucose in the blood and aid the body in resisting long-term stress.
Glucose (gloo′kōs) The principal blood sugar; the main sugar used by cells for energy.
Glycogen (gli′ko-jen) A long chain of glucose molecules; the main form in which sugar is stored in animal cells; glycogen takes the form of dense granules in the cytoplasm.
Goblet cells Individual mucus-secreting cells of the respiratory and digestive tracts.
Gonad (go′nad) Primary reproductive organ: the testis of the male or the ovary of the female.
Gonadotropins (go-nad″o-trōp′ins) Gonad-stimulating hormones secreted by the anterior pituitary: follicle-stimulating hormone and luteinizing hormone.
Graded potential A local change in membrane potential that varies directly with the strength of the stimulus, declines with distance, and is not conducted along axons.
Gray matter Gray area of the central nervous system; contains neuron cell bodies and unmyelinated processes of neurons.
Ground substance The viscous, spongy part of the extracellular matrix of connective tissue; its large molecules attract water and hold tissue fluid.
Growth hormone Hormone that stimulates growth of the body; secreted by the anterior pituitary; also called somatotropin and somatotropic hormone (SH).
Gustation (gus-ta′shun) Taste.
Gyrus (ji-rus) A ridge on the surface of the cerebral cortex.
Hair follicle Tube-like invagination of the epidermis of the skin from which a hair grows.
Haustra (hos′trah) Pouches (sacculations) of the colon.
Heart block Impaired transmission of impulses from atria to ventricles.
Heart murmur Abnormal heart sound (usually resulting from valve problems).
Helper (CD 4^+) T cell A type of T lymphocyte that participates in the activation of other lymphocytes by secreting chemicals that stimulate newly activated lymphocytes to multiply.
Hematocrit (he-mat′o-krit) The percentage of total blood volume occupied by erythrocytes.
Hematoma (he″mah-to′mah) A mass of blood that has bled from blood vessels into the tissues.

Hematopoiesis (hem″ah-to-poi-e′sis) Blood cell formation; hemopoiesis.
Hemoglobin (he′mo-glo″bin) Oxygen-transporting protein in erythrocytes.
Hemopoiesis *See* Hematopoiesis.
Hemorrhage (hem′ŏ-rij) Bleeding.
Hepatic portal system (hĕ-pat′ik) The part of the circulation in which veins receive nutrients from capillaries in the stomach and intestines and carry these nutrients to capillaries in the liver; the liver cells then process the nutrients.
Hepatitis (hep″ah-ti′tis) Inflammation of the liver.
Hernia (her′ne-ah) Abnormal protrusion of an organ or body part through the containing wall of its cavity.
Hilus (hi′lus) A slit on the surface of an organ through which the vessels and nerves enter and leave; the spleen, lungs, kidneys, lymph nodes, and ovaries have prominent hiluses.
Histamine (his′tah-mēn) Chemical substance that increases vascular permeability in the initial stages of inflammation.
Histology (his-tol′o-je) Branch of anatomy dealing with the microscopic structure of tissues, cells, and organs.
Holocrine gland (hol′o-krin) A gland in which entire cells break up to form the secretion product; sebaceous (oil) glands of the skin are the only example.
Hormones Messenger molecules that are released by endocrine glands and travel in the blood to regulate specific body functions.
Hypertension (hi″per-ten′shun) High blood pressure.
Hypertrophy (hi-per′tro-fe) Enlargement of an organ or tissue due to an increase in the size of its cells.
Hypodermis (hi″po-der′mis) The fatty layer deep to the skin; consists of adipose and areolar connective tissue; also called the *subcutaneous layer* and *superficial fascia.*
Hypophysis (hi-pof′ĭ-sis) The pituitary gland.
Hypothalamus (hi″po-thal′ah-mus) Inferior region of the diencephalon; visceral control center of the brain.
Ileum (il′e-um) Coiled terminal part of the small intestine, located between the jejunum and the cecum of the large intestine.
Immune response Antigen-specific defenses mounted by activated lymphocytes (T and B cells).
Immunity (ĭ-mu′nĭ-te) Ability of the body to develop resistance to specific foreign agents (both living and nonliving) that can cause disease.
Immunocompetence Ability of the body's immune system to recognize specific antigens.
Infarct (in′farkt) Region of dead, deteriorating tissue resulting from a lack of blood supply.
Inferior (caudal) Below; toward the feet.
Inflammation (in″flah-ma′shun) A physiological response of the body to tissue injury; includes dilation of blood vessels and an increase in capillary permeability; indicated by redness, heat, swelling, and pain in the affected area.
Inguinal (ing′gwĭ-nal) Pertaining to the groin region.
Inner cell mass Accumulation of cells in the blastocyst from which the body of the embryo derives.
Innervation (in″er-va′shun) Supply of nerves to a body part.
Insertion Movable part or attachment of a muscle, as opposed to the muscle's origin.
Inspiration Drawing of air into the lungs; inhalation.
Insulin (in′su-lin) Hormone secreted by beta cells in pancreatic islets; it decreases blood glucose levels.
Integumentary system (in-teg″u-men′tar-e) The skin and its appendages (hairs, nails, and skin glands).
Intercalated discs (in-ter′kah-la″ted) Complex junctions that interconnect cardiac muscle cells in the wall of the heart.
Intercellular Between body cells.
Internal Deep to.
Internal capsule Band of white matter in the brain, between the basal nuclei and the thalamus.
Internal respiration Exchange of gases between blood and tissue fluid and between tissue fluid and cells.
Interneuron (in″ter-nu′ron) (1) Nerve cell that lies between a sensory neuron and a motor neuron in a reflex arc; (2) any nerve cell that is confined entirely within the central nervous system.
Interoceptor (in″ter-o-sep′tor) Nerve ending situated in a visceral organ; responds to changes and stimuli within the body's internal environment; also called *visceroceptor.*
Interstitial fluid (in″ter-stish′al) *See* Tissue fluid.
Intervertebral discs (in″ter-ver′tĕ-bral) The discs between the vertebrae of the spinal column; each consists of fibrous rings surrounding a springy core.
Intervertebral foramina (fo-ra′min-ah) Openings between the dorsal projections of adjacent vertebrae through which the spinal nerves pass.
Intracellular Within a cell.
Invaginate To form an inpocketing or to grow inward. For example, an epithelium pushes inward, pocket-like, into the underlying connective tissue to form a gland during development.
Ion (i′on) Atom or molecule with a positive or negative electrical charge.
Ipsilateral (ip″sĭ-lat′er-al) Situated on the same side of the body; opposite of *contralateral.*
Ischemia (is-ke′me-ah) Local decrease in blood supply.
Jejunum (jĕ-joo′num) The coiled part of the small intestine that is located between the duodenum and ileum.
Joint Junction of two or more elements of the skeleton; an articulation.
Keratin (ker′ah-tin) Tension-resisting protein found in the epidermis, hair, and nails; keratin makes these structures tough and able to resist friction.
Labium (la′be-um) Lip.
Labyrinth (lab′ĭ-rinth) Bony cavities and membranes of the inner ear.
Lacrimal (lak′rĭ-mal) Pertaining to tears.
Lactation (lak-ta′shun) Production and secretion of milk.
Lacteal (lak′te-al) Lymphatic capillaries in the small intestine that take up lipids.
Lacuna (lah-ku′nah) Little depression or cavity; in bone and cartilage, each lacuna is occupied by a cell.
Lamina (lam′ĭ-nah) (1) A thin layer or flat plate; (2) the portion of a vertebra between the transverse process and the spinous process.
Laparoscopy (lap″ah-ros′ko-pe; "observing the flank") Examination of the peritoneal cavity and the associated organs with a laparoscope, a viewing device (endoscope) at the end of a thin tube that is inserted through the abdominal wall.
Larynx (lar′ingks) Cartilaginous organ located between the pharynx and trachea; contains the vocal cords; the voice box.
Lateral Away from the body midline.
Leukocyte (loo′ko-sīt) White blood cell; the five types of leukocytes are all involved in the defense against disease.
Ligament (lig′ah-ment) Band of dense regular connective tissue that connects bones.
Limbic system (lim′bik) Functional brain system involved in emotional and visceral responses; structurally, it includes a part of the cerebrum (limbic lobe) and the hypothalamus of the diencephalon.
Lumbar (lum′bar) Region of the back between the thorax and the pelvis.
Lumen (lu′men) The cavity inside a tube, blood vessel, or hollow organ.
Luteinizing hormone (LH) (loo′te-in-īz″ing) A hormone secreted by the anterior pituitary; in females, it aids maturation of follicles in the ovary and triggers ovulation; in males, it signals the interstitial cells of the testis to secrete testosterone.

Lymph (limf) The clear fluid transported by the lymphatic vessels.

Lymph node Bean-shaped lymphoid organ that filters and cleanses the lymph.

Lymphatic system (lim-fat'ik) Organ system consisting of lymphatic vessels, lymph nodes, and the lymphoid organs and tissues; drains excess tissue fluid and fights disease.

Lymphatics General term used to designate lymphatic vessels.

Lymphocyte (lim'fo-sīt) Agranular white blood cell that arises from bone marrow and becomes functionally activated in the lymphoid organs of the body; the main cell type of the immune system; each lymphocyte recognizes a specific antigen.

Lymphoid organs The organs of the lymphatic system that house lymphocytes and function in immunity; spleen, lymph nodes, tonsils, and thymus are the main examples.

Lymphoid tissue The main tissue of the immune system; a reticular connective tissue that houses and activates many lymphocytes.

Lysosome (li'so-sōm) A membrane-bound, sac-like cytoplasmic organelle that contains a wide variety of digestive enzymes.

Macrophages (mak'ro-fāj-es) The general phagocytic cells of the body, capable of engulfing and digesting a wide variety of foreign cells, particles, and molecules; present throughout the connective tissues of the body and especially abundant in lymphoid tissues of the immune system.

Malignant (mah-lig'nant) Life-threatening; pertains to neoplasms such as cancer that spread and lead to death.

Mammary glands (mam'ar-e) The breasts.

Mastication (mas″tĭ-ka'shun) Chewing.

Meatus (me-a'tus) A canal or opening.

Mechanoreceptor (mek″ah-no-re-sep'tor) Receptor sensitive to mechanical forces, such as touch, stretch, pressure, or vibration.

Medial Toward the midline of the body.

Median (me'de-an) In the midline of the body; midsagittal.

Mediastinum (me″de-ah-sti'num) Region of the thoracic cavity between the lungs; contains the heart, thoracic aorta, esophagus, and other structures.

Medulla (me-dul'ah) Middle or internal region of certain organs.

Medulla oblongata (ob″long-gah'tah) Inferior part of the brain stem.

Meiosis (mi-o'sis) A process of nuclear division that occurs during the production of the sex cells and reduces the chromosome number by half; results in the formation of haploid (n) cells.

Melanin (mel'ah-nin) Dark pigment formed by cells called melanocytes; imparts color to the skin and hair.

Memory cells T and B lymphocytes that provide for immunologic memory (acquired, long-term immunity from diseases).

Meninges (mĕ-nin'jēz) Protective coverings around the brain and spinal cord; from external to internal, they are the dura mater, arachnoid mater, and pia mater.

Meningitis (men″in-ji'tis) Inflammation of the meninges.

Menstrual cycle (men'stroo-al) The changes in the female reproductive organs that occur every month (28 days, on the average).

Menstruation (men″stroo-a'shun) The periodic, cyclic discharge of blood and tissue from the lining of the female uterus in the absence of pregnancy.

Mesencephalon (mes″en-sef'ah-lon) Midbrain.

Mesenchyme (mes'eng-kīm) The type of embryonic tissue from which connective tissues and muscle tissues arise.

Mesenteries (mes'en-ter″ēz) Double-layered sheets of peritoneum that support most organs in the abdominopelvic cavity.

Mesoderm (mez'o-derm) The embryonic germ layer that gives rise to most structures in the body, including the skeleton, muscles, dermis, connective tissues, kidneys, and gonads.

Metabolic rate (mĕ-tah-bol'ik) Energy expended by the body per unit time.

Metabolism (mĕ-tab'o-lizm) Sum total of all the chemical reactions occurring in the cells of the body.

Metastasis (mĕ-tas'tah-sis) The spread of cancer from one body part or organ to another not directly connected to it.

Microvilli (mi″kro-vil'i) Immotile, cellular projections on the free surface of most epithelia; microvilli anchor sheets of mucus or increase surface area for absorption.

Micturition (mik″tu-rish'un) Urination, or voiding; emptying the bladder.

Midbrain Region of the brain stem that lies between the diencephalon and the pons.

Mineralocorticoids (min″er-al-o-kor'tĭ-koids) Steroid hormones secreted by the adrenal cortex that increase the reabsorption of sodium and water by the kidneys; the main mineralocorticoid is aldosterone.

Mitochondrion (mi″to-kon'dre-on) Cytoplasmic organelle that generates most adenosine triphosphate (ATP) for cellular activities; mitochondria are the cell's "power plants."

Mitosis (mi-to'sis) Division of the nucleus during the typical process of cell division, during which the chromosomes are distributed to the two daughter nuclei.

Mixed nerve A nerve containing fibers of both sensory and motor neurons; most nerves are mixed nerves.

Monocyte (mon'o-sīt) An agranular white blood cell, with a large nucleus that is often bent into the shape of a C; the largest of all blood cells; develops into a macrophage.

Motor neuron Nerve cell that signals muscle cells to contract or gland cells to secrete; also called *efferent neuron.*

Motor unit A motor neuron and all of the muscle cells it stimulates.

Mucosa (mu-ko'sah) *See* Mucous membrane.

Mucous membranes Moist membranes that line all tubular organs and body cavities that open to the exterior (digestive, respiratory, urinary, and reproductive tracts).

Mucus (mu'kus) A sticky, viscous fluid that covers many internal surfaces in the body; it consists of the protein mucin and a large amount of water.

Multipolar neuron (mul″tĭ-po'lar) A nerve cell that has more than two processes; most neurons are multipolar, having several dendrites and an axon.

Muscle fiber Muscle cell.

Muscle spindle Complex, spindle-shaped receptor in skeletal muscles that senses muscle stretch.

Muscle tone Continuous, low levels of contractile force produced by muscles that are not actively shortening.

Myelencephalon (mi″el-en-sef'ah-lon) Caudal part of the hindbrain; the medulla oblongata.

Myelin sheath (mi'ĕ-lin) Fatty insulating sheath that surrounds all but the thinnest nerve fibers; formed of the plasma membrane of supporting cells wrapped in concentric layers around the nerve fiber.

Myocardial infarction (mi″o-kar'de-al) Condition characterized by dead tissue areas in the myocardium of the heart; caused by interruption of blood supply to the area; also called *heart attack.*

Myocardium (mi″o-kar'de-um) Layer of the heart wall composed of cardiac muscle.

Myofibril (mi″o-fi'bril) Rod-like bundle of contractile myofilaments in the cytoplasm of a skeletal muscle cell; made of repeating segments called sarcomeres.

Myofilament (mi″o-fil'ah-ment) The contractile filaments in muscle cells; the two varieties are thick (myosin) filaments and thin (actin) filaments.

Myometrium (mi″o-me′tre-um) The thick layer of smooth muscle in the wall of the uterus.

Myosin (mi′o-sin) A contractile protein in cells, especially abundant in muscle cells.

Nares (na′rēz) The nostrils.

Necrosis (nĕ-kro′sis) Death of a cell or tissue caused by disease or injury.

Neoplasm (ne′o-plazm) *See* Tumor.

Nephron (nef′ron) A major division of the uriniferous tubule in the kidney (the other division is the collecting tubule).

Nerve A collection of nerve fibers (long axons) in the peripheral nervous system.

Nerve fiber Any long axon of a neuron.

Neural crest Embryonic tissue derived from ectoderm that migrates widely within the embryo and gives rise to sensory neurons, all nerve ganglia, melanocytes, and other structures.

Neuroglia (nu-rog′le-ah) The supporting cells in the central nervous system; they support, protect, and insulate neurons.

Neurohypophysis (nu″ro-hi-pof′ĭ-sis) The part of the pituitary gland that derives from the brain; contains the stalk-like infundibulum and the posterior lobe of the pituitary.

Neuron (nu′ron) Cell of the nervous system specialized to generate and transmit electrical signals; a nerve cell.

Neurotransmitter (nu″ro-trans′mit-er) Chemical released by neurons that may, upon binding to receptors on neurons or effector cells, stimulate or inhibit them.

Neutrophil (nu′tro-fil) Most abundant type of white blood cell; a granulocyte specialized for destroying bacteria.

Nucleic acid (nu-kle′ik) The class of organic molecules that includes deoxyribonucleic acid (DNA) and ribonucleic acid (RNA).

Nucleolus (nu-kle′o-lus) A small, dark-staining body in the cell nucleus; represents parts of several chromosomes and manufactures the basic subunits of ribosomes.

Nucleus (nu′kle-us) (1) Control center of a cell; contains genetic material; (2) a cluster of neuron cell bodies in the brain.

Occipital (ok-sip′ĭ-tal) Pertaining to the area at the back of the head.

Occlusion (ŏ-kloo′zhun) Closure or obstruction.

Olfaction (ol-fak′shun) Smell.

Olfactory epithelium (ol-fak′to-re) A sensory receptor region in the superior lining of the nasal cavity; this epithelium contains olfactory neurons that respond to odors in the air.

Oocyte (o′o-sīt) Immature egg undergoing the process of meiosis.

Oogenesis (o″o-jen′ĕ-sis) Process of ovum (female gamete) formation.

Ophthalmic (of-thal′mik) Pertaining to the eye.

Optic (op′tik) Pertaining to the eye.

Optic chiasma (ki-az′mah) A cross-shaped structure anterior to the diencephalon of the brain, representing the point of crossover of half the axons of the optic nerve.

Organ A part of the body formed of two or more tissues and adapted to carry out a specific function; the stomach and biceps brachii muscle are examples of large organs, but many organs are smaller and simpler (sweat gland, hair follicle, muscle spindle).

Organ system A group of organs that work together to perform a vital body function; e.g., the nervous system.

Organelles (or″gah-nelz′) Small structures in the cytoplasm (ribosomes, mitochondria, and others) that perform specific functions for the cell. Nucleus is also an organelle.

Origin Attachment of a muscle that remains relatively fixed during muscular contraction.

Osmosis (oz-mo′sis) Diffusion of a solvent (water molecules) through a membrane from a dilute solution into a more concentrated one.

Ossicles (os′ĭ-k′lz) The three tiny bones in the middle ear: malleus (hammer), incus (anvil), and stapes (stirrup).

Ossification (os″ĭ-fĭ-ka′shun) Bone formation. *See* Osteogenesis.

Osteoblast (os′te-o-blast″) A bone-forming cell.

Osteoclast (os′te-o-klast″) Large cell that reabsorbs or breaks down the bone matrix.

Osteocyte (os′te-o-sīt″) Mature bone cell, shaped like a spider with a body and long processes, that occupies a lacuna in the bone matrix.

Osteogenesis (os″te-o-jen′ĕ-sis) Process of bone formation.

Osteon (os′te-on) Tube-shaped unit in mature, compact bone; consists of concentric layers of bone lamellae surrounding a central canal; also called *Haversian system.*

Osteoporosis (os″te-o-po-ro′sis) Age-related condition (affects many elderly women) in which bones weaken as bone reabsorption outpaces bone deposition; the weakened bones break easily.

Ovarian cycle (o-va′re-an) Monthly cycle of follicle development in the ovaries, ovulation, and formation of the corpus luteum; the menstrual cycle as it involves the ovaries.

Ovary (o′var-e) Female sex organ in which ova (eggs) are produced; the female gonad in the pelvis.

Ovulation (o″vu-la′shun) Ejection of an egg (oocyte) from the ovary.

Ovum (o′vum) (1) General meaning: the female germ cell, or egg; (2) specific meaning: female germ cell after a sperm has entered it but before the sperm nucleus and the egg nucleus have fused.

Oxytocin (ok″sĭ-to′sin) Hormone produced by the hypothalamus and released by the posterior pituitary; it stimulates contraction of the uterus during childbirth and the ejection of milk during nursing.

Palate (pal′at) Roof of the mouth.

Palpation (pal-pa′shun) Using one's fingers to feel deep organs through the skin of the body surface.

Pancreas (pan′kre-as) Tadpole-shaped gland posterior to the stomach; produces both exocrine and endocrine secretions.

Pancreatic juice Bicarbonate-rich secretion of the pancreas containing enzymes for digestion of all food categories.

Parasympathetic (par″ah-sim″pah-thet′ik) The division of the autonomic nervous system that oversees digestion, elimination, and glandular function; the resting and digesting division.

Parathyroid glands (par″ah-thi′roid) Small endocrine glands located on the posterior aspect of the thyroid gland.

Parathyroid hormone (PTH) Hormone secreted by the parathyroid glands; it increases the concentration of calcium ions in the blood.

Parenchyma (pah-reng′kĭ-mah) The functional tissue of an organ, as opposed to its supporting tissue (stroma); in a gland, the parenchyma is the secretory cells, rather than the connective-tissue capsule.

Parietal (pah-ri′ĕ-tal) Pertaining to the walls of a cavity.

Pars distalis (parz dis-tal′is) The main division of the adenohypophysis of the pituitary gland; the anterior lobe of the pituitary.

Pars nervosa (ner-vo′sah) The region of the neurohypophysis of the pituitary gland from which hormones are secreted; the posterior lobe of the pituitary.

Pectoral (pek′tor-al) Pertaining to the chest.

Pectoral girdle Bones that attach an upper limb to the axial skeleton: clavicle and scapula.

Pelvic girdle The paired hip bones that attach the lower limbs to the axial skeleton.

Pelvis (pel′vis) Inferior region of the body trunk; contains the basin-shaped, bony structure called the bony pelvis.

Pepsin (pep′sin) Protein-digesting enzyme secreted by the stomach lining.

Pericardium (per″ĭ-kar′de-um) Double-layered sac that encloses the heart and forms its superficial layer.

Perichondrium (per″ĭ-kon′dre-um) Membrane of fibrous connective tissue that covers the external surface of cartilages.

Perimysium (per″ĭ-mis′e-um) Connective tissue that surrounds and separates fascicles (bundles) of muscle fibers within a skeletal muscle.

Perineum (per″ĭ-ne′um) The region of the trunk containing the anus, vulva (females), and the posterior of the scrotum (males).

Periosteum (per″e-os′te-um) Membrane of fibrous connective tissue that covers the external surface of bones of the skeleton.

Peripheral nervous system (PNS) Portion of the nervous system consisting of nerves and ganglia that lie outside the brain and spinal cord.

Peristalsis (per″ĭ-stal′sis) Progressive, wave-like contractions that squeeze foodstuffs through the alimentary canal (or that move other substances through other body organs).

Peritoneum (per″ĭ-to-ne′um) Serous membrane that lines the interior of the abdominopelvic cavity and covers the surfaces of the organs in this cavity.

Peritonitis (per″ĭ-to-ni′tis) Infection and inflammation of the peritoneum.

Peritubular capillaries Capillaries in the kidney that surround the proximal and distal convoluted tubules; active in reabsorption.

Phagocytosis (fag″o-si-to′sis) The process by which a cell forms cytoplasmic extensions to engulf foreign particles, cells, or macromolecules and then uses lysosomes to digest these substances.

Pharynx (far′ingks) Muscular tube extending from the region posterior to the nasal cavity to the esophagus; the "throat" part of the digestive tube.

Photoreceptors (fo″to-re-sep′tors) Specialized receptor cells that respond to light energy: rod cells and cone cells.

Pia mater (pi′ah ma′ter) Most internal and most delicate of the three membranes (meninges) covering the brain and spinal cord.

Pinocytosis (pin″o-si-to′sis) The process by which cells engulf extracellular fluids.

Pituitary gland (pĭ-tu′ĭ-tār″e) A hormone-secreting, golf-club-shaped structue that hangs inferiorly from the brain and performs a variety of endocrine functions, such as regulating the gonads, thyroid gland, adrenal cortex, lactation, and water balance. Also called the *hypophysis*.

Placenta (plah-sen′tah) Temporary organ formed from both fetal and maternal tissues that provides nutrients and oxygen to the developing fetus, carries away fetal waste molecules, and secretes the hormones of pregnancy; shed as the afterbirth when labor is over.

Plasma (plaz′mah) The nonliving, fluid component of blood, within which the blood cells are suspended.

Plasma cell Cell formed from the division of an activated B lymphocyte; secretes antibodies.

Plasma membrane Membrane that encloses cell contents; the external, limiting membrane of the cell.

Plasmalemma (plaz″mah-lem′ah) *See* Plasma membrane.

Platelet Cell fragment found in blood; plugs small tears in blood vessels and helps initiate clotting.

Pleura (ploo′rah) Serous membrane that lines the pleural cavity in the thorax and covers the external surface of the lung.

Plexus (plek′sus) A network of converging and diverging nerves or veins.

Plica (pli′kah) A fold.

Podocytes (pod′o-sīts) Octopus-shaped epithelial cells that surround the glomerular capillaries; they help produce and maintain the basement membrane (a filtration membrane in the kidney).

Pons The part of the brain stem between the midbrain and the medulla oblongata.

Portal system A system of vessels in which two capillary beds, rather than one, lie between the incoming artery and the outgoing vein of a body region; a portal vein lies between the two capillary beds; examples are the hepatic portal system and the hypophyseal portal system.

Posterior Toward the back; dorsal.

Posterior pituitary *See* Pars nervosa.

Postganglionic neuron (pōst″gang-gle-on′ik) Autonomic motor neuron that has its cell body in a peripheral ganglion and projects its axon to an effector.

Preganglionic neuron (pre″gang-gle-on′ik) Autonomic motor neuron that has its cell body in the central nervous system and projects its axon to a peripheral ganglion.

Prime mover Muscle that bears the major responsibility for a particular movement; agonist.

Process (1) Prominence or projection; (2) series of actions for a specific purpose.

Progesterone (pro-jes′tĕ-rōn) Hormone that prepares the uterus to receive the implanting embryo.

Pronation (pro-na′shun) Medial rotation of the forearm that causes the palm to face posteriorly.

Prone Refers to a body lying horizontally with the face downward.

Proprioceptor (pro″pre-o-sep′tor) Receptor that senses movement in the musculoskeletal system; more specifically, proprioceptors sense stretch in muscles, tendons, and joint capsules.

Protein (pro′tēn) A long chain of amino acids or several linked chains of amino acids; the amino acid chains have bent and folded (and often coiled) to give each protein a distinct shape.

Proximal (prok′sĭ-mal) Toward the attached end of a limb, or near the origin of a structure.

Pseudostratified (soo″do-strat′ĭ-fīd) Pertaining to an epithelium that appears to be stratified (consisting of more than one layer of cells) but is not; the cells vary in height, but all touch the base of the epithelium.

Puberty (pu′ber-te) Period of life when reproductive maturity is reached.

Pulmonary (pul′mo-ner″e) Pertaining to the lungs.

Pulmonary circuit System of blood vessels that serve gas exchange in the lungs: the pulmonary arteries, capillaries, and veins.

Pulse Rhythmic expansion and recoil of arteries resulting from the contraction of the heart; can be felt from outside the body.

Pupil Opening in the center of the iris through which light enters the eye.

Purkinje fibers (pur-kin′je) Long rows of modified cardiac muscle cells of the conduction system of the heart; also called *conduction myofibers*.

Pus Fluid product resulting from the defense response to bacterial infection; composed of dead white blood cells, bacteria, and a thin fluid.

Pyloric region (pi-lor′ik) A funnel-shaped region of the stomach, just proximal to the pylorus.

Pylorus The distal, ring-shaped portion of the stomach that joins the small intestine and contains the pyloric sphincter muscle.

Ramus (ra′mus) Branch of a nerve, artery, or bone.

Raphe (ra′fe) A seam in the midline.

Receptor Peripheral nerve ending, or complete cell, that responds to particular types of stimulus.

Receptor molecule A cell component, usually a membrane protein, that binds to specific molecules to signal a certain response.

Reduction Restoring broken bone ends or dislocated bones to their original positions.

Reflex Automatic response to a stimulus.

Relay nucleus Any nucleus in the brain whose neurons receive signals from one region of the central nervous system and relay this information to another region; within the relay nucleus, the information is organized and edited.

Renal (re′nal) Pertaining to the kidney.

Renin (re′nin) Hormone released by the kidneys that is involved with raising blood pressure, blood volume, and the sodium concentration in blood.

Respiratory system Organ system that carries out gas exchange; includes the nose, pharynx, larynx, trachea, bronchi, and lungs.

Rete (re′te) A network, often composed of nerve fibers or blood vessels.

Reticular cell (rĕ-tik′u-lar) A fibroblast in reticular connective tissue (in bone marrow, spleen, lymph nodes, and so on).

Reticular formation Functional system that runs through the core of the brain stem; involved in alertness, arousal, and sleep; also contains visceral centers that control heart rate, breathing rate, and vomiting; controls some body movements as well.

Reticulocyte (rĕ-tik′u-lo-sīt) Immature or young erythrocyte.

Retina (ret′ĭ-nah) Neural tunic of the eyeball; contains the photoreceptor cells for vision.

Rhinencephalon (ri″nen-sef′ah-lon) The part of the cerebrum that receives and integrates olfactory (smell) impulses.

Ribonucleic acid (RNA) (ri″bo-nu-kle′ik) Nucleic acid that contains the sugar ribose; acts in protein synthesis.

Ribosome (ri′bo-sōm) Cytoplasmic organelle on which proteins are synthesized.

Rod cell One of the two types of photoreceptor cells in the retina of the eye.

Rostral Toward the nasal region or higher brain centers.

Rugae (roo′ge) Elevations or ridges, as in the mucosa of the stomach.

Sacral (sa′kral) Pertaining to the sacrum; the region in the midline of the buttocks.

Sagittal plane (saj′ĭ-tal) A vertical plane that divides the body or a body part into right and left portions.

Sarcolemma (sar″ko-lem′ah) The plasma membrane of a muscle cell.

Sarcomere (sar′ko-mēr) The smallest contractile unit of skeletal and cardiac muscle; the part of a myofibril between two Z discs; contains myofilaments composed mainly of contractile proteins (actin, myosin).

Sarcoplasm (sar′ko-plazm) The cytoplasm of a muscle cell.

Sarcoplasmic reticulum (sar′ko-plaz-mik rĕ-tik′u-lum) Specialized smooth endoplasmic reticulum of muscle cells; stores calcium ions.

Sclera (skle′rah) Outer fibrous tunic of the eyeball.

Scrotum (skro′tum) The external sac that encloses the testes.

Sebaceous gland (se-ba′shus) Gland in the skin that produces an oily secretion called sebum.

Sebum (se′bum) The oily secretion of sebaceous glands.

Secretion (se-kre′shun) (1) The passage of material formed by a cell to its exterior; (2) cell product that is transported to the exterior of a cell.

Section A cut through the body (or an organ) along a particular plane; a thin slice of tissue prepared for microscopic study.

Semen (se′men) Fluid mixture containing sperm and secretions of the male accessory reproductive glands.

Semilunar valves (sem″ĭ-lu′nar) Valves at the base of the aorta and the pulmonary trunk that prevent blood from returning to the heart ventricles after ventricular contraction.

Seminiferous tubules (sem″ĭ-nif′er-us) Highly convoluted tubules within the testes that form sperm.

Sensory nerve Nerve that contains only sensory fibers.

Sensory neuron Nerve cell that carries information received from sensory receptors; also called *afferent neuron.*

Serosa (se-ro′sah) *See* Serous membrane.

Serous cell Exocrine gland cell that secretes a watery product containing digestive enzymes.

Serous fluid (se′rus) A clear, watery lubricant secreted by cells of a serous membrane.

Serous membrane Moist, slippery membrane that lines internal body cavities (pleural, pericardial, and peritoneal cavities) and covers visceral organs within these cavities; also called *serosa.*

Serum (se′rum) Amber-colored fluid that exudes from clotted blood plasma as the clot shrinks and then no longer contains clotting factors.

Sinoatrial (SA) node (si″no-a′tre-al) A collection of specialized cardiac muscle cells in the superior wall of the right atrium; pacemaker of the heart.

Sinus (si′nus) (1) Mucous-membrane-lined, air-filled cavity in certain bones of the face; (2) dilated channel for the passage of blood or lymph.

Sinusoid (si′nu-soid) An exceptionally wide, twisted leaky capillary; large protein molecules or whole blood cells pass easily through the walls of sinusoids.

Smooth muscle Musculature consisting of spindle-shaped unstriped (nonstriated) muscle cells; present in the walls of most visceral organs.

Somatic (so-mat′ik) Pertaining to the region of the body that lies external to the ventral body cavity, including the skin, skeletal muscles, and the skeleton; opposite of *visceral.*

Somite (so′mīt) A mesodermal segment of the body of the embryo.

Special Pertaining to sensory inputs or motor outputs that are *localized* rather than being widespread through the body; opposite of *general.*

Special senses The senses whose receptors are confined to a small region rather than distributed widely through the body: taste, smell, vision, hearing, and equilibrium.

Spermatogenesis (sper″mah-to-jen′e-sis) The process by which sperm (male gametes) form in the testes; involves meiosis.

Sphincter (sfingk′ter) A muscle surrounding an opening; acts as a valve to close and open the orifice.

Spinal nerves The 31 pairs of nerves that attach to the spinal cord.

Spinal reflex A reflex mediated through the spinal cord.

Squamous (skwa′mus) Flat, plate-like; pertaining to flat epithelial cells that are wider than they are tall.

Stenosis (stĕ-no′sis) Constriction or narrowing.

Steroids (ste′roids) Group of lipid molecules containing cholesterol and some hormones.

Stimulus (stim′u-lus) An excitant or irritant; a change in the environment that evokes a response.

Striated muscle (stri′āt-ed) Muscle consisting of cross-striated (striped) muscle fibers: skeletal and cardiac muscle.

Stroke Condition in which brain tissue is deprived of its blood supply, as in blockage of a cerebral blood vessel; also called *cerebrovascular accident.*

Stroma (stro′mah) The connective tissue framework of an organ.

Subcutaneous (sub″ku-ta′ne-us) Deep to the skin.

Sulcus (sul′kus) A groove.

Superficial Located close to or on the body surface; outer; external; opposite of *deep.*

Superficial fascia The hypodermis; the fatty layer just below the skin.

Superior Closer to the head; above.

Supination (soo″pĭ-na′shun) Lateral rotation of the forearm that causes the palm to face anteriorly.

Supine (soo′pīn) Refers to a body lying horizontally with the face upward.
Supporting cells Cells in nervous tissue that support, nourish, and insulate neurons.
Suprarenal gland (soo″prah-re′nal) Another name for an adrenal gland.
Surfactant (ser-fak′tant) Detergent-like fluid secreted by certain cells lining the respiratory alveoli in the lungs; reduces the surface tension of water molecules, thus preventing collapse of the alveoli after each breath.
Suture (soo′cher) An immovable, fibrous joint; except at the jaw joint, all bones of the skull are united by sutures.
Sweat gland Tubular gland in the skin that secretes sweat (which cools the body).
Sympathetic division (sim″pah-thet′ik) Division of the autonomic nervous system that prepares the body to cope with danger or excitement; the fight, fright, or flight division.
Symphysis (sim′fĭ-sis) A joint in which the bones are connected by fibrocartilage.
Synapse (sin′aps) Specialized cell junction between two neurons, at which the neurons communicate.
Synaptic cleft Fluid-filled space at a synapse between neurons; also called *synaptic gap.*
Synarthrosis (sin″ar-thro′sis) Any immovable joint.
Synchondrosis (sin″kon-dro′sis) A joint in which bones are united by hyaline cartilage.
Syndesmosis (sin″des-mo′sis) A joint in which bones are united only by a ligament.
Synergist (sin′er-jist) Muscle that aids the action of a prime mover by contributing to the same movement or by stabilizing joints to prevent undesirable movements.
Synostosis (sin″os-to′sis) A completely ossified joint; a joint fused by bone.
Synovial fluid (sĭ-no′ve-al) Fluid secreted by the synovial membranes of the freely movable joints of the body; lubricates the joint surfaces and nourishes the articular cartilages.
Synovial joint Freely movable joint with a cavity and a capsule. *See* diarthrosis.
Systemic (sis-tem′ik) Pertaining to the whole body.
Systemic circuit System of blood vessels that carries oxygenated blood to the tissues throughout the body.
Systole (sis′to-le) Period during which the ventricles or the atria of the heart contract.
T cells Lymphocytes that mediate cellular immunity; include cytotoxic and helper T cells; also called *T lymphocytes.*
T tubule Extension of the muscle cell plasmalemma (sarcolemma) that protrudes deeply into the muscle cell.
Target cell A cell that is capable of responding to a hormone because it bears receptors to which the hormone can bind.
Taste buds Bulb-shaped sensory organs on and around the tongue that house the receptor cells for taste.
Telencephalon (tel″en-sef′ah-lon) Rostral division of the embryonic forebrain that develops into the cerebrum.
Tendon (ten′don) Cord of dense regular connective tissue that attaches muscle to bone.
Testis (tes′tis) Male primary sex organ that produces sperm; male gonad in the scrotum.
Testosterone (tes-tos′tĕ-rōn) *See* Androgen.
Thalamus (thal′ah-mus) An egg-shaped mass of gray matter in the diencephalon of the brain; consists of nuclei through which information is relayed to the cerebral cortex.
Thermoreceptor (ther″mo-re-sep′tor) Receptor sensitive to temperature changes.
Thoracic (tho-ras′ik) Pertaining to the chest or thorax.
Thoracic duct Large lymphatic duct that ascends anterior to the vertebral column; drains the lymph from up to three-fourths of the body (all except the body's superior right quarter).
Thorax (tho′raks) That portion of the body superior to the diaphragm and inferior to the neck.
Thymus (thi′mus) Organ of the immune system that is essential for the production of T cells (T lymphocytes); located in the anterior thorax.
Thyroid gland (thi′roid) Butterfly-shaped endocrine gland in the anterior neck; its main hormone (thyroid hormone) increases metabolic rate.
Tight junction A type of cell junction that closes off the intercellular space; also called a *zonula occludens.*
Tissue A group of similar cells (and extracellular material) that perform similar functions; primary tissues of the body are epithelial, connective, muscle, and nervous tissue.
Tissue fluid Watery fluid that, along with the molecules of ground substance, occupies the extracellular matrix of connective tissue; it is a filtrate of the blood containing all the small molecules of blood plasma; also called *interstitial fluid.*
Trabecula (trah-bek′u-lah) (1) Any one of the fibrous bands extending from the capsule to the interior of an organ; (2) a piece of the bony network in spongy bone.
Trachea (tra′ke-ah) Windpipe; the cartilage-reinforced airtube that extends from the larynx to the bronchi.
Tract A collection of nerve fibers in the central nervous system having the same origin, destination, and function.
Transverse process One of a pair of projections that extend laterally from the neural arch of a vertebra.
Trauma (traw′mah) A wound, injury, or shock, usually caused by external forces.
Trochanter (tro-kan′ter) A large, somewhat blunt process on a bone.
Trophoblast (trof′o-blast) External layer of cells in the blastocyst (early embryo); forms the embryo's contribution to the placenta.
Tropic hormone (tro′pik) A hormone that regulates the function of another endocrine organ; tropic hormones signal endocrine glands to secrete their own hormones.
Tubercle (too′ber-k′l) A nodule or small rounded process on a bone.
Tuberosity (too″bĕ-ros′ĭ-te) A broad process on a bone, larger than a tubercle.
Tumor An abnormal growth of cells; a swelling; a neoplasm; can be cancerous.
Tunica (too′nĭ-kah) A covering or coat; a layer or membrane of tissue.
Tympanic membrane (tim-pan′ik) The eardrum located between the outer and middle ear.
Ulcer (ul′ser) Erosion of the surface of an organ or tissue, such as a peptic ulcer in the wall of the stomach or small intestine.
Umbilical cord (um-bil′ĭ-kal) A cord that attaches to the navel before birth and connects the fetus to the placenta; contains the umbilical arteries and vein.
Umbilicus (um-bil′ĭ-kus) Navel; belly button.
Unipolar neuron (u″nĭ-po′lar) A sensory neuron in which a single short process projects from the cell body but divides like a T into two long processes (central process and peripheral process).
Urea (u-re′ah) The main nitrogen-containing waste excreted in urine.
Ureter (u-re′ter) Tube that carries urine from kidney to bladder.
Urethra (u-re′thrah) Tube that carries urine from the bladder to the exterior.
Uterine tube (u′ter-in) Tube through which the ovum travels to the uterus; also called *fallopian tube* and *oviduct.*

Uterus (u′ter-us) Hollow, thick-walled pelvic organ that receives the developing embryo; site where embryo/fetus develops; the womb.

Vasa recta (va′sah rek′tah) Capillary-like blood vessels that supply the loops of Henle and collecting tubules in the medulla of the kidney.

Vascularized (vas′ku-lar-īzd″) Having a blood supply; containing blood vessels.

Vasoconstriction (vas″o-kon-strik′shun) Narrowing of blood vessels, normally through the contraction of smooth muscle cells in the vessel walls.

Vasodilation (vas″o-di-la′shun) Relaxation of smooth muscle cells in the walls of blood vessels, causing the vessels to dilate (widen).

Vasomotor fibers (vas″o-mo′tor) Sympathetic nerve fibers that regulate the contraction of smooth muscle in the walls of blood vessels, thereby regulating the diameter of the vessels.

Vein Vessel that carries blood toward the heart.

Ventilation (ven″tĭ-la′shun) Breathing; consists of inspiration and expiration.

Ventral (ven′tral) Toward the front of the body; anterior.

Ventricles (1) Paired, inferiorly located heart chambers that function as the major blood pumps; (2) fluid-filled cavities of the brain.

Venule (ven′ūl) A small vein.

Vertebral column (ver′tĕ-bral) The spine or spinal column; formed of a number of bones called vertebrae, the discs between these vertebrae, and two composite bones (sacrum and coccyx).

Vesicle (ves′ĭ-k′l) A small, liquid-filled sac; also refers to the urinary bladder.

Vesicular follicle (vĕ-sik′u-lar fol′ĭ-k′l) Mature ovarian follicle; formerly called *Graafian follicle.*

Villus (vil′us) One of many finger-like projections of the internal surface of the small intestine that together increase the surface area for nutrient absorption.

Viscera (visceral organs) (vis′er-ah) The organs within the ventral body cavity, including the stomach, bladder, heart, lungs, spleen.

Visceral Pertaining to the organs and structures within the ventral body cavity and to all smooth muscle and glands throughout the body; opposite of *somatic.*

Visceral muscle Smooth muscle and cardiac muscle.

Vitamins Organic compounds required by the body in minute amounts that generally must be obtained from the diet.

Vulva (vul′vah) The external genitalia of the female.

White matter White substance of the central nervous system; contains tracts of myelinated nerve fibers.

Yolk sac Embryonic sac that stores a tiny quantity of yolk and gives rise to the lining of the digestive tube; also gives rise to the primordial germ cells and the blood cells.

Zygote (zi′gōt) Fertilized egg.

PHOTOGRAPHY AND ILLUSTRATION CREDITS

Photography Credits

Cover (from left to right): © Scott Camazine/Photo Researchers, Inc. © GJLP/CNRI/Phototake. © Mehau Kulyk/Science Photo Library. Back cover (from left to right): © Mehau Kulyk/Science Photo Library. © Scott Camazine/Photo Researchers, Inc. © Scott Camazine/Photo Researchers, Inc.
Chapter 1 Chapter Opener: © GJLP/CNRI Phototake. 1.1: © 1997 PhotoDisc. 1.3: © Benjamin/ Cummings Publishing Company. Photo by Paul Waring/Biomed Arts Associates, Inc. 1.5 top: © Jenny Thomas/Addison Wesley Longman. 1.5a and c: © Howard Sochureck. 1.5b: © Petit Format/Photo Researchers, Inc. 1.13a: © Ed Reschke. 1.13b: © R. Rodewald, University of Virginia/Biological Photo Service. 1.13c: © Science Photo Library/Photo Researchers, Inc. 1.14: © Vanessa Vick/Photo Researchers, Inc. 1.15: © Clinique Ste Catherine/CNRI/Science Photo Library/Photo Researchers, Inc. 1.16: © Custom Medical Stock Photography. All rights reserved. 1.17: © Howard Sochureck. All rights reserved. 1.18: Courtesy of Diane Roth. 1.19a: © Geoff Thompkins/Science Photo Library/ Photo Researchers, Inc. 1.19b: Photo by Jeff Shaw, Vanderbilt University, Knoxville, TN. Courtesy of Sun Microsystems.

Chapter 2 Chapter Opener: © Gopal Murti Phototake. 2.2b: Courtesy of J. David Robertson. 2.3b: © Dr. R. Alexley/Peter Arnold Inc. 2.6b: © Don Fawcett/Visuals Unlimited. 2.7b: Courtesy of Nicolai Simionescu. 2.8b: © Dr. R. Alexley/Peter Arnold, Inc. 2.9b: Courtesy of Dr. Barry King, University of California, Davis School of Medicine. 2.11b: R.Rodewald, University of Virginia/Biological Photo Service. 2.13: R. Rodewald, University of Virginia/Biological Photo Service. 2.14c: © John Heuser. 2.15c: © W. L. Dentler, University of Kansas/Biological Photo Service. 2.16c: © Dr. Leonard H. Rowl/UCLA School of Medicine. 2.17b: © CNRI/SPL/Photo Researchers, Inc. 2.19a: © Ada Olins/Biological Photo Services. 2.19b: Courtesy of GF Bahr, Armed Forces Institute of Pathology. 2.21: © Ed Reschke.

Chapter 3 Chapter Opener: Professor P.M. Motta et al/Science Photo Library. 3.11a and b: © Lennart Nillson. From *A Child is Born*. © 1990 Dell Publishing, New York, NY. A Closer Look: Courtesy of Fetal Alcohol & Drug Unit, University of Washington, Seattle. From Streissguth, A. P., Clarren, S. K., & Jones, K. L. (1985, July). "Natural History of the Fetal Alcohol Syndrome: A 10-year follow-up of eleven patients." *The Lancet*. 2, 85–91.

Chapter 4 Chapter Opener: © Dr. Dennis Kunkel/Phototake. 4.3a: © Dennis Strete. 4.3b and c: © Benjamin/Cummings Publishing Company. Photo by Allen Bell, University of New England. 4.3d: © R. Calentine/Visuals Unlimited. 4.3e–h: © Benjamin/Cummings Publishing Company. Photo by Allen Bell, University of New England. 4.5b and c: W. L. Dentler, University of Kansas/Biological Photo Service. 4.7: © David M. Phillips/The Population Council/Science Source/Photo Researchers, Inc. 4.8: L. Orci and A. Perrelet, Feeze-Etch Histology. Heidelberg: Springer-Verlag, 1975. 4.9: © David M. Phillips/The Population Council/Science Source/Photo Researchers, Inc. 4.10a: © CNRI/Science Photo Library. 4.15a and b: © Ed Reschke. 4.15c and d: © Benjamin/ Cummings Publishing Company. Photo by Allen Bell, University of New England. 4.15e: © Ed Reschke. 4.15f: © Benjamin/ Cummings Publishing Company. Photo by Allen Bell, University of New England. 4.15g: © Ed Reschke. 4.15h–j: © Benjamin/ Cummings Publishing Company. Photo by Allen Bell, University of New England. 4.15k: © Ed Reschke. 4.17a: © Eric Graves/Photo Researchers, Inc. 4.17b: © Ed Reschke. 4.17c: © Benjamin/Cummings Publishing Company. Photo by Allen Bell, University of New England. 4.18: © Ed Reschke. A Closer Look: © John D. Gearhart, John Hopkins Medical School.

Chapter 5 Chapter Opener: © Jean Claude Revy/Phototake. 5.3: © Courtesy of *Gray's Anatomy*, Churchill Livingstone Publishers. 5.4b: © Carolina Biological Supply/Phototake NYC. 5.4d: © Manfred Kage/Peter Arnold, Inc. 5.5: © CNRI/Science Photo Library/ Photo Researchers, Inc. 5.7a: © Cabisco/Visuals Unlimited. 5.7b: © John D. Cunningham/Visuals Unlimited. 5.9b.1: © Myrleen Ferguson/PhotoEdit. 5.9b.2: Dr. P. Marazzi/Science Photo Library/Photo Researchers, Inc. 5.9b.3: © 1997 Jon Meyer/Custom Medical Stock Photography. All rights reserved. 5.10a and b: © Dr. P. Marazzi/Science Photo Library/Photo Researchers, Inc. 5.10c: © Zeva Oelbaum/Peter Arnold, Inc. A Closer Look: © Rachel Epstein/Stuart Kenter Associates.

Chapter 6 Chapter Opener: © Michael Abbey/Photo Researchers, Inc. 6.3d: © Cabisco/Visuals Unlimited. 6.5b: © 1999 The Bergman Collection. 6.6c: © Benjamin/Cummings Publishing Company. Photo by Allen Bell, University of New England. 6.10: © Ed Reschke. 6.13: © Professor P. Motta, Department of Anatomy, University "La Sapienza," Rome/Science Photo Library/Photo Researchers, Inc. 6.14: © Carolina Biological Supply. Table 6.1, 1: © 1992 Michael English, MD. Custom Medical Stock Photography. All rights reserved. Table 6.1, 2: © SIU/Peter Arnold, Inc. Table 6.1, 3: Courtesy of San Francisco General Hospital. Table 6.1, 4: Courtesy of San Francisco General Hospital. Table 6.1, 5: Courtesy of San Francisco General Hospital. Table 6.1, 6: © 1997 Bates M. D. Custom Medical Stock Photography. All rights reserved. A Closer Look: © Gary M. Prior/Allsport.

Chapter 7 Chapter Opener: D. Young Riess, M.D./Tony Stone Images. 7.3c: © R. T. Hutchings. 7.4b: © R.T. Hutchings. 7.5: © R.T. Hutchings. 7.6a–c: © R.T. Hutchings. 7.7: © R.T. Hutchings. 7.9a: © R.T. Hutchings. 7.14d: © L. Bassett/Visuals Unlimited. 7.22a–c: Courtesy of Center for Cranialfacial Anomalies, University of California, San Francisco.

Chapter 8 Chapter Opener: © CNRI/Phototake NYC. 8.1d and e: © R. T. Hutchings. 8.3c: © L.V. Bergman & Associates, Inc. NY. 8.6c: Department of Anatomy and Histology, University of California, San Francisco. 8.9c: © 1999 The Bergman Collection. 8.10c and d: © R. T. Hutchings. Table 8.2: From *A Stereoscopic Atlas of Human Anatomy* by David L. Bassett. A Closer Look: Courtesy of Norian Corporation.

Chapter 9 Chapter Opener: © GJLP/CNRI/Phototake. 9.3b: © Dr. Gilbert Faure/Science Photo Library/Photo Researchers, Inc. 9.5a–f: © John Wilson White/Addison Wesley Longman, Inc. 9.6a–f: © John Wilson White/Addison Wesley Longman, Inc. 9.9b: © Addison Wesley Longman, Inc. Photo by Mark Neilsen, University of Utah. 9.9e: © VideoSurgery/Photo Researchers, Inc. 9.10d: From *A Stereoscopic Atlas of Human Anatomy* by David L. Bassett. 9.11d: From *A Stereoscopic Atlas of Human Anatomy* by David L. Bassett. 9.12f: © L. Bassett/Visuals Unlimited. 9.16: © SIU/Peter Arnold, Inc.

Chapter 10 Chapter Opener: © Albert Tousson/Phototake. 10.1b: © John D. Cunningham/ Visuals Unlimited. 10.2: Courtesy of Thomas S. Leeson. 10.4a: © Marian Rice. 10.5: © K. G. Murti/Visuals Unlimited. 10.7: Courtesy of Dr. H. E. Huxley, Brandeis University, Waltham, MA. 10.9a and b: From *Functional Histology* 2e by Paul Wheater and H. George Burkitt. Churchill-Livingstone. Copyright © 1987. 10.10c: © Manfred Kage/Peter Arnold, Inc. 10.11: © Dr. Dennis Kunkel/Phototake NYC. 10.12c: © Fred E. Hossler/Visuals Unlimited. 10.12f: © Fawcett/Cooke/Photo Researchers, Inc. Table 10.1, 4: © Eric Graves/Photo Researchers, Inc. Table 10.1, 5: © Marian Rice. Table 10.1, 6: Department of Anatomy and Histology, University of California, San Francisco. A Closer Look: © Tony Stone Images/David Madison.

Chapter 11 Chapter Opener: © Corbis Images. 11.9c: From *A Stereoscopic Atlas of Human Anatomy* by David L. Bassett. 11.10c: From *A Stereoscopic Atlas of Human Anatomy* by David L. Bassett. 11.16c: © Benjamin/ Cummings Publishing. Photo by Stephen Spector.

Chapter 12 Chapter Opener: © Dr. Dennis Kunkel/Phototake. 12.5: © David M. Phillips/Photo Researchers, Inc. 12.8c: © Dennis Kunkel/CNRI/Phototake NYC. 12.16: © Don Fawcett/Photo Researchers, Inc. 12.17b: From *Tissues and Organs: A text-Atlas of Scanning Electron Microscopy* by R. Kessel and R. Kardon. © 1979 W. H. Freeman. 12.17c: © Benjamin/Cummings Publishing Company. Photo by Victor Eroschenko, University of Idaho.

Chapter 13 Chapter Opener: © Alfred Pasieka/Science Photo Library. 13.1: From *A Stereoscopic Atlas of Human Anatomy* by David L. Bassett. 13.7d: © Dr. Robert A. Chase. 13.9: © Volker Steger/Peter Arnold, Inc. 13.14b: © Pat Lynch/Photo Researchers, Inc. 13.19: © Leonard Lessin/Peter Arnold, Inc. 13.22b and c: © L. Bassett/Visuals Unlimited. 13.25b: From *A Stereoscopic Atlas of Human Anatomy* by David L. Bassett. 13.28: © Carroll H. Weiss. All rights reserved. 13.29b–d: From *A Stereoscopic Atlas of Human Anatomy* by David L. Bassett. 13.35: *The New England Journal of Medicine*, 1996, 334:12

page 775. 13.36: © Biophoto Associates/Science Source/Photo Researchers, Inc. A Closer Look: © Lawrence Migdale/Photo Researchers, Inc.

Chapter 14 Chapter Opener: © Ron Boardman. Table 14.3: © Benjamin/Cummings Publishing Company. Photo by William Thompson. A Closer Look: Courtesy of William E. Collins, Inc.

Chapter 15 Chapter Opener: © Nancy Kedersha/UCLA/Science Photo Library.

Chapter 16 Chapter Opener: © Dr. G. Oran Bredberg/Science Photo Library. 16.1d: © Carolina Biological Supply Company/Phototake NYC. 16.3c: © John D. Cunningham/Visuals Unlimited. 16.4: © Addison Wesley Longman, Inc. Photo by Richard Tauber. 16.7b: From *A Stereoscopic Atlas of Human Anatomy* by David L. Bassett. 16.9b: Dr. Jerry Kuszak, Rush-Presbyterian St. Luke's Hospital, Chicago, IL. 16.10b: © Ed Reschke/Peter Arnold, Inc. 16.12: © 1990 Custom Medical Stock Photography. All rights reserved. 16.13: © NMSB/Custom Medical Stock Photography. All rights reserved. 16.15b: © Benjamin/ Cummings Publishing Company. Photo by Stephen Spector.

Chapter 17 Chapter Opener: © Dr. Dennis Kunkel. 17.2a: National Cancer Institute/Science Photo Library/Photo Researchers, Inc. 17.2b: © Benjamin/Cummings Publishing Company. Photo by Victor Eroschenko, University of Idaho. 17.4a–e: J.O. Ballard, M.D. 17.7: © Lennart Nilsson/Boehringer Ingelheim International GmbH. 17.8b: © Ed Reschke/Peter Arnold, Inc. 17.8c: © Benjamin/Cummings Publishing Company. Photo by Allen Bell, University of New England. A Closer Look: Courtesy of Cbr Systems, Inc./Cord Blood Registry.

Chapter 18 Chapter Opener: © CNRI/Phototake. 18.2d: © L. Bassett/Visuals Unlimited. 18.6: © L. Bassett/Visuals Unlimited. 18.8b: From *A Stereoscopic Atlas of Human Anatomy* by David L. Bassett. 18.8c: © Lennart Nilsson. From *The Body Victorious.* © Boehringer Ingelheim International GmbH.

Chapter 19 Chapter Opener: © GJLP/CNRI/Phototake. 19.1b: © Benjamin/Cummings Publishing Company. 19.2: © Benjamin/ Cummings Publishing Company. 19.9b: © Science Photo Library/Photo Researchers, Inc. 19.18: © L. Bassett/Visuals Unlimited. A Closer Look, a: GCA/CNRI/Phototake, NYC., b: © Howard Sochurek., c: © Howard Sochurek.

Chapter 20 Chapter Opener: © Biology Media. 20.4b: © Biophoto Associates/Photo Researchers, Inc. 20.5b: © L. Bassett/Visuals Unlimited. 20.6b: © Don Fawcett/Photo Researchers, Inc. 20.7c: © Benjamin/Cummings Publishing Company. Photo by Victor Eroschenko, University of Idaho. 20.7d: © Francis Leroy, Biocosmos/Science Photo Library/Photo Researchers, Inc. 20.9c: © Benjamin Cummings Publishing Company. Photo by Mark Neilsen. 20.9d: Courtesy of LUMEN Histology, Loyola University Medical Education Network, http://www.lumen.luc.edu/lumen/MedEd/Histo/frames/histo_frames.html. 20.10: © Michael Abbey/Photo Researchers, Inc. 20.11: © John Cunningham/Visuals Unlimited. 20.12: © Biophoto Associates/Science Source/Photo Researchers, Inc. A Closer Look: © Montagnier/Institut Pasteur/SPL/Science Source/Photo Researchers, Inc.

Chapter 21 Chapter Opener: © Prof. P. Motta/Dept. of Anatomy/University "La Sapienza," Rome/Science Photo Library. 21.3b: From *A Stereoscopic Atlas of Human Anatomy* by David L. Bassett. 21.4a: © Biophoto Associates/Photo Researchers, Inc. 21.4b: © Science Photo Library/Photo Researchers, Inc. 21.5c and d: From *A Stereoscopic Atlas of Human Anatomy* by David L. Bassett. 21.7b: Courtesy of the University of San Francisco. 21.8b: © Martin Dohm, Royal College of Surgeons/Science Source. 21.8c: © Ed Reschke/Peter Arnold, Inc. 21.9b: © Carolina Biological Supply/Phototake. 21.10a: © R. Kessel and R. Kardon. 21.12a: © Benjamin/Cummings Publishing Company. Photo by Richard Tauber. 21.13a: © Martin Rotker/Phototake. 21.13b: © L. Bassett/Visuals Unlimited. 21.13c: From *A Stereoscopic Atlas of Human Anatomy* by David L. Bassett. 21.17a and b: Kenneth Siegesmund, Ph.D., Department of Anatomy, Medical College of Wisconsin. A Closer Look: © Martin Rotker/Phototake.

Chapter 22 Chapter Opener: © Scott Camazine/Photo Researchers. 22.9: From *Tissues and Organs* by R. Kessel and R. Kardon. © 1979 W. H. Freeman. 22.10b: © Science Photo Library/Photo Researchers, Inc. 22.13a: © Biophoto Associates/Photo Researchers, Inc. 22.13b: From *Color Atlas of Histology* by Leslie P. Garner and James L. Hiatt © 1990 Williams and Wilkins. 22.14c: From *Color Atlas of Histology* by Leslie P. Garner and James L. Hiatt © 1990 Williams and Wilkins. 22.15d: © Biophoto Associates/Photo Researchers, Inc. 22.17d: Courtesy of LUMEN Histology, Loyola University Medical Education Network, http://www.lumen.luc.edu/lumen/MedEd/Histo/frames/histo_frames.html. 22.19a: © Ed Reschke/Peter Arnold, Inc. 22.19b: © Biophoto Associates/Photo Researchers, Inc. 22.20a and b: From *A Stereoscopic Atlas of Human Anatomy* by David L. Bassett. 22.22b: From *A Stereoscopic Atlas of Human Anatomy* by David L. Bassett. 22.23: © L. Bassett/Visuals Unlimited. 22.24b: © M. Walker/Photo Researchers, Inc. 22.25b: © Carolina Biological Supply. 22.26a: From *A Stereoscopic Atlas of Human Anatomy* by David L. Bassett. A Closer Look: © Carroll H. Weiss/Camera MD Studios.

Chapter 23 Chapter Opener: © Alfred Pasieka/Science Photo Library. 23.2b: Courtesy of Ruedi Thoeni. M.D., University of California, San Francisco. 23.3a: © L. Bassett/Visuals Unlimited. 23.5: © Biophoto Associates/Photo Researchers, Inc. 23.7b: © Professor P. M. Motta and M. Castellucci/Science Photo Library/Photo Researchers, Inc. 23.8c: From *Tissues and Organs,* by R. Kessel and R. Kardon © 1979 W. H. Freeman. 23.11: © L. Bassett/Visuals Unlimited. 23.12: © Biophoto Associates/Photo Researchers, Inc. 23.16: © John D. Cunningham/Visuals Unlimited. A Closer Look: A. Glauberman/Photo Researchers, Inc.

Chapter 24 Chapter Opener: © Yorgos Nikas. 24.3b: © Benjamin/Cummings Publishing Company. Photo by Victor Eroschenko, University of Idaho. 24.3c: From *A Stereoscopic Atlas of Human Anatomy* by David L. Bassett. 24.4a: From *Tissues and Organs,* by R. Kessel and R. Kardon © 1979 W. H. Freeman. 24.5b: © Manfred Kage/Peter Arnold, Inc. 24.6a: © Benjamin/ Cummings Publishing Company. Photo by Victor Eroschenko, University of Idaho. 24.6b: © Biophoto Associates/Photo Researchers, Inc. 24.9c: © Allen Bell/Benjamin Cummings. 24.12b: © Ed Reschke. 24.14: © C. Edelman/La Vilette/Photo Researchers, Inc. 24.16: From *A Stereoscopic Atlas of Human Anatomy* by David L. Bassett. 24.17: © L. Bassett/Visuals Unlimited. 24.18a: © Benjamin/ Cummings Publishing Company. Photo by Victor Eroschenko, University of Idaho. 24.23b: © Lennart Nilsson. From *A Child is Born.* © 1990 Dell Publishing, New York, New York. 24.24b: Photographed for the Carnegie Collection by Dr. Allen C. Enders, University of California, Davis. 24.26: From *A Stereoscopic Atlas of Human Anatomy* by David L. Bassett.

Chapter 25 Chapter Opener: © Michael Siegel/Phototake. 25.5: © Biophoto Associates/Photo Researchers, Inc. 25.8b: © Ed Reschke. 25.9b: *Color Histology,* by Leslie Garner and James Hiatt. © 1990 Williams and Wilkins. 25.10b: © Ed Reschke. 25.11b: © Don Fawcett/Photo Researchers, Inc. 25.12a: © Carolina Biological Supply/Phototake NYC. 25.13a and b: Courtesy of Dr. Charles B. Wilson, Neurological Surgery, University of California Medical Center, San Francisco. A Closer Look: © 1991 Andy Levin. All rights reserved/Photo Researchers, Inc.

Chapter 26 Chapter Opener: © GJLP/CNRI/Phototake. 26.5a and b: © Benjamin/Cummings Publishing Company. Photo by Paul Waring/Biomed Arts Associates, Inc. 26.6: © Benjamin/Cummings Publishing Company. Photo by Paul Waring/Biomed Arts Associates, Inc. 26.7a: © Benjamin/Cummings Publishing Company. Photo by Paul Waring/Biomed Arts Associates, Inc. 26.8: © Benjamin/Cummings Publishing Company. Photo by Paul Waring/Biomed Arts Associates, Inc. 26.10a and b: © Benjamin/ Cummings Publishing Company. Photo by Paul Waring/Biomed Arts Associates, Inc. 26.11: © Benjamin/ Cummings Publishing Company. Photo by Paul Waring/Biomed Arts Associates, Inc. 26.15a and b: © Benjamin/ Cummings Publishing Company. Photo by Paul Waring/Biomed Arts Associates, Inc. 26.17: © Benjamin/Cummings Publishing Company. Photo by Paul Waring/Biomed Arts Associates, Inc. 26.18a: © Benjamin/ Cummings Publishing Company. Photo by Paul Waring/Biomed Arts Associates, Inc. 26.19: © Benjamin/Cummings Publishing Company. Photo by Paul Waring/Biomed Arts Associates, Inc. 26.20a and b: © Benjamin/ Cummings Publishing Company. Photo by Paul Waring/Biomed Arts Associates, Inc. 26.21: © Benjamin/Cummings Publishing Company. Photo by Paul Waring/Biomed Arts Associates, Inc. 26.22: © Benjamin/Cummings Publishing Company. Photo by Paul Waring/Biomed Arts Associates, Inc. 26.23: © Benjamin/Cummings Publishing Company. Photo by Paul Waring/Biomed Arts Associates, Inc. 26.24: © Benjamin/Cummings Publishing Company. Photo by Paul Waring/Biomed Arts Associates, Inc. 26.25a–d: © Benjamin/ Cummings Publishing Company. Photo by Paul Waring/Biomed Arts Associates, Inc. 26.26:

Illustration Credits

Chapter 1 1.1: Jeanne Koelling/Kristin Mount. 1.2a–l: Vincent Perez/Wendy Hiller Gee. 1.4a and b: Imagineering. 1.6: Imagineering. 1.7a–c: Imagineering. 1.8a and b: Imagineering. 1.9a–d: Imagineering. 1.10: Imagineering. 1.11a and b: Imagineering. 1.12: Imagineering. Table 1.1: Imagineering.

Chapter 2 2.1: Tomo Narashima. 2.2a: George Klatt/Kristin Mount. 2.3a: Imagineering. 2.4a and b: Imagineering. 2.5: Imagineering. 2.6a: Imagineering. 2.7a: Tomo Narashima. 2.8a: Imagineering. 2.9a: Imagineering. 2.10: Imagineering. 2.11a: Tomo Narashima. 2.12: Tomo Narashima. 2.14a and b: Imagineering. 2.15b: Tomo Narashima. 2.16a and b: Imagineering. 2.17a: Tomo Narashima. 2.18: Imagineering. 2.20: Imagineering. 2.21: Precision Graphics/Kristin Mount. 2.22: Imagineering. A Closer Look: Imagineering.

Chapter 3 3.1: Imagineering. 3.2: Imagineering. 3.3: Imagineering. 3.4: Imagineering. 3.5: Imagineering. 3.6: Imagineering. 3.7: Imagineering. 3.8: Imagineering. 3.9a–c: Imagineering. 3.10: Imagineering. Table 3.2: Imagineering.

Chapter 4 4.1: Imagineering. 4.4: Imagineering. 4.5a: Precision Graphics. 4.6: Precision Graphics. 4.7: Imagineering. 4.8: Imagineering. 4.9: Imagineering. 4.11: Imagineering. 4.13: Imagineering. 4.14: Kristin Otwell. 4.16a–c: Imagineering. 4.19a–c: Imagineering.

Chapter 5 5.1: Tomo Narashima. 5.2: Imagineering. 5.6: Imagineering. 5.7: Imagineering. 5.8a and b: Imagineering. 5.9a: Imagineering.

Chapter 6 6.1: Precision Graphics. 6.2: Imagineering. 6.3a–c: Carla Simmons. 6.4: Imagineering. 6.5a: Imagineering. 6.6a and b: Carla Simmons. 6.7: Imagineering. 6.8: Imagineering. 6.9: Carla Simmons. 6.11: Imagineering. 6.12: Carla Simmons.

Chapter 7 7.1a and b: Laurie O'Keefe. 7.2a and b: Kristin Mount. 7.3a and b: Kristin Mount. 7.4a and c: Kristin Mount. 7.5: Kristin Mount. 7.6a and b: Kristin Mount. 7.7: Kristin Mount. 7.8a–c: Kristin Mount. 7.9b: Kristin Mount. 7.10a and b: Kristin Mount. 7.11a and b: Vincent Perez. 7.12: Kristin Mount. 7.13: Kristin Otwell. 7.14a–c: Laurie O'Keefe/Kristin Mount. 7.15: Laurie O'Keefe. 7.16a–c: Laurie O'Keefe. 7.17a–c: Laurie O'Keefe. 7.18a and b: Laurie O'Keefe. 7.19a and b: Imagineering. 7.20a and b: Laurie O'Keefe. 7.21a and b: Nadine Sokol. Table 7.1: Imagineering.

Chapter 8 8.1: Kristin Mount. 8.1a–c: Laurie O'Keefe/Kristin Mount. 8.2a–c: Laurie O'Keefe. 8.3a and b: Laurie O'Keefe. 8.4a and b: Laurie O'Keefe. 8.5a–c: Kristin Mount. 8.6a and b: Kristin Mount. 8.7a–c: Laurie O'Keefe. 8.8a and b: Kristin Mount. 8.9b: Laurie O'Keefe. 8.10a and b: Laurie O'Keefe. 8.11a–c: Kristin Mount/Kristin Mount. 8.12: Imagineering. Table 8.1: Laurie O'Keefe. Table 8.2: Kristin Otwell.

Chapter 9 9.1a–c: Imagineering. 9.2a–c: Carla Simmons/Kristin Mount. 9.3a and c: Imagineering. 9.4a and b: Carla Simmons/Kristin Mount. 9.7a–f: Imagineering. 9.8a and b: Imagineering. 9.9a, c and d: Carla Simmons/Kristin Mount. 9.10a–c: Carla Simmons. 9.11a–c: Carla Simmons. 9.12a: Carla Simmons/Kristin Mount. 9.12b: Kristin Mount. 9.12c–e: Carla Simmons/Kristin Mount. 9.13a and b: Imagineering. 9.14: Imagineering. 9.15a–c: Imagineering. Table 9.2: Imagineering.

Chapter 10 10.1a: Raychel Ciemma. 10.3: Imagineering. 10.4b–d: Imagineering. 10.6a–b: Imagineering. 10.8: Imagineering. 10.10a, b, d and e: Imagineering. 10.12a, b, d and e: Imagineering. Table 10.1: Imagineering. Table 10.2: Imagineering.

Chapter 11 11.2a–c: Imagineering. 11.3: Imagineering. 11.4a–e: Imagineering. 11.5a and b: Raychel Ciemma. 11.6: Raychel Ciemma. 11.7a–c: Raychel Ciemma. 11.8a and b: Raychel Ciemma. 11.9a, b, d and e: Raychel Ciemma. 11.10: Raychel Ciemma. 11.11a–c: Raychel Ciemma. 11.12a–c: Raychel Ciemma. 11.13a and b: Raychel Ciemma. 11.14a–d: Raychel Ciemma. 11.15a–c: Raychel Ciemma. 11.16a and b: Raychel Ciemma. 11.17: Imagineering. 11.18a–d: Laurie O'Keefe. 11.19a–c: Raychel Ciemma. 11.20a–c: Wendy Hiller Gee. 11.21a–d: Raychel Ciemma. 11.22a–c: Raychel Ciemma. 11.23a–f: Raychel Ciemma. 11.24: Imagineering. 11.25a–e: Laurie O'Keefe.

Chapter 12 12.1: Imagineering. 12.2: Imagineering. 12.3: Imagineering. 12.4: Imagineering. 12.6: Imagineering. 12.7: Imagineering. 12.8a and b: Imagineering. 12.9: Imagineering. 12.10a–c: Imagineering. 12.11: Imagineering. 12.12a–e: Imagineering. 12.13a–e: Imagineering. 12.14a and b: Imagineering. 12.15a–c: Imagineering. 12.17a: Charles W. Hoffman/Kristin Mount. 12.18a and b: Imagineering. 12.19: Imagineering. 12.20a–c: Imagineering.

Chapter 13 13.2a–e: Imagineering. 13.3a–d: Imagineering. 13.4: Imagineering. 13.5: Imagineering. 13.6a and b: Imagineering. 13.7a, b and e: Imagineering. 13.8: Imagineering. 13.10a and b: Imagineering. 13.11: Imagineering. 13.12: Imagineering. 13.13a and b: Imagineering. 13.14a and b: Imagineering. 13.15: Imagineering. 13.16a and b: Imagineering. 13.17: Imagineering. 13.18: Imagineering. 13.19: Imagineering. 13.20a–d: Imagineering. 13.21a–c: Imagineering. 13.22: Imagineering. 13.23: Imagineering. 13.24: Imagineering. 13.25a: Imagineering. 13.26: Imagineering. 13.27a and b: Imagineering. 13.29a: Imagineering. 13.30a and b: Stephanie McCann. 13.31: Imagineering. 13.32: Imagineering. 13.33a and b: Imagineering. 13.34a and b: Imagineering. Table 13.1: Imagineering.

Chapter 14 14.1: Imagineering. 14.2: Imagineering. 14.3: Imagineering. 14.4: Imagineering. 14.5a and b: Imagineering. 14.6a and b: Imagineering. 14.7: Imagineering. 14.8: Imagineering. 14.9a and b: Stephanie McCann. 14.10: Imagineering. 14.11a, c and d: Imagineering. 14.12a and b: Imagineering. 14.13a and b: Imagineering. 14.14a and b: Imagineering. Table 14.1: Imagineering. Table 14.3: Imagineering.

Chapter 15 15.1: Imagineering. 15.2: Imagineering. 15.3: Imagineering. 15.4: Imagineering. 15.5: Imagineering. 15.6: Imagineering. 15.7: Imagineering. 15.8: Imagineering. 15.9: Imagineering. 15.10: Imagineering. 15.11: Imagineering. 15.12: Imagineering. 15.13: Imagineering. 15.14: Imagineering. 15.15: Imagineering. 15.16: Imagineering.

Chapter 16 16.1a–c: Imagineering. 16.2: Imagineering. 16.3a and b: Imagineering. 16.5a and b: Charles W. Hoffman. 16.6a and b: Wendy Hiller Gee/Kristin Mount. 16.6c: Kristin Mount. 16.7a: Imagineering. 16.8: Imagineering. 16.9a: Imagineering. 16.10a and c: Imagineering. 16.11: Imagineering. 16.14: Imagineering. 16.16a–e: Imagineering. 16.17a and b: Imagineering/Kristin Mount. 16.18: Kristin Mount. 16.19: Precision Graphics/Kristin Mount. 16.20: Precision Graphics/Kristin Mount. 16.21a–c: Imagineering. 16.22a and b: Imagineering. 16.23a–c: Imagineering. 16.24: Imagineering. 16.25: Imagineering. 16.26a–d: Imagineering. A Closer Look: Imagineering.

Chapter 17 17.1: Imagineering. 17.3: Imagineering. 17.5: Imagineering. 17.6a and b: Imagineering. 17.8a: Imagineering. 17.9: Imagineering. Table 17.1: Imagineering.

Chapter 18 18.1: Imagineering. 18.2a–c: Wendy Hiller Gee/Kristin Mount. 18.3: Imagineering. 18.4: Carla Simmons/Kristin Mount. 18.5a–d: Imagineering. 18.7: Kristin Mount. 18.8a: Imagineering. 18.9a and b: Imagineering. 18.10a and b: Imagineering. 18.11: Imagineering. 18.12: Carla Simmons/Kristin Mount. 18.13: Imagineering. 18.14a and b: Imagineering. 18.15: Imagineering. 18.16: Imagineering. 18.17a–f: Imagineering. A Closer Look: Imagineering.

Chapter 19 19.1a: Imagineering. 19.2: Imagineering. 19.3a and b: Imagineering. 19.4a–c: Imagineering. 19.5a–b: Imagineering. 19.6: Carla Simmons. 19.7: Imagineering. 19.8: Imagineering. 19.9a and c: Imagineering. 19.10: Imagineering. 19.11: Imagineering. 19.12: Imagineering. 19.13: Imagineering. 19.14a and b: Imagineering. 19.15: Imagineering. 19.16: Imagineering. 19.17: Imagineering. 19.19a and b: Imagineering. 19.20a and b: Imagineering. 19.21: Imagineering.

19.22: Imagineering.
19.23: Imagineering.
19.24a and b: Imagineering.
19.25: Imagineering.
19.26: Imagineering.

Chapter 20 20.1: Imagineering.
20.2a and b: Imagineering.
20.3: Imagineering.
20.4a: Imagineering.
20.5a: Imagineering.
20.6a: Imagineering.
20.7a and b: Imagineering.
20.8: Imagineering.
20.9a and b: Imagineering.

Chapter 21 21.1: Imagineering.
21.2: Imagineering.
21.3a: Imagineering.
21.5a and b: Imagineering.
21.6a–d: Imagineering.
21.7a: Imagineering.
21.8a: Imagineering.
21.9a and b: Imagineering.
21.10b–d: Imagineering.
21.11: Imagineering.
21.12a and b: Imagineering.
21.14: Imagineering.
21.15a–d: Imagineering.
21.16: Imagineering.
21.18a–c: Imagineering.

Chapter 22 22.1: Kristin Mount. 22.2: Imagineering. 22.4a–c: Imagineering. 22.6: Tomo Narashima. 22.7a and b: Cyndie Wooley. 22.8: Imagineering.
22.9: Imagineering.
22.10a: Imagineering.
22.11: Imagineering.
22.12: Imagineering.
22.14a–b: Imagineering.
22.15a–c: Imagineering.
22.16: Imagineering.
22.17a and b: Imagineering.
22.18b: Precision Graphics.
22.21: Imagineering.
22.22c and d: Imagineering.
22.24a: Imagineering.
22.25a: Imagineering.
22.26b–d: Imagineering.
22.27a and b: Imagineering.

Chapter 23 23.1a and b: Imagineering.
23.2a: Imagineering.
23.3b: Imagineering.
23.4: Imagineering.
23.6: Imagineering.
23.7a and c: Imagineering.
23.8a and b: Imagineering.
23.9: Imagineering.
23.10: Imagineering.
23.14: Imagineering.
23.15a and b: Linda McVay.
23.17: Imagineering.
23.18a–d: Linda McVay/Kristin Mount.

Chapter 24 24.1: Martha Blake/Kristin Mount.
24.2: Imagineering.
24.3a: Imagineering.
24.4c: Kristin Mount.
24.5a: Martha Blake/Kristin Mount.
24.7: Imagineering.
24.8a and b: Martha Blake/Kristin Mount. 24.9a and b: Imagineering.
24.10a and b: Martha Blake/Kristin Mount.
24.11: Imagineering.
24.12a: Martha Blake.
24.13: Imagineering.
24.15: Imagineering.
24.18b: Martha Blake/Kristin Mount. 24.19a–d: Imagineering.
24.20: Carla Simmons. 24.21: Imagineering. 24.22a and b: Martha Blake. 24.23a: Martha Blake/Kristin Mount. 24.24a: Imagineering. 24.25a–f: Martha Blake. 24.27a–d: Martha Blake/Kristin Mount. 24.28: Imagineering. 24.29a and b: Imagineering. 24.30a–c: Imagineering.

Chapter 25 25.1: Imagineering.
25.2a–c: Imagineering. 25.3a–c: Imagineering. 25.4: Imagineering.
25.6: Imagineering.
25.7: Imagineering.
25.8a and c: Imagineering.
25.9a and c: Imagineering.
25.10: Imagineering.
25.14a–d: Imagineering.

Chapter 26 26.1a and b: Laurie O'Keefe. 26.2: Raychel Ciemma/Kristin Mount. 26.3: Kristin Mount. 26.4: Imagineering.
26.7b: Imagineering.
26.9a and b: Imagineering.
26.12a and b: Imagineering.
26.13: Imagineering.
26.14a and b: Imagineering.
26.16a–c: Imagineering.
26.18b: Imagineering.

INDEX

NOTE: Page numbers in **boldface** type indicate a major discussion. A *t* following a page number indicates tabular material and an *f* following a page number indicates a figure.

A bands, sarcomere, **248,** 249*f*
ABCD rule, **124**
ABCD(E) rule, **124**
Abdomen, 15–16
  arteries of, 559–61
  bony landmarks of, 779*f,* 781
  muscles and other surface features of, 779*f,* 781–83
  quadrants of, **16,** 17*f*
  regions of, **16,** 17*f*
  veins of, 569–71
Abdominal aorta, **553,** 554*f,* 559
Abdominal cavity, **13,** 14*f*
Abdominal organs, sympathetic pathways to, 456
Abdominal wall
  hernia in, 290
  innervation of, 426*f,* 427*f,* 428
  muscles of, 289–90*t,* 291*f*
  regions and quadrants of, 636*f,* 637
Abdominopelvic cavity, 13, 14*f*
Abducens nerve (VI), **418,** 419*t,* 422*t*
Abducens nerve paralysis, 422
Abduction movement, **221,** 222*f*
Abductor digiti minimi muscle, 309*t,* 310*f,* 329*t,* 330*f*
Abductor hallucis muscle, 329*t,* 330*f*
Abductor pollicis brevis muscle, 309*t,* 310*f*
Abductor pollicis longus muscle, 305*t,* 306*f*
Absorption
  of food, **636**
  function of epithelia in, 76
  by small intestine, 653–54
Absorptive cells
  of large intestine, 657, 658
  of small intestine, **653,** 654
Accessory digestive organs, **635**
Accessory glands, male, 710–12
Accessory hemiazygos vein, 568, 569
Accessory nerve (XI), 381, **419,** 425*t,* 777
Accessory nerve damage, 425
Accessory pancreatic duct, **663**
Accessory sex organs, **702**
Accessory structure of eye, 470–73
Accommodation of eye, **481**
Acetabular labrum, **229,** 230*f*
Acetabulum, **194,** 195*f*
Acetylcholine (ACh), 447
Achalasia, of the cardia, **460**
Achilles (calcaneal) tendon, **332,** 791, 793*f*
Achondroplasia, **147**
Acid hydrolases, 38
Acidic stains, **17**
Acidophils, **749**
Acinar cells, **663,** 664*f,* 758
Acne, 121, 124
Acquired immune deficiency syndrome (AIDS), **592–93**
Acquired umbilical hernias, **781**
Acromegaly, **760**
Acromial end, clavicle, **186,** 187*f*
Acromioclavicular joint, 783, 784*f*
Acromion, scapula, **188,** 783, 784*f*
Acrosomal reaction, fertilization and, **727**
Acrosome, **705**
Actin, **40**
Actin filaments, **40,** 100, **248,** 249*f*
Action potentials. *See* Nerve impulses (action potentials)
Adam's apple, **610,** 774
Addison's disease, **760**
Adduction movement, **221,** 222*f*
Adductor brevis muscle, 313*t,* 315*f*
Adductor hallucis muscle, 330*t,* 331*f*
Adductor longus muscle, 312*t,* 314*f,* 791, 792*f*
Adductor magnus muscle, 313*t,* 315*f*
Adductor muscles, thigh, 312–13*f,* 314*f,* 791, 792*f*
Adductor pollicis muscle, 309*t,* 310*f*
Adductor strain, muscle, **332**
Adductor tubercle, **199,** 201*f*
Adenine (A), 43
Adenocarcinoma, **626**
Adenocorticotropic hormone (ACTH), **748,** 749*t*
Adenohypophysis, **746, 747–51**
  aging and, 762
  cells and hormones of, 747, 748, 749*t*
  hypothalamic control of hormone secretion from, 750–51
  relationship between hypothalamus and, 750*f*
Adenoidectomy, **629**
Adenoids (pharyngeal tonsil), 608
Adenoma, **106**
Adenosine triphosphate (ATP), **34**
Adhesions, **105**
Adipocytes, 97
Adipose capsule, 677*f,* **678**
Adipose tissue, 93*f,* **97–98**
Adluminal compartment, **705,** 706*f*
ADPKD (autosomal dominant polycystic kidney disease), **695, 696–97**
Adrenal cortex, 744, **755–57**
  disorders of, 760, 761*f*
Adrenal (suprarenal) gland, **456,** 744, **754–57**
  cortex of (*see* Adrenal cortex)
  embryonic development of, 460*f,* 762*f*
  medulla of (*see* Adrenal medulla)
  role in sympathetic division, 456, 457*f*
Adrenaline (epinephrine), **457**
Adrenal medulla, **456,** 457*f,* 744, 754, **755**
Adrenergic fibers, 447
Adrenocorticotropic hormone (ACTH), 749*t*
Adulthood, blood vessels in, 578
Adult hypothyroidism, **760**
Adventitia, **612, 687,** 688*f*
Afferent arteriole, **686,** 687*f*
Afferent lymphatic vessels, **586**
Afferent signals. *See* Sensory (afferent) signals
Age-related macular degeneration (AMD), **484–85**
Ageusia, **497**
Aggregated lymphoid nodules (Peyer's patches), 595*f,* **598,** 599*f,* 654
Aging
  autonomic nervous system and, 460
  blood and, 517
  bones and, 147
  brain, spinal cord, and, 404
  cell development and theories of, 52–53
  digestive system and, 668
  endocrine organs and, 762
  growth hormone as treatment for, 752
  heart and, 539
  joints and, 238–39
  lymphatic and immune systems and, 599–600
  peripheral nervous system and, 437
  respiratory system and, 629
  skin and, 124–25
  tissues and, 106
  urinary system and, 695
  vertebral column and, 169, 181–82
  vision, hearing, and, 497
Agnosia, 367*f,* **371**
Agonist muscles, **268**
Agranulocytes, **506,** 507*f,* **508–10,** 511*t*
AIDS (acquired immune deficiency syndrome), **592–93**
  growth hormone as aid for, 752
Air-blood barrier, 614
Ala
  of ilium, **194,** 195*f*
  of nose, **770,** 775*f*
  of sacrum, **176,** 177*f*
Alar plate, **354,** 355*f*
Albumin, 505
Alcoholism, 572, 662
Aldosterone, **756**
Alimentary canal, **634–37**
  digestive processes in, 635–36
  as endocrine structure, 759
  tissues of, 638–40
Allantois, **667,** 694
Allergies, 91, 508
Alopecia, **125**
Alpha (A) cells, pancreatic, **758**
α (alpha) efferent neurons, **415**
Alveolar ducts, **614,** 615*f*
Alveolar exocrine glands, **87,** 88*f*
  sebaceous glands as, 120–21
Alveolar macrophages, **614,** 616*f*
Alveolar margin, mandible, **163,** 164*f*
Alveolar margin, maxillary bones, **163,** 164*f*
Alveolar pores, **614,** 616*f*
Alveolar sacs, **614**
Alveoli, **614,** 615*f,* 621*t*
Alzheimer's disease, **402–3**
AMD (age-related macular degeneration), **484–85**
Amino acid-based hormones, **745**
Amniocentesis, **72**
Amnion, **61**
Amniotic sac cavity, **61**
Amphiarthroses, **212**
Ampulla
  of ductus deferens, **709**
  of semicircular canals, 490*f,* **491**
  of uterine tube, 714*f,* **719**
Amygdala, **372,** 375*f,* 386–87
  control of autonomic nervous system by, 458
Amyloid-beta peptide (Abeta), 403
Amyloid precursor protein (APP), 403
Amyotrophic lateral sclerosis (ALS), **404**
Anabolic steroid use, effects of, 260

Anal canal, 655*f,* 656, **657**, 658
Anal columns, 655*f,* **657**
Anal fissure, **668**
Anal sinuses, 655*f,* **657**
Anal sphincters, **292**, 293*f,* 655*f,* **657**
Anal valves, 655*f,* **657**
Anaplasia, **53**
Anastomoses
portal-systemic, 572–73
vascular, 550–51
Anatomical neck, humerus, 189*f,* **190**
Anatomical position, **4**, 8*f*
Anatomy, **2–23**
clinical, 19–23
gross, 4–16
microscopic, 16–18
overview of, 2–4
surface (*see* Surface anatomy)
Anconeus muscle, 299*f,* 300*t*
Androgen-binding protein, **707**
Androgens, 142, **708**, 717, 749, **759**. *See also* Testosterone
hair and, 119, 120
muscle strength and, 261
sebum secretion and, 121
Anemia, **515**
Anencephaly, **403**
Aneurysm, **573–74**
Angina pectoris, **535**
Angiogram, **578**
Angiosarcoma, **578**
Angle, rib, **179**, 180*f*
Angles (canthi), **470**, 471*f*
Angular movements, **220–21**
Ankle joint, 233–34, 235*f*
muscles acting on, 319–28
Ankyloglossia, **641**
Ankylosing spondylitis, **239**
Ankylosis, **237**
Annular ligament, **227**, 229*f*
Annular ring, **472**, 473*f*
Annulospiral sensory endings, **415**
Annulus (anulus) fibrosus, 130, **168**, 169*f*
Anosmia, 420, **470**
Antagonist muscles, **268**
Anterior, defined, 8, 9*t*
Anterior arch, atlas vertebrae, **173**
Anterior axillary fold, 779*f,* **783**
Anterior cardiac vein of heart, 534*f,* **535**
Anterior cerebral artery, **555**, 556*f*
Anterior chamber of eye, 475*f,* **480**
Anterior circumflex humeral arteries, **557**, 558*f*
Anterior communicating artery, **555**, 556*f*
Anterior cranial fossa, **154**, 158*f,* 159*f*
Anterior crest, tibia, **202**, 203*f*
Anterior cruciate ligaments, **231**, 232*f,* 233*f*
Anterior cutaneous branches of intercostal nerves, 427*f,* **428**
Anterior division of brachial plexus trunk root, **429**, 430*f*
Anterior fontanels, **179**, 181*f*
Anterior funiculus, 394*f,* **396**
Anterior gluteal lines, **194**, 195*f*
Anterior horns, spinal cord, 394*f,* **395**
Anterior inferior iliac spine, **194**, 195*f*
Anterior intercostal arteries, **557**, 558*f*
Anterior interventricular artery, heart, **534**
Anterior interventricular sulcus, heart, **525**, 526*f*
Anterior lateral neck muscles, 283*t,* 284*f*
Anterior lobe, cerebellar hemisphere, **385**, 386*f*
Anterior longitudinal ligaments, **168**, 169*f*
Anterior median fissure, 394*f,* **395**
Anterior muscle compartment of leg, 791, 792*f*
Anterior pole, eyeball, **474**
Anterior semicircular canals, 490*f,* **491**
Anterior spinocerebellar pathway, **397**, 398*f,* 399*t*
Anterior spinothalamic tract, **397**, 398*f,* 399*t*
Anterior superior iliac spine, **194**, 195*f,* 779*f,* **781**
Anterior talofibular ligaments, **234**, 235*f*
Anterior tibial artery, 562*f,* 563
Anterior tibial vein, **571**, 572*f*
Anterior tibiofibular ligaments, **234**, 235*f*
Anterior triangle of neck, **777**
Anteverted uterus, 714*f,* **720**
Antibodies, 91, **509**
Antidiuretic hormone (ADH), 749*t,* **752**
Antigen, **508**, 589, 592
Antigen challenge, 592
Antigen presenting cell, 592
Antrum, 717*f,* **718**
Anucleate cells, 42
Anus, 635, 656, 657
Aorta, **553**, 554*f*
Aortic aneurysm, 573–74
Aortic bodies, **623**, 624*f*
Aortic lymph nodes, 586, 587*f*
Aortic plexus, **456**
Aortic valve of heart, 528, **529**, 530*f*
Apex
of heart, **522**, 523*f*
of lung, **617**, 618*f*
of nose, **770**, 775*f*
Apical foramen, **645**
Apocrine glands, **121**
Aponeurosis, 98, 247, 248
Apoptosis, 50, **53**
Appendectomy, **656**
Appendicitis, **656**, 781
Appendicular region, **8**
Appendicular skeleton, **152**, **185–210**, 771*f*
bone markings for, 153*t*
disorders of, 191, 193, 205–6
lower limb, 199–205
pectoral girdle, 186–89
pelvic girdle, 194–99
upper limb, 189–94
Appendix, 595*f,* **598**, 638, 656
Appositional growth
of bone, **142**
of cartilage, **130**
Aqueous humor, eye, 474, **480**
Arachnoid mater, **389**
Arachnoid villi (arachnoid granulations), **389**
Arches, foot, 205
Arcuate arteries, 562*f,* **563**, **678**, 679*f,* 715*f,* **722**
Arcuate popliteal ligament, **231**, 232*f*
Arcuate veins, **679**
Areola, **726**
Areolar connective tissue, 89–91, 92*f,* 97
fibers providing support for, 90
functions of, 89–90
Arm. *See also* Forearm
arteries of, 557, 558*f*
bones of, 189–90, 191*f,* 192*f,* 196*t*
innervation of, 429–32
lymphedema of, 586
muscles acting on, 297–98*t,* 299*f,* 307*t*
surface anatomy of, 783, 784*f,* 786*f*
veins of, 567, 568*f*
Arrector pili muscle, **119**
sympathetic pathways to, 454, 455*f*
Arterial anastomoses, 550–51
Arterial pulse, points on body surface for feeling, 774*f*
Arteries, 504, 544, **545–47**
abdominal, 559–61
of adrenal glands, 754
aorta, 553, 554*f*
arterioles, 546–47
coronary heart, 534–35, 551
disorders of, 534–35, 550, 551, 573–76
elastic, 546
of head and neck, 553–57, 774, 776, 777*f*
kidney (*see* Renal arteries)
muscular, 545*f,* 546
nutrient, in bones, 134, 135*f*
ovarian, 559*f,* 560, 715
of pelvis and lower limb, 561–63
of penis, 713
of pituitary gland, 746, 748*f*
pulmonary, 552–53
of retina, 474, 479
of small intestine, 652
of stomach, 648
structure of, 544, 545*f*
systemic, 553–63, 564*t*
of testes, 704
of thorax, 557–59
of upper limb, 557, 558*f*
of urinary bladder, 689
uterine, 715, 722
Arterioles, 544, **546–47**
Arteriosclerosis, **574**
Arteriovenous malformation, **575–76**
Arthritis, **236–37**
Arthroplasty, **239**
Arthroscopic surgery, **236**
Articular capsule, synovial joint, **215**
Articular cartilage, 130, 131*f*
synovial joint, **214**, 215*f*
Articular disc, synovial joint, 215*f,* **217**
Articular surfaces, synovial joint stability and, 219
Articular tubercle, **226**
Articulations, **152**, **212**. *See also* Joint(s)
Artifacts, **18**
Artificial joints, **238**
Arytenoid cartilages, 609*f,* **610**, 611*f*
Ascending colon, 655*f,* **656**
Ascending (sensory) pathways, spinal cord, 397, 398*f,* 399*t*
Ascites, **668**
Association areas, cerebral cortex, 370–71
Association fibers, **372**, 374*f*
Asthma, 624
Astigmatism, **482**
Astrocytes, 346*f,* **347**
Asystole, **540**
Atelectasis, **629**
Atheromas, **574**
Atherosclerosis, 32, 402, 551, 573, **574–75**
sex differences and, 578
Atherosclerotic plaques, **574**
Athlete's foot, **125**
Atlas vertebrae, **173**, 175*f,* 176*f*
Atonic bladder, **461**
ATP (adenosine triphosphate), **34**
ATPase enzymes, 248
Atria, **522**
Atrial fibrillation, **537**
Atrial natriuretic peptide (ANP), **759**
Atrioventricular bundle, 532, **533**
Atrioventricular (AV) node, 532, **533**
Atrioventricular valves, 529
function of, 531*f*
left, **529**, 530*f,* 531*f*
right, 525, **529**, 530*f*
Atrium of heart, **525–28**
development of, **538**
left, **525**, 526*f,* 527*f,* 528
right, **525**, 526*f,* 527*f*

Atrium of respiratory zone, **614,** 615*f*
Auditory areas, cerebral cortex, 367*f*, **370**
Auditory association area, 367*f*, **370**
Auditory canal, **486,** 487*f*
Auditory pathway, **494,** 495*f*
Auricle, ear, **486,** 487*f*, 774, 775*f*
Auricular surface, ilium, **194,** 195*f*
Auscultation, triangle of, 779
Autograft (skin graft), 122
Autoimmune diseases, **237,** 353
Autologous cartilage implantation, **236**
Autonomic ganglion, **445,** 450*f*
Autonomic nerve plexuses, 449, 450*f*
Autonomic nervous system (ANS), 338, 410, **443–64**
 central control of, 457–58
 defined, **444**
 disorders of, 459–60, 461
 divisions of, 445–47
 thoughout human lifespan, 460
 hypothalamus and control of, 377
 innervation of heart and, 533, 534*f*
 parasympathetic division, 447–57
 somatic motor system compared to, 444–45
 visceral reflexes and, 459
 visceral sensory neurons of, 458, 459*f*
Autosomal dominant polycystic kidney disease (ADPKD), **695, 696–97**
Axial muscles, 271*f*, **272**
Axial region, **8**
Axial skeleton, **152,** 771*f*
 bone markings for, 153*t*
 bony thorax (thoracic cage), 177–79
 disorders of, 179–81
 throughout human lifespan, 179–81
 skull, 154–67
 vertebral column, 168–77
Axilla, surface anatomy of, 779*f*, 783
Axillary folds, 779*f*, **783**
Axillary lymph nodes, 586, 587*f*, 726, 783
Axillary nerve, brachial plexus, **431**
Axillary vein, **567,** 568*f*
Axis vertebrae, **173,** 175*f*, 176*f*
Axodendritic synapses, **341**
Axon(s), 338, 339*f*, **340**
 connection of, to reticular activating system, 388
 motor, 416–17
 nerve impulses (action potentials) on, 342, 343*f*
 post ganglionic, 445
 preganglionic, 445
 tracts as bundles of, 353
Axonal transport, **340**
Axon collaterals, 339*f*, **340**
Axoneme, **83**
Axon hillock, **340**
Axon terminals, 339*f*, **340,** 342*f*
Axosomatic synapses, **341**
Azurophilic granules, 507
Azygos system, 568*f*
Azygos vein, 568, 569

Back
 innervation of, 428
 intrinsic muscles of, 284*f*, 285*t*, 286*f*
 pain in, 169, 172, 261
 surface anatomy of, 777–79
Baldness, 120
Ball-and-socket joint, 225*f*, **226**
Balloon angioplasty, **574**
Band cells, **513,** 514*f*
Bare area of liver, **658,** 659*f*, 660*f*
Basal body, **83,** 84*f*
Basal cell(s), special senses and, **466, 468**
Basal cell carcinoma, **123,** 124*f*
Basal compartment, **705,** 706*f*
Basal lamina, 77, **86,** 547
Basal nuclei (basal ganglia), 364*f*, **365, 372–74,** 375*f*
Basal plate, **354,** 355*f*
Base
 of axilla, 779*f*, **783**
 of heart, **522,** 523*f*
 of lung, **617,** 618*f*
Basement membrane, 77, **86**
Bases, DNA nucleotide, 43
Basic stains, **17**
Basilar arteries, **555,** 556*f*
Basilar membrane, **493**
Basilic vein, **567,** 568*f*, 786*f*, 787
Basioccipital bone, **157**
Basophils, 506, 507*f*, **508,** 511*t*, **749**
B cells (lymphocytes), **508,** 509*f*, 589, 599
Behavior, hypothalamus and control of, 377
Bell's palsy, **423**
Benign prostatic hyperplasia (BPH), **711**
Benign tumor, **50**
Beta (B) cells, pancreatic, 758*f*, **759**
Biaxial movement, **226**
Biceps brachii muscle, 299*f*, 300*t*, 783, 784*f*
Biceps femoris muscle, 316*f*, 318*t*, 791, 793*f*
Bilaminar embryonic disc, **61,** 537*f*
Bilateral symmetry, **12**
Bile, **658**
Bile canaliculi, 661*f*, **662**
Bile duct, 652*f*, **660,** 661*f*, **662**
Biopsy, **50**
Bipennate arrangement of fascicles, **268,** 269*f*
Bipolar cells, retina, 477*f*, **478**
Bipolar neurons, **343,** 344*f*
Birth, premature, 68
Birth defects, 68, **70–71**
Bladder. *See* Urinary bladder
Bladder cancer, **693**
Blastocyst, **60**
 implantation of, 60, 61*f*, 728*f*, 730*f*
Blastomeres, **59**
Blister, **113**
Blood, 97*f*, **99, 503–20**
 blood plasma, 97*f*, 99, 504–5
 bone marrow transplants, 516
 chemistry of, 623
 clotting of, 103
 composition of, 504
 disorders of, 515–16, 517
 formation of, 133, 512–15
 formed elements in, 505–11
 glucose levels in, and stress, 756
 throughout human lifespan, 516–17
 pathway of, through heart, 528–29
 supply of, to heart, 534–35
 supply of, to lungs, 619
Blood-brain barrier, **391,** 549
Blood cells (formed elements), **505–11**
Blood islands, 516, 537
Blood plasma, 97*f*, 99, **504–5**
Blood pressure, 382
 vasoconstriction of arterioles and rising, 446, 546–47
Blood sinusoids, 513
Blood stem cell, **513,** 514*f*
Blood-testis barrier, **705,** 706*f*
Blood-thymus barrier, 597
Blood vessels, 504, **543–82.** *See also* Arteries; Capillaries, blood; Vein(s)
 bone markings for depressions/openings allowing, 153*t*
 disorders of, 550, 573–76
 throughout human lifespan, 576–78
 of large intestine, 657
 microscopic, associated with uriniferous tubules, 685–86
 of pulmonary circulation, 552–53
 in retina, 479
 in skeletal muscles, 247
 structure of walls of, 544, 545*f*
 sympathetic pathway to peripheral, 454, 455*f*
 of systemic circulation, 553–73
 in synovial joints, **216**
 types of, 544–50
 vasa vasorum of, 546*f*, 551
 vascular anastomoses of, 550–51
Blue baby, **578**
Body
 of hyoid bone, **167**
 of ilium, **194,** 195*f*
 of ischium, **194,** 195*f*
 of mandible, **163,** 164*f*, 774, 775*f*
 of nail, **121,** 122*f*
 of penis, 710*f*, **712**
 of pubis, 195*f*, **198**
 of sphenoid bone, **160,** 161*f*
 of sternum, **177,** 779
 of stomach, **647,** 648*f*
 of uterus, 714*f*, **720**
 of vertebrae, **172**
Body, human. *See also* Female body; Male body
 abdominal regions and quadrants of, 15–16, 17*f*
 anatomical position of, 4, 8*f*
 body plan of, 12–13, 59*f*
 cavities and membranes of, 13–15, 59
 directional and regional terms applied to, 8, 9*t*, 10*f*
 hierarchy of structural organization in, 3–4
 movement of, 220–24, 385
 organ systems of, 4, 5–7*f* (*see also* name of specific system, e.g., Cardiovascular system)
 planes and sections of, 10, 11*f*, 12
 skeleton, 152*f* (*see also* Skeletal system)
 temperature (*see* Temperature regulation)
 surface anatomy (*see* Surface anatomy)
 upper-lower body ratio, 205, 206*f*
Body periphery, sympathetic pathways to, 454, 455*f*
Body stalk, **728,** 729, 730*f*
Boils, **125**
Bolus of food, 641
Bone(s), **132–47, 152**
 anatomy of epiphyseal growth areas, 140–41
 of appendicular skeleton (*see* Appendicular skeleton)
 of axial skeleton (*see* Axial skeleton)
 of back, surface anatomy of, 777–79
 classification of, 133–34
 development and growth of, 139–42
 disorders of, 143–47
 fractures of (*see* Fractures, bone)
 functions of, 132–33
 gross anatomy of, 134–36
 throughout human lifespan, 147
 postnatal growth of endochondral, 141–42
 relationship of muscles to, 266–68

Bone(s), *continued*
  repair of, 143, 146*f*, **206–7**
  of thorax, surface anatomy of, 779–80
  tissue of (*see* Bone tissue)
Bone deposition, **142**
Bone graft, **147**
Bone markings, **153*t***, 247
Bone marrow, formation of blood and, 133, **512–13**
Bone marrow biopsy, **517**, 790
Bone marrow transplant, **516**
Bone reabsorption, **142**, 143*f*
Bone remodeling, **142–43**
Bone substitutes, **206**
Bone tissue, 96*f*, **99**, 136–38
  chemical composition of, 137
  microscopic structure of, 136–37, 138*f*
  ossification of, 139–40
Bony callus formation, bone fracture repair and, **143**, 146*f*
Bony labyrinth, inner ear, **489**, 490*f*
Bony pelvis, **194**, 195*f*
Bony spur, **147**
Bony thorax (thoracic cage), **177–79**, 181
  ribs, 178–79
  sternum, 177–78
Bowel sounds, **783**
Bowman's capsule (glomerular capsule), **682**, 683*f*, 687*f*
Brachial artery, 557, 783, 787
Brachialis muscle, 299*f*, 300*t*
Brachial plexus, **429–32**, 777
  branches of, 431*t*
  injuries of, 431
  roots, trunks, divisions, and cords of, 430*f*
Brachial pulse, 783
Brachial vein, **567**, 568*f*
Brachiocephalic trunk, **553**, 555*f*
Brachiocephalic veins, 564, 565*f*, 568
Brachiomeric muscles, 271*f*, **272**
Brachioradialis muscle, 299*f*, 300*t*, 304*t*, 306*f*, 786*f*, 787
Brain, 66, 336, **360–91**
  arteries of head and, 553–57
  basic parts and organization of, 362
  brain stem, 362, 379–82, 383*f*, 384*f*
  cerebellum, 362, 382–85
  cerebral hemispheres, 362, 363–74
  diencephalon, 362, 374–79, 380*f*, 383*f*
  dysfunction and disease of, 371, 372, 374, 385, 401–3
  embryonic development of, 360, 361*f*, 403
  functional systems of, 385–88
  functions of major parts of, 390*t*
  myelin sheaths in, 348
  protection of, 388–91
  sex-related differences in structure of, 373
  venous drainage of, 566, 567*f*
  ventricles of, 363
Brain stem, 362, **379**–82, 383*f*, 384*f*
  control of autonomic nervous system by, 457–58
  embryonic development of, **360**, 361*f*
  functions of, 390*t*
  medulla oblongata, 381–82
  midbrain, 379–81
  pons, 381
Brain tumors, 347
Branchial groove, **496**
Branchial motor area of nervous system, **338**, 410*f*, 444*f*
Breast cancer, 726, **734**
Breasts (mammary glands), 725–26, 738
Breathing, **621–23**
  mechanisms of, 621–23
  muscles of, 287*t*, 288*f*
  neural control of, 623
  rhythm and rate of, 382
Broad ligament, 714*f*, **715**
Broca's area, brain, 367*f*, **368–69**
Brodmann areas, **365**, 367*f*
Bronchi, 613–14
Bronchial arteries, 559, 619
Bronchial asthma, **624**
Bronchial tree, 613–14, 621*t*
Bronchial veins, 569, 619
Bronchioles, **614**
Bronchitis, 625
Bronchomediastinal (lymph) trunks, **589**
Bronchopulmonary dysplasia, 623
Bronchopulmonary segments of lungs, **617**, 620*f*
Bronchoscopy, **629**
Brown adipose tissue, **98**
Bruises, 116
Brunner's glands (duodenal glands), 653*f*, 654
Bruxism, **669**
Buccinator muscle, 277*t*, 278*t*, 279*t*
Buffy coat, blood, 504
Bulb of the penis, 710*f*, **713**
Bulbospongiosus muscle, 292*f*, 293*f*
Bulbourethral glands, 702, 710*f*, **712**
Bulbs of the vestibule, 725
Bulbus cordis, **538**
Bulk transport, 30
Bulla, **113**
Bundle branches of heart, 532, **533**
Bunion, **206**
Burns, **122–23**
Bursa, **219**
Bursitis, **236**
Buttocks, prominences of, 789*f*, 790
Cadherins, **84**
Calcaneal (Achilles) tendon, 791, 793*f*
  rupture of, **332**
Calcaneofibular ligament, **234**, 235*f*
Calcaneus, **202**, 204*f*
Calcarine sulcus, 367*f*, **369**
Calcified cartilage, **131**
Calcitonin, **754**
Calcium
  in bones, 137, 142, 143, 144–45
  muscle contraction and role of, 251, 254, 257
  parathyroid gland and concentrations of, 745, 746*f*, 754
  storage of, in endoplasmic reticulum, 36
  thyroid gland, hormone calcitonin, and blood levels of, 754
  vitamin D and uptake of, 117, 146, 759–60
Calculus, **645**
Callus, **113**
Calorie restriction, slowed aging linked to, 52
Canaliculi, **137**, 138*f*
Cancer, **50–51**
  adenomas and carcinomas, 106, 626
  angiosarcoma, 578
  bone, 146–47
  bone marrow, 515
  breast, 726, 734
  cervical, 734
  colon, 51*f*
  diagnosis, staging, and treatment of, 50–51 (*see also* Chemotherapy)
  endometrial, 733–34
  genesis of, 50
  gliomas, 347
  Kaposi's sarcoma, 592
  leukemia, 515
  lung, 623, 626–27
  of lymphoid tissues, 599
  ovarian, 733
  prevalence and causes of, 50
  prostate, 711, 733
  sarcomas, 106
  skin, 117, 123–24
  testicular, 733
  thyroid, 763
  of urinary organs (bladder, kidney), 693
Canines (teeth), **644**, 645*f*
Capillaries, blood, 504, 544, **547–50**
  beds of, 547
  permeability of, 547–49
  relationship of lymph capillaries to, 585*f*
  structure of, 544, 545*f*, 548*f*
Capillaries, lymph. *See* Lymph capillaries
Capillary beds, **547**
Capillary network, skeletal muscle, 247*f*
Capitate bone, **191**, 193*f*
Capitulum, humerus, 189*f*, **190**
Capsular space, glomerular, 681*f*, **682**, 683*f*
Capsule, lymph node, **586**, 588*f*
Carbuncles, **125**
Carcinogenesis, 50
Carcinoma, **106**
Cardiac catheterization, **540**
Cardiac center in reticular formation of medulla, 381–82, 457, 534
Cardiac muscle tissue, **100**, 101*f*, **244**, **254**, 256*f*
  cardiac muscle bundles, 525*f*
  conducting system of, 532–33
  fibrosis of, 539
  smooth and skeletal muscle compared to, 245–46*t*
Cardiac notch, **617**, 618*f*
Cardiac orifice, **646**, 648*f*
Cardiac plexus, **449**
Cardiac reserve, decline in, 539
Cardiac sphincter, 646
Cardiac tamponade, **524**
Cardiac veins of heart, 534*f*, **535**
Cardia (cardiac region) of stomach, **647**, 648*f*
Cardioacceleratory center, **534**
Cardioinhibitory center, **534**
Cardiomyopathy, **540**
Cardiovascular system, 4, 6*f*, **504**, 522–42
  blood vessels of, 552–78
  heart and, 522–42
Caries (cavities), **645**
Carina of trachea, **612**
Carotene, **116**
Carotid arteries, 553, 555*f*, 623, 624*f*, 776, 777*f*
Carotid bodies, **623**, 624*f*
Carotid canal, **160**
Carotid endarterectomy, **578**
Carpal arch, 557
Carpals, **191**
Carpal tunnel syndrome, 191, **193**
Carpus, **190–91**, 193*f*
Cartilage(s), **99**, **130–32**, **152**
  arytenoid, corniculate, and cuneiform, 610, 611*f*
  cricoid, 610
  elastic, 95*f*, 99, 130, 131*f*
  fibrocartilage, 96*f*, 99, 130, 131*f*
  growth of, 130–31
  hyaline, 95*f*, 99, 130, 131*f*
  location and basic structure of, 130, 131*f*
  organization of, within epiphysis of growing bone, 140–41, 142*f*
  properties of, 132
  thyroid, 609–10
  torn, 236
Cartilage bones, **139**

Cartilaginous joints, **212, 213**
characteristics of, 216–18*t*
summary of, 212*t*
symphyses, **213,** 214*f*
synchondroses, 213, 214*f*
CAT (computed axial tomography), **19,** 20*f*
Catalases, 39
Cataract, **480**
Catheter, **692**
Cauda equina, **395**
Caudal, defined, **360**
Caudate lobe of liver, 659*f*, **660**
Caudate nucleus, **372,** 375*f*
Caveolae, **33,** 257, 549
Cavernous sinuses, **566,** 567*f*
Cavities (caries), **645**
CBC (complete blood count), **510**
$CD8^+$ T lymphocyte, **509,** 589
Cecum, 655*f*, **656**
Celiac ganglion, 452, 453*f*, **456**
Celiac plexus, **449**
Celiac trunk, 559, 560*f*
Cell(s), 4, **27–56**
cancer as uncontrolled division of, 50–51
cytoplasm of, 33–41, 44*t*
developmental aspects of, 52–53
diversity of, 48, 49*f*, 52
introduction to, 28–29
life cycle of, 45–48
nucleus of, 41–45
plasma membrane of, 29–33, 44*t*
structural organization at level of, 3*f*, **4**
structure of generalized, 28*f*
Cell body of neuron, **338,** 339*f*
graded potentials on, 342
gray matter as clustered, 352
Cell differentiation, **52**
Cell division, 40, **45–48**
Cell junctions, **84–86**
Cell life cycle, **45–48**
Cellular defenses, 91
Cellular organelles, 4. *See also* Organelles
Cementum, **645**
Centimeter, **4**
Central arteries, spleen, **595,** 596*f*
Central artery of retina, 474*f*, **479**
Central canal (Haversian canal), **137,** 138*f*
Central canal, spinal cord, 394*f*, **395**
Central chemoreceptors, 623, 624*f*
Central nervous system (CNS), **336,** 337*f*
brain, 360–91 (*see also* Brain)
disorders of, 401–3
embryonic development of, 354, 355*f*, 403
gray matter and white matter of, 352–53
throughout human lifespan, 403–4
mylelin sheaths in, 348, 349*f*
neuron cell bodies in, 340*f* (*see also* Neuron(s) (nerve cells))
simplified design of, 351–53
spinal cord, 391–401 (*see also* Spinal cord)
supporting cells in, 346–47
Central process, **343,** 345
Central sulcus, cerebral hemisphere, **363,** 364*f*
Central tendon, **292**
Central vein of liver, **660,** 661*f*
Centrioles, **40,** 41*f*, 44*t*
Centrosome, **40,** 41*f*
Centrosome matrix, **40**
Centrum, vertebrae, **172**
Cephalic vein, **567,** 568*f*, 786*f*, 787
Cerebellar hemispheres, **385,** 386*f*
Cerebellar peduncles, **379, 381,** 382*f*, 383*f*, **385,** 386*f*
Cerebellum, 362, **382–85**
embryonic development of, **360,** 361*f*, 403
functions of, 390*t*
regions of, 385, 386*f*
Cerebral aqueduct, **363,** 379
Cerebral arterial circle (circle of Willis), **555,** 556*f*
Cerebral arteries, 555, 556*f*
Cerebral cortex, 364*f*, **365–72**
association areas of, 370–71
control of autonomic nervous system by, 458
functional and structural areas of, 367*f*
lateralization of, 371–72
motor areas of, 366–69, 385
relationship of, to thalamus, 375, 378*f*
sensory areas of, 369–70
visual pathway to, 481–83, 484*f*
Cerebral hemispheres, 362, **363–74**
embryonic development of, **360,** 361*f*, 403
functions of, 390*t*
grooves of, 363, 364*f*
lateralization of function in, 371–72
lobes of, 363, 364*f*
Cerebral palsy, **404**
Cerebral peduncles, 379
Cerebral white matter, 364*f*, **365, 372**
Cerebrospinal fluid (CSF), 347, 388, **391**
formation and circulation patterns of, 392*f*
Cerebrovascular accidents (CVAs), **402**
Cerebrum, **360**
Cerumin, 486
Ceruminous glands, 121, 486
Cervical canal, 714*f*, **720**
Cervical cancer, **734**
Cervical curvature, 168*f*, **172**
Cervical enlargement, 393*f*, **395**
Cervical ganglia, **452,** 453*f*
Cervical glands, 720
Cervical lymph nodes, 586, 587*f*
Cervical plexus, **428–29,** 777
branches of, 429*t*
Cervical vertebrae, 168*f*, **172, 173,** 174*t*, 175*f*, 176*f*
Cervix, 714*f*, **720**
Chagas' disease, **461**
Chalazion, **471**
Chancre, **712**
Charley horse, **332**
Cheeks, **640**
Chemical digestion, **636**
Chemical level, structural organization at, 3–4
Chemical senses, 466–70
disorders of, 470
smell, 468–70
taste, 466–68
Chemoreceptors, **413, 466**
for blood chemistry, 623, 624*f*
for smell, 468–70
for taste, 466–68
Chemotherapy, 48, 51
hair loss and, 120
Chewing. *See* Mastication
Chief (zymogenic) cells
of gastric glands, 649*f*, **650**
of parathyroid glands, **754,** 755*f*
Childbirth, 729–32
episiotomy in, 725
oxytocin and, 752
Childhood polycystic kidney, **695**
Chlamydia, **713,** 720
Chlamydia trachomatis, 485, 713
Cholecystokinin, 654
Cholesterol
gallstones and, 662
hypercholesterolemia, 32
in plasma membrane, 29*f*, 30
Cholinergic fibers, 447
Chondroblasts, 99, 142*f*
Chondrocyte, **99,** 142*f*
Chondromalacia patella, **239**
Chorion, **728,** 730*f*
Chorionic villi, **728,** 730*f*
Choroid, 474*f*, **476**
Choroid plexus, 381, **391,** 392*f*
Chromaffin, **755**
Chromatin, **42–45**
arrangement of DNA and histones in, 43*f*
in neutrophils, 507
Chromatophilic (Nissl) bodies, neuron, **338,** 339*f*
Chromophobes, **750**
Chromosomes, **45**
telomeres on ends of, 52–53
Chronic bronchitis, **625**
Chronic lung disease, 623
Chronic obstructive pulmonary disease (COPD), **624–25**
Chyle, 585, 598
Chylomicrons, 654
Chylothorax, 598
Chyme, **646**
Cilia, **83,** 84*f*
Ciliary body, 474*f*, **476**
Ciliary ganglion, **448**
Ciliary muscle, 475*f*, **476**
Ciliary processes, **476**
Ciliary zonule, 475*f*, **476**
Cineradiography, **19**
Cingulate gyrus, 386, **387**
Circadian rhythms, **377,** 758
Circle of Willis (cerebral arterial circle), **555,** 556*f*
Circular folds (plicae circulares) of small intestine, **653**
Circular layer, smooth muscle, **254**
Circular pattern of fascicles, **268,** 269*f*
Circulatory system, **504.** *See also* Blood; Cardiovascular system; Lymphatic system
Circumcision, **712–13**
Circumduction movement, **221,** 222*f*
Circumferential lamellae, **137,** 138*f*
Circumflex artery, heart, 534*f*, **535**
Circumvallate papillae, 641, **642**
Cirrhosis of liver, **662**
Cis-Golgi network, 37*f*, **38**
Cisterna chyli, **589**
Cisternae, rough ER, **35**
Clathrin, **32**
Clavicles, **186,** 187*f*, 780, 783, 784*f*
Cleavage, **59,** 60*f*
Cleft palate, **179–80,** 182*f*
Clinical anatomy, **19–23**
Clitoris, 724*f*, **725**
Cloaca, **667,** 694
Cloacal membrane, **667,** 668*f*
Closed reduction, bone fracture treatment by, **143**
Clot, blood, **510**
Clotting
blood, 510
protein, 103
Clubfoot, **205**
Coated pit, **32**
Coated vesicle, **32**
Coccygeus muscle, 292*t*, 293*f*
Coccyx, 168*f*, **172, 177**
Cochlea, inner ear, 489, **491–94**
anatomy of, 493*f*
role of, in hearing, 494*f*
Cochlear duct, 489, **493**
Cochlear implants, **495**
Cochlear nerve, **491,** 493*f*
Cochlear nuclei, **381,** 384*f*, **494,** 495*f*
Coelom, **62**
Cognition, **370–71,** 385
Cold sores, **125**
Colic arteries, 560, 561*f*
Collagen fibers, **90,** 115
in compact bone osteons, 137
Collarbones (clavicles), 186, 187*f*

Collateral channels for blood flow, 550
Collecting tubules, nephron and, 680*f*, **684–85**
Colles' fracture, **787**
Colloid in thyroid gland, 753*f*, **754**
Colon, 655*f*, **656**
Colon cancer, development of, 51*f*
Colony-forming units (CFUs), 513
Columnar epithelia, 77*f*, **78**, 81*f*
  pseudostratified, 78, 80*f*
  simple, 78, 80*f*
Comminuted bone fracture, 144*t*
Commissural fibers, **372**, 374*f*
Commissures, **372**, 374*f*, 395
Common fibular nerve, 432, **434**, 791
Common hepatic artery, **560**
Common hepatic duct, **660**
Common iliac arteries, 561
Common iliac vein, **571**, 572*f*
Common interosseous artery, **557**
Communicating arteries, 555, 556*f*
Compact bone, **134**, 135*f*
  microscopic structure of, 137, 138*f*, 139*f*
Complete blood count (CBC), **510**
Composite tissues, 99
Compound fracture, bone, **143**
Compound glands, **87**, 88*f*
Compression bone fracture, 144*t*
Computed axial tomography (CAT), **19**, 20*f*
Computed tomography (CT), **19**, 20*f*
Conception, **59**, 60*f*
Concha of ear, **774**, 775*f*
Concussion, brain, **402**
Condensed chromatin, **42**
Conducting arteries, 546
Conducting system of heart, **532**, 533*f*
  disorders of, 537
Conducting zone, bronchi in, 613–14
Conduction deafness, **495**
Conduction myofibers, 533
Condylar process, mandible, 163
Condyles
  femur, **199**, 791, 792*f*
  tibia, **202**, 791, 792*f*
Condyloid joints, **225–26**
Cone cells, **478**
Congenital abnormality, **70**
  of heart, 537, 539*f*
Congenital deafness, **497**
Congenital inguinal hernias, **737**
Congenital megacolon, **460**
Congestive heart failure, **535**
Conjoined (Siamese) twins, **72**
Conjunctiva, **471**
Conjunctival sac, **471**
Conjunctivitis, **471**
Connective tissue, **88–99**
  blood, 97*f*, 99 (*see also* Blood)
  bone tissue, 96*f*, 99 (*see also* Bone(s))
  cartilage, 95*f*, 96*f*, 99
  classes of, 89*f*
  connective tissue proper, 88–98
  skeletal muscle held by, 244, 246*f*, 247
Connective tissue papilla, **117**, 119
Connective tissue root sheath, 118*f*, **119**
Conoid tubercle, clavicle, **186**
Continuous capillary, 548*f*
Contrast medium, **19**
Contusion, brain, **402**
Conus medullaris, 393*f*, **395**
Convergent pattern of fascicles, **268**, 269*f*
Coracobrachialis muscle, **297**, 298*t*, 299*f*
Coracohumeral ligament, **227**, 228*f*
Coracoid process, scapula, **188**
Cordae tendineae of heart, **525**, 526*f*, 529
Cord-blood transplants, **516**
Cords of brachial plexus, **429**, 430*f*
Cornea, 474*f*, **475**, 547
Corneal transplant, **476**
Corniculate cartilages, 609*f*, **610**, 611*f*
Cornified (horny) cells, 115
Coronal (frontal) plane, **10**, 11*f*
Coronal suture, **154**, 156*f*
Corona radiata, **372**, 716*f*, **718**
Coronary arteries of heart, 534–35, 551
Coronary artery disease, **535**
Coronary bypass surgery, 558, 572, **574**
Coronary sinus of heart, 525, 534*f*, **535**
Coronary sulcus of heart, **525**, 526*f*
Coronoid fossa, humerus, 189*f*, **190**
Coronoid process
  mandible, **162**, 164*f*
  ulna, **190**, 192*f*
Corpora cavernosa, 710*f*, **713**, 725
Corpora quadrigemina, **380**, 383*f*
Cor pulmonale, **535**
Corpus albicans, 717*f*, **718**
Corpus callosum, **372**, 374*f*
Corpus luteum, 717*f*, **718**
Corpus spongiosum, 710*f*, **713**
Corpus striatum, **372**, 375*f*
Corrugator supercilii muscle, 276*t*, 277*f*
Cortex
  brain, 362, 365–72, 385
  hair, **117**, 118*f*
  lymph nodes, 594*f*, **595**
Cortex of thymus, **597**
Cortical nephrons, **684**
Cortical reaction, fertilization and, **727**
Corticospinal tracts. *See* Pyramidal (corticospinal) tracts
Corticosteroids, **756**
Corticotropic cells, **748**
Corticotropin-releasing hormone (CRH), 750, 756
Cortisol, **756**
Costal cartilages, **130**, 131*f*
Costal groove, rib, **179**, 180*f*
Costal margin, **178**, 780
Costocervical trunk, 553, 556*f*, **557**
Covering membranes, **99**, 100*f*
Coverings of heart, 524
Coxal bone, **194**, 195*f*
Coxal (hip) joint, **229–31**
Cranial base (floor), 154
Cranial bones, **154–62**, 170–71*t*
Cranial cavity, **13**, 14*f*
Cranial fossa, **154**, **157**, 158*f*, 159*f*, **160**
Cranial nerve(s), 154, 336, **418–26**, 566
  I (olfactory), 370, 418, 419*t*, 420*t*
  II (optic), 418, 419*t*, 420*t*, **481**, 484*f*
  III (oculomotor), 380, 418, 419*t*, 420*t*, 448
  IV (trochlear), 380, 418, 419*t*, 421*t*
  V (trigeminal), 418, 419*f*, 421–22*t*
  VI (abducens), 418, 419*t*, 422*t*
  VII (facial), 160, 418, 419*t*, 423*t*, 448, 466–68, 643
  VIII (vestibulocochlear), 160, 418, 419*t*, 424*t*
  IX (glossopharyngeal), 381, 419, 424*t*, 448, 468
  X (vagus), 381, 419, 425*t*, 449, 450*f*
  XI (accessory), 381, 419, 425*t*, 777
  XII (hypoglossal nerve), 157, 381, 419, 426*t*
  summarized by functional group, 419*t*
Cranial nerve nuclei, 381, 383*f*
Cranial outflow of parasympathetic division, 448–49, 450*f*
Cranial sensory ganglia, **419**
Cranial vault (calvaria), 154
Craniosacral (parasympathetic) divsion of autonomic nervous system, **446**
Craniotomy, **182**
Cranium, **154**. *See also* Cranial bones
  surface anatomy of, 770, 775*f*
Creatinine, 676
Cremaster muscles, **703**
Cretinism, **760**
Cribriform plates, **162**
Cricoid cartilage, 609*f*, **610**, 611*f*, 774
Cricothyroid ligament, 774
Crista ampullaris, **491**, 492*f*
Cristae, **33**, 34*f*
Crista galli, **162**
Crista terminalis of heart, **525**, 526*f*
Crohn's disease, 666
Croup, **629**
Crown of tooth, **644**, 645*f*
Cruciate ligaments, 231, 232*f*, 233*f*
Crura of the penis, 710*f*, **713**
Cryptochidism, **737**
Crypts, tonsil, **598**
Crypts of Lieberkuhn, 654
CT (computed tomography), **19**, 20*f*
Cubital fossa, 786*f*, **787**
Cubital veins, 567, 786*f*, 787
Cuboidal epithelia, 77*f*, **78**
  simple, 78, 79*f*
Cuboid tarsal bone, **202**, 204*f*
Cuneiform cartilages, 609*f*, **610**
Cuneiform tarsal bones, **202**, 204*f*
Cupula of crista ampullaris, **491**, 492*f*
Cushing's disease (Cushing's syndrome), **760**, 761*f*
Cutaneous branches of intercostal nerves, 427*f*, **428**
Cutaneous membrane, **99**, 100*f*
Cutaneous nerves, **429**
Cuticle, **117**, 119*f*
Cyanosis, **117**
Cystic duct, 652*f*, **662**
Cystic fibrosis (CF), **626–27**, **667**
Cystitis, **692**
Cystocele, **695**
Cystoscopy, **695**
Cytokinesis, 40, 46–47*f*, **48**
Cytomegalovirus retinitis, 592
Cytoplasm, 29, **33–41**, 44*t*
Cytosine (C), 43
Cytoskeleton, 39*f*, **40**
Cytosol (cytoplasmic matrix), **33**
Cytotoxic T cells, **508–9**, 589
Cytotrophoblast, **728**

Dartos muscle, **703**
Decidua basalis, **729**, 730*f*
Decidua capsularis, **729**, 730*f*
Deciduous teeth, **644**
Decubitus ulcer, **125**
Decussation of the pyramids, **381**
Deep, defined, **8**, 9*t*
Deep artery of the arm, **557**, 558*f*
Deep artery of the thigh, 562*f*, **563**
Deep cerebellar nuclei, 385, 386*f*
Deep fibular (peroneal) nerve, **434**
Deep inguinal ring, 709*f*, **710**
Deep palmar arch, **557**, 558*f*

Deep palmar venous arches, **567,** 568*f*
Deep transverse perineus muscle, 292*t*, 293*f*
Deep veins
of pelvis and lower limbs, 571, 572*f*
of upper limbs, 567, 568*f*
Deep vein thrombosis of the lower limb, **574**
Defecation, **636,** 657
Defecation reflex, 459, 657
Defense(s)
cellular, 91
keratinocytes in skin and, 113
Degenerative conditions. *See also* Disease
bone, 143–45
brain, 402–3
joint, 236–38
muscular, 258
nervous system, 353–54
Dehydroepiandrosterone (DHEA), 756–57
Delta (D) cells, pancreatic, **759**
Deltoid muscle, **297,** 298*t*, 299*f*, 783, 784*f*, 785*f*
Deltoid tuberosity, humerus, 189*f*, **190**
Demifacets, thoracic vertebrae, **173,** 176*f*
Dendrite(s), 338, 339*f*, **340,** 342*f*, 345
graded potentials on, 342
Dendritic cells, 114, **592**
Dens, axis vertebrae, **173,** 175*f*, 176*f*
Dense bodies, smooth muscle, **257**
Dense connective tissue, 88, **98**
irregular, 94*f*, **98**
regular, 94*f*, **98**
Dense plaques, smooth muscle, **257**
Dental formula, 644
Dental plaque, **645**
Denticulate ligaments, 393*f*, **395**
Dentin (dentine), **645**
Dentinal tubules, **645**
Dentition, 644
Deoxyribonucleic acid. *See* DNA (deoxyribonucleic acid)
Depressed bone fracture, 144*t*
Depression movement, **222,** 223*f*
Depressor anguli oris muscle, 276*t*, 277*f*
Depressor labii inferioris muscle, 276*t*, 277*f*
Dermal papillae, **115**
Dermal ridges, 115
Dermatitis, **124**
Dermatomes, **67,** 68*t*, **435–36**
map of, 436*f*
Dermis, 112, 113*f*, 114*f*, 115
tatoos and, 116
Descending aorta, **553**
Descending colon, 655*f*, **656**
Descending limb, loop of Henle, **682,** 684*f*
Descending (motor) pathways, spinal cord, 397, 399, 400*f*, 401*t*
Desmosomes, **84,** 85*f*
capillary permeability and, 547–49
Detached retina, **479**
Detrusor muscle, **689**
Developmental anatomy, **2**
Development aspects of cells, 52–53
Deviated septum, **629**
DHEA (dehydroepiandrosterone), 756–57
Diabetes insipidus, **760**
Diabetes mellitus, **761**
effects of, on basement membranes, 86
microangiopathy of, 575
Dialysis, **695**
Diapedesis, **506**
Diaphragm, breathing and action of, 622
Diaphragmatic surface of liver, **658,** 659*f*, 660*f*
Diaphragm muscle, 287*t*, 288*f*
Diaphysis, **134,** 135*f*
Diarthroses, **212**
Diastole, **528–29**
Diencephalon, 362, **374–79,** 380*f*, 383*f*
embryonic development of, **360,** 361*f*
epithalamus, 376*f*, 379
functions of, 390*t*
hypothalamus, 376*f*, 377, 379*f*
thalamus of, 375–77, 378*f*
Diet. *See* Nutrition
Diethylstilbestrol (DES), 70
Differential white blood cell (WBC) count, **510**
Diffuse neuroendocrine system (DNES), **759**
Diffusion, 30
Digastric muscles, **278,** 280*t*, 282*f*
Digestive system, 7*f*, **633–73**
anatomy of, 637–66
disorders of, 643, 645, 646, 650–51, 656, 657, 662, 663, 666–67, 668, 669
overview of, 634–37
Digestive tube, 59, 66
Digital arteries, **557,** 558*f*
Digital rectal exam, **711**
Digital subtraction angiography (DSA), **20,** 21*f*
Dilation (first) stage of labor, **732**
Diploe, **135**
Disease. *See also* Degenerative conditions; names of individual diseases, e.g. AIDS; Cancer
autoimmune, 237, 353
bone, 143–47
infectious (*see* Infectious disease)
sexually transmitted, 592–93, 712–13, 720
Dislocation (luxation), joint, 205, 227, 231, **234**
Dissection, **2**
Distal convoluted tubule, 680*f*, **682**
Distal radioulnar joint, 190
Diverticulosis, **656**
Divertivulitis, **656**
DNA (deoxyribonucleic acid), **41**
double helix structure of, 43*f*
Dorsal, defined, **8,** 9*t*
Dorsal body cavity, **13,** 14*f*
Dorsal column pathway (ascending sensory pathway), **397,** 398*f*, 399*t*
Dorsal hollow nerve cord, **12**
Dorsal interossei muscle, 310*t*, 311*f*
Dorsalis pedis artery, 562*f*, **563,** 793
Dorsalis pedis vein, **571,** 572*f*
Dorsal motor nucleus, 449
Dorsal nerves of penis, 713
Dorsal ramus, **427,** 428
Dorsal root ganglia, 452
Dorsal sacral foramina, **176,** 177*f*
Dorsal spinal root, **426,** 427*f*
Dorsal stream, visual association area of cerebral cortex, 369, 370*f*
Dorsal venous arch, **571,** 572*f*, 792*f*, 793
Dorsal venous network, 788*f*, 789
Dorsiflexion movement, **222,** 223*f*
Dorsum nasi, **770,** 775*f*
Drugs
anabolic steroids, 260
birth defects due to some, 70
phenobarbitol, tolerance to, 36
for treatment of AIDS, 593
for treatment of heart disease, 536
for treatment of stomach ulcers, 650–51
vinblastine, for cancer therapy, 48
DSA (digital subtraction angiography), **20,** 21*f*
DSR (dynamic spatial reconstruction), **20**
Duchenne muscular dystrophy, **258**
Ductless glands. *See* Endocrine glands
Duct of the epididymis, 704*f*, **708**
Ducts of exocrine glands, **87,** 88*f*
Ductus arteriosus, 553, **576**
Ductus deferens, 702, 708*f*, **709**
Ductus venosus, **576,** 577*f*
Duodenal glands, 653*f*, **654**
Duodenal ulcers, 650
Duodenum, **652**
Dural sinuses, 388, 389*f*, 563, 566, 567*f*
Dura mater, **388,** 389*f*, 390*f*
Dynamic spatial reconstruction (DSR), **20**
Dynein, 40, 83
Kartagener's syndrome and faulty formation of, 83
Dyskinesia, **372**
Dysplasia, **53**
Dyspnea, **624**
Dystrophin, 258

Ear, **486–96**
embryonic development of, 496
disorders affecting, 486, 488, 495, 497
equilibrium and auditory pathways, 494–95
inner ear, 487*f*, 488*f*, 489–94
middle ear, 486–89
outer (external), 486, 487*f*
"popping" in, 487–88
surface anatomy of, 774
Eardrum (tympanic membrane), 486, 487*f*
Early erythroblasts, **513,** 514*f*
Ear ossicles, 171*t*
Eccrine glands, 120*f*, **121**
Echocardiography, **540**
Ectoderm, **62,** 63*f*
derivatives of, 66, 68*t*
development of nervous tissue from, 354, 355*f*
Ectodermal placodes, **437**
Ectopic pregnancy, **72**
Edema, **103**
burns and, 122
lymph vessels and, 584, 586
Effector, reflex, 350, 352*f*
Effector lymphocytes, 593
Efferent arteriole, **686,** 687*f*
Efferent ductules, 704
Efferent lymphatic vessels, **586**
Efferent signals. *See* Motor (efferent) signals
Effort, lever systems and, **266**
Ejaculation, 714–15
Ejaculatory duct, 702, **709**
Elastic arteries, **546**
Elastic cartilage, 95*f*, 99, **130,** 131*f*
Elastic connective tissue, **98**
Elastic fibers, **90,** 115
Elastin, **90**
Elbow joint, **227,** 229*f*
muscles crossing, 299*f*, 300*t*
tennis elbow, **301**
Elbow region, surface anatomy of, 783, 784*f*, 786*f*, 787
Electrical stimulation of fracture sites, **206**
Electroencephalography (EEG), **366**
Electromyography, **332**
Elephantiasis, **586**
Elevation movement, **222,** 223*f*
EM (electron microscopy), **16,** 18*f*
Embolus (emboli), **511, 517**

Embryo, human, 13*f,* 58. *See also* Embryonic period
developmental abnormalities in, 70–71
fate of, 71
fertilization of egg and formation of, 727–29
stem cells of, 105
Embryology, **2–3, 57–74**
defined, **58**
embryonic period in (*see* Embryonic period)
fetal period in (*see* Fetal period)
Embryonic period, **58, 59–68.** *See also* Fetal period
autonomic nervous system development in, 460
blood formation in, 516–17
body formation in (week 4), 66–68
bone formation in, 139
brain and spinal cord development in, 360, 361*f,* 403–4
digestive system development in, 667, 668*f*
ear development in, 496
endocrine organ development in, 762*f*
eye development in, 485–86
heart development in, 537–39
joint formation in, 238
lymphatic system development in, 599
muscle development in, 259–60, 270, 271*f,* 272
nervous tissue development in, 354, 355*f*
peripheral nervous system in, 437
respiratory system development in, 628
second month of development in (weeks 5–8), 68
sex organ development in, 734–37
special sense development in, 470, 485, 496
three-layered embryo in (week 3), 62–65
two-layered embryo in (week 2), 61
urinary system development in, 693, 694*f,* 695
from zygote to blastocyst in (week 1), 59–61
Embryonic stem cells, **105**
Emmetropia, **482,** 483*f*
Emotional responses
hypothalamus and control of, 377
limbic system and, 386–87
Emphysema, 624, 625*f*
Enamel of tooth, **644,** 645*f*
Encapsulated dendritic endings, **413–16**
Encephalitis, **389**
Encephalopathy, **404**
Endemic goiter, **760**
Endocarditis, **540**
Endocardium, 524, **525**
Endochondral bones, **139**
postnatal growth of, 141–42
Endochondral ossification, **139–40,** 141*f*
Endocrine glands, **86,** 87–88. *See also* Hormone(s)
Endocrine system, 4, 6*f,* **743–67**
disorders of, 760–62, 763
throughout human lifespan, 762–63
hypothalamus and control of, 377
major organs of, 744*f,* 745–59
other endocrine structures of, 759–60
overview of, 744–45
Endocrinology, **744**
Endocytosis, **30,** 31*f,* 32*f,* 33*f*
Endoderm, **62,** 63*f*
derivatives of, 66, 68*t*
Endolymph, **489**
Endometrial cancer, **733–34**
Endometriosis, **724**
Endometrium, 714*f,* **721,** 722*f*
Endomysium, 246*f,* **247**
Endoneurium, **348**
Endoplasmic reticulum (ER), **35–36,** 44*t*
Endoscopy, **669**
Endosome, **32**
Endosteum, **134**
Endothelium, **78,** 544
Enteric nervous system, **640**
Enteric neurons, 639
Enteritis, **669**
Enterocytes, 653
Enteroendocrine cells, 651, 654, 759
Enzymes, **29**
in lysosomes, 38–39
in peroxisomes, 39
Eosinophils, **91,** 506, 507*f,* **508,** 511*t*
Ependymal cells, 346*f,* **347**
Epiblast, **61**
Epicardium, **524, 525**
Epicondyles
femur, 199, 201*f*
humerus, 189*f,* **190,** 783, 784*f*
Epicranius muscles, 276*t,* 277*f*
Epidermal ridges (friction ridges), 115, 789
Epidermis, 66, 112, **113–15**
Epididymis, 702, 704*f,* **708–9**
Epidural space, **391,** 394*f*
Epigastric region of abdominal wall, 636*f,* **637**
Epiglottis, 130, **610,** 611*f,* 628
Epiglottitis, **628**
Epimysium, **246**
Epinephrine (adrenaline), **457**
Epineurium, **348**
Epiphyseal arteries and veins, 134
Epiphyseal bone fracture, 145*t*
Epiphyseal line, **134,** 135*f*
Epiphyseal plates (epiphyseal discs, growth plates), **140**
anatomy of bone growth and, 140–41, 142*f*
closure of, 141
synchondrosis cartilaginous joint and, 214*f*
Epiphyses, **134,** 135*f*
bone growth at, 140–41, 142*f*
Epiploic appendages of large intestine, 655*f,* **656**
Episiotomy, **725**
Epistaxis, **627**
Epithalamus, **379**
Epithelia, 62, **76–86**
alimentary canal, **638,** 639*f*
classification of, 77–83
endocrine cells derived from, 744–45
sensory receptors in, 412*t,* 413
special characteristics of, 76*f,* 77
surface features of, 83–86
Epithelial cell, 48, 49*f*
Epithelial-connective tissue interactions, **106**
Epithelial reticular cells, 597
Epithelial root sheath, 118*f,* **119**
Epitympanic recess, middle ear, **486,** 487*f*
Eponychium (cuticle), **121,** 122*f*
Epstein-Barr virus, 599
Equilibrium, **337,** 381, 385
disorders of, 495
head rotation and, 491, 492*f*
pathway of, 494
static, 489, 490
vestibular cortex, 367*f,* **370**
Equilibrium pathway, **494**
ER (endoplasmic reticulum), **35–36,** 44*t*
Erectile bodies, 710*f,* **713**
Erection, penis, 713
Erector spinae muscle, **283,** 285*t,* 286*f,* 778*f,* 779
Erythroblasts, 513, 514*f*
Erythrocytes, 42, 48, 50*f,* 97*f,* 504, **505–6,** 511*t*
disorders of, 515
formation of, 513
Erythropoietin, 759
Esophageal hiatus, 646
Esophageal plexus, **449**
Esophagus, 634, 635*f,* **646**
achalasia of the cardia in, 460
gross anatomy of, 634*f,* 646, 648*f*
microscopic anatomy of, 646, 647*f*
Estrogens, 142, **718, 759**
brain structure and, 373
heart disease and, 578
osteoporosis and deficiency of, 144
Ethmoid bone, **162,** 171*t*
Ethmoid sinuses, **162**
Eustachian tube. *See* Pharyngotympanic tube
Eversion movement, **222,** 223*f*
Excitatory synapses, **343**
Exocrine glands, **86,** 87
multicellular, 87, 88*f*
unicellular, 87*f*
Exocytosis, **30,** 31*f*
Expiration, breathing, **622–23**
Expulsion (second) stage of labor, **732**
Extended chromatin, **43**
Extension movement, **221**
Extensor carpi radialis brevis muscle, 304*t,* 306*f*
Extensor carpi radialis longus muscle, 304*t,* 306*f,* 787
Extensor carpi ulnaris muscle, 305*t,* 306*f,* 787
Extensor digitorum brevis muscle, 321*f,* 329*t*
Extensor digitorum longus muscle, 319*t,* 320*f,* 792*f,* 793
Extensor digitorum muscle, 304*t,* 306*f,* 788*f,* 789
Extensor hallucis longus muscle, 319*t,* 320*f,* 792*f,* 793
Extensor indicis muscle, 305*t,* 306*f*
Extensor pollicis brevis and longus muscle, 305*t,* 306*f,* 788*f,* 789
Extensor retinacula muscle, **301**
External acoustic meatus, **159**
External anal sphincter, **292,** 293*f,* 655*f,* **657**
External auditory canal, **486,** 487*f,* 774, 775*f*
External auditory meatus, **159,** 160*f*
External carotid arteries, **553,** 556*f,* 776, 777*f*
External genitalia, female, **724–25,** 736*f*
External iliac artery, 562
External iliac vein, **571,** 572*f*
External intercostal muscles, 287*t,* 288*f*
External jugular vein, 567, 776, 777*f*
External nose, **604,** 605*f*
External oblique muscle, 289*t,* 290*f,* 783
External occipital protuberance, **157**
External occipital protuberance of cranium, 770, 775*f*
External os, 714*f,* **720**
External respiration, **604**
External root sheath, hair, **119**
External urethral orifice, **690, 691**
External urethral sphincter, **690**
Exteroceptors, **411**
Extracapsular ligaments, 216
Extracellular matrix, **88,** 89*f*
Extracorporeal shock wave therapy, 693
Extraembryonic membrane, **728,** 730*f*
Extrafusal muscle fibers, **415**

Extrinsic eye muscles, **472,** 473*f*
Extrinsic (accessory) glands, 638
Extrinsic tongue muscles, 278*t,* 279*f,* 641
Eye, **470–86**
  accessory structures of, 470–73
  anatomy of eyeball, 474–80
  disorders of, 473, 479, 480, 482–83, 484–85
  embryonic development of, 485–86
  as optical device, 480–81
  visual pathways of, 481–84
Eyeball, **470**
Eyebrows, **470**
Eyelashes, **471**
Eyelids, **470,** 471*f*

Face
  cavernous sinus and danger zone of, 566
  muscles of, 276–77*t*
  surface anatomy of, 770, 774, 775*f*
Facial artery, **553,** 556*f,* 774
Facial bones, 154, 155*f,* 156*f,* **162–64,** 171*t*
Facial expressions
  muscles of, 276–77*t,* 774
  recognition of, 387
Facial nerves (VII), 160, **418,** 419*t,* 423*t,* 643
  cranial outflow via, 448
  gustatory pathway and, 466, 468
Falciform ligament, **658,** 659*f,* 660*f,* **664,** 665*f*
Fallopian tubes, **718**. *See also* Uterine tubes
False pelvis, 198, **199**
False ribs, **178**
False vocal cords, **610,** 611*f*
Falsiform ligament, **664,** 665*f*
Falx cerebelli, **388,** 390*f*
Falx cerebri, 162, **388,** 390*f*
Familial hypercholesterolemia, 32
Fascia, **98**
  superficial, 116
Fasciae adherans, **254**
Fascicles, **246,** 247
  arrangement of, in muscles, **268,** 269*f*
Fasciculus cuneatus, **396,** 398*f,* 399*t*
Fasciculus gracilis, **396,** 398*f,* 399*t*
Fat cell, 48, 49*f,* **97**
Fauces, 606*f,* **608**
F (PP) cells, pancreatic, **759**
Feedback loops, 745
Female body
  brain structure of, 373
  heart disease in, 578
  menopause in, 738
  ovarian arteries of, 559*f,* 560–61
  ovarian veins of, 569
  pelvic structure of, and childbearing, 199, 200*t*
  perineum of, 724*f,* **725,** 782*f*
  puberty in, 738
Female reproductive system, 7*f,* **715–34**
  disorders of, 712–13, 720, 724, 733–34, 738, 739
  embryonic development of, 59, 60*f,* 734–37
  exernal genitalia and perineum, 724–25
  internal reproductive organs, *f*
  mammary glands, 725–26
  ovaries, 715–18
  pregnancy and childbirth and, 726–32
  uterine tubes, 718–20
  uterus, 720–24
  vagina, 724
Femoral artery, 562–63, 791
Femoral hernia, **791**
Femoral nerve, **432,** 433*f*
Femoral triangle, 785*f,* **791,** 792*f*
Femoral vein, **571,** 572*f,* 791
Femoropatellar joint, 231, 232*f*
Femur, **199,** 201*f,* 230, 231
Fenestrated capillary, 548*f*
Fertilization, **728**
Fetal alcohol effect, 70
Fetal alcohol syndrome, **70,** 71*f*
Fetal period, **58, 68–71**
  bone development in, 139, 140, 179–81
  brain and spinal cord development in, 360, 361*f,* 403–4
  developmental events of, 69*t*
  development of special senses in, 497
  heart development in, 537–39
  respiratory system development in, 628–29
Fetus, 58*f,* **59,** 71*f*
  blood vessels and circulation in, 576–78
  cleft palate in, 179–80, **182**
  development of (*see* Fetal period)
  fontanels in skull of, 179, 181*f*
  skin of, 124
Fever blisters, **125**
Fibers of areolar connective tissue, **90**
Fibrinogen, 505
Fibroblasts, 48, 49*f,* **90,** 91, 98
Fibrocartilage, 96*f,* 99, **130,** 131*f*
Fibrocartilaginous callus, bone fracture repair and formation of, **143,** 146*f*
Fibrocytes, 90
Fibroids, **738**
Fibromuscular stroma, **711**
Fibromyalgia, **259**
Fibrosis, **103**
Fibrous capsule, synovial joint, **215**
Fibrous joints, **212,** 213*f*
  characteristics of, 216–18*t*
  gomphoses, 213
  summary of, 212*t*
  sutures, 212, 213*f*
  syndesmoses, 213–14
Fibrous pericardium, **524**
Fibrous skeleton of heart, 530*f,* **532**
Fibrous tunic, **474–76**
Fibula, **202,** 203*f,* 791, 792*f*
Fibular (peroneal) artery, 562*f,* **563**
Fibular collateral ligament, **231,** 232–33*f*
Fibularis (peroneus) brevis muscle, 321*f,* 322*t,* 790*f,* 791
Fibularis (peroneus) longus muscle, 320*f,* 322*t,* 790*f,* 791
Fibularis (peroneus) tertius muscle, 319*t,* 320*f,* 790*f,* 791
Fibular nerves, **434,** 435*t*
Fibular (peroneal) vein, **571,** 572*f*
Filaments of olfactory nerve, **469**
Filiform papillae, 641, **642**
Filtration, kidney, **681**
Filtration membrane (filtration barrier), **682,** 683*f*
Filtration slits, **682,** 683*f*
Filum terminale, 393*f,* **395**
Fimbriae, 714*f,* **718**
Finger(s)
  bones, 193*f,* **194**
  muscles of, 309–10*t,* 311*f*
  popping sounds in joints of, 218
Fingerprints, 115
First-class levers, **267**
First-degree burns, **122,** 123*f*
Fissures
  cerebellar hemisphere, **385,** 386*f*
  cerebral hemisphere, **363,** 364*f*
  spinal cord, 394*f,* 395
  on visceral surface of liver, **658,** 660*f*
Fixator muscles, **269**
Fixed specimens, **17**
Flaccid paralysis, **403**
Flagellum, 83
Flat bones, **134**
  structure of, 134–35, 136*f*
Flexion creases of hand, 789
Flexion movement, **221**
Flexor accessorius (quadratus plantae) muscle, 329*t,* 330*f*
Flexor carpi radialis muscle, 301*t,* 302*f,* 787, 788*f*
Flexor carpi ulnaris muscle, 302*t,* 787, 788*f,* 789
Flexor digiti minimi brevis muscle, 309*t,* 310*f,*
330*t,* 331*f*
Flexor digitorum brevis muscle, 329*t,* 330*f*
Flexor digitorum longus muscle, 323*t,* 325*f*
Flexor digitorum profundus muscle, 303*t*
Flexor digitorum superficialis muscle, 302*t,* 303*f*
Flexor hallucis brevis muscle, 329*t,* 331*f*
Flexor hallucis longus muscle, 323*t,* 325*f*
Flexor pollicis brevis muscle, 309*t,* 310*f*
Flexor pollicis longus muscle, 303*t*
Flexor retinacula muscle, **301**
Flexure lines, 115
Flexures, brain, 360, 361*f*
Floating ribs (vertebral ribs), **178**
Flocculonodular lobe, cerebellar hemisphere, **385,** 386*f*
Flower spray sensory endings, **415**
Fluid mosaic model of membranes, 29–30
Fluoroscope, **19**
Focusing disorders of eye, 482–83
Folding, embryonic body formation by, 66
Folia, **385,** 386*f*
Folic acid, 180
Follicles
  lymphoid nodules, 594*f,* **595**
  ovarian, **716,** 717, 718
  thyroid gland, 753*f,* **754**
Follicle-stimulating hormone (FSH), **705,** 717, **748,** 749*t*
Follicular cells
  ovarian, **717**
  thyroid gland, 753*f,* **754**
Follicular phase of ovarian cycle, **717**
Fontanels, **179,** 181*f*
Foot
  arches of, 205
  bones of, 197*t,* 202, 204*f,* 205
  clubfoot, 205
  muscles of, 319–31
  surface anatomy of, 791, 792*f,* 793*f*
Footdrop, **435**
Foot processes of podocytes, **682,** 683*f*
Foramen lacerum, **160**
Foramen magnum, **157,** 158*f*
Foramen ovale, 161*f,* **162,** 538, **576**
Foramen rotundum, 161*f,* **162**
Foramen spinosum, 161*f,* **162**
Forearm
  arteries of, 557, 558*f*
  bones of, 190, 191*f,* 192*f,* 196*t*
  innervation of, 429–32
  muscles acting on, 299*f,* 300*t,* 301–5*t,* 306*f,* 307*t,* 308*t*
  surface anatomy of, 786*f,* 787, 788*f,* 789
  veins of, 567, 568*f*
Forebrain (prosencephalon), **360,** 361*f*
  internal structure of, 365*f*

Foregut, **667**, 668*f*
Formed elements of blood, **505–11**
Fornix, **386**, 387*f*, 714*f*, **724**
Fossa ovalis of heart, **525**, 526*f*
Fourchette, 724*f*, **725**
Fourth ventricle, **363**
Fovea capitis, **199**, 201*f*
Fovea centralis, **479**
Fractures, bone, **143**
Colles', of radius styloid process, 787
clavicle, 186
mandible, 163
metatarsal stress, 202
repair of, 143, 146*f*, **206–7**
types of, 144–45*t*
Free dendritic endings, **413**, 414*f*, 416
Free edge, nail, **121**, 122*f*
Free radicals, **39**, 52
Free ribosomes, 34*f*, **35**
Free vascular fibular graft technique, **206**
Friction, skin's response to, 113
Fright and flight reactions, 380
Frontal bone, **154**, 155*f*, 170*t*
Frontal eye field, 367*f*, **368**
Frontalis muscle, 276*t*, 277*f*, 770, 775*f*
Frontal lobe, cerebral hemisphere, **363**, 364*f*
Frontal (coronal) plane, **10**, 11*f*
Frontal processes, maxillary bones, **163**, 164*f*
Frontal sinuses, **154**, 156*f*
Fulcrum, **266**
Full-thickness burns, **122**
Functional anatomy, 2
Functional morphology, **3**
Functional MRI, **23**
Functional neuroimaging techniques, **366**
Fundus
of stomach, **647**, 648*f*
of uterus, 714*f*, **720**
Fungiform papillae, 641, **642**
Funiculi, 394*f*, **396**

G1 (growth 1) phase, **45**, 46–47*f*
G2 (growth 2) phase, cell cycle, **45**, 46–47*f*
Galea aponeurotica, 277*f*, 770
Gallbladder, 635, 652*f*, 654, 660*f*, **662**, 781
Gallstones, **662**
Gametes, **702**
g gamma efferent fibers, **415**
Ganglia (ganglion), 338, 339*f*, **410**. *See also* Basal nuclei (basal ganglia)
Ganglion cells of retina, 477*f*, **478**
Gap junction, **86**
Gas transport, erythrocytes and, 506
Gastric arteries, 559, 560*f*
Gastric glands, 649*f*, **650–51**
Gastric intrinsic factor, **651**
Gastric pits, 649*f*, **650**
Gastric ulcers, 650
Gastrin, 651
Gastrocnemius muscle, 322*t*, 324*f*, 791, 793*f*
Gastroduodenal artery, **560**
Gastroenteritis, **668**
Gastroepiploic artery, **559**, 560*f*
Gastroesophageal reflex disease (GERD), **646**
Gastrointestinal tract (GI), **759**. *See also* Alimentary canal
Gastrulation, **62**
Gemellus muscle, 316*f*, 318*t*
Gene, 52
General interpretation area, brain, 367*f*, **371**
General sensory receptors, **413–16**
encapsulated, 412*t*, 413–16
proprioceptors, 413*t*, 414–16
structure of, 414*f*
unencapsulated, 412*t*, 413
General somatic motor area of nervous system, **337**, 338, 366–69, 410*f*, 444*f*
autonomic nervous system compared to, 444–45
General visceral motor fibers, 426
General visceral motor system, 444. *See also* Autonomic nervous system (ANS)
General visceral senses, **337**, **444**
neurons, 458, 459*f*
reflexes, 459
Genetic theory of aging, 52
Genicular artery, 563
Geniculate nuclei, **377**, **481**, 484*f*, **494**, 495*f*
Genioglossus muscle, 278*t*, 279*f*
Geniohyoid muscles, 281*t*, 282*f*
Genital herpes, **713**
Genitalia, **702**
embryonic development of, 736*f*
female, 724–25
male, 703–8, 712–15
Genital tubercle, **736**
Genital warts, **713**
Germinal centers, **595**
Germinal epithelium, **716**
Germ layers, embryonic, 62, 63*f*, 64*f*
derivatives of, 66–67, 68*t*
Gigantism, **760**
Gingiva, **645**
Gingivitis, **645**
Glabella, **154**, 155*f*
Glands, **86–88**. *See also* Endocrine glands; Exocrine glands
in integumentary system, 120–21
innervation of, 417–18
Glans penis, 710*f*, **712**
Glaucoma, **480**
Glenhumeral (shoulder) joint, **227**, 228*f*
Glenohumeral ligaments, **227**, 228*f*
Glenoid cavity, scapula, **186**, 188*f*
Glenoid labrum, **227**, 228*f*
Glial cells, **346–47**
Gliding movement, **220**
Globulins, 505
Globus pallidus, **372**, 375*f*
Glomerular capsule (Bowman's capsule), **682**, 683*f*
Glomeruli, **469**
Glomerulus, 681*f*, **682**, 683*f*, 687*f*
Glossary, 800–811
Glossopharyngeal nerve (IX), 381, **419**, 424*t*
cranial outflow via, 448
gustatory pathway and, 468
Glossopharyngeal nerve damage, 424
Glottis, **610**
Glucagon, 663, **758**
Glucocorticoids, **756**
Gluteal arteries, **561**, 563*f*
Gluteal cleft, 789*f*, 790
Gluteal fold, 789*f*, **790**
Gluteal lines, **194**, 195*f*
Gluteal nerves, **434**, 435*t*
Gluteal region, surface anatomy of, 789–91
Gluteal tuberosity, **199**, 201*f*
Gluteus maximus muscle, **312**, 316*f*, 317*t*, 785*f*, 789*f*, 790–91
Gluteus medius muscle, **312**, 316*f*, 317*t*
Gluteus minimus muscle, **312**, 316*f*, 317*t*
Glycocalyx (cell coat), 30
Glycolipids
in plasma membrane, 29*f*, 30
Tay-Sachs disease and undigested, 38–39
Glycosaminoglycans, 91
Glycosomes, **41**
Goblet cell, **87**
in large intestine, 657, 658*f*
in small intestine, 653*f*, 654
Golgi apparatus, **36–38**, 44*t*
Golgi tendon organs, 413*t*, **415**, 416
Gomphosis, **213**
Gonadal arteries, 559*f*, 560–61
Gonadal ridges, **734**, 735*f*
Gonadal veins, 569
Gonadotropic cells, **748**
Gonadotropin-releasing hormone, 750
Gonadotropins, **748–49**
Gonads (primary sex organs), 59, **702**, 744, **759**
descent of, 737
sex hormones secreted by, 759
Gonorrhea, **712**, 720
Gouty arthritis (gout), **237**
Gracilis muscle, 313*t*, 314*f*
Graded potentials, 340, **342**
Graft-versus-host disease, **516**
Gram, **4**
Granulation tissue, **103**, 104*f*
Granulocytes, **506**, 507–8, 511*t*
Granulosa cells, **717**
Graves' disease, **760**
Gray commissure, 394*f*, **395**
Gray matter
brain, 362, 379–80, 385
central nervous system, **352**, 353*f*, 362*f*
spinal cord and spinal roots, 394*f*, 395, 396*f*
Gray rami communicates, **452**, 454*f*
Great cardiac vein of heart, 534*f*, **535**
Greater curvature of stomach, **647**, 648*f*
Greater omentum, 665*f*, **666**
Greater sciatic notch, **194**, 195*f*
Greater trochanter of femur, **199**, 201*f*, 790*f*
Greater tubercle of humerus, 189*f*, **190**, 783, 784*f*
Greater vestibular glands, 725
Greater wings, sphenoid bone, **161**
Great saphenous vein, **572**, 793
Greenstick bone fracture, 145*t*
Gross anatomy, **2**, **4–16**
of bones, 134–36
medical imaging techniques in study of, 19–23
Ground substance in connective tissue, 90*f*, **91**
Growth hormone, 142, **747**, 749*t*, 760
potential medical uses of, 752
Growth hormone-releasing hormone, 750
Guanine (G), 43
Gubernaculum, **737**
Gum disease, **645**
Gustation. *See* Taste (gustation)
Gustatory cells, **466**
Gustatory (taste) cortex, 367*f*, **370**
Gustatory pathway, 466–68
Gynecology, **738**
Gyri (gyrus), cerebral hemisphere, **363**, 364*f*

Hair, **117–20**
color, 117
nose, 606
structure of, 117, 118*f*, 119
thinning and baldness, 120
types and growth of, 119–20
Hair bulb, **117**, 118*f*
Hair cells, inner ear, **489**, 491*f*, **493**, 494
Hair follicles, **117–19**
Hair matrix, **119**
"Hair raising," 119
Hallux valgus, **332**
Hamate bone, **191**, 193*f*
Hamstring muscle, **312**, 316*f*, 318*t*, 791, 793*f*

Hand
- arteries of, 557, 558*f*
- bones of, 190–94, 196*t*
- muscles of, 307*t*, 308*f*, 309–10*t*, 311*f*
- surface anatomy of, 787, 788*f*, 789

Hard keratin, **117**
Hard palate, **163**, 166, **605**, 606*f*, 641
Haustra of large intestine, 655*f*, **656**
Haversian canal, **137**, 138*f*
Haversian system, **137**
Head
- of femur, **199**, 201*f*
- of fibula, **202**, 203*f*, 791, 792*f*
- of humerus, 189*f*, **190**
- of radius, **190**, 191*f*, 192*f*
- of sperm, **705**
- of ulna, **190**, 192*f*, 787

Head of body
- arteries of, 553–57
- bones of (*see* Skull)
- equilibrium and rotation of, 491, 492*f*
- muscles of mastication and tongue movement, 278*t*, 279*f*, 641
- muscles of movement of, 283*t*, 284*f*, 286*f*
- muscles of neck and throat, 280–83*t*
- muscles of scalp and face, 276–77*t*, 770
- surface anatomy of, 770, 774, 775*f*
- sympathetic pathways to, 454, 455*f*
- veins of, 566–67

Healing by first and second intention, **106**
Hearing
- auditory pathways, 494–95
- disorders of, 486, 488, 495, 497
- ear anatomy and, 486–94
- embryonic development of ear and, 496

Heart, **522–42**, **759**
- in adulthood and old age, 539
- blood supply to, 534–35
- cardiac center in brain, 381–82, 457, 534
- cardiac muscle tissue of, **100**, 101*f*, **244**, 245*t*, **254**, 256*f*, 525, 539
- conducting system and innervation of, 532–34, 537
- development of, 537–38, 539*f*
- disorders of, 524, 532, 533, 535–37, 538, 539*f*, 540, 578
- as endocrine structure, 759
- fibrous skeleton of, 530*f*, 532
- location and orientation with thorax, 522–24
- pathway of blood through, 522*f*, 526–27*f*, 528–29
- structure of, 524–28
- sympathetic pathways to, 455, 456*f*
- treating diseases of, 536
- valves of, 528, 529–32, 539

Heart attacks, 458, 511
Heartbeat, 381, 457, **528–29**
- arrhythmia of, 537
- cardiac muscle and rhythmicity of, 254
- cardiac reserve decline and, 539
- conducting system, heart innervation and, 532–34
- irregularities of, 237

Heart block, **533**
Heart chambers, 525–28
Heart failure, **535–36**
Heart transplants, 536
Heart valves, 528, 529–32
- hardening and thickening of, 539

Heart wall, layers of, 524–25
Helicobacter pylori bacteria, 651
Helicotrema, **493**
Helix, ear, **486**, 487*f*
Helper ($CD4^+$) T lymphocyte, **592**, 595
Hematocrit, **504**
Hematoma, **116**
Hematoma formation, bone fracture repair and, **143**, 146*f*
Hematopoietic stem cells, 143
Hemiazygos vein, 568, 569
Hemochromatosis, **517**
Hemoglobin
- in erythrocytes, 505–6, 515
- skin color and, **116**, 117

Hemophilus influenzae, 628
Hemopoiesis, **512–15**
Hemorrhage, **517**
Hemorrhoids, **550**, 655*f*, **657**
Heparin, 91
Hepatic artery proper, **560**, 660
Hepatic ducts, **660**
Hepatic (right colic) flexure, 655*f*, **656**
Hepatic macrophages (Kupffer cells), 661*f*, 662
Hepatic portal system, 563, 570*f*, 571*f*
Hepatic portal vein, **570**, 571, 660
Hepatic veins, 570, 660
Hepatitis, 662, **667**
Hepatocytes, **658**, 662
Hepatopancreatic ampulla, **652**
Hernia, **290**
- congenital, 737
- femoral, 791
- hiatal, 646
- inguinal, 710, 781
- umbilical, 781

Herniated discs, **169**, 172, 432
Hiatal hernia, **646**
Hilton's law, **435**
Hilus
- of kidney, **678**
- of lungs, **617**, 618*f*
- of lymphatic vessels, **586**, **595**, 596*f*

Hindbrain (rhombencephalon), **360**, 361*f*
Hindgut, **667**, 668*f*
Hinge joints, **224–25**
Hip bones, **194**, 195*f*
Hip dislocation, 231
Hip dysplasia (congenital dislocation of the hip), **205**
Hip (coxal) joint, **229–31**
- muscles of, 312–13*t*, 314*f*

Hippocampal formation, 386
Hippocampus, **387**, 403
Hip pointer injury, **194**
Hirschsprung's disease, **460**
Histamine, 91
Histology (microscopic anatomy), **2**, **16–18**
Histones, **43**
Hodgkin's disease, **599**
Holocrine secretion, **120**
Hooke, Robert, 28
Horizontal fissure of lung, **617**, 619*f*
Horizontal (transverse) plane, **10**, 11*f*
Horizontal plates, palatine bones, **163**, 164*f*
Hormonal stimuli, control of hormonal release by, **745**, 746*f*
Hormone(s), **87**, **744**, **745**
- basic action of, 745
- bone growth regulated by, 142
- bone remodeling and, 143
- brain structure and, 373
- classes of, 745
- control of secretions of, 745, 746*f*
- obesity and, 98
- of pituitary gland, 749*t*
- sex (*see* Sex hormones)

Horner's syndrome, **461**
Horns, hyoid bone, **167**
Horseshoe kidney, **695**
Housemaid's knee, **236**
Human immunodeficiency virus (HIV), 592
Humerus, 189*f*, **190**
- muscles affecting movement of, 297–98*t*, 299*f*
- surface anatomy of, 783, 784*f*

Humoral stimuli, control of hormone release by, **745**, 746*f*
Hunger, hypothalamus and control of, 377
Huntington's disease, **374**
Hyaline cartilage, 95*f*, 99, **130**, 131*f*, 612*f*
Hydrocele, **738**
Hydrocephalus, **391**, 392*f*
Hydrogen peroxide, 39
Hydronephrosis, **678**
Hydroxyapatites, 137
Hymen, 721*f*, **724**
Hyoglossus muscle, 278*t*, 279*f*
Hyoid bone, **167**, 774, 776*f*
Hypercalcemia, **763**
Hypercholesterolemia, 32
Hyperextension, **221**
Hyperopia, **482**, 483*f*
Hyperplasia, **53**
Hypertension, **459**
Hypertrophic cardiomyopathy, **540**
Hypertrophy, **53**
Hypoblast, **61**
Hypochondriac regions of abdominal wall, 636*f*, **637**
Hypodermis, 97, 112, **115–16**
Hypogastric plexuses, **456**
Hypogastric regions of abdominal wall, 636*f*, **637**
Hypoglossal canal, **157**
Hypoglossal nerve (XII), 157, 381, **419**, 426*t*
Hypoglossal nerve damage, 426
Hypophyseal arteries, 746, 748*f*
Hypophyseal portal veins, **751**
Hypophyseal pouch, **762**
Hypophysectomy, **763**
Hypophysis, **745**
Hypospadias, **737**
Hypothalamic-hypophyseal tract, **751**
Hypothalamic neurons, 750, 751*f*
Hypothalamus, **377**, 744
- control of autonomic nervous system by, 458
- control of hormone secretion from adenohypophysis, 750–51
- median eminence of, 746
- nuclei of, 379*f*

Hypothenar eminence, 789
Hypothyroidism, 760
Hysterectomy, **738**
H zone, sarcomere, **248**, 249*f*

I bands, sarcomere, **248**, 249*f*
IgE antibody, 91
Ileocecal valve, 655*f*, **656**
Ileocolic artery, **560**, 561*f*
Ileum, **652**
Iliac arteries, 561, 562*f*, 563*f*, 689
Iliac crest, **194**, 195*f*, 777, 779*f*, **781**
Iliac fossa, **194**, 195*f*
Iliac lymph nodes, 586, 587*f*
Iliac region of abdominal wall, 636*f*, **637**
Iliac spine, **194**, 195*f*, 779*f*, **781**, 789
Iliacus muscle, 312*t*, 314*f*
Iliocostalis muscle, 285*t*, 286*f*
Iliofemoral ligament, **229**, 230*f*
Iliopsoas muscle, 312*t*
Ilium, **194**, 195*f*
Immune system and immunity, 91, 102, 114, **589–98**
- AIDS and, 592–93
- antigens and, 508, 589, 592
- disorders of, 598–99
- effects of glucocorticoids on, 756
- throughout human lifespan, 599–60
- lymphocytes, activation of, 589–93

Immune system, *continued*
lymphocytes and, 508, 509*f*, 589, 593
lymphoid organs, 595–98
lymphoid tissue, 589, 593–95
Immunocompetence, 589
Immunoglobulin-like adhesion proteins, 84
Impetigo, **125**
Implantation of blastocyst, **60**, 61*f*, **728**, 730*f*
Incisors (teeth), **644**
Inclusions, cytoplasmic, 40–41
Incompetent heart valves, **532**
Incus, **488**, 489*f*
Infants
birth defects in, 404
blood vessels and circulation in, 576, 577*f*
bone development in, 147
brown adipose tissue in, 98
cleft palate in, 179–80, 182*f*
growth of endochondral bones in, 141–42
hydrocephalus in, 391, 392*f*
respiratory distress syndrome in, 623, 628
skin of, 124
skull of, 179
special senses in, 497
sudden infant death syndrome in, 629
Infectious disease
as cause of some birth defects, 70
cellular defenses against, 91
opportunistic, and AIDS, 592
parasitic, 508
tissue response to, 104
Inferior, defined, **8**, 9*t*
Inferior angle, scapula, **186**, 188*f*, 777
Inferior articular process, vertebral arch, **172**, 174*f*
Inferior cerebellar peduncles, **381**, 382*f*, **385**, 386*f*
Inferior cervical ganglia, **452**, 453*f*
Inferior colliculi, **380**, 384*f*, **494**, 495*f*
Inferior gluteal lines, **194**, 195*f*
Inferior gluteal nerve, **434**, 435*t*
Inferior hypogastric ganglia, 452, 453*f*, **456**
Inferior hypogastric plexus, **450**
Inferior hypophyseal artery, **746**, 748*f*
Inferior mesenteric ganglia, 452, 453*f*, **456**
Inferior mesenteric plexus, **456**
Inferior mesenteric vein, **570**, 571*f*
Inferior nasal conchae, **163**, 166*f*, 171*t*, 606*f*, 607
Inferior nuchal lines, **157**, 158*f*
Inferior oblique muscle, **472**, 473*f*
Inferior olivary nucleus, **381**, 384*f*
Inferior orbital fissure, **163**, 165*f*
Inferior phrenic arteries, 559–60
Inferior rami, pubis, 195*f*, **198**
Inferior rectus muscle, **472**, 473*f*
Inferior sagittal sinus, **566**, 567*f*
Inferior thyroid artery, 556*f*, 557
Inferior vena cava vein, 525, 526*f*, 564, 565*f*, 566*f*, 660
Inferior vesicle arteries, 689
Inflammation, 102–3, 550
joint, 236–38
of lymph vessel, 598–99
mast cells and, 91
Inflammatory bowel disease, **666**
Inflammatory chemicals, 102, 103
Influenza, 628, 629
Infrahyoid muscles, **280**, 281*t*, 282*f*, 777
Infraorbital foramen, **163**, 165*f*
Infraspinatus muscle, **297**, 298*t*, 299*f*
Infraspinous fossae, scapula, **189**
Infrasternal angle of thoracic wall, **780**
Infundibular stalk of neurohypophysis, **746**, 751*f*
Infundibulum
pituitary gland, **745**, 747*f*
uterine tubes, 714*f*, **718**
Ingestion of food, **635**
Ingrown toenail, 121–22
Inguinal canal, **709–10**
Inguinal hernia, 709*f*, **710**, 791
diagnosis of, 781
Inguinal ligament, 779*f*, **781**
Inguinal lymph nodes, 586, 587*f*, 791
Inguinal regions of abdominal wall, 636*f*, **637**
Inhibin, **707**
Inhibiting hormones, **750**
Inhibitory synapses, **343**
Initial segment of axon, **340**
Injuries
ankle, 234
bone, 186, 194
brain, 402
burns, 122–23
elbow, **227**
head and neck traumas, 173, 182
hip dislocation, 231
joint, 234–36, 239
knee, 232–33
muscle, 252, 332
nerve, 420, 421, 424, 425, 426, 431, 432, 434
neuron, 355, 356
repetitive stress, 193
shoulder dislocation, 227
spinal cord, 356, 436
tissue response to, **102–6**
Inner cell mass, blastocyst, **60**
Inner (internal) ear, **489–94**
Inner ear cavities, 154
Innominate bone, **194**, 195*f*
Insertion, muscle attachment, **247**
Inspiration (breathing), **621–22**
Insufficiency, heart, **532**
Insula, 364*f*, **365**, **371**
Insulin, 32, 663, **759**, 761
Insulin-dependent diabetes, **761**
Insulin-like growth factor-1 (somatomedin), 747
INTEGRA Artificial Skin, 122
Integral proteins, **30**
Integration, nervous system function of, **336**
Integration center, reflex, 350, 352*f*
Integrins, 91
Integumentary system, 4, 5*f*, **111–28**
defined, **112**
disorders of, 122–25
throughout human lifespan, 124–25
hypodermis, 97, 112, 115–16
skin (epidermis, dermis), 112–17 (*See* also Skin (integument))
skin appendages (hair, glands, nails), 117–22
Interatrial septum, heart, **525**, 526*f*
Intercalated disc, **254**
Intercellular clefts, **548**, 549
Interchromosomal domains, 45
Intercondylar eminence, **202**, 203*f*
Intercondylar notch, **199**, 201*f*
Intercostal arteries, 557, 558*f*
Intercostal muscles, 287*t*, 288*f*
breathing and action of, 622
Intercostal nerves, 426*f*, 427*f*, **428**
Intercostal spaces, 177
Intercostal veins, 569
Interlobar arteries, **678**, 679*f*
Interlobar veins, **679**
Interlobular arteries, **678**, 679*f*
Interlobular veins, **679**
Intermediate cuneiform bone, **202**, 204*f*
Intermediate fast-twitch fibers, 252, 253*f*
Intermediate filaments, 39*f*, **40**, 44*t*
Intermediate mesoderm, **62**, 65*f*
derivatives of, 67, 68*t*
Internal acoustic (auditory) meatus, 158*f*, **160**
Internal anal sphincter, 655*f*, **657**
Internal capsule, cerebral white matter, **372**, 374*f*
Internal carotid artery, **555**, 556*f*
Internal iliac arteries, 561, 562*f*, 563*f*
Internal iliac vein, **571**, 572*f*
Internal intercostal muscles, 287*t*, 288*f*
Internal jugular vein, 564, **566**, 567*f*, 776, 777*f*
Internal oblique muscle, **289**, 290*t*, 291*f*
Internal os, 714*f*, **720**
Internal pudendal artery, **561**
Internal respiration, **604**
Internal root sheath, hair, **119**
Internal thoracic artery, **557**, 558*f*
Internal urethral sphincter, **689**, 690
Interneurons, 344*f*, **345**
location of, in nervous system, 351, 353*f*
in spinal gray matter, 395, 396*f*
Internodal pathway, heart, 532, **533**
Interoceptors, **411**
Interossei muscles, **309**, 310*t*, 311*f*
Interosseous arteries, 557, 558*f*
Interosseous membrane, **190**, **202**, 213
Interphase, cell cycle, 45, 46–47*f*
Interstitial cells (Leydig cells) between seminiferous tubules, 703*f*, **707–8**
secretion of steroid hormones by, 757*f*
Interstitial connective tissue, **681**
in kidneys, 686
Interstitial growth in cartilage, **130**
Interstitial lamellae, **137**, 138*f*
Intertrochanteric crest, 199
Intertrochanteric line, 199
Intertubercular groove, humerus, 189*f*, **190**
Interventricular arteries, 534–35
Interventricular foramen, **363**
Interventricular septum, heart, **525**, 526*f*
Intervertebral discs, **168–72**
symphyses joints and, 214*f*
Intervertebral foramina, 168*f*, **172**
Intestinal arteries, **560**, 561*f*
Intestinal crypts, **654**
Intestinal flora, 654
Intestinal juice, 654
Intestinal obstruction, **666**
Intestinal (lymph) trunk, **587**
Intra-articular disc, synovial joint, 215*f*, **217**
Intracapsular ligaments, 216, 231, 232*f*
Intrafusal muscle fibers, **414**
Intramembranous ossification, **139**, 140*f*
Intramural ganglia, 449
Intramuscular injections
gluteal, 790–91
sites of, 783, 785*f*
Intraocular pressure, 480
Intraperitoneal organs, **638**, 666*t*
Intrastromal corneal implants, 482
Intrinsic glands, 638
Intrinsic tongue muscles, **278**, 641

Inversion movement, **222,** 223*f*
Involuntary nervous system. *See* Autonomic nervous system (ANS)
Iodine, 760
Ion transport, epithelia and, 76
Ipsilateral fibers, cerebellum, **385**
Iris (eye), 475*f,* **476–77**
Iron in hemoglobin, 505–6
Irregular bones, **134**
  structure of, 134–35
Ischemia, **402**
  silent, 535
Ischial spine, **194,** 195*f*
Ischial tuberosity, 195*f,* **198,** 789*f,* 790
Ischiocavernosus muscle, 292*t,* 293*f*
Ischiofemoral ligament, **229,** 230*f*
Ischium, **194,** 195*f,* 198
Islets of Langerhans (pancreatic islets), **663,** 664*f,* **758**
Isthmus
  of thyroid gland, **753,** 774
  of uterine tube, 714*f,* **719**
  of uterus, 714*f,* **720**
Itch receptors, 413

Jejunum, **652**
Joint(s), **152, 211–42**
  arterial anastomoses around, 551
  bone markings for, 153*t*
  cartilaginous, 212*t,* 213, 214*f*
  characteristics of body, 216–18*t*
  classification of, 212
  defined, **212**
  development of artificial, 238
  disorders of, 227, 234–38
  "double-jointed," 220
  fibrous, 212–13
  thoughout human lifespan, 238–39
  synovial, 212*t,* 214–34
Joint cavity (synovial cavity), synovial joint, **214,** 215*f*
Joint kinesthetic receptors, 413*t,* **415–16**
Jugular foramen, **160**
Jugular (suprasternal) notch, bony thorax, **178,** 774, 776*f*
Jugular (lymph) trunks, **589**
Jugular vein, 564, 566, 567*f,* 776, 777*f*
Juxtaglomerular apparatus, **686,** 687*f*
Juxtaglomerular cells, **686,** 687*f*

Kaposi's sarcoma, 592, **600**
Kartagener's syndrome, 83
Keratin, **113,** 114
  hard and soft forms of, 117
Keratinocytes, **113,** 114, 124
Keratohyaline granules, 114
Kidney cancer, **693**
Kidneys, 59, **676–86, 759**
  arteries of, 559*f,* 560, 678, 679*f,* 686*f*
  as endocrine structure, 759
  gross anatomy of, 676–81
  microscopic anatomy of, 681–86
  veins of, 569, 678, 679*f,* 686*f*
Kidney transplant, **693**
Killer T cells (lymphocyte), 114, **508–9**
Kilogram, **4**
Kinesins, 40
Kneecap. *See* Patella (kneecap)
Knee joint, **231–33**
  muscles of, 312–18
Knock-knee, **207**
Krause's end bulbs, 413, **414**
Kupffer cells, 661*f,* **662**
Kyphosis, **179,** 181

Labia (lips), **640**
Labia frenulum, **641**
Labia majora, 724*f,* **725**
Labia minora, 724*f,* **725**
Labioscrotal swellings, **736**
Labor in childbirth, **729–32**
  oxytocin and, 752
Labyrinth (inner ear), **489–94**
Lacrimal apparatus, **471–72**
Lacrimal bones, **163,** 165*f,* 171*t*
Lacrimal canaliculi, **472**
Lacrimal caruncle, **470,** 471*f*
Lacrimal fossa, 770, 775*f*
Lacrimal gland, **471,** 472*f*
Lacrimal nucleus, 448
Lacrimal puncta, **471–72**
Lacrimal sac, 163, **472**
Lacteals, 585, **654**
Lactiferous ducts, **726**
Lactiferous sinus, **726**
Lacunae, **137,** 138*f,* **728,** 730*f*
Lamboid suture, **154,** 155*f*
Lamella, **137,** 138*f,* 139*f*
Laminae, vertebral arch, **172**
Lamina propria, 99
  alimentary canal, **638,** 639*f*
Laminated granules, 114
Laminectomy, **182**
Landmarks, anatomical, 2
Langerhans cells, 113*f,* **114**
Language area, brain, 367*f,* **371**
Lanugo, 124
Large intestine, 634, **654–58**
  congenital megacolon, 460
  gross anatomy of, 655*f,* 656–57
  microscopic anatomy and tissues of, 657–58
Laryngeal prominence (Adam's apple), **610,** 774
Laryngitis, **611**
Laryngopharynx, 606*f,* **608–9,** 640*f,* **645**
Laryngotracheal bud, **628**
Larynx, 130, 167, 425, 553, **609–12,** 621*t*
  innervation of, 611–12
  sphincter functions of, 611
  voice production and, 610, 611*f*
Late erythroblast, **513,** 514*f*
Lateral, defined, **8,** 9*t*
Lateral angle, scapula, **186,** 188*f*
Lateral apertures, **363**
Lateral border, scapula, **186,** 188*f*
Lateral cervical (cardinal) ligaments, 714*f,* **720**
Lateral circumflex femoral artery, 562*f,* **563**
Lateral columns, 394*f,* **395**
Lateral condyles
  femur, **199,** 201*f,* 791, 792*f*
  tibia, **202,** 203*f,* 791, 792*f*
Lateral cuneiform bone, **202,** 204*f*
Lateral cutaneous branches of intercostal nerve, 427*f,* **428**
Lateral epicondyles
  femur, **199,** 201*f*
  humerus, 189*f,* **190,** 783, 784*f*
Lateral funiculus, 394*f,* **396**
Lateral geniculate nucleus of thalamus, **377, 481,** 484*f*
Lateral head of gastrocnemius muscle, 791, 793*f*
Lateral lemniscus, **494,** 495*f*
Lateral ligament
  ankle, **234,** 235*f*
  temporomandibular joint, **226**
Lateral longitudinal arch, **205**
Lateral malleolus of fibula, **202,** 203*f,* 791, 792*f*
Lateral mass
  atlas vertebrae, **173**
  ethmoid bone, **162**
Lateral menisci, knee, **231,** 232*f*
Lateral nuclear group, **387,** 388*f*
Lateral patellar retinacula, **231,** 232*f*
Lateral plantar arteries, 562*f,* **563**
Lateral plate, mesoderm, **62,** 65*f*
Lateral pterygoid muscle, 278*t,* 279*f*
Lateral rectus muscle, **472**
Lateral rotation movement, **222**
Lateral sacral crest, **176,** *177*
Lateral semicircular canals, 490*f,* **491**
Lateral spinothalamic tract, **397,** 398*f,* 399*t*
Lateral sulcus, cerebral hemispheres, 364*f,* **365**
Lateral supracondylar lines, **199,** 201*f*
Lateral supracondylar ridges, 189*f,* **190**
Lateral thoracic artery, **557,** 558*f*
Lateral ventricles, **363**
Latissimus dorsi muscle, **294,** 296*f,* 297*t,* 299*f,* 778*f,* 779
Law of levers, **266**
Left atrioventricular valve of heart, **529,** 530*f,* 531*f*
Left atrium of heart, **525,** 526*f,* 527*f,* 528
Left brachiocephalic veins, **564,** 565*f*
Left colic artery, **561**
Left colic (splenic) flexure, 655*f,* **656**
Left common carotid artery, **553,** 555*f,* 776, 777*f*
Left coronary artery, **534**
Left gastric artery, **559,** 560*f*
Left gastroepiploic artery, **559,** 560*f*
Left hepatic duct, **660**
Left hypochondriac region, 636*f,* **637**
Left iliac region, 636*f,* **637**
Left lobe of liver, **658,** 659*f,* 660*f*
Left lumbar region, 636*f,* **637**
Left pulmonary artery, **552**
Left subclavian arteries, **553,** 555*f*
Left ventricle of heart, 526*f,* 527*f,* **528**
Left ventricular assist devices (LVADS), 536
Leg
  arteries of, 561–63
  bones of, 197*t,* **202,** 203*f*
  innervation of, 432–35
  muscles of, 319–28
  surface anatomy of, 791, 792*f,* 793*f*
Lens, eye, 474*f,* 475*f,* **480,** 547
Lens epithelium, **480**
Lens fiber, **480**
Lens placode, 485*f,* **486**
Lens vesicle, 485*f,* **486**
Lentiform nucleus, **372,** 375*f*
Leptin, 98
Lesion, **106**
Lesser curvature of stomach, **647,** 648*f*
Lesser omentum, **664,** 665*f*
Lesser sciatic notch, **194,** 195*f*
Lesser trochanter, **199,** 201*f*
Lesser tubercle, humerus, 189*f,* **190**
Lesser wings, sphenoid bone, **161**
Leukemia, **515**
  treatment for, 516
Leukocytes, 504, **506–10,** 511*t*
  agranulocytes, 506, 508–10
  disorders of, 515
  formation of, 513–15
  granulocytes, 506, 507–8
  percentage of different types of, 508*f*
Leukocytosis, **506**
Levator ani muscle, 292*t,* 293*f,* 690
Levator labii superioris muscle, 276*t,* 277*f*
Levator palpebrae superioris, **471**
Levator scapulae muscle, 295*t,* 296*f*
Lever, **266**
Lever systems, 266–68
Leydig cells (interstitial cells), 704*f,* **707–8**
Ligament(s), 98, 130, **152**
  affect of, on synovial joint stability, 219
  bone markings for attachment of, 153*t*

Ligament(s), *continued*
at joints, 216, 219, 226, 227, 229, 230, 231, 232, 233*f*, 234
spinal, 168, 169*f*
supporting uterus, 720–21
vocal, 610
Ligament of the head of the femur, **230**
Ligamentum arteriosum, **553**, 555*f*
Ligamentum flavum, 98, **168**, 169*f*
Ligamentum nuchae, 98
Ligamentum teres, **576**, 577*f*, 659*f*, **660**
Ligamentum venosum, **576**, 577*f*, 659*f*, **660**
Light microscope (LM), **16**
Limb(s), 59
bones of lower, 199–205
bones of upper, 189–94
embryonic formation of, 68*f*
muscles of, 271*f*, **272**
muscles of lower, 312–31
muscles of upper, 297–311
Limb buds, **68**
Limbic system, 370, **385–87**, 390*t*, 469
Limbus, 474*f*, **475**
Linea alba, **289**, 779*f*, 781
Linea aspera, **199**, 201*f*
Linea semilunaris, 779*f*, **781**
Lingual artery, **553**, 556*f*
Lingual frenulum, 640*f*, **641**
Lingual tonsil, **598**, 606*f*, **608**, 642
Lining membranes, **99**, 100*f*
Lipid cells, 97
Lipid droplets, **41**
Lipid metabolism, 36
Lipofuscin, 338
Lips (mouth), **640–41**
Liquid ventilation, 623
Lisfranc injury, **207**
Liter, **4**
Liver, 635, **658–62**
gross anatomy of, 658–60
microscopic anatomy and tissue of, 660–62
Liver biopsy, **669**
Liver lobes, **660**, 661*f*
Liver sinusoids, 570, **660**, 661*f*, 662
Liver stem cells, 662
Load, lever systems and, **266**
Lobar arteries, **552**, **678**, 679*f*
Lobar bronchi, **614**
Lobes
of kidney, 678, 679*f*
of lung, **617**, 618*f*, 619*f*
of mammary gland, **726**
of thyroid gland, **753**, 774
Lobule(s)
of breasts, **726**
of ear, **486**, 487*f*, 774, 775*f*
of lungs, **617**, 619*f*
of testes, **704**
of thymus, **597**
Long bones, **134**
endochondral ossification of, 141*f*
structure of typical, 134, 135*f*
Longissimus muscle, 285*t*, 286*f*
Longitudinal fissure, cerebral hemisphere, **363**, 364*f*
Longitudinal layer, smooth muscle, **254**
Longitudinal ligaments, **168**, 169*f*
Loop of Henle, 680*f*, **682**, 684*f*
Loose connective tissue, **88–98**
Lordosis, **179**
Low-density lipoproteins (LDLs), 32
Lower back pain, **261**
Lower limb
arteries of, 561–63
bones of, 197*t*, 199–205
deep vein thrombosis of, 574
innervation of, 432–35
muscles of, 312–31
surface anatomy of, 789–93
veins of, 571, 572*f*
Lower lobe of lung, **617**, 619*f*
Lumbar arteries, 559*f*, 561
Lumbar curvature, 168*f*, **172**
Lumbar enlargement, 393*f*, **395**
Lumbar plexus, **432**, 433*f*
branches of, 433*t*
Lumbar puncture (spinal tap), **395**, 779
Lumbar regions of abdominal wall, 636*f*, **637**
Lumbar spine, stenosis of, 179
Lumbar (lymph) trunks, **587**
Lumbar veins, 569
Lumbar vertebrae, 168*f*, **172**, 174*t*, **175**
Lumbosacral plexus, 432
Lumbosacral trunk, **432**, 433*f*
Lumbrical muscles, **309**, 310*t*, 311*f*, 329*t*, 330*f*
Lumen, **544**, 545*f*
Lumpectomy, **734**
Lunate bone, **191**, 193*f*
Lungs, **617–21**
blood supply and innervation of, 619–21
cancer of, 623, 626–27
collapsed, 623
gross anatomy of, 617–19
pleurae, 615–17
sounds of, 779
Lung volume reduction surgery, 625
Lunula, nail, **121**, 122*f*
Luteal phase of ovarian cycle, 717*f*, **718**, 723*f*
Luteinizing hormone (LH), **708**, 717, **748**, 749*t*
Lyme disease, **237–38**
Lymph, **584**
movement of, by muscle contractions, 586
Lymphadenopathy, **600**
Lymphangiography, **586**
Lymphangitis, **598–99**
Lymphatic collecting vessels, 584, **585–86**
Lymphatic duct, 587*f*, **589**
Lymphatic sacs, 599
Lymphatic system, 4, 6*f*, **504**, **584–89**
capillaries of, 584, 585
collecting vessels, 584, 585–86
disorders of, 586, 598–99, 600
ducts, 584, 587*f*, 589, 590*f*
throughout human lifespan, 599–600
nodes, 584, 586–87, 588*f*
trunks, 584, 587–89, 590*f*
Lymphatic vessels, **584**
afferent, and efferent, 586
draining of breast by, 726
edema and, 586
Lymph capillaries, 584, **585**
Lymph duct, 584, **589**
Lymphedema, **600**
Lymph nodes, 584, **586**, 587*f*, 726
cortex and medulla of, 594*f*, 595
immune system and, 595
structure of, 586, 588*f*
swollen, 587
Lymphocytes, **91**, 97*f*, 506, 507*f*, **508–9**, 511*t*, **584**, **589**
activation of, 589–93
B cell, 508–9, 589, 599
effector, 593
glucocorticoids and redirection of, 756
memory, 593, 599
natural killer cells, 509
T cell, 508–9, 589, 592, 595, 597
Lymphoid connective tissue, 508
Lymphoid nodules (follicles), **595**
Lymphoid organs, **584**, **595–98**
Lymphoid stem cells, **513**
Lymphoid tissue, **584**, **593–95**
Lymphoma, **600**
Lymph sinuses, **586**
Lymph trunks, 584, **587–89**
Lysosomes, 31, **38–39**, 44*t*
Lysozyme, **472**

McBurney's point, 779*f*, **781**
Macromolecules, transport of, across membranes, 30–33
Macrophage(s), 48, 49*f*, 90*f*, **91**, **510**
alveolar, 614, 616*f*
Macula, inner ear, **489–90**, 491*f*
Macula densa, **686**, 687*f*
Macula lutea, eye, **479**
Magnetic resonance imaging (MRI), **22**, 23*f*, **366**
Magnetic resonance spectroscopy (MRS), **23**
Main bronchi, **613**
Main pancreatic duct, 652*f*, **663**
Major calices, **678**, 679*f*
Major duodenal papilla, **652**
Male body
brain structure of, 373
heart disease in, 578
muscular strength of, 261
pelvic structure of, 199, 200*t*
perineum of, **715**, 782*f*
puberty in, 738
reproductive system of (*see* Male reproductive system)
testicular arteries of, 559*f*, 560–61
testicular veins of, 569
thyroid cartilage of, 610
urinary bladder and urethra of, 689*f*, 690*f*, 691, 710
Male pattern baldness, **120**
Male reproductive system, 7*f*, **702–15**
accessory glands, 710–12
disorders of, 710, 711, 712–13, 733, 737, 738, 739
embryonic development of, 734–37
penis, 712–15
perineum, 715
reproductive duct system, 708–10
testes, 703–8
Malignant neoplasm, **50**
Mallet finger, **332**
Malleus, **488**, 489*f*
Mammary glands (breasts), 121, 715, **725–26**, 738
Mammillary bodies, **377**
Mammogram, **734**
Mammotropic cells, **748**
Mandible, **162–63**, 164*f*, 171*t*, 774, 775*f*
Mandibular angle, **162**, 164*f*
Mandibular arch, **163**, 164*f*
Mandibular condyle, **163**, 164*f*
Mandibular foramen, **163**, 164*f*
Mandibular fossa, **159**, 160*f*, **226**
Mandibular symphysis, **163**, 164*f*
Manubrium, **177**, 779
Marginal artery of heart, 534*f*, **535**
Masseter muscle, 278*t*, 279*f*, 774
Mass peristaltic movements, **656**
Mass reflex reaction, **459–60**
Mast cells, **91**, 508
Mastication
muscles of, 278*t*, 279*f*, 641
teeth and, 643–45
Mastoid air cells, 486
Mastoid antrum, **486**, 487*f*
Mastoid fontanels, **179**, 181*f*
Mastoid process, **159**, 160*f*, 770
Mastoid region, temporal bone, **159**, 160*f*
Matrix, mitochondria, **34**
Matrix granules, mitochondria, **34**
Maxillary artery, **553**, 556*f*

Maxillary bones (maxillae), **163,** 164*f,* 171*t*
Maxillary sinuses, **163,** 167*f*
Mechanical advantage, **266**
Mechanical digestion, **636**
Mechanical disadvantage, **266**
Mechanoreceptors, **411,** 413
Medial, defined, **8,** 9*t*
Medial bicipital furrow (groove), **783,** 784*f*
Medial border, scapula, **186,** 188*f,* 777
Medial circumflex femoral artery, 562*f,* **563**
Medial condyles
  femur, **199,** 201*f,* 791, 792*f*
  tibia, **202,** 203*f,* 791, 792*f*
Medial cuneiform bone, **202,** 204*f*
Medial epicondyles
  femur, **199,** 201*f*
  humerus, 189*f,* **190,** 783, 784*f*
Medial geniculate nuclei, **377, 494,** 495*f*
Medial head of gastrocnemius muscle, 791, 793*f*
Medial lemniscus tract, **397,** 398*f*
Medial (deltoid) ligament, ankle, **234,** 235*f*
Medial longitudinal arch, **205**
Medial malleolus, **202,** 203*f*
Medial menisci, knee, **231,** 232
Medial nuclear group, **387,** 388*f*
Medial patellar retinacula, **231,** 232*f*
Medial plantar arteries, 562*f,* **563**
Medial pterygoid muscle, 278*t,* 279*f*
Medial rectus muscle, **472,** 473*f*
Medial rotation movement, **221,** 222*f*
Medial supracondylar lines, **199,** 201*f*
Medial supracondylar ridges, 189*f,* **190**
Medial umbilical ligaments, **576,** 577*f*
Median aperature, **363**
Median cubital vein, **567,** 568*f,* 786*f,* 787
Median eminence of hypothalamus, **746**
Median nerve, brachial plexus, **431,** 789
Median plane, 11*f,* **12**
Median sacral artery, 561
Median sacral test, **176,** 177*f*
Median vein of the forearm, **567,** 568*f*
Mediastinum, **13,** 14*f*
Mediastinum testis, 704
Medical imaging techniques, **19–23**
Medulla
  hair, **117,** 118*f*
  lymph nodes, 594*f,* **595**
  thymus, **597**
Medulla oblongata, **360,** 361*f,* **381–82,** 384*f,* 390*t,* 457, 623
Medullary cavity (marrow cavity), **134**
Medullary pyramids (renal pyramids), **678,** 679*f*
Medullary respiratory center in reticular formation of medulla, 382
Megakaryoblasts, 514, **515**
Megakaryocyte, 514*f,* **515**
Meiosis, sperm formation and, **705,** 706*f*
Meissner's corpuscle, 413, **414**
Melanin, **114, 116,** 381
Melanocytes, 113*f,* **114,** 117
Melanocyte-stimulating hormone (MSH), **748,** 749*t*
Melanoma, **124**
Melatonin, **379, 758**
Membrane(s)
  basement, 77, **86**
  on bone surface, 134
  covering and lining, **99,** 100*f*
  mucous (mucosa), **99,** 100*f*
  plasma (*see* Plasma membrane (plasmalemma))
  serous (serosa), **14,** 15*f,* **99,** 100*f*
Membrane bones, **139**
Membranous ampulla, 490*f,* **491**
Membranous labyrinth, inner ear, **489,** 490*f*
Membranous urethra, 690*f,* **691,** 710
Memory
  hypothalamus and formation of, 377
  limbic system and retrieval of, 387
Memory lymphocytes, 593, 599
Meniere's syndrome, **495**
Meningeal layer, **388,** 389*f*
Meninges, **388–89,** 390*f*
Meningitis, **389**
Meningocele, **403**
Menisci, knee joint, 130, **231,** 232*f*
Meniscus, synovial joint, 215*f,* **217**
Menopause, **738**
Menstrual cycle, **715,** 723*f,* 749
  ovarian cycle and, 716–18
Menstrual phase of uterine cycle, 723*f,* **724**
Mental foramen, mandible, **163,** 164*f*
Mentalis muscle, 277*t*
Merkel cells, 113*f,* **114**
Merkel discs, **413**
Mesencephalon (midbrain), **360,** 361*f*
Mesenchyme tissue, 62, 66
  connective tissue originating from, **88,** 89*f,* 92*f,* 106
Mesenteric arteries, 559*f,* 560, 561*f*
Mesenteric veins, 570, 571*f*
Mesenteries, 637*f,* **638, 663–66**
  support for uterus by, 720–21
Mesentery proper, 665*f,* **666**
Mesoderm, **62,** 63*f*
  derivatives of, 67, 68*t*
  differentiation of, 62–65
  muscle development from, 270, 271*f,* 272
Mesometrium, 714*f,* **720**
Mesonephric (Wolffian) duct, **693,** 694*f,* **734,** 735*f*
Mesonephros, **693,** 694*f*
Mesosalpinx, 714*f,* **719**
Mesothelium, **78**
Mesovarium, 714*f,* **715**
Messenger RNA (mRNA), **35**
Metabolism, **29**
Metacarpals, **193**
Metacarpus, **193**–**94**
Metamyelocyte, **513,** 514*f*
Metanephros, **693,** 694*f*
Metarteriole, **547**
Metastasis, **50**
Metatarsals, **202,** 204*f*
Metatarsal stress fracture, **202**
Metatarsus, **202,** 204*f*
Metencephalon, **360,** 361*f*
Meter, **4**
Metric system, **4,** 797
Microangiopathy of diabetes, **575,** 763
Microcephaly, **404**
Microfilaments, 39*f,* **40,** 44*t*
Microglia, 346*f,* **347**
Micrometer, **4**
Microscopic anatomy (histology), **2, 16–18**
Microscopy, **16–18**
Microtubules, 39*f,* **40,** 41*f,* 44*t*
  arrangement of, in cilium, 83, 84*f*
Microvilli of epithelial cells, **83**
  in small intestine, 653*f,* **654**
Micturition, 459, **691–92**
Midaxillary line of thoracic wall, **780**
Midbrain (mesencephalon), **360,** 361*f,* **379–81,** 382–83*f,* 384*f,* 390*t*
Midclavicular line of thoracic wall, **780**
Middle cardiac vein of heart, 534*f,* **535**
Middle cerebellar peduncles, **381,** 383*f,* **385,** 386*f*
Middle cerebral artery, **555,** 556*f*
Middle cervical ganglia, **452,** 453*f*
Middle colic artery, **560,** 561*f*
Middle cranial fossa, 159*f,* **160**
Middle ear, **486–89**
Middle ear cavities, **15,** 16*f,* 154, 159
Middle lobe of lung, **617,** 619*f*
Middle meningeal artery, 553, 555
Middle nasal conchae, **162,** 166*f,* 606*f,* 607
Middle suprarenal arteries, 559*f,* **560**
Midgut, **667,** 668*f*
Midpiece of sperm, **705**
Midsagittal plane, 11*f,* **12**
Migraine headaches, **437**
Milk lines, 725
Milliliter, **4**
Mineral crystals in compact bone osteons, 137
Mineralocorticoids, **756**
Minor calices, **678,** 679*f*
Mitochondria, **33,** 34*f,* 44*t*
  free radicals and, 52
Mitochondrial theory of aging, 52
Mitosis, 46–47*f,* **48**
Mitral cells, **469**
Mitral valve of heart, 528, **529,** 530*f*
Mitral valve prolapse, **532**
Mixed (motor and sensory) nerves, 410, **419–20**
M line, sarcomere, **248,** 249*f*
Modiolus of cochlea, **491,** 493*f*
Molars (teeth), **644**
Monoblasts, 514*f,* **515**
Monocytes, 347, 506, 507*f,* **509–10,** 511*t,* 514*f,* **515**
Mononucleosis, **599**
Monosynaptic reflex, **350,** 352*f*
Mons pubis, 724*f,* **725**
Morphology, **2.** *See also* Anatomy
Morula, **59,** 60*f*
Motion sickness, **495**
Motor areas in cerebral cortex, 366–69
Motor axons, innervation of skeletal muscles by, 416–17
Motor endings, peripheral, **410, 416–18**
  innervation of skeletal muscle, 416–17
  innervation of visceral muscle and glands, 417–18
Motor end plates, 416
Motor homunculus, **366,** 368*f*
Motor nerves, **419**
Motor (efferent) neurons, 344*f,* **345**
  in autonomic nervous system, 444
  locations of, in nervous system, 351, 353*f*
  postganglionic, 445, 449, 452, 453*f,* 455–57*f,* 460
  preganglionic, 445, 452, 453*f,* 455–57*f,* 460
  reflexes and, 350, 352*f*
Motor nuclei, 381, 383*f*
Motor output, **336**
Motor proteins, 40, 83
Motor (efferent) signals, **336,** 337*f*
Motor unit, **417**
Mouth (oral cavity), 14, 16*f,* 634, **640–41**
  lips and cheeks of, 640–41
  salivary glands of, 642–43
  teeth of, 643–45

Mouth, *continued*
tissues of, 640
tongue of, 641–42
Movement(s)
brain and coordination of body, 385
ciliary, 83, 84*f*
equilibrium and head rotation, 491, 492*f*
synovial joints and human body, 220–24
venous blood movement through human body, 550, 551*f*
M (mitotic) phase, cell cycle, 45*f*, **48**
MRI (magnetic resonance imaging), **22,** 23*f*, **366**
MRS (magnetic resonance spectroscopy), **23**
Mucin, **87**
Mucosa-associated lymphoid tissue (MALT), **593,** 594*f*, 598, 638
Mucosa of alimentary canal wall, **638,** 639*f*
Mucosa of ureters, **687**
Mucous cells of salivary glands, 643
Mucous membrane (mucosa), **99,** 100*f*
alimentary canal wall, **638,** 639*f*
olfactory, 606
respiratory, 606, **607**
in small intestine, 653–54
stomach, 648, 649*f*, 650, 651
trachea, **612**
ureter, **687**
Mucous neck cells of gastric glands, 649*f*, **650**
Mucus, **87**
respiratory, 607
Multiaxial movement, **226**
Multicellular glands, **87**
Multinucleate cells, 42
Multipennate arrangement of fascicles, **268,** 269*f*
Multiple sclerosis (MS), **353–54**
Multipolar neurons, **343,** 344*f*
Multiunit innervation, 257
Mumps, **643**
Muscle(s). *See also* Skeletal muscle(s); and names of individual muscles
arrangement of fascicles in, 268, 269*f*
bone markings for attachment of, 153*t*
contraction of, 248, 250*f*, 251*f*
development-based organization of, 270–72
extension of, 248–50
eye, 472, 473*f*
functions of, 244
injuries, 252, 332
relationship of bones to, 266–68
superficial, 274–75*f*, 772*f*, 773*f*
tissue of (*see* Muscle tissue)
Muscle attachments, 247–48
Muscle fiber, **99–100, 244**
skeletal muscle, 248, 249*f*, 250
Muscle spindles, 413*t*, **414,** 415
Muscle tissue, **99–102, 243–64**
cardiac, 254, 256*f* (*See* also Cardiac muscle tissue)
disorders of, 252, 257–59
effect of anabolic steroids on, 260
throughout human lifespan, 259–61
overview of, 244
skeletal, 244–53, 255*t* (*See* also Skeletal muscle tissue)
smooth, 254, 257 (*See* also Smooth muscle tissue)
types of, 244
Muscle tone, affect of, on synovial joint stability, 220
Muscular (distributing) arteries, 545*f*, **546**
Muscular dystrophy, **258**
Muscularis externa, **638,** 639*f*
Muscularis mucosae, **638,** 639*f*
Muscularis of ureters, **687**
Muscular system, 4, 5*f*
Musculocutaneous nerve, brachial plexus, **431**
Myalgia, **261**
Myasthenia gravis, **437**
Mycobacterium tuberculosis, 625
Myelencephalon, **360,** 361*f*
Myelin, **347,** 348*f*
Myelin sheaths, **347,** 348*f*, 350*f*
multiple sclerosis and breakdown of, 353–54
Myelitis, **404**
Myelocyte, **513,** 514*f*
Myelogram, **405**
Myeloid stem cells, **513**
Myelomeningocele, **403**
Myenteric nerve plexus, **639,** 648
Mylohyoid muscles, **278,** 280*t*, 282*f*
Myoblasts, 259, 260, **513,** 514*f*
Myocardial infarction, **535**. *See also* Heart attacks
Myocarditis, **540**
Myocardium, 524, **525**
Myofascial pain syndrome, **258–59**
Myofibrils, skeletal muscle, **248,** 250*f*
Myofilaments in muscle tissue, 100, **244,** 257, 259*f*
Myoglobin in red slow-twitch muscle fibers, 252, 253*f*
Myoid cells, 704*f*, **707**
Myometrium, 714*f*, **720**
Myopathy, **261**
Myopia, **482,** 483*f*
Myosin, unconventional, 40
Myosin filaments, 100, **248,** 249*f*
Myosin heads (cross bridges), 248, 249*f*
Myotome, **67,** 68*t*, 270, 271*f*
Myotonic dystrophy, **258**
Myringotomy, **488**
Myxedema, **760**

Nail(s), **121,** 122*f*
Nail bed, **121,** 122*f*
Nail folds, **121,** 122*f*
Nail matrix, **121,** 122*f*
Nares, **604–5**
Nasal bones, **163,** 171*t*
Nasal cavity, **15,** 16*f*, 154, **165,** 166*f*, 167, **604–8**
Nasal conchae, **162, 163,** 166*f*, 606*f*, **607**
Nasal septum, **605**
Nasolacrimal duct, **472**
Nasopharynx, 606*f*, **608**
Natal cleft, 789*f*, **790**
Natural killer cells (lymphocytes), **509**
Navicular tarsal bone, **202,** 204*f*
Neck
arteries of, 553–57, 776, 777*f*
femur, **199,** 201*f*
innervation of, 428–29
muscles of, 280–81*t*, 282*f*, 774–76, 777*f*
rib, **179,** 180*f*
skeletal landmarks, 774, 776*f*
surface anatomy of, 774–77
of tooth, **644,** 645*f*
triangles of, 777
Necrosis, **53**
Neoplasm, **50**
Nephron, **682–84**
classes of, 684
renal corpuscle, 682, 683*f*, 687*f*
tubular section of, 682–84
vessels around, 684–85*f*
Nerve(s), 336, **348–50, 410**
bone markings for depressions/openings allowing, 153*t*
cranial, 418–26
heart innervation and, 533, 534*f*
of joints, **216,** 435
of kidneys, 679
of large intestine, 657
of larynx, 611–12
of lower limb, 432–35
of lungs, 619–21
of penis, 713
renal, 679
in skeletal muscle, 247
of skin, 435–36
of small intestine, 652
spinal, 426–36
of stomach, 648
structure of, 351*f*
of testes, 704
of upper limb, 429–32
Nerve cells. *See* Neuron(s) (nerve cells)
Nerve fascicles, **348**
Nerve fiber, **340**
Nerve impulses (action potentials), 338, 340, **342,** 343*f*
Nerve injuries, **439**
Nerve plexus, 426*f*, **428**. *See also* name of specific plexus, e.g. Pampiniform plexus
of alimentary canal, 639–40
Nervous system, 4, 5*f*, **335–58**
basic divisions of, 336–38 (*See* also Autonomic nervous system (ANS); Central nervous system (CNS); Peripheral nervous system (PNS))
enteric, of alimentary canal, 639–40
throughout human lifespan, 354–56
nerves in, 348–50
nervous tissue of, 338–48, 353–54
reflex arcs in, 350, 352*f*
simplified design of, 351–53
simplified functions of, 336*f*
Nervous tissue, **102,** 103*f*, **338–48**
disorders of, 353–54
throughout human lifespan, 354–56
neurons, 336, 338–45 (*See* also Neuron(s) [nerve cells])
supporting cells, 338, 346–48
Neural crest, **62,** 65*f*, 66, 354
Neuralgia, **439**
Neural groove, **62,** 65*f*, 403–4
Neural layer of retina, **477–78**
Neural plasticity, **402**
Neural plate, **62,** 65*f*
Neural stimuli, control of hormone release by, **745,** 746*f*
Neural tube, 62, 354, 355*f*
Neural tube defects, **403**
Neurilemma, **347,** 349*f*
Neuritis, **439**
Neuroblastoma, **356**
Neuroendocrine organ, 744
Neuroepithelial cells, **354,** 355*f*
Neurofibril, **338,** 339*f*
Neurofibrillar tangles, 403
Neurofilaments, 338, 339*f*
Neuroglia. *See* Glial cells
Neurohypophyseal bud, **762**
Neurohypophysis, **746,** 747*f*, **751–53**
cells and hormones of, 749*t*
Neurolemmocytes, 347
Neurologist, **356**
Neuromuscular junction, 247, **416–17**
Neuron(s) (nerve cells), 48, 49*f*, **102,** 103*f*, 336, **338–45**
cell body of, 338, 339*f*, 352
classification of, 343–45
embryonic formation of, 354, 355*f*
enteric, 639
hypothalamic, 750, 751*f*
injuries to, 355, 356
processes of, 338–40
signals carried by, 342–43

Neuron(s), *continued*
synapses of, 340, 341*f*
visceral sensory, 395, 396*f*, 458, 459*f*
Neuron processes (axons, dendrites), **338–40**
Neuropathy, **356**
Neurotoxin, **356**
Neurotransmitters, 340, 341, 343, 416
in sympathetic and parasympathetic divisions of autonomic nervous system, 446–47
Neurotrophins, 354
Neurulation, **62,** 65*f*
Neutrophils, **91,** 97*f*, 506, **507–8,** 511*t*
Nipple, **726,** 780
Nociceptors, **413**
Nodes of Ranvier (neurofibral nodes), **348,** 349*f*
Nonaxial movements, **224**
No-new-neurons doctrine, 355
Non-Hodgkin's lymphoma, **599**
Norepinephrine, 447
Norian SRS bone cement, 207
Normoblast, **513,** 514*f*
Nose, **604,** 605*f*, 621*t*
surface anatomy of, 770, 775*f*
Nosebleed, 627
Notochord, **12, 62,** 64*f*
derivatives of, 67, 68*t*
Nuchal lines, **157,** 158*f*
Nuclear envelope, **42,** 44*t*
Nuclear pores, **42**
Nucleoli (nucleolus), 42*f*, 44*t*, **45**
Nucleoplasm, 42
Nucleosomes, **43**
Nucleotides, 43
Nucleus, 29, **41–45**
Nucleus cuneatus, **381,** 398*f*
Nucleus gracilis, **381,** 398*f*
Nucleus pulposus, 67, **168,** 169*f*
Nutrient arteries in bones, 134, 135*f*
Nutrient foramen, 134
Nutrient veins in bones, 134, 135*f*
Nutrition
antioxidants in diet, 52
for disease prevention, 145
diseases related to inadequate, 91, 145–46
for healthy tissue, 106
for prevention of birth defects, 180, 404

Obdurator externus muscle, 316*f*, 317*t*
Obdurator internus muscle, 316*f*, 318*t*
Obesity, 98
Oblique fissure of lung, **617,** 619*f*
Oblique muscles, 289–90*t*, 291*f*, **472,** 473*f*
Oblique popliteal ligament, **231,** 232*f*
Oblique sections, **12**
Obstructive emphysema, **624,** 625*f*
Obturator artery, **561,** 563*f*
Obturator foramen, **198**
Obturator nerve, **432,** 433*f*
Occipital artery, **553,** 556*f*
Occipital bone, 155*f*, **157,** 170*t*
Occipital condyles, **157,** 158*f*
Occipitalis muscle, 276*t*, 277*f*
Occipital lobe, cerebral hemispheres, 364*f*, **365**
Occipitofrontalis muscles, 276*t*, 277*f*
Occipitomastoid sutures, 156*f*, 157
Ocular (bulbar) conjunctiva, **471,** 472*f*
Oculomotor nerve (III), 380, 418, 419*t*, 420*t*
cranial outflow via, 448
Oculomotor nerve paralysis, 420
Oculomotor nucleus, 448
Odontoblast, 645
Odontoid process, **173**
Olecranon bursitis, **236**
Olecranon fossa, humerus, 189*f*, **190**
Olecranon process of ulna, **190,** 191*f*, 783
Olfaction (smell), 468–70
Olfactory bulb, 367*f*, **370, 469**
Olfactory cilia, **468,** 469*f*
Olfactory (smell) cortex, 367*f*, **370**
Olfactory epithelium, **468,** 608
Olfactory foramina, 162
Olfactory mucosa, 606
Olfactory nerve (I), 370, **418,** 419*t*, 420*t*
filaments of, 469
Olfactory pits, 628
Olfactory placode, **628**
Olfactory receptor cells, **468,** 469*f*
Olfactory tract, 367*f*, **370**
Oligodendrocytes, 346*f*, **347,** 348
Olive, **381**
Omohyoid muscles, 281*t*, 282*f*
Oncogenes, **50,** 51
Oocyte (egg), 48, 49*f*, 59, 60*f*, **716**
Oogenesis, **718,** 719*f*
Oogonia, **718,** 719*f*
Open reduction, bone fracture treatment by, **143**
Opponens digiti minimi muscle, 309*t*, 310*f*
Opponens pollicis muscle, 309*t*, 310*f*
Opposition movement, **222,** 223*f*, 224
Opthalmic artery, **555,** 556*f*
Opthalmic vein, **566,** 567*f*
Opthalmology, **497**
Optical device, eye as, 480, 481*f*
Optic chiasma, **377,** 380*f*, **481,** 484*f*, 746
Optic cups, 485*f*, **486**
Optic disc, **479**
Optic fissure, 485*f*, **486**
Optic foramen, **161**
Optic nerve (II), **418,** 419*t*, 420*t*, **481,** 484*f*
Optic nerve damage, 420
Optic radiation, **481,** 484*f*
Optic stalks, 485*f*, **486**
Optic tract, **481,** 484*f*
Optic vesicles, **485**
Oral cavity (mouth), **14,** 16*f*, **640–41**
Oral cavity proper, **640**
Oral membrane, **667,** 668*f*
Oral orifice, **640**
Ora serrata retinae, 476*f*, **479**
Orbicularis oculi muscle, 276*t*, 277*f*
Orbicularis oris muscle, 277*t*
Orbital cavity, **15,** 16*f*
Orbital fissures, 161*f*, **162, 163,** 165*f*
Orbital plates, ethmoid bone, **162**
Orbits, skull, 154, **165**
Orchitis, **738**
Organ(s). *See also* Organ system(s)
capillaries of, 544
digestive, 634, 635, 637–66
effects of parasympathetic and sympathetic divisions on
select, 451*t*
endocrine, 744*f*, 745–59
lymphoid, 595–98
respiratory, 604–21
stem cells for treatment of failing, 105, 106
target, of hormones, 87
urinary, 676–91
visceral (*see* Viscera (visceral organs))
Organelles, 4, **28, 33–41,** 44*t*
Organismal level, structural organization at level of, 3*f*, **4**
Organization, formation of granulation tissue, **103,** 104*f*
Organ system(s). *See also* Viscera (visceral organs); names of specific systems, e.g. Integumentary system
overview of, 4, 5–7*f*
structural organization at level of, 3*f*, **4**
Organ transplants
heart, 536
kidney, 693
Origin, muscle attachment, **247**
Oropharynx, 606*f*, **608,** 640*f*, **645**
Orthopedist, **182**
Os coxae, **194,** 195*f*
Osmosis, 30
Ossicles, middle ear, 487*f*, **488,** 489*f*
Ossification, **139–40**
Ossification centers, **179**
Ostealgia, **147**
Osteoarthritis (OA), **236–37,** 238
Osteoblasts, 99, **139,** 140*f*, 142
Osteoclasts, **142,** 143*f*
Osteocytes, **99, 137,** 138*f*
Osteogenesis, **139–40**
Osteoid, **139,** 140*f*
Osteomalacia, **145**
Osteomyelitis, **147**
Osteon, **137,** 138*f*, 139*f*
Osteoporosis, **143–45,** 146*f*, 179
Osteosarcoma, **146–47**
Otic ganglion, **448**
Otitis externa, **497**
Otitis media, **488**
Otolithic membrane, **489–90,** 491*f*
Otoliths, **490,** 491*f*
Otorhinolaryngology, **497**
Otosclerosis, **488**
Outer body wall, 59
Outer (external) ear, **486,** 487*f*
Oval window, middle ear, **486,** 487*f*
Ovarian arteries, 559*f*, **560,** 715
Ovarian cancer, **733**
Ovarian cortex, **716**
Ovarian cycle, **716–18,** 723*f*
Ovarian ligament, 714*f*, **715**
Ovarian medulla, **716**
Ovarian veins, 569
Ovaries, 714*f*, **715–18**
descent of, 737–38
ovarian cycle and, 716–18
structure of, 716*f*
Overflow incontinence, **692**
Oviducts, **718.** *See also* Uterine tubes
Ovulation, 717, 718*f*, 723*f*
Ovum (egg), 48, 49*f*, 59, 60*f*, **718,** 719*f*
Oxidases, 39
Oxygen, transport of, by erthrocytes, 506
Oxyphil cells, **754,** 755*f*
Oxytocin, 731, 749*t*, **752**
therapeutic uses of, 753

Pacinian corpuscles, 413, **414,** 415
Pagetic bone, 146
Paget's disease, **146**
Pain, 103
back, 169, 172, 261
brain and, 380
heart-related, 524, 535
muscle, 258, 259
peptic ulcers and stomach, 650
referred, 458, 459*f*
visceral, 458
Palate, **641**
cleft, 179–80, 182*f*
hard, **163,** 166, **605,** 606*f*, 641
soft, 605, 606*f*, 641
Palatine bones, **163,** 164*f*, 171*t*
Palatine processes, maxillary bones, **163,** 164*f*
Palatine tonsils, **597,** 598*f*, 606*f*, **608**
Palatoglossal arches, **641,** 642*f*
Palatopharyngeal arches, **641,** 642*f*

Palmar arches, 557, 558*f*
Palmar interossei muscle, 310*t*, 311*f*
Palmaris longus muscle, 301*t*, 302*f*, 787, 788*f*
Palmar venous arches, 567, 568*f*
Palpation, **770**
Palpebrae, **470**, 471*f*
Palpebral conjunctiva, **471**
Palpebral fissure, **470**, 471*f*
Pampiniform plexus, 703*f*, **704**
Pancreas, 635, **663**, 664*f*, 744, **758–59**
Pancreatic islets (islets of Langerhans), **663**, 664*f*, **758**
Pancreatic polypeptide, **759**
Pancreatic transplantation, **763**
Pancreatitis, **663**
Paneth cells, 654
Papanicolaou (Pap) smear, **734**
Papilla, kidney, **678**, 679*f*
Papilla ducts, nephron and, 680*f*, **684**
Papillary layer of dermis, **115**
Papillary muscles of heart, **525**, 527*f*, 529
Parafollicular (C) cells, 753*f*, **754**
Parahippocampal gyrus, **387**
Parallel arrangement of fascicles, **268**, 269*f*
Paramesonephric (Mullerian) ducts, **734**, 735*f*
Paranasal sinuses, **167**, **608**
Paraplegia, **403**
Pararenal fat, **678**
Parasagittal plane, **12**
Parasternal lymph nodes, 726
Parasympathetic division of peripheral nervous system, 410, **446**, **447–57**
  cranial outflow, 448–49, 450*f*
  effects of, on select organs, 451*t*
  sacral outflow, 449–50
  sympathetic division versus, 447*f*, 448*t*
Parathyroid glands, 744, **754**, 755*f*
  calcium concentration monitored by, 745, 746*f*, 754
  embryonic development of, 762*f*
Parathyroid hormone (PTH) (parathormone), 143, **754**, 755*f*
Paraxial mesoderm, **62**, 65*f*
Paresthesia, **439**
Parietal bones, **154**, 155*f*, 156*f*, 170*t*
Parietal (oxyntic) cells of gastric glands, 649*f*, **650–51**
Parietal layer
  of glomerular capsule, **682**, 683*f*
  of serous pericardium, **524**
Parietal lobe, cerebral hemisphere, **363**, 364*f*
Parietal peritoneum, 637*f*, **638**
Parietal pleura, **615**, 616, 617*f*
Parietal serosa, **14**, 15*f*
Parieto-occipital sulcus, cerebral hemisphere, 364*f*, **365**
Parkinson's disease, **372–74**, 381
Parotid duct, **643**
Parotid gland, **643**
Pars distalis (anterior lobe) of adenohypophysis, **746**, 747*f*
  cells and hormones of, 749*t*
  tissue of, 748*f*
Pars intermedia of adenohypophysis, **746**, 747*f*, 750
Pars nervosa (posterior lobe) of neurohypophysis, **746**, 751*f*, 760
Pars tuberalis of adenohypophysis, **746**, 747*f*, 750
Partial-thickness burns, **122**
Parturition, **729–32**. *See also* Childbirth
Patella (kneecap), 197*t*, **199**, 201*f*, 231, 232*f*, 239, 791, 792*f*
Patellar ligament, **231**, 232*f*, 791, 792*f*
Patellar retinacula, **231**, 232*f*
Patellar surface, **199**, 201*f*
Patellofemoral pain syndrome, **239**
Pathological anatomy, **3**
Pathologic fracture, **147–48**
Pathology, **3**
Pectinate muscles of heart, **525**, 527*f*
Pectineal line, 198
Pectineus muscle, **312**, 313*t*, 314*f*
Pectoral girdle (shoulder), **186–89**
  clavicles, 186, 187*f*
  scapulae, 186, 187*f*, 188
  surface anatomy of, 783–89
Pectoralis major muscle, **294**, 296*f*, 297*t*, 299*f*, 779*f*, 780
Pectoralis minor muscle, 294*t*, 296*f*
Pedicles
  of podocytes, **682**, 683*f*
  vertebral arch, **172**
Pelvic brim, **198**
Pelvic cavity, **14**
Pelvic diaphragm, muscles of, 292*t*, 293*f*
Pelvic girdle, **194–99**. *See also* Pelvis
  surface anatomy of, 782*f*, 783
Pelvic inflammatory disease (PID), **720**
Pelvic inlet, 198*f*, 199, 200*t*
Pelvic kidney, **695**
Pelvic organs, sympathetic pathways to, 456, 457*f*
Pelvic outlet, 198*f*, 199, 200*t*
Pelvic plexus, 450
Pelvic splanchnic nerves, 449*f*, **450**, 456
Pelvic structure and childbearing, 199, 200*t*
Pelvimetry, **207**
Pelvis
  arteries of, 561–63
  bony, **194**, 195*f*
  false, 198, **199**
  muscles of, 312*t*, 314–15*f*
  surface anatomy of, 783
  true, 198, **199**
  veins of, 571, 572*f*
Penis, 702, 710*f*, **712–15**
  circumcision of, 712–13
  external anatomy of, 712
  internal anatomy of, 713
  sexual response and erection of, 713–15
Pennate pattern of fascicles, **268**, 269*f*
Pepsin, **646**
Pepsinogen, **651**
Peptic ulcers, 650–51
Percussion, **540**
Perforated eardrum, **486**
Periaqueductal gray matter, **379–80**, 384*f*
Pericardial cavity, **13**, 14*f*, **524**
Pericarditis, **524**
Pericardium of heart, **14**, 15*f*, **524**
Perichondrium, 130
Pericytes, **548**
Perikaryon, 338, 339*f*
Perilymph, **489**
Perimetrium, 714*f*, **721**
Perimysium, **246–47**
Perineum
  female, 724*f*, **725**, 782*f*
  male, **715**, 782*f*
  muscles of, 292*t*, 293*f*
  surface anatomy of, 782*f*, 783
Perineurium, **348**
Periodontal ligament, 213, **645**
Periodontitis, **645**
Periodontium, **645**
Periosteal bud, **140**
Periosteal layer, **388**, 389*f*
Periosteum, **134**
Peripheral chemoreceptors, 623, 624*f*
Peripheral nervous system (PNS), **336**, **409–42**
  cranial nerves, 418–26 (*See* also Cranial nerve[s])
  defined, **410**
  disorders of, 420, 421, 422, 423, 424, 425, 426, 437, 438
  functional organization of, 410*f*
  general visceral motor part of (*see* Autonomic nervous system [ANS])
  throughout human lifespan, 437
  myelin sheaths in, 347–48, 349*f*
  peripheral motor endings of, 416–18
  peripheral sensory receptors of, 411–16
  simplified design of, 351–53
  spinal nerves, 426–36 (*See* also Spinal nerves)
  structural components of, 410, 411*f*
  supporting cells in, 346*f*, 347
Peripheral neuropathy, **439**
Peripheral process, **343**, 345
Peripheral proteins, **30**
Peripheral reflexes, 459
Peripheral sensory receptors, **411–16**
  classification of, by location, 411
  classification of, by stimulus detected, 411–13
  classification of, by structure, 412, 413–16
  in epithelia, 76
Perirenal fat, **678**
Peristalsis as digestive process, 635*f*
Peritoneal cavity, **14**, 15*f*, 637*f*, **638**
Peritoneal organs (intraperitoneal organs), **638**, 666*t*
Peritoneum, **14**, 15*f*, **637–38**
Peritonitis, **638**
Peritubular capillaries, 684*f*, **686**
Permanent teeth, **644**
Peroxisomes, **39**, 44*t*
Perpendicular plate, ethmoid bone, **162**
Perpendicular plates, palatine bones, **163**, 164*f*
Perphorating canals (Volkmann's canals), **137**, 138*f*
PET (positron emission tomography), **21**, **366**
Petrous region, temporal bone, 158*f*, **159**, 160
Peyer's patches (aggregated lymphoid nodules), 595*f*, 598, 599*f*, 654
Phagocytosis, **31**
  microglia as, 346
  performed by blood cells, 507, 508
Phagosome, **31**
Phalanges
  fingers, 193*f*, **194**
  toes, 204*f*, 205
Phantom limb pain, **405**
Pharyngeal arch muscles, 271*f*, **272**, 338
Pharyngeal constrictor muscles, **280**, 281*t*, 282*f*, 646
Pharyngeal pouches, **12–13**, 496, 667
Pharyngeal tonsils, **598**, 606*f*, **608**
Pharyngotympanic tube, **486**, 487*f*, 488*f*
Pharynx, **12**, 606*f*, **608–9**, 621*t*, 634, 640*f*, **645–46**
Phenobarbitol, 36

Pheochromocytoma, **763**
Philtrum of face, **770,** 775*f*
Phlebitis, **578**
Phlebotomy, **578**
Phosphate ions in bones, 142, 143
Photoreceptor cells, **413,** 477*f,* **478–79**
Photorefractive keratectomy, 482
Phrenic arteries, 558, 559–60
Phrenic nerve, **429,** 777
Physiology, **2**
Pia mater, **389**
Pigmented layer of retina, **477**
Pigments
  hair color and, 117
  iris of eye and, 476–77
  skin color and, 114, **116–17**
Pineal gland (pineal body), **379,** 744, **758**
Pinealocytes, 758
Pinkeye, **471**
Pinna, ear, **486,** 487*f*
Pinocytosis, **31–32**
Pinocytotic vesicle, **32**
Piriformis muscle, 316*f,* 317*t*
Piriform lobe, 370, 469–70
Pisiform bone, **191,** 193*f*
Pituitary dwarfs, **760**
Pituitary gland, 377, 744, **745–53,** 754
  adenohypophysis of, 746, 747–51
  disorders of, 760
  embryonic development of, 762
  neurohypophysis of, 746, 747*f,* 751–53
Pivot joints, **225**
Placenta, **729, 759**
  anatomy of, 729, 731*f*
  blood vessels and circulation through, 576, 577*f*
  displaced, 729
  as endocrine structure, 759
  formation of, 728–29, 730–31*f*
Placental abruption, **729**
Placental barrier, 729
Placental-blood transplants, **516**
Placental (third) stage of labor, **732**
Placenta previa, **729**
Plane joint, **224**
Plantar and dorsal interossei muscles, 330*t,* 331*f*
Plantar arch, 562*f,* **563**
Plantar arteries, 562*f,* 563
Plantar flexion movement, **222,** 223*f*
Plantaris muscle, 323*t,* 324*f*
Plantar nerves, **434**
Plantar veins, **571**
Plasma, blood. *See* Blood plasma
Plasma cells, **91, 509**
Plasma membrane (plasmalemma), **29–33,** 44*t*
  functions of, 30–33
  structure of, 29–30
Platelets, 504, **510,** 511*t*
  disorders of, 515–16
  formation of, 513–15
Platysma muscle, 277*t*
Pleura, **14,** 15*f*
  of lungs, **615,** 616, 617*f,* 621*t*
Pleural cavity, **13,** 14*f,* **615,** 617*f,* 780
Pleural effusion, **617**
Pleural fluid, 615
Pleurisy (pleuritis), **617**
Plicae circulares (circular folds) of small intestine, **653**
Pluripotential hematopoietic stem cell, **513,** 514*f*
Pneumocystis carinii, pneumonia caused by, 592, 625
Pneumonia, 592, **625**
Pneumothorax, **623**
Podiatry, **207**
Podocytes, **682,** 683*f*
Polar bodies, **718,** 719*f*
Polarity of epithelia, 77
Polio, 438
Polycythemia, **515**
Polymorphonuclear cells, 507
Polyribosomes, **35**
Polysynaptic reflexes, **350,** 352*f*
Pons, **360,** 361*f,* **381,** 382*f,* 383*f,* 384*f,* 390*t,* 691
Pontine nuclei, 381
Popliteal artery, 562, 563, 791
Popliteal fossa, **791,** 793*f*
Popliteal ligaments, **231,** 232*f*
Popliteal vein, **571,** 572*f,* 791
Popliteus muscle, 323*t,* 325*f*
Pore, sweat, 115, **121**
Porta hepatis, 559*f,* **660**
Portal hypertension, **572**
Portal systemic anastomoses, 572–73
Portal triad (portal tract), **660,** 661*f*
Positron emission tomography (PET), **21, 366**
Postcentral gyrus, cerebral hemisphere, 364*f,* **365**
Posterior, defined, **8,** 9*t*
Posterior arch, atlas vertebrae, **173**
Posterior axillary fold, 779*f,* **783**
Posterior cerebral arteries, **555,** 556*f*
Posterior chamber of eye, 475*f,* **480**
Posterior circumflex humeral arteries, **557,** 558*f*
Posterior cranial fossa, **157**
Posterior cruciate ligament, **231,** 232*f,* 233*f*
Posterior division of brachial plexus root trunk, **429,** 430*f*
Posterior fontanels, **179,** 181*f*
Posterior funiculus, 394*f,* **395**
Posterior gluteal lines, **194,** 195*f*
Posterior horns, spinal cord, 394*f,* **395**
Posterior inferior iliac spine, **194,** 195*f*
Posterior intercostal arteries, **558**
Posterior intercostal veins, 568*f,* **569**
Posterior interventricular artery of heart, 534*f,* **535**
Posterior interventricular sulcus, heart, **525,** 526*f*
Posterior lobe, cerebellar hemisphere, **385,** 386*f*
Posterior longitudinal ligaments, **168,** 169*f*
Posterior median furrow, **777,** 778*f*
Posterior median sulcus, 394*f,* **395**
Posterior nasal apertures, **605**
Posterior pole, eyeball, **474**
Posterior segment of eye, 474*f,* 475*f,* **480**
Posterior semicircular canals, 490*f,* **491**
Posterior spinocerebellar pathway, **397,** 398*f,* 399*t*
Posterior superior iliac spine, **194,** 195*f,* 789
Posterior talofibular ligaments, **234,** 235*f*
Posterior tibial artery, 562*f,* 563, 791
Posterior tibial vein, **571,** 572*f*
Posterior tibiofibular ligaments, **234,** 235*f*
Posterior triangle of neck, **777**
Postganglionic axon, **445**
  neurotransmitters released by, 446–47
Postganglionic neuron, **445,** 449, 452, 453*f,* 455–57*f,* 460
Postpolio syndrome (PPS), 438
Postsynaptic densities, **341**
Postsynaptic neuron, **341**
Power stroke/recovery stroke of cilium, 83, 84*f*
Pre-Bozinger complex, 382, 623
Precapillary sphincters, **547**
Precentral gyrus, cerebral hemispheres, **363,** 364*f*
Prefrontal cortex, 367*f,* **370–71**
Preganglionic axon, **445**
Preganglionic neuron, **445,** 452, 453*f,* 455–57*f,* 460
Pregnancy, 726–32
  development of embryo and fetus during, 59–69
  ectopic, 720
  events leading to fertilization and, 727–29
  formation and anatomy of placenta, 728–29, 730–31*f*
  implantation of blastocyst in, 728*f,* 730*f*
Premature birth, **68**
Premolars (teeth), **644**
Premotor cortex, 367*f,* **368**
Prenatal period, **58.** *See also* Embryonic period; Fetal period
Preoptic nucleus, **377,** 379*f*
Prepuce (foreskin), 710*f,* **712**
Prepuce of the clitoris, 724*f,* **725**
Presbycusis, **497**
Presbyopia, **482**
Presynaptic densities, **341**
Presynaptic neuron, **341**
Pretectal nuclei, **484**
Prevertebral ganglia, **452,** 453*f*
Primary auditory cortex, 367*f,* **370**
Primary brain vesicles, **360,** 361*f*
Primary capillary plexus, **751**
Primary follicle, **717**
Primary motor cortex, **366,** 367*f,* 368*f*
Primary oocytes, **718,** 719*f*
Primary ossification center, **140,** 147*f*
Primary sex organs (gonads), 59, **702**
Primary somatosensory cortex, 367*f,* 368*f,* **369**
Primary spermatocytes, **705,** 706*f*
Primary visual (striate) cortex, 367*f,* **369**
Prime mover muscle, **268**
Primitive gut, **66,** 667
Primitive intestinal loop, **667,** 668*f*
Primitive node, **62,** 63*f*
Primitive streak, **62,** 63*f*
Primordial follicles, **717**
Proctodeum, **667,** 668*f*
Proerythroblasts, **513,** 514*f*
Progenitor cells, blood, 513
Progesterone, **718, 759**
Projection fibers, **372,** 374*f*
Prolactin (PRL), **748,** 749*t*
Prolactin-inhibiting hormone, 750
Prolactinoma, **763**
Prolapsed disc, **169**
Prolapse of the uterus, 714*f,* **715**
Proliferative phase of uterine cycle, 723*f,* **724**
Promonocytes, 514*f,* **515**
Promyelocytes, **513,** 514*f*
Pronation movement, **222,** 223*f*
Pronator quadratus muscle, **301,** 302*f,* 303*f,* 304*t*
Pronator teres muscle, 301*t,* 302*f,* 786*f,* 787
Pronephric duct, **693**
Pronephros, **693,** 694*f*
Proprioceptive senses (proprioception), **336**
Proprioceptors, **411,** 413*t,* 414–16
  structure of, 415*f*
Propulsion of food, **635**
Prosencephalon (forebrain), **360**
Prostaglandins, 711
Prostate cancer, 711, **733**
Prostate gland, 702, 689, **711–12**
  benign prostatic hyperplasia of, 711

Prostate-specific antigen (PSA), 711
Prostatic urethra, 690*f*, **691**, 710
Prostatitis, **712**
Protease inhibitors, 593
Proteases, 91
Protein(s)
  in blood plasma, 505
  in plasma membrane, 29*f*, 30
  as receptors, 30, 32
Protein synthesis
  DNA and instructions for, 41–42
  ribosomes and, 34–35
  at rough endoplasmic reticulum, 36*f*
Proteoglycans, 91
Protraction movement, **222**, 223*f*
Proximal convoluted tubule, 680*f*, 681*f*, **682**
Proximal radioulnar joint, 190
Pseudopods, 40
Pseudostratified columnar epithelium, 78, 80*f*
Psoas muscle, 312*t*, 314*f*
Psoriasis, **125**
Psychosocial dwarfism, **763**
Pterygoid processes, sphenoid bone, **161**
Pterygolid muscles, **278**, 279*f*
Pterygopalatine ganglion, **448**
Puberty, **738**
Pubic (subpubic) arch, 195*f*, **198**
Pubic crest, 195*f*, **198**, 779*f*, **781**
Pubic region of abdominal wall, 636*f*, **637**
Pubic symphysis, 195*f*, **198**, 779*f*, 781
Pubic tubercle, 195*f*, **198**, 779*f*, 781
Pubis (pubic bone), 195*f*, **198**
Pubofemoral ligament, **229**, 230*f*
Pudendal artery, 561
Pudendal nerve, **434**, 435*t*
Pudendal nerve block, **434**
Pudendum, **724**
Pulmonary arteries, **552**, 618*f*, **619**
Pulmonary capillary networks, 616*f*, **619**
Pulmonary circuit, **522**, 552
  shunts away from, 576–78
Pulmonary circulation, **552–53**
Pulmonary plexus, **449**
Pulmonary trunk, **525**, 527*f*, **552**
Pulmonary (semilunar) valve of heart, 525, **529**, 530*f*, 531*f*
Pulmonary veins, **525**, 527*f*, **552**, 618*f*, **619**
Pulmonary ventilation, **604**, **621–23**
Pulp cavity of tooth, **645**
Pulp of tooth, **645**
Pulse oxymeter, 774
Pulse points, 774*f*, 783, 788*f*, 789
Pupil, eye, 476
Pupillary light reflex, **476**
Purkinje fibers of heart, 532, **533**
Purkinje myocytes, 533
Pus, **508**
Putamen, **372**, 375*f*
Pyelitis, **692**
Pyelogram, **687**
Pyelonephritis, **678**, **692**
Pyloric antrum, **647**, 648*f*
Pyloric canal, **647**, 648*f*
Pyloric region of stomach, **647**, 648*f*
Pyloric sphincter, **647**, 648*f*
Pyloric stenosis, 650, **669**
Pylorus, **647**, 648*f*
Pyramid(s), **381**, 382*f*
Pyramidal cells, 366
Pyramidal (corticospinal) tracts, 366, **397**, 400*f*, 401*t*

Quadrants of abdominal wall, 636*f*, **637**
Quadrate lobe of liver, 659*f*, **660**
Quadratus femoris muscle, 316*f*, 318*t*
Quadratus lumborum muscle, 285*t*, 286*f*
Quadriceps femoris muscle, **312**, 313*t*, 314*f*, 791, 792*f*
Quadriplegia, **403**

Rabies, **356**
Radial arteries, 557, 558*f*, **722**, 788*f*, 789
Radial collateral ligament, **227**, 229*f*
Radial fossa, humerus, 189*f*, **190**
Radial groove, humerus, 189*f*, **190**
Radial keratectomy, 482
Radial nerve, brachial plexus, 431*f*, **432**
Radial notch, ulna, **190**, 192*f*
Radial tuberosity, radius, **190**, 192*f*
Radial vein, **567**, 568*f*
Radiation therapy, 51
Radical mastectomy, **734**
Radiographic anatomy, **3**
Radius, 190, 191*f*, 192*f*, 787
Rami communicantes, 427
  white, and gray, 452, 453*f*, 454*f*
Ramus
  dorsal and ventral, 427–36
  ischium, **194**, 195*f*
  mandible, **162**, 164*f*, 774, 775*f*
  pubis, 195*f*, **198**
Raphe, 248
Raphe nuclei, **387**, 388*f*
Rathke's pouch, 762
Raynaud's disease, **459**
Reabsorption, kidney, **681**
Receptive sites, dendrites as, 340
Receptor(s)
  proteins as, **30**
  reflex, 350, 352*f*
Receptor cells of special senses, **466**
  olfactory, 468, 469*f*
  photoreceptor, 477*f*, 478–79
  taste, 466, 467*f*
Receptor-mediated endocytosis, **32**, 33*f*
Rectal arteries, 561
Rectal valves, 655*f*, **657**
Rectocele, **669**
Rectouterine pouch, 721
Rectum, 655*f*, **656–57**
  evacuation of, 611, 657
Rectus abdominis muscle, 289*t*, 291*f*, 779*f*, 781
Rectus femoris muscle, **312**, 313*t*, 314*f*
Rectus muscles, **472**, 473*f*
Recurrent laryngeal nerves, 611–12
Red blood cells. *See* Erythrocytes
Red marrow, **512**
Red nucleus, **381**, 384*f*
Red pulp of spleen, **596**
Red slow-twitch muscle fibers, 252, 253*f*
Reduction, bone fracture treatment by, **143**
Reed-Sternberg cells, 599
Referred pain, **458**, 459*f*
Reflex arcs, **350**, 352*f*
  visceral, 459
Reflexes, **350**, 352*f*, 391
  visceral, 459
Refractory media, **481**
Regeneration of tissue, 77, **103**, 104
  spinal cord, 356
Regional anatomy, **2**
Regional terms, **8**, 10*f*
Reinforcing ligaments, synovial joint, 215*f*, **216**
Releasing hormones (releasing factors), **750**
Renal arteries, 559*f*, 560, 677*f*, **678**, 679*f*
Renal calculi, **693**
Renal capsule, 677*f*, **678**
Renal columns, **678**, 679*f*
Renal corpuscle, 682*f*, **683**, 684*f*, 687*f*
Renal cortex, **678**, 679*f*
Renal fascia, 677*f*, **678**
Renal infarct, **695**
Renal medulla, **678**, 679*f*
Renal pelvis, **678**, 679*f*
Renal plexus, 679
Renal pyramids (medullary pyramids), **678**, 679*f*
Renal sinus, 678, 679*f*
Renal veins, 569, 677*f*, **679**
Renin, **686**, 759
Repair of tissues, 103–6
Repetitive stress injuries, **193**
Reproductive duct system, male, 708–10
Reproductive system, 7*f*, **701–42**
  disorders of, 732–34
  female, 715–34
  throughout human lifespan, 734–38
  male, 702–15
  pregnancy and childbirth, 726–32
Respiration, **604**
Respiratory bronchioles, **614**, 615*f*
Respiratory distress syndrome (RDS), **623**, 628
Respiratory membrane, **614**, 616*f*
Respiratory mucosa, 606, **607**
Respiratory system, 4, 6*f*, **603–32**
  disorders of, 608, 617, 623–28
  functional anatomy of, 604–21
  throughout human lifespan, 628–29
  principal organs of, summary of, 621*t*
  ventilation and, 621–23
Respiratory zone, **614**, 615*f*, 616*f*
Response to injury hypothesis, **574**
Rete testis, **704**
Reticular activating system (RAS), 388*f*
Reticular cells, **98**, 513
Reticular connective tissue, 93*f*, **98**
Reticular fibers, **90**, 115
Reticular formation, 381–82, 384*f*, **387–88**, 390*t*, 623
Reticular layer of dermis, **115**
Reticulocyte, **513**, 514*f*
Reticulocyte count, **513**
Reticulospinal tract, 388, **397**, 401*t*
Retina, **477–79**
  blood supply in, 479
  detached, 479
  photoreceptors of, 478*f*, 479
  regional specialization of, 479
Retinacula muscles, **319**
Retinal detachment, **479**
Retinopathy of prematurity, **485**
Retraction movement, **222**, 223*f*
Retroperitoneal organs, **638**, 666*t*
Retroverted uterus, 714*f*, **720**
Reverse transcriptase inhibitors, 593
Rhabdomyolysis, **252**
Rheumatoid arthritis, **237**
Rhinencephalon, 370, 386
Rhinitis, **607**
Rhombencephalon (hindbrain), **360**, 361*f*
Rhomboid muscles, 295*t*, 296*f*
Rib(s), 176, **178–79**, 180*f*
  joint between sternum and, 214*f*
Ribosomal RNA, **34**, 45
Ribosomes, **34–35**, 44*t*
RICE (rest, ice, compression, elevation), **261**

Rickets, **145–46**
Right atrioventricular (tricuspid) valve of heart, 525, **529**, 530*f*, 531*f*
Right atrium of heart, **525**, 526*f*, 527*f*
Right auricle of heart, **525**, 527*f*
Right brachiocephalic veins, **564**, 565*f*
Right colic artery, **560**, 561*f*
Right colic flexure (hepatic flexure), 655*f*, **656**
Right coronary artery, 534*f*, **535**
Right gastric artery, **560**
Right gastro-epiploic artery, **560**
Right hepatic duct, **660**
Right hypochondriac region, 636*f*, **637**
Right iliac region, 636*f*, **637**
Right lobe of liver, **658**, 659*f*, 660*f*
Right lumbar regions, 636*f*, **637**
Right lymphatic duct, 587*f*, **589**
Right pulmonary artery, **552**
Right ventricle of heart, **525**, 526*f*, 527*f*
Rima glottidis, **610**, 611*f*
Risorius muscle, 276*t*, 277*f*
Rod cells, **478**
Root
  of hair, **117**, 118*f*
  of lung, **617**, 618*f*, 619*f*
  of nail, **121**, 122*f*
  of penis, 710*f*, **712**
  of tooth, **644**, 645*f*
Root canal, **645**
Root canal therapy, **645**
Root hair plexus, **117**, **413**
Rootlets, spinal **426**, 427*f*
Roots of brachial plexus, 428, **429**, 430*f*
Rostral, defined, **360**
Rostral ventrolateral medulla (RVLM), 623
Rotation movement, **221**, 222*f*
Rotator cuff, **227**, 228*f*
Rough endoplasmic reticulum (rough ER), **35**, 36*f*, 44*t*
Round ligament (ligamentum teres), 659*f*, **660**
Round ligaments of the uterus, 714*f*, **721**
Round window, middle ear, **486**, 487*f*
Rubrospinal tract, **397**, 400*f*, 401*t*
Ruffini's corpuscles, 413, **414**, 415–16
Rugae of stomach mucosa, **648**
Rule of nines, **122**, 123*f*

Saccule, **489**, 490*f*
Sacral artery, 561
Sacral canal, **176**, 177*f*
Sacral crest, **176**, 177*f*
Sacral curvature, 168*f*, **172**
Sacral foramina, 176
Sacral hiatus, **177**
Sacral outflow of parasympathetic division, 449–50
Sacral plexus, **432–35**
Sacral promontory, **176**, 177*f*
Sacroiliac joint, 176, 789
Sacrospinalis muscle, 285*t*, 286*f*
Sacrospinous ligament, 194
Sacrotuberous ligament, 198
Sacrum, 168*f*, **172**, **175–77**, 779
Saddle joints, 225*f*, **226**
Sagittal planes, 11*f*, **12**
Sagittal sinuses, 566, 567*f*
Sagittal suture, **154**, 155*f*
Saliva, 642
Salivary glands, 448, 635, **642**, 643*f*
Salivary nucleus, 448
Salpingectomy, **720**
Salpingitis, **720**
Saphenous veins, 572, 793
Sarcolemma, **244**, 249*f*
Sarcoma, **106**
Sarcomeres, **248**, 249*f*, 250*f*, 251*f*
Sarcopenia, **261**
Sarcoplasm, **244**, 250*f*
Sarcoplasmic reticulum, **251**, 252*f*
Sartorius muscle, 312*t*, 314*f*
Satellite cells, 261, 346*f*, **347**
Satorius muscle, 791, 792*f*
Scala media, **493**
Scala tympani, **493**
Scala vestibuli, **493**
Scale, 4
Scalenes muscle, 283*t*, 284*f*
Scalp, muscles of, 276*t*, 277*f*, 770
Scanning electron microscopy, **18**
Scaphoid bone, **191**, 193*f*
Scapula, **186–89**
  muscles affecting movement of, 294, 296*f*
  spine of, 777
Scapular winging, **439**
Scar tissue, **103**, 104
Schleiden, Matthias, 28
Schwann, Theodor, 28
Schwann cells, 346*f*, **347**
Sciatica, **434–35**
Sciatic nerve, **432**, 434*f*
Sciatic nerve injury, 434
Sciatic notch, **194**, 195*f*
Sclera, **474**
Sclera venous sinus, **475**
Sclerotome, **67**, 68*t*
Scoliosis, **179**
Scotoma, **497**
Scrotum, **703**
Scurvy, **91**
Sea bone, 206
Sebaceous glands, **120–21**, 124
Sebum, **120–21**
Secondary brain vesicles, **360**, 361*f*
Secondary capillary plexus, **751**
Secondary (antral) follicle, 716*f*, **718**
Secondary oocyte, **718**, 719*f*
Secondary ossification centers, **140**
Secondary sex characteristics, **738**
Secondary spermatocytes, **705**, 706*f*
Second-class levers, **267**
Second-degree burns, **122**, 123*f*
Secretin, 654
Secretion
  epithelia and, 76
  of glands, 86–88
  kidney, **682**
Secretory phase of uterine cycle, 723*f*, **724**
Secretory vesicles (secretory granules), 37*f*, **38**
Sections, tissue, **17**
Segmental arteries, **678**, 679*f*
Segmental bronchi, **614**
Segmentation, body, **12**
Segmentation as digestive process, 635*f*, **636**
Selectively permeable barrier, plasma membrane as, 30
Sella turcica, **161**
Semen, **710**
Semibiceps femoris, 791, 793*f*
Semicircular canals, inner ear, 489, 490*f*, **491**
Semicircular duct, 489, 490*f*, **491**
Semilunar valve of heart, 525, **529**, 530*f*, 531*f*
Semimembranous muscle, 316*f*, 318*t*, 791, 793*f*
Seminal vesicles (seminal glands), 702, **710–11**
Seminiferous tubules, **704**, **705–8**
  spermatogenic cells of, 705, 706*f*
  sustentacular cells of , 705–7
Semispinalis muscle, 285*t*, 286*f*
Semitendinosus muscle, 316*f*, 318*t*, 791, 793*f*
Senile plaques, 403
Senses
  general visceral, 337, 444, 458, 459
  nervous system and, 336–38
  special (hearing, vision, taste, smell) (*see* Special senses)
Sensorineural deafness, **495**
Sensory areas, cerebral cortex, 367*f*, 368*f*, 369–70
Sensory ganglia, **351**, 353*f*
Sensory homunculus, 368*f*, **369**
Sensory input, **336**
Sensory nerves, **419**
Sensory (afferent) neurons, **343**, 344*f*
  embryonic formation of, 354, 355*f*
  locations of, in nervous system, 351, 353*f*
  reflexes and, 350, 352*f*
Sensory nuclei, 381, 383*f*
Sensory (ascending) pathways, spinal cord, 397, 398*f*, 399*t*
Sensory receptors, 113, 336, **410**
  peripheral, **411–16**
Sensory (afferent) signals, **336**, 337*f*
Sensory tunic (retina), **477–79**
Sentinel lymph node in cancer, **600**
Sentinel node mapping, **734**
Septal cartilage, 167
Septal nuclei, 386
Septum pellucidum, 363, 376*f*
Seromucous cells of salivary glands, 643
Seromucous glands, 612
Serosa of alimentary canal, **638**, 639*f*
Serous body cavities, 14, 15*f*
Serous cells of salivary glands, 643
Serous fluid, **14**, 99
Serous membrane (serosa), **14**, 15*f*, **99**, 100*f*
Serous pericardium, **524**
Serratus anterior muscle, 294*t*, 296*f*, 779*f*, 780
Sertoli cells (sustentacular cells), **705**, 706*f*
Sesamoid bones, **134**
Sex hormones, 702, 756. *See also* Estrogens; Testosterone
  brain structure and, 373
  gonads and secretion of, 759
Sex organs
  embryonic development of, 734–37
  female, 715–34
  male, 702–15
  primary, and accessory, 702
Sexually indifferent stage, **734**, 735*f*
Sexually transmitted diseases, 592–93, **712–13**, **720**
Sexual response, male, 713–15
Shaft
  of hair, **117**, 118*f*, 119*f*
  of penis, 710*f*, **712**
  of rib, **179**, 180*f*
Sharpey's fibers, **134**
Shingles (herpes zoster), 436, **437**
Shin splints, **332**
Short bones, **134**
  structure of, 134–35
Shoulder, surface anatomy of, 783, 784*f*, 785*f*
Shoulder dislocation, 227
Shoulder girdle. *See* Pectoral girdle (shoulder)
Shoulder (glenohumeral) joint, **227**, 228*f*
  muscles affecting, 297–98*t*, 299*f*
Sickle cell disease, **515**
Sigmoidal artery, **561**
Sigmoid colon, 655*f*, **656**
Sigmoid mesocolon, 665*f*, **666**
Sigmoid sinus, **566**, 567*f*
Silent ischemia, **535**

Simple columnar epithelium, 78, 80*f*
Simple cuboidal epithelilum, 78, 79*f*
Simple epithelia, 77*f*, **78**, 79*f*
Simple fracture, bone, **143**
Simple glands, **87**, 88*f*
Simple squamous epithelium, 78, 79*f*
Single-unit innervation, 257
Sinoatrial (SA) node, 532, **533**
Sinuses. *See* names of specific sinuses
Sinusitis, **608**
Sinusoids (sinusoidal capillaries), **549–50**
Sinus venosus, development of, **537**
Skeletal muscle(s), **244–53**, **266**. *See also* names of individual muscles
  of abdominal wall, 289–91
  arrangement of fascicles in, 268, 269*f*
  back, intrinsic, 284*f*, 285*t*, 286*f*
  back, superficial, 778*f*, 779
  development-based organization of, 270–72
  of head, face, and tongue, 276–79, 774–76, 777*f*
  innervation of, 416–17
  interactions of, in body, 268–70
  naming, 270
  of neck and throat, 280–82
  of neck and vertebral column, 283–86
  of pelvic floor and perineum, 292, 293*f*
  relationship of bones to, 266–68
  of shoulder, arm, forearm, and hands, 297–311
  structure and organizational levels of, 255*t*
  summary of major, 272–73
  superficial, 274–75*f*
  of thigh, leg, and foot, 312–31
  of thorax, 287–88, 294–96, 779*f*, 780
Skeletal muscle cell, 48, 49*f*
Skeletal muscle fiber, 248, 249*f*, 255*t*
  contraction and extension of, 248, 250*f*, 251*f*
  fascicles of, 246–47, 268, 269*f*
  intracellular tubules of, 251, 252*f*
  length of, and force of contraction, 250
  types of, 252–53
Skeletal muscle pump, 550, 551*f*
Skeletal muscle tissue, **100**, 101*f*, **244–53**, 255*t*
  anatomy of, 248–53
  basic features of, 244–48
  cardiac and smooth muscle compared to, 245–46*t*
Skeletal system, 4, 5*f*, 152–53
  appendicular, 152*f*, 185–210
  axial, 152*f*, 154–82
  bones of, 132–47, 771*f* (*See* also Bone[s])
  cartilage of, 130, 131*f*
  throughout human lifespan, 147
Skin (integument), 59, **112–17**
  cancer of, 117, 123–24
  dermis, 112, 113*f*, 114*f*, 115
  disorders of, 121, 122–25
  epidermis, 112, 113–15
  grafts, and artificial, 122–23
  throughout human lifespan, 124–25
  innervation of, 435–36
  itch receptor in, 413
  pigments and color of, 114, 116–17
  response of, to friction, 113
  sensory receptors in, 412*t*, 413
  stretch marks, 115
  structure of, 112*f*
  tatoos, 116
  tissue repair in wound to, 104*f*
Skin appendages, **117–22**
  hair and hair follicles, 117–20
  nails, 121, 122*f*
  sebaceous glands, 120–21
  sweat glands, 120*f*, 121
Skull, **154–67**
  bones and markings of, 155–60*f*, 170–71*t*
  cranial bones, 154–62, 170–71*t*
  facial bones, 162–64, 171*t*
  fetal and infant, 179, 181*f*
  flat bones of, 136*f*
  overview of, 154
  special parts of, 165–67
Sleep-wake cycles, 379
  hypothalamus and control of, 377
Sliding filament theory, **248**, 250*f*, 251*f*
Slit diaphragm, **682**
Small cardiac vein of heart, 534*f*, **535**
Small cell carcinoma, **626**
Small intestine, 634, **652–54**
  gross anatomy of, 652
  microscopic anatomy and tissues of, 653–54
  muscles of, 258–59*f*
Small saphenous vein, **572**
Smell (olfaction), 468–70
Smell (olfactory) cortex, 367*f*, **370**
Smoking-related lung disease, 624, 625, 626–27, 629
Smooth endoplasmic reticulum (smooth ER), 35*f*, **36**, 44*t*
Smooth muscle cell, 48, 49*f*
Smooth muscle tissue, **102**, **244**, **254–57**
  cardiac and skeletal muscle compared to, 245–46*t*
Soft keratin, **117**
Soft palate, **605**, 606*f*, 641
Soleus muscle, 322*t*, 324–25*f*, 791, 793*f*
Solitary nucleus, 468
Somatic body region, 336
Somatic mesoderm, **62**, 65*f*, 67
  derivatives of, 67, 68*t*
Somatic motor area of nervous system, **337**, 338, 410*f*, 444*f*
  autonomic nervous system compared to, 444–45
  cerebral cortex and control of, 366–69
  skeletal muscle innervation and, 416–17
Somatic motor (SM) neuron, **395**, 396*f*
Somatic reflexes, 350
Somatic sensory area of nervous system, **336**, 337*f*
Somatic sensory (SS) neuron, **395**, 396*f*
Somatosensory association area, brain, 367*f*, **369**
Somatostatin, **759**
Somatotopy, **366**
Somatotropic cells, **747**
Somatotropic hormones, **747**
Somatotropin, **747**
Somites, **62**, 65*f*
Somitomeres, **270**, 271*f*
Sonography (ultrasound imaging), **21**, 22*f*
Sounds
  bowel, 783
  heart, 530, 532*f*
  lung, 779
  vocal speech, 610, 611*f*
Spasm, muscle, **261**
Spastic paralysis, **403**
Spatial discrimination, **369**
Special senses, 465–99
  chemical (taste, smell), 466–70
  ear (hearing, equilibrium), 486–96
  eye and vision, 470–86
  throughout human lifespan, 496–97
Special sensory receptors, **466**
Special somatic senses, **337**
Special visceral senses, **337**
Specific granules, 507
Speech, 610, 611*f*
Sperm
  formation of, 705–8
  movement of, through duct of epididymis, 708–9
Spermatic cord, **709–10**
Spermatids, **705**
  transformation of, into sperm, 707*f*
Spermatocytes, **705**, 706*f*
Spermatogenesis, **705**, 706*f*
Spermatogonia, **705**, 706*f*
Spermatozoa, 705
Spermatozoon, **705**
Sperm cell, 49*f*, 52
  flagellum of, 83
Spermiogenesis, **705**, 706*f*, 707*f*
S (synthetic) phase, cell cycle, **45**, 46–47*f*
Sphenoidal fontanels, **179**, 181*f*
Sphenoid bone, 158*f*, 159*f*, **160–62**, 170*t*
Sphenoid sinuses, **161**
Sphincter muscle, 268
Sphincter urethrae muscle, 292*t*, 293*f*, 690
Spinal bifida, **403–4**
Spinal bifida cystica, **403**
Spinal column. *See* Vertebral column
Spinal cord, 66, 336, **391–401**
  anatomy of, 394*f*
  control of autonomic nervous system by, 457–58
  embryonic development of, 360, 361*f*, 403–4
  gray matter and spinal roots of, 394*f*, 395, 396*f*
  gross structure of, 393*f*
  injuries and damage to, 356, 403
  male versus female, 373
  myelin sheaths in, 348
  sensory and motor pathways of, 397–401
  white matter of, 395–97
Spinal dural sheath, **391**, 394*f*
Spinal fusion, **182**
Spinalis muscle, 285*t*, 286*f*
Spinal nerves, 336, 394*f*, 395, **426–36**
  brachial plexus and innervation of upper limb, 429–32
  cervical plexus and innervation of neck, 428–29, 777
  innervation of anterior thoracic and abdominal wall, 428
  innervation of back, 427*f*, 428
  innervation of body joints, 435
  innervation of skin (dermatomes), 435–36
  lumbar plexus and innervation of lower limb, 432, 433*t*
  nerve plexuses and, 426*f*, 428
  posterior view of, 426*f*
  sacral plexus and innervation of lower limb, 432–35
Spinal roots, 394*f*, 395, 396*f*, **426**, 427*f*
Spinal tap (lumbar puncture), **395**, 779
Spinal tracts, sensory and motor pathways and, 397–401
Spine, scapula, **188**
Spinocerebellar pathways, **397**, 398*f*, 399*t*
Spinothalamic pathway, **397**, 398*f*, 399*t*
Spinothalamic tracts, **397**, 398*f*, 399*t*
Spinous process, **172**, 174*f*, 173, 176*f*, 774, 777
Spiral (coiled) arteries, **722**

Spiral bone fracture, 145*t*
Spiral ganglion, 493*f*, **494**
Spiral lamina of cochlea, **491**, 493*f*
Spiral organ of Corti, **493**, 495*f*
Splanchnic mesoderm, **62**, 65*f*, 67
derivatives of, 67, 68*t*
Spleen, **595**, 596*f*
Splenectomy, **596**
Splenic artery, **559**, 560*f*
Splenic cords, **596**
Splenic (left colic) flexure, 655*f*, **656**
Splenic vein, **570**, 571*f*
Splenius muscles, **283**, 285*t*, 286*f*
Splenomegaly, **600**
Spongy bone (cancellous bone), **134**, 135*f*
microscopic structure of, 137
Spongy (penile) urethra, 690*f*, **691**, 710
Sprains, joint, **234**
Squamous cell carcinoma, **124**, **626**
Squamous epithelia, 77*f*, **78**
simple, 78, 79*f*
Squamous region, temporal bone, **159**, 160*f*
Squamous suture, **155**, 156*f*
Stained specimens, **17**–18
Stapedius, **489**
Stapes, **488**, 489*f*
Stellate ganglion, 450*f*, **452**
Stem cells, **105**, **106**
hematopoietic, 143, 513
liver, 662
Stenosis of the lumbar spine, **179**
Stenotic heart valves, **532**
Stereocilia, **708**
Sternal angle, bony thorax, **178**, 779–80
Sternal end, clavicle, **186**, 187*f*
Sternocleidomastoid muscle, 283*t*, 284*f*, 774–76, 777*f*
Sternohyoid muscles, 281*t*, 282*f*
Sternothyroid muscles, 281*t*, 282*f*
Sternum, **177–78**, 779
joint between rib and, 214*f*
Steroids, **745**
effects of anabolic, 260
secretion of, by adrenal cortex, 755–57
Stomach, 634, **646–51**
gross anatomy of, 646–48
microscopic anatomy and tissues of, 648–51
peptic ulcers in, 650–61
Stomodeum, **667**, 668*f*
Strabismus, **473**
Straight arteries (basal arteries), **722**
Straight sinuses, **566**, 567*f*
Strain, muscle, **261**
Stratified columnar epithelium, 82*f*, 83
Stratified cuboidal epithelium, 81*f*, 83
Stratified epithelia, 77*f*, **78**, 81*f*, 82*f*, 83
Stratified squamous epithelium, 78, 81*f*, 83
Stratum basale, 113*f*, **114**
Stratum basel (basal layer) of endometrium, **722**
Stratum corneum, 114*f*, **115**
Stratum functionalis (functional layer) of endometrium, **721**, 722*f*
Stratum germinativum, 114
Stratum granulosum, 113*f*, **114**
Stratum lucidum, 114*f*, **115**
Stratum spinosum, 113*f*, **114**, 124
Streptokinase, **575**
Stress
bone design and, 135, 136*f*
glucocorticoids, high blood glucose levels and, 756
hypertension and, 459
peptic ulcers and, 650
response to emotional, 387
Stress incontinence, **692**
Stretch reflexes, 350
Striations, skeletal muscle fiber, 248, 249*f*, 251*f*
Stroke (ischemia), 402, 483
Stroma, 619
Student's elbow, **236**
Sty, **471**
Styloglossus muscle, 278*t*, 279*f*
Stylohyoid ligaments, 167
Stylohyoid muscles, 280*t*, 282*f*
Styloid process
radius, **190**, 192*f*, 787*f*
skull, **159**, 160*f*
ulna, **190**, 192*f*, 787*f*
Stylomastoid foramen, 158*f*, **159**
Subarachnoid space, 363, **389**
Subclavian arteries, 553, 555*f*, 557, 558*f*, 777
Subclavian (lymph) trunks, **589**
Subclavian vein, 564, **567**, 568*f*
Subclavius muscle, 295*t*, 296*f*
Subcostal arteries, 558
Subcostal plane, 637
Subcutaneous layer, integumentary system, 116, 770
Subcutaneous prepatellar bursa, **231**
Subdural space, **389**
Subendothelial layer, blood vessel walls, **544**, 545*f*
Sublingual gland, **643**
Subluxation, **234**
Submandibular ganglion, **448**
Submandibular (submaxillary) gland, **643**, 777
Submucosa
alimentary canal, **638**, 639*f*
trachea, **612**
Submucosal nerve plexus, **639**
Subscapular artery, **557**, 558*f*
Subscapular fossa, **189**
Subscapularis muscle, **297**, 298*t*, 299*f*
Substantia nigra, **380**, 384*f*
Subthalamic nucleus, 375*f*, **377**
Sudden infant death syndrome (SIDS), **629**
Sudoriferous glands, 120*f*, **121**
Sulci (sulcus), cerebral hemisphere, **363**, 364*f*
Sulcus terminalis, **642**
Sunlight, effect of UV rays and, on skin, 113, 114, 117, 124, 125
Superciliary arches, 770, 775*f*
Superficial, defined, **8**, 9*t*
Superficial fascia, **116**
Superficial fibular (peroneal) nerve, **434**
Superficial inguinal ring, 709*f*, **710**
Superficial palmar arch, **557**, 558*f*
Superficial palmar venous arch, **567**, 568*f*
Superficial perineal space, muscles of, 292*t*, 293*f*
Superficial temporal artery, **553**, 556*f*, 774
Superficial transverse perineus muscle, 292*t*, 293*f*
Superficial veins, 563, 567, 568*f*, 571, 572*f*
Superior, defined, **8**, 9*t*
Superior angle, scapula, **186**, 188*f*
Superior articular process
sacrum, **176**, 177*f*
vertebral arch, **172**, 174*f*
Superior border, scapula, **186**, 188*f*
Superior cerebellar peduncles, **379**, 382*f*, **385**, 386*f*
Superior cervical ganglia, **452**, 453*f*
Superior colliculi, **380**, 384*f*, **484**
Superior gluteal nerves, **434**, 435*t*
Superior hypophyseal artery, **746**, 748*f*
Superior mesenteric artery, 560, 561*f*
Superior mesenteric ganglia, 452, 453*f*, **456**
Superior mesenteric plexus, **449**
Superior mesenteric vein, **570**, 571*f*
Superior nasal conchae, **162**, 166*f*, 606*f*, 607
Superior nuchal lines, **157**, 158*f*, 770
Superior oblique muscle, **472**, 473*f*
Superior olivary nuclei, **494**, 495*f*
Superior orbital fissure, 161*f*, **162**
Superior rami, pubis, 195*f*, **198**
Superior rectal arteries, **561**
Superior rectus muscle, **472**, 473*f*
Superior sagittal sinus, 388, 389*f*, **566**, 567*f*
Superior tarsal muscle, 454, 455*f*
Superior thyroid artery, **553**, 556*f*
Superior vena cava vein, 525, 526*f*, 564, 565*f*
Superior vesicle arteries, 689
Supination movement, **222**, 223*f*
Supinator muscle, **301**, 305*t*, 306*f*
Supporting cells of nervous tissue, **102**, 103*f*, **346–48**
of special senses, **466**, **468**, **489**, 491*f*
Suprachiasmatic nucleus, **377**, 379*f*, 484
Supracondylar lines, femur, **199**, 201*f*
Supracondylar ridges, humerus, 189*f*, **190**
Supracristal line of the back, **777**, 778*f*
Suprahyoid muscles, 280–81*t*, 282*f*, 777
Supraorbital foramen, **154**, 155*f*
Supraorbital margin, **154**, 155*f*
Suprarenal arteries, 559*f*, 560, 754
Suprarenal veins, 569, 754
Suprascapular notch, **189**
Supraspinatus muscle, **297**, 298*t*, 299*f*
Supraspinous fossae, scapula, **189**
Sural nerve, **434**
Surface anatomy, **2**, **769–97**
bones of human skeleton, 771*f*
head, 770–74, 775*f*
lower limb and gluteal region, 789–93
muscles of human body, 772*f*, 773*f*
neck, 774–77
trunk, 777–83
upper limb and shoulder, 783–89
Surfactant, 614, **623**
Surgical neck, humerus, 189*f*, **190**
Suspensory ligament of the breast, **726**
Suspensory ligament of the ovary, 714*f*, **715**
Sustentacular cells (Sertoli cells), **705**, 706*f*
Sustentaculum tali, **202**, 204*f*
Sutural bones, 155*f*, 157
Sutures
fibrous joint, **212**, 213*f*
skull, 154, 155*f*, 156*f*
Sweat glands, 120*f*, **121**
sympathetic pathways to, 454, 455*f*
Sweat pores, 115, **121**
Sympathetic chain ganglia, 427, 452, 454*f*
Sympathetic division of autonomic nervous system, 410, **445**, 446*f*
basic organization of, 452, 453*f*

Sympathetic division of autonomic nervous system, *continued*
blood pressure and adjustment of arterioles by, 546–47
effects of, on select organs, 451*t*
parasympathetic division versus, 447*f*, 448*t*
pathways of, 452–56
role of adrenal medulla in, 456–57
Sympathetic pathways, 452–56
to body periphery, 454, 455*f*
to head, 454, 455*f*
to pelvic organs, 456, 457*f*
to thoracic organs, 455, 456*f*
Sympathetic trunk ganglia, **452**, 454*f*
Sympathetic trunks, **452**, 454*f*
Symphysis, **213**, 214*f*
Synapse, **340**, 341*f*, 342*f*
excitatory, and inhibitory, 343
Synaptic cleft, **341**
Synaptic potentials, 342–43
Synaptic vesicles, **341**
Synarthroses, 212
Synchondroses, **213**, 214*f*
Syncytiotrophoblast, **728**
Syndesmoses, **212**, 213*f*
Syndesmosis ankle sprains, 234
Synergist muscles, **268–69**
Synovial cavities, **15**, 16*f*
Synovial fluid, **215**–16
Synovial joints, **214–34**
ankle, 233–34, 235*f*
bursae and tendon sheaths, 219
characteristics of, 216–17*t*
classification of, by shape, 224–26
elbow, 227, 229*f*
factors influencing stability of, 219–20
function of, 217–18
general structure of, 214–17
hip, 229–31
knee, 231–34
movements allowed by, 220–24
shoulder, 227, 228*f*
summary of, 212*t*
temporomandibular, 226–27, 774, 775*f*
Synovial membrane, **215**
Synovitis, **239**
Syphilis, **712–13**
Systemic anatomy, **2**
Systemic circuit, **522**, 552*f*
Systemic circulation, **553–73**
arteries of, 553–63, 564*t*
veins of, 563–73
Systole, **528–29**

Tail of sperm, **705**
Talofibular ligaments, **234**, 235*f*
Talus, **202**, 204*f*
Target cells of hormones, **745**
Target organs of hormones, 87
Tarsal glands, **471**, 472*f*
Tarsal plates, **470**, 472*f*
Tarsals, **202**, 204*f*
Tarsus, **202**, 204*f*
Taste (gustation), 466–68
Taste buds, **466**, 467*f*, 641, 642
Taste (gustatory) cortex, 367*f*, **370**
Taste pore, **466**
Tatoos, 116
Tay-Sachs disease, 38–39
T cells (lymphocytes), **508**, 509*f*, 589, 592, 595, 597
Tears, 163, 471–72
Tectorial membrane, 493*f*, **494**
Tectospinal tract, **397**, 400*f*, 401*t*
Teeth, 635, **643–45**
dentition and dental formula, 644
tooth structure, 644–45
Telencephalon, **360**, 361*f*
Telomerase, 53
Telomeres, 52–53
TEM (transmission electron microscope), **16**, 18*f*
Temperature regulation
dermis and, 115
hypothalamus and control of, 377
sweat glands and, 121
Temporal bones, **157–60**, 170*t*
Temporalis muscle, 278*t*, 279*f*, 770, 775*f*
Temporal lobe, cerebral hemisphere, 364*f*, **365**
Temporomandibular disorders, **227**
Temporomandibular joint (TMJ), 159, **226–27**, 774, 775*f*
Tendinous insertions, 781
Tendon(s), 98, **247**, 248
central, 292
Tendonitis, **236**
Tendon sheath, **219**
Teniae coli of large intestine, 655*f*, **656**
Tenosynovitis, **239**
Tensegrity structure, 39*f*, 40
Tensor fasciae latae muscle, **312**, 313*t*, 314*f*
Tensor tympani, **489**
Tentorium cerebelli, **388**, 390*f*
Teratogens, **70**
Teratology, **72**
Teres major muscle, **297**, 298*t*, 299*f*
Teres minor muscle, **297**, 298*t*, 299*f*
Terminal, hair, **119**
Terminal branches of neuron, 339*f*, **340**
Terminal bronchioles, **614**
Terminal cisternae, **251**, 252*f*
Terminology, anatomical, 2. *See also* Glossary
Testes, 702, **703–8**
descent of, 737
gross anatomy of, 703–5
seminiferous tubules and spermatogenesis in, 705–8
Testicular arteries, 559*f*, 560
Testicular cancer, **733**
Testicular fluid, **707**
Testicular torsion, **738–39**
Testicular veins, 569
Testosterone, 260, 261, **705**, **759**
brain structure and, 373
Thalamus, **375–77**
geniculate nucleus of, **377**, **481**, 484*f*, **494**, 495*f*
relationship of, to cerebral cortex, 375, 378*f*
Thalassemia, **517**
Thalidomide, 70
Theca folliculi, **717**
Thenar eminence, 789
Thermoreceptors, **413**
Thick (myosin) filaments, **248**, 249*f*
Thick segment of loop of Henle, **682**, 684*f*
Thick skin, **114**
Thigh
arteries of, 561–63
bones of, 197*t*, **199**, 201*f*
innervation of, 432–35
muscles of, 312–13*t*, 314–16*f*, 317–18*t*, 326–27*t*, 328*f*
surface anatomy of, 790*f*, 791, 792*f*, 793*f*
Thin (actin) filaments, **248**, 249*f*
Thin segment of loop of Henle, **682**, 684*f*
Thin skin, 113*f*, **114**
Third-class levers, **267**
Third-degree burns, **122**, 123*f*
Third ventricle, **363**
Thirst, hypothalamus and control of, 377
Thoracic aorta, **553**, 554*f*, 555*f*
Thoracic arteries, 557, 558*f*
Thoracic cage. *See* Bony thorax (thoracic cage)
Thoracic cavity, **13**, 14*f*
division of, by pleurae, 617
lungs and pleural sacs in, 617, 618*f*
Thoracic curvature, 168*f*, **172**
Thoracic duct, **589**, 590*f*
Thoracic splanchnic nerves, **456**
Thoracic vertebrae, 168*f*, **172**, **173**, 174*t*, 176*f*
Thoracic volume, breathing and changes in, 622*f*
Thoracoacromial artery, **557**, 558*f*
Thoracolumbar (sympathetic) division of autonomic nervous system, **446**
Thorax
arteries of, 557–59
bones of, surface anatomy of, 779–80
deep muscles of (breathing), 287*t*, 288*f*
innervation of, 426*f*, 427*f*, 428
location and orientation of heart within, 522–24
muscles, 294–95*t*, 296*f*, 779*f*, 780
sympathetic pathways to organs of, 455, 456*f*
sympathetic trunks and trunk ganglia in, 454*f*
veins of, 568–69
Thoroughfare channel, **547**
Throat, 608
Thrombocytes. *See* Platelets
Thrombocytopenia, **515–16**
Thrombophlebitis, 574
Thrombus, **510**
Thumb, muscles of ball of, 309*t*, 311*f*
Thymic (Hassall's) corpuscles, **597**
Thymic hormones, **759**
Thymine (T), 43
Thymus, **597**, 599, 744, **759**
embryonic development of, 762*f*
Thyrocervical trunk, 553, 556*f*, **557**
Thyroglobulin, **754**
Thyrohyoid muscles, 281*t*, 282*f*
Thyroid arteries, 556*f*, 557
Thyroid cancer, **763**
Thyroid cartilage, **609**, 611*f*
Thyroid gland, 744, **753–54**, 774
disorders of, 760
embryonic development of, 762*f*
Thyroid hormone, 142, **754**
Thyroid-stimulating hormone (TSH), **748**, 749*t*, 754
Thyroid storm, **763**
Thyrotropic cells, **748**
Tibia, **202**, 203*f*, 791, 792*f*
Tibial arteries, 562*f*, 563, 791
Tibial collateral ligaments, **231**, 232–33*f*
Tibialis anterior muscle, 319*t*, 320*f*
Tibialis posterior muscle, 323*t*, 325*f*
Tibial nerve, 432, **434**, 435*t*
Tibial tuberosity, **202**, 203*f*, 791, 792*f*
Tibial veins, 571, 572*f*
Tibiofibular joints, 202
Tibiofibular ligaments, **234**, 235*f*
Tic douloureux, **346**, **421**
Tight junctions, **85**
capillary permeability and, 547–49
Tinnitus, **497**
Tissue(s), **75–100**
of alimentary canal, 638–40
capillary beds in, 547
connective, 88–99
covering and lining membranes, 99, 100*f*
defined, and types of, **76**
epithelia, 62, 76–86
glands, 86–88
throughout human lifespan, 106
injury of, responses and repair mechanisms and, 102–6
mesenchyme, 62, 66, 88, 92*f*

Tissue(s), *continued*
of mouth, 640
muscle, 99–102
nervous, 102, 103*f*
preparation of, for microscopy, 17–18
of small intestine, 653*f*, 654
stem cells for treatment of failing, 105, 106
of stomach, 648–51
structural organization at level of, 3*f*, **4**
Tissue engineering, 105
Tissue fluid (interstitial fluid), **91**, 584, 585*f*
Tissue plasminogen activator (t-PA), **575**
Tissue-specific stem cells, **105**
Tissue-type plasminogen activator, 402
Titin, 249*f*, **250–51**
Toes
bones of, 204*f*, 205
muscles acting on, 319–31
Tongue, 635, **641–42**
Tongue muscles, **278**, 641
Tonofilaments, 114, 115
Tonsilitis, **600**
Tonsils, **597**, 598*f*, 608
lingual, **598**, 606*f*, **608**, 642
pharyngeal (adenoids), 608
tubal, **598**, 606*f*, **608**
Torticollis, **283**
Trabeculae, 134
lymph node, **586**, 588*f*
Trabeculae carneae of heart, **525**, 527*f*
Trachea, **612**, 621*t*
Trachealis muscle, **612**
Tracheobronchial lymph nodes, 586, 587*f*
Tracheotomy, **629**
Trachoma, **485**
Traction, **148**
Tracts, axon bundles as, 353
Tragus of ear, **774**, 775*f*
Transcranial magnetic stimulation, **366**
Transcription, **43**
Trans-Golgi network, 37*f*, **38**
Transitional epithelium, 82*f*, 83
Translation, 35
Transmission electron microscope (TEM), **16**, 18*f*
Transmyocardial revascularization, 536
Transport mechanisms across plasma membrane, 30–33
Transport of respiratory gases, **604**
Transport vesicles, 37*f*, **38**
Transtubercular plane, 637
Transverse arch, **205**
Transverse colon, 655*f*, **656**, 665*f*, **666**
Transverse fissure, cerebral hemisphere, **363**, 364*f*
Transverse foramen, **173**, 174*f*
Transverse lines, sacrum, **176**, 177*f*
Transverse (horizontal) plane, **10**, 11*f*
Transverse process, vertebral arch, **172**, 174*f*
Transverse sinuses, **566**, 567*f*
Transversus abdominis muscle, **289**, 290*t*, 291*f*
Trapezium bone, **191**, 193*f*
Trapezius muscle, 295*t*, 296*f*, 776, 777*f*, 778*f*, 779
Trapezoid bone, **191**, 193*f*
Trapezoid line, clavicle, **186**
Triad, T tubule complex, **251**
Triangle
of auscultation, **779**
of neck, **777**
Triceps brachii, 299*f*, 300*t*
Triceps surae muscle, 322*t*, 324*f*
Tricuspid valve of heart, 525, **529**, 530*f*, 531*f*
Trigeminal nerve (V), **418**, 419*f*, 421–22*t*
Trigone, **689**, 690*f*
Triquetral bone, **191**, 193*f*
Trochanter, **199**, 201*f*
Trochlea
eye, **472**, 473*f*
humerus, 189*f*, **190**
Trochlear nerve (IV), 380, **418**, 419*t*, 421*t*
Trochlear nerve damage, 421
Trophoblast, **60**, 61*f*, **728**
Tropic hormones, **749**
True pelvis, 198, **199**
True ribs (vertebrosternal ribs), **178**
Trunk, surface anatomy of, 777–83
abdomen, 781–83
back, 777–79
pelvis and perineum, 782*f*, 783
thorax, 779–80
Trunk movements, muscles of, 289–90*t*
Trunks of brachial plexus, **429**, 430*f*
T tubules, **251**, 252*f*
Tubal ligation, **739**
Tubal tonsils, **598**, 606*f*, **608**
Tuber calcanei, **202**, 204*f*
Tubercle
humerus, 189*f*, 190
rib, **179**, 180*f*
Tuberculosis (TB), **625–26**
Tube-within-a-tube body plan, **12**
Tubular exocrine glands, **87**, 88*f*
Tubuloalveolar glands, **87**, 646
Tubulus rectus, **704**
Tumors, **50**. *See also* Cancer
Tumor-suppressor genes, **50**, 51
Tunica albuginea, **704**, **716**
Tunica externa, **544**, 545*f*, 550
Tunica intima, **544**, 545*f*
Tunica media, **544**, 545*f*, 550
Tunica vaginalis, **704**
Tunics
blood vessel walls, 544, 545*f*
eyeball, **474–79**
Tympanic cavity. *See* Middle ear
Tympanic membrane (eardrum), **486**, 487*f*
Tympanic region, temporal bone, **159**, 160*f*
Type I cells, alveolus, **614**, 615*f*
Type I diabetes, **761**
Type II cells, alveolus, **614**, 616*f*
Type II (non-insulin-dependent) diabetes, **761**
Type A daughter cells, **705**, 706*f*
Type B daughter cells, **705**, 706*f*

Ulcerative colitis, 666
Ulcers, esophageal and stomach, 650–51
Ulna, 190, 191*f*, 192*f*, 787
Ulnar artery, 557*f*, 558, 789
Ulnar collateral ligament, **227**
Ulnar nerve, 431*t*, **432**, 783
Ulnar notch, **190**, 192*f*
Ulnar vein, **567**, 568*f*
Ultimobranchial bodies, 762
Ultrasound imaging (sonography), **21**, 22*f*
bone repair using, **206**
Ultrastructure, **17**
Ultraviolet light (UV), skin and, 113, 114, 117, 124, 125
Umbilical arteries, **576**, 577*f*
Umbilical hernias, 781
Umbilical region of abdominal wall, 636*f*, **637**
Umbilical vein, **576**, 577*f*
Umbilicus (navel), 781
Uncinate fits, **470**
Unconventional myosin, **40**
Uncus, **367**, **370**
Uniaxial joints, **225**
Unicellular glands, **87**
Unipennate pattern of fascicles, **268**, 269*f*
Unipolar neurons, **343**, 344*f*
Unmyelinated axons, **348**, 349*f*
Upper limb
arteries of, 557, 558*f*
bones of, 189–94, 196–98*t*
innervation of, 429–32
muscles of, 297–311
surface anatomy of shoulder and, 783–89
veins of, 567, 568*f*
Upper lobe of lung, **617**, 619*f*
Upper-lower (UL) body ratio, **205**, 206*f*
Urachus, **689**
Urea, 676
Ureteric bud, **694**
Ureters, 676, **686–87**, 688*f*
gross anatomy of, 686–87
microscopic anatomy of, 687, 688*f*
Urethra, 676, **689–91**
female, 690–91
male, 690*f*, 691, 702, 710
Urethral folds (genital folds), **736**
Urethral gland, 710
Urethral groove, **736**
Urge incontinence, **692**
Uric acid, 676
Urinary bladder, 676, **687–89**
Urinary incontinence, **691**
Urinary retention, **692**
Urinary system, 7*f*, **675–700**
disorders of, 692–93
throughout human lifespan, 693–95
kidneys, 676–86
micturition and, 691–92
organs of, 676*f*
ureters, 686–87
urethra, 689–91
urinary bladder, 687–89
Urinary tract infections, **692**
Urine, production of, 681–82, 694
Uriniferous tubule, 680*f*, **681**
microscopic blood vessels associated with, 685–86
Urogenital diaphragm, 690
muscles of, 292*t*, 293*f*
Urogenital ridges, 693
Urogenital sinus, **694**
Urologist, **695**
Uterine arteries, 715, **722**
Uterine cycle, **722–24**
Uterine glands, **722**
Uterine tubes, 59, 60*f*, 714*f*, 715, **718–20**
Uterine wall, 721–22
Uterus, 714*f*, 715, **720–24**
cycle of, 722–24
supports of, 720–21
wall of, 721–22
Utricle, **489**, 490*f*
Utrophin, 258
Uvula, 606*f*, **608**

Vagina, 714*f*, 715, 724
Vaginal orifice, 714*f*, **724**
Vaginal process, **737**
Vaginitis, **713**
Vagotomy, **461**
Vagus nerve (X), 381, 419, 425*t*
cranial outflow via, 449, 450*f*
Vagus nerve damage, 425
Valgus and varus injuries, **239**
Vallate papillae, 641, **642**
Valsalva's maneuver, **611**
Varicocele, **739**
Varicose veins, **550**, 575
Vasa recta, 684*f*, **686**
Vasa vasorum, 546*f*, **551**
Vascular anastomoses, 550–51
Vascular endothelial growth factor (VEGF), 536
Vascular tunic, 474*f*, **476–77**
Vasectomy, **709**
Vasoconstriction, **446**, 459, 544, 547
Vasomotor center in reticular formation of medulla, 382, 457
Vasomotor nerve fibers, 544
Vasopressin, 752
Vastus intermedius muscle, 313*t*, 314*f*
Vastus lateralis muscle, 313*t*, 314*f*, 785*f*, 791, 792*f*

Vastus medialis muscle, 313*t*, 314*f*
Vaults, **40**, 41*f*, 44*t*
Vein(s), 504, 544, **550**
abdominal, 569–71
of adrenal glands, 754
disorders of, 550, 573–76
of head and neck, 566–67
of heart, 525
hemorrhoids and varicose, 550
nutrient, in bones, 134, 135*f*
of pelvis and lower limbs, 571–72
of penis, 713
of pituitary gland, 746, 748*f*
pulmonary, 552–53
of small intestine, 652
of stomach, 648
structure of, 544, 545*f*
systemic, 563–73
of testes, 704
of thorax, 568–69
of upper limbs, 567, 568*f*
of urinary bladder, 689
valves in, and muscular pump, 550, 551*f*
venae cavae and major tributaries, 564, 565*f*
Vein of the retina, 474*f*, **479**
Vellus, hair, **119**
Vena cava veins, 564, 565*f*, 566*f*, 660
Venous anastomoses, 551
Venous disease, **574–75**
Venous plexus, 563
Venous sinuses, spleen, **596**
Ventilation, **604**, **621–23**
expiration, 622–23
inspiration, 621–22
muscles of, 287*t*, 288*f*
neural control of, 623
Ventral, defined, **8**, 9*t*
Ventral body cavity, **13**, 14*f*
Ventral gluteal site, 785*f*, **790**
Ventral posterior lateral nucleus, **375**
Ventral ramus, **427**, 428
Ventral sacral foramina, **176**, 177*f*
Ventral spinal root, **426**, 427*f*
Ventral stream, visual association area of cerebral cortex, 369, 370*f*
Ventricles, brain, **363**, 376*f*
embryonic development of, **360**, 361*f*
Ventricles, heart, **522**
development of, **538**
heartbeat and, 529
left, 526*f*, 527*f*, **528**
right, **525**, 526*f*, 527*f*
Ventricular fibrillation, **537**
Ventricular reduction, 536
Ventricular septal defect, **538**, 539*f*
Venules, 544, **550**
Vermiform appendix, 655*f*, **656**
Vermis, **385**, 386*f*
Vernix caseosa, 124
Vertebrae, **12**, **168**, 777
characteristics of regional, 173–77
general structure of, 172
Vertebral arch, **172**
Vertebral arteries, 553, **555**, 556*f*
Vertebral cavity, **13**, 14*f*
Vertebral column, **168–77**
curvatures of, 168*f*, 172, 179, 180–81
general structure of vertebrae, 172
throughout human lifespan, 180–81
intervertebral discs of, 168–72
muscles of, 284*f*, 285*t*, 286*f*
regional vertebral characteristics, 173–77
Vertebral foramen, **172**
Vertebral spine, **172**
Vertebral vein, 567
Vertebra prominens, **173**
Vertebrates, body plan of, 12, 13*f*
Vesicle, membrane transport and, 30
Vesicoureteric reflex, **695**
Vesicouterine pouch, 721
Vesicular (Graafian) follicle, 716*f*, **718**
Vestibular (equilibrium) cortex, 367*f*, **370**
Vestibular folds (false vocal cords), **610**, 611*f*
Vestibular ganglia, 490
Vestibular membrane, **493**
Vestibular nerve, **489**, 491*f*
Vestibule
of female genitalia, 724*f*, **725**
inner ear, **489–90**
mouth, **640**
nasal cavity, **606**
Vestibulocochlear nerve (VIII), 160, 418, 419*t*, 424*t*, 489
Vestibulocochlear nerve damage, 424
Vestibulospinal tract, **397**, 401*t*
Vesticular nuclei, **381**, 384*f*
Villi of small intestine, **653**, 654
Vinblastine, 48
Viral hepatitis, 662, **667**
Virchow, Rudolf, 28
Viscera (visceral organs), **13**, 14*f*
adhesions of, 105–6
arteries of thoracic, 559
of digestive system, 637–40, 646–66
effects of sympathetic and parasympathetic divisions on, 451*t*
muscles for compression of abdominal, 289–90*t*, 291*f*
musculature of, 272
sympathetic pathways to, 455, 456*f*, 457*f*
Visceral body region, 336
Visceral layer
of glomerular capsule, **682**
of serous pericardium, **524**
Visceral motor area of nervous system, **338**, 410*f*, 444*f*. *See also* Autonomic nervous system (ANS)
Visceral motor (VM) neuron, **395**, 396*f*
Visceral muscle, 244
innervation of, 417–18
Visceral peritoneum, **637–38**
Visceral pleura, **615**, 616, 617*f*
Visceral reflex arcs, **459**
Visceral reflexes, 350
Visceral sensory area of nervous system, **337**, 444, 458–59
Visceral sensory (VS) neuron, **395**, 396*f*, 458, 459*f*
Visceral serosa, **14**, 15*f*
Visceral surface of liver, **658**, 659*f*, 660*f*
Vision
accessory structures of eye, 470–73
anatomy of eyeball, 474–80
disorders of, 484–85
embryonic development of eye and, 485–86
light and, 480–81
visual pathways, 481–84
Visual areas, brain, 367*f*, **369**
Visual association area, brain, 367*f*, **369**, 370*f*
Visual cortex, **481**, 484*f*
Visual pathways, **481–84**
Visual pigments, 479
Vitamins
A, 106
B, 404
C, 52, 91
D, 117, 145, 146, 759–60
E, 52
Vitelline duct, **667**
Vitiligo, **125**
Vitreous humor, eye, 474*f*, 475*f*, **480**
Vocal cords, 609*f*, **610**, 611*f*
Vocal folds, 609*f*, **610**, 611*f*
Voice production, 610, 611*f*
Volkmann's canals (perforating canals), **137**, 138*f*
Voluntary nervous system, 337
Vomer, **163**, 166*f*, 171*t*
Vulva, 715, **724**
Water, osmosis of, across membranes, 30
Weeping lubrication, synovial joint, **218**
Weight lifting, muscles and, 253
Wernicke's area, brain, 367*f*, **370**
Whiplash, **182**
White adipose tissue, 98
White blood cells. *See* Leukocytes
White columns (funiculi), 394*f*, **396**
White fast-twitch muscle fibers, **252**, 253*f*
White matter
brain, 372, 385
central nervous system, **352**, 353*f*, 362*f*
spinal cord, 395, 396*f*, 397
White pulp of spleen, **596**
White rami communicantes, **452**, 453*f*, 454*f*
Wisdom teeth, 644
Withdrawal reflexes, 350
Woven bone tissue, **139**, 140*f*
Wristdrop, **432**

Xenon CT, **20**
Xiphisternal joint, 780
Xiphisternal joint, bony thorax, **178**
Xiphoid process, **177**, 779
X ray imaging techniques, 19–20

Yellow marrow, **512**
Yolk sac, **61**
Youth
autonomic nervous system in, 460
cell development in, 52
muscle development in, 261
smoking and respiratory development in, 629

Z discs (Z lines), **248**, 249*f*
Zona fasciculata, **755**
Zona glomerulosa, **755**
Zona pellucida, **717**
Zona reticularis, **756**
Zonula occludens, 85
Zygomatic arch, 159, 774, 775*f*
Zygomatic bones, **163**, 171*t*
Zygomatic process
maxillary bones, **163**, 164*f*
temporal bone, **159**, 160*f*
Zygomaticus muscle, **276**, 277*f*
Zygote, **59**, 60*f*
fertilization of egg and formation of, 728
Zymogen granules, **663**, 664*f*

Dear Student:

As the authors of ***Human Anatomy,* Third Edition**, we hope that this textbook has been a useful learning tool for your course. With the vast amounts of factual material that you must learn, our goal is to make the material as logical and accessible as possible. We would appreciate hearing about your experiences with this textbook and its multimedia support, and we invite your suggestions for improvements!

Many thanks,

Elaine N. Marieb — Jon Mallatt

1. Did you use the **Study Partner CD-ROM**? ☐ Yes ☐ No
If so, which features were most useful to you? Did you use it on your own, or did your instructor require you to use it? ______________________________

Did the **Study Partner CD-ROM** help you to study? Was it easy to use, or did you experience problems? How can we improve the **Study Partner CD-ROM**? ______________________________

2. Did you use **The A&P Place**? ☐ Yes ☐ No
If so, which features were useful to you? Did you use it on your own, or did your instructor require it?

Did **The A&P Place** help you to study? Was it easy to use, or did you experience problems? How can we improve **The A&P Place**? ______________________________

3. Did you use **Adam Online Anatomy (AOA)**? ☐ Yes ☐ No
If so, which features were most useful to you? Did you use it on your own, or did your instructor require it?

4. Which study tool(s) — e.g. Chapter Summary, Review Questions, "Disorders" sections — did you find most useful?

6. What did you like most about ***Human Anatomy,* Third Edition**? Please provide three examples, in order of priority. ______________________________

____

____

____

7. Do you have any ideas about how this book can be improved? If so, please write your specific suggestions, in order of priority, citing page numbers as appropriate. ______________________________

____

____

____

____

NO POSTAGE NECESSARY IF MAILED IN THE UNITED STATES

BUSINESS REPLY MAIL

FIRST-CLASS MAIL PERMIT NO. 275 SAN FRANCISCO CA

POSTAGE WILL BE PAID BY ADDRESSEE

BENJAMIN/CUMMINGS SCIENCE
PEARSON EDUCATION
1301 SANSOME STREET
SAN FRANCISCO CA 94111-9328

School: ______________________________

Optional:

Your name: ______________________________

Email: ______________________________

Date: ______________________________

May Benjamin Cummings have permission to quote your comments in promotions for ***Human Anatomy***? ☐ Yes ☐ No